COLD SPRING HARBOR SYMPOSIA ON QUANTITATIVE BIOLOGY

VOLUME L

COLD SPRING HARBOR SYMPOSIA ON QUANTITATIVE BIOLOGY

VOLUME L

Molecular Biology of Development

COLD SPRING HARBOR LABORATORY
1985

COLD SPRING HARBOR SYMPOSIA ON QUANTITATIVE BIOLOGY
VOLUME L

© 1985 by The Cold Spring Harbor Laboratory
International Standard Book Number 0-87969-050-X (cloth)
International Standard Book Number 0-87969-051-8 (paper)
International Standard Serial Number 0091-7451
Library of Congress Catalog Card Number 43-8174

Printed in the United States of America
All rights reserved

COLD SPRING HARBOR SYMPOSIA ON QUANTITATIVE BIOLOGY

Founded in 1933 by
REGINALD G. HARRIS
Director of the Biological Laboratory 1924 to 1936

Previous Symposia Volumes

I (1933) Surface Phenomena
II (1934) Aspects of Growth
III (1935) Photochemical Reactions
IV (1936) Excitation Phenomena
V (1937) Internal Secretions
VI (1938) Protein Chemistry
VII (1939) Biological Oxidations
VIII (1940) Permeability and the Nature of Cell Membranes
IX (1941) Genes and Chromosomes: Structure and Organization
X (1942) The Relation of Hormones to Development
XI (1946) Heredity and Variation in Microorganisms
XII (1947) Nucleic Acids and Nucleoproteins
XIII (1948) Biological Applications of Tracer Elements
XIV (1949) Amino Acids and Proteins
XV (1950) Origin and Evolution of Man
XVI (1951) Genes and Mutations
XVII (1952) The Neuron
XVIII (1953) Viruses
XIX (1954) The Mammalian Fetus: Physiological Aspects of Development
XX (1955) Population Genetics: The Nature and Causes of Genetic Variability in Population
XXI (1956) Genetic Mechanisms: Structure and Function
XXII (1957) Population Studies: Animal Ecology and Demography
XXIII (1958) Exchange of Genetic Material: Mechanism and Consequences

XXIV (1959) Genetics and Twentieth Century Darwinism
XXV (1960) Biological Clocks
XXVI (1961) Cellular Regulatory Mechanisms
XXVII (1962) Basic Mechanisms in Animal Virus Biology
XXVIII (1963) Synthesis and Structure of Macromolecules
XXIX (1964) Human Genetics
XXX (1965) Sensory Receptors
XXXI (1966) The Genetic Code
XXXII (1967) Antibodies
XXXIII (1968) Replication of DNA in Microorganisms
XXXIV (1969) The Mechanism of Protein Synthesis
XXXV (1970) Transcription of Genetic Material
XXXVI (1971) Structure and Function of Proteins at the Three-dimensional Level
XXXVII (1972) The Mechanism of Muscle Contraction
XXXVIII (1973) Chromosome Structure and Function
XXXIX (1974) Tumor Viruses
XL (1975) The Synapse
XLI (1976) Origins of Lymphocyte Diversity
XLII (1977) Chromatin
XLIII (1978) DNA: Replication and Recombination
XLIV (1979) Viral Oncogenes
XLV (1980) Movable Genetic Elements
XLVI (1981) Organization of the Cytoplasm
XLVII (1982) Structures of DNA
XLVIII (1983) Molecular Neurobiology
XLIX (1984) Recombination at the DNA Level

Authorization to photocopy items for internal or personal use, or the internal or personal use of specific clients, is granted by Cold Spring Harbor Laboratory for libraries and other users registered with the Copyright Clearance Center (CCC) Transactional Reporting Service, provided that the base fee of $1.00 per article is paid directly to CCC, 21 Congress St., Salem MA 01970. [0-87969-050-X/85 $1.00 + .00] This consent does not extend to other kinds of copying, such as copying for general distribution, for advertising or promotional purposes, for creating new collective works, or for resale.

All Cold Spring Harbor Laboratory publications may be ordered directly from Cold Spring Harbor Laboratory, Box 100, Cold Spring Harbor, New York 11724. Phone: 1-800-843-4388   In New York (516)367-8423

Symposium Participants

ACKERMAN, ERIC, NIADDK, National Institutes of Health, Bethesda, Maryland
AKAM, MICHAEL, Dept. of Genetics, University of Cambridge, England
ALEXANDER, STEPHEN, Scripps Clinic and Research Foundation, La Jolla, California
ALPERT, SUSAN, Cold Spring Harbor Laboratory, New York
AMATI, PAOLO, Dept. of Cell Biology, University of Sapienza, Rome, Italy
ANDERSON, DAVID, Institute of Cancer Research, Columbia University, New York, New York
ANDREWS, MATTHEW, Dept. of Embryology, Carnegie Institution of Washington, Baltimore, Maryland
ARTAVANIS-TSAKONAS, SPRYOS, Dept. of Biology, Yale University, New Haven, Connecticut
ARTZT, KAREN, Memorial Sloan-Kettering Cancer Center, New York, New York
ASHBURNER, MICHAEL, Dept. of Genetics, University of Cambridge, England
AVIV, HAIM, Bio-Technology Genetic Corp., New York, New York
AWGNLEWITSCH, ALEXANDER, Dept. of Biology, Yale University, New Haven, Connecticut
AXEL, RICHARD, Institute for Cancer Research, Columbia University, New York, New York
BALTIMORE, DAVID, Whitehead Institute, Cambridge, Massachusetts
BARGMANN, CORNELIA, Whitehead Institute, Cambridge, Massachusetts
BEACH, DAVID, Cold Spring Harbor Laboratory, New York
BEER, MICHAEL, Dept. of Biophysics, Johns Hopkins University, Baltimore, Maryland
BELOTE, JOHN, Dept. of Biology, University of California, San Diego, La Jolla
BENDER, WELCOME, Dept. of Biological Chemistry, Harvard Medical School, Boston, Massachusetts
BENZER, SEYMOUR, Dept. of Biology, California Institute of Technology, Pasadena
BERNSTEIN, SANFORD, Dept. of Biology, San Diego State University, California
BINGHAM, PAUL, Dept. of Biochemistry, State University of New York, Stony Brook
BIRCHMEIER, CARMEN, Cold Spring Harbor Laboratory, New York
BIRNSTEIL, MAX, Dept. of Microbiology, University of Zurich, Switzerland
BISHOP, MICHAEL, University of California, San Francisco
BOHN, MARTHA, Dept. of Neurobiology and Behavior, State University of New York, Stony Brook
BONCINELLI, E., I.I.G.B.-CNR, Naples, Italy
BOTSTEIN, DAVID, Dept. of Biology, Massachusetts Institute of Technology, Cambridge
BRASH, DOUGLAS, NCI, National Institutes of Health, Bethesda, Maryland
BREEDEN, LINDA, Laboratory of Molecular Biology, MRC, Cambridge, England
BRENNAN, SEAN, Dept. of Molecular Embryology, University of Cambridge, England
BRENNER, SYDNEY, MRC, Cambridge, England
BROEK, DANIEL, Cold Spring Harbor Laboratory, New York
BROWN, DONALD, Dept. of Embryology, Carnegie Institution of Washington, Baltimore, Maryland
BRULET, PHILIPPE, Dept. of Molecular Biology, Institut Pasteur, Paris, France
BRYDOLF, BARBARA, Dept. of Biology, University of California, San Diego, La Jolla
BUCK, LINDA, Institute for Cancer Research, Columbia University, New York, New York
CAPPELLO, JOSEPH, Whitehead Institute, Cambridge, Massachusetts
CARROLL, SEAN, Dept. of Biology, University of Colorado, Boulder
CEDAR, HOWARD, Dept. of Molecular Biology, Hebrew University, Jerusalem, Israel
CHAMPER, ROBERT, Dept. of Molecular Biology, Albert Einstein College of Medicine, Bronx, New York
CHUNG, SU-YUN, NCI, National Institutes of Health, Bethesda, Maryland
CLINE, THOMAS, Dept. of Biology, Princeton University, New Jersey
COHEN, STEPHEN, Whitehead Institute, Cambridge, Massachusetts
COLBERG-POLEY, ANAMARIS, Molecular Biology Center, University of Heidelberg, Federal Republic of Germany
COOLEY, LYNN, Dept. of Embryology, Carnegie Institution of Washington, Baltimore, Maryland
COOPER, ASHLEY, Dept. of Biochemistry, Imperial College of Science and Technology, London, England
CORDEN, JEFFERY, Dept. of Molecular Genetics, Johns Hopkins University, Baltimore, Maryland
COSTANTINI, FRANK, Dept. of Human Genetics, Columbia University, New York, New York
COVARRUBIAS, LUIS, Institute for Cancer Research, Philadelphia, Pennsylvania
CREWS, STEPHEN, Dept. of Biological Sciences, Stanford University, California
DALDAL, FEVZI, Cold Spring Harbor Laboratory, New York

DAVIDSON, ERIC H., Dept. of Biology, California Institute of Technology, Pasadena
DAWID, IGOR, NICHD, National Institutes of Health, Bethesda, Maryland
DAWSON, ADAM, Amersham International plc, Buckingham, England
DE ROBERTIS, EDWARD, Dept. of Cell Biology, University of Basel, Switzerland
DEVREOTES, PETER, Dept. of Biological Chemistry, Johns Hopkins University, Baltimore, Maryland
DJIAN, PHILIPPE, Dept. of Physiology and Biophysics, Harvard Medical School, Boston, Massachusetts
DOUBLE, DENNIS, Laboratoire de Génétiqui Moléculaire des Eucaryotes CNRS, Strausbourg, France
EDELMAN, GERALD, Rockefeller University, New York, New York
EMMONS, SCOTT, Dept. of Molecular Biology, Albert Einstein College of Medicine, Bronx, New York
ETKIN, LALURENCE, Dept. of Genetics, M.D. Anderson Hospital, Houston, Texas
EVANS, MARTIN, Dept. of Genetics, University of Cambridge, England
EVANS, RONALD, Salk Institute, San Diego, California
FEDOROFF, NINA, Dept. of Embryology, Carnegie Institution of Washington, Baltimore, Maryland
FIRTEL, RICHARD, Dept. of Biology, University of California, San Diego, La Jolla
FONG, DUNNE, Dept. of Biology, University of Alabama, Birmingham
FRISCHAUF, A.-M., Dept. of Molecular Biology, European Molecular Biology Laboratory, Heidelberg, Federal Republic of Germany
FULLER, MARGARET, Dept. of Biology, University of Colorado, Boulder
GALLIE, BRENDA L., Dept. of Opthamology, Ontario Cancer Institute, Toronto, Canada
GALLIN, WARREN, Rockefeller University, New York, New York
GASIC, GREGORY, Rockefeller University, New York, New York
GEHRING, WALTER, Dept. of Cell Biology, University of Basel, Switzerland
GELBART, WILLIAM, Dept. of Cell and Developmental Biology, Harvard University, Cambridge, Massachusetts
GERISCH, GUNTHER, Dept. of Cell Biology, Max-Planck Institute for Biochemistry, Munich, Federal Republic of Germany
GILBERT, WALTER, Biogen, Cambridge, Massachusetts
GIULIANO, GIOVANNI, Cold Spring Harbor Laboratory, New York
GOUSTIN, SCOTT A., Dept. of Cell Biology, Mayo Clinic, Rochester, Minnesota
GURDON, JOHN, Dept. of Zoology, University of Cambridge, England
HAFEN, ERNST, Dept. of Biochemistry, University of California, Berkeley
HAMILTON, BARBARA, Duke University Medical Center, Durham, North Carolina
HAMMER, ROBERT, Dept. of Reproductive Physiology, University of Pennsylvania School of Veterinary Medicine, Philadelphia
HANAHAN, DOUGLAS, Cold Spring Harbor Laboratory, New York
HARDISON, ROSS, Dept. of Molecular and Cell Biology, Pennsylvania State University, University Park
HARPER, MARY I., Roche Institute of Molecular Biology, Nutley, New Jersey
HARRELSON, ALLEN L., Rockefeller University, New York, New York
HART, CHARLES, Dept. of Biology, Yale University, New Haven, Connecticut
HAWLEY, ROBERT, Institute of Cancer Research, Philadelphia, Pennsylvania
HEILMANN, LARRY, National Institutes of Health, Bethesda, Maryland
HENNING, WOLFGANG, Dept. of Genetics, Catholic University, Nijmegen, The Netherlands
HERBOMEL, PHILIPPE, Dept. of Molecular Biology, Institut Pasteur, Paris, France
HIGASHINAKAGAWA, TORU, Dept. of Biology, Tokyo Metropolitan University, Japan
HIRSH, DAVID, Dept. of Biology, University of Colorado, Boulder
HODGKIN, JONATHAN, Laboratory of Molecular Biology, MRC, Cambridge, England
HOGNESS, DAVID, Dept. of Biochemistry, Stanford University Medical Center, California
HORSCH, ROBERT, Monsanto Corporation, St. Louis, Missouri
HORVITZ, ROBERT, Dept. of Biology, Massachusetts Institute of Technology, Cambridge
HOWARD, KENNETH, Imperial Cancer Research Fund, London, England
HOWE, CHIN, Wistar Institute, Philadelphia, Pennsylvania
HSU, Y.C., Johns Hopkins University, Baltimore, Maryland
INGHAM, P.W., Dept. of Developmental Genetics, Imperial Cancer Research Fund, London, England
JACKLE, HERBERT, Dept. of Biochemistry, Max-Planck Institute, Tubingen, Federal Republic of Germany
JACOB, F., Dept. of Molecular Biology, Institut Pasteur, Paris, France
JAENISCH, RUDOLF, Whitehead Institute, Cambridge, Massachusetts
JOYNER, ALEXANDRA, Dept. of Anatomy, University of California, San Francisco
KAFATOS, FOTIS C., The Biological Laboratories, Harvard University, Cambridge, Massachusetts
KAISER, DALE, Dept. of Biochemistry, Stanford University, California
KASAI, MASATAKA, Dept. of Molecular Immunology, Dana Farber Cancer Institute, Boston, Massachusetts
KATAOKA, TOHRU, Cold Spring Harbor Laboratory, New York
KAUVAR, LAWRENCE, Dept. of Biochemistry, University of California, San Francisco

SYMPOSIUM PARTICIPANTS

KEMLER, ROLF, Friedrich Miescher Laboratory, Max-Planck Institute, Tubingen, Federal Republic of Germany
KENNISON, JAMES, Dept. of Genetics, University of Alberta, Edmonton, Canada
KHILLAN, J.S., NICHD, National Institutes of Health, Bethesda, Maryland
KIBERSTIS, PAULA, *Cell*, Cambridge, Massachusetts
KLAR, AMAR, Cold Spring Harbor Laboratory, New York
KLEIN, PETER, Dept. of Biological Chemistry, Johns Hopkins University School of Medicine, Baltimore, Maryland
KORN, LAURENCE, Dept. of Genetics, Stanford University, California
KORNBERG, T., Dept. of Biochemistry and Biophysics, University of California, San Francisco
KROOS, LEE, Dept. of Biochemistry, Stanford University Medical Center, California
KRUMLAUF, ROBB, Institute for Cancer Research, Philadelphia, Pennsylvania
LACY, ELIZABETH, Memorial Sloan-Kettering Cancer Center, New York, New York
LANG-UNNASCH, N., Dept. of Molecular Biology, Massachusetts General Hospital, Boston
LASKEY, RON, Laboratory of Molecular Biology, MRC, Cambridge, England
LAUER, JOYCE, Genetics Institute, Cambridge, Massachusetts
LAUGHON, ALLEN, Dept. of Biology, University of Colorado, Boulder
LAWRENCE, PETER, Laboratory of Molecular Biology, MRC, Cambridge, England
LEDER, PHILIP, Dept. of Genetics, Harvard Medical School, Boston, Massachusetts
LEE, C.S., Dept. of Zoology, University of Texas, Austin
LEMKE, GREG E., Institute of Cancer Research, New York, New York
LENNARZ, WILLIAM, M.D. Anderson Hospital, Houston, Texas
LEVINE, ARNOLD, Dept. of Molecular Biology, Princeton University, New Jersey
LEVINE, MICHAEL, Dept. of Biological Sciences, Columbia University, New York, New York
LEWIN, BENJAMIN, *Cell*, Cambridge, Massachusetts
LEWIS, E., Dept. of Biology, California Institute of Technology, Pasadena
LIPSHITZ, HOWARD, Dept. of Biochemistry, Stanford University, California
LOGAN, SUSAN, Brandeis University, Waltham, Massachusetts
LOOMIS, WILLIAM, Dept. of Biology, University of California, San Diego, La Jolla
LOSICK, RICHARD, The Biological Laboratories, Harvard University, Cambridge, Massachusetts
LOUIS, CHRISTOS, Dept. of Biology, University of Crete, Greece
LOVELL-BADGE, ROBIN, Mammalian Development Unit, MRC, London, England
LYON, M.F., Radiobiology Unit, MRC, Oxon, England
MACWILLIAMS, HARRY, Dept. of Zoology, University of Munich, Federal Republic of Germany
MAELICKE, A., Max-Planck Institut, Dortmund, Federal Republic of Germany
MAHON, KATHLEEN, NICHD, National Institutes of Health, Bethesda, Maryland
MAINI, CLAUDE, Brandeis University, Waltham, Massachusetts
MALMBERG, RUSSELL, Cold Spring Harbor Laboratory, New York
MANIATIS, THOMAS, Dept. of Biochemistry and Molecular Biology, Harvard University, Cambridge, Massachusetts
MANN, RICHARD, Whitehead Institute, Cambridge, Massachusetts
MARTIN, GAIL, Dept. of Anatomy, University of California, San Francisco
MASON, IVOR, National Institute for Medical Research, London, England
MAXWELL, GERALD D., Dept. of Anatomy, University of Connecticut Health Center, Farmington
MCGHEE, JAMES, Dept. of Medical Biochemistry, University of Calgary, Canada
MCGINNIS, WILLIAM, Dept. of Molecular Biophysics and Biochemistry, Yale University, New Haven, Connecticut
MCLAREN, ANNE, Mammalian Development Unit, MRC, London, England
MELTON, DOUGLAS, Dept. of Biochemistry and Molecular Biology, Harvard University, Cambridge, Massachusetts
MEYEROWITZ, ELLIOT, Dept. of Biology, California Institute of Technology, Pasadena
MILLER, KATHY, Dept. of Biochemistry and Biophysics, University of California, San Francisco
MINTZ, BEATRICE, Fox Chase Cancer Center, Philadelphia, Pennsylvania
MITSIALIS, ALEX, The Biological Laboratories, Harvard University, Cambridge, Massachusetts
MITSIS, PAUL, Brandeis University, Waltham, Massachusetts
MOHIER, ELIANE, University of Strasbourg, France
MONMANEY, TERRENCE, *Science*, Washington, DC
MOORE, ROBERT, Monsanto Corp., Chesterfield, Missouri
MORATA, G., Dept. of Molecular Science, University of Madrid, Spain
MOSCHONAS, NIKOS, Dept. of Biology, University of Crete, Greece
MUGLIA, LISA, Dept. of Cell Biology and Anatomy, Northwestern University, Chicago, Illinois
MUGLIA, LOUIS, University of Chicago, Illinois
MULLIGAN, RICHARD, Dept. of Biology, Whitehead Institute, Cambridge, Massachusetts
MURPHY, DAVID, Dept. of Molecular Embryology, National Institute for Medical Research, London, England
NASSER, DELILL, National Science Foundation, Washington, DC

NEWPORT, JOHN, Dept. of Biology, University of California, San Diego, La Jolla
NGUYEN, HANH, The Biological Laboratories, Harvard University, Cambridge, Massachusetts
NGUYEN-HUU, CHI, Dept. of Microbiology, Columbia University, New York, New York
NICOLAS, J.F., Dept. of Molecular Biology, Institut Pasteur, Paris, France
NOLL, MARKUS, Dept. of Cell Biology, University of Basel, Switzerland
NÖTHIGER, ROLF, Zoological Institute, University of Zurich, Switzerland
NUSSLEIN-VOLHARD, J., Federal Research Institute for Animal Viruses, Tubingen, Federal Republic of Germany
O'FARRELL, PATRICK, Dept. of Biology, University of California, San Francisco
ORNITZ, DAVID, Dept. of Biochemistry, University of Washington, Seattle
OVERBEEK, PAUL, National Institutes of Health, Bethesda, Maryland
OVERTON, CHRISTIAN, Wistar Institute, Philadelphia, Pennsylvania
PARKS, SUKI, Dept. of Embryology, Carnegie Institution of Washington, Baltimore, Maryland
PARO, RANATO, Dept. of Biochemistry, Stanford University, California
PEARSON, MARK, E.I. Dupont de Nemours, Wilmington, Delaware
PETERSON, MARTHA, Institute for Cancer Research, Philadelphia, Pennsylvania
PHILLIPS, R.A., Ontario Cancer Institute, Toronto, Canada
PIERANDREI-AMALDI, PAOLA, Institute for Cellular Biology, CNR, Rome, Italy
POIRIER, FRANCOIS, Imperial College of Science and Technology, London, England
POWERS, SCOTT, Cold Spring Harbor Laboratory, New York
RASSOULZADEGAN, M., Dept. of Biochemistry, University of Nice, France
REED, STEVE, Dept. of Biological Sciences, University of California, Santa Barbara
REINHARDT, SIGRID, University of Essen, Federal Republic of Germany
REYMOND, C., Dept. of Biology, University of California, San Diego, La Jolla
RICHARDS, G., Laboratoire de Génétique Moléculaire des Eucaryotes CNRS, Strasbourg, France
RIGBY, PETER, Dept. of Biochemistry, Imperial College of Science, London, England
ROBERTSON, M., *Nature*, London, England
ROBINOW, STEVEN, Dept. of Biology, Brandeis University, Waltham, Massachusetts
ROSS, SUSAN, Dept. of Biochemistry, University of Illinois, Chicago
RUAN, KE-SAN, Dept. of Molecular Biology, Albert Einstein College of Medicine, Bronx, New York
RUBEN, MARTHA, Dept. of Medicine, Ottawa, Canada

RUBIN, G., Dept. of Biochemistry, University of California, Berkeley
RUDDLE, FRANK, Dept. of Biology, Yale University, New Haven, Connecticut
RUDERMAN, JOAN, Dept. of Anatomy and Cell Biology, Harvard Medical School, Boston, Massachusetts
RUSSELL, M.A., Dept. of Genetics, University of Alberta, Edmonton, Canada
SASSONE-CORSI, PAOLO, Dept. of Chemical Biology, University of Strasbourg, France
SCHEDL, PAUL, Dept. of Biology, Princeton University, New Jersey
SCHELL, JEFF, Max-Planck Institute, Cologne, Federal Republic of Germany
SCHIERENBERG, EINHARD, Max-Planck Institute for Experimental Medicine, Gottingen, Federal Republic of Germany
SCOLNIK, PABLO, Cold Spring Harbor Laboratory, New York
SCOTT, MATTHEW, Dept. of Biology, University of Colorado, Boulder
SCOTT, RICHARD, E.I. Dupont de Nemours Company, Wilmington, Delaware
SEDIVY, JOHN M., Massachusetts Institute of Technology, Cambridge
SHAPIRO, LUCY, Dept. of Molecular Biology, Albert Einstein College of Medicine, Bronx, New York
SHIN, HEE-SUP, Memorial Sloan-Kettering Cancer Center, New York, New York
SINGH, HARINDER, Massachusetts Institute of Technology, Cambridge
SIU, CHI-HUNG, University of Toronto, Canada
SMOLIK-UTLAUT, SARAH, Dept. of Biology, California Institute of Technology, Pasadena
SOBIESKI, DONNA A., National Institutes of Health, Bethesda, Maryland
SOELLER, WALTER, University of California, San Francisco
SOLTER, DAVOR, Wistar Institute, Philadelphia, Pennsylvania
SORIANO, PHILIPPE, Whitehead Institute, Cambridge, Massachusetts
SPADARO, ANTONELLA, Dana Farber Cancer Institute, Boston, Massachusetts
SPRADLING, ALLAN, Dept. of Embryology, Carnegie Institution of Washington, Baltimore, Maryland
STEIN, REUBEN, Institute for Cancer Research, Columbia University, New York, New York
STELLER, HERMAN, Dept. of Biochemistry, University of California, Berkeley
STERN, PETER, Dept. of Immunology, University of Liverpool, England
STEVENS, LESLIE, Dept. of Neurobiology, Harvard Medical School, Boston, Massachusetts
STEWARD, RUTH, Dept. of Biology, Princeton University, New Jersey
STILLMAN, BRUCE, Cold Spring Harbor Laboratory, New York

STRAUS, DONALD, Massachusetts General Hospital, Boston
STRUHL, KEVIN, Dept. of Biological Chemistry, Harvard Medical School, Boston, Massachusetts
SWAROOP, ANAND, Dept. of Molecular Biochemistry and Biophysics, Yale University, New Haven, Connecticut
SWENSON, KATHERINE, Dept. of Anatomy and Cell Biology, Harvard Medical School, Boston, Massachusetts
TAKETO, MAKOTO, Jackson Laboratory, Bar Harbor, Maine
TAMRICH, MILAN, NCI, National Institutes of Health, Bethesda, Maryland
TERAO, MINEKO, Institute for Cancer Research, Fox Chase, Philadelphia, Pennsylvania
TERHORST, COX, Dana Farber Cancer Institute, Boston, Massachusetts
TILGHMAN, SHIRLEY, Fox Chase Cancer Center, Philadelphia, Pennsylvania
TJIAN, ROBERT, Dept. of Biochemistry, University of California, Berkeley
TODE, TAKASHI, Cold Spring Harbor Laboratory, New York
TURNER, DAVID, Dept. of Neuroscience, Harvard Medical School, Boston, Massachusetts
VAN DYKE, TERRY, Dept. of Molecular Biology, Princeton University, New Jersey
VERMA, INDER, Salk Institute, San Diego, California
VOGT, THOMAS, Institute for Cancer Research, Philadelphia, Pennsylvania
WAGNER, ERWIN, European Molecular Biological Laboratory, Heidelberg, Federal Republic of Germany
WASSARMAN, PAUL, Roche Institute for Molecular Biology, Nutley, New Jersey
WATSON, CHRISTINE, Dept. of Biochemistry, Imperial College London, England
WEINHEIMER, STEVEN, Dept. of Embryology, Carnegie Institution of Washington, Baltimore, Maryland
WEINTRAUB, H., Fred Hutchison Cancer Center, Seattle, Washington
WEIR, MICHAEL, Dept. of Biochemistry, University of California, San Francisco
WENSINK, PEITER, Brandeis University, Waltham, Massachusetts
WESTPHAL, HEINDER, NICHD, National Institutes of Health, Bethesda, Maryland
WIESCHAUS, ERIC, Dept. of Biology, Princeton University, New Jersey
WIGLER, MICHAEL, Cold Spring Harbor Laboratory, New York
WILLISON, KEITH R., Institute of Cancer Research, London, England
WILSON, K., Dept. of Biochemistry and Biophysics, University of California, San Francisco
WOLGEMUTH, DEBRA, Dept. of Genetics and Development, Columbia University College of Physicians & Surgeons, New York, New York
WOOD, WILLIAM, Dept. of Biology, University of Colorado, Boulder
WOOD, JOHN, Lilly Research Laboratories, Indianapolis, Indiana
WRIGHT, PAUL, Dept. of Biochemistry, University of Iowa, Iowa City
WU, CARL, NCI, National Institutes of Health, Bethesda, Maryland
WYLIE, C.C., Dept. of Anatomy, St. George's Hospital Medical School, London, England
YOUNG, MICHAEL, Rockefeller University, New York, New York
ZACHAR, ZUZANA, Dept. of Biochemistry, State University of New York, Stony Brook

First row: J.D. Watson; 50th anniversary cake; S. Brenner
Second row: D. Hanahan; F. Jacob, M. Birnsteil, S. Brenner; D. Hogness
Third row: A. Spradling; H. Westphal, A. Hershey; P. Sassone-Corsi
Fourth row: Wine and cheese party

First row: G. Martin; F. Ruddle; I. Verma
Second row: J. Gurdon, A. McLaren; J. Schell, N. Fedoroff; D. Kaiser
Third row: 50th anniversary portrait

First row: M. Ashburner; D. Baltimore, R. Jaenisch; F. Jacob
Second row: L. Shapiro, R. Losick; S. Benzer being interviewed for documentary on Cold Spring Harbor Laboratory
Third row: D. Baltimore, G. Rubin; R. Axel, R. Mulligan; W. Gehring, H. Westphal

First row: M. Bishop; M.F. Lyon, R. Jaenisch; R. Steward
Second row: F. Jacob, W. Gilbert, L. Shapiro, D. Hogness; G. Rubin, K. Struhl
Third row: E. Lewis; Coffee break; S. Brenner, D. Botstein
Fourth row: T. Higashinakagawa; Cocktail party

First row: M. Birnsteil; Lunch break; B. Mintz
Second row: W. Gilbert; End of a session; S. Benzer
Third row: Coffee break; A. McLaren; Picnic
Fourth row: Between sessions; P. O'Farrell; Lunch on Blackford lawn

Foreword

Developmental biologists are an audacious breed. Their ambition is to provide an intellectually satisfying account of the forces that guide the development of multicellular organisms around the circle from fertilized egg to embryo to adult to gamete. These processes, however, lie close to the border beyond which genetics may not be feasible nor necessarily useful; furthermore, meaningful biochemical studies pose extraordinary challenges because of the small amounts of material that are available and the general rapidity of events in the early embryo. In the face of these difficulties it is not surprising that for many years developmental biology was essentially a branch of anatomy, in that it was exclusively descriptive in nature. Accurate and elegant though these descriptions were, they led to an understanding of developmental processes less frequently than to an amazement at their beauty.

In recent years, however, developmental biology has undergone a dramatic change and has matured from a descriptive to an analytical science. There is no doubt in my mind that this change stems almost entirely from two technical advances—the ability to use molecular cloning to isolate and characterize wild-type and mutant versions of genes that control or are expressed at specific developmental stages and, second, the ability to generate transgenic organisms in which the expression of the introduced gene(s) is correct both spatially and temporally. In consequence, the developmental biologist has not only the capacity to describe but also now to analyze and influence the events that guide a fertilized egg to its destiny. Our choice of the topic molecular embryology for this year's Symposium—the 50th—celebrates this new-found freedom.

The breadth of the program meant that I had to seek much outside help. In particular, I wish to thank John Gurdon, Gerry Rubin, Sydney Brenner, David Ish-Horowicz, Doug Hanahan, Don Brown, Paul Bingham, Tom Maniatis, and Paul Wassarman. The traditional Introduction, given on the first evening by John Gurdon, was an appropriate mixture of philosophy, history, and foresight. We then heard talks from Seymour Benzer, Sydney Brenner, and Anne McLaren who reminisced about earlier Cold Spring Harbor Symposia and spoke about aspects of developmental biology of particular interest to them. The 50th Symposium therefore started on a high intellectual plane that was sustained through the 101 formal presentations given during the next 6 days. Despite this length, most of the 254 people attending the Symposium stayed for the entire meeting. The final talk was given by Gerry Rubin who provided an excellent summary of the highlights of the meeting.

Much-needed financial support for this year's meeting was provided in part by the National Science Foundation, the Department of Energy, and the National Cancer Institute, National Institutes of Health. These funds are used to help defray travel expenses and meeting costs for our invited speakers and overseas guests.

Contributions from the following Corporate Sponsors provide core support for the Cold Spring Harbor meetings program: Agrigenetics Corporation; American Cyanamid Company; Amersham International plc; Becton Dickinson and Company; Biogen S.A.; Cetus Corporation; Ciba-Geigy Corporation; CPC International, Inc.; E.I. du Pont de Nemours & Company; Genentech, Inc.; Genetics Institute; Hoffmann-La Roche Inc.; Johnson & Johnson; Eli Lilly and Company; Mitsui Toatsu Chemicals, Inc.; Monsanto Company; Pall Corporation; Pfizer Inc.; Schering-Plough Corporation; Smith Kline & French Laboratories; and Upjohn Company.

We wish to thank the Meetings Office staff—Gladys Kist, Barbara Ward, Maureen Berejka, and Micki McBride—for making everyone feel welcome and for overseeing the registration and housing of Symposium participants. Herb Parsons arranged the audiovisual setups for the meetings, and Mike Ockler, Dave Greene, and Sue Zehl provided art and photography services, including our 50th anniversary portrait. Our especial thanks go to Marilyn Goodwin for handling the large volume of correspondence involved in organizing the meeting. Rapid publication of this volume is due to the efforts of Nancy Ford, Director of Publications, and of editors Judy Cuddihy and Dorothy Brown.

Joe Sambrook
June 1985

Contents

Symposium Participants	v
Foreword	xv

Introduction

J.B. Gurdon — 1

Nuclear/Cytoplasmic Interactions in Early Development

The Mouse Egg's Receptor for Sperm: What Is It and How Does It Work? *P.M. Wassarman, J.D. Bleil, H.M. Florman, J.M. Greve, R.J. Roller, G.S. Salzmann, and F.G. Samuels* — 11

Localized Maternal mRNAs in *Xenopus laevis* Eggs *D.L. Weeks, M.R. Rebagliati, R.P. Harvey, and D.A. Melton* — 21

Altered Morphogenesis and Its Effects on Gene Activity in *Xenopus laevis* Embryos *M. Jamrich, T.D. Sargent, and I.B. Dawid* — 31

Germ Plasm and Germ Cell Determination in *Xenopus laevis* as Studied by Cell Transplantation Analysis *C.C. Wylie, S. Holwill, M. O'Driscoll, A. Snape, and J. Heasman* — 37

Nuclear Transfer in Mouse Embryos: Activation of the Embryonic Genome *D. Solter, J. Aronson, S.F. Gilbert, and J. McGrath* — 45

Molecular Analysis of the First Differentiations in the Mouse Embryo *P. Brûlet, P. Duprey, M. Vasseur, M. Kaghad, D. Morello, P. Blanchet, C. Babinet, H. Condamine, and F. Jacob* — 51

Cell Determination during Early Embryogenesis of the Nematode *Caenorhabditis elegans* *E. Schierenberg* — 59

Genes Affecting Early Development in *Caenorhabditis elegans* *D. Hirsh, K.J. Kemphues, D.T. Stinchcomb, and R. Jefferson* — 69

Studies on the Cytoplasmic Organization of Early *Drosophila* Embryos *K.G. Miller, T.L. Karr, D.R. Kellogg, I. Mohr, M. Walter, and B.M. Alberts* — 79

Developmental Expression of Cell-surface (Glyco)Proteins Involved in Gastrulation and Spicule Formation in Sea Urchin Embryos *S.R. Grant, M.C. Farach, G.L. Decker, H.D. Woodward, H.A. Farach, and W.J. Lennarz* — 91

Lineage and Segmentation/Pattern Formation

Genes That Affect Cell Fates during the Development of *Caenorhabditis elegans* *W. Fixsen, P. Sternberg, H. Ellis, and R. Horvitz* — 99

Bicaudal Mutations of *Drosophila melanogaster*: Alteration of Blastoderm Cell Fate *J. Mohler and E.F. Wieschaus* — 105

Pattern Formation in the Muscles of *Drosophila* *P.A. Lawrence* — 113

The Decapentaplegic Gene Complex in *Drosophila melanogaster* *W.M. Gelbart, V.F. Irish, R.D. St. Johnston, F.M. Hoffmann, R. Blackman, D. Segal, L.M. Posakony, and R. Grimaila* — 119

Isolation and Structural Analysis of the Extra Sex Combs Gene of *Drosophila* *E. Frei, D. Bopp, M. Burri, S. Baumgartner, J.-E. Edström, and M. Noll* — 127

Molecular and Genetic Analysis of the hairy Locus in *Drosophila* *D. Ish-Horowicz, K.R. Howard, S.M. Pinchin, and P.W. Ingham* — 135

Genes Affecting the Segmental Subdivision of the *Drosophila* Embryo *C. Nüsslein-Volhard, H. Kluding, and G. Jürgens* — 145

Homeotic Mutants

Regulation of the Genes of the Bithorax Complex in *Drosophila* E.B. Lewis	155
Anatomy and function of the Bithorax Complex of *Drosophila* E. Sánchez-Herrero, J. Casanova, S. Kerridge, and G. Morata	165
Domains of *Cis*-interaction in the Bithorax Complex W. Bender, B. Weiffenbach, F. Karch, and M. Peifer	173
Regulation and Products of the *Ubx* Domain of the Bithorax Complex D.S. Hogness, H.D. Lipshitz, P.A. Beachy, D.A. Peattie, R.B. Saint, M. Goldschmidt-Clermont, P.J. Harte, E.R. Gavis, and S.L. Helfand	181
Function and Expression of Ultrabithorax in the *Drosophila* Embryo M.E. Akam, A. Martinez-Arias, R. Weinzier, and C.D. Wilde	195
Genetic Control of the Spatial Pattern of Selector Gene Expression in *Drosophila* P.W. Ingham	201
Expression of the Homeo Box Gene Family in *Drosophila* M. Levine, K. Harding, C. Wedeen, H. Doyle, T. Hoey, and H. Radomska	209
Expression of the Dorsal Gene R. Steward, L. Ambrosio, and P. Schedl	223
The engrailed Locus of *Drosophila melanogaster*: Genetic, Developmental, and Molecular Studies Z. Ali, B. Drees, K.G. Coleman, E. Gustavson, T.L. Karr, L.M. Kauvar, S.J. Poole, W. Speller, M.P. Weir, and T. Kornberg	229
Embryonic Pattern in *Drosophila*: The Spatial Distribution and Sequence-specific DNA Binding of engrailed Protein P.H. O'Farrell, C. Desplan, S. DiNardo, J.A. Kassis, J.M. Kuner, E. Sher, J. Theis, and D. Wright	235

Homeo Boxes

Homeotic Genes, the Homeo Box, and the Genetic Control of Development W.J. Gehring	243
Common Properties of Proteins Encoded by the Antennapedia Complex Genes of *Drosophila melanogaster* A. Laughon, S.B. Carroll, F.A. Storfer, P.D. Riley, and M.P. Scott	253
Homeo Box Sequences of the Antennapedia Class Are Conserved Only in Higher Animal Genomes W. McGinnis	263
The *Xenopus* Homeo Boxes E.M. DeRobertis, A. Fritz, J. Goetz, G. Martin, I.W. Mattaj, E. Salo, G.D. Smith, C. Wright, and R. Zeller	271
Mammalian Homeo Box Genes F.H. Ruddle, C.P. Hart, A. Awgulewitsch, A. Fainsod, M. Utset, D. Dalton, N. Kerk, M. Rabin, A. Ferguson-Smith, A. Fienberg, and W. McGinnis	277
Expression of Murine Genes Containing Homeo Box Sequences during Visceral and Parietal Endoderm Differentiation of Embryonal Carcinoma Stem Cells A.M. Colberg-Poley, S.D. Voss, and P. Gruss	285
Structure and Expression of Two Classes of Mammalian Homeo-box-containing Genes A. Joyner, C. Hauser, T. Kornberg, R. Tjian, and G. Martin	291
Human cDNA Clones Containing Homeo Box Sequences E. Boncinelli, A. Simeone, A. LaVolpe, A. Faiella, V. Fidanza, D. Acampora, and L. Scotto	301

Tissue Specificity/Position Effects

The *Ac* and *Spm* Controlling Element Families in Maize J. Banks, J. Kingsbury, V. Raboy, J.W. Schiefelbein, O. Nelson, and N. Fedoroff	307
Regulation of Tc*1* Transposable Elements in *Caenorhabditis elegans* S.W. Emmons, K. Ruan, A. Levitt, and L. Yesner	313
Lineage-specific Gene Expression in the Sea Urchin Embryo E.H. Davidson, C.N. Flytzanis, J.J. Lee, J.J. Robinson, S.J. Rose III, and H.M. Sucov	321
Germ Line Specificity of P-element Transposition and Some Novel Patterns of Expression of Transduced Copies of the white Gene G.M. Rubin, T. Hazelrigg, R.E. Karess, F.A. Laski, T. Laverty, R. Levis, D.C. Rio, F.A. Spencer, and C.S. Zuker	329
On the Molecular Basis of Transvection Effects and the Regulation of Transcription Z. Zachar, C.H. Chapman, and P.M. Bingham	337

The 68C Glue Puff of *Drosophila* E.M. Meyerowitz, M.A. Crosby, M.D. Garfinkel, C.H. Martin, P.H. Mathers, and K. VijayRaghavan	347
Hybrid Genes in the Study of Glue Gene Regulation in *Drosophila* M. Bourouis and G. Richards	355

Expression of Genes Introduced into Transgenic Mice

Developmental Regulation of Human Globin Genes in Transgenic Mice F. Costantini, G. Radice, J. Magram, G. Stamatoyannopoulos, T. Papayannopoulou, and K. Chada	361
Regulated Expression of α-Fetoprotein Genes in Transgenic Mice R. Krumlauf, R.E. Hammer, R. Brinster, V.M. Chapman, and S.M. Tilghman	371
Use of Gene Transfer to Increase Animal Growth R.E. Hammer, R.L. Brinster, and R.D. Palmiter	379
Inducible and Developmental Control of Neuroendocrine Genes R.M. Evans, C. Weinberger, S. Hollenberg, L. Swanson, C. Nelson, and M.G. Rosenfeld	389
Elastase I Promoter Directs Expression of Human Growth Hormone and SV40 T Antigen Genes to Pancreatic Acinar Cells in Transgenic Mice D.M. Ornitz, R.D. Palmiter, A. Messing, R.E. Hammer, C.A. Pinkert, and R.L. Brinster	399
Promoter Sequences of Murine αA Crystallin, Murine α2(I) Collagen or of Avian Sarcoma Virus Genes Linked to the Bacterial Chloramphenicol Acetyl Transferase Gene Direct Tissue-specific Patterns of Chloramphenicol Acetyl Transferase Expression in Transgenic Mice H. Westphal, P.A. Overbeek, J.S. Khillan, A.B. Chepelinsky, A. Schmidt, K.A. Mahon, K.E. Bernstein, J. Piatigorsky, and B. de Crombrugghe	411
Studies of Immunodifferentiation using Transgenic Mice D. Baltimore, R. Grosschedl, D. Weaver, F. Costantini, and T. Imanishi-Kari	417
Transfer and Regulation of Expression of Chimeric Genes in Plants J. Schell, H. Kaulen, F. Kreuzaler, P. Eckes, S. Rosahl, L. Willmitzer, A. Spena, B. Baker, L. Herrera-Estrella, and N. Fedoroff	421
Transgenic Plants R.B. Horsch, S.G. Rogers, and R.T. Fraley	433

Induced Developmental Defects

Retroviruses and Insertional Mutagenesis R. Jaenisch, M. Breindl, K. Harbers, D. Jähner, and J. Löhler	439
Early Developmental Mutations Due to DNA Rearrangements in Transgenic Mouse Embryos L. Covarrubias, Y. Nishida, and B. Mintz	447
An Insertional Mutation in a Transgenic Mouse Line Results in Developmental Arrest at Day 5 of Gestation W.H. Mark, K. Signorelli, and E. Lacy	453
Molecular Analysis of Krüppel, a Segmentation Gene of *Drosophila melanogaster* H. Jäckle, U.B. Rosenberg, A. Preiss, E. Seifert, D. Knipple, A. Kienlin, and R. Lehmann	465
Genetics of Polyamine Synthesis in Tobacco: Developmental Switches in the Flower R.L. Malmberg, J. McIndoo, A.C. Hiatt, and B.A. Lowe	475

Control of Gene Expression

spo0H: A Developmental Regulatory Gene for Promoter Utilization in *Bacillus subtilis* P. Zuber and R. Losick	483
Constitutive and Coordinately Regulated Transcription of Yeast Genes: Promoter Elements, Positive and Negative Regulatory Sites, and DNA Binding Proteins K. Struhl, W. Chen, D.E. Hill, I.A. Hope, and M.A. Oettinger	489
A Simple Gene with a Complex Pattern of Transcription: The Alcohol Dehydrogenase Gene of *Drosophila melanogaster* C. Savakis and M. Ashburner	505
Identification of DNA Sequences Required for the Regulation of *Drosophila* Alcohol Dehydrogenase Gene Expression J.W. Posakony, J.A. Fischer, and T. Maniatis	515
Developmental Control of *Drosophila* Yolk Protein 1 Gene by cis-Acting DNA Elements P.C. Wensink, B. Shepherd, M.J. Garabedian, and M.-C. Hung	521

Localization of Sequences Regulating *Drosophila* Chorion Gene Amplification and Expression *L. Kalfayan, T. Orr-Weaver, S. Parks, B. Wakimoto, D. de Cicco, and A. Spradling* 527

Studies on the Developmentally Regulated Expression and Amplification of Insect Chorion Genes *F.C. Kafatos, S.A. Mitsialis, N. Spoerel, B. Mariani, J.R. Lingappa, and C. Delidakis* 537

The Molecular Basis of Differential Gene Expression of Two 5S RNA Genes *D.D. Brown and M.S. Schlissel* 549

Symbiotic Nitrogen Fixation: Developmental Genetics of Nodule Formation *N. Lang-Unnasch, K. Dunn, and F.M. Ausubel* 555

Sex Determination

Master Regulatory Loci in Yeast and Lambda *I. Herskowitz* 565

Aspects of Dosage Compensation and Sex Determination in *Caenorhabditis elegans* *W.B. Wood, P. Meneely, P. Schedin, and L. Donahue* 575

Sex Determination Pathway in the Nematode *Caenorhabditis elegans*: Variations on a Theme *J. Hodgkin, T. Doniach, and M. Shen* 585

Sex-lethal, A Link between Sex Determination and Sexual Differentiation in *Drosophila melanogaster* *E.M. Maine, H.K. Salz, P. Schedl, and T.W. Cline* 595

Control of Sexual Differentiation in *Drosophila melanogaster* *J.M. Belote, M.B. McKeown, D.J. Andrew, T.N. Scott, and B.S. Baker* 605

A Single Principle for Sex Determination in Insects *R. Nöthiger and M. Steinmann-Zwicky* 615

Sex Determination in Mice *A. McLaren* 623

Cell-cycle Effects

Genetic and Molecular Analysis of Division Control in Yeast *S.I. Reed, M.A. de Barros Lopes, J. Ferguson, J.A. Hadwiger, J.-Y. Ho, R. Horwitz, C.A. Jones, A.T. Lörincz, M.D. Mendenhall, T.A. Peterson, S. Richardson, and C. Wittenburg* 627

Control of Sexual Differentiation by the *ran1$^+$* Gene in Fission Yeast *D. Beach* 635

Regulation of the Yeast *HO* Gene *L. Breeden and K. Nasmyth* 643

The Role of Mitotic Factors in Regulating the Timing of the Midblastula Transition in *Xenopus* *J. Newport, T. Spann, J. Kanki, and D. Forbes* 651

Chromosome Replication in Early *Xenopus* Embryos *R.A. Laskey, S.E. Kearsey, M. Mechali, C. Dingwall, A.D. Mills, S.M. Dilworth, and J. Kleinschmidt* 657

Regulation of Histone Gene Expression *M. Busslinger, D. Schümperli, and M.L. Birnstiel* 665

Pluripotent Cells/Oncogenes

A Comparison of Several Lines of Transgenic Mice Containing the SV40 Early Genes *T. Van Dyke, C. Finlay, and A.J. Levine* 671

Persistence and Transmission in the Mouse of Autonomous Plasmids Derived from Polyoma Virus *M. Rassoulzadegan and J. Vailly* 679

The Ability of EK Cells to Form Chimeras after Selection of Clones in G418 and Some Observations on the Integration of Retroviral Vector Proviral DNA into EK Cells *M.J. Evans, A. Bradley, M.R. Kuehn, and E.J. Robertson* 685

Gene Transfer into Murine Stem Cells and Mice using Retroviral Vectors *E.F. Wagner, G. Keller, E. Gilboa, U. Rüther, and C. Stewart* 691

The Regulation of Gene Expression in Murine Teratocarcinoma Cells *C.M. Gorman, D.P. Lane, C.J. Watson, and P.W.J. Rigby* 701

Transformation of Embryonic Stem Cells with a Human Type-II Collagen Gene and Its Expression in Chimeric Mice *R.H. Lovell-Badge, A.E. Bygrave, A. Bradley, E. Robertson, M.J. Evans, and K.S.E. Cheah* 707

Introduction of Genes into Embryonal Carcinoma Cells and Preimplantation Embryos by Retroviral Vectors *J.F. Nicolas, J.L.R. Rubenstein, C. Bonnerot, and F. Jacob* 713

Conservation and Divergence of *RAS* Protein Function during Evolution C. Birchmeier, D. Broek, T. Toda, S. Powers, T. Kataoka, and M. Wigler	721
Proto-oncogenes of *Drosophila melanogaster* J.M. Bishop, B. Drees, A.L. Katzen, T.B. Kornberg, and M.A. Simon	727
Proto-oncogene *fos* Is Expressed during Development, Differentiation, and Growth J. Deschamps, R.L. Mitchell, F. Meijlink, W. Kruijer, D. Schubert, and I.M. Verma	733
Viral Enhancer Activity in Teratocarcinoma Cells P. Sassone-Corsi, D. Duboule, and P. Chambon	747
Coordinate Expression of Myogenic Functions and Polyoma Virus Replication A. Felsani, R. Maione, L. Ricci, and P. Amati	753

Cellular Differentiation

Structure and Regulated Transcription of DIRS-1: An Apparent Retrotransposon of *Dictyostellium discoideum* J. Cappello, K. Handelsman, S.M. Cohen, and H.F. Lodish	759
Regulation of Cell-type-specific Differentiation in *Dictyostelium* W.F. Loomis	769
Two Feedback Loops May Regulate Cell-type Proportions in *Dictyostelium* H. MacWilliams, A. Blaschke, and I. Prause	779
cAmp Receptors Controlling Cell-Cell Interactions in the Development of *Dictyostelium* P. Klein, D. Fontana, B. Knox, A. Theibert, and P. Devreotes	787
Regulation of Cell-type-specific Gene Expression in *Dictyostelium* R. Gomer, S. Datta, M. Mehdy, T. Crowley, A. Sivertson, W. Nellen, C. Reymond, S. Mann, and R.A. Firtel	801
Early *Dictyostelium* Development: Control Mechanisms Bypassed by Sequential Mutagenesis G. Gerisch, J. Hagmann, P. Hirth, C. Rossier, U. Weinhart, and M. Westphal	813
Cell Interactions Govern the Temporal Pattern of *Myxococcus* Development D. Kaiser, L. Kroos, and A. Kuspa	823
Temporal and Spatial Control of Flagellar and Chemotaxis Gene Expression during *Caulobacter* Cell Differentiation R. Champer, R. Bryan, S.L. Gomes, M. Purucker, and L. Shapiro	831

Developmental Neurobiology

Molecular Genetics of *Drosophila* Neurogenesis B. Yedvobnick, M.A.T. Muskavitch, K.A. Wharton, M.E. Halpern, E. Paul, B.G. Grimwade, and S. Artavanis-Tsakonas	841
Gene Expression in Differentiating and Transdifferentiating Neural Crest Cells D.J. Anderson, R. Stein, and R. Axel	855
A Biological Clock in *Drosophila* M.W. Young, F.R. Jackson, H.S. Shin, and T.A. Bargiello	865
Cell Adhesion Molecule Expression and the Regulation of Morphogenesis G.M. Edelman	877
Neurogenesis in Grasshopper and fushi tarazu *Drosophila* Embryos C.Q. Doe and Corey S. Goodman	891

Summary

G.M. Rubin	905
Author Index	909
Subject Index	911

COLD SPRING HARBOR SYMPOSIA ON QUANTITATIVE BIOLOGY

VOLUME L

Introductory Comments

J.B. GURDON

CRC Molecular Embryology Unit, Department of Zoology, University of Cambridge, Cambridge CB2 3EJ, England

THE PAST

For many centuries, a central question in developmental biology has been how a fertilized egg with little apparent internal organization is so quickly converted, without external instructions, into a highly organized functional larva. This question only became accessible to concerted experimental analysis toward the end of the 19th century, when, as in most areas of science, progress was limited much more by available methods than by concepts. Until about 100 years ago, analysis was largely restricted to descriptive cytology; as can be seen from the justly celebrated volume of E.B. Wilson (1925), the precision of descriptive detail at the light microscope level has not been surpassed today. Since 1900, methods of embryological intervention have been much the same as those that are used today with greater sophistication. These include egg centrifugation (or rotation), hair-loop constriction, blastomere separation and their culture in isolation, blastomere or tissue transplantation, and the study of species hybrids. Even the in vitro culture of specialized cells was initiated by Harrison as long ago as 1910.

Early attempts to understand development in molecular terms were limited by available biochemical procedures. As can be seen from J. Needham's three-volume treatise on chemical embryology published in 1931, all that could be done at that time was to measure gross changes in the embryonic content of total carbohydrates, proteins, lipids, etc., and of certain poorly characterized enzymes. In 1957, J. Brachet's *Biochemical Cytology* was published, a work of major importance and one that prepared the scene for the application of macromolecular analysis to problems of development. At that time the most important biochemical information relevant to development had come from histochemistry. As J. Brachet says in the introduction to his book, it is remarkable that RNase was used as a reagent by histologists some 15 years before it started to be used by biochemists.

From about 1960, methods of macromolecular analysis improved dramatically, and these were applied, as they became available, to eggs and embryos. The isolation of genes by cloning, the rapid sequencing of DNA, and the development of sensitive transcript assays, all leading to the characterization of developmentally important genes and their activities, have made an immense contribution to the molecular analysis of development. However technological traffic has not been in only one direction. For example, amphibian oocytes, with which I happen to be familiar, have yielded several procedures of general value in cell and molecular biology. The first eukaryotic chromosomal gene to be purified was amphibian ribosomal DNA by Wallace and Birnstiel (1966), before the days of cloning. A eukaryotic chromosomal gene-specific transcription factor was first isolated from extracts of oocytes added to an egg cell-free system (Engelke et al. 1980). In situ hybridization of nucleic acids was developed on oocyte lampbrush chromosomes (Gall and Pardue 1969), as was the electron microscopic analysis of transcription complexes (Miller and Beatty 1969). The injection of purified macromolecules, such as mRNA and DNA, into oocytes has become a useful procedure in connection with gene cloning, as for example when very sensitive mRNA assays are required (reviewed by Gurdon and Melton 1981).

Some General Concepts

Over the years, a number of concepts concerning mechanisms of development have gained acceptance. Those that are most important for the purposes of my subsequent comments are the following.

It is useful to distinguish differentiation from determination. It is now clear, most particularly from the serial transplantation of imaginal discs in *Drosophila* (Hadorn 1968), that cells can retain an ability to differentiate in only one way (such as wing) for 100 generations or more, even though there are no known morphological or biochemical criteria by which they differ from other cells only able to differentiate in a different direction (such as leg). This is the state of determination and it is remarkably stable, transdetermination being a rare event (Hadorn 1968). The transition from a determined to an overtly differentiated state can be induced by many different conditions, so that the determined cells will now display their commitment, adopting a cell-type-specific morphology and synthesizing cell-type-specific gene products. The transition between a determined and an overtly differentiated state is readily reversible, and it is the acquisition of the determined rather than differentiated state that we most need to understand.

It has also become clear that the stability of the determined state is only true of a complete cell. In nuclear transfer or cell hybrid experiments, dramatic changes in gene expression take place. For example, it is impossible to convert a terminal skin cell or an amnion cell into a muscle cell. But a skin cell nucleus transplanted to an egg will yield muscle cells in the resulting embryo (Gurdon et al. 1975), and an amnion cell nucleus will express muscle genes in an amnion-myoblast heterokaryon (Blau et al. 1983). Therefore the kinds of genes expressed in a nucleus can be rapidly changed by its cytoplasmic environment, and it may well be that nuclear genes are under continual regulation by cytoplasmic components.

Lastly, it is now generally believed that genomic DNA does not undergo irreversible changes in most cases of cell differentiation. A range of experiments argue in this direction, including cell transformation (lens regeneration from other cells; Eguchi 1979), DNA sequences as far as they are known at present (Tsujimoto and Suzuki 1984), and particularly nuclear transplantation. Nuclear transfers in amphibians, where the technique works best, have yielded fully formed, feeding-stage tadpoles from the nuclei of different adult cell types (e.g., Gurdon et al. 1975). Therefore not only cell-type differentiation but also morphogenesis to the rather substantial extent seen in swimming larvae that have reached the feeding stage do not depend on irreversible genetic changes. It is not very surprising that antibody-forming cells are an exception to this rule, since the aim of this particular type of differentiation is to generate as many different proteins as possible, whereas a characteristic of most other types of cell differentiation is to produce the least differences among cells of the same type. Gene amplification has been demonstrated for ribosomal DNA in the oocytes of some amphibian species (Brown and Dawid 1968), for chorion genes in the ovary of certain insects (Kalfayan et al.; Kafatos et al.; both this volume), and in a few other cases. It is extraordinary that gene amplification as a mechanism has been developed, and could therefore have been used to provide a general means of yielding differential gene expression; yet gene amplification is used only rarely, illustrating the general principle that the evolution of a mechanism does not ensure its widespread use. The general lack of irreversible genetic changes observed in somatic cells does not exclude *reversible* changes, such as DNA methylation.

THE PRESENT—HOW DOES DEVELOPMENT START?

Most routes to an understanding of early development, at all levels including that of morphogenesis, pass through an analysis of gene activity or expression and its control. I will not discuss the control of genes which fall into the "housekeeping" category, such as 5S RNAs (Brown 1984), 28 and 18S rRNAs, histones (Busslinger et al., this volume), and ribosomal proteins, interesting as the network of controls operating on these genes certainly is. Rather I will discuss, from the gene control point of view, a central question of development, that of how differences between cells first arise at the beginning of development, and how the spatial arrangements of cells are subsequently established. I believe that there may be only two fundamental mechanisms that account for cell differentiation and morphogenesis, and these seem to me to represent current problems of development in special need of a molecular solution.

Localization and Asymmetric Cell Division

If a fertilized amphibian egg or very early embryo is marked at a point on its surface by a colored dye or by other means, this appears reproducibly within a certain region of the resulting larva. This means that under normal conditions of development a particular region of egg cytoplasm will be included in, or give rise to, certain kinds of cells or embryonic structures. This does not at all imply a causal connection; for example, those ribosomes that are present in one part of an egg will always end up, by lineage, in the equivalent position in an embryo.

A more meaningful association between a region of egg cytoplasm and a particular type of development is achieved by removing parts of an egg. This was done with great success by Crampton in 1896, a student at that time in Wilson's laboratory. He took advantage of the fortuitous formation of a polar lobe in the cleaving eggs of the mollusc *Dentalium*. The removal of this lobe results in a larva deficient for the posttrochal and other regions, an effect not observed when other regions of equivalent size are removed (Wilson 1925). The best known example of a localization in eggs is the pole plasm of insects or germ plasm of amphibia. This histologically recognizable material of insects can be removed or destroyed, yielding flies that are sterile but in all other respects normal; fertility can be restored by injecting polar cytoplasm into the posterior region (Mahowald 1979). In a very important experiment, Illmensee and Mahowald (1974) were able to transform animal pole cells of a *Drosophila* embryo into gametes by transferring polar cytoplasm from the posterior to the anterior end of an egg. These experiments clearly establish a causal relationship between one highly localized region of an egg and the differentiation of certain cell types, in this case gametes. They have given direct support to the concept of cytoplasmic determinants, already set out by Morgan (1934).

In the most widely applicable form of this concept, determinants may be thought of as macromolecules, which cannot pass from cell to cell. They are asymmetrically distributed in an egg, and are consequently distributed in very different concentrations to cells formed during cleavage. The same principle is involved whether material is localized (as in pole plasm near one end of an egg), distributed in a continuous gradient throughout an egg, or differentially segregated into daughter cells during cleavage, even though evenly distributed in an egg (Fig. 1A,B); the latter condition seems to apply in ctenophores and may also take place in mammalian development when outside and inside cells are formed (Graham and Wareing 1984). The unequal distribution of materials between daughter cells can occur at any cell division, including those that take place in adults. In some cases, as in nematode development, the two daughter cells of a division are of markedly different size, and this could contribute to

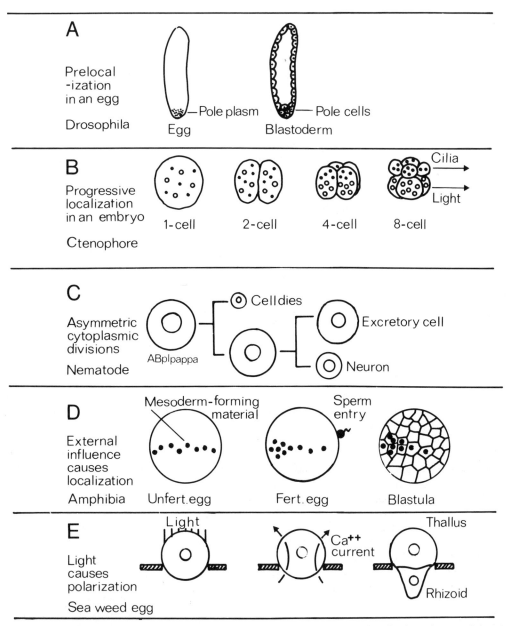

Figure 1. The principle of cytoplasmic localization and asymmetric cell division. (*A*) Prelocalization of *Drosophila* pole plasm (review by Mahowald 1979). (*B*) Progressive localization of cytoplasmic materials required for the formation of ciliated or light-producing cells in ctenophores (Freeman 1979); cilia and light cells are seen several hours after the eight-cell stage. (*C*) An example of asymmetric cell divisions in *Caenorhabditis elegans*, associated with divergent cell differentiation; the lineage shown starts with a cell designated ABplpappa (Sulston et al. 1983). (*D*) The point of sperm entry in frog eggs determines the dorsoventral polarity of the embryo; the black spots represent determinants for muscle formation and mark the dorsal (left-hand as shown) side of the embryo (from Gurdon et al. 1985b) (*E*) Seaweed eggs are placed on a grid; the cell leading to root (rhizoid) formation forms opposite the direction of light, and the future shoot (thallus) on the same side as light (from Robinson and Jaffe 1975).

the partitioning of cytoplasmic determinants (Fig. 1C). If certain components are located in a small area on the surface of a cell, and if 90% of these are partitioned at mitosis into the smaller of two daughter cells, a very large concentration difference between the daughter cells in respect of this component could arise. The concept is that the macromolecular determinants will reach a high enough concentration in one daughter cell of an asymmetric division to activate a gene or cause some other change in cell properties, whereas this effective concentration is not reached in a symmetric division.

It is unlikely that eggs have a large number of determinants. Apart from pole plasm there are a few other examples of substances localized in a fertilized, but uncleaved egg. One is the yellow cytoplasm of ascidian eggs, first described by Conklin (1905) and now known to be related, under several experimental conditions, to muscle formation in the larval tail (Whittaker 1982; Jeffery 1984). All components required for mesoderm (including muscle) formation in amphibia are localized in a subequatorial region of a newly fertilized egg (Gurdon et al. 1985b). Various maternal-effect mutations in *Drosophila* are known to affect the anterior-posterior and the dorsoventral development of embryos, and this implies the existence of determinants for these axes (see Anderson and Nusslein-Volhard 1984).

In no case has a morphogenetic substance or determinant in animal eggs been identified, and until this is achieved it is unclear whether there is a direct connection between the substances present in an egg and subsequent cell differentiation. A direct connection is exemplified by a gene-specific transcription factor; an example of an indirect connection would be a raised concentration of a metabolite that leads through several sequential steps, possibly including cell interactions, to the end result.

How is the localization of morphogenetic substances or determinants brought about? The great merit of localization and asymmetric cell division as a developmental mechanism is that one is not forced endlessly backwards, arguing that each step is itself determined by some previous event of a similar or different kind. Localization within an egg or other cell is usually induced by asymmetric external influences on that cell (Fig. 1D,E). These may be of a very unspecific kind; for example, the stimulus due to the site of sperm entry which causes the appearance of dorsoventral symmetry in amphibian and other eggs can be replaced by penetration with a glass needle and many other disturbances to one side of the egg. Light normally causes polarity in the eggs of *Pelvetia* (Robinson and Jaffe 1975). Nearly all cells arranged in sheets, like epithelia, have an asymmetric external environment, such as the luminal side of cells constituting a tube. The concept therefore exists that unspecific external influences can generate the asymmetric distribution of (specific) macromolecules within a cell. It is not necessary to explain the synthesis of these macromolecules, since the products of constitutive genes could be unequally distributed to daughter cells after an asymmetric division, but equally distributed in other dividing cells when the external stimulus is absent.

Two other general points concerning localization are worth mentioning. One is that the materials subject to localization are at first freely mobile in the cell, and subsequently fixed in position. At least this is the case with amphibian eggs, in which Schultze showed in 1894 (see Wilson 1925) that twinning occurs if a newly fertilized egg is inverted; halfway through the first cell cycle, the dorsal symmetry is no longer changed by inversion or centrifugation (Gerhart et al. 1981). There is no evidence that substances are localized in the cortex of the egg; the experiments of Curtis that appeared to show this have now been reinvestigated and reinterpreted (Gerhart et al. 1981).

The last point is that whatever substances are localized seem not to *commit* a cell to the type of differentiation which they promote. In this Symposium, we see that even germ cells carrying germ plasm can be made to differentiate into other cell types, if moved as single cells to an environment of cells of a different kind (Wylie et al., this volume). We must therefore think of localized determinants as the first step in a series of events that lead ultimately to overt cell differentiation or gene expression. Determinants cause a bias, not a commitment.

Embryonic Induction and Cell Interactions

The single most widely used mechanism in vertebrate development is embryonic induction. The best known example is primary induction in amphibians, when mesoderm induces the overlying ectoderm of a gastrula to become neural tissue, an effect discovered by Mangold and Spemann in 1924 (see Spemann 1938), when they found that the dorsal lip of a gastrula induced a secondary axis after it had been grafted to the ventral side of another gastrula. Cell interactions of the same kind as embryonic induction occur throughout vertebrate development, almost all vertebrate organs depending on one or more inductive steps, typically between endoderm and mesoderm, or ectoderm and mesoderm. The micromeres of a sea urchin embryo have inductive effects on other cells (Horstadius 1973), and even in the lineage-related development of a nematode, cell interactions are involved in the development of the vulva (Sternberg and Horvitz 1984). The organization center in experiments with *Euscelis* (Sander 1975) and the grafted apical ectodermal ridge of developing chick wing (Saunders and Gasseling 1968) both have developmental and morphogenetic effects resembling that of an amphibian organizer graft, and may therefore involve similar processes.

The unifying characteristic of embryonic induction and equivalent cell interactions is the transfer of information from one cell to another. If the mechanism involves the passage of molecules between cells, these are likely to be small, since molecules of less that 1000 daltons are readily transferred between cells, whereas this happens only rarely with larger molecules. The following paragraphs summarize some of the principal characteristics of induction processes, as background to the kinds of analysis that may prove useful for a molecular understanding of this important mechanism.

Inductions usually amplify a type of differentiation for which there is already a bias. In the case of amphibian muscle formation, a subequatorial localization of substances in the egg appears to cause some degree of muscle differentiation in the absence of cell contact, but this is greatly enhanced by an induction of animal cells by equatorial or vegetal cells (Nieuwkoop 1977; Gurdon et al. 1985a). The same is true for several other inductive systems (for review, see Saxen and Wartiovaara 1984). This phenomenon by which a developmental step depends on both localization with asymmetric division as well as induction may turn out to be a general characteristic of development. It would provide a means of progressively building up a structure, and of correcting its formation during development.

An important question in the analysis of induction

is whether the inducing signal is instructive or permissive. A totally permissive induction is one in which the responding tissue can only respond in one way whatever the stimulus, whereas an instructive induction is one where the type of response is determined by the properties of the inducer. An instructive induction can only be established if at least three possible directions of differentiation exist for the responding tissue. For example, early gastrula ectoderm will differentiate into epidermis if it receives no induction at all. If placed on top of mesoderm it becomes nerve, but the same ectoderm will become muscle if placed in the same way over endoderm (Fig. 2A). The fact that the two inducing tissues lead to different responses, both of which differ

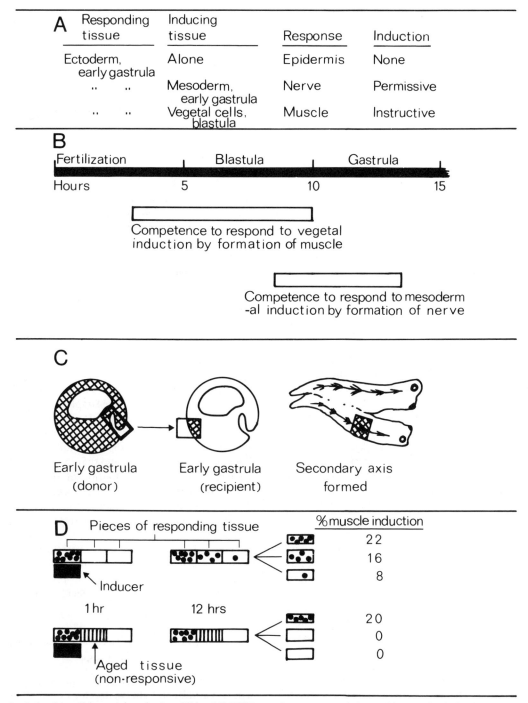

Figure 2. Inductive cell interactions in Amphibia. (*A*) Difference between permissive and instructive inductions (see text). (*B*) Changing competence of the same tissue (animal or ectoderm cells) as development proceeds; based on Gurdon et al. (1985a). (*C*) The dorsal lip region of one species (pigmented) is grafted to the ventral side of another (unpigmented) gastrula, and causes a second axis to be formed largely from host (unpigmented) tissue (Spemann 1938). (*D*) An inductive signal is transmitted from tissue in contact with an inducer to other tissue not in contact; the inducer was removed after 1 hr, and the responding tissue divided at 12 hr; the inductive signal cannot pass through normal living (but nonresponsive) cells (Kurihara and Sasaki 1981).

from the response in the absence of induction, shows that these inducers are to some degree instructive. To be sure of this, it is necessary to eliminate the possibility that a different subpopulation of cells in the ectoderm responds to each inducer. In some cases, though not in this, interpretation is further complicated by extensive cell proliferation, which could amplify different subpopulations of responding cells. It is, in fact, very hard to eliminate decisively the idea that there may be functionally distinct but morphologically indistinguishable subpopulations of cells in the responding tissue. In several other cases there is also evidence of instructive induction; a case often quoted is that chick epidermis will form different structures according to the origin of the inducing mesoderm (Sengel et al. 1969).

Even when there is evidence of instructive induction, it is clear that competence to respond is a far more precisely controlled feature of inductive cell interactions than the nature of the inducer. This applies to both the kind and the timing of the response. Most competent tissues can respond to natural inducers in only a limited number of ways. Thus, chick epidermis can never be induced to form muscle or blood. The time of response is also strictly limited (Fig. 2B). Competence typically appears just before the normal induction time and disappears soon after this (Saxen and Toivonen 1962; Saxen and Wartiovaara 1984).

Soon after induction was discovered, attempts were initiated to identify the chemical nature of an inducing substance. This was seen in the 1930s as the most accessible approach to a molecular understanding of development. Neural induction in amphibians was intensely investigated and it was soon found that pure chemical substances had the ability to induce ectoderm to form nerve. However, the range of agents found to have this ability became increasingly diverse and unspecific, such as high pH, low pH, and even distilled water (Holtfreter 1947); it became clear that newt or salamander ectoderm is so far predisposed to form nerve that almost any, even sublethal, stimulus will push it in that direction. The importance of this work was to emphasize the crucial nature of competence. To some extent this experimental cul-de-sac has been circumvented by the discovery of extracts that can induce the same test tissue to form nonneural structures, such as muscle or endoderm. An example of such a vegetalizing factor is the boiled swim bladder of a carp (Saxen and Toivonen 1962). It is not suggested that such "heterologous" inducers include the normal inducing substance, but they provide a source of material which is purifiable. Indeed Tiedemann and colleagues (1976) have partially purified a 32-kD protein from 12-day chick embryos as a vegetalizing factor and low-molecular-weight substances from various sources as neuralizing inducers. There is, however, some doubt whether the identification of heterologous inducers is informative; after all, lithium chloride is also an effective vegetalizing inducer in many organisms.

Somewhat more success has been achieved in attempts to characterize the normal mechanism of induction. In particular, filters have been used to show that some types of induction can be effected by diffusible molecules which can pass through 0.1-μm pores, and that in these cases, which include neural induction by mesoderm, cell contact is not required. However, there are also cases, such as kidney tubule induction, in which the same test indicates a need for physical contact between cells (see Saxen and Wartiovaara 1984). There is no simple relationship between the need for cell contact and instructive or permissive inductions, all combinations being found.

One of the most interesting aspects of induction is that its effect can be transmitted from one responsive cell to another. Some of the cells that respond to the induction have never been in contact with the inducing cells. The vegetal induction of muscle cells in amphibian embryos results in a group of muscle cells mostly some distance from the nearest vegetal cells. Since there is very little cell movement within the responding animal cells, this situation has not arisen by cell rearrangement. The same is true of the various organizer graft experiments referred to above.

The last general point worth emphasizing about induction is that there is often a gradient response. In the amphibian organizer grafts of Spemann, a second axis containing many cell types is induced in the responding cells of the embryo (Fig. 2C). Inductive responses can be arranged in order of the strength of induction required for their formation, as was demonstrated very clearly in the blastomere recombination experiments of Horstadius (1973). The kind of response obtained is influenced by the amount of inducing tissue, the duration of the contact time between inducing and responding tissues, and the distance of the responding cells from the inducing tissue. The simplest interpretation is that a diffusible molecule is released either by the inducing tissue or by the responding cells closest to it, that this passes through the responding tissue becoming more dilute as it goes, and that the type of response elicited is related to the dilution of the inducer (Fig. 2D). It is unclear whether the differential responses to one inducing tissue depend on different concentrations of the same molecule, or on different molecules that diffuse over greater or lesser distances through the responding tissue. Embryological experiments aimed at these questions favor the simplest (first) interpretation, but are not decisive, and will probably never be so without the aid of current molecular analysis of the kind discussed at this meeting.

The Cooperative Effects of Developmental Mechanisms

The single most remarkable property of a fertilized egg is that it nearly always develops normally without external instructions; 10^7 sea urchin eggs from one animal can nearly all form normal larvae of very complex construction compared to that of the egg. It is hard to imagine a way of achieving this except by gradually amplifying and continually correcting differences between, and arrangements among, cells. A current view of development is that a small number of morphogenetic substances are localized, rather roughly, in the fertilized egg. As soon as these have been segregated into separate cells as a result of cleavage, these cells interact with others to form new kinds of cells, and these in turn will induce further differences. These intercellular interactions result not only in new cell types, but seem also to be an essential means of generating shapes and patterns. Development may therefore be thought of as requiring the continuous operation of the two fundamental mechanisms of development. Cells and their daughters probably pass through a series of states in which they are temporarily responsive to the next effect, whether an intracellular component concentrated by localization and unequal division, or an inductive stimulus. An important consequence of such a series of events, especially inductive interactions which depend on the approximation of cell layers by folding, is that cell differentiation and morphogenesis can be coordinated and continuously corrected in different parts of an embryo. The series of states through which a cell and its daughters pass reflects progressive determination, as it becomes increasingly hard to divert a cell from its ultimate developmental fate.

FUTURE PROSPECTS

To what extent are the experimental strategies and methods discussed at this Symposium capable of providing molecular explanations of developmental processes, including localization and inductive cell interactions? Are the prospects any better now than they have been in the past? The great strength of genetic and molecular analysis, that is, the collection and biochemical characterization of mutants, is evident from the succeeding papers and summary chapter (Rubin, this volume). Rather than trying to anticipate this, it seems to me more helpful to indicate some current limitations on progress and some nongenetic types of analysis likely to be useful.

Cell-type-specific Markers of Gene Expression

Until recently, much embryological work has been analyzed by histological assessment, often of embryos cultured for many days after an experiment, and often based on an arrangement of cells not themselves individually definable; examples include patterns on insect cuticle and amphibian neural structures, such as archencephalon, resulting from mesodermal induction. The absence of a molecular assay for insect and amphibian pole plasm (the only assays now available being to count germ cells or score adults for fertility) has probably contributed more than anything else to the failure to identify the biologically important components. There seems to me to be a need for a more widespread use of in situ hybridization using gene-specific probes, as described in *Drosophila* (Akam et al., this volume) and in sea urchins (Davidson et al., this volume), or of cell-type-specific antibodies. Eventually it will be desirable to have probes that recognize the expression of *single genes* at the *single cell* level. It would also be valuable to have probes that recognize all levels at which gene expression is regulated (e.g., different splicing patterns), and which can be used with single cell resolution.

Cell-type-specific markers of gene expression could give valuable information concerning the following questions. Are the cells that first activate a gene in a particular position, such as adjacent to another structure? Which genes are activated first, and which secondarily, in the same cells?

Analysis of Mutants

Mutations most likely to be helpful in analyzing the earliest steps in development include those with maternal effects. The first maternal effect mutation to be analyzed was one causing sinistral coiling in snails (Boycott et al. 1930), and the idea of injecting extracts of wild-type eggs or oocytes to rescue the defective development of mutant eggs was pioneered by Briggs for the axolotl *o* mutation (Briggs and Cassens 1966). Recently this experimental approach has given very encouraging results, in that some *Drosophila* maternal-effect mutants such as dorsal can be rescued by the injection of poly(A)$^+$ RNA (Anderson and Nusslein-Volhard 1984). This could lead directly to a cDNA, and hence to a gene and its product, required for the action of a developmental determinant. In *Drosophila* and nematodes, it may be efficient to isolate a gene such as dorsal directly, but in other organisms without the same genetic advantages, analysis "by rescue" can offer an important way forward.

Maternal-effect mutations are not easy to obtain, and many of them have their effect for embryologically uninteresting reasons, such as an inability of the mother to deposit eggs (Gans et al. 1975). Even if attention is restricted to mutations that cause defects in early development, not all of these are of interest. For example, the first maternal-effect mutation whose gene product has been identified was rudimentary in *Drosophila*; this turned out to be a defect in pyrimidine-synthesizing enzymes, such that rudimentary embryos are rescued by an injection of pyrimidines (Okada et al. 1974). It seems likely that pyrimidines are general cell metabolites of which a particularly high rate of synthesis is required in the early stages of development, but whose availability is in no way determinative in wild-type embryos.

It is not always a simple matter to collect mutations affecting particular processes, especially those that operate at the beginning of development and that affect mechanisms such as induction, competence, and the localization of determinants. A very hopeful alternative is the use of antisense RNA, especially for species in which the selection of appropriate mutants is not practicable, and for processes dependent on maternal mRNA. Antisense RNA prepared from cDNAs cloned into an SP6 vector and injected into eggs (Melton 1985; Jackle et al., this volume) or introduced as DNA coding for it (Izant and Weintraub 1984) seems to cut out individual kinds of gene expression, as does a mutation.

Even when mutants have been collected and characterized, or when one species of antisense mRNA has an interesting developmental effect, a potential problem lies ahead. It should be possible, given time, biotechnology, and a DNA transformation assay, to identify the DNA sequence, and hence protein(s), associated with the mutant effect. Progress this far should proceed fast. However, a much more difficult problem is to find out the function of a gene product. There could even be a problem of unemployment among genes; their products can be identified, but this does not expose a direct route to their function. Hopefully the localization of gene products by antibodies will give a clue.

cis- and trans-acting Elements

It is a reasonable assumption that molecules which regulate genes include ones that interact directly with them. The success of DNA transformation in the mouse, and especially in *Drosophila* via P elements, has yielded several instances in which genes injected as DNA are correctly regulated; this leads, in turn, to the mapping of *cis*-acting sequences required for the temporal or regional regulation of genes.

The same design of experiment is potentially useful for finding sequences bound by *trans*-acting factors. The transcription of injected genes is commonly assayed when they have integrated into the host cells' chromosomal DNA (e.g., Posakony et al.; Hammer et al.; both this volume). However, a more useful situation would be realized if the injected genes would come under correct regulation when present as nonintegrated minichromosomes; this happens, for example, in amphibians for the first 2 days of development (Gurdon and Melton 1981). Apart from eliminating the uncertain effects of neighboring chromosomal DNA on the expression of injected genes, this would open up good prospects for experiments in which cloned segments of DNA in or around a gene are coinjected with a gene but in great excess. If the sequence of DNA injected is one that binds a *trans*-acting factor, it ought to sequester most of the cells' supply of this factor and hence alter activity of the injected gene.

Another route toward the identification of *trans*-acting factors involves gene injection into amphibian oocytes (see Gurdon and Melton 1981). If crude cell or nuclear extracts are injected into oocyte cytoplasm, molecules normally resident in a nucleus will migrate to, and accumulate in, the oocyte nucleus. A day later, when this has taken place, genes, in the form of cloned DNA, can be injected into the same oocyte nucleus. The DNA is then assembled into chromatin incorporating histones and other DNA binding proteins. This provides a situation where cloned DNA is assembled into a normal chromatin structure, under in vivo conditions and in the presence of transcription factors such as TFIIIA which may need to bind to DNA before histones (Brown 1984). Ideally the cellular material injected into oocyte cytoplasm would be mRNA, since one mRNA molecule is translated into 10^4 proteins per day, and identification of the mRNA involved could proceed via cDNA.

It is hard to see how development can be fully understood at the molecular level without the use of cell-free systems. Ones that simulate developmental regulation may be difficult to achieve since even the most thoroughly understood genes (5S, globin) cannot be regulated in their expression in any cell-free system as yet. If it proves possible to obtain cells with nonintegrated, but regulated, genes as mentioned above, a lysate of such cells could be an appropriate starting point for the removal and replacement of components.

We may ask how rare gene-controlling molecules are likely to be, and whether there is a hope that they can be isolated. It is generally found to be impracticable, even with a good assay, to undertake a purification of more than 10^5 times. If gene-specific molecules are rare, would the search for these or other developmentally important molecules seem like looking for a needle in a haystack, without the ability to see anything smaller than the farmer's daughter? The very few gene-specific molecules so far identified and quantitated are present at about 10^4 molecules per nucleus (Table 1), a concentration equivalent to that of the *lac* repressor in a bacterium. It is a reasonable prediction that a protein

Table 1. Abundance and Concentration of Nuclear Molecules

	Number of molecules		Concentration by volume (in somatic nucleus)
Component	per amphibian oocyte nucleus	per somatic cell nucleus	
Core histones	2×10^{12}	2×10^8	10^{-2}
TFIIIA	10^{11}	10^4	2×10^{-6}
Steroid receptor	—	10^4	2×10^{-6}
Needle in haystack	—	—	10^{-11}

Numbers of molecules: Core histones (Woodland and Adamson 1977); transcription factor IIIA (refs. from Brown 1984); steroid receptor (Eriksson 1983). Concentration by volume assumes a nucleus of 250 μm^3, a 15K histone of 16 nm^3, a TFIIIA of 60 nm^3, a receptor of 60 nm^3, and a needle of 10 mm^3.

could not find a specific DNA sequence in a eukaryotic cell nucleus at less than this concentration, in view of the lower but significant affinity that proteins have for all other DNA (Lin and Riggs 1975). A typical eukaryotic cell, such as a mouse teratocarcinoma cell, contains $\sim 10^9$ proteins (assuming a cell of 15 μm diameter, and total proteins [average size 60 kD] at 100 mg/ml). Of these, 10% will be nuclear and 1% of these DNA binding. A nucleus could therefore contain 100 different proteins each at 10^4 per nucleus, more than sufficient for gene regulation.

Cell-type Transformation

As an increasing variety of molecules associated with development comes to be characterized, the question will always arise whether they are a cause or consequence of gene activation and morphogenetic events. A useful route to the solution of this problem involves the redirection of cell differentiation; if a molecule is truly causative in its developmental effect, it should be capable of changing the direction of development. Of course this requires that it is present at an appropriate concentration and at the appropriate time in that cell's developmental history.

Conditional mutants of the homeotic type would be invaluable. Alternatively or in addition, it would be useful to be able to redirect the differentiation of cells by introducing known kinds of mRNA or protein. An important step in this direction was achieved by Illmensee and Mahowald (1974) who transformed anterior blastoderm cells of *Drosophila* (which would normally give epidermis or brain) into gametes by the injection of pole plasm. To some extent, an equivalent effect is achieved by altering the cleavage plane in embryonic ascidian cells (Whittaker 1982), or by removing a high proportion of the cytoplasm of embryonic nematode cells (Schierenberg, this volume). The prospect of redirecting the differentiation of cells by the introduction of purified macromolecules is highly desirable, and, though not yet achieved, may not be as difficult as it may sound. It is impressive that injected mRNA is stable and so successfully translated in the wrong cell type; proteins are correctly modified and assembled; and both RNA and proteins tend to take up their normal intracellular locations in inappropriate cell types (Gurdon 1974). These comments apply to cells of the same species, and are certainly not always true for prokaryotic molecules in eukaryotic cells. Ingenious means exist, and others will surely follow, of causing introduced DNA to be expressed in foreign cells at the time required. Once the products of developmentally important genes have been identified, it seems realistic to try to redirect cell differentiation with them, and hence to provide proof of their causal connection with some aspect of development. From this design of experiment, it should be possible to determine the effective concentration of a constituent and the exact time at which this concentration is required for a particular developmental step to proceed.

CONCLUSION

The overall impression from work presented at this Symposium is that analysis involving nucleic acid technology can proceed with great precision, and, as far as this can be used in the analysis of development, progress should be fast. Analysis at the protein level is less advanced. Antibodies cannot yet be used to isolate rare proteins from a complex mixture as effectively as cDNA can for species of mRNA, and sequencing is much more laborious for proteins than is currently true for nucleic acids. Biochemical methods for the purification of other macromolecules such as carbohydrates and lipids are much less satisfactory. Those developmental processes, which depend on concentration differences in ions or other small molecules, may be even harder to analyze.

A major limitation on progress at present appears to be that of identifying molecules that interact specifically with a DNA sequence. Looking further ahead, it is not clear that the currently favored direction of nucleic acid–protein analysis will lead to a molecular understanding of complex developmental processes such as color patterns of lepidopteran wings, birds, and fish, but this is already looking toward an area likely to be covered at the next Cold Spring Harbor Symposium on development. In the more immediate future, readers of the ensuing pages will, I believe, have every reason to be optimistic about progress toward a molecular understanding of how differences first arise among cells of an embryo, at least at the level of gene activation.

ACKNOWLEDGMENTS

I am most grateful to R.A. Laskey, P.A. Lawrence, M.E. Akam, L.J. Korn, and J.O. Thomas for valuable discussion and criticism.

REFERENCES

ANDERSON, K. and C. NUSSLEIN-VOLHARD. 1984. Information for the dorsal-ventral pattern of the *Drosophila* embryo is stored as maternal mRNA. *Nature* 311: 223.

BLAU, H.M., C.-P. CHIU, and C. WEBSTER. 1983. Cytoplasmic activation of human nuclear genes in stable heterokaryons. *Cell* 32: 1171.

BOYCOTT, A.E., C. DIVER, S.L. GARSTANG, and F.M. TURNER. 1930. The inheritance of sinistrality in *Limnaea peregra*. *Philos. Trans. R. Soc. Lond. B* 219: 51.

BRACHET, J. 1957. *Biochemical cytology*. Academic Press, New York.

BRIGGS, R. and G. CASSENS. 1966. Accumulation in the oocyte nucleus of a gene product essential for embryonic development beyond gastrulation. *Proc. Natl. Acad. Sci.* 55: 1103.

BROWN, D.D. 1984. The role of stable complexes that repress and activate eukaryotic genes. *Cell* 37: 359.

BROWN, D.D. and I.B. DAWID. 1968. Specific gene amplification in oocytes. *Science* 160: 272.

CONKLIN, E.G. 1905. The organization and cell lineage of the ascidian egg. *J. Acad. Natl. Sci.* 13: 1.

EGUCHI, G. 1979. Transdifferentiation in pigmented epithelial cells of vertebrate eyes *in vitro*. In *Mechanisms of change* (ed. J.D. Ebert and T.S. Okada), p. 273. Wiley, New York.

ENGELKE, D.R., S.Y. NG, B.S. SHASTRY, and R.G. ROEDER. 1980. Specific interaction of a purified transcription factor with an internal control region of 5S RNA genes. *Cell* **19**: 717.

ERIKSSON, H. 1983. Regulation of estrogen receptor concentration in target organs of the rat. *Nobel Symp.* **57**: 389.

FREEMAN, G. 1979. The multiple roles that cell division can play in the localization of developmental potential. In *Determinants of spatial organization* (ed. S. Subtelny and I.R. Konigsberg), p. 53. Academic Press, New York.

GALL, J.G. and M.L. PARDUE. 1969. Formation and detection of RNA-DNA hybrid molecules in cytological preparations. *Proc. Natl. Acad. Sci.* **63**: 378.

GANS, M., C. AUDIT, and M. MASSON. 1975. Isolation and characterization of sex-linked female-sterile mutants in *D. melanogaster*. *Genetics* **81**: 683.

GERHART, J., G. UBBELS, S. BLACK, K. HARA, and M. KIRSCHNER. 1981. A reinvestigation of the role of the grey crescent in axis formation in *Xenopus laevis*. *Nature* **292**: 511.

GRAHAM, C.F. and P.F. WAREING. 1984. *Developmental control in animals and plants*. Blackwell, Oxford, England.

GURDON, J.B. 1974. *The control of gene expression in animal development*. Oxford University Press, England.

GURDON, J.B. and D.A. MELTON. 1981. Gene transfer in amphibian eggs and oocytes. *Annu. Rev. Genet.* **15**: 189.

GURDON, J.B., R.A. LASKEY, and O.R. REEVES. 1975. The developmental capacity of nuclei transplanted from keratinized skin cells of adult frogs. *J. Embryol. Exp. Morphol.* **34**: 93.

GURDON, J.B., S. FAIRMAN, T.J. MOHUN, and S. BRENNAN. 1985a. The activation of muscle-specific actin genes in *Xenopus* development by an induction between animal and vegetal cells of a blastula. *Cell* **41**: 913.

GURDON, J.B., T.J. MOHUN, S. FAIRMAN, and S. BRENNAN. 1985b. All components required for the eventual activation of muscle-specific actin genes are localized in the subequatorial region of an uncleaved amphibian egg. *Proc. Natl. Acad. Sci.* **82**: 139.

HADORN, E. 1968. Transdetermination in cells. *Sci. Am.* **219**(5): 110.

HARRISON, R.G. 1910. The outgrowth of the nerve fiber as a mode of protoplasmic movement. *J. Exp. Zool.* **9**: 787.

HOLTFRETER, J. 1947. Neural induction in explants which have passed through a sublethal cytolysis. *J. Exp. Zool.* **106**: 197.

HORSTADIUS, S. 1973. *Experimental embryology of the echinoderms*. Oxford University Press, England.

ILLMENSEE, K. and A.P. MAHOWALD. 1974. Transplantation of posterior polar plasm in *Drosophila*. Induction of germ cells at the anterior pole of the egg. *Proc. Natl. Acad. Sci.* **71**: 1016.

IZANT, J. and H. WEINTRAUB. 1984. Inhibition of thymidine kinase gene expression by anti-sense RNA: A molecular approach to genetic analysis. *Cell* **36**: 1007.

JEFFERY, W.R. 1984. Pattern formation by ooplasmic segregation in ascidian eggs. *Biol. Bull.* **166**: 277.

KURIHARA, K. and N. SASAKI. 1981. Transmission of homoiogenetic induction in presumptive ectoderm of newt embryo. *Dev. Growth Differ.* **23**: 361.

LIN, S. and A.D. RIGGS. 1975. The general affinity of *lac* repressor for *E. coli* DNA: Implications for gene regulation in prokaryotes and eukaryotes. *Cell* **4**: 107.

MAHOWALD, A.P. 1979. Genetic control of oogenesis in *Drosophila*. In *Mechanisms of change* (ed. J.D. Ebert and T.S. Okada), p. 101. Wiley, New York.

MELTON, D.A. 1985. Injected anti-sense RNAs specifically block mRNA translation *in vivo*. *Proc. Natl. Acad. Sci.* **82**: 144.

MILLER, O.L. and B.R. BEATTY. 1969. Visualization of nucleolar genes. *Science* **164**: 955.

MORGAN, T.H. 1934. *Embryology and genetics*. Columbia University Press, New York.

NEEDHAM, J. 1931. *Chemical embryology*. Cambridge University Press, England.

NIEUWKOOP, P.D. 1977. Origin and establishment of embryonic polar axes in amphibian development. *Curr. Top. Dev. Biol.* **11**: 115.

OKADA, M., I.A. KLEINMAN, and H.A. SCHNEIDERMAN. 1974. Repair of a genetically-caused defect in oogenesis in *D. melanogaster* by transplantation of cytoplasm from wild-type eggs and by injection of pyrimidine nucleosides. *Dev. Biol.* **37**: 55.

ROBINSON, K.R. and L.F. JAFFE. 1975. Polarizing fucoid eggs drive a calcium current through themselves. *Science* **187**: 70.

SANDER, K. 1975. Pattern specification in the insect embryo. Cell patterning. *CIBA Found. Symp.* **29**: 241.

SAUNDERS, J.W. and M.T. GASSELING. 1968. Ectodermal-mesenchymal interactions in the origin of limb symmetry. In *Epithelial-mesenchymal interactions* (ed. R. Fleischmajer and R.E. Billingham), p. 78. Williams and Wilkins, Baltimore, Maryland.

SAXEN, L. and S. TOIVONEN. 1962. *Primary embryonic induction*. Academic Press, London.

SAXEN, L. and J. WARTIOVAARA. 1984. Embryonic induction. In *Developmental control in animals and plants* (ed. C.F. Graham and P.F. Wareing), p. 176. Blackwell, Oxford, England.

SENGEL, P., D. DHOUAILLY, and M. KIENY. 1969. Aptitude des constituants cutanes de l'apterie médioventrale du polet à former des plumes. *Dev. Biol.* **19**: 436.

SPEMANN, H. 1938. *Embryonic development and induction*. Yale University Press, New Haven, Connecticut.

STERNBERG, P.W. and H.R. HORVITZ. 1984. The genetic control of cell lineage during nematode development. *Annu. Rev. Genet.* **18**: 489.

SULSTON, J.E., E. SCHIERENBERG, J.G. WHITE, and J.N. THOMSON. 1983. The embryonic cell lineage of the nematode *C. elegans*. *Dev. Biol.* **100**: 64.

TIEDEMANN, H. 1976. Pattern formation in early developmental stages of amphibian embryos. *J. Embryol. Exp. Morphol.* **35**: 437.

TSUJIMOTO, Y. and Y. SUZUKI. 1984. Natural fibroin genes purified without using cloning procedures from fibroin-producing and -nonproducing tissues reveal indistinguishable structure and function. *Proc. Natl. Acad. Sci.* **81**: 1644.

WALLACE, H. and M.L. BIRNSTIEL. 1966. Ribosomal cistrons and the nucleolus organizer. *Biochim. Biophys. Acta* **114**: 296.

WHITTAKER, J.R. 1982. Muscle lineage cytoplasm can change the developmental expression in epidermal lineage cells of ascidian embryos. *Dev. Biol.* **93**: 463.

WILSON, E.B. 1925. *The cell in development and heredity*, 3rd edition. Macmillan, New York.

WOODLAND, H.R. and E.D. ADAMSON. 1977. The synthesis and storage of histones during the oogenesis of *Xenopus laevis*. *Dev. Biol.* **57**: 118.

The Mouse Egg's Receptor for Sperm: What Is It and How Does It Work?

P.M. WASSARMAN,* J.D. BLEIL,* H.M. FLORMAN, J.M. GREVE,* R.J. ROLLER,* G.S. SALZMANN,* AND F.G. SAMUELS*

Department of Biological Chemistry, Harvard Medical School, Boston, Massachusetts 02115

Fertilization is the process by which sperm and eggs unite to form a zygote, the true beginning of a new individual. In mammals, sperm first make contact with eggs at the surface of the egg's extracellular coat, or zona pellucida (Fig. 1a). This contact can lead to binding of sperm to eggs via species-specific sperm receptors present in the zona pellucida. Bound sperm then undergo changes that enable them to penetrate the zona pellucida and to fuse with the egg's plasma membrane. This fusion results in activation of eggs, and development of the organism ensues (for review, see Gwatkin 1977; Yanagimachi 1981; Bedford 1982; Wassarman 1983a; Wassarman et al. 1984a).

Here, we describe experiments from our laboratory that have led to the identification and characterization of the mouse egg's sperm receptor, ZP3, and have provided clues to its mechanism of action. Overall, we suggest that sperm recognize and bind to a specific class of O-linked oligosaccharides present on ZP3, an 83-kD M_r glycoprotein present in mouse egg zonae pellucidae, and that binding is sufficient to induce those changes in sperm that permit them to penetrate the zona pellucida and to fuse with the egg's plasma membrane. Subsequent to sperm–egg fusion, ZP3 is altered such that it no longer serves as a sperm receptor.

RESULTS

Early Events in Mouse Gamete Interaction

In all of our experiments, we use gametes and embryos isolated from randomly bred, Swiss albino mice (CD-1) obtained from Charles River Breeding Laboratories. Detailed procedures for isolation and culture of mouse oocytes, eggs, sperm, and embryos have appeared (Bleil and Wassarman 1980a, 1983; Florman et al. 1984; Florman and Wassarman 1985).

To study the sperm receptor, we examine early events of mouse gamete interaction in vitro. Ovulated eggs and two-cell embryos are added to sperm and, within seconds, zonae pellucidae of both eggs and embryos are covered with motile sperm. These sperm are loosely associated with zonae pellucidae and can be removed by gentle pipetting with a broad-bore micropipet; this state of adhesion is referred to as attachment. Shortly thereafter, contact between sperm and egg zonae pellucidae becomes more tenacious, such that gentle pipetting no longer dissociates the gametes; this state of adhesion is referred to as binding. Although the initial, reversible attachment of sperm to embryo zonae pel-

*Present address: Department of Cell Biology, Roche Institute of Molecular Biology, Nutley, New Jersey 07110.

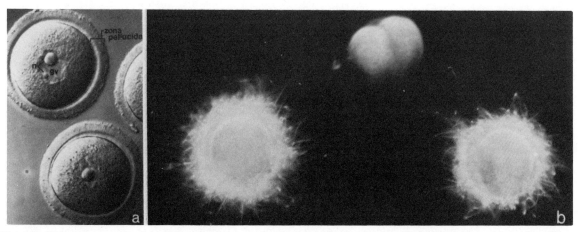

Figure 1. Photomicrographs of mouse oocytes, eggs, and embryos. (*a*) Light micrograph of fully grown oocytes, about 85 μm in diameter, observed with Nomarski differential interference contrast optics. The zona pellucida, about 7 μm thick, is indicated. (gv) Germinal vesicle; (n) nucleoli. (*b*) Light micrograph of unfertilized eggs and a two-cell embryo in the presence of sperm, obtained using a Planachromat (Zeiss) objective with illumination through a Phase 3 condenser. Sperm are bound to zonae pellucidae of unfertilized eggs, but not to the zona pellucida of the two-cell embryo.

lucidae is virtually indistinguishable from that observed with eggs, in the former case attachment does not proceed to the binding state. This pronounced difference in behavior provides an operational definition of bound sperm as those adhering to egg zonae pellucidae under conditions that result in complete removal of sperm from embryo zonae pellucidae (Fig. 1b). In the mouse, binding of sperm to egg zonae pellucidae occurs at plasma membrane overlying the anterior region of the sperm head. Shortly after binding, sperm undergo the acrosome reaction. This is a membrane fusion event, wherein the contents of the acrosome, a large, lysosome-like vesicle overlying the sperm nucleus, are exposed following vesiculation of the anterior region of sperm plasma membrane and outer acrosomal membrane. Completion of the acrosome reaction is required for sperm penetration of zonae pellucidae, as well as for fusion of plasma membrane toward the posterior region of the sperm head with egg plasma membrane (fertilization).

Nature of Mouse Egg Zonae Pellucidae

The zona pellucida is a relatively thick extracellular coat that surrounds all mammalian eggs and plays important roles during both fertilization and preimplantation development. For example, zonae pellucidae contain species-specific sperm receptors that function during fertilization, participate in a secondary block to polyspermy following fertilization, and protect early embryos as they pass down the oviduct into the uterus. It is not until the expanded blastocyst stage of development that embryos hatch from their zonae pellucidae, thereby enabling them to implant in the uterine wall.

Mouse egg zonae pellucidae are approximately 7 μm thick and contain about 3 ng of protein (Fig. 1a). Transmission electron micrographs of sectioned material indicate that the zona pellucida is an amorphous network ("sponge-like"), not unlike a number of other types of extracellular coats in appearance. As such, zonae pellucidae do not present a permeability barrier to large macromolecules or even small viruses. Zonae pellucidae are dissolved by a variety of agents that either do (e.g., proteinases and reducing agents) or do not (e.g., heat, low pH, and SDS) break covalent bonds, with egg zonae pellucidae being more susceptible to these agents than embryo zonae pellucidae. The latter state is thought to reflect a "hardening" of zonae pellucidae following fertilization of eggs (Inoue and Wolf 1974a,b; Schmell and Gulyas 1980; Bleil et al. 1981).

Results of recent structural studies carried out in our laboratory (Greve and Wassarman 1985) demonstrate that the zona pellucida is a highly organized, three-dimensional matrix of interconnected (branched) filaments. Individual filaments, in turn, resemble "beads-on-a-string," with each bead (9.4 ± 2.2 nm in diameter) located every 17.2 ± 4.6 nm (center-to-center distance) along the filament. ZP1, one of three glycoproteins that make up the zona pellucida (discussed below), apparently serves as a cross-linker of individual filaments, giving rise to the three-dimensional matrix.

Glycoproteins of Mouse Egg Zonae Pellucidae

The 3 ng or so of zona pellucida protein is distributed among only three glycoprotein species, ZP1, ZP2, and ZP3 having apparent M_r of 200, 120, and 83 kD, respectively (Bleil and Wassarman 1978, 1980a,b,c). These three glycoproteins together account for virtually all of the mass of the zona pellucida. ZP2 and ZP3 appear to be present in approximately equimolar amounts, accounting for as much as 80% of zona pellucida protein. Results of amino acid analysis, high-resolution two-dimensional gel electrophoresis, electrophoretic analysis of proteolytic digests, and immunological analysis all strongly suggest that each glycoprotein represents a unique polypeptide chain. All three zona pellucida glycoproteins are acidic (pI < 5.5), in part due to the presence of sialic acid, and only ZP1 consists of more than one polypeptide chain (linked by intermolecular disulfides). These are the mature forms of the glycoproteins found in zonae pellucidae.

Identification of the Mouse Egg's Sperm Receptor

As described above, sperm bind to zonae pellucidae of unfertilized eggs, but not to zonae pellucidae of embryos. We (Bleil and Wassarman 1980a) and others (Gwatkin and Williams 1976) have found that pretreatment of sperm with solubilized egg zonae pellucidae prevents binding of these sperm to unfertilized eggs. On the other hand, pretreatment with embryo zonae pellucidae has no effect on sperm binding. These observations suggest that sperm receptors present in solubilized egg zona pellucida preparations bind to appropriate sites on sperm, thereby preventing subsequent binding of these sperm to zonae pellucidae of unfertilized eggs. This forms the basis of a competition assay used to identify the mouse egg's sperm receptor.

Each of the three zona pellucida glycoproteins, purified to homogeneity, has been tested for sperm receptor activity in this competition assay (Bleil and Wassarman 1980a; Wassarman and Bleil 1982; Florman et al. 1984; Florman and Wassarman 1985). In each case, the extent of binding of treated and untreated sperm to unfertilized eggs was compared (number of sperm bound per egg). Whereas neither purified ZP1 nor ZP2 have a significant effect on sperm binding, ZP3 is nearly as effective as solubilized egg zonae pellucidae in reducing the number of sperm bound to eggs ($ID_{50} \approx 1$ zona pellucida equivalent/μl) (Fig. 2). Comparisons of results obtained with purified ZP3 and solubilized zona pellucida preparations indicate that ZP3 alone accounts for all sperm receptor activity present in egg zonae pellucidae. As expected, purified ZP3 from embryo zonae pellucidae has no effect on sperm binding at concentrations at which egg ZP3 inhibits binding by 80% or more. These results suggest that ZP3 serves as receptor for sperm in unfertilized eggs and that ZP3 is

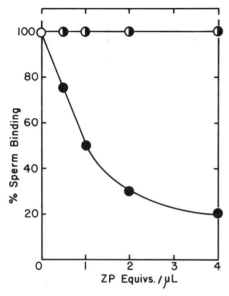

Figure 2. Binding of sperm to eggs in the presence of various concentrations of glycoproteins purified from unfertilized egg zonae pellucidae. Binding of sperm to eggs was compared after exposure of sperm to purified ZP1 (◐), ZP2 (◐), or ZP3 (●), or to culture medium alone (○). In each case, eggs were transferred to a microscope slide, fixed in 1% glutaraldehyde, and the number of sperm bound to egg zonae pellucidae counted. In each case, the concentration of the glycoprotein tested is expressed as zona pellucida equivalents per microliter (Bleil and Wassarman 1980a; Florman et al. 1984; Florman and Wassarman 1985).

altered as a result of fertilization ($ZP3_f$) such that it no longer exhibits sperm receptor activity.

Consequences of Sperm Receptor Binding

Mouse sperm that have undergone the acrosome reaction fail to bind to unfertilized eggs; only acrosome-intact sperm bind (Saling and Storey 1979; Saling et al. 1979; Florman and Storey 1982; Bleil and Wassarman 1983). In this context, autoradiograms of whole-mount preparations of sperm incubated with radiolabeled ZP3 demonstrate that ZP3 binds to the head of acrosome-intact, but not of acrosome-reacted, sperm (Fig. 3; J.D. Bleil and P.M. Wassarman, unpubl.). These observations suggest that plasma membrane overlying the anterior region of the sperm head interacts with the zona pellucida via ZP3 and that a zona pellucida component induces bound sperm to undergo the acrosome reaction.

Each of the three zona pellucida glycoproteins, purified to homogeneity, has been tested for acrosome reaction-inducing activity (Wassarman and Bleil 1982; Bleil and Wassarman 1983; Florman et al. 1984; J.D. Bleil and P.M. Wassarman, unpubl.). Three different methods, transmission electron microscopy, immunofluorescence microscopy, and Nomarski differential interference contrast microscopy, were used to score acrosome-intact and -reacted sperm in the presence or absence of zona pellucida glycoproteins. In each case,

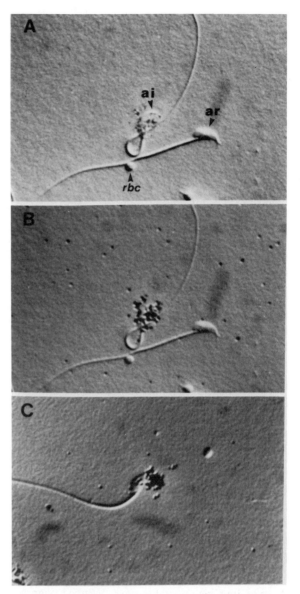

Figure 3. Binding of purified unfertilized egg ZP3 to sperm. Sperm were incubated in the presence of radiolabeled ZP3 and then processed for and subjected to whole-mount autoradiography. Shown are photomicrographs of autoradiograms observed by Nomarski differential interference contrast optics. (*A*) Light micrograph of an acrosome-intact (ai) and an acrosome-reacted (ar) sperm. Note the "ridge" overlying the acrosomal cap region of the head of the acrosome intact sperm that is not present on the acrosome-reacted sperm. rbc, Residual body component. (*B*) Light micrograph of the same two sperm shown in *A*, but focusing on grains overlying the acrosomal cap region of the head of the acrosome-intact sperm. No grains are found associated with the head of the acrosome-reacted sperm. (*C*) Light micrograph of an acrosome-intact sperm with grains associated with the sperm head.

it is clear that only ZP3 induces sperm to undergo the acrosome reaction. Comparisons of results obtained with purified ZP3 and solubilized zona pellucida preparations indicate that ZP3 alone accounts for all acrosome reaction-inducing activity present in egg zonae pellucidae. Furthermore, ZP3 is as effective as ionophore A23187 in inducing the acrosome reaction. Purified $ZP3_f$ from embryo zonae pellucidae has little, if any, acrosome reaction-inducing activity. These results suggest that binding of acrosome-intact sperm to the sperm receptor ZP3 is sufficient to induce sperm to undergo the acrosome reaction.

Structure of the Mouse Egg's Sperm Receptor

Much of our information about the structure of ZP3 has come from biosynthetic studies carried out with growing mouse oocytes cultured in vitro (for review, see Wassarman et al. 1984b). It is during this 2- to 3-week period of oogenesis that the zona pellucida first appears (Wassarman and Josefowicz 1978; Wassarman 1983b).

We have demonstrated that growing mouse oocytes themselves synthesize and secrete all three zona pellucida glycoproteins (Bleil and Wassarman 1980c). By using polyclonal antisera directed against zona pellucida glycoproteins, together with metabolic radiolabeling of intracellular precursors of these glycoproteins, we have learned a great deal about the structures of ZP1 (Wassarman et al. 1984b; G.S. Salzmann and P.M. Wassarman, unpubl.), ZP2 (Greve et al. 1982; Roller and Wassarman 1983), and ZP3 (Salzmann et al. 1983; Roller and Wassarman 1983). Results of pulse-chase experiments, experiments utilizing either enzymes or chemical agents that remove N- and/or O-linked oligosaccharides from glycoproteins (Fig. 4A,B), and experiments carried out in the presence of tunicamycin to prevent N-linked glycosylation of nascent proteins (Fig. 4C) have enabled us to construct the biosynthetic pathway for ZP3 shown in Figure 4D. ZP3 (≈ 83 kD M_r [range, 74–90 kD]; pI≈ 4.7 [range, 4.3–5.2]) is synthesized as a 44-kD M_r polypeptide chain to which either three or four high-mannose-type, N-linked oligosaccharides are added, giving rise to 53-kD and 56-kD M_r (pI≈ 6.7) intermediates, respectively. The N-linked oligosaccharides are subsequently processed to complex-type, and O-linked oligosaccharides are added to nascent ZP3 prior to its secretion into the zona pellucida.

To obtain information about the primary structure of ZP3, we have prepared and screened a mouse oocyte cDNA library for clones containing ZP3 coding sequences (R.J. Roller and P.M. Wassarman, unpubl.). cDNAs, synthesized by using RNA from about 16,000, individually isolated, growing oocytes as template, were inserted into the expression vector λgt11. The resulting library contains about 80,000 clones. The library has been screened by using rabbit antisera directed against zona pellucida glycoproteins, and clear, reproducible, positive clones have been obtained. One of these contains a 2.5-kb insert that apparently includes most or all of the ZP3 coding sequence, since the *lacZ* gene-insert fusion encodes a protein about 45 kD larger than native β-galactosidase.

Role of Polypeptide Chain in Sperm Receptor Activity

The sperm receptor activity of ZP3 is extremely stable (Bleil and Wassarman 1980a; Wassarman et al. 1984a). It is virtually unaffected by lyophilization, freezing and thawing, boiling, or by exposure to reducing agents (dithiothreitol or 2-mercaptoethanol), detergents (0.1% SDS or 1% Triton X-100), or denaturants (7 M urea). This suggests that protein conformation plays little, if any, role in the sperm receptor function of ZP3.

To assess directly the role of polypeptide chain, purified ZP3 has been digested extensively with insoluble carboxymethyl cellulose (CMC)-conjugated Pronase and tested for sperm receptor activity in the competition assay (Florman et al. 1984). Glycopeptides, 1.5–6 kD M_r, are virtually as effective as intact ZP3 (83 kD M_r) in inhibiting binding of sperm to eggs (Fig. 5), and the effect is not simply attributable to an inhibition of sperm motility that could affect the frequency of gamete contact. This result demonstrates that the integrity of the polypeptide chain of ZP3 is not essential for its sperm receptor activity.

Role of Polypeptide Chain in Acrosome Reaction-inducing Activity

The acrosome reaction-inducing activity of ZP3 is also extremely stable, being unaffected by exposure to denaturants, detergents, and other agents. However, small glycopeptides (1.5–6 kD M_r) obtained by extensive digestion of ZP3 with CMC-conjugated Pronase, although functional in the sperm receptor assay, fail to induce sperm to undergo the acrosome reaction (Fig. 6; Florman et al. 1984). The extent of loss of acrosome reaction-inducing activity is a function of the extent of digestion of ZP3 by Pronase, such that glycopeptides larger than about 45 kD M_r retain the ability to induce the acrosome reaction, albeit at a diminished efficiency as compared with intact ZP3 (Fig. 6). These results suggest that the polypeptide chain of ZP3 does play a role in its function as inducer of the acrosome reaction.

Role of Oligosaccharides in Sperm Receptor Activity

The stability of the sperm receptor activity of ZP3, as well as the ability of small ZP3 glycopeptides to act as sperm receptor, suggests that carbohydrate components, rather than polypeptide chain, are involved. We have examined directly the role of carbohydrate in sperm receptor function by selective removal of N- and/or O-linked oligosaccharides from ZP3 (Florman and Wassarman 1985; F.G. Samuels and P.M. Wassarman, unpubl.).

Extensive deglycosylation of ZP3 with trifluoromethanesulfonic acid (TFMS), a reagent that removes both N- and O-linked oligosaccharides from glycoproteins (Edge et al. 1981), converts ZP3 to a species with

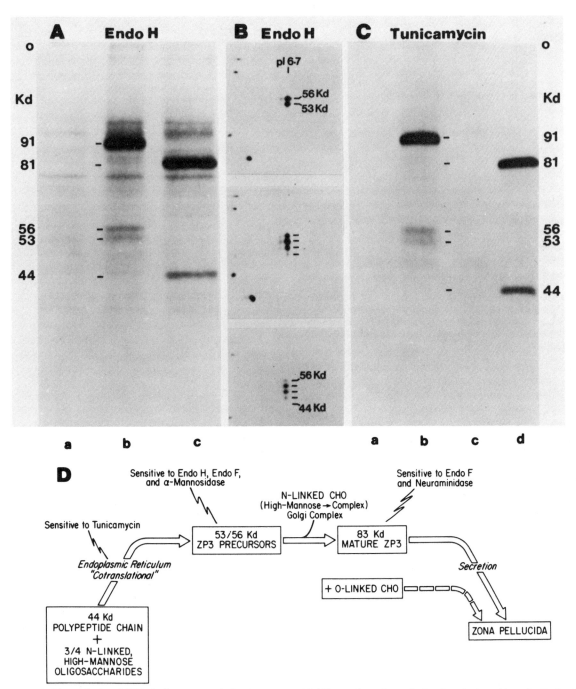

Figure 4. Biosynthesis of ZP3 during oogenesis in the mouse. (*A*) Electrophoretic analysis of endoglycosidase H (Endo H) digests of [^{35}S]methionine-labeled ZP3 precursors from growing mouse oocytes. Zona pellucida-free oocytes were treated with rabbit antisera directed against either mouse IgG (lane *a*) or ZP2/ZP3 (lanes *b* and *c*). Shown is a fluorogram in which anti-ZP2/ZP3 immunoprecipitates were either analyzed immediately (lane *b*) or digested with Endo H for 48 hr (lane *c*), and then processed for and subjected to SDS-PAGE and fluorography. ZP2 (91 kD) and ZP3 (53 and 56 kD) precursors and their respective Endo H digestion products (ZP2, 81 kD; ZP3, 44 kD) are indicated. (*B*) Electrophoretic analysis of Endo H partial digests of [^{35}S]methionine-labeled ZP3 precursors. Shown are portions of fluorograms of two-dimensional gels in which anti-ZP2/ZP3 immunoprecipitates were either not exposed to Endo H (*top*) or were exposed to Endo H for 1 hr (*middle*) or 2 hr (*bottom*) and were then processed for and subjected to two-dimensional gel electrophoresis and fluorography. (*C*) Electrophoretic analysis of the effect of tunicamycin on the biosynthesis of ZP3. Growing oocytes were labeled with [^{35}S]methionine for 3 hr in the presence or absence of tunicamycin. Shown is a fluorogram in which anti-IgG (lanes *a* and *c*) and anti-ZP2/ZP3 (lanes *b* and *d*) immunoprecipitates of oocytes cultured either in the presence (lanes *c* and *d*) or absence (lanes *a* and *b*) of tunicamycin were analyzed. ZP2 (91 kD) and ZP3 (53 and 56 kD) precursors and their respective unglycosylated forms (ZP2, 81 kD; ZP3, 44 kD) are indicated. (*D*) Biosynthetic pathway for ZP3 during oogenesis in the mouse (Greve et al. 1982; Roller and Wassarman 1983; Salzmann et al. 1983; Wassarman et al. 1984b).

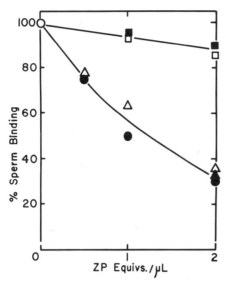

Figure 5. Binding of sperm to eggs in the presence of modified forms of mouse egg ZP3. Binding of sperm to eggs was compared after exposure of sperm to untreated ZP3 (●), Pronase-treated ZP3 (△), trifluoromethanesulfonic acid-treated ZP3 (■), endoglycosidase F-treated ZP3 (▲), and alkali-treated ZP3 (□), or to culture medium alone (○). Samples were examined as described in the legend to Fig. 2 (Florman et al. 1984; Florman and Wassarman 1985).

a relative molecular weight approximating that of the polypeptide chain (44 kD) that is inactive as a sperm receptor (Fig. 5). On the other hand, selective removal of N-linked oligosaccharides from ZP3 by digestion with endo-β-N-acetyl-D-glucosaminidase F (Endo F;

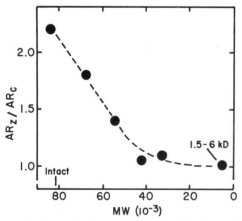

Figure 6. Relationship between the size of ZP3 glycopeptides and their acrosome reaction-inducing activity. ZP3 purified from unfertilized egg zonae pellucidae was treated with CMC-conjugated Pronase and subjected to SDS-PAGE, and glycopeptides of differing molecular weights were electroeluted and tested for acrosome reaction-inducing activity. The number of acrosome-reacted sperm in the presence of ZP3 glycopeptides (AR_Z) is compared with the number in the absence of ZP3 glycopeptides (AR_C). The effects of intact ZP3 (≈83 kD M_r) and the smallest ZP3 glycopeptides (1.5–6 kD M_r) are indicated (Florman et al. 1984; H.M. Florman and P.M. Wassarman, unpubl.).

Elder and Alexander 1982) has no effect on its sperm receptor activity (Fig. 5). These results suggest that O-linked oligosaccharides are responsible for the sperm receptor activity of ZP3.

ZP3 has been subjected to mild alkaline hydrolysis (β-elimination; Sharon 1975) under various conditions to selectively remove O-linked oligosaccharides without breakage of peptide bonds. Following β-elimination of ZP3 in the presence of 5 mM NaOH and 1 M $NaBH_4$, glycosidically linked serine and threonine residues are converted to alanine and α-aminobutyric acid, respectively, N-acetyl-D-galactosamine at the reducing termini of released oligosaccharides is converted to the sugar alcohol, and the polypeptide chain remains intact. After such treatment, sperm receptor activity is found associated with released oligosaccharides (≈3.4–4.6 kD apparent M_r) having N-acetyl-D-galactosaminitol at the reducing terminus (Fig. 7A), but not with the N-glycosylated polypeptide chain (Fig. 5). Therefore, the sperm receptor activity of ZP3 is attributable to a specific size class of O-linked oligosaccharides.

Binding of Oligosaccharides with Receptor Activity to Sperm

Our results strongly suggest that a specific size class of O-linked oligosaccharides on ZP3 accounts for the glycoprotein's sperm receptor activity. To determine whether or not these oligosaccharides bind to sperm, radiolabeled oligosaccharides, released from ZP3 during mild alkaline reduction in the presence of NaB^3H_4, were incubated with sperm, the sperm centrifuged through dibutyl phthalate into sucrose–Triton X-100, and the associated oligosaccharides subjected to gel filtration analysis (Florman and Wassarman 1985).

Under the conditions just described, there is a significant enrichment of oligosaccharides in a region of the elution profile found to possess sperm receptor activity (≈3.4–4.6 kD apparent M_r) (Fig. 7B). Although other size classes of ZP3 oligosaccharides are associated with sperm, the extent of their association simply reflects their relative abundance in the total population. Binding of ZP3 oligosaccharides to sperm appears to be specific, since analogous experiments using mouse adipocytes do not reveal any selective binding of oligosaccharides to these cells. It should be noted that neither ZP3 oligosaccharides nor ZP3 glycopeptides bind as tightly to sperm as intact ZP3 (J.D. Bleil et al., unpubl.). These results provide further experimental evidence for the role of O-linked oligosaccharides of ZP3 in sperm receptor function.

DISCUSSION

We have briefly described the identification and characterization of a glycoprotein responsible for binding of sperm to eggs, as well as for certain subsequent events that lead to fertilization of mouse eggs. This gly-

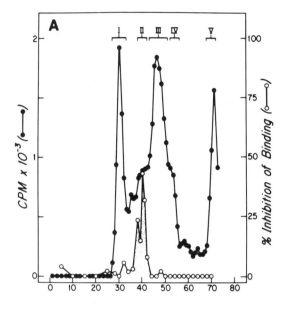

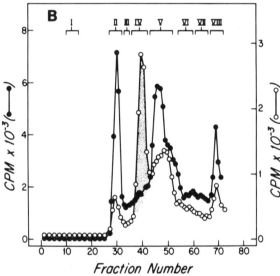

Figure 7. Gel filtration analysis of sperm receptor activity and sperm binding of O-linked oligosaccharides released from purified unfertilized egg ZP3. (*A*) O-linked oligosaccharides, released from purified ZP3 during mild alkaline hydrolysis in the presence of NaB³H₄, were subjected to gel filtration on Bio-Gel P6. Shown are the elution profile for radiolabeled oligosaccharides (●) and the sperm receptor activity profile (○). The region exhibiting sperm receptor activity is stippled. Region I represents the void volume and [³H]borohydride eluted in region V. (*B*) O-linked oligosaccharides were obtained as in *A* and an aliquot was incubated with sperm. Sperm were centrifuged through dibutyl phthalate into sucrose–Triton X-100, and oligosaccharides in the detergent phase were subjected to gel filtration on Bio-Gel P6. Shown are the elution profiles for oligosaccharides not incubated with sperm (●) and associated with sperm after a 1-hr incubation (○). The region of the elution profile that was significantly enriched following incubation of oligosaccharides with sperm (region IV) is stippled. Only region IV exhibited sperm receptor activity in the competition assay. Region II represents the void volume and [³H]borohydride eluted in region VIII (Florman and Wassarman 1985).

coprotein, called ZP3, is found in more than a billion copies in zonae pellucidae, the mouse egg's extracellular coat. ZP3 is recognized by acrosome-intact sperm, apparently in a relatively species-specific manner, and interacts with a component of plasma membrane overlying the anterior region of the sperm head. A specific size class of O-linked oligosaccharides solely accounts for the receptor activity of ZP3, whereas the polypeptide chain of ZP3 is also involved in its acrosome reaction-inducing activity. As a consequence of fertilization, ZP3 is altered such that it no longer exhibits either sperm receptor or acrosome reaction-inducing activity.

Based on evidence gathered to date, it is tempting to suggest that binding of sperm to eggs in mice is mediated by lectin-like interactions. Especially germane to this suggestion is the identification of O-linked oligosaccharide as the "recognition" moiety on ZP3 (Florman and Wassarman 1983, 1985). Taken together with observations that various lectins, monosaccharides, and glycoconjugates inhibit binding of sperm to mammalian eggs (Oikawa et al. 1973; Ahuja 1982; Huang et al. 1982; Shur and Hall 1982; Wassarman et al. 1984a), and that various monosaccharides and polysaccharides inhibit sperm–egg interaction in several invertebrate and plant species (Bolwell et al. 1979, 1980; Rosati and De Santis 1980; Glabe et al. 1982; Barnum and Brown 1983), a relatively strong case emerges in support of such a mechanism. In this context, several lines of evidence suggest that gamete adhesion in sea urchins is mediated by "bindin," a lectin-like sperm protein associated with acrosomes, and by a carbohydrate-containing sperm receptor in the egg's vitelline envelope (Vacquier and Moy 1977; Glabe and Vacquier 1978; Glabe et al. 1982; Rossignol et al. 1984). It remains for a bindin-like molecule to be identified on plasma membrane overlying the head of mammalian sperm.

ZP3 plays a multifaceted role in regulating fertilization in mice. It mediates binding of sperm to eggs, induces bound sperm to undergo the acrosome reaction, and participates in a secondary block to polyspermy. Whereas O-linked oligosaccharides alone account for the receptor activity of ZP3, a role for polypeptide chain is indicated for its acrosome reaction-inducing activity. In view of the harsh procedures used to purify ZP3, it seems unlikely that anything other than polypeptide chain primary structure is important to acrosome reaction-inducing activity. In this context, we suggest that there are multiple O-linked oligosaccharide binding sites on each ZP3 molecule, and these sites must be joined by a polypeptide chain (polyvalent) for ZP3 to induce sperm to undergo the acrosome reaction. The location, number, and/or spacing of O-linked oligosaccharides along the polypeptide chain may influence the receptor and acrosome reaction-inducing activities of intact ZP3, as well as of glycopeptides derived from ZP3. Following binding of acrosome-intact sperm to ZP3, it is possible that a "capping-like" reaction occurs over plasma membrane at the anterior region of the sperm head; preliminary results from our laboratory support this possibility (J.D. Bleil and P.M. Wassarman, unpubl.). This, in turn, could alter the affected region of plasma membrane such that it is rendered capable of fusing with outer acrosomal membrane. Since calcium is involved in the acrosome reaction, and ionophores such as A23187 induce the acrosome reaction, it seems likely that the alteration could be to ion channels. Further experiments are needed to sort out this particular aspect of sperm-egg interaction in mammals.

Sperm do not bind to zonae pellucidae of fertilized mouse eggs. Furthermore, $ZP3_f$ purified from embryo zonae pellucidae does not exhibit either receptor or acrosome reaction-inducing activity at concentrations at which ZP3 from unfertilized egg zonae pellucidae exhibits both activities. At much higher concentrations (three- to fivefold), receptor and acrosome reaction-inducing activities are detected in preparations of embryo $ZP3_f$ due to the presence of residual ZP3 (J.D. Bleil and P.M. Wassarman, unpubl.). In other words, not every molecule of ZP3 is converted to $ZP3_f$ following fertilization of eggs; a similar situation applies to the proteolytic conversion of ZP2 to $ZP2_f$ following fertilization (Bleil et al. 1981).

Both inactivation of ZP3 and proteolysis of ZP2, which occur in fertilized eggs, are probably attributable to cortical granule components released into zonae pellucidae as a consequence of the cortical reaction. In sea urchins, it has been suggested that proteases that originate from the egg's cortical granules release sperm receptors from the vitelline envelope following fertilization (Vacquier et al. 1972, 1973; Glabe and Vacquier 1978). However, there is no evidence as yet to suggest that proteolysis is involved in conversion of ZP3 to $ZP3_f$ (Bleil and Wassarman 1980a; Bleil et al. 1981). Whatever the nature of the alteration of ZP3, it is subtle. In view of the role of O-linked oligosaccharides in sperm receptor activity, the possibility that a cortical granule glycosidase modifies ZP3 should be considered seriously.

Finally, there is evidence to suggest that mammalian sperm receptors exhibit a certain degree of species specificity (Bedford 1981; Wassarman 1983a; Wassarman et al. 1984a; Yanagimachi 1984). For example, in vitro fertilization of zona pellucida-free eggs by heterologous sperm is quite common, whereas hybrid fertilization of zona pellucida-intact eggs is rare (Adams 1974; Barros and Leal 1980; Gulyas and Schmell 1981; Yanagimachi 1981, 1984). Whether or not ZP3 and/or O-linked oligosaccharides derived from ZP3 exhibit any species specificity in the in vitro competition assay described here has not been determined as yet. Certainly, the diversity of oligosaccharide structures is compatible with generation of relative species specificity for sperm receptors (Kobata 1984; Sadler 1984). It will be of interest to compare the structures of ZP3 O-linked oligosaccharides possessing receptor activity with functionally analogous oligosaccharides from other mammalian species.

ACKNOWLEDGMENTS

This research was supported in part by grants awarded to P.M.W. by the National Institute of Child Health and Human Development and the National Science Foundation. H.M.F. was a National Institutes of Health postdoctoral fellow, J.M.G. and F.G.S. were Rockefeller Foundation postdoctoral fellows, and J.D.B., R.J.R., and G.S.S. were predoctoral fellows supported by National Research Service Awards.

Finally, P.M.W. is delighted to acknowledge the many tangible and intangible contributions of his colleague, collaborator, and friend, Dr. Melvin DePamphilis, to this research.

REFERENCES

ADAMS, C.E. 1974. Species specificity in fertilization. In *Physiology and genetics of reproduction* (ed. E.N. Coutinho and F. Fuchs), p. 69. Plenum Press, New York.

AHUJA, K.K. 1982. Fertilization studies in the hamster. The role of cell-surface carbohydrates. *Exp. Cell Res.* **140:** 353.

BARNUM, S.R. and G.G. BROWN. 1983. Effects of lectins and sugars on primary sperm attachment in the horseshoe crab, *Limulus polyphemus*. *Dev. Biol.* **95:** 352.

BARROS, C. and J. LEAL. 1980. *In vitro* fertilization and its use to study gamete interactions. In *In vitro fertilization and embryo transfer* (ed. E. Hafez and K. Semm), p. 37. A.R. Liss, New York.

BEDFORD, J.M. 1981. Why mammalian gametes don't mix. *Nature* **291:** 286.

―――. 1982. Fertilization. In *Reproduction in mammals: Germ cells and fertilization* (ed. C.R. Austin and R.V. Short), vol. 1, p. 128. Cambridge University Press, Cambridge, England.

BLEIL, J.D. and P.M. WASSARMAN. 1978. Identification and characterization of the proteins of the zona pellucida. *J. Cell Biol.* **79:** 173a.

―――. 1980a. Mammalian sperm-egg interaction: Identification of a glycoprotein in mouse egg zonae pellucidae possessing receptor activity for sperm. *Cell* **20:** 873.

―――. 1980b. Structure and function of the zona pellucida: Identification and characterization of the proteins of the mouse oocyte's zona pellucida. *Dev. Biol.* **76:** 185.

―――. 1980c. Synthesis of zona pellucida proteins by denuded and follicle-enclosed mouse oocytes during culture in vitro. *Proc. Natl. Acad. Sci.* **77:** 1029.

―――. 1983. Sperm-egg interactions in the mouse: Sequence of events and induction of the acrosome reaction by a zona pellucida glycoprotein. *Dev. Biol.* **95:** 317.

BLEIL, J.D., C.F. BEALL, and P.M. WASSARMAN. 1981. Mammalian sperm-egg interaction: Fertilization of mouse eggs triggers modification of the major zona pellucida glycoprotein. *Dev. Biol.* **86:** 189.

BOLWELL, G.P., J.A. CALLOW, and L.V. EVANS. 1980. Fertilization in brown algae. III. Preliminary characterization of putative gamete receptors from eggs and sperm of *Fucus serratus*. *J. Cell. Sci.* **43:** 209.

BOLWELL, G.P., J.A. CALLOW, M.W. CALLOW, and L.V. EVANS. 1979. Fertilization in brown algae. II. Evidence for lectin-sensitive complementary receptors involved in gamete recognition in *Fucus serratus*. *J. Cell. Sci.* **36:** 19.

EDGE, A.S.B., C.R. FALTYNEK, L. HOF, L.E. REICHERT, and P. WEBER. 1981. Deglycosylation of glycoproteins by trifluoromethanesulfonic acid. *Anal. Biochem.* **118:** 131.

ELDER, J.H. and S. ALEXANDER. 1982. Endo-β-N-acetylglucosaminidase F: Endoglycosidase from *Flavobacterium meningosepticum* that cleaves both high-mannose and complex glycoproteins. *Proc. Natl. Acad. Sci.* **79:** 4540.

FLORMAN, H.M. and B.T. STOREY. 1982. Mouse gamete interactions: The zona pellucida is the site of the acrosome reaction leading to fertilization *in vitro*. *Dev. Biol.* **91:** 121.

FLORMAN, H.M. and P.M. WASSARMAN. 1983. The mouse egg's receptor for sperm: Involvement of O-linked carbohydrate. *J. Cell. Biol.* **97:** 26a.

———. 1985. O-Linked oligosaccharides of mouse egg ZP3 account for its sperm receptor activity. *Cell* **41:** 313.

FLORMAN, H.M., K.B. BECHTOL, and P.M. WASSARMAN. 1984. Enzymatic dissection of the functions of the mouse egg's receptor for sperm. *Dev. Biol.* **106:** 243.

GLABE, C.B. and V.D. VACQUIER. 1978. Egg surface glycoprotein receptor for sea urchin sperm bindin. *Proc. Natl. Acad. Sci.* **75:** 881.

GLABE, C.B., L.B. GRABEL, V.D. VACQUIER, and S.D. ROSEN. 1982. Carbohydrate specificity of sea urchin bindin: A cell surface lectin mediating sperm-egg interaction. *J. Cell. Biol.* **94:** 123.

GREVE, J.M. and P.M. WASSARMAN. 1985. Mouse extracellular coat is a matrix of interconnected filaments possessing a structural repeat. *J. Mol. Biol.* **181:** 253.

GREVE, J.M., G.S. SALZMANN, R.J. ROLLER, and P.M. WASSARMAN. 1982. Biosynthesis of the major zona pellucida glycoprotein secreted by oocytes during mammalian oogenesis. *Cell* **31:** 749.

GULYAS, B.J. and E.D. SCHMELL. 1981. Sperm-egg recognition and binding in mammals. In *Bioregulators of reproduction* (ed. G. Jagiello and H.J. Vogel), p. 499. Academic Press, New York.

GWATKIN, R.B.L. 1977. *Fertilization mechanisms in man and mammals*. Plenum Press, New York.

GWATKIN, R.B.L. and D.T. WILLIAMS. 1976. Receptor activity of the solubilized hamster and mouse zona pellucida before and after the zona reaction. *J. Reprod. Fertil.* **49:** 55.

HUANG, T.T.F., E. OHZU, and R. YANAGIMACHI. 1982. Evidence suggesting that L-fucose is part of a recognition signal for sperm-zona pellucida attachment in mammals. *Gamete Res.* **5:** 355.

INOUE, M. and D.P. WOLF. 1974a. Solubility properties of the murine zona pellucida. *Biol. Reprod.* **10:** 512.

———. 1974b. Comparative solubility properties of the zona pellucida of fertilized and unfertilized mouse ova. *Biol. Reprod.* **11:** 558.

KOBATA, A. 1984. The carbohydrates of glycoproteins. In *Biology of carbohydrates* (ed. V. Ginsburg and P.W. Robbins), vol. 2, p. 87. Wiley, New York.

OIKAWA, T., R. YANAGIMACHI, and G.L. NICOLSON. 1973. Wheat germ agglutinin blocks mammalian fertilization. *Nature* **241:** 256.

ROLLER, R.J. and P.M. WASSARMAN. 1983. Role of asparagine-linked oligosaccharides in secretion of glycoproteins of the mouse egg's extracellular coat. *J. Biol. Chem.* **258:** 13243.

ROSATI, F. and R. DESANTIS. 1980. Role of surface carbohydrates in sperm-egg interaction in *Ciona intestinalis*. *Nature* **283:** 762.

ROSSIGNOL, D.P., B.J. EARLES, G.L. DECKER, and W.J. LENNARZ. 1984. Characterization of the sperm receptor on the surface of the eggs of *Strongylocentrotus purpuratus*. *Dev. Biol.* **104:** 308.

SADLER, J.E. 1984. Biosynthesis of glycoproteins: Formation of O-linked oligosaccharides. In *Biology of carbohydrates* (ed. V. Ginsburg and P.W. Robbins), vol. 2, p. 199. Wiley, New York.

SALING, P.M. and B.T. STOREY. 1979. Mouse gamete interactions during fertilization *in vitro*. Chlortetracycline as a fluorescent probe for the mouse acrosome reaction. *J. Cell Biol.* **83:** 544.

SALING, P.M., J. SOWINSKI, and B.T. STOREY. 1979. An ultrastructural study of epididymal mouse spermatozoa binding to zonae pellucidae *in vitro*: Sequential relationship to the acrosome reaction. *J. Exp. Zool.* **109:** 229.

SALZMANN, G.S., J.M. GREVE, R.J. ROLLER, and P.M. WASSARMAN. 1983. Biosynthesis of the sperm receptor during oogenesis in the mouse. *EMBO J.* **2:** 1451.

SCHMELL, E.D. and B.J. GULYAS. 1980. Ovoperoxidase activity in ionophore treated mouse eggs. II. Evidence for the enzyme's role in hardening of the zona pellucida. *Gamete Res.* **3:** 279.

SHARON, N. 1975. *Complex carbohydrates: Their chemistry, biosynthesis, and functions*. Addison-Wesley, Reading, Massachusetts.

SHUR, B.D. and G. HALL. 1982. A role for mouse sperm galactosyltransferases in sperm binding to the egg zona pellucida. *J. Cell Biol.* **95:** 574.

VACQUIER, V.D. and G.W. MOY. 1977. Isolation of bindin: The protein responsible for adhesion of sperm to sea urchin eggs. *Proc. Natl. Acad. Sci.* **74:** 2456.

VACQUIER, V.D., D. EPEL, and L.A. DOUGLAS. 1972. Sea urchin eggs release protease activity at fertilization. *Nature* **237:** 34.

VACQUIER, V.D., M.J. TEGNER, and D. EPEL. 1973. Protease released from sea urchin eggs at fertilization alters the vitelline layer and aids in preventing polyspermy. *Exp. Cell Res.* **80:** 111.

WASSARMAN, P.M. 1983a. Fertilization. In *Cell interactions and development: Molecular mechanisms* (ed. K. Yamada), p. 1. Wiley, New York.

———. 1983b. Oogenesis: Synthetic events in the developing mammalian egg. In *Mechanism and control of animal fertilization* (ed. J.F. Hartmann), p. 1. Academic Press, New York.

WASSARMAN, P.M. and J.D. BLEIL. 1982. The role of zona pellucida glycoproteins as regulators of sperm-egg interactions in the mouse. In *Cellular recognition* (ed. W.A. Frazier et al.), p. 845. A.R. Liss, New York.

WASSARMAN, P.M. and W.J. JOSEFOWICZ. 1978. Oocyte development in the mouse: An ultrastructural comparison of oocytes isolated at various stages of growth and meiotic competence. *J. Morphol.* **156:** 209.

WASSARMAN, P.M., H.M. FLORMAN, and J.M. GREVE. 1984a. Receptor mediated sperm-egg interactions in mammals. In *Biology of fertilization* (ed. C.B. Metz and A. Monroy), vol. 2, p. 341. Academic Press, New York.

WASSARMAN, P.M., J.M. GREVE, R.M. PERONA, R.J. ROLLER, and G.S. SALZMANN. 1984b. How mouse eggs put on and take off their extracellular coat. In *Molecular biology of development* (ed. E.H. Davidson and R.A. Firtel), p. 213. A.R. Liss, New York.

YANAGIMACHI, R. 1981. Mechanisms of fertilization in mammals. In *Fertilization and embryonic development in vitro* (ed. L. Mastroianni and J.D. Biggers), p. 82. Plenum Press, New York.

———. 1984. Zona-free hamster eggs: Their use in assessing fertilizing capacity and examining chromosomes of human spermatozoa. *Gamete Res.* **10:** 187.

Localized Maternal mRNAs in *Xenopus laevis* Eggs

D.L. WEEKS, M.R. REBAGLIATI, R.P. HARVEY, AND D.A. MELTON
Department of Biochemistry and Molecular Biology, Harvard University, Cambridge, Massachusetts 02138

Cells that divide to give rise to nonidentical progeny must be able to assign each daughter cell its own identity. The most dramatic example of this problem is the development of an embryo from an egg. In 1925 Wilson suggested that one solution to this problem during embryogenesis might be the differential distribution of maternal factors in the egg. Maternal factors or cytoplasmic determinants that are segregated into particular cells during cleavage can, in principle, specify the fates of the recipient cells.

Most of the evidence suggesting that localized maternal factors have a determinative role in development comes from experiments in which a particular developmental potential is correlated with a cytoplasmic region of the egg. For example, the formation of germ cells during frog development is prevented by the irradiation of the egg's vegetal pole and restored by the reintroduction of vegetal pole cytoplasm from nonirradiated eggs. Irradiation of the opposite end of the egg or attempts to restore germ cell formation with cytoplasm from other egg locations have no effect on germ cell development (Smith 1966). Germ cell formation in insects has been examined by similar experiments (Illmensee and Mahowald 1974; Okada et al. 1974) with the conclusion that localized maternal factors are responsible for their proper development. Other events best explained by the localization of maternal factors include the formation of comb plates and photocytes in ctenophores (Freeman 1976), body axis formation in amphibians and insects (Nieuwkoop 1977; Nusslein-Volhard 1979; Gerhart et al. 1983; Kalthoff 1983), and the induction of embryonic muscle actin genes in frog embryos (Gurdon et al. 1985).

Xenopus laevis provides a convenient system for further studies on the role of localized maternal components in development. *Xenopus* eggs are large and some of their molecular components are well characterized. The embryos are transcriptionally silent from the time the egg matures until just prior to gastrulation (Newport and Kirschner 1982). This situation allows for the examination of a relatively constant pool of maternal RNA and prevents having to differentiate between localized maternal RNA and localized transcription of embryonic DNA. The pigment covering the animal pole of the oocyte, egg, and early embryo reliably distinguishes the animal and vegetal poles of the organism. Fate-mapping experiments have shown that the animal pole and the vegetal pole give rise to different cell types. In addition to these differences in developmental potential, the molecular composition is not identical in opposite ends of *Xenopus* eggs. The distribution of total protein (Moen and Namenwirth 1977) and RNA (Brachet 1977; Sagata et al. 1981; Carpenter and Klein 1982; Phillips 1982) along the animal-vegetal axis of the egg has been examined. These studies indicate that there is an unequal distribution of both of these molecular components and suggest that there are individual members of these general classes of molecules that reside at one pole or the other.

In this paper we review our studies showing that RNAs localized to either the animal or vegetal poles of *Xenopus* eggs exist and have been isolated as cDNA clones. We characterize the size, location, and transcription pattern of four localized RNAs during frog development using cloned cDNA probes. The isolation of these cDNA clones provides specific examples of RNAs that are localized within a single cell, the frog egg. As a class, these RNAs are candidates for the maternal factors that may be involved in specifying cell fate. Moreover, they provide substrates for studies to determine how informational molecules are localized within a frog egg.

EXPERIMENTAL PROCEDURES

Materials

RNase-free DNase and SP6 RNA polymerase were prepared as described previously (Melton et al. 1984) and SP6 RNA polymerase was obtained from Promega-Biotec. Reverse transcriptase was purchased from Life Sciences Inc; mare serum gonadotropin (No. G-4877) and human chorionic gonadotropin (No. CG-10) were purchased from Sigma; *X. laevis* specimens were purchased from Xenopus I (Ann Arbor, Michigan); and radionucleotides were purchased from Amersham.

Oocyte cDNA Library Construction

Total poly(A)$^+$ RNA was isolated from defolliculated oocytes (all stages) and copied into double-stranded cDNA. Following digestion with S1 nuclease, the cDNA was methylated with *Eco*RI methylase and ligated to *Eco*RI linkers. The cDNA was then digested with *Eco*RI and purified from the linkers by Sepharose CL-4B (Pharmacia) chromatography. The cDNA was then ligated into *Eco*RI-digested λGT10. Details of these steps can be found in Toole et al. (1984) and Huynh et al. (1985). The initial packaging gave a library of about 2×10^6 independent clones. Assays on 24 random clones show that the average cDNA insert is 2.5 kb.

Extraction of RNA from Egg and Embryo Sections

X. laevis females were primed with mare serum gonadotropin and induced to lay eggs with an injection of human chorionic gonadotropin. Embryos were obtained by artificial insemination and staged according to Nieuwkoop and Faber (1967). Eggs and embryos were oriented, frozen on dry ice, and sectioned with a scalpel under a dissecting microscope to cut off either one-third to one-fifth of the animal pole or the one-third to one-fifth of vegetal pole. Jelly coats were not removed prior to sectioning. Sections were homogenized in a buffer containing Proteinase K and total nucleic acids were extracted and ethanol-precipitated, as described previously (Krieg and Melton 1984). To prepare total RNA free from polysaccharides, the ethanol precipitate was dissolved in diethyl pyrocarbonate (DEPC)-treated water and an equal volume of 8 M LiCl was added. After incubation at $-20°C$, the RNA was recovered by centrifugation.

Probe Synthesis

To prepare cDNA probes for library screening, animal (An) or vegetal (Vg) poly(A)$^+$ RNA was copied into ^{32}P-labeled, single-stranded cDNA to yield probes with a specific activity of about 5×10^8 cpm/μg. To synthesize single-stranded RNA probes for Northern blots and RNase mapping, An and Vg cDNAs were subcloned into the *Eco*RI site of pSP65 (Melton et al. 1984). In vitro transcription with SP6 RNA polymerase was performed as described previously (Melton et al. 1984).

Library Screening

Duplicate nitrocellulose filters (Benton and Davis 1977) were used to lift plaques from plates containing the library. Filters were air-dried and then stacked between sheets of Whatmann 3MM paper. The stack was then autoclaved for 2 minutes under low steam pressure at 100°C. Autoclave pressure was released by fast exhaust, the dry cycle run for 10 seconds, and the filters removed. After baking for 1 or 2 hours at 80°C, the filters were prehybridized at 42°C in 50% formamide, 5× SSC, 5× Denhardt's, 50 mM sodium phosphate (pH 7), and 200 μg/ml denatured salmon sperm DNA. Filters were hybridized in a similar solution with 0.6× Denhardt's at 42°C using 8×10^6 to 15×10^6 cpm of cDNA probe per 15-cm filter. Filters were washed with 0.5× SSC and 0.1% SDS at 65°C and exposed to preflashed X-ray film with intensifying screens.

Northern Blots and RNase Mapping

RNAs were electrophoresed in formaldehyde agarose gels as described (Goldberg 1980), blotted to Gene Screen (New England Nuclear), and cross-linked for 5 minutes at 600 μW/cm^2 with a short-wave UV lamp (UV Products, Inc., bulb #34000801), as described (Church and Gilbert 1984). The blot was baked for 1 hour at 80°C in vacuo. For a nick-translated DNA probe, the hybridization was done at 42°C in 50% formamide, 1% SDS, 5× SSPE, 5× Denhardt's, 50 mM sodium phosphate (pH 7), 7% dextran sulfate, and 100 μg/ml denatured salmon sperm DNA. Filters were washed in 0.2× SSPE and 0.5% SDS at 65°C. To reuse the blot, bound probe was stripped by washing at 70–75°C for 30–60 minutes in 50% formamide, 1 mM EDTA, 25 mM sodium phosphate (pH 7), and 0.1% SDS. For SP6 RNA probes, hybridizations were done at 58–60°C in 50% formamide, 5× SSPE, 5% SDS, 100 μg/ml denatured salmon sperm DNA, and 5 μg/ml poly(A). Filters were washed at 65°C in 0.1× SSPE and 0.5% SDS. Densitometric analysis of autoradiographs was performed on a Helana Quick-Scan R and D densitometer. RNase mapping was performed as previously described (Melton et al. 1984).

Testing for Mitochondrial or Ribosomal RNA Sequences

The cDNA inserts of pAn1, pAn2, pAn3, and pVg1 were blotted onto nitrocellulose and hybridized with nick-translated mitochondrial or ribosomal DNA fragments of pX1M31 (Rastl and Dawid 1979) or pX1r101 (Dawid and Wellauer 1976), respectively. Although we estimate that about 3.2% of the sequences cloned in the library are mitochondrial and 0.9% represent ribosomal RNA sequences, there was no detectable hybridization of either probe to the An or Vg cDNA sequences under conditions in which positive controls were detected.

Identification of Localized RNA as mRNA

Polysomes were prepared as outlined previously (Richter et al. 1984). EDTA release was accomplished by treating homogenates prior to centrifugation with 100 mM EDTA for 10 minutes on ice. Cloning of An2 *Eco*RI fragment into pSP64T and its subsequent use as a translation template was carried out as previously described (Krieg and Melton 1984). In vitro translations were performed with a wheat germ extract from BRL.

RESULTS

Selection of cDNA Coding for Localized Maternal RNAs

The known differences in developmental potential between the animal and vegetal poles of the unfertilized egg encouraged us to look for maternal RNAs localized along this axis. To identify these sequences we constructed a cDNA library in λGT10 (Huynh et al. 1985). Poly(A)$^+$ RNA was isolated from defolliculated *Xenopus* oocytes and served as the cDNA template. The library contains about 2×10^6 independent clones with an average insert size of 2.5 kb and therefore should contain representatives of nearly all the transcripts present in oocytes and eggs.

Unfertilized eggs were frozen on dry ice and manually sectioned under a dissecting microscope. The top one-third to one-fifth of the egg was used as a source of animal RNA and a comparable portion of the bottom of the egg was the source for vegetal pole RNA (Fig. 1). Radioactive cDNA probes were synthesized from poly(A)$^+$ RNAs isolated from 400–1000 egg sections. Replica filters of the oocyte cDNA library were screened with these [^{32}P]cDNA probes. The autoradiographs resulting from this type of screening show that it is possible to identify plaques containing sequences that are differentially represented in the animal or vegetal RNA preparations (Fig. 1).

Following an overnight exposure of a typical library screening, only about 5% of the plaques are detected. With longer exposures of the autoradiograms, 60% of the plaques can be detected, but we were never able to detect a signal from all of the recombinant phage on a filter. Independent tests show that >90% of the recombinant phage contain a cDNA insert. Together, these results suggest that many (40%) of the recombinant phage contain sequences representing RNAs that are present in amounts too low to allow for detection of their corresponding phage with our [^{32}P]cDNA probes.

In these studies we have screened in excess of 10^6 recombinant phage and find that most of the sequences in the library are equally represented in both ends of the unfertilized egg. A small subset of recombinant phage, however, contain sequences that hybridize only

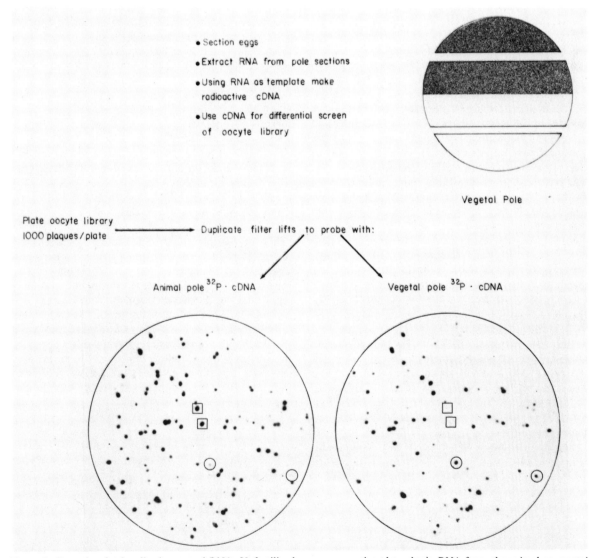

Figure 1. Screening for localized maternal RNA. Unfertilized eggs were sectioned to obtain RNA from the animal or vegetal pole. About one-third to one-fifth of the egg height was cut from the ends of the eggs as shown. [^{32}P]cDNA synthesized from this RNA was used to probe replica filters of portions of a λGT10 cDNA library. Representative autoradiographs of two such filters are shown. Putative animal pole- or vegetal pole-specific sequences are in boxes or circles, respectively.

to probe made from one end or the other (Fig. 1). From our primary screening we estimate that as many as 1.2% of the library represents sequences localized to the animal pole of the egg and 0.2% represents sequences localized to the vegetal pole. However, secondary screening revealed that a large number of these candidates were false positives.

Further rounds of screening eliminated many candidates and finally we selected four cDNA clones for further study. Three of these cDNA clones, An1, An2, and An3, represent RNAs localized in the animal pole. One clone, Vg1, corresponds to a vegetal-specific RNA.

Characterization of cDNA Clones for Localized RNAs

The inserts contained in the λGT10 recombinants were subcloned into SP6 vectors. In these vectors, the inserted DNA can serve as a template for the synthesis of single-stranded RNA in vitro, using SP6 RNA polymerase (Melton et al. 1984). This feature simplifies the generation of probes used for some of the remaining studies. Schematics of the constructions that allow the synthesis of single-stranded RNA complementary to the endogenous RNA are shown in Figure 2b.

The sizes of homologous transcripts present in unfertilized eggs have been determined for the four sequences by RNA blot analysis. An1, An2, and Vg1 all are present predominately as single transcripts of 2.9, 1.9, and 2.7 kb, respectively. An3 is present as a major transcript of 5.1 kb and a minor transcript of 3.5 kb. The relationship between these two transcripts is not currently known. We estimate that An2 is roughly 0.5% of the total poly(A)$^+$ RNA in the egg and that the other three are each between 0.1% and 0.25% (Rebagliati et al. 1985). We have not detected any cross-hybridization between these sequences and other abundant egg RNAs. This suggests that the localized RNAs are not members of that class of transcripts containing interspersed repeat elements that has been characterized by Davidson, Smith, and their colleagues (Anderson et al. 1982; Richter et al. 1984).

Northern blots performed with RNA isolated from animal or vegetal poles dramatically demonstrate that An1, An2, An3, and Vg1 are localized (Fig. 2a). This assay also provides a good measurement of the extent to which these sequences are localized. Each of the four localized sequences is present in at least 10-fold excess in one end of the egg.

RNA blots have been reprobed with Xenopus histone sequences to examine the amounts of histone RNA in the animal and vegetal tracks. The reanalysis of a blot first used to examine Vg1 localization shows that histone RNAs are about evenly distributed in the animal and vegetal sections (Fig. 2a). Careful measurements reveal that there is in fact a slight (about twofold) concentration gradient of histone RNA, with the high concentration at the animal pole. This is consistent with previous measurements of total poly(A)$^+$ distribution that demonstrated the presence of more RNA in the nonyolky animal pole (Brachet 1977; Sagata et al. 1981; Phillips 1982) and with estimates made for histone distribution in stage-VI oocytes as judged by in situ hybridization (Jamrich et al. 1984). Random cDNA clones from the oocyte cDNA library show a similar twofold difference in the distribution of their corresponding RNA along the animal-vegetal axis in this type of blot assay (data not shown). Thus, comparison of the distribution of histone or other randomly selected RNAs with An1, An2, An3, or Vg1 highlights the extent to which the latter are localized.

The An and Vg RNAs Remain Localized during Development

It is known that there are extensive cytoplasmic movements in frog eggs following fertilization (Phillips 1985) and these movements could potentially rearrange components like the An and Vg RNAs. The position of localized RNAs must be maintained if blastomeres are to receive different maternal information. We have therefore tested the distribution of An and Vg RNAs at later developmental stages. In these experiments (Fig. 3) the position of the RNAs was examined using a sensitive RNase protection assay (Melton et al. 1984). Single-stranded [^{32}P]RNA probes were synthesized in vitro with SP6 RNA polymerase. Assays performed on early blastula embryos, nine rounds of cell division later after fertilization, give the same result found for four cell embryos (Fig. 3). Thus, not only are the An and Vg RNAs localized in unfertilized eggs, they remain localized during early development.

We have also examined the expression pattern for these localized RNAs from early oogenesis (Dumont 1972) through the swimming tadpole stage (Fig. 4). These transcripts are first detectable in stage-II oocytes (early in oogenesis) and reach their maximum maternal level in stage-VI oocytes. There may be a reduction in localized RNA levels during maturation to eggs after which time they remain at a constant level throughout early embryogenesis until shortly after the midblastula transition. Two of the localized RNAs are present only up to the gastrula stages and are not found later in development. An1 transcripts disappear after the early gastrula stage (stage 10) and Vg1 RNAs are not detectable after late gastrula (stage 12). In contrast, the genes encoding An2 and An3 seem to be reactivated later in embryogenesis. Although this type of assay only measures the accumulation of stable RNAs, the results presented in Figure 4 suggest that the An2 gene is transcriptionally active at neurula (stage 18) and An3 at gastrula (stage 12). These results allow one to classify An1 and Vg1 transcripts as strictly maternal in their expression, whereas An2 and An3 are reexpressed and may have a role later in development.

The An and Vg RNAs Are mRNAs

Several properties of the An and Vg RNAs are consistent with their being mRNAs. First, the cDNA li-

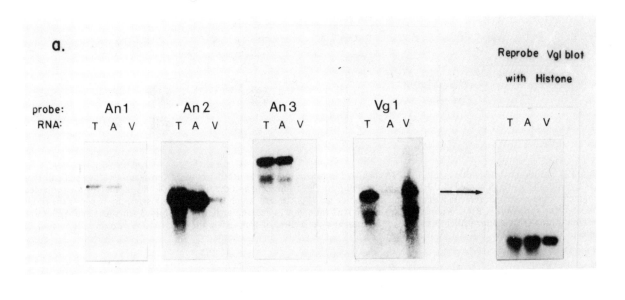

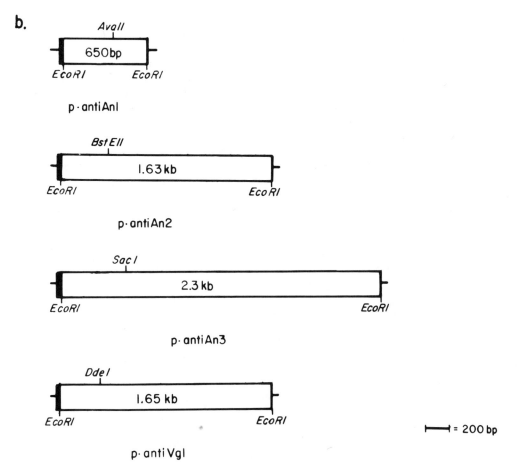

Figure 2. Identification and cloning of localized sequences. (*a.*) Northern blot analysis of localized RNA. RNAs were electrophoresed in a denaturing agarose gel and blotted onto Gene Screen. Blots were hybridized with ^{32}P-labeled nick-translated probes as indicated above each set of three lanes. Lane designations are: (T) total egg RNA; (A) animal pole egg RNA; (V) vegetal pole egg RNA. The Northern blot initially hybridized to Vg1 was washed free of probe and then rehybridized to histone probe. (*b.*) The structure of each cDNA clone is indicated. cDNA inserts from λGT10 were cloned into pSP65. Shown here are the antisense (p·anti) clones. The black box at the left represents the phage SP6 promoter. Internal restriction sites noted were used for linearizing plasmids for in vitro RNA synthesis (see Fig. 3).

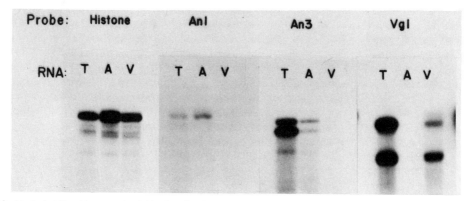

Figure 3. Solution hybridization assay for RNA localized in four-cell embryos. RNA isolated from sectioned four-cell embryos was hybridized to a single-stranded [^{32}P]RNA probe. Following hybridization and digestion with RNase A and T1 the protected fragments were electrophoresed in a denaturing acrylamide gel, a portion of which is shown. The histone probe shows the result obtained for nonlocalized RNA. Results obtained with An2 (not shown) are similar to those obtained with An1 and An3. Lane designations: (T) total four-cell embryo RNA; (A) animal pole section of four-cell embryo RNA; (V) vegetal pole section of unfertilized egg RNA.

brary and the probes used to screen the library were constructed with poly(A)$^+$ RNA, so that these RNAs are likely to be polyadenylated. Second, we have shown that these RNAs are not homologous to ribosomal DNA (nor to mitochondrial DNA). However, as mentioned earlier, eggs contain an unusual class of poly(A)$^+$ RNA, so-called interspersed repeat RNA, that does not code for protein (Richter et al. 1984). It is difficult to prove conclusively that the An and Vg RNAs do not belong to that class. Instead, we have done two experiments that suggest these RNAs do indeed code for proteins.

We have isolated RNA from eggs and embryos in order to look for the presence of localized RNAs in polysomes (Woodland 1974; Richter et al. 1984). Free RNA was separated from RNA bound in a RNA:protein complex and assayed by Northern blots and solution hybridization for the presence of localized transcripts. All four localized species were found in both free and protein-bound fractions of RNA (Fig. 5). This result indicates that some portion of these RNAs is found on polysomes during early development. In addition, the protein-coding capacity of one of the clones, An2, was demonstrated directly by using the cDNA to synthesize a functional mRNA in vitro. The An2 cDNA, which is nearly full length, was cloned into pSP64T (Krieg and Melton 1984) and used as a template for SP6 in vitro transcription. This vector provides a 5' leader and a 3' trailer including a poly(A) tail from a *Xenopus* β-globin gene. The resulting hybrid synthetic mRNA was used to direct protein synthesis in a wheat germ extract. Figure 6 shows that this synthetic An2 mRNA directs the synthesis of a 50-kD protein in vitro. A final proof that An2 and the other localized RNAs are in fact mRNAs will require isolation of the proteins they encode from developing eggs.

DISCUSSION

There is a long history of embryological research, including those studies cited in the introductory section, that supports the view that localized maternal factors can direct blastomeres to assume particular cell fates in development. In general, these data show that there is some factor or cytoplasmic determinant which is segregated to an embryonic cell and thereby affects the fate of that cell. There is obviously great interest in extending these studies by purifying the determinants and explaining how they specify cell fate. Although the molecules involved have never been identified or purified, significant preliminary progress in this area has been made.

When the existence of a determinant has been shown by a functional test, several studies implicate RNA involvement. For example, Kalthoff has shown that the determinants for body axis in midges contain an RNA component (Kalthoff 1983). Similarly, in *Drosophila*, RNA injections can rescue mutants that fail to form a correct dorsal-ventral axis (Anderson and Nusslein-Volhard 1984). Finally, with respect to germ cell formation in amphibians (Smith 1966) and insects (Togashi and Okada 1982), a localized RNA component has been implicated as a critical agent.

Experiments designed to look for localized molecules have also been performed. This approach assumes that molecules localized in eggs are unlikely to be involved in the growth and development of all cell types and are therefore candidates for cytoplasmic determinants. Previous studies have examined the concentration of both total protein (Moen and Namenwirth 1977) and RNA (Sagata et al. 1981; Carpenter and Klein 1982; Phillips 1982) along the animal-vegetal axis of the *Xenopus* egg. More recently, King and her colleagues have provided evidence for the existence of localized mRNAs using in vitro translation of isolated RNAs as an assay (M.L. King, pers. comm.). These surveys have shown that at least some maternal proteins and RNAs are unevenly distributed in unfertilized eggs or oocytes. Our present study confirms and extends these earlier observations by obtaining clones for RNAs that are strikingly localized along the animal-vegetal axis. Having identified molecules that are

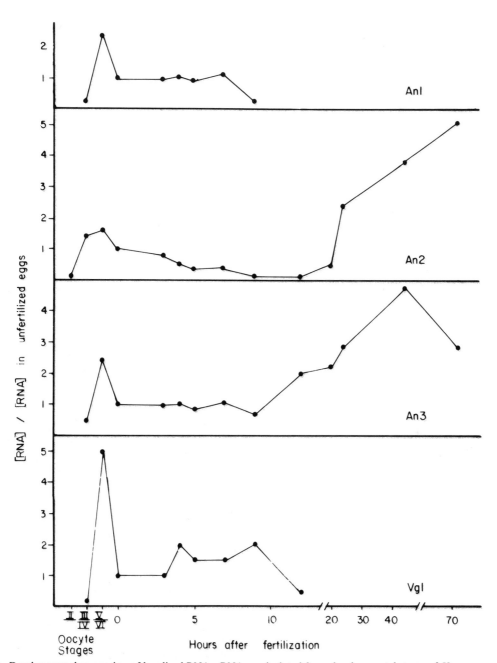

Figure 4. Developmental expression of localized RNAs. RNA was isolated from developmental stages of *Xenopus* ranging from stage-II oocytes to swimming tadpoles (stage 41). RNA was electrophoresed in denaturing agarose gels and Northern blots were prepared. Each blot was hybridized with ^{32}P-labeled probe from one of the localized cDNA clones. The graphs show the amount of RNA present at each stage as judged by densitometric tracing of the autoradiographs. RNA amounts are given relative to the amount of RNA present in the unfertilized egg. RNA from the following oocyte and embryonic stages was used: oocyte stages II, III/IV, V/VI, unfertilized egg, stages 6, 7, 8, 9, 10, 12, 18/20, 24, 34, and 41.

localized in eggs, the next step is to establish whether any of these molecules play a role in early determinative events.

As yet, there is no direct evidence that the localized An and Vg RNAs are developmentally important. However, their isolation and cloning expands our ability to address this issue. First, we can examine the function of these RNAs by preventing the expression

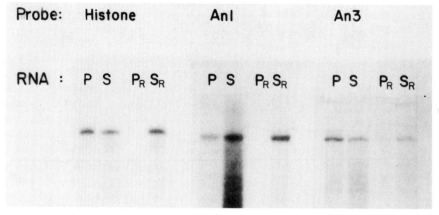

Figure 5. Localized RNAs are found on polysomes. Polysomal pellets were isolated from eggs and assayed for the presence of localized RNAs by solution hybridization with ^{32}P-labeled single-stranded RNA probes. After treatment with RNase A and T1, RNA was electrophoresed in denaturing acrylamide gels. Lane designations are: (P) pellet (i.e., RNA associated with protein, including polysomal RNA), (S) supernatant, unbound RNA, (PR) EDTA release, pellet, (SR) EDTA release supernatant. Each pellet/supernatant pair corresponds to the material from 15 eggs. The results shown for the histone, An1, and An3 RNA are representative of results seen for An2 and Vg1. Similar results were obtained with polysomes isolated from stage-4 embryos.

of their protein products. This can be accomplished by the injection of antisense (complementary) RNA into oocytes or eggs. Previous studies have shown that injected antisense RNAs hybridize to their complementary mRNA in vivo and thereby block translation (Melton 1985). In early developmental stages when there is no transcription, the antisense RNAs must be provided by the direct injection of in vitro-synthesized RNAs. At later stages, the antisense RNAs can be supplied by transcription from a suitably constructed DNA template (Izant and Weintraub 1984). A second way to investigate the role of these RNAs during development is to place them in regions of embryos where they are not normally found. For example, a sense copy of the Vg1 RNA can be injected into the animal pole, and thereby provide presumptive ectoderm cells with a RNA normally found only in presumptive endoderm. The sense copy of Vg1 mRNA can be synthesized in vitro (Krieg and Melton 1984) and injected at the eight-cell stage so that the cell membranes will keep the Vg RNA in the "wrong" cells. This type of mislocalization experiment may allow us to test whether particular maternal molecules can direct or, in the case of the mislocalized RNA, redirect the developmental fate of a cell. A third approach to the problem of function relies on an identification of the protein product of the An and Vg RNAs. Antibodies directed against these proteins can be used in attempts to block function (Boucaut et al. 1984) and identify the normal site of action of the proteins.

We are also investigating the mechanism by which RNA becomes localized. These studies involve synthesizing copies of An1, An2, An3, and Vg1 in vitro and reinjecting them into areas of oocytes and unfertilized eggs where they are not usually found. If the RNAs relocalize, as is suggested by the studies of Capco and Jeffery (Capco and Jeffery 1981), it may be possible to identify the cis-acting signals that specify location in the cytoplasm. The mechanism that exists to localize maternal RNA is apparently resistant to the cytoplasmic movements that occur during fertilization (Phillips 1985). It is possible, for instance, that the RNA is retained in particular locations in the egg and

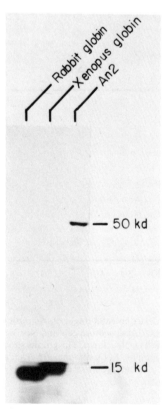

Figure 6. In vitro translation of synthetic mRNA. SP6 transcripts of Xenopus β-globin and An2 cDNA were used to direct protein synthesis in a wheat germ extract. Natural rabbit β-globin mRNA was included as a control. Translation products were labeled using [^{35}S]methionine during synthesis and fractionated on an SDS acrylamide gel. Size markers (not shown) were used to estimate the sizes of the proteins shown.

embryo via attachment to the cellular cytoskeleton (Lenk et al. 1977). By examination of the molecules that become associated with the reinjected RNAs both during and after relocalization we may be able to identify the cellular components that are involved.

ACKNOWLEDGMENTS

We would like to thank Paul Krieg and Jan Fassler for comments on the manuscript, Mary Lou King for sharing her unpublished data, and Kate Breakey for help with the figures. M.R. and D.W. acknowledge support from an NSF Predoctoral Fellowship and a Damon Runyon-Walter Winchell Postdoctoral Fellowship, respectively. This work was supported by grants from the National Institutes of Health and The Chicago Community Trust/Searle Scholars Program.

REFERENCES

ANDERSON, D.M., J.D. RICHTER, M.E. CHAMBERLIN, D.H. PRICE, R.J. BRITTEN, L.D. SMITH, and E.H. DAVIDSON. 1982. Sequence organization of the poly(A) RNA synthesized and accumulated in lampbrush stage *Xenopus laevis* oocytes. *J. Mol. Biol.* **155**: 281.

ANDERSON, K.V. and C. NUSSLEIN-VOLHARD. 1984. Information for the dorsal-ventral pattern of the *Drosophila* embryo is stored as maternal mRNA. *Nature* **311**: 223.

BENTON, W.D. and R.W. DAVIS. 1977. Screening λGT recombinant clones by hybridization to single plaques in situ. *Science* **196**: 180.

BOUCAUT, J., T. DARRIBERE, H. BOULEKLOACHE, and J.P. THEIRY. 1984. Prevention of gastrulation but not neurulation by antibodies of fibronectin in amphibian embryos. *Nature* **307**: 364.

BRACHET, J. 1977. An old enigma: The grey crescent of amphibian eggs. *Curr. Top. Dev. Biol.* **11**: 133.

CARPENTER, C.D. and W.H. KLEIN. 1982. A gradient of poly A$^+$ RNA sequences in *Xenopus laevis* eggs and embryos. *Dev. Biol.* **91**: 43.

CAPCO, D. and W.R. JEFFERY. 1981. Regional accumulation of vegetal pole poly(A)$^+$ RNA injected into fertilized *Xenopus* eggs. *Nature* **294**: 255.

CHURCH, G. and W. GILBERT. 1984. Genomic sequencing. *Proc. Natl. Acad. Sci.* **81**: 1991.

DAWID, I.B. and P.K. WELLAUER. 1976. A reinvestigation of 5'-3' polarity in 40S ribosomal RNA precursor of *Xenopus laevis*. *Cell*. **8**: 443.

DUMONT, J.N. 1972. Oogenesis in *Xenopus laevis* (Daudin). I. Stages of oocyte development in laboratory maintained animals. *J. Morphol.* **136**: 153.

FREEMAN, G. 1976. The role of cleavage in the localization of developmental potential in the ctenophore *Mnemiopsis leidyi*. *Dev. Biol.* **49**: 143.

GERHART, J., S. BLACK, R. GIMLICH, and S. SCHARF. 1983. Control of polarity in the amphibian egg. In *Time, space and pattern in embryonic development* (ed. W.R. Jeffery and R.A. Raff), p. 261. A.R. Liss, New York.

GOLDBERG, D.A. 1980. Isolation and partial purification of the *Drosophila* alcohol dehydrogenase gene. *Proc. Natl. Acad. Sci.* **77**: 5794.

GURDON, J.B., T.J. MOHUN, S. FAIRMAN, and S. BRENNAN. 1985. All components required for the eventual activation of muscle-specific actin genes are localized in the subequatorial region of an uncleaved egg. *Proc. Natl. Acad. Sci.* **82**: 139.

HUYNH, T., R.A. YOUNG, and R.W. DAVIS. 1985. Constructing and screening cDNA libraries in lambda gt10 and lambda gt11. In *DNA cloning: A practical approach* (ed. D.M. Glover). IRL Press, Oxford. (In press.)

ILLMENSEE, K. and A.P. MAHOWALD. 1974. Transplantation of posterior pole plasm in *Drosophila*. Induction of germ cells at the anterior pole of the egg. *Proc. Natl. Acad. Sci.* **71**: 1016.

IZANT, J. and H. WEINTRAUB. 1984. Inhibition of thymidine kinase gene expression by anti-sense RNA: A molecular approach to gene analysis. *Cell* **36**: 1007.

JAMRICH, M., K.A. MAHON, E.R. GAVIS, and J. GALL. 1984. Histone RNA in amphibian oocytes visualized by in situ hybridization to methacrylate-embedded tissue sections. *EMBO J.* **3**: 1939.

KALTHOFF, K. 1983. Cytoplasmic determinants in Dipteran eggs. In *Time, space and pattern in embryonic development* (ed. W.R. Jeffery and R.A. Raff), p. 313. A.R. Liss, New York.

KRIEG, P. and D. MELTON. 1984. Functional messenger RNAs are produced by SP6 in vitro transcription of cloned DNAs. *Nucleic Acids Res.* **12**: 7057.

LENK, R., L. RANSOM, Y. KAUFMANN, and S. PENMANN. 1977. A cytoskeletal structure with associated polyribosomes obtained from HeLa cells. *Cell* **10**: 67.

MELTON, D.A. 1985. Injected anti-sense RNAs specifically block messenger RNA translation in vivo. *Proc. Natl. Acad. Sci.* **82**: 144.

MELTON, D.A., P.A. KRIEG, M.R. REBAGLIATI, T. MANIATIS, K. ZINN, and M.R. GREEN. 1984. Efficient in vitro synthesis of biologically active RNA and RNA hybridization probes from plasmids containing a bacteriophage SP6 promoter. *Nucleic Acids Res.* **12**: 7035.

MOEN, T.L. and M. NAMENWIRTH. 1977. The distribution of soluble proteins along the animal-vegetal axis of frog eggs. *Dev. Biol.* **58**: 1.

NEWPORT, J. and M. KIRSCHNER. 1982. A major developmental transition in early *Xenopus* embryos. I. Characterization and timing of cellular changes at the mid-blastula stage. *Cell* **30**: 675.

NIEUWKOOP, P.D. 1977. Origin and establishment of embryonic polar axes in amphibian development. *Curr. Top. Dev. Biol.* **11**: 115.

NIEUWKOOP, P.D. and J. FABER. 1967. *Normal table of* Xenopus laevis *(Daudin)*. Elsevier/North-Holland, Amsterdam.

NUSSLEIN-VOLHARD, C. 1979. Maternal effect mutations that alter the spatial coordinates of the embryo of *Drosophila melanogaster*. *Symp. Soc. Dev. Biol.* **37**: 185.

OKADA, M., I.A. KLEINMAN, and H.A. SCHNEIDERMAN. 1974. Restoration of fertility in sterilized *Drosophila* eggs by transplantation of polar cytoplasm. *Dev. Biol.* **37**: 43.

PHILLIPS, C.R. 1982. The regional distribution of poly(A) and total RNA concentrations during early *Xenopus* development. *J. Exp. Zool.* **223**: 265.

———. 1985. Spatial changes in poly A concentrations during early embryogenesis in *Xenopus laevis*: Analysis by in situ hybridization. *Dev. Biol.* **109**: 299.

RASTL, E. and I.B. DAWID. 1979. Expression of the mitochondrial genome in *Xenopus laevis*: A map of transcripts. *Cell* **18**: 501.

REBAGLIATI, M.R., D.L. WEEKS, R.P. HARVEY, and D.A. MELTON. 1985. Identification and cloning of localized maternal RNAs from *Xenopus* eggs. *Cell* (in press).

RICHTER, J.D., L.D. SMITH, D.M. ANDERSON, and E.H. DAVIDSON. 1984. Interspersed poly(A) RNAs of amphibian oocytes are not translatable. *J. Mol. Biol.* **173**: 227.

SAGATA, N., K. OKUYAMA, and K. YAMANA. 1981. Localization and segregation of maternal RNAs during early cleavage of *Xenopus laevis* embryos. *Dev. Growth Differ.* **23**: 23.

SMITH, L.D. 1966. The role of a germinal plasm in the formation of primordial germ cells in *Rana pipiens*. *Dev. Biol.* **14**: 330.

TOGASHI, S. and M. OKADA. 1982. Restoration of pole cell forming ability to UV-sterilized *Drosophila* embryos by injection of an RNA fraction extracted from eggs. In *The ultrastructure and functioning of insect cells* (ed. H. Akai et al.), p. 41. Society for Insect Cells, Tokyo.

TOOLE, J.J., J.L. KNOPF, J.M. WOZNEY, L.A. SULTZMAN, J.L. BUECKER, D.D. PITTMAN, R.J. KAUFMAN, E. BROWN, C. SHOEMAKER, E.C. ORR, G.W. AMPHLETT, W.B. FOSTER, M. COE, G.L. KNUTSON, D.N. FASS, and R.M. HEWICK. 1984. Molecular cloning of a cDNA encoding human antihaemophilic factor. *Nature* **312:** 342.

WILSON, E.B. 1925. *The cell in development and heredity.* Macmillan, New York.

WOODLAND, H.R. 1974. Changes in the polysome content of developing *Xenopus laevis* embryos. *Dev. Biol.* **40:** 90.

Altered Morphogenesis and Its Effects on Gene Activity in *Xenopus laevis* Embryos

M. JAMRICH, T. D. SARGENT, AND I. B. DAWID

Laboratory of Molecular Genetics, National Institute of Child Health and Human Development, Bethesda, Maryland 20205

During many weeks of development, the giant amphibian oocytes accumulate large amounts of RNA and protein (for review, see Davidson 1976; Smith and Richter 1985). In *Xenopus* these components are sufficient to support normal embryonic development for several hours after fertilization, since 13 cell divisions take place without any measurable transcription. The proteins inherited from the oocyte are distributed among the arising embryonic cells and are supplemented by new synthesis on polysomes using maternal mRNAs. About 7 hours after fertilization, during a period called the midblastula transition (MBT), the zygotic genome of *Xenopus* embryos is activated for the first time (Brown and Littna 1964; Bachvarova and Davidson 1966; Shiokawa et al. 1981; Newport and Kirschner 1982a,b). At this time, one set of genes is expressed whose members have not been transcribed during oogenesis and whose products are therefore not represented in the maternal RNA population (Sargent and Dawid 1983). Expression at this critical period of development makes this set of genes good candidates for having a significant role in the formation of germ layers and tissue anlagen, either as regulatory signals or as structural components required in the process of differentiation itself.

In this paper, we report experiments in which we tested the activity of several early expressed genes in embryos raised under normal and altered developmental conditions. We used culture of embryos in high salt media and UV irradiation of embryos as experimental tools. Both of these procedures have been reported previously to alter the developmental pattern of *Xenopus* embryos. High salt concentration in the culture media is known to cause exogastrulation, a process in which the normal movement of mesodermal cells into the area between the ectoderm and endoderm is prevented (Gurdon et al. 1984). In such embryos differentiation of the mesoderm, as well as the ectoderm and endoderm, does occur, but the germ layers do not assume their normal positions in the embryo. UV irradiation produces effects on development that are, in some respect, more severe than exogastrulation. Such irradiation of embryos prior to first cleavage interferes with, and in extreme cases prevents, the formation of dorsal mesodermal and neural structures, leading to embryos that lack an axis entirely (Grant and Wacaster 1972; Malacinski et al. 1975, 1977; Youn and Malacinski 1981; Scharf and Gerhart 1980, 1983; Gimlich and Gerhart 1984).

With the aim of generating probes for the study of early developmental phenomena in *Xenopus*, we have prepared a subtracted library of cDNA clones that represent genes expressed for the first time between MBT and the gastrula stage (Sargent and Dawid 1983). These clones, as assayed by RNA dot blot procedures, do not hybridize to maternal RNA or to RNA of pre-MBT embryos, but do show significant homology to RNA from gastrulae; they have been named DG clones. The developmental profiles of several DG clones have been studied and will be presented below. Among these we have identified markers for the activity of two of the germ layers: Clones DG81 and DG70 have been identified as epidermal keratin genes by sequence analysis (Jonas et al. 1985; J. Winkles et al., in prep.) and have been used in our study as ectodermal markers. Gene DG42 is specifically expressed in the endoderm (M. Jamrich and T.D. Sargent, unpubl.), and thus was used as a endodermal marker. We used a cardiac α-actin clone as a mesodermal marker. In addition, we used several other DG cDNAs in this study whose developmental behavior in the normal embryo is known but whose location of expression has not yet been determined.

MATERIALS AND METHODS

Exogastrulation. Embryos were cultured in full-strength Barth's medium as described by Gurdon et al. (1984), except that the vitelline membrane was not removed.

UV treatment of embryos. Embryos were placed in a Petri dish in which the bottom was replaced by plastic film. Embryos were placed on top of a Transilluminator TM-38 (Ultra-Violet Products, Inc., San Gabriel, California) and irradiated at 400 μW/cm^2 for 1–5 min. This treatment yields, in our hands, embryos with a range of malformations (grades 1–5 according to Malacinski et al. 1977; see also Gimlich and Gerhart 1984). Embryos classified as grade 5 on the scale of severity of malformations were selected for further study.

Molecular methods. Isolation of RNA and RNA dot and gel blots were carried out as described by T.D. Sargent et al. (in prep.).

RESULTS AND DISCUSSION

Accumulation of Embryo-specific RNAs during Normal *Xenopus* Development

Seventeen clones representing genes expressed for the first time in *Xenopus* development after the MBT were selected for this study. These genes carrying the prefix DG (*d*ifferentially expressed in *g*astrula), were analyzed for their expression by dot blot hybridization to RNAs of different developmental stages. Eggs derived from a single female were fertilized in vitro and allowed to develop normally. At the intervals indicated in Figure 1, samples were collected and RNAs were isolated. The RNAs were spotted onto nitrocellulose filters, and the filters were hybridized with ^{32}P-labeled nick-translated cDNA clones, washed, and exposed. The resulting series of dots indicates the dynamics of the accumulation of RNA complementary to each clone (Fig. 1). It is clear that the accumulation profiles for individual clones vary, but they all are activated no later than midgastrula stage, as required by the method of preparation of the cDNA library. In most of the cases, the transcription (or the stability of the transcripts) declines after neurulation, although several clones persist for longer periods. We have shown that these latter RNAs are, nevertheless, restricted in their developmental expression: DG81, for example, is abundant throughout most of tadpole development but decreases in concentration by at least 100-fold during metamorphosis (Jonas et al. 1985). In Figure 1 we have arranged the clones according to their identification numbers rather than by developmental pattern. It appears that there is a common but not invariant relationship between the time of first accumulation of any RNA and its eventual time of decay. Thus DG4, DG42, and DG56 RNAs accumulate and decay early, DG70 and DG72 RNAs somewhat later, and DG81 and DG76 RNAs later still. There are, however, exceptions to this pattern, including DG39 and DG83 RNAs, which begin accumulation early but persist through later tadpole stages. Clone r5, at the bottom of Figure 1, does not belong to the DG class of genes and was used as a positive control for hybridization of pre-MBT RNA; r5 RNA is present in oocytes and all other stages and tissues of *Xenopus* that we have examined.

In the experiments described below we have used several of the DG cDNA clones as markers for developmental gene expression. In addition, we have monitored the expression of a cardiac α-actin gene, represented here by clone H2, which was originally selected from a *Xenopus* tadpole library (Dworkin and Dawid 1980). This clone was positively identified as encoding cardiac α-actin by nucleotide sequence analysis and by hybridization to gel blots of RNA prepared from different tissues, including heart and skeletal muscle (T.D. Sargent, unpubl.). The α-actin mRNA has been shown by Mohun et al. (1984) to appear first at stage 13 in presomite mesoderm, making this gene the latest to be

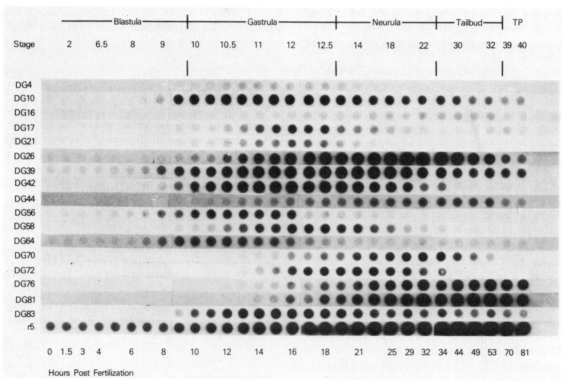

Figure 1. Time course of accumulation and decay of DG genes. RNA samples prepared from embryos at the stages indicated were applied as dots on filters and hybridized with labeled probes from the DG cDNA clones shown. Clone r5 is a control sequence; the r5 RNA is present in eggs and all other stages and tissues tested.

activated and the only known mesodermal marker in our selection of clones.

Expression of Developmentally Regulated Genes in UV-irradiated Embryos

Xenopus embryos were UV-irradiated prior to first cleavage at a dose empirically determined to yield embryos showing deformations ranging from grade 1 to 5, as defined by Malacinski et al. (1977). Embryos with the highest degree of deformity, grade 5, were selected for these experiments. Such embryos have no axial structures, forming almost radially symmetrical "belly pieces," yet survive in this form for days or weeks. Irradiated and normal control embryos were collected at several times after fertilization and the expression of several genes was analyzed as described above. Twelve DG clones were analyzed, and none showed a significant reduction of expression under these conditions, as shown for six examples in Figure 2. In contrast, the accumulation of α-actin RNA was completely eliminated in the irradiated embryos (Fig. 3). The elimination of α-actin mRNA accumulation is consistent with prior observations that UV treatment of embryos prevents the formation of dorsal mesodermal structures, including muscle cells, which are the sites of α-actin synthesis (Malacinski et al. 1977; Scharf and Gerhart 1980). The fact that none of the DG RNAs was negatively affected by this treatment suggests that the DG clones that are included in our sample are not related to the formation of dorsal mesodermal structures. Since the mesoderm develops during gastrulation, it has not fully established itself at midgastrula, the time at which the embryos for DG library construction had been collected. Thus, it might be expected that mesoderm-specific sequences are not represented with high frequency in this library, possibly explaining the fact that none of the DG genes tested were inhibited in UV-irradiated embryos. Among the DG clones whose expression is unaffected by UV irradiation are both ectoderm-specific (DG81, DG70, DG76) and endoderm-specific (DG42) sequences. This result is not entirely surprising since UV treatment does not prevent formation of these germ layers and possibly does not affect their initial differentiation at all. There is a slight increase in the abundance of all DG transcripts in the total RNA of UV-irradiated embryos that might be due, at least in part, to a larger proportion of ectodermal and endodermal cells in the absence of mesoderm.

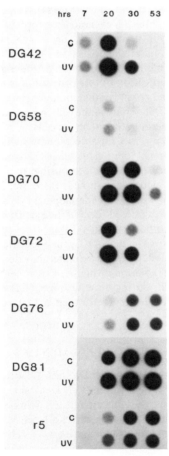

Figure 2. Activity of DG genes in UV-irradiated embryos. Fertilized eggs were irradiated before first cleavage and cultured for different periods of time, shown at the top of the figure. Control embryos from the same clutch reached midblastula (st 8 +) at 7 hr, neurula (st 13/14) at 20 hr, tailbud (st 21) at 30 hr, and tadpole (st 32) at 53 hr. At 20, 30, and 53 hr, highly affected individuals (grade 5, see text) were selected from among the treated embryos. RNA was analyzed by dot blot hybridization.

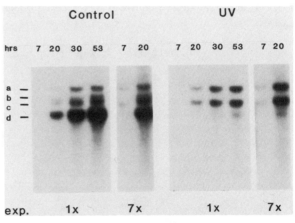

Figure 3. Failure of UV-treated embryos to accumulate α-actin mRNA. Treated and control embryos collected at the same times as in Fig. 2 were analyzed by gel blot hybridization. Four actin bands were seen: Bands a and c are cytoskeletal actins; band d is α-actin homologous to the probe used; band b is not definitely characterized but appears to be a distinct α-actin mRNA. For the 7-hr and 20-hr times, points to exposure levels are shown to illustrate the point that blastulae (and unfertilized eggs) contain the cytoskeletal actin mRNAs but no detectable α-actin RNA, and, further, that the cytoskeletal actin mRNAs accumulate during development in both normal and treated embryos, which constitutes a control for survival and metabolic activity in the experimental embryos.

Exogastrulation Does Not Affect the Expression of DG Genes

High salt concentration in the incubation medium leads to an alteration of normal spatial relationships between the germ layers in the developing frog embryo. This treatment inhibits the involution of cells during gastrulation, preventing most of the mesodermal cells from having correct contact with ectodermal and endodermal cells. Although normal development of the embryo is prevented, all three germ layers are generated. Figure 4 shows the analysis of the levels of several DG RNAs and of the ubiquitously expressed gene r5 in these embryos. It is obvious that the accumulation of these RNAs is not affected. Similarly, expression of α-actin is unchanged, as has been shown previously by Gurdon et al. (1984). These results indicate that correct shape of the embryo is not required for the expression of any of the developmentally regulated genes studied in this work.

CONCLUSION

We have analyzed the dependence of activation of several developmentally regulated genes on normal morphological relationships during *Xenopus* embryogenesis by experimentally interfering with the establishment of these relationships. Exogastrulation, induced by culture in medium of high salt content, allows the establishment of all three germ layers but in an abnormal spatial relation. None of the genes tested, including several DG genes and the gene for α-actin, is inhibited under these conditions. Thus, expression of the genes that we have analyzed is not dependent on the correct shape of the embryo, as long as all three germ layers are fully established. UV irradiation of *Xenopus* zygotes before the first cleavage prevents the differentiation of all dorsal mesodermal and neural structures in highly affected individuals. Such embryos nevertheless express all tested DG genes at the normal time and at normal levels. In contrast, α-actin mRNA fails completely to accumulate in these embryos. The inhibition of expression of this mesodermal marker correlates with the suppression of dorsal mesoderm formation in the UV-treated embryos.

The conclusions drawn from these experiments support the view that inductive interactions are required for the formation of the mesoderm in *X. laevis* (reviewed by Gerhart 1980). UV irradiation is known not to destroy a cytoplasmic determinant physically, since its effect can be reversed entirely by turning the irradiated eggs by 90° before first cleavage (Scharf and Gerhart 1980). Thus, UV irradiation affects the position of substances in the egg rather than their chemical integrity. The inability of irradiated embryos to develop a dorsal axis is most likely the result of their failure to establish inductive capacity in dorsal/vegetal blastomeres, as shown by the transplantation experiments of Gimlich and Gerhart (1984). The correlation between the inhibition of mesoderm formation and accumulation of α-actin mRNA suggests that the inductive interactions implicated in axis determination are also responsible for the activation of the α-actin gene. A further implication of this conclusion is that the α-actin gene, and possibly other mesoderm-specific genes as well, is activated only in the cell lineage that actually forms mesoderm rather than in cells in which these genes had been preset for activation by virtue of a specific cytoplasmic component inherited from the egg. The regulatory substances responsible for the activation of the α-actin gene should, therefore, be formed in the embryo during early development rather than be present in a sequestered form in the egg. In contrast, it could be hypothesized that substances responsible for the activation of certain DG genes specific for ectoderm and endoderm might be maternally derived, since neither UV irradiation or high salt treatment, nor dispersion of cells during blastula stages (T.D. Sargent et al., unpubl.), interferes with their activity in the embryo.

ACKNOWLEDGMENTS

We would like to thank Dr. Jeff Winkles for help with the experiment in Figure 1, Drs. Mark Dworkin, Erszebeth Jonas, Brian Kay, and Jeff Winkles for sharing unpublished results, and Dr. Kathleen Mahon for a critical reading of this manuscript.

REFERENCES

BACHVAROVA, R. and E.H. DAVIDSON. 1966. Nuclear activation at the onset of amphibian gastrulation. *J. Exp. Zool.* **163:** 285.

BROWN, D.D. and E. LITTNA. 1964. RNA synthesis during development of *Xenopus laevis*, the South African clawed toad. *J. Mol. Biol.* **8:** 669.

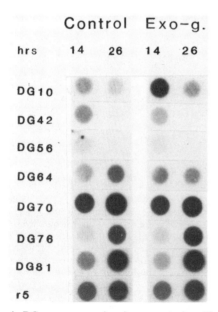

Figure 4. DG gene expression in exogastrulae. Two time points were collected in this experiment. See legends to Figs. 1 and 2 for other details.

DAVIDSON, E.H. 1976. *Gene activity in early development*. Academic Press, New York.

DWORKIN, M.B. and I.B. DAWID. 1980. Construction of a cloned library of expressed embryonic gene sequences from *Xenopus laevis*. *Dev. Biol.* **76:** 435.

GERHART, J.C. 1980. Mechanisms regulating pattern formation in the amphibian egg and early embryo. In *Biological regulation and development* (ed. R.F. Goldberger), vol. 2, p. 133. Plenum Press, New York.

GIMLICH, R.L. and J.C. GERHART. 1984. Early cellular interactions promote embryonic axis formation in *Xenopus laevis*. *Dev. Biol.* **104:** 117.

GRANT, P. and J.F. WACASTER. 1972. The amphibian grey crescent — A site of developmental information? *Dev. Biol.* **28:** 454.

GURDON, J.B., S. BRENNAN, S. FAIRMAN and T.J. MOHUN. 1984. Transcription of muscle-specific actin gene in early *Xenopus* development: Nuclear transplantation and cell dissociation. *Cell* **38:** 691.

JONAS, E., T.D. SARGENT, and I.B. DAWID. 1985. Epidermal keratin gene expressed in embryos of *Xenopus laevis*. *Proc. Natl. Acad. Sci.* **82:** 5413.

MALACINSKI, G.M., H. BENFORD, and H.-M. CHUNG. 1975. Association of an ultraviolet irradiation sensitive cytoplasmic localization with the future dorsal side of the amphibian egg. *J. Exp. Zool.* **191:** 97.

MALACINSKI, G.M., A.J. BROTHERS, and H.-M. CHUNG. 1977. Destruction of components of the neural induction system of the amphibian egg with ultraviolet irradiation. *Dev. Biol.* **56:** 24.

MOHUN, T.J., S. BRENNAN, S. FAIRMAN, N. DATHAN, and J.B. GURDON. 1984. Cell-specific activation of actin genes in the early amphibian embryo. *Nature* **311:** 716.

NEWPORT, J. and M. KIRSCHNER. 1982a. A major developmental transition in early *Xenopus* embryos. I. Characterization of timing of cellular changes at the midblastula stage. *Cell* **30:** 675.

———. 1982b. A major developmental transition in early *Xenopus* embryos. II. Control of the onset of transcription. *Cell* **30:** 687.

SARGENT, T.D. and I.B. DAWID. 1983. Differential gene expression in the gastrula of *Xenopus laevis*. *Science* **222:** 135.

SCHARF, S.R. and J.C. GERHART. 1980. Determination of the dorsal-ventral axis in eggs of *Xenopus laevis*: Complete rescue of UV-impaired eggs by oblique orientation before first cleavage. *Dev. Biol.* **79:** 181.

———. 1983. Axis determination in eggs of *Xenopus laevis*: A critical period before first cleavage, identified by the common effects of cold, pressure, and ultraviolet irradiation. *Dev. Biol.* **99:** 75.

SHIOKAWA, K., K. TASHIRO, Y. MISUMI, and K. YAMANA. 1981. Non-coordinated synthesis of RNAs in pregastrular embryos of *Xenopus laevis*. *Dev. Growth Differ.* **23:** 589.

SMITH, L.D. and J.D. RICHTER. 1985. Synthesis, accumulation, and utilization of maternal macromolecules during oogenesis and oocyte maturation. In *Biology of fertilization* (ed. A. Monroy and C. Metz), vol. 1, p. 141. Academic Press, New York.

YOUN, B.W. and G.M. MALACINSKI. 1981. Axial structure development in ultraviolet-irradiated (notochord-defective) amphibian embryos. *Dev. Biol.* **83:** 339.

Germ Plasm and Germ Cell Determination in *Xenopus laevis* As Studied by Cell Transplantation Analysis

C.C. Wylie, S. Holwill, M. O'Driscoll, A. Snape, and J. Heasman

Department of Anatomy, St. George's Hospital Medical School, London, SW17 ORE, England

This paper will review separate but related groups of data on the ontogeny and possible roles of germ plasm in the establishment of the germ line in anuran amphibian embryos. We shall discuss the paradoxical situation that germ plasm is apparently necessary for the normal formation and/or the migration of primordial germ cells (PGCs), but that at the same time PGCs remain pluripotent, at least until they reach the gonad, and are not determined in the classical sense of being restricted in potential to form only germ line cells.

Germ plasm refers to a collection of small islands of basophilic cytoplasm found in the vegetal cortex of eggs of anuran amphibians such as *Rana pipiens* and *Xenopus laevis* (for review, see Eddy 1975). These masses of cytoplasm aggregate during the first few cleavage cycles of development in *Xenopus* until several (usually four) large masses of germ plasm are formed at the 32-cell stage, each occupying one large vegetal pole blastomere (Fig. 1). Until the early gastrula stage, these germ plasm-bearing cells divide unequally, so that only one daughter cell retains the germ plasm. The other is assumed to enter the embryonic endoderm. After the gastrula stage, the germ plasm-bearing cells divide equally, so that their numbers now increase. They are incorporated into the early gut endoderm, and during gut differentiation they migrate dorsally until they occupy a dorsal crest on the embryonic gut tube. As the gut mesentery forms by splitting of the lateral plate mesoderm, the germ plasm-bearing cells, now morphologically identifiable as PGCs, migrate into the dorsal mesentery, and then to the dorsal body wall where the gonad develops. By this time there is a variable number of PGCs, which can be as low as 15 or as many as 50 PGCs per embryo in any clutch of embryos. There is also considerable variation in numbers within individual clutches of embryos which probably indicates variation in the initial number of germ plasm-bearing cells and the number of divisions they undergo during migration.

The Ontogeny of Germ Plasm

Recently, the origin of germ plasm in the oocyte of *X. laevis* has been established using a combination of light and electron microscopical analysis of oocyte stages (Heasman et al. 1984a).

One of the most obvious cytoarchitectural features of the previtellogenic *Xenopus* oocyte is a large phase-dense cytoplasmic mass known as the mitochondrial cloud (Fig. 2). This has been thought in the past to

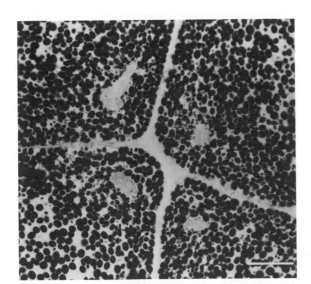

Figure 1. Horizontal section through the vegetal pole of a *Xenopus* embryo between the 16- and 32-cell stages. Profiles of the four most vegetal blastomeres are shown. Each contains a pale, granular mass of germ plasm. Scale bar, 50 μm.

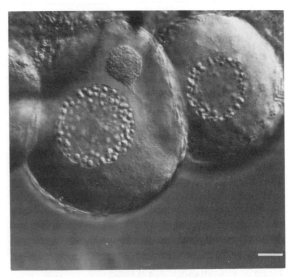

Figure 2. Differential interference contrast image of a living previtellogenic oocyte. The mitochondrial cloud can be seen as a large dense round cytoplasmic mass next to the nucleus. Scale bar, 25 μm.

provide the maternal store of mitochondria (Dawid and Blackler 1972), suggested by its breakdown and apparent dispersion throughout the oocyte cytoplasm at the beginning of vitellogenesis (Balinsky and Devis 1963; Billett and Adam 1976). However, the study of Heasman et al. (1984a) showed several new and unexpected aspects of both mitochondrial biogenesis and mitochondrial cloud function. First, the oocyte mitochondria are not solely provided by the mitochondrial cloud. During the previtellogenic stage, a shell of smaller masses of mitochondria appears and grows around the nucleus. These are connected together by strands of cytoplasm that also contain mitochondria. This changing distribution of mitochondria was shown in living oocytes using the vital dye for mitochondria, Rhodamine 123 (Johnson et al. 1980). Figure 3 shows the distribution of mitochondria at the middle previtellogenic stage. The mitochondrial cloud and the smaller perinuclear mitochondrial masses can be seen.

Second, the breakdown products of the mitochondrial cloud do not disperse throughout the oocyte, but remain in small discrete clumps in one sector of the oocyte cortex. This region of the cortex becomes the vegetal pole. Furthermore, the mitochondrial cloud and its cortical breakdown products contain not only mitochondria, but also masses of granulo-fibrillar material, which under the electron microscope are identical to the germinal granules seen at the vegetal pole of the fertilized egg (Fig. 4a,b). Thus, it would seem that at least one important role of the mitochondrial cloud is as the source, and the distribution mechanism, of germ plasm during *Xenopus* oogenesis.

So the ontogeny of germ plasm is as follows. It appears from the nucleus and accumulates in the mitochondrial cloud during early previtellogenesis. This breaks down at the beginning of vitellogenesis and the fragments spread in a broad arc to one pole of the oo-

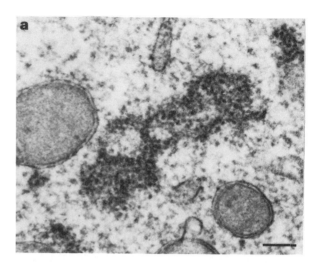
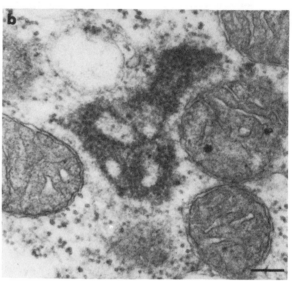

Figure 4. Electron micrographs of granulo-fibrillar material found in the mitochondrial cloud (*a*) and in the germ plasm of the fertilized egg (*b*). Scale bars, 137 nm.

cyte cortex. Here they remain until fertilization, after which they aggregate (Smith and Williams 1979) during the first few cleavage cycles to form approximately four large masses of mitochondria and germinal granules contained in the vegetal pole blastomeres. Clearly the cytoskeleton of the oocyte and early embryo plays an important role in these changes in germ plasm distribution.

Evidence for the Importance of Germ Plasm in Normal Germ Cell Formation and Migration

It has been convincingly shown in several species that the germ line progenitor cells cannot be replaced if destroyed or removed (for review, see Nieuwkoop and Sutasurya 1979). Therefore, presumptive somatic cells cannot be reprogrammed to enter the germ line. There is also much experimental evidence that in some species, including anuran amphibians, the germ plasm is a

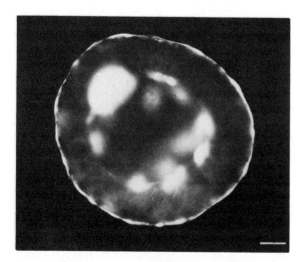

Figure 3. Fluorescence image of living previtellogenic oocyte in which the mitochondria are stained with Rhodamine 123. The mitochondrial cloud is the largest mass. It is connected to many smaller mitochondrial aggregates arranged in a shell around the nucleus. Scale bar, 25 μm.

requirement for normal emergence of the germ line cells. The germ plasm can be traced microscopically to the PGCs that migrate to the gonad (Bounoure 1939). Its destruction, either by physical removal (Buehr and Blackler 1970) or by UV light irradiation (for review, see Smith and Williams 1979), reduces the germ cell population and delays migration. In some cases, damage due to UV irradiation can be reversed by injection of unirradiated vegetal cytoplasm (Smith 1966). Furthermore, increased numbers of germ cells can be generated by injecting extra vegetal pole cytoplasm into unirradiated eggs (Wakahara 1978). Such results have led to the idea that a germ cell determinant exists at the vegetal pole, which causes the cells that inherit it to enter the germ line. Such a hypothesis is supported by the existence of morphologically similar structures in the egg of *Drosophila* (for review, see Mahowald and Boswell 1983) and *Caenorhabditis* (Strome and Wood 1983).

In some species, however, there is little evidence that germ cells are determined early in development, or that a specific region of cytoplasm is inherited by germ line cells. Single-cell transplantation experiments in mouse embryos have shown that all or most of the primary ectoderm cells of the early gastrula are pluripotent, forming germ cells as well as other lineages. Therefore it is most unlikely that a population of cells is set aside to become only germ cells very early in mouse embryogenesis (Gardner 1978; Gardner et al. 1985). These paradoxical results have led to the idea that two distinct mechanisms of germ cell determination exist in vertebrates: preformationist and epigenetic (Nieuwkoop and Sutasurya 1979).

However, the real problem is one of methodology. No cell transplantation studies like those in mammals have been carried out in frog embryos to test the capacity of early germ line cells to differentiate into somatic cell types.

The Use of Cell Transplantation to Test the State of Commitment of *Xenopus* Germ Cells

Cell transplantation analysis of cell potency has been made possible in amphibian embryos by the development of lineage markers that are not diluted out during early development and are cell autonomous. The best known of these are horseradish peroxidase (Weisblat et al. 1978), fluoresceinated lysine dextran (Gimlich and Cooke 1983), and tetramethyl rhodamine isothiocyanate (TRITC, Heasman et al. 1984b).

The only studies so far on single-cell transplantation in *Xenopus* have used the TRITC label (Heasman et al. 1984b, 1985a,b). In these studies, several facts concerning cell commitment in *Xenopus* embryos have been established. First, early blastula cells from both the animal and vegetal poles are pluripotent when inserted into the blastocoel cavities of unlabeled late blastula hosts. Their progeny are found in all germ layer derivatives in most or all host embryos. The blastocoel of the late blastula therefore offers a wide (if not complete) range of developmental options to a donor cell. However, vegetal pole cells from later blastula donors show decreasing pluripotency, until by the gastrula stage all the donor cells enter only derivatives of the embryonic endoderm. Thus, on a single-cell basis, the vegetal pole cells become committed to form cells of their normal fate (Heasman et al. 1984b). This method of analysis of cell commitment is different from classical grafting experiments (for review, see Spemann 1938) in one significant respect. In grafting experiments, a small piece of tissue is sealed by the grafting technique in a single new location, so the range of developmental options open to it are limited. There are only two results possible (apart from the demise of the graft)—either the graft differentiates according to its origin, or to its new position. When separated cells are transplanted into the blastocoel, however, each cell has a very wide choice of developmental options, and so a more complete analysis of potency is possible.

The cell transplantation method therefore offers the possibility of testing whether or not germ plasm-bearing cells are "determined" in the classical sense, i.e., whether they will only form germ line cells when exposed to a range of developmental signals. There is a philosophical conundrum here, of course. The end result of the female germ line, the egg, is by definition totipotent, since during normal development all cell types appear amongst its descendants. However, there is no experimental evidence to suggest whether or not cells of the developing germ line are totipotent. Indeed there must be some mechanism in the embryo that restricts them to germ line differentiation. It has been widely assumed in the past that this happens very early in anuran embryos and is caused by the germ plasm.

To test the potency of germ plasm-bearing cells, we carried out two cell transplantation experiments (Wylie et al. 1985b). In the first, we selected vegetal pole cells from the early blastula (stage 7 according to the Nieuwkoop and Faber [1956] normal table) that were shown by histology of sibling embryos to have a >50% chance of containing germ plasm. These were labeled with TRITC and transplanted singly into unlabeled late blastula hosts. The hosts were allowed to develop to the early swimming tadpole stage and analyzed by serial sectioning. Out of 32 embryos examined, and which contained TRITC-labeled cells, none contained fluorescent PGCs. Instead, labeled cells were found in a variety of other tissues of the tadpoles, derived from various combinations of germ layers (see Table 1 for details). This result suggests that the presence of germ plasm does not inevitably cause some descendents of the donor blastomere to enter the germ line.

One possible objection to this interpretation, however, is that blastomeres were taken before the germ plasm had its action, and the still pluripotent blastomeres were committed to other pathways before it could act. In the second experiment, therefore, migrating PGCs were isolated from the dorsal gut mesentery of stage-45 early swimming tadpoles (see Heasman et al. 1981 for details of method). These were labeled with

Table 1. Distribution of TRITC-labeled Cells in *Xenopus* Embryos

Number of tadpoles	Number with TRITC	Distribution of labeled cells at stage 46[a]						
		end	end + ect	end + mes	end + ect + mes	mes + ect	mes	ect
49	32	18	1	7	4	0	2	0

Single vegetal pole blastomeres were isolated from stage-7 embryos, TRITC labeled, and inserted into unlabeled host embryos at late blastula stage. Each host received one donor blastomere. Embryos were allowed to grow to stage 46, fixed, embedded, sectioned, and examined for TRITC-labeled cells. (Reprinted, with permission, from Wylie et al. 1985b.)

[a]End, endoderm; mes, mesoderm; ect, ectoderm.

TRITC and inserted into the blastocoel cavities of unlabeled late blastula hosts. About 15 PGCs were inserted into each host embryo. The hosts were allowed to grow to early swimming tadpole stage and analyzed by serial sections. Once again, the labeled cells were found in derivatives of all three germ layers (see Table 2). No labeled cells were found in the PGC population of the host embryos.

The first conclusion to be drawn from this experiment is that PGCs from later embryos cannot migrate to the correction position of earlier host embryos, and traverse their normal migratory route a second time. However, the most important result came from a close examination of the labeled cells in the various tissues containing them. In many cases, these cells obviously were differentiated according to their new position.

Table 2. Number of TRITC-labeled Cells Found at Stage 46 in the Following Tissues

Ectoderm	
neural tube	22
epidermis	5
neural crest	4
Endoderm	
gut lining	13
pharynx	1
liver	2
esophagus	1
Mesoderm	
notochord	2
somite	5
mesothelium	3
loose connective tissue	3
pronephros	1

Primordial germ cells were isolated from stage-45 *Xenopus* embryos. They were labeled with TRITC, and injected into unlabeled host embryos at the late blastula stage (stage 9). PGCs (15) were injected into each host embryo. Hosts were allowed to develop to stage 46, fixed, embedded, sectioned, and examined for the presence of TRITC-labeled cells. The above data are aggregated from 16 host embryos. In no case was a labeled cell found amongst the host PGC population. (Reprinted, with permission, from Wylie et al. 1985b.)

Figures 5 and 6 give examples of these in somite (Fig. 5a-d), notochord (Fig. 5e-i), gut lining (Fig. 6a-d), and pronephric tubule (Fig. 6e-g). The use of polarizing microscopy allowed a further test of the degree of differentiation of the somite and notochord cells. Somitic muscle cells contain organized arrays of actin and myosin filaments. These cause form birefringence, arranged with characteristic periodicity. Figure 5c shows the labeled PGC-derived cell containing birefringent cytoplasm. The inset shows a single myofibril, with periodic A and I bands, running across the cell. The notochord also exhibits form birefringence that is not periodic. This is due to organized arrays of cytokeratin filaments (S.F. Godsave et al., in prep.). Figure 5g shows that the PGC-derived cell exhibits the same birefringent pattern as the rest of the notochord. Labeled cells in the skin also showed nonperiodic birefringence due to the high content of aligned cytokeratin filaments (data not shown). Thus, in these cases, the PGC-derived cells had synthesized and organized different cytoskeletal molecules in concert with their new position. It should be mentioned that PGCs are not normally birefringent. The labeled gut epithelial cells also showed convincing evidence of their differentiation, since their apical surfaces were covered in microvilli (Fig. 6d) in common with the other unlabeled gut lining cells. In many cases, the labeled cells were not as obviously differentiated as the ones selected in these figures. New differentiation markers are now becoming available in *Xenopus* embryos that will allow more complete analysis of their differentiated state. One interesting feature of these labeled cells was that they often showed one characteristic feature of PGC morphology, despite their new differentiated state. Figure 5d and 6g show two examples of the lobed nucleus that is characteristic of PGCs.

These results show the *Xenopus* germ line cells remain pluripotent, at least until they reach the gonad. This resolves a longstanding paradox over the data (derived from different techniques) from mammalian and anuran embryos. When their potency is tested by injection into the blastocoel, progenitors of the germ line in both vertebrate groups are pluripotent. This makes it likely that similar mechanisms of germ cell differentiation act in most, if not all, vertebrates. It remains unknown, of course, what normally restricts the germ

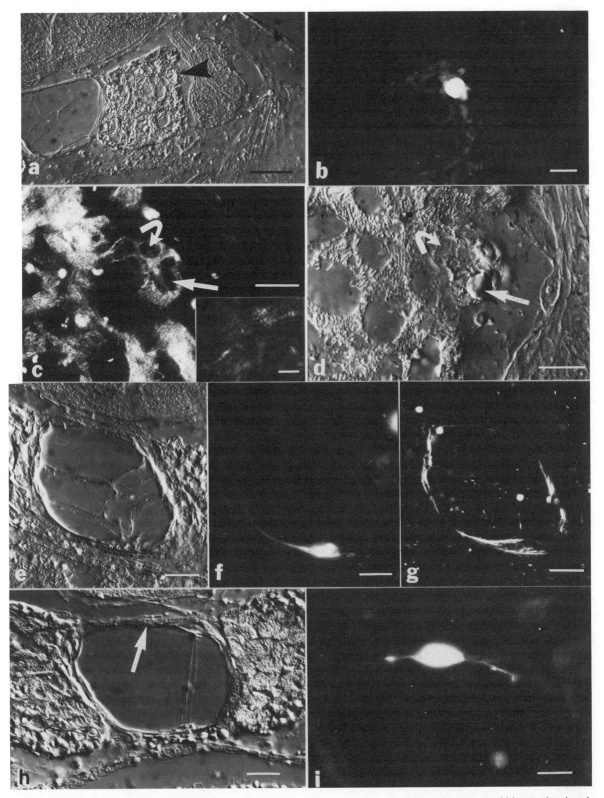

Figure 5. A TRITC-labeled cell in developing striated muscle of the myotome. (*a*) A low-power Nomarski image showing the labeled cell (arrowed) in the lateral border of the myotome. (*b*) The TRITC labeling of the cell arrowed in *a*. (*c*) Form birefringence in the cytoplasm of the labeled cell, in common with the other cells of the myotome. In places, striated myofibrils can be seen passing in and out of the plane of section (inset to *c*), in which alternating isotropic and anisotropic bands are seen. Areas of negative birefringence in the cell are due to yolk platelets (straight arrow) and lobes of the multilobed nucleus (curved arrow). These features can also be seen in *d*. (*e–i*) Labeled cells found in the host notochord. (*e–g*) Nomarksi, TRITC labeling, and birefringence images of the same field of view. The labeled cell in the notochord sheath shows form birefringence, in common with the other notochord cells. (*h,i*) Nomarski and TRITC labeling of a notochord cell that contributes to the septae in the center of the notochord. Scale bars: *a*, 100 μm; *b*, 50 μm; *c*, 20 μm (*inset*, 4 μm); *d*, 20 μm; *e–h*, 40 μm; *i*, 20 μm. (Reprinted, with permission, from Wylie et al. 1985b.)

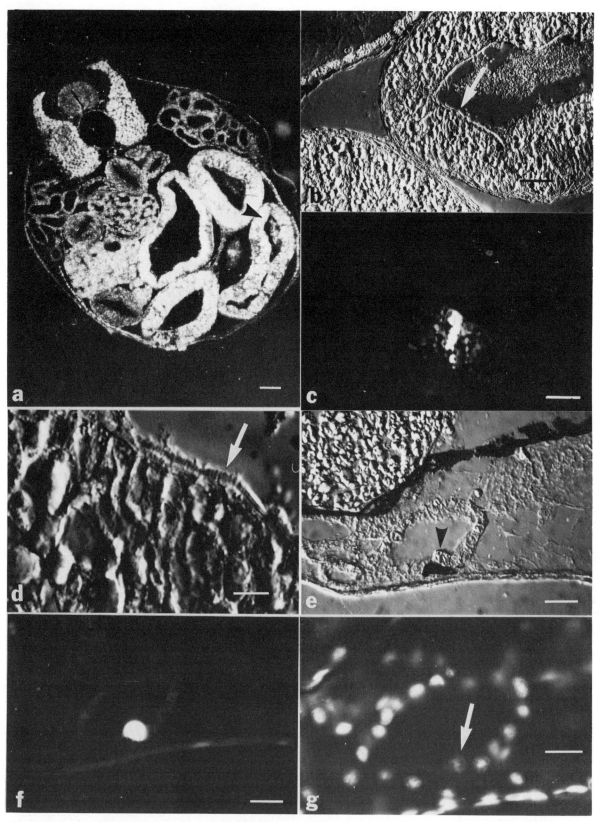

Figure 6. (*a–d*) Labeled cell forming part of the gut lining epithelium. The position of the labeled cell is arrowed in the low-power dark-ground image of the unstained section shown in *a*. (*b* and *c*) A Nomarski and fluorescence pair showing the position of the TRITC-labeled cell (arrowed in *b*). The labeled cell shows a brush border typical of the gut epithelial cells, arrowed in the high-power Nomarski image shown in *d*. (*e–g*) Labeled cell in the pronephric duct. Once again the TRITC-labeled cell (arrowed in *e*, with TRITC fluorescence shown in *f*) conforms in shape and morphology to the host tissue. This section was stained with 4,6-diamidine-2-phenylindole dihydrochloride (DAPI) to show nuclear shape (shown in *g*). In this site too, the nucleus of the labeled cell was lobed (arrowed in *g*). One lobe lies in the plane of focus, the other curves away below it. Scale bars: *a*, 10 μm; *b,c*, 20 μm; *d*, 8 μm; *e,f*, 50 μm; *g*, 20 μm. (Reprinted, with permission, from Wylie et al. 1985b.)

plasm-bearing cells to the germ line, but a position effect is suggested. This concept can be tested by grafting labeled PGCs into the correct region of the early embryo (the vegetal pole).

The most interesting problem now is to establish the role of the germ plasm in anuran embryos. There are several possibilities. It may yet be a germ line determinant, but either in a reversible fashion or at a later stage than that tested here; or it may be a maternally synthesized store of molecules required for migration and/or targetting of the PGCs, a speculation previously made (Smith et al. 1983). The latter role might explain the absence of germ plasm in mammalian embryos. Development to the larval stage in *Xenopus* is very rapid and relies on maternal stores of many molecules. Mammalian early development is very slow, and synthesis of the same molecules is carried out in the developing embryo. A cytoskeletal role of germ plasm is made more attractive by the fact that both the mitochondrial cloud and its breakdown products in the oocyte, as well as germ plasm in the egg, stain strongly with anti-tubulin as well as several different anti-vimentin antibodies (Wylie et al. 1985a). The intermediate filaments of *Xenopus* are not yet characterized, but this result suggests that germ plasm may contain maternally synthesized cytoskeletal material.

Whatever the role of germ plasm, it clearly does not irreversibly restrict the developmental capacity of the cells that inherit it. It will be interesting to test its role further, and to analyze the potency of germ line progenitor cells in other species.

ACKNOWLEDGMENTS

We are grateful to the Wellcome Trust and the Cancer Research Campaign for financial support, to Claire Varley and Lynn Albert for expert technical assistance, and to Melanie Coulton for typing the manuscript.

REFERENCES

Balinsky, B.I. and R.J. Devis. 1963. Origin and differentiation of cytoplasmic structures in the oocytes of *Xenopus laevis*. *Acta Embryol. Morphol. Exp.* **6:** 55.

Billett, F.S. and E. Adam 1976. The structure of the mitochondrial cloud of *Xenopus laevis* oocytes. *J. Embryol. Exp. Morphol.* **33:** 697.

Bounoure, L. 1939. *L'origine des cellules reproductive et le problem de la lignée germinale.* Gauthier-Villars, Paris.

Buehr, M. and A.W. Blackler. 1970. Sterility and partial sterility in the South African clawed toad following pricking of the egg. *J. Embryol. Exp. Morphol.* **23:** 375.

Dawid, I.B. and A.W. Blackler. 1972. Maternal and cytoplasmic inheritance of mitochondrial DNA in *Xenopus*. *Dev. Biol.* **29:** 152.

Eddy, E.M. 1975. Germ plasm and the differentiation of the germ cell line. *Int. Rev. Cytol.* **43:** 229.

Gardner, R.L. 1978. Developmental potency of normal and neoplastic cells of the early mouse embryo. In *Proceedings of the 5th International Conference on Birth Defects* (ed. J. Littlefield and J. de Grouchy), p. 153. Elsevier/North-Holland, Amsterdam.

Gardner, R.L., M.F. Lyon, E.P. Evans, and M.D. Burtenshaw. 1985. Clonal analysis of X-chromosome inactivation and the origin of the germ line in the mouse embryo. *J. Embryol. Exp. Morphol.* (in press).

Gimlich, R.L. and J. Cooke. 1983. Cell lineage and the induction of second nervous systems in amphibian development. *Nature* **306:** 471.

Heasman, J., J. Quarmby, and C.C. Wylie. 1984a. The mitochondrial cloud of *Xenopus* oocytes: The source of germinal granule material. *Dev. Biol.* **105:** 458.

Heasman, J., A. Snape, J.C. Smith, and C.C. Wylie. 1985a. Single cell analysis of commitment in early embryogenesis. *J. Embryol. Exp. Morphol.* (in press).

———. 1985b. Cell lineage and commitment in early amphibian development. *Philos. Trans. R. Soc. Lond. B* (in press).

Heasman, J., C.C. Wylie, P. Hausen, and J.C. Smith. 1984b. Fates and states of determination of single vegetal pole blastomeres of *Xenopus laevis*. *Cell* **37:** 185.

Heasman, J., R.O. Hynes, A.P. Swan, V.A. Thomas, and C.C. Wylie. 1981. Primordial germ cells of *Xenopus* embryos: The role of fibronectin in their adhesion during migration. *Cell* **27:** 437.

Johnson, L.V., M.L. Walsh, and L.B. Chen. 1980. Localization of mitochondria in living cells with Rhodamine 123. *Proc. Natl. Acad. Sci.* **77:** 990.

Mahowald, A.P. and R.E. Boswell. 1983. Germ plasm and germ cell development in invertebrates. In *Current problems in germ cell differentiation* (ed. A. McLaren and C.C. Wylie), p. 3. Cambridge University Press, Cambridge, England.

Nieuwkoop, P.D. and J. Faber. 1956. *A normal table of* Xenopus laevis. Elsevier/North-Holland, Amsterdam.

Nieuwkoop, P.D. and L.A. Sutasurya. 1979. Primordial germ cells in the chordates, embryogenesis and phylogenesis. Cambridge University Press, Cambridge, England.

Smith, L.D. 1966. The role of 'germinal plasm' in the formation of primordial germ cells in *Rana pipiens*. *Dev. Biol.* **14:** 330.

Smith, L.D. and M. Williams. 1979. Germinal plasm and germ cell determinants in anuran amphibians. In *Maternal effects in development* (ed. D. Newth and M. Balls), p. 167. Cambridge University Press, Cambridge, England.

Smith, L.D., P. Michael, and M. Williams. 1983. Does a pre-determined germ line exist in amphibians? In *Current problems in germ cell differentiation* (ed. A. McLaren and C.C. Wylie), p. 19. Cambridge University Press, Cambridge, England.

Spemann, H. 1938. *Embryonic development and induction* (reprinted, 1967). Hafner, New York.

Strome, S. and W.B. Wood. 1983. Generation of asymmetry and segregation of germ-line granules in early *C. elegans* embryos. *Cell* **35:** 15.

Wakahara, M. 1978. Induction of supernumerary primordial germ cells by injecting vegetal pole cytoplasm into *Xenopus* eggs. *J. Exp. Zool.* **203:** 159.

Weisblat, D.A., R.T. Sawyer, and G.S. Stent. 1978. Cell lineage analysis by intracellular injection of a tracer enzyme. *Science* **202:** 1295.

Wylie, C.C., D. Brown, S.F. Godsave, J. Quarmby, and J. Heasman. 1985a. The cytoskeleton of *Xenopus* oocytes and its role in development. *J. Embryol. Exp. Morphol.* (in press).

Wylie, C.C., J. Heasman, A. Snape, M. O'Driscoll, and S. Holwill. 1985b. Primordial germ cells of *Xenopus laevis* are not irreversibly determined early in development. *Dev. Biol.* (in press).

Nuclear Transfer in Mouse Embryos: Activation of the Embryonic Genome

D. SOLTER,* J. ARONSON,* S. F. GILBERT,† AND J. McGRATH*

*The Wistar Institute, Philadelphia, Pennsylvania 19104; †Department of Biology, Swarthmore College, Swarthmore, Pennsylvania 19081

Following fertilization, normal development requires a strictly defined inventory of cellular components. These components, the products of male and female gametogenesis, reflect two distinct processes of differentiation. It is apparent that the transcriptional activity of the respective gametic nuclei must meet different morphogenetic and functional requirements. In the male, transcriptional activity persists into the haploid phase of spermatogenesis (Distel et al. 1984), whereas synthesis of poly(A)$^+$ RNA in oocytes in the female continues beyond the completion of oocyte growth (Wassarman 1983). Thus, the male and female genomes at the moment of fertilization reflect functionally different germ cell lineages that must undergo a coordinated transition for the specification of embryo-specific gene products. This transition requires complex interactions among the pronuclei, the ovum cytoplasm, and their environment. The nature of these interactions and the individual contributions of each cellular component may be assessed by nuclear and cytoplasmic transfer. We present experiments in which such transfers were performed in order to define functional differences between the maternal and paternal genomes, the ability of exogenous nuclei (either from later embryonic stages or pronuclei from another species) to support development, and the specific timing of gene activation in the mammalian embryo.

Analysis of nuclear function during embryogenesis cannot occur without consideration of the cytoplasmic environment in which the nucleus resides. The egg cytoplasm contains elements necessary for the support of early development. The length of time the embryo can develop without nuclear input varies between species but is probably very short in mammals (see Magnuson and Epstein 1981; Johnson 1981). Despite its brevity, the autonomous expression of an intrinsic developmental program by the mammalian ovum cytoplasm does occur following fertilization (Waksmundzka et al. 1984; Longo and Chen 1985; Van Blerkom 1985). Thus, the ovum cytoplasm contains elements essential for early developmental events, including proper gene activation within the embryonic nucleus. Besides being unique in its ability to perform these predesigned developmentally essential functions, the egg cytoplasm is also unique in its ability to accept the foreign genome of the sperm and to mold its activity. Most of the presented experiments test the genomic potential of various nuclei in the cytoplasm of the fertilized egg. In an initial experimental series, we considered possible nonequivalence of the maternal and paternal genomes as a cause for death of mammalian parthenogenetic embryos. Extensive evidence for restricted development of diploid parthenogenones (Graham 1974; Kaufman et al. 1977) exists; however, the cause of death was complicated by the possibility of extragenomic sperm contribution and/or homozygosity for lethal genes. The experiments presented here circumvent these problems and conclusively demonstrate the requirement of both male and female genomes for successful development (McGrath and Solter 1984a). In addition, we present preliminary evidence using two-dimensional gel analysis showing the maternal and paternal genomes express generally equivalent information during preimplantation development but with differential expression in a few instances.

The mammalian embryonic genome undergoes progressive differentiation which is reflected in a continuously changing profile of gene products (Epstein and Smith 1974; Van Blerkom and Brockway 1975; Levinson et al. 1978; Howe and Solter 1979; Bolton et al. 1984). Nuclear transfer permits a test of the degree of reversibility of these genomic changes. In amphibians, the reversibility is substantial, allowing extensive development of zygotes whose nuclei were replaced by nuclei of somatic cells of tadpoles (Gurdon 1962). Our results (McGrath and Solter 1984b) indicate that the ability of embryonic nuclei in mammals to support normal development is substantially reduced very early. The data presented here suggest that severe reduction in developmental potential already exists in nuclei from mid- to late preimplantation-stage embryos. Similarly, pronuclei obtained from one species when transplanted into the cytoplasm of a second species undergo early preimplantation-stage developmental arrest. Taken together these results underscore the absolute necessity for appropriate transcriptional activity in response to cytoplasmic signals in the early mammalian embryo.

In addition to analyses of developmental potential of the whole or a large part of the genome, nuclear transfer may also be used to investigate the onset of activation of a particular gene. Most notably, since the maternal genome can be separated from the original egg cytoplasm, nuclear transfer is uniquely suitable for monitoring the onset of expression of maternally derived genes. We have used nuclear transfer to demonstrate the simultaneous activation of the maternal and paternal alleles coding for glucose phosphate isomerase by, or prior to, day 4 of embryogenesis.

METHODS

Nuclear transfer and embryo culture. Nuclear transfer procedures have been described in detail elsewhere (McGrath and Solter 1983a,b). Briefly, fertilized ova from spontaneously ovulating females were exposed to 5 µg/ml of cytochalasin B (Sigma) and 0.1 µg/ml of demecolcine (Sigma) in Whitten's medium (Whitten 1971). Removal of one or both pronuclei within small, plasma membrane-bound, cytoplasmic vesicles was achieved without penetrating the plasma membrane of the egg. The resulting karyoplast was then fused with a previously enucleated zygote by placing it under the zona pellucida together with a small amount of inactivated Sendai virus. These virus-fused nuclear transfer embryos were cultured in Whitten's medium containing 100 µM EDTA (Abramczuk et al. 1977) or transferred to the oviducts of pseudopregnant foster mothers.

Two-dimensional gel electrophoresis. Nuclear transfer and control embryos grown in vitro for an appropriate length of time were exposed to [^{35}S]methionine (4 mCi/ml; sp. act. 1000 Ci/mM) in Whitten's medium for 4 hours. Two-dimensional (18 × 18 cm) gel electrophoresis, using isoelectric focusing followed by SDS-PAGE was performed as described (Garrels 1979).

Glucose phosphate isomerase analysis. Glucose phosphate isomerase (GPI-1) isozymes were analyzed by a modification of the method of Eppig et al. (1977), using the citrate-based staining procedure of Peterson et al. (1978). Embryos at various developmental stages were placed into 1-µl droplets of lysis buffer (50 mM Tris, 1mM EDTA, 1 mM 2-mercaptoethanol, pH 7.5) and lysed by freeze-thawing three times. The lysate was spotted onto cellulose-acetate plates (Titan III, Helena Laboratories) and run at 200 V for 3–3.5 hours in 43 mM Tris and 46 mM glycine buffer (pH 8.6). Stain was applied in a 0.67% agarose overlay and plates incubated at 37°C.

RESULTS

Development of Diploid Androgenones and Gynogenones

Diploid biparental androgenones (embryos containing two male pronuclei) and gynogenones (embryos containing two female pronuclei) were produced by removing one of the pronuclei from the fertilized egg and replacing it with the pronucleus of the opposite parental derivation from another zygote. The maternal and paternal origin of pronuclei was determined on the basis of their size and position within the egg. The smaller female pronucleus is typically located beneath the second polar body. In addition, the zygotes used in these experiments were derived from reciprocal crosses of two different inbred strains which varied in coat color and GPI-1 phenotype so that the genotype of each nuclear transplant embryo could be deduced at term. Control embryos, which were produced by replacing one pronucleus with a pronucleus of the same parental sex from another zygote, developed to term (Table 1). Analysis of development of a large series of putative androgenones and gynogenones revealed that the phenotypes of the animals born were those in which visual assignment of the maternal/paternal origin was incorrect, i.e., they were derived from zygotes containing both maternal and paternal pronuclei (Table 1). On the basis of the number of control progeny born, we would have expected 19 gynogenones and 14 androgenones; however, no progeny with the appropriate phenotype was observed ($X^2 = 27.15$; $p < 0.001$). Therefore, these results indicate that diploid embryos containing two maternal or two paternal genomes do not complete development. Since all experimental embryos were derived from fertilized zygotes, we conclude that neither the absence of an extragenomic sperm contribution nor excessive homozygosity can explain the death of these embryos. Our previous investigation has shown that preimplantation development and the rate of implantation of androgenones and gynogenones is comparable to that of control embryos; however, following implantation, these embryos soon display signs of growth retardation and degeneration, and all die by midgestation (McGrath and Solter 1984a).

Table 1. Development of Biparental Androgenones, Gynogenones, and Control Embryos

Genotype of the zygote	Number of embryos	Number of animals born with the appropriate gynotype
Androgenones (paternal/paternal)	328	0[a]
Gynogenones (maternal/maternal)	339	0[b]
Control (maternal/paternal)	348	18

[a]Two animals born in this group were genotypically maternal/paternal.
[b]Three animals born in this group were genotypically maternal/paternal.

Table 2. Development of Enucleated Zygotes Containing Nuclei from Preimplantation Embryos

Nuclear donor	Total number of nuclear transfer embryos	Percent of embryos reaching	
		four-cell stage	morula/blastocyst
Zygote	25	4	96
Two-cell	191	19	20
Four-cell	87	3	5
Eight-cell	120	1	0
Inner cell mass cell	84	1	0

Hemizygous Gene Expression in Androgenones and Gynogenones

On the basis of results obtained by single pronuclear transfer (Table 1), we concluded that maternal and paternal genomes are functionally different and that each contributes uniquely to the developing embryo. This assumption is supported by data demonstrating that the T^{hp} deletion is lethal when present on the maternal chromosome 17 and not when present on paternal chromosome 17 and that this phenomenon is due to nuclear postfertilization gene activity (McGrath and Solter 1984c). These results are consistent with the hypothesis (McLaren 1979) that some part of the chromosome 17 (corresponding to the T^{hp} deletion) is expressed when this chromosome is maternally, but not when it is paternally, derived. Several similar cases have been also reported (see Discussion). If there is differential gene expression by the maternal and paternal genomes during embryogenesis, one might expect to observe differences in proteins synthesized by androgenetic and gynogenetic embryos. Our preliminary results (Fig. 1) demonstrate that one can indeed observe infrequent differences when two-dimensional protein gels of in vitro-grown androgenones and gynogenones are compared. No extensive evaluation of these data has been completed, but these preliminary findings suggest that gene products derived from only one set of parental chromosomes can be identified.

Developmental Potential of Nuclei from Preimplantation Embryos

To assess the ability of preimplantation-stage nuclei to support development, karyoplasts from zygote, two-, four-, and eight-cell-stage embryos, and single inner cell mass (ICM) cells were introduced into enucleated zygotes as previously described (McGrath and Solter 1983b, 1984b). The resultant embryos were subsequently cultured for 5 days in vitro. The results (Table 2) show that whereas a high proportion of enucleated zygotes receiving pronuclei from a second zygote successfully complete preimplantation development, relatively few of the embryos receiving later-stage nuclei do so. Although these latter embryos frequently divide to the two-cell stage, subsequent development is significantly impaired. Approximately 20% of embryos receiving two-cell-stage nuclei developed to the morula–blastocyst stages. A small proportion of embryos (5%) receiving four-cell stage nuclei developed to the early morula stage; blastocoelic cavity formation by these embryos, however, was impaired and resulted in embryos with multiple, small, fluid-filled cavities. Embryos receiving eight-cell-stage or ICM cell nuclei only infrequently developed beyond the two-cell stage and never achieved the morula–blastocyst stages.

Control experiments (McGrath and Solter 1984b) demonstrate that the transfer of cytoplasts from cleavage-stage embryos or transfer of corresponding karyoplasts into nonenucleated zygotes does not inhibit development. Thus, we can conclude that impaired development in nuclear transfer embryos is not due to the inhibitory factors present in transplanted material but due to the failure of transplanted nuclei to support development.

In a similar manner, we have performed interspecies pronuclear transfers between *Mus caroli* and *M. musculus* zygotes. The results (Table 3) show that of 35 enucleated *M. musculus* embryos receiving both pronuclei from a *M. caroli* zygote, none completed preimplantation development. Similarly, all of the 26 enucleated *M. caroli* zygotes receiving *M. musculus* pro-

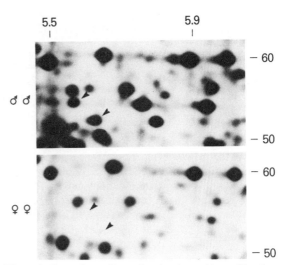

Figure 1. Part of two-dimensional polypeptide profile of androgenones (♂ ♂) and gynogenones (♀ ♀) grown in vitro for 3 days and labeled with [^{35}S]methionine. The polypeptide spots present in androgenones and absent in gynogenones are marked with arrowheads.

Table 3. In Vitro Development of *M. musculus* ↔ *M. caroli* Nuclear Transplant Embryos

	Total number of embryos	One-cell[a]	Two-cell[a]	Three-cell[a]	Four-cell[a]	Eight-cell[a]	Morula[a]	Blastocyst[a]
M. caroli$_N$→*M. musculus*$_C$[b]	35	3	29		3			
M. musculus$_N$→*M. caroli*$_C$	26		17	7	2			
M. caroli$_N$→*M. caroli*$_C$	13						1	12
M. musculus$_N$→*M. musculus*$_C$	22			1			1	20
M. caroli$_C$→nonenucleated *M. musculus* zygote	10						3	7
M. caroli control	73		2		1		5	65
M. musculus control	29	1					1	27

[a]Numbers refer to the most advanced stage of development achieved by successful nuclear transplant embryos during 5 days of in vitro culture.

[b]N, Species origin of the nucleus; C, species origin of the cytoplasmic recipient subsequent to enucleation.

nuclei arrested at the two- to four-cell stages. Complete preimplantation development, however, was observed when both pronuclei were removed from either *M. caroli* or *M. musculus* zygotes and returned to an enucleated embryo of the same species (Table 3). Inhibition of development resulting from the transfer of foreign cytoplasm in the karyoplast can be excluded as a possible cause for the poor development observed, since the transfer of *M. caroli* membrane-bound cytoplasm into nonenucleated *M. musculus* zygotes was compatible with preimplantation development (Table 3). Although these data are preliminary, they once again underscore the extreme requirement in the early mammalian embryo for correct and synchronous nuclear–cytoplasmic interactions.

Onset of Activation of Embryonic Genes

In order to determine the developmental stage when a particular embryonic gene is activated, one must distinguish between the gene product inherited via egg cytoplasm (either in the form of protein or maternal mRNA) from the newly transcribed and translated gene product. In a case when the gene products of two different alleles are distinguishable, activation of the paternal allele is easily demonstrated in F_1 embryos; however, activation of maternal allele can be masked by the presence of gene product in egg cytoplasm. We used nuclear transfer techniques to demonstrate the simultaneous activation of paternal and maternal genes for GPI-1 (*Gpi-1*). At the time of activation of embryonic genes, there is no indication that functional maternal mRNA is still present (Gilbert and Solter 1985). Activation of *Gpi-1* occurs at the 8- to 16-cell stage, at which time there is still a substantial amount of enzyme inherited via egg cytoplasm. By the blastocyst stage (Fig. 2) the total enzyme is probably derived from newly transcribed mRNA since the isozyme originally present in the egg cytoplasm is no longer detectable.

DISCUSSION

The fertilized embryo contains all of the information necessary for successful embryogenesis. The results summarized in this report, however, indicate that few alterations in the nuclear content of the mammalian embryo are compatible with normal development. This is most obvious in the experiments involving androgenetic and gynogenetic embryos. From our results (McGrath and Solter 1984a) and those presented by others (Barton et al. 1984; Mann and Lovell-Badge 1984; Surani et al. 1984), it is now quite certain that some functional nuclear information contributed to the zygote by the male and female gametes differs and that both are indispensable for normal development. These data are supported by genetic studies that have demonstrated developmental differences for certain chromosomes or chromosomal regions depending upon whether these chromosomes were derived from the male or from the female gamete (Johnson 1975; Lyon and Glenister 1977; Searle and Beechey 1978; McGrath and Solter 1984c; Cattanach and Kirk 1985). The difference between male and female contributions to the embryonic genome is thus the most likely explanation for the developmental failure of mammalian parthenogenones. Two questions, however, remain: one dealing with the mechanism(s) by which this different functional status is achieved and the other dealing with the purpose of this feature, probably unique for mammalian development. The mechanisms presently are not clear, but reports demonstrating unique methylation

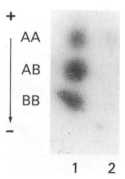

Figure 2. GPI-1 pattern of control F_1 embryos (lane *1*) and nuclear transfer embryos (lane *2*). Nuclear transfer embryos were produced by transferring nuclei from *Gpi-1*^{a}/*Gpi-1*^{a} zygotes into enucleated *Gpi-1*^{b}/*Gpi-1*^{b} zygotes. Embryos were grown in culture for 6 days. Only two nuclear transfer embryos were used, thus the enzyme activity is rather faint.

patterns occurring during spermatogenesis in avian species (Groudine and Conkin 1985) and the increased frequency of meiotic gene conversion during oogenesis in mice (Loh and Baltimore 1984) suggest that processes unique for male and female gametogenesis might occur. Differential parental contribution to the functioning genome may be of physiological significance in early development, but its real role is at present unclear. Analysis of X-linked genes coding for various enzymes (Epstein 1983) indicates that gene dosage regulates the level of the gene products resulting in a double amount of a given enzyme in the female when compared with the male embryos. This differential level of gene product in male and female embryos is subsequently eliminated by the inactivation of one X chromosome in female embryos. We can speculate that differential parental genomic contribution might represent an alternative to the gene dosage compensation. If this is correct, we should be able to detect the presence of some gene product in androgenetic but not in gynogenetic embryos and vice versa. Our very preliminary data using two-dimensional gels support such a possibility. Another, not mutually exclusive, possibility is that such differential genomic activity is required for normal functions of extraembryonic tissues, thus representing a specific mammalian adoption to viviparity. While pondering the purpose of differential parental contributions to the functioning genome, it is of interest that homeo box-containing sequences have been mapped to mouse chromosomes 6 and 11 (McGinnis et al. 1984; Joyner et al. 1985; Rabin et al. 1985), which also contain genes with differential activity depending on the parental origin of the chromosome (Cattanach and Kirk 1985). Sequences corresponding to the *engrailed* locus homeo box have recently been mapped to mouse chromosome 1 (Joyner et al., this volume), whose maternal or paternal disomy does not affect development (Searle and Beechey 1978). Further mapping of other homeo box sequences should reveal whether observed chromosomal colocalizations are spurious or instead indicative of a functional relationship.

Developmental failure of nuclear transfer embryos containing nuclei from cleavage-stage blastomeres was unexpected. Chimera experiments have indicated that blastomeres from eight-cell-stage mouse embryos are very likely totipotent (Kelly 1977). In addition, in amphibian species, nuclear transfer from much older embryos resulted in normal development, whereas transfer of adult nuclei yielded extensive if not complete development (Gurdon 1974). It is therefore probable that the reason for developmental failure of nuclear transfer mouse embryos lies in some specific characteristic of mammalian development. Although the early development of nonmammalian species is largely independent of the embryonic genome (Davidson 1976), activation of the embryonic genome in mammals occurs very early, possibly in zygote and certainly in two-cell-stage embryos (Sawicki et al. 1981; Clegg and Piko 1982, 1983; Flach et al. 1982; Bolton et al. 1984). It is possible that stage-specific genes required for the transition from one embryonic stage to another are not appropriately reactivated in nuclear transplant embryos receiving exogenous nuclei. Another possibility is that genes active in later-stage nuclei continue to be expressed after transfer and that such expression is deleterious for development. This need for very precise matching of nucleus and cytoplasm is also emphasized by our recent data, which demonstrate that interspecies pronuclear transfer embryos do not complete preimplantation development. Possible solutions to these problems might include the transfer of nuclei into developmentally more advanced cytoplasm, the serial transplantation of nuclei as has been employed in amphibian nuclear transplant embryos (King and Briggs 1957; Gurdon 1974), or the introduction of the donor nucleus at a time coincident with cytoplasmic activation (Czolowska et al. 1984).

The experiments presented point to unique aspects of mammalian development and hopefully will provide information as to the precise timing of gene activation and the requirements for correct nuclear–cytoplasmic interactions. This background information should prove invaluable in identifying genes that control development and in deciphering the mechanism of their action.

ACKNOWLEDGMENTS

This work was supported in part by grants HD-12487 and HD-17720 from the National Institute of Child Health and Human Development and by grants CA-10815 and CA-25875 from the National Cancer Institute. S.F.G. was supported by grant SG-134 from The American Cancer Society.

REFERENCES

Abramczuk, J., D. Solter, and H. Koprowski. 1977. The beneficial effect of EDTA on development of mouse one-cell embryos in chemically defined medium. *Dev. Biol.* **61:** 378.

Barton, S.C., M.A.H. Surani, and M.L. Norris. 1984. Role of paternal and maternal genomes in mouse development. *Nature* **311:** 374.

Bolton, V.N., P.G. Oades, and M.H. Johnson. 1984. The relationship between cleavage, DNA replication, and gene expression in the mouse 2-cell embryo. *J. Embryol. Exp. Morphol.* **79:** 139.

Cattanach, E.M. and M. Kirk. 1985. Differential activity of maternally and paternally derived chromosome regions in mice. *Nature* **315:** 496.

Clegg, K.B. and L. Piko. 1982. RNA synthesis and cytoplasmic polyadenylation in the one-cell mouse embryo. *Nature* **295:** 342.

———. 1983. Quantitative aspects of RNA synthesis and polyadenylation in 1-cell and 2-cell mouse embryos. *J. Embryol. Exp. Morphol.* **74:** 169.

Czolowska, R., J.A. Modlinski, and A.K. Tarkowski. 1984. Behavior of thymocyte nuclei in non-activated and activated mouse oocytes. *J. Cell Sci.* **69:** 19.

Davidson, E.M. 1976. *Gene activity in early development*, 2nd edition. Academic Press, New York.

Distel, R.J., K.C. Kleene, and N.B. Hecht. 1984. Haploid expression of a mouse testis alpha-tubulin gene. *Science* **224:** 68.

EPPIG, J.J., L.P. KOZACK, E.M. EICHER, and L.C. STEVENS. 1977. Ovarian teratomas in mice are derived from oocytes that have completed the first meiotic division. *Nature* **269:** 517.

EPSTEIN, C.J. 1983. The X chromosome in development. In *Cytogenetics of the mammalian X chromosome, part A: Basic mechanisms of X chromosome behavior* (ed. A.A. Sandberg), p. 51. A.R. Liss, New York.

EPSTEIN, C.J. and S.A. SMITH. 1974. Electrophoretic analysis of proteins synthesized by preimplantation mouse embryos. *Dev. Biol.* **40:** 233.

FLACH, G., M.H. JOHNSON, P.R. BRAUDE, R.A.S. TAYLOR, and V.N. BOLTON. 1982. The transition from maternal to embryonic control in the 2-cell mouse embryo. *EMBO J.* **1:** 681.

GARRELS, J.I. 1979. Two-dimensional gel electrophoresis and computer analysis of proteins synthesized by clonal cell lines. *J. Biol. Chem.* **254:** 7961.

GILBERT, S.F. and D. SOLTER. 1985. Onset of paternal and maternal Gpi-1 expression in preimplantation mouse embryos. *Dev. Biol.* **109:** 515.

GRAHAM, C.F. 1974. The production of parthenogenetic mammalian embryos and their use in biological research. *Biol. Rev.* **49:** 399.

GROUDINE, M. and K.F. CONKIN. 1985. Chromatin structure and *de novo* methylation of sperm DNA: Implications for activation of the paternal genome. *Science* **228:** 1061.

GURDON, J.B. 1962. The developmental capacity of nuclei taken from intestinal epithelium cells of feeding tadpoles. *J. Embryol. Exp. Morphol.* **10:** 622.

———. 1974. *The control of gene expression in animal development*. Clarendon Press, Oxford.

HOWE, C.C. and D. SOLTER. 1979. Cytoplasmic and nuclear protein synthesis in preimplantation mouse embryos. *Dev. Biol.* **52:** 209.

JOHNSON, D.R. 1975. Further observations on the hairpintail (T^{hp}) mutation in the mouse. *Genet. Res.* **24:** 207.

JOHNSON, M.H. 1981. The molecular and cellular basis of preimplantation mouse development. *Biol. Rev.* **56:** 463.

JOYNER, A.L., R.V. LEBO, Y.W. KANT, R. TJIAN, D.R. COX, and G.R. MARTIN. 1985. Comparative chromosome mapping of a conserved homeo box region in mouse and human. *Nature* **314:** 173.

KAUFMAN, M.H., S.C. BARTON, and M.A.H. SURANI. 1977. Normal postimplantation development of mouse parthenogenetic embryos to the forelimb bud stage. *Nature* **265:** 53.

KELLY, S.J. 1977. Studies of the developmental potential of 4- and 8-cell stage mouse blastomeres. *J. Exp. Zool.* **200:** 365.

KING, T.J. and R. BRIGGS. 1957. Serial transplantation of embryonic nuclei. *Cold Spring Harbor Symp. Quant. Biol.* **21:** 271.

LEVINSON, J., P. GOODFELLOW, M. VADEBONCOEUR, and H. MCDEVITT. 1978. Identification of stage-specific polypeptides synthesized during murine preimplantation development. *Proc. Natl. Acad. Sci.* **75:** 3332.

LOH, D.Y. and D. BALTIMORE. 1984. Sexual preference of apparent gene conversion events in MHC genes of mice. *Nature* **309:** 639.

LONGO, F.J. and D.-Y. CHEN. 1985. Development of cortical polarity in mouse eggs: Involvement of the meiotic apparatus. *Dev. Biol.* **107:** 382.

LYON, M.F. and P.H. GLENISTER. 1977. Factors affecting the observed number of young resulting from adjacent-2 disjunction in mice carrying a translocation. *Genet. Res.* **29:** 83.

MAGNUSON, T. and C.J. EPSTEIN. 1981. Genetic control of very early mammalian development. *Biol. Rev.* **56:** 369.

MANN, J.R. and R.H. LOVELL-BADGE. 1984. Inviability of parthenogenones is determined by pronuclei, not egg cytoplasm. *Nature* **310:** 66.

MCGINNIS, W., C.P. HART, W.J. GEHRING, and F.H. RUDDLE. 1984. Molecular cloning and chromosome mapping of a mouse DNA sequence homologous to homeotic genes of *Drosophila*. *Cell* **38:** 675.

MCGRATH, J. and D. SOLTER. 1983a. Nuclear transplantation in the mouse embryo by microsurgery and cell fusion. *Science* **220:** 1300.

———. 1983b. Nuclear transplantation in mouse embryos. *J. Exp. Zool.* **228:** 355.

———. 1984a. Completion of mouse embryogenesis requires both the maternal and paternal genomes. *Cell* **37:** 179.

———. 1984b. Inability of mouse blastomere nuclei transferred to enucleated zygotes to support development in vitro. *Science* **226:** 1317.

———. 1984c. Maternal T^{hp} lethality in the mouse is a nuclear, not cytoplasmic, defect. *Nature* **308:** 550.

MCLAREN, A. 1979. The impact of pre-fertilization events on post-fertilization development in mammals. In *Maternal effects in development* (ed. D.R. Newth and M. Ball), p. 287. Cambridge University Press, Cambridge, England.

PETERSON, A.C., P.M. FRAIR, and G.G. WONG. 1978. A technique for detection and relative quantitative analysis of glucose phosphate isomerase isozymes from nanogram tissue samples. *Biochem. Genet.* **16:** 681.

RABIN, M., C.P. HART, A. FERGUSON-SMITH, W. MCGINNIS, M. LEVINE, and F.H. RUDDLE. 1985. Two homeo box loci mapped in evolutionarily related mouse and human chromosomes. *Nature* **314:** 175.

SAWICKI, J., T. MAGNUSON, and C.J. EPSTEIN. 1981. Evidence for the expression of the paternal genome in the two-cell mouse embryo. *Nature* **294:** 450.

SEARLE, A.G. and C.V. BEECHEY. 1978. Complementation studies with mouse translocations. *Cytogenet. Cell Genet.* **20:** 282.

SURANI, M.A.H., S.C. BARTON, and M.L. NORRIS. 1984. Development of reconstituted mouse eggs suggests imprinting of the genome during gametogenesis. *Nature* **308:** 548.

VAN BLERKOM, J. 1985. Extragenomic regulation and autonomous expression of a developmental program in the early mammalian embryo. *Ann. N.Y. Acad. Sci.* (in press).

VAN BLERKOM, J. and G. BROCKWAY. 1975. Qualitative patterns of protein synthesis in the preimplantation mouse embryo. I. Normal pregnancy. *Dev. Biol.* **44:** 148.

WAKSMUNDZKA, M., E. KRYSIAK, J. KARASIEWICZ, R. CZOLOWSKA, and A.K. TARKOWSKI. 1984. Autonomous cortical activity in mouse eggs controlled by a cytoplasmic clock. *J. Embryol. Exp. Morphol.* **79:** 77.

WASSARMAN, P.M. 1983. Oogenesis: Synthetic events in the developing mammalian egg. In *Mechanism and control of animal fertilization* (ed. J.F. Hartman), p. 1. Academic Press, New York.

WHITTEN, W.K. 1971. Nutrient requirements for the culture of preimplantation embryos *in vitro*. *Adv. Biosci.* **6:** 129.

Molecular Analysis of the First Differentiations in the Mouse Embryo

P. Brûlet, P. Duprey, M. Vasseur, M. Kaghad, D. Morello,* P. Blanchet, C. Babinet,*
H. Condamine, and F. Jacob

*Unité de Génétique cellulaire du Collège de France et de l'Institut Pasteur and *Unité de Génétique des Mannifères, Institut Pasteur, 75724 Paris Cedex 15, France*

The transition from the mouse eight-cell morula to the blastocyst stage is characterized by the appearance of two different cell types, the inner cell mass (ICM), which is composed of pluripotent cells, and the trophectoderm, which contains the first epithelial cells (Ziomek and Johnson 1982). To analyze the mechanisms involved in the regulation of gene expression during this first differentiation, we have isolated two types of markers that are expressed in these two tissues. Selection and characterization of such markers have been performed in the teratocarcinoma in vitro system which provides cell lines equivalent to the embryonic pluripotent cells (the embryonal carcinoma or EC cells) and differentiated derivatives equivalent to the trophectoderm cells (the trophoblastoma cells).

The trophoblastoma cells are characterized by the expression of cytokeratin intermediate filaments which are not found in EC cells (Brûlet et al. 1980). One of the major components of the intermediate filaments of the trophectoderm is the cytokeratin Endo A, also referred to as cytokeratin A (Brûlet et al. 1980; Oshima 1982). In the blastocyst, Endo A has been detected only in the trophectoderm but not in the ICM, and later, in extraembryonic endoderm cells but not in pluripotent embryonic ectoderm (Kemler et al. 1981). To analyze the regulation of Endo A expression during the early steps of embryogenesis, we recently have isolated genomic clones encoding this protein (Vasseur et al. 1985).

The EC cells contain abundant transcripts from a family of repetitive elements that present a retroviral-like structure, which we named E.Tn, for early transposon. The E.Tn family is only poorly expressed in the trophoblastoma cells (Brûlet et al. 1983). In vivo, this family is transcribed in the pluripotent cell lineage from the blastocyst stage to the primitive streak stage (Brûlet et al. 1985).

We have used these two markers to investigate the regulation of transcription during the first stages of mouse embryogenesis. We describe here some characteristic features of the structure and expression of the E.Tn family and report results which suggest that although Endo A mRNA expression is likely to be regulated at the transcriptional level both in vitro and in vivo, the promoter of the Endo A gene does not present a strict tissue specificity of transcription in transient transfection experiments.

RESULTS

Analysis of the Regulation of a Trophectoderm-specific Marker

Expression of Endo A mRNA in EC cells, in differentiated cells, and during early embryogenesis. We studied Endo A mRNA expression by S1 mapping using a 295-bp-long SmaI–SmaI fragment (Fig. 1) which encompasses the 5' end of the mRNA. The use of such

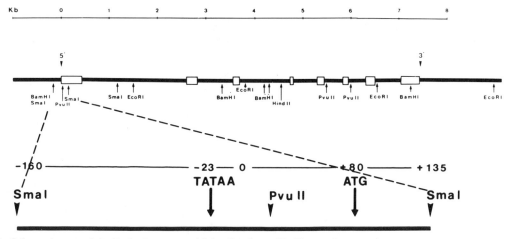

Figure 1. Schematic map of the Endo-A α_1 gene with its 5' end, used in S1 mapping experiments.

a fragment provides both a qualitative and quantitative probe that allows us to measure the amount of mRNA and to analyze its 5' end. Moreover, this fragment does not hybridize to any of the other members of the cytokeratin multigene family and is strictly specific for Endo A. This fragment was subcloned into M13mp8 in order to prepare single-stranded probes which were hybridized to RNA extracted from TDM1, 3T3, PCC3, F9, and retinoic acid-treated F9 cells (cell lines used in this work are referenced in Kemler et al. 1981). Hybridization with RNA from the trophoblastoma cell line TDM1 yields a major 130-bp-long fragment, which corresponds to the major cap site, and a minor 153-bp band, which reveals an initiation in the TATAA box (Fig. 2). Very low levels of Endo A transcripts were found in F9, but a 96-hour retinoic acid treatment of these cells increases the amount of Endo A mRNA by more than 50-fold, as evaluated by densitometric scanning (Fig. 2). A low amount of Endo A mRNA was also found in the PCC3 EC cell line. In contrast, we did not detect any Endo A mRNA expression in 3T3 cells, even after prolonged exposure of the autoradiogram. Thus, the block for Endo A expression appears to be completely stringent in fibroblasts, which do not express any cytokeratin in their cytoskeleton. In contrast EC cells, which are likely to be committed to differentiate into epithelial endodermic cells, express a low basic level of Endo A mRNA. This low level probably reflects the spontaneous endodermal differentiation that occurs during the in vitro culture of EC cell lines.

Since Endo A protein has been detected in vivo during mouse embryogenesis from the eight-cell stage, we studied the Endo A mRNA expression from the two-cell stage to the blastocyst stage. Total RNA was extracted from two-cell embryos, eight-cell morulae, and blastocysts, and was analyzed by the S1 mapping procedure. No transcript was detected in two-cell-stage embryos, even when using as many as 1200 embryos for one hybridization experiment. At the eight-cell stage, we detected Endo A transcripts after prolonged exposure of the autoradiogram; 1000 morulae of this stage were used for each hybridization. In blastocysts, the expression of Endo A mRNA was significantly enhanced and detected in hybridizations performed with 250 embryos (Fig. 2). Because of the scarcity of material available, we did not measure the amounts of RNA in these different experiments. However taking into account the measurements of RNA content of early mouse embryos reported in the literature (Piko and Clegg 1982) and the number of embryos used for the different experiments, the increase of Endo A mRNA from the eight-cell to the blastocyst stage can be evaluated to about 10-fold.

Analysis of the Endo A gene promoter activity in transfection experiments. The results described

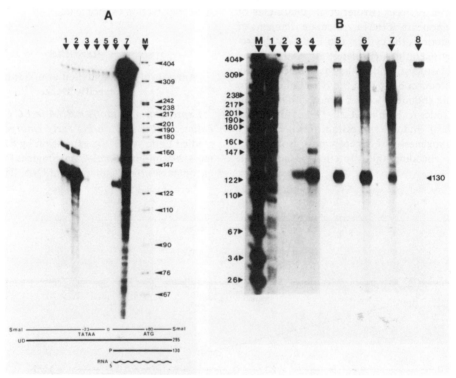

Figure 2. S1 mapping experiments using the single-stranded *Sma*I–*Sma*I probe schematized below the gel in *A*. (UD) Undigested probe; (P) protected probe. (*A*) Analysis of RNA extracted from various cell lines. (Lane *1*) TDM1; (lane *2*) F9 96 hr after induction of differentiation by retinoic acid; (lane *3*) F9; (lane *4*) 3T3; (lane *5*) SV T2; (lane *6*) PCC3; (lane *7*) nondigested probe. Exposure of the autoradiogram was 18 hr at −80°C. (*B*) Same as *A* with RNAs from: (lane *3*) liver; (lane *4*) TDM1; (lane *5*) 250 blastocysts; (lane *6*) 500 blastocysts; (lane *7*) 1000 eight-cell morula; (lane *8*) 1200 two-cell embryo. (Lanes *1* and *2*) Probe before and after S1 digestion. Exposures of autoradiograms were 15 hr (lanes *1–4*) and 120 hr (lanes *5–7*).

above, obtained with a highly sensitive S1 procedure, suggest that the expression of Endo A could be regulated at the transcriptional level during early embryogenesis and EC cell differentiation. To study the promoter of the Endo A gene, and to analyze its efficiency in different cell types, we used the chloramphenicol acetylase (CAT) system of Gorman et al. (1982) in which the expression of the bacterial CAT gene, linked to an eukaryotic promoter, provides a measure of the transcriptional activity of this promoter. This system has been used to compare the efficiencies of different promoters and to analyze the effect of enhancer sequences on transcription (Yaniv 1984). We have designed constructions (1) to test the efficiency of the Endo A promoter and (2) to investigate the enhancing effect of sequences located upstream or downstream from the mRNA cap site and to analyze their potential role in the tissue-specific expression of the Endo A mRNA.

We used a PvuII–EcoRI DNA fragment which contains 900 bp upstream and 20 nucleotides downstream from the cap site of the Endo A mRNA (see Fig. 1) (Vasseur et al. 1985). This fragment was fused to the coding region of the CAT gene, followed by the usual SV40 DNA fragments providing the RNA processing signals (Fig. 3). This recombinant vector was then transfected into epithelial (PYS or TDM1), fibroblastic (3T3 or 3T6), or EC cells (F9) and the expression of the CAT activity measured after 48 hours. To compare the Endo A promoter activity with the efficiencies of other well-characterized promoters, we transfected the same cells with CAT plasmids linked to either the β-actin or the α2(I) collagen promoters (gift of P. Herbomel, Institut Pasteur). Moreover, to achieve accurate comparisons between the different transfection assays, we introduced, as an internal marker, the pCH110 plasmid that codes for a β-galactosidase fusion protein, expressed under the control of the SV40 early promoter. We cotransfected every construction with this plasmid in a 4:1 molar ratio. Then 48 hours later, each cell extract was tested for both CAT and β-galactosidase activities.

The Endo A gene promoter sequence provides initiation signals allowing efficient expression of the CAT gene (Fig. 4). The level of CAT activity was in the range of the results observed with the collagen promoter and 50 times lower than the activity measured with the actin promoter. Moreover, the Endo A promoter was efficient in all the different cell lines tested, as well in 3T3 or 3T6 (which do not express any Endo A mRNA) as in PYS or TDM1 (which express high level of Endo A mRNA). In PYS cells, the Endo A promoter was only twofold more efficient than in 3T6. The in vitro activity of the Endo A promoter in a fibroblastic cell line such as 3T6, which does not express any Endo A mRNA, is surprising and shows that transient expression assays do not reflect exactly the normal in vivo gene activity. This discrepancy could perhaps be explained by the Endo A gene extrachromosomal location which might suppress the effect of cis-acting factors.

To test if any tissue-specific enhancer element, acting in cis, was located in the vicinity of the promoter, we inserted Endo A gene fragments upstream from the

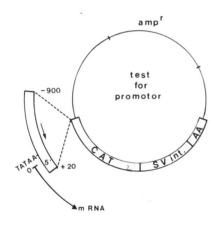
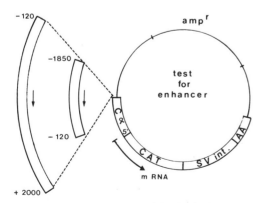

Figure 3. CAT plasmids containing Endo A gene sequences.

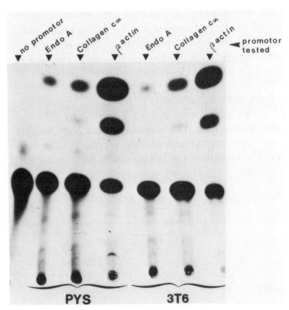

Figure 4. CAT assays using the promoter constructions shown in Fig. 3. Endo A promoter was compared with the α2(I)-collagen and β-actin promoters in PYS and 3T6 cells.

α2-collagen promoter, itself fused to the coding region of the CAT gene (referred to as COL-CAT). We used for these constructions a minimum version of the α-collagen promoter, spanning from nucleotide −280 to nucleotide +110. Upstream from these elements, we inserted a *Bam*HI-*Bam*HI −120/−1850 fragment (Fig. 1) containing 1850 bp upstream from the Endo A mRNA cap site, or a *Bam*HI-BamHI −120/+2000 fragment containing a part of the coding region and the first larger intron. These constructions were tested as described above in the different cells, but did not allow the detection of any tissue-specific enhancing effect of any of the sequences tested.

Although it is difficult to draw conclusions from negative results, it seems that the tissue specificity of Endo A transcription is not provided by a single DNA region located around the promoter elements. Tissue-specific controlling elements have been identified between a hundred and several hundred nucleotides preceding the 5′ end of genes, or even inside introns (for review, see Yaniv 1984). It has to be pointed out that such elements have been detected around, or in genes encoding specialized, terminally differentiated functions, i.e., immunoglobulin heavy chains, insulin, chymotrypsin, and crystallin protein. In contrast, Endo A is expressed in several epithelial cell types localized in various organs of the embryo and adult mouse, from mesodermal or endodermal origins (Kemler et al. 1981). Thus, the regulation of Endo A may be, at least in part, dependent on other mechanisms requiring a proper localization in the chromatin, and common to all cell lineages committed to the epithelial phenotype.

Early Transcription of E.Tn in Pluripotential Cells

The mouse genome harbors several families of long repeated sequences with retroviral-like structure. Recently, we have characterized one of them (Brûlet et al. 1983), which we call E.Tn. One interesting characteristic of the E.Tn family is that its transcription peaks in the pluripotent cell lineage from blastocyst to primitive streak stage (Brûlet et al. 1985). In vitro an abundant RNA species is detected in EC cell lines.

Retroviral characteristics of E.Tn. E.Tn sequences are about 5.5 kb long and are bordered by direct repeats 322 bp long. These sequences are 200–400 scattered copies per mouse haploid genome and are not found in other mammalian genomes. As in other long repeat families, the integrity of the sequence is partially conserved as evidenced by a polymorphism at the restriction site level. One of these E.Tn sequences was cloned, the so-called pMAC-2 plasmid.

A structural analysis has established that the two long terminal repeats (LTRs) bordering the long moderately repeated E.Tn sequences are identical to retroviral LTRs. By a combination of nucleotide sequencing, primer extension, and S1 mapping experiments, the E.Tn LTR can be subdivided into three domains U_3, R, and U_5, which are 192, 12, and 118 bp long, respectively. The central 12-base R sequence is duplicated at both ends of E.Tn RNA whereas the U_3 and U_5 sequences are found at the 3′ and 5′ ends of the RNA. A previous electron microscope analysis of heteroduplexes had shown that E.Tn RNA is colinear with a randomly selected genomic E.Tn sequence. The E.Tn LTR structure gives further support to the notion that E.Tn sequences are organized in a way similar to retroviral sequences.

An inverted repeat, ending in TCT....ACA, borders the E.Tn LTR. Inverted repeats that border retroviral LTRs are involved in the integration of the provirus. Duplication of cellular DNA at the cellular DNA junction has been recognized as a feature of the integration process of transposable elements and retroviruses. A 6-bp duplication brackets the genomic E.Tn sequence in pMAC-2 (Fig. 5).

Two base pairs 3′ from the E.Tn's 5′ LTR is a sequence homologous to $tRNA_3^{Lys}$. Retroviruses are known to have a tRNA binding site at this position which serves as a primer for the synthesis of the minus DNA strand (Temin 1981; Varmus 1982; Weiss et al. 1982).

Pattern of early embryonic transcription. We have investigated the presence of E.Tn transcripts in the early mouse embryo by in situ RNA-DNA hybridization. Single-strand, uniformly labeled probes were synthesized that covered various portions of the E.Tn sequences. For instance, we have used a probe covering the 5′ LTR or 3′ LTR as well as probes from the E.Tn internal domains. Controls were always run with a probe of opposite polarity. By this technique we have monitored the amount of E.Tn RNA inside the cells.

The results of in situ hybridization experiments show that the E.Tn sequences are transcribed at a high rate in early pluripotent cell lineages, from the blastocyst

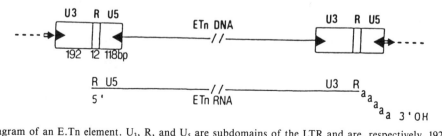

Figure 5. Diagram of an E.Tn element. U_3, R, and U_5 are subdomains of the LTR and are, respectively, 192, 12, and 118 bp long. Opposed arrows at the end of the LTR: a 9-bp inverted repeat with two mismatches. A 6-bp direct repeat in the immediate flanking genomic DNA is indicated by smaller arrows. The RNA transcript is indicated with its 5′ and 3′ ends within the 5′ and 3′ LTRs.

ICM to embryonic ectoderm of the late primitive-streak stage (between 3.5 and 7.5 days of embryonic development) but also in extraembryonic ectoderm layers (Fig. 6). Low or very low levels of hybridization were seen in early cleavage stages, endoderm layers of egg cylinder stages, and tissues of 8.5- to 10.5-day and older embryos.

The fact that E.Tn RNA is abundant in embryonic and extraembryonic ectoderm tissues indicates that E.Tn transcription does not occur in a single cell lineage in the early embryo. Gardner and Papaioannou (1975) have shown that the embryonic and extraembryonic ectoderms derive from the two distinctly committed lineages in the blastocyst.

The low amount of E.Tn transcription detected by the in situ hybridization technique in tissue from 9.5- to 13.5-day embryos was correlated with the presence of a small amount of correctly initiated E.Tn RNA by S1 mapping experiments. The general picture that emerges from those experiments is that E.Tn transcription peaks in the pluripotent cell lineages in 4- to 7.5-day embryos and also in extraembryonic ectoderm layers; then transcription drops slowly down to a few percent of its peak value in tissues of older embryos. The precise onset of E.Tn transcription in preimplantation embryos is not yet known.

E.Tn transcription in an heterologous system. We do not yet know which molecular components are responsible for the E.Tn transcription during early embryogenesis. As a first step in analyzing the regulation of E.Tn transcription, we have used a transfection assay.

As recipient cells we have chosen Rat 3T3 fibroblast cells because we could not detect E.Tn sequences in the genome of those cells and therefore did not have to distinguish the introduced E.Tn sequence from any endogenous copies.

The plasmid pMAC2, containing an entire E.Tn sequence, was cointroduced into the 3T3 cells by the calcium phosphate precipitate method, with a plasmid pAG60 carrying the gene coding for the resistance to the antibiotic g418 (a gift from Axel Garapin). After 48 hours, selections were applied with the antibiotic g418 and resistant clones were grown. Figure 7 shows a Southern analysis of the genomic DNA extracted from a few clones. The DNA was restricted with the endonuclease *Eco*RI which yields the entire E.Tn-sequence as a 6-kb fragment from plasmid pMAC-2 (Brûlet et al. 1983). Most of the analyzed clones contained E.Tn sequences and in a few of them no rearrangement was detected after integration. Three clones were further analyzed for the E.Tn transcription. Clones A and J were estimated to contain 4 and 10 E.Tn copies, respectively, whereas clone H contained many more copies.

Figure 8 shows the results of a dot blot analysis of RNA extracted from those three clones. E.Tn RNA is clearly detected in each of them with a particularly high amount of transcription occurring in clone A. This relatively high level of transcription might be due to E.Tn integration taking place in "open" regions of fibroblast chromatin or downstream from a strong promoter. To analyze further the transcription in those clones, S1 nuclease mapping experiments were performed with a probe covering the 5' LTR and upstream region (Brûlet et al. 1985). Using this assay, only in clone A could we detect some correctly initiated and terminated E.Tn transcripts (data not shown).

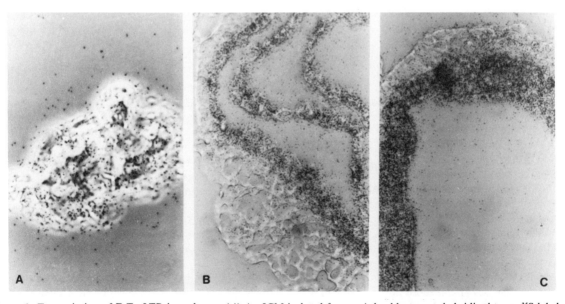

Figure 6. Transcription of E.Tn LTR in embryos. (*A*) An ICM isolated from a 4-day blastocyst, hybridized to an ^{35}S-labeled single-strand probe covering the 5' LTR (Brûlet et al. 1985) (4×10^5 cpm/slide, 4 weeks of exposure). (*B*) Enlargement of the extraembryonic ectoderm region of a 7.5-day embryo showing the strongly labeled tissues in contrast to the weakly labeled extraembryonic endoderm (2×10^6 cpm/slide, 4.5 weeks of exposure). (*C*) Enlargement of the labeled embryonic ectoderm region of a 7.5-day embryo. The mesoderm and endoderm are less labeled. Same experimental conditions as in *B*.

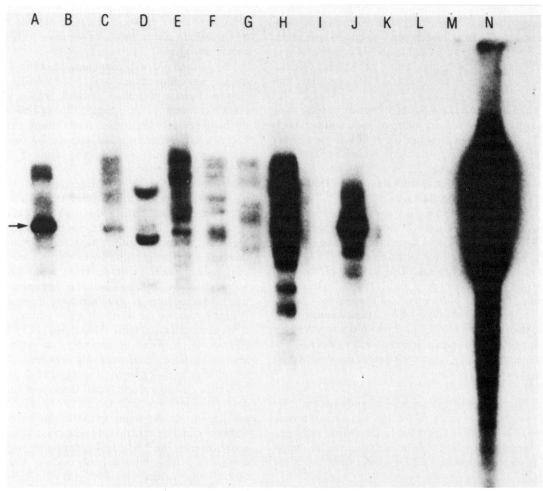

Figure 7. Southern blot analysis of EcoRI-restricted DNA extracted from Rat 3T3 clones transfected with the plasmid pMAC-2. The arrow indicates the 6-kb EcoRI–EcoRI fragment containing the E.Tn sequence. (Lanes *A–L*) The 12 analyzed clones; (lane *M*) the recipient Rat 3T3 cells; (lane *N*) a mouse EC cell line PCC4. The probe is a nick-translated PstI–PstI internal E.Tn fragment (Brûlet et al. 1983). Calibration of intensity in the bands was obtained by appropriate dilution of EcoRI-restricted pMAC-2.

Clone A cells were grown to confluence to analyze if the introduction of E.Tn sequences in clone A had any easily detectable physiological impact on the Rat 3T3 cells. No loss of contact inhibition could be detected by this assay.

CONCLUDING REMARKS

The experiments reported here show that cloned mouse genes corresponding to sequences that are activated in the early embryonic lineages are available: The Endo A gene is first expressed in the morula and remains active on trophectoderm and endoderm cells, but not in pluripotent undifferentiated cells thereafter, whereas the moderately repeated E.Tn sequences have a different pattern of expression, becoming mainly active in blastocyst ICM cells and their pluripotential derivatives. It is probable that both Endo A gene and E.Tn sequence expression is regulated at the transcriptional level.

These tools should help us to understand the molecular mechanism underlying the initial appearance of cell heterogeneities in the mouse embryo. It is generally believed that at least up to the eight-cell stage, all the blastomeres are developmentally equivalent (Kelly 1975; however, see Wassarman et al. 1984) and that no cytoplasmic determinants or prepatterning preexist the emergence of two cell states in the blastocyst. How purely extrinsic cues brought by the environment might be converted into positional information responsible for such differentiations remains an entirely unsolved problem. Clearly, the molecular markers described here should be useful since ultimately one will have to understand how some positional information triggers differential gene activation in trophectoderm and ICM cells. The results have not made it possible to identify *cis* sequences specifically involved in the regulation of Endo A sequences. This leaves open the possibility that somewhat unusual mechanisms are at work in the control of gene expression of very early mammalian embryonic stages.

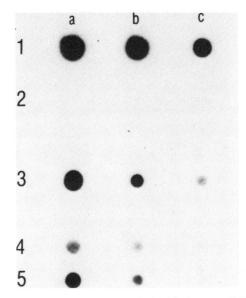

Figure 8. RNA dot blots probed with the same E.Tn fragment as in Fig. 7. *a, b,* and *c* refer to three amounts, 25, 15, and 5 μg, of total RNA. *1* and *2* refer, respectively, to RNA extracted from the EC PCC4 and Rat 3T3 cell line. *3, 4,* and *5* refer to RNA extracted from clones A, H, and J.

ACKNOWLEDGMENTS

We would like to thank D. Boullier and M. Maury for excellent technical assistance and P. Herbomel for the gift of CAT plasmids containing β-actin and α-collagen promoters. This work was supported by the Centre National de la Recherche Scientifique (Grant UA 269 and ATP 955 189), The Fondation pour la Recherche Médicale, the Ligue Nationale contre le Cancer, the Ministère de l'Industrie et de la Recherche (Grant 82V1388), and the Fondation André Meyer.

REFERENCES

BRÛLET, P., H. CONDAMINE, and F. JACOB. 1985. Spatial distribution of transcripts of the long repeated E.Tn sequence during early mouse embryogenesis. *Proc. Natl. Acad. Sci.* **82:** 2054.

BRÛLET, P., C. BABINET, R. KEMLER, and F. JACOB. 1980. Monoclonal antibodies against trophectoderm-specific markers during mouse blastocyst formation. *Proc. Natl. Acad. Sci.* **77:** 4113.

BRÛLET, P., M. KAGHAD, Y.S. XU, O. CROISSANT, and F. JACOB. 1983. Early differential tissue expression of transposon-like repetitive DNA sequences of the mouse. *Proc. Natl. Acad. Sci.* **80:** 5641.

GARDNER, R.G. and U.E. PAPAIOANNOU. 1975. Differentiation in the trophectoderm and inner cell mass. In *The early development of mammals* (ed. M. Balls and A.E. Wild), p. 107. Cambridge University Press, Cambridge, England.

GORMAN, C.M., L.F. MOFFAT, and B.H. HOWARD. 1982. Recombinant genomes which express chloramphenicol acetyl-transferase in mammalian cells. *Mol. Cell. Biol.* **2:** 1044.

KELLY, S.J. 1975. Studies of the potency of the early cleavage blastomeres of the mouse. In *The early development of mammals* (ed. M. Balls and A.E. Wild), p. 97. Cambridge University Press, Cambridge, England.

KEMLER, R., P. BRÛLET, M.T. SCHNEBELEN, J. GAILLARD, and F. JACOB. 1981. Reactivity of monoclonal antibodies against intermediate filament proteins during embryonic development. *J. Embryol. Exp. Morphol.* **64:** 45.

OSHIMA, R.G. 1982. Developmental expression of murine extraembryonic endodermal cytoskeletal proteins. *J. Biol. Chem.* **257:** 3414.

PIKO, L. and K.B. CLEGG. 1982. Quantitative changes in total RNA, total poly(A) and ribosomes in early mouse embryos. *Dev. Biol.* **89:** 362.

TEMIN, H.M. 1981. Structure, variation and synthesis of retrovirus long terminal repeat. *Cell* **27:**1.

VARMUS, H.E. 1982. Form and function of retroviral proviruses. *Science* **216:** 812.

VASSEUR, M., P. DUPREY, P. BRÛLET, and F. JACOB. 1985. One gene and one pseudogene for the cytokeratin endo A. *Proc. Natl. Acad. Sci.* **82:** 1155.

WASSARMAN, P.M., J.M. GREVE, R.M. PERONA, R.J. ROLLER, and G.S. SALZMANN. 1984. How mouse eggs put on and take off their extracellular coat. *UCLA Symp. Mol. Cell. Biol. New Ser.* **19:** 213.

WEISS, R., N. TEICH, H. VARMUS, and J. COFFIN, eds. 1982. *Molecular biology of tumor viruses*, 2nd edition: *RNA tumor viruses*. Cold Spring Harbor Laboratory, Cold Spring Harbor, New York.

YANIV, M. 1984. Regulation of eukaryotic gene expression by transactivating proteins and cis acting DNA elements. *Biol. Cell.* **50:** 203.

ZIOMEK, C.A. and M.H. JOHNSON. 1982. The roles of phenotype and position in guiding the fate of 16-cell mouse blastomeres. *Dev. Biol.* **91:** 440.

Cell Determination during Early Embryogenesis of the Nematode *Caenorhabditis elegans*

E. SCHIERENBERG

Max-Planck-Institut für experimentelle Medizin Abteilung Chemie, D-3400 Göttingen, Federal Republic of Germany

Investigations in many different organisms indicate that early embryonic development is controlled essentially by maternal gene products, which are deposited in the unfertilized egg (cf. Davidson 1976). Nevertheless, how a specific developmental pathway for a cell is determined, including eventual functional differentiation, as well as decisions such as the number of progeny a cell will produce and its final position in the organism remain open questions.

Some cases in which determination of cell fate is obviously mediated by differential segregation of cytoplasmic components have been reported for somatic (van Dam et al. 1982; Whittaker 1982) and germ line cells (Illmensee and Mahowald 1974; Wakahara 1978). On the other hand, models for the asymmetric segregation of developmental potential via chromosome lineage have been discussed (Cairns 1975).

The nematode *Caenorhabditis elegans* is a suitable organism for studying the question of cell determination during development. On the basis of classical studies (e.g., Boveri 1910), nematodes are thought to be the best example for "mosaic" development. Laser-induced cell ablations in embryos of *C. elegans* (Sulston et al. 1983), in fact, support the notion that here embryogenesis proceeds essentially in a cell-autonomous way. This concept allows an easier interpretation of experiments, because in the first approximation no intercellular regulation phenomena must be taken into consideration.

A second advantage of *C. elegans* is that eggs are small and transparent. They can develop from the zygote to hatching outside the mother. Embryonic cell lineages have been completely analyzed in developing eggs under the microscope and the fate of each cell has been determined (Sulston et al. 1983). Development was found to be invariant from individual to individual. During the first hour after the onset of cleavage a series of asymmetric divisions of the germ line cells P_0, P_1, P_2, and P_3 leads to the formation of five somatic founder cells designated AB, MS, E, C, D and the primordial germ cell P_4 (Fig. 1). Characteristic of the cleavage pattern of the germ line is the anterior-posterior orientation of newly formed daughter cells and an apparent inversion of polarity. The division of P_0 and P_1 each produces a pair of cells with the smaller germ line cell posterior. The division of P_2 and P_3 each produces a pair of cells with the smaller germ line cell anterior (Fig. 2). This pattern is more obvious in partial than in normal embryos (Laufer et al. 1980;

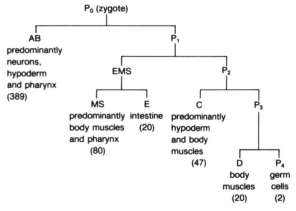

Figure 1. Schematic early cell lineage pattern in embryos of *C. elegans*. A series of asymmetric divisions of the germ line cells (P_0-P_3) leads to the formation of five somatic founder cells (AB, MS, E, C, D) and the primordial germ cell P_4. The tissues to which each founder cell contributes and the number of surviving cells produced in each lineage during embryogenesis are given below cell names. (Reprinted, with permission, from Schierenberg 1984.)

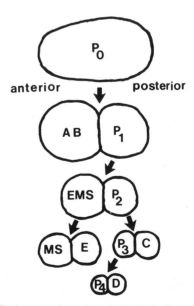

Figure 2. Apparent inversion of polarity in the germ line. The division of P_0 and P_1 each produces a pair of cells with the smaller germ line cell posterior. The division of P_2 and P_3 each produces a pair of cells with the smaller germ line cell anterior.

E. Schierenberg, unpubl.) where cellular topogenesis is less restricted by neighboring cells and the egg shell. The cell lineage derived from each founder cell produces a fixed number of cells, which contribute to one or several tissues (Fig. 1). In the early embryo the progeny of each of the five somatic founder cells divide synchronously and rhythmically, with cell cycle periods different for each lineage (Deppe et al. 1978). This results in a specific sequence of cleaving cell groups that appears to be important in establishing the correct spatial patterns (Schierenberg et al. 1980; Denich et al. 1984; Schierenberg 1984).

Two strategies have been used to study the question of cell determination in early embryos of *C. elegans*. The genetic approach includes the analysis of temperature-sensitive mutants that require maternal gene expression. With this approach the question of maternal versus zygotic contribution to early development is studied. Although the nematode embryo is not very amenable to manipulation from outside due to its protective eggshell and the crucial microenvironment inside, in the second approach a more direct way to interfere with normal development has been chosen. Using a laser microbeam coupled to a microscope, techniques have been developed to remove parts of the embryo and to transfer cytoplasm from one cell to another (Laufer and von Ehrenstein 1981; Schierenberg 1984; Schierenberg and Wood 1985). This approach explores the question of cytoplasmic versus nuclear contribution, which may not be necessarily a synonym for maternal versus zygotic influence. Altogether, the experiments done so far document the importance of maternal/cytoplasmic components for various aspects of embryonic cell determination.

RESULTS

Maternal Control of Early Development

In different laboratories, sets of *ts*-embryonic arrest mutants of *C. elegans* have been isolated after ethylmethanesulfonate-induced mutagenesis (Hirsh and Vanderslice 1976; Miwa et al. 1980; Cassada et al. 1981). Their modes of gene expression have been tested, temperature-sensitive periods have been determined, and early cellular development has been analyzed (Vanderslice and Hirsh 1976; Hirsh et al. 1977; Schierenberg et al. 1980; Wood et al. 1980; Isnenghi et al. 1983; Denich et al. 1984; E. Schierenberg, unpubl.).

Here, only a subclass of these mutants, those that require maternal gene expression, will be considered. The vast majority of the more than 30 mutants of this kind shows abnormalities during embryogenesis before the 50-cell stage, if the mother has been transferred to the restrictive temperature (25°C). Some of them do not cleave at all or arrest after a few cell divisions. Some seem to suffer from membrane defects. However, the most prominent kind of defect, visible in the majority of these mutants, is an abnormal cell cycle period (CCP). The time from one mitosis to the next is either too long or too short in one or several cell lineages. This leads to an alteration of the normally invariant sequence of cell divisions. Subsequently abnormal cell positioning can be observed and the embryos eventually arrest as "monsters," usually without undergoing a visible morphogenesis. It has been suggested that abnormal cell cycle timing is directly responsible for cell pattern defects and embryonic death (Schierenberg et al. 1980). A common defect in this mutant subclass is the premature division of the two gut precursor (E) cells after the 24-cell stage (Fig. 3d–f). In normal embryos these two cells migrate into the interior of the egg before they divide (Fig. 3a–c). Either genes controlling cell cycle timing are particularly sensitive to this kind of mutagenesis, or a large number of genes are involved in this process. In addition it seems likely that the selection method favors the isolation of this mutant type.

Cellular analysis has not been performed in enough detail to decide whether, in addition to or as a consequence of abnormal cell cycle timing, the differentiation program of cells has been affected in these mutants. A candidate for such a defect is the mutant *zyg-9* which inhibits pronuclear migration and shows abnormal orientation of the first cleavage spindle perpendicular to the long axis of the egg (Albertson 1984). This orientation leads to the equal distribution of germ line-specific granules to both daughter cells (Strome and Wood 1983). Recently K. Kemphues (pers. comm.) isolated an absolute lethal embryonic arrest mutant, which cleaves with its first division into cells of similar size. They both behave like AB cells regarding their further symmetric cleavage, their short and (initially) synchronous cell cycle periods, and the number of cells they produce.

A number of zygotic mutants with postembryonic lineage defects have been studied extensively, some of them representing homeotic-like mutations. Their study led to the identification of developmental control genes and the formulation of a binary switch model for the control of cell fates (Greenwald et al. 1983; Sternberg and Horvitz 1984).

Cytoplasmic Control of Cell Cycle Timing

Because the analysis of mutants indicates a strong maternal influence on early embryonic cell cycle timing and maternal gene products are assumed to be present in the cytoplasm, the importance of the cytoplasm for cell cycle behavior was tested under various conditions (Schierenberg and Wood 1985).

Enucleated one-cell embryos cycle with periods close to those of nuclei in cleavage-blocked embryos. Periodic waves of surface contraction in enucleated egg fragments have been described in several species (for review, see Satoh 1982), and similar contractions can be observed in enucleated embryos of *C. elegans*. When both pronuclei are extruded from an embryo prior to first cleavage (for early events during embryogenesis of

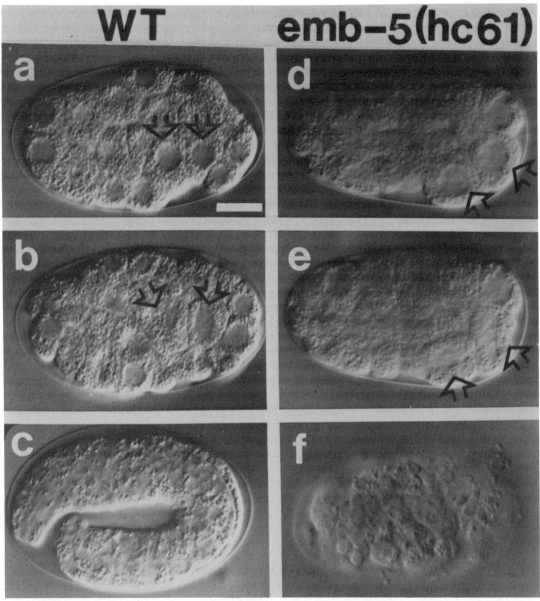

Figure 3. Abnormal cell cycle timing in the mutant *emb-5*. (*a*) Wild-type, 28-cell stage. Arrows point to both gut precursor (E) cells, which have migrated from the ventral side into the interior of the embryo. Time: ~110 min after fertilization (af). (*b*) Wild-type, 44-cell stage. Arrows point to both E cells in division. Time: ~122 min af. (*c*)Wild-type, "pretzel-stage" embryo has shaped into a worm and moves. Time: ~8 hr af. (*d*) Temperature-sensitive embryonic arrest mutant *emb-5* (allele *hc 61*), 26-cell stage. Arrows point to both E cells. Time: ~107 min af. (*e*) *emb-5*, 26-cell stage. Both E cells (arrows) have started mitosis prematurely before migration. Time: ~112 min af. (*f*) *emb-5*, terminal phenotype. Time: ~14 hr af. Nomarski optics; orientation, anterior is left, dorsal is top. Bar, 10 μm.

C. elegans, see von Ehrenstein and Schierenberg 1980; Strome and Wood 1983) the remaining cytoplast (enucleate cell) undergoes cycles of contraction and relaxation (Fig. 4). The periods and phasing of these cytoplasmic cycles are almost identical to those of the synchronous nuclear division cycles in one-cell embryos whose cell division has been blocked by cytochalasin D.

Enucleated blastomeres from different lineages cycle with different CCPs. The cycling behavior of enucleated cells from different lineages (which normally express different CCPs) was examined in two-cell embryos. Cytoplasts of both the AB and P_1 cells were extruded simultaneously through a laser-induced hole in the side of the eggshell and separated from the egg. AB cytoplasts were found to cycle with shorter periods than P_1 cytoplasts.

CCPs are only marginally influenced by changes of cell size. Cell volume was reduced by extruding considerable amounts of cytoplasm. In a two-cell embryo, about 50% of the cytoplasm of the AB cell was removed, so that AB was definitely smaller than P_1 (for

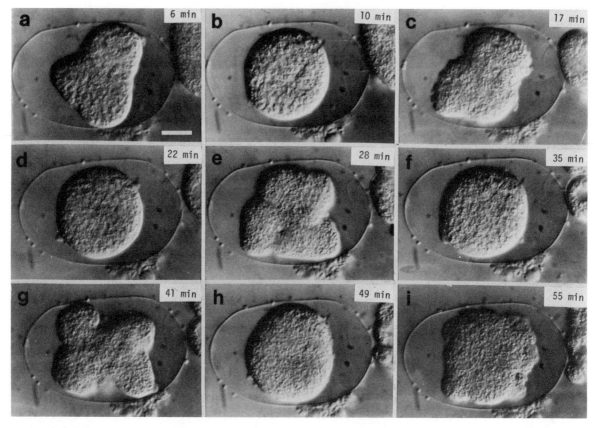

Figure 4. Periodical cortical contractions in an enucleated one-cell embryo. Maternal and paternal pronuclei were extruded together, just prior to pronuclear fusion, through a perforation at the posterior pole. Inserts indicate time after pronuclear fusion would have occurred. Nomarski optics; orientation, anterior is left. Bar, 10 μm. (Reprinted, with permission, from Schierenberg and Wood 1985.)

normal size ratios, see Fig. 2). Nevertheless, CCPs of all cells in the AB lineage remained faster than those of any P_1 descendant.

The volume of a cell can be increased by fusion of a cell to a neighboring cytoplast from the same lineage. ABa (anterior daughter of AB) was enucleated and two cell cycles later was fused to a granddaughter of ABp (posterior daughter of AB). In this manner the cytoplasmic volume of the recipient cell was increased severalfold. Nevertheless, the cell cycles of the giant cell and its descendants were found to be only slightly shorter than those of the untouched AB cells.

CCPs are only marginally influenced by DNA content per cell. Haploid embryos were generated by extruding the maternal pronucleus from fertilized eggs before zygote formation. Early cleavage pattern and cell cycle timing remained unaffected but the embryos arrested at a stage of several hundred cells with no initiation of visible morphogenesis, which may require maternal and paternal gene action (McGrath and Solter 1984). Embryos of a tetraploid variant of *C. elegans* (sp343; Madl and Herman 1979) have uniformly 10–15% longer CCPs, but the relative order of cleavages remains unaltered.

CCPs can be considerably altered by addition of cytoplasm from a different lineage. Using a combination of nuclear extrusion followed by fusion of the resulting cytoplast with an adjacent cell, cytoplasm can be added from one lineage to a cell from another lineage. In this way the CCPs of the recipient cell and its descendants can be lengthened or shortened in a manner that depends upon the CCP of the donor cell. Cytoplasm from a faster-dividing cell (e.g., ABp) can shorten the period of a cell from a slower-dividing lineage, (e.g., P_2) and vice versa. However, the CCP of the recipient cell is always closer to that of the more rapidly dividing cell.

Our observations suggest the presence of a cytoplasmic oscillator whose components are partitioned at different concentrations to early blastomeres. Their periods of oscillation depend upon the concentrations of these components. Thereby they regulate the activity of mitosis-inducing and/or mitosis-inhibiting factors and thus the duration of the cell cycle.

Mechanism of cell cycle control. Several molecular mechanisms have been proposed for the determination of the CCP (for review, see Fantes and Nurse 1981). In most of them, cell size, nuclear:cytoplasmic ratio, or protein content of a cell are involved as critical parameters. A fundamental difference between those experimental systems and ours is that we are looking at cleaving embryos. Here, divisions subdivide the cytoplasm without cellular growth. Cell cycles are short, probably because of simultaneous initiation of DNA synthesis at many sites (Kriegstein and Hogness 1974) and the absence of a G_1 period (Hinegardner et al. 1964; Graham and Morgan 1966; Brookbank 1970; van den Biggelaar 1971).

Cytoplasmic factors that change periodically in concentration or activity during the cell cycle have been identified as possible components of the cell-cycle control mechanism. One such factor, termed maturation-promoting factor (MPF), promotes maturation in oocytes and mitosis in fertilized eggs. Its activity changes with time in the cell cycle and may be controlled, in turn, by the cycling activity of an inactivating antagonist. MPF is a protein and has been extracted in active form from oocytes and eggs of various species and cultured cells, even from yeast (see Gerhart et al. 1984). MPF from all species promotes maturation of frog oocytes, suggesting long evolutionary conservation. Further experiments have to show whether these findings may point to a universal mechanism for cell cycle control.

Cytoplasmic Control of Early Cleavage Pattern

Nematodes are classical examples for the early appearance of the germ line and its early separation from the soma (Boveri 1887). Typical for the germ line is a series of asymmetric divisions always generating a larger somatic cell with a faster CCP and a smaller germ line cell with a slower CCP (Figs. 1 and 2). Somatic cells pass, in the first approximation, through symmetric divisions, always producing two cells of similar size whose descendants all cleave synchronously. Using the laser extrusion technique, various portions of the fertilized one-cell embryo were extruded to localize the region that carries the potential for asymmetric cleavage and to get a better idea about the mechanism of cell determination.

Posterior localization of potential for asymmetric cleavage. Up to about 40% of total cytoplasm can be extruded from the anterior pole of a one-cell embryo without removing nuclear material or centrioles (Fig. 5a). The remainder of the zygote nevertheless divides asymmetrically (Fig. 5b). The two daughter cells behave like AB and P_1 regarding their early cleavage pattern (symmetric vs. asymmetric cell divisions and different CCPs). However, abnormal cell positioning occurs, probably as a consequence of ample free space inside the eggshell, leading to altered directions of cell division or atypical cellular rearrangements. The embryo cleaves into several hundred cells, certain indices of cell differentiation can be detected (typical autofluorescence of E-cell descendants, muscle twitching), but no worm is formed. Removal of more than approximately 25% of the cytoplasm from the posterior pole abolishes the potential of a one-cell embryo to cleave asymmetrically (Fig. 5c,d). Even a large piece of approximately 75% of total cytoplasm extruded from the anterior pole, including nucleus and centrioles, is not able to divide in an asymmetric fashion (Fig. 5e,f). But even a small piece of cytoplasm extruded from the posterior pole (approximately 20% of total cytoplasm), if it includes nucleus and centrioles, is able to pass through the typical series of asymmetric cleavages (Fig. 5g,h).

These experiments indicate that the potential for asymmetric cleavage is localized in the posterior region of the uncleaved zygote.

Two developmental options for some partial embryos. Partial embryos generated by extrusion of cytoplasm plus nucleus and centrioles from the posterior pole can develop in two different ways. The first possibility is to cleave like a normal one-cell embryo, as described above. The second possibility is to make a symmetric division first (Fig. 5i,j). Then both daughters pass through asymmetric cleavages (Fig. 5k,l). In early cleavage patterns, each of the two initial cells behaves like a P_1 cell, formally producing two posterior half-twins. Cell positioning is variably abnormal. A similar result had been obtained by Boveri (1910) after the centrifugation of uncleaved *Ascaris* zygotes. The culture media used (Schierenberg and Wood 1985; E. Schierenberg, unpubl.) to allow cleavage of extruded cells is not sufficient so far to promote development to a stage where cells have differentiated visibly.

No influence of nuclear and centriolar history on early cleavage pattern. The experiments described above suggest that cytoplasmic quality determines a cell either to divide in a symmetric fashion (like a somatic cell) or to divide in an asymmetric fashion (like a germ line cell). The following experiment tests whether nucleus and/or centriole carry any imprint that binds them to a once-chosen alternative or whether the decision for one of the basic cleavage patterns is newly made with each division according to cytoplasmic environment.

The majority of cytoplasmic volume of a one-cell embryo was extruded from the anterior pole, including nucleus and centrioles (Fig. 5m). The cytoplasmic connection to the cytoplast remaining in the eggshell was not disrupted but was interrupted temporarily by moderately shaking the extruded part (repeatedly pressing on the coverslip). The first division of the exovulate generated two cells of similar size (Fig. 5n). During early anaphase of the simultaneously started mitoses, the still-existing connection between one of the cells and the cytoplast was reopened. Either this could occur accidentally without any manipulation or it could be induced by pressure on the embryo or by a laser pulse on the clogged cytoplasmic bridge. While the unfused cell completed its symmetric cleavage, one of the newly formed daughter nuclei of the fused cell slipped

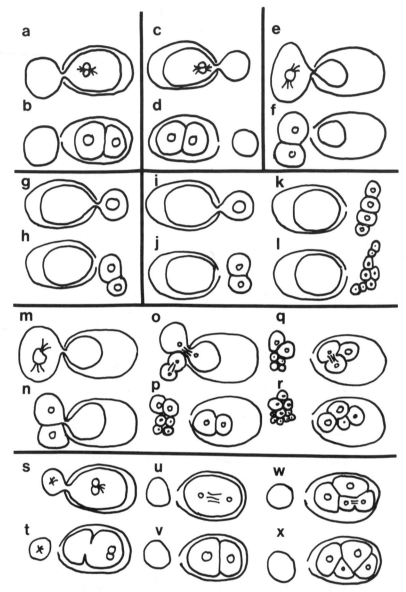

Figure 5. Cytoplasmic control of early cleavage pattern. This figure diagrams seven different experiments. Each box represents one experiment. (*a*) Cytoplasm is extruded from the anterior pole of uncleaved zygote. (*b*) Zygote has cleaved asymmetrically. (*c*) Cytoplasm is extruded from the posterior pole of uncleaved zygote. (*d*) Zygote has cleaved symmetrically. (*e*) Cytoplasm is extruded from the anterior pole of the uncleaved zygote including nucleus and centrioles. (*f*) The extruded part has cleaved symmetrically. (*g*) Cytoplasm is extruded from the posterior pole of the uncleaved zygote including nucleus and centrioles. (*h*) The extruded part has cleaved asymmetrically. (*i*) Same as *g*. (*j*) The extruded part has cleaved symmetrically. (*k*) Both cells have cleaved asymmetrically. (*l*) First the two larger of the four cells (located in the center) have cleaved symmetrically, then the two smaller cells (on opposite ends) have divided asymmetrically, so that their smaller daughters lie toward the center. (*m*) Cytoplasm is extruded from the anterior pole of the uncleaved zygote including nucleus and centrioles. (*n*) The extruded part cleaves symmetrically. (*o*) The lower cell has passed through another symmetric cleavage. The upper cell has started mitosis, one daughter nucleus has slipped into the cytoplast inside the eggshell. No cell division occurs. (*p*) The extruded part has been separated from the egg. The cell inside the eggshell has cleaved asymmetrically. (*q*) The larger, anterior cell inside the eggshell cleaves symmetrically. (*r*) The smaller, posterior cell inside the eggshell has cleaved asymmetrically, so that its smaller daughter lies anteriorly. (*s*) Cytoplasm is extruded from the anterior pole of the uncleaved zygote including one centrosome. (*t*) A pseudocleavage occurs, and nuclear material is located in the posterior region of the egg. (*u,v*) After having passed through a cell cycle without cell division, the zygote cleaves asymmetrically. (*w,x*) First the larger anterior cell has divided symmetrically. Then the smaller posterior cell divides asymmetrically, so that its smaller daughter lies anteriorly. Orientation, anterior is left. For further description, see text.

into the cytoplast (Fig. 5o). No actual cell division occurred here, only a separation of nuclei. The extruded part of the embryo was then separated from the egg to prevent any further interaction. The first division of the newly formed cell inside the eggshell produced a larger anterior and a smaller posterior daughter (Fig. 5p). This demonstrates that the potential of this cell to divide asymmetrically is not lost by the temporary stay (and replication) of nucleus and centriole in a cytoplasmic environment which can only promote symmetric cleavage. Next the anterior cell divided symmetrically (Fig. 5q). Shortly afterwards the posterior cell cleaved into a smaller anterior and a larger posterior daughter (Fig. 5r), like a P_2 cell (Fig. 1,2). In favorable cases, the disc-shaped centriolar region, which is typical for the germ line cell (Schierenberg and Cassada 1983), can be detected in the smaller cell.

Judged by the further cleavage pattern of its descendants, the cell generated (Fig. 5o) behaves similarly to a P_1 cell. However, the differences in CCPs between the presumptive MS and E cells are less pronounced than normal and the presumptive P_4 cell divides prematurely. Similar deviations from the typical sequence of cleavages have been observed in early embryos after fusion of AB and P_1 (Schierenberg 1984). They are probably due to a remixing of cytoplasm caused by experimental manipulation. It is still an open question whether a cell that behaves like a certain somatic founder cell with respect to its cleavage pattern will necessarily differentiate like such a cell. If one assumed that in the experiment just described the posterior part of the egg remaining in the eggshell contained the potential for asymmetric cell division, one could have predicted that the transfer of a nucleus into that cytoplast would have generated a P_0-like cell, because its cytoplasm has not yet passed through a single cell division. Alternatively, one could have predicted that this nuclear transfer would have generated a P_2-like cell because its nucleus has already passed through two divisions. The actually observed P_1-like behavior suggests a third alternative, namely that the number of cytoplasmic cycles (which run independently of nucleus and centriole, Fig. 4) through which this cytoplast has already passed is involved in the determination of its future cleavage behavior after addition of a nucleus.

Partial embryos generated in this way arrest several hours later with many cells but without visible initiation of morphogenesis. Attempts to achieve nuclear transfer one cycle later (in expectation of a cell that cleaves like P_2) have been unsuccessful so far.

Cell cycle and cell determination. To test further the idea of the cytoplasmic cycle as a critical parameter for early cell determination, a different experiment was performed. Just prior to pronuclear fusion, one centrosome was extruded with some cytoplasm from the anterior pole of a one-cell embryo (Fig. 5s). Mitosis that normally occurs immediately after pronuclear fusion could not be completed successfully because of the missing centrosome. A temporary constriction occurred (Fig. 5t) and the undivided zygote passed through the cell cycle generating a tetraploid configuration. The first division (Fig. 5u) generated a larger anterior and a smaller posterior cell (Fig. 5v). The division of the anterior cell was symmetric (Fig. 5w). The posterior cell divided asymmetrically, with the smaller daughter lying anterior (Fig. 5x), thus behaving like P_2. The further cleavage pattern reveals that (as in the preceding experiment) the initial cell behaves like P_1 (with the same reservations as above). Embryos manipulated in this way arrest several hours later with many cells but without visible morphogenesis. However, muscle contractions can be observed. How close cleavage potential is correlated to differentiation potential also needs to be tested here. Efforts to achieve the extrusion of another centrosome during the second attempt of the egg to cleave (expecting that it would behave like a P_2 cell) have so far been unsuccessful.

Determination of Cell Fate

One main question of animal development is how the fates of individual cells are determined during embryogenesis. In the early embryo of *C. elegans*, cell patterns are identical from individual to individual. The invariance in position and function of cells demonstrates that the developmental program is tightly controlled. To make a distinction between positional and ancestral influences, one must interrupt the normal chain of events experimentally. Using a laser microbeam to perform microsurgery on embryos, we have investigated whether fusion of a cell from one lineage with cytoplasm from a second lineage can lead to expression of a differentiation marker that is specific to the second lineage (Wood et al. 1984).

Cytoplasmic localization of gut cell potential. The eggshell of a two-cell embryo was punctured at the posterior pole and the P_1 nucleus was extruded with a small amount of cytoplasm. This extruded P_1 fragment was separated from the embryo, and the P_1 cytoplast remaining inside the shell was fused to a descendant of AB (Fig. 6d). After development overnight to a terminal phenotype of several hundred cells (Fig. 6e), many embryos showed localized autofluorescence characteristic of differentiated gut cells in a normal embryo (Fig. 6f). In control experiments with no fusion following P_1 enucleation (Fig. 6a), usually no autofluorescence was observed (Fig. 6c). Some embryos in both types of experiments showed diffuse low-level fluorescence, perhaps resulting from cell degeneration. Preliminary experiments indicate that developmental potential for nerve and muscle cells may also be segregated with the cytoplasm.

DISCUSSION

The data presented here indicate that cytoplasmic components are involved in the determination of cell cycle periods, asymmetric cleavage pattern, and cell

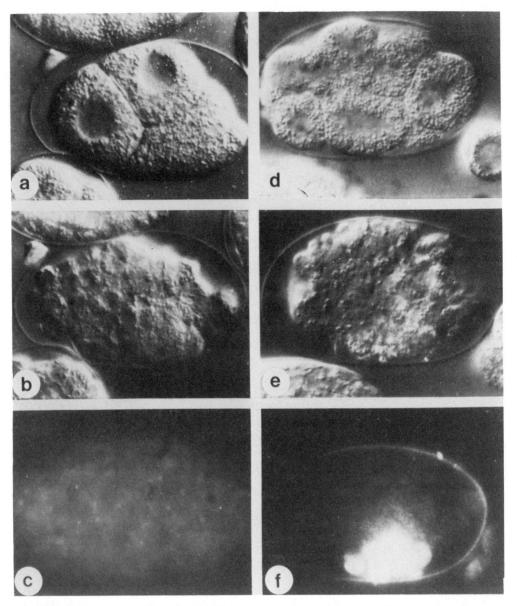

Figure 6. Cytoplasmic localization of gut cell potential. (*a*) Embryo after extrusion of the P_1 nucleus at the two-cell stage and subsequent division of the AB cell. (*b*) Several hours later, the AB cell has given rise to many progeny cells while the P_1 cytoplast is still visible in the lower right region of the embryo. (*c*) After development overnight no localized autofluorescence can be detected with UV epi-illumination. (*d*) Embryo after extrusion of the P_1 nucleus at the two-cell stage and subsequent fusion of the P_1 cytoplast with one of the four AB granddaughter cells. The resulting hybrid cell has divided once to give the two large cells in the lower right region of the embryo. (*e*) Same embryo as in panel *d* after development overnight. (*f*) Same embryo as in panel *e* photographed with UV epi-illumination to show the localized autofluorescence typical of the developing gut. (Reprinted, with permission, from Wood et al. 1984.)

fate in early embryos of *C. elegans*. The only cytoplasmic structures detected so far in embryos of this nematode that behave like determinants are germ line-specific granules (Strome and Wood 1982, 1983; Wolf et al. 1983; Yamaguchi et al. 1983). As visualized by antibodies, these granules are segregated from one germ line cell to the succeeding one. During interphase they seem to be freely diffusible in the cell, but prior to mitosis they become concentrated at one pole. The mechanism of their segregation depends on microfilaments (Strome and Wood 1983).

Several questions can be asked in this context: Are P granules a visible expression of how germ line determination works? Can they serve as a paradigm for the determination of somatic cells? Are the different levels of cell determination (i.e., cell cycle timing, cleavage pattern, cell fate) correlated? Are they based on the same underlying mechanism? No definite answer to any of these questions is available yet, but the information we have seems to exclude some possibilities and may thereby help us to find a working hypothesis to start with.

Preliminary experiments suggest that normal early cleavage pattern does not require the strict segregation of P granules into the germ line cell. Therefore, cleavage pattern is probably not determined by them. These results do not exclude that the mechanism per se on which segregation of P granules is based must be functional for correct early cleavage pattern. Nor do they exclude that P granules (or structures they are associated with) are necessary for later germ cell differentiation.

The analysis of mutants (Schierenberg et al. 1980; Denich et al. 1984) and of cell fusion experiments (Schierenberg 1984) shows that cell cycle periods can be altered considerably without necessarily preventing cell-specific differentiation (autofluorescence of gut-precursor cells). Therefore, cell cycle timing and cell differentiation are at least not strictly correlated. The addition of "fast" AB cytoplasm to a "slow" germ line cell can accelerate the future cell cycle periods of the germ line cell and its descendants (Schierenberg and Wood 1985) without interfering with the pattern of asymmetric cleavages. So, cell cycle timing and cleavage potential are not strictly correlated either.

Efforts to isolate antibodies against structures that could be candidates for somatic cell determinants in *C. elegans* embryos have been unsuccessful so far. But several new monoclonal antibodies against P granules have been isolated this way (S. Strome, pers. comm.). Either one must assume the marked germ line structures to be much more antigenic than any comparable hypothetic somatic structures or one must consider the possibility that no such structures exist.

Some of the results reported here suggest that the number of cytoplasmic cycles (with or without actual cell division; see Fig. 4) a cell has passed through is involved in cell determination. If that were true not only for early cleavage pattern but also for later differentiation of a cell, a simple model for early cell determination could be formulated. It needs only two things, polarity and a counting mechanism.

With each division of the germ line cell, a binary decision is made between "germ line" and "non-germ line." This requires the existence of a polarity that could be present in the form of a gradient along the anterior-posterior axis of the egg or could be some absolute difference between one end of the egg and the other (e.g., localized germ line determinants). Which pair of cells is generated with the division of the germ line cell, $AB + P_1$, $EMS + P_2$, $C + P_3$, or $D + P_4$, depends on the number of cycles through which the cytoplasm has already passed. Cell cycles could be counted for instance via a substance that is progressively degraded with each cycle.

The analysis of the *lin-12* locus indicates that during postembryonic development cell fate can be selected by "binary switch" genes. A high level of gene activity appears to specify one cell fate and a low level specifies an alternative fate (Greenwald et al. 1983). It remains to be tested whether a similar mechanism may be active in early embryos, where different concentrations of a maternally inherited cytoplasmic factor may cause a blastomere to be shunted into one or another developmental pathway.

ACKNOWLEDGMENTS

I thank Mechthild Ziemer for expert photographic work and Randy Cassada for critical reading of the manuscript. Part of the work reported here was done in the laboratory of W.B. Wood, University of Colorado, Boulder, and was supported by a grant of the Deutsche Forschungsgemeinschaft.

REFERENCES

ALBERTSON, D.G. 1984. Formation of first cleavage spindle in nematode embryos. *Dev. Biol.* **101:** 61.

BOVERI, T. 1887. Über die Differenzierung der Zellkerne während der Furchung des Eies von *Ascaris megalocephala*. *Anat. Anz.* **2:** 668.

———. 1910. Die Potenzen der Ascaris-Blastomeren bei abgeänderter Furchung. In *Festschrift für R. Hertwig*, p. 133. Fischer, Jena.

BROOKBANK, J.W. 1970. DNA synthesis and development in reciprocal interordinal hybrids of a sea urchin and a sand dollar. *Dev. Biol.* **21:** 29.

CAIRNS, J. 1975. Mutation selection and the natural history of cancer. *Nature* **255:** 197.

CASSADA, R., E. ISNENGHI, M. CULOTTI, and G. VON EHRENSTEIN. 1981. Genetic analysis of temperature-sensitive embryogenesis mutants in *Caenorhabditis elegans*. *Dev. Biol.* **84:** 193.

DAVIDSON, E.H. 1976. *Gene activity in early development*, 2nd edition. Academic Press, New York.

DENICH, K.T.R., E. SCHIERENBERG, E. ISNENGHI, and R. CASSADA. 1984. Cell-lineage and developmental defects of temperature-sensitive embryonic arrest mutants of the nematode *Caenorhabditis elegans*. *Wilhelm Roux's Arch. Dev. Biol.* **193:** 164.

DEPPE, U., E. SCHIERENBERG, T. COLE, C. KRIEG, D. SCHMITT, B. YODER, and G. VON EHRENSTEIN. 1978. Cell lineages of the embryo of the nematode *Caenorhabditis elegans*. *Proc. Natl. Acad. Sci.* **75:** 376.

FANTES, P.A. and P. NURSE. 1981. Division timing: Controls, models and mechanisms. In *The cell cycle* (ed. P.C.L. John), p. 11. Cambridge University Press, Cambridge, England.

GERHART, J., M. WU, and M. KIRSCHNER. 1984. Cell cycle dynamics of an M-phase-specific cytoplasmic factor in *Xenopus laevis* oocytes and eggs. *J. Cell Biol.* **98:** 1247.

GRAHAM, C.F. and R.W. MORGAN. 1966. Changes in the cell cycle during early amphibian development. *Dev. Biol.* **14:** 439.

GREENWALD, I.S., P.W. STERNBERG, and H.R. HORVITZ. 1983. The *lin-12* locus specifies cell fates in *Caenorhabditis elegans*. *Cell* **34:** 435.

HINEGARDNER, R.T., B. RAO, and D.E. FELDMAN. 1964. The DNA synthetic period during early development of the sea urchin egg. *Exp. Cell. Res.* **36:** 53.

HIRSH, D. and R. VANDERSLICE. 1976. Temperature-sensitive developmental mutants of *C. elegans*. *Dev. Biol.* **19:** 220.

HIRSH, D. W.B. WOOD, R. HECHT, S. CARR, and R. VANDERSLICE. 1977. Expression of genes essential for early development in the nematode *C. elegans*. In *Molecular approaches to eukaryotic genetic systems* (ed. G. Wilcox et al.), p. 347. Academic Press, New York.

ILLMENSEE, K. and A.P. MAHOWALD. 1974. Transplantation of posterior polar plasm in *Drosophila*. Induction of germ cells at the anterior pole of the egg. *Proc. Natl. Acad. Sci.* **71:** 1016.

ISNENGHI, E., R. CASSADA, K. SMITH, K. DENICH, K. RADNIA, and G. VON EHRENSTEIN. 1983. Maternal effects and temperature-sensitive period of mutations affecting embryogenesis in *Caenorhabditis elegans*. *Dev. Biol.* **98:** 465.

KRIEGSTEIN, H.J. and D.S. HOGNESS. 1974. Mechanism of DNA replication in *Drosophila* chromosomes: Structure of replication forks and evidence for bidirectionality. *Proc. Natl. Acad. Sci.* **71:** 135.

LAUFER, J.S. and G. VON EHRENSTEIN. 1981. Nematode development after removal of egg cytoplasm: Absence of localized unbound determinants. *Science* **211:** 402.

LAUFER, J.S., P. BAZZICALUPO, and W.B. WOOD. 1980. Segregation of developmental potential in early embryos of *Caenorhabditis elegans*. *Cell* **19:** 569.

MADL, J.E. and R.K. HERMAN. 1979. Polyploids and sex determination in *C. elegans*. *Genetics* **93:** 393.

MCGRATH, J. and D. SOLTER. 1984. Completion of mouse embryogenesis requires both the maternal and paternal genomes. *Cell* **37:** 179.

MIWA, J., E. SCHIERENBERG, S. MIWA, and G. VON EHRENSTEIN. 1980. Genetics and mode of expression of temperature-sensitive mutations arresting embryonic development in *Caenorhabditis elegans*. *Dev. Biol.* **76:** 160.

SATOH, N. 1982. Timing mechanisms in early embryonic development. *Differentiation* **22:** 156.

SCHIERENBERG, E. 1984. Altered cell division rates after laser induced cell fusion in nematode embryos. *Dev. Biol.* **101:** 240.

SCHIERENBERG, E. and R. CASSADA. 1983. Cell division patterns and cell diversification in the nematode *Caenorhabditis elegans*. In *Identification and characterization of stem cell populations* (ed. C.S. Potten), p. 67. Churchill-Livingstone, Edinburgh, England.

SCHIERENBERG, E. and W.B. WOOD. 1985. Control of cell-cycle timing in early embryos of *Caenorhabditis elegans*. *Dev. Biol.* **107:** 337.

SCHIERENBERG, E., J. MIWA, and G. VON EHRENSTEIN. 1980. Cell lineage and developmental defects of temperature-sensitive embryonic arrest mutants in *Caenorhabditis elegans*. *Dev. Biol.* **76:** 141.

STERNBERG, P.W. and H.R. HORVITZ. 1984. The genetic control of cell lineage during nematode development. *Annu. Rev. Genet.* **18:** 489.

STROME, S. and W.B. WOOD. 1982. Immunofluorescence visualization of germ-line-specific cytoplasmic granules in embryos, larvae, and adults of *Caenorhabditis elegans*. *Proc. Natl. Acad. Sci.* **79:** 1558.

———. 1983. Generation of asymmetry and segregation of germ-line granules in early *C. elegans* embryos. *Cell* **35:** 15.

SULSTON, J.E., E. SCHIERENBERG, J. WHITE, and N. THOMSON. 1983. The embryonic cell lineage of the nematode *Caenorhabditis elegans*. *Dev. Biol.* **100:** 64.

VAN DAM, W.I., M.R. DOHMEN, and N.H. VERDONK. 1982. Localization of morphogenetic determinants in a special cytoplasm present in the polar lobe of *Bithynia tentaculata* (Gastropoda). *Wilhelm Roux's Arch. Dev. Biol.* **191:** 371.

VAN DEN BIGGELAAR, J.A.M. 1971. Timing of the phases of the cell cycle with tritiated thymidine and Feulgen cytophotometry during the period of synchronous division in Lymnea. *J. Embryol. Exp. Morphol.* **26:** 351.

VANDERSLICE, R. and D. HIRSH. 1976. Temperature-sensitive zygote defective mutants of *Caenorhabditis elegans*. *Dev. Biol.* **49:** 236.

VON EHRENSTEIN, G. and E. SCHIERENBERG. 1980. Cell lineages and development of *Caenorhabditis elegans* and other nematodes. In *Nematodes as biological models* (ed. B. Zuckerman), p. 1. Academic Press, New York.

WAKAHARA, M. 1978. Induction of supernumerary primordial germ cells by injecting vegetal pole cytoplasm into *Xenopus* eggs. *J. Exp. Zool.* **203:** 159.

WHITTAKER, J.R. 1982. Muscle lineage cytoplasm can change the developmental expression in epidermal lineage cells of ascidian embryos. *Dev. Biol.* **93:** 463.

WOLF, N., J. PRIESS, and D. HIRSH. 1983. Segregation of germ-line granules in early embryos of *Caenorhabditis elegans*: An electron microscopic analysis. *J. Embryol. Exp. Morphol.* **73:** 297.

WOOD, W.B., E. SCHIERENBERG, and S. STROME. 1984. Localization and determination in early embryos of *Caenorhabditis elegans*. In *Molecular biology of development* (ed. E.H. Davidson and R.A. Firtel), p. 37, A.R. Liss, New York.

WOOD, W.B., R. HECHT, S. CARR, R. VANDERSLICE, N. WOLF, and D. HIRSH. 1980. Parental effects and phenotypic characterization of mutations that affect early development in *Caenorhabditis elegans*. *Dev. Biol.* **74:** 446.

YAMAGUCHI, Y., K. MARAKAMI, M. FURUSAWA, and J. MIWA. 1983. Germ line-specific antigens identified by monoclonal antibodies in the nematode *Caenorhabditis elegans*. *Dev. Growth Differ.* **25:** 121.

Genes Affecting Early Development in *Caenorhabditis elegans*

D. HIRSH, K.J. KEMPHUES,* D.T. STINCHCOMB, † AND R. JEFFERSON
Department of Molecular, Cellular and Developmental Biology, University of Colorado, Boulder, Colorado 80309

We have identified several genes that appear to control the early events in the development of *Caenorhabditis elegans* (Hirsh and Vanderslice 1976; Wood et al. 1980). We continue to study them genetically and characterize them morphologically. We describe here three of these genes that appear to be involved in establishing the organization of the embryonic cytoplasm and in positioning the early cleavage furrows (K.J. Kemphues et al., in prep.). In addition, we review our continuing studies on the characterization of the *C. elegans* genome, the goal of which has been to facilitate the study of the molecular basis of the role of these genes in early development. Our earlier physical studies on the nematode genome provided an understanding of the sequence organization of the DNA and revealed considerable DNA sequence polymorphism between different strains of *C. elegans* (Emmons et al. 1979, 1980; Hirsh et al. 1982). Most of the DNA polymorphisms were found to be due to the transposable element Tc1 (Emmons et al. 1983; Liao et al. 1983). Polymorphism mapping combined with a knowledge of the structure of Tc1 facilitated the linking of the genetic and physical maps and the biochemical isolation of developmentally important genetic loci. The ability to isolate developmentally significant loci has made it possible to analyze and compare their sequences, as exemplified in this volume by Horvitz et al. We are continuing our studies on the *C. elegans* DNA by developing methods for DNA transformation in order to be able to reintroduce DNA sequences into the nematode (D.T. Stinchcomb et al., in prep.). Hopefully, by using such techniques, we will soon begin to understand the molecular basis for the activities of these developmental genes and gain insights into the detailed mechanisms by which they affect early embryogenesis.

C. elegans is especially favorable for these kinds of developmental studies. The genome is relatively small, facilitating the cloning of individual genes. The genetics are well understood, making the isolation and characterization of developmental mutants feasible (Brenner 1974). Microscopy on living embryos is straightforward and a map of the somatic cell lineages has been charted, making it now possible to analyze the course of development on a cell-by-cell basis (Sulston and Horvitz 1977; Kimble and Hirsh 1979; Sulston et al. 1983).

Before describing the developmental mutants, it is worthwhile to review the sequence of events in early *C. elegans* development. Immediately after fertilization, the *C. elegans* embryo completes meiosis and forms two polar bodies. The embryo then undergoes dramatic cytoplasmic streaming and reorganization that includes the formation and disappearance of a pseudocleavage furrow and the segregation of germ line-specific granules to the posterior of the embryo (Nigon 1960; Strome and Wood 1982, 1983; Wolf et al. 1983). The maternal pronucleus then migrates to the male pronucleus, the two join, move to the center of the egg, and rotate 90°. The first cleavage then ensues.

The first embryonic cleavage in *C. elegans* is asymmetric, forming a large AB and a small P1-blastomere (Fig. 1). The germ line granules segregate into the P1 blastomere, and the AB blastomere then divides symmetrically to form the more anterior daughter Aba and the more posterior daughter ABp. The AB lineage continues dividing both synchronously and symmetrically. The AB descendants form mostly neurons and hypodermal cells and some pharyngeal muscle and gland cells. The P1 blastomere divides asymmetrically to form the smaller P2 and the larger EMS blastomere. The germ line-specific granules located in the cytoplasm once again segregate to the posterior where P2 forms, and then into each P cell thereafter (Strome and Wood 1982; Wolf et al. 1983). EMS divides to form the E and MS blastomeres. All of the E descendants form the intestine. MS descendants mainly form muscle and gland cells, a few neurons, and the precursors to the somatic cells of the gonad. P2 subsequently divides asymmetrically forming the smaller P3 and the larger C blastomeres, and P3 divides to form the D and P4 blastomeres; P4 forms Z2 and Z3, which give rise to all of the germ cells.

As in other invertebrates with an asymmetric first cleavage, early development in *C. elegans* is mosaic. Blocking cleavage as early as the two-cell stage results in blastomeres that display different fates. For example, in a cleavage-arrested two-cell embryo, the P1 blastomere which is the embryonic precursor of the intestinal lineage displays characteristic intestinal markers, such as fluorescent granules, refractile bodies, and a specific esterase activity (Laufer et al. 1980; Edgar and McGhee 1985). Also, ablation of early blastomeres has little effect on the subsequent proliferation and

---

*Present address: Section of Genetics and Development, Cornell University, Ithaca, New York 14850.
†Present address: Department of Cellular and Developmental Biology, Harvard University, Cambridge, Massachusetts 02138.

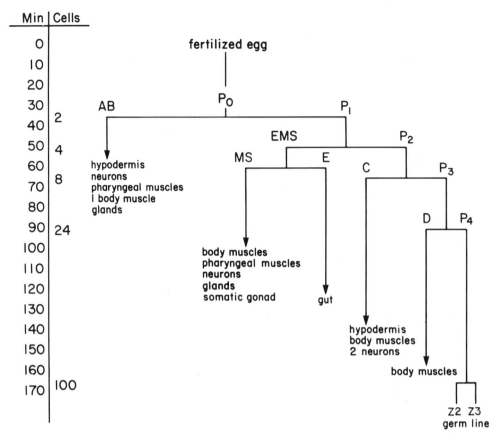

Figure 1. Embryonic lineages of *C. elegans*. P₀ represents the formation of the zygote nucleus. Time of development and number of cells proceeds along the vertical axis. Cell divisions are designated by the horizontal lines connecting the letters that designate the names of the blastomeres. The major cell types derived from each lineage are indicated. (Adapted from Strome and Wood 1982 and Sulston et al. 1983.)

differentiation of the remaining blastomeres (Priess and Hirsh 1985). In contrast, manipulating or blocking cleavage of the one-cell embryo invariably disrupts further morphogenesis and differentiation. Thus, early embryogenesis in *C. elegans* can be characterized by cytoplasmic reorganization, asymmetric first cleavage, and highly determined fates of blastomeres. We are interested in the genes that control these events.

ZYGOTE-DEFECTIVE GENES

We, as well as others, have identified a number of mutations that disrupt embryogenesis in *C. elegans* (Hirsh and Vanderslice 1976; Wood et al. 1980; Cassada et al. 1981; Denich et al. 1984). Most of these mutations have maternal effects. These maternal effect mutations were of special interest to us because they represent maternally expressed functions essential for early embryogenesis. Half of the maternal effect mutations that we isolated showed strict maternal phenotypes; a paternal wild-type allele failed to support development of an egg derived from a homozygous mutant mother. A few of these strict maternal mutations including some amber alleles, had no effects on any other phases of development, implying that they represented functions required only during embryogenesis. Furthermore, the mutant embryos displayed cellular phenotypes that reflected disturbances in the organization of the embryonic cytoplasm and early cleavage patterns.

Genetic complementation tests placed several of these mutations with interesting properties into three genes, designated *zyg-9*, *-11*, and *-14*, where *zyg* is an abbreviation for "zygote defective" (K.J. Kemphues et al., in prep. and unpubl.). All of the mutations in these three genes show strict maternal effect behavior. Temperature-sensitive alleles have temperature-sensitive periods (TSPs) around the time of fertilization. All three genes are essential for proper cytoplasmic organization and completion of a normal first cleavage in the embryo. Mutations disrupt no other stages in the life cycle besides early embryogenesis. Only limited differentiation occurs when these gene products are defective. We describe these three genes in greater detail here, beginning with *zyg-14*, which seems to have the most straightforward and intriguing effects on development.

zyg-14

Two mutant alleles have been isolated in this gene on chromosome V, and both alleles show the same mutant phenotypes. Mutant embryos cleave symmetrically and synchronously until the fourth or fifth cleavage (Fig. 2). P granules that normally segregate to the posterior of the one-cell embryo and then to the germ line precursor P cells fail to segregate and are located in all of the blastomeres. This is particularly evident in the four-cell stage. Asynchrony sets in after the fifth cleavage. Cleavages continue until a ball of 1000 cells forms containing no obvious signs of morphogenesis. Gut-specific granules do not form, indicating failure of differentiation in the intestinal lineage. Thus, it appears likely that the *zyg-14* gene encodes a protein involved in establishing the asymmetrical position of the first cleavage furrow. Presumably the protein is a component of the cytoskeletal system. The failure of the P-granule segregation would imply that the protein interacts either directly or indirectly with the thin filaments that are involved in P-granule localization (Strome and Wood 1983). It is clear, however, that *zyg-14* is not an actin gene because it does not map to the position of any of the four actin genes in *C. elegans* (Hirsh et al. 1982; D. Albertson, pers. comm.).

Zyg-11

Six alleles of this gene on chromosome II have been isolated. One is suppressible by the amber suppressor *sup-7(st5)X* (Waterston 1981). Temperature-sensitive alleles have TSPs around the time of fertilization. All of the mutants show the same phenotypes: 33% of the embryos with symmetric first division, 31% with formation of a cytoplast instead of a first cleavage, 24% with an asymmetric first cleavage but with reversed polarity, and 12% with asymmetric divisions with normal polarity (Fig. 3).

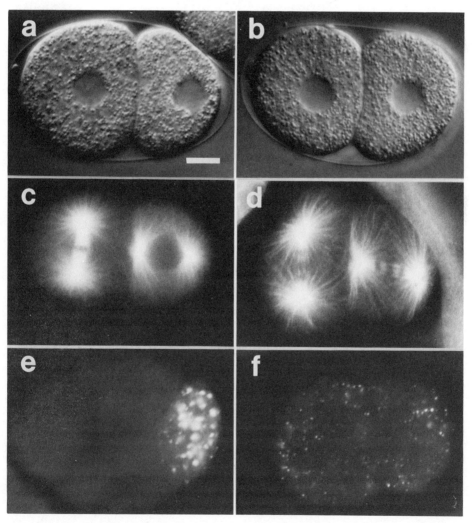

Figure 2. Comparison of wild-type and *zyg-14(b274)* mutant embryos. (*a* and *b*) Normarski images. (*c–f*) Immunofluorescence images. Anterior is to the left in all panels. (*a*) Wild-type two-cell embryo showing the large AB blastomere and small P1 blastomere. (*b*) *b274* two-cell embryo showing the symmetrical mutant cleavage. (*c*) Wild-type two-cell embryo stained with anti-tubulin antibody showing asynchronous mitoses. (*d*) *b274* two-cell embryo showing synchronous mitoses. (*e*) Wild-type four-cell embryo stained with anti-P-granule antibody. (*f*) *b274* four-cell embryo stained with anti-P-granule antibody showing the P granules present in all blastomeres. Bar in *a*, 10 μm.

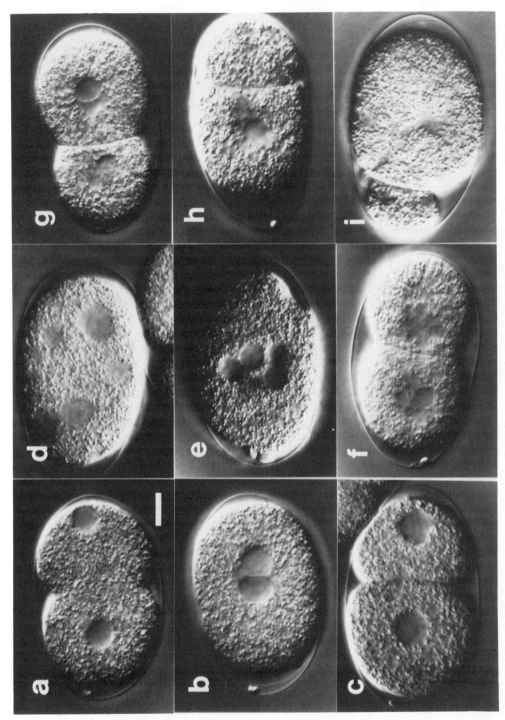

Figure 3. Phenotypes of *zyg-11(mn40)* mutant embryos. (*a–c*) Wild-type embryos: (*a*) early pronuclear stage, (*b*) pronuclear fusion, and (*c*) end of first cleavage. (*d–i*) *zyg-11(mn40)* embryos: (*d*) pre-pronuclear stage containing pools of clear cytoplasm, (*e*) pre-pronuclear stage containing multiple pronuclei, (*f*) even cleavage, (*g*) cleavage with reversed polarity, (*h*) cleavage with normal polarity, and (*i*) cytoplast formation. Bar in *a*, 10 μm.

In wild-type development, the maternal pronucleus undergoes two meiotic divisions following fertilization. Normally, the end of meiosis II is accompanied by extensive and complex cytoplasmic movements culminating in formation of a pseudocleavage that precedes the migration of the maternal pronucleus to the paternal pronucleus. The first noticeable abnormal phenotype in *zyg-11* mutants appears during meiosis II which lasts four times longer than meiosis II in wild-type embryos. Furthermore, mutant *zyg-11* embryos fail to complete meiosis II and do not form second polar bodies. Mutant *zyg-11* embryos also undergo abnormal cytoplasmic movements and fail to form pseudocleavages. The mutant embryos fail to localize P granules and usually do not form gut granules characteristic of intestinal differentiation. Thus, this gene appears to encode a protein that is required for properly completing meiosis II. Whether this is its primary function or whether there are earlier events not yet detected is unclear. The failure of the premeiotic and/or meiotic events then has pleiotropic effects on the subsequent steps involving cytoplasmic reorganization and cleavages.

Zyg-9

There are six alleles of this gene on chromosome II; one is amber-suppressible. Temperature-sensitive alleles have TSPs around the time of fertilization. Mutant embryos have first cleavages perpendicular to the direction of a normal cleavage. The first cleavage runs along the length instead of the width of the egg (Fig. 4). Cytoplasts also form at the anterior end of the egg. The perpendicular cleavage pattern results from the failure of the maternal pronucleus to migrate to the paternal pronucleus. Because the pronuclei never join, the 90° rotation of the mitotic spindle of the zygotic nucleus never occurs (Albertson 1984a). Consequently, the spindle remains perpendicular to the position it normally would assume for the first cleavage. Other earlier abnormal phenotypes have also been observed in certain *zyg-9* mutants including multiple maternal pronuclei, abnormal polar bodies, and disorganized meiotic spindles. These phenotypes strongly suggest a failure in the functioning of the proteins participating in the cytoskeletal network required for meiosis and the subsequent cytoplasmic reorganization.

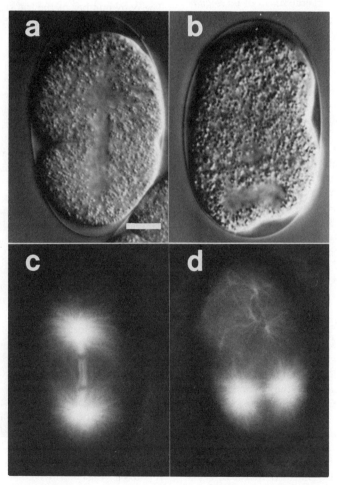

Figure 4. Comparison of wild-type and *zyg-9* mutant early embryos. (*a* and *b*) Nomarski optics images. (*c* and *d*) Immunofluorescence images using antitubulin antibody. (*a*) Wild-type embryo in late anaphase of the first cleavage, (*b*) *b301* embryo in anaphase of the first cleavage, (*c*) wild-type embryo at anaphase of first cleavage, and (*d*) *b301* embryo during the pseudocleavage stage containing two maternal nuclei. Bar in *a*, 10 μm.

STRATEGIES FOR ISOLATING GENES AFFECTING EARLY DEVELOPMENT

We are currently attempting to isolate these genes affecting early development in order to study them and their gene products biochemically. The *C. elegans* genome consists of 8×10^7 bp of DNA, of which 17% is repetitive and 83% is single-copy DNA (Sulston and Brenner 1974). Unique sequences 1–5 kb in length are interspersed among repetitive DNA sequences that are a few hundred base pairs long (Emmons et al. 1980). Various strains and species of *Caenorhabditis* differ dramatically in the content and genomic distribution of certain classes of repetitive DNA sequences (Emmons et al. 1979). This variation in distribution of repetitive DNAs provides a rich array of interstrain polymorphisms, which have been particularly useful for mapping cloned genes onto the genetic map (Hirsh et al. 1982; Rose et al. 1982).

The best studied class of repetitive sequences that differs in abundancy and distribution among different *Caenorhabditis* strains is that of the transposon, Tc*1* (Emmons et al. 1983; Liao et al. 1983). The positions of several Tc*1* polymorphisms have been mapped genetically. A Tc*1* from the Bergerac strain of *C. elegans* has been cloned and sequenced (Rosenzweig et al. 1983). Therefore, Tc*1* DNA has been used as the starting point to isolate adjacent fragments of genomic DNA in order to "walk" to a nearby gene of interest (Fixsen et al., this volume). We currently are using these physical methods to isolate *zyg-9* and *zyg-11* on chromosome II (P. Carter and D. Hirsh, unpubl.). These two genes fortunately reside in an area of chromosome II in which there are numerous deletions that accelerate the walking (Sigurdson et al. 1984).

Tc*1* transposition should now make it possible to isolate developmental genes by generating Tc*1* insertional mutations and then directly cloning the genes via the inserted Tc*1* (Moerman and Waterston 1984; Eide and Anderson 1985). Indeed, this method has been used to isolate *unc-54* mutants and the *unc-22* gene (P. Anderson and D. Moerman, pers. comm.). We are currently pursuing the use of this technique to isolate *zyg* genes, in particular *zyg-14*.

DNA TRANSFORMATION

We have begun developing DNA transformation in *C. elegans* as another strategy for isolating *zyg* genes. In addition, transformation should serve a more general purpose, such as a means for studying in vivo those genes that have been altered in vitro, and as a tool for correlating cell lineages with specific gene expression (D.T. Stinchcomb et al., in prep.). We have introduced cloned DNA into young adult worms by injecting a few picograms of DNA into the ovary. The ovary is a syncytium of approximately 1000 nuclei (Hirsh et al. 1976). Therefore, several germ nuclei are presumably exposed to the exogenous DNA after each injection. Progeny from the injected worm were collected individually and allowed to propagate through self-fertilization. After three generations, each population of worms was harvested and the total DNA isolated. This DNA was analyzed for the presence of exogenous DNA by hybridization to the bacterial DNA sequences in the plasmids used as cloning vectors. Approximately 10% of the injected animals had at least one progeny that carried the exogenous DNA. The transformed progeny propagated the DNA for several generations.

When either supercoiled or linear DNA molecules were injected, the transformed DNA was found in high-molecular-weight arrays. When the cloned DNA was cut with a restriction enzyme that cleaved only once, fragments were released having the length of the injected linear plasmid. Therefore, the majority of the transformed DNA was present as a head-to-tail tandem array. Minor amounts of larger and smaller DNA fragments were also present. When the transformed DNA was reisolated, recloned in *Escherichia coli*, and restriction mapped, the majority of the reisolated DNA molecules had the same structure as the injected DNA but some molecules contained partial deletions or duplications. These altered molecules must have given rise to the minor class of different sized fragments on gels. The copy number of the transformed DNA was determined by hybridization to plasmid DNA using the four actin genes of *C. elegans* as internal standards. Transformants contained from 80 to 300 copies per haploid genome of the exogenous DNA.

Remarkably, no worm DNA was required for achieving DNA transformation. Several different bacterial and yeast vectors have been used in the transformation experiments, and the behavior of the DNA has been the same irrespective of whether nematode sequences were present. The only common element thus far has been 2.2 kb of pBR322 DNA present in all of the plasmids. The transformed DNA was propagated from generation to generation albeit unstably. Stability was determined by picking individual worms at succeeding generations, propagating them by self-fertilization, and analyzing individual clones of worms for the presence of the exogenous DNA by hybridization. The results were that 30–90% of the worms in a particular generation carried the foreign DNA. The numbers varied from one population of transformants to another. These frequencies correspond approximately to a 5% probability of loss of the foreign sequences per cell division, since there are approximately 1000 germ cells generated through about 14 cell divisions. This degree of instability of the transformed DNA is similar to that observed for radiation-induced free duplications in *C. elegans*; about 50% of the worms lose the duplication at each generation (Herman et al. 1976). The instability of segregation of transformed DNA persisted even when a segment of nematode DNA that displayed segregator activity in yeast was included in the transforming DNA. This worm segregator DNA had been isolated on the basis of stabilizing mitotic and meiotic segregation of a plasmid in yeast (Stinchcomb et al. 1985).

The instability of the transformed DNA led us to ask whether the DNA was segregating inefficiently as an autonomous extrachromosomal array. Indeed, after staining the DNA with 4,6-diamidine-2-phenylindole dihydrochloride (DAPI), one and occasionally two or three fluorescent bodies could be seen in addition to the six meiotic chromosome figures. These bodies were present only in worms that proved to be transformants on the basis of positive hybridization of plasmid DNA.

We carried out in situ hybridization experiments to demonstrate directly the presence of the exogenous DNA sequences in the transformants. ^{35}S-Labeled plasmid DNA was hybridized to worms and to embryos at various stages of development using adaptations of the method of Albertson (1984b). These experiments revealed that about half of the embryos derived from a population of transformants contained foreign DNA in all of their cells. The remaining embryos displayed a mosaic distribution of the exogenous DNA. This mosaicism revealed that the unstable segregation that had been apparent in the germ line from generation to generation also occurred in the somatic lineages. Many of the adult worms contained a mosaic distribution of in situ hybridization in both their somatic and germ tissues. However, worms could also be found with transformed DNA in their somatic nuclei but none in their ovaries. Conversely, some worms contained the transformed DNA in their germ cells and in only some of the somatic tissues. In some mosaic worms, the distribution of the transformed DNA could be correlated with the pattern of embryonic cell lineages to distinguish where the loss of exogenous DNA occurred. For example, one worm appeared to contain exogenous DNA in germ tissues and in all somatic tissues except the intestine, as if loss occurred at the EMS cleavage.

Expression of the Transformed DNA

It is essential that the transformed DNA be expressed to maximize the usefulness of the transformation technique in carrying out biologically significant experiments in the future. For example, expression will be necessary for isolating genes by complementation of mutations and for studying the in vivo effects of mutations generated in vitro. At this time, we have observed expression of the transformed DNA, but the expression appears to be constitutive and tissue-nonspecific. We assume that this unregulated behavior is because the DNA is present in such high copy number that any *trans*-acting regulatory molecules are overwhelmed.

To assay expression, we have constructed gene fusions between nematode and bacterial genes. Gene fusions are useful because they can provide easy, sensitive assays for the expression of genes whose proteins are otherwise difficult to assay, such as in the case of structural proteins. They are especially useful for studying expression of a eukaryotic gene that is one member of a multigene family. The fusion allows the expression of the gene of interest to be distinguished from that of the other family members.

We have constructed fusions between the flanking regions of nematode genes and the coding region of the *E. coli* β-glucuronidase gene, which is designated *uidA*. The *uidA* gene serves as a useful indicator of gene expression for several reasons. The gene has been cloned and sequenced, thus facilitating the construction of gene fusions (R.A. Jefferson et al., in prep.). There are spectrophotometric and extremely sensitive fluorescent assays for the enzyme activity. There are histochemical colorimetric assays for localizing enzyme activity in situ. Finally, the *C. elegans* β-glucuronidase gene has been identified genetically and mapped and null mutations have been isolated (P. Bazzicalupo, in prep.). Thus, no interfering endogenous background enzyme activity is present in these mutant strains.

We have fused a nematode collagen gene to the *uidA* gene in a plasmid vector (R.A. Jefferson et al., in prep.). There are between 50 and 150 collagen genes in *C. elegans* (Cox et al. 1984). These genes encode 30,000-molecular weight collagens that make up the cuticle. Expression of the genes is developmentally regulated to coincide with the synthesis of new cuticles during molting (Cox and Hirsh 1985). Several of the genes and their flanking regions have been sequenced, thus facilitating the construction of gene fusions with *E. coli uidA*. We fused approximately 600 bp of upstream sequences and the first five amino acids of the *col-1* gene in-frame to the *uidA* gene. The *col-1* gene is expressed strongly in embryos when the first cuticle is synthesized and much less so later in development (Cox and Hirsh 1985; Kramer et al. 1985). The cloned chimeric DNA was microinjected into the mutant strain *gus-1 (b408)I* of *C. elegans* lacking any endogenous β-glucuronidase activity. Transformants containing the exogenous DNA were obtained. The transformed DNA had all of the physical properties and stability characteristics of the transformants described above. Extracts of populations of transformants were assayed fluorometrically and β-glucuronidase activity was detected (Fig. 5). The activity was present only in worms containing the exogenous DNA. No activity was found in untransformed worms or in worms transformed with a vector carrying an out-of-frame gene fusion between *col-1* and *uidA*. Embryos from populations of transformants stained bright red when assayed histochemically using a reaction that couples the hydrolysis of naphthol-ASBI-glucuronide to pararosaniline. Nontransformed embryos remained colorless in the histochemical assay.

The specific activity of β-glucuronidase was significantly higher in embryos than in later-stage worms. Whereas this is consistent with the pattern of *col-1* expression, this increase in specific activity can be accounted for simply by the 100-fold increase in protein that occurs during worm development along with a constant total amount of β-glucuronidase activity. Therefore, expression of the transformed DNA may not be regulated in a manner expected for the *col-1* gene.

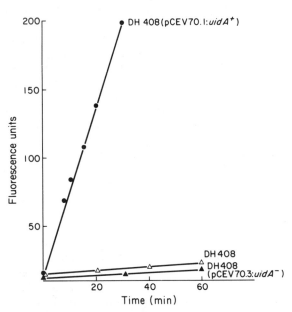

Figure 5. β-Glucuronidase expression in transformants. Extracts were made from a population of worms derived from the transformant DH408 (pCEV70.1 *uidA*⁺) containing the in-frame fusion of the nematode *col-1* promoter to the *E. coli uidA* gene encoding β-glucuronidase. Half of the worms in the population contained transformed DNA, as determined by hybridization to labeled plasmid DNA. Extracts were assayed for enzyme activity using methyl umbelliferyl glucuronide as substrate. As controls, extracts from the untransformed worms (DH408) were assayed as were extracts from DH408(pCEV70.3 *uidA*⁻) worms that were transformed with an out-of-frame fusion. DH408 is the strain designation for the mutant, *gus-1(b408)I*, lacking endogenous β-glucuronidase activity.

Similar results have been obtained with a gene fusion between the nematode major sperm protein gene (MSP) and *uidA* (R.A. Jefferson et al., in prep.). MSP is an abundant 15,000-molecular weight protein present only in sperm (Klass and Hirsh 1981; Burke and Ward 1983). The MSP genes are a multigene family of closely related sequences (Burke and Ward 1983; Klass et al. 1984). MSP transcription occurs only during spermatogenesis in the fourth larval stage of hermaphrodite development. The promoter region of the p3L4 MSP gene was fused in-frame to *uidA* and the MSP transcriptional terminator region was placed downstream of *uidA*. The chimeric DNA was transformed into the β-glucuronidase null strain of *C. elegans*. The transformants expressed β-glucuronidase in all developmental stages and in many tissues besides the germ cells. Therefore, it appears again that expression is constitutive.

SUMMARY AND FUTURE PROSPECTS

We have described three maternal effect genes that specify essential functions for cytoplasmic organization in the early embryo. *zyg-14* seems to be the most fundamental to embryonic cytoplasmic organization because in contrast to *zyg-9* and *zyg-11*, it does not appear to affect earlier meiotic events. The isolation and characterization of the maternal effect genes in *C. elegans* have established the basic criteria for searching for more genes in this organism that specifically affect early embryogenesis. We continue to search for more maternal effect genes, particularly those controlling cytoplasmic organization and early cleavages. We also continue to search for more alleles of the existing maternal effect genes, including amber mutations or deletions. These null mutations will be critical for verifying that there are no other phenotypes in the absence of the gene products and that the mutations are indeed maternal effect. Furthermore, small deletions will be valuable for identifying the genes in cloning experiments.

In addition to identifying genes that grossly affect cytoplasmic organization, it will be interesting to find those genes with subtler controls on the organization of the embryonic cytoplasm. The classical grandchildless phenotype in *Drosophila* is one example in which there appears to be a limited disruption of a cytoplasmic component, presumably that specifying germ plasm differentiation. Whether the P granules of *C. elegans* are part of a similar system of regional organization of specific germ plasm determinants within the embryonic cytoplasm remains unclear.

Studies on maternal effect genes at a molecular level hopefully will open new vistas on understanding the mechanisms of early embryogenesis. The cloning methods that we have described should permit the isolation of these genes and their gene products. The two proven methods thus far for isolating genetic loci in *C. elegans* have both depended on the use of Tc1. Genomic walking from a nearby Tc1 to the gene of interest has succeeded as one approach, and insertion of a Tc1 into the gene of interest has succeeded as another (Fixsen et al., this volume; D. Moerman and R. Waterston, pers. comm.). Both methods are currently being explored for isolating *zyg* genes.

Genetic complementation by DNA transformation currently is being explored as a third method for isolating developmentally important genes, and improvements in the transformation are still being sought. Stable segregation and regulated expression of the transformed DNA will be required to use transformation for complementation and gene isolation. Stable segregation might be attainable by adding centromeric and telomeric sequences to the vectors, thereby creating new linkage groups or by integrating the transformed DNA into the nematode chromosome. So far, we have found integrants to be extremely rare, even when using linear DNA, which has been shown in yeast to be more recombinogenic than circular DNA (Orr-Weaver et al. 1981). However, a few cases of integration of the transformed DNA have been identified by A. Fire and J. Hodgkin (pers. comm.) using suppression of an amber mutation by transformed *sup-7* DNA as the assay. The factors that led to the integration or whether there is regulated transcription are not understood at this time. We assume that the unregulated gene expression observed thus far is due to the high copy number of the transformed DNA. Regulated expression might be achieved along with stable segregation, and low copy number of exogenous DNA sequences.

By analogy with the P elements in *Drosophila*, the best hope at this time for integrating foreign DNA into the chromosome seems to be with Tc*1* (Rubin and Spradling 1983). To that end, we have constructed vectors with multiple cloning sites within Tc*1*. Selectable drug resistance genes, such as the *E. coli* hygromycin phosphotransferase, have been fused to nematode promoters and are being inserted into these sites (Kaster et al. 1983). Coinjecting these vectors with intact Tc*1* might facilitate insertion of exogenous DNA.

Biochemical analysis of the activity of developmental genes, such as the *zyg* genes that we have described, should shed light on several long-standing questions. For example, which DNA sequences temporally and spatially regulate the developmental expression, which sequences specify maternal transfer of the gene products to the egg, which sequences localize the gene products in the appropriate cytoplasmic compartment, and in which embryonic functions do these gene products participate?

The ability to insert and express foreign DNA in *C. elegans* is of utility even in the absence of regulated expression and stable segregation. The instability of the concatemers causes loss of the DNA frequently enough such that about half the animals have a mosaic distribution. The constitutive expression of the β-glucuronidase then marks each cell containing the DNA. Loss of the enzyme activity thereby distinguishes the descendants of any cell that has lost the foreign DNA during development. This ability to mark lineages might be especially convenient in studying some of the more aberrant patterns of embryogenesis in the mutants. If developmentally significant genes can be included in the same vector, then mosaic analysis might also be carried out to determine the effects of certain cloned genes on specific lineages, particularly since all of the wild-type lineages have been described.

ACKNOWLEDGMENTS

We are grateful to our many colleagues who contributed to these experiments and to the ideas behind them. In particular, we thank Jocelyn Shaw, Bill Wood, Mike Krause, Steve Carr, and Nurit Wolf. This work was supported by Public Health Service grants GM19851 and GM26515.

REFERENCES

ALBERTSON, D.G. 1984a. Formation of the first cleavage spindle in nematode embryos. *Dev. Biol.* **101**: 61.

———. 1984b. Localization of the ribosomal genes in *Caenorhabditis elegans* chromosomes by *in situ* hybridization using biotin-labeled probes. *EMBO J.* **3**: 1227.

BRENNER, S. 1974. The genetics of *C. elegans. Genetics* **77**: 71.

BURKE, D.J. and S. WARD. 1983. Identification of a large multigene family encoding the major sperm protein of *C. elegans. J. Mol. Biol.* **171**: 1.

CASSADA, R., E. ISNENGHI, M. CULOTTI, and G. VON EHRENSTEIN. 1981. Genetic analysis of temperature-sensitive embryogenesis mutants in *Caenorhabditis elegans. Dev. Biol.* **84**: 193.

COX, G.N. and D. HIRSH. 1985. Stage-specific patterns of collagen gene expression during development of *C. elegans. Mol. Cell. Biol.* **5**: 363.

COX, G.N., S. CARR, J.M. KRAMER, and D. HIRSH. 1984. Genetic mapping of *C. elegans* collagen genes using DNA polymorphisms as phenotypic markers. *Genetics* **109**: 513.

DENICH, K.T.R., E. SCHIERENBERG, E. ISNENGHI, and R. CASSADA. 1984. Cell lineage and developmental defects of temperature sensitive embryonic arrest mutants of the nematode *C. elegans. Wilhelm Roux's Arch. Dev. Biol.* **193**: 164.

EDGAR, L.G. and J.D. MCGHEE. 1985. Embryonic expression of a gut specific esterase in *Caenorhabditis elegans. Dev. Biol.* (in press).

EIDE, D.J. and P. ANDERSON. 1985. Transposition of Tc1 in the nematode *C. elegans. Proc. Natl. Acad. Sci.* **82**: 1756.

EMMONS, S.W., M.R. KLASS, and D. HIRSH. 1979. Analysis of the constancy of DNA sequences during development and evolution of the nematode *Caenorhabditis elegans. Proc. Natl. Acad. Sci.* **76**: 1333.

EMMONS, S.W., B. ROSENZWEIG, and D. HIRSH. 1980. The arrangement of repeated sequences in the DNA of the nematode *Caenorhabditis elegans. J. Mol. Biol.* **144**: 481.

EMMONS, S.W., L. YESNER, K. RUAN, and D. KATZENBERG. 1983. Evidence for a transposon in *C. elegans. Cell* **32**: 55.

HERMAN, R.K., D.G. ALBERTSON, and S. BRENNER. 1976. Chromosome rearrangements in *Caenorhabditis elegans. Genetics* **83**: 91.

HIRSH, D. and R. VANDERSLICE. 1976. Temperature sensitive developmental mutants of *Caenorhabditis elegans. Dev. Biol.* **49**: 220.

HIRSH, D., J. FILES, and S. CARR. 1982. Isolation and genetic mapping of actin genes in *C. elegans*. In *Muscle development: Molecular and cellular control* (ed. M.L. Pearson and H.F. Epstein), p. 77. Cold Spring Harbor Laboratory, Cold Spring Harbor, New York.

HIRSH, D., D. OPPENHEIM, and M. KLASS. 1976. Development of the reproductive system of *C. elegans. Dev. Biol.* **49**: 200.

KASTER, K.R., S.G. BURGETT, R.N. RAO, and T.D. INGOLIA. 1983. Analysis of a bacterial hygromycin B resistance gene by transcriptional and translational fusions and by DNA sequencing. *Nucleic Acids Res.* **11**: 6895.

KIMBLE, J. and D. HIRSH. 1979. The post-embryonic cell lineages of the hermaphrodite and male gonads in *Caenorhabditis elegans. Dev. Biol.* **70**: 396.

KLASS, M. and D. HIRSH. 1981. Sperm isolation and biochemical analysis of the major sperm protein from *Caenorhabditis elegans. Dev. Biol.* **84**: 299.

KLASS, M.R., S. KINSLEY, and L.C. LOPEZ. 1984. Isolation and characterization of a sperm-specific gene family in the nematode *C. elegans. Mol. Cell. Biol.* **4**: 529.

KRAMER, J.M., G.N. COX, and D. HIRSH. 1985. Expression of the *C. elegans* collagen genes *col-1* and *col-2* is developmentally regulated. *J. Biol. Chem.* **260**: 1945.

LAUFER, J.S., P. BAZZICALUPO, and W.B. WOOD. 1980. Segregation of developmental potential in early embryos of *Caenorhabditis elegans. Cell* **19**: 569.

LIAO, L.W., B. ROSENZWEIG, and D. HIRSH. 1983. Analysis of a transposable element in *Caenorhabditis elegans. Proc. Natl. Acad. Sci.* **80**: 3585.

MOERMAN, D.G. and R.H. WATERSTON. 1984. Spontaneous unstable *unc-22 IV* mutations in *C. elegans* var. Bergerac. *Genetics* **108**: 859.

NIGON, V., P. GUERRIER, and H. MONIN. 1960. L'architecture

polaire de l'oeuf et les mouvements des constituants cellulaires au cours des prémières étapes du development chez quelques nematodes. *Bull. Biol. Fr. Belg.* **93:** 131.

ORR-WEAVER, T., J.W. SZOSTAK, and R.J. ROTHSTEIN. 1981. Yeast transformation: A model system for the study of recombination. *Proc. Natl. Acad. Sci.* **78:** 6354.

PRIESS, J. and D. HIRSH. 1985. Development of displaced blastomeres in the embryo of *Caenorhabditis elegans. Dev. Biol.* (in press).

ROSE, A.M., D.L. BAILLIE, E.P.M. CANDIDO, K.A. BECKENBACH, and D. NELSON. 1982. The linkage mapping of cloned restriction fragment length differences in *C. elegans. Mol. Gen. Genet.* **188:** 286.

ROSENZWEIG, B., L.W. LIAO, and D. HIRSH. 1983. Sequence of the *C. elegans* transposable element Tc1. *Nucleic Acids Res.* **11:** 4201.

RUBIN, G.M. and A.C. SPRADLING. 1983. Genetic transformation of *Drosophila* with transposable element vectors. *Science* **218:** 348.

SIGURDSON, D.C., G.J. SPANIER, and R.K. HERMAN. 1984. *Caenorhabditis elegans* deficiency mapping. *Genetics* **108:** 331.

STINCHCOMB, D.T., D. MELLO, and D. HIRSH. 1985. *Caenorhabditis elegans* DNA that directs segregation in yeast cells. *Proc. Natl. Acad. Sci.* **82:** 4167.

STROME, S. and W.B. WOOD. 1982. Immunofluorescence visualization of germ-line-specific cytoplasmic granules in embryos, larvae, and adults of *Caenorhabditis elegans. Proc. Natl. Acad. Sci.* **79:** 1558.

———. 1983. Generation of asymmetry and segregation of germ-line granules in early *C. elegans* embryos. *Cell* **35:** 15.

SULSTON, J.E. and S. BRENNER. 1974. The DNA of *Caenorhabditis elegans. Genetics* **77:** 95.

SULSTON, J.E. and H.R. HORVITZ. 1977. Post-embryonic cell lineages of the nematode *Caenorhabditis elegans. Dev. Biol.* **56:** 110.

SULSTON, J.E., E. SCHIERENBERG, J.G. WHITE, and J.N. THOMSON. 1983. The embryonic cell lineage of the nematode *C. elegans. Dev. Biol.* **106:** 64.

WATERSTON, R.W. 1981. A second informational suppressor, *sup-7* (X), in *C. elegans. Genetics* **97:** 307.

WOLF, N., J. PRIESS, and D. HIRSH. 1983. Segregation of germline granules in early embryos of *C. elegans*: An electron microscopic analysis. *J. Embryol. Exp. Morphol.* **73:** 297.

WOOD, W.B., R. HECHT, S. CARR, R. VANDERSLICE, N. WOLF, and D. HIRSH. 1980. Parental effects and phenotypic characterization of mutations that affect early development in *C. elegans. Dev. Biol.* **74:** 446.

Studies on the Cytoplasmic Organization of Early *Drosophila* Embryos

K.G. MILLER, T.L. KARR, D.R. KELLOGG, I.J. MOHR, M. WALTER, AND B.M. ALBERTS
Department of Biochemistry and Biophysics, University of California, San Francisco, San Francisco, California 94143

We are attempting to use the early, single-cell *Drosophila* embryo as a model system to study two central biological problems. The first relates to the high degree of organization found in the intracellular compartment termed the cytosol (defined as the cytoplasm outside of membrane-bounded organelles). Cells apparently have the ability to position specific soluble proteins in defined locations, so that (for example) the cytosol that surrounds the Golgi apparatus and the cytosol located just beneath the plasma membrane contain different mixtures of proteins. What is the biochemical basis for this positioning? Although the cell cytoskeleton is almost certainly involved, our rudimentary knowledge of its structure and function is insufficient to explain the amount of cytoplasmic order that is observed.

We would also like to use the early *Drosophila* embryo to study the problem of positional information. Generating the proper pattern of cell specialization requires that each cell in a developing multicellular organism obtain information on its relative position, but there is very little data on how this information is encoded or transmitted. What is the molecular basis for positional information?

We believe that the two problems just described are closely related to each other. In a number of systems, evidence is accumulating that the cell cytoskeleton plays a central role in the process of pattern formation. For example, in *Xenopus*, there is good evidence that positional information for dorsal-ventral polarity is associated with changes in the structural framework of the early embryonic cytoplasm (Gerhardt et al. 1981; Sharf and Gerhardt 1983; Ubbels et al. 1983). But the most striking findings have been made in experiments in which the position of the P granules associated with the germ cell lineage in the nematode *Caenorhabditis elegans* have been followed by antibody staining. These P granules may or may not be determinants (positional information molecules). However, the observation that they are segregated in an orderly manner into one of two daughter cells during the early cell cleavages reveals that cells could, in principle, sort an initially non-localized determinant correctly at each division, allowing the progeny cells to acquire components that specify the formation of the appropriate cell type (Strome and Wood 1982). Because P-granule segregation is dependent on actin filaments and independent of the spindle, special cytoskeletal machinery must be involved in the segregation process (Strome and Wood 1983). Thus, by studying the cytoskeletal filament systems in an embryo, one would hope to be able to elucidate details of cytoplasmic organization that reveal how positional cues are generated and transmitted in cells.

Drosophila is a convenient organism for a detailed study of both of the problems just discussed. Its embryology has been well studied morphologically (Turner and Mahowald 1976; Zalokar and Erk 1976; Foe and Alberts 1983) and genetically (see, for example, Lewis 1979; Nusslein-Volhard 1979; Nusslein-Volhard et al. 1980; Nusslein-Volhard and Wieschaus 1980). The embryo initially develops as a single giant cell (a syncytium): it requires 13 nuclear divisions to reach the cellular blastoderm stage, when membranes form that divide the approximately 6000 nuclei present into separate cells. Differential gene expression occurs in the syncytium (Hafen et al. 1984; other papers, this volume), so that the initially unspecified nuclei must be instructed to begin transcribing genes according to their position during the divisions that precede cellularization. This means that neighboring nuclei can become different in appropriate ways even without cell membranes to separate them. This paper describes some of our preliminary results that aim at understanding how the cytoplasm of these embryos is organized so as to make such differences possible.

MATERIALS AND METHODS

Fixation and staining of **Drosophila** *embryos.* Embryos were collected in timed intervals from population cages of *Drosophila melanogaster* Oregon-R. The procedure for fixation and staining of whole embryos is reported elsewhere in detail (T.L. Karr and B.M. Alberts, in prep.); in general, we have used the procedure of Mitchison and Sedat (1983), with the addition of a low concentration of taxol during the fixation to preserve microtubules.

Preparation of F-actin affinity columns. The procedure for column construction will be reported in detail elsewhere (K.G. Miller and B.M. Alberts, in prep.). Briefly, F-actin affinity columns were prepared using rabbit skeletal muscle actin purified by the method of Spudich and Watt (1971). Columns were prepared in sterile syringes fitted with polypropylene filter discs (Ace glass, Vineland, N.J.) as bed supports. Columns with total bed volumes between 3 and 20 ml have been used. To make the matrix, 0.5 column bed volume of 2 mg/ml F actin in 75 mM HEPES (pH 7.5), 0.2 mM ATP, 0.2 mM $CaCl_2$, 0.1 M KCl, 6 mM $MgCl_2$, and 5 µg/ml phalloidin (Sigma) was added to one column volume of packed resin, consisting of a 1:1 mixture of Affi-Gel 10 (Bio-Rad) and Sepharose CL-6B (Pharmacia), and gently mixed with a spatula. Coupling was allowed to proceed for 6-15 hours at 4°C in the syringe. To inactivate unreacted groups on the resin, 3 M ethanolamine (redistilled and adjusted to pH 8) was then added to a final concentration of 80 mM. After the solution containing actin and ethanolamine was recirculated through the column bed by pumping at a flow rate of 3-10 ml/hour for 4 hours, the excess actin solution was drained and saved for quantitation. The column was washed with 50 mM HEPES (pH 7.5), 50 mM KCl, 10 mM $MgCl_2$, and 0.2 mM $CaCl_2$ until all the unbound protein was washed off. The column was then washed with 50 mM HEPES (pH 7.5), 1 M KCl, 10 mM $MgCl_2$ (3-5 column volumes), followed by washing into 50 mM HEPES (pH 7.5), 50 mM KCl, 10 mM $MgCl_2$, and 0.02% NaN_3 for storage at 4°C. Phalloidin (5 µg/ml) was added in the final wash. The total actin bound is most easily quantitated by subtraction of the actin removed in the washes from the total input. The accessible F actin on the matrix can be quantitated, if desired, by the binding of a saturating amount of heavy meromyosin followed by its elution with pyrophosphate. As controls, G actin and albumin columns were prepared in an identical manner, except that actin was dialyzed against depolymerization buffer (5 mM HEPES [pH 7.5], 0.2 mM $CaCl_2$, 0.2 mM ATP) and coupled in this buffer; the albumin concentration was 3 mg/ml in F-actin coupling buffer without phalloidin.

Preparation of the Drosophila *extract for F-actin chromatography.* Collections of embryos were obtained by leaving plates containing cornmeal agar in population cages of *D. melanogaster* Oregon-R for 4-10 hours. The embryos were harvested by rinsing the plates with 0.03% Triton-X-100 and 0.4% NaCl onto a Nitex screen. Embryos were dechorionated in 50% chlorox for 1.5-2 minutes, washed extensively in Triton-NaCl, and then rinsed with extract buffer (5 mM HEPES [pH 7.5], 1 mM Na_3EDTA, 0.05% NP-40, plus four protease inhibitors: 10 µg/ml leupeptin, 10 µg/ml aprotinin, 10 µg/ml pepstatin, 1 mM phenylmethylsulfonyl fluoride [PMSF]).

For preparation of extract A, embryos were resuspended in 5 volumes of extract buffer and homogenized in a Teflon-glass homogenizer with a motor-driven pestle. All operations were carried out at 4°C. A low-speed supernatant was prepared by centrifugation at 10,000g for 20 minutes. Lipid was removed by filtration through two layers of H_2O-washed cheese cloth. Dithiothreitol (DTT) was added to 0.5 mM. The extract was then centrifuged at 80,000g for 60 minutes, KCl was added to 20 mM, and the extract was again centrifuged at 80,000g for 60 minutes. The resultant supernatant is extract A. Using a flow rate of 1 ml/hour, this extract was loaded directly onto F-actin, G-actin, and albumin columns which were previously washed with column buffer (5 mM HEPES [pH 7.5], 20 mM KCl, 1 mM Na_3EDTA, 1 µg/ml leupeptin, 1 µg/ml aprotinin, 1 µg/ml pepstatin, 0.05% NP-40, 0.5 mM DTT). Each column was then rinsed extensively with column buffer at a flow rate of 5 ml/hour for about 2 hours. The column was then eluted in steps of about three column bed volumes each with column buffer containing 0.5 mM ATP, 3 mM $MgCl_2$; followed by column buffer containing 0.1 M KCl, 0.5 M KCl, and 1 M KCl (final concentration), in succession. Flow rates during the elution were about 5 ml/hour.

Extract B was derived by solubilization of proteins from the above high- and low-speed pellets with 1% NP-40 detergent at low ionic strength. First, the low-speed pellet from extract A was homogenized in five volumes (original embryo weight) of a low-salt rinse buffer (1 mM HEPES [pH 7.6], 1 mM Na_3EDTA, 0.5 mM Na_3EGTA, 0.05% NP-40, plus the four protease inhibitors). The homogenized pellet was then centrifuged at 20,000g for 15 minutes and the supernatant discarded. The new pellet was reextracted by homogenization in 5 volumes of a buffer containing 1% NP-40, 0.1 mM Na_3EDTA, plus the four protease inhibitors; after 1 hour the homogenate was mixed with the high-speed pellet from extract A (rinsed free of salt by a quick H_2O overlay) and rehomogenized briefly. This combined extract of high- and low-speed pellets was then centrifuged at 20,000g for 15 minutes. HEPES (pH 7.6) and KCl were added to the supernatant to 5 mM and 20 mM, respectively, and the pH was adjusted to 7.6 with 2 N KOH. A high-speed supernatant was then made by centrifuging at 80,000g for 1 hour. After removal of the floating lipid layer, the supernatant is extract B. This extract was loaded onto actin columns and eluted with ATP and salt as described for extract A, with the exception that 0.3% NP-40 was included in all wash and elution buffers. Proteins bound and eluted from these columns were analyzed on 7.5 or 8.5% SDS-polyacrylamide gels (Laemmli 1970).

Affinity chromatography of embryo extracts on immobilized microtubules. For preparation of a column matrix containing attached microtubules, bovine brain tubulin was purified by two successive cycles of polymerization followed by chromatography on phosphocellulose. The tubulin (2 mg/ml), in 80 mM PIPES (pH 6.8), 1 mM Na_3EGTA, 1 mM $MgCl_2$, and 1 mM GTP, was polymerized by the stepwise addition of taxol to 20 µM at 37°C. After adjusting the pH to 7.5 with KOH, the microtubules were coupled to the column bed by following the procedure described above for constructing F-actin affinity columns.

To prepare the extract, embryos were homogenized in 100 mM Tris (pH 7.4), 20 mM KCl, 1 mM Na_3EGTA, 1 mM $MgCl_2$, 0.05% NP-40, 1 mM PMSF, 10 mM benzamidine-HCl, 10 µg/ml leupeptin, 10 µg/ml aprotinin, and 10 µg/ml pepstatin A. The extract was then clarified according to the F-actin procedure. All steps in extract preparation were performed at 4°C to ensure depolymerization of endogenous *Drosophila* microtubules.

The column buffer used in this case contained 100 mM Tris-HCl (pH 7.4), 20 mM KCl, 1 mM Na_3EGTA, 1 mM $MgCl_2$, 1 μM taxol, 1 mM PMSF, 1 mM benzamidine-HCl, 1 μg/ml leupeptin, 1 μg/ml aprotinin, and 1 μg/ml pepstatin A. After loading the extract, the column was rinsed extensively with this buffer and then eluted with the same buffer containing in addition either 100 mM, 250 mM, or 500 mM KCl (final concentration). Columns were stored in the column buffer plus 0.02% NaN_3 and 1 μM taxol. Proteins bound and eluted from this column were analyzed on 7.5% SDS-polyacrylamide gels (Laemmli 1970).

Production of monoclonal antibodies. Standard procedures were used for all steps (see Oi and Herzenberg 1980, for typical protocol). Each antigen injection was with approximately 100 μl of whole fixed embryos (1:1 mixture with Freund's adjuvant or phosphate-buffered saline [PBS]) or with 100–400 μg of pooled actin-binding proteins. Mice were injected intraperitoneally at least three times at 2-week intervals prior to testing their sera for response to the antigen by indirect immunofluorescence and immunoblotting. Mice with positive sera were then boosted by intraperitoneal injections of antigen on each of the 5 days preceding the fusion. Spleen cells from these mice were fused to the SP 2/0 (mouse myeloma) cell line. Hybridomas obtained were screened by indirect immunofluorescence on whole fixed *Drosophila* embryos prepared as described above (T.L. Karr and B.M. Alberts, in prep.). Protein blotting for screening of sera and monoclonal antibodies was done by standard procedures (Towbin et al. 1979).

RESULTS AND DISCUSSION

It requires about 3 hours for the fertilized *Drosophila* egg to develop to the cellular blastoderm stage. During this period, the nuclei divide nearly synchronously, with the first six nuclear divisions occurring at about 8-minute intervals in the yolky interior of the embryo. The nuclei then begin migrating outward toward the periphery, the first of them reaching the posterior pole of the embryo during nuclear cycle 9. During the next nuclear cycle, these nuclei form the pole cells, which are the precursors of the germ cells. The rest of the migrating nuclei reach the cortex during nuclear cycle 10. Once there, they divide four more times as a monolayer (nuclear cycles 11–14) before membranes grow inward around the nuclei to segregate them into cells. Gastrulation commences almost immediately thereafter, so that a true cellular blastoderm exists only transiently (for details, see Fig. 1 in Foe and Alberts 1983).

There are Two Types of "Structured Cytoplasm" in the Early Embryo

Despite the absence of membranes that separate different portions of the embryo during the early nuclear divisions, it is highly organized, containing at least two distinct types of "structured cytoplasm." When sectioned embryos are examined in the light microscope, these special regions of the cytoplasm are detected as areas from which the yolk and other visible particles are excluded (Rabinowitz 1941; Scriba 1964). One type of structured cytoplasm (the cortical cytoplasm) underlies the entire plasma membrane in early embryos. This cortical cytoplasm extends to a depth of about 3 μm, and it is highly enriched in actin filaments, microtubules, and intermediate filaments (Walter and Alberts 1984; T.L. Karr and B.M. Alberts, in prep.). Because the cortical cytoplasm has been implicated as the carrier of positional information (Vogel 1982), detailed knowledge of its structure is likely to be important for an understanding of the early pattern formation process in *Drosophila*.

A second type of structured cytoplasm occurs around each of the nuclei in the embryo as a distinct cytoplasmic domain. These domains also appear to be organized by cytoskeletal elements, and they accompany each nucleus in the interior of the embryo during its outward migration to the cortex (Scriba 1964; Foe and Alberts 1983; T.L. Karr and B.M. Alberts, in prep.). When the nuclei arrive at the periphery, their individual cytoskeletal elements mix with those already present in the cortical cytoplasm, leading to a global reorganization of the cytoplasm in the cortical region. These changes have been followed by indirect immunofluorescence, using antibodies against actin, tubulin, and intermediate filaments to stain each of these three types of filaments (Walter and Alberts 1984; T.L. Karr and B.M. Alberts, in prep.); the distribution of filamentous actin has also been followed by staining *Drosophila* embryos with rhodamine-labeled phalloidin (Warn et al. 1984). Selected micrographs that show aspects of the organization of each of the three major filament types around the postmigration nuclei are presented in Figures 1 and 2.

Figure 1 shows triple-label staining of a small region of the surface of a cycle-10 embryo in interphase. At the top, the nuclei have been visualized by staining with the dye Hoechst 33258; in the center panel, the microtubules are seen, stained with a fluorescein-labeled secondary antibody; at the bottom, actin has been stained with a rhodamine-labeled secondary antibody. Comparison of these three panels reveals that each nucleus in the monolayer at the embryo surface is surrounded by an astral array of microtubules, which seems to help separate the adjacent nuclei from each other while delineating a distinct domain of cytoplasm for each nucleus. An actin "cap" overlies each of these cytoplasmic domains; from sections of embryos this actin can be seen to be located just beneath the plasma membrane. As development proceeds, the actin and microtubules undergo major changes in their organization as

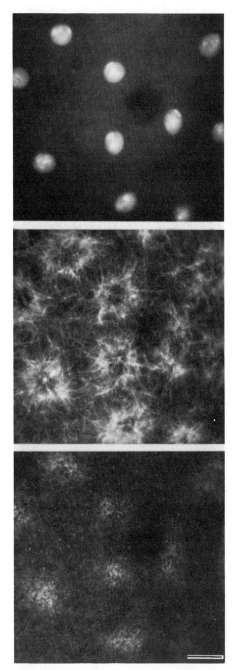

Figure 1. Nuclei, microtubule, and actin structures present during interphase of nuclear cycle 10. Shown is a small area (approximately 65 × 64 μm of the blastoderm surface) of an embryo that was triply stained to reveal the DNA (*top*), microtubule (*middle*), and actin (*bottom*) structures surrounding nuclei during early interphase of nuclear cycle 10. The focal distance from the surface is 3, 2, and 0.5 μm for top, middle, and bottom panels, respectively, and indicates the relative spatial distributions of these structures to one another. Bar, 12.5 μm.

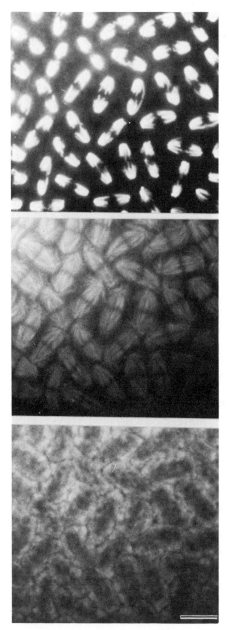

Figure 2. Intermediate filament and tubulin patterns in *Drosophila* embryos during mitosis of nuclear cycle 12. Embryos were stained for intermediate filaments with the monoclonal antibody Ah6 (*bottom*, visualized with rhodamine), for tubulin with an antitubulin antibody (*middle*, visualized with fluorescein), and for DNA (*top*, Hoechst dye 33258). Intermediate filament staining is in domains around each spindle, excluding the spindle and chromosomes, yet leaving a space in between the neighboring spindles. Bar, 12.5 μm.

the nuclei enter mitosis during each nuclear cycle; however, the subdivision of the cytoplasm into domains around each nucleus is maintained through mitosis. In each interphase one observes that every nucleus is surrounded by a microtubule network and overlaid with an actin cap, the nuclei becoming more and more closely spaced as their number doubles with each nuclear cycle (T.L. Karr and B.M. Alberts, in prep.).

The intermediate filaments are also organized into domains by each nucleus during nuclear cycles 10–14 (Walter and Alberts, 1984). An interesting aspect of their organization is shown in Figure 2, which shows an embryo in anaphase of nuclear cycle 12 (note separating chromosomes in the DNA-stained panel at the top). In contrast to the interphase microtubule network, which undergoes a drastic rearrangement during mitosis to form the mitotic spindle (middle panel), the interphase network of intermediate filaments remains essentially unchanged during mitosis. As shown in the bottom panel, the intermediate filaments seem to surround the spindle, being excluded from a central region during mitosis just as they are excluded from the nucleus during interphase.

A Working Hypothesis for the Role of the Cytoskeleton in Pattern Formation in *Drosophila*

To study a problem as wide-ranging and complicated as the organization of the cell cytoplasm, one needs a well-defined entry point as a focus for experiments. Our study of the cytoskeletal proteins of the precellular *Drosophila* embryo focuses on their function as a scaffold on which a great deal of further cytoplasmic order is based. Our working hypothesis is outlined in Figure 3, where we have arbitrarily chosen the layer of structured cytoplasm located just beneath the plasma membrane (the cortical cytoplasm) as an example. Prior to cycle 9, when the first nuclei arrive at the embryo surface (the pole cell nuclei), the plasma membrane of the entire embryo is underlaid with an apparently uniform, filament-rich network that is highly enriched for the three major types of cytoskeletal proteins: microtubules, actin filaments, and intermediate filaments (Walter and Alberts 1984; T.L. Karr and B.M. Alberts, in prep.). The major cytoskeletal protein diagramed in Figure 3 could therefore be any one of these three major filaments; however, for the purpose of simplicity, we will focus our discussion on actin.

Figure 3 incorporates our previous observation that the actin is evenly distributed over the entire embryo cortex; as might have been anticipated, there appears to be no pattern information in the bulk distribution of the major cytoskeletal proteins. From studies in other organisms, a large number of actin-binding proteins are known to exist (for review, see Korn 1982). Like actin, many of these accessory proteins of the cytoskeleton will be distributed evenly across the embryo, because they are required for the general organization and stability of actin filaments. In addition, we anticipate the existence of additional, more minor species of actin-binding proteins that are not evenly distributed in the embryo. Such proteins will create a regional heterogeneity in the cell cytoplasm, as schematically illustrated in the middle of Figure 3. At the next level of organization, we propose that the accessory proteins in turn bind specific cytosolic proteins, acting as "micro-chromatography columns" that partition these proteins into specific cortical regions (Fig. 3, bottom). In the extreme version of this view, a great deal of the cytosol could become highly organized in this way, being built up layer by layer through a branching series of protein-protein interactions that eventually fill the cell.

It is, of course, obvious that the cytoplasm is far from being a static structure, no matter how highly organized it may be. Proteins must be free to move rapidly from place to place, rapidly readjusting their locations as the nuclei divide and the embryo changes with development. For this reason, one would anticipate that many of the important protein-protein interactions in a cell will be weak ones; such weak interactions will usually be missed when a cell is broken open and a crude extract prepared. Thus, many (if not most) of the soluble proteins that remain in the supernatant fraction of a biochemist's extract are likely to be specifically bound to the networks that are created by cytoskeletal proteins inside a cell, where very high protein concentrations make even weak interactions significant (Fulton 1982).

In Figure 3, we have made the distinction between interactions of proteins with the cytoskeleton that are secondary, involving the direct attachment of proteins to either actin filaments, microtubules, or intermediate filaments, and those that are tertiary, involving the less direct binding of proteins to the cytoskeleton via an accessory protein linker. Although we visualize the tertiary interactions as in general being weaker than the secondary interactions, this need not always be the case: for example, there is no a priori reason to rule out weak secondary interactions as also being important to the cell.

The simple model just discussed has been important to us in suggesting the design of experiments, as we shall describe below. It is not an original idea: a major microtubule-binding protein found in nerve cells, microtubule-associated protein 2 (MAP 2), is asymmetrically distributed in developing neurons (Bernhardt and Matus 1982) and in the mature cell it ends up on the microtubules of dendrites, while being excluded from the axon that emanates from the same cell (DeCamilli et al. 1984). This MAP 2 protein in turn binds the R_{II} subunit of the cAMP-dependent protein kinase, which has been suggested to be localized in dendrites as well (Miller et al. 1982). Therefore, cells

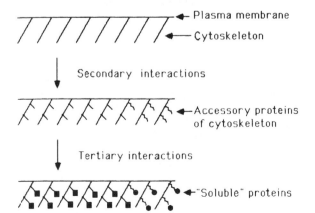

Figure 3. Schematic diagram of a simple general model for cytoplasmic organization. See text for details.

clearly know how to segregate proteins that are attached to cytoskeletal elements (see also Strome and Wood 1983). How they do it is a puzzle that we would like to help solve.

Detection of Accessory Proteins of the Cytoskeleton by Protein-Affinity Chromatography

Because we expect many of the most interesting cytoskeletally bound proteins to be only weakly bound to the cytoskeleton and/or present in small amounts, we sought to develop methodology that would allow us to isolate and identify such proteins reproducibly and with relative ease. In previous work from this laboratory, DNA replication proteins of interest were attached covalently to a solid support and the matrix then used for fractionation of a crude cell extract (Alberts et al. 1983; Formosa et al. 1983). If proteins are immobilized on such a matrix at concentrations of a milligram per milliliter or higher, protein–protein interactions with a K_D as weak as 10^{-5} M can usually be detected (Alberts et al. 1983). Such protein affinity chromatography methods exploit a continuous partitioning process. Therefore, they have a high resolving power compared with the more conventional batch purifications, which rely on cosedimentation following cycles of polymerization and depolymerization of actin filaments or microtubules.

We began by devising a method that allows stable columns containing filamentous actin (F actin) to be used for chromatography. Due to the viscosity of highly concentrated F-actin solutions, producing a column of high F-actin content that had good flow properties turned out to be difficult. However, the successful protocol outlined in Materials and Methods was eventually devised. The technique relies on the use of phalloidin, a mushroom toxin that stabilizes F actin against depolymerization, to allow the actin on the column to remain polymerized under the low-ionic-strength conditions used to depolymerize most of the actin in the crude extract. Therefore, when high-speed supernatants of extracts prepared from Drosophila embryos (extract A) are applied to F-actin columns, those actin-binding proteins that interact with actin filaments (as opposed to the actin monomer) will bind. Columns were also run that contained immobilized monomeric actin (G-actin) for comparison, and columns with bovine serum albumin coupled to the resin served as controls for nonspecific binding. After loading identical aliquots of the same extract onto F-actin, G-actin, and albumin columns, each column was washed extensively with column buffer and then eluted with the same buffer containing ATP and Mg^{++}. The majority of the actin-binding proteins were then eluted with stepwise increases of KCl concentration in the column buffer. Much more protein was recovered from the actin filament column (approximately 0.2% of the protein in the extract) than from either of the two other columns (less than 0.04% of the protein in the extract). In addition, other proteins were specifically retarded by the F-actin column; these relatively weakly binding proteins appear in the early wash fractions that follow the fractions containing the extract that passes through the column. Specific proteins are present in these fractions that are not retarded on either the G-actin (not shown) or the albumin columns.

It occurred to us that many actin-binding proteins may not be solubilized under the conditions just described. For instance, proteins bound to other cytoskeletal systems or to the plasma membrane would end up in the pellet after centrifugation of the extract. Hence, we sought to find conditions that would solubilize additional actin-binding proteins from Drosophila embryos. The procedure we have used involves resuspending the pellets from the centrifugation steps in a buffer of very low ionic strength (less than 0.5 mM) containing high concentrations of a nonionic detergent. The extract (extract B) made in this manner was loaded onto F-actin and albumin columns and eluted with ATP and increasing concentrations of KCl as described for extract A. Compared with extract A, there was approximately 10-fold less protein in extract B. Approximately 0.5% of this protein bound and specifically eluted from F-actin columns.

An analysis of these results by SDS-polyacrylamide gel electrophoresis is presented in Figure 4. Lanes labeled "wash" contain the proteins that bind most weakly to the F-actin columns, and appear in the wash after loading (lane A, actin column; lane C, albumin column). The major proteins are two species in the molecular weight range of approximately 230,000 daltons that migrate as a doublet which resembles spectrin (compare to molecular weight markers). The same proteins are seen in both extract A and extract B (not shown) washes. Lanes labeled "ATP" contain the proteins that appear to interact with actin in an ATP-dependent manner, being eluted from the column when ATP and Mg^{++} are added to the column buffer. (Mg^{++} alone elutes no protein.) Four major bands are observed in the ATP-eluting fraction from extract A (lane A, ATP), the most abundant of which is of an appropriate size for the myosin heavy chain. The others have molecular weights of about 90,000, 125,000, and 130,000 and their nature is unknown. There are approximately 10 major species eluted with buffers of increasing ionic strength from extract A (lanes A; 0.1 M, 0.5 M, 1.0 M), several of which are of very high molecular weight. These proteins are in the molecular weight range expected for the actin-binding proteins spectrin (fodrin), filamin, and talin; however, these proteins have not yet been characterized from Drosophila. Also seen are smaller proteins of about 100,000 daltons, which are of the approximate size of α-actinin, a known actin-binding protein characterized in other systems. In total, about 15 major protein species plus many other more minor bands are detected as binding specifically to actin filaments in these experiments. Most of

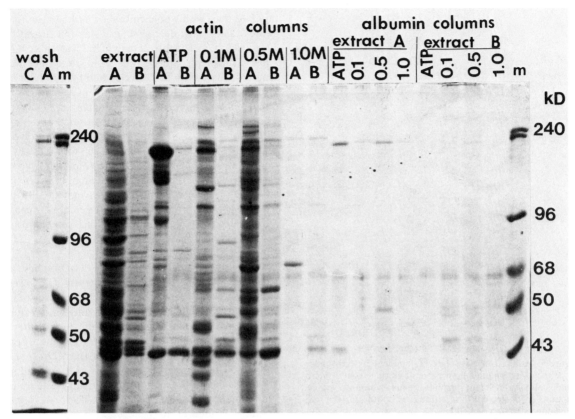

Figure 4. Analysis by SDS-polyacrylamide gel electrophoresis of proteins eluting from F-actin columns or from control (albumin) columns. Each lane shows Coomassie Blue-stained proteins eluted in the steps indicated and described in the text. Elutions from columns loaded with extract A (labeled A) have approximately 15% of the total eluate loaded. Elutions from columns loaded with extract B (labeled B) have approximately 30% of the total eluate loaded. The molecular weight standards (labeled m) include a doublet at about 240,000 daltons that is human red cell spectrin and actin at 43,000 daltons. Note that actin is present in all eluates from the F-actin columns.

these proteins are not detected in the eluates from either the G-actin (not shown) or the albumin columns (Fig. 4, lanes labeled albumin columns) run in parallel. However, on the albumin columns there are occasionally small amounts of proteins that comigrate with the putative myosin heavy chain and actin as seen in ATP and 0.5 M KCl lanes. We believe this is an aggregate of myosin and actin (and perhaps other proteins) which sometimes forms during the loading of the column.

Comparison of the proteins that elute from columns loaded with extract A (lanes A) and those that elute from columns loaded with extract B (lanes B) shows that although some proteins are in common between the two extracts, a number of unique actin-binding proteins appear to be solubilized under the conditions used to prepare extract B. These include proteins of approximate molecular weights 210 kD, 85 kD, and two proteins in the 110-kD range. These are candidates for proteins that mediate attachment of actin filaments to the plasma membrane or to other components of the embryo.

Accessory proteins of the cytoskeleton can also be isolated on affinity columns that contain immobilized microtubules. In this case, the extract is prepared under conditions that depolymerize the endogenous microtubules, and the column is otherwise loaded, washed, and eluted as described for the actin columns. The proteins eluted from this column bind specifically to microtubules, since they do not bind to a control column similarly constructed with albumin (data not shown). The proteins that elute in the three steps of increasing KCl concentration are analyzed by polyacrylamide gel electrophoresis in Figure 5. There are approximately seven major proteins that elute in these washes. Many minor proteins are also present. Some of the proteins have high molecular weights in the range of known microtubule-associated proteins (for example, MAPs 1 and 2 from mammalian brain). The remainder of the proteins, however, span a broad molecular weight range and may be responsible for mediating a wide variety of functions involving microtubules. Their surprising diversity is likely to contribute to microtubule heterogeneity in early embryos. Thus, some may be components of the electron-dense centrosomal cloud responsible for microtubule nucleation, some may be specific spindle components, whereas others may be molecules that mediate intracellular motility. In future studies, it will be interesting to use this technique to identify molecules that can be eluted from the column with ATP, or Ca^{++}/calmodulin, and to obtain monoclonal antibodies directed against the various microtubule-binding proteins detected in Figure 5.

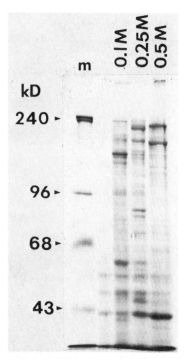

Figure 5. Analysis by SDS-polyacrylamide gel electrophoresis of the *Drosophila* embryo proteins eluting from a microtubule column. Each lane contains the same volume of pooled fractions from step elutions with buffer containing the indicated concentration of KCl.

Production of Monoclonal Antibodies to *Drosophila* Actin-binding Proteins

To further characterize the cytoskeleton in *Drosophila* embryos, we plan to generate monoclonal antibodies against the proteins that associate with the cytoskeleton. We have used two approaches to generate such monoclonal antibodies. In one approach, we immunized mice with whole embryos that had been fixed and extracted with detergent in a way that leaves the cytoskeleton reasonably intact (T.L. Karr and B.M. Alberts, in prep.). Spleen cells from these mice were removed, fused with SP2/0 cells, and the resultant hybridomas were screened for antibody production by indirect immunofluorescence on the same fixed embryos. The other approach we have used is to immunize mice with the proteins that elute from actin affinity columns with KCl. Hybridomas derived from these mice were similarly screened by indirect immunofluorescence staining of fixed embryos. Our hope is to find antibodies that recognize proteins asymmetrically distributed in the *Drosophila* embryo resembling some of the proteins postulated in Figure 3. Such proteins could be involved in providing positional information to the nucleus during early development.

We have generated a wide variety of antibodies that recognize *Drosophila* embryo proteins. The results from five fusions, three using whole fixed embryos and two using actin-binding proteins as antigen, are summarized in Table 1. We have placed these antibodies in seven general categories according to the kind of pattern they show when used to stain fixed *Drosophila* embryos. These categories are rather arbitrary, and not all antibodies in the same category show exactly the same immunofluorescence pattern. There are several notable features of these results. First, hybridomas from three different fusions using mice immunized with whole embryos produced the same limited range of staining patterns. Some of these antibodies no doubt recognize the most abundant and/or immunogenic proteins in fixed embryos. Although others are likely to represent antibodies directed against antigens present in the preimmune mouse, it is notable that antibodies with an intermediate filament-like staining pattern were only obtained using fixed embryos as antigen.

The limited range of the immune response to whole

Table 1. Classification of the Monoclonal Antibodies Produced Following Immunization with Different Antigens

A. Antigen	Colonies screened	Positives (first screen)
Whole fixed embryos	1124	13%
Actin-binding fraction (extract A)	1000	42%
Actin-binding fraction (extract B)	576	41%

B. Immunofluorescence pattern	Percent of monoclonals producing each immunofluorescence class		
	whole embryo antigen	extract A antigen	extract B antigen
Cytoplasmic	25	50	38
Actin-like	9	7	8
Microtubule-like	7	13	11
Intermediate filament-like	13	0	0
Chromatin[a]	9	11	9
Nuclear/cytoplasmic[b]	13	4	12
Other	24	15	22

[a]Stain condensed chromatin at metaphase. Some of these also stain cytoplasmic regions, or components of the spindle.

[b]Associated with the nucleus at some stages of the cell cycle, cytoplasmic at other stages.

embryos convinced us of the importance of using purified fractions as the antigen for the purpose of obtaining antibodies that recognize proteins associated with the cytoskeleton. Hybridomas from mice immunized with pooled proteins that elute from actin affinity columns produced a number of new staining patterns not seen in fusions using mice immunized with whole embryos. To determine whether these antibodies recognize actin-binding proteins, we probed protein blots of whole embryo extracts and of salt-eluting proteins from actin affinity columns with some of these antibodies. Of 82 hybridoma supernatants tested, 30 recognized proteins on the blot. Of these, 10 recognized proteins that elute from actin affinity columns. These 10 antibodies recognize at least six different putative actin-binding proteins of molecular weights 100 kD, 60 kD, 53 kD, 30 kD, and two different proteins in the 200-kD range.

Some examples of embryos stained with the monoclonal antibodies obtained after immunization with actin-binding proteins are presented in Figure 6. We show patterns for four monoclonal antibodies that react with antigens that are enriched from a crude extract by their preferential affinity for the actin filament column. The fluorescence micrographs represent the results of indirect immunofluorescent staining of whole fixed Drosophila embryos, with both interphase and metaphase staining patterns presented. The top panel of Figure 6 (A and B) shows an example of an antibody that recognizes a 60-kD protein from extract B. Its distribution mimics actin, which forms a cap over each nucleus during interphase. When prophase of cycles 11–13 begins, this actin cap spreads out and begins to move into furrows between the forming spindles. At metaphase, actin appears as a bright line forming the border between neighboring mitotic spindles. Like actin, the 60-kD protein also forms a cap over each nucleus at interphase, while at metaphase, it becomes concentrated in the border between neighboring domains.

The middle panels of Figure 6 (C and D; E and F) show two examples of antibodies that react with actin-binding proteins isolated from extract A. On immunoblots (not shown), the antibody shown in Figure 6, C and D, reacts with a protein of approximately 200 kD. This antigen is concentrated in the cytoplasmic area around each nucleus during interphase but does not appear to be concentrated in a cap over each nucleus (Fig. 6C). However, at metaphase (Fig. 6D), it forms a thin line separating neighboring mitotic spindles, apparently mimicking the distribution of actin. The antibody shown in the third set of panels (Fig. 6E,F) reacts with a protein of approximately 30 kD. This antigen is concentrated over each nucleus like actin during interphase (Fig. 6E); however, its distribution does not mimic actin at metaphase (Fig. 6F) since it is excluded from the region where actin would appear most concentrated (between neighboring domains). The distribution of proteins recognized by the first antibody (Fig. 6A,B) is consistent with its being bound to all actin filaments. The distribution of the proteins recognized by antibodies shown in the middle two panels of Figure 6 (C and D; E and F) is consistent with their being actin-binding proteins that bind to some actin filaments but not to others.

The last antibody to be discussed reacts with an approximately 53-kD protein that elutes from the actin columns loaded with extract B. Its distribution is unexpected for an actin-binding protein. As seen in Figure 6G, during interphase the perinuclear region stains brightly with more diffuse staining in the rest of the cytoplasmic domain. As nuclei enter mitosis, the nuclear periphery still stains but the spindle poles stain most brightly (Fig. 6H). Further experiments are needed to determine if there is an in vivo association of this protein with actin.

CONCLUSIONS AND FUTURE PROSPECTS

Even though this project is still in its early stages, it seems worthwhile to look back and consider what we have learned that might be of general interest. First of all, the visualization of the major cytoskeletal filaments in early Drosophila embryos reveals that each nucleus in the syncytial blastoderm embryo organizes its own domain of cytoplasm, in which specific molecules can be held by cytoskeletal attachments. Presumably, different molecules are localized in this way in the vicinity of different nuclei, some of which may carry positional information. From whence could this positional information arise? It is tempting to speculate that important factors that act to inform nuclei of their relative position in an embryo (determinants) first become asymmetrically distributed in the thin layer of cortical cytoplasm prior to nuclear migration, attached to the cytoskeleton there. When the nuclei reach the cortex in cycle 10, these would be incorporated automatically into the individual nuclear domains, with different nuclei picking up different factors according to their position. This could explain how preblastoderm embryos of the leaf-hopper Eusclis can be highly deformed with respect to their anterior–posterior axis and still develop normally, inasmuch as the two-dimensional continuity of the cortical cytoplasm should be left relatively intact in these embryos (Vogel 1982).

Among the monoclonal antibodies screened to date (Table 1), there have been none that stain early embryos in an obvious gradient, patchy, or striped pattern. The failure to find antigens distributed in this way could mean that antigens of this type are not attached to actin filaments, or that they are only very minor proteins compared with the relatively abundant proteins that are the predominant species isolated on our actin filament and microtubule affinity columns. However, because of problems with immunodominance, we have not yet come close to our goal of obtaining antibodies directed against every major Drosophila actin-binding protein, and alternate methods may be required to increase the power of our approach. Moreover, we have not yet produced monoclonal anti-

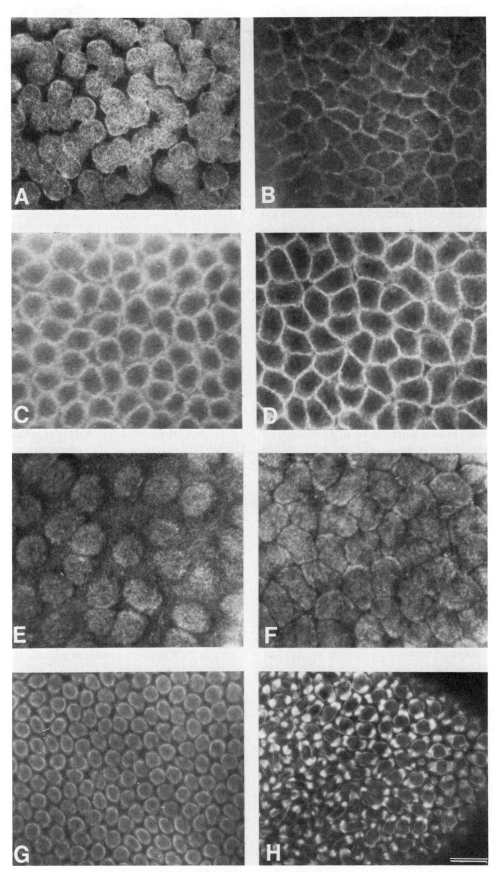

Figure 6. (*See facing page for legend.*)

bodies using either microtubule-binding proteins or intermediate filament-binding proteins as the antigen. Therefore, it is still much too early to know if our experiments will provide an efficient pathway to studies of pattern formation in these embryos.

Our results to date are much more encouraging if we ignore the above issues and concentrate instead on the second major problem that we set out to study in the early *Drosophila* embryo—the organization of the cell cytosol. Our protein-affinity chromatography methods have revealed a rich variety of actin filament and microtubule-binding proteins that can be reproducibly purified from early embryos in large amounts. We suspect that the large number of different accessory proteins in the *Drosophila* cell cytoskeleton reflects a complexity and subtlety of cytoplasmic ordering processes that we can only dimly perceive at present. The results shown in Figure 6 are especially promising because they demonstrate that new cytoskeletal proteins that are distributed in interesting patterns in *Drosophila* embryos can be readily isolated and characterized. For example, two of the antigens identified in Figure 6 (C and D; E and F) would seem to represent proteins that bind preferentially to a subset of the actin filaments inside of the cell. It should be a relatively simple matter to purify these proteins further to homogeneity, so that their biochemical properties can be studied in detail. Moreover, both the purified proteins and the monoclonal antibodies that bind to them can be readily injected into living embryos, making possible more refined studies of intracellular locations, as well as functional tests. Current technologies also enable the gene for any especially interesting cytoskeletal protein to be isolated, which in turn makes possible various functional studies of the protein that exploit the power of *Drosophila* genetics. There is much to be done in this area, for the cell cytoskeleton is an enormously rich structure, whose nature we as yet only barely comprehend.

ACKNOWLEDGMENTS

We would like to thank Roger Cooke and Kathy Franks for advice and protocols for actin purification, Terri Burgess for preliminary experiments on actin column construction, David Gard for help with tubulin purification, and Christina Thaller for help in the early phases of the monoclonal work. We are especially grateful to Frank McKeon for advice and instruction on procedures for generating monoclonal antibodies. We are also grateful to Kathleen Rañeses for preparation of the manuscript. K.G.M. was supported by a National Institutes of Health postdoctoral fellowship (GM08740) during this work. This work was supported by National Institutes of Health grant GM23928 to B.M.A.

REFERENCES

ALBERTS, B.M., J. BARRY, P. BEDINGER, T. FORMOSA, C.V. JONGENEEL, and K.N. KREUZER. 1983. Studies on DNA replication in the T4 bacteriophage *in vitro* system. *Cold Spring Harbor Symp. Quant. Biol.* **47:** 655.

BERNHARDT, R. and A. MATUS. 1982. Initial phase of dendrite growth: Evidence of the involvement of high molecular weight microtubule-associated proteins (HMWP) before the appearance of tubulin. *J. Cell Biol.* **92:** 589.

DECAMILLI, P., P.E. MILLER, F. NAVONE, W.E. THEURKAUF, and R.B. VALLEE. 1984. Distribution of microtubule-associated protein 2 in the nervous system of the rat studied by immunofluorescence. *Neuroscience* **11:** 817.

FOE, V.E. and B.M. ALBERTS. 1983. Studies of nuclear and cytoplasmic behavior during the five mitotic cycles that precede gastrulation in *Drosophila* embryogenesis. *J. Cell Sci.* **61:** 31.

FULTON, A.B. 1982. How crowded is the cytoplasm? *Cell* **30:** 345.

FORMOSA, T., R.L. BURKE, and B.M. ALBERTS. 1983. Affinity purification of T4 bacteriophage proteins essential for DNA replication and genetic recombination. *Proc. Natl. Acad. Sci.* **80:** 2442.

GERHARDT, J., G. UBBELS, S. BLACK, K. HARA, and M. KIRSCHNER. 1981. A reinvestigation of the role of the grey crescent in axis formation in *Xenopus laevis*. *Nature* **292:** 511.

HAFEN, E., A. KURIOWA, and W.J. GEHRING. 1984. Spatial distribution of transcripts from the segmentation gene *fushi tarazu* during *Drosophila* embryonic development. *Cell* **37:** 833.

KORN, E.D. 1982. Actin polymerization and its regulation by proteins from non-muscle cells. *Physiol. Rev.* **62:** 672.

LAEMMLI, U.K. 1970. Cleavage of the structural proteins during the assembly of the head of bacteriophage T4. *Nature* **227:** 680.

LEWIS, E.B. 1979. A gene complex controlling segmentation in *Drosophila*. *Nature* **276:** 565.

MILLER, P., U. WALTER, W.E. THEURKAUF, R.B. VALLEE, and P. DECAMILLI. 1982. Frozen tissue sections as an experimental system to reveal specific binding sites for the regulatory subunit of type II cAMP-dependent protein kinase in neurons. *Proc. Natl. Acad. Sci.* **79:** 5562.

MITCHISON, T.J. and J. SEDAT. 1983. Localization of antigenic determinants in whole *Drosophila* embryos. *Dev. Biol.* **99:** 261.

NUSSLEIN-VOLHARD, C. 1979. Maternal effect mutations that alter the spatial coordinates of the embryo of *Drosophila*

Figure 6. Indirect immunofluorescent staining of whole fixed *Drosophila* embryos with monoclonal antibodies generated against proteins that bind to actin affinity columns. (*A* and *B*) Embryos stained with a monoclonal antibody that reacts with a *Drosophila* protein of 60 kD that binds to the actin filament column and elutes with KCl. The distribution of this antigen in interphase of cycle 12 is shown on the left (*A*), whereas the panel on the right (*B*) shows this antigen at metaphase of cycle 12. (*C* and *D*) Embryos stained with a monoclonal antibody that reacts with an approximately 200-kD protein that binds to actin columns and is eluted with KCl. The left panel (*C*) shows this antibody staining an embryo in interphase of cycle 12; the right panel (*D*) shows the same antibody staining of an embryo in metaphase of cycle 12. The third antibody shown (*E* and *F*) reacts with a protein of 30 kD present in eluates from actin columns. On the left (*E*) an embryo in interphase of cycle 11 stained with this antibody is shown; an embryo stained with this antibody in metaphase of cycle 11 is shown on the right (*F*). The last antibody shown (*G* and *H*) reacts with a protein of 53 kD present in the KCl eluates from actin columns used to fractionate extract B. On the left (*G*) is an embryo stained with this antibody in interphase of cycle 13; distribution of this antibody at metaphase of cycle 13 is shown on the right (*H*). In each panel only a small portion of the embryo surface is shown. Bar, 15 μm.

melanogaster. In *Determinants of spatial organization* (ed. S. Subtelney and I.R. Konigsberg), p. 185. Academic Press, New York.

NUSSLEIN-VOLHARD, C. and E. WIESCHAUS. 1980. Mutations affecting segment number and polarity in *Drosophila*. *Nature* **287:** 795.

NUSSLEIN-VOLHARD, C., M. LOHS-SCHARDIN, K. SANDER, and C. CREMER. 1980. A dorsoventral shift of embryonic primordia in a new maternal-effect mutant of *Drosophila*. *Nature* **283:** 474.

OI, V.T. and L.A. HERZENBERG. 1980. Immunoglobulin-producing hybrid cell lines. In *Selected methods in cellular immunology* (ed. B.B. Mishell and S.M. Shiigi), p. 351. W.H. Freeman, San Francisco.

RABINOWITZ, M. 1941. Studies on the cytology and early embryology of the egg of *Drosophila melanogaster*. *J. Morphol.* **69:** 1.

SCRIBA, M.E.L. 1964. Beeinflussung der fruhen Embryonalentwicklung von *Drosophila melanogaster* durch chromosomenaberrationen. *Zool. Jb. Anat. Bd.* **81:** 435.

SHARF, S.R. and J.C. GERHARDT. 1983. Axis determination in eggs of *Xenopus laevis*: A critical period before cleavage identified by the common effects of cold, pressure, and ultraviolet irradiation. *Dev. Biol.* **99:** 75.

SPUDICH, J.A. and S. WATT. 1971. The regulation of rabbit skeletal muscle contraction. *J. Biol. Chem.* **246:** 4866.

STROME, S. and W.B. WOOD. 1982. Immunofluorescence visualization of germ-line-specific cytoplasmic granules in embryos, larvae, and adults of *Caenorhabditis elegans*. *Proc. Natl. Acad. Sci.* **79:** 1558.

———. 1983. Generation of asymmetry and segregation of germ-line specific granules in early *C. elegans* embryos. *Cell* **35:** 15.

TOWBIN, H., T. STAEHLIN, and J. GORDON. 1979. Electrophoretic transfer of proteins from polyacrylamide gels to nitrocellulose sheets: Procedures and some applications. *Proc. Natl. Acad. Sci.* **76:** 4350.

TURNER, F.R. and A.P. MAHOWALD. 1976. Scanning electron microscopy of *Drosophila* embryogenesis. I. The structure of the egg envelope and the formation of the cellular blastoderm. *Dev. Biol.* **50:** 95.

UBBELS, G.A., K. HARA, C.H. KOSTER, and M.W. KIRSCHNER. 1983. Evidence for a functional role of the cytoskeleton in determination of the dorsoventral axis in *Xenopus laevis* eggs. *J. Embryol. Exp. Morphol.* **77:** 15.

VOGEL, O. 1982. Development of complete embryos in drastically deformed leaf hopper eggs. *Wilhelm Roux's Arch. Dev. Biol.* **191:** 134.

WALTER, M. and B.M. ALBERTS. 1984. Intermediate filaments in tissue culture cells and early embryos of *Drosophila melanogaster*. *UCLA Symp. Mol. Cell. Biol. New Ser.* **19:** 263.

WARN, R.M., R. MAGRATH, and S. WEBB. 1984. Distribution of F-actin during cleavage of the *Drosophila* syncytial blastoderm. *J. Cell Biol.* **98:** 156.

ZALOKAR, M. and I. ERK. 1976. Division and migration of nuclei during early embryogenesis of *Drosophila melanogaster*. *J. Microbiol. Cell.* **25:** 97.

Developmental Expression of Cell-Surface (Glyco)proteins Involved in Gastrulation and Spicule Formation in Sea Urchin Embryos

S.R. Grant, M.C. Farach, G.L. Decker, H.D. Woodward, H.A. Farach, Jr., and W.J. Lennarz

Department of Biochemistry and Molecular Biology, The University of Texas System Cancer Center, M.D. Anderson Hospital and Tumor Institute, Houston, Texas 77030

Gastrulation in the sea urchin embryo begins shortly after the embryo emerges from the fertilization envelope (Gustafson and Wolpert 1967). This process, which eventually produces the germ layers, involves a series of cell migration and differentiation events. First, the descendants of the micromeres enter the blastocoel where they become the primary mesenchyme cells. Second, the blastoderm invaginates, beginning to form the archenteron. Secondary mesenchyme cells at the tip of the archenteron extend filopodia to the inner surface of the blastocoel, helping to elongate the archenteron into a tube. During invagination of the archenteron, the filopodia of the primary mesenchyme cells fuse to form a syncytial cable-like structure that later will be the site for deposition of the skeletal matrix of the spicule. This cable system, which is initially triradiate, seems to predetermine the shape of the spicule (Okazaki 1975). During the late stages of gastrulation, the primary mesenchyme cells initiate deposition of $CaCO_3$ to form the spicule. Over the next 2 days of development, the spicules continue to grow in size and complexity and define the overall morphological features of the prism- and pluteus-stage embryos. This process is accompanied by further growth and development of a functional gut system. At the point where the tip of the archenteron finally makes contact with the wall of the blastocoel, the mouth will form. The pluteus embryo, now possessing a functional gut, can begin to feed.

We have previously shown that the onset of gastrulation is preceded by a marked increase in the synthesis of N-linked glycoproteins (Lennarz 1983). Inhibition of glycosylation by addition of tunicamycin, a specific inhibitor of N-glycosylation, prevents gastrulation (Schneider et al. 1978; Heifetz and Lennarz 1979); the effect of tunicamycin on gastrulation has been corroborated by ultrastructural studies (Akasaka et al. 1980). In addition, at the gastrula stage of development microinjection of concanavalin A (Con A) into the blastocoel also inhibits gastrulation (Spiegel and Burger 1982). The glycoproteins binding to Con A have been reported to be localized at the roof of the blastocoel at the animal pole (Katow and Solursh 1982; Spiegel and Burger 1982). The hypothesis that N-linked glycoprotein synthesis is required for gastrulation is also supported by the finding that normal gastrulation does not occur when de novo synthesis of dolichylphosphate (Dol-P) is inhibited by the drug compactin (Carson and Lennarz 1979, 1981).

We have also found that spicule formation is inhibited by tunicamycin, suggesting a role for N-linked glycosylation in skeleton development. Initially we studied spicule formation in a mixed cell culture system (Mintz et al. 1981). It was found that although inhibitors of DNA synthesis had no effect on spicule formation, actinomycin D (an inhibitor of transcription) blocked spiculogenesis. A marked increase in incorporation of [^{3}H]glucosamine, [^{3}H]lysine, and $^{45}Ca^{++}$ was detected at the time that formation of spicules was observed. In contrast, incorporation of [^{3}H]leucine and [^{3}H]proline increased progressively from the time the cells were plated. As mentioned, tunicamycin inhibited the incorporation of [^{3}H]glucosamine into macromolecules and blocked skeleton formation.

In addition, we found that one or more novel hydroxyproline-containing proteins apparently must be synthesized by the primary mesenchyme cells during skeleton formation (Mintz et al. 1981). This protein(s) does not appear to have the properties of collagen since it does not contain pepsin-resistant domains and is not degraded by collagenase. However, its synthesis and processing seems to be essential for spicule formation since the presence of inhibitors of prolyl hydroxylase, lysyl hydroxylase, and lysyl oxidase, three enzymes that are involved in posttranslational modification of amino acid residues in collagen-like proteins, blocked spicule formation. In this context it has recently been proposed that collagen-like proteins are components of an extracellular matrix necessary for spicule formation (Blankenship and Benson 1984). These results provide evidence that during differentiation of the primary mesenchyme cells, the biosynthesis of N-linked glycoproteins and one or more hydroxyproline-containing proteins are prerequisites for $CaCO_3$ deposition and skeleton formation. In addition to the above findings, in subsequent studies it was found that the level of cAMP controls some component essential for $CaCO_3$ formation in the spicule-forming cells, since agents that elevate cAMP levels inhibit spiculogenesis (Mintz and Lennarz 1982).

Recently, our objective has been to extend these studies using cultures that are highly enriched in primary mesenchyme cells. In the past, in vitro studies have been carried out by using either isolated micromeres, which are the predecessors of the primary mesenchyme cells (Okazaki 1975), or isolated primary mesenchyme cells contained within a basal laminar sac (Harkey and Whiteley 1980). Using the observation that primary mesenchyme cells form stable attachments on plastic dishes whereas ectodermal cells do not, we have devised a culture system consisting of 90–95% primary mesenchyme cells obtained from dissociated mesenchyme blastula-stage embryos (Carson et al. 1985). These cultures consistently produce spicules over a 3-day culture period.

In this paper, studies on the regulation of the synthesis of cell-surface glycoproteins that are believed to participate in the process of gastrulation and on the function of a cell-surface protein that is involved in spicule formation are reported.

METHODS

The experimental procedures employed are cited in Mintz et al. (1981), Carson et al. (1985), and in the figure legends.

RESULTS AND DISCUSSION

Regulation of N-Linked Glycoprotein Synthesis During Gastrulation

Given the fact that increased incorporation of mannose (Man) and glucosamine (GlcN) into N-linked glycoproteins was observed at the onset of gastrulation (Lennarz 1983), it was of interest to study the possible factors regulating glycoprotein synthesis in the developing sea urchin embryo. Based on studies in a number of higher biological systems, the pathway for dolichol-mediated synthesis of N-linked glycoproteins has been elucidated (Struck and Lennarz 1980). As shown in Figure 1, the site of inhibition by tunicamycin (which blocks gastrulation) is the first step in assembly of the oligosaccharide chain whereby GlcNAc-1-P is transferred from UDP-GlcNAc to Dol-P. As shown, compactin, which also inhibits normal gastrulation, blocks an even earlier step, namely the synthesis of the isoprenoid precursor, mevalonic acid.

Polyisoprenoid biosynthesis. Given these findings with compactin, it was of interest to examine the regulation of dolichol synthesis in the developing sea urchin embryo. Therefore, cells were labeled for 2-hour intervals over the course of development with [^{3}H]acetate, a precursor of polyisoprenoids (Carson and Lennarz 1981). At each time point dolichol and cholesterol were isolated and the level of radioactivity in each was determined. The results of these studies revealed that there was little or no synthesis of dolichol or cholesterol in the egg, but following fertilization there was an essentially linear increase in the rate of dolichol and cholesterol synthesis up to the gastrula stage. Since in most isoprenoid-synthesizing systems β-hydroxy-β-methyl glutaryl (HMG)CoA reductase, the enzyme that converts HMGCoA to mevalonic acid, is the rate-limiting enzyme, it was of interest to see if this activity was developmentally regulated in the sea urchin embryo. Therefore, cell-free extracts were prepared from eggs and embryos over the course of development. Assays of these homogenates for HMGCoA reductase revealed that the enzyme activity was virtually absent in eggs, but upon fertilization its level increased, first relatively slowly and then more rapidly until the gastrula stage, where it plateaued (H.D. Woodward and W.J. Lennarz, unpubl.).

It has been shown that a CTP-dependent dolichol kinase is present in a number of higher eukaryotic organisms. In addition, a phosphatase, which converts Dol-P to dolichol, has been described (see Fig. 1). Given the fact that other evidence from our laboratory sug-

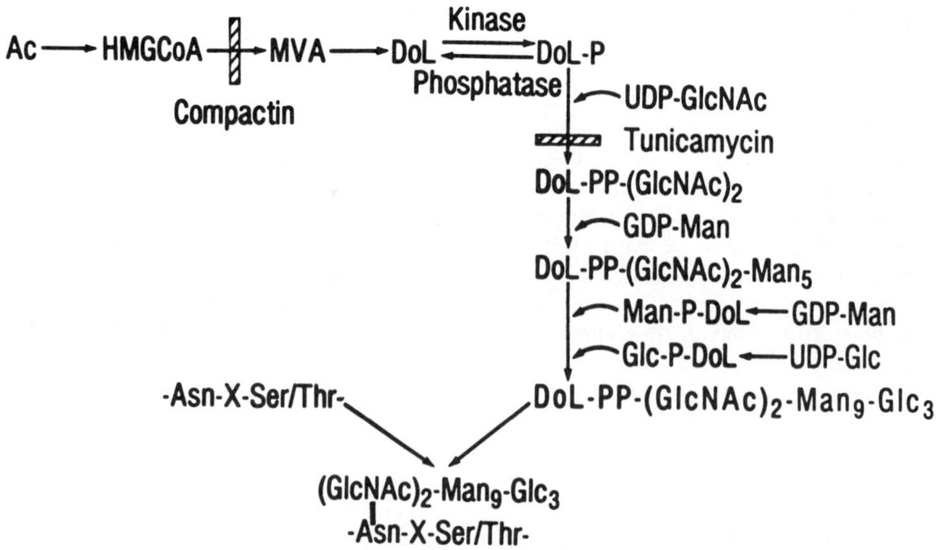

Figure 1. Pathway for the biosynthesis of Dol-P and N-linked oligosaccharide chains.

gests that the end product of the polyisoprenoid pathway in the urchin embryo is dolichol and not Dol-P, it was of interest to examine homogenates of sea urchin embryos for the presence of these two enzymes. Indeed, both enzymes were found to be present and to have properties similar to those described earlier in a variety of higher organisms (Rossignol et al. 1981, 1983). When the enzyme activities were examined in cell-free membrane preparations over the course of development, the results shown in Figure 2 were obtained. It is apparent that dolichol kinase activity is present in the egg, but increases markedly until late blastula, where it begins to decline. In contrast, the phosphatase activity is very low in the egg and increases relatively slowly until early gastrula stage, when it increases rapidly. These results suggest that these two enzymes may control the relative levels of Dol-P and dolichol which, in turn, would be expected to be a controlling factor in the rate of glycosylation.

Oligosaccharide assembly. As is apparent from examination of Figure 1, another potential site of regulation of *N*-linked glycoprotein synthesis is the assembly of the oligosaccharide chain linked to dolichylpyrophosphate (Dol-PP). Consequently, we undertook to study several steps in this pathway (Welply et al. 1985). Conditions were optimized for measurement of the in vitro activities of one enzyme early in the assembly process, namely that which catalyzes synthesis of $GlcNAc_2$-PP-Dol. In vitro assays of this enzyme system revealed that, consistent with its role as the first committed step in the pathway, this activity appeared to be under developmental control. Thus, its activity was low in the egg and in early stages of development and then increased markedly at the early gastrula stage. In contrast, two later enzymes in the pathway, Man-P-Dol synthase and Glc-P-Dol synthase, did not show the same patterns of expression. Although the level of Glc-P-Dol synthase activity was relatively low in the egg, it increased progressively from the onset of fertilization rather than remaining low during early development. Interestingly, Man-P-Dol synthase was high in the egg and showed a progressive decline in activity. It should be noted, however, that because this enzyme activity is 100-fold higher than that of the other enzymes in the pathway, the observed decline probably has relatively little impact on the overall rate of oligosaccharide chain assembly.

Glycoprotein mRNAs. As shown in Figure 1, another potential site of control of glycosylation is the availability of polypeptide chains containing -Asn-X-Ser/Thr- sequences. The availability of glycosylatable proteins could potentially be controlled by new transcription or by the activation of silent mRNA encoding for such proteins. To approach this question, we adopted the strategy of assessing, in a qualitative fashion, the level of these mRNAs by measuring the glycoproteins translated and glycosylated in vitro as a function of sea urchin embryo RNA (Lau and Lennarz 1983). Subsequently, the glycosylated proteins were selected out of the mixture of the total translation products by use of Con A agarose. The bound [^{35}S]methionine-labeled proteins were eluted with α-methyl-mannoside and then analyzed by SDS-PAGE. Using this technique, we examined membrane-bound mRNAs for their ability to encode for glycoproteins. When this was done over the course of development, only a very low level of translation products was detected using membrane-bound RNA from stages prior to the late blastula stage. At late blastula and gastrula stages, four major and many minor glycoproteins were detected. This result, showing that mRNAs were absent from membrane-bound polysomes at earlier stages of development, raised the question of whether these mRNAs were present at all. To answer this question, we isolated total mRNA from eggs and compared the in vitro translated glycoproteins with those formed using gastrula-stage total mRNA. The results of these experiments indicated that, of the four prominent translation products found at gastrula, only two were found in the egg. These results suggest that the mRNAs for two of the four glycoproteins are stored in silent form until gastrulation. In contrast, it is likely that the other two mRNAs are probably new transcripts made just prior to the onset of gastrulation. Clearly, this method of assessing mRNA levels is qualitative. Hopefully in the future we can study this in a more quantitative fashion once cDNA probes for these mRNAs are available.

It is clear, then, from the results presented so far that the control of glycoprotein synthesis at gastrulation is a complex event, regulated by many factors, including the availability of Dol-P, the level of activity of the enzymes involved in oligosaccharide assembly, and the presence of translatable mRNAs encoding for the glycoproteins. In addition, recently we have found that a

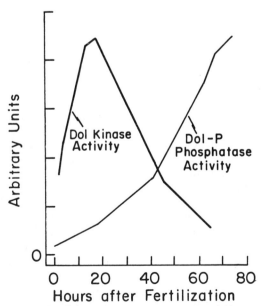

Figure 2. Developmental expression of dolichol kinase and Dol-P phosphatase.

low level of de novo synthesis of N-linked glycoproteins occurs prior to gastrulation, perhaps as a result of utilization of stored oligosaccharide-lipid made during oogenesis (S.R. Grant and W.J. Lennarz, unpubl.).

Function of N-Linked Glycoproteins in Gastrulation

At present we are concentrating our efforts on the perhaps more interesting and certainly more difficult question of the function of these N-linked glycoproteins in gastrulation. Because these newly synthesized glycoproteins may well be a very minor subset of glycoproteins made earlier in development or during oogenesis, we decided to utilize monoclonal antibodies as specific probes for the glycoproteins that are important at the gastrula stage. With such monoclonal antibodies, it should be possible to localize the newly synthesized glycoproteins and to test the hypothesis that they are involved in cell recognition events during the gastrulation process. To prepare a monoclonal antibody library to N-linked glycoproteins, we isolated the total Con A-bindable glycoproteins from gastrula-stage embryos. This yielded a mixture of 20 major and many minor N-linked glycoproteins that were used as the immunogens. At present, 12 monoclonal antibodies have been prepared. These antibodies are being screened for binding to cells of dissociated embryos by an ELISA assay. They are also being screened for surface binding to live cells from dissociated embryos by indirect immunofluorescence. In addition, we are checking that they are indeed surface binders by utilizing electron microscopy and gold-conjugated antibody. Finally, we are in the process of identifying the antigens by Western blots. Although at present this work is incomplete, Figures 3 and 4 illustrate the results with one antibody DB-1 directed toward a potentially very interesting an-

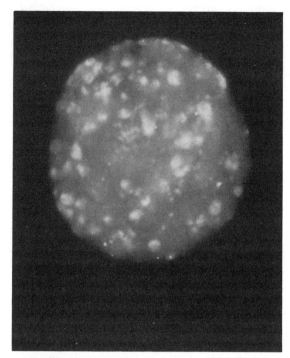

Figure 4. Binding of monoclonal antibody DB-1 to the gastrula-stage embryo.

tigen. When cells at the 16-cell stage are examined by indirect immunofluorescence, it is obvious that there is highly specific binding to one cell type, either the mesomeres or the micromeres (Fig. 3). Interestingly, by the gastrula stage, when these embryos are treated with the same antibody, they exhibit a punctate cell-surface distribution of antigen throughout the embryo (Fig. 4). Preliminary results of Western blots using the antibody indicate that the glycoprotein antigen is absent from the egg, present at low and relatively constant levels early in development, and markedly increased at gastrulation. Hopefully, these antibodies will be powerful tools in localizing the glycoproteins and gaining a better understanding of their function in gastrulation. Additionally, they may be useful in screening cDNA libraries.

In Vitro Spicule Formation

As gastrulation is nearing completion, another dramatic morphogenetic event occurs: spicule formation. As discussed in the introductory remarks, we have developed a simple in vitro culture system to study spicule formation that consists of 90–95% spicule-producing primary mesenchyme cells (Carson et al. 1985). This procedure takes advantage of the selective attachment of primary mesenchyme cells to tissue culture plates in the presence of seawater supplemented with 4% horse serum. The procedure is shown in Figure 5. Two days after the cells obtained from dissociated embryos are placed in culture, the unattached blastomeres are removed, leaving an essentially pure population of

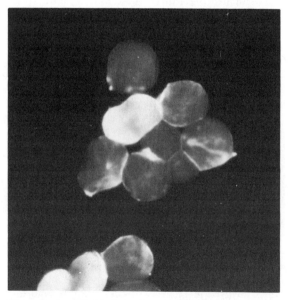

Figure 3. Specific binding of monoclonal antibody DB-1 to one cell type in the 16-cell-stage embryo.

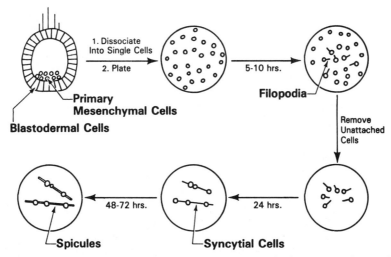

Figure 5. Protocol for the preparation of enriched cultures of primary mesenchyme cells.

spicule-forming cells attached to the dishes. These cells will consistently produce spicules that elongate rapidly on the third day in culture. In this culture system, as well as the other in vitro systems mentioned in the introductory remarks, it is clear that the ability to produce spicules is developmentally controlled and occurs over a characteristic time interval, just as is the case in the intact embryo. Thus, the spicule system is ideally suited to study the mineralization process in general because: (1) This process occurs during a very well-defined period of development, and (2) studies can be carried out on highly enriched cultures of primary mesenchyme cells cultured in vitro.

Antibody that blocks spicule formation. In connection with other studies, a monoclonal antibody (MAb 1223) has been prepared that binds strongly to some of the cells of the gastrula-stage embryo. As will be apparent, this antibody may be very useful in understanding some of the basic aspects of spicule formation. Indirect immunofluorescence of cells from dissociated gastrula-stage embryos first suggested that this antibody might be primary mesenchyme cell specific. It was found that about 5% of the cells in a large population were illuminated by the antibody. Since primary mesenchyme cells represent about 5% of the total cell population (Okazaki 1975), this suggested that, in fact, this antibody might be specifically directed toward an antigen localized to these cells. This finding was further supported by immunofluorescence studies on gastrula-stage embryos that had been depleted of most of their ectoderm. It was found that the antibody was specifically localized to what appeared to be primary mesenchyme cells and their syncytia. The most interesting effect of MAb 1223 was its effect on spicule growth in vitro. Shown in Figure 6A and B is the appearance of control cultures after 2 and 3 days in culture. It is apparent that during this 24-hour period considerable spicule growth occurs. In Figure 6C is shown a culture that was treated at day 2 with MAb 1223 and then photographed on day 3. The Fab fragment of MAb 1223 has the same effect as the intact antibody. It is apparent that there was very little spicule growth between days 2 and 3 in the presence of the antibody. The other striking effect of the antibody is that if it is added at day 0, it prevents attachment of the cells to the culture dish (Fig. 6D). However, the antibody is not toxic and has no effect on viability staining nor on polypeptide synthesis. It also does not prevent reaggregation of dissociated embryos. Given the possibility that the 1223 antigen may interact with Ca^{++}, and that Ca^{++} is involved in cell substratum attachment, it may be that the effect of antibody in this instance is not related to its effect on spicule formation.

Given the effect of this antibody on spicule growth, it seemed likely that if spicule growth were monitored by $^{45}Ca^{++}$ incorporation, similar inhibition would be observed. That this is the case is shown in Figure 7A. In the top panel, the effect of 1223 addition is seen to be immediate upon addition of the antibody. In Figure 7B is shown the effect of the 1223 antibody and another inhibitor of spicule growth, acetazolamide, which blocks deposition of $CaCO_3$ by inhibiting the cytosolic enzyme carbonic anhydrase. Like the effect of acetazolamide, the effect of the 1223 antibody is rapid and, in fact, the inhibitory effect of both together is additive. The basic question that we then sought to answer was the mode of action of the antibody.

Specifically, we hoped to determine whether MAb 1223 binding inhibited a function permitting Ca^{++} uptake into the cells or whether it affected the actual deposition of Ca^{++} into alkaline hypochlorite-insoluble $CaCO_3$. We previously found that the processes of Ca^{++} uptake and conversion to $CaCO_3$ appear to be very tightly coupled in primary mesenchyme cells undergoing spicule growth. For example, when cultures on day 2, such as those in Figure 6A, are labeled with $^{45}Ca^{++}$ for 2 hours, approximately 75% of the total radioactivity accumulated is found to be in alkaline

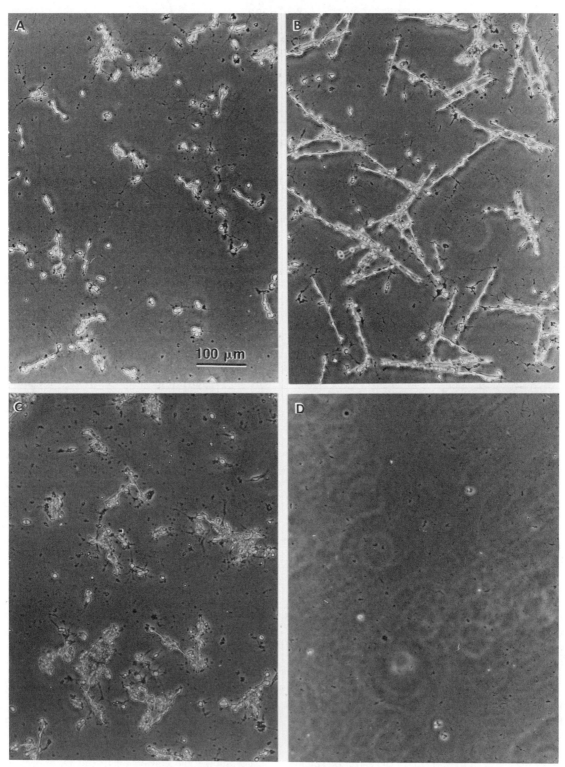

Figure 6. Effects of 1223 antibody on spicule formation in spicule-forming cell cultures. (*A*) Culture after 2 days of incubation in the absence of antibody. (*B*) Culture after 3 days in the presence of 10 µg/ml normal mouse IgG. (*C*) Culture after 2 days of incubation in the absence of antibody, followed by 1 day in the presence of 10 µg/ml 1223 antibody. Note that spicule growth has been retarded relative to control cultures such as that shown in *B*. (*D*) Culture after 3 days in the presence of 10 µg/ml 1223 antibody. Few cells are evident, since all unattached cells are routinely removed after 48 hr in culture, and the 1223 antibody inhibits cell attachment.

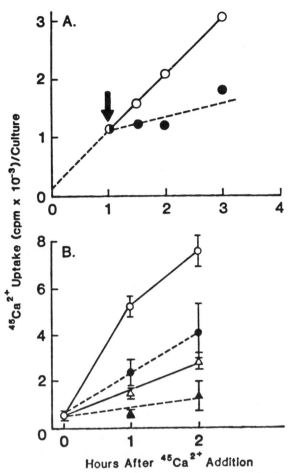

Figure 7. Effects of 1223 antibody on Ca^{++} accumulation by spicule-forming cell cultures. In both A and B [^{45}Ca]Cl$_2$ was added at time zero, and Ca^{++} accumulation was assessed at the indicated times thereafter by measuring the total cell-associated radioactivity in washed cell pellets. (A) Cultures were incubated in the absence of antibody for 2 days. On day 2 of culture, unattached cells were removed, and Ca^{++} accumulation into attached cells was assessed. At time 0, 10 μCi of [^{45}Ca]Cl$_2$ was added and 10 μg/ml of 1223 antibody was added to some cultures (●) after 1 hr (arrow). (B) Day 2 cultures consisting only of attached cells were pretreated for 10 min with inhibitor, with antibody, or with both. The acetazolamide was added from a fresh 1 M stock solution prepared in dimethylsulfoxide (DMSO). Control cultures received DMSO and preimmune IgG. At time 0, 20 μCi of [^{45}Ca]Cl$_2$ was added, and accumulation was measured at the time intervals indicated. (○) Control (preimmune); (●) preimmune IgG and acetazolamide; (△) monoclonal 1223 and DMSO; (▲) 1223 and acetazolamide. The end points for ^{45}Ca^{++} accumulation for these four samples measured at 24 hr were 20,344, 7706, 2034, and 794 cpm, respectively.

hypochlorite-insoluble material. This rapid conversion makes it very difficult to differentiate between uptake and deposition under these conditions. However, recently we found that detachment of cells from the culture dish on day 2 apparently uncouples deposition from Ca^{++} uptake. As shown in Table 1, only 4.7% of the ^{45}Ca^{++} accumulated by detached cells is found to be alkaline hypochlorite insoluble. These cells remain viable as assessed by Trypan Blue exclusion and can be replated to produce long spicules. The limited deposition that occurs in these cells is consistent with the modest inhibitory effect of acetazolamide on ^{45}Ca^{++} accumulation (Table 1 and Fig. 8), given the fact that this drug blocks CaCO$_3$ formation. In contrast, MAb 1223 inhibits the level of total ^{45}Ca^{++} that becomes cell-associated in a 40-minute assay by approximately 50% (Table 1 and Fig. 8). This level of inhibition is much greater than could be expected if the effect of MAb 1223 were to inhibit completely the very limited deposition of Ca^{++} into the spicules under these conditions. This finding leads us to conclude that the antigen recognized by MAb 1223 is not directly involved in deposition of CaCO$_3$ into the spicule and is likely to be involved in the Ca^{++} uptake process.

The 1223 antigen has, in fact, been identified by immunoaffinity chromatography of leucine-labeled late-gastrula-stage proteins. The major protein that binds to and is subsequently recovered from the immunoaffinity column is a 130-kD protein. Occasionally, 250- and 50-kD proteins are also observed. The relationship among these three proteins remains to be established. Future work should provide further understanding of

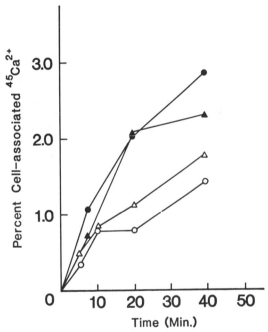

Figure 8. Accumulation of ^{45}Ca^{++} by detached primary mesenchyme cells. Cells were detached and demineralized from tissue culture dishes in Ca^{++}/Mg^{++}-free sea water containing 2 mM EDTA. At time 0, ^{45}Ca^{++} was added to cells concentrated to 2×10^7 cells/ml in Ca^{++}-free isotonic sea urchin saline containing 10 mM Tris-HCl (pH 8.0). Uptake was monitored using a glass fiber filter disk assay and sequential washes (Farach et al. 1984). The percent of the added ^{45}Ca^{++} that became cell-associated during the labeling period was determined at the indicated time intervals following addition of 0.5 mM LaCl$_3$ to halt uptake. (●) Control receiving preimmune mouse IgG; (▲) uptake in the presence of 1 mM acetazolamide; (△) cultures receiving MAb 1223; and (○) cultures receiving 1 mM acetazolamide and MAb 1223.

Table 1. Effect of MAb 1223 and Acetazolamide on $^{45}Ca^{++}$ Accumulation and Conversion of $^{45}Ca^{++}$ to Alkaline Hypochlorite-insoluble Form in Detached Primary Mesenchyme Cells

	Total $^{45}Ca^{++}$ accumulated (cpm/40 min)	Percent inhibition	Alkaline hypochlorite insoluble $^{45}CaCO_3$ (cpm/40 min)	Percent inhibition
Control (pI)	94,015	—	4383	—
MAb 1223 1 mM	48,410	48.5	3171	27.6
Acetazolamide	80,189	14.7	349	92.0
0.5 mM LaCl$_3$	<background	100.0	ND[a]	ND

[a]ND, Not determined.

the developmental regulation of this antigen, as well as its function in spiculogenesis.

ACKNOWLEDGMENTS

This work was supported by a National Institutes of Health grant HD-18600 (to W.J.L.). Dr. William J. Lennarz, who is a Robert A. Welch Professor of Chemistry, acknowledges with gratitude the Robert A. Welch Foundation.

REFERENCES

AKASAKA, K., S. AMEMIYA, and H. TERAYAMA. 1980. Scanning electron microscopical study of the inside of sea urchin embryos (*Pseudocentrotus depressus*): Effects of aryl β-xyloside, tunicamycin and deprivation of sulfate ions. *Exp. Cell Res.* **129:** 1.

BLANKENSHIP, J. and S. BENSON. 1984. Collagen metabolism and spicule formation in sea urchin micromeres. *Exp. Cell Res.* **152:** 98.

CARSON, D.D. and W.J. LENNARZ. 1979. Inhibition of polyisoprenoid and glycoprotein biosynthesis causes abnormal embryonic development. *Proc. Natl. Acad. Sci.* **76:** 5709.

———. 1981. Relationship of dolichol synthesis to glycoprotein synthesis during embryonic development. *J. Biol. Chem.* **256:** 4679.

CARSON, D.D., M.C. FARACH, D.S. EARLES, G.L. DECKER, and W.J. LENNARZ. 1985. A monoclonal antibody inhibits calcium entry and skeleton formation in cultured embryonic cells of the sea urchin. *Cell* **41:** 639.

FARACH, H.A., D.K. MUNDY, G.L. DECKER, W.J. LENNARZ, and W.J. STRITTMATTER. 1984. A possible role for metalloendoproteases in sea urchin fertilization. *J. Cell Biol.* **99:** 263a.

GUSTAFSON, T. and L. WOLPERT. 1967. Cellular movement and contact in sea urchin morphogenesis. *Biol. Rev. Camb. Philos. Soc.* **42:** 442.

HARKEY, M.A. and A.H. WHITELEY. 1980. Isolation, culture and differentiation of echinoid primary mesenchyme cells. *Wilhelm Roux's Arch. Dev. Biol.* **189:** 111.

HEIFETZ, A. and W.J. LENNARZ. 1979. Biosynthesis of N-glycosidically linked glycoproteins during gastrulation of sea urchin embryos. *J. Biol. Chem.* **254:** 6119.

KATOW, H. and M. SOLURSH. 1982. In situ distribution of concanavalin A-binding sites in mesenchyme blastulae and early gastrulae of the sea urchin *Lytechinus pictus*. *Exp. Cell Res.* **139:** 171.

LAU, J.T.Y. and W.J. LENNARZ. 1983. Regulation of sea urchin glycoprotein mRNAs during embryonic development. *Proc. Natl. Acad. Sci.* **80:** 1028.

LENNARZ, W.J. 1983. Glycoprotein synthesis and embryonic development. *CRC Crit. Rev. Biochem.* **14:** 257.

MINTZ, G.R. and W.J. LENNARZ. 1982. Spicule formation by cultured embryonic cells from the sea urchin. *Cell Differ.* **11:** 331.

MINTZ, G.R., S. DE FRANCESCO, and W.J. LENNARZ. 1981. Spicule formation by cultured embryonic cells from the sea urchin. *J. Biol. Chem.* **256:** 13105.

OKAZAKI, K. 1975. Spicule formation by isolated micromeres of the sea urchin embryo. *Am. Zool.* **15:** 567.

ROSSIGNOL, D.P., W.J. LENNARZ, and C.J. WAECHTER. 1981. Induction of phosphorylation of dolichol during embryonic development of the sea urchin. *J. Biol. Chem.* **256:** 10538.

ROSSIGNOL, D.P., M. SCHER, C.J. WAECHTER, and W.J. LENNARZ. 1983. Metabolic interconversion of dolichol and dolichyl phosphate during development of the sea urchin embryo. *J. Biol. Chem.* **258:** 9122.

SCHNEIDER, E.G., H. NGUYEN and W.J. LENNARZ. 1978. The effect of tunicamycin, an inhibitor of protein glycosylation, on embryonic development of sea urchins. *J. Biol. Chem.* **253:** 2348.

SPIEGEL, M. and M.M. BURGER. 1982. Cell adhesion during gastrulation. *Exp. Cell Res.* **139:** 377.

STRUCK, D.K. and W.J. LENNARZ. 1980. The role of saccharide lipids in glycoprotein synthesis. In *The biochemistry of glycoproteins and proteoglycans* (ed. W.J. Lennarz), p. 35. Plenum Press, New York.

WELPLY, J.K., J.T. LAU, and W.J. LENNARZ. 1985. Developmental regulation of glycosyl transferase activities for synthesis of N-linked glycoproteins in sea urchin embryos. *Dev. Biol.* **107:** 252.

Genes That Affect Cell Fates during the Development of *Caenorhabditis elegans*

W. FIXSEN, P. STERNBERG,* H. ELLIS,† AND R. HORVITZ
Department of Biology, Massachusetts Institute of Technology, Cambridge, Massachusetts 02139

The development of a multicellular organism involves the generation of many different cell types from a single-celled fertilized egg. An understanding of the molecular mechanisms that cause cells to become different from one another is a fundamental goal of developmental biology. One organism well suited for the study of this problem at the level of single cells and individual cell divisions is the nematode *Caenorhabditis elegans*.

The complete pattern of cell divisions, migrations, and differentiations that occur during the development of *C. elegans* is known (Sulston and Horvitz 1977; Kimble and Hirsh 1979; Sulston et al. 1980, 1983). Two factors have facilitated the elucidation of this detailed knowledge of *C. elegans* development: (1) It is relatively easy to observe individual cells in living nematodes directly and continuously, using Nomarski differential interference contrast light microscopy. (2) The *C. elegans* cell lineage (i.e., the pattern of cell divisions and the fates of the cells generated by those divisions) is essentially invariant. Many cell fates appear to be autonomously (intrinsically) determined during the development of *C. elegans*; in addition, some cell fates are specified by cell interactions and therefore must be controlled extrinsically (Sulston and Horvitz 1977, 1981; Sulston and White 1980; Kimble 1981; Sulston et al. 1983). To understand the molecular basis of the generation of cellular diversity, it will be important to define, isolate, and characterize molecules responsible for both the intrinsic and extrinsic determination of cell fates. We have taken a genetic approach toward the identification of such molecules. Specifically, mutations that cause cells that normally express different cell fates to express instead the same cell fate should define genes (and therefore molecules) that are involved (either intrinsically or extrinsically) in the specification of cell fates.

Specification of Cell Fates during Normal Development

We regard every cell that is generated during the development of *C. elegans* as having a "fate" (for discussion, see Sternberg and Horvitz 1984). The fate of some cells is to differentiate terminally, whereas the fate of other cells is to undergo a particular pattern of cell divisions and to generate a particular complement of terminally differentiated cell types. As mentioned above, many cell fates appear to be determined intrinsically during *C. elegans* development. Specifically, the ablation of all possible neighboring cells does not appear to have an effect on the fate of the remaining cell in a number of cases that have been tested (Sulston and Horvitz 1977; Sulston and White 1980; Sulston et al. 1983; P. Sternberg and R. Horvitz, in prep.). However, the fates of some cells are clearly specified by factors extrinsic to those cells. For example, certain pairs of cells can adopt either of two positional configurations, with the fate of each cell then depending on its position. The cells P5 and P6, which normally have different fates, are one such pair (Sulston and Horvitz 1977). Two cells in the newly hatched larva, called P5/6L and P5/6R, are located symmetrically on the left and right sides, respectively. During postembryonic development, these cells adopt positions along the ventral midline such that one is anterior to the other, with either configuration (i.e., P5/6L anterior or P5/6R anterior) occurring with equal frequency (Fig. 1). The subsequent fate of each cell (the more anterior cell is now P5 and the more posterior cell is now P6) depends on its position and not on its ancestry. Thus, the two cells P5/6L and P5/6R appear to be of equivalent developmental potential. Such sets of equipotential cells in nematodes are said to define "equivalence groups" (Kimble et al. 1979; Sulston and White 1980).

A number of other members of equivalence groups have been revealed by the limited natural variability that exists during nematode development (Sulston and Horvitz 1977; Kimble and Hirsh 1979). Additional equivalence group members have been defined by cell ablation experiments utilizing a laser microbeam (Kimble et al. 1979; Sulston and White 1980; Kimble 1981; Sulston et al. 1983). Specifically, if the ablation of one cell results in its replacement by another cell, the two cells must share at least the potential to express the fate of the first cell. In this way, many equivalence groups have been defined; e.g., it has been found that P5 and P6 are members of a larger equivalence group that includes both the more anterior cells P3 and P4 and the more posterior cells P7 and P8. Furthermore, the posterior daughters of these six cells, P(3-8).p, themselves constitute an equivalence group known as the vulval equivalence group (Sulston and White 1980). During

---

Present addresses: *Department of Biochemistry and Biophysics, University of California, School of Medicine, San Francisco, California 94143; †Department of Biology, University of California, San Diego, La Jolla, California 92093.

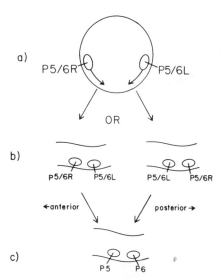

Figure 1. (a) Schematic cross section of a newly hatched animal as viewed from the anterior shows the bilaterally symmetrical nuclei of P5/6L and P5/6R. (b) During postembryonic development, these nuclei migrate ventrally and adopt positions at the ventral midline; each anteroposterior configuration is equally likely. (c) The cell that adopts the more anterior position is named P5, and the cell that adopts the more posterior position is named P6.

normal development, three of these cells, P(5–7).p, divide to generate the vulva. If any of these three cells are ablated by laser microsurgery, other cells of the vulval equivalence group can replace them, allowing a normal vulva to be formed. This replacement occurs in a strict hierarchical fashion; e.g., if any of the P(3,4,8).p cells are ablated, the normal vulval precursor cells will not replace them. This regulation results in a net loss in cell number, since it occurs at the level of cell fates and not at the level of cell number. Since the determination of cell fates in equivalence groups involves cell interactions, the isolation of mutations that alter these fates should define genes that are normally involved in the extrinsic specification of cell fates.

Although the fate expressed by each member of the vulval equivalence group is determined by cell interactions, the lineage subsequently generated by each cell appears to be determined autonomously. For example, cell ablation experiments in which the vulval precursor cells have been destroyed at different times have defined a time after which the subsequent patterns of cell divisions and cell fates are not perturbed by ablating neighboring cells (Sulston and White 1980; P. Sternberg and R. Horvitz, in prep.).

Since, in many cases, cell lineages appear to be determined autonomously, the developmental history of a cell must play an important role in determining its fate. Generally, at each cell division, daughter cells become different from their mothers and also from each other. The autonomy of most cell fates indicates that this asymmetry, at least for most cell divisions, is likely to be caused by factors intrinsic to the cells involved. Isolation of mutations that affect the differences between sister and sister, or between mother and daughter, should identify genes that are normally involved in the intrinsic specification of cell fates.

During *C. elegans* development, most cells of any particular cell type are not generated clonally. Instead, such cells are in many cases generated by identical cell lineages derived from a set of morphologically indistinguishable precursor cells. For example, the six VC neurons are generated by identical lineages from the six cells P(3–8).a (Sulston 1976) (see Fig. 5). Because certain cell lineages are expressed multiple times, it has been proposed that these lineages are generated as integral units (for discussion, see Chalfie et al. 1981; Sternberg and Horvitz 1982; Sulston et al. 1983). Specifically, each of these "sublineages" is thought to result from the action of the same set of genes in a number of different precursor cells. In some instances, almost identical sublineages differ in the fates of a single progeny cell. The isolation of mutations that eliminate the differences between such almost identical sublineages should identify genes that normally act to modify cell fates within sublineages.

Mutations Affecting the Specification of Cell Fates

Many mutations have been isolated in *C. elegans* that affect cell lineages and cell fates (Horvitz and Sulston 1980; Chalfie et al. 1981; Sulston and Horvitz 1981; Greenwald et al. 1983; Ambros and Horvitz 1984; Ferguson and Horvitz 1985). Some of these mutations cause cells that normally express different cell fates to express instead the same cell fate. Thus, these mutations appear to identify genes that are normally involved in making certain cells different from each other. We discuss below a number of these genes, some of which affect intrinsic factors and others of which probably affect extrinsic factors necessary for normal development. We have recently prepared a more comprehensive review of genes that affect cell lineages and cell fates during nematode development (Sternberg and Horvitz 1984).

Genes That Affect Sister-Sister Decisions

During the development of *C. elegans*, the fates of sister cells are almost always different from each other (Fig. 2a). We have identified mutations in several genes that disrupt the normal segregation of cell fates between certain sister cells. These genes were initially defined on the basis of their effects on the vulval cell lineages (Ferguson and Horvitz 1985). Mutations in each of these genes cause sisters that are normally different to be instead the same. Each gene affects only certain pairs of sister cells, and different genes affect different pairs of sister cells.

We consider first the mutation *lin-26(n156)*, which was identified because it prevents the formation of a vulva (Ferguson and Horvitz 1985). As described above, vulval development involves a set of hypodermal P

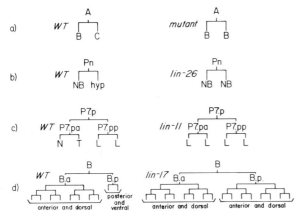

Figure 2. (a) In many cell divisions, sister cells are different from each other. Mutations have been isolated that cause some normally asymmetric cell divisions instead to be symmetric. A, B, and C represent canonical cell fates that differ from one another. (b) The first P cell division normally generates a neuroblast (NB) and a hypodermal cell or hypodermal blast cell (hyp). In *lin-26* mutant animals, the fates of both sister cells are to become neuroblasts. (c) In the hermaphrodite, the cell P7.p normally generates two daughters, P7.pa and P7.pp, that have different fates. The fate of P7.pa is to divide to generate two cells, each of which has different fates (called N and T); the fate of P7.pp is to divide to generate two cells, both of which have the same fate (called L). In *lin-11* mutant hermaphrodites, the sisters P7.pa and P7.pp have the fate normally expressed by P7.pp. (d) In the male, the B cell divides to generate two daughters, B.a and B.p. The fate of B.a normally is to produce eight blast cells that are dorsal and anterior; the fate of B.p normally is to produce two blast cells that are ventral and posterior. In *lin-17* mutant males, the sisters B.a and B.p have the fate normally expressed by B.a.

cells, which divide to produce the members of the vulval equivalence group P(3-8).p. In fact, each of the 12 P cells divides to generate an anterior daughter that is a neuroblast and a posterior daughter that is a hypodermal cell or a hypodermal blast cell (Sulston and Horvitz 1977). In *lin-26(n156)* animals, both daughters are instead neuroblasts (Fig. 2b). This transformation prevents the generation of the cells of the vulval equivalence group, and for this reason results in a "vulvaless" phenotype.

Mutations in the gene *lin-11* also result in a vulvaless phenotype and also eliminate sister-sister differences. *lin-11* appears to be specific for one of the vulval sublineages, that expressed by both P5.p and P7.p. In each of these lineages, two sisters that normally differ in their fates (P5.pa and P5.pp or P7.pa and P7.pp) instead express the same fate (that of P5.pp or P7.pa) (Fig. 2c).

Mutations in the gene *lin-17* were also isolated on the basis of their effects on the vulval cell lineages. *lin-17* animals have a "multivulva" phenotype (Ferguson and Horvitz 1985). Although the nature of the effect on the vulval cell lineages is unclear, *lin-17* mutations affect a number of other sets of cells so that sister cells that normally have different fates instead have the same fate. For example, in *lin-17* mutant males, the anterior and posterior daughters of the B cell, B.a and B.p, both have the same fate as the normal B.a cell (Fig. 2d). Thus, mutations in *lin-26*, *lin-11*, and *lin-17* cause specific cell divisions that are normally asymmetric to be instead symmetric.

Genes That Affect Mother-Daughter Decisions

At most cell divisions, daughter cells have fates that differ from the fates of their mothers (Fig. 3a). Mutations have been isolated that apparently affect some of the decisions to make daughters different from mothers. For example, mutations in the gene *unc-86* alter at least seven neural lineages. In most cases, specific mother cells in *unc-86* animals divide to produce one daughter with its normal cell fate and the other daughter with the same cell fate as the mother (Chalfie et al. 1981; Horvitz et al. 1983). For example, in the wild type, the cell V5.paa divides to generate a dopaminergic neuron and a cell that divides again to generate a nondopaminergic neuron and a cell that undergoes programmed cell death (Fig. 3b). In mutant *unc-86* animals, V5.paa divides to generate a dopaminergic neuron and a cell that, like V5.paa, divides to generate a dopaminergic neuron. However, mutations in *unc-86* also result in other defects; e.g., the HSN neuron fails to undergo the decision to differentiate, which occurs independently of cell division (Horvitz et al. 1983; M. Finney and R. Horvitz, unpubl.). Therefore, the *unc-86* gene may encode a product necessary both for the generation of normal mother-daughter differences and for the expression of other developmental decisions.

Like mutations in *unc-86*, certain mutations in the genes *lin-4* and *lin-14* also cause the reiterative expression of specific cell types within certain cell lineages (Chalfie et al. 1981; Ambros and Horvitz 1984). How-

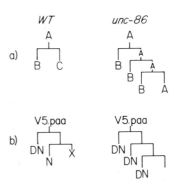

Figure 3. (a) In many cell divisions, mother cells are different from both of their daughters. In *unc-86* mutant animals, some mother cells instead divide to produce one daughter different from the mother and another daughter the same as the mother. A, B, and C represent canonical cell fates that differ from one another. (b) Normally, the cell V5.paa is a neuroblast that divides to generate a dopaminergic neuron (DN) and another neuroblast that divides to generate a nondopaminergic neuron (N) and a cell that undergoes programmed cell death (X). In *unc-86* mutant animals, the neuroblast V5.paa instead divides to generate a dopaminergic neuron and a neuroblast that, like V5.paa, again divides to produce a dopaminergic neuron and a neuroblast (adapted from Chalfie et al. 1981).

ever, in these cases, both parental and grandparental lineage reiterations occur; i.e., in some cases, cells acquire the fates not of their mother cells but rather of their grandmother cells. Current evidence suggests that in these mutants, mother-daughter developmental decisions are normal, but abnormalities in the specification of development timing ("heterochronies") cause certain cells to express fates normally expressed one, two, or three larval stages earlier (Ambros and Horvitz 1984).

Genes That Affect Cell Fates Specified by Cell Interactions

As discussed above, P5/6L and P5/6R form an equivalence group that can be observed during normal development because of the variability that occurs in the relative positioning of these cells along the ventral midline. These cells generate two members of the larger vulval equivalence group. Many genes have been identified that affect the determination of cell fates within the vulval equivalence group. For example, mutations in the genes *lin-2*, *lin-3*, *lin-7*, *lin-10*, and *let-23* (see below) transform three cells of the vulval equivalence group so that the cells that normally generate vulval cell lineages (P[5-7].p) instead act like the cells that normally do not generate vulval cell lineages (P[3,4,8].p) (Sternberg and Horvitz 1984; Ferguson and Horvitz 1985). Conversely, mutations in the genes *lin-1*, *lin-13*, *lin-15* (see below), *lin-31*, and *lin-34*, as well as in certain pairs of genes (e.g., *lin-8* and *lin-9*), transform the three other cells of the vulval equivalence group so that the cells that usually do not generate vulval cell lineages (P[3,4,8].p) act like the cells that normally do generate vulval cell lineages (P[5-7].p) (Sternberg and Horvitz 1984; Ferguson and Horvitz 1985). Therefore, mutations in these two sets of genes cause reciprocal transformations in cell fates, and in both cases, cells that normally express different extrinsically determined fates instead express the same cell fates.

The gene *lin-12*, which was also identified on the basis of its effects on the vulval cell lineages, has been studied in detail (Greenwald et al. 1983; Horvitz et al. 1983; Ferguson and Horvitz 1985). There exist two opposite allelic states of *lin-12*, *lin-12(d)*, and *lin-12(0)*, which reciprocally transform certain sets of cells in *C. elegans*. For example, the fates of the cells W and G2 are affected by the state of *lin-12*. In *lin-12(d)* mutant animals, both cells have the normal G2 cell fate; in *lin-12(0)* mutant animals, both cells have the normal W cell fate (Greenwald et al. 1983). W and G2 are members of an equivalence group (Sulston et al. 1983); other equivalence groups also are affected by the state of the *lin-12* gene. In each case, it appears that the level of *lin-12* activity determines which of the alternative cell fates is expressed. Specifically, genetic analyses have indicated that the *lin-12(d)* mutations cause increased *lin-12* activity, and the *lin-12(0)* mutations cause a loss of *lin-12* activity (Greenwald et al. 1983). Since during normal development, the fates of most cells affected by *lin-12* are determined by cell interactions, it seems likely that the level of *lin-12* activity normally is specified extrinsically and controls the activity of other genes responsible for the differences in cell fates. When the level of *lin-12* activity in different cells becomes fixed by mutation, cells that normally express different fates express the same fate instead.

Like *lin-12*, the genes *let-23* and *lin-15* also affect other extrinsically determined cell fates in addition to those involved in vulval development. In particular, these two genes affect the fates of an equivalence group consisting of P11 and P12. Just as in the case of P5/6L and P5/6R (see above), the bilaterally symmetric pair of P cells (P11/12L and P11/12R) adopt positions along the ventral midline. For this pair of P cells, however, the relative anteroposterior order appears to be fixed and not random (Sulston and Horvitz 1977). The cell from the left adopts the anterior position and becomes P11, and the cell from the right adopts the posterior position and becomes P12 (Fig. 4a). If P11/12R (P12) is ablated before these cells adopt their positions at the ventral midline, P11/12L (P11) replaces P12 and expresses its fate (Sulston and White 1980). Thus, P11/12L and P11/12R are members of an equivalence group, and their fates must be determined extrinsically. Mutations in the genes *lin-15* and *let-23*, isolated because of their effects on the cells of the vulval equivalence group, also affect the P11/P12 equivalence group. *lin-15* mutations cause both P11 and P12 to adopt the P12 cell fate (Fig. 4b). Conversely, a *let-23* mutation causes both P11 and P12 to adopt the P11 cell fate (Fig. 4c). Thus, mutations in these two genes, which cause reciprocal transformations in cell fates in the vulval equivalence group (see above), also cause reciprocal

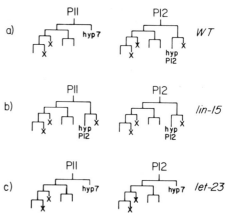

Figure 4. (*a*) Cell lineages of P11 and P12 in the wild-type hermaphrodite. The posterior daughter of P11 is a hypodermal cell that fuses with the large "hyp7" syncytium, whereas the posterior daughter of P12 is a blast cell that divides to generate a hypodermal cell of a different type (called hypP12) and a cell that undergoes programmed cell death (X). Unlabeled cells are neurons. (*b*) In *lin-15* mutant hermaphrodites, the cell lineages of P11 and P12 are both like that normally observed for P12. (*c*) In *let-23* mutant hermaphrodites, the cell lineages of P11 and P12 are both like that normally observed for P11.

transformations in the P11/P12 equivalence group. Therefore, *lin-15* and *let-23* appear to encode molecules that are involved in the cellular interactions of, or responses to, those interactions responsible for a number of extrinsically determined cell fates. Interestingly, mutations in *lin-12* (which affect at least 11 sets of cells) do not affect the P11/P12 equivalence group, although mutations in *lin-12*, *lin-15*, and *lin-23* all affect cell fates within the vulval equivalence group.

Genes That Affect the Fates of Cells Generated by Repeated Sublineages

As described above, certain sets of identical cell types are generated by the repeated expression of the same cell lineages by a number of morphologically indistinguishable precursor cells. There is evidence that, in general, the execution of such sublineages appears to utilize the same set of genes wherever or whenever a particular sublineage is utilized. For example, the 12 sublineages generated by the P cells are all affected by the mutation *lin-26(n156)* (Fig. 2b). Thus, each P cell utilizes a developmental pathway that is affected by *lin-26(n156)*.

If the same genes are utilized each time a particular sublineage is executed, then identical patterns of cell fates should always be generated. The sublineages generated by the 12 P cells, although strikingly similar in most respects, do differ slightly. Therefore, there may be genes that can modify sublineages such that cells that would otherwise be the same instead differ. One such difference involves P11 and P12 (Fig. 4). As discussed above, these cells are members of an equivalence group and differences between them must be determined extrinsically. At least two genes, *lin-15* and *let-23*, are involved in making P11 and P12 different.

The generation of the VC neurons, in contrast, might involve an intrinsically determined modification of the P cell sublineages, since cell fates within these sublineages appear to be autonomously determined (Sulston and Horvitz 1977, 1981; Sulston and White 1980). As noted above, the six VC cells are generated from P(3–8).a. The corresponding cells derived from the other six P cell lineages (i.e., from P[1,2].a and P[9–12].a) undergo programmed cell death (Sulston 1976; Horvitz et al. 1982) (Fig. 5a). The mutation *lin-39(n709)* causes the six cells that normally differentiate into VC neurons to undergo programmed cell death (Fig. 5b). Therefore, the *lin-39* gene may encode a product that is capable of modifying the P cell lineage intrinsically and is involved in making cells different that would otherwise be the same.

Mutations that affect the utilization of sublineages have also been isolated. For example, in *lin-22(n372)* mutant animals, certain anterior blast cells of the lateral hypodermis (V[1–4]) express the same sublineage as a more posterior blast cell (V5) (Horvitz et al. 1983; W. Fixsen and R. Horvitz, in prep.). Similarly, in *lin-28* mutant animals, blast cells (e.g., V5) that normally express different sublineages in the two sexes instead express the same sublineage (Ambros and Horvitz 1984). In both of these cases, cells that normally express different fates instead express the same fate.

SUMMARY AND PROSPECTS

We have identified a number of mutations that cause cells that are normally different to be the same. Some of these mutations affect cells with fates that are specified intrinsically; these mutations prevent the normal asymmetry generated between sister and sister cells or mother and daughter cells. Others of these mutations affect cells with fates that are specified extrinsically; these mutations probably alter interactions with, or responses to, neighboring cells or other positional signals. The genes defined by the mutations we have isolated may normally function in specifying cell fates. Genes that affect intrinsically specified cell fates may encode products that are necessary for the asymmetric segregation of developmental potential. Genes that affect extrinsically specified cell fates may encode products that are necessary for the production or interpretation of intercellular signals. The identification of such genes and the isolation and characterization of their products provide one approach toward an understanding of the molecular basis of the specification of cell fates.

Recently, methods have been developed for the molecular cloning of *C. elegans* genes that have been identified only by mutations. These methods include transposon-induced mutagenesis and "walking" from genetically mapped restriction fragment polymorphisms. Using these approaches, the genes *lin-12* (I.

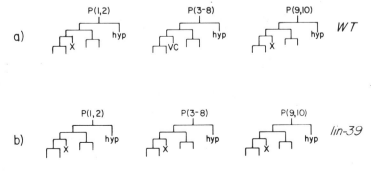

Figure 5. (*a*) First larval stage sublineages of P(1–10) in the wild-type hermaphrodite. The cells that undergo programmed cell death (X) in the P(1,2) and the P(9,10) sublineages instead survive and become VC neurons in the P(3–8) sublineages. (*b*) In *lin-39* mutant hermaphrodites, the cells that normally become VC neurons instead undergo programmed cell death. In all cases, unlabeled cells are other (non-VC) neurons and "hyp" cells are hypodermal or hypodermal blast cells.

Greenwald, pers. comm.) and *lin-14* (G. Ruvkun et al., unpubl.) have been cloned and are now being analyzed molecularly. It should now be possible to clone many of the other genes that we have identified as involved in making cells different from each other in *C. elegans*. Once molecular probes are obtained, we should be able to study at the molecular level when, where, and how these genes act during development.

ACKNOWLEDGMENTS

We thank L. Avery, C. Desai, M. Finney, J. Thomas, and E. Wolinsky for comments on the manuscript and M.E. Davis for help in its preparation. The research in our laboratory was supported by U.S. Public Health Service grants GM-24663 and GM-24943 and Research Career Development Award HD-00369 to H.R.H.

REFERENCES

Ambros, V. and R. Horvitz. 1984. Heterochronic mutants of the nematode *Caenorhabditis elegans*. *Science* **226**: 409.

Chalfie, M., R. Horvitz, and J. Sulston. 1981. Mutations that lead to reiterations in the cell lineages of *Caenorhabditis elegans*. *Cell* **24**: 59.

Ferguson, E. and R. Horvitz. 1985. Identification and characterization of 22 genes that affect the vulval cell lineages of the nematode *Caenorhabditis elegans*. *Genetics* **110**: 17.

Greenwald, I., P. Sternberg, and R. Horvitz. 1983. The *lin-12* locus specifies cell fates in *Caenorhabditis elegans*. *Cell* **34**: 435.

Horvitz, R. and J. Sulston. 1980. Isolation and genetic characterization of cell-lineage mutants of the nematode *Caenorhabditis elegans*. *Genetics* **96**: 435.

Horvitz, R., H. Ellis, and P. Sternberg. 1982. Programmed cell death in nematode development. *Neurosci. Comment.* **1**: 56.

Horvitz, R., P. Sternberg, I. Greenwald, W. Fixsen, and H. Ellis. 1983. Mutations that affect neural cell lineages and cell fates during the development of the nematode *Caenorhabditis elegans*. *Cold Spring Harbor Symp. Quant. Biol.* **48**: 453.

Kimble, J. 1981. Alterations in cell lineage following laser ablation of cells in the somatic gonad of *Caenorhabditis elegans*. *Dev. Biol.* **86**: 286.

Kimble, J. and D. Hirsh. 1979. Post-embryonic cell lineages of the hermaphrodite and male gonads in *Caenorhabditis elegans*. *Dev. Biol.* **70**: 396.

Kimble, J., J. Sulston, and J. White. 1979. Regulative development in the postembryonic lineages of *Caenorhabditis elegans*. *INSERM Symp.* **10**: 59.

Sternberg, P. and R. Horvitz. 1982. Postembryonic nongonadal cell lineages of the nematode *Panagrellus redivivus*: Description and comparison with those of *Caenorhabditis elegans*. *Dev. Biol.* **93**: 181.

———. 1984. The genetic control of cell lineage during nematode development. *Annu. Rev. Genet.* **18**: 489.

Sulston, J.E. 1976. Post-embryonic development in the ventral cord of *Caenorhabditis elegans*. *Philos. Trans R. Soc. Lond. B.* **275**: 287.

Sulston, J. and R. Horvitz. 1977. Post-embryonic cell lineages of the nematode *Caenorhabditis elegans*. *Dev. Biol.* **56**: 110.

———. 1981. Abnormal cell lineages in mutants of the nematode *Caenorhabditis elegans*. *Dev. Biol.* **82**: 41.

Sulston, J. and J. White. 1980. Regulation and cell autonomy during postembryonic development of *Caenorhabditis elegans*. *Dev. Biol.* **78**: 577.

Sulston, J., D. Albertson, and N. Thomson. 1980. The *Caenorhabditis elegans* male: Post-embryonic development of non-gonadal structures. *Dev. Biol.* **78**: 542.

Sulston, J., E. Schierenberg, J. White, and N. Thomson. 1983. The embryonic cell lineage of the nematode *Caenorhabditis elegans*. *Dev. Biol.* **100**: 64.

Bicaudal Mutations of *Drosophila melanogaster*: Alteration of Blastoderm Cell Fate

J. MOHLER AND E.F. WIESCHAUS
Department of Biology, Princeton University, Princeton, New Jersey 08540

One of the most intriguing questions of developmental biology is how the organization of the developing embryo is regulated. In *Drosophila melanogaster*, a number of mutations have been isolated that disrupt the organization of the early embryo. Some of these mutations are expressed after fertilization in the developing zygote and have only a limited effect on the overall organization of the developing embryo (see, e.g., Nüsslein-Volhard and Wieschaus 1980). Given the limited transcription in *Drosophila* during cleavage and nuclear migration, the global organization of the developing embryo, defined during the cellular blastoderm stage, is more likely to be under maternal control during oogenesis. The maternal-effect mutations that appear to have the greatest effect on the global organization of the embryo are those that affect one of the two major axes of the embryo: the anterior-posterior axis or the dorsal-ventral axis. Mutations, such as dorsal (*dl*), cause the loss of the most ventral pattern elements along the dorsal-ventral axis (Anderson and Nüsslein-Volhard 1984; Steward et al., this volume). Among mutations that affect the anterior-posterior axis of the embryo, the bicaudal mutations have one of the most severe effects on embryonic pattern: Most of the anterior portions of the embryo are lost from the differentiated embryo and replaced by duplicated posterior structures.

Mutations at three loci will give rise to embryos with a bicaudal phenotype, *bic* (Bull 1966), *BicC*, and *BicD* (J. Mohler and E. Wieschaus, in prep.). Mutations at the *bic* locus are recessive; the female must be homozygous for the embryo to appear mutant. Mutations at the *BicC* and *BicD* loci are semidominant; heterozygous females produce some phenotypically mutant embryos. Homozygotes of *BicC* and *BicD* behave differently: homozygous *BicC* females are blocked in oogenesis at about stage 8, whereas homozygous *BicD* females show a severe increase in the penetrance of the bicaudal phenotype. Mutations at all of these loci produce the same range of phenotypes and are suspected to act via the same or similar mechanisms to give rise to the bicaudal phenotypes. Because the *BicD* locus gives the highest penetrance of the bicaudal phenotype, we have concentrated our studies on this locus. Genetic analysis of the *BicD* mutations suggest that these semidominant mutations are antimorphic gain-of-function mutations, whereby the altered gene product induces the production of bicaudal embryos and can be outcompeted by the wild-type gene product (J. Mohler and E. Wieschaus, in prep.).

The Bicaudal Phenotype

Females mutant at one of the bicaudal loci produce three types of differentiated embryos: (1) those with normal cuticular morphology that may or may not hatch, (2) those with normal polarity throughout the animal, but lacking head and sometimes thoracic structures, and (3) those with duplicated posterior structures of reversed polarity in the anterior region of the embryo (double-abdomen embryos). For a more complete description of the classes of bicaudal embryos, see Nüsslein-Volhard (1977) and J. Mohler and E. Wieschaus (in prep.). Figure 1 shows an example of an embryo from the most predominant class of each of the three types of differentiated bicaudal embryos (wild type, Fig. 1a,d; mouthparts reduced, Fig. 1b; symmetric double abdomen, Fig. 1c,e). The range of defects in bicaudal embryos suggests that the final differentiated pattern of bicaudal embryos reflects deletions of anterior pattern elements extending a variable distance from the anterior end of the embryo toward the posterior end. In embryos where the pattern deletions extend deeply into the posterior half of the embryo (i.e., past the thorax in the final differentiated pattern), the organization of the embryo along the anterior-posterior axis becomes reorganized to produce a mirror-image pattern with the duplication of the posterior end (J. Mohler and E. Wieschaus, in prep.).

BicD Acts during Oogenesis

The production of bicaudal embryos depends strictly on the genotype of the mother and is not altered by the number of wild-type alleles in the embryo. Because the expression of both alleles of *BicD* is temperature-dependent, it is possible to use temperature shifts to determine whether the altered *BicD* gene product is necessary during early embryogenesis or oogenesis. 85% of the embryos produced by homozygous *BicD* females at 18°C and 50% of the embryos produced at 25°C develop as double-abdomen embryos, whereas at 29°C, the frequency is always less than 5%. No change in the frequencies is observed when embryos are shifted between restrictive and permissive temperatures immediately after the egg is laid (Fig. 2). This argues that the

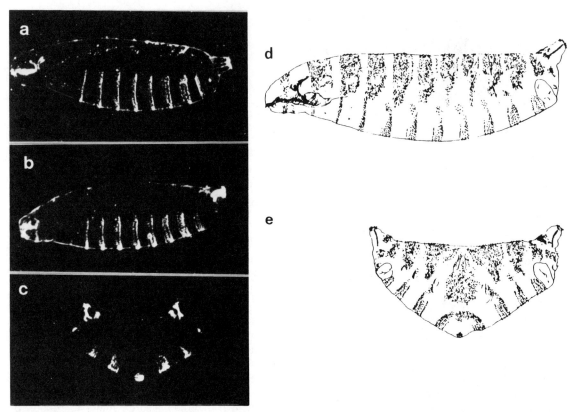

Figure 1. Phenotypes of embryos produced by females mutant for a bicaudal mutation. (*a–c*) Dark-field micrographs; (*a*) normal (unhatched normal class); (*b*) head defect (mouthparts-reduced class); (*c*) double abdomen (symmetric double abdomen, S2.5). (*d,e*) Camera lucida drawings; (*d*) normal (hatched larva); (*e*) double abdomen (symmetric, S4.5). Magnifications: (*a–c*) 80×; (*d,e*) 110×.

critical events controlling double-abdomen production occur during oogenesis. When the temperature of the *BicD* females is shifted, the frequency of double-abdomen embryos laid 6 hours later is altered. Therefore, the stage in which the production of bicaudal embryos is temperature-dependent occurs late in oogenesis, during the final 6 hours of oogenesis, which corresponds roughly to between stages 11 and 14 of oocyte maturation.

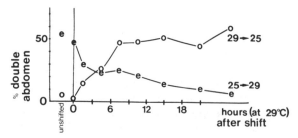

Figure 2. Frequency of double-abdomen embryos produced by homozygous *BicD* females ($BicD^{71.34}/BicD^{IIIE48}$) among embryos at different times following a temperature shift between 25°C and 29°C. (o) Embryos from females shifted from 29°C to 25°C; (e) embryos from females shifted from 25°C to 29°C. The leftmost datapoints represent the frequency of double-abdomen embryos from unshifted cultures (o = 29°C, e = 25°C). Shifts between 18°C and 29°C produce similar curves (E. Wieschaus, unpubl.).

Gastrulation of Bicaudal Embryos

In the development of normal, wild-type *Drosophila* embryos, the zygote nucleus initially divides syncytially without cell division for 13 nuclear divisions. After the 9th nuclear division, the majority of the nuclei migrate to the egg cortex to form the syncytial blastoderm. After the 13th nuclear division, the cell membrane of the egg invaginates around the nuclei, segregating the nuclei into separate cells and forming the cellular blastoderm. Following the cellular blastoderm, gastrulation begins with the formation of a ventral furrow at the ventral midline of the embryo. The posterior midgut invagination forms soon thereafter beneath the pole cells. As the ventral portion of the embryo (the germ band) elongates, the posterior midgut invagination is pushed anteriorly along the dorsal side of the embryo. During germ band elongation, the cephalic furrow pinches in on the lateral sides of the embryo about one third of the way from the anterior end of the embryo. More anteriorly, the anterior midgut invagination forms on the ventral midline. For a more complete review of gastrulation in *Drosophila*, see Turner and Mahowald (1977). Figure 3 indicates the major morphogenetic movements in normal *Drosophila* gastrulation.

Gastrulation in symmetric double-abdomen embryos differs from the wild-type gastrulation pattern in

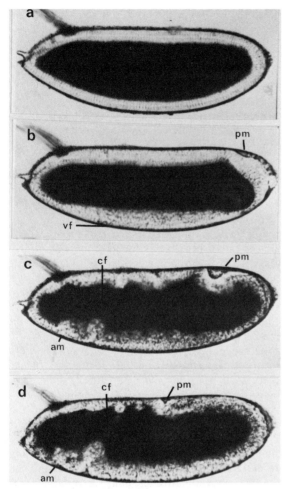

Figure 3. Early development of a normal embryo. (*a*) Cellular blastoderm; (*b,d*) early gastrula. (pm) Posterior midgut invagination; (cf) cephalic furrow; (am) anterior midgut invagination; (vf) ventral furrow. Magnification, 140×.

a manner consistent with the final differentiation pattern. In double-abdomen embryos, a normal syncytial and cellular blastoderm is formed. Gastrulation initiates with a ventral furrow formed along the entire length of the ventral midline. A posterior midgut invagination is formed at the normal position beneath the pole cells, and a novel invagination forms at the anterior end of the embryo. The formation of the novel anterior invagination is often delayed slightly (about 5 min) relative to the posterior midgut invagination. As the germ band elongates, the posterior midgut invagination is pushed anteriorly along the dorsal side while the novel anterior invagination is pushed posteriorly along the dorsal side (Fig. 4). However, pole cells are not duplicated at the anterior end, indicating that only the somatic tissues are reorganized in the double-abdomen embryo.

Embryos from bicaudal females that differentiate into animals with head defects, but no polarity reversals, also gastrulate abnormally. Gastrulae of these embryos have the normal ventral furrow and posterior midgut invaginations but are lacking the cephalic furrow and anterior midgut invaginations (Fig. 5). In

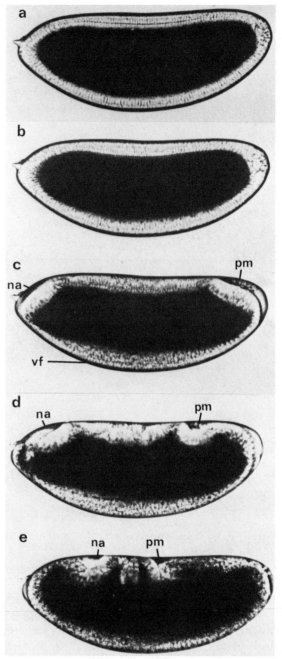

Figure 4. Early development of a symmetric double-abdomen embryo. (*a*) Syncytial blastoderm; (*b*) cellular blastoderm; (*c–e*) early gastrula. (pm) Posterior midgut invagination; (vf) ventral furrow; (na) novel anterior invagination. Magnification, 140×.

some embryos that differentiate with weak head defects (some mouthparts-reduced embryos), a cephalic furrow is formed but is displaced anteriorly relative to its normal position in the embryo.

The analysis presented above indicates that the *BicD* mutation alters the gastrulation patterns of the embryo in a manner consistent with the final differentiated pattern. Although the deletions and duplications of the pattern in the final differentiated embryos could be formally described as resulting from cell death

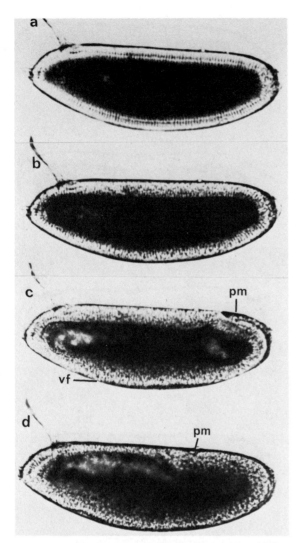

Figure 5. Early development of a headless embryo. (*a*) Syncytial blastoderm; (*b*) cellular blastoderm; (*c*,*d*) early gastrula. (pm) Posterior midgut invagination; (vf) ventral furrow. Magnification, 140×.

and subsequent regeneration, no evidence for either of these phenomena was detected in these observations of early development. Instead, it appears that the bicaudal mutation alters the spatial pattern of cell determination events at the blastoderm stage, at least with respect to cells undergoing specific morphogenetic movements during gastrulation.

Spatial Expression of the *ftz* Gene in Bicaudal Embryos

To determine whether the bicaudal mutation altered specifically only the pattern of gastrulation or whether it altered other more general aspects of the spatial pattern at the blastoderm stage, we investigated the effect of the *BicD* mutation on the spatial expression of the fushi tarazu (*ftz*) gene transcript. As shown by Hafen et al. (1984), in wild-type embryos, the *ftz* gene is transcribed heavily in the cellular blastoderm in the region fated to give rise to the head, thoracic, and abdomen segments. The bulk of this transcription occurs in seven stripes that correspond to the regions of the embryo affected by mutations of the *ftz* locus; these mutations result in a final differentiated embryo with defects spaced at distances of every other segment.

The transcription pattern of *ftz* in an early double-abdomen gastrula, as detected by in situ hybridization of ^{3}H-labeled *ftz* DNA to embryonic RNA, is shown in Figure 6b. This embryo has two invaginations: the posterior midgut invagination and the novel anterior invagination. Each invagination is flanked by three stripes of *ftz* hybridization. The *ftz* transcription that flanks the novel anterior invagination is occurring in cells that in the wild type would not normally express the *ftz* gene. A similar alteration in the spatial expression of the *ftz* gene can be found in some blastoderm-stage embryos (Fig. 6d), where stripes of hybridization are spaced from the anterior end to the posterior end. In these individuals, the anterior stripes of hybridization are angled obliquely to the posterior stripes, resulting in a hybridization pattern that is visually mirror-image symmetric.

Although the *ftz* transcription pattern in many blastoderm-stage embryos from bicaudal mutant females is different from the wild-type pattern, some blastoderm-stage embryos show a wild-type pattern of *ftz* gene expression. This is related to the fact that the *BicD* mutation is rarely fully penetrant, so presumably the blastoderm-stage embryos with normal *ftz* gene expression correspond to the normal differentiated embryos from bicaudal females. In confirmation of this fact, the frequency of blastoderm embryos with the normal *ftz* transcription pattern in a given population is roughly correlated with the frequency of normal pattern differentiated embryos: Seven out of eight blastoderm-stage embryos had mirror-image *ftz* transcription patterns from a strain where 96% of the embryos developed as symmetric double-abdomen embryos (the eighth was also visibly different from the wild-type pattern). Figure 7 shows a composite representation of the pattern of *ftz* transcription generated from serial reconstructions of these seven blastoderm-stage embryos; 70% of the population from which these seven embryos were culled differentiated as symmetric double-abdomen embryos, in which the plane of mirror-image symmetry intersected the ventral midline in the anterior portion of the sixth abdominal segment or the posterior portion of the fifth abdominal segment (as is a common phenomenon among double-abdomen embryos, more segments are found on the ventral side of the embryo than on the dorsal side). These blastoderm-stage embryos have two anterior and two posterior stripes of *ftz* gene transcription, which extend around the circumference of the embryo. Also, in at least two of these embryos, there is an additional ring of transcription, displaced to one side of the embryo, located at the midline of the anterior-posterior axis. If this ring is placed on the ventral side of the blastoderm, the spatial transcription pattern of *ftz* in blastoderm reflects the final differentiated pattern of the

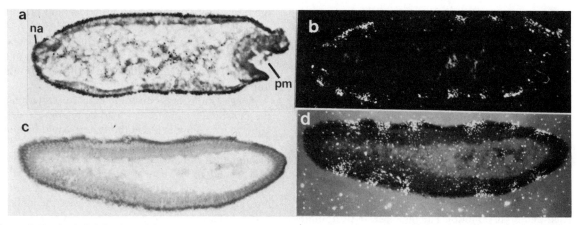

Figure 6. In situ hybridization of ³H-labeled DNA (pDm439H3.2), containing the 3.2-kb HindIII fragment including the *ftz* locus, to RNA in embryos from females carrying a bicaudal mutation. (*a,b*) Early gastrula embryo from a female raised at 25°C, homozygous for BicD⁷¹·³⁴ (exposed 12 days); hybridization was done according to procedure of Hafen et al. (1983). (*a*) Bright-field illumination; (*b*) dark-field illumination. (*c,d*) Cellular blastoderm embryo from a BicD⁷¹·³⁴/Df(2L)TW119 female raised at 25°C (exposed 21 days); hybridization was done according to the procedure of Ambrosio and Schedl (1984). (*c*) Bright-field illumination; (*d*) dark-field illumination. Magnification, 150×.

embryo. Therefore, it appears that the maternal expression of the *BicD* mutation alters the spatial expression of the *ftz* gene in the cellular blastoderm.

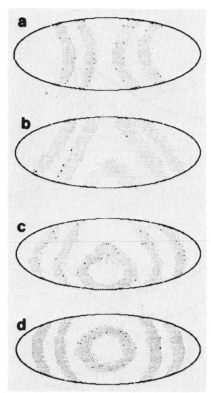

Figure 7. Rotated views of *ftz* transcription patterns in a composite blastoderm embryo constructed from serial reconstructions of *ftz* transcription patterns of seven embryos from females (BicD⁷¹·³⁴/E(2)Bic) that yield 96% double-abdomen embryos (70% are symmetric with either 2.5 or 3 segments in mirror-image symmetry). Assuming that the view with the fewest stripes corresponds to the dorsal side (which has the fewest segments in a differentiated double-abdomen embryo), these views represent *ftz* transcription patterns on the dorsal (*a*), lateral (*b*), ventral-lateral (*c*), and ventral (*d*) surfaces of the blastoderm.

This altered pattern of *ftz* gene expression appears to correspond to the altered pattern in the fully differentiated embryo. Because the cellular expression of the *ftz* gene is related to final fates of those cells, these results indicate that much, if not all, of the blastoderm fate plan is altered by the *BicD* mutation and that the *BicD* mutation alters the global pattern of cell-determination events in the blastoderm.

Role of *BicD* in Patterning the *Drosophila* Embryo

The results presented above indicate that females bearing a bicaudal mutation lay eggs with an altered blastoderm fate plan. The final pattern of the differentiated bicaudal embryo is due to and directly reflects an alteration in the programming of individual blastoderm cells. Whether this effect is due to the inability of the cell to interpret a properly positioned morphogenetic determinant or whether it is due to the improper positioning of the morphogenetic determinant has not been resolved. Although the effects of the mutation vary from one embryo to another, the pattern within any given embryo is usually integrated, rather than patchy. This global effect is difficult to explain as altered interpretation of a determinant, because to achieve it, each blastoderm cell would have a varied and unequal defect in its capacity to interpret the morphogenic determinant. This observation, coupled with our observation of a prefertilization temperature-dependent period, suggests that the bicaudal mutations affect the proper positioning of an anterior-posterior morphogenetic determinant.

The phenotypes of bicaudal embryos indicate that the anterior-posterior morphogenetic determinant is continuously variable. The most anterior structure in a symmetric double-abdomen embryo (always on the ventral midline) can be at any place within any segment between the posterior part of the third abdominal segment and the anterior portion of the sixth abdominal segment, with no discrete steps. Similar continuously variable defects can be observed in the head-defective embryos. Other experiments on *Drosophila* embryos (such as the embryo-ligation experiments of Schubiger et al. 1977) also suggest a continuously varying anterior-posterior determinant. The conceptually simplest model for a continuously varying determinant is a concentration gradient of morphogen along the anterior-posterior axis, where a particular concentration of morphogen will determine a particular cell fate.

The best-developed model for the generation of a concentration gradient of morphogen is the activator-inhibitor model of Meinhardt (1982). This model postulates an activator, which stimulates the synthesis of itself and of an inhibitor. The inhibitor, in turn, inhibits the synthesis of the activator. Thus, whenever and wherever the activity of the activator surpasses the activity of the inhibitor, both are synthesized, otherwise neither is synthesized. Both the activator and inhibitor are degraded throughout the morphogenetic field. Because the activator diffuses less well than the inhibitor, a stable gradient of activator and inhibitor can be formed, with synthesis of both activator and inhibitor in a single "activator peak." Because the inhibitor is more readily diffusible, the concentration of the inhibitor serves as the morphogenetic determinant. In the case of the *Drosophila* embryo, the activator peak would normally be localized at the posterior end of the embryo. The resultant gradients of activator and inhibitor are graphed in Figure 8a.

If such gradients do form the basis for patterning in the *Drosophila* embryo, it would be relevant to determine which components in the system could be modified to yield the bicaudal phenotypes. Symmetric double-abdomen embryos might result from the generation of a second activator peak at the anterior end of the animal. A second activator peak can be generated by creating an activator activity that exceeds the inhibitor activity. This could be accomplished by lowering the inhibitor concentration at the anterior end (by decreasing the diffusion of the inhibitor) or by increasing the activator concentration at the anterior end (by increasing the activator diffusion or decreasing the degradation rate of the activator). Decreased diffusion of the inhibitor is unlikely to be the defect caused by bicaudal mutations, because it cannot be used to explain the generation of headless embryos produced under less-penetrant conditions. A decrease in the diffusion of the

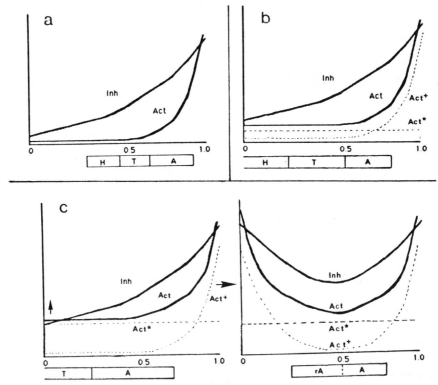

Figure 8. (*a*) Concentration gradients of activator and inhibitor hypothesized to exist along the anterior-posterior axis of the *Drosophila* embryo according to the activator-inhibitor model of Meinhardt (1982). Different threshold levels of the inhibitor induce cells to differentiate into head (H), thorax (T), and abdomen (A). (Act) Total activator; (Inh) inhibitor. (*b*) Concentration gradients in an embryo in which a low level of the mutant nondegradable activator (Act*) is synthesized, relative to a wild-type, degradable form (Act+). Because the mutant activator is stable, a very shallow gradient of mutant activator will accumulate. The total activator concentration (Act* plus Act+) will be increased, resulting in slightly increased inhibitor levels and the loss of anterior structures. (*c*) Concentration gradients in which a higher level of the mutant activator is synthesized. In this case, the total activator activity exceeds the inhibitor activity at the anterior end, inducing a second activator peak and a symmetric double-abdomen embryo. (These graphs represent our expectations of how the activator-inhibitor model would react and have not been verified by computer simulations.)

inhibitor (insufficient to generate a second peak) would generate a steeper gradient of inhibitor from the posterior end and would result in an embryo in which the entire pattern is compressed toward the posterior end of the embryo. In contrast, an increase in the activator concentration (e.g., by a decrease in its rate of degradation) would result in a concomitant increase in inhibitor levels. A second peak would form if and where the activator level locally exceeds that of the increased inhibitor. In embryos where a second peak does not form, the increased inhibitor would result in the elimination of head structures.

Our favorite hypothesis for the function of the *BicD* locus in an activator-inhibitor model is that the *BicD* locus consists of a number of tandem copies of the gene encoding the activator substance. Bicaudal mutations at the *BicD* locus would cause one of these copies to be converted to a nondegradable version of the activator. Depending on the relative utilization of the gene encoding the nondegradable form as compared with the wild-type gene copies, embryos from mothers with a bicaudal mutation would develop either normally, with head defects (as diagramed in Fig. 8b), or as a symmetric double-abdomen embryo (Fig. 8c). Such a hypothesis would account for much of the genetics of the *BicD* locus: (1) the bicaudal mutation would be enhanced by a deletion in *trans* of the *BicD* locus (increasing the ratio of nondegradable to wild-type gene copies), (2) the bicaudal mutation would be suppressed by a duplication of a wild-type *BicD* locus (decreasing the ratio of nondegradable to wild-type gene copies), (3) new recessive, loss-of-function alleles of the *BicD* locus would be difficult to obtain on wild-type chromosomes, due to the existence of multiple copies, and (4) revertants of a dominant allele would be expected to be wild type if they knock out only the nondegradable gene copy, whereas they might lack any anterior-posterior axis if the entire *BicD* locus were deleted. The latter phenotype might correspond to a known revertant, $BicD^{71.34.R26}$, which produces no oocyte.

Because the phenotypes of mutations at the other bicaudal loci are so similar to *BicD* and because all the bicaudal mutations appear to enhance each other (J. Mohler and E. Wieschaus, in prep.), it seems likely that all the bicaudal loci are affecting the same general process. Under the hypothesis proposed above, we would expect the other bicaudal loci to increase diffusion of the activator, to decrease degradation rates of the normal activator, or to cause a minor constitutive level of activator synthesis throughout the morphogenetic field of the embryo.

ACKNOWLEDGMENTS

We thank Linda Ambrosio, Kate Harding, and Michael Levine for technical advice and expert assistance; Michael Shedd for technical assistance; Paul Schedl and Michael Levine for the generous gift of laboratory space and supplies; Matthew Scott for the pDm439H3.2 (*ftz*) clone; and Pam Hoppe and Helen Salz for assistance in preparing the manuscript. This work was supported by a National Institutes of Health postdoctoral fellowship to J.M. and an NIH research grant (PHS-HD-15587) to E.F.W.

REFERENCES

AMBROSIO, L. and P. SCHEDL. 1984. Gene expression during *Drosophila melanogaster* oogenesis: Analysis by in situ hybridization to tissue sections. *Dev. Biol.* **105:** 80.

ANDERSON, K.V. and C. NUSSLEIN-VOLHARD. 1984. Information for the dorsal-ventral axis is stored as maternal RNA. *Nature* **311:** 223.

BULL, A. 1966. Bicaudal, a genetic factor which affects the polarity of the embryo of *Drosophila melanogaster*. *J. Exp. Zool.* **161:** 221.

HAFEN, E., A. KUROIWA, and W. GEHRING. 1984. Spatial distribution of transcripts from the segmentation gene *fushi tarazu* during *Drosophila* development. *Cell* **37:** 833.

HAFEN, E., M. LEVINE, R.L. GARBER, and W.J. GEHRING. 1983. An improved *in situ* hybridization method for the detection of cellular RNAs in *Drosophila* tissue sections and its application for localizing transcript of the homeotic *Antennapedia* gene complex. *EMBO J.* **2:** 617.

MEINHARDT, H. 1982. *Models of biological pattern formation.* Academic Press, New York.

NUSSLEIN-VOLHARD, C. 1977. Genetic analysis of pattern formation in the embryo of *Drosophila melanogaster*: Characterization of the maternal-effect mutant *bicaudal*. *Wilhelm Roux's Arch. Dev. Biol.* **183:** 249.

SCHUBIGER, G., R.C. MOSELEY, and W.J. WOOD. 1977. Interaction of different egg parts in determination of different body regions in *Drosophila melanogaster*. *Proc. Natl. Acad. Sci.* **74:** 2050.

TURNER, F.R. and A. MAHOWALD. 1977. Scanning electron microscopy in *Drosophila melanogaster* embryogenesis. II. Gastrulation and segmentation. *Dev. Biol.* **57:** 403.

Pattern Formation in the Muscles of *Drosophila*

P.A LAWRENCE
MRC Laboratory of Molecular Biology, Cambridge, CB2 2QH, England

In the *Drosophila* thorax, there are about 60 individual muscles arranged in an intricate three-dimensional pattern that varies little from individual to individual (Miller 1950). There is no immediate prospect of our understanding the mechanisms used in the construction of this complex pattern; all I can do here is to describe some recent experiments aimed at identifying genes involved in the segmental diversification of muscles. Since we know most about pattern formation in the epidermis, it makes sense to investigate how far principles identified in that tissue can be applied directly to the developing muscles and also to identify those aspects of muscle development that present special problems.

Selector Genes and the Mesoderm

In 1975, Garcia-Bellido proposed the selector gene hypothesis. Because the epidermis is constructed in precise compartments that are defined by the lineage of their constituent cells (Garcia-Bellido et al. 1973; Crick and Lawrence 1975) and because some mutations transform one compartment into the pattern appropriate to another, he suggested that compartments are coextensive with the realms of action of a special class of genes. He also proposed that these "selector" genes act together to define the pattern made by the group of cells that found the compartment so that the combination of active and inactive selector genes makes a binary code word that is unique in each compartment; we call this code word the genetic address (Garcia-Bellido et al. 1979). Evidence in favor of this hypothesis has increased considerably since 1975, and selector genes so far studied include the elements of the bithorax complex (Lewis 1978), engrailed (Morata and Lawrence 1975; Kornberg 1981), and the elements of the Antennapedia complex (Kaufman 1983).

Here, I examine how far this hypothesis can be applied to the somatic muscles, and to do this we first must compare their development with that of the epidermis. At the extended germ band stage (~5 hr of development), the *Drosophila* embryo consists of a superficial ectoderm (forming epidermis and nervous system) and an internal mesoderm divided into an inner visceral mesoderm and an outer somatic mesoderm. Our present view is that the majority of the embryonic ectoderm consists of a chain of 28 compartments extending from the presumptive mouthparts to the anal region (see Martinez-Arias and Lawrence 1985). Mapping of engrailed transcription gives some support, since there are 14 stripes of active transcription in the main body of the embryo corresponding to 14 posterior (P) compartments, and 14 interstripes corresponding to the 14 anterior (A) compartments (Fjose et al. 1985; Kornberg et al. 1985). At this stage, there are surface grooves in the embryo that seem to coincide with the A/P boundaries and divide the ectoderm into parasegments: units consisting of the P compartment of one segment and the A compartment of the neighboring segment. The mesoderm is divided into cell masses that, because they are in register with those in the epidermis, are also thought to be parasegments (Martinez-Arias and Lawrence 1985).

Cell lineage analysis of the adult somatic muscles shows that the muscles of each segment seem to have an independent primordium that is established within the first few hours of development, but there is no evidence for an A/P subdivision of that primordium (Lawrence 1982). Furthermore, there appears to be no requirement for the engrailed gene in the mesoderm, as mesodermal cells that are homozygous for lethal alleles of engrailed develop quite normally (Lawrence 1982; Lawrence and Johnston 1982, 1984a). Together, these results suggest that the mesoderm is divided into single metameric units that are not subdivided into A and P compartments (Lawrence and Johnston 1984a), and evidence summarized below indicates that the segmental units defined by cell lineage are actually the parasegments seen in the embryo.

If this is the case, there must be a relative shift between the ectoderm and mesoderm such that the mesoderm that originates in register with one parasegment (e.g., parasegment 6, consisting of the P compartment of the third thoracic segment [T3] and the A compartment of the first abdominal segment A1) produces the muscles that span the entire A1. Evidence of such a shift was provided by Akam (1983), who studied, in the third larval stage, the pattern of transcription of the *Ubx* gene, an element in the bithorax complex (Lewis 1978); Akam observed that transcription in the muscles was out of register with that in the epidermis. More recently, Akam and Martinez-Arias (1985) have shown, by in situ hybridization with a *Ubx* probe, that the shift occurs in the embryo. They have followed the mesoderm that originates in parasegment 6 and have found that by germ band shortening (~9–11 hr), most of the cells seem to move caudally with respect to the epidermis and become the muscles of A1. In addition, they have made the important observation that Ubx^+ seems to be transcribed in parasegmental units. In the mesoderm, the pattern of *Ubx* transcription is simpler than that in the ectoderm; parasegments 6–12 express *Ubx* evenly, the limits of expression coinciding with parasegmental boundaries as defined by the grooves in the mesoderm (Akam and Martinez-Arias 1985). Antennapedia$^+$ is also active in a defined portion of the mesoderm, but it is not clear whether the boundaries of expression are parasegmental or segmental (Martinez-Arias and Lawrence 1985).

In conclusion, it seems that the selector-gene hypothesis could apply to the mesoderm. Like the ectoderm, the mesoderm is divided into lineage units, and at least one selector gene (Ubx^+) is transcribed in a subset of those units. In the simplest view, one might expect that the muscle pattern of any one segment could depend entirely on its genetic address as in the epidermis; however, there are complications that apply specifically to the muscles.

Some Complications

Experiments on embryos have shown that although the ectoderm can develop autonomously, the mesoderm cannot. When the ectoderm is isolated it develops almost normally, showing that patterning does not depend on any information coming from the mesoderm. However, isolated mesoderm fails to form a proper pattern, and when it is moved relative to the ectoderm, the mesoderm's pattern of differentiation is altered appropriately. These experiments have indicated that information passes from ectoderm to mesoderm in the embryo (Bock 1942; Haget 1953).

Later in development, more experiments speak for a dependence of muscle pattern on the ectoderm. The muscle pattern is invariant because muscles attach reliably to the epidermal cells in particular places. These muscle-attachment sites are registered in the epidermal pattern and behave as do other features of that pattern. For example, if a piece of epidermis containing an attachment site is removed, rotated, and reimplanted, the polarity and position of the new site are defined by the altered segmental gradient (Williams and Caveney 1980a), just as the polarity and position of other epidermal landmarks are determined (Locke 1959; Lawrence 1973). Moreover, during development and metamorphosis, muscles change their position and orientation; they do so by "walking" from one epidermal coordinate to another. Working with the beetle *Tenebrio*, Williams and Caveney (1980b) proved this by transplanting a marked graft into the place where at metamorphosis one end of a muscle was expected to move to. After metamorphosis, they found that the muscle was still attached to the host at one end but was now attached to the grafted epidermis at the other end. These experiments show that both early and later in development, the muscles interact with the epidermal cells.

Another important difference between the ectoderm and muscles is that muscles form by the fusion of cells. According to Ho et al. (1983), who studied the developing locust, a muscle is founded by a large "pioneer" mesodermal cell and later smaller myoblasts fuse with it. This observation relates to the observation that when genetically marked myoblasts from the wing disk are implanted into the mature larva of *Drosophila*, they fuse indiscriminately with developing adult muscles. Even when the muscles contain many donor-derived cells, they develop appropriate to their position in the host. For example, the abdominal muscle fibers develop normally when they contain cells from the wing imaginal disk (Lawrence and Brower 1982). One explanation for this might be that the fate of the muscle cells is completely controlled by the associated epidermal cells. Another explanation, which I currently prefer, is that the fusing nuclei become entrained by the cytoplasm of the forming myotube (Lawrence and Johnston 1984b): The myoblasts of the wing disk (T2) do not express Ubx^+ (Lewis 1978; Akam 1983; White and Wilcox 1984; Beachy et al. 1985). Imagine one of these cells fusing with either a pioneer myoblast or a developing myotube in the abdomen, where Ubx^+ is being expressed. Since the amount of cytoplasm injected into the developing myotube by each cell would be relatively small and *if* the state of Ubx activity is regulated by diffusible molecules, the arriving nucleus would soon start transcribing Ubx^+. If, in normal development, the pattern of muscle development were influenced by Ubx^+ and other selector genes, the eventual differentiation of a myotube would depend on the genetic address of the pioneer cell or cells and not on the source of any smaller myoblasts that contribute to it later. Such a model removes the requirement for rigid lineage segregation (which is so important in the epidermis), since it would not matter if small numbers of myoblasts were to wander from one segment to another. Indeed, studies of muscle lineage in the abdomen indicated that some myoblasts do wander from segment to segment (Lawrence and Johnston 1982). We now have evidence for some errant myoblasts even in the thorax: By transplanting genetically marked nuclei that make wild-type succinate dehydrogenase to hosts that make a temperature-sensitive form of the enzyme (Lawrence 1981), mosaics can be made and donor cells can be traced in the adult. Although colonization of muscle compartments is generally all or none, there seem to be a small number of stray cells that fuse with nearby compartments late in development (P.A. Lawrence and P. Johnston, in prep.).

Role of Selector Genes in Muscle Development

The three experiments described below attempt to define the role of selector genes in the muscles themselves. In these studies, it is important to use a muscle that can be recognized out of context. Examples of such muscles would be the indirect flight muscles in T2, which can be recognized by their fibrillar structure, and the abdominal muscle found in A5 of males, which can be easily distinguished from other abdominal muscles (Lawrence and Johnston 1984b).

In all three experiments, the aim is to make mosaics in which muscle cells are mutant for a homeotic gene.

The mutants that were chosen clearly transform part of the fly (cuticle and muscles) to a pattern normally found elsewhere. For example, in the first experiment, we used *Abd-B* (Sanchez-Herrero et al. 1985), a mutation in the bithorax complex that transforms A6 and A7 to A5 and therefore forms three pairs of male muscles (Fig. 1). We wanted to determine whether the development of this muscle depends on the genotype of the epidermis to which the muscle attaches or on the genotype of the muscle cells themselves or on something else. Gynandromorphs were made in which the male tissue was *Abd-B* and the female tissue was wild type. Although male and female parts of the cuticle could be distinguished, other male and female cells could not. Gynandromorphs (183) of this type were made and examined and the results were clear (Lawrence and Johnston 1984b): In A5, differentiation of the muscle was independent of the sex of the epidermis to which it attached; thus, male muscles could attach to female epidermis and vice versa. Moreover, ectopic male muscles were formed in A6 and A7, and these could form when the epidermis was female, showing that the homeotic transformation of muscles from A6 and A7 to A5 also occurred independently of the epidermal cells. So where are the cells that determine the differentiation of the ectopic muscles? Using statistical mapping methods requiring a large sample of gynandromorphs (Garcia-Bellido and Merriam 1969; Hotta and Benzer 1972), these cells were located to a region in each segment that was ventral and near or identical to the expected origin of the muscle cells themselves. The experiment did not prove that the genotype of the muscle cells (whether male and *Abd-B*) was responsible, since other cells (such as primordia of the nervous system) could come from that location (Lawrence and Johnston 1984b).

More recently, we have used another mutation in the bithorax complex called *Mcp* (Lewis 1978) in an attempt to identify these cells. We made mosaics by nuclear transplantation in which the donor nuclei were *Mcp/Mcp* and wild type for succinate dehydrogenase; these nuclei were transplanted from donors at the syncytial blastoderm stage into hosts that carried only a temperature-sensitive form of the enzyme. This method has the advantage that almost every donor-derived cell can be identified, since it will stain blue in the mosaic. Host-derived cells, which are almost invariably in a large majority, do not stain. *Mcp* transforms A4 into A5, giving an ectopic male muscle in A4 (Fig. 1). Among the 111 mosaics, there were only four cases where a muscle formed in A4, and of those, two cases were uninformative, since both the cuticle and the muscle itself were derived from the donor and therefore mutant. In one case, the muscle was donor-derived, but the cuticle was formed by host cells, which illustrates again that the cuticle and muscles can develop independently. In the last case, the muscle that formed in A4 was unstained and therefore came from the host. This is evidence of nonautonomy, since something other than the muscle cells themselves must have induced that muscle in A4. However, it is unwise to build an argument on only one case because, however clear, who knows what unusual combination of circumstances may have contributed to it? Only further experiments can resolve this point.

Other evidence for nonautonomy comes from an analysis of A5 in gynandromorphs produced by nuclear transplantation in the *Mcp* experiment described above. In these gynandromorphs, the donor- and host-derived cells can also be distinguished by their staining for succinate dehydrogenase. Where the donor is female and the host male, there were 11 cases of muscles in A5 and 8 of these were entirely unstained and therefore derived from the host. However, in 3 of these cases, they were stained, which means that the male muscle can be entirely or partially derived from female cells, i.e., cells that would normally never contribute to such a muscle.

These experiments indicate that the development of the male muscle is influenced by other factors apart from the genotype of the muscle cells themselves and confirm that these factors are not determined by the

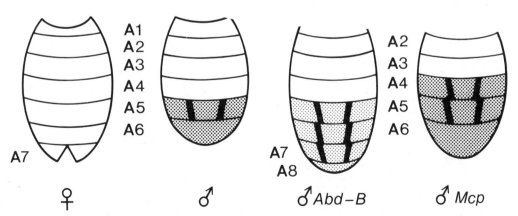

Figure 1. Male muscle in flies of different genotypes. The dorsal abdomen is sketched; the black stripe indicates a muscle that spans abdominal segment A5 in the wild-type male (Lawrence and Johnston 1984b). In *Abd-B* homozygotes, male muscles occur in segments A5, A6, and A7, whereas in *Mcp* homozygous males, the muscles occur in A4 and A5. Stippling indicates the darker pigmentation of the cuticle found in A5 and A6 of wild-type males.

adult epidermis that forms the muscle-attachment sites.

Finally, in another experiment, we have tested the role of the bithorax complex in the muscle cells. Again using nuclear transplantation, we have used a donor cross in which one fourth of the donor embryos were $Dp\ bxd^{100}Df115/Df\ Ubx^{109}$, a combination deficient for a large section of the bithorax complex, including bxd^+ and $abd\text{-}A^+$ (Lewis 1978; Sanchez-Herrero et al. 1985). In the cuticle, the abdominal segments are transformed into parasegment 5, i.e., the P compartment of T2 plus the A compartment of T3. With such a severe transformation, it is not surprising that the majority of experimental mosaics died; however, 6 survived in a total of 79 mosaics. None of these included mesodermal cells in the region anterior to A5 (Fig. 2), and this provides a weak argument in favor of a direct role for the bithorax complex in the abdominal mesoderm: If like engrailed the bithorax complex were not required in muscle cells, then mosaics that colonize the mesoderm should be as common in this experiment as when engrailed-lethal mosaics were made (Lawrence and Johnston 1984a). In the engrailed experiment, 3 flies containing engrailed-lethal cells survived from a total of 28 mosaics. All three contained mesodermal organs that were donor-derived and two of the three colonized the midabdominal mesoderm extensively; yet in the bithorax experiment, not one of the six experimental mosaics did so. In the extreme posterior region of the body (apparently unaffected by the mutant genotype we have used), in five of six experimental mosaics, mesodermal structures such as muscles of the genitalia and the visceral muscles surrounding the hindgut were derived from the donor. None of these included more anterior mesodermal structures. It must be reiterated that the argument is weak, but it does suggest that the bxd^- $abd\text{-}A^-$ mosaics that included areas of mutant mesoderm in the abdomen were lethal.

The results of the three experiments summarized above do not contradict each other but do not provide a clear or simple picture. They and the in situ data (Akam 1983; Akam and Martinez-Arias 1985) make it likely that features of muscle development are determined autonomously by the genetic state of the muscle cells themselves. Williams and Caveney (1980a,b) had proved that the epidermis specifies the attachment sites, and the present experiments show that, in both sexes, several segments of the abdomen (A4–A7) are competent to form sites for a muscle that is normally found only in A5 of males. The single anomalous case in the *Mcp* experiment raises the possibility that the patterning of adult muscles is also influenced by something that is neither the muscle cells themselves nor the epidermal attachments; it could be the arrangement of preexisting larval muscles, it could be the genotype of the innervating neurons or of the most ventral part of the larval epidermis, or even a combination of these. One experiment still remains to be done properly, i.e., the removal of an appropriate selector gene from *only* the muscle cells to determine what the effect on muscle pattern would be.

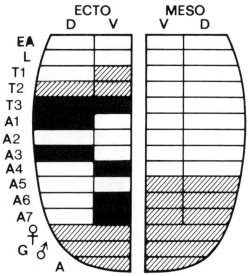

Figure 2. Distribution of marked territory of the *mwh jv Dp bxd^{100} red Df115 e^{11}/mwh jv red Df(3R)Ubx^{109}* genotype in six adult mosaics. (*Left*) Ectoderm; (*right*) mesoderm. Hatched area indicates that the corresponding part was *sdh$^+$* and therefore of the mutant genotype in at least one mosaic; black area indicates that the cuticular pattern was transformed toward parasegment 5. (V) Ventral; (D) dorsal; (EA) head; (L) proboscis; (T1–T3) thoracic segments; (A1–A7) abdominal segments; (G) genitalia; (A) analia. Other parts mutant in the larva were parts of A4, A6, and A7 (transformed to parasegment 5, giving either a T3a/T2p leg or haltere/wing imaginal disks and Keilin's organs in the pupal case). Mutant parts in the adult were fat body in the posterior abdomen, the hindgut, the malpighian tubules (showing the *red* phenotype), the visceral muscle of both the hindgut and posterior regions of the midgut, most of the internal genitalia of both male and female, the posterior region of the heart and a few posterior pericardial cells, the visceral muscle lining the male internal genitalia, part of the posterior midgut, and the germ cells in male and female. All these tissues seemed normal and could not be distinguished from patches in control mosaics. Donor nuclei were placed near the middle of the egg in approximately the position of A1 (Lohs-Schardin et al. 1979). (Results from P.A Lawrence and P. Johnston, unpubl.)

ACKNOWLEDGMENTS

I thank Paul Johnston for much help and my colleagues at the LMB, Michael Akam, Gines Morata, and Gary Struhl, for advice. Alfonso Martinez-Arias and Michael Akam have generously shared their results and ideas with me. Fly stocks were kindly provided by Alain Ghysen, Ed Lewis, Gines Morata, and Gary Struhl.

REFERENCES

Akam, M. 1983. The location of *Ultrabithorax* transcripts in *Drosophila* tissue sections. *EMBO J.* **2:** 2075.

Akam, M. and A. Martinez-Arias. 1985. *Ultrabithorax* expression in *Drosophila* embryos. *EMBO J.* (in press).

Beachy, P.A., S.L. Helfand, and D.S. Hogness. 1985. Segmental distribution of bithorax complex protein during *Drosophila* development. *Nature* **313:** 545.

Bock, E. 1942. Wechselbeziehung zwischen den Keimblattern bei der Organbildung von *Chrysopa perla* (L.). I. Die En-

twicklung des Ektoderms in mesodermdefekten Keimteilen. *Wilhelm Roux's Arch. Dev. Biol.* **141:** 159.
CRICK, F.H.C. and P.A. LAWRENCE. 1975. Compartments and polyclones in insect development. *Science* **189:** 340.
FJOSE, A., W. MCGINNIS, and W.J. GEHRING. 1985. Isolation of a homoeobox-containing gene from the *engrailed* region of *Drosophila* and the spatial distribution of its transcripts. *Nature* **313:** 284.
GARCIA-BELLIDO, A. 1975. Genetic control of wing disc development in *Drosophila*. *Ciba Found. Symp.* **29:** 161.
GARCIA-BELLIDO, A. and J.R. MERRIAM. 1969. Cell lineage of the imaginal discs in *Drosophila* gynandromorphs. *J. Exp. Zool.* **170:** 61.
GARCIA-BELLIDO, A., P.A. LAWRENCE, and G. MORATA. 1979. Compartments in animal development. *Sci. Am.* **241:** 102.
GARCIA-BELLIDO, A., P. RIPOLL, and G. MORATA. 1973. Developmental compartmentalization of the wing disc of *Drosophila*. *Nat. New Biol.* **245:** 251.
HAGET, A. 1953. Analyse experimentale des facteurs de la morphogenèse embryonnaire chez le coleoptère *Leptinotarsa*. *Bull Biol. Fr. Belg.* **87:** 123.
HO, R.K., E.E. BALL, and C.S. GOODMAN. 1983. Muscle pioneers: Large mesodermal cells that erect a scaffold for developing muscles and motoneurones in grasshopper embryos. *Nature* **301:** 66.
HOTTA, Y. and S. BENZER. 1972. Mapping of behaviour in *Drosophila* mosaics. *Nature* **240:** 527.
KAUFMAN, T.C. 1983. The genetic regulation of segmentation in *Drosophila melanogaster*. In *Time, space and pattern in embryonic development* (ed. W.R. Jeffery and R.A. Raff), p. 365. A.R. Liss, New York.
KORNBERG, T. 1981. *engrailed*: A gene controlling compartment and segment formation in *Drosophila*. *Proc. Natl. Acad. Sci.* **78:** 1095.
KORNBERG, T., I. SIDEN, P. O'FARRELL, and P. SIMON. The *engrailed* locus of *Drosophila*: In situ localization of transcripts reveals compartment-specific expression. *Cell* **40:** 45.
LAWRENCE, P.A. 1973. The development of spatial patterns in the integument of insects. In *Development systems: Insects* (ed. S.J. Counce and C.H. Waddington), vol. 2, p. 157. Academic Press, London.

———. 1981. A general cell marker for clonal analysis of *Drosophila* development. *J. Embryol. Exp. Morphol.* **64:** 321.
———. 1982. Cell lineage of the thoracic muscles of *Drosophila*. *Cell* **29:** 493.
LAWRENCE, P.A. and D.L. BROWER. Myoblasts from *Drosophila* wing disks can contribute to developing muscles throughout the fly. *Nature* **295:** 55.
LAWRENCE, P.A. and P. JOHNSTON. 1982. Cell lineage of the *Drosophila* abdomen: The epidermis, oenocytes and ventral muscles. *J. Embryol. Exp. Morphol.* **72:** 197.
———. 1984a. On the role of the *engrailed*[+] gene in the internal organs of *Drosophila*. *EMBO J.* **3:** 2839.
———. 1984b. Genetic specification of pattern in a *Drosophila* muscle. *Cell* **36:** 775.
LEWIS, E.B. 1978. A gene complex controlling segmentation in *Drosophila*. *Nature* **276:** 565.
LOCKE, M. 1959. The cuticular pattern in an insect *Rhodnius prolixus* Stål. *J. Exp. Biol.* **36:** 459.
LOHS-SCHARDIN, M., C. CREMER, and C. NÜSSLEIN-VOLHARD. 1979. A fate map of the larval epidermis of *Drosophila melanogaster*: Localized defects following irradiation of the blastoderm with an ultraviolet laser microbeam. *Dev. Biol.* **73:** 239.
MARTINEZ-ARIAS, A. and P.A. LAWRENCE. 1985. Parasegments and compartments in the *Drosophila* embryo. *Nature* **313:** 639.
MILLER, A. 1950. The internal anatomy and histology of the imago of *Drosophila melanogaster*. In *The biology of Drosophila* (ed. M. Demerec), p. 420. Wiley, New York.
MORATA, G. and P.A LAWRENCE. 1975. Control of compartment development by the *engrailed* gene in *Drosophila*. *Nature* **255:** 614.
SANCHEZ-HERRERO, E., I. VERNOS, R. MARCO, and G. MORATA. 1985. Genetic organisation of the *Drosophila* bithorax complex. *Nature* **313:** 108.
WHITE, R.A.H. and M. WILCOX. 1984. Protein products of the bithorax complex in *Drosophila*. *Cell* **39:** 163.
WILLIAMS, G.J.A. and S. CAVENEY. 1980a. A gradient of morphogenetic information involved in muscle patterning. *J. Embryol. Exp. Morphol.* **58:** 35.
———. 1980b. Changing muscle patterns in a segmental epidermal field. *J. Embryol. Exp. Morphol.* **58:** 13.

The Decapentaplegic Gene Complex in *Drosophila melanogaster*

W.M. Gelbart, V.F. Irish, R.D. St. Johnston, F.M. Hoffmann,* R.K. Blackman,
D. Segal,† L.M. Posakony, and R. Grimaila

*Department of Cellular and Developmental Biology, Harvard University,
Cambridge, Massachusetts 02138-2097*

We have been studying a genetic region termed the decapentaplegic gene complex (DPP-C) in *Drosophila melanogaster*. Our interest in the complex was triggered by the phenotypes engendered by the first series of DPP-C mutations we encountered. These highly pleiotropic decapentaplegic (*dpp*) mutations give rise to defects in the larval imaginal disks. After metamorphosis, these defects are transformed into multiple pattern deletions and duplications in the adult epidermis. In particular, those structures derived from the central regions of most or all the imaginal disks of the animal are missing. On the basis of information from studies of imaginal disk regeneration (Bryant 1978), it seemed that the decapentaplegic phenotypes could be rationalized as effects of *dpp* mutations on the elaboration or interpretation of an element of positional information common to all imaginal disks. Recent studies indicate that the DPP-C is more extensive, with another region being required for normal epidermal development in the embryo. As described in this paper, the total removal of the DPP-C function leads to the complete ventralization of the epidermis in the embryo. One central focus of our work remains the elucidation of the developmental role(s) of the DPP-C.

The other major aim of our work on the complex is to understand the structure and regulation of the functions within the DPP-C. By genetic and phenotypic criteria, the complex can be divided into three regions: shortvein, Haplo-insufficient, and decapentaplegic. Allelic interactions, limited in severity, are seen in some shortvein (*shv*)/*dpp* heterozygous genotypes. These interactions are never as severe as interactions between pairs of *shv* or pairs of *dpp* alleles, indicating some level of functional autonomy of these regions. In contrast, mutations in the Haplo-insufficient region appear to inactivate *shv* and *dpp* functions, suggesting that the entire complex is a single integrated functional unit. The interpretation of allelic interactions has been additionally muddied by the fact that most alleles of *shv* and *dpp* are associated with gross chromosomal rearrangements. Some of the *shv-dpp* interactions can be viewed as polar inactivations of functions in the complex due to rearrangement position effects. We are in the process of dissecting the molecular nature of these interactions.

In this paper, we review and preview several aspects of our work on the DPP-C. We first review the basic phenotypes elicited by mutations in the DPP-C and then briefly summarize our progress in the overall molecular description of the organization of the DPP-C. Finally, we focus on a more detailed description of the *Hin-d* (Haplo-insufficient near decapentaplegic) region, which has only been tangentially described in previous reports. We present our preliminary information on the structure and transcriptional activities within the *Hin-d* region and then consider how these observations help to define the next set of questions to be addressed in attempting to understand the contribution of DPP-C to development.

RESULTS

Recovery and General Characteristics of DPP-C Mutations

We have generated over 100 lesions in the DPP-C. With the exception of a very few alleles generated in F_2 lethal screens, all induced mutations in the complex have been recovered among the F_1 of mutagenized individuals by virtue of an effect on adult morphology. These mutations are recessive and are allelic to at least one of three tester mutations, conferring either abnormally oriented (heldout) wings (*dpp*^{ho}; Spencer et al. 1982), wings lacking normal venation patterns (shortvein) (*shv*; Segal and Gelbart 1985), or smaller than normal (blink) eyes (*dpp*^{blk}; kindly provided by J. Sparrow). Because their recovery depended on the inactivation of a function within the imaginal disk-specific *dpp* region of the complex, alleles identified on the basis of heldout wings or blink eyes have been termed *dpp* mutations. Similarly, alleles recovered because of a shortvein phenotype have been classified as *shv* mutations. In addition to the previously reported types of recessive mutations falling into the *shv* and *dpp* regions of the complex, our more recent procedures have permitted the efficient recovery of Haplo-lethal alleles, called *Hin-d* mutations. These *Hin-d* mutations are described in more detail below.

Present addresses: *McArdle Laboratory, University of Wisconsin, Madison, Wisconsin 53706; †Department of Neurobiology, Weismann Institute, Rehoveth, Israel.

Mutant Phenotypes in the DPP-C

Mutations in the DPP-C fall into four major phenotypic categories. Three of these (shortvein, embryonic lethal, and decapentaplegic) have been described in previous reports; here, we describe the properties of the fourth class (Haplo-lethal alleles) for the first time.

shv **alleles.** These alleles engender wing venation defects. Most also lead to defects in the maxillary palps and vibrissae. For most *shv* mutations, shv^x/shv^y heteroallelic combinations are inviable, and in the three cases that have been examined carefully, death occurs after an extended larval period; dying animals remain the size of newly hatched first-instar larvae (Segal and Gelbart 1985).

Imaginal disk-specific **dpp** *alleles.* These mutations are characterized by imaginal disk defects, in which pattern elements derived from the centers of the imaginal disks are missing and somewhat more peripherally derived pattern elements are duplicated. Previously, we had separated these mutations into five phenotypic classes, of which *dpp*-I (heldout wings) is the mildest and *dpp*-V is the most severe. All of these mutations are allelic to heldout. Mutations in a more severe class have all of the phenotypic defects of the next milder class and lack additional pattern elements as well. It is as if in each more extreme mutant class, a larger core of each affected imaginal disk is being removed. The alleles in classes *dpp*-I through *dpp*-III are adult-viable, whereas alleles in classes *dpp*-IV and *dpp*-V are pupal-lethal. *dpp*-III through *dpp*-V alleles appear to affect all major disks, whereas *dpp*-II alleles only affect the wing, halter, genital, and, to a small degree, eye disks. The blink (*blk*) mutation is novel in that it displays a phenotype that does not readily fit in any of these classes. Blink has the eye defect of a strong *dpp*-III allele but is otherwise wild type. Alleles generated using blink tester chromosomes are of the standard *dpp*-III and *dpp*-V types (and *Hin-d* mutations are recovered as well). Heterozygotes for two mutations of different decapentaplegic classes generally have a phenotype close to that of the milder allele (Spencer et al. 1982).

Embryonic-lethal **EL** *alleles.* Most of these alleles were recovered in F_2 recessive-lethal screens (for a description of one such screen, see Spencer et al. 1982). Most of these alleles behave as if they are leaky versions of the *Hin-d* lesion (see below). As their name implies, homozygotes bearing these mutations die during embryogenesis. *EL/EL* embryos exhibit defects in dorsally derived cephalic and caudal structures of the developing larvae. From analyzing heterozygotes of these mutations with mutations of other classes, we can infer that *EL* mutations appear to be partially inactive for all functions within the DPP-C.

We first encountered transvection effects at decapentaplegic in the synapsis-dependent complementation of EL/dpp^{ho} genotypes (Gelbart 1982). Complementation is seen in such genotypes under conditions in which synapsis of DPP-C homologs can occur, but noncomplementation results when structural heterozygosity obviates synapsis of distal chromosome arm 2L. Transvection effects have also been noted in heterozygotes for *EL* and for the other cytologically normal alleles in other parts of the DPP-C (Gelbart and Wu 1982).

Haplo-lethal **(Hin-d)** *alleles.* The existence of a Haplo-lethal region just distal to *dpp* was inferred from our inability to recover adults monosomic for this region (Spencer et al. 1982; Segal and Gelbart 1985). However, we were not able to follow up on this inference immediately by determining if there was a class of DPP-C mutations that were themselves Haplo-insufficient (i.e., *Hin-d*$^-$) or if the *Hin-d* gene was functionally unrelated to the complex. To make such a determination, we needed to recover *Hin-d*$^-$ mutations that were not gross deletions.

Our strategy was to search for mutations allelic to the dpp^{ho} or dpp^{blk} mutations, but utilizing tester genotypes in which a trisomic copy of *Hin-d*$^+$ was present. This extra copy was defective in *dpp* function so that it would not obscure the heldout or blink phenotypes of newly induced recessive mutations in the mutagenized homolog. This extra copy then compensated for the loss of function of the new *Hin-d*$^-$ mutation. This strategy proved successful and generated several *Hin-d*$^-$ mutations that were recovered as *ho* or *blk* allelic lesions.

In some variations of this paradigm, we included heterozygous rearrangements known to disrupt transvection at *dpp* (Gelbart 1982) in the tester genotypes. As it turned out, most of the cytologically normal *Hin-d*$^-$ alleles would not have been recovered were it not for the presence of such rearrangements. The heldout phenotype is elicited by such genotypes (e.g., $Hin\text{-}d^-Dp[2;2]Hin\text{-}d^+/dpp^{ho}$) only when apposition of homologous DPP-C regions is prevented by the presence of an appropriate heterozygous rearrangement involving chromosome arm 2L.

By definition, the *Hin-d*$^-$ mutations are inviable as *Hin-d*$^-$/*Hin-d*$^+$ heterozygotes but are viable in the presence of an extra dose of *Hin-d*$^+$. Therefore, the dominant lethality of these mutations is due to the insufficiency of one copy of *Hin-d*$^+$ to fulfill its role. Insofar as we have been able to ascertain, by using *Hin-d*$^+$ duplications deleted for other parts of the DPP-C, and by analyzing rare *Hin-d*$^-$/*Hin-d*$^+$ escapers, *Hin-d*$^-$ alleles are inactivated for all functions within the complex. In some cases, the inactivation is transvection-sensitive; i.e., it is only apparent when synapsis of DPP-C homologs is prevented.

Both *Hin-d*$^-$/*Hin-d*$^+$ and *Hin-d*$^-$/*Hin-d*$^-$ animals die during embryogenesis. *Hin-d*$^-$/*Hin-d*$^+$ embryos are very similar to *EL/EL* embryos, with defects in dorsally derived cephalic and caudal structures of the developing larvae. The phenotype of *Hin-d*$^-$/*Hin-d*$^-$ homozygotes (or heterozygotes for two different *Hin-d*$^-$ alleles) is characterized by a spectacular loss of dorsal pattern elements (compare the wild-type embryo in Fig. 1 with the *Hin-d*$^-$/*Hin-d*$^-$ embryo in Fig.

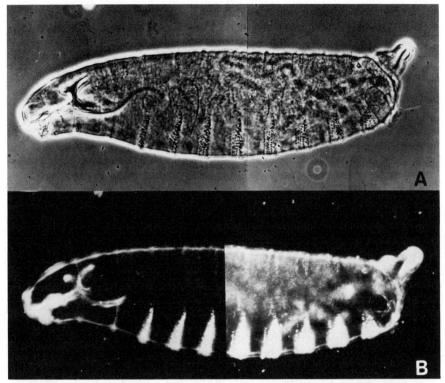

Figure 1. Phase-contrast (*A*) and dark-field (*B*) photomicrographs of the cuticle of a mature wild-type embryo. (*Left*) Anterior; (*bottom*) ventral. Note the presence of the thick denticle belts on the ventral side of each thoracic and abdominal segment.

2). These embryos display a "ventralization" of the normal body plan, with the denticle belts, normally ventral epidermal structures, completely enwrapping the embryo. Embryos of this genotype gastrulate abnormally. They are defective by 4 hours of development, displaying defects in germ band extension and in the cephalic furrow.

Organization of the DPP-C

The number of mutations in each class within the complex are summarized in Table 1. Within the imaginal disk-specific *dpp*-I through *dpp*-V region, the bulk of the mutant alleles are of the *dpp*-III and the *dpp*-V types. Classes I and IV are represented only by one allele each, leading us to question whether it is proper to put them into separate classifications.

Delineation of the genetic organization of the DPP-C has been facilitated by the cytogenetic properties of most mutations in the complex (Table 1). Virtually all of the mutations of the decapentaplegic (imaginal disk-specific) and shortvein types are associated with gross chromosomal rearrangements involving the polytene doublet 22F1-2. Several of these alleles are rearranged in a manner such that it has been possible to use them to generate synthetic deletions or duplications for individual portions of the complex. By utilizing such rearrangements (Spencer et al. 1982; Segal and Gelbart 1985), we have been able to infer that the order of functions within the complex is (telomere of chromosome

Table 1. Distribution of DPP-C Mutations According to Their Cytogenetic Properties

	Number of mutant alleles		
Mutational class	cytologically normal	cytologically abnormal	
		deletions	rearrangements
dpp-I	1	0	0
dpp-II	1	0	4[a]
dpp-III	2[b]	3	27[a]
dpp-IV	0	0	1
dpp-V	0	3	18
blk	1	0	0
EL	4	0	1[c]
Hin-d	6	12[d]	2
shv	7[e]	0	22
Total	22	18	75

The cytological definition of mutant alleles is determined by analysis of polytene chromosomes in larval salivary gland squashes of mutant/wild-type heterozygotes. The "rearrangement" category includes inversions, translocations, and transpositions having a breakpoint in 22F1-2. The "deletions" either all have one deletion endpoint within 22F1-2 or their endpoints span 22F1-2, completely removing this polytene doublet.

[a]One of these alleles has been lost.

[b]One allele from this group behaves genetically as a deletion: In addition to eliminating class I through III activities, it is also inactivated for *1(2)ND1*, the nearest vital gene proximal to DPP-C.

[c]Classification of this allele as a member of the embryonic-lethal (*EL*) class is tentative.

[d]The 12 deletions remove most, if not all, sequences of DPP-C.

[e]Four of these mutations, selected on the basis of allelism to an adult viable shortvein mutant are tentatively termed *shv* alleles. These alleles are inviable in heterozygotes with *EL* alleles, and three of them display mutant interactions with some *dpp* alleles. The possibility exists that they are misclassified and should be more appropriately considered as *EL* alleles.

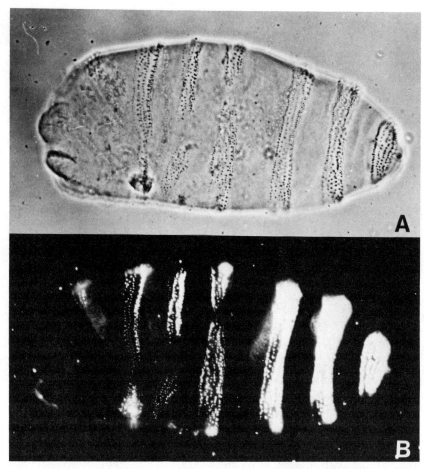

Figure 2. Phase-contrast (*A*) and dark-field (*B*) photomicrographs of a mature *Hin-d⁻/Hin-d⁻* embryo (*dpp³⁷/Df[2L]DTD2*). As in Fig. 1, anterior is to the left and ventral is down in each frame. Note that the normally ventral denticle belts wrap around many of the abdominal segments. Although cephalic structures are highly disrupted in these embryos, it has not yet been possible to describe these defects in detail.

arm 2L)−*shv*−*Hin-d*−V−III−I−(centromere of chromosome 2).

These rearrangements have also aided greatly in the molecular analysis of the DPP-C. We have cloned the DPP-C by the method of "transposon tagging" utilizing a strain bearing a copy of the F element (for a description of this transposon, see Di Nocera et al. 1983) in polytene band 22F1-2. A genomic library prepared from this strain, when probed with a subclone containing that mobile element, yielded a bacteriophage λ insertion, in which the mobile element resided next to single-copy DNA derived from 22F1-2. This single-copy DNA was used to initiate a 140-kbp walk within 22F1-2. The walk was terminated when it was clear that we had encountered breakpoints of mutations spanning the complex. (A description of the chromosome walk and the detailed structure of the DPP-C will be presented elsewhere [F.M. Hoffmann et al., in prep.].)

By using a combination of polytene chromosome in situ hybridization analysis and whole-genome Southern blotting, we have begun to determine the positions of rearrangement breakpoints associated with mutational disruption of the DPP-C. A summary of our progress in this mapping is presented in Table 2. The cogent points that derive from this analysis are that (1) the entire complex appears to reside in approximately 40–45 kbp of DNA (from coordinates 68–74 through 112–114 on our molecular map), (2) the mutations cluster according to phenotype, and (3) the clusters fall in the order predicted from our genetic analysis. The possibility exists that a clean separation of mutant classes will not occur in the class II through IV region.

We have begun to analyze the cytologically normal mutations in the *dpp* region of the complex. Of the five cytologically normal alleles we have identified (*dpp^ho^*, *dpp^blk^*, one *dpp*-II allele, and two *dpp*-III alleles), we have initiated molecular mapping of four of these. We have not yet examined one of the *dpp*-III alleles. Indeed, all four of these mutations are associated with altered restriction maps in the anticipated regions of the complex. The *dpp*-II allele has a breakpoint within the 109.5–111.0 region, whereas the *dpp*-III allele is a deletion beginning at approximately 101–102 and deleting all DPP-C sequences proximal to it. Interestingly, the only two disk-specific mutations, *dpp^ho^* and *dpp^blk^*, are associated with deletions of approximately 2.8 and 5.2 kbp of DNA, respectively. The *dpp^blk^* deletion is adjacent to the insertion of a 1.6-kbp mobile element se-

Table 2. Distribution of the DPP-C Mutations on the Molecular Map of the Region

Type of mutation	Number of alleles mapped	Maximum limits to the locations of mutant breakpoints
shv	13	68–82.5
Hin-d	5	83.5–89
dpp-V	7	91.5–99
dpp-IV	1	99.5–100.5
dpp-III	13	99.5–110
dpp-II	3	109.5–113
dpp-I	1	111.5–114.5

The molecular map of the *dpp* region includes a chromosome walk of about 140 kbp of DNA. The distal end of the walk has arbitrarily been set at position 0 and the proximal end is position 140 (in units of kilobase pairs). The locations of the mutant breakpoints are given in kilobase pair coordinates on this standard DNA map of the region. Although at present we cannot rule out the possibility of a small amount of overlap of breakpoint positions in mutations of different classes, it is quite possible that these potential overlaps will disappear as we refine the mapping of breakpoints near the boundaries of the several mutant classes.

quence and resides in the 106–111 region of our map, whereas dpp^{ho} deletes most of the 111.5–114.5 interval. Thus, within the imaginal disk-specific *dpp* region, only one candidate remains (by virtue of not having been examined yet) for a true point mutation.

Similarly, within the *shv* region, most alleles are associated with gross chromosomal rearrangements. The mildest *shv* alleles appear to be those furthest from the *Hin-d* region. We have not yet initiated a molecular analysis of the cytologically normal *shv* alleles.

Molecular Organization of the *Hin-d* Region

Because it is associated with the earliest and most severe mutational disruptions within the complex, we have focused particular attention on the organization of the *Hin-d* region. Only our most recent mutational screens have permitted identification of *Hin-d*⁻ alleles, and thus we have fewer representatives of this class than of *shv* or *dpp*. The spectrum of *Hin-d* mutations differs in its cytogenetic properties from those of *shv* and *dpp* lesions. As discussed above, the vast majority of *dpp* and *shv* mutations are associated with translocation or inversion breakpoints in 22F1-2. Most of the Haplo-lethal mutations turn out to be deletions of the entire DPP-C. Of the remainder, only two are associated with gross chromosomal rearrangements. Even the cytologically normal *Hin-d*⁻ mutations appear to inactivate the other activities of the complex.

Our present state of information concerning the *Hin-d* region is summarized in Figure 3 (the data on which this figure is based will be presented elsewhere). This region has been delimited to the 82–91 interval on the basis of the locations of Haplo-sufficient *shv* breakpoints (shv^{S2} and shv^{S18}) on the distal side and a *dpp* breakpoint (dpp^{24}) proximal to *Hin-d*. Within this region, alterations in the restriction maps of five γ-ray-induced *Hin-d* mutations have been identified. Three of these mutations (dpp^{46}, dpp^{47}, and dpp^{61}) are small deletions, and two are associated with rearrangement breakpoints in the *Hin-d* region (dpp^{37} and dpp^{45}). One other *Hin-d*⁻ mutation, dpp^{48}, shows no restriction map alterations.

Transcriptional Activities within the *Hin-d* Region

Our initial studies of transcripts within the DPP-C have been aimed at defining the *Hin-d*⁺ transcriptional product(s). At present, the transcriptional complexity of this region has prevented assignment of a *Hin-d*⁺ role to a particular transcript. An example of a developmental Northern blot, using several staged poly(A)⁺ RNA preparations, is presented in Figure 4. Such blots have identified three sizes of transcripts: 3.5, 4.0, and 4.5 kb. Using single-stranded M13 probes, we have determined that the three transcripts are each homolo-

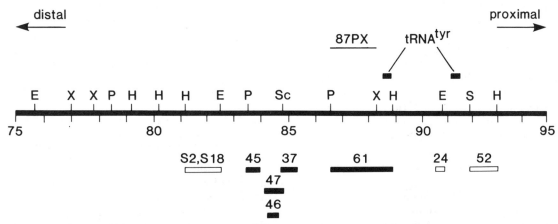

Figure 3. A map of the Haplo-lethal, or *Hin-d*, region of the DPP-C. Coordinates 75 through 95 indicate the positions of this region (in kbp) on our 140-kbp chromosomal walk. Arrow marked distal points in the direction of the telomere of chromosome arm 2L, and arrow marked proximal points toward the centromere of chromosome 2. Approximate positions of the two tRNA^{Tyr} sequences are indicated. The line marked 87PX indicates the extent of the insertion used to probe the Northern blot in Fig. 4. The restriction map indicates the positions of the following restriction-enzyme-sensitive hexanucleotide sequences: (E) *Eco*RI; (H) *Hin*dIII; (P) *Pst*I; (S) *Sal*I; (Sc) *Sac*I; (X) *Xho*I. Below the line, the positions of Haplo-lethal (■) and Haplo-viable (□) mutations in the region are shown. (See text for a description of these mutations.)

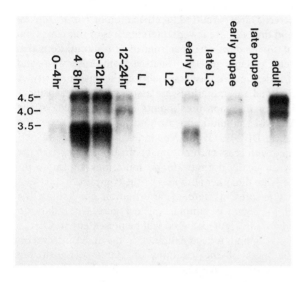

Figure 4. Developmental Northern blot identifying poly(A)+ transcripts homologous to a Haplo-lethal region probe. The RNAs in this blot were prepared from an isogenic *dp cn bw* strain (wild type for the DPP-C) from four embryonic time points (0–4 hr, 4–8 hr, 8–12 hr, and 12–24 hr), from four larval stages (L1, first-instar larva; L2, second-instar larva; early L3, first day of the third larval instar; and late L3, second day of the third larval instar), from the first two days of pupation (early pupae), and from the last two days of this stage (late pupae), as well as from adults. Note the developmental profiles of the three RNA size fractions identified (3.5, 4.0, and 4.5 kb in length) with a single-strand probe of the 87PX fragment (Fig. 3) cloned in M13. This probe identifies transcripts running 5′ to 3′ in the distal-to-proximal direction on the map in Fig. 3. (See text for further discussion of transcripts in the Haplo-lethal region.)

gous to sequences contained in the 84–85 interval and in the 87–89 interval. They are each transcribed from distal (5′) to proximal (3′) on the map in Figure 3. Consistent with our preliminary mapping of embryonic cDNAs (data not shown), the exons in the 84–85 and 87–89 intervals appear to represent the more 3′ exons of these RNAs. Although our evidence is incomplete, we have no indications that any of these transcripts contain DPP-C sequences proximal to 89. Probes from the 76–81 region show differential hybridization to the three transcript size classes. These studies are not yet complete, but they already suggest that the transcripts differ in splicing toward the 5′ ends of the messages and/or are initiated from different promoters. We have not yet been able to determine whether these transcript sizes represent individual species of RNAs or whether one or more of them are composites of multiple RNAs indistinguishable in size.

The different transcripts vary in their developmental profiles. The 3.5- and 4.5-kb transcripts peak early in embryonic development (4–8 hr), whereas the embryonic peak of the 4.0-kb transcript is later (12–24 hr). The 3.5-kb transcript reappears in early third-instar larvae, 4.0-kb RNA reappears in early pupae and may be present in adults, and 4.5-kb RNA weakly reappears in early third-instar larvae and in early pupae and is relatively abundant in adults. Preliminary evidence suggests that the adult RNA is primarily ovarian.

Finally, two tRNA^{Tyr} genes reside in the vicinity of the *Hin-d* region. Our search for these sequences was prompted by the observations of Dudler et al. (1981) and Hayashi et al. (1982) that radiolabeled tRNA^{Tyr} showed in situ hybridization to 22F1-2, the decapentaplegic polytene chromosome doublet, as well as to several other sites in the genome. At present, our evidence points to two and only two sites of tRNA hybridization within our 140-kbp chromosome walk. This has been determined by Southern blots of restriction-enzyme-digested cloned DNA sequences from the walk, which were probed in one set of experiments with purified tRNA^{Tyr} (kindly proved by E. Kubli) and probed in another set with a mixed *Drosophila* tRNA tracer (kindly provided by S. Sharp and D. Soll). These sites (Fig. 3) fall in the interval between the *Hin-d* region and the *dpp* region (one between coordinates 88 and 89 and the other between 91 and 92). Restriction fragments containing the entire tRNA^{Tyr}-homologous regions have been sequenced (B. Suter and E. Kubli, pers. comm.), and they completely match the known tRNA^{Tyr} sequence. However, since there are approximately 25 copies of tRNA^{Tyr} per haploid genome, we were unable to determine if the copies in the DPP-C are transcriptionally active.

With the possible exception of dpp^{61}, all of our *Hin-d* mutations fall distal to the more distal of the two tRNA^{Tyr} genes. The more proximal tRNA^{Tyr} gene is clearly distal to the most distal *dpp*-V allele (dpp^{52}). The only mutation that falls between the two tRNA^{Tyr} genes is dpp^{24}, a rearrangement-associated allele with novel properties. It behaves as a *dpp*-V allele except in tests with *EL* mutations. Unlike other *dpp*-V alleles, dpp^{24}/*EL* animals die as embryos or larvae. (Other genotypes of *dpp*-V/*EL* are characterized by pupal lethality.) Hence, we have tentatively classified dpp^{24} as the only rearrangement-associated *EL* allele. Although we are not yet in a position to determine the contribution, if any, of its tRNA^{Tyr} genes to DPP-C activities, we find it curious that they seem to reside at a boundary, with quite different phenotypes being conferred by mutations distal to, proximal to, or in between the two tRNA^{Tyr} genes.

DISCUSSION

Organization and Regulation of the DPP-C

The molecular analysis of mutations in the DPP-C has been of great help in clarifying our picture of the complex. On the basis of both the transcriptional activities within it and our ability to generate point and pseudopoint mutations within it, we anticipate that the Haplo-lethal region will encode one or more protein products from its several poly(A)+ transcripts. It is likely that many or all of the *shv* rearrangements will interrupt one or more of these transcripts, but it is curious that so few *shv* alleles are potential point mutations. One attractive and testable possibility is that the region defined by *shv* breakpoints does not have protein-coding capacity.

The striking lack of point mutations in the *dpp* region argues that very little, if any, of this region is protein-coding. Of the 60 mutations in the *dpp* region that have been surveyed, all are associated with restriction-map-visible alterations. Most of these are gross chromosomal rearrangements that fragment the complex into two disconnected pieces. Although the fragment containing the DPP-C material distal to the breakpoint appears to remain functional, we always find that the DPP-C material in the proximal fragment is inactivated (at the phenotypic level). Indeed, it is likely that mutations without such polar effects could not have been recovered in our F_1 screens. Our data are consistent with the possibility that the *dpp* region is composed of a number of *cis*-acting elements that regulate the expression of DPP-C genes distal to them.

We have not demonstrated that the small deletions present in the *dpp^{ho}* and *dpp^{blk}* strains cause the respective heldout and blink mutant phenotypes. However, we have not encountered such deletions in wild-type strains, and thus we expect that these deletions are responsible for the mutant phenotypes. We speculate that these deletions remove *cis*-acting sequences required for expression of some chromosomally more distal DPP-C product in the wing and eye, respectively. Thus, all DPP-C breakpoints distal to the location of the *dpp^{ho}*-associated deletion separate the *cis*-regulatory element necessary to ensure normal wing posture development from the function it controls. Similarly, all breakpoints distal to the site of the *dpp^{blk}*-associated deletions separate the *cis*-regulatory element necessary for normal eye development from the function it controls. Consistent with this is the observation that all *dpp* breakpoints distal to 106 have severe eye defects, whereas those proximal to 107 have normal or near-normal eyes (data not shown).

If these speculations are correct, then we must begin asking if the remainder of the *dpp* region is a cluster of such *cis*-regulatory elements. Furthermore, we will have to determine the location and nature of the function(s) these elements control. Are they regulating the transcripts in the Haplo-lethal and *shv* region? Or are they regulating the expression of the tRNA^{Tyr} genes? Finally, are there protein-coding transcripts in the *dpp* region itself that are being regulated? In addition to understanding the *cis*-regulation of the DPP-C, hopefully, a determination of the contribution of the *dpp* region will aid in unraveling the mysteries of transvection.

Developmental Contributions of the DPP-C

We are unable at this time to pinpoint the developmental role(s) of the DPP-C. Both its pattern-specific effects on imaginal disk development and the dramatic ventralization phenotype observed in Hin-d$^-$/Hin-d$^-$ embryos, which is unique among zygotic mutations (C. Nüsslein-Volhard and K. Anderson, pers. comm.), lead us to speculate that the DPP-C will play an important role in morphogenesis or growth regulation in developing epidermal tissues. We do not as yet have evidence of any nonepidermal tissues that are affected by mutations in the DPP-C. Indeed, we have found (Spencer 1984) that the imaginal disk phenotypes of *dpp* mutations are disk-autonomous. How the imaginal and embryonic activities within the complex relate to one another remains to be determined.

ACKNOWLEDGMENTS

We are grateful for the technical help of Lee Kerkhof, Mark van Doren, Pablo Figueroa, and Ruth Catavalo. This work was supported by an award to W.M.G. from the National Institutes of Health. F.M.H. was supported by postdoctoral fellowships from the National Institutes of Health and from the Massachusetts Medical Society. R.K.B. is supported by a postdoctoral fellowship from the National Institutes of Health. D.S was supported by a postdoctoral fellowship from the Massachusetts branch of the American Cancer Society. V.F.I., L.M.P., and R.G. have been supported by predoctoral training grants from the National Institutes of Health.

REFERENCES

BRYANT, P.J. 1978. Pattern formation in imaginal discs. In *The genetics and biology of* Drosophila (ed. M. Ashburner and T.R.F. Wright), vol. 2C, p. 229. Academic Press, London.

DI NOCERA, P.P., M.E. DIGAN, and I.B. DAWID. 1983. A family of oligo-adenylate-terminated transposable sequences in *Drosophila melanogaster*. *J. Mol. Biol.* **168:** 715.

DUDLER, R., T. SCHMIDT, M. BIENZ, and E. KUBLI. 1981. The genes coding for tRNA^{Tyr} of *Drosophila melanogaster*: Localization and determination of the gene numbers. *Chromosoma* **84:** 49.

GELBART, W.M. 1982. Synapsis-dependent allelic complementation at the decapentaplegic gene complex in *Drosophila melanogaster*. *Proc. Natl. Acad. Sci.* **79:** 2636.

GELBART, W.M. and C.-T. WU. 1982. Interactions of zeste mutations with loci exhibiting transvection effects in *Drosophila melanogaster*. *Genetics* **102:** 179.

HAYASHI, S., I.C. GILLIAM, T.A. GRIGLIATTI, and G.M. TENER. 1982. Localization of tRNA genes of *Drosophila melanogaster* by in situ hybridization. *Chromosoma* **86:** 279.

SEGAL, D. and W.M. GELBART. 1985. Shortvein: A new component of the decapentaplegic gene complex in *Drosophila melanogaster*. *Genetics* **109:** 119.

SPENCER, F.A. 1984. "The decapentaplegic gene complex and adult pattern formation in *Drosophila*." Ph.D. thesis. Harvard University, Cambridge, Massachusetts.

SPENCER, F.A., F.M. HOFFMANN, and W.M. GELBART. 1982. Decapentaplegic: A gene complex affecting morphogenesis in *Drosophila melanogaster*. *Cell* **28:** 451.

Isolation and Structural Analysis of the extra sex combs Gene of *Drosophila*

E. Frei, D. Bopp, M. Burri, S. Baumgartner, J.-E. Edström,* and M. Noll
*Department of Cell Biology, Biocenter of the University, CH-4056 Basel, Switzerland; *European Molecular Biology Laboratory, D-6900 Heidelberg, Federal Republic of Germany*

During *Drosophila* embryogenesis, the developmental pathways of the abdominal, thoracic, and the posterior head segments are determined by genes of the bithorax complex (BX-C) (Lewis 1978; Sánchez-Herrero et al. 1985) and Antennapedia complex (ANT-C) (Kaufman et al. 1980; Lewis et al. 1980a,b). Whereas each of these genes determines the identity of only a single or a few neighboring compartments, other genes affect the identity of all or most segments (Lewis 1978; Struhl 1981; Duncan 1982; Duncan and Lewis 1982; Ingham 1983, 1984; G. Jürgens, pers. comm.). Genes of this second type appear to play a central role in the overall regulation of the activity of the gene products of the BX-C and ANT-C genes, as well as of genes determining the head segments. Therefore, to understand the regulatory circuits active during early development, it seems important to identify the molecular targets of these genes and gene products. One of these genes is the extra sex combs (*esc*) gene. Its importance in determining segment identity was first demonstrated in a series of elegant papers by Struhl (1981, 1983; Struhl and Brower 1982). The *esc* gene shows a maternal effect: Homozygous *esc*$^-$ embryos from heterozygous *esc*$^+$/*esc*$^-$ mothers develop normally, whereas *esc*$^-$ embryos derived from homozygous *esc*$^-$ mothers die as late embryos, in which all segments exhibit a pattern resembling that of the normal eighth abdominal segment (Struhl 1981). The gene product might thus be a regulatory protein that activates or represses genes whose products are essential during development in the determination of segment identity.

We have recently cloned a large region containing the *esc* gene. Subsequently, the sequences that harbor the gene and are responsible for its proper expression in the female germ line have been limited to about 12 kb (Frei et al. 1985). Here, we summarize our strategy to clone and identify the gene; such a technique for identifying genes whose patterns of temporal or spatial expression are known is of potential wide application. In addition, we identify the *esc* transcript and present data on the structure of the gene.

METHODS

Most experimental procedures were carried out according to standard methods (Maniatis et al. 1982) or have been described in a recent publication (Frei et al. 1985). A description of the expression vector pTRB0 will be published elsewhere (T.R. Bürglin, in prep.).

Northern analysis of *esc* transcripts after electrophoresis of the RNA in agarose gels containing formaldehyde (Lehrach et al. 1977) and transfer to nitrocellulose filters (Southern 1975) have been described previously. Preparation (Ullmann 1984) and analysis of β-Gal-*esc* fusion proteins by SDS-PAGE (Laemmli 1970) again followed standard procedures.

RESULTS

Cloning of the *esc*^{10} Deficiency

Since the only cytologically visible mutation of *esc* was a deletion (*Df(2L)esc*^{10}) removing the chromosomal band 33B1,2 on the left arm of the second chromosome (Struhl 1981), we decided first to clone the DNA deleted in the *esc*^{10} mutant by microdissection and subsequent chromosomal walking. As shown in Figure 1 (small horizontal bars), the region cloned by screening a *Drosophila* library with the subcloned microdissected DNA essentially covered the proximal part of the *esc*^{10} deletion and two small regions to the left and right of the deficiency. The location and orientation of the cloned DNA were determined by in situ hybridizations to polytene chromosomes (not shown) of *CyO esc*^{2}/*Df(2L)esc*^{10} *Dp(2;2)GYL* larvae that carried, on the right arm of the second chromosome, a large duplication (*Dp(2;2)GYL*) of a region containing the *esc*$^+$ gene (Ashburner 1982; Yannopoulos et al. 1982). Chromosomal walking soon united the DNA clones within the deficiency *esc*^{10} with those to the right of it. However, two long marches from either side (F and C) were required to fill in the 400-kb gap between their starting points (Fig. 1). Mapping the proximal and distal breakpoints of *Dfesc*^{10} by whole-genome Southern analysis and by in situ hybridization to salivary gland chromosomes (not shown) revealed that 380 kb had been deleted in *esc*^{10} (Fig. 1). This appeared to be a rather large amount of DNA for the single chromosomal band 33B1,2 (Struhl 1981). Yet, upon close inspection of *Dfesc*^{10} polytene chromosomes, we found that in addition to 33B1,2, a large portion of 33A was also missing.

Reduction of the Region Containing *esc* by *Df(2L)prd*$^{1.7.20}$ and *Dp(2;2)GYL*

To narrow down the region containing *esc*, we analyzed two mutants by whole-genome Southern analysis and by in situ hybridization to polytene chromosomes

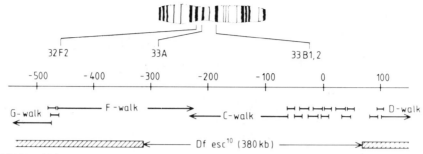

Figure 1. Cloned DNA within polytene chromosome segments 32F to 33B. About 700 kb of cloned DNA from chromosomal bands 32F and 33B are shown. Only those bands that were easily visible in a light microscope are indicated in the polytene chromosome segments drawn schematically at the top. Cloned DNA of recombinant phages isolated from the Canton S library by screening with microdissected DNA is represented by horizontal lines, whereas the regions cloned by, and the directions of, chromosomal walking are indicated by arrows (scale refers to DNA lengths in kb). At the bottom, the breakpoints and the extent of the deletion Df(2L)esc[10] are shown. (Reprinted, with permission, from Frei et al. 1985.)

(not shown). One carries a deficiency of the paired gene, Df(2L)prd[1.7.20], located to the right of esc (C. Nüsslein-Volhard, pers. comm.) and hence sets a proximal limit to the region comprising esc (Fig. 2). The other is the Dp(2;2)GYL mutant (see above) that carries a functional esc+ gene. The distal breakpoint of this duplication is located about 200 kb to the right of the left breakpoint of Dfesc[10] (Fig. 2). Therefore, the distal boundary of the region that must contain esc is moved considerably to the right and thus reduces this region to about 160 kb (Fig. 2).

Screening for esc by Differential Transcript Mapping

Because whole-genome Southern analysis did not reveal any convincing differences between wild-type DNA and a number of esc⁻ mutants, another criterion had to be applied to reduce the 160 kb containing esc even further. The known developmental profile of esc expression provides one such criterion. Since esc exhibits a maternal effect (Struhl 1981), one expects to find esc transcripts in follicles. On the other hand, the late requirement for esc in the meso- and metathoracic leg disks probably ends during late or early pupal life (Tokunaga and Stern 1965). Since females transcribe esc in their ovaries, esc transcripts were not expected to be found in adult males. Hence, we compared the patterns of transcripts in follicles and adult males hybridizing within the entire 160-kb region. "Differential transcript mapping" was performed by hybridizing radioactive cDNA of poly(A)+ RNA from follicles or adult males to identical Southern blots of EcoRI digests of

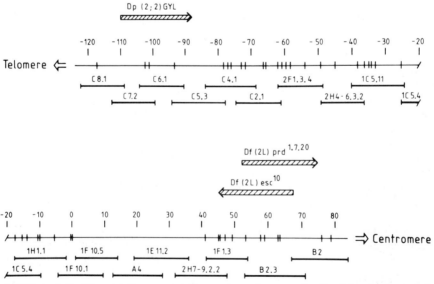

Figure 2. Cloned DNA and EcoRI restriction map at the esc locus between the distal breakpoints of Dp(2;2)GYL and Df(2L)prd[1.7.20]. A map of EcoRI sites (scale in kb) and the locations of corresponding cloned DNA segments of recombinant phages isolated from the Canton S library (horizontal lines) are shown. The coordinates and the vertical marks indicate the distances of EcoRI sites from the distal end of the 41-kb EcoRI fragment. The names of recombinant phages isolated by screening with microdissected DNA start with the number 1 or 2, and those of phages isolated by chromosomal walking start with the letters A, B, or C. The breakpoints and orientations of the duplication Dp(2;2)GYL, the deficiency Df(2L)esc[10], and the deficiency Df(2L)prd[1.7.20] are indicated by hatched arrows. (Reprinted, with permission, from Frei et al. 1985.)

all recombinant phage DNAs shown in Figure 2. As evident from Figure 3, transcripts are detected in follicles only in two DNA regions that appear to remain silent in adult males. One region, which is contained entirely within phage C2.1 (Fig. 3, lane 6), comprises three contiguous *Eco*RI fragments of 5.3 kb, 0.65 kb, and 3.7 kb, and the other region consists of the 2.9-kb fragment of phage C6.1. However, this fragment is also weakly labeled by adult male cDNA (not detectable in Fig. 3) and hence is unlikely to code for *esc*.

Identification of the *esc* Gene by P-element-mediated Transformation

Differential transcript mapping suggests that the *esc* gene is part of, or entirely within, the DNA cloned in phage C2.1 (at −70 kb in Fig. 2). To confirm this conclusion, we decided to rescue the *esc*⁻ phenotype by P-element-mediated gene transfer (Rubin and Spradling 1982; Spradling and Rubin 1982), using the insert of phage C2.1. Accordingly, most of the inserted DNA of phage C2.1 was subcloned into the *Eco*RI site of Carnegie 4, a nonautonomous P-element vector (Rubin and Spradling 1983), as shown by the restriction map of subclone C2.1.5(Car4) in Figure 4.

Subclone C2.1.5(Car4) was coinjected with DNA of the helper P-element plasmid pπ25.7wc (Karess and Rubin 1984) into the posterior pole of cleavage-stage embryos where the primordial germ cells form. Of 30 injected embryos, six developed into adult flies. Only one of them could be shown in subsequent crosses to produce females that were *trans*-heterozygous *esc*⁻ on their second chromosome but exhibited an *esc*⁺ phenotype as they were able to support normal development of their *esc*⁻ offspring (for details, see Frei et al. 1985).

Proof for Transformation with C2.1.5(Car4)

To test directly whether the observed *esc*⁺ phenotype was the result of transformation with C2.1.5(Car4), two experiments were carried out. First, it was shown that both sequences of C2.1 subcloned in the Carnegie 4 vector and the flanking terminal repeats of the P element were present at the same ectopic chromosomal site, 7A, on the X chromosome (Fig. 5a,b). Hence, we concluded that only sequences flanked by the terminal repeat of the P element in C2.1.5(Car4) had integrated at 7A and were solely responsible for the *esc*⁺ phenotype. The possibility that P-element sequences had integrated together with sequences from 33B comprising more than the DNA inserted in C2.1 was ruled out because no hybridization signal appeared at 7A when sequences flanking the insert of C2.1 on either side were hybridized to salivary gland chromosomes (Fig. 5c,d).

The second proof that C2.1.5(Car4) was responsible for transformation to the *esc*⁺ phenotype came from a comparison, based on whole-genome Southern analysis, of transformed stocks and their corresponding mutant and wild-type stocks (Fig. 6). The *Eco*RI fragment at the right end of the insert in C2.1.5(Car4) is bounded at its right limit by a synthetic *Eco*RI site originating from the C2.1 phage DNA. Hence, its size of 1.0 kb differs from that of the overlapping genomic fragment, which is either 1.3 kb or 2.5 kb, depending on the presence or absence of an additional *Eco*RI site in various fly stocks (Fig. 4). This polymorphism permits us to distinguish this fragment in *CyO*, *CyO esc²*, or *esc¹*

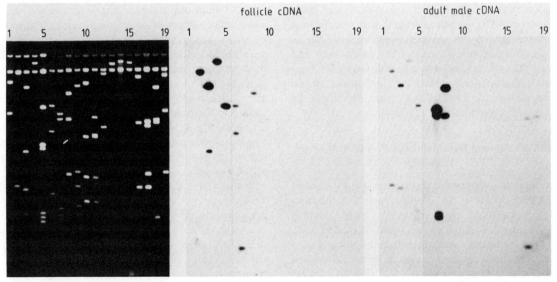

Figure 3. Differential transcript mapping of follicles and adult males in the DNA region shown to contain the *esc* gene. All recombinant phage DNAs shown in Fig. 2 were digested with *Eco*RI and run in a 0.6% agarose gel in TBE buffer (Maniatis et al. 1982). (*1–19*) C8.1 to B2 DNAs loaded in the order of their inserts along the chromosome (Fig. 2). The gel was stained with ethidium bromide (*left* panel), and two Southern blots, obtained by a bidirectional transfer of the gel, were hybridized with labeled cDNA of poly(A)⁺ RNA from RNA from follicles or adult males as indicated. (Reprinted, with permission, from Frei et al. 1985.)

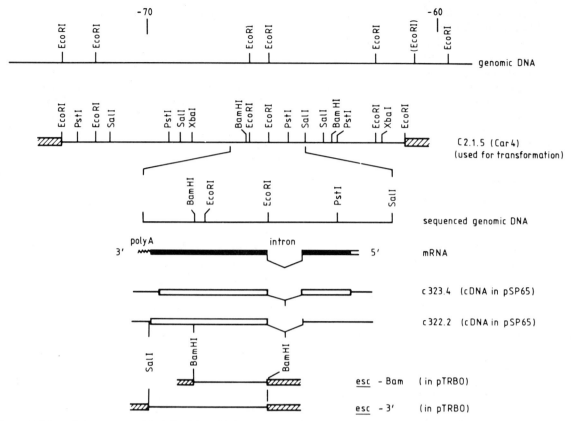

Figure 4. Map of the *esc* gene. (*Top*) EcoRI restriction map of the genomic region between −73 kb and −60 kb (cf. Fig. 2). Below this is a more detailed restriction map of the insert in the Carnegie 4 vector used for transformation. (Note that the positions of the *Sal*I and *Xba*I site at −68.8 kb and −68.4 kb, respectively, had to be slightly corrected with respect to a previous map [Frei et al. 1985].) Part of the insert that codes for the *esc* transcript has been sequenced and is depicted in an enlarged scale underneath. The region transcribed and processed into *esc* mRNA shown below has been determined from the two cDNA clones c323.4 and c322.2. The precise location of the 5′ end of the *esc* transcript is not yet known, and additional small introns may be present in this region. (*Bottom*) Two cDNA subclones in the expression vector pTRB0, *esc*-Bam and *esc*-3′.

chromosomes (2.5 kb) from that in *b pr* or esc^6 chromosomes (1.3 kb), as well as from that in transformed chromosomes (1.0 kb). As clearly shown in Figure 6, the transformed stocks, esc^+; $CyO\ esc^2/Dfesc^{10}$ and esc^+; $CyO\ esc^2/esc^6$, exhibited in addition to the fragments characteristic for their location on the second chromosomes, the 1.0-kb *Eco*RI fragment diagnostic for transformation with C2.1.5(Car4).

Identification and Developmental Profile of the esc^+ Transcript

We conclude from the transformation experiment that all sequences of the *esc* gene, including those required for its proper expression in the female germ line, are contained in the 12 kb subcloned in C2.1.5(Car4). Only one transcript was detectable by Northern analysis in this entire region in early embryos. As evident from Figure 7, this poly(A)$^+$ RNA of about 1.8 kb exhibits a developmental profile consistent with that predicted from clonal analysis (Tokunaga and Stern 1965) and a temperature-sensitive mutant of *esc* (Struhl and Brower 1982) and hence most likely represents the mRNA of the *esc* gene. It is most abundant in follicles and 0–4-hour-old embryos but is drastically diminished at later stages, although small amounts appear to be present throughout development up to the early pupal stage at 7–8 days after egg deposition. Older pupae have not been analyzed, and in adult males, no *esc* poly(A)$^+$ RNA was detected, which is in contrast to its abundance in adult females, where it is presumably localized in the ovaries. The orientation of the *esc* transcript (see Fig. 4) was determined by hybridization with single-stranded probes. These were obtained as labeled transcripts of subclones in the pSP65 vector (Melton et al. 1984).

Within C2.1.5(Car4), the limits of the esc^+ transcript were determined by Northern analysis and transcript mapping (not shown). Northern analysis did not reveal any poly(A)$^+$ RNAs in 0–4-hour-old embryos to the left of the *Sal*I site at −68.8 kb or to the right of the *Bam*HI site at −63.6 kb (Fig. 4). This result was confirmed by transcript mapping. Hybridization of labeled cDNA of poly(A)$^+$ RNA from 0–4-hour-old embryos to restriction fragments of C2.1.5(Car4) showed that the 1.8-kb *esc* mRNA lies entirely within a 3.2-kb region between the *Hinc*II site at −67.8 kb and the *Sal*I site at −64.6 kb.

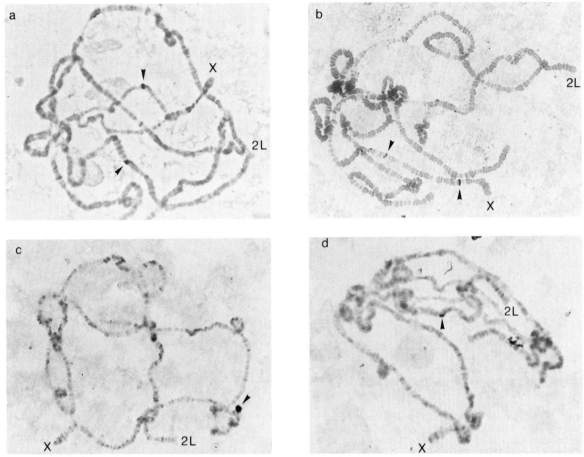

Figure 5. In situ hybridizations to salivary gland chromosomes of esc^6/esc^6 or $esc^6/Dfesc^{10}$ larvae transformed with C2.1.5(Car4). Salivary gland chromosomes of transformed esc^6/esc^6 or $esc^6/Dfesc^{10}$ female larvae were hybridized in situ with the following biotinylated DNAs: one half of a salivary gland with pAT153 containing the 5.3-kb EcoRI fragment of C2.1 DNA (*a*) and the other half with the Carnegie 4 vector (*b*). Alternatively, one half of a salivary gland was hybridized with a subclone in pAT153 of the 15-kb EcoRI fragment of phage C5.3 (*c*) or with a mixture of subclones in pAT153 of the 6.6-kb EcoRI fragment of phage 2H4-6.3.2 and of the 4.3-kb and 4.7-kb EcoRI fragments of phage 2F1.3.4 (*d*). For *c* and *d*, the other half of the salivary gland showed a signal at 7A after hybridization with biotinylated C2.1.5(Car4) DNA. Only female larvae were used for the squashes to facilitate detection of a possible transposition of *DpGYL* into the X chromosome. The left telomere of the second chromosome (2L) and the telomere of the X chromosome (X) are indicated. Signals (arrowheads) are at 33B1,2 and 7A (*a*), 7A and 3C (*b*), and 33B1,2 (*c,d*). The signal at 3C (*b*) is due to the presence of white sequences in the Carnegie 4 vector (Rubin and Spradling 1983). (*a* and *b* were reprinted, with permission, from Frei et al. 1985.)

Comparison of Genomic and cDNA Sequences and Synthesis of a β-Gal-*esc* Fusion Protein

A number of *esc* cDNA clones were isolated from follicle and 0-4-hour-old embryo libraries and sequenced. Comparison of the sequences of the c322.2 cDNA clone shown in Figure 4 with that of the corresponding genomic DNA revealed the position of the 3' end (site of poly[A] addition) and an intron of 364 bp. The 5' end has not yet been determined precisely by S1-nuclease or primer-extension mapping. Its approximate location is inferred from the length of the c323.4 cDNA, which places it between the PstI and SalI site. This position of the 5' end is consistent with the transcript mapping (see above) that labels the PstI-SalI fragment about half as intensely as the neighboring EcoRI-PstI fragment containing most of the intron sequence (Fig. 4). It is possible that additional small introns exist close to the 5' end between PstI and SalI.

It was convenient that upon splicing, a new BamHI site was generated in the cDNAs (Fig. 4). This allowed us to subclone the entire (*esc-3'*) or most of the 3' domain (*esc-Bam*) into the expression vector pTRB0 (Fig. 4). The two β-Gal-*esc* fusion proteins were analyzed as shown in Figure 8. Whereas the *esc-3'* fusion protein exhibited considerable degradation, little or no breakdown was detected for the *esc-Bam* fusion protein. The lengths of the two fusion proteins are consistent with the sequencing data. Sufficiently large amounts have been prepared for immunization of mice to obtain mono- and polyclonal antibodies. Using this obvious approach, we hope to identify the molecules interact-

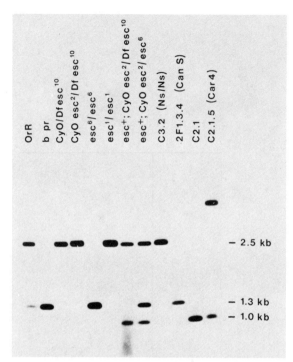

Figure 6. Whole-genome Southern analysis of transformed, esc^-, and esc^+ stocks. EcoRI digests of DNA from esc^+ (Oregon R [OrR] or $b\ pr$), esc^- ($CyO\ esc^2/Dfesc^{10}$, esc^6/esc^6, or esc^1/esc^1), or transformed fly stocks (esc^+; $CyO\ esc^2/Dfesc^{10}$ or esc^+; $CyO\ esc^2/esc^6$) were analyzed according to the method of Southern (1975) after electrophoresis in a 0.6% agarose gel in TBE buffer (Maniatis et al. 1982). The hybridization probe was a subclone in pAT153 of the 1.3-kb EcoRI fragment of phage 2F1.3.4 that includes the 1.0-kb EcoRI fragment at the right end of the insert in phage C2.1 and in the Carnegie 4 subclone C2.1.5(Car4) used for transformation (cf. Fig. 4). For comparison, EcoRI digests of the phages 2F1.3.4 and C2.1, of the subclone C2.1.5(Car4) in the Carnegie 4 vector, of a phage containing esc from a Ns/Ns library (C3.2), and of the CyO chromosome ($CyO/Dfesc^{10}$) are shown.

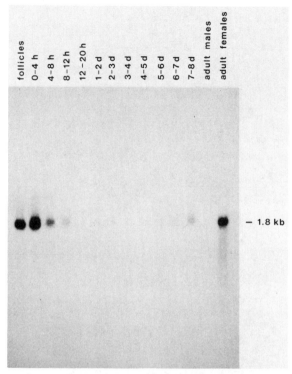

Figure 7. Developmental profile of the esc^+ transcript. Poly(A)$^+$ RNAs (2 μg each) isolated from the various stages indicated at the top of each lane were run in a 1.1% agarose gel containing formaldehyde. RNA was transferred to a nitrocellulose filter and hybridized with labeled RNA transcribed from the c322.2 cDNA subclone in pSP65 (cf. Fig. 4). The length of the esc transcript has been calibrated with fragments of a HindIII digest of bacteriophage λ DNA. No signals were detectable after the filter had been hybridized with transcripts from the opposite strand.

ing with the esc gene product and thus shed light on its regulatory role in the determination of segment identity.

DISCUSSION

To clone the esc gene, we first cloned a large region containing the gene by microdissection and chromosomal walking. This strategy was adopted because no esc mutants (insertion of a known transposable element or inversion) were known that would have permitted a simpler approach. In a second step, the region containing esc was narrowed down to a DNA size that could be used for transformation in a P-element vector. Particularly useful in this step was the technique of differential transcript mapping. The criterion that a specific mRNA is present at some developmental stage (e.g., in young embryos), yet not at another (e.g., in larvae or adults), proved to be rather stringent in the case of esc mRNA: Only one transcript within 160 kb of DNA satisfied its partly known developmental profile. The region comprising this transcript was finally shown by P-element-mediated transformation to contain the esc gene.

Since we relied on a single esc^+ transformant, it was crucial to prove that rescue of the phenotype was indeed a consequence of transformation, rather than caused by reversion to wild type or by transposition of DNA carrying esc^+ sequences from its original locus at 33B to another chromosomal site. The possibility of reversion was ruled out by the observation that replacing, in genetic crosses, the second chromosome with second chromosomes carrying other esc^- alleles did not eliminate the esc^+ phenotype. Positive evidence for transformation came from two independent experiments. First, in situ hybridizations with C2.1 or P-element sequences to corresponding halves of salivary glands showed that both sequences were present at the new location, 7A (Fig. 5a,b). Second, whole-genome Southern analysis of transformants revealed a 1.0-kb EcoRI fragment that is characteristic for the sequence subcloned in C2.1.5(Car4) and whose size differs in nontransformed genomes (Fig. 6). Therefore, rescue of the esc^- phenotype occurred by transformation of the germ line with C2.1.5(Car4) DNA and not by any other mechanism.

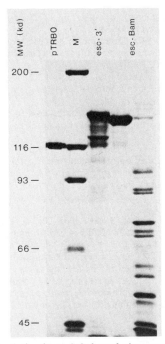

Figure 8. Analysis of two β-Gal-*esc* fusion proteins by SDS-PAGE. The two β-Gal-*esc* fusion proteins *esc*-3' and *esc-Bam* (cf. Fig. 4) were analyzed by electrophoresis in a 7.5% SDS-polyacrylamide gel (Laemmli 1970) and subsequently stained with Coomassie brilliant blue. The two lanes to the far right show strongly and weakly induced samples of the *esc-Bam* fusion protein. For comparison, β-Gal obtained from the expression vector alone (pTRB0) and the molecular-weight markers (M) myosin (200,000), β-Gal (116,350), phosphorylase b (92,500), bovine serum albumin (66,200), and ovalbumin (45,000) are shown.

Since the *esc* gene also has a late function (Tokunaga and Stern 1965), reflected in the transformation of second and third legs into first legs in homozygous *esc*⁻ flies (Hannah-Alava 1958), the question arises as to whether this late function has also been rescued by the transformation. Therefore, esc^+/Y; esc^-/esc^- males, obtained from the cross esc^+ (transposon)/$FM6$; CyO $esc^2/esc^6 \times FM6/Y$; CyO esc^2/esc^6, were examined for the presence of sex combs on their second and third legs; in no case were extra sex combs observed. In addition, when fertile females of the esc^+-transformed CyO esc^2/esc^6 stock were analyzed, none exhibited a transformation of the second leg into the first leg, which would have been recognized most easily by the appearance of tarsal and tibial transverse rows of bristles on the second legs (Hannah-Alava 1958). In contrast, most of the sterile females showed this homeotic transformation. Therefore, in all likelihood, the late function of the *esc* gene is included on the C2.1.5(Car4) subclone as well. As esc^+ transformant males were viable, integration at 7A did not induce a lethal mutation on the X chromosome.

We have no rigorous proof that the poly(A)⁺ RNA found within the DNA segment used for transformation is the sole transcript of the *esc* gene. Since it is clear that a gene product of *esc* exists (Struhl and Brower 1982), only another transcript, coded by the insert of C2.1.5(Car4), that remained undetected could be the true *esc* mRNA. This possibility seems unlikely for the following reason. The DNA used for transformation was the only region within 160 kb in which a transcript exhibited the expected developmental profile, as judged from differential transcript mapping, and it was the RNA assigned to the *esc* gene that was responsible for the signals observed in the differential transcript mapping (cf. Figs. 3 and 4). Furthermore, Northern analysis of this RNA (Fig. 7) revealed a developmental profile entirely consistent with that expected for the *esc* transcript. However, final proof that this RNA is the *esc* transcript will have to await the identification of the structural lesions caused by *esc* mutations.

ACKNOWLEDGMENTS

We thank François Kilchherr for excellent technical assistance, Thomas Bürglin for his pTRB0 expression vector, Marek Mlodzik for providing bio-dUTP, and Hans Noll for comments on the manuscript. This work was supported by grants (3.180-0.82 and 3.600-1.84) from the Swiss National Science Foundation and by the Canton of Basel.

REFERENCES

ASHBURNER, M. 1982. The genetics of a small autosomal region of *Drosophila melanogaster* containing the structural gene for alcohol dehydrogenase. III. Hypomorphic and hypermorphic mutations affecting the expression of hairless. *Genetics* **101**: 447.

DUNCAN, I. 1982. Polycomblike: A gene that appears to be required for the normal expression of the bithorax and Antennapedia gene complexes of *Drosophila melanogaster*. *Genetics* **102**: 49.

DUNCAN, I. and E.B. LEWIS. 1982. Genetic control of body segment differentiation in *Drosophila*. In *Developmental order: Its origin and regulation* (ed. S. Subtelny), p. 533. A.R. Liss, New York.

FREI, E., S. BAUMGARTNER, J.-E. EDSTRÖM, and M. NOLL. 1985. Cloning of the *extra sex combs* gene of *Drosophila* and its identification by P-element-mediated gene transfer. *EMBO J.* **4**: 979.

HANNAH-ALAVA, A. 1958. Developmental genetics of the posterior legs in *Drosophila melanogaster*. *Genetics* **43**: 878.

INGHAM, P.W. 1983. Differential expression of *bithorax complex* genes in the absence of the *extra sex combs* and *trithorax* genes. *Nature* **306**: 591.

———. 1984. A gene that regulates the bithorax complex differentially in larval and adult cells of *Drosophila*. *Cell* **37**: 815.

KARESS, R.E. and G.M. RUBIN. 1984. Analysis of P transposable element functions in *Drosophila*. *Cell* **38**: 135.

KAUFMAN, T.C., R. LEWIS, and B. WAKIMOTO. 1980. Cytogenetic analysis of chromosome 3 in *Drosophila melanogaster*: The homeotic gene complex in polytene chromosome interval 84A-B. *Genetics* **94**: 115.

LAEMMLI, U.K. 1970. Cleavage of structural proteins during the assembly of the head of bacteriophage T4. *Nature* **227**: 680.

LEHRACH, H., D. DIAMOND, J.M. WOZNEY, and H. BOEDTKER. 1977. RNA molecular weight determinations by gel electrophoresis under denaturing conditions, a critical reexamination. *Biochemistry* **16**: 4743.

Lewis, E.B. 1978. A gene complex controlling segmentation in *Drosophila*. *Nature* **276:** 565.

Lewis, R.A., T.C. Kaufman, R.E. Dennell, and P. Tallerico. 1980a. Genetic analysis of the Antennapedia gene complex (ANT-C) and adjacent chromosomal regions of *Drosophila melanogaster*. I. Polytene chromosome segments 84B-D. *Genetics* **95:** 367.

Lewis, R.A., B.T. Wakimoto, R.E. Dennell, and T.C. Kaufman. 1980b. Genetic analysis of the Antennapedia gene complex (ANT-C) and adjacent chromosomal regions of *Drosophila melanogaster*. II. Polytene chromosome segments 84A-84B1,2. *Genetics* **95:** 383.

Maniatis, T., E.F. Fritsch, and J. Sambrook, eds. 1982. *Molecular cloning: A laboratory manual*. Cold Spring Harbor Laboratory, Cold Spring Harbor, New York.

Melton, D.A., P.A. Krieg, M.R. Rebagliati, T. Maniatis, K. Zinn, and M.R. Green. 1984. Efficient *in vitro* synthesis of biologically active RNA and RNA hybridization probes from plasmids containing a bacteriophage SP6 promoter. *Nucleic Acids Res.* **12:** 7035.

Rubin, G.M. and A.C. Spradling. 1982. Genetic transformation of *Drosophila* with transposable element vectors. *Science* **218:** 348.

———. 1983. Vectors for P element-mediated gene transfer in *Drosophila*. *Nucleic Acids Res.* **11:** 6341.

Sánchez-Herrero, E., I. Vernós, R. Marco, and G. Morata. 1985. Genetic organization of *Drosophila* bithorax complex. *Nature* **313:** 108.

Southern, E. 1975. Detection of specific sequences among DNA fragments separated by gel electrophoresis. *J. Mol. Biol.* **98:** 503.

Spradling, A.C. and G.M. Rubin. 1982. Transposition of cloned P elements into *Drosophila* germ line chromosomes. *Science* **218:** 341.

Struhl, G. 1981. A gene product required for correct initiation of segmental determination in *Drosophila*. *Nature* **293:** 36.

———. 1983. Role of the esc^+ gene product in ensuring the selective expression of segment-specific homeotic genes in *Drosophila*. *J. Embryol. Exp. Morphol.* **76:** 297.

Struhl, G. and D. Brower. 1982. Early role of the esc^+ gene product in the determination of segments in *Drosophila*. *Cell* **31:** 285.

Tokunaga, C. and C. Stern. 1965. The developmental autonomy of extra sex combs in *Drosophila melanogaster*. *Dev. Biol.* **11:** 50.

Ullmann, A. 1984. One-step purification of hybrid proteins which have β-galactosidase activity. *Gene* **29:** 27.

Yannopoulos, G., A. Zacharopoulou, and N. Stamatis. 1982. Unstable chromosome rearrangements associated with male recombination in *Drosophila melanogaster*. *Mutat. Res.* **96:** 41.

Molecular and Genetic Analysis of the hairy Locus in *Drosophila*

D. Ish-Horowicz, K.R. Howard, S.M. Pinchin, and P.W. Ingham
*Laboratory of Developmental Genetics, Imperial Cancer Research Fund,
Mill Hill Laboratories, London NW7 1AD, England*

A universal feature of the insect body pattern is its segmental (metameric) organization (for review, see Lawrence 1981). The *Drosophila* embryo has three thoracic (T1–T3) segments and eight abdominal (A1–A8) segments that, although homologous, show distinct segment-specific cuticular patterns (Fig. 1a). This metameric organization extends into the larval head, which is composed of five or six segmental units, although extreme morphological differences obscure their segmental nature. The diversity of the segment characters is controlled by sets of selector (homeotic) genes that are activated in different regions of the em-

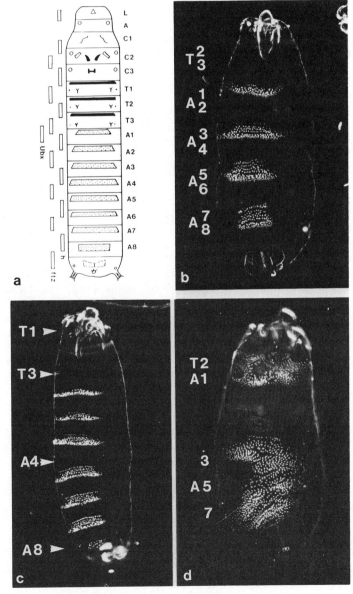

Figure 1. (*a*) Diagrammatic representation of the pattern elements of the *Drosophila* larva indicating the deletion phasing of *h* and *ftz* and the parasegmental *Ubx* domain. Continuous lines across the larvae indicate the segment boundaries. Segments: (L) labral; (A) antennal; (C1) mandibular; (C2) maxillary; (C3) labial; (T1–T3) thoracic; (A1–A8) abdominal. Black areas are the thoracic ventral denticle belts, and the stippled areas are the abdominal ventral denticles. The phenotypic description of *ftz* is according to P.W. Ingham and A. Martinez-Arias (in prep.), which differs in several respects from the description of Wakimoto et al. (1984) and Nüsslein-Volhard et al. (1982), and is based on detailed studies of several *ftz* alleles and combinations with other genes. The *Ubx* domain indicated is the region of major accumulation of *Ubx* gene products (White and Wilcox 1984; Akam and Martinez-Arias 1985; Beachy et al. 1985), although it corresponds to the genetically defined *bxd* domain (Lewis 1978; Akam 1983). The phenotypic description of *h* is according to analysis of *h Ubx* embryos (Fig. 3) and other unpublished observations (P.W. Ingham et al., in prep.). (*b–d*) Ventral aspect of the larval cuticle of *h* mutants, viewed under dark-field illumination. (*b*) $Df(3L)h^{i22}$ homozygote exhibiting the "ideal" pair-rule phenotype. The composite character of each segmental unit is indicated by the designations at the left. (*c*) h^1/h^{5HO7} trans-heterozygote exhibiting the weak *h* phenotype. Arrowheads indicate the segments that are affected. (*d*) h^{5HO7} homozygote exhibiting an extreme phenotype in which the posterior three abdominal composite segments have fused and differentiated a single mass of denticles.

bryo. The identity of a particular segment is specified by a unique combination of active selector genes (Garcia-Bellido 1975).

The establishment of metameric pattern requires a distinct set of genes, the segmentation loci, that affect the number and organization of segmental units, without necessarily affecting their identity (Nüsslein-Volhard and Wieschaus 1980). There are two classes of segmentation genes: maternally acting genes, which encode products deposited in the developing oocyte, and zygotic genes, which are expressed after fertilization. Mutations in the zygotic genes fall into three phenotypic classes: (1) "gap" genes, which lead to the deletion of contiguous sets of segments; (2) "segment-polarity" genes, which disturb polarity within each segment without affecting segment number; and (3) "pair-rule" genes, which cause pattern abnormalities in alternate segments. The latter class demonstrates that an underlying double-segment periodicity precedes the establishment of overt unisegmental organization (Nüsslein-Volhard and Wieschaus 1980).

Mutations in selector genes have no effect on segment number, but they cause homeotic transformations, i.e., the differentiation of homologous structures characteristic of an alternative segmental identity. Thus, for example, the Ultrabithorax (Ubx) mutations transform parts of the metathorax (T3) and first abdominal segment (A1) into mesothorax (T2) (Lewis 1978). One striking aspect of Ubx activity is that the domains of its activity are not segmental (Kerridge and Morata 1982; Hayes et al. 1984; Struhl 1984). Rather, its expression appears to define an alternative segment-wide unit, the parasegment, that lies about one third of a segment anterior to the morphological unit (Fig. 1a) (Martinez-Arias and Lawrence 1985). The parasegmental domain is bounded by lines of lineage restriction (compartments) defined in the adult cuticle (Garcia-Bellido et al. 1973).

The initial nuclear cleavages of the developing Drosophila embryo are not accompanied by cell division. The nuclei divide nine times before migrating to the periphery of the egg, forming the syncytial blastoderm. They divide a further four times before cellularization (cellular blastoderm). Several lines of evidence indicate that the reiterated pattern that underlies segmentation appears to have been established at this time or shortly thereafter. Segmental restrictions of cell lineage are established by blastoderm (Wieschaus and Gehring 1976; Szabad et al. 1979; Simcox and Sang 1983). In addition, spatially periodic transcription of the pair-rule gene fushi tarazu (ftz) is detectable before blastoderm cellularization (Hafen et al. 1984). ftz transcription is initially uniform, suggesting that establishment of periodicity occurs during the syncytial blastoderm stage. Moreover, spatially restricted patterns of Antennapedia and Ubx transcription are detectable at blastoderm (Levine et al. 1983; Akam and Martinez-Arias 1985). Finally, the segmental organization of the embryo is reflected by the transcription of the engrailed gene, which is already reiterated by the onset of gastrulation (Fjose et al. 1985; Kornberg et al. 1985).

To understand how the spatial organization of the embryo is achieved, we are using genetic and molecular techniques to study the hairy (h) locus, a member of the pair-rule class of genes (Nüsslein-Volhard and Wieschaus 1980). h embryos suffer deletions in alternate segments such that, in ideal cases, the embryos develop with only half the normal number of segments. h is also interesting because it is genetically complex; it encodes a second function required later in development for the establishment of the pattern of adult sensory organs (microchaete, bristles) (see Ingham et al. 1985). This latter phenotype has led to the suggestion that h encodes a negative regulator of the achaete function (Falk 1963; Botas et al. 1982).

We have cloned the h locus and attempted to relate h expression to the embryological effects of h mutations. We show that h encodes several overlapping transcripts, some of which correlate with the two different h functions. h transcription in the early embryo is spatially periodic, correlating with defects in h mutant embryos. We have also studied the relative domains of h, ftz, and Ubx transcription in order to define possible metameric phasings in the blastoderm.

RESULTS

h Is Involved in Two Distinct Developmental Processes

h mutations can affect two apparently unrelated developmental processes: embryonic segmentation and adult bristle patterning. The two requirements for h are recessive and can be analyzed independently because there are h mutations that affect one phenotype while largely complementing the other (Ingham et al. 1985).

Lack of h during embryogenesis results in a series of anteroposterior pattern deletions. The simplest phenotype is the deletion of alternate segment-wide regions, the limits of which do not correspond to morphological segment boundaries (see below). h embryos showing a pair-rule phenotype (Fig. 1b) have half the normal number of segments, each originating from a pair of adjacent segments. Anterior T2 is fused with posterior T3, anterior A1 with posterior A2, etc. (Fig. 1a). This deletion pattern is approximately complementary to the phenotype of ftz embryos, where anterior T3 is fused with posterior A1, anterior A2 with posterior A3, etc. (Fig. 1a).

Head structures are also affected in h embryos, with loss of labial (C3) and mandibular (C1) structures and lack of the medial (labral) tooth (P.W. Ingham et al., in prep.). In total, pair-rule h embryos suffer eight sets of deletions (Fig. 1a). Some h alleles have weaker phenotypes with occasional partial or complete segment fusions, usually between T2 and T3, A3 and A4, and A7 and A8 (Fig. 1c). Homozygotes for these alleles often survive as adults, which sometimes show segment defects, particularly of A4.

The most severe h mutations delete more pattern elements than the idealized pair-rule phenotype. The naked cuticle between double segments is deleted, leading to further fusions, especially in the posterior abdomen. In extreme cases, the three posterior double segments fuse together into a single band of denticles (Fig. 1d). The "null" phenotype, associated with a complete lack of h activity, ranges from pair-rule embryos to these extreme fusions (Ingham et al. 1985).

Most h alleles cause the development of extra bristles throughout the adult epidermis, particularly on the wing blades, scutellum, and pleura. This phenotype also shows considerable allele variation, weaker alleles displaying extra bristles only on the wing blade, especially the L2 vein (Ingham et al. 1985). This h function is required during late larval development (Garcia-Bellido and Merriam 1971), whereas the function affecting segmentation must act early.

We have shown that the two h phenotypes are not equally affected in different alleles. Figure 2 shows that we can define four classes of mutations according to their relative effects on segmentation and bristle development. This suggests that h encodes two different but overlapping functions that are expressed or required at different times (Ingham et al. 1985).

Fused h Double Segments Preserve Two Segmental Identities

Homeotic genes and segmentation genes appear to have independent affects on segment identity and segment pattern, respectively. Thus, h embryos have an altered pattern but no obvious segmental transformations. However, the abdominal denticle bands are sufficiently similar that transformations between abdominal identities would be hard to detect. Therefore, we have examined h embryos in which a Ubx mutation transforms the A1 denticle belt into a T2 belt. The morphologies of thoracic and abdominal ventral denticles are clearly distinguishable, the former being much smaller and finer than the latter. Figure 3 shows that the fused A1/A2 band in h Ubx embryos displays both types of denticles, confirming that both segmental identities can be present in the double segment (Nüsslein-Volhard and Wieschaus 1980).

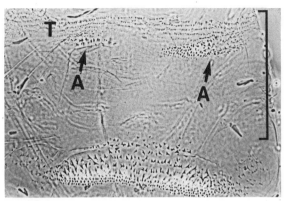

Figure 3. Detail of the ventral larval cuticle of a h^{79K} Ubx^1 homozygote. The region shown is the unit considered to be composed of cells of both A1 and A2 identity. The structures differentiated by the A1 cells are typical of segment T2 as a result of the Ubx mutation. However, cells determined to be A2 should be largely unaffected by Ubx and differentiate typically abdominal denticles. The composite nature of the unit is thus revealed by the presence of both thoracic (T) and abdominal (A; arrows) denticles in a single denticle belt.

A second inference from this experiment is that the deletions in h pair-rule embryos extend into two adjacent segments. The phenotype of h Ubx embryos indicates that the deletion boundaries lie within the denticle belts. If the regions deleted in h embryos are an accurate reflection of the site of h activity, the domains of h expression would correspond to a unit other than the segment. However, we cannot distinguish primary and secondary effects on the pattern of h mutations by looking at morphological phenotypes. Thus, we have used molecular probes to examine h expression during early embryogenesis.

Cloning and Characterization of the h Locus

We have isolated DNA from the h locus by making insertional h mutations using hybrid dysgenesis. We generated several h mutations containing P elements at 66D9-11, the site of the h locus (Ingham et al. 1985), and used these to clone the h region from a wild-type Oregon-R stock. Our 50-kb cloned region joins the distal end of the region cloned by Spradling (1981) in the chromosome walk from the 66E chorion-protein locus and includes the region cloned by Holmgren (1984). Wild-type revertants of the dysgenic h^{d4} mutation confirm that it is indeed due to the P element at 66D and that the segmentation and bristle phenotypes are both due to the insertion (P.W. Ingham et al., in prep.).

Figure 4 shows a restriction map of the h region and the sites of chromosomal lesions associated with a number of h mutations. It is striking that there is physical segregation between the h mutations in the different phenotypic classes. Distal mutations affect only or predominantly the bristle function, whereas proximal mutations affect both h functions. This suggests that h encodes two overlapping functions, the distal domain being required only for the bristle function.

Functions

Class	Segmentation	Bristle
I	+	−
II	±	−
III	−	±
IV	−	−

Figure 2. Genetic complementation of h alleles. The ability of different h mutations to complement the segmentation or the bristle phenotypes was tested in *trans* to appropriate h alleles. (+) Complete complementation, i.e., wild-type activity for this function; (±) partial complementation; (−) lack of complementation (Ingham et al. 1985).

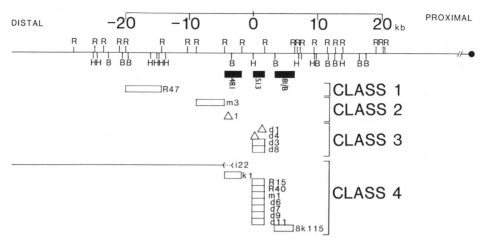

Figure 4. Restriction map of the *h* region. The cloning of the *h* region and its characterization are described by P.W. Ingham et al. (in prep.). (□) Regions within which chromosome aberrations associated with *h* mutations map. (△) Cloned insertional mutations. $Df(3L)h^{i22}$ deletes the proximal *h* region as indicated. (■) Sequences used as probes in Fig. 5. The phenotypic classes of *h* mutations are described in Fig. 2. (R) *Eco*RI; (H) *Hin*dIII; (B) *Bam*HI.

Two Classes of *h* Transcripts Correlate with the Two *h* Functions

The pattern of transcription from this region is consistent with the above interpretation. The proximal region encodes a transcript of about 2.1 kb (α_L) that is expressed at high levels in embryos 2–4 hours old (Fig. 5a), the time at which the segmentation function should act. Expression of this transcript is much reduced in embryos older than 4 hours and is almost undetectable in 0–2-hour-old embryos and ovaries, but resumes during larval and adult stages. A 2.0-kb transcript (α_S) is present at low levels in embryos 2–4 hours old and at an amount equal to α_L in larva and adults. Hybridization with different probes (see Fig. 5) shows that these RNAs are encoded within a 6.4-kb region as indicated in Figure 4.

Two other transcripts, β_S (1 kb) and β_L (3 kb), appear only during larval development. Hybridization with chromosomal probes (Fig. 5) shows that these RNAs overlap the α transcripts but specifically extend into a more distal domain (Fig. 4). The time of expression and chromosomal origin of the β RNAs are those expected for the *h* bristle function. The reexpression of the α transcripts suggests that they may be involved in both *h* functions.

Since all RNAs are transcribed in a distal-to-proximal direction, they must differ at their 5′ ends. Together with the genetic data, which implicates the upstream domain exclusively in bristle development, this suggests that the different *h* functions are encoded by overlapping transcripts from at least two promoters that are differentially utilized during embryogenic and larval development.

Spatial Distribution of *h* Transcription during Early Embryogenesis

The reiterated nature of the pattern deletions in *h* mutant embryos suggests a spatially periodic requirement for h^+ in the early embryo. We have analyzed the expression of the *h* transcript by visualizing the spatial distribution of α_L transcripts with in situ hybridization to tissue sections of *Drosophila* embryos. At cellular blastoderm, *h* transcripts localize to eight domains (Fig. 6e,f), the number of regions that suffer deletions in *h* embryos. For convenience, the domains are numbered from 0 to 7 (anterior to posterior). The labeling in domains 1 through 7 is periodic, with alternating labeled and unlabeled regions, three to four nuclei wide. Thus, the zones of *h* expression at blastoderm are approximately one segment wide. The labeled regions circumscribe the egg, including both ectodermal and mesodermal progenitor cells. Hybridization is seen over the peripheral cytoplasm, not in the nucleus or inner cytoplasm.

Domain 0 lies at the anterior of the embryo, between 85% and 95% the length of the egg. It differs from the other domains, being restricted to the dorsal surface of the embryo. This region corresponds to the blastoderm primordium of the clypeolabrum (see Technau and Campos-Ortega 1985), from which develops the medial tooth, one of the head structures deleted in *h* embryos.

Although the number and position of the *h* domains approximately correspond to the anlagen of the structures deleted in *h* embryos (Fig. 1a), the lack of landmarks at blastoderm makes it hard to define the exact relationship between the *h* domains and the larval fatemap. The first anteroposterior distinction arises at the beginning of gastrulation when two groups of lateral cells at about 67% egg length invaginate to form the cephalic furrow. We have mapped *h* expression in early gastrulae and find that the furrow, which maps within the maxillary primordium (Underwood et al. 1980; Technau and Campos-Ortega 1985), lies between *h* domains 1 and 2 (Fig. 6g,h). *h* domain 1 lies just anterior to the furrow, i.e., in the mandibular/maxillary region, again corresponding to a region affected in *h* embryos. Thereafter, the segmental periodicity of the labeling implies that the more posterior *h* domains correspond

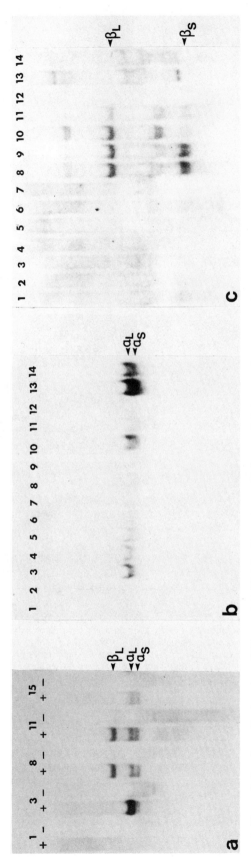

Figure 5. Transcription from the *h* region. RNA was hybridized with antisense single-stranded RNA probes as described by P.W. Ingham et al. (in prep). (*a*) Polyadenylated RNA homologous to the 5.1.3 probe; (*b*) total RNA probed with the distal region, 48.1; (*c*) total RNA probed with the proximal fragment 81/B. The probes are described in Fig. 4. RNA samples: (*1*) ovaries; Embryos: (*2*) 0–2 hr; (*3*) 2–4 hr; (*4*) 4–6 hr; (*5*) 6–9 hr; (*6*) 9–13 hr; (*7*) 13–20 hr. Larvae: (*8*) first instar; (*9*) second instar; (*10*) early third instar; (*11*) late third instar; (*12*) pupae; (*13*) adult males; (*14*) adult females; (*15*) mixed adults. Each track contains an equivalent amount of RNA. (+) Poly(A)+ fraction; (−) poly(A)− fraction.

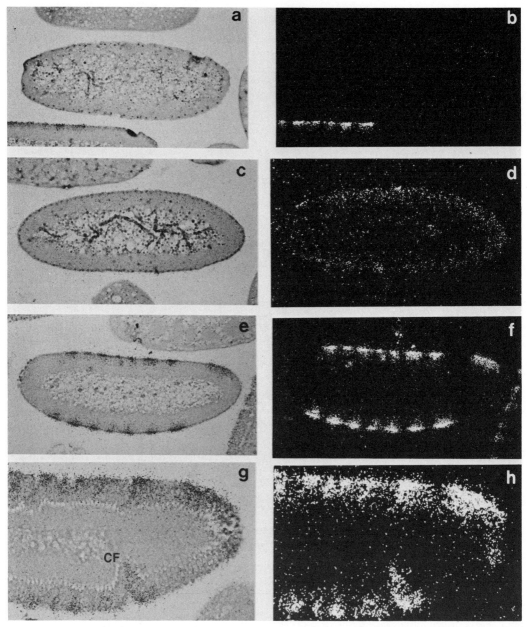

Figure 6. Sections (6 μm) of early stage wax-embedded *Drosophila* embryos hybridized with ^{35}S-labeled *h* probe 5.1.3 (see Fig. 4). The photographs are pairs of bright-field and dark-field images of sections of embryos of successively later stages. (*a,b*) Frontal section of a stage-10 embryo. *h* transcripts are not detectable at this stage. (*c,d*) Frontal section of a stage 12 embryo showing uniform expression of *h* (compare labeling with absence of signal over earlier cleavage-stage embryo in *a*). (*e,f*) Sagittal section of a midstage 14 embryo showing eight distinct domains of *h* expression; (*g,h*) sagittal section of an embryo at the onset of gastrulation indicating the position of *h* domain 1 relative to the cephalic furrow (CF). All sections are oriented with their anterior pole to the right. The sagittal sections have their dorsal surface uppermost.

to primordia deleted in *h* embryos, although the localization is necessarily approximate.

In contrast to the periodic expression at late blastoderm (stage 14; Foe and Alberts 1983), the initial appearance of *h* transcription appears uniform. We have not yet detected either *h* or *ftz* expression prior to stage 12 (not shown). At stages 12 and 13, *h* transcripts are detectable all around the embryo (Fig. 6a–d). At this time, *ftz* transcription is found only between 15% and 65% the length of the egg (Hafen et al. 1984). The first signs of spatially restricted *h* transcription is at late stage 13 or early stage 14. By the onset of stage 14, the localization of *h* transcripts closely resembles that shown in Figure 6 g and h.

Expression of *h* decays rapidly following the onset of gastrulation so that no labeling of the early *h* domains is detectable by mid-germ-band extension. At this time, *ftz* expression is still detectable (not shown).

h and *ftz* Domains May Overlap at Blastoderm

The structures deleted in *ftz* embryos suggest that the deletion frames of *ftz* and *h* are roughly complementary (Fig. 1a). If the deletions were a direct indication of the cells that express each gene, one might have expected that the patterns of *h* and *ftz* would be exactly out of phase. We have shown that this is not the case at late blastoderm by hybridizing a mixture of *h* and *ftz* probes to sections and visualizing the regions that express either of the two genes.

In the central region, wide bands of hybridizing cells are separated by unlabeled regions, about one cell wide (Fig. 7). This shows that one cell per double segment expresses neither *h* nor *ftz*. Although we cannot directly visualize cells that express both *h* and *ftz*, we can infer that there should be about one such cell per double segment, since the domains of *h* and *ftz* labeling (Hafen et al. 1984) are of equal size.

The mixed probes reveal nine hybridizing domains, the anterior *h* domain 0 plus eight stripes. The anterior stripe lies anterior to the cephalic furrow (not shown) and must correspond to *h* domain 1; the posterior zone must be the posteriormost *ftz* domain. The central six stripes thus represent regions of overlapping *h* and *ftz* hybridization. Indeed, these zones are about five to six cells wide, in contrast to the three to four cell-wide domains that hybridize to each probe alone.

h, *ftz*, and *Ubx* Define Three Different Metameric Phases

Although *Ubx* acts to determine segmental identity, the domains of its activity appear to be parasegmental, not segmental. Both *Ubx* transcription and translation appear to be parasegmentally organized (White and Wilcox 1984, 1985; Akam and Martinez-Arias 1985; Beachy et al. 1985). Akam and Martinez-Arias (1985) have shown that *ftz* expression is not parasegmental, being expressed one cell anterior to the *Ubx* domain at blastoderm. Thus, the spatial domain of *Ubx* activity cannot be regulated solely by the *ftz* gene.

Figure 7. In situ hybridization with mixed ^{3}H-labeled *h* and *ftz* probes to a late stage 14 embryo.

We have analyzed the relative phasing of *h* and *Ubx* by hybridizing to sections with a mixture of both probes. The two probes can be distinguished because *h* labels the outer cytoplasm, whereas *Ubx* labeling is nuclear (Akam 1983). Our results suggest that by the time the cellularization of the blastoderm is complete, the anterior limit of the domain of *Ubx* activity is one cell anterior to *h* domain 4 (Fig. 8b). At the same stage, the *Ubx* domain is one cell posterior to the third *ftz* domain (Akam 1985) (see also Fig. 8c). Thus, neither *h* nor *ftz* transcription is parasegmental and, together with *Ubx*, they define three alternative metameric phases.

DISCUSSION

The early expression of *h* is clearly required for the establishment of segmentation. Our studies of *h* transcription have demonstrated a major 2.1-kb RNA (α_L) whose temporal and spatial patterns of expression are those expected of the *h* segmentation function: The transcript is present at high levels only during early embryogenesis (Fig. 5); its spatial distribution at late blastoderm is periodic, roughly corresponding to the primordia of structures affected in *h* mutant embryos (Fig. 6). We also find that the domains of *h* transcription are out of register with those of *ftz* (Fig. 7), indicating that the spatial distributions of *h* and *ftz* expression are not determined merely by each other, but must involve the action of other patterning genes.

The pattern of *h* expression confirms the double-segment periodicity first suggested by Nüsslein-Volhard and Wieschaus (1980) and demonstrated directly by Hafen et al. (1984). The pattern of *h* expression differs from that of *ftz* in several significant respects. Although both show reinterated expression, only *h* labels the very anterior of the embryo (domain 0) (Fig. 6). In addition, their patterns of labeling in the overtly metameric region of the embryo are more than one segment out of phase so that *h* is not expressed in the posteriormost *ftz* domain and *ftz* is not expressed in *h* domain 1. Together, *h* and *ftz* define a region of double-segment periodicity between 15% and 70% the length of the egg. *h* domain 1 shows that the region extends anterior of the cephalic furrow (Fig. 6c). Domain 0 demonstrates homology between the labral region of the embryo and the central metameric region. Homologies of the antennal segment to labial (C3) and thoracic segments have been demonstrated previously (Postlethwait and Schneiderman 1971; Struhl 1981a), but the metameric origin of the labral segment has been a matter of some dispute (for discussion, see Struhl 1981b).

The mechanism of spatial restriction of domain 0 may differ somewhat from those of the other *h* domains. We have observed a separate anterior *h* domain before the reiteration of domains 1 through 7 is apparent (D. Ish-Horowicz et al., unpubl.). Moreover, only domain 0 is restricted to the dorsal surface of the em-

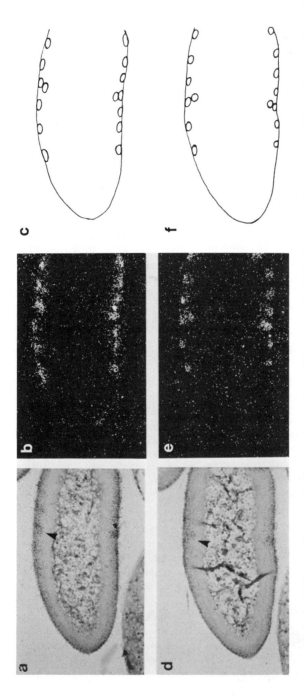

Figure 8. Relative phasing of *h*, *ftz*, and *Ubx*. (*a–c*) In situ hybridization of mixed *h* and *Ubx* probes; (*d–f*) hybridization with mixed *ftz* and *Ubx* probes. The *Ubx* probe is pBM012 (Akam and Martinez-Arias 1985). Both probes are labeled with ^{35}S. (*a,d*) Bright field; (*b,e*) dark field; (*c,f*) diagrammatic representation showing the nuclear *Ubx* localization and the out-of-phase peripheral *h* and *ftz* hybridizations.

bryo, presumably as a response to the maternal dorsalizing genes (Anderson and Nüsslein-Volhard 1984).

Why does lack of h^+ lead to pattern deletions? We find that h mutant embryos have reiterated zones of cell death, whose number and position roughly correspond to the deleted regions (P.W. Ingham et al., in prep.). This would explain how the double segments in h embryos maintain two segmental identities, viz, segmental character is established normally and the fusions arise later due to death of intervening cells. Such cell death may be restricted to the cells that normally express h as a direct (autonomous) consequence of the lack of h activity. However, it could also be a secondary consequence of the disruption of patterning caused by absence of h^+. This nonautonomy of h action is implied by the phenotype of the strongest h alleles, where more structures are lacking than correspond to the h domains; i.e., nonexpressing cells are also lost in h embryos. Similarly, extreme ftz phenotypes also appear nonautonomous (Wakimoto et al. 1984; P.W. Ingham and A. Martinez-Arias, in prep.).

The most striking result is that h expression defines a reiterated unit that differs from those associated with either ftz or Ubx transcription. The partial overlap of h and ftz domains is clearly illustrated in Figure 7a. It demonstrates that some blastoderm cells express neither h nor ftz and suggests that some cells express both. This contrasts with the apparently complementary phenotypes of h and ftz embryos (Fig. 1a) (P.W. Ingham and A. Martinez-Arias, in prep.) and is further evidence of nonautonomy of the cuticular phenotypes.

The relationship between h and Ubx domains is also unexpected. Although Ubx transcription is initially restricted to a single domain at T3/A1, later Ubx expression shows clear metameric periodicity (Akam 1983; White and Wilcox 1984; Akam and Martinez-Arias 1985; Beachy et al. 1985). Thus, the Ubx domain defines a metameric register that differs from both the h and ftz phasing, h domain 4 being about one blastoderm cell anterior to the Ubx domain (Fig. 8). The only overt embryonic unit that could correspond to this h domain would be the A1 segment, although we cannot infer a direct connection between h domains and segments. Nonautonomy of h action could explain the discrepancy between an approximately segmental h domain and the boundary of h deletion in h Ubx embryos, which lies within a denticle band (Fig. 3).

Three reiterative phasings are sufficient to define a polar periodic pattern (Meinhardt 1982). However, we would not wish to imply that either h or ftz acts to define cell states. One would expect such cell labels to be cell-autonomous and to persist during early development. h transcripts decay very rapidly following the onset of gastrulation, and ectodermal ftz transcripts are detectable only to the end of germ band extension (Hafen et al. 1984). Moreover, the cell autonomy of h and ftz is unproven, perhaps even doubtful. Indeed, nonautonomy is to be expected of genes involved in establishing a pattern that, of necessity, requires spatial communication. Thus, either or both h and ftz might be regionalizing the embryo, whereas other genes act to maintain cell identity.

Of course, the major question is how the reiterated expression itself arises. As h and ftz expression is initially uniform, metamerism must be arising between stage 13 and midstage 14, a period of about 30-40 minutes (Foe and Alberts 1983). Are h and ftz responding passively to maternally coded components that interact among themselves to give the primary pattern, or do h and ftz actively participate in establishing reiterated organization? We are currently analyzing transcription in embryos mutant for h or ftz to resolve these alternatives. Preliminary results (not shown) suggest that ftz expression is disturbed in h embryos, i.e., that h is involved in the establishment of ftz periodicity.

The other aspect of h is its complex genetic and biological effects. The evidence strongly suggests that the two h functions correspond to two different, overlapping transcripts. The early segmentation function correlates with the α_L transcript and the late bristle function correlates with the β transcripts. Genetic analysis implicates the upstream region only in bristle function, suggesting that the two classes have different promoters. We are currently analyzing h cDNA clones to determine the exact structures and coding capacities of the transcripts. By combining genetic, embryological, and biochemical approaches in our studies, we should be able to analyze the exact roles that h plays in segmentation and bristle development.

ACKNOWLEDGMENTS

We thank Drs. C.A. Rushlow, J.G. Williams, J.M.W. Slack, and B. Hogan for critically reading the manuscript. We also thank Drs. M.A. Akam and A. Martinez-Arias for many helpful discussions of their unpublished observations and for providing us with the Ubx probes. D.I.H. also thanks Dr. C. Nüsslein-Volhard who first encouraged him toward studying the molecular genetics of embryonic pattern formation.

REFERENCES

AKAM, M.E. 1983. The location of *Ultrabithorax* transcripts in *Drosophila* tissue sections. *EMBO J.* **2:** 2075.
———. 1985. Spatially restricted expression of homoeotic genes. *Proc. R. Soc. Lond. B.* (in press).
AKAM, M.E. and A. MARTINEZ-ARIAS. 1985. *Ultrabithorax* expression in *Drosophila* embryos. *EMBO J.* **4:** 1689.
ANDERSON, K.V. and C. NÜSSLEIN-VOLHARD. 1984. In *Pattern formation* (ed. G.M. Malacinski and S. Bryant), p. 269. Macmillan, New York.
BEACHY, P.A., S.L. HELFAND, and D.S. HOGNESS. 1985. Segmental distribution of bithorax complex proteins during *Drosophila* development. *Nature* **313:** 545.
BOTAS, J., J. MOSCOSO DEL PRADO, and A. GARCIA-BELLIDO. 1982. Gene-dose titration analysis in the search of trans-regulatory genes in *Drosophila*. *EMBO J.* **1:** 307.
FALK, R. 1963. A search for a gene control system in *Drosophila*. *Am. Nat.* **97:** 129.
FJOSE, A., W.J. MCGINNIS, and W.J. GEHRING. 1985. Isolation

of a homoeo box-containing gene from the *engrailed* region of *Drosophila* and the spatial distribution of its transcripts. *Nature* **313**: 284.

FOE, V.E. and B.M. ALBERTS. 1983. Studies of nuclear and cytoplasmic behavior during the five mitotic cycles that precede gastrulation in *Drosophila* embryogenesis. *J. Cell Sci.* **61**: 31.

GARCIA-BELLIDO, A. 1975. Genetic control of wing disc development in *Drosophila*. *Ciba Found. Symp.* **29**: 161.

GARCIA-BELLIDO, A. and J. MERRIAM. 1971. Genetic analysis of cell heredity in imaginal discs of *Drosophila melanogaster*. *Proc. Natl. Acad. Sci.* **68**: 2222.

GARCIA-BELLIDO, A., P. RIPOLL, and G. MORATA. 1973. Developmental compartmentalisation of the wing disk of *Drosophila*. *Nat. New Biol.* **245**: 251.

HAFEN, E., A. KUROIWA, and W.G. GEHRING. 1984. Spatial distribution of transcripts from the segmentation gene *fushi tarazu* during *Drosophila* embryonic development. *Cell* **37**: 833.

HAYES, P.H., T. SATO, and R.E. DENELL. 1984. Homoeosis in *Drosophila*: The *Ultrabithorax* larval syndrome. *Proc. Natl. Acad. Sci.* **81**: 545.

HOLMGREN, R. 1984. Cloning sequences from the *hairy* gene of *Drosophila*. *EMBO J.* **3**: 569.

INGHAM, P.W., S.M. PINCHIN, K.R. HOWARD, and D. ISH-HOROWICZ. 1985. Genetic analysis of the *hairy locus in Drosophila melanogaster*. *Genetics* (in press).

KERRIDGE, S. and G. MORATA. 1982. Developmental effects of some newly induced *Ultrabithorax* alleles of *Drosophila*. *J. Embryol. Exp. Morphol.* **68**: 211.

KORNBERG, T., I. SIDEN, P. O'FARRELL, and M. SIMON. 1985. The *engrailed* locus of *Drosophila*: In situ localisation of transcripts reveals compartment-specific expression. *Cell* **40**: 45.

LAWRENCE, P.A. 1981. The cellular basis of segmentation in insects. *Cell* **26**: 3.

LEVINE, M., E. HAFEN, R.L. GARBER, and W.J. GEHRING. 1983. Spatial distribution of *Antennapedia* transcripts during *Drosophila* development. *EMBO J.* **2**: 2037.

LEWIS, E.B. 1978. A gene complex controlling segmentation in *Drosophila*. *Nature* **276**: 565.

LOHS-SCHARDIN, M., C. CREMER, and C. NUSSLEIN-VOLHARD. 1979. A fate map of the larval epidermis of *Drosophila melanogaster*: Localised cuticle defects following irradiation of the blastoderm with an ultraviolet laser microbeam. *Dev. Biol.* **73**: 239.

MARTINEZ-ARIAS, A. and P.A. LAWRENCE. 1985. Parasegments and compartments in the *Drosophila* embryo. *Nature* **313**: 639.

MEINHARDT, H. 1982. *Models of biological pattern formation*. Academic Press, London.

NÜSSLEIN-VOLHARD, C. and E. WIESCHAUS. 1980. Mutations affecting segment number and polarity in *Drosophila*. *Nature* **287**: 795.

NÜSSLEIN-VOLHARD, C., E. WIESCHAUS, and G. JURGENS. 1982. Segmentation bei *Drosophila* — Eine genetische Analyse. *Verh. Dtsch. Zool. Ges.*: 91.

POSTLETHWAIT, J. and H.A. SCHNEIDERMAN. 1971. Pattern formation and determination in the antenna of the homoeotic mutant *Antennapedia* of *Drosophila melanogaster*. *Dev. Biol.* **25**: 606.

SIMCOX, A.A. and J.H. SANG. 1983. When does determination occur in *Drosophila* embryos? *Dev. Biol.* **97**: 212.

SPRADLING, A.C. 1981. The organisation and amplification of two chromosomal domains containing *Drosophila* domain genes. *Cell* **27**: 193.

STRUHL, G. 1981a. Anterior and posterior compartments in the proboscis of *Drosophila*. *Dev. Biol.* **84**: 372.

———. 1981b. A blastoderm fate map of compartments and segments of the *Drosophila* head. *Dev. Biol.* **84**: 386.

———. 1984. Splitting the bithorax complex of *Drosophila*. *Nature* **308**: 454.

SZABAD, J., T. SCHUPBACH, and E. WIESCHAUS. 1979. Cell lineage and development in the larval epidermis of *Drosophila melanogaster*. *Dev. Biol.* **73**: 265.

TECHNAU, G.M. and J.A. CAMPOS-ORTEGA. 1985. Fate-mapping in wild-type *Drosophila melanogaster*. II. Injections of horseradish peroxidase in cells of the early gastrula stage. *Wilhelm Roux's Arch. Dev. Biol.* **194**: 196.

UNDERWOOD, E.M., F.R. TURNER, and A.P. MAHOWALD. 1980. Analysis of cell movements and fate mapping during early embryogenesis in *Drosophila melanogaster*. *Dev. Biol.* **74**: 286.

WAKIMOTO, B.T., F.R. TURNER, and T.C. KAUFMAN. 1984. Defects in embryogenesis in mutants associated with the *Antennapedia* gene complex of *Drosophila melanogaster*. *Dev. Biol.* **102**: 147.

WHITE, R.H. and M. WILCOX. 1984. Protein products of the bithorax complex in *Drosophila*. *Cell* **39**: 163.

———. 1985. Distribution of *Ultrabithorax* proteins in *Drosophila*. *EMBO J.* **4**: 2035.

WIESCHAUS, E. and W.G. GEHRING. 1976. Clonal analysis of primorial disc cells in the early embryo of *Drosophila melanogaster*. *Dev. Biol.* **50**: 249.

WIESCHAUS, E., C. NÜSSLEIN-VOLHARD, and H. KLUDING. 1984. *Kruppel*, a gene whose activity is required early in the zygotic genome for normal embryonic segmentation. *Dev. Biol.* **104**: 172.

Genes Affecting the Segmental Subdivision of the *Drosophila* Embryo

C. NÜSSLEIN-VOLHARD, H. KLUDING, AND G. JÜRGENS
Friedrich Miescher Laboratorium der Max-Planck Gesellschaft, 7400 Tübingen, Federal Republic of Germany

Segmentation is the process that subdivides an initially uniform blastoderm into a series of repeating homologous units, the segments. In *Drosophila*, a number of genes have been identified that are involved in segmentation (Nüsslein-Volhard and Wieschaus 1980; Nüsslein-Volhard et al. 1982, 1984; Jürgens et al. 1984; Wieschaus et al. 1984a). For a complete and comprehensive understanding of the process, genetic analysis must proceed basically in two directions: (1) identification of all the relevant genes by mutations that lead to an altered segmentation pattern and (2) evaluation of the role of each individual gene in the process by studying its phenotype, its developmental effects, and finally its molecular biology. While the screens for maternal mutants have not yet reached saturation, most, if not all, of the genes required specifically in the zygote are known. The initial phenotypic analysis grouped these zygotic genes, according to the level of spatial organization affected in mutant embryos, into only three classes with less than ten genes each. Two of these classes are characterized by mutants with periodic patterns of defects that occur in a single segment repeat in segment-polarity genes, whereas homologous regions in every other segment are affected in pair-rule genes. In the third mutant class of the three gap genes, essentially one large region is affected while the rest of the embryo develops normally. No mutant that abolishes segmentation completely throughout the embryo has yet been reported (Nüsslein-Volhard and Wieschaus 1980; Nüsslein-Volhard et al. 1982).

Some of the segmentation genes have been studied in much detail both phenotypically and molecularly; the choice of the genes studied depended largely on historical or technical factors (Gloor 1950; Wieschaus et al. 1984b; Wakimoto and Kaufman 1981; Hafen et al. 1984; Ish-Horowicz et al., this volume). In these cases, the spatial pattern of gene expression in the blastoderm stage, as revealed by in situ hybridization of cloned probes to tissue sections, was found to correspond to the pattern of defects in the mutant phenotype. However, since not all genes of a given class are known in as much detail, it is not possible to evaluate the individual roles of the genes or to decide whether a similarity in phenotype reflects analogous functions or whether the role of some of the genes within a class is unique. Of particular importance is the exact description of the mutant phenotype in the complete absence of a functional gene product (the amorphic condition) that defines the realm of action of a particular gene, as well as the time of requirement—be it in the zygote only or during oogenesis. In this paper we describe relevant genetic and phenotypic parameters of the three pair-rule genes odd-skipped (*odd*), paired (*prd*), and even-skipped (*eve*) and compare them with other segmentation genes described earlier.

EXPERIMENTAL PROCEDURES

The general methods for growing flies, collecting eggs, making mutants, and scoring mutant phenotypes were described previously (see Nüsslein-Volhard et al. 1984). EMS-induced mutant alleles of the three loci described here were also described previously (Nüsslein-Volhard and Wieschaus 1980; Jürgens et al. 1984; Nüsslein-Volhard et al. 1984). Marker mutants and balancers were described by Lindsley and Grell (1968).

X-ray mutagenesis. Males (1-4 days old) were irradiated with 5000 rad (330 rad/min; 1-mm aluminum filter, 100 kV, Philips X-ray MG 102) and mated with the same number of virgin females. F_1 females or males were mated individually with males or females, respectively, that were heterozygous for odd^{IIID}, prd^{IIB}, and eve^{ID} in egg-laying blocks. Eggs were collected on agar plates in two 24-hour intervals, and developed unhatched embryos were inspected in place under oil for the appearance of a mutant phenotype (Nüsslein-Volhard 1977). Putative mutants were recovered from the progeny of matings yielding mutant embryos, using standard crossing procedures. The genotypes and the crossing protocol of the experiments are outlined in Figure 1.

Pole cell transplantation. Donor embryos for pole cell transplantations were obtained from suitable crosses of two alleles of the respective locus. For the three pair-rule genes, the alleles odd^{IIID} (strong), $prd^{2.45}$ (strong), and prd^{6L} (strong) were used, and for *eve*, the allele $eve^{3.77}$ (medium) and the deficiency $Df(2R)eve^{1.27}$ (strong) were used. Host embryos were obtained from wild-type females mated to $ovo^{D1}v^{24}$ males (Busson et al. 1983). Pole cells were removed from 2-3-hour-old embryos by inserting the needle (15-20 μm dia) at the anterior pole. Three to five pole cells each were injected into the posterior end of host embryos of the same age. The injected embryos were kept on agar plates covered with oil. Hatched larvae trapped in drops of fresh yeast suspension on agar plates were allowed to develop until the late pupal stage and then trans-

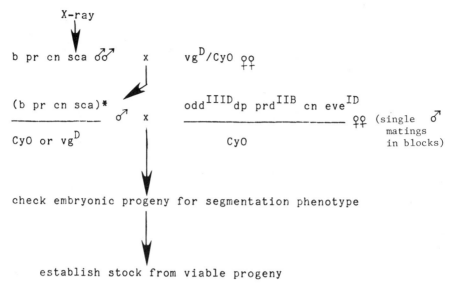

Figure 1. Crossing scheme used for screening and recovery of X-ray-induced aberrations uncovering the embryonic phenotypes of *odd*, *prd*, or *eve* (for details, see Experimental Procedures).

ferred individually into fly tubes. Eclosed females were mated individually with appropriate heterozygous mutant males, and eggs were collected over several days from fertile females. Hatch rates and frequency of mutant embryos were determined as described previously (Nüsslein-Volhard 1977). In addition, a sample of unhatched embryos was prepared for microscopic inspection.

Analysis of mutant embryos. For cuticle preparations, dechorionated eggs were placed on a dry plastic petri dish and covered with water. The vitelline membrane was pricked with a needle, and the embryos were helped out of the membrane. After fixation in glycerol and acetic acid (1:4) for 15 minutes at 60°C, the embryos were mounted in Hoyer's medium diluted 1:1 with lactic acid and cleared by incubation at 60°C for 1 hour. Photomicrographs were taken with a Zeiss photomicroscope.

RESULTS

Isolation of Chromosomal Deficiencies by Means of Embryonic Phenotypes

For the genes described in this paper, a number of point mutants have been collected in a series of systematic screens for mutations that alter the morphology of the larval cuticle. All genes are defined by alleles with various strengths (see below). Without deletion mutants that physically eliminate the gene, however, it is not possible to decide whether the point mutant with the strongest known phenotype represents the lack of function phenotype or whether this is modified by some residual gene activity (hypomorphic phenotype). Therefore, it was necessary to collect deficiencies for the segmentation genes and compare their phenotypes in homozygous embryos with those of the point mutants.

In *Drosophila*, deficiencies covering about 30% of the genome have been collected and described. These have been isolated traditionally by means of one of the following criteria: (1) uncovering of an adult visible recessive phenotype, (2) reversion of a dominant gain-of-function mutant, or (3) a haplo-insufficient dominant phenotype. The usefulness of these screening criteria for the isolation of a deletion of a particular segmentation gene is dependent on the close proximity of a suitable "marker" gene.

For the second chromosomal pair-rule genes, *odd*, *prd*, and *eve*, suitable marker genes were not available. Therefore, instead of screening for an adult visible phenotype of a nearby gene, we screened directly for the embryonic visible phenotype of the segmentation gene. Males heterozygous for an X-ray-treated chromosome were crossed individually with females heterozygous for *odd*, *prd*, and *eve*. Eggs were collected from these crosses on agar plates (Nüsslein-Volhard 1977), and the embryonic progeny were inspected for a pair-rule-type segmentation defect in a fraction of the eggs. When mutant embryos were detected, a stock was established from the viable siblings of the matings. Of 11,000 single F_1 male matings, 7,000 produced fertilized eggs that could be scored for the embryonic phenotype; 56 mutants were recovered, of which 29 uncovered *odd*, 5 uncovered both *odd* and *prd*, 9 uncovered *prd*, and 13 uncovered *eve*.

In contrast to conventional screens for mutations uncovering adult visible phenotypes, in this screen, the detectability of the mutant did not depend on the survival of the mutants to the adult stage. Thus, the screening procedure permitted, in addition to point mutants and small deficiencies, the detection of translocations as very large deficiencies and their isolation in combination with the corresponding duplication. Translocations could be distinguished from simple deficiencies or point mutants by the fact that in the F_2

generation, *trans*-heterozygotes of the mutant and tester chromosome survived.

Of the 56 mutants, 6 were lethal over a mutant allele of one of the three loci. Chromosomal analysis showed that three of these lethal alleles are cytologically normal, one each for *odd*, *prd*, and *eve*. Two of the lethal *prd* mutants, as well as one *eve* mutant, are deletions. All other mutants are translocations. To determine the approximate extent of the translocated chromosome piece, complementation tests were performed with a number of embryonic visible mutants (Nüsslein-Volhard et al. 1984). Because the segregants haploid for large chromosomal regions die as embryos, often showing poor differentiation and failure in some late morphogenetic movements, the usefulness of the embryonic mutants for mapping the extent of the deficiencies depends on the detectability of the mutant phenotype superimposed on the aneuploid phenotype. The embryonic visible markers we found useful in this screen were odd-skipped (8), sloppy-paired (8), midline (16), wingless (30), paired (45), and crinkled (52) on chromosome 2L and patched (59), even-skipped (59), lines (59), filzig (59), engrailed (62), shavenoid (62), Posterior sex combs (67), mastermind (71), and twisted (100) on chromosome 2R (Nüsslein-Volhard et al. 1984). The deficiency mapping showed that in several cases, the deficiency was very large, including up to an entire chromosome arm. Some of the translocations allowed an ordering of genes mapping close to each other (see also Nüsslein-Volhard et al. 1984). For deficiencies and smaller translocations, the cytology was determined by chromosomal analysis (Table 1). The results indicate placement of *odd* into 23E-24B, *prd* into 33B6,7-E2,3, and *eve* into 46C3,4;C9,11.

Do the Segmentation Genes Have a Maternal Effect?

The isolation of the chromosomal aberrations described above allowed the construction of zygotes that lack the respective segmentation gene altogether and thus the determination of the amorphic mutant phenotypes (see below). However, homozygous mutant embryos are usually obtained from heterozygous parents; if the gene were already transcribed during oogenesis, the functional gene product would be present in the egg and would thus weaken the zygotic amorphic phenotype. To test whether maternal gene expression takes place, it is necessary to obtain homozygous and heterozygous mutant embryos from a homozygous mutant female germ line and compare their phenotypes with those obtained from heterozygous mutant females (Frohnhöfer 1982; Lawrence et al. 1983; Lehmann 1985; E. Wieschaus and E. Noell, in prep.).

Chimeric flies with a homozygous mutant germ line surrounded by wild-type somatic tissue have been constructed by transplanting pole cells from blastoderm-stage mutant embryos into embryos heterozygous for the dominant female sterile mutant ovo^{D1} (Busson et al. 1983). $ovo^{D1}/+$ females are agametic and produce eggs only when they have received functional female pole cells by transplantation; in such chimeras, all of the progeny are derived from the implanted pole cells. Transplants of homozygous mutant germ cells were identified by the frequency of occurrence of mutant embryos in matings to heterozygous males. Figure 2 shows the results of such an experiment in which the parental genotype of the donor embryos was *odd*/+. The progeny from a homozygous mutant germ line was indistinguishable from that of a heterozygous germ line, both showing the typical strong odd-skipped phenotype. Furthermore, the nonmutant progeny developed completely normally, yielding hatching larvae. Thus, *odd* is not required during oogenesis for normal development of the embryos, and its presence or absence in the female germ line has no influence on the mutant phenotype.

Pole cell transplantation experiments have been carried out with all three segmentation genes described in this paper, with very similar results: In all three cases, the absence of a functional gene in the female germ line has no influence on the normal development of the heterozygous progeny, and thus zygotic expression is sufficient for normal development. In *prd* and in weak alleles of *eve*, as in the case of *odd*, homozygous mutant embryos derived from a homozygous mutant germ line look no different than those derived from a heter-

Table 1. Cytology of Chromosomal Aberrations Uncovering *odd*, *prd*, or *eve*

Mutation	Deficiency breakpoints	Dupl.	Deficiency uncovers	Other information
$T(1;2)odd^{1.10}$	tip-24B	4A3,4	odd	reciprocal translocation
$T(Y;2)odd^{2.31}$	tip-25C	Y	odd,slp	associated with $In(2L)25Ci,38B$
$T(2;3)odd^{3.29}$	tip-25C	100F	odd,slp	
$T(Y;2)odd^{4.13}$	22A-25F	Y	odd,slp,mid	
$T(Y;2)odd^{4.25}$	tip-24D	Y	odd,slp	
$T(2;?)odd^{5.1}$	23E-24EF	?	odd,slp	
$Df(2L)prd^{1.7}$	33B2,3-34A1,2	—	prd	
$Df(2L)prd^{1.25}$	33B6,7-E2,3	—	prd	in $T(Y;2L)21-40$
$T(2;3)prd^{2.27}$	31B;33D	97CD	prd,bsk	
$T(Y;2)prd^{5.12}$	33A;35B	Y	prd	associated with $T(2R;3R)$
$T(2;3)eve^{1.18}$	43F;46E	96A	ptc,flz,lin,eve	
$Df(2R)eve^{1.27}$	46C3,4;C9,11	—	eve	

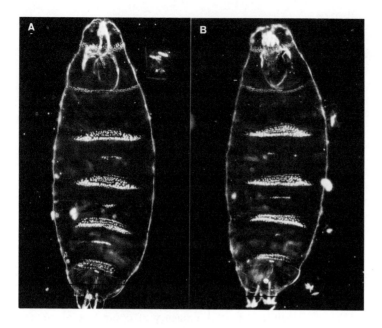

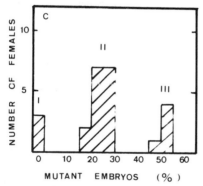

Figure 2. Assay for maternal effect of *odd* by pole cell transplantation into agametic flies. (*Top*) *odd* embryos derived from pole cells homozygous (*A*) or heterozygous (*B*) for *odd*. (*Bottom*) Proportion of host females that received +/+ (class I), *odd*/+ (class II), or *odd*/*odd* (class III) pole cells as evidenced by the proportion of *odd* embryos among their offspring.

ozygous germ line; thus, the genes are probably not expressed, or very weakly expressed, during oogenesis (or the expression is irrelevant for embryonic development).

Mutant Phenotypes

odd-skipped. In the X-ray mutant screen described above, no deletion for *odd* was isolated. The one X-ray-induced allele, whose phenotype is identical to that of several EMS-induced alleles, proved to be cytologically normal. In mutant embryos, the posterior part of the denticle band and adjacent naked cuticle in every other segment (mesothorax and the odd-numbered abdominal segments) are replaced by a mirror-image duplication of the anterior part of the denticle band and adjacent naked cuticle (Fig. 4). This is the same type of duplication seen in the segment-polarity mutant patched in every segment and in the pair-rule mutant runt in every other segment (with a deletion of the alternate segments; Nüsslein-Volhard and Wieschaus 1980; Gergen and Wieschaus 1985). This pattern is also observed in a synthetic deficiency of 23E-24B created by crossing $T(1;2)odd^{1.10}$ with $T(2;3)odd^{5.1}$, although the differentiating quality of embryos heterozygous for any of the two deficiencies is rather poor, indicating that it reflects the amorphic condition of the gene. As mentioned above, the construction of germ line chimeras to obtain homozygous mutant embryos from a homozygous germ line did not yield an increase in the size of the deletions. Thus, despite the similarity in name, odd-skipped in phenotype is quite different from the pair-rule gene even-skipped, described below.

paired. The amorphic phenotype of *prd* as represented by deletion homozygotes is identical to that of strong point mutants (Nüsslein-Volhard and Wieschaus 1980; Sander et al. 1980), although in embryos mutant for strong alleles, there is a tendency of irregular double-segment fusion, especially on the dorsal and lateral sides of the larva. Strong *prd* embryos have half the normal number of segmental units, which are somewhat larger than normal, the increase in size predominantly affecting the denticle band, rather than the naked cuticle of the "double segment." Thus, the size of the deleted region approaches one segmental unit per segment pair. The determination of the exact deletion frame and the segmental identity of the remaining regions followed two strategies: recording the deletion pattern in weak alleles and double-mutant combinations with the homeotic mutant Ultrabithorax (*Ubx*), which specifically transforms the anterior part of A1 into thorax. The deletions are smaller and less regular in a weak *prd* allele than in strong alleles, starting in the naked cuticle between the denticle bands and, with increasing size, including the anterior band of the posterior segment of the pair (Fig. 3). The exact boundary of segment identity is best seen in the first abdominal segment of embryos mutant for both *prd* and *Ubx*. In such embryos, a large anterior part of this denticle band shows thoracic morphology, whereas the posterior margin and the naked cuticle that follow posteriorly are abdominal (presumably A2) in character (Fig. 4).

In antero-posterior order, the deletions in the *prd* pattern start with derivatives of the mandibular segment, the ventral arms, and the Lateralgräten. The mouthhook is somewhat reduced in size, whereas the other maxillary derivatives (maxillary sense organ and cirri) are normal. The labial sense organs and the anterior belt of the prothorax are missing, and the deletion pattern proceeds posteriorly in a very regular pattern, deleting posterior mesothorax and anterior metathorax, posterior A1 and anterior A2, and so forth. Of the structures of the posterior terminal region (the telson), the posterior lateral sense organs are missing, the tuft is reduced to about five denticles, and the filzkörper (the tracheal endings) are not stretched.

even-skipped. In the initial screens for segmentation mutants (Nüsslein-Volhard and Wieschaus 1980), only two *eve* alleles were isolated, of which one is rather weak and the other is temperative-sensitive. At room temperature, the latter allele has a typical and regular double segmental phenotype similar to the paired phenotype (Nüsslein-Volhard and Wieschaus 1980). Although it was clear that neither of these alleles could be amorphic, it came as a surprise that a small deletion (including the *eve* locus), isolated in the X-ray screen described above, had a phenotype that was qualitatively different from that of the point mutants: Embryos homozygous for $Df(2R)eve^{1.27}$ do not produce segmental subdivisions, but instead their ventral region is covered with a lawn of blunt denticles pointing toward the ventral midline (Figs. 4 and 5). Subsequently, an EMS-induced point mutant, eve^{R13}, which shows the same apparently amorphic phenotype, was fortuitously isolated. At the anterior margin of this lawn, the denticles are fine and unpigmented like thoracic denticles, suggesting that different segmental identities are present throughout the field. The entire lack of segmentation in these embryos is also evident in the terminal regions, head and telson, of the embryo. All identifiable derivatives of the gnathal segments are missing, such as maxillary sense organs, cirri, mouthhooks, labial sense organs, and the mandibular parts of the cephalopharyngeal skeleton. The labrum, the antennal sense organs, and a rudimentary skeleton are the only remains of the larval head. Of the structures of the telson, the anal plates and some of the sensory organs are present, whereas the spiracles, the filzkörper, and the tuft are very much reduced in size.

The weakest *eve* phenotype is seen in larvae heterozygous for the deficiency, in which frequently the denticle band of the sixth abdominal segment and sometimes that of the second abdominal segment and the metathorax are partly or completely deleted. This haplo-insufficient dominant phenotype in certain backgrounds can be strong enough to reduce the viability of the heterozygotes to about 10% of the normal value.

All three hypomorphic alleles in *trans* to the deficiency show the regular pair-rule phenotype described for even-skipped above, the remaining denticle bands being those of the odd-numbered abdominal and even-numbered thoracic segments. The deletions span a region of the size of one segment and extend from the middle of the naked region, which in the thorax is defined by the Keilin's organs, to the middle of the naked region of the next segment. The frame of deletions is thus shifted by a third of a segment's length compared with paired (Figs. 4 and 7).

With the increasing strength of the allele combination ($eve^{ID}/Df(2R)eve^{1.27}$) at higher temperature, an irregular fusion of the double segments is observed, accompanied by substantial enlargements of the denticle bands, finally leading to the lawn phenotype expressed in the deficiency homozygote. This series of stronger phenotypes involves not only deletion patterns, but also enlargements of remaining pattern elements. The enlargements only concern the regions covered with denticle bands, and as the phenotypes get stronger, the naked regions get smaller.

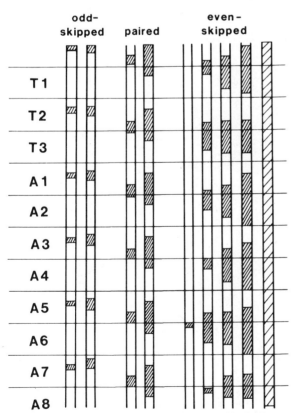

Figure 3. Schematic representation of weak and strong pair-rule phenotypes, showing pattern deletions (hatched areas). The rightmost column represents the "lawn" phenotype caused by the amorphic allele eve^{R13}.

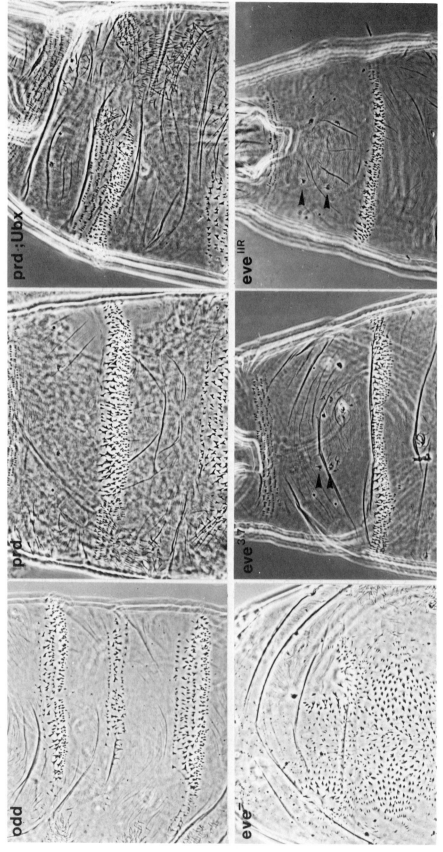

Figure 4. Phase-contrast photographs of cuticle pattern of pair-rule mutant embryos. *odd*: denticle bands of A2 to A4 are shown with polarity reversal in A3. *prd*: first abdominal denticle band composite of anterior A1 and posterior A2 as revealed by homeotic transformation in *prd;Ubx*. *eve*: no segmented pattern in *eve*[R13], tandem array of Keilin's organs in T2/T3 double segment of *eve*[3.77] and *eve*[IIR] (arrowheads).

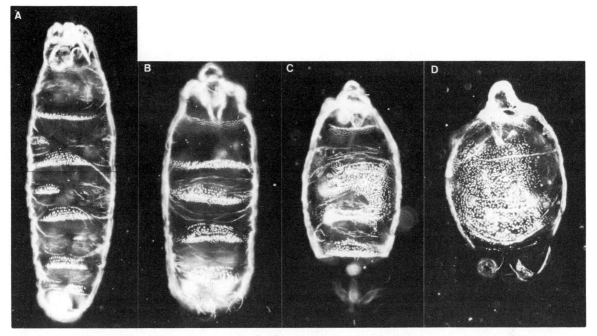

Figure 5. Dark-field photographs of a phenotypic series of *eve* mutant embryos showing progressive loss of segments from left to right. Genotypes: eve^{IIR} (A); $eve^{ID}/Df(2R)eve^{1.27}$ at 25°C (B) and at 27°C (C); $Df(2R)eve^{1.27}$ (D).

The phenotypes of even-skipped are shown in Figures 4 and 5, and the phenotypic series is illustrated in Figure 3. This description also shows that the frequency and size of the deletions in the weak and in the strong phenotypes are not equal throughout the larva. In the weak phenotypes, the first affected is the denticle band of A6 and the last affected is that of A8; in the strong phenotypes, the fusion first affects the double segments derived from A1 and A3, and last to fuse is the thoracic region.

DISCUSSION

Among the pair-rule genes, the *eve* gene is unique because in the complete absence of gene function, segmentation is abolished completely throughout the embryo. Instead, a lawn of denticles without polarity covering the ventral side of the larva is formed. In weak *eve* alleles, a rather regular pair-rule-type phenotype is observed that superficially resembles that of other pair-rule genes, *prd*, fushi tarazu (*ftz*), odd-paired (*opa*), and hairy (*h*), showing half the normal number of segmental units. It has been shown that the pair-rule genes *ftz* and *h* are transcribed during the blastoderm stage in a pattern of seven stripes the width of one segmental anlage, interspersed with equally sized stripes of cells that do not express the gene (Hafen et al. 1984; Ish-Horowicz et al., this volume). Whether the unsegmented "lawn" phenotype seen in *eve*⁻ homozygous embryos reflects an expression of the gene in all cells of the segmented region of the embryo will be seen when cloned probes of the gene become available. It is also equally possible that *eve* is expressed in stripes (as suggested by the weak phenotypes) but that the gene function is not autonomous and not limited to the cells in which the gene is transcribed.

Although *eve* is unique among the pair-rule genes, the lawn phenotype is not unique, since it is also expressed in double-mutant combinations of other segmentation genes. Figure 6 shows an embryo homozygous for the gap mutants hunchback (*hb*) and knirps (*kni*) as well as one homozygous for the pair-rule mutants *h* and *ftz*. The component mutants in these double-mutant combinations have almost complementary phenotypes: what is lacking in one is present in the other and vice versa. In *hb* embryos, the thorax and

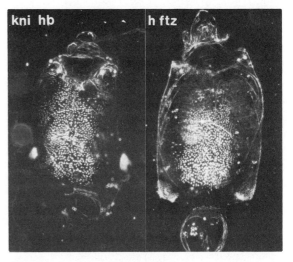

Figure 6. Dark-field photographs of the unsegmented lawn phenotype caused by the gap gene double-mutant *kni hb* (*left*) and by the pair-rule double-mutant *h ftz* (*right*).

the eighth abdominal segment are deleted (Lehmann 1985), whereas in *kni*, abdominal segments A1 through A7 do not develop the normal segmented pattern. The pair-rule genes *h* and *ftz* show alternating deletion patterns in a metasegmental or parasegmental frame (see below). The lawn phenotype, in contrast to the phenotypes of most of the segmentation mutants, cannot be described as a deletion pattern only, since the denticle region is larger than expected from a simple addition of the remaining regions in the single mutants. It is also larger than the denticle regions remaining in weak alleles of *eve*. We interpret the lawn phenotype as a failure of the embryo to segment at all. Why a homogeneous field of denticles, rather than a naked cuticle, remains is not clear. No zygotic mutants that form only the unsegmented naked cuticle have been found.

Deletion of exactly one segmental equivalent per double-segment unit is the most frequent general phenotype among the pair-rule loci. It has been pointed out previously (Nüsslein-Volhard and Wieschaus 1980; Nüsslein-Volhard et al. 1982) that the borders of the deleted regions never coincide with the segment boundary. Instead, they appear to be defined by either of two lines within the segment, and examples of both are given in the hypomorphic phenotype of *eve* and the amorphic phenotype of *prd* as described in this paper.

Figure 7 indicates the positions of the two lines as well as the segment boundary in the thorax and anterior abdomen of the larva. For the determination of the homologous positions in thorax and abdomen, the "spiral" larva shown in Figure 8, in which different segmental identities are expressed in the right and left sides, is helpful. The one line runs through the Keilin's organs in the thorax and homologous positions in the abdominal segments. This line corresponds to the anterior-posterior compartment boundary within the larval segment. It also defines the realm of action of mutants in the bithorax complex (Struhl 1984); the units bounded by this line have been termed parasegments (Martinez-Arias and Lawrence 1985). The pair-rule genes *eve* and *ftz* have a parasegmental deletion pattern, whereby the *ftz* pattern is shifted by one parasegment relative to the *eve* pattern (Fig. 7). The other line runs through the posterior margin of the denticle bands in the abdomen and homologous positions in the thoracic segments, separating the anterior compartment into two parts. In addition to *prd*, the pair-rule genes *opa* and probably *h* (Ingham et al. 1985) show a deletion pattern defined by this line, which we propose to call the metasegmental pattern. Whether this line also defines a compartment boundary is not clear and cannot be readily assessed. Meinhardt (1982, 1984) postulates the existence of at least three compartments per segment, termed S (separation), A (anterior), and P (posterior), with the segment boundary forming at the juxtaposition of P and S. Although Meinhardt's view is based primarily on theoretical arguments, support can be found in several transplantation experiments of Bohn (1974) and Wright and Lawrence (1981). Furthermore, in transplantation experiments within the larval segments of the bug *Oncopeltus*, D.A. Wright (pers. comm.) found evidence for two clonal restriction lines located within the segment that might well correspond to the two lines defined by the deletion pattern of the pair-rule mutants. It is interesting to note that in the larval segment from anterior to posterior, three muscle-attachment sites are present in the ventral side (Szabad et al. 1979). The most prominent site defines the segment boundary, and the two minor sites may well correspond to the lines defining the meta- and parasegmental frames.

Although most of the pair-rule genes show mutant phenotypes deleting alternating parasegments (*eve*, *ftz*) or metasegments (*prd*, *opa*, *h*), which in both frames cover the entire segmented region, the odd-skipped phenotype is different. Only a small region within every other segment is deleted in embryos that are deficient for the gene and also when the gene is already absent during oogenesis. In *odd*, the deletions are found in the mesothorax and the odd-numbered abdominal segments; a gene with a complementary phenotype (i.e., deleting homologous regions in the even-numbered segments) has not been identified. Another gene with a singular pair-rule phenotype is *runt*, in which the deletions are larger than one segment and accompanied by mirror-image duplications. These unique phenotypes indicate a nonequivalence of alternating segments and a hierarchy in gene function in the process of segmental subdivisions.

The construction of germ line chimeras has shown

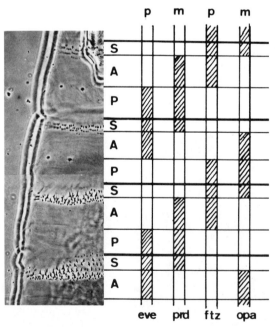

Figure 7. Schematic representation of repetitive subdivisions of the larval pattern shown here for thoracic segments T2 and T3 and abdominal segments A1 and A2. Two repeating units, each one segment long, are defined by the complementary pattern defects (hatched areas) of pair-rule mutants, parasegments (p) by *eve* and *ftz*, and metasegments (m) by *prd* and *opa*. Their boundaries (thin lines) and the intersegmental boundaries (thick lines) define three subdivisions, or compartments, in each segment, called S, A, and P (see text).

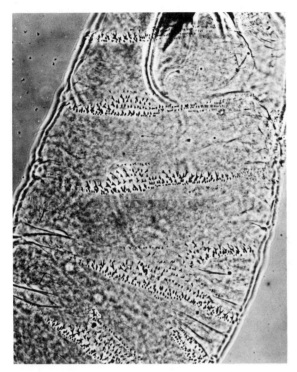 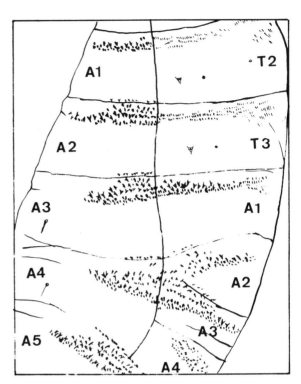

Figure 8. Phase-contrast photograph (*left*) and drawing (*right*) of a "spiral" larva revealing homology of positions between thoracic and abdominal segments. Vertical line separates the left half and the right half of the segments; horizontal lines indicate segment boundaries. This pattern abnormality was caused by UV-laser irradiation of wild-type embryo (Lohs-Schardin et al. 1979).

that of the three genes *odd*, *prd*, and *eve*, none are required during oogenesis; the same results have been obtained for runt (E. Wieschaus and E. Noell, in prep.), *h* (Ingham et al. 1985), and *ftz* (C. Nüsslein-Volhard, unpubl.). In both *prd* and *eve* mutant embryos, a deviation from normal development can be observed at the onset of gastrulation, indicating gene expression during the blastoderm stage. The temperature-sensitive periods for both *prd* and *eve* occur for a short time around the cellular blastoderm stage. For runt, a very similar temperature-sensitive period has been determined (E. Wieschaus, pers. comm.). These observations and the time course of gene expression determined by in situ hybridization for the *h* and *ftz* genes mentioned above suggest that all the pair-rule gene functions are required for a very short period of development, the cellular blastoderm stage.

ACKNOWLEDGMENTS

We thank Eric Wieschaus for his valuable contributions during earlier periods of this work, Ulrike Schier for expert technical assistance in the chromosomal analysis, and Ruth Lehmann for stimulating discussions.

REFERENCES

BOHN, H. 1974. Extent and properties of the regeneration field in the larval legs of cockroaches (*Leucophaea maderae*). *J. Embryol. Exp. Morphol.* **31:** 557.

BUSSON, D., M. GANS, K. KOMITOPOULOU, and M. MASSON. 1983. Genetic analysis of three dominant female-sterile mutations located in the X-chromosome of *Drosophila melanogaster*. *Genetics* **105:** 309.

FROHNHÖFER, H.G. 1982. "Abgrenzung maternales and zygotischer Anteile bei der Genetischen Controlle der Musterbildung in *Drosophila melanogaster*." Ph. D. thesis, Tübingen University.

GERGEN, J.P. and E.F. WIESCHAUS. 1985. The localized requirements for a gene affecting segmentation in *Drosophila*: Analysis of larvae mosaic for *runt*. *Dev. Biol.* **109:** 321.

GLOOR, H. 1950. Schädigungsmuster eines Letalfaktors (*Kr*) in *Drosophila melanogaster*. *Arch. Julius Klaus-Stift.* **29:** 277.

HAFEN, E., A. KUROIWA, and W.J. GEHRING. 1984. Spatial distribution of transcripts from the segmentation gene *fushi tarazu* during *Drosophila* embryonic development. *Cell* **37:** 833.

INGHAM, P.W., S.M. PINCHIN, K.R. HOWARD, and D. ISH-HOROWICZ. 1985. Genetic analysis of the *hairy* locus in *Drosophila melanogaster*. *Genetics* (in press).

JÜRGENS, G., E. WIESCHAUS, C. NÜSSLEIN-VOLHARD, and H. KLUDING. 1984. Mutations affecting the pattern of the larval cuticle in *Drosophila melanogaster*. II. Zygotic loci on the third chromosome. *Wilhelm Roux's Arch. Dev. Biol.* **193:** 283.

LAWRENCE, P.A., P. JOHNSON, and G. STRUHL. 1983. Different requirements for homeotic genes in the soma and germ line of *Drosophila*. *Cell* **35:** 27.

LEHMANN, R. 1985. "Regions spezifische Segmentierungsmutanten bei *Drosophila melanogaster*." Ph. D. thesis, Tübingen University.

LINDSLEY, D.L. and E.H. GRELL. 1968. Genetic variations of *Drosophila melanogaster*. *Carnegie Inst. Wash. Publ.* no. 627.

LOHS-SCHARDIN, M., C. CREMER, and C. NÜSSLEIN-VOL-

HARD. 1979. A fate map for the larval epidermis of *Drosophila melanogaster*: Localized cuticle defects following irradiation of the blastoderm with an UV-laser microbeam. *Dev. Biol.* **73:** 239.

MARTINEZ-ARIAS, A. and P.A. LAWRENCE. 1985. Parasegments and compartments in the *Drosophila* embryo. *Nature* **313:** 639.

MEINHARDT, H. 1982. *Models of biological pattern formation.* Academic Press, London.

———. 1984. Models for positional signalling, the threefold subdivision of segments and the pigmentation patterns of molluscs. *J. Embryol. Exp. Morphol.* (suppl.) **83:** 289.

NÜSSLEIN-VOLHARD, C. 1977. A rapid method for screening eggs from single *Drosophila* females. *Drosophila Inform. Serv.* **52:** 166.

NÜSSLEIN-VOLHARD, C. and E. WIESCHAUS. 1980. Mutations affecting segment number and polarity in *Drosophila*. *Nature* **287:** 795.

NÜSSLEIN-VOLHARD, C., E. WIESCHAUS, and G. JÜRGENS. 1982. Segmentierung bei *Drosophila*—Eine genetische Analyse. *Verh. Dtsch. Zool. Ges.* **91:** .

NÜSSLEIN-VOLHARD, C., E. WIESCHAUS, and H. KLUDING. 1984. Mutations affecting the pattern of the larval cuticle in *Drosophila melanogaster*. I. Zygotic loci on the second chromosome. *Wilhelm Roux's Arch. Dev. Biol.* **193:** 267.

SANDER, K., M. LOHS-SCHARDIN, and M. BAUMANN. 1980. Embryogenesis in a *Drosophila* mutant expressing half the normal segment number. *Nature* **287:** 841.

STRUHL, G. 1984. Splitting of the *Bithorax* complex of *Drosophila*. *Nature* **308:** 454.

SZABAD, J., T. SCHÜPBACH, and E. WIESCHAUS. 1979. Cell lineage and development of the larval epidermis of *Drosophila melanogaster*. *Dev. Biol.* **73:** 265.

WAKIMOTO, B.T. and T.C. KAUFMAN. 1981. Analysis of larval segmentation in lethal genotypes associated with the *Antennapedia* gene complex in *Drosophila melanogaster*. *Dev. Biol.* **81:** 51.

WIESCHAUS, E., C. NÜSSLEIN-VOLHARD, and G. JÜRGENS. 1984a. Mutations affecting the pattern of the larval cuticle in *Drosophila melanogaster*. III. Zygotic loci on the X-chromosome. *Wilhelm Roux's Arch. Dev. Biol.* **193:** 296.

WIESCHAUS, E., C. NÜSSLEIN-VOLHARD, and H. KLUDING. 1984b. *Krüppel*, a gene whose activity is required early in the zygotic genome for normal embryonic segmentation. *Dev. Biol.* **104:** 172.

WRIGHT, D.A. and P.A. LAWRENCE. 1981. Regeneration of segment boundary in *Oncopeltus*. *Dev. Biol.* **85:** 317.

Regulation of the Genes of the Bithorax Complex in *Drosophila*

E.B. LEWIS
Division of Biology, California Institute of Technology, Pasadena, California 91125

The bithorax complex (BX-C) is a set of master control genes that regulates other genes outside the complex to program much of the development of the thorax and abdomen of the fly. The emphasis in this paper will be on how the genes of the complex are themselves regulated, rather than on how they regulate other genes.

The BX-C is regulated in *cis* and in *trans*. Only a brief review of its regulation by *trans*-acting loci will be given, since our primary concern in this paper is to identify the rules that govern regulation in *cis* and to outline a model for *cis*-regulation based on these rules.

Regulation of the BX-C in *trans* appears to be chiefly negative in the sense that loss-of-function alleles at several loci outside the complex are known to result in derepression of BX-C genes in regions of the body where the latter are normally repressed. Examples are Polycomb, *Pc* (Lewis 1978; Duncan and Lewis 1981; Denell and Frederick 1983); Polycomb-like, *Pcl* (Duncan 1982); extra sex combs, *esc* (Struhl 1981); super sex combs, *sxc* (Ingham 1984); and polyhomeotic, *ph* (Dura et al. 1985).

Only one gene outside the BX-C has been found to act as a positive regulator of the complex in the sense that loss-of-function alleles fail to activate the wild-type functions of BX-C. For example, animals heterozygous for a dominant Regulator-of-bithorax (*Rg-bx*), have slightly reduced pigmentation of the tergite of the fifth abdominal segment (A5); a similar phenotype is seen in deficiencies, which include *Rg-bx* and are deleted for several bands in 88B, namely, $Df(3)red^{P52}/+$ and $Df(3)red^{P93}/+$ animals (see Lewis 1968, 1982). Garcia-Bellido and Capdevila (1978) and Capdevila and Garcia-Bellido (1981) have shown that *Rg-bx* acts maternally, as well as zygotically, in repressing BX-C genes. A recessive mutant, trithorax (*trx*), has been shown by Ingham and Whittle (1980) and Ingham (1981) to have strong maternal and zygotic repression of many gene functions of the BX-C and to fail to complement with *Rg-bx* or with the two small red deficiencies in 88B cited above. Thus, *Rg-bx* and *trx* evidently occupy a single locus that codes for a regulatory product; *trx* might have inducer-like properties or it could operate as an activator of the BX-C genes. However, Ingham (1983) has shown that the BX-C genes are still differentially expressed in animals that are doubly mutant for *trx* and *esc*.

In the egg, a gradient in the concentration of an inducer (Lewis 1963) or of a repressor (Lewis 1978) has been invoked to account for sequential derepression of the BX-C genes. Specifically, in the latter case, an antero-posterior gradient in the concentration of repressor along the body axis is assumed to be coupled with a proximo-distal gradient along the chromosome in the affinities for repressor of *cis*-regulatory elements of the BX-C genes. The proximo-distal gradient has been proposed to account for the preliminary finding that the order of the genes in the chromosome is colinear with the order of their expression along the body axis of the organism. Evidence that this type of colinearity extends to the distal half of the complex is reviewed below.

The remainder of this paper deals with *cis*-regulation of the BX-C. Two highly unusual genetic phenomena have been shown to characterize such regulation in the proximal portion of the complex: (1) cisvection, or *cis*-inactivation of one gene function by a mutant allele of a neighboring gene; and (2) transvection, or synapsis-dependent complementation of neighboring gene functions (for review, see Lewis 1955, 1982). In this paper evidence is summarized which indicates that cisvection and transvection also occur in the distal portion of the BX-C.

METHODS

Two genetic screens have been devised for the purpose of identifying as many as possible of the functional units within the BX-C. The first, or MCP screen, involves searching for revertants of the dominant gain-of-function mutant Miscadastral pigmentation (*Mcp*). The *Mcp* phenotype consists of a strong transformation of the fourth abdominal segment (A4) toward the fifth abdominal segment (A5) and is readily scored in males as they have solid black pigmentation on A4, A5, and A6, whereas wild-type males have such pigmentation only on A5 and A6. M. Crosby, who found and mapped *Mcp* (see Lewis 1978), showed that several types of revertants can be induced, one of which results from a loss of function in an adjoining infra-abdominal-5 (*iab-5*) gene. In this latter case, the revertant proved to be a second mutant, $iab-5^{C7}$, very close to the right of *Mcp* (M. Crosby, unpubl.); this and several other revertants have been mapped by molecular methods (Karch et al. 1985). To apply the MCP screen, one mutagenizes *Mcp* homozygotes (either males or females, as desired) and mates to wild type. The F_1 males are then scored for partial or complete loss of the male-type pigmentation on A4.

The second, or global rearrangement (GR) screen, is designed to detect virtually all gross chromosomal rearrangements that have one breakage point within the BX-C. This screen employs a *cis*-arrangement of two dominant mutants, Contrabithorax (*Cbx*), a dominant gain-of-function mutant that transforms the second thoracic segment (T2) toward the third thoracic segment (T3), and Ultrabithorax (*Ubx*), a dominant loss-of-function mutant that transforms T3 toward T2. The *cis*-heterozygote, *Cbx Ubx*/ + +, has nearly, but never quite entirely, complete suppression of the *Cbx* phenotype when the third chromosomes are structurally homozygous for the wild-type sequence, but it does have complete suppression of the *Cbx* phenotype when those chromosomes are structurally heterozygous for a transvection-suppressing rearrangement (TSR) of the type that suppresses transvection of the bx^{34e}/*Ubx* genotype (Lewis 1954).

The GR screen involves mutagenizing the wild-type males and mating to females homozygous for *Cbx Ubx* and for a duplication, *Dp(3;1)68*, which is an insertion of region 89D to 89E into the proximal portion of the X chromosome. Such females have wild-type wings and halteres. Thus, this duplication fully rescues the otherwise lethal *Cbx Ubx* homozygote. These tester-strain females are also homozygous for a recessive marker in X (yellow-2) and one in the third chromosome (glass-3). The F_1 flies that are *Dp(3:1)68; Cbx Ubx*/ + + have weak traces of the *Cbx* phenotype, consisting of spread wings and partially reduced alulae; however, when flies of this genotype are structurally heterozygous for a TSR, their wings are normal (Lewis 1955). It should be noted that the extra copy of *Ubx*⁺ is without influence on the wing phenotype for two reasons: (1) The slight *Cbx* effect is a dominant gain of function not suppressible by an extra dose of wild type, and (2) the *Ubx*⁺ gene in the duplication is apparently too remotely located to be affected by the *Cbx* mutant. F_1 males with normal wings are therefore selected and backcrossed to the tester strain to check for transmission of suppression of transvection, or, since the vast majority of normal-winged flies in the F_1 of the screen do show such transmission, the F_1 males are often mated directly to a homozygous bithoraxoid (*bxd*) line. From this latter mating, male larvae that have eight rows of abdominal setal bands (see Fig. 1) are heterozygous for the mutagenized wild-type chromosome, whereas those with seven rows have a *bxd/Cbx Ubx* genotype; the latter are therefore discarded and the former are analyzed cytologically by examining their salivary gland chromosomes.

Rearrangements detected in the GR screen obey the same rules as those found in the original study of TSRs (Lewis 1954). Although the breakage points of such rearrangements may occur anywhere in a region from the centromere to the BX-C and slightly beyond, a distance of over 500 bands in the salivary gland chromosomes, only those having breakages within the 89D–89F region are generally saved for further study; they, in turn, are tested in a variety of ways, including determining their phenotypes when opposite deletions for portions or all of the BX-C.

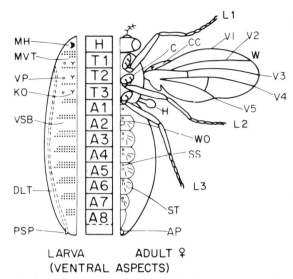

Figure 1. Comparison of ventral cuticular patterns in the larval first-instar and adult stages of *Drosophila*. (MH) Mandibular hooks; (MVT) midventral tuft; (VP) ventral pits; (KO) Keilin's organ; (VSB) ventral setal belts; (DLT) dorsal longitudinal (tracheal) trunk; (L) leg; (W) wing; (H) haltere; (C) coxa; (CC) costal cell (of wing); (V) vein; (WO) Wheeler's organ; (SS) sensillum (on segments A1 to A7, inclusive); (ST) sternite; (AP) anal plate. For a detailed account of larval cuticular patterns, see Lohs-Schardin et al. (1979). (Modified from Lewis 1982.)

The MCP screen detects in the F_1 both lethal and nonlethal mutants that weaken the dominance of *Mcp*. The GR method detects in the F_1 generation not only lethal and nonlethal mutants, but also dominant sterile mutants that are made fertile by covering with *Dp(3;1)68*; in addition, the GR method can detect breakages within the BX-C that have little or no detectable effect on any of the BX-C genes.

The GR screen detects the chromosomal rearrangements, almost exclusively. Point or pseudopoint mutations are not detected, except in the case of *Ubx* mutants. The latter are detected as would be expected on the basis that substitution of such a mutation for *Ubx*⁺ in the *Dp-68; Cbx Ubx*/ + + genotype would be expected to suppress the remaining slight *Cbx* phenotype. The GR method also detects tandem duplications of the BX-C, evidently because the pairing of the two BX-C regions within the duplication-bearing chromosome tends to exclude their interaction with the homologous *Cbx Ubx* chromosome.

RESULTS

Relevant cuticular patterns of the wild-type organism are shown in Figure 1. The current status of the correlation of cytological, genetic, and molecular maps of the BX-C is shown in Figure 2. The entire complex

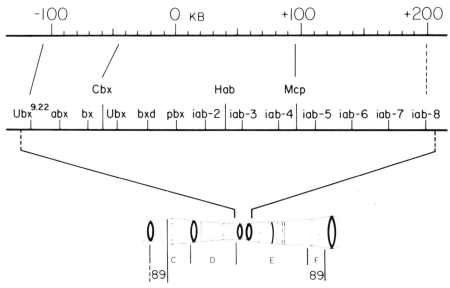

Figure 2. Correlation of the cytological, genetic, and molecular maps of the BX-C. On the DNA map, the location of *pbx* is 0.0, since it was the starting point of the walk (Bender et al. 1983); for reference points, only the approximate locations of the *9.22* allele of *Ubx*, *Cbx* insertion point, and a pseudopoint small deletion mutant *Mcp* are shown. For details of the genetic mapping of the region, see Lewis (1982) and for the molecular maps see Karch et al. (1985).

appears to be contained within the first two doublets of section 89E.

Several revisions and additions are shown in the genetic map since the last map was published (Bender et al. 1983). The location of $Ubx^{9.22}$ (Kerridge and Morata 1982) on the genetic and molecular maps is based on the work of Akam et al. (1984), who found it to be a pseudopoint mutation in the major leftmost, or 3', exon of the *Ubx* transcripts. The location of Hyperabdominal (*Hab*) has been found to be to the right of an *iab-2* allele, *iab-2*^{C53} (I. Duncan, pers. comm.); previously, its position was only known to be at the right of *pbx*. Evidence for the existence of additional loci beyond *iab-2* is based on genetic and recombinant DNA studies summarized by Karch et al. (1985) and results reported in this paper.

From a screen for detecting lethals, Sanchez-Herrero et al. (1985) reported finding two lethal complementation groups in the distal half of the complex. They have suggested that this implies only two functional domains in that half. From the work of Karch et al. (1985), one of their lethal complementation groups is confined to the *iab-2* region, whereas the other includes one or more loci distal to *iab-5*. A lethal complementation screen fails to detect lesions in *iab-3*, *iab-4*, and *iab-5*, since inactivations of these genes, including deletions thereof, are viable as adults. However, each has a specific function, and *iab-4*$^+$ turns out to control whether gonads will form. The utility of using survival to adulthood as a criterion for defining higher functional units with the BX-C is open to question, since animals totally lacking the BX-C complete embryonic development and die as first-instar larvae (Lewis 1978); however, Sanchez-Herrero et al. (1985) have done an important service in identifying what may turn out to be a significant correlation between adult lethality and the presence of homeo boxes within the complex (see below).

Cisvection in the Proximal Half of BX-C

Cisvection in the region from anterobithorax (*abx*) to postbithorax (*pbx*), inclusive, has been discussed elsewhere (Lewis 1952, 1955, 1982). Examples shown in Figure 3 are those cases of cisvection in which the direction of the effect can be deduced. Consider the two mutants *bxd* and *pbx*. The *cis*-heterozygote *bxd pbx/ + +* is wild type, whereas the *trans*-heterozygote *bxd +/+ pbx* has a *pbx* phenotype. Thus, in the latter case, the first abdominal segment (A1) fails to transform toward a thoracic one, as it does in homozygous *bxd* genotypes, whereas the posterior compartment of the third thoracic segment (T3) partially transforms toward that of the second (T2), but to a slightly lesser degree than it does in *bxd* or *pbx* homozgyotes. In other words, *bxd* partially *cis*-inactivates *pbx*$^+$, but *pbx* does not show detectable *cis*-inactivation of *bxd*$^+$.

Cisvection in the region from the position of the $Ubx^{9.22}$ mutation to that of Ubx^1 is of special interest, since these mutants occupy different exons in the same transcription unit, as described above. The two mutants are indistinguishable from one another phenotypically in a large number of combinations tested, including those shown in Figure 3. Ubx^1 and $Ubx^{9.22}$ each shows partial complementation with either *abx* or *bx* mutants. Since these four types of mutants express the same type of transformation (anterior T3 toward anterior T2), direction of *cis*-inactivation cannot be unambiguously determined; therefore, these cases of cisvection have been omitted from Figure 3. Although the

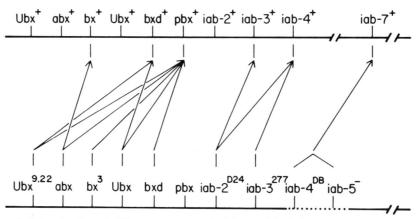

Figure 3. Direction of *cis*-inactivation of wild-type genes (top line) imposed by individual mutant genes whose locations are shown on the bottom line is indicated by means of sloping lines ending in an arrow and is from a proximal to a distal direction in every case. Only cases in which the direction of *cis*-inactivation has been established are indicated, as explained in the text. For descriptions of mutant symbols, see Fig. 2 and text.

abx and *bx* mutants show a very similar type of transformation of the anterior compartment T3 toward that of T2, *abx* differs qualitatively in that it causes, with variable expressivity, a more complete transformation in the anteriormost region of that compartment than does bx^3. In $abx+/+bx^3$ animals, there is no trace of these effects of *abx*; however, such animals show only very weak and variable transformations of T3 toward T2. The direction of *cis*-inactivation can nevertheless be deduced indirectly in the sense that the bx^3 homozygote acts as if it causes overactivity of the adjoining abx^+ gene, rather than inactivation of that gene, as described in more detail elsewhere (Lewis 1981). Hence, in Figure 3, *abx* is shown as *cis*-inactivating bx^+.

Cisvection in the Distal Half of BX-C

Analysis of cisvection in the right half of the complex has been complicated by the lack of dramatic differences in some of the segments, such as A3 and A4, and by the finding that hemizygotes for a wild-type third chromosome express slight dominant loss-of-function phenotypes for many of the loci, especially *iab-5* and *iab-6*, as judged from partial loss of male pigmentation on A5 (A5 transforming toward A4) and extra bristles on the male sternite of A6 (A6 transforming toward A5). The analysis has also been handicapped by a shortage of point or pseudopoint mutations that would permit *cis-trans* tests to be performed. Suggestive evidence for cisvection can nevertheless be inferred from analysis of chromosomal rearrangements having breakages in that portion of the complex. The first example involved the *iab-2* and *iab-3* loci.

Previously, only one example of a recessive loss of function corresponding to that predicted for an *iab-3* gene had been found, namely, Ultra-abdominal-4 (Uab^4) (Lewis 1978). Although this mutant was originally detected as a dominant transformation of A1 toward A2, it exhibits, when hemizygous, a very strong transformation of the predicted phenotype for an *iab-3* mutant, namely, transformation of A3 toward A2.

A second *iab-3* mutant, $iab-3^{277}$ (found by D. Baker, California Institute of Technology), is a revertant of *Mcp* and has not been separated from *Mcp*. It is associated with a transposition of a region from 94 to 96 of the right arm of the third chromosome into the BX-C approximately at +65 kb (Karch et al. 1985), presumably within or adjacent to the *iab-3* gene. This mutant is associated with a strong transformation of A3 toward A2, not only when hemizygous, but also when in *trans* with *iab-2* mutants, including $iab-2^{D24}$ that Karch et al. (1985) find may be a true point mutation. Thus, *iab-2* is inferred from these results to *cis*-inactivate $iab-3^+$.

The $iab-3^{277}$ mutant lacks the strong, dominant *Uab* effect associated with the Uab^4 mutant; however, when hemizygous, it expresses a very weak transformation of A1 toward A2, or, in other words, a very weak *Uab* effect. This transformation was first noticed by B. Weiffenbach and W. Bender (pers. comm.).

Five *iab-4* mutants have been identified. Two, $iab-4^{45}$ and $iab-4^{302}$ (found by R. Baker, California Institute of Technology), are from the GR screen and three are from the MCP screen. Two of the latter, $iab-4^{125}$ (found by J. von der Ahe, California Institute of Technology), and $iab-4^{166}$ (found by R. Baker), are partial revertants of *Mcp*, and the third (found by D. Baker) is a complete revertant of *Mcp* and has a deletion extending from +80 to +118 kb on the DNA map (Fig. 2); it is designated $iab-4,5^{DB}$ to indicate its dual loss of function in the *iab-4* and *iab-5* regions.

All five *iab-4* mutants are viable when hemizygous. Externally, in males hemizygous for the *45* and *302* alleles, the black band bordering the rear margin of segments A3 and A4 fails to reach the edge in A4 and in A3, whereas in wild-type males, it reaches the margin in A4 but not in A3 (the latter subtle difference was noticed by I. Duncan [pers. comm.]; for further details and illustration of this effect, see Karch et al. [1985]). Since the *125* and *166* alleles still have residual *Mcp* effects, they are not useful for analyzing loss-of-function effects in the A4 tergite.

Internally, hemizygotes for four of the *iab-4* mutants show loss of gonads in both sexes, whereas the *iab-4⁴⁵* hemizygote shows only partial loss, or reduction in size, of the gonads. Since loss-of-function mutants of *iab-5*, such as *iab-5³⁰¹* (described below), do develop gonads, the absence of gonads in *iab-4* mutants is taken to mean that *iab-4⁺* has the important function of controlling the initiation of gonadal development.

Partial to complete loss of gonads has been reported for heterozygotes involving the P10 allele of *iab-2* and the *Uab⁴* mutant (Lewis 1978). Such loss has also been observed in the case of *iab-3²⁷⁷* when opposite several *iab-2* mutants, including *iab-2^{D24}*. Finally, *iab-3* mutants in *trans* with several *iab-4* mutants tested also show partial to complete loss of gonads. Thus, *iab-2* and *iab-3* mutants *cis*-inactivate *iab-4⁺*. The *iab-4³⁰²* mutant is viable when homozygous, and in addition to showing strongly reduced gonad development, the ventral sternite of segment A2 is converted to one resembling that of A3. A weak transformation of this type in hemizygotes for the *iab-4³⁰²* allele was first noted by F. Karch and W. Bender (pers. comm.). The *iab-4⁴⁵* and *iab-4³⁰²* rearrangement-associated mutants show only partial *cis*-inactivation of *iab-5⁺* in the sense that in the hemizygotes, A5 transforms much more strongly toward A4 than do wild-type hemizygotes (*Df-P9/+*), whereas in the homozygote, which is viable only in the case of *iab-4³⁰²*, there is little or no effect on A5. (Since the 125 and 166 alleles of *iab-4* still carry the *Mcp* mutant, they are not useful for cisvection analysis.)

Cis-inactivation beyond *iab-5* is most clearly seen in the case of *iab-4,5^{DB}*. It is quite viable when homozygous and expresses a strong transformation of A4 toward A3 (already described above as an *iab-4* mutant effect) and of A5 toward A4 (an *iab-5* mutant effect expected because of the associated deletion of DNA in the *iab-5* region). In addition, the homozygote has partial transformation of A6 toward A5, principally dorsally in the middle of the tergite (a weak *iab-6* mutant effect). The hemizygote (*iab-4,5^{DB}/Df-P9*) is also quite viable and differs from the homozygote in having a much more extreme transformation of A6 toward A5 and in having a small tergite on A7 that is significantly larger than the tiny streak of A7 tergite tissue found in *Df-P9/+*. Thus, *iab-4,5^{DB}* shows partial *cis*-inactivation of *iab-6⁺* and *iab-7⁺* functions.

Two rearrangements from the GR screen, *iab-5³⁰¹* and *iab-5⁸⁴³* (found by R. Baker), are viable as hemizygotes and express a strong transformation of A5 toward A4 but show only a weak transformation of A6 toward A5, detectable in the pattern of trichomes. The breakage points of these rearrangements within the BX-C occur in the region to the right of *iab-4* breakages and near the middle of the *iab-4,5^{DB}* deletion that extends from approximately 85 kb to 115 kb (F. Karch and W. Bender; cited in Karch et al. 1985).

In the case of the *iab-5³⁰¹* rearrangement, the homozygote and the hemizygote are viable. The former shows a transformation of the pigmentation and bristle pattern at the lateral edge of the tergite of A3 toward that found in A4. The direction of this transformation is that expected for a gain-of-function phenotype (Lewis 1978). Again, as in the case of *iab-3* and certain *iab-4* mutants, there is gain of function for a gene just proximal to the gene exhibiting *cis*-inactivation of gene(s) immediately distal. Results indicating that *iab-5^{DB}* and another pseudopoint mutation, *iab-5^{C7}* (M. Crosby; cited in Lewis 1981), show *cis*-inactivation extending to *iab-7⁺* are discussed below.

Transvection

The known instances of transvection are diagrammed in Figure 4. An example of transvection between *Ubx* and *bx³⁴ᵉ* has been described previously (Lewis 1964), as well as one between *Ubx* and *pbx²* (Lewis 1982).

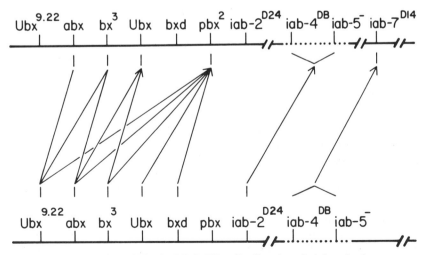

Figure 4. Established cases of transvection within the BX-C. When the direction of *cis*-inactivation accompanying the transvection phenomenon is known, it is indicated by an arrow and is always from a proximal to a distal direction. For a description of mutant symbols, see Fig. 2 and text.

Transvection between *iab-2* and *iab-4* has been assayed in *trans*-heterozygotes for the pseudopoint mutations, *iab-2^{D24}* and *iab-4,5^{DB}* (E.B. Lewis and R. Baker, in prep.). For example, among a total of 35 males of this genotype that were structurally homozygous for normal third chromosomes, only 3 had no trace of gonadal tissue, 16 had incompletely developed testes, and the remaining 16 each had one or both seminal vesicles pigmented, suggesting (see Stern and Hadorn 1939) that traces of gonadal tissue may have been present. When the *iab-2^{D24}* + / + *iab-4,5^{DB}* genotype carried a TSR, of the 65 males dissected, 49 had no trace of gonadal tissue, 1 had a testis that was one-half normal size, and the remaining 15 had one or both seminal vesicles pigmented. Hence, the TSR significantly reduced gonadal development.

Transvection involving the region from *iab-5* to *iab-7*, inclusive, has been detected by comparing *trans*-heterozygotes for *iab-5^{C7}* and *iab-7^{D14}* in the presence and absence of structural heterozygosity (E.B. Lewis and I. Duncan, in prep.). A significantly larger seventh tergite develops when a TSR is present than when the third chromosomes are structurally homozygous for the wild-type sequence. In this comparison, tergite size was measured by counting the number of bristles on that tergite on slides that had been coded and scored blindly. The results indicate that *iab-5^{C7}* *cis*-inactivates *iab-7$^+$* and that the TSR when heterozygous enhances that effect. The results, however, are complicated by the unexpected finding that the *iab-5^{C7}* chromosome acts as if it carries a dominant and closely linked maternal modifier of the *Rg-bx* type. Thus, when the mother carries that chromosome and the father contributes the *iab-7^{D14}* chromosome, the *iab-5^{C7}* + / + *iab-7^{D14}* male progeny have a significantly larger number of bristles on the seventh tergite than do males of the same genotype from the reciprocal mating. Nevertheless, transvection was demonstrable regardless of the maternal genotype. Care has therefore been taken to carry out transvection experiments with the mother of the same genotype; i.e., the TSR is always introduced through the male parent. Parallel analyses with *iab-5^{DB}* and *iab-7^{D14}* also indicate a significant effect of a TSR in enhancing tergite development in A7.

DISCUSSION

Rules for *Cis*-regulation

From the above results, three rules governing *cis*-regulation of the BX-C can be recognized: (1) The colinearity (COL) rule states that there is colinearity between the order of the genes in the chromosome and the order in which they show their effects along the body axis of the organism; (2) the *cis*-inactivation (CIN) rule states that a mutant lesion in a given gene tends to *cis*-inactivate one or more neighboring genes lying distal to that gene; and (3) the *cis*-overexpression (COE) rule states that certain lesions in a given gene tend to lead to an overexpression of the gene lying immediately proximal. Examples upon which these three rules are based are discussed below.

The COL rule continues to apply for much, and possibly all, of the distal half of the complex (Figs. 2 and 3). This conclusion is based on extensive molecular mapping studies of that region (Karch et al. 1985), as well as a limited number of genetic mapping and deletion experiments. The only known exception to this rule is the *pbx* locus, which would have been expected to lie proximal to *bxd*, since *pbx* affects T3 while *bxd* affects A1 (as well as T3).

The first examples of the CIN rule involved the *bx*, *Ubx*, and *bxd* mutants (Lewis 1952). Additional examples, reported in this paper, are summarized in Figure 3. Exceptions to the CIN rule occur in several cases. The known *pbx* mutant alleles, *pbx* and *pbx^2*, are pseudopoint mutations involving deletions of 17 kb and 15 kb, respectively, yet they fail to show detectable *cis*-inactivation of *iab-2$^+$*, the next most distally known gene in the complex (see Struhl 1984). A critical test of whether the distal half of the complex can function independently of the proximal half requires analysis of rearrangements having breakages in the intervening region. The GR screen is capable of detecting such rearrangements even though they might lack any detectable phenotype, other than their ability to acts as TSRs. None have yet been found in the GR screen; however, only nine TSRs that have breakages within the distal-half complex have been recovered. Whether it is chance that has kept the BX-C intact or whether the two halves are functionally interrelated cannot be determined from the existing data.

The first known example of the COE rule involves the *bxd* mutant. Thus, accumulation of the *Ubx$^+$* product has been specifically invoked (Lewis 1955) to explain the finding that *bx^3* + + *bxd*/*bx^{34e}* + + has a less extreme transformation of T3 toward T2 in the notal area than does the corresponding genotype in which *bxd* is replaced by *bxd$^+$*, namely, *bx^3* + + /*bx^{34e}* + + . (To make explicit that *Ubx$^+$* is homozygous, a plus sign is included as the middle symbol in these genotypes and in others that involve the *bxd* comparisons.)

That *Ubx$^+$* overexpression is actually present both in T3 and in T2 is strongly suggested by two other comparisons in which the possibility for such an effect has been sensitized by the use of gain-of-function mutants (E.B. Lewis, unpubl.). *Cbx* + *bxd*/ + + + has a more extreme transformation of T2 toward T3 than does *Cbx* + + / + + + . That the effect involves *cis*-regulation is shown by the finding that the *trans*-heterozygote, *Cbx* + + / + + *bxd*, fails to show such an enhancement of the *Cbx* phenotype. Second, a very weak *Cbx*-like wing effect that occurs in *Pc^3*/ + animals is intensified slightly but significantly when there is heterozygosity for a *bxd* mutant (whether in the *Pc* or + chromosome). Thus, in these two sets of comparisons involving, in the one case, *Cbx* as the sensitizer and, in the other, *Pc^3*, the *bxd* mutant effects a stronger transformation of T2 toward T3. Hence, *bxd* acts as if it causes overexpression of *Ubx$^+$* in T2, which is one segment anterior to that in which the latter allele normally becomes fully expressed. Using indirect immunofluorescence, Beachy et al. (1985) have obtained direct evidence for this inferred effect of a *bxd* mutant on *Ubx+* expression by showing that in embryos homozygous for an extreme *bxd* mutant, *bxd113*, specific staining of a part of the *Ubx+* protein could first be detected in nervous tissue of T2, compared with T3.

In the *bx3* case, weak transformation of extreme anterior T2, especially in the homozygous mutant, is consistent with there being overexpression of *abx+* in T2, one segment anterior to that (T3) in which *abx+* is normally assumed to be first expressed. The utility of applying the CIN and COE rules to these mutants was demonstrated with the finding that the homozygous double mutant, *abx bx3*, has the effect of such overexpression largely blocked in T2 and in T3, so that the triple mutant, *abx bx3 pbx*, when homozygous, produces a much more nearly perfect four-winged fly (see Lewis 1982) than does the *bx3 pbx* homozygote (see Lewis 1964).

In the case of *iab-3277* (see above), there is evidence for weak transformation of the dominant gain-of-function type that implies overexpression of *iab-2+* in A1, one segment ahead of the segment in which that gene normally becomes fully expressed. Whether that overexpression occurs in A2, where that gene normally becomes fully expressed, has not been studied.

Strong overexpression of *iab-2+* has been invoked to account for the strong transformation of A1 toward A2 in the case of the *Uab4* mutant, which is associated with a recessive loss of function of the *iab-3* type, as already discussed above (Lewis 1978); however, caution must be used in interpreting such examples because it was the dominant gain of function that led to the detection of that mutant, rather than the recessive loss of function in *iab-3* that was only shown later. Such strong dominant mutants of the gain-of-function type tend to occur extremely rarely compared with mutants of the recessive loss-of-function type, as if the former require special types of DNA association. For a discussion of the behavior of numerous *cis*-dominant mutants of the overexpression type involving genes in the BX-C, see Lewis (1978). Still other dominant gain-of-function mutants show gross misregulation of certain genes of the BX-C (see, e.g., Gausz et al. 1981; Celniker and Lewis 1984; and other examples cited in Karch et al. 1985).

In the case of *iab-4*, the *iab-4302*, *iab-445*, and *iab-4125* mutants transform the sternite of A2 toward that of A3 and therefore are examples of mutants that obey the COE rule; i.e., they show an overexpression of the *iab-3+* gene. *iab-4302* shows much greater overexpression in the homozygote than in the hemizygote. The remaining two *iab-4* alleles, *iab-4,5DB* and *iab-4166*, do not show this overexpression. In the case of *iab-5*, the *iab-5301* allele obeys the COE rule since, as reported above, the homozygote survives and shows a transformation of A3 toward A4. This result is, of course, consistent with overexpression of *iab-4+* in A3.

The CIN and COE rules have been found to apply not only to pseudopoint mutants, including deletions and gypsy insertion sequences, but also to gross chromosomal rearrangements, including wholly euchromatic and euchromatic-heterochromatic ones. Too few mutants, however, have been characterized at each locus within the complex to detect patterns related to specific types of DNA lesions.

A Model for *Cis*-regulation

The CIN and COE rules suggest a model for explaining cisvection and transvection in the case of the BX-C. It is sufficient to consider three closely linked genes, *a*, *b*, and *c*, to illustrate the main features of the model. (Gene will be used here in the broad sense of embracing an entire transcription unit, including all noncoding and coding regions.) These genes would make three specific *trans*-regulatory products (TRPs), designated A, B, and C, respectively. Each gene would code for at least one TRP (or a group, if more than one spliced message were involved). TRPs in some cases would be translated into proteins. Thus, White and Wilcox (1984) and Beachy et al. (1985) have shown that *Ubx* codes for a protein, whereas the discovery of homeo boxes by McGinnis et al. (1984) and Scott and Weiner (1984) combined with the molecular mapping data of Karch et al. (1985) indirectly implicates the *iab-2* gene and either the *iab-6* or the *iab-7* gene as coding for proteins. However, for the TRP coming from a gene that is transcribed but appears not to be translated, as may be the case with *bxd+* (Peattie and Hogness 1984), it is assumed that the RNA transcript still functions, albeit in an unknown way, as a *trans*-regulator of other genes.

The model invokes a *cis*-regulatory entity (E) that is assumed to be an essential ingredient in the machinery needed by a gene to produce its final transcript(s). The interrelationships of the three genes with E is depicted in a series of diagrams, starting with the wild-type haploid complement (the chromosome is represented by the horizontal line at the bottom of the diagram):

(1)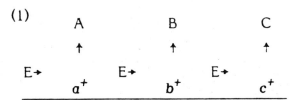

The source of E is not specified and could be pretranscriptional or cotranscriptional. In either case, it would be desirable to have a mechanism that would account for the finding that the direction of *cis*-inactivation appears to be opposite of that of transcription (based on analyses of *Ubx* transcripts [see Beachy et al. 1985] and of a cDNA from the *bxd* region [H. Lipshitz and D. Hogness, pers. comm.], as well as indirect inferences based on molecular studies of *iab-2* and *iab-6* or *iab-7* regions that code for the two known homeo box sequences in the distal half of the complex [W. Bender, pers. comm.]). The most obvious mechanism would be one that could track the other DNA strand from that used by DNA-dependent RNA polymerase. If E is made pretranscriptionally, then an analogy for such tracking would be the case of the hypothetical primosome of prokaryotes (Arai and Kornberg 1981). If E is produced cotranscriptionally, then E might be (or contain) antisense RNA either from some portion of the coding or noncoding regions of the BX-C genes or from one or more *trans*-regulatory genes that would code for E. Such speculation is invoked here only to suggest that there may be a plausible basis at a molecular level for the direction of the CIN effect.

At this stage of the analysis, it seems more profitable to pursue a model in which E is assumed to be a substance, as opposed to some type of construct that would involve, for example, structural deformation of the chromosome. To account for cisvection and transvection, it is necessary to assume that E has a nonrandom distribution in the nucleus and that its effective radius of action is very limited, suggesting that E is unstable or is produced in extremely minute amounts, or both. If cotranscriptionally produced, then E, or the effective part of E, is likely to be RNA rather than protein, since the latter should act in *trans*, not in *cis*.

In case a mutant lesion in one of the genes, such as *b*, obeys the CIN rule, the new pattern of relationships among the TRPs and E can be diagramed as follows:

(2)

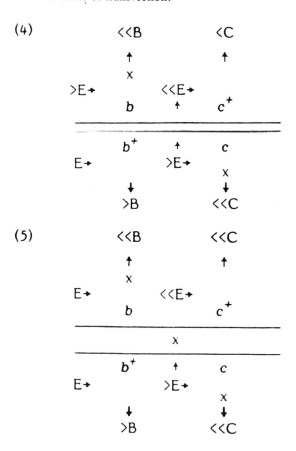

The DNA lesion in the *b* mutant is assumed to have two effects: (1) It leads to either a defective TRP or lowered quantity thereof (either possibility being symbolized by "< < B") and (2) it tends to block the transport of E so that the effective concentration of E that reaches c^+ is greatly reduced ("< < E") relative to that found in the case (Diagram 1) of wild type. The postulated block in the transport of E to c^+ would be responsible for the *cis*-inactivation of c^+ leading to less C ("< < C"): the CIN rule. To account for the COE rule, it is further assumed that E accumulates ahead of *b* sufficiently in amount to cause overexpression of the a^+ gene and hence an increase in concentration of A (">A") relative to that in the wild type.

A corollary of the CIN rule is that the *cis*-inactivation tends to spread over several genes located distally to the mutant lesion; however, the farther the gene is from that lesion, the less is the CIN effect. Such a result implies that the concentration of E gradually builds up the more distal the gene from the lesion. Some type of restart mechanism for E production from either outside or within the complex is also needed in the case

of iab-2^+, since neither *pbx* nor *bxd* mutant lesions have induced detectable *cis*-inactivation of that gene. Similarly, for the case of a *c* mutant:

(3)

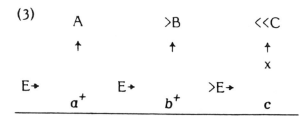

Combining Diagrams 2 and 3 to give a *trans*-heterozygote, $b+/+c$, with closely paired chromosomes, or with pairing reduced or eliminated by a TSR (Diagrams 4 and 5, respectively), yields the following two genotypes that are needed to examine the application of the model to cases of transvection:

(4)

(5)

In Diagrams 4 and 5, the concentration of C is expected to be lower than that in the *cis*-heterozygote, $bc/++$ (not diagramed), in agreement with the observed findings that the *trans*-heterozygotes in both cases express the "c" phenotype rather than the wild-type phenotype (or that of a slight dominant effect that arises whenever one of the mutants is incompletely dominant, as in the case of a mutant such as *Ubx*); i.e., *Ubx* mutants, deletions that include *Ubx* and such double mutants as *bx Ubx*, *Ubx bxd*, and *Ubx pbx*, all are virtually identical in their causing a slightly enlarged haltere when tested opposite a wild-type chromosome.

To account for transvection, it is assumed that when the chromosomes are paired, as in Diagram 4, E diffuses from the lower "c"-bearing chromosome to a sufficient extent to permit the c^+ in the upper two chromosomes to function more efficiently, thereby producing more C, than when the chromosomes are unpaired, as in Diagram 5 in which E is assumed not to be able to so diffuse. The nature of the restriction placed on the ability of E to diffuse readily over such short distances might reside in its being produced in very small amounts, in being unstable, or both. As a result, Diagram 4 is expected to have a less extreme "c" phenotype than that of Diagram 5: the observed result in the known cases of transvection involving such *trans*-heterozygotes.

To account for the COE rule, it is necessary to postulate that there is a gradient in the expression of many, if not all, of the BX-C genes. The application of this principle to the most relevant genotypes is shown in Table 1. To illustrate, let n, $n-1$, and $n-2$ represent abdominal segments A4, A3, and A2, respectively, then the gene symbols a, b, and c correspond, respectively, to *iab-2*, *iab-3*, and *iab-4*. To explain why a lesion in *iab-4* can cause an overexpression of *iab-3*$^+$ in relevant cells of segment A2, it is necessary to assume that in wild type, *iab-4*$^+$ is very weakly expressed in cells of A2, or incipiently expressed in the sense that E would be assumed to have prepared it to be so expressed.

In a number of instances, it has been necessary to assume that a given gene is weakly expressed one segment ahead of the segment in which it is normally active in order to account for certain phenotypes (Lewis 1978). Evidence that can be interpreted as pointing to an expression of this kind extending to two segments ahead has been found in one instance by Duncan (1982). Recently, evidence has been obtained (E.B. Lewis, unpubl.) pointing to the conclusion that an *iab-3*$^+$ function in the embryo (namely, promotion of development of special denticles of a triangular shape in the dorsal posterior half of the abdominal segments [Lohs-Schardin et al. 1979]) appears to be weakly manifested in A1 of wild type and increases in strength posteriorly in proportion to the number of copies of the wild-type BX-C present in the embryo.

The present model, which is quite different from one recently proposed by Beachy et al. (1985), has a number of testable predictions. It would, for example, predict that each gene, including *iab-3*, *iab-4*, and *iab-5*, would make a transcript that would be expected to appear in appropriate segments in accord with the COE rule. Specifically, a *bxd*$^+$ would be expected to be primarily active in A1 and segments beyond, especially in genotypes such as *Dp-P10*; *Df-P9*, where other transcripts from genes to the right of *bxd* would be absent and therefore not likely to interfere with the expression of the *bxd*$^+$ function (namely, suppression of sense organs in the abdominal segments and promotion of ventral setal bands). As pointed out above, molecular results obtained by Beachy et al. (1985) are consistent with the COE rule in showing that an extreme *bxd* mutant causes overexpression of *Ubx*$^+$ function in T2, one segment anterior to T3, i.e., the segment in which that function normally becomes fully expressed.

SUMMARY

The BX-C is a set of master control genes that *trans*-regulate other genes and thereby control much of the segmentation pattern of the fly. The BX-C genes are themselves regulated in *cis* and *trans*. Three rules governing *cis*-regulation of BX-C are applicable over a region extending from *Ubx* to at least *iab-7*, a distance of nearly 300 kb on the DNA map: (1) The colinearity (COL) rule: genes are colinear with respect to map location and order of expression along the body axis, the only exception thus far being *pbx*$^+$; (2) the *cis*-inactivation (CIN) rule: a mutant lesion in one gene tends to *cis*-inactivate the wild-type gene(s) immediately distally; and (3) the *cis*-overexpression (COE) rule: certain mutant lesions in a given gene cause the next most proximal gene to overexpress one segment more anterior to the one in which the latter gene normally expresses. A model is proposed that attempts to account for these rules by invoking a special *cis*-regulatory entity (E) that diffuses more efficiently along the chromosome than between chromosomes.

Table 1. *Cis*-overexpression or COE Rule

	Trans-regulatory products					
	genotype (1)			genotype (3)		
	--a^+--	--b^+--	--c^+--	--a^+--	--b^+--	--c--
	↓	↓	↓	↓	↓	↓
Segment number	A	B	C	A	>B	<<C
$n-2$	+++	++	+	+++	+++	±
$n-1$	+++	+++	++	+++	++++	±
n	+++	+++	+++	+++	++++	±

Levels of expression of TRPs coming from the chromosome genotypes (1) and (2) in progressively more anterior body segments; e.g., if n, $n-1$, and $n-2$ correspond to abdominal body segments A4, A3, and A2, respectively, then a^+, b^+, and c^+ correspond to *iab-2*$^+$, *iab-3*$^+$, and *iab-4*$^+$, respectively. (±) Little or none; (+) very weak or incipient; (++) weak; (+++) wild type; (++++) overexpression.

ACKNOWLEDGMENTS

I am indebted to Welcome Bender, Ian Duncan, Patrick O'Farrell, and Sue Celniker for helpful discussion; to David Baker, Rollin Baker, and Josephine Macenka of this laboratory for excellent technical assistance; and to Renee Thorf, Connie Katz, Rollin Baker, and Pamela Lewis for assistance in preparing the manuscript. I thank Ian Duncan for use of his iab-2^{D24} and iab-7^{D14} mutants before their publication. This work was supported by U.S. Public Health Service grant HD-06331.

REFERENCES

AKAM, M., H. MOORE, and A. COX. 1984. *Ultrabithorax* mutations map to distant sites within the bithorax complex of *Drosophila*. *Nature* **309:** 635.

ARAI, K. and A. KORNBERG. 1981. Unique primed start of phage ϕX174 DNA replication and mobility of the primosome in a direction opposite chain synthesis. *Proc. Natl. Acad. Sci.* **78:** 69.

BEACHY, P.A., S.L. HELFAND, and D.S. HOGNESS. 1985. Segmental distribution of bithorax complex proteins during *Drosophila* development. *Nature* **313:** 545.

BENDER, W., M. AKAM, F. KARCH, P.A. BEACHY, M. PEIFER, P. SPIERER, E.B. LEWIS, and D.S. HOGNESS. 1983. Molecular genetics of the bithorax complex in *Drosophila melanogaster*. *Science* **221:** 23.

CAPDEVILA, M.P. and M. GARCIA-BELLIDO. 1981. Genes involved in the activation of the bithorax complex of *Drosophila*. *Wilhelm Roux's Arch. Dev. Biol.* **190:** 339.

CELNIKER, S.E. and E.B. LEWIS. 1984. *Transabdominal*: A new dominant mutant in the bithorax gene complex (BX-C) of *Drosophila melanogaster*. *Genetics* **107:** S17.

DENELL, R.E. and R.D. FREDERICK. 1983. Homoeosis in *Drosophila*: A description of the Polycomb lethal syndrome. *Dev. Biol.* **97:** 34.

DUNCAN, I. 1982. Localization of bithorax complex (BX-C) and Antennapedia complex (ANT-C) gene activities along the body axis of *Drosophila*. *Genetics* **100:** S20.

DUNCAN, I. and E.B. LEWIS. 1981. Genetic control of body segment differentiation in *Drosophila*. *Symp. Soc. Dev. Biol.* **40:** 3.

DURA, J.M., H.W. BROCK, and P. SANTAMARIA. 1985. *Polyhomeotic*: A gene of *Drosophila melanogaster* required for correct expression of segmental identity. *Mol. Gen. Genet.* **198:** 213.

GARCIA-BELLIDO, A. and M.P. CAPDEVILA. 1978. The initiation and maintenance of gene activity in a developmental pathway of *Drosophila*. *Symp. Soc. Dev. Biol.* **36:** 3.

GAUSZ, J., H. GYURKOVICS, and A. PARDUCZY. 1981. A new homeotic mutation, SGA62, in *Drosophila melanogaster*. *Acta Biol. Acad. Sci. Hung.* **32:** 219.

INGHAM, P.W. 1981. *trithorax*: A new homeotic mutation of *Drosophila melanogaster*. II. The role of trx^+ after embryogenesis. *Wilhelm Roux's Arch. Dev. Biol.* **190:** 365.

———. 1983. Differential expression of bithorax complex genes in the absence of the *extra sex combs* and *trithorax* genes. *Nature* **306:** 591.

———. 1984. A gene that regulates the bithorax complex differentially in larval and adult cells of *Drosophila*. *Cell* **37:** 815.

INGHAM, P.W. and J.R.S. WHITTLE. 1980. *trithorax*: A new homeotic mutation of *Drosophila melanogaster* causing transformations of abdominal and thoracic imaginal segments. *J. Mol. Gen. Genet.* **179:** 607.

KARCH, F., B. WEIFFENBACH, M. PEIFER, W. BENDER, I. DUNCAN, S. CELNIKER, M. CROSBY, and E.B. LEWIS. 1985. The abdominal region of the bithorax complex. *Cell* (in press).

KERRIDGE, S. and G. MORATA. 1982. Developmental effects of some newly induced *Ultrabithorax* alleles of *Drosophila*. *J. Embryol. Exp. Morphol.* **68:** 211.

LEWIS, E.B. 1952. Pseudoallelism and gene evolution. *Cold Spring Harbor Symp. Quant. Biol.* **16:** 159.

———. 1954. The theory and application of a new method of detecting chromosomal rearrangements in *Drosophila melanogaster*. *Am. Nat.* **88:** 225.

———. 1955. Some aspects of position pseudoallelism. *Am. Nat.* **89:** 73.

———. 1963. Genes and developmental pathways. *Am. Zool.* **3:** 33.

———. 1964. Genetic control and regulation of developmental pathways. In *Role of chromosomes in development* (ed. M. Locke), p. 231. Academic Press, New York.

———. 1968. Genetic control of developmental pathways in *Drosophila melanogaster*. *Proc. Int. Congr. Genet.* **2:** 96.

———. 1978. A gene complex controlling segmentation in *Drosophila*. *Nature* **276:** 565.

———. 1981. Developmental genetics of the bithorax complex in *Drosophila*. *ICN-UCLA Symp. Mol. Cell. Biol.* **23:** 189.

———. 1982. Control of body segment differentiation in *Drosophila* by the bithorax gene complex. In *Proceedings of the 9th Congress of the International Society of Developmental Biologists* (ed. M.M. Burger), p. 269. A.R. Liss, New York.

LOHS-SCHARDIN, M., C. CREMER, and C. NÜSSLEIN. 1979. A fate map for the larval epidermis of *Drosophila melanogaster*: Localized cuticle defects following irradiation of the blastoderm with an ultraviolet laser microbeam. *Dev. Biol.* **73:** 239.

MCGINNIS, W., M.S. LEVINE, E. HAFEN, A. KUROIWA, and W. GEHRING. 1984. A conserved DNA sequence in homeotic genes of the *Drosophila* Antennapedia and bithorax complexes. *Nature* **308:** 428.

PEATTIE, D.A. and D.S. HOGNESS. 1984. Molecular analysis of the *bxd* locus within the bithorax complex of *Drosophila*. *Genetics* **107:** S81.

SANCHEZ-HERRERO, E., I. VERNOS, R. MARCO, and G. MORATA. 1985. Genetic organization of *Drosophila* bithorax complex. *Nature* **313:** 108.

SCOTT, M.P. and A.J. WEINER. 1984. Structural relationships among genes that control development: Sequence homology between the *Antennapedia*, *Ultrabithorax*, and *fushi tarazu* loci of *Drosophila*. *Proc. Natl. Acad. Sci.* **81:** 4115.

STERN, C. and E. HADORN. 1939. The relation between the color of testes and vasa efferentia in *Drosophila*. *Genetics* **24:** 162.

STRUHL, G. 1981. A gene product required for the correct initiation of segmental determination in *Drosophila*. *Nature* **293:** 36.

———. 1984. Splitting the bithorax complex of *Drosophila*. *Nature* **308:** 454.

WHITE, R.A.H. and M. WILCOX. 1984. Protein products of the bithorax complex in *Drosophila*. *Cell* **39:** 163.

Anatomy and Function of the Bithorax Complex of *Drosophila*

E. SÁNCHEZ-HERRERO,* J. CASANOVA,* S. KERRIDGE,*† AND G. MORATA*
Centro de Biología Molecular, Facultad de Ciencias, Universidad Autónoma de Madrid, Canto Blanco, 28049 Madrid, Spain; †Centre Universitaire, Marseille-Luminy, Marseille 90, France

The bithorax complex (BX-C) comprises a number of functionally related, adjacent genes specifying the development of two thoracic (T2-T3) segments and the abdominal (A1-A8) segments of *Drosophila*. The genetic and developmental aspects of the BX-C have been elaborated in a lifetime of research by Lewis (1955, 1963, 1978, 1981, 1982). From the developmental side, two critical points emerge from his work:

1. Elimination of all BX-C genes results in the segmental pattern being profoundly modified in both embryos and larvae: Posterior to T2, all segments develop as thoracic ones. Lewis (1978) described these segments as mesothoracic (T2), but further work (Morata and Kerridge 1981; Kerridge and Morata 1982; Hayes et al. 1984; Struhl 1984; Sánchez-Herrero et al. 1985) showed that the anterior compartments develop as the anterior one of T2 and all the posterior compartments develop as the posterior one of T1. This is the basis for the BX-C being a segmental unit with a mosaic pattern T1p-T2a (a for anterior and p for posterior compartments). The phenotype of these embryos defective in the entire BX-C function defines the realm of action of the complex: Anatomically, it is limited by the anteroposterior boundary in T2 and by the posterior end of A8. The BX-C does not have a function outside this area.

2. Many mutations within the BX-C show phenotypes that are part of the total deficiency (Lewis 1963, 1978). The effects of these mutations are restricted to specific segments or groups of segments within the entire BX-C domain, indicating that the BX-C is functionally complex and comprises a number of discrete elements.

The genetic and developmental data suggested (Lewis 1978) a model in which the development of each segment was determined by a particular combination of active BX-C genes. The model implied that the morphological characteristics of each segment are established by the activity of a specific gene that acts on a ground pattern corresponding in each case to the segment located immediately anterior.

During the last few years, we have performed (Kerridge and Morata 1982; Sánchez-Herrero et al. 1985) a comprehensive mutagenesis screen aimed at recovering mutations in all BX-C genes. We have found that the BX-C is made up of three adjacent genes that function in concert to establish the appropriate pattern in each segment. Further genetic and developmental analyses of these genes revealed a functional and genetic diversity within each locus.

MATERIALS AND METHODS

Mutagenesis. To facilitate the genetic and developmental analyses, our mutations were induced on a chromosome (*mm*) carrying the markers *mwh jv st red sbd^2 e^{11} ro ca* (for details, see Lindsley and Grell 1968). Ethyl methane sulfonate (EMS; 0.025%) and X-rays (3500-4000 rad) were used as mutagens. We have used two methods of mutagenesis. In the first method (Kerridge and Morata 1982), X-rayed *mm* males were crossed to *bx^{34e}* females, and the progeny were inspected for mutations or deletions that fail to complement *bx^{34e}*. This method is quick and effective, but it has the obvious limitation that it will not detect any mutation complementing *bx^{34e}*. In addition, many of the F$_1$ flies selected for their *bx* phenotype turned out to be sterile due to a dominant haplo-insufficient factor located distally in the BX-C.

The second method (Sánchez-Herrero et al. 1985) is an F$_2$ procedure that permits the isolation of any BX-C lethal or visible mutation. Essentially, it consists of mutagenizing males homozygous for the *mm* chromosome and checking for complementation of the treated *mm* chromosomes with *DfP9*, a chromosome (Lewis 1978) where the entire BX-C is deleted. Chromosomes that are lethal or show visible mutant traits over *DfP9* are isolated and stocks are established.

Embryonic phenotypes. To study the effect of mutations on larval segments, embryos and larvae were mounted in Hoyer's medium after fixation in a mixture of glycerol and acetic acid according to the method of Van der Meer (1977). Embryos were first gently dechorionated.

Adult phenotypes. The effect of the different mutations on the adult cuticular structures was studied under the compound microscope. Pertinent cuticular pieces were cut, washed in alcohol, treated in 10% KOH to digest soft parts, and mounted in Eurapal. In the case of visible mutations, flies homo- or hemizygous could be studied directly. However, the majority of mutations were lethal, and we used methods to generate mosaic flies bearing patches of mutant tissue. The genetic mosaics were produced by X-ray-induced mitotic recombination, a technique that permits interchange of genetic material between homologous chromosomes. Two main methods were used: (1) mitotic recombination in the first chromosome (Fig. 1A) and (2) mitotic recombination in the third chromosome (Fig. 1B). In the first method, we produced female flies of genotype $M(1)o^{Sp}DpP115/ywf^{36a}$; $Ubx/DfP9$ (the position of the different mutations on the first chromosome in Fig. 1A). In these flies, the mutant lethal combination $Ubx/DfP9$ in the third chromosome is covered by the presence in the first chromosome of the $DpP115$ containing a full dose of the BX-C$^+$. The event of recombination gave rise to two daughter cells. One cell had lost DpP115 and was marked genetically with yellow (y) and forked (f^{36a}); because of the loss of the retarding $M(1)o^{Sp}$ mutation (Morata and Ripoll 1975), it had a proliferation advantage over neighboring cells. The other sister cell died because it was homozygous for $M(1)o^{Sp}$ (Stern and Tokunaga 1971; Morata and Ripoll 1975). The advantage of this method is that the frequency of mitotic recombination in the first chromosome is the highest of all chromosomes (García-Bellido 1972); hence, a large number of mutant clones can be easily obtained. A disadvantage is that since it requires two X chromosomes, it can only be used in females. Also significant is the frequency of the mitotic recombination (~20%) between the $DpP115$ and f^{36a} (Fig. 1A) that gives rise to clones that are marked yf^{36a} but retain the $DpP115$ and are therefore not mutant for the BX-C.

The second method is shown in Figure 1B. As exemplified for the abdominal-A (abd-A) mutation, the recombination is induced in the left arm of the third chromosome. The mutant combination abd-A/Df-$P115$ ($DfP115$ is a deletion for the entire BX-C) is covered by the $Dp146$ carrying a full dose of the BX-C. A chromosomal interchange proximal to the $M(3)i^{55}$ results in cell clones marked with the mutants y, mwh, and jv that lose $Dp146$ and $M(3)i^{55}$. They are mutant

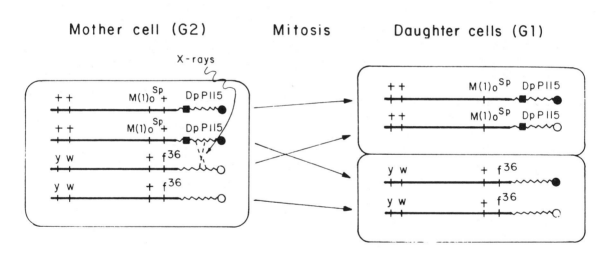

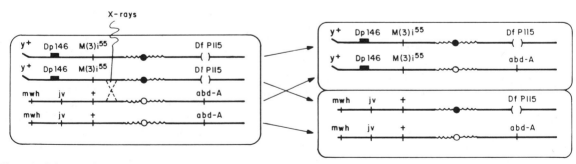

Figure 1. Schemes of mitotic recombination to generate marked clones deficient in BX-C activity. In A, the recombination event in the first chromosome gives rise to a clone homozygous for y, w, f^{36a}, and M^+, which, because of the loss of $DpP115$, will express the BX-C mutant genotype of the third chromosome (not shown). In B, after recombination in the left arm of the third chromosome, the mutant clone for the BX-C (exemplified in abd-A) is marked with y, mwh, jv, and M^+ (see Materials and Methods).

for BX-C and have a proliferation advantage. This method gives rise to a lower yield of clones than the first method, but the marking is better and the method can be used for both males and females. All irradiations were performed using a Philips 151Be X-ray (150 kV; 15 mA) at a dose of 300 rad/min.

RESULTS

Complementation Groups within DfP9

By using deficiency mapping and complementation analysis, we have identified five complementation groups within *DfP9*. Each locus is defined by a number of alleles; the number is high enough that we can be confident that all genes within the chromosomal fragment that mutate to lethality or to a visible trait have been found. In a similar study, Tiong et al. (1985) also found the same number of complementation groups, and preliminary tests indicate a one-to-one correspondence between their loci and ours. By using different overlapping deletions (Fig. 2), we have determined the genetic order of the five loci. Only three of these, Ultrabithorax (*Ubx*), abdominal-A (*abd-A*), and abdominal-B (*Abd-B*) (see below), affect the segmental pattern. The other two genes, which map at opposite ends, lethal left of bithorax (*llb*) and lethal right of bithorax (*lrb*), appear not to have any effect on the segments either in larvae or in adult flies (Sánchez-Herrero et al. 1985; Tiong et al. 1985). We therefore believe that they are not related to the BX-C.

The Ubx Gene

There are many mutant alleles of the *Ubx* gene. In different experiments (Kerridge and Morata 1982; Sánchez-Herrero et al. 1985), we have isolated 21 new alleles, but many more are available (Lindsley and Grell 1968), as well as physical deletions of the gene (Lewis 1978). Most *Ubx* alleles are recessive lethals, but some that are EMS-induced are viable (Sánchez-Herrero et al. 1985).

The developmental consequences of the elimination of *Ubx*+ activity in the larval and adult segments have been studied by examining *Ubx*− larvae (Lewis 1978; Struhl 1981; Sánchez-Herrero et al. 1985) or by generating *Ubx*− cell clones in the cuticle of the adult segments (Morata and García-Bellido 1976; Morata and Kerridge 1981; Kerridge and Morata 1982; Struhl 1982). Essentially, the same effect is observed in larvae and in adults and can be described as a transformation of the pattern of compartments T2p-T3a-T3p-A1a (which we call the *Ubx* domain) into that of T1p-T2a-T1p-T2a.

Anterior to T2p, no effect is observed either in the thorax or in cephalic segments. Posterior to A1a, *Ubx*− clones develop normally in the adult cuticle, suggesting that *Ubx*+ has no effective role there. However, in *Ubx*− larvae, abdominal segments posterior to A1 develop ventral pits, a characteristic thoracic structure, thus indicating a slight thoracic transformation of A2-A7 and therefore a requirement for *Ubx*+ activity in these segments. The possible role of *Ubx*+ in abdominal segments is discussed below. The phenotype of viable *Ubx* alleles resembles in a mild form that of the lethal alleles. Lethality is probably not an essential feature of *Ubx* mutations but depends on the degree with which the normal body pattern is altered.

The temporal and spatial requirements of *Ubx*+ activity have been investigated by the genetic mosaic method (Morata and Kerridge 1981; Kerridge and Morata 1982; Struhl 1982). The elimination of *Ubx* early in development, up to 7 hours after fertilization, results in the compartment series T2p-T3a-T3p-A1a developing as T1p-T2a-T1p-T2a. However, if *Ubx*+ is eliminated after 7 hours, only the compartments T3a-T3p-A1a are affected and are transformed into T2a-T2p-T2a; compartment T2p remains normal and T3p becomes T2p (not T1p as in early induced mosaics). These observations indicate that the *Ubx* gene contains discrete functions with different temporal requirements. In particular, there is an early function, termed postprothorax (*ppx*) (Morata and Kerridge 1981), that is necessary to prevent T1p development of T2p and T3p and that is active only in the embrionary period. In contrast, other *Ubx* functions necessary for normal

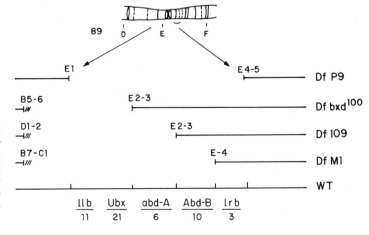

Figure 2. Deficiency mapping and complementation analysis of mutations within *DfP9*. The number of alleles is indicated under the name of each locus. (*llb*) Lethal left of bithorax; (*Ubx*) Ultrabithorax; (*abd-A*) abdominal-A; (*Abd-B*) abdominal-B; (*lrb*) lethal right of bithorax. For a description of the different deficiencies, see Lewis (1978), Struhl (1981), and Sánchez-Herrero et al. (1985).

T3a-T3p-A1a development are active during the embrionary and larval periods.

A considerable genetic complexity of the *Ubx* locus is suggested by the existence of a number of viable recessive mutations with phenotypes that are part of *Ubx*. These are mutations anterobithorax (*abx*), bithorax (*bx*), bithoraxoid (*bxd*), and postbithorax (*pbx*) (for a description of these mutations, see Lewis 1981, 1982). These mutations all fail to complement *Ubx*, and the *trans*-heterozygote shows the phenotype of the mutant in question, suggesting that *Ubx* contains a number of discrete genetic subunits specific for different parts of the entire *Ubx* domain.

We have recently carried out a detailed study of the phenotypes of several of these mutations (Casanova et al. 1985) to establish their respective contributions to the entire *Ubx* syndrome. In addition, there was the problem that the *ppx* transformation described in *Ubx*⁻ clones or embryos had not been associated with any individual mutation. There remained the possibility that this transformation resulted not from the loss of a particular function within *Ubx* but from the elimination of the entire *Ubx* activity (Lewis 1982).

Our results indicate that the phenotypes of *abx* and *bx* mutations are essentially the same; compartments T2p and T3a become transformed toward T1p and T2a, respectively. One exception is the mutation bx^1, which does not have any effect in T2p. The effect of *bx* and *abx* mutants in T2p is the same *ppx* transformation observed in *Ubx*⁻ clones, which is now shown to be the result of the loss of function of the *abx* locus. The reason why the *ppx* transformation went unnoticed after many years of work with *bx* mutations (Lewis 1963, 1978; Morata and Kerridge 1980) is that in most combinations, it has a low penetrance at 25°C, the temperature at which *Drosophila* cultures are normally kept. At 17°C, the penetrance of the *ppx* transformation is much higher and can be readily detected. We believe that *abx* and *bx* alleles are the same class of mutation. They partially inactivate the abx^+ subunit of the *Ubx* gene. However, even though the transformations at T2p (toward T1p) and at T3a (toward T2a) are genetically linked, they are different developmentally (Casanova et al. 1985). The elimination of the normal abx^+ gene by mitotic recombination results in a *ppx* phenotype only when the gene is eliminated before 7 hours of development; thereafter it is dispensable. Nevertheless, the elimination of abx^+ by the same method results in a transformation of T3a into T2a even when the mutant clones are generated at the end of the third larval period.

The phenotypes of the *bxd* and *pbx* mutants define a second genetic subunit contained within the *Ubx* gene. They produce a transformation of compartments T3p and A1a into T2p and T3a, respectively (Lewis 1963; Miñana and García-Bellido 1982; Hayes et al. 1984), the same phenotype shown by *Ubx*⁻ cells or embryos in this area.

As a general rule, mutations at the abx^+ subunit (*abx* and *bx*) complement those at the bxd^+ subunit (*bxd* and *pbx*), although abx^1 and bx^3 produce a slight inactivation of the pbx^+ function encoded by the *bxd* subunit (Lewis 1963, 1981), indicating that abx^+ and bxd^+ are not totally independent.

The *abd-A* Gene

Mutations at this locus are lethal, with the exception of the allele $abd\text{-}A^{M3}$. Some alleles are induced by EMS and others are induced by X-rays (Sánchez-Herrero et al. 1985). The breakpoint in the IIIR of $T(2;3)P10$ (Lewis 1978) also behaves as a mutant allele of this locus. The effect of *abd-A* mutations has been studied in both larval and adult segments. A synthetic deletion of the gene $DpP10;Df109$ has been observed and has been characterized both genetically and developmentally (Morata et al. 1983).

The loss of function of $abd\text{-}A^+$ does not alter the cephalic region or the normal *Ubx* domain. The limit of the morphological effect of *abd-A* appears to be in the anteroposterior compartment boundary at A1. In the posterior region of A1, there is frequently a monohair, interpreted (Struhl 1984; Sánchez-Herrero et al. 1985) to be a rudiment of a Keilin's organ and indicative of thoracic transformation of the posterior part of the segment. All the abdominal segments posterior to A1 show the monohair in a homologous position, indicating that they are transformed toward a posterior thoracic segment. The anterior parts of all segments posterior to A1 show a pattern of denticle belts resembling that of A1. The transformation is complete up to A4 or A5, whereas A6, A7, and A8 present patterns intermediate between A1 and the more posterior segments (Fig. 3).

In the adult cuticle, the phenotypic effect of $abd\text{-}A^-$ has been studied by mosaic analysis (see Materials and Methods), generating $abd\text{-}A^-$ cell clones in flies with a normal segmental pattern. The results parallel those seen in the larval epidermis (Fig. 4). Clones in segments A2–A4 show a complete transformation toward A1, but posterior to A5, they show different pattern elements intermediate between A1 and the more posterior segments. We refer to the A1p–A4 anatomical region of the body as the *abd-A* domain.

Like the *Ubx* gene, *abd-A* appears to be complex and possibly contains several genetic subunits that specify the development of particular areas within the *abd-A* domain. The main argument in favor of this view is the existence of the allele $iab\text{-}2^K$ (Kuhn et al. 1981) that produces a transformation of A2 into A1 but has no effect on any other segment. All of the *abd-A* mutations fail to complement $iab\text{-}2^K$, but the effect is restricted to A2 exclusively (Sánchez-Herrero et al. 1985), indicating that *iab-2* is part of the *abd-A* gene, and suggests that the latter may contain several subunits of this type.

The *Abd-B* Gene

The locus is defined genetically by a number of EMS-induced and X-ray-induced alleles. In our laboratory, we have isolated ten *Abd-B* mutations, and Tiong et al. (1985) have reported ten more. Preliminary comple-

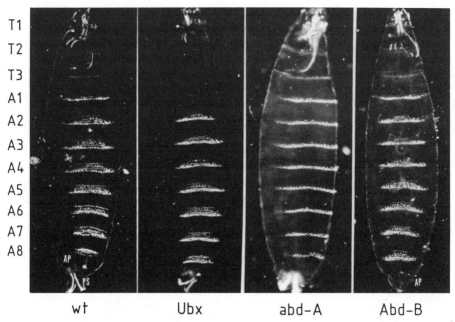

Figure 3. Ventral view of segments in wild type, *Ubx*, *abd-A*, and *Abd-B* larvae. (AP) Anal pads; (PS) posterior spiracles. Note that in *Abd-B*, PS are missing and there is also a region of naked cuticle between the denticle belt of the eighth segment and the AP. Mutant larvae shown are hemizygous over *DfP9*.

mentation tests indicate that the two sets of mutants are allelic. All of our *Abd-B* mutations show a dominant phenotype consisting of a slight transformation of segments A5, A6, and A7 toward a more anterior one. In addition, *Abd-B*⁻ males present a rudimentary A7 and the genitalia are often rotated. The same phenotype is seen in *DfP9/+* or *DfC4/+*, indicating that the *Abd-B* locus is haplo-insufficient. The phenotype of hemizygous *Abd-B* has been defined both in larval and adult segments. The study of the adult phenotype has been facilitated by the fact that several *Abd-B* alleles, although lethal, occasionally produce "escapers" (homo- or hemizygous flies that survive to adulthood) in which the phenotype can be inspected directly. Nevertheless, we have performed a mosaic analysis of two alleles that do not produce escapers and are presumably the strongest mutations.

In the *Abd-B* escapers, the cephalic, thoracic, and abdominal segments (including the *Ubx* and *abd-A* domains) are normal, but segments A5–A8 are transformed toward A4. This can be visualized better in males that normally do not develop A7 and A8 in the adult cuticle, having only six abdominal segments. The transformation of A5–A8 into A4 results in males carrying eight abdominal segments, in addition to analia and genitalia. Posterior to A4, all of the abdominal segments develop as A4. The genitalia disappear or are reduced in females, whereas they are sometimes present in males. Analia almost always remains normal in both sexes. This is consistent with the current view of terminalia development (Nöthiger et al. 1977) in which female genitalia derive from A8, male genitalia from A9, and analia from A10. However, the *Abd-B*^{MX2} allele produces escapers in which segments A5–A7 are transformed to A4, but genitalia and analia are normal. Some exceptions to this description are considered below.

The larval segments of *Abd-B* mutants exhibit a transformation consistent with that seen in adult segments (Fig. 3), although the lack of specific landmarks makes it impossible to adjudicate the segment pattern of intermediate abdominal segments. However, A7 and A8 can be seen clearly transformed toward a more anterior abdominal segment.

Like that of *abd-A*, the genetic structure of the *Abd-B* locus reveals a high degree of complexity that has not yet been clarified. There is one allele, *Abd-B*^{M1}, that does not produce escapers, but its effects on abdominal segments and terminalia have been studied using genetic mosaics (Sánchez-Herrero et al. 1985 and unpubl.). In addition to the effect of A5–A8 described above, *Abd-B*^{M1} clones transform female genitalia (probably into a sternite) but do not appear in male genitalia (A9), whereas clones in the analia (A10) are usually normal.

The development of terminalia is also affected, being reduced or absent, in a recessive mutation *x23-1* (J. Casanova et al., unpubl.). It complements regular *Abd-B* mutations, but not *Abd-B*^{M1} nor a new EMS-induced *Abd-B* allele (J. Casanova and G. Morata, unpubl.); *x23-1/Abd-B*^{M1} flies show the same phenotype of hemizygous *x23-1* flies, and there is no transformation toward A4 other than the slight dominant effect of *Abd-B* mutations. *x23-1* also fails to complement *tuh-3* and Ultra-abdominal^1 (*Uab*^{1}) for *x23-1/tuh-3*, and *x23-1/Uab*^{1} flies show a reduction or elimination of genitalia and analia. Intriguingly, all of these combinations affecting the entire terminalia exhibit in embryos and larvae an extra-abdominal segment posterior to A8. The appearance of this extra segment has also been ob-

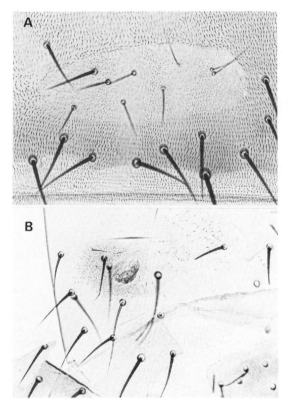

Figure 4. $abd\text{-}A^-$ clones in the adult cuticle (marked with y, mwh, and jv) generated in the same experiment in segments A4 (*A*) and A6 (*B*). In *A*, the clone presents a clean transformation toward A1, as indicated by the size and aspect of the bristles and the pigmentation of the cuticle. In *B*, the clone is transformed toward a segment intermediate between A1 and A6, as indicated by the presence of trichomes and the size of bristles. Typically, it is also associated with abnormal differentiations of the cuticle such as foldings and areas of uneven pigmentation.

served by Tiong et al. (1985) in two of their *Abd-B* alleles.

This extra segment A9 does not appear in embryos deficient for the entire BX-C (Lewis 1978; Struhl 1981), pointing to the possibility that the extra A9 may be the result of a gain rather than a loss of some BX-C function and suggesting a regulatory role of this locus on the expression of the rest of the BX-C on the A9 segment (E. Sánchez-Herrero et al., unpubl.). On the basis of the phenotype of $Pc^3Df(3R)C4$ homozygous embryos, Duncan and Lewis (1982) suggested the possibility that genes at the end of the BX-C are necessary to suppress the appearance of this extra segment.

DISCUSSION

Number of BX-C Genes and Their Anatomical Expression

Our studies (Sánchez-Herrero et al. 1985) demonstrate that the BX-C is made up of three adjacent genes, *Ubx*, *abd-A*, and *Abd-B*, that are independent and complement one another. Ubx^+ has been shown to function isolated from *abd-A* and *Abd-B* (Struhl 1984); thus, structural integrity of the complex is not a prerequisite for normal function.

The activity of Ubx^+ is limited anteriorly by the anteroposterior compartment boundary on the second thoracic segment. Anterior to this line there is no requirement of Ubx^+, and its absence does not cause any alteration in the development of those parts. From the anteroposterior boundary at T2 there is a requirement of Ubx^+ activity in T3 and in all the abdominal segments down to at least A7. However, there is a sharp distinction between the effect of Ubx^- in the compartments T2p-T3a-T3p-A1a that are completely transformed into T1p-T2a-T1p-T2a (Morata and Kerridge 1981; Kerridge and Morata 1982; Hayes et al. 1984; Struhl 1984) and the effect on A2-A7, which is a slight thoracic transformation only manifested in the presence of ventral pits, a regular thoracic structure. In the adult segments (Kerridge and Morata 1982), there is no detectable effect of the loss of Ubx^+ in abdominal segments posterior to A1. Thus, although there is an unquestionable requirement for Ubx^+ in A2-A7, its principal role is in T2p-T3a-T3p-A1a. We call these four compartments the *Ubx* domain (Sánchez-Herrero et al. 1985).

Mutations of the *abd-A* class also affect a specific anatomical region of the body. The phenotypes of individual mutants (Sánchez-Herrero et al. 1985) and of a synthetic deletion of the gene (Morata et al. 1983) indicate that there is no requirement of $abd\text{-}A^+$ anterior to the anteroposterior boundary of A1; thus, there is no function of $abd\text{-}A^+$ in the *Ubx* domain. From A1p down to A8, the loss of $abd\text{-}A^+$ results in a clear alteration in both the larval and adult segments (Morata et al. 1983; Sánchez-Herrero et al. 1985). The effect is more complete in the more anterior abdominal segments up to A4 or A5. Posterior to these, segments appear to be the intermediate type between A1 and the more posterior ones. This is clearly seen in the effect of $abd\text{-}A^-$ clones in the adult segments A4 and A6 (Fig. 4). In A4, the clone is completely transformed toward A1, whereas in A6, it is transformed in a structure that is clearly not A1, but may be A3 or A4.

The effect of *Abd-B* mutations and deletions extends to A5-A8, which develop like A4 (Sánchez-Herrero et al. 1985) (Fig. 4), and no effect is seen in any other segment.

The three major BX-C genes therefore have different, but overlapping, realms of action. However, each defines a specific area of the body where it is exclusively responsible for the morphological characteristics. The compartment series T2p-T3a-T3p-A1a defines the *Ubx* domain (Sánchez-Herrero et al. 1985). Neither $abd\text{-}A^+$ nor $Abd\text{-}B^+$ has any role in this area, which in the absence of Ubx^+ is transformed into T1p-T2a-T1p-T2a; hence, the appropriate development of these compartments depends exclusively on Ubx^+.

The area including the compartments A1p-A4 defines the *abd-A* domain. There is no function of $Abd\text{-}B^+$ in this area, and the elimination of $abd\text{-}A^+$ results in the whole area developing an A1a-T3p series. Since the acquisition of the A1a-T3p pattern requires Ubx^+ activity, this indicates that this gene is active in the abd-A domain but that the morphological diversity is conferred by abd-A^+. The simultaneous elimination of both Ubx^+ and abd-A^+ as in $Df109$ (Lewis 1978; Struhl 1981; Morata et al. 1983) results in both domains developing equally as T1p-T2a units. The same phenotype is observed in the double-mutant combination $Ubx^{Xa7}abd$-A^{M1} (J. Casanova and G. Morata, unpubl.).

The area including the segments A5–A8 defines the Abd-B domain. On the basis of the phenotypes of $Antp^{Scx}Df109$ embryos, Duncan (1982) concluded that the BX-C genes not deleted by $Df109$ are active in A4p. Since Abd-B mutations complement $Df109$, it suggests that the most anterior limit of the Abd-B domain may be the anteroposterior boundary at A4. The Abd-B^+ gene acts exclusively in this area, and in its absence, segments A5–A8 develop equally as A4. The normal A4 pattern requires Ubx^+ and abd-A^+ activity, indicating that these two genes are active in the Abd-B domain but that the morphological diversity of the domain is established exclusively throughout Abd-B^+. The simultaneous elimination of abd-A^+ and Abd-B^+ (as in $DpP10;DfP9$) results in the two domains developing equally as T3p-A1a (Lewis 1978; Morata et al. 1983), a pattern resulting from Ubx^+ activity alone. When there is a total lack of BX-C function as in $DfP9$ (Lewis 1978; Struhl 1981), the three domains develop identically as T1p-T2a, the ground pattern upon which the BX-C acts and establishes the identity of thoracic and abdominal segments.

Spatial Deployment of BX-C Functions

The entire anatomical area under the control of the BX-C is subdivided into three domains. In each domain, a particular BX-C gene establishes the appropriate segment pattern. However, each domain is made up of a number of compartments that have separate lineages from the blastoderm period (Steiner 1976; Lawrence and Morata 1977) and also display a particular morphology. Somehow, each BX-C gene must account for the morphological diversity within its domain. The observation that the Ubx gene includes two subunits, abx^+ and bxd^+ (Casanova et al. 1985), affecting compartments T2p-T3a and T3p-A1a, respectively, suggests that each BX-C gene may contain individual subunits that control specific subdomains. This is supported by the phenotype of the iab-2^K allele being part of that of abd-A mutants and also by the different types of mutations found at the Abd-B locus. Probably, the morphological diversity within the domains matches the genetic diversity within the controlling genes.

Lewis (1978, 1981, 1982) put forth a model describing how BX-C genes may control segment development. In each segment, there is activity of a particular combination of BX-C genes, but the precise morphology of the segment would be established principally or exclusively by one specific gene. The other BX-C-acting genes in the segment provide the ground pattern on which the specific gene acts. The model implies that there is one gene per segment. Although we and other investigators (Tiong et al. 1985) find only three major BX-C genes, our results do not contradict the Lewis model, since it appears that each BX-C gene contains individual subunits (abx^+, bxd^+, iab-2^+) with the properties of the Lewis genes. However, our results clearly indicate that the activity of the BX-C is not deployed spatially on the basis of segments but on larger areas that appear to be demarcated by anteroposterior compartment boundaries. Within these domains, there are subdomains that in well-defined cases appear to be anatomical units of the "parasegment" type (Martínez-Arias and Lawrence 1985); these seem to be the primary areas of BX-C activity. At this level, the activity of BX-C genes would be deployed as proposed by Lewis (1978), with the modification suggested by Struhl (1984) and Hayes et al. (1984) that the areas of gene activity be shifted by one half-segment. It is important to emphasize, however, that the genetic structures of the abd-A and Abd-B genes are still by and large unknown, and it remains to be seen how many individual genetic elements they contain and how their domains are subdivided.

ACKNOWLEDGMENTS

The experimental work described here has been supported by the Comisión Asesora de Investigación Científica y Técnica (CAICYT). S.K. was supported by an EMBO fellowship.

REFERENCES

CASANOVA, J., E. SÁNCHEZ-HERRERO, and G. MORATA. 1985. Prothoracic transformation and functional structure of the *Ultrabithorax* gene of *Drosophila*. *Cell* **42:** 663.

DUNCAN, I. 1982. Localization of bithorax complex (BX-C) and Antennapedia complex (ANT-C) gene activities along the body axis of *Drosophila*. *Genetics* **100:** s20.

DUNCAN, I. and E.B. LEWIS. 1982. Genetic control of body segment differentiation in *Drosophila*. In *Developmental order: Its origin and regulation* (ed. S. Subtelny), p. 533. A.R. Liss, New York.

GARCÍA-BELLIDO, A. 1972. Some parameters of mitotic recombination in *Drosophila melanogaster*. *Mol. Gen. Genet.* **115:** 54.

HAYES, P.H., T. SATO, and R.E. DENELL. 1984. Homoeosis in *Drosophila*: The *Ultrabithorax* larval syndrome. *Proc. Natl. Acad. Sci.* **81:** 545.

KERRIDGE, S. and G. MORATA. 1982. Developmental effects of some newly induced *Ultrabithorax* alleles of *Drosophila*. *J. Embryol. Exp. Morphol.* **68:** 211.

KUHN, D.T., D.F. WOODS, and J.L. COOK. 1981. Analysis of a new homeotic mutation (iab-2) within the bithorax complex in *Drosophila melanogaster*. *Mol. Gen. Genet.* **181:** 82.

LAWRENCE, P.A. and G. MORATA. 1977. The early development of mesothoracic compartments in *Drosophila*. *Dev. Biol.* **56:** 40.

LEWIS, E.B. 1955. Some aspects of position pseudoalleles m. *Am. Nat.* **89:** 73.

———. 1963. Genes and developmental pathways. *Am. Zool.* **3:** 33.

———. 1978. A gene complex controlling segmentation in *Drosophila*. *Nature* **276:** 565.

———. 1981. Developmental genetics of the bithorax complex of *Drosophila*. *ICN-UCLA Symp. Mol. Cell. Biol.* **23:** 189.

———. 1982. Control of body segment differentiation in *Drosophila* by the bithorax gene complex. In *Embryonic development: Genes and cells* (ed. M. Burger), p. 1. A.R. Liss, New York.

LINDSLEY, D.L. and E.H. GRELL. 1968. Genetic variations of *Drosophila melanogaster*. *Carnegie Inst. Wash. Publ.*, no. 627.

MARTINEZ-ARIAS, A. and P.A. LAWRENCE. 1985. Parasegments and compartments in the *Drosophila* embryo. *Nature* **313:** 639.

MIÑANA, F.J. and A. GARCÍA-BELLIDO. 1982. Preblastoderm mosaics of mutants of the bithorax-complex. *Wilhelm Roux's Arch. Dev. Biol.* **191:** 331.

MORATA, G. and A. GARCÍA-BELLIDO. 1976. Developmental analysis of some mutants of the bithorax system of *Drosophila*. *Wilhelm Roux's Arch. Dev. Biol.* **179:** 125.

MORATA, G. and S. KERRIDGE. 1980. An analysis of the expressivity of some bithorax transformations. In *Development and neurobiology of Drosophila* (ed. O. Sidiqqi et al.), p. 141. Plenum Press, New York.

———. 1981. Sequential functions of the bithorax complex of *Drosophila*. *Nature* **290:** 778.

MORATA, G. and P. RIPOLL. 1975. Minutes: Mutants of *Drosophila* autonomously affecting cell division rate. *Dev. Biol.* **42:** 211.

MORATA, G., J. BOTAS, S. KERRIDGE, and G. STRUHL. 1983. Homeotic transformations of the abdominal segments of *Drosophila* caused by breaking or deleting a central portion of the bithorax complex. *J. Embryol. Exp. Morphol.* **78:** 319.

NÖTHIGER, R., A. DÜBENDORFER, and F. EPPER. 1977. Gynandromorphs reveal two separate primordia for male and female genitalia in *Drosophila melanogaster*. *Wilhelm Roux's Arch Dev. Biol.* **181:** 367.

SÁNCHEZ-HERRERO, E., I. VERNÓS, R. MARCO, and G. MORATA. 1985. Genetic organization of *Drosophila* bithorax complex. *Nature* **313:** 108.

STEINER, E. 1976. Establishment of compartments in the developing leg imaginal discs of *Drosophila melanogaster*. *Wilhelm Roux's Arch. Dev. Biol.* **180:** 9.

STERN, C. and C. TOKUNAGA. 1971. On cell lethals in *Drosophila*. *Proc. Natl. Acad. Sci.* **68:** 329.

STRUHL, G. 1981. A gene product required for correct initiation of segmental determination in *Drosophila*. *Nature* **293:** 36.

———. 1982. Genes controlling segmental specification in the *Drosophila* thorax. *Proc. Natl. Acad. Sci.* **79:** 7380.

———. 1984. Splitting the bithorax complex of *Drosophila*. *Nature* **308:** 454.

TIONG, S., L.M. BONE, and T.R.S. WHITTLE. 1985. Recessive lethal mutations within the bithorax complex in *Drosophila melanogaster*. *Mol. Gen. Genet.* **200:** 335.

VAN DER MEER, J.M. 1977. Optical clean and permanent whole mount preparation for phase-contrast microscopy of cuticular structures of insect larvae. *Drosophila Inform. Serv.* **52:** 160.

Domains of *cis*-Interaction in the Bithorax Complex

W. BENDER, B. WEIFFENBACH, F. KARCH, AND M. PEIFER
Department of Biological Chemistry, Harvard Medical School, Boston, Massachusetts 02115

The bithorax complex (BX-C) in *Drosophila melanogaster* specifies the proper determination of segments in the fly. The third thoracic segment and the first through the eighth abdominal segments are affected by various mutations within the complex. Recessive loss-of-function mutations transform a segment or segments to resemble a more anterior segment; dominant gain-of-function alleles transform a segment to resemble a more posterior one. The BX-C is the largest cluster of homeotic mutations in *Drosophila*, and thus it is a particularly appealing locus for a molecular study of developmental decisions.

DNA of the BX-C was first isolated 6 years ago (Bender et al. 1983a), but we still have only preliminary information on one group of gene products (White and Wilcox 1984; Beachy et al. 1985). We have no idea how many other products exist, or what the phenotypic consequences are for the loss of any one product. The genetic analysis is complicated by surprising patterns of noncomplementation between mutations of different phenotypic classes; it is difficult to enumerate separable functions. Another difficulty is that the BX-C is very large, covering 300 kb or more. The left half, which includes the anterobithorax (*abx*), bithorax (*bx*), Ultrabithorax (*Ubx*), bithoraxoid (*bxd*), and postbithorax (*pbx*) alleles, is spread out over about 125 kb (Bender et al. 1983b) and the right half, which covers infra-abdominal-2 (*iab-2*) through infra-abdominal-8 (*iab-8*) alleles, has been recently mapped out over the adjacent 175 kb (Karch et al. 1985). The mutant lesions for any one phenotypic class are typically spread over many kilobases, and two noncomplementing classes are typically separated by tens of kilobases. Many of the available mutants in the left half are insertions of mobile elements, which somehow affect sequences distant from the site of the insertion (Bender et al. 1983b).

Point mutations exist for three phenotypic classes (*Ubx*, *iab-2*, and *iab-7*), but they have a variety of phenotypes within each class, and none have yet been located on the DNA. Small deletions, which are most useful for assigning phenotypic functions to particular DNA sequences, are rare among the available alleles.

We have recently focused on mutations associated with chromosomal rearrangements splitting the BX-C. We can be more certain of the molecular consequences of rearrangement breakpoints than of deletion or insertion lesions. A break will disrupt the function of a gene not only when it interrupts the actual coding sequence, but also when it falls within an intron, or when it splits between the coding region and any *cis*-acting regulatory regions. We have located several new rearrangement breakpoints throughout the *bxd* region and have correlated their DNA map positions with their phenotypes. The progression of phenotypes suggests that there are multiple elements in the *bxd* region that interact with the *Ubx* region. The correlation of breaks and phenotypes in the abdominal half of the complex can be interpreted with similar models.

RESULTS

Figure 1 shows the 400-kb map of the chromosomal walk in the BX-C, compiled from Bender et al. (1983b) and Karch et al. (1985). The map is marked with the sites for about 50 rearrangement breakpoints that split the complex; these breaks span 300 kb. Two breakpoints are also indicated that lie outside the complex to the left (TE77) and right (S485). The breaks all have loss-of-function phenotypes (excepting the Contrabithorax [Cbx^3 and Cbx^{Twt}] breaks at the left); the different phenotypic classes are indicated. The breakpoint

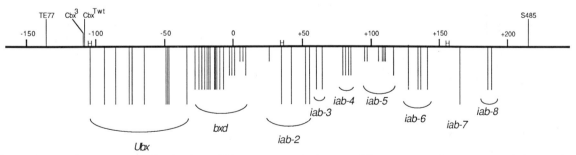

Figure 1. Map of rearrangement breakpoints within the BX-C. The horizontal line represents the DNA map of the complex, marked in kilobases. Vertical lines below the DNA line indicate the positions of breakpoints. The names of the different phenotypic classes are indicated below the breaks; longer lines denote more extreme phenotypes. Above the DNA line are shown breaks with no apparent loss-of-function phenotype.

lines are drawn in different lengths to indicate differences in severity of the mutant transformation. There is an even progression of severity in the *bxd* region and in the *iab-3* through *iab-6* region; these will be discussed more fully below.

The BX-C can be considered to have two separable "halves," the left or thoracic half spanning roughly from −110 to +20 kb on the molecular map and the right or abdominal half spanning from +20 to +190 kb. The separation of the two halves is based on complementation tests. Failures of complementation are seen among mutant alleles in the left half, and comparable interactions among mutations in the right half, but no discernible interactions are seen across the boundary at about +20 kb. The separation has been confirmed by constructing a fly homozygous for *Dp(2;3)P10* (containing DNA of the left half through +35 kb) and for *Df(3)bxd^{100}* (lacking DNA of the left half through −15 kb); such flies resemble the wild type (Struhl 1984). Both tests rule out *cis*-interaction between the left and right halves of the complex, but it is still possible that thoracic and abdominal functions overlap in the boundary region. It seems likely that each half is confined to a separate chromosome band (89E1,2 and 89E3,4), although the constriction at 89E makes it difficult to do precise cytology.

Breakpoints spanning about 73 kb at the left side of the complex all have extreme *Ubx* phenotypes (dominant enlargement of the haltere, recessive lethal with transformation of the third thoracic and first abdominal segments toward the second thoracic segment [Lewis 1978; Hayes et al. 1984; Struhl 1984]). The DNA domain defined by these breakpoints corresponds exactly to the transcribed region defined by two cDNA clones, which have exons spread out over the same 73 kb (Beachy et al. 1985; R. Saint, M. Akam, P. Beachy, and D.S. Hogness, unpubl.). These cDNAs are thus presumed to code for *Ubx* products. The cDNA sequences include the conserved homeo box region near their 3′ ends (Scott and Weiner 1984; McGinnis et al. 1984). The homology between the predicted amino acid sequence of the homeo box and the amino acid sequences of known DNA-binding proteins (Laughon and Scott 1984) suggests that the *Ubx* products may also be DNA-binding proteins.

Ubx alleles show a striking *cis*-inactivation of all functions in the left half of the BX-C. *Ubx* mutations fail to complement with *abx*, *bx*, *bxd*, or *pbx* mutations. It is possible that the *abx* and *bx* mutations, which map within the *Ubx* transcript, affect exons that are used by alternate patterns of splicing of the 73-kb transcript. This would not explain the *Ubx* inactivation of the *bxd* and *pbx* regions, unless there are transcripts yet undiscovered that begin in the *pbx* region and extend over more than 100 kb to the left end of the BX-C.

Breakpoints to the right of the *Ubx* region give recessive *bxd* transformations; their map positions are diagrammed in Figure 2. The most extreme *bxd* transformations include transformation of posterior haltere toward wing, loss of first abdominal tergite, and appearance of one or two legs from the first abdominal

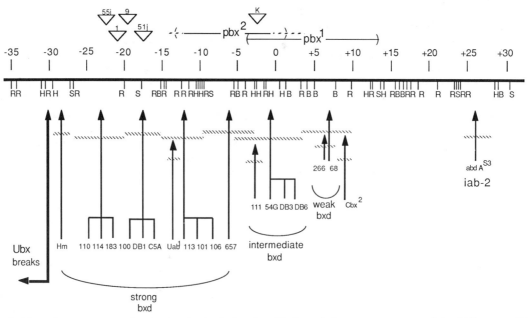

Figure 2. Map of rearrangement breakpoints in the *bxd* region. Horizontal line represents part of the DNA map of Fig. 1, marked in kilobases. Restriction sites are indicated for the enzymes *Eco*RI (R), *Bam*HI (B), *Hin*dIII (H), and *Sal*I (S). Vertical arrows show the positions of *bxd* breaks; hatched bars indicate the uncertainty of their positions. The breakpoint positions for the *bxd* alleles *Hm*, *110*, *100*, *Uab^1*, *113*, *111*, and *Cbx^2* have been confirmed by in situ hybridizations or by recloning; the remainder are preliminary assignments based on genomic blots. The breaks are grouped in three phenotypic classes, by severity. The triangles above the map indicate the insertion sites of gypsy mobile elements in spontaneous *bxd* alleles; the lines in parentheses show the extents of the two *pbx* deletions. The leftmost *iab-2* break is also shown; it has no *bxd* phenotype.

segment (Lewis 1963). In the late embryo and the larva, extreme *bxd* animals have a ventral setal belt of the first abdominal segment resembling that of the third thoracic segment, and they have thoracic-type ventral pits and Keilin organs on the abdominal segments (Lewis 1981a). Animals heterozygous for a breakpoint close to *Ubx*, such as *Tp(3)bxd^{100}* at about −17 kb, show the full spectrum of *bxd* transformations; these animals die as pharate adults. Other breaks between −28 and −10 kb have similar phenotypes. Breakpoints further to the right, such as *T(2;3)bxd^{DB3}* at about −3 kb, have a less-severe phenotype in the larvae. The ventral setal belt on the first abdominal segment is intermediate between the normal third thoracic belt and the normal first abdominal belt (Fig. 3). This intermediate transformation was first described for *T(1;3)bxd^{111}* (Lewis 1981a). The adult phenotype of *bxd^{DB3}* hemizygotes is also less severe; the flies are able to hatch from the pupal case, and they rarely have extra legs (Fig. 4).

We have recently recognized that two breakpoints further to the right (*bxd^{266}* and *bxd^{1068}*) have yet milder *bxd* phenotypes. Larvae of *bxd^{266}* hemizygotes have a first abdominal setal belt that is nearly wild type, although they still show the thoracic pits on the first abdominal segment (Fig. 3). Hemizygous adults look nearly wild type; the first abdominal tergite is slightly shortened laterally (Fig. 4). Homozygotes of *bxd^{266}* look wild type and can be kept as a healthy stock. The *Cbx^2* break is in the same region as *bxd^{266}*, but it has an intermediate phenotype like that of *bxd^{DB3}*. There is a striking dominant phenotype of *Cbx^2* (Lewis 1981b) that may be due to abnormal expression of bithorax functions, directed by the new DNA sequences to which bithorax is fused. We suspect that such a position effect could disrupt normal functions in the *bxd* region at some distance from the *Cbx^2* break. A break at about +26 kb (the *abd-A^{s3}* break in the *iab-2* group) has no phenotypic effect on the first abdominal segment, and no apparent interaction with *Ubx* alleles. We are currently examining the other *bxd* breakpoints for additional gradations in the spectrum of *bxd* transformations.

The positions and phenotypes of a series of breaks in the abdominal half of the BX-C have recently been reported (Karch et al. 1985). There are nested series of progressively less severe transformations with successive breaks, analogous to the progression of *Ubx* and *bxd* breaks. The *iab-2*, *iab-3*, and *iab-4* regions form the clearest progression. Four rearrangements breaking between +35 and +56 kb have severe *iab-2* phenotypes, as do small deficiencies at either end of this region (Fig. 1) (Karch et al. 1985). In *iab-2* hemizygous embryos, the second through the eighth abdominal segments are transformed toward the first abdominal segment (Morata et al. 1983; Sanchez-Herrero et al. 1985). We have isolated a cDNA clone made from embryonic RNA that has exons spanning this same region (F. Karch, unpubl.), and we presume it represents a major *iab-2* product. There are two breakpoints with *iab-3* phenotypes at about +60 and +64 kb. *iab-3* hemizygous adults show transformation of the third through sixth abdominal segments toward the second abdominal segment; this can be considered a subset of the transformation due to loss of *iab-2* function. The four breaks with *iab-4* phenotypes fall between +76 and +86 kb (Fig. 1). *iab-4* hemizygous adults show transformation of the fourth and fifth segments toward the third segment. Again, this *iab-4* transformation can be considered a subset of the *iab-3* transformation.

The complementation pattern of the *iab-2*, *iab-3*, and *iab-4* regions is also analogous to the *Ubx*, *bxd* region. Heterozygotes between *iab-4* and either *iab-3* or *iab-2* appear similar to *iab-4* hemizygotes, and *iab-3/iab-2* heterozygotes appear similar to *iab-3* hemizygotes. Thus, *iab-2* mutations *cis*-inactivate the *iab-3* and *iab-4* regions, just as *Ubx* mutations *cis*-inactivate the *bxd* region.

A similar phenotypic progression is seen in the *iab-7*, *iab-8* region. *iab-7* hemizygotes transform the fifth, sixth, and seventh segments toward the fourth or third segment and the eighth segment toward the sixth or fifth segment (Karch et al. 1985). *iab-8* hemizygotes transform the eight abdominal segment toward the seventh abdominal segment; this can be considered a subset of the *iab-7* transformation. We have only a single *iab-7* break at about +165 kb, and there are two *iab-8* breaks at +185 and +187 kb. *iab-7/iab-8* heterozygotes resemble *iab-8* hemizygotes, and so, again, the *iab-7* lesion seems to *cis*-inactivate the *iab-8* region.

The *iab-6* breakpoints have a phenotype that can also be considered a subset of the *iab-7* transformation; *iab-6* hemizygotes transform the sixth and seventh segments toward the fifth segment. However, the *iab-6* breaks, between +126 and +142 kb, fall to the left of *iab-7*. There is a strong failure of complementation between *iab-6* and *iab-7* alleles, but we cannot be certain of the direction of *cis*-inactivation (Karch et al. 1985).

The *iab-5* phenotype (transformation of fifth abdominal segment toward the fourth or third) can be considered a subset of either the *iab-2* or the *iab-7* phenotypes. The *iab-5* breaks fall between the *iab-4* and *iab-6* breaks (Fig. 1), and so, by map position or by phenotype, *iab-5* could be included either in the *iab-2,3,4* series or in the *iab-7,6* series. However, the complementation pattern of *iab-5* alleles is curious; they show partial complementation with both *iab-2* and *iab-7* alleles. It may be proper to consider *iab-5* as part of both series.

DISCUSSION

Figure 5 summarizes the examples of nested *cis*-inactivation described above. In the *bxd* regions, we can distinguish at least three gradations in the phenotypes of *bxd* breaks. The functions of the *bxd* region must depend on continuity with the *Ubx* region for proper function, since the phenotypes of all these *bxd* breaks over *Ubx^1* are nearly as severe as over a deficiency for

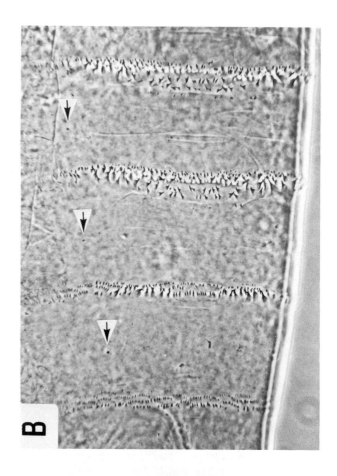

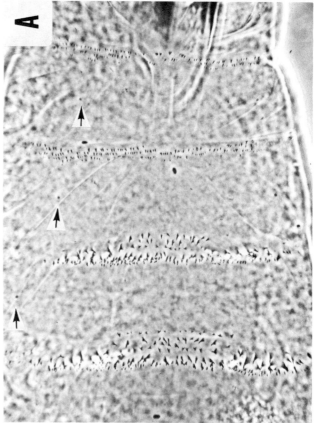

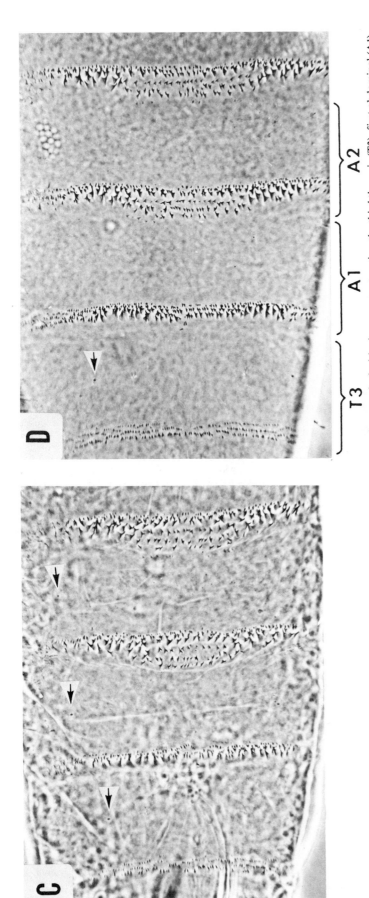

Figure 3. Progression of *bxd* larval transformations. The ventral surfaces of first-instar larvae were photographed with phase-contrast optics; the third thoracic (T3), first abdominal (A1), and second abdominal (A2) segments are shown. Arrows indicate the ventral pits. (*A*) *bxd¹⁰⁰/DfUbx¹⁰⁹*. The A1 setal belt resembles the T3 belt in the size and number of denticles, and ventral pits appear on abdominal segments. (*B*) *bxd^{DB3}/DfUbx¹⁰⁹*: The A1 denticle belt is intermediate between the normal T3 and A1 belts, and ventral pits appear on abdominal segments. (*C*) *bxd²⁶⁶/DfUbx¹⁰⁹*. The A1 denticle belt is near wild type, but ventral pits are still present on the abdominal segments. (*D*) *iab-2(abd-A^{S3})/DfUbx¹⁰⁹*: These larvae are indistinguishable from +/*DfUbx¹⁰⁹* larvae; ventral pits are present only on the thoracic segments.

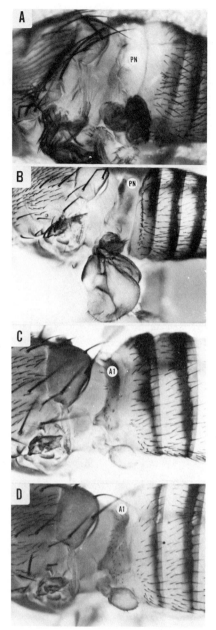

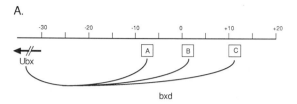

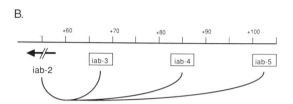

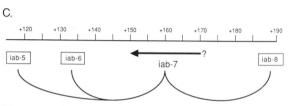

Figure 5. Diagram of the interaction of elements in the BX-C. Horizontal lines indicate the DNA map with coordinates marked as in Figs. 1 and 2. (*A*) *Ubx* region is postulated to interact with at least three separable elements in the *bxd* region. *bxd²⁶⁶* splits away only element C, *bxd^DB3* splits away B and C, and *bxd¹⁰⁰* splits away A, B, and C. (*B*) *iab-2* region interacts with *iab-3* and *iab-4* elements; there is also a partial interaction with *iab-5*. (*C*) *iab-7* region interacts with one or more elements for *iab-8*, and there may be a similar interaction with elements in the *iab-5* and *iab-6* regions.

Figure 4. Progression of *bxd* adult transformations. The dorsal cuticles of adult flies are shown. The wings have been removed from the animals in *B*, *C*, and *D*. (*A*) *bxd¹⁰⁰/DfUbx¹⁰⁹*: This pharate adult was dissected from its pupal case. The normal first abdominal tergite is absent, but a band of postnotal tissue (PN) is present. The posterior haltere is transformed into wing tissue, but it remains folded. This animal also had eight legs. (*B*) *bxd^DB3/DfUbx¹⁰⁹*. The first abdominal tergite is replaced by postnotal tissue as in *A*, and the posterior haltere is transformed to wing tissue. This animal had six legs. (*C*) *bxd²⁶⁶/DfUbx¹⁰⁹*: The first abdominal tergite (A1) is reduced, but there are no extra legs, no transformation of the haltere, and no postnotal tissue. (*D*) *+/DfUbx¹⁰⁹*. This animal has a wild-type first abdominal tergite (A1).

the whole region (*DfUbx¹⁰⁹*). Thus, we can postulate at least three elements that interact with *Ubx*. With the same logic, we can postulate at least two elements (*iab-3* and *iab-4*) that interact with *iab-2*, and one or two elements (*iab-8* and maybe *iab-6*) that interact with *iab-7*. The *iab-2* and *iab-7* regions both contain transcripts that include sequences homologous to the homeo box (Regulski et al. 1985); like *Ubx*, they may encode DNA-binding proteins. Four models can be proposed for these *cis*-interactions: alternate domains of transcription, *trans*-splicing, controlling RNAs, and enhancers.

The different elements of the *bxd* region could be different initiation sites for transcription; all transcripts would then proceed through the *Ubx* region and be spliced to include *Ubx* exons, including the homeobox-coding region. Alternate transcription domains were first seen for the α-amylase gene in different tissues of the mouse (Young et al. 1981). In *Drosophila*, the alcohol dehydrogenase gene has different initiation sites for different times in development (Benyajati et al. 1983). Such alternate initiation sites in bithorax could be used in different segments (abdominal elements) or different tissues within a segment (*bxd* elements). If the alternate exons included protein-coding regions, there might be a family of *bxd* products, for example, with variable (*bxd*) and constant (*Ubx*) protein domains. The *Ubx*, *iab-2*, and *iab-7* homeo boxes are all in the same orientation on the chromosome, so that the *bxd*, *iab-3*, *iab-4*, and *iab-8* elements all lie upstream of the respective homeo boxes. If the interaction of *iab-6* with *iab-7* utilizes the same mechanism, then its position downstream from the *iab-7* homeo box presents a problem. There is a similar problem with the *cis*-inactivation of the *bx* region by *Ubx* mutations to the right of *bx*.

The elements in the *bxd* region might encode separate transcripts that are joined together with the *Ubx* transcript after both are transcribed. Such a model has been proposed for the variable surface glycoprotein mRNAs in trypanosomes, where the 5' exon comes from a separately transcribed RNA molecule (Nelson et al. 1983). Such RNAs might be transcribed from either strand in the *bxd* region, and then be *trans*-spliced to *Ubx* RNA products. In the case of *iab-7*, its RNA could interact with separate RNA products from regions to the left (*iab-6*) or right (*iab-8*) on the chromosome. Since the *bxd* elements must be close to the *Ubx* region for proper function, we must also postulate that such *trans*-spliced RNAs would be rapidly degraded. The RNAs would then be in sufficient concentration only near the site of their transcription. A difficulty with this and the preceding model is that none of the cDNAs examined by us or reported by others (Beachy et al. 1985) include exons from both the *bxd* and *Ubx* regions or from any pair of *cis*-interacting abdominal regions.

The elements of the *bxd* region might make RNA products that interact with the *Ubx* RNA or with other small RNA molecules involved in processing the *Ubx* RNA. In *Escherichia coli*, a small antisense RNA from the *ompC* locus binds to the mRNA from the *ompF* gene and blocks its expression (Mizuno et al. 1984). RNAs from the *bxd* region could block expression of a subset of RNA products of the *Ubx* region, or they might direct processing of a *Ubx* transcript. As with the previous model, such *bxd* RNAs would have to be rapidly degraded to explain the requirement for chromosomal proximity. Similar models for *cis*-interaction in the BX-C have been proposed by D.S. Hogness and by P. O'Farrell (both pers. comm.).

A final proposal for the *bxd* elements is that they are tissue-specific enhancers controlling the transcription of *Ubx*. Enhancer sequences were first identified in SV40 viral DNA (Banerji et al. 1981), and they have been shown to be tissue-specific in other systems (Gillies et al. 1983). Such sequences can "enhance" the level of transcription of a neighboring transcription initiation site independent of the orientation of the enhancer sequence and at a distance of as much as a few thousand base pairs away. The molecular nature of the interaction between an enhancer and an initiation site remains unknown. An enhancer model has been proposed for the *cis*-acting sequences at the white locus in *Drosophila* and has been suggested, by analogy, for the BX-C (Davison et al. 1985). The most distant *bxd* elements lie 40 kb or more from the *Ubx* region, a far greater distance than has yet been seen for enhancer action. Our present ignorance about enhancers makes such a model difficult to evaluate.

The *cis*-interactions like those between *Ubx* and the *bxd* elements seem pervasive in the BX-C. The alternate molecular models predict different numbers of products from the complex, and the number of potential products is fundamental to theories on how the products might act to direct segmental development.

ACKNOWLEDGMENTS

We are grateful to E.B. Lewis for providing the BX-C mutations and for describing their phenotypes and cytology. E.B. Lewis and I. Duncan have helped with the interpretation of mutant interactions. Most of the new *bxd* alleles were isolated by David Baker and Rollin Baker.

REFERENCES

BANERJI, J., S. RUSCONI, and W. SCHAFFNER. 1981. Expression of a beta-globin gene is enhanced by remote SV40 DNA sequences. *Cell* 27: 299.

BEACHY, P.A., S.L. HELFAND, and D.S. HOGNESS. 1985. Segmental distribution of bithorax complex proteins during *Drosophila* development. *Nature* 313: 545.

BENDER, W., P. SPIERER, and D.S. HOGNESS. 1983a. Chromosomal walking and jumping to isolate DNA from the *Ace* and *rosy* loci and the bithorax complex in *Drosophila melanogaster*. *J. Mol. Biol.* 168: 17.

BENDER, W., M. AKAM, F. KARCH, P.A. BEACHY, M. PEIFER, P. SPIERER, E.B. LEWIS, and D.S. HOGNESS. 1983b. Molecular genetics of the Bithorax complex in *Drosophila melanogaster*. *Science* 221: 23.

BENYAJATI, C., N. SPOEREL, H. HAYMERLE, and M. ASHBURNER. 1983. The messenger RNA for alcohol dehydrogenase in *Drosophila melanogaster* differs in its 5' end in different developmental stages. *Cell* 33: 125.

DAVISON, D., C.H. CHAPMAN, C. WEDEEN, and P.M. BINGHAM. 1985. Genetic and physical studies of a portion of the *white* locus participating in transcriptional regulation and in synapsis-dependent interactions in *Drosophila* adult tissue. *Genetics* 110: 479.

GILLIES, S.D., S.L. MORRISON, V.T. OI, and S. TONEGAWA. 1983. A tissue-specific transcription enhancer element is located in the major intron of a rearranged immunoglobin heavy chain gene. *Cell* 33: 717.

HAYES, P.H., S. SATO, and R.E. DENELL. 1984. Homoeosis in *Drosophila*: The Ultrabithorax larval syndrome. *Proc. Natl. Acad. Sci.* 81: 545.

KARCH, F., B. WEIFFENBACH, M. PEIFER, W. BENDER, I. DUNCAN, S. CELNIKER, M. CROSBY, and E.B. LEWIS. 1985. The abdominal region of the Bithorax complex. *Cell* (in press).

LAUGHON, A. and M.P. SCOTT. 1984. Sequence of a *Drosophila* segmentation gene: Protein structure homology with DNA-binding proteins. *Nature* 310: 25.

LEWIS, E.B. 1963. Genes and developmental pathways. *Am. Zool.* 3: 33.

———. 1978. A gene complex controlling segmentation in *Drosophila*. *Nature* 276: 565.

———. 1981a. Developmental genetics of the bithorax complex in *Drosophila*. *ICN-UCLA Symp. Mol. Cell. Biol.* 23: 189.

———. 1981b. Control of body segment differentiation in *Drosophila* by the bithorax gene complex. In *Embryonic development: Genes and Cells* (ed. M. Burgher), p. 269. A.R. Liss, New York.

McGINNIS, W., M.S. LEVINE, E. HAFEN, A. KUROIWA, and W. GEHRING. 1984. A conserved DNA sequence in homoeotic genes of *Drosophila* Antennapedia and bithorax complexes. *Nature* 308: 428.

MIZUNO, T., M. CHOU, and M. INOUYE. 1984. A unique mechanism regulating gene expression: Translational inhibition by a complementary RNA transcript (micRNA). *Proc. Natl. Acad. Sci.* 81: 1966.

Morata, G., J. Botas, S. Kerridge, and G. Struhl. 1983. Homeotic transformations of the abdominal segments of *Drosophila* caused by breaking or deleting a central portion of the bithorax complex. *J. Embryol. Exp. Morphol.* **78:** 319.

Nelson, R.G., M. Parsons, P.J. Barr, K. Stuart, M. Selkirk, and N. Agabian. 1983. Sequences homologous to the variant antigen mRNA spliced leader are located in tandem repeats and variable orphons in *Trypanosoma brucei*. *Cell* **34:** 901.

Regulski, M., K. Harding, R. Kostriken, F. Karch, M. Levine, and W. McGinnis. 1985. Homeo box genes of the Antennapedia and Bithorax complexes of *Drosophila*. *Cell* (in press).

Sanchez-Herrero, E., I. Vernos, R. Marco, and G. Morata. 1985. Genetic organization of the *Drosophila* bithorax complex. *Nature* **313:** 108.

Scott, M.P. and A.J. Weiner. 1984. Structural relationships among genes that control development: Sequence homology between the Antennapedia, Ultrabithorax, and fushi tarazu loci of *Drosophila*. *Proc. Natl. Acad. Sci.* **81:** 4115.

Struhl, G. 1984. Splitting the bithorax complex of *Drosophila*. *Nature* **308:** 454.

White, R.A.H. and M. Wilcox. 1984. Protein products of the bithorax complex in *Drosophila*. *Cell* **39:** 163.

Young, R.A., O. Hagenbuchle, and U. Schibler. 1981. A single mouse alpha-amylase gene specifies two different tissue specific mRNAs. *Cell* **23:** 451.

Regulation and Products of the *Ubx* Domain of the Bithorax Complex

D.S. HOGNESS, H.D. LIPSHITZ, P.A. BEACHY, D.A. PEATTIE, R.B. SAINT,* M. GOLDSCHMIDT-CLERMONT,† P.J. HARTE, E.R. GAVIS, AND S.L. HELFAND‡
Department of Biochemistry, Stanford University School of Medicine, Stanford, California 94305

The cluster of homeotic genes known as the bithorax complex (BX-C) (Lewis 1978) is divisible into three complementation groups, or functional domains (Sánchez-Herrero et al. 1985). Each domain provides the major developmental determinants for segment identity in one of three adjacent metameric regions that together extend from the anterior/posterior compartment boundary (a/p) of the second thoracic segment (T2) into the eighth abdominal segment (A8) (Fig. 1). Thus, the Ultrabithorax (*Ubx*) domain controls the T2p-A1a region (Lewis 1963, 1978, 1981, 1982; Morata and Kerridge 1981; Hayes et al. 1984; Struhl 1984; Sánchez-Herrero 1985 and this volume), and the abdominal-A (abd-A) and Abdominal-B (*Abd-B*) domains control the A1p-A4 and A5-A8 regions, respectively (Sánchez-Herrero et al. 1985).

These positional assignments of domain functions derive from transformations of segmental identities in the larval and adult epidermis caused by mutations in the respective domains. Recent evidence indicates that the BX-C also provides metameric identity functions for the central nervous system (CNS) and musculature—evidence derived both from mutational studies (Lawrence and Johnston 1984; Teugels and Ghysen 1985) and from the metameric distribution of BX-C RNAs (Akam 1983; Akam and Martinez-Arias 1985 and this volume; Harding et al. this volume) and proteins (White and Wilcox 1984; Beachy et al. 1985).

Present addresses: *Division of Entomology, CSIRO, P.O. Box 1700, Canberra City, ACT 2601, Australia; †Department of Molecular Biology, University of Geneva, 30 quai Ernest Ansermet, 1211 Geneva 4, Switzerland; ‡Department of Biology, Yale University, New Haven, Connecticut.

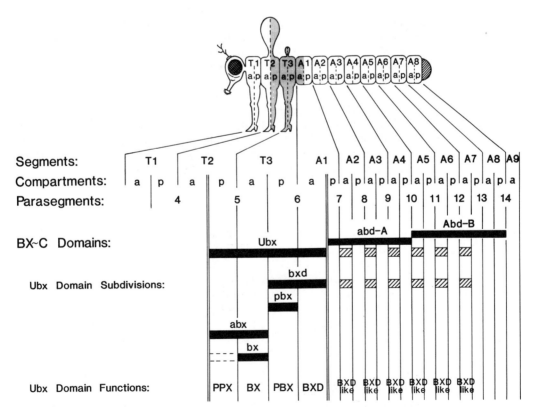

Figure 1. Mutations and functions of the BX-C domains. The diagram of the fly is in the style of Morata (Sánchez-Herrero et al. 1985). The stippled portion of the fly represents the region where the *Ubx* domain functions are the primary determinants of metameric identity.

Taken together, these observations indicate that the BX-C genes act as general transducers of positional information for the specification of metameric identities in quite different tissues. An understanding of the role played by the BX-C in metameric development therefore requires the solution of two basic problems: How is the expression of the BX-C genes regulated in a position-specific manner? How do the products of that expression specify the morphologies and functions that differentiate the metameric units within each of several different tissues?

A more precise molecular definition of these problems is required for their solution. To that end, we have concentrated on the molecular analysis of the *Ubx* domain, both because it is the best defined of the three BX-C domains and because we think it will serve as a model for the other two. The *Ubx* domain occupies the proximal or left one third of the BX-C DNA and contains two long transcription units (Fig. 2) (Bender et al. 1983; Beachy et al. 1985). In this paper we focus on the problem of how these two units can interact in *cis* so that one regulates the expression of the other in a position-specific manner.

Definition of the *cis*-Regulatory Problem

Genetic definition. The *Ubx* domain is defined by five classes of recessive mutations (Lewis 1963, 1978, 1981, 1982; Morata and Kerridge 1981; Hayes et al. 1984; Struhl 1984; Sánchez-Herrero et al., this volume). Four of these (anterobithorax [*abx*], bithorax [*bx*], bithoraxoid [*bxd*], and postbithorax [*pbx*]) serve to define the individual functions of the domain, whereas the fifth (*Ubx*), *cis*-inactivates the set of functions defined by the other four and thereby defines the genetic determinants of the set as a single functional domain.

The definition of these functions derives from the transformations of identity effected by these mutations in the T2p-A1a region of the larval and adult epidermis. These effects can best be described in parasegmental units, where a parasegment consists of the posterior compartment of a given segment plus the adjacent anterior compartment of the next segment (Fig. 1). According to the convention adopted by Martinez-Arias and Lawrence (1985), the region controlled by the *Ubx* domain therefore consists of parasegments 5 (T2p, T3a) and 6 (T3p, A1a). Four compartment-specific functions can be defined according to the rule that mutational inactivation of a given function transforms the compartment in which its expression is required for normal development to the identity of the homologous compartment in the anteriorly adjacent parasegment. These functions are referred to here as postprothorax (PPX), BX, PBX, and BXD and are required, respectively, for the normal development of T2p, T3a, T3p, and A1a (Fig. 1). (The PPX function was previously referred to as ABX [Beachy et al. 1985] and is changed here to conform with an earlier designation [Morata and Kerridge 1981] and with usage elsewhere in this volume [Akam and Martinez-Arias; Sánchez-Herrero et al.].)

The *abx* and *bxd* mutations effect a division of these four functions into two subsets required for the normal development of parasegments 5 and 6, respectively (Fig. 1). Thus, *abx* mutations induce transformations in both compartments of parasegment 5 and are therefore considered to inactivate both members of the parasegment 5 subset (PPX, BX), whereas *bxd* mutations are considered to inactivate both members of the parasegment 6 subset (PBX, BXD) because they transform both compartments of this parasegment (Lewis 1963, 1978, 1981, 1982; Sánchez-Herrero et al., this volume; Casanova et al. 1985). The division of each subset into two functions results from the more restricted transformations of the adult epidermis observed for *bx* and *pbx* mutations, which affect, respectively, T3a (inactivation of BX, but not PPX) and T3p (inactivation of PBX, but not BXD). *Ubx* mutations are considered to *cis*-inactivate both subsets because they transform parasegments 5 and 6 to the "ground state" identity of parasegment 4 and because they fail to complement any of the other four mutational classes (Lewis 1963, 1978; Morata and Kerridge 1981; Kerridge and Morata 1982; Hayes et al. 1984; Struhl 1984; Sánchez-Herrero et al. 1985).

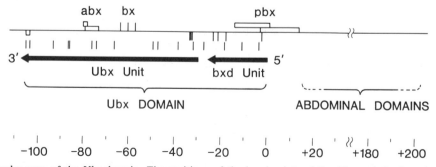

Figure 2. Molecular map of the *Ubx* domain. The positions of the breakpoints of the *Ubx* and the four *bxd* chromosomal rearrangements that define the *Ubx* and *bxd* regions are indicated by the vertical lines just above the solid arrows denoting the orientations and extents of the *Ubx* and *bxd* transcription units. The *Ubx* and *bxd* pseudopoint mutations are indicated just below the long horizontal line representing the chromosomal DNA. The *abx* and *pbx* deletions and the sites of the *bx* insertional mutations are indicated above this line. The scale is in kilobases. (Modified from Beachy et al. 1985.)

The above division of the parasegment 5 subset into the compartment-specific PPX and BX functions is supported by two additional lines of evidence. First, these two functions, or, more precisely, their genetic determinants, are dispensable after quite different times of development: PPX after only 7 hours of embryogenesis (Morata and Kerridge 1981) and BX after completion of embryonic and most of larval development (Lewis 1964; Morata and Garcia-Bellido 1976). Second, *abx* mutations are cold-sensitive with respect to the PPX, but not the BX, function (Casanova et al. 1985).

The mutational subdivision of this subset may not be as simple as we have supposed, however, since Casanova et al. (1985) have found that at low temperature (17°C), certain *bx* mutants exhibit a slightly defective PPX function in addition to their loss of the BX function. We have not attempted to account for such secondary complexities in the scheme given in Figure 1. Rather, this scheme is viewed as a useful first approximation that is most accurate with respect to the following conclusions that are particularly relevant to this paper. (1) The *Ubx* domain provides *multiple* identity functions for the T2p-A1a region. (2) These functions are divisible into two subsets that are specific for parasegments 5 and 6 and are inactivated by *abx* and *bxd* mutations, respectively. (3) Each subset consists of more than one function. (4) *Ubx* mutations inactivate in *cis* all functions of the domain. In addition, a set of functions that are expressed in the anterior portions of A2-A7 and inactivated by *Ubx* and *bxd* mutations are shown in Figure 1 (BXD-like functions) to account for the partial transformations induced by both mutations in these regions of the larval epidermis (Lewis 1978, 1981; Bender et al., this volume).

Molecular definition. Figure 2 shows a molecular map of the *Ubx* domain that emphasizes the relationship between the sites of mutations that define the domain (Bender et al. 1983) and two transcription units (~75 kb and ~25 kb) found within it (Beachy et al. 1985; R. Saint et al.; M. Goldschmidt-Clermont et al.; both in prep.). The sites of the *Ubx* and *bxd* mutations are not interspersed, rather, they occupy regions (*Ubx* and *bxd* regions) that are closely correlated with those defined by the 75-kb and 25-kb transcription units, which are therefore called the *Ubx* and *bxd* units respectively. The *abx* and *bx* mutations occupy closely linked but nonoverlapping regions within the *Ubx* unit, whereas the *pbx* mutations overlap the *bxd* unit.

The most curious aspect of this distribution is that mutations in either of two closely linked regions, each of which is closely correlated with a transcription unit, can *cis*-inactivate the parasegment 6 and BXD-like subsets of domain functions. These subsets therefore depend on the integrity of the DNA in each of two long regions and on the close linkage of these regions within the same DNA molecule. The dependence of the parasegment 6 functions on this close linkage is emphasized by the observation that these functions can be expressed only if the DNA spanning both regions is contiguous (Lewis 1978; Bender et al. 1983). In contrast, the *Ubx* unit is sufficient for the expression of the parasegment 5 functions defined by the *abx* and *bx* mutations located within that unit (Morata and Kerridge 1981; Bender et al. 1983). The *Ubx* unit therefore appears to be required for all domain functions, subsets of which also depend on the close linkage of the DNA in the *bxd* region.

The Domain Functions Depend on the *Ubx* Unit Because They Require Proteins Encoded by That Unit

An explanation for the dependence of all functions on the *Ubx* unit derives from an analysis of its RNAs (*Ubx* RNAs) and from the structure and location of the *Ubx* mutations. Figure 3 shows the three major size classes of the mature *Ubx* RNAs, along with their temporal profiles of expression. We have defined the structure of the 3.2-kb-size class by nucleotide sequence analyses of two cloned cDNAs and of homologous regions in the genomic DNA and by S1-protection and primer-extension analyses of its 5' end (R. Saint et al., in prep.). These analyses indicate that the 3.2-kb RNAs have a common 5' end and consist of four exons, of which the 5' and 3' exons account for 97% of the RNA sequence, the two internal 51-bp microexons accounting for the remainder. The two cDNA clones differ with respect to the donor splice site for the 5' exon, one being 27 bp removed from the other. This heterogeneity in the 3.2-kb *Ubx* RNAs can be translated into a heterogeneity among the proteins encoded by the *Ubx* unit (*Ubx* proteins) because the 27 bp are in phase with a long open reading frame of 380 codons (ORF Ia; inclusion of the 27 bp yields ORF Ib of 389 codons) that is initiated by an AUG codon in the 5' exon and terminated by a UAG codon in the 3' exon (Fig. 4).

There is little doubt that these ORFs are translated to yield *Ubx* proteins in *Drosophila melanogaster* cells. The strongest evidence derives from experiments with antibodies directed against the *Ubx* determinants of β-galactosidase (β-Gal)-*Ubx* hybrid proteins encoded by a hybrid gene in which a sequence of codons from ORF Ia has been fused in-phase with the coding region of the *Escherichia coli lacZ* gene (White and Wilcox 1984; Beachy et al. 1985). These antibodies bind proteins that are present in wild-type but not in *Ubx* mutant embryos. These *Ubx* proteins exhibit metameric distributions in embryos and larval imaginal disks (see below) that are consistent with those observed for *Ubx* RNAs (Akam 1983; Akam and Martinez-Arias 1985), and they migrate in SDS gels with electrophoretic mobilities approximating those expected for these ORFs. The conclusion that these ORFs encode functional proteins is also supported by the following observations: The codon usage in these ORFs fits well with the codon frequencies observed for known *D. melanogaster* structural genes (Beachy et al. 1985); their 3'-exon regions contain a "homeo box" (Fig. 4) (McGinnis et al. 1984; Scott and Weiner 1984), the only one found in the *Ubx*

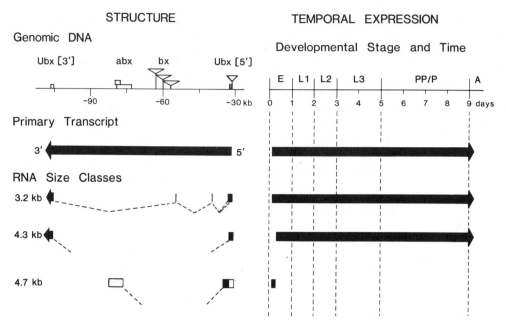

Figure 3. Structural and temporal expression of RNAs from the *Ubx* unit. The exon-containing regions are indicated by the filled boxes below the representation of the primary transcript. In the case of the 3.2-kb RNA size class, the dashed lines indicate the structure of the transcripts as defined by the cDNAs. Note the alternative donor splice site of the 5' exon. The 3' exon-containing region of the 4.3-kb size class can be seen to extend beyond that of the 3.2-kb class. Pseudopoint mutations in the genomic DNA are indicated by open boxes or vertical lines (deletions) and inverted triangles (transposable-element insertions). The gray-level of the boxes representing the 4.7-kb class indicates the relative strength of hybridization from these regions. The solid arrows and box on the right indicate the qualitative aspects of the temporal expression of each of the RNA size classes. (E) Embryonic; (L1, L2, L3) first, second, and third larval instar; (PP/P) prepupal and pupal stages; (A) adult.

domain (R. Saint et al., in prep.); and the 5'-exon sequence common to both ORFs is highly conserved in *D. melanogaster*, *D. funebris*, and *Musca domestica* (D. Wilde and M. Akam, pers. comm.; R. Saint et al., in prep.). Finally, the 3.2-kb RNAs in cytoplasmic extracts cosediment with polysomes of the appropriate size (D. Peattie, unpubl.).

The 4.3-kb-size class provides a likely additional source of mRNAs encoding *Ubx* proteins. They exhibit the same 5' ends as the 3.2-kb mRNAs, and Northern

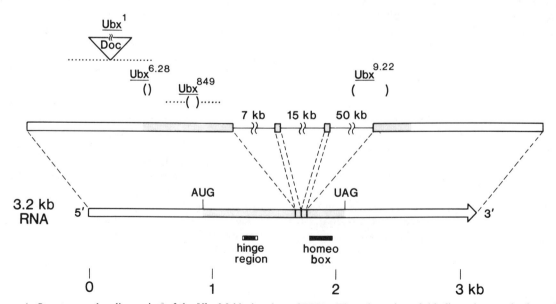

Figure 4. Structure and coding region of the *Ubx* 3.2-kb size class of RNAs. The orientation of this figure is opposite from that in Figs. 2, 3, 7, and 8. The locations of the *Ubx* pseudopoint mutations are shown above the horizontal bar representing the genomic DNA. The three introns are indicated by the thin lines. Stippled areas in the DNA and in the arrow below it representing the processed transcripts indicate the extent of the ORFs. The locations of the hinge region and homeo box are also shown. (Modified from Beachy et al. 1985.)

analyses with probes from the *Ubx* unit are indicative of strong sequence homologies between the 5' and 3' exons of the 3.2-kb mRNAs and the corresponding regions of the 4.3-kb RNAs; such analyses are not, however, sensitive enough to detect internal microexons. The two size classes differ in the times of their appearance during early embryogenesis and in their 3' ends, that for the 4.3-kb class lying downsteam from that of the 3.2-kb class (Fig. 3) (R. Saint et al., in prep.). Although the exon structure of the 4.3-kb RNAs has not yet been determined by nucleotide sequence analyses of cloned cDNAs, the above characteristics suggest that they encode *Ubx* proteins with amino acid sequences akin to those encoded by ORF Ia and Ib.

What effects do the *Ubx* mutations have on the synthesis and/or structure of these two size classes of mRNAs and the proteins they encode? The *Ubx* mutations shown in Figure 2 form two classes. The "pseudopoint" class consists of three small deletions and a 4.3-kb insertion that are located within the *Ubx* unit near either of its ends. The other class consists of 12 "gross rearrangements" (translocations, inversions, and deficiencies), each with a single break within the *Ubx* unit. These breaks are scattered throughout the unit and are sealed by the fusion of the *Ubx* DNA with DNA from loci that are separated from the *Ubx* unit by large genomic distances.

Figure 4 shows the structure and distribution of the four pseudopoint mutations relative to the exon structure of the 3.2-kb mRNA and, more particularly, to its coding region (ORF Ia or Ib). Three of these mutations are deletions, one of which, $Ubx^{6.28}$, deletes only 32 bp from the coding region near its 5' end (M. Akam, pers. comm.). The resulting shift in reading frame yields a stop codon shortly downstream from the deletion and should prevent the translation of all but a small fraction of ORF Ia and Ib. This conclusion has been verified by the observation that $Ubx^{6.28}$ homozygotes lack all *Ubx* proteins detectable in wild-type strains by anti-*Ubx* antibodies directed against that part of a β-Gal-*Ubx* hybrid protein encoded by codons 9–300 in ORF Ia (Beachy et al. 1985).

The other two pseudopoint deletions also eliminate sequences common to ORF Ia and Ib; Ubx^{849} deletes approximately 110 bp of the 5' exon downstream from $Ubx^{6.28}$, and $Ubx^{9.22}$ deletes approximately 1.6 kb that includes part of the homeo box of ORF Ia and Ib and extends upstream to include the 3'-exon splice site and adjacent sequences in the last intron (Bender et al. 1983; R. Saint et al., in prep.; M. Akam, pers. comm.). Thus, all three pseudopoint deletions exhibit the common characteristic of altering the structure of the *Ubx* proteins encoded by ORF Ia and Ib. Furthermore, each will similarly alter any member of a family of *Ubx* proteins whose coding sequence begins with the AUG codon that initiates ORF Ia and Ib, extends downstream to include part or all of the approximately 110 bp deleted by Ubx^{849}, and by splicing, incorporates the homeo box. Such a family would thus be characterized by two constant regions that include, respectively, an amino-proximal sequence and the homeo box sequence. Members of this family would be distinguished by a variable region connecting the two constant regions (e.g., the proteins encoded by ORF Ia and Ib) and perhaps by another variable region downstream from the homeo box. Additional members of this family could be encoded by other ORFs derived from the 3.2-kb and/or the 4.3-kb mRNAs by differing patterns of splicing. Indeed, a hint of such additional members is given by the observation that SDS-gel electrophoresis of proteins extracted from embryos or haltere imaginal disks yields at least three bands capable of binding the anti-*Ubx* antibodies (White and Wilcox 1984; Beachy et al. 1985).

Because the only obvious common characteristic of these *Ubx* deletions is that each will alter the structure of all members of this family, and because each deletion is defective with respect to both the parasegment 5 and parasegment 6 functions (Kerridge and Morata 1982; Bender et al. 1983), we think that each of these functions requires one or more proteins from this family—a conclusion that explains why each function is dependent on the *Ubx* unit.

The structures of the other *Ubx* mutations shown in Figures 2 and 4 are consistent with this conclusion, since each can be interpreted as inhibiting the synthesis, or altering the structure, of all members of the above family of *Ubx* proteins. This is certainly the case for 10 of the 12 gross rearrangements whose breakpoints clearly lie between the 5' end of the *Ubx* unit and its homeo box and may also be true for the two breakpoints nearest the 3' end of the unit, although they have not been mapped at sufficient resolution to be sure (Bender et al. 1983; R. Saint et al., in prep.). The remaining mutation, Ubx^1, results from the insertion of a 4.3-kb transposable element called "Doc" that has been mapped to a region within the 5' exon that overlaps adjacent parts of its untranslated and coding regions, as shown in Figure 4. Ubx^1 homozygotes, like $Ubx^{6.28}$ homozygotes, lack *Ubx* proteins detectable by anti-*Ubx* antibody (S.L. Helfand, unpubl.). In sum, 14 of the 16 *Ubx* mutations clearly alter the structure or inhibit the synthesis of all members of the family, and the data for the other two are consistent with their having the same effect. We therefore propose that each *Ubx* mutation inactivates all domain functions by inactivating all proteins in this family.

This proposition does not exclude the possibility that one or more of the domain functions also require other products of the *Ubx* unit, particularly the 4.7-kb *Ubx* RNA (Fig. 3). This poly(A)⁻ RNA has not been sufficiently characterized, however, to give more than a hint as to its possible functions. Northern analyses of this poly(A)⁻ RNA indicate that it cannot encode proteins belonging to the above family since it lacks detectable sequences from the 3' end of the *Ubx* unit, including the homeo box (R. Saint et al., in prep.). Although it contains some sequences from the 5' exon of the 3.2-kb mRNA, it differs from this and the 4.3-kb mRNA in containing sequences just downstream from that exon, and preliminary evidence indicates that it contains sequences from a genomic region that overlaps the two *abx* deletions (E. Gavis, unpubl.). The transient expression of the 4.7-kb RNA during embryogenesis indicates that its processing is controlled differently from that for the 3.2-kb and 4.3-kb mRNAs and suggests the possibility that it may act as an early determinant of metameric identity, as has been suggested previously in respect to parasegment 6 (Akam and Martinez-Arias 1985).

The Distribution of *Ubx* Proteins Correlates with the Distribution of Domain Functions

The proposition that all functions of the *Ubx* domain require members of the *Ubx* protein family predicts the presence of one or more of these proteins in each metameric unit to which a domain function has been assigned by the mutational data (Fig. 1). We have tested this prediction by an immunofluorescence assay of the *Ubx* proteins present in the CNS of approximately 12-hour-old embryos and in the larval imaginal disks (Beachy et al. 1985). Because the polyclonal antibody used in these assays is directed against that part of a hybrid protein encoded by codons 9-300 of ORF Ia, it should detect any *Ubx* protein containing the amino-proximal constant region of the family. It does not, however, effectively bind to the homeo box constant region, since it does not detect proteins encoded by other homeotic genes containing that box, as witnessed by the observation, noted above, that Ubx^1 and $Ubx^{6.28}$ homozygotes lack proteins detectable by this antibody. The *Ubx* proteins exhibit a strong nuclear localization in all tissues where they have been detected by this antibody. Close examination of antibody-labeled nuclei indicates that the *Ubx* proteins are localized within the nucleus rather than being bound to the nuclear membrane (Beachy et al. 1985), as do preliminary results, indicating that these proteins exhibit sequence-specific binding to DNA (P.A. Beachy, unpubl.).

Figure 5 shows the results obtained when the CNS of an approximately 12-hour-old wild-type embryo was labeled with this antibody (see Fig. 6 for a schematic of this labeling). The distribution of label is in good agreement with the distribution of the domain functions given in Figure 1 and, consequently, with the above prediction. The most intense labeling is in parasegment 6, where virtually all nuclei are labeled at the same intensity. A single wave of label extends anteriorly with a trough in T3a and a peak in T2p, specific subsets of nuclei in each of these presumptive compartments of parasegment 5 being labeled. Seven waves of decreasing amplitude of label extend posteriorly from parasegment 6, with troughs in the posterior parts of A1-A7 and peaks in the anterior parts of A2-A8, each wave again corresponding to a parasegment. The number and arrangement of the subset of nuclei labeled in each of these abdominal peaks are similar, with the exception of the last, where the nuclei in A8 are too faintly labeled for comparison. These results have been confirmed by White and Wilcox (1984), using a mono-

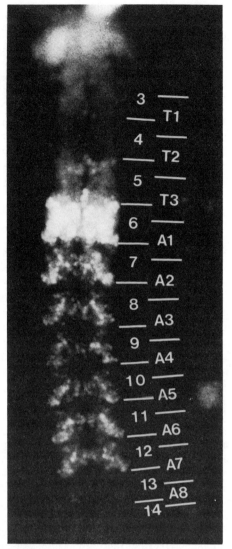

Figure 5. *Ubx* proteins in the CNS of an approximately 12-hr wild-type embryo. The parasegments and segments are indicated (for details, see text).

clonal antibody induced by a β-Gal-*Ubx* hybrid protein in which the *Ubx*-encoded part was somewhat shorter than that used here.

The distribution of *Ubx* proteins in the imaginal disks overlaps that for the mid-embryonic CNS at the qualitative level, although the disk distribution does not include A1-A8, since these adult epidermal segments arise from small larval histoblasts that were not examined (White and Wilcox 1984; Beachy et al. 1985). Both the T3a and T3p compartments of the third leg and haltere disks contain relatively high levels of the *Ubx* proteins in most if not all of their nuclei, although the levels are generally higher in the T3p than T3a nuclei of the third leg disk, which also exhibits localized heterogeneities along the proximo-distal axis. More anteriorly, the levels drop precipitously. In T2, the second leg disk exhibits localized low levels, and no *Ubx* proteins were detected in the wing disk proper, although

White and Wilcox (1984) found them in its peripodial membrane, which is of uncertain metameric origin. Similarly, no *Ubx* proteins were detected in T1 (first leg) and head (eye-antennal) disks, nor in the genital disk that contributes to the most posterior epidermal segment of the adult.

The low and high levels in the T2 and T3 disks, respectively, are consistent with clonal analyses which demonstrate that the abx^+ and Ubx^+ determinants of the PPX function of T2p are dispensable after a critical embryonic period, whereas the bx^+, pbx^+, and Ubx^+ determinants of the BX(T3a) and PBX(T3p) functions are not dispensable until after most of larval development (Lewis 1964; Morata and Garcia-Bellido 1976; Morata and Kerridge 1981; Kerridge and Morata 1982; Casanova et al. 1985). They are also consistent with the distribution of *Ubx* transcripts in the disks, as determined by in situ hybridization (Akam 1983) and by Northern analyses, which revealed the 3.2-kb and 4.3-kb mRNAs in isolated haltere disks but not in wing disks (R. Saint et al., in prep.). Akam (1983) also failed to detect *Ubx* transcripts in the wing disk by in situ hybridization, whereas low levels were obvious in the second leg disk. The presence of even low levels of *Ubx* transcripts and proteins in the second leg disk is curious at this late stage of larval development, particularly since temperature-shift experiments based on the cold sensitivity of PPX in *abx* mutants indicate that the PPX function is not required for second leg development after embryogenesis (Casanova et al. 1985). Although these *Ubx* products may represent residual ineffective expression of PPX proteins, their absence from the wing disk, whose development also requires PPX during embryogenesis (Lewis 1981, 1982; Kerridge and Morata 1982; Casanova et al. 1985), suggests that they may reflect another domain function not yet defined by mutation.

The *Ubx* protein and RNA distributions in the mid-embryonic CNS are also in good agreement (Akam 1983). Both these and the imaginal disk distributions fail, however, to resolve the important question of whether different members of the respective protein and mRNA families exhibit different distributions. Akam and Martinez-Arias (1985) have approached this problem by constructing hybridization probes that distinguish the 4.3-kb mRNAs from the sum of the mRNAs. Although these authors did not examine the mid-embryonic CNS (~12 hr), they did use these probes to determine that the 3.2-kb mRNA, but not the 4.3-kb mRNA, is expressed at 7-8 hours in the ectodermal cell layer containing CNS precursors with a pattern resembling that for the *Ubx* proteins in the mid-embryonic CNS; at a later stage, when the CNS is partially condensed, the probes for the 4.3-kb mRNA and for the sum of the two yield the same pattern. Hence, the only differential determinant for mRNA expression detected thus far is time (Fig. 3). The preparation of antibodies specific to individual proteins of the *Ubx* family may offer a more productive approach, particularly as such antibodies can be prepared on the basis of the nucleotide sequence of the respective cDNAs, e.g., antibodies directed against the sequence of nine amino acids that differentiates the proteins encoded by ORF Ia and Ib.

The *bxd* Unit Regulates the Metameric Distribution of *Ubx* Proteins

The conclusion that all domain functions, including those inactivated by mutations in the *bxd* region, require one or more proteins in the family encoded by the *Ubx* unit induces the proposition that the *bxd* region acts as a *cis*-regulator of the *Ubx* unit, particularly as the long *bxd* transcript does not appear to yield any mRNAs (see below). We have tested this proposition by examining the distribution of *Ubx* proteins in the mid-embryonic CNS of embryos homozygous for certain *bxd* mutations (Beachy et al. 1985). Teugels and Ghysen (1985) observed that extreme *bxd* mutations alter CNS development of both presumptive compartments of parasegment 6, in keeping with their effects on epidermal development. According to the above proposition, we should therefore expect such *bxd* mutations to alter the distribution of *Ubx* proteins in this parasegment and perhaps in more posterior parasegments where *bxd* mutations effect partial transformations in the larval epidermis (Lewis 1978, 1981; Bender et al., this volume).

Like the *Ubx* mutations, the *bxd* mutations divide into pseudopoint mutations and gross rearrangements with a breakpoint in the *bxd* region; the pseudopoint mutations that have been mapped each result from the insertion of a 7.3-kb transposable element of the "gypsy" family, although the orientation of insertion varies (Fig. 2) (Bender et al. 1983). We examined the *Ubx* protein distribution in homozygotes of two pseudopoint mutations (bxd^1, bxd^{55i}) located near the 3′ end of the *bxd* unit and one gross rearrangement, bxd^{113}, an inversion with a breakpoint near the center of the unit associated with a small deletion of DNA at or near the breakpoint (Fig. 7). This gross rearrangement exhibits a more extreme transformation of parasegment 6 in the adult epidermis than do the two pseudopoint mutations (Lewis 1981), one of which, bxd^1, was examined by Teugels and Ghysen (1985) and found to induce only minor alterations of the development of the CNS.

Although no obvious changes in the *Ubx* protein distribution were observed in homozygotes for the weaker pseudopoint mutations, the bxd^{113} homozygote produced a striking change as shown in Figure 6. As expected, parasegments 6–12 exhibit conspicuous changes in their *Ubx* protein distributions. The level of the *Ubx* proteins in the parasegment 6 nuclei is markedly reduced from that in the wild type, as is the level in parasegments 7–12, which also suffer a regionally specific approximately twofold reduction in the number of nuclei containing detectable *Ubx* proteins. However, the effect of bxd^{113} is not limited to these parasegments; both presumptive compartments of parasegment 5 exhibit an increase in the number of nuclei containing

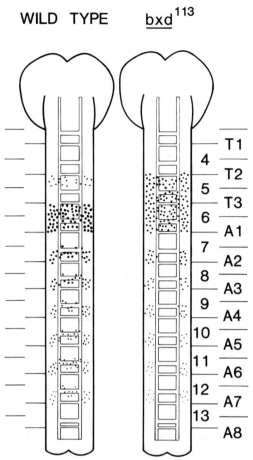

Figure 6. Spatial distribution of *Ubx* proteins in the CNS of wild-type and *bxd*[113] mutant embryos. The parasegments and segments are indicated. The size of the dots correlates with the relative staining intensity of the nuclei, while the number of dots approximate the relative number of nuclei that are stained. In the wild type, the weak staining in T3a and A8 discussed in the text is not shown. (Modified from Beachy et al. 1985.)

Ubx proteins and in the amount of these proteins per nucleus. This increase appears to be coordinated with the decreased nuclear levels in parasegment 6 so that the two parasegments produce a uniform zone in which most of the nuclei contain equivalent amounts of the *Ubx* proteins. Evidently, position-specific controls on the expression of the *Ubx* unit are relaxed or overridden by the *bxd*[113] mutation.

The change observed in parasegment 5 is curious since its epidermal identity does not appear to be affected by *bxd*[113] or by *bxd* mutations generally, although they are known to modify the effects of other mutations in this region (Lewis 1955 and pers. comm.). Since other *bxd* gross rearrangements have recently been shown to cause similar changes in the *Ubx* protein distribution of parasegment 5 (S.L. Helfand, unpubl.), it is likely that this effect results from the loss of *bxd* DNA and not from the juxtaposition of novel sequences with the *bxd* DNA that remains linked to the *Ubx* unit. In any case, the important question for our immediate purpose is not whether, or how, a new distribution of *Ubx* proteins might cause a change in the morphological identity of a given parasegment in the CNS, but rather, whether *bxd* mutations can cause a change in the metameric distribution of *Ubx* proteins. Clearly, these mutations do cause such a change, and the fact that they do indicates that the *bxd* region regulates the expression of the adjacent *Ubx* unit in a position-specific manner.

Transcripts of the *bxd* Unit

Given that the *bxd* region acts as a *cis*-regulator of the *Ubx* unit, it is reasonable to suppose that its transcription plays a role in that regulation. This supposition implies that the transcripts of the *bxd* unit (*bxd* RNAs) function in another capacity than that of mRNAs. The peculiar properties of the early class of *bxd* RNAs described here indicate that they are not mRNAs, although the properties of the late *bxd* RNA strongly suggest it is an mRNA (M. Goldschmidt-Clermont et al., in prep.; H. Lipshitz; D. Peattie; both unpubl.).

The early class is first detected between 1.5 hours and 3 hours of embryogenesis and disappears by 6 hours (Fig. 7). Northern analyses with DNA probes specific for leftward transcription reveal two distinct sets of RNAs: a poly(A)+ set with lengths in the 1.2 ± 0.1-kb range, and a heterogeneous smear of large poly(A)− RNAs that extends to lengths greater than 10 kb. The 1.2-kb poly(A)+ RNAs are detectable with probes overlapping either end of the unit or a central region (Fig. 7), whereas the poly(A)− smear is detectable with all probes from the unit. No transcripts were detected with adjacent probes on the left or right or with probes specific for rightward transcription.

Members of the 1.2-kb poly(A)+ set exhibit the same 5′ terminus and 5′ exon, as determined by S1-protection and primer-extension analyses at the nucleotide level. Similar S1-protection analyses covering the entire unit detect a central exon and a complex set of five exons near the 3′ end of the unit, of which four represent an overlapping series and the other lies downstream from this series without overlapping any of its members. The stoichiometry of the exon lengths suggests that the common 5′ exon is spliced to the central exon in all of the 1.2-kb RNAs, which vary according to which 3′ exon is spliced to the central exon. Figure 7 shows the structure of one of these RNAs in relation to sites of the eight *bxd* mutations given in Figure 2. This structure derives from the nucleotide sequence of a cloned cDNA that includes and extends from the 3′ poly(A) tail to within 50 nucleotides of the common 5′ terminus defined above and from sequence analyses of the *bxd* unit. The 1144 nucleotides in this RNA (exclusive of its poly[A] tail) derive from the common 5′ exon, the central exon, and the furthest downstream of the 3′ exons. Southern blot analysis of a second cloned cDNA establishes a structure like that of the first, except that its 3′ exon belongs to the set of overlapping exons located

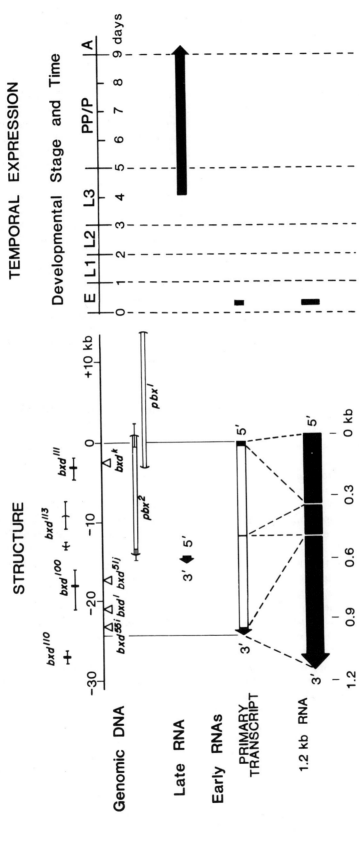

Figure 7. Structure and temporal expression of RNAs from the *bxd* unit. The extent of the *bxd* region DNA and the locations of the mutations discussed in the text are shown. The structures of the 0.8-kb late RNA and one of the 1.2-kb early RNAs are shown below. The exons contributing to the latter are shown as solid boxes within the arrow representing the early primary transcript.

just upstream of that in the first cDNA clone. Both cDNAs are therefore consistent with the prediction from the S1-protection data.

One reason for doubting that the 1.2-kb *bxd* RNAs are mRNAs is that the above 1144-nucleotide sequence does not contain any significantly long AUG-initiated ORFs. The longest ORF contains only 46 codons and is initiated in the 3′ exon, some 630 nucleotides from the 5′ end. Furthermore, it does not fit the codon usage frequencies for known structural genes in *D. melanogaster*, as do the ORFs in the *Ubx* mRNAs. Another reason for this doubt is that the 1.2-kb *bxd* RNAs exhibit an anomalous bimodal sedimentation pattern when embryonic extracts are zone-sedimented in a sucrose gradient. One mode overlaps polysomes small enough to represent the translation of a 46-codon ORF, whereas the other sediments as rapidly as the polysomes that translate the 3.2-kb *Ubx* mRNAs, whose ORFs would occupy the entire sequence of the above *bxd* RNA. It is difficult to account for this sedimentation pattern if the 1.2-kb *bxd* RNAs are mRNAs, particularly as preliminary sequence data indicate that the overlapping series of 3′ exons also lack ORFs of appreciable length. The sedimentation data suggest instead that these *bxd* RNAs may be packaged as RNP particles of two different-size classes.

The effects of four *bxd* gypsy insertion mutations (Fig. 7) on the 1.2-kb RNAs emphasize the questionable nature of their function. Northern analyses of the RNA in 0–6-hour embryos homozygous for bxd^K, bxd^{51j}, or bxd^{55i} indicate that neither the amount nor the lengths of the 1.2-kb RNAs are significantly altered by these mutations. In contrast, the bxd^1 insertion causes a conspicuous increase in their lengths to yield a peak value of about 1.4 kb, again without change in amount. However, a spontaneous revertant of bxd^1 (bxd^{1rev}) (Lewis stock 3-125), in which the gypsy element is replaced by a 0.5-kb insert presumed to be a gypsy terminal repeat (Bender et al. 1983), produces the same length distribution as bxd^1, although its adult phenotype is wild type. Taken together, these results indicate either that the 1.2-kb *bxd* RNAs are not required for the expression of the functions inactivated by gypsy insertions or that these insertions modify other characteristics of these RNAs, such as their spatial distribution, that have not been measured. In the latter case, one would also have to assume that the altered *bxd* RNAs in bxd^{1rev} retain the wild-type activity.

The late *bxd* RNA appears to be transcribed from a small central portion of this unit (Fig. 7) during a period extending from mid-third larval instar through the pupal and into the adult phases. Between the early and late periods, there is a quiescent period when no transcripts are detected by Northern analyses with probes covering the entire *bxd* unit. A single approximately 0.8-kb poly(A)⁺ RNA is detected during the late period with a probe from the center of the unit that is specific for leftward transcription. None of the other probes from the unit detect this late RNA, nor do any of the probes detect the smear of high-molecular-weight poly(A)⁻ RNAs that characterize the early class. Furthermore, S1-protection experiments with DNA probes covering the central region to which the 0.8-kb RNA sequences were localized by Northern analyses yield a single exon of approximately 630 nucleotides located at about −14 kb on the molecular map (Fig. 7). These data strongly suggest that the late primary transcript is not spliced and is initiated and terminated within the second intron for the 1.2-kb early RNA.

We have found the 0.8-kb RNA in the abdomen, but little or none in the thorax of dissected pupae. Although this result needs confirmation at the higher resolution afforded by in situ hybridization, it suggests that the 0.8-kb RNA may play a role in the BXD function of A1a of the adult. More specifically, it may encode a protein that acts in cooperation with proteins of the *Ubx* family to execute this BXD function—a possibility that is enhanced by cDNA sequence data demonstrating that the 0.8-kb RNA contains a significantly long AUG-initiated ORF that exhibits a good fit with the codon usage frequencies of known *D. melanogaster* structural genes (H. Lipshitz, unpubl.). In this respect, it should be emphasized that the preceding arguments demand only that *Ubx* proteins are necessary for the BXD function, but do not demand that these proteins are sufficient for this function. In the case imagined above, the adult epidermal BXD function would require a protein encoded by the late *bxd* RNAs in addition to the requisite *Ubx* protein(s).

Sequence Complementarity between the *Ubx* and Early *bxd* Transcripts

The peculiar properties of the early 1.2-kb *bxd* RNAs induced consideration of the possibility that the primary transcript of the *bxd* unit might be the *cis*-regulatory element by which the expression of the *Ubx* unit is controlled. More specifically, we imagined that the primary transcripts of the adjacent units might form a complex that would facilitate or inhibit the processing pathways of the *Ubx* transcript and thereby modulate the ratios among the processed *Ubx* RNAs.

We therefore used Southern hybridization techniques at low stringency to search the DNA of the two units for homologous sequences that might facilitate a specific interaction between their transcripts (H. Lipshitz, unpubl.). Two regions in the *bxd* unit (X^b and Y^b) were found that contain sequences homologous to those in two regions of the *Ubx* unit (X^U and Y^U, respectively; Fig. 8). The regions are intronic and lie in complementary order within the units (5′-$Y^U X^U$-3′ and 3′-$Y^b X^b$-5′). Interaction between the two transcripts could therefore occur in an inverted linear order by base-pairing in both regions if their transcript sequences are complementary. We have shown that this is the case for the X homology region (Fig. 8) (H. Lipshitz, unpubl.) but have not yet examined the Y region in this detail.

The nucleotide sequence of the 400-bp X^U region was determined and compared to that of the larger X^b region, whose sequence was known from previous studies (D. Peattie and L. Prestidge, unpubl.). The only sig-

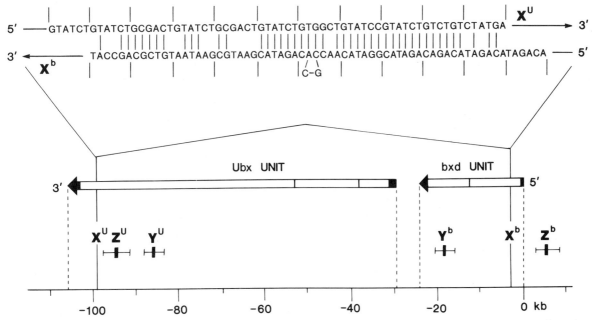

Figure 8. Homology between the *Ubx* and *bxd* regions. Homologies X, Y, and Z (see text) are shown as vertical bars above the horizontal line representing the genomic DNA of the *Ubx* domain. The extent of uncertainty in the location of these homologies is indicated. Note that only the X^b and Y^b homologies are within the *bxd* unit. The arrows represent the *Ubx* and *bxd* primary transcripts with the solid portions indicating the exons contributing to the 3.2-kb *Ubx* and 1.2-kb *bxd* RNAs. Details of the X^U and X^b homology are shown at the top. In each case, the sequence of the sense strand of the DNA is shown and base complementarity is indicated. The hexanucleotide repeats are also indicated.

nificant homology between the two is represented by an inverted repeat that provides the sequence complementarity between the two transcripts. In the configuration shown in Figure 8, the frequency of base pairing is 77% for a 57-base sequence, and within this region, 91% and 100% pairing is observed for 32- and 19-base sequences, respectively. Other configurations of significant pairing are possible because the sequences consist of an array of imperfectly repeated hexanucleotides. The consensus sequence for the 11 hexanucleotides in the sense strand of X^U is $5'-G_{10}T_9A_7T_8C_9T_9-3'$, whereas that for the complementary 10 hexanucleotides in X^b is $3'-C_6A_8T_5A_8G_5A_6-5'$, where the subscripts indicate the number of times the consensus base appears in the 11 or 10 hexanucleotides, respectively. (X^b is interrupted at its center by an octanucleotide.) Sequences homologous to those in the arrays appear to be present at a dozen or so other euchromatic loci, as assayed by whole-genome Southern hybridization with probes containing the arrays and by in situ hybridization to polytene chromosomes with genomic DNA clones identified by these probes.

Puzzling Aspects of Early *bxd* Transcription

Mutational analyses place certain constraints on the above processing model, or any model based on early *bxd* transcription. Since clonal analyses demonstrate that the *bxd*+ DNA is not dispensable until the end of larval development (Morata and Garcia-Bellido 1976), it is clear that such models can account for only a part of all *bxd*+ DNA functions. The putative *bxd* mRNA appearing late in development may account for some of this late requirement for *bxd*+ DNA, but probably not all. Among the multiple domain functions associated with the *bxd* region, one might then expect that disruption of early *bxd* transcription would inactivate only a subset. Consequently, only functions that are commonly inactivated by mutations that disrupt early *bxd* transcription could possibly belong to the subset.

Among the four gross rearrangements shown in Figure 7, three (*bxd*[100], *bxd*[113], *bxd*[111]) result from breaks that clearly map within the *bxd* unit and therefore will disrupt its early transcription. Direct evidence that the *pbx*[1] and *pbx*[2] deletions prevent early *bxd* transcription derives from our finding that 0-6-hour embryos of the *pbx*[1]/*pbx*[2] heterozygote do not contain *bxd* RNAs detectable by Northern analysis (M. Goldschmidt-Clermont et al., in prep.). This result was expected from the mapping data for *pbx*[1], but not necessarily from that for *pbx*[2], since the data for this deletion are not sufficiently precise to be sure it includes the 5' terminus of the *bxd* unit (Fig. 7).

A comparison of the phenotypes of this group of five mutations indicates that the BXD and BXD-like functions required for early larval A1-A7 development and the BXD function for adult A1a epidermal adult development do not require early *bxd* transcription because they are not commonly inactivated by these mutations. The PBX function for adult T3p epidermal development is commonly inactivated and consequently could require this transcription. Thus, the two rearrangements with breakpoints nearest the *Ubx* unit (*bxd*[100], *bxd*[113]) inactivate the largest number of functions. Each exhibits the T3p(PBX) and A1a(BXD) transformations of the adult epidermis and the three early larval transformations, namely, (1) appearance of thoracic ventral pits on A1-A7 (BXD and BXD-like), (2) partially developed thoracic Keilin organs on A1 (BXD), and (3) transformation of the A1a ventral setal belts to the T3a identity (BXD) (Lewis 1981). The bxd^{III} rearrangement exhibits the same transformations, with the exception of that for the larval ventral setal belts and a reduced adult A1a transformation represented by a lower frequency of extra legs in A1 (Lewis 1981 and pers. comm.). Finally, pbx^1 exhibits only the adult T3p transformation (pbx^2 transformations are similarly limited in the adult but have apparently not been examined in larvae; E.B. Lewis, pers. comm.).

Models based on early bxd transcription are further constrained by the metameric distribution of bxd transcription during its transient expression between approximately 2.5 and 6 hours of embryogenesis. The processing model is additionally restricted by the overlap of the Ubx transcription distribution with that of bxd transcription during this period, since this model requires coincident transcription of both units.

Detailed in situ hybridization analyses of the Ubx distribution at late cellular blastoderm (i.e., during the period of bxd transcription) indicate that virtually all transcripts (~95%) are restricted to parasegments 6-12, half of these forming a high, narrow peak at parasegment 6 (Akam and Martinez-Arias 1985). This distribution was obtained with a probe from the 5' end of the unit that registers all nascent transcripts, as well as the two processed RNAs present at this time (3.2-kb and 4.7-kb RNAs). Given that the time required to transcribe the long Ubx unit (~75 min) represents a large fraction of the period of transcription initiation preceding the time at which this distribution was measured, we think that it is determined primarily by the nascent transcripts. Distributions for the early bxd transcript are reported in this volume (Akam and Martinez-Arias). These distributions were generated with a cDNA probe to the 1.2-kb bxd RNA shown in Figure 7. For this reason and because of the shorter length of the bxd unit, such distributions are probably not dominated by the bxd nascent transcripts. These distributions are relatively constant over time and do not exhibit a peak at parasegment 6. To the contrary, the level of bxd transcripts increases posteriorly from a low, probably insignificant, level in parasegment 6 to exhibit significant overlap with the Ubx distribution only in the seventh and more posterior parasegments.

This is not what would be expected were early bxd transcription required for the PBX function, the only allowed candidate from the mutational analyses, particularly since PBX is expressed only in the anterior portion of parasegment 6 (i.e., T3p). Given the apparent absence of bxd transcription in parasegment 6 of the wild type, how could the general loss of bxd transcription caused by the pbx^1 deletion account for the inactivation of the parasegment 6 PBX function? Such a disruption of bxd transcription by pbx^1 would be expected to have effects in more posterior parasegments

where the levels are normally high, yet pbx^1 has no effect on development in this region. The mutation and distribution data therefore argue against the relevance of early bxd transcription to the cis-regulatory functions of the bxd region. One can work around this argument by assumptions regarding, for example, the bias of the bxd probe for the processed bxd RNAs, but these are ad hoc assumptions that have no simple, attractive basis. Evidently, the interest in the striking sequence complementarity between the transcripts of the two units (Fig. 8) should be switched to an interest in the inverted repeat between the bxd and Ubx DNAs and its possible function.

Transcription Control by the bxd Unit and Concluding Remarks

The existing data on the early and late transcripts of the bxd region invite the surprising conclusion that the early RNAs, with all their intriguing properties, are not required for cis-regulation of the Ubx unit. A corollary to this conclusion is that the cis-regulatory elements of the bxd region consist of the bxd DNA. We assume that these DNA elements regulate the transcription of the adjacent Ubx unit; we call them transcription-control elements to avoid implication that they have the properties of enhancers or silencers. Given the different combinations of functions inactivated by the different bxd rearrangements and pbx mutations shown in Figure 7, one might expect that the bxd region is divisible into transcription-control elements, or combinations of these elements, that can be differentially regulated.

A striking correlation between the position of the breakpoints of bxd gross rearrangements and their functional inactivations is presented in this volume by Bender et al. in their study of an extensive set of rearrangements with breaks that extend from almost −30 kb to approximately +10 kb on the molecular map. That correlation can be approximated by the rule that the greater the length of bxd DNA between the 5' end of the Ubx unit and breakpoint, the lower the level of inactivation—where level of inactivation is loosely defined by a mixture of degree of inactivation of a given function and the number of functions affected. Thus, breaks in the region of bxd^{110} and bxd^{100}, which lie closest to Ubx (Fig. 7), exhibit severe inactivation of all functions (see above), breaks lying at the other end of the bxd region inactivate functions to a lesser degree or not at all, and breaks in the central region caused graded inactivations, particularly of the A1a ventral setal band of early larvae. Curiously, the larval ventral pit transformation of A1-A7 is retained by all rearrangements, although it is absent in pbx^1, as noted earlier. A similar correlation between phenotypic effect and breakpoint position has been observed for rearrangements over an approximately 50-kb region in the achaete-scute gene complex of *D. melanogaster* by Campuzano et al. (1985).

It is not clear from the morphogenetic effects of these rearrangements whether the bxd region can be simply divided into regions specific to the domain functions, particularly if the effects of the *pbx* deletions are to be accounted. Of course, it is a long molecular jump from a change in arrangement of the *bxd* DNA to a change in the arrangement of the setae in a ventral setal band. To shorten that jump, we have been examining the effects of this array of *bxd* rearrangements on the amount and arrangement of *Ubx* proteins in the mid-embryonic CNS, using the immunofluorescent techniques described earlier (Fig. 5). Although these studies are in the initial stage, the differences of the effects of rearrangements near the *Ubx* unit (bxd^{110}, bxd^{100}) and in the middle of the *bxd* region (bxd^{113}, bxd^{106}) on *Ubx* protein expression are striking (Beachy et al. 1985; S.L. Helfand, unpubl.). The differences between wild type and bxd^{113}, which appears to have a somewhat stronger effect on *Ubx* protein expression than bxd^{106}, are shown in Figure 6 and have been discussed. In the bxd^{110} homozygote (and bxd^{100}, which is similar to it), parasegments 5 and 6 also form a block of similarly stained nuclei (i.e., with the anti-*Ubx* antibody) but at a much lower level than in bxd^{113} and bxd^{106}, while the abdominal pattern of staining is more dramatically changed. No nuclei are stained in A7 and A8, and only a few nuclei are stained in A2-A6. Thus, the loss of the *bxd* DNA that is present in bxd^{113} (and bxd^{106}), but not in bxd^{100} (Fig. 7), results in a decrease of *Ubx* protein expression and, perhaps as a consequence of this decrease, a change in the abdominal pattern of nuclei containing *Ubx* proteins. Similarly, the loss of *bxd* DNA to the right of the bxd^{113} appears to cause the aberrant expression of *Ubx* proteins in parasegment 5 (particularly in T3a), as well as a decreased expression of these proteins in more posterior parasegments.

These interpretations of the data depend on the assumption that the imported DNA fused to the *bxd* DNA that remains linked to the *Ubx* unit has little or no effect, a reasonable assumption in view of the similar effects observed for different members of a given pair (i.e., bxd^{110}, bxd^{100}, or bxd^{113}, bxd^{106}), and the ordered morphological effects observed by Bender et al. (this volume). The similarity between the effects of bxd^{110} and bxd^{100} is not surprising, since rearrangements with breakpoints more proximal to the *Ubx* unit than the most proximal transcription-control element will have the same effect as those that break that element, as they will separate the *bxd* transcription-control region from the *Ubx* unit by long genomic distances, thus preventing its *cis*-regulatory functions. Consequently, it is difficult to define the *Ubx* proximal end of the transcription-control region with these rearrangements.

The linkage between the *Ubx* unit and the Y^b homology region is not broken by bxd^{113}, but may be in bxd^{100}; consequently, Y^b may be relevant to the difference of *Ubx* protein expression between these two rearrangements (Figs. 7 and 8). The X^b inverted repeat is also located in what appears to be a critical part of the *bxd* transcription-control region, as does another region of homology (Z^b; Fig. 8) located in the rightmost part of the *bxd* region. Its counterpart in the *Ubx* unit (Z^U) is located between X^U and Y^U. The position of these homology regions in the *Ubx* unit induces the question of whether members of the homology pairs may become coupled by proteins specific to each pair, thereby producing a chromatin loop that includes most of the *Ubx* unit and, most importantly, its 5' end and upstream sequences. Such a coupling might alter the coiling properties of the chromatin within the loop to inhibit or facilitate *Ubx* transcription, different combinations of the coupling proteins in different cells thereby effecting cell-specific differential control of *Ubx* expression.

Finally, two unresolved questions deserve comment. The first is whether the different functions of the *Ubx* domain depend on different combinations of the *Ubx* proteins, as proposed by Beachy et al. (1985), and the second concerns the relevance of the early transcription of the *bxd* unit. The presumption that the *cis*-regulatory function of the *bxd* DNA consists in the control of *Ubx* transcription affects the first question only in denying a *cis*-regulatory role of that DNA in the regulation of the different processing pathways for the *Ubx* mRNAs. Clearly, different combinations of *trans*-regulatory factors for these pathways in the different metameres that require the *bxd* DNA for their development could lead to different combinations of *Ubx* proteins in those metameres. It is equally clear, however, that there is no need to assume differences in *Ubx* protein combinations to account for the different domain functions. For example, if the combination of *Ubx* proteins resulting from *Ubx* transcription were invariant, the different domain functions could result from the cooperative action of those proteins with others that are differently distributed among the metameres. There is, indeed, a rapidly expanding list of genes whose products may have *trans*-regulatory effects on the functions of the *Ubx* domain, including Polycomb (*Pc*) (Lewis 1978; Beachy et al. 1985), extra sex combs (*esc*) (Struhl 1981), Polycomb-like (*Pcl*) (Duncan 1982), super sex combs (*scx*), trithorax (*trx*) and certain segmentation genes (Ingham, this volume), and several others (Jürgens 1985) that probably include the *abd-A* and *Abd-B* domains of the BX-C. However, in none of these cases is it known whether the observed effect is direct or indirect, and as yet they are of little help in resolving the above question. Nor, as we have seen, are there adequate data regarding the distribution of the different *Ubx* mRNAs and proteins to resolve this question.

The early *bxd* transcription unit remains an enigma, despite a considerable knowledge of its products and its temporal and spatial expression during development. It would be most puzzling if a transcription unit with the properties of the *bxd* unit were not relevant to any BX-C function. Recall that it is an exceptionally long unit by *Drosophila* standards, that it occupies a majority of the *bxd* region, that its primary transcript is multiply processed in a specific manner, that its expression is limited to a crucial time during development, and that this expression is restricted to metameres whose identities are specified by the BX-C. Indeed, we think that it is relevant, but not to the *cis*-regulatory functions of the *bxd* DNA. Rather, we suppose that early *bxd* transcription, as well as the late transcription, yields *trans*-acting products and that the defects in the *cis*-regulatory functions caused by mutations in the *bxd* region so dominate the phenotype that defects in the *trans*-functions resulting from both early and late *bxd* transcriptions are not visible. Thus, if a mutation adversely affects the expression of the required *Ubx* proteins via a *cis*-regulatory defect, its effect on the expression of the *trans*-acting product of *bxd* transcription will be inconsequential to the phenotype. Clearly, we need mutations that affect the expression of these *trans*-acting products but not the *cis*-regulatory function. These may be difficult to obtain since their target size is apt to be small. However, the effect of such *trans*-acting products may be detectable when P-element-mediated insertions of the *bxd* DNA are tested against the pbx^2 deletion, which exhibits a lesser *cis*-regulatory defect than other mutations in the *bxd* region (Lewis 1982). If so, alterations in the *bxd* DNA could be made in vitro and tested in vivo. We are proceeding along these lines with the constructions of appropriate P-element cosmids (Steller and Pirotta 1985) with which we also plan to test the effects of the *bxd* DNA on the transcription of the *Ubx* unit.

ACKNOWLEDGMENTS

We thank Kenneth Burtis for his generous help in the analysis of the nucleotide sequence data and Ed Lewis, Welcome Bender, and Michael Akam for sharing their unpublished data and for many rewarding discussions. This work was supported by grants from the National Institutes of Health and the National Science Foundation to D.S.H. and by fellowships from the Helen Hay Whitney Foundation (H.D.L. and D.A.P.), the Jane Coffin Childs Foundation (M.G.-C. and P.J.H.), the National Institutes of Health (P.A.B., E.R.G., P.J.H., and S.L.H.), Damon Runyon–Walter Winchell Cancer Fund (R.A.S.), and the National Science Foundation (P.A.B.). Computer resources used in these studies included the BIONET™ National Computer Resource for Molecular Biology.

REFERENCES

Akam, M.E. 1983. The location of Ultrabithorax transcripts in *Drosophila* tissue sections. *EMBO J.* **2:** 2075.

Akam, M.E. and A. Martinez-Arias. 1985. The distribution of Ultrabithorax transcripts in *Drosophila* embryos. *EMBO J.* **4:** 1689.

Beachy, P.A., S.L. Helfand, and D.S. Hogness. 1985. Segmental distribution of bithorax complex proteins during *Drosophila* development. *Nature* **313:** 545.

Bender, W., M. Akam, F. Karch, P.A. Beachy, M. Pfeifer, P. Spierer, E.B. Lewis, and D.S. Hogness. 1983. Molecular genetics of the bithorax complex in *Drosophila melanogaster*. *Science* **221:** 23.

Campuzano, S., L. Carramolino, C.V. Cabrera, M. Ruiz-Gomez, R. Zillares, A. Boronat, and J. Modelell. 1985. Molecular genetics of the achaete-scute gene complex of *D. melanogaster*. *Cell* **40:** 327.

Casanova, J., E. Sanchez-Herrero, and G. Morata. 1985. Prothoracic transformation and functional structure of the Ultrabithorax gene in *Drosophila*. *Cell* **39:** 663.

Duncan, I.M. 1982. Polycomblike: A gene that appears to be required for the normal expression of the bithorax and antennapedia complexes of *Drosophila melanogaster*. *Genetics* **102:** 49.

Hayes, P.H., T. Sato, and R.E. Dennell. 1984. Homoeosis in *Drosophila*: The Ultrabithorax larval syndrome. *Proc. Natl. Acad. Sci.* **81:** 545.

Jürgens, G. 1985. A group of genes controlling the spatial expression of the bithorax complex in *Drosophila*. *Nature* **316:** 153.

Kerridge, S. and G. Morata. 1982. Developmental effects of some newly induced Ultrabithorax alleles of *Drosophila*. *J. Embryol. Exp. Morphol.* **68:** 211.

Lawrence, P.A. and P. Johnston. 1984. The genetic specification of pattern in a *Drosophila* muscle. *Cell* **36:** 775.

Lewis, E.B. 1955. Some aspects of position pseudoallelism. *Am. Nat.* **89:** 73.

———. 1963. Genes and developmental pathways. *Am. Zool.* **3:** 33.

———. 1978. A gene complex controlling segmentation in *Drosophila*. *Nature* **276:** 565.

———. 1981. Developmental genetics of the bithorax complex in *Drosophila*. *ICN-UCLA Symp. Mol. Cell. Biol.* **23:** 189.

———. 1982. Control of body segment differentiation in *Drosophila* by the bithorax gene complex. In *Embryonic development: Genes and cells* (ed. M. Burger), p. 269. A.R. Liss, New York.

———. 1984. Genetic control and regulation of developmental pathways. In *Role of chromosomes in development* (ed. M. Locke), p. 231. Academic Press, New York.

Martinez-Arias, A. and P.A. Lawrence. 1985. Parasegments and compartments in the *Drosophila* embryo. *Nature* **313:** 639.

McGinnis, W., M.S. Levine, E. Hafen, A. Kuroiwa, and W.J. Gehring. 1984. A conserved DNA sequence in homeotic genes of the *Drosophila* Antennapedia and bithorax complexes. *Nature* **308:** 428.

Morata, G. and A. Garcia-Bellido. 1976. Developmental analysis of some mutants of the bithorax system of *Drosophila*. *Wilhelm Roux's Arch. Dev. Biol.* **179:** 125.

Morata, G. and S. Kerridge. 1981. Sequential functions of the bithorax complex of *Drosophila*. *Nature* **290:** 778.

Sánchez-Herrero, E., I. Vernós, R. Marco, and G. Morata. 1985. Genetic organization of *Drosophila* bithorax complex. *Nature* **313:** 108.

Scott, M.P. and A.J. Weiner. 1984. Structural relationships among genes that control development: Sequence homology between the Antennapedia, Ultrabithorax and fushi tarazu loci of *Drosophila*. *Proc. Natl. Acad. Sci.* **81:** 4115.

Steller, H. and V. Pirotta. 1985. A transposable P vector that confers selectable G418 resistance to *Drosophila* larvae. *EMBO J.* **4:** 167.

Struhl, G. 1981. A gene product required for correct initiation of segmental determination in *Drosophila*. *Nature* **293:** 36.

———. 1984. Splitting the bithorax complex of *Drosophila*. *Nature* **308:** 454.

Teugels, E. and A. Ghysen. 1985. The domains of action of Bithorax genes in *Drosophila* central nervous system. *Nature* **314:** 558.

White, R.A.H. and M. Wilcox. 1984. Protein products of the Bithorax complex in *Drosophila*. *Cell* **39:** 163.

Function and Expression of Ultrabithorax in the *Drosophila* Embryo

M.E. AKAM, A. MARTINEZ-ARIAS,* R. WEINZIERL, AND C.D. WILDE

Department of Genetics, University of Cambridge, Cambridge, CB2 3EH, England; *MRC Laboratory of Molecular Biology, Cambridge, England

We discuss here the functions of the Ultrabithorax domain of the bithorax complex (BX-C) of *Drosophila* and the role in development of the two major transcription units within this region. We use the term Ultrabithorax domain to refer to a region of the BX-C that is approximately 100 kb long. This region is not a single gene, but by genetic criteria, it forms a single functional unit (Lewis 1963, 1978; Bender et al. 1983; Beachy et al. 1985; Sanchez-Herrero et al. 1985). Within this domain, there are at least two transcription units. We follow Beachy et al. (1985) in referring to one of these as the Ultrabithorax (*Ubx*) transcription unit. This transcription unit spans the region of the chromosome where most mutations inactivate all functions of the Ultrabithorax domain; such mutations are described as *Ubx* alleles (Lewis 1963). A second transcription unit lies 5' to the *Ubx* unit (Beachy et al. 1985; M. Goldschmidt-Clermont and D. Hogness, unpubl.) in a region of DNA where mutations affect some, but not all, functions of the Ultrabithorax domain. We refer to this simply as the upstream transcription unit. As discussed below, it is far from clear how transcripts from this unit relate to the various functions of the Ultrabithorax domain.

Ubx mutations transform parasegments 5 (T2p and T3a) and 6 (T3p and A1a) into parasegment 4 (for a definition of these terms and abbreviations, see Fig. 3). Other mutations within the Ultrabithorax domain transform only parts of this region. Such mutations fall into two major classes. The "*abx*" class (anterobithorax [*abx*] and bithorax [*bx*] alleles) affects functions that are required primarily in parasegment 5 (Lewis 1981; Casanova et al. 1985). These mutations map within the *Ubx* transcription unit (Bender et al. 1983). The *bxd* class (bithoraxoid [*bxd*] and postbithorax [*pbx*] alleles) affects functions that are required primarily in parasegment 6 (Lewis 1978; Bender et al. 1983). These map in the upstream region, spanning the 30 kb of DNA 5' to the *Ubx* transcription unit. Mutations in either of these two classes complement those in the other, but none complements *Ubx* alleles.

The genetic relationship between *Ubx* alleles and mutations of these two subclasses originally suggested a multicistronic organization. *Ubx* alleles were assumed to identify operator-like mutations exerting regulatory effects on the "genes" *bx*, *bxd*, and *pbx* (Lewis 1963). It now seems likely that this view of the complex should be inverted; *Ubx* is, without doubt, a gene. One long *Ubx* open reading frame is generated by the splicing of exons located at the 5' and the 3' extremes of the *Ubx* transcription unit, across the intervening 70 kb (Beachy et al. 1985). Proteins that derive from at least a part of this reading frame have been identified (White and Wilcox 1984; Beachy et al. 1985). It seems likely that many mutations of the *abx* and *bxd* classes are affecting the expression of these *Ubx* proteins (see below) (Ingham 1984; Beachy et al. 1985), but it is not clear whether additional proteins are encoded by other regions of the Ultrabithorax domain. Nor is it clear how *abx* and *bxd* mutations affect *Ubx* expression.

We discuss below the characteristics of some *Ubx* mutations, which suggest that proteins including the 5' exon of the *Ubx* transcription unit implement those functions of the Ultrabithorax domain that are specific to parasegment 6 and some, but possibly not all, of those required in parasegment 5. We also summarize the spatial distribution of transcripts from the *Ubx* and upstream transcription units. During embryogenesis, *Ubx* transcripts are most prominent in parasegment 6, supporting the contention that *Ubx* transcripts play a unique role in this metamere. In contrast, embryonic transcripts of the upstream unit are not prominent in parasegment 6, even though the *bxd* mutations that map in this transcription unit exert their effects primarily in parasegment 6.

Characteristics of *Ubx* Mutations

We have examined the structure of *Ubx* mutations to determine what range of molecular lesions can disrupt the function of the entire Ultrabithorax domain. The *Ubx* alleles examined include gross chromosomal rearrangements, small deletions, and a group of chemically induced mutations that are almost certainly true point alleles (Bender et al. 1983; Akam et al. 1984; R. Weinzierl and M. Akam, unpubl.).

Chromosomal breaks associated with *Ubx* mutations are spread throughout the 70 kb of the *Ubx* transcription unit. All of these show very similar phenotypes, suggesting that the continuity of sequences at both ends of the unit is required for the normal expression of most *Ubx* subfunctions. However, not all breaks have identical effects. Kerridge and Morata (1982) assayed the extent to which different *Ubx* alleles inactivate one function of the Ultrabithorax domain required in posterior T2 (the postprothorax [*ppx*] function). These authors divided *Ubx* alleles into two classes: those that have an effect similar to a deletion of the *Ubx* unit and those that leave some residual function. We now know that mutations of both these classes can arise from breaks in the *Ubx* transcription unit (Bender et al. 1983). Moreover, we notice a correlation between the location of different *Ubx* breakpoints and their effect on this *ppx* function (Fig. 1A). Rearrangements that occur within the 5' region of the *Ubx* transcription unit result in a weak *ppx* transformation, but those distal to the loci of *bx* and *abx* mutations have a much stronger effect. This suggests that some aspect of the function of the Ultrabithorax domain can be expressed even when the *bx-abx* locus and 3' *Ubx* sequences are detached from the 5' end of the *Ubx* transcription unit. In contrast to this, all *Ubx* rearrangements result in similar extreme *bxd* and *pbx* phenotypes when made homozygous in clones of cells (Kerridge and Morata 1982; G. Morata, pers. comm.), even though the distance of these rearrangements from the *bxd* "locus" varies from less than 10 kb to more than 75 kb.

We have examined in detail two small deletions associated with *Ubx* mutations. One of these, $Ubx^{6.28}$, is located in the *Ubx* 5' exon; the other, $Ubx^{9.22}$, is located in the 3' exon. These two mutations are indistinguishable from each other in their interactions with *bx*, *bxd*, and *pbx* mutations (Morata and Kerridge 1980; E. Lewis, pers. comm.). From sequence analysis, it is clear that both mutations inactivate the major protein-coding sequence of the *Ubx* embryonic RNAs (R. Weinzierl and M. Akam, unpubl.). $Ubx^{9.22}$ deletes about 1.3 kb, including the splice acceptor site and most of the homeo box from the *Ubx* 3' exon; $Ubx^{6.28}$ is a 32-base deletion located just downstream from the initiator methionine of the open reading frame. It leads to premature termination of this reading frame with only 27 correct amino acids (Fig. 1B).

Sequences removed by the $Ubx^{9.22}$ deletion include approximately 1 kb close to the 3' end of the long *Ubx* intron, a region that we might expect to serve in the processing of *Ubx* RNAs, as well as for protein coding. We would therefore not argue that the complete spectrum of phenotypes associated with the $Ubx^{9.22}$ mutation results from the inactivation of the homeo-box-containing open reading frame. However, we have no reason to suspect that the few bases removed by the $Ubx^{6.28}$ deletion are important for any function other than protein coding. Therefore, we suspect that all of the effects of this mutation, including the inactivation of *bxd* and *pbx* functions, result from the elimination of proteins that contain the *Ubx* 5' exon.

Spatial Domains and Temporal Phases of *Ubx* Expression

Transcription of *Ubx* generates a set of different, processed RNA products that can be distinguished by

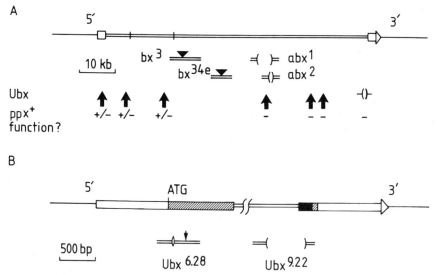

Figure 1. Characteristics of *Ubx* mutations. (*A*) Chromosome rearrangements (arrows) distributed throughout the 70-kb *Ubx* transcription unit disrupt all subfunctions of the *Ubx* domain (*ppx*, *bx*, *bxd*, and *pbx*), giving rise to typical *Ubx* alleles. However, chromosome breaks in the 5' region of the transcription unit inactivate the *ppx* function only weakly. Classification of the *ppx* effect of *Ubx* alleles was described by Kerridge and Morata (1982): (+/−) Partial inactivation; (−) complete inactivation. *Ubx* alleles tested were from 5' to 3': 5.12, 8.8, 6.26, 130, 5.2326, 12.5 (all rearrangement breaks), and 9.22 (a deletion; see panel *B*). The locations of these were taken from Bender et al. (1983) except Ubx^{130} (M. Akam, unpubl.). The locations of insertions and deletions associated with some *bx* and *abx* alleles are also shown (Bender et al. 1983). (*B*) The structure of two *Ubx* alleles that block the major embryonic open reading frame. 5' and 3' exons of the 3.2-kb *Ubx* embryonic transcript are shown as boxes (Beachy et al. 1985). Hatched area shows the major protein-coding open reading frame. The initiator ATG indicated is not the first in the message, but the evolutionary conservation of protein-coding sequence begins at this point (D. Wilde and M. Akam, unpubl.). Homeo box in the 3' exon is shaded. Two microexons lying between the 5' and 3' exons are not shown on this diagram. $Ubx^{6.28}$ deletes 32 bases, introducing a frameshift that terminates the open reading frame prematurely (arrow). $Ubx^{9.22}$ deletes the 3' splice acceptor site and most of the homeo box sequences (R. Weinzierl and M. Akam, unpubl.).

their sizes (4.7, 4.3, 3.2 kb) and by their developmental kinetics (Beachy et al. 1985). A 4.7-kb RNA appears only early in development (3-6 hr after fertilization; 25°C). A 3.2-kb RNA appears at about the same time and is the predominant species between 6 and 9 hours, and a 4.3-kb RNA is predominant in the later stages of embryogenesis (Akam and Martinez-Arias 1985; R. Saint and D. Hogness, unpubl.). Other bands on the gels have not yet been characterized.

The spatial pattern of *Ubx* expression is complex. One element of this complexity is rapid change with time. Using in situ hybridization to monitor the abundance of *Ubx* transcripts in embryos, we distinguish three embryonic phases of *Ubx* expression (Figs. 2 and 3) (Akam and Martinez-Arias 1985). The first phase extends from the beginning of cellularization in the blastoderm to the onset of gastrulation. During this time, the pattern evolves rapidly. Initially, *Ubx* transcripts show a broad distribution throughout the posterior part of the egg (20-50% egg length, measured from the posterior pole). Shortly after, they accumulate predominantly at the anterior margin of this region. This phase of expression culminates in the very late cellular blastoderm, when a single metameric primordium has accumulated much higher levels of *Ubx* transcripts than any other. We believe this to be the primordium for parasegment 6 (T3p/A1a; Akam and Martinez-Arias 1985).

At or shortly after the onset of gastrulation, the second phase of *Ubx* expression is initiated by rapidly increasing concentrations of *Ubx* transcripts throughout a large region of the presumptive abdomen. At this time, a "pair-rule" modulation of *Ubx* expression can be detected (Akam 1985; Akam and Martinez-Arias 1985). Four metameric stripes of cells that accumulate *Ubx* transcripts (the primordia for parasegments 6, 8, 10, and 12) are separated by three stripes with somewhat lower levels. This pair-rule modulation is transient, and by the completion of germ band extension, *Ubx* transcripts are abundant throughout the entire block of seven primordia, parasegments 6-12. Most or all cells in this region of the germ band contain *Ubx* transcripts, including both ectoderm and mesoderm.

From the genetics of the *Ubx* domain, we would expect parasegment 5 to be a major site of *Ubx* transcription. In these first two phases of expression, we can detect *Ubx* transcripts in only a few cells of parasegment 5. In the blastoderm, these are located in the dorsal quarter of the eggs circumference, and in the extended germ band, they are located laterally in the ectoderm, around the tracheal pit.

In the third phase, which begins shortly after the completion of the germ band, the relatively uniform expression of *Ubx* in parasegments 6-12 gives place to a much more complex pattern (Akam 1983; Akam and Martinez-Arias 1985). Within each metamere, the abundance of *Ubx* transcripts becomes very different in different tissue primordia, and in different regions of the segmental repeat. Differences also appear between metameres. In the nervous system, in the epi-

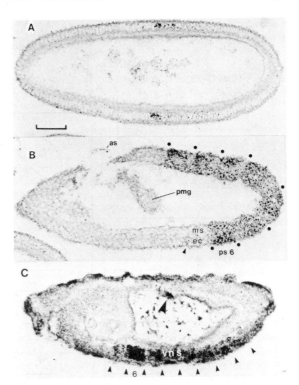

Figure 2. In situ hybridization with a probe for *Ubx* transcripts, illustrating three embryonic phases of *Ubx* expression. (*A*) Cellular blastoderm (3 hr at 25°C). *Ubx* transcripts are most abundant in a zone three to four cells wide that girdles the embryo between 45% and 50% egg length. (Egg lengths are measured from the posterior pole.) We believe that this zone defines the primordium for parasegment 6. (*B*) Extended germ gand (5-6 hr). *Ubx* transcripts are now abundnant throughout a block of 7 metameric units in the germ band, parasegments 6-12. Ectoderm (ec) and mesoderm (ms) both contain *Ubx* transcripts. A small group of ectodermal cells located laterally in parasegment 5 also contain *Ubx* transcripts at this stage. The section shown here just grazes the edge of this cell cluster (arrowhead). (*C*) Embryo at 12-14 hr, after the germ band has shortened. In the ventral nervous system (vns), parasegment 6 is strongly and uniformly labeled. A repeating motif of differential expression is visible within each of the parasegments 7-12, and a different set of cells are labeled in parasegment 5 of the vns. Strong hybridization is also visible to a single region of the visceral mesoderm, which surrounds the midgut (arrowhead). Panel *A* is a nearly horizontal section through dorsolateral regions of the blastoderm. Panels *B* and *C* are near-sagittal sections. In each case, anterior is to the left, and in *B* and *C*, dorsal is at the top. The probe used here is a 5' fragment common to all or most embryonic transcripts of the *Ubx* unit, including the open reading frame shown in Fig. 1. For details of probe and hybridization conditions, see Akam and Martinez-Arias (1985). Scale, 50 µm.

dermis, and probably also in the visceral mesoderm, parasegment 6 is delimited by the accumulation of uniquely high levels of *Ubx* transcripts in most or all cells. In more posterior regions, the distribution of *Ubx* products appears as a repeating motif expressed at progressively lower levels in more posterior segments, with virtually no expression detectable in the eighth abdominal segment. *Ubx* products are now found in both the epidermal and neural structures of parasegment 5, but

198　AKAM ET AL.

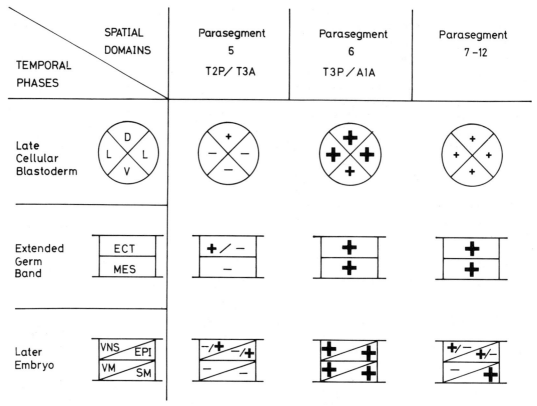

Figure 3. Spatial domains and temporal phases of *Ubx* expression. The distribution of *Ubx* transcripts in embryonic tissues is shown schematically for three stages in development (see Fig. 2). Within the region of the embryo that expresses *Ubx*, three spatial domains are distinguished. These domains are parasegments (Martinez-Arias and Lawrence 1985), i.e., metameric units running from the presumed antero/posterior compartment boundary in one segment to that in the next. Thus, parasegment 6 includes the posterior compartment of the third thoracic segment (T3p) and the anterior compartment of the first abdominal segment (A1a). Each illustration represents the major divisions of one metameric unit of the embryo at the indicated stages. Circles, for the blastoderm, are transverse sections. (D) Dorsal; (V) ventral; (L) lateral. Dorsal-lateral regions will give rise to ectoderm, and the ventral region will give rise to mesoderm. In the extended germ band, one metameric unit is shown divided into ectoderm (ECT) and mesoderm (MES). In the later embryo, the ectoderm gives rise to ventral nervous system (VNS) and epidermis (EPI), and the mesoderm gives rise to somatic (SM) and visceral (VM) musculature. Different-size plus (+) signs indicate the abundance of *Ubx* transcripts in different regions. Expression in a subset of cells within any one region is indicated by +/−. (Data abstracted from Akam and Martinez-Arias 1985.)

the "motif" of *Ubx* expression here is quite different from that seen in parasegment 6, or parasegments 7–12.

To some extent, we can correlate this changing spatial and temporal pattern of *Ubx* expression with the transcript species observed. The early concentration of transcripts in parasegment 6 coincides with the period of expression of the 4.7-kb *Ubx* transcript. A probe derived from an exon specific to this transcript hybridizes strongly, but not uniquely, to parasegment 6 (Akam and Martinez-Arias 1985). The 3.2-kb *Ubx* transcript is predominant in the extended germ band phase, when *Ubx* expression is nearly uniform throughout parasegments 6–12. A *Ubx* probe containing the 3′ sequences of this transcript (but not the sequences of the 4.7-kb RNA) shows the same broad distribution throughout parasegments 6–12 not only in the extended germ band, but also throughout blastoderm stages. It fails to detect the peak of *Ubx* expression in parasegment 6. We conclude from this that transcripts which span the *Ubx* unit are distributed throughout the wider *Ubx* domain (parasegments 6–12) but that additional transcripts derived

from 5′ sequences accumulate preferentially in parasegment 6; among these is the 4.7-kb RNA. We cannot yet determine whether specific transcript(s) characterize parasegment 5.

Expression of the Upstream Transcription Unit

Transcripts homologous to the upstream unit can first be detected in the syncytial blastoderm, a little earlier than those from the *Ubx* unit itself (Fig. 4) (M. Akam, unpubl.). From this time on, until gastrulation, they are widely distributed in a region of the abdomen from 10% to 45% egg length. The anterior boundary of this domain lies posterior to the peak of *Ubx* expression in the cellular blastoderm, with little or no transcription being detected in the primordium for parasegment 6, where we might expect to see maximal transcription of a *bxd* function. When we can first identify these early transcripts of the upstream unit, they are located in nuclei, but from the cellular blastoderm stage onwards, it is clear that they are present

Figure 4. Section of an embryo shortly after the last nuclear division in the syncytial blastoderm (2.5 hr) is shown in bright-field (*A*) and dark-field (*B*) illumination. The hybridization probe was derived from a cDNA clone homologous to exons in the upstream transcription unit of the Ultrabithorax domain (M. Goldschmidt et al., pers. comm.), which lies upstream of the *Ubx* promoter and is probably an independent transcription unit (Beachy et al. 1985). Transcripts homologous to this probe are detected in a region of the embryo from 10% to 45% egg length, spanning the primordia for the abdominal segments. Details of probe preparation and hybridization conditions are similar to those in Fig. 2 (see Akam and Martinez-Arias 1985).

in both nuclei and cytoplasm. They become less abundant after gastrulation, although they can be detected throughout germ band elongation. At no time are they abundant in parasegment 6 (M. Akam, unpubl.).

DISCUSSION

Functions of the Ultrabithorax Domain

The phenotypes of different mutations in the Ultrabithorax domain have been used to define subfunctions of Ultrabithorax: a bithorax function required in T3a, a postbithorax function required in T3p, a bithoraxoid function required in A1a (Lewis 1978), and, most recently, a postprothorax function required in T2p (Morata and Kerridge 1981), which is partially inactivated by *abx* alleles (Casanova et al. 1985). Each of these subfunctions is assumed to be required for the normal development of a single compartment and to play a role exclusively in that compartment. A model of this type is strongly suggested by the effects of *bx* and *pbx* mutant alleles, which completely transform the pattern of the adult epidermis, but only in single compartments.

A rather different model is suggested by our observations of the spatial and temporal patterns of *Ubx* transcription in the embryo. Within each compartment, it is not a single *Ubx* function that acts, but a series of different products, each expressed during a particular phase of development. We do not know the precise molecular structure of these *Ubx* products in each compartment, but the available data suggest that at early stages in development, a given product of the *Ubx* unit is expressed not just in a single compartment, but in parasegments or even in spatial domains spanning several segments. Thus, although the final epidermal pattern of the adult fly may depend on the compartment-specific expression of certain *Ubx* products, the development of that compartment is not controlled exclusively by one *Ubx* product but is dependent on the expression of a set of products, of which some may be expressed sequentially, and others simultaneously, in overlapping morphological domains of varying extent.

These different products of the Ultrabithorax domain are expressed from overlapping DNA sequences, and thus many mutant lesions are likely to affect the expression of several different products to greater or lesser extents. This is probably why such mutations as *bx* and *pbx* may have extreme and compartment-specific effects on the development of the adult epidermis while having weak and variable effects on the development of larval structures throughout a wider morphological domain.

Domains of Ultrabithorax Expression

The pattern of *Ubx* transcript accumulation in the embryo can be considered in three spatial domains, within each of which a qualitatively different pattern of *Ubx* expression is observed. The most anterior of these domains is parasegment 5. From genetic data, we know that the functions of the Ultrabithorax domain that are expressed here (conventionally, the *bx* and *ppx* subfunctions) are dependent only on sequences within the *Ubx* transcription unit itself and not on the upstream sequences of the Ultrabithorax domain. The surprising feature of our results is that *Ubx* transcript accumulation is so limited in parasegment 5; we see little transcription in the blastoderm and, in later development, detectable transcripts only in certain cells of the ectoderm. One possible reason for this is that *Ubx* transcripts specific for parasegment 5 may derive from regions of the *Ubx* unit that are not included in our probes. Since *Ubx* 5′ rearrangements fail to inactivate the *ppx* function completely, such transcripts may initiate at a second promoter within the *Ubx* unit.

The second domain, parasegment 6, is marked by the abundant and ubiquitous accumulation of *Ubx* transcripts. We conclude that these *Ubx* transcripts mediate functions of the Ultrabithorax domain that are required uniquely in parasegment 6. From genetic data again, we expect that these are the *bxd* functions, which for their expression require that the *Ubx* transcription unit be contiguous with DNA sequences throughout the 30-kb upstream region.

The third domain, parasegments 7–12, includes those abdominal segments where both *Ubx* and other, abdominal, genes of the BX-C are active. Within each metamere of this domain, we see essentially the same repeating pattern of *Ubx* expression, suggesting that the role of *Ubx* is similar in each. The expression of *Ubx* in this region has little obvious effect on morphology—the only external structures altered by *Ubx* mutations are the ventral pits on the larval cuticle—but in view of the abundant *Ubx* transcripts, particularly in the mesoderm of the abdominal region, we suspect that thorough examination will reveal other effects of *Ubx* in these abdominal segments.

When abdominal genes of the BX-C are not active, the activity of *Ubx* in abdominal segments becomes the dominant determinant of segment morphology, and each segment resembles the normal parasegment 6 (Lewis 1978). This suggests that at least some of the differences in *Ubx* expression between parasegment 6 and the more posterior domain result from the repression of the Ultrabithorax domain by abdominal genes of the BX-C. One effect of this interaction is the reduction in *Ubx* transcript levels observed in parasegments 7–12 of the nervous system after germ band shortening (R. Marco and M. Akam, unpubl.).

Role of the Upstream Element

We have used data on the structure of *Ubx* mutations and on the distribution of *Ubx* transcripts to argue that *bxd* functions in parasegment 6 are mediated by transcripts and proteins of the *Ubx* transcription unit. Several other lines of evidence support this view. Beachy et al. (1985) have shown that *bxd* mutations alter the distribution of *Ubx* proteins in the ventral nerve cord, and we have found that similar mutations have a complex effect on the distribution of *Ubx* transcripts in earlier embryonic stages (M.E. Akam and A. Martinez-Arias, unpubl.). It then becomes of some interest to determine why *bxd* mutations in the upstream element of the Ultrabithorax domain, which may be as much as 30 kb from the start of *Ubx* transcription, affect this subset of Ultrabithorax functions.

Much of the upstream element is transcribed in early embryos. It gives rise to a set of spliced polyadenylated RNAs that do not appear to encode proteins (M. Goldschmidt-Clermont et al., unpubl.). Beachy et al. suggest that these transcripts elicit *bxd* functions of the *Ubx* unit by altering the splicing of *Ubx* precursor RNAs, and hence the spectrum of *Ubx* proteins. Thus, we would expect them to be active in parasegment 6 and in more posterior segments but not in parasegment 5. The actual distribution of processed transcripts from the upstream unit does not conform to this expectation, nor does it match that of any function attributed to the Ultrabithorax domain. We can barely detect them in parasegment 6 but find that they are relatively abundant posterior to this.

We are left with two possibilities. If transcripts of the upstream element are regulating the expression of *Ubx*, then the distribution of these effector transcripts cannot be the same as that of the stable RNAs that we observe by in situ hybridization. The effectors may be labile species, acting at or close to their site of transcription. This would be consistent with genetic evidence that the upstream element is active only when linked in *cis* to an intact *Ubx* unit (Lewis 1963). Alternatively, the early transcripts of the upstream element may be irrelevant to the regulation of *Ubx* in parasegment 6, but they may serve some as yet undefined function in more posterior segments. We would then invoke other features of the upstream element (e.g., enhancer activity) to explain the *cis*-limited regulation of *Ubx*. This latter view is certainly not excluded by the available genetic evidence.

ACKNOWLEDGMENTS

We thank W. Bender, P. Beachy, M. Goldschmidt-Clermont, D. Hogness, G. Morata, D. Peattie, and R. Saint for communicating their results prior to publication and M. Goldschmidt-Clermont for the cDNA clone of the upstream unit. Helen Moore assisted with many aspects of this work, which was supported by the Medical Research Council of Great Britain.

REFERENCES

AKAM, M.E. 1983. The location of *Ultrabithorax* transcripts in *Drosophila* tissue sections. *EMBO J.* **2**: 2075.

———. 1985. Segments, lineage boundaries and the domains of expression of homeotic genes. *Proc. R. Soc. Lond. B* (in press).

AKAM, M.E. and A. MARTINEZ-ARIAS. 1985. The distribution of *Ultrabithorax* transcripts in *Drosophila* embryos. *EMBO J.* **4**: 1689.

AKAM, M.E., H. MOORE, and A. COX. 1984. *Ultrabithorax* mutations map to distant sites within the bithorax complex of *Drosophila*. *Nature* **309**: 635.

BEACHY, P.A., S.L. HELFAND, and D.S. HOGNESS. 1985. Segmental distribution of bithorax complex proteins during *Drosophila* development. *Nature* **313**: 545.

BENDER, W., M. AKAM, F. KARCH, P.A. BEACHY, M. PEIFER, P. SPIERER, E.B. LEWIS, and D.S. HOGNESS. 1983. Molecular genetics of the bithorax complex in *Drosophila melanogaster*. *Science* **221**: 23.

CASANOVA, J., E. SANCHEZ-HERRERO, and G. MORATA. 1985. The prothoracic transformation and functional structure of the *Ultrabithorax* gene of *Drosophila*. *Cell* (in press).

INGHAM, P.W. 1984. A gene that regulates the bithorax complex differentially in larval and adult cells of *Drosophila*. *Cell* **37**: 815.

KERRIDGE, S. and G. MORATA. 1982. Developmental effects of some newly induced *Ultrabithorax* alleles of *Drosophila*. *J. Embryol. Exp. Morphol.* **68**: 211.

LEWIS, E.B. 1963. Genes and developmental pathways. *Am. Zool.* **3**: 33.

———. 1978. A gene complex controlling segmentation in *Drosophila*. *Nature* **276**: 565.

———. 1981. Developmental genetics of the bithorax complex in *Drosophila*. In *Developmental biology using purified genes* (ed. D.D. Brown and C.F. Fox), p. 189. Academic Press, New York.

MARTINEZ-ARIAS, A. and P.A. LAWRENCE. 1985. Parasegments and compartments in the *Drosophila* embryo. *Nature* **313**: 639.

MORATA, G. and S. KERRIDGE. 1980. An analysis of the expressivity of some bithorax transformations. In *Development and neurobiology of* Drosophila (ed. O. Siddiqi et al.), p. 141. Plenum Press, New York.

———. 1981. The role of position in determining homeotic gene function in *Drosophila*. *Nature* **300**: 191.

SANCHEZ-HERRERO, E., I. VERNOS, R. MARCO, and G. MORATA. 1985. The genetic organization of the bithorax complex of *Drosophila*. *Nature* **313**: 108.

WHITE, R. and M. WILCOX. 1984. Protein products of the bithorax complex of *Drosophila*. *Cell* **39**: 163.

Genetic Control of the Spatial Pattern of Selector Gene Expression in *Drosophila*

P.W. INGHAM

Developmental Genetics Laboratory, Imperial Cancer Research Fund
Mill Hill, London NW7 1AD, England

The body of the adult insect is an assemblage of highly specialized and widely differing structures and appendages derived from a series of homologous segments. In *Drosophila*, this diversity of form depends on the differential expression in each body segment of a specific set of genes known as selector genes (Garcia-Bellido 1975). Extensive genetic analysis has shown that most selector genes are organized in two clusters known as the Antennapedia complex (ANT-C) and bithorax complex (BX-C). The ANT-C genes are involved in controlling the differentiation of anteriorly located body segments (in the head and thorax), whereas the BX-C is required for the differentiation of the posterior part of the animal (the thorax and abdomen).

The requirements for different selector genes in the development of particular body segments have been inferred from studies of their mutant phenotypes. For example, in the absence of the wild-type Antennapedia (*Antp*) gene, the mesothoracic (T2) leg is transformed into an antenna, whereas the antenna itself develops normally (Struhl 1981a). This implies that the expression of the wild-type *Antp* gene is restricted to the thorax, where its function is to promote leg development, rather than antennal development.

In a similar manner, mutations of the BX-C gene Ultrabithorax (*Ubx*) cause transformations of segment T3 structures to their T2 equivalents but have no effect on the differentiation of prothoracic (T1) or head structures. This suggests that *Ubx* is specifically required to promote the T3 level of development (see Lewis 1978).

Confirmation that selector genes are differentially expressed in the various body segments has recently been obtained at the molecular level. Studies of the spatial patterns of expression of the *Antp* (Levine et al. 1983) and *Ubx* (Akam 1983) genes have demonstrated that both genes are expressed only in particular regions of the body. An intriguing question raised by the selector gene model concerns the way in which differential expression of selector genes is initiated and maintained. One approach to this problem is to identify other genes required for the correct spatial pattern of expression of the ANT-C and BX-C. Mutations in putative regulatory genes are expected to cause homeotic transformations that mimic those of ANT-C and BX-C mutants. Of the several genes described so far, most display characteristics implicating them in the negative regulation of the BX-C and ANT-C (Lewis 1978; Struhl 1981b; Duncan 1982; Duncan and Lewis 1982; Ingham 1984), whereas mutations of the trithorax locus suggest a role for this gene in the positive regulation of the two complexes (Ingham and Whittle 1980; Ingham 1983).

In this paper, I describe the effects of loss-of-function mutations of the trithorax (*trx*) and super sex combs (*sxc*) loci and show that some of the transformations they cause in the adult body segments can be directly attributed to changes in the spatial regulation of *Ubx* gene expression. In addition, I discuss the complexity of the *Ubx* locus and provide further evidence that expression of the *Ubx* transcript is modulated by the postbithorax (*pbx*) function.

MATERIALS AND METHODS

Genetics. All flies were grown at 25°C. The trx^1 mutation is temperature-sensitive (Ingham and Whittle 1980), but other mutations employed do not exhibit significant variations with temperature. Hemizygous trx^1 larvae were generated by the cross ♀♀ *st trx¹* × ♂♂ *Df(3R)red^{P52}/Sb Tb* and identified by their wild-type (not tubby) appearance. *Df(3R)red^{P52}* lacks the *trx* gene, and in this paper it is referred to as *Df trx*.

sxc^3/sxc^4 trans-heterozygous larvae were distinguished by virtue of the pigment mutations *cn* and *bw* in *cis* with both alleles. The combination of these two mutations results in a complete absence of pigment from the Malpighian tubules. In principle, this allows distinction between *cn bw* homozygotes and *cn* homozygotes (all *sxc* heterozygotes are homozygous for *cn*, since all available second-chromosome balancers carry *cn*). In practice, this distinction was greatly enhanced by growing larvae on medium containing 50 μg/ml riboflavin. Larvae homozygous for the *pbx¹ e* chromosome were identified by the dark pigmentation of the anterior spiracles due to the *e* mutation. For descriptions of these mutations (other than *trx* and *sxc*), see Lindsley and Grell (1968).

Molecular probes. Transcripts deriving from the *Ubx* transcription unit were detected using the A113 probe described by Akam (1983). Protein products of the *Ubx* transcripts were detected by the FP3.38 monoclonal antibody described by White and Wilcox (1984).

In situ hybridization of radioactively labeled RNA probes to tissue sections. For the preparation of tissue sections, third-instar larvae were fixed for 1 hour at room temperature in Carnoys fixative, dehydrated through graded ethanols (80%, 90%, 95%, 100%, 100%, 30 min each) and xylene (2×, 30 min), infiltrated with a 1:1 mixture of paraffin wax and xylene for 30 minutes at 58°C, and then by two further 20-minute incubations in pure wax (58°C). The larvae were then embedded in wax, and 6-μm sections were cut on a Cambridge Rotary Microtome. Sections were floated on drops of distilled H_2O on poly-D-lysine-coated glass microscope slides and dried at 40°C overnight. Prior to hybridization, the wax was removed by two 10-minute washes in xylene.

Details of the preparation and hybridization of ^{35}S-labeled RNA probes will be reported elsewhere (P.W. Ingham et al., in prep.).

For autoradiography, slides were coated with Kodak NTB^{-2} emulsion and exposed in light-tight boxes at 4°C for the stated times. Development was carried out with a 1:1 dilution of Kodak D19 for 2 minutes at 20°C, followed by fixing in 30% sodium thiosulfate for 2 minutes.

Detection of *Ubx* antigen was achieved using the monoclonal antibody FP3.38 essentially as described by White and Wilcox (1984).

RESULTS

Effect of *sxc* Mutations on Segment Differentiation

The *sxc* gene is one of a number of loci that are apparently required for the negative control of genes of both the ANT-C and BX-C.

Absence of *sxc*$^+$ during embryogenesis has a major effect on the differentiation of larval segments (Ingham 1984). If *sxc*$^+$ is absent from both the female parental germ line (thereby eliminating the *sxc*$^+$ gene product from the egg) and the zygotic genome, most of the larval body segments differentiate structures characteristic of the eighth abdominal segment. According to the model proposed by Lewis (1978), this is the segment in which all of the elements of the BX-C are expected to be expressed. Such transformations therefore suggest that *sxc*$^+$ is required to regulate expression of BX-C genes, suppressing their expression in more anterior body segments. This interpretation is supported by the finding that the somewhat variable phenotype is significantly enhanced by an additional wild-type copy of the entire BX-C in the zygotic genome (Ingham 1984).

The *sxc* gene continues to be required during the larval stages for the correct differentiation of adult body segments. However, the homeotic transformations that ensue when *sxc*$^+$ is removed from adult cells differ considerably from those of their larval counterparts. In the head, the maxillary palps are reduced or absent, whereas the antennae are partially transformed toward the T1 leg. This latter effect is indicative of a change in the spatial control of the expression of two ANT-C genes. Normal T1 leg development requires the activity of the ANT-C genes *Sex combs reduced* (*Scr*) and *Antp*.

In the absence of *Scr*$^+$, the T1 leg is transformed toward its T2 equivalent (Lewis et al. 1980), whereas the absence of *Antp*$^+$ from T2, in turn, results in the leg being transformed toward the antenna (Struhl 1981a). The transformation of the antenna to the prothoracic leg (typical of *sxc*$^-$ animals) can therefore be viewed as the consequence of the inappropriate expression of both the *Scr*$^+$ and *Antp*$^+$ genes in those imaginal cells that normally give rise to the antenna.

Further evidence that *sxc*$^+$ regulates *Scr* expression is provided by the homeotic transformations of the thoracic legs caused by *sxc* mutations. In *sxc*^{3}/*sxc*^{4} flies, the anterior compartments of all three legs are identical, and prothoracic in character (Fig. 1b). This implies that the *Scr*$^+$ gene is being expressed inappropriately in segments T2 and T3. In contrast, the posterior compartments of all three legs are not similarly prothoracic, but instead exhibit characteristics of the posterior compartment of the wild-type T2 and T3 legs. This suggests that in the posterior compartments of each leg, ectopic expression of the *Scr* either does not occur or is overridden by the expression of other genes, in par-

Figure 1. Adult homeotic transformations caused by the *sxc* mutation. (*a*) Dorsal thorax of an *sxc*^{3}/*sxc*^{4} animal. The appendages associated with segment T2 are drastically reduced in size and closely resemble the normal T3 appendage, the haltere (H), although they are somewhat larger. Only a few bristles characteristic of the proximal region of the wing remain. This phenotype is relatively consistent in flies of this genotype. Note that the rest of segment T2, the mesonotum, is essentially wild type in appearance, as are the halteres; however, an ectopic ridge of cuticle (arrow) is associated with the metanotum. (*b*) Thoracic legs of the *sxc*^{3}/*sxc*^{4} adult male. The legs of the three segments T1, T2, and T3 differentiate almost identical patterns, which in the anterior compartment is characteristic of segment T1. Particularly striking is the presence of a T1-specific sex comb (arrow) on all three legs.

ticular, *Ubx*, which controls the T3 level of development (Lewis 1978).

Ectopic expression of *Ubx* is also suggested by the transformation of the wing toward the haltere caused by *sxc* mutations (see Fig. 1a), since development of the normal haltere requires the activity in segment T3 of the *Ubx* gene (Lewis 1978). Interestingly, the transformation caused by *sxc* is limited to the T2 appendage. The mesonotum develops quite normally, whereas expression of *Ubx*⁺ in the homologous region of segment T3 results in the differentiation of the much smaller metanotum.

Differentiation of the haltere, on the other hand, is not affected by *sxc* mutations, although a small ridge of cuticle does develop ectopically in the metanotum; the identity of this structure is unclear. These transformations of adult thoracic segments contrast with those of their larval counterparts. Although segments T1, T2, and T3 are all transformed toward A8 in *sxc*⁻ larvae, there is no evidence for transformation of any adult thoracic segment toward the abdomen. This implies a varying requirement for the *sxc* gene at different developmental stages.

The *Ubx* Gene Is Ectopically Expressed in *sxc*⁻ Animals

To test directly the effect on the expression of *Ubx* inferred from the *sxc*⁻ mutant phenotype, the spatial distribution of *Ubx* gene products has been analyzed in the nervous system and imaginal disks of the *sxc*⁻ third-instar larvae.

Expression of the *Ubx* gene in wild-type imaginal disks as detected either by hybridization of nucleic acid probes (Akam 1983) or by immunofluorescence (White and Wilcox 1984; Beachy et al. 1985) is most prominent among those of the third thoracic segment. In the haltere disk, expression appears almost uniform, with no detectable differences between anterior and posterior compartments; in the third leg disk, however, appreciably higher levels of labeling are found in the posterior compartment.

As might be expected from the phenotype of the adult, the pattern of expression of *Ubx* in the haltere disks of sxc^3/sxc^4 larvae is essentially the same as that in wild type. Similarly, the third leg disks of sxc^3/sxc^4 larvae label strongly with anti-*Ubx* antibody, indicating the expression of *Ubx* protein in both compartments of the disk (Fig. 3).

In the wild-type wing imaginal disk, no significant levels of *Ubx* transcription can be detected using a probe complementary to the known major *Ubx* transcription products (Akam 1983). Broadly consistent with this finding is the absence of *Ubx*-specific antigen in most of the cells of the wild-type wing disk (Fig. 2) (White and Wilcox 1984). In contrast, wing disks from sxc^3/sxc^4 third-instar larvae exhibit high levels of *Ubx* expression, as assayed both at the RNA and protein levels (Fig. 2). The adventitious expression of *Ubx* revealed in both cases is highly region-specific and correlates well with the transformation of T2 structures toward their T3 homolog, typical of sxc^3/sxc^4 flies. Thus, it is the presumptive wing region of the disk that expresses *Ubx*, whereas those cells that give rise to the mesonotum remain unlabeled. Note that the labeled region includes cells from both anterior and posterior compartments of the disk.

In contrast to the wild-type wing disk, low levels of *Ubx* products are detectable in its ventral counterpart, the second leg disk (Akam 1983; White and Wilcox 1984). In sxc^3/sxc^4 larvae, the levels of *Ubx* expression detected in this disk by immunofluorescence are significantly higher than those detected in wild type (Fig. 3).

Concomitant with these changes in the imaginal disks, the pattern of expression of *Ubx* in the central nervous system is also altered in sxc^3/sxc^4 larvae. In the wild type, there is one major region of *Ubx* expression, corresponding to the first abdominal neuromere (Akam 1983). In *sxc* mutants, however, three additional regions of strong labeling are apparent in the central nervous system, corresponding to the three thoracic neuromeres (Fig. 4).

Relationship between the *Ubx* Transcript and *Ubx* Subfunctions

The *Ubx* products detected by the A113 nucleic acid probe and the FP3.38 antibody derive from a single transcription unit, within which all known *Ubx* mutations map (Bender et al. 1983). The *Ubx* mutations are, however, only a subset of the *Ubx* pseudo-allelic series that together define *Ubx* as a genetic unit. Mutations of two of the other allelic classes, namely, *abx* and *bx*, map within the *Ubx* transcription unit (Bender et al. 1983; Beachy et al. 1985), whereas *bxd* and *pbx* mutations map within a second more distal transcription unit termed the *bxd* unit (Bender et al. 1983; Beachy et al. 1985). Whereas *bx* mutations are limited in their effect to the anterior compartment of the haltere, *pbx* mutations affect only its posterior compartment. Moreover, *bx* and *pbx* mutations more or less fully complement one another; however, neither are complemented by *Ubx* mutations. These findings led to the proposition that *bx* and *pbx* represent two genes required for the development of the anterior and posterior compartments, respectively, of segment T3, whereas *Ubx* defines a third gene whose expression is required in both compartments of T3 (Lewis 1964, 1978).

Molecular analyses of the structure and expression of the *Ubx* region have cast some doubt on this interpretation. The *Ubx* mutations do map within a single transcription unit, which is expressed in both the anterior and posterior compartments of the imaginal disks of segment T3 (Akam 1983; White and Wilcox 1984; Beachy et al. 1985). However, there is no evidence for the existence of a separate *bx* gene, since all *bx* mutations map within the *Ubx* transcription unit. This raises the possibility that *Ubx* expression alone directs the development of anterior T3. Since *bx* mutations

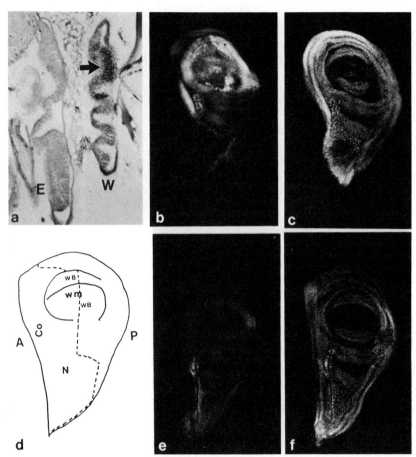

Figure 2. (a) Section through an sxc^3/sxc^4 third-instar larva showing the wing (W) disk and eye antennal (E) disk. The section was hybridized with a ^{35}S-labeled A113 probe (see Materials and Methods). Arrow indicates a region of strong labeling, demonstrating ectopic expression of the Ubx gene in the presumptive wing region of the disk. Note that the rest of the disk is unlabeled, as is the eye antennal disk (exposure: 5 days). (b) Whole mount of an sxc^3/sxc^4 wing imaginal disk labeled with monoclonal antibody FP3.38, followed by Texas Red conjugated sheep anti-mouse Ig. The labeled regions correspond to the presumptive wing region of the disk (see d for fate map). There is some variation in the pattern of labeling between disks, but it is always confined to the wing region, with the presumptive costa invariably labeled. Note that both anterior and posterior compartments are labeled. (c) Same disk as in b with Hoescht 33258. (d) Fate map of the wild-type wing disk, indicating the anterior/posterior compartment border as defined by clonal analysis (based on Brower et al. 1981). (A) Anterior; (P) posterior; (N) notum; (Co) costa; (WM) wing margin; (WB) wing blade. (e) Wild-type ($sxc^3/+$) wing imaginal disk labeled with FP3.38 and Texas Red conjugated sheep anti-mouse Ig and photographed under conditions identical to those of the disk shown in b. Only the cells associated with the trachea (arrow) exhibit significant labeling. (f) Same disk as in e stained with Hoescht 33258.

map within the Ubx transcription unit, it is not surprising that they result in a T3-to-T2 transformation. But what is the relationship between Ubx and pbx? Lewis (1964) suggested that Ubx mutations exert a polar effect on the pbx locus, thus explaining the lack of complementation between Ubx and pbx. However, an alternative possibility is that pbx cis-regulates expression of the Ubx gene.

Evidence for this comes from the analysis of the differentiation of segment T3 in sxc^- animals that lack either or both wild-type bx and pbx alleles. The data presented above demonstrate that the absence of sxc^+ results in the inappropriate expression of the Ubx unit in part of segment T2, causing the transformation of both compartments of the wing to haltere. In sxc^- animals also homozygous or hemizygous for mutant bx alleles, such as bx^{34e} or bx^3, this transformation of the wing to haltere is suppressed in both compartments,

suggesting that the bx mutants have inactivated the ectopically expressed Ubx gene (Ingham 1984). In $sxc^3/sxc^4;pbx^1$ animals, however, both compartments of the wing remain transformed to haltere. Moreover, the transformation of the posterior haltere normally caused by the pbx^1 mutation is almost completely suppressed (Ingham 1984). This finding is consistent with the notion that posterior T3 development requires expression of the Ubx transcript and that this expression is modulated by the pbx function. In the absence of pbx^+, Ubx expression would be suppressed in posterior T3, but this requirement for pbx can be overridden by the derepression of Ubx caused by the absence of sxc^+.

This interpretation is supported by the phenotype of $sxc^3/sxc^4;bx^3pbx^1/Ubx^{130}$ flies. In such animals, the effect of ectopic expression of Ubx caused by the absence of sxc^+ is expected to be suppressed by the in-

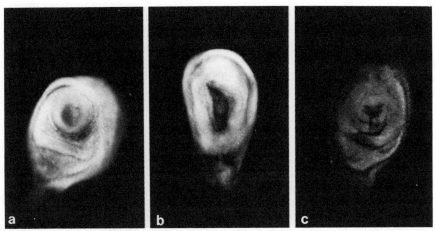

Figure 3. Leg imaginal disks labeled with FP3.38 and Texas Red conjugated sheep anti-mouse Ig. (*a*) sxc^3/sxc^4 third leg disk. Cells in both compartments show significant labeling, with higher levels in the posterior compartment (*left*). (*b*) sxc^3/sxc^4 second leg disk. Cells in both compartments exhibit levels of labeling significantly higher than those of the wild-type second leg disk shown in *c*.

activation of *Ubx* by the bx^3 mutation. Hence, the suppression of the pbx^1 mutation by *sxc* should itself be suppressed. This is indeed the case. In $sxc^3/sxc^4;bx^3 pbx^1/Ubx^{130}$ flies, both compartments of the haltere are transformed to wing, whereas the wing develops normally.

A strong prediction of this interpretation is that *Ubx* will not be expressed in posterior T3 of animals homozygous for *pbx* mutations. This has been tested by a direct analysis of *Ubx* expression in the imaginal disks of pbx^1 larvae. In the haltere disk of pbx^1 homozygotes, *Ubx* transcript is detectable only in that part of the disk that gives rise to structures of the anterior compartment (Fig. 5). Thus, as predicted, the pbx^1 mutation suppresses *Ubx* expression in the posterior compartment of the haltere.

Effect of *trx* Mutations on Segment Differentiation

In contrast to *sxc*, the *trx* gene appears to be required for the positive regulation of ANT-C and BX-C

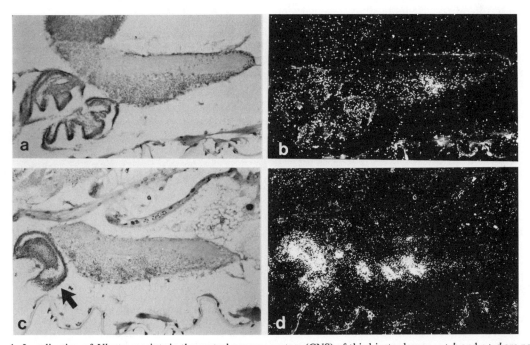

Figure 4. Localization of *Ubx* transcripts in the central nervous system (CNS) of third-instar larvae. *a + b* and *c + d* are pairs of bright-field and dark-field micrographs of sections through wild-type and sxc^3/sxc^4 larvae, respectively. In the wild-type larva, there is one region of strong labeling in the CNS corresponding to the first abdominal neuromere (Akam 1983). In the mutant larva, three additional regions of strong labeling anterior to the first abdominal neuromere are present. These appear to correspond to the three thoracic neuromeres. Note also the strong labeling of the second leg disks (arrow) in the mutant larva.

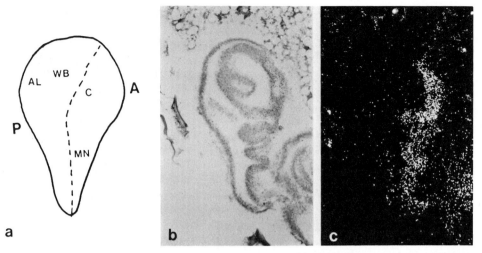

Figure 5. Expression of *Ubx* in a *pbx¹* haltere disk. (*a*) Fate map of the *pbx¹* haltere disk showing the approximate boundary between the anterior (A) compartment (which differentiates haltere structures) and the posterior (P) compartment (which differentiates wing structures). (AL) Alula; (WB) wing blade; (C) capitellum; (MN) metanotum (see also Fig. 1; based on Adler 1978). (*b*) Bright-field and dark-field (*c*) micrographs of a section through a *pbx¹* homozygous larva showing the haltere disk. The section was hybridized with a ^{35}S-labeled A113 probe. The labeling is restricted to the right-hand region of the disk, which would give rise to the anterior compartment structures. The posterior compartment (which is transformed to wing by *pbx¹*) is unlabeled (exposure, 5 days).

genes. Thus, *trx* mutations result in homeotic transformations of adult body structures that correspond to a loss of function of ANT-C and BX-C genes (Ingham and Whittle 1980).

Amorphic alleles of *trx* are homozygous lethal, but their effect on adult segment differentiation can be assessed in clones of mutant cells produced by X-ray-induced mitotic recombination. Such clones reveal that *trx*⁺ is required in the head, thorax, and abdomen. The common feature of *trx*⁻ clones in all of these parts of the body is that they are transformed to structures typical of the mesothorax (Ingham 1985).

Less extreme homeotic transformations also occur in flies homozygous or hemizygous for the viable allele *trx¹*, although because of its "leaky" nature, they are somewhat variable. However, in *trx¹/Df trx* animals, several transformations are consistently observed. First, the T1 leg is transformed partially to its mesothoracic equivalent. Since the differentiation of T1, rather than T2, is controlled by expression of the *Scr* gene of the ANT-C (Lewis et al. 1980), this transformation suggests that *trx*⁺ may normally be required for *Scr* expression.

Second, both dorsal and ventral derivatives of the T3 segment are transformed to their T2 equivalents. Thus, the haltere is replaced by structures typical of the wing, an extra mesonotum develops posterior to the normal mesonotum (Fig. 6), and the third leg bears specific bristles characteristic of the second leg. In both the haltere and the third leg, the anterior compartment is transformed more frequently than the posterior compartment. These transformations suggest that the absence of *trx*⁺ results in a loss of expression of the *Ubx* gene, primarily in the anterior compartment of T3.

Finally, various abdominal segments are partially transformed to their more anterior counterparts. Transformation toward segment A1 is indicated by a change in the size and distribution of the tergite bris-

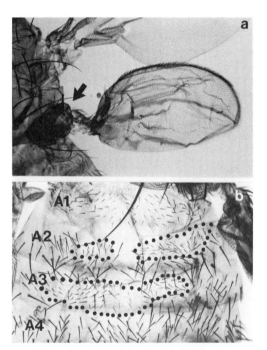

Figure 6. Adult transformations caused by the *trx¹* mutation. (*a*) Partial transformation of dorsal T3 toward T2. This is characterized by the replacement of the haltere by wing tissue, although such transformations are normally imperfect as illustrated here: Only the anterior part of the haltere is transformed. In addition, the metanotum is partially transformed to mesonotum (arrow), although again this effect is imperfect. (*b*) Partial transformation of abdominal segments A2 and A3 toward A1. This is characterized by regions of the tergites of A2 and A3 that differentiate small bristles characteristic of A1 (regions are marked by dotted boundaries).

tles and occurs most frequently in segments A2, A3, and A4 (Fig. 6). In addition, segments A5, A6, and A7 are frequently transformed to A4. These transformations are identical to those caused by mutations of the *Abd-A* and *Abd-B* genes, respectively, of the BX-C (Sánchez-Herrero et al. 1985) and thus suggest that *trx+* is required for the normal expression of elements of the BX-C other than *Ubx*.

Changes in *Ubx* Expression in *trx¹* Hemizygotes

The pattern of expression of *Ubx* in *trx¹* hemizygotes differs from that of wild type in two major respects. First, expression in the T3 imaginal disks is altered. The patchy expression in the haltere disk shown in Figure 7 directly parallels the patchy transformation of the metathorax typical of *trx¹* hemizygotes. Similarly, the absence of *Ubx* labeling in the anterior region only of the third leg disk shown in Figure 7 is consistent with the finding that the posterior compartment of the third leg is only infrequently transformed. Second, significantly higher levels of *Ubx* labeling are observed in the abdominal ganglia of the central nervous system (Fig. 7). The implications of this increased expression of *Ubx* in the abdomen are discussed below.

DISCUSSION

Mutations of the *sxc* and *trx* genes cause homeotic transformations of different body segments, which suggest that both genes are required for the correct spatial pattern of expression of genes of the ANT-C and BX-C (Ingham 1981, 1984, 1985). The results of this analysis show that these transformations are accompanied by changes in the spatial distribution of products of the *Ubx* gene, thus confirming the role of both *sxc* and *trx* in the regulation of selector gene expression.

In *sxc* mutations, *Ubx* is ectopically expressed in both anterior and posterior compartments of the wing disk, resulting in a transformation of wing to haltere. Since this transformation is independent of the presence or absence of a *pbx+* allele, it follows that expression of the *Ubx* transcription unit is itself sufficient to direct cells along the T3 developmental pathway in both anterior and posterior compartments. The wild-type function of *pbx* would thus be to control expression of *Ubx* in posterior T3 (Ingham 1984). This interpretation is supported by the finding that *Ubx* expression is suppressed in the posterior compartment of haltere disks of *pbx* homozygotes. Other evidence that the *bxd* transcription unit (which is partially deleted by the *pbx* mutations) regulates expression of the *Ubx* unit has recently been presented by Beachy et al. (1985).

Despite the ectopic expression of *Ubx* in both the dorsal and ventral imaginal disks of segment T2 (i.e., the wing disk and second leg disk), the structures they differentiate are not typical of the same segment. Thus, whereas the wing is transformed to haltere (segment T3), the anterior compartment of the second leg is transformed to first leg (T1). Moreover, although *Ubx* is expressed in both compartments of the T3 leg disk in *sxc* mutants, the anterior compartment of this leg also differentiates structures characteristic of the T1 leg. Since the T1 developmental pathway is controlled by the *Scr* gene, these transformations of T2 and T3 to T1 have been interpreted as reflecting the ectopic expression of *Scr+* in *sxc* mutants. In the absence of suitable probes for *Scr* expression, it has not been possible to test this hypothesis directly. However, if correct, it follows that expression of *Scr* results in anterior compartment cells following the T1 developmental pathway, irrespective of the level of expression of *Ubx*. In contrast, the posterior compartments of all three legs differentiate some structures characteristic of T3. The explanation for this difference in the developmental levels followed by anterior and posterior compartments remains obscure.

The transformation of the T3 derivatives to their T2 counterparts typical of *trx¹/Df trx* flies is reflected in

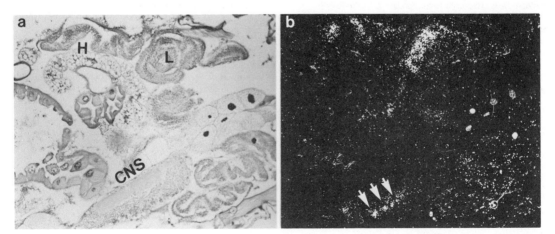

Figure 7. Bright-field (*a*) and dark-field (*b*) micrographs of a section through a *trx¹/Df trx* third-instar larva hybridized with a ^{35}S-labeled A113 probe. Note the patchy distribution of signal in the haltere (H) disk. In the third leg disk (L), the labeling is predominantly localized in the posterior compartment. In the CNS, significant labeling is present in the abdominal neuromeres (compare with the wild-type CNS shown in Fig. 3).

a reduction or absence of *Ubx* expression in particular regions of the T3 imaginal disks. This implies that *trx*$^+$ is required for the expression of *Ubx*. In contrast, levels of *Ubx* expression appear increased in the abdominal ganglia of *trx/Df trx* larvae. This apparently paradoxical finding is directly in line with two characteristics of *trx* mutants. First, as described here, the adult epidermal derivatives of the abdominal segments are frequently transformed toward A1. Second, the abdominal neuromeres of *trx* mutants exhibit morphological changes indicative of a transformation to A1 (Ghysen et al. 1985). Since in the wild type, maximal expression of *Ubx* in the central nervous system is limited to the A1 neuromere, it is not inconsistent that increased levels of *Ubx* expression occur in the abdominal neuromeres of *trx* mutants.

The simplest explanation of these apparently opposing effects on *Ubx* expression in different segments is to consider the elevation in *Ubx* levels to be an indirect consequence of the requirement for *trx*$^+$ for the expression of other BX-C genes. Evidence that the distal genes of the BX-C suppress *Ubx* expression in the abdominal segments has recently been presented by Beachy et al. (1985). Thus, the increase in *Ubx* expression in the abdominal neuromeres can be attributed to the reduction in the expression of *Abd-A* and *Abd-B* genes in *trx* mutants.

ACKNOWLEDGMENTS

I am grateful to Rob White and Mike Wilcox for performing the immunofluorescence staining, to Michael Akam for providing the *Ubx* clone, and to Ken Howard for making the ^{35}S-labeled *Ubx* probe.

REFERENCES

ADLER, P.N. 1978. Positional information in imaginal discs transformed by homoeotic mutations. *Wilhelm Roux's Arch. Dev. Biol.* **185:** 271.

AKAM, M.E. 1983. The location of *Ultrabithorax* transcripts in *Drosophila* tissue sections. *EMBO J.* **2:** 2075.

BEACHY, P.A., S.L. HELFAND, and D.S. HOGNESS. 1985. Segmental distribution of bithorax complex proteins during *Drosophila* development. *Nature* **313:** 545.

BENDER, W., M. AKAM, F. KARCH, P.A. BEACHY, M. PFEIFER, P. SPIERER, E.B. LEWIS, and D.S. HOGNESS. 1983. Molecular genetics of the bithorax complex in *Drosophila melanogaster*. *Science* **221:** 23.

BROWER, D.L., P.A. LAWRENCE, and M. WILCOX. 1981. Clonal analysis of the undifferentiated wing disk of *Drosophila*. *Dev. Biol.* **86:** 448.

DUNCAN, I. 1982. *Polycomblike*: A gene that appears to be required for the normal expression of the Bithorax and Antennapedia Complexes of *Drosophila melanogaster*. *Genetics* **102:** 49.

DUNCAN, I. and E.B. LEWIS. 1982. Genetic control of body segment differentiation in *Drosophila*. In *Developmental order: Its origin and regulation* (ed. S. Subtelny), p. 533. A.R. Liss, New York.

GARCÍA-BELLIDO, A. 1975. Genetic control of wing disc development in *Drosophila*. *Ciba Found. Symp.* **29:** 161.

GHYSEN, A., L.Y. JAN, and Y.N. JAN. 1985. Segmental determination in *Drosophila* central nervous system. *Cell* **40:** 943.

INGHAM, P.W. 1981. *Trithorax*: A new homoeotic mutation of *Drosophila melanogaster*. II. The role of *trx*$^+$ after embryogenesis. *Wilhelm Roux's Arch Dev. Biol.* **190:** 365.

———. 1983. Differential expression of bithorax complex genes in the absence of the *extra sex combs* and *trithorax* genes. *Nature* **306:** 591.

———. 1984. A gene that regulates the bithorax complex differentially in larval and adult cells of *Drosophila*. *Cell* **37:** 815.

———. 1985. The role of the *trithorax* gene in the diversification of segments in *Drosophila melanogaster*. *J. Embryol. Exp. Morphol.* **89:** (in press).

INGHAM, P.W. and R. WHITTLE. 1980. *Trithorax*: A new homoeotic mutation of *Drosophila melanogaster* causing transformations of abdominal and thoracic imaginal segments. *Mol. Gen. Genet.* **179:** 607.

LEVINE, M., E. HAFEN, R.L. GARBER, and W.J. GEHRING. 1983. Spatial distribution of *Antennapedia* transcripts during *Drosophila* development. *EMBO J.* **2:** 2037.

LEWIS, E.B. 1964. Genetic control and regulation of developmental pathways. In *Role of chromosomes in development* (ed. M. Locke), p. 231. Academic Press, New York.

———. 1978. A gene complex controlling segmentation in *Drosophila*. *Nature* **276:** 565.

LEWIS, R.A., B.T. WAKIMOTO, R.E. DENELL, and T.C. KAUFMAN. 1980. Genetic analysis of the Antennapedia gene complex (ANT-C) and adjacent chromosomal regions in *Drosophila melanogaster*. II. Polytene chromosome segments 84A-84B1,2. *Genetics* **95:** 383.

LINDSLEY, D.L. and E.L. GRELL. 1968. Genetic variations of *Drosophila melanogaster*. *Carnegie Inst. Wash. Publ.* no. 627.

SÁNCHEZ-HERRERO, E., I. VERNOS, R. MARCO, and G. MORATA. 1985. Genetic organisation of *Drosophila* bithorax complex. *Nature* **313:** 108.

STRUHL, G. 1981a. A homoeotic mutation transforming leg to antenna in *Drosophila*. *Nature* **292:** 635.

———. 1981b. A gene product required for the correct initiation of segmental determination in *Drosophila*. *Nature* **293:** 36.

WHITE, R.A.H. and M. WILCOX. 1984. Protein products of the bithorax complex in *Drosophila*. *Cell* **39:** 163.

Expression of the Homeo Box Gene Family in *Drosophila*

M. LEVINE, K. HARDING, C. WEDEEN, H. DOYLE, T. HOEY, AND H. RADOMSKA
Department of Biological Sciences, Columbia University, New York, New York 10027

The fruit fly is composed of eight abdominal, three thoracic, and four to six head segments (Ferris 1950). Several of the constituent tissues of a given segment acquire morphological properties unique to that segment. For example, the larval and adult epidermis elaborate cuticular structures, such as legs and antennae, that are unique to a particular body segment. In addition, recent histologic and genetic studies have revealed morphologic differences between some of the mesodermal (Lawrence and Johnson 1984) and neural (Kankel et al. 1980; Jimenez and Campos-Ortega 1981) tissues of each segment.

Homeotic genes are thought to establish the diverse developmental pathways by which each embryonic and larval segment acquires a distinct adult phenotype (Lewis 1978; for review, see Ouweneel 1976). Mutation of a homeotic gene can result in the partial or complete transformation of the epidermal tissues of one segment into those of another segment. For example, several dominant mutations of the Antennapedia (*Antp*) locus result in a second pair of mesothoracic legs in place of antennal structures. It appears that at least some homeotic transformations might include the mesodermal and neural tissues of the affected segment as well (Jimenez and Campos-Ortega 1981; Teugels and Ghysen 1983; Lawrence and Johnson 1984).

Many homeotic genes are clustered within one of two regions of the *Drosophila* genome, the bithorax complex (BX-C) (Lewis 1978; Sanchez-Herrero et al. 1985) or the Antennapedia complex (ANT-C) (Denell 1973; Kaufman et al. 1980). Genes of the BX-C are required for the establishment of morphologically diverse segments in the posterior regions of the fly (Lewis 1963, 1978, 1982). Lewis has identified a number of homeotic loci within the BX-C on the basis of adult mutant phenotypes (Karch et al. 1985; Lewis 1978) More recently, on the basis of lethal complementation analyses, the BX-C has been shown to contain three essential domains of homeotic function: Ultrabithorax (*Ubx*), abdominal-A (*abd-A*), and Abdominal-B (*Abd-B*; Sanchez-Herrero et al. 1985). *Ubx* is most critically required for proper development of the posterior compartment of the mesothorax (T2p) through the anterior compartment of the first abdominal segment (A1a). The *abd-A* function is required for proper embryonic development of A1p through A8, and *Abd-B* is involved in correct morphogenesis of A5 through A8 (Sanchez-Herrero et al. 1985).

In contrast to the BX-C, genes of the ANT-C are necessary for the establishment of morphologically diverse segments in the anterior regions of the fly (Wakimoto and Kaufman 1981; Wakimoto et al. 1984). Several homeotic lethal complementation groups have been identified for the ANT-C (Kaufman et al. 1980; Wakimoto and Kaufman 1981; Hazelrigg and Kaufman 1983; Wakimoto et al. 1984), including the *Antp*, Sex combs reduced (*Scr*), and Deformed (*Dfd*) loci. The *Antp* locus has been shown to be required for proper segment morphogenesis of the thorax (Struhl 1980, 1982; Wakimoto and Kaufman 1981). Embryos that lack *Antp*$^+$ activity display an epidermal transformation of the meso- and metathorax (T2 + T3) into the homologous tissues of the prothorax (T1; Wakimoto and Kaufman 1981). Analyses of *Scr*$^-$ and *Dfd*$^-$ mutant embryos suggest that these genes are required for differentiation of the prothorax and posterior head regions (Wakimoto and Kaufman 1981; Wakimoto et al. 1984). Additional lethal homeotic functions might occur within the ANT-C that map to the left of *Dfd* (see Fig. 1; Hazelrigg and Kaufman 1983).

The regions of developing embryos that accumulate transcripts specified by *Antp* and *Ubx* have been identified by in situ hybridization to tissue sections (Akam 1983; Hafen et al. 1983; Levine et al. 1983). The principal sites of *Antp*$^+$ and *Ubx*$^+$ expression were found to correspond roughly to the embryonic segments and segment primordia that are most severely disrupted in *Antp*$^-$ and *Ubx*$^-$ mutants. *Antp* transcripts principally accumulate in T1/T2 and T2/T3 (See Fig. 2) (Hafen et al. 1983; Levine et al. 1983); *Ubx* transcripts are detected in T3/T1 (Akam 1983). Despite these differences in transcript distribution, the overall developmental patterns of *Antp* and *Ubx* expression are remarkably similar. *Antp* and *Ubx* transcripts are detected in epidermal, neural, and mesodermal tissues at approximately the same times during development (Akam 1983; Levine et al. 1983). This observation is consistent with previous proposals that ANT-C and BX-C genes are members of a diverse gene family and have evolved from a common ancestral gene or gene complex (Lewis 1978; McGinnis et al. 1984b). Further support for this proposal is the demonstration that the *Antp* and *Ubx* loci share direct nucleotide sequence homology within a conserved protein-coding region, designated the homeo box (Laughon and Scott 1984; McGinnis et al. 1984a,b; Scott and Weiner 1984).

Seven genomic DNA fragments have been shown to cross-hybridize strongly with both the *Antp* and *Ubx*

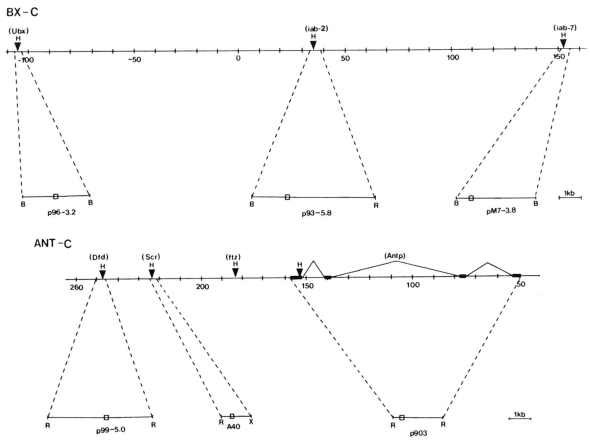

Figure 1. Genetic and molecular maps of the BX-C and the ANT-C. Horizontal lines correspond to overlapping cloned genomic DNAs that span the two complexes (units in kilobase pairs) (for details, see Bender et al. 1983; Garber et al. 1983; Scott et al. 1983). Arrowheads indicate the positions of homeo box copies within the two complexes. Homeo-box-containing cloned genomic DNAs used as probes for in situ hybridizations are indicated below the horizontal lines; dashed lines indicate the positions within the chromosome walk from which these cloned DNAs are derived. Open boxes within each cloned DNA used as probe correspond to the position of the homeo box within each fragment. The *ftz* homeo box region was not used in this analysis. p99-5.0 is a 5-kb *Eco*RI genomic DNA fragment that derives from the *Dfd* locus. A40 is a 1.4-kb *Hpa*I/*Xho*I genomic fragment that appears to derive from the homeo box region of the *Scr* locus (the exact position of the homeo box within the fragment is unknown); p903 is a 2.3-kb *Antp* cDNA; p96-3.2 is a 3.2-kb *Bam*HI genomic DNA fragment that has homology with *Ubx* transcripts; p93-5.8 is 5.8-kb *Bam*HI/*Eco*RI genomic fragment that has homology with *iab-2* transcripts; pM7-3.8 is 3.8-kb *Bam*HI genomic fragment that has homology with *iab-7* transcripts (Karch et al. 1985; Regulski et al. 1985).

homeo boxes (Regulski et al. 1985). These seven regions correspond to the Antennapedia class of the homeo box gene family and are all located within either the BX-C or ANT-C (Regulski et al. 1985). The three BX-C loci that contain a homeo box copy correspond to *Ubx*, infra-abdominal-2 (*iab-2*), and infra-abdominal-7 (*iab-7*) (see Fig. 1) (Regulski et al. 1985; Karch et al. 1985). The four homeo box copies in the ANT-C are associated with the *Antp*, fushi tarazu (*ftz*), *Scr*, and *Dfd* loci. In this paper we show that six of these seven homeo-box-containing loci specify transcripts that accumulate in largely discrete regions of the embryonic central nervous system (CNS). To a close approximation, the regions of the CNS that contain transcripts encoded by each of these loci correspond to the embryonic segments that are disrupted in mutants for these genes.

A central problem in homeosis is how different ANT-C and BX-C homeotic loci come to be expressed in largely nonoverlapping regions of the embryonic and larval CNS. We propose that spatially restricted expression of each ANT-C and BX-C locus involves hierarchical, cross-regulatory interactions that are mediated by the homeo box protein domains. Support for this model is based on analysis of homeo box transcript distribution patterns in various homeotic mutant embryos.

Since the patterns of ANT-C/BX-C homeo box gene expression might involve cross-regulatory interactions, it is of interest to determine the nature of any other regions of the *Drosophila* genome that contain homeo box copies. In this paper we describe the isolation and characterization of additional loci that contain the homeo box. One of these (S67) is expressed during oogenesis and displays a localized transcript distribution pattern at preblastoderm periods of embryogenesis.

MATERIALS AND METHODS

In situ hybridization to tissue sections. Frozen tissue sections were prepared with a cryostat as described previously (Hafen et al. 1983). Embryos were not treated with fixative prior to sectioning. Tritiated hybridization probes were prepared by nick translation as described by Hafen et al. (1983), who used total recombinant DNA (insert + vector) in the nick-translation reaction. However, in this study, the insert DNA fragment was gel-purified prior to labeling.

Homeo box screen of recombinant DNA library. A total of 4×10^4 recombinants (approximately four genome equivalents) from the Charon 4/*Drosophila* DNA library of Maniatis et al. (1978) were screened using homeo box probes derived from the *Scr* and *ftz* loci of the ANT-C (see Fig. 1). The conditions for reduced stringency hybridization were carried out exactly as described previously (McGinnis et al. 1984b).

In situ hybridization to polytene chromosomes. Recombinant DNA was biotinylated by nick translation and hybridized to chromosome spreads as described previously (Langer-Safer et al. 1982). The hybridization signal was visualized with a strepavidinbiotinylated horseradish peroxidase polymer as described by the supplier (ENZO Biochemicals).

RESULTS

Transcripts Encoded by Different Homeo-box-containing Regions of the BX-C and ANT-C Accumulate in Discrete Regions of the Embryonic CNS

The identities of tissues that contain *Dfd*, *Scr*, *iab-2*, and *iab-7* RNAs were determined by a previously described in situ hybridization technique (Hafen et al. 1983). Serial tissue sections of wild-type embryos at various stages of development were separately hybridized to cloned homeo-box-containing genomic DNA fragments after radiolabeling with tritiated deoxynucleotides. A summary of the probes used in this analysis is shown in Figure 1. A primary site of hybridization for each of these probes corresponds to neural tissues.

The embryonic CNS is composed of a dorsally located brain that is connected to a ventral nerve cord by a subesophageal ganglion. The ventral cord is a composite of 11 repeating ganglia, each of which contains approximately 300 nerve cells (Poulson 1950). The epidermal portions of each of the three thoracic and eight abdominal segments come to be innervated by a corresponding ganglion of the ventral cord.

Figure 2, a and b, shows an autoradiogram of a 12–14-hour (midstage) embryo after hybridization to the *Dfd* probe, p99-5.0. *Dfd* transcripts are detected in a central portion of the subesophageal ganglion, S1/S2. Hybridization signals above background levels are not observed in other regions of the embryonic CNS. Figure 2, c and d, shows an autoradiogram of a midstage embryo after hybridization to the *Scr* probe, A40 (see Fig. 1). *Scr* transcripts are detected in the S3 region of the subesophageal ganglion. A second focus of embryonic hybridization is observed over epidermal and mesodermal tissues of a posterior head segment, probably the labium (Fig. 2c,d, arrowhead). As previously described, *Antp* transcripts are found in a more posterior portion of the CNS, including the T1 and T2 ventral ganglia (Hafen et al. 1983; Levine et al. 1983); *Ubx* transcripts are detected in T3/A1 (Akam 1983).

Figure 2, e and f, shows an autoradiogram of a midstage embryo after hybridization with the *iab-2* probe, p93-5.8. Strong hybridization signals are detected in the region of the ventral cord that contains the A1/A2 through A6/A7 ventral ganglia. Lower levels of labeling are observed in the posterior-most ventral ganglia. Labeling above background levels is not detected in portions of the embryonic CNS anterior to the A1/A2 ventral ganglia.

RNAs homologous to the *iab-7* hybridization probe, pM7-3.8, are detected primarily in the posterior-most ventral ganglia, including at least a posterior portion of the A7 ganglion and all of the composite A8/A9 ganglion (Fig. 2g,h). Less-intense hybridization signals are observed over the A5 and A6 ganglia; no labeling above background levels is detected in the CNS anterior to the A4/A5 region of the ventral cord.

Altered Distribution of *Antp*⁺ Transcripts in BX-C⁻ Embryos

BX-C⁻ embryos (*Df[3R]P9* homozygotes) die during the terminal stages of embryogenesis (Lewis 1978). Lewis has shown that the epidermal tissues of embryonic segments T3→A7 acquire features characteristic of the mesothorax (T2). More recently, it has been suggested that the transformed segments might acquire both T1 and T2 characteristics (Duncan 1981; Struhl 1983). In advanced-stage BX-C⁻ embryos, *Antp*⁺ transcripts persist at high steady-state levels in the region of the ventral cord that encompasses the T1/T2, T3, and first seven abdominal ganglia (Hafen et al. 1984).

Figure 3, a and b, shows an autoradiogram of a BX-C⁻ embryo (P9/P9) at the cellular blastoderm period of development after hybridization to the *Antp* 903 cDNA probe. Segment primordia T1/T2 and T2/T3 display intense hybridization signals (Fig. 3, arrows). Substantially lower levels of hybridization are observed in other segment primordia. Similar discrete hybridization signals are detected in gastrulating P9/P9 embryos (Fig. 3c,d).

In wild-type embryos undergoing germ band retraction, *Antp* transcripts are broadly distributed over the progenitors of the ventral cord. However, labeling of the presumptive abdominal ganglia is substantially less intense as compared with future thoracic ganglia (Levine et al. 1983). In contrast, P9/P9 embryos of this developmental stage show almost uniformly intense *Antp* hybridization signals from the future T1/T2 primordium posteriorly through A8 (Fig. 3e,f, arrows).

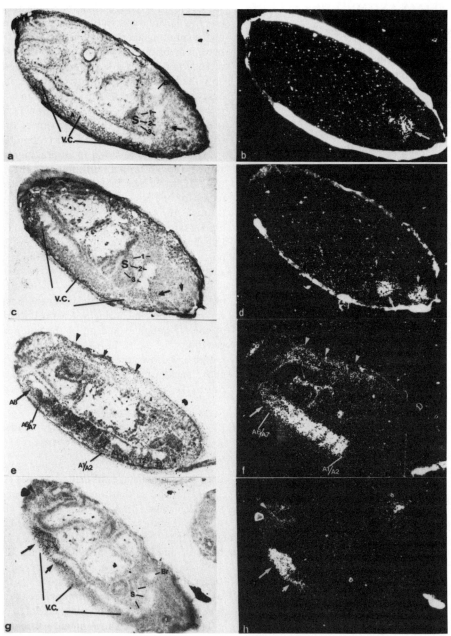

Figure 2. Localization of *Dfd*, *Scr*, *iab-2*, and *iab-7* in tissue sections of 12–14-hr-old embryos. All sections are sagittal and are oriented such that anterior is to the right and dorsal is up. The luminescence of the chorion in dark-field photomicrographs is due to refraction of light and does not indicate nonspecific or specific hybridization. Bar in *a* represents 0.1 mm. Our in situ localizations of transcripts could be in error by half a segment. The term "T1/T2" indicates that the region of hybridization could be in tissues of T1 and/or T2. Similarly, A1/A2 indicates labeling in all or portions of the first and/or second abdominal ganglia, etc. (*a*) Localization of transcripts homologous to the *Dfd* probe (p99-5.0). Hybridization signals are seen over the cells of the S1/S2 region of the subesophageal ganglion (arrow). No signals over background are detected in regions either anterior or posterior to S1/S2. (*b*) Dark-field photomicrograph of *a*. Arrow indicates the principal site of hybridization. (*c*) Localization of transcripts homologous to the putative *Scr* probe, A40. Hybridization signals are seen over the S2/S3 region of the subesophageal ganglion and possibly some cells of the T1 ventral ganglion (arrow); hybridization is also seen to labial tissues (arrowhead). No other sites of hybridization are detected. (*d*) Dark-field photomicrograph of *c*. Arrow indicates S3/T1 labeling (see above), arrowhead indicates labeling of the labium. (*e*) Localization of transcripts homologous to the *iab-2* probe. The strongest hybridization signals occur over A1/A2 through A6/A7 ventral ganglia. Weaker signals are observed in the A8 ganglion (arrow) as well as in the hypodermal and mesodermal regions of segments A2 through A6/A7 (arrowheads). (*f*) Dark-field photomicrograph of *e*. Arrow indicates A8 labeling; arrowheads indicate hypodermal and mesodermal labeling. (*g*) Localization of transcripts homologous to the *iab-7* probe. The strongest hybridization signals are seen in the A7 and A8 ventral ganglia (large arrow); weaker hybridization is observed in A6 (small arrow) and in hypodermal tissues (arrowhead). (*h*) Dark-field photomicrograph of *g*. Large arrow indicates the A7 and A8 ganglia; the small arrow shows the A5/A6 region of the ventral cord, and the arrowhead indicates hypodermal labeling. (A) Abdominal region of the ventral cord; (Br) brain; (S) subesophageal ganglion; (T) thoracic region of the ventral cord; (V.C.) ventral cord.

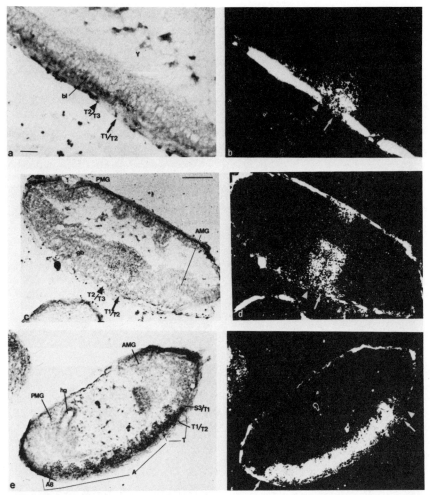

Figure 3. Localization of *Antp* transcripts in BX-C⁻ embryos. All sections are oriented such that anterior is to the right and dorsal is up. Parallel sections through the same embryos failed to show hybridization to either the *Ubx* or *iab-2* probes. The luminescence of the chorion in dark-field photomicrographs is due to refraction of light and does not indicate specific or nonspecific hybridization. (*a*) High magnification of a section through a BX-C⁻ embryo at the cellular blastoderm stage (bar, 0.025 mm). Hybridization signals are seen over cells in the regions of the embryo that give rise to T1/T2 (arrow) and T2/T3 (arrowhead). (*b*) Dark-field photomicrograph of *a*. The arrow indicates T1/T2, and the arrowhead, T2/T3. (*c*) A section through a gastrulating BX-C⁻ embryo (bar, 0.1 mm). Hybridization signals are detected in regions giving rise to T1/T2 (arrow) and T2/T3 (arrowhead). (*d*) Dark-field photomicrograph of *c*. The arrow indicates T1/T2 and the arrowhead, T2/T3. (*e*) A section through a BX-C⁻ embryo undergoing germ band retraction. Uniformly intense signals are detected in T1/T2 through A8. Weaker signals are detected in cells of S3/T1. (*f*) Dark-field photomicrograph of *e*. Arrows demarcate T1/T2 to A8, and the arrowhead indicates S3/T1 labeling. (A) Abdominal region of the germ band; (AMG) anterior midgut invagination; (bl) blastoderm; (hg) hindgut; (PMG) posterior midgut invagination; (T) thoracic region of the germ band.

Distribution of ANT-C and BX-C Transcripts in Young (9–11 Hour) Polycomb⁻ Embryos

Embryos that lack Polycomb (*Pc*) function display gross homeotic transformations of most body segments (Duncan and Lewis 1982). The anterior segments of such mutant embryos acquire features normally charcteristic of more posterior segments. Lewis has postulated that these transformations result from the derepression in anterior segment primordia of BX-C homeotic loci that are normally expressed only in posterior regions (Lewis 1978; Duncan and Lewis 1982). Support for this proposal has been obtained by localizing ANT-C and BX-C RNAs in tissue sections of *Pc⁻* embryos, as described below. The *Antp*, *Ubx*, *iab-2*, and *iab-7* hybridization probes used in this analysis are shown in Figure 1.

Figure 4, a and b, shows a wild-type distribution pattern of *Antp* RNAs in a 10–11-hour embryo. The most intense hybridization signals correspond to the T1/T2 and T2/T3 regions of the ventral cord; weaker hybridization is detected in more posterior ventral ganglia. Deviation from this wild-type hybridization pattern is observed in *Pc⁻* embryos of the same developmental stage, as shown in Figure 4 c and d. In contrast to wild-type embryos, *Pc⁻* embryos show equally intense *Antp* hybridization signals throughout the CNS, including the T1→A7 ventral ganglia as well as the subesophageal ganglion and brain.

A similar expansion of the normal *Ubx* transcript

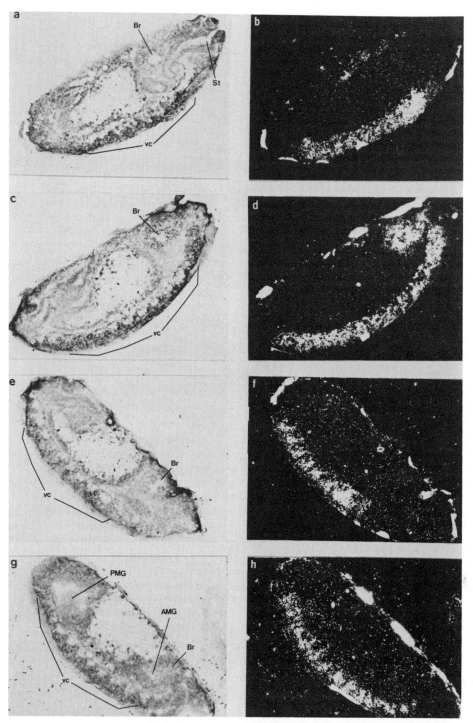

Figure 4. Localization of *Antp* and *Ubx* transcripts in 9–11-hr Pc^- and wild-type embryos. All sections are sagittal and are oriented such that anterior is to the right and dorsal is up. The luminescence of the chorion in dark-field photomicrographs is due to refraction of light and does not indicate nonspecific hybridization. The magnification of each photomicrograph is identical to that used for the panels shown in Fig. 2. (*a*) Tissue autoradiogram of a wild-type embryo after hybridization with the *Antp* (903) probe. The strongest hybridization signals are seen over the T1/T2 and T2/T3 ventral ganglia; weaker labeling is observed in posterior ventral ganglia through A7. (*b*) Dark-field photomicrograph of *a*. (*c*) Localization of transcripts homologous to the *Antp* probe in a Pc^3/Pc^3 embryo. Hybridization is broadly distributed throughout most of the CNS. (*d*) Dark-field photomicrograph of *c*. The embryos shown in *a* and *c* were subjected to identical conditions of hybridization and autoradiography. (*e*) Localization of transcripts homologous to the *Ubx* probe, p96. The strongest site of hybridization within the ventral cord corresponds to the T3/A1 ganglia. Weaker signals are observed in more posterior ventral ganglia through A7. (*f*) Dark-field photomicrograph of *e*. (*g*) Distribution of transcripts homologous to the *Ubx* probe, p96, in a Pc^3/Pc^3 embryo. Hybridization is detected throughout most of the ventral cord, including the subesophageal ganglion. (*h*) Dark-field photomicrograph of *g*. The embryos shown in *e* and *g* were subjected to identical conditions of hybridization and autoradiography. (AMG) Anterior midgut; (Br) brain; (PMG) posterior midgut invagination; (St) stomadeum; (VC) ventral cord.

distribution pattern is observed in Pc^- embryos. Figure 4, e and f, shows the wild-type Ubx pattern in the CNS of a 10–11-hour embryo. The strongest hybridization signals are detected in the T3/A1 region of the ventral cord; weaker signals are observed in more posterior ventral ganglia. Pc^- embryos of a similar developmental stage show the transcripts in the T1→A7 ganglia of the ventral cord as well as the subesophageal ganglion (Fig. 4g,h). The Pc mutation results in a similar expansion of the CNS transcript distribution pattern for the iab-2 and iab-7 loci of the BX-C as well (data not shown).

Distribution of ANT-C and BX-C Transcripts in Advanced-stage (14–16 Hour) Pc^- Embryos

During Pc^- embryogenesis, there is a successive decline in the levels of $Antp$ and Ubx RNAs in the CNS. By 14–16 hours following fertilization, Pc^- embryos display less than one-fifth the level of autoradiographic signal than that observed in 9–11-hour Pc^- embryos after hybridization with the same $Antp$ and Ubx probes. In contrast, iab-2 and iab-7 RNAs persist at high steady-state levels in advanced-stage Pc^- embryos.

Figure 5, a and b, shows a tissue autoradiogram of a 14–16-hour wild-type embryo following hybridization with the $Antp$ 903 probe. The strongest labeling is detected over ganglion cells in the T1/T2 region of the ventral cord. Hybridization of a 14–16-hour Pc^- embryo with the same probe (Fig. 5c,d) reveals that $Antp$ RNAs are present at substantially reduced steady-state levels as compared with wild-type embryos (Fig. 5a,b) or 9–11-hour Pc^- embryos (Fig. 4c,d). Similarly, a comparison of panels e and f (wild-type) with panels g and h (Pc^-) (Fig. 5) shows that Ubx RNAs are present at sharply lower levels in advanced-stage Pc^- embryos. In contrast to the successive decline of $Antp$ and Ubx RNAs, iab-2 and iab-7 transcripts are maintained at high steady-state levels over the course of Pc^- development (data not shown).

Isolation and Cytogenetic Mapping of New Homeo Box Loci

Approximately four genome equivalents of a recombinant $Drosophila$ DNA library were screened with homeo box probes as described in Materials and Methods. More than 60 cross-hybridizing phage were plaque-purified. Recombinants that derive from previously isolated homeo-box-containing regions of the $Drosophila$ genome were identified by high-stringency hybridizations with the homeo box probes listed in Table 1 (data not shown). In this way, six of the cross-hybridizing clones (~10%) were tentatively identified as new homeo-box-containing loci. Figure 6 shows the locations of hybridization within polytene chromosome spreads for five of the six clones. Table 1 lists the cytogenetic map positions for each of the six clones.

Transcript Distribution Patterns of New Homeo Box Cross-hybridizing Clones

To help determine whether the new homeo box cross-hybridizing clones derive from loci involved in the elaboration of positional information during $Drosophila$ development, the transcript distribution patterns for these sequences have been determined by in situ hybridization to tissue sections. The results for two of the six new clones are described below. Genomic DNA fragments that encompass the homeo box cross-hybridizing regions (abbreviated HBC) from each clone were used as probes in this analysis.

Figure 7 shows the hybridization pattern of the S67 probe (see Table 1). Transcripts homologous to the S67

Table 1. Cytogenetic Locations of Homeo Box Copies in *Drosophila*

(a) *Locations of homeo boxes isolated in previous studies*

Gene	Cytogenetic map position	References
Deformed (*Dfd*), p99-5.0	84A	McGinnis et al. (1984b); Regulski et al. (1985)
Sex combs reduced (*Scr*), A40	84B	Harding et al. (1985)
fushi tarazu (*ftz*), p523B	84B	McGinnis et al. (1984b)
Antennapedia (*Antp*), 903	84B	Garber et al. (1983); McGinnis et al. (1984b); Regulski et al. (1985)
Ultrabithorax (*Ubx*), p96-3.2	89E	Bender et al. (1983); McGinnis et al. (1984b)
infra-abdominal-2 (*iab*-2), p93-5.8	89E	McGinnis et al. (1984b); Regulski et al. (1985)
infra-abdominal-7 (*iab*-7) pM7-3.8	89E	McGinnis et al. (1984b); Regulski et al. (1985)
engrailed	48A	Fjose et al. (1985); Poole et al. (1985)

(b) *Putative homeo box clones isolated in current study*

Clone	Cytogenetic map position
F90-2	84A (3R)
S-59	93E1,2 (3R)
S-60	84A/B (3R)
S-63	88B (3R)
S-67	38F (2L)
S-8	46C (2R)

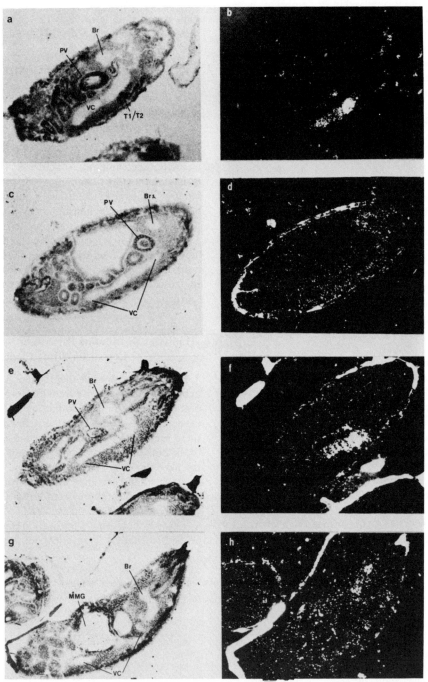

Figure 5. Localization of *Antp* and *Ubx* transcripts in advanced-stage (14–16 hr) Pc^- and wild-type embryos. All sections are sagittal and are oriented such that anterior is to the right and dorsal is up. The luminescence of the chorion in dark-field photomicrographs is due to refraction of light and does not indicate nonspecific or specific labeling by hybridization probes. The magnification of each photomicrograph is identical to that used in Figs. 2 and 4. (*a*) Tissue autoradiogram of a wild-type embryo after hybridization with the *Antp* probe, 903. The strongest labeling corresponds to the T1/T2 position of the ventral cord. Weaker hybridization signals are detected over more posterior ventral ganglia. (*b*) Dark-field photomicrograph of *a*. (*c*) Pc^3/Pc^3 embryo after hybridization with the *Antp* probe. Weak labeling of the CNS over background levels can be discerned. (*d*) Dark-field photomicrograph of *c*. The embryos shown in *a* and *c* were subjected to identical conditions of hybridization and autoradiography. Consecutive tissue sections through the embryo shown in *c* were hybridized with *iab-2* and *iab-7* probes. In both cases, intense labeling was observed throughout the CNS (data not shown). (*e*) Localization of transcripts homologous to the *Ubx* probe, p96. The strongest site of hybridization corresponds to the T3/A1 region of the ventral cord; hybridization is also observed over more posterior ventral ganglia through A7. (*f*) Dark-field photomicrograph of *e*. (*g*) Tissue autoradiogram of a Pc^3/Pc^3 embryo after hybridization with the *Ubx* probe. Labeling over background is detected over a portion of the brain. (*h*) Dark-field photomicrograph of *g*. The embryos shown in *e* and *g* were subjected to identical conditions of hybridization and autoradiography. Moreover, serial sections through the Pc^- embryo shown in *g* hybridized strongly with *iab-2* and *iab-7* probes (data not shown). (Br) Brain; (MMG) middle midgut; (PV) proventriculus; (VC) ventral cord.

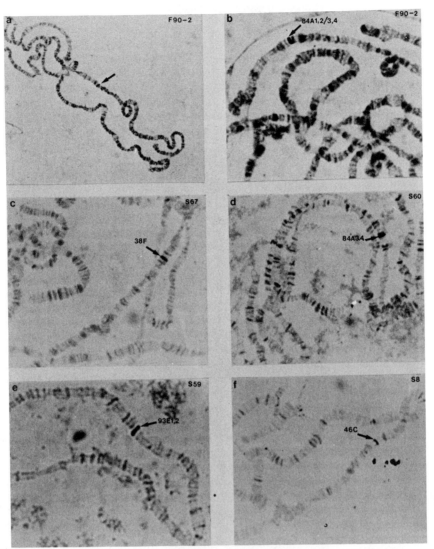

Figure 6. Cytogenetic locations of new homeo box cross-hybridizing sequences. Polytene chromosome spreads were prepared as described in Materials and Methods. Charon 4/*Drosophila* genomic DNA recombinants were biotinylated by nick translation using DNA polymerase I and a biotinylated deoxyuridine triphosphate (Langer-Safer et al. 1982). Hybridization and subsequent histochemical detection were done as described in Materials and Methods. The chromosome spreads shown in *a* and *b* were hybridized with the F90-2 recombinant; labeling is observed to the 84A region on the right arm of chromosome 3 (arrows). (*c*) Chromosome spread hybridized with the S67 recombinant; the arrow shows labeling to the 38F region on the left arm of chromosome 2. (*d*) Chromosome spread hybridized with S60, which labels the 84A/B region of 3R (arrow) B. (*e*) Chromosome spread hybridized with S59; (*f*) site of hybridization with the S8 probe.

HBC are uniformly distributed throughout the ooplasm of developing oocytes (data not shown). Following fertilization, these transcripts persist through the early cleavage stages of embryogenesis. A nonuniform S67 hybridization pattern is first observed in the sixth or seventh cleavage stage of embryogenesis (Fig. 7a,b). At this time, more than 70% of the S67 hybridization signal is detected in the posterior half of the embryo. Over the course of development, S67 hybridization signals gradually become more biased toward the posterior end of the embryo such that by the cellular blastoderm stage, virtually all of the detectable S67 transcripts are localized within a discrete band near the posterior pole (Fig. 7c,d). According to fate map studies (Poulson 1950; Lohs-Schardin et al. 1979), this region of the embryo will give rise to the malpighian tubules and portions of the posterior midgut (PMG). S67 hybridization signals are detected over these structures for the duration of embryogenesis. Figure 7, e and f, shows a 6–7-hour-old embryo after hybridization with the S67 probe. Strong hybridization is detected in the proximal portion of the PMG; weaker labeling is observed to more distal portions of the PMG as well. Figure 7, g and h, shows a tissue autoradiogram of an advanced-stage embryo (16–18 hr). Hybridization is observed to the malpighian tubules, portions of the posterior midgut, as well as to hindgut derivatives.

Figure 8 shows a series of embryo tissue sections after

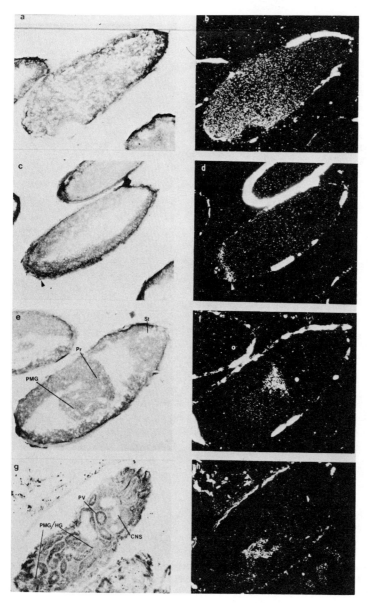

Figure 7. Localization of S67 transcripts in wild-type embryos. All sections are sagittal and are oriented such that anterior is to the right and dorsal is up. The luminescence of the chorion in dark-field photomicrographs is due to refraction of light and does not indicate nonspecific or specific labeling by hybridization probes. The magnification of each photomicrograph is identical to that used in Figs. 2, 4, and 5. The hybridization probe used in this analysis is a 3.7-kb *Eco*RI/*Bam*HI genomic DNA fragment that contains the entire S67 homeo box cross-hybridizing region. (*a*) Tissue autoradiogram of a wild-type embryo at cleavage stage 7 or 8 after hybridization with the S67 probe. Stronger labeling is observed to the posterior half of the embryo as compared with the anterior half. (*b*) Dark-field photomicrograph of *a*. (*c*) Localization of S67 transcripts in an embryo undergoing cellularization of the blastoderm. Specific hybridization signal is confined to a region near the posterior pole. (*d*) Dark-field photomicrograph of *c*. (*e*) Hybridization to a wild-type embryo at the germ band extension stage. Transcripts are localized within distal and proximal regions of the posterior midgut invagination. (*f*) Dark-field photomicrograph of *e*. (*g*) Localization of transcripts within a 16-18-hr-old wild-type embryo. Labeling is observed over the malpighian tubules and regions of the posterior midgut and hindgut. (*h*) Dark-field photomicrograph of *g*. (CNS) Central nervous system; (HG) hindgut; (PMG) posterior midgut; (Pr) proctodeum; (PV) proventriculus; (St) stomadeum.

hybridization with the HBC region of the F90-2 clone. Following germ band retraction (at 8-9 hr postfertilization), F90-2 transcripts are detected primarily within a discrete region of the midgut rudiment (Fig. 8a-d, arrows). A second site of hybridization is observed just posterior to the developing brain (Fig. 8a,b).

F90-2 transcripts are observed to persist in derivatives of the anterior midgut through the terminal stages of embryogenesis. Figure 8, e-f, shows that the F90-2 probe labels the anterior midgut and the portion of the middle midgut that is derived from the anterior midgut invagination (AMG). Figure 8, g and h, shows that the labeling over a region of the CNS just posterior to the brain also persists through embryonic development.

DISCUSSION

By mid-embryonic periods of development, each of the homeo-box-containing homeotic genes of the ANT-C and BX-C comes to specify transcripts that appear to accumulate in largely discrete, nonoverlapping regions of the mature embryonic CNS. These foci of transcript accumulation within the CNS roughly correspond to the embryonic segments that are most severely disrupted in mutants for the corresponding genes. The CNS is the most easily visualized, but it is not the only focus of homeotic gene expression. During gastrulation and germ band extension, strong expression is also detected in epidermal and mesodermal tissues (Akam 1983; Levine et al. 1983; Harding et al. 1985). By advanced stages of embryonic development, however, much of the labeling over the hypodermal and mesodermal tissues is no longer detectable. Because expression is seen to persist in the CNS during embryonic and larval development, it serves as a convenient model for analyzing the spatial regulation of homeotic gene expression.

A striking feature of ANT-C and BX-C homeo box

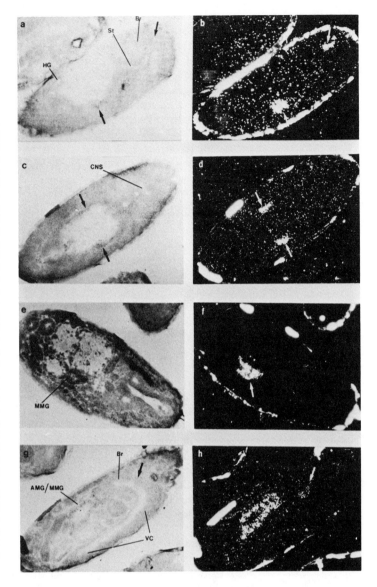

Figure 8. Distribution of F90-2 transcripts during embryogenesis. With the exception of panels c and d, all sections are sagittal and are oriented such that anterior is to the right and dorsal is up. The luminescence of the chorion in dark-field photomicrographs is due to refraction of light and does not indicate nonspecific or specific hybridization. The magnification used for each photomicrograph is identical to that used in Figs. 2, 4, 5, and 7. The hybridization probe used in this analysis is a 2.7-kb *Bam*HI genomic DNA fragment that contains the entire F90-2 homeo box cross-hybridizing region. (*a*) Tissue autoradiogram of a wild-type embryo undergoing germ band retraction (8–9 hr) after hybridization with the F90-2 probe. Two foci of hybridization are observed: a discrete region within the midgut rudiment, and a site within the CNS just posterior to the brain. (*b*) Dark-field photomicrograph of *a*. (*c*) Horizontal section through the medial region of an 8–9-hr wild-type embryo after hybridization with the F90-2 probe. Labeling to a region within the midgut rudiment is observed. (*d*) Dark-field photomicrograph of *c*. (*e*) Tissue autoradiogram of a 12–14-hr wild-type embryo after hybridization with the F90-2 probe. Strong labeling is observed to the region of the middle midgut where derivatives of the posterior midgut and anterior midgut fuse. (*f*) Dark-field photomicrograph of *e*. (*g*) Tissue autoradiogram of 15–16-hr wild-type embryo. Labeling is observed over the anterior and middle midgut and to a region within the CNS just posterior to the brain. (*h*) Dark-field photomicrograph of *g*. (AMG) Anterior midgut; (Br) brain; (CNS) central nervous system; (HG) hindgut; (MMG) middle midgut; (St) stomadeum; (VC) ventral cord.

gene expression is the colinearity between the physical order of the genes along the chromosome and their sites of expression along the body axis. Lewis (1978) first demonstrated such colinearity for the BX-C loci and their corresponding domains of function. F90-2 appears to be the proximal-most (closest to the centromere) homeo-box-containing region of the ANT-C. Transcripts homologous to F90-2 are detected within the CNS, probably anterior to the site of *Dfd* expression. By mid-embryogenesis, *Dfd* transcripts are principally detected in the S1/S2 region of the subesophageal ganglion (see Fig. 2). The *Scr* locus is distal to *Dfd*, and transcripts homologous to the A40 clone accumulate in a region of the mature embryonic CNS (S2/S3) that is just posterior to the site of *Dfd* expression. *Antp* transcripts are mostly confined to the T1/T2 region of the ventral cord. Similarly, the proximal-most BX-C homeo box locus (*Ubx*) specifies the most anteriorly localized BX-C products within the CNS (T3/A1; Akam 1983; White and Wilcox 1984; Beachy et al. 1985). The more distal *iab-2* and *iab-7* loci specify transcripts that are detected primarily in ventral ganglia A1/A2→A6/A7 and A7/A8, respectively. It is not clear whether such colinearity between the genetic map and spatial domains of expression is critical for the normal regulation of homeotic gene function.

It has been suggested that diverse pathways of segment morphogenesis are established by different combinations of homeotic gene products (Garcia-Bellido 1977; Lewis 1978; Struhl 1982). In particular, Lewis (1978, 1982) has proposed that each embryonic segment primordium in the posterior half of the embryo contains a unique combination of BX-C gene products. According to the Lewis model, there is a sequential activation of BX-C gene expression along the anterioposterior axis of the embryo. Thus, the most anterior segment anlage that is acted upon by the BX-C (T2p/T3a) would contain products encoded by only one BX-C locus (*Ubx*), whereas the posterior-most segment primordium (A8) is expected to contain all BX-C gene products (i.e., *Ubx*, *iab-2*, and *iab-7*).

During extension and retraction of the germ band,

the transcript distribution patterns for the *Ubx, iab-2,* and *iab-7* loci are consistent with the Lewis model (Harding et al. 1985). Progenitors of the metathorax and first abdominal segment contain *Ubx* transcripts but do not display detectable levels of either *iab-2* or *iab-7* transcripts. In contrast, progenitors of the seventh (and possibly eighth) abdominal segment contain transcripts encoded by each of the homeo-box-containing BX-C loci.

Over the course of embryonic development, there is a successive restriction in the spatial limits of transcript accumulation for at least some of the ANT-C and BX-C homeo-box-containing loci. As compared with earlier periods of embryogenesis, advanced-stage embryos show relatively discrete transcript accumulation patterns for each of the ANT-C and BX-C genes that contains a homeo box (Harding et al. 1985). Although not specifically predicted by the Lewis model, this observation is consistent with the demonstration that each ANT-C and BX-C homeotic locus has a discrete primary domain of function.

A central problem in the homeotic control of segment morphogenesis is how each homeo-box-containing ANT-C and BX-C locus comes to be expressed within a spatially restricted region of the developing embryo. At least some homeotic transformations are associated with the stable accumulation of a particular homeotic gene product within inappropriate embryonic segment primordia (see below) (Hafen et al. 1984). It has been previously suggested that hierarchical interactions among homeotic genes might play a role in this spatial restriction of expression (Struhl 1982). In particular, there is genetic evidence that the *Ubx* locus might exert negative control over *Scr* expression during early periods of embryogenesis (Morata and Kerridge 1982; Struhl 1982).

More recent evidence for homeotic cross-regulatory interactions has been obtained by analysis of homeo box transcript distribution patterns in homeotic mutant embryos. *Antp* transcripts have been shown to accumulate stably in the posterior ventral ganglia of advanced-stage embryos that lack all known genes of the BX-C. On this basis, it was suggested that one or more BX-C gene products either directly or indirectly inhibit *Antp* expression (Hafen et al. 1984). Here, we present further support for this proposal.

The altered distribution of *Antp* transcripts in advanced-stage BX-C embryos could be either a direct or an indirect consequence of deleting the BX-C. An example of an indirect effect is that the absence of BX-C allows for the activation of *Antp* expression by a common *trans*-regulatory component that in wild-type embryos preferentially initiates BX-C gene expression. The data presented in Figure 3 are not obviously consistent with this model. BX-C$^-$ embryos display a wild-type *Antp*$^+$ transcript distribution pattern during early developmental stages. Only after the sixth hour of embryogenesis do BX-C$^-$ individuals show a deviation from wild type in the *Antp*$^+$ transcript pattern. It is therefore unlikely that initiation of *Antp* expression is altered in BX-C$^-$ embryos. The altered *Antp*$^+$ pattern observed in Figure 3 (e and f) more likely results from the absence of products encoded by the BX-C, rather than from the physical removal of the complex per se.

The molecular basis for possible cross-regulatory interactions among *Drosophila* homeotic loci is not known. Such interactions might be mediated by the different homeo box protein domains encoded by the various ANT-C and BX-C loci. There appears to be weak structural homology between the homeo domain and known (or suggested) DNA-binding proteins (Laughon and Scott 1984; Shepherd et al. 1984). It is possible that each homeo-box-containing locus inefficiently represses its own expression, thereby permitting relatively high steady-state levels of product to accumulate within its primary domain of function. In contrast, the homeo domain encoded by one locus might more efficiently repress the expression of other homeo-box-containing genes. For example, perhaps only high levels of *Antp* protein products repress *Antp* expression. This would result in the maintenance of high levels of *Antp* product within the primary domain of *Antp* function (i.e., T1/T2 ventral ganglion cells). In embryonic ventral ganglia posterior to T1/T2, *Antp* product accumulation might be hindered by high-affinity binding of BX-C homeo-box-containing proteins to the *Antp* promoter. As a result, *Antp* products do not accumulate at high steady-state levels in the posterior ventral ganglia of wild-type embryos.

Consistent with this proposal is the demonstration that *Antp* RNAs stably accumulate in posterior portions of the developing ventral cord in embryos that lack all genes of the BX-C (Hafen et al. 1984). Moreover, we have shown that the altered distribution of *Antp* transcripts in BX-C$^-$ embryos probably does not result from initiation of *Antp* expression within inappropriate segment primordia (see Fig 3). Rather, the absence of products encoded by one or more BX-C loci appears to be either directly or indirectly responsible for the stable accumulation of *Antp* products in posterior ventral ganglia.

Analysis of homeo box transcript distribution patterns in Polycomb mutant embryos (Pc^-) provides further support for the occurrence of inhibitory cross-regulatory interactions among homeotic loci. Pc^3 homozygous embryos display grossly altered *Antp, Ubx, iab-2,* and *iab-7* transcript distribution patterns such that each locus is ectopically expressed throughout most or all regions of the early embryonic CNS (see Fig. 4; Beachy et al. 1985). The basis for this altered pattern of homeotic gene expression is not known. However, despite the alteration in the *Antp, Ubx, iab-2,* and *iab-7* transcript distribution patterns in Pc^- embryos, it is possible that putative cross-regulatory interactions between homeotic loci occur as for wild type. On the basis of analyzing the *Antp* transcript distribution pattern in BX-C$^-$ embryos (discussed above), it was predicted that ectopic expression of *Ubx, iab-2,* and/or *iab-7* would result in a successive reduction in the levels of *Antp* transcripts during Pc^- embryonic development. By advanced periods of Pc^- embryonic development, *Antp* RNAs are detected at less than one-fifth the level observed in young Pc^- embryos or in the T1/T2 ganglia of advanced-stage wild-type embryos (cf. Figs. 4 and 5).

In a manner similar to the putative negative control of *Antp* expression by one or more BX-C loci, it is possible that *Ubx* expression is influenced by the *iab-2* locus (and possibly *iab-7*). This proposal was based on the observation that *Ubx* transcripts are present at higher levels in the A2→A6 ventral ganglia of $iab\text{-}2^-$ embryos as compared with wild type (K. Harding and M. Levine, unpubl.). Further support for this proposal is presented in Figures 4 and 5. Comparison of these figures reveals that the levels of *Ubx* transcript gradually diminish in the CNS of Pc^- embryos over the course of development. In contrast, *iab-2* and *iab-7* RNAs are maintained at high levels. It is possible that this decline in *Ubx* transcript accumulation results from inhibition of *Ubx* expression by *iab-2* and/or *iab-7* products.

A detailed understanding of possible cross-regulatory interactions among homeotic loci depends on the identification and characterization of all loci that contain a homeo box. Six new homeo box cross-hybridizing genomic DNA sequences have been isolated (see Table 1). It is not known at this time whether any of these loci encode products that interact with ANT-C or BX-C genes. However, the transcript distribution pattern for at least one of these new loci (S67) is consistent with a role in the elaboration of positional information during embryonic development (see Fig. 7).

ACKNOWLEDGMENTS

We thank E.B. Lewis and Rob Denell for fly stocks. This work was funded by grants from the National Institutes of Health (GM-34431-02) and the Searle Scholars Program.

REFERENCES

AKAM, M. 1983. The location of *Ultrabithorax* transcripts in *Drosophila* tissue sections. *EMBO J.* **2:** 2075.

BEACHY, P., D. HELFAND, and D.S. HOGNESS. 1985. Segmental distribution of bithorax complex proteins during *Drosophila* development. *Nature* **313:** 541.

BENDER, W., M. AKAM, F. KARCH, P.A. BEACHY, M. PFEIFER, P. SPIERER, E.B. LEWIS, and D.S. HOGNESS. 1983. Molecular genetics of the bithorax complex in *Drosophila melanogaster*. *Science* **221:** 23.

DENELL, R.E. 1973. Homeosis in *Drosophila*. I. Complementation studies with revertants of *Nasobemia*. *Genetics* **75:** 279.

DUNCAN, I. 1981. Localization of bithorax complex (BX-C) and Antennapedia complex (ANT-C) gene activities along the body axis of *Drosophila*. *Genetics* **100:** 520.

DUNCAN, I. and E.B. LEWIS. 1982. Genetic control of body segment differentiation in *Drosophila*. In *Developmental order: Its origin and regulation*. p. 533. A.R. Liss, New York.

FERRIS, G.F. 1950. External mophology of the adult. In *Biology of* Drosophila (ed. M. Demerec), p. 368. Wiley, New York.

FJOSE, A., W. MCGINNIS, and W.J. GEHRING. 1985. Isolation of a homeo box-containing gene from the *engrailed* region of *Drosophila* and the spatial distribution of its transcripts. *Nature* **313:** 284.

GARBER, R.L., A. KUROIWA, and W.J. GEHRING. 1983. Genomic and cDNA clones of the homeotic locus *Antennapedia* in *Drosophila*. *EMBO J.* **2:** 2027.

GARCIA-BELLIDO, A. 1977. Homeotic and atavic mutations in insects. *Am. Zool.* **17:** 613.

HAFEN, E., M. LEVINE, and W.J. GEHRING. 1984. Regulation of Antennapedia transcript distribution by the bithorax complex in *Drosophila*. *Nature* **307:** 287.

HAFEN, E., M. LEVINE, R.L. GARBER, and W.J. GEHRING. 1983. An improved *in situ* hybridization method for the detection of cellular RNAs in *Drosophila* tissue sections and its application for localizing transcripts of the homeotic *Antennapedia* gene complex. *EMBO J.* **2:** 617.

HARDING, K., C. WEDEEN, W. MCGINNIS, and M. LEVINE. 1985. Spatially regulated expression of homeotic genes in *Drosophila*. *Science* **229:** 1236.

HAZELRIGG, T. and T.C. KAUFMAN. 1983. Revertants of dominant mutations associated with the Antennapedia gene complex of *Drosophila melanogaster*: Cytology and genetics. *Genetics* **105:** 581.

JIMENEZ, F. and J.A. CAMPOS-ORTEGA. 1981. A cell arrangement specific to thoracic ganglia in the central nervous system of the *Drosophila* embryo: Its behavior in homeotic mutants. *Wilhelm Roux's Arch. Dev. Biol.* **190:** 370.

KANKEL, D.R., A. FERRUS, S.H. GAREN, P.J HARTE, and P.E. LEWIS. 1980. The structure and development of the nervous system. In *The genetics and biology of* Drosophila (ed. M. Ashburner and T.S. Wright), vol. 2, p. 295. Academic Press, New York.

KARCH, F., B. WEIFFENBACH, W. BENDER, M. PEIFER, I. DUNCAN, S. CELNECKER, M. CROSBY, and E.B. LEWIS. 1985. The abdominal region of the bithorax complex. *Cell* (in press).

KAUFMAN, T.C., R. LEWIS, and B. WAKIMOTO. 1980. Cytogenetic analysis of chromosome 3 in *Drosophila melanogaster*: The homeotic gene complex in polytene chromosome interval 84 A-B. *Genetics* **91:** 115.

LANGER-SAFER, P.R., M. LEVINE, and D. WARD. 1982. Immunological method for mapping genes on *Drosophila* polytene chromosomes. *Proc. Natl. Acad. Sci.* **79:** 4381.

LAUGHON, A. and M.P. SCOTT. 1984. Sequence of a *Drosophila* segmentation gene: Protein structure homology with DNA-binding proteins. *Nature* **310:** 25.

LAWRENCE, P.A. and P. JOHNSON. 1984. The genetic specification of pattern in a *Drosophila* muscle. *Cell* **36:** 775.

LEVINE, M., E. HAFEN, R.L. GARBER, and W.J. GEHRING. 1983. Spatial distribution of *Antennapedia* transcripts during *Drosophila* development. *EMBO J.* **2:** 2037.

LEWIS, E.B. 1963. Genes and developmental pathways. *Am. Zool.* **3:** 33.

———. 1978. A gene complex controlling segmentation in *Drosophila*. *Nature* **276:** 565.

———. 1982. Control of body segment differentiation in *Drosophila* by the bithorax gene complex. In *Embryonic development*, part 2: *Genetic aspects*, p. 269. A.R. Liss, New York.

LOHS-SCHARDIN, M., C. CREMER, and C. NÜSSLEIN-VOLHARD. 1979. A fate map for the larval epidermis of *Drosophila melanogaster*: Localized cuticle defects following irradiation of the blastoderm with an ultraviolet laser microbeam. *Dev. Biol.* **73:** 239.

MANIATIS, T., R.C. HARDISON, E. LACY, J. LAUER, C. O'CONNELL, D. QUON, D.K. SIM, and A. EFSTRATIADIS. 1978. The isolation of structural genes from libraries of eucaryotic DNA. *Cell* **15:** 687.

MCGINNIS, W., R.L. GARBER, J. WIRZ, A. KUROIWA, and W.J. GEHRING. 1984a. A homologous protein-coding sequence in *Drosophila* homeotic genes and its conservation in other metazoans. *Cell* **37:** 403.

MCGINNIS, W., M.S. LEVINE, , E. HAFEN, , A. KUROIWA, and W.J. GEHRING. 1984b. A conserved DNA sequence in homeotic genes of the *Drosophila* Antennapedia and bithorax complexes. *Nature* **308:** 428.

MORATA, G. and S. KERRIDGE. 1982. The role of position in

determining homeotic gene function in *Drosophila*. *Nature* **300:** 191.

OUWENEEL, W.H. 1976. Developmental genetics of homeosis. *Adv. Genet.* **18:** 179.

POOLE, S., L. KAUVAR, B. DREES, and T. KORNBERG. 1985. The engrailed locus of *Drosophila*: Structural analysis of an embryonic transcript. *Cell* **40:** 37.

POULSON, D.F. 1950. Histogenesis, organogenesis, and differentiation in the embryo of *Drosophila melanogaster* Meigen. In *Biology of* Drosophila (ed. M. Demerec), p. 168. Wiley, New York.

REGULSKI, M., K. HARDING, R. KOSTRIKEN, F. KARCH, M. LEVINE, and W. MCGINNIS. 1985. Homeo box genes of the Antennapedia and Bithorax complexes of *Drosophila*. *Cell* (in press).

SANCHEZ-HERRERO, I. VERNOS, R. MARCO, and G. MORATA. 1985. Genetic organization of *Drosophila bithorax* complex. *Nature* **313:** 108.

SCOTT, M.P. and A.J. WEINER. 1984. Structural relationships among genes that control development: Sequence homology between the *Antennapedia*, *Ultrabithorax* and *fushi tarazu* loci of *Drosophila*. *Proc. Natl. Acad. Sci.* **81:** 4115.

SCOTT, M.P., A.J. WEINER, T.I. HAZELRIGG, B.A. POLISKY, V. PIROTTA, F. SCALENHE, and T.C. KAUFMAN. 1983. The molecular organization of the *Antennapedia* locus of *Drosophila*. *Cell* **35:** 763.

SHEPHERD, J.C.W., W. MCGINNIS, A.E. CARRASCO, M. DE ROBERTIS, and W.J. GEHRING. 1984. Fly and frog homeo domains show homologies with yeast mating type regulatory proteins. *Nature* **310:** 70.

STRUHL, G. 1980. A homeotic mutation transforming leg to antenna in *Drosophila*. *Nature* **292:** 635.

———. 1982. Genes controlling segmental specification in the *Drosophila* thorax. *Proc. Natl. Acad. Sci.* **79:** 7380.

———. 1983. Role of the *esc*$^+$ gene product in ensuring the selective expression of segment-specific homeotic genes in *Drosophila*. *J. Embryol. Exp. Morphol.* **76:** 297.

TEUGELS, E. and A. GHYSEN. 1983. Independence of the numbers of legs and leg ganglia in *Drosophila bithorax* mutants. *Nature* **304:** 440.

WAKIMOTO, B.T. and T.C. KAUFMAN. 1981. Analysis of larval segmentation on lethal genotypes associated with the Antennapedia gene complex in *Drosophila melanogaster*. *Dev. Biol.* **81:** 51.

WAKIMOTO, B.T., F.R. TURNER, and T.C. KAUFMAN. 1984. Defects in embryogenesis in mutants associated with the Antennapedia gene complex of *Drosophila melanogaster*. *Dev. Biol.* **102:** 147.

WHITE, R. and M. WILCOX. 1984. Protein products of the bithorax complex in *Drosophila*. *Cell* **39:** 163.

Expression of the Dorsal Gene

R. STEWARD, L. AMBROSIO, AND P. SCHEDL
Department of Biology, Princeton University, Princeton, New Jersey 08544

The establishment of polarity within the *Drosophila* embryo provides the basis for the elaboration of complex morphogenetic patterns during embryogenesis. This process is initiated during oogenesis and is completed by the blastoderm stage of embryogenesis. The establishment of polarity within the embryo appears to be largely dependent on genes that are expressed during oogenesis, whereas the interpretation of polarity seems to be dependent on genes that are activated in the zygote.

Two groups of maternal-effect genes involved in the establishment of polarity have been identified. The earliest-acting genes seem to be essential for the polarity of the ovarian chamber, and mutations in these genes, e.g., K10 (see Wieschaus 1979) and gurken (T. Schüpbach, pers. comm.), show defects in both eggshell and embryonic polarity. The second group of maternal-effect genes appears to be essential only for the polarity of the embryo. This group contains two classes of genes that appear to function in the establishment of the two major embryonic axes: anterior-posterior and dorsal-ventral. Mutants in one class, the "bicaudal" genes, disrupt the anterior-posterior axis (Nüsslein-Volhard 1977; Mohler and Wieschaus, this volume), whereas lesions in the other class disrupt dorsal-ventral polarity (Nüsslein-Volhard 1979; Anderson and Nüsslein-Volhard 1984b). Once polarity has been established in the early embryo, zygotic genes become activated along the two embryonic axes (cf. Mohler and Wieschaus, this volume). These zygotically active genes appear to be primarily involved in the interpretation of polarity (e.g., anterior-posterior axis: Krüppel [Wieschaus et al. 1984] and runt [Gergen and Wieschaus 1985]; dorsal-ventral axis: snail and twist [Simpson 1983]). Although null mutations in these zygotically active genes also show pattern abnormalities along the two embryonic axes, the overall polarity of the embryo appears to be maintained.

Even though anterior-posterior and dorsal-ventral polarity is clearly evident in the outside envelope of the mature *Drosophila* egg, the organization of components within the oocyte appears to show at most only limited polarity. The oocyte nucleus is located at an anterior-dorsal position, whereas the posterior tip contains a specialized subcellular structure, the polar granules (Mahowald 1972). Hence, it is not clear what signals are ultimately responsible for determining anterior-posterior and dorsal-ventral polarity in the early embryo, and at which point in development (during oogenesis or embryogenesis) the steps in the establishment of these two axes are initiated. To learn more about the mechanisms involved in determining polarity, we have begun studies on dorsal (*dl*), a maternal-effect gene that is essential for the establishment of dorsal-ventral polarity in the embryo (Nüsslein-Volhard 1979).

Females that are homozygous mutant for *dl* produce embryos that fail to establish normal dorsal-ventral polarity, irrespective of the genotype of the father. In the most severe alleles, the embryo completely lacks dorsal-ventral polarity, and the mature embryo consists of a tube of dorsal cuticle. Weaker alleles develop structures derived from more lateral-ventral positions in the blastoderm, e.g., filzkörper, and occasionally bands of ventral setae (Nüsslein-Volhard 1979). The *dl* mutation does not appear to cause readily apparent lesions during either oogenesis or the very beginning of embryogenesis, and the *dl* phenotype is first observed during the formation of the cellular blastoderm about 2.5–3 hours after fertilization. In normal embryos, ventral cell walls are completed 5 minutes before the dorsal cell walls; however, in the *dl* embryos, the cells on both sides are formed at the same time. Interestingly, Nüsslein-Volhard and co-workers (Santamaria and Nüsslein-Volhard 1983; Anderson and Nüsslein-Volhard 1984a) have shown that the *dl* phenotype can be partially rescued by injection of wild-type cytoplasm from cleavage-stage embryos into *dl* hosts of the same age. These findings, as well as injection studies on other maternal-effect dorsalizing mutations (Anderson and Nüsslein-Volhard 1984b), suggest that although the *dl* gene is expressed during oogenesis, the dorsal-ventral polarity may only be fixed after fertilization.

RESULTS

Cloning of the *dl* Region

Genetic studies (R. Steward and C. Nüsslein-Volhard, in prep.) place *dl* within a small deficiency interval on the left arm of the second chromosome. This deficiency interval is bordered on the distal side by the distal breakpoints of *Df(2L)TW119* and *Df(2L)TW137* (36C2-C4) (Wright et al. 1976) and on the proximal side by the distal breakpoint of three deficiencies: *Df(2L)VA18* (36C3, C4-D1), *Df(2L)MH^{55}* (36D1-E1), and *Df(2L)TW330* (Wright et al. 1976 and T.R.F. Wright, pers. comm.). Two inversions, *In(2L)dl^H* and *In(2L)dl^T*, with breakpoints within this 36C interval have been isolated. Both of these inactivate the *dl* gene and presumably have DNA lesions either within or in close proximity to the *dl* locus. The proximal breakpoint of *In(2L)dl^H* maps just to the distal side of 37C1, 2, which contains the gene for dopa decarboxylase (*Ddc*). Since the *Ddc* gene and a surrounding ~100-kb region have been cloned by Hirsh and Davidson (1981), we were able to isolate the *dl* region by first "walking" (Bender et al. 1983) distally from *Ddc* to the *In(2L)dl^H* breakpoint. We then screened a bacteriophage λ library constructed from DNA prepared from flies carrying the *In(2L)dl^H* chromosome and recovered several recombinants that contain the inversion breakpoint. As expected, these recombinants were found to hybridize in situ to two distinct sites, 36C and 37B-C, in wild-type chromosomes. A restriction fragment from one of these recombinants that was derived from the 36C side of the inversion was used to identify a wild-type recombinant (phage D1) from the *dl* region. Starting from this wild-type recombinant, we isolated a series of overlapping λ recombinants spanning an ~100-kb DNA segment surrounding the *In(2L)dl^H* breakpoint.

To localize DNA sequences within this ~100-kb segment (which are essential for normal *dl* gene function), we mapped the breakpoints of the two inversions. As might be expected from our cloning strategy, the *dl^H* breakpoint was located within the first wild-type recombinant (phage D1) isolated in the *dl* region. To identify the *dl^T* breakpoint, we probed *dl^T* DNA with the λ recombinants from the 36C region. These experiments showed that the *dl^T* breakpoint was also contained within recombinant D1. In subsequent experiments, we used restriction fragment probes isolated from the phage D1 to compare the DNA sequence organization of the inversions with that of the wild-type parental chromosome. As shown in Figure 1, these experiments place both inversion breakpoints within a small 2.3-kb *Hin*dIII-*Sst*I restriction fragment.

To identify the *dl* transcription unit, we probed Northern blots of total and poly(A)+ ovarian and early embryonic RNAs with the recombinant phage from our walk. These experiments revealed that the phage D1 contains sequences homologous to a 2.8-kb poly(A)+ RNA. More detailed studies using isolated restriction fragments from D1 as probes indicated that this 2.8-kb poly(A)+ RNA is encoded by a transcription unit more than 8.0 kb in length, which is transcribed proximal-to-distal as indicated in Figure 1. This transcription unit includes the 2.3-kb *Hin*dIII-*Sst*I fragment that contains both inversion breakpoints (see Fig. 1). In addition, we found that the 2.8-kb poly(A)+ RNA is missing in ovaries prepared from females carrying either of the inversion chromosomes (over a deficiency). These results indicate that this transcription unit corresponds to the *dl* gene and that it encodes a 2.8-kb poly(A)+ RNA (Steward et al. 1984).

To define the *dl*-coding region further, we have used the 2.3-kb *Hin*dIII-*Sst*I fragment to screen cDNA libraries constructed from 0-3-hour embryo poly(A)+ RNA (Poole et al. 1985). Several cDNA clones containing inserts of varying sizes were isolated. Using cross-hybridization to genomic recombinants, we used these cDNAs to map the approximate location of introns and exons. As indicated in Figure 1, these experiments have led to the identification of two introns, one of about 1 kb and one of about 5 kb, and have revealed at least one exon upstream (proximal) of the larger intron, which was not detected in our original Northern experiments. Hence, the *dl* transcription unit is likely to be more than 12 kb in length.

Developmental Profile of the *dl* RNA

To examine the developmental profile of the 2.8-kb *dl* transcript, we prepared poly(A)+ RNA from dissected ovaries and staged embryos: 0-2.5-hour (cleavage and early blastoderm), 2.5-5-hour (blastoderm and gastrula), 8-10.5-hour (postgastrula), and 20-24-hour embryos collected at 22°C. Northern blots of these RNAs were then probed with a fragment from the *dl*-coding region or with a recombinant containing the 5C actin gene (Fyrberg et al. 1983). As shown in Figure 2b, high and very nearly constant levels of actin mRNA can be detected at all developmental stages examined.

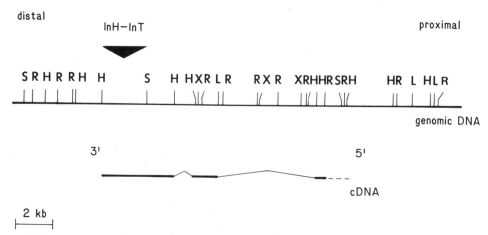

Figure 1. Restriction map of the *dl* region, RNA-coding region, and chromosomal breakpoints. (R) *Eco*RI; (H) *Hin*dIII; (L) *Sal*I; (S) *Sst*I; (X) *Xho*I. (InH) *In(2L)dl^H*; (InT) *In(2L)dl^T*.

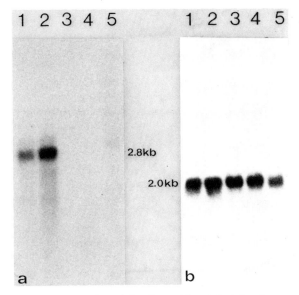

Figure 2. Transcripts of the *dl* gene (*a*) and the 5C actin gene (*b*) during embryonic development. All lanes contain 5 μg of poly(A)⁺ RNA. (*1*) Ovary; (*2*) 0–2.5-hr embryo; (*3*) 2.5–5-hr embryo; (*4*) 8–10.5-hr embryo; (*5*) 20–24-hr embryo.

In contrast, the 2.8-kb *dl* RNA shows a much more restricted developmental profile. Relatively high levels of this RNA are observed in isolated ovaries (Fig. 2a, lane 1) and immediately after fertilization in 0–2.5-hour cleavage and early syncytial blastoderm embryos (Fig. 2, lane 2). However, this RNA cannot be detected in 2.5–5-hour embryos or at the later embryonic stages (Fig. 2, lanes 3–5). These findings suggest that *dl* is probably not expressed in the zygote (at least after 2.5 hr) and that the preexisting *dl* RNA is very rapidly degraded at an early stage of embryogenesis. In other studies, we were unable to detect *dl* transcripts in adult males, and only very low levels were found in total poly(A)⁺ RNA from adult females. This latter finding would suggest that the expression of the *dl* gene in adult females may be restricted to the ovaries.

Expression of the *dl* Gene during Oogenesis

Since genetic studies, as well as our Northern analysis, suggest that *dl* is expressed in the ovaries of adult females, it was of interest to examine the pattern of *dl* RNA accumulation and its localization during oogenesis. The ovary of *Drosophila* consists of about 16 ovarioles (for ovarian development, see King 1970). In the upper part of the ovariole, the free stem cell divides into 16 cytocytes, 15 of which become nurse cells and the 16th, the oocyte. The nurse cells are connected to each other, and four are connected to the oocyte via cytoplasmic bridges, resulting from incomplete cytokinesis during cell division. This group of cells is surrounded by follicle cells and together these form the egg chamber. The subsequent development of the egg chamber has been subdivided into 14 stages (King 1970). Up to stage 6, the egg chamber grows in volume with little change in morphology. At the onset of vitellogenesis (stage 7), the oocyte is localized to the posterior end of the chamber and begins to enlarge, whereas the nurse cells occupy the anterior portion of the chamber. Vitellogenesis lasts from stage 7 to stage 11, and during this period the oocyte enlarges more than a thousandfold in volume. Since the oocyte nucleus is transcriptionally inactive throughout most of oogenesis (Muckenthaler and Mahowald 1966), much of this growth is due to the deposition of yolk proteins, which are produced in the follicle cells and the fat body (Gelti-Douka et al. 1974). In addition, the nurse cells that are polyploid (and in some cases, contain more than 1000 genome equivalents by stage 10a) exhibit quite high levels of RNA and protein synthesis. Much of this newly synthesized RNA and protein is ultimately transported from the nurse cells into the oocyte via the cytoplasmic bridges. At stage 10b-11, the nurse cells empty virtually all of their cytoplasm into the oocyte and then degenerate.

To examine the expression of the *dl* gene during oogenesis, we hybridized in situ a *dl* cDNA recombinant to tissue sections of wild-type *Drosophila* ovaries (Ambrosio and Schedl 1984). The hybridization pattern of this probe to *dl* RNA in several examples of egg chambers at different developmental stages is shown in Figure 3. To analyze the expression and accumulation of *dl* mRNA during oogenesis, we undertook a detailed analysis of the grain distribution in the three different cell types of the ovarian chamber (follicle, nurse, and oocyte) throughout the course of oogenesis. The grain density as a function of the cell type and the developmental stage are plotted in Figure 4B. Taking into account the changes in volume of each cell type as oogenesis proceeds, we then estimated the total amount of *dl* mRNA in the entire ovarian chamber—the nurse cell complex, the oocyte, and the follicle cell complex—at each developmental stage (Fig. 4A) (Ambrosio and Schedl 1984).

As shown in Figure 4B, there is little or no *dl* RNA in the follicle cells throughout oogenesis, and we observe only a low level of grains over these cells, which probably represents background hybridization. In contrast, *dl* RNA can be detected as early as stage 5 in the nurse cells. Although the grain density over the nurse cells remains nearly constant (at least up to about stage 10b; see Figs. 3 and 4B), there is, in fact, a quite substantial accumulation of *dl* RNA in the nurse cells from stage 5 up to stage 11. This accumulation can be seen from our estimates of the relative number of *dl* RNAs in the nurse cells (see Fig. 4A). We have also calculated the number of *dl* RNAs per genome equivalent in the nurse cells at each developmental stage. As shown in Figure 5, this calculation suggests that there is a relatively low and constant level of *dl* RNA transcription from stage 5 to stage 7. The number of *dl* RNAs per genome equivalent then begins to increase gradually between stages 8 and 9. This increase becomes even more pronounced at stages 10a and 10b, and by the end of stage 10b, the number of *dl* RNAs per genome

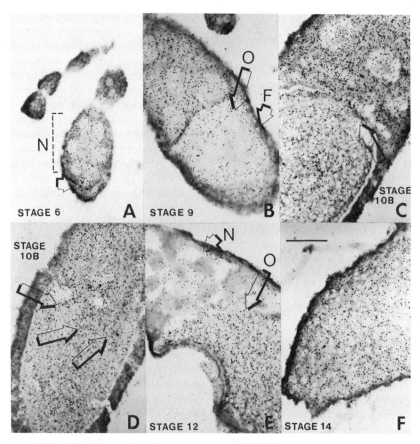

Figure 3. In situ hybridization of *dl* RNA at different stages of egg chamber development. (*A*) Series of early egg chambers; the largest one represents a stage 6 or 7 egg chamber. Arrow indicates intense hybridization over oocyte. (*B*) Late-stage-9 egg chamber. (*C,D*) Two egg chambers at stage 10b. Arrow indicates large quantities of *dl* RNA transported from the nurse cells into the oocyte. Note the high level of hybridization in the nurse cells. (*E*) Stage-12 egg chamber. Note the low level of hybridization left over the disintegrating nurse cell complex and the high grain density over the oocyte. (*F*) Oblique section through stage 14, mature oocyte. (N) Nurse cells; (F) follicle cells; (O) oocyte. Sections are 7 μm thick. Probe was cDNA subcloned into pBR; exposure time was 5 weeks. Bar, 40 μm.

equivalent is more than six times higher than that observed at stage 7. This finding would suggest that there may be a transition in the transcriptional activity of the *dl* gene that begins around stage 8 and reaches a maximum at stage 10b. Interestingly, a similar "induction" around stage 8 has also been observed for a variety of other maternal RNAs (Ambrosio and Schedl 1984).

As shown in Figure 4A, the newly synthesized *dl* RNA appears to be transported from the nurse cells into the growing oocyte. The transport and accumulation of *dl* RNA in the oocyte can first be detected between stages 6 and 7, and this continues through stage 11. At this point, there is a precipitous drop in the number of grains over the nurse cells (see Figs. 3 and 4), and all of the remaining *dl* RNA is extruded from the nurse cells into the oocyte. Moreover, our calculations (see Fig. 4A) suggest that virtually all of the *dl* RNA synthesized in the nurse cells during oogenesis is ultimately transported into the oocyte.

Since the *dl* gene is essential for the establishment of dorsal-ventral polarity in the embryo, it was clearly of interest to determine whether there was an asymmetric distribution of the *dl* transcript in the ovarian chamber. During the period when the nurse cells are expressing the *dl* gene, we were unable to detect any differences in the patterns of *dl* RNA accumulation among the 15 nurse cells. Moreover, analysis of cross sections of mature stage-14 oocytes also failed to reveal an unequal distribution of the *dl* RNA. Since the distribution of *dl* RNA in the mature oocyte is essentially uniform, it seems unlikely that the establishment of dorsal-ventral polarity is dependent on the asymmetric localization of this RNA, at least prior to fertilization. In this regard, it should be pointed out that we have found another maternal RNA species that is specifically localized in the oocyte along the anterior-posterior axis (L. Ambrosio, unpubl.). Hence, although the *dl* RNA is apparently distributed uniformly throughout the egg, mechanisms exist for localizing at least some RNAs to specific regions within the developing oocyte.

DISCUSSION

The in situ hybridization studies presented here indicate that the *dl* gene is expressed in the nurse cells of the ovarian chamber, and this newly synthesized RNA

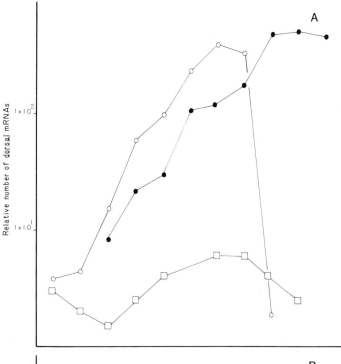
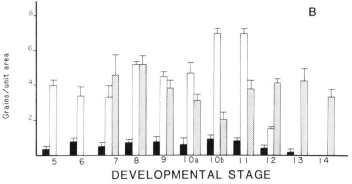

Figure 4. (*A*) Relative number of *dl* mRNAs that accumulate in the nurse cell complex (○), the follicle cell complex (□), the oocyte (■) at stages 5 to 14 of egg chamber development. The numbers are calculated from the mean grains per unit area (see below) multiplied by the estimates of total volume of each complex at the different developmental stages. (*B*) Actual numbers of grains counted per unit area over nurse cells, follicle cells, and the oocyte at stages 5 to 14 of egg chamber development. Each bar represents the mean number of grain counts over four different sections. The error represents standard error of the mean.

is transported from the nurse cells into the growing oocyte. The pattern of *dl* gene expression appears to be divided into two phases. In the first phase, between stages 5 and 7, we observe a relatively low and nearly constant level of *dl* RNA accumulation in the nurse cells. In the second phase, which begins at stage 8, there is a gradual increase in the transcriptional activity of the gene, and *dl* expression appears to reach its maximum at stage 10b. Virtually all of the newly synthesized *dl* RNA is rapidly deposited into the oocyte, where it appears to be stored in a stable form until the end of oogenesis.

Our Northern analysis indicates that substantial amounts of the *dl* RNA are also found in early embryos. It seems likely that this embryonic *dl* RNA represents the maternal RNA synthesized during oogenesis, rather than newly transcribed message in the zygote. First, genetic studies have shown that *dl* is not zygotically rescued by wild-type sperm (Nüsslein-Volhard 1979). Second, molecular studies indicate that there is very little zygotic transcription until the syncytial blastoderm stage (McKnight and Miller 1976;

Zalokar 1976). In fact, by the time high levels of zygotic gene activity are observed, the bulk of the *dl* RNA appears to have been degraded and can no longer be detected.

Although these arguments would suggest that the *dl* gene is not transcribed during embryogenesis, the presence of high levels of maternally derived *dl* RNA in early embryos raises the possibility that this RNA may be translated after fertilization. Zalokar (1976) has shown that translation of maternally derived mRNA commences just after fertilization. If the *dl* RNA is translated at this time, this process must be completed by the time of cellularization, when the *dl* phenotype can first be observed and when the *dl* RNA can no longer be detected.

Injection studies have shown that the *dl* phenotype can be partially rescued by cytoplasm from wild-type cleavage stage and syncytial blastoderm stage embryos. Santamaria and Nüsslein-Volhard (1983) have shown that this rescuing activity is uniformly distributed in wild-type embryos immediately after fertilization but subsequently, during the cleavage stage, becomes local-

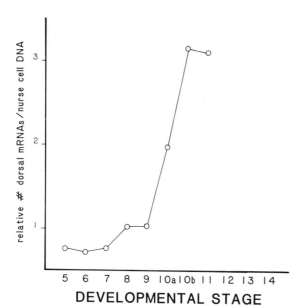

Figure 5. Relative number of *dl* mRNAs in nurse cells as a function of the nurse cell ploidy level at stages 5 to 11 of egg chamber development. The total number of *dl* mRNAs in the nurse cell complex is divided by the number of genome equivalents.

ized to the ventral side. Consistent with this finding, our studies have shown that there is a uniform distribution of *dl* RNA in the mature oocyte. However, it remains to be determined whether this RNA or the *dl* protein product becomes asymmetrically localized (or activated) during early embryogenesis.

ACKNOWLEDGMENTS

We thank Le Nguyen for technical help and Larry Kauvar for sending us his cDNA libraries. This work was supported by grants from the National Institutes of Health and the March of Dimes Birth Defect Foundation.

REFERENCES

AMBROSIO, L. and P. SCHEDL. 1984. Gene expression during *Drosophila melanogaster* oogenesis; analysis by *in situ* hybridization to tissue sections. *Dev. Biol.* **105**: 80.

ANDERSON, K. and C. NÜSSLEIN-VOLHARD. 1984a. Genetic analysis of dorsal-ventral embryonic pattern in *Drosophila*. In *Primers in developmental biology* (ed. G. Malacinski and S. Bryant), p. 269. MacMillan, New York.

―――. 1984b. Information for the dorsal-ventral pattern of *Drosophila* is stored as maternal mRNA. *Nature* **311**: 223.

BENDER, W., P. SPIERER, and D.S. HOGNESS. 1983. Chromosomal walking and jumping to isolate DNAs from the *Ace* and *rosy* loci and the bithorax complex in *Drosophila melanogaster*. *J. Mol. Biol.* **168**: 17.

FYRBERG, E.A., J.W. MAHAFFEY, B.J. BOND, and N. DAVIDSON. 1983. Transcripts of the six *Drosophila* actin genes accumlate in a stage- and tissue-specific manner. *Cell* **33**: 115.

GELTI-DOUKA, H., T.R. GINGERAS, and M.P. KAMBYSELLIS. 1974. Yolk proteins in *Drosophila*: Identification and site of synthesis. *J. Exp. Zool.* **187**: 167.

GERGEN, J.P. and E.F. WIESCHAUS. 1985. Localized requirements for a gene affecting segmentation in *Drosophila*: Analysis of larvae mosaic for *runt*. *Dev. Biol.* **109**: 321.

HIRSH, J. and N. DAVIDSON. 1981. Isolation and characterization of the *dopa decarboxylase* gene of *Drosophila melanogaster*. *Mol. Cell. Biol.* **6**: 475.

KING, R.C., ed. 1970. *Ovarian development in* Drosophila melanogaster. Academic Press, New York

MAHOWALD, A.P. 1972. Oogenesis. In *Development systems: Insects* (ed. S.J. Counce and C.H. Waddington), vol. 1, p. 1. Academic Press, New York.

MCKNIGHT, S. and O.L. MILLER. 1976. Ultrastructural patterns of RNA synthesis during early embryogenesis of *Drosophila melanogaster*. *Cell* **8**: 305.

MUCKENTHALER, F.A. and A.P. MAHOWALD. 1966. DNA synthesis in the ooplasm of *Drosophila melanogaster*. *J. Cell. Biol.* **28**: 199.

NÜSSLEIN-VOLHARD, C. 1977. Genetic analysis of pattern formation in the embryo of *Drosophila melanogaster*. *Wilhelm Roux's Arch Dev. Biol.* **183**: 249.

―――. 1979. Maternal effect mutations that alter the spatial coordinates of the embryo of *Drosophila melanogaster*. In *Determination of spatial organization* (ed. S. Subtelney and I.R. Koenigsberg), p. 185. Academic Press, New York.

POOLE, J.S., L.M. KAUVAR, B. DREES, and T. KORNBERG. 1985. The *engrailed* locus of *Drosophila*: Structural analysis of an embryonic transcript. *Cell* **40**: 37.

SANTAMARIA, P. and C. NÜSSLEIN-VOLHARD. 1983. Partial rescue of *dorsal*, a maternal effect mutation affecting the dorsal-ventral pattern of the *Drosophila* embryo, by the injection of wild-type cytoplasm. *EMBO J.* **2**: 1693.

SIMPSON, P. 1983. Maternal zygotic gene interactions involving the dorsal-ventral axis in *Drosophila* embryos. *Genetics* **105**: 615.

STEWARD, R., F. MCNALLY, and P. SCHEDL. 1984. Isolation of the *dorsal* locus in *Drosophila*. *Nature* **311**: 262.

WIESCHAUS, E. 1979. *fs(1)K10*, a female sterile mutation altering the pattern of both the egg coverings and the resultant embryos in *Drosophila*. In *Cell lineage, stem cell, and cell differentiation* (ed. N. Le Douarin), p. 291. Elsevier/North-Holland, New York.

WIESCHAUS, E., C. NÜSSLEIN-VOLHARD, and H. KLUDING. 1984. *Krüppel*, a gene whose activity is required early in the zygotic genome for normal embryonic segmentation. *Dev. Biol.* **104**: 172.

WRIGHT, T.R.F., R.B. HODGETTS, and A.F. SHERALD. 1976. The genetics of dopa decarboxylase in *Drosophila melanogaster* I, isolation and characterization of deficiencies that delete the dopa-decarboxylase-dosage-sensitive region and the α-methyl-dopa-hypersensitive locus. *Genetics* **84**: 267.

ZALOKAR, M. 1976. Autoradiographic study of protein synthesis in *Drosophila*. *Exp. Cell Res.* **19**: 184.

The engrailed Locus of *Drosophila melanogaster*: Genetic, Developmental, and Molecular Studies

Z. ALI, B. DREES, K.G. COLEMAN, E. GUSTAVSON, T.L. KARR, L.M. KAUVAR, S.J. POOLE, W. SOELLER, M.P. WEIR, AND T. KORNBERG
Department of Biochemistry and Biophysics, University of California, San Francisco, California 94143

A fertilized egg divides and differentiates into distinct cell types arranged in recognizable and reproducible patterns. How this spatial organization is achieved is not known, but recent studies in *Drosophila* have been able to relate such patterns of spatial organization to patterns of gene expression. Morphologically, *Drosophila* is organized as a series of metameric segments. Although not obvious from their morphology, each segment is functionally subdivided into an anterior and posterior developmental compartment (García-Bellido et al. 1973; Steiner 1976; Weischaus and Gehring 1976; Lawrence et al. 1979; Kornberg 1981a; Struhl 1981). Thus, the animal is constructed as a series of bands of cells that alternate in terms of compartment type (anterior-posterior-anterior-posterior) from the head to the tail.

The engrailed (*en*) gene of *Drosophila* is essential for the normal development of cells of the posterior compartments but not of their anterior neighbors (Lawrence and Morata 1976; Kornberg 1981a,b; Lawrence and Struhl 1982). Therefore, the cells of the anterior and posterior compartments differ not only with respect to their location, but also with respect to their requirement for the *en* gene. This discontinuous pattern of *en* function correlates precisely with its pattern of transcription. Transcripts of the *en* gene are detectable in only the posterior compartments, in a banded, zebra-like pattern of *en* expressing cells and *en* nonexpressing cells (Kornberg et al. 1985). The metameric organization of the insect obvious in later stages is reflected at early stages in the striped array of *en* gene expression.

en is one of a number of genes that, on the basis of their mutant phenotype, have been implicated in the regulation of key developmental decisions. Three main categories of such regulatory genes have been described: (1) maternal-effect genes, which are expressed during oogenesis and are essential for the normal spatial organization of the egg (see, e.g., Nüsslein-Volhard 1979); (2) segmentation genes, which are involved in determining the number and polarity of the body segments (Jurgens et al. 1984; Nüsslein-Volhard et al. 1984; Wieschaus et al. 1984); and (3) homeotic genes, which specify segment identity (see, e.g., Lewis 1978). Mutations in all of these genes cause severe developmental defects and, in most cases, embryonic lethality.

The *en* gene is essential for normal segmentation and thus can be considered a segmentation gene. Animals lacking *en* function die during embryogenesis; abnormal segment morphology is evident during gastrulation (Kornberg 1981b). During the past few years, the *en* gene has been isolated by recombinant DNA technology (Kuner et al. 1985). Fifteen independently isolated mutations with chromosomal rearrangements in the *en* locus have been localized on the cloned DNA; they define a 70-kbp region of DNA within which chromosome breaks inactivate the locus. In contrast to other large *Drosophila* genes such as Antennapedia (*Antp*) and Ultrabithorax (*Ubx*), which each have transcription units longer than 70 kbp (Garber et al. 1983; Scott et al. 1983; Hogness et al., this volume), the *en* transcription unit is located only in the centromere-proximal third of the locus and is derived from only about 4 kbp (B. Drees and T. Kornberg, in prep.). A full-length cDNA copy of the *en* transcript has been sequenced, and this study has revealed the presence of a 1700-bp potential open reading frame that includes a new class of homeo box (Poole et al. 1985).

RESULTS AND DISCUSSION

The *en* Embryonic Phenotype

During the first hours of embryogenesis, the *Drosophila* embryo undergoes a rapid series of nuclear divisions, generating within 2 hours of fertilization a blastoderm composed of several thousand nuclei distributed evenly over the embryo surface, 25–45 pole cells and 100–200 yolk nuclei. Patterns of transcription are already present in these young embryos and provide the earliest indication of segmental patterning. In situ hybridization has revealed that the transcripts of several genes, including fushi tarazu (*ftz*) and hairy (*h*), are restricted to alternating bands of cells even in precellular blastoderm embryos (Hafen et al. 1984; Ish-Horowicz et al., this volume). How the embryo becomes subdivided into the segmental units is not known, nor is the relationship between the formation of segments and compartments and the role of the *en* locus understood.

Two lines of experimental evidence have been considered pertinent to hypotheses of *en* function: (1) studies of the behavior of *en* mutant clones that had been induced by mitotic recombination have indicated that the *en* gene is essential to maintain the compartmental (and segmental) organization once it has been generated (Lawrence and Morata 1976; Kornberg 1981a) and (2) marked clones of cells induced by mitotic recombination are restricted to either the anterior or posterior compartments as early as such clones can be made (Weischaus and Gehring 1976). For technical reasons, these experiments cannot demonstrate compartment formation earlier than gastrulation and thus have only implicated the *en* locus in maintenance of the compartmental organization, not in its formation.

To understand better the possible role of the *en* locus in the organizational events of the pregastrulating embryo, the phenotypes of wild-type and *en* mutant embryos have been examined. Various hemizygous, homozygous, and *trans*-heterozygous combinations of *en* alleles were examined with a dissecting microscope and with fluorescent probes following formaldehyde fixation. Using both methods, abnormalities of morphogenesis were visible prior to cellularization.

First, the clearing of the peripheral yolk-plasm that results as the nuclei reach the cortical region of the embryo was quite uniform in wild-type embryos, but in *en* mutant embryos, it was visibly uneven (not shown). With this dissecting microscope assay, *en* embryos could be selected with greater than 90% accuracy, as judged from their subsequent development. Second, when blastoderm-stage embryos with nonuniform peripheral cytoplasm were fixed in formaldehyde and stained with the fluorescent DNA-binding dye DAPI, more details of their morphological abnormalities were revealed. In contrast to wild-type embryos, the nuclei on the periphery of the mutant embryo were nonuniformly distributed (Fig. 1), and the internal yolk nuclei were abnormal in terms of their number and location (Fig. 2). Among selected precellular embryos, only the *en* embryos were abnormal. As expected of a phenotype associated with a recessive zygotically expressed gene, such precellular defects were observed in one fourth of the embryos in a cross of heterozygous *en* parents. Defects were never observed among the progeny of wild-type parents or in crosses in which only one parent carried a heterozygous *en* mutation.

The precellular blastoderm phenotype of *en* embryos was unexpected; only maternal-effect mutations have hitherto been observed to cause preblastoderm defects (Rice and Garen 1975). Since the precellular blastoderm phenotype was observed only in genetically *en* embryos and *en* mutations have no maternal effect (Lawrence et al. 1983; Z. Ali and T. Kornberg, unpubl.), it is clear that functioning of the *en* locus is essential for normal morphogenesis in the zygote during precellular stages. We conclude that the *en* function is involved not only in maintaining the developmental compartments once they are established, but also in the initial processes that organize them.

A requirement for early zygotic expression of the *en* locus suggests that maternal information alone is not sufficient to organize the spatial patterns of the early embryo. Indeed, if a gene such as *en* functions by regulating other genes, then it is expected that the zygotic expression of other genetic functions will also be required for normal pattern formation. A preliminary

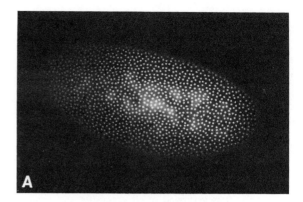

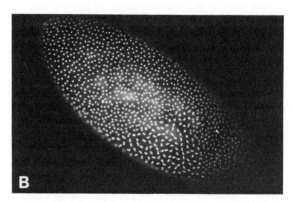

Figure 1. Blastoderm nuclei of wild-type and *en* mutant embryos. Wild-type (*A*) and *en* mutant (*B*) embryos were prepared according to the method of T. Karr and B. Alberts (pers. comm.). Embryos were washed thoroughly with 0.4% NaCl and 0.03% Triton X-100 and transferred to 2.6% sodium hypochlorite (50% clorox) for 90 sec to remove the chorion. Dechorionated embryos were washed with the NaCl/Triton solution and transferred into 5 ml of PBS, an equal volume of heptane, and 1 ml of 20% formaldehyde. After shaking vigorously for 30 sec, the embryos were collected and transferred to a test tube containing, at −70°C, 5 ml of heptane, 4.5 ml of methanol, and 0.5 ml of 0.05 M EGTA. Agitation was carried out for 10 min, whereupon rapid warming to room temperature under a stream of hot water removed the vitelline membrane (Mitchison and Sedat 1983). Devitellinized embryos were rinsed with 90% methanol, rehydrated by passing through methanol/PBS (1:1), stained for 5 min with 1 μg/ml DAPI (4,6 diamino-2-phenylindole), rinsed with PBS for 15 min, and viewed with epi-fluorescence optics. Note the uneven distribution of nuclei in the mutant embryo.

search among several of the mutants known to affect segmentation has revealed that preblastoderm phenotypes were not detected among *ftz* embryos but that *h* embryos had morphological defects similar to, and possibly more extreme than, *en* (data not shown). We conclude that there exists a class of genes that must function during the nuclear cleavage stages in order for the normal segmentation pattern to be formed.

Expression of the *en* Locus

RNA transcripts from the *en* locus have been detected by probing Northern blots of RNA isolated from

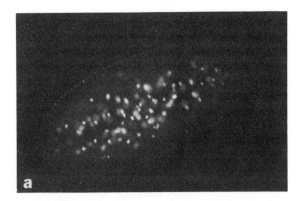
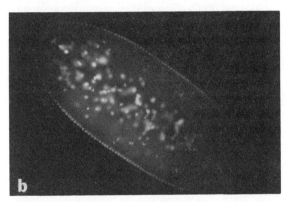

Figure 2. Yolk nuclei of wild-type and *en* mutant embryos. Wild-type (*a*) and *en* mutant (*b*) embryos were prepared as described in Fig. 1 and were photographed under fluorescent epi-illumination. Note the greater number, lower brightness, and abnormal distribution of the nuclei in the mutant embryo.

animals at various developmental periods (B. Drees and T. Kornberg, unpubl.). This analysis has revealed a predominant 2.7-kb transcript. cDNA copies of this transcript have been isolated and sequenced, revealing that the RNA had been processed to remove two intervening sequences (Poole et al. 1985). Nick-translated restriction fragment probes are sufficiently sensitive to detect this transcript in preparations of poly(A)$^+$ RNA during the postblastoderm embryonic periods, during the three larval periods, and during the early pupal period. However, such probes did not detect *en* transcripts in RNA isolated from the blastoderm or preblastoderm stages. Using more sensitive single-stranded probes, a low level of the 2.7-kb *en* transcript was detected in RNA prepared from blastoderm and precellular blastoderm embryos. To date, this transcript has been detected as early as nuclear division stage 10. This result is consistent with the observation that mutant *en* embryos are morphologically abnormal prior to cellularization.

Localization of the cells that express the *en* locus by in situ hybridization has revealed that only the cells of the posterior compartments contain *en* RNA (Kornberg et al. 1985). Nick-translated probes have been unable to detect a pattern of hybridization that can be distinguished above background in embryos prior to cellularization. However, once cellularization of the blastoderm begins, a striking pattern of hybridization is observed (Fig. 3). A band of hybridizing cells one cell wide was found at the region of the future cephalic furrow. Autoradiographic grains were observed in the nuclear and cytoplasmic regions of these cells. With the

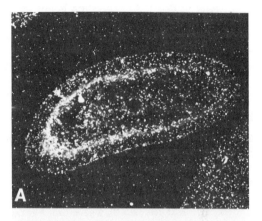
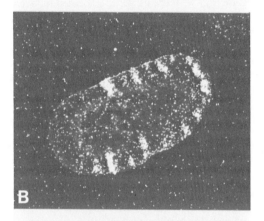

Figure 3. Localization of *en* transcripts in sections of blastoderm stage embryos. Dark-field micrographs of sections of embryos that are cellularizing (*A,B*) or have initiated gastrulation (*C*). Note the single band of hybridization in *A*, the bands of alternating intensity in *B*, and the bands of equal intensity in *C*. In situ hybridization was carried out according to the method of Hafen et al. (1984), except that the embryos were prefixed and devitellinized as described in Fig. 1.

completion of cellularization and the onset of gastrulation, more bands of hybridizing cells were seen, each one cell wide and occurring at regular intervals posterior to the cephalic furrow invagination. The number of autoradiographic grains was not equal among these hybridizing bands, rather they alternate in intensity. Shortly thereafter, as the elongation of the germ band proceeds, the bands of hybridizing cells in each segment were of equal intensity. Although the significance of the nonuniformity of *en* expression is not clear at this time, mutant phenotypes with a periodicity of two segments (Nüsslein-Volhard and Weischaus 1980; Kornberg 1981b) and expression of the *ftz* and *h* genes in two-segment intervals (Hafen et al. 1984; Ish-Horowicz et al., this volume) have been observed.

en-related Genes

DNA sequences related to the *en* gene have been found in *Drosophila* and in other organisms. By hybridizing the full-length *en* cDNA or selected portions of it at lowered stringency to genomic DNA, homologous sequences were detected. In *Drosophila*, invected (*inv*, formerly called *en*-related), a gene that has extensive homology with the *en* gene, is located adjacent to it on the centromere-proximal side (Poole et al. 1985).

Isolation and sequence analyses of several *inv* cDNA clones have indicated that (1) the *inv* gene contains a homeo box that is highly homologous in sequence to the one in the *en* gene, thus defining a class of homeo box distinct from the *Ubx* and *Antp* class; (2) similar to the *en* homeo box, the *inv* homeo box is split by an intervening sequence; and (3) a 29-kbp intron separates the regions that have sequence homology with the *en* gene. In situ hybridization has revealed that expression of the *inv* gene is also restricted to the posterior compartments of the gastrulating embryo (not shown). We conclude that two genes closely related in sequence and pattern of expression are located together at the site of the *en* locus. Evaluation of their relative contributions to the establishment and maintenance of the posterior compartment developmental pathway must await the isolation and characterization of *inv*-gene-defective mutants.

Related sequences in other species were isolated by using probes generated from the homeo-box-containing region of the *en* cDNA. In collaboration with Joyner and Martin (Martin et al., this volume), *en* homologous sequences from mouse DNA were isolated and analyzed. A region of extensive homology with the *en* homeo box and sequences 3' to the homeo box were found. In collaboration with D. Darnell and C. Ordahl (unpubl.), a similarly homologous region was isolated from chicken DNA. Although any conclusions regarding the function of these related genes must await further analysis of their pattern of expression and developmental role, the presence of these related sequences suggests that their function is similarly conserved. If so, then the *en* gene plays a fundamental role in a widely utilized process that subdivides an embryo into smaller developmental units.

ACKNOWLEDGMENTS

This work was supported by a research grant from the National Institutes of Health to T.K., by National Institutes of Health graduate student and postdoctoral training grant support to B.D., K.G.C., E.G., and T.L.K., by a Weingart Foundation fellowship to L.M.K., by National Institutes of Health postdoctoral fellowships to S.J.P. and W.S., and by a fellowship from the Jane Coffin Childs Memorial Fund to M.P.W.

REFERENCES

Garber, R.C., A. Kuroiwa, and W.J. Gehring. 1983. Genomic and cDNA clones of the homeotic locus *Antennapedia* in *Drosophila*. *EMBO J.* **2:** 2027.

Garcia-Bellido, A., P. Ripoll, and G. Morata. 1973. Developmental compartmentalization of the wing disc of *Drosophila*. *Nat. New. Biol.* **245:** 251.

Hafen, E., A. Kuroiwa, and W. Gehring. 1984. Spatial distribution of transcripts from the segmentation gene *fushi tarazu* during *Drosophila* development. *Cell* **37:** 833.

Jurgens, G., E. Weischaus, C. Nüsslein-Volhard, and H. Kluding. 1984. Mutations affecting the pattern of the larval cuticle in *Drosophila melanogaster*. II. Zygotic loci on the third chromosome. *Wilhelm Roux's Arch. Dev. Biol.* **193:** 283.

Kornberg, T. 1981a. engrailed: A gene controlling compartment and segment formation in *Drosophila*. *Proc. Natl. Acad. Sci.* **78:** 1095.

———. 1981b. Compartments in the abdomen of *Drosophila* and the role of the *engrailed* locus. *Dev. Biol.* **86:** 363.

Kornberg, T., I. Sidèn, P. O'Farrell, and M. Simon. 1985. The *engrailed* locus of *Drosophila*: In situ localization of transcripts reveals compartment-specific expression. *Cell* **40:** 45.

Kuner, J., M. Nakanishi, Z. Ali, B. Drees, E. Gustavson, J. Thies, L. Kauvar, T. Kornberg, and P. O'Farrell. 1985. The *engrailed* locus of *Drosophila melanogaster*: Molecular cloning. *Cell* **42:** (in press).

Lawrence, P. and G. Morata. 1976. Compartments in the wing of *Drosophila*: A study of the *engrailed* gene. *Dev. Biol.* **50:** 321.

Lawrence, P. and G. Struhl. 1982. Further studies of the *engrailed* phenotype in *Drosophila*. *EMBO J.* **1:** 827.

Lawrence, P., P. Johnston, and G. Struhl. 1983. Different requirements for homeotic genes in the soma and germ line of *Drosophila*. *Cell* **40:** 37.

Lawrence, P., G. Struhl, and G. Morata. 1979. Bristle patterns and compartment boundaries in the tarsi of *Drosophila*. *J. Embryol. Exp. Morphol.* **51:** 195.

Lewis, E.B. 1978. A gene complex controlling segmentation in *Drosophila*. *Nature* **276:** 565.

Mitchison, T.J. and J. Sedat. 1983. Localization of antigenic determinants in whole *Drosophila* embryos. *Dev. Biol.* **99:** 261.

Nüsslein-Volhard, C. 1979. Maternal effect mutations that alter the spatial coordinates of the embryo of *Drosophila melanogaster*. In *Determinants of spatial organization* (ed. S. Subtelny and I.R. Konigsberg), p. 185. Academic Press, New York.

Nüsslein-Volhard, C. and E. Wieschaus, 1980. Mutations affecting segment number and polarity. *Nature* **287:** 795.

Nüsslein-Volhard, C., E. Weischaus, and H. Kluding. 1984. Mutations affecting the pattern of the larval cuticle in *Drosophila melanogaster* I. Zygotic loci on the second chromosome. *Wilhelm Roux's Arch. Dev. Biol.* **193:** 267.

Poole, S., L.M. Kauvar, B. Drees, and T. Kornberg. 1985. The *engrailed* locus of *Drosophila*: Structural analysis of an embryonic transcript. *Cell* **40:** 37.

Rice, T.B. and A. Garen. 1975. Localized defects of blastoderm formation in maternal effect mutants of *Drosophila*. *Dev. Biol.* **43:** 277.

Scott, M.P., A.J. Weiner, T.I. Hazelrigg, B.A. Polisky, V. Perotta, F. Scalenghe, and T.L. Kaufman. 1983. The molecular organization of the *Antennapedia* locus of *Drosophila*. *Cell* **35:** 763.

Steiner, E. 1976. Establishment of compartments in the developing leg imaginal discs of *Drosophila melanogaster*. *Wilhelm Roux's Arch. Dev. Biol.* **178:** 233.

Struhl, G. 1981. Anterior and posterior compartments in the proboscis of *Drosophila*. *Dev. Biol.* **84:** 372.

Wieschaus, E. and W. Gehring. 1976. Clonal analysis of primordial disc cells in the early embryo of *Drosophila melanogaster*. *Dev. Biol.* **50:** 249.

Weischaus, E., C. Nüsslein-Volhard, and G. Jurgens. 1984. Mutations affecting the pattern of the larval cuticle in *Drosophila melanogaster III*. Zygotic loci on the X-chromosome and fourth chromosome. *Wilhelm Roux's Arch Dev. Biol.* **193:** 296.

Embryonic Pattern in *Drosophila*: The Spatial Distribution and Sequence-specific DNA Binding of engrailed Protein

P.H. O'FARRELL, C. DESPLAN, S. DiNARDO, J.A. KASSIS, J.M. KUNER,*
E. SHER, J. THEIS, AND D. WRIGHT
Department of Biochemistry and Biophysics, University of California, San Francisco, California 94143

The engrailed gene (*en*) is a regulator of pattern formation in *Drosophila melanogaster*. Here, we summarize immunofluorescence studies that reveal an elaborate spatial and temporal program of expression of the *en* gene product in the early embryo. These results suggest that *en* expression is controlled by a number of other genes. Likely candidates for these genes are other known pattern-forming loci. We propose that a large proportion of the extensive *en* locus (70 kb) contains target sequences for such controlling gene products. We also present results which demonstrate that *en* encodes a sequence-specific DNA-binding activity. This observation suggests that *en* may control selection of developmental pathways by acting as a specific regulator of transcription.

Background

Genetic analysis has lead to the concept that key developmental regulators (the selector genes) are expressed in spatially restricted patterns and act to direct cells toward particular developmental pathways (Garcia-Bellido 1975; Lewis 1978). Two somewhat different classes of selector genes have been identified: the segmentation and homeotic genes. Embryonic expression of at least 22 segmentation genes is required to establish the number and polarity of segments (Nüsslein-Volhard and Wieschaus 1980; Jurgens et al. 1984; Wieschaus et al. 1984; Nüsslein-Volhard et al. 1984). According to mutant phenotypes, these genes are classified into "gap" loci, responsible for a large contiguous region of the embryonic pattern; "pair-rule" loci, responsible for pattern elements spaced at paired-segment intervals; and "segment polarity" loci, responsible for setting up pieces of each segment along the body axis (we consider *en* a member of this last class). The homeotic loci establish identity within each of the segments (e.g., *Ubx* within the bithorax complex; Lewis 1978).

The "selector" gene concept has received dramatic support from recent determinations of the spatial patterns of distribution of specific RNAs (Akam 1983;

Levine et al. 1983; Hafen et al. 1984; Kornberg et al. 1985). For example, consider the genetic predictions and molecular analysis for the *en* gene. Genetic studies demonstrated that embryonic cells are divided into two distinct populations: those contributing to anterior compartments of adult segments and those contributing to posterior compartments of adult segments (Garcia-Bellido et al. 1973; Garcia-Bellido 1975; Morata and Lawrence 1975). The *en* gene is required for posterior cells to maintain their commitment (Morata and Lawrence 1975; Kornberg 1981). The selector gene hypothesis predicts that selective expression of *en* in posterior cells specifies these cells as posterior. Molecular analysis confirmed the predicted spatial pattern of *en* product; *en* RNA and *en* protein accumulate in the posterior part of every segment in the early embryo (Fjose et al. 1985; Kornberg et al. 1985; DiNardo et al. 1985).

Although the above studies have been seminal, they leave many unanswered questions about *en* function in embryonic pattern formation. In particular, how is the pattern of *en* gene expression laid down and how does the *en* gene product direct the developmental pathway taken by *en*-expressing cells? Answers to these questions may contribute to a general understanding of the mechanisms of selector gene function. In the next four sections, we describe spatial and temporal aspects of *en* gene expression and discuss possible mechanisms responsible for this control. In the later sections, we describe and discuss results implicating sequence-specific DNA binding in the functioning of the *en* protein.

Spatial Information in the Early Embryo

The *en* gene is expressed in a periodic pattern of segmental stripes at 2.5 hours of development (see below). Here, we attempt to put this patterned expression into the context of the developing pattern of the whole embryo.

Some aspects of pattern are evident very early. The following examples illustrate that the cellular blastoderm embryo (2.5 hr) has a detailed spatial organization, although it is not visually evident. Transplantation experiments show that cells from different regions of the blastoderm have distinct developmental potentials (Simcox and Sang 1983). Although not specified

---

*Present address: Synergen, 1885 33rd Street, Boulder, Colorado 80301.

to a particular cell type, a blastoderm cell will contribute only to a limited area of the body, i.e., it is positionally specified (Simcox and Sang 1983). Additionally, although all are in their fourteenth cell cycle, the cells of the blastoderm are on different mitotic schedules. During germ band elongation (4-7 hr), they will enter mitosis according to a spatially ordered pattern (Hartenstein and Campos-Ortega 1985; V. Foe, pers. comm.). Because regulation of cell-cycle length involves events that long precede mitosis, this highly programmed pattern of cell division is presumably set in motion much earlier (Pardee et al. 1978). Finally, within minutes of completing cellularization (the beginning of cellular blastoderm), the complex and highly ordered processes of gastrulation and neuroblast formation begin. Clearly, the blastoderm is not a shell of uniform properties.

Molecular patterns are presumed to precede and direct these early events. What is the time lag between molecular specification of pattern information and its readout in terms of morphological pattern? The elaborate pattern in the blastoderm presumably requires an equivalently elaborate molecular pattern at an earlier stage. How elaborate and how early are presently not known.

Molecular probes have detected localized accumulation of zygotic segmentation gene products prior to cellular blastoderm. A gap gene, Krüppel, is expressed in the central portion of the embryo two or three cell cycles prior to the cellular blastoderm (see Knipple et al. 1985). The striped distributions of two pair-rule gene products, fushi tarazu (*ftz*) and hairy (*h*), develop prior to and during cellularization (Hafen et al. 1984; Ish-Horowicz et al., this volume). About 1 hour after localized expression of Krüppel is detected, the *en* gene product begins to accumulate in stripes. We envision the initiation of *en* expression as another step in a continuous process whereby the code of spatial information is further refined. The molecular pattern must be intricate and its development must be rapid if it is to keep pace and direct the extraordinarily rapid and complex development of morphological pattern that ensues.

en Protein Accumulates in a Pattern of Segmental Stripes

Two nonoverlapping sements of the *en*-coding region (391 and 505 bp) were expressed as fusion proteins in *Escherichia coli* (Fig. 1). Rabbit polyclonal antibodies were prepared and affinity-purified. In immunofluorescent studies, these sera specifically detect an *en* gene product. The *en* protein is nuclear and accumulates to high levels in a subset of cells in the embryo (DiNardo et al. 1985).

The most dramatic feature of the pattern of *en* expression is the early accumulation of *en* protein in a series of stripes that transect the anterior-posterior axis of the embryo. We see the first indications of these stripes immediately after the beginning of gastrulation

Figure 1. Representation of major embryonic coding region. On the basis of analysis of a cDNA (Poole et al. 1985), the major *en*-coding exons have been defined and are represented as horizontal line segments with the spacing of the introns indicated. (■) Nonoverlapping segments of the same open reading frame. These segments were used in the construction of fusion proteins and generation of antisera specific to the *en* protein. (□) Homeo box sequence split by the second intron.

(Fig. 2). The stripes, each a row of intensely staining nuclei, are predominantly one cell wide at this stage. The stripes are separated by about three weakly staining cells. This periodic pattern of *en* expression (15 stripes corresponding to 3 oral, 3 thoracic, and 9 abdominal segments) is established during the 30 minutes following ventral furrow formation. However, these stripes do not appear synchronously but rather in a complex fashion. First, a single stripe, that of the maxillary segment, develops just posterior to the forming cephalic furrow (not shown). The next step reveals a provocative pattern. Stripes representing every second segment (first thoracic, third thoracic, and second abdominal segments) develop an intense fluorescence (Fig. 2a). Only later do the alternating segments (labial, second thoracic, first abdominal, and third abdominal segments) develop a comparable immunofluorescent signal (cf. a and b in Fig. 2). As discussed below, we think it likely that this pattern is a consequence of the regulation of *en* expression by the pair-rule genes. Further complexities in the development of the striped pattern suggest that expression of *en* must respond to additional regulators. For example, the appearance of more posterior stripes (fourth through ninth abdominal segments) lags behind the central region of the embryo (Fig. 2a,b). Finally, each stripe is initially intense ventrally and faint dorsally (Fig. 2b).

By mid-germ band elongation, the 15 positive stripes are more highly fluorescent and appear uniformly intense. The stripes have widened to about three cells and are separated by five to six cells (Fig. 2c). The change in segment width is the combined result of one cell division and cell movements associated with germ band formation and extension. Embryos with an almost fully extended germ band exhibit new domains of *en* expression in the head anlagen (Fig. 2c,d). Rather than simple stripes, four patches of expression are seen anterior to the stripe of the mandibular segment. The spatial program of expression in the head is distinguished by timing and pattern from the earlier establishment of segmental stripes.

During germ band shortening, we can clearly see that the stripes correspond to cells in the posterior region of the now evident segments (Fig. 3a). Furthermore, complex rearrangements have transformed the initially simple segmental divisions of the oral region and of the terminal segments (eighth, ninth, and tenth abdominal segments). Such transformations can mask the segmental origins of the final structures.

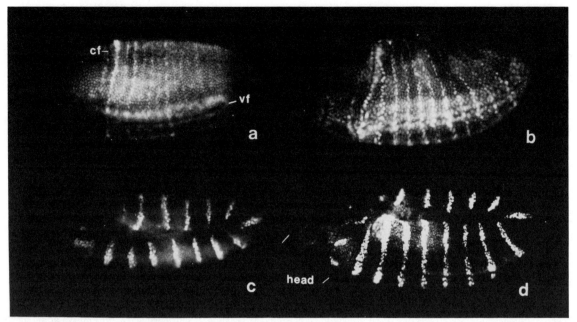

Figure 2. Immunofluorescent detection of *en* protein in whole-mount embryos. Embryos are oriented anterior to the left and more dorsal surface upward. (*a*) Embryo just beginning gastrulation. Ventro-lateral view: (cf) cephalic furrow; (vf) ventral furrow. *en* protein is accumulating in stripes transecting the anterior-posterior axis of the embryo. The strongest signal is from the maxillary stripe, just posterior to the cephalic furrow. Antigen is more prevalent in alternate stripes (maxillary, first and third thoracic, and second abdominal). *en* accumulation posterior to second abdominal segment is delayed. (*b*) Embryo about 15 min older than in *a*. The signal in maxillary through second abdominal segments is equalizing in intensity. Additionally, *en* antigen is now beginning to accumulate posterior to A2, still in an alternate-segment fashion. Note also that there is a stronger signal in the more ventrally located nuclei. (*c*) Embryo at full germ band extension, about 5-6 hr after egg laying. Most ectodermal cells have completed the fourteenth cell cycle. There are nine abdominal, three thoracic, two oral (labial and maxillary), and two head signals evident (the mandibular signal cannot be seen in this view). (*d*) Embryo slightly older than that in *c*. Most ectodermal cells have completed the fifteenth cell cycle. There are a total of four signals in the head region (all showing bilateral symmetry).

Expression of *en* in Neuronal Cells

In the interior of the germ-band-shortened embryo, a few nuclei of the segmental nervous system stain intensely (Fig. 3b). Prior to contraction of the nerve cord, the segmentally repeated pattern of neuronal cells staining for *en* is aligned with the periodic pattern of staining in the ectoderm (Fig. 3c). However, unlike the positively staining segmental stripes in the ectoderm, in the ventral nerve cord, a few scattered brightly staining cells are intermixed with a large number of faint or nonstaining cells. We think that the expression in the nervous system serves functions other than marking cells as members of posterior or anterior compartments. Perhaps expression of *en* in a few specific cells of the nervous system plays a role in specifying their developmental fate and thus the specific interactions among neurons. Detailed studies have described the lineages and interactions of the neuronal cells (Bate 1976; Bate and Grunewald 1981; Thomas et al. 1984). Perhaps identification of *en*-positive cells will reveal commonalities in their genesis and behavior. Here, we emphasize an obvious conclusion. After segmentation is complete, new patterns of *en* expression are established; these indicate additional levels of regulation of *en* gene expression.

A new *en* staining feature on the lateral surface of each abdominal segment of the germ-band-shortened embryo suggests respecification of *en* expression (Fig. 3d). A small spot of *en* staining appears within an area that is otherwise negative for *en* expression. The spots, which correspond to one or a few closely spaced nuclei, are likely to represent peripheral sensor neurons (A. Ghysen et al., in prep.). Since we have not seen movement of *en*-positive cells into the nonstaining region of the segments, we presume that these spots result from induction of new expression.

Regulation of the *en* Gene

Physical mapping of *en* mutations has shown that the genetic complementation unit is at least 70 kb, whereas the coding sequences appear to be included in a primary transcription unit of less than 4.5 kb of sequences (Kuner et al. 1985; Poole et al. 1985). Phenotypes of *en* mutants that disrupt sequences outside the coding unit suggest that much of the flanking sequence has regulatory functions (Kuner et al. 1985; O'Farrell et al. 1985). The pattern and extent of evolutionary conservation of these flanking sequences suggest that numerous regulatory interactions are involved (J. Kassis et al., in prep. and unpubl.). In the preceding section, we described the complex spatial and temporal pattern of *en* gene expression. This overall pattern can be seen as being made up of a number of individual

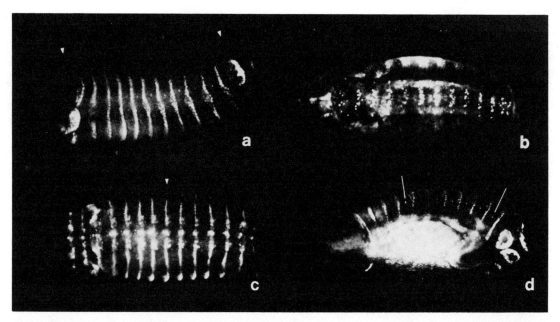

Figure 3. Immunofluorescent detection of *en* protein in embryos at the germ-band-shortened stage. (*a*) Embryo at about 9 hr after egg laying, after germ band shortening. Segmentation is apparent at this stage. The *en* signal is clearly localized to a strip of nuclei at the posterior border of each segment (first thoracic through seventh abdominal). The width of the *en* stripe is not constant along the dorsal-ventral axis, but bulges at the lateral midline. The oral and terminal regions of the embryo have undergone gross rearrangement (arrowheads) and no longer appear as simple segmental divisions. (*b*) Embryo after head involution, about 10–11 hr after egg laying, ventral view of the internal segmental ventral nervous system. Particular nuclei accumulate *en* protein. Due to curvature of the embryo, the focal plane shifts to different depths along the length of the nerve cord. The ectodermal signal has decayed. (*c*) Ventral view of an embryo at a stage similar to *a*. Focal plane is just below surface, so signals in ectodermal nuclei are slightly out of focus near the midline. The expression in the ventral nerve cord aligns with that in the ectoderm (arrowhead marks a particular segment, A2). (*d*) Embryo at same stage as those in *a* and *c*, dorsal view. Displaced dorsally from the lateral midline in segments A1 through A7, one or two nuclei just below the ectoderm accumulate *en* protein. These lateral spots may be in precursors to peripheral neurons.

programs or "subroutines" of expression, i.e., early expression of segmental stripes, patches of expression in the head, and expression in particular neurons of the developing nervous system. We suggest that the regulatory interactions required to execute each subroutine of *en* expression occur in distinct regions of the *en* locus.

What factors direct the pattern of *en* expression? The initial appearance of *en* stripes in an alternating segmental period has suggested to us that *en* expression may be controlled by pair-rule loci (e.g., *ftz*) that are known to be expressed at paired segment intervals along the embryonic axis. This predicts altered patterns of *en* expression in pair-rule mutants. Preliminary results show alterations in even-skipped (*eve*), *ftz*, paired gene (*prd*), and *runt* mutant embryos (S. DiNardo, unpubl.; see also Ingham, this volume). Because the pair-rule genes are expressed in stripes that are a few cells wide, control of *en* expression to produce a stripe a single cell wide must involve additional specificity. We suggest that the domains of expression of pair-rule gene products overlap to uniquely specify bands of cells around the blastoderm. *en* expression would be induced by one combination of pair-rule gene products in odd segments and a different combination in even segments (Fig. 4). We hope to identify developmental genes that act as transcriptional regulators of *en* to test these ideas about the regulatory circuitry and the organization of upstream control sequences.

Genetic studies have indicated that an early embryonic decision to express or not express *en* must be stably transmitted to some cells. However, these studies have focused on requirements for *en* only in adult cuticle. Molecular experiments suggest another level of complexity. We have shown that new patterns of *en* expression develop subsequent to the early subdivision into segmental units. These later patterns of expression must require that some cells reverse an earlier commitment to express or not express the gene. We suggest that the requirement for stable control of *en* expression is confined to the ectoderm and that in other parts of the embryo (such as the nervous system), *en* is free to adopt alternative patterns of expression and to play other regulatory roles at different stages.

Sequence-specific Binding of *en* Fusion Proteins

How might the specifier genes act to execute their control over cell fate? A prevalent notion is that they act by regulating expression of a bank of genes—the cytodifferentiation genes (García-Bellido 1975). Most simply, this could be achieved if the proteins acted as transcriptional regulators. This leads to the further prediction that the selector gene products will be se-

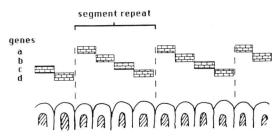

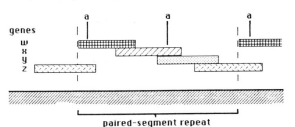

Figure 4. Model for segmental pattern information and the derivation of this pattern from a pair-segment pattern. (*a*) Different segmentation genes are expressed in spatially restricted patterns. Here, it is proposed that at cellular blastoderm, a segmental division is four cells across in the anterior-posterior direction. Expression (indicated by bars) of four genes, *a*, *b*, *c*, and *d*, localized in stripes each a single cell wide could uniquely specify four rings of cells per segment repeat. This is the pattern of expression that we have seen for the *en* gene, which might be considered analogous to *d* in the above schematic. (*b*) Pair-rule genes are also expressed in a spatially restricted pattern. Four pair-rule genes, *w*, *x*, *y*, and *z*, expressed in an overlapping pattern (differently shaded bars in the above schematic) could uniquely specify eight rings of cells per paired segment division. The coarser repeat of the paired-segment pattern could be divided into the segmental repeat if the pair-rule gene products acted in specific combinations to induce expression of particular segmentation genes. Thus, if pair-rule gene products *w* and *z* acted together to induce segmentation gene *a*, it would be expressed in a strip one cell wide in every other segment. To achieve a segmental repeat pattern of expression, gene *a* would also have to be induced by a different combination of pair-rule gene products, here *x* and *y*.

quence-specific DNA-binding proteins. These ideas are supported by analyses of an evolutionarily conserved protein domain, the "homeo domain" (Laughon and Scott 1984; McGinnis et al. 1984b). Several homeotic and segmentation gene products include a homeo domain. This domain is conserved in evolution, even in vertebrate species (Carrasco et al. 1984; McGinnis et al. 1984a). In yeast—it is present in a slightly more divergent but extremely illuminating form—it is within proteins that determine cell fate (mating-type) through transcriptional regulation (Laughon and Scott 1984; Shepherd et al. 1984; Johnson and Herskowitz 1985). On the basis of the weak homology to a protein structural motif that characterizes prokaryotic DNA-binding proteins, it is argued that the homeo domain functions in DNA binding (Laughon and Scott 1984).

We tested the DNA-binding activity of *en*-encoded proteins. Guided by DNA sequence data (Poole et al. 1985; J. Kassis et al., in prep.), we fused the bacterial *lacZ* gene to all or part of the *en* gene (Fig. 5) and expressed the encoded fusion proteins in *E. coli*. To assay the DNA-binding activity of these fusions, we incubated labeled restriction fragments with *E. coli* extracts, precipitated the fusion protein with β-galactosidase antibody, and identified bound DNA fragments as those fragments coprecipitating with the fusion protein (McKay 1981; Johnson and Herskowitz 1985). Extracts from *E. coli* making either β-galactosidase or a fusion of β-galactosidase with the aminoterminal 347 amino acids of the *en* protein (total of 552 amino acids) did not show any DNA-binding activity. In contrast, extracts containing fusions of β-galactosidase to the entire *en*-coding sequence or to the carboxyterminal 143 amino acid residues showed DNA-binding activity. The active fusions contained the entire homeo domain (amino acids 453–513), whereas the inactive fusion lacked these sequences. When the binding is done in low salt and at low DNA concentrations, these fusions bound all DNA fragments nonselectively. However, addition of competitor DNA or high salt eliminated this nonspecific binding and revealed sequence-specific interactions. Figure 6 shows the binding behavior of restriction fragments from a plasmid (p615) carrying a 5-kb *en* insert that is about half coding and half 5′ upstream sequences. Both binding and nonbinding fragments are seen. Using a number of different restriction enzyme digests, we have roughly located five binding sites within the insert sequences of the plasmid. Three of these sites have been localized within 1.2 kb upstream of *en*-coding sequences (C. Desplan et al., in prep.) and preliminary observations suggest that one other site is located further upstream and one site is within the first *en* intron (C. Desplan, unpubl.). Binding sites are also found in some DNAs other than that from *Drosophila*. Several restriction fragments of bacteriophage λ DNA bind with varying efficiency (Fig. 6). Nevertheless, whether the binding is tested with λ DNA or cloned *en* sequences, it is clear that the interaction is specific; binding selects a small subset of the fragments (Fig. 6). Although competition experiments show that the affinity of the *en* fusion proteins for the various sites differs, the range of affinities seen for λ and *en* sites overlap (C. Desplan, unpubl.).

We conclude from these results that *en* encodes a sequence-specific DNA-binding protein. Our binding data do not distinguish the *en* sites from fortuitous sites. Nonetheless, because of their clustering and their location, we propose that the binding sites upstream of the *en* locus are functional and involved in the regulation of expression of *en* transcription. Clearly, functional assays will be required to test this speculation.

A Diverse Family of Regulatory Proteins

A large number of protein-coding sequences containing homology with the homeo domain have re-

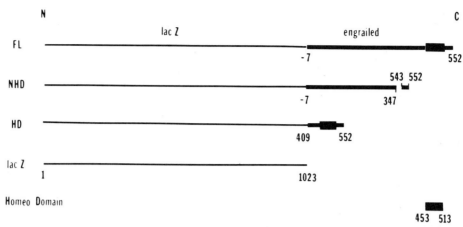

Figure 5. Constructs used to produce fusion proteins for tests of DNA binding. DNA fragments from *en* cDNA and genomic clones (Kuner et al. 1985; Poole et al. 1985) were inserted into the polylinker of expression vector plasmids pUR290, 291, or 292 (Ruther and Muller-Hill 1983). These different plasmids allow inserted DNA to be expressed as a carboxyterminal extension of β-galactosidase. In the full-length fusion protein (FL), the extension includes seven amino acids that precede the first methionine of the *en* protein as well as the entire *en* protein. In the homeo domain construct (HD), a *Bam*HI fragment was spliced out of the FL construction, keeping only the last quarter of the *en*-coding sequence from amino acid 409 to the end. This fusion protein contains the entire homeo domain (amino acids 453–513). Amino acids 347–542, containing the homeo domain, have been deleted in the non-homeo-domain protein (NHD) by removing a *Xho*I fragment from the FL construct. In the absence of insertion, the *lacZ* gene is expressed as the complete β-galactosidase.

cently been identified. Proposals for functional relationships among this family of sequences have been based on either biological or physical roles. According to one proposal, based on shared "biological roles," the related proteins define the developmental fate of segmental units (Lewis 1978; McGinnis et al. 1984b; Scott and Weiner 1984). Another proposal suggests that the related proteins are involved in similar physical interactions, i.e., they all bind to DNA (Laughon and Scott 1984; McGinnis et al. 1984b).

The number of homeo-domain-containing proteins is dependent on criteria used to define homology. At relaxed criteria, the family of related proteins includes yeast proteins, a large number of *Drosophila* proteins, and undefined vertebrate sequences. This "extended family" encompasses many more genes than the original family of related genes whose homology defined the homeo domain sequence (McGinnis et al. 1984c; Scott and Weiner 1984). In the earlier experiments, relatively stringent hybridization conditions identified highly homologous sequences. These were eventually mapped to seven loci, Antennapedia (*Antp*), *ftz*, Sex combs reduced (*Scr*), and Deformed (*Dfd*) in the Antennapedia complex and Ultrabithorax (*Ubx*), abdominal A (*abd-A*), and abdominal B (*Abd-B*) in the bithorax complex (Harding et al. 1985). In addition, a number of sequences identified in vertebrate DNA share extraordinary homology with the defining homeo box sequences. Thus, a "core family" would include the closely related *Drosophila* and vertebrate sequences.

McGinnis and colleagues (1984b,c) noted that the *Drosophila* members of the core family are clustered at two major complexes encoding selector genes. Apparently, it is not only the sequence that is conserved among these genes—they also play similar "biological roles." Thus, the name homeo box was proposed, and the hypothesis was advanced that the vertebrate sequences that were members of the core family would act as specifiers of developmental pathways in vertebrate development.

Whatever biological role a homeo-box-containing gene might serve, the homeo domain must be involved in similar physical interactions since its sequence is so highly conserved. It has been proposed that one of these interactions is DNA binding. It should be emphasized that the data and the arguments favoring this proposal are derived from members of the "extended family." As we describe here, *en* fusion proteins bind DNA. Previous work (Johnson and Herskowitz 1985) had shown that the yeast protein α2 is a sequence-specific DNA-binding protein and a regulator of transcription. Both *en* and α2 have unambiguous homology with the homeo box sequences of the core family but are more diverged than the core members themselves (Laughon and Scott 1984; Shepherd et al. 1984; O'Farrell et al. 1985; Poole et al. 1985). Thus, we propose that all the members of the extended family possess a sequence that has approximate homology with the homeo domain and that this domain specifies DNA binding. Although a physical interaction is presumed to be conserved among these diverse proteins, we would expect that they might adopt many biological roles.

The core family of homeo-domain-containing sequences is a subset of this extended family. Perhaps as originally proposed, the regulatory roles for this core family will be restricted to related biological roles. Although it is clear that conservation of sequence implies conservation of physical interactions that the sequence is involved in, it is not clear that a conserved regulator would serve the same biological role in very diverse organisms. The selective forces are not known, but it is nevertheless generally found that the regulators, and to some extent their biological roles, are strongly conserved (Tomkins 1975). Perhaps complex interdepend-

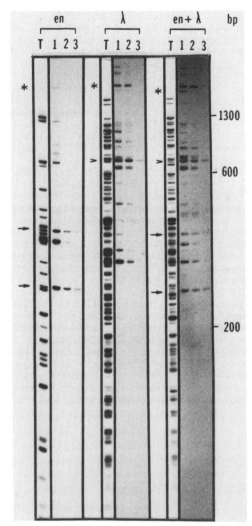

Figure 6. End-labeled restriction fragments of DNA are selectively bound by an *en* fusion protein. DNA from a 10-kb plasmid with a 5-kb insert of *en* DNA (P615; see text) was restricted with *Cla*I and *Hinf*I and end-labeled (en). *Hinf*I and *Cla*I restricted bacteriophage λ DNA was similarly labeled (λ). Lanes designated T show the total pattern of restriction fragments from en, λ, and a mixture of en and λ. The remaining lanes show fragments selectively bound by the HB *en* fusion protein (see Fig. 5 and text). Binding assays displayed in lanes *1*, *2*, and *3* included 0.03, 0.25, and 2 μg of competitor DNA (unlabeled restricted equimolar mixture of en and λ), respectively. Because of the selective displacement of those fragments having lower affinity, the relative efficiency of recovery for a fragment in lanes *1*, *2*, and *3* provides an indication of the binding affinity of the fusion protein for that fragment. The arrows and arrowhead indicate the en fragments and λ fragments, respectively, that have high affinity binding. Some bands (asterisks) are seen in the binding reactions but not in the starting material. These are the result of an uncharacterized modification of DNA fragments.

ence of vital processes results in fixation of fundamental regulatory steps (Tomkins 1975). In any case, further analysis will indicate whether we can properly regard the group of homeo-domain-containing proteins as a diverse group of DNA-binding proteins that is further divisible into close-knit families within which both sequence and biological role are strongly conserved (O'Farrell et al. 1985).

ACKNOWLEDGMENTS

We thank Tim Karr and Marika Walter for advice on immunofluorescence, Sandy Johnson for valuable discussions, Bruce Alberts for use of facilities, and Judy Piccini for her help with preparation of the manuscript. This work was supported by a National Science Foundation grant (DCB-8418016) and fellowship support from Damon-Runyon (E.S.), National Institutes of Health Fogarty Fellowship (C.D.), Helen Hay Whitney (S.D.), and National Institutes of Health training grants (J.T. and J.A.K.).

REFERENCES

AKAM, M.E. 1983. The location of *Ultrabithorax* transcripts in *Drosophila* tissue sections. *EMBO J.* **2**: 2075.

BATE, C.M. 1976. Embryogenesis of an insect nervous system. A map of the thoracic and abdominal neuroblasts in *Locusta migratioria*. *J. Embryol. Exp. Morphol.* **35**: 107.

BATE, C.M. and E.B. GRUNEWALD. 1981. Embryogenesis of an insect nervous system. II. A second class of neuron precursor cells and the origin of the intersegmental connectives. *J. Embryol. Exp. Morphol.* **61**:317.

CARRASCO, A.E., W. MCGINNIS, W.J. GEHRING, and E.M. DE ROBERTIS. 1984. Cloning of an *X. laevis* gene expressed during early coding for a peptide region homologous to *Drosophila* homeotic genes. *Cell* **37**: 409.

DINARDO, S., J. KUNER, J. THEIS, and P.H. O'FARRELL. 1985. Development of embryonic pattern in *D. melanogaster* as revealed by accumulation of the nuclear *engrailed* protein. *Cell* (in press).

FJOSE, A., W.J. MCGINNIS, and W.J. GEHRING. 1985. Isolation of a homoeo box-containing gene from the *engrailed* region of *Drosophila* and the spatial distribution of its transcripts. *Nature* **313**: 284.

GARCÍA-BELLIDO, A. 1975. Genetic control of wing disc development in *Drosophila*. *Ciba Found. Symp.* **29**: 161.

GARCÍA-BELLIDO, A., P. RIPOLL, and G. MORATA. 1973. Developmental compartmentalization of the wing disc of *Drosophila*. *Nat. New Biol.* **245**: 251.

HAFEN, E., A. KUROIWA, and W.J. GEHRING. 1984. Spatial distribution of transcripts from the segmentation gene *fushi tarazu* during *Drosophila* embryonic development. *Cell* **37**: 833.

HARDING, K., C. WEDEEN, W. MCGINNIS, and M. LEVINE. 1985. Spatially regulated expression of homeotic genes in *Drosophila*. *Science* **229**: 1236.

HARTENSTEIN, V. and J.A. CAMPOS-ORTEGA. 1985. Fate-mapping in wild-type *Drosophila melanogaster*. I. The spatiotemporal pattern of embryonic cell divisions. *Wilhelm Roux's Arch. Dev. Biol.* **194**: 181.

JOHNSON, A. and I. HERSKOWITZ. 1985. A repressor (*Mat a2* product) and its operator control expression of a set of cell-type specific genes in yeast. *Cell* **42**: 237.

JURGENS, G., E. WIESCHAUS, C. NÜSSLEIN-VOLHARD, and M. KLUDING. 1984. Mutations affecting the pattern of the larval cuticle in *Drosophila melanogaster*. II. Zygotic loci on the third chromosome. *Wilhelm Roux's Arch. Dev. Biol.* **193**: 283.

KNIPPLE, D.C., E. SEIFERT, U.B. ROSENBERG, A. PREISS, and H. JÄCKLE. 1985. Spatial and temporal patterns of *Krüppel* gene expression in early *Drosophila* embryos. *Nature* **317**: 40.

KORNBERG, T. 1981. *engrailed*: A gene controlling compartment and segment formation in *Drosophila*. *Proc. Natl. Acad. Sci.* **78**: 1095.

KORNBERG, T., I. SIDEN, P. O'FARRELL, and M. SIMON. 1985. The *engrailed* locus of *Drosophila*: In situ localization of transcripts reveals compartment-specific expression. *Cell* **40**:45.

KUNER, J.M., M. NAKANISHI, Z. ALI, B. DREES, E. GUSTAVson, J. Theis, L. Kauvar, T. Kornberg, and P.H. O'Farrell. 1985. Molecular cloning of *engrailed*: A gene involved in the development of pattern in *Drosophila melanogaster*. *Cell* **42:**309.

Laughon, A. and M.P. Scott. 1984. Sequence of a *Drosophila* segmentation gene: Protein structure homology with DNA binding proteins. *Nature* **310:** 25.

Levine, M., E. Hafen, R.L. Garber, and W.J. Gehring. 1983. Spatial distribution of *Antennapedia* transcripts during *Drosophila* development. *EMBO J.* **2:** 2037.

Lewis, E.B. 1978. A gene complex controlling segmentation in *Drosophila*. *Nature* **276:** 565.

McGinnis, W., C.P. Hart, W.J. Gehring, and F. Ruddle. 1984a. Molecular cloning and chromosome mapping of a mouse DNA sequence homologous to homeotic genes of *Drosophila*. *Cell* **38:** 675.

McGinnis, W., R.L. Garber, J. Wirz, A. Kuroiwa, and W.J. Gehring. 1984b. A homologous protein-coding sequence in *Drosophila* homeotic genes and its conservation in other metazoans. *Cell* **37:** 403.

McGinnis, W., M.S. Levine, E. Hafen, A. Kuroiwa, and W.J. Gehring. 1984c. A conserved DNA sequence in homeotic genes of the *Drosophila* Antennapedia and bithorax complexes. *Nature* **308:** 428.

McKay, R. 1981. Binding of a simian virus 40 T-antigen related protein to DNA. *J. Mol. Biol.* **145:** 471.

Morata, G. and P.A. Lawrence. 1975. Control of compartment development by the *engrailed* gene in *Drosophila*. *Nature* **255:** 614.

Nüsslein-Volhard, C. and E. Wieschaus. 1980. Mutations affecting segment number and polarity in *Drosophila*. *Nature* **287:** 795.

Nüsslein-Volhard, C., E. Wieschaus, and H. Kluding. 1984. Mutations affecting the pattern of the larval cuticle in *Drosophila melanogaster*. I. Zygotic loci on the second chromosome. *Wilhelm Roux's Arch. Dev. Biol.* **193:** 267.

O'Farrell, P.H., C. Desplan, S. DiNardo, J. Kassis, J. Kuner, E. Lim, E. Sher, J. Theis, and D. Wright. 1985. Molecular analysis of the involvement of the *Drosophila engrailed* gene in embryonic pattern formation. *Mol. Cell. Biol. New Ser.* **31:** (in press).

Pardee, A.B., R. Dubrow, J.L. Hamlin, and R.F. Kletzein. 1978. Animal cell cycle. *Annu. Rev. Biochem.* **47:** 715.

Poole, S.J., L.M. Kauvar, B. Drees, and T. Kornberg. 1985. The *engrailed* locus of *Drosophila*: Structural analysis of an embryonic transcript. *Cell* **40:** 37.

Ruther, U., and B. Muller-Hill. 1983. Easy identification of cDNA clones. *EMBO J.* **2:** 1791.

Scott, M.P. and A.J. Weiner. 1984. Structural relationships among genes that control development: Sequence homology between the Antennapedia, Ultrabithorax, and fushi tarazu loci of *Drosophila*. *Proc. Natl. Acad. Sci.* **81:** 4115.

Shepherd, J.C.W., W. McGinnis, A.E. Carrasco, E.M. De Robertis, and W.J. Gehring. 1984. Fly and frog homoeo domains show homologies with yeast mating type regulatory proteins. *Nature* **310:** 70.

Simcox, A.A. and J.H. Sang. 1983. When does determination occur in *Drosophila* embryos? *Dev. Biol.* **97:** 212.

Thomas, J.B., M.J. Bastiani, M. Bate, and C.S. Goodman. 1984. From grasshopper to *Drosophila*: A common plan for neuronal development. *Nature* **310:** 203.

Tomkins, G.M. 1975. The metabolic code. *Science* **189:** 760.

Wieschaus, E., C. Nüsslein-Volhard, and C. Jurgens. 1984. Mutations affecting the pattern of the larval cuticle in *Drosophila melanogaster*. III. Zygotic loci on the X-chromosome. *Wilhelm Roux's Arch. Dev. Biol.* **193:** 296.

Homeotic Genes, the Homeo Box, and the Genetic Control of Development

W.J. GEHRING
Biozentrum, University of Basel, CH-4056 Basel, Switzerland

Genomic DNA contains three kinds of basic information: (1) genetic information, which is passed on from generation to generation; (2) a precise developmental program, according to which an organism develops and functions; and (3) historical information, reflecting the evolution of the organism. How the linear information, stored in the sequence of the nucleotides in the DNA, is translated into the three-dimensional structure of an organism (or into four dimensions if we also include time as an additional dimension) is the enigma of developmental genetics. Organisms have their own history, going back millions of years, and develop along historical lines. They are not constructed in the most rational and economic way, although reproduction and development have been optimized to a large extent in the course of evolution. Some aspects of development can only be understood on the basis of their evolutionary history. This historical component sometimes makes it difficult to distinguish the essential aspects, reflecting the basic mechanisms of development, from phenomena resulting from historical serendipity. However, it also provides the opportunity to use DNA segments from one organism to identify and isolate the homologous sequences from another organism, in trying to understand the generalities of different developmental programs.

As early as 1934, T.H. Morgan proposed the *theory of differential gene activity in development*, which assumes that "different batteries of genes come into action as development proceeds." Morgan also proposed that "the initial differences in the protoplasmic regions of the egg are affecting the activity of the genes." The theory of differential gene activity has gained considerable support over the last 50 years, but the factors localized in different regions of the egg cytoplasm have proved to be difficult to identify. They are either considered to be qualitatively different cytoplasmic determinants or graded distributions of morphogens specifying positional information, but their molecular nature remains to be elucidated. Considering the facts that the various developmental processes are interdependent and that the genomes of higher organisms contain on the order of 10^4 to 10^5 genes, it is inconceivable that each gene is regulated individually and independently of the others: Gene activity must be coordinated. Therefore, we have to assume that there are key controlling genes that coordinate and regulate the activity of sets of other genes involved in the construction and maintenance of the organism. The isolation of homeotic genes in *Drosophila* has identified one family of such key controlling genes involved in the spatial organization of the embryo. A DNA sequence element, called the homeo box, which forms a highly conserved part of these genes, has been identified. Using the homeo box as a probe, related genes have been isolated from a variety of higher organisms including vertebrates and man. The homeo box may provide the key to the understanding of the genetic control of development.

Genes Involved in the Control of Development

In *Drosophila*, at least three classes of genes involved in the spatial organization of the embryo have been identified by mutations: maternal-effect genes, segmentation genes, and homeotic gene loci. The egg structure, with its cytoplasmic determinants and positional information, is largely elaborated under the control of nuclear genes during oogenesis in the mother. Therefore, mutations in such genes can be detected by their maternal effect. The structure, polarity, and the spatial coordinates of the egg are controlled by such maternal-effect genes (Gehring 1973; Nüsslein-Volhard 1979; Anderson and Nüsslein-Volhard 1984). Mutagenesis experiments have also identified a large number of segmentation genes that are expressed after fertilization in the zygote and affect the number and polarity of the body segments (see Nüsslein-Volhard and Wieschaus 1980; Jürgens et al. 1984; Nüsslein-Volhard et al. 1984; Wieschaus et al. 1984). Finally, homeotic genes are thought to specify segmental identity, i.e., the kind of segment that is formed. The term homeosis (originally spelled homoeosis) was proposed by Bateson (1894) and refers to the replacement of one structure of the body by a homologous structure from another body segment. Homeotic mutations transform one body part into another, e.g., an antenna into a leg in the case of the Antennapedia (*Antp*) mutation. The antenna belongs to a head segment and is thought to be homologous to a leg, but in some other mutants, it is difficult to prove the homology between the two structures replacing each other. The fact that in homeotic mutants like *Antp*, the antennae are replaced by a pair of middle legs indicates that the normal $Antp^+$ gene ensures that the legs are formed in the proper place, i.e., in the second thoracic segment. This leads to the conclusion that homeotic genes are involved in the control of the spatial organization of the embryo, specifying the basic "architecture."

There are two main clusters of homeotic genes in *Drosophila*, the Antennapedia complex (ANT-C), which appears to control the head and anterior thoracic segments, and the bithorax complex (BX-C), which specifies the more posterior thoracic and abdominal segments. The genetic aspects of the BX-C have been studied with great dedication by Lewis (1964, 1978), who has proposed a model based on his genetic data and on evolutionary considerations. Insects are thought to have evolved from milliped-like ancestors whose body is subdivided into a large number of similar segments, each one carrying a pair of legs. In insects, the legs are confined to the three thoracic segments (T1–T3), and most orders of insects have two pairs of wings, except diptera (including *Drosophila*), which have only one pair of wings on T2, the second pair of wings being reduced to small halteres (on T3). Lewis proposed that the mesothorax (T2) represents the prototype segment, with a pair of legs and a pair of wings, and that the activity of at least one additional homeotic gene of the BX-C is needed for each of the consecutive posterior segments. Inactivation of the genes involved exclusively in the specification of T3 leads to the transformation of T3 to T2, with a second pair of wings replacing the halteres, giving rise to a four-winged fly. Since the ancestors of insects possessed two pairs of wings, this amounts to a "reversal" of the evolutionary process. Deletion of the entire BX-C leads to the formation of mesothoracic structures in all segments posterior to T2, i.e., a reversion to the prototype segment or ground state. Since the genes in the BX-C are arranged in a linear order in the chromosome similar to the order of the segments in which they are expressed, Lewis proposed that homeotic genes may have arisen from a prototype gene by tandem duplication.

The ANT-C is thought to form the extension of the BX-C into the anterior thoracic and the head region (Kaufmann et al. 1980). The anterior segments are more difficult to study, since during evolution, considerable "remodeling" has occurred in the head region. When the ANT-C is also included in the Lewis model, difficulties arise concerning the ground state (Struhl 1982), which remain to be solved. The classic genetic approach has proved highly useful in identifying the relevant genes, but it has serious limitations as far as explaining the underlying mechanisms.

Cloning of the Homeotic Genes and Discovery of the Homeo Box

The molecular approach to the study of homeotic genes became possible when methods for gene cloning were developed that did not require any biochemical information about the nature of the gene products. One of these procedures, chromosomal walking, was pioneered by D.S. Hogness and his group (Bender et al. 1983) in order to clone the BX-C. It involves the progressive isolation of a series of overlapping DNA segments proceeding along the chromosome until the desired locus has been reached. Using similar procedures, we have cloned approximately 300 kb of DNA in the ANT-C (Fig. 1) (Garber et al. 1983; A. Kuroiwa et al., in prep.). The various genes were first mapped on this stretch of DNA by means of deletions and inversion breakpoints. Chromosomal DNA segments were then used as probes for the isolation of cDNA clones from cDNA libraries, and the transcripts were mapped on the chromosomal DNA. *Antp* turned out to be a complex locus spanning over 100 kb, with two major transcripts of 3.4 and 5 kb and at least eight exons (Fig. 1). When mapping the exons represented by the cDNAs on the chromosomal DNA, a faint but significant cross-homology was detected between the 3'-most exon of *Antp* and sequences outside the *Antp* locus, as defined by *Df(3R)9A99*, which leaves *Antp* intact. The cross-hybridizing sequences were found to be associated with a different transcription unit, identified as the fushi tarazu (*ftz*) gene (Hafen et al. 1984a; Kuroiwa et al. 1984). *ftz* has been identified previously by mutations (Wakimoto et al. 1984). In Japanese it means "not enough segments" and refers to the fact that it is a segmentation mutant, lacking portions of alternate body segments. To test whether *Antp* and *ftz* are members of a multigene family, the *Antp* cDNA clone was cut into small fragments, which were subcloned separately and used as probes for whole-genome Southern hybridization experiments under reduced-stringency conditions (McGinnis et al. 1984c). Two kinds of repetitive sequences were detected, the M repeat, which is present several hundred times per genome, and the H repeat, which yields a pattern of seven strong bands and some weaker bands on whole-genome Southern blots. The same bands were detected by both *Antp* and *ftz*, and in both cases, it is the 3' exon that includes the homology region. Next, we performed experiments to test for the presence of the H repeat in other homeotic genes. The 3' exon of the Ultrabithorax (*Ubx*), a homeotic gene of the BX-C kindly provided by P. Spierer (University of Geneva) and D.S. Hogness (Stanford University), turned out to contain another copy of the H repeat. The cross-homology between *Antp*, *ftz*, and *Ubx* was detected independently by Scott and Weiner (1984). The significance of this cross-homology was tested by probing a genomic library with the H repeats of both *Antp* and *ftz*. More than ten additional genes were isolated on the basis of cross-homology, and all of the genes isolated so far are either homeotic or at least involved in the spatial organization of the embryo. Since the H repeat appears to be a characteristic feature of homeotic genes, we named it the homeo box. The term box refers to the fact that it is a short DNA segment of approximately 180 bp (McGinnis et al. 1984b; see below). At least six homeo-box-containing genes have been found in the ANT-C, of which *Antp*, *ftz*, Sex combs reduced (*Scr*) (A. Kuroiwa et al., in prep.), and Deformed (*Dfd*) (McGinnis et al. 1984c and in prep.) have been identified (Fig. 1). Another three

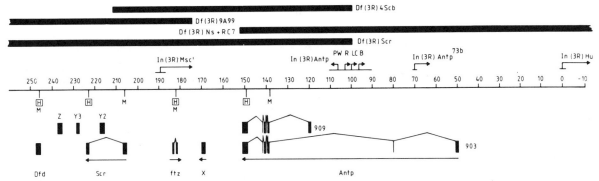

Figure 1. Genomic organization of the ANT-C and localization of homeo boxes. The coordinates are given in kilobases with 0 corresponding to the breakpoint of the *Humeral* inversion, which marks the starting point of the "chromosomal walk." The extent of various deletions (Df) are indicated by the black bars. Arrows indicate the inversion breakpoints (In). (H) Homeo box; (M) M repeat. The genes are indicated at the bottom of the figure, their exons are indicated by closed boxes, the introns by connecting lines. The direction of transcription is indicated. (*Dfd*) Deformed; (*Scr*) Sex combs reduced; (*ftz*) fushi tarazu; (*Antp*) Antennapedia. x, y, and z are transcripts, which remain to be identified. 903 and 909 are two *Antp* cDNA clones representing two different transcripts (Garber et al. 1983; A. Kuroiwa et al., in prep; S. Schneuwly et al., unpubl.). Two more homeo-box-containing genes have been identified within the ANT-C outside the chromosomal segment shown here.

genes, *Ubx*, infra-abdominal-2 (*iab-2*), and infra-abdominal-7 (*iab-7*), have been mapped in the BX-C, corresponding to the three lethal complementation groups identified by Sanchez-Herrero et al. (1985). Besides the two major homeotic gene clusters, engrailed (Fjose et al. 1985; Poole et al. 1985) and caudal, a homeotic gene that is expressed both maternally and zygotically (M. Mlodzik et al., in prep.), have been cloned on the basis of the homeo box homology; several other cloned genes with a homeo box remain to be characterized. The significance of the homeo box cross-homology is supported by the fact that homeo-box-containing genes have subsequently been detected in a variety of higher metazoa, including vertebrates and man (see below).

Comparative DNA Sequence Analysis

DNA sequence data have been obtained for six homeo box regions in *Drosophila*, two in *Xenopus*, one in the mouse, and two in humans. A representative sample is shown in Figure 2a. The homology is confined to approximately 180 bp and ranges from 66% to 79%. In the flanking sequences, the homology drops off rapidly to random levels. In most cases, the homeo box is located in the most distal exon toward the 3' end of the gene, close to an intron-exon boundary. This DNA sequence arrangement is compatible with the gene duplication hypothesis of Lewis (1978), but it fits even better with the concept of exon shuffling.

All known homeo box sequences have the same open reading frame, indicating that the homeo box codes for a domain of a protein, designated as homeo domain. The conceptual translation of the DNA into the protein sequence is shown in Figure 2b, which indicates that, in general, the degree of conservation is higher at the protein than at the DNA level. The most astonishing homology is found between the homeo domain of the frog gene MM3 (Müller et al. 1984) and *Antp*, in which 59 out of 60 amino acid residues are identical.

Considering the fact that vertebrates and invertebrates have evolved separately for more than 500 million years, this high degree of conservation is remarkable.

The homeo domain forms part of much larger proteins and is highly basic, since more than 30% of its amino acids are either lysine or arginine, which is compatible with the idea that the homeotic proteins are DNA-binding or chromatin proteins. This is supported by antibody studies showing that the Ubx^+ proteins are localized in the cell nuclei (White and Wilcox 1984; Beachy et al. 1985). A further hint as to the function of these proteins comes from a computer search in protein sequence data banks. A small, but apparently significant, degree of homology between the homeo domain and amino acid sequences encoded by the *MAT* genes of two species of yeast has been detected (Shepherd et al. 1984). The *MAT* genes of yeast have been shown to control a set of genes involved in mating-type differentiation and sporulation. They code for sequence-specific DNA-binding proteins that bind to those genes they regulate (Johnson and Herskowitz 1985; Miller et al. 1985). The partial homology between the homeo domain and amino acid sequences in the *MAT* proteins suggests that the homeotic proteins might have a similar gene regulatory function. This assumption is supported by the genetic data and some preliminary biochemical data on DNA binding of homeotic fusion proteins. A small amount of homology has also been reported for the *MAT* proteins and prokaryotic gene regulatory proteins (see Pabo and Sauer 1984). It is interesting to note that the highest degree of conservation (Fig. 2b, boxed area) within the homeo box sequences corresponds to the α-helix, which in prokaryotic gene regulatory proteins provides the sequence-specific interaction with the DNA. The only homeo boxes (known so far) that have an amino acid substitution (Arg→Ala at position 43) in this highly conserved region are engrailed and engrailed-related (Fjose et al. 1985; Poole et al. 1985). These two genes

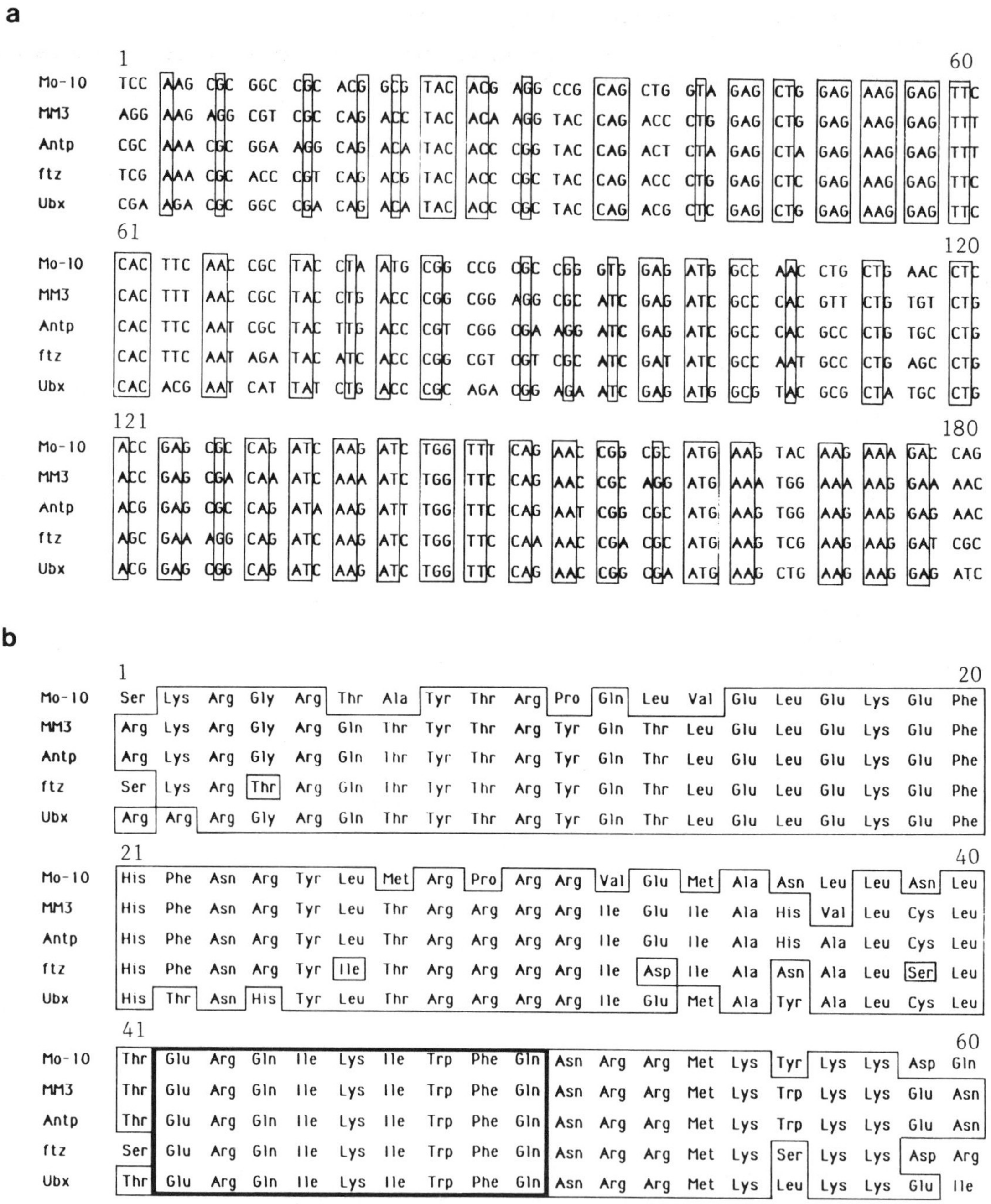

Figure 2. (*a*) DNA sequences of five homeo boxes from mouse (Mo-10), *Xenopus* (MM3), and *Drosophila* (*Antp, ftz, Ubx*). Nucleotides shared by all five sequences are marked. (*b*) Protein sequences of the homeo domains derived by conceptual translation of the common open reading frame. Boxed areas enclose those amino acids that are invariant, taking *Antp* as the standard. (After McGinnis et al. 1984a,b; Müller et al. 1984.)

also differ from the others by having an intron within the homeo box.

Localized Expression of Homeo-box-containing Genes during Embryogenesis

Following fertilization, the *Drosophila* embryo develops as a syncytium of rapidly dividing nuclei. At the 512 nuclear stage, a few nuclei reach the posterior pole of the egg, and a group of pole cells is budded off, which later give rise to the germ cells. After 13 synchronous divisions, most of the remaining nuclei (~6000) have entered the cortical egg cytoplasm, where they become enclosed by plasma membranes, and a monolayer of cells (the blastoderm) is formed. During the early divisions, the nuclei are totipotent and equivalent (Sturtevant 1929), but during cellularization at the blastoderm stage, the cells become determined (Chan and Gehring 1971; Wieschaus and Gehring 1976; Morata and Lawrence 1977). Clones derived from single blastoderm cells are confined to single body segments, indicating that the segmentation pattern is determined during cellularization at the blastoderm stage.

One of the genes involved in the process of segmentation is *ftz* (Wakimoto et al. 1984). Homozygous mutant embryos lack portions of alternate body segments, leading to an embryo with only half the number of segments (Fig. 3). The mutant embryos die prior to hatching from the egg shell. Improved procedures for in situ hybridization of cloned DNA probes to RNA in tissue sections allow us to localize homeotic transcripts and to find out when and where a given gene is expressed (Hafen et al. 1983). The application of these techniques to *ftz*$^+$ gave spectacular results (Hafen et al. 1984a). As shown in Figure 4, the *ftz*$^+$ transcripts accumulate in a pattern of seven belts, three to four cells wide, corresponding in width to the body segments as determined by laser ablation experiments (Lohs-Schardin et al. 1979). Alignment of the pattern of the in situ hybridization data with the blastoderm fate map indicates that the segmental portions missing in the mutant embryo correspond to those regions in the wild-type embryo that express the *ftz*$^+$ gene (Figs. 3 and 6). It should be pointed out that the in situ hybridization pattern is out of phase with respect to the segmental pattern. The domains of *ftz*$^+$ expression have been called parasegments (Martinez-Arias and Lawrence 1985), but at present, we do not know in how many phases the various segmentation genes are expressed, and there may be several overlapping parasegments in the embryo. Examination of the phenotypes of other segmentation mutants suggests that there may be at least three different phases (Nüsslein-Volhard and Wieschaus 1980).

Closer examination of the pattern of in situ hybridization of the *ftz*$^+$ transcripts (see Hafen et al. 1984a) shows that the first distinct labeling is found after the eleventh nuclear division when the nuclei have arrived in the cortical cytoplasm on the surface of the egg. At

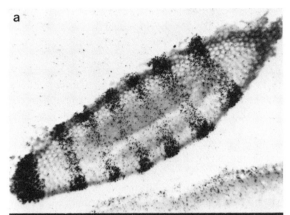
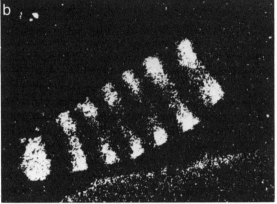

Figure 4. In situ hybridization of a cloned *ftz*$^+$ probe (p523B) to a syncytial wild-type blastoderm (tangential section). (*a*) Bright field: The nuclei appear as white circles in the cortical cytoplasm, the *ftz*$^+$ transcripts are represented by the black silver grains. (*b*) Dark field: The *ftz*$^+$ transcripts form a pattern of seven evenly spaced bands around the embryo. (After Hafen et al. 1984a.)

this stage, the region from 15% to 65% the length of the egg (0% is defined as the posterior pole) is fairly uniformly labeled, and the silver grains are localized over the nuclei, suggesting that these transcripts mark the beginning of transcription. After the following twelfth division, the grains appear mostly over the cytoplasm, forming larger patches in the region from 15% to 65% egg length, and after the thirteenth division, the well-defined pattern of seven stripes is observed. Since this segmental pattern is formed prior to completion of the cell membranes, we propose that the determination of the blastoderm cells involves a direct interaction between the cortical cytoplasm and the immigrating nuclei. The cortical cytoplasm is thought to contain determinants or positional information signals that allow the nuclei to "read" their position within the egg and to respond by activating controlling genes such as *ftz*$^+$ in a position-dependent segmental pattern (Gehring 1984). I have therefore proposed that genes like *ftz*$^+$ might serve as sensors that allow the nuclei to determine their position in the egg and to respond by establishing a segment-specific pattern of gene activity. To examine the mechanisms of this localized gene expression, germ line transformation experiments and

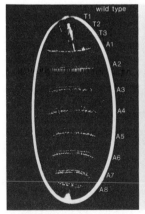
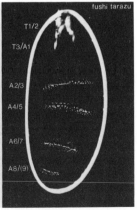

Figure 3. Comparison of wild-type and *ftz* mutant embryos. Segmentation is indicated by the denticle belts shown in dark field. (T1–T3) Thoracic segments; (A1–A8) abdominal segments. The *ftz* mutant embryo lacks portions of alternate segments, and the remaining segments are fused (e.g., A2/3), resulting in only half the number of denticle belts.

gene-fusion studies have been carried out (Y. Hiromi et al., in prep.). DNA segments including the *ftz*⁺ gene were inserted into a P-vector plasmid and injected into fertilized eggs in order to obtain germ line transformation. The transformants were tested for their capacity to complement the *ftz*⁻ mutant phenotype of the recipient. To rescue the mutant phenotype and to obtain viable adults with normal segmentation, a surprisingly large segment of DNA is required. As much as 6 kb of 5'-flanking sequences are needed for complete rescue, indicating that *ftz*⁺ has an unusually large controlling region. When only 3.1-kb of 5'-flanking sequences are used for transformation, the segmentation pattern of the embryos is partially normalized, but the transformed embryos still fail to hatch.

Segment-specific gene expression was analyzed further by constructing a fusion gene consisting of the 5'-flanking sequences of *ftz*⁺ fused to the protein-coding part of the β-galactosidase gene of *Escherichia coli*. The fusion gene was introduced into *ftz*⁺ recipients by P-factor-mediated germ line transformation. Transformants carrying the fusion gene express the bacterial β-galactosidase in a *ftz*⁺-specific segmental pattern of seven stripes. This was shown both at the RNA level by in situ hybridization using β-galactosidase DNA as a probe and at the protein level by means of fluorescent antibodies or activity staining of the enzyme. The protein pattern is detected later than the RNA, and the protein appears to be quite stable. These experiments show that a foreign gene can be expressed under *ftz*⁺ control and that the 5' sequences of *ftz*⁺ contain a morphogenetic DNA element controlling the pattern of gene expression in space and time.

The expression of *ftz*⁺ is confined to early embryonic stages, and by the time segments become morphologically detectable, the gene is no longer active. Studies of genetic mosaics indicate that *ftz*⁺ is not required at later stages of development (Wakimoto et al. 1984).

At late blastoderm and early gastrula stages, another homeo-box-containing gene, engrailed (*en*), becomes expressed (Fjose et al. 1985; Kornberg et al. 1985). Earlier genetic experiments had indicated that the *en*⁺ is specifically required in the posterior compartment of each segment (Morata and Lawrence 1975, 1977; Kornberg 1981). In situ hybridization experiments strongly support this conclusion. During germ band extension, when *en*⁺ is most strongly expressed, a segmental pattern of 14 stripes is observed in that section of the embryo that includes the seven stripes of *ftz*⁺ (Fig. 5). The localizations of the *en*⁺ transcripts as compared with those of *ftz*⁺ are shown in Figure 6. Taken together, these studies indicate that after fertilization, the egg is first subdivided into a series of metameric units corresponding in size to the segments. In the next step, each segment is subdivided into compartments, reflecting the process of stepwise determination.

Whereas *ftz*⁺ and *en*⁺ are involved in the general process of segmentation and compartment formation, a large group of homeotic genes is thought to specify the identity of individual segments. These genes are

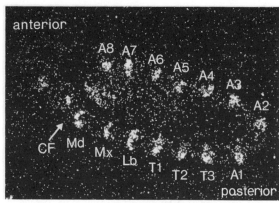

Figure 5. In situ hybridization of a cloned homeo box probe from the *en* locus (pF7036) to a sagittal section through a 6-hr-old embryo. The germ band is extended around the posterior pole. The arrow indicates the cephalic furrow (CF). Segments: (Md) mandibulary; (Mx) maxillary; (Lb) labial segment; (T1–T3) thoracic segments; (A1–A8) abdominal segments. Two additional sites of labeling are detected in the anterior head region and one over the posterior midgut area (Fjose et al. 1985).

preferentially expressed in one or a few segments only. The segment-specific expression is illustrated in Figure 7, showing the localization of the Deformed (*Dfd*⁺) transcripts that at early gastrula stages accumulate in a single belt on either side of the cephalic furrow, presumably corresponding to the mandibulary and maxillary head segments on the fate map. At later stages of development, the segmental localization can be seen most clearly in the ventral nervous system. In an anterior-to-posterior sequence, *Dfd*⁺ is followed by *Scr*⁺, which is most strongly expressed in the labial and first thoracic segments (A. Kuroiwa et al., in prep.), and *Antp*⁺, which is expressed in the second thoracic segment (Fig. 8) (Levine et al. 1983). These three genes of the ANT-C are followed in turn by three homeo-box-containing genes of the BX-C: *Ubx*⁺, which is preferentially expressed in the third thoracic segment and the

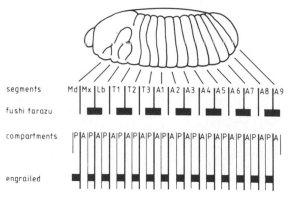

Figure 6. Segmentation and compartment formation in the *Drosophila* embryo and localization of the transcripts of *ftz* and *en*. The segments are labeled as in Fig. 5. (A) Anterior compartment; (P) posterior compartment. (■) Localization of the transcripts.

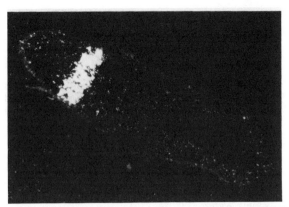

Figure 7. In situ hybridization of a cloned Dfd^+ probe (p93) to an early gastrula embryo (dark field). This tangential section shows that the Dfd^+ transcripts are localized in a single band over the cephalic furrow. (After McGinnis et al. 1984c.)

first abdominal segments (T3 + A1) (Akam 1983), *iab-2* in A2 to A6, and *iab-7* in A7 and A8 (McGinnis et al. 1984c; W. Bender and M. Levine, pers. comm.). On the basis of homeo box homology, we also have identified an additional gene called caudal (*cad*), which is expressed both during oogenesis and zygotically (M. Mlodzik et al., in prep.). This gene has been mapped to the second chromosome, i.e., outside the major homeotic gene clusters, and its transcripts accumulate in the posterior-most abdominal segments. Other genes in this series remain to be identified. In most cases, the transcripts accumulate strongly in one segment and to a lesser extent in the more posterior segments. This is consistent with a model where it is the combination of the various homeotic genes that are expressed in a given segment, which specifies the segmental identity.

The mutual interaction of homeotic genes in determining segmental identity has been studied by localizing the $Antp^+$ transcripts in mutant embryos lacking the entire BX-C or parts of it (Hafen et al. 1984b).

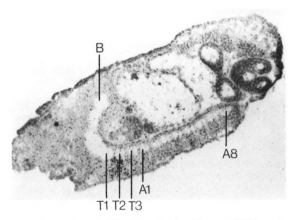

Figure 8. In situ hybridization of a cloned *Antp* cDNA probe (903) to a wild-type embryo (sagittal section). The cells of the ventral nervous system in the second thoracic segment (T2) are preferentially labeled. (B) Brain; (T1-T3) thoracic segments; (A1-A8) abdominal segments. (After Levine et al. 1983.)

Deletion of the BX-C leads to the transformation of all of the segments posterior to T2 into additional second thoracic (T2) segments, except A8. Under these conditions, *Antp* is strongly expressed in all the second thoracic segments, indicating that in the normal embryo, the genes of the BX-C directly or indirectly repress $Antp^+$ in the segments posterior to T2. In contrast to ftz^+, which is expressed only during early embryogenesis, the expression of genes like *Ubx* and *Antp* is required all through larval development at least until metamorphosis, in order to maintain the state of segmental determination (Lewis 1964; Akam 1983; Levine et al. 1983).

Evolutionary Considerations

Using *Drosophila* homeo box sequences as probes, we screened the genomes of a variety of organisms for the presence of homeo-box-homologous sequences by Southern blotting experiments. Under conditions of reduced stringency, similar hybridization patterns were found in other *Drosophila* species and in other insects, such as beetles and the silk worm, where homeotic mutants are known to occur (McGinnis et al. 1984b; A. Kuroiwa, unpubl.). Homologous sequences were also detected in annelids (*Lumbricus* spec.), which are considered to be the ancestors of arthropods and insects. More surprising is the observation that homologous sequences were also detected in vertebrates, including frog, chicken, mouse, and man (McGinnis et al. 1984b). The first vertebrate homeo-box-containing gene was cloned in collaboration with De Robertis and his group (Carrasco et al. 1984), and the sequencing data eliminate the possibility of a hybridization artifact. Subsequently, homeo box sequences were cloned from mouse (McGinnis et al. 1984a) and man (Levine et al. 1984). A more complete evolutionary survey is given by McGinnis (this volume).

The reduced stringency conditions of hybridization used to identify the homologous sequences do not detect homologies of less than approximately 60%. The homology of *MAT* genes of yeast found by computer search (Shepherd et al. 1984) would not be sufficient to be detected by hybridization. The small degree of homology between the homeo domain, the *MAT* proteins, and the gene regulatory proteins of prokaryotes probably reveals some common features of sequence-specific DNA-binding proteins.

DNA-sequencing data, and the conceptual translation into protein sequences, show a remarkable degree of homology among the genes from different species, which reaches over 90% for two human genes and 98% for MM3 of *Xenopus*, when compared with *Antp* in *Drosophila*. This indicates that these sequences are highly conserved during evolution and thus subject to considerable selective pressure, suggesting that the corresponding genes in vertebrates are likely to serve a similar function as those of *Drosophila*. There are in fact mutants in the mouse that might be homeotic in nature, such as *Rachiterata*, which transforms the seventh cervical vertebra into a first thoracic vertebra with a rib (Theiler et al. 1974). The homology is not confined to the homeo box but extends also into the flanking sequences of a pair of genes in mouse and man (Joyner et al. 1985; Rabin et al. 1985). Also, gene organization seems to be similar in *Drosophila* and vertebrates. In many cases, the 5' end of the homeo box is located in the most distal exon, close to the intro-exon boundary, and in both mouse and man, clusters of homeo-box-containing genes similar to those in *Drosophila* have been found. It remains to be shown whether the homeo-box-containing genes in vertebrates serve the same or a similar regulatory function to those of *Drosophila*.

CONCLUDING REMARKS

The homeo box has identified a family of developmental controlling genes specifying the general architecture of the *Drosophila* embryo. DNA sequence comparison suggests that the homeotic genes code for proteins that bind to specific DNA sequences and thereby regulate the activity of those genes which they control. The circuits of genetic control, the interactions between homeotic genes, and the positional information in the egg cytoplasm can now be elucidated. By means of these controlling circuits, the linear information contained in the base sequence of the DNA is converted into the three-dimensional structure of the embryo. Therefore, the homeo box may well provide a key to the understanding of the genetic control mechanisms of development.

ACKNOWLEDGMENTS

I thank all the members of my research group who over many years have contributed to this work. In particular Rick Garber, Atsushi Kuroiwa, Bill McGinnis, Ernst Hafen, Michael Levine, Anders Fjose, Yash Hiromi, Stephan Schneuwly, Johannes Wirz, and Marek Mlodzik, who have been involved in the discovery and analysis of the homeo box. My thanks also go to Eddy De Robertis, Iain Mattaj, John Shepherd, and David Hogness for their friendship and helpful collaboration. The technical assistance of U. Weber, U. Kloter, P. Baumgartner, and E. Wenger-Marquardt is gratefully acknowledged. This work was supported by the Kanton of Basel-Stadt and grants from the Swiss National Science Foundation.

REFERENCES

AKAM, M.E. 1983. The localization of *Ultrabithorax* transcripts in *Drosophila* tissue sections. *EMBO J.* **2**: 2075.
ANDERSON, K.V. and C. NÜSSLEIN-VOLHARD. 1984. Information for the dorsal-ventral pattern of the *Drosophila* embryo is stored as maternal mRNA. *Nature* **311**: 223.
BATESON, W. 1894. *Materiais for the study of variation treated with especial regards to discontinuity in the origin of species.* MacMillan, London.
BEACHY, P.A., S.L. HELFAND, and D.S. HOGNESS. 1985. Segmental distribution of bithorax complex proteins during *Drosophila* development. *Nature* **313**: 545.
BENDER, W., M. AKAM, F. KARCH, P.A. BEACHY, M. PFEIFER, P. SPIERER, E.B. LEWIS, and D.S. HOGNESS. 1983. Molecular genetics of the biothorax complex in *Drosophila melanogaster. Science* **221**: 23.
CARRASCO, A.E., W. MCGINNIS, W.J. GEHRING, and E.M. DE ROBERTIS. 1984. Cloning of an *X. laevis* gene expressed during early embryogenesis coding for a peptide region homologous to *Drosophila* homeotic genes. *Cell* **37**: 409.
CHAN, L.-N. and W.J. GEHRING. 1971. Determination of blastoderm cells in *Drosophila melanogaster. Proc. Natl. Acad. Sci.* **68**: 2217.
FJOSE, A., W.J. MCGINNIS, and W.J. GEHRING. 1985. Isolation of a homeo box-containing gene from the *engrailed* region of *Drosophila* and the spatial distribution of its transcripts. *Nature* **313**: 284.
GARBER, R.L., A. KUROIWA, and W.J. GEHRING. 1983. Genomic and cDNA clones of the homeotic locus Antennapedia in *Drosophila. EMBO J.* **2**: 2027.
GEHRING, W.J. 1973. Genetic control of determination in the *Drosophila* embryo. In *Genetic mechanisms of development* (ed. F.H. Ruddle), p. 103. Academic Press, New York.
———. 1984. Homeotic genes and the control of cell determination. *UCLA Symp. Mol. Cell. Biol. New Ser.* **19**: 3.
HAFEN, E., A. KUROIWA, and W.J. GEHRING. 1984a. Spatial distribution of transcripts from the segmentation gene *fushi tarazu* during *Drosophila* embryonic development. *Cell* **37**: 833.
HAFEN, E., M. LEVINE, and W.J. GEHRING. 1984b. Interaction of homeotic genes in *Drosophila*: Altered pattern of Antennapedia⁺ expression in bithorax mutant embryos. *Nature* **307**: 287.
HAFEN, E., M. LEVINE, R.L. GARBER, and W.J. GEHRING. 1983. An improved *in situ* hybridization method for the detection of cellular RNAs in *Drosophila* tissue sections and its application for localizing transcripts of the homeotic Antennapedia complex. *EMBO J.* **2**: 617.
JOHNSON, A. and I. HERSKOWITZ. 1985. A repressor (MAT α2 product) and its operator control: Expression of a set of cell-type specific genes in yeast. *Cell* (in press).
JOYNER, A.L., R.V. LEBO, Y.W. KAN, R. TJIAN, D.R. COX, and G.R. MARTIN. 1985. Comparative chromosome mapping of a conserved homeo box region in mouse and human. *Nature* **314**: 173.
JÜRGENS, G., E. WIESCHAUS, C. NÜSSLEIN-VOLHARD, and H. KLUDING. 1984. Mutations affecting the pattern of the larval cuticle in *Drosophila melanogaster*. II. Zygotic loci on the third chromosome. *Wilhelm Roux's Arch. Dev. Biol.* **193**: 283.
KAUFMAN, T.C., R. LEWIS, and B. WAKIMOTO. 1980. Cytogenetic analysis of chromosome 3 in *Drosophila melanogaster*. The homeotic gene complex in polytene chromosome interval 84A-B. *Genetics* **94**: 115.
KORNBERG, T. 1981. *engrailed*: A gene controlling compartment and segment formation in *Drosophila. Proc. Natl. Acad. Sci.* **78**: 1095.
KORNBERG, T., I. SIDEN, P. O'FARRELL, and M. SIMON. 1985. The *engrailed* locus of *Drosophila*: *In situ* localization of transcripts reveals compartment-specific expression. *Cell* **40**: 45.
KUROIWA, A., E. HAFEN, and W.J. GEHRING. 1984. Cloning and transcriptional analysis of the segmentation gene *fushi tarazu* of *Drosophila. Cell* **37**: 825.
LAWRENCE, P.A. and G. MORATA. 1977. The early development of mesothoracic compartments in *Drosophila. Dev. Biol.* **56**: 40.
LEVINE, M., G. RUBIN, and R. TJIAN. 1984. Human DNA sequences homologous to a protein coding region conserved between homeotic genes of *Drosophila. Cell* **38**: 667.
LEVINE, M., E. HAFEN, R.L. GARBER, and W.J. GEHRING. 1983. Spatial distribution of Antennapedia transcripts during *Drosophila* development. *EMBO J.* **2**: 2037.
LEWIS, E.B. 1964. Genetic control and regulation of devel-

opmental pathways. In *The role of chromosomes in development* (ed. M. Locke), p. 231. Academic Press, New York.
———. 1978. A gene complex controlling segmentation in *Drosophila*. *Nature* **276:** 565.
LOHS-SCHARDIN, M., C. CREMER, and C. NÜSSLEIN-VOLHARD. 1979. A fate map for the larval epidermis of *Drosophila melanogaster*: Localized cuticle defects following irradiation of the blastoderm with an ultraviolet laser microbeam. *Dev. Biol.* **73:** 239.
MARTINEZ-ARIAS, A. and P.A. LAWRENCE. 1985. Parasegments and compartments in the *Drosophila* embryo. *Nature* **313:** 639.
MCGINNIS, W., C.P. HART, W.J. GEHRING, and F.H. RUDDLE. 1984a. Molecular cloning and chromosome mapping of a mouse DNA sequence homologous to homeotic genes of *Drosophila*. *Cell* **38:** 675.
MCGINNIS, W., R.L. GARBER, J. WIRZ, A. KUROIWA, and W.J. GEHRING. 1984b. A homologous protein-coding sequence in *Drosophila* homeotic genes and its conservation in other metazoans. *Cell* **37:** 403.
MCGINNIS, W., M.S. LEVINE, E. HAFEN, E. KUROIWA, and W.J. GEHRING. 1984c. A conserved DNA sequence in homoeotic genes of the *Drosophila* Antennapedia and bithorax complexes. *Nature* **308:** 428.
MILLER, A.M., V.L. MACKAY, and K.A. NASMYTH. 1985. Identification and comparison of two sequence elements that confer cell-type specific elements in yeast. *Nature* **314:** 598.
MORATA, G. and P.A. LAWRENCE. 1975. Control of compartment development by the *engrailed* gene in *Drosophila*. *Nature* **255:** 614.
———. 1977. Homeotic genes, compartments and cell determination in *Drosophila*. *Nature* **265:** 211.
MORGAN, T.H. 1934. *Embryology and genetics*. Columbia University Press, New York.
MÜLLER, M., A.E., CARRASCO, and E.M. DE ROBERTIS. 1984. A homeo-box-containing gene expressed during oogenesis in *Xenopus*. *Cell* **39:** 157.
NÜSSLEIN-VOLHARD, C. 1979. Maternal effect mutations that alter the spatial coordinates of the embryo. In *Determinants of spatial organization* (ed. S. Subtelny and I. Konigsberg), p. 185. Academic Press, New York.
NÜSSLEIN-VOLHARD, C. and E. WIESCHAUS. 1980. Mutations effecting segment number and polarity in *Drosophila*. *Nature* **287:** 795.
NÜSSLEIN-VOLHARD, C., E. WIESCHAUS, and H. KLUDING. 1984. Mutations affecting the pattern of the larval cuticle in *Drosophila melanogaster*. I. Zygotic loci on the second chromosome. *Wilhelm Roux's Arch. Dev. Biol.* **193:** 267.
PABO, C.O. and R.T. SAUER. 1984. Protein-DNA recognition. *Annu. Rev. Biochem.* **53:** 293.
POOLE, S.J., L.M. KAUVAR, B. DREES, and T. KORNBERG. 1985. The *engrailed* locus of *Drosophila*: Structural analysis of an embryonic transcript. *Cell* **40:** 37.
RABIN, M., C.P. HART, A. FERGUSON-SMITH, W. MCGINNIS, M. LEVINE, and F.H. RUDDLE. 1985. Two homoeo box loci mapped in evolutionarily related mouse and human chromosomes. *Nature* **314:** 175.
SANCHEZ-HERRERO, E., I. VERNOS, R., MARCO, and G. MORATA. 1985. Genetic organization of *Drosophila* bithorax complex. *Nature* **313:** 108.
SCOTT, M.P. and A.J. WEINER. 1984. Structural relationships among genes that control development: Sequence homology between Antennapedia, Ultrabithorax and fushi tarazu loci of *Drosophila*. *Proc. Natl. Acad. Sci.* **81:** 4115.
SHEPHERD, J.C.W., W. MCGINNIS, A.E. CARRASCO, E.M. DE ROBERTIS, and W.J. GEHRING. 1984. Fly and frog homoeo domains show homologies with yeast mating type regulatory proteins. *Nature* **310:** 70.
STRUHL, G. 1982. Genes controlling segmental specification in the *Drosophila* thorax. *Proc. Natl. Acad. Sci.* **79:** 7380.
STURTEVANT, A.H. 1929. The claret mutant type of *Drosophila simulans*: A Study of chromosome elimination and cell-lineage. *Z. Wiss. Zool.* **135:** 323.
THEILER, K., D. VARNUM, and L.C. STEVENS. 1974. Development of Rachiterata, a mutation in the house mouse with 6 cervical vertebrae. *Z. Anat. Entwicklungsgesch.* **145:** 75.
WAKIMOTO, B.T., F.R. TURNER, and T.C. KAUFMAN. 1984. Defects in embryogenesis in mutants associated with the Antennapedia gene complex of *Drosophila melanogaster*. *Dev. Biol.* **102:** 147.
WHITE, R.A.H. and M. WILCOX. 1984. Protein products of the bithorax complex in *Drosophila*. *Cell* **39:** 163.
WIESCHAUS, E. and W. GEHRING. 1976. Clonal analysis of primordial disc cells in the early embryo of *Drosophila melanogaster*. *Dev. Biol.* **50:** 249.
WIESCHAUS, E., C. NÜSSLEIN-VOLHARD, and G. JÜRGENS. 1984. Mutations affecting the pattern of the larval cuticle in *Drosophila melanogaster*. III. Zygotic loci on the X-chromosome and fourth chromosome. *Wilhelm Roux's Arch. Dev. Biol.* **193:** 296.

Common Properties of Proteins Encoded by the Antennapedia Complex Genes of *Drosophila melanogaster*

A. LAUGHON, S.B. CARROLL, F.A. STORFER, P.D. RILEY, AND M.P. SCOTT
Department of Molecular, Cellular, and Developmental Biology, University of Colorado, Boulder, Colorado 80309

Homeotic mutations cause one part of an organism to develop into the likeness of another part (Bateson 1894; Ouweneel 1976). Evidence available to date, primarily from the study of *Drosophila melanogaster*, indicates that homeotic genes act to control development by selecting among alternative developmental pathways or programs. Many of the homeotic genes of *Drosophila* are clustered in two complexes called the bithorax complex (BX-C) (Lewis 1952, 1963, 1978, 1982) and the Antennapedia complex (ANT-C) (Kaufman et al. 1980). Genes of the BX-C regulate pattern formation in part of the thorax and in the abdomen, whereas different genes of the ANT-C function in the head, thorax, and abdomen. Genes in both complexes are commonly described as regulating "segmental identity."

Molecular analysis of the two homeotic gene complexes (Bender et al. 1983; Garber et al. 1983; Scott et al. 1983) has already revealed some of the fine structure of the genes. Certain recurrent features stand out: (1) The genes often have large transcription units and give rise to multiple RNA species (Akam 1983a; Bender et al. 1983; Garber et al. 1983; Scott et al. 1983). (2) Many of the mutations are due to insertions of transposable elements or to chromosome rearrangements, and the character of the mutations may be dependent not only on the location of the mutation, but also on the activity of the transposon or of the newly juxtaposed DNA (Bender et al. 1983; Scott et al. 1983). (3) Many of the homeotic genes share two repetitive sequences, the homeo box (McGinnis et al. 1984c; Scott and Weiner 1984) and a GC-rich repeat that has been named strep, M, and opa (Kidd et al. 1983; McGinnis et al. 1984c; Scott and Weiner 1984; Wharton et al. 1985). Both repeats are found in protein-coding regions of at least some of the ANT-C and BX-C genes. The homeo box encodes a 60-amino-acid structure, the homeo domain, that has some homology with bacterial DNA-binding proteins (Laughon and Scott 1984) and excellent homology with genes of unknown function in higher vertebrates (Carrasco et al. 1984; Levine et al. 1984; McGinnis et al. 1984a,b; Müller et al. 1984). The existence of homologous sequences in the genes of the ANT-C and BX-C can be taken as evidence that the two homeotic complexes constitute related "families" of genes.

Recent work has provided details about the patterns of expression of homeotic genes and has yielded some information about the nature of their encoded products. Localization of transcripts by in situ hybridization of DNA probes has demonstrated that the timing and position of expression of homeotic genes are correlated with their realms of activity as defined by mutant phenotypes (Akam 1983b; Hafen et al. 1983, 1984; Levine et al. 1983; Kornberg et al. 1985). Such results have been further substantiated in the case of the Ultrabithorax (*Ubx*) gene by detection of *Ubx*-encoded protein using immunofluorescent staining of tissues (White and Wilcox 1984; Beachy et al. 1985). *Ubx* protein is located primarily in nuclei, a property consistent with the proposed regulatory role of homeotic gene products. Recent results indicate that a protein encoded by the engrailed (*en*) gene (a homeotic locus unlinked to the ANT-C or BX-C) is localized in the nuclei of embryos undergoing gastrulation (S. DiNardo and P. O'Farrell, pers. comm.).

DNA sequences for two homeo-box-containing genes, fushi tarazu (*ftz*) and *en*, have been reported (Laughon and Scott 1984; Poole et al. 1985). The *ftz* protein contains a high percentage of glutamine, proline, and tyrosine residues, whereas the *en* protein contains runs of glutamine, alanine, serine, and glutamic or aspartic acid. The *Ubx*-encoded protein contains a glycine-rich "hinge" region (Beachy et al. 1985). The significance of stretches of protein sequences rich in single amino acids is unclear.

Previous studies have focused on the organization and expression of the genes of the ANT-C. Here, we present information on the structure and expression of ANT-C proteins. These results are discussed in the context of current models for the function of homeotic genes.

MATERIALS AND METHODS

A full report of the immunochemical methods used in the preparation of antisera and affinity-purified antibodies and in the fixation and immunofluorescent staining of tissues and chromosomes appears in Carroll and Scott (1985) and in Carroll et al. (1985).

Chain-terminator dideoxynucleotide sequencing of DNA was performed using pEMBL plasmid vectors as described previously (see Laughon and Scott 1984). pDmG1981 was isolated from the 1–5-hour embryonic cDNA library of M. Goldschmidt-Clermont by virtue of its hybridization to the Antennapedia (*Antp*) and *ftz* homeo box sequences.

RESULTS

Molecular Organization of the ANT-C

Previous work has defined the positions of the *Antp* and *ftz* genes on the molecular map of the ANT-C (Garber et al. 1983; Scott et al. 1983; Weiner et al. 1984). The newly mapped positions of the Deformed (*Dfd*) and Sex combs reduced (*Scr*) genes are shown in Figure 1.

Dfd maps between zerknüllt and Sex combs reduced; the original dominant *Dfd* allele results in defects in head development (Kaufman et al. 1980). The 1.65-kb *Dfd* cDNA clone, pDmG1981, hybridizes to exons at −5 and +3 on the ANT-C molecular map as defined by Scott et al. (1983). The assignment of the cDNA clone to the *Dfd* gene is based on three facts: (1) The insertion of a large piece of the third chromosome into the "0" location reverts the dominant *Dfd* allele (Hazelrigg and Kaufman 1983; Scott et al. 1983). (2) The Ns^{+R17} deletion (Kaufman et al. 1980) removes DNA distal to "0" (B. Baker, pers. comm.) and deletes the function of the *Dfd* locus and loci to the right but not the zen locus that is left of *Dfd*. (3) A homeo-box-containing clone from the putative *Dfd* region hybridizes in situ to embryonic thoracic and head region cells (p99 in McGinnis et al. 1984c), in which *Dfd* is believed to function (Lewis et al. 1980; Wakimoto et al. 1984). A comparison of cDNA and genomic sequences (data not shown) maps the *Dfd* homeo box to a 290-bp exon at position +3. The orientation of the homeo box in genomic clones indicates that transcription is from proximal to distal (left to right as the ANT-C is usually represented).

The extent of the *Scr* transcription unit is uncertain. Several cDNA clones hybridize to genomic DNA between positions 28 and 32. In situ hybridization of a genomic DNA fragment from the putative *Scr* region to tissue sections of late embryos reveals that homologous RNAs are expressed in the nervous system of the labial and prothoracic segments (M. Levine, pers. comm.). These results suggest that the cDNAs are from the *Scr* gene. However, *Scr* mutation breakpoints extend from position 30 to 85 (55 kb), leading us to suspect that the *Scr*-transcribed region extends further distally than is indicated in Figure 1. Some mutations affecting *Scr* are associated with chromosome rearrangements broken in or to the right of the *ftz* locus (Scott et al. 1983), suggesting that sequences on both sides of the *ftz* gene are required for normal *Scr* function.

Homeo Domain Structure

We have determined the sequences of the *Dfd* and *Scr* homeo boxes (Fig. 2). This brings the number of sequenced *Drosophila* homeo boxes to 7, and the total number of published homeo box sequences to 13 (Fig. 2). The sequences of the ANT-C and *Ubx* homeo boxes are closely related and constitute a group distinct from the *en* and *en*-related (*en-r*) homeo boxes (Poole et al. 1985). Of the 60 amino acids encoded by the homeo boxes, 24 are perfectly conserved among the 13 sequences (Fig. 2, arrowheads). The homology between part of the homeo domain (the 60-amino-acid protein sequence encoded by homeo boxes) and two products of the yeast mating-type locus (Laughon and Scott 1984; Shepherd et al. 1984) is also shown. It was noted previously that homeo domains contain a region of 20 amino acids that resemble the sequences of DNA-binding domains in certain bacterial proteins (Laughon and Scott 1984). This observation was originally based only on the sequences of the homeo domains for *Antp*, *ftz*, and *Ubx*.

The residues of the homeo domain thought to be important for the formation of the helix-turn-helix DNA-binding structure occur at positions 35, 36, 39, 40, 41, 46, and 49 (Fig. 2). Positions 35, 39, 41, and 49 correspond to hydrophobic residues in the bacterial proteins and are perfectly conserved except for position 35, which is somewhat variable but hydrophobic in every case. In the bacterial proteins, an alanine in the first helix (position 36) comes into close contact with a conserved isoleucine or valine (position 46) in the second helix. In the homeo domains, an alanine always occurs at position 36, except in *en* and *en-r*, where serine is found instead, and an isoleucine is found in every homeo domain at position 46. Position 40 corresponds to the β-turn of the bacterial proteins, in which glycine, serine, or cysteine connects the two helices (Pabo and Sauer 1984). With the exception of *Dfd* and Mo10, a glycine, cysteine, or serine appears at position 40 in all of the homeo domains. Therefore, the sequence data for the growing list of homeo domains continue to agree quite well with the proposed DNA-binding model.

A striking degree of sequence conservation occurs in the region corresponding to the "recognition" helix of bacterial DNA-binding proteins (residues 43–51). The only substitution in this region is alanine for arginine in *en* and *en-r*. If the DNA-binding model is correct, this level of conservation in the recognition helix suggests that all of the homeo domains (except for *en* and *en-r*) bind to the same sets of DNA sequences.

Partial Sequence of a Deformed Protein Product

In contrast to the homeo domain, less is known about the entire sequences of homeotic gene protein

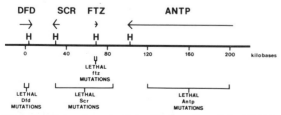

Figure 1. Molecular map of the distal portion of the ANT-C. Arrows show transcribed regions. (H) Homeo box.

products. On the basis of DNA sequences, two segmentation gene protein sequences have been published (of fushi tarazu [Laughon and Scott 1984] and of engrailed [Poole et al. 1985]), and two homeotic gene protein sequences have been completed (Ultrabithorax [P.A. Beachy et al., pers. comm.] and Antennapedia [A. Laughon and M.P. Scott, unpubl.]). From the sequence of a cDNA clone, we have determined the partial amino acid sequence of a protein encoded by another homeotic locus, the *Dfd* gene of the ANT-C (see Fig. 3).

The *Dfd* DNA sequence reveals a protein sequence with unusual characteristics, encoded by a single large open reading frame (Fig. 3). The *Dfd* homeo box starts 594 bp from the beginning of the 1280-bp open reading frame that extends from the 5' end of the cDNA. Presumably, the *Dfd*-coding region continues further upstream in the mRNA, since McGinnis et al. (1984a) found that a 5-kb *Eco*RI fragment containing the *Dfd* homeo box hybridized to a 2.8-kb mRNA, 1150 bp longer than pDmG1981. McGinnis and co-workers also mapped an "M" repetitive sequence to the same 5-kb *Eco*RI fragment containing the *Dfd* homeo box. The M or opa element consists of GC-rich repeats (Wharton et al. 1985; A. Laughon and M.P. Scott, unpubl.); such a sequence occurs in the *Dfd* cDNA starting 100 bp 3' to the end of the homeo box (and is also in a 290-bp homeo-box-containing exon; data not shown). The CAG repeat in *Dfd* encodes 16 glutamines in a stretch of 20 amino acids. Near the 3' end of the *Dfd* open reading frame, repeating AAT and AAC triplets encode 23 asparagines in a stretch of 28 amino acids.

Characteristics of Homeo-domain-containing Proteins

The sequences of *Antp* cDNA and genomic clones demonstrate that *Antp* contains a GC-rich repeat (A. Laughon and M.P. Scott, in prep.). However, in contrast to *Dfd*, in *Antp*, the repetitive sequence occurs upstream of the homeo box and is part of a separate exon. The *Antp* protein contains an abundance of glutamines that are encoded by CAG (and CAA) repeats; the protein overall is about 18% glutamine residues. In one 46-amino-acid section, the *Antp* protein has 31 glutamine residues. The *Dfd* protein (Fig. 3) has an interval in which 16 out of 20 residues are glutamine, another interval in which 23 out of 28 residues are asparagine, a third interval in which 19 out of 32 residues are glycine, and a fourth interval in which 14 out of 29 residues are aspartate or glutamate. The *Ubx* protein has been reported to have a "hinge" region in which 26 out of 31 residues are glycine (Beachy et al. 1985). The *en* protein is the most striking case: It includes an interval where 22 of 35 residues are glutamine, one in which 18 of 20 residues are alanine, one in which 27 of 49 residues are serine, and one in which 13 of 16 residues are glutamate or aspartate (Poole et al. 1985). The *ftz* protein is the least dramatic, having only a rather pedestrian region in which 12 of 23 residues are gluta-

mine (Laughon and Scott 1984). McGinnis et al. (1984a) reported that M repeat sequences also occur in BX-C DNA, possibly at the infra-abdominal-2 locus. The occurrence of repeating amino acids in *Antp*, *ftz*, *Dfd*, and *en* is illustrated in Figure 4.

These types of repeating amino acids are unusual; the fact that they are found in all of the homeo-domain-containing proteins for which sequences are known suggests that such protein structures may perform a common function in the regulation of development by homeotic genes.

Nuclear Localization of Proteins Encoded by Homeotic and Segmentation Genes

Homeo-domain-containing proteins appear to be localized in the nucleus. This was first shown to be the case by using immunofluorescence to localize the *Ubx* product (White and Wilcox 1984; Beachy et al. 1985). Here we show that the *ftz* protein is also localized in nuclei. A portion of the *ftz* cDNA clone pDmG20 (Weiner et al. 1984) was ligated into the λ vector gt11, resulting in the synthesis of a 175-kD β-Gal-*ftz* hybrid protein (Carroll and Scott 1985). Antiserum raised against the fusion protein was affinity-purified and used to stain whole embryos that had been dechorionated, permeabilized in heptane, fixed in formalhyde, and devitellinized in heptane/methanol. A complete account of these experiments appears elsewhere (Carroll and Scott 1985). The *ftz* protein first appears in seven transverse bands of nuclei in blastoderm-stage embryos undergoing cellularization (Fig. 5). This pattern persists through gastrulation until about the time of germ band extension, when it gradually disappears.

We have examined the distribution of the set of proteins that contain homeo domains using an antiserum prepared against a synthetic oligopeptide. The oligopeptide corresponds to an 11-amino-acid part of the homeo domain. For a detailed report of the results, see Carroll et al. (1985). The oligopeptide used, Arg-Gln-Thr-Tyr-Thr-Arg-Tyr-Gln-Thr-Leu-Glu, was coupled to bovine serum albumin (BSA) and injected into rabbits. The antibodies were affinity-purified on a column of oligopeptide cross-linked to Sepharose 4B. The final antibody preparation reacts on protein blots with homeo-domain-containing *ftz*, *Antp*, and *Ubx* proteins produced in bacteria (data not shown) but not with BSA or any detectable *Escherichia coli* proteins. The antibodies readily detect proteins in fly tissues using indirect immunofluorescence techniques. The detected proteins are localized in the nuclei of all tissues and are present at all developmental stages examined. The tissues examined include whole embryos at all stages, imaginal disks, trachea, muscle, salivary gland, brain, ventral ganglia, fat body, testes, and ovaries. Haploid germ line cells have not yet been examined. An example is shown in Figure 6, in which a portion of a stained whole-mount salivary gland preparation is visible. The nuclei that contain the polytene chromosomes are the major sites of fluorescence.

	1	2	3	4	5	6	7	8	9	10	11	12	13	14	15	16	17	18	19	20
ANTP	GLU	ARG	LYS	ARG	GLY	ARG	GLN	THR	TYR	THR	ARG	TYR	GLN	THR	LEU	GLU	LEU	GLU	LYS	GLU
UBX	LEU	ARG	ARG	ARG	GLY	ARG	GLN	THR	TYR	THR	ARG	TYR	GLN	THR	LEU	GLU	LEU	GLU	LYS	GLU
Scr	GLU	THR	LYS	ARG	GLN	ARG	THR	SER	TYR	THR	ARG	TYR	GLN	THR	LEU	GLU	LEU	GLU	LYS	GLU
Dfd	GLU	PRO	LYS	ARG	GLN	ARG	THR	ALA	TYR	THR	ARG	HIS	GLN	ILE	LEU	GLU	LEU	ASP	LYS	GLU
ftz	ASP	SER	LYS	ARG	THR	ARG	GLN	THR	TYR	THR	ARG	TYR	GLN	THR	LEU	GLU	LEU	GLU	LYS	GLU
en	ASP	GLU	LYS	ARG	PRO	ARG	THR	ALA	PHE	SER	SER	GLU	GLN	LEU	ALA	ARG	LEU	LYS	ARG	GLU
en-r	GLU	ASP	LYS	ARG	PRO	ARG	THR	ALA	PHE	SER	GLY	THR	GLN	LEU	ALA	ARG	LEU	LYS	HIS	GLU
Hu1	ASP	GLY	LYS	ARG	ALA	ARG	THR	ALA	TYR	THR	ARG	TYR	GLN	THR	LEU	GLU	LEU	GLU	LYS	GLU
Hu2	PRO	THR	ALA	GLY	GLY	ARG	GLN	THR	TYR	THR	ARG	TYR	GLN	THR	LEU	GLU	LEU	GLU	LYS	GLU
Mo10	SER	SER	LYS	ARG	GLY	ARG	THR	ALA	TYR	THR	ARG	PRO	GLN	LEU	VAL	GLU	LEU	GLU	LYS	GLU
FROG AC1	ASP	ARG	ARG	ARG	GLY	ARG	GLN	ILE	TYR	SER	ARG	TYR	GLN	THR	LEU	GLU	LEU	GLU	LYS	GLU
FROG MM3	ASP	ARG	LYS	ARG	GLY	ARG	GLN	THR	TYR	THR	ARG	TYR	GLN	THR	LEU	GLU	LEU	GLU	LYS	GLU
MOUSE M6	ASP	ARG	LYS	ARG	GLY	ARG	GLN	THR	TYR	THR	ARG	TYR	GLN	THR	LEU	GLU	LEU	GLU	LYS	GLU

▲ (6)   ▲ (12)   ▲ (15)   ▲ (17)   ▲ (19)

HELIX (32–39)

	21	22	23	24	25	26	27	28	29	30	31	32	33	34	35	36	37	38	39	40
ANTP	PHE	HIS	PHE	ASN	ARG	TYR	LEU	THR	ARG	ARG	ARG	ARG	ILE	GLU	ILE	ALA	HIS	ALA	LEU	CYS
UBX	PHE	HIS	THR	ASN	HIS	TYR	LEU	THR	ARG	ARG	ARG	ARG	ILE	GLU	MET	ALA	HIS	ALA	LEU	CYS
Scr	PHE	HIS	PHE	ASN	ARG	TYR	LEU	THR	ARG	ARG	ARG	ARG	ILE	GLU	ILE	ALA	HIS	ALA	LEU	CYS
Dfd	PHE	HIS	TYR	ASN	ARG	TYR	LEU	THR	ARG	ARG	ARG	ARG	ILE	GLU	ILE	ALA	HIS	THR	LEU	VAL
ftz	PHE	HIS	PHE	ASN	ARG	TYR	ILE	THR	ARG	ARG	ARG	ILE	ASP	ILE	ALA	ASN	ALA	LEU	SER	
en	PHE	ASN	GLU	ASN	ARG	TYR	LEU	THR	GLU	ARG	ARG	ARG	GLN	GLN	LEU	SER	SER	GLU	LEU	GLY
en-r	PHE	ASN	GLU	ASN	ARG	TYR	LEU	THR	GLU	LYS	ARG	ARG	GLN	GLN	LEU	SER	GLY	GLU	LEU	GLY
Hu1	PHE	HIS	PHE	ASN	ARG	TYR	LEU	THR	ARG	ARG	ARG	ARG	ILE	GLU	ILE	ALA	HIS	ALA	LEU	CYS
Hu2	PHE	HIS	TYR	ASN	ARG	TYR	LEU	THR	ARG	ARG	ARG	ARG	ILE	GLU	ILE	ALA	HIS	ALA	LEU	CYS
Mo10	PHE	HIS	PHE	ASN	ARG	TYR	LEU	MET	ARG	PRO	ARG	ARG	VAL	GLU	MET	ALA	ASN	ALA	LEU	ASN
FROG AC1	PHE	HIS	PHE	ASN	ARG	TYR	LEU	THR	ARG	ARG	ARG	ARG	ILE	GLU	ILE	ALA	HIS	ALA	LEU	CYS
FROG MM3	PHE	HIS	PHE	ASN	ARG	TYR	LEU	THR	ARG	ARG	ARG	ARG	ILE	GLU	ILE	ALA	HIS	VAL	LEU	CYS
MOUSE M6	PHE	HIS	PHE	ASN	ARG	TYR	LEU	THR	ARG	ARG	ARG	THR	LEU	GLU	ILE	ALA	HIS	ALA	LEU	CYS
YEAST a1										LYS	GLU	LYS	GLU	GLU	VAL	ALA	LYS	LYS	CYS	GLY
YEAST α2										LYS	GLY	LEU	GLU	ASN	LEU	MET	LYS	ASN	THR	SER

▲ (21)   ▲ (23)   ▲ (25)   ▲ (30) ▲ (31)   ▲ (37)

TURN — **RECOGNITION HELIX** (42–51)

	41	42	43	44	45	46	47	48	49	50	51	52	53	54	55	56	57	58	59	60
ANTP	LEU	THR	GLU	ARG	GLN	ILE	LYS	ILE	TRP	PHE	GLN	ASN	ARG	ARG	MET	LYS	TRP	LYS	LYS	GLU
UBX	LEU	THR	GLU	ARG	GLN	ILE	LYS	ILE	TRP	PHE	GLN	ASN	ARG	ARG	MET	LYS	LEU	LYS	LYS	GLU
Scr	LEU	THR	GLU	ARG	GLN	ILE	LYS	ILE	TRP	PHE	GLN	ASN	ARG	ARG	MET	LYS	TRP	LYS	LYS	GLU
Dfd	LEU	SER	GLU	ARG	GLN	ILE	LYS	ILE	TRP	PHE	GLN	ASN	ARG	ARG	MET	LYS	TRP	LYS	LYS	ASP
ftz	LEU	SER	GLU	ARG	GLN	ILE	LYS	ILE	TRP	PHE	GLN	ASN	ARG	ARG	MET	LYS	SER	LYS	LYS	ASP
en	LEU	ASN	GLU	ALA	GLN	ILE	LYS	ILE	TRP	PHE	GLN	ASN	LYS	ARG	ALA	LYS	ILE	LYS	LYS	SER
en-r	LEU	ASN	GLU	ALA	GLN	ILE	LYS	ILE	TRP	PHE	GLN	ASN	LYS	ARG	ALA	LYS	LEU	LYS	LYS	SER
Hu1	LEU	SER	GLU	ARG	GLN	ILE	LYS	ILE	TRP	PHE	GLN	ASN	ARG	ARG	MET	LYS	TRP	LYS	LYS	ASP
Hu2	LEU	THR	GLU	ARG	GLN	ILE	LYS	ILE	TRP	PHE	GLN	ASN	ARG	ARG	MET	LYS	TRP	LYS	LYS	GLU
Mo10	LEU	THR	GLU	ARG	GLN	ILE	LYS	ILE	TRP	PHE	GLN	ASN	ARG	ARG	MET	LYS	TYR	LYS	LYS	ASP
FROG AC1	LEU	THR	GLU	ARG	GLN	ILE	LYS	ILE	TRP	PHE	GLN	ASN	ARG	ARG	MET	LYS	TRP	LYS	LYS	GLU
FROG MM3	LEU	THR	GLU	ARG	GLN	ILE	LYS	ILE	TRP	PHE	GLN	ASN	ARG	ARG	MET	LYS	TRP	LYS	LYS	GLU
MOUSE M6	LEU	THR	GLU	ARG	GLN	ILE	LYS	ILE	TRP	PHE	GLN	ASN	ARG	ARG	MET	LYS	TRP	LYS	LYS	GLU
YEAST a1	ILEU	THR	PRO	LEU	GLN	VAL	ARG	VAL	TRP	PHE	ILE	ASN	LYS	ARG	MET	ARG	SER	LYS	STOP	
YEAST α2	LEU	SER	ARG	ILE	GLN	ILE	LYS	GLN	TRP	VAL	SER	ASN	ARG	ARG	ARG	LYS	GLU	LYS	THR	ILE

▲ (42) ▲ (44) ▲ (45) ▲ (46) ▲ (47) ▲ (48) ▲ (49) ▲ (50) ▲ (51) ▲ (53) ▲ (55) ▲ (57) ▲ (59)

Figure 2. (*See facing page for legend.*)

It is difficult to rule out the possibility that the antibodies cross-react with a ubiquitous nuclear protein that contains part or all of the 11-amino-acid sequence but is not the product of a known homeotic gene. In fact, such a protein (and gene) could be of substantial interest. It is clear that the antigen detected is not a ubiquitous chromatin component such as a histone, since nucleoli are not stained by the antiserum (data not shown).

DISCUSSION

From the data presented, it appears that genes of the ANT-C and BX-C and at least one unlinked gene (*en*) encode proteins with the following properties: (1) They contain a homeo domain; (2) they contain regions enriched in, or with continuous stretches of, single amino acids; and (3) they are localized within the nucleus. Repeated amino acid sequences in the proteins could serve solely as structural components of the proteins. Alternatively, repeated amino acids could reflect translational control of synthesis or could affect rates of degradation of the proteins. The proposed DNA-binding functions of the homeo domain and the nuclear localization of homeotic proteins are consistent with the hypothesis that homeotic genes regulate other genes (García-Bellido 1977). The structure of the proposed DNA-binding region of the homeo domain has implications for the mechanism of gene regulation by homeotic proteins. Before discussing possible regulatory mechanisms, it is important to summarize the genetic evidence for combinatorial action by homeotic genes.

Combinatorial Regulation by Genes Containing the Homeo box

The combinatorial nature of homeotic gene activity has been emphasized in several studies (Lewis 1978; Duncan and Lewis 1982; Struhl 1982; Kaufman and Abbott 1984). Combinatorial regulation may involve the expression of several homeotic genes within the same cell, with the resulting interactions deciding the fate of that cell. We briefly review here the genetic and molecular data that several homeo-box-containing genes are (or can be) active in the same cells, using the ventral, posterior third thoracic segment as an example. *Ubx*, *Antp*, and *en* are all required for normal development of (and, therefore, must be active in) the ventral, posterior third thoracic cells. Clones of cells lacking any one of the three genes do not develop normally (Morata and Kerridge 1981; Struhl 1981, 1982; Lawrence and Struhl 1982; Kaufman and Abbott 1984).

The genetic and developmental data demonstrating combinatorial effects have now been supplemented with molecular data on the distribution of transcripts and proteins. There is no a priori certainty that a gene is functioning whenever its gene products are found, but the genetic and molecular data together make the strongest, and most precise, case for combinatorial effects. In situ hybridization (Akam 1983b) and protein localization (White and Wilcox 1984; Beachy et al. 1985) have shown that RNA and protein products of the *Ubx* gene are located in posterior T3 cells (and other cells). Transcripts from *en* are found in posterior T3 cells in blastoderm-stage embryos and at later stages (Kornberg et al. 1985); *Antp* transcripts are observed in blastoderm cells that are precursors to T3 cells and transiently in the T3 ventral ganglion cells (Levine et al. 1983). Here we have used the T3 segment as an example, but in general, the molecular cytology results reveal multiple instances in which the RNA products of different homeotic genes coexist within the same cells. Often, the distribution of gene products changes during development (see, e.g., Levine et al. [1983] for the *Antp* case); some encounters between homeotic gene products may well be transient.

The description above neglects some of the subtleties of the spatial and temporal specificities of gene expression. However, we merely wish to describe some of the data that support the idea that regulation by homeotic genes is combinatorial. The activities of multiple genes *within each cell* are required for normal pattern formation.

Molecular Basis for Determinative Switching

A striking feature of the homeo domain is the resemblance of its carboxyterminal half to the helix-turn-helix bacterial DNA-binding domain (Laughon and Scott 1984). Out of 60 amino acids in the homeo domain, the most conserved region is what we have predicted to be the "recognition helix" (Fig. 2) that, in current models for the DNA-binding proteins λcro, repressor, and CAP, binds to the DNA major groove of an operator site (Ptashne et al. 1980; Ohlendorf et al. 1983; Pabo and Sauer 1984).

Suggestive evidence for DNA-binding properties of homeo-domain-containing proteins comes from biochemical studies of yeast mating-type fusion proteins. Johnson and Herskowitz (1985) have demonstrated that the yeast *MATα2* protein (in the form of a β-galactosidase hybrid protein) binds in vitro to specific DNA sequences upstream of genes regulated by *MATα2*. *MATα2* and *MATa1* (also believed to be a DNA-bind-

---

Figure 2. Tabulation of homeo domain sequences. Positions of sequences related to the helix-turn-helix structure of bacterial DNA-binding proteins are shown above the sequences (Laughon and Scott 1984). Conserved positions are indicated by arrowheads. References for the homeo domain sequences are Scott and Weiner (1984) and McGinnis et al. (1984b) for *Antp*, *Ubx*, and *ftz*, Poole et al. (1985) and Fjose et al. (1985) for *en*, Poole et al. (1985) for *en-r*, Levine et al. (1984) for Hu1 and Hu2, McGinnis et al. (1984a) for Mo10, Carrasco et al. (1984) for frog AC1, Müller et al. (1984) for frog MM3, and Colberg-Poley et al. (1985) for mouse M6. Yeast *MATα2* and *MATa1* sequences are from Nasmyth et al. (1981) and Miller (1984). Positions in *MATα2* and *MATa1* that are homologous to the homeo domain are underscored.

```
                    EcoRI
            IleSerAlaGlyAlaValHisSerAspProThrAsnGlyTyrGlyProAlaAlaAsnValProAsnThrSerAsnGlyGlyGlyGlyGly
5' GAATTCCATATCCGCAGGAGCGGTGCACTCCGATCCCACGAACGGATACGGACCGGCGGCAAACGTTCCAAATACGAGCAATGGCGGTGGTGGCGGAGGA

   SerGlyAlaValLeuGlyGlyGlyAlaValGlyGlySerAlaAsnGlyTyrTyrGlyGlyTyrGlyGlyTyrGlyGlyTyrGlyGlyThrAlaAsnGlySerValGly
   AGTGGCGCAGTGCTCGGAGGTGGCGCAGTGGGGGGATCGGCAAATGGATATTACGGCGGGTACGGTGGGGGTTATGGGACGGCGAACGGAAGTGTGGGCA  200
                                                                                                   PstI
   IleThrHisSerGlnGlyHisSerProHisSerGlnMetMetAspLeuProLeuGlnCysSerSerThrGluProProThrAsnThrAlaLeuGlyLeuGln
   TCACCCACTCCCAAGGACACTCGCCGCACTCCCAGATGATGGATCTGCCCCTTCAGTGCAGCTCCACGGAACCACCGACGAACACGGCGCTGGGACTGCA
                                                                                              BamHI
   GluLeuGlyLeuLysLeuGluLysArgIleGluGluAlaValProAlaGlyGlnGlnLeuGlnGluLeuGlyMetArgLeuArgCysAspAspMetGly
   GGAATTAGGCCTAAAACTAGAAAAACGCATAGAAGAAGCTGTACCTGCCGGACAACAACTGCAAGAGCTAGGGATGCGATTGCGGTGTGATGATATGGGA  400

   SerGluAsnAspAspMetSerGluGluAspArgLeuMetLeuAspArgSerProAspGluLeuGlySerAsnAspAsnAspAspAspLeuGlyAspSer
   TCCGAAAACGACGATATGTCAGAGGAGGATCGGCTCATGCTGGATCGATCTCCAGACGAGCTGGGCTCCAACGACAACGACGACGACCTCGGGGACTCAG

   AspSerAspGluAspLeuMetAlaGluThrThrAspGlyGluArgIleIleTyrProTrpMetLysLysIleHisValAlaGlyValAlaAanGlySerTyr
   ACAGCGATGAGGATCTGATGGCGGAGACGACCGATGGCGAACGGATCATCTACCCCTGGATGAAGAAGATCCATGTGGCGGGAGTTGCGAACGGGTCTTA  600

   GlnProGlyMetGluProLysArgGlnArgThrAlaTyrThrArgHisGlnIleLeuGluLeuAspLysGluPheHisTyrAsnArgTyrLeuThrArg
   CCAGCCGGGAATGGAGCCAAAACGCCAACGCACCGCCTACACACGCCATCAGATCCTGGAACTGGAAAAGGAGTTCCACTACAACCGCTACCTGACGCGT
                                                          BglII
   ArgArgArgIleGluIleAlaHisThrLeuValLeuSerGluArgGlnIleLysIleTrpPheGlnAsnArgArgMetLysTrpLysLysAspAsnLys
   CGGCGGCGCATCGAGATTGCCCATACGTTAGTTCTCTCGGAGCGGCAGATCAAGATCTGGTTCCAGAACAGGCGCATGAAGTGGAAGAAGGACAACAAGC  800

   LeuProAsnThrLysAsnValArgLysLysThrValAspAlaAsnGlyAsnProThrProValAlaLysLysProThrLysArgAlaAlaSerLysLysGln
   TGCCCAACACCAAGAACGTGCGCAAGAAGACGGTGGACGCCAACGGCAAACCAACACCGGTAGCGAAGAAACCCACCAAGCGGGCCGCCTCCAAAAAGCA

   GlnGlnAlaGlnGlnGlnGlnGlnSerGlnGlnGlnGlnThrGlnGlnThrGlnGlnThrProValMetAsnGluCysIleArgSerAspSerLeuGlu
   GCAGCAAGCGCAGCAGCAGCAGCAGTCGCAGCAGCAGCAGACGCAGCAGACGCAGCAGACTCCGGTGATGAATGAGTGCATTCGTTCCGACAGTTTGGAG  1000

   SerIleGlyAspValSerSerSerLeuGlyAsnProProTyrPheProAlaAlaProGluThrThrSerSerTyrProGlySerGlnGlnHisLeuSer
   AGTATCGGTGACGTCAGCTCGTCCCTGGGCAATCCGCCCTATATACCGGCGGCACCTGAGACGACCAGCTCTTATCCGGGATCTCAGCAGCACCTCAGCA

   AsnAsnAsnAsnGlySerGlyAsnAsnAsnAsnAsnAsnAsnAsnAsnSerAsnLeuAsnAsnAsnAsnAsnAsnGlnMetGlyHisThrAsn
   ATAACAACAACAATGGCAGCGGCAATAACAATAATAACAACAACAACAACAGCAACCTCAATAACAATAACAATAACAATCAAATGGGTCACACGAA  1200

   LeuHisGlyHisLeuGlnGlnGlnGlnSerAspLeuMetThrAsnLeuGlnLeuHisIleLysGlnAsnTyrAspLeuThrAlaLeu
   TCTGCATGGGCACCTTCAACAGCAACAATCCGATCTCATGACCAATCTTCAGCTACACATCAAGCAGGACTACGATCTGACGGCCCTGTAGAATCAGCAG

   GGATATCTTCAAGATGACTTCAGCATTGTAAATTAAATACCCCTATCATTCTATGTAGATTTTAGTTTTTCATTTTGAATCAATGACCTTGAACTTCAGG  1400

   CAAGGTTAAAGACAAATCGGAGTATACAGTGAACGTAGTTAGTTAAGGACCGACCATTACCTAGACGACTCCCATTTTATAATGTTGTAGTCCCAATTCC

   CCACAAGTGTTGTATATGAGTTTTACTTAAGCACCCTAGGATTCAGGTTTAAGAGTTTGTATAAAGCCTAGAACGATTCGTCATAATAATATATCATTAA  1600
                         EcoRI
                           ↓
   TAATATTATCATTATTATCATTAATAATATTTAATGATAACTATAAATGGAATTC  3'
```

Figure 3. Sequence of the *Dfd* cDNA clone, pDmG1981. The translated sequence of the 973-bp open reading frame is shown above the DNA sequence. The homeo box is underscored. *Eco*RI sites at the ends of the sequence are from synthetic DNA linkers and do not correspond to genomic sequences.

ing regulatory protein) contain regions related in sequence to the homeo domain (Fig. 2) (Laughon and Scott 1984; Shepherd et al. 1984). Preliminary results indicate that a *ftz*-β-Gal fusion protein binds to DNA-cellulose in vitro and can be eluted with 0.25–0.5 M NaCl. This property is consistent with a DNA-binding role for *ftz* but does not prove it (data not shown).

Each of the genes mentioned in the previous section, *Ubx*, *Antp*, and *en*, contains a homeo box (McGinnis et al. 1984b; Scott and Weiner 1984; Poole et al. 1985;

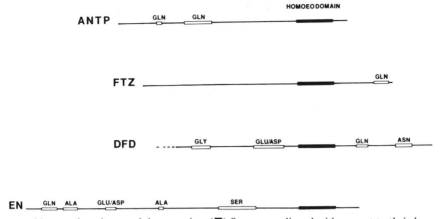

Figure 4. Structure of homeo-domain-containing proteins. (■) Sequences aligned with respect to their homeo domains; (□) blocks of repeated amino acids. Sequences are from *Antp* (A. Laughon and M.P. Scott, unpubl.), *ftz* (Laughon and Scott 1984), *Dfd* (this paper), and *en* (Poole et al. 1985).

F.A. Storfer and M.P. Scott, unpubl.), and it appears that all three genes are expressed in the same T3 cells. If one function of the homeo domain structure is DNA binding, then the problem arises as to how several different proteins, containing closely related homeo domains, can act in a combinatorial way and give gene-specific effects. In particular, the proposed recognition helix structure, which is identical in *Ubx* and *Antp*, is thought to be, in large part, responsible for DNA sequence recognition in bacterial DNA-binding proteins. How then could regulatory specificity be achieved by the homeotic gene products? A possible answer comes from recent studies of yeast mating-type regulation, in which different sets of genes are the targets of different mating-type gene products: Some haploid-specific genes are repressed by diploid **a**/α regulation, whereas other genes are expressed only in α cells or only in **a** cells. As was originally proposed by Strathern et al. (1981), the *MAT*α*2* product represses **a**-specific genes; in diploids, *MAT*α*2* and *MAT***a***1* products act together to repress haploid-specific genes. Therefore, regulation is, or can be, combinatorial. The *MAT* protein products are believed to act as dimers to bind to specific DNA sequences; in vitro, monomers of *MAT*α*2* protein are not sufficient for sequence-specific binding (Johnson and Herskowitz 1985; Miller et al. 1985).

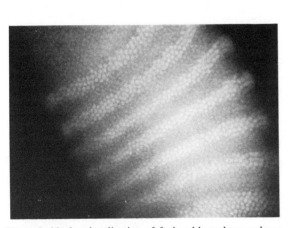

Figure 5. Nuclear localization of *ftz* in a blastoderm embryo. Immunofluorescence staining of fixed whole embryos was carried out with affinity-purified anti-*ftz* antibody, followed by secondary staining with affinity-purified goat anti-rabbit IgG fluorescein conjugate (Cappel). Anterior end of the embryo is at the top of the photograph.

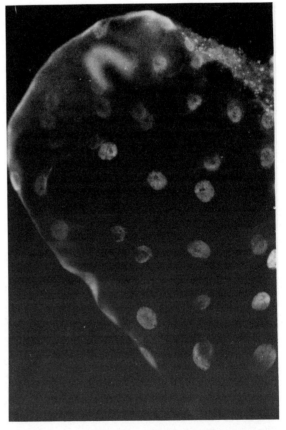

Figure 6. Immunofluorescent staining of salivary gland nuclei with affinity-purified anti-homeo-domain peptide antibody. Secondary antibody staining was carried out with affinity-purified goat anti-rabbit fluorescein conjugate (Cappel).

The following are the key points of the *MAT* model: (1) DNA-binding proteins can positively or negatively regulate a set of target genes by binding to DNA sequences near the promoters. (2) Binding sites are roughly symmetrical, and a protein monomer binds to each half-site (which is also the case for viral and bacterial proteins such as *cro*). (3) The spacing between the two halves of a binding site varies. (4) *MATα2* can recognize one set of binding sites when it acts as a homodimer (i.e., in α haploids) and a different set of sites (or perhaps additional ones) when *MATa1* is present, i.e., in a/α diploids, due to heterodimerization of the *MATα2* and *MATa1* proteins. (5) The heterodimers could recognize different target sites because the spacing of half-sites is different (even if the half-sites have the same sequence) or because the α2 and a1 proteins recognize different half-sites and the heterodimer will bind only to correctly juxtaposed half-sites.

There is as yet no direct evidence that the combinatorial action of homeotic genes is due to heterodimerization of their products. However, the analogy with yeast *MAT* function does help to resolve some questions about how DNA binding could be a viable way for homeotic proteins to regulate an array of other genes. The first question is how do proteins with identical amino acids in the DNA-binding parts of the proteins regulate only certain target genes? One explanation would be that other parts of each protein transmit structural influences to the DNA-binding part of the protein, and by distorting its structure, bias its DNA sequence specificity. This ad hoc postulate is unnecessary if the selection of binding sites is dependent on the formation of homodimers or heterodimers (or higher-order associations) of different proteins and on pairing of the appropriate half-binding sites. As in the yeast model, certain homo- or heterodimers could bind to pairs of half-sites that are properly spaced. The non-DNA-binding parts of the protein monomers would, among other things, set the spacing of the two homeo domains. Spacing does not, of course, mean only linear distance, since the heterodimer must wrap around the DNA more as the distance between the half-sites increases. Thus, the angle between two monomer proteins has to vary, as well as the distance between them.

It is possible that interactions between homeo-domain-containing proteins, and the spacing between homeo domains, are in part dependent on the runs of polyamino acids described above. A spacing or linking role has been suggested for the glycine-rich "hinge" region of the *Ubx* proteins (Beachy et al. 1985). Unequal crossing-over within repetitive DNA-coding regions would lead to changes in the number of polyamino acids, providing a mechanism for fine variation in the spacing of DNA-binding protein monomers. Since homeotic proteins are expressed in a position-specific way, target genes are regulated differently in different parts of the embryo. A change in one of the homeotic protein structures could lead to a particular target gene being turned on or off in a novel location, leading to a phenotypic change.

Why should the putative DNA-recognition part of the homeo domain be so conserved in the proteins examined thus far? Variation in DNA-binding specificity might be expected to add to the flexibility of a regulatory system. However, if a single homeo-domain-containing protein regulated the expression of a set of genes in an ancestral organism, the DNA-binding specificity of the homeo domain would have become fixed. A change in the specificity would have necessitated simultaneous compensatory alterations of multiple binding sites for the regulatory circuit to remain functional. The constraints on recognition helix structure would be especially stringent if the target DNA sequences vary somewhat; any change in the helix sequence could lead to loss of recognition of a whole subset of target sequences. Such a system might have increased in complexity through the evolution of duplications of the homeo box, together with alterations in other protein structural features or in the spatial and temporal expression of homeo-domain-containing proteins. There may well be other classes of proteins with homeo-domain-like sequences that have different DNA-binding specificities; such protein families may comprise regulatory circuits, analogous to the homeotic system, for the control of other aspects of development.

ACKNOWLEDGMENTS

We thank Drs. Steve DiNardo, Pat O'Farrell, Alexander Johnson, Ira Herskowitz, Philip Beachy, and David Hogness for communication of unpublished results and stimulating discussions, and Dr. James Kennison for comments on the manuscript. We are especially grateful to Cathy S. Inouye for her help with the manuscript. This research was supported by grants from the National Institutes of Health to M.P.S. (HD-18163 and KO4-HD00539) and by postdoctoral fellowships from the American Cancer Society and the Damon Runyon-Walter Winchell Cancer Fund to A.L. and S.B.C., respectively.

REFERENCES

Akam, M. 1983a. Decoding the *Drosophila* complexes. *Trends Biochem. Sci.* **8**: 173.

———. 1983b. The location of *Ultrabithorax* transcripts in *Drosophila* tissue sections. *EMBO J.* **2**: 2075.

Bateson, W. 1894. *Materials for the study of variation.* MacMillan, London.

Beachy, P.A., S.L. Helfand, and D.S. Hogness. 1985. Segmental distribution of bithorax complex proteins during *Drosophila* development. *Nature* **313**: 545.

Bender, W., M. Akam, F. Karch, P. Beachy, M. Peifer, P. Spierer, E.B. Lewis, and D.S. Hogness. 1983. Molecular genetics of the bithorax complex in *Drosophila melanogaster*. *Science* **221**: 23.

Carrasco, A.E., W. McGinnis, W.J. Gehring, and E.M. De Robertis. 1984. Cloning of an *X. laevis* gene expressed during early embryogenesis coding for a peptide region homologous to *Drosophila* homeotic genes. *Cell* **37**: 409.

CARROLL, S.B. and M.P. SCOTT. 1985. Localization of the *fushi tarazu* protein during *Drosophila* embryogenesis. *Cell* (in press).

CARROLL, S.B., P.D. RILEY, M. KLYMKOWSKY, J. VAN BLERKOM, J. STEWARD, and M.P. SCOTT. 1985. Localization of homoeodomain-containing proteins using antisera against synthetic oligopeptides. *Symp. Soc. Dev. Biol.* **44:** (in press).

COLBERG-POLEY, A.M., S.D. VOSS, K. CHOWDHURY, and P. GRUSS. 1985. Structural analysis of murine genes containing homoeo box sequences and their expression in embryonal-carcinomal cells. *Nature* **314:** 713.

DUNCAN, I.M. and E.B. LEWIS. 1982. Genetic control of body segment differentiation in *Drosophila*. *Symp. Soc. Dev. Biol.* **40:** 533.

FJOSE, A., W.J. MCGINNIS, and W.J. GEHRING. 1985. Isolation of a homoeobox-containing gene from the *engrailed* region of *Drosophila* and the spatial distribution of its transcript. *Nature* **313:** 284.

GARBER, R.L., A. KUROIWA, and W.J. GEHRING. 1983. Genomic and cDNA clones of the homoeotic locus *Antennapedia* in *Drosophila*. *EMBO J.* **2:** 2027.

GARCÍA-BELLIDO, A. 1977. Homoeotic and atavistic mutations in insects. *Am. Zool.* **17:** 613.

HAFEN, E., A. KUROIWA, and W.J. GEHRING. 1984. Spatial distribution of transcripts from the segmentation gene *fushi tarazu* during *Drosophila* embryonic development. *Cell* **37:** 833.

HAFEN, E., M. LEVINE, R.L. GARBER, and W.J. GEHRING. 1983. An improved *in situ* hybridization method for the detection of cellular RNAs in *Drosophila* tissue sections and its application for localizing transcripts of the homoeotic Antennapedia gene complex. *EMBO J.* **4:** 617.

HAZELRIGG, T. and T.C. KAUFMAN. 1983. Revertants of dominant mutations associated with the Antennapedia gene complex of *Drosophila melanogaster*: Cytology and genetics. *Genetics* **105:** 581.

JOHNSON, A.D. and I. HERSKOWITZ. 1985. A repressor (*MATα2 Product*) and its operator control expression of a set of cell-type specific genes in yeast. *Cell* **42:** 237.

KAUFMAN, T.C. and M. ABBOTT. 1984. Homoeotic genes and the specification of segmental identity in the embryo and adult thorax of *Drosophila melanogaster*. In *Molecular aspects of early development* (ed. G.M. Malacinski and W.H. Klein), p. 189. Plenum Press, New York.

KAUFMAN, T.C., R. LEWIS, and B. WAKIMOTO. 1980. Cytogenetic analysis of chromosome 3 in *Drosophila melanogaster*: The homoeotic gene complex in polytene chromosome interval 84A,B. *Genetics* **94:** 115.

KIDD, S., T.J. LOCKETT, and M.W. YOUNG. 1983. The *Notch* locus of *Drosophila melanogaster*. *Cell* **34:** 421.

KORNBERG, T., I. SIDEN, P. O'FARRELL, and F. SIMON. 1985. The *engrailed* locus of *Drosophila*: In situ localization of transcripts reveals compartment-specific expression. *Cell* **40:** 45.

LAUGHON, A. and M.P. SCOTT. 1984. Sequence of a *Drosophila* segmentation gene: Protein structure homology with DNA-binding proteins. *Nature* **310:** 25.

LAWRENCE, P.A. and G. STRUHL. 1982. Further studies of the *engrailed* phenotype in *Drosophila*. *EMBO J.* **1:** 827.

LEVINE, M., G.M. RUBIN, and R. TJIAN. 1984. Human DNA sequences homologous to a protein coding region conserved between homoeotic genes of *Drosophila*. *Cell* **38:** 667.

LEVINE, M., E. HAFEN, R.L. GARBER, and W.J. GEHRING. 1983. Spatial distribution of *Antennapedia* transcripts during *Drosophila* development. *EMBO J.* **2:** 2037.

LEWIS, E.B. 1952. Pseudoallelism and gene evolution. *Cold Spring Harbor Symp. Quant. Biol.* **16:** 159.

———. 1963. Genes and developmental pathways. *Am. Zool.* **3:** 33.

———. 1978. A gene complex controlling segmentation in *Drosophila*. *Nature* **276:** 565.

———. 1982. Control of body segment differentiation in *Drosophila* by the bithorax gene complex. In *Embryonic development: Genes and cells* (ed. M. Burger), p. 269. A.R. Liss, New York.

LEWIS, R.A., B.T. WAKIMOTO, R.E. DENELL, and T.C. KAUFMAN. 1980. Genetic analysis of the Antennapedia gene complex (ANT-C) and adjacent regions of *Drosophila melanogaster*. II. Polytene chromosome segments 84A-84B1,2. *Genetics* **95:** 383.

MCGINNIS, W., C.P. HART, W.J. GEHRING, and F.H. RUDDLE. 1984a. Molecular cloning and chromosome mapping of a mouse DNA sequence homologous to homoeotic genes of *Drosophila*. *Cell* **38:** 675.

MCGINNIS, W., R.L. GARBER, J. WIRZ, A. KUROIWA, and W.J. GEHRING. 1984b. A homologous protein-coding sequence in *Drosophila* homoeotic genes and its conservation in other metazoans. *Cell* **37:** 403.

MCGINNIS, W., M. LEVINE, E. HAFEN, A. KUROIWA, and W.J. GEHRING. 1984c. A conserved DNA sequence found in homoeotic genes of the *Drosophila* Antennapedia and bithorax complexes. *Nature* **308:** 428.

MILLER, A.M. 1984. The yeast *MATa*1 gene contains two introns. *EMBO J.* **3:** 1061.

MILLER, A.M., V.L. MACKAY, and K.A. NASMYTH. 1985. Identification and comparison of two sequence elements that confer cell-type specific transcription in yeast. *Nature* **314:** 598.

MORATA, G. and S. KERRIDGE. 1981. Sequential functions of the bithorax complex of *Drosophila*. *Nature* **290:** 778.

MÜLLER, M.M., A.E. CARRASCO, and E.M. DE ROBERTIS. 1984. A homoeo-box-containing gene expressed during oogenesis in *Xenopus*. *Cell* **39:** 157.

NASMYTH, K.A., K. TATCHELL, B.D. HALL, C. ASTELL, and M. SMITH. 1981. Physical analysis of mating-type loci in *Saccharomyces cerevisiae*. *Cold Spring Harbor Symp. Quant. Biol.* **45:** 961.

OHLENDORF, D.H., W.F. ANDERSON, and B.W. MATTHEWS. 1983. Many gene-regulatory proteins appear to have a similar α-helical fold that binds DNA and evolved from a common precursor. *J. Mol. Evol.* **19:** 109.

OUWENEEL, W. 1976. Developmental genetics of homoeosis. *Adv. Genet.* **16:** 179.

PABO, C.O. and R.T. SAUER. 1984. Protein-DNA recognition. *Annu. Rev. Biochem.* **53:** 293.

POOLE, S.J., L.M. KAUVAR, B. DREES, and T. KORNBERG. 1985. The *engrailed* locus of *Drosophila*: Structural analysis of an embryonic transcription. *Cell* **40:** 37.

PTASHNE, M., A. JEFFREY, A.D. JOHNSON, R. MAURER, B.J. MEYER, C.O. PABO, T.M. ROBERTS, and R.T. SAUER. 1980. How the repressor and cro work. *Cell* **19:** 1.

SCOTT, M.P. and A.J. WEINER. 1984. Structural relationships among genes that control development: Sequence homology between the *Antennapedia*, *Ultrabithorax*, and *fushi tarazu* loci of *Drosophila*. *Proc. Natl. Acad. Sci.* **81:** 4115.

SCOTT, M.P., A.J. WEINER, T.I. HAZELRIGG, B.A. POLISKY, V. PIRROTTA, F. SCALENGHE, and T.C. KAUFMAN. 1983. The molecular organization of the *Antennapedia* complex of *Drosophila*. *Cell* **35:** 763.

SHEPHERD, J.C.W., W. MCGINNIS, A.E. CARRASCO, E.M. DE ROBERTIS, and W.J. GEHRING. 1984. Fly and frog homoeo domains show homologies with yeast mating type regulatory proteins. *Nature* **310:** 70.

STRATHERN, J.N., J.B. HICKS, and I. HERSKOWITZ. 1981. Control of cell type by the mating type locus: The α1-α2 hypothesis. *J. Mol. Biol.* **147:** 357.

STRUHL, G. 1981. A homoeotic mutation transforming leg to antenna in *Drosophila*. *Nature* **292:** 635.

———. 1982. Genes controlling segmental specification in the *Drosophila* thorax. *Proc. Natl. Acad. Sci.* **79:** 7380.

WAKIMOTO, B.T., F.R. TURNER, and T.C. KAUFMAN. 1984. Defects in embryogenesis in mutants associated with the Antennapedia gene complex of *Drosophila melanogaster*. *Dev. Biol.* **102:** 147.

WEINER, A.J., M.P. SCOTT, and T.C. KAUFMAN. 1984. A molecular analysis of *fushi tarazu*, a gene in *Drosophila melanogaster* that encodes a product affecting embryonic segment number and cell fate. *Cell* **37:** 843.

WHARTON, K.A., B. YEDVOBNICK, V.G. FINNERTY, and S. ARTAVANIS-TSAKONAS. 1985. *opa*: A novel family of transcribed repeats shared by the *Notch* locus and other developmentally regulated loci in *D. melanogaster*. *Cell* **40:** 55.

WHITE, R.A.H. and M. WILCOX. 1984. Protein products of the bithorax complex in *Drosophila*. *Cell* **39:** 163.

Homeo Box Sequences of the Antennapedia Class Are Conserved Only in Higher Animal Genomes

W. MCGINNIS

Department of Molecular Biophysics and Biochemistry, Yale University, New Haven, Connecticut 06511

On the basis of mutant phenotypes and localized gene expression, homeotic gene products are believed to be the key elements in determining the morphogenetic fates of cells in the *Drosophila* body plan (Garcia-Bellido 1977; Lewis 1978; Struhl 1982; Akam 1983; Levine et al. 1983; Hafen et al. 1984b; White and Wilcox 1984; Beachy et al. 1985). Many of these homeotic genes, as judged from the above phenotypic criteria, appear to assign differential cell fates along the longitudinal body axis, thus specifying one region of the body to develop into a specific element of the metameric anatomy of the fly. For example, in the absence of the gene products for the homeotic gene Antennapedia (*Antp*), no second thoracic segment is formed; however, when the *Antp* gene products are ectopically expressed, multiple copies of the second thoracic segment are formed along the embryonic body axis (Wakimoto and Kaufman 1981; Hafen et al. 1984a). Although proof of a direct involvement of the *Antp* gene products in triggering second-thoracic development versus other pathways has not yet been obtained, all genetic and molecular data so far gathered are consistent with this role.

Antp is a member of a multigene family within the *Drosophila* genome. All of the closely conserved members of this family appear to be capable of mutating to phenotypes in which segmentation or segmental diversity is disrupted. The signal homology for this multigene family is the homeo box, a protein-coding sequence of 180–200 bp (McGinnis et al. 1984b; Scott and Weiner 1984). Originally detected in the *Antp* and fushi tarazu (*ftz*) genes of the Antennapedia complex (ANT-C) and the Ultrabithorax (*Ubx*) gene of the bithorax complex (BX-C), it is now known to occur in well-conserved form in at least four other genes within the ANT-C and BX-C (Fig. 1) (F. Karch et al.; W. McGinnis et al.; both in prep.). The seven copies within the two gene clusters are located within the Deformed (*Dfd*), Sex combs reduced (*Scr*), *ftz*, *Antp*, *Ubx*, infra-abdominal-2 (*iab-2*), and infra-abdominal-7 (*iab-7*) loci. On the basis of their cross-homology to the founder member of the family at *Antp*, these genes are grouped as the Antennapedia class of homeo box gene family (W. McGinnis et al., in prep.). As shown in Figure 1, this class of homeo box gene family includes all those genes most homologous to the *Antp* and *Ubx* homeo box regions.

Slight homology can be seen in Figure 1 between the *Ubx* homeo box sequence and members of another class (the engrailed class) of the homeo box gene family. The members of the engrailed class appear to be closely conserved in inter se comparisons, but in toto, they are rather dissimilar to the members of the Antennapedia class (Fjose et al. 1985; Poole et al. 1985). Weak signals detected with other homeo box probes on low-stringency Southern blots may represent other classes of the homeo box gene family (W. McGinnis, unpubl.). Although the homeo box is the best-characterized region of cross-homology between these loci, many also share the M repeat, which corresponds to the *opa* repetitive element originally identified in the Notch locus (Poole et al. 1985; Wharton et al. 1985; W. McGinnis et al., in prep.). The *opa* element is a trinucleotide repeat of CA (A or G), also found in engrailed, and might code for stretches of polyglutamine. The *opa* element is found in one form or another at approximately 300 sites in the *Drosophila* genome.

The restriction of highly conserved copies of the homeo box to homeotic loci of *Drosophila*, its potentially interesting structure, and its strong evolutionary conservation suggest that it has a crucial role in the function of many homeotic gene products. Although this crucial function need not be part of the homeotic function of the genes, its limited distribution provides circumstantial evidence for this hypothesis. At least part of this crucial function is believed to involve DNA binding. The homeo domain is basic and shows weak structural homology with the **a**1 and α2 proteins from the *MAT* locus of yeast (Shepherd et al. 1984). Both are sequence-specific transcriptional repressors, and **a**1 has been proposed to have weak homology with repressor proteins from prokaryotes (Ohlendorf et al. 1983). The region of this weak structural homology between homeo domains, yeast **a**1 and α2 and prokaryotic repressors, lies in the carboxyl half of the homeo domain (Laughon and Scott 1984; Shepherd et al. 1984).

The entire homeo domain, not just the carboxyl half, has been conserved in genes of unknown function in a variety of species of higher metozoa, including frogs, mice, and humans (Carrasco et al. 1984; Levine et al. 1984; McGinnis et al. 1984a). The strong conservation of the entire domain from the evolutionary branch point that separates vertebrates from arthropods ($\sim$600 million years) argues that the homeo domains of the Antennapedia class serve a very similar or identical function or functions in whatever animal they reside. As the results of this paper show, whatever these Antennapedia class homeo domain functions are, they

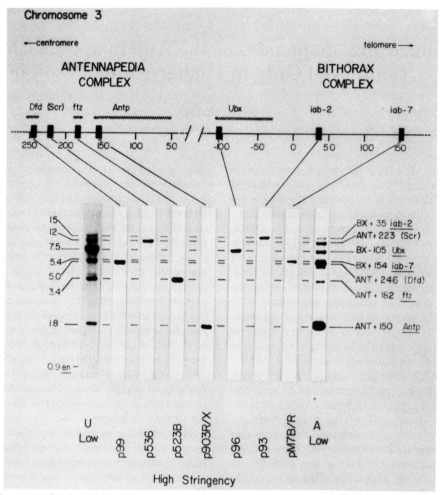

Figure 1. The Antennapedia class of the homeo box gene family. (*Top*) Molecular maps of the ANT-C and BX-C DNAs. The numbering system in kilobases for each of the complexes is depicted below the lines representing the DNA. The ANT-C map was derived from Garber et al. (1983), and the BX-C map was derived from Bender et al. (1983) and F. Karch et al. (in prep.). (■) Sites of homeo box homology. Gene names and their associated transcription units are indicated above the map. (*Bottom*) Hybridization signals resulting from *Drosophila* genomic blots probed under high-stringency conditions with each of the ANT-C and BX-C homeo box regions. These signals are aligned with low-stringency hybridization signals derived from probes of replicate genomic blots with *Antp* (A) and *Ubx* (U) homeo boxes. The sizes in kilobases of hybridizing fragments resulting from *Eco*RI digestion are indicated at the left, and the genomic locations in ANT-C and BX-C are indicated at the right. Lanes were hybridized with the following probes: (U) homeo box probe, see Fig. 2; (p99) 5.0-kb genomic *Eco*RI fragment from *Dfd* that includes the homeo box (W. McGinnis et al., in prep.); (p523b) *ftz* genomic clone (Kuroiwa et al. 1984); (p903R/X) 300-bp *Eco*RI-*Xba*I fragment from the 3' end of the 903 *Antp* cDNA (contains the homeo box exon only; McGinnis et al. 1984c); (p96) *Ubx* genomic clone containing *Ubx* 3' exon with homeo box (McGinnis et al. 1984c; Beachy et al. 1985); (p93) *iab-2* 5.8-kb *Bam*HI-*Eco*RI genomic fragment (W. McGinnis et al., in prep.); (pM7B/R) *iab-7* *Bam*HI-*Eco*RI fragment containing the *iab-7* region homeo box (W. McGinnis et al., in prep.); (A) same as 903 *Eco*RI-*Xba*I, only hybridized and washed under nonstringent conditions (McGinnis et al. 1984c). Near the bottom of the leftmost lane is also indicated the hybridization signal corresponding to the engrailed (*en*) homeo box fragment.

have apparently been evolutionarily fixed in the biological systems used by many higher animal phyla and are missing in other branches of the phylogenetic tree.

METHODS

Genomic DNA isolation. Total genomic DNA was isolated from *Drosophila* (Oregon-R Munchen) and other animals according to the procedure of Frei et al. (1985). This procedure sufficed for the isolation of high-molecular-weight DNA in all cases except that of sponge DNA, which was invariably severely degraded and thus not tested.

Southern blot analysis. Southern blots were performed as described by Meinkoth and Wahl (1984) with the following modifications. The gels used for unidirectional transfer to nitrocellulose were 3 mm thick or less, which seemed to result in increased signal. Blots were hybridized with nick-translated DNA fragments described in the figure legends. These probes were hybridized and washed in nonstringent and stringent buffers (see McGinnis 1984c) for 36 hours. The average specific activity of the probes was 4×10^8 cpm/µg, the average single-strand length was 250 bp, and the average concentration was 2×10^6 cpm/ml of hybridization buffer.

RESULTS

To detect homeo box homology comparable to that exhibited by the genes of the Antennapedia class, whole genomic Southern blots were hybridized with homeo box probes from the *Antp*- and *Ubx*-coding regions (Fig. 2). Restriction fragments that hybridized both probes were considered to contain highly homologous homeo box copies. This method has been previously used to detect homeo box copies in genomic DNA from chicken, mouse, human, and frog sources (McGinnis et al. 1984b). Some of these copies have since been isolated and sequenced and have been found to encode protein domains that have striking homology with those found in *Antp* and *Ubx*. By these empirical standards, it appears that clear hybridization bands on whole genomic blots are a fairly reliable criteria for homeo-domain-coding sequences in a genome.

The copies detected by this method are likely to have considerable homology with the Antennapedia homeo box class, but probably less with other *Drosophila* genes with homeo box homology. For example, applying the above empirical standards to the *Drosophila* genome itself, the engrailed fragment at 0.9 kb would be considered as lacking homeo box homology, since it hybridizes the *Ubx* probe but not the *Antp* probe (Fig. 1). As Fjose et al. (1985) point out, the *Ubx* homeo box shows only 5% more homology (58% vs. 53%) than *Antp* with the engrailed homeo box region. So it is likely that in genomes the size of that of *Drosophila*, 55% is the lower limit of homology for a positive signal in this test. In genomes with a greater complexity, this limit is likely to be higher.

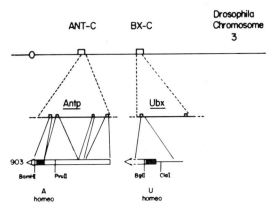

Figure 2. Antennapedia and Ultrabithorax homeo box probes. (*Top*) Relative locations of the ANT-C and BX-C gene complexes on chromosome 3. (*Middle*) Expansion of the *Antp* and *Ubx* region of each chromosome showing a simplified schematic of the exons (□) at each locus. (*Bottom*) Regions of *Antp* and *Ubx* coding DNA that were used as homeo box probes. Exons are bars, introns are lines, and homeo box homologies are cross-hatched bars. Arrows indicate the direction of the transcription.

The ambiguity associated with the described test is dependent on the likelihood of false positive and false negative signals. To minimize the possibility of false positives, two probe fragments have been used whose only cross-homology is in the homeo box region. Nevertheless, false positives do occasionally occur, as shown below. All seem due to slight homology between probes and tandem repeat sequences, which in some cases correspond to rDNA repeats. The possibility of false negatives is more difficult to assess, since there are no well-conserved homeo-domain-coding regions in the *Drosophila* genome that are known to go undetected with *Antp* and *Ubx* hybridization probes. However, by artificially manipulating codon usage, one can design a homeo-box-coding sequence that has less than 50% homology with *Antp* at the nucleotide level while retaining 80% homology at the amino acid level. Thus, if average codon usage in a particular species is substantially different than in *Drosophila*, then relatively well-conserved homeo boxes can go undetected by genomic blot analysis. However, it is clear from previous results that well-conserved homeo box sequences can be detected in a variety of different animal DNAs, with presumably a variety of different codon preferences. The genomes tested in the experiments described here were designated as positive for homeo box homology if they contained one or more fragments that hybridized with both *Antp* and *Ubx* homeo box probes, *and* if the hybridizing fragment(s) did not correspond to repeated fragments, as revealed by prominent bands on the ethidium bromide stain of the electrophoresed genomic DNA.

Homeo Box Homology in Animal Genomes

Each blot was prepared from a gel that had duplicate lanes of *Drosophila* (positive control) and test DNAs. Results of the first "phylo" blot are shown in Figure 3a. The *Drosophila* control lanes (left two panels) exhibit the extent to which the Antennapedia class copies are visible at the exposure level shown. The other genomes that tested positive on this blot were *Amphioxus* (*Branchiostoma lancelatum*), a cephalochordate; sea urchin (*Sphaerechinus granularis*), an echinoderm; and leech (*Erpobdella octoculata*), an annelid. DNA from a flatworm (*Planaria gonocephala*) had no hybridizing copies of homeo box homology. At least ten copies are present in the *Amphioxus* genome, five or more in the sea urchin genome, and four or more in the leech genome. The signal intensity of the *Amphioxus* fragments is even greater than cross-hybridization in the fly genome. This is presumably due to a small genome size for this cephalochordate and the fact that four times as much *Amphioxus* DNA as *Drosophila* DNA was loaded per lane. The relative weakness of the sea urchin signals is presumably due in part to the large size of the sea urchin genome as compared with that of *Drosophila* (Graham et al. 1974). The sea urchin copies, in fact, hybridize with an intensity roughly equal to the copies in frog or mouse genomes, when all are present on the same blot (W. McGinnis, unpubl.).

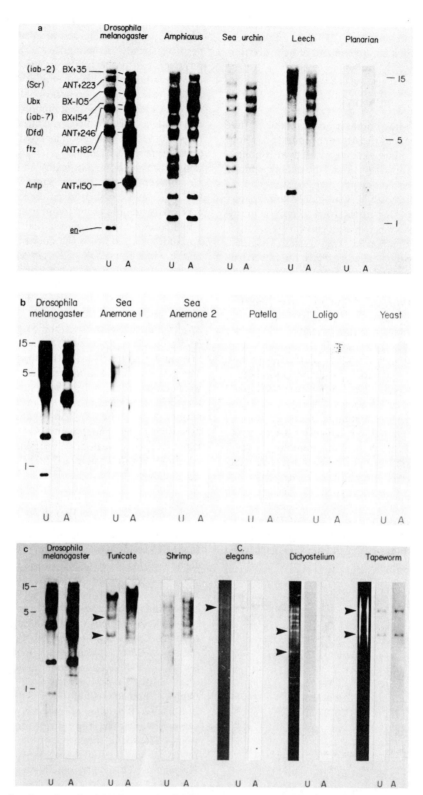

Figure 3. Homeo box homology in animal genomic DNA. Duplicate genomic Southern blots with DNA from the indicated animals were hybridized under reduced stringency conditions (see Methods) with either the A (*Antp* 600-bp *Bam*HI-*Pvu*II fragment in Fig. 2) or the U (*Ubx* 450-bp *Bgl*I-*Cla*I fragment in Fig. 2) homeo box probes. Each *Drosophila*, *C. elegans*, *Dictyostelium*, and yeast lane contained 2.5 μg of DNA digested with *Eco*RI. All the rest contained 10 μg of DNA digested with *Eco*RI. In panel *a*, the hybridizing *Drosophila* fragments are labeled with their genetic origin and location in BX-C or ANT-C DNA. Migration distances of 15-, 5-, and 1-kb size standards are marked in all panels. In panel *c*, the *C. elegans*, *Dictyostelium*, and tapeworm lanes are aligned with a photograph of the ethidium-bromide-stained DNA from which they were derived. Arrows indicate the position of genomic repeats.

The second phylo blot results are shown in Figure 3b. The positive control lanes with *Drosophila* DNA show strong signals; however, all other lanes had no signal. These contained DNA from two species of sea anemone (*Aequorea foscalea* and *Sarsia* sp.), both coelenterates. The genomes from two species of mollusks, *Patella coerula*, a gastropod, and *Loligo* sp., a squid, were also negative with the homeo box probes, as was the yeast (*Saccharomyces cerevisiae*) genome. All of the experiments reported here have been done at least twice, and in not one of the attempts has the mollusk or coelenterate genomes hybridized the homeo box probes. On one occasion, the yeast genome showed weak hybridization to two bands. These signals correspond to the repeated *Eco*RI fragments that contain part of the rDNA genes of yeast (Phillipsen et al. 1978). This spurious hybridization is also seen in the rDNA repeat fragments in other genomes, as shown below.

Results from phylo blot number 3 are shown in Figure 3c. The genomes hybridizing the homeo box probes are tunicate (*Phallusia mammillata*), an urochordate, and shrimp (*Leander squilla*), a crustacean. The other three genomes on this blot showed hybridization to one or more of the probes; however, in all cases, the hybridization signal corresponded to a position in which a genomic repeat was migrating (see arrows to stained gel lanes in Fig. 3c). In the case of *Caenorhabditis elegans* DNA, the genomic repeat that hybridizes the probe migrates at 7 kb, which is the size expected for the *C. elegans* rDNA repeats when digested with *Eco*RI (Files and Hirsh 1981). A re-probe of this blot with an rDNA probe from *Drosophila* (DmrY22C) (Dawid et al. 1978) strongly hybridized the same position. *Dictyostelium* DNA also contained two repeats that weakly hybridized the *Ubx* probe; these also apparently correspond to rDNA fragments, since both hybridized the same *Drosophila* rDNA probe, which contained both the 18S and 28S rDNA genes. The two bands hybridized in the tapeworm (*Hymenolepsis nana*, a platyhelminth) also corresponded to genomic repeat positions, one of which strongly hybridized the *Drosophila* rDNA probe on a re-probe of the blot.

Since it was obvious that some spurious signal could be obtained from genomic repeats, all blots, after probing with homeo boxes, were re-probed with an rDNA probe from *Drosophila*. Although all genomes showed hybridization, providing a positive control for effective transfer and binding of genomic DNA in the various lanes, in only one case did the signals correspond to presumptive homeo box fragments. This occurred in the tunicate genomic blot (Fig. 3c) at the positions marked with arrows. This left only one position in the tunicate genomic blot that strongly hybridized both probes.

DISCUSSION

The results of these and previous hybridization analyses demonstrate that detectable copies of the homeo box sequence are restricted to a few different phyla of higher metazoa. These include arthropods, annelids, echinoderms, urochordates (ascidians), and chordates. No homology was detected after repeated attempts in genomes of aschelminths, platyhelminths, coelenterates, mollusks, fungi, or bacteria.

Although only one or a few species from each phylum have been tested, the strong conservation across phyletic boundaries indicates the likelihood of a considerable conservation within a phylum. In addition, unpublished results with other species from the mollusk and echinoderm phyla are all consistent with the above generalizations. A summary of the results is shown in Table 1, and Figure 4 shows the distribution of homeo box homology on one of the proposed phylogenetic trees for the animal kingdom. The experiments have also been confirmed with homeo box probes from other loci in the Antennapedia class (e.g., *Scr* and *Dfd*). In every experiment, the phylogenetic distribution of hybridization was consistent with the results obtained with the *Antp* and *Ubx* probes. In addition, a systematic search of plant genomes for homeo box homology has yielded no positive signals (J. Medford, unpubl.). Proof of the postulated distribution shown in Figure 4 will come from cloning and sequencing analyses, which, in most of the phyla shown, is either completed or in progress.

The potential interest of genes containing this conserved coding element lies in its preferential association with genes in *Drosophila* that influence pattern formation in the developing animal, in particular the process of segmentation (the *ftz* locus) and regional specification (*Dfd*, *Scr*, *Antp*, *Ubx*, *iab-2*, and *iab-7*) (McGinnis et al. 1984b,c; Scott and Weiner 1984; W. McGinnis et al., in prep.). These genes contain the most conserved (relative to *Antp* and *Ubx*) copies in the *Drosophila* genome and have been termed the Antennapedia class. Other copies of the homeo box with diverged versions of this coding sequence exist and, extrapolating from the limited evidence available in the case of engrailed, may form other classes of homeo box genes with strong inter se homology (Poole et al. 1985). This immediately brings up the possibility that since engrailed class homeo boxes would not be classified as positive using the genomic hybridization experiments employed in this paper, could there not be engrailed class homeo boxes in the phyla classified as negative? Experiments similar to those described here with engrailed class homeo box sequences are in progress and should resolve this question.

Although the functions of the homeo domains(s) are a subject of very active research, on the basis of weak structural homology, the most likely subfunction is that of DNA binding. The structural homology argument would place this subfunction in the carboxyl half of the homeo domain. The characterized genes of the Antennapedia homeo box class conserve both this subdomain and approximately another 30 amino acids to the aminoterminal side. In all probability, this aminoterminal subdomain is also well conserved in all of the homeo-domain-coding regions detected by the *Antp*

Table 1. Summary of Phylogenetic Distribution of Homeo Box Homology of Antennapedia Class

Phyla/subphyla	Genome tested	Homeo box homology
Eubacteria	*E. coli*	not detected
Fungi	*D. discoideum*	not detected
	S. cerevisiae	not detected
Coelenterata	*A. foscalea* (sea anemone)	not detected
	Sarsia sp. (sea anemone)	not detected
Platyhelminthes	*P. gonocephala* (planarian)	not detected
	H. nana (tapeworm)	not detected
Aschelminthes	*A. lumbricoides* (nematode)	not detected
	C. elegans (nematode)	not detected
Mollusca	*P. coerula*	not detected
	Loligo sp. (squid)	not detected
Annelida	*Lumbricus sp.* (earthworm)	multiple copies
	H. octoculata (leech)	multiple copies
Arthropoda	*D. melanogaster* (fly)	multiple copies
	D. hydei (fly)	multiple copies
	T. molitor (beetle)	multiple copies
	L. squilla (shrimp)	multiple copies
Echinodermata	*S. granularis* (sea urchin)	multiple copies
Urochordata	*P. mammillata* (tunicate)	one or few copies
Cephalochordata	*B. lancelatum* (amphioxus)	multiple copies
Vertebrata	*X. laevis* (frog)	multiple copies
	G. domesticus (chicken)	multiple copies
	M. musculus (mouse)	multiple copies
	H. sapiens (human)	multiple copies

Results are derived from those reported in this paper and in McGinnis et al. (1984a,b), Levine et al. (1984), and Carrasco et al. (1984).

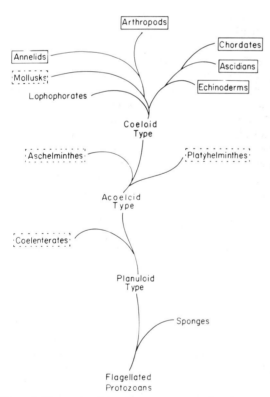

Figure 4. Homeo box homology on a proposed evolutionary tree. The proposed phylogenetic relationships (Hyman 1940) of a variety of metazoans are shown in simplified form. Each phylum, including organisms that contain detectable homeo box homology of the Antennapedia class, is boxed; those phyla enclosed by dotted lines showed no detectable homology. Lophophorates and sponges were not tested.

and *Ubx* probes on genomic blots. Indeed, this aminoterminal region is very well conserved in the sequenced homeo boxes from frog, mouse, and human (Carrasco et al. 1984; Levine et al. 1984; McGinnis et al. 1984a). Such extreme conservation of primary amino acid sequence is often noted in proteins that are required to form intimate protein/protein contacts to effect their function; i.e., the genes encoding these interacting proteins must evolve in a concerted, and thus very gradual, fashion. The most striking example of this occurs in the extremely well-conserved histone proteins of the nucleosome core particle (McGhee and Felsenfeld 1980). It thus seems most plausible that the extreme conservation of the homeo domain over the past 600 million years might be explained by the necessity for it to cooperatively interact with other highly conserved protein domains. These interactive domains might even be other homeo domains from a particular class, which could explain the conservation of a class of homeo box genes. Cooperative interactions between some sequence-specific DNA-binding proteins are a crucial element to their proper function (e.g., those modulating the developmental switch controlling lysis/lysogeny in the bacteriophage λ life cycle [Ptashne et al. 1980] or those controlling mating-cell type in yeast [Miller et al. 1985]) and could be of equal or perhaps more importance in gene regulatory proteins imposing stable cellular fates. Thus, the molecular details of the interactions of the domains with DNA could be complex variations on a theme simply elaborated in the λ life cycle.

The high-level conservation of the homeo box in a few genes of *Drosophila* (all believed to be involved in patterning the fly body plan) and the restriction of this coding region to a few phyla of higher metazoa make it impossible not to speculate whether the homeo domain could be associated with a common developmental process in these disparate phyla. The early embryology of the variety of animals that conserve the homeo box is profoundly different; indeed, this is an important part of the evidence that has been used to phylogenetically classify these groups. Although this does not rule out the homeo domain encoding similar developmental functions in all the phyla, it is a strong argument against it. Since one of the genes with an Antennapedia class homeo box, *ftz*, is necessary for proper segmentation (Nüsslien-Volhard and Wieschaus 1980; Wakimoto et al. 1984), and others function in the determination of segmental diversity (see Lewis 1978; Kaufman 1983), one can propose a function evolutionarily fixed to the process of metameric pattern elaboration. However, the presence of well-conserved copies in a sea urchin genome disputes this hypothesis, since even a broadest definition of metamerism excludes the sea urchin body plan (unless one designates, as some do [Hadzi 1963], the early subdivisions of the echinoderm coelom as metameric or oligomeric body planning gone bizarre). Yet it is intriguing that the annelids and mollusks, whose early embryology is similar, are apparently distinct in that the annelids (metameric) conserve multiple homeo box copies, while the mollusks (nonmetameric) have no detectable copies. On the other branch of the presumed protostome/deuterostome split (Hyman 1940) (shown diagrammatically in Fig. 4), the cephalochordate *Amphioxus* (strong metamerism) has multiple copies, whereas the closely related urochordate *Phallusia* (weak metamerism, only in larval stage) shows one or a few detectable copies.

Although the functions effected by the gene family marked by the homeo box are not yet known, the evidence from the genetics and from structural comparisons to other better-characterized gene products strongly favors a genetic regulatory role in the determination of cell fate. The ability to study these genetic functions in a variety of different organisms, with a wide spectrum of experimental advantages, should advance our knowledge about homeotic gene functions and subfunctions and determine which of these can be generalized among animal developmental systems.

ACKNOWLEDGMENTS

My thanks go to Walter Gehring, Marco Totolli, and Judy Varner who helped acquire and identify many of the organisms. Thanks also go to Bill Loomis for *Dictyostelium* DNA, Markus Noll for yeast DNA, Maria Salvato for *C. elegans* DNA, and Igor Dawid for the *Drosophila* rDNA clone. I am indebted to June Medford for allowing me to mention her unpublished results, to Nadine McGinnis for help with experiments, to Michael Levine for constructive criticism of both the experiments and the manuscript, and to Erika Wenger for typing the manuscript.

REFERENCES

AKAM, M.E. 1983. The location of *Ultrabithorax* transcripts in *Drosophila* tissue sections. *EMBO. J.* **2**: 2075.

BEACHY, P.A., S.L. HELFAND, and D.S. HOGNESS. 1985. Segmental distribution of bithorax complex proteins during *Drosophila* development. *Nature* **313**: 545.

BENDER, W., M. AKAM, F. KARCH, P.A. BEACHY, M. PFEIFER, P. SPIERER, E.B. LEWIS, and D.S. HOGNESS. 1983. Molecular genetics of the bithorax complex in *Drosophila melanogaster*. *Science* **221**: 23.

CARRASCO, A.E., W. MCGINNIS, W.J. GEHRING, and E.M. DE ROBERTIS. 1984. Cloning of a *Xenopus laevis* gene expressed during early embryogenesis coding for a peptide region homologous to *Drosophila* homeotic genes. *Cell* **37**: 409.

DAWID, I.B., P.K. WELLAWER, and E.O. LONG. 1978. Ribosomal DNA in *Drosophila melanogaster*. I. Isolation and characterization of cloned fragments. *J. Mol. Biol.* **126**: 749.

FILES, J.G. and D. HIRSH. 1981. Ribosomal DNA of *Caenorhabditis elegans*. *J. Mol. Biol.* **149**: 223.

FJOSE, A., W. MCGINNIS, and W.J. GEHRING. 1985. Isolation of a homeo box-containing gene from the *engrailed* region of *Drosophila* and the spatial distribution of its transcripts. *Nature* **313**: 284.

FREI, E., S. BAUMGARTNER, J.E. EDSTROM, and M. NOLL. 1985. Cloning of the *extra sex combs* gene of *Drosophila* and its identification by P-element mediated gene transfer. *EMBO J.* **4**: 979.

GARBER, R.L., A. KUROIWA, and W.J. GEHRING. 1983. Genomic and cDNA clones of the homeotic locus *Antennapedia* in *Drosophila*. *EMBO J.* **2**: 2027.

GARCIA-BELLIDO, A. 1977. Homeotic and atavic mutations in insects. *Am. Zool.* **17**: 613.

GRAHAM, D.E., B.R. NEUFELD, E.H. DAVIDSON, and R.J. BRITTEN. 1974. Interspersion of repetitive and non-repetitive sequences in the sea urchin genome. *Cell* **1**: 127.

HADZI, J., ed. 1963. *The evolution of the metazoa*. MacMillan, New York.

HAFEN, E., A. KUROIWA, and W.J. GEHRING. 1984a. Spatial distribution of transcripts from the segmentation gene *fushi tarazu* during *Drosophila* embryonic development. *Cell* **37**: 833.

HAFEN, E., M. LEVINE, and W.J. GEHRING. 1984b. Regulation of *Antennapedia* transcript distribution by the bithorax complex in *Drosophila*. *Nature* **307**: 287.

HYMAN, L., ed. 1940. *The invertebrates: Protozoa through Ctenophora*. Vol. 1. McGraw-Hill, New York.

KAUFMAN, T.C. 1983. The genetic regulation of segmentation in *Drosophila melanogaster*. In *Time, space and pattern in embronic development* (ed. W.R. Jeffery and R.A. Raff), p. 365. A.R. Liss, New York.

KUROIWA, A., E. HAFEN, and W.J. GEHRING. 1984. Cloning and transcriptional analysis of the segmentation gene *fushi tarazu* of *Drosophila*. *Cell* **37**: 825.

LAUGHON, A. and M.P. SCOTT. 1984. Sequence of a *Drosophila* segmentation gene: Protein structure homology with DNA-binding proteins. *Nature* **310**: 25.

LEVINE, M., G.M. RUBIN, and R. TJIAN. 1984. Human DNA sequences homologous to a protein coding region conserved between homeotic genes of *Drosophila*. *Cell* **38**: 667.

LEVINE, M., E. HAFEN, R.L. GARBER, and W.J. GEHRING. 1983. Spatial distribution of *Antennapedia* transcripts during *Drosophila* development. *EMBO J.* **2**: 2037.

LEWIS, E.B. 1978. A gene complex controlling segmentation in *Drosophila*. *Nature* **276**: 565.

MCGHEE, J.D. and G. FELSENFELD. 1980. Nucleosome structure. *Annu. Rev. Biochem.* **49**: 1115.

MCGINNIS, W., C.P. HART, W.J. GEHRING, and F.H. RUDDLE. 1984a. Molecular cloning and chromosome mapping of a mouse DNA sequence homologous to homeotic genes of *Drosophila*. *Cell* **38**: 675.

McGinnis, W., R.L. Garber, J. Wirz, A. Kuroiwa, and W.J. Gehring. 1984b. A homologous protein-coding sequence in *Drosophila* homeotic genes and its conservation in other metazoans. *Cell* **37**: 403.

McGinnis, W., M. Levine, E. Hafen, A. Kuroiwa, and W.J. Gehring. 1984c. A conserved DNA sequence found in homeotic genes of the *Drosophila* Antennapedia and bithorax complexes. *Nature* **308**: 428.

Meinkoth, J. and G. Wahl. 1984. Hybridization of nucleic acids immobilized on solid supports. *Anal. Biochem.* **138**: 267.

Miller, A.M., V.L. MacKay, and K.A. Nasmyth. 1985. Identification and comparison of two sequence elements that confer cell-type specific transcription in yeast. *Nature* **314**: 598.

Nüsslein-Volhard, C. and E. Wieschaus. 1980. Mutations affecting segment number and polarity in *Drosophila*. *Nature* **287**: 795.

Ohlendorf, D.H., W.F. Anderson, and B.F. Matthews. 1983. Many gene-regulatory proteins appear to have a similar helical fold that binds DNA and evolved from a common precursor. *J. Mol. Evol.* **1983**: 109.

Phillippsen, P., M. Thomas, R.A. Kramer, and R.W. Davis. 1978. Unique arrangements of coding sequences for 5S, 5.8S, 18S, and 25S ribosomal RNA in *Saccharomyces cerevisiae* as determined by R-loop and hybridization analysis. *J. Mol. Biol.* **123**: 387.

Poole, S.J., L.M. Kauver, B. Drees, and T. Kornberg. 1985. The *engrailed* locus of *Drosophila*: Structural analysis of an embryonic transcript. *Cell* **40**: 37.

Ptashne, M., A. Jeffrey, A.D. Johnson, R. Maurer, B.J. Meyer, C.O. Pabo, T.M. Roberts, and R.T. Sauer. 1980. How the repressor and *cro* work. *Cell* **19**: 1.

Scott, M.P. and A.J. Weiner. 1984. Structural relationships among genes that control development: Sequence homology between the *Antennapedia*, *Ultrabithorax*, and *fushi tarazu* loci of *Drosophila*. *Proc. Natl. Acad. Sci.* **81**: 4115.

Shepherd, J.C.W., W. McGinnis, A.E. Carrasco, E.M. De Robertis, and W.J. Gehring. 1984. Fly and frog homeo domains show homologies with yeast mating type regulatory proteins. *Nature* **310**: 70.

Struhl, G. 1982. Genes controlling segmental specification in the *Drosophila* thorax. *Proc. Natl. Acad. Sci.* **79**: 7380.

Wakimoto, B.T. and T.C. Kaufman. 1981. Analysis of larval segmentation in lethal genotypes associated with the Antennapedia gene complexes in *Drosophila melanogaster*. *Dev. Biol.* **81**: 51.

Wakimoto, B.T., F.R. Turner, and T.C. Kaufman. 1984. Defects in embryogenesis in mutants associated with the Antennapedia gene complex of *Drosophila melanogaster*. *Dev. Biol.* **102**: 147.

Wharton, K.A., B. Yedvobnick, V.G. Finnerty, and S. Artavanis-Tsakonas. 1985. *opa*: A novel family of transcribed repeats shared by the *Notch* locus and other developmentally regulated loci in *D. melanogaster*. *Cell* **40**: 55.

White, R.A.H. and M. Wilcox. 1984. Protein products of the bithorax complex in *Drosophila*. *Cell* **39**: 163.

The *Xenopus* Homeo Boxes

E.M. DE ROBERTIS, A. FRITZ, J. GOETZ, G. MARTIN, I.W. MATTAJ,
E. SALO, G.D. SMITH, C. WRIGHT, AND R. ZELLER
Biocenter, University of Basel, CH-4056 Basel, Switzerland

Embryologists have been searching for "cytoplasmic determinants" in egg cells for many decades. These elusive substances are thought to be located in the cytoplasm of most eggs and to become localized in specific blastomeres of the embryo where they then act upon the genes to determine the initial steps in the differentiation of the embryo (Conklin 1905; Wilson 1928; Davidson 1979; Gurdon 1977). In 1934, T.H. Morgan wrote, "It is known that the protoplasm of different parts of the embryo is somewhat different. The initial differences in the protoplasmic regions may be supposed to affect the activity of genes. The genes will then in turn affect the protoplasm, which will start a new series of reciprocal reactions. In this way we can picture to ourselves the gradual elaboration and differentiation of the various regions of the embryo."

The task of identifying and isolating the molecules that control development out of the myriad of unknown components that constitute a vertebrate egg is daunting. Although a few maternal-effect mutants have been studied in vertebrates, there is little evidence to suggest that they encode the molecules that determine the initial developmental decisions. The difficulties, however, have not deterred modern embryologists from trying, and the advent of new and powerful techniques (such as monoclonal antibodies, differential screening of cDNA libraries, and antisense RNA) gives hope that the problem can be attacked even in the absence of conventional genetics. Our laboratory has been interested in the early development of the frog *Xenopus laevis* for a number of years. We initially chose to study how macromolecules migrate from the cytoplasm into the nucleus in early embryos, because this was one aspect of the expected behavior of cytoplasmic determinants that we felt was amenable to analysis at the time (for discussion, see De Robertis and Black 1981; De Robertis 1983).

We have recently discovered what may turn out to be a new way to identify and isolate molecules that may be involved in establishing the early developmental decisions in vertebrate embryos (Carrasco et al. 1984; Muller et al. 1984). The finding that *X. laevis*, a vertebrate, has homeo-box-containing genes was possible because we shared the same floor with W. Gehring at the Biozentrum in Switzerland at the time the *Drosophila* homeo box was discovered (for review, see Gehring 1985). Our close scientific interaction in a department dedicated to the embryology of fruit flies and frogs led to the unusual, but eventually productive, idea of screening a *Xenopus* library with an Antennapedia probe. The initial finding of homeo boxes in frogs was rapidly followed by the finding of similar sequences in a variety of other vertebrates, including man, by Southern blots (McGinnis et al. 1984b) and sequencing (Levine et al. 1984; McGinnis et al. 1984a). Several lines of evidence (see below) suggest that homeo-box-containing genes may control vertebrate cell differentiation, although definite proof is still lacking.

One of the *Xenopus* genes we isolated (Muller et al. 1984) has the interesting feature of being transcribed in fully mature *Xenopus* oocytes. This raises the possibility that the protein product of this maternal gene may be located in egg cytoplasm for the control of the initial steps of embryonic differentiation.

RESULTS AND DISCUSSION

Frog Homeo Boxes

The homeo box codes for a region of about 60 amino acids, usually located in the last exon of much longer mRNAs, which at the protein level is called the homeo domain (McGinnis et al. 1984b,c; Scott and Weiner 1984). The homeo domain, which starts and ends abruptly at about the same point in all fly and vertebrate genes, thus represents only a short segment of the complete homeotic protein. The selective pressure to keep the primary structure constant applies to a very defined region of the protein.

Figure 1 compares the sequences of XlH-1 (for *X. laevis* homeo box 1, previously called AC1; Carrasco et al. 1984) and XlH-2 (*X. laevis* homeo box 2, previously called MM3; Muller et al. 1984). The codon usage suggests that the frog homeo boxes must be translated into protein at some stage during the lifetime of the organism, because in 180 nucleotides, there are 23 silent nucleotide changes that do not change the amino acid composition (Fig. 1).

Table 1 shows that the two frog homeo boxes are more conserved at the amino acid level than at the protein level. The homeo domains of frogs (as well as those in mice and humans; McGinnis et al. 1984a; Levine et al. 1984) are strikingly conserved when compared with those of *Drosophila*, to a degree comparable to the homology shared by Antennapedia (*Antp*), Ultrabithorax (*Ubx*), and fushi tarazu (*ftz*) among themselves. The record is held by XlH-2, which shares 59 out of 60 amino acids with the Antennapedia homeo box (Muller et al. 1984).

We have recently been able to isolate a large collec-

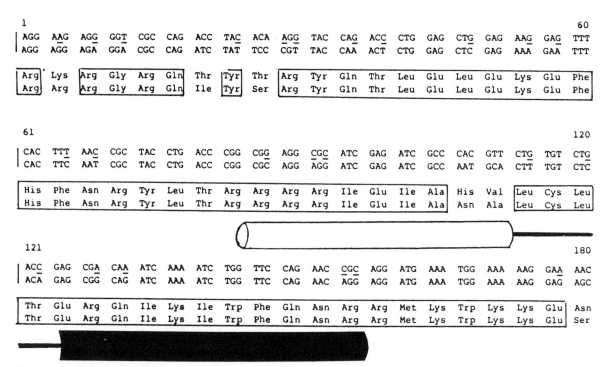

Figure 1. Sequence comparison of two *X. laevis* homeo boxes. The sequence of XlH-2 is shown above that of XlH-1. Silent nucleotide changes that do not change amino acids are underlined. The position of the helix-turn-helix motif present in prokaryotic proteins (and presumed to exist in the homeo box, only on the basis of the indirect argument that homeo boxes, yeast mating-type proteins, and prokaryotic repressors share homologies) is indicated by a white cylinder and a black cylinder under the sequence. Note that the sequence of amino acids in the black cylinder is identical in most homeo boxes sequenced to date.

tion of homeo-box-containing clones from a cDNA library. The characterization of these cDNA clones, which include six different homeo-box-containing genes, will be described elsewhere (A. Fritz et al., unpubl.).

The Functional Significance of Homeo Boxes in Frog Development

The developmental role of the homeo domain in frogs and other vertebrates is not yet known. If we could show that these transcripts are regionally localized in early embryos by in situ hybridization, we could then ascertain whether these genes have a function similar to those of *Drosophila*, which are in general involved in segmentation. Unfortunately, in situ hybridization has not been successful in vertebrate material up to now, but since it is being pursued vigorously in a number of laboratories (including our own), we should soon know the distribution of these transcripts in embryos. In the meantime, however, four lines of indirect evidence suggest that the vertebrate homeo-box-containing genes may indeed be developmentally important.

Transcription in embryos. The two *Xenopus* genes thus far studied by Northern blots have a tightly controlled pattern of transcription during early development. XlH-1 produces three transcripts starting at gastrulation, each one of which is under specific temporal control (Carrasco et al. 1984). Although clearly different, the situation is reminiscent of the temporal control

Table 1. Sequence Comparison of Frog and Fly Homeo Boxes

Compared homeo boxes	XlH-2 XlH-1	XlH-2 Antp	XlH-1 Antp	XlH-2 ftz	XlH-1 ftz	XlH-2 Ubx	XlH-1 Ubx	Antp ftz	ftz Ubx	Ubx Antp
Identical nucleotides (out of 180)	148	141	140	133	130	131	131	138	133	143
Percentage of homology (180 bp = 100%)	82	78	78	74	72	73	73	77	74	79
Identical amino acids (out of 60)	54	59	55	49	48	51	51	50	45	51
Percentage of homology (60 amino acids = 100%)	90	98	92	82	80	85	85	83	75	85

Note that the degree of conservation between fly and frog is similar to that of fly genes among themselves. The *Drosophila* sequences were determined by McGinnis et al. (1984b).

of *Ubx* and *Antp* transcripts in *Drosophila* (Bender et al. 1983; Garber et al. 1983; Beachy et al. 1985).

XlH-2 is predominantly expressed in oocytes, which, as mentioned above, makes it of considerable embryological interest. Its transcripts disappear during cleavage, and two transcripts become detectable again after the midblastula transition (Muller et al. 1984).

Transcript structure. The *Xenopus* XlH-1 and XlH-2 transcripts resemble those of *Drosophila* in several respects: (1) The homeo box is immediately preceded by an intron; (2) the homeo box is in the last exon of the transcripts; and (3) all transcripts have long 3'-untranslated regions (A. Fritz et al., unpubl.).

Genomic Southern blots. In *Drosophila*, of about ten genomic restriction fragments that contain detectable homeo boxes (McGinnis et al. 1984c), at least eight correspond to homeotic or segmentation genes (Gehring 1985). Thus, in *Drosophila*, the homeo box seems to be associated specifically with the genes that control development. Vertebrates have a similar number of homeo boxes: Southern blots of chicken, mouse, human (McGinnis et al. 1984b), and frog (Carrasco et al. 1984) genomic DNAs also show about ten detectable hybridizing restriction fragments. In mammals, homeo boxes are clustered in the genome *(Hox-1* and *Hox-2* gene complexes; Rabin et al. 1985), as are the genes of the bithorax and Antennapedia complexes in *Drosophila*.

Homologies to the yeast mating-type genes. Soon after the discovery of *Drosophila* and *Xenopus* homeo boxes, their sequences were used in a computer search of the Dayhoff protein library. Significant homologies were found only with the *Saccharomyces cerevisiae* mating-type proteins **a**1 and α2 (Shepherd et al. 1984). As shown in Figure 2, the closest homology between the homeo domain and the mating-type genes occurs in the last third of the homeo box. This is also the region that is conserved in the *Drosophila* gene engrailed (Fjose et al. 1985; Poole et al. 1985), which has a homeo box that has diverged considerably from those of *Antp*, *Ubx*, and *ftz*. A mating-type gene from the fission yeast *Schizosaccharomyces pombe*, which is very different from its *S. cerevisiae* counterpart, is nevertheless conserved at the residues shared with the homeo box (M. Smith and D. Beach, pers. comm.).

This homology with yeast genes is remarkable, because the yeast mating-type cassette genes are known to mediate the stable determination of cell types in yeast, i.e., a type of cell differentiation (Herskowitz and Oshima 1981; Nasmyth 1982). Current understanding on how mating-type genes work in yeast is progressing rapidly, and hopefully this will also provide insights into how homeo domains might work.

Only One Type of Homeo Box?

The yeast mating-type proteins **a**1 and α2 are known to share weak homologies with prokaryotic DNA-binding proteins (Matthews et al. 1983); thus, it was proposed (Laughon and Scott 1984; Shepherd et al. 1984) that the homeo domain might be DNA-binding. Prokaryotic repressors have a conserved helix-turn-helix structure (Pabo and Sauer 1984); the regions of the homeo box (residues 29–52) corresponding to this structure are shown in Figure 1.

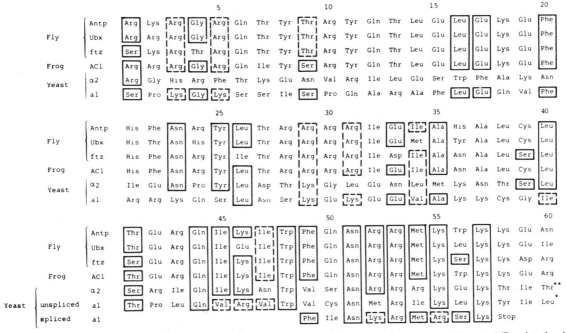

Figure 2. The *Drosophila* and frog homeo domains share homologies with yeast mating-type cassette genes. (Reprinted, with permission, from Shepherd et al. 1984.)

Studies with prokaryotic repressors have shown that the sequence specificity is determined by the amino acid sequence of the helical region (Fig. 1, black cylinder) which fits into the major groove of DNA (Ebright et al. 1984; Wharton et al. 1984). The linear sequence of amino acids in this region is identical in two frog, two human, one mouse, and two *Drosophila* homeo boxes (Carrasco et al. 1984; Levine et al. 1984; McGinnis et al. 1984a,b; Muller et al. 1984; Scott and Weiner 1984). Thus, if we accept the reasoning that the region is a sequence-specific DNA-binding helix, the logical consequence is that all of these proteins should bind to the same operator sequence on DNA. On the other hand, a "simple" organism like yeast has at least two different such sequences in the a1 and α2 proteins. Thus, it seems unlikely that there would be only one type of homeo box in higher organisms. Perhaps the problem stems from the fact that most studies have used the Antennapedia homeo box as the main probe, and consequently most genes isolated are similar to it. Future screening will be carried out using probes of the homeo boxes that have diverged most, in the hope of identifying a new, but still related, homeo box family.

An important concept that has emerged from the yeast system (Miller et al. 1985) is that the spacing between operator sequences on DNA might be very important in obtaining combinatorial effects between different DNA-binding proteins. We cannot examine this hypothesis in depth here, but it could explain the complex interactions known to exist between homeotic genes (Hafen et al. 1984), even if there were only a single or a few DNA-binding specificities (see Miller et al. 1985).

Although it does seem likely that the homeo domain has, at least in part, a DNA-binding function, its extreme conservation over a linear stretch of 60 amino acids (far more than what is required for DNA binding in prokaryotes) suggests that it might also recognize or be recognized by other components. Perhaps it interacts with an essential cellular component present in all cells (e.g., RNA polymerase or nucleosomes). Since there are multiple homeo-box-containing genes that must interact with this component, it may be difficult to introduce new mutations. In this way, the sequence may have become fixed in evolution, providing a way of isolating the genes that contain it.

Are Homeo Boxes in Vertebrates Related to Segmentation?

Until the results from the in situ hybridizations are available, we will not know to what extent vertebrate homeo-box-containing genes are functionally related to their *Drosophila* counterparts. We would like to put forth the argument here that, even if they do not turn out to hybridize in stripes, the family of developmentally regulated *Xenopus* genes we have discovered deserves further study.

Although on genomic Southern blots the presence of homeo boxes correlates in general with segmented metazoans (McGinnis et al. 1984b), yeast cells do have genes for homologous proteins (Shepherd et al. 1984), but they cannot hybridize at the nucleic acid level. Thus, the correlation between homeo boxes and segmentation might not be complete.

In a recent study, Gruss and colleagues (Colberg-Poley et al. 1985) have shown that embryonal carcinoma F9 stem cells induced to differentiate with retinoic acid will switch on transcription of a mouse homeo-box-containing gene. In this case, the switch cannot be related to segmentation, because F9 cells under these culture conditions differentiate into extraembryonic parietal endoderm cells.

If we assume that homeo-box-containing genes code for proteins that control the transcription of batteries of other genes, then we cannot necessarily expect their behavior in other species to be identical to that in *Drosophila*. In the fruit fly, the earliest developmental decisions are made at the blastoderm stage and are connected with segmentation. In the amphibian embryo, on the other hand, segmentation of the mesoderm does not occur until after gastrulation. In the mammalian embryo, the earliest decisions are connected with the determination of extraembryonic tissues. Thus, in other organisms, the transcription-controlling proteins involved in early embryogenesis may have very different patterns of expression from those in *Drosophila*.

One possibility we would like to suggest is that vertebrate homeo-box-containing proteins could control batteries of other genes by binding to DNA within gene promoters and interacting, directly or indirectly, with tissue-specific gene enhancers. They may even turn out to be the enhancer-binding proteins.

We can envisage that in the future, most of the mouse, human, and frog homeo-box-containing genes will be cloned and characterized, and their counterparts in each species will be established by sequence comparisons and hybridization using gene-specific probes (lacking the homeo box region). Thus, the findings concerning the function or tissue distribution of a gene in frogs might be immediately applicable to mammals, and vice versa. In this context, it will be interesting to study the transcription of the counterparts of the frog maternally expressed gene in mouse embryos, because in mice, the body derives from a group of very few cells from the inner cell mass, whereas the majority of the blastocyst cells give rise to extraembryonic membranes. Thus, in mammals, some developmental decisions occur much later than in *Xenopus*, making it less likely that the molecules involved will be maternally inherited. The *Xenopus* studies might have something to contribute to our understanding of the strategy of early development in higher vertebrates as well.

We cannot say at this point whether the vertebrate homeo-box-containing genes discovered initially in *Xenopus* are directly involved in the control of vertebrate development. With the matter now being pursued vigorously in many laboratories throughout the world, it is likely that we shall soon know. We believe that further study of these intriguing *Xenopus* genes, which are subject to a very specific developmental regulation, should provide valuable insights into the strategy of early vertebrate development.

REFERENCES

BEACHY, P.A., S.L. HELFLAND, and D.S. HOGNESS. 1985. Segmental distribution of bithorax complex proteins during *Drosophila* development. *Nature* **313**: 545.

BENDER, W., M. AKAM, F. KARCH, P. BEACHY, M. PFEIFER, P. SPIERER, E.B. LEWIS, and D.S. HOGNESS. 1983. Molecular genetics of the bithorax complex in *Drosophila melanogaster*. *Science* **221**: 23.

CARRASCO, A.E., W. MCGINNIS, W.J. GEHRING, and E.M. DE ROBERTIS. 1984. Cloning of a *X. laevis* gene expressed during early embryogenesis coding for a peptide region homologous to *Drosophila* homeotic genes. *Cell* **37**: 409.

COLBERG-POLEY, A.M., S.D. VOSS, K. CHOWDHURY, and P. GRUSS. 1985. Structural analysis of murine genes containing homeobox sequences and their expression in embryonal carcinoma cells. *Nature* **314**: 713.

CONKLIN, E.G. 1905. The organization and cell lineage of the ascidian egg. *J. Natl. Acad. Sci.* **13**: 1.

DAVIDSON, E.H. 1976. *Gene activity in early development*, 2nd edition. Academic Press, New York.

DE ROBERTIS, E.M. 1983. Nucleocytoplasmic segregation of proteins and RNAs. *Cell* **32**: 1021.

DE ROBERTIS, E.M. and P. BLACK. 1981. Frog oocyte nuclear proteins and the analysis of developmental determinants. *Fortschr. Zool.* **26**: 49.

EBRIGHT, R.H., P. COSSART, B. GICQUEL, and J. BECKWITH. 1984. Mutations that alter the DNA sequence specificity of the catabolite gene activator protein of *E. coli*. *Nature* **311**: 232.

FJOSE, A.W., W.J. MCGINNIS, and W.J. GEHRING. 1985. Isolation of a homeobox-containing gene from the engrailed region of *Drosophila* and the spatial distribution of its transcripts. *Nature* **313**: 284.

GARBER, R.L., A. KUROIWA, and W.J. GEHRING. 1983. Genomic and cDNA clones of the homeotic locus Antennapedia in *Drosophila*. *EMBO J.* **2**: 2027.

GEHRING, W.J. 1985. The homeobox: A key to the understanding of development? *Cell* **40**: 3.

GURDON, J.B. 1977. Egg cytoplasm and gene control in development. *Proc. R. Soc. Lond. B* **198**: 211.

HAFEN, E., M. LEVINE, and W.J. GEHRING. 1984. Interaction of homeotic genes in *Drosophila*: Altered pattern of Antennapedia + expression in bithorax mutant embryos. *Nature* **307**: 287.

HERSKOWITZ, I. and Y. OSHIMA. 1981. Control of cell-type in *Saccharomyces cerevisiae*: Mating type and mating type interconversion. In *The molecular biology of the yeast Saccharomyces cerevisiae: Life cycle and inheritance* (ed. J.N. Strathern et al.), p. 181. Cold Spring Harbor Laboratory, Cold Spring Harbor, New York.

LAUGHON, A. and M.P. SCOTT. 1984. Sequence of a *Drosophila* segmentation gene: Protein structure homology with DNA-binding proteins. *Nature* **310**: 25.

LEVINE, M., G.M. RUBIN, and R. TJIAN. 1984. Human DNA sequences homologous to a protein coding region conserved between homeotic genes of *Drosophila*. *Cell* **38**: 667.

MATTHEWS, B.W., D.H. OHLENDORF, W.F. ANDERSON, R.G. FISHER, and Y. TAKEDA. 1983. Cro repressor protein and its interaction with DNA. *Cold Spring Harbor Symp. Quant. Biol.* **47**: 427.

MCGINNIS, W., C.P. HART, W.J. GEHRING, and F.H. RUDDLE. 1984a. Molecular cloning and chromosome mapping of a mouse DNA sequence homologous to homeotic genes of *Drosophila*. *Cell* **38**: 675.

MCGINNIS, W., R.L. GARBER, H. WIRZ, A. KUROIWA, and W.J. GEHRING. 1984b. A homologous protein-coding sequence is conserved in *Drosophila* homeotic genes: Evidence for the conservation of a similar sequence in other metazoans. *Cell* **37**: 403.

MCGINNIS, W., M. LEVINE, E. HAFEN, A. KUROIWA, and W.J. GEHRING. 1984c. A conserved DNA sequence in homeotic genes of the *Drosophila* Antennapedia and bithorax complexes. *Nature* **308**: 428.

MILLER, A.M., V.L. MACKAY, and K.A. NASMYTH. 1985. Identification and comparison of two sequence elements that confer cell-type specific transcription in yeast. *Nature* **314**: 598.

MORGAN, T.H. 1934. *Embryology and genetics*. University Press, New York.

MULLER, M.M., A.E. CARRASCO, and E.M. DE ROBERTIS. 1984. A homeobox-containing gene expressed during oogenesis in *Xenopus*. *Cell* **39**: 157.

NASMYTH, K. 1982. Molecular genetics of yeast mating type. *Annu. Rev. Genet.* **16**: 439.

PABO, C.O. and R.T. SAUER. 1984. Protein-DNA recognition. *Annu. Rev. Biochem.* **53**: 293.

POOLE, S.J., L.M. KANVAR, B. DREES, and T. KORNBERG. 1985. The engrailed locus of *Drosophila*: Structural analysis of an embryonic transcript. *Cell* **40**: 37.

RABIN, M., C.P. HART, A. FERGUSON-SMITH, W. MCGINNIS, M. LEVINE, and F.H. RUDDLE. 1985. Two homeobox loci mapped in evolutionarily related mouse and human chromosomes. *Nature* **314**: 175.

SCOTT, M.P. and A.J. WEINER. 1984. Structural relationships among genes that control development: Sequence homology between the Antennapedia, Ultrabithorax, and fushi tarazu loci of *Drosophila*. *Proc. Natl. Acad. Sci.* **81**: 4115.

SHEPHERD, J.C.W., W. MCGINNIS, A.E. CARRASCO, E.M. DE ROBERTIS, and W.J. GEHRING. 1984. Fly and frog homeo domains show homologies with yeast mating-type regulatory proteins. *Nature* **310**: 70.

WHARTON, R.P., E.L. BROWN, and M. PTASHNE. 1984. Substituting an α-helix switches the sequence specific DNA interactions of a repressor. *Cell* **38**: 361.

WILSON, E.B. 1928. *The cell in development and heredity*. Macmillan, New York.

Mammalian Homeo Box Genes

F.H. RUDDLE,*† C.P. HART,* A. AWGULEWITSCH,* A. FAINSOD,* M. UTSET,†
D. DALTON,† N. KERK,* M. RABIN,* A. FERGUSON-SMITH,* A. FIENBERG,† AND W. MCGINNIS‡

Departments of *Biology, †Human Genetics, and ‡Molecular Biophysics and Biochemistry, Yale University, New Haven, Connecticut 06511

Drosophila homeo box sequences are 180-bp exonic elements residing within genes involved in the determination of body segment number and identity. These genes include members of the Antennapedia and bithorax complexes as well as others involved in *Drosophila* morphogenesis, such as engrailed (Laughon and Scott 1984; McGinnis et al. 1984b,c; Scott and Weiner 1984; Fjose et al. 1985; Poole et al. 1985). The functions of the homeo box sequences themselves are not yet known. It has been established that they exist in open reading frames and that they are located in genes whose protein products are nuclear in location in the few instances examined (White and Wilcox 1984; Beachy et al. 1985; Scott et al., this volume). Deduced amino acid sequences indicate a high level of conservation between organisms as distantly related as *Drosophila* and man (Levine et al. 1984; McGinnis et al. 1984a; Colberg-Poley et al. 1985), high levels (~30%) of the basic amino acids lysine and arginine (McGinnis et al. 1984b), and a 3' region that exhibits the properties of a DNA-binding domain (Laughon and Scott 1984; Shepherd et al. 1984). These features have suggested to some that the genes possessing homeo box sequences may perform their developmental function by controlling the expression of ensembles of genes through transcriptional regulation. However, other possible mechanisms of action are possible and cannot yet be ruled out.

Homeo box sequences have been highly conserved in a broad array of deuterostomes and protostomes, suggesting their origins in a common ancestor to these lineages (Carrasco et al. 1984; McGinnis et al. 1984b; McGinnis, this volume). Vertebrate and insect homeo boxes have been shown to have DNA sequence conservation at the level of about 70%, whereas the deduced amino acid conservation is even higher at a level of around 85%. Many of the nucleotide differences can be explained by third-position changes in codons that specify identical amino acids or amino acids of a similar biochemical type. Such patterns of conservation in gene structure are usually associated with conservation in genetic function, suggesting that mammalian genes containing homeo box sequences may be involved in some aspect of morphogenesis, although their specific roles may have diverged over the course of evolutionary time.

In this paper we attempt to shed some light on this general problem of homeo box function by describing our experimental studies on the homeo box sequences in mammals, using the laboratory mouse and man as model systems. Both species are well described from a genetic point of view, and the mouse has been an object of developmental analysis for at least 75 years. We have been able to show that many of the homeo box sequences are organized into complexes, at least superficially similar to the Antennapedia and bithorax complexes of *Drosophila*. We also find a high degree of conservation in the organization of one of these complexes between man and mouse, suggestive of the same functional activity. We also show that the complexes encode RNA transcripts produced during mouse embryogenesis and, moreover, that the expression of the transcripts is likely to possess spatial and temporal specificity. All of these findings are consistent with the view that the homeo box sequences in mammals are genetically functional, and may contribute to the regulation of the developmental process.

MATERIALS AND METHODS

Genomic library screening. The genomic library (BALB/c mouse embryo DNA cloned into Charon 28) was provided by P. Leder (Harvard Medical School) and screened as described previously (McGinnis et al. 1984a). Plaques (10^6) were screened in duplicate with the *Drosophila* Antennapedia and Ultrabithorax homeo box probes (for a description, see McGinnis et al. 1984b).

Southern blot hybridization. Genomic DNA (15 μg) or phage DNA (0.25 μg) was digested with restriction endonuclease and electrophoresed in 0.85% agarose Tris-borate gels. DNA was transferred to nitrocellulose according to the procedure of Southern (1975). Hybridizations were done at 65°C in 6×SSC, 2×Denhardt's, 0.1% SDS, and 150 μg/ml of denatured sonicated salmon sperm DNA, and in the case of genomic blots, 10% dextran sulfate. The hybridization probes were nick-translated in the presence of [^{32}P]dCTP to a specific activity of approximately 10^8 cpm/μg and added at 10 ng/ml of hybridization mix. Following a 12–16-hour incubation, the filters were first washed at 65°C in 2×SSC and 0.1% SDS for 30 minutes and then washed in 0.1×SSC and 0.1% SDS again at 65°C for 30 minutes.

Chromosome mapping. The somatic-cell hybrids used were formed between the Chinese hamster cell line

E36 and various types of mouse cells (for a summary, see McGinnis et al. 1984a). The murine chromosomal content of the individual hybrid cell lines was determined by isozymes, karyotypes, and/or DNA hybridization.

RNA isolation. Immediately after dissection, the tissue was transferred to 5 ml of an ice-cold guanidinium thiocyanate solution, and the tissue was homogenized for 1 minute at the highest speed with an N8 blade of an Ultra Turrax tissuemizer (Janke and Kunkel). After the addition of 2.2 g of solid CsCl to the lysate, the RNA was isolated by centrifugation through a 5.7 M CsCl step gradient in an SW 50.1 rotor (Beckman) at 35,000 rpm and 20°C for 12-15 hours. The poly(A)$^+$ RNA fraction was selected by passing the total RNA through an oligo(dT)-cellulose (Collaborative Research) column.

Gel electrophoresis of RNA and Northern blotting. The RNAs were separated by electrophoresis in a formaldehyde-containing 0.8% agarose gel; 8 µg of RNA was loaded per lane. *Escherichia coli* RNA (5 µg; Boehringer) provided molecular weight markers (16S and 23S rRNAs), in addition to the 18S and 28S rRNAs in the poly(A)$^-$ fraction of 13-day-old embryos. After electrophoresis, the lanes containing the molecular size markers were cut off and stained with ethidium bromide, and the RNAs from the rest of the gel were transferred to nitrocellulose.

RNA filter hybridization. The RNA filters were prehybridized for 8-15 hours in 50% formamide, 5×SSC, 50 mM sodium phosphate (pH 6.6), 0.1% SDS, 5×Denhardt's solution, and 100-200 µg/ml of denatured sonicated salmon sperm DNA (Sigma). In the hybridization mixture, the Denhardt's concentration was reduced to 1× and 10% dextran sulfate was added. The specific activity of the ^{32}P-labeled DNA probe ranged from 5×10^7 to 3×10^8 cpm/µg, and the concentration was about 3 ng/ml. After hybridization for 20-24 hours at 42°C, the filters were washed in 2×SSC and 0.1% SDS at room temperature for about 10 minutes, followed by three washes in 0.2×SSC and 0.1% SDS at 65°C for 20 minutes. The filters were autoradiographed with XAR-5 film (Kodak) and an intensifying screen for 3 days at -70°C.

RESULTS

Linkage Relationships

We have attempted to define the map positions of the individual homeo box sequences (1) to provide an identity and a system of classification for the individual sequences, (2) to detect patterns of homeo box clustering, and (3) to demonstrate proximity of location or possibly allelic identity with other known genes. We have concentrated on establishing gene map positions in the mouse, but in addition, we have located homeo box sequences in man, especially as a means of determining genetic homologies between the two species.

Four loci containing homeo box sequences have been identified in the mouse. We propose that these be designated *Hox-1*, *Hox-2*, *Hox-3*, and *Hox-4*. *Hox-1*, the first to be mapped, was assigned to mouse chromosome 6, using mouse/Chinese hamster somatic-cell, hybrid-gene-mapping panels, and to the region proximal to the IgK locus, using hybrid cells containing subfragments of chromosome 6. This region is estimated to be approximately 30 cM in length or equivalent to about 30,000 kb of DNA (Fig. 1). We first assigned one homeo box segment, having the trivial plasmid name Mo-10, to this region (McGinnis et al. 1984a). In this paper we report a second homeo box that maps in this region (see Table 1 and Figs. 2 and 3).

The human cognates of Mo-10 and Mo-11 (*Hox-1*) have not yet been defined, nor has a homologous map position been established. Since genetic linkage groups are often conserved in humans and the mouse, it is often possible to predict the location of a particular gene in one species given its chromosomal assignment

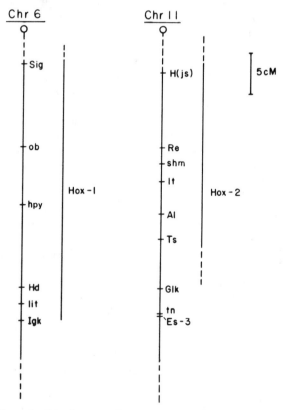

Figure 1. Genetic maps of the murine chromosomal regions containing the *Hox-1* and *Hox-2* homeo loci. (*Sig*) Sightless; (*ob*) obese; (*hyp*) hydrocephalic-polydactyl; (*hd*) hypodactyly; (*lit*) little; (Igk) immunoglobulin kappa; (H[js]) histocompatibility (js); (*Re*) rex; (*shm*) shambler; (*lt*) lustrous; (*Al*) alopecia; (*Ts*) tail short; (*Glk*) galactokinase; (*tn*) teetering; (Es-3) esterase-3.

Table 1. Segregation of Mo-11 in Mouse-Hamster Hybrids
(Presence of Chromosome/Blot Hybridization)

Mouse chromosome	Concordant			Discordant		
	+/+	−/−	total	+/−	−/+	total
1	12	8	20	2	1	3
2	10	5	15	5	2	7
3	11	9	20	1	2	3
4	12	8	20	2	1	3
5	5	6	11	3	8	11
6	11	10	21	0	0	0
7	13	3	16	7	0	7
8	7	7	14	3	6	9
9	8	6	14	3	6	9
10	10	6	16	4	3	7
11	1	8	9	0	9	9
12	12	0	12	10	1	11
13	2	3	5	3	1	4
14	3	6	9	4	10	14
15	3	1	4	5	0	5
16	0	1	1	5	3	8
17	11	4	15	6	2	8
18	2	9	11	1	5	6
19	6	3	9	7	4	11
X	7	3	10	7	2	9

Summary of the blot hybridization results of Mo-11 with 23 mouse/hamster cell line DNAs. There were zero discordancies for the presence or absence of the Mo-11 hybridization and the presence or absence of the proximal portion of mouse chromosome 6 (cen → B2).

in the other. Human chromosome 2 has been implicated in this regard since it carried IgK, the murine homolog of which maps close to the *Hox-1* locus on mouse chromosome 6. However, an equally strong association with mouse chromosome 6 may be made between human chromosomes 7 and 12, respectively, since linkage relationships among homologous genes in mouse and man are also conserved on these chromosomes.

It will of course be extremely important to establish identity between a known genetic mutation and any of the homeo-box-containing genes. This could provide an immediate clue to the function of the homeo box gene. There are at least three morphogenetic mutations that map into the interval between IgK and the centromere. These are *sightless, hydrocephalic polydactyl,* and *hypodactyly*. The genes *obese* and *little* also map in this region, and they have been described as affecting the differentiation of the endocrine function of the pancreas and the pituitary, respectively (Green 1981). It will be necessary to obtain high-resolution mapping data to fix the position of *Hox-1* more precisely in order to establish its genetic map position relevant to these genes. As we shall see, more progress has been made along these lines in the case of *Hox-2*.

Hox-2 was mapped to mouse chromosome 11 again by means of mouse/Chinese hamster hybrid cell-mapping panels. At first, the subregional localization of the locus was deduced from information on the mapping of homologous homeo box sequences in man. Independently, Rabin et al. (1985) and Joyner et al. (1985), using techniques of in situ hybridization and chromosome sorting, respectively, showed that human *Hox-2* sequences mapped to the middle of the long arm of human chromosome 17. This region contains the gene thymidine kinase (*tk*) and galactosekinase (*glk*) in close linkage (Kozak and Ruddle 1977). It had also been established earlier by means of recombinant inbred strain mapping that the mouse *glk* gene mapped to the distal region of mouse chromosome 11 (Mishkin et al. 1976). Therefore, by homology, one could argue that the mouse *Hox-2* locus was also located in that region of the chromosome. Subsequent Mendelian linkage tests making use of a restriction-fragment-length polymorphism (RFLP) for the murine *Hox-2* locus have provided direct experimental evidence to support that prediction. On the basis of these experiments (D. Dalton and F.H. Ruddle, unpubl.), it is likely that the murine *Hox-2* locus will map in the vicinity of the gene *rex* on mouse chromosome 11.

As in the case of *Hox-1*, it is interesting to inspect the *Hox-2* region for previously reported genes that might be relevant to the possible function of the *Hox-2* locus. The relevant genes in the vicinity of the *rex* region are *rex* itself, *trembler, lustrous, alopecia,* and *tail short*. *rex, lustrous,* and *alopecia* all affect different aspects of hair differentiation, whereas *trembler* and *shambling* are neurological. *Tail short* (*Ts*) is semidominant and morphogenetic in its expression. In the *Ts* heterozygotes, the tail is typically kinked, and there are numerous skeletal abnormalities, including vertebral fusions and additional pairs of ribs. The homozygote animals show abnormalities as early as 3.5 days of development and are dead by 5.5 days (Green 1981). Using the RFLP variant, it will now be possible to test for allelism between *Hox-2* and the loci described above. These experiments are currently under way.

Evidence at the molecular level has been obtained showing that the murine *Hox-2* locus contains at least four homeo boxes over a relatively short distance of approximately 30 kb (C.P. Hart et al., in prep.). This evidence is presented in detail below. These homeo boxes have been designated *Hox-2.1, Hox-2.2, Hox-2.3,* and *Hox-2.4*. In the cognate human *Hox-2* locus, three homeo boxes have been described; two, designated Hu-1 and Hu-2 (Levine et al. 1984), are homologous to the murine *Hox-2.1* and *Hox-2.2* homeo boxes. These relationships are thoroughly discussed below. Thus, both *Hox-1* and *Hox-2* show the clustering of multiple homeo box sequences.

We have recently mapped a homeo box sequence (trivial name: EA) to mouse chromosome 15. We propose that this locus be designated *Hox-3*.

A fourth homeo box region has recently been described by G. Martin (University of California, San Francisco) and her associates. They have isolated a murine analog of the *Drosophila* engrailed homeo box. A detailed account of these findings are presented in these proceedings. The murine, engrailed-like homeo box maps to mouse chromosome 1 in the vicinity of the *dominant hemimelia* (*Dh*) gene. We suggest that this locus be termed *Hox-4*.

Structure of the Hox-2 Locus

Three phage clones containing murine homeo box sequences were isolated by cross-hybridization to *Drosophila* homeo box sequences. These were designated Mo-1, Mo-3, and Mo-4. Each was found to carry two homeo boxes, and restriction mapping showed three overlapping inserts. As shown in Figure 2, restriction analysis defines four different homeo boxes in tandem array over a distance of approximately 30 kb together with flanking sequences. We termed these homeo boxes *Hox-2.1*, *Hox-2.2*, *Hox-2.3*, and *Hox-2.4* (from right to left). Phage restriction fragments were isolated in order to identify unique genomic sequences free of mouse middle-repetitive sequences. Surprisingly, individual hybridization probes covering more than 80% of the cloned region were all found to be single-copy sequences by the Southern transfer technique. To confirm and generalize this result, we tested for hybridization homology between our entire phage inserts and total genomic DNA. These tests confirmed the absence of middle-repetitive elements throughout the *Hox-2* domain as defined above. We regard this finding significant given the ubiquity of middle-repetitive DNA in the mouse genome. It suggests that the entire *Hox-2* region is engaged in essential genetic activity. We will see this viewpoint confirmed in regard to other attributes of the locus, such as transcriptional expression and the conservation of sequence homology in comparison with the human *Hox-2* locus.

As described above, gene-mapping data showed homology between homeo box sequences located on mouse chromosome 11 and human chromosome 17. This suggested that the human homeo box inserts Hu-1 and Hu-2, which had been shown to map to chromosome 17, might be homologs of the mouse series. It was also known that the Hu-1 and Hu-2 homeo boxes were in tight linkage, separated by a distance of 5 kb, approximately the same distance separating mouse homeo boxes 2.1 and 2.2. We therefore decided to test for homology between these mouse and human sequences (1) by hybridization analysis and (2) by a comparison of nucleotide sequences in the regions flanking the homeo boxes proper. Using selected cloned genomic segments containing homeo box sequences and their flanks, we demonstrated that strong hybridization homology existed between the homeo box and adjacent sequences of *Hox-2.1* and Hu-1, on the one hand, and between *Hox-2.2* and Hu-2, on the other. Strong homology attributable to the flanking sequences was found only with the above combinations. This result was confirmed by the direct examination of flanking nucleotide sequences that showed more than 90% homology in flanking sequences between cognate regions. This suggests that this homeo box gene is part of a genetic function very similar to that shared by these mammalian species, which are separated by over 80 million years of evolutionary time.

The results of the homology tests described above provide the basis for a revised nomenclature for the human *Hox-2* homeo boxes. We recommend that Hu-1 and Hu-2 henceforth be termed *Hox-2.1* and *Hox-2.2*, respectively. The homology relationships between the human and mouse *Hox-2* loci also predict that the human *Hox-2* locus will possess homeo box sequences equivalent to the murine homeo boxes *Hox-2.3* and *Hox-2.4*. Taking all of the mouse and human *Hox-2* data into account, it is likely that there is a minimum of four homeo boxes at this locus in both species.

Hox-1 and Hox-2 Loci Expression

Northern blotting and in situ hybridization analysis has provided evidence for the transcriptional activity of homeo-box-associated genes. Our most extensive data stem from an analysis of the murine *Hox-2* locus (Fig. 3). A total of 12 different DNA probes covering this locus have been used individually for Northern blot hybridizations with poly(A)$^-$ and poly(A)$^+$ RNAs from 13-day-old embryos, as well as poly(A)$^+$ RNA from adult liver. With the probes containing a homeo box sequence, we were able to identify five major transcripts of different molecular weights in embryonal poly(A)$^+$ RNA but not in poly(A)$^+$ RNA from adult liver. The size of these transcripts ranges from about 1.2 kb to 2.2 kb. In addition to these lower-molecular-weight species of rather high signal intensity, we detected transcripts of higher molecular weight in the range of about 3–6 kb, which gave signals of considerably lower intensity. The high-molecular-weight species also appear, with one exception, only in the em-

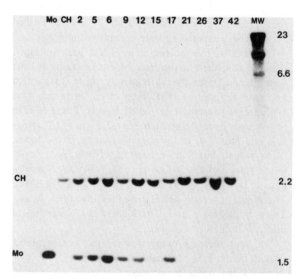

Figure 2. Chromosomal mapping of λMo-11. The 1.5-kb *Hin*dIII fragment from λMo-11 was nick-translated to a specificity activity of $10^8/\mu g$ and hybridized to a Southern blot of mouse/Chinese hamster cell line DNAs digested with *Hin*dIII. (Mo) Mouse genomic DNA; (CH) Chinese hamster genomic DNA. (Tracks 2–42) Mouse/Chinese hamster cell line hybrids of the If-4 series; (MW) λDNA, all digested with *Hin*dIII. The probe hybridizes itself in the mouse control track and the mouse/hamster hybrids retaining chromosome 6. Note the strong cross-hybridization to the Chinese hamster 2.2-kb *Hin*dIII band.

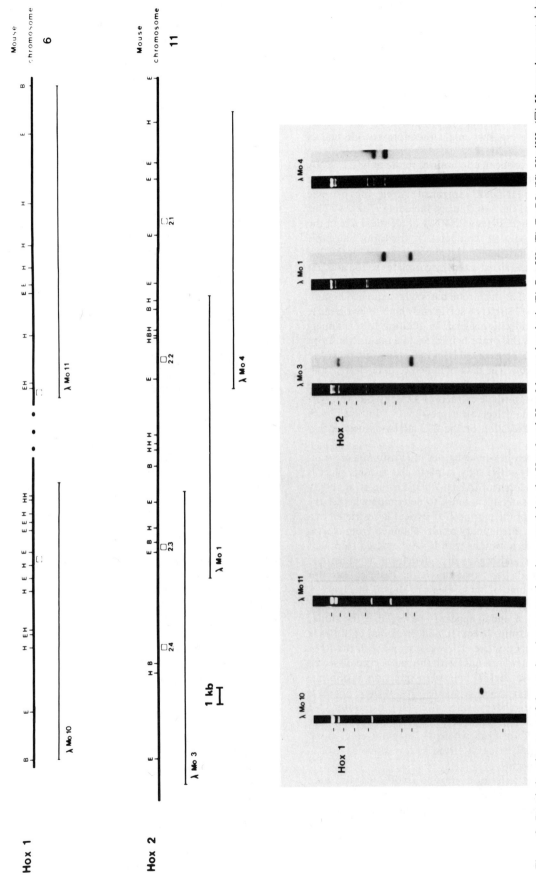

Figure 3. Restriction site maps of the murine chromosomal regions containing the *Hox-1* and *Hox-2* homeo box loci. (B) *Bam*HI; (E) *Eco*RI; (H) *Hind*III. (□) Homeo-box-containing restriction fragments. At the bottom are shown the recombinant phage genomes digested with *Eco*RI and *Hind*III and hybridized to a *Drosophila* Antennapedia homeo box probe (see McGinnis et al. 1984b). For molecular-weight determinations, the migration of DNA digested with *Hind*III is shown by the marks at the left of the ethidium-bromide-stained DNA tracks. The relative orientation and distance separating λMo-10 and λMo-11 are unknown at this time.

bryonal poly(A)+ RNA fraction but not in the poly(A)+ RNA from adult liver. It is possible that the high-molecular-weight transcripts might represent the unprocessed precursor molecules of the abundant transcripts represented by the hybridizing RNAs in the size range of 1.2–2.2 kb. This idea is supported by the fact that some probes, containing only homeo-box-flanking sequences, hybridize exclusively to the high-molecular-weight RNAs but not to the putative mature transcripts. These probes might therefore encode mainly intronic sequences.

The only probe that hybridizes to the poly(A)+ RNA from adult liver was a rather large DNA fragment (a 5.6-kb BamHI-EcoRI fragment) from the extreme "left" end of the Hox-2 locus, including Hox-2.4. This probe also hybridizes with RNA of identical size in the embryonic poly(A)+ and poly(A)- fractions. Since this common hybridization signal comigrates with the 28S rRNA and the signal intensity correlates to the amount of 28S rRNA present in each of these three fractions, we assume that this probe has slight sequence homology with the 28S rRNA sufficiently high to give a fairly strong hybridization signal. In addition to this apparent artifact, this probe hybridizes to a major transcript of about 2.2 kb exclusively in the embryonic poly(A)+ RNA fraction. The data obtained to date clearly indicate transcriptional expression of all the Hox-2 homeo box regions in whole extracts of 13.5-day-old embryos; they show no detectable expression in adult liver, with the single exception of the 28S rRNA-associated hybridization.

The homeo-box-linked genes in Drosophila are transcriptionally active only in particular subsets of cells during development (Akam 1983; Levine et al. 1983). It is of considerable interest to determine whether the expression of the murine Hox-2 locus is restricted in a similar way. Preliminary results obtained from Northern blot hybridizations with RNA extracted from various tissues and organs dissected from newborn and adult animals show that this is the case, at least for Hox-2.1. Some of these data are shown in Figure 4, where the thymus, brain, and liver of newborn animals (tracks 1, 3, 4, and 6) appear to be negative, the kidney is weakly positive (track 2), and the spinal cord (track 5) is strongly positive. The weak signal with the RNA from total newborn mice with the spinal cord dissected away (Fig. 4, track 7) indicates expression also in one or more other tissues or organs. One of these organs is the newborn kidney, since after long exposures (>1 week), a faint signal is detected (Fig. 4, track 2). The poly(A)+ RNA from whole 13.5-day-old embryo extracts (Fig. 4, track 8) served as a positive control in this experiment.

A limited series of expression studies has also been carried out using adult tissues. It was found that both adult spinal cord (Fig. 4, track 10) and kidney tissues (track 13) continue to express Hox transcripts; the adult adrenal gland (track 9) and spleen (track 14) were negative.

Northern blot analysis has also been carried out us-

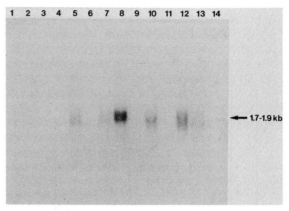

Figure 4. Northern blot with RNAs from different tissues and organs from newborn and adult mice probed with a ^{32}P-labeled 2.8-kb DNA fragment from the Hox-2 locus, containing the Hox-2.1 homeo box sequence. Tracks 1 through 7 contain the following RNAs from tissues of newborn mice: (1) total RNA from thymus; (2) total RNA from kidney; (3) poly(A)- RNA from brain; (4) poly(A)+ RNA from brain; (5) poly(A)+ RNA from the spinal column; (6) poly(A)+ RNA from liver; (7) poly(A)+ RNA whole extracts of newborn mice without spinal column; (8) poly(A)+ RNA from whole extracts of 13-day-old embryos. Tracks 9 through 14 contain RNAs from adult tissues; (9) total RNA from adrenal gland; (10) total RNA from spinal cord; (11) poly(A)- from brain; (12) poly(A)+ RNA from brain; (13) poly(A)+ RNA from kidney; (14) poly(A)+ RNA from spleen. The rather diffuse hybridization signals correspond most likely to two distinct transcripts of about 1.7–1.9 kb in size, which has been shown in separate experiments (data not shown).

ing Mo-10-cloned DNA as a probe. As stated above, this homeo box maps to the Hox-1 locus. The hybridization signals obtained with poly(A)+ RNA from 13-day-old embryos correspond to at least three different transcripts in the size range of 3.2–5.0 kb. These transcripts have been detected in the spinal cords of newborn and adult mice. In this respect, they evince a pattern of expression similar to Hox-2.1.

It is now clear from these results that there is good evidence to support tissue specificity of expression of Hox transcripts. Preliminary evidence for temporal patterns of differential patterns of transcript expression for particular tissues has recently been obtained for Hox-2.1 transcripts. In this instance, the Hox-2.1 probe detects a poly(A)+ transcript in adult brain, whereas the newborn brain is negative.

DISCUSSION

Homeo-box-mapping studies in the mouse have provided evidence for two separate homeo-box-containing loci, Hox-1 and Hox-2. A minimum of six homeo boxes can be accounted for at these two loci. Southern transfer analyses of genomic DNA indicate a total of at least 10 homeo box sequences in the murine genome. Some of these sequences will undoubtedly map to the already identified loci, but it is likely that others will map to one or several homeo box loci yet to be defined, assuming that there are homeo box sequences sufficiently homologous to the *Drosophila* Antennapedia or Ultrabithorax homeo boxes to be detected by blot hybridization. More divergent homeo-box-like sequences that do not cross-hybridize with these defining sequences may also exist. The engrailed sequence would be an example of a homeo box sequences belonging to such a category.

Our analysis of *Hox-2* shows that homeo boxes can occur in tightly linked groups at least superficially reminiscent of the clustering of *Drosophila* homeo box sequences in the Antennapedia and bithorax gene complexes. This is of particular interest since it supports the possibility of a functional relationship between the *Drosophila* complexes and those defined in the mammals. One clear difference between the mammalian and *Drosophila* homeo box cluster is size. The intervals between homeo boxes in *Drosophila* (McGinnis; Gehring; both this volume) are approximately tenfold longer than those in mammals. The significance (if any) of this difference is not clear.

The biological significance of the tight tandem arrangement of the homeo boxes in the mammals is supported by the apparent high conservation of this pattern of organization between mouse and man. With respect to *Hox-2.1* and *Hox-2.2*, there exists a high degree of homology between the two species based on hybridization and nucleotide sequence analysis. These data show that cognate homeo boxes as well as their flanking sequences are highly conserved. The lack of repetitive elements of *Hox-2* is consistent with the notion that the entire *Hox-2* locus is functional and cannot tolerate their insertion. The arguments supporting a functional role for *Hox-2* based solely on structural considerations are reinforced by an examination of the transcriptional activity of the locus.

We have tested for the transcriptional expression of the *Hox-2* locus using the Northern transfer technique applied to whole-body extracts of 13-day-old embryos. A number of separate transcripts are encoded by the genomic sequences of the complex. We are currently isolating cDNA clones from this region in order to define clearly the transcriptional units of the *Hox-2* complex. The complexity of the patterns of gene expression at this locus is again at least superficially reminiscent of the situation that obtains in the case of *Drosophila* homeotic gene complexes. One of the attributes of the *Drosophila* loci is temporal and tissue specificity of expression. Recent data as described above suggest that similar patterns of expression are also associated with the mouse homeo box loci.

Tissue-specific expression has been demonstrated using the Northern transfer technique applied to 13-day-old embryos, newborn animals, and adults. It can be shown that *Hox-2.1* expression is primarily restricted to the spinal cord from 13-day-old embryos and newborn animals and to the spinal cord and to the brain of adults. This suggests another parallel with the *Drosophila* system, since Antennapedia transcripts are strongly expressed in the ventral nerve cord during larval development (Levine et al. 1984). In the mouse system, it can also be shown that *Hox-2.1* is weakly expressed in the kidney tissues of newborns and adults. The significance of this finding is not yet clear.

One objective of this paper has been to review some of the similarities between the mammalian homeo box system and that of *Drosophila*. We believe the similarities are striking, although quite possibly superficial. We record a high level of conservation in the homeo box nucleotide sequence itself, a roughly similar number of homeo box elements in insects and in mammals, organization of homeo boxes into gene complexes, and finally, similarities with regard to complex patterns of expression of homeo box transcripts that evince features of tissue and temporal specificity.

It has been argued that the structure of the homeo box sequence suggests a role in the regulation of ensembles of genes. One might suspect that such a general function might be retained over the course of evolution, but adapted to different specific functions. This line of reasoning suggests that the mammalian homeo box genes may be used in a different way than they are in *Drosophila*. Although it is interesting to argue these different positions, we can be confident that the evolutionary relationships will only finally become clear when the actual function of the mammalian homeo box system is determined through direct experimentation. Toward that end, we are pursuing a number of experimental approaches in the emerging discipline of reverse genetics. These include tests for allelism between extant mouse mutants and various *Hox* loci, tests for tissue- or region-specific expression during mouse development, and the inactivation of the endogenous genes by antisense gene methodologies.

ACKNOWLEDGMENTS

We acknowledge the secretarial assistance of Marie Siniscalchi and the photographic skills of Maria Pafka. We also thank Dimi Pravtcheva for cell hybrid DNAs and their analysis. A.F. is a Fellow of the National Cancer Cytology Center. M.U. is supported by Medical Scientist training program grant T32-GM-07205-10. These studies were supported by a grant from the Research Development Program of the American Cancer Society and National Institutes of Health (GM-09966).

REFERENCES

AKAM, M.E. 1983. The location of *Ultrabithorax* transcripts in *Drosophila* tissue sections. *EMBO J.* **2:** 2075.

BEACHY, P.A., S.L. HELFAND, and D.S. HOGNESS. 1985. Segmental distribution of bithorax complex proteins during *Drosophila* development. *Nature* **313:** 545.

CARRASCO, A.E., W. McGINNIS, W.J. GEHRING, and E.M. DE ROBERTIS. 1984. Cloning of an *X. laevis* gene expressed during early embryogenesis coding for a peptide region homologous to *Drosophila* homeotic genes. *Cell* **37:** 409.

COLBERG-POLEY, A.M., S.D. VOSS, K. CHOWDHURY, and P. GRUSS. 1985. Structural analysis of murine genes containing homeo box sequences and their expression in embryonal carcinoma cells. *Nature* **314:** 713.

FJOSE, A., W.J. McGINNIS, and W.J. GEHRING. 1985. Isolation

of a homeo box-containing gene from the *engrailed* region of *Drosophila* and the spatial distribution of its transcripts. *Nature* **313:** 284.

GREEN, M.C. 1981. *Genetic variants and strains of the laboratory mouse.* Fischer, New York.

JOYNER, A.L., R.V. LEBO, Y.W. KAN, R. TJIAN, D.R. COX, and G.R. MARTIN. 1985. Comparative chromosome mapping of a conserved homeo box region in mouse and human. *Nature* **314:** 173.

KOZAK, C.A. and F.H. RUDDLE. 1977. The assignment of the genes for thymidine kinase and galactokinase to *Mus musculus* chromosome 11 and the preferential segregation of this chromosome in Chinese hamster/mouse somatic cell hybrids. *Somatic Cell Genet.* **3:** 121.

LAUGHON, A. and M.P. SCOTT. 1984. Sequence of a *Drosophila* segmentation gene: Protein structure homology with DNA binding proteins. *Nature* **310:** 25.

LEVINE, M., G.M. RUBIN, and R. TJIAN. 1984. Human DNA sequences homologous to a protein coding region conserved between homeotic genes of *Drosophila*. *Cell* **38:** 667.

LEVINE, M., E. HAFEN, R.L. GARBER, and W. GEHRING. 1983. Spatial distribution of *Antennapedia* transcripts during *Drosophila* development. *EMBO J.* **2:** 2037.

McGINNIS, W., C.P. HART, W.J. GEHRING, and F.H. RUDDLE. 1984a. Molecular cloning and chromosome mapping of a mouse DNA sequence homologous to homeotic genes of *Drosophila*. *Cell* **38:** 675.

McGINNIS, W., R.L. GARBER, J. WIRZ, A. KUROIWA, and W.J. GEHRING. 1984b. A homologous protein-coding sequence in *Drosophila* homeotic genes and its conservation in other metazoans. *Cell* **37:** 408.

McGINNIS, W., M. LEVINE, E. HAFEN, A. KUROIWA, and W.J. GEHRING. 1984c. A conserved DNA sequence in homeotic genes of *Drosophila* Antennapedia and bithorax complexes. *Nature* **308:** 428.

MISHKIN, J.D., B.A. TAYLOR, and W.J. MELLMAN. 1976. *Glk*: A locus controlling galactokinase activity in the mouse. *Biochem. Genet.* **14:** 635.

POOLE, S.J., L.M. KAUVER, B. DREES, and T. KORNBERG. 1985. The engrailed locus of *Drosophila*: Structural analysis of an embryonic transcript. *Cell* **40:** 37.

RABIN, M., C.P. HART, A. FERGUSON-SMITH, W. McGINNIS, M. LEVINE, and F.H. RUDDLE. 1985. Two homeo box loci mapped in evolutionarily related mouse and human chromosomes. *Nature* **314:** 175.

SCOTT, M.P. and A.J. WEINER. 1984. Structural relationships among genes that control development: Sequence homology between the Antennapedia, Ultrabithorax, and fushi tarazu loci of *Drosophila*. *Proc. Natl. Acad. Sci.* **78:** 1095.

SHEPHERD, J.C.W., W. McGINNIS, A.E. CARRASCO, E.M. DE ROBERTIS, and W.J. GEHRING. 1984. Fly and frog homoeo domains show homologies with yeast mating type regulatory proteins. *Nature* **310:** 70.

SOUTHERN, E.M. 1975. Detection of specific sequences among DNA fragments by gel electrophoresis. *J. Mol. Biol.* **98:** 503.

WHITE, R.A.H. and M. WILCOX. 1984. Protein products of the bithorax complex in *Drosophila*. *Cell* **39:** 163.

Expression of Murine Genes Containing Homeo Box Sequences during Visceral and Parietal Endoderm Differentiation of Embryonal Carcinoma Stem Cells

A.M. COLBERG-POLEY, S.D. VOSS, AND P. GRUSS

Zentrum für Molekulare Biologie der Universität Heidelberg D-6900 Heidelberg, Federal Republic of Germany

The first visible event that occurs during mouse embryogenesis, and which may be defined as a differentiation event, is the segregation and establishment of the trophectodermal and inner cell mass cell lineages. The subsequent delineation of ectodermal and endodermal lineages from the inner cell mass is distinguished by the appearance, late on the fourth day of development, of the primitive or primary endoderm along the blastocoelic surface of the inner cell mass. Although very little is understood about the steps leading to the establishment of the primitive endoderm, Gardner (1982) has shown that this primitive endoderm is capable of giving rise to two types of extraembryonic endoderm: parietal endoderm, which lies adjacent to cells of the trophoblast, and visceral endoderm, which surrounds the egg cylinder. The extremely small number of primitive endoderm "founder cells" present on the blastocoelic surface, however, has restricted their direct analysis, resulting in limited available information (Hogan et al. 1983).

The ability of F9 embryonal carcinoma (EC) cells to differentiate into either visceral or parietal endoderm from an apparent stem-cell population provides an interesting and reliable model system (Martin 1980) to investigate the early events during embryogenesis that lead to the establishment of the committed, although undifferentiated, primitive endoderm and to its subsequent differentiation into either visceral or parietal endoderm (Strickland and Mahdavi 1978; Strickland et al. 1980; Hogan et al. 1981). In differentiating along these alternative pathways, new genes in these cells are transcribed while transcripts of other genes are no longer present. In our laboratory, we are analyzing these early murine embryonic events in several ways, including the correlation of the transcriptional activity of specific cellular *onc* genes with EC cell differentiation, the construction of complementary DNA libraries specific for differentiating EC cells and various-stage mouse embryos, and, finally, by the use of the sequences conserved in some of the developmental genes of *Drosophila* (homeo boxes) as probes to isolate analogous murine sequences. The latter approach has recently proved to be the most fruitful in providing access to genes expressed during development, which may play a role in the decision-making processes in early embryos and in their in vitro EC cell counterparts. The homeo boxes present in *Drosophila* DNA are small stretches (~180 bp) of highly conserved nucleotides common to at least five genes in the Antennapedia complex (ANT-C), three genes in the bithorax complex (BX-C), and more than five genes outside these two major clusters (McGinnis et al. 1984c; Scott and Weiner 1984; Gehring 1985). These sequences are not only conserved in *Drosophila* developmental genes, but also widely distributed throughout the animal kingdom, including mouse and human (McGinnis et al. 1984b), suggesting that these vertebrates could possess the equivalent of the homeotic genes of *Drosophila* that play a key role in regulating its development.

Murine DNA fragments containing homeo box homology have been isolated and characterized by sequencing and mapping on chromosomes (McGinnis et al. 1984a; Colberg-Poley et al. 1985; Rabin et al. 1985). Expression of the sequences that flank a murine homeo box, m6-12, was observed in EC stem cells together with the induction of a specific transcript that appears to contain the homeo box sequences and is only present during differentiation of the F9 cells of parietal endoderm (Colberg-Poley et al. 1985). We wished to examine in further detail the arrangement of the homeo box sequences in this region of murine DNA and to determine whether the expression of the induced RNA is specifically associated with the determination of parietal endoderm from its stem-cell precursor or whether it is common to the establishment of both visceral and parietal endoderm tissue types.

MATERIALS AND METHODS

Cells. F9 cells were grown in cell culture flasks pretreated with 0.1% gelatin in Dulbecco's modified Eagle's medium (DMEM) containing 10% fetal bovine serum and antibiotics. For differentiation into parietal endoderm, cells were treated with 5×10^{-7} M retinoic acid (RA) and dibutyryl cAMP (10^{-3} M), and for differentiation into visceral endoderm, they were treated with 5×10^{-7} M RA and aggregated in bacterial petri dishes. PC13 cells were grown and differentiated into visceral-endoderm-like cells in a manner similar to that of F9 cells, except that they were maintained in cell culture flasks throughout differentiation (Rees et al. 1979; D. Barlow, pers. comm.). 3TDM cells were grown in DMEM with 10% fetal bovine serum.

Isolation of DNA, RNA, and hybridizations. Phage DNAs were isolated as described previously

(Umene and Enquist 1981). RNA samples were isolated using the guanidinium thiocyanate method (Chirgwin et al. 1979) and treated as described recently (Colberg-Poley et al. 1985). The probes were produced by nick translation of purified fragments or recombinant plasmids containing the sequences of interest or by primer extension of M13 cloned DNAs. Hybridizations of Southern blots under stringent and nonstringent conditions and of Northern blots were carried out as described recently (Colberg-Poley et al. 1985).

RESULTS

Organization of Murine Homeo Boxes

Screening of a murine genomic library cloned in λEMBL3A (Frischauf et al. 1983) for murine homeo boxes was performed using two *Drosophila* probes, Antennapedia (*Antp*; Garber et al. 1983) and fushi tarazu (*ftz*; Kuroiwa et al. 1984). Of the eight independent isolates obtained, six shared partially overlapping sequences that permitted the construction of a restriction map for this region (Fig. 1A). DNAs from these phage were examined by hybridization with several homeo box probes: *Antp* and *ftz* from *Drosophila* DNA and a probe generated from one region of the murine sequences contained in λm6 (Fig. 1B,C). This murine homeo box probe is the 206-bp *Eco*RI-*Pvu*II fragment contained in the 1.9-kbp *Eco*RI fragment, m6-12 (Colberg-Poley et al. 1985), and has been subcloned into pBR322. Two *Eco*RI fragments (1.9 kbp and 4 kbp) contain sequences that hybridize with all of the probes tested. The approximate positions of these sequences are shown in Figure 1A. Both of these homeo boxes have been sequenced, are homologous to, but not identical with, known homeo boxes, and lie in the same orientation about 10 kbp apart on the murine DNA (Colberg-Poley et al. 1985 and in prep.). Outside of these regions are other sequences that hybridize to one or more homeo box probes. The sequences in the 5.8-kbp *Eco*RI fragment hybridize under nonstringent conditions with the *Eco*RI-*Pvu*II 206-bp probe and very weakly to *Antp*, suggesting the presence of an additional homeo box. The two remaining regions hybridize exclusively with the *ftz* probe. Whether these latter hybridizations are spurious or whether they detect parts of or complete authentic homeo boxes remains to be determined. Thus, there appears to be a clustering of homeo boxes in this region of murine DNA.

Expression of a Gene Containing Murine Homeo Box Sequences

The transcripts detected by m6-12 DNA are encoded by the same strand of DNA (Colberg-Poley et al. 1985). The sequences contained in these transcripts were further examined by hybridization of probes containing homeo box sequences, m6-35, and the *Eco*RI-*Pvu*II 206-bp fragment of m6-12 (Fig. 2B). Each probe de-

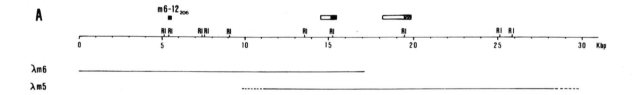

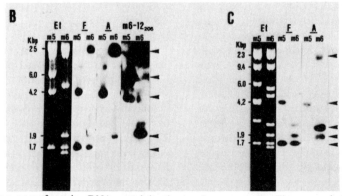

Figure 1. (*A*) Restriction map of murine DNA containing homeo box sequences. The mouse inserts from this region in two representative recombinant phage (λm5 and λm6) are indicated by the lines below the map. Dashed lines represent uncertainty in the identity of the fusion fragments. Closed boxes represent DNA that hybridizes to all of the homeo box probes tested: *Antp*, *ftz*, and the *Eco*RI-*Pvu*II 206-bp fragment from m6-12. Hatched box represents a region that hybridizes weakly to *Antp* and strongly to the homeo box of m6-12; the open box represents regions that hybridize only to *ftz*. (*B,C*) Hybridization of λm5 and λm6 DNAs with homeo box probes. λm5 and λm6 DNAs were cleaved with *Eco*RI (*B*) or with *Eco*RI and *Sal*I (*C*) and hybridized with *ftz* (F), *Antp* (A), or the homeo box probe from m6-12 (*Eco*RI-*Pvu*II, 206 bp; m6). Arrowheads indicate the positions of DNA fragments that hybridize to the homeo box probes.

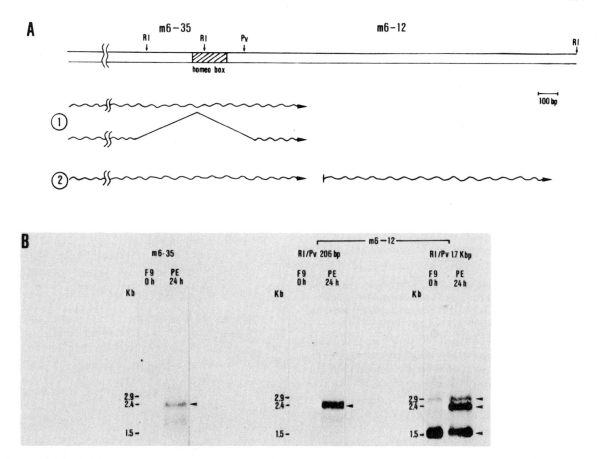

Figure 2. (*A*) Models of the transcription of murine DNA containing the homeo box. The restriction map shows the three probes tested for hybridization with F9 stem cells (0 hr) and F9 parietal endoderm (24 hr) poly(A)⁺ RNAs. m6-35, an *Eco*RI 280-bp fragment, contains 60 bp of the homeo box and the two subfragments from m6-12 DNA: the *Eco*RI-*Pvu*II 206-bp fragment (contains the remaining 120 bp of the homeo box) and the 1.7-kbp *Pvu*II-*Eco*RI flanking sequence probe. (*B*) Hybridization of poly(A)⁺ RNA from F9 stem cells (0 hr) and F9 cells treated with RA and dibutyryl cAMP for 24 hr (PE) with the probes represented in *A*. Arrowheads indicate the positions of the hybridizing RNAs encoded by this region.

tects the presence of two RNAs. The larger of these (2.4 kb) maps in this region as shown by hybridization of the flanking unique sequences (1.7-kbp *Pvu*II-*Eco*RI). The RNAs that hybridize weakly and exclusively to the homeo box probes appear to map outside of the unique sequences contained in the 1.7-kbp *Pvu*II-*Eco*RI probe.

Either of two simple models could explain the pattern of hybridization observed for this region (Fig. 2A). The first model postulates that the three transcripts are colinear and that the 2.9- and 1.5-kb RNAs are spliced, removing the homeo box sequences. The second model places the 5' end of the 2.4-kb RNA upstream of the homeo box and the 5' ends of the 2.9-kb and 1.5-kb RNAs downstream from the homeo box. The three transcripts would have sequences contained in the unique flanking DNA probe (1.7-kbp *Eco*RI-*Pvu*II). Both of these models are still plausible.

Specificity of the Induced Homeo Box Transcript

Following the identification of a homeo-box-containing transcript in RNA from F9 cells differentiated into parietal endoderm, we wished to evaluate the tissue specificity of the transcript by examining its expression in visceral endoderm (see Fig. 3A). Poly(A)⁺ RNAs from visceral endoderm (7 days) were blotted and hybridized to the m6-12 probe. Comparison of poly(A)⁺ RNAs from parietal endoderm (24 hr) and from F9 stem cells (0 hr) showed the presence in the visceral endoderm poly(A)⁺ RNAs of an induced transcript similar in size to that observed in parietal endoderm. The apparent lower levels of the transcripts present in visceral endoderm (7 days) are similar in quantity to those obtained at late times (5 days), following differentiation into parietal endoderm (Colberg-Poley et al. 1985). To demonstrate that differentiated F9 cells were present in these cultures, poly(A)⁺ RNA from the visceral endoderm cells was also hybridized with DNA of a marker specifically expressed in the visceral endoderm, α-fetoprotein (AFP) (Dziadek and Adamson 1978; Janzen et al. 1982). As expected, AFP hybridized with a 2.2-kb RNA present only in the visceral endoderm (Fig. 3B). The sequences contained in the visceral endoderm transcripts that hybridized with m6-12 were examined using the two *Eco*RI-*Pvu*II fragments present in m6-12 (Fig. 4A,B). The smaller probe contained

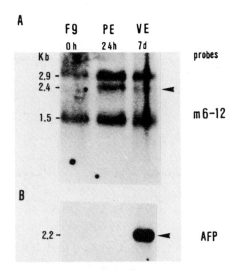

Figure 3. Hybridization of poly(A)+ RNA of visceral endoderm (VE) with murine sequences containing homeo box sequences (m6-12) and AFP DNA. Poly(A)+ RNAs were isolated from F9 stem cells (0 hr), F9 cells treated with RA and dibutyryl cAMP for 24 hr (PE, 24h), and F9 cells treated with RA and aggregation for 7 days (VE, 7d). Duplicate blots were hybridized with ^{32}P-labeled m6-12 DNA (A) or AFP DNA (B). Arrowheads indicate the positions of the induced RNA (2.4 kb) and of AFP RNA (2.2 kb).

the homeo box sequences hybridized to a 2.4-kb RNA in the visceral endoderm sample as did the flanking unique sequences (1.7 kbp). In addition, the larger probe detected 2.9-kb and 1.5-kb RNAs seen in both stem-cell RNA and differentiated F9 cell RNA.

To rule out the possibility that the presence of homeo box transcript(s) in F9-induced visceral endoderm was an F9-associated (rather than tissue-related) phenomenon, poly(A)+ RNAs from another EC cell line (PC13) and from 3TDM cells were also examined using the m6-12 probe (Fig. 5A,B). Hybridization of 2.4-kb RNA was observed in poly(A)+ RNA from visceral-endoderm-like cells produced by differentiation of PC13 (Rees et al. 1979; D. Barlow, pers. comm.) but not in the undifferentiated PC13 RNA or in RNA from

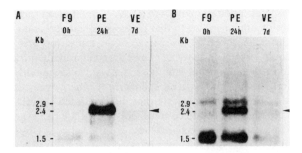

Figure 4. Hybridization of poly(A)+ RNA from VE with subfragments of m6-12 DNA. Poly(A)+ RNAs were separated, blotted, and hybridized to a ^{32}P-labeled EcoRI-PvuII 206-bp fragment (A) or a PvuII-EcoRI 1.7-kbp fragment (B). Poly(A)+ RNAs were obtained from F9 stem cells (F9, 0h), F9 cells treated with RA and cAMP for 24 hr (PE, 24h), and F9 cells treated with RA and aggregation for 7 days (VE, 7d). Arrowheads indicate the position of the induced 2.4-kb RNA.

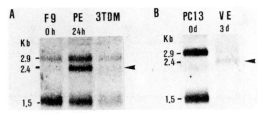

Figure 5. Hybridization of poly(A)+ RNA from 3TDM cells and VE with m6-12 DNA. Poly(A)+ RNAs were obtained from F9 stem cells (F9, 0h), F9 cells treated with RA and cAMP for 24 hr (PE, 24h), and 3TDM cells (A) and from PC13 stem cells (PC13, 0d) and PC13 cells treated with RA for 3 days (VE, 3d) (B). Arrowheads indicate the position of the induced RNA (2.4 kb).

3TDM cells, a cell line resembling trophoblast cells (Nicolas et al. 1976; Brûlet and Jacob 1982).

DISCUSSION

We have shown here that a cluster of murine homeo boxes is contained within a 30-kbp region of the murine genome and that transcription of the coding strand of one of the homeo box genes, m6-12 (Colberg-Poley et al. 1985), produces a 2.4-kb RNA that appears during differentiation of F9 cells into both visceral endoderm and parietal endoderm. Our initial hypothesis was that the use of Drosophila homeo box sequences could select, by virtue of their conservation, murine developmental genes analogous to those of Drosophila. The role, if any, of homeo box genes in murine development remains to be established. However, Rabin and co-workers (1985) have suggested, on the basis of chromosomal mapping of a murine homeo box (Hox-2), the possibility that this gene is identical to Tail-short (Ts), a dominant mutation that is viable in heterozygous form but, when homozygous, causes death in the embryo (Green 1981). On the other hand, we have documented the expression of murine homeo box sequences in EC cells following the differentiation of these cells.

The presence of at least two homeo boxes in this region of cloned murine DNA confirms a clustering of homeo boxes similar to that reported by other investigators (McGinnis et al. 1984a; Rabin et al. 1985). These clusters, however, appear to map in different regions of murine DNA, since their restriction maps cannot be aligned. Although clustering of these sequences has been found, chromosomal mapping shows that they are present on at least two murine chromosomes, 6 and 11 (McGinnis et al. 1984a; Joyner et al. 1985; Rabin et al. 1985). Similarly, clustering of homeo boxes has also been observed in human and Drosophila genomes (Levine et al. 1984; Gehring 1985; Joyner et al. 1985). In addition, such a clustering of homeo boxes suggests a linkage of murine genes potentially analogous to the homeotic genes of Drosophila. In this context, examination of two homeo boxes in this cluster shows that they exhibit stage-specific expression during embryogenesis (A.M. Colberg-Poley et al., in prep.). From the results of hybridization studies, we conclude that there is at least one additional homeo box in this cluster. The sequences contained in these regions, however, must be determined in order to exclude spurious homology with the homeo box probes that might be responsible for the observed hybridizations. The isolation of such a high proportion of the recombinant phage that contain inserts mapping in the same region of murine DNA from a genomic library suggests that, by using two *Drosophila* homeo box probes, we selected a family of related homeo box sequences.

The pattern of hybridization observed for the m6-12 DNA region led us to propose two alternative models for the transcription of this region. Using hybridization with single-stranded probes, we have determined that all three transcripts detected by m6-12 are produced by transcription of the coding strand (Colberg-Poley et al. 1985). Additional analysis of the induced RNA (2.4 kb) using the two probes containing both parts of the m6-12 homeo box sequences suggests strongly that, of the transcripts mapping in this region, only the 2.4-kb RNA contains homeo box sequences. Two other transcripts (2.9 kb and 1.5 kb), however, map in close proximity as they also hybridize with the same probe containing unique flanking DNA sequences. Other RNAs (~1.7 and 1.9 kb), detected only by the homeo box probes, appear to map elsewhere on the murine genome, since they are not detected with the unique DNA probe. The three RNAs encoded by this region could be products of differential splicing, which would remove homeo box sequences from both the 2.9- and 1.5-kb RNAs but not from the 2.4-kb RNA. Alternatively, the initiation of transcription within the unique flanking DNA sequences (1.7-kbp *Pvu*II-*Eco*RI) could result in production of the 2.9- and 1.5-kb RNAs, whereas the 2.4-kb RNA would result from the transcription of sequences upstream of and including the m6-12 homeo box.

Although weak hybridization of the homeo box probe (fragment isolated from *Eco*RI-*Pvu*II-cleaved m6-12 DNA) to the 1.5-kb RNA is seen in Figure 4A, it seems that the 1.5-kb RNA does not contain homeo box sequences. This was demonstrated by hybridization of the cloned *Eco*RI-*Pvu*II 206-bp probe (Fig. 2B) that did not detect the 1.5-kb RNA. The hybridization to the 1.5-kb RNA must be attributed then to use of the 206-bp probe slightly contaminated with the larger flanking sequence (1.7-kbp) fragment.

Since the 2.4-kb RNA is induced in visceral endoderm as well as parietal endoderm, we conclude that the expression of this information is neither tissue-specific nor an F9-associated phenomenon; i.e., its expression is not sufficient to direct a cell into a specific differentiation program. Thus, transcription of these sequences may be complex and not necessarily confined to specific regions of the developing embryo, suggesting a role in a much wider decision-making process. This is confirmed by the presence of additional homeo box transcripts induced by the differentiation of F9 cells into parietal endoderm, as well as by differentiation of P19 cells into myogenic cells (Edwards et al. 1983; Dony et al. 1985; A.M. Colberg-Poley et al., in prep.). Thus, the interaction of several gene products carrying homeo box homology with other cellular factors might be necessary to determine precisely the specificity of a differentiation pathway. We are currently trying to identify transcripts carrying homeo box homology that are specific for one cell type. Subsequent reintroduction of complementary DNAs containing such homeo box sequences into EC cells will facilitate the study of the function of putative proteins encoded by these transcripts in differentiating cells.

ACKNOWLEDGMENTS

We thank Erwin Wagner and Colin Stewart for their comments on the manuscript and for providing the PC13 and 3TDM cells, Shirley Tilghman for the AFP DNA probe, Carola Dony for poly(A)$^+$ RNA from the F9 visceral endoderm, Rosemary Franklin for typing the manuscript, E. Vosshans-Bosbach for technical assistance, and S. Mähler for photography. A.M.C.-P. initially held a fellowship from the Alexander von Humboldt Stiftung. S.D.V. holds a Fulbright Fellowship. This research was supported by the Deutsche Forschungsgemeinschaft (DFG Ba 384/18-4).

REFERENCES

BRÛLET, P. and F. JACOB. 1982. Molecular cloning of a cDNA sequence encoding a trophectoderm-specific marker during mouse blastocyst formation. *Proc. Natl. Acad. Sci.* **79:** 2328.

CHIRGWIN, J.M., A.E. PRZYBYLA, R.J. MACDONALD, and W.J. RUTTER. 1979. Isolation of biologically active ribonucleic acid from sources enriched in ribonuclease. *Biochemistry* **18:** 5294.

COLBERG-POLEY, A.M., S.D. VOSS, K. CHOWDHURY, and P. GRUSS. 1985. Structural analysis of murine genes containing homoeo box sequences and their expression in embryonal carcinoma cells. *Nature* **314:** 713.

DONY, C., M. KESSEL, and P. GRUSS. 1985. An embryonal carcinoma cell line as a model system to study developmentally regulated genes during myogenesis. *Cell. Differ.* **15:** 275.

DZIADEK, M. and E. ADAMSON. 1978. Localization and synthesis of alphafetoprotein in post-implantation mouse embryos. *J. Embryol. Exp. Morphol.* **43:** 289.

EDWARDS, M.K.S., J.F. HARRIS, and M.W. MCBURNEY. 1983. Induced muscle differentiation in an embryonal carcinoma cell line. *Mol. Cell. Biol.* **3:** 2280.

FRISCHAUF, A.-M., H. LEHRACH, A. POUSTKA, and N. MURRAY. 1983. Lambda replacement vectors carrying polylinker sequences. *J. Mol. Biol.* **170:** 827.

GARBER, R.L., A. KUROIWA, and W.J. GEHRING. 1983. Genomic and cDNA clones of the homeotic locus *Antennapedia* in *Drosophila*. *EMBO J.* **2:** 2027.

GARDNER, R.G. 1982. Investigation of cell lineage and differentiation in the extraembryonic endoderm of the mouse embryo. *J. Embryol. Exp. Morphol.* **68:** 175.

GEHRING, W.J. 1985. The homeo box: A key to the understanding of development? *Cell* **40:** 3.

GREEN, M.C., ed. 1981. *Genetic variants and strains of the laboratory mouse.* Fischer, New York.

HOGAN, B.L.M., D.P. BARLOW, and R. TILLY. 1983. F9 teratocarcinoma cells as a model for the differentiation of parietal and visceral endoderm in the mouse embryo. *Cancer Surv.* **2:** 115.

HOGAN, B.L.M., A. TAYLOR, and E. ADAMSON. 1981. Cell interactions modulate embryonal carcinoma cell differentiation into parietal or visceral endoderm. *Nature* **291:** 235.

JANZEN, R.G., G.K. ANDREWS, and T. TAMOAKI. 1982. Synthesis of secretory proteins in the developing mouse yolk sac. *Dev. Biol.* **90:** 18.

JOYNER, A.L., R.V. LEBO, Y.W. KAN, R. TJIAN, D.R. COX, and G.R. MARTIN. 1985. Comparative chromosome mapping of a conserved homeo box region in mouse and human. *Nature* **314:** 173.

KUROIWA, A., E. HAFEN, and W.J. GEHRING. 1984. Cloning and transcriptional analysis of the segmentation gene *fushi tarazu* of *Drosophila*. *Cell* **37:** 825.

LEVINE, M., G.M. RUBIN, and R. TJIAN. 1984. Human DNA sequences homologous to a protein coding region conserved between homeotic genes of *Drosophila*. *Cell* **38:** 667.

MARTIN, G.R. 1980. Teratocarcinoma and mammalian embryogenesis. *Science* **209:** 768.

MCGINNIS, W., C.P. HART, W.J. GEHRING, and F.H. RUDDLE. 1984a. Molecular cloning and chromosome mapping of a mouse DNA sequence homologous to homeotic genes of *Drosophila*. *Cell* **38:** 675.

MCGINNIS, W., R.L. GARBER, J. WIRZ, A. KUROIWA, and W.J. GEHRING. 1984b. A homologous protein coding sequence in *Drosophila* homeotic genes and its conservation in other metazoans. *Cell* **37:** 403.

MCGINNIS, W., M.S. LEVINE, E. HAFEN, A. KUROIWA, and W.J. GEHRING. 1984c. A conserved DNA sequence in homeotic genes of the *Drosophila* Antennapedia and bithorax complexes. *Nature* **308:** 428.

NICOLAS, J.F., P. AVNER, J. GAILLARD, J.L. GUENET, H. JAKOB, and F. JACOB. 1976. Cell lines derived from teratocarcinomas. *Cancer Res.* **36:** 4224.

RABIN, M., C.P. HART, A. FERGUSON-SMITH, W. MCGINNIS, M. LEVINE, and F.H. RUDDLE. 1985. Two homoeo box loci mapped in evolutionarily related mouse and human chromosomes. *Nature* **314:** 175.

REES, A.R., E.D. ADAMSON, and C.F. GRAHAM. 1979. Epidermal growth factor receptors increase during the differentiation of embryonal carcinoma cells. *Nature* **281:** 309.

SCOTT, M.P. and A.J. WEINER. 1984. Structural relationships among genes that control development: Sequence homology between the Antennapedia, Ultrabithorax and fushi tarazu loci of *Drosophila*. *Proc. Natl. Acad. Sci.* **81:** 4115.

STRICKLAND, S. and V. MAHDAVI. 1978. The induction of differentiation in teratocarcinoma stem cells by retinoic acid. *Cell* **15:** 393.

STRICKLAND, S., K. SMITH, and K.R. MAROTTI. 1980. Hormonal induction of differentiation in teratocarcinoma stem cells: Generation of parietal endoderm by retinoic acid and dibutyryl cAMP. *Cell* **21:** 347.

UMENE, K. and L.W. ENQUIST. 1981. A deletion analysis of lambda hybrid phage carrying the U_s region of herpes simplex virus type I (Patton). I. Isolation of deleted derivatives and identification of *chi*-like sequences. *Gene* **13:** 251.

Structure and Expression of Two Classes of Mammalian Homeo-box-containing Genes

A. JOYNER,* C. HAUSER,† T. KORNBERG,‡ R. TJIAN,† AND G. MARTIN*
*Department of Anatomy, ‡Department of Biochemistry and Biophysics, University of California, San Francisco, California 94143; †Department of Biochemistry, University of California, Berkeley, California 94720

Genetic experiments in the fly, Drosophila melanogaster, have identified a number of loci that play a basic role in the control of pattern formation during embryonic development. The Drosophila genes that regulate these processes each fall into one or more of three general categories: maternal-effect genes, which are expressed during oogenesis and specify the structure and spatial coordinates of the egg (Nüsslein-Volhard 1979); segmentation genes, which determine the number and polarity of the body segments (Nüsslein-Volhard and Wieschaus 1980); and homeotic genes, which specify segment identity (Ouweneel 1976).

Identification of the 180-nucleotide homeo box sequence as a constituent of several segmentation and homeotic genes has led to the hypothesis that this 60-amino-acid protein domain has been highly conserved in these genes because it serves an important function in the control of pattern formation during embryonic development (McGinnis et al. 1984b,c; Scott and Weiner 1984). Thus far, homeo-box-containing genes have been found clustered in three distinct regions of the Drosophila genome, known as the Antennapedia complex (ANT-C) (Kaufman et al. 1980), bithorax complex (BX-C) (Lewis 1978), and a region that contains the engrailed (en) gene (Garcia-Bellido and Santamaria 1972).

The Drosophila homeo-box-containing genes thus far identified can be grouped into two classes. The larger class (ANT-C/BX-C) includes at least seven genes in the Antennapedia and bithorax complexes (Gehring 1985). The smaller class (engrailed complex, EN-C) is represented only by the two genes in the en genomic region (Fjose et al. 1985; Poole et al. 1985). There are several features that distinguish the two classes of genes. The ANT-C and BX-C genes are on the right arm of chromosome 3 and the EN-C is located on the right arm of chromosome 2. The homeo boxes found in the ANT-C/BX-C genes have continuous open reading frames, whereas those in the EN-C genes are each interrupted by an intervening sequence. The polypeptides encoded by the homeo boxes found in ANT-C/BX-C genes share 75% or more amino acid homology. The polypeptides encoded by the two homeo boxes found in the EN-C also share approximately 87% amino acid homology; however, there is only approximately 50% amino acid homology between homeo boxes from genes of the two classes. Thus, homeo boxes themselves can be generally classified as ANT-C/BX-C or EN-C on the basis of their degree of amino acid sequence homology with members of a given class. Finally, the two genes in the EN-C show DNA sequence and putative amino acid homology outside their homeo boxes. This unique region of homology encodes 31 amino acids immediately 3′ to the homeo box (Poole et al. 1985). In contrast, no homology outside the homeo box has been found among any of the ANT-C/BX-C homeo-box-containing genes.

The finding that all vertebrate genomes examined contain 8–12 copies of DNA sequences that cross-hybridize with members of the Drosophila ANT-C/BX-C class of homeo box raises the possibility that the homeo box has been conserved not only in the evolution of Drosophila genes, but throughout vertebrate evolution as well (McGinnis et al. 1984b). This hypothesis has recently been substantiated by sequencing studies of frog (Carrasco et al. 1984; Muller et al. 1984), mouse (McGinnis et al. 1984a; Colberg-Poley et al. 1985), and human homeo-box-containing genomic clones (Levine et al. 1984). These studies showed that vertebrate genomes contain highly conserved homeo boxes that encode polypeptides that are between 70% and 92% homologous to the peptides encoded by the ANT-C/BX-C class of homeo boxes. One inference drawn from these observations is that the presence of conserved homeo box sequences provides a means of identifying vertebrate genes with functions similar to those of Drosophila homeo-box-containing genes, i.e., the control of pattern formation and the specification of developmental pathways. Consistent with this idea are the findings that certain homeo-box-containing sequences are expressed during Xenopus development (Carrasco et al. 1984; Muller et al. 1984) and during differentiation of mouse teratocarcinoma cells (Colberg-Poley et al. 1985). However, the concept that inclusion of a homeo box sequence in a vertebrate gene signifies its involvement in the regulation of developmental processes remains to be validated.

Our long-term goal is to determine the function of homeo-box-containing genes in mammals. We have begun by studying genes containing representatives of both classes of homeo boxes in humans and mice. First, we are characterizing a region of human chromosome 17 that contains a cluster of at least three homeo boxes of the ANT-C/BX-C class. Our previous hybridization studies suggested that this region has been highly conserved during mammalian evolution and that its homolog exists on mouse chromosome 11 (Joyner et al. 1985). To confirm this, we have isolated a portion of this homolog from the mouse genome and have carried out a comparative sequencing study of a homeo box region in the two species. Second, we have used a *Drosophila en* gene cDNA clone as a probe to identify homologous sequences in mammalian genomes. Two regions of the mouse genome and one region of the human genome were found to share homology with the *Drosophila en* homeo box and/or sequences 3' to it, and one of these mouse sequences, Mo-en.1, has been cloned and further analyzed. In addition, as a first step toward determining the function of these mammalian homeo-box-containing sequences, we have shown that they are indeed genes expressed during normal mouse embryogenesis.

MATERIALS AND METHODS

DNA probes. All probes were double-stranded and radiolabeled by nick translation unless otherwise specified.

Mammalian ANT-C/BX-C homeo box region probes. The H.1 probe is a 1.7-kb HindIII restriction fragment that contains the human Hu1 homeo box; the RF.1 probe is a HindIII-SacI restriction fragment flanking Hu1 and is derived from the λHu1 clone (Joyner et al. 1985). The rs probe, used in the Northern blot analysis, is an EcoRI-SalI restriction fragment derived from the Mu1 recombinant phage that extends approximately 1.5 kb 5' from the EcoRI site in the Mu1 homeo box.

engrailed and Mo-en.1 probes. The 1.4-kb *Drosophila* cDNA fragment probe extends from the EcoRI site that is approximately 300 bp 3' to the homeo box to a site approximately 900 bp 5' of the homeo box (Poole et al. 1985). The λMo-en.1-derived probes included hbr, a single-stranded probe complementary to the antisense strand of the Mo-en.1 homeo box region. This probe was approximately 140 bp in length, extending 3' from the SstI site in the homeo box, and was derived from an M13 clone according to the method of Church and Gilbert (1984) by primer extension under conditions of limiting radiolabeled [^{32}P]dCTP, followed by electrophoresis of the denatured M13 DNA on a 5% acrylamide-7 M urea gel and isolation of the labeled fragment of appropriate size. The single-copy fragment probe derived from Mo-en.1 that lacks homeo box sequences, mp, is a 3.6-kb EcoRI-SstI fragment that extends 3' from an EcoRI site 360 bp 3' of the homeo box. The probe used in the Northern blot analysis, eb, is a 700-bp EcoRI-BamHI λMo-en.1 fragment that contains the homeo box.

Isolation of the λMu1 and λMo-en.1 clones. A recombinant λ EMBL 3A library (kindly provided by J. Vogel and R. Goodenow, University of California, Berkeley) was screened with radiolabeled RF.1 or the 1.4-kb *en* cDNA fragments. The library contains fragments (12–20 kb) of C3Hf mouse genomic DNA partially digested with Sau3A. Replica nitrocellulose filters were prepared according to the method of Maniatis et al. (1982) and hybridized to radiolabeled RF.1 or *en* cDNA fragments for 16 hours at 64°C in 6 × SSC, 10 × Denhardt's, 0.1% SDS, and 0.1% sodium pyrophosphate. The filters were washed two times for 30 minutes each in 2 × SSC and 0.1% SDS at 50°C, followed by one wash in 2 × SSC at 50°C.

DNA sequence analysis. All DNA sequences were obtained by the dideoxynucleotide chain terminating technique (Sanger et al. 1977), after fragments were cloned into M13 vectors (Messing 1983). All of the sequences discussed here were determined by independently sequencing both strands.

Southern blot analysis. Recombinant λ DNAs (1 μg) or total genomic DNAs (10 μg) isolated from human and mouse tissues were cleaved with restriction enzymes, electrophoretically fractionated in 1% agarose gels, and transferred to nitrocellulose or Gene Screen (New England Nuclear) in 10 × SSC. Subsequent to transfer, the nitrocellulose filters were baked according to the method of Southern (1975), whereas the Gene Screen nylon membranes were exposed to UV irradiation according to the method of Church and Gilbert (1984). The Southern blots of mouse genomic DNA or recombinant λ clone DNA on nitrocellulose filters were hybridized to the *Drosophila en* probe or H.1 using the following "low-stringency" conditions: Hybridizations were done at 37°C for 48 hours in 5 × SSC, 5 × Denhardt's solution, 250 μg/ml sonicated boiled salmon sperm DNA, 50 mM NaPO$_4$ (pH 7), 0.1% SDS, and 45% deionized formamide, and the filters were washed as described above for the phage replica filters. The Southern blots of mouse and human genomic DNAs on Gene Screen membranes were hybridized with the Mo-en.1-derived probes using the following "moderate-stringency" conditions: Hybridizations were carried out as described by Church and Gilbert (1984), with the addition of 15% formamide, and the filters were washed twice at 50°C for 30 minutes in 2 × SSC and 1% SDS, followed by one wash in 0.2 × SSC at 50°C.

Northern blot analysis. Total cellular RNA was isolated from mouse embryos of various strains according to the method of Chirgwin et al. (1979). The day on which the vaginal plug was detected was termed 0.5 days of gestation. Embryos at the designated stages were dissected from the implantation sites and separated from the extraembryonic membranes. Cellular RNA was dissolved in H$_2$O and loaded onto oligo(dT)-cellulose columns (Collaborative Research), and poly(A)$^+$ was eluted as described by Maniatis et al. (1982). RNA was separated on 1% agarose-formaldehyde gels and transferred to Gene Screen in 20 × SSC as described by Maniatis et al. (1982), except that the running buffer contained 0.5 M HEPES (pH 7.0), 10 mM EDTA, and 50 mM Na-acetate, and the Gene Screen membranes were UV-irradiated according to the method of Church and Gilbert (1984). Hybridizations were carried out at 65°C for 24 hours in 0.1% bovine serum albumin, 0.2 M NaPO$_4$ (pH 7.5), 1 mM EDTA, 7% SDS, and 45% formamide. The filters were washed twice in 2 × SSC and 1% SDS at 65°C for 30 minutes and once in 30% formamide and 0.2 × SSC at 65°C for 30 minutes.

RESULTS

Comparison of Human and Homologous Mouse Clones That Contain Homeo Boxes of the ANT-C/BX-C Class

We reported previously that human chromosome 17 contains two homeo box sequences, Hu1 and Hu2, that are contained within a 6-kb fragment (Levine et al. 1984; Joyner et al. 1985). By using restriction hybridization analysis of recombinant λ clones containing human homeo box sequences, we have now found that a third homeo box region, Hu5, is situated within 15 kb of Hu1. A composite map of these three clones, λHu1, λHu2, and λHu5, is shown in Figure 1. This 25-kb region on the long arm of chromosome 17 contains three homeo boxes.

In the course of chromosome mapping studies, it was shown that sequences flanking the Hu1 and Hu2 homeo boxes hybridize to single-copy fragments on mouse chromosome 11 (Joyner et al. 1985). These data suggested that a region including and extending beyond the homeo box might be conserved between man and mouse. To determine the degree of homology as well as to facilitate examination of the developmental expression and functional role of these homeo-box-containing sequences in mammals, we have isolated part of the homologous region from the mouse genome. A mouse genomic library was screened by hybridization to a radiolabeled RF.1 probe, a HindIII-SacI restriction fragment that flanks the Hu1 homeo box (Fig. 1). One cross-hybridizing genomic clone, λMu1, was isolated and found to contain a single homeo box, Mu1.

To examine directly the degree of homology in the region surrounding the Hu1 and Mu1 homeo boxes, the DNA sequences of a 384-bp segment, including 126 bp 5' to the homeo box and 75 bp 3' to the homeo box, were determined. The data show that the putative translation products of these homeo box sequences from the two species share a very high degree of homology: Although there are 13 base differences between the Hu1 and Mu1 homeo boxes, none of these lead to an amino acid change (Fig. 2). Figure 3 shows the amino acid sequence of these two mammalian homeo boxes in comparison with the amino acid sequences encoded by three Drosophila homeo boxes of the ANT-C/BX-C class. When only identical amino acids are considered, the Hu1/Mu1 homeo box is 72–90% homologous to each member of the Drosophila ANT-C/BX-C class of homeo box described thus far. Furthermore, of the 45 amino acids that are per-

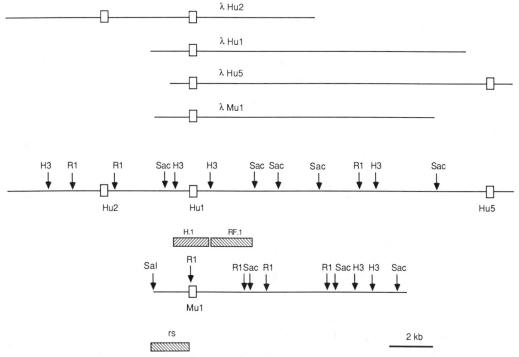

Figure 1. Restriction map of the regions surrounding the Hu1 and Mu1 homeo boxes. The genomic regions contained in recombinant phage are shown above, with the Hu1 and Mu1 homeo boxes aligned. Below is a composite restriction map of the human and mouse genomic regions. The probes derived from the genomic clones and used for screening the mouse genomic library (RF.1), for the Southern blot analysis of mouse genomic DNA (H.1), and for the Northern blot analysis (rs) are indicated by hatched rectangles beneath the relevant restriction maps. Restriction endonucleases used for mapping were EcoRI (R1), HindIII (H3), SacI (Sac), and SalI (Sal).

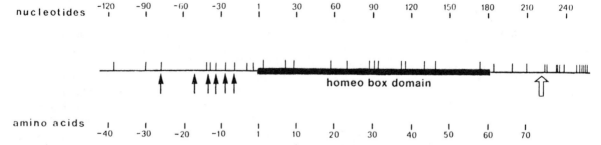

Figure 2. Graphic representation of nucleotide and amino acid sequence differences between Mu1 and Hu1. Each vertical line above the horizontal line indicates one nucleotide difference, and each closed arrow below the line indicates an amino acid difference. The homeo box domain is the bold portion of the horizontal line. The open arrow indicates the end of the open reading frame containing the homeo box. The nucleotide and amino acid designated 1 are each at the start of the homeo box sequence.

fectly conserved in these three *Drosophila* ANT-C/BX-C homeo boxes, 43 have been conserved in Hu1/Mu1.

In addition to the sequence homology within the Hu1 and Mu1 homeo boxes, the nucleotide and amino acid sequences on both sides of the mammalian homeo boxes are highly conserved. The 126 bp 5' of the Hu1 and Mu1 homeo boxes have 91% DNA sequence homology and 86% amino acid homology. Additional sequences on the 5' side of these homeo boxes, which have not been confirmed by determining sequences of both strands, indicate that the homology ends about 150 bp 5' to the homeo boxes. The nucleotide sequences on the 3' side of both the Mu1 and Hu1 homeo boxes encode an identical 15-amino-acid sequence followed by a stop codon. These sequence data, summarized in Figure 2, specifically show that there are conserved coding regions directly flanking the homeo boxes. These results suggest that the genes containing Hu1 and Mu1 have been conserved through evolution and that they may serve a similar function in man and mouse.

Isolation of a Mouse Genomic Region That Shows Sequence Homology with the *Drosophila* en Gene

To determine whether the mouse genome contains sequences homologous to those in the *Drosophila* en gene, Southern blot analysis was carried out using mouse genomic DNA digested separately with four dif-

Hu1/Mu1	Gly	Lys	Arg	Ala	Arg	Thr	Ala	Tyr	Thr	Arg	Tyr	Gln	Thr	Leu	Glu
ANTP	Arg	Lys	Arg	Gly	Arg	Gln	Thr	Tyr	Thr	Arg	Tyr	Gln	Thr	Leu	Glu
FTZ	Ser	Lys	Arg	Thr	Arg	Gln	Thr	Tyr	Thr	Arg	Tyr	Gln	Thr	Leu	Glu
UBX	Arg	Arg	Arg	Gly	Arg	Gln	Thr	Tyr	Thr	Arg	Tyr	Gln	Thr	Leu	Glu

Hu1/Mu1	Leu	Glu	Lys	Glu	Phe	His	Phe	Asn	Arg	Tyr	Leu	Thr	Arg	Arg	Arg
ANTP	Leu	Glu	Lys	Glu	Phe	His	Phe	Asn	Arg	Tyr	Leu	Thr	Arg	Arg	Arg
FTZ	Leu	Glu	Lys	Glu	Phe	His	Phe	Asn	Arg	Tyr	Ile	Thr	Arg	Arg	Arg
UBX	Leu	Glu	Lys	Glu	Phe	His	Thr	Asn	His	Tyr	Leu	Thr	Arg	Arg	Arg

Hu1/Mu1	Arg	Ile	Glu	Ile	Ala	His	Ala	Leu	Cys	Leu	Ser	Glu	Arg	Gln	Ile
ANTP	Arg	Ile	Glu	Ile	Ala	His	Ala	Leu	Cys	Leu	Thr	Glu	Arg	Gln	Ile
FTZ	Arg	Ile	Asp	Ile	Ala	Asn	Ala	Leu	Ser	Leu	Ser	Glu	Arg	Gln	Ile
UBX	Arg	Ile	Glu	Met	Ala	Tyr	Ala	Leu	Cys	Leu	Thr	Glu	Arg	Gln	Ile

Hu1/Mu1	Lys	Ile	Trp	Phe	Gln	Asn	Arg	Arg	Met	Lys	Trp	Lys	Lys	Asp	Asn
ANTP	Lys	Ile	Trp	Phe	Gln	Asn	Arg	Arg	Met	Lys	Trp	Lys	Lys	Glu	Asn
FTZ	Lys	Ile	Trp	Phe	Gln	Asn	Arg	Arg	Met	Lys	Ser	Lys	Lys	Asp	Arg
UBX	Glu	Ile	Trp	Phe	Gln	Asn	Arg	Arg	Met	Lys	Leu	Lys	Lys	Glu	Ile

Figure 3. Comparison of the putative amino acid sequences of the Hu1/Mu1 homeo boxes with those of three *Drosophila* homeo boxes of the ANT-C/BX-C class. Shaded portions highlight the 43 amino acids that have been perfectly conserved in all five homeo box domains. The Antennapedia (ANTP), fushi tarazu (FTZ), and Ultrabithorax (UBX) sequences are those reported by McGinnis et al. (1984b).

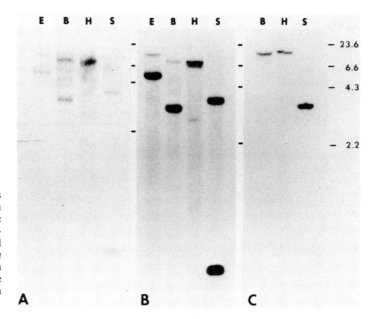

Figure 4. Mouse and human genomic DNAs contain conserved EN-C sequences. Southern blots of mouse (*A,B*) or human (*C*) genomic DNA (10 μg) digested with various endonucleases (E, *Eco*RI; B, *Bam*HI; H, *Hin*dIII; and S, *Sst*I) were hybridized to the *Drosophila en* cDNA probe (*A*) or to the homeo box region (hbr) probe derived from Mo-en.1 (*B,C*). The sizes indicated at the right of each lane are in kilobase pairs.

ferent restriction endonucleases. Under conditions of low-stringency hybridization and washing, the Southern blot was probed with a 1.4 kb *Drosophila en* cDNA clone. Two prominent restriction fragment bands hybridized to this probe in each of the four mouse DNA digests, suggesting that the mouse genome contains two different sequences homologous to sequences within the *en* cDNA (Fig. 4A). When the same Southern blot was rehybridized to the H.1 probe that contains the human Hu1 homeo box (Fig. 1), six to eight restriction fragments were detected in each DNA digest; however, none of these fragments comigrated with the two detected with the *en* probe (data not shown), indicating that the *en*-homologous sequences are distinct from those containing ANT-C/BX-C homeo boxes.

A clone representing one of these two *en*-like sequences was then isolated from a mouse genomic library by screening with the 1.4-kb *en* cDNA probe under conditions of low stringency. Southern blot analysis, using the *en* cDNA probe, of restriction endonuclease digests of this clone, Mo-en.1, indicated that it represents one of the two genomic regions containing *en* homology. The *en*-homologous region in Mo-en.1 was found to localize to a 700-bp *Bam*HI-*Eco*RI fragment. A partial restriction map of the Mo-en.1 clone, which was further analyzed as described below, is shown in Figure 5.

The results of sequencing studies of the 700-bp *Bam*HI-*Eco*RI fragment indicated that the Mo-en.1 clone contains a homeo box that can encode a peptide considerably more homologous to the putative translation products of homeo boxes of the EN-C class than the ANT-C/BX-C class. When only identical amino acids are considered, the Mo-en.1 homeo box segment shares 75% homology with the homeo boxes of each of the EN-C genes but is only 45–53% homologous to each member of the *Drosophila* ANT-C/BX-C class of homeo box described thus far. A similar degree of reduced homology (45–53%) exists between Mo-en.1 and each member described thus far of the ANT-C/BX-C class of homeo box from vertebrates. As illustrated in Figure 6, of the 45 out of 60 amino acids that are perfectly conserved in three *Drosophila* ANT-C/BX-C homeo boxes, only 25 are also conserved in the homeo boxes of the two *Drosophila* EN-C genes. The Mo-en.1 homeo box conserves the same 25 amino acids. In addition to the homology between the homeo boxes of

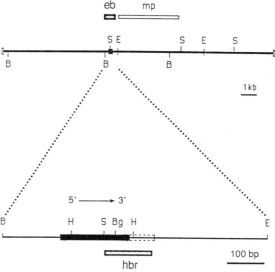

Figure 5. Restriction mapping of the region surrounding the Mo-en.1 homeo box. The genomic region contained in λMo-en.1 is shown above. The lower horizontal line is an enlargement of the *Eco*RI-*Bam*HI restriction fragment containing the homeo box. The arrow indicates the direction of transcription of the homeo box, which is shown as a solid rectangle. The extended region of homology between *en* and Mo-en.1 is delineated by a hatched rectangle. The two probes used for the Southern blot analyses (homeo box region probe, hbr, and mp probe) and the Northern blot analysis (eb probe) are diagramed. (B) *Bam*HI; (S) *Sst*I; (E) *Eco*RI; (H) *Hae*III; (Bg) *Bgl*II.

296   JOYNER ET AL.

Mo-en.1	Asp	Lys	Arg	Pro	Arg	Thr	Ala	Phe	Thr	Ala	Glu	Gln	Leu	Gln	Arg
EN	Glu	Lys	Arg	Pro	Arg	Thr	Ala	Phe	Ser	Ser	Glu	Gln	Leu	Ala	Arg
INV	Asp	Lys	Arg	Pro	Arg	Thr	Ala	Phe	Ser	Gly	Thr	Gln	Leu	Ala	Arg
ANTP	Arg	Lys	Arg	Gly	Arg	Gln	Thr	Tyr	Thr	Arg	Tyr	Gln	Thr	Leu	Glu
FTZ	Ser	Lys	Arg	Thr	Arg	Gln	Thr	Tyr	Thr	Arg	Tyr	Gln	Thr	Leu	Glu
UBX	Arg	Arg	Arg	Gly	Arg	Gln	Thr	Tyr	Thr	Arg	Tyr	Gln	Thr	Leu	Glu

Mo-en.1	Leu	Lys	Ala	Glu	Phe	Gln	Ala	Asn	Arg	Tyr	Ile	Thr	Glu	Gln	Arg
EN	Leu	Lys	Arg	Glu	Phe	Asn	Glu	Asn	Arg	Tyr	Leu	Thr	Glu	Arg	Arg
INV	Leu	Lys	His	Glu	Phe	Asn	Glu	Asn	Arg	Tyr	Leu	Thr	Glu	Lys	Arg
ANTP	Leu	Glu	Lys	Glu	Phe	His	Phe	Asn	Arg	Tyr	Leu	Thr	Arg	Arg	Arg
FTZ	Leu	Glu	Lys	Glu	Phe	His	Phe	Asn	Arg	Tyr	Ile	Thr	Arg	Arg	Arg
UBX	Leu	Glu	Lys	Glu	Phe	His	Thr	Asn	His	Tyr	Leu	Thr	Arg	Arg	Arg

Mo-en.1	Arg	Gln	Thr	Leu	Ala	Gln	Glu	Leu	Ser	Leu	Asn	Glu	Ser	Gln	Ile
EN	Arg	Gln	Gln	Leu	Ser	Ser	Glu	Leu	Gly	Leu	Asn	Glu	Ala	Gln	Ile
INV	Arg	Gln	Gln	Leu	Ser	Gly	Glu	Leu	Gly	Leu	Asn	Glu	Ala	Gln	Ile
ANTP	Arg	Ile	Glu	Ile	Ala	His	Ala	Leu	Cys	Leu	Thr	Glu	Arg	Gln	Ile
FTZ	Arg	Ile	Asp	Ile	Ala	Asn	Ala	Leu	Ser	Leu	Ser	Glu	Arg	Gln	Ile
UBX	Arg	Ile	Glu	Met	Ala	Tyr	Ala	Leu	Cys	Leu	Thr	Glu	Arg	Gln	Ile

Mo-en.1	Lys	Ile	Trp	Phe	Gln	Asn	Lys	Arg	Ala	Lys	Ile	Lys	Lys	Ala	Thr
EN	Lys	Ile	Trp	Phe	Gln	Asn	Lys	Arg	Ala	Lys	Ile	Lys	Lys	Ser	Thr
INV	Lys	Ile	Trp	Phe	Gln	Asn	Lys	Arg	Ala	Lys	Leu	Lys	Lys	Ser	Ser
ANTP	Lys	Ile	Trp	Phe	Gln	Asn	Arg	Arg	Met	Lys	Trp	Lys	Lys	Glu	Asn
FTZ	Lys	Ile	Trp	Phe	Gln	Asn	Arg	Arg	Met	Lys	Ser	Lys	Lys	Asp	Arg
UBX	Glu	Ile	Trp	Phe	Gln	Asn	Arg	Arg	Met	Lys	Leu	Lys	Lys	Glu	Ile

Figure 6. Comparison of the putative amino acid sequences of the Mo-en.1 homeo box with those of the two homeo boxes of the *Drosophila* EN-C class and with three of the ANT-C/BX-C class. The amino acid sequences of the three members of the EN-C class of homeo box, Mo.en.1, EN (engrailed), and INV (invected, previously termed engrailed-related [Poole et al. 1985]) are shown above. The amino acid sequences of three members of the ANT-C/BX-C class of homeo box are shown below. Boxed areas indicate the amino acids that have been perfectly conserved among all three members of each class. Shaded portions indicate the 25 amino acids that are perfectly conserved among all six homeo boxes. The sequences of EN and INV were determined Poole et al. (1985). The Antennapedia (ANTP), fushi tarazu (FTZ), and Ultrabithorax (UBX) sequences are those reported by McGinnis et al. (1984b).

Mo-en.1 and the EN-C genes of *Drosophila*, we found sequence homology outside of the homeo box. Of the 31 amino acids 3' to the homeo box that are highly conserved between the two EN-C genes, 17 of the first 21 amino acids immediately 3' to the homeo box are also conserved in Mo-en.1. These data are summarized in Figure 7. In contrast, there is no such homology between vertebrate and *Drosophila* sequences in regions flanking any homeo box of the ANT-C/BX-C class.

Conservation of the EN-C Homeo Box Regions in the Mouse and Human Genomes

As shown in Figure 4A, hybridization of the 1.4-kb *en* cDNA probe to Southern blots of mouse genomic DNA indicated that the mouse genome contains two regions of homology with this probe. The restriction map of λMo-en.1 indicates that it represents one of these two regions. To determine whether the second region detected with the *en* cDNA probe also contains sequences homologous to those in the conserved EN-C homeo box region, Southern blots of mouse genomic DNA were hybridized to a Mo-en.1-derived homeo box region (hbr) probe that extends from the *Sst*I site in the homeo box to approximately 50 bp 3' of the homeo box (see Fig. 5). Using moderate stringency conditions, we found that this probe hybridized to two (or three) restriction fragments in mouse genomic DNA that had been digested with any of the enzymes tested (Fig. 4B). In each case, the bands of hybridization observed were indistinguishable from those detected with the 1.4-kb *en* cDNA probe, and one set represents Mo-en.1. Thus,

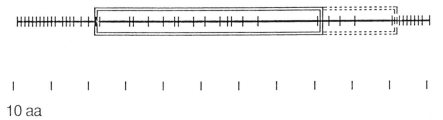

Figure 7. Graphic representation of putative amino acid sequence differences between Mo-en.1 and the *en* homeo box regions. The horizontal line represents the Mo-en.1 genomic region. The rectangle outlined by a solid double line delineates the homeo box and the rectangle outlined by the dashed double line delineates the 3' region of homology between Mo-en.1 and *en*. Each vertical line represents one amino acid difference between the Mo-en.1 genomic and *en* cDNA sequences.

the remaining band(s) of hybridization represents a second region, Mo-en.2, that is homologous to the conserved EN-C region found in Mo-en.1.

Our studies (described above) of the structure of a cluster of ANT-C/BX-C homeo boxes on human chromosome 17 have demonstrated a strong conservation of this region in mouse DNA, including sequences flanking the homeo boxes. To determine whether there is similar conservation of sequences in the EN-C homeo box region, restriction-enzyme-digested human genomic DNA was probed with the Mo-en.1-derived hbr probe as described above. Human DNA was found to contain at least one region that hybridized to the Mo-en.1 hbr probe (Fig. 4C). To determine whether this human region contains additional homology with Mo-en.1, a similar Southern blot was hybridized to a single-copy probe, derived from Mo-en.1, that lacks any of the sequences conserved in *Drosophila* (data not shown). This probe (mp), which extends from an *Eco*RI site 360 bp 3' of the homeo box to a *Sst*I site approximately 4 kb 3' of the homeo box, detected a single band in human DNA that had been digested with any of three different restriction enzymes, and these bands were found to comigrate with the most prominent bands detected with the hbr probe. Thus, the human genome contains a region homologous to Mo-en.1 that includes both a homeo box region as well as 3'-flanking sequences.

Studies of the Expression of Mu1 and Mo-en.1 Genomic Regions in Mouse Embryos

The strong conservation of coding capacity in the homeo box regions of Mu1 and Mo-en.1 relative to their respective *Drosophila* counterparts suggests a conservation of function of the proteins encoded by these regions. We would therefore predict that these regions are transcribed and translated during mouse embryogenesis, as they are during embryonic development in *Drosophila*. As a first step in testing this hypothesis, poly(A)+ RNAs from mouse embryos at 11.5, 12.5, and 13.5 days of gestation were analyzed for Mu1 and Mo-en.1 sequences. Figure 8 shows the results of a Northern blot experiment in which the RNA from embryos at these stages was hybridized to the rs probe (derived from Mu1) that contains approximately 60 bases of the homeo box and approximately 1.5 kb of 5'-flanking

DNA or the eb probe (derived from Mo-en.1) that contains the homeo box and approximately 150 bp 5' and 350 bp 3' to it. A transcript of approximately 2.3 kb from the Mu1 genomic region and two transcripts (the most prominent being a 3.1-kb mRNA) from the Mo-en.1 genomic region were detected in the embryo RNA. These data indicate that these two homeo-box-containing regions are contained in genes expressed during the mid-gestation stages of mouse embryogenesis.

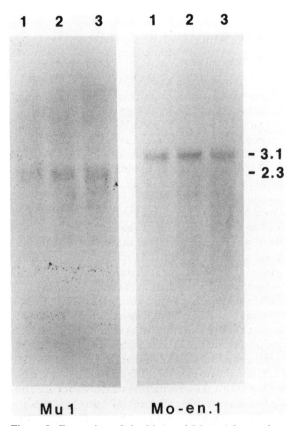

Figure 8. Expression of the Mu1 and Mo-en.1 homeo-box-containing genes in mouse embryos. Northern blots of 5 μg of poly(A)+ RNA extracted from mouse embryos at 11.5 (lane *1*), 12.5 (lane *2*), and 13.5 (lane *3*) days of gestation were hybridized with the rs probe derived from the Mu-1 genomic clone and containing 60 nucleotides of the Mu1 homeo box (*left*) and the eb probe derived from the Mo-en.1 genomic clone and containing the Mo-en.1 homeo box (*right*). The sizes indicated at the right are in kilobases.

DISCUSSION

The data reported here, taken in conjunction with the results of an earlier study (Joyner et al. 1985), demonstrate that a cluster of at least three ANT-C/BX-C-like homeo box domains is located on human chromosome 17 and that the region around at least one homeo box in the cluster, Hu1, has been highly conserved on mouse chromosome 11. The concept that the remainder of the cluster, including the regions around Hu2 and Hu5, is also conserved is supported, at least in part, by results of other investigators showing that a mouse genomic clone that appears to overlap λMu1, by restriction site comparison, contains a second homeo box 5' to Mu1 (Rabin et al. 1985). The location of this second homeo box with respect to the one we have termed Mu1 suggests that it is the Hu2 homolog and that the homology between the human and mouse clusters extends well beyond the region around Hu1/Mu1. An intriguing possibility is that this conserved cluster is analogous to either the ANT-C or BX-C cluster of homeo-box-containing genes in *Drosophila*. In the absence of sequence homology outside the homeo boxes, it will be difficult to determine whether this is indeed the case until it is known whether the functions of the genes in this mammalian homeo box cluster are in any way analogous to the functions of the genes in the ANT-C or BX-C.

In the case of the mouse gene Mo-en.1, which contains sequence homology with the *Drosophila en* gene, sequences outside the homeo box have been conserved. This gene can code for a protein containing 81 amino acids that shares 75% homology with sequences in the two genes of the *Drosophila* EN-C. These 81 amino acids include a 60-amino-acid homeo box domain that is significantly more homologous to homeo boxes of the EN-C genes than to the other homeo boxes described thus far. The remaining 21 amino acids that lie immediately 3' to the homeo box are also highly conserved between Mo-en.1 and the two genes of the *Drosophila* EN-C. Mo-en.1 thus represents the first example of a vertebrate gene that can be identified with a specific complex of *Drosophila* homeo-box-containing genes. Interestingly, our data provide evidence that an *en* homolog may also exist in the human genome.

One problem that arises when considering the possible ways in which *Drosophila* and vertebrate homeo-box-containing genes might have similar functions is that the fundamental similarities between development in higher invertebrates and vertebrates are not obvious. In defining these parallels, it is possible that the vertebrate somite is the developmental homolog of the *Drosophila* segment. One prediction of this hypothesis is that the vertebrate homologs of homeo-box-containing genes should be expressed in the vertebrate somites in a manner analogous to the position-dependent arrays that characterize the expression of the *Drosophila* homeo-box-containing genes (Akam 1983; Levine et al. 1983; Hafen et al. 1984a; White and Wilcox 1984; Beachy et al. 1985; Fjose et al. 1985; Kornberg et al. 1985). The results of in situ RNA hybridization experiments and antigen-localization studies using antibodies against the putative Mu1 and Mo-en.1 protein products should help to determine whether this is indeed the case.

A second problem that arises when trying to compare the functions of different homeo-box-containing genes is that the biochemical function has not been elucidated for any of the homeo-box-containing gene products, and in particular, the function of the peptide sequence specified by the homeo box has not been determined. Genetic and molecular studies of *Drosophila* homeotic genes suggest that not only do they regulate other genes involved in development, but they also regulate the expression of other homeotic genes (Duncan and Lewis 1982; Struhl 1983; Hafen et al. 1984b; Ingham 1984). The inferred structure of the amino acid domain encoded by the homeo box and its resemblance to DNA-binding proteins (Laughon and Scott 1984; Shepherd et al. 1984), along with the nuclear localization of homeotic gene products (White and Wilcox 1984; Beachy et al. 1985), suggest that these proteins may recognize and bind to specific DNA sequences. It is an attractive model that homeo-box-containing gene products bind to specific DNA sequences and regulate transcription of various classes of genes in a manner similar to that of eukaryotic sequence-specific DNA-binding proteins (for review, see Dynan and Tjian 1985). One reason we have chosen to examine the expression of mammalian homeo-box-containing genes, in spite of the difficulty of investigating the stage-specific and position-specific expression of these genes, is the availability of fractionated in vitro transcription systems from human cells (Matsui et al. 1980; Dynan and Tjian 1983). To begin to examine interactions between these proteins and various templates, purified proteins from homeo-box-containing genes can be tested in in vitro transcription reactions.

Finally, in view of the biological insights that have been gained by studying mutant alleles of the homeotic and segmentation genes in *Drosophila*, it is clear that the study of mutant alleles of the mammalian homeo-box-containing genes would help to illuminate both their function and the developmental processes in which they are involved. As a first step toward determining whether mutant alleles of these genes exist, we have mapped the cluster of ANT-C/BX-C homeo-box-containing genes described here to human chromosome 17 and mouse chromosome 11 (Joyner et al. 1985). In addition, it has been shown that Mo-en.1 maps to mouse chromosome 1 (A. Joyner et al., unpubl.) and therefore is physically separate from all other mouse homeo-box-containing genes that have been mapped to date. The knowledge that the cluster of ANT-C/BX-C class homeo boxes maps to mouse chromosome 11 and that Mo-en.1 maps to chromosome 1 has enabled us to focus our attention on several existing mouse mutations that might be alleles of these homeo-box-containing genes. Mutations such as tail-short on chromosome 11 and dominant hemimelia on chromosome 1, which are lethal when homozygous, have phenotypes that are suggestive of the types of developmental abnormalities that might result from lesions in functions that control embryonic development (Green 1981). Mapping studies currently in progress should help to determine whether any of these mutations might be alleles of the homeo-box-containing genes under investigation. In addition, the results of studies aimed at determining the spatial and temporal expression of both the ANT-C/BX-C classes of homeo-box-containing genes we have isolated, as well as studies of the possible role of the homeo box in the control of transcription, should help to provide new insights into the function of homeo-box-containing genes in mammals as well as the relationship between the processes that control development in higher vertebrates and invertebrates.

ACKNOWLEDGMENTS

We thank T. Learned for help with the sequencing studies, Kevin Coleman for providing the engrailed cDNA probe, George Church for helpful discussions, and Karen Ronan for preparing the manuscript. This work was funded by grants from the National Foundation for Cancer Research, a National Institute of Environmental Health Sciences Center grant, and grants from the National Institutes of Health. C.A.H. was supported by a Jane Coffin Childs fellowship and A.L.J. was supported by a fellowship from the M.R.C. of Canada.

REFERENCES

AKAM, M.E. 1983. The localization of *Ultrabithorax* transcripts in *Drosophila* tissue sections. *EMBO J.* **2:** 2075.

BEACHY, P.A., S.L. HELFAND, and D.S. HOGNESS. 1985. Segmental distribution of bithorax complex proteins during *Drosophila* development. *Nature* **313:** 545.

CARRASCO, A.E., W. MCGINNIS, W.J. GEHRING, and E.M. DE ROBERTIS. 1984. Cloning of an *X. laevis* gene expressed during early embryogenesis coding for a peptide region homologous to *Drosophila* homeotic genes. *Cell* **37:** 409.

CHIRGWIN, J.M., A.E. PRZYBYLA, R.J. MCDONALD, and W.J. RUTTER. 1979. Isolation of biologically active ribonucleic acid from sources enriched in ribonuclease. *Biochemistry* **18:** 5294.

CHURCH, G.M. and W. GILBERT. 1984. Genomic sequencing. *Proc. Natl. Acad. Sci.* **81:** 1991.

COLBERG-POLEY, A.M., S.D. VOSS, K. CHOWDHURY, and P. GRUSS. 1985. Structural analysis of murine genes containing homeo box sequences and their expression in embryonal carcinoma cells. *Nature* **314:** 713.

DUNCAN, I. and E.B. LEWIS. 1982. Genetic control of body segment differentiation in *Drosophila*. In *Developmental order: Its origin and regulation* (ed. S. Subtelny), p. 533. A.R. Liss, New York.

DYNAN, W.S. and R. TJIAN. 1983. Isolation of transcription factors that discriminate between different promoters recognized by RNA polymerase II. *Cell* **32:** 669.

———. 1985. Control of eukaryotic messenger RNA synthesis by sequence-specific DNA binding proteins. *Nature* **316:** 774.

FJOSE, A., W.J. MCGINNIS, and W.J. GEHRING. 1985. Isolation of a homeo box-containing gene from the engrailed region of *Drosophila* and spatial distribution of its transcripts. *Nature* **313:** 284.

GARCIA-BELLIDO, A. and P. SANTAMARIA. 1972. Developmental analysis of the wing disc in the mutant *engrailed* of *Drosophila melanogaster*. *Genetics* **72:** 87.

GEHRING, W. 1985. The homeo box: A key to the understanding of development? *Cell* **40:** 3.

GREEN, M., ed. 1981. *Genetic variants in strains of laboratory mouse*. Gustave Fischer Verlag, Stuttgart.

HAFEN, E., A. KUROIWA, and W.J. GEHRING. 1984a. Spatial distribution of transcripts from the segmentation gene *fushi tarazu* during *Drosophila* embryonic development. *Cell* **37:** 833.

HAFEN, E., M. LEVINE, and W.J. GEHRING. 1984b. Regulation of *Antennapedia* transcript distribution by the bithorax complex in *Drosophila*. *Nature* **307:** 287.

INGHAM, P.W. 1984. A gene that regulates the bithorax complex differentially in larval and adult cells of *Drosophila*. *Cell* **37:** 815.

JOYNER, A.L., R.V. LEBO, Y.W. KAN, R. TJIAN, D.R. COX, and G.R. MARTIN. 1985. Comparative chromosome mapping of a conserved homeo box region in mouse and human. *Nature* **314:** 173.

KAUFMAN, T.C., R. LEWIS, and R. WAKIMOTO. 1980. Cytogenetic analysis of chromosome 3 in *Drosophila melanogaster*: The homeotic gene complex in polytene chromosome interval 84A-B. *Genetics* **94:** 115.

KORNBERG, T., I. SIDEN, P. O'FARRELL, and M. SIMON. 1985. The engrailed locus of *Drosophila*: In situ localization of transcripts reveals compartment-specific expression. *Cell* **40:** 45.

LAUGHON, A. and M.P. SCOTT. 1984. Sequence of a *Drosophila* segmentation gene: Protein structure homology with DNA binding proteins. *Nature* **310:** 25.

LEVINE, M., G.M. RUBIN, and R. TJIAN. 1984. Human DNA sequences homologous to a protein coding region conserved between homeotic genes of *Drosophila*. *Cell* **38:** 667.

LEVINE, M., E. HAFEN, R.I. GARBER, and W.J. GEHRING. 1983. Spatial distribution of *Antennapedia* transcripts during *Drosophila* development. *EMBO J.* **2:** 2037.

LEWIS, E.B. 1978. A gene complex controlling segmentation in *Drosophila*. *Nature* **276:** 565.

MANIATIS, T., E.F. FRITSCH, and J. SAMBROOK. 1982. *Molecular cloning: A laboratory manual*. Cold Spring Harbor Laboratory, Cold Spring Harbor, New York.

MATSUI, T., J. SEGALL, P.A. WEIL, and R.G. ROEDER. 1980. Multiple factors required for accurate initiation of transcription by purified RNA polymerase II. *J. Biol. Chem.* **255:** 11992.

MCGINNIS, W., C.P. HART, W.J. GEHRING, and F.H. RUDDLE. 1984a. Molecular cloning and chromosome mapping of a mouse DNA sequence homologous to homeotic genes of *Drosophila*. *Cell* **38:** 675.

MCGINNIS, W., R.L. GARBER, J. WIRZ, A. KUROIWA, and W.J. GEHRING. 1984b. A homologous protein-coding sequence in *Drosophila* homeotic genes and its conservation in other metazoans. *Cell* **37:** 403.

MCGINNIS, W., M.J. LEVINE, E. HAFEN, A. KUROIWA, and W.J. GEHRING. 1984c. A conserved DNA sequence in homeotic genes of the *Drosophila* Antennapedia and bithorax complexes. *Nature* **308:** 428.

MESSING, J. 1983. New M13 vectors for cloning. *Methods Emzymol.* **101:** 28.

MULLER, M.M., A.E. CARRASCO, and E.M. DE ROBERTIS. 1984. A homeo box containing gene expressed during oogenesis in *Xenopus*. *Cell* **39:** 157.

NÜSSLEIN-VOLHARD, C. 1979. Maternal effect mutations that alter the spatial coordinates of the embryo of *Drosophila melanogaster*. In *Determinants of spatial organization* (ed. S. Subtelny and I.R. Konisberg), p. 185. Academic Press, New York.

NÜSSLEIN-VOLHARD, C. and E. WIESCHAUS. 1980. Mutations affecting segment number and polarity in *Drosophila*. *Nature* **287:** 795.

OUWENEEL, W.J. 1976. Developmental genetics of homeosis. *Adv. Genet.* **18:** 179.

Poole, S.J., L.M. Kauvar, B. Drees, and T. Kornberg. 1985. The *engrailed* locus of *Drosophila*: A structural analysis of an embryonic transcript. *Cell* **40:** 37.

Rabin, M., C.P. Hart, A. Ferguson-Smith, W. McGinnis, M. Levine, and F.H. Ruddle. 1985. Two homeobox loci mapped in evolutionarily related mouse and human chromosomes. *Nature* **314:** 175.

Sanger, F., S. Nicklen, and A.R. Coulson. 1977. DNA sequencing with chain-terminating inhibitors. *Proc. Natl. Acad. Sci.* **74:** 5463.

Scott, M.P. and A.J. Weiner. 1984. Structural relationships among genes that control development: Sequence homology between the *Antennapedia*, *Ultrabithorax*, and *fushi tarazu* loci of *Drosophila*. *Proc. Natl. Acad. Sci.* **81:** 4115.

Shepherd, J.C., W. McGinnis, A.E. Carrasco, E.M. De Robertis, and W. Gehring. 1984. Fly and frog domains show homology with yeast mating type regulatory proteins. *Nature* **310:** 70.

Southern, E.M. 1975. Detection of specific sequences among DNA fragments separated by gel electrophoresis. *J. Mol. Biol.* **98:** 503.

Struhl, G. 1983. Role of esc^+ gene product in ensuring the selective expression of segment specific homeotic genes in *Drosophila*. *J. Embryol. Exp. Morphol.* **76:** 297.

White, R.A.H. and M. Wilcox. 1984. Protein products of the bithorax complex in *Drosophila*. *Cell* **39:** 163.

Human cDNA Clones Containing Homeo Box Sequences

E. BONCINELLI, A. SIMEONE, A. LA VOLPE, A. FAIELLA, V. FIDANZA,
D. ACAMPORA, AND L. SCOTTO*
International Institute of Genetics and Biophysics, CNR, Naples, Italy;
*Immunology Department, II Medical School, University of Naples, Naples, Italy

One of the central aims of developmental biology is to understand the molecular mechanisms governing the precise control of differentiation processes. Early in the development of many metazoans, cells in different regions of the embryo are committed to separate fates based on their position. These separate fates result eventually in a differential pattern of morphogenesis in different regions of the embryo. A conserved basic pattern in animal embryogenesis is that of metamerism, in which groups of cells are organized into morphologically distinct segments. It is of great interest to identify the genetic loci in vertebrates that determine pattern formation and developmental pathways. Unfortunately, the genetics of most vertebrates does not easily lend itself to this type of analysis. In contrast, a number of loci in *Drosophila* have been defined that are implicated in directing the formation of distinct body segments (Ouweneel 1976). These include the homeotic genes (Lewis 1978), which specify separate identities to different elements of the segmental pattern, and the segmentation genes (Nüsslein-Volhard and Wieschaus 1980), which determine the number and polarity of body segments. In *Drosophila*, many homeotic genes are grouped in two gene clusters on the right arm of the third chromosome, the Antennapedia complex (ANT-C) and the bithorax complex (BX-C) (Lewis 1978; Kaufman et al. 1980). More than ten *Drosophila* genes, mapping mainly in these complexes, share a highly conserved 183-bp sequence, designated the homeo box (McGinnis et al. 1984b,c; Gehring 1985). The ANT-C has at least five genes containing the homeo box, including Antennapedia (*Antp*) (McGinnis et al. 1984b), fushi tarazu (*ftz*) (McGinnis et al. 1984b; Scott and Weiner 1984), Deformed (*Dfd*), and Sex combs reduced (*Scr*), whereas the BX-C contains at least three homeo boxes, corresponding to three lethal complementation groups, namely, in Ultrabithorax (*Ubx*) (McGinnis et al. 1984b), infra-abdominal-2 (*iab-2*), and infra-abdominal-7 (*iab-7*). Genes containing the homeo box, in addition to the two main homeotic clusters in *Drosophila*, have also been cloned. Among them, a homeo box has been identified within the engrailed (*en*) gene and a closely linked gene, designated engrailed-related (*en-r*) (Fjose et al. 1985; Poole et al. 1985).

Using the *Drosophila* homeo boxes as probes, homeo box sequences have been detected by Southern blotting in many organisms, including vertebrates and man (McGinnis et al. 1984b), and homeo-box-containing genomic sequences have been isolated in *Xenopus* (Carrasco et al. 1984; Mueller et al. 1984), in the mouse (McGinnis et al. 1984a; Colberg-Poley et al. 1985), and in humans (Levine et al. 1984). The estimated number of genes containing the homeo box in the mouse and in humans is about ten and may be higher. From this point of view, it would be relevant to determine which of these genes are transcribed in which tissues in determined periods in the embryo and/or in adults. On the other hand, some of these genes could represent pseudogenes frequently found associated with gene families. With this in mind, we decided to look for cDNA clones containing the homeo box, and we report here on the isolation and characterization of six human cDNA clones obtained by screening with *Drosophila* homeo boxes as probes, using a cDNA library prepared from poly(A)$^+$ RNA of human fibroblasts transformed by SV40.

RESULTS AND DISCUSSION

Isolation of Human cDNA Clones Containing Sequences Homologous to the *Drosophila* Homeo Box

Using the *Antp* and *ftz* homeo boxes as probes, we screened approximately 5×10^5 recombinant colonies of a cDNA library prepared from SV40-transformed human fibroblasts (Okayama and Berg 1983). The hybridization and washing conditions were those adopted to detect human DNA sequences that share homology with the *Drosophila* homeo boxes (McGinnis et al. 1984b). Six positive cDNA clones with an insert length ranging from 1 kb to 1.7 kb were isolated. Four of these clones (HHO.c1, HHO.c2, HHO.c3, and HHO.c10) showed a comparatively strong hybridization signal, whereas two clones (HHO.c8 and HHO.c13) showed a weaker hybridization signal.

Figure 1 shows the restriction map of these clones. Restriction mapping analysis and cross-hybridization experiments revealed that clones c1, c2, and c3 derived presumably from the same polyadenylated transcript, although they differ in their 5' truncation point and/or in the length of the poly(A) tail. Clones c1, c10, c8, and c13 do not cross-hybridize. The homeo box region lies proximal to the 5' terminus of each clone with the exception of c13.

Using high-stringency conditions of hybridization and washing, these clones were used to detect specific

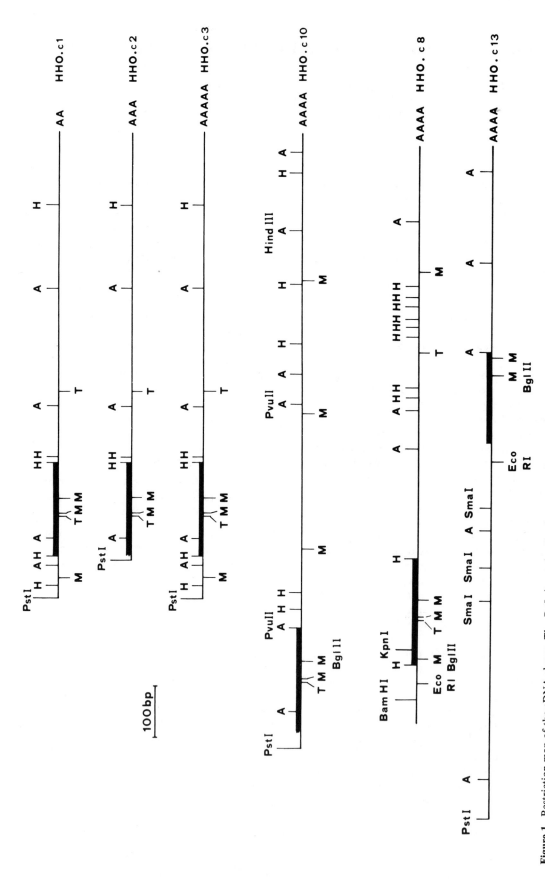

Figure 1. Restriction map of the cDNA clones. The *PstI* site at the 5' end corresponds to the *PstI* site of the vector, which is just upstream of the oligo(dG) tail (Okayama and Berg 1983). Restriction sites for *AluI* (A), *HaeIII* (H), *TaqI* (T), and *MboI* (M) are shown. (■) Regions containing sequences homologous to the *Drosophila* homeo box.

hybridizing bands in total human Southern blot analysis (Fig. 2). Clones c1, c2, and c3 detect a single EcoRI restriction fragment of approximately 9 kb, strengthening the hypothesis that clones c1, c2, and c3 derive from the same transcripts. In the same type of experiment, clone c10 detected a single EcoRI restriction fragment of 11 kb, whereas clones c8 and c13, which contain an EcoRI-recognition site, detected two EcoRI restriction bands as described in Figure 2. These data indicate that our cDNA clones represent transcripts deriving from four different single-copy genes that include sequences homologous to the homeo box region of Drosophila.

Homeo Boxes Contained in Our cDNA Clones Show a High Degree of Conservation

To determine the extent of homology between the homeo box regions contained in our clones and the homeo box regions found in Drosophila and in other systems, we determined the nucleotide sequence of a region around the homeo box contained in clones c8, c10, and c13 and in clone c1, as a representative of the group of clones including c1, c2, and c3. Figure 3 shows the conceptual translation of the homeo boxes present in these clones and in other published clones of Drosophila, Xenopus, mice, and humans. In every clone, the reported peptide sequence is included within the longest open reading frame. Figure 4 summarizes the homology between different homeo boxes at the nucleotide level and predicted peptide sequence level. A high degree of conservation is apparent throughout all the homeo boxes, with the exceptions of the en and en-r sequences (Fjose et al. 1985; Poole et al. 1985).

The c1 homeo box shares 85% direct sequence homology with the Antp homeo box. The homology is even higher in the peptide sequence derived from the conceptual translation of both regions (60 amino acid residues out of 61). The c1 homeo box shares 86.9% nucleotide sequence homology and 57 amino acids with the human Hu2 homeo box. The homeo boxes present in the human clone c1 and in the Xenopus clone MM3 share 59 amino acid residues. The similarity of these two clones extends to their predicted carboxy termini, which are 20 amino acids downstream from the homeo domain, ending with six glutamic acid residues in a row (Mueller et al. 1984).

The insert of clone c10 starts within the homeo box region from its seventh codon. This human homeo box region also shares high (80%) nucleotide sequence homology with the Antp homeo box region. The nucleotide sequence of clone c10 is identical to the determined region from the human Hu1 genomic sequence (Levine et al. 1984), with a single nucleotide substitution at the third position of the sixth codon downstream from the homeo box. The cDNA clone c10 appears to represent a transcript from the Hu1 human gene sequence.

The predicted peptide sequence of the homeo domain present in clone c8 is identical to that of the homeo domain of the Xenopus clone AC1 (Carrasco et al. 1984). The open reading frame present in clone c8 terminates 34 amino acid residues downstream from the homeo domain. The peptide sequence of this terminal region shares a high degree of homology with the corresponding region of clone AC1. Furthermore, the peptide sequence derived from the short region of clone c8 upstream of the homeo domain shares 20 amino acid residues out of 21 with the corresponding region present in a Xenopus cDNA clone homologous to clone AC1 (De Robertis et al., this volume). The homology between the human gene corresponding to clone c8 and the frog gene corresponding to clone AC1 appears to be more extended than the homeo box region. The observation that two human genes show such a high degree of homology in the peptide sequence with two frog genes suggests a common function of their products in these species, in addition to the specific role played by the homeo domain. Comparison of human cDNA clones c1, c8, and c10 with homologous genomic sequences suggests the presence of an intron just upstream of the homeo box sequence in each case. A potential glycosylation site (Asn-X-Ser/Thr) is present at the 60th amino acid residue of c1 and Antp homeo domains and at the 61st amino acid residue of c8 and AC1 homeo domains.

The homeo box contained in clone c13 appears to be uniformly related to most homeo boxes, since it shares a 70% nucleotide homology with every homeo box analyzed with the exclusion of en and en-r. The nucleotide sequence analysis of clone c13 reveals that this cDNA clone may represent the almost full-length complementary copy of a polyadenylated transcript, due to the presence of a 200-bp 5'-untranslated region (data not shown).

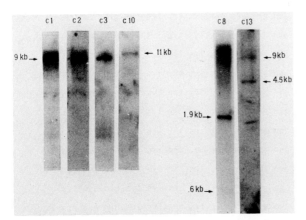

Figure 2. Southern blot analysis of human genomic DNA. Each lane contains 15 µg of human genomic DNA isolated from peripheral lymphocytes, digested with endonuclease EcoRI, electrophoresed on a 1% agarose gel, transferred to nitrocellulose, and hybridized to the indicated cDNA probes.

Figure 3. Putative amino acid sequence of the common open reading frame found in the homeo box region of human clones c1, c8, c10, c13, Hu1, and Hu2; of *Drosophila* genes *Antp*, *ftz*, and *Ubx*; of *Xenopus* clones MM3 and AC1; and of mouse clones Mo10 and m6. The amino acid changes relative to the *Antp* domain are underscored.

```
                                         10                                              20
c1     arg lys arg gly arg gln thr tyr thr arg tyr gln thr leu glu leu glu lys glu phe
c8     arg arg arg gly arg gln ile tyr ser arg tyr gln thr leu glu leu glu lys glu phe
c10                            ala tyr thr arg tyr gln thr leu glu leu glu lys glu phe
c13    pro lys arg ser arg thr ala tyr thr arg gln gln val leu glu leu glu lys glu phe
Antp   arg lys arg gly arg gln thr tyr thr arg tyr gln thr leu glu leu glu lys glu phe
ftz    ser lys arg thr arg gln thr tyr thr arg tyr gln thr leu glu leu glu lys glu phe
Ubx    arg arg arg gly arg gln thr tyr thr arg tyr gln thr leu glu leu glu lys glu phe
Hu1    gly lys arg ala arg thr ala tyr thr arg tyr gln thr leu glu leu glu lys glu phe
Hu2    thr ala gly gly arg gln thr tyr thr arg tyr gln thr leu glu leu glu lys glu phe
MM3    arg lys arg gly arg gln thr tyr thr arg tyr gln thr leu glu leu glu lys glu phe
AC1    arg arg arg gly arg gln ile tyr ser arg tyr gln thr leu glu leu glu lys glu phe
m6     arg lys arg gly arg gln thr tyr thr arg tyr gln thr leu glu leu glu lys glu phe
Mo10   ser lys arg gly arg thr ala tyr thr arg pro gln leu val glu leu glu lys glu phe

                                         30                                              40
c1     his tyr asn arg tyr leu thr arg arg arg arg ile glu ile ala his ala leu cys leu
c8     his phe asn arg tyr leu thr arg arg arg arg ile glu ile ala asn ala leu cys leu
c10    his phe asn arg tyr leu thr arg arg arg arg ile glu ile ala his ala leu cys leu
c13    his phe asn arg tyr leu thr arg arg arg arg ile glu ile ala his thr leu cys leu
Antp   his phe asn arg tyr leu thr arg arg arg arg ile glu ile ala his ala leu cys leu
ftz    his phe asn arg tyr ile thr arg arg arg arg ile asp ile ala asn ala leu ser leu
Ubx    his thr asn his tyr leu thr arg arg arg arg ile glu met ala tyr ala leu cys leu
Hu1    his phe asn arg tyr leu thr arg arg arg arg ile glu ile ala his ala leu cys leu
Hu2    his tyr asn arg tyr leu thr arg arg arg arg ile glu ile ala his ala leu cys leu
MM3    his phe asn arg tyr leu thr arg arg arg arg ile glu ile ala his val leu cys leu
AC1    his phe asn arg tyr leu thr arg arg arg arg ile glu ile ala asn ala leu cys leu
m6     his phe asn arg tyr leu thr arg arg arg thr leu glu ile ala his ala leu cys leu
Mo10   his phe asn arg tyr leu met arg pro arg arg val glu met ala asn leu leu asn leu

                                         50                                              60
c1     thr glu arg gln ile lys ile trp phe gln asn arg arg met lys trp lys lys glu asn lys
c8     thr glu arg gln ile lys ile trp phe gln asn arg arg met lys trp lys lys glu ser asn
c10    ser glu arg gln ile lys ile trp phe gln asn arg arg met lys trp lys lys asp asn lys
c13    ser glu arg gln ile lys ile trp phe gln asn arg arg met lys trp lys lys asp his lys
Antp   thr glu arg gln ile lys ile trp phe gln asn arg arg met lys trp lys lys glu asn lys
ftz    ser glu arg gln ile lys ile trp phe gln asn arg arg met lys ser lys lys asp arg thr
Ubx    thr glu arg gln ile lys ile trp phe gln asn arg arg met lys leu lys lys glu ile gln
Hu1    ser glu arg gln ile lys ile trp phe gln asn arg arg met lys trp lys lys asp asn lys
Hu2    thr glu arg gln ile lys ile trp phe gln asn arg arg met lys trp lys lys glu ser lys
MM3    thr glu arg gln ile lys ile trp phe gln asn arg arg met lys trp lys lys glu asn lys
AC1    thr glu arg gln ile lys ile trp phe gln asn arg arg met lys trp lys lys glu arg asn
m6     thr glu arg gln ile lys ile trp phe gln asn arg arg met lys trp lys lys glu his lys
Mo10   thr glu arg gln ile lys ile trp phe gln asn arg arg met lys tyr lys lys asp gln lys
```

	Antp	ftz	Ubx	EN	ER	AC1	MM3	Mouse10	Hu1	Hu2		
c1	85.2%	78%	80%	54.6%	55.2%	78.7%	82%	72.7%	82.5%	86.9%		
	60	49	52	30	30	54	59	44	54	57		
c10	80.3%	78.7%	76.5%	59%	57.4%	74.3%	81.4%	73.8%	100%	79.2%		
	55	50	46	33	32	50	54	46	61	51	c1	
c8	77.6%	75.4%	75.4%	55%	52%	82%	83.6%	72.6%	78%	78%	83.6%	
	55	50	53	30	29	61	54	44	51	54	54	
c13	72%	68.3%	65.6%	51%	53.6%	68.3%	72.6%	71%	74.3%	73.2%	72%	72%
	51	47	43	32	31	47	51	46	55	47	50	47

Figure 4. Comparison of the nucleotide sequence and predicted peptide sequence of several homeo boxes. (*Top*) Percentage of nucleotide homology; (*bottom*) number of identical amino acid residues out of 61.

Expression of Human cDNA Clones in Transformed Fibroblasts

In *Drosophila melanogaster*, homeotic genes are expressed at defined developmental periods in the embryo and when the determination of adult structures takes place. Frequently, these genes appear to be complex transcriptional units with multiple transcripts (Bender et al. 1983; Garber et al. 1983; Scott et al. 1983; Kuroiwa et al. 1984; Beachy et al. 1985). Similar results have been reported in *Xenopus* (Carrasco et al. 1984; Mueller et al. 1984) and in mouse embryonal carcinoma cells (see Colberg-Poley et al. 1985), although expression analysis of one human clone indicated that there was no detectable mRNA corresponding to the homeo box, expressed in some normal and transformed primate tissue-culture cells (Levine et al. 1984). We isolated total RNA from the cell line GM0637 of SV40-transformed human fibroblasts, from which the screened cDNA library was prepared, electrophoresed under denaturing conditions, and transferred onto nitrocellulose membranes. Hybridization at high stringency to our cDNA clones is shown in Figure 5. The insert of clone c1 reveals four major transcripts of approximately 4.8, 4.4, 2.5, and 2.2 kb, with the 2.5-kb band representing about ten molecules per cell. Lower-abundance transcripts are detected by clones c10 and c13, whereas clone c8 hybridizes to barely detectable transcripts 2.2 kb and 2.4 kb long. Intermediate and low-abundance transcripts are detected by all four clones in normal fibroblasts and peripheral blood lymphocytes (data not shown). It is rewarding to investigate which transcript contains which portions of these cDNA clones and in particular the homeo box regions. The presence in some adult human tissues of transcripts complementary to our cDNA clones suggests that different transcripts of the same complex transcriptional unit containing the homeo box might have different functions in embryonic and adult tissues.

ACKNOWLEDGMENTS

We are greatly indebted to Prof. W.J. Gehring for providing the *Drosophila* clones and to Dr. H.K. Okayama for providing the cDNA library. We thank Drs. S. Boast, P.P. Di Nocera, and L. Lania for helpful discussions. This work was supported by Progetti Finalizzati CNR Ingegneria Genetica e Basi Molecolari delle Malattie Ereditarie and Oncologia.

REFERENCES

BEACHY, P.A., S.L. HELFAND, and D.S. HOGNESS. 1985. Segmental distribution of bithorax complex proteins during *Drosophila* development. *Nature* **313:** 545.

BENDER, W., M. AKAM, F. KARCH, P.A. BEACHY, M. PFEIFER, P. SPIERER, E.B. LEWIS, and D.S. HOGNESS. 1983. Molecular genetics of the bithorax complex in *Drosophila melanogaster*. *Science* **221:** 23.

CARRASCO, A.E., W. MCGINNIS, W.J. GEHRING, and E.M. DE ROBERTIS. 1984. Cloning of an *X. laevis* gene expressed during early embryogenesis coding for a peptide region homologous to *Drosophila* homeotic genes. *Cell* **37:** 409.

COLBERG-POLEY, A.M., S.D. VOSS, K. CHOWDHURY, and P. GRUSS. 1985. Structural analysis of murine genes containing homoeo box sequences and their expression in embryonal carcinoma cells. *Nature* **314:** 713.

FJOSE, A., W.J. MCGINNIS, and W.J. GEHRING. 1985. Isolation

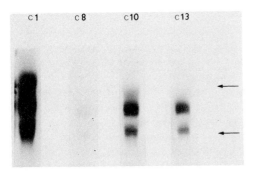

Figure 5. Northern blot analysis of transcripts homologous to our cDNA clones in a cell line (GM0637) of human fibroblasts transformed by SV40. Total RNA (25 μg) has been loaded in every lane. Arrows indicate the position of human 28S and 18S ribosomal RNAs.

of a homeo box-containing gene from the engrailed region of *Drosophila* and the spatial distribution of its transcripts. *Nature* **313:** 284.

GARBER, R.L., A. KUROIWA, and W.J. GEHRING. 1983. Genomic and cDNA clones of the homeotic locus *Antennapedia* in *Drosophila*. *EMBO J.* **2:** 2027.

GEHRING, W.J. 1985. The homeo box: A key to the understanding of development? *Cell* **40:** 3.

KAUFMAN, T.C., R. LEWIS, and B. WAKIMOTO. 1980. Cytogenetic analysis of chromosome 3 in *Drosophila melanogaster*: The homeotic gene complex in polytene chromosome interval 84A-B. *Genetics* **94:** 115.

KUROIWA, A., E. HAFEN, and W.J. GEHRING. 1984. Cloning and transcriptional analysis of the segmentation gene *fushi tarazu* of *Drosophila*. *Cell* **37:** 825.

LEVINE, M., G.M. RUBIN, and R. TJIAN. 1984. Human DNA sequences homologous to a protein coding region conserved between homeotic genes of *Drosophila*. *Cell* **38:** 667.

LEWIS, E.B. 1978. A gene complex controlling segmentation in *Drosophila*. *Nature* **276:** 565.

MCGINNIS, W., C.P. HART, W.J. GEHRING, and F.H. RUDDLE. 1984a. Molecular cloning and chromosome mapping of a mouse DNA sequence homologous to homeotic genes of *Drosophila*. *Cell* **38:** 675.

MCGINNIS, W., R.L. GARBER, J. WIRZ, A. KUROIWA, and W.J. GEHRING. 1984b. A homologous protein-coding sequence in *Drosophila* homeotic genes and its conservation in other metazoans. *Cell* **37:** 403.

MCGINNIS, W., M.L. LEVINE, E. HAFEN, A. KUROIWA, and W.J. GEHRING. 1984c. A conserved DNA sequence in homeotic genes of the *Drosophila* Antennapedia and bithorax complex. *Nature* **308:** 428.

MUELLER, M.M., A.E. CARRASCO, and E.M. DE ROBERTIS. 1984. A homeo box-containing gene expressed during oogenesis in *Xenopus*. *Cell* **39:** 157.

NÜSSLEIN-VOLHARD, C. and E. WIESCHAUS. 1980. Mutations affecting segment number and polarity in *Drosophila*. *Nature* **287:** 795.

OKAYAMA, H. and P. BERG. 1983. A cDNA cloning vector that permits expression of cDNA inserts in mammalian cells. *Mol. Cell. Biol.* **3:** 280.

OUWENEEL, W.H. 1976. Developmental genetics of homeosis. *Adv. Genet.* **18:** 179.

POOLE, S.J., L.M. KAUVER, B. DREES, and T. KORNBERG. 1985. The engrailed locus of *Drosophila*: Structural analysis of an embryonic transcript. *Cell* **40:** 37.

SCOTT, M.P. and A.J. WEINER. 1984. Structural relationships among genes that control development: Sequence homology between the Antennapedia, Ultrabithorax, and fushi tarazu loci of *Drosophila*. *Proc. Natl. Acad. Sci.* **81:** 4115.

SCOTT, M.P., A.J. WEINER, T.I. HAZELRIGG, B.A. POLISKY, V. PIRROTTA, F. SCALENGHE, and T.C. KAUFMAN. 1983. The molecular organization of the Antennapedia locus of *Drosophila*. *Cell* **35:** 763.

The *Ac* and *Spm* Controlling Element Families in Maize

J. BANKS,* J. KINGSBURY,* V. RABOY,† J.W. SCHIEFELBEIN,† O. NELSON,† AND N. FEDOROFF*
Department of Embryology, Carnegie Institution of Washington, Baltimore, Maryland 21210; †Laboratory of Genetics, University of Wisconsin, Madison, Wisconsin 53706

Several of the maize transposable elements first identified and designated "controlling elements" by Barbara McClintock have recently been isolated. The controlling elements described here belong to what we now designate as different "families" of transposable elements (Fedoroff 1983). These are the *Activator-Dissociation* (*Ac-Ds*) and *Suppressor-mutator* (*Spm*) element families. As evidenced below, a transposable element family comprises a structurally heterogeneous collection of elements, some of which are capable of autonomous transposition and others of which are not. The members of a given family of elements interact genetically; they either promote the transposition or transpose in the presence of other elements belonging to the same family.

The first elements McClintock understood to transpose, and later the first elements to be isolated, belong to the *Ac-Ds* element family. The *Ds* element was initially identified as a locus of chromosome dissociation or breakage (McClintock 1947, 1948, 1949, 1952a). McClintock reported that chromosome breakage at the *Ds* locus required the simultaneous presence of a second locus, designated the *Ac* locus for its ability to activate chromosome breakage at *Ds*. In subsequent investigations, McClintock showed that both *Ac* and *Ds* loci were transposable elements, but that transposition of *Ds* was dependent on the simultaneous presence of *Ac* (McClintock 1952a). She also showed that both *Ac* and *Ds* could cause insertion mutations. *Ac* insertion mutations were inherently unstable, reverting at a high frequency, while *Ds* mutations were unstable only in the presence of *Ac*. These observations, taken together with the observation that a *Ds* mutation could be derived directly from an *Ac* insertion mutation (McClintock 1955, 1957a, 1962), strongly suggested that *Ds* elements could be functionally defective *Ac* elements. Several *Ac* and many *Ds* elements have now been isolated and studied in detail.

The *Spm* family of controlling elements, like the *Ac-Ds* element family, contains both transposition-competent and transposition-defective elements. Its name derives from the genetic interaction between a transposition-competent element and certain mutations caused by transposition-defective family members (McClintock 1954). A transposition-defective *Spm* element occasionally inserts into a locus in such a way that it does not completely prevent gene expression, giving a phenotype that is wild-type or intermediate between wild-type and null in the absence of a fully functional *Spm* element. When a nondefective *Spm* element is present in the same genome, however, expression of the affected locus is suppressed completely, except in cell lineages within which the defective element is excised, restoring normal gene function. Thus the *Spm* element appears to have two *trans*-acting functions. The "suppressor" function inhibits expression of the gene with an insertion mutation, and the "mutator" function promotes excision of the inserted element. The first member of this element family isolated genetically was the *Enhancer* (*En*) element, identified and named by Peterson (Peterson 1953, 1960, 1961). The *En* element has recently been cloned (Pereira et al. 1985).

Ac and *Ds* Elements

Ac and *Ds* elements were first isolated from the *sh* and *wx* loci (Geiser et al. 1982; Fedoroff et al. 1983). The *Ac* element, which is capable of autonomous transposition and which can *trans*-activate *Ds* elements to transpose, resembles other small transposons structurally. The *Ac* element is 4.5 kb in length, and the general features of its structure are shown in Figure 1. Two *Ac* elements have been cloned from the *wx* locus; both have been sequenced and are identical (Fedoroff et al. 1983; Behrens et al. 1984; Pohlman et al. 1984a,b; Muller-Neumann et al. 1985). A third *Ac* element has been isolated from the *bz* locus and resembles the others (Fedoroff et al. 1984a). Both of the elements isolated from the *wx* locus have an imperfect terminal repetition of 11 bp (Fig. 1). The *Ac* element has several open reading frames, the three largest of which are depicted in Figure 1. Two of the open reading frames are in the same orientation and overlap; these may comprise a single transcription unit (Pohlman et al. 1984b).

Unlike *Ac* elements, *Ds* elements have been found to be rather heterogeneous in structure. The first evidence that a *Ds* element can be derived directly from an *Ac* element by mutation came from the isolation and analysis of the *Ds* element at the *wx* locus in the *wx-m9* allele (Fedoroff et al. 1983). McClintock derived this allele from the original *Ac* mutation, designated *Ac wx-m9*, that was used as the source of the first cloned *Ac* (McClintock 1963). The *wx-m9* allele proved to have an insertion at the same site as the *Ac* insertion in the *Ac wx-m9* allele (Fedoroff et al. 1983). The element was inserted in the same orientation as the *Ac* element and differed from it by an internal deletion of 194 bp in one of the open reading frames (Fig. 1; Pohlman et al. 1984a). Since the genetic consequence of the mutation had been the conversion of an autonomously transposable Ac element into a Ds element, it appears likely that the deletion was the cause of the mutation and had the effect of disrupting expression of a *trans*-acting transposition function, a transposase.

Several rather different Ds elements have been isolated in various laboratories (Geiser et al. 1982; Fedoroff et al. 1983; Döring et al. 1984a,b; Sutton et al. 1984). A 2-kb Ds element isolated from the wx-m6 allele of the wx locus comprises the ends of the Ac element (Fig. 1). The wx-m6 allele was derived, albeit indirectly, from the original Ds element studied by McClintock (B. McClintock, pers. comm.). Although the Ds6 element, too, may have arisen by a deletion from an Ac element, its origin cannot be traced genetically to an Ac element and remains conjectural. Three additional Ds elements that had an origin similar to that of the wx-m9 allele have been studied at the bz locus (McClintock 1955, 1956, 1962). Two of the alleles have 4-kb insertions, one of which has been cloned and found to be a defective Ac element with an internal deletion of 0.9 kb (Fig. 1). The structure of the insertion element in the third allele is rather surprising. The original Ac element is no longer present, but the locus contains an 0.4-kb Ds element structurally similar to the Ds1 element recovered from the Adh locus (Sutton et al. 1984; J. Schiefelbein and O. Nelson, unpubl.). The resemblance between this Ds element and the Ac element is confined to the terminal inverted repetitions. This allele appears to have arisen, therefore, by the removal of the Ac element and the introduction of a new type of Ds element. The observation that there is a group of Ds elements whose structural homology to the Ac element is confined to the terminal inverted repetitions suggests that all or virtually all of the structural information required for transposition resides in the termini. Thus, it may be that any sequence flanked by the appropriate inverted repetition can transpose.

A rather complex Ds mutation at the sh locus has been investigated (Geiser et al. 1982; Courage-Tebbe et al. 1983; Döring et al. 1983). McClintock reported the isolation of several Ds mutations of the sh locus with the unusual property that revertants still showed evidence of a Ds element at the locus (McClintock 1952b, 1953, 1954, 1955, 1956). The presence of Ds in revertants was detected by the persistence of Ds-mediated chromosome breakage at the locus. Analysis of the mutation designated sh-m5933 and several Sh revertants derived from it showed the mutation to be quite complex (Courage-Tebbe et al. 1983). The sh locus in this allele is interrupted by an insertion of about 30 kb with inverted terminal repetitions of about 10 kb. The ends of the insertion comprise internally duplicated Ds elements having a unit structure resembling that of the 2-kb Ds element cloned from the wx locus (Courage-Tebbe et al. 1983; Fedoroff et al. 1983; Döring et al. 1984a). At one end of the large insertion there is a double Ds element consisting of one 2-kb Ds element inserted in inverted orientation into the center of another identical element. At the other end of the large insertion, there is a 3-kb Ds element resembling the double Ds element, but lacking one half of one of the elements. In addition, the mutant chromosome carries a nearby duplication of part of the insertion and the portion of the sh locus on the 5′ side of the insertion site. Revertants of this mutation lack the large insertion in the sh locus but retain the duplication with its associated Ds elements. The persistence of the duplication explains the continued presence of Ds at the locus.

Differences in the structure of the several Ds elements that have been isolated provide some insight into the "chromosome breakage" phenomenon. Through detailed genetic and cytogenetic studies, McClintock established that chromosome breakage at the site of Ds insertion involves the formation of dicentric and acentric chromosome fragments (McClintock 1948, 1949, 1952a). It appears likely that the formation of the fragments occurs at the time of chromosome replication and may well represent aberrant transposition events. The propensity for acentric-dicentric formation appears to be a property of some, but not all, Ds elements. Ds elements have been cloned from two alleles of the sh locus that show a high frequency of chromosome breakage. These are the sh-m5933 allele described above and the sh-m6233 allele (Döring et al. 1984a; Weck et al. 1984). Both have yielded the complex double Ds element. The sh-m6233 allele, however, contains a simple insertion of the double Ds element in the sh transcription unit. The Ds elements isolated from the wx locus are not internally duplicated and do not show a high frequency of chromosome breakage. These observations suggest that some feature of the in-

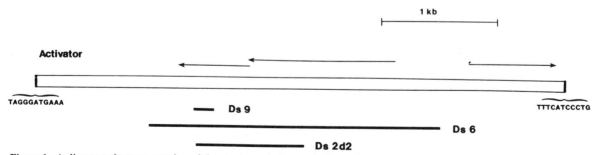

Figure 1. A diagramatic representation of the Ac element. Arrows indicate the extent and polarity of the element's three largest open reading frames (Pohlman et al. 1984a,b). Heavy lines indicate the Ac element sequence that is not present in several Ds elements, designated Ds6, Ds9 (Fedoroff et al. 1983), and Ds2d2.

ternally duplicated *Ds* element promotes acentric-dicentric formation. Since acentric-dicentric chromosome formation at a *Ds* insertion site is dependent on the presence of an *Ac* element, it appears likely that acentric-dicentric formation is not a simple consequence of the internal inverted duplication within the element. In view of the evidence that elements of this family transpose by a "cut-and-paste" mechanism (for review, see Fedoroff 1983), it appears plausible that the availability of several specific cleavage sites or element ends in close proximity promotes incorrect cleavage and ligation reactions.

Sequence information about the insertion site and the termini of *Ac* and *Ds* elements has accumulated from several laboratories and has recently been reviewed (Döring and Starlinger 1984). Many but not all *Ds* elements have perfect terminal inverted repetitions, unlike the two *Ac* elements whose sequence is known. Both *Ac* and *Ds* elements generate an 8-bp duplication upon insertion. Curiously, when an *Ac* or *Ds* element excises, the former insertion site is often marked by the presence of either part or all of the original duplication. In some cases, one or more of the central nucleotides of the duplication is also altered. These residual duplications undoubtedly reflect the excision mechanism and a hypothesis has recently been proposed in explanation (Saedler and Nevers 1985).

Spm Elements

Several elements belonging to the *Spm* family of elements have been isolated from the *wx, bz,* and *a* loci (Fedoroff et al. 1984b; Schwarz-Sommer et al. 1984; Pereira et al. 1985). Although we refer to this element family as *Spm* here because we are working with McClintock's materials, it should be kept in mind that the same element family was independently identified and studied by Peterson (1953, 1960, 1961). Peterson reported in 1965 that the *En* element had the same genetic properties as McClintock's *Spm* element, and it is now apparent that the elements are also structurally similar or identical (Peterson 1965; Pereira et al. 1985; N. Fedoroff, unpubl.). It is now also apparent that the *Spm* element family comprises both defective and nondefective elements analogous to the *Ds* and *Ac* elements. McClintock did not give the defective elements a separate name, but eventually designated the *Spm* element a "regulator" and transposition-defective family members as "operators" (McClintock 1965). Other investigators have used the designation "receptor" or *Rs* for defective elements (Klein and Nelson 1983). Peterson (1953) gave them the specific designation *Inhibitor* (*I*). Because the elements that have been isolated so far bear a clear structural resemblance to the nondefective *Spm* element, we refer to them simply as defective *Spm* elements (*dSpm*).

The first nondefective element of this family to be isolated was an *En* element inserted at the *wx* locus (Pereira et al. 1985). Using an *a* locus probe kindly provided by K. O'Reilly, S. Schwarz-Sommer, and H. Saedler (O'Reilly et al. 1985), we have isolated an element designated *Spm-s* by McClintock from the *a-m2* allele of the *a* locus. The structure of the element is shown in Figure 2 and has so far proved indistinguishable from the structure of the *En* element isolated from the *wx* locus (Pereira et al. 1985; N. Fedoroff, unpubl.). Both elements are 8.4 kb in length. McClintock identified and studied several genetically distinguishable nondefective *Spm* elements (for review, see Fedoroff 1983). One type of element, designated *Spm-weak* or *Spm-w* to distinguish it from the standard *Spm-s*, is characterized by its tendency to transpose and *trans*-activate transposition of defective elements late in the development of the plant (McClintock 1956, 1957b, 1961, 1963). We have isolated an *Spm-w* derivative of the *Spm-s* element from another of the *a-m2* alleles and its structure is shown in Figure 2. The element is inserted at the same position within the *a* locus as the *Spm-s* element and appears to have been derived from the original *Spm-s* by an internal deletion of about 1.7 kb near the center of the element. The deletion is located near a portion of the element that is transcribed (Gierl et al. 1985) and

Figure 2. A diagramatic representation of the *Spm* element. Heavy lines indicate the *Spm* sequence missing from various *dSpm* elements cloned from the *bz, wx,* and *a* loci. The *Spm-s* element was cloned from the *a-m2-7991* allele, and the *Spm-w* element was cloned from the *a-m2-8011* allele. The *dSpm-8* element was cloned from the *wx-m8* allele, and the *dSpm-13* element was cloned from the *bz-m13* allele (Fedoroff et al. 1984b). The remaining *dSpm* elements were cloned from derivatives of the original *a-m2* allele and bear the number of the original allele designation (*a-m2-7995, a-m2-7977,* and *a-m2-8004*). The arrow shows the approximate extent of the transcript believed to encode the element's transposase; its length has been estimated from the data of Gierl et al. (1985) together with that presented in the text.

may well affect the quantity or developmental time of expression of an element-encoded transposition function. A striking genetic property of this and other *Spm-w* elements is the frequent occurrence of mutations that restore the *Spm-s* phenotype. We are now engaged in the isolation and analysis of such mutations.

Defective *Spm* elements have been isolated from several loci, including the *wx*, *bz*, and *a* loci (Fedoroff et al. 1984b; Schwarz-Sommer et al. 1984). A 2-kb defective element comprising the ends of the *Spm* element has been cloned from the *wx-m8* allele of the *wx* locus and, more recently, the *bz-m13* allele of the *bz* locus (Fedoroff et al. 1984b; Schwarz-Sommer et al. 1984). These elements have been designated *dSpm-8* and *dSpm-13*, respectively; they appear to be identical. Their structural relationship to the intact element is shown in Figure 2. Like the *Ds6* element of the *Ac-Ds* family, they comprise the ends of the *Spm* element. McClintock described the isolation, initially at the *a* locus, of alleles containing defective *Spm* elements that showed different somatic reversion characteristics, as well as different levels of *a*-locus expression in the absence of a nondefective *Spm* element (McClintock 1955, 1957b, 1958, 1968). She designated such derivatives "states" of a locus (McClintock 1955). Similar derivatives have been selected from the *bz-m13* allele (Schiefelbein et al. 1985). The original *bz-m13* allele resembles the *wx-m8* allele in its ability to revert early in the development of the kernel in the presence of an *Spm* element. The several derivatives that have been analyzed genetically and structurally differ in the developmental time of element excision. These derivatives appear to have arisen by internal deletions within the inserted element, rather than by changes in the position of the element within the *bz* locus (Schiefelbein et al. 1985). All of the derivatives respond to the "mutator" or transposition function of the *Spm* element, albeit substantially later in development than did the original *bz-m13* allele. The observation that the various deletions remove virtually all of the *dSpm* sequence except possibly 0.2 kb at each end bears the implication that only the ends of the element are required for excision, but that additional sequences within the larger *dSpm* element influence the developmental time of excision. Since all of the derivatives respond to the "suppressor" function of *Spm* as did the original *bz-m13* allele, it follows that suppression of *bz* gene expression is also mediated by the extreme ends of the *Spm* element.

The *a* locus has been cloned from several additional derivatives of the *a-m2* allele. These are designated *a-m2-7995*, *a-m2-7977*, and *a-m2-8004*. In each of these alleles, the locus contains a *dSpm* element at the original *Spm* insertion site in the *a-m2* allele. Each of the *dSpm* elements is missing an internal sequence of the *Spm* element and therefore arose by a deletion mutation that removed a sequence necessary for expression of the element's "mutator" or transposition function. The sequence of the *Spm* element missing from each of these *dSpm* elements is indicated in Figure 2. The *dSpm* elements range in length from 1.2 kb to 3.4 kb. All retain at least 0.2–0.3 kb at each end of the element. The portion of the element deleted in the *dSpm-7995* allele includes the region known to encode a 2.7-kb element transcript (Gierl et al. 1985). Since all of these elements are transposition defective, but the *Spm-w* is not, it appears likely that the sequence encoding the element's transposition function is located near the right end of the element, as indicated in Figure 2.

DISCUSSION

Over the past several years, we and other investigators have succeeded in isolating the DNA molecules responsible for the unstable mutations in maize identified and studied by maize geneticists for a period of almost 80 years. We have found that the maize controlling elements, the first transposable elements to be identified and studied genetically, are similar in size and structure to the transposons that were subsequently identified and studied in a variety of prokaryotic and eukaryotic organisms. The maize *Ac* and *Spm* elements are small transposons with the short terminal inverted repetitions that have been found to be characteristic of transposable elements in all organisms. Like other transposons, the maize elements generate short duplications on insertion. These observations suggest that there are universal principles employed by all transposable elements to move from place to place within the genome. The results of these studies also suggest that there are features of the transposition mechanism that are, so far, unique to plant elements. It has been found that the excision of elements belonging to both the *Ac-Ds* and *Spm* element families is imprecise in an unexpected way, generally leaving behind an imperfect version of the short duplication generated on insertion of the element. The imprecision of the excision event gives us clues about the way the DNA molecules are cut and rejoined by transposition enzymes (Saedler and Nevers 1985). The sequence changes that mark the former site of transposon insertion enlarge our knowledge of the sources of genomic sequence change associated with transposable elements (Schwarz-Sommer et al. 1985).

What remains most intriguing about maize controlling elements is their instability. Maize controlling elements mutate at an extremely high frequency to elements that are either incapable of autonomous transposition or are altered in the timing or frequency of transposition. We are just beginning to understand the structural correlates of such changes. Almost all of the altered elements that have been investigated appear to have arisen by internal deletions. Although the number of mutant elements investigated in detail is still quite small, their analysis has already yielded some insights into element function. Analysis of certain derivatives of the *Ac* and *Spm* elements has allowed us to identify regions that are likely to encode *trans*-acting transposition functions. Further structural analysis of elements, combined with studies on transcripts and element-encoded proteins, should give us a better understanding not only of transposition, but also of the molecular mechanisms underlying changes in the developmental timing of transposition.

REFERENCES

Behrens, U., N. Fedoroff, A. Laird, M. Muller-Neumann, P. Starlinger, and J. Yoder. 1984. Cloning of *Zea mays* controlling element *Ac* from the *wx-m7* allele. *Mol. Gen. Genet.* **194:** 346.

Courage-Tebbe, U., H.P. Döring, N. Fedoroff, and P. Starlinger. 1983. The controlling element *Ds* at the *shrunken* locus in *Zea mays*: Structure of the unstable *sh-m5933* allele and several revertants. *Cell* **34:** 383.

Döring, H.P. and P. Starlinger. 1984. Barbara McClintock's controlling elements: Now at the DNA level. *Cell* **39:** 253.

Döring, H.P., E. Tillmann, and P. Starlinger. 1984a. DNA sequence of the maize transposable element *Dissociation*. *Nature* **307:** 127.

Döring, H.P., M. Geiser, E. Weck, W. Werr, U. Courage-Tebbe, E. Tillmann, and P. Starlinger. 1983. Comparison of genomic clones derived from the *Sh* gene in *Zea mays* L. and of two mutants of this gene which are caused by the insertion of the controlling element *Ds*. In *Genetic engineering of plants* (ed. T. Kosuge et al.), p. 203. Plenum Press, New York.

Döring, H.P., M. Freeling, S. Hake, M.A. Johns, R. Kunze, A. Merckelbach, F. Salamini, and P. Starlinger. 1984b. A *Ds*-mutation of the *Adh1* gene in *Zea mays* L. *Mol. Gen. Genet.* **193:** 199.

Fedoroff, N. 1983. Controlling elements in maize. In *Mobile genetic elements* (ed. J.A. Shapiro), p. 1. Academic Press, New York.

Fedoroff, N., D. Furtek, and O. Nelson. 1984a. Cloning of the *Bronze* locus in maize by a simple and generalizable procedure using the transposable controlling element *Ac*. *Proc. Natl. Acad. Sci.* **81:** 3825.

Fedoroff, N., S. Wessler, and M. Shure. 1983. Isolation of the transposable maize controlling elements *Ac* and *Ds*. *Cell* **35:** 235.

Fedoroff, N., M. Shure, S. Kelly, M. Johns, D. Furtek, J. Schiefelbein, and O. Nelson. 1984b. Isolation of *Spm* controlling elements from maize. *Cold Spring Harbor Symp. Quant. Biol.* **49:** 339.

Geiser, M., E. Weck, H.P. Döring, W. Werr, U. Courage-Tebbe, E. Tillmann, and P. Starlinger. 1982. Genomic clones of a wild type allele and a transposable element-induced mutant allele of the sucrose synthetase gene of *Zea mays* L. *EMBO J.* **1:** 1455.

Gierl, A., Z. Schwarz-Sommer, and M. Saedler. 1985. Molecular interactions between the components of the *En-I* transposable element system of *Zea mays*. *EMBO J.* **4:** 579.

Klein, A.S. and O.E. Nelson. 1983. Biochemical consequences of the insertion of a *suppressor-mutator* (*Spm*) receptor at the *bronze-1* locus in maize. *Proc. Natl. Acad. Sci.* **80:** 7591.

McClintock, B. 1947. Cytogenetic studies of maize and *Neurospora*. *Carnegie Inst. Wash. Year Book* **46:** 146.

———. 1948. Mutable loci in maize. *Carnegie Inst. Wash. Year Book* **47:** 155.

———. 1949. Mutable loci in maize. *Carnegie Inst. Wash. Year Book* **48:** 142.

———. 1952a. Chromosome organization and genic expression. *Cold Spring Harbor Symp. Quant. Biol.* **16:** 13.

———. 1952b. Mutable loci in maize. *Carnegie Inst. Wash. Year Book* **51:** 212.

———. 1953. Mutation in maize. *Carnegie Inst. Wash. Year Book* **52:** 227.

———. 1954. Mutations in maize and chromosomal aberrations in *Neurospora*. *Carnegie Inst. Wash. Year Book* **53:** 254.

———. 1955. Controlled mutation in maize. *Carnegie Inst. Wash. Year Book* **54:** 245.

———. 1956. Mutation in maize. *Carnegie Inst. Wash. Year Book* **55:** 323.

———. 1957a. Controlling elements and the gene. *Cold Spring Harbor Symp. Quant. Biol.* **21:** 197.

———. 1957b. Genetic and cytological studies of maize. *Carnegie Inst. Wash. Year Book* **56:** 393.

———. 1958. The suppressor-mutator system of control of gene action in maize. *Carnegie Inst. Wash. Year Book* **57:** 415.

———. 1961. Some parallels between gene control systems in maize and in bacteria. *Am. Nat.* **95:** 265.

———. 1962. Topographical relations between elements of control systems in maize. *Carnegie Inst. Wash. Year Book* **61:** 448.

———. 1963. Further studies of gene-control systems in maize. *Carnegie Inst. Wash. Year Book* **62:** 486.

———. 1965. The control of gene action in maize. *Brookhaven Symp. Biol.* **18:** 162.

———. 1968. The states of a gene locus in maize. *Carnegie Inst. Wash. Year Book* **66:** 20.

Muller-Neumann, M., J. Yoder, and P. Starlinger. 1985. The sequence of the *Ac* element of *Zea mays*. *Mol. Gen. Genet.* (in press).

O'Reilly, C., N.S. Shepherd, A. Pereira, Z. Schwarz-Sommer, I. Bertram, D.S. Robertson, P.A. Peterson, and H. Saedler. 1985. Molecular cloning of the *a1* locus of *Zea mays* using the transposable elements *En* and *Mu1*. *EMBO J.* **4:** 877.

Pereira, A., Z. Schwarz-Sommer, A. Gierl, I. Bertram, P.A. Peterson, and M. Saedler. 1985. Genetic and molecular analysis of the enhancer (*En*) transposable element system of *Zea mays*. *EMBO J.* **4:** 17.

Peterson, P.A. 1953. A mutable pale green locus in maize. *Genetics* **38:** 682.

———. 1960. The pale green mutable system in maize. *Genetics* **45:** 115.

———. 1961. Mutable a_1 of the *En* system in maize. *Genetics* **46:** 759.

———. 1965. A relationship between the *Spm* and *En* control system in maize. *Am. Nat.* **99:** 391.

Pohlman, R.F., N.V. Fedoroff, and J. Messing. 1984a. The nucleotide sequence of the maize controlling element *Activator*. *Cell* **37:** 635.

———. 1984b. Correction nucleotide sequence of *Ac*. *Cell* **39:** 417.

Saedler, H. and P. Nevers. 1985. Transposition in plants: A molecular model. *EMBO J.* **4:** 585.

Schiefelbein, J.W., V. Raboy, N.V. Fedoroff, and O.E. Nelson. 1985. Deletions within a defective *Suppressor-mutator* element in maize affect the frequency and developmental timing of its excision from the *bronze* locus. *Proc. Natl. Acad. Sci.* (in press).

Schwarz-Sommer, Z., A. Gierl, H. Cuypers, P.A. Peterson, and H. Saedler. 1985. Plant transposable elements generate the DNA sequence diversity needed in evolution. *EMBO J.* **4:** 591.

Schwarz-Sommer, Z., A. Gierl, R.B. Klosgen, U. Wienand, P.A. Peterson, and H. Saedler. 1984. The *Spm* (*En*) transposable element controls the excision of a 2-kb DNA insert at the *wx-m8* allele of *Zea mays*. *EMBO J.* **3:** 1021.

Sutton, W.D., W.L. Gerlach, D. Schwartz, and W.J. Peacock. 1984. Molecular analysis of *Ds* controlling element mutations at the *Adh1* locus of maize. *Science* **223:** 1265.

Weck, E., U. Courage, H.-P. Döring, N. Fedoroff, and P. Starlinger. 1984. Analysis of *sh-m6233*, a mutation induced by the transposable element *Ds* in the sucrose synthetase gene of *Zea may*. *EMBO J.* **3:** 1713.

Regulation of Tc1 Transposable Elements in *Caenorhabditis elegans*

S.W. EMMONS, K.S. RUAN,* A. LEVITT, AND L. YESNER

Department of Molecular Biology, Division of Biological Sciences, Albert Einstein College of Medicine, Bronx, New York 10461

The genome of the nematode *Caenorhabditis elegans* contains multiple copies of a transposable genetic element that, although it has been known for only a few years, has already proven to be of great utility to workers in the field. This element, denoted Tc1, is inserted at approximately 30 sites in the genomes of many *C. elegans* strains. In other strains the copy number has amplified to 300 or more (Emmons et al. 1983; Liao et al. 1983). It has been possible to take advantage of the resulting Tc1-induced DNA strain polymorphisms to determine the genetic map location of cloned DNA fragments and to isolate genetically defined genes (Files et al. 1983; Cox et al. 1985; Fixsen et al., this volume). In certain strain backgrounds, the element undergoes germ line transposition and makes a significant contribution to the spontaneous mutation rate (Moerman and Waterston 1984 and pers. comm.; Eide and Anderson 1985b). These strains make it possible to use Tc1 as an insertional mutagen. To date, two genes have been cloned using Tc1 as a molecular tag (D. Moerman and R. Waterston; I. Greenwald; both pers. comm.).

Transposable elements, first identified genetically in maize (McClintock 1952), have since been found in a wide range of eukaryotes and prokaryotes. Eukaryotic elements have a great variety of structures and characteristics but can be subdivided into a small number of types based on structural properties that probably reflect their mode of transposition. Representatives of the various types have been found in species as diverse as yeast and mammals, suggesting universal distribution in fungi, plants, and animals. One large class that encompasses copia-like elements of *Drosophila*, Ty elements of yeast, and retroviruses of vertebrates has a structure that correlates with a mode of transposition involving an RNA intermediate. These elements have been dubbed "retrotransposons" (Baltimore 1985). A second class, including FB elements of *Drosophila* and Tu elements of sea urchin, have an overall inverted repeat structure arising from reiteration of a short sequence (Potter et al. 1980; Lieberman et al. 1983). Several plant transposons have a somewhat similar reiterated substructure (Rhodes and Vodkin 1985). F elements of *Drosophila* and many sequences of vertebrate genomes have a structure suggesting they may represent a third general class consisting of sequences that arise by reverse transcription of cellular polyadenylated RNAs or RNA polymerase III products. (Van Arsdell et al. 1981; Schmid and Jelenik 1982; Di Nocera et al. 1983).

Tc1 elements of *C. elegans* fall into a fourth structural class. This class consists of elements that contain open reading frames bounded by short inverted repeats (in the range of 10–50 bp). Tc1 elements are 1.6 kb in length, have 54-bp perfect terminal inverted repeats, and contain an open reading frame capable of encoding a polypeptide of 273 amino acids (Rosenzweig et al. 1983a). They thus bear a structural resemblance to IS elements of bacteria (Iidu et al. 1983). Other elements in this class include P elements of *Drosophila* and Ac elements of maize (Fedoroff et al. 1983; O'Hare and Rubin 1983). Unlike most elements in the other classes, all the elements of this class manifest multiple genetic regulatory phenomena that control excision or transposition of the element. This was first and most strikingly demonstrated for the maize elements. Mutations caused by insertion of these elements into genes are unstable due to frequent excision of the element, and both the time during development and the tissue and location within a tissue where excision occurs are strictly regulated (see Fedoroff 1983). In *Drosophila*, P elements repress their own transposition in the germ line. Relief of repression, which can occur following certain crosses, results in rapid transposition of the element in a phenomenon known as hybrid dysgenesis (Kidwell et al. 1977). During hybrid dysgenesis, transposition of the elements occurs only in germ tissue; the elements remain quiescent in somatic tissues (Engels 1979).

The activity of Tc1 elements similarly shows regulation. This regulation affects the frequency and tissue specificity of two activities of the element: transposition and excision. Transposition and excision of Tc1 both occur in the germ line at frequencies that vary over a thousandfold range in various strains (Moerman and Waterston 1984 and pers. comm.; Eide and Anderson 1985a,b). Excision also occurs in somatic tissues at a thousandfold higher frequency than in germ tissues (Emmons and Yesner 1984; Eide and Anderson 1985b). In this latter respect, Tc1 elements are similar to many plant transposable elements, which also undergo frequent excision from their insertion sites. We are presently attempting to understand these regulatory phe-

---

*Permanent address: Institute of Biophysics, Academy of Sciences of China, Beijing, China.

nomena. The restriction of high-frequency excision to somatic tissues suggests that there may be some general difference between germ and somatic tissues with respect to the expression of functions that affect the rate of excision. We present evidence that is consistent with this view. Excision could be the first step of a transposition pathway, and we describe results of experiments aimed at characterizing in detail the products of the reaction. Finally, we present preliminary results that may point the way to an understanding of the differences in transposition and excision frequencies among strains.

Somatic Excision of Tc1

Tc1 elements at specific genomic sites are, by and large, stable genetic components of the *C. elegans* genome. Long-term nonselective propagation of nematodes failed to reveal a change in Tc1 arrangement (Emmons and Yesner 1984), and Tc1 elements can be used as stable genetic markers in interstrain crosses (e.g., Fixsen et al., this volume). Therefore, excision or transposition of Tc1 in the germ line is rare. In somatic tissues, however, Tc1 elements undergo frequent excision from their insertion sites. These events can be detected in genomic Southern hybridization experiments by the appearance of a restriction fragment 1.6 kb smaller than a known restriction fragment carrying a Tc1 element (Emmons et al. 1983). Such fragments arise by religation of the target sequence after excision of the element (Fig. 1). "Empty" target sites arising by excision are present at a level of a few percent in nematode DNA preparations.

That excision does not occur at high frequency in the germ line was demonstrated by showing that Tc1 elements at specific sites are not lost during propagation of worms over many generations (Emmons and Yesner 1984). Direct evidence of somatic localization of the excision process is provided by experiments in which the amount of excision that has occurred at a particular Tc1 site is analyzed in staged populations of worms as they develop. An example is shown in Figure 2a. In this experiment, a HindIII fragment of 2.65 kb is used as a probe of HindIII-digested genomic DNAs. The probe fragment was isolated by T. Blumenthal and coworkers from a region of the X chromosome of the Bristol strain encoding vitellogenin genes (Blumenthal et al. 1984). In the Bergerac strain it hybridizes to a HindIII fragment of 4.25 kb instead of 2.65 kb, due to the presence of a Tc1 element in this region of the Bergerac genome. However, a small amount of the 2.65-kb fragment is also present in Bergerac DNA, due to excision of the Tc1 element followed by religation of the target site. When DNA of synchronously developing populations of nematodes is examined, the amount of the 2.65-kb fragment is seen to increase as the worms develop. The resulting accumulated empty sites, however, are not passed into the next generation. Schwarz-Sommer et al. (1984) have provided a similar demonstration of somatic excision of a plant transposable element. Using Southern hybridizations to detect empty target sites, they showed that the element *Spm-I8* of maize undergoes excision in leaves when another element, *En*, is also present in the plant genome.

In this manner, somatic excision has been demonstrated for Tc1 elements at five separate sites in the genome of the Bergerac strain, a strain that has about 300 Tc1 elements in all (Emmons and Yesner 1984; Eide and Anderson 1985b). Two of these five elements were identified by Eide and Anderson (1985b) as newly inserted elements interrupting the *unc-54* gene on linkage group I. The product of the *unc-54* gene is a myosin heavy chain, and mutations at this locus result in disorganized myofibrils in body-wall muscle cells. For these two Tc1 elements, somatic excision could also be seen as mosaic reversion of the mutation, observable as patches of wild-type muscle tissue in homozygous mutant animals (Eide and Anderson 1985b). In these two cases, where excision can be scored genetically, germ line excision can also be detected as stably revertant wild-type animals. It occurs at a frequency of around 10^{-5}, or at a rate three orders of magnitude lower than somatic excision.

The Mechanism of Tissue-specific Regulation

The difference in the observed rate of excision of Tc1 elements in germ cells and somatic cells cannot be due to a difference in the nature of these lineages per se (e.g., a difference in the number of rounds of cell division in each). Both germ and somatic lineages are of roughly the same length and give rise to a similar number of cells (Hirsh et al. 1976; Sulston and Horvitz 1977; Kimble and Hirsh 1979; Sulston et al. 1983). Rather, there must be a difference in the rate of the excision reaction itself between the tissues.

We imagine two alternative explanations for this dif-

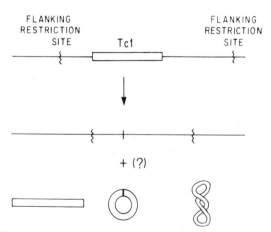

Figure 1. The excision of a Tc1 element. Excision is detected on a Southern hybridization as the appearance of a restriction fragment 1.6 kb smaller than the fragment containing the Tc1 element. Cloned sequences flanking the Tc1 element are used as the probe. Hypothetical reciprocal products of the reaction (extrachromosomal copies of Tc1) are shown. These can be detected in some strains.

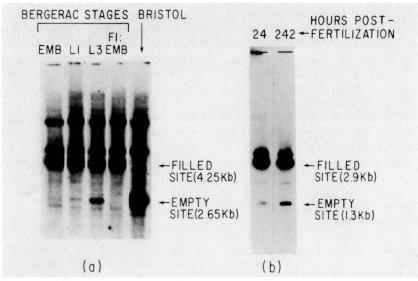

Figure 2. (*a*) Excision during development. Somatic excision is demonstrated by hybridization of a 2.65-kb *Hin*dIII fragment of the Bristol strain to DNA of Bergerac worms of various stages. In Bergerac, this region of the genome contains a Tc1 element (Blumenthal et al. 1984). Nematode DNAs are digested with *Hin*dIII. The fragment carrying the Tc1 element in Bergerac (4.25 kb) and the fragment lacking Tc1 equivalent to the probe (2.65 kb) are indicated. The amount of the latter fragment increases in the Bergerac strain as the worms develop. Other fragments that appear carry additional vitellogenin genes that cross-hybridize with the probe. (EMB) DNA of embryos; (L1) DNA of L1 larvae; (L3) DNA of L3 larvae; (F1:EMB) DNA of embryos of the next generation. (*b*) Excision during developmental arrest. A restriction fragment flanking a Tc1 element is used to demonstrate continuing excision of the Tc1 element during developmental arrest. The Tc1 element and surrounding DNA were cloned and kindly provided by V. Ambrose. DNAs of L1 larvae of the Bergerac strain arrested for the times indicated were digested with *Eco*RI. The fragment carrying the Tc1 element (2.9 kb) and the fragment that arises by excision of the element (1.3 kb) are indicated. The other bands are due to incomplete digestion of the genomic DNA.

ference in rate. One possibility is that *trans*-acting factors required for Tc1 excision, or that can repress excision, are under tissue-specific regulation. Thus, for example, a repressor could be present in the germ line, or excision could require factors that are present at higher levels in some or all somatic tissues than in the germ line. An alternative possibility is that excision of individual Tc1 elements at various genomic sites is regulated by *cis*-acting sequences. Necessary *trans*-acting factors might be generally present, but the accessibility of a given Tc1 element at a particular genomic site to their action might depend on local chromatin structure, which would be governed in turn by DNA sequences flanking the element, giving rise to a position effect. In this second case, elements at different genomic sites would fall under the control of different flanking sequences and could undergo excision at distinct times during development or in distinct tissues. Such a hypothesis, the position hypothesis, has been put forward to explain the variable activity of certain plant transposable elements when inserted at different sites within a locus (Peterson 1977; Muller-Neumann et al. 1984).

Under the second hypothesis, Tc1 elements at some sites might well undergo high-frequency excision in the germ line. The observation that Tc1 elements at all sites so far studied undergo high-frequency excision only in somatic tissues might simply mean that no element activated in the germ line has yet been encountered. However, we note that any element that is activated to undergo high-frequency excision in the germ line is destined to be lost rather rapidly. Thus, high-frequency excision of all presently existing genomic Tc1 elements might be restricted to somatic tissues, wihout implying the presence of any general germ-soma difference.

Under the hypothesis of control via *trans*-acting factors, all elements in the genome would undergo excision simultaneously within a given tissue, whereas *cis*-acting regulation would result in differences between elements at different genomic sites. We therefore sought to distinguish these hypotheses by comparing the rate of excision of elements at different sites. The rates of excision of five separate Tc1 elements residing at distinct sites were compared in the L1 larval stage of the Bergerac strain. The L1 was chosen for these studies because nematodes can be arrested at this stage by hatching embryos into buffer. Without food, L1 larvae survive for a period from days to weeks depending upon temperature. No postembryonic cell division or other development occurs until food is provided. The L1 larva consists of 550 somatic cells and two germ cells (Sulston and Horvitz 1977).

All five elements were found to undergo excision during the L1 stage at similar rates. During L1 arrest, excision of Tc1 elements continues, as can be seen by an increasing level of empty sites. An example of such an increase is shown for one Tc1 element in Figure 2b. This increase in the level of empty sites must be due to new excision events. It cannot be due to replication of DNA strands where excision had previously occurred, since no cell division or chromosome replication is taking place in the arrested worms. By quantitative analysis of Southern hybridizations similar to the one shown in Figure 2b, the rate of excision of a particular Tc1 element can be fairly accurately determined. Results for one element are shown in Figure 3. Excision occurs continuously for as long as the worms are viable. When subpopulations are withdrawn and fed and the amount of excision that has occurred during the resulting period of normal development is compared with that taking place in the starved population during the same period, no difference is seen. Thus the observed excision is not induced by starvation, and the rate of excision appears to be the same through several larval stages. The largest amount of excision obtained was 11%, with no indication of a decrease in rate suggestive of a plateau. Excision of this element is evidently taking place in well over 20% of the somatic cells, since otherwise a decline in the excision rate with time would have been observable.

Similar results were obtained for the elements at all five sites studied (S.W. Emmons, in prep.). Hence, we can find no evidence for sharp differences between elements at different sites that would suggest regulation primarily by *cis*-acting sequences. Of course, regulation could occur at both the level of *trans*-acting factors and *cis*-acting sequences, and there is some indication in our data for slight differences in the rates of excision among the five elements. However, the overall data are consistent with the hypothesis that all Tc1 elements in the genome are activated in some if not all somatic tissues, and we have been unable to obtain evidence for the restriction of excision to a particular time during development or to a particular somatic lineage. Hence, the data favor developmental regulation of Tc1 elements via developmental regulation of *trans*-acting factors rather than via control in *cis* by flanking sequences. They suggest a general distinction between the germ line and the soma in the expression of these factors.

Extrachromosomal Tc1 DNA

Does excision of Tc1 represent the first step of the transposition pathway? If so, then extrachromosomal copies of Tc1 must be present at some level as transposition intermediates. Extrachromosomal Tc1 elements can indeed be detected in DNA preparations of the Bergerac strain (Fig. 4a) (Rose and Snutch 1984; Ruan and Emmons 1984). We have isolated and studied these molecules, which are present at a level corresponding to about 0.1 copy per cell. Three molecular species are present: a monomer linear, a monomer relaxed circle, and a monomer supercoil. The ends of the linear molecule are nonpermuted and are located at positions corresponding approximately or exactly to the ends of an integrated element. The structure of circular molecules is consistent with circularization of a genomic element, but definitive studies on cloned material have not yet been carried out to determine the exact sequence at the joint.

Presumably, these extrachromosomal molecules are products of the somatic excision reaction taking place at most or all of the 300 Tc1 insertion sites in the genome of the Bergerac strain. If this is the case, then they are unstable and are metabolized to other forms subsequent to excision. This was shown by measuring the level of extrachromosomal elements in DNA of arrested L1 larvae in which excision had been going on for a number of days. As shown in Figure 4b, the amount of extrachromosomal Tc1 DNA does not increase with time, unlike the level of chromosomal excision products (empty chromosomal insertion sites; see Fig. 2b). In the case of the L1 larvae starved for 242 hours and shown in the figure, the probable average amount of excision at each Tc1 site is approximately 10%. Therefore, an average of 60 excision events per diploid cell will have taken place in these worms. But the level of extrachromosomal Tc1 molecules is fewer than one per cell.

Tc1 Excision Can Be Imprecise

We have begun an analysis of the DNA sequences of religated target sites produced after Tc1 excision to determine whether Tc1 excision is precise or imprecise. In one case so far examined, a religated target frag-

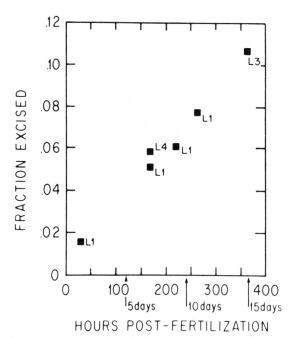

Figure 3. Accumulation of chromosomal fragments lacking Tc1 during developmental arrest. Continuous excision of a Tc1 element in L1 larvae arrested by starvation is demonstrated by quantitative analysis of a Southern hybridization similar to the one shown in Fig. 2b. The Tc1 site here is the same as that shown in Fig. 2a. The fraction of the DNA strands at this particular insertion site that lack the element is plotted as a function of the time of starvation (points labeled L1). Two subpopulations were withdrawn, fed, allowed to develop, and analyzed at later stages (points labeled L3 and L4).

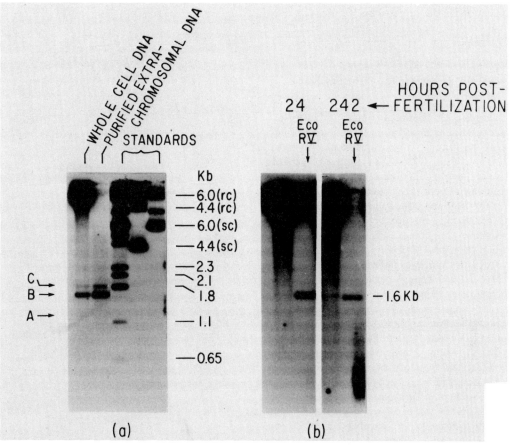

Figure 4. Extrachromosomal Tc1 elements in Bergerac DNA are demonstrated by hybridizing Tc1 to a Southern filter of undigested genomic DNA fractionated on a 1.2% agarose gel. A, B, and C indicate, respectively, supercoiled (visible on the autoradiogram), linear, and relaxed circular monomer molecules. Extrachromosomal molecules were purified on a sucrose gradient for further analysis of their structure with restriction endonucleases. Supercoiled (sc), relaxed circular (rc), and linear length standards on the same filter are shown. (b) Extrachromosomal Tc1 elements do not accumulate during prolonged starvation of L1 larvae. The amount of extrachromosomal Tc1 DNA in undigested genomic DNA of L1 larvae starved for the times shown is quantitated by comparison with the level of hybridization to genomic elements in the same DNA preparation (lanes EcoRV). The latter is determined in an EcoRV digestion of a 1:300 dilution of the sample. After EcoRV digestion, each genomic Tc1 element gives rise to a single fragment of 1.6 kb due to an EcoRV site in the inverted repeats at the ends of the element. Since there are approximately 300 genomic elements per haploid complement in this strain, the level of hybridization to the EcoRV lanes is a signal equivalent to two Tc1 copies per diploid cell.

ment was cloned from the Bergerac strain, sequenced, and compared with the corresponding sequence from the Bristol strain, which lacks a Tc1 element at this site (Rosenzweig et al. 1983b). Four extra nucleotides were found in the Bergerac fragment relative to the Bristol fragment at the site of excision. The two sequences were identical otherwise. Although other interpretations of this result are possible, it nevertheless suggests that Tc1 excision can be imprecise. Analysis of products of additional excision events will be required to confirm this conclusion.

Examples of both precise and imprecise excision of eukaryotic transposable elements have been found previously. FB elements and P elements of *Drosophila* have been shown to undergo precise excision (Collins and Rubin 1983; O'Hare and Rubin 1983). Flanking duplications of target sequences of 9 bp and 8 bp, respectively, are excised with the element. In the experiments with P elements, excision events were detected as genetic revertants, and hence precise excision events might have been selectively recovered in this case. In other experiments with P elements, imprecise events have been observed (Searles et al. 1982). The *Ds* element of maize can undergo imprecise excision, leaving behind an 8-bp target site duplication (Sutton et al. 1984).

The four extra nucleotides found at the site of Tc1 excision in the one example studied are not present at that site before excision and hence do not represent a target site duplication created by the element upon insertion and left behind upon excision. Despite sequencing results at several Tc1 insertion sites, it remains undetermined whether Tc1 elements create a target site duplication on insertion (Rosenzweig et al. 1983b; P. Anderson, pers. comm.). A duplication of a TA dinucleotide flanks all sequenced Tc1 elements, and all sequenced Tc1 elements have inserted at a TA dinucleotide site. The duplicated bases could therefore be a duplication of the target dinucleotide, but they could equally well be a part of the element itself; these two alternatives cannot be distinguished at present.

A Novel Polymorphic Repetitive Family

The activity of Tc1 elements differs not only between tissues, but also between strains. This was initially suggested by the observation that the number of elements in the genomes of various strains differed widely (Emmons et al. 1983; Liao et al. 1983). Genetic analyses subsequently demonstrated strain differences in the frequency of germ-line transposition (Moerman and Waterston 1984 and pers. comm. Eide and Anderson 1985a,b). Recently, our own analysis of somatic excision has shown that somatic excision still occurs in strains where germ line transposition is not detectable, but at a frequency that is significantly lower than in strains where Tc1 is transpositionally active in the germ line (S.W. Emmons, in prep.).

There are of course many possible explanations for these observations. Among these, one is that Tc1 could be part of a two-element system. Differing levels of activity of Tc1 in various strains would, under this hypothesis, reflect the presence of a larger or smaller number (or absence) of a second element in those strains. Such an explanation has precedence in other organisms. In both maize and *Drosophila*, *Ac/Ds* and P transposable elements are members of heterogeneous transposon families (Fedoroff et al. 1983; O'Hare and Rubin 1983). Individual family members may or may not encode all of the *trans*-acting functions required for transposition. The small size and low coding capacity of the Tc1 element (1 open reading frame) suggest that here too some functions required for one or more of the various phenomena involving Tc1 might be encoded elsewhere. Experiments to detect expression of this open reading frame have not been carried out. In maize, the frequency of excision of *Ds*, the nonautonomous component of a two-element system, is affected by the dosage of the autonomous component *Ac* (McClintock 1952).

In maize and *Drosophila*, the various elements of heterogeneous transposon families have some sequences in common. Accordingly, we asked whether sequences homologous to Tc1 but of a different structure were present in the genomes of several *C. elegans* strains. Figure 5 presents evidence for such sequences. In some strains, an *Eco*RV fragment of 4.8 kb is detectable in a Southern hybridization experiment using a Tc1 probe. Normal Tc1 elements give rise to an *Eco*RV fragment of 1.6 kb due to *Eco*RV sites in the inverted repeats at each end of the element. A fragment of 4.8 kb can therefore only arise from a novel sequence involving Tc1. Interestingly, the presence or absence of the 4.8-kb fragment correlates with the activity or inactivity of germ line transposition of Tc1 in the various strains examined, with one exception. Germ line transposition occurs at a frequency at least several orders of magnitude higher in Bergerac (BO) than in

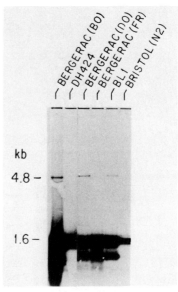

Figure 5. Evidence for novel sequences homologous to Tc1. Genomic DNA of various strains is digested with *Eco*RV and hybridized to a Tc1-specific probe. Most genomic elements give rise to a single *Eco*RV fragment of 1.6 kb due to *Eco*RV sites in the inverted repeats at the ends of the element. The 4.8-kb fragment in Bergerac (BO) was cloned and studied as described. Bands of molecular weight lower than 1.6 kb have not been analyzed further.

DH424, Bergerac (FR), BL1, or Bristol (N2) (Moerman and Waterston 1984; Eide and Anderson 1985a,b). The correlation between the presence of the 4.8-kb *Eco*RV fragment and germ line transposition holds with the exception of the BL1 strain. The frequency of germ line transposition has not been determined in Bergerac (DO). The correlation appeared to justify further investigation, and we therefore isolated the 4.8-kb *Eco*RV fragment for further analysis.

The 4.8-kb *Eco*RV fragment contains a truncated Tc1 element as well as sequences that are not homologous to Tc1. When the non-Tc1 sequences are used as a hybridization probe in a genomic Southern, a new, polymorphic repetitive family is revealed (Fig. 6). Like Tc1, and unlike any of the many other repetitive families in the *C. elegans* genome we have studied to date, the number of copies of this sequence differs widely between strains, from 20–30 in the Bergerac (BO) strain (containing 300 Tc1 elements) to 4–6 in the Bristol (N2) strain (containing 30 Tc1 elements). This is a characteristic that might be expected for a transposon family regulating Tc1 activity. However, it is also quite possible that these correlations are merely a coincidence or that this new family consists of transposable elements that respond to the same factors that also govern Tc1 activity or, perhaps, respond to Tc1 itself. In *Drosophila*, copia elements appear to be activated along with P elements in P-element-induced dysgenic crosses, even though the two elements are structurally unrelated (Rubin et al. 1982). Our current experiments are aimed at elucidating the structure of elements of this new

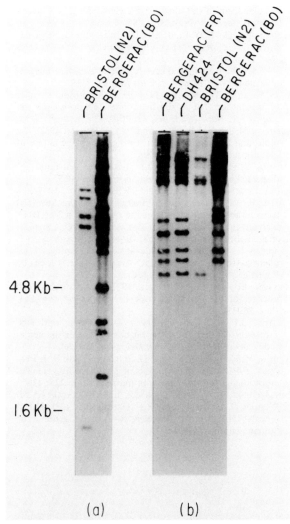

Figure 6. A novel polymorphic repetitive family. Genomic DNAs of various strains are digested with EcoRV (a) or BamHI (b) and hybridized to the non-Tc1 portion of the 4.8-kb EcoRV fragment shown in Fig. 5.

family and at determining what relationship, if any, they have with Tc1.

SUMMARY

C. elegans strains contain variable numbers of a 1.6-kb transposable genetic element. Activity of this element, which is denoted Tc1, shows regulation at at least two levels. At one level, excision of Tc1 elements occurs in somatic cells at a frequency several orders of magnitude higher than in germ cells. Evidence is presented suggesting that this results from regulation at the level of trans-acting functions that are required for excision or that repress excision. At the second level, germ line transposition of Tc1 occurs at greater frequency in some strains than in others. The hypothesis is proposed that this is because Tc1 is one component of a two-element system, the second element of which differs between strains. Evidence for a second putative transposable element family in C. elegans is presented. This family has properties that suggest a relationship to Tc1. This possibility is currently under investigation.

ACKNOWLEDGMENTS

We thank T. Blumenthal and V. Ambrose for providing Tc1 elements or insertion sites cloned by them. We thank Drs. Lucy Shapiro, Leslie Leinwand, and Scott Hawley for their comments on the manuscript. This work was supported by grant PCM 8316338 from the National Science Foundation. S.E. is supported by ACS Salary Award FRA-227. A.L. is supported by ACS Fellowship PF-2537. Some nematode strains used in this work were provided by the *Caenorhabditis* Genetics Center, which is supported by a grant from the National Institutes of Health to the University of Missouri.

REFERENCES

BALTIMORE, D., 1985. Retroviruses and retrotransposons: The role of reverse transcription in shaping the eukaryotic genome. *Cell* **40:** 481.

BLUMENTHAL, T., M. SQUIRE, S. KIRTLAND, J. CANE, M. DONEGAN, J. SPIETH, and W. SHARROCK. 1984. Cloning of a yolk protein gene family from *Caenorhabditis elegans*. *J. Mol. Biol.* **174:** 1.

COLLINS, M. and G.M. RUBIN. 1983. High-frequency precise excision of the *Drosophila* foldback transposable element. *Nature* **303:** 259.

COX, G.N., S. CARR, J.M. KRAMER, and D. HIRSH. 1985. Genetic mapping of *Caenorhabditis elegans* collagen genes using DNA polymorphisms as phenotypic markers. *Genetics* **109:** 513.

DI NOCERA, P.P., M.E. DIGAN, and I.B. DAWID. 1983. A family of oligoadenylate-terminated transposable sequences in *Drosophila melanogaster*. *J. Mol. Biol.* **168:** 715.

EIDE, D. and P. ANDERSON. 1985a. The gene structures of spontaneous mutations affecting a *Caenorhabditis elegans* myosin heavy chain gene. *Genetics* **109:** 67.

———. 1985b. Transposition of Tc1 in the nematode *Caenorhabditis elegans*. *Proc. Natl. Acad. Sci.* **82:** 1756.

EMMONS, S.W. and L. YESNER. 1984. High-frequency excision of transposable element Tc1 in the nematode *Caenorhabditis elegans* is limited to somatic cells. *Cell* **36:** 599.

EMMONS, S.W., L. YESNER, K.-S. RUAN, and D. KATZENBERG. 1983. Evidence for a transposon in *Caenorhabditis elegans*. *Cell* **32:** 55.

ENGELS, W.R. 1979. Extrachromosomal control of mutability in *Drosophila melanogaster*. *Proc. Natl. Acad. Sci.* **76:** 4011.

FEDOROFF, N.V. 1983. Controlling elements in maize. In *Mobile genetic elements* (ed. J.A. Shapiro), p. 1. Academic Press, New York.

FEDOROFF, N.V., S. WESSLER, and M. SHURE. 1983. Isolation of the transposable maize controlling elements Ac and Ds. *Cell* **35:** 125.

FILES, J.G., S. CARR, and D. HIRSH. 1983. Actin gene family of *Caenorhabditis elegans*. *J. Mol. Biol.* **164:** 355.

HIRSH, D., D. OPPENHEIM, and M. KLASS. 1976. Development of the reproductive system of *Caenorhabditis elegans*. *Dev. Biol.* **49:** 200.

IIDU, S., J. MEYER, and W. ARBER. 1983. Prokaryotic IS elements. In *Mobile genetic elements* (ed. J.A. Shapiro), p. 159. Academic Press, New York.

KIDWELL, M.G., J.F. KIDWELL, and J.A. SVED. 1977. Hybrid dysgenesis in *Drosophila melanogaster*: A syndrome of aberrant traits including mutation, sterility, and male recombination. *Genetics* **86:** 813.

KIMBLE, J. and D. HIRSH. 1979. The postembryonic cell lineages of the hermaphrodite and male gonads in *Caenorhabditis elegans*. *Dev. Biol.* **70:** 396.

LIAO, L.W., B. ROSENZWEIG, and D. HIRSH. 1983. Analysis of a transposable element in *Caenorhabditis elegans*. *Proc. Natl. Acad. Sci.* **80:** 3585.

LIEBERMAN, D., B. HOFFMAN-LIEBERMAN, J. WEINTHAL, G. CHILDS, R. MAXSON, A. MAURON, S.N. COHEN, and L. KEDES. 1983. An unusual transposon with long terminal inverted repeats in the sea urchin *Stronglylocentrotus purpuratus*. *Nature* **306:** 342.

MCCLINTOCK, B. 1952. Chromosome organization and genic expression. *Cold Spring Harbor Symp. Quant. Biol.* **16:** 13.

MOERMAN, D.G. and R.H. WATERSTON. 1984. Spontaneous unstable *Unc-22* IV mutations in *C. elegans* var. Bergerac. *Genetics* **108:** 859.

MULLER-NEUMANN, M., J.I. YODER, and P. STARLINGER. 1984. The DNA sequence of the transposable element *Ac* of *Zea mays* L. *Mol. Gen. Genet.* **198:** 19.

O'HARE, K. and G.M. RUBIN. 1983. Structures of P transposable elements and their sites of insertion and excision in the *Drosophila melanogaster* genome. *Cell* **34:** 25.

PETERSON, P.A. 1977. The position hypothesis for controlling elements in maize. In *DNA insertion elements, plasmids, and episomes* (ed. A.I. Bukhari et al.), p. 429. Cold Spring Harbor Laboratory, Cold Spring Harbor, New York.

POTTER, S.S., M. TRUETT, M. PHILLIPS, and A. MAHER. 1980. Eukaryotic transposable genetic elements with inverted terminal repeats. *Cell* **20:** 639.

RHODES, P.R. and L.O. VODKIN. 1985. Highly structured sequence homology between an insertion element and the gene in which it resides. *Proc. Natl. Acad. Sci.* **82:** 493.

ROSE, A.M. and T.P. SNUTCH. 1984. Isolation of the closed circular form of the transposable element Tc1 in *Caenorhabditis elegans*. *Nature* **311:** 485.

ROSENZWEIG, B., L.W. LIAO, and D. HIRSH. 1983a. Sequence of the *C. elegans* transposable element Tc1. *Nucleic Acids Res.* **11:** 4201.

———. 1983b. Target sequences for the *C. elegans* transposable element Tc1. *Nucleic Acids Res.* **11:** 7137.

RUAN, K. and S.W. EMMONS. 1984. Extrachromosomal copies of transposon Tc1 in the nematode *Caenorhabditis elegans*. *Proc. Natl. Acad. Sci.* **81:** 4018.

RUBIN, G.M., M.G. KIDWELL, and P.M. BINGHAM. 1982. The molecular basis of P-M hybrid dysgenesis: The nature of induced mutations. *Cell* **29:** 987.

SCHMID, C.W. and W.R. JELINEK. 1982. The *Alu* family of dispersed repetitive sequences. *Science* **216:** 1065.

SCHWARTZ-SOMMER, Z., A. GIERL, R.B. KLOSSEN, U. WIENAND, P.A. PETERSON, and H. SAEDLER. 1984. The *Spm* (*En*) transposable element controls the excision of a 2-kb DNA insert at the *wx m-8* allele of *Zea mays*. *EMBO J.* **3:** 1021.

SEARLES, L.L., R.S. JOKERST, P.M. BINGHAM, R.A. VOELKER, and A.L. GREENLEAF. 1982. Molecular cloning of sequences from a *Drosophila* RNA polymerase II locus by P element transposon tagging. *Cell* **31:** 585.

SULSTON, J.E. and H.R. HORVITZ. 1977. Post-embryonic cell lineages of the nematode, *Caenorhabditis elegans*. *Dev. Biol.* **56:** 110.

SULSTON, J.E., E. SCHIERENBERG, J.G. WHITE, and J.N. THOMSON. 1983. The embryonic cell lineage of the nematode *Caenorhabditis elegans*. *Dev. Biol.* **100:** 64.

SUTTON, W.D., W.L. GERLACH, D. SCHWARTZ, and W.J. PEACOCK. 1984. Molecular analysis of *Ds* controlling element mutations at the *Adh1* locus of maize. *Science* **223:** 1265.

VAN ARSDELL, S.W., R.A. DENISON, L.B. BERNSTEIN, A.M. WEINER, T. MANSER, and R.F. GESTELAND. 1981. Direct repeats flank three small nuclear RNA pseudogenes in the human genome. *Cell* **26:** 11.

Lineage-specific Gene Expression in the Sea Urchin Embryo

E.H. DAVIDSON, C.N. FLYTZANIS, J.J. LEE, J.J. ROBINSON, S.J. ROSE III, AND H.M. SUCOV
Division of Biology, California Institute of Technology, Pasadena, California 91125

Within a few days of fertilization, the sea urchin embryo develops into a small differentiated organism consisting of about 1800 cells and capable of feeding, swimming, and the further ontogenic transformations required in the succeeding weeks of larval growth. A number of distinct cell lineages that are clearly specialized at the morphological and functional levels can be discerned in the advanced embryo, and many of these can be traced back to particular sets of early blastomeres. Classical cell lineage and experimental studies (Hörstadius 1939; for review, see Angerer and Davidson 1984) have shown that certain of these lineages appear to be specified, at least in part, in consequence of the maternal components inherited in those regions of egg cytoplasm occupied by their progenitor cells. Specification of others among the early cell lineages clearly depends on inductive interactions that occur between blastomeres during cleavage. For the molecular biologist, as for his predecessors, this rapidly developing and simply constructed embryo offers the advantages of experimental accessibility. Thus, in respect to direct molecular-level analyses of gene activity in the embryo, for both specific genes and overall transcript populations and their protein products, the sea urchin is at present the best known embryonic system (e.g., reviews of Hentschel and Birnstiel 1981; Davidson et al. 1982; Angerer and Davidson 1984).

A focus of recent efforts in our laboratory has been the acquisition of a library of cloned genes that are expressed in the early embryo in a lineage-specific manner, for use in examination of the molecular processes by which these genes are differentially specified for activity early in development. This is, of course, the fundamental and general problem in understanding how the zygote gives rise to a functionally differentiated embryo, even given the well-established cytoplasmic anisotropy of most eggs (reviewed by Davidson 1976). It is a problem that is far from being solved for any embryo. Furthermore, diverse solutions may well be utilized in different modes of development. In the following, we review current progress on the isolation and characterization of lineage-specific sea urchin embryo genes, and then describe briefly recent studies that demonstrate apparently correct ontogenic expression of such genes after microinjection into the unfertilized egg.

The Sea Urchin Actin Genes

There are eight actin genes per haploid genome in *Strongylocentrotus purpuratus*, of which two are probably pseudogenes (Lee et al. 1984). All of the remaining six are expressed in the embryo. Of these, one is a muscle actin gene (M) and the others code for cytoskeletal (Cy) actin proteins (Durica et al. 1980; Scheller et al. 1981; Lee et al. 1984; Shott et al. 1984). Contrary to the implication of the cliché that cytoskeletal actin genes are "housekeeping" genes, our analysis of the patterns of expression of the individual members of this gene family has shown that each gene is expressed according to a specific ontogenic program. The genomic linkages of these genes, the nomenclature by which they are designated, and their patterns of activity in embryo and adult cell types, are summarized in Figure 1. Although their protein-coding regions are largely homologous, these genes differ greatly in the 3' nontranslated trailer sequences of their mRNAs (Scheller et al. 1981; Lee et al. 1984), and this affords a means of preparing gene-specific molecular probes to identify their transcripts individually. Such probes have been utilized in RNA gel blot and in situ hybridization studies for determining the spatial and temporal patterns of actin gene expression in the embryo (Angerer and Davidson 1984; Shott et al. 1984; K.H. Cox et al., in prep.). As indicated in Figure 1 the expression of several of the actin genes is strictly lineage specific, and none is active ubiquitously. For example, it is shown in Figure 2 that transcripts of the M actin gene are confined to two bilateral patches, each containing 10-20 cells that are associated with the newly formed coelomic pouches in the late embryo. These cells are of the secondary mesenchyme cell lineage, and their role is the construction of the pharyngeal muscle required for larval feeding. Expression of the CyIIIa and CyIIIb cytoskeletal actin genes is also confined to a single embryonic cell lineage, although CyIIIa messages are present as well in maternal RNA at the low level of about 1200 molecules per egg (J.J. Lee et al., unpubl.). They are not spatially localized in the egg (K.H. Cox et al., in prep.). After fertilization, newly synthesized transcripts of the CyIIIa and CyIIIb actin genes appear only in aboral ectoderm cells, although on different schedules. An in situ hybridization experiment carried out with a CyIIIa probe that demonstrates the early localization of these transcripts is shown in Figure 3 (Angerer and Davidson 1984; K.H. Cox et al., in prep.).

Expression of the CyIIa gene is restricted to mesenchyme cells, and in late embryos to parts of the gut. The CyI and CyIIb genes are regulated similarly to one another, and their transcripts are more widely distributed than those of the other actin genes. These messages are also found in small quantities in the unfertil-

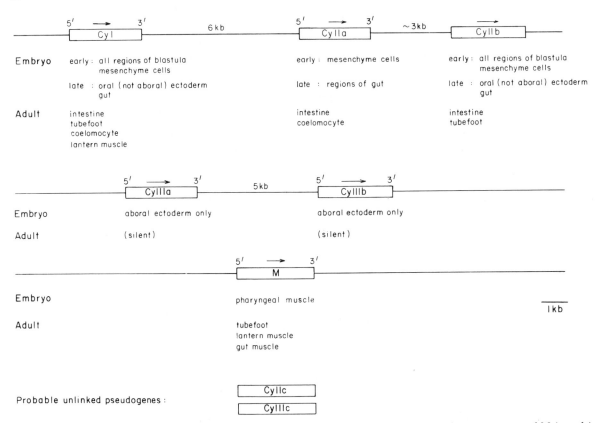

Figure 1. Organization and expression of actin gene family in the sea urchin *S. purpuratus*. Actin genes are named M (muscle) or Cy (cytoskeletal, i.e., expressed in nonmuscle cell types). Roman numerals designate the three nonhomologous 3' nontranslated trailer sequences found in the cytoskeletal actin genes. a, b, and c designate different though homologous trailer sequence variants. Linkage data are from analyses of cloned genes (Lee et al. 1984; R. Akhust, F. Calzone, R. Britten, and E. Davidson, in prep.). Expression patterns in adult tissues were determined by RNA gel blot hybridizations, as reported by Shott et al. (1984). Expression of actin genes in embryonic cell types is summarized from the in situ hybridization study of K.H. Cox et al. (in prep.).

ized egg, and both genes are then expressed in all regions of the early embryo. However, in the ectoderm of pluteus-stage embryos, the pattern of expression is exactly complementary to that of CyIIIa, as their messages are located in the oral but not in the aboral ectoderm. Thus, except for CyI and CyIIb, which function similarly, each of the actin genes is utilized in a particular set of cells at particular times during embryogenesis and it may be supposed that each possesses its own unique *cis*-regulatory genomic control apparatus.

The number of mRNA molecules produced during embryonic development by five of the six functional actin genes has recently been measured by J.J. Lee et al. (unpubl.). These estimates were obtained by a single-strand probe excess titration method (Wallace et al. 1977; Scheller et al. 1978; Lev et al. 1980) using RNA transcripts synthesized in vitro from an Sp6 promoter (Butler and Chamberlin 1982; Melton et al. 1984). Since the cell types in which the various actin genes are expressed are known, the number of molecules of each mRNA species per active cell can be calculated from these data. These results are summarized in Table 1.

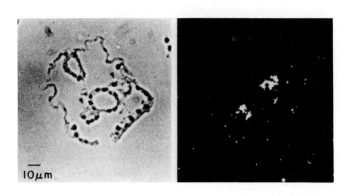

Figure 2. Expression of M actin gene in bilateral pharyngeal muscle anlage cells, visualized by in situ hybridization. Phase photomicrograph of embryo, an 82-hr pluteus, is shown at the left. (Reprinted, with permission, from K.H. Cox, L.M. Angerer, J.J. Lee, E.H. Davidson, and R.C. Angerer, in prep.)

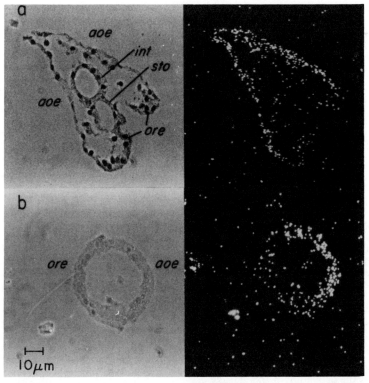

Figure 3. Expression of CyIIIa actin gene in aboral ectoderm of pluteus-stage embryo (*a*) and of blastula (*b*) visualized by in situ hybridization. Labeling in other regions is at background level, as shown by control experiments with probes that are not represented in sea urchin RNA. Phase photomicrographs (*left*): *aoe*, aboral ectoderm; *int*, intestine; *sto*, stomach; *ore*, oral ectoderm. (Reprinted, with permission, from Angerer and Davidson 1984.)

Gene transfer experiments described briefly in a following section of this report have focused on the CyIIIa actin gene. This is the single most intensely expressed of all the actin genes in the embryo and its activity is confined to embryonic and larval stages. Thus, it is not utilized at all in the postmetamorphosis sea urchin (Shott et al. 1984). As shown in Table 1, late in embryogenesis there are about 200 molecules of CyIIIa mRNA per aboral ectoderm cell. The time course of CyIIIa gene expression during early development, as established by titration measurements, is shown in Figure 4. Within a few hours beginning at the very early blastula stage, new transcripts of this gene appear, and they accumulate dramatically to a level of over 80,000 molecules per embryo. Nuclear run-off experiments (S. Johnson, R. Britten, and E. Davidson, unpubl.) indicate that this accumulation is regulated primarily at the transcription level.

Other Lineage-specific Cloned Genes Active in the Sea Urchin Embryo

Two other sea urchin genes that, like the CyIIIa and CyIIIb actin genes, are expressed specifically in aboral ectoderm have been characterized by Dr. William Klein and his associates (Bruskin et al. 1981, 1982; Lynn et al. 1983; Carpenter et al. 1984; S.L. Houck, C.D. Carpenter, P.E. Hardin, A.M. Bruskin, and W.H. Klein, unpubl.). These genes, called Spec1 and Spec2 (Spec, *S. purpuratus* ectoderm), code for intracellular Ca^{++}-binding proteins related to troponin C. The aboral ec-

Table 1 Transcripts per Cell for Five Actin Genes Expressed in the Embryo of *S. purpuratus*

Gene	Approximate number of cells transcribing actin mRNA per embryo	Number of actin transcripts per embryo	Number of actin transcripts per expressing cell
CyI	1000	9×10^4	90
CyIIa	160	1.4×10^4	90
CyIIb	1000	5.8×10^4	60
CyIIIa	470	8.8×10^4	190
M	20–40	2.5×10^4	650–1300

Data are from J.J. Lee, F.J. Calzone, R.J. Britten, R.C. Angerer, and E.H. Davidson (unpubl.); measurements refer to 65-hr pluteus-stage embryos.

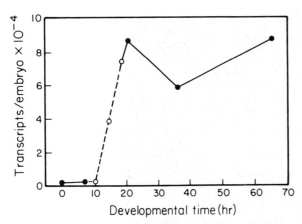

Figure 4. Quantities of CyIIIa actin gene transcripts accumulated during embryonic development. Transcripts were measured by solution titration with excess single-strand complementary probes. Various quantities of whole-embryo RNA were reacted with a gene-specific RNA probe synthesized in vitro. The shape of the curve for the dashed line is derived from relative RNA gel blot data (Shott et al. 1984) (open circles). Titration measurements are designated by closed circles.

toderm descends initially from one animal pole quadrant of the cleavage-stage blastomeres, and its differentiation thus involves the activation in this lineage of a whole battery of diverse genes. This battery may include many other genes in addition to the examples already known, i.e., Spec1, Spec2, CyIIIa, and CyIIIb.

Another battery of genes active in the embryo of which diverse representatives are already in hand is that functional in the muscle precursor cells. In addition to the M actin genes (Fig. 2), we have isolated the muscle myosin heavy-chain gene, using a *Drosophila* probe kindly provided by Dr. Charles Rozek (Case Western University). A significant homology between a region of the *Drosophila* and the sea urchin gene sequences is shown in Figure 5. As stressed in early considerations of the mechanisms underlying specification of cell type, pleiotropic control of batteries of distinct genes is an essential aspect of developmental gene regulation (e.g., see Morgan 1934; Britten and Davidson 1969, 1971).

Two additional genes that display strict lineage specificity in the early embryo have recently been cloned in our laboratory. In collaboration with Dr. William Lennarz (M.D. Anderson Hospital and Tumor Institute) we have recovered from a λgt11 cDNA clone library several probes for the hyalin gene (J.J. Robinson et al., in prep.). Hyalin is the major component of the tough extracellular coat surrounding the sea urchin embryo, the hyaline layer. This protein can be solubilized by removal of Ca^{++} from the medium, as initially discovered by Herbst (1900). It is stored in the cortical granules of the unfertilized egg, released by exocytosis on fertilization, and from the gastrula stage on is a prominent synthetic product of the embryonic ectoderm cells (Hylander and Summers 1982; McClay and Fink 1982). Its developmental localization is illustrated by indirect immunocytofluorescence in Figure 6, using the anti-hyalin antibody applied in the detection of hyalin cDNA clones in the λgt11 expression library. Initial experiments with these probes reveal the expected pattern of appearance of hyalin mRNA during development and show that this protein probably consists of cross-linked subunits of about 50-kD mass. We have also isolated from the same λgt11 cDNA library a gene for the major spicule matrix protein. The spicules are the $CaCO_3$ "bones" that ultimately provide a structural armature for the larva, and they are synthesized exclusively by primary mesenchyme cells and their descendants. These cells in turn derive from four vegetal pole blastomeres (the micromeres) formed at fourth cleavage. S. Benson, N. Crise-Benson, and F. Wilt (unpubl.) have shown that the protein matrix remaining when spicules are demineralized consists of nine proteins, one of which accounts for 60% of the total. The gene for this protein has now been cloned and sequenced (H. Sucov, S. Benson, J. Robinson, R. Britten, F. Wilt, and E. Davidson, in prep.). This gene is expressed exclusively in primary mesenchyme cells, according to in situ hybridization (S. Benson, H. Sucov, L. Stevens, E. Davidson, and F. Wilt, in prep.). The fate of the micromere-primary mesenchyme cell lineage appears to be determined by localized maternal cytoplasmic factors, and hence the spicula matrix protein gene is of particular interest. Thus, it affords an opportunity to test experimentally the hypothesis that activation of this gene may be caused by interaction with maternal regulators localized initially in the micromeres.

Though knowledge of lineage-specific gene expression in the sea urchin embryo is as yet largely descriptive, an interesting generality has already emerged. This is that specification of the first embryonic cell lineage occurs long before lineage-specific gene expression can

```
            Glu         Asp Val         Arg Glu
SU      GCC GAG AAG GAC GTC CGT CGA CAG
        •••  •• •••  ••  •  ••
DROS    GCC CAG AAG AAC CTG CGC AAG TCC
            Gln         Asn Leu         Lys Ser

                                    Ile Ile
SU      GAA CGT CGC GTC AAG GAA ATC ATC
        ••  ••• ••• ••• ••• ••   •  • •
DROS    GAG CGT CGC GTC AAG GAG CTG AGC
                                    Leu Ser

            Asn
SU      CTT AAC CAG TCC GAG GAG GAC CGC
        ••  • • ••• ••• ••• ••• ••• •••
DROS    CTG AGC CAG TCC GAG GAG GAC CGC
            Ser

                            Ile
SU      AAA AAT CAC GAA CGA ATC CAG GAT
        ••  ••  ••• ••  ••  ••  ••• •••
DROS    AAG AAC CAC GAG CGC ATG CAG GAT
                            Met
```

Figure 5. Sequence homology between a region of the sea urchin and the *Drosophila* myosin heavy-chain genes. The region shown derives from a genomic *S. purpuratus* recombinant. Dots indicate identical nucleotides, and where changes result in coding differences the respective amino acids are indicated. In the sequence shown 23/32 amino acids are identical. *Drosophila* sequence kindly provided by Dr. Sanford Bernstein (pers. comm.). Data from S. Rose, M. Rosenberg, D. Chen, R. Britten, and E. Davidson (unpubl.).

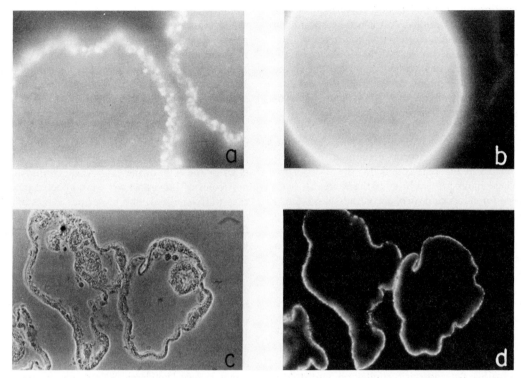

Figure 6. Localization of hyalin displayed by indirect immunofluorescence. Pluteus-stage (74-hr) embryos, fertilized eggs, and unfertilized eggs were fixed, embedded, and sectioned, then reacted with anti-hyalin rabbit antibody. Bound antibody was visualized with fluorescein isothiocyanate (FITC)-conjugated goat antirabbit IgG. (*a*) Unfertilized eggs. (*b*) Fertilized eggs. (*c*) Phase-contrast micrograph of 5-μm sections of pluteus-stage embryos. (*d*) Immunofluorescence from same pluteus-stage embryos as in *c*. (From J.J. Robinson, N. Ruiz-Bravo, H.M. Sucov, B.R. Hough-Evans, R.J. Britten, W.J. Lennarz, and E.H. Davidson, in prep.)

be detected. By specification is meant the process by which the differentiative fate of the descendants of given blastomeres is first established (whether irreversibly or not is, for this argument, irrelevant). There are only a few "target" nuclei when specification takes place, while expression occurs after a number of divisions have intervened, and a meaningful number of properly situated blastomeres have been produced. Different kinds of mechanisms can be envisioned. For example, genes such as those described here might be activated secondarily by the products of pleiotropically active master regulator genes, which might be the only genes directly affected in the initial specification events. Alternatively, such upper-level hierarchy control genes might not be involved at all in the early development of this embryo, and immediate regulators for the lineage-specific structural genes could be stored in the egg or released by appropriate interblastomere contacts. Specification would then consist of the direct interaction of such regulators with *cis*-regulatory sequences of lineage-specific structural genes, such as those described in this article.

Gene Transfer: Expression of Microinjected CyIIIa Genes

We recently described a method for introducing genes into the sea urchin egg by microinjection, and provided evidence that the exogenous DNA is replicated and stably retained during development (Flytzanis et al. 1985; McMahon et al. 1985). Unfertilized eggs are bound by electrostatic attraction to protamine-coated dishes, and several thousand linearized DNA molecules are injected into the cytoplasm of each. Sperm is added, and the eggs develop in situ until hatching, when they secrete an enzyme that dissolves the fertilization envelope, releasing the swimming blastulae. We have shown that within an hour of injection the exogenous DNA has formed random end-to-end concatenates, and during cleavage these replicate an average of 30- to 100-fold. Incorporation of the exogenous DNA into early blastomere nuclei apparently occurs with good efficiency, and the replicated sequences subsequently persist throughout the several weeks of larval life. In about 30% of larvae, the quantity of exogenous DNA continues to increase. Presumably in these larvae the DNA had originally been incorporated in nuclei of those cell lineages that remain mitotically active during larval growth. In several experiments (see Flytzanis et al. 1985), the genomic DNA of postmetamorphosis juveniles descendant from the injected eggs was examined, and 6-15% were found to bear integrated sequences. This fraction probably reflects the frequency with which the exogenous DNA is incorporated into cells that are ancestral to the imaginal rudiment from which the postmetamorphosis juvenile derives. The integrated DNA has been analyzed in cloned isolates from such juveniles, and can be recovered in the DNA of their sperm when sexual maturity has been attained. Such a gene transfer system could thus be utilized to produce transgenic lines of sea urchins. In the experiments described here, however, we have focused on expression of the exogenous DNA in the embryos deriving immediately from the injected eggs.

Initial studies on the expression of DNA injected into sea urchin eggs were reported by McMahon et al. (1984). In this work a construct containing the regulatory elements of the gene coding for the 70-kD *Drosophila* heat shock protein fused to the bacterial structural gene for chloramphenicol acetyltransferase (CAT) (Di Nocera and Dawid 1983) was injected into eggs by the method described, and the developed embryos were subjected to heat stress. It was found that at 25°C, a temperature at which the endogenous heat shock response is elicited in this sea urchin species, CAT enzyme synthesis was induced in the host embryos. Since in *Drosophila*, from which the exogenous heat shock gene sequences derive, this gene is silent at 25°C, transcription of the injected construct must respond to the diffusible signals produced or activated in the heat-stressed sea urchin cells. A similar result has been obtained in other heterospecific studies on *Drosophila* heat shock gene expression (e.g., Corces et al. 1981). It follows that the exogenous DNA is present in the sea urchin embryo in an intracellular location, and in a form that permits regulated transcriptional expression.

We were thus encouraged to investigate the expression of sea urchin genes that in normal embryos display an easily identified ontogenic pattern of expression. An initial choice was the CyIIIa actin gene, which, as shown in Figures 3 and 4, is expressed in a spatially and temporally specific way. Figure 7 displays the stucture of the in-frame fusion between the CyIIIa gene and the CAT gene that was utilized for the following experiments (C. Flytzanis, R. Britten, and E. Davidson, in prep.). The injected DNA contains several kilobases of upstream CyIIIa sequence, plus a large (2.2-kb) intron wholly included in the 5' leader of the CyIIIa transcript and coding sequence for only a few of the aminoterminal amino acids of the actin molecule. The remainder of the actin structural gene has been replaced by the CAT gene. Translation could start at either the actin or the CAT ATG codon. About 3000 molecules of this construct were injected per egg, the eggs were fertilized, and at various stages of development aliquots of 20–50 embryos were harvested and the CAT enzyme activity present in the embryos measured. Results from two such experiments are shown in Figure 8a and b. CAT activity appears essentially on the schedule expected for CyIIIa transcripts (cf. Fig. 4). Thus, by 20 hours the amount of CAT enzyme has attained maximum value. Comparison to an absolute standard for CAT enzyme activity (McMahon et al. 1984) suggests that about 10^6 molecules of CAT enzyme are produced by the CyIIIa-CAT genes per embryo, on the average, and assuming the usual translational parameters (see Davidson 1976) it may be roughly estimated that in the 20-hour embryo the quantity of fusion gene mRNA is on the order of about one-fourth of the natural quantity of CyIIIa message. An important additional observation is shown in Figure 8c. Here is presented a parallel series of experiments in which the upstream regulatory sequences of the *S. purpuratus* early H2a histone gene (for review of sea urchin histone gene regulation see Hentschel and Birnstiel 1981) have replaced those of the CyIIIa gene in a similar CAT

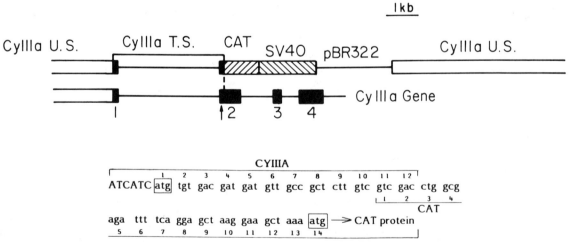

Figure 7. CyIIIa-CAT fusion constructs. (T.S.) Transcribed sequence; (U.S.) upstream sequence. A diagram of the linearized construct as injected into sea urchin eggs is shown in the top line. The plasmid was linearized at an *Sph*I site 2.5 kb upstream of the 5' end of the CyIIIa transcript. Since the molecules form a random end-to-end concatenate after injection (McMahon et al. 1985), in half the cases the original 8-kb upstream sequence included in the plasmid will be regenerated. In the second line is a diagram of the CyIIIa gene, from data of R.J. Akhurst et al. (in prep.). Introns are denoted as thin lines and exons are numbered. The third and fourth lines show the sequence of the junction region (shown by the arrow in the second line) between the CyIIIa and the bacterial CAT genes, with the two in-frame ATG translation start codons, the first deriving from the CyIIIa and second from the CAT gene.

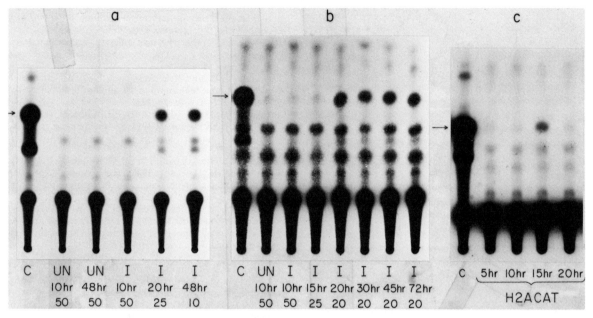

Figure 8. CAT expression from fusion genes injected into sea urchin eggs. The genes were injected into unfertilized eggs which were then allowed to develop until collected for assay of CAT activity, as described by McMahon et al. (1984, 1985; see text). (*a* and *b*) Two independent experiments using the CyIIIa-CAT fusion gene depicted in Fig. 7. (C) Control assay containing bacterial CAT enzyme. The arrow indicates the monoacetylated CAT product. Other samples contained sea urchin embryo extracts. (UN) Embryos derived from uninjected eggs; (I) embryos derived from injected eggs. The number of hours postfertilization and the number of embryos in the sample analyzed are indicated, respectively, in the bottom two rows of numerals. (*c*) CAT activity in embryos injected with an α-H2a-CAT construct. This contained all sequence upstream of the H2a gene to the preceding H3 histone gene, but only 30 nucleotides of coding sequence.

fusion construct. In embryos injected with this construct the CAT enzyme appears at 15 hours, just after the amount of endogenous H2a mRNA reaches its peak value in normal embryos (Mauron et al. 1982).

In preliminary experiments so far available, the CAT activity then disappears, just as does the endogenous α-histone message. This result would imply that within the sea urchin embryo the newly synthesized CAT protein is unstable, and also that the mRNA produced by the fusion gene is unstable. The same conclusion follows from the constant amount of CAT activity observed after the blastula stage in embryos injected with the CyIIIa-CAT fusion. Most significantly for the results shown in Figure 8a and b, it is clear from Figure 8c that the absence of CyIIIa-CAT expression at 15 hours cannot be due to insufficiency of exogenous DNA or inaccessibility of this DNA for transcriptional activation in 15-hour (as opposed to 20-hour) embryos. Some time is required for accumulation of the CAT protein once the message has appeared, and it must also be taken into consideration that injected embryos are delayed in their development by about one division cycle (McMahon et al. 1985). These factors account for the minor retardation, in regard to the respective endogenous transcripts in control embryos, in the appearance of CAT enzyme in both the CyIIIa-CAT and H2a-CAT experiments shown. We may conclude that both the exogenous CyIIIa gene sequences and the exogenous α-H2a sequences promote expression of their fusion constructs according to the predicted ontogenic schedule.

It should thus be possible to explore directly the mechanism by which lineage-specific genes are activated in the early sea urchin embryo. This is not necessarily the same problem as determining what activates genes in the terminal differentiation processes of later development, or in the physiologically induced genes of adult organisms. The early embryo utilizes spatial regulatory information that originates maternally; it relies to some extent on oriented interblastomere interactions for lineage determination; and, unlike the case in advanced developmental systems, its nuclei are equivalent and totipotent, i.e., until the initial events of lineage specification have taken place. Furthermore, the results summarized by Solter et al. elsewhere in this volume show that some genes destined to function early in mouse development may be "imprinted" at the chromatin level during gametogenesis. The qualitative results illustrated in Figure 8 are already instructive in this respect, however, in that they suggest that genes activated in the early sea urchin embryo respond to *trans*-activators that are at least initially present in excess. At the level of genomic function, the key to the mechanisms underlying the initial specification of embryonic cell lineages may lie in the nature, origins, and spatial disposition or release of such *trans*-acting embryonic regulatory molecules.

ACKNOWLEDGMENTS

This work was supported by National Institutes of Health grants HD-05753 and GM-20927. C.N.F. was supported by a Lieve Fellowship (S-11-83) from the American Cancer Society, California Division; J.J.R. by a Proctor & Gamble Fellowship; S.J.R. by a Muscular Dystrophy Association Fellowship; and H.M.S. and J.J.L. by a National Institutes of Health training grant (GM-07616).

REFERENCES

Angerer, R.C. and E.H. Davidson. 1984. Molecular indices of cell lineage specification in sea urchin embryos. *Science* 226: 1153.

Britten, R.J. and E.H. Davidson. 1969. Gene regulation for higher cells: A theory. *Science* 165: 349.

———. 1971. Repetitive and nonrepetitive DNA sequences and a speculation on the origins of evolutionary novelty. *Q. Rev. Biol.* 46: 111.

Bruskin, A.M., A.L. Tyner, D.E. Wells, R.M. Showman, and W.H. Klein. 1981. Accumulation in embryogenesis of five mRNAs enriched in the ectoderm of the sea urchin pluteus. *Dev. Biol.* 87: 308.

Bruskin, A.M., P.-A. Bedard, A.L. Tyner, R.M. Showman, B.P. Brandhorst, and W.H. Klein. 1982. A family of proteins accumulating in ectoderm of sea urchin embryos specified by two related cDNA clones. *Dev. Biol.* 91: 317.

Butler, E.T. and M.J. Chamberlin. 1982. Bacteriophage SP6-specific RNA polymerase. I. Isolation and characterization of the enzyme. *J. Biol. Chem.* 257: 5772.

Carpenter, C.D., A.M. Bruskin, P.E. Hardin, M.J. Keast, J. Anstrom, A.L. Tyner, B.P. Brandhorst, and W.H. Klein. 1984. Novel proteins belonging to the troponin C superfamily are encoded by a set of mRNAs in sea urchin embryos. *Cell* 36: 663.

Corces, V., A. Pellicer, R. Axel, and M. Meselson. 1981. Integration, transcription and control of a *Drosophila* heat shock gene in mouse cells. *Proc. Natl. Acad. Sci.* 78: 7038.

Davidson, E.H. 1976. *Gene activity in early development.* Academic Press, New York.

Davidson, E.H., B.R. Hough-Evans, and R.J. Britten. 1982. Molecular biology of the sea urchin embryo. *Science* 217: 17.

Di Nocera, P.P. and I.B. Dawid. 1983. Transient expression of genes introduced into cultured cells of *Drosophila*. *Proc. Natl. Acad. Sci.* 80: 7095.

Durica, D.S., J.A. Schloss, and W.R. Crain, Jr. 1980. Organization of actin gene sequences in the sea urchin: Molecular cloning for a cytoplasmic actin. *Proc. Natl. Acad. Sci.* 77: 5683.

Flytzanis, C.N., A.P. McMahon, B.R. Hough-Evans, K.S. Katula, R.J. Britten, and E.H. Davidson. 1985. Persistence and integration of cloned DNA in postembryonic sea urchins. *Dev. Biol.* 108: 431.

Hentschel, C.C. and M.L. Birnstiel. 1981. The organization and expression of histone gene families. *Cell* 25: 301.

Herbst, C. 1900. Uber das Auseinandergehen von Forchurgs- und Gewebezellen in Kalkfreiem Medium. *Wilhelm Roux' Arch. Entwicklungsmech. Org.* 9: 424.

Hörstadius, S. 1939. The mechanics of sea urchin development studied by operative methods. *Biol. Rev. Camb. Philos. Soc.* 14: 132.

Hylander, B.L. and R.G. Summers. 1982. An ultrastructural immunocytochemical localization of hyalin in the sea urchin egg. *Dev. Biol.* 93: 368.

Lee, J.J., R.J. Shott, S.J. Rose III, T.L. Thomas, R.J. Britten, and E.H. Davidson. 1984. Sea urchin actin gene subtypes. Gene number, linkage and evolution. *J. Mol. Biol.* 172: 149.

Lev, Z., T.L. Thomas, A.S. Lee, R.C. Angerer, R.J. Britten, and E.H. Davidson. 1980. Developmental expression of two cloned sequences coding for rare sea urchin embryo messages. *Dev. Biol.* 76: 322.

Lynn, D.A., L.M. Angerer, A.M. Bruskin, W.H. Klein, and R.C. Angerer. 1983. Localization of a family of mRNAs in a single cell type and its precursor in sea urchin embryos. *Proc. Natl. Acad. Sci.* 80: 2656.

Mauron, A., L. Kedes, B.R. Hough-Evans, and E.H. Davidson. 1982. Accumulation of individual histone mRNAs during embryogenesis of the sea urchin *Strongylocentrotus purpuratus*. *Dev. Biol.* 94: 425.

McClay, D.R. and R.D. Fink. 1982. Sea urchin hyalin: Appearance and function in development. *Dev. Biol.* 92: 285.

McMahon, A.P., T.J. Novak, R.J. Britten, and E.H. Davidson. 1984. Inducible expression of a cloned heat shock fusion gene in sea urchin embryos. *Proc. Natl. Acad. Sci.* 81: 7490.

McMahon, A.P., C.N. Flytzanis, B.R. Hough-Evans, K.S. Katula, R.J. Britten, and E.H. Davidson. 1985. Introduction of cloned DNA into sea urchin egg cytoplasm: Replication and persistence during embryogenesis. *Dev. Biol.* 108: 420.

Melton, D.A., P.A. Krieg, M.R. Rebagliati, T. Maniatis, K. Zinn, and M.R. Green. 1984. Efficient in vitro synthesis of biologically active RNA and RNA hybridization probes from plasmids containing a bacteriophage SP6 promoter. *Nucleic Acids Res.* 12: 7035.

Morgan, T.H. 1934. *Embryology and genetics*. Columbia University Press, New York.

Scheller, R.H., F.D. Costantini, M.R. Kozlowski, R.J. Britten, and E.H. Davidson. 1978. Representation of cloned interspersed repetitive sequences in sea urchin RNAs. *Cell* 15: 189.

Scheller, R.H., L.B. McAllister, W.R. Crain, D.S. Durica, J.W. Posakony, T.L. Thomas, R.J. Britten, and E.H. Davidson. 1981. Organization and expression of multiple actin genes in the sea urchin. *Mol. Cell. Biol.* 1: 609.

Shott, R.J., J.J. Lee, R.J. Britten, and E.H. Davidson. 1984. Differential expression of the actin gene family of *Strongylocentrotus purpuratus*. *Dev. Biol.* 101: 295.

Wallace, R.B., S.K. Dube, and J. Bonner. 1977. Localization of the globin gene in the template active fraction of chromatin of Friend leukemia cells. *Science* 198: 1166.

Germ Line Specificity of P-element Transposition and Some Novel Patterns of Expression of Transduced Copies of the white Gene

G.M. Rubin, T. Hazelrigg, R.E. Karess,* F.A. Laski, T. Laverty, R. Levis, D.C. Rio, F.A. Spencer, and C.S. Zuker
Department of Biochemistry, University of California, Berkeley, California 94720

P TRANSPOSABLE ELEMENTS

P elements, the family of mobile genetic elements that are responsible for the phenomenon of P-M hybrid dysgenesis in *Drosophila melanogaster*, are of particular interest because their mobility has been shown to be under genetic control (for reviews, see Bregliano and Kidwell 1983; Engels 1983). When the elements are quiescent they are said to be in the P cytotype (Engels 1979a), the cellular environment of P-strain flies. P cytotype is apparently determined by the P factors themselves (Engels 1979b). P factors, which are defined by genetic assays, are presumed to correspond to a fully functional subset of the biochemically defined P elements (O'Hare and Rubin 1983). Flies lacking functional P elements are called M-strain flies, and are said to possess the M cytotype. Hybrid dysgenesis occurs when P-strain males are crossed to M-strain females, thereby introducing functional P elements into the M cytotype. The offspring of such a cross (dysgenic hybrids) show a series of genetic aberrations, all of which are confined to the germ line of the dysgenic hybrid. These may include chromosomal rearrangements, visible and lethal mutations, male recombination, and a high level of gonadal sterility. In the reciprocal cross, between an M-strain male and P-strain female, or in a P × P cross, the P elements are maintained in the P cytotype, and no dysgenic traits are observed.

A number of P elements have been isolated (Rubin et al. 1982) and characterized (O'Hare and Rubin 1983). About one-third of the 50 elements found in a typical strain share a conserved 2.9-kb structure; the others are heterogeneous in size and smaller, missing sequences internal to the 31-bp terminal inverted repeats that flank the element. The 2.9-kb element has been shown to supply a *trans*-acting function required both for its own transposition (Spradling and Rubin 1982) and for the transposition of defective nonautonomous P elements (Rubin and Spradling 1982). These observations led to the development of the P-element-mediated gene-transfer system (Rubin and Spradling 1982; Spradling and Rubin 1982).

Strategy for the Analysis of P Transposable Element Functions

A genetic approach to dissecting the functions encoded by P elements requires that one be able to examine a single element of defined structure, rather than the heterogeneous population of defective and nondefective P elements found in the genomes of natural P strains. Additionally, the element needs to be marked genetically, so that its presence and location in the genome of a living fly can be easily detected, independent of its ability to produce the symptoms of hybrid dysgenesis.

We have made a P-element derivative which carries the wild-type rosy gene, but which acts in several respects like the nondefective 2.9-kb P element (Karess and Rubin 1984; see Fig. 1). When introduced into an M-strain fly, this element, called Pc[*ry*], continues to transpose autonomously within the genome, and is able to destabilize other nonautonomous P elements. Our strategy has been to mutagenize the Pc[*ry*] element in vitro, and to assay its activity in vivo, both singly and in combination with other P elements to identify the regions of the element encoding the transposase function.

A Frameshift Mutation in Any of the P-element Open Reading Frames Abolishes Transposase Activity

The most sensitive assay of P transposase activity involves measuring the rate of destabilization of the singed weak mutation (Engels 1984). singed weak (sn^w) is a hypermutable allele of the singed bristle locus on the X chromosome. It arose in the offspring of a dysgenic hybrid (Engels 1979a) and its phenotype results from the presence of two small, defective P elements at the *sn* locus (H. Roiha et al., unpubl.). In the offspring of a P male by M female dysgenic cross in which one parent carries sn^w, up to 50% of the gametes of the F_1 dysgenic hybrids no longer carry the parental sn^w allele (Engels 1979a, 1984). One or the other of the two defective P elements at sn^w excises, generating one of two new phenotypes in the F_2 offspring: a much more extreme singed bristle (sn^e), and an apparently wild-type bristle (sn^+) (H. Roiha et al., unpubl.). However, the sn^w allele is essentially stable when it is maintained in a genome devoid of autonomous P elements. Thus the

---
*Present address: Department of Biochemistry, Imperial College of Science and Technology, London SW72AZ, England.

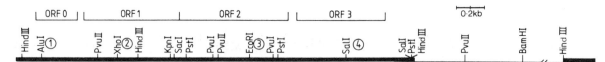

Figure 1. Map of prototype Pc[(ry)] element and derivatives. The heavy line represents P-element sequences. A portion of the 7.2-kb HindIII fragment containing the rosy sequences (thin line) is shown. Numbered restriction sites have been altered in one or more derivatives of the element. The four frameshift mutations were introduced onto this element: in P[(ry)Alu] (ORF 0 frameshift mutant), site 1 has become an EcoRI site by the addition of an 8-bp linker. (The positions of other AluI sites are not shown.) P[(ry)Xho] (ORF 1 mutant), P[(ry)R1] (ORF 2 mutant), and P[(ry)Sal] (ORF 3 mutant) are 4-bp insertions at sites 2, 3, and 4, respectively. The four potential open reading frames (ORF 0–ORF 3) of the P element, as determined by O'Hare and Rubin (1983), are indicated above the map.

destabilization of sn^w is an extremely sensitive assay for the presence of transposase activity provided by functional P elements. A single autonomous P element is sufficient to induce sn^w instability (Spradling and Rubin 1982).

The wild-type Pc[ry] element can destabilize sn^w, but mutant derivatives of Pc[ry] containing frameshift mutations in any of the four open reading frames (ORFs) cannot (Karess and Rubin 1984; see Fig. 1). Furthermore, no complementation is observed between mutations in different ORFs, suggesting that each of the four ORFs contributes to a single polypeptide required for transposition.

Why is P-element Transposition Limited to the Germ Line?

P elements transpose only in the germ line and not in somatic tissues (see Engels 1983 for a review of data supporting this statement). Two classes of models can be proposed for the mechanism of this tissue specificity. First, transposase production might be limited to the germ line. For example, the transcriptional promoter for the transposase gene might only be active in germ line cells. Alternatively, transposase might be made in all cells, but other components (polymerases, ligases) or cofactors needed for the transposition reaction might be germ line limited. We have been able to distinguish between these classes of models by the series of in vitro mutagenesis experiments described below.

We first asked whether the limitation of transposition to the germ line could be overcome by substituting for the natural P promoter, a transcriptional promoter known to be highly active in somatic cells. P-element constructs were made which contained the P-element coding sequence under the control of the heat-inducible hsp70 promoter (Lis et al. 1983). Although a high level of transcription was observed in somatic cells following heat induction, transposition was still limited to the germ line (F. Laski et al., in prep.).

A clue to the mechanism of germ line specificity was provided by the structure of the major P-element transcript. As described above, the results of in vitro mutagenesis experiments suggested that the four ORFs present in the 2.9-kb element are joined by RNA splicing into a single continuous sequence encoding transposase. Examination of the P-element DNA sequence (O'Hare and Rubin 1983) reveals that potential 5′ and 3′ splice sites are indeed located in appropriate positions near the ends of the ORFs. An analysis of the major P-element transcript by S1 nuclease mapping, however, indicates that one of the postulated introns was still present in the mature polyadenylated transcript isolated from embryos and other sources (F. Laski et al., in prep.; see Fig. 2). This result was puzzling in two respects. First, while there are many examples of differential or alternative RNA splicing patterns, no case has been reported of a stable RNA in which some, but not all, of the introns have been removed. Second, given that the sequences of the last ORF are required (Karess and Rubin 1984), such a transcript should be incapable of encoding active transposase. A way to reconcile this apparent contradiction is to propose that the RNA whose structure we have analyzed is not in fact a functional mRNA for transposase. We isolated RNA from whole embryos in which only about 1% of the cells are of the germ line lineage. Perhaps removal of this last intron, and thus the production of functional transposase mRNA, only occurs in the germ line. Given the very low abundance of P transcripts (Karess and Rubin 1984) such a small fraction of fully spliced RNA might have escaped our detection.

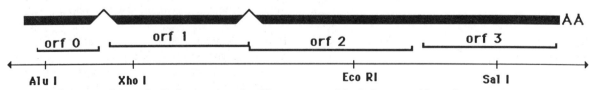

Figure 2. Structure of the major P-element transcript. The structures of the P element and its major transcript and the position of the four ORFs are diagramed. The four restriction enzyme sites at which frameshift mutations were created (AluI, XhoI, ExoXI, and SalI) are shown. The heavy line represents the structure of the major polyadenylated P-element transcript as determined by S1 nuclease mapping. The introns between ORF 0 and ORF 1 and between ORF 1 and ORF 2 have been spliced but the putative intron between ORF 2 and ORF 3 is present in this transcript.

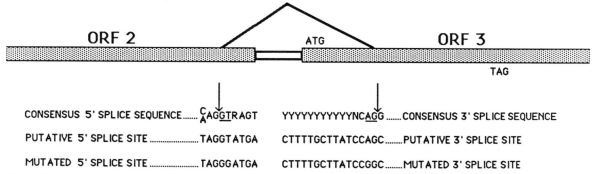

Figure 3. Analysis of the putative ORF 2 to ORF 3 splice junctions by in vitro mutagenesis. A diagram of the region of the P element containing the ORF 2–ORF 3 junction is shown. The shaded bars represent the positions of the ORFs as determined from the DNA sequence (O'Hare and Rubin 1983). An ATG codon found at the beginning of ORF 3 is shown. The position of the proposed splice is indicated and the DNA sequences at the putative 5' and 3' splice junctions are shown and compared with those of the consensus splice junctions (Mount 1982). The DNA sequences of the mutated splice junctions are also shown. In addition, a nonsense mutation was made in ORF 3 at the position marked by the TAG. P elements containing this mutation do not produce transposase.

We tested the hypothesis that germ line-specific splicing of the third intron is the basis for the tissue specificity of transposase production by constructing a number of Pc[ry] elements in which this intron was either removed or altered (F. Laski et al., in prep.; see Fig. 3). These elements were introduced into *Drosophila* and assayed for sn^w destabilization as described by Karess and Rubin (1984).

Consistent with the hypothesis that removal of the intron is a prerequisite for transposase production, mutations that alter the consensus 5' or 3' splice sites of this last intron abolish transposase production as assayed by sn^w destabilization. The mutation that proved to be most informative was one that removed the intron, thereby joining the coding sequences of the last two ORFs without the need for RNA splicing. Flies carrying such a mutant element, which we call Pc[ry, Δ2-3], produce comparable, if not more, transposase in their germ lines as do those with wild-type Pc[ry] elements. Most importantly, however, they now produce transposase in their somatic cells as well. This can most easily be seen by the ability of Pc[ry, Δ2-3] elements to induce somatic mosaicism by causing transposition and excision of P elements in somatic cells.

Two simple assays for detecting somatic transposase activity are outlined in Figure 4. In the first assay (Fig. 4A), which measures somatic P-element excision, flies carrying a Pc[ry, Δ2-3] element are crossed to flies carrying a nonautonomous transposon, P[w^+], as their only normal copy of the white locus. Such flies have red eyes by virtue of the wild-type white gene contained within the P[w^+] transposon. The F₁ offspring of such a cross will carry both the P[w^+] and Pc[ry, Δ2-3] elements. Autonomous P elements, such as Pc[ry, Δ2-3], are able to induce the excision of nonautonomous elements, such as Pc[w^+]. Normally this activity is limited to the germ line and in such a case all of the F₁ offspring would have red eyes, as only their germ lines would be mosaic. Since the white gene is cell autonomous, somatic excisions of P[w^+] would produce clones of nonpigmented cells in the eye. Such clones are observed in nearly all flies carrying both the P[w^+] and Pc[ry, Δ2-3] transposons. An example is shown in Figure 5a.

The second assay is designed to measure transposition of a P[w^+] element from one chromosome site to another (Fig. 4B). In this assay the P[w^+] transposon used confers a pale yellow eye color due to its location at a site in the genome that precludes normal white gene expression (Fig. 5b; Hazelrigg et al. 1984). Transposition of this P[w^+] to a new genomic site or transposase-induced local rearrangements can restore a red eye color (Levis et al. 1985). Evidence of such somatic events, which produce red clones on the yellow background, is seen in the eyes of nearly all flies carrying both this P[w^+] transposon and the Pc[ry, Δ2-3] ele-

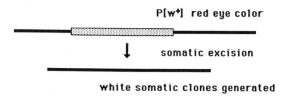

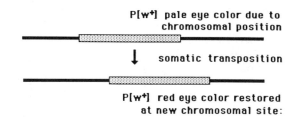

Figure 4. Transposase assays. Schematic representations of the excision (*A*) and transposition (*B*) assays used to measure somatic transposition rates of P[w^+] transposons are shown (see text for details).

ment (Figs. 5c,d). The size and number of clones observed suggest that multiple events occur in the cell lineage leading to each eye and that these events can occur both early and late in development.

Taken together these data indicate that the limitation of P-element transposition to the germ line results from the failure to remove a particular intron from P transcripts in somatic cells. Clearly, all other requirements for P-element transposition are met by somatic as well as germ line cells. Further experiments will be required to determine whether this type of control is used by other genes expressed specifically in the germ line or is peculiar to P elements. In addition to these biological questions, the ability to induce somatic mosaics at high frequency may be useful in marking cells for lineage studies. Moreover, it is possible that removal of the intron between ORFs 2 and 3 will not only overcome the germ line limitation, but also broaden the host range of P-element transposition to include other organisms.

NOVEL, HERITABLE PATTERNS OF WHITE GENE EXPRESSION IN THE EYE

The relationship between the chromosomal location of a gene and its expression has been studied extensively in *Drosophila* using chromosomal rearrangements such as inversions and translocations. Rearrangements can result in the inactivation of genes located at some distance from the rearrangement breakpoints. This inactivation commonly occurs in only a portion of the cells in which a gene is normally expressed, causing the fly to have a mosaic phenotype. Such variegating position effects are usually associated with rearrangements that bring euchromatic genes into the proximity of centromeric heterochromatin (for review, see Spofford 1976).

The development of the technique of P-element-mediated DNA transformation in *Drosophila* has expanded our ability to study genomic position effects. A gene, borne within a P transposable element, can be introduced at many genomic positions and its expression at these different positions compared (Spradling and Rubin 1983). Several genes that have been reintegrated into the germ line by this method show proper developmentally regulated expression at a variety of euchromatic positions (Goldberg et al. 1983; Scholnick et al. 1983; Spradling and Rubin 1983).

The white (*w*) gene confers an essentially wild-type (red-eyed) phenotype when introduced to >30 different chromosomal locations by P element vectors (Gehring et al. 1984; Hazelrigg et al. 1984; Levis et al. 1985). In contrast, in approximately one in 10 cases a transduced white gene confers a mutant mosaic eye color. For example, flies carrying one copy of the transduced white gene A^R4-3 are more darkly pigmented in the anterior portion of their eyes than in the posterior portion (Fig. 5e; Hazelrigg et al. 1984). This transduced gene is located near the centromeric heterochromatin of chromosome arm 2L.

We have demonstrated that the mutant phenotype of A^R4-3 is the result of a genomic position effect by moving the transduced white gene in A^R4-3 to new chromosomal positions (Levis et al. 1985). To relocate the transduced white gene in A^R4-3 we mobilized the P-white transposon in this strain. Embryos were injected with plasmid DNA containing the "wings-clipped" P element. This P-element derivative can induce the transposition of P-element transposons, but does not itself integrate into the chromosomes (Karess and Rubin 1984). Therefore, the P-white transposon should transpose in the germ cells of some of the injected embryos but should be stable in subsequent generations. The fact that relocating the gene resulted in a wild-type eye color at most new positions demonstrates that its mutant phenotype is due to a position effect, and not due to a mutation intrinsic to the gene.

The mobilization of the P-white transposon of A^R4-3 produced, in addition to 12 red-eyed derivatives, a derivative having a heritable, novel mosaic pattern of eye pigmentation. The transduced white gene of this derivative is located in the 24CD cytogenetic region and we will refer to it as A^R4-24. The ventral halves of the eyes of A^R4-24 flies are more darkly pigmented than the dorsal halves (Fig. 5f). This pattern is like that of A^R4-3 flies in that one part of the eye is reproducibly more darkly pigmented than the other. However, the eye pigmentation patterns of A^R4-24 and A^R4-3 are oriented perpendicular to each other.

A^R4-3 and A^R4-24 are fundamentally different from other white position-effect mutants in which patches of pigmented and unpigmented tissue vary, more or less at random, in their locations in the eyes of flies carrying the same mutant allele. The reproducible patterns of pigmentation in A^R4-3 and A^R4-24 indicate that the phenotypic expression of white, at least in these mutants, must be sensitive to positional information in the developing eye. Moreover, the pattern of pigmentation in these alleles can be altered to produce new reproducible patterns by the action of the *trans*-acting gene zeste. For example, Figure 5g and h, shows the pigmentation produced by A^R4-3 flies heterozygous or hemizygous, respectively, for the $zeste^1$ gene. Additional examples of reproducible, heritable pigmentation patterns produced in flies carrying transduced copies of the white locus and the $zeste^1$ allele are shown in Figure 5i, l, and m. The specifics of their genotypes are given in the legend to Figure 5.

Why is the spectrum of position effects on white expression we observe with transduced white genes so different from that seen with classical chromosomal rearrangements which tend to produce clonal, nonreproducible variegated phenotypes? In most classical position effects the white gene is brought near to centric heterochromatin but the chromosomal breakpoint near white is often many kilobases or tens of kilobases from the gene. A transduced gene may be more sensitive to position effects than a gene in a chromosomal rearrangement, due to the small size of the transduced segment, leading to the observation of position effects

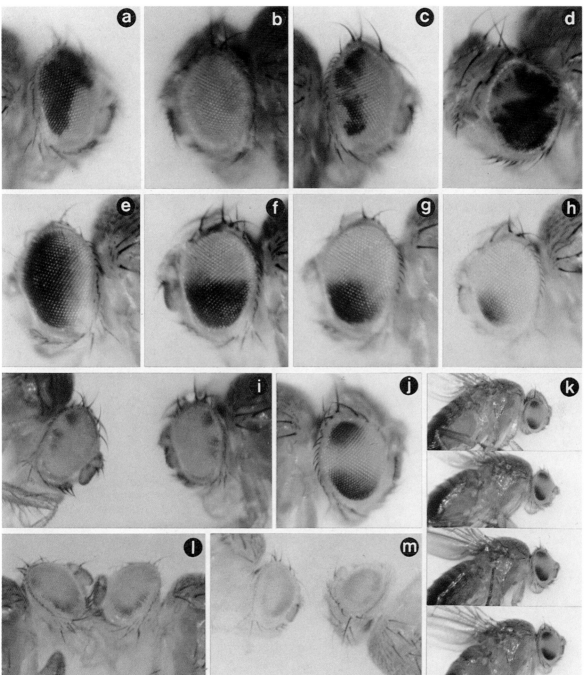

Figure 5. Examples of eye color phenotypes. (*a*) P[w^+], Pc[*ry*, Δ2-3], showing examples of somatic excision. (*b*) P[(*w, ry*)A]4-4, showing pale color due to the chromosomal position of the white transposon (Hazelrigg et al. 1984). (*c*) P[(*w, ry*)A]4-4; Pc[*ry*, Δ2-3], showing examples of somatic events that restore full gene activity to clones of cells in the eye. (*d*) Other eye of the fly shown in *c*. (*e*) w^{1118}/w^{1118}; P[(*w, ry*)A^R]4-3/+. (*f*) w^{1118}/w^{1118}; P[(*w, ry*)A^R]4-24/P[(*w, ry*)A^R]4-24. (*g*) $z^1w^{1118}cv/w^{1118}$; P[(*w, ry*)A^R]4-24/+. (*h*) $z^1w^{1118}cv$/Y; P[(*w, ry*)A^R]4-24/+. (*i*) Two females of the same genotype, homozygous for $z^1w^{1118}cv$ X chromosomes, *trans*-heterozygotes for two P[(*w, ry*)A^R] transposons both inserted at 92B (see Levis et al. 1985 for details). (*j*) Male carrying a white allele derived spontaneously from the w^{DZL} revertant w^{rD320} (see Levis and Rubin 1982) in combination with z^1, which is required for expression of the mosaic pattern. (*k*) Four males, same genotype as *j*. (*l*) z^1w^{1118}, P[w^{dl}]9.3/z^1w^{1118}, P[w^{dl}]9.3. (*m*) z^1w^{1118}/Y; P[w^{dB}]21/P[w^{dB}]21.

at a variety of euchromatic as well as heterochromatic sites. Indeed, the white-zeste-halo mutation (Judd 1975) which, unlike the vast majority of chromosomal rearrangements, produces a reproducible pigmentation pattern not unlike that shown in Figure 5m, results from a chromosomal translocation with two euchromatic breaks, one of which is within a few kilobases of the white gene (Levis et al. 1982). Another example of a heritable reproducible pattern is shown in Figure 5j and k. This pattern results from a small inversion with two euchromatic breakpoints, one very near the white locus (R. Levis, unpubl.).

The patterns of pigmentation within the eye produced by mutant copies of the white gene would a priori be thought to represent one of the simplest examples of pattern formation. The white gene itself is nonessential for the viability of the cells that express it. Moreover, these cells are located in a regular array, the formation of which does not require white gene function. Yet we cannot easily explain the various pigmentation patterns we observe by a coherent model of pattern formation based on cell lineage, gradients of morphogens, or zones of inhibition.

ACKNOWLEDGMENTS

T.H. was supported by an American Cancer Society California Division Senior Postdoctoral Fellowship. D.R. and C.Z. are Fellows of the Jane Coffin Childs Memorial Fund for Medical Research. F.L. is an Exxon Education Fellow of the Life Sciences Research Foundation. F.S. is an American Cancer Society Postdoctoral Fellow. R.K. was a National Institutes of Health Postdoctoral Fellow. This work was supported by the American Cancer Society, the Life Sciences Research Foundation, the Jane Coffin Childs Memorial Fund for Medical Research, and the National Institutes of Health.

REFERENCES

Bregliano, J.C. and M.G. Kidwell. 1983. Hybrid dysgenesis determinants. In *Mobile genetic elements* (ed. J.A. Shapiro), p. 363. Academic Press, New York.

Engels, W.R. 1979a. Extrachromosomal control of mutability in *Drosophila melanogaster. Proc. Natl. Acad. Sci.* **76:** 4011.

———. 1979b. Hybrid dysgenesis in *Drosophila melanogaster*: Rules of inheritance of female sterility. *Genet. Res.* **33:** 219.

———. 1983. The P family of transposable elements in *Drosophila. Annu. Rev. Genet.* **17:** 315.

———. 1984. A *trans*-acting product needed for P factor transposition in *Drosophila. Science* **226:** 1194.

Gehring, W.J., R. Klemenz, U. Weber, and U. Kloter. 1984. Functional analysis of the $white^+$ gene of *Drosophila* by P-factor-mediated transformation. *EMBO J.* **3:** 2077.

Goldberg, D.A., J.W. Posakony, and T. Maniatis. 1983. Correct developmental expression of a cloned alcohol dehydrogenase gene transduced into the *Drosophila* germ line. *Cell* **34:** 59.

Hazelrigg, T., R. Levis, and G.M. Rubin. 1984. Transformation of *white* locus DNA in *Drosophila*: Dosage compensation, *zeste* interaction, and position effects. *Cell* **36:** 469.

Judd, B.H. 1975. Genes and chromosomes of *Drosophila*. In *The eukaryote chromosome* (ed. W.J. Peacock and R.D. Brock), p. 169. Australian University Press, Canberra.

Karess, R.E. and G.M. Rubin. 1984. Analysis of P transposable element functions in *Drosophila. Cell* **38:** 135.

Levis, R. and G.M. Rubin. 1982. The unstable w^{DZL} mutation of *Drosophila* is caused by a 13 kilobase insertion that is imprecisely excised in phenotypic revertants. *Cell* **30:** 543.

Levis, R., P.M. Bingham, and G.M. Rubin. 1982. Physical map of the *white* locus of *Drosophila melanogaster. Proc. Natl. Acad. Sci.* **79:** 564.

Levis, R., T. Hazelrigg, and G.M. Rubin. 1985. Effects of genomic position on the expression of transduced copies of the *white* gene in *Drosophila. Science* **229:** 558.

Lis, J.T., J.A. Simon, and C.A. Sutton. 1983. New heat shock puffs and β-galactosidase activity resulting from transformation of *Drosophila* with an hsp70-lacZ hybrid gene. *Cell* **35:** 403.

Mount, S.M. 1982. A catalogue of splice junction sequences. *Nucleic Acids Res.* **10:** 459.

O'Hare, K. and G.M. Rubin. 1983. Structures of P transposable elements and their sites of insertion and excision in the *Drosophila melanogaster* genome. *Cell* **34:** 25.

Rubin, G.M. and A.C. Spradling. 1982. Genetic transformation of *Drosophila* with transposable element vectors. *Science* **218:** 348.

Rubin, G.M., M.G. Kidwell, and P.M. Bingham. 1982. The molecular basis of P-M hybrid dysgenesis: The nature of induced mutations. *Cell* **29:** 987.

Scholnick, S.B., B.A. Morgan, and J. Hirsh. 1983. The cloned *dopa decarboxylase* gene is developmentally regulated when reintegrated into the *Drosophila* genome. *Cell* **34:** 37.

Spofford, J.B. 1976. Position-effect variegation in *Drosophila*. In *The Genetics and Biology of* Drosophila, *1c* (ed. M. Ashburner and E. Novitski), p. 955. Academic Press, London.

Spradling, A.C. and G.M. Rubin. 1982. Transposition of cloned P elements into *Drosophila* germ line chromosomes. *Science* **218:** 341.

———. 1983. The effect of chromosomal position on the expression of the *Drosophila* xanthine dehydrogenase gene. *Cell* **34:** 47.

On the Molecular Basis of Transvection Effects and the Regulation of Transcription

Z. ZACHAR, C.H. CHAPMAN, AND P.M. BINGHAM
Department of Biochemistry, State University of New York, Stony Brook, New York 11794

Since the original cloning of sequences from the white locus of *Drosophila* (Bingham et al. 1981), white has been subjected to extensive molecular analysis (Bingham et al. 1982; Rubin et al. 1982; Zachar and Bingham 1982; O'Hare et al. 1983; Levis et al. 1984; Pirrotta and Brockl 1984; Bingham and Zachar 1985; Davison et al. 1985). We describe here selected portions of our study at the white locus of a class of phenomena referred to as transvection effects. Our results suggest that a single, tissue-specific enhancer is ultimately responsible for all the superficially distinct transvection effects at white. On the basis of our analysis, we propose a new, general model for the molecular basis of transvection effects. We discuss the relevance of this proposal to the detailed mechanistic basis of enhancer function and transvection. We further discuss selected technical and evolutionary implications of our model.

Several lines of cytological evidence argue that interphase chromosomes in *Drosophila* are normally synapsed in precise register in many polyploid and diploid somatic tissues (for recent discussions, see Lifschytz and Hareven 1982; Agard and Sedat 1983; Hammond and Laird 1985). The two alleles at a locus in the diploid organism are thus not randomly arranged within the nuclear volume but rather are intimately associated with one another during the stage of the cell cycle in which transcription occurs. This normally occurring synapsis is disrupted in the vicinity of chromosome rearrangement break points in rearrangement heterozygotes, allowing the impact of synapsis on the interaction of the two alleles at a locus in the diploid organism to be experimentally assessed.

The impact of synapsis of alleles on the phenotypes produced by those alleles was first assessed by Lewis (1954) with a surprising result. Lewis observed that certain pairs of mutant alleles at the bithorax locus would substantially complement (based on morphological phenotype) when they were allowed to synapse but would not complement when synapsis between them was blocked. Lewis designated these effects of synapsis "transvection effects."

After a long hiatus in which transvection effects were regarded as arcane peculiarities by most non-*Drosophila* biologists and as important inscrutabilities by most *Drosophila* biologists, several new observations reopened the study of these phenomena (Ashburner 1970; Jack and Judd 1979; Bingham 1980; Korge 1981; Gelbart and Wu 1982). These genetic and cytogenetic studies revealed and extensively documented synapsis-dependent genetic interactions at several loci and suggested that most or all transvection effects had some common mechanistic basis. Although it was pointed out that various mechanisms might account for synapsis-dependent genetic interactions (including those assuming these interactions to be mediated by diffusion of unstable RNAs produced at the interacting loci; Ashburner 1970, 1971; Jack and Judd 1979; Bingham 1980), these fundamentally genetic experiments were inherently incapable of establishing the detailed molecular mechanism of these effects.

EXPERIMENTAL PROCEDURES

Fly strains and methods for purification and analysis of RNAs have been described previously (Bingham and Zachar 1985; Davison et al. 1985). Flies were reared at 24.5–25.5°C. DNA sequence probes used in Figure 2 were single stranded and were generated as described by Hu and Messing (1982) and used as described by Bingham and Zachar (1985).

RESULTS

The white locus is dispensable (allowing routine, extensive genetic manipulation) and produces a conspicuous, cell-autonomous pigment deposition phenotype (allowing the convenient isolation of mutations exerting cell-type-specific or tissue-specific effects on white expression). As a result of its properties, the locus has long been a favorite tool of geneticists and molecular biologists. Early studies of the white locus led to the discoveries of sex-chromosome mechanics (Morgan 1910) and dosage compensation (Muller 1950a), and its more recent analysis has led to the discoveries of the P-element transposon (Bingham et al. 1982; Rubin et al. 1982) and the unexpectedly large role of transposon insertion in spontaneous mutation in *Drosophila* (Collins and Rubin 1982; Zachar and Bingham 1982). Seventy years of genetic analysis of white by a very large number of investigators has produced a quite substantial collection of mutant alleles and genetic phenomena relevant to the analysis of transvection effects.

Evidence That white Transcription in Eye Pigment Cells Is Controlled by a Tissue-specific Regulatory Element

Among white mutant alleles are members of a set referred to as the white-spotted (w^{sp}) alleles. The w^{sp}

mutant alleles have a variety of exotic genetic properties (for a review, see Green 1959) and the original molecular analysis of these alleles led to the proposal that they inactivated a genetic element (the w^{sp} genetic element) mapping outside of and 5' to the white transcription unit (Zachar and Bingham 1982; Fig. 1). Subsequent physical analysis of the white transcription unit supported this proposal (Levis et al. 1984; Pirrotta and Brockl 1984; Davison et al. 1985) and analysis of transcription in selected deletion alleles confirmed that the major w^+ transcript is produced by white alleles in which the entire w^{sp} region is deleted (Davison et al. 1985). The results of Davison et al. (1985) place the w^{sp} genetic element outside of and 5' to the primary transcription unit producing the major w^+ transcript by at least 440 bases and probably by 1100-1300 bases (Fig. 1).

The w^{sp} mutant alleles were originally isolated on the basis of their producing a dramatic reduction in adult eye pigmentation (Lindsley and Grell 1968 and references therein). In spite of this, the w^{sp1} and w^{sp2} mutations have no discernible effect on pigment deposition in the other major tissues showing white-dependent pigmentation—larval and adult Malpighian tubules and adult male testis sheath (Davison et al. 1985; results not shown). Moreover, analysis of whole-organism white transcript levels produced by the w^{sp1} insertion mutation (Levis et al. 1984; Pirrotta and Brockl 1984; Davison et al. 1985) and several w^{sp} region deletion mutations (Davison et al. 1985) failed to detect an effect of these mutations on transcription. These various observations led us to propose that the w^{sp} genetic element exerts tissue-specific effects on white locus transcription.

We have tested this hypothesis directly by examining white transcript levels in anterior pupal tissues (adult head precursor tissues) (Fig. 2). (Cell division in immature eye tissue ceases around puparium formation and eye pigment cells are proceeding through terminal differentiation during the pupal stages examined in our experiments [for a recent discussion of eye development, see Campos-Ortega et al. 1979]. white-dependent eye pigment deposition occurs during this terminal differentiation process.) We observe high levels of white transcripts in this tissue fraction from w^+ individuals, and this high level is dependent on an intact w^{sp} genetic element (Fig. 2). These results demonstrate that the w^{sp} genetic element exerts tissue-specific effects on white transcript levels.

The question arises as to the nature of the w^{sp} genetic element. The tissue-specific, apparently long-range cis effects of the element are reminiscent of the enhancers first recognized in vertebrate systems (Banerji et al. 1981, 1983; Gillies et al. 1983; Queen and Baltimore 1983). Our sequence analysis of the w^{sp} region and the w^{sp} mutations revealed a region with statistically significant homology to the functional portions of two vertebrate enhancers in the interval of about 150 bases containing all or a portion of the w^{sp}

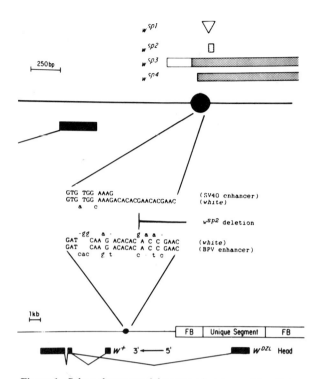

Figure 1. Selected structural features of various white alleles and their transcripts. Solid bars represent exons; attached diagonal lines represent intervening sequences. Transcription is right to left (centromere to telomere). (*Top*) The white locus region immediately surrounding the apparent start site for the major adult white transcript. The horizontal line represents the DNA sequences of the wild-type allele. The stippled triangle indicates the point of insertion of the B104 transposon responsible for the w^{sp1} allele, and the stippled bars indicate the white sequences deleted in the w^{sp2}, w^{sp3}, and w^{sp4} alleles. (The uncertainty in placement of the w^{sp3} break point is indicated by the open portion of the corresponding bar.) The approximate minimal extent of the w^{sp} genetic element is indicated by the solid circle. (*Bottom*) The structure of the w^{DZL} allele. The horizontal line represents white locus sequences, and the open bar represents the w^{DZL} transposon. The segment of *Drosophila* DNA from the second chromosome (unique segment) and FB termini responsible for the transposition of this segment are indicated. The positions of the w^{sp} genetic element (solid circle), the major adult white transcript (initiated to the left of the w^{sp} genetic element), and the composite w^{DZL} head transcript (initiated in the w^{DZL} transposon) are shown. (Note that the single, large exon derived from unique segment sequences could consist of two smaller exons separated by an intervening sequence not detected by our experiments [Bingham and Zachar 1985].) (*Middle*) The sequence homology between the w^{sp} region and two papovavirus enhancers. The white DNA sequence shown is 5' to 3' and is written centromere to telomere. The position of this region of homology in white locus sequences is indicated by the connecting lines. Superposed on the white DNA sequence segment is the position of the 5'-most (centromere proximal) portion of the w^{sp2} deletion.

genetic element (Davison et al. 1985; Fig. 1). Although the interpretation of sequence homologies of this sort is inevitably ambiguous, the various properties of the w^{sp} genetic element (see results) certainly corroborate the hypothesis that the element is an enhancer.

Evidence That the w^{sp} Element Can Activate white Transcription in *trans* in the Case of Somatically Synapsed white Genes

The w^{sp} mutations substantially (though not completely) complement all white mutations that damage the white structural gene without damaging the w^{sp} element (Green 1959; Zachar and Bingham 1982; Levis et al. 1984; Pirrotta and Brockl 1984; Davison et al. 1985). We reinterpret these observations to indicate that in such a complementation test, the w^{sp} element (on the white structural gene mutant chromosome) can act in *trans* to activate the structurally normal white transcription unit (on the w^{sp} mutant chromosome).

This complementation test examines the interaction of somatically synapsed white alleles (on nonrearranged chromosomes). Thus, we interpret these observations to indicate that the w^{sp} element can activate white transcription in *trans* in the case of synapsed chromosomes (Fig. 3).

Evidence That the white Transvection Effect Potentiating the zeste-white Interaction Is Transcriptional and Involves the w^{sp} element

The z^l mutant allele at the zeste locus causes extreme reduction in adult eye pigment deposition in individuals with two intact, synapsed white alleles but does not produce this reduction when an asynapsed white allele is present or when one of the two synapsed white alleles is mutant for the w^{sp} site (Gans 1953; Green 1959; Jack and Judd 1979).

We interpret these earlier observations as follows: Two copies of the w^{sp} element must be present for the extreme z^l mutant pigment deposition phenotype to occur; in the presence of one copy, a wild-type eye pigment deposition phenotype is observed. These two copies of the w^{sp} element must be synapsed with one another when in *trans* to potentiate this z^l mutant phenotype, and we interpret the analysis of small duplications of portions of white (Judd 1961; Green 1963) to indicate that two tandemly repeated copies of the w^{sp} element potentiate the z^l mutant pigment deposition phenotype as effectively as a synapsed *trans* pair of copies. (Structural analysis of the w^l, w^{ch}, and w^e mutations suggests that genetic elements around the presumptive transcription start site might also play a role in potentiating the z^l mutant phenotype; however, ambiguity remains in interpretation of the significance of these mutant alleles [Zachar and Bingham 1982; Levis et al. 1984; Pirrotta and Brockl 1984].)

We find that z^l produces a substantial reduction in white transcript levels in the adult head precursor tissue fractions in which white expression is dependent on the w^{sp} element; this reduction is most extreme in individuals with white genotypes potentiating the z^l mutant eye pigment deposition phenotype (Fig. 2; Bingham and Zachar 1985). The white transcript levels in all tissue fractions examined other than adult head and adult head precursor tissues are unaffected by the z^l mutation (Fig. 2).

Our results indicate that the product of the z^l mutant allele reduces or eliminates the tissue-specific burst of white transcription dependent on the w^{sp} element. (The z^l mutant allele is a neomorphic allele at zeste—i.e., an allele producing a functionally altered product rather than no product [Jack and Judd 1979; Bingham 1980; Gelbart and Wu 1982].) The transcriptional effects of the z^l product are strongest in the presence of two or more copies of the w^{sp} element intimately associated in *trans* (synapsed) or in *cis* (locally, tandemly duplicated). Although the precise molecular rationale for this requirement for multiple, intimately associated copies of the w^{sp} element is unclear (see Discussion), it is presumably this requirement that is largely or entirely responsible for the white transvection effect for potentiation of the z^l mutant phenotype (Fig. 3).

Evidence That w^{DZL} Transvection Effects Are Transcriptional and Involve the w^{sp} Element

The properties of the w^{DZL} mutant allele at white suggested that the allele was a transcriptional regulatory mutation resulting from the insertion of a transposon in or to the right of the rightmost portion of the white locus (Bingham 1980, 1981). This mutant allele is unlike most white mutant alleles in being dominant to w^+, and the dominance of w^{DZL} to w^+ requires that the two alleles be synapsed (Bingham 1980, 1981).

Subsequent physical analysis (Levis et al. 1982; Zachar and Bingham 1982; Fig. 1) demonstrated that w^{DZL} results from insertion to the right of the white locus of a complex composite transposon consisting of a segment of normally single-copy *Drosophila* DNA (referred to below as the unique segment) mobilized by two copies of the FB transposon family. However, early attempts to demonstrate transcriptional effects of this insertion failed (O'Hare et al. 1983).

We have characterized the transcripts produced by the w^{DZL} allele (Bingham and Zachar 1985; Figs. 1 and 2) and find that the w^{DZL} transposon insertion produces complex transcriptional effects in adult head and head precursor tissues. First, we observe a novel transcription unit apparently beginning in the w^{DZL} transposon unique segment, proceeding through white locus sequences and yielding a composite, mature transcript (called the w^{DZL} head transcript) containing both white and w^{DZL} transposon unique segment sequences (Figs. 1 and 2; for the detailed analysis of the structure of the w^{DZL} head transcript, see Bingham and Zachar 1985). Further, when unique segment sequences are used as probe, we detect only the composite transcript in w^{DZL} individuals and no transcript in w^+ individuals (Fig. 2; Bingham and Zachar 1985).

This last observation demonstrates that the unique segment transcription start site does not promote the formation of detectable quantities of a discrete, polyadenylated RNA in tested tissues in its original loca-

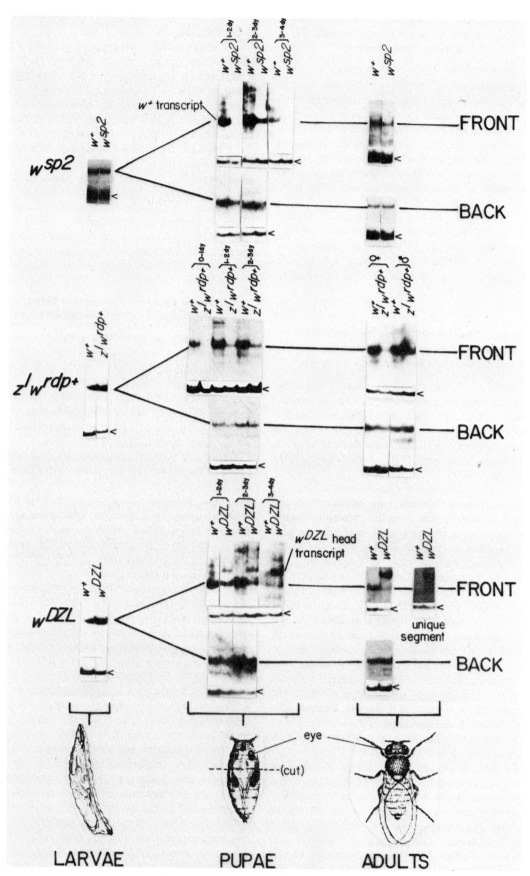

Figure 2. *(See facing page for legend.)*

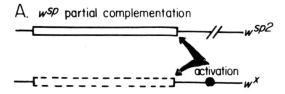

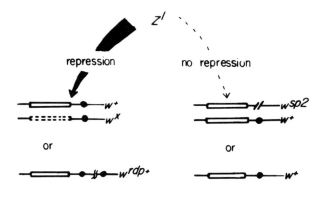

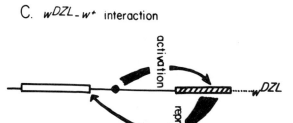

Figure 3. Proposed interpretations of three processes occurring in synapsed *trans* and involving the w^{sp} genetic element. In each panel the major white primary transcription unit is indicated by the open boxes, and the direction of transcription is right to left; a solid boundary for an open box indicates a w^+ allele, and a dashed boundary indicates a mutation altering sequences within the transcription unit. In each panel the solid circle represents the w^{sp} genetic element; deletion of the w^{sp} element in the w^{sp2} allele is indicated by a gap in the line at the position of the w^{sp} element. (All of the w^{sp} mutant alleles [Fig. 1] behave as diagrammed for w^{sp2} in this figure.) The hatched box indicates the w^{DZL} transposon (Fig. 1). Close juxtaposition of pairs of alleles indicates synapsis. The w^{rdp+} allele carries a tandem duplication covering the w^{sp} element; portions of this duplicated segment (at the point of the line break) are not shown. w^x is a generic allele designation indicating any deletion, duplication, base substitution, or transposon insertion mutation functionally damaging white-transcription-unit sequences without disturbing the w^{sp} genetic element. The relevant processes or actions are indicated by the bold arrows and are discussed in detail in the text. In *A*, the activation of the w^{sp2} transcription unit in *trans* presumably also occurs when w^x is replaced by w^+. In *B*, not only the w^+/w^x combination is repressed by z^l but also w^+/w^+ and all tested combinations of w^x/w^x. In *C*, w^x alleles are generally repressed in *trans* just as is the w^+ allele shown. In *B* and *C*, the processes labeled "repression" are equivalently well interpreted as prevention of activation by the w^{sp} genetic element (see text).

Figure 2. Northern gel analysis of the levels and structures of transcripts from various white alleles during late larval, pupal, and adult stages. For all panels (except the single panel labeled "unique segment" at the lower right), the sequence probe was a white locus fragment containing the second white exon and the first 650 bases of the third white exon (Fig. 1). For the unique segment panel, the probe consisted of a fragment of the w^{DZL} transposon with about 1100 bases of homology to the w^{DZL} head transcript (see Fig. 1). At the bottom of each panel (indicated by the caret) is superposed a control experiment for the total amount of polyadenylated RNA. In most cases, this control experimental consisted of running 5% of each sample to be analyzed on a separate gel (in strict parallel with the experimental samples) and probing a transfer of this gel with DNA sequences from the ribosomal protein gene, *rp49* (O'Connell and Rosbash 1984). In a few cases the control experiments consisted of reprobing the original experimental transfer with *rp49* sequences. The major white transcript and the w^{DZL} head transcript are indicated. The results of analysis of each of the three experimental genotypes are organized in the appropriately labeled row. The pupal and adult rows are divided into two subrows. The upper subrow corresponds to the front portion of the organism (including mature or immature eye tissue), and the lower subrow corresponds to the back portion of the organism. For pupae, the boundary between front and back is indicated by the cut line diagramed at the bottom of the figure. For adults, fronts correspond to purified heads, backs to purified thoraxes and abdomens. The leftmost panel in each row shows white transcripts in 4- to 6-day-old larvae (third instar larvae). The central set of panels in each row shows white transcripts in pupae of various ages (indicated in days). For pupal stages, all panels corresponding to the same age group are organized in a column extending through all three rows. The rightmost set of panels shows white transcripts in adult flies. (The panel at the lower right labeled "unique segment" shows adult head transcripts homologous to the unique segment probe.) The $z^l w^{rdp+}$ and w^{DZL} genotypes produce transcriptional effects persisting into mature adulthood. In contrast, the w^{sp2} mutation does not produce reproducible adult effects under our conditions. We note that small effects of the w^{sp2} mutation on adult white transcript levels would not have been detected in these experiments.

tion on the second chromosome (all strains analyzed carry the second chromosome copy of the unique segment). The w^{sp} element apparently activates the white promoter from a substantial distance, and an attractive mechanistic interpretation of the pattern of transcripts promoted from the w^{DZL} transposon is that the w^{sp} genetic element activates transcription from the juxtaposed unique segment transcription start site. Strong circumstantial support for this proposal is found in the details of the tissue and stage specificity of production of the composite w^{DZL} head transcript. Specifically, this transcript begins to accumulate at the same time and in the same tissue fraction as the w^{sp}-dependent burst of white transcription in w^+ individuals (Fig. 2).

The second transcriptional effect of the w^{DZL} transposon insertion that we observe in adult head and head precursor tissues is a substantial reduction in levels of the w^+ transcript (Fig. 2; Bingham and Zachar 1985). This reduction occurs in the tissue fraction and at the time when the w^{sp}-dependent burst of white transcription occurs and when the composite w^{DZL} head transcript is being made (Fig. 2). Equivalently, in all tested tissue fractions in which the w^{sp} element is not implicated in white transcription, the w^{DZL} unique segment promoter is silent and white is transcribed normally.

Collectively, we interpret these results as follows: The w^{sp} element activates transcription from the w^{DZL} transposon unique segment promoter and, for some reason, this interaction prevents or antagonizes activation of the white promoter by the w^{sp} element (Fig. 3). The mutant eye color phenotype presumably results from depression in the levels of authentic white message (presumably, the composite w^{DZL} head transcript does not produce a functional white protein product). (We note that the complex interaction between white and the w^{DZL} transposon unique segment might be a molecular paradigm for the effects of juxtaposed sequences on white expression suspected on indirect genetic grounds [Judd 1974; Hazelrigg et al. 1984].)

As above, our results allow us to reinterpret the results of earlier genetic analysis (Bingham 1980, 1981). We interpret the synapsis-dependent dominance of w^{DZL} over w^+ to indicate that the w^{sp}-dependent activation of the unique segment promoter is associated with prevention of activation of the white promoter not just in *cis*, but also in *trans* when the interacting alleles are synapsed.

The detailed molecular mechanism of the complex interaction between the white transcription unit, the w^{DZL} transposon, and the w^{sp} genetic element in *cis* and in synapsed *trans* is obscure (see Discussion). However, it is clear that the w^{sp} element is likely to be the pivotal element in the w^{DZL} transvection effect.

DISCUSSION

Our results demonstrate that the w^{sp} genetic element is likely to be the central actor in all known transvection effects at the white locus. Moreover, several of our observations strongly suggest that the w^{sp} element is a tissue-specific enhancer. Based on several items of evidence to be discussed below, we propose that our observations can be generalized and that enhancers or enhancer-like elements (tissue-specific, long-range, *cis*-acting regulatory elements) are responsible for all transvection effects at all loci. For economy, we refer to this generalized hypothesis as the "enhancer hypothesis" for the molecular basis of transvection.

Evidence That the white Locus Case Can Be Generalized to Account for All Transvection Effects

Several sets of observations suggest to us that the molecular basis of transvection is the same at all loci and, thus, that our analysis at white can be generalized to account for all transvection effects.

First, the zeste locus participates in transvection at all three loci where the appropriate genetic analysis has been attempted.

1. white: The allelic state of the zeste locus determines whether the w^{sp} partial complementation reaction (Fig. 3; Results) will occur (Babu and Bhat 1980). Moreover, the z^l allele at zeste causes the repression of white transcription that is apparently responsible for the synapsis-dependent genetic phenomenon known as the zeste-white interaction (see Results). Further, the w^{DZL} allele (see Results) interacts specifically and synergistically with the z^l allele as assessed by adult eye pigmentation phenotype (Bingham 1980).

2. bithorax: Synapsis-dependent complementation between selected mutant alleles at the bithorax locus (Lewis 1954) is prevented in the presence of one class of mutant alleles at the zeste locus (Kaufman et al. 1973).

3. decapentaplegic: Synapsis-dependent complementation between selected mutant alleles at the decapentaplegic locus is prevented in the presence of various mutant alleles at the zeste locus (Gelbart and Wu 1982). (We note that currently available information about transvection effects at bithorax and decapentaplegic is consistent with our proposal for the mechanistic basis of transvection. The detailed molecular analysis of both these loci should provide important future tests of our proposal.)

Second, recent molecular analyses of the *Sgs-4* locus can be used to reinterpret the results of earlier genetic analysis to suggest a situation strikingly analogous to one of the cases we have characterized at the white locus. Specifically, Korge (1981 and references therein) isolated and characterized a special group of alleles at the *Sgs-4* locus that did not normally produce detectable levels of *Sgs-4* protein product (a salivary gland glue protein) but could be induced to produce protein by wild-type alleles at the locus in *trans*; however, this *trans* restoration of expression appeared to require somatic synapsis, and Korge (1981) recognized the possibility that this represented a transvection effect. Recent molecular analysis of Korge's transvecting mutant alleles has shown them to be deletion mutations affecting sequences 300-500 bases 5' to the Sgs-4 transcription start site and to cause depressed levels of Sgs-4 transcripts (Muskavitch and Hogness 1982; McGinnis et al. 1983). We interpret Korge's results to suggest that the effects of the absence of a long-range, cis-acting regulatory element at Sgs-4 (a presumptive enhancer-like element) can be partially overcome by providing that element in trans in the synapsed (but not in the asynapsed) configuration. This interpretation is precisely analogous to our interpretation of the partial complementation reaction involving the presumptive w^{sp} enhancer at the white locus (see Results; Fig. 3).

Mechanistic Implications of the Enhancer Hypothesis

Significant evidence exists suggesting that the widely recognized capacity of enhancers to increase steady-state transcript levels results from their capacity to increase the frequency of initiation of transcription (e.g., see Treisman and Maniatis 1985; Weber and Schaffner 1985); however, the detailed mechanism of this increase in initiation is still quite obscure. Our proposal that enhancers represent the elements responsible for synapsis-dependent genetic interactions (transvection effects) suggests that transvection effects might provide unique insights to the molecular basis of enhancer function. Although rigorously informative attempts to capitalize on the potential power of this approach have yet to be undertaken, several reflections may be of value at this early stage in the analysis.

First, one model for enhancer function supposes that enhancers are sites at which the degree of supercoiling of a domain of DNA sequence surrounding the enhancer can be controlled. Changes in the degree of supercoiling initiated at these sites are supposed to be responsible for the activation of promoters in the vicinity. This model is motivated in large part by the compelling evidence that the degree of supercoiling can influence promoter function in E. coli (Sternglanz et al. 1981). This model for enhancer function has received some additional circumstantial support from the observations that in one case a topoisomerase is a conspicuous nuclear structural element (Berrios et al. 1985) and that in another case a different topoisomerase is apparently associated with some transcribing genes (Fleishmann et al. 1984). Although the degree of supercoiling may vary when genes are activated, our proposal that enhancers are responsible for synapsis-dependent genetic interactions argues against the hypothesis that enhancers are the cis-acting elements responsible for changes in the degree of supercoiling. Specifically, with this model the enhancer hypothesis for transvection would require, in the simplest case, that both members of a synapsed pair of alleles be members of the same domain with respect to degree of supercoiling. This seems implausible to us.

A second class of models for the molecular basis of enhancement proposes that enhancers represent sites of facilitated binding of one or more limiting transcription factors. Subsequent migration of bound factor(s) along the DNA strand away from such facilitated entry sites is energetically neutral, allowing local accumulation of the factor. For a model of this sort to be consistent with our proposal, it is required that both members of synapsed allele pairs be part of the same binding surface for the transcription factor in question. Though this is probably conceivable (depending on the unknown nature of the contacts between synapsed allele pairs), it seems unlikely to us.

A third class of models supposes that enhancers function by controlling nuclear localization of themselves and, thus, of appended sequences. Several lines of evidence support this hypothesis circumstantially, including the observations that the folding patterns of polytene chromosomes are substantially controlled in a tissue-specific fashion (Agard and Sedat 1983) and that apparently expressed chromosome sets in certain plant hybrids are localized in the nuclear periphery, whereas apparently silent chromosome sets are centrally localized (Bennet 1984). This class of models efficiently predicts our proposal that enhancers exert synapsis-dependent trans effects as well as long-range cis effects (see the next paragraph). (A ~6.5-kb gene like white is about 600 Å long as a 300-Å-diameter chromatin fiber, and the diameter of a diploid Drosophila nucleus is on the order of 10,000-50,000 Å. Thus, genes like white can undergo highly significant spatial displacement in an interphase nucleus.)

According to this third model the capacity of the presumptive w^{sp} enhancer to activate white transcription in synapsed-trans as well as in cis is explained by assuming that a synapsed allele pair undergoes spatial displacement as a unit so that an intact w^{sp} element on one of the two synapsed alleles can direct that migration. The rationale for the w^{DZL} and z^l effects on white transcription with this model is less clear; however, we note that these effects can be explained by assuming that these mutations (in concert with the presumptive w^{sp} enhancer) actively antagonize positioning of the white locus appropriate for efficient expression of the white promoter. (Note that both w^{DZL} and z^l "repression" of white transcription [Figs. 2 and 3] are equivalently well interpreted as prevention of activation of white transcription by the presumptive w^{sp} enhancer.)

There are, of course, numerous other models for enhancer function. It seems likely that the analysis of transvection will contribute significantly to the definition of the basis of enhancer function whatever that basis ultimately proves to be.

Selected Technical Implications of the Enhancer Hypothesis

Our analysis of the molecular basis of transvection at the white locus has an important technical implication for the isolation of regulatory mutations and the recognition of transvection effects in Drosophila. Our results suggest that conventional genetic analysis systematically excludes mutations inactivating enhancer-like regulatory elements for at least three reasons.

Firstly and most importantly, mutations inactivating enhancer-like regulatory elements are systematically excluded because of their extensive complementation with alleles inactivating structural genetic information at the same locus. Available evidence suggests that spontaneous and X-ray-induced mutations inactivating the presumptive w^{sp} enhancer occur at only about 3–10% of the rate of mutations affecting structural genetic information at the locus — proportions expected from the relative target sizes of these segments of the locus (Zachar and Bingham 1982). Thus, when a mutant allele is isolated at a previously unrecognized locus, this allele will usually result from inactivation of structural gene sequences. When this reference allele is used to screen for new alleles at the locus, mutations inactivating enhancer-like regulatory elements will tend to be excluded because of complementation with the reference allele. (Some mutant screens have been carried out by selecting new mutations allelic to multilocus deletions. Such studies should escape this bias against regulatory mutations. However, such mutant screens account for a relatively small fraction of the mutant alleles that have been subjected to the genetic experimentation appropriate to detect transvection effects.)

Secondly, the presumptive enhancerless white alleles (w^{sp} alleles) are slightly leaky. They allow continued expression of white in eye pigment cells sufficient to allow deposition of several percent of the wild-type level of pigment. Some genetic screens fail to identify mutant alleles that do not more completely prevent expression.

Thirdly, our analysis suggests that white is under the control of the presumptive w^{sp} enhancer in only one tissue fraction (adult head precursor tissues) and is expressed independently of the w^{sp} element in various other tissues (see Results; Davison et al. 1985). A number of genes do not produce cell-autonomous phenotypes (e.g., rosy and vermilion; Beadle and Ephrussi 1936); inactivation of tissue-specific regulatory elements at such loci would generally elude experimental detection.

Collectively, these considerations suggest that the apparent rarity of loci showing recognizable transvection effects may be an artifact of the genetic properties of enhancers. The enhancer hypothesis predicts that transvection effects will be quite common when the appropriately refined approaches are used to detect them.

The Enhancer Hypothesis and the Evolution of Somatic Synapsis

Muller (1950b) originally showed that "recessive" lethal mutations are generally not strictly recessive but rather have a slight dominant effect resulting in reduction of viability of the heterozygous carrier by an average of about 2%. Our analysis of transvection effects suggests that this small dominant effect on fitness would be partially or entirely ameliorated for the subset of mutant alleles inactivating enhancer-like elements in individuals possessing somatic synapsis. This partial rescue of viability for enhancerless alleles would not occur in individuals lacking somatic synapsis and represents an individual selective advantage of somatic synapsis. In finite populations (where newly arising, rare alleles are possibly often lost by stochastic processes), the individual advantage of synapsis (above) represents a reduction in the total number of genetic deaths.

We speculate that the synapsis-dependent properties of enhancer-like regulatory elements account for the evolution and persistence of somatic synapsis. It will be of interest to carry out a detailed theoretical analysis of the implications of synapsis-dependent genetic interactions for the evolution of somatic synapsis.

On the Occurrence of Transvection in Vertebrates and Other Nonarthropods

Our analysis suggests that transvection effects, resulting from the properties of a generally occurring class of regulatory elements, should occur in all organisms showing somatic synapsis. Moreover, the evolutionary arguments made in the preceding section suggest that somatic synapsis may be a much more generally occurring phenomenon than has been generally appreciated.

Somatic synapsis is recognized in *Drosophila* primarily on the basis of two classes of observations. The first is the presence of polytene chromosomes. In practice, the clear definition of polyteny requires polyploidy on the order of 500-fold to 1000-fold. This level of polyploidy is not encountered in nonarthropods commonly studied, and polytene chromosomes would not be expected to be observable in these nonarthropods even if they exhibit somatic synapsis. (One case in which very high levels of polyploidy are achieved outside of Arthropoda is in ciliates during amplification of the sexual nucleus in the early stages of macronucleus formation; under these conditions apparently polytene chromosomes are observed [Spear and Lauth 1976].)

The second principal observation leading to the recognition of somatic synapsis in *Drosophila* is the existence of transvection effects themselves. For the reasons discussed above, most conventional approaches to genetic analysis tend to exclude the mutations that reveal transvection effects. The initial recognition of these effects in *Drosophila* is entirely attributable to the existence of a very large collection of mutations (arising spontaneously) that were recognized serendipitously as unselected homozygous segregants in the small populations represented by the vial stocks in which most *Drosophila* strains are carried. The existence of a large number of mutations so isolated occurs in no experimental genetic system other than *Drosophila*. Thus, genetic evidence for somatic synapsis is not currently expected in most organisms even if these organisms undergo somatic synapsis.

Collectively, these considerations suggest that somatic synapsis would not have been detected in vertebrates and other nonarthropods even if it occurred. It will be of interest to learn whether these groups show transvection effects. Perhaps this issue will be resolved as our understanding of the basis of these effects in *Drosophila* is refined.

ACKNOWLEDGMENTS

We are grateful to M. Ashburner, W. Eanes, W. Gelbart, P. Hearing, A. Levine, and H. Weintraub for helpful discussions and to J. Goland for help in carrying out the experiments described. This work was supported by grant GM32003 from the National Institutes of Health to P.M.B.

REFERENCES

AGARD, D.A. and J.W. SEDAT. 1983. Three-dimensional architecture of a polytene nucleus. *Nature* **302:** 676.

ASHBURNER, M. 1970. The genetic analysis of puffing in polytene chromosome of *Drosophila*. *Proc. R. Soc. Lond. B.* **176:** 319.

———. 1971. A prodromus to the genetic analysis of puffing in *Drosophila*. *Cold Spring Harbor Symp. Quant. Biol.* **35:** 533.

BABU, P. and S.G. BHAT. 1980. Effect of *zeste* on *white* complementation. In *Development and neurobiology of Drosophila* (ed. O. Siddiqi et al.), p. 35. Plenum Press, New York.

BANERJI, J., L. OLSON, and W. SCHAFFNER. 1983. A lymphocyte-specific cellular enhancer is located downstream of the joining region in immunoglobulin heavy chain genes. *Cell* **33:** 729.

BANERJI, J., S. RUSCONI, and W. SCHAFFNER. 1981. Expression of a β-globin gene is enhanced by remote SV40 DNA sequences. *Cell* **27:** 299.

BEADLE, G. and B. EPHRUSSI. 1936. Development of eye colors in *Drosophila*: Transplantation experiments with supressor of vermilion. *Proc. Natl. Acad. Sci.* **22:** 536.

BENNET, M.D. 1984. Nuclear architecture and its manipulation in plant breeding. In *Gene manipulation in plant improvement* (ed. J.P. Gustafson), p. 469. Plenum Press, New York.

BERRIOS, M., N. OSHEROFF, and P.A. FISHER. 1985. In situ localization of DNA topoisomerase II, a major polypeptide component of the *Drosophila* nuclear matrix fraction. *Proc. Natl. Acad. Sci.* **82:** 4142.

BINGHAM, P.M. 1980. The regulation of white locus expression: A dominant mutant allele at the white locus of *Drosophila*. *Genetics* **95:** 341.

———. 1981. A novel, dominant allele at the *white* locus of *Drosophila* is mutable. *Cold Spring Harbor Symp. Quant. Biol.* **45:** 519.

BINGHAM, P.M. and Z. ZACHAR. 1985. Evidence that two mutations, w^{DZL} and z^{l}, that alter the synapsis-dependent genetic behavior of *white* are transcriptional regulatory mutations. *Cell* **40:** 819.

BINGHAM, P.M., M.G. KIDWELL, and G.M. RUBIN. 1982. The molecular basis of PM hybrid dysgenesis: The role of the P element, a P strain-specific transposon family. *Cell* **29:** 995.

BINGHAM, P.M., R. LEVIS, and G.M. RUBIN. 1981. Cloning of DNA sequences from the *white* locus of *D. melanogaster* by a novel and general method. *Cell* **25:** 693.

CAMPOS-ORTEGA, J.A., G. JURGENS, and A. HOFBAUER. 1979. Cell clones and pattern formation: Studies on *sevenless*, a mutant of *Drosophila melanogaster*. *Wilhelm Roux's Arch. Dev. Biol.* **186:** 27.

COLLINS, M. and G.M. RUBIN. 1982. Structure of the *Drosophila* mutable allele, *white-crimson*, and its *white-ivory* and wild-type derivatives. *Cell* **30:** 71.

DAVISON, D., C.H. CHAPMAN, C. WEDEEN, and P.M. BINGHAM. 1985. Genetic and physical studies of the portion of the white locus involved in transcriptional regulation and in synapsis-dependent interactions in *Drosophila* adults. *Genetics* **110:** 479.

FLEISCHMANN, G., G. PFLUGFELDER, E.K. STEINER, K. JAVAHERIAN, G.C. HOWARD, J.C. WANG, and S.C.R. ELGIN. 1984. *Drosophila* DNA topoisomerase I is associated with transcriptionally active region of the genome. *Proc. Natl. Acad. Sci.* **81:** 6958.

GANS, M. 1953. Étude génétique et physiologique du mutant z de *Drosophila melanogaster*. *Bull. Biol. Fr. Belg.* (suppl.) **38:** 1.

GELBART, W.M. and C.-T. WU. 1982. Interactions of zeste mutations with loci exhibiting transvection effects in *Drosophila melanogaster*. *Genetics* **102:** 179.

GILLIES, S.D., S.L. MORRISON, V.T. OI, and S. TONEGAWA. 1983. A tissue-specific transcription enhancer element is located in the major intron of a rearranged immunoglobulin heavy chain gene. *Cell* **33:** 717.

GREEN, M.M. 1959. Spatial and functional properties of pseudoalleles at the white locus in *Drosophila melanogaster*. *Heredity* **13:** 302.

———. 1963. Unequal crossing over and the genetical organization of the white locus of *Drosophila melanogaster*. *Z. Indukt. Abstammungs. Vererbungsl.* **94:** 200.

HAMMOND, M.P. and C.D. LAIRD. 1985. Chromosome structure and DNA replication in nurse and follicle cells of *Drosophila melanogaster*. *Chromosoma* **91:** 267.

HAZELRIGG, T., R. LEVIS, and G.M. RUBIN. 1984. Transformation of *white* locus DNA in *Drosophila*: Dosage compensation, zeste interaction and position effects. *Cell* **36:** 469.

HU, N. and J. MESSING. 1982. The making of strand-specific M13 probes. *Gene* **17:** 271.

JACK, J.W. and B.H. JUDD. 1979. Allelic pairing and gene regulation: A model for the zeste-white interaction in *Drosophila melanogaster*. *Proc. Natl. Acad. Sci.* **76:** 1368.

JUDD, B.H. 1961. Formation of duplication-deficiency products by asymmetrical exchange within a complex locus of *Drosophila melanogaster*. *Proc. Natl. Acad. Sci.* **47:** 545.

———. 1974. Genes and chromosomes of *Drosophila*. In *The eukaryotic chromosome* (ed. W.J. Peacock and R.D. Brock), p. 169. Australian National University Press, Canberra.

KAUFMAN, T.C., S.E. TASAKA, and D.T. SUZUKI. 1973. The interaction of two complex loci, zeste and bithorax, in *Drosophila melanogaster*. *Genetics* **75:** 299.

KORGE, G. 1981. Genetic analysis of the larval secretion gene Sgs-4 and its regulatory chromosome sites in *Drosophila melanogaster*. *Chromosoma* **84:** 373.

LEVIS, R., M. COLLINS, and G.M. RUBIN. 1982. FB elements are the common basis of the instability of the w^{DZL} and w^c *Drosophila* mutations. *Cell* **30:** 551.

LEVIS, R., K. O'HARE, and G.M. RUBIN. 1984. Effects of transposable element insertions on RNA encoded by the *white* gene of *Drosophila*. *Cell* **36:** 471.

LEWIS, E.B. 1954. The theory and application of a new method of detecting chromosomal rearrangement in *Drosophila melanogaster*. *Am. Nat.* **88:** 225.

LIFSCHYTZ, E. and D. HAREVEN. 1982. Heterochromatin markers: Arrangement of obligatory heterochromatin, histone genes and multisite gene families in the interphase nucleus of *D. melanogaster*. *Chromosoma* **86:** 443.

LINDSLEY, D.L. and E.H. GRELL. 1968. Genetic variations of *Drosophila melanogaster*. *Carnegie Inst. Wash. Publ.* no. 627.

MCGINNIS, W., A. SHERMOEN, J. HAEMSKERK, and S. BECKENDORF. 1983. DNA sequence changes in an upstream DNase-I hypersensitive region correlates with reduced gene expression. *Proc. Natl. Acad. Sci.* **80:** 1063.

MORGAN, T.H. 1910. Sex-limited inheritance in *Drosophila*. *Science* **32:** 120.

MULLER, H.J. 1950a. Evidence of the precision of genetic adaptation. *Harvey Lect.* **18:** 165.

———. 1950b. Our load of mutations. *Am. J. Hum. Genet.* **2:** 111.

MUSKAVITCH, M.A.T. and D.S. HOGNESS. 1982. An expandable glue gene that encodes a *Drosophila* glue protein is not expressed in variants lacking remote upstream sequences. *Cell* **29:** 1041.

O'CONNELL, P. and M. ROSBASH. 1984. Sequence, structure and codon preference of the *Drosophila* ribosomal protein 49 gene. *Nucleic Acid Res.* **12:** 5495.

O'HARE, K., R. LEVIS, and G.M. RUBIN. 1983. Transcription of the *white* locus in *Drosophila melanogaster*. *Proc. Natl. Acad. Sci.* **80:** 6917.

PIRROTTA, V. and C. BROCKL. 1984. Transcription of the *Drosophila white* locus and some of its mutants. *EMBO J.* **3:** 563.

QUEEN, C. and D. BALTIMORE. 1983. Immunoglobulin gene transcription is activated by downstream sequence elements. *Cell* **33:** 741.

RUBIN, G.M., M.G. KIDWELL, and P.M. BINGHAM. 1982. The molecular basis of PM hybrid dysgenesis: The nature of induced mutation. *Cell* **29:** 987.

SPEAR, B.B. and M.R. LAUTH. 1976. Polytene chromosomes of *Oxytricha*: Biochemical and morphological changes during macronuclear development in a ciliated protozoan. *Chromosoma* **54:** 1.

STERNGLANZ, R., S. DINARDO, D. VOELKEL, Y. NISHIMURA, Y. HIROTA, K. BECHERER, L. ZUMSTEIN, and J.C. WANG. 1981. Mutations in the gene coding for *Escherichia coli* DNA topoisomerase I affect transcription and transposition. *Proc. Natl. Acad. Sci.* **78:** 2747.

TREISMAN, R. and T. MANIATIS. 1985. Simian virus 40 enhancer increases number of RNA polymerase II molecules on linked DNA. *Nature* **315:** 72.

WEBER, F. and W. SCHAFFNER. 1985. Simian virus 40 enhancer increases RNA polymerase density with the linked gene. *Nature* **315:** 75.

ZACHAR, Z. and P.M. BINGHAM. 1982. Regulation of *white* locus expression: The structure of mutant alleles at the *white* locus of *Drosophila melanogaster*. *Cell* **30:** 529.

The 68C Glue Puff of *Drosophila*

E.M. Meyerowitz, M.A. Crosby, M.D. Garfinkel, C.H. Martin, P.H. Mathers,
and K. VijayRaghavan

Division of Biology, California Institute of Technology, Pasadena, California 91125

Drosophila melanogaster begins its life as a fertilized egg, which undergoes embryogenesis within an egg shell for about a day and then hatches from the egg as a first instar larva. The wormlike larva eats yeast for a day, then sheds its first instar skin and crawls out as a second instar larva. One day later another molt occurs, resulting in a third instar larva. The third instar stage lasts approximately 40 hours at 25°C; toward the end of this stage the larva crawls out of its medium and onto a dry surface. After a few hours of wandering, it stops, contracts, and secretes a protein glue from its salivary glands that hardens and attaches the larva to its substrate (Fraenkel and Brookes 1953). The affixed larva then tans its third larval instar cuticle into a puparial case, and some hours later pupates within this case. Complete metamorphosis follows pupation, and after several days the pupal case is opened and a new adult emerges.

The third instar salivary glands that are the source of the secreted glue consist of a pair of separate lobes connected by a common anterior duct. Each lobe is a blind sac with a simple lumen; the lumen communicates with the anterior duct, and through the duct with the larval mouth, and the outside world. Each lobe contains approximately 130 giant cells, which were set aside during embryogenesis, and have enlarged without cell division since that stage. The enlargement of the cells parallels an increase in the DNA content of each cell: As the gland develops the chromosomes of its cells replicate up to 11 times, with the replicated chromatids staying aligned in register, to form the familiar polytene chromosomes that exist in many dipteran tissues (Berendes and Ashburner 1978). These chromosomes are enormous, with a diameter of around 5 μm, and lengths of hundreds of micrometers. They show an abundance of cytological detail. One prominent cytological feature is the pattern of bands: Bands are dark-staining transverse stripes on the chromosomes that sit in a constant relation to the genetic map, and thus to the DNA sequence. They are unchanging features, with the same banding pattern recognizable in all stages of salivary gland development in which the chromosomes are large enough to allow the banding pattern to be resolved. Individual bands can be identified, and each is named according to a standard system (Bridges 1935). In addition to bands, polytene chromosomes have puffs, transient structures which are local regions of chromatin decondensation visible in the light microscope as areas of expanded diameter and indistinct banding. Puffs are normally sites of highly active transcription (Ashburner and Berendes 1978). The puffs seen in midthird instar larvae, when the salivary chromosome banding pattern is first distinct enough to allow individual puff locations to be ascertained, are called the intermolt puffs. There are about 10 of them, with the most prominent found in band region 3C on the X chromosome, 25B on the left arm of the second chromosome, 68C on the left arm of the third chromosome, and 90BC on the right arm of the third chromosome. About five hours before puparium formation these puffs regress, with restoration of the normal diameter and sharp banding pattern of the previously puffed regions. Simultaneous with the regression of the intermolt puffs is the appearance of a new set of puffs called the early puffs; these regress by the time of puparium formation, and are followed by a set of puffs called the late puffs. The changes in puffing pattern reflect changing patterns of gene expression in the developing salivary glands (Ashburner 1967, 1973).

The best characterized set of puffs is the intermolt set. Both genetic and molecular experiments have shown that these puffs contain the structural genes for the polypeptides that comprise the salivary gland glue (Korge 1975, 1977; Akam et al. 1978; Muskavitch and Hogness 1980; Velissariou and Ashburner 1980, 1981; Crowley et al. 1983; Gautam 1983; Guild and Shore 1984). This glue consists of at least eight polypeptides that start to be synthesized in the salivary glands in early to midthird instar, and stop synthesis at the time of puparium formation (Beckendorf and Kafatos 1976). The signals that induce these puffs are not known; they regress in response to a steroid hormone, ecdysterone. This response occurs both in vivo, as a result of release of the hormone from the larval ring gland, and in vitro, when midthird instar larval salivary glands are cultured in a medium containing ecdysterone (Ashburner 1973). If protein synthesis inhibitors are added to cultured intermolt salivary glands along with ecdysterone, the regression of at least three of the intermolt puffs (3C, 25B, and 68C) still occurs, indicating that the regression is a direct response to the hormone, and does not require intermediate steps of gene expression (Ashburner 1974).

The intermolt puffs thus represent a set of developmentally and hormonally controlled genes that express in a tissue-specific manner and code for known products of known function. In addition, their expression is accompanied by visible and striking changes in chromatin structure. To understand better the mechanisms of tissue-specific and developmentally specific gene expression, and the relation of puffing to gene expression, we are pursuing molecular and genetic studies of the 68C puff.

DNA Sequence Organization of the 68C Puff

The cloning of the 68C puff DNA started with isolation of cDNA clones representing RNAs specific to and abundant in midthird instar salivary glands. These were screened by in situ hybridization to salivary gland polytene chromosomes and five major classes of cDNA inserts were found. Clones within each class cross-hybridize, clones in different classes do not. One class hybridizes to the 3C glue puff (which codes for the sgs-4 glue polypeptide) in in situ hybridization experiments, and one to the 90BC glue puff (the site of the structural gene of the sgs-5 glue protein). The remaining three classes of clones hybridize in situ to 68C (Wolfner 1980). Each of the three 68C cDNA clone classes hybridizes to a different third instar salivary gland RNA; the three RNAs are 360, 320, and 1120 nucleotides in length. The RNAs are all polyadenylated and polysomal (Meyerowitz and Hogness 1982). By comparing conceptual translations of the cDNA sequences with partial amino acid sequences of different glue polypeptides, it was found that each of the 68C RNAs codes for a different glue component (Crowley et al. 1983). The 1120-nucleotide RNA is the mRNA for the sgs-3 glue polypeptide, which had previously been shown by genetic mapping to be coded in or near the 68C puff (Korge 1975; Akam et al. 1978). The other two RNAs coded for two previously undescribed glue fractions: the 320-base RNA is the mRNA for the sgs-7 polypeptide, and the 360-base RNA is for the sgs-8 polypeptide (Crowley et al. 1983).

The amino acid sequences of the three 68C glue proteins are related: All three have a 23-residue aminoterminal signal peptide that is cleaved off before secretion of the mature protein, and all of the proteins have carboxyterminal regions of about 50 amino acids that show considerable homology with each other. In addition to these two modules, the sgs-3 protein has a third, central, module that is 234 amino acids long and largely comprised of tandem repeats of the five amino acids Pro-Thr-Thr-Thr-Lys. Thus, the secreted sgs-7 and sgs-8 peptides are small and similar to each other, whereas the sgs-3 protein is much larger, and contains over 40% threonine (Garfinkel et al. 1983). sgs-3 is also different from its two relatives in its high degree of glycosylation, with sugar molecules likely attached to threonine residues in the repetitive central module (Beckendorf and Kafatos 1976; Crowley et al. 1983).

Clones representing the chromosomal DNA of the 68C puff were collected by screening a genomic library with the cDNA clones and then with the newly obtained genomic clones. Overlapping clones containing over 50,000 bp of puff DNA were eventually characterized (Meyerowitz and Hogness 1982; Fig. 1). This DNA represents all of the sequences found in the 68C puff: If the leftmost fragment (the 2057 fragment in Fig. 1) from this chromosomal walk is biotinylated, hybridized in situ to surface-spread polytene chromosomes, then bound by anti-biotin antibody which is in turn bound with gold sphere-labeled anti-antibody, the gold spheres are seen in the transmission electron microscope to be positioned over the 68C4 band, which is just to the left of the puff. Similar experiments show that the right end of the 68C chromosomal walk (the 2058 fragment in Fig. 1) is in the 68C7 band, just to the right of the puff (Kress et al. 1985). The only region of the 68C puff that contains genes coding for abundant salivary gland RNAs is the central 5000 bp.

Hybridization and DNA sequencing studies show the structure of this region (Fig. 2). The *Sgs7* and *Sgs8* genes are transcribed as a divergent pair, with less than 500 bp separating their 5' ends. *Sgs3* is transcribed from the same DNA strand as *Sgs7*, with almost 2000 bp separating the poly(A) addition site of the *Sgs7* RNA from the 5' end of the *Sgs3* gene. The *Sgs7* and *Sgs3* genes each have their own promoter, and their RNA products are not processed from a common precursor. Each of the genes has a single, small intervening sequence. The introns are in comparable positions in the three genes, all of them occurring between the second

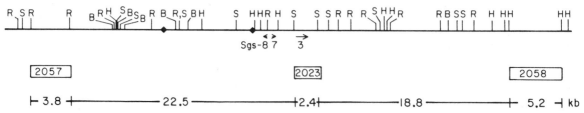

Figure 1. Restriction map of 68C puff DNA. The 68C puff is bounded on the left by the 3.8-kbp 2057 *Eco*RI fragment, and on the right by the 5.2-kbp 2058 *Hin*dIII fragment. The only genes in the puff DNA known to express in salivary glands are the *Sgs3*, *Sgs7*, and *Sgs8* structural genes. The restriction endonuclease recognition sites shown are: R, *Eco*RI; S, *Sal*I; B, *Bam*HI; H, *Hin*dIII. The diamonds on the map represent the termini of a transposable element of the *roo* family (Meyerowitz and Hogness 1982) that is present at this location in some strains, but not others. As drawn, the centromere is to the right and the tip of the left arm of the third chromosome to the left.

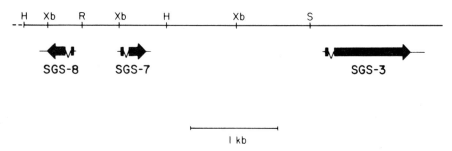

Figure 2. Salivary gland transcripts at 68C. The central portion of the 68C puff DNA contains the three 68C glue genes, transcribed in the directions shown. Translated portions of each glue RNA are shown as heavy lines, the upstream and downstream untranslated regions and the intervening sequences are shown as lighter lines. Above the transcripts is a restriction map of the genomic region containing the genes. The restriction endonuclease recognition sites shown are: H, HindIII; Xb, XbaI; R, EcoRI; S, SalI; Xh, XhoI.

and third bases of the tenth codon. The *Sgs8* intron is 69 nucleotides, the one in *Sgs7* 66 nucleotides, and the *Sgs3* intervening sequence 73 bases long (Garfinkel et al. 1983). There is no detectable homology between the three introns, despite the similarity of the protein-coding portions of the mRNAs.

RNA hybridization experiments show that the three 68C glue RNAs are coordinately expressed. They are only found in third instar larvae, and only in salivary glands (Meyerowitz and Hogness 1982). Gel blots of RNA from salivary glands of individual midthird instar larvae show that any individual larva has either none or all three of the 68C RNAs present at any time; no individual (out of 25) had only one or two of the RNAs present (J. Kendall et al., unpubl.). Thus, the RNAs begin accumulating simultaneously. The three RNAs disappear together as well, just at the time of puparium formation. Therefore, in addition to the question of tissue- and time-specific regulation of each of the RNAs, there is a question of the mechanism of the precise coordination of expression of the three genes in the 68C cluster.

Factors Required in *trans* for 68C Regulation

Knowledge of the sequence organization of the 68C puff has allowed analysis of the sequences required in *cis*, and the factors required in *trans*, for the normal regulation of the 68C genes. Two *trans*-acting factors are known. The first is ecdysterone, which not only causes puff regression, as described above, but is also required for the initiation of accumulation of the 68C RNAs, and for the later cessation of their synthesis.

The evidence that ecdysterone induces RNA accumulation at 68C comes largely from work with a temperature-sensitive X-chromosomal mutation *l(1)su(f)^{ts67g}*. At 18°C flies homozygous or hemizygous for this mutation are normal. At 30°C, they do not synthesize ecdysterone. If mutant larvae are raised to the restrictive temperature near the time of the second to third larval instar molt, they fail to accumulate any of the 68C RNAs in their salivary glands. If such larvae are then fed ecdysterone, accumulation of the RNAs begins (Hansson and Lambertsson 1983). Thus, the aspect of the *l(1)su(f)^{ts67g}* phenotype responsible for failure of 68C glue gene expression is the steroid, and the hormone is required for initiation of 68C gene expression. In these ecdysteroneless mutants, the 68C puff is present at the restrictive temperature, despite the absence of accumulation of the glue RNAs (Hansson et al. 1981).

Ecdysterone is also the signal that stops transcription of the three genes. If midthird instar salivary glands are dissected from larvae and cultured in a simple medium, they continue to synthesize the 68C glue RNAs for hours. This is demonstrated in pulse-labeling experiments, where salivary glands are cultured for 2 hours without label, then for 10 or 15 minutes in the presence of tritiated uridine or adenosine. RNA is then extracted from the glands and purified, and each of the individual 68C glue mRNAs is isolated from this mixture by hybrid selection (Crowley and Meyerowitz 1984). Scintillation spectrometry shows that each of the three RNAs is labeled in such an experiment, and thus is still being synthesized at high levels in the cultured glands. If ecdysterone is added to the culture medium to a concentration of 10^{-5} M for as little as 15 minutes just before the pulse labeling, no significant radioactivity can be detected in the 68C RNAs after pulse-labeling and hybrid selection. Thus, the accumulation of newly synthesized 68C glue mRNA is very rapidly stopped by a short exposure to ecdysterone.

The hormonal control of the 68C RNAs in cultured salivary glands reflects the control mechanisms that operate in vivo. At about the time of the second to third instar molt, and again 5 hours before puparium formation, there are transient increases in the ecdysterone titer in the larval hemolymph (Berreur et al. 1979). The first of these ecdysterone peaks corresponds approximately to the time when 68C glue RNA begins accumulating, and the second corresponds to the time when the RNAs cease synthesis and when the 68C puff regresses. One might ask how the same hormone could both activate and inactivate the same genes in the same tissue. One possibility is that the initial hormone response of the salivary gland is to activate both the 68C genes and other genes, whose products eventually accumulate and cause a reversal in the cellular response to the hormone. Another possibility is that the levels of hormone involved in the two responses are different. These possibilities, and others, are currently being tested in our lab.

The second factor required in *trans* for initiation of expression of the three 68C glue RNAs is the product of a different X-chromosomal gene, the *l(1)npr-1* gene. This mutation was first discovered as a late larval lethal that is not rescued by addition of ecdysterone to the larval medium (Kiss et al. 1978). It was afterward recognized that the 68C puff is present in late third instar in mutant larvae (Belyaeva et al. 1981), and demonstrated that in cultured salivary glands the 68C puff of these mutant larvae appears normal, but will not regress in response to the addition of exogenous ecdysterone (Crowley et al. 1984). Thus, it seemed possible that the 68C RNAs might be produced constitutively in the mutant. This is not the case. When *l(1)npr-1* mutant salivary gland RNA is subjected to electrophoresis and blotted to a nitrocellulose filter, and the filter-bound RNA hybridized with labeled 68C RNA-specific DNA probes, no 68C RNA is detected. Furthermore, if *l(1)npr-1* mutant salivary gland RNAs are pulse-labeled in organ culture experiments, no radioactivity is incorporated into the 68C glue RNAs, despite normal levels of incorporation into total salivary gland RNA (Crowley et al. 1984). Thus, the *l(1)npr-1* mutation prevents accumulation of newly synthesized RNA from the 68C puff, without affecting the appearance of the puff. Puffing and RNA accumulation are thus again shown to be separable. The *l(1)npr-1* mutation and a deletion of the *l(1)npr-1* locus both behave identically in their effect on 68C RNA levels, showing that the effect of the mutation is to eliminate the normal function of the locus (Crowley et al. 1984). It can therefore be concluded that the normal product of the *l(1)npr-1* locus is required in *trans* for initiation of expression of the 68C glue genes.

What is the *l(1)npr-1* product? One possibility is that it is a component of the ecdysterone receptor, which binds ecdysterone when it enters cells. This binding is believed to be the first step in the action of the hormone. One piece of evidence that indicates that the gene product may not be a crucial component of the receptor is that the *l(1)npr-1* mutation only reduces, and does not eliminate, accumulation of RNA from the 3C glue puff, which codes for the sgs-4 polypeptide (Crowley et al. 1984). This RNA, like the 68C RNAs, requires ecdysterone to trigger the initiation of its accumulation (Hansson and Lambertsson 1983). Thus, if the *l(1)npr-1* gene product is a component of the ecdysterone receptor, it is a component that is only required for some, and not other, ecdysterone responses.

Sequences Required in *cis* for 68C Regulation

What DNA or RNA sequences might the *trans*-acting factors interact with? Classical genetic experiments show that no more than 20,000 bp of contiguous DNA sequence is required for normal puffing and expression of the 68C glue genes: *Df(3L)vin³* is a deficiency that removes a large chromosomal segment normally found to the right of the 68C puff. The left deficiency breakpoint is just over 1000 bp downstream of the 3' end of the *Sgs3* gene. This deficiency has no effect on expression of the adjacent *Sgs3* gene or on the puffing of the 68C DNA. *In(3L)HR15* is an inversion of a chromosomal region that includes a substantial part of the left arm of the third chromosome. The right inversion breakpoint is about 15 kbp to the left of the 3' end of *Sgs8*, and the effect of the inversion is to move sequences normally found adjacent to the 68C puff to a distant location. This inversion has no effect on expression of any of the 68C glue genes or on 68C puffing (M.A. Crosby and E.M. Meyerowitz, in prep.).

Thus, the sequences that control the 68C glue genes are near the genes. These classical experiments do not show whether there is only one set of sequences that regulate all three of the genes, thus explaining why these coordinately regulated genes may be found in a gene cluster, or whether there are separate regulatory regions for each of the genes. The DNA sequence of the genes is no help in this respect either: There are no recognizable stretches of similar sequences either upstream or downstream of the three genes that are obvious candidates for control sequences (Garfinkel et al. 1983). To localize the regulatory sequences that interact with *trans*-acting regulatory factors, P-element transformation experiments have been performed. In these experiments various fragments of the 68C DNA have been cloned into P-element transformation vectors (Rubin and Spradling 1982; Goldberg et al. 1983; Spradling and Rubin 1983) adjacent to genes whose product is selectable or easily recognized in adult flies (alcohol dehydrogenase, which confers resistance to high levels of ethanol, or xanthine dehydrogenase, which affects fly eye color). These constructs are then injected into *Drosophila* embryos in a way that allows reintegration of the cloned genes into random locations in the chromosomes of the primordial germ cells of the embryos. Transformed lines are recognized by the phenotype that results from expression of the selected marker, and the expression of the unselected 68C fragments is assayed in the salivary glands of larvae from these lines.

Most of the experiments have been aimed at determining the extent of *Sgs3* regulatory sequences. To assay expression of introduced *Sgs3* genes, it is necessary to distinguish the RNA and protein products of the introduced genes from those of endogenous genes. This has been done in our laboratory by injecting genes from the Oregon-R wild-type strain into embryos whose *Sgs3* genes derived either from the Hikone-R or Formosa wild strains. While the *Sgs3* RNA in the Oregon-R strain is 1120 nucleotides in length, those from the other strains are smaller, and are easily separated from the Oregon-R type by agarose gel electrophoresis. If an *Sgs3* gene is reintroduced to the fly genome, along with

2.27 kb of upstream sequence and some downstream flanking sequence as well, it expresses RNA normally: The RNA is only found in third larval instar and only in salivary glands. It accumulates to an approximately normal level, and is translated to a normal amount of sgs-3 protein (M.A. Crosby and E.M. Meyerowitz, in prep.). The fact that sgs-3 protein is made demonstrates that RNA from the introduced gene is spliced, as failure of splicing would lead to an RNA that would translate to a short peptide easily distinguished from sgs-3. Despite the proper expression of RNA and protein from the introduced gene, in at least one case where careful measurements have been made, there is no puff at the site of integration of the new gene. That an *Sgs3* gene with 2.7 kb of upstream sequence expresses RNA normally without puffing has also been shown by Richards and his collaborators, in experiments in which the *Sgs3* gene of the Formosa strain gene has been integrated into the genome of a strain that produces a 1120-nucleotide *Sgs3* RNA (Richards et al. 1983.) If only 130 nucleotides of sequence are left upstream of the *Sgs3* gene, our laboratory and Richards' obtain somewhat different results. We find, in experiments using an alcohol dehydrogenase vector and involving introduction of an Oregon-R gene into a Formosa background, that the introduced genes express at the correct time and in the correct tissue, but at a level 10- to 40-fold lower than normal (M.A. Crosby and E.M. Meyerowitz, in prep.). They find that a Formosa *Sgs3* gene introduced with a xanthine dehydrogenase vector into a strain with an Oregon-R-sized endogenous gene gives a level of RNA more than 400-fold lower than normal (Bourouis and Richards 1985). The basis of difference between the results of the two labs is not yet known; differences in transformation vector, genetic background of transformed flies, and strain from which the introduced gene derived are all possibilities. Since the Richards lab has transformants with 1.0 and 1.5 kb of upstream *Sgs3* sequence that express the RNA at a reduced level, but in appropriate tissue and stage, all of the work agrees in the conclusion that the upstream regulatory sequences of the *Sgs3* gene contain two separable elements. One is responsible for the level of expression of the RNAs and includes sequences from -130 to -2270 from the start of transcription; the other controls the tissue and time of expression and is closer to or in the gene.

The *Sgs3* experiments demonstrate that this gene has its own regulatory sequences nearby but do not reveal if the *Sgs7* and *Sgs8* genes respond to the same sequences as the *Sgs3* gene, or have their own regulatory regions. To find out, we have fused the *Sgs8* gene to an *Escherichia coli* β-galactosidase gene, creating a fusion gene whose RNA and protein products are easily distinguished from the products of the endogenous *Sgs8* genes. This construct, with upstream sequences derived from the *Sgs8* gene, and an *Sgs8* 3' untranslated region and poly(A) addition site added downstream of the β-galactosidase gene, has been cloned into a xanthine dehydrogenase P-element vector. Larvae homozygous for the newly introduced construct have been tested both for the fusion RNA and for *E. coli* β-galactosidase activity. Both the RNA and the activity are present in third instar larval salivary glands, and in no other third instar tissues (E.M. Meyerowitz, unpubl.). Since this construct shares no 68C sequences with the *Sgs3* constructs, the *Sgs8* gene must have its own regulatory sequences, and need not be near an *Sgs3* gene to express appropriately. Therefore we can exclude models that explain coordinate expression of the 68C genes by postulating that all three of the genes rely on the same regulatory sequences. We are also led to the conclusion that the clustering of the three genes at 68C is a consequence of their evolution by local duplication followed by divergence, and not a functional necessity.

Interaction of *trans*-acting Factors and *cis*-acting Sequences

By crossing the *trans*-acting mutations into strains containing introduced modified glue genes, we can learn about the sequences that interact with the *trans*-acting factors, and about the factors themselves. An example of such an experiment is one in which the *l(1)npr-1* mutation was crossed into a transformed strain bearing the Hikone-R *Sgs3* variant as its endogenous *Sgs3* genes and, closer to the tip of the left arm of the third chromosome, an introduced *Sgs3* construct with 2.27 kb of upstream sequence, the Oregon-R variant of the *Sgs3* gene, and approximately 2.5 kb of downstream sequence. In the presence of a wild-type *l(1)npr-1* locus, this introduced gene expresses *Sgs3* RNA and protein at approximately normal levels and in the appropriate developmental stage and tissue. When this wild-type locus is replaced with the *l(1)npr-1* mutation, neither the endogenous nor the introduced genes express (Crowley et al. 1984). Thus, the normal product of the *l(1)npr-1* locus, or something induced by the product of this locus, interacts with sequences in or near the *Sgs3* gene to allow its expression. The sequences could be DNA upstream, in, or downstream of the gene, or sequences of the RNA transcript of the gene. Just which sequences are involved is being determined with constructs that separate upstream, genic, and downstream sequences. Additional experiments may reveal whether the product of the wild-type *l(1)npr-1* gene is a factor that directly binds to DNA or RNA and allows its expression, or is a product that causes a repressor to leave the DNA or RNA. If the former is true, constructs that express *Sgs3* normally will always fail to express when the *l(1)npr-1* mutation is present. If the latter is the case, some constructs that lack the repressor binding site may express the *Sgs3* gene despite the absence of the wild-type *l(1)npr-1* product.

Similar experiments with strains containing introduced *Sgs8* β-galactosidase fusion genes show that the *l(1)npr-1* product acts, either directly or indirectly, on sequences in or near the *Sgs8* gene in addition to sequences near the *Sgs3* gene. When the *l(1)npr-1* mutation is crossed into one of the strains carrying the fusion construct, the fusion gene no longer directs synthesis of β-galactosidase in third instar larvae (K. VijayRaghavan and E.M. Meyerowitz, unpubl.). The *l(1)npr-1* product thus acts independently on the *Sgs3* and *Sgs8* genes.

One further approach to identifying the 68C glue puff sequences recognized by *trans*-acting factors involves analysis of regions of the genomes of other *Drosophila* species that are homologous with the *D. melanogaster* 68C puff. It is known that several species of the *melanogaster* species subgroup of *Drosophila* have puffs at their equivalent of polytene chromosome location 68C (Ashburner and Lemeunier 1972; Ashburner and Berendes 1978). These regions have been cloned and characterized (Meyerowitz and Martin 1984). All of them contain genes homologous with the 68C glue genes of *D. melanogaster* that exist in coordinately regulated clusters and express abundant polyadenylated RNAs in third instar salivary glands. *D. simulans* and *D. erecta* both have three genes in the 68C-homologous clusters, with two small genes that cross-hybridize with *Sgs8* or *Sgs7*, and one large gene that has homology with *Sgs3*. *D. yakuba* and *D. teissieri* have four genes, three small genes, and one *Sgs3*-like large gene. Examination of the DNA sequences of these genes is the converse of a mutagenesis experiment. In a mutagenesis experiment, one seeks the smallest changes in a DNA sequence that alter its function. In these experiments, we are looking at the largest amount of divergence in a DNA sequence that leaves its function intact. The sequences that have not diverged are expected to be those whose function is critical, including those that interact with diffusible regulatory factors. Those sequences that are bound by regulatory factors are expected to diverge more slowly than others, because a change in an important base in such a sequence would have to be accompanied by a corresponding change in the regulatory factor if function is to be retained. We are currently analyzing sequences of the *Sgs3*-equivalent genes, and their flanking regions, from *D. simulans*, *D. erecta*, and *D. yakuba*. There are conserved sequences upstream of the genes, separated by regions of low sequence homology (C.H. Martin and E.M. Meyerowitz, unpubl.). When the conserved regions have been thoroughly analyzed, interspecies transformation experiments will allow us to determine if these sequences do indeed serve the same function in all of the species. In addition, the *Sgs3* gene sequences will enable us to see how the repetitive central protein module of the gene evolves, and perhaps to understand its origin.

Conclusions and Further Questions

The *Drosophila* 68C glue gene cluster is a chromosomal region whose transcription is regulated hormonally and in a tissue-specific manner. In wild-type larvae transcription of the genes of the cluster and visible chromatin decondensation are coincident. Analysis of *trans*-acting mutations has revealed two diffusible factors required by the 68C genes for proper RNA expression, and has allowed a rough localization of the sequences with which one of them interacts. In vitro mutagenesis of the *Sgs3* gene has shown that the sequences required in *cis* for normal gene activity contain two functional regions, one controlling the level of gene transcription and one directing the time and tissue of gene expression. The *Sgs8* gene has been demonstrated to have its own regulatory sequences, independent of the similar *Sgs3* sequences. Nonetheless, many questions remain. The identity of the molecules that interact with the 68C sequences is unknown, as is their exact site of interaction, and their precise effects. A major question raised by our work, and not yet answered, regards the relation of gene expression and puffing at 68C. Puffing in wild-type salivary glands always corresponds with high-level transcription of the puffed DNA, implying that puffing is either a precondition to high-level gene expression, or a result of it. In mutant and transformed larvae, however, puffing and high-level expression of the 68C glue mRNAs have been separated in several ways. The puff can be present and normal in appearance without concomitant accumulation of the puff RNAs, as in $l(1)su(f)^{ts678}$ and *l(1)npr-1* larvae, and at least one of the 68C genes (*Sgs3*, which accounts for more than half of the transcription in the 68C gene cluster) can express at an approximately normal level without giving rise to a recognizable puff. In these abnormal situations, puffing is neither necessary nor sufficient for abundant transcription, and high-level transcription is neither necessary nor sufficient for puffing. Additional experiments designed to reveal the relation of gene transcription and polytene chromosome puffing, and to define better the *trans*-acting factors and *cis*-acting sequences that control the 68C glue genes, are in progress.

ACKNOWLEDGMENTS

Our 68C puff work has been funded by National Institutes of Health grant GM28075 to E.M.M., M.D.G., and P.H.M. are supported by National Research Service Award T32 GM07616; C.H.M. is supported by a National Science Foundation predoctoral fellowship; and K.V. is supported by a Procter and Gamble postdoctoral fellowship.

REFERENCES

AKAM, M.E., D.B. ROBERTS, G.P. RICHARDS, and M. ASHBURNER. 1978. *Drosophila*: The genetics of two major larval proteins. *Cell* **13**: 215.

ASHBURNER, M. 1967. Patterns of puffing activity in the salivary gland chromosomes of *Drosophila*. I. Autosomal puffing patterns in a laboratory stock of *Drosophila melanogaster*. *Chromosoma* **21**: 398.

———. 1973. Sequential gene activation by ecdysone in polytene chromosomes of *Drosophila melanogaster*. I. Dependence upon ecdysone concentration. *Dev. Biol.* **35**: 47.

———. 1974. Sequential gene activation by ecdysone in polytene chromosomes of *Drosophila melanogaster*. II. The effects of inhibitors of protein synthesis. *Dev. Biol.* **39:** 141.

ASHBURNER, M. and H.D. BERENDES. 1978. Puffing of polytene chromosomes. In *The genetics and biology of Drosophila* (ed. M. Ashburner and T.R.F. Wright), vol. 2B, p. 315. Academic Press, London.

ASHBURNER, M. and F. LEMEUNIER. 1972. Patterns of puffing activity in the salivary gland chromosomes of *Drosophila*. VII. Homology of puffing patterns on chromosome arm 3L in *D. melanogaster* and *D. yakuba*, with notes on puffing in *D. teissieri*. *Chromosoma* **38:** 283.

BECKENDORF, S.K. and F.C. KAFATOS. 1976. Differentiation in the salivary glands of *Drosophila melanogaster*: Characterization of the glue proteins and their developmental appearance. *Cell* **9:** 365.

BELYAEVA, E.S., I.E. VLASSOVA, Z.M. BIYASHEVA, V.T. KAKPAKOV, G. RICHARDS, and I.F. ZHIMULEV. 1981. Cytogenetic analysis of the 2B3-4–2B11 region of the X-chromosome of *Drosophila melanogaster*. II. Changes in 20-OH ecdysone puffing caused by genetic defects of puff 2B5. *Chromosoma* **84:** 207.

BERENDES, H.D. and M. ASHBURNER. 1978. The salivary glands. In *The genetics and biology of Drosophila* (ed. M. Ashburner and T.R.F. Wright), vol. 2B, p. 453. Academic Press, London.

BERREUR, P., P. PORCHERON, J. BERREUR-BONNEFANT, and P. SIMPSON. 1979. Ecdysteroid levels and pupariation in *Drosophila melanogaster*. *J. Exp. Zool.* **201:** 347.

BOUROUIS, M. and G. RICHARDS. 1985. Remote regulatory sequences of the *Drosophila* glue gene *Sgs-3* as revealed by P element transformation. *Cell* **40:** 349.

BRIDGES, C.B. 1935. Salivary chromosome maps. *J. Hered.* **26:** 60.

CROWLEY, T.E. and E.M. MEYEROWITZ. 1984. Steroid regulation of RNAs transcribed from the *Drosophila* 68C polytene chromosome puff. *Dev. Biol.* **102:** 110.

CROWLEY, T.E., M.W. BOND, and E.M. MEYEROWITZ. 1983. The structural genes for three *Drosophila* glue proteins reside at a single polytene chromosome puff locus. *Mol. Cell. Biol.* **3:** 623.

CROWLEY, T.E., P.H. MATHERS, and E.M. MEYEROWITZ. 1984. A *trans*-acting regulatory product necessary for expression of the *Drosophila melanogaster* 68C glue gene cluster. *Cell* **39:** 149.

FRAENKEL, G. and V.J. BROOKES. 1953. The process by which the puparia of many species of flies become fixed to a substrate. *Biol. Bull.* **105:** 442.

GARFINKEL, M.D., R.E. PRUITT, and E.M. MEYEROWITZ. 1983. DNA sequences, gene regulation and modular protein evolution in the *Drosophila* 68C glue gene cluster. *J. Mol. Biol.* **168:** 765.

GAUTAM, N. 1983. Identification of two closely located larval salivary protein genes in *Drosophila melanogaster*. *Mol. Gen. Genet.* **189:** 495.

GOLDBERG, D.A., J.W. POSAKONY, and T. MANIATIS. 1983. Correct developmental expression of a cloned alcohol dehydrogenase gene transduced into the *Drosophila* germ line. *Cell* **34:** 59.

GUILD, G.M. and E.M. SHORE. 1984. Larval salivary gland secretion proteins in *Drosophila*: Identification and characterization of the *Sgs-5* structural gene. *J. Mol. Biol.* **179:** 289.

HANSSON, L. and A. LAMBERTSSON. 1983. The role of *su(f)* gene function and ecdysterone in transcription of glue polypeptide mRNAs in *Drosophila melanogaster*. *Mol. Gen. Genet.* **192:** 395.

HANSSON, L., K. LINERUTH, and A. LAMBERTSSON. 1981. Effects of the *l(1)su(f)^{ts67g}* mutation of *Drosophila melanogaster* on glue protein synthesis. *Wilhelm Roux's Arch. Dev. Biol.* **190:** 308.

KISS, I., J. SZABAD, and J. MAJOR. 1978. Genetic and developmental analysis of puparium formation in *Drosophila*. *Mol. Gen. Genet.* **164:** 77.

KORGE, G. 1975. Chromosome puff activity and protein synthesis in larval salivary glands of *Drosophila melanogaster*. *Proc. Natl. Acad. Sci.* **72:** 4550.

———. 1977. Larval saliva in *Drosophila melanogaster*: Production, composition and relation to chromosome puffs. *Dev. Biol.* **58:** 339.

KRESS, H., E.M. MEYEROWITZ, and N. DAVIDSON. 1985. High resolution mapping of *in situ* hybridized biotinylated DNA to surface spread *Drosophila* polytene chromosomes. *Chromosoma* (in press).

MEYEROWITZ, E.M. and D.S. HOGNESS. 1982. Molecular organization of a *Drosophila* puff site that responds to ecdysone. *Cell* **28:** 165.

MEYEROWITZ, E.M. and C.H. MARTIN. 1984. Adjacent chromosomal regions can evolve at very different rates: Evolution of the *Drosophila* 68C glue gene cluster. *J. Mol. Evol.* **20:** 251.

MUSKAVITCH, M.A.T. and D.S. HOGNESS. 1980. Molecular analysis of a gene in a developmentally regulated puff of *Drosophila melanogaster*. *Proc. Natl. Acad. Sci.* **77:** 7362.

RICHARDS, G., A. CASSAB, M. BOUROUIS, B. JARRY, and C. DISSOUS. 1983. The normal developmental regulation of a cloned *Sgs-3* "glue" gene chromosomally integrated in *Drosophila melanogaster* by P element transformation. *EMBO J.* **2:** 2137.

RUBIN, G.M. and A.C. SPRADLING. 1982. Genetic transformation of *Drosophila* with transposable element vectors. *Science* **218:** 348.

SPRADLING, A.C. and G.M. RUBIN. 1983. The effect of chromosomal position on the expression of the *Drosophila* xanthine dehydrogenase gene. *Cell* **34:** 47.

VELISSARIOU, V. and M. ASHBURNER. 1980. The secretory proteins of the larval salivary glands of *Drosophila melanogaster*. *Chromosoma* **77:** 13.

———. 1981. Cytogenetic and genetic mapping of a salivary gland secretion protein in *Drosophila melanogaster*. *Chromosoma* **84:** 173.

WOLFNER, M. 1980. "Ecdysone-responsive genes of the salivary gland of *Drosophila melanogaster*." Ph.D. thesis, Stanford University, Stanford, California.

Hybrid Genes in the Study of Glue Gene Regulation in *Drosophila*

M. BOUROUIS AND G. RICHARDS

Laboratoire de Génétique Moléculaire des Eucaryotes du CNRS, Unité 184 de Biologie Moléculaire et de Génie Génétique de l'INSERM, Institut de Chimie Biologique, Faculté de Médecine, 67085 Strasbourg Cedex, France

Gene transfer experiments in *Drosophila* have led to important advances in our understanding of gene structure in higher eukaryotes. Since the introduction of germ line transformation using P-element vectors (Rubin and Spradling 1982), we have learned a good deal regarding the regulation of a small number of carefully chosen genes and there have been confirmatory data from a larger number of genes whose detailed analysis is more difficult. In the case of genes with relatively short transcription units, spanning 2 or 3 kb of DNA, normal developmental regulation is possible with less than a 10-kb genomic fragment and this in the majority of chromosomal insertion sites (Goldberg et al. 1983; Richards et al. 1983; Scholnick et al. 1983; Spradling and Rubin 1983). This result is important for our concepts of genome structure and function because classical studies in *Drosophila*, largely based on polytene chromosome banding patterns, had suggested that genes might be regulated in chromosomal domains containing an average of 30 kb of DNA. Even in cases where related genes are clustered, and therefore good candidates for coordinated regulation, individual transcription units can be expressed in isolation (e.g., hsps, Lis et al. 1983, Hoffman and Corces 1984; *Sgs3*, Richards et al. 1983; chorion genes, Kalfayan et al. 1984).

The second general result that is now emerging is that normal regulation requires sequences much further from the transcribed regions than was envisioned from earlier transcription studies in which *Drosophila* genes were expressed in heterologous systems (e.g., Mirault et al. 1982; Pelham 1982). These recent results show similarities to those that were first obtained in the analysis of viral gene regulation and led to the discovery of enhancer sequences (Grosschedl and Birnstiel 1980; Banerji et al. 1981; Moreau et al. 1981). Of particular importance in *Drosophila* studies is the demonstration by Garabedian et al. (1985) of the existence of separable sequences showing different tissue-specific effects on the synthesis of the mRNA for yolk polypeptides YP1 and YP2.

In the current phase of our research, we are trying to discover the number and nature of the control elements necessary in *cis* for the normal developmental regulation of the *Drosophila melanogaster* larval salivary gland glue gene *Sgs3* and are investigating the extent of similarities between our findings and those defining regulatory sequences in other eukaryotic systems (see Chambon et al. 1984 for recent review). Our approach has been to transform a short *Sgs3* allele (Formosa) into a host strain carrying a long allele (ry^{506}) and then to compare the activity of the inserted gene to that of the resident gene. Our previous results (Bourouis and Richards 1985) have suggested that it may be experimentally useful to consider three regions involved in the regulation of *Sgs3* (Fig. 1). Region A contains the *Sgs3* transcribed sequence as well as 127 bp 5' to the presumed cap site and 320 bp of 3' sequence. In transformation experiments this region alone rarely gives rise to *Sgs3* transcripts ($\leq 1\%$ of normal levels). Constructs containing the 850-bp region B in addition to A result in consistently detectable stage and tissue-specific transcripts but at levels averaging 5–10% of wild type. If region C (1.75 kb) is now added, normal levels of activity are restored, suggesting that this region contains an element essentially regulating levels of *Sgs3* expression. Finally, if the combined B and C regions are inverted with respect to region A we again find normal levels of activity (g71I; Fig. 1). This last result suggested that region A is a transcription unit capable of responding to remote regulatory elements and that regions B and/or C contained a bidirectionally active element. The experiments we will discuss derive initially from these observations.

MATERIALS AND METHODS

Transposon Construction

Sgs3 **Formosa genomic segments and general conventions for transposons.** The designation of DNA regions of the *Sgs3* glue gene as A, B, and C in the text results from our analysis of transformants containing various Formosa *Sgs3* DNA fragments. Details of plasmids not described here will be found in Bourouis and Richards (1985). Figure 1 shows schematic representations of the DNA contained in transposons g71 and g71I, the former being the normal genomic arrangement. The restriction enzyme sites given are those that define the limits of the different regions; those not reconstituted in a transposon construct are shown in parentheses. All fragments, including those described below, were inserted in the same orientation in the polylinker region of the Carnegie 20 vector (Rubin and Spradling 1983) which carries the wild-type rosy (*ry*) gene. Where *Sgs3* sites are denoted by nucleotide num-

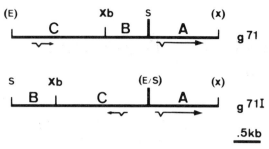

Figure 1. The *Sgs3* Formosa genomic segments contained in the Carnegie 20 transposons g71 and g71I. The large and small arrows represent the transcribed sequences of *Sgs3* and *Sgs7*, respectively. A, B, and C designate regions that are discussed in the text. See Materials and Methods for conventions. (E) *Eco*RI; (Xb) *Xba*I; (S) *Sal*I; (X) *Xho*I; (K) *Kpn*I throughout.

bering, this refers equally to the Oregon-R sequence (Garfinkel et al. 1983) or the Formosa sequence (Mettling et al. 1985).

β-Globin constructs. The rabbit β-globin gene was obtained on a 2.1-kb *Xho*I fragment from the plasmid pEP1 (Rautmann et al. 1984) derived from pSXβ+ (Banerji et al. 1981) and inserted in the *Sal*I site of the Carnegie 20 vector to give C20β. βES and βESI were obtained by inserting the Formosa *Eco*RI(1713)–*Sal*I(4327) fragment by blunt-end ligation into the adjacent *Hpa*I polylinker site 5' to the globin gene and selecting for the two orientations. βESX is derived from βES by deletion of the 1.3-kb *Xba*I fragment which removes the right-hand portion of region C leaving the 420-bp fragment normally upstream of *Sgs7* as shown in Figure 2. βES170 was obtained by first substituting the *Sal*I(4327)–*Alu*I(4490) *Sgs3* promoter fragment for the −424 to −9 fragment of the globin promoter fragment in pEP1. The fusion was then isolated as a 1.87-kb *Sal*I–*Xho*I fragment and inserted in the unique *Sal*I site of the plasmid C20g1 ΔSal. The plasmid C20g1 ΔSal was constructed by inserting the Formosa 6.7-kb *Sgs3 Eco*RI fragment into the *Eco*RI site of the polylinker of Carnegie 20 and then deleting the sequences from *Sal*I(4327) to the 3' *Sal*I site in the Carnegie 20 polylinker. In the fusion gene the leader and first two codons of *Sgs3* are fused to the −9 (*Pvu*II site) of the β-globin sequence which results in an out-of-frame fusion relative to the *Sgs3* initiation codon.

Construction of an Sgs3–alcohol dehydrogenase (Adh) fusion gene (g1Ad). The *Sal*I(4327)–*Alu*I(4490) fragment of *Sgs3* Formosa was subcloned into the *Sal*I–*Bam*HI sites of pEMBL19+ (Mettling et al. 1985), the latter site having been filled in so as to maintain the *Bam*HI site. To remove the *Sgs3* initiation codon, the subclone was then reopened at this site, digested briefly with BAL-31, filled in, a *Hin*dIII linker (CAAGCTTG) added, and the plasmid recircularized by blunt-end ligation. The extent of the BAL-31 digest was monitored by the remaining restriction sites in the polylinker as nucleotides were removed in both directions from the *Bam*HI site. One clone, which had lost the *Kpn*I site but retained the *Eco*RI site, was selected for sequencing and proved to have the *Hin*dIII linker at +10 on the *Sgs3* leader. The 143-bp *Sal*I–*Hin*dIII fragment was isolated from this plasmid and substituted for the *Sal*I–*Hin*dIII heat shock promoter fragment in pR3XA (a plasmid carrying an hsp-*Adh* fusion gene; Dudler and Travers 1984). This *Hin*dIII site is at +9 in the *Adh* larval leader and was created by the insertion of a *Hin*dIII linker (Bonner et al. 1984). The 2.6-kb *Sal*I–*Xho*I fragment was then isolated from this construct and inserted in the *Sal*I site of C20g1 ΔSal so as to restore the *Sgs3* genomic sequence from *Eco*RI(1713) via *Sal*I(4327) to the *Hin*dIII linker leader fusion (4467).

Insertion of the SV40 enhancer upstream of Sgs3. The plasmid C20g6, which contains regions A and B of *Sgs3*, was modified so as to contain a *Kpn*I linker 5' and adjacent to the genomic *Xba*I site, which was retained. The SV40 enhancer sequence contained in the 180-bp *Kpn*I–*Bam*HI fragment of the plasmid pHS102 (Benoist and Chambon 1981) was subcloned into the polylinker of pEMBL19+ and recovered as a *Kpn*I–*Xba*I fragment, which was inserted into the modified C20g6 to give the plasmid C20g672.

Recipient Strains and Production of Transformed Lines

The general scheme for the isolation of transformed lines was as previously published (Bourouis and Richards 1985). All transformants, including g1Ad, were selected as ry+ flies and then their progeny analyzed for the segregation of the ry+ phenotype. In addition, the

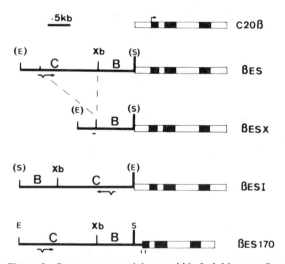

Figure 2. Constructs containing a rabbit β-globin gene (box segments) fused to *Sgs3* remote regulatory regions (line segments). In the globin gene, exons are denoted by filled box segments, the cap site is shown by an arrow for C20β. B and C regions of *Sgs3* are as in Fig. 1. The *Xba*I segment deleted from βES to form βESX is denoted by broken lines. In βESI the combined B and C regions were inverted as in g71I. The two vertical arrows in βES170 denote the position of the *Sgs3* and globin ATG codons, respectively.

β-globin series of constructs were analyzed by genomic blots and in situ hybridizations to confirm that lines carried a single insertion. These detailed analyses, with those undertaken previously, have established that the genetic criterion of a 1:1 segregation in the G_2 or subsequent subline is usually sufficient to indicate that a single insertion is involved. We have found this to be the case in particular for lines derived from G_1 progenies where there are few ry^+ flies. For each construct several insertions were obtained and are denoted as previously βES:1, βES:2, etc., to indicate independent transformed lines. In this report data are presented for selected homozygous representative lines for each construct, with the exception of g1Ad transformants. The β-globin-containing series of constructs were injected into ry^{506} using pπ25.1 as helper plasmid. The fusion gene g1Ad was injected into a strain b Adh^{n4}, ry^{506} constructed from w, b Adh^{n4} (supplied by K. Moses, Cambridge University) and ry^{506} (W. Bender, Harvard University), using the helper plasmid pπ25.7WC (Karess and Rubin 1984). The Adh^{n4} strain is a phenotype null allele that expresses an Adh message at a few percent of wild type (K. Moses, pers. comm.). The construct g672 was injected into ry^{506}, C-S 75% (gift of A. Spradling), a strain with improved viability when compared to the original ry^{506} strain, with pπ25.7WC as helper.

Analysis of Gene Expression

RNA analysis by Northern blotting was as previously described using nick-translated probes. Probes were as follows: pEP1 (see Fig. 3), which contains a 2.1-kb fragment from the rabbit β-globin gene in addition to pBR322 and SV40 sequences; pG3 (see Fig. 5), which is a pBR322 containing a 294-bp Sgs3 cDNA insert (Richards et al. 1983).

Staining for alcohol dehydrogenase activity was essentially that of Bonner et al. (1984). Glands were dissected in insect Ringer's, stained 10 minutes, and destained 10 minutes before mounting for photography using a transmitted light source. Prolonged staining of dissected larvae from the Adh^{n4} strain results in coloration in several tissues, notably the gut, but as in the wild type there is little or no staining of the salivary glands.

RESULTS

Fusions of Sgs3 Upstream Sequences to a Heterologous Transcription Unit

Our results with Sgs3 constructs had suggested that the combined regions B and C (Fig. 1) might be capable of activating a transcription unit in a stage- and tissue-specific manner. On the basis of this idea, we constructed a series of transposons (βES, βESI, and βESX) in which region A, containing the Sgs3 transcription unit, was replaced by a rabbit β-globin gene. A Carnegie 20-based transposon containing only the β-globin sequences (C20β) was also constructed as a control

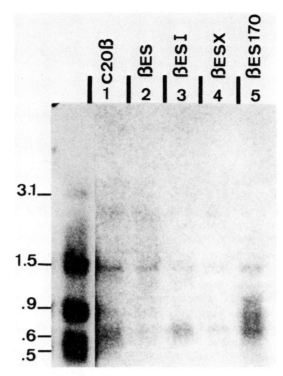

Figure 3. Northern analysis of total RNA from transformants carrying rabbit β-globin gene constructs. Each lane contains the RNA obtained from five late third instar larvae. The blot was probed with pEP1 and exposed 15 days. Under similar conditions a comparable signal for Sgs3 RNA would be obtained in 3-6 hr. Transformed strains are C20β:4, βES:4, βESI:3, βESX:1, and βES170:3. Size markers (kb) are from DNA fragments of pEP1 run in parallel.

(Fig. 2). The rabbit β-globin unit consists of 420 bp of 5'-flanking sequence, its three exonic sequences, and 350 bp of 3'-flanking sequence carried on a 2.1-kb genomic segment. This unit has been shown to include at least three promoter elements that are needed for efficient and accurate transcription of the gene either in HeLa cells (Grosveld et al. 1982a,b) or transformed mouse cells (Dierks et al. 1981). These are the -30 ATA box, the -80 homology (Benoist et al. 1980) or CCAAT box, and the -100 homology (McKnight and Kingsbury 1982) within which the sequence CACCC is crucial for β-globin expression (Dierks et al. 1983). For convenience we will refer to these as the proximal promoter elements.

The analysis of RNA from third instar larvae of a strain carrying each of the above constructs is shown in Figure 3. The β-globin probe reveals two faint bands at 1.5 and 0.6 kb, the latter, from its size, being most probably the processed rabbit β-globin message whereas the former corresponds to transcription from the anti-sense strand (data not shown). For these first four constructs (lanes 1-4), the slight differences in the levels of these bands are no greater than those due to position effects, which, as we have seen for Sgs3, can be of the order of two- to threefold at the level of RNA. From the length of exposure of the autoradiogram, we can estimate that the levels of RNA we detect are approximately 100-fold lower than those of the accumulated *Sgs3* message in third instar larvae. This low level of expression is most likely a consequence of the activity of the β-globin promoter alone, as the presence of the B and C regions does not stimulate RNA levels significantly when compared with C20β (lane 1). This may result from an inability of the proximal β-globin promoter to interact with *Drosophila* cellular transcription factors in the same way as the *Sgs3* promoter. To test this hypothesis we constructed a gene fusion, βES170, in which the β-globin proximal promoter was substituted by a *Sgs3* promoter fragment (see Materials and Methods for details). In this transformant (Fig. 3, lane 5) we do, in fact, find changes in the RNA pattern. There is clearly both an increase in the overall amount of RNA and the appearance of larger RNA species of unknown origin. The quantity of RNA detected remains very low compared with the *Sgs3* RNA, although in this construct we have effectively exchanged the coding sequences while retaining 2.7 kb of *Sgs3* 5′-flanking sequences as well as the leader and the first two codons of the *Sgs3* protein. We conclude that this relatively low level of expression may result either from an instability of the rabbit globin message in the *Drosophila* cellular environment or that the presence of the *Sgs3* gene itself is necessary for elevated levels of transcription.

The Alcohol Dehydrogenase Gene Can be Specifically Expressed in the Third Larval Instar Salivary Gland

In a construction similar to βES170, we have made a fusion gene using the upstream sequences, promoter, and first transcribed bases of *Sgs3* fused to the leader sequence of the larval *Adh* transcript (Fig. 4). The *Adh*

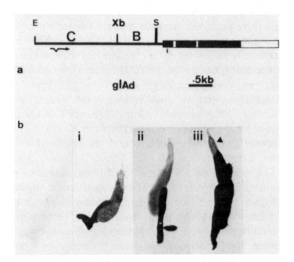

Figure 4. (*a*) The g1Ad construct (see Materials and Methods). Symbols are as in Fig. 2, except that box segments represent the gene for larval alcohol dehydrogenase. (*b*) *Adh* activity in larval salivary glands of (*i*) Adh^+, ry^{506}; (*ii*) Adh^{n4}, ry^{506}; (*iii*) Adh^{n4}, ry^{506} P[g1Ad,ry^+]. *Adh* activity results in a deep violet staining which appears black here. In *iii* the arrow indicates proximal cells that do not show *Adh* activity.

gene is amenable to rapid histochemical analysis and, because of the powerful selection techniques that have been developed for this gene in *Drosophila*, may prove useful for the genetic dissection of *Sgs3* regulation in a manner similar to that recently proposed by Bonner et al. (1984). We have recovered several lines carrying the glue-*Adh* fusion gene and have tested them for the presence of *Adh* activity in the third instar salivary gland. Neither ry^{506} nor the derived strain Adh^{n4}, ry^{506} show significant staining in the salivary gland, although the fat body of the former is rapidly stained (Fig. 4b, i,ii). In the transformed strains, coloration is rapid in the salivary gland cells, is limited to those cells normally accumulating the salivary gland secretion (Berendes and Ashburner 1978), and is not found in the proximal cells nor the duct cells (Fig. 4b, iii). This contrasts not only with the wild-type distribution of *Adh* activity but also with that seen when the *Adh* gene is transformed with its own regulatory elements (Goldberg et al. 1983). From this we conclude that the *Adh* gene in this construct is regulated by the stage- and tissue-specific elements of *Sgs3*. Preliminary analyses of the transcript from the fusion gene in third instar larvae confirm that there is specific synthesis of an RNA of the expected size in the salivary gland (data not shown). The level of this transcript is rather less than that seen for *Adh* in the rest of the larva in wild-type strains and is intermediate between that seen for the globin message in βES170 and that of the normal *Sgs3* gene.

The Effect of a Viral Enhancer on *Sgs3* Transcription

As discussed in the introduction, the formal distinction between elements contained in the B and C regions is that the latter contains the regulatory elements necessary for wild-type levels of *Sgs3* transcription, whereas the former is sufficient for stage- and tissue-specific regulation but at reduced levels (5–10% of wild type). We have asked whether or not we can stimulate the level of transcription by substituting a viral enhancer for region C, as these promoter elements are known to stimulate transcription on heterologous promoters over large distances. For this experiment we have fused the SV40 enhancer element obtained from the plasmid pHS102 (Benoist and Chambon 1981) to sequences −980 bp from the *Sgs3* cap site (Fig. 5). This enhancer is active in a number of host-type cells and has previously been used by Pelham (1982) to induce transcription of the *Drosophila* hsp70 gene in COS cells in the absence of heat shock.

We have analyzed RNA from male and female third instar larvae from two lines carrying this construct (g672:1 and g672:2) and compared them with lines carrying the *Sgs3* A and B regions (g6) or the A, B, and C regions (g71). From the Northern analysis (Fig. 5), it appears that the relative abundance of RNA from the transformed Formosa *Sgs3* sequence (lower band) compared with the resident gene (upper band) is not increased by the presence of the SV40 enhancer and is

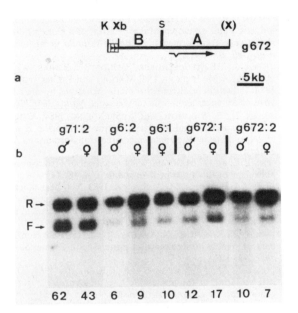

Figure 5. (a) The g672 construct. The SV40 sequences, shown by box segments, are immediately 5' to region B. (b) Northern analysis of Sgs3 RNA from individual male and female larvae from transformed lines probed with pG3. R indicates transcripts from the resident Sgs3 gene and F those from the introduced Formosa gene. Values under each lane indicate the amount of the F transcript relative to the R transcript, in percentage, obtained by densitometric scanning. Different transformed lines are denoted as g672:1, g672:2, etc.

clearly different from that seen when the region C is present. Therefore we conclude that the SV40 element cannot substitute for the sequences contained in region C of g71.

DISCUSSION

These experiments were designed to further our analysis of the number, location, and nature of the regulatory elements of Sgs3. Our recent findings, using deletion analysis of cis-acting DNA sequences, distinguished two kinds of control of Sgs3 transcription: one (contained in regions A and B) sufficient for low levels but cell-specific expression, the other (contained in region C) regulating levels of expression. As this regulation extends over 2 kb of 5' sequence, we are currently attempting to define more precisely the limits of these elements.

Our constructs with the β-globin gene have shown that there is no evidence for stimulation by regions B and C of the low basal activity seen with the β-globin gene alone. After the substitution of the 170-bp Sgs3 proximal promoter sequence, there is an increase, suggesting that this sequence is necessary in addition to elements in B and C, but the transcript levels remain low compared with those seen with Sgs3. Taken together with previous results and those of the Sgs3-Adh fusion, it seems probable that an element situated between −127 bp and the cap site acts in concert with element(s) in the region B to produce cell-specific transcription. Whether or not this specificity is conferred by elements in A or B or both can now be tested by regarding the tissue distribution of RNA in the globin constructs.

We have shown with the Sgs3-Adh leader fusion that we can change the tissue-specific control of the Adh gene to that of a glue gene. This result derives primarily from the rapid and specific histochemical staining of the salivary gland. In contrast, we do not detect the abundant transcripts we would have expected if the Sgs3 sequence had remained in place. This suggests that 3' of the Sgs3 cap site there are further elements necessary either for the rate of initiation of transcription or for the stability of Sgs3 mRNA in the salivary gland. Such a kind of regulation has been described recently for another major product of a specialized tissue as, in Xenopus, vitellogenin mRNA is selectively stabilized in the liver by estrogen (Brock and Shapiro 1983).

We had postulated the existence of an element that activates the Sgs3 transcription unit at a distance. This activation is a property of the eukaryotic regulatory elements known as enhancers (see Gluzman and Shenk 1983). The molecular mechanism of enhancer function is not fully understood. However, there is evidence for two components of its activity. First, it may contain a DNA sequence that prevents regular nucleosome assembly (Wasylyk et al. 1979; Jongstra et al. 1984) and, second, it functions as a target binding site for factors that stimulate transcription (Sassone-Corsi et al. 1984, 1985; Wildeman et al. 1984). Using the SV40 strong enhancer, we have been unable to substitute for the normal Drosophila element. Of course we do not know if the Drosophila cellular environment provides the factors that normally interact with the SV40 enhancer, but this result does, however, exclude the hypothesis that the DNA sequence alone of the enhancer is able to stimulate transcription by disturbing chromatin structure. In addition, some enhancer elements are associated with tissue-specific transcription (Gillies et al. 1983), whereas our element in region C is apparently not involved in determining tissue specificity. For these reasons it is perhaps premature to assume an absolute correspondence between this element and enhancers. In consequence, we prefer to think of the remote element in region C as an "activator" sequence until the experimental data are more complete.

ACKNOWLEDGMENTS

We thank B. Wasylyk for his comments on a draft of this manuscript. This manuscript was prepared with the help of B. Boulay. The work was supported by grants from the CNRS (ATP 3582), the INSERM, the Foundation pour la Recherche Medicale, the Association pour le Développement de la Recherche sur le Cancer, and the Fondation Simone et Cino del Duca. G.R. was supported by a Research Fellowship of the University Louis Pasteur, Strasbourg.

REFERENCES

Banerji, J., S. Rusconi, and W. Schaffner. 1981. Expression of a globin gene is enhanced by remote SV40 sequences. *Cell* 27: 299.

Benoist, C. and P. Chambon. 1981. *In vivo* sequence requirements of the SV40 early promoter region. *Nature* 290: 304.

Benoist, C., K. O'Hare, R. Breathnach, and P. Chambon. 1980. The ovalbumin gene-sequence of putative control regions. *Nucleic Acids Res.* 8: 127.

Berendes, H.D. and M. Ashburner. 1978. The salivary glands. In *The genetics and biology of* Drosophila (ed. M. Ashburner and T.R. Wright), vol. 2B, p. 453. Academic Press, New York.

Bonner, J.J., C. Parks, J. Parker-Thornburg, M.A. Mortin, and H.R.B. Pelham. 1984. The use of promoter fusions in *Drosophila* genetics: Isolation of mutants affecting the heat shock response. *Cell* 37: 979.

Bourouis, M. and G. Richards. 1985. Remote regulatory sequences of the *Drosophila* glue gene sgs3 as revealed by P-element transformation. *Cell* 40: 349.

Brock, M.L. and D.J. Shapiro. 1983. Estrogen stabilizes vitellogenin mRNA against cytoplasmic degradation. *Cell* 34: 207.

Chambon, P., A. Dierich, M.-P. Gaub, S. Jakowlev, J. Jongstra, A. Krust, J.-P. LePennec, P. Oudet, and T. Reudelhuber. 1984. Promoter elements of genes coding for proteins and modulation of transcription by estrogens and progesterone. *Recent Prog. Horm. Res.* 40: 1.

Dierks, P., A. van Ooyen, N. Mantei, and C. Weissman. 1981. DNA sequences preceding the rabbit β-globin gene are required for formation in L cells of β-globin RNA with correct 5' terminus. *Proc. Natl. Acad. Sci.* 78: 1411.

Dierks, P., A. van Ooyen, M.D. Cochran, C. Dobkin, J. Reiser, and C. Weissmann. 1983. Three regions upstream from the cap site are required for efficient and accurate transcription of the rabbit β-globin gene in mouse 3TG cells. *Cell* 32: 695.

Dudler, R. and A.A. Travers. 1984. Upstream elements necessary for optimal function of the *hsp70* promoter in transformed flies. *Cell* 38: 391.

Garabedian, M.J., M.-C. Hung, and P.C. Wensink. 1985. Independent control elements that determine yolk protein gene expression in alternative *Drosophila* tissues. *Proc. Natl. Acad. Sci.* 82: 1396.

Garfinkel, M.D., R.E. Pruitt, and E.M. Meyerowitz. 1983. DNA sequences, gene regulation and modular protein evolution in the *Drosophila* 68C glue gene cluster. *J. Mol. Biol.* 168: 765.

Gillies, S.D., S.L. Morrison, V.T. Oi, and S. Tonegawa. 1983. A tissue-specific transcription enhancer element is located in the major intron of a rearranged immunoglobulin heavy chain gene. *Cell* 33: 991.

Gluzman, Y. and T. Shenk, eds. 1983. *Current communications in molecular biology: Enhancers and eukaryotic gene expression*. Cold Spring Harbor Laboratory, Cold Spring Harbor, New York.

Goldberg, D.A., J.W. Posakony, and T. Maniatis. 1983. Correct developmental expression of a cloned alcohol dehydrogenase gene transduced into the *Drosophila* germ line. *Cell* 34: 59.

Groschedl, R. and M.L. Birnstiel. 1980. Identification of regulatory sequences in the prelude sequences of an H2A histone gene by the study of specific deletion mutants *in vivo*. *Proc. Natl. Acad. Sci.* 77: 1432.

Grosveld, G.C., A. Rosenthal, and R.A. Flavell. 1982a. Sequence requirements for the transcription of the rabbit β-globin gene *in vivo*. The −80 region. *Nucleic Acids Res.* 10: 4951.

Grosveld, G.C., E. de Boer, C.K. Shewmaker, and R.A. Flavell. 1982b. DNA sequences necessary for the transcription of the rabbit β-globin gene *in vivo*. *Nature* 295: 120.

Hoffman, E.P. and V.G. Corces. 1984. Correct temperature induction and developmental regulation of a cloned heat shock gene transformed into the *Drosophila* germ line. *Mol. Cell. Biol.* 4: 2883.

Jongstra, J., T.L. Reudelhuber, P. Oudet, C. Benoist, C.-B. Chae, J.-M. Jeltsch, D.J. Mathis, and P. Chambon. 1984. Induction of altered chromatin structures by simian virus 40 enhancer and promoter elements. *Nature* 307: 708.

Kalfayan, L., B. Wakimoto, and A. Spradling. 1984. Analysis of transcriptional regulation of the s38 chorion gene of *Drosophila* by P element-mediated transformation. *J. Embryol. Exp. Morphol.* (suppl.) 83: 137.

Karess, R.E. and G.M. Rubin. 1984. Analysis of P transposable element functions in *Drosophila*. *Cell* 38: 135.

Lis, J.T., J.A. Simon, and C.A. Sutton. 1983. New heat shock puffs and β-galactosidase activity resulting from transformation of *Drosophila* with an hsp70-*lacZ* hybrid gene. *Cell* 35: 403.

McKnight, S.L. and R. Kingsbury. 1982. Transcriptional control signals of a eukaryotic protein-coding gene. *Science* 217: 316.

Mettling, C.M., M. Bourouis, and G. Richards. 1985. Allelic variation at the nucleotide level in *Drosophila* glue genes. *Mol. Gen. Genet.* (in press).

Mirault, M.-E., R. Southgate, and E. Delwart. 1982. Regulation of heat-shock genes: A DNA sequence upstream of *Drosophila* hsp70 genes is essential for their induction in monkey cells. *EMBO J.* 1: 1279.

Moreau, P., R. Hen, B. Wasylyk, R. Everett, M.P. Gaub, and P. Chambon. 1981. The SV40 72 base pair repeat has a striking effect on gene expression both in SV40 and other chimeric recombinants. *Nucleic Acids Res.* 9: 6047.

Pelham, H.R.B. 1982. A regulatory upstream promoter element in the *Drosophila* hsp70 heat-shock gene. *Cell* 30: 517.

Rautmann, G., H.W.D. Matthes, M.J. Gait, and R. Breathnach. 1984. Synthetic donor and acceptor sites function in an RNA polymerase B (II) transcription unit. *EMBO J.* 3: 2021.

Richards, G., A. Cassab, M. Bourouis, B. Jarry, and C. Dissous. 1983. The normal developmental regulation of a cloned sgs3 "glue" gene chromosomally integrated in *Drosophila melanogaster* by P element transformation. *EMBO J.* 2: 2137.

Rubin, G.M. and A.C. Spradling. 1982. Genetic transformation of *Drosophila* with transposable element vectors. *Science* 218: 348.

———. 1983. Vectors for P element mediated gene transfer in *Drosophila*. *Nucleic Acids Res.* 11: 6341.

Sassone-Corsi, P., A.G. Wildeman, and P. Chambon. 1985. A *trans*-acting factor is responsible for the simian virus 40 enhancer activity *in vitro*. *Nature* 313: 458.

Sassone-Corsi, P., J.P. Dougherty, B. Wasylyk, and P. Chambon. 1984. Stimulation of *in vitro* transcription from heterologous promoters by the SV40 enhancer. *Proc. Natl. Acad. Sci.* 81: 308.

Scholnick, S.B., B.A. Morgan, and J. Hirsh. 1983. The cloned dopa decarboxylase gene is developmentally regulated when reintegrated into the *Drosophila* genome. *Cell* 34: 37.

Spradling, A.C. and G.M. Rubin. 1983. The effect of chromosomal position on the expression of the *Drosophila* xanthine dehydrogenase gene. *Cell* 34: 47.

Wasylyk, B., P. Oudet, and P. Chambon. 1979. Preferential *in vitro* assembly of nucleosome cores on some AT-rich regions of SV40 DNA. *Nucleic Acids Res.* 7: 705.

Wildeman, A.G., P. Sassone-Corsi, T. Grundström, M. Zenke, and P. Chambon. 1984. Stimulation of *in vitro* transcription from the SV40 early promoter by the enhancer involves a specific trans-acting factor. *EMBO J.* 3: 3129.

Developmental Regulation of Human Globin Genes in Transgenic Mice

F. Costantini, G. Radice, J. Magram, G. Stamatoyannopoulos,*
T. Papayannopoulou,* and K. Chada

*Department of Human Genetics and Development, College of Physicians and Surgeons, Columbia University, New York, New York 10032; *Department of Medicine, University of Washington School of Medicine, Seattle, Washington 98195*

The globin gene family represents a particularly interesting system for the study of developmental gene regulation. Not only are the globin genes expressed specifically in erythroid cells, but different members of the gene family are utilized at sequential stages of development, a phenomenon known as "hemoglobin switching." Of the genes encoding β-like chains in the human, the first to be expressed is the ε-globin gene, which is active primarily in the primitive erythroid cells that develop in the blood islands of the yolk sac. Between the 5th and 10th weeks of gestation, there is a shift to "definitive" fetal liver erythropoiesis, and during this time the ε gene is switched off and the two γ-globin genes begin to be expressed. The γ-globin genes are maximally active through most of fetal life and are switched off around the time of birth. In contrast, the "adult" genes, δ and β, are active at only a low level in the fetal liver erythroid cells and are turned on fully only after birth, as the site of erythropoiesis shifts from liver to bone marrow (Weatherall and Clegg 1981). All of the human β-like globin genes are located in a single cluster (Fig. 1) and are arranged 5' to 3' in the order of their expression (Fritsch et al. 1980).

Recently, there has been some progress in understanding the mechanism by which transcription of the adult β-globin gene is activated during erythroid cell differentiation. This appears to be a multistep process that is initiated before the gene is actually transcribed (Stalder et al. 1980a; Cohen and Sheffery 1985; Yu and Smith 1985). The earliest events, which may involve changes in chromatin structure, in methylation patterns, or in as yet undefined properties of the gene, appear to be necessary but not sufficient for transcription (Groudine and Weintraub 1982; Charnay et al. 1984; Yu and Smith 1985). The later events, which permit transcription, probably involve the interaction of the β-globin gene with *trans*-acting factors specific to differentiated erythroid cells (Chao et al. 1983; Wright et al. 1983; Emerson et al. 1985). Through gene transfer experiments using cultured erythroid cell lines, such as murine erythroleukemia (MEL) cells, it has been possible to begin to define the DNA sequences that appear to interact with *trans*-acting regulatory factors in adult erythroid cells. The available data indicate that multiple sequences, both 5' and 3' to the initiation codon, are involved in regulating transcription during the terminal events of erythroid differentiation (Charnay et al. 1984; Wright et al. 1984). Virtually nothing is known about the control of the earlier steps in gene activation, and since cloned genes transfected into cultured cells apparently bypass these early regulatory events (Robins et al. 1982; Charnay et al. 1984), the question cannot be readily addressed using MEL or other cultured cells.

The mechanism of hemoglobin switching also remains a mystery. Although alterations in chromatin structure and DNA methylation patterns accompany hemoglobin switching (Stalder et al. 1980a; Groudine

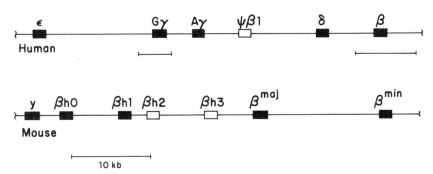

Figure 1. Organization of the human and mouse β-like globin gene clusters. The top line shows the human β-like globin gene cluster (Fritsch et al. 1980), which contains one embryonic gene (ε), two fetal genes (Gγ and Aγ), and two adult genes (δ and β); ψβ1 is a pseudogene. The bars below the human gene map show two human DNA fragments that were introduced into the mouse germ line; these are a 4.0-kb BglII-SphI fragment containing the Gγ gene, and a 7.7-kb HindIII-HindIII fragment containing the β gene. The bottom line shows the organization of the "diffuse" haplotype mouse β-like globin gene cluster (Jahn et al. 1980). The y, βh0, and βh1 genes are embryonic globin genes (Hansen et al. 1982; Farace et al. 1984), whereas the β-major (β^{maj}) and β-minor (β^{min}) genes serve as both fetal and adult globin genes; βh2 and βh3 are pseudogenes.

et al. 1983), it is not known whether these represent a cause or an effect of changes in gene expression. One popular although vaguely defined model is that the chromatin structure of a globin gene cluster or "domain," established early in the differentiation of an erythroid cell, dictates whether the embryonic, fetal, or adult genes will be accessible for transcription (Stalder et al. 1980b). An alternative model is that switching is mediated by the *trans*-acting factors that function during the later stages of erythroid cell differentiation (Chao et al. 1983; Wright et al. 1983), and that transcriptional factors specific for embryonic, fetal, or adult genes are present in erythroid cells at different stages of development. Experiments in which fetal globin genes were introduced into cultured adult erythroid cells by DNA-mediated gene transfer (Wright et al. 1983; Anagnou et al. 1985; M. Chao, pers. comm.) or by chromosomal transfer (Willing et al. 1979; Papayannopoulou et al. 1985) have been interpreted to support both types of mechanisms. In any case, it is likely that the process of switching is mediated by *cis*-acting DNA sequences within the globin gene clusters; these might interact with factors that modulate chromatin organization, with factors that more directly activate (or repress) transcription, or perhaps with both kinds of factors.

Our approach to these problems has involved the introduction of cloned globin genes into the mouse germ line by microinjection of DNA into the zygote (Gordon et al. 1980; Brinster et al. 1981; Costantini and Lacy 1981; Wagner et al. 1981) and the analysis of gene expression during the development of the transgenic mice. In the transgenic mouse, an introduced globin gene is present in the complete embryonic, fetal, and adult erythroid cell lineages, from the multipotential stem cell to the erythrocyte, as well as in all nonerythroid cells. If globin gene expression is regulated by the interaction of *trans*-acting factors with *cis*-acting DNA sequences in the globin gene clusters, then cloned globin genes introduced into the mouse germ line together with the *cis*-acting control elements should be expressed appropriately. As described in this paper, we find that cloned adult β-globin genes are indeed expressed in a tissue-specific (Chada et al. 1985; Townes et al. 1985) as well as a stage-specific pattern (Magram et al. 1985) after transfer into the mouse germ line. In addition, we find that a human fetal globin gene is subject to developmental regulation in transgenic mice; however, it behaves as an embryonic rather than a fetal globin gene. This finding can be explained by considering the evolutionary history of the mouse and human globin gene families, and provides insight into the possible mechanisms by which patterns of hemoglobin switching have evolved in mammals.

EXPERIMENTAL PROCEDURES

Production of transgenic mice. The DNA fragments described in Figure 1 were excised from plasmid clones of the human beta and Gγ-globin genes, purified by agarose gel electrophoresis, and eluted. After further purification by banding in cesium chloride and dialysis (Hogan et al. 1986), the DNAs were microinjected at concentrations of 1-2 µg/ml into the pronuclei of mouse zygotes, as described (Costantini and Lacy 1982; Hogan et al. 1985). Transgenic mice were identified by Southern blot analysis of DNA extracted from the tail (Grosschedl et al. 1984).

Analysis of DNase I hypersensitive sites. Mice were injected intraperitoneally on 4 consecutive days with 0.1 ml of 0.9% phenylhydrazine and were sacrificed on the sixth day. The spleens were removed and gently homogenized in anticoagulation buffer (2% acid citrate dextrose, 14% citrate phosphate dextrose, 84% phosphate-buffered saline [PBS], 1 mM phenylmethylsulfonyl fluoride [PMSF]; Wintrobe et al. 1981) in a Teflon glass homogenizer. Nucleated cells were isolated from the spleen homogenate by centrifugation over Lymphocyte Separation Medium (Bionetics), and nuclei were prepared according to the procedure of Larsen and Weintraub (1982). Brains were removed and homogenized in PBS, and nuclei were similarly isolated from total brain cells. DNase I digestions were performed according to Sheffery et al. (1982), using DNase concentrations ranging from 2 to 16 µg/ml.

Detection of human β-globin chains by direct immunofluorescence. Blood cells were collected from the tails of adult mice or from the umbilical cord of embryos; fetal livers were dissected and cell suspensions were prepared. The cells were washed and blood films were prepared, fixed, stained with a fluorescein-conjugated monoclonal antibody against human β-globin chains, and examined by fluorescence microscopy as described (Stamatoyannopoulos et al. 1983).

Analysis of human β- and Gγ-globin mRNA in transgenic mouse tissues. Peripheral blood (0.3-0.5 ml) was collected from the tails of adult mice into PBS containing 10 units/ml heparin (PBS/heparin), the cells were collected by centrifugation, and the pellet was resuspended in guanidine thiocyanate buffer (Chirgwin et al. 1979). RNA was isolated by centrifugation through cesium chloride and reprecipitation from guanidine hydrochloride buffer as described (Chirgwin et al. 1979). Transgenic embryos and fetuses were obtained by mating a transgenic male mouse with a normal female and sacrificing the female at the appropriate day of gestation (the morning the copulation plug was observed was designated day 0). Blood was collected by allowing the embryos to bleed through the umbilical cord into PBS/heparin, and the cells were recovered, washed by centrifugation, and lysed in guanidinium thiocyanate. Fetal livers were dissected and homogenized in guanidinium thiocyanate. Total RNA was prepared from the blood or livers of an entire litter, as described above for adult blood. To determine how many embryos in each litter had inherited the foreign globin gene, DNA was isolated from the remaining portion of each embryo or fetus and screened by dot blot hybridization. RNA concentrations were determined by absorbance at 260 nm, and the integrity of all RNA preparations was confirmed by formaldehyde-agarose gel electrophoresis and staining with ethidium bromide (Maniatis et al. 1982).

A 770-nucleotide EcoRI–PstI DNA fragment containing the 3' end of the human β-globin gene was cloned into the polylinker of the plasmid pSP64, and the SP6 polymerase was used to transcribe a ^{32}P-labeled single-stranded RNA probe (Melton et al. 1984). Similarly, a 560-nucleotide EcoRI–HindIII DNA fragment containing the 3' end of the human Gγ-globin gene was cloned into the same vector and a probe was prepared. Of each probe, 4 ng was hybridized with 1–10 μg of RNA in 15 μl of 80% formamide buffer, at 45°C (β probe) or 49.5°C (γ probe) as described by Chada et al. (1985). The hybrids were digested with RNase A (40 μg/ml) and RNase T1 (2 μg/ml) for 1 hour at 15°C, and analyzed by electrophoresis on a 6% polyacrylamide gel containing 7 M urea.

RESULTS AND DISCUSSION

Tissue-specific Expression of Adult β-Globin Genes

Before attempting to study the regulation of hemoglobin switching using transgenic mice, we asked the simpler question whether an introduced adult β-globin gene could be expressed in a tissue-specific pattern in the adult mouse. Cloned adult globin genes had been readily expressed when transfected into MEL cells (Chao et al. 1983; Wright et al. 1983). However, in several studies in which similar globin genes in plasmid or bacteriophage vectors were introduced into the mouse germ line, no expression in erythroid tissues was observed (e.g., Stewart et al. 1982; Lacy et al. 1983). It is now apparent that the failure of these early experiments was due to an inhibitory effect of prokaryotic vector sequences on globin gene expression in transgenic mice (Chada et al. 1985; Townes et al. 1985). The mechanism of this inhibition remains to be determined. It is not specific to globin genes, as inhibitory effects on other genes such as the mouse α-fetoprotein gene (R. Krumlauf and S. Tilghman, pers. comm.) and fusion genes containing the elastase promoter (R. Palmiter, pers. comm.) have been observed.

When an adult β-globin gene is excised and purified from vector sequences before being injected into mouse zygotes, the resulting transgenic mice express the gene specifically in erythroid cells. This was first demonstrated using a hybrid gene containing the 5' portion of the mouse β-major adult globin gene fused to the remaining 3' portion of the human adult β-globin gene, and including 1.2 kb of 5'- and 1.9 kb of 3'-flanking DNA (Chada et al. 1985). Animals in four out of seven transgenic lines carrying the hybrid gene expressed the gene exclusively or predominantly in erythroid tissues. This demonstrated that regulatory sequences closely linked to the β-globin gene are sufficient to specify a correct pattern of tissue-specific expression in a developing mouse, when the gene is integrated at a subset of foreign chromosomal positions. In this experiment, the steady-state levels of hybrid globin mRNA in erythroid cells ranged from 0.02% to 2% of the endogenous β-globin mRNA level (Chada et al. 1985), and this appears to result from a correspondingly low rate of transcription as measured by nuclear runoff analysis (G. Radice, unpubl.).

To ask whether the mouse DNA sequences in the hybrid gene were essential for regulated expression during mouse erythropoiesis, we have introduced an intact human β-globin gene into the mouse germ line. Similar experiments have been reported by Townes et al. (1985). We injected a 7.7-kb HindIII fragment containing the human adult β-globin gene (Fig. 1) into mouse zygotes, and obtained five transgenic mice. As an initial screen for expression of the human gene, we stained peripheral blood cells from each founder transgenic animal with a fluorescent monoclonal antibody specific for human β-globin chains (Stamatoyannopoulos et al. 1983). Several examples of the results are shown in Figure 2. Mouse Hβ56 (Fig. 2A) shows bright staining of all erythrocytes, while only about 5% of the erythrocytes from mouse Hβ66 (Fig. 2B) appear to contain human β-globin. The mosaic staining of cells from mouse Hβ66 may be due to a mosaic distribution of the foreign gene, and in accord with this explanation the animal failed to transmit the gene to any of its first 28 progeny. In all, four of the five transgenic mice appeared to contain human β-globin chains in some or all erythrocytes by this analysis (Table 1). In addition, blood from mouse Hβ56 was shown by isoelectric focusing to contain a globin chain comigrating with normal human β-globin chains (G. Stamatoyannopoulos and T. Papayannopoulou, unpublished data).

Human β-globin mRNA in the mouse blood cells was measured by hybridization with a single-stranded RNA probe, ribonuclease digestion, and analysis of the products on a polyacrylamide gel (Fig. 3). β-Globin mRNA from human cord blood protects a 212-nucleotide fragment of the probe from digestion, and the same protected fragment is seen when several of the transgenic mouse RNAs are analyzed. By this analysis, blood cells from two of the transgenic mice (Hβ56 and Hβ58) contain human β-globin mRNA at a level of 2000 pg/μg total RNA, which is comparable to the level of endogenous β-globin mRNA (Chada et al. 1985; Townes et al. 1985). A third mouse (Hβ66) contains a tenfold lower level (when corrected for mosaicism), and two mice contain no detectable human β-globin mRNA (Table 1). This range of expression levels is similar to that observed by Townes et al. (1985). A cloned mouse β-major globin gene has also been expressed at a relatively high level in transgenic mice; of two transgenic mice carrying a 7-kb EcoRI fragment including this gene, one expressed the gene at about 30% of the endogenous level (F. Costantini, unpubl.). It is not clear why the mouse/human hybrid β-globin gene used previously was expressed at much lower levels than either

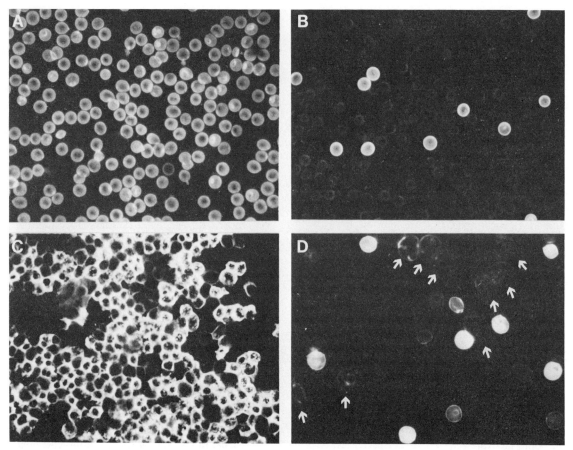

Figure 2. Detection of human β-globin chains in transgenic mouse erythroid cells by immunofluorescence labeling. (*A*) Peripheral red blood cells from adult founder transgenic mouse Hβ56 show virtually pancellular and homogeneous accumulation of human β-globin chains. (*B*) A similar preparation from founder mouse Hβ66 shows labeling of only a small proportion of red cells, presumably due to mosaicism for the foreign gene (see Table 1). (*C*) Positive staining of definitive fetal liver erythroblasts in a cytocentrifuge smear of cells from the liver of a 13.5-day fetus from line Hβ56. (*D*) Circulating blood cells pooled from four 13.5-day fetal progeny of mouse Hβ56. Some of the enucleated, fetal red cells are positive for human β-globin chains, whereas those that are negative probably derive from one or more fetuses that did not inherit the human gene. However, all of the larger, nucleated embryonic erythroblasts present in this field (arrows) are negative.

Table 1. Expression of the Human β-Globin Gene in Blood Cells

Founder mouse	Estimated gene copy number[a]	mRNA level[b] (pg/μg)	Anti-human β-globin fluorescent staining
Hβ56	50–100	2000	all cells strong positive
Hβ58	10–20	2000	strong, heterogeneous[c]
Hβ64	5–15	<1	medium, homogeneous[d]
Hβ66	50–100	10	5% of cells strong positive[e]
Hβ81	1	<1	negative

[a]Gene copy number, per diploid genome, was estimated from the intensity of bands on a Southern blot.

[b]The mRNA levels in total blood RNA were measured as described in the text and in Chada et al. (1985) using human cord blood RNA as a standard; the levels are expressed as picograms of human β-globin mRNA per microgram of total blood RNA. For comparison, the level of β-globin mRNA in mouse reticulocytes has been estimated to be 5000 (Chada et al. 1985) or 1500 (Townes et al. 1985).

[c]The reason for the heterogeneous staining of blood cells from this founder mouse is unclear; blood from a transgenic progeny in this line showed essentially homogeneous staining. The founder did not appear to be a mosaic as judged by germ line transmission frequency.

[d]Although blood cells from this founder animal stained moderately positive for human β-globin, no human β-globin mRNA was detected in one experiment.

[e]This founder mouse appeared to be a mosaic for the presence of the foreign gene. It transmitted the gene to less than 10% of its progeny. The level of human β-globin mRNA in positive cells, if corrected for the apparent degree of mosaicism, is approximately 200 pg/μg.

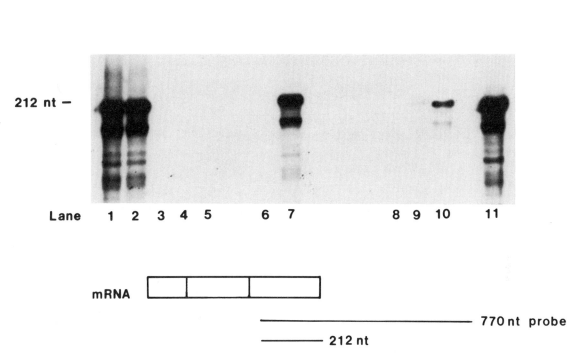

Figure 3. Analysis of human β-globin mRNA in transgenic mice. The amount of human β-globin mRNA in RNA samples from transgenic mice carrying the human β-globin gene (*Hin*dIII fragment shown in Fig. 1) was measured by hybridization with a uniformly labeled single-stranded RNA probe, followed by ribonuclease digestion, electrophoresis on a urea-polyacrylamide gel, and autoradiography. A 212-nucleotide fragment of the probe is protected from digestion by hybridization to human β-globin mRNA, as shown in the diagram. (Lanes *1–5*) Hybridization with 0.5 μg of total blood RNA from the indicated founder transgenic mouse. On a longer exposure, a 212-nucleotide band is also visible in lane *3* (mouse Hβ66). (Lane *6*) 0.7 μg of total embryonic blood RNA from a litter of eight 11.5-day embryos sired by mouse Hβ56, six of which inherited the human β-globin gene. (Lane *7*) 0.8 μg of total fetal liver RNA from a litter of nine 14.5-day embryos sired by mouse Hβ56, six of which inherited the human β-globin gene. (Lanes *8–11*) Human cord blood RNA containing the indicated number of picograms of β-globin mRNA (Chada et al. 1985).

the mouse or human gene alone. One possibility is a partial inhibition of expression by the small (350 nucleotide) segment of plasmid DNA introduced with the hybrid gene (Chada et al. 1985).

DNase I Hypersensitive Sites in a Foreign Globin Gene

Transcriptionally active globin genes have been shown to assume an altered chromatin conformation, one feature of which is an extreme sensitivity to nuclease cleavage at specific sites within or flanking the gene (Stalder et al. 1980b; Sheffery et al. 1982; Groudine et al. 1983). To ask whether an introduced globin gene in a foreign chromosomal position could establish a normal chromatin structure during erythroid cell differentiation, we analyzed several transgenic mouse lines carrying the mouse/human hybrid β-globin gene for the presence of DNase I hypersensitive sites. If the hybrid gene develops DNase hypersensitivity at the same sites where the normal mouse and human β-globin genes are hypersensitive (Sheffery et al. 1982; Groudine et al. 1983), one would expect to see DNase I cleavage at the three sites designated by arrows in Figure 4.

Figure 4 shows an analysis of one transgenic mouse from a line (46) that carries a single copy of the hybrid β-globin gene and contains hybrid β-globin mRNA specifically in erythroid tissues. Nuclei isolated from adult erythroblasts (from the spleen of an anemic animal), or from brain cells, were digested with varying amounts of DNase I. DNA was then isolated, digested with *Pst*I, and analyzed by Southern blotting using a probe for the large intron of the human β-globin gene (Fig. 4). In DNA from erythroid cell nuclei, subbands of 2.4 and 2.7 kb were generated, consistent with DNase cleavage at two of the predicted sites in the promoter region and in the third exon. Double digestion of the DNA with *Bam*HI and *Pst*I yielded a single subband of 1.0 kb (data not shown), confirming the location of these two DNase cleavage sites (Fig. 4). By a similar analysis, DNase cleavage at the third predicted hypersensitive site in the 3′-flanking region was also detected (data not shown). None of the three DNase hypersen-

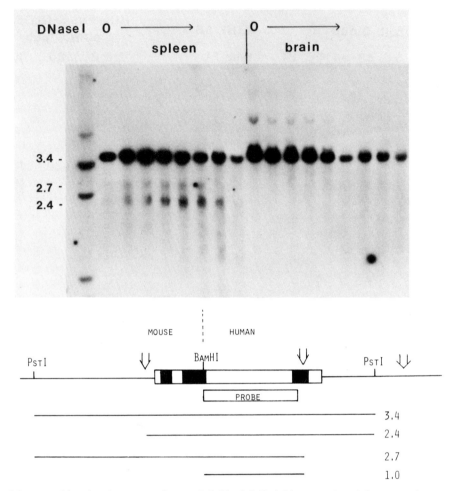

Figure 4. DNase I hypersensitive sites in a mouse/human hybrid adult β-globin gene. An adult transgenic mouse carrying a single copy of a hybrid mouse/human β-globin gene (line 46, Chada et al. 1985) was treated with phenylhydrazine to induce splenic erythropoiesis. Nuclei isolated from spleen cells or brain cells were digested with varying amounts of DNase I, and the DNA was isolated, digested with PstI, fractionated by agarose gel electrophoresis, transferred to nitrocellulose, and hybridized to a probe for the large intron of the human β-globin gene (BamHI-EcoRI fragment). In the absence of DNase I cleavage, the probe hybridizes to a 3.4-kb PstI fragment. The arrows in the diagram show the location of a DNase I hypersensitive site seen in the promoter region of the mouse β-major globin gene (Sheffery et al. 1982) and two hypersensitive sites in the third exon and 3'-flanking region of the human β-globin gene (Groudine et al. 1983) in erythroid cells. In the transgenic mouse spleen DNA, but not brain DNA, two subbands of 2.4 and 2.7 kb are seen, consistent with DNase cleavage at the normal hypersensitive sites in the 5'-flanking DNA and in the third exon. When the same DNA was digested with BamHI and PstI and hybridized with the same probe, a single 1.0-kb subband was observed (data not shown), confirming the location of these two hypersensitive sites.

sitive sites could be detected in brain cell nuclei from the same mouse (Fig. 4).

A summary of the data obtained with three different transgenic lines is presented in Table 2. In two lines that express the foreign gene specifically in erythroid tissues (at the mRNA level), the predicted DNase I hypersensitive sites were observed in erythroid cell but not brain cell nuclei. In a third transgenic line that does not express the foreign gene, no hypersensitive sites were detected in erythroid or brain cell nuclei.

Table 2. DNase I Hypersensitive Sites in a Mouse/Human Hybrid β-Globin Gene, in Erythroid, and in Nonerythroid Cells

Mouse line	Gene copy number	Expression (mRNA)		DNase I hypersensitivity	
		erythroid	brain	erythroid	brain
46	1	+	−	+	−
77	3	+	−	+	−
47	1	−	−	−	−

"Erythroid" cells were total nucleated cells isolated from the spleens of mice made anemic with phenylhydrazine. DNase I hypersensitive sites were detected as described in the text and Fig. 4. Expression data and gene copy numbers are from Chada et al. (1985).

Several conclusions can be drawn from these experiments. First, as transcriptionally active endogenous globin genes always contain DNase I hypersensitive sites (Stalder et al. 1980b; Sheffery et al. 1982; Groudine et al. 1983), our observations support the interpretation that the tissue-specificity of hybrid globin mRNA accumulation in the transgenic mice (Chada et al. 1985) is due at least in part to transcriptional regulation. Second, although the level of synthesis of the hybrid β-globin mRNA in the mice is only 1-2% of the endogenous β-globin mRNA level, the hybrid gene contains all the normal DNase I hypersensitive sites. If *trans*-acting factors in erythroid cells are responsible for the generation of hypersensitive sites, the hybrid gene appears to be appropriately recognized by these factors. The low level of expression may result from a failure of some other step in the process of gene activation. These results are consistent with the observation that the human adult β-globin gene contains a normal complement of DNase I hypersensitive sites in fetal liver erythroid cells, in which it is expressed at only 5% of its maximal level (Groudine et al. 1983).

Third, the apparently inactive chromatin conformation of the foreign globin gene in brain cell nuclei contrasts with the results obtained when globin genes are transfected into nonerythroid cells in culture. In the in vitro experiments, transfected genes appear to assume an active conformation independent of the recipient cell type, and to circumvent the suppression that normally occurs during development (Robins et al. 1982; Weintraub 1983; Charnay et al. 1984). In contrast, during the development of a transgenic mouse the normal suppression mechanisms appear to operate in nonerythroid cell types. This implies that the DNA sequences responsible for this early event in globin gene regulation should be amenable to analysis using transgenic mice.

Stage-specific Expression of Adult β-Globin Genes

The expression of the hyrid mouse/human β-globin gene in erythroid cells of adult transgenic mice allowed us to begin to approach the issue of hemoglobin switching. To ask whether the hybrid adult gene was subject to appropriate stage-specific regulation, we analyzed the expression of the gene in erythroid cells of the transgenic mice at various stages of development. Circulating blood cells, consisting of primitive nucleated erythrocytes derived from the embryonic yolk sac, were collected from transgenic embryos at 10.5 or 11.5 days of gestation. Mouse erythroid cells at this stage synthesize the embryonic β-like chains y and z, but do not make significant amounts of the adult β-globin chains (Craig and Russell 1964; Fantoni et al. 1967; Gilman and Smithies 1968). The hybrid mouse/human globin mRNA was not detected in the embryonic blood cells in any of three transgenic lines examined (Magram et al. 1985). The fetal liver is the major erythropoietic organ between days 12 and 17 of gestation (Kovach et al. 1967; Rifkind et al. 1969), and these fetal erythroid cells are the first to express the mouse adult β-globin genes at high levels (Craig and Russell 1964; Fantoni et al. 1967). The mouse/human hybrid gene was found to be expressed in the fetal livers of all four transgenic lines that express the gene in adult erythroid tissues (Magram et al. 1985).

A similar developmental analysis was conducted for one transgenic line (Hβ56) carrying the human β-globin gene, and the same result was obtained. Figure 2 shows that while human β-globin chains are found in adult red cells (Fig. 2A), in nucleated fetal liver erythroblasts (Fig. 2C), and in circulating enucleated fetal red cells (Fig. 2D), they are not detected in circulating nucleated cells derived from the embryonic yolk sac (Fig. 2D). Figure 3 shows that at the mRNA level, expression of the human β-globin gene in this line is also confined to fetal and adult erythroid cells.

Thus, a cloned adult β-globin gene introduced into the mouse germ line is first activated at the same stage of development as the endogenous adult β-globin genes. The human and mouse/human hybrid genes, which include only a few kilobases of 5'- or 3'-flanking DNA, contain the information to remain inactive in primitive erythroid cells which are synthesizing embryonic hemoglobins, and to be specifically activated later in development in fetal and adult erythroid cells.

Embryonic Expression of a Human Fetal Globin Gene in Transgenic Mice

In humans, the Gγ- and Aγ-globin genes (Fig. 1) serve as fetal globin genes; that is, they are inactive in the earliest yolk sac erythroid cells, they are expressed maximally through the remainder of gestation, and are almost entirely switched off after birth (Weatherall and Clegg 1981). To ask how a fetal globin gene would behave in the mouse, which has no separate fetal globin genes, we produced transgenic mice carrying the human Gγ-globin gene (a 4-kb *Bgl*II–*Sph*I DNA fragment shown in Fig. 1). The stage-specificity of expression of this gene was analyzed as described above for the adult β-globin gene, and the results for one transgenic line (G 9) are shown in Figure 5. Human Gγ-globin mRNA was detected in 11.5-day embryonic blood but not in 14-day fetal liver (Fig. 5) or adult blood (data not shown). Similar results were obtained with two other independent transgenic lines, while in three other lines no expression was observed at any stage (data not shown). As seen in Figure 5, the level of Gγ mRNA in the embryonic mouse blood cells is similar to that observed in the human cell line K562 after induction with hemin. K562 is a leukemic cell line that can be induced to undergo erythroid differentiation, synthesizing embryonic and fetal hemoglobins (Lozzio and Lozzio 1975; Benz et al. 1980). We conclude that in the developing mouse, a human fetal globin gene behaves as an embryonic rather than a fetal gene (K. Chada et al., in prep.).

To interpret this result, it is necessary to consider the evolutionary history of the β-like globin gene clusters in mouse and humans. The human Gγ and Aγ genes, and the mouse βh0 and βh1 genes (Fig. 1), are thought

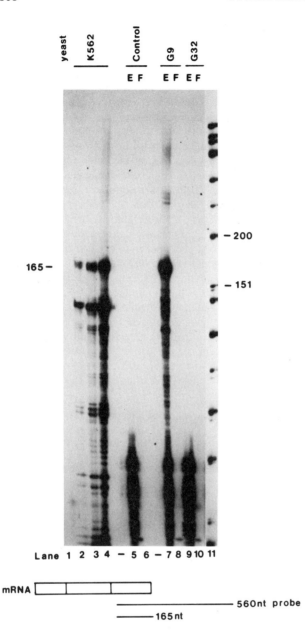

Figure 5. Analysis of human Gγ-globin mRNA in transgenic mice. Human Gγ-globin mRNA was measured by hybridization with a 560-nucleotide RNA probe, ribonuclease digestion, urea-polyacrylamide gel electrophoresis, and autoradiography; Gγ mRNA protects a 165-nucleotide fragment of the probe from digestion. (Lane 1) Hybridization with yeast RNA; (lanes 2–4) hybridization with 1, 2, or 10 μg, respectively, of total RNA from human K562 cells induced with 20 μM hemin (a gift of M. Donovan-Peluso, Columbia University). (Lanes 5 and 6) 10 μg of total RNA from normal 11.5-day mouse embryonic blood or normal 14.5-day mouse fetal liver, respectively. (Lanes 7 and 9) 10 μg of total blood RNA from a litter of 11.5-day embryos sired by mouse G9 and mouse G32, respectively. (Lanes 8 and 10) 10 μg of total fetal liver RNA from a litter of 14.5-day fetuses sired by mice G9 and G32, respectively. The fraction of transgenic embryos or fetuses in each litter were: lane 7, 5/12; lane 8, 2/7; lane 9, 3/11; lane 10, 2/6.

to descend from a single ancestral gene termed proto-γ (Czelusniak et al. 1982; Hardison 1984; Hill et al. 1984). The mouse βh0 and βh1 genes are embryonic rather than fetal in their pattern of expression (Farace et al. 1984), and there is reason to believe that the ancestral proto-γ gene was also an embryonic globin gene (Hill et al. 1984). Thus, the proto-γ gene was "recruited" for fetal expression during the evolution of the primates, while it continued to function as an embryonic gene in most other mammalian orders.

When a human Gγ gene is introduced into the mouse germ line, we find that it reverts to an embryonic pattern of expression. This suggests that the gene has retained its ability to respond specifically to signals produced in embryonic erythroid cells in the mouse; the shift in its pattern of expression during the evolution of primates appears, therefore, to have been accomplished by a shift in the timing of the *trans*-acting regulatory signals specific for the γ genes. An alternate mechanism of fetal recruitment might have been a change in *cis*-acting elements associated with the γ genes, causing them to respond to a different fetal regulatory signal and no longer to an embryonic signal (Hill et al. 1984). However, such a model predicts that a human γ-globin gene would have lost the ability to be expressed in mouse embryonic blood cells, and might either behave as a fetal gene in the mouse or simply fail to be expressed. Our results therefore argue in favor of an evolutionary mechanism of the former type.

Aside from these evolutionary considerations, our studies indicate that the developmental activation and inactivation of a fetal globin gene, like that of the adult β-globin gene, can be controlled by closely linked *cis*-

acting sequences. The gross organization of the globin gene cluster appears not to be functionally important for the control of hemoglobin switching. At least some of the *trans*-acting regulatory factors that interact with the γ- and β-globin genes appear to be different, because the two genes show opposite patterns of developmental stage-specific expression in transgenic mice. The number of regulatory factors involved in switching, the stage of erythroid cell differentiation at which they act, and the relationship of these factors to those defined by other experimental approaches (Chao et al. 1983; Wright et al. 1983; Emerson et al. 1985) remain to be established.

The ability of cloned globin genes to undergo switching during mouse development opens several new avenues of experimentation. First, by generating "fusion genes" incorporating different portions of the β- and Gγ-globin genes, and testing their developmental regulation in transgenic mice, it should be possible to define the DNA sequences that interact with stage-specific regulatory factors. Second, the transgenic mouse may provide an experimental system in which to analyze the various mutations that cause heritary persistence of fetal hemoglobin in humans. If these mutations occur in regulatory sequences involved in the postnatal inactivation of fetal globin synthesis, as has been proposed (for review, see Collins and Weissman 1985), the mutant γ-globin genes may show an altered pattern of stage-specific expression in the developing mouse.

ACKNOWLEDGMENTS

K.C. was supported by a fellowship from NATO and F.C. by an Irma T. Hirschl Career Scientist Award. This work was funded by grants from the National Institutes of Health (HD17704 and HL20899) and the National Foundation–March of Dimes (MDBDF 5-410).

REFERENCES

Anagnou, N.P., S. Karlsson, A.D. Moulton, G. Keller, and A.W. Neinhuis. 1985. Promoter sequences required for function of the human gamma-globin gene. *Proc. Natl. Acad. Sci.* (in press).

Benz, E.J., M.J. Murname, B.L. Tonkonow, B.W. Berman, E.M. Mazur, C. Cavellesco, T. Jenko, E.L. Snyder, B.G. Forget, and R. Hoffman. 1980. Embryonic-fetal erythroid characteristics of a human leukemic cell line. *Proc. Natl. Acad. Sci.* 77: 3509.

Brinster, R.L., H.Y. Chen, M. Trumbauer, A.W. Senear, R. Warren, and R.D. Palmiter. 1981. Somatic expression of a herpes thymidine kinase gene in mice following injection of a fusion gene into eggs. *Cell* 27: 223.

Chada, K., J. Magram, K. Raphael, G. Radice, E. Lacy, and F. Costantini. 1985. Specific expression of a foreign beta-globin gene in erythroid cells of transgenic mice. *Nature* 314: 337.

Chao, M.V., P. Mellon, P. Charnay, T. Maniatis, and R. Axel. 1983. The regulated expression of β-globin genes introduced into mouse erythroleukemia cells. *Cell* 32: 483.

Charnay, P., R. Treisman, P. Mellon, M. Chao, R. Axel, and T. Maniatis. 1984. Differences in human alpha- and beta-globin gene expression in mouse erythroleukemia cells: The role of intragenic sequences. *Cell* 38: 251.

Chirgwin, J., A. Przybyla, R. MacDonald, and W.J. Rutter. 1979. Isolation of biologically active ribonucleic acid from sources enriched in ribonuclease. *Biochemistry* 18: 5294.

Cohen, R.B. and M. Sheffery. 1985. Nucleosome disruption precedes transcription and is largely limited to the transcribed domain of globin genes in murine erythroleukemia cells. *J. Mol. Biol.* 182: 109.

Collins, F.S. and S.M. Weissman. 1985. The molecular genetics of human hemoglobin. *Prog. Nucleic Acid Res.* 31: 315.

Costantini, F. and E. Lacy. 1981. Introduction of a rabbit beta-globin gene into the mouse germ line. *Nature* 294: 92.

———. 1982. Gene transfer into the mouse germ line. *J. Cell. Physiol.* (suppl.) 1: 219.

Craig, M.L. and E.S. Russell. 1964. A developmental change in hemoglobins correlated with an embryonic red cell population in the mouse. *Dev. Biol.* 18: 191.

Czelusniak, J., M. Goodman, D. Hewett-Emmett, M.L. Weiss, P.J. Venta, and R.E. Tashian. 1982. Phylogenetic origins and adaptive evolution of avian and mammalian haemoglobin genes. *Nature* 298: 297.

Emerson, B.M., C.D. Lewis, and G. Felsenfeld. 1985. Interaction of specific nuclear factors with the nuclease-hypersensitive region of the chicken adult β-globin gene: Nature of the binding domain. *Cell* 41: 21.

Fantoni, A., A. Bank, and P.A. Marks. 1967. Globin composition and synthesis of hemoglobins in developing fetal mice erythroid cells. *Science* 157: 1327.

Farace, M.G., B.A. Brown, G. Raschella, J. Alexander, R. Gambari, A. Fantoni, S.C. Hardies, C.A. Hutchison III, and M.H. Edgell. 1984. The mouse βh1 gene codes for the z chain of embryonic hemoglobin. *J. Biol. Chem.* 259: 7123.

Fritsch, E.F., R.M. Lawn, and T. Maniatis. 1980. Molecular cloning and characterization of the human β-like globin gene cluster. *Cell* 19: 959.

Gilman, J.G. and O. Smithies. 1968. Fetal hemoglobin variants in mice. *Science* 160: 885.

Gordon, J.W., G.A. Scangos, D.J. Plotkin, J.A. Barbosa, and F.H. Ruddle. 1980. Genetic transformation of mouse embryos by microinjection of purified DNA. *Proc. Natl. Acad. Sci.* 77: 7380.

Grosschedl, R., D. Weaver, D. Baltimore, and F. Costantini. 1984. Introduction of a μ immunoglobulin gene into the mouse germ line: Specific expression in lymphoid cells and synthesis of functional antibody. *Cell* 38: 647.

Groudine, M.A. and H. Weintraub. 1982. Propagation of globin DNase I-hypersensitive sites in absence of factors required for induction: A possible mechanism for determination. *Cell* 30: 131.

Groudine, M., T. Kohwi-Shigematsu, R. Gelinas, G. Stamatoyannopoulos, and T. Papayannopoulou. 1983. Human fetal to adult hemoglobin switching: Changes in chromatin structure of the β-globin gene locus. *Proc. Natl. Acad. Sci.* 80: 7551.

Hansen, J.N., D.A. Konkel, and P. Leder. 1982. The sequence of a mouse embryonic β-globin gene. *J. Biol. Chem.* 257: 1048.

Hardison, R.C. 1984. Comparison of the β-like globin gene families of rabbits and humans indicates that the gene cluster 5'-ε-γ-δ-β-3' predates the mammalian radiation. *Mol. Biol. Evol.* 1: 390.

Hill, A., S.C. Hardies, S.J. Phillips, M.G. Davis, C.A. Hutchison III, and M.H. Edgell. 1984. Two mouse early embryonic β-globin gene sequences: Evolution of the nonadult β-globins. *J. Biol. Chem.* 259: 3739.

Hogan, B., F. Costantini, and E. Lacy. 1986. *Manipulating the mouse embryo: A laboratory manual.* Cold Spring Harbor Laboratory, Cold Spring Harbor, New York. (In press.)

Jahn, C.L., C.A. Hutchison III, S.J. Phillips, S. Weaver, N.L. Haigwood, C.F. Voliva, and M.H. Edgell. 1980.

DNA sequence organization of the β-globin complex in the BALB/c mouse. *Cell* **21:** 159.

Kovach, J.S., P.A. Marks, E.S. Russell, and H. Epler. 1967. Erythroid cell development in fetal mice: Ultrastructural characteristics and hemoglobin synthesis. *J. Mol. Biol.* **25:** 131.

Lacy, E., S. Roberts, E.P. Evans, M.D. Burtenshaw, and F. Costantini. 1983. A foreign beta-globin gene in transgenic mice: Integration at abnormal chromosomal positions and expression in inappropriate tissues. *Cell* **34:** 343.

Larsen, A. and H. Weintraub. 1982. An altered DNA conformation detected by S1 nuclease occurs at specific regions in active chick globin chromatin. *Cell* **29:** 609.

Lozzio, C.B. and B.B. Lozzio. 1975. Human chronic myelogenous leukemia with positive Philadelphia chromosome. *Blood* **45:** 321.

Magram, J., K. Chada, and F. Costantini. 1985. Developmental regulation of a cloned adult beta-globin gene in transgenic mice. *Nature* **315:** 338.

Maniatis, T., E.F. Fritsch, and J. Sambrook. 1982. *Molecular cloning: A laboratory manual.* Cold Spring Harbor Laboratory, Cold Spring Harbor, New York.

Melton, D.A., P.A. Krieg, M.R. Rebagliati, T. Maniatis, K. Zinn, and M.R. Green. 1984. Efficient in vitro synthesis of biologically active RNA and RNA hybridization probes from plasmids containing a bacteriophage SP6 promoter. *Nucleic Acids Res.* **12:** 7035.

Papayannopoulou, T., D. Lindsley, S. Kurachi, K. Lewison, T. Hemenway, M. Melis, N.P. Anagnou, and V. Najfeld. 1985. Adult and human fetal globin genes are expressed following chromosomal transfer into MEL cells. *Proc. Natl. Acad. Sci.* **82:** 780.

Rifkind, R.A., D.H.K. Chui, and H. Epler. 1969. An ultrastructural study of early morphogenetic events during the establishment of fetal hepatic erythropoiesis. *J. Cell Biol.* **40:** 343.

Robins, D.M., I. Paek, P.H. Seeburg, and R. Axel. 1982. Regulated expression of human growth hormone genes in mouse cells. *Cell* **29:** 623.

Sheffery, M., R.A. Rifkind, and P.A. Marks. 1982. Murine erythroleukemia cell differentiation: DNaseI hypersensitivity and DNA methylation near the globin genes. *Proc. Natl. Acad. Sci.* **79:** 1180.

Stalder, J., M. Groudine, J.B. Dodgson, J.D. Engel, and H. Weintraub. 1980a. Hb switching in chickens. *Cell* **19:** 973.

Stalder, J., A. Larsen, J.D. Engel, M. Dolan, M. Groudine, and H. Weintraub. 1980b. Tissue-specific cleavages in the globin domain introduced by DNase I. *Cell* **20:** 451.

Stamatoyannopoulos, G., M. Farquhar, D. Lindsley, M. Brice, T. Papayannopoulou, and P.E. Nute. 1983. Monoclonal antibodies specific for globin chains. *Blood* **61:** 530.

Stewart, T.A., E.F. Wagner, and B. Mintz. 1982. Human β-globin gene sequences injected into mouse eggs, retained in adults, and transmitted to progeny. *Science* **217:** 1046.

Townes, T.M., J.B. Lingrel, R.L. Brinster, and R.D. Palmiter. 1985. Erythroid specific expression of human beta-globin genes in transgenic mice. *EMBO J.* **4:** 1715.

Wagner, E.F., T.A. Stewart, and B. Mintz. 1981. The human beta-globin gene and a functional thymidine kinase gene in developing mice. *Proc. Natl. Acad. Sci.* **78:** 5016.

Weatherall, D.J. and J.B. Clegg. 1981. *The thalassemia syndromes*, 3rd edition. Blackwell, Oxford.

Weintraub, H. 1983. A dominant role for DNA secondary structure in forming hypersensitive structures in chromatin. *Cell* **32:** 1191.

Willing, M.C., A.W. Neinhuis, and W.F. Anderson. 1979. Selective activation of human beta- but not gamma-globin gene in human fibroblast × mouse erythroleukaemia cell hybrids. *Nature* **277:** 534.

Wintrobe, M., G. Lee, D. Boggs, T. Bithell, J. Foerster, J. Athens, and J. Lukens. 1981. Transfusion of blood and blood components. Transfusion of blood and blood components. *Clinical hematology*, 8th edition, p. 492. Lea and Febiger, Philadelphia.

Wright, S., E. deBoer, F.G. Grosveld, and R.A. Flavell. 1983. Regulated expression of the human β-globin gene family in murine erythroleukemia cells. *Nature* **305:** 333.

Wright, S., A. Rosenthal, R. Flavell, and F. Grosveld. 1984. DNA sequences required for regulated expression of β-globin genes in murine erythroleukemia cells. *Cell* **38:** 266.

Yu, J. and R.D. Smith. 1985. Sequential alterations in globin gene chromatin during erythroleukemia cell differentiation. *J. Biol. Chem.* **260:** 3035.

Regulated Expression of α-Fetoprotein Genes in Transgenic Mice

R. KRUMLAUF,* R.E. HAMMER,† R. BRINSTER,† V. M. CHAPMAN, ‡
AND S.M. TILGHMAN*

*Institute for Cancer Research, Philadelphia, Pennsylvania 19111; †Laboratory of Reproductive Physiology, School of Veterinary Medicine, University of Pennsylvania, Philadelphia, Pennsylvania 19104; ‡Roswell Park Memorial Institute, Buffalo, New York 14263

Eukaryotic organisms have evolved multiple ways in which to generate variation in the tissue-specific expression of cellular genes. In some instances, gene duplication followed by divergence of the regulatory sequences has been used to accomplish this, as best typified by the actin gene families from sea urchins (Angerer and Davidson 1984) to humans (Gunning et al. 1983). Alternative splicing of the primary transcripts of a single-copy gene in the calcitonin gene (Rosenfeld et al. 1984) and tissue-specific use of duplicated promoters in the mouse α-amylase gene (Schibler et al. 1983) represent ways in which diversity is built into single-copy genes.

None of these mechanisms can be invoked to explain the diverse pattern of expression of the murine α-fetoprotein (AFP) and albumin genes. These genes, which arose as the products of a gene duplication 500 million years ago (Kioussis et al. 1981), form a small multigene family on chromosome 5 of the mouse (D'Eustachio et al. 1981; Ingram et al. 1981). Both genes are activated in concert during embryonic development in the visceral endoderm of the yolk sac, the fetal liver, and the fetal gastrointestinal tract (Tilghman and Belayew 1982; Young and Tilghman 1984). However the levels of AFP and albumin mRNA are markedly different in these tissues, with AFP mRNA constituting 20% of the mRNA in the visceral endoderm, 10% in fetal liver, and less than 0.1% in the gut (Andrews et al. 1982; Janzen et al 1982; Tilghman and Belayew 1982). Albumin mRNA, on the other hand, is present at high levels only in liver.

The AFP gene is differentially regulated in liver relative to the albumin gene in that its rate of transcription undergoes a 10,000-fold decline shortly after birth (Tilghman and Belayew 1982). This decline is under genetic control by at least one *trans*-acting regulatory locus, termed *raf*, which determines the adult basal level of AFP mRNA (Olsson et al. 1977; Belayew and Tilghman 1982). We wish to understand the biochemical basis for the common tissue distribution of the expression of these two genes, as well as the means by which their rates of transcription in each tissue are established.

Insight into the nature of the *cis*-acting DNA sequences required for the tissue-specific activation of the AFP in visceral endoderm has recently been obtained from experiments using F9 teratocarcinoma cells (Scott et al. 1984) which are capable of differentiation into either visceral or parietal endoderm under the appropriate stimuli (Hogan et al. 1983). Modified copies of a cloned AFP gene that were introduced into F9 cells by DNA-mediated transformation were activated only upon formation of visceral endoderm, thereby displaying a tissue-specific pattern of expression qualitatively similar to the endogenous AFP gene. These results established that the introduced DNA, which included the entire 14-kb intergenic DNA between the 3' end of the albumin gene and the 5' end of the AFP gene (see Fig. 1), contained sufficient sequence information to confer tissue-specific expression, at least in visceral endoderm. However, the response of the gene in the other tissues and during development could not be ascertained.

Recent studies have demonstrated that transgenic mice that carry microinjected genes integrated into nonhomologous chromosomal locations can express the gene appropriately and transmit the expression to progeny (Brinster et al. 1983; Grosschedl et al 1984; Storb et al. 1984; Swift et al. 1984). This approach provides an ideal experimental system for identifying the molecular mechanisms underlying the diverse patterns of expression of the AFP gene during development and to distinguish whether the mechanisms for activation and postnatal repression in liver are the same or different. For this reason, modified AFP minigenes have been introduced into the germ line of mice via the microinjection of fertilized eggs (Krumlauf et al. 1985).

One of the other important consequences of introducing DNA into heterologous locations in the mouse germ line is that it provides an opportunity to examine the effect that chromosomal position plays in the

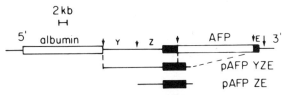

Figure 1. Structure of mouse AFP minigenes. The AFP minigenes YZE and ZE were prepared using cloned Y, Z, and E *Eco*RI restriction fragments, indicated by the arrows, from the albumin-AFP locus. YZE contains 14 kb of 5'-flanking DNA and ZE contains 7 kb of 5'-flanking DNA.

expression of the genes. For example, what chromosomal position effects would be exerted on the expression of an autosomal gene integrated on the inactivated X chromosome? The identification of transgenic animals carrying X-linked AFP genes allows us to examine whether the inactivation process can extend to an autosomal gene that is normally expressed at high levels in the cell.

EXPERIMENTAL PROCEDURES

Minigene constructs and microinjection. The AFP minigenes YZE and ZE were constructed (Scott et al. 1984; Krumlauf et al. 1985) by joining in pBR322 two (Z,E) or three (Y,Z,E) EcoRI fragments from the albumin/AFP locus, as shown in Figure 1. Transgenic mice were produced using the YZE and ZE minigene plasmids linearized at the unique SalI site in pBR322. The male pronucleus of F_2 hybrid eggs (obtained by mating C57Bl/6 × SJL hybrid adults) were microinjected with 2 pl containing approximately 260 copies of YZE or 360 copies of ZE. The injected eggs were reimplanted into pseudopregnant mice as previously described (Brinster et al. 1981, 1983) and allowed to develop for 18–19 days, at which time the fetuses and their yolk sacs were removed by caesarean section and frozen.

Isolation and analysis of nucleic acids. Tissues were dissected from frozen embryos rapidly thawed in 0.9% NaCl. DNA and RNA were isolated from the fetal tissues by phenol extraction according to Iynedjian and Hanson (1977) and Krumlauf et al. (1985). DNA was obtained from sections of mouse tails as described by Palmiter et al. (1982). DNAs were digested with restriction enzymes and following electrophoresis in agarose gels transferred to nitrocellulose. Poly(A)$^+$ RNA was isolated by oligo(dT)-cellulose chromatography (Aviv and Leder 1972), electrophoresed in formaldehyde agarose gels, and transferred to nitrocellulose (Thomas 1980). Filters were hybridized in 50% formamide containing dextran sulfate, as described (Wahl et al. 1979), except that hybridizations with the single-stranded RNA probe were performed at 57°C.

A hybridization probe specific for AFP was prepared by subcloning into pUC9 a 440-bp HincII fragment containing the entire first exon and a portion of the first intron of the AFP gene. Purified insert was isolated by digestion with HincII, followed by electrophoresis and electroelution, and the fragments were labeled by nick-translation. For RNA analysis an SP6 single-stranded RNA probe (Melton et al. 1984) was prepared by subcloning the 440-bp HincII fragment into pSP65 (Promega Biotec).

Tissue dissections and enzyme assays. C3H/HeHa females congenic for the variant isozymes phosphoglycerate kinase (PGK)-A and hypoxanthine-guanine phosphoribosyl transferase (HPRT)-A were crossed to male transgenics which expressed PGK-B, HPRT-B. At 14–16 days of gestation, fetal livers were harvested and the yolk sac was separated into its endodermal and mesodermal components by enzymatic digestion (West et al. 1977). PGK and HPRT assays were performed as previously described (West et al. 1977; Kratzer et al. 1983). The sex of fetuses was determined by hybridizing their DNAs to a retroviral probe (M720) which is able to detect Y-chromosome-specific sequences (Phillips et al. 1982).

RESULTS

Microinjection of Modified Copies of the AFP Gene

Two cloned AFP minigenes that differ only in the amount of 5'-flanking DNA were chosen for microinjection into mouse eggs. The YZE minigene has 14 kb of 5'-flanking DNA, which includes the entire intergenic region between the albumin and AFP genes, while ZE has 7 kb of 5' flanking sequence (Fig. 1). These constructs, termed pAFPYZE and pAFPZE, are comprised of the first three and last two coding blocks of the 15-coding-block AFP gene (Ingram et al. 1981; Kioussis et al. 1981), with 0.4 kbp of 3'-flanking DNA. This five-exon minigene, when introduced into cells, produces a fully processed, correctly initiated poly(A)$^+$ transcript 600 nucleotides long (Scott et al. 1984; Krumlauf et al. 1985) which can be readily distinguished from the endogenous 2.2-kb AFP mRNA by gel electrophoresis.

In our initial experiments all progeny that resulted from the microinjection of linearized pAFPYZE or pAFPZE were sacrificed by caesarean section at 18–19 days of gestation. Genomic DNA was isolated from the yolk sacs and was used to identify those animals that carried the microinjected DNA. Approximately 25% of the mice at term carried exogenous YZE or ZE DNA, with the copy number varying from 1 to >70 copies per diploid genome, as summarized in Table 1 for the nonmosaic animals. The DNAs were typically integrated into single chromosomal locations and in breeding studies segregated as Mendelian markers, as has been described for other genes (Brinster et al. 1981; Palmiter et al. 1982; Lacy et al. 1983).

Tissue-specific Expression of the AFP Minigenes

Total poly (A)$^+$ RNA was isolated from the yolk sacs of animals identified as carrying integrated DNA to ask whether the minigenes were being expressed. Significant levels of a 600-nucleotide RNA that comigrated with authentic minigene RNA and initiated at the AFP cap site (Krumlauf et al. 1985) were observed in 15 of 34 yolk sacs examined (Table 1). The data in Table 1 serve to illustrate that not all transgenic animals express the minigene and there is variability in the levels of minigene RNA in those that do. This variability in the level of minigene mRNA, which ranged from 0.1% to 25% of the endogenous AFP mRNA representing 40 to 10,500 copies of RNA per cell, does not correlate with the copy number of the DNA.

Table 1. Analysis of Transgenic Mice Carrying AFP Minigenes

A. Minigene construct		5'-Flanking DNA	DNA-positive mice	Mice expressing minigene RNA
YZE		14 kb	8/51	3/8 (37%)
ZE		7 kb	26/109	12/26 (46%)

B. Minigene construct	Mouse	Minigene copies per cell	Level of minigene mRNA in yolk sac (percent of endogenous AFP)	Minigene mRNA in yolk sac (molecules/cell)
YZE	5-2	4	8	3,100
YZE	39-6	12	1	420
YZE	42-1	0.5	0.2	80
ZE	50-2	10	25	10,500
ZE	28-4	2	10	4,200
ZE	47-3	12	8	3,400
ZE	35-7	8	4	1,700
ZE	47-4	40	3.4	1,400
ZE	47-2	50	2.2	920
ZE	32-3	6	2	840
ZE	55-6	1	1.7	710
ZE	54-4	70	1.5	630
ZE	54-3	70	0.9	380
ZE	55-5	50	0.6	250
ZE	47-1	15	0.1	40

The number of DNA-positive mice does not include 12 mosaic animals also observed.

DNA copy numbers and relative levels of minigene mRNA were derived from autoradiographs by densitometric analysis of different exposures, using the endogenous AFP gene and mRNA levels as internal standards. The endogenous AFP mRNA comprises 20% of the yolk sac poly(A)+ equal to 42,000 molecules/cell.

Recently, the source of the variable expression of the AFP minigene in the yolk sac of transgenic mice was attributed largely to the presence of the prokaryotic vector sequence on the DNA fragment injected. As shown in Figure 2, when the levels of minigene RNA in yolk sacs were examined in 12 fetuses that had developed from eggs microinjected with the ZE minigene freed of pBR322 DNA, all 12 exhibited high-level expression. Chada et al. (1985) suggested the possibility of plasmid sequences affecting gene expression and Townes et al. (1985) demonstrated a dramatic increase in the expression of human β-globin genes in transgenic mice following removal of prokaryotic DNA (Hammer et al., this volume). The explanation for this repression is at present unknown. The rest of the experiments described in this paper were conducted with animals that had been generated earlier with plasmid-bearing fragments.

To determine whether the YZE and ZE minigenes were being expressed in other tissues, those 19-day fetuses producing minigene mRNA in yolk sac were dissected into liver, gut, kidney, heart, and brain. Genomic DNA was isolated from the fetal tissues at the same time to ensure that each contained the same number of copies of exogenous genes originally observed for the corresponding yolk sac (data not shown). Total poly(A)+ RNAs from the fetal tissues were fractionated by electrophoresis and hybridized to an AFP probe (see Fig. 3). In addition to the yolk sac, authentic AFP mRNA was detected in fetal liver, and to a lesser extent fetal gut, but not in kidney, heart, or brain. These results, which demonstrate a strict observance by the minigene of appropriate tissue-specific expression, have been reproduced with 10 independent transgenic lineages.

The levels of exogenous transcripts were generally lower than AFP mRNA in the liver and gut as well as in the yolk sac. In most cases, the yolk sac, liver, and gut in a single animal exhibited levels of minigene mRNA that represented similar percentages of the endogenous AFP mRNA, as shown for mice 28-4 and 35-7 in Figure 3. However, occasionally a discordancy was observed, where the levels of minigene mRNA in fetal

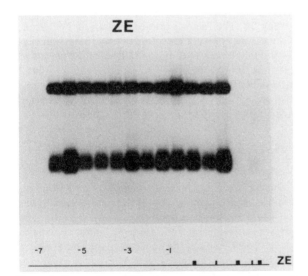

Figure 2. Expression of ZE minigene mRNA in yolk sacs of transgenic mice containing no vector sequences. Twelve fetuses which developed from eggs microinjected with gel-purified fragments of the ZE minigene were sacrificed at 19 days. Total poly(A)+ RNA was isolated from the individual yolk sacs and analyzed by hybridization after gel electrophoresis for expression of AFP mRNA (*upper band*) and ZE mRNA (*lower band*).

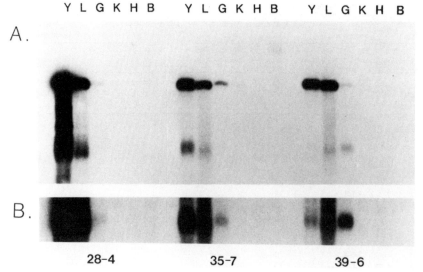

Figure 3. Analysis of minigene expression in fetal tissues of three transgenic mice. Three fetuses at days 18–19 of gestation were dissected into five tissues. Total poly(A)+ mRNA was isolated from each tissue and 15–30 ng of RNA from the yolk sac (Y), liver (L), gut (G), kidney (K), heart (H), and brain (B) was hybridized to an AFP-specific single-stranded RNA probe as detailed in Experimental Procedures. A 3-hr (*A*) exposure and 36-hr (*B*) exposure of bottom half the same filter. The upper band represents the authentic 2.2-kb AFP mRNA, and the lower band the 600-nucleotide minigene mRNA.

gut more closely approximated that of the endogenous AFP mRNA in that tissue. For example, in mouse 39-6 the relative levels of YZE mRNA were 250% of the endogenous level in fetal gut, but only 1% in the yolk sac and 4% in the liver. We have also looked for minigene expression in the tissues of the transgenic mice that did not express the minigene in our original yolk sac screen. None of these expressed YZE or ZE in any tissue. In summary, 44% (15/34) of the transgenic mice carrying the AFP minigenes expressed significant levels of mRNA, always in a tissue-specific manner. The observation that the YZE and ZE constructs behaved identically argues that the intergenic region between 7 kb and 14 kb upstream of the AFP gene is not required for its tissue-specific expression.

Developmental Regulation of the AFP Minigene in Liver

We next wished to examine whether the ZE minigene contained sequences sufficient for its transcriptional repression in neonatal liver. Fertilized eggs microinjected with ZE constructs were allowed to develop to term, delivered by caesarean section, and given to foster mothers. Hybridization analyses of the yolk sac DNA and poly(A)+ RNA were used to identify those progeny that carried and expressed the minigene, and were therefore of interest. For the developmental studies, we first confirmed that the tissue-specific pattern of expression in a given line was stably inherited in all offspring. For this purpose F_1 animals from five separate transgenic lines were sacrificed 1 day after birth and total poly(A)+ RNA was isolated from several tissues. In each of three lines all progeny that inherited the minigene expressed its mRNA in a qualitatively and quantitatively identical manner as shown for one line in Figure 4. In the two other lines none of the offspring produced minigene transcripts. This apparently represents an inability of some integration sites to transmit expression through the germ line as has been observed for other genes (Palmiter et al. 1982; Lacy et al. 1983; Wagner et al. 1983).

Liver poly(A)+ RNA was isolated at day 18 of gestation and 3, 7, 14, and 28 days after birth from the three transgenic lines that were transmitting the expression of the minigene. As shown for two lines in Figure 5, the minigene and AFP mRNA levels decreased over 100-fold between days 3 and 14 and were not detectable by day 28. The relative rate of decline of both transcripts as measured by densitometry was identical, suggesting that their transcription rates were decreasing at the same time and to a similar extent. From this data, we conclude that the ZE minigene contains the regions required not only for tissue specificity, but also for the postnatal transcriptional repression in liver.

The Expression of an AFP Minigene Integrated on the X Chromosome

During breeding experiments with the 47-3-13 line, we observed that segregation of the AFP minigene was consistent with its being X linked. That is, in over 300 progeny derived from five transgenic males, all females but no males were transgenic. In contrast, females were able to transmit the introduced AFP genes equally to males and females.

Previous studies had shown that genes on the paternally derived X chromosome were preferentially inactivated in the extraembryonic visceral and parietal endoderm of the yolk sac (West et al. 1977; Papaioannou and West 1981). To determine whether the X-linked AFP genes were also inactivated on the paternal X, an-

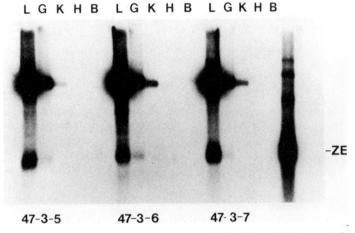

Figure 4. Analysis of minigene expression in F$_1$ progeny. Tissues were isolated from day-1 neonatal transgenic litter mates derived from 47-3-13. Total poly(A)$^+$ RNAs were prepared from liver (L), gut (G), kidney (K), heart (H), and brain (B). The RNAs were electrophoresed in an agarose gel, blotted, and hybridized with an AFP-specific single-stranded RNA probe.

imals were obtained by caesarean section from matings between (C57BL/6 × SJL)F$_1$ mice with either male or female transgenic mice at days 14–16 of fetal development. Total yolk sac poly(A)$^+$ RNA from transgenic progeny was isolated, fractionated by electrophoresis, and hybridized to an AFP probe. The results shown in Figure 6 indicate that the AFP minigene was expressed on both the paternal and maternal X chromosomes in the yolk sac. There was at most a two- to threefold higher level of AFP minigene expression on the maternal X chromosome.

To ensure that the paternal X chromosome was actually inactivated, male transgenic animals carrying the X-linked $Pgk-1^b$ and $Hprt^b$ loci were crossed to C3H/He female mice carrying the $Pgk-1^a$ and $Hprt^a$ alleles. Fetuses were harvested at days 14–16 of gestation and the visceral endoderm was isolated from the yolk sacs by enzymatic digestion. Assays for PGK, HPRT, and the AFP minigene mRNA were performed on both fetal liver and visceral endoderm. In liver both the A and B isozymes of PGK and HPRT were expressed, while only the A isozymes from the maternal-derived

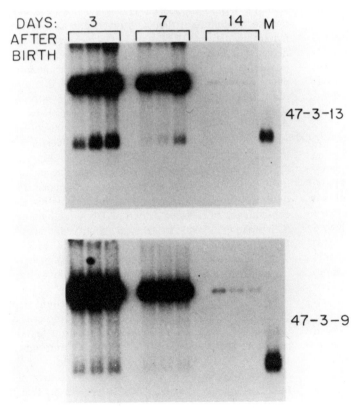

Figure 5. Developmental regulation of the levels of minigene and AFP mRNA in neonatal liver. Transgenic males 47-3-9 and 47-3-13 were mated to females and progeny were sacrificed at 3, 7, and 14 days after birth. Mice carrying minigenes were identified by DNA blot analysis and poly(A)$^+$ mRNA was isolated from the corresponding livers. RNAs (100 ng) from three animals at each of the time points were electrophoresed in an agarose gel, blotted, and hybridized with a labeled, single-stranded RNA probe homologous to the first AFP coding block.

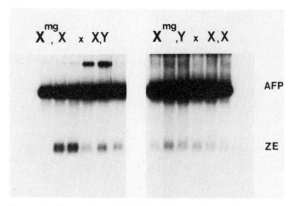

Figure 6. Expression of AFP minigenes located on the X chromosome. Male (*right panel*) and female (*left panel*) transgenic mice, denoted by "mg" were mated to (C57BL/6 × SJL) F_1 and progeny were obtained by caesarean section at days 14–16 of gestation. The yolk sacs were removed and total poly(A)$^+$ RNAs derived from them were electrophoresed on an agarose gel, blotted, and hybridized with an AFP-specific probe.

X chromosome were expressed in the visceral endoderm of the same animals (data not shown). As before, the AFP minigene mRNA was expressed in both tissues. Therefore, although the paternal X is preferentially inactivated in the yolk sac, the minigenes can escape this inactivation process.

DISCUSSION

Transgenic mice provide an ideal approach to investigate the requirements for diverse patterns of gene expression of a single-copy gene in different tissues. In the case of the AFP gene, the interaction between tissue-specific factors and their response to humoral signals has been reproduced with the minigenes in mice in that the pattern of expression of the minigenes was qualitatively identical to the endogenous AFP gene. Indeed, we have never generated an animal in which the minigene was expressed in an inappropriate tissue.

Appropriate expression of the ZE minigene was transmitted to progeny in three of five lines, allowing us to ask whether the minigene mRNA was also developmentally regulated in neonatal liver. The data in Figure 5 clearly demonstrate that this was the case, in that the reductions in AFP and minigene RNAs followed a parallel course. This is the first instance in which a cloned gene introduced into the mouse genome has been shown not only to express in the proper tissues but also to respond to developmental signals that modify its pattern of expression.

From this we conclude that the DNA signals necessary for developmental repression as well as tissue-specific activation are contained in the DNA that has been introduced into mice. The precise location of these sequences and their degree of overlap, if any, must now be addressed. More recently we have introduced into mice constructs that contain only 1 kb of 5'-flanking DNA. In all cases, no expression of the AFP minigene has been detected, suggesting that sequences between 1 and 7 kb are essential for expression (R.E. Hammer et al., unpubl.). It remains to be established whether sequences further upstream, including those in the closely linked albumin gene, have any influence on the expression of the AFP gene.

We have observed several instances in which the relative levels of minigene RNA in each of the three tissues in a single animal are not identical. A particularly striking example of this is shown in Figure 3, where the relative level of minigene RNA in the gut of 39-6 is significantly higher than in either liver or yolk sac. Likewise, several animals exhibited higher levels in the liver than in the yolk sac, in contrast to the endogenous mRNA. These unusual relationships might be explained by differences in the activity of chromosomal domains adjacent to the foreign gene in the different tissues (Jaenisch et al. 1981). However another intriguing possibility is that these fluctuations are indicative of different regulatory mechanisms used by these tissues to modulate the transcription of the gene. If that is so, then it may be possible, with the appropriate constructions in transgenic mice, to dissociate the expression of the minigene in each of the three cell lineages and examine the DNA sequence requirements for both the activation and the developmental modulation of AFP in each tissue. Now that we can obtain consistent high-level expression of the minigene relatively free of position effects by injecting only the eukaryotic DNA, such experiments are feasible and in progress.

One striking example where the AFP minigene was able to overcome a predicted chromosome position effect was on the paternally derived inactive X chromosome in visceral endoderm. This could not be attributed to the failure of that X to inactivate, as only the maternally derived HPRT-A and PGK-A isozymes were detected in the tissue. What seemed more likely was the possibility that either the establishment or maintenance of X inactivation in visceral endoderm does not result in a stably condensed heterochromatic chromosome. Functional and biochemical differences between the inactive X chromosome in visceral endoderm versus other somatic tissues include the relatively early time of replication of the former early in development (Takagi 1974; Sugawara et al. 1983), and the accessibility of X-linked genes like *hprt* for gene transfer into tissue culture cells (Kratzer et al. 1983), possibly the consequence of a lesser degree of DNA methylation (Chapman et al. 1982; Lester et al. 1982; Venolia et al. 1982). It is thought that these differences lie at the level of the maintenance rather than the establishment of the inactive state, as by all markers tested, both chromosomes are equally inactive.

To ask, then, whether the AFP minigene could also escape inactivation on a nonvisceral endoderm-derived inactive X, we crossed a transgenic male to a female carrying a Searle's translocation between chromosomes 16 and X. In all females that carry the translocated X, the normal X (in this case derived from the transgenic father) is inactive (Takagi 1980; Disteche et al. 1981); otherwise partial monosomy of chromosome 16 would result. In the fetal livers of such females, the minigene was inactive (R. Krumlauf et al., in prep.). Thus the transcriptional activity of the minigene on the inactive X chromosome in visceral endoderm is specific to the state of that inactive X, and not due to the integration of the minigene into a region that normally escapes inactivation. The exact biochemical nature of that difference is of intense interest, but is, as yet, unknown.

ACKNOWLEDGMENTS

We thank Beverly Hazel for technical assistance, Myrna Trumbauer for experimental assistance and R. Palmiter, R. Scott, and T. Vogt for valuable discussions. This work was supported by grants HD17321, CA06927, CA28050, and RR05539 from the National Institutes of Health and an appropriation from the Commonwealth of Pennsylvania. R.E.H. was supported by National Institutes of Health postdoctoral training grant HD07155; R.K. was supported by National Institutes of Health postdoctoral training grant CA09035.

REFERENCES

ANDREWS, G.K., R.G. JANZEN, and T. TAMAOKI. 1982. Stability of α-fetoprotein messenger RNA in mouse yolk sac. *Dev. Biol.* **89:** 111.

ANGERER, R.C. and E.H. DAVIDSON. 1984. Molecular indices of cell lineage specification in sea urchin embryos. *Science* **226:** 1153.

AVIV, H. and P. LEDER. 1972. Purification of biologically active globin mRNA by chromatography on oligothymidylic acid-cellulose. *Proc. Natl. Acad. Sci.* **69:** 1408.

BELAYEW, A. and S.M. TILGHMAN. 1982. Genetic analysis of α-fetoprotein synthesis in mice. *Mol. Cell. Biol.* **2:** 1427.

BRINSTER, R.L., H.Y. CHEN, M. TRUMBAUER, A.W. SENEAR, R. WARREN, and R.D. PALMITER. 1981. Somatic expression of herpes thymidine kinase in mice following injection of a fusion gene into eggs. *Cell* **27:** 223.

BRINSTER, R.L., K.A. RITCHIE, R.E. HAMMER, R.L. O'BRIEN, B. ARP, and U. STORB. 1983. Expression of a microinjected immunoglobulin gene in the spleen of transgenic mice. *Nature* **306:** 332.

CHADA, K., J. MAGRAM, K. RAPHAEL, G. RADICE, E. LACY, and F. COSTANTINI. 1985. Specific expression of a foreign β-globin gene in erythroid cells of transgenic mice. *Nature* **314:** 377.

CHAPMAN, V.M., P.G. KRATZER, L.D. SIRACUSA, B.A. QUARANTILLO, R. EVANS, and R.M. LISKAY. 1982. Evidence for DNA modification in the maintenance of X-chromosome inactivation of adult mouse tissue. *Proc. Natl. Acad. Sci.* **79:** 5357.

D'EUSTACHIO, P., R.S. INGRAM, S.M. TILGHMAN, and F.H. RUDDLE. 1981. Murine α-fetoprotein and albumin: Two evolutionarily linked proteins encoded on the same mouse chromosome. *Somatic Cell Genet.* **7:** 289.

DISTECHE, C.M., E.M. EICHER, and S.A. LATT. 1981. Late replication patterns in adult and embryonic mice carrying Searle's X-autosome translocation. *Exp. Cell Res.* **133:** 357.

GROSSCHEDL, R., D. WEAVER, D. BALTIMORE, and F. COSTANTINI. 1984. Introduction of a μ immunoglobulin gene into the mouse germ line: Specific expression in lymphoid cells and synthesis of functional antibody. *Cell* **38:** 647.

GUNNING, P., P. PONTE, H. BLAU, and L. KEDES. 1983. α-Skeletal and α-cardiac actin genes are coexpressed in adult human skeletal muscle and heart. *Mol. Cell. Biol.* **3:** 1985.

HOGAN, B.L.M., D.P. BARLOW, and R. TILLY. 1983. F9 teratocarcinoma cells as a model for the differentiation of parietal and visceral endoderm in the mouse embryo. *Cancer Surv.* **2:** 115.

INGRAM, R.S., R.W. SCOTT, and S.M. TILGHMAN. 1981. α-Fetoprotein and albumin genes are in tandem in the mouse genome. *Proc. Natl. Acad. Sci.* **78:** 4694.

IYNEDJIAN, P.B. and R.W. HANSON. 1977. Increase in level of functional messenger RNA coding for phosphoenolpyruvate carboxykinase (GTP) during induction by cyclic adenosine 3':5'-monophosphate. *J. Biol. Chem.* **252:** 655.

JAENISCH, R., D. JAHNER, P. NOBIS, I. SIMON, J. LOHLER, K. HARBERS, and D. GROTKOPP. 1981. Chromosomal position and activation of retroviral genomes inserted into the germ line of mice. *Cell* **24:** 519.

JANZEN, R.G., G.K. ANDREWS, and T. TAMAOKI. 1982. Synthesis of secretory proteins in developing mouse yolk sac. *Dev. Biol.* **90:** 18.

KIOUSSIS, D., F. EIFERMAN, P. VAN DE RIJN, M.B. GORIN, R.S. INGRAM, and S.M. TILGHMAN. 1981. The evolution of α-fetoprotein and albumin II. The structures of the α-fetoprotein and albumin genes in the mouse. *J. Biol. Chem.* **256:** 1960.

KRATZER, P.G., V.M. CHAPMAN, H. LAMBERT, R.E. EVANS, and R.M. LISKAY. 1983. Differences in the DNA of the inactive X chromosomes of fetal and extraembryonic tissues of mice. *Cell* **33:** 37.

KRUMLAUF, R., R.E. HAMMER, S.M. TILGHMAN, and R.L. BRINSTER. 1985. Developmental regulation of α-fetoprotein genes in transgenic mice. *Mol. Cell. Biol.* **5:** 1639.

LACY, E., S. ROBERTS, E.P. EVANS, M.D. BURTENSHAW, and F.D. COSTANTINI. 1983. A foreign β-globin gene in transgenic mice: Integration at abnormal chromosomal positions and expression in inappropriate tissues. *Cell* **34:** 343.

LESTER, S.C., N.J. KORN, and R. DEMARS. 1982. Derepression of genes on the human inactive X chromosome: Evidence for differences in locus-specific rates of derepression and rates of transfer of active and inactive genes after DNA-mediated transfection. *Somatic Cell Genet.* **8:** 265.

MELTON, D.A., P.A. KRIEG, M.R. RABAGLIATI, T. MANIATIS, K. ZINN, and M.R. GREEN. 1984. Efficient *in vitro* synthesis of biologically active RNA and RNA hybridization probes from plasmids containing a bacteriophage SP6 promoter. *Nucleic Acids Res.* **12:** 7025.

OLSSON, M., G. LINDAHL, and E. RUOSHLAHTI. 1977. Genetic control of alphafetoprotein synthesis in the mouse. *J. Exp. Med.* **145:** 819.

PALMITER, R.D., R.L. BRINSTER, R.E. HAMMER, M.E. TRUMBAUER, M.G. ROSENFELD, N.C. BIRNBERG, and R.M. EVANS. 1982. Dramatic growth of mice that develop from eggs microinjected with metallothionein-growth hormone fusion gene. *Nature* **300:** 611.

PAPAIOANNOU, V.E. and J.D. WEST. 1981. Relationship between the parental origin of the X chromosomes, embryonic cell lineage and X chromosome expression in mice. *Genet. Res.* **37:** 183.

PHILLIPS, S.J., E.H. BIRKENMEIER, R. CALLAHAN, and E.M. EICHER. 1982. Male and female mouse DNA's can be discriminated using retroviral probes. *Nature* **297:** 241.

ROSENFELD, M.G., S.G. AMARA, and R.M. EVANS. 1984. Alternative RNA processing: Determining neuronal phenotype. *Science* **225:** 1315.

SCHIBLER, U., O. HAGENBUCHLE, P.K. WELLAUER, and A.C. PITTET. 1983. Two promoters of different strengths control the transcription of the mouse alpha-amylase gene amy-1^a in the parotid gland and the liver. *Cell* **33:** 501.

SCOTT, R.W., T.F. VOGT, M.E. CROKE, and S.M. TILGHMAN. 1984. Tissue-specific activation of a cloned α-fetoprotein gene during differentiation of a transfected embryonal carcinoma cell line. *Nature* **310:** 562.

STORB, U., R.L. O'BRIEN, M.D. MCMULLEN, K.A. GOLLAHON, and R.L. BRINSTER. 1984. High expression of cloned immunoglobulin κ gene in transgenic mice is restricted to B lymphocytes. *Nature* **310:** 238.

SUGAWARA, O., N. TAKAGI, and M. SASAKI. 1983. Allocyclic

early replicating X chromosome in mice: Genetic inactivity and shift into a late replicator in early embryogenesis. *Chromosoma* **88**: 133.

Swift, G.H., R.E. Hammer, R.J. MacDonald, and R.L. Brinster. 1984. Tissue-specific expression of the rat pancreatic elastase I gene in transgenic mice. *Cell* **38**: 639.

Takagi, N. 1974. Differentiation of X chromosomes in early female mouse embryos. *Exp. Cell Res.* **86**: 127.

———. 1980. Primary and secondary nonrandom X chromosome inactivation in early female mouse embryos carrying Searle's translocation T(X;16)16H. *Chromosoma* **81**: 439.

Thomas, P.S. 1980. Hybridization of denatured RNA and small DNA fragments transferred to nitrocellulose. *Proc. Natl. Acad. Sci.* **77**: 5201.

Tilghman, S.M. and A. Belayew. 1982. Transcriptional control of the murine albumin/α-fetoprotein locus during development. *Proc. Natl. Acad. Sci.* **79**: 5254.

Townes, T.M., J.B. Lingrel, H.Y. Chen, R.L. Brinster, and R.D. Palmiter. 1985. Erythroid specific expression of human β-globin genes in transgenic mice. *EMBO J.* (in press).

Venolia, L., S.M. Gartler, E.R. Wassman, P. Yen, T. Mohandas, and L.J. Shapiro. 1982. Transformation with DNA from 5-azacytidine-reactivated X chromosomes. *Proc. Natl. Acad. Sci.* **79**: 2352.

Wagner, E.F., L. Covarrubias, T.A. Stewart, and B. Mintz. 1983. Prenatal lethalities in mice homozygous for human growth hormone gene sequences integrated in the germ line. *Cell* **35**: 647.

Wahl, G.M., M. Stern, and G.R. Stark. 1979. Efficient transfer of large DNA fragments from agarose gels to diazobenzylorymethyl-paper and rapid hybridization by using dextran sulfate. *Proc. Natl. Acad. Sci.* **76**: 3683.

West, J.D., W.I. Freis, V.M. Chapman, and V. Papaioannou. 1977. Preferential expression of the maternally derived X chromosome in the mouse yolk sac. *Cell* **12**: 873.

Young, P.R. and M.S. Tilghman. 1984. Induction of α-fetoprotein synthesis in differentiating F9 teratocarcinoma cells is accompanied by a genome-wide loss of DNA methylation. *Mol. Cell. Biol.* **4**: 989.

Use of Gene Transfer to Increase Animal Growth

R.E. HAMMER,* R.L. BRINSTER,* AND R.D. PALMITER†
Laboratory of Reproductive Physiology, University of Pennsylvania, School of Veterinary Medicine, Philadelphia, Pennsylvania 19104; †Department of Biochemistry, Howard Hughes Medical Institute, University of Washington, Seattle, Washington 98195

In mammals, somatic growth is controlled by a cascade of circulating hormones emanating from the hypothalamus, pituitary gland, and liver. Growth hormone (GH), an intermediate in this cascade, is a 22,000-dalton, single-chain polypeptide produced in acidophiles of the anterior pituitary gland. The regulation of GH secretion is mediated, in part, by two hypothalamic hormones that reach the pituitary via the hypothalamo-hypophyseal portal blood system. Somatostatin, a neuropeptide of 14 or 28 amino acids inhibits the release of GH while growth hormone-releasing factor (GRF) stimulates both GH synthesis and secretion. GH stimulates the production of somatomedin C or insulin-like growth factor I (IGF-I) which is thought to mediate postnatal somatic growth by binding to membrane receptors of peripheral mescenchymal cells.

The genes coding for polypeptides involved in the regulation of growth have been cloned, including GRF (Mayo et al. 1983, 1985), somatostatin (Shen et al. 1982; Goodman et al. 1983), GH (Barta et al. 1981; Seeburg 1982; Woychik et al. 1982; Barsh et al. 1983; Gordon et al. 1983), and IGF-I and -II (Jansen et al. 1983; Bell et al. 1984; Ullrich et al. 1984). To determine if expression of integrated foreign GH genes accelerated growth of eggs (Hammer et al. 1984a), we introduced rat (r) or human (h) GH genes into mice by microinjection. Transgenic mice containing these constructs had very low levels of foreign GH mRNA, no foreign serum GH, and did not show enhanced growth, in agreement with the results of Wagner et al. (1983). Because a fusion gene composed of the mouse metallothionein I (MT-I) promoter/regulator and herpes virus thymidine kinase structural gene was expressed in transgenic mice (Brinster et al. 1981), we decided to use this promoter to direct the expression of GH genes.

Transgenic mice harboring either MTrGH or MThGH fusion genes commonly exhibit high, metal-inducible levels of fusion mRNA, substantial quantities of foreign GH in their sera, and enhanced growth (Palmiter et al. 1982b, 1983). In addition, the gene is stably incorporated into the germ line, making the gene and large phenotype heritable. Here we report the results from experiments in transgenic mice examining the growth-promoting action of MT-bovine (b) GH constructs, the steroid and metal responsiveness of the human MT-II$_A$ promoter in mice and rabbits, and the consequences of MT-human GRF gene expression in mice.

EXPERIMENTAL PROCEDURES

Plasmid Constructions

The 2.6-kb MTbGH DNA fragment includes the mouse MT-I promoter and 5'-flanking sequences to the BstEII site (-350) fused to the bGH structural gene (Gordon et al. 1983).

The 2.7-kb hMTII$_A$hGH fusion gene includes 500 bp of the human MT-II (hMT) promoter and 5'-flanking sequence (Karin et al. 1984b) fused to the hGH structural gene (Barsh et al. 1983).

The 2.5-kb MThGRF construct combines a 750-bp fragment of the mouse MT-I gene fused to a human GRF minigene that includes the entire coding region of the GRF precursor protein (Mayo et al. 1983, 1985). The human GRF gene includes both cDNA and genomic clones such that 10 kb of human GRF has been reduced to less than 1 kb and retains a single intron of 230 bp.

Microinjection

The results of several gene transfer experiments in our laboratory suggested that prokaryotic vector sequences could adversely affect gene expression. To test this possibility further, we constructed two groups of identical genes with and without plasmid DNA (Table 1). In both groups, removal of the vector DNA resulted in a dramatic increase in both the number of fetuses that express the gene and the level of expression. Interference of vector DNA on the expression of the human β-globin gene (Townes et al. 1985) and the mouse α-fetoprotein gene (Krumlauf et al., this volume) has also been observed. On the basis of these results we now routinely remove prokaryotic DNA from every gene we introduce into animals.

Purified linearized preparations of plasmid without prokaryotic DNA were diluted to about 2 ng/μl and microinjected into the male pronuclei of F$_2$ hybrid eggs of C57Bl/6 × SJL parents (Brinster et al. 1985) or fertilized rabbit eggs from superovulated New Zealand White rabbits, as previously described (Hammer et al. 1985b). The injected eggs were transferred into the reproductive tracts of pseudopregnant mice or rabbits and allowed to develop to term. Animals that retained the foreign DNA were detected by DNA dot hybridization to nucleic acids isolated from tail biopsies (Brinster et al. 1985). Those animals that retained the foreign DNA sequences were analyzed for expression.

Table 1. Effect of Prokaryotic DNA on Expression of MThGH Fusion Genes in Liver of Transgenic Fetuses

Group[a]	Prokaryotic DNA (presence)	Fetuses[b] (number with MThGH genes)	hGH mRNA[c] (number)	(molecules/cell)
1	+	10	4	10 ± 3
	−	18	11	112 ± 46
2	+	11	0	0
	−	17	5	105 ± 64

[a]MThGH constructs in group 1 contained the mouse MT-I promoter (−185 to +6) fused to +2 of the hGH gene with (+) and without (−) ~1.5 kb of prokaryotic cloning vector. Constructs in group 2 contained MT-I (−185 to −46) fused to −90 of the hGH gene with (+) and without (−) ~2.6 kb of cloning vector. Transgenic mice were made as described in Experimental Procedures.

[b]The presence of MThGH sequences was determined by dot hybridization to liver nucleic acids isolated from 13-day-old fetuses.

[c]MThGH-mRNA was measured by solution hybridization as described in Experimental Procedures. Values are means ± SEM.

Animal Treatments

To induce MT expression, transgenic animals were either injected twice with $CdSO_4$ (1 mg/kg) 18 hours and 4 hours before partial hepatectomy, or their drinking water was supplemented with 25 mM $ZnSO_4$ for 2 weeks and then a partial hepatectomy was performed. To determine the effect of glucocorticoids on hMThGH expression, dexamethasone (10 mg/kg) was administered as equal amounts of a soluble and insoluble preparation to adrenalectomized or intact animals 18 hours and 6 hours before partial hepatectomy.

mRNA Analysis

Tissues (~70 mg) were homogenized in 4 ml of SET (1% SDS, 1 mM EDTA, 10 mM Tris, pH 7.5) with 100 μg of proteinase K. Total nucleic acids were then isolated by phenol/chloroform extraction and ethanol precipitation (Durnam and Palmiter 1983). MTbGH and hMTGH mRNAs were measured by solution hybridization with ^{32}P-labeled oligonucleotide probes (Ornitz et al. 1985). MThGRF mRNA was analyzed by Northern blotting and quantitated by densitometry of autoradiograms (Hammer et al. 1985a).

GH and GRF Analysis

For detection of hGH, mouse serum was assayed in duplicate using a hGH radioimmunoassay kit provided by Dr. Raiti (National Hormone and Pituitary Program) (Hammer et al. 1985b). Animals with values less than 2 ng/ml were designated negative for hGH. Procedures used for the detection of mouse GH and hGRF in plasma from MThGRF mice have been described (Hammer et al. 1985a).

RESULTS AND DISCUSSION

Expression of MTbGH Genes

The linear DNA fragment containing the mouse MT-bGH fusion gene (MTbGH) that was microinjected into F_2 hybrid mouse eggs is shown in Figure 1. Out of the 65 animals that developed from injected eggs, 10

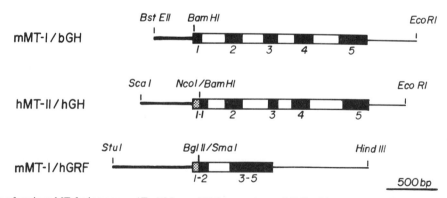

Figure 1. Diagram of various MT fusion genes. (*Top*) Mouse MT-I promoter and 5'-flanking sequences back to the *Bst*EII site (−350) fused to bGH structural gene. The fusion is between an artificial *Bam*HI site at +9 of mMT-I and +6 of bGH. (*Middle*) Human MT-II$_A$ promoter and 5'-flanking sequence back to *Sca*I (−500) fused to hGH gene. The fusion is between an *Nco*I site (+71) of hMT-II and *Bam*HI (+6) of hGH. (*Bottom*) Mouse MT-I promoter and 5'-flanking sequence back to *Stu*I (−750) fused to a hGRF minigene. The fusion is between a *Bgl*II site (+64) of mMT-I and a *Sma*I site in exon 2 of hGRF. The hGRF minigene has the introns between exons 3 and 5 deleted. In all diagrams, the exons are shown as solid or stippled boxes and they are numbered; introns are open boxes; the heavy solid line represents the MT 5'-flanking sequences and the stippled boxes represent the first exon of MT gene. The diagrams show the DNA fragments that were excised from bacterial plasmids, purified by agarose gel electrophoresis, and microinjected into fertilized eggs.

mice contained from 1 to 50 copies of the fusion gene and 7 grew significantly larger than control littermates (Table 2). As observed in other MT-GH mice (Palmiter et al. 1982b, 1983), there was no correlation between growth rates and MTbGH copy number. Transgenic mice exhibiting this enhanced growth had substantial quantities of bGH mRNA in their liver. In one animal, 899-2, bGH mRNA was induced about 20-fold by supplementing drinking water with $ZnSO_4$ (Table 2).

Two of the largest bGH males (896-3, 897-7) were bred to verify that this fusion gene and the large phenotype were heritable (Table 2). Offspring from these matings were analyzed for the presence of the MTbGH gene and for their rate of growth. In one pedigree (896-3), the gene was transmitted to about half of the offspring (8/21) and all of the transgenic progeny grew to about twice the size of normal littermates. In the other pedigree (897-7), none of the 13 offspring inherited the gene. Such incidences of germ line mosaicism occur frequently and current estimates are that ~30% of transgenic founder animals exhibit some degree of mosaicism (Palmiter and Brinster 1985). The three MTbGH females that exhibited enhanced growth (Table 2) were mated to determine the effects of MTbGH expression on fertility. After being housed with fertile males for more than 3 months, only one of the three females bred and she gave birth to six pups. Five of the six progeny inherited the fusion gene and all of them had substantial quantities of bGH mRNA in their liver.

These results indicate that mice bearing MTbGH inserts exhibit expression and growth characteristics similar to other MT-GH mice (Palmiter et al. 1982b, 1983). The rat, human, and bovine GH genes appear to be functionally equivalent when expressed in the mouse under the control of the MT promoter. Although the effects of MTbGH expression on female fertility have not been fully explored, it is apparent that fertility is impaired, as was observed with other MT-GH fusion genes (Hammer et al. 1984a).

Effects of Increased GH on Organ Weights

The ability to generate lines of mice containing and expressing MT-GH genes permits the systematic evaluation of the effects of chronic exposure to elevated GH on such parameters as skeletal and visceral growth. Such studies have previously been conducted in rats bearing GH-secreting pituitary tumors of GH_3 cell lines (Prysor-Jones and Jenkins 1980) or animals receiving daily injection of large amounts of GH for a relatively short period (Bates et al. 1964). We have used sixth generation progeny of a MTrGH mouse (MGH 10) containing eight copies of the fusion gene in a single chromosomal location to examine the degree of enlargement of various organs following chronic exposure to rGH (Palmiter et al. 1982b). In MT-GH mice, foreign mRNA can be detected as early as day 13 of fetal life, yet the mice apparently do not respond to the foreign GH until approximately 3 weeks of age when the first evidence of increased body weights is detectable (Palmiter et al. 1983). This approach allows chronic exposure to high levels of GH from fetal life onward and hence the effects may not be the same as with injections of GH or secretion of GH from tumors.

The mean body weight of unstimulated MTrGH mice was increased about twofold over control littermates (Table 3). This enhanced somatic growth was not increased by elevating mean serum GH levels from 1250 ng/ml in unstimulated animals to 12,500 ng/ml with zinc treatment (Hammer et al. 1984b). The constitutive activity of this gene provides enough GH to maximize growth. Of all the organs examined, the liver and spleen showed the greatest enlargement, ~3.5-fold and 3.0-fold increases, respectively. Other organs, notably kidney, heart, lung, and testis showed weight increases proportional to the total elevation in body weight. Surprisingly, the brains of transgenic male and female mice did not differ in weight from control animals. In MTrGH females, mean ovarian weight was significantly decreased ($p < 0.05$). MT-GH females commonly exhibit impaired fertility and this reduction in ovarian size may be related to such impairments (Hammer et al. 1984a).

In MT-GH mice, the disproportionate increase in the size of the liver as compared with body weight and the proportionate increase in the size of other organs could result from increases in cell number and/or cell size. If the organ weights of transgenic mice are elevated strictly as a result of hyperplasia, then the DNA content of these organs should be increased proportionately. In contrast, if the organ weights are increased as a result of marked hypertrophy, then the change in DNA content would be less than the increase in weight. The increase in DNA content of livers from MT-GH mice as compared with control littermates (3.1-fold) is only slightly less than the increase in organ weight (3.7-fold), indicating that there is marked hyperplasia and some hypertrophy (Table 4). In contrast, the increase in DNA content of kidneys from transgenics is equal to the increase in weight suggesting hyperplasia. These changes in the size of organs are very similar to those

Table 2. Expression of MTbGH in Transgenic Mice

Mouse	Gene Copy[a] (number per cell)	Liver bGH mRNA[b] (molecules/cell) control	+Zn	Relative growth[c] (ratio)
893-8 ♀	1	0	0	0.91
896-3 ♂	2	108	102	1.86
897-7 ♂	5	132		1.62
898-1 ♀	1	0	0	1.10
899-2 ♀	3	396	9840	1.73
899-5 ♂	6	4164		1.39
900-1 ♂	1	1344		1.65
900-5 ♀	1	ND		1.91
900-6 ♀	2	696		1.78
900-8 ♂	1	0	0	1.08

[a]The presence of MTbGH sequences was determined by dot hybridization to tail nucleic acids with a nick-translated probe.

[b]MTbGH mRNA was measured by solution hybridization as described in Experimental Procedures. ND, Not determined.

[c]The relative weights of transgenic mice compared with sex-matched littermates at 9 weeks are shown.

Table 3. Body and Organ Weights of Normal and MTrGH Transgenic Mice at 17 Weeks of Age

Group	Number of animals	Body weight (g)	Organ weights[a] (mg)						
			liver (g)	kidney	spleen	lung	heart	testis/ovary	brain
Transgenic ♂	5	53.8 ± 5.9 (1.8)	4.9 ± 0.28 (3.7)	437 ± 61 (2.0)	307 ± 46 (3.3)	271 ± 30 (1.8)	308 ± 32 (1.7)	144 ± 18 (1.5)	527 ± 21 (1.1)
Transgenic ♀	4	43.3 ± 2.7 (1.9)	3.4 ± 0.07 (3.1)	319 ± 30 (2.0)	258 ± 50 (2.4)	252 ± 20 (1.7)	279 ± 20 (2.0)	5.1 ± 1.0 (0.76)	502 ± 30 (1.1)
Normal ♂	3	29.4 ± 1.4	1.3 ± 0.11	215 ± 16	91 ± 18	151 ± 16	197 ± 24	96 ± 4	467 ± 11
Normal ♀	3	22.6 ± 3.2	1.1 ± 0.17	158 ± 20	106 ± 3	145 ± 8	139 ± 5	6.7 ± 0.4	470 ± 8

[a] The mean values ± S.D. are given. Mice are 17-week-old, sixth-generation progeny of an MTrGH male (MGH10) containing eight copies of the fusion gene per cell (Palmiter et al. 1982b). Values in parentheses are the relative body or organ weights of transgenic mice compared with sex-matched littermates.

observed in acromegaly and although details of the structure and function of these organs have not been explored, these MT-GH mice might serve as models for the study of tissue changes in gigantism and acromegaly.

Expression of hMThGH Genes

In mammals, MT gene expression is regulated by heavy-metal ions and glucocorticoid hormones. These inductions are due to increased transcriptional activity of the MT gene (Durnam and Palmiter 1981; Hager and Palmiter 1981). When mouse MT genes are introduced into cells, the genes retain transcriptional regulation by cadmium but not by glucocorticoids (Mayo et al. 1982). In contrast, cells transfected with the human MT-II$_A$ (hMT) gene maintain responsiveness to both heavy-metal ions and glucocorticoids (Karin et al. 1984a).

A number of fusion genes utilizing the mouse MT-I promoter/regulator region have now been introduced into mice and in most cases MT-I has provided transcriptional regulation of the fused foreign structural gene by heavy metals (Palmiter et al. 1982a; R.E. Hammer and N.C. Birnberg, unpubl.). Whereas MTrGH and MThGH expression results in metal-regulatable production of foreign GH, we were never able to observe glucocorticoid regulation (data not shown). Karin et al. (1984b) have shown that the human MT-II$_A$ promoter has a glucocorticoid receptor binding site and confers steroid regulation to fusion genes after transfection into tissue culture cells (Karin et al. 1984a). Therefore, it was of interest to determine if the hMT promoter when fused to the hGH gene would confer responsiveness to both metals and glucocorticoids when introduced into mice.

Six of the nine transgenic mice produced had detectable levels of fusion transcripts in their liver and all of these exhibited metal-regulatable expression (Table 5). Of the six mice with hGH mRNA transcripts four contained hGH in their sera (data not shown) and showed enhanced growth. One mouse 886-5, which exhibited increased body weight, lacked detectable liver hGH mRNA. Such findings are unusual and suggest that in this line other organs are primary sites of fusion gene

Table 4. Concentration of DNA in Various Organs of Transgenic Mice Bearing Growth-promoting Genes

Gene	Line	Number of animals	DNA concentration[a] (μg/mg wet weight)			
			liver	kidney	brain	pituitary
MTrGH[b]	45-3	6	2.98 ± 0.20	4.02 ± 0.36	1.14 ± 0.15	ND
Control		5	3.59 ± 0.13	4.17 ± 0.34	1.16 ± 0.08	ND
MTbGH[c]	896-3	2	3.18 ± 0.06	4.39 ± 0.57	1.11 ± 0.08	ND
Control		3	2.97 ± 0.19	4.24 ± 0.20	1.09 ± 0.10*	ND
MThGRF[d]	803-4	4	ND	ND	1.15 ± 0.07	4.42 ± 0.12*
Control		4	ND	ND	1.15 ± 0.16	10.17 ± 0.55

[a] The mean values ± S.D. are given. ND, Not determined. Organs were weighed and pieces removed, weighed, and homogenized in SET buffer with 100 μg/ml of proteinase K (see Experimental Procedures). The DNA concentration was determined by measuring the fluorescence of bisbenzimide H33258 (Laborca and Paigen 1980) using a Perkin Elmer fluorimeter. Calf thymus DNA was used as a standard.

[b] Mice were 17-week-old, sixth-generation transgenic and control progeny of MGH10 (Table 2). Mean liver and kidney weights were 3.7 and 2.0 greater than normal, respectively (See Table 3). Mean brain weights were not significantly different.

[c] Mice were 12-week-old, first-generation transgenic and control male progeny of 896-3 male bearing 2.0 copies of MTbGH (Table 1). Mean liver and kidney weights of these transgenic progeny were 2.8- and 2.1-fold greater than normal, respectively. Mean brain weights were not significantly different. *Group contains two samples.

[d] Mice were 21-week-old, first-generation transgenic and control progeny of 803-4 female bearing 10 copies of MThGRF (Table 6). Pituitaries from these transgenic offspring were ~4.8-fold larger than control littermates. Mean brain weights were not significantly different. *Group contains three samples.

Table 5. Expression of hMThGH in Transgenic Mice

Mouse	Gene copy[a] (number per cell)	Liver hGH mRNA[b] (molecules/cell) control	Liver hGH mRNA[b] (molecules/cell) + Zn	Relative growth[c] (ratio) + Zn
883-2 ♀	7	570	2700	1.38
884-2 ♂	3	42	312	0.89
884-5 ♂	1	0	720	1.18
884-9 ♂	7	0	0	0.98
886-4 ♂	4	222	156	1.08
886-5 ♀	1	0	0	1.33
887-5 ♀	4	18	180	1.47
890-3 ♂	1	84	792	1.48
891-6 ♀	5	108	180	1.56

[a]The presence of hMThGH sequences was determined by dot hybridization to tail nucleic acids.

[b]hMThGH mRNA was measured by solution hybridization as described in the Experimental Procedures.

[c]The relative weights of transgenic mice compared with sex-matched littermates at 10 weeks are shown. Animals had their drinking water supplemented with 25 mM ZnSO$_4$ for 3 weeks prior to weighing.

expression. Expression of MT constructs in the intestine is frequently high especially when mice are treated with zinc.

To test glucocorticoid regulation in transgenic animals, we bred several hMThGH mice and utilized the transgenic progeny. To eliminate the influence of endogenous glucocorticoids on hMT expression, two mice from each line were adrenalectomized 1 week prior to initial partial hepatectomy. Mice in one line (884-5, Table 6) had very little hMT mRNA and did not exhibit any steroid responsiveness (data not shown). In contrast, in the 884-2 line, both the intact (control) and adrenalectomized animals showed about a fivefold induction in hMThGH expression following dexamethasone treatment. This is the first demonstration of a glucocorticoid response of an MT fusion gene in transgenic mice.

Introduction and Transmission of hMThGH Genes in Rabbits

MT-GH fusion constructs have been introduced and are expressed at substantial levels in a high frequency of transgenic mice (Palmiter et al. 1982b, 1983). Because of this success, we decided to develop the transgenic rabbit as a model to study further the effects of GH gene expression and to pilot gene transfer experiments in large domestic animals. Rabbits are the most convenient large laboratory animal available and methods for the collection, culture, and transfer of fertilized eggs are well established. In addition, nuclear structures in one-cell and two-cell rabbit eggs are easily seen with interference-contrast microscopy (Fig. 2), unlike eggs from other domestic species where the opacity of the cytoplasm interferes with nuclear visualization.

We initiated our studies in rabbits with MThGH constructs (Hammer et al. 1985b), but because of the low frequency of expressing MThGH-bearing rabbits (25%) as compared with mice (70%) we decided to try the human MT-II$_A$ promoter to regulate hGH expression.

Linear DNA fragments of hMThGH (Fig. 1) were injected into rabbit eggs (Fig. 2) and transferred to pseudopregnant recipients. Of 27 animals that developed from these eggs, eight (29%) contained intact multiple copies of the gene and two had detectable levels of hGH mRNA in their liver (Table 7). Two of the

Table 6. Regulation of hMThGH Expression by Glucocorticoids

Transgenic mouse[a]	hGH mRNA[b] (molecules/cell)			
	control		ADX	
	– Dex	+ Dex	– Dex	+ Dex
884-2-1	252	858		
884-2-6			378	2382
884-1-27			440	2310

[a]Mice are progeny of a hMThGH male (Table 2).

[b]Liver hMThGH mRNA was determined by solution hybridization as described in the Experimental Procedures. Initial, partial hepatectomies were performed on intact (control) and adrenalectomized (ADX) hGH mice. Two weeks following the first partial hepatectomy, dexamethasone (Dex 10 mg/kg; i.p.) was administered 18 hr and 6 hr before a final hepatectomy was performed.

Table 7. Characteristics of hMThGH Transgenic Rabbits

Rabbit	Gene copy[a] (number per cell)	hGH mRNA[b] (molecules/cell) liver	hGH mRNA[b] (molecules/cell) testis
251-1	16	17	
266-2	2	0	
266-4	24	0	
266-5	4	0	
266-11	47	0	
273-4 ♂	26	0	105
273-8 ♂	24	0	72
273-9	62	42	

[a]The presence of hMThGH sequences was determined by dot hybridization to tail nucleic acids.

[b]hMThGH mRNA was measured by solution hybridization as described in Experimental Procedures.

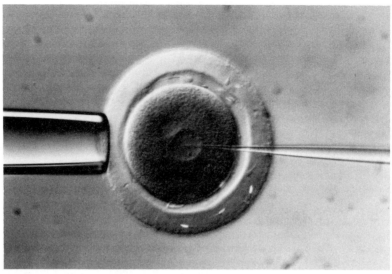

Figure 2. Interference-contrast photomicrograph of a one-cell, fertilized rabbit ova. The egg is being held by a blunt, holding pipette and an injecting pipette (diameter ∼1.5 μm) is seen within the male pronucleus immediately following injection of buffer containing DNA. Injection was monitored by observing the diameter of the pronucleus expand by ∼50%.

animals (273-4 and 273-8) were sacrificed and mRNA preparations from liver, kidney, heart, intestine, and testis were tested for the presence of hGH mRNA. In both animals, hGH mRNA was present only in testicular tissue. In addition, liver hMT was not inducible by acute cadmium treatment. The reasons for the low frequency and levels of expression of mMT and hMT fusion genes in rabbits as compared with mice are unknown but could be related to species differences in MT gene regulation.

Nevertheless, it was important to determine if the gene would be transmitted. Male 273-4 was bred and 27-day-old fetuses examined for the presence of the fusion gene and hGH mRNA. The gene was transmitted in a Mendelian fashion, indicative of integration on a single chromosome, but like their father, none of the offspring had detectable hGH mRNA in their liver (Fig. 3). These results demonstrate that the genes can be introduced at high frequencies and are heritable in transgenic rabbits.

Expression of MThGRF Genes

Our success in controlling growth in mice by ectopic production of foreign GH raised the possibility of regulating growth by stimulating the production of endogenous GH by introducing a GRF gene. For this experiment, the mouse MT-I promoter/regulator was fused to a hGRF minigene that included the entire coding region of the GRF precursor protein (Hammer et al. 1985a).

When this construct was introduced into the germ line of mice, 10 of 14 transgenic animals showed a significant increase in body weight (Hammer et al. 1985a). The characteristics of five representative mice from this group are shown in Table 8. This enhanced growth was correlated with the presence of hGRF mRNA in the liver and intestine, detectable levels of plasma hGRF, and increased plasma GH. Apparently many tissues are capable of synthesizing the mRNA, however, it is not certain which are capable of correctly processing this precursor polypeptide to generate functional GRF.

To determine if the fusion gene and phenotype were heritable, founder mice were bred and progeny examined for the gene and large phenotype. In one pedigree, 803-4, the gene has been transmitted to 8 of 15 progeny and all transgenic offspring show enhanced growth equal to or greater than the parent (Fig. 4). In the other pedigree, 765-2, the gene has now been passed into the second generation and the large phenotype is stably in-

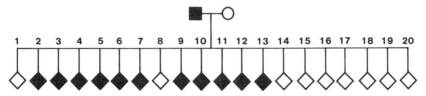

Figure 3. Inheritance of GH genes in progeny of transgenic rabbits 273-4 (Table 5). Male is shown as a square, female as a circle, and fetuses of unknown sex as diamonds. Solid symbols signify the inheritance of hMThGH sequences as determined by dot hybridization (see Experimental Procedures). Numbers above symbols designate the animal number.

Table 8. Expression of MThGRF in Transgenic Mice

Mouse	Gene copy[a] (number per cell)	Liver hGRF mRNA[b] (relative amount)	Plasma hGRF[c] (ng/ml)	Plasma mGH[c] (ng/ml)	Growth[d] (ratio)
762-5 ♀	4	63	26	141	1.33
765-2 ♀	2	102	207	415	1.24
801-5 ♀	8	38	99	809	1.35
803-4 ♀	10	118	263	1095	1.41
803-6 ♂	10	16	50	302	1.24

[a] The presence of MThGRF sequences was determined by dot hybridization to tail nucleic acids.

[b] Liver MThGRF mRNA was analyzed by Northern blotting. Autoradiograms were densitometrically scanned to determine relative RNA amounts.

[c] Plasma GRF and GH were determined by radioimmunoassay as described in Experimental Procedures. The GH assay was able to detect 16 ng/ml (two control animals had 16 and 46 mg/ml GH) and the GRF assay was able to detect 10 ng/ml (two control animals had < 10 ng/ml GRF).

[d] The relative weights of transgenic mice compared with sex-matched littermates at 9 weeks are shown.

herited. All of the MThGRF founder females were mated to determine if fertility was impaired. Significantly, all six females bred and delivered viable litters, unlike MT-GH females, which are generally infertile. This suggests that the enhanced growth resulting from stimulation of endogenous GH may be more physiological and less deleterious to reproductive function than MT-GH gene expression. Such results may have important implications for the application of such growth-promoting techniques to other mammalian species.

The enhanced growth of mice bearing MThGRF genes is a clear demonstration of a direct role for this releasing factor in the positive regulation of growth. The long-term effects of chronic overstimulation of the somatotrophs are just beginning to be explored; however, one result is a two- to sevenfold increase in pituitary size (Fig. 5). While the sizes of pituitaries in MThGRF progeny of one founder animal, 803-4, were 4.8 times normal size, the DNA content was only increased twofold, indicating that this enlargement is a result of marked hypertrophy and hyperplasia (Table 4). This condition is similar to clinical findings in patients having pancreatic or bronchial carcinomas which secrete GRF; therefore these mice will provide an excellent model for the study of such disorders.

CONCLUSIONS

Currently, growth can be enhanced by introducing DNA fragments consisting of MT promoters fused to GH and GRF structural genes into the germ line of mice. This method has been utilized to enhance normal growth as well as to restore growth to a mutant strain of mice with a genetically inherited growth disorder (Hammer et al. 1984a). In addition, such germ line manipulations have been extended to other mammalian species with the ultimate hope of enhancing growth and

Figure 4. Growth of mice expressing MThGRF genes. A mouse (803-4-4; 44 g) carrying approximately 10 copies of the MThGRF gene compared with a normal littermate (20 g) at age 15 weeks.

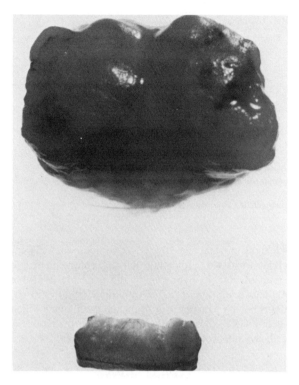

Figure 5. Growth of the pituitary gland in mice expressing MThGRF fusion genes. A pituitary gland (28 mg) from a 1-year-old MThGRF transgenic female (803-4) carrying 10 copies of the fusion gene compared with a pituitary gland (2 mg) from a normal animal.

improving the feed efficiency of various farm livestock (Hammer et al. 1985b).

Our first approach to regulating growth has been to utilize the mouse MT-I promoter fused to GH genes from three different species (rat, human, and bovine). Such constructs provide metal-regulatable control over GH transcription in several organs resulting in the production of the foreign peptide and animals twice normal size. Founding animals containing each of these constructs have transmitted the gene and large phenotype to their offspring. Although the mouse MT-I fusion genes are transcriptionally controlled by heavy metals, a glucocorticoid response has not been obtained. In contrast, the human MT-II$_A$ promoter appears to provide both metal- and steroid-regulated control over GH expression.

Another approach to increasing somatic growth was to fuse the MT-I promoter to a human GRF minigene (Mayo et al. 1985). In transgenic mice harboring such constructs, the presence of this biologically active peptide elevates endogenous GH secretion such that animals grow larger than normal (Hammer et al. 1985a). One consequence of this chronic stimulation of somatotrophs is a dramatic increase in the size of the pituitary gland due to hypertrophy and hyperplasia.

Gene insertion by microinjection has now been extended successfully to other mammalian species (Hammer et al. 1985b). Both mMT and hMT constructs have been introduced into the germ line of rabbits. While we are not certain of the effects of foreign GH on the growth of this species, it is now clear that this approach to gene introduction can be utilized to explore a number of fundamental questions concerning gene expression and regulation in other species.

ACKNOWLEDGMENTS

We thank Richard Maurer and Michael Karin for the bGH and the human MT-II$_A$ genes, respectively. We are grateful to Douglas Bolt for performing the radioimmunoassays of hGH. We thank Myrna Trumbauer and Mary Yagle for their expert technical assistance and Paul Fallon and Stephanie Mengel for exceptional animal care. The work was supported by research grants from USDA (Section 1433 formula funds) and NIH (HD-19018) to R.L.B. and NIH (HD-09172) to R.D.P.

REFERENCES

Barsh, G.S., P.H. Seeburg, and R.E. Gelinas. 1983. The human growth hormone gene family: Structure and evolution of the chromosomal locus. *Nucleic Acids Res.* **11:** 3939.

Barta, A., R.I. Richards, J.D. Baxter, and J. Shine. 1981. Primary structure and evolution of rat growth hormone gene. *Proc. Natl. Acad. Sci.* **78:** 4867.

Bates, R.W., S.M. Kovic, and M.M. Garrison. 1964. Effects of prolactin, growth hormone and ACTH alone and in combination, upon organ weights and adrenal function in normal rats. *Endocrinology* **74:** 714.

Bell, G., J.P. Merryweather, R. Sanchez-Pescador, M.M. Stempien, L. Priestly, J. Scott, and L.B. Rall. 1984. Sequence of a cDNA clone encoding human preproinsulin-like growth factor II. *Nature* **310:** 775.

Brinster, R.L., H.Y. Chen, M.E. Trumbauer, M. Yagle, and R.D. Palmiter. 1985. Factors affecting the efficiency of introducing foreign DNA into mice by microinjecting eggs. *Proc. Natl. Acad. Sci.* **82:** 4438.

Brinster, R.L., H.Y. Chen, M.E. Trumbauer, A. Senear, R. Warren, and R.D. Palmiter. 1981. Somatic expression of herpes thymidine kinase in mice following injection of a fusion gene into eggs. *Cell* **27:** 223.

Durnam, D.M. and R.D. Palmiter. 1981. Transcriptional regulation of the mouse metallothionein-I gene by heavy metals. *J. Biol. Chem.* **256:** 5717.

―――. 1983. A practical approach for quantitating specific mRNAs by solution hybridization. *Anal. Biochem.* **131:** 385.

Goodman, R.H., D.C. Aron, and B.A. Ross. 1983. Rat preprosomatostatin: Structure and processing by microsomal membranes. *J. Biol. Chem.* **258:** 5570.

Gordon, D.G., D.P. Quick, C.R. Erwin, J.E. Donelson, and R.A. Maurer. 1983. Nucleotide sequence of the bovine GH chromosomal gene. *Mol. Cell. Endocrinol.* **33:** 81.

Hager, L.H. and R.D. Palmiter. 1981. Transcriptional regulation of mouse liver metallothionein-I gene by glucocorticoids. *Nature* **291:** 340.

Hammer, R.E., R.D. Palmiter, and R.L. Brinster. 1984a. Partial correction of a murine hereditary growth disorder by germ-line incorporation of a new gene. *Nature* **311:** 65.

―――. 1984b. The introduction of metallothionein-growth hormone fusion genes into mice. In *Advances in gene technology: Human genetic disorders* (ed. S. Black et al.), vol. 1, p. 52). ICSU Press, Miami.

Hammer, R.E., R.L. Brinster, M.G. Rosenfeld, R.M. Evans, and K.E. Mayo. 1985a. Expression of human growth hormone-releasing factor in transgenic mice results in increased somatic growth. *Nature* 315: 413.

HAMMER, R.E., V.G. PURSEL, C.E. REXROAD, JR., R.J. WALL, D.J. BOLT, K.M. EBERT, R.D. PALMITER, and R.L. BRINSTER. 1985b. Production of transgenic rabbits, sheep and pigs by microinjection. *Nature* 315: 680.

JANSEN, M., F.M.A. VAN SCHAIK, A.T. RICKER, B. BULLOCK, D.E. WOODS, K.H. GABBAY, A.L. NUSSBAUM, J.S. SUSSENBACH, and J.L. VAN DEN BRANDE. 1983. Sequence of cDNA encoding human insulin-like growth factor I precursor. *Nature* 306: 609.

KARIN, M., A. HASLINGER, H. HOLTGREVE, G. CATHALA, E. SLATER, and J.D. BAXTER. 1984a. Activation of a heterologous promoter in response to dexamethasone and cadmium by metallothionein gene 5′-flanking DNA. *Cell* 36: 371.

KARIN, M., A. HASLINGER, H. HOLTGREVE, R.I. RICHARDS, P. KRAUTER, H.M. WESTPHAL, and M. BEATO. 1984b. Characterization of DNA sequences through which cadmium and glucocorticoid hormones induce human metallothionein-II$_A$ gene. *Nature* 308: 513.

LABORCA, C. and K. PAIGEN. 1980. A simple and sensitive DNA assay procedure. *Anal. Biochem.* 102: 344.

MAYO, K.E., R. WARREN, and R.D. PALMITER. 1982. The mouse metallothionein-I gene is transcriptionally regulated by cadmium following transfection into human or mouse cells. *Cell* 29: 99.

MAYO, K.E., W. VALE, J. RIVIER, M.G. ROSENFELD, and R.M. EVANS. 1983. Expression-cloning and sequence of a cDNA encoding human growth hormone-releasing factor. *Nature* 306: 86.

MAYO, K.M., G.M. CERELLI, R.V. LEBO, B.D. BRUCE, M.G. ROSENFELD, and R.M. EVANS. 1985. Gene encoding human growth hormone-releasing factor precursor: Structural sequence and chromsomal assignment. *Proc. Natl. Acad. Sci.* 82: 63.

ORNITZ, D.M., R.D. PALMITER, R.E. HAMMER, R.L. BRINSTER, G.H. SWIFT, and R.J. MACDONALD. 1985. Specific expression of an elastase-human growth hormone fusion gene in pancreatic acinar cells of transgenic mice. *Nature* 313: 600.

PALMITER, R.D. and R.L. BRINSTER. 1985. Transgenic mice. *Cell* 41: 343.

PALMITER, R.D., H.Y. CHEN, and R.L. BRINSTER. 1982a. Differential regulation of metallothionein-thymidine kinase fusion genes in transgenic mice and their offspring. *Cell* 29: 710.

PALMITER, R.D., G. NORSTEDT, R.E. GELINAS, R.E. HAMMER, and R.L. BRINSTER. 1983. Metallothionein-human growth hormone fusion genes stimulate growth of mice. *Science* 222: 809.

PALMITER, R.D., R.L. BRINSTER, R.E. HAMMER, M.E. TRUMBAUER, M.G. ROSENFELD, N.C. BIRNBERG, and R.M. EVANS. 1982b. Dramatic growth of mice that develop from eggs microinjected with metallothionein-growth hormone fusion genes. *Nature* 300: 611.

PRYSOR-JONES, R.A. and J.S. JENKINS. 1980. Effect of excessive secretion of growth hormone on tissues of the rat, with particular reference to the heart and skeletal muscle. *J. Endocrinol.* 85: 75.

SEEBURG, P.H. 1982. The human growth hormone gene family: Nucleotide sequences show recent divergence and predict a new polypeptide hormone. *DNA* 1: 239.

SHEN, L.P., R.L. PICTET, and W.J. RUTTER. 1982. Human somatostatin-I: Sequence of the cDNA. *Proc. Natl. Acad. Sci.* 79: 4575.

TOWNES, T.M., J.B. LINGREL, H.Y. CHEN, R.L. BRINSTER, and R.D. PALMITER. 1985. Erythroid specific expression of human β-globin genes in transgenic mice. *EMBO J.* 4: 1715.

ULLRICH, A., C.H. BERMAN, T.J. HULL, A. GRAY, and J.M. LEE. 1984. Isolation of the human insulin-like growth factor I gene using a single synthetic DNA probe. *EMBO J.* 3: 361.

WAGNER, E.F., L. COVARRUBIAS, T.A. STEWART, and B. MINTZ. 1983. Prenatal lethalities in mice homozygous for human growth hormone gene sequences integrated in the germ line. *Cell* 35: 647.

WOYCHIK, R.P., S.A. CAMPER, R.H. LYONS, S. HOROWITZ, E.C. GOODWIN, and F.M. ROTHMAN. 1982. Cloning and nucleotide sequencing of the bovine growth hormone gene. *Nucleic Acids Res.* 10: 7197.

Inducible and Developmental Control of Neuroendocrine Genes

R.M. Evans,* C. Weinberger,* S. Hollenberg,* L. Swanson,* C. Nelson,†
AND M.G. Rosenfeld†

Howard Hughes Medical Institute Laboratory and Gene Expression Laboratory, The Salk Institute, San Diego, California 92138; †Howard Hughes Medical Institute Laboratory and Regulatory Biology Program, School of Medicine, University of California, San Diego, California 92037

In animals, the endocrine and the nervous systems produce a diverse set of cells and molecules which interact to exert profound effects of developmental physiology and behavior. Development of the neuroendocrine system, like other tissue and organs, includes proliferation, migration, cell-cell interaction, and the establishment of phenotypic diversity which is ultimately based on precise temporal and spatial patterns of differential gene expression. Accordingly, the sequential production of regulatory proteins capable of activating gene expression will lead to progressive phenotypic change. These proteins are expected to be sequence-specific DNA-binding proteins capable of binding high-affinity sites on chromatin and regulating transcription of a limited set of genes. Such proteins may be expressed transiently and thus function during discrete phases of development or may be expressed indefinitely and thus promote expression of the differentiated phenotype. These factors would accordingly be *trans*-acting and exert their transcriptional effects through the recognition of *cis*-acting regulatory sequences.

Nucleotide sequences necessary for eukaryotic gene transcription (Breathnach and Chambon 1981; Grosveld et al. 1982; McKnight and Kingsbury 1982; Dierks et al. 1983) and for specific regulation of inducible genes (Pelham 1982; Raag and Weissman 1983; Karin et al. 1984) have been demonstrated by in vitro mutagenesis and gene transfer analysis. Furthermore, the transfer of a variety of intact and fusion genes into differentiated cells that express related endogenous genes has demonstrated the potential of enhancer sequences to direct tissue- and cell-specific expression (Renkawitz et al. 1982; Spandidos and Paul 1982; Gillies et al. 1983; Oi et al. 1983; Queen and Baltimore 1983; Walker et al. 1983). However, this approach is limited by the availability of appropriate cultured cell types, and to genes that do not require sequential activation by factors produced in earlier lineages of the recipient cell.

The recent development of methods for introducing foreign genes into the germ line of mice provides a complementary approach for studying mechanisms underlying inducible and developmental gene regulation (Gordon et al 1980; Brinster et al. 1981; Costantini and Lacy 1981). Transgenic animals expressing foreign genes can thus be used to test models of the role played by specific DNA sequences in determining cell-specific expression. This approach has led to the demonstration that *Drosophila* genes along with their flanking chromosomal regions can be expressed appropriately when inserted into the germ line of mutant strains of flies (Goldberg et al. 1983; Scholnick et al. 1983; Spradling and Rubin 1983.) Similarly, recombined immunoglobulin genes and flanking sequences have been shown to be expressed appropriately in the spleen but not in the liver of transgenic mice, and the rat elastase I promoter and the rat insulin promoter are expressed selectively in the exocrine and endocrine pancreas of transgenic animals, respectively (Brinster et al. 1983; Hanahan 1985; Ornitz et al., this volume). Taken together, these results suggest that the expression of some genes may be the consequence of a *cis*-acting, tissue-specific regulatory sequence. We have initiated an analysis of the molecules' sequence and the events associated with developmental and homeostatic regulation of gene expression.

Cloning of the Glucocorticoid Receptor — An Enhancer Protein

The regulation of eukaryotic gene expression in response to intercellular signals, such as hormones, represents a critical strategy by which development and homeostatic regulation are achieved in eukaryotic organisms. Such regulation is modulated by compounds that bind intracellular receptors and those that interact with plasma membrane receptors. For example, steroid hormones regulate the transcription of certain genes as a consequence of binding specific nuclear receptors. The steroid hormone–receptor interaction initiates a transformation of the complex, after which it is capable of binding high-affinity receptor sites on chromatin and, by an unknown process, regulating transcription of a limited number of genes. For instance, the rates of transcription of mouse mammary tumor virus (MMTV), mouse metallothionein (MT), and rat growth hormone (GH) genes are stimulated by one class of steroid hormones, known as glucocorticoids, in cultured cell lines (Ringold et al. 1977; Mayo and Palmiter 1981; Evans et al. 1982; Spindler et al. 1982). Purified rat liver glucocorticoid receptor complexes will bind a specific region of cloned MMTV DNA in vitro, suggesting that steroid receptors modulate transcription by binding specific regulatory sequences near promoters (Payvar et al. 1981; Pfahl et al. 1982). Furthermore, deletion analysis of the human metallothionein II (MT-II) gene and of the MMTV promoter has delimited regions that identify upstream control elements necessary for steroid response (Hynes et al. 1983; Karin et al. 1984).

Several studies have led to a model that suggests the existence of distinct steroid binding and DNA binding domains in the receptor polypeptide as well as a major immunogenic region. Although sufficient protein for a direct structural analysis has not been available, it has been possible to prepare polyclonal and monoclonal antibodies to partially purified glucocorticoid receptor (Okret et al. 1981). The receptor has been characterized as a 94,000-dalton polypeptide on the basis of biochemical and immunological criteria as well as covalent binding studies using labeled steroid analogs. We have attempted to isolate human glucocorticoid receptor cDNA clones in the absence of amino acid sequence information using a polyclonal antiserum that is reactive with several proteins in addition to the receptor.

To screen efficiently a large number of clones, a λgt11 cDNA library was prepared with size-fractionated poly(A)$^+$ mRNA from the human IM-9 B-cell line which expresses approximately 10^5 receptors per cell (Harmon et al. 1984). Four immunopositive phage were identified and further characterized using a novel technique called "epitope selection." In this procedure (Fig. 1) fusion proteins from phage plaques are adsorbed on to nitrocellulose filters and incubated for 1 hour with the polyclonal antiserum. Each of these isolates presumably expressed one or a small set of epitopes that was recognized by the polyclonal antiserum. Affinity-purified antibodies immobilized on individual filters were eluted and used to detect their cognate IM-9 proteins by Western blot analysis. Three of the four clones selectively bound antibody specific for 94- and 79-kD proteins expressed in human lymphoid cells (Fig. 2A, lanes 2 and 4). Antibody purified from one clone (GR80) specifically recognized a protein doublet complex of approximately 38 kD (Fig. 2A, lane 5). On the basis of these results, the fusion protein from the HGR2.9C lysogen was partially purified and covalently linked to an activated agarose resin to permit large-scale purification of epitope-specific antibody. Antibody retained on this affinity column was eluted and assayed for its ability to bind receptor. This antibody bound only the 94- and 79-kD proteins in extracts from human lymphoid cells (Fig. 2B, lane 1).

If the transformed receptor becomes a DNA-binding protein with high affinity for chromatin, cytoplasmic receptor levels should decrease and nuclear levels should decrease, and nuclear levels should increase in response to steroid administration. In accord with this model, treatment of lymphoid cells with glucocorticoid resulted in a decrease of the 94- and 79-kD proteins in the cytoplasm (Fig. 2B, lanes 2 and 3). Combined with the appearance of receptor in nuclear extracts following a 2-hour steroid treatment (Fig. 2A,

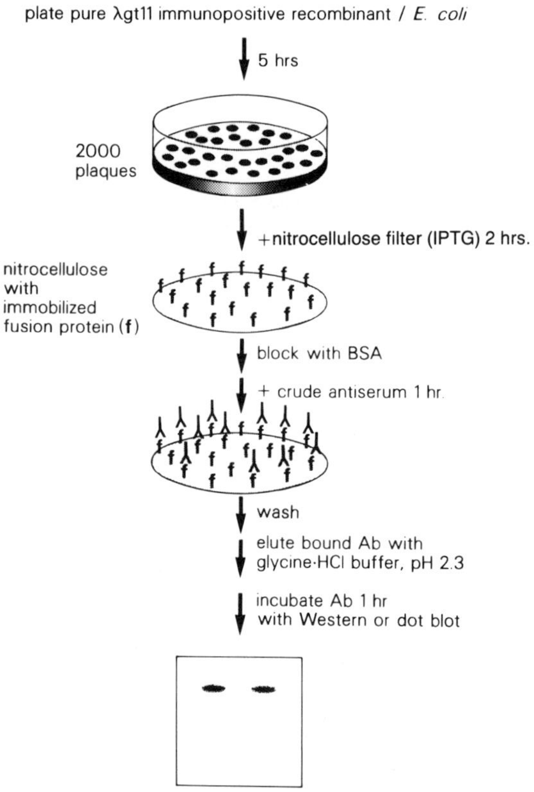

Figure 1. Protocol used for antibody purification and epitope identification using immunopositive recombinants. Immunopositive phage (2000 plaque-forming units) were used to infect Y1090 at 42°C for 5 hr, nitrocellulose filters were overlaid, and plates were transferred to 37°C for 2 hr. The filters were processed as described in detail (Weinberger et al. 1985). Briefly, crude antiserum was diluted 1:100 and incubated with the filters for 1 hr. Filters were washed three times, blotted dry, and antibody-eluted separately from each filter with three 1-min acid washes. These antibodies were then reacted with Western blots as described in Fig. 2.

lanes 5 and 6), these results provide evidence that the fusion protein encodes a specific epitope representing the physiological glucocorticoid receptor.

Further evidence on the identity of the receptor clones was provided by chromosomal mapping studies. Somatic cell fusion analysis has previously localized the human glucocorticoid receptor (GR) gene to chromosome 5 (Gehring et al. 1985). Using the insert from clone HGR1.2A as a hybridization probe, CHO cells carrying human chromosome 5 were shown to be positive for the receptor gene (Fig. 3). The presence of appropriately hybridizing fragments absolutely correlates with the presence of chromosome 5. Dot-blot analysis of DNA from high-resolution, dual laser-sorted human chromosomes confirms the presence of GR sequences on chromosome 5 (data not shown). In this technique a custom-modified fluorescence-activated cell sorter was used to sort mitotic chromosome suspensions stained with 4',6-bis(2''-imidazolinyl-4H, 5H)-2-phenylindole (DIPI)/chromomycin in conjunction

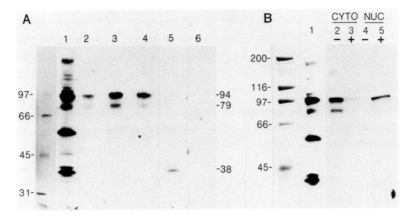

Figure 2. (*A*) Identification of recombinant clones by epitope selection. Four immunopositive recombinants and λgt11 were used to select antibody with the protocol outlined in Fig. 1. Each selected antibody was then used to probe Western blots of IM-9 cytoplasmic extracts. Antibodies used are as follows: crude polyclonal antiserum (lane *1*) and epitope-selected antibodies from clones HGR1.2A (lane *2*), HGR2.9C (lane *3*), HGR5.16A (lane *4*), GR80 (lane *5*), and λgt11 (lane *6*). (*B*) Specificity of selected antibody. IM-9 cells were grown in suspension culture in the absence (−) or presence (+) of triamcinolone acetonide (10^{-6} M) for 2 hr prior to harvest. One hundred micrograms of protein from cytoplasmic and nuclear fractions was prepared and subjected to Western blot analysis with epitope-specific antibody which had been affinity-purified against clone HGR2.9C. (Lane *1*) Cytoplasm from control cells; (lanes *2* and *3*) cytoplasm (CYTO) from control and steroid-treated cells, respectively; (lanes *4* and *5*) nuclear (NUC) fraction from control and steroid-treated cells, respectively. Lane *1* was incubated with crude (nonselected) polyclonal receptor antibody; lanes *2-5* were incubated with epitope-specific antibody that had been affinity-purified against clone HGR2.9C.

with Hoechst 33258/chromomycin (Lebo et al. 1984). Chromosomes are directly spotted onto nitrocellulose and denatured prior to hybridization with the hGR probe. A schematic of the receptor structure is shown in Figure 4.

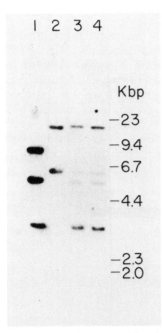

Figure 3. Chromosome mapping. DNA (10 mg) from hybrid CHO cell lines (Dana and Wasmuth 1982) was probed with DNA (10 mg) from human placenta; CHO-human hybrid and CHO cell lines were digested to completion with *Eco*RI or *Hin*dIII. Products from these reactions were separated on a 0.8% agarose gel and transferred to nitrocellulose paper by the method of Southern. The bound DNA fragments were detected by hybridization using a nick-translated 0.75-kbp fragment from the receptor cDNA coding region.

Developmental Regulation of the Growth Hormone Gene

The anatomical and physiological interaction of the hypothalamus and the pituitary provides a central paradigm to investigate the nature of the coordination of the nervous system and endocrine system in regulating homeostasis in higher metazoan organisms. We have studied the GH gene with particular interest in understanding the developmental and hormonal mechanisms regulating its expression. The GH gene is expressed at high levels only in the anterior pituitary. Growth hormone is the principal gene product of the somatotropic cells necessary for normal growth and development. Immunohistochemical studies suggest that GH is the only pituitary hormone secreted from somatotropes and thus represents a cell-specific gene product. To investigate the mechanisms involved in cell-specific gene expression, we have designed experiments to test for regulatory elements in the 5'-flanking region of the GH gene. Accordingly, promoter-proximal sequences were linked to the coding sequence of the bacterial enzyme chloramphenicol acetyl transferase (CAT) to serve as a reporter for transcriptional activity. GH recombinants were introduced into pituitary cells as well as into other nonpituitary cells and the relative expression of the CAT gene was measured. Initial constructions utilized are shown in Figure 5 and the results of the first series of experiments are shown below in Figure 5b. These data reveal the presence of an element in the GH gene capable of activating transcription in endocrine pituitary cells but not in heterologous cells. To determine the location of these elements, deletions of 5'-flanking sequences were produced. For the GH promoter, a drastic reduction was observed in deleting a region from −230 to −110 (Fig. 5). This suggests that a reg-

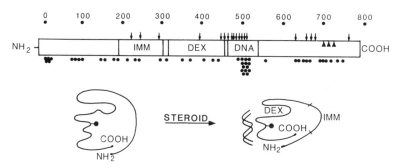

Figure 4. Schematic of the predicted structure of the human glucocorticoid receptor.

ulatory sequence (or a 5' border for such a sequence) necessary for efficient and cell-specific expression is located in this 120-bp fragment. We next wish to determine whether this region acts in an orientation-specific fashion or whether it can activate transcription independent of position and orientation. Accordingly, a fragment containing the putative GH regulatory element was inserted upstream of a promoter lacking enhancer elements. The first 450 nucleotides of the rat prolactin promoter fulfill these conditions and thus serve as a particularly useful construction to assay the effects of the GH regulatory sequence. As shown in Figure 6, the GH regulatory sequences activate transcription of the Prl promoter in an orientation-independent manner. However, these constructions express only in pituitary cells and not in fibroblasts. These results led to the proposal that the GH promoter-proximal sequence contains a pituitary cell-specific transcriptional enhancer.

A Role for Downstream Sequences

The presence of a pituitary-specific enhancer sequence in the 5'-flanking region of the GH gene does not exclude a role for other regulatory sequences in the downstream region of these genes. To begin to examine this possibility, we have fused structural portions of the GH gene to a heterologous promoter. Because this fusion gene represents a novel chimeric transcription unit combining potential regulatory elements from two different genes, it is not possible to predict "a priori" in what cell type such a hybrid gene would be expressed. Therefore, to assay all possible tissue types, these fusion genes were introduced into mice by injection of DNA into fertilized eggs. Thus, stably integrated fusion genes will be present throughout the normal development of the mouse and have the opportunity to be expressed in every cell in the body.

mGH Fusion Gene Expression in the Brain

We have previously described the generation of transgenic mice that develop from eggs microinjected with GH fusion genes (Palmiter et al. 1982). In these experiments the structural part of the GH gene was fused to the promoter region of the mouse MT-I gene (Fig. 7A), and it is thus expected that the fusion gene would be expressed in tissues where endogenous MT levels are high. It was therefore unexpected to find in transgenic animals exceptionally high levels of the fusion gene expressed in the brain, a tissue normally expressing low levels of MT and no GH (see Fig. 7C).

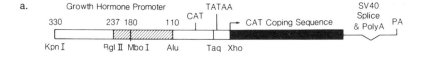

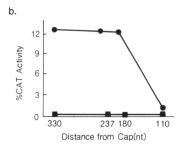

Figure 5. Pituitary-specific control sequences. Relative activity of plasmids carrying various lengths of the GH promoter fused to the reporter enzyme CAT (Gorman et al. 1982). Deletions from the 5' end were produced at the indicated restriction sites. Activity was determined following transfection into appropriate cells.

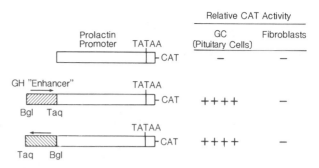

Figure 6. Enhancer activity of the GH promoter element. The *Bgl*II–*Taq*I fragment from the GH promoter was fused in both orientations to a plasmid carrying the promoter region of the rat prolactin gene (Supowit et al. 1984) linked to the CAT coding sequence. Relative CAT activities of the various constructions in each cell type are indicated.

These findings led us to examine in more detail the cellular distribution of fusion gene expression, and to determine whether this specificity is a consequence of sequences within the gene or the site of fusion gene insertion in the genome.

GH Distribution in Brain

Initial exploration of these results led to the surprising observation that expression of the fusion gene was highly restricted to several anatomically defined areas of the brain. Using GH-specific antisera, the most obvious pattern of cell staining in transgenic mice was observed in the hypothalamus. A detailed examination of eight animals representing offspring from five generations of a single transgenic parent (MGH-10) (Palmiter et al. 1982) showed an identical distribution of GH-stained neuronal cell bodies in seven of the animals. GH-stained cells were concentrated in three parts of the hypothalamus, including the large neurons in the paraventricular and supraoptic nuclei (Fig. 8A,B), which almost certainly synthesize oxytoxin or vasopressin, project to the posterior pituitary, and are involved in the integration of neuroendocrine and autonomic responses (Swanson and Sawchenko 1983). In addition, small cells in the suprachiasmatic nucleus

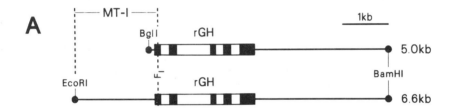

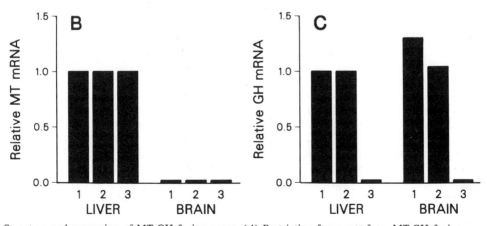

Figure 7. Structure and expression of MT-GH fusion genes. (*A*) Restriction fragments from MT-GH fusion genes that were injected into the male pronucleus of fertilized eggs. Terminal restriction enzyme sites are indicated. The genes are aligned according to the MT-GH fusion boundary. Solid boxes are exons, open boxes are intervening sequences, and solid lines are flanking chromosomal sequences. (*B*) Relative amount of MT RNA in livers and brains of transgenic (1 and 2) and normal control (3) mice as detected by slot-blot hybridization with nick-translated MT-I cDNA as probe. (*C*) Relative levels of GH RNA in livers and brains of transgenic (1 and 2) and normal (3) mice. rGH cDNA was used as probe.

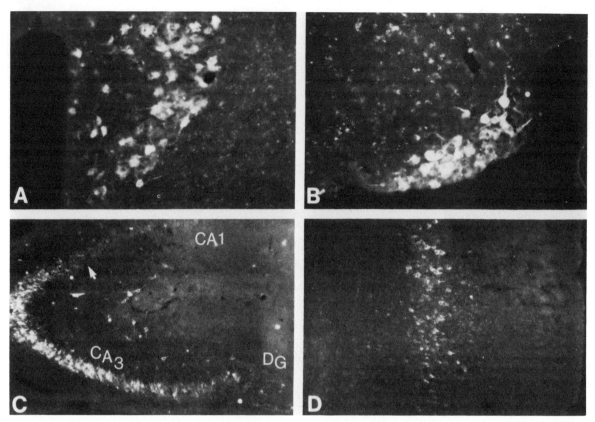

Figure 8. Frontal sections through the hypothalamus of an MGH-10 animal. Large neurons in the paraventricular (*A*) and supraoptic (*B*) nuclei stained with an antiserum to MGH. Small bright patches of staining outside of the nuclei were autofluorescent, and cannot be distinguished from specific immunofluorescence in these black and white photomicrographs. Magnification, 150×. (*C*) GH immunostaining in cell bodies in the hippocampus are confined to the pyramidal cell layer in field CA3, and to the region of interneurons in the hilus of the dentate gyrus (DG), and in fields CA1 and CA3. The arrowhead shows the boundary between fields CA1 and CA2; the fimbria is in the lower left-hand corner. Magnification, 75×. (*D*) Stained pyramidal cell bodies in layer V of the junctional region between the somatic sensory (upper half) and motor (lower half) cortex on the left side of the brain. The corpus callosum is to the far left and the pial surface is to the right. Magnification, 75×. (Reprinted, with permission, from Swanson et al. 1985.)

were even more brightly stained, and scattered large neurons were seen throughout many parts of the hypothalamus. In all cases, neuronal staining was blocked by the addition of rGH to both of the anti-GH sera used. No glial or vascular cell types were stained.

The prominent but highly restricted expression of the GH fusion gene in the hypothalamus led us to examine possible expression in other areas of the brain as well. Unexpectedly, specific immunostaining was observed in several areas, including regions of the neocortex, hippocampus, and olfactory cortex in the MGH-10 pedigree. Each animal examined showed a similar pattern of GH immunostaining that was greatly enhanced by colchicine pretreatment, which blocks axonal transport and leads to accumulation of peptides in cell bodies. Immunoreactive pyramidal cells restricted to layer V of the neocortex (Fig. 8D) were found in all animals expressing the fusion gene in liver and hypothalamus. The cortex is broadly divided into six layers with unique cell types that are morphologically and physiologically distinct. Bright immunostaining in this region is highly localized and does not appear to enter other layers.

The hypothalamus and neocortex are extensively interconnected with a ring of forebrain structures called the limbic region, which is broadly concerned with the maintenance of homeostasis. In the brains of transgenic mice, specific immunostaining is seen in the pyramidal cells of hippocampal field CA3 (Fig. 8C), which is a region of the limbic system that is dominated by mossy fiber afferents from the adjacent dentate gyrus. Although pyramidal cells in the other hippocampal fields (CA1, 2, and 4) were negative, some interneurons were stained here, as well as in the hilus of the dentate gyrus (Fig. 8C).

Neurons in several other areas of the brain consistently express the fusion gene; these include cells in layer II of the olfactory cortex, the central nucleus of the amygdala, the pars compacta of the substantia nigra and adjacent parts of the ventral tegmental area, and finally, the dorsal motor nucleus of the vagus nerve.

To determine whether the site of chromosomal integration plays a critical role in determining the pattern of mGH expression in neurons, six other integrations of the MT-rGH gene (Palmiter et al. 1982; Hammer et al. 1984), and two integrations of the MT-hGH gene (Palmiter et al. 1983), were examined in a total of 11 animals. In each brain, the pattern of GH immunostaining in the hypothalamus was examined in detail, and was identical to that described for the MGH-10 pedigree (Fig. 9), although the intensity of staining was variable. For example, immunostaining in the dwarf little (*Lit*) mouse was generally weak, while in two different MT-hGH integrations it was markedly more intense. In situ hybridization studies were initiated to examine independently the expression of the mGH gene. The regions of most intense immunoreactivity (paraventricular nucleus and suprachiasmatic nucleus) are also the most intensely reactive by hybridization analysis with growth hormone probes (Fig. 9).

In summary, 16 of 18 transgenic mice analyzed, representing nine different integrations of the rGH and hGH fusion genes, showed the same cellular distribution of GH in the brain (see Fig. 9). An identical pattern of GH immunostaining was seen in males and females, as well as in albino, agouti, and black strains of mice. These data strongly suggest that the precise site of integration into the genome may not play a critical role in determining which neurons in the brain express the fusion gene, although it may influence the relative level of gene expression. No immunostaining was observed in the brains of two animals, and here immunostaining in the liver was extremely weak.

MT Distribution in Brain

If sequences in the MT promoter determine the specific pattern of the mGH fusion gene expression in the brain of transgenic mice, then the endogenous MT-I gene should be expressed with an identical distribution. Antiserum to rat MT-I was used to identify the pattern of immunostaining in normal and transgenic mice. This antiserum consistently stained small cells with the appearance of astrocytes in all parts of the white and gray matter, both in normal and transgenic animals. This tentative identification was confirmed in double immunostaining experiments with a monoclonal antibody (Haan et al. 1982) to the astrocyte marker S-100 and a polyclonal antiserum (raised in rabbit) to MT-I. Essentially all cells were doubly labeled in this material (Fig. 10), and no evidence for MT staining of neurons was obtained. These results indicate that al-

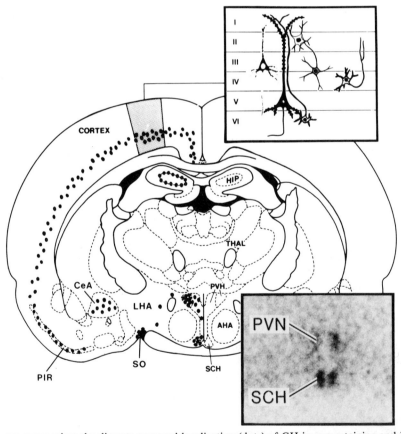

Figure 9. This diagram summarizes the discrete neuronal localization (dots) of GH immunostaining and in situ hybridization in a frontal section through the forebrain of a rodent. The upper insert shows that while the cerebral cortex contains many cell types that are distributed through six layers, unequivocal GH staining is confined to large pyramidal cells (dark) in layer V. The lower insert shows discrete hybridization of a single-stranded GH probe to a coronal section of the hypothalamus of a transgenic mouse. Abbreviations: AHA, anterior hypothalamic area; CeA, central nucleus of the amygdala; HIP, hippocampus; LHA, lateral hypothalamic area; PIR, piriform cortex; PVH, paraventricular nucleus; SCH, suprachiasmatic nucleus; SO, supraoptic nucleus; THAL, thalamus. (Reprinted, with permission, from Swanson et al. 1985.)

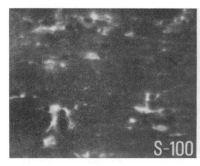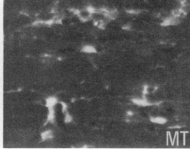

Figure 10. Colocalization of MT and the astrocyte-specific marker protein S-100 in the same section through the corpus callosum (a fiber tract) of an MGH10 animal. The section was incubated in a mixture of a rabbit antiserum to MT and a mouse monoclonal antibody to S-100. The MT was shown with a fluorescein-conjugated secondary antiserum and the S-100 was shown with a rhodamine-conjugated secondary antiserum. The control (CON) panel on the right shows the corpus callosum after incubation in anti-MT serum that was blocked by the addition of MT. Magnification, 400×. (Reprinted, with permission, from Swanson et al. 1985.)

though the fusion gene and MT-I appear to be coexpressed in hepatocytes, their expression is independently and differently regulated in the brain.

Implications for a Developmental Code

Because MT expression in the brain fails to colocalize with expression of the fusion gene, it is tempting to propose a model that requires sequences within both the 5′ flanking and the 3′ structural regions of the fusion gene for the observed cellular specificity. According to this hypothesis, sequences present in downstream regions of the GH gene are capable of promoting expression outside of the anterior pituitary. This would further suggest that pituitary-specific expression of the endogenous GH gene may require an interaction between developmental regulatory sequences in 5′ and 3′ regions of the gene. Integration site of the fusion genes does not appear to control the neuronal specificity because the pattern of fusion gene expression was identical in the brains of unrelated transgenic animals. We have recently examined the potential role of the MT-I promoter to direct neuronal specific expression by fusion with a structural gene encoding the precursor for the hypothalamic peptide human GH-releasing factor (Hammer et al. 1985). Although this fusion gene is efficiently expressed in visceral tissues, no evidence for discrete neuronal expression has yet been found. This suggests that neuronal expression is not simply a consequence of intrinsic specificity of the MT promoter.

Whatever structural elements are finally identified, the results presented here suggest a model in which developmentally specific expression of some genes may be determined by specific combinations of regulatory sequences. In this model, the combination of sequences provide a code that would be deciphered by the cell such that alterations of the combination of sequences could alter the pattern of specificity of gene expression. Similarly, their linkage to sequences from other genes could generate novel specificities such as those observed with the mGH fusion gene. One advantage of such combinatorial regulation would be that it would allow for adaptation, acquisition, and expansion of tissue specificity by events that alter the combination of developmental sequences. Such a model would imply that a relatively limited number of such sequences and factors could enforce the large and diverse range of tissues specificities characteristic of eukaryotic development.

ACKNOWLEDGMENTS

We thank Marijke ter Horst for her secretarial and administrative assistance. This work was supported in part by grants from the National Institutes of Health. Dr. Evans acknowledges generous support from the Fritz B. Burns and Mathers Foundations. Stan Hollenberg and Chris Nelson are graduate students in the Department of Biology at the University of California, San Diego, and receive support from a U.S. Public Health Service predoctoral training grant.

REFERENCES

Breathnach, R. and P.A. Chambon. 1981. Organization and expression of eukaryotic split genes coding for proteins. *Annu. Rev. Biochem.* **50:** 349.

Brinster, R.L., H.Y. Chen, M. Trumbauer, A.W. Senear, R. Warran, and R.D. Palmiter. 1981. Somatic expression of herpes thymidine kinase in mice following injection of a fusion gene into eggs. *Cell* **27:** 223.

Brinster, R.L., K.A. Richie, R.E. Hammer, R. O'Brian, B. Arp, and U. Storb. 1983. Expression of a micro-injected immunoglobulin gene in the spleen of transgenic mice. *Nature* **306:** 332.

Costantini, F. and E. Lacy. 1981. Introduction of a rabbit β-globin gene into the mouse germ line. *Nature* **294:** 92.

Dana, S.L. and J.J. Wasmuth. 1982. Linkage of the *leuS*, *emtB* and *chr* genes on chromosome 5 in humans and expression of human genes encoding protein synthetic components in human-Chinese hamster hybrids. *Somatic Cell Genet.* **8:** 245.

Dierks, P., A. van Ooyen, M.D. Cochran, C. Dobkin, J. Reiser, and C. Weissman. 1983. Three regions upstream from the cap site are required for efficient and accurate transcription of the rabbit β-globin gene in mouse 3T6 cells. *Cell* **32:** 692.

Evans, R.M., N. Birnberg, and M.G. Rosenfeld. 1982. Glucocorticoid and thyroid hormones transcriptionally regulate growth hormone gene expression. *Proc. Natl. Acad. Sci.* **79:** 7659.

GEHRING, U., B. SEGNITZ, B. FOELLMER, and U. FRANKE. 1985. Assignment of the human gene for the glucocorticoid receptor to chromosome 5. *Proc. Natl. Acad. Sci.* **82:** 3751.

GILLIES, S.D., S.L. MORRISON, V.T. OI, and S. TONEGAWA. 1983. A tissue-specific transcription enhancer element is located in the major intron of a rearranged immunoglobulin heavy chain gene. *Cell* **33:** 717.

GOLDBERG, D.A., J.W. POSAKONY, and T. MANIATIS. 1983. Correct developmental expression of a cloned alcohol dehydrogenase gene transduced into the *Drosophila* germ line. *Cell* **34:** 59.

GORDON, J.W., G.A. SCANGOS, D.J. PLOTKIN, J.A. BARBOSA, and F.H. RUDDLE. 1980. Genetic transformation of mouse embryos by microinjection of purified DNA. *Proc. Natl. Acad. Sci.* **77:** 7380.

GORMAN, C., L. MOFFAT, and B. HOWARD. 1982. Recombinant genomes which express chloramphenicol acetyltransferase in mammalian cells. *Mol. Cell. Biol.* **2:** 1044.

GROSVELD, G.C., A. ROSENTHAL, and R.A. FLAVELL. 1982. Sequence requirements for the transcription of the rabbit β-globin gene *in vivo*: The −80 region. *Nucleic Acids Res.* **10:** 4951.

HAAN, E.A., B.D. BOSS, and W.M. LOWAN. 1982. Production and characterization of monoclonal antibodies against the "brain-specific" proteins 14-3-2 and S-100. *Proc. Natl. Acad. Sci.* **79:** 7585.

HAMMER, R.E., R. PALMITER, and R.L. BRINSTER. 1984. Partial correction of murine hereditary growth disorder by germ-line incorporation of a new gene. *Nature* **311:** 65.

HAMMER, R.E., R. BRINSTER, M.G. ROSENFELD, R.M. EVANS, and K.E. MAYO. 1985. Expression of human growth hormone-releasing factor in transgenic mice results in increased somatic growth. *Nature* **315:** 413.

HANAHAN, D. 1985. Heritable formation of pancreatic beta-cell tumors in transgenic mice expressing recombinant insulin/simian virus 40 oncogenes. *Nature* **315:** 115.

HARMON, J.M., H.J. EISEN, S.T. BROWER, S.S. SIMONS, C.L. LANGLEY, and E.B. THOMPSON. 1984. Identification of human leukemic glucocorticoid receptors using affinity labeling and anti-human glucocorticoid receptor antibodies. *Cancer Res.* **44:** 4540.

HYNES, N., A.J.J. VAN OOYEN, N. KENNEDY, P. HERRLICH, H. PONTA, and B. GRONER. 1983. Subfragments of the large terminal repeat cause glucocorticoid-responsive expression of mouse mammary tumor virus and of an adjacent gene. *Proc. Natl. Acad. Sci.* **80:** 3637.

KARIN, M., A. HASLINGER, H. HOLTGREVE, R. RICHARDS, P. KRAUTER, H. WESTPHAL, and M. BEATO. 1984. Characterization of DNA sequences through which cadmium and glucocorticoid hormones induce human metallothionein II gene. *Nature* **308:** 513.

LEBO, R., F. GOREN, R. FLETTERICK, F.-T. KAO, M.-C. CHEUNG, B. BRUCE, and Y. KAN. 1984. High resolution chromosome sorting and DNA spot-blot analysis assign McArdle's Syndrome to chromosome 11. *Science* **225:** 57.

MAYO, K.E. and R.D. PALMITER. 1981. Glucocorticoid regulation of metallothionein-I mRNA synthesis in cultured mouse cells. *J. Biol. Chem.* **256:** 2621.

MCKNIGHT, S.L. and R. KINGSBURY. 1982. Transcriptional control signals of a eukaryotic protein-coding gene. *Science* **217:** 316.

OI, V., S.L. MORRISON, L.A. HERZENBERG, and P. BERG. 1983. Immunoglobulin gene expression in transformed lymphoid cells. *Proc. Natl. Acad. Sci.* **80:** 825.

OKRET, S., J. CARLSTEDT-DUKE, O. WRANGE, K. CARLSTROM, and J.-A. GUSTAFSSON. 1981. Immunochemical analysis of the glucocorticoid receptor: Identification of a third domain separate from the steroid-binding and DNA-binding domains. *Biochim. Biophys. Acta* **677:** 205.

PALMITER, R.D., G. NORSTEDT, R.E. GELINAS, R.E. HAMMER, and R.L. BRINSTER. 1983. Metallothionein human GH fusion genes stimulate growth of mice. *Science* **222:** 809.

PALMITER, R.D., R.L. BRINSTER, R.E. HAMMER, M.E. TRUMBAUER, M.G. ROSENFELD, N.C. BIRNBERG, and R.M. EVANS. 1982. Dramatic growth of mice that develop from eggs microinjected with metallothionein growth hormone fusion genes. *Nature* **300:** 611.

PAYVAR, F., O. WRANGE, J. CARLSTEDT-DUKE, S. OKRET, J.-A. GUSTAFSSON, and K.R. YAMAMOTO. 1981. Purified glucocorticoid receptors bind selectively *in vitro* to a cloned DNA fragment whose transcription is regulated by glucocorticoids *in vivo*. *Proc. Natl. Acad. Sci.* **78:** 6628.

PELHAM, H. 1982. A regulatory upstream promoter element in the *Drosophila* Hsp 70 heat-shock gene. *Cell* **30:** 517.

PFAHL, M. 1982. Specific binding of the glucocorticoid-receptor complex to the mouse mammary tumor proviral promoter region. *Cell* **31:** 475.

QUEEN, C. and D. BALTIMORE. 1983. Immunoglobulin gene transcription is activated by downstream sequence elements. *Cell* **33:** 741.

RAAG, H. and C. WEISSMAN. 1983. Not more than 117 base pairs of 5′-flanking sequence are required for inducible expression of a human 1FN-gene. *Nature* **303:** 439.

RENKAWITZ, R., H. BEUG, T. GRAF, P. MATTHAIS, M. GREG, and G. SCHUTZ. 1982. Expression of a chicken lysozyme recombinant gene is regulated by progesterone and dexamethasone after microinjection into oviduct cells. *Cell* **31:** 167.

RINGOLD, G.M., K.R. YAMAMOTO, J.M. BISHOP and H.E. VARMUS. 1977. Glucocorticoid-stimulated accumulation of mouse mammary tumor virus RNA: Increased rate of synthesis of viral RNA. *Proc. Natl. Acad. Sci.* **74:** 2879.

SCHOLNICK, S.B., B.A. MORGAN, and J. HIRSH. 1983. The cloned dopa decarboxylase gene is developmentally regulated when reintegrated into the *Drosophila* genome. *Cell* **34:** 37.

SPANDIDOS, D.A. and J. PAUL. 1982. Transfer of human globin genes to eythioleukemia mouse cells. *EMBO J.* **1:** 15.

SPINDLER, S.R., S.H. MELLON, and J.D. BAXTER. 1982. Growth hormone gene transcription is regulated by thyroid and glucocorticoid hormones in cultured rat pituitary cells. *J. Biol. Chem.* **257:** 11627.

SPRADLING, A.C. and G.M. RUBIN. 1983. The effect of chromosomal position on the expression of the *Drosophila* xanthine dehydrogenase gene. *Cell* **34:** 47.

SUPOWIT, S.C., E. POTTER, R.M. EVANS, and M.G. ROSENFELD. 1984. Polypeptide hormone regulation of gene transcription: Specific 5′ genome sequences are required for epidermal growth factor and phorbol ester regulation of prolactin gene expression. *Proc. Natl. Acad. Sci.* **81:** 2975.

SWANSON, L.W. and P.E. SAWCHENKO. 1983. Hypothalamic integration: Organization of the paraventricular and supraoptic nuclei. *Annu. Rev. Neurosci.* **6:** 269.

SWANSON, L.W., D.M. SIMMONS, J. ARRIZA, R. HAMMER, R. BRINSTER, M.G. ROSENFELD, and R.M. EVANS. 1985. Novel developmental specificity in the nervous system of transgenic animals expressing growth hormone fusion gene. *Nature* **317:** 363.

WALKER, M., T. EDLUND, A. BOULET, and W. RUTTER. 1983. Cell specific expression controlled by the 5′ flanking region of insulin and chymotrypsin genes. *Nature* **306:** 557.

WEINBERGER, C., S.M. HOLLENBERG, E.S. ONG, J.M. HARMON, S.T. BROWER, J. CIDLOWSKI, E.B. THOMPSON, M.G. ROSENFELD, and R.M. EVANS. 1985. Identification of human glucocorticoid receptor cDNA clones by epitope selection. *Science* **228:** 740.

Elastase I Promoter Directs Expression of Human Growth Hormone and SV40 T Antigen Genes to Pancreatic Acinar Cells in Transgenic Mice

D.M. ORNITZ,* R.D. PALMITER,* A. MESSING,† R.E. HAMMER,‡ C.A. PINKERT,‡ AND R.L. BRINSTER‡

*Howard Hughes Medical Institute Laboratory, Department of Biochemistry, University of Washington, Seattle, Washington 98195; †Department of Pathobiological Science, School of Veterinary Medicine, University of Wisconsin, Madison, Wisconsin 53706; ‡Laboratory of Reproductive Physiology, School of Veterinary Medicine, University of Pennsylvania, Philadelphia, Pennsylvania 19104

The exocrine pancreas, which differentiates from the endothelial cells of the foregut, functions to synthesize and store digestive enzymes that are released into the intestinal tract. The pancreas also contains endocrine cells probably of ectodermal origin, which produce metabolic regulatory hormones, including insulin, glucagon, and somatostatin. During mouse development the pancreatic acinar cells can be detected by about day 10 and begin to synthesize low levels of digestive enzymes between days 11 and 14. Shortly before birth the levels of these enzymes increase dramatically and adult levels are attained within the first several weeks of life (Rutter et al. 1972). Expression of pancreatic serine protease genes is several orders of magnitude higher in the acinar cell than in any other cell type (Swift et al. 1984a). Because expression of these digestive enzymes in inappropriate tissues could be deleterious to an organism, there may be strong selective pressure to maintain this very precise cell-specific expression. To gain an understanding of the regulatory mechanisms governing this family of digestive enzymes, we have begun studying the regulation of the rat elastase I gene. This gene is one of at least nine serine protease genes expressed exclusively in pancreatic acinar cells.

Regulation of cell-specific gene expression is being studied by introducing various genes into tissue culture cells that maintain a differentiated phenotype or by producing transgenic mice carrying these genes. Although cis-acting, cell-specific elements have been identified for a variety of genes by DNA-mediated transfection into cells (Banerji et al. 1983; Chao et al. 1983; Gillies et al. 1983; Kondoh et al. 1983; Stafford and Queen 1983; Walker et al. 1983), transgenic mice have the distinct advantage that developmentally regulated genes can be assayed in every possible cell type throughout the normal development of the mouse. Furthermore, normal levels of expression are observed for many genes in transgenic mice, whereas in tissue culture the level of expression of differentiated genes is often very low. In transgenic mice, tissue-specific expression has been demonstrated for immunoglobulin heavy- and light-chain genes (Brinster et al. 1983; Grosschedl et al. 1984; Storb et al. 1984; Rusconi and Kohler 1985), the elastase I gene (Swift et al. 1984a;

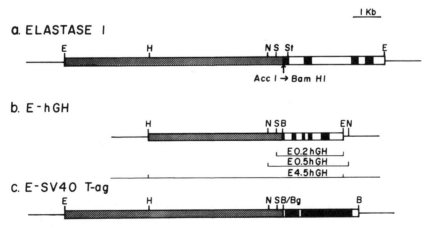

Figure 1. Elastase-hGH and elastase-SV40 fusion gene constructions. (*a*) A portion of the rat elastase I gene with 7.2 kb of 5' sequence and 3.6 kb of structural gene containing the first four exons. The *Acc*I site at +8 was converted to a *Bam*HI site, as described in Experimental Procedures. (*b*) The three elastase-hGH fusion genes introduced into transgenic mice. (*c*) The elastase-SV40 T antigen fusion gene introduced into transgenic mice. Black bars represent exons, stippled regions represent elastase 5'-flanking sequence, and the thin line represents pBR322 vector sequences. (Bg) *Bgl*II; (E) *Eco*RI; (H) *Hin*dIII; (N) *Nde*I; (S) *Sal*I; (St) *Stu*I.

Ornitz et al. 1985), the insulin gene (Hanahan 1985), β-globin genes (Chada et al. 1985; Townes et al. 1985), the myosin light-chain gene (Shani 1985), and the α-fetoprotein gene (Krumlauf et al. 1985).

Previous studies in transgenic mice have demonstrated that the entire rat elastase I gene, containing 7 kb of 5' sequence, 11 kb of coding sequence and introns, and 5 kb of 3' flanking sequence, is expressed higher in the pancreas than in any other tissue by three to four orders of magnitude (Swift et al. 1984a). To identify sequences required for pancreas-specific expression in the 23-kb rat elastase I genomic clone, we tested fusion genes containing 5' elastase I sequence and the human growth hormone (hGH) structural gene in transgenic mice (Ornitz et al. 1985). These studies revealed that no more than 213 bp of elastase 5'-flanking sequence are required for acinar cell-specific expression.

The hGH structural gene has several advantages for use in transgenic mice. The gene is small, the five exons span 1.6 kb of genomic DNA (Seeburg 1982, Fig. 1b), and the mRNA appears to be spliced normally and it is reasonably stable in many tissues (Palmiter et al. 1983). The protein product is easily detected by radioimmune assay (Palmiter et al. 1983) or indirect immunofluorescence (Ornitz et al. 1985). Furthermore, if hGH is synthesized in a cell that secretes into the bloodstream, it can stimulate the growth of transgenic mice, a parameter that can be monitored easily (Palmiter et al. 1983). However, if hGH is produced in an exocrine cell type, and consequently secreted externally, then no enhanced growth would be expected.

In this paper we demonstrate that 213 bp of elastase 5' sequence are sufficient for normal levels of pancreas-specific expression, that the tissue-specificity is manifested at the transcriptional level, and that expression of an elastase-hGH fusion gene in transgenic mice correlates with a tissue-specific DNase I hypersensitive site corresponding to the region of the elastase regulatory sequence. We further demonstrate that the elastase regulatory region can direct expression of SV40 T antigen in the pancreatic acinar cells resulting in characteristic pancreatic tumors in transgenic mice.

EXPERIMENTAL PROCEDURES

Construction of fusion genes. As a first step in constructing pancreatic acinar cell-specific expression vectors for use in transgenic mice, we introduced a unique restriction site in the first exon of the rat elastase I gene. To accomplish this, we subcloned a 400-bp SalI-StuI fragment that spans the mRNA initiation site and contains 205 bp of elastase 5'-flanking sequence from plasmid Elastase I (containing 7.2 kb of elastase 5' sequence and 3.6 kb of the structural gene; Fig. 1a) into pUC13. A BamHI linker was placed at the AccI site (position +8) in this subclone (E0.2). An elastase-hGH fusion gene (E0.2hGH; Fig. 1b) was made by cloning the hGH structural gene (on a 2.1-kb BamHI-EcoRI fragment) into E0.2. Elastase 5'-flanking sequence was added to this construction to generate E4.5hGH (Fig. 1b), a plasmid containing 4.5 kb of elastase 5'-flanking sequence. The E7.2 SV40 T antigen fusion gene (Fig. 1c) was made in a similar fashion. The fusion in this construction is at +8 of elastase and at the StuI site at +34 (converted to a BglII site) in SV40. The resulting mRNA should contain 8 bp of elastase 5' untranslated sequence, a synthetic linker fusion (BamHI/BglII), and 27 bp of SV40 5' untranslated mRNA, followed by T antigen coding sequence.

Production of transgenic mice. All plasmids were linearized prior to microinjection. E4.5hGH (Fig. 1b) was linearized at the EcoRI site prior to injection, E0.2hGH was linearized at SalI, and E7.2SV40 T antigen (Fig. 1c) was linearized at an XhoI site in the plasmid vector. The E0.5hGH (a 2.8-kb NdeI fragment) and some of the E0.2hGH (a 2.3-kb SalI-EcoRI fragment) mice were produced with genes that were separated from vector sequences prior to injection. The restricted fragments were isolated from 1% agarose gels by a perchlorate elution procedure (Chen and Thomas 1980), and 100-300 copies were microinjected into the male pronucleus of fertilized mouse eggs as described (Brinster et al. 1981, 1985). The injected eggs were introduced into the oviduct of pseudopregnant foster mothers and allowed to develop to term. Mouse pups were screened by tail DNA dot hybridization for the presence of the injected gene. Growth rates of elastase-hGH transgenic mice were monitored until they were about 10 weeks old, at which time the mice were either bred to establish transgenic lines or sacrificed for tissue analysis. Elastase SV40 mice were bred and monitored for signs of pathology.

Nucleic acid isolation and RNA analysis. Fifty milligrams of each tissue were homogenized in 4 ml of SET (1% SDS, 1 mM EDTA, 10 mM Tris, pH 7.5) with 100 µg/ml proteinase K. Total nucleic acids were purified as previously described (Durnam and Palmiter 1983) and stored frozen in 0.2 × SET. mRNA levels were analyzed by a solution hybridization procedure utilizing a complementary end-labeled 21-base oligonucleotide probe. Hybridization and S1 nuclease digestion were carried out as previously described (Ornitz et al. 1985).

Isolation of nuclei. Nuclei were isolated from about 100 mg of pancreas or liver by Dounce homogenization followed by centrifugation essentially as described (Mulvihill and Palmiter 1977), except that 1 mM EGTA was added to buffers NA, NB, and NC. Nuclei were stored in buffer NC at -70°C and thawed on ice for transcription and DNase I hypersensitivity assays.

Nuclear transcription assays. Transcription assays were performed as previously described (McKnight and Palmiter 1979), except that nitrocellulose filter discs for hybridization contained about 4 µg of plasmid DNA and hybridizations were incubated at 45°C for 36 hours. Individual filters were washed as previously described (McKnight and Palmiter 1979) and counted for

10 minutes in a Packard Liquid Scintillation spectrometer. Transcription rates (expressed in ppm) were calculated by dividing hybridized cpm minus background by the length of the cloned probe in kilobases and by the input counts of ^{32}P-labeled RNA.

DNase I hypersensitivity assays. DNase I hypersensitivity studies were carried out by adding 0.1 μg/ml to 5 μg/ml DNase I in 5 mM MgCl$_2$ and 1 mM CaCl$_2$ to approximately 20 μg of nuclear DNA in buffer NC. Digestion was carried out at 37°C for 10 minutes with pancreas nuclei, and at 25°C for 5 minutes with liver nuclei. Nuclei were then incubated in 1% SDS and 100 μg/ml proteinase K for 1 hour at 37°C followed by extraction with phenol/chloroform and then chloroform. Total nucleic acids were precipitated in 70% ethanol and 0.1 M NaCl. Nucleic acids were dissolved in 10 mM Tris-HCl, 1 mM EDTA (pH 7.5), treated with RNase A, and digested with the indicated restriction enzymes and analyzed by Southern blotting.

Histology. Tissue samples were fixed overnight in Bouin's fixative. Tissue samples were imbedded in paraffin by standard procedures, and 6-μm sections were stained with hematoxylin and eosin.

Immunoperoxidase staining. Tissues were fixed overnight in Bouin's fixative, washed in 70% ethanol, and embedded in paraffin. Six-micrometer sections were mounted on glass slides and stained for SV40 large T antigen by an indirect immunoperoxidase procedure, using the monoclonal antibody 412 (Gurney et al. 1980) and the ABC avidin-biotin system (Vector Laboratories). Sections were exposed to a 1:100 dilution of 412 culture supernatant (gift from A.J. Levine, Princeton University) overnight at 4°C, treated with the ABC reagents, and then incubated for 5 minutes in 0.3 mg/ml diaminobenzidine, 0.01% H$_2$O$_2$, and 2% nickel ammonium sulfate.

RESULTS AND DISCUSSION

Pancreas-specific Expression of Elastase-hGH Fusion Genes

Transgenic mice containing elastase-hGH fusion genes with either 4.5 kb, 0.5 kb, or 0.2 kb of elastase 5'-flanking sequence (Fig. 1b) expressed hGH mRNA only in the pancreas. In most cases we examined eight tissues: intestine, kidney, liver, pancreas, parotid gland, spleen, submandibular gland, and testes or ovary. We have never observed significant expression in a nonpancreatic tissue. Analysis of 21 transgenic mice containing these three fusion genes (Table 1) reveals that hGH mRNA levels are up to three to four orders of magnitude higher in the pancreas than in other tissues in 15 out of 21 mice. Five mice failed to express the foreign genes in any tissue and one mouse had very low-level expression in the pancreas. Some mice in each group expressed more hGH mRNA than endogenous elastase mRNA (about 10,000 molecules/cell; Swift et al. 1984a), suggesting that elastase sequences can function in both a quantitatively and qualitatively normal fashion when inserted into random chromosomal positions.

Table 1. Expression of Elastase-hGH Genes in Tissues of Transgenic Mice

Plasmid	Mouse[a]	Genes/cell[b]	hGH mRNA[c] pancreas (molecules/cell)	other tissues (molecules/cell)
E4.5hGH	37-9	6.5	<10	<10
	31-7	2.2	35	<10
	34-4	1.5	1,170	<10
	34-9	3.2	2,560	<10
	34-2	1.4	4,000	<10
	35-1	1.2	8,490	<10
	33-6	6.2	15,100	<10
	34-10	6.2	28,100	<10
E0.5hGH	43-4*	1.0	<10	<10
	43-10*	6.0	<10	<10
	42-6*	1.2	1,460	<10
	44-6*	199.0	9,760	<10
	44-4*	136.0	11,900	<10
	43-5*	2.3	18,000	<10
	40-2*	4.2	39,400	<10
E0.2hGH	47-4	3.3	<10	<10
	48-4	39.4	<10	<10
	49-3	2.1	3,340	<10
	136-2*	3.4	16,100	<10
	132-5*	3.4	33,100	<10
	129-3*	3.4	51,000	<10

[a]*, These mice received elastase-hGH genes lacking vector sequences.
[b]Determined by quantitative DNA dot hybridization (Ornitz et al. 1985).
[c]Determined by solution hybridization with a 21-base oligonucleotide probe. (See Experimental Procedures.)

Our previous data indicated that elastase-hGH fusion genes yield an hGH mRNA of normal size which is translated into an immunoreactive hGH protein that is secreted into the pancreatic ducts along with the digestive enzymes (Ornitz et al. 1985). Immunohistology was used to demonstrate that hGH was located exclusively in all acinar cells of the pancreas (Ornitz et al. 1985). Furthermore, all transgenic mice containing elastase-hGH fusion genes grow normally. This observation indicates that these fusion genes are not expressed in any cells that secrete into the bloodstream. If hGH were expressed in such cells, then we would expect to observe an increased growth rate in these mice as is observed when metallothionein (MT)-hGH fusion genes are expressed in transgenic mice (Palmiter et al. 1983). Thus, the lack of enhanced growth, the lack of hGH mRNA in nonpancreatic cells, and the exclusive immunofluorescence of hGH in the acinar cells of the pancreas argue that the 213 bp of elastase sequences present in E0.2hGH are sufficient to direct expression of hGH to pancreatic acinar cells.

Analysis of additional deletion mutants and constructs in which the elastase region is inverted or moved further away will be necessary to establish the precise cis-acting sequences involved in pancreas-specific expression and to determine if this region has enhancer-like properties. The 213-bp elastase I sequence contains a 37-bp region that is conserved between the elastase I and II genes; it is also homologous to similarly positioned sequences in the chymotrypsin gene and two trypsin genes (Swift et al. 1984b). Walker et al. (1983) have shown that this region is essential for expression of chymotrypsin-CAT fusion genes in cultured acinar cells.

Elastase hGH Pedigrees

We have established two elastase-hGH lines to determine whether tissue-specific expression is transmissible and whether the level of expression is relatively stable from one offspring to another. Pedigree 43-5 shows three generations of mice containing the E0.5hGH genes (Fig. 2a). The founder mouse transmitted the E0.5hGH genes to only 2 out of 10 of its progeny, whereas in the F_1 generation these genes were transmitted normally (6/12). The average amount of hGH mRNA in the offspring was 63,000 molecules/cell compared with 18,000 in the founder. These observations suggested that the founder was probably mosaic. Consistent with this interpretation, quantitative DNA blot hybridization revealed that the founder mouse contained an average of 2.3 copies of E0.5hGH per cell,

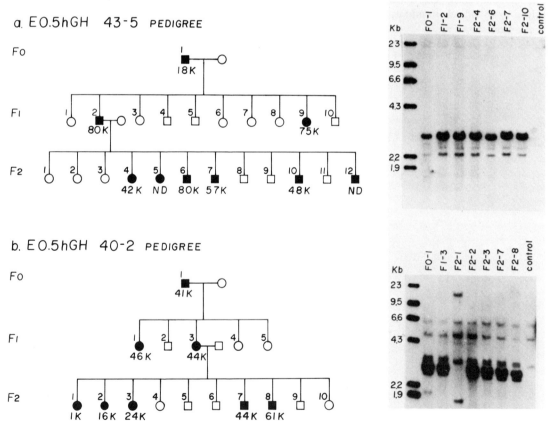

Figure 2. Elastase-hGH pedigrees (*left*). Squares represent males, circles represent females; solid symbols represent animals containing E0.5hGH fusion genes. The numbers below the symbols indicate hGH mRNA levels in thousands of molecules per cell. The Southern blots (*right*) are *Sst*I digests of kidney DNA. In *a* the blot was probed with a nick-translated hGH structural gene fragment. In *b* the blot was probed with a MT-hGH fusion gene fragment (the bands at 6.0, 4.5, and 2.9 kb represent endogenous MT gene fragments).

whereas its progeny contained about seven genes per cell (data not shown). Furthermore, a Southern blot of DNA digested with SstI, an enzyme that cuts once within the 2.8-kb piece of DNA that was introduced into this mouse, yielded a prominent band of 2.8 kb, indicative of a tandem head-to-tail array, plus fainter junction fragments (Fig. 2a). The intensity of the bands in the offspring is significantly greater than the founder, in agreement with the quantitative dot analysis, but the pattern is identical in all of the mice.

A second pedigree, 40-2, is shown in Figure 2b. These mice carry about four copies of E0.5hGH genes and the genes appear to be transmitted in a normal Mendelian manner as though all the genes were closely linked on one chromosome. Southern blot analysis of SstI-digested DNA (Fig. 2b) indicates a major 2.8-kb fragment plus several junction fragments; the pattern and intensity of the bands are similar in all of the mice except one mouse in the F_2 generation (F2-1) that is missing the tandem repeat and has acquired several new bands. This mouse also has an unusually low level of hGH mRNA in the pancreas. Apparently, a DNA rearrangement occurred during meiosis of the egg that gave rise to this mouse. (The bands at 6.0, 4.5, and 2.9 kb represent endogenous MT DNA fragments that were probed as an internal control.)

Both pedigrees reveal that elastase-hGH fusion genes are in the germ line and can be stably (with one notable exception) transmitted through several generations. Expression of the fusion genes is also reasonably constant. Mice from these lines were used to study transcription and chromatin structure of the elastase fusion genes (see below). These lines will also be useful for studying developmental regulation of elastase-hGH gene expression.

Transcriptional Activity of the Elastase Fusion Genes

To ascertain whether the elastase regulatory region is capable of conferring quantitatively normal transcription as well as tissue specificity, we measured the rate of elastase-hGH transcription relative to the endogenous elastase genes. This approach avoids problems associated with mRNA and/or protein accumulation which might be subject to differential stability.

Nuclei were isolated from pancreas and liver of offspring from transgenic line 43-5 containing seven tandem copies of the E0.5hGH gene. These nuclei were allowed to continue transcriptional elongation in the presence of [α-^{32}P]UTP in a so called nuclear "run-on" assay. Specific ^{32}P-labeled RNA transcripts were quantitated by hybridizing them to nitrocellulose-bound plasmid DNA containing homologous cloned sequences of the hGH gene, the mouse elastase I gene, the liver albumin gene, or pBR322 plasmid DNA. The slope from a plot of input ^{32}P-labeled RNA versus hybridized specific RNAs (Fig. 3) gives a value that represents relative transcription rates. These values are ex-

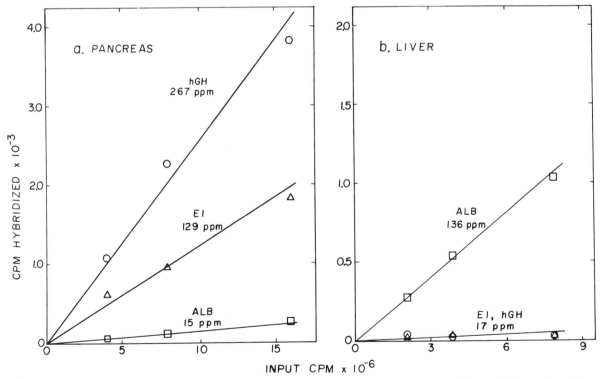

Figure 3. Relative rate of transcription of elastase, E0.5-hGH, and albumin genes in pancreas and liver nuclei from mice of the 43-5 line. Nuclei were isolated and endogenous RNA polymerases were allowed to elongate products started in vivo as described under Experimental Procedures. Labeled RNA transcripts were hybridized to immobilized plasmid DNA containing sequences corresponding to hGH (○), elastase (△), and albumin genes (□). (a) Hybridization of RNA synthesized in pancreas nuclei; (b) a similar experiment using liver nuclear RNA. The relative rate of transcription in ppm was calculated as described in Experimental Procedures.

pressed in ppm after correcting for probe lengths and background, as described in Experimental Procedures.

We calculate that in the pancreas of mice from line 43-5 there are about two times more RNA polymerase molecules engaged in E0.5hGH gene transcription than in endogenous elastase I gene transcription (Fig. 3a). This apparent increased transcription rate may be attributable to the seven tandem copies of the E0.5hGH gene in this transgenic line compared with the two endogenous elastase I genes. Because there is little relationship between gene copy number and mRNA accumulation in transgenic mice (see Table 1), we do not know whether all seven copies of the elastase-hGH fusion genes are transcribed equally well. Nevertheless, we can set some limits: If only one gene is functional, then it is transcribed at about four times the rate of an endogenous elastase gene, and if all seven are transcribed, then they are transcribed at about half the rate of the endogenous elastase genes. Either way, the results suggest that the foreign elastase promoter is recognized about as efficiently as the endogenous promoters, despite being in a completely foreign chromosomal environment.

Transcription of the E0.5hGH gene and the mouse elastase I gene was very low in liver nuclei (Fig. 3b), indicating that the tissue-specific expression is due predominantly, if not exclusively, to transcriptional specificity. Conversely, the albumin gene is transcribed at a high level in the liver and not in the pancreas. This gene serves as a control for functional liver nuclei and also demonstrates a background transcription rate in the pancreas, similar to that of elastase and hGH in the liver. We believe that the background rate of 15-17 ppm seen for all three genes is due to nonspecific binding to our plasmid probes.

The Elastase Sequence Is Associated with a DNase I Hypersensitive Site

DNase I hypersensitive sites are frequently associated with actively transcribed structural genes (Stalder et al. 1980; Sweet et al. 1982; Senear and Palmiter 1983; Becker et al. 1984; Kunnath and Locker 1985). These sites are often at the 5' end of the gene and may reflect perturbations of chromatin structure due to sequence-specific DNA binding proteins. These hypersensitive sites have been shown to correspond with tissue-specific and developmental regulatory elements. For example, the immunoglobulin genes have hypersensitive sites in the intron between the VDJ and C regions in B-cell lines (Parslow and Granner 1982), the insulin gene has a hypersensitive site in its 5' region specifically in β cells (Wu and Gilbert 1981), and the β-globin gene develops a hypersensitive site in its 5' region when mouse erythroleukemia cells are induced to differentiate (Miller et al. 1978). The hypersensitive sites in the murine mammary tumor virus (MMTV) long terminal repeat (LTR) correlate with transcriptional activation of the MMTV promoter by the activated glucocorticoid receptor (Zaret and Yamamoto 1984).

Here we show that a hypersensitive site is established near the elastase regulatory element in the chromatin from pancreas of mice from the 43-5 line. This mouse line carries seven tandem copies of E0.5hGH (NdeI restriction fragment from Fig. 1b). The map of this integrant is shown in Figure 4a. Digestion with BglII yields a 2.84-kb internal BglII fragment and a 12-kb band including the junction between mouse DNA and the insert. These bands are shown in Figure 4b (lane 1, no DNase I treatment or incubation). To detect hypersensitive sites near the 5' end of these genes, we purified DNA from DNase I-digested nuclei, restricted with BglII, and hybridized a Southern blot of this DNA with a probe derived from the middle region of the hGH gene (as shown in Fig. 4a). Digestion of pancreas nuclei with DNase I revealed a band of about 1.4 kb (lanes 2-9) that was not present when liver nuclei were digested (lanes 2-5). The band of approximately 1.4 kb maps a hypersensitive site to the vicinity of the elastase promoter (see Fig. 4a). Additional experiments with more precise size markers (SalI-BglII- and BamHI-BglII-digested DNA) indicate that the hypersensitive site maps to the region between -150 and $+50$ of the E0.5hGH fusion gene. Liver nuclei show no hypersensitive sites in the region of the elastase promoter element, but they do show a uniform 1-kb band present in all lanes that is probably due to digestion by endogenous nucleases during nuclei isolation.

This experiment demonstrates that the elastase sequence generates a DNase I-sensitive region in mouse pancreatic chromatin in close proximity to the tissue-specific regulatory region and promoter. Because the integration site of the E0.5hGH genes is probably random in transgenic mice, this experiment also demonstrates that the ability to form a tissue-specific hypersensitive site is independent of precise chromosomal location. Furthermore it is independent of the elastase I structural gene. It remains to be demonstrated whether the elastase regulatory element can yield a hypersensitive site in the absence of a functional transcription unit.

Pancreatic Carcinomas Induced by Elastase-SV40 T Antigen

The early region of the SV40 virus codes for two proteins: large and small T antigens. These proteins affect the expression of viral and cellular genes and can lead to cell transformation (Tooze 1980). Introduction of SV40 early-region genes into mice resulted in a high frequency of choroid plexus tumors (Brinster et al. 1984). Further experiments demonstrated that large T antigen was sufficient for tumorigenesis and that the SV40 enhancer (72-bp repeats) was required for development of choroid plexus tumors (Palmiter et al. 1985). When the SV40 early region was fused to the 5'-flanking region of the rat insulin gene and introduced into mice, the transgenic mice expressed T antigen only in the β cells and ultimately developed β-cell tumors (Hanahan 1985). Thus, it appears that the site of tu-

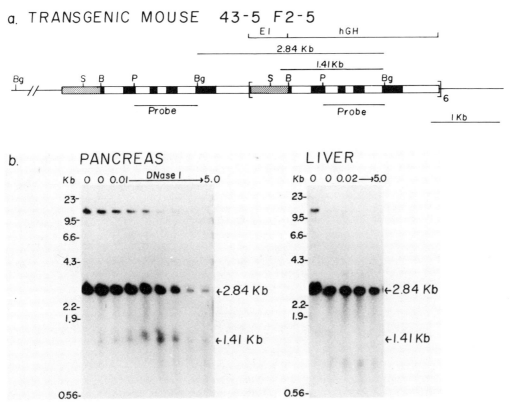

Figure 4. Analysis of DNase I hypersensitivity sites in E0.5hGH genes from line 43-5. (*a*) Map of the insertion containing seven tandem copies of the E0.5hGH gene. The elastase sequences are stippled; the hGH exons are solid. The inserted DNA was 2.84 kb and gives a prominent band of this size when cut with *Bgl*II, an enzyme that cut once within the insert. The 1.41-kb line marks the location of the band that appears after nuclease digestion. (B) *Bam*HI; (Bg) *Bgl*II; (P) *Pvu*II; (S) *Sal*I. (*b*) Southern blot analysis of DNase I-treated nuclei. Pancreas nuclei were digested at 37°C for 10 min and liver nuclei were digested at 25°C for 5 min; DNase I concentrations ranged from 0 to 5 μg/ml; one sample (next to markers) was not incubated. After digestion, the DNA was purified, digested with *Bgl*II, electrophoresed on an agarose gel, transferred to nitrocellulose, and probed with a nick-translated *Pvu*II–*Bgl*II DNA fragment shown in *a*.

morigenesis can be directed to specific cell types by using different transcriptional regulatory elements.

The foreign gene that we injected contained 7.2 kb of elastase 5'-flanking sequence fused to the SV40 T antigen structural gene (Fig. 1c). Five mice containing this fusion gene were produced. These mice were bred to start transgenic lines and then observed for the development of any pathological symptoms. Transgenic line 177-5 was studied in detail (Fig. 5, Table 2). All of the mice in the F_1 generation developed large abdominal masses at 3–7 months of age and died within a few weeks of obvious abdominal swelling. Mice in the F_2 generation are approaching the age when tumors developed in their parent.

The abdominal masses that originated in the pancreas were firm and lobular and generally grew rapidly to enormous size. The mouse in Figure 6A had a 9.1-g tumor when sacrificed. These tumors occasionally spread by transcoelomic seeding. Rare metastases to liver and lung were also observed. Histological examination revealed nearly complete replacement of normal pancreatic tissue (Fig. 6B) by sheets of pleomorphic epithelial cells (Fig. 6C), many of which formed pseudoacinar structures, with numerous mitoses. SV40 large T antigen was detected immunocytochemically in the nucleus of most tumor cells (Fig. 6D).

To assess the relationship between T antigen gene expression and tumorigenesis, we analyzed pancreatic sections from mice of different ages that inherited the elastase T antigen genes. Sections taken between 4 and 28 days after birth revealed a few isolated cells in the exocrine pancreas that were positive for large T antigen (Fig. 6E); tissue morphology at this stage was essentially normal. Between 4 and 8 weeks the isolated foci increased in size, possibly by clonal expansion (Fig. 6F). The earliest pathological changes evident by hematoxylin and eosin staining were an increase in size and irregularity of the acinar cell nuclei. The first histologic evidence of actual tumor formation was an expanding growth of disorganized acinar epithelial cells with numerous mitoses (Fig. 6G). These cells showed more intense staining for T antigen (Fig. 6H), but there was marked cell-to-cell variation even within such a focus.

We also measured the SV40 mRNA in pancreas from transgenic mice of different ages (Fig. 5, Table 2). In young mice, SV40 mRNA was either undetectable or very low. By 11 weeks, SV40 mRNA was generally detectable but the amount increased sharply in overt tu-

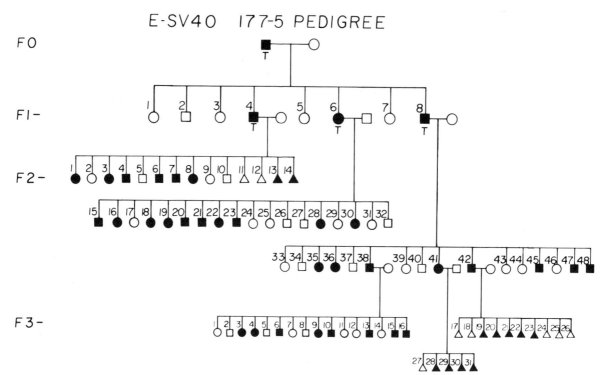

Figure 5. Pedigree of the 177-5 line of mice carrying E7.2SV40 T antigen fusion genes. Symbols are the same as in Fig. 2 legend. T signifies that a pancreatic tumor developed; triangles indicate mice where sex was not determined.

Table 2. Pathology and Gene Expression in Pancreatic Tumors of Transgenic Mice

Mouse[a]	Age[b] (days)	Histology[c]	Pancreatic mRNA content		T antigen[e] (molecules/cell)
			elastase I[d]	amylase[d]	
			(% normal pancreas)		
F1-4	210	+++	3.4	3.8	269
F2-4	140	+++	59.0	70.3	25
F2-38	108	+++	27.8	44.0	13
F2-41	108	+++	39.6	59.0	16
F2-47	87	++	75.7	87.1	12
F2-48	87	++	93.4	100.0	14
F3-19	26	+	ND	ND	<5
F3-20	26	+	ND	ND	<5
F3-28	19	+	ND	ND	<5
F3-29	19	+	ND	ND	<5
F3-30	19	+	ND	ND	<5
F3-31	19	+	ND	ND	<5
F3-21	4	−	ND	ND	<5
F3-22	4	+	ND	ND	<5
F3-23	4	−	ND	ND	<5
261-4	80	+++	9.1	23.2	11
264-4	140	+++	34.1	20.1	22
266-5	200	+++	3.6	3.1	12
266-6	112	+++	10.5	15.7	40

[a]Individual mice in the F_1, F_2, and F_3 generations from pedigree 177-5 are identified in Fig. 5.

[b]The age when mice either died of pancreatic tumors or were sacrificed for tissue analysis.

[c]+++, Large tumors; ++, focal tumors; +, isolated cells or very small T antigen-positive foci; −, no T antigen-positive cells.

[d]mRNA levels were determined by solution hybridization of total nucleic acid (TNA) with ^{32}P-labeled oligonucleotides specific for amylase and elastase mRNA. The values are presented relative to normal pancreas.

[e]T antigen mRNA was quantitated by solution hybridization with an oligonucleotide specific to the 3′ end of the T antigen gene. Molecules per cell were calculated by using an M13 standard, estimating the size of T antigen mRNA, and determining the RNA/DNA ratio of the TNA samples (Durnam and Palmiter 1983).

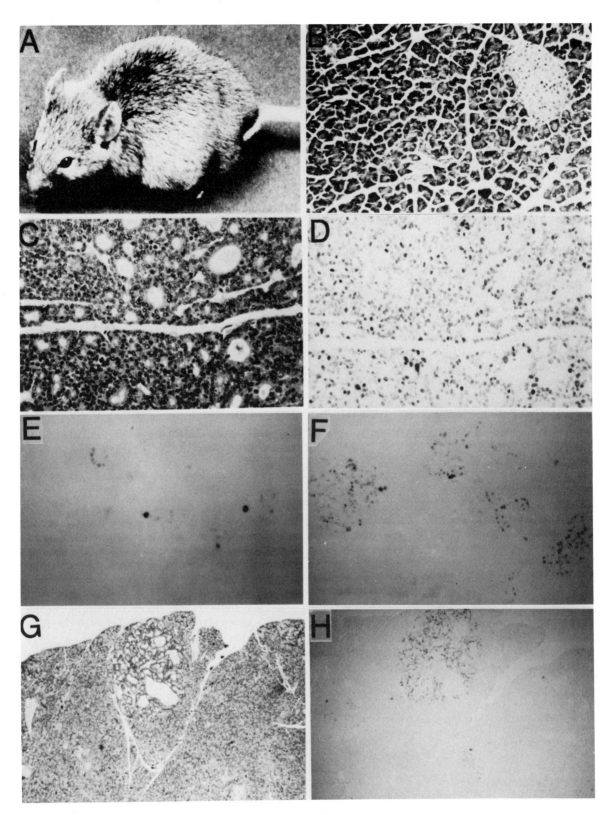

Figure 6. Histology of mice containing elastase-SV40 T antigen fusion genes. (*a*) Transgenic mouse with abdominal swelling (this mouse had a 9.1-g pancreatic tumor). (*b*) Hematoxylin and eosin-stained section of normal pancreas. (*c*) Hematoxylin and eosin-stained section of a pancreatic tumor (mouse F1-4; Fig. 5). (*d*) Indirect immunoperoxidase staining for T antigen of the pancreatic tumor shown in *c* (280-day-old mouse). (*e*) Immunoperoxidase staining for T antigen in pancreas of a 26-day-old mouse (F3-22). (*f*) Immunoperoxidase staining for T antigen in pancreas of a 87-day-old mouse (F3-16). (*g*) A focal tumor in the pancreas of mouse F3-16 visualized with hematoxylin and eosin. (*h*) Immunoperoxidase staining for T antigen in the focal tumor shown in *g*.

mors. These results corroborate the immunological staining for SV40 T antigen and suggest that the increased expression is due to proliferation of a subpopulation of cells expressing T antigen. We also noted that as the expression of T antigen increased there was a progressive decrease in the amount of elastase and amylase mRNA (Table 2).

The results obtained with elastase SV40 T antigen genes are quite different from those observed with elastase-hGH. In the latter case, all the acinar cells expressed hGH and at very high levels, whereas T antigen expression appears to commence in a small number of acinar cells and then increase presumably by clonal expansion. These results also contrast with those obtained by Hanahan (1985) with a comparable insulin T antigen construct. He observed relatively uniform expression of T antigen in all β-cells that was associated with hypertrophy of those cells; eventually some of those cells proliferated and formed overt tumors.

We do not know what causes the activation of SV40 T antigen gene expression in selected pancreatic acinar cells. The availability of lines of mice that routinely develop pancreatic tumors (Fig. 5) will allow a more detailed analysis of the molecular events associated with tumor development.

Pancreatic tumor tissue from mouse 266-6 was used to start a cell line. This cell line has been maintained in culture for several months. The cells grow in clumps (resembling acini) and they express high levels of SV40 mRNA (450 molecules/cell); this value is considerably higher than in the primary tumor (Table 2). This cell line retained about 2% of normal elastase mRNA, but amylase mRNA was barely detectable. Because these cells contain an elastase-SV40 fusion gene, we suspect that there is selective pressure to maintain some degree of differentiation. Failure to maintain the proteins involved in elastase promoter function would presumably lead to cessation of the transformed phenotype.

CONCLUSIONS

To identify the *cis*-acting DNA sequence elements required for pancreas-specific expression of the rat elastase I gene, the 5'-flanking region of the elastase I gene was joined to the hGH structural gene. Elastase-hGH fusion genes with 4.5, 0.5, and 0.2 kb of elastase 5'-flanking sequence were introduced into mice. Most of the mice carrying these three constructs expressed high levels of hGH mRNA in the pancreas but undetectable levels in other tissues. The data indicate that the sequence between −205 and +8 bp of the elastase promoter is sufficient to direct expression of hGH exclusively to the pancreatic acinar cells. Two lines of mice were established that transmit tissue-specific expression of hGH in a stable manner. Nuclei isolated from pancreas or liver of these mice were used to demonstrate that transcription of the elastase-hGH gene is tissue specific and quantitatively similar to the endogenous elastase genes. These nuclei were also used to demonstrate that there is a DNase I hypersensitive site located in the elastase regulatory element in pancreatic nuclei but not in liver nuclei. The results suggest that sequences in the close proximity to the elastase promoter bind a specific protein(s) and this leads to transcriptional activation exclusively in the acinar cells of the pancreas. The function of this sequence is apparently independent of chromosomal location.

When the elastase 5' sequences were fused to SV40 T antigen genes, large pancreatic tumors routinely developed at 3–7 months of age in transgenic founder mice and their offspring. Development of these tumors appears to reflect activation of SV40 T antigen expression in a small number of acinar cells, followed by their proliferation into overt tumors rather than high-level expression in all acinar cells, as was observed when this regulatory element was fused to the hGH gene.

ACKNOWLEDGMENTS

We thank Brian Davis, Galvin Swift, and Raymond MacDonald for providing mouse elastase I cDNA clones and synthetic oligomers for quantitating mouse elastase and amylase mRNA, Mary Yagle and Kathy DePaul for technical assistance, and Myrna Trumbauer for DNA microinjection. D.M.O. was supported by the Medical Scientist Training Program at the University of Washington. This research was supported by the National Institutes of Health grants HD-09172, CA-38635, and NS-00956.

REFERENCES

Banerji, J., L. Olson, and W. Schaffner. 1983. A lymphocyte-specific cellular enhancer is located downstream of the joining region in immunoglobulin heavy chain genes. *Cell* **33:** 729.

Becker, P., R. Renkawitz, and G. Schutz. 1984. Tissue-specific DNase I hypersensitive sites in the 5'-flanking sequences of the tryptophan oxgenase and the tyrosine aminotransferase genes. *EMBO J.* **3:** 2015.

Brinster, R.L., H.Y. Chen, M. Trumbauer, M.K. Yagle, and R.D. Palmiter. 1985. Factors affecting the efficiency of introducing foreign DNA into mice by microinjecting eggs. *Proc. Natl. Acad. Sci.* **82:** 4438.

Brinster, R.L., H.Y. Chen, A. Messing, T. van Dyke, A.J. Levine, and R.D. Palmiter. 1984. Transgenic mice harboring SV40 T-antigen genes develop characteristic brain tumors. *Cell* **37:** 367.

Brinster, R.L., H.Y. Chen, M. Trumbauer, A.W. Senear, R. Warren, and R.D. Palmiter. 1981. Somatic expression of herpes thymidine kinase in mice following injection of a fusion gene into eggs. *Cell* **27:** 223.

Brinster, R.L., K.A. Ritchie, R.E. Hammer, R.L. O'Brien, B. Arp, and U. Storb. 1983. Expression of a microinjected immunoglobulin gene in the spleen of transgenic mice. *Nature* **306:** 332.

Chada, K., J. Magram, K. Raphael, G. Radice, E. Lacy, and F. Costantini. 1985. Specific expression of a foreign β-globin gene in erythroid cells of transgenic mice. *Nature* **314:** 377.

Chao, M.V., P. Mellon, P. Charnay, T. Maniatis, and R. Axel. 1983. The regulated expression of β-globin genes introduced into mouse erythroleukemia cells. *Cell* **32:** 483.

Chen, C.W. and C.A. Thomas, Jr. 1980. Recovery of DNA fragments from agarose gels. *Anal. Biochem.* **101:** 339.

Durnam, D. and R.D. Palmiter. 1983. A practical approach

for quantitating specific mRNAs by solution hybridization. *Anal. Biochem.* **131:** 385.
GILLIES, S.D., S.L. MORRISON, V.T. OI, and S. TONEGAWA. 1983. A tissue-specific transcription enhancer element is located in the major intron of a rearranged immunoglobulin heavy chain gene. *Cell* **33:** 717.
GROSSCHEDL, R., D. WEAVER, D. BALTIMORE, and F. COSTANTINI. 1984. Introduction of a μ immunoglobulin gene into the mouse germ line: Specific expression in lymphoid cells and synthesis of functional antibody. *Cell* **38:** 647.
GURNEY, E.G., R.O. HARRISON, and J. FENNO. 1980. Monoclonal antibodies against simian virus T-antigen: Evidence for distinct subclasses of large T-antigen and for similarities among non-viral T-antigens. *J. Virol.* **34:** 752.
HANAHAN, D. 1985. Cell specific expression of recombinant insulin/SV40 oncogenes produces β cell tumors in transgenic mice. *Nature* **315:** 115.
KONDOH, H., K. YASUDA, and T.S. OKADA. 1983. Tissue-specific expression of a cloned chick δ-crystallin gene in mouse cells. *Nature* **301:** 440.
KRUMLAUF, R., R.E. HAMMER, S.M. TILGHMAN, and R.L. BRINSTER. 1985. Developmental regulation of α-fetoprotein genes in transgenic mice. *Mol. Cell. Biol.* **5:** 1639.
KUNNATH, L. and J. LOCKER. 1985. DNase I sensitivity of the rat albumin and α-fetoprotein genes. *Nucleic Acids Res.* **13:** 115.
MCKNIGHT, G.S. and R.D. PALMITER. 1979. Transcriptional regulation of the ovalbumin and conalbumin genes by steroid hormones in chick oviduct. *J. Biol. Chem.* **254:** 9050.
MILLER, D.M., P. TURNER, A.W. NIENHUIS, D.E. AXELROD, and T.V. GOPALAKRISHNAN. 1978. Active conformation of the globin genes in uninduced and induced mouse erythroleukemia cells. *Cell* **14:** 511.
MULVIHILL, E.R. and R.D. PALMITER. 1977. Relationship of nuclear estrogen receptor levels to induction of ovalbumin and conalbumin mRNA in chick oviduct. *J. Biol. Chem.* **252:** 2060.
ORNITZ, D.M., R.D. PALMITER, R.E. HAMMER, R.L. BRINSTER, G.H. SWIFT, and R.J. MACDONALD. 1985. Specific expression of an elastase-human growth hormone fusion gene in pancreatic acinar cells of transgenic mice. *Nature* **313:** 600.
PALMITER, R.D., H.Y. CHEN, A. MESSING, and R.L. BRINSTER. 1985. The SV40 enhancer and large T-antigen are instrumental in development of choroid plexus tumors in transgenic mice. *Nature* **316:** 457.
PALMITER, R.D., G. NORSTEDT, R.E. GELINAS, R.E. HAMMER, and R.L. BRINSTER. 1983. Metallothionein-human GH fusion genes stimulate growth of mice. *Science* **222:** 809.
PARSLOW, T.G. and D.K. GRANNER. 1982. Chromatin changes accompany immunoglobulin K gene activation: A potential control region within the gene. *Nature* **299:** 449.

RUSCONI, S. and G. KOHLER. 1985. Transmission and expression of a specific pair of rearranged immunoglobulin μ and ϰ genes in a transgenic mouse line. *Nature* **314:** 330.
RUTTER, W.J., J.D. KEMP, W.S. BRADSHAW, W.R. CLARK, R.A. RONZIO, and T.G. SANDERS. 1972. Regulation of specific protein synthesis in cytodifferentiation. *J. Cell. Physiol.* (suppl.) **1:** 1.
SEEBURG, P.H. 1982. The human growth hormone gene family: Nucleotide sequences show recent divergence and predict a new polypeptide hormone. *DNA* **1:** 239.
SENEAR, A.W. and R.D. PALMITER. 1983. Expression of the mouse metallothionein-I gene alters the nuclear hypersensitivity of its 5' regulatory region. *Cold Spring Harbor Symp. Quant. Biol.* **47:** 539.
SHANI, M. 1985. Tissue-specific expression of rat myosin light-chain 2 gene in transgenic mice. *Nature* **314:** 283.
STAFFORD, J. and C. QUEEN. 1983. Cell-type specific expression of a transfected immunoglobulin gene. *Nature* **306:** 77.
STALDER, J., A. LARSEN, J.D. ENGEL, M. DOLAN, M. GROUDINE, and H. WEINTRAUB. 1980. Tissue-specific DNA cleavages in the globin chromatin domain introduced by DNase I. *Cell* **20:** 451.
STORB, U., R.L. O'BRIEN, M.D. MCMULLEN, K.A. GOLLAHON, and R.L. BRINSTER. 1984. High expression of cloned immunoglobulin K gene in transgenic mice is restricted to B lymphocytes. *Nature* **310:** 238.
SWEET, R.W., M.V. CHAO, and R. AXEL. 1982. The structure of the thymidine kinase gene promoter: Nuclease hypersensitivity correlates with expression. *Cell* **31:** 347.
SWIFT, G.H., R.E. HAMMER, R.J. MACDONALD, and R.L. BRINSTER. 1984a. Tissue-specific expression of the rat pancreatic elastase I gene in transgenic mice. *Cell* **38:** 639.
SWIFT, G.H., C.S. CRAIK, S.J. STARY, C. QUINTO, R.G. LAHAIE, W.J. RUTTER, and R.J. MACDONALD. 1984b. Structure of the two related elastase genes expressed in the rat pancreas. *J. Biol. Chem.* **250:** 14271.
TOOZE, J., ed. 1980. *The molecular biology of tumor viruses*, 2nd edition: *DNA tumor viruses*. Cold Spring Harbor Laboratory, Cold Spring Harbor, New York.
TOWNES, T.M., J.B. LINGREL, H.Y. CHEN, R.L. BRINSTER, and R.D. PALMITER. 1985. Erythroid specific expression of human β-globin genes in transgenic mice. *EMBO J.* **4:** 1715.
WALKER, M.D., T. EDLUND, A.M. BOULET, and W.J. RUTTER. 1983. Cell-specific expression controlled by the 5'-flanking region of insulin and chymotrypsin genes. *Nature* **306:** 557.
WU, C. and W. GILBERT. 1981. Tissue-specific exposure of chromatin structure at the 5' terminus of the rat preproinsulin II gene. *Proc. Natl. Acad. Sci.* **78:** 1577.
ZARET, K.S. and K.R. YAMAMOTO. 1984. Reversible and persistent changes in chromatin structure accompany activation of a glucocorticoid-dependent enhancer element. *Cell* **38:** 29.

Promoter Sequences of Murine αA Crystallin, Murine α2(I) Collagen or of Avian Sarcoma Virus Genes Linked to the Bacterial Chloramphenicol Acetyl Transferase Gene Direct Tissue-specific Patterns of Chloramphenicol Acetyl Transferase Expression in Transgenic Mice

H. Westphal,* P.A. Overbeek,* J.S. Khillan,* A.B. Chepelinsky,† A. Schmidt,‡
K.A. Mahon,* K.E. Bernstein,* J. Piatigorsky,† and B. de Crombrugghe‡

*Laboratory of Molecular Genetics, National Institute of Child Health and Human Development; †Laboratory of Molecular and Developmental Biology, National Eye Institute; ‡Laboratory of Molecular Biology, Division of Cancer Biology and Diagnosis, National Cancer Institute, National Institutes of Health, Bethesda, Maryland 20205

Our study deals with the spatial and temporal expression of chimeric genes in transgenic mice. In three separate sets of experiments, we have made insertions into mice gene constructs in which 5'-flanking control sequences of the mouse αA crystallin gene, mouse α2(I) collagen gene, or avian sarcoma virus (Rous sarcoma virus, RSV) were fused to the bacterial sequence encoding chloramphenicol acetyl transferase (CAT). The chimeric genes also contained SV40 splicing and polyadenylation signals at their 3' ends. The DNA constructs were introduced into the mouse germ line by zygote microinjection (Gordon and Ruddle 1981), and the transgenic strains were assayed for CAT activity in individual tissues at various stages of pre- and postnatal development.

Expression of genes inserted into the genomes of mice, often seemingly independent of their position within the genome, is amply documented in a recent review article by Gordon and Ruddle (1985) and in a number of articles in this volume. The rationale of our present experiments is based on the expectation that in many viral and cellular genes, sequences important for host cell-specific activation of expression are located upstream of the coding region and that individual control sequences are recognized by factors present in cells that are able to transcribe the corresponding gene (for review, see Emerson et al. 1985). With the help of suitable cell cultures transfected in vitro with cloned gene constructs, upstream sequences indicating cell-specific gene activation have been identified. In this type of experiment, the upstream sequence to be mapped is often placed 5' of a marker gene coding for an easily recognizable gene product. The bacterial CAT enzyme has been utilized by Gorman et al. (1982) as a useful indicator of chimeric gene activity in mammalian cells. CAT assays are simple and sensitive. We noted that enzyme activity can be measured in crude extracts by preparing low-speed supernatants of tissue homogenates. There is no comparable activity in any tissue we examined. Two further considerations prompted us to use CAT in all our gene constructs. First, we wanted to avoid autoregulation of the inserted gene by its own product. The bacterial CAT enzyme has no corollary in eukaryotes and can therefore be considered neutral in this regard. Second, chimeric genes can be constructed in such a way that the CAT enzyme is active in *Escherichia coli*. It can thus be utilized as a selection marker if one wishes to isolate the inserted chimeric gene with flanking mouse sequences from a genomic library of a transgenic animal. We will cite one experiment where this becomes an important consideration.

Each of our three chimeric gene constructs has its own intrinsic interest, which we will discuss under separate headings. As a whole, our work is a logical succession of experiments. Beginning with the αA crystallin-CAT construct, we demonstrate that a short, highly specialized control sequence directs CAT activity at the correct time of development to the correct cells in the ocular lens of the transgenic animal and its progeny. The next construct, α2(I) collagen-CAT, has a broader tissue distribution. Therefore, we decided to produce a large number of independent transgenic strains with this construct in the hope that matching patterns of CAT distribution in the various tissues would point out to us the types of cells that contain activation factors able to recognize the α2(I) collagen control region. Finally, with RSV-CAT we were not sure whether we could establish any patterns of CAT distribution at all because the control region of this construct is from an avian virus and foreign to the mouse. Again, we had to generate a larger number of independent strains. The distribution we found is revealing with respect to the disease specificity of RSV.

αA Crystallin-CAT

Crystallins are the major structural proteins of the ocular lens. Each of the three major classes of mouse crystallins, α, β, and γ, is induced at a specific time of lens development and has a characteristic distribution among the different lens tissues. One of our laboratories has isolated and characterized several members of the vertebrate crystallin gene family and has identified regulatory sequences in transfection assays (for review, see Piatigorsky 1984). In particular, a 409-bp region of the mouse αA crystallin gene, containing 364 bp of 5'-flanking region and 45 bp of 5' untranslated region of exon 1, was fused to a CAT coding sequence and SV40 splicing and polyadenylation signals and shown to express CAT activity in chicken lens explants (Chepelinsky et al. 1985). A 2102-bp linear DNA fragment containing this chimeric gene construct was injected into the male pronucleus of mouse FVB/N × FVB/N zygotes. Embryos were brought to term in foster mothers, and two strains of transgenic mice expressing CAT activity selectively in the eye were obtained. One of these strains was characterized in detail (Overbeek et al. 1985). The F_0 animal of this strain contained one copy of the transferred gene construct, which was integrated intact at a site different from the αA crystallin locus. The gene was transmitted stably and without apparent rearrangement to progeny, and a homozygous strain was obtained by inbreeding.

Of several organs tested, only the eye contained RNA hybridizing with a CAT-specific probe. A control hybridization of this RNA to a αA crystallin-specific cDNA probe allowed us to estimate that the lens of this transgenic strain has about 400 times more αA crystallin mRNA than CAT RNA. The low yield of CAT RNA and the limited availability of lens tissue has made it difficult for us to characterize the transcript more thoroughly. Thus, although we predict that the RNA is initiated within the 409-bp αA crystallin control region upstream of the CAT sequence, a firm statement in this regard will have to await further experimentation.

CAT enzyme assays were performed by the standard procedure of Gorman et al. (1982) with some useful changes. We learned (Mercola et al. 1985) that factors in crude tissue extracts that interfere with the assay can be inactivated by brief incubation at elevated temperature without affecting CAT activity. Also, if acetyl-coenzyme A (CoA) is resupplied frequently, the linear range of transacetylation can be increased to at least 8–10 hours at 37°C. This means that activities as low as 1 μU/mg of extracted protein can be determined with confidence. This is an important consideration because, in a given organ screen, CAT activities may vary more than 10^5-fold from one tissue to the next. We dissected an F_1 progeny animal of the αA crystallin-CAT transgenic strain and prepared crude extracts of 10 organs and tissues, including the eye, tongue, muscle, gonads, heart, kidney, spleen, liver, lung, and brain (Fig. 1). The only positive organ, the eye, contained about 1.1 mU of enzyme/mg of extracted protein (for definition of enzyme units, see Shaw 1975). All other organs were negative, that is, they contained either no CAT activity or else less than 0.001 times the amount found in the eye.

In accordance with Mendelian transmission of the trait, approximately one-half of the offspring of each of a number of F_0 × wild type or F_1 × wild type matings contained comparable levels of CAT activity in total eye extracts. Further breeding of F_1 animals resulted in a homozygous strain expressing doubly as much CAT activity. The correct spatial and temporal control of

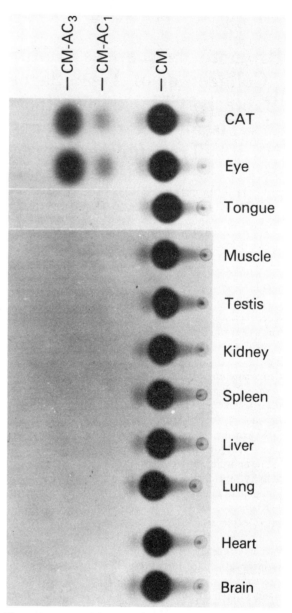

Figure 1. CAT activity in various organs of a 3.5-month-old αA crystallin-CAT transgenic mouse. Protein (10 μg) from the homogenates of the indicated organs was assayed. For details of the assay, see Overbeek et al. (1985). (CAT) *E. coli* CAT enzyme (purchased from P-L Biochemicals); (CM) unacetylated chloramphenicol; (CM-Ac$_1$ and CM-AC$_3$) chloramphenicol acetylated at the 1-hydroxyl and the 3-hydroxyl positions, respectively.

CAT expression from the chimeric gene at its unknown chromosomal site of integration was demonstrated in the following series of experiments (Overbeek et al. 1985). CAT activity in the eye was found only in tissues that are specialized in αA crystallin production, namely lens fiber and lens epithelia. No activity was detectable in retina or any other tissue of the eye. Immunoblots of proteins extracted from transgenic and wild-type lenses showed comparable distributions of the crystallins, indicating that CAT enzyme production does not interfere with α-crystallin synthesis.

With respect to temporal control, we examined embryo eye extracts for the presence of α-crystallin and of CAT activity. The α-crystallins appear as early as day 10.75 of gestation (Zwaan 1983). Accordingly, we prepared eye extracts from 10.5-, 11.5-, 12.5-, and 13.5-day embryos and analyzed them simultaneously for the presence of α-crystallins in immunoblots and for CAT activity in our enzyme assay. The 10.5- and 11.5-day extracts were negative for both signals, but a clear band of α-crystallins coincided with detectable levels of CAT activity in the day-12.5 extract. The specific activity of CAT increased strongly in the interval between days 12.5 and 13.5 of gestation and reached peak levels after birth.

Summarizing our results with αA crystallin-CAT, we can state that a 409-bp 5′ control sequence of a highly specialized mouse gene confers correct temporal and spatial control of expression to the chimeric gene construct. Since both CAT-specific RNA and CAT enzymatic activity were detected selectively in the lens of the transgenic mice, it is attractive to postulate that expression of the inserted chimeric gene construct is controlled at the level of transcriptional activation by factors present in the appropriate lens tissues. Further experimentation may reveal whether these putative lens factors recognize one or more distinct structural elements within the 409-bp control sequence contained in our gene construct and whether temporal and spatial control of expression is exerted by the same or by distinct factors.

Short sequences in the vicinity of the RNA start site have in other instances been shown to mediate tissue-specific regulation of gene expression. A good example is a 213-bp 5′-flanking region of the rat elastase gene fused to sequences encoding the human growth hormone. Expression of the human hormone in transgenic mice carrying this chimeric gene construction was selectively directed to acinar cells of the pancreas (Ornitz et al. 1985). Although this underscores the importance of promoter proximal sequences for tissue-specific control of gene expression in vivo, contributions from regulating sequences elsewhere in the gene are, of course, by no means excluded. In this context, we like to point out that, on a quantitative basis, expression of our αA crystallin-CAT chimeric gene is no match for that of the genomic αA crystallin gene in the mouse. One may evoke lack of appropriate gene control sequences, incorrect chromosomal location of the chimeric gene, differential RNA or protein stabilities, or any of a number of other possible reasons to explain this discrepancy. Let it suffice to conclude that chimeric gene constructs of the type presented here can be used to analyze the control of gene expression in the mammalian organism in considerable molecular detail.

α2(I) Collagen-CAT

The collagens constitute the most abundant proteins of the extracellular matrix. In higher vertebrates, there are at least 10 different types of collagen, each with its own characteristic tissue distribution (for review, see Prockop and Kivirikko 1984). We have focused our attention on the gene encoding the α2 chain of type-I collagen. The gene is large, with ~50 introns interrupting the coding sequence and at least two separate control sequences present in the region upstream of the RNA start site (A. Schmidt et al., in prep.). While it is known that, in vivo, type-I collagen may preferentially be found in tissues such as tendon, bone, skin, and smooth muscle (Focht and Adams 1984), the signals directing gene expression to these or other tissues have not been determined. Given the complex structure of the α2(I) collagen gene, it appeared appropriate to replace its coding region by the CAT sequence in an effort to study selectively the tissue specificity conferred by the 5′-flanking regions of the gene. In a transfection experiment, this chimeric gene construct was expressed at significantly higher levels in fibroblasts than in myeloma cells (A. Schmidt et al., in prep.). Encouraged by this indication of cell specificity, we generated a total of eight transgenic mouse strains carrying the α2(I) collagen-CAT construct pAZ1003, i.e., a 5′-flanking region, −2000 to +54, of the mouse collagen gene fused to the CAT coding sequence and to SV40 splicing and polyadenylation signals (J.S. Khillan et al., in prep.). Mating patterns indicated that the gene had integrated at a single locus in all eight strains. Strains containing more than one copy contained multimeric head-to-tail arrangements of the gene at the site of integration. In seven of the eight strains, Southern analysis indicated that the integrated chimeric gene unit was contiguous, whereas in strain 8, the gene was found rearranged. In all eight strains integration patterns of parents and offspring were indistinguishable, i.e., there was no evidence for rearrangement during transmission.

Were the α2(I) collagen 5′-flanking sequences sufficient to direct tissue-specific CAT expression in our transgenic mice? The answer came from a correlation of CAT enzyme activities, determined in an extensive tissue screen, with the levels of genuine α2(I) collagen mRNA found in several of these tissues. CAT activities are listed in Table 1. Enzyme activity was detected in strains 1–7, i.e., in all strains in which the chimeric α2(I) collagen-CAT gene had integrated intact. The highest levels of activity were found in extracts of the tail, but some of the strains showed appreciable levels of CAT activity in other tissues as well, for example, in skin, intestine, lung, and brain. Tail CAT activity varied from a low value of 2.6 μU/mg in strain 5 to a high value of 651 μU/mg in strain 7. There was no correlation between the specific activity of CAT and gene copy number in the various strains. In an effort to compare CAT expression with that of genuine α2(I) collagen, we determined α2(I) collagen mRNA levels in various organs. We found much more of this RNA in the tail, an organ rich in connective tissue, than in skin, brain, or liver. The correlation between levels of CAT activity and of endogenous α2(I) collagen mRNA suggests that at least in this organ, our chimeric gene responds to the same tissue-specific activation signals that mediate expression of the genuine collagen gene. The same

Table 1. CAT Activity (μU/mg Protein) in Various Tissues of Eight Mouse Strains Carrying the α2(I) Collagen-CAT Chimeric Gene

	Strain							
	1	2	3	4	5	6	7	8
Tail	12.5	15.4	6.4	17.3	2.6	462.0	651.3	<0.6
Skin	<0.6	<0.6	<0.6	<0.6	2.5	10.9	139.6	<0.6
Intestine	1.6	2.6	<0.6	<0.6	<0.6	48.5	<0.6	<0.6
Kidney	<0.6	<0.6	<0.6	<0.6	1.3	<0.6	<0.6	<0.6
Lung	<0.6	<0.6	<0.6	<0.6	<0.6	<0.6	29.8	<0.6
Brain	<0.6	<0.6	<0.6	<0.6	<0.6	33.7	20.1	<0.6
Testis/ovary	<0.6	<0.6	<0.6	<0.6	2.3	<0.6	<0.6	<0.6

No CAT activity was found in serum, thymus, eye, muscle, tongue, heart, spleen, or sternum.

holds true for the timing of expression during embryonic life. Adamson and Ayers (1979) and Leivo et al. (1980) have detected type-I collagen as early as day 8 of embryonic life. Correspondingly, working with our two high-expressor strains 6 and 7, we found appreciable levels of CAT activity in total embryo extracts at day 8.5 of gestation.

Organ screens are obviously a very crude way of determining tissue specificity. We are presently exploiting techniques of in situ hybridization for the purpose of detecting CAT transcripts in methacrylate-embedded tissue sections (Jamrich et al. 1984). Not only would this approach constitute a more faithful account of the cell specificity of a collagen 5′ control region in the living animal, but it may also help us to detect more refined cell specificities when we begin to dissect the large α2(I) collagen control region prior to integrating the chimeric gene in the mouse.

RSV-CAT

RSVs have been implicated in the etiology of a variety of connective tissue tumors such as fibromas, osteomas, myxomas, or chondromas, and their malignant equivalents, the sarcomas (for review, see Purchase and Burmester 1978). Svet-Moldavsky (1958) was first to show that RSV can cause sarcomas not only in birds but in rodents as well. Thus it appears that, in a broad range of vertebrate organisms, RSV selects cells or tissues of mesodermal origin, presumably because it finds there conditions favorable for viral gene expression. Control sequences that respond to cell or tissue-specific stimuli of expression are located in the U_3 region within the long terminal repeat (LTR) of RSV (Luciw et al. 1983; Laimins et al. 1984; Cullen et al. 1985). Corresponding LTR regions in other retroviruses have been shown to be responsible for distinct disease specificities (Chatis et al. 1984; DesGroseillers and Jolicoeur 1984; Lenz et al. 1984; Cullen et al. 1985; Davis et al. 1985). If it is indeed the LTR that determines tissue tropism, we felt that it would be to our advantage to dissociate this structure from the rest of the viral genome and analyze its expression in transgenic mice using the familiar pRSV-CAT construct of Gorman et al. (1982). This chimeric gene contains the LTR of the Schmidt-Ruppin strain of avian sarcoma virus fused to the CAT coding sequence and to SV40 splicing and polyadenylation signals. The transfection experiments of Gorman et al. (1982) had shown that the RSV sequence contained in pRSV-CAT is recognized as a powerful promoter by the mammalian cell.

We generated a total of nine transgenic mouse strains carrying the pRSV-CAT gene (P.A. Overbeek et al., in prep.) and we have begun to characterize these strains by measuring CAT enzyme activity in crude tissue extracts. Results obtained with adult mice of four strains are shown in Table 2. There is a pronounced specificity for muscle, tendon, and bone tissue, but low activities were detected in almost all tissues. These data allow us to conclude that the avian sarcoma virus LTR integrated in the mouse directs CAT expression preferentially to tissues that are also targets for attack by the intact sarcoma viruses. This stresses the role of the LTR in retroviral disease specificity. We would like to point out in this context recent papers by Brinster et al. (1984) and by Stewart et al. (1984) that also deal with tissue specificities exerted by viral control sequences in transgenic mice. The way is now open to identify the target tissue(s) of many other viral or cellular control sequences fused to a suitable marker gene and inserted in the mouse.

CAT activities in mice carrying the RSV-CAT construct are among the highest observed in any of our transgenic strains. In comparable target tissues, the activities are, on average, more than 100-fold higher than in strains containing the α2(I) collagen CAT construct. Therefore, the RSV LTR would appear to be a good choice of 5′-flanking sequence in constructs containing genes intended for expression in muscle or connective tissues.

One of the RSV-CAT strains (not listed in Table 2) was peculiar because small litter sizes resulted from $F_0 \times$ wild type or $F_1 \times$ wild type matings. We examined pregnant females surgically and noted that 50-60% of the implanted embryos had arrested development prior to day 8 of gestation. Surviving offspring again consistently produced small litters as a result of embryonic lethality. There is no sex bias in the heritability of this trait since both males and females are equally affected. The case is different from reported insertional inactivation of essential genes (Wagner et al. 1983; Löhler et al. 1984) in that the observed embryonic lethality is not

Table 2. Cat Activity (μU/mg protein) in Various Tissues of Four Mouse Strains Carrying the RSV-CAT Chimeric Gene

	Strain			
	1	2	3	4
Leg muscle	1,800	27,000	120,000	5,500
Abdominal muscle	740	13,000	62,000	1,500
Foot	160	3,000	20,000	12,000
Tail	32	625	160	22,000
Sternum	240	11,000	25,000	620
Heart	2	5,700	280	2,900
Ear	1	100	1,100	240
Serum	<0.6	184	7	<0.6
Spleen	<0.6	300	56	1
Lung	1	46	58	1
Brain	<0.6	150	270	3
Thymus	<0.6	12	50	14
Eye	<0.6	16	154	12
Kidney	2	50	110	<0.6
Tongue	2	155	190	35
Intestine	<0.6	170	13	46
Liver	<0.6	38	37	6
Testis	<0.6	32	206	10

recessive but dominant, and yet, some offspring survive to pass the trait on to subsequent generations. The unusual features of this strain could be explained by chromosomal rearrangements and these have indeed been detected. In addition, we have isolated integrated RSV-CAT sequences from a genomic cosmid library of an F_1 mouse, a task simplified by the fact that cosmids containing the RSV-CAT insert render their *E. coli* host cell chloramphenicol resistant. A molecular analysis of mouse sequences flanking the RSV-CAT insert in combination with detailed karyotyping and in situ chromosomal hybridization should allow us to shed more light on the genetics of this RSV-CAT strain.

In summary, the temporal and spatial control of CAT expression in mice carrying the αA crystallin-CAT or the α2(I) collagen CAT construct reflects that of the genuine mouse genes from which the 5'-flanking sequences of the chimeric genes were derived. In mice carrying the RSV-CAT construct, CAT expression is preferentially directed to muscle and connective tissue. This reflects the disease specificity of sarcoma viruses. Finally, we report on one RSV-CAT transgenic strain which is characterized by a dominant trait of embryonic lethality.

ACKNOWLEDGMENTS

We thank Sing-Ping Lai, Kurtis Van Quill, and Barbara Norman for dedicated technical assistance and Dawn Sickles for expert typing of the manuscript.

REFERENCES

ADAMSON, E.D. and S.E. AYERS. 1979. The localization and synthesis of some collagen types in developing mouse embryos. *Cell* **4**: 953.

BRINSTER, R.L., H.Y. CHEN, A. MESSING, T. VAN DYKE, A.J. LEVINE, and R.D. PALMITER. 1984. Transgenic mice harboring SV40 T-antigen genes develop characteristic brain tumors. *Cell* **37**: 367.

CHATIS, P.A., C.A. HOLLAND, J.E. SILVER, T.N. FREDERICKSON, N. HOPKINS, and J.W. HARTLEY. 1984. A 3' end fragment encompassing the transcriptional enhancers of nondefective Friend virus confers erythroleukemogenicity on Moloney leukemia virus. *J. Virol.* **52**: 248.

CHEPELINSKY, A.B., C.R. KING, P.S. ZELENKA, and J. PIATIGORSKY. 1985. Lens-specific expression of the chloramphenicol acetyltransferase gene promoted by 5' flanking sequences of the murine αA crystallin gene in explanted chicken lens epithelia. *Proc. Natl. Acad. Sci.* **82**: 2334.

CULLEN, B.R., K. RAYMOND, and G. JU. 1985. Transcriptional activity of avian retroviral long terminal repeats directly correlates with enhancer activity. *J. Virol.* **53**: 515.

DAVIS, B., E. LINNEY, and H. FAN. 1985. Suppression of leukaemia virus pathogenicity by polyoma virus enhancers. *Nature* **314**: 550.

DESGROSEILLERS, L. and P. JOLICOEUR. 1984. Mapping the viral sequences conferring leukemogenicity and disease specificity in Moloney and amphotropic murine leukemia viruses. *J. Virol.* **52**: 448.

EMERSON, B.M., C.D. LEWIS, and G. FELSENFELD. 1985. Interaction of specific nuclear factors with the nuclease-hypersensitive region of the chicken adult β-globin gene: The nature of the binding domain. *Cell* **41**: 21.

FOCHT, R.J. and S.L. ADAMS. 1984. Tissue specificity of type I collagen gene expression is determined at both transcriptional and post-transcriptional levels. *Mol. Cell. Biol.* **9**: 1843.

GORDON, J.W. and F.H. RUDDLE. 1981. Integration and stable germ line transmission of genes injected into mouse pronuclei. *Science* **214**: 1244.

———. 1985. DNA-mediated genetic transformation of mouse embryos and bone marrow—A review. *Gene* **33**: 121.

GORMAN, C.M., G.T. MERLINO, M.C. WILLINGHAM, I. PASTAN, and B.H. HOWARD. 1982. The Rous sarcoma virus long terminal repeat is a strong promoter when introduced into a variety of eukaryotic cells by DNA-mediated transfection. *Proc. Natl. Acad. Sci.* **79**: 6777.

JAMRICH, M., K.A. MAHON, E.R. GRAVIS, and J.G. GALL. 1984. Histone RNA in amphibian oocytes visualized by in situ hybridization to methacrylate-embedded tissue sections. *EMBO J.* **3**: 1939.

LAIMINS, L.A., P. GRUSS, R. POZZATTI, and G. KHOURY. 1984.

Characterization of enhancer elements in the long terminal repeat of Moloney murine sarcoma virus. *J. Virol.* **49:** 183.

LEIVO, I., A. VAHERI, and R. TIMPL. 1980. Appearance and distribution of collagens and laminin in the early mouse embryo. *Dev. Biol.* **76:** 100.

LENZ, J., D. CELANDER, R.L. CROWTHER, R. PATARCA, D.W. PERKINS, and W.A. HASELTINE. 1984. Determination of the leukemogenicity of a murine retrovirus by sequences within the long terminal repeat. *Nature* **308:** 467.

LÖHLER, J., R. TIMPL, and R. JAENISCH. 1984. Embryonic lethal mutation in mouse collagen I gene causes rupture of blood vessels and is associated with erythropoietic and mesenchymal cell death. *Cell* **38:** 597.

LUCIW, P.A., J.M. BISHOP, H.E. VARMUS, and M.R. CAPECCHI. 1983. Location and function of retroviral and SV40 sequences that enhance biochemical transformation after microinjection of DNA. *Cell* **33:** 705.

MERCOLA, M., J. GOVERMAN, C. MIRELL, and K. CALAME. 1985. Immunoglobulin heavy-chain enhancer requires one or more tissue-specific factors. *Science* **227:** 266.

ORNITZ, D.M., R.D. PALMITER, R.E. HAMMER, R.L. BRINSTER, G.H. SWIFT, and R.J. MACDONALD. 1985. Specific expression of an elastase-human growth hormone fusion gene in pancreatic acinar cells of transgenic mice. *Nature* **33:** 600.

OVERBEEK, P.A., A.B. CHEPELINSKY, J.S. KHILLAN, J. PIATIGORSKY, and H. WESTPHAL. 1985. Lens-specific expression and developmental regulation of the bacterial chloramphenicol acetyltransferase gene driven by the murine αA-crystallin promoter in transgenic mice. *Proc. Natl. Acad Sci.* **82:** (in press).

PIATIGORSKY, J. 1984. Lens crystallins and their gene families. *Cell* **38:** 620.

PROCKOP, D.J. and K.I. KIVIRIKKO. 1984. Heritable diseases of collagen. *N. Engl. J. Med.* **311:** 376.

PURCHASE, J.T. and B.R. BURMESTER. 1978. Neoplastic diseases: Leukosis/sarcoma group. In *Disease of poultry* (ed. M.S. Hofstad et al.), p. 418. Iowa State University Press, Ames.

SHAW, W.V. 1975. Chloramphenicol acetyltransferase from chloramphenicol-resistant bacteria. *Methods Enzymol.* **43:** 737.

STEWART, T.A., P.K. PATTENGALE, and P. LEDER. 1984. Spontaneous mammary adenocarcinomas in transgenic mice that carry and express MTV/myc fusion genes. *Cell* **38:** 627.

SVET-MOLDAVSKY, G.J. 1958. Sarcoma in albino rats treated during the embryonic stage with Rous virus. *Nature* **182:** 1452.

WAGNER, E.F., L. COVARRUBIAS, T.A. STEWART, and B. MINTZ. 1983. Prenatal lethalities in mice homozygous for human growth hormone gene sequences integrated in the germ line. *Cell* **35:** 647.

ZWAAN, J. 1983. The appearance of α-crystallin in relation to cell cycle phase in the embryonic mouse lens. *Dev. Biol.* **96:** 173.

Studies of Immunodifferentiation Using Transgenic Mice

D. Baltimore,* R. Grosschedl,* D. Weaver,* F. Costantini,† and T. Imanishi-Kari‡

*Whitehead Institute for Biomedical Research, Cambridge, Massachusetts 02142 and Department of Biology, Massachusetts Institute of Technology, Cambridge, Massachusetts 02139; †Department of Human Genetics and Development, College of Physicians and Surgeons, Columbia University, New York, New York 10032; ‡Center for Cancer Research, Massachusetts Institute of Technology, Cambridge, Massachusetts 02139

The development of an organism from a zygote involves an enormous number of decisions, each of which differentiates one body cell from another. In the terminal stages of this differentiation tree, cells develop their definitive characteristics. This process of terminal differentiation turns a relatively nondescript but multipotential cell into one with very particular characteristics but with only a limited range of further differentiation. Understanding the processes involved in terminal differentiation represents a major challenge in developmental biology.

A very informative system for the study of terminal differentiation is differentiation along the B lymphoid lineage. B lymphoid cells ultimately become cells that secrete immunoglobulins, but they go through many intermediate stages of differentiation before becoming the plasma cell that is the ultimate immunoglobulin factory. These intermediate stages involve some very well-defined changes, particularly changes in DNA structure. The B lymphoid lineage is one of the few that incorporates rearrangement of DNA. Rearrangement is particularly useful to this lineage because DNA

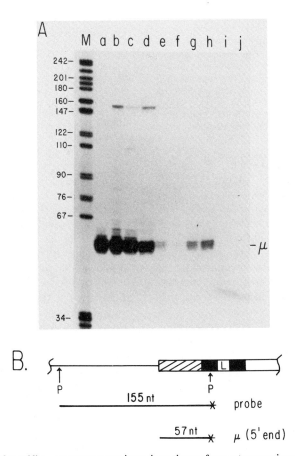

Figure 1. S1 nuclease analysis of specific μ gene sequences in various tissues from a transgenic mouse. (*A*) Analysis of the 5' end of specific μ transcripts. The probe used is described below. The fragment marked μ represents the 57-nucleotide protected fragment indicative of appropriately initiated transcripts. Lanes represent: (*M*) size marker (pBR322 DNA cleaved with *Hpa*II); (*a*) specific hybridoma RNA; (*b–j*) RNA from lymph nodes, spleen, thymus, kidney, brain, heart, lung, liver, and cultured fibroblasts. (*B*) Structure of the probe and protected fragments. The 5' end of the gene is represented above with leader (L) sequences in black, intron sequences as an open box, and 5' noncoding sequences hatched. The probe goes between *Pvu*II (P) sites and is labeled at its 5' end (indicated by *). (Reprinted, with permission, from Grosschedl et al. 1984.)

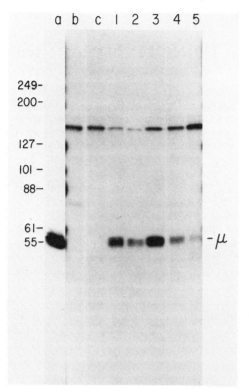

Figure 2. S1 nuclease analysis of specific RNA from Ab-MLV transformants of bone marrow cells from transgenic mice. Cytoplasmic RNA was hybridized to the probe described in Fig. 1B and the 57-nucleotide characteristic fragment was displayed (marked μ). (Lanes a–c) Specific hybridoma RNA as marker and MPC11 myeloma and 70Z RNAs as negative controls, respectively; (lanes 1–5) individual Ab-MLV transformants. (Reprinted, with permission, from Weaver et al. 1985.)

processing is the key to the development of immunoglobulin diversity which gives the immune system its tremendous range of potential reactivity.

B-cell development has been studied for many years, mainly in cell culture. An important adjunct has been the study of B lymphoid tumor cells, cells that are apparently frozen in their differentiation at particular stages along the B lymphoid lineage. To understand the full range of this terminal differentiation pathway, however, it is necessary to follow the myriad events that occur in an animal. One way to transfer studies from cell culture into the animal would be to make transgenic mice (Gordon et al. 1980; Brinster et al. 1985) that incorporate specific immunoglobulin genes. Such transgenic mice would only be useful if they expressed the genes at the right time in the right tissues and in the right amounts.

To this end, we began a number of years ago to produce mice transgenic for a rearranged immunoglobulin heavy-chain gene (Grosschedl et al. 1984). Four lineages of such mice have now been established, each of which transmits multiple copies of a rearranged heavy-chain gene in Mendelian fashion to its offspring.

Specific Synthesis of the Heavy Chain

To examine whether the transgenic heavy chain was being expressed specifically in lymphoid tissues, we prepared RNA from numerous tissues of a transgenic mouse and examined whether the RNA contained transcripts specific to the transgene (Grosschedl et al. 1984). Such transcripts could be assayed using an S1 nuclease procedure with a single-stranded DNA probe that spans the initiation site for transcription. Protection of this probe by authentic mRNA gave rise to a 57-nucleotide signature fragment (Fig. 1B). This fragment was obvious when RNA from spleen, lymph nodes, and thymus was used but was either not evident or faint with RNA from any other tissues (Fig. 1A). The faint signal is evident in a number of tissues, but in all cases except the heart, that faint signal is clearly due to contamination with blood cells (data not shown). The amount of specific RNA in these tissues is very high, being comparable to the amount found in a hybridoma that expresses the specific variable region incorporated into the transgene.

The ability of the transgene to be expressed in a tissue-specific manner was demonstrated in a number of different mice from a number of different lineages. It would therefore appear that a rearranged immunoglobulin heavy-chain gene is expressed in an autonomous fashion, independently of the surrounding DNA. Although we do not know the sites of integration of the transgenes, they each behave as a Mendelian unit and it is known from numerous investigations that integration of transgenes is to a first approximation a random process (Palmiter and Brinster 1985).

Knowing that the transgene is expressed in the appropriate tissues of transgenic mice, we examined whether each B lymphoid cell expresses the transgene. To carry out this analysis, rather than investigating each individual cell, we chose to immortalize a large number of clones of B lymphoid cells using Abelson murine leukemia virus (Ab-MLV). This virus transforms early cells of the B lymphoid lineage into immortal clones (Baltimore et al. 1979).

When a number of individual Ab-MLV bone marrow cell transformants were investigated, all were found to express the transgene. This is evident in Figure 2 where the S1 nuclease procedure was used to investigate the expression of the transgene in different Ab-MLV transformants. In all cases, the 57-nucleotide signature frag-

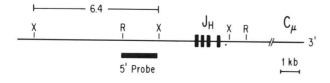

Figure 3. Probe used for analyzing rearrangements. The structure of J_H-associated DNA is shown with the four J_H regions marked by solid vertical bars. The EcoRI (R)–XbaI (X) ^{32}P-labeled probe is indicated. It hybridizes to a 6.4-kb XbaI fragment. (Reprinted, with permission, from Weaver et al. 1985.)

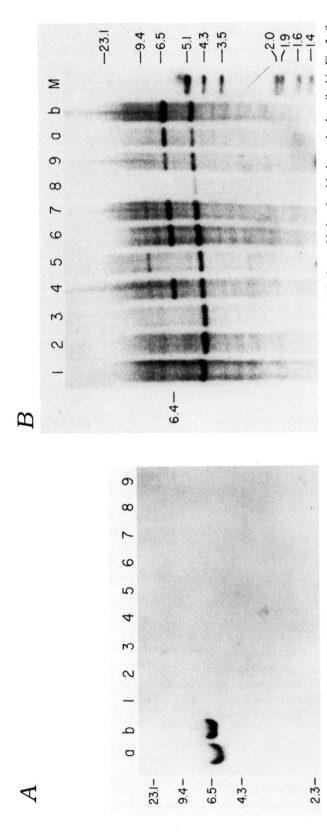

Figure 4. Germ line J_H alleles in Ab-MLV transformants. (*A*) Transformants from normal C57BL/6 mice. (Lanes *1–9*) No detectable bands with the probe described in Fig. 3; (lanes *a* and *b*) positive controls. (*B*) Transformants from transgenic mice. (Lanes *1–9*) Different transformants; (lanes *a* and *b*) positive controls; (lane *M*) markers. The 5.1-kb band is an artifact present in all of the DNA. (Reprinted, with permission, from Weaver et al. 1985.)

ment was protected by RNA from the transformants. It was not, however, protected by RNA from transformants of normal, nontransgenic C57BL/6 bone marrow.

Does the Transgene Prevent Endogenous Rearrangement?

The lymphoid lineage is characterized by a series of DNA rearrangements. Having incorporated an expressed, rearranged heavy-chain gene into an animal, we could examine whether there is regulation of the rearrangement process. Our specific question was: Does the presence of the rearranged transgene inhibit rearrangement of endogenous heavy-chain genes? Earlier work from this laboratory had predicted that there is regulation of heavy-chain gene rearrangement and that the expression of a rearranged heavy-chain gene should prevent rearrangement of endogenous heavy-chain genes (Alt et al. 1984). We investigated this question by examining the gene structure in the various Ab-MLV transformants that are known to express the transgene (Weaver et al. 1985). The most informative experiment was to probe the DNA with a probe that hybridizes to a fragment found upstream of the J_H regions. Because immunoglobulin gene rearrangement involves deletion of DNA upstream of the J_H regions (Alt et al. 1984), a germ line chromosome should be evident by the presence of the upstream DNA whereas rearranged chromosomes would give no signal (Fig. 3). When Ab-MLV transformants from nontransgenic mice were examined, no hybridization of this probe to the transformant DNA was evident (Fig. 4A). The 6.4-kb germ line fragment was evident, however, in the DNA from a number of Ab-MLV transformants from the transgenic mice (Fig. 4B; the 5.1-kb band evident in this figure was not related to immunoglobulin gene rearrangement). From these as well as other data, we can conclude that 26% of the chromosomes in Ab-MLV transformants from transgenic animals are in the germ line configuration while 0% of germ line chromosomes occur in transformants from normal animals. Thus, the predicted inhibition of rearrangement is evident although it is incomplete. We are presently continuing these studies to examine whether it is the transgene itself or its product that is responsible for the inhibition of rearrangement. Ritchie et al. (1984) found that transgenic, rearranged $\varkappa$-chain genes can also suppress rearrangement of endogenous genes and Rusconi and Kohler (1985), who introduced both heavy- and light-chain rearranged genes, found a partial suppression.

CONCLUSIONS

This study shows that mice transgenic for rearranged immunoglobulin heavy-chain genes can help us to understand the process of immunodifferentiation. We already have learned that immunoglobulin genes are expressed in a position-independent manner and that all of the controlling elements for the activation and expression of the gene in the B lymphoid lineage are present within a small distance either side of the heavy-chain gene. Further studies using deleted heavy-chain genes are planned to examine the role of individual sequence elements. The ability of the rearranged gene to prevent the rearrangements of the endogenous heavy-chain genes shows that there is feedback regulation in the B lymphoid pathway and suggests that the transgenic animal represents a very powerful way to examine feedback pathways during terminal differentiation.

ACKNOWLEDGMENTS

D.W. was supported by a Helen Hay Whitney Foundation postdoctoral fellowship. F.C. is a recipient of the Irma T. Hirschl Career Scientist Award. This work was supported by grants from the National Institutes of Health (to D.B., F.C., and T.I.-K.) and a grant from the March of Dimes (to F.C.).

REFERENCES

ALT, F.W., G.D. YANCOPOULOS, K. BLACKWELL, C. WOOD, E. THOMAS, M. BOSS, R. COFFMAN, N. ROSENBERG, S. TONEGAWA, and D. BALTIMORE. 1984. Ordered rearrangement of immunoglobulin heavy chain variable region segments. *EMBO J.* **3:** 1209.

BALTIMORE, D., N. ROSENBERG, and O.N. WITTE. 1979. Transformation of immature lymphoid cells by Abelson murine leukemia virus. *Immunol. Rev.* **48:** 3.

BRINSTER, R.L., H.Y. CHEN, M.E. TRUMBAUER, M.K. YAGLE, and R.D. PALMITER. 1985. Factors affecting the efficiency of introducing foreign DNA into mice by microinjecting eggs. *Proc. Natl. Acad. Sci.* **82:** 4438.

GORDON, J., G. SCANGOS, D. PLOTKIN, J. BARBOSA, and F. RUDDLE. 1980. Genetic transformation of mouse embryos by microinjection of purified DNA. *Proc. Natl. Acad. Sci.* **77:** 7380.

GROSSCHEDL, R., D. WEAVER, D. BALTIMORE, and F. COSTANTINI. 1984. Introduction of a μ immunoglobulin gene into the mouse germ line: Specific expression in lymphoid cells and synthesis of functional antibody. *Cell* **38:** 647.

PALMITER, R.D. and R.L. BRINSTER. 1985. Transgenic mice. *Cell* **41:** 343.

RITCHIE, K., R. BRINSTER, and U. STORB. 1984. Allelic exclusion and control of endogenous immunoglobulin gene rearrangement in ϰ transgenic mice. *Nature* **312:** 517.

RUSCONI, S. and G. KOHLER. 1985. Transmission and expression of a specific pair of rearranged immunoglobulin μ and ϰ genes in a transgenic mouse line. *Nature* **314:** 330.

WEAVER, D., F. COSTANTINI, T. IMANISHI-KARI, and D. BALTIMORE. 1985. A transgenic immunoglobulin mu gene prevents rearrangement of endogenous genes. *Cell* **42:** 117.

Transfer and Regulation of Expression of Chimeric Genes in Plants

J. SCHELL,*† H. KAULEN,* F. KREUZALER,* P. ECKES,* S. ROSAHL,* L. WILLMITZER,*
A. SPENA,* B. BAKER,* L. HERRERA-ESTRELLA,† AND N. FEDOROFF‡

*Max-Planck-Institut für Züchtungsforschung, 5000 Köln 30, Federal Republic of Germany; †Laboratorium voor Genetika, Rijksuniversiteit Gent, 9000 Gent, Belgium; ‡Carnegie Institution of Washington, Baltimore, Maryland

Plant gene vectors derived from the Ti plasmid of *Agrobacterium tumefaciens* (Zambryski et al. 1983) were used to introduce a number of chimeric genes into tobacco plants. The actual genes involved in these studies were chosen to serve as models for investigating the involvement of 5' upstream sequences and 3' downstream sequences in the regulation of gene expression in plants.

In particular we wanted to study induction by light, tissue-specific expression of genes, and induction by elevated temperatures. Finally we wanted to test whether controlling elements, such as the *Activator* from maize, would be active after transfer to heterologous plants.

THE LIGHT-INDUCIBLE CHALCONE SYNTHASE

Light is one of the most important effectors of differentiation and development in higher plants (Mohr 1972; Mohr and Schäfer 1983). For our investigations we have selected the light-inducible enzyme chalcone synthase (CS), one of the key enzymes involved in flavonoid biosynthesis. Flavonoids constitute one of the most abundant classes of phenolic compounds in higher plants and serve important functions as flower pigments, antimicrobial agents (phytoalexins), and complete UV-protective compounds (Hahlbrock and Grisebach 1979). CS catalyzes the stepwise condensation of three acetate units from malonyl CoA with 4-coumaryl-CoA to give naringenin chalcone (Heller and Hahlbrock 1980). This chalcone is the central intermediate in the biosynthesis of flavones, flavonols, and various other flavonoids. CS from parsley consists of two M_r 42,000 subunits that are coded for by a mRNA of approximately 1700 nucleotides (Kreuzaler et al. 1979). In higher plants three photoreceptors are known to be involved in the absorption and transduction of light. The most important and probably the best analyzed system is phytochrome. Phytochrome is a water-soluble chromo-protein which at physiological temperatures can exist in two defined photointerconvertible forms. One form (Pr) absorbs maximally near 665 nm and is considered to be inactive, whereas the other form (Pfr) absorbs maximally near 730 nm and is morphogenically active (Butler et al. 1959; Shropshire and Mohr 1983).

The second system involved in photomorphogenesis is the blue light-absorbing photoreceptor. Although its function is well documented, e.g., in the development of chloroplasts, its chemical structure is still a matter of discussion. It seems to contain flavin derivatives as prosthetic groups (Senger 1980, 1982).

The third receptor involved in photomorphogenesis is activated by UV light. The chemical nature of the UV-absorbing-photoreceptor is not yet understood, but the inducing effect of UV light on the transcription of specific genes is well documented (Schröder et al. 1979; Kreuzaler et al. 1983).

Induced Expression of the CS Gene in *Antirrhinum majus* and in Tobacco Induced by Light of Different Wavelengths

The induction pattern of the endogenous CS in tobacco and *Antirrhinum majus* was determined using monospecific affinity-purified antibodies against the CS from parsley (provided by Dr. J. Schröder, Freiburg). In tobacco, expression of the CS is induced by UV light at 305–310 nm. The most effective way to induce expression of the CS is to take young plants grown without UV light and illuminate them with normal sunlight for about 2 days. Induction can also be obtained by illuminating young plants with a UV lamp at 310 nm. In young seedlings of *A. majus*, expression of CS can be induced by normal light activation of phytochrome alone. In older green plants, expression can be enhanced by UV light of about 320–350 nm (unpublished data, obtained in collaboration with Dr. E. Wellmann, University of Freiburg).

To identify the DNA sequences involved in the regulation of the CS genes by different wavelengths of light, chimeric genes were constructed consisting of various lengths of the 5' upstream sequences of the CS gene isolated from *A. majus*. A promoter fragment of 3.9 kb of the CS gene was cloned from a genomic library of *A. majus* by Dr. H. Sommer using the cDNA-CS from *Petroselinum hortense* (Kreuzaler et al. 1983) as a probe.

A chimeric gene was constructed, containing the CS 5' upstream promoter region and the neomycin phosphotransferase II (NPTII) from Tn5 as a reporter protein. A 600-bp-long DNA fragment from CS gene No. 1 from *P. hortense* containing the termination region was subsequently ligated in the correct orientation to the 3' end of the chimeric NPT. The translation product of the chimeric genes should yield a fusion protein with 18 amino acid residues added to the aminoterminal end of the NPT protein. When analyzed by PAGE (Reiss et al. 1984), the fusion protein was expected to show a shift of the active enzyme band to a position of reduced electrophoretic mobility.

To analyze the location of potential sites required for regulation of transcription, the 3.9-kb 5' upstream promoter was shortened, in various constructs, to the sizes 1.2 kb, 0.47 kb, and 0.15 kb. Internal deletions upstream from the TATA box have also been introduced into this promoter region (see Fig. 1).

In the deletion KpnI–KpnI (FHEI74 ICR), a fragment of 158 bp was removed. This deletion starts 5 bp upstream from the TATA box. In the deletion KpnI–HincII (FHEI 714 3CR), 314 bp were removed, again starting from the 5 bp upstream from the TATA box. The TATA box was kept intact in all these chimeric gene constructions.

Transforming Plants with Chimeric Genes

The different constructs were inserted into the plasmid pGV710 (R. Deblaere, pers. comm.), mobilized to *A. tumefaciens*, and integrated into the Ti plasmid pGV3850 or pGV3851 (Zambryski et al. 1983, 1984) by homologous recombination. Tobacco plants were transformed with these *Agrobacterium* strains and either intact plants or teratoma tissues were obtained from the transformations (Zambryski et al. 1983; De Block et al. 1984). The transformed state of a tissue or a plant was revealed by the presence of nopaline synthase activity as a cotransferred marker and confirmed by DNA/DNA hybridizations.

Regulation of Expression

Expression of the various CS-NPTII constructs in teratoma tissue was analyzed using the NPT assay of Reiss et al. (1984, modified by Schreier et al. 1985). For each test about 10 calli were pooled to minimize any possible insert position effects on expression. The results show that the genes with either the 3.9-kb or the 1.2-kb promoter fragment are expressed well in teratoma tissue, whereas the expression of the gene containing the 0.47-kb promoter and the internal deletions is markedly reduced. The promoter regions from -39 to -353 and from -594 to -661 must therefore be important for the expression of these chimeric genes. It is, however, important to note that the construction devoid of all sequences upstream from the TATA box shows no NPTII activity.

When the CS promoter from *A. majus* is analyzed carefully for possible sequences that might play a role in regulating of expression (sequence data kindly provided by H. Sommer, Max-Planck-Institute, Köln) the following features can be noted:

1. Close to the TATA box, the sequence TACCAT is present twice and separated by only 6 bp.
2. Between positions -272 and -304, a stretch of 12 As and 20 Ts is followed by a stretch of 14 Gs and 1 C (positions -320 to -336).
3. Upstream from the HincII site at positions -564 to -661, two direct repeat sequences are observed each with a length of 47 bp. The homology between the repeats is 85%.
4. At positions -601 to -608, a sequence GTGGTAG is observed that is identical to the consensus core sequence for enhancers in animal systems:

$$\text{GTGG} \begin{matrix} \text{AAA} \\ \text{G} \\ \text{TTT} \end{matrix}$$

5. A 22-bp sequence at positions -591 through -613 in the CS 5' upstream region has homology to a similar sequence in the 5' upstream region of the light-inducible SS Rubisco gene from pea (at positions -156 through -177) (Herrera-Estrella et al. 1984). The homology is 82%.

Light Inducibility of the CS-NPTII Gene (pFHEI714CR) in Teratoma Tissue

A teratoma callus was grown in sterile culture under light conditions (16 hr white light, 8 hr darkness). After a period of growth time, the callus was divided into two parts, one was kept in the light, the other was put in the dark. These experiments demonstrated that the CS-NPTII gene with the 1.2-kb promoter fragment was expressed under light conditions but not in the dark. The activity was restored when the material kept in the dark is subsequently illuminated. We would like to mention, however, that in some transformants we were not able to completely switch off the expression in the dark or in some other transformants to switch it on again after a dark period. It could be that this was due to an unfavorable developmental state of the callus (it is known that cells have to be in a phase of competence to react to the regulatory effect of photoreceptors) or to an interfering effect of cytokinins on the regulation of expression because transcript 4 from the T-DNA (Joos et al. 1983) is present in the vector pGV3851. We are presently analyzing the regulation of transcription of the CS-NPTII chimeric genes in young intact plants grown from seeds (F_1 generation). The seeds were obtained by selfing transformed tobacco plants. The first results show that UV light can induce the transcription of the CS-NPTII gene. A detailed analysis is in progress.

ISOLATION AND CHARACTERIZATION OF DEVELOPMENTALLY CONTROLLED GENES FROM *SOLANUM TUBEROSUM*

The differentiated state of cells in a higher plant is characterized by the developmentally controlled expression of specific gene pools. Polysomal RNA se-

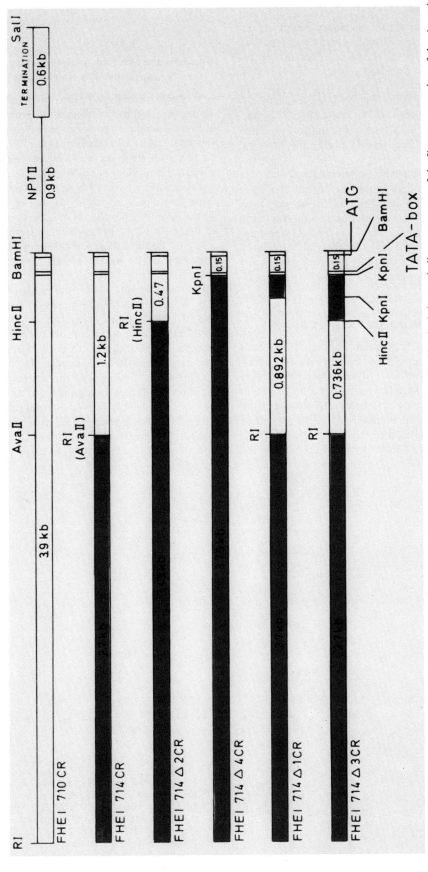

Figure 1. Schematic drawing illustrating the structure of the different CS chimeric genes used in this study. The shaded areas indicate segments of the 5' upstream region of the *A. majus* CS that were detected in the respective constructs.

quences present in the different vegetative (i.e., leaf, root, and stem) and floral (i.e., petal, anther, and ovary) organ systems of tobacco plants contain at least 6000–11,000 different mRNAs (representing 25–40% of the total sequence complexity) detected also in the polysomes of other organs. However, 10–40% of the nuclear RNA complexity is organ-specific. Comparison of polysomal and nuclear RNA sequences of different organs indicates that the organ-specific expression must be controlled at the transcriptional as well as at the posttranscriptional level (Kamalay and Goldberg 1984).

Here we describe experiments aiming to elucidate the mechanism underlying developmentally controlled gene expression in higher plants. Potato (*S. tuberosum*) was chosen for this project because it is an important crop plant and because it is possible to regenerate whole potato plants from isolated single cells and Ti-plasmid-mediated gene transfer is readily achieved. Leaves and tubers were used as a source for organ-specific genes.

Isolation of Organ-specific cDNA Clones

cDNA libraries were prepared by standard methods using either poly(A)$^+$ RNA from leaves or tubers. About 5000 clones of each library subsequently were screened for organ-specific clones in a colony hybridization experiment using ^{32}P-labeled cDNA probes from both corresponding homologous (i.e., leaf as well as tuber) and heterologous (i.e., root, tuber as well as leaf, stem, root) organs. Clones hybridizing only to the cDNAs from the homologous organ were characterized further.

Northern and Dot Blot Analysis Demonstrates Organ-specific Expression of a Number of cDNA Clones

To verify the organ-specific expression of the cDNA clones obtained by the differential screening described above, the inserts were isolated from a number of leaf-specific cDNA clones (pcL 600, pcL 700, and pcL 900; cL is cDNA from leaf) as well as from some tuber-specific clones (i.e., pcT 700, pcT 800, and pcT 1500) and hybridized against RNAs of different organs isolated from field- or greenhouse-grown potato plants. The results of a Northern-type experiment are shown in Figure 2. Clone pcL 900 hybridizes strongly to an RNA of about 900 nucleotides, which is predominantly present in leaves. A weak hybridization is also seen with stem poly(A)$^+$ RNA whereas no hybridization is visible with RNA from either roots or tubers (Fig. 2, lane b). A similar picture is seen for pcL 600 and pcL 700: a strong hybridization to RNA of about 600–700 nucleotides, no hybridization with RNA from either root or tuber, and a weaker though evident hybridization with RNA from stem (Fig. 2, lanes a and c). The results obtained with the different tuber cDNAs are shown in Figure 2 (lanes d–f). In all cases hybridization is visible only with RNA from tubers; no signal is seen with RNA from leaves, roots, or stems. Semiquantitative experiments performed using dot blot assays show that the level of RNAs present in the homol-

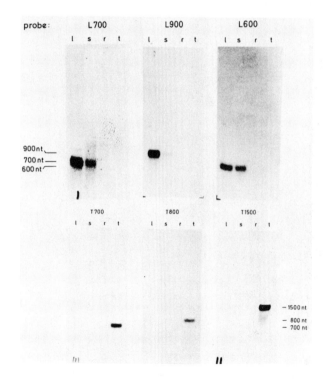

Figure 2. Northern gel analysis of leaf-specific clones (pcL 700, 600, 900) and tuber-specific clones (pcT 700, 800, 1500). Poly(A)$^+$ RNA (10 μg) of leaf (l), stem (s), root (r), and tuber (t) tissue were separated on 1.5% denaturing formaldehyde gel, blotted on DBM filters, and hybridized to ^{32}P-nick-translated cDNA inserts of pcL 700, 600, 900 and pcT 700, 800, 1500. Sizes of the specific RNAs are indicated in nucleotides (nt).

ogous organs (i.e., leaf as well as tuber) is at least several hundredfold higher than the levels detected in heterologous organs (data not shown).

Organ-specific Expression Is Controlled at Both the Transcriptional as Well as the Posttranscriptional Level

Organ-specific gene expression can result from both transcriptional as well as posttranscriptional control. To distinguish between these possibilities nuclei were isolated from different organs of the potato plant under conditions retaining the transcriptional activity. After isolation and purification the nuclei were incubated for transcription to proceed in the presence of ^{32}P-labeled RNA precursors. After a pulse of 30 minutes, ^{32}P-labeled newly synthesized RNA was isolated and probed against single-stranded M13 clones of the different cDNAs immobilized on nitrocellulose filters. As is evident in Figure 3, the RNA synthesized in the nuclei hybridizes only to the coding strand, indicating that the run-off transcription experiments reflect the normal transcription process. Concerning the transcriptional activity, it is evident from Figure 3 that in the case of the leaf-specific cDNA clones pcL 600, pcL 700, and pcL 900 hybridization is seen with RNA pulse-labeled in nuclei isolated from leaves, whereas no signal is seen with RNA pulse-labeled in nuclei from root or tuber. A reciprocal picture is obtained with the tuber-specific clones pcT 700 and pcT 1500 (i.e., transcriptional activity can be detected only in nuclei from tubers). These results therefore indicate that the developmental (organ)-specific expression of these five clones is mainly (if not exclusively) controlled at the level of transcription.

A different picture emerges for the tuber-specific clone pcT 800. Hybridization toward this cDNA is seen not only with RNA pulse-labeled in nuclei from tuber (the homologous organ) but also with RNA pulse-labeled in leaf nuclei. This therefore raises the possibility that the organ-specificity of the steady-state RNA homologous to pcT 800 is due mainly to posttranscriptional processes involving either differential processing and/or stability.

Run-off experiments performed in the presence of different concentrations of α-amanitin support the assumption that transcription of all six cDNA clones is performed by RNA polymerase II (data not shown).

Proteins Encoded by the Different cDNA Clones

Hybrid selection and in vitro translation experiments as well as nucleotide sequence data (data not shown) have identified the product of pcL 900 to be the small subunit of the ribulose-1,5-bisphosphate carboxylase, that of pcT 800 to be the major 20-kD protein of potato tubers, and that of pT 1500 to be the major 40-kD protein of potato tubers, called "patatin." The products of pcL 600, pcL 700, and pcT 700 have not yet been identified.

All cDNA Clones Except pcL 700 Are Encoded by Small Multigene Families

Southern blot analysis of a haploid potato line (HH5793) using the cDNA inserts of the different

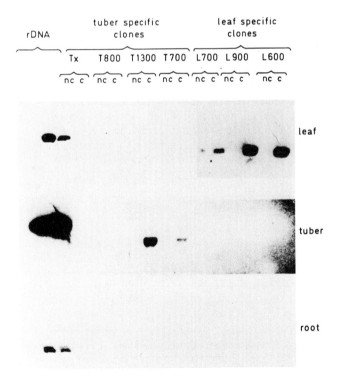

Figure 3. Comparison of runoff transcripts of leaf- and tuber-specific clones in nuclei from leaves, tubers, and roots. A total of 500 ng of the coding strand (c) and the noncoding strand (n) of pcL 700, 900, 600 (all leaf specific) and of pcT 700, 800, 1500 (all tuber specific) were run on 0.8% agarose gels and subsequently blotted on nitrocellulose filters. Filters were hybridized against RNA, pulse-labeled in isolated nuclei from leaves (top), tubers (middle), and roots (bottom). Two clones containing rDNA sequences were always included as an internal control.

clones as probes showed that all clones, except pcL 700, hybridized to a number of bands (ranging from 7 to 12 using different restriction enzymes), indicating that they are encoded by small multigene families. The only exception is pcL 700, which hybridized predominantly to a single band (data not shown).

Analysis of Genomic Clones Homologous to a Leaf-specific (pcL 700) as Well as a Tuber-specific (pcT 1500) Clone

A genomic library of a haploid potato line (HH 5793) was established using the λ vector EMBL 4 and screened for genomic clones hybridizing to the different cDNA clones by plaque hybridization. Genomic clones corresponding to pcL 700 and pcT 1500 were used for further characterization.

Characterization of a Genomic Clone Homologous to pcL 700

The structure of the genomic clone (Fig. 4) pgL 700 was deduced from a comparison of the nucleotide sequences of cDNA and genomic DNA as well as by S1-type mapping experiments. pgL 700 contains four introns with lengths varying from 96 to 838 nucleotides. The largest exons with 248 and 186 nucleotides are found at the 3' and 5' termini; the three internal exons are all rather small since they consist of less than 60 nucleotides each. The sequence GATAAA is found about 20 nucleotides in front of the polyadenylation signal and probably functions as the polyadenylation signal. All intron–exon boundaries obey the GT-AG rule. pgL 700 contains an open reading frame of 414 nucleotides bordered by a 32-bp-long untranslated 5' leader sequence and a 167-bp-long 3' untranslated region. The amino acid sequence as deduced from nucleotide sequence data indicates that the protein contains a fairly high amount of hydrophobic amino acids (70%). The cDNA cloning has indicated that the mRNA encoded by pgL 700 is abundant (14 out of 5000 cDNA clones screened were homologous to pgL 700). To see whether the cloned 4.5-kb genomic fragment contains all the necessary information for a high level of expression and before engaging in extensive sequencing, pgL 700 was earmarked by inserting a 470-bp DNA fragment derived from the coding region of the T-DNA gene 2 of the octopine-type plasmid pTi ACH 5 (Gielen et al. 1984). This DNA fragment was inserted into the last exon, as indicated in Figure 4, and transferred back to potato using *Agrobacterium*-derived vectors. If this gene were actively expressed in potato, it should give rise to a chimeric RNA, 470 nucleotides longer than its normal counterpart. In addition, this RNA should hybridize specifically to the 470-bp fragment. The analysis of transformed potato shoots is shown in Figure 5. The reintroduced gene is definitely transcribed, giving rise to an RNA species of about 1200 nucleotides. Furthermore, the amount of RNA made is about the same as the amount made from the endogenous gene. We take this as evidence that the 4.5-kb genomic fragment pgL 700 contains all the signals necessary for a high level of expression.

Characterization of a Genomic Clone (pgpat 1) Homologous to pcT 1500 (Encoding the Patatin)

Figure 6 shows the structure of a genomic clone homologous to the patatin-encoding cDNA clone pcT 1500. This structure was derived from a comparison of nucleotide sequences of different cDNA clones and pgpat 1. The gene is composed of six introns and seven exons of varying length, and contains an open reading frame coding for a protein of 386 amino acids. The exon–intron boundaries again obey the classical GT-AG rule. An AATAA sequence is located 20 nucleotides upstream of the polyadenylation site.

pgpat1 was earmarked by exon modification (cf. Fig. 6) and was introduced in potato using *Agrobacterium*-derived vectors. The analysis of the activity of this modified gene, however, has to await the formation of tubers.

Conclusion

cDNA and genomic clones have been isolated and characterized from two genes of potato that show an organ-specific expression. The further functional analysis of these genes by modification and Ti plasmid-mediated gene transfer into both potato and tobacco plants should lead to a deeper understanding of the mechanisms leading to developmentally controlled

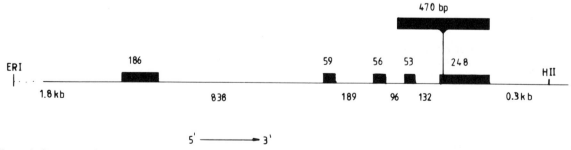

Figure 4. Structure of pgL 700 (*S. tuberosum*). Sizes and locations of the five exons (boxed) and the four intervening sequences (lined) are shown. To distinguish the activity of the endogenous pgL 700 from the reintroduced pgL 700, the last exon was modified by insertion of a 470-bp fragment of the T-DNA.

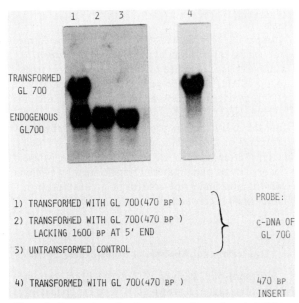

Figure 5. Expression of endogenous and transferred gL700 in potato leaves. cDNA of GL 700: (lane *1*) transformed with GL 700 (470 bp); (lane *2*) transformed with GL 700 (470 bp) lacking 1600 bp at 5' end; (lane *3*) untransformed control. Insert of 470 bp: (lane *4*) transformed with GL 700 (470 bp).

expression of plant genes. Our preliminary results indicate that 5' upstream sequences of these genes play an essential role in regulating the expression of these genes.

HEAT-INDUCIBLE EXPRESSION OF A CHIMERIC GENE UNDER CONTROL OF A PROMOTER DERIVED FROM THE HSP70 GENE FROM *DROSOPHILA*

The induction of heat shock proteins has been studied in several organisms including bacteria, fungi, animals, and plants (for review, see Schlesinger et al. 1982). The induction mechanism involves, in almost all of these studies, the transcriptional activation of heat shock genes and this mechanism is apparently highly conserved through evolution.

The evolutionary conservation of the heat shock response was illustrated even more convincingly when hsp70 genes were shown to be expressed in heterologous systems such as mouse cells, monkey cells, and *Xenopus* oocytes (Pelham 1982; Pelham and Bienz 1982). The DNA sequence responsible for the control of the heat induction was mapped to a stretch of 66 bp (Pelham 1982) or 97 bp (Dudler and Travers 1984) di-

rectly upstream of the initiation of transcription site. A short, imperfect inverted repeat, located about 11 nucleotides upstream of the TATA box, was first noted by Ingolia et al. (1980) in a number of heat shock genes. Sequence comparison of different *Drosophila* genes suggested a consensus sequence CT-GAA--TTC-AG for the heat shock element. When an element upstream of the TATA box of the *tk* promoter was substituted with a synthetic sequence CTAGAAGCTTCTAG sequence, the promoter became inducible both in COS cells and in *Xenopus* oocytes (Pelham and Bienz 1982). The DNA sequence analysis of heat shock genes from several eukaryotes has shown the presence in their promoter regions of consensus sequences very closely related to the *Drosophila* one (Farrelly and Finkelstein 1984; Schöffl et al. 1984). In view of this remarkable degree of conservation, we decided to test whether a *Drosophila* heat shock promoter would be able to trigger heat induction of chimeric genes in plants.

Experimental Design

The question whether the promoter of the heat shock p70 gene of *Drosophila* (Ingolia et al. 1980) would be activated by a temperature shift in tobacco tissues was tested by using a 457-bp *Xba*I–*Xmn*I fragment from the hsp70 gene, containing the promoter and 199 nucleotides of untranslated leader sequence (Steller and Pirrotta 1984) to construct a chimeric gene (*hsneo*) having the coding sequence of the Tn5 neomycin phosphotransferase II (NPTII) as a reporter protein (Fig. 7). A polyadenylation site and transcription terminator were derived from the octopine synthase gene. The construction was transferred via bacterial conjugation (Van Haute et al. 1983) from *Escherichia coli* to the Ti plasmid-derived vector pGV3850 (Zambryski et al. 1983). The transfer of the T-DNA region of pGV3850hsneo was obtained by cocultivation of *A. tumefaciens* bacteria harboring the construction with tobacco protoplasts. Plant cell clones were selected on the basis of kanamycin resistance.

Neomycin Phosphotransferase Assays

Callus material obtained by transformation of SR1 protoplasts by pGV3850hsneo and found to be resistant to 50 µg/ml of kanamycin was assayed for the enzymatic activity of the NPTII gene (Reiss et al. 1984). No NPTII activity was detected in normal SR1 tobacco tissue (data not shown), in contrast to the enzymatic activity detected on transformed calli grown at 24°C

Figure 6. Structure of pgT 1500. The figure shows size and location of the seven exons (boxed) and six introns (line) of the patatin genomic clone pgT 1500. The insertion of the 470-bp fragment into a *Hin*dIII site in the third exon is indicated.

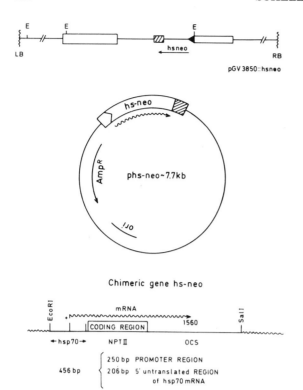

Figure 7. Schematic drawing of the *hs-neo* gene and of this gene as is is integrated in the pGV3850::hs-neo vector.

(Fig. 8, lane 1). When callus material was exposed for 45 minutes at 40°C and then tested, a heat-induced increase in NPTII activity was detected. The heat induction is already evident soon after heat shock, but its maximum is reached after approximately 1-2 hours of recovery at 24°C.

The level of NPTII enzymatic activity elicited by heat treatment varies in different calli, ranging from 5 to 40 times the basal level. Despite this difference, the pattern of induction is the same, with maximum activity between 1-2 hours after heat shock. Not all the calli are heat inducible; out of 17 independent clones tested, four showed constitutive low expression of the NPTII gene.

In parallel (Fig. 8, lanes 6-10) NPT assays performed on tobacco callus material transformed by *nosneo1103*, a chimeric gene obtained by fusion of the *nos* promoter to the NPTII coding region (Hain et al. 1985). The NPTII activity remains practically constant, since only a slight decrease is detectable after heat shock. Therefore the results hereby presented show that the hsp70 promoter of *Drosophila* elicits an increase of NPTII activity after heat shock in tobacco tissue, whereas the activity of plant material transformed by a similar construction under the control of a constitutive promoter (*nos*) is constant.

Transcriptional Analysis of Transformed Tissue

Northern blot analysis of poly(A) RNA, extracted from callus material grown at 24°C or exposed at 40°C for 45 minutes, shows the heat inducibility of mRNA when hybridized with probes containing all of the NPTII coding region (Fig. 9A, lane 2). The heat-induced transcripts are very rapidly degraded during the recovery period (Fig. 9A, lanes 3, 4, and 5). The same analysis, performed on material transformed by a chimeric gene under the control of a constitutive promoter (*nos*), shows that the steady-state level of the NPTII transcripts is not affected by heat shock (Fig. 9B, lanes 1-5).

Conclusion

Our results indicate that DNA sequences derived from the 5' upstream end of the *Drosophila* hsp90 genes are functional in plants and control the heat inducible expression of chimeric *hs-neo* genes. To our knowledge, the hsp70 is the only animal promoter thus far to conserve its functional features after transfer into the plant genome.

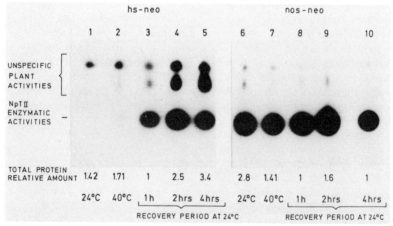

Figure 8. Assay of NPT activity in transformed calli. (Lanes *1-5*) NPTII activity in calli containing the *hs-neo* gene; (lanes *6-10*) NPTII activity in control calli containing the *nos-neo* gene.

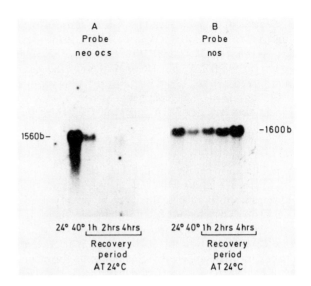

Figure 9. Analysis by Northern blotting of poly(A) RNA from callus material subjected to heat shock induction. (*A*) Probe specific for NPTII and 3′ *ocs* sequences as coded for by the *hs-neo* chimeric gene. (*B*) Probe specific for the transcript of the nopaline synthase gene.

TRANSPOSITION OF MAIZE CONTROLLING ELEMENT *ACTIVATOR* IN A HETEROLOGOUS PLANT

Activator (*Ac*) is a maize transposable genetic element capable of autonomous excision from a specific chromosomal site and insertion at another new site within the maize genome. *Ac* was first identified and studied by McClintock using standard genetic means (for review, see Fedoroff 1983). The DNA segment comprising *Ac* has been recently isolated (Fedoroff et al. 1983; Behrens et al. 1984) and sequenced (Müller-Neumann et al. 1984; Pohlmann et al. 1984).

Transposable elements have been identified and studied in numerous prokaryotic and eukaryotic organisms. Studies to discern the mechanism and biological impact of a transposable element have best been accomplished by complementation experiments using mutants constructed in vitro. The capability to transfer exogenous genetic information to cells is a requirement for the rapid accumulation of information about the structure, function, and regulation of any gene or genetic element such as *Ac*. In this context we have been applying exogenous gene transfer techniques to the study of the maize controlling elements.

At present, efficient transfer of exogenous genetic information and subsequent regeneration of mature maize plants has not proven possible. Due to this limitation we chose to begin our investigation of *Ac* function in tobacco, a plant proven amenable to exogenous gene transfer experiments. If *Ac* transposition occurs in tobacco it will be possible to exploit this system to study the function and regulation of maize transposable elements using techniques similar to those successfully applied to transposable elements of bacteria, yeasts and *Drosophila* (Shapiro 1983).

To introduce *Ac* into tobacco we used Ti plasmid-derived plant gene vectors. The *Ac9* element isolated from the *waxy* (*wx*) locus (Federoff et al. 1983) is contained on a 4.8-kbp *Pst*I DNA fragment (see Fig. 10). The element itself is 4.56 kbp in length and flanked by, respectively, 121 bp and 128 bp on either side by *wx* gene sequences. A revertant 247-bp *Wx Pst*I fragment, isolated from a maize strain following *Ac* excision from *waxy*, was also used in control experiments described below.

The 4.8-kb *Ac9 Pst*I fragment and the 247-bp *Wx Pst*I fragment were cloned into the *E. coli* plasmid pLGV1103neo. Conjugation between *E. coli* and *A. tumefaciens* (strain C58C1 containing pGV3850 [Zambryski et al. 1983]) was performed using the intermediate *E. coli* strain, GJ23 (Van Haute et al. 1983). Homologous recombination between (1) a control unaltered pLGV1103neo, (2) pBL1103-30 (containing *Ac* flanked by *wx*), (3) pBL1103-7 (containing the *wx* revertant segment) resulted in the insertion of the desired DNA segments into the T-DNA of the cointegrate plasmids. The resulting *Agrobacterium* strains were then used to transform protoplasts of *Nicotiana tabacum* cv. Petit Havana SR1 by cocultivation (Hain et al. 1985). Tobacco transformants were detected by screening for the presence of nopaline in the transformed tissue. Seven independent tobacco-transformed lines were propagated in tissue culture. Genomic DNA of several of these lines was analyzed to determine the structure of integrated T-DNA as well as the structure of the

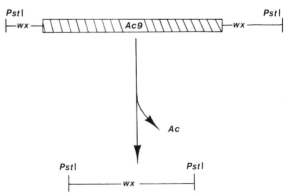

Figure 10. Transposition of the *Ac9* element from the *wx* locus.

Ac element and wx sequences. DNA from tobacco lines transformed with either pGV3850::pLGV1103neo (no Ac, no wx), pGV3850::pBL1103-30 (with Ac flanked by wx), or pGV3850::pBL1103-7 (with Wx) and nontransformed control tissue was digested with various restriction enzymes and analyzed by Southern blotting. The 4.8-kbp PstI Ac9 DNA fragment, the 247-bp Wx DNA fragment, the plasmid pLGV1103neo DNA, and a Tn5-specific DNA fragment, isolated from pLGV1103neo were used as probes in succession on the same filter or in parallel on identical filters.

Two specific results from this analysis indicate that Ac transposed in at least three out of seven tobacco lines from its original site of integration (within the introduced T-DNA) to numerous new sites within the tobacco genome. (1) In two cell lines there is an obvious increase in the number of Ac homologous fragments when compared with the copy number of flanking wx and vector T-DNA sequences detected by appropriate probes. (2) In two of these three lines, a 247-bp PstI fragment was detected by hybridization to the Wx-specific probe. The presence of the 247-bp fragment suggests that Ac has excised from its original position within wx leaving behind an uninterrupted Wx fragment. Tobacco lines transformed with only the Wx construction contained the 247-bp Wx-specific genomic DNA fragment as expected. Tobacco lines transformed with pGV3850::pLGV1103 and nontransformed tobacco tissue did not hybridize to the Ac probe nor to the wx probe. Further evidence for transposition of Ac within these lines will be obtained by isolation and sequence analysis of the Ac and Wx components.

These results, accompanied by an appropriate functional test system for transposition of Ac in tobacco cells, will fulfill the requirements for further detailed analysis of the mechanism of transposition and applied work on gene isolation and vector development.

ACKNOWLEDGMENTS

The authors are indebted to Dr. H. Sommer, Prof. H. Saedler, and Prof. K. Hahlbrock for their contributions to the work on chalcone synthase. Dr. H. Sommer and Prof. H. Saedler provided us with the cloned CS promoter from *A. majus* and with the relevant sequence data along with considerable advice. Prof. K. Hahlbrock allowed us to use the cDNA of CS isolated from parsley that was isolated in his laboratory by one of us (F. Kreuzaler), and it is his pioneering work on the CS that induced us to perform these experiments. We are very much indebted also to our many colleagues and friends of the Laboratory of Genetics, Gent, for their help and advice with Ti plasmid vectors and selectable marker genes.

REFERENCES

Behrens, U., N. Federoff, A. Laird, M. Müller-Neumann, P. Starlinger, and J. Yoder. 1984. Cloning of *Zea mays* controlling element *Ac* forms the *wx-m7* allele. *Mol. Gen. Genet.* **194**: 346.

Butler, W.L., K.H. Norris, H.W. Siegelmann, and S.B. Hendrichs. 1959. Detection assay and preliminary purification of the pigment controlling photoresponsive development of plants. *Proc. Natl. Acad. Sci.* **45**: 1703.

De Block, M., L. Herrera-Estrella, M. Van Montagu, J. Schell, and P. Zambryski. 1984. Expression of foreign genes in regenerated plants and in their progeny. *EMBO J.* **3**: 1681.

Dudler, R. and A.A. Travers. 1984. Upstream elements necessary for optimal function of the hsp 70 promoter in transformed flies. *Cell* **38**: 391.

Farrelly, F.W. and D.B. Finkelstein. 1984. Complete sequence of the heat shock-inducible *hsp90* gene of *Saccharomyces cerevisiae*. *J. Biol. Chem.* **259**: 5745.

Fedoroff, N. 1983. Controlling elements in maize. In *Mobile genetic elements* (ed. J. Shapiro), p. 1. Academic Press, New York.

Fedoroff, N., S. Wessler, and M. Share. 1983. Isolation of the transposable maize controlling elements *Ac* and *Ds*. *Cell* **35**: 235.

Gielen, J., M. de Beuckeleer, J. Seurinck, F. Deboeck, H. de Greve, M. Lemmers, M. Van Montagu, and J. Schell. 1984. The complete nucleotide sequence of the TL-DNA of the *Agrobacterium tumefaciens* plasmid pTiAch5. *EMBO J.* **3**: 835.

Hahlbrock, K. and H. Grisebach. 1979. Enzymic controls on the biosynthesis of lignin and flavonoids. *Annu. Rev. Plant Physiol.* **30**: 105.

Hain, R., P. Stabel, A.P. Czernilofky, H.H. Steinbiss, L. Herrera-Estrella, and J. Schell. 1985. Uptake, integration, expression and genetic transmission of a selectable chimaeric gene by plant protoplasts. *Mol. Gen. Genet.* **199**: 161.

Heller, W. and K. Hahlbrock. 1980. Highly purified flavanone synthase from parsley catalyses the formation of naringenin chalcone. *Arch. Biochem. Biophys.* **200**: 617.

Herrera-Estrella, L., G. Van den Broeck, R. Maenhaut, M. Van Montagu, and J. Schell. 1984. Light-inducible and chloroplast-associated expression of a chimaeric gene introduced into *Nicotiana tabacum* using a Ti plasmid vector. *Nature* **310**: 115.

Ingolia, T.D., E.A. Craig, and B.J. McCarthy. 1980. Sequence of three copies of the gene for the major *Drosophila* heat shock induced protein and their flanking regions. *Cell* **21**: 669.

Joos, H., D. Inze, A. Caplan, M. Sorman, M. Van Montagu, and J. Schell. 1983. Genetic analysis of T-DNA transcripts in nopaline crown galls. *Cell* **32**: 1057.

Kamalay, J.C. and R.B. Goldberg. 1984. Organ specific nuclear RNAs in tobacco. *Proc. Natl. Acad. Sci.* **81**: 2801.

Kreuzaler, F., H. Ragg, E. Fautz, D.N. Kuhn, and K. Hahlbrock. 1983. UV-induction of chalcone synthase mRNA in cell suspension cultures of *Petroselinum hortense*. *Proc. Natl. Acad. Sci.* **80**: 2591.

Kreuzaler, F., H. Ragg, W. Heller, R. Teich, I. Witt, and D. Hammer. 1979. Flavanone synthase from *Petroselinum hortense*. *Eur. J. Biochem.* **99**: 89.

Mohr, H. 1972. *Lectures on photomorphogenesis*. Springer-Verlag, Berlin.

Mohr, H. and H. Schäfer. 1983. Photoreception and de-etiolation. *Philos. Trans. R. Soc. Lond.* B **303**: 489.

Müller-Neumann, M., I. Yoder, and P. Starlinger. 1984. The DNA sequence of the transposable element *Ac* of *Zea mays* L. *Mol. Gen. Genet.* **198**: 19.

Pelham, H.R.B. 1982. A regulatory upstream promoter element in the *Drosophila* hsp 70 heat-shock gene. *Cell* **30**: 517.

Pelham, H.R.B. and M. Bienz. 1982. A synthetic heat-shock promoter element confers heat-inducibility on the herpes simplex virus thymidine kinase gene. *EMBO J.* **1**: 1473.

Pohlmann, R.F., N. Fedoroff, and J. Messing. 1984. The nucleotide sequence of the maize controlling element activator. *Cell* **37**: 635.

Reiss, B., R. Sprengel, H. Will, and H. Schaller. 1984. A

new sensitive method for qualitative and quantitative analysis of neomycin phosphotransferase in crude cell extracts. *Gene* **30:** 217.

Schlesinger, M.I., M. Ashburner, and A. Tissières, eds. 1982. *Heat shock: From bacteria to man.* Cold Spring Harbor Laboratory, Cold Spring Harbor, New York.

Schöffl, F., E. Raschke, and R.T. Nagao. 1984. The DNA sequence analysis of soybean heat-shock genes and identification of possible regulatory promoter elements. *EMBO J.* **3:** 2491.

Schreier, P.H., E.A. Seftor, J. Schell, and H.J. Bohnert. 1985. The use of nuclear-encoded sequences to direct the light-regulated synthesis and transport of a foreign protein into plant chloroplasts. *EMBO J.* **4:** 25.

Schröder, J., F. Kreuzaler, E. Schäfer, and K. Hahlbrock. 1979. Concomitant induction of phenylalanine ammonia-lyase and flavanone synthase mRNA's in irradiated plant cells. *J. Biol. Chem.* **254:** 57.

Senger, H., ed. 1980. *The blue light syndrome.* Springer-Verlag, Berlin.

———. 1982. The effect of blue light on plants and microorganisms. *Photochem. Photobiol.* **35:** 911.

Shapiro, J., ed. 1983. *Mobile genetic elements.* Academic Press, New York.

Shropshire, W. and H. Mohr. 1983. Photomorphogenesis. In *Encyclopedia of Plant Physiology* (ed. A. Pirsan and M.H. Zimmermann), vol. 16a, p. 24. Springer-Verlag, Berlin.

Steller, H. and V. Pirrotta. 1984. Regulated expression of genes injected into early *Drosophila* embryos. *EMBO J.* **3:** 165.

Van Haute, E., H. Joos, M. Maes, G. Warren, M. Van Montagu, and J. Schell. 1983. Intergenic transfer and exchange recombination of restriction fragments cloned in pBR322: A novel strategy for the reversed genetics of the Ti-plasmids of *Agrobacterium tumefaciens*. *EMBO J.* **2:** 411.

Zambryski, P., L. Herrera-Estrella, M. De Block, M. Van Montagu, and J. Schell. 1984. The use of the Ti plasmid of *Agrobacterium* to study the transfer and expression of foreign DNA in plant cells: New vectors and methods. In *Genetic engineering, principles and methods* (ed. A Hollaender and J. Setlow), vol. 6, p. 253. Plenum Press, New York.

Zambryski, P., H. Joos, C. Genetello, J. Leemans, M. Van Montagu, and J. Schell. 1983. Ti plasmid vector for the introduction of DNA into plant cells without alterations of their normal regeneration capacity. *EMBO J.* **2:** 2143.

Transgenic Plants

R.B. HORSCH, S.G. ROGERS, AND R.T. FRALEY
Biological Sciences, Monsanto Company, St. Louis, Missouri 63198

Plants are relatively simple organisms, but they contain a diversity of cell types, tissues, and organs. Many species have remarkable regeneration capacity, including the ability to form embryos from somatic cells cultured in vitro. Since there is no cell movement during plant development, the position of cells reflects their origin. The process of development continues indefinitely at the growing points, and major changes in development (e.g., commitment to flower) can be triggered by simple environmental signals, such as day length.

The ability to introduce foreign genes into eukaryotic organisms has permitted experimental investigation of the function of genes at every level: DNA sequences such as promoters and enhancers, posttranscriptional and posttranslational control mechanisms, protein structure/function relationships, and the effects of specific gene products in physiological or developmental processes.

Although such experiments and our overall understanding of molecular genetics are less advanced in higher plant systems than in mammalian systems, yeast, or *Drosophila*, technology that should facilitate rapid progress has recently been developed. It is now possible easily and rapidly to produce large numbers of transgenic plants of species such as petunia, tobacco, or tomato. This powerful gene transfer technology, coupled with the developmental competence of somatic cells of these plants, offers promising systems for studying the processes and regulatory mechanisms that are involved in growth and development.

Agrobacterium tumefaciens causes crown gall disease in many plants by transferring a defined piece of DNA into the nuclear genome of infected cells (for recent reviews, see Bevan and Chilton 1982; Depicker et al. 1983; Nester et al. 1984). The transferred DNA (T-DNA) is found on a large tumor-inducing (Ti) plasmid and contains genes that code for phytohormone biosynthetic enzymes (Akiyoshi et al. 1984; Barry et al. 1984; Schröder et al. 1984; Thomashow et al. 1984). The phytohormones produced in the transformed cells cause them to proliferate, forming a gall. In addition, another gene or set of genes found on T-DNA codes for enzymes that create one or more unique metabolites called opines. One such opine, nopaline, is a conjugate of arginine and α-ketoglutarate and is diagnostic for plant cells transformed by an *A. tumefaciens* strain that carries the nopaline synthase gene. Plant cells do not degrade opines, but *A. tumefaciens* has catabolic genes corresponding to the synthetic genes found in their T-DNA. The T-DNA is defined by a 25-bp sequence present as a direct repeat at each end of the T-DNA (Simpson et al. 1982; Yadav et al. 1982; Zambryski et al. 1982; Barker et al. 1983). Another region of the Ti plasmid contains a set of genes that are essential for virulence (*vir* genes) (Klee et al. 1983; Hille et al. 1984). Presumably the virulence gene products cause transfer of the T-DNA into plant cells.

The most useful feature of the DNA transfer capability of *A. tumefaciens* is the complete separation of the transfer mechanism from the transferred DNA. The *vir* genes and 25-bp border sequences are necessary but are not themselves found integrated in transformed plant cells, whereas the T-DNA genes are completely unnecessary for the transfer process (Zambryski et al. 1983). Thus any DNA located between the border sequence repeat will be transferred: All the disease-related sequences can be removed and any desired genes can be added. Once in a plant chromosome, the T-DNA has no inherent capacity to move again and usually remains as a permanent addition to the genome.

Plant cells can be transformed in vitro by coculture with *A. tumefaciens*. Initial studies with nonengineered strains used tissue such as inverted stem sections of tobacco or carrot root slices and depended on tumorous proliferation of cells caused by the phytohormone biosynthetic genes in the T-DNA.

A more refined technique, cocultivation, was developed where single plant cells were transformed in vitro by *A. tumefaciens* (Mártón et al. 1979; Wullems et al. 1981; Fraley and Horsch 1983). This permitted production of hundreds of independent transformed colonies in a single experiment (Fraley et al. 1984).

Cocultivation of cells with *A. tumefaciens* provided an ideal means to test the functioning of chimeric gene constructions for antibiotic resistance (Bevan et al. 1983; Fraley et al. 1983a,b; Herrera-Estrella et al. 1983) and then, with improved vectors, permitted regeneration of the first transgenic plants (DeBlock et al. 1984; Horsch et al. 1984). Recently, we have reported a simple and rapid leaf disk transformation procedure that produces transgenic plants within 4–6 weeks for species such as petunia or tobacco (Horsch et al. 1985). Vectors, selectable markers, and nontumorigenic strains of *A. tumefaciens* have been constructed that permit routine production of transgenic plants.

RESULTS

Vector System

The split end vector (SEV) system (Fraley et al. 1985) is shown in Figure 1. The wild-type *A. tumefaciens*

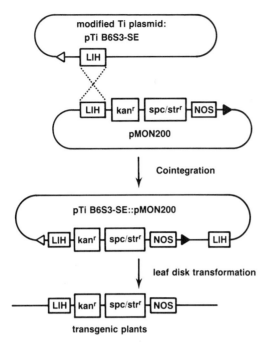

Figure 1. SEV system for plant transformation. The disarmed Ti plasmid pTiB6S3-SE contains the left border sequence (arrow) and a region of T-DNA (LIH) but none of the phytohormone biosynthetic genes. The pMON200 plasmid contains a region of T-DNA homology (LIH) for homologous recombination, a chimeric kanamycin-resistance gene (kan^r), a bacterial spectinomycin–streptomycin-resistance determinant (spc/str^r) for selection of cointegrates, a nopaline synthase gene (NOS), and the right border sequence (solid arrow). Reciprocal recombination within LIH leads to a cointegrate plasmid, pTiB6S3-SE::pMON200, which contains a complete T-DNA. During leaf disk transformation the T-DNA located between the borders is transferred into the plant genome, resulting in selectable kanamycin resistance and production of nopaline, which is easily assayed.

the vector. This plasmid can be transferred to *A. tumefaciens* by conjugation where it must integrate via homologous recombination at LIH to be maintained.

Leaf Disc System

The leaf disc system is diagrammed in Figure 2. *A. tumefaciens* transforms cells at the edge of the disc, transferring the engineered T-DNA containing a selectable marker and other components of the vector.

The medium used during this initial phase contains a combination of phytohormones that induces both cell growth and development of adventitious (de novo) shoots. After 2 days' coculture of the leaf discs with the bacteria, the discs are transferred to similar media containing carbenicillin to kill the bacteria and kanamycin to inhibit the wild-type plant cells. The transformed plant cells are able to grow and develop into plantlets at the edges of the discs within 2–4 weeks.

The shoots are cut from the discs and transferred to rooting medium also containing 100 μg/ml kanamycin to screen for nonresistant survivors of the initial selection. Wild-type shoots will not root in the presence of 100 μg/ml kanamycin whereas transformants root normally. Despite the fact that no growth or shoot development occurred from control leaf discs on selective medium, many of the shoots that grew from transformed discs would not root in the presence of kanamycin. These shoots may have initially been resistant to kanamycin but became sensitive during development because of loss of the T-DNA or loss of its expression. Alternatively they may have been able to escape selection by cross-protection of some kind from nearby resistant cells. Several such escapes were found to contain DNA homologous with the pMON200 vector.

Transgenic Plants

The plants that formed roots in the presence of kanamycin were transplanted to soil and grown to maturity. About half of the transgenic plants exhibited predictable, stable expression of the foreign genes they contained. Samples from any tissue or organ (e.g., roots, stems, leaves, petals, anthers) contained nopaline and explants from these sources were resistant to kanamycin in culture. However, many of the transgenic plants exhibited aberrant expression of one or more of the foreign genes. The nopaline synthase gene, at the outside edge of the T-DNA, gave the most frequent abnormalities. In one sample of 25 plants that had rooted in the presence of kanamycin, only 12 produced nopaline in leaves while growing in soil, whereas 20 were resistant to kanamycin (including all 12 nopaline-positive plants) when leaf explants were tested in culture. Of the eight nopaline-negative, kanamycin-resistant plants, five showed induction of nopaline synthesis upon transfer of leaves to culture medium. Similar results were obtained in tobacco and tomato. This loss of expression of the foreign gene(s) after the

strain GV3111 contains plasmid pTiB6S3 and causes crown gall disease on many plants, including tobacco, petunia, and tomato. This strain has two different T-DNAs, T_R, which contains several genes for opine synthesis, and T_L, which contains phytohormone biosynthetic genes and octopine synthase. A deletion was created that removed all of T_R and most of T_L, removing the phytohormone biosynthetic genes but leaving the region designated LIH (left inside homology). This avirulent plasmid, pTiB6S3-SE, is used as the recipient for homologous recombination of intermediate vectors such as pMON200. The intermediate vectors are designed so that they create a functional T-DNA when they integrate into the disarmed Ti plasmid, pTiB6S3-SE.

The features of the intermediate vector include the T-DNA right border sequence, a region of DNA homology (LIH) for directing integration into pTiB6S3-SE, a chimeric plant kanamycin resistance gene (kan^R), an intact nopaline synthase gene (NOS), and a polylinker region to facilitate the addition of other genes to

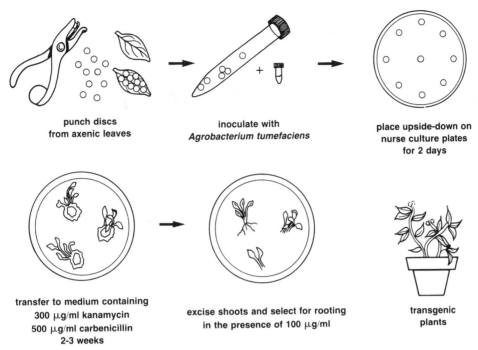

Figure 2. Leaf disc transformation system. Details of the procedure are described in the text.

rooting assay may be due to the same problem(s) that resulted in the initial escapes. In any event, it is a simple matter to screen for those plants that constitutively express both markers.

More troublesome is the quantitative variation in level of expression of genes between independent transformants. We have observed large differences in neomycin-phosphotransferase (NPTII) activity between plants containing identical gene constructions. These differences do not correlate with copy number of the inserted T-DNA. They are most likely due to influences of the surrounding plant DNA—chromatin structure, methylation, or other "position effects."

Again, these position effects must be dealt with by use of sufficiently large numbers of independent transformants to ensure representative averages in expression experiments. Future-generation vectors might be devised to reduce the variability of position effects if DNA structures or signals are discovered that can override the influence of surrounding chromatin.

Inheritance Studies

The T-DNA is usually transmitted to progeny in a normal Mendelian fashion. To facilitate careful, detailed analysis where infrequent abnormalities could be detected and evaluated, we have begun genetic mapping of the sites of T-DNA insertion in a special hybrid petunia created at the Department of Genetics, University of Amsterdam. This hybrid (V23 × R51) is heterozygous for markers on all seven chromosome pairs. Thus the T-DNA will be inserted in one of the 14 unique chromosomes and can be mapped in a single backcross generation to each of the parental lines (V23 and R51). Thus far we have localized eight different T-DNAs to four different chromosomes. No unusual behavior of the T-DNA has yet been observed in this small sample.

While most of our transgenic petunia, tobacco, and tomato plants show the expected 3:1 ratio in selfed progeny, one unusual transgenic plant of *Nicotiana plumbaginifolia* has been produced. The plant, NPK7, gives progeny in a 2:1 ratio with nearly all seeds germinating. In reciprocal backcrosses to wild-type, it gives a 1:1 ratio as expected. Among the transgenic progeny, no homozygotes have been found. One simple explanation is that the T-DNA is linked to or has caused a recessive lethal mutation that acts very early in embryo development, preventing production of a seed. There is also a competitive disadvantage in growth rate in transgenic progeny compared with non-transgenic progeny, indicating that the mutation is deleterious in the heterozygous stage. This property might permit an assay for analysis of the function of the mutation and for cloning the gene by complementation.

The T-DNA places selectable and scoreable markers into chromosomes that can be used to facilitate genetic manipulations and plant breeding. We are exploring the potential for use of a comprehensive collection of mapped transgenic plants for chromosomal manipulations and molecular genetic studies such as chromosome walking or insertional mutagenesis.

Additional Markers and Vectors

We have constructed a variety of new vectors and selectable markers to increase the utility of gene transfer for molecular and genetic studies. A binary vector has been made that will permit shotgun transformation of gene libraries into *A. tumefaciens* for subsequent bulk transformation of plant cells by cocultivation or the leaf disc method.

The cointegrate plasmids such as pTiB6S3-SE:pMON200 are inefficient for such mass transfers between *Escherichia coli* and *A. tumefaciens*, since the frequency of cointegrate formation ranges between 10^{-5} and 10^{-7}. Nonintegrating, "binary" vectors permit very efficient mobilization into *A. tumefaciens* (10–100%). The binary vector need only contain the 25-bp border sequence to be transferred into plant cells by the virulence gene products on the resident Ti plasmid (which has had its own T-DNA deleted).

The pMON200 plasmid was converted to a binary vector by replacement of the LIH region with a 3.8-kb RK2 replicon. This binary vector, pMON505, transforms plant cells at normal frequency and retains all of the advantages of the pMON200 vector (H. Klee and R.B. Horsch, unpubl.). Cosmid vectors and cDNA expression vectors could be constructed that would then permit shotgun complementation, selection, or screening for the desired transformed cell or plant followed by recloning of the vector from the plant to recover the gene of interest.

Several other selectable markers have been used to obtain transformed plant cells (Herrera-Estrella et al. 1983), however none are as effective as kanamycin. We recently have constructed a chimeric gene containing the cauliflower mosaic virus 35s promoter, a methotrexate-resistant mouse dihydrofolate reductase (*dhfr*) gene, and the NOS polyadenylation signals. The gene confers high-level methotrexate resistance to plant cells and promises to be a useful selectable marker. A binary vector, pMON809, was constructed that contains this methotrexate-resistance gene and the gene for octopine synthase (OCS) (H. Klee et al., unpubl.). This vector can be used to transfer a second T-DNA into transgenic plants containing the kanamycin-resistance marker and the NOS gene. Thus the two selectable markers can be observed separately and the two scoreable markers (OCS and NOS) can be observed simultaneously. Genetic analysis of two gene or two component systems is now simple in hybrids between plants containing each of the vectors.

The technology for transferring functional foreign genes into plants is now routine for species such as petunia, tobacco, or tomato. A variety of versatile vectors are available for many types of gene transfer experiments. This powerful technology is now being applied to studies of gene expression, biochemistry, physiology, and developmental biology of plants.

ACKNOWLEDGMENTS

This work was a team effort with valuable contributions from many people, including C.L. Fink, J.S. Flick, J.S. Fry, M.B. Hayford, N. Hoffmann, H. Klee, A. Lloyd, S. McCormick, J. Niedermeyer, P.R. Sanders, and M. Wallroth. We thank B. Schiermeyer for typing the manuscript and L. Kraus for the drawings. We also thank Dr. E.G. Jaworski for encouragement and support during the course of this work.

REFERENCES

AKIYOSHI, D., H. KLEE, R. AMASINO, E. NESTER, and M. GORDON. 1984. T-DNA of *Agrobacterium tumefaciens* encodes an enzyme of cytokinin biosynthesis. *Proc. Natl. Acad. Sci.* **81:** 5994.

BARKER, R.F., K.B. IDLER, D.V. THOMPSON, and J.D. KEMP. 1983. Nucleotide sequence of the T-DNA region from the *Agrobacterium tumefaciens* octopine Ti plasmid PTi15955. *Plant Mol. Biol.* **2:** 335.

BARRY, G., S. ROGERS, R. FRALEY, and L. BRAND. 1984. Identification of a cloned cytokinin biosynthetic gene. *Proc. Natl. Acad. Sci.* **81:** 4776.

BEVAN, M. and M.-D. CHILTON. 1982. T-DNA of the *Agrobacterium* Ti and Ri plasmids. *Annu. Rev. Genet.* **16:** 357.

BEVAN, M., R.B. FLAVELL, and M.-D. CHILTON. 1983. A chimeric antibiotic resistance gene as a selectable marker for plant cell transformation. *Nature* **304:** 184.

DEBLOCK, M., L. HERRERA-ESTRELLA, M. VAN MONTAGU, J. SCHELL, and P. ZAMBRYSKI. 1984. Expression of foreign genes in regenerated plants and their progeny. *EMBO J.* **3:** 1681.

DEPICKER, A., M. VAN MONTAGU, and J. SCHELL. 1983. In *Genetic engineering of plants: An agricultural perspective* (ed. T. Kosuge et al.), p. 143. Plenum Press, New York.

FRALEY, R.T. and R.B. HORSCH. 1983. In vitro plant transformation systems using liposomes and bacterial co-cultivation. In *Genetic engineering of plants: An agricultural perspective* (ed. T. Kosuge et al.), p. 177. Plenum Press, New York.

FRALEY, R., S. ROGERS, and R. HORSCH. 1983a. Use of a chimeric gene to confer antibiotic resistance to plant cells. In *Advance in gene technology: Molecular genetics of plants and animals* (ed. K. Downey et al.), p. 211. Academic Press, New York.

FRALEY, R., R. HORSCH, A. MATZKE, M.-D. CHILTON, W. CHILTON, and P. SANDERS. 1984. *In vitro* transformation of petunia cells by an improved method of cocultivation with *A. tumefaciens* strains. *Plant Mol. Biol.* **3:** 371.

FRALEY, R., S. ROGERS, R. HORSCH, D. EICHHOLTZ, J. FLICK, C. FINK, N. HOFFMANN, and P. SANDERS. 1985. The SEV system: A new disarmed Ti plasmid vector for plant transformation. *Biotechnology* **3:** 629.

FRALEY, R.T., S.G. ROGERS, R.B. HORSCH, P. SANDERS, J. FLICK, S. ADAMS, M. BITTNER, L. BRAND, C. FINK, J. FRY, G. GALLUPPI, S. GOLDBERG, N. HOFFMANN, and S. WOO. 1983b. Expression of bacterial genes in plant cells. *Proc. Natl. Acad. Sci.* **80:** 4803.

HERRERA-ESTRELLA, L., M. DEBLOCK, E. MESSENS, J.P. HERNALSTEENS, M. VAN MONTAGU, and J. SCHELL. 1983. Chimeric genes as dominant selectable markers in plant cells. *EMBO J.* **2:** 987.

HILLE, J., J. VAN KAN, and R. SCHILPEROORT. 1984. trans-Acting virulence functions of the octopine Ti plasmid from *Agrobacterium tumefaciens*. *J. Bacteriol.* **158:** 754.

HORSCH, R., R. FRALEY, S. ROGERS, P. SANDERS, A. LLOYD, and N. HOFFMANN. 1984. Inheritance of functional foreign genes in plants. *Science* **223:** 496.

HORSCH, R.B., J.E. FRY, N.L. HOFFMANN, M. WALLROTH, D. EICHHOLTZ, S.G. ROGERS, and R.T. FRALEY. 1985. A simple and general method for transferring genes into plants. *Science* **227:** 1229.

KLEE, H.J., F.F. WHITE, V.N. IYER, M.P. GORDON, and E.W. NESTER. 1983. Mutational analysis of the virulence region of an *Agrobacterium tumefaciens* Ti plasmid. *J. Bacteriol,* **153:** 878.

Martón, L., G.J. Wullems, L. Molendijk, and R.A. Schilperoort. 1979. *In vitro* transformation of cultured cells from *Nicotiana tabacum* by *Agrobacterium tumefaciens*. *Nature* 227: 129.

Nester, E.W., M.P. Gordon, R.M. Amasino, and M.F. Yanofsky. 1984. Crown gall: A molecular and physiological analysis. *Annu. Rev. Plant Physiol.* 35: 387.

Schröder, G., S. Waffenschmidt, E.W. Weiler, and J. Schröder. 1984. The T-region of Ti plasmids codes for an enzyme synthesizing indole-3-acetic acid. *Eur. J. Biochem.* 138: 387.

Simpson, R., P. O'Hara, C. Lichtenstein, A.L. Montoya, W. Kwok, M.P. Gordon, and E.W. Nester. 1982. The DNA from A6S/2 tumor contains scrambled Ti plasmid sequence near its junction with plant DNA. *Cell* 29: 1005.

Thomashow, L.S., S. Reeves, and M.F. Thomashow. 1984. Crown gall oncogenesis: Evidence that a T-DNA gene from the *Agrobacterium* Ti plasmid pTiA6 encodes an enzyme that catalyzes synthesis of indoleacetic acid. *Proc. Natl. Acad. Sci.* 81: 5071.

Wullems, G.J., L. Molendijk, G. Ooms, and R.A. Schilperoort. 1981. Differential expression of crown gall tumor markers in transformants obtained after *in vitro Agrobacterium tumefaciens*-induced transformation of cell wall regenerating protoplasts derived from *Nicotiana tabacum*. *Proc. Natl. Acad. Sci.* 78: 4344.

Yadav, N.S., J. Vanderleyden, D. Bennet, W.M. Barnes, and M.-D. Chilton. 1982. Short direct repeats flank the T-DNA on a nopaline Ti plasmid. *Proc. Natl. Acad. Sci.* 79: 6322.

Zambryski, P., A. Depicker, K. Kruger, and H. Goodman. 1982. Tumor induction by *Agrobacterium tumefaciens*: Analysis of the boundaries of T-DNA. *J. Mol. Appl. Genet.* 1: 361.

Zambryski, P., H. Joos, C. Genetello, J. Leemans, M. Van Montagu, and J. Schell. 1983. Ti plasmid vector for the introduction of DNA into plant cells without alteration of their normal regeneration capacity. *EMBO J.* 2: 2143.

Retroviruses and Insertional Mutagenesis

R. JAENISCH,*† M. BREINDL,† K. HARBERS,† D. JÄHNER,†‡
and J. LÖHLER†

*Whitehead Institute for Biomedical Research and Department of Biology at Massachusetts Institute of Technology, Cambridge, Massachusetts 02142; †The Heinrich-Pette-Institut für Experimentelle Virologie und Immunology an der Universitat Hamburg, 2000 Hamburg 20, Federal Republic of Germany

The phenotypic analysis of experimentally induced or spontaneous mutations has long been the subject of developmental genetics in the mouse. Lethal mutations have had a significant role in identifying pleiotropic effects of presumably single genes on complex developmental processes. Examples of such genetic systems are the well-studied T locus (Silver 1981), the albino deletions (Glueksohn-Waelsch 1979), or mutations affecting hematopoiesis (Russell 1979). This descriptive approach, however, has its limitations in elucidating the underlying developmental defect on a molecular basis, because a randomly mutated gene or its gene product is difficult to identify. An alternative approach is to induce mutations by insertional mutagenesis and to use the inserted foreign DNA element as a tag to molecularly clone the mutated gene.

In mammals, the experimental insertions of retroviruses and of recombinant DNA are two approaches that have been successful in inducing developmental mutations by insertional mutagenesis. In these experiments, either mouse zygotes were microinjected with DNA (Wagner et al. 1983; Palmiter et al. 1984) or mouse embryos were exposed to infectious retroviruses at different stages of development (Jaenisch 1976; Jähner and Jaenisch 1980; Jaenisch et al. 1981) to derive animals that carried the foreign genetic information in their germ line.

In this report, we will review the derivation of a mouse strain with a recessive lethal embryonic mutation that was induced by provirus insertion into the α1(I) collagen gene. Furthermore, our studies to characterize the mutation on a molecular level and the use of the mutant to investigate the role of type-I collagen in embryonic development will be summarized. Finally, we will compare the known insertional mutations in mice induced by retroviruses as opposed to insertion of recombinant DNA.

RESULTS AND DISCUSSION

Insertion of Moloney Leukemia Viruses into the Germ Line of Mice

Mouse strains, termed Mov substrains, carrying a single Moloney leukemia proviral copy (Mo-MLV) as a Mendelian determinant, were derived by exposing mouse embryos to infectious virus at different developmental stages (Table 1). The great majority of mouse strains were derived from virus-infected preimplantation embryos (Jaenisch et al. 1981) or from a zygote that was microinjected with proviral DNA (Stewart et al. 1983), whereas the Mov-13 strain was derived from an embryo microinjected with infectious virus at midgestation (Jaenisch 1980; Jaenisch et al. 1981). Although the frequency of germ line integrations is high in animals derived from infected preimplantation embryos, infection of primordial germ cells with virus at midgestation is a rare event, and so far Mov-13 is the only strain obtained by this infection protocol.

Infectious virus is activated at different stages of development in some, but not in other Mov substrains (see Table 1). We have shown that virus inactivity in Mov-2, Mov-7, and Mov-10 mice is due to a mutation in the proviral genome (Schnieke et al. 1983b). The different timing of virus activation in Mov-1 and Mov-13 (Jaenisch et al. 1981; Fiedler et al. 1982) and in Mov-3, Mov-9, and Mov-14 mice (T. Berleth et al., in prep.), however, suggests that the chromosomal position is important for the developmental activation of the respective proviral genomes (Jaenisch 1983).

Provirus Insertion into the α1(I) Collagen Gene Caused a Recessive Lethal Mutation in Mov-13 Mice

Heterozygosity at any of the Mov loci did not interfere with normal embryonic development or result in a detectable phenotype in the postnatal animals. To detect possible recessive mutations caused by insertion of the virus, offspring were derived from heterozygous parents and analyzed to identify animals homozygous at the respective Mov locus (Table 2). With the exception of Mov-13, all crosses resulted in homozygous, healthy offspring, indicating that proviral insertion at none of the different loci disrupted essential gene functions. The Mov-13 substrain, as the other strains, carries a single provirus in its germ line. In contrast to the other strains, however, homozygosity at the Mov-13 locus leads to death of the embryos between days 13 and 14 of gestation (Jaenisch et al. 1983). The host sequences flanking the insertion site were molecularly cloned by using the virus as a probe. Hybridization of these sequences to RNA extracted from normal embryos or various tissue culture lines detected two spe-

‡Present address: Salk Institute, MBVL, P.O. Box 85800, San Diego, California 92138.

Table 1. Insertion of Mo-MLV into the Germ Line of Mice

Stage of exposure to virus	Genetic locus	Location	Virus expression	Time of virus activation
Zygote (microinjection of DNA)	Mov-14	X chromosome	+	day 15 of gestation
Four- to-eight-cell preimplantation stage	Mov-1	chromosome 6	+	7 days postnatal
	Mov-2		−	
	Mov-3		+	4 days postnatal
	Mov-4		−	
	Mov-5		−	
	Mov-6		−	
	Mov-7		−	
	Mov-8		−	
	Mov-9		+	day 18 of gestation
	Mgpt-1[a]		−	
	Mov^s-15[b]	X, Y chromosome	+	
Blastocyst	Mov-10		−	
	Mov-11		−	
	Mov-12		−	
Midgestation	Mov-13	chromosome 11	+	day 16 of gestation

[a] Jähner et al. 1985.
[b] K. Harbers et al., in prep.

cies of RNA. The developmentally regulated and cell-specific pattern of *Mov-13* locus expression and its sensitivity to transformation by sarcoma viruses led us to screen genes coding for proteins of the extracellular matrix and thus to identify the α1(I) collagen gene as being interrupted by virus insertion in Mov-13 mice (Schnieke et al. 1983a).

Molecular Analysis of Mutated Collagen Gene

Sequence analysis of the host DNA flanking the virus showed that provirus insertion had occurred in the first intron of the α1(I) collagen gene with the transcriptional orientation of the provirus opposite to that of the host gene (Harbers et al. 1984). Furthermore, S1 mapping experiments using the first exon of the collagen gene as a probe demonstrated that virus insertion into these noncoding host sequences completely blocked collagen transcription in vivo as well as in tissue culture cells. In an effort to understand the virus-induced inactivation of the collagen gene on a molecular level, we have analyzed the chromatin structure and the methylation status of the α1(I) collagen in Mov-13 mice. Both, DNA hypomethylation (Jaenisch and Jähner 1984; Jähner and Jaenisch 1984) and an open chromatin conformation (Elgin 1981) have been correlated with transcriptional activity of genes.

Virus insertion changes the chromatin conformation of the α1(I) collagen gene. The chromatin structures in the 5′-flanking region that are implicated in transcriptional control of eukaryotic genes often display an unusual sensitivity to digestion with nucleases (Weisbrod 1982). To analyze whether virus insertion into the α1(I) collagen gene in Mov-13 mice induced a change on the chromatin level, the chromatin structures of the mutant and wild-type alleles were compared (Breindl et al. 1984). Our results localized three hypersensitive sites in the chromatin structure at the 5′ end of the α1(I) collagen gene (Fig. 1). Two of these, located 3′ of the first exon, are present, regardless of whether the gene is transcribed. This was shown for F9 cells and for brain and liver of normal adult mice, which do not produce type-I collagen. In contrast, the presence of a third hypersensitive site located 5′ of the mRNA start point, correlated strictly with active transcription and was found in chromatin of fibroblastic cell lines and of day-15 normal mouse embryos. This suggests that this

Table 2. Derivation of Mouse Strains Homozygous for Mo-MLV Insertion (♀ Mov/wt × ♂ Mov/wt)

Strain	Ratio of offspring with genotype		
	Mov/Mov	Mov/wt	wt/wt
Mov-1 to Mov-12, and Mov-14, Mov-15, Mgpt-1	1	2	1
Mov-13			
adults, embryos day 16–19	0	2	1
embryos (day 11–14)	1[a]	2	1

[a] Most homozygous embryos at day 13 or older have stopped development or are dead.

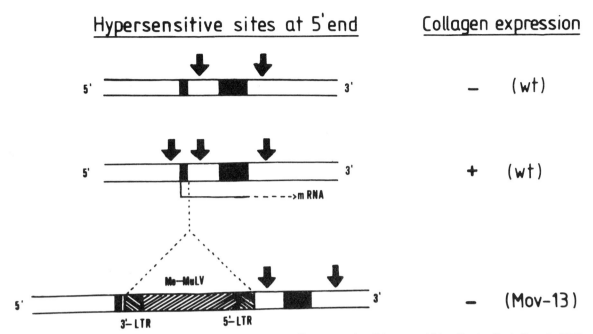

Figure 1. DNase-hypersensitive regions at 5' end of the α1(I) collagen gene in wild-type and Mov-13 mice (Breindl et al. 1984). The black boxes represent exons, the open areas introns, and the hatched areas in the lowest panel represent the Mo-MLV genome with the 3' and 5' LTRs. DNase I-hypersensitive areas are marked by arrows and collagen transcription is indicated by the arrow in the middle panel. (*Upper panel*) Wild-type cells that do not express collagen have two hypersensitive sites. (*Middle panel*) Wild-type cells that express collagen display a third hypersensitive site approximately 200 bp 5' of the cap site. (*Lower panel*) In Mov-13 mice, this transcription-associated site never appears and its absence correlates with gene inactivity.

site appears during embryonic development in cells that activate the type-I collagen gene.

The comparison of the chromatin structure of the wild-type and mutant α1(I) collagen genes showed that virus insertion is associated with a change in chromatin structure as well as with gene inactivity. Restriction enzyme analysis showed that the two transcription-independent hypersensitive sites are present, whereas the transcription-associated site 5' of the gene is absent in the mutant allele. This site is not present in embryonal carcinoma cells and presumably not in early embryos. It must therefore appear some time during normal development, and its absence in the mutant allele suggests that the provirus insertion prevents its developmentally regulated appearance in Mov-13 embryos and thereby interferes with proper activation of the gene.

Virus insertion causes de novo methylation of flanking host sequences. Proviruses carried in the germ line of Mov substrains of mice are highly methylated and not expressed in the tissues analyzed (Stuhlmann et al. 1981; Jaenisch et al. 1985; for review see Jähner and Jaenisch 1984). A causal relationship between hypermethylation and gene inactivity has been established for retroviral genomes by molecular cloning and transfection experiments (Harbers et al. 1981; Simon et al. 1983). Although cells of midgestation embryos or somatic cells are highly permissive for retrovirus expression (Jaenisch 1980), viruses cannot replicate in cells of the preimplantation embryo (Jaenisch et al. 1975) or in embryonic carcinoma cells (Teich et al. 1977). Nonpermissivity of preimplantation mouse embryos (Jähner et al. 1982) or of embryonal carcinoma cells (Stewart et al. 1982; Gautsch and Wilson 1983; Niwa et al. 1983) correlates with the presence of a de novo methylation activity that is a characteristic of early embryonic cells and that efficiently methylates exogenous DNA de novo after genomic integration. De novo methylation is not observed after introduction of retroviruses into cells of the postimplantation embryo or into embryonal carcinoma cells that have been induced to differentiate. Furthermore, it has been shown using a transient expression assay, that the enhancer of the viral long terminal repeat (LTR) does not function in embryonic carcinoma cells but functions efficiently in somatic cells (Linney et al. 1984). Thus de novo methylation and a different requirement for gene transcription represent two parameters that distinguish embryonic from somatic cells and that appear to be involved in nonpermissivity of embryonic cells for retrovirus replication.

To investigate whether proviral insertion can change the methylation pattern of flanking host sequences, we have compared the methylation status of the normal α1(I) collagen with that of the mutant allele in Mov-13 mice. Figure 2 summarizes the results (Jähner and Jaenisch 1985a). All testable CpG sites of the wild-type α1(I) collagen gene are unmethylated in sperm, in day-12 embryos, and in different tissues of the adult animal. Provirus integration into the first intron resulted in complete methylation of all sites within 1 kb of sequences flanking the insertion sites when somatic cells

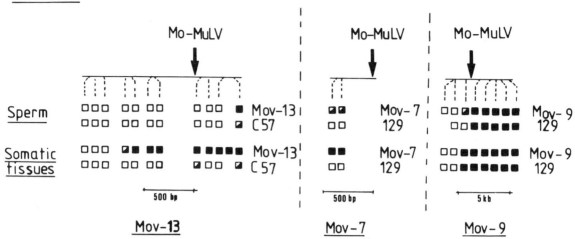

Figure 2. Retrovirus-induced de novo methylation of adjacent cellular sequences (Jähner and Jaenisch 1985a). The methylation status of host sequences flanking the provirus in Mov-7, Mov-9, and Mov-13 and of the sequences of the preintegration site in the respective parental strains (129 and C57B1/6, respectively) was analyzed by restriction mapping. The DNA was isolated from sperm and from different tissues of day-12 embryos, day-17 embryos, and adult mice. Filled, half-filled, or open boxes indicate greater than or equal to 80%, 20–80%, or less than 20% methylation, respectively, of the individual sites analyzed.

were analyzed. The methylation status of sites that were further distant than 1 kb was not affected by provirus integration. In sperm, however, the sequences flanking the Mov-13 provirus were completely unmethylated, indicating that virus insertion induced de novo methylation of unique cellular sequences during embryogenesis. So far, de novo methylation has been observed only in repetitive sequences (Chapman et al. 1984; Ponzetto-Zimmermann and Wolgemuth 1984) or with proviral (Jähner and Jaenisch 1985b) or cellular sequences (Palmiter et al. 1982) introduced into transgenic mice. The lower level of methylation in sperm than in somatic cells of Mov-13 mice may either indicate germ line-specific lack of de novo methylation or specific demethylation during spermatogenesis at these sites. The relative hypomethylation of satellite DNA in extraembryonic versus the embryo tissues (Chapman et al. 1984) is consistent with the notion that de novo methylation occurs after the blastocyst stage, when these cell lineages have separated. Similarly to Mov-13, provirus insertion at the *Mov-7* locus induced de novo methylation of flanking host sequences, which are unmethylated in the parental strain (Fig. 2). In contrast to Mov-7 and Mov-13, the provirus in Mov-9 mice has integrated into a hypermethylated region, the methylation status of which was not detectably affected by the insertion. Similar results were found for Mov-2, Mov-3, and Mov-10 mice (D. Jähner, unpubl.).

De novo methylation of retroviral sequences introduced into mouse zygotes, early mouse embryos, or embryonic carcinoma cells correlates with transcriptional inactivity, whereas lack of de novo methylation in somatic cells correlates with gene expression (Jaenisch and Jähner 1984). Our results show that a functional provirus becomes repressed even when integrated into a chromosomal region which, as in Mov-13 mice, has all characteristics of active gene expression, i.e., hypomethylation and the potential to develop an open chromatin configuration (Breindl et al. 1984). Thus the integration of retroviruses may lead to altered chromatin structures that allow de novo methylation. The molecular basis for both provirus and host gene suppression may be the maintenance of these structures due to their methylation. Nonretroviral genes that have been introduced into the germ line of mice are often not expressed (Palmiter et al. 1982; Lacy et al. 1983; Stewart et al. 1984). It remains to be seen whether this correlates with similar changes in chromatin structure and methylation pattern, as described here for Mov mice.

Role of Type-I Collagen in Early Embryonic Development

Type-I collagen, as the major protein of connective tissues, is likely to have a variety of functions in the organization of the vertebrate body. That it is crucial for the mechanical stability of bones is illustrated in patients with osteogenesis imperfecta, a severe genetic disease caused by mutations in type-I collagen genes (McKusick 1983). Furthermore, as a major component of the extracellular matrix, type-I collagen has been proposed to have an important function in mesenchymal–epithelial interactions during organogenesis (Bernfield 1980; Hay 1981). Experimental evidence for such a function includes studies in which morphogenesis in organ culture was shown to be disturbed by collagenase (Grobstein and Cohen 1965) or by drugs that interfere with collagen as well as protein synthesis (Spooner and Faubion 1980). The availability of a molecularly defined mouse mutant provides the opportunity to study functionally the role of type-I collagen in mammalian development without having to rely on the use of inhibitors that are likely to have a wide variety of effects on embryo development. Therefore, we analyzed homozygous embryos for pathological events that precede their sudden death (Löhler et al. 1984).

The majority of homozygous embryos die between day 13 and day 14 of gestation. Most organ rudiments are already laid down on or before day 12 of gestation, and growth and histomorphogenesis, rather than determinative events, are characteristic of the subsequent days of mouse development (Rugh 1968). While type-I collagen is already detected in day-8 embryos by immunohistochemical techniques, abundant transcription, which may be related to the formation of mesenchyme, begins at day 12 (Shnieke et al. 1983a) and thus coincides with the death of homozygous embryos. This suggests that type-I collagen has no important role in organ determination but rather is essential for the growth phase subsequent to day 12 of gestation, a period during which its expression increases substantially in normal embryos. Immunohistochemical techniques to detect other components of the extracellular matrix (ECM), such as collagen types II, III, IV, laminin, and fibronectin, demonstrated that the blockage of type-I collagen synthesis does not affect the synthesis or distribution of the other ECM components in the mutant embryo (Löhler et al. 1984).

Histologically, homozygous embryos at day 12 showed two distinct alterations: a progressive necrosis of erythropoietic and mesenchymal cells followed by rupture of major blood vessels, which causes sudden death by breakdown of the circulatory system (Löhler et al. 1984). The active hematopoietic proliferation at day 12 of gestation leads to a rapid expansion of the blood volume, and it is likely that the blood vessels in homozygous embryos that lack type-I collagen cannot resist the ensuing increase of blood pressure. Whereas type-III collagen as the major collagen present in the cardiovascular system is thought to provide strength and elasticity, the Mov-13 phenotype ascribes also to type-I collagen an important role for establishing and maintaining the mechanical stability of the circulatory system.

To study the role of type-I collagen in epithelial branching morphogenesis during the formation of embryonic organs, organ rudiments from homozygous embryos were explanted in organ culture. In vitro development of all organ types tested, including lung, kidney, salivary gland, mammary buds, and skin, was perfectly normal despite the absence of type-I collagen (K.D. Kratochwil et al., in prep.). These observations indicate that type-I collagen has either no essential role for early morphogenesis or that its function can be replaced by other interstitial collagens.

Insertional Mutagenesis by Retroviruses and Recombinant DNA

Table 3 summarizes currently known mutations in mice that were caused by insertional mutagenesis. For comparison, some retrovirus-induced mutations obtained in tissue culture cells are included in the table. The published information allows the conclusion that DNA microinjected into the zygote pronucleus is highly mutagenic and mutations can be estimated realistically to occur in 10–20% of all transgenic mouse strains generated. In contrast, none of the 15 Mov mouse strains that were derived from virus-exposed preimplantation

Table 3. Insertional Mutagenesis in Mice

Insertional mutagen	Developmental stage of exposure	Frequency of mutation induced	Affected gene	Insertion of provirus	Mutant phenotype	References[a]
Retrovirus	preimplant embryo	0/15	—	—	—	1
	midgestation embryo	1/1	α1(I) collagen	first intron	embryonic lethal	1
	spontaneous mutation	?	?	noncoding	coat color (d)	2
	tissue culture cells	low	src	first intron	reversion of transformation	3
		?	p-53	first intron	tumor rejection	4
		?	x-light chain	first intron	decreased IgG synthesis	5
		low	β-microglobulin	first exon or intron	no protein expression	6
		low	hprt	intron	HPRT⁻	7
DNA	zygote pronucleus (microinjection)	10–20%	?		embryonic lethal	8
			?		transmission distortion	9
			?		limb deformity	10

[a]1. Jaenisch et al. 1983; 2. Hutchison et al. 1984; 3. Varmus et al. 1981; 4. Wolf and Rotter 1984; 5. Kuff et al. 1983; 6. T.V. Rajan, pers. comm.; 7. King et al. 1985; 8. Wagner et al. 1983; 9. Palmiter et al. 1984; 10. P. Leder, pers. comm.

embryos showed evidence for a mutant phenotype. The only mutant obtained by experimental manipulation, the Mov-13 mouse strain, was obtained by infecting a primordial germ cell at the midgestation stage. The mode of infection by which the *dilute* mutation (Jenkins et al. 1981) was generated is not known.

Information on the possible mechanism(s) involved in insertional mutagenesis has so far been obtained only for retrovirus-induced mutations. In the most extensively studied mutant, the Mov-13 mouse strain, the virus has inserted into noncoding sequences in the first intron of the collagen gene, which results in a block of gene transcription. Molecular analyses demonstrated two virus-induced alterations of the mutated gene: (1) the prevention of a DNase I-hypersensitive and transcription-associated site to appear during embryonic development (Fig. 1); (2) de novo methylation of sequences flanking the proviral insertion (Fig. 2). Both of these changes are associated with gene inactivity rather than with active gene transcription.

It is remarkable that retroviruses have inserted at the 5' end of a given gene, most frequently into the first intron in almost all known cases of insertional mutagenesis. A Mo-MLV proviral copy was found in the "intron" of the Rous sarcoma virus (RSV) genome in two revertants of RSV-transformed cells (Varmus et al. 1981). Likewise, a provirus copy has mutated the p53 gene (Wolf and Rotter 1984), the IgG gene (Kuff et al. 1983), and the β-microglobulin gene (T.V. Rajan, pers. comm.) by insertion into sequence at the 5' end of the respective gene. Insertional mutation of the *hprt* gene involved insertion of a Mo-MLV provirus into an intron in the body of this gene (King et al. 1985). Furthermore, the provirus that caused the *dilute* mutation has integrated into noncoding sequences (Hutchison et al. 1984). It thus appears that proviral copies of spontaneously induced mutations most frequently cluster at the 5' end of the affected gene, which may suggest that retroviruses do not randomly integrate but have a preference for chromosomal regions with an open chromatin conformation.

The molecular anatomy of genes that have been mutated by recombinant DNA has not yet been elucidated in a single instance. The remarkably high frequency of mutations induced by DNA injection into the zygote pronucleus (Wagner et al. 1983; Palmiter et al. 1984), as opposed to retrovirus-induced mutations (Table 3), may suggest, however, that introduction of DNA may induce major rearrangements or deletions that cause mutations in the host cell. The use of insertional mutagenesis by retroviruses or by DNA injection is of great potential for dissecting genetic mechanisms of mammalian development, and elucidating the underlying molecular events remains an important challenge.

ACKNOWLEDGMENTS

This work was supported by grants from the Stiftung Volkswagenwerk, the Deutsche Forschungemeinschaft, the National Institutes of Health, and the National Cancer Institute.

REFERENCES

BERNFIELD, M. 1980. Organization and remodeling of the extracellular matrix in morphogenesis. In *Development and pattern formation* (ed. Conelly et al.), p. 139. Raven Press, New York.

BREINDL, M., K. HARBERS, and R. JAENISCH. 1984. Retrovirus-induced lethal mutation in collagen I gene of mice is associated with altered chromatin structure. *Cell* 38: 9.

CHAPMAN, V., L. FORRESTER, J. SANFORD, N. HASTIE, and J. ROSSANT. 1984. Cell lineage-specific undermethylation of mouse repetitive DNA. *Nature* 307: 284.

ELGIN, S. 1981. DNase I-hypersensitive sites of chromatin. *Cell* 27: 413.

FIEDLER, W., P. NOBIS, D. JAHNER, and R. JAENISCH. 1982. Differentiation and virus expression in BALB/Mo mice: Endogenous Moloney leukemia virus is not activated in hematopoietic cells. *Proc. Natl. Acad. Sci.* 79: 1874.

GAUTSCH, J.W. and M.C. WILSON. 1983. Delayed *de novo* methylation in teratocarcinoma suggests additional tissue-specific mechanisms for controlling gene expression. *Nature* 301: 32.

GLÜECKSOHN-WAELSCH, S. 1979. Genetic control of morphogenetic and biochemical differentiation: Lethal albino deletions in the mouse. *Cell* 16: 225.

GROBSTEIN, C. and J. COHEN. 1965. Collagenase: Effect on the morphogenesis of embryonic salivary epithelium *in vitro*. *Science* 150: 626.

HARBERS, K., M. KUEHN, H. DELIUS, and R. JAENISCH. 1984. Insertion of retrovirus into the first intron of alpha 1(I) collagen gene leads to embryonic lethal mutation in mice. *Proc. Natl. Acad. Sci.* 81: 1504.

HARBERS, K., A. SCHNIEKE, H. STUHLMANN, D. JÄHNER, and R. JAENISCH. 1981. DNA methylation and gene expression: Endogenous retroviral genome becomes infectious after molecular cloning. *Proc. Natl. Acad. Sci.* 78: 7609.

HAY, E. 1981. Collagen and embryonic development. In *Cell biology of the extracellular matrix* (ed. E. Hay), p. 379. Plenum Press, New York.

HUTCHISON, K., N. COPELAND, and N. JENKINS. 1984. Dilute-coat-color locus of mice: Nucleotide sequence analysis of the d+23 and d+Ha revertant allele. *Mol. Cell. Biol.* 4: 2899.

JAENISCH, R. 1976. Germ line integration and Mendelian transmission of the exogenous Moloney leukemia virus. *Proc. Natl. Acad. Sci.* 73: 1260.

———. 1980. Retroviruses and embryogenesis: Microinjection of Moloney leukemia virus into midgestation mouse embryos. *Cell* 19: 181.

———. 1983. Endogenous retroviruses. *Cell* 32: 5.

JAENISCH, R. and D. JÄHNER. 1984. Methylation expression and chromosomal position of genes in mammals. *Biochim. Biophys. Acta* 782: 1.

JAENISCH, R., H. FAN, and B. CROKER. 1975. Infection of preimplantation mouse embryos and of newborn mice with leukemia virus: Tissue distribution of viral DNA and RNA and leukemogenesis in the adult animal. *Proc. Natl. Acad. Sci.* 72: 4008.

JAENISCH, R., A. SCHNIEKE, and K. HARBERS. 1985. Treatment of mice with 5-azacytidine efficiently activates silent retroviral genomes in different tissues. *Proc. Natl. Acad. Sci.* 82: 1451.

JAENISCH, R., D. JÄHNER, P. NOBIS, I. SIMON, J. LÖHLER, K. HARBERS, and D. GROTKOPP. 1981. Chromosomal position and activation of retroviral genomes inserted into the germ line of mice. *Cell* 24: 519.

JAENISCH, R., K. HARBERS, A. SCHNIEKE, J. LÖHLER, I. CHUMAKOV, D. JÄHNER, D. GROTKOPP, and E. HOFFMANN. 1983. Germline integration of Moloney murine leukemia virus at the Mov13 locus leads to recessive lethal mutation and early embryonic death. *Cell* 32: 209.

JÄHNER, D. and R. JAENISCH. 1980. Integration of Moloney leukemia virus into the germ line of mice: Correlation between site of integration and virus activation. *Nature* 287: 456.

———. 1984. DNA methylation in early mammalian development. In *DNA methylation* (ed. A. Razin et al.), p. 189. Springer-Verlag, New York.

———. 1985a. Retrovirus induced *de novo* methylation of flanking host sequences correlates with gene inactivity. *Nature* 315: 594.

———. 1985b. Chromosomal position and specific demethylation of enhancer sequences in germ line transmitted genomes during mouse development. *Mol. Cell. Biol.* (in press).

JÄHNER, D., K. HAASE, R. MULLIGAN, and R. JAENISCH. 1985. Insertion of the bacterial gpt gene into the germ line of mice via retrovirus infection. *Proc. Natl. Acad. Sci.* (in press).

JÄHNER, D., H. STUHLMANN, C.L. STEWART, K. HARBERS, J. LÖHLER, I. SIMON, and R. JAENISCH. 1982. De novo methylation and expression of retroviral genomes during mouse embryogenesis. *Nature* 298: 623.

JENKINS, N., N. COPELAND, B. TAYLOR, and B. LEE. 1981. Dilute (d) coat colour mutation of DBA/2J mice is associated with the site of integration of an ecotropic MuLV genome. *Nature* 293: 370.

KING, W., M. PATEL, L. LOBEL, S. GOFF, and C. NGUYEN-HUU. 1985. Insertion mutagenesis of embryonal carcinoma cells by retroviruses. *Science* 228: 554.

KUFF, E.L., A. FEENSTRA, K. LUEDERS, L. SMITH, R. HAWLEY, N. HOZUMI, and M. SHULMAN. 1983. Intracisternal A-particle genes as movable elements in the mouse genome. *Proc. Natl. Acad. Sci.* 80: 1992.

LACY, E., S. ROBERTS, E.P. EVANS, D. BURTENSHAW, and F.D. COSTANTINI. 1983. A foreign beta-globin gene in transgenic mice: Integration at abnormal chromosomal positions and expression in inappropriate tissues. *Cell* 34: 343.

LINNEY, E., B. DAVIS, J. OVERHAUSER, E. CHAO, and H. FAN. 1984. Non-function of a Moloney murine leukemia virus regulatory sequence in F9 embryonal carcinoma. *Nature* 308: 470.

LÖHLER, J., R. TIMPL, and R. JAENISCH. 1984. Embryonic lethal mutation in mouse collagen I gene causes rupture of blood vessels and is associated with erythropoietic and mesenchymal cell death. *Cell* 38: 597.

MCKUSICK, V. 1983. *Mendelian inheritance in man*. Johns Hopkins University Press, Baltimore, Maryland.

NIWA, O., Y. YOKOTA, H. ISHIDA, and T. SUGAHARA. 1983. Independent mechanisms involved in suppression of the Moloney leukemia virus genome during differentiation of murine teratocarcinoma cells. *Cell* 32: 1105.

PALMITER, R.D., H.Y. CHEN, and R.L. BRINSTER. 1982. Differential regulation of metallothionein-thymidine kinase fusion genes in transgenic mice and their offspring. *Cell* 29: 701.

PALMITER, R.D., T.M. WILKIE, H.Y. CHEN, and R.L. BRINSTER. 1984. Transmission distortion and mosaicism in an unusual transgenic mouse pedigree. *Cell* 36: 869.

PONZELTO-ZIMMERMAN, C. and D.J. WOLGEMUTH. 1984. Methylation of satellite sequences in mouse spermatogenic and somatic DNAs. *Nucleic Acids Res.* 12: 2807.

RUGH, R. 1968. *The mouse: Its reproduction and development*. Burgess, Minneapolis, Minnesota.

RUSSELL, E.S. 1979. Hereditary anemias of the mouse: A review for geneticists. *Adv. Genet.* 20: 357.

SCHNIEKE, A., K. HARBERS, and R. JAENISCH. 1983a. Embryonic lethal mutation in mice induced by retrovirus insertion into the alpha 1(I) collagen gene. *Nature* 304: 315.

SCHNIEKE, A., H. STUHLMANN, K. HARBERS, I. CHUMAKOV, and R. JAENISCH. 1983b. Endogenous Moloney leukemia virus in nonviremic Mov substrains of mice carries defects in the proviral genome. *J. Virol.* 45: 505.

SILVER, L.M. 1981. Genetic organization of the mouse *t* complex. *Cell* 27: 239.

SIMON, D., H. STUHLMANN, D. JÄHNER, H. WAGNER, E. WERNER, and R. JAENISCH. 1983. Retrovirus genomes methylated by mammalian but not bacterial methylase are non-infectious. *Nature* 304: 275.

SPOONER, B. and J. FAUBION. 1980. Collagen involvement in branching morphogenesis of embryonic lung and salivary gland. *Dev. Biol.* 77: 84.

STEWART, T.A., P.K. PATTENGALE, and P. LEDER. 1984. Spontaneous mammary adenocarcinomas in transgenic mice that carry and express MTV/myc fusion genes. *Cell* 38: 627.

STEWART, C., K. HARBERS, D. JÄHNER, and R. JAENISCH. 1983. X-chromosome-linked transmission and expression of retroviral genomes microinjected into mouse zygotes. *Science* 221: 760.

STEWART, C.L., H. STUHLMANN, D. JÄHNER, and R. JAENISCH. 1982. De novo methylation, expression, and infectivity of retroviral genomes introduced into embryonal carcinoma cells. *Proc. Natl. Acad. Sci.* 79: 4098.

STUHLMANN, H., D. JÄHNER, and R. JAENISCH. 1981. Infectivity and methylation of retroviral genomes is correlated with expression in the animal. *Cell* 26: 221.

TEICH, N.M., R.A. WEISS, G.R. MARTIN, and D.R. LOWY. 1977. Virus infection of murine teratocarcinoma stem cell lines. *Cell* 12: 973.

VARMUS, H.E., N. QUINTRELL, and S. ORTIZ. 1981. Retroviruses as mutagens: Insertion and excision of a nontransforming provirus alter expression of a resident transforming provirus. *Cell* 25: 23.

WAGNER, E.F., L. COVARRUBIAS, T.A. STEWART, and B. MINTZ. 1983. Prenatal lethalities in mice homozygous for human growth hormone gene sequences integrated in the germ line. *Cell* 35: 647.

WEISBROD, S. 1982. Active chromatin. *Nature* 297: 289.

WOLF, D. and V. ROTTER. 1984. Inactivation of p53 gene expression by an insertion of Moloney murine leukemia virus-like sequences. *Mol. Cell. Biol.* 4: 1402.

Early Developmental Mutations Due to DNA Rearrangements in Transgenic Mouse Embryos

L. COVARRUBIAS, Y. NISHIDA, AND B. MINTZ
Institute for Cancer Research, Fox Chase Cancer Center, Philadelphia, Pennsylvania 19111

The integration of exogenous DNA into the genome of a mammalian cell would be expected—despite limited knowledge of the recombinatorial mechanisms—to have an intrinsic potential for generating change in the donor or host DNA sequences. The changes might entail tandem duplications, deletions, other rearrangements, such as inversions or translocations, or even base substitutions.

When the recipient cell is the fertilized egg rather than a somatic cell, the consequences of such a mutation may be far-reaching, especially if the change is deleterious. Not only can the new genotype be replicated and found in all somatic cells of the animal derived from that egg; it can be transmitted to the next generation through the germ cells. A mutation initially present in the heterozygous state could appear in the homozygous form of a percentage of progeny descended from a pair of heterozygotes. Thus, a transgenic individual derived from an egg with integrated exogenous (proviral or recombinant) DNA would have at risk any cell type(s) in which the relevant genetic region is ordinarily expressed, if the mutation is a dominant or codominant one, and could generate handicapped or inviable homozygotes, if the mutation is recessive.

There is abundant evidence that DNA that is transfected or injected, with or without viral sequences, into cultured mammalian somatic cells undergoes mutation at a high frequency, up to approximately 1% per gene (Lebkowski et al. 1984; Miller et al. 1984; Mounts and Kelly 1984). The mutations are chiefly base substitutions and deletions in the donor DNA, but they also include insertions from the host genome. Cotransfected DNA species can themselves become covalently linked by homologous recombination or by blunt-end ligation, and may experience rearrangement (Wigler et al. 1978, 1979; Winocour and Keshet 1980; Folger et al. 1982).

After microinjection of plasmid DNA into a pronucleus of the fertilized mouse egg, donor DNA isolated from the resultant transgenic mice has often been found, in Southern blot analyses, to have undergone tandemization (e.g., some cases in Brinster et al. 1981 and in Wagner et al. 1981). It is also apparent from Southern blots in virtually all published studies of transgenic mice, irrespective of the type of DNA injected, that there are instances of rearrangements involving host flanking regions. This can be deduced partly from the presence of extra fragments, in addition to the expected number of flanking sequences, that hybridize to the plasmid probe. We therefore undertook an investigation of the occurrence of insertional mutagenesis following injection of recombinant DNA into the mouse egg (Wagner et al. 1983); the developmental and molecular characterizations of two independent insertional mutations are presented here. There have been only two previous germ line cases of mutations examined in the mouse; both were due to integrated retroviruses. One resulted from an experimental early infection with murine leukemia virus whose disruption of the native α1 collagen gene led to lethality of homozygous fetuses (Jaenisch et al. 1983; Schnieke et al. 1983). The other has apparently occurred spontaneously by retroviral integration near the *dilute* pigment gene locus (Jenkins et al. 1981; Hutchison et al. 1984).

RESULTS

Prenatal Lethalities among Transgenic Mice Homozygous for Human Growth Hormone Gene Sequences

The phGH plasmid used in these experiments contained the human growth hormone gene subcloned as a 2.6-kb *Eco*RI fragment into the *Eco*RI site of the 4.3-kb pBR322 vector (Fiddes et al. 1979). Of 20 animals born from (C3H × C57BL/6)F$_1$ fertilized eggs injected in the male pronucleus with approximately 6000 copies of the plasmid DNA, six mice (five males and one female) were positive for the donor sequences in tissue samples tested in Southern blots (Wagner et al. 1983). All also contained donor DNA in their germ cells. The six positive animals, each heterozygous for the foreign sequences (*HGH*/+), became the founders of six new strains, designated HUGH/1 through HUGH/6. These were derived by backcrossing to wild-type (+/+) C57BL/6 followed by brother-sister *HGH*/+ matings. The founder of one of the strains (HUGH/5) subsequently proved to have had two separate integrations of the plasmid DNA and HUGH/5 was ultimately replaced by two newer strains, each based on segregation of one or the other integration site. Copy numbers varied from a single-copy integration (in HUGH/2) to approximately 20 (in HUGH/1).

Homozygotes (*HGH*/*HGH*) were distinguished from heterozygotes in Southern blots by inclusion of an internal single-copy mouse gene as a standard in the hybridization mixture. Viable and apparently healthy *HGH/HGH* homozygotes were identified in most of the strains, based on band intensities and comparison with the constant internal controls.

In striking contrast to the other strains, matings between *HGH/+* heterozygotes failed to produce any *HGH/HGH* postnatal homozygotes among 31 progeny tested in the HUGH/3 strain and 52 animals tested in the HUGH/4 strain. Although the DNA tests were conducted on spleen or tail samples taken at 4-6 weeks of age, the low incidence of deaths between birth and sampling meant that those losses could not account for the missing homozygotes. DNA tests of some of the few dead animals confirmed the absence of homozygotes among them. Homozygosity was therefore lethal at some time before birth in each of these two strains.

The average litter size at birth was reduced in *HGH/+ × HGH/+* matings (2.9 in HUGH/3 and 5.4 in HUGH/4), in comparison with *HGH/+ × +/+* matings (4.7 in HUGH/3 and 6.9 in HUGH/4). This was expected in view of the absence of the homozygous class. However, there is no obvious explanation for the fact that the average litter size in HUGH/3 was distinctly below normal, even from *HGH/+ × +/+* matings. Inasmuch as virtually all these matings involved heterozygous males and wild-type females, the small litter size cannot be attributed to gestational problems of *HGH/+* females. In both HUGH/3 and HUGH/4, the normal viability of heterozygotes was indicated by the normal incidence of this class in backcross or heterozygous matings.

Interstrain matings between heterozygotes of HUGH/3 and HUGH/4 produced healthy individuals with the distinctive hybridization patterns of both strains present concurrently (as well as animals with the separate strain patterns) in Southern blots after *Eco*RI digestion. The double heterozygotes therefore confirm that there are probably different defects caused by different integrations in HUGH/3 and HUGH/4.

Early Postimplantation Developmental Arrests in Homozygous Mutant Embryos

Homozygous lethality in HUGH/3. On day 3 of gestation (counting the vaginal plug date as day 0), all 23 embryos obtained from matings between *HGH/+* heterozygotes of the HUGH/3 strain appeared to be normal blastocysts. Dissection and histological examination of a series of later stages revealed that the primary defect, accounting at first for developmental arrest and death of approximately 25% of the embryos, occurred in the day-4 to -5 period, shortly after implantation. Through at least day 12 of gestation, the number of implantation sites (conspicuous decidual swellings) continued to be normal, with an average of 8.3 per pregnant female, based on 166 implantation sites found in a total of 20 litters. This attested to the success of implantation, even in *HGH/HGH* homozygotes. However, upon further inspection, dead or dying embryos were observed in some of the implantation sites on days 4-5; and embryos were largely absent on days 6-8 in 17 (24%) out of the 70 decidual swellings. Thus, the primary defect caused by the phGH integration in the HUGH/3 strain was manifest in the egg cylinder stage, the time when many morphogenetic changes ordinarily begin to be seen.

While losses through day 8 account for the deaths of the expected 25% homozygous mutant segregants, they do not explain the fact (noted above) that the litter size at birth in these matings is approximately three rather than six, as would have been the case if only the *HGH/HGH* embryos died. The basis for the disparity was found at subsequent stages of gestation: Further losses of roughly a similar magnitude occurred, starting chiefly on day 9, bringing the total subsequent deaths up to 59% (33 of 56) by day 12. From these data, a litter size of three would in fact be expected at birth. These "second-stage" deaths were characterized by a set of gross morphological abnormalities. The *HGH* genotype of the dead embryos has not been analyzed and the phenomenon remains unexplained. It is unlikely that an unusually high incidence of the *HGH/HGH* type, and therefore an aberrant *HGH* transmission ratio, is at issue, because, as already indicated, an apparently normal incidence of *HGH/+* was obtained from backcrosses of HUGH/3 heterozygotes to wild-type.

Homozygous lethality in HUGH/4. Early postimplantation lethality of homozygotes also occurred in HUGH/4 and proved to be remarkably similar to the result in HUGH/3, although the respective mutations are attributable to entirely different patterns of donor-DNA integration and probably also to insertion on different chromosomes (data not shown). In HUGH/4, 23/68 (34%) of the implantation sites visible through day 12 of pregnancy had dead or missing embryos. Deaths again occurred in the egg cylinder stage. However, no further losses were found at later stages.

DNA Integration Patterns in the Mutant Strains

We have shown previously that the integration patterns of phGH sequences in HUGH/3 and HUGH/4 are quite complex (Wagner et al. 1983). Information on the physical state of the vector sequences was obtained from restriction enzyme analyses of single and double digests together with the use of different hybridization probes. In addition to the entire phGH plasmid probe, the isolated 2.6-kb *Eco*RI fragment containing the human growth hormone gene or pBR322 DNA alone were used to define the origin of particular fragments. Genomic DNA from these strains digested with *Eco*RI hybridized to the phGH DNA probe in eight fragments of different sizes from HUGH/3 and four from HUGH/4. The 2.6-kb and 4.3-kb *Eco*RI fragments that hybridize to phGH most intensely contained multiple (presumably tandemly arranged) copies of the plasmid sequences. The remaining fragments would be expected to contain the junctions between mouse and plasmid sequences. The possibility of changes in these mouse flanking DNAs, as well as the overall configuration of the insert, were further examined in clonal isolates.

Clones of the Flanking Regions

EcoRI-digested fractions of HUGH/3 and HUGH/4 genomic DNAs containing the putative flanking regions were ligated to λgtWES EcoRI arms and a library was prepared and screened for clones hybridizing with a nick-translated phGH probe. From HUGH/3, four different positive clones were isolated (λHUGH/3-1, 3-3, 3-4, and 3-5). From HUGH/4, one positive clone was isolated (λHUGH/4-1). The one EcoRI fragment from each clone that hybridized to phGH was subcloned into the EcoRI site of pUC9. The resultant subclones are designated pHUGH/3-1, pHUGH/3-3, etc. Each fragment included murine DNA sequences as well as either hGH or pBR322 sequences bordering an EcoRI site. The flanking regions and their tentative arrangement in the genomic DNA of HUGH/3 and HUGH/4 are depicted in Figure 1.

Under hybridization conditions (Steinmetz et al. 1980) that revealed sequences present at more than 100 copies per genome, repetitive sequences were found to be present in some of the clones (data not shown). The pHUGH/3-1 fragment lacked repetitive sequences in its 5–6 kb of mouse DNA. pHUGH/3-3 contained a member of the B1 family (Krayev et al. 1982) of repetitive DNA. The remaining fragments (pHUGH/3-4, 3-5, and 4-1) included unidentified repetitive sequences. The 3.2-kb StuI–EcoRI fragment from pHUGH/3-3 and the 2.2-kb HindIII–EcoRI fragment from pHUGH/4-1, both of which lacked repetitive sequences, were subcloned into pUC9, yielding the pHUGH/3-3Δ and pHUGH/4-1Δ clones, respectively.

Wild-type Mouse Sequences Were Rearranged during Integration of Plasmid DNAs

Evidence in the HUGH/3 strain. Approximately three to five total copies of plasmid DNA were integrated. The region of integration in this strain has an unusual structure characterized by the presence of approximately 5 kb of mouse DNA interrupting tandem arrays of the phGH insert in each of two locations (Fig. 1). Further analyses of the mouse DNA-flanking regions in the designated subclones was carried out by probing Southern blots of restricted wild-type genomic DNA with these flanking probes. The data revealed a number of discrepancies between the predicted and the actual results.

An example of a disparity concerns the fact that the mouse DNA in pHUGH/3-1, which has only single internal HindIII and PvuII sites, should hybridize to two of the correspondingly digested fragments from wild-type DNA of the original parental (C3H and C57BL/6) strains. However, only one strongly hybridizing band was found for each enzyme (Fig. 2). Moreover, pHUGH/3-1, which has three internal PstI sites and should therefore hybridize to four restricted wild-type fragments, in fact contains only two hybridizing bands of 1.9 kb and 4.8 kb that are different from the ones expected. When the same filter was exposed for a longer period (as shown in Fig. 2), additional weakly hybridizing bands were seen. Some of these varied between C3H and C57BL/6 DNAs and may have been due to restriction fragment length polymorphisms in the parental strains. It is also possible that some translocation of DNA pieces from other regions or other chromosomes may have been inserted into this segment of mouse DNA flanking the insert, or that there are some other chromosomal regions with sequences homologous to some of those in the probe.

Evidence in the HUGH/4 strain. Approximately three to five copies of plasmid DNA were also integrated in HUGH/4. In this strain there appears to be a single multicopy insertion and no evidence for internal interruptions of plasmid by mouse DNA, based on release of two flanking fragments of 3.4 kb and 2.8 kb, of which 3–5 kb is mouse DNA, after EcoRI digestion (Fig. 1). If HUGH/4 were actually generated by a simple insertion, however, the pHUGH/4-1Δ probe should hybridize to a single 3- to 5-kb EcoRI fragment in wild-type DNA. In fact, this probe hybridizes to two EcoRI fragments (8.6 and 1.6 kb) in wild-type DNA, although the probe has no internal EcoRI site (Fig. 3). Similarly, the same probe hybridizes to three HindIII fragments in wild-type DNA despite the absence of a HindIII site

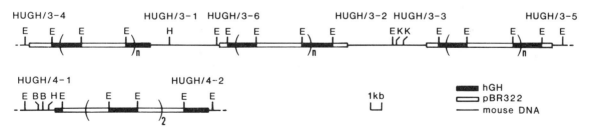

Figure 1. phGH integration maps for the HUGH/3 and HUGH/4 mouse strains. The maps are based on Southern blot analyses of single and double digests hybridized with hGH or pBR322 probes. In HUGH/3 (*above*), up to approximately five copies of phGH sequences are located in three regions (in parentheses), with the flanking EcoRI fragments indicated. In HUGH/4 (*below*), there are at least two tandem intact copies of the phGH plasmid (in parentheses), with the two flanking EcoRI fragments shown. (E), EcoRI; (Bg), BglII; (H), HindIII; (K), KpnI.

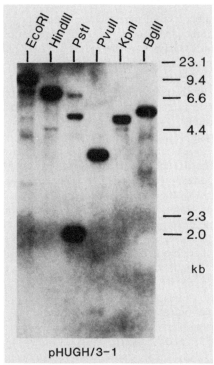

Figure 2. Southern blot hybridization analyses of wild-type mouse genomic DNA after digestion with the restriction enzymes indicated and hybridization with the pHUGH/3-1 flanking sequence probe. Autoradiographic exposure has been prolonged in order to reveal faint bands. The observed numbers and sizes of hybridizable fragments differ from those predicted from the restriction map of the λHUGH/3-1 clone.

in the probe. (There was no hybridization of the phGH probe to the mouse growth hormone gene under the conditions of the experiments, thereby ruling that out as a possible artifact.)

DISCUSSION

The results just described imply that integration of foreign DNA after injection into a pronucleus of the mouse egg was a highly complex event during which some mouse DNA sequences in the target region were deleted and/or otherwise rearranged—apart from any rearrangements that may have occurred in the donor DNA itself. Among the changes, some DNA pieces around the insert that were originally separated by at least 10 kb were brought closer together (data not shown), thereby disrupting an endogenous gene. It seems unlikely that these changes are attributable to the specific plasmid DNA employed here. On the contrary, Southern blot patterns in the literature on transgenic mice, whose donor DNAs come from a wide variety of sources, suggest that rearrangements in host DNA have been relatively commonplace during integration of exogenous genetic material.

A possible causal factor may be damage to the introduced DNA (Miller et al. 1984) and activation of DNA repair mechanisms. In addition, plasmid sequences may

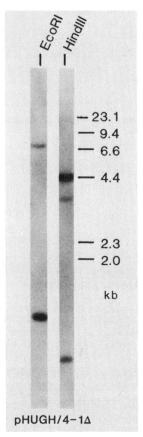

Figure 3. Southern blot hybridization analyses of wild-type mouse genomic DNA after digestion with EcoRI or HindIII and hybridization with the pHUGH/4-1Δ flanking sequence probe. The patterns of hybridizable fragments do not conform to those predicted from the restriction map of the λHUGH/4-1 clone.

become characterized by an unstable structure, for which the commonly observed tandem integrants provide some evidence; this early instability may implicate both insert and target sequences in such a way as to generate changes tending toward a more stable conformation. A question of interest is whether some chromosome regions are more susceptible to exposed DNA ends (of linearized plasmid molecules), and thus are "hot spots" of integration as well as rearrangement; such regions could be inherently more recombinatorial, or even fragile.

In the present study, based on six transgenic founder mice, complex integration patterns occurred not only in the two strains (HUGH/3 and HUGH/4) with insertional recessive lethal mutations, but also in HUGH/5 and HUGH/6 (Wagner et al. 1983) without obvious, if any, phenotypic consequences in the latter cases. If dominant early lethals had occurred in any of the injected embryos, they would not have been distinguishable among the other (technical) losses.

The two independent lethals caused by insertional mutagenesis in HUGH/3 and HUGH/4 constitute the first reported early-stage lethalities due to the introduction of cloned DNA. The sole previously described recessive lethal, which resulted from integration of proviral DNA from a murine leukemia retrovirus (administered after implantation) into the α1 collagen gene (Jaenisch et al. 1983; Schnieke et al. 1983) caused death of homozygotes at a much later stage (day 12), after all the organ primordia are established. Possibly, late actively transcribing genetic regions were more exposed, and more susceptible to DNA integration, in this late-stage viral infection, in contrast to early-stage introduction of recombinant DNA in the present report. The separate occurrence of the HUGH/3 and HUGH/4 recessive lethalities in the early postimplantation period (days 4-5) may also signify that a great many changes dependent upon many newly activated genes are occurring at that time, thereby defining a period of exceptionally great vulnerability to deleterious effects of many sorts of mutations. There are in normal development numerous morphological manifestations of change following implantation. These include the delineation of embryonic and extraembryonic ectoderm, and of proximal and distal endoderm, and the appearance of a proamniotic cavity; the so-called embryonic "ectoderm" is of particular importance because the cells are developmentally totipotent (Diwan and Stevens 1976). It is unlikely that preexisting maternal transcripts (Mintz 1964) can supply all the proteins necessary for these events. An early lethal mutation may create an immediate problem in only a limited area or aspect of embryogenesis, but there are probably many such problems that would entrain secondary consequences due to the interrelations and scaffolding of development, and would ultimately cause death. It should, in fact, be strongly emphasized that an anomaly first seen or detected in a given tissue at a given time during embryonic development does not necessarily signify that the primary site of gene action is in that tissue; less obvious defects may have been expressed sooner and/or in other tissues. Moreover, the expression of the primary defect may have been very limited in time. Despite the great interest in understanding gene control of mammalian development at these early stages, the possibility of very localized or very transitory expression of a given gene, even before gross abnormalities are apparent, and the extremely small amounts of available material thus continue to pose practical problems, e.g., in identification of specific transcripts. Nevertheless, insertional mutagenesis by exogenous DNA integration in the mammalian germ line may be expected to provide an important new tool for isolating and characterizing genes active in development. Some of these genes may well turn out merely to encode de novo products needed for newly involved essential metabolic pathways which, while needed for survival, are not implicated in any peculiarly "developmental" processes. The special promise of experimental insertional mutagenesis, however, is that it may also reveal some genes responsible for the very developmental phenomena that have remained most elusive.

The relative ease with which genetic rearrangements seem to occur under the experimental conditions described here raises the question to what extent such rearrangements occur spontaneously in the mammalian genome. It seems entirely admissible that some (or many) of the known specific allelic series of mutations already catalogued for the mouse, for example, simply represent different rearrangements and/or deletions of DNA sequences within a relatively localized region.

ACKNOWLEDGMENTS

This work was supported by grants HD-10646, CA-06927, and RR-05539 from the U.S. Public Health Service and by an appropriation from the Commonwealth of Pennsylvania.

REFERENCES

Brinster, R.L., H.Y. Chen, M. Trumbauer, A.W. Senear, R. Warren, and R.D. Palmiter. 1981. Somatic expression of herpes thymidine kinase in mice following injection of a fusion gene into eggs. *Cell* **27**: 223.

Diwan, S.B. and L.C. Stevens. 1976. Development of teratomas from the ectoderm of mouse egg cylinders. *J. Natl. Cancer Inst.* **57**: 937.

Fiddes, J.C., P.H. Seeburg, F.M. De Noto, R.A. Hallewell, J.D. Baxter, and H.M. Goodman. 1979. Structure of genes for human growth hormone and chorionic somatomammotropin. *Proc. Natl. Acad. Sci.* **76**: 4294.

Folger, K.R., E.A. Wang, G. Wahl, and M.R. Capecchi. 1982. Patterns of integration of DNA microinjected into cultured mammalian cells: Evidence for homologous recombination between injected plasmid DNA molecules. *Mol. Cell. Biol.* **2**: 1372.

Hutchison, K.W., N.G. Copeland, and N.A. Jenkins. 1984. Dilute-coat-color locus of mice: Nucleotide sequence analysis of the d^{+2J} and d^{+Ha} revertant alleles. *Mol. Cell. Biol.* **4**: 2899.

Jaenisch, R., K. Harbers, A. Schnieke, J. Löhler, I. Chumakov, D. Jähner, D. Grotkopp, and E. Hoffman. 1983. Germline integration of Moloney murine leukemia virus at the *Mov13* locus leads to recessive lethal mutation and early embryonic death. *Cell* **32**: 209.

Jenkins, N.A., N.G. Copeland, B.A. Taylor, and B.K. Lee. 1981. Dilute (d) coat colour mutation of DBA/2J mice is associated with the site of integration of ectotropic MuLV genome. *Nature* **293**: 370.

Krayev, A.S., T.V. Markusheva, D.A. Kramerov, A.P. Ryskov, K.G. Skryabin, A.A. Bayev, and G.P. Georgiev. 1982. Ubiquitous transposon-like repeats B1 and B2 of the mouse genome: B2 sequencing. *Nucleic Acids Res.* **10**: 7461.

Lebkowski, J.S., R.B. Du Bridge, E.A. Antell, K.S. Greisen, and M.P. Calos. 1984. Transfected DNA in mutated in monkey, mouse, and human cells. *Mol. Cell. Biol.* **4**: 1951.

Miller, J.H., J.S. Lebkowsi, K.S. Greisen, and M.P. Calos. 1984. Specificity of mutations induced in transfected DNA by mammalian cells. *EMBO J.* **3**: 3117.

Mintz, B. 1964. Synthetic processes and early development in the mammalian egg. *J. Exp. Zool.* **157**: 85.

Mounts, P. and T.J. Kelly, Jr. 1984. Rearrangements of host and viral DNA in mouse cells transformed by simian virus 40. *J. Mol. Biol.* **177**: 431.

Schnieke, A., K. Harbers, and R. Jaenisch. 1983. Embryonic lethal mutation in mice induced by retrovirus insertion into the α1(I) collagen gene. *Nature* **304**: 315.

Steinmetz, M., J. Höchtl, H. Schnell, W. Gebhard, and H.G. Zachau. 1980. Cloning of V region fragments from mouse liver DNA and localization of repetitive DNA sequences in the vicinity of immunoglobulin gene segments. *Nucleic Acids Res.* **8:** 1721.

WAGNER, E.F., T.A. STEWART, and B. MINTZ. 1981. The human β-globin gene and a functional viral thymidine kinase gene in developing mice. *Proc. Natl. Acad. Sci.* **78:** 5016.

WAGNER, E.F., L. COVARRUBIAS, T.A. STEWART, and B. MINTZ. 1983. Prenatal lethalities in mice homozygous for human growth hormone gene sequences integrated in the germ line. *Cell* **35:** 647.

WIGLER, M., A. PELLICER, S. SILVERSTEIN, and R. AXEL. 1978. Biochemical transfer of single-copy eucaryotic genes using total cellular DNA as donor. *Cell* **14:** 725.

WIGLER, M., R. SWEET, G.K. SIM, B. WOLD, A. PELLICER, E. LACY, T. MANIATIS, S. SILVERSTEIN, and R. AXEL. 1979. Transformation of mammalian cells with genes from prokaryotes and eukaryotes. *Cell* **16:** 777.

WINOCOUR, E. and I. KESHET. 1980. Indiscriminate recombination in simian virus 40-infected monkey cells. *Proc. Natl. Acad. Sci.* **77:** 4861.

An Insertional Mutation in a Transgenic Mouse Line Results in Developmental Arrest at Day 5 of Gestation

W.H. MARK, K. SIGNORELLI, AND E. LACY
Graduate Program in Molecular Biology, Memorial Sloan-Kettering Cancer Center, New York, New York 10021

Over the past 50 years many spontaneous and radiation-induced mutations have been identified that disrupt the development of the early mouse embryo (McLaren 1976; Green 1981). For many of these mutations, such as those in the T/t complex (Bennett 1975; Silver 1981), yellow (A^y; Pedersen 1974; McLaren 1976), albino (*c*; Gluecksohn-Waelsch 1979), and oligosyndactyly (*Os*; Van Valen 1966; Magnuson and Epstein 1984), the mutant phenotype has been well characterized and the chromosomal location of the mutated gene(s) genetically mapped. However, in none of these cases, has it yet been possible to analyze molecularly the mutated gene and to identify its protein or RNA product, whose abnormal expression results in developmental arrest. A major impediment to the molecular analysis of these existing mouse mutants has been the relatively low resolution of genetic mapping in the mouse. Although in *Drosophila* several developmentally important genes have been cloned through a combined genetic and molecular approach, the large variety of genetic variants and the detailed chromosomal maps that made this approach feasible in the fly do not exist for most loci in the mouse. Consequently, the potential of such spontaneous and radiation-induced mutations to provide insights into the normal mechanisms of mammalian development has been severely restricted. Recently, several mutations, also affecting embryogenesis in the mouse, have been generated by a new approach, which has the advantage of allowing the molecular cloning and analysis of the mutated loci (Jaenisch et al. 1983; Schnieke et al. 1983; Wagner et al. 1983). This approach involves the production of transgenic mice and exploits the ability of the introduced DNA molecules to serve both as an insertional mutagen and as a probe for molecular cloning.

It is well established that spontaneous mutations in prokaryotes (Bukhari et al. 1977), yeast (Roeder and Fink 1980), *Drosophila* (Zachar and Bingham 1982), and maize (McClintock 1952; Fedoroff 1983) often arise from the insertion of DNA in the form of a transposable element. An altered phenotype results from either the activation (Hayward et al. 1981) or inactivation (Shortle et al. 1982) of a cellular gene by the integrated element. In mice, to date, only one spontaneous mutation has been definitively shown to result from the insertion of DNA; this is the *dilute* (*d*) coat color mutation of DBA/2J mice, which results from the integration of an ecotropic murine leukemia virus (MLV) provirus (Jenkins et al. 1981; Copeland et al. 1983). One unique feature of transposon-induced mutations is that the disrupted genes are effectively tagged by the DNA sequences of the mobile element. Therefore the element can be used as a probe to clone the DNA sequences that flank the site of insertion and that presumably contain the affected gene. Such an approach has been successfully employed to isolate mutant genes and their normal counterparts from bacteria (Calos and Miller 1980), yeast (Roeder et al. 1980), and *Drosophila* (Bingham et al. 1981). With the aid of a MLV-specific probe, sequences from the mouse *dilute* coat color locus have similarly been cloned (Copeland et al. 1983; Hutchison et al. 1984).

Over the past 4 years, methods have been developed by which new mutations can be generated in the mouse by the experimental insertion of exogenous DNA into the germ line (Gordon et al. 1980; Brinster et al. 1981; Costantini and Lacy 1981; Jaenisch et al. 1981; Wagner et al. 1981). These insertional mutations are similar to the transposon-induced mutations in that they will be tagged by the newly introduced exogenous DNA sequences, which can then be used as a molecular probe to clone the disrupted host gene. DNA can be efficiently introduced into the germ line by either of two methods: microinjection of DNA into the pronuclei of fertilized mouse eggs (Costantini and Lacy 1981; Gordon and Ruddle 1981; Palmiter et al. 1982; Stewart et al. 1982) or viral infection of preimplantation embryos (Jahner and Jaenisch 1980; Jaenisch et al. 1981). The exogenous DNA integrates randomly into the host chromosomes and is transmitted through the germ line as a Mendelian trait (Lacy et al. 1983). Mice that contain new genetic information as a result of these manipulations are called transgenic mice (Gordon and Ruddle 1981). Initially, transgenic mice are hemizygous for the integrated foreign DNA. Therefore, a new insertional mutation will be recognized only if its phenotype is dominant and nonlethal. To identify recessive mutations, intercrosses must be set up between heterozygous transgenic mice and their offspring screened for homozygosity with respect to the foreign DNA. If the inserted DNA has disrupted the function of a gene essential for development, no viable homozygotes will be produced. On the other hand, if the inserted DNA has caused a nonlethal recessive mutation, it will only be detected if the homozygous progeny exhibit a readily scorable mutant phenotype. Through such analyses, several new mutations have been identified in transgenic mice. The first was discovered by Jaenisch et al. (1983) in one of 13 transgenic mouse lines carrying exogenous copies of the Moloney murine leukemia provirus. The homozygous embryos in this line become arrested in development between days 11 and 12 of gestation due to the inactivation of the α1(I) collagen gene (Schnieke et al. 1983). Two other recessive prenatal lethal mutations have been detected in transgenic mouse lines generated by the microinjection of the human growth hormone gene (Wagner et al. 1983; Covarrubias et al., this volume). Recently, a recessive nonlethal mutation has been identified in a transgenic line bearing copies of the mouse *myc* gene; this mutation is apparently allelic to a previously defined locus carrying the spontaneous mutation limb deformity (ld^j) (Leder et al., this volume). A dominant nonlethal mutation was discovered in a transgenic line carrying an insertion of a head-to-head dimer of a metallothionein–thymidine kinase fusion gene. Males in this line, though fertile, do not transmit the foreign DNA to progeny; it has been proposed that the foreign DNA insert has disrupted a gene required during the haploid stages of spermatogenesis (Palmiter et al. 1984).

We have recently identified an additional recessive prenatal lethal mutation induced in a transgenic mouse line. This line, referred to as line 4, was generated by the microinjection of a bacteriophage λ recombinant, λRβG2, containing the rabbit adult β-globin gene (Costantini and Lacy 1981; see Fig. 1). In these mice, three to four copies of λRβG2 are integrated as a head-to-tail tandem array on chromosome 3 (Lacy et al. 1983), and the locus defined by this insertion has been designated the Rβ3 locus. An interesting characteristic of the heterozygous mice in line 4 is that they transcribe the rabbit β-globin gene inappropriately but specifically in skeletal muscle. We have previously proposed that this abnormal pattern of expression results from a chromosomal position effect that has placed the rabbit β-globin gene under the control of host cell-specific *cis*-acting regulatory mechanisms operating at the Rβ3 locus (Lacy et al. 1983).

In this paper we report the experiments that identified a recessive prenatal lethal mutation at the Rβ3 locus in line 4 mice. In addition, we establish the stage in development at which the homozygous embryos in line 4 are arresting.

EXPERIMENTAL PROCEDURES

Isolation of DNA from the tail. Animals were anesthetized with ether or by an intraperitoneal injection of Avertin and a 3- to 4-cm piece of tail was cut off with a razor blade. The tail tips were then incubated overnight, with continuous rocking, at 55°C in 0.6 ml of 0.1 M EDTA, 0.05 M Tris-HCl (pH 8), 0.5% SDS, and 1 mg/ml Proteinase K. Following extraction with phenol and phenol/chloroform/iso-amyl alcohol (1:0.96:0.04) the DNA was precipitated by adding NaOAc to 0.3 M and an equal volume of EtOH at room temperature. The precipitated DNAs were washed with 70% EtOH, dried, and then resuspended in 100–400 μl of 0.01 M Tris (pH 8) and 0.001 M EDTA.

Quantitative dot blot hybridization. Duplicate aliquots of 125, 250, and 500 ng of tail DNA from each animal were spotted onto nitrocellulose in the presence of 2 μg of salmon sperm DNA following the procedures of Kafatos et al. (1979). One set of duplicates was hybridized to nick-translated λRβG2 DNA and the other set to nick-translated plivs-1, a cDNA clone of a mouse urinary protein gene (Derman 1981; Derman et al. 1981). The hybridizations were performed as previously described (Costantini and Lacy 1981). The amount of radioactivity hybridized to each dot was determined by scintillation counting and the ratio of hybridized λRβG2/plivs-1 counts calculated for each animal.

The isolation and culture of mouse embryos. Preimplantation and postimplantation embryos were isolated according to the procedures described by Hogan et al. (1986). The nominal age of the embryos was established by the vaginal plug method and the morning the plug was found is defined as day 0.5 of gestation or day 0.5 post coitus (p.c.; Bronson et al. 1968).

Preimplantation embryos were cultured in vitro following the methods of Whittingham (1971) and Hogan et al. (1986). Briefly, mouse embryos were flushed from the oviduct at days 1.5 and 2.5 of gestation using M2 culture medium (Quinn et al. 1982). The embryos were rinsed in M2 and then cultured individually in microdrops of M16 medium (Whittingham 1971) under paraffin oil (Hsu et al. 1974) at 37°C in a CO_2 incubator. When the embryos reached the blastocyst stage, the M16 medium was replaced with Dulbecco's modified minimal essential medium (MEM) (GIBCO) supplemented with 2 mM glutamine, 1 mM pyruvate, 0.1 mM nonessential amino acids (GIBCO), and 10% fetal bovine serum and, subsequently, the culture medium was changed daily. Mouse blastocysts were flushed from the uterus at day 3.5 of gestation using M2 medium and then immediately placed in Dulbecco's MEM for culture.

The embryos were examined and photographed us-

Figure 1. Schematic diagram of the clone λRβG2. Thick lines represent the bacteriophage λCH4A vector sequences and the thin line represents the 19-kb rabbit DNA insert. The solid boxes labeled β1 and ψβ2 are the adult β-globin gene and a β-like pseudogene, respectively.

ing a Nikon Diaphot inverted microscope equipped with Hoffman Modulation Contrast optics.

RESULTS

The Insertion of λRβG2 at the Rβ3 Locus Generated a Recessive Prenatal Lethal Mutation

Five different transgenic mouse lines, carrying λRβG2 sequences at different chromosomal loci, were tested to determine whether viable adult offspring could be produced that were homozygous for the integrated foreign DNA. Intercrosses were set up between heterozygous male and female mice in each of the five lines and their progeny screened by a quantitative dot blot assay to identify animals that were either hemizygous or homozygous for the λRβG2 DNA insert. For this assay, duplicate samples of tail DNA from each progeny were spotted onto a nitrocellulose filter (Kafatos et al. 1979); one set of duplicates was hybridized to ^{32}P-labeled λRβG2 DNA and the other set to a radiolabeled cDNA clone (plivs-1) of a mouse urinary protein gene (MUP) (Derman 1981; Derman et al. 1981). The amount of radioactivity hybridized to each dot was measured by scintillation counting and the ratio of λRβG2/plivs-1 counts determined for each embryo. In this assay, a homozygous mouse will have twice the ratio of hybridized counts compared with a known heterozygous animal.

The results from quantitative dot blot assays on 2- to 5-month-old progeny from lines 3, 4, and 23 are summarized in the histogram shown in Figure 2. The 28 offspring analyzed from line 23 consisted of 19 heterozygous animals, 6 homozygous animals, and 3 wild-type or nontransgenic animals. Of the 56 progeny screened from line 3, 25 were heterozygous, 12 homozygous, and 19 wild type. Similarly, the two lines, which are not included in Figure 2 (lines 38 and 4-12), also yielded viable homozygous offspring. The genotype of the homozygous animals in each of these four lines was subsequently confirmed genetically by mating them to wild-type mice and screening their progeny for λRβG2 sequences. Each animal tested transmitted the foreign DNA to 100% of its progeny, thus establishing the validity of the quantitative dot blot assay.

In contrast, of the 64 progeny screened from line 4, 38 were heterozygous, 26 were wild type, but none were homozygous. As a significant number of deaths did not occur between birth and weaning (only 3 deaths out of 67 births), the integration of λRβG2 at the Rβ3 locus must have caused a recessive prenatal lethal mutation.

Homozygous Embryos in Line 4 Die before 7.5 Days of Gestation

To characterize this prenatal lethal mutation further, a set of experiments was carried out to determine the stage of development at which the line 4 homozygous embryos were arresting. These experiments involved setting up heterozygous intercrosses and examining the resulting embryos at different stages of postimplantation development between 7.5 and 11.5 days p.c. At each day of gestation the uteri were removed from the pregnant females and the implantation sites dissected open to look for the presence or absence of an embryo. Each embryo that was found was photographed and its de-

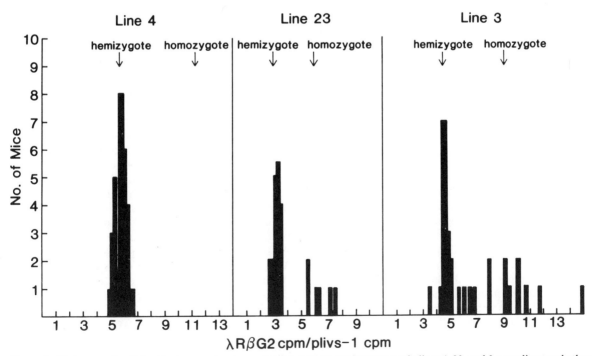

Figure 2. Histogram classifying the transgenic progeny from heterozygous intercrosses in lines 4, 23, and 3 according to whether they are hemizygous or homozygous for the λRβG2 DNA insert.

velopmental age confirmed by comparison to the morphology of normal embryos, which is well documented for all gestational stages (Snell and Stevens 1968; Theiler 1972). In addition, DNA was prepared from each embryo so that its genotype at the Rβ3 locus could subsequently be determined in a quantitative dot blot assay. Matings between nontransgenic litter mates were also set up to determine the frequency at which abnormal development of wild-type embryos occurs between 7.5 and 11.5 days of gestation. The results obtained from these experiments are summarized in Table 1.

A total of 142 implantation sites were examined from the control matings; of these, 135 contained normal embryos, 3, retarded or abnormal embryos, and 4, resorbed embryos. Thus, approximately 5% of the embryos generated in the matings between nontransgenic litter mates developed abnormally. Of the 232 implantation sites analyzed in the heterozygous intercrosses, 87 were classified as containing retarded, abnormal, or resorbed embryos. After correcting for the fraction of embryos that developed abnormally in the control matings, the frequency of abnormal development between 7.5 and 11.5 days of gestation in the experimental matings is 34%. A similarly high frequency of abnormal development, which is primarily due to the resorbed embryos, is seen at each day of gestation in the heterozygous intercrosses. Even as early as 7.5 days, approximately one-quarter of the embryos produced in the heterozygous intercrosses have been resorbed. Therefore most, if not all, of the homozygous embryos in line 4 must be arresting in development before day 8 of gestation. In support of this conclusion, none of the 32 10.5- to 12.5-day-old embryos that were analyzed by a quantitative dot blot assay were homozygous for the λRβG2 insert at the Rβ3 locus (data not shown). (Currently, we are developing a quantitative Southern blotting procedure that will allow us to genotype the embryos between 7.5 and 9.5 days of development.)

The Homozygous Embryos Are Viable during the Preimplantation Stages of Development

Early mouse development can be divided into two stages: preimplantation and implantation. The preimplantation stage takes place within the oviduct between 0.5 and 2.5 days p.c.; during this time the egg cleaves five times to produce a compacted 32-cell morula. By 3.5 days p.c. the embryo has entered the uterus and formed a 64- to 128-cell blastocyst, containing the first two distinct cell lineages: the trophectoderm, which will give rise only to extraembryonic tissues, and the inner cell mass (ICM), which will give rise to the three germ layers of the fetus as well as to extraembryonic tissues (Snell and Stevens 1968; Gardner 1983). On day 4.5 of gestation, the blastocyst hatches from the zona pellucida and begins to implant into the uterine wall. In response to the implanting embryo, the uterus forms a specialized structure known as the decidua (Enders and Schlafke 1969). The observation that approximately one-quarter of the decidua (or implantation sites) produced in the heterozygous intercrosses contained resorbed embryos suggests that the homozygotes in line 4 are capable of triggering a decidual response in the uterus and, thus, that they are viable through the preimplantation stage. As a first step toward testing this prediction, we have examined embryos obtained from heterozygous intercrosses and control matings during the first 4 days of gestation.

For this experiment, embryos were flushed from the oviducts on days 1.5 and 2.5 p.c., when normal embryos should be two-cell and morulae, respectively. On day 3.5 p.c., when normal embryos should be at the blastocyst stage, they were flushed from the uterus. All embryos were photographed and then classified as normal or abnormal. The results from this study are summarized in Table 2. The fraction of day-1.5 embryos that are at the two-cell stage is essentially the same in both the control and experimental matings. At day 2.5 p.c., a larger percentage of the embryos are abnormal in the heterozygous intercrosses; however, we do not believe that the line 4 homozygotes are arresting between day 1.5 and 2.5 of development. If this were the case, one would expect to find an accumulation of two-cell and four-cell embryos at day 2.5 p.c.; this was not observed. At day 3.5 p.c. over 90% of the isolated embryos were morphologically normal blastocysts in both the control and experimental matings. Thus, as predicted, most of the homozygous line 4 embryos do not appear to arrest before the fourth day of gestation.

Homozygous Preimplantation Embryos in Line 4 Develop Abnormally in Culture

To determine whether the homozygous embryos in line 4 develop normally and remain viable up to the time of implantation, we monitored the development of isolated two-cell embryos and morulae in vitro. Culture conditions that can support the preimplantation development of a one-cell mouse embryo into a hatching blastocyst (a stage of development equivalent to implantation in vivo) are well established (Whittingham 1971; Hsu et al. 1974). Figure 3 shows the stages of preimplantation development that can be observed in vitro after placing two-cell embryos in culture. Briefly, the individual cells of the embryo divide to give rise to the eight-cell morula (Fig. 3-1). The morula then undergoes compaction (Fig. 3-2) and begins to secrete fluid internally to form the blastocoelic cavity (Fig. 3-3 and 3-4). The blastocoel, which gradually enlarges, is largely surrounded by an outer layer of trophectoderm cells and contains the ICM to one side. After reaching the late or enlarged blastocyst stage (Fig. 3-5 and 3-6), the embryo goes through a series of contractions (Fig. 3-7) and hatches out of the zona pellucida (Fig. 3-8). Upon hatching, the trophectoderm cells spread out to form an adherent layer of cells on the tissue culture dish, and the ICM, which continues to proliferate, often appears as a ball of cells on top of the trophectoderm cells (Fig. 3-9). In our assay of in vitro development, embryos are classified as normal

Table 1. Development between 7.5 and 11.5 Days of Gestation

	Day of gestation	Number of litters	Total number of implantation sites	Number of normal embryos	Number of retarded or abnormal embryos	Number of resorbed embryos	Number of retarded + abnormal + resorbed embryos/total number implantation sites (%)	
Control matings	7.5–11.5	15	142	135	3	4	7/142	(5%)
Heterozygous intercrosses	7.5	3	33	22	2	9	11/33	(30%)
	8.5	5	45	25	2	18	20/45	(42%)
	9.5	7	55	36	6	13	19/55	(31%)
	10.5	5	52	31	3	18	21/52	(37%)
	11.5	5	47	31	0	16	16/47	(31%)
total		25	232	145	13	74	87/232	(34%)

In the heterozygous intercrosses, the percent of the implantation sites that contain retarded, abnormal, or resorbed embryos is corrected for the 5% abnormally developing embryos observed in the control matings.

Table 2. Preimplantation-stage Embryos

A. Day-1.5 embryos

	Number of litters	Total number of embryos	Number at two-cell stage	Number of unfertilized eggs	Number of dead eggs	Number of empty zona	Number at one-cell stage	Number of irregular two-cell embryos	Percent retarded or abnormal
Control matings	7	72	43	1	13	3	10	2	40
Heterozygous intercrosses	6	64	39	4	11	4	5	1	39

B. Day-2.5 embryos

	Number of litters	Total number of embryos	Number of morulae	Number of unfertilized eggs	Number of dead eggs	Number of empty zona	Number of one-cells	Number of two-cells	Percent retarded or abnormal
Control matings	8	90	71	1	16	1	1	0	21
Heterozygous intercrosses	10	128	82	1	31	3	10	1	36

C. Day-3.5 embryos

	Number of litters	Total number of embryos	Number of blastocysts	Number of morulae	Number of collapsed or lysed blastocysts	Percent retarded or abnormal
Control matings	11	81	74	3	4	9
Heterozygous intercrosses	7	50	46	1	3	8

In the heterozygous intercrosses, the percentages at days 1.5, 2.5, and 3.5 p.c. are not corrected for the fraction of embryos that were abnormal in the control matings.

only if they completely hatch out of the zona and form an adherent layer of trophectoderm cells.

Table 3 summarizes our observations on the in vitro development of two-cell embryos and morulae that were obtained both from control matings between nontransgenic litter mates and from experimental matings between heterozygous animals in line 4. After correcting for the fraction of embryos that developed abnormally in the control matings, the frequency of abnormally developing two-cell embryos and morulae from the heterozygous intercrosses is 23% and 30%, respectively. Thus, although the homozygous preimplantation embryos in line 4 are viable up to 3.5 days p.c. in vivo, they are apparently exhibiting their mutant phenotype upon placement in culture, as approximately one-quarter of the embryos derived from experimental matings do not develop normally in vitro.

A further analysis of this data revealed that the abnormally developing embryos, in both the control and experimental matings, could be divided into three classes: A, incomplete hatching of the blastocyst out of the zona; B, incomplete formation of the blastocyst; and C, the collapse and consequent failure of the late blastocyst to hatch (see Fig. 4). In Table 4 we have combined the data on the development of the two-cell embryos and morulae in culture and have categorized the abnormally developing embryos according to the three classes shown in Figure 4. Both the control and experimental matings generated a similar percentage of abnormally developing embryos in class A. In contrast, classes B and C are enriched for the abnormally developing embryos produced in the heterozygous intercrosses, with most of these falling into class C. Thus, in culture, the homozygous embryos appear to be ar-

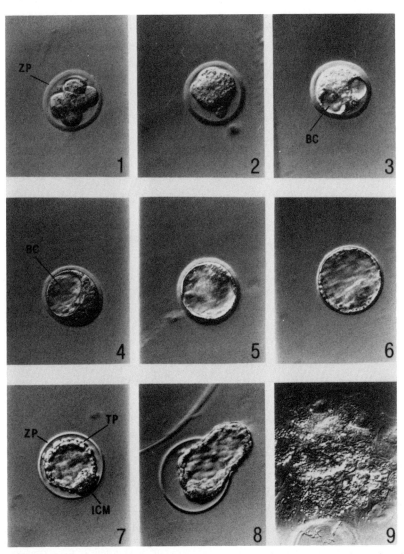

Figure 3. In vitro development of preimplantation mouse embryos. A two-cell-stage mouse embryo placed in culture develops into a morula (*1*) which then undergoes compaction (*2*) and differentiation to give rise to a blastocyst (*3-6*). The blastocyst, containing a recognizable inner cell mass (ICM), trophectoderm (TP), and blastocoel (BC), contracts and hatches out of the zona pellucida (ZP) (*7-8*). The cells of the trophectoderm adhere to the bottom of the culture vessel while cells from the ICM continue their development (*9*).

Table 3. In Vitro Development of Preimplantation Embryos

A. Day-1.5 embryos	Number of litters	Total number of two-cell embryos	In vitro development		Frequency of abnormal development
			normal	abnormal	
Control matings	7	43	33	10	23%
Heterozygous intercrosses	6	39	23	16	41% (23%)[a]

B. Day-2.5 embryos	Number of litters	Total number of morula	In vitro development		Frequency of abnormal development
			normal	abnormal	
Control matings	8	70	62	8	11%
Heterozygous intercrosses	10	82	51	31	38% (30%)[a]

Embryos are classified as normal only if they completely hatch out of the zona pellucida and form an adherent layer of trophectoderm cells on the tissue culture dish.

[a] The number in parentheses is the frequency of abnormal development observed in the heterozygous intercrosses, after correcting for the fraction of embryos that develop abnormally in the control matings.

resting in development over a period of time between the early and late blastocyst stages, with a majority of the homozygotes arresting at the late blastocyst stage. It is at this time of development at which the embryo, in vivo, would normally begin the process of implantation into the uterine wall.

DISCUSSION

In this paper, we present the identification and partial characterization of a recessive prenatal lethal mutation that was generated in the mouse by the integration of λRβG2 DNA on chromosome 3, at a locus now designated as Rβ3. An examination of pre- and post-

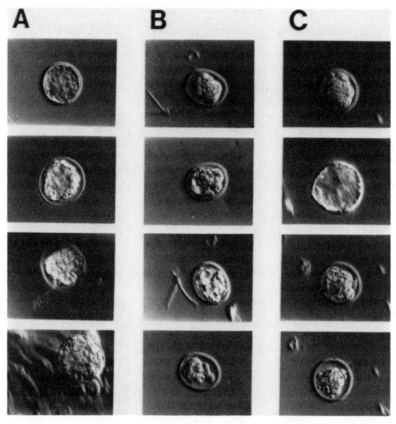

Figure 4. Three classes of abnormal in vitro development of preimplantation mouse embryos. Each set of photographs (four panels) depicts the development of one individual embryo. (*A*) Incomplete hatching: The mouse embryo develops in vitro but does not completely escape from the zona pellucida during hatching (*bottom panels*). (*B*) Incomplete blastocyst formation: The developing embryo reaches the morula stage (*top panel*) and begins vacuolating but fails to form a complete blastocoelic cavity (*middle panels*). These embryos then collapse and degenerate (*bottom panel*). (*C*) Failure to hatch: The embryo develops to the late blastocyst stage (*top panels*) but fails to hatch out of the zona pellucida. These embryos, again, collapse and eventually degenerate (*bottom panels*).

Table 4. Abnormal Development In Vitro

	Total number of embryos tested	Normal development	Arrest at two-cell/morula stage	A[a]	B[a]	C[a]
Control matings	113	95	2	12	3	1
Heterozygous intercrosses	121	74	1	15	9	22

[a]Classes A, B, and C refer to the categories of abnormal development depicted in Fig. 4: (A) Incomplete hatching; (B) incomplete blastocyst formation; (C) failure to hatch.

implantation embryos (Tables 1 and 2) revealed that the homozygotes are viable and morphologically normal up to the blastocyst stage at 3.5 days p.c. and that they can trigger a decidual response in the uterus on day 4.5 p.c. However, by 7.5 days p.c. they have been resorbed. Thus, the homozygous embryos appear to arrest in development about the time of implantation into the uterine wall. Our analysis of the preimplantation development of embryos in culture supports this conclusion. Approximately one-quarter of the embryos obtained from heterozygous intercrosses and cultured from the two-cell or eight-cell stage were unable to hatch from the zona pellucida and form an adherent layer of trophectoderm cells (Tables 3 and 4). This process of hatching observed in vitro normally occurs at a stage in development equivalent to that at which implantation occurs in utero. Although the homozygous mutant embryos flushed from the uterus at 3.5 days p.c. were morphologically normal blastocysts (Table 2), between one-quarter to one-third of the homozygous two-cell embryos and morulae followed in culture failed to develop to the blastocyst stage (Table 4). This discrepancy between the developmental potential of the homozygotes in vivo and in vitro may reflect the inadequacy of conditions for embryo culture.

To define more precisely the phenotypic defect of the homozygous embryos, we plan to construct chimeras between mutant and wild-type embryos. By reconstituting blastocysts from wild-type ICMs and mutant trophoblastic vesicles (and by performing the reciprocal reconstitution as well), it will be possible to establish whether the mutation at the Rβ3 locus results in a defect specific to either ICM or trophectoderm cells (Gardner et al. 1973). Similarly the formation of aggregation chimeras between wild-type and mutant morulae will allow us to determine whether the homozygous embryos can be rescued by the introduction of wild-type cells. At least two mechanisms by which wild-type cells could rescue mutant cells in a chimera can be envisioned. First, rescue could be achieved if the wild-type cells provide an essential secreted protein lacking in the homozygous embryos, or if they can cooperate metabolically with the mutant cells. Second, if the mutation affects only a small subset of cells in the implanting embryo, then the replacement of this specific cell type by wild-type cells may also rescue the homozygous embryos.

The most likely cause of the recessive lethal mutation is that the λRβG2 DNA inserted into and inactivated a gene essential for development. Lesions in several different types of genes might be expected to cause developmental arrest at the blastocyst stage. For example, the inactivated gene might encode a protein product that is normally synthesized in the oocyte as well as in the developing embryo. In this case, the egg cytoplasm might provide enough of the maternally derived protein to support the embryo to the early stages of implantation, after which the homozygous embryo would arrest due to its inability to synthesize the protein. Consistent with this model is the observation that the activity of several maternally derived enzymes persists up to the 8- to 16-cell stage (Johnson 1981; Magnuson and Epstein 1981) or, in the case of glucose phosphate isomerase, to the late blastocyst stage (West and Green 1983). Alternatively, the inactivated gene might encode a product that is synthesized by the embryonic genome and that is first required by the embryo at implantation. In either case, the normal expression of the inactivated gene need not be limited to embryonic cells; adult tissues might express this gene as well. In fact the chromosomal position effect resulting in transcription of the rabbit β-globin gene specifically in skeletal muscle in heterozygous mice suggests that the Rβ3 locus does include genes expressed in the adult mouse.

Because the recessive lethal mutation at the Rβ3 locus is an insertional mutation, the inactivated gene at this locus can be isolated and characterized following the strategy established for transposon-induced mutations. First, using sequences from λRβG2 as a hybridization probe, the mouse DNA sequences flanking the foreign DNA insert can be cloned; subsequently, unique regions in the flanking mouse DNA can be used as a probe to isolate the wild-type Rβ3 locus. The mechanism by which the essential gene at the Rβ3 locus was inactivated (deletion, rearrangement, or interruption) should be revealed by comparing the cloned wild-type and mutant alleles. With respect to the mechanism of inactivation, it is interesting to note that metaphase chromosome spreads revealed no chromosomal abnormalities in line 4 mice and Southern blot hybridization experiments provided no evidence for rearrangements in the integrated λRβG2 molecules (Lacy et al. 1983). The identity and function of the inactivated gene may be determined by screening different tissues from adults and embryos for RNA transcripts derived from the Rβ3 locus.

In conclusion, an insertional mutation at the Rβ3 locus has identified a gene essential for early mouse development. By combining the methods of molecular biology with the large repertoire of techniques for manipulating mouse embryos, it should be possible to identify this gene and elucidate its role in the early embryogenesis of the mouse.

ACKNOWLEDGMENTS

We thank Nils Lonberg and Frank Costantini for useful discussions and critical reading of the manuscript, Eva Derman for the cDNA clone plivs-1, Maureen Del Re and Dianna Rynkiewicz for excellent technical assistance, and Mary Wentzler for expert secretarial assistance. This work was supported by a grant from the G. Harold and Leila Y. Mathers Charitable Foundation.

REFERENCES

Bennett, D. 1975. The t-locus of the mouse. *Cell* **6:** 441.

Bingham, P., R. Lewis, and G.M. Rubin. 1981. Cloning of DNA sequences from the white locus of *D. melanogaster* by a novel and general method. *Cell* **25:** 693.

Brinster, R.L., H.Y. Chen, M. Trumbauer, A.W. Senear, R. Warren, and R.D. Palmiter. 1981. Somatic expression of herpes thymidine kinase in mice following injection of a fusion gene into eggs. *Cell* **27:** 223.

Bronson, F.H., C.P. Dagg, and G.P. Snell. 1968. Reproduction. In *Biology of the laboratory mouse* (ed. E.L. Green), p. 187. Dover, New York.

Bukhari, A., J. Shapiro, and S. Adhya, eds. 1977. *DNA insertion elements, plasmids and episomes.* Cold Spring Harbor Laboratory, Cold Spring Harbor, New York.

Calos, M.P. and J.H. Miller. 1980. Transposable elements. *Cell* **20:** 579.

Copeland, N.G., K.W. Hutchison, and N.A. Jenkins. 1983. Excision of the DBA ecotropic provirus in dilute coat-color revertants of mice occurs by homologous recombination involving the viral LTRs. *Cell* **33:** 379.

Costantini, F. and E. Lacy. 1981. Introduction of a rabbit β-globin gene into the mouse germ line. *Nature* **294:** 92.

Derman, E. 1981. Isolation of a cDNA clone for mouse urinary proteins: Age- and sex-related expression of mouse urinary protein genes is transcriptionally controlled. *Proc. Natl. Acad. Sci.* **78:** 5425.

Derman, E., K. Krauter, L. Walling, C. Weinberger, M. Ray, and J.E. Darnell. 1981. Transcriptional control in the production of liver-specific mRNAs. *Cell* **23:** 731.

Enders, A.C. and S. Schlafke. 1969. Cytological aspects of trophoblast-uterine interaction in early implantation. *Am. J. Anat.* **125:** 1.

Fedoroff, N. 1983. Controlling elements in maize. In *Mobile genetic elements* (ed. J.A. Shapiro), p. 1. Academic Press, New York.

Gardner, R.L. 1983. Origin and differentiation of extraembryonic tissues in the mouse. *Int. Rev. Exp. Pathol.* **24:** 64.

Gardner, R.L., V.E. Papaioannou, and S.C. Barton. 1973. Origin of the ectoplacental cone and secondary giant cells in mouse blastocysts reconstituted from isolated trophoblast and inner cell mass. *J. Embryol. Exp. Morphol.* **30:** 561.

Gluecksohn-Waelsch, S. 1979. Genetic control of morphogenetic and biochemical differentiation: Lethal albino deletions in the mouse. *Cell* **16:** 225.

Gordon, J.W. and F.H. Ruddle. 1981. Integration and stable germline transmission of genes injected into mouse pronuclei. *Science* **214:** 1244.

Gordon, J.W., G.A. Scangos, D.J. Plotkin, J.A. Barbosa, and F.H. Ruddle. 1980. Genetic transformation of mouse embryos by microinjection of purified DNA. *Proc. Natl. Acad. Sci.* **77:** 7380.

Green, M.C., ed. 1981. Catalog of mutant genes and polymorphic loci. In *Genetic variants and strains of the laboratory mouse*, p. 8. Gustav Fischer Verlag, New York.

Hayward, W., B. Neel, and S. Astrin. 1981. Activation of a cellular onc gene by promoter insertion in ALV-induced lymphoid leukosis. *Nature* **290:** 475.

Hogan, B., F. Costantini, and E. Lacy. 1986. *Manipulating the mouse embryo: A laboratory manual.* Cold Spring Harbor Laboratory, Cold Spring Harbor, New York. (In press.)

Hsu, Y.-C., J. Baskar, L.C. Stevens, and J.E. Rash. 1974. Development in vitro of mouse embryos from the two-cell egg stage to the early somite stage. *J. Embryol. Exp. Morphol.* (suppl. 1) **31:** 235.

Hutchison, K.W., N.G. Copeland, and N.A. Jenkins. 1984. Dilute-coat-color locus of mice: Nucleotide sequence analysis of the d^{+2} and d^{+a} revertant alleles. *Mol. Cell. Biol.* **4:** 2899.

Jaenisch, R., D. Jahner, P. Nobis, I. Simon, J. Lohler, K. Harbers, and D. Grotkopp. 1981. Chromosomal position and activation of retroviral genomes inserted into the germ line of mice. *Cell* **24:** 519.

Jaenisch, R., K. Harbers, A. Schnieke, J. Lohler, I. Chumakov, D. Jahner, D. Grotkopp, and E. Hoffmann. 1983. Germline integration of Moloney murine leukemia virus at Mov13 locus leads to recessive lethal mutation and early embryonic death. *Cell* **32:** 209.

Jahner, D. and R. Jaenisch. 1980. Integration of Moloney leukemia virus into the germ line of mice: Correlation between site of integration and virus activation. *Nature* **287:** 456.

Jenkins, N.A., N.G. Copeland, B.A. Taylor, and B.K. Lee. 1981. Dilute (d) coat colour mutation of DBA/2J mice is associated with the site of integration of an ecotropic MuLV genome. *Nature* **293:** 370.

Johnson, M.H. 1981. The molecular and cellular basis of preimplantation mouse development. *Biol. Rev. Camb. Philos. Soc.* **56:** 463.

Kafatos, F.C., C.W. Jones, and A. Efstratiadis. 1979. Determination of nucleic acid sequence homologies and relative concentrations by a dot hybridization procedure. *Nucleic Acids Res.* **7:** 1541.

Lacy, E., S. Roberts, E.P. Evans, M.D. Burtenshaw, and F. Costantini. 1983. A foreign β-globin gene in transgenic mice: Integration at abnormal chromosomal positions and expression in inappropriate tissues. *Cell* **34:** 343.

Magnuson, T. and C.J. Epstein. 1981. Genetic control of very early mammalian development. *Biol. Rev. Camb. Philos. Soc.* **56:** 369.

——— 1984. Oligosyndactyly: A lethal mutation in the mouse that results in mitotic arrest very early in development. *Cell* **38:** 823.

McClintock, B. 1952. Chromosome organization and genetic expression. *Cold Spring Harbor Symp. Quant. Biol.* **16:** 13.

McLaren, A. 1976. Genetics of the early mouse embryo. *Annu. Rev. Genet.* **10:** 361.

Palmiter, R.D., H.Y. Chen, and R.L. Brinster. 1982. Differential regulation of metallothionein-thymidine kinase fusion genes in transgenic mice and their offspring. *Cell* **29:** 701.

Palmiter, R.D., T.M. Wilkie, H.Y. Chen, and R.L. Brinster. 1984. Transmission distortion and mosaicism, in an unusual transgenic mouse pedigree. *Cell* **36:** 869.

Pedersen, R.A. 1974. Development of lethal yellow (A^y/A^y) mouse embryos in vitro. *J. Exp. Zool.* **188:** 307.

QUINN, P., C. BARROS, and D.G. WHITTINGHAM. 1982. Preservation of hamster oocytes to assay the fertilizing capacity of human spermatozoa. *J. Reprod. Fertil.* **66:** 161.

ROEDER, G.S. and G.R. FINK. 1980. DNA rearrangements associated with a transposable element in yeast. *Cell* **21:** 239.

ROEDER, G.S., P.J. FARABAUGH, P.T. CHALEFF, and G.R. FINK. 1980. The origins of gene instability in yeast. *Science* **209:** 1375.

SCHNIEKE, A., K. HARBERS, and R. JAENISCH. 1983. Embryonic lethal mutation in mice induced by retrovirus insertion into the α1 (I) collagen gene. *Nature* **304:** 315.

SHORTLE, P., J. HABER, and D. BOTSTEIN. 1982. Lethal disruption of the yeast actin gene by integrative DNA transformation. *Science* **217:** 371.

SILVER, L.M. 1981. Genetic organization of the mouse t complex. *Cell* **27:** 239.

SNELL, G.P. and L.C. STEVENS. 1968. Early embryology. In *Biology of the laboratory mouse* (ed. E.L. Green), p. 205. Dover, New York.

STEWART, T.A., E.F. WAGNER, and B. MINTZ. 1982. Human β-globin gene sequences injected into mouse eggs, retained in adults, and transmitted to progeny. *Science* **217:** 1046.

THEILER, K. 1972. *The house mouse*. Springer-Verlag, New York.

VAN VALEN, P. 1966. Oligosyndactylism, an early embryonic lethal in the mouse. *J. Embryol. Exp. Morphol.* **15:** 119.

WAGNER, E.F., T.A. STEWART, and B. MINTZ. 1981. The human β-globin gene and a functional viral thymidine kinase gene in developing mice. *Proc. Natl. Acad. Sci.* **78:** 5016.

WAGNER, E.F., L. COVARRUBIAS, T.A. STEWART, and B. MINTZ. 1983. Prenatal lethalities in mice homozygous for human growth hormone gene sequences integrated in the germ line. *Cell* **35:** 647.

WEST, J.D. and J.F. GREEN. 1983. The transition from oocyte-coded to embryo-coded glucose phosphate isomerase in the early mouse embryo. *J. Embryol. Exp. Morphol.* **78:** 127.

WHITTINGHAM, D.G. 1971. Culture of mouse ova. *J. Reprod. Fertil.* (suppl.) **14:** 11.

ZACHER, Z. and P.M. BINGHAM. 1982. Regulation of white locus expression: The structure of mutant alleles at the white locus of *Drosophila melanogaster*. *Cell* **30:** 529.

Molecular Analysis of Krüppel, a Segmentation Gene of *Drosophila melanogaster*

H. Jäckle, U.B. Rosenberg, A. Preiss, E. Seifert,
D.C. Knipple, A. Kienlin, and R. Lehmann
Max-Planck-Institut für Entwicklungsbiologie, 7400 Tübingen, Federal Republic of Germany

Mutations of the three "gap" genes, a class of segmentation genes of *Drosophila melanogaster*, cause contiguous groups of adjacent segments of the embryo to fail to develop (Nüsslein-Volhard and Wieschaus 1980). In hunchback, the thoracic segments are deleted, whereas in knirps only two rather than eight denticle bands are formed in the abdomen of the embryo (Nüsslein-Volhard and Wieschaus 1980). The gap in the segmentation pattern of homozygous Krüppel (*Kr*) embryos overlaps the borders seen in the segmentation pattern of the other two gap mutants (Nüsslein-Volhard and Wieschaus 1980; Wieschaus et al. 1984).

The 26 *Kr* alleles obtained by either X-ray or chemical mutagenesis can be ordered into an allelic series where the size of the gap is proportional to the strength of the mutant allele (Wieschaus et al. 1984; Preiss et al. 1985). Strong *Kr* alleles lack all thoracic and five anterior abdominal segments, which are replaced by a mirror-image duplication of the normal posterior abdominal segments (see Fig. 1). Embryos homozygous for the weakest allele isolated so far develop a normal head, prothorax, and seven abdominal segments while the meso- and metathorax and one abdominal segment are deleted. Weak and intermediate *Kr* alleles lack mirror image duplications in the segment pattern except for Filzkörper, the posterior tracheal endings. These frequently appear at ectopic locations adjacent to the head region of *Kr* intermediate phenotypes (Wieschaus et al. 1984, and Fig. 1).

The requirement for Kr^+ gene activity is strictly zygotic and appears to be confined to early stages of embryogenesis (Wieschaus et al. 1984). Maternal dosage of Kr^+ activity has no apparent effect on the embryonic phenotype, nor does homozygosity for *Kr* prevent germ cells from making eggs capable of normal development when fertilized by wild-type sperm (Wieschaus et al. 1984). Since homozygous embryos can be distinguished from wild-type or heterozygous embryos shortly after the beginning of gastrulation (Wieschaus et al. 1984), Kr^+ activity is required for the normal formation of thorax and anterior abdomen after egg deposition and earlier than gastrulation.

In addition to the segmentation defect, Kr^+ gene activity might be involved in other functions such as the formation of Malpighian tubules and the organization of neural tissue in the thoracic and anterior abdominal segments. This can be deduced from the strong *Kr* phenotype which lacks Malpighian tubules and shows severe distortion of the nervous system (Gloor 1954). The present paper describes the molecular analysis of Kr^+ gene activity during embryogenesis. Our data suggest that the Kr^+ gene is polyfunctional and that its expression is both under temporal and spatial control.

EXPERIMENTAL PROCEDURES

Stock keeping of *Drosophila*, handling of the embryos, preparation of larval cuticles, analysis of injected embryos, and genetic markers have been described in detail (Preiss et al. 1985; Rosenberg et al. 1985). Procedures concerning molecular cloning, DNA preparations, handling, and analysis of cloned DNA and RNA techniques were described by Preiss et al. (1985). Injections of embryos with cloned DNA, RNA, or cytoplasm were described by Preiss et al. (1985), Rosenberg et al. (1985), and Müller-Holtkamp et al. (1985), respectively.

In situ hybridization of the *Kr*-antisense RNA (Rosenberg et al. 1985) to tissue sections of embryos was carried out according to the procedure of Akam and Martines-Arias (1985), as described by D. Knipple et al. (in prep.). Histological sections of wild-type and *Kr* embryos were prepared as described by Zalokar and Erk (1976). After fixation with glutaraldehyde followed by osmium tetroxide, embryos were embedded in Durcupan (Fluka) prior to sectioning. Semithin sections (2 μm) were stained with an aqueous solution of 0.05% Methylene Blue, 0.01% Toluidine Blue, and 0.05% sodium tetraborate decahydrate at 60°C.

RESULTS

Molecular Cloning of Kr^+ Gene Sequences

Kr maps to the tip of the right arm of the second chromosome (Gloor 1954; Wieschaus et al. 1984). Using seven chromosomal rearrangements, we mapped the *Kr* locus into polytene chromosome band 60F3 (Preiss et al. 1985). This band was microdissected from salivary gland chromosome squashes, and its DNA was cloned using the microcloning approach (Pirrotta et al. 1983). One of the microclones was used as a starting point for chromosomal walking (Bender et al. 1983) to collect overlapping recombinant DNA clones covering the entire *Kr* region (Preiss et al. 1985). The relevant portion of the *Kr* region is shown in Figure 2.

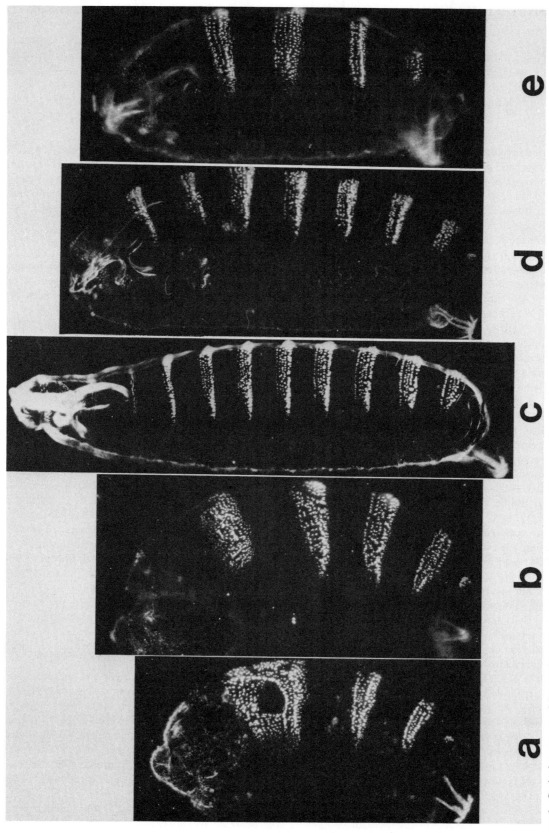

Figure 1. Cuticular pattern of *Drosophila* larvae: (*a*) strong Kr phenotype, (*b*) intermediate Kr phenotype, (*c*) newly hatched wild-type larva, (*d*) weak Kr phenotype, and (*e*) Kr phenocopy obtained after injection of *Kr* antisense RNA into wild-type embryos (see legend of Table 2 for details). For a detailed description of the different Kr phenotypes, see Wieschaus et al. (1984) and Preiss et al. (1985).

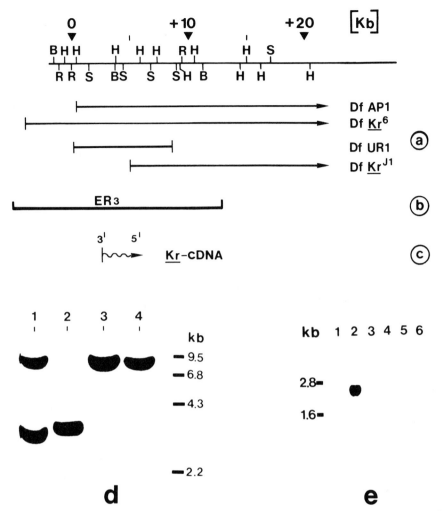

Figure 2. Physical map of the DNA of the *Kr* region (R, *Eco*RI; B, *Bam*HI; H, *Hin*dIII; S, *Sal*I). (*a*) Deletion breakpoints of four *Kr* alleles. Solid line represents the deleted DNA. Note that Kr^{J1}/Kr UR1 embryos contain a DNA deletion smaller than 4 kb. For a detailed map see Preiss et al. (1985). (*b*) Position of the clone ER3-DNA used for rescue experiments shown in Table 2. (*c*) Position of the *Kr* cDNA clone used for the *Kr* antisense RNA studies (Table 1). (*d*) Breakpoint analysis of DNA by Southern analysis. Southern blots containing *Eco*RI-digested: (lane *1*) *AP1/SM1*; (lane *2*) homozygous AP1; (lane *3*) the homozygous parental *If*; and (lane *4*) *If/SM1* DNA was probed with the DNA fragment of region 0–9.5 (see physical map). For designation of the deletions and the marker chromosomes see Preiss et al. (1985). Note the 9.5-kb DNA in lanes *1, 3,* and *5,* and the breakpoint DNA segment in lanes *1* and *2*. The precise position of the *AP1* DNA breakpoint is indicated in *b*. (*e*) Developmental profile of Kr^+ poly(A)$^+$ RNA on Northern blots containing about 10 µg of poly(A)$^+$ RNA from 0- to 30-min (lane *1*), 2- to 5-hr (lane *2*), 8- to 24-hr (lane *3*)-old embryos; larvae (lane *4*); pupae (lane *5*); and adult (lane *6*). RNAs were probed with a 9.5-kb DNA fragment (see above). Note a broad band of about 2.5 kb in 2- to 5-hr-old embryos that is a duplet of two poly(A)$^+$ RNAs of similar size (U.B. Rosenberg, unpubl.).

Nine out of 26 *Kr* alleles that we analyzed were associated with DNA deletions. Four of the alleles contained DNA breakpoints in the *Kr* region (Fig. 2a). They allowed us to map a small DNA deletion present in transheterozygous Kr^{J1}/KrUR1 mutants (Fig. 2a), so indicating that the corresponding DNA segments present in the wild-type represent an essential part of the Kr^+ gene.

Mapping of Functional Parts of the Kr^+ Gene

We injected DNA from various recombinant phages encompassing the *Kr* region shown in Figure 2 into embryos of Ddc^{n7}Kr^1/SM1 parents. The Dopa-decarboxylase mutation allows Kr^1/Kr^1 embryos to be unequivocally distinguished from their siblings as it renders the cuticle and mouth parts of homozygous Kr^1 larvae unpigmented (Wright et al. 1976; Wieschaus et al. 1984). The homozygous Kr^1 embryos have DNA from the entire *Kr* region deleted (Preiss et al. 1985). After injection of DNA from clone ER3 (Fig. 2b), segments anterior to the sixth abdominal segment of homozygous Kr^1 embryos often showed normal polarity (Table 1). Such embryos are significantly different from the uninjected Kr^1 embryos possessing the strong Kr phenotype. The rescue was not observed after injection of other DNA segments.

The rescued Kr^1 embryos can be ordered into the al-

Table 1. Phenotypic Rescue of Kr Embryos Injected with ER3-DNA

Site of injection (% of egg length from posterior)[a]	Number of injected homozygous Ddc-marked Kr¹ embryos scored[b]	Percent of Kr¹ embryos without rescue response[c]	Percent of Kr embryos showing rescue response[d]
0–40	71	100	0
50–60	115	55	45
60–80	90	88	12
80–100	59	98	2

About 300 pl of ER3 DNA (130 μg/ml) was injected into embryos of Ddc^{n7} Kr/SM1 parents. For details on the injection procedure, handling, and scoring of the injected embryos see Preiss et al. (1985).

[a] Embryos were injected at about pole cell stage into a lateral region along their longitudinal axis.
[b] Homozygous Kr¹ embryos were scored by their Ddc phenotype which renders mouthhooks and denticle belts unpigmented (details in Preiss et al. 1985). Only uninjured embryos were scored.
[c] Kr¹ embryos expressing the extreme Kr phenotype (see Fig. 1a).
[d] Kr¹ embryos with normal polarity in at least four abdominal segments resembling an intermediate or weak Kr phenotype (see Fig. 1b,c).

lelic series of Kr mutants (see introductory section). This response suggests weak Kr^+ activity derived from the injected ER3-DNA in Kr¹ embryos. Although the rescue effect of injected DNA is far from being sufficient to provide the level of Kr^+ activity required for a normal embryonic development, the ER3-DNA should contain essential sequences of the Kr gene. It should be emphasized that this weak rescue response was dependent on both, the period of development when the DNA was injected and the site of injection. The best rescue response was obtained when DNA was injected just after the pole cell stage in the middle of the Kr embryos (Table 1). These results already indicate that the middle region of the Kr embryo is most responsive to the presence of weak Kr^+ activity (see below).

Kr Poly (A)⁺ RNA Is Transcribed Early in Embryogenesis

Deletion mapping and the phenotypic rescue of the Kr mutant embryos with cloned DNA both localized Kr^+ function in the region 0 to +10 on the physical map shown in Figure 2. The corresponding DNA was used to probe Northern blots loaded with equal amounts of poly (A)⁺ RNA from various stages of Drosophila development. As shown in Figure 2e, the EcoRI fragment (0 to +9.5) hybridized to a poly(A)⁺ RNA transcript of about 2.5 kb present only between 2 and 5 hours postfertilization. Using this fragment as a probe, a cDNA clone was isolated from cDNA library constructed from poly(A)⁺ RNA from wild-type embryos of the corresponding stage, but no cDNA was found in the library prepared from poly(A)⁺ RNA of earlier embryos (U.B. Rosenberg et al., in prep.).

The Kr-cDNA clone, 2.3 kb in size, was used to determine the number of Kr^+ transcripts in wild-type embryos being at blastoderm to germ band extension stages (2–5 hr postfertilization at 25°C). Titration experiments using actin mRNA as a reference on Northern blots showed that the Kr^+ transcript amounts to less than 0.01% of total poly(A)⁺ RNA. We have to note, however, that the broad band shown in Figure 2e corresponds to a doublet of bands of similar size. In addition, we observed three longer transcripts which amount to less than 1% of the doublet band in intensity. This suggests that more than one poly(A)⁺ RNA is transcribed from the Kr^+ gene (U.B. Rosenberg et al., in prep.). Using the Kr-cDNA as a probe, we obtained eight positive clones from a freshly prepared, nonamplified cDNA-library (see above) containing 230,000 recombinant phages. This number corresponds to one in about 40,000 poly(A)⁺ RNA sequences, or some 10^5 Kr^+ transcripts per embryo, which is about 10 Kr^+ transcripts per average blastoderm cell.

Production of Kr Phenocopies by Injection of Kr-antisense RNA into Wild-type Embryos

To demonstrate that the 2.3-kb cDNA clone contained sequences of the Kr gene, we injected Kr-antisense RNA into wild-type embryos to specifically inactivate Kr^+ transcripts in vivo.

Kr-antisense RNA was prepared by in vitro transcription from a plasmid construct containing the SP6 promoter linked to the Kr-cDNA in 3'–5' orientation (Rosenberg et al. 1985). Wild-type embryos, injected at syncytial blastoderm stage with a thousandfold excess of Kr antisense RNA (4 μg/μl) over endogenous Kr^+ transcripts, developed in high-frequency lethal embryos strongly resembling the weak and intermediate but not the strong Kr phenotype (Table 2). This suggests that Kr-antisense RNA hybridized to Kr^+ transcripts in vivo and blocked directly the translation of Kr mRNA. The lack of extreme Kr phenocopies indicates that under the condition applied, Kr^+ activity was not abolished completely.

Kr-antisense RNA was injected without any kind of protection and may be target of subsequent degradation by endogenous RNases. In this case, Kr^+ transcripts may be inactivated during a short developmental period following the injection at syncytial blastoderm stage. This implies that Kr has an early function at about blastoderm stage, as expected from the genetic studies (see introductory section). It was then important to know the precise temporal and spatial characteristics of Kr^+ gene expression during early embryogenesis.

Table 2. *Kr* Phenocopy Production Resulting from Injection of *Kr* Sense and Antisense RNA into *Drosophila* Wild-type Embryos

Injected RNA[a]	RNA concentration (μg/μl)	Number of injected embryos that developed a larva[b]	Hatched larvae (%)	Weak Kr phenocopies[c] (%)	Strong Kr phenocopies[d] (%)	Injection artefacts[e] (%)
Kr sense	3	263	73	0	0	27
Kr antisense	0.7	226	72	7	2	19
	4	174	41	22	15	22

[a]*Kr* sense and antisense RNA were prepared from an SP6 promoter construct as described in Rosenberg et al. (1985) and were injected into the middle region of preblastoderm-stage Oregon-P2 embryos in about 300-pl volumes.

[b]Only embryos that developed a complete cuticle were scored except for hatched embryos.

[c]Weak Kr phenocopies developed prothorax and up to seven abdominal segments. They cannot be distinguished morphologically from the corresponding phenotype shown in Fig. 1d.

[d]Strong Kr phenocopies (Fig. 1e) developed ectopic Filzkörper in the head region and were indistinguishable from intermediate Kr phenotypes (Fig. 1b).

[e]Embryos resembling weak Kr phenocopies in the presence of segmental defects due to the injection.

Temporal and Spatial Expression of the *Kr*+ Gene

The SP6-plasmid construct (see above) was used to prepare ^{3}H-labeled *Kr*-antisense RNA probes (D. Knipple et al., in prep.). These were hybridized to frozen sections of *Drosophila* embryos, and sites of hybridization corresponding to *Kr*+ transcript accumulation were analyzed after autoradiography.

Transcripts of the *Kr* gene are first detected at syncytial blastoderm stage (after the eleventh nuclear division) in a circumferential band with a width of about 10 nuclei in a region of 45–55% egg length from the posterior pole. By completion of the thirteenth nuclear division, at early cellular blastoderm, a much stronger hybridization signal is over the cytoplasm surrounding about 14 nuclei in the same region of the embryo. A strong hybridization signal is also seen over the cytoplasm surrounding yolk nuclei that are in this region, indicating that *Kr*+ gene expression occurs throughout the embryo in a defined region rather than in the peripheral layer of energids (Fig. 3a). We refer to this region as the *Kr*+ central domain.

During the blastoderm stage, *Kr*+ transcripts continued to be present in the *Kr*+ central domain. In addition, a new region of *Kr*+ gene expression appears late at this stage in a band of about 10 cells adjacent to the pole cells (Fig. 3b). This *Kr*+ posterior domain persists during the germ band extension period marking the posterior end of the germ band. At the beginning of gastrulation, a third zone of *Kr*+ gene expression, the *Kr*+ anterior domain, appears in a band of cells about six to nine cell diameters anterior and parallel to the cephalic furrow (Fig. 3c). At the same time, the *Kr*+ central domain broadens posteriorly, starting from its initial position, which is about six to nine cell diameters posterior to the cephalic furrow. At the end of the germ band extension stage, this domain extends from the posterior part of the cephalic furrow to the posterior midgut invagination; the *Kr*+ gene is now expressed in ectoderm and mesoderm as well as ventral neural tissue. When the germ band is fully extended (Fig. 3d), *Kr*+ transcripts are associated with the segmental ganglia of the brain and the somatogastric nervous system (*Kr*+ anterior domain), the mesoderm including the ventral nervous system (*Kr*+ central domain), and the Malpighian tubules (*Kr*+ posterior domain). At later stages, *Kr*+ transcript hybridization was detected slightly above background in most cells of the developing embryo.

The in situ hybridization experiments revealed that *Kr*+ gene expression is under temporal control and constitutes a complex spatial pattern of transcription in three regions of the embryo. To follow the fate of the corresponding cells during development, we analyzed histological sections of *Kr* mutant and wild-type embryos.

The *Kr*+ Central Domain Is Associated with Cell Death in *Kr* Mutant Embryos

Histological sections of *Kr*^{1} mutant embryos revealed significant cell death in the thoracic and anterior abdominal region during germ band extension. The region of cell death in 6- to 7-hr-old embryos is rather small and covers a mesodermal region smaller than the gap in segmentation later seen in the *Kr*^{1} mutant embryo (Fig. 4). After germ band shortening, dead cells are found in a region that roughly corresponds to the maximal gap in the segmentation pattern, including both mesoderm and neural tissue. Cell death was not observed in the head region and not in or adjacent to the posterior midgut rudiment (Fig. 4).

Injection of Wild-Type Cytoplasm Partially Rescues the Strong Kr Phenotype

We have injected cytoplasm from various regions of blastoderm stage wild-type embryos into the middle region of homozygous *Kr*^{1} embryos (Table 3) to assay for their response to *Kr*+ activity. Only the cytoplasm from the middle region of the donor embryo would rescue the *Kr*^{1} phenotype similar to injected ER3 DNA (see above). Cytoplasm taken from outside the 70–30% region (placing the zero point at the posterior pole) of wild-type embryos had no effect (see Table 3). The rescue response was only seen when the cytoplasm was injected into this 70–30% region of the *Kr*^{1} recipient

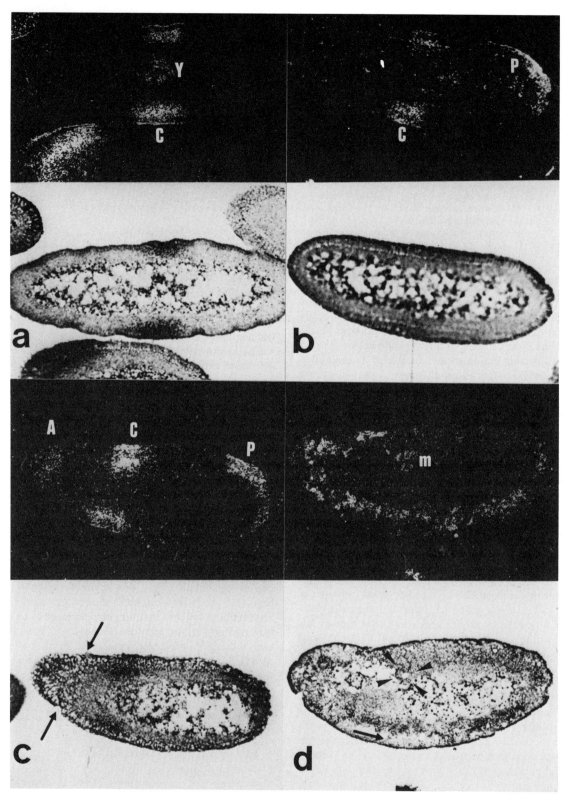

Figure 3. Localization of *Kr* transcripts in *Drosophila* wild-type embryos. Dark-field micrograph and corresponding photomicrograph (*below*) of tissue sections of embryos (anterior, *left*; posterior, *right*), hybridized in situ with the *Kr*+ probe. (*a*) Horizontal section of an early blastoderm stage embryo, after 13th nuclear division. Note the *Kr*+ central domain (C) including yolk energids (Y) at about 45-55% egg length from posterior. (*b*) Sagittal section of a late blastoderm-stage embryo (elongated nuclear stage). Note the appearance of the *Kr*+ posterior domain (P; 0-10% egg length). (*c*) Tangential section of an embryo at early gastrulation stage. Note the appearance of the *Kr*+ anterior domain (A) six to nine cells anterior and parallel to the cephalic furrow (arrows). (*d*) Oblique section of an embryo at extended germ band stage. Note *Kr*+ transcript accumulation in mesoderm and its absence in ectoderm (arrow). Note *Kr*+ transcripts in Malpighian tubules (m and arrowheads). For a detailed analysis of *Kr*+ gene expression see D. Knipple et al. (in prep.).

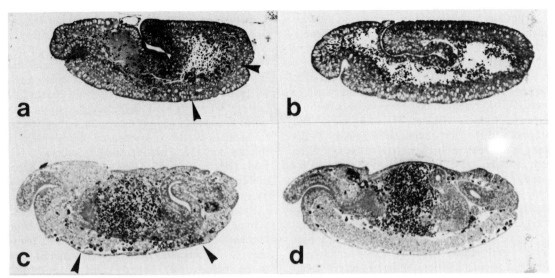

Figure 4. Histological sections (sagittal) of 6- to 7-hr (*a,b*) and 11- to 13-hr (*c,d*)-old wild-type and *Kr* embryos. Note an accumulation of dead cells (delimited by arrowheads) in *Kr* (*a,c*) but not in wild-type embryos (*b,d*). Dead cells accumulate at the yolk-mesoderm border in a small region at germ band extension stage (*a*) and throughout mesoderm and neural tissue in a larger region after germ band shortening (*c*). Photomicrographs, in contrast to stained sections under the microscope, do not allow distinguishing yolk particles from dead cells unambiguously.

embryo. Furthermore, when cytoplasm taken from any region of the wild-type embryo was injected into the posterior 0–10% egg length of *Kr* mutant embryos, rescue of the Malpighian tubules was never observed (Table 3 and E. Seifert, unpubl.). These results demonstrate that weak *Kr*+ activity is highest in the middle of the wild-type embryos, and that this region is most sensitive to the absence of *Kr*+ gene activity as suggested by both the *Kr* mutant phenotype (see introductory section) and the spatial rescue response pattern of *Kr* mutant embryos. Futhermore, Malpighian tubule formation cannot be restored under similar experimental conditions.

DISCUSSION

Our results demonstrate the use of *Drosophila* mutant embryos as an in vivo test system for specific biological response both to cytoplasm of wild-type embryos and to molecular probes injected. Our principal finding is a complex temporal and spatial pattern of *Kr*+ gene expression during early stages of embryogenesis.

Kr+ gene requirement is strictly zygotic. This suggests a cell-specific or regulatory, rather than a general housekeeping, function in all cells of the developing embryo. That *Kr*+ activity is essential for the establish-

Table 3. Phenotypic Rescue of *Kr* Embryos Injected with Wild-type Cytoplasm

Region of donor cytoplasm[a] (% of egg length from posterior)	Site of injection of recipient Kr^1 embryos[b] (% of egg length from posterior)	Number of injected homozygous *Ddc*-marked Kr^1 embryos scored[c]	Percent of *Kr* embryos with phenotypic rescue [d]	Percent of Kr^1 embryos developing Malpighian tubules[e]
0–10	40–60	59	0	0
10–30	40–60	53	0	0
30–40	40–60	52	12	0
40–60	40–60	68	29	0
60–70	40–60	54	9	0
70–90	40–60	57	0	0
40–60	0–10	62	0	0
40–60	10–30	52	0	0
40–60	30–40	53	8	0
40–60	40–60	68	26	0
40–60	80–100	56	0	0
0–10	0–10	51	0	0

[a]Cytoplasm was taken from a region of Oregon-P2 embryos at blastoderm stage and injected into the region of embryos (see footnote 3) from *Ddc Kr¹/SM1* parents (see general footnote, Table 1). *Ddc Kr¹/Ddc Kr¹* embryos, injected just after pole cell formation, were scored. For details on the technical parts, handling, and scoring of embryos see Müller-Holtkamp et al. (1985).

[b-d]See footnotes a–c in Table 1.

[e]Malpighian tubules were scored prior to embedding, but they were never observed.

ment of normal segmentation of thorax and anterior abdomen of the *Drosophila* embryo is best demonstrated by the series of different *Kr* alleles (Wieschaus et al. 1984; Preiss et al. 1985).

On a molecular level, the initial *Kr+* gene expression is in a zone in the middle region of blastoderm-stage embryos. This *Kr+* central domain corresponds to (1) the anlagen of three segments, including meso- and metathorax and the first abdominal segment; (2) the gap seen in the weakest Kr phenotype so far identified (Wieschaus et al. 1984; Preiss et al. 1985); (3) the region of *Kr+* activity in the cytoplasm of the wild-type embryo; (4) the region of *Kr* embryos most responsive to injection of both cloned DNA and wild-type cytoplasm; and (5) a region of cell death seen in 6- to 7-hr-old amorphic *Kr* mutant embryos. However, the pattern of *Kr+* gene expression has no direct correspondence in the segmental deletions of the strong, amorphic Kr phenotype. In this respect, *Kr+* gene expression is very different from other segmentation genes such as fushi tarazu (Hafen et al. 1984) and hairy (D. Ish-Horowiz, pers. comm.), where the spatial pattern of gene expression coincides with the deletion pattern later seen in the embryo. The absence of additional flanking segments to the zone of the initial *Kr+* central domain in most *Kr* mutants suggests that *Kr+* gene expression exerts a secondary effect on the formation of these segments or that their formation could be mediated by the spread of *Kr+* gene expression into these anlagen, beginning at early gastrula (see D. Knipple et al., in prep. for details) and possibly involving control by other segmentation genes.

The specific role of *Kr+* activity in establishing the normal segment pattern was interpreted as transfer of positional values (of a maternally established gradient reaction) to individual blastoderm cells in the middle of the embryo (Wieschaus et al. 1984). In this interpretation, the lack of proper transfer of positional information results in a distortion of positional values which, in turn, cause the formation of a second, reversed posterior abdomen in place of eight deleted segments. The new posterior identity of these cells is due to genetic as well as morphological transformation which depends on a change in cell fate in *Kr* embryos (Wieschaus et al. 1984). The localized appearance of the *Kr+* central domain (including *Kr+* gene expression in yolk energids) is compatible with a switch on of *Kr+* gene activity in response to the proposed positional values of maternal origin (for details, see Wieschaus et al. 1984). We note, however, that the lack of *Kr+* gene expression in amorphic *Kr* alleles, at least partially, results in cell lethality. This phenomenon and the observation that the blastoderm fate map of *Kr* embryos is not altered with respect to the segments that develop normally (R. Lehmann, unpubl.) suggest that the mirror image duplication may result from secondary effects resulting in a change of fate of the nonlethal cells in the region. In this view, injected *Kr+* DNA (Table 1) or *Kr+* cytoplasm may each rescue a critical number of normally lethal cells which then survive to form a normal abdominal segment. Residual *Kr+* activity present in weaker *Kr* alleles may have a similar effect.

Cell lethality in response to the absence of normal *Kr+* activity is not a general phenomenon in all regions of the embryo. It was not observed in anlagen related to the *Kr+* anterior and *Kr+* posterior domain, which correspond on the blastoderm fate map to the anlagen of the procephalic neurogenic region, optical lobes, and the esophagus, and to the anlagen of both Malpighian tubules and the posterior midgut rudiment, respectively (Hartenstein et al. 1985). There are no apparent defects seen in the corresponding tissues of *Kr* alleles, except for the lack of Malpighian tubules.

Rescue experiments involving both cloned DNA and cytoplasm, and *Kr+*-antisense injection experiments (Tables 1–3) were only effective in the middle region of the embryo, indicating that this region is most sensitive to the absence of *Kr+* activity consistent with the pattern of cell lethality observed. This implies that *Kr+* gene activity acts differently in different embryonic regions: Cells of the middle regions require *Kr+* activity for viability whereas cells in the posterior region survive to a different cell fate or stop dividing rather than forming Malpighian tubules.

Northern blot analysis and library screening allow a rough estimation of the number of *Kr+* transcripts present in 2- to 5-hour embryos. Based on 2 ng of poly(A)$^+$ RNA (about 2×10^9 transcripts; average length of 2 kb) per embryo (Anderson and Lengyel 1979), we expect some 0.5×10^5 to 1×10^5 *Kr+* transcripts accumulating in a blastoderm to extended germ band-stage embryo. This number relates to about 10 transcripts per average blastoderm cell or about 100 *Kr+* transcripts per *Kr+*-expressing cell within each domain. However, the transcripts of each domain may be different in function, quality, and quantity (U.B. Rosenberg et al., unpubl.). We hope that further molecular analysis of the *Kr* gene in combination with extended genetic studies will allow an understanding of the biological function of *Kr*, a developmentally regulated segmentation gene that appears to be more complex, both in function and regulation, as initially expected on the basis of genetic studies.

ACKNOWLEDGMENTS

We thank our colleagues in the lab and Dr. D. Glover for critically reading this manuscript, Ms. B. Hieber for typing, and Mrs. R. Groemke-Lutz for preparing the photographs.

REFERENCES

AKAM, M. and A. MARTINEZ-ARIAS. 1985. Ultrabithorax expression in *Drosophila* embryos. *EMBO J.* (in press).

ANDERSON, K.V. and J.A. LENGYEL. 1979. Rates of synthesis of major classes of RNA in *Drosophila* embryos. *Dev. Biol.* **70:** 217.

BENDER, W., P. SPIERER, and D.S. HOGNESS. 1983. Chromosomal walking and jumping to isolate DNA from the *Ace* and *rosy* loci and the *bithorax* complex in *Drosophila melanogaster*. *J. Mol. Biol.* **168:** 17.

GLOOR, H. 1954. Phänotypus der Heterozygoten bei der unvollständig dominanten, homozygot letalen Mutante *Kr* (= Krüppel) von *Drosophila melanogaster. Arch. Julius Klaus-Stift. Vererbungsforsch. Sozialanthropol. Rassenhyg.* **29**: 277.

HAFEN, E., A. KUROIWA, and W.J. GEHRING. 1984. Spatial distribution of transcripts from the segmentation gene *fushi tarazu* during *Drosophila* embryonic development. *Cell* **37**: 833.

HARTENSTEIN, V., G.M. TECHNAU, and J. CAMPUS-ORTEGA. 1985. Fate mapping in wild-type *Drosophila* melanogaster. III. A fate map of the blastoderm. *Wilhelm Roux's Arch. Dev. Biol.* **194**: 213.

MÜLLER-HOLTKAMP, F., D.C. KNIPPLE, E. SEIFERT, and H. JÄCKLE. 1985. An early role of maternal mRNA in establishing the dorsoventral pattern in *pelle* mutant *Drosophila* embryos. *Dev. Biol.* **110**: 238.

NÜSSLEIN-VOLHARD, C. and E. WIESCHAUS. 1980. Mutations affecting segment number and polarity in *Drosophila. Nature* **287**: 795.

PIROTTA, V., H. JÄCKLE, and J.-E. EDSTRÖM. 1983. Microcloning of microdissected chromosome fragments. In *Genetic engineering: Principles and methods* (ed. A. Hollaender and J.K. Setlow), vol. 5, p. 1. Plenum Press, New York.

PREISS, A., U.B. ROSENBERG, A. KIENLIN, E. SEIFERT, and H. JÄCKLE. 1985. Molecular genetics of *Krüppel*, a gene required for segmentation of the *Drosophila* embryo. *Nature* **313**: 27.

ROSENBERG, U.B., A. PREISS, E. SEIFERT, H. JÄCKLE, and D.C. KNIPPLE. 1985. Production of phenocopies by *Krüppel* antisense RNA injection into *Drosophila* embryos. *Nature* **313**: 703.

WIESCHAUS, E., C. NÜSSLEIN-VOLHARD, and H. KLUDING. 1984. *Krüppel*, a gene whose activity is required early in the zygotic genome for normal embryonic segmentation. *Dev. Biol.* **104**: 172.

WRIGHT, T.R.F., G.C. BEWLEY, and A.F. SHERALD. 1976. The genetics of Dopa decarboxylase in *D. melanogaster*. Isolation and characterization of dopa decarboxylase mutants and their relationship to the alpha-methyl dopa hypersensitive mutants. *Genetics* **84**: 287.

ZALOKAR, M. and I. ERK. 1976. Division and migration of nuclei during early embryogenesis of *Drosophila melanogaster. J. Microsc. Biol. Cell.* **25**: 97.

Genetics of Polyamine Synthesis in Tobacco: Developmental Switches in the Flower

R. L. MALMBERG,* J. MCINDOO, A.C. HIATT, AND B.A. LOWE
Cold Spring Harbor Laboratory, Cold Spring Harbor, New York 11724

Tobacco Flowers

The study of the induction and regulation of flowering is one of the great themes of plant developmental biology. One particular motif in this research area has been the study of the control of sex in the flower. This is a wonderfully complex subject in higher plants because of the many reproductive strategies that exist: complete hermaphrodism with both sexes in one flower; sexes on separate plants; mixed male and female flowers on the same plant; and virtually all possible combinations of these. A variety of genetic and hormonal mechanisms have been discovered that govern these patterns of floral differentiation (for review, see Frankel and Galun 1977).

Nicotiana tabacum (tobacco) has perfect, or complete, flowers, with both male and female differentiation normally occurring in the same flower. In a vegetatively growing tobacco plant, the terminal apical meristem gives rise to the body of the plant; periodically, lateral initials are generated that differentiate into leaves. After the signal to flower has been received, the lateral initials develop into the parts of the flower—sepals, petals, stamen, carpel—instead of leaves. This sequence is both temporal and spatial, with the sepals developing first and occupying the outermost position in the completed flower and the male stamen and female carpel occupying the innermost and last developing positions. The tobacco flower has a fivefold symmetry; there are normally five of each flower part, except for the carpel. The sepals and petals each develop as fusions of their lateral initials, whereas the carpel develops as a composite fusion of the central meristem and the last developing initials (Steeves and Sussex 1972).

In this paper we describe a collection of mutants that alter the development of the tobacco flower; many of them seem to alter the expression of the initials that generate the flower parts, so that a variety of developmental switches are seen. They were isolated using the powerful somatic cell culture system of tobacco that allows one to perform microbial-like selections on defined media, and then regenerate whole plants from the mutated or transformed cells (for review, see Chaleff 1981). We obtained these mutants by selecting for lesions in the polyamine biosynthetic pathway (Fig. 1). We will therefore also describe studies on the normal developmental regulation of the polyamine synthesis enzymes in wild-type tobacco.

Polyamines in Higher Plants

Recently, a great deal of interest has developed in the study of the polyamine pathway in higher plants. Correlations have been discovered between polyamines and response of the plant to a variety of stress conditions (Young and Galston 1983), and also to the application of hormones to the plant (Dai et al. 1982). These data have suggested that polyamines may be an important growth regulatory substance in higher plants (Galston 1983). An additional stimulus to polyamine research has been the development of a variety of inhibitors of the enzymes in the pathway (for review, see Tabor and Tabor 1983) (Fig. 1). Difluoromethylornithine (DFMO) and difluoromethylarginine (DFMA) block ornithine decarboxylase (OrnDC) and arginine decarboxylase (ArgDC), respectively. Methylglyoxal-bis(guanylhydrazone) (MGBG) and dicyclohexylammoniumsulfate (DCHA) inhibit S-adenosylmethionine decarboxylase (SamDC) and spermidine synthase, respectively. Other inhibitors have also been developed for both OrnDC and SamDC. These compounds offer a potent tool for probing the pathway.

Two pathways to the synthesis of putrescine have been documented in higher plants: from arginine via ArgDC through agmatine (Smith 1981), and directly from ornithine via OrnDC (Heimer and Mizrahi 1982). Spermidine and spermine are then synthesized by the combined actions of SamDC, spermidine synthase, and spermine synthase. The relative importance of the two paths to putrescine has been resolved partially by consideration of tissue specificity and phylogeny. ArgDC is the enzyme found in most plants tested, and in particular is found in leaves (Smith 1979). OrnDC has been found in the rapidly proliferating tissues of tobacco cell cultures and tomato ovaries (Heimer and Mizrahi 1982); thus, in addition to the link to division, it is possible that OrnDC is preferentially found in the Solonaceae. Cohen et al. (1982) demonstrated that exogenous application of DFMO could block tomato ovary development, whereas Berlin and Forche (1981) and Bagni et al. (1983) have similarly demonstrated that DFMO can inhibit growth of tobacco cell cultures. Hence, there is evidence for a requirement for OrnDC in these tissues.

An important feature in plants is the conjugation of polyamines to phenolic compounds (Smith and Best

---

*Present address: Botany Department, University of Georgia, Athens, Georgia 30602.

1978); Smith et al. (1983) have demonstrated that the majority of polyamines in tobacco flowers may be bound up in these polyamine conjugates, which therefore may act in a buffering capacity for the free polyamines. There is some tissue specificity in the whole plant in the distribution of conjugates across tissues, particularly in the flower (Smith et al. 1983). Dumas and colleagues (Dumas et al. 1981, 1982) have found low polyamine-conjugate levels in a nonflowering line of tobacco.

Several researchers have noted an effect of polyamine depletion on embryogenesis. The model somatic cell embryogenic system is carrot, which can be switched from growth as a suspension cell culture to an embryogenic pathway by media and hormone manipulations. Montague et al. (1978) have shown an increase in ArgDC activity following the switch from cell culture to embryogenesis. More recently, Feirer et al. (1984) have shown that DFMA blocks the transition to embryogenesis, while it does not alter the simple growth of suspension cultures. Putrescine added at the same time as DFMA allows the transition to proceed. Z.R. Sung (pers. comm.) has shown that a mutant carrot defective in the switch to embryogenesis can be partially rescued by exogenous putrescine.

The reports just discussed find intriguing correlations between polyamines and a variety of developmental processes in plants. We felt that a mutational analysis of the pathway would be a useful approach to sorting this out.

RESULTS

Developmental Regulation of the Enzymes

As an initial characterization of the polyamine pathway in tobacco, we measured enzyme specific activities for OrnDC, ArgDC, and SamDC in the different organ systems and in cell cultures. These data are summarized in Figure 2. OrnDC activity is found preferentially in flowers and cell cultures. ArgDC activity is found in all tissues, although to a somewhat lesser extent in flowers and cultures. SamDC activity is preferentially, although not exclusively, found in flowers.

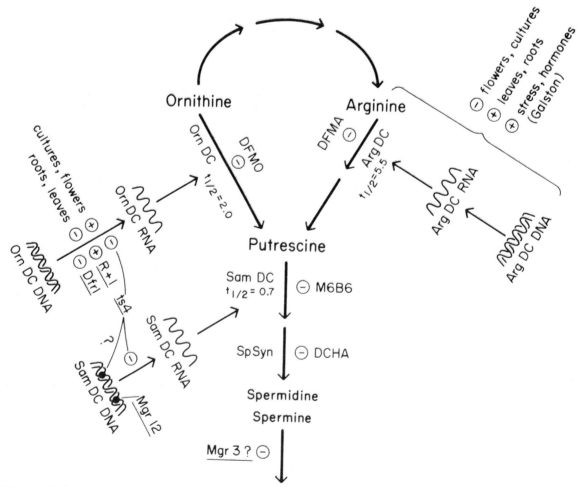

Figure 1. Polyamine pathway in tobacco. (OrnDC) Ornithine decarboxylase; (ArgDC) arginine decarboxylase; (SamDC) S-adenosylmethionine decarboxylase; (SpSyn) spermidine or spermine synthase; (DFMO) difluoromethylornithine; (DFMA) difluormethylarginine; (MGBG) methylglyoxal-bis(guanylhydrazone); (DCHA) dicyclohexylammonium sulfate.

	OrnDC	ArgDC	SamDC	O/A ratio
Flower	79	20	100	2.24
Leaf	8	59	22	0.08
Stem	9	79	21	0.07
Root	20	100	24	0.11
Culture	100	31	31	1.83

Figure 2. Developmental regulation of enzyme activities. (OrnDC) Ornithine decarboxylase; (ArgDC) arginine decarboxylase; (SamDC) S-adenosylmethionine decarboxylase. The data shown in the first three columns are the specific activities of the enzymes relative to the tissue in which the specific activity is the maximum. The fourth column (O/A ratio) is the ratio of the specific activities of OrnDC to ArgDC.

Thus, there is considerable developmental regulation of the enzyme activities with respect to reproductive versus nonreproductive organs. Cell cultures are theoretically undifferentiated, although in this particular instance, the state of the pathway in culture resembles flowers more than it does other tissues of the plant.

In tobacco cell cultures, we have found that DFMO inhibits growth, whereas DFMA does not (Hiatt et al. 1985), implying that OrnDC is the more vital pathway in cell cultures. One mode of regulation of the tobacco cell culture enzymes is rapid protein turnover (Hiatt et al. 1985); we measured the half-lives of OrnDC, ArgDC, and SamDC as 2 hours, 5.5 hours, and 0.7 hour, respectively. The end products of the pathway, when added to a culture, rapidly eliminate ArgDC and SamDC activities, with kinetics of shutdown that are the same as that of cycloheximide. That suggests this regulation may operate during some phase of protein synthesis. OrnDC showed only partial activity reduction when cells were grown in various polyamines, and maintained a basal, constitutive level of activity. Since we are studying putrescine, we felt it necessary to examine decay.

Recently, molecular probes have become available for OrnDC and SamDC. McConlogue and Coffino (1984) have obtained a full-length molecular clone for the mouse cDNA that encodes OrnDC. They obtained this clone from a mouse cell line that had a gene amplification for OrnDC, due to a stepwise selection for resistance to DFMO. D. Morris (pers. comm.) has obtained a partial clone for the bovine SamDC by construction of an oligonucleotide probe from the protein sequence data. P. Coffino (University of California, San Francisco) and D. Morris (University of Washington) have generously given us their molecular probes for mammalian OrnDC and SamDC, respectively. Under conditions of moderate stringency washes, we were able to detect hybridization of both probes to RNA from various tobacco tissues and cell lines (Fig. 3). The degree of hybridization to the RNA from different tissue types accurately reflects the enzyme activity profile given above, thus helping to confirm the preferential expression of these genes in flowers. This also suggests that the developmental regulation of OrnDC and SamDC in tobacco occurs at a pretranslational stage.

Isolation and Characterization of Mutants

Over the last few years we have isolated tobacco cell lines that are temperature sensitive, or that are resistant to either MGBG, DFMO, or DCHA. Part of our success in mutant isolation derives from an UV light mu-

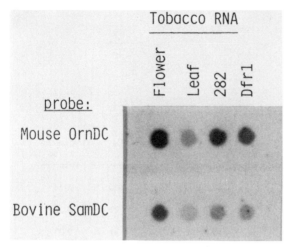

Figure 3. Heterologous probing of tobacco RNA with the mouse OrnDC and bovine SamDC clones.

tagenesis scheme (Malmberg and McIndoo 1984), coupled with a plating method that gave a 12.5% plating efficiency of resistant cell lines under actual selection conditions. In the case of MGBG resistance, we obtained 31 lines from mutagenized cultures and 0 lines from nonmutagenized control cultures; although the mutagenesis was successful, the mutation rate was still extremely low, being slightly less than 10^{-7} per cell division. This suggests that MGBG resistance is, biochemically, not an easy trait to obtain. When we regenerated these cultures into whole plants, every one that had flowered had abnormal flowers (Malmberg 1980; Malmberg and McIndoo 1983, 1984). Some showed simple male and female sterilities without gross morphological changes, and others showed one or another developmental switch. Some of these cell lines and their current characterization are (see Fig. 4 and Table 1 also):

ts4. This was our original variant blocked in polyamine synthesis. This cell line was isolated by a bromodeoxyuridine-negative selection scheme (Malmberg 1979), is pleiotropic, and has 4% of the wild-type levels of OrnDC and SamDC. On SDS-polyacrylamide gels, an extra band of 35,000 daltons is seen; wild-type cell cultures grown on sublethal levels of MGBG also overproduce this 35,000-dalton polypeptide (Malmberg and McIndoo 1983). We have recently demonstrated that this extra polypeptide is SamDC (Hiatt et al. 1985), so that *ts4* produces large amounts of an inactive SamDC. This cell line regenerates into a weak, light green plant that is dwarfed, and that never flowers (Malmberg 1980).

Rt1. A revertant of *ts4*, that is partially blocked in polyamine synthesis, this line has near wild-type levels of OrnDC, but reduced levels of SamDC. It does not overproduce the 35,000-dalton polypeptide. Plants regenerated from *Rt1* are nearly normal in color and size. The flowers are both male and female sterile, but with the abnormality that the anthers are partially turned into petals (petaloid anthers) (Malmberg 1980).

Dfr1. This line is resistant to DFMO. Enzyme assays of this line indicate that it has extremely low levels of OrnDC, and the residual activity that it does have is sensitive to DFMO. *Dfr1* regenerates into an extremely weak dwarf plant that never flowers. Physically it resembles the *ts4* mutant just described. Both *ts4* and *Dfr1* took an unusually long time to regenerate into plantlets, over a year for each, whereas wild type normally regenerates in 2 months.

Mgr3. This line is resistant to MGBG and overproduces spermidine and spermine relative to wild type. Plants regenerated from *Mgr3* have the abnormality that their ovules are switched into stamens (stamenoid ovules), a female-to-male switch (Malmberg and McIndoo 1983). This mutant is male sterile with defective anthers as well. In one case, we were able to obtain nine seeds from a cross of wild type to *Mgr3*. The drug resistance and floral phenotype cosegregated as a nuclear dominant trait within this small F_1 generation. Other lines that display the stamenoid ovule phenotype are *Mgr1*, *Mgr4*, *Mgr9*, and *Mgr31*.

Mgr9. This line is resistant to MGBG. In addition to stamenoid ovules, this line has regenerated into plants with stigmoid anthers, a male-to-female switch, and with simple defective anthers. Other lines that display the stigmoid anther phenotype are *Mgr1*, *Mgr12*, and *Mgr27*.

Mgr12. *Mgr12* is resistant to MGBG. In regenerated plants *Mgr12* is male sterile, and is a nuclear dominant mutation by meiotic genetic analysis (Malmberg and McIndoo 1984). Biochemical analysis of crude extracts of cultures established from heterozygous F_1 plants showed that approximately 50% of the SamDC activity was resistant to the presence of MGBG in the assay cocktail. This suggests that the mutation may be in the structural gene for SamDC. In preliminary experiments, approximately one-third of the MGBG-resistant mutants show similar kinetic evidence of being a structural gene mutation.

Mgr21. Resistant to MGBG, this plant differs from the bulk of the MGBG-resistant plants in that it is extremely dwarfed. It flowers with a "puzzle box" or "nested doll" phenotype. When one dissects a normal tobacco flower, the sequence of structures is sepal, petal, stamen, and carpel. In *Mgr21* the carpel is a hollow cylinder with the stigma surface having a hole down the middle. Inside this hole, a partial floral sequence is repeated, including stamen, and carpel. The mutant seems to alter partially the determinate character of the normal tobacco flower; the terminal meristem is no longer committed to developing into part of the carpel, but rather dedifferentiates into an earlier phase of floral development, i.e., it stutters. These flowers also show a partial conversion of sepals into leaves. *Mgr15* is similar to *Mgr21*.

Mgr27. Resistant to MGBG, the flowers on this plant have combined many of the features of the other mutants listed. They frequently have stigmoid and petaloid anthers in a single flower, and also frequently have petaloid sepals, and extra petals.

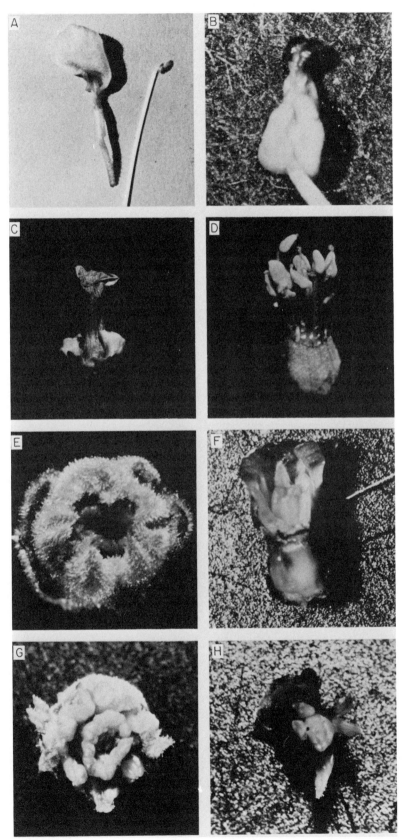

Figure 4. Abnormal floral development in regenerated plants. (*A*) Petaloid and wild-type anthers; (*B*) stigmoid anther; (*C*) petaloid sepals; (*D*) flower with both petaloid and stigmoid anthers; (*E–H*) progressive dissection of a *Mgr2l* puzzle box flower.

Table 1. Whole Plant Phenotypes of Polyamine Synthesis Variants

Strain	Coral like	Very dwarf	Petal sepal	Extra petal	Petal anther	Stigma anther	Stamen ovule	Puzzle box
Class I								
Mgr5-RO[a]	X							
Mgr6-RO	X							
Mgr23-RO	X							
Class II								
Dfr1-RO		X						
ts4-RO		X						
Class III								
Mgr15-RO		X						X
Mgr21-RO		X						X
Class IV								
Mgr25-RO				X				
Rt1-RO					X			
Mgr3-RO							X	
Mgr3-F1[b]							X	
Mgr31-RO			X				X	
Mgr9-RO						X	X	
Mgr12-RO								
Mgr12-F1				X	X	X		
Mgr1-RO		X				X	X	
Mgr27-RO		X	X	X	X			

[a] RO, Plants regenerated from culture.
[b] F1, Progeny from a cross of an RO plant to wild type.

Within a given plant, not all flowers will show the same degree of abnormality. Generally, the most terminal, earliest opening flowers on a given branch will have the most extreme phenotype. We exploited this phenomenon to obtain the small seed samples from crosses to Mgr3 and Mgr12, although other plants have so far proven recalcitrant and fully sterile. Cultures can be made from leaves of mutant plants, and these prove to be fully resistant when tested. Secondary regenerants, made from cultures that had been leaves, generally show the same phenotype as the primary regenerant; however, there is frequently some phenotypic drift to the abnormal floral structures seen on another plant. This is only true for the class-IV plants of Table 1 that usually display the phenotype shown, but that with a low frequency will display the phenotype of another class-IV plant. There is no phenotypic drift across class lines.

The possibility exists that these polyamine lines may have double mutations. They were isolated from mutagenized cultures, and simple passage of some plants through cell culture can also generate variation (somaclonal variation, Larkin and Scowcroft 1981); hence, the important test of correlation between polyamine lesions and flower development is a genetic analysis to analyze genetic linkage of the floral trait with the pathway phenotype. Because of the sterilities in these abnormal flowers, we have been able to perform meiotic crosses in only two cases. Mgr3 and Mgr12 are nuclear dominant traits, and the abnormal flowers cosegregate with the altered polyamine synthesis. For our many other lines, we have no evidence yet for a linkage between the polyamine mutant selection and the abnormal floral development, beyond the statistical argument that all the polyamine selections resulted in abnormal flowers, whereas plants regenerated from unselected cultures do not show these effects on morphology.

DISCUSSION AND SPECULATIONS

The various floral and plant phenotypes of the polyamine mutants fit into four categories as summarized in Table 1. Class I includes lines that have been difficult to regenerate. These often give the appearance of short stems studded with tiny leaves, but will regenerate no further than this. The class-II plants are ts4 and Dfr1; these are both dwarf and have not yet flowered. Both also took a long time to regenerate at all. Biochemically, ts4 has low levels of OrnDC and SamDC activities, and Dfr1 also has low OrnDC activity. The revertant of ts4 has increased levels of ODC, and also flowers, albeit with a developmental switch. These genetic data imply that OrnDC is not only floral specific in its expression, but that it is also required for flowering. This may not be a profound conclusion, in the sense that many biochemical functions are probably required for flowering and not for vegetative development.

The class-IV mutants are the most numerous. These lines have a wide range of abnormalities in the flower, including partial sex changes. There is some evidence that all the phenotypes in this group actually represent a single phenotype that might better be described as an instability after floral induction. Specifically, as secondary and tertiary regenerants are formed by the cycle of putting leaves into culture and pulling them back into plants, we observe that many plants show other developmental switches characteristic of class IV. This is also apparent in the F_1 MGBG-resistant segregants of the cross $Mgr12 \times$ wild type. That is, at a low frequency the flowers of this F_1 display virtually all the characteristic class-IV switches. This has also been observed in the F_1 progeny of the similar cross $Mgr3 \times$ wild type. The one constant phenotype in all the class-IV plants is male sterility. Whether or not a developmental switch is found in a flower; it is always true that the anthers are sterile. The base phenotype is that of a shrunken anther with no pollen. Biochemically, class IV includes all those MGBG-resistant lines for which we have preliminary indications of a SamDC structural gene alteration. It seems unlikely that all of class IV is a SamDC mutation.

Developmentally the class-III puzzle box phenotype is one of the most interesting we have seen. The floral phenotype is coincident with an extreme dwarfness, and also seems to be mutually exclusive of the class-IV switches. In a sense it has parallels with the heterochronic mutants of Caenorhabditis (Ambros and Horvitz 1984). We have diagramed floral lineage development of wild type and $Mgr21$ in Figure 5. This puzzle box tobacco mutant is unable to complete normal carpel development, instead it repeats the previous developmental step and starts over with a row of stamens. The shift back to the previous developmental step is also seen in the occasional leafy sepals found on these flowers. We have no biochemical or genetic analysis of these lines, yet, although the analogy to the heterochronic mutants suggests they might be dominant hypermorphs.

One philosophy of a mutational analysis is to attempt to gain insight into the normal function of the genes being studied. The question thus remains as to what the normal role of polyamines is in a higher plant. Clearly our developmental and mutational data indicate some role for polyamines in floral development. OrnDC may be required for initiation of flowering, and SamDC may be required for proper anther development, or may be part of a male-female balancing system. Neither of these ideas necessarily implies more than that these enzymes supply one of the many functions that are required for these developmental events. As mentioned in the introductory section, there has been speculation that polyamines are involved in the hormone response of plants. Our data are compatible with the notion that they are specifically part of the gibberellic acid response, since gibberellic acid has been shown to control internodal length (dwarfness), the ability to flower, and the male-female balance.

A very interesting set of tobacco lines has been developed by Burke, Gerstel, and colleagues (for review, see Gerstel 1980). A series of recurrent back-crosses were performed with N. tabacum always as the male parent, and with a wild Nicotiana as the initial female parent. This led to lines that had the cytoplasm of the wild species, and with a nucleus mostly derived from N. tabacum. In these lines many of the same sort of floral abnormalities were found as we see in class-IV

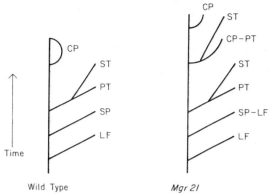

Figure 5. Floral lineage diagrams of wild type and $Mgr21$ "puzzle box" mutants. (LF) Leaf; (SP) sepal; (PT) petal; (ST) stamen; (CP) carpel; (SP-LF) leafy sepal. Since plants develop from a terminal meristem with a small number of cells, the developmental patterns are both temporal and spatial, with the earliest-forming tissue occupying the outermost position in the flower.

polyamine lesions. This parallel suggests that nuclear-cytoplasmic interactions may be contributing to floral development as well.

Higher plants, particularly tobacco, are wonderful organisms for developmental genetic studies. They can be manipulated in cell culture, transformed with exogenous DNA, and regenerated into plants. Also available are the classic manipulations of grafting, fate mapping, and application of exogenous compounds at critical stages. Thus, the potential exists for intriguing insights into eukaryotic development.

ACKNOWLEDGMENTS

This research has been supported by National Science Foundation grants PCM-8203213 (Genetic Biology), DMB-8415322 (Metabolic Biology), by a McKnight Foundation grant (Plant Biology), and by a Damon Runyan–Walter Winchell fellowship to A.C.H.

REFERENCES

Ambros, V. and R.H. Horvitz. 1984. Heterochronic mutants of the nematode Caenorhabditis elegans. Science **226:** 409.

Bagni, N., P. Torrigiani, and P. Barbieri. 1983. In vitro and in vivo effects of ornithine and arginine decarboxylase inhibitors in plant tissue culture. Adv. Polyamine Res. **4:** 409.

Berlin, J. and E. Forche. 1981. Difluoromethylornithine causes enlargement of cultured tobacco cells. Z. Pflanzenphysiol. **101:** 277.

Chaleff, R.S. 1981. Genetics of higher plants: Applications of cell culture. Cambridge University Press, Cambridge, England.

Cohen, E., S. Malis-Arad, Y.M. Heimer, and Y. Mizrahi. 1982. Participation of ornithine decarboxylase in early stages of tomato fruit development. Plant Physiol. **70:** 540.

Dai, Y., R. Kaur-Sawhney, and A.W. Galston. 1982. Promotion of gibberellic acid of polyamine biosynthesis in internodes of light grown dwarf peas. Plant Physiol. **69:** 103.

Dumas, E., E. Perdrizet, and J.-C. Vallee. 1981. Evolution

quantitative des acides amines et amines libres au cours du developpement de diverses especes de *Nicotiana*. *Physiol. Veg.* **19:** 155.

Dumas, E., J.-C. Vallee, and E. Perdrizet. 1982. Étude comparée du metabolisme des acides amines et amines libres chez deux especes de Tabac. *Physiol. Veg.* **20:** 505.

Feirer, R., G. Mignon, and J.D. Litvay. 1984. Arginine decarboxylase and polyamines required for embryogenesis in the wild carrot. *Science* **223:** 1433.

Frankel, R. and E. Galun. 1977. *Pollination mechanisms, reproduction, and plant breeding.* Springer-Verlag, New York.

Galston, A.W. 1983. Polyamines as modulators of plant development. *Bioscience* **33:** 382.

Gerstel, D. 1980. Cytoplasmic male sterility in *Nicotiana*. *North Carolina Agricultural Research Service Technical Bulletin*, No. 263.

Heimer, Y. and Y. Mizrahi. 1982. Characterization of ornithine decarboxylase of tobacco cells and tomato ovaries. *Biochem. J.* **201:** 373.

Hiatt, A.C., J. McIndoo, and R.L. Malmberg. 1985. Regulation of polyamine biosynthesis in tobacco: Effects of inhibitors and exogenous polyamines on arginine decarboxylase, ornithine decarboxylase, and S-adenosylmethionine decarboxylase. *J. Biol. Chem.* (in press).

Larkin, P. and W.R. Scowcroft. 1981. Somaclonal variation—A novel source of variability from cell cultures for plant improvement. *Theor. Appl. Genet.* **60:** 197.

Malmberg, R.L. 1979. Temperature sensitive variants of *Nicotiana tabacum* isolated from somatic cell culture. *Genetics* **92:** 215.

———. 1980. Biochemical, cellular, and developmental characterization of a temperature sensitive mutant of *Nicotiana tabacum* and its second site revertant. *Cell* **22:** 603.

Malmberg, R.L. and J. McIndoo. 1983. Abnormal floral development of a tobacco mutant with elevated polyamine levels. *Nature* **305:** 623.

———. 1984. Ultraviolet mutagenesis and genetic analysis of resistance to methylglyoxal-bis(guanylhydrazone) in tobacco. *Mol. Gen. Genet.* **196:** 28.

McConlogue, L., M. Gupta, L. Wu and P. Coffino. 1984. Molecular cloning and expression of the mouse ornithine decarboxylase gene. *Proc. Natl. Acad. Sci.* **81:** 540.

Montague, M., J. Koppenbrink, and E. Jaworski. 1978. Polyamine metabolism in embryogenic cells of *Daucus carota*. *Plant Physiol.* **62:** 430.

Smith, T.A. 1979. Arginine decarboxylase of oat seedlings. *Phytochemistry* **18:** 1447.

———. 1981. Biosynthesis and metabolism of polyamines in plants. In *Polyamines in biology and medicine* (ed. D.R. Morris and L.J. Marton), p. 77. Marcel-Dekker, New York.

Smith, T.A. and G.R. Best. 1978. Distribution of the hordatines in barley. *Phytochemistry* **17:** 1093.

Smith, T.A., J. Negrel, and C.R. Bird. 1983. The cinnamic acid amides of the di-and polyamines. *Adv. Polyamine Res.* **4:** 347.

Steeves, T.A. and I.M. Sussex. 1972. *Patterns in plant development.* Prentice-Hall, Englewood Cliffs, New Jersey.

Tabor, H. and C.W. Tabor, eds. 1983. Polyamines. *Methods Enzymol.*, vol. 94.

Young, N.D. and A.W. Galston. 1983. Putrescine and acid stress, induction of arginine decarboxylation activity and putrescine accumulation by low pH. *Plant Physiol.* **71:** 767.

spo0H: A Developmental Regulatory Gene for Promoter Utilization in *Bacillus subtilis*

P. ZUBER AND R. LOSICK

Department of Cellular and Developmental Biology, The Biological Laboratories, Harvard University, Cambridge, Massachusetts 02138

Cells of the gram-positive bacterium *Bacillus subtilis* are induced to enter an elaborate cycle of morphological differentiation in response to conditions in which certain essential nutrients, such as carbon, nitrogen, or phosphorous, become limiting for growth (for reviews, see Losick and Youngman 1984; Youngman et al. 1985). This developmental process culminates in the formation of a dormant cell type known as the spore or more precisely the endospore, a complex multilayered structure that is able to resist extremes of environmental conditions. Under conditions that support germination, the endospore is rapidly converted back to a vegetative cell. Figure 1 is an electron micrograph of a germinating spore in which a vegetative cell is emerging from the broken shell of the spore casing (the protein layers that comprise the spore coat).

The formation of the endospore and its transformation into a growing cell is governed by a large number of genes known as sporulation (*spo*), germination (*ger*), and outgrowth (*out*) genes, which are located at many sites on the *B. subtilis* chromosome (Piggot et al. 1981). At least 50 *spo*, *ger*, and *out* genes have been identified, but recent studies based on new methods for identifying, mapping, and analyzing developmental mutations, using transposon-mediated insertional mutagenesis, suggest that the list of developmental loci may ultimately grow to over 100 genes (Losick and Youngman 1984; P. Youngman et al., unpubl.).

The initiation phase of the sporulation process is dependent upon a class of genes known as the *spo0* loci (Hoch 1976; Hoch et al. 1978). Mutations in *spo0* genes do not interfere with vegetative growth but block sporulation prior to any of its earliest morphological events under conditions that would induce sporulation in Spo+ bacteria. One or more of the *spo0* gene products may be regulatory proteins that are involved in initiating the program of sporulation gene expression in response to conditions of nutrient deprivation (Losick

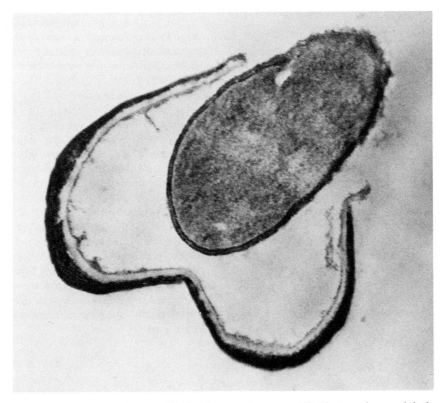

Figure 1. A germinating spore of *B. subtilis* magnified by electronmicroscopy. (The electronmicrograph is the gift of C. Robinow of the University of Western Ontario.)

1981). Approximately eight *spo0* genes are known, and these are located at widely scattered sites on the chromosome.

We are interested in the role of *spo0* genes in the initiation phase of sporulation. To investigate this problem, we have employed as a model system a sporulation gene known as *spoVG* whose transcription is switched on almost immediately after *B. subtilis* cells are induced to sporulate by nutrient deprivation (Segall and Losick 1977; Ollington et al. 1978). Figure 2 is a genetic map of the region of the *B. subtilis* chromosome within which the *spoVG* gene is located (Trowsdale et al. 1979; Rosenbluh et al. 1981). Transcription of *spoVG* is significantly impaired by mutations in the *spo0* loci *spo0A* and *spo0B*, and is almost completely abolished by deficiencies in a third *spo0* locus known as *spo0H* (also located in the *spoVG* chromosomal region, Fig. 2; Segall and Losick 1977; Ollington et al. 1978; Zuber and Losick 1983; Zuber 1985). We have taken advantage of the strong dependence of *spoVG* transcription on *spo0H* gene product as a tool for investigating the role of *spo0H* protein in developmental gene regulation. Here we discuss data that indicate that *spo0H* protein exerts its effect on *spoVG* transcription at a site within the *spoVG* promoter region, and that the dependence of *spoVG* expression on *spo0H* protein is highly sensitive to the (superhelical?) state of *spoVG* DNA. We also report on an extragenic suppressor mutation that allows *spoVG* expression in a *spo0A* mutant but not in a *spo0H* mutant.

Structure of the *spoVG* Promoter Region

The transcription initiation region of *spoVG* is composed of two overlapping promoters whose start points of RNA synthesis are separated by 10 bp (Moran et al. 1981; Johnson et al. 1983a). The upstream promoter is designated P1 and the downstream promoter P2. Transcription from P1 and P2 in vitro is catalyzed separately by two different forms of RNA polymerase holoenzyme that contain different species of RNA polymerase σ-factor. σ-Factors are subunits of RNA polymerase that confer on the core moiety of polymerase the ability to recognize and initiate transcription from promoter sites. *Bacillus* and certain other genera of bacteria contain multiple species of σ-factor that confer on polymerase the capacity to recognize different classes of promoters (Losick and Pero 1981; Grossman et al. 1984; Westpheling et al. 1985). Transcription from the P1 promoter is initiated by a holoenzyme form containing the σ^{37} species of σ-factor ($E\sigma^{37}$), whereas P2 RNA synthesis is carried out by σ^{32}-containing RNA polymerase ($E\sigma^{32}$) (Johnson et al. 1983a).

$E\sigma^{37}$ and $E\sigma^{32}$ are thought to direct transcription initiation from their corresponding promoters through recognition of cognate nucleotide sequences centered approximately 35 and 10 bp upstream from their respective transcription start sites. Putative recognition sequences for $E\sigma^{37}$ are indicated by the lines above the nucleotide sequence of Figure 3; underlined bases designate putative $E\sigma^{32}$ recognition sequences (Johnson et al. 1983a).

A striking structural feature of the *spoVG* promoter is the presence of an upstream region that is highly enriched for A:T base pairs, as identified by the stretch of boldface bases in Figure 3. This so-called "AT-box" is largely organized in alternating stretches of As and Ts. Interestingly, transcription from both the P1 and P2 start sites is strongly dependent upon AT-box sequences; deletion-mutated promoter DNAs which lack the upstream AT box are extremely poor templates for in vitro RNA synthesis by $E\sigma^{37}$ or $E\sigma^{32}$ RNA polymerases (Banner et al. 1983).

spo0H-dependent Promoter Utilization

To monitor conveniently the utilization of the *spoVG* promoter region in vivo, we (Zuber and Losick 1983) constructed a specialized transducing phage (a derivative of the *B. subtilis* temperate phage SPβ) bearing an in-frame fusion of the 5'-proximal region of the *spoVG* gene to the *lacZ* gene of *E. coli*. This SPβ::*spoVG-lacZ* phage contains only 157 bp of the *spoVG* transcription and translation initiation region and can be conveniently introduced in single copy into a variety of Spo⁻ mutants simply by specialized transduction. Using this SPβ derivative, we (Zuber and Losick 1983; Zuber 1985) examined the expression of the *spoVG-lacZ* fusion in a large number of Spo⁻ mutants, including the principal Spo0 mutants. This survey indicated that *spoVG* expression was most severely impaired in *spo0A*, *spo0B*, and *spo0H* cells. Of these three mutants, *spo0H* cells are of special interest because *spoVG*-directed β-galactosidase synthesis was almost totally eliminated in these bacteria. Table 1 shows that the level of *spoVG*-directed β-galactosidase synthesis in two *spo0A* mutants, including a strain bearing a deletion of *spo0A*, was severalfold reduced from that observed in Spo⁺ bacteria, but two independent *spo0H* mutants produced nearly undetectable levels of enzyme (less than 1% of the level in Spo⁺ cells). We conclude from these experiments that the *spo0H* gene plays an essential role in the induction of the *spoVG* promoters and that *spo0H* product or a protein under its

Figure 2. Genetic map of the origin of replication region of the *B. subtilis* chromosome.

```
                                                                              P1 →
ATCCTATTTTTTCAAAAAATATTTTAAAAACGAGCAGGATTTCAGAAAAAATCGTGGAATTGATACACTAATGCTTTTATATAGG
                                                                                P2 →
```

Figure 3. Nucleotide sequence of the nontranscribed strand of the *spoVG* promoter region. The lines designate putative "−10" and "−35" regions for the overlapping P1 and P2 promoters. The upstream AT-rich box is highlighted with bold letters.

control acts at or near the *spoVG* promoter region (that is, within the 157-bp segment contained in the *spoVG-lacZ* fusion). *spo0H* product could act to stimulate transcription initiation from P1 and P2 either by direct activation of the *spoVG* promoters (positive control) or by antagonism of a repressor of *spoVG* transcription.

An Extragenic Suppressor of *spo0A* Mutations Does Not Relieve the *spo0H* Requirement

spo0 mutations are highly pleiotropic, blocking not only spore formation at its earliest stage but also several of the properties of vegetative cells (for reviews, see Hoch et al. 1978; Losick 1981). Mutations in *spo0A*, the most highly pleiotropic locus, prevent sporulation-associated antibiotic and protease production, interfere with the acquisition of genetic competence and confer sensitivity during growth to surface-active, antibiotics and certain phages that are otherwise unable to grow in Spo+ bacteria. Mutations at a suppressor locus known as *abrB*, which maps in the replication origin region of the chromosome (Fig. 2), reverse many of the pleiotropic effects of *spo0A* mutations, although these *abrB* mutations do not enable *spo0A* mutants to sporulate (Trowsdale et al. 1977, 1979; Hoch et al. 1978). We wondered whether *abrB* mutations would relieve the dependence of *spoVG* expression on the *spo0A* gene. Table 1 shows that expression of *spoVG-lacZ* in an *abrB*, *spo0A* double mutant was as high as that observed in Spo+ bacteria. Thus, like antibiotic and protease production, the requirement for *spo0A* in *spoVG-lacZ* expression is not indispensable and can be relieved by an *abrB* mutation.

We also examined the effect of *abrB* on the requirement for *spo0H* in *spoVG* expression (Table 1). Interestingly, in this case, *abrB* had little or no effect in restoring active β-galactosidase synthesis, a finding that may indicate that *spo0A* is less directly involved than *spo0H* in *spoVG* expression.

Localizing the Site of *spo0H*-dependent Regulation

To localize more precisely the region within which the *spo0H* product exerts its effect, we (Zuber and Losick 1983; Zuber 1985) constructed deletion mutations in vitro, that extended into the promoter region of *spoVG* contained in a *spoVG-lacZ* fusion. One such mutant gene fusion was deleted for sequences upstream to the "−35" region of P1 and, as a consequence, entirely lacked the upstream AT box (Zuber 1985). This mutant promoter displayed only about 2% of the activity of the parental, wild-type promoter. Interestingly, however, the *pattern* of induction of β-galactosidase synthesis by the mutant promoter was indistinguishable from that of the wild type. Moreover, this residual mutant promoter-directed enzyme synthesis was completely dependent upon *spo0H* product, as little or no β-galactosidase could be detected in *spo0H* cells bearing *spoVG-lacZ* DNA lacking the AT box. We conclude that the AT box serves as a general enhancer of *spoVG* transcription, and that the target of *spo0H*-dependent developmental regulation is downstream

Table 1. Genetic Regulation of *spoVG-lacZ*

Strain	Genotype	Enzyme activity[a] (percent maximum)
ZB308	(SPβc2de12::Tn917::*spoVG-lacZ*)[b] *trpC2*	100
ZB233	(SPβc2de12::Tn917::*spoVG-lacZ*) *spo0A12, trpC2, pheA1*	12–26
ZB383	(SPβc2de12::Tn917::*spoVG-lacZ*) *spo0AΔ204, trpC2, pheA1*	9
ZB326	(SPβc2de12::Tn917::*spoVG-lacZ*) *spo0H17, trpC2, rpoB2*	0.3
ZB269	(SPβc2de12::Tn917::*spoVG-lacZ*) *spo0HΔHindD, trpC2, pheA1*	0.1
ZB369	(SPβc2de12::Tn917::*spoVG-lacZ*) *spo0AΔ204, abrB, trpC2, pheA1*	118
ZB416	(SPβc2de12::Tn917::*spoVG-lacZ*) *spo0AΔ204, spo0HΔHinD, abrB, trpC2, pheA1*	0.1

[a] Enzyme activity is expressed as a percent of the β-galactosidase-specific activity observed in Spo+ cells at 1–2 hr after the end of exponential phase growth (approximately 400 Miller units).

[b] An SPβ prophage bearing the *spoVG-lacZ* gene fusion.

from the AT box. This limits the region within which *spo0H* exerts its effect to a segment of 95 bp.

A Position Effect in *spo0H*-dependent Regulation

We (Zuber and Losick 1983; and Table 2) have unexpectedly discovered that the requirement for *spo0H* product in *spoVG* induction can be partially suppressed by propagation of the *spoVG-lacZ* fusion on a multicopy plasmid. This observation was made by examining the expression of the gene fusion contained in the *B. subtilis* plasmid pBD64 (Gryczan et al. 1980), which had been introduced by transformation into Spo⁻ cells bearing a mutation of the *spo0H* gene known as *spo0H17*. Table 2 (line 3) shows that the plasmid-borne gene fusion produced high levels of β-galactosidase in the asporogenous bacteria. This high level of enzyme synthesis was not simply a consequence of "readthrough" from a plasmid promoter, as deletion-mutated *spoVG-lacZ* DNA that had been cloned into the plasmid vector was inactive in directing β-galactosidase synthesis (line 4). This effect of propagating *spoVG-lacZ* on a plasmid was also not a consequence of titrating a repressor by having increased dosage of *spoVG* promoter DNA, because suppression of *spo0H17* was only observed in cis; that is, propagation of the *spoVG* promoter region on a multicopy plasmid did not permit the induction of *spoVG-lacZ* contained in single copy in the chromosome (data not shown). Moreover, amplification of *spoVG-lacZ* in the chromosome (achieved by multiple Campbell-like integration events) to a copy number of approximately 10 failed to bypass the requirement for *spo0H* product (line 2). We tentatively conclude that the state of *spoVG* DNA when borne on a plasmid (for example, its degree of superhelicity) is different from that of chromosomal DNA, and that this difference in DNA structure somehow bypasses the requirement for *spo0H* product in promoter utilization. Conceivably, enhanced superhelicity of plasmid-borne DNA could facilitate transcription initiation by Eσ^{37} and Eσ^{32} in the absence of fully functional *spo0H* protein.

Interestingly, however, suppression of *spo0H17* does not represent a complete bypass of the requirement for *spo0H* product. Plasmid-borne *spoVG-lacZ* propagated in cells containing an in vitro-constructed deletion mutation of the *spo0H* gene (*spo0HΔHinD*; Weir et al. 1984) failed to direct significant levels of β-galactosidase synthesis (line 7). Evidently, *spo0H17* produces a partially inactive product whose defect can be compensated by an alteration in the (superhelical?) state of the promoter DNA. Nevertheless, this defective *spo0H17* product must perform an indispensable function in Eσ^{37}- and Eσ^{32}-directed promoter utilization, as the *spoVG* promoters are apparently inactive in the entire absence of *spo0H* product.

DISCUSSION

Figure 4 summarizes our working model for the organization of the *spoVG* promoter region and its regulation by *spo0H* protein. The transcription initiation region of *spoVG* is composed of two overlapping promoters (P1 and P2), which are separately recognized by the Eσ^{37} and Eσ^{32} forms of *B. subtilis* RNA polymerase holoenzyme. As indicated in the diagram each holoenzyme (conceivably each σ-factor) is thought to contact its cognate "−10" and "−35" nucleotide sequences, which are arranged in the alternating order: Eσ^{37} "−35," Eσ^{32} "−35," Eσ^{37} "−10," Eσ^{32} "−10." Transcription from both overlapping promoters is strongly enhanced by an upstream AT box, which is

Table 2. A Position Effect for *spo0H*-dependent Regulation

Strain	Plasmid	Genotype	Enzyme activity (Miller units)
ZB366	—	*spoVG-lacZ*,[a] *spo0H17* *trpC2*, chr::Tn917HU146[b]	0.3
ZB337	—	(*spoVG-lacZ*)$_{10}$,[c] *spo0H17 trpC2*	0.6
ZB323	pZL205[d] (*spoVG-lacZ*)	*spo0H17, trpC2,* chr::Tn917HU146	183
ZB360	pZΔ303[e] (*spoVGΔ303-lacZ*)	*trpC2, pheA1*	1.1
ZB310	—	(SPβc2de12::Tn917::*spoVG-lacZ*) *spo0HΔHindD, trpC2, pheA1*	1.6
ZB339	—	(*spoVG-lacZ*)$_{10}$, *spo0H HindD, trpC2, pheA1*	1.8
ZB317	pZL205 (*spoVG-lacZ*)	*spo0HΔHindD, trpC2, pheA1*	3.3

[a]The *spoVG-lacZ* fusion was contained in pZL207 and inserted into the chromosome in single copy by Campbell recombination as described previously.
[b]A silent insertion of transposon Tn917 that was isolated by K. Sandman and P. Youngman (pers. comm.).
[c]The *spoVG-lacZ* fusion was contained in pZL207 and inserted and amplified in the chromosome to a copy number of about 10 by Campbell recombination and selection to a high level of plasmid-borne drug resistance (Cm^r).
[d]An autonomously replicating *B. subtilis* plasmid containing the *spoVG-lacZ* fusion.
[e]The same as pZL205 except that the gene fusion harbors an 86-bp deletion in the *spoVG* promoter.

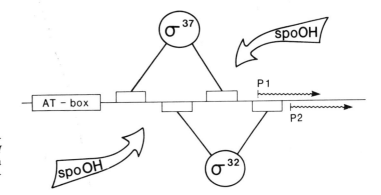

Figure 4. A speculative cartoon of the interaction of *spoVG* with Eσ^{37}, Eσ^{32}, and *spo0H* protein. The figure is not meant to imply a direct interaction of *spo0H* protein with promoter DNA.

largely organized in alternating stretches of As and Ts. This enhancement could be mediated by direct interaction of the AT box with the RNA polymerase, as it is known from "DNA footprinting" experiments that Eσ^{37} and Eσ^{32} partially protect AT box sequences from digestion by DNase I (Johnson et al. 1983b).

Efficient utilization of the *spoVG* promoters in vivo requires in addition to Eσ^{37} and Eσ^{32} the products of several *spo0* genes, putative regulatory proteins upon which the initial events of the *B. subtilis* differentiation cycle depend. Genetic experiments reported here (Table 1) and elsewhere indicate that the product of the *spo0H* gene plays a key role in the induction and expression of *spoVG*. Indeed, in the presence of the suppressor mutation *abrB*, *spo0H* protein may be the only *spo0* gene product required for *spoVG* induction.

How does the *spo0H* influence transcription from the *spoVG* promoters? From earlier work we know that the product of this regulatory gene does not determine significantly the abundance or apparently the activity of the minor holoenzyme forms (e.g., Eσ^{37} is present in extracts of Spo0H cells [C. Johnson, unpubl. results]; also, an Eσ^{37}- and Eσ^{32}-transcribed gene *ctc* does not depend on *spo0H* product for its transcription, and is actively transcribed in *spo0H* mutant cells in which *spoVG* is not expressed [Ollington et al. 1978; M. Igo, unpubl.]). We therefore suppose that the *spo0H* gene product acts in *conjunction* with the RNA polymerases to stimulate transcription from P1 or P2, either as a positive regulator or by antagonism of a repressor. Our deletion analysis shows that the target of *spo0H* action is downstream of the AT box within the region of the overlapping promoter recognition sequences. Conceivably, *spo0H* protein (or a protein under its control) interacts directly with promoter DNA or with the polymerases themselves. The recent cloning by I. Smith and colleagues (Dubnau et al. 1981; Weir et al. 1984) of the *spo0H* gene from *B. licheniformis* and *B. subtilis* should facilitate the isolation of *spo0H* product and make possible an examination of the effect of this protein on *spoVG* transcription in vitro. (The cloning of the *spo0H* gene has also led to the interesting discovery [Smith et al. 1985] that *spo0H* mutations cause enhanced transcription of the *spo0H* gene itself. Thus, *spo0H* may be autoregulatory in addition to its role in promoting *spo* gene transcription.)

Finally, we note that *spoVG* expression must be coupled somehow to environmental signals (nutrient deprivation) that trigger the onset of sporulation. It is appealing to suppose that this coupling is achieved by the products of one or more *spo0* genes, which could in turn regulate the synthesis or activity of *spo0H* protein.

ACKNOWLEGMENTS

This work was supported by a grant from the National Institutes of Health (GM18568). We thank C. Robinow for the electronmicrograph.

REFERENCES

BANNER, C.D.B., C.P. MORAN, and R. LOSICK. 1983. Deletion analysis of a complex promoter for a developmentally regulated gene from *Bacillus subtilis*. *J. Mol. Biol.* **168:** 351.

DUBNAU, E., N. RAMAKRISHNA, K. CABANE, and I. SMITH. 1981. Cloning of an early sporulation gene in *Bacillus subtilis*. *J. Bacteriol.* **147:** 622.

GROSSMAN, A.D., J.W. ERICKSON, and C.A. GROSS. 1984. The *htpR* gene product of *E. coli* is a sigma factor for heat-shock promoters. *Cell* **38:** 383.

GRYCZAN, T., A.G. SHIVAKUMAR, and D. DUBNAU. 1980. Characterization of chimeric plasmid cloning vehicles in *Bacillus subtilis*. *J. Bacteriol.* **141:** 246.

HOCH, J.A. 1976. Genetics of bacterial sporulation. In *Advances in genetics* (ed. E.W. Caspari), p. 67. Academic Press, New York.

HOCH, J.A., M. SHIFLETT, J. TROWSDALE, and S.M.A. CHEN. 1978. State 0 genes and their products. In *Spores VII* (ed. G. Chambliss and J.C. Vary), p. 127. American Society for Microbiology, Washington, D.C.

JOHNSON, W.C., C.P. MORAN, JR., and R. LOSICK. 1983a. Two RNA polymerase sigma factors from *Bacillus subtilis* discriminate between overlapping promoters for a developmentally regulated gene. *Nature* **302:** 800.

JOHNSON, W.C., C.P. MORAN, JR., C.D. BANNER, P. ZUBER, and R. LOSICK. 1983b. Anatomy of a complex procaryotic promoter under developmental regulation. In *Gene expression* (ed. D. Nierlich et al.), p. 235. A.R. Liss, New York.

LOSICK, R. 1981. Sigma factors, stage 0 genes and sporulation. In *Spores VIII* (ed. H.S. Levinson et al.), p. 48. American Society for Microbiology, Washington, D.C.

LOSICK, R. and J. PERO. 1981. Cascades of sigma factors. *Cell* **25:** 582.

LOSICK, R. and P. YOUNGMAN. 1984. Endospore formation in *Bacillus*. In *Microbial development* (ed. R. Losick and L.

Shapiro), p. 63. Cold Spring Harbor Laboratory, Cold Spring Harbor, New York.

Moran, C.P., Jr., N. Lang, C.D.B. Banner, W.G. Haldenwang, and R. Losick. 1981. Promoter for a developmentally regulated gene in *Bacillus subtilis*. *Cell* **25:** 783.

Ollington, J.F., W.G. Haldenwang, J.V. Huynh, and R. Losick. 1978. Developmentally regulated transcription in a cloned segment of the *Bacillus subtilis* chromosome. *J. Bacteriol.* **147:** 432.

Piggot, P.J., A. Moir, and D.A. Smith. 1981. Advances in the genetics of *Bacillus subtilis* differentiation. In *Spores VIII* (ed. H.S. Levinson et al.), p. 29. American Society for Microbiology, Washington, D.C.

Rosenbluh, A., C.D.B. Banner, R. Losick, and P.C. Fitz-James. 1981. Identification of a new developmental locus in *Bacillus subtilis* by construction of a deletion mutation in a cloned gene under sporulation control. *J. Bacteriol.* **148:** 341.

Segall, J. and R. Losick. 1977. Cloned *Bacillus subtilis* DNA containing a gene that is activated early during sporulation. *Cell* **11:** 751.

Smith, I., E. Dubnau, J. Weir, J. Oppenheim, N. Ramakrishna, and K. Cabane. 1985. Regulation of the *Bacillus spo0H* gene. In *Spores IX*. American Society for Microbiology, Washington, D.C. (In press.)

Trowsdale, J., S.M.H. Chen, and J.A. Hoch. 1977. Genetic analysis of phenotypic revertants of *spo0A* mutants in *Bacillus subtilis*. In *Spores VII* (ed. G. Chambliss and J.C. Vary), p. 131. American Society for Microbiology, Washington, D.C.

———. 1979. Genetic analysis of a class of polymyxin resistant partial revertants of stage 0 sporulation mutants of *Bacillus subtilis*: Map of the chromosome region near the origin of replication. *Mol. Gen. Genet.* **173:** 61.

Weir, J., E. Dubnau, N. Ramakrishna, and I. Smith. 1984. *Bacillus subtilis spo0H* gene. *J. Bacteriol.* **157:** 405.

Westpheling, J., M. Ranes, and R. Losick. 1985. RNA polymerase heterogeneity in *Streptomyces coelicolor*. *Nature* **313:** 22.

Youngman, P.J., P. Zuber, J.B. Perkins, K. Sandman, M. Igo, and R. Losick. 1985. New ways to study developmental genes in spore-forming bacteria. *Science* **228:** 285.

Zuber, P. 1985. Localizing the site of *spo0*-dependent regulation in the *spoVG* promoter of *Bacillus subtilis*. In *Spores IX*. American Society for Microbiology, Washington, D.C. (In press.)

Zuber, P. and R. Losick. 1983. Use of a *lacZ* fusion to study the role of *spo0* genes in developmental regulation in *Bacillus subtilis*. *Cell* **35:** 275.

Constitutive and Coordinately Regulated Transcription of Yeast Genes: Promoter Elements, Positive and Negative Regulatory Sites, and DNA Binding Proteins

K. STRUHL, W. CHEN, D.E. HILL, I.A. HOPE, AND M.A. OETTINGER
Department of Biological Chemistry, Harvard Medical School, Boston, Massachusetts 02115

The yeast genome specifies approximately 5000 protein coding genes that are densely clustered on 16 linear chromosomes. In the form of nuclear chromatin, these genes are transcribed by RNA polymerase II from discrete initiation sites. The average yeast gene is transcribed about five to ten times during each cell cycle which results in a steady-state mRNA level of one to two molecules per cell. However, some genes are transcribed constitutively at considerably different rates, whereas other genes are transcribed at variable rates depending on the physiological circumstances. Such regulated expression is achieved either by positive factors that increase transcription above a basal level or by negative factors that repress transcription below a basal level (for a general review of yeast promoters, see Struhl 1985a).

The transcriptional properties of the yeast genes investigated in these experiments are illustrated in Figure 1. *pet56*, *his3*, and *ded1* are adjacent but unrelated genes located on chromosome XV (Struhl and Davis 1981; Struhl 1985b). Under normal growth conditions, *his3* and *pet56* are transcribed at average levels, whereas *ded1* is transcribed at a fivefold higher level. Under conditions of amino acid starvation, *his3* transcription increases by a factor of 5, whereas the other genes are not affected (Struhl and Davis 1981). This physiological circumstance also results in the coordinate increase in the expression of many amino acid biosynthetic genes which are scattered around the genome (for review, see Jones and Fink 1982). The coordinately regulated *gal1-gal10* genes are closely linked on chromosome II, and they are transcribed divergently (St. John and Davis 1981). In glucose medium, the *gal* genes are not expressed, whereas in galactose medium, they each account for about 1% of the steady-state mRNA (40 molecules/cell). Thus, these genes provide representative examples of different kinds of yeast promoters.

Experimentally, mutant and hybrid promoters are created by a variety in vitro manipulations of cloned DNA, structurally characterized by DNA sequencing, and introduced back into yeast cells such that there is one copy per cell exactly at the normal chromosomal location; in many experiments, the DNAs to be tested directly replace the normal chromosomal sequence. In this way, mutations constructed in vitro are examined under true in vivo conditions for their effects on transcription and chromatin structure (for a review of this methodology, see Struhl 1983b). The transcriptional properties of these promoters are assayed by a quantitative 5′ mapping procedure which provides information concerning the level and the initiation sites.

This paper describes the properties of *cis*-acting elements that determine the accuracy, level, and regulation of transcription as well as the DNA binding properties of the *gcn4* positive regulatory protein. It also provides evidence suggesting that constitutive and regulatory promoters may operate by different molecular mechanisms.

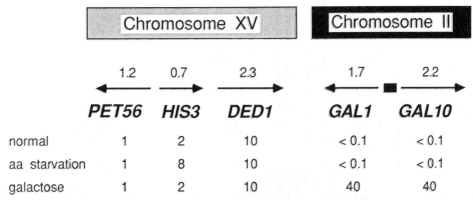

Figure 1. Structural and transcriptional properties of the *pet56-his3-ded1* and *gal1-gal10* genes regions. The *pet56-his3-ded1* region is located on chromosome XV (gray bar) and the *gal1-gal10* region is located on chromosome II (black bar). The size (kb), directionality, and relative order of the transcripts are indicated. The transcription levels observed during growth in normal conditions, amino acid starvation conditions, or growth using galactose instead of glucose as a sole source of carbon are indicated (in mRNA molecules/cell at the steady state).

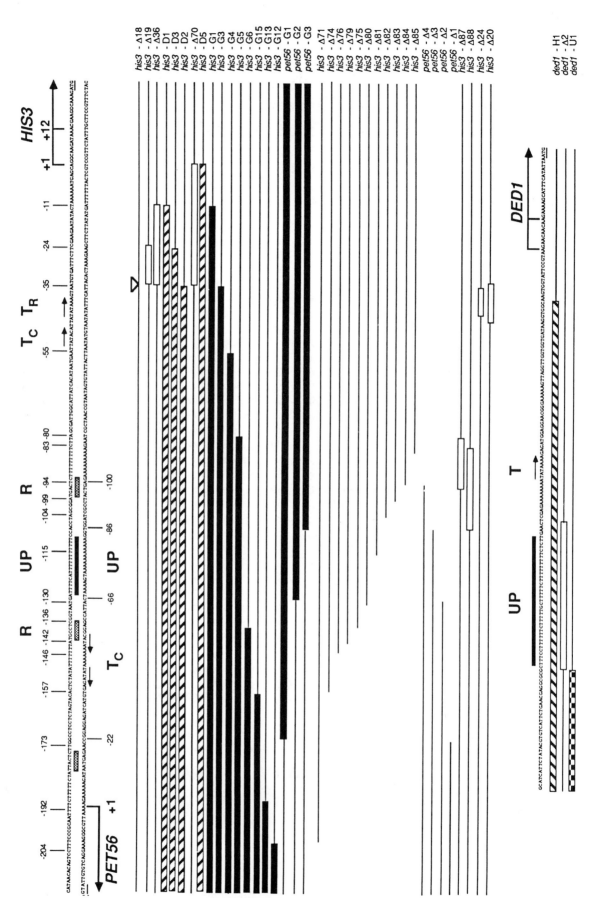

Figure 2. (See facing page for legend.)

ACCURATE INITIATION

Transcription of the wild-type *his3* gene is initiated with equal efficiency at two sites (defined as +1 and +12). At least two distinct elements, called the upstream and TATA sequences, are necessary for wild-type levels of transcription (Struhl 1981, 1982a). Deletion of either promoter element markedly reduces the level of expression. On the other hand, deletions that remove sequences between or adjacent to these elements do not significantly alter the basal expression level.

Which, if any, of these promoter elements determines where transcription begins? Analysis of mutant and hybrid promoters strongly suggests that unlike the situation in higher eukaryotes, accurate yeast mRNA initiation is determined primarily by an "initiator" element, not by the distance from the TATA sequence (Chen and Struhl 1985; Fig. 2; Table 1).

First, when the spacing between the *his3* TATA element and initiation sites is altered by an 8-bp insertion (Δ18), or by 3- and 16-bp deletions (Δ19, Δ36), *his3* transcription is indistinguishable from the wild-type gene. This indicates that the precise distance is relatively unimportant, and that something else must impart initiation specificity.

Second, hybrids between the *his3* and *ded1* promoter regions indicate that accurate initiation occurs even when transcription depends on "foreign" elements that are located at different positions from the "normal" elements. When the *ded1* initiation region downstream from −15 is fused to the *his3* TATA and upstream elements (*ded1*-H1), the initiation pattern is indistinguishable from the wild-type *ded1* gene. In this case, the *his3* TATA element is only 33 bp from the *ded1* +1 site as compared with the 60-bp separation in the wild-type *ded1* gene. Conversely, hybrid promoters (*his3*-D1 through D4) consisting of the *ded1* upstream and TATA elements fused to a series of *his3* segments containing the initiation region all initiate transcription at the +1 site. The distances between the *ded1* TATA element and *his3* initiation site range from 69 to 93 bp whereas the spacing in the wild-type *his3* promoter is 45 bp. These experiments indicate that the signal for accurate initiation depends on specific sequences downstream of the TATA element. In particular *his3* sequences upstream of −11 and *ded1* sequences upstream of −15 are unnecessary for accurate initiation.

Third, when the *his3* or *ded1* promoter elements are fused to *his3* nucleotide +2 (*his3*-Δ70 and *his3*-D5), transcription initiated from a position that is equivalent to +1 is not observed. This cannot be explained by a "closeness" effect because in *his3*-D5, the TATA element is 51 bp from the +1 equivalent site. Moreover, both derivatives produce the same initiation pattern, even though the TATA elements are located at different distances from the observed +12 and +22 mRNA start sites.

Although there is no precise spacing relationship between the TATA element and RNA start sites, there are distance limits over which a TATA element can act. The maximum distance appears to be about 90–100 bp. The best evidence for this comes from hybrid promoters (*his3*-D2, D3, D4) in which transcription from +12 is reduced compared with that from +1, and aberrant initiation from −3 to −6 is observed. The simplest interpretation for these observations is that the *ded1* TATA element is too far from the +12 site for efficient transcription. In *his3*-D2, the distance is 103 bp from the +12 site whereas it is only 92 bp from the +1 site, which is efficiently used. In *his3*-D4, the relevant distances are 92 and 81 bp. The minimal distance is more difficult to define. In *his3*-Δ36, the distance from the "best" sequence (TATAAA) is only 28 bp, similar sequences (TATATA and TATACA) are 30 and 37 bp away and lie within the region defined by functional criteria to be the TATA element (Struhl 1982a, 1984). Thus, it seems that the TATA element can be located anywhere between 30 and 90 bp from the initiation site with minimal functional consequences.

POSITIVE AND NEGATIVE REGULATORY ELEMENTS MEDIATE TRANSCRIPTIONAL CONTROL OF THE *gal1,10* GENES

The basic experiment is to fuse a 365-bp DNA fragment encoding the *gal* upstream regulatory region (Guarente et al. 1982; Guarente and Mason 1983) to a set of *his3* derivatives that systematically delete sequences upstream from the mRNA initiation sites (Struhl 1984, 1985c; Fig. 2). This DNA segment is responsible both for extremely high transcription in galactose medium as well as lowered levels in medium containing glucose and galactose (catabolite repression). Because the wild-type *his3* gene is always expressed but is not regulated as a function of growth on glucose or galactose, these *gal-his3* fusions are useful for distinguishing between positive and negative regulatory control mechanisms. The phenotypes of these fusions are summarized in Table 2.

Figure 2. The *his3-pet56* and *ded1* promoter regions. The nucleotide sequence of both strands of the divergent *his3-pet56* promoter region is shown at the top of the figure. *his3* deletion breakpoints are indicated as vertical lines above the top strand and *pet56* deletion breakpoints are indicated below the bottom strand. The positions of the poly(dA:dT) sequence for the constitutive expression of both genes (UP, black bar), the TGACTC regulatory sequences (R, gray bars), TATA elements (T_C, T_R, thin arrows), RNA initiation sites (vertical lines and thick arrows), and AUG initiation codons (underlined) are indicated. The structures of many of the mutant derivatives are shown below the DNA sequence. For each derivative, the horizontal line indicates the normal DNA sequences that are present, open boxes indicate internally deleted regions, striped boxes indicate the *ded1* promoter region (from −447 to −15), and the black boxes indicate the 365-bp *gal* regulatory region. The nucleotide sequence of the *ded1* promoter region is shown at the bottom of the figure. Symbols are as described above except that the striped bar indicates the *his3* promoter region (from −2500 to −35) and the checkered bar indicates *ura3* sequences.

Table 1. Requirements for Accurate Initiation

Alleles	End points		Spacing				Patterns of initiation			
			his3		ded1		his3		ded1	
			+1	+12	+1	+10	+1	+12	+1	+10
Wild type	–	–	45	56	65	74	+ +	+ +	+ +	+ +
Spacing mutants										
his3-Δ18	–35	–35	53	64	65	74	+ +	+ +	+ +	+ +
his3-Δ19	–35	–24	42	53	65	74	+ +	+ +	+ +	+ +
his3-Δ36	–35	–11	29	40	65	74	+ +	+ +	+ +	+ +
"Swap" mutants	ded1	his3								
his3-D1	–15	–11	69	80	65	74	+ +	+ +	+ +	+ +
his3-D3	–15	–24	82	93	65	74	+ +	+[a]	+ +	+ +
his3-D4	–15	–31	89	100	65	74	+ +	+[a]	+ +	+ +
his3-D2	–15	–35	93	104	65	74	+ +	+[a]	+ +	+ +
	his3	ded1								
ded1-H1	–35	–15	–	–	33	42	–	–	+ +	+ +
"Initiator" mutants										
his3-Δ70	–35	+2	18[b]	29	65	74	–	+ +[c]	+ +	+ +
	ded1	his3								
his3-D5	–15	+2	58[b]	69	65	74	–	+ +[c]	+ +	+ +

Spacing mutants represent deletions and insertions between the *his3* TATA element and the normal initiation sites; swap mutants represent hybrids between *his3* and *ded1* promoter and initiation regions; initiator mutants represent deletions that remove part of the initiation region (Chen and Struhl 1985; see Fig. 2). The end points are measured with respect to the normal *his3* or *ded1* +1 initiation site, and they refer to the last nucleotide that is present before the EcoRI linker joint. The spacing indicates the distance between the most upstream T of the relevant TATA element and the *his3* or *ded1* initiation sites. For each normal initiation site, transcriptional levels are indicated as follows: (+ +) the level that is expected from the upstream promoter element; (+) detectable, but lower than expected levels; and (–) very low or undetectable RNA levels.

[a]There are also aberrant initiations at –3 to –6.

[b]Since the +1 nucleotide is removed in these two derivatives, these numbers refer to the distances corresponding to the equivalent +1 position (i.e., to the C residue of the EcoRI site).

[c]The +22 site, normally a minor *his3* initiation site, is utilized at equal or nearly equal efficiency as the +12 site.

Positive Control is Mediated by an Enhancer-like Sequence

The *gal* regulatory region possesses many properties of enhancer elements (Struhl 1984). When appropriately fused to the *his3* mRNA coding region, it confers extremely high transcriptional activity, but only in galactose medium. In terms of mRNA molecules per cell, *his3* expression in such *gal-his3* fusion strains occurs at equivalent levels to those observed for the wild-type *gal1* and *gal10* genes. Although the *gal* region is clearly necessary for this activation, the region between –35 and –55, which coincides precisely with the *his3* TATA element, is also required (compare *his3*-G3 and *his3*-G4). The *gal* element activates transcription at various distances from the *his3* initiation site, even as far as 600 bp away (*his3*-G11). Finally, the *gal* element functions equally well in either orientation with respect to the *his3* coding sequences (compare *his3*-G4 and *his3*-G16), and it initiates transcription in an indistinguishable manner. Fine-scale deletion analysis and inversion of smaller fragments indicate that the *gal* segment does not contain two separate sites that independently regulate *gal1* and *gal10* expression (Johnston and Davis 1984; West et al. 1984; Giniger et al. 1985). Thus, a bidirectional element coordinately regulates the divergently transcribed *gal1* and *gal10* genes.

One way in which the *gal* regulatory fragment apparently differs from enhancer sequences is that it does not activate transcription even when placed only 82 bp downstream from the *his3* initiation region (Struhl 1984). Trivial explanations for this result, such as distance effects and unstable RNA species due to the insertion of the *gal* element into the *his3* structural gene, have been excluded. However, one possible explanation is that the region including the TATA and initiator elements may "block" galactose activation in a manner analogous to effects mediated by the bacterial *lexA* repressor protein binding to its cognate operator site (Brent and Ptashne 1984). Indeed, in both experiments that demonstrated the inability of upstream elements to function in a downstream location, this possibility was addressed and blocking effects were observed (Guarente and Hoar 1984; Struhl 1984). Thus, it may be that the properties of yeast upstream elements are not inherently different from enhancer sequences, but rather that these effects are due to apparent differences in the mechanism of transcriptional initiation, such as described in the previous section.

The *gal* element differs from the SV40 (Jongstra et al. 1984) and glucocorticoid receptor (Zaret and Yamamoto 1984) enhancers in terms of nuclease hypersensitivity of chromatin. Although the *gal* regulatory region is hypersensitive to DNase I digestion, this structural feature does not correlate with transcriptional activation (Struhl 1984). It is observed both in glucose and in galactose medium, and in the presence or absence of a TATA element. These results implicate

Table 2. Analysis of *gal-his3* Fusions

Allele	End point	Glucose	Galactose
his3-G1	−8	−	−
his3-G2	−23	−	−
his3-G3	−35	−	−
his3-G4	−55	−	+++
his3-G16	−55	−	+++
his3-G5	−80	−	+++
his3-G6	−136	±	+++
his3-G15	−157	±	+++
his3-G14	−173	±	+++
his3-G13	−192	±	+++
his3-G12	−204	±	+++
his3-G7	−253	+	++
his3-G8	−330	+	++
his3-G9	−357	+	++
his3-G11	−389	+	++
pet56-G1	−86	±	±
pet56-G2	−66	±	±
pet56-G3	−22	−	−

For each allele, the 365-bp DNA segment containing the *gal* upstream regulatory sequences (indicated as black boxes in Figs. 1 and 2) was fused to the *his3* or to the *pet56* promoter region at the positions listed (Struhl 1984; 1985c; see also Fig. 2). The difference between *his3*-G4 and *his3*-G16 is the orientation of the *gal* segment with respect to the *his3* transcription unit. For each allele, appropriate strains were grown in broth containing 2% glucose or galactose as the sole source of carbon. *his3* or *pet56* transcription levels were assayed by hybridizing to completion 50 μg of RNA with an excess of the relevant ^{32}P-end-labeled, single-stranded DNA probe, digesting the reaction mixture with S1 nuclease, and separating the denatured products by polyacrylamide gel electrophoresis (see Fig. 3). The transcription levels are indicated as follows: (+) normal constitutive levels of *his3* or *pet56* RNA; (−) no detectable RNA; (±) low RNA levels, roughly 20% of normal; (+++) RNA levels that are equivalent to those of the normal *gal1-gal10* genes (i.e., complete induction); and (++) RNA levels that are approximately fivefold above the normal *his3* level but below the completely induced *gal* levels (Struhl 1984, 1985b).

a protein bound to this region which requires a conformational change or another protein for galactose induction. Presumably, this effect is not due to the *gal4* regulatory protein which interacts directly with four related 17-bp sequences located within the 365-bp DNA fragment (Bram and Kornberg 1985; Giniger et al. 1985).

The Catabolite Repression Regulatory Site Exerts Its Effects When Upstream of an Intact Promoter Region

Catabolite repression is a phenomenon whereby the expression of genes involved in carbon metabolism is reduced when glucose is present in the growth medium (Magasanik 1962). In principle, the molecular mechanism for catabolite repression could involve positive and/or negative control. In other words, the lowered levels in glucose medium could reflect either a failure to activate transcription (positive control) or actual repression of transcription that would otherwise occur (negative control). In *Escherichia coli*, catabolite repression, despite its name is a positive control mechanism because it depends on transcriptional activation by the cAMP:CAP protein complex (for review, see deCrombrugghe et al. 1984).

The properties of several *gal-his3* fusions strongly suggest that catabolite repression in yeast occurs by a true repression mechanism (Struhl 1985c; Table 2). When the 365-bp *gal* regulatory site is fused to *his3* derivatives that contain the entire promoter region (*his3*-G12 through G15), transcription in glucose medium is repressed below the wild-type basal level. This effect is clearly due to catabolite repression because it is abolished in raffinose medium, and the observed repression affects both the normal and the induced *his3* levels (Struhl 1985c). Therefore, the behavior of alleles *his3*-G12 through G15 in glucose medium cannot be explained by the lack of a functional activator protein, because if this were the case, wild-type *his3* expression levels should have been observed.

Two other points are worth noting. First, catabolite repression and galactose activation are separable properties of the *gal* regulatory region. In particular, catabolite repression of *his3* expression is observed in the absence of galactose. This strongly suggests that these two regulatory effects are mediated, at least in part, by different DNA sequences, a conclusion supported by deletion mutants of the *gal* segment described by West et al. (1984). Second, the ability of the *gal* region to exert its effects even when located at variable positions upstream from an intact promoter is markedly different from standard repression sites. In *E. coli*, the repression sites overlap or are between the required promoter elements (for review, see Reznikoff and McClure 1985), and in yeast, *lexA* repression of galactose induction occurs only if the operator site is between the *gal* upstream regulatory site and the TATA element (Brent and Ptashne 1984). Thus, it seems probable that the repression mechanism does not involve steric competition between the presumptive regulatory protein and the transcriptional apparatus. Instead, it seems possible that repression and enhancer-like activation represent opposite sides of the same basic mechanism.

POLY(dA:dT) SEQUENCES ACT AS UPSTREAM ELEMENTS FOR CONSTITUTIVE EXPRESSION

The *his3*, *pet56*, and *ded1* upstream promoter elements necessary for wild-type levels of transcription were determined by analyzing the phenotypes of deletion mutants that successively remove DNA sequences upstream from the respective structural genes (Struhl 1985d; Fig. 2; Table 3). Such sequential 5′ deletion analysis defines the minimum contiguous sequence necessary for wild-type transcription levels. For all three genes, the critical sequences are naturally occurring stretches of poly(dA:dT) located upstream of the TATA element. Deletion mutants that retain a particular stretch of dA:dT residues are transcribed equally efficiently as the wild-type gene, whereas related derivatives that lack this region are transcribed significantly below the normal level.

Table 3. Deletions of the *his3*, *pet56*, and *ded1* Promoter Regions

Allele	End point	Normal	Starvation
HIS3+	—	+	+++
his3-Δ71	−204	+	+++
his3-Δ72	−192	+	+++
his3-Δ73	−173	+	+++
his3-Δ74	−157	+	+++
his3-Δ76	−146	+	+++
his3-Δ79	−142	+	+++
his3-Δ75	−136	+	++
his3-Δ80	−130	+	++
his3-Δ81	−115	±	++
his3-Δ82	−104	±	++
his3-Δ83	−99	±	++
his3-Δ84	−94	−	−
his3-Δ85	−83	−	−
his3-Δ87	−95/−80	+	+
his3-Δ88	−109/−83	+	+
his3-Δ24	−44/−35	+	+
his3-Δ20	−46/−34	+	+
his3-142	−100/−98	+	+
PET56+	—	+	+
pet56-Δ4	−100	+	+
pet56-Δ3	−86	+	+
pet56-Δ2	−66	±	NT
pet56-Δ1	−22	−	NT
DED1+	—	+++	+++
ded1-U1	−123	+++	+++
ded1-Δ2	−123/−78	+	NT

Alleles with one listed end point represent deletions that remove all sequences upstream of that position; those with two end points indicate internal deletions that remove sequences between these positions (Struhl, 1982b, 1985c,d; D.E. Hill and K. Struhl, in prep.; Fig. 2). For each allele, wild-type and *gcd1*⁻ strains were grown in broth containing 2% glucose as the sole source of carbon; *gcd1*⁻ strains cause induced levels of amino acid biosynthetic genes under all conditions, and hence provide a convenient method to simulate amino acid starvation (Wolfner et al. 1975; Struhl 1982b; Donahue et al. 1983). *his3* transcription levels were assayed as described in Table 2 and Fig. 3; a *ded1* hybridization probe was included to provide an internal standard (Struhl 1985d; D.E. Hill and K. Struhl, in prep.). The transcription levels are indicated as follows: (+) normal constitutive levels of *his3* RNA; (−) no detectable RNA; (±) low RNA levels, roughly 20% of normal; (+++) levels equivalent to complete induction of the wild-type *his3* gene; and (++) partial induction (see text).

For the *his3* gene, deletions that retain as few as 130 bp upstream from the mRNA coding region behave indistinguishably from the wild-type gene, whereas related deletions that extend to −115 or further are transcribed at about 20% of the wild-type level. Earlier analyses of the *his3* promoter region indicate that besides the TATA element, sequences downstream from −115 are not important for normal expression levels (Struhl 1982a). Thus, the region between −115 and −130, which coincides with a 17-bp region containing 15 dT residues in the coding strand, is necessary and sufficient for constitutive expression.

Similar analysis of the *pet56* promoter indicates that this same 17-bp poly(dA:dT) region defines the upstream promoter element. In particular, sequences more than 86 bp upstream from the *pet56* initiation site are unimportant for full promoter function, whereas the region between −66 and −86 is necessary.

In other words, although *his3* and *pet56* encode genes of unrelated function, the same poly(dA:dT) region located between these divergently transcribed genes is responsible for the constitutive transcription of both. This suggests that poly(dA:dT) sequences activate transcription in a bidirectional manner.

The *ded1* gene contains a typical TATA sequence located between −60 and −65 as well as a 34-bp region between −88 and −121 that contains 28 dT residues in the coding strand. Deletion of this region (*ded1*-Δ2) reduces *ded1* transcription to about 20% of the normal level. On the other hand, insertion of a *ura3* DNA fragment at position −121 (*ded1*-U1) does not affect *ded1* RNA levels, suggesting that sequences upstream of −125 are relatively unimportant. The fact that the *ded1* poly(dA:dT) region is longer than the *his3*-*pet56* region may explain why constitutive *ded1* RNA levels are roughly five times higher than *his3* or *pet56* levels.

The influence of long poly(dA:dT) sequences was initially observed in constitutive up-promoter mutants of the *adr2* gene (Russell et al. 1983). In two such mutants, the expansion of a normal 20-bp dA:dT sequence to a 54- or 55-bp homopolymer stretch presumably causes high levels of constitutive expression. Presumably, these rare mutants reflect functions of poly(dA:dT) tracts that are actually used in wild-type genes.

These are two mechanisms by which poly(dA:dT) sequences might confer their effects. One possibility is that a transcription factor binds to such sequences and activates transcription. A second and particularly attractive suggestion is that the RNA polymerase II transcription machinery recognizes the unusual structure of poly(dA:dT). Such sequences have a helix repeat of 10.0 bp instead of the normal 10.6 (Peck and Wang 1981; Rhodes and Klug 1981), and they are associated with kinks in DNA (Marini et al. 1983). Of particular interest is the observation that dA:dT regions prevent nucleosome formation in vitro (Kunkel and Martinson 1981; Prunell 1982). An attractive feature of this second suggestion is that because specific transcription factors are not invoked, it explains why poly(dA:dT) sequences behave as constitutive upstream elements.

his3 UPSTREAM REGULATORY ELEMENTS

When yeast cells are subjected to conditions of amino acid starvation, transcription of *his3* and other amino acid biosynthetic genes is induced over the basal level. Deletion mutations analyzed previously indicate that this regulation of *his3* expression depends on a sequence between −83 and −103 (Struhl 1982b). Deletion mutations that lack this region express *his3* at the normal basal level, but are unable to induce transcription in response to starvation conditions. Thus, the −83 to −103 region encodes a positive regulatory site necessary for transcriptional induction.

The TGACTC sequence (−99 to −94) within the *his3* regulatory region is of particular interest because it is repeated, with minor variations, several times in promoters of coregulated genes (Struhl 1982b; Donahue et al. 1983; Hinnebusch and Fink 1983a). In the *his3* gene, perfect TGACTC sequences are found at −99 to −94 and −258 to −263, and imperfect variants containing five out of six matches are found at −142 to −137, −181 to −176, −216 to −221, and −225 to −230 (Struhl 1985b).

Two TGACTC Sequences Are Necessary for Full *his3* Induction

Sequential 5'-deletion mutants similar to those described in the section on constitutive expression were analyzed for their abilities to induce *his3* expression to the level observed for the wild-type gene (D.E. Hill and K. Struhl, in prep.). A fortuitous feature of *his3* transcription provides a very sensitive method for distinguishing between constitutive and regulated expression. As will be discussed in detail in a later section, when wild-type strains are subjected to conditions of amino acid starvation, transcription from the +12 site is induced fivefold whereas transcription from the +1 site is unaffected. Since the relative levels of the +1 and +12 transcripts are directly determined, even twofold changes in the ratio (which correspond to a total induction of only 50%) are easily observed.

Deletion mutants that encroach as far downstream as −142 behave indistinguishably from the wild-type gene (Table 3). Thus the four copies of the TGACTC repeat that lie upstream from this point are unimportant for regulation. However, deletions that extend to −136 or as far downstream as −99 fail to induce *his3* expression to the maximal level, although partial induction is observed. Because the only difference between the −142 and −136 derivatives is the imperfect TGCCTC sequence, this result indicates that two copies of the conserved sequence are necessary for full induction.

Several lines of evidence indicate that the proximal TGACTC sequence located between −99 and −94 is essential for induction (D.E. Hill and K. Struhl, in prep.). First, small deletions (*his3*-Δ87, *his3*-Δ88) that remove this region but retain all the other repeat sequences are uninducible (Table 3). Second, *his3*-Δ83, which removes all nucleotides upstream from the proximal TGACTC sequence shows the typical partial induction phenotype, whereas a derivative that extends 5 bp further and hence retains only the last C residue is transcriptionally defective under all conditions. This result also indicates that the proximal regulatory site is sufficient to confer partial levels of induction. Third and most compelling, a single base pair deletion of the dT residue at position −99 (*his3*-142) abolishes *his3* inducibility (Table 3).

Nucleotide Requirements of the TGACTC Regulatory Site

By using a set of synthetic oligonucleotides corresponding to the region between −102 and −91, mutants that differ by single base pair changes in the TGACTC regulatory site have been analyzed (D.E. Hill and K. Struhl, in prep.). Specifically, these oligonucleotides were joined via a SacI linker to a *his3* test fragment that contains all sequences downstream of nucleotide −83. Yeast strains containing the test fragment are defective in *his3* transcription and consequently are unable to grow in the absence of histidine. However, when the wild-type oligonucleotide is fused to the test fragment, the resulting strains do grow. Thus, this growth phenotype provides a simple and qualitative assay for the effects of such point mutations with respect to *his3* transcription.

There are 18 possible base pair substitutions of the TGACTC regulatory sequences, of which 13 have been examined. As shown in Table 4, 12 out of these 13 do not permit growth in the absence of histidine, and hence are functionally defective. The sole exception is a C:T change resulting in the sequence TGACTT (*his3*-158). The fact that each nucleotide of the consensus can be mutated to produce a nonfunctional regulatory site indicates that each nucleotide is of some importance.

Although the TGACTC sequence is clearly essential, several lines of evidence suggest that the nine dT residues immediately downstream from the regulatory site are also important. First and most important, when the wild-type oligonucleotide is joined to the test fragment (*his3*-145), the resulting strains do not grow in the presence of aminotriazole, a competitive inhibitor of IGP dehydratase, the *his3* gene product. In contrast, a deletion mutant that simply removes all sequences upstream of −102 (listed as HIS3+ in Table 4) permits growth under these circumstances. The only difference between these two derivatives is that the six downstream-most dT residues after the TGACTC sequence are replaced by a SacI linker (GAGCTC). A derivative containing six dT residues (*his3*-143) is fully functional, whereas a derivative containing four dT residues (*his3*-144) is only partially functional. Second, the only functional point mutation (*his3*-158) produces a stretch of five dT residues; in comparison, the wild type and all the other derivatives contain only three dT residues. Third, other coregulated genes such as *his4*, *arg5*, and *trp5* all have dT tracts just downstream of the TGACTC sequence. Fourth, enhanced DNase I cleavage at the dT residues is observed when the GCN4 protein is bound to the regulatory sequences (Hope and Struhl 1985; see below).

gcn4 PROTEIN BINDS TO THE *his3* REGULATORY SEQUENCES

his3 transcription is regulated coordinately with many other genes involved in amino acid biosynthesis. *Trans*-acting factors involved in this general control phenomenon have been identified by mutations that fail to regulate properly the biosynthetic genes (Schurch et al. 1974; Wolfner et al. 1975; Penn et al. 1983; Hinnebusch and Fink 1983b). Epistatic relationships

Table 4. Point Mutations of the *his3* Regulatory Site

Allele	Sequence	Expression
HIS3+	GGAT<u>GACTC</u>TTTTTTTTTC	+
his3-143	GGATGACTCTTTTTTGAGCTC	+
his3-144	GGATGACTCTTTTGAGCTC	±
his3-145	GGATGACTCTTTGAGCTC	±
his3-146	GGATGAC<u>A</u>CTTTGAGCTC	–
his3-147	GGATGAC<u>C</u>CTTTGAGCTC	–
his3-148	GGATGAC<u>G</u>CTTTGAGCTC	–
his3-149	GGA<u>A</u>GACTCTTTGAGCTC	–
his3-150	GGA<u>C</u>GACTCTTTGAGCTC	–
his3-151	GGA<u>T</u>GA<u>A</u>TCTTTGAGCTC	–
his3-152	GGATGA<u>T</u>TCTTTGAGCTC	–
his3-153	GGATC<u>A</u>CTCTTTGAGCTC	–
his3-154	GGAT<u>A</u>ACTCTTTGAGCTC	–
his3-155	GGAT<u>T</u>ACTCTTTGAGCTC	–
his3-156	GGATG<u>T</u>CTCTTTGAGCTC	–
his3-157	GGATG<u>C</u>CTCTTTGAGCTC	–
his3-158	GGATGACT<u>T</u>TTTTGAGCTC	+

For each allele, the DNA sequence between positions -102 and -83 is shown; in all cases, sequences between -102 and -447 are deleted (D.E. Hill and K. Struhl, in prep.; see text). For the wild-type derivative, the TGACTC sequence is underlined; in the point mutations, the altered base is underlined. *his3* expression was determined by the ability of strains to grow in the absence of histidine. (+) Growth equivalent to the wild-type derivative; (±) slower growth; and (−) no growth.

among these mutations suggest that the *gcn4* gene product has the most direct role in the transcriptional regulation process (Hinnebusch and Fink 1983b).

Recent experiments demonstrate that the *gcn4* gene encodes a specific DNA binding protein. This was accomplished by using a new and general method for analyzing protein-DNA interactions (Hope and Struhl 1985). Specifically, the *gcn4* protein-coding sequences were cloned into a vector containing a promoter for SP6 RNA polymerase, mRNA was synthesized by transcribing the template with this enzyme, and *gcn4* protein was synthesized as a pure ^{35}S-labeled species by in vitro translation of this mRNA. DNA binding activity was detected by incubating the labeled protein with specific DNA fragments, and separating protein-DNA complexes from free protein by electrophoresis in native acrylamide gels. By this assay, protein-DNA complexes are observed when *gcn4* protein is incubated with *Taq*I-cleaved pUC8-*his3* DNA (contains the 1.7-kb *Bam*HI fragment) (Hope and Struhl 1985). Such specific complexes are not observed with the vector DNA, although faint bands on the autoradiogram indicate that the protein does possess some nonspecific binding activity.

Four lines of evidence indicate that *gcn4* protein binds specifically to *his3* regulatory sequences and to promoter regions of other coregulated genes (Hope and Struhl 1985). First, analysis of deletion DNAs described previously indicates that the 20 bp region between -85 and -104 is necessary and sufficient for *gcn4* binding. This corresponds precisely with the proximal TGACTC regulatory sequences. Second, *gcn4* protein protects a 10-bp region including the TGACTC nucleotides from DNase I cleavage. This directly demonstrates an interaction between the critical regulatory sequences and the *gcn4* protein. Third, *gcn4* mutant proteins lacking various amounts of the carboxyl terminus, which were generated after restriction endonuclease cleavage of the original template, have no specific or nonspecific DNA binding activity. In particular, the mutant protein lacking the carboxyterminal 40 amino acids is almost identical to the predicted protein produced by strains carrying the *gcn4*-Δ1306 allele of Hinnebusch (1984). Thus, the fact that such strains are unable to induce the transcription of the co-regulated genes is probably due to the inability of the mutant *gcn4* protein to bind DNA. Fourth, *gcn4* protein binds to the promoter regions of three other coregulated genes (*his4*, *trp5*, *arg4*), whereas it does not bind to analogous regions of four unregulated genes (*ded1*, *gal1*, *10*, *ura3*, *trp1*). All these results indicate that at the level of protein and DNA sequences, there is a direct correlation between DNA binding in vitro and transcriptional activation in vivo.

CONSTITUTIVE AND REGULATORY *his3* EXPRESSION

The results presented in previous sections indicate that the *his3* gene contains two distinct upstream promoter elements. Constitutive expression depends on the poly(dA:dT) sequence between -130 and -115, whereas induced expression in response to conditions of amino acid starvation depends on the TGACTC regulatory sequences. Thus, constitutive and regulated *his3* expression is most simply explained by a core promoter being activated by two independent upstream elements with different transcriptional specificity. However, the observations reported below strongly suggest that the constitutive and regulatory promoters for *his3* expression are qualitatively different.

Regulated *his3* Transcription Is Initiated at the +12 Site

Under normal growth conditions, *his3* transcription is initiated at equal efficiency from positions +1 and +12. Surprisingly, under conditions of amino acid starvation, transcription from +1 remains at the normal basal level, whereas transcription from +12 is induced about fivefold (K. Struhl, in prep.; see Fig. 3). In addition, transcription from +22, normally a minor initiation site, is also induced. Thus, constitutive and regulated modes of *his3* expression are distinguished not only by their required upstream sequences, but also by their utilization of transcriptional initiation sites.

The same selectivity is observed in strains containing *gal-his3* fusion promoters in which the *his3* upstream region is replaced by the *gal1,10* enhancer-like sequence (Struhl 1984; Fig. 3). When such strains are grown in galactose medium, essentially all the transcripts are initiated at +12 and +22. This result does not depend on the location of the *gal* regulatory site because the same transcription pattern is observed in many different *gal-his3* fusions (Fig. 3).

A third example of this phenomenon is represented by revertants of *his3*-Δ13, a promoter mutation that contains the TATA element but lacks the entire upstream promoter region (Oettinger and Struhl 1985; Fig. 3). These revertants are all due to recessive suppressor mutations in three different genes, *ope1*, *ope2*, and *ope3*. In strains containing *his3*-Δ13 and any one of the *ope* suppressor mutations, *his3* transcripts initiated preferentially from the +12 site are observed in minimal medium, whereas no transcripts from +1 or +12 are detected in rich broth. The suppressor mutations presumably cause their transcriptional effects by activating cryptic upstream promoter elements.

Thus, in these three examples, the same initiation pattern is observed even though *his3* transcription depends on different upstream regulatory elements. This

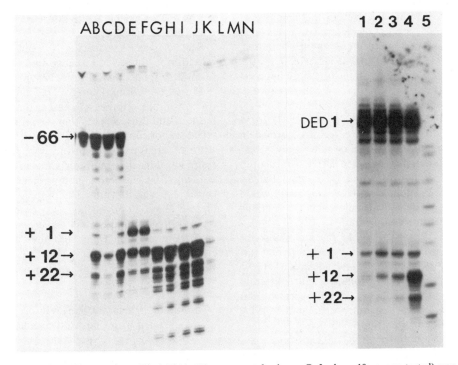

Figure 3. 5' mapping of RNA transcripts. Total RNA (50 μg except for lanes G–J where 10 μg was tested) was hybridized to completion with an excess of a single-stranded *his3* probe (end-labeled with ^{32}P at position +174 and extending to −136), and in some cases with a *ded1* probe (end-labeled at position +262 and extending to −75). The reaction products were treated with S1 nuclease, denatured, and subjected to electrophoresis in a 6% polyacrylamide gel containing 7 M urea (Chen and Struhl 1985; Struhl 1985b). In lanes A–D, RNAs were prepared from cells grown in minimal medium from strains of the following genotypes: *his3*-Δ13 (lane A); *his3*-Δ13, *ope1* (lane B); *his3*-Δ13, *ope2* (lane C), and *his3*-Δ13, *ope3* (lane D). The *his3*-Δ13 allele, which deletes all *his3* sequences upstream of −66, does not express *his3*; the recessive *ope* suppressor mutations restore wild-type levels of expression (Oettinger and Struhl 1985). Lanes E and F contain RNA from a wild-type strain grown in broth containing 2% glucose (lane E) or 2% galactose (lane F) as the sole carbon source. RNAs from the following *gal-his3* fusions are analyzed in lanes G–J (galactose medium) and K–N (glucose medium): *his3*-G13 (lanes G,K); *his3*-G15 (lanes H,L); *his3*-G6 (lanes I,M); and *his3*-G4 (lanes J,N). In the autoradiogram on the right side of the figure, RNAs were examined from strains of the following genotypes: *gcn4*-1 (lane *1*); *gcn4*-2 (lane *2*); wild type (lane *3*); *gcd1*-1 (lane *4*); molecular weight standards (lane *5*). As described in the text, *gcn4* encodes the positive regulatory protein that binds the *his3* regulatory sequences; *gcd1*-1 is a mutation that causes induced *his3* transcription under all conditions, and hence is equivalent to amino acid starvation. The positions representing *ded1* transcripts and *his3* transcripts that initiate from nucleotides +1, +12, and +22 as well as read-through transcripts (RT) initiating upstream of nucleotide −66 (only relevant for strains containing *his3*-Δ13) are indicated.

suggests that although different proteins interact with these different regulatory sequences, the basic mechanism of transcriptional activation is similar. In contrast, the observed initiation pattern during constitutive *his3* expression is qualitatively different. This suggests that the poly(dA:dT) upstream element behaves in a functionally distinct manner from the regulatory elements.

pet56 Transcription Is Uninducible by the *his3* or *gal* Elements

The promoter region for the divergently transcribed *his3* and *pet56* genes appears symmetrical. Constitutive transcription of both genes is mediated by a centrally located dA:dT region that acts as a bidirectional upstream promoter element. Copies of the TGACTC regulatory elements are located between this common dA:dT region and the *his3* and *pet56* TATA elements. However, even though the TGACTC regulatory sequence functions in both orientations (Hinnebusch et al. 1985), *pet56* transcription is not increased when wild-type cells are subject to amino acid starvation (Struhl and Davis 1981).

To determine whether this lack of inducibility was due to properties of the *his3* regulatory sequences (e.g., the *his3* proximal copy is identical to the consensus, whereas the *pet56* proximal copy is imperfect), the *gal* regulatory element was fused to the *pet56* promoter region at several positions. In striking contrast to *gal-his3* and *gal-cyc1* fusions, which used the identical *gal* DNA segment (Guarente et al. 1982; Struhl 1984), *pet56* transcription is not induced when cells are grown in galactose medium (Table 2). This result cannot be explained by failure to include *pet56* TATA sequences in the fusions. Both *pet56*-G2 and G3 contain the sequence TATACA implicated as a functional *pet56* TATA element (Struhl 1985d), and *pet56*-G3 contains the entire *pet56* promoter region.

Thus, the *pet56* transcript, like the *his3* transcript initiating at +1, is not activated by two different upstream regulatory elements. This indicates that the constitutive and regulatory promoters must differ in ways other than their upstream promoter elements. In principle, such differences could occur at the TATA element, the initiator element, or other sequences that have not yet been defined. From the experiments reported below, I suggest that there are two distinct classes of TATA elements.

Evidence for Two Classes of TATA Elements

The basic model is that TATA elements (overall consensus sequence TATAAA) can be divided into two classes, constitutive and regulatory. The functional distinction between these hypothetical classes is that regulatory TATA elements (T_R) are active in the presence of any upstream regulatory site, whereas constitutive TATA elements (T_C) are not. By this proposal, the *his3* promoter contains both kinds of TATA elements, whereas the *pet56* promoter contains only the constitutive type (see Fig. 2). Two independent lines of genetic evidence support this model.

First, derivatives with small deletions in the TATA region (*his3*-Δ24 and *his3*-Δ20) do not affect the basal level of *his3* expression, but they do prevent induction (Struhl 1982b; Table 3). Moreover, in contrast to the wild-type gene, equal levels of both the +1 and +12 transcripts are observed even during starvation conditions (K. Struhl, in prep.). Formally, this result indicates that differences in the TATA region can account for differences between constitutive and regulatory expression. Presumably, these deletion mutants remove the T_R element without affecting the T_C element. Both of the mutations that delete the *his3* T_R element destroy the only perfect TATAAA sequence in the promoter region (nucleotides −45 to −40). However, they retain the sequence TATACA (nucleotides −54 to −49) that is also located in the functionally defined TATA region. The *pet56* gene does not contain any perfect TATAAA sequences, although both TATAGA (nucleotides −40 to −35) and CATAAA (nucleotides −50 to −45) are found in the region implicated as being functionally important. Thus the consensus sequence may act as a T_R element, while imperfect sequences may constitute T_C elements.

Second, when the spacing between the TATA and initiation region is increased by 8 bp (*his3*-Δ19), transcriptional induction is observed equally at the +1 and +12 sites (K. Struhl, in prep). This shows directly that transcription from the +1 site has the potential to be induced and thus suggests that the initiation region itself does not confer any specificity with regard to constitutive versus regulatory expression. In addition, this result suggests that the preferential utilization of the +12 initiation site occurs simply because T_R is too close to the +1 initiation site. The proposed T_C element is located about 10 bp further away from the initiation region, which would explain why it is capable of activating transcription from both sites with equal efficiency. Thus, the distinct initiation patterns may reflect the activities of the different TATA elements.

Chromatin Structural Changes in the TATA Region Associated with Constitutive and Regulatory Expression

In nuclear chromatin, the TATA region of the wild-type *his3* gene is preferentially cleaved by micrococcal nuclease (Struhl 1983a). This hypersensitivity is seen in *his3* derivatives lacking all sequences upstream of −155, whereas it is not observed in deletion mutants that lack the upstream promoter region (Struhl 1983a). For example, *his3*-Δ13 strains, which contain the TATA element but lack all sequences upstream of −66, do not show nuclease hypersensitivity and they are transcriptionally defective. However, in the presence of *ope* suppressor mutations, the *his3*-Δ13 allele is transcribed at wild-type levels but only under certain growth conditions (Oettinger and Struhl 1985). Nevertheless, these strains do not have a wild-type chromatin structure under transcriptionally active or inactive conditions (Oettinger and Struhl 1985). Similarly, when the *gal* upstream regulatory site replaces the *his3* upstream promoter region (*his3*-G4), nuclease sensitivity is not observed in galactose medium, conditions causing extremely high levels of *his3* expression (K. Struhl, in prep.).

Thus, nuclease sensitivity at the TATA region is not correlated with transcription per se, but rather is associated with normal levels of transcription initiating at +1. In other words, these experiments provide direct evidence for a structural change at the TATA region that distinguishes constitutive expression from regulated expression; thus, they support the proposal that there are two different classes of TATA elements.

MOLECULAR MECHANISMS: INFERENCES AND SPECULATIONS

The properties of yeast promoter/regulatory elements are summarized schematically in Figure 4, and the conclusions derived from them are listed below.

1. Transcriptional initiation is not a simple enzyme-substrate interaction between RNA polymerase and DNA because there is no precise spacing arrangement of any of the promoter elements.
2. The facts that the distance between the TATA element and the initiation site is variable and that purified RNA polymerase II does not initiate at specific sequences in vitro suggest that a protein distinct from RNA polymerase II recognizes the TATA element. This is supported by TATA binding proteins isolated from higher eukaryotic organisms (Davison et al. 1984; Parker and Topol 1984).
3. Promoter specificity is determined primarily by the upstream element. The regulatory properties of the *his3* and *gal1,10* genes are determined by different DNA binding proteins that recognize different DNA sequences. The constitutive expression of the apparently unrelated *pet56*, *his3*, and *ded1* genes is mediated by poly(dA:dT) sequences, and the length of such homopolymer tracts influences the level of transcription.
4. It seems unlikely that transcriptional activation by upstream promoter elements is mediated by specific protein-protein interactions between the activator proteins and the transcriptional machinery. All these elements function bidirectionally, and in the case of the *gal* enhancer-like element, act at long and variable distances from the TATA element and initiation site. In addition, a given TATA element can function with different upstream elements, and presumably the cognate activator proteins.
5. Negative control of transcription cannot occur simply by steric competition between a repressor protein and the transcription apparatus. In contrast to prokaryotic repression models, the *gal* catabolite repression site exerts its effects when upstream from an intact promoter region.
6. The full induction of *his3* expression in response to conditions of amino acid starvation requires two copies of the TGACTC regulatory sequence, although the promoter-proximal copy is sufficient to confer partial induction. Each nucleotide of the TGACTC sequence is important for induction, as is the stretch of dT residues immediately downstream. *gcn4* protein, which is required for induction in vivo, binds specifically to the *his3* proximal TGACTC regulatory sequences and to promoter regions of other coregulated genes.
7. Two classes of TATA elements, constitutive (T_C) and regulatory (T_R), can be distinguished by their ability to respond to upstream regulatory elements and by their physical structure in nuclear chromatin. The arrangement of such elements in the divergent *his3-pet56* promoter region explains why constitutive expression of both genes depends on a common dA:dT region whereas only the *his3* gene is inducible.
8. Constitutive and regulated *his3* expression is mediated by interlaced promoters. Regulated transcription is initiated primarily at the +12 site, and it depends on the TGACTC upstream sequences and

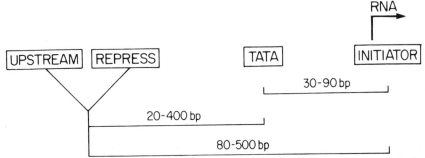

Figure 4. Promoter/regulatory elements in yeast. *cis*-acting elements of a hypothetical yeast gene are indicated as boxes. The initiator element, which is located near the RNA start (arrow) is important for determining where transcription begins. The TATA element, located 30-90 bp away from the RNA start, is required for transcription. The upstream promoter element, which can be located at variable distances away from the other elements, is important for transcription and also for regulation. Repressor sites, which are important for negative control, are also located at variable positions upstream of the TATA element. (See text for details.)

the T_R element. On the other hand, constitutive transcription is initiated equally at the +1 and +12 sites, and it depends on the poly (dA:dT) tract and the T_C element.

Transcriptional Initiation

A highly speculative model for constitutive and regulated transcription is shown in Figure 5. The basic assumption is that standard chromatin structure represents an inert form of DNA which is not recognized by RNA polymerase II. Thus, this structure must be disrupted for transcription to occur. Two mechanisms are proposed, both of which involve the upstream promoter element. Nucleosomes could be excluded from the promoter region either by the binding of specific activator proteins to their cognate regulatory sequences, or alternatively by the unusual structure of

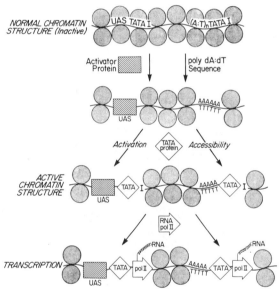

Figure 5. Molecular models for transcription. As described in the text, these models are highly speculative and are presented mainly to summarize the data. The top part of the figure shows a region of the yeast genome coated with nucleosomes (pairs of shaded circles). The promoter sequences of two genes are indicated. The gene on the left contains an upstream promoter element (UAS) typical for a regulated gene, whereas the gene on the right contains poly(dA:dT) tracts typical of a constitutively expressed gene. Both genes contain TATA and initiator (I) elements. The first step of transcriptional activation is diagramed as a disruption in chromatin structure mediated by an activator protein (striped box) or by the unusual properties of the poly(dA:dT) region. The second step involves interaction of the TATA protein (open diamond) with its cognate promoter element. This is pictured either as activation mediated by the particular protein that binds to the upstream element (for the regulated gene) or as accessibility due to nucleosome exclusion (for the constitutively expressed gene). By either of the proposed mechanisms, the result is an active chromatin structure. The final step is shown as the recognition of this active structure by RNA polymerase II followed by transcription initiation. The precise start point is mediated in some manner by the initiator element.

poly(dA:dT) sequences implicated as constitutive upstream elements. It seems less likely that the initial structural change would occur at the TATA element, because nuclease sensitivity of the *his3* region, which presumably measures a specific protein-DNA interaction, requires an upstream element.

Such structural changes, although necessary, are insufficient for transcription because the TATA element is also required for transcription. In addition, R. Brent and M. Ptashne (pers. comm.) have provided direct evidence for an activation step that is distinct from DNA binding. Specifically, a *lexA-gal4* fusion protein, but not the *lexA* protein itself, activates transcription from a promoter containing the *lexA* operator as an upstream element. The fact that different regulatory elements can be functionally associated with a given TATA and initiation region suggests that this activation mechanism must be somewhat general.

The properties of the upstream elements described here and elsewhere suggest that a signal initiated at the upstream element must be transmitted downstream to the TATA element. This signal could be the movement of a protein, the obvious candidates being the activator protein, the TATA protein, or RNA polymerase II. Several considerations discussed in more detail elsewhere (Struhl 1985a) favor the TATA protein. Alternatively, the signal could represent a structural change induced at the upstream element and propagated in both directions. If activation represents nucleosome exclusion, the critical promoter sequences would be more accessible to the TATA protein and RNA polymerase II. Alternatively, if activation represents a change such as local supercoiling, the TATA protein could bind and/or be activated by recognizing such a change.

The basic proposal is that by any of these specific suggestions, an active chromatin structure is created which is recognized by RNA polymerase II such that the enzyme can bind nearby. The size of this binding region would correspond to the variability in spacing between the TATA element and initiation site, and the initiator element would correspond to the particular sequences within the binding region that are preferred by RNA polymerase II. This specificity could be due to intrinsic preferences of the polymerase or to an initiation factor that positions the enzyme.

This model provides a simple way to understand the basis of regulation. Regulated expression is mediated by specific binding proteins, whereas constitutive transcription depends on the unusual structural properties of poly(dA:dT) sequences. Positive control is achieved by transcription factors, which are functional only in association with cofactors that exist under specific environmental or developmental circumstances. In their active form, these proteins disrupt the normal chromatin structure, whereas in their inactive form, the chromatin is inert. Negative control is carried out by repressor proteins that also are affected by cofactors. Functional repressors could either alter the chromatin such that the transcription process does not begin, or they could block the activation process that was begun by a positive factor. Thus, complex regulation can be viewed as a competition between activator and repressor proteins, each recognizing a specific DNA sequence and each subject to particular physiological controls, to determine the activity state of chromatin.

General Control of Amino Acid Biosynthesis

The *gcn4* gene, whose product is necessary for the coordinate induction of amino acid biosynthetic genes, encodes a protein that binds specifically to DNA sequences that are critical for this regulation (Hope and Struhl 1985). Thus, the simplest hypothesis is that *gcn4* protein is a specific transcription factor for genes that contain the TGACTC binding sites (Fig. 6). More complex hypotheses, such as *gcn4* protein interacting with a separate transcription factor, cannot be excluded.

This proposed mechanism, however, does not explain how yeast knows when to induce coordinately the transcription of amino acid biosynthetic genes. A cell must interpret the physiological state of starvation to produce a molecular signal, and it must transmit this signal to the effector molecule, *gcn4* protein, such that transcription of the coregulated genes is induced under appropriate conditions. The starvation signal is almost certainly produced during the process of translation because this is the only situation when all 20 amino acids are used for the same purpose, and because starvation for single amino acids results in the same transcriptional induction. Although the molecular nature of the starvation signal is unknown, the unusual structure and regulation of the *gcn4* gene (Hinnebusch 1984; Thireos et al. 1984) suggest how this signal is transmitted to the effector molecule, *gcn4* protein. The most revealing observation is that the level of a *gcn4-lacZ* fusion protein (and by inference the *gcn4* protein itself) is extremely low under normal growth conditions. Moreover, *gcn4* protein levels are increased dramatically under conditions of amino acid starvation, whereas the level of mRNA is minimally affected (Hinnebusch 1984; Thireos et al. 1984). Therefore, transcriptional induction of amino acid biosynthetic genes is mediated, at least in part, by the intracellular level of the specific DNA binding protein that recognizes the TGACTC regulatory sequences in the relevant promoter regions. Indeed, strains containing multiple copies of the *gcn4* gene cause high expression levels of amino acid biosynthetic genes even under normal growth conditions; presumably, this reflects an overproduction of the *gcn4* binding protein.

The structure of the *gcn4* gene strongly suggests a mechanism for its translational control (Hinnebusch 1984; Thireos et al. 1984). Unlike typical yeast mRNAs that contain short 5' leaders before the AUG initiation codon, the leader for the *gcn4* mRNA is almost 600 bp in length and it includes four AUG codons that cannot initiate *gcn4* protein synthesis. Extensive studies of translation in yeast indicate that initiation begins at the 5' proximal AUG codon and that it cannot be efficiently reinitiated at more downstream AUG codons (Sherman et al. 1980). Thus, the normal rules of translation preclude the synthesis of *gcn4* protein, thereby providing a simple explanation for why the coregulated genes are normally expressed at a basal level. However, under conditions of amino acid starvation, the basic translation rules must be altered in some manner such that *gcn4* protein is synthesized from its unusual mRNA.

An appealing feature of this model (Fig. 6) is that both the production of the molecular signal and its mode of transmission to the ultimate effector, *gcn4* binding protein, are associated with the basic translation process. It is tempting to speculate that these as-

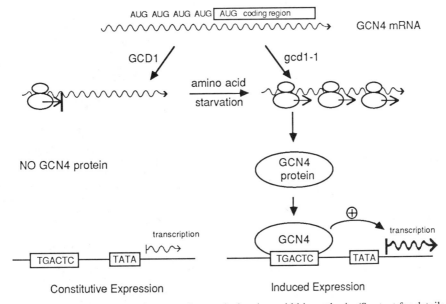

Figure 6. Molecular model for the general control of amino acid biosynthesis. (See text for details.)

pects of general control are related mechanistically and that the *gcd1* gene product is involved in both. *gcd1* mutant strains are defective in general control of amino acid biosynthetic genes (Wolfner et al. 1975), in protein synthesis (D.E. Hill and K. Struhl, in prep.), and in the translational control of *gcn4* protein levels (Hinnebusch 1984). For example, *gcd1* protein could sense the translationally stalled ribosomes that undoubtedly occur during starvation due to the lack of available charged tRNA species, and respond by releasing the block to translational reinitiation at the critical AUG codon of *gcn4* protein (Fig. 6). This provides a molecular mechanism for the translational specificity that ultimately determines the coordinate induction of amino acid biosynthetic genes in response to starvation conditions.

ACKNOWLEDGMENTS

This work was supported by grants from the Royal Society (I.A.H.), the Damon Runyon-Walter Winchell Cancer Fund (D.E.H.), the National Institute of Health (K.S.), and the Searle Scholars Program (K.S.).

REFERENCES

BRAM, R. and R. KORNBERG. 1985. Specific protein binding to far upstream activating sequences in polymerase II promoters. *Proc. Natl. Acad. Sci.* **82:** 43.

BRENT, R. and M. PTASHNE. 1984. A bacterial repressor protein or a yeast transcriptional terminator can block upstream activation of a yeast gene. *Nature* **312:** 612.

CHEN, W. and K. STRUHL. 1985. Yeast mRNA initiation sites are determined primarily by specific sequences, not by the distance from the TATA element. *EMBO J.* (in press).

DAVISON, B.L., J.M. EGLY, E.R. MULVIHILL, and P. CHAMBON. 1983. Formation of stable preinitiation complexes between eukaryotic class b transcription factors and promoter sequences. *Nature* **301:** 680.

DECROMBRUGGHE, B., S. BUSBY, and H. BUC. 1984. Activation of transcription by the cyclic AMP receptor protein. In *Biological regulation and development* (ed. K. Yamamoto), vol. 3B, p. 129. Plenum Press, New York.

DONAHUE, T.F., R.S. DAVES, G. LUCCHINI, and G.R. FINK. 1983. A short nucleotide sequence required for the regulation of *HIS4* by the general control system of yeast. *Cell* **32:** 89.

GINIGER, E., S.M. VARNUM, and M. PTASHNE. 1985. Specific DNA binding of GAL4, a positive regulatory protein of yeast. *Cell* **40:** 767.

GUARENTE, L. and E. HOAR. 1984. Upstream activation sites of the *cyc1* gene of *Saccharomyces cerevisiae* are active when inverted but not when placed downstream of the "TATA box." *Proc. Natl. Acad. Sci.* **81:** 7860.

GUARENTE, L. and T. MASON. 1983. Heme regulates transcription of the *cyc1* gene of *S. cerevisiae* via an upstream activation site. *Cell* **32:** 1279.

GUARENTE, L., R.R. YOCUM, and P. GIFFORD. 1982. A *GAL10-CYC1* hybrid yeast promoter identifies the *GAL4* regulatory region as an upstream site. *Proc. Natl. Acad. Sci.* **79:** 7410.

HINNEBUSCH, A.G. 1984. Evidence for translational regulation of the activator of general amino acid control in yeast. *Proc. Natl. Acad. Sci.* **81:** 6442.

HINNEBUSCH, A.G. and G.R. FINK. 1983a. Repeated DNA sequences upstream from *HIS1* also occur at several other co-regulated genes in *Saccharomyces cerevisiae*. *J. Biol. Chem.* **258:** 5238.

———. 1983b. Positive regulation in the general control of *Saccharomyces cerevisiae*. *Proc. Natl. Acad. Sci.* **80:** 5374.

HINNEBUSCH, A., G. LUCCHINI, and G.R. FINK. 1985. A synthetic *HIS4* regulatory element confers general amino acid control on the cytochrome C gene (*CYC1*) of yeast. *Proc. Natl. Acad. Sci.* **82:** 498.

HOPE, I.A. and K. STRUHL. 1985. GCN4 protein, synthesized in vitro, binds *H153* regulatory sequences: Implications for general control of amino acid biosynthetic genes in yeast. *Cell* (in press).

JOHNSTON, M. and R.W. DAVIS. 1984. Sequences that regulate the divergent GAL1-GAL10 promoter in *Saccharomyces cerevisiae*. *Mol. Cell. Biol.* **4:** 1440.

JONES, E.W. and G.R. FINK. 1982. Regulation of amino acid biosynthesis and nucleotide biosynthesis in yeast. In *The molecular biology of the yeast* Saccharomyces: *Metabolism and gene expression* (ed. J.N. Strathern et al.), p. 181. Cold Spring Harbor Laboratory, Cold Spring Harbor, New York.

JONGSTRA, J., T.L. REUDELHUBER, P. OUDET, C. BENOIST, C.B. CHAE, J.M. JELTSCH, D.J. MATHIS, and P. CHAMBON. 1984. Induction of altered chromatin structures by simian virus 40 enhancer and promoter elements. *Nature* **307:** 708.

KUNKEL, G.R. and H.G. MARTINSON. 1981. Nucleosomes will not form on double-stranded RNA or over poly (dA):poly(dT) in recombinant DNA. *Nucleic Acids Res.* **9:** 6869.

MAGASANIK, B. 1962. Catabolite repression. *Cold Spring Harbor Symp. Quant. Biol.* **26:** 249.

MARINI, J.C., S.D. LEVENE, D.M. CROTHERS, and P.T. ENGLUND. 1983. A bent helix in kinetoplast DNA. *Cold Spring Harbor Symp. Quant. Biol.* **47:** 279.

OETTINGER, M.A. and K. STRUHL. 1985. Suppressors of *Saccharomyces cerevisiae* promoter mutations lacking the upstream element. *Mol. Cell. Biol.* **5:** 1901.

PARKER, C.S. and J. TOPOL. 1984. A *Drosophila* RNA polymerase II transcription factor contains a promoter-region specific DNA binding activity. *Cell* **36:** 357.

PECK, L.J. and J.C. WANG. 1981. Sequence dependence of the helical repeat of DNA in solution. *Nature* **292:** 375.

PENN, M.D., B. GALGOCI, and H. GREER. 1983. Identification of *AAS* genes and their regulatory role in general control of amino acid biosynthesis in yeast. *Proc. Natl. Acad. Sci.* **80:** 2704.

PRUNELL, A. 1982. Nucleosome reconstitution on plasmid inserted poly(dA):(dT). *EMBO J.* **1:** 173.

REZNIKOFF, W. and W. MCCLURE. 1985. Prokaryotic promoters. In *From gene to protein: Steps dictating the maximal level of gene expression* (ed. W. Reznikoff and L. Gold). Benjamin/Cummings, California. (In press.)

RHODES, D. and A. KLUG. 1981. Sequence dependent helical periodicity of DNA. *Nature* **292:** 378.

RUSSELL, D.W., M. SMITH, D. COX, V.M. WILLIAMSON, and E.T. YOUNG. 1983. DNA sequences of two yeast promoter-up mutants. *Nature* **304:** 652.

SCHURCH, A., G. MIOZZARI, and R. HUTTER. 1974. Regulation of tryptophan biosynthesis in *Saccharomyces cerevisiae*: Mode of action of 5-methyl-tryptophan and 5-methyl-tryptophan sensitive mutants. *J. Bacteriol.* **117:** 1131.

SHERMAN, F., J.W. STEWART, and A.M. SCHWEINGRUBER. 1980. Mutants of yeast initiation translation of iso-1-cytochrome c within a region spanning 37 nucleotides. *Cell* **20:** 215.

ST. JOHN, T.P. and R.W. DAVIS. 1981. The organization and transcription of the galactose gene cluster of *Saccharomyces*. *J. Mol. Biol.* **152:** 285.

STRUHL, K. 1981. Deletion mapping a eukaryotic promoter. *Proc. Natl. Acad. Sci.* **78:** 4461.

———. 1982a. The yeast *his3* promoter contains at least two distinct elements. *Proc. Natl. Acad. Sci.* **79:** 7385.

———. 1982b. Regulatory sites for *his3* expression in yeast. *Nature* **300:** 284.

———. 1983a. Promoter elements, regulatory elements, and chromatin structure of the yeast *his3* gene. *Cold Spring Harbor Symp. Quant. Biol.* **47:** 901.

———. 1983b. The new yeast genetics. *Nature* **305:** 391.

———. 1984. Genetic properties and chromatin structure of the yeast *gal* regulatory element: An enhancer-like sequence. *Proc. Natl. Acad. Sci.* **81:** 7865.

———. 1985a. Yeast promoters. In *From gene to protein: Steps dictating the maximal level of gene expression* (ed. W. Reznikoff and L. Gold). Benjamin/Cummings, California. (In press.)

———. 1985b. Nucleotide sequence and transcriptional mapping of the yeast *pet56–his³–ded 1* gene region. *Nucleic Acids Res.* (in press).

———. 1985c. The regulatory site that mediates catabolite repression in yeast exerts its effects when upstream of an intact promoter region. *Nature* (in press).

———. 1985d. Naturally occurring polydA:dT sequences are upstream promoter elements for constitutive transcription in yeast. *Proc. Natl. Acad. Sci.* (in press).

STRUHL, K. and R.W. DAVIS. 1981. Transcription of the *his3* gene region in *Saccharomyces cerevisiae*. *J. Mol. Biol.* **152:** 535.

THIREOS, G., M.D. PENN, and H. GREER. 1984. 5′ Untranslated sequences are required for the translational control of a yeast regulatory gene. *Proc. Natl. Acad. Sci.* **81:** 5096.

WEST, R.W., R.R. YOCUM, and M. PTASHNE. 1984. Yeast GAL1-GAL10 divergent promoter region: Location and function of the upstream activating sequence-UAS_G. *Mol. Cell. Biol.* **4:** 2467.

WOLFNER, M., D. YEP, F. MESSENGUY, and G.R. FINK. 1975. Integration of amino acid biosynthesis into the cell cycle of *Saccharomyces cerevisiae*. *J. Mol. Biol.* **96:** 273.

ZARET, K.S. and K.R. YAMAMOTO. 1984. Reversible and persistent changes in chromatin structure accompany activation of a glucocorticoid-dependent enhancer element. *Cell* **38:** 29.

A Simple Gene with a Complex Pattern of Transcription: The Alcohol Dehydrogenase Gene of *Drosophila melanogaster*

C. SAVAKIS* AND M. ASHBURNER
Department of Genetics, University of Cambridge, Cambridge, England

The dimeric enzyme alcohol dehydrogenase (ADH) is coded for by a single, relatively simple, gene in *Drosophila melanogaster*. Natural populations of this species are commonly polymorphic for two different classes of *Adh* allele that code for electrophoretically distinguishable enzymes. This polymorphism allowed Grell et al. (1965) to map the responsible genetic locus and, subsequently, to induce mutations that lacked ADH activity (Grell et al. 1968). These mutant *Adh* alleles revealed the function of the gene; flies lacking ADH activity were found to be very susceptible to ethanol. Moderate (e.g., 6%) concentrations of ethanol that are quite harmless to wild-type *D. melanogaster* kill genotypes that are homozygous for *Adh*-null mutations (Vigue and Sofer 1976). Subsequently Sofer and his colleagues (Sofer and Hatkoff 1972; O'Donnell et al. 1975) were able to use chemical selection to recover flies with functional *Adh* alleles from those that are mutant, since ADH will oxidize relatively nontoxic secondary alcohols to toxic ketones. The relatively small size of ADH (its monomeric molecular weight is about 27 kD) and its abundance in the fly, encouraged, in the years B.C., studies of its protein chemistry which rigorously proved that *Adh* was indeed the only structural gene for this enzyme, and that the major electrophoretic polymorphism in natural populations was due to a single amino acid substitution (Thatcher 1980).

The availability of homozygous viable *Adh* deletions (O'Donnell et al. 1977; Woodruff and Ashburner 1979) allowed Goldberg (1980) to double-screen a genomic library of *Drosophila* DNA with two populations of RNA, one from a wild-type strain and one from a homozygous *Adh* deletion, and recover genomic *Adh* clones. At about the same time Benyajati and her colleagues (1980) recovered a partial cDNA clone, corresponding to the 3' of the *Adh* mRNA. At first it seemed as if the structure of the *Adh* gene was as simple as had been hoped, a single coding region interrupted by two small introns (of 65 and 70 bp), with modest 5' and 3' nontranslated ends and corresponding to a single abundant mRNA. Two groups then began to define the 5' end of the mRNA at the nucleotide level. By chance they used RNA prepared from different developmental stages, adult flies and third instar larvae. Confusing results were found when the 5' end of the larval RNA was mapped using primers extended with chain-terminating 2-deoxyribonucleotides. Extending from a primer internal to the coding region, the sequence was clearly colinear with the genomic sequence (which had then been independently determined by Benyajati et al. [1981] and by Haymerle [1983]) up to 36 bases upstream of the AUG codon. Beyond that point the sequence was ambiguous, indeed it appeared to be a mixture of two quite different sequences. Luckily, for all concerned, the two groups were able to compare their data at the Storrs Drosophila Research Conference in March 1982, and it immediately became clear that the *Adh* gene has two different 5' ends and that the most abundant mRNAs of larvae and adult flies were transcribed from different promoters. The confusing data from the primer extension experiments with larval RNA resulted from the fact that, in larvae, both types of mRNA are present. These data were published jointly by the two laboratories (Benyajati et al. 1983). Almost simultaneously, and independently of either group, Henikoff had been using the *Adh* gene as a model system to develop a method for exon mapping. He too discovered two different 5' exons in *Adh* mRNA (Henikoff 1983).

The structural model for the *Adh* gene proposed by Benyajati et al. (1983) has stood the test of time, although Henikoff has evidence for an alternative splice acceptor junction for the most 5' intron (this was not seen in S1 nuclease protection experiments by C. Savakis et al., in prep.). This model is illustrated in Figure 1. It will be seen that transcription can begin downstream of either of two different promoters but that the distal transcript (which is that most abundant in adult flies) is processed by the removal of a 654-base intron. This intron overlaps the 5' untranslated region of the proximal transcript (that most abundant in larvae), but from a position of 36 bases 5' to the AUG the two transcripts are identical. The two transcripts must, therefore, code for identical polypeptides.

The general structure of the *Adh* gene of *D. melanogaster* is also found in related species. Bodmer and Ashburner (1984) have sequenced this gene from three sibling species of *D. melanogaster*, that is to say *D. simulans*, *D. mauritiana*, and *D. orena*, and all show identical structural features. There are, of course, differences at the nucleotide level, affecting both coding

---

Present address: Institute for Molecular Biology and Biotechnology, Research Center of Crete, Heraklion, Crete, Greece.

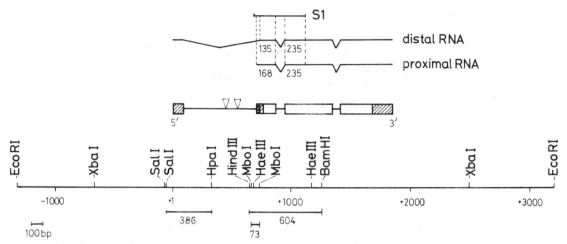

Figure 1. The structure of the *Adh* gene of *Drosophila melanogaster*. Exons are represented as boxes and introns as single lines. The untranslated regions of the exons are hatched. The two major transcripts are indicated above the representation of the gene as is the DNA fragment used in the S1 nuclease protection experiments (S1). The various DNA probes used to analyze the RNAs are shown, with their sizes in base pairs, below the gene. The numbering assumes that the cap nucleotide of the distal transcript is +1. The sequence data are from a 7423-bp sequence of the *Adh-Af:S* allele (bases 1–4293), the Canton-S *Adh-S* allele (bases 4294–6833), and the *Adh-Fl:2S* allele (bases 6834–7423) (Benyajati et al. 1981; Haymerle 1983; Kreitman 1983 and unpubl.). The EcoRI site at −1370 is at position 3447 on this consensus sequence. The positions of insertional polymorphism within the first intron of most *Adh-F* alleles are indicated as open triangles (from Kreitman 1983; see also Fig. 6a). Only relevant restriction enzyme cleavage sites are shown.

and noncoding regions of the gene. Indeed, *D. orena*, the most distant relative to *D. melanogaster* of these species, has very considerable differences in the long intron of the distal transcript. It is interesting that even in *D. melanogaster* this intron is quite variable. Kreitman (1983) has sequenced 11 different *Adh* alleles of *D. melanogaster*, five coding for the electrophoretically slow allozyme and six for the electrophoretically fast allozyme. Most fast alleles differ from all slow alleles by one or two insertions (of 34 and 36 bp) within this intron (see Figs. 1 and 6a, below).

The discovery of a technique to transform *Drosophila* (Rubin and Spradling 1982), and the successful transformation of the *Adh* gene (Goldberg et al. 1983), meant that a far more detailed study of the developmental expression of the two *Adh* mRNAs was required, in order to judge the fidelity (or not) of expression of transformed genes. This study has shown that the transcriptional expression of *Adh* is rather more complex, and even more interesting, than we had previously thought (Benyajati et al. 1983). For reasons that will become clear, we now prefer to refer to the two transcripts of *Adh* as the distal and proximal transcripts, rather than, as before, the adult and larval transcripts. Some of the results to be discussed are published elsewhere (C. Savakis et al., in prep.).

RESULTS AND DISCUSSION

Adh mRNAs in Larvae and Adults

Using DNA probes that will specifically hybridize to either the distal or the proximal transcript, as well as a probe that will hybridize to both (Fig. 1), we surveyed a range of developmental stages of *D. melanogaster* for transcript expression. The result of this survey is shown in Figure 2 (a–c). It is clear that there are two discrete periods in development during which *Adh* transcripts accumulate to relatively high levels—in the early third instar larva and in the adult fly. These correspond to the maxima of ADH enzyme activity, measured spectrophotometrically in extracts of flies. Two peaks of specific activity are seen, the first in the late third instar larva and the second a few days after eclosion of the adult fly (Ursprung et al. 1970; Maroni and Stamey 1983). However, from the Northern blot hybridizations we see that the most abundant *Adh* transcripts differ in larvae and adult flies. In larvae the proximal transcript is abundant, the distal is either absent or rarer. In the adult the distal transcript is abundant, the proximal is much rarer.

The changes in steady-state level of *Adh* transcripts in larvae are remarkable. Quite suddenly, at a time in the middle of the third instar, the proximal transcript disappears. Due to the difficulty in synchronizing larval development, we cannot state just how sudden this disappearance of the proximal transcript is. At most, the time from maximum to minimum levels would appear to be about 10 hours (at 25°C). This abrupt transition is not characteristic of all RNAs. It is not seen, for example, for the major transcripts of the copia transposable element (C. Savakis et al., in prep.) nor for the transcript of the rosy gene, which codes for the enzyme xanthine dehydrogenase and is, like *Adh*, expressed in both the fat body and Malpighian tubules (see Fig. 5c, below). This transition occurs too early in development for it to be due to the surge in ecdysteroid hormone titer that characterizes the late third instar larvae. It could, however, correlate with a minor peak in ecdysteroid titer that may initiate larval wandering

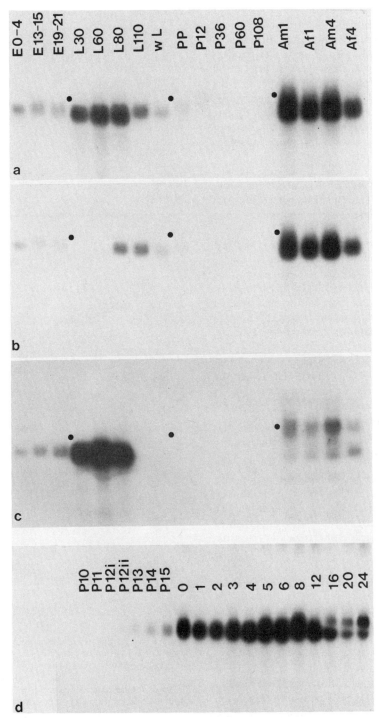

Figure 2. Northern blot hybridizations to RNA prepared from different stages of development of *D. melanogaster* probed for *Adh* sequences. In *a–c* the stages are: (E) embryos, ages after egg laying in hours; (L) larvae, ages after egg laying in hours; (wL) wandering third instar larvae; (PP) white prepupae; (P) pupae, ages after pupariation in hours; and (A) adults, ages after eclosion in days (m, males; f, females). The probes used were the 604-bp *Hin*dIII–*Bam*HI fragment (gel-purified from pBR322 and nick-translated) that will hybridize to both distal and proximal transcripts (*a*), the 386-bp *Sal*I–*Hpa*I fragment (gel-purified from the *Eco*RI–*Eco*RI [Fig. 1] fragment and nick-translated), that will hybridize only to the distal transcript and the 73-base *Mbo*I–*Mbo*I fragment (a uniformly labeled single-stranded probe was made by primer extension from a clone in m13mp7) that will hybridize to the proximal transcript and to the precursor to the processed distal transcript (*c*). (*d*) A developmental series around the time of eclosion of the adult flies. P10–P15 represent the standard stages of pharate adult development (Bainbridge and Bownes 1981) and 0–24 the ages of adult flies, in hours after eclosion. Stage P13 is about 18 hr before eclosion. Probed with the 604-bp *Hin*dIII–*Bam*HI fragment. Equal amounts of total RNA (5 μg) were loaded onto each slot. The intensities of the signals seen with different probes cannot be directly compared, due to the differences in specific activities and length of the probes.

behavior in the mid-third instar (see C. Savakis et al., in prep, for discussion).

The expression of the distal *Adh* transcript during the third larval instar, previously noted by Benyajati et al. (1983), can be clearly seen from the experiment shown in Figure 2 (a–b). It is now clear that this distal transcript begins to accumulate just at the time the sudden drop in the level of the proximal transcript occurs. No proximal transcript can be detected in the late third instar larvae. The amount of distal transcript slowly falls so that, midway during "pupal" development, it too is undetectable.

Since the function of ADH is to detoxify ingested alcohols, the absence of *Adh* mRNA, and the low specific activity of the ADH enzyme (Ursprung et al. 1970; Maroni and Stamey 1983), during metamorphosis is not too surprising. It was surprising, however, to see the very sudden accumulation of *Adh* transcript at the time the adult fly hatches from the pupal case. As can be seen in more detail in Figure 2d, the reappearance of *Adh* mRNA at eclosion occurs in two discrete steps. About 18 hours before eclosion, a slow and gradual increase in the level of *Adh* mRNA begins. Then, quite suddenly, within 30 minutes of the hatching of the fly, there is a massive accumulation of this transcript. The amount of transcript then remains high, and is approximately the same in male and female flies (when the dilution of *Adh* mRNA by the large amounts of RNA in the ovaries is taken into account).

The Transition from Proximal to Distal Transcripts Occurs in Individual Tissues

The metamorphosis of *Drosophila*, in common with all holometabolous insects, involves the destruction by histolysis of most of the tissues of the larva and the development of the tissues of the adult fly. Only three larval tissues escape the fate of destruction, and for one of these this escape is only temporary. These tissues are the central nervous system, the Malpighian tubules (the fly's excretory organ), and the larval fat body. The larval fat body persists into the adult fly, though it is histolysed within the first few days of adult life. Two of these tissues, the Malpighian tubules and the fat body, express ADH (Ursprung et al. 1970; Maroni and Stamey 1983). By S1 nuclease protection experiments we have analyzed the *Adh* transcripts of these tissues in early and late third instar larvae and in adult flies. The results of these experiments (Fig. 3) are clear, in both cases organs from young third instar larvae express *Adh* from the proximal promoter but organs from late third instar larvae, or adult flies, express *Adh* from the distal promoter. Both the Malpighian tubules and fat body consist of a small number of highly polyploid (indeed polytene) cells that do not divide at all subsequent to organogenesis in the embryo. These tissues stain uniformly for ADH when histochemical reagents are used to detect enzyme activity: Each cell expresses this gene. (The cellular autonomy

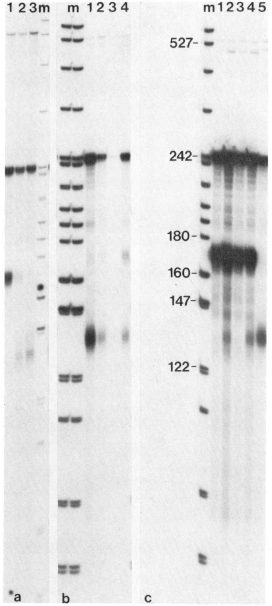

Figure 3. (*a*) S1 nuclease protection experiments of the 489-base *Hae*III–*Hae*III fragment (Fig. 1) by RNA from Malpighian tubules of 80-hr third instar larvae (track *1*), from wandering third instar larvae (track *2*), and from adult flies (track *3*). (*b*) A similar series of experiments, but with RNA from adult fat bodies (track *1*), the residual larval fat bodies of young adult flies (track *2*), male reproductive organs (track *3*), and ovaries and associated tissues (track *4*). (*c*) S1 nuclease protection experiments with RNA from second instar larvae (58–62 hr) (track *1*), 78- to 82-hr third instar larvae (track *2*), 84- to 88-hr third instar larvae (track *3*), 90- to 94-hr third instar larvae (track *4*), and wandering third instar larvae (track *5*). The tracks labeled *m* are *Hpa*II fragments of the plasmid pBR322, used as size markers. Three different fragments of the *Hae*III–*Hae*III DNA will be protected, a 236-base fragment by both distal and proximal mRNAs, a 169-base fragment by the proximal mRNA, and a 135-base fragment by the distal mRNA (see Fig. 1).

of *Adh* expression in the larval fat body has been proved by P. Martin and W. Sofer, pers. comm.). This means that the switch from proximal to distal transcript occurs within single cells without the necessity of cell division.

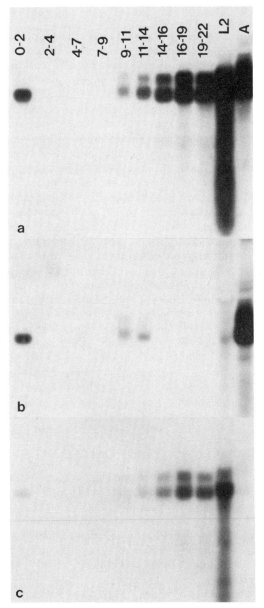

Figure 4. Northern blot hybridizations to RNA extracted from embryos (ages, in hours, after egg laying, the eggs being collected from a rapidly laying population for 2 or 3 hr). Hybridized with the 604-bp *Hin*dIII–*Bam*HI probe for all *Adh* RNAs in *a*, with the 386-bp *Sal*I–*Hpa*I probe, for distal transcripts in *b*, and with the 73-base *Mbo*I–*Mbo*I probe, for proximal transcripts in *c*. The embryos were F$_1$ hybrids between wild-type (Canton-S) mothers and homozygous *Adh*^{nLA248} fathers. This mutant allele codes for a transcript that is longer than that from the wild-type (see text). The lanes marked L2 and A are RNAs from second instar larvae and adult flies, respectively, showing the specificity of the hybridization with the *Sal*I–*Hpa*I and *Mbo*I–*Mbo*I probes. The RNA in the L2 lane is partially degraded.

Adh Transcripts in Embryos

We saw, in our first survey of *Adh* transcripts (Fig. 2a–c), that the youngest stage then analyzed, 0- to 4-hour-old embryos, contained both proximal and distal transcripts. This was a rather surprising observation, and to study the embryonic expression of *Adh* in more detail we have analyzed an accurately timed series of embryonic stages. For this we wanted a way to distinguish between the maternally and paternally derived alleles. We used an X-ray-induced *Adh*-null allele, *Adh*^{nLA248}, since we had already discovered that this allele makes a longer *Adh* mRNA, and that the levels of this RNA are only slightly lower than those of the Canton-S wild type. The *Adh*^{nLA248} mutation is a 251-bp duplication of parts of exon 2 and exon 3 (and of intron 3 which separates them), and its transcript is normally processed to remove both copies of the third intron. Nevertheless, this mRNA is not translated, *Adh*^{nLA248} flies have no ADH activity detectable spectrophotometrically, immunologically, or on 2-D gels (Kelley et al. 1985; W. Chia et al., in prep.).

Our embryos were, therefore, F$_1$ hybrids between Canton-S (i.e., wild type) mothers and homozygous *Adh*^{nLA248} fathers. The Northern blot hybridizations with the three different *Adh* probes are illustrated in Figure 4. It is clear that both *Adh* transcripts are inherited maternally. If so, they must also be present in ovaries. To check this, an S1 nuclease protection experiment, with RNA from adult ovaries, was done. This showed (Fig. 3b) that both proximal and distal transcripts are indeed present in ovaries, and in approximately equal amounts. Surprisingly, these maternal transcripts are unstable; by 2–4 hours after egg laying they can no longer be detected. The significance of these transcripts is not known. We presume that they function during oogenesis, and are, so to speak, inadvertently stabilized, along with other maternal RNAs, until fertilization.

Transcription of the zygotic genome begins at about 9 hours after egg laying. Unsurprisingly, it occurs from both maternal and paternal genomes at the same time. But we did not expect to see that the initial transcription of *Adh* leads to the accumulation of both distal and proximal RNAs. Between 9 and 14 hours after egg laying, both transcripts are clearly seen. However, the distal transcript soon disappears, whereas the proximal continues to increase in level until the mid-third larval instar.

Is the Pattern of *Adh* Transcription Dependent on the Genetic Position of the Gene or Its Functional Expression?

Transformation of *Drosophila* embryos with exogenous DNA allows the study of the effect on genetic position on expression to be studied. From the earliest data with the rosy, *Adh*, and *Ddc* genes (Goldberg et al. 1983; Scholnick et al. 1983; Spradling and Rubin

1983), the rather surprising general result—that expression, assayed at the level of the gene's protein product, was relatively position independent—was found. Although position-dependent expression of transformants has now been found (see, e.g., Bourouis and Richards 1985; Simon et al. 1985), it is nevertheless striking that *Adh* genes occupying quite novel genetic positions can be expressed, at the level of transcription, quite normally. Figure 5a illustrates a rather coarse developmental profile of a strain transformed with the 3161-bp *Xba*I–*Xba*I fragment that includes *Adh* (see Fig. 1). This gene has integrated into a position on the X chromosome (at 15A) and shows a very similar transcriptional profile to the wild-type gene. Interestingly, as Goldberg et al. (1983) also found, this X-linked *Adh* gene shows a degree of dosage compensation. At the transcriptional level the ratio of *Adh* mRNA (in adult flies) in males and females is 1.3 (data not shown).

The autoregulation of gene activity is now well known (Cove and Pateman 1969; Goldberger 1974) and probably also occurs in *Drosophila* at both some heat shock genes (DiDomenico et al. 1982) and at the early salivary gland ecdysteroid-induced genes E74 and E75 (Walker and Ashburner 1981). Were it to occur at *Adh* we would expect an altered transcriptional profile in mutant strains that lack any translational product. We have used the Adh^{nLA248} mutation to see whether or not this is so. The fact that the developmental profiles of *Adh* mRNAs are very similar in this mutant strain and wild-type (compare Fig. 5d with Fig. 5b), despite the complete absence of any ADH protein in the mutant, clearly rules out autoregulation, from protein product to gene, for *Adh*.

Species Differences in *Adh* Gene Expression

Not surprisingly, since they inhabit different environments, different species of *Drosophila* express their *Adh* genes in different ways. Perhaps the most dramatic of these differences discovered so far is in a complex of species that breeds exclusively in the giant cacti of the southwestern deserts of the United States and in Mexico, the *mulleri* species subgroup of the *repleta* species group. Phylogenetically these species are quite distant from *D. melanogaster*. They belong to a different subgenus (the subgenus *Drosophila*, rather than the subgenus *Sophophora*). These two subgenera have been distinct for at least 40 million years (Throckmorton 1975; Beverley and Wilson 1985) and, in general terms, we can consider the subgenus *Sophophora* to be more primitive than the subgenus *Drosophila*. As was first discovered from studies of the electrophoretic mobilities of ADH the *mulleri* species subgroup, but not all other members of the *repleta* species group, possess two linked *Adh* genes (see Batterham et al. 1984). These two genes are differentially expressed during development, in a way that recalls the differential promoter use of *D. melanogaster*. That is to say, in general, one is expressed in adult flies and the other in larvae (Batterham et al. 1983). These genes have been cloned from three members of the subgroup, *D. mohavensis*, *D. arizonensis* (D. Sullivan, pers. comm.), and *D. mulleri* (J. Fischer and T. Maniatis, pers. comm.) and are very closely linked and transcribed from the same DNA strand. Both genes show a structural organization that resembles the proximal "gene" of *D. melanogaster*, that is to say they both lack an intron within the 5' untranslated region of the gene.

We have made a preliminary study of the expression of the *Adh* gene of a species much closer to *D. melanogaster* than the *mulleri* species subgroup. This is *D. orena*, a species only known from one mountain in the West Cameroons of equatorial Africa. *D. orena* is one of the eight species that comprises the *melanogaster* species subgroup. However, by polytene chromosome inversion analysis (Lemeunier and Ashburner 1984) and by DNA sequence studies of its *Adh* gene (Bodmer and Ashburner 1984), it is a quite distant relative of *melanogaster*, the two species having been separated by perhaps 15 million years. Its *Adh* gene has the same general structure to that of *D. melanogaster*, the coding region is identical in length, and the three introns in identical positions. However, the sequence of the first intron is very different from that of *D. melanogaster* at its 5' end (Fig. 6a). This correlates with a rather different developmental expression. As can be seen from the Northern blot hybridization shown in Figure 6b, the *Adh* gene of *D. orena* is expressed to very different levels in larvae and adults, in *D. melanogaster* the ratio of larval to adult expression is of the order of 1 or 2. In *D. orena* it is less than 0.1. We have now constructed "recombinants" between the *Adh* genes of these species, in an attempt to dissect, by transformation, the DNA sequences responsible for this difference (K. Moses, unpubl.).

CONCLUDING REMARKS

This account of the developmental changes in transcriptional pattern of the *Adh* gene of *D. melanogaster* raises as many problems as it solves. The major unresolved problem is the extent to which the changes in *Adh* mRNA concentrations reflect changes in the rates of transcription and/or changes in the stabilities of the mRNAs. Whole *Drosophila* are inconvenient for the study of transcriptional rate by the method of run-off assays from isolated nuclei (e.g., Carneiro and Schibler 1984). They are also inconvenient for the determination of mRNA stability. A possible method to determine whether or not changes in mRNA concentration reflect changes in transcriptional rate is to compare the levels of mRNA precursors with those of the mRNAs themselves. Assuming that the turnover of precursors is rapid in comparison to the turnover of mRNAs, and assuming that the stabilities of precursor and product are independent, then any change in mRNA concentration will be reflected in a change in precursor concentration, if the control is transcriptional, but not otherwise. Preliminary experiments suggest that indeed the precursor concentrations do parallel their respective

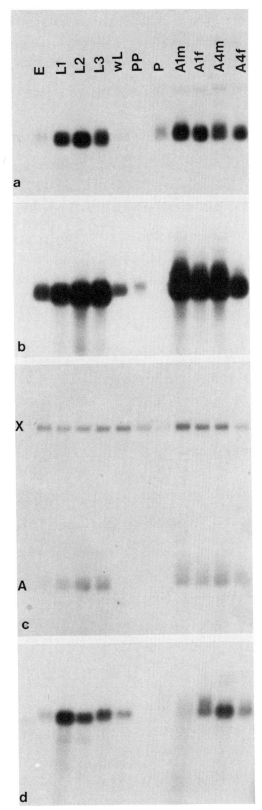

Figure 5. (*a*) A developmental profile of *Adh* transcription in the transformed strain *w; Tr[Adh]AX9.3; b Adh^{n4}*. An *Adh^{n4}* strain was transformed with the *Xba*I–*Xba*I fragment (see Fig. 1) in the vector CY4. The transformants were selected by the resistance of flies to 6% ethanol. Three integrations of *Adh* occurred in the original transformant (*AX9*), at 3C, 15A, and region 23. In the subline *Tr[Adh]AX9.3* only the site at 15A, on the X chromosome remains, the others having been removed by recombination. The *Adh^{n4}* mutation is a nonsense mutation in codon 83 (W. Chia, unpubl.) which leads to the accumulation of very low levels of *Adh* mRNAs (see C. Savakis et al., in prep.). (*b*) A similar developmental profile to that seen in *a* but with RNA from wild-type (Canton-S) larvae. The 604-bp *Hin*dIII–*Bam*HI probe was used for both experiments. The autoradiograph shown in *b* is overexposed relative to that shown in *a*. (*c*) The same filter as in *b*, but hybridized, after most of the *Adh* signal had decayed, with the 7.2-kb *Hin*dIII–*Hin*dIII fragment from the vector CY20 that includes the rosy (xanthine dehydrogenase) structural gene; the *Xhd* (X) and residual *Adh* (A) transcripts are indicated. (*d*) A developmental profile of *Adh* expression in flies homozygous for the *Adh^{nLA248}* mutation. (E) 0- to 24-hr embryos; (L1) first instar larvae; (L2) second instar larvae; (L3) 80-hr third instar larvae; (wL) wandering third instar larvae; (WP) white prepupae; (P) pupae, 90-hr after puparium formation; (A1) 0- to 1-day-old adult males (m) and females (f); and (A4) 4-day-old adult males (a) and females (f).

a

```
GTAAGTAGCAAAAGGGCACCCAATTAAAGGAAATTCTTGTTTAATTGAATTTATTATGCAAGTGCGGAAATAAAATGACA
  * **      ** *  * ************* ** * ****************************************
GTAAGT-GGGAAAGGGACCTCATTATGCAATGTCGAATAGTAAGAGATC-------------------------------


GTATTAATTAGTAAATATTTTGTAAAATCATATATAATCAAATTTATTCAATCAGAACTAATTCAAGCTGTCACAAGTAG
 *** * * ** ******  * **  **   *     * ** *     **  ** **  *    * *** *       *
--ACT-ATCACTAAT---GGTGGAGCAT-A-ATAAAATCAATTGCATGCAATCGAAATGAATGCAAACCGGCACAAGCAG


TGCGAACTCAATTAATTGGCATCGAATTAAAATTTGGAGGCCTGTGCCGC-----ATATTCGTCTTGGAAAATCACCTGT
***  ** *  *   ****** * *                        ******      **
TAGCAAACCTACTAA-----A-CAAATTAAAATTTGGAGGCCTGTGCCGTGGCGAATATTTGACTTGGAAAATCACCTGT


TAGTTAACTTCTAAAAATAGGAATTTTAACATAACTCGTCCCTGTTAATCGGCGCCGTGCCTTCGTTAGCTATCTCAAAA
 **   **        *      *** **
TGTTTAACCGCTAAAAATAGGAATTTTAACATTAAGCACCCCTGTTAATCGGCGCCGTGCCTTCGTTAGCTATCTCAAAA


GCGAGCGCGTGCAGACGAGCAGTAATTTTCCAAGCATCAGGCAT▽AGTTGGGCATAAATTATAAACATACAAACC]GAATAC
 *                                          ⌊                            *  ⌉
GCGCGCGCGTGCAGACGAGCAGTAATTTTCCAAGCATCAGGCATAGTTGGGCATAAATTATAAACATAGAAACTGAATAC


TAATATAGAAAAGCTTTGCCGGTACAAAATCCCAAACAAAAACAAACCGTGTGTGCCGAAAAAT-----AAA▽AATAAAC
          *                                    * ***  *   *****          ⌊
TAATATAGAAAAGCTTTGCCGGCACAAAATCCCAAACAAAAACAAA--GAGAGTGCC-AAAAATAAAACAAAAATAAAC


CATAAACTAGGC]AGCGCTGCCGTC---GCCGGCTGAGCAGCCTGCGTACATAGCC--GAGATCGCGTAACGGTAGATAAT
    *  ***   ⌋  *      ***    * **    **   **       **    * **               *
CGTAAACCGAGCAGCGTTGCCGTCGTTGCGGGCTGTGAAGCTTACGTGAATAGCCGAGAGATCGCGTAATGATAGATAAA


GAAAAGCTCTACGTAACCGAAGCTTCTGCTGTACGGATCTTCCTATAAATACGGGGCCGACACGAACTGGAAACCAACAA▼
 *         *      *   ** ** ***                          *   *           *
G-AAAGCTCTACGTAAGCGAAGCTTCTGGGGGATAGATCTTCCTATAAATACGGGACCGACGCGAACTGGAAACGAACAA


CTAACGGAGCCCTCTTCCAATTGAAACAG
           *
CTAACGGAGCCCTCTTCCCATTGAAACAG
```

b

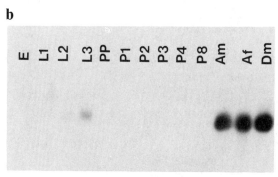

Figure 6. (*See facing page for legend.*)

mRNA concentrations. In the following discussion we will assume that this means that transcriptional control is the major determinant of *Adh* mRNA level, though we give the caution that this assumption has yet to be rigorously tested.

Although the functional significance of the gross changes in the level of *Adh* mRNA is clear, they correlate well with the changes in enzyme-specific activity, and these correlate with the ingestion of dietary alcohols. Why should this gene have two different promoters, coding for identical polypeptides? As we have suggested before (Benyajati et al. 1983; C. Savakis et al., in prep.), there are two general explanations of this observation. Since a necessary consequence of two promoters is mRNAs that differ in their 5' ends, this may be the functional reason for this organization of the *Adh* gene. A number of speculative suggestions may be made as to why the major mRNAs of larvae and adult have different 5' ends: they may differ in their translational efficiencies, in their requirements for translational factors, or in their stabilities. The distal regions of the two 5' untranslated ends of the mRNAs show some small sequence similarities (C. Savakis et al., in prep.). On the other hand, the fact that the major larval and adult mRNAs have different 5' ends may be quite irrelevant, simply being a necessary consequence of initiating transcription at different sites. Perhaps *Adh* is included within a hierarchical set of genes with respect to its control, one subset of which is active in larvae and the other in adults. Then, to be able to activate this gene, within the context of its control hierarchy, different signals (factors?) may be used at different stages in development, and the possession of two promoters may simply be a way to achieve this end.

Why is the control of transcription of the major larval and adult mRNAs not more precise? That is to say, why is the distal (i.e., the major adult) mRNA expressed, albeit to a low level, in the 9- to 14-hour embryo and then in the third instar larva? Indeed is this expression of the distal transcript biologically relevant? To take the last question first, we should emphasize that we do not know whether or not the distal transcript that accumulates in the 9- to 14-hour embryo and in the late larvae is functional. Since distal and proximal mRNAs code for identical proteins, this is not a trivial question to answer. It could possibly be done by the in vitro mutagenesis of the proximal leader, creating a new AUG codon so that the translational products of the two mRNAs could be distinguished by virtue of their size or antigenic properties. But to return to the first question: Why is there distal transcription in embryos and late larvae? It may be significant that the distal promoter appears to be used whenever the proximal promoter is either turned "on" or turned "off." Changes in the conformation of chromatin that accompany the proximal promoter transients may, for their duration, permit access of RNA polymerase II (and other factors) to the distal promoter, and hence, willy nilly, result in distal transcription. Another, but not mutually exclusive, explanation is suggested by the discovery of promoter exclusion in genes carried by a retroviral vector (Emerman and Temin 1984). In the embryo both promoters may indeed be activated but the proximal promoter subsequently excludes the distal. The distal remains inactive until such time as the proximal promoter is inactivated, in the third instar. When this happens the dominance of the active proximal promoter over distal is lost, and the latter is again expressed. Of course, this would demand another mechanism to inactivate the distal promoter after puparium formation.

ACKNOWLEDGMENTS

Original work reported in this paper was financed by a Programme Grant from the Medical Research Council, London, to M.A. C.S. was funded by a long-term Fellowship from EMBO. We thank our colleagues W. Chia, H. Haymerle, R. Karp, J. Luchessi, K. Moses, and N. Spoerel for data, and B. Durrant for technical help and for drawing the figures. We thank Martin Bishop for his computer programs. Some of the experiments reported here were done in collaboration with J. Willis. We thank other workers in this field for generously sharing unpublished information, material and ideas, especially C. Benyajati, J. Fischer, D. Goldberg, T. Maniatis, P. Martin, J. Posakony, W. Sofer, and D. Sullivan. This paper was written whilst M.A. was a guest of the Institute for Molecular Biology and Biotechnology and of the Department of Biology, University of Crete, Iraklio, Crete, Greece. M.A. thanks Professor Fotis Kafatos and his colleagues for their hospitality.

Figure 6. (*a*) A comparison of the sequence of the first *Adh* intron, within the 5' untranslated region of the distal transcript, from *D. melanogaster* and its sibling species *D. orena* (from Bodmer and Ashburner 1984, with corrections). The sequence of the *D. melanogaster Adh-S* allele is shown, with the insertions characteristic of most *Adh-F* alleles indicated by open arrows (from Kreitman 1983). The most 5' insertion is associated with a 29-bp deletion and the 3' insertion with a tandem duplication of 19 bp that immediately follows the insertion site. These sequences are enclosed in square brackets. The sequences were aligned with the ALIGNDNA program of M.J. Bishop which uses Dijkstra's algorithm. The top sequence is that of the *Adh-S* allele, the bottom that of *D. orena*. Nonsimilarities in sequence are indicated by asterisks. The consensus splice donor and acceptor sequences and the proximal TATA box are in bold type. The cap site of the proximal transcript is indicated by the closed arrow above the *Adh-S* sequence. (*b*) A developmental profile of *Adh* transcription in *D. orena* (strain 188.1), hybridized with the *D. melanogaster* 604-bp HindIII–BamHI fragment. *D. orena* was cultured at 18°C, rather than at 25°C, as was the case for *D. melanogaster*. Abbreviations: (E) 0- to 19-hr embryos; (L1) 26- to 28-hr first instar larvae; (L2) 48- to 72-hr second instar larvae; (L3) 8- to 9-day-old third instar larvae; (PP) white prepupae; (P1, P2, P3, P4, P8) pupae and pharate adults, 1, 2, 3, 4, and 8–9 days after puparium formation; (Am) adult males; and (Af) adult females. The final track (Dm) is from 4-day-old adult *D. melanogaster* females (grown at 25°C).

REFERENCES

Bainbridge, S.P. and M. Bownes. 1981. Staging the metamorphosis of *Drosophila melanogaster*. *J. Embryol. Exp. Morphol.* **66**: 57.

Batterham, P., G.K. Chambers, W.T. Starmer, and D.T. Sullivan. 1984. Origin and expression of an alcohol dehydrogenase gene duplication in the genus *Drosophila*. *Evolution* **38**: 644.

Batterham, P., J.A. Lovett, W.T. Starmer, and D.T. Sullivan. 1983. Differential regulation of duplicate alcohol dehydrogenase genes in *Drosophila mojavensis*. *Dev. Biol.* **96**: 346.

Benyajati, C., A.R. Place, D.A. Powers, and W. Sofer. 1981. Alcohol dehydrogenase of *Drosophila melanogaster*: Relationships of intervening sequences to functional domains in the protein. *Proc. Natl. Acad. Sci.* **78**: 2717.

Benyajati, C., N. Spoerel, H. Haymerle, and M. Ashburner. 1983. The messenger RNA for alcohol dehydrogenase in *Drosophila melanogaster* differs in its 5' ends in different developmental stages. *Cell* **33**: 125.

Benyajati, C., N. Wang, A. Reddy, E. Weinberg, and W. Sofer. 1980. Alcohol dehydrogenase in *Drosophila melanogaster*: Isolation and characterization of messenger RNA and a cDNA clone. *Nucleic Acids Res.* **8**: 5649.

Beverley, S. and A.C. Wilson. 1985. Ancient origin for Hawaiian Drosophilinae inferred from protein comparisons. *Proc. Natl. Acad. Sci.* **82**: 4753.

Bodmer, M. and M. Ashburner. 1984. Conservation and change in the DNA sequences coding for alcohol dehydrogenase in sibling species of *Drosophila*. *Nature* **309**: 425.

Bourouis, M. and G. Richards. 1985. Remote regulatory sequences of the *Drosophila* glue gene *sgs-3* as revealed by P-element transformation. *Cell* **40**: 349.

Carneiro, M. and U. Schibler. 1984. Accumulation of rare and moderately abundant mRNAs in mouse L-cells is mainly post-transcriptionally regulated. *J. Mol. Biol.* **178**: 869.

Cove, D.J. and J.A. Pateman. 1969. Autoregulation of the synthesis of nitrate reductatase in *Aspergillus nidulans*. *J. Bacteriol.* **97**: 1374.

DiDomenico, B.J., G.E. Bugaisky, and S.L. Lindquist. 1982. The heat shock response is self-regulated at both transcriptional and posttranscriptional levels. *Cell* **31**: 593.

Emerman, M. and H. Temin. 1984. Genes with promoters in retrovirus vectors can be independently suppressed by an epigenetic mechanism. *Cell* **39**: 459.

Goldberg, D.A. 1980. Isolation and partial characterization of the *Drosophila* alcohol dehydrogenase gene. *Proc. Natl. Acad. Sci.* **77**: 5794.

Goldberg, D.A., J.W. Posakony, and T. Maniatis. 1983. Correct developmental expression of a cloned alcohol dehydrogenase gene transduced into the *Drosophila* germ line. *Cell* **34**: 59.

Goldberger, R.F. 1974. Autogenous regulation of gene expression. *Science* **183**: 810.

Grell, E.H., K.B. Jacobson, and J.B. Murphy. 1965. Alcohol dehydrogenase of *Drosophila melanogaster*: Isozymes and genetic variants. *Science* **149**: 80.

———. 1968. Alterations of genetic material for analysis of alcohol dehydrogenase isozymes of *Drosophila melanogaster*. *Ann. N.Y. Acad. Sci.* **151**: 441.

Haymerle, H. 1983. "The alcohol dehydrogenase gene of *Drosophila melanogaster*: DNA sequence, in vivo and in vitro transcripts." Ph.D. thesis, University of Cambridge, Cambridge, England.

Henikoff, S. 1983. Cloning exons for mapping transcription: Characterization of the *Drosophila melanogaster* alcohol dehydrogenase gene. *Nucleic Acids Res.* **11**: 4735.

Kelley, M.R., I.P. Mims, C.F. Farnet, S.A. Dicharry, and W.R. Lee. 1985. Molecular analysis of X-ray-induced alcohol dehydrogenase (*Adh*) null mutations in *Drosophila melanogaster*. *Genetics* **109**: 365.

Kreitman, M. 1983. Nucleotide polymorphism at the alcohol dehydrogenase locus of *Drosophila melanogaster*. *Nature* **304**: 412.

Lemeunier, F. and M. Ashburner. 1984. Studies on the evolution of the *melanogaster* species subgroup of the genus *Drosophila (Sophophora)*. IV. The chromosomes of two new species. *Chromosoma* **89**: 343.

Maroni, G. and S.C. Stamey. 1983. Developmental profile and tissue distribution of alcohol dehydrogenase. *Drosophila Inform. Serv.* **59**: 77.

O'Donnell, J., L. Gerace, F. Leister, and W. Sofer. 1975. Chemical selection of mutants that affect alcohol dehydrogenase in *Drosophila melanogaster*: Use of 1-pentyn-3-ol. *Genetics* **79**: 73.

O'Donnell, J., H.C. Mandel, M. Krauss, and W. Sofer. 1977. Genetic and cytogenetic analysis of the *Adh* region in *Drosophila melanogaster*. *Genetics* **86**: 533.

Rubin, G.M. and A.C. Spradling. 1982. Genetic transformation of *Drosophila* with transposable element vectors. *Science* **218**: 348.

Scholnick, S.B., B.A. Morgan, and J. Hirsh. 1983. The cloned dopa decarboxylase gene is developmentally regulated when reintegrated into the *Drosophila* genome. *Cell* **34**: 37.

Simon, J.A., C.A. Sutton, R.B. Lobell, R.L. Glaser, and J.T. Lis. 1985. Determinants of heat shock-induced chromosome puffing. *Cell* **40**: 805.

Sofer, W. and M.A. Hatkoff. 1972. Chemical selection of alcohol dehydrogenase negative mutations of *Drosophila melanogaster*. *Genetics* **72**: 545.

Spradling, A.C. and G.M. Rubin. 1983. The effect of chromosomal position on the expression of the *Drosophila* xanthine dehydrogenase gene. *Cell* **34**: 47.

Thatcher, D. 1980. The complete amino-acid sequence of three alcohol dehydrogenase alloenzymes (Adh n-11, Adh-S and Adh-uf) from the fruit-fly *Drosophila melanogaster*. *Biochem. J.* **187**: 875.

Throckmorton, L.H. 1975. The phylogeny, ecology, and geography of *Drosophila*. In *Handbook of Genetics* (ed. R.C. King), vol. 3, p. 421. Plenum Press, New York.

Ursprung, H., W.H. Sofer, and N. Burroughs. 1970. Ontogeny and tissue distribution of alcohol dehydrogenase in *Drosophila melanogaster*. *Wilhelm Roux's Arch. Dev. Biol.* **164**: 201.

Vigue, C. and W. Sofer. 1976. Chemical selection of mutants that effect alcohol dehydrogenase activity in *Drosophila*. III. Effects of ethanol. *Biochem. Genet.* **14**: 127.

Walker, V. and M. Ashburner. 1981. The control of ecdysterone regulated puffs in *Drosophila* salivary glands. *Cell* **26**: 269.

Woodruff, R.C. and M. Ashburner. 1979. The genetics of a small autosomal region of *Drosophila melanogaster* containing the structural gene for alcohol dehydrogenase. I. Characterization of deletions and mapping of *Adh* and visible mutations. *Genetics* **92**: 117.

Identification of DNA Sequences Required for the Regulation of *Drosophila* Alcohol Dehydrogenase Gene Expression

J.W. POSAKONY,* J.A. FISCHER, AND T. MANIATIS
Harvard University, Department of Biochemistry and Molecular Biology, Cambridge, Massachusetts 02138

Drosophila alcohol dehydrogenase (*Adh*) genes display a complex pattern of developmental expression. The level of ADH enzyme activity varies significantly during *Drosophila* development, and the expression of the gene is limited to specific tissues (Ursprung et al. 1970; Batterham et al. 1983). In *D. melanogaster* there is one copy of the *Adh* gene per haploid genome (Goldberg 1980), and this gene is transcribed from two different promoters that are utilized at different times during *Drosophila* development (Benyajati et al. 1983). One promoter ("adult" or distal) is active transiently in embryos, at a moderate level in third instar larvae, and at a high level in adults, while the other promoter ("larval" or proximal) is used principally in late embryos and all larval stages, and at a low level in adults (C. Savakis et al., in prep.).

In previous work in this laboratory, we have studied the expression of an 11.8-kb fragment of the *D. melanogaster Adh* gene (Goldberg 1980), which was introduced into the germ line of ADH null flies by P-element-mediated transformation (Goldberg et al. 1983). The expression of the transduced *Adh* gene was found to be regulated in a temporal and tissue-specific pattern indistinguishable from that of the endogenous *Adh* gene.

We have since taken two approaches to the identification of *cis*-acting DNA sequences required for the normal pattern of *Adh* gene expression. In the first approach, we have introduced deletions into the cloned *D. melanogaster Adh* gene and then determined the effect of these mutations on *Adh* expression in vivo using P-element transformation. These studies have identified three distinct DNA sequences containing *cis*-acting regulatory elements. In the second approach, we have analyzed the expression of *Adh* genes from *D. mulleri*, a member of a species complex in which the *Adh* gene has been duplicated (Oakeshott et al. 1982; Batterham et al. 1983, 1984). In contrast to the *D. melanogaster Adh* system, where two different promoters are utilized at different stages of development, *D. mulleri* has two different *Adh* genes, one of which is expressed only in larvae, and the other of which is expressed in both larvae and adults (a pattern similar to the differential utilization of the two promoters in *D. melanogaster*). A comparison of the larval-to-adult switch in *Adh* gene expression in these two species should identify common regulatory components, and possibly reveal changes in gene regulation that have occurred during evolution.

cis-Acting Regulatory Elements of the *D. melanogaster Adh* Gene

As noted above, all of the *cis*-acting DNA sequences required for normal *Adh* gene expression in vivo are contained within an 11.8-kb *Sac*I restriction fragment which includes 5.5 kb and 4.5 kb of 5'- and 3'-flanking sequences, respectively (Goldberg et al. 1983). Here we describe experiments in which P-element-mediated germ line transformation (Rubin and Spradling 1982; Spradling and Rubin 1982) has been used to locate and characterize the regulatory elements within this fragment.

First, a series of P-element transformation vectors was constructed, which contain P-element and flanking DNA sequences from pπ25.1 (Spradling and Rubin 1982; O'Hare and Rubin 1983) cloned into the bacterial plasmid pUC9 (Vieira and Messing 1982; Yanisch-Perron et al. 1985). A polylinker from the plasmid πVX (Seed 1983) was also added to provide additional cloning sites within the P-element termini. The 4.8-kb *Eco*RI or 3.2-kb *Xba*I fragments of *Adh* (Fig. 1A) were inserted into these vectors and introduced into ADH null recipients (e.g., Adh^{fn4} pr cn; Benyajati et al. 1982); transformants were identified by ethanol selection of adult G_1 flies (Vigue and Sofer 1976; Goldberg et al. 1983). In other experiments, the wild-type xanthine dehydrogenase (XDH) gene (rosy$^+$; Spierer et al. 1983) was also included in the P-element construction as a visible (eye color) marker, and recipient strains null for both ADH and XDH (e.g., Adh^{fn6} cn; ry^{502}) were constructed and transformed. To permit segregation analysis and the establishment of homozygous transformant lines, a stock of the genotype b Df(2L)A47 cn bw/CyO^{nB}; MKRS/TM2, ry was constructed; it carries an ADH-null second chromosome balancer and an XDH-null third chromosome balancer. (Descriptions and references for mutations and balancer chromosomes may be found in Lindsley and Grell 1968; Goldberg et al. 1983.)

Transformed strains homozygous for one trans-

---
*Present address: Department of Biology B-022, University of California, San Diego, La Jolla, California 92093.

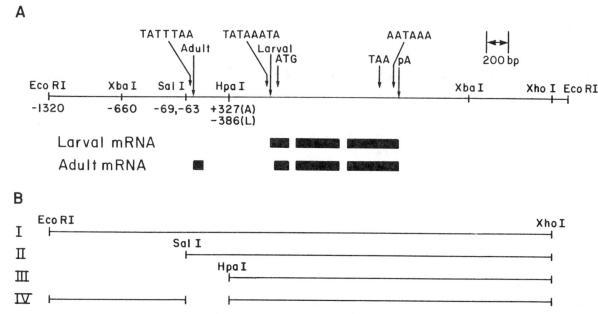

Figure 1. A restriction map of the *D. melanogaster Adh* gene, showing the location of in vitro deletions used to identify regulatory sequences. (*A*) The top line shows a map of restriction sites adjacent to and within the *Adh* gene. The numbers below the line indicate the number of base pairs upstream (−) or downstream (+) from the distal and proximal transcription start sites, which are indicated by the arrows labeled Adult and Larval, respectively. The initiation and termination sites of translation are indicated by ATG and TAA, respectively. The distal and proximal TATA box sequences are indicated by the arrows labeled TATTTAA and TATAAATA, respectively. The arrows labeled AATAAA and pA indicate the polyadenylation signal sequence and polyadenylation site, respectively. The exons present in larval and adult mRNA (initiated at the proximal and distal promoters, respectively) are indicated by the black boxes below the line, while the spaces between these boxes represent introns. (*B*) Maps of in vitro deletion mutants, the in vivo expression phenotypes of which are discussed in the text. Restriction sites corresponding to those in *A* are shown.

duced copy of either the 4.8-kb or the 3.2-kb fragment of *Adh* exhibited the same expression phenotype, and this was independent of the method of selection (ethanol resistance or eye color): (1) Correct tissue specificity of *Adh* expression was observed in both third instar larvae and adults, as assayed by histochemical staining of dissected tissues (Ursprung et al. 1970; Goldberg et al. 1983). Thus, ADH activity was detected in larval fat body, midgut, and Malpighian tubules; and in adult fat body, crop, midgut, hindgut, rectum, and Malpighian tubules, as well as in accessory genital organs (vasa deferentia, seminal vesicles, and sperm pump) of adult males. These results show that the 3.2-kb *Xba*I fragment of *Adh* (Fig. 1A), which has 0.66 kb and 0.64 kb of 5′- and 3′-flanking sequences, respectively, contains all of the *cis*-acting DNA sequences necessary for the major tissue specificities of *Adh* expression. (2) The levels of ADH activity (Table 1) and mRNA (J. Posakony and T. Maniatis, in prep.) in adult flies were nearly normal, being reduced an average of less than twofold from the wild-type. In contrast, third instar larval levels of activity (Table 1) and mRNA were reduced to a much greater extent, approximately 10-fold. Thus, DNA sequences contained within the 11.8-kb *Sac*I fragment but not the 4.8-kb *Eco*RI fragment of *Adh* are required for normal levels of larval *Adh* expression. Subsequent experiments, which will not be described here, have shown that these sequences are located in a 3.4-kb region on the 5′ side of the gene, and lie at least 2 kb upstream of the proximal promoter.

To localize further the *Adh* regulatory sequences within the *Eco*RI fragment, we analyzed the effect of a series of in vitro deletions on the in vivo expression of the gene (Fig. 1B). In all of these experiments, a rosy⁺ gene included in the P-element plasmid construction was used as a marker to identify germ line transformants. The phenotypic descriptions that follow re-

Table 1. ADH Enzyme Activity (nmoles/min/organism) in Transformed Strains Carrying *Eco*RI or *Xba*I Fragments of *Adh*

Strain[a]	Larvae[b]	Adults[c]
tAP-9	1.0	23.0
tAP-11	4.8	39.4
tAP-14	9.2	41.0
tAP-16	9.5	37.9
tAP-18	1.6	28.9
tAP-19	6.1	49.0
tAP-21	3.6	48.9
tAP-23	2.3	30.5
Average	4.8	37.3
w; *Adh*^{F} (wild type)	49.9	70.0

[a] tAP-9 through tAP-19 carry the *Eco*RI fragment of *Adh*; tAP-21 and tAP-23 carry the *Xba*I fragment (see Fig. 1A).
[b] Late third instar.
[c] Aged 4–7 days.

fer to transformed strains homozygous for one insertion of the P-element transposon.

When the DNA sequences 5' to the SalI site (−69) are deleted (Fig. 1), detectable expression from the distal promoter is lost (J. Posakony and T. Maniatis, in prep.). This was determined by nuclease protection analysis of the RNA (Zinn et al. 1983; Melton et al. 1984), using labeled SP6 RNA probes that distinguish between transcripts initiated at the two promoters. The proximal promoter, however, continues to function in both larvae and adults at the low level observed with the EcoRI and XbaI fragments, giving rise to a small amount of ADH activity at both stages (Table 2). This result shows that sequences between −660 and −69 upstream of the distal promoter are required for its activity.

Truncation of the gene to the HpaI site (Fig. 1), which lies 386 bp 5' to the proximal promoter transcription start, preserves a similar low level of proximal promoter activity in both larvae and adults (Table 2). This expression exhibits appropriate tissue specificity in larvae, indicating that the cis-acting regulatory elements responsible are located 3' to the HpaI site.

The phenotype of an internal deletion of the sequences between the SalI and HpaI sites (Fig. 1) provides evidence for a regulatory element upstream of, but separable from, the distal promoter. The deletion removes a 390-bp fragment which includes the distal TATA box and transcription initiation point, and thus fuses sequences upstream of the distal promoter to the −386 truncation just described. Adult flies carrying this deletion exhibit a high level of ADH activity, elevated approximately 10-fold above that observed with the −386 truncation itself (Table 2). RNA analysis reveals that the Adh mRNA in these adults is similarly increased in quantity, and is accurately initiated at the proximal promoter (J. Posakony and T. Maniatis, in prep.). Thus, upstream of the distal promoter there appears to be a positive regulatory element that can act on the proximal promoter in adults when the distal promoter is deleted. Remarkably, high levels of ADH activity (and mRNA initiated at the proximal promoter) are also observed in larvae carrying this same deletion (Table 2). This effect may be a reflection of the onset of distal promoter transcription in the third larval instar normally observed with the wild-type Adh gene (C. Savakis et al., in prep.).

Evidence that the distal upstream sequence has the properties of a tissue-specific enhancer element has been obtained by analyzing P-element transformants carrying a construction in which the region between the EcoRI and SalI sites 5' to the distal promoter of Adh (−1320 to −69; Fig. 1) are fused to the dopadecarboxylase (Ddc) gene at a position 210 nucleotides upstream of its transcription start point. It has been shown that the −210 truncation of Ddc exhibits essentially normal expression in transformed flies (J. Hirsh, pers. comm.). Transformants carrying the Adh-Ddc fusion, by contrast, have five- to 10-fold elevated levels of DDC activity in adults, and their Ddc mRNA is correctly initiated (J. Posakony and T. Maniatis, in prep.). This quantitative effect of the Adh distal upstream sequences on Ddc is very similar to their effect on the proximal promoter of Adh, described above. Thus, these sequences are capable of activating a high level of expression in adult flies, even from a heterologous promoter. Moreover, the Adh sequences appear to contain a cis-acting element involved in controlling at least part of the tissue specificity of Adh expression. In the transformed lines carrying the Adh-Ddc fusion, DDC activity is easily detectable in the accessory genital apparatus of adult males, a tissue that normally expresses ADH but not DDC (data not shown).

On the basis of these observations, we propose that the expression of the Adh gene in D. melanogaster is regulated by cis-acting sequence elements in at least three different regions upstream of and within the transcription unit (Fig. 2). One region lies over 2000 bp upstream from the proximal promoter, and is necessary for normal levels of Adh transcription in larvae. A second region is located between −69 and −660 from the distal transcription start point, and contains an element or elements that confer temporal and tissue specificities characteristic of the distal promoter. The third region is located 3' to position −386 from the proximal transcription start point, and is responsible for the tissue-specific activity of the proximal promoter in larvae.

Structure and Transcription of the D. mulleri Adh Locus

An alternative approach to the analysis of temporal switching of Adh gene expression is suggested by the observation that members of the mulleri subgroup (distant relatives of D. melanogaster) produce two different ADH proteins, ADH-1 (larval) and ADH-2 (adult), in a temporal pattern similar to the expression from the D. melanogaster proximal and distal promoters (Oakeshott et al. 1982; Batterham et al. 1983, 1984). Genetic analyses of electrophoretic variants of ADH-1 and ADH-2 in D. buzzatii (a species of the mulleri subgroup) provided evidence for the hypothesis that ADH-1 and ADH-2 are encoded by two closely linked genes that are differentially expressed during development (Oakeshott et al. 1982). However, other possibilities, such as alternate splicing of a single transcript to

Table 2. Average ADH Enzyme Activity (nmoles/min/organism) in Transformed Strains Carrying Mutant Adh Genes

Construction[a]	Larvae[b]	Adults[c]
I D(−1320)	5.4	36.5
II D(−69)	2.8	2.1
III P(−386)	3.8	1.7
IV D(−69) to P(−386)	26.3	20.6
w; Adh^F (wild type)	55.4	61.4

[a] See Fig. 1B.
[b] Late third instar.
[c] Aged 4–7 days.

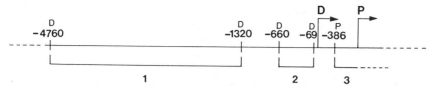

Figure 2. A restriction map of the 5' end and flanking sequences of the *D. melanogaster Adh* gene, showing the location of *cis*-acting regulatory elements. The long horizontal line represents gene sequences. The distal and proximal transcription initiation sites are indicated by arrows labeled D and P, respectively. Numbers above the line show the distance in base pairs upstream (5') of the distal (D) or proximal (P) transcription starts. Numbered brackets below the line show the location of sequence regions containing *cis*-acting regulatory elements, as discussed in the text.

produce the two transcripts, could not be ruled out. Whatever the processes involved, a comparison of the structure and transcriptional regulation of the *Adh* genes in these distantly related species could lead to an understanding of the mechanism and biological significance of the temporal switch in *Adh* promoter selection. Toward this goal, we have cloned and sequenced the *Adh* locus of *D. mulleri*, a species of the *mulleri* subgroup (J. Fischer and T. Maniatis, in prep.).

The *Adh* genes of *D. mulleri* were cloned from a phage λ library of *D. mulleri* genomic DNA by hybridization with a *D. melanogaster Adh* gene probe. All of the DNA sequences in the *D. mulleri* genome homologous to *Adh* were contained within one 7-kb region. The nucleotide sequence of this 7-kb region of DNA was determined. Figure 3 summarizes schematically the structural information revealed by the DNA sequence. The *D. mulleri Adh* locus consists of three *Adh* genes which are all transcribed from the same DNA strand. Conceptual translation of the two downstream genes revealed that they differ by 10 amino acids. These differences identified the upstream gene as *Adh-2* and the downstream gene as *Adh-1*, since ADH-2 is the more basic protein (Oakeshott et al. 1982). Approximately 1.2 kb upstream of *Adh-2* is an *Adh*-like gene which has accumulated numerous nucleotide substitutions, deletions, and additions, so that it no longer encodes an ADH protein. For these reasons, we have designated the upstream gene a pseudogene.

Using SP6 RNA probes specifically complementary to putative transcripts from each of the three *Adh* genes (Zinn et al. 1983; Melton et al. 1984), we have shown that each gene is transcribed (J. Fischer and T. Maniatis, in prep.). The 5' ends of *Adh-1* and *Adh-2* each map 28 bp downstream from a TATA-box sequence. There is no DNA sequence evidence for a distal promoter and splice junction sequences, resembling the organization of the *D. melanogaster Adh* gene (Benyajati et al. 1983) upstream of either *Adh-1* or *Adh-2*. Although there is no TATA box at an appropriate distance upstream of the 5' end of the transcript, the pseudogene is transcribed. The pseudogene transcripts are found at approximately one-tenth the level of the *Adh-1* and *Adh-2* transcripts.

Expression of the *D. mulleri Adh* Genes

Figure 4 shows an analysis of the transcription of *Adh-1* and *Adh-2* during the larval stages of development and in adults of *D. mulleri*. *Adh-1* transcripts are present during all larval stages and at extremely low levels in adults. *Adh-2* transcripts are present at low levels in first instar larvae, at steadily increasing levels through the third instar, and at levels in adults similar to the level in third instar larvae. These results correlate with a previous developmental analysis of the ADH-1 and ADH-2 proteins (Batterham et al. 1984). The temporal pattern of expression of the pseudogene is not shown. The pseudogene is transcribed in late (5-day-old) third instar larvae, and in adults.

Thus, the switch from predominantly *Adh-1* expression in early larvae, to expression of both *Adh-1* and *Adh-2* in later larval stages, to expression of predominantly *Adh-2* in adults, occurs at the level of transcrip-

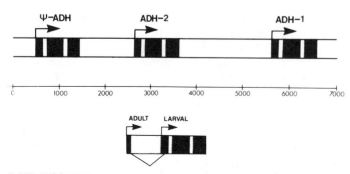

Figure 3. Schematic diagram of the *D. melanogaster* and *D. mulleri Adh* genes. The genes are indicated by rectangles; the black regions represent exons, the dotted regions represent introns. The direction of transcription of each gene is indicated by an arrow.

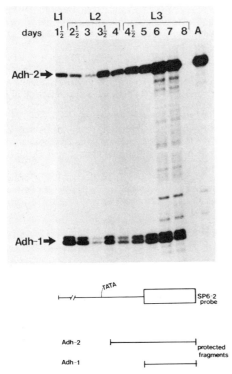

Figure 4. Transcription of *Adh-1* and *Adh-2* in *D. mulleri* larval stages and adults. One microgram of RNA from *D. mulleri* first (L1), second (L2), and third instar (L3) larvae and adults (A) was hybridized with the RNA probe SP6-2. SP6-2 protects an approximately 130-nucleotide fragment of the *Adh-2* transcript, from the 5′ end through most of the first coding exon. Because *Adh-1* and *Adh-2* have the same nucleotide sequence for 62 nucleotides in each of their first exons, SP6-2 also protects a 62-nucleotide fragment of the *Adh-1* transcript. The numbers above the lanes refer to the approximate number of days following oviposition that the larvae were harvested. The flies and larvae were grown at 25°C.

tion. In addition, the expression of *Adh-1* and *Adh-2* during larval development and in adults is quite similar to the pattern of *D. melanogaster Adh* expression from the proximal and distal promoters. C. Savakis et al. (in prep.) have shown that the proximal promoter is active throughout larval development and is expressed at an extremely low level in adults. They have also shown that the distal promoter becomes active in late second instar larvae, increases its activity through the third instar, and is used at high levels in adult flies. Thus, the major difference in the temporal pattern of *Adh* expression between the two species is the time at which *Adh-2* or the distal promoter becomes active; *Adh-2* is transcribed at low levels as early as the first instar, whereas no transcripts from the *D. melanogaster* adult promoter can be detected until the late second instar.

In summary, we have shown that although the transcriptional organization of the *D. mulleri Adh* genes is quite different from that of the *D. melanogaster Adh* gene, *Adh-1* and *Adh-2* share a similar temporal pattern of transcription with the proximal and distal promoters. Because the tissue-specific pattern of expression is also quite similar in the two species (Ursprung et al. 1970; Batterham et al. 1983), we thought it might be possible that the *D. mulleri Adh* genes would be expressed and regulated if introduced into *D. melanogaster*. Thus, a 9.3-kb *SacI* fragment containing the *D. mulleri Adh* locus (1.5 kb upstream of the pseudogene to 1.5 kb downstream of *Adh-1*) was introduced into *D. melanogaster* via P-element transformation using the rosy gene as a selectable marker. Seven independent transformed lines were isolated. Remarkably, all show the proper temporal and tissue-specific regulation of *Adh-1* and *Adh-2* (J.A. Fischer and T. Maniatis, in prep.). Thus, the factors present in *D. melanogaster* which regulate the temporal and tissue specific expression of the *Adh* gene from both of its promoters can also regulate *Adh-1* and *Adh-2*, making it possible to use P-element transformation of *D. melanogaster* to investigate the *Adh* regulatory elements of *D. mulleri*.

ACKNOWLEDGMENTS

We acknowledge the excellent technical assistance of Abby Telfer. We thank Charalambos Savakis, Michael Ashburner, and Marty Kreitman for communication of unpublished results, Art Chovnick for providing the ry⁻ TM2 chromosome, and Welcome Bender for a subclone of the rosy⁺ gene. J.W.P. was supported by an American Cancer Society postdoctoral fellowship. J.A.F. acknowledges support from a National Science Foundation predoctoral fellowship. This work was funded by a grant to T.M. from the National Institutes of Health.

REFERENCES

BATTERHAM, P., J.A. LOVETT, W.T. STARMER, and D.T. SULLIVAN. 1983. Differential regulation of duplicate alcohol dehydrogenase genes in *Drosophila mojavensis*. *Dev. Biol.* **96:** 346.

———. 1984. Origin and expression of an alcohol dehydrogenase gene duplication in the genus *Drosophila*. *Evolution* **38:** 644.

BENYAJATI, C., N. SPOEREL, H. HAYMERLE, and M. ASHBURNER. 1983. The messenger RNA for alcohol dehydrogenase in *Drosophila melanogaster* differs in its 5′ end in different developmental stages. *Cell* **33:** 125.

BENYAJATI, C., A.R. PLACE, N. WANG, E. PENTZ, and W. SOFER. 1982. Deletions at intervening sequence splice sites in the alcohol dehydrogenase gene of *Drosophila*. *Nucleic Acids Res.* **10:** 7261.

GOLDBERG, D.A. 1980. Isolation and partial characterization of the *Drosophila* alcohol dehydrogenase gene. *Proc. Natl. Acad. Sci.* **77:** 5794.

GOLDBERG, D.A., J.W. POSAKONY, and T. MANIATIS. 1983. Correct developmental expression of a cloned alcohol dehydrogenase gene transduced into the *Drosophila* germ line. *Cell* **34:** 59.

LINDSLEY, D.L. and E.H. GRELL. 1968. *The genetic variations of Drosophila melanogaster*. Carnegie Inst. Wash. Publ. no. 627.

MELTON, D.A., P.A. KRIEG, M.R. REBAGLIATI, T. MANIATIS, K. ZINN, and M.R. GREEN. 1984. Efficient *in vitro* synthesis of biologically active RNA and RNA hybridization probes from plasmids containing a bacteriophage SP6 promoter. *Nucleic Acids Res.* **12:** 7035.

OAKESHOTT, J.G., G.K. CHAMBERS, P.D. EAST, J.B. GIBSON,

and J.S.F. Barker. 1982. Evidence for a genetic duplication involving alcohol dehydrogenase genes in *Drosophila buzzatii* and related species. *Aust. J. Biol. Sci.* **35:** 73.

O'Hare, K. and G.M. Rubin. 1983. Structures of P transposable elements and their sites of insertion and excision in the *Drosophila melanogaster* genome. *Cell* **34:** 25.

Rubin, G.M. and A.C. Spradling. 1982. Genetic transformation of *Drosophila* with transposable element vectors. *Science* **218:** 348.

Seed, B. 1983. Purification of genomic sequences from bacteriophage libraries by recombination and selection *in vivo*. *Nucleic Acids Res.* **11:** 2427.

Spierer, P., A. Spierer, W. Bender, and D.S. Hogness. 1983. Molecular mapping of genetic and chromomeric units in *Drosophila melanogaster*. *J. Mol. Biol.* **168:** 35.

Spradling, A.C. and G.M. Rubin. 1982. Transposition of cloned P elements into *Drosophila* germ line chromosomes. *Science* **218:** 341.

Ursprung, H., W. Sofer, and N. Burroughs. 1970. Ontogeny and tissue distribution of alcohol dehydrogenase in *Drosophila melanogaster*. *Wilhelm Roux's Arch. Dev. Biol.* **164:** 201.

Vieira, J. and J. Messing. 1982. The pUC plasmids, an M13mp7-derived system for insertion mutagenesis and sequencing with synthetic universal primers. *Gene* **19:** 259.

Vigue, C. and W. Sofer. 1976. Chemical selection of mutants that affect ADH activity in *Drosophila*. III. Effects of ethanol. *Biochem. Genet.* **14:** 127.

Yanisch-Perron, C., J. Vieira, and J. Messing. 1985. Improved M13 phage cloning vectors and host strains: Nucleotide sequences of the M13mp18 and pUC19 vectors. *Gene* **33:** 103.

Zinn, K., D. DiMaio, and T. Maniatis. 1983. Identification of two distinct regulatory regions adjacent to the human β-interferon gene. *Cell* **34:** 865.

Developmental Control of *Drosophila* Yolk Protein 1 Gene by *cis*-acting DNA Elements

B. SHEPHERD, M.J. GARABEDIAN, M.-C. HUNG, AND P.C. WENSINK
Department of Biochemistry and The Rosenstiel Center, Brandeis University, Waltham, Massachusetts 02254

In *Drosophila melanogaster* the three yolk proteins and their mRNAs occur in a specific developmental pattern. The mRNAs and the precursors for yolk proteins are synthesized by two tissues of the adult female, the ovarian follicle cells and the fat bodies (Gelti-Douka et al. 1974; Bownes and Hames 1977; Warren and Mahowald 1979; Barnett and Wensink 1981; Brennan et al. 1982; Isaac and Bownes 1982). One of these tissues, the fat body, is scattered throughout the organism, in the head, the abdomen, and the thorax (Miller 1965). In this tissue the yolk protein mRNAs and precursors are first detected shortly after the female hatches. They steadily increase in concentration to reach a plateau which is maintained from approximately 1 day to 1 week after hatching (Barnett and Wensink 1981; Garabedian et al. 1985). In contrast to the fat bodies, the follicle cells contain yolk protein mRNA for a short portion of their life span, in only two of the 14 stages of oocyte development (Brennan et al. 1982). The yolk proteins are secreted from both of these tissues and then sequestered in the egg. They are the most abundant proteins of the egg and are among the most abundant female proteins. They are undetectable in normal males.

Each of the three proteins is encoded by a different single-copy gene, *yp1*, *yp2*, or *yp3* (Barnett et al. 1980). None of the genes is amplified or rearranged in the tissue where they are most actively transcribed, the fat bodies (Barnett and Wensink 1981). The transcripts from each gene have the same 5' end in both tissues. The arrangement and structure of the *yp1* and *yp2* genes are shown in Figure 1A. Each of the two genes has a single small intron. They are divergently transcribed and separated by 1225 bp (Hung and Wensink 1981, 1983; Hung et al. 1982). The *yp3* gene is approximately 1000 kb proximal to the other *yp* genes on the X chromosome and has two small introns (Barnett et al. 1980; Hung et al. 1982).

In this paper we summarize our results describing the yolk protein 1 gene (*yp1*) and the DNA elements near it that control its developmental pattern of expression. These results lead us to conclude that two different *cis*-acting DNA elements control the tissue specificity of *yp1* expression. One is necessary for expression in fat bodies and the other is necessary for expression in ovarian follicle cells. We have localized the fat body element to a 125-bp DNA segment. Furthermore, we have shown that this fat body element can control a heterologous *Drosophila* promoter to express in the *yp1* fat body expression pattern. We conclude that the developmental expression pattern of this gene is controlled by at least two independent tissue-specific elements and that another element or elements are responsible for the sex and time specificity of its expression.

METHODS

DNA constructions. The vector plasmids for transformation were pRPL1 (Garabedian et al. 1985) and pCP20.1 (Simon et al. 1985). The *yp* genes in the series of *yp1* 5' deletions were marked by insertion of the 676-bp M13mp8 *Bgl*II/*Xmn*I fragment containing gene II (van Wezenbeek et al. 1980; Messing et al. 1981). It was inserted (Garabedian et al. 1985) into the *Xho*I site of *yp1* (Hung and Wensink 1981) and replaced a 393-bp fragment between the *Stu*I and *Bgl*II sites of *yp2* (Hung and Wensink 1983). The resulting constructions were inserted into pRPL1. The series of *yp1* 3' deletions was constructed using two plasmids that contain the hsp70-*lacZ* fusion gene (Lis et al. 1983; Costlow et al. 1985). These plasmids were generously given to us by John Lis (Cornell University) and his colleagues. One of them has a *Xho*I site 196 bp upstream of the hsp70 cap site, the other (a deletion derivative of the first) has an *Xho*I site 43 bp upstream of the hsp70 cap site. We placed *Xho*I linkers at different locations upstream of the *yp1* promoter and then joined upstream regions of *yp1* to the appropriate hsp70-*lacZ* fusion via the *Xho*I linkers. The *yp1*-hsp70-*lacZ* fusion genes were inserted into pCP20.1 (Figs. 1 and 5, below). The methods of construction are described elsewhere (Schleif and Wensink 1981; Maniatis et al. 1982; Garabedian et al. 1985).

Germ line transformation. The method of Spradling and Rubin (1982) was used to transform ry⁻ embryos (ry$^{\Delta 506}$, a gift from Welcome Bender, Harvard University Medical School) with DNA constructs that included a ry⁺ gene. Adults derived from injected embryos were mated to ry⁻ flies. The ry⁺ progeny were selected and mated to ry⁻ flies. The ry⁺ progeny of this mating were parents of the resulting transformed line.

Analysis of germ line transformants. Total and poly(A)⁺ RNA was prepared (Barnett et al. 1980) from hand-dissected adult (3- to 5-day-old) tissues, adults, or the developmental stages described in the Results section. DNA was prepared (Schleif and Wensink 1981)

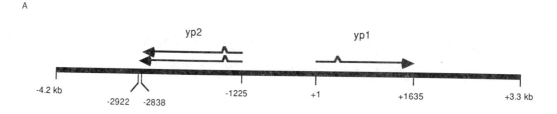

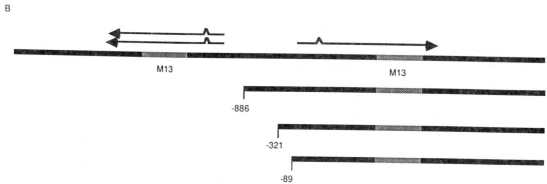

Figure 1. The *yp1* and *yp2* gene structure and the series of *yp1* 5' deletions. (*A*) Arrangement of *yp1* and *yp2* genes. The *yp1* transcript and the two *yp2* transcripts are represented by arrows. The transcript termini are indicated by their nucleotide position relative to the capping site of *yp1*. (*B*) Four 5' deletions of *yp1* DNA. The M13 DNA segment introduced into *yp1* and *yp2* is indicated by the stippled portion of the lines.

from 3- to 5-day-old adults. DNA and RNA electrophoresis, blotting, hybridization, radiolabeling of nucleic acids, and autoradiography have been described (Kalfayan and Wensink 1982; Garabedian et al. 1985). The X-gal assay for β-galactosidase activity has been described by Lis et al. (1983).

RESULTS

Different DNA Elements Control *yp1* Transcription in Different Tissues

The approximate 5' boundary of elements controlling the in vivo pattern of *yp1* transcription were determined by constructing altered forms of the gene and then examining the transcription pattern of these altered genes after they had been introduced into the *Drosophila* germ line. A series of deletions (Fig. 1) approaching the 5' end of *yp1* was constructed. In these constructions a fragment of gene II from M13mp8 was inserted into the *yp1* structural gene so that transcripts from the altered gene could be distinguished from those of the normal endogenous gene. The altered *yp1* genes were placed in a vector DNA which allows P-element-mediated transformation of the germ line (Spradling and Rubin 1982).

After introducing these reconstructed *yp1* genes into the germ line, the DNA of each resulting line of flies was examined by Southern blots to determine the number of gene copies present in the genome. All of the lines described in this paper contain a single copy of the introduced gene. An example of copy number determination is shown in Figure 2.

As a first step in locating the 5' boundary of elements controlling *yp1*, we identified a large DNA fragment which, when introduced into the germ line, permitted the *yp1*-M13 gene to be expressed in the pattern of the endogenous *yp1* gene. Transcripts from this DNA fragment (which extends from 3 kb upstream of the 5' end of *yp1* to 1.5 kb downstream of the 3' end) occur in the normal *yp1* pattern. This DNA fragment includes all of the *yp2* gene as well as the *yp1* gene (Fig. 1). Northern gel blots of transcripts from transformed flies indicate that transcripts from both of the introduced genes are undetectable in males, but occur at high concentration in females (Fig. 3). Unlike the simple pattern of transcripts from the normal gene, there are multiple transcripts from the introduced genes. The extra transcripts are due to the inserted M13 DNA. As described elsewhere (Garabedian et al. 1985), this M13 fragment contains several sequences either identical to or very similar to the sequence (AATAAA) that determines the 3' end of many eukaryotic transcripts (Proudfoot and Brownlee 1976; Montell et al. 1983). These sequences are fairly effective in the *yp1* gene and cause most of the transcripts to terminate within the M13 sequence. This artifact does not obsure the fact that the introduced *yp1* gene is expressed only in females. Its transcripts also occur in the normal *yp* tissue-specific pattern, that is, in both fat bodies and fol-

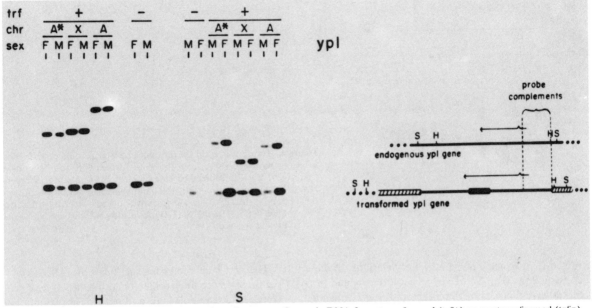

Figure 2. Genomic copy number of three *yp1* transformants. Genomic DNA from transformed (trf+) or nontransformed (trf−) adult males (M) or females (F) was digested with either *Hin*dIII (H) or *Sac*I (S) restriction enzymes, electrophoresed in agarose gels, and blotted onto nitrocellulose. Transformed flies are heterozygous for the introduced gene. The introduced gene is on either the X chromosome (chr X) or an autosome (A). In one case (A*) the transformed line is female-sterile. A scale diagram of the transforming and endogenous *yp1* genes and the radiolabeled probe hybridized to the blots is shown to the right.

licle cells (Fig. 3). We conclude that this 7.5-kb fragment contains all of the *cis*-acting DNA necessary for the normal expression pattern of *yp1*.

The next deletion of DNA upstream of *yp1* removed all of the *yp2* gene and a large portion of the intergenic region, leaving only 886 nucleotides of DNA normally found upstream of the *yp1* capping site (Fig. 1). This

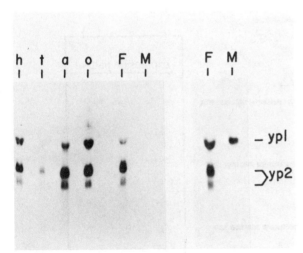

Figure 3. Transcripts from the introduced *yp1-yp2* gene region. RNA from male (M) and female (F) flies or hand-dissected ovaries (o), heads (h), and fat bodies from the thorax (t) and abdomen (a) were electrophoresed in formaldehyde/agarose gels and blotted onto nitrocellulose paper. The left panel was hybridized to radiolabeled M13 DNA and the right panel to radiolabeled M13 and α1-tubulin gene DNA (Garabedian et al. 1985). The tubulin gene transcript is present in both males and females and has an electrophoretic mobility similar to the *yp1* transcript.

deletion separates *yp1* from *cis*-acting DNA necessary for one of its tissue specificities. A Northern gel blot of RNA purified from ovarian and from several fat body tissues of the transformed flies was hybridized to a radiolabeled DNA complementary to the M13 fragment inserted into the *yp1* gene. This blot (Fig. 4) demonstrates that the *yp1* DNA in this structure includes all of the *cis*-acting DNA necessary for the appearance of transcripts in fat bodies of adult females (hereafter termed the fat body expression pattern). However, this DNA structure is missing an element or elements necessary for the appearance of ovarian transcripts. The radiolabeled DNA used in this experiment also includes the α1-tubulin gene (Kalfayan and Wensink 1981, 1982). This α1 DNA detected α1 mRNA in all RNA samples (Fig. 4), thereby indicating that the absence of a *yp1*-M13 autoradiographic signal in the ovarian RNA lanes of this gel blot is not due to the absence of RNA in the lanes. We conclude that there are at least two tissue-specific DNA elements controlling *yp1* gene expression. One is necessary for expression in ovaries and the other is necessary for expression in fat bodies.

Northern blot analysis of RNAs from other deletions (Fig. 1) demonstrated that the fat body expression pattern does not require DNA upstream of −321, but does require DNA upstream of −89. These results indicate that the 5′ boundary of DNA necessary for proper fat body expression is between −321 and −89.

A 125-bp Fragment from *yp1* Confers the Fat Body Expression Pattern on a Heterologous Promoter

To determine the 3′ boundary of DNA necessary for the *yp1* female-, adult-, and fat body-specific expres-

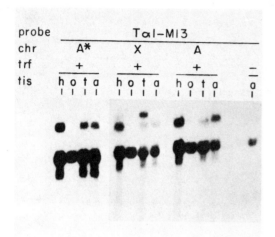

Figure 4. Transcripts from the introduced −886 deletion of *yp1*. Blots of electrophoresed RNA from flies transformed with the −886 deletion of *yp1* were hybridized to radiolabeled M13 and α1-tubulin gene DNA. In this autoradiogram, the band shared by all lanes contains the α1 gene transcript. The bands above and below the α1 transcript are complementary to the M13 DNA introduced into the *yp1* gene. The abbreviations are as in Fig. 3.

sion, we constructed a second set of deletions. In these constructions (Fig. 5) the *yp1* structural gene and promoter were replaced by the 5′ region of the *D. melanogaster* hsp70 heat shock gene fused in frame to the *E. coli lacZ* structural gene (Lis et al. 1983). This construction strategy allows transcription from the hsp70 promoter to be assayed by a test for activity of the *lacZ* gene product, β-galactosidase. Staining for this activity gives a rapid and fairly precise assay for the tissue and time of gene activity because the hsp70-*lacZ* mRNA and protein appear to be stable and active throughout development and in almost all tissues (Lis et al. 1983). All staining described below was performed under non-heat shock conditions.

The first 3′ deletion construct fused the −886 to −47 region of *yp1* to nucleotide −43 of the hsp70-*lacZ* gene (Fig. 5). When adult male and female flies from a line with a single genomic copy of this construct are stained using the β-galactosidase-sensitive dye X-gal (Lis et al. 1983), the dye turns dark blue in female fat bodies, but is colorless in ovaries and males (Fig. 6). Adult males and females of the host ry− strain remain colorless when they are stained similarly (Fig. 6). Since flies containing the hsp70-*lacZ* gene deleted to −195 (Lis et al. 1983) or the hsp70 gene deleted to −44 (Cohen and Meselson 1984) also do not express in female fat bodies, it appears that the −886 to −47 region of *yp1* controls the hsp70 promoter to produce the *yp1* fat body

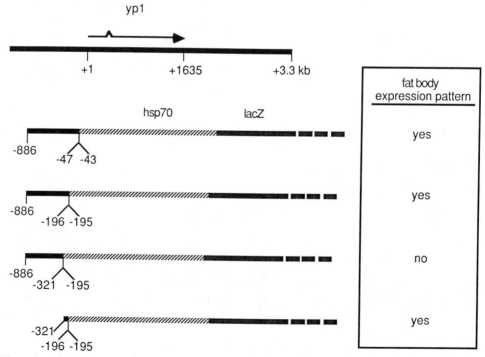

Figure 5. The series of *yp1* 3′ deletions. The *yp1* (heavy black line), hsp70 (hatched line), and *lacZ* (gray line) DNA in the introduced DNA are diagramed. The numbers represent the breakpoints in DNA relative to the capping site of the gene in that DNA. The table at the right summarizes the transcription pattern of the introduced DNAs.

expression pattern. Since the hsp70 DNA in this construction is missing the DNA elements (Pelham boxes I and II) known to be necessary for normal heat shock induction of the hsp70 gene (Pelham 1982; Parker and Topol 1984; Wu 1984), we conclude that these hsp70 elements are not necessary for the fat body expression pattern caused by the addition of *yp1* upstream DNA. However, it is possible that this transcription initiates within the *yp1* DNA and is processed to have a 5' end at the normal 5' end of the hsp70 mRNA. This possibility has been eliminated by our detection of fat body-specific transcripts from a construct in which the *yp1* upstream region was placed downstream of the hsp70-*lacZ* gene (data not shown). We conclude that transcription begins at the hsp70 promoter and is controlled by the *yp1* upstream region.

Two other constructions in this deletion series indicate that the 3' boundary of the fat body expression element lies between nucleotides −321 and −196 (Fig. 5). A *yp1* upstream fragment from −886 to −196 fused to nucleotide −195 of hsp70 has the same expression pattern as the −47 fusion. However, no transcripts could be detected from a construction in which a *yp1* fragment from −886 to −321 was fused to −195 of hsp70-*lacZ*.

The results from these 5' and 3' deletions indicate

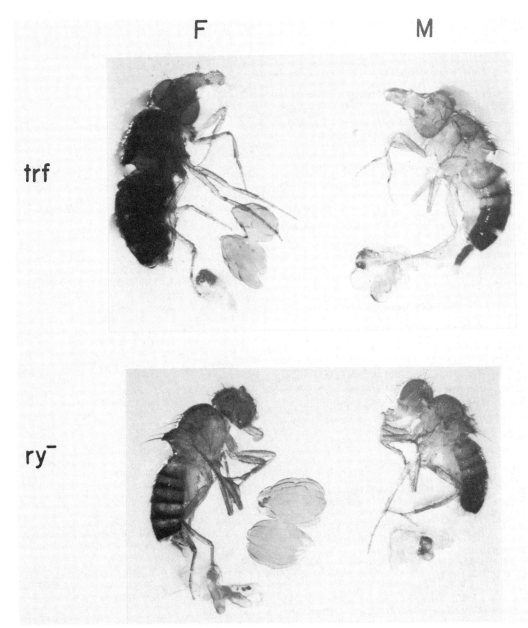

Figure 6. Tissue-specific *lacZ* gene expression controlled by *yp1* upstream DNA. The upper panel (trf) shows female (F) and male (M) flies containing a single germ line copy of *yp1* DNA from nucleotides −886 to −47 joined at nucleotide −43 of the hsp70-*lacZ* gene. The lower panel shows flies from the untransformed host strain (ry⁻). The flies were dissected in a solution of X-gal and allowed to stain for 2 hr prior to photography.

that *yp1* DNA between −321 and −196 may be sufficient for the *yp1* fat body expression pattern. To test this possibility, we constructed a fusion between this upstream region of *yp1* and the hsp70-*lacZ* gene. When a single copy of this construction was introduced into the germ line of *Drosophila*, staining was observed in adult female fat bodies, but not in ovarian tissue. Males had light staining only in a few intestinal tissues known to produce heat shock RNA without heat shock induction (R.L. Glaser and J.T. Lis, pers. comm.). This male staining was not observed in the −886 to −47 *yp1* fusion described above. We presume that this is because the hsp70 DNA elements necessary for this transcription have been eliminated in the latter construction. We conclude that the 125 bp between −321 and −196 are sufficient to cause a heterologous promoter to express in the *yp1* fat body pattern.

SUMMARY

In this paper we have demonstrated that at least two tissue-specific *cis*-acting DNA elements are necessary for the normal developmental pattern of *Drosophila yp1* gene expression. One of these elements is necessary for expression in the adult female follicle cells and the other is necessary for expression in adult female fat bodies. We have localized the fat body expression element to a 125-bp fragment that lies between nucleotides −196 and −321 of the *yp1* gene. This small fragment has at least one of the characteristics of enhancer sequences because it can direct a heterologous *Drosophila* promoter to be transcribed with the *yp1* fat body expression pattern.

ACKNOWLEDGMENTS

We thank John Lis and his colleagues for several hsp70-*lacZ* fusions given to us before they had been described in publications. We also thank the National Institutes of Health for financial support (GM21626). B.M.S. was supported by a EMBO long-term fellowship. We are especially grateful for stimulating discussions with our colleagues, Howard Baum, Claude Maina, Paul Mitsis, and Bill Theurkauf.

REFERENCES

Barnett, T. and P.C. Wensink. 1981. Transcription and translation of yolk protein mRNA in the fat bodies of *Drosophila*. *ICN-UCLA Symp. Mol. Cell. Biol.* **23:** 97.

Barnett, T., C. Pachl, J.P. Gergen, and P.C. Wensink. 1980. The isolation and characterization of *Drosophila* yolk protein genes. *Cell* **21:** 729.

Bownes, M. and B.D. Hames. 1977. Accumulation and degradation of three major yolk proteins in *Drosophila melanogaster*. *J. Exp. Zool.* **200:** 149.

Brennan, M.D., A.J. Weiner, T.J. Goralski, and A.P. Mahowald. 1982. The follicle cells are a major site of vitellogenin synthesis in *Drosophila melanogaster*. *Dev. Biol.* **89:** 225.

Cohen, R.S. and M. Meselson. 1984. Inducible transcription and puffing in *Drosophila melanogaster* transformed with hsp70-phage λ hybrid heat shock genes. *Proc. Natl. Acad. Sci.* **81:** 5509.

Costlow, N.A., J.A. Simon, and J.T. Lis. 1985. A hypersensitive site in hsp70 chromatin requires adjacent not internal DNA sequence. *Nature* **313:** 147.

Garabedian, M.J., M.-C. Hung, and P.C. Wensink. 1985. Independent control elements that determine yolk protein gene expression in alternative *Drosophila* tissues. *Proc. Natl. Acad. Sci.* **82:** 1396.

Gelti-Douka, H., T.R. Gingeras, and M.P. Kambysellis. 1974. Yolk proteins in *Drosophila*: Identification and site of synthesis. *J. Exp. Zool.* **187:** 167.

Hung, M.-C. and P.C. Wensink. 1981. The sequence of the *Drosophila melanogaster* gene for yolk protein 1. *Nucleic Acids Res.* **9:** 6407.

———. 1983. Sequence and structure conservation in yolk proteins and their genes. *J. Mol. Biol.* **164:** 481.

Hung, M.-C., T. Barnett, C. Woolford, and P.C. Wensink. 1982. Transcript maps of *Drosophila* yolk protein genes. *J. Mol. Biol.* **154:** 581.

Isaac, P.G. and M. Bownes. 1982. Ovarian and fat-body vitellogenin synthesis in *Drosophila melanogaster*. *Eur. J. Biochem.* **123:** 527.

Kalfayan, L. and P.C. Wensink. 1981. α-Tubulin genes of *Drosophila*. *Cell* **24:** 97.

———. 1982. Developmental regulation of *Drosophila* α-tubulin genes. *Cell* **29:** 91.

Lis, J.T., J.A. Simon, and C.A. Sutton. 1983. New heat shock puffs and β-galactosidase activity resulting from transformation of *Drosophila* with an hsp70-lacZ hybrid gene. *Cell* **35:** 403.

Maniatis, T., E.F. Fritsch, and J. Sambrook. 1982. *Molecular cloning*. Cold Spring Harbor Laboratory, Cold Spring Harbor, New York.

Messing, J., R. Crea, and P.H. Seeburg. 1981. A system for shotgun DNA sequencing. *Nucleic Acids Res.* **9:** 309.

Miller, A. 1965. The internal anatomy and histology of the imago of *Drosophila melanogaster*. In *The biology of Drosophila* (ed. M. Demerec), p. 420. Hafner, New York.

Montell, C., E.F. Fisher, M.H. Caruthers, and A.J. Berk. 1983. Inhibition of RNA cleavage but not polyadenylation by a point mutation in mRNA 3′ consensus sequence AAUAAA. *Nature* **305:** 600.

Parker, C.S. and J. Topol. 1984. A *Drosophila* RNA polymerase II transcription factor binds to the regulatory site of an hsp 70 gene. *Cell* **37:** 273.

Pelham, H.R.B. 1982. A regulatory upstream promoter element in the *Drosophila* hsp 70 heat-shock gene. *Cell* **30:** 517.

Proudfoot, N.J. and G.G. Brownlee. 1976. 3′ non-coding region sequences in eukaryotic messenger RNA. *Nature* **263:** 211.

Schleif, R.F. and P.C. Wensink. 1981. *Practical methods in molecular biology*. Springer-Verlag, New York.

Simon, J.A., C.A. Sutton, R.B. Lobell, R.L. Glaser, and J.T. Lis. 1985. Determinants of heat shock-induced chromosome puffing. *Cell* **40:** 805.

Spradling, A. and G.M. Rubin. 1982. Transposition of cloned P elements into *Drosophila* germ line chromosomes. *Science* **218:** 341.

van Wezenbeek, P.H.G.F., T.J.M. Hulsebos, and J.G.G. Schoenmakers. 1980. Nucleotide sequence of the filamentous bacteriophage M13 DNA genome; comparison with phage fd. *Gene* **11:** 129.

Warren, T.G. and A.P. Mahowald. 1979. Isolation and partial chemical characterization of the three major yolk polypeptides from *Drosophila melanogaster*. *Dev. Biol.* **68:** 130.

Wu, C. 1984. Activating protein factor binds *in vitro* to upstream control sequences in heat shock gene chromatin. *Nature* **311:** 81.

Localization of Sequences Regulating *Drosophila* Chorion Gene Amplification and Expression

L. Kalfayan, J. Levine, T. Orr-Weaver, S. Parks, B. Wakimoto, D. de Cicco, and A. Spradling

Department of Embryology, The Carnegie Institute of Washington, Baltimore, Maryland 21210

Recently there has been considerable progress in determining the structure of numerous eukaryotic genes. Sequences important for the initiation, processing, and termination of gene transcripts have been mapped in detail, and the biochemical characterization of regulatory factors is progressing rapidly. The elucidation of control mechanisms acting on genes within multicellular organisms during development has proved to be a more difficult problem, however. Experimental systems that unambiguously reproduce developmental controls are rare. During the last 3 years, the development of gene transfer methods for several higher organisms has facilitated a purely genetic approach to this problem. Modified genes are introduced into the germ line of multicellular organisms such as *Drosophila*, mouse, or tobacco, where their activity can be studied during the development of succeeding generations. The chief limitation with such experiments has now become simply the time that elapses during the generations required to establish sufficient numbers of genetically uniform transformed organisms.

Drosophila Chorion Genes

In this paper we summarize and discuss gene transfer studies examining the regulation of the *Drosophila* chorion genes, a family of genes expressed during *Drosophila* oogenesis. Oogenesis in *Drosophila* (for review, see Mahowald and Kambysellis 1980) requires only about 4 days. This rapid transition from oogonial cell to mature egg is part of a biological strategy which makes it possible for the adult female to lay more than one-half of her weight in eggs each day. One of the final steps in oogenesis is the secretion by the polyploid ovarian follicle cells of a protein-rich eggshell around the oocyte. The shell is produced as a series of layers which are laid down sequentially—first the vitelline membrane, followed by the two chorionic layers, the endochorion and exochorion. Perhaps because the process of eggshell synthesis occurs so rapidly (the chorion layers are produced in about 5 hr), genes encoding the major chorion proteins exhibit several regulatory mechanisms which make them useful models for study. These include specific gene amplification and temporally regulated transcription.

Two clusters of tandemly organized genes (Fig. 1) contain between them single copies of genes encoding all six major chorion proteins as well as some putative minor eggshell components (Petri et al. 1976; Waring and Mahowald 1979; Spradling 1981; Griffin-Shea et al. 1982; S. Parks and A.C. Spradling, in prep.). Commitment to complete oogenesis occurs at the beginning of stage 8, apparently as the result of a hormonal signal (see Mahowald and Kambysellis 1980). At this time both chorion gene clusters begin to amplify in the follicle cells, resulting in a 15-fold (X chromosome) or 60-fold (chromosome III) increase in copy number by stage 13 (Fig. 1A). The extra DNA is lost during the breakdown of the follicle cells that occurs subsequent to stage 14, hence it makes no contribution to future generations.

Individual chorion genes are expressed during specific subintervals of the final 5 hours of oogenesis comprising stages 11–14, as illustrated in Figure 1B for genes s38-1 and s15-1. During these short periods, an mRNA encoding an abundant chorion protein may increase until it constitutes more than 10% of the total poly(A)-containing RNA, reflecting extraordinary temporally controlled rates of synthesis and decay. Each gene is subject to individual temporal controls, since profiles of expression differ (Fig. 1B), but all appear to be ovary specific. Studies in which major chorion genes were directly visualized in the electron microscope demonstrated that the control of chorion gene expression occurs largely at the transcriptional level (Osheim and Miller 1983). Despite high transcription rates, amplification is probably necessary to allow a single germ line gene copy to produce the large amounts of the major chorion proteins that accumulate in this brief period of time. One goal of our studies has been to elucidate the sequences that are necessary for the particular program of transcription of each individual chorion gene.

Mechanism of Amplification

Each chorion gene cluster is amplified as a unit along with 80–100 kb of flanking chromosomal DNA. Studies of the structure of amplified chorion DNA (Spradling 1981) and observation of amplified molecules in the electron microscope (Osheim and Miller 1983) support a simple reinitiation model as the mechanism of differential replication (see Fig. 2). Normal controls preventing reinitiation within a replicated region must be suppressed within replicons containing the two amplified regions. The gradient structure of the amplified

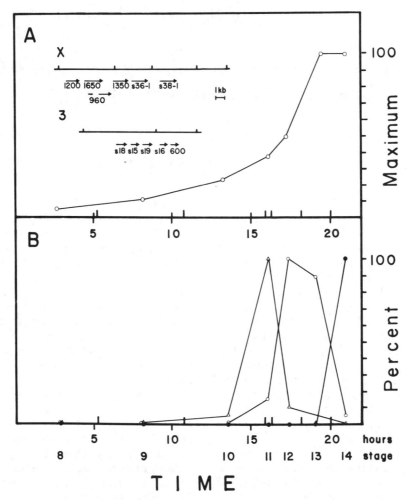

Figure 1. Developmental regulation of chorion gene amplification and expression. The inset in A shows the structure of 11 transcription units encoding six identified and five presumptive chorion genes within the X chromosome and third chromosome gene clusters (see Spradling 1981; Griffin-Shea et al. 1982; S. Parks et al., in prep.). The time course of amplification (A) and the profile of poly(A)+ RNA accumulation (B) for s15-1 (●), s38-1 (○), and the 1650-bp transcript (△) is shown. The control point at the junction of stage 8 is taken as time zero (see Mahowald and Kambysellis 1980). The data are plotted as percent of the maximum values (data in A from A. Spradling, unpubl.; see also Orr et al. 1984; data in B from S. Parks and A. Spradling, in prep.).

region, and its behavior following chromosomal rearrangement (Spradling and Mahowald 1981) imply that origin sequences for replication must be located within the maximally amplified central "plateau" region of each amplified domain. Amplification occurs uniformly throughout these 12- to 15-kb segments, which contain the entire chorion gene clusters. We refer to these putative origin sequences and to the sequences that control the tissue and temporal specificity of amplification as amplification control elements (ACEs). Previous studies suggested that sequences controlling the developmental specificity of amplification were not widely separated from the sites of replication initiation, justifying our treatment of them as structurally simple elements (de Cicco and Spradling 1984). The second major goal of our studies has been to learn more about the location, structure, and regulatory mechanisms governing these elements.

RESULTS

P-Element-mediated Gene Transfer

The process of P-element-mediated gene transfer (Rubin and Spradling 1982; Spradling and Rubin 1982) was used to investigate the organization of regulatory sequences within the chorion gene clusters. The method is summarized in Figure 3. A defective P-element transposon vector (Rubin and Spradling 1983) was modified to contain the chorion gene of interest, as well as a marker gene to allow transformants to be identified. The wild-type rosy gene (ry^+) is a particularly useful marker, since it complements the brown eye color of rosy mutants such as ry^{506}. To carry out gene transfer, the transposon was microinjected into host embryos from the ry^{506} strain, along with a disabled helper transposon, $p\pi25.7wc$ (Karess and Rubin 1984), which cannot itself insert into host chromosomes. Phenotyp-

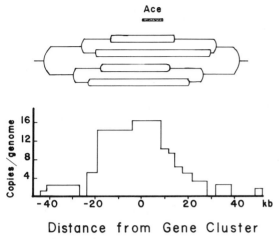

Figure 2. Disproportionate replication model of chorion gene amplification. Below is shown the extent of amplification along 100 kb of genomic DNA surrounding the X chromosome chorion gene cluster at 7EF in stage-13 egg chamber DNA (see Spradling 1981). Nested bidirectional replication forks are illustrated above. A large number of such forks are believed to give rise to the observed pattern of amplification. According to the model, an ACE is expected to reside somewhere within the maximally amplified "plateau" region.

tion, DNA was also isolated from ovariectomized females. Digestion with appropriate restriction enzymes allowed the introduced and endogenous chorion and rosy sequences to be distinguished. In the example shown, about 16-fold transposon amplification occurs, as measured by the increase in the ratio of the transposon-specific band (Tc), compared with the nonamplifying normal rosy gene (Hr). Transposon amplification occurs with the same tissue and temporal specificity as normal amplification (Hc1). The rosy DNA within the transposon (Tr) also undergoes differential replication. Thus transposon amplification probably occurs by the initiation of new rounds of bidirectional replication within the chorion DNA, which then spreads into the flanking sequences.

The differential replication of insertions containing chorion sequences within 123 different lines has been analyzed (Fig. 5). All transposons that amplified retained their normal pattern of developmental specificity. However, differences were observed in the amplification levels induced with different constructs, and

ically wild-type individuals from the G_1 generation were mated singly to ry^{506} partners, and DNA from each individual isoline was tested to determine if it had acquired a single, unrearranged copy of the transposon. In situ hybridization to polytene chromosomes from the lines allowed accurate cytogenetic mapping of the insertion site. P-element-mediated gene transfer resulted in insertions at sites throughout the entire euchromatic genome, although strong site preferences probably occur on a local level (O'Hare and Rubin 1983). Only lines containing single, nonidentical insertions of intact transposons were retained for further study. Viable insertions were kept as homozygous stocks; lethal or sterile inserts were balanced, or maintained by selection.

Mapping Sequences Controlling Gene Amplification

To identify and localize amplification control elements, a series of transposons was constructed incorporating different segments of the gene clusters. Each construct was transformed into germ line chromosomes at a variety of insertion sites. Lines containing single insertions were tested by quantitative Southern blotting to determine if the introduced transposon underwent tissue-specific amplification within the ovarian follicle cells during oogogesis.

An example of such a test is shown in Figure 4A for a transformed line containing the R7.7 transposon (de Cicco and Spradling 1984). DNA was isolated from preamplification (stages 1–8), midamplification (stages 9–10), and postamplification (stages 12–13) egg chambers. To control for the tissue specificity of amplifica-

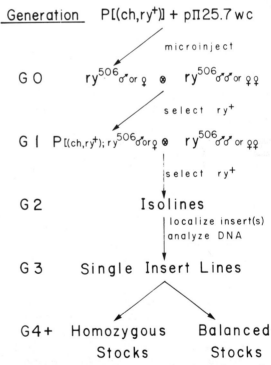

Figure 3. Experimental plan for mapping *cis*-regulators using P-element-mediated gene transfer. Transposons are constructed containing modified chorion gene genomic fragments marked with the wild-type rosy gene (ry^+), and denoted P[(ch,ry$^+$)]. Transformation is initiated by microinjecting the transposon of interest along with the helper transposon pπ25.7wc (Karess and Rubin 1984) in host ry^{506} embryos. The strategy then involves the establishment of sublines from individual transformants of the G_1 generation. Testing of DNA from the sublines allows the identification of multiple lines, each of which contains a single, unrearranged copy of the initial transposon. (See text for further details.) The entire procedure may require five or more generations (G_0–G_4).

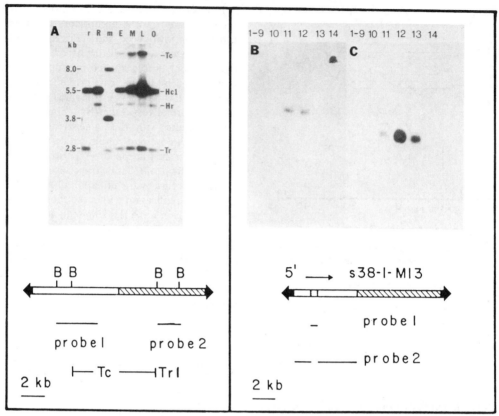

Figure 4. Transformed chorion sequences undergo developmentally regulated amplification and transcription. (*A*) DNA from early (E), middle (M), and late (L) egg chambers or from nonovarian tissues (0), from a strain containing transposon P[(ch,ry⁺)R7.7] was analyzed for amplification (see de Cicco and Spradling 1984, for details). The structure of the transposon, including the chorion sequences (open bars) and rosy sequences (hatched bars), is diagrammed below. DNA was digested with *Bam*HI and transfers were probed with a mixture of the two probes shown below. Comparison of the intensity of the transposon-specific fragments (Tc and Tr) to the control fragments derived from the normal chorion region (Hc1) and host rosy gene (Hr) indicate that transposon amplification occurs in late-stage egg chambers. (*B* and *C*) RNA was isolated from egg chambers staged as shown above from a strain containing a modified s38-1 gene (see Kalfayan et al. 1984 for details). (*C*) Blot was hybridized with probe 2, which labels predominantly the normal 1.4-kb s38-1 mRNA (compare Fig. 1A). In *B*, the same blot was probed with probe 1 which is specific for the phage M13 insertion within the transformed s38-1 gene. The activity of the transformed gene is regulated normally.

between different insertions of the same transposon. Only rarely did the level of transposon amplification approach normal levels. Thus the ability to induce amplification was frequently subject to quantitative position effects.

These results verify the existence of amplification control elements within both gene clusters, are consistent with the presence of only a single element per cluster, and allow the approximate position of the elements to be determined. For example, only chromosome 3 fragments containing a central 3.85-kb *Sal*I fragment underwent amplification following transformation. The location of this fragment (ACE3) is indicated in Figure 5A. ACE1 appeared to map within a 4.7-kb RI fragment containing the s38-1 gene (Fig. 5B). Transposons containing segments from the X chromosome gene cluster amplified less frequently and at lower levels than chromosome 3 inserts. This suggests that the mechanisms that normally cause the X chromosome cluster to amplify fourfold less than the third chromosome cluster are mediated through the corresponding control elements, and are not an effect of the sequences surrounding these regions.

Transposon amplification in virtually all the lines tested was less than normal. Furthermore, a larger proportion of the single insert lines containing ACE3 within small genomic fragments failed to amplify, compared with insertions having greater amounts of normal flanking DNA. Although nonessential, the extra flanking sequences may normally function to enhance amplification. Alternatively, specific inhibitory sequences may be found frequently throughout most of the genome, but act over only a limited distance. When located in the sequences immediately flanking an insertion site, they might inhibit replication of the smaller transposons.

To allow more detailed mapping of the ACE elements despite the problems of position effects, a different strategy was employed (T. Orr-Weaver and A. Spradling, in prep.). Deletions of 0.1–0.5 kb were constructed within a large genomic fragment (R7.7) which consistently induced amplification. This allowed se-

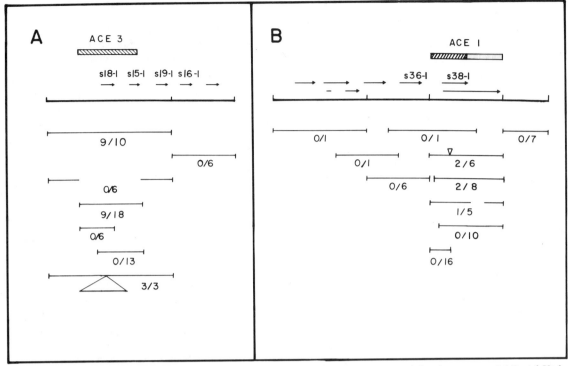

Figure 5. Localization of *ACE3* and *ACE1* by transformation. *Eco*RI restriction maps of the chromosome 3 (*A*) and X chromosome (*B*) chorion gene clusters are indicated. Horizontal bars below indicate genomic segments that were incorporated into P-element transposons and tested for their ability to induce specific gene amplification. The number of single insert lines in which the transposon underwent amplification compared with the total number of single insert lines tested is shown for each region. The locations of the amplification control elements (*ACE3* and *ACE1*) are indicated by hatched and stippled bars (data from de Cicco and Spradling 1984; T. Orr-Weaver and A. Spradling, in prep.; B. Wakimoto et al., unpubl.).

quences essential for amplification within the 3.85-kb region shown to contain *ACE3* (Fig. 5A) to be determined. These experiments demonstrated that neither s18-1 nor s15-1 transcription was required for amplification. Essential sequences were identified within an A/T-rich 428-bp *Bal*I fragment located upstream (−187 to −615) from s18-1 (see Fig. 6). Deletion analysis within the 4.7-kb RI fragment containing *ACE1* (B. Wakimoto et al., unpubl.) implicated a region upstream from s38-1 (−266 to −654) as playing an essential role for amplification.

Mapping Sequences Regulating Chorion Gene Transcription

Transformed chorion genes can also undergo developmentally regulated expression in the egg chambers of transformed strains (Wakimoto et al. 1983, 1985). This provides an assay for localizing *cis*-acting sequences that are necessary for tissue-specific and temporal regulation of gene expression. An example of such an experiment is illustrated in Figure 4B. The s38-1 gene within the transposon was modified by insertion of a 0.6-kb fragment of the *Escherichia coli* phage M13 (see diagram). Transcription of the modified gene should yield a 2.0-kb RNA that could be distinguished from the normal 1.4-kb s38-1 mRNA. Figure 4B demonstrates that an RNA of the expected size is specifically labeled with a probe specific for the M13 insert in RNA from egg chambers of the transformed strain. Expression of the transformed s38-1 gene occurred with normal stage specificity. Additional experiments demonstrated that expression of the transformed gene could not be detected in other tissues. However, the level of gene activity was subject to position effects which caused the amount of RNA accumulated per gene to vary over a 10-fold range (Wakimoto et al. 1983, 1985). Unlike the case of amplification-competent transposons, all the insertions of transposons that contained active chorion genes produced detectable product at all sites of insertion. Furthermore, no alterations in the developmental specificity were observed at different sites of insertion, except for very small changes in the temporal profile of mRNA accumulation, which probably resulted from changes in mRNA stability (Wakimoto et al. 1985).

Regions required for tissue- and time-specific s38-1 expression were mapped by BAL-31 digestion experiments. Regulated expression was maintained when sequences between −487 and +74 relative to the site of transcription initiation were present. However, further deletion from either the 5′ (to −266) or 3′ (to +9) direction resulted in the reduction of transposon-derived RNA below detectable levels (B. Wakimoto et al.; L.

```
       -650
       CGTCTTCTGG CTACTGGATG CTGGTACCCT GAGCCTGGCC AACATCTAAA
       TTATATGGTA CTTTAAACTG ATGGTTTAAT CA-TTACATG GATTTTCTAA

       -600
       TAATTTATTA TTCATTATTA AATGTTTGCG CCCACCCATA AGCCATTCAC
       ATTAAAAATG GTCATGTGAA GATAGCCACT CTTCTAACAA TCTAATCACA

       -550
       AATTTGTGTA GCGCCAATTG AATGTTATAA AAAGCTTAGT GCGGCAGTTT
       TTTATAGTAA GAAATACAAT ACAATACAAT ACAATACAAT ACAATACAAT

       -500
       GGAAAGTGGA ACGGTTGTGT TTATAATTTT ATTGTAATTT TATCTCAATT
       ACAATAGAAA GACAATCGAA TCTGCGC-AT CCGTGTGAAA TTCAAGGACT

       -450
       TTTTTTGCTT TTGTATATAA ATTCTACCAA CGCAGCAGAA TTTTCAGGCC
       ACAGCTGGGT GGCTAATCAT TTCCCCCTAT CCA-TTACAC CTCGGATTAC

       -400
       ACTGCCTTGA CTTCACTGTG TCACTGAAAA ATCGGTGTCA AGCTCTCGGC
       CTCTTATTCC GACTCCCGGA GTCTTGTGTC TGCCAATGCG GAACTATTTT

       -350
       ACCGTGGGGC AAAGCAACTG CAATACTGAT CGAAACTATG CGGATCCGGA
       CGCTATCTGA ACAGACGTTC GGACCTCGAT ATGCGGCAAA GATTCACAGC

       -300
       GCACGAAGAG TCATGCGGTC GGAATCTTAC GTAATGGGTC TCGTCTCTGG
       CCGGCTGTTG ATTCCGATTC GGTGGCAATG TGTTCGTTGT TATTGTAAAA

       -250
       TAGACGATGG CGTAAGCACA GACGCCTGCT ATCTGGACCG GCCCGAATTG
       CGGGCAATGG CAACTGGGCA GTGGGCAGTG GGGTTTTCGG GTTGTGGCTT

       -200
       AGAGCCAGCA TTTTGGCCA
       CTACGTAAGT GGAAGAG
```

Figure 6. Nucleotide sequence of region required for *ACE3* function. The nucleotide sequence of a 420-bp region upstream from the s18-1 gene is shown (*upper sequence*) (Levine and Spradling 1985). Deletion of a *Bal*I fragment containing this region eliminates amplification (T. Orr-Weaver and A. Spradling, in prep.). The lower sequence shows the comparable region upstream from s38-1. Deletion analysis (B. Wakimoto et al., unpubl.) suggests that the region between −266 and −654 may be required for X chromosome amplification (see Fig. 5). Small arrows show direct repeats, while larger arrows indicate the conserved sequence mentioned in the text.

DISCUSSION

Sequences Controlling *Drosophila* Chorion Gene Amplification and Expression Are Localized within the Gene Clusters

Both chorion gene clusters contain a series of structurally similar chorion genes in a tandem head-to-tail arrangement. We have mapped within each cluster a single amplification control element responsible for the differential replication in follicle cells of all the genes and their flanking sequences. A replication origin activated by the control element must also be present in the 12- to 15-kb region undergoing maximal amplification surrounding each cluster (Fig. 2). As discussed below, we believe it is likely that the replication origins utilized during amplification are located within the control elements that have been identified. However, since it remains possible that the essential *cis*-regulatory sequences we have mapped could activate distant origins, their exact location remains to be determined.

ACE1 and *ACE3* are located at different positions in the two gene clusters. *ACE3* is located near the 5′ proximal gene, in contrast to the 3′ proximal site of *ACE1*. The functional significance of these divergent locations remains unclear, however. Since replication forks involved in amplification all elongate sufficiently to produce a 12- to 15-kb "plateau" region of maximal amplification surrounding their initiation site, a single

```
              -150
       s38-1  ATCTATCAAC ATTTGGCCAT GTTTCGTTCG CATCGCGTGG GCCCGGAGCG
       s18-1  CGGCGTCTCA AGATTGCTGG ACAAAGAGGC GAGGCCTGGA ACTGCGTCTC
       s15-1  ACAAGTACAT AAATCAAATG TGAGTATATT CCAGCCGGGC AATTATGAAA
       s19-1  AAAGCTAGAA ATCCACAGAA AGTTCCCCAA CAAACTGGCC GAGAAGAGAC

              -100
       s38-1  GAACAGCCGG CACCCGGAGTT GGCATCAATC CAAATGTCAC GTACCCGGAG
       s18-1  CGGGAACCCG GAGAGCCGAA ACTTCGCATCA TATTCGTCAC GTAAGAGTTG
       s15-1  TGCCATTTCT GGGCTGAAAC AGAACAATTA GTGTATATAG GTCACGTAAA
       s19-1  GGCGAAGCCA GCTCTTGAGC CGTGATAAAT TTCTGGGCGA GATCACGTTT

              -50
       s38-1  CCGGAGACGC GTCCGGAGCA TATTTAAAGT AGTCGGCCAC CAATGGAGGG
       s18-1  GGCCTCTGCC TGGATCTGGT ATAAAAACAA AACATTGCGC CAGAATAAGA
       s15-1  TGTCCAGGCT AAAATTTGCG TATAAAGCG AGCGTTTCTG GTCGGTAAAT
       s19-1  CGAGTGCAAC AATAAATTTG CTTATATAAA GAAGTGTGCT TGGCCATTTA

              +1
       s38-1  CAGCAGAAGA CAGCAGACAG TCCAAGCGGG AGCACACCAG AAGCCGAAGA
       s18-1  CATTAGTTAC CTTCGCATCG ATCAACTAAC CAACTCAGCC TCAGAATGAT
       s15-1  CATAGTTTGA TTGATTACCC CAAACCAACA AAACTAAGCA CTCACCATGA
       s19-1  ATATGTTAAT TCAGCCAACT GTGCCAAAAC CCATACATCA TAGCCATGAA
```

Figure 7. Nucleotide sequences of four chorion genes. The sequences from −150 to +50 of four chorion genes are compared. The initiation site, TATAA box, conserved sequence (G/A)TCACGT(A/T), and proposed binding sites of putative regulatory factors at several inverted repeat sequences are indicated.

Kalfayan et al.; both in prep.). Additional experiments have shown that a 5′ boundary for regulated expression of s18-1 is located between positions −189 and −3300 (T. Orr-Weaver and A. Spradling, in prep.). No more than 385 bp 5′ to s15-1 were required for correct temporal regulation (T. Orr-Weaver and A. Spradling, in prep.). These results demonstrate that a region of no more than 561 bp (−487 to +74) is required for the precise developmental regulation of s38-1. They suggest that similar control regions may exist within the spacer region upstream from each chorion gene. Figure 7 compares the nucleotide sequence of a portion of the s38-1 control region with the corresponding sequences upstream from three other chorion genes (Levine and Spradling 1985; Wong et al. 1985).

origin located anywhere within one of the 8- to 12-kb gene clusters should insure equal replication of all its component genes. Nevertheless, the location of the control elements does correspond to the location of the most abundantly expressed genes within each cluster. The s38-1 and s18-1 genes, which have been associated with *ACE1* and *ACE3*, respectively, encode two of the five major eggshell constituents. Although the chorion genes within each cluster are equally amplified, genes distant from the control elements are expressed at lower levels, particularly in the case of the X chromosome.

While all the genes within a cluster share a single amplification control element, the s38-1 gene contains its own transcriptional control sequences located predominantly in the region 5′ to the initiation site. At least two other chorion genes, s18-1 and s15-1, also contain independent regulatory sequences located in the same general region, although they have not yet been mapped in detail. All three of these genes are expressed normally when separated from the remainder of the gene cluster on an integrated transposon. The configuration of *cis*-acting sequences controlling gene amplification and expression correlates with the observed behavior of chorion genes during development. All the genes in a cluster are amplified coordinately, while each gene shows similar, but temporally and quantitatively distinct, patterns of expression during oogenesis.

Sequences sufficient for correct s38-1 developmental regulation extend less than 500 bp upstream from the initiation site. In contrast, the spacer region between each gene is larger, varying between about 800 and 1400 bp. Some of these spacer sequences may mediate the quantitative regulation of different chorion genes, whose mRNAs vary in abundance more than 100-fold. Transcription terminators are also expected to reside in these regions (Osheim and Miller 1983). Perhaps reflecting this putative differentiation in spacer function, the A/T content of the spacers that have been sequenced is greater in the region upstream from the control sequences identified in these experiments.

Nucleotide Sequence of Amplification Control Elements

The nucleotide sequence of much of the third chromosome chorion gene cluster was reported recently (Levine and Spradling 1985; Wong et al. 1985). Sequencing of the X chromosome cluster surrounding s36-1 and s38-1 has also been carried out (B. Wakimoto et al., in prep.). Comparison of the sequences within the control regions revealed some insights into potential regulatory mechanisms. Extensive homology was not found between the essential regions of the two amplification control elements (Fig. 6), or between either element and previously characterized replication origins. These regions behave as single-copy DNA when hybridized with whole genomic DNA (Spradling 1981). In the yeast *Saccharomyces cerevisiae*, autonomously replicating sequences (*ARS*) are essential for maintenance of extrachromosomal plasmids (Stinchcomb et al. 1979). *ARS* elements, which have been proposed to represent replication origins, also lack extensive sequence homology, but do share an 11-bp "core" consensus sequence (T/A)TT(T/A)AT(A/G)TTT(A/T) (Broach et al. 1983; Celniker et al. 1984; Kearsey 1984). One copy of this consensus sequence is found 801 bp upstream from the initiation site of s18-1 (Levine and Spradling 1985), within a region whose functional significance has yet to be tested by deletion. *ARS* consensus sequences are absent from the *ACE1* region or from any portion of the X chromosome gene cluster yet studied. However, detailed comparison of the sequences in Figure 6 revealed the presence of a related sequence TT(T/C)TATTGT(T/A)(T/A)T, shared between both control elements (see Fig. 6). This sequence is located between −474 and −462 in *ACE3*, and in reverse orientation between −503 and −491 in *ACE1*. This consensus or its complement is not found elsewhere within the 9.8 kb of chorion DNA for which sequence is available, or in any other DNA in the GENBANK invertebrate sequence library. Furthermore, in both cases the consensus is associated with unusual direct repeats (Fig. 6) which are themselves only present at these sites. These sequences are therefore candidates as essential components of *Drosophila* amplification control elements and may represent the core sequences of replication origins used during amplification.

Nucleotide Sequences Regulating Chorion Gene Transcription

The regulation of gene transcription has frequently been associated with sequences located near the 5′ start site (see Guarente 1984). Small direct or inverted repeat consensus sequences have been identified in these regions and in some cases shown to be necessary for regulation (see, for example, Pelham 1982). Conserved regulatory sequences may represent binding sites for regulatory proteins (Tjian 1978; Parker and Topol 1984) Consequently, we searched the 561-bp region containing sequences regulating s38-1 transcription, and compared it with regions upstream of three other sequenced chorion genes. Several putative regulatory sequences were identified which were shared between these chorion genes. The most striking similarities were found near the 5′ start site, and are illustrated in Figure 7.

All four chorion genes contain the sequence (G/T)TCACGT(A/T) about 60 bp upstream from the start of transcription. This sequence is not found at this location in any other invertebrate gene to which it could be compared. Another shared sequence motif is located near the putative 5′ junction of the s38-1 control region. It consists of variants of the hairpin sequence GAAC NNNN GTTC. This sequence is found in one or two copies in the −340 to −380 region of all four genes. It could function as part of a positive regulatory sequence required for chorion gene expression in follicle cells. However, the *YP2* gene, which is also expressed in follicle cells, but at stages 8-9, does not contain either of these sequence features (Hung and Wensink 1983).

All four chorion genes contain hairpin sequences at about position −100. If these serve as binding sites for regulatory factors, then different factors may bind to each site, since the sequences vary between the genes (Fig. 7). Since the genes are activated at different times during oogenesis, these sites are candidates as targets for the factors regulating temporal expression.

Model of *ACE* Structure

These observations suggest similarities between the control of the s18-1 and s38-1 chorion genes and the regulatory region of polyoma virus. The polyoma genome contains a transcriptional enhancer element required for early transcription located near the viral origin of replication. The enhancer is also required in *cis* for viral DNA replication (Tyndall et al. 1981; de Villiers et al. 1984). We speculate that a transcriptional enhancer is located 300-400 bp upstream from s38-1 and s18-1, and that the conserved TT(T/C)TATTGT(T/A)(T/A)T sequences, located about 100 bp distant, are replication origins. The tissue-specific activation of replication at these origins, as well as the activation of s18-1 and s18-1 transcription, is postulated to be dependent on this enhancer sequence. The enhancer appears to be required directly, since deletion experiments demonstrated that transcription from the s18-1 mRNA initiation site was not required for amplification (T. Orr-Weaver and A. Spradling, in prep.).

This model makes several predictions which are currently being tested. First, substitution of an enhancer with a different tissue specificity for the putative s18-1 enhancer should result in the amplification of the entire third chromosome chorion gene cluster in the tissue in which the enhancer is active. However, only the s18-1 gene would be transcribed in this tissue. Second, deletion of the putative origin sequence should eliminate amplification, but not s18-1 expression. The ability to rescue the amplification of such a deletion mutation might provide an assay for *Drosophila* replication origins.

Position Effects Frequently Reduce the Quantitative Expression of Transformed Chorion Genes

Drosophila genes introduced into the germ line at diverse chromosomal positions on P-element transposons are frequently able to complement the corresponding mutant phenotype. However, the expression of transformed genes is subject to position effects that may reduce gene activity below wild-type levels (Goldberg et al. 1983; Spradling and Rubin 1983; Bourouis and Richards 1985; Wakimoto et al. 1985) and occasionally alter regulation (Hazelrigg et al. 1984). Chorion gene amplification is also subject to position effects (de Cicco and Spradling 1984). The mechanisms that interfere with amplification at a particular site are not always identical to those limiting chorion gene expression, since levels of amplification and expression are not strongly correlated (Wakimoto et al. 1985).

The fact that significant quantitative position effects are observed on both replication and transcription of inserted genes has important ramifications for using P-element-mediated transformation to map *cis*-regulatory elements. If no means of reducing these variations could be found, the ability to map elements exerting quantitative effects of less than two- or threefold would require averaging so many independent insertions as to become impractical. However, the effects of flanking sequences can be reduced by the inclusion of moderate amounts of flanking DNA sequences beyond those absolutely required for normal regulation (Spradling and Rubin 1983; T. Orr-Weaver and A. Spradling, in prep.). Analyzing the effects of small alterations within a large construct held otherwise constant, may provide a generally useful approach to mitigating these limitations.

In addition to providing technical complications for the design of transformation experiments, position effects represent a potentially significant unanswered question concerning the structure and regulation of genes in complex organisms. Do position effects result from interactions of biological significance between genes and larger units within the chromosome, for example, replicons? Do they merely indicate that regulatory circuitry can easily be perturbed when normally distant elements are juxtaposed? Despite the clarity with which we can now examine the genes of higher organisms and their *cis*-regulatory sequences, a more complete picture must still await resolution of this question.

ACKNOWLEDGMENTS

This research was supported by a grant from the National Institutes of Health to A.C.S. B.T.W. is a recipient of a fellowship from the Helen Hay Whitney Foundation. L.J.K. is a recipient of a fellowship from the National Institutes of Health. T.O.-W. was supported by a fellowship from the Jane Coffin Childs Fund.

REFERENCES

BOUROUIS, M. and G. RICHARDS. 1985. Remote regulatory sequences of the *Drosophila* glue gene sgs3 as revealed by P-element transformation. *Cell* 40: 349.

BROACH, J.R., Y.-Y. LI, J. FELDMAN, M. JAYARAM, J. ABRAHAM, K.A. NASMYTH, and J.B. HICKS. 1983. Localization and sequence analysis of yeast origins of DNA replication. *Cold Spring Harbor Symp. Quant. Biol.* 47: 1165.

CELNIKER, S.E., K. SWEDER, F. SRIENC, J.E. BAILEY, and J. CAMPBELL. 1984. Deletion mutations affecting autonomously replicating sequence *ARS1* of *Saccharomyces cerevisiae*. *Mol. Cell. Biol.* 4: 2455.

DE CICCO, D. and A. SPRADLING. 1984. Localization of a *cis*-acting element responsible for the developmentally regulated amplification of *Drosophila* chorion genes. *Cell* 38: 45.

DE VILLIERS, J., W. SCHAFFNER, C. TYNDALL, S. LUPTON, and R. KAMEN. 1984. Polyoma virus DNA replication requires an enhancer. *Nature* **312:** 242.

GOLDBERG, D.A., J.W. POSAKONY, and T. MANIATIS. 1983. Correct developmental expression of a cloned alcohol dehydrogenase gene transduced into the *Drosophila* germ line. *Cell* **34:** 59.

GRIFFIN-SHEA, R., G. THIREOS, and F.C. KAFATOS. 1982. Organization of a cluster of four chorion genes in *Drosophila* and its relationship to developmental expression and amplification. *Dev. Biol.* **91:** 325.

GUARENTE, L. 1984. Yeast promoters: Positive and negative elements. *Cell* **36:** 799.

HAZELRIGG, T.I., R. LEVIS, and G.M. RUBIN. 1984. Transformation of white locus DNA in *Drosophila*, dosage compensation, zeste interaction, and position effects. *Cell* **36:** 469.

HUNG, M.-C. and P. WENSINK. 1983. Sequence and structure conservation in yolk proteins and their genes. *J. Mol. Biol.* **164:** 481.

KALFAYAN, L., B. WAKIMOTO, and A.C. SPRADLING. 1984. Analysis of transcriptional regulation of the s38 chorion gene of *Drosophila* by P element-mediated transformation. *J. Embryol. Exp. Morphol.* **83:** 137.

KARESS, R.E. and G.M. RUBIN. 1984. Analysis of P transposable element functions in *Drosophila*. *Cell* **38:** 135.

KEARSEY, S. 1984. Structural requirements for the function of a yeast chromosomal replicator. *Cell* **37:** 299.

LEVINE, J. and A.C. SPRADLING. 1985. DNA sequencing of 3.8 kb region controlling *Drosophila* chorion gene amplification. *Chromosoma* **92:** 136.

MAHOWALD, A.P. and M.P. KAMBYSELLIS. 1980. Oogenesis. In *The genetics and biology of* Drosophila (ed. M. Ashburner and T.R.F. Wright), p. 141. Academic Press, New York.

O'HARE, K. and G.M. RUBIN. 1983. Structure of P transposable elements and their sites of insertion and excision in the *Drosophila melanogaster* genome. *Cell* **34:** 25.

ORR, W., K. KOMITOPOULOU, and F.C. KAFATOS. 1984. Mutants suppressing in *trans* chorion gene amplification in *Drosophila*. *Proc. Natl. Acad. Sci.* **81:** 3773.

OSHEIM, Y.N. and O.L. MILLER. 1983. Novel amplification and transcriptional activity of chorion genes in *Drosophila melanogaster* follicle cells. *Cell* **33:** 543.

PARKER, C.S. and J. TOPOL. 1984. A *Drosophila* RNA polymerase II transcription factor that binds to the regulatory site of an hsp70 gene. *Cell* **37:** 273.

PELHAM, H.R.B. 1982. A regulatory upstream promoter element in the *Drosophila* hsp70 heat shock gene. *Cell* **30:** 517.

PETRI, W.H., A.R. WYMAN, and F.C. KAFATOS. 1976. Specific protein synthesis in cellular differentiation. III. The eggshell proteins of *Drosophila melanogaster* and their program of synthesis. *Dev. Biol.* **49:** 185.

RUBIN, G.M. and A.C. SPRADLING. 1982. Genetic transformation of *Drosophila* with transposable element vectors. *Science* **218:** 348.

———. 1983. Vectors for P element-mediated transformation in *Drosophila*. *Nucleic Acids Res.* **11:** 6341.

SPRADLING, A.C. 1981. The organization and amplification of two clusters of *Drosophila* chorion genes. *Cell* **27:** 193.

SPRADLING, A.C. and A.P. MAHOWALD. 1981. A chromosome inversion alters the pattern of specific DNA replication in *Drosophila* follicle cells. *Cell* **27:** 203.

SPRADLING, A.C. and G.M. RUBIN. 1982. Transposition of cloned P elements into *Drosophila* germ line chromosomes. *Science* **218:** 341.

———. 1983. The effect of chromosomal position on the expression of the *Drosophila* xanthine dehydrogenase gene. *Cell* **34:** 47.

STINCHCOMB, D.T., K. STRUHL, and R.W. DAVIS. 1979. Isolation and characterization of a yeast chromosomal replicator. *Nature* **282:** 39.

TJIAN, R. 1978. The binding site on SV40 DNA for a T antigen-related protein. *Cell* **13:** 165.

TYNDALL, C., G. LA MANTIA, C.M. THACKER, J. FAVALORO, and R. KAMEN. 1981. A region of the polyoma virus genome between the replication origin and late protein coding sequences is required in *cis* for both early gene expression and viral DNA replication. *Nucleic Acids Res.* **9:** 6231.

WAKIMOTO, B., L. KALFAYAN, and A. SPRADLING. 1983. Analysis of transcriptional regulation of the s38 gene by transformation. *Carnegie Inst. Wash. Year Book* **82:** 198.

———. 1985. Developmentally regulated expression of *Drosophila* chorion genes introduced at diverse chromosomal positions. *J. Mol. Biol.* (in press).

WARING, G. and A.P. MAHOWALD. 1979. Identification and time of synthesis of chorion proteins in *Drosophila melanogaster*. *Cell* **16:** 599.

WONG, Y.C., J. PUSTELL, N. SPOEREL, and F.C. KAFATOS. 1985. Coding and potential regulatory sequences of a cluster of chorion genes in *Drosophila melanogaster*. *Chromosoma* **92:** (in press).

Studies on the Developmentally Regulated Expression and Amplification of Insect Chorion Genes

F.C. Kafatos,*† S.A. Mitsialis,* N. Spoerel,* B. Mariani,* J.R. Lingappa,* and C. Delidakis*

*Department of Cellular and Developmental Biology, Harvard University, Cambridge, Massachusetts 02138; †Institute of Molecular Biology and Biotechnology and University of Crete, Heraklio, Crete, Greece

The insect eggshell or chorion has been studied extensively as a model system of gene regulation during development (for reviews, see Kafatos 1983; Goldsmith and Kafatos 1984; Kafatos et al. 1985). Before we summarize our recent results, a brief introduction to the system is necessary.

In both silk moths and fruit flies, the ovary consists of multiple strings of progressively more mature follicles. Each follicle, in turn, consists of three cell types: a single oocyte, a small number of nutritive nurse cells connected to the oocyte, and a monolayer of follicular epithelial cells which surround the oocyte-nurse cell complex. The follicular epithelial cells number ~1000 per follicle in Drosophila, or up to 10,000 in moths. At the end of oogenesis, they synthesize sequentially a complex mixture of structural proteins, and secrete them onto the surface of the oocyte to form the chorion. The programmed, sequential production of the proteins is probably related to their complex morphogenetic functions. The composition, morphogenesis, and structure of the chorion differ substantially in moths and flies (Fig. 1), which is not surprising, given the large evolutionary distance between these insect groups (more than 200 million years, or nearly as much as that between birds and mammals; see Mitsialis and Kafatos 1985).

While the fly chorion is comparatively simple and consists of six major and approximately 14 minor proteins, produced over 5 hours, in the moths the number of chorion genes and the duration of choriogenesis are

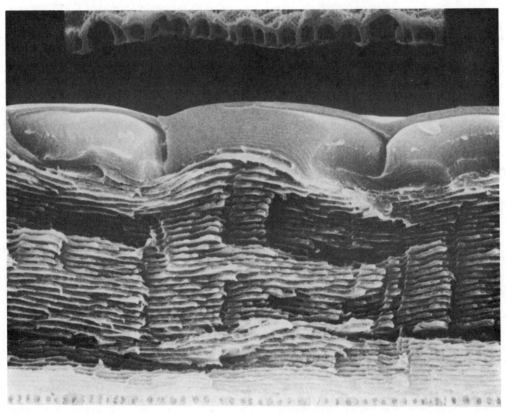

Figure 1. Comparison of fruit fly and silk moth chorions. Scanning electron micrographs of transverse rips through mature eggshells of *D. melanogaster* (*upper*) and *B. mori* (*lower*) are shown at approximately the same scale. In both cases the vitelline membrane encompassing the oocyte would be toward the bottom of the picture. (Courtesy Dr. G.D. Mazur, Harvard University.)

each approximately an order of magnitude higher. The moth genes can be classified into a small number of multigene families, the members of which have arisen during evolution by gene reduplication and diversification, so that they now range from nearly identical "copies" to those that differ by as much as 50% in sequence. Three families, A, B, and C, have been studied in the wild Saturniid silk moths, such as *Antheraea polyphemus*; the same families, plus two additional ones, Hc-A and Hc-B, have been characterized in the moth *Bombyx mori*. The fly genes are probably also related by evolutionary origin but have become so distinct that they score as single-copy sequences in hybridization reactions. It is uncertain whether fly and moth chorion genes are evolutionarily homologous, since their sequence similarities are quite limited.

Moth and fly chorion genes differ in organization as well as sequence. Moth genes are largely organized as divergently transcribed, coordinately expressed pairs, each embedded in a long and variable DNA segment that is tandemly repeated along the chromosome. In the moth *B. mori*, the chorion locus maps on chromosome 2 and appears to consist of two segments (Goldsmith and Kafatos 1984) that probably total more than 1000 kb of DNA. Indeed, contiguous sequences of 270 kb have been obtained thus far from one of these segments by chromosomal walking (Eickbush and Kafatos 1982). In *Drosophila melanogaster*, two short 5- to 10-kb clusters have been defined, each encoding tandemly oriented chorion genes; one cluster is located on the X chromosome and one on the third (Griffin-Shea et al. 1980, 1982; Spradling et al. 1980; Spradling 1981). We have concentrated on studying the chromosome III cluster, at 66D12-15, which encodes four low-molecular-weight proteins known as s15-1, s16-1, s18-1, and s19-1.

In *Drosophila*, the demand for large amounts of chorion proteins produced in a very short time is met by differential gene amplification (Spradling and Mahowald 1980). Amplification begins several hours before chorion formation (in stage 8-9 follicles), and by the end of choriogenesis (stages 13-14) reaches maximal levels of 20-fold for the X and more than 80-fold for the chromosome III chorion gene cluster. Amplification results from extra rounds of replication, which initiate at a central origin or origins and extend bidirectionally for variable distances up to 40-50 kb, generating a multiforked "onion-skin" structure (Spradling 1981; Osheim and Miller 1983). In moths, where choriogenesis lasts longer and gene multiplicity is high, amplification does not occur.

In summary, insect choriogenesis entails tissue- and temporally specific developmental phenomena: amplification in *Drosophila* and differential gene expression in both *Drosophila* and moths. *cis*-regulatory elements implicated in these phenomena can be studied by reverse genetic procedures that are becoming standard: DNA cloning and sequence analysis, in vitro engineering, and functional testing of the modified constructs after germ line transformation. *Drosophila* is especially attractive because it should also permit identification of *trans*-acting elements by direct genetics. The *trans*-acting elements are of special biological interest, for they are likely to reveal previously unsuspected functions and ultimately illuminate the logic of the developmental program as a whole.

METHODS

Vector construction. *Drosophila* transformation vectors were derivatives of pCar20 (Rubin and Spradling 1983). Cloned genomic fragments from the *D. melanogaster* chromosome III chorion gene cluster or from the *B. mori* chorion gene locus were inserted into the vector polylinker, and were thus flanked by ry^+ sequences and P-element terminal sequences. The detailed structure of vectors bearing *B. mori* chorion genes has been reported (Mitsialis and Kafatos 1985).

The hybrid s15-poly gene was constructed by the insertion of a 288-bp *Hae*II fragment from the *A. polyphemus* cDNA plasmid pc408 (Jones et al. 1979) into the unique *Hae*II site of a *Bgl*II-*Sal*I genomic subclone encompassing the *D. melanogaster* s15-1 gene (see Fig. 3, below). The moth sequence, representing part of the coding region of a member of the B family of silk moth chorion genes, was inserted in frame with the s15-1 gene coding sequence in a position 14 codons upstream from the carboxyl terminus. The resulting 1.86-kb fragment was inserted into the *Hpa*I site of the polylinker in pCar20 or was used to substitute the analogous *Bgl*II-*Sal*I fragment in pCtc201. The latter plasmid is a pCar20 derivative harboring a 10-kb segment of the chromosome III chorion locus (see text).

Deletions in the 5'-flanking region of s15-1 were made by linearizing a plasmid harboring the *Bgl*II-*Sal*I s15-poly gene construct at the unique *Xba*I site at position -386 with respect to the putative s15-1 cap site (defined as $+1$). After BAL-31 nuclease digestion, the ends were filled with the Klenow fragment of DNA polymerase I and ligated to *Xba*I linkers and the plasmids were recircularized. The original and truncated inserts were excised as *Xba*I-*Sal*I fragments and inserted between the *Xba*I and *Sal*I sites of a version of pCar20 carrying a modified polylinker. In parallel, they were used to substitute the analogous fragment in pCtc201.

Transformation. *Drosophila* transformation was carried out essentially as described (Rubin and Spradling 1982; Spradling and Rubin 1982). Embryos of the *D. melanogaster* cn;ry strain were injected with a DNA mixture of vector plasmid and helper plasmid pπ25.7WC (G. Rubin, pers. comm.). Resulting G_0 adults were mated with the parental strain and ry^+ transformants in the G_1 progeny were used to establish lines. Genomic DNA isolated from adult flies was subjected to Southern analysis (Southern 1975) to select lines that contained single inserts at unique chromosomal positions (as judged by the presence of line-specific junction fragments).

Quantitation of chorion gene amplification levels. DNA isolated from late-stage follicles of transformants was subjected to Southern analysis utilizing a mixture of probes from the chorion gene cluster on chromosome III and from single-copy, nonamplified regions (Kafatos et al. 1985). The grain density in bands corresponding to fragments derived from the endogenous chorion gene cluster, the transduced chorion cluster regions in the insert, and the single-copy regions was determined by densitometry. Conditions for autoradiography were carefully optimized in order to assure that a linear relationship between band intensity and quantity of corresponding fragment was valid over a wide range of DNA concentrations. The values for the transduced and the endogenous chorion gene sequences were normalized with the single-copy control sequence. The ratio of these normalized values was used to calculate the amplification level of the insert relative to the amplification level of the endogenous chorion gene sequences.

RNA analyses. Total RNA was isolated from ovaries of female flies or from isolated follicles representing the various stages of *Drosophila* oogenesis (King 1970). RNA preparations were glyoxylated (McMaster and Carmichael 1977), fractionated on agarose gels, and blotted on nylon membrane filters. The RNA was cross-linked to filters by UV irradiation and hybridized with appropriate probes essentially under the conditions described by Church and Gilbert (1984). The cRNA probes (Melton et al. 1984) and the exact conditions of hybridization utilized in the analysis of silk moth chorion gene transcripts have been described (Mitsialis and Kafatos 1985).

Quantitation of silk moth chorion gene transcript levels. Stage-specific RNA was isolated from follicles of transformants and analyzed as described above. Filters were hybridized with probes specific for the *B. mori* chorion genes and the *D. melanogaster* s15-1 and α1-tubulin genes (Kalfayan and Wensink 1981). Hybridization intensities were determined by densitometry, and the values obtained for the chorion gene transcripts in each developmental stage were normalized with the appropriate value for the α1-tubulin transcript.

Sequence analysis. The sequence of the segment of *D. melanogaster* chromosome III encompassing chorion genes s18-1, s15-1, and s19-1 has been reported (Wong et al. 1985). A 3.1-kb genomic *Eco*RI fragment encompassing *B. mori* chorion genes A/B.L12 was subcloned in M13mp8 and sequenced in both orientations (N. Spoerel et al., in prep.) using the systematic sequencing strategy of Hong (1982). Sequences were analyzed using a comprehensive package of sequence analysis programs (Pustell and Kafatos 1984, 1986).

RESULTS AND DISCUSSION

cis- and trans-regulation of Amplification

cis-regulatory elements involved in chorion gene amplification have been studied by both the Spradling lab (Carnegie Institution of Washington) and ours, using the technique of P-element-mediated transformation (Rubin and Spradling 1982; Spradling and Rubin 1982).

de Cicco and Spradling (1984) showed that the minimal chromosome III sequence that amplifies autonomously in transformants is a 3.8-kb *Sal*I–*Sal*I fragment, called amplification control element 3 (*ACE3*), encompassing chorion genes s18-1 and s15-1 (fragment S1S2 in Fig. 2). Although amplification of this fragment is developmentally normal, it occurs sporadically, and its level is much reduced relative to that of the endogenous chorion locus. Thus, although *ACE3* contains cis-acting sequences that are essential for developmentally regulated amplification, the activity of these sequences in the transformants is apparently affected by chromosomal elements abutting the different insertion sites.

In our own experiments, which are summarized in Figure 2, we have found low levels of amplification even when *ACE3* is extended by ~2 kb, to encompass the additional chorion gene, s19-1 (fragment S1R2); even with that segment, amplification occurs in only 50% of the transformant lines, and then only up to 10% of the endogenous level. By contrast, a still larger seg-

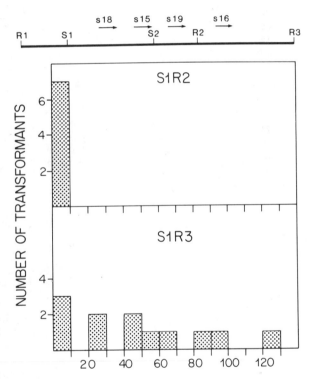

Figure 2. cis-regulation of chorion gene amplification. (*Top*) Schematic representation of the *D. melanogaster* chromosome III chorion gene cluster. The positions and the direction of transcription of the four major chorion gene transcripts are indicated by arrows. R1, R2, and R3 are *Eco*RI sites; S1 and S2 are *Sal*I sites. In the constructs used for transformation, the ry^+ gene would be on the left. The cluster (R1R3) is 12 kb long. (*Bottom*) Quantitation of insert amplification levels in transformants. The number of independent lines containing the S1R2 or the S1R3 insert is indicated on the ordinate and the extent of insert amplification, expressed relative to the amplification of the endogenous chorion gene cluster, is indicated on the abscissa.

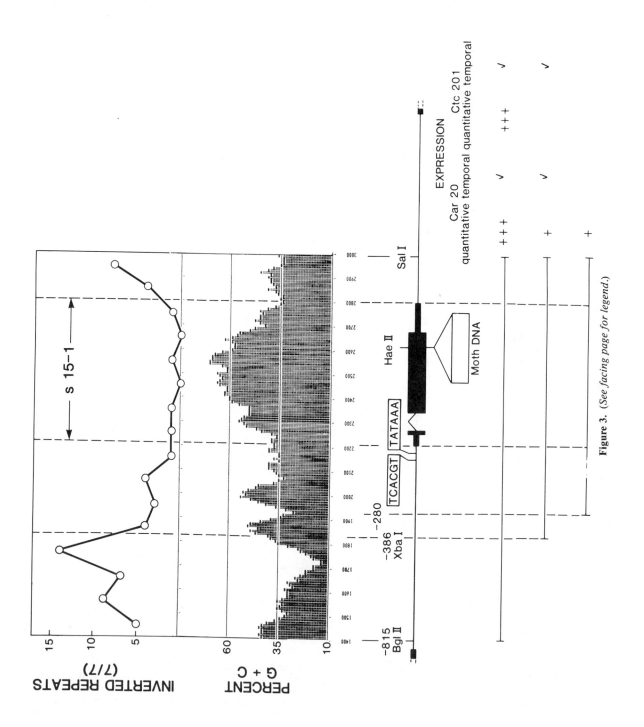

Figure 3. (See facing page for legend.)

ment (S1R3), which encompasses ~10 kb and all the chorion genes in this locus, amplifies in nearly every transformant line; the mean amplification is approximately 50% of the endogenous level, and transformants amplifying at wild-type levels are not infrequent. Even with that segment, chromosomal position effects are prominent, since amplification levels in different transformants vary from 0% to 130% of endogenous. The molecular basis of these effects is as yet unknown. They might indicate, as mentioned before, sensitivity of amplification to flanking DNA or local chromatin states. Alternatively, and not mutually exclusive with the above, they might imply that additional positively acting *cis* elements participate in the amplification mechanism augmenting the essential ones located within *ACE3*. These hypothetical elements would be redundant, occurring in multiple chromosomal locations. In that connection, it is interesting that the extra DNA present in S1R3 (beyond R2), contains an autonomously replicating sequence (*ARS*) capable of autonomous replication in yeast (Kafatos et al. 1985; G. Thireos, unpubl.). Studies utilizing internal deletions within S1R3 are underway to discriminate between these possibilities.

trans-acting functions that affect amplification have been revealed by genetic analysis. Our genetic efforts have taken as a point of departure mutants with a visible effect on chorion structure. In the X chromosome, eight complementation groups have been identified with prominent ultrastructural defects in the chorion (Komitopoulou et al. 1983 and in prep.). Similarly, in chromosome III, four complementation groups with similar effects have been identified (P. Snyder and V. Galanopoulos, unpubl.).

Two of the X-linked chorion mutants, K451 and K1214, were subjected to detailed molecular analysis (Orr et al. 1984). Their prominent ultrastructural defects are accompanied by substantial underproduction of all six major chorion proteins (K. Komitopoulou et al., in prep.), which was traced to a corresponding deficiency of their mRNAs. The temporal specificities of the mRNAs were not affected, however, clearly indicating that the mutants do not have a trivial explanation, such as cell death. In turn, the reduced mRNA levels were attributed to a correspondingly reduced gene dosage, resulting from a substantial suppression of amplification in both the X and the third chromosome. Since the loci in question map far from or even in a different chromosome than the amplification sites that they affect, they must operate in *trans*. More recently, four additional chorion-defective mutants, two from the X (K. Komitopoulou and S. Kouyanou, unpubl.) and two from chromosome III (P. Snyder and V. Galanopoulos, unpubl.), were shown to reduce amplification in *trans*. Since three of the six complementation groups that are known to affect amplification exist as single alleles, it appears that many genetic functions are necessary for normal amplification. This observation may be related to the remarkable degree to which amplification is subject to position effects (see above).

In general, the functions necessary for normal amplification may be quite diverse, covering a very wide spectrum from nonspecific to specific. At one extreme might be mutants in the general replication machinery: It is likely that amplification places especially heavy demands on the replication process, so that leaky and therefore viable general replication mutants are scored as amplification defective. At the other extreme would be mutants of highly specific diffusible factors that act directly on the amplification-control elements. In between these extremes would be mutants in functions that are indirect but developmentally interesting. For example, mutants may affect functions involved in intercellular coordination of development between oocyte, nurse cells, and follicle cells, or intracellular functions that relate the program of amplification to other aspects of follicular cell development. Other functions might affect amplification through their influence on local chromosomal conditions, such as the state of chromatin condensation. Although each type of mutant would be of some intrinsic interest, we are continuing our analysis of chorion mutants, by both genetic and molecular procedures, in the direction of identifying and characterizing those that are of special developmental significance.

cis-regulation of Chorion Gene Expression in *Drosophila*

We have chosen chorion gene s15-1, which is predominantly expressed in very late choriogenesis (stage 14), for analysis of the *cis*-acting elements involved in differential expression of the chorion genes during development. The approach is to define the minimal DNA segment that permits normal expression of the gene, when incorporated into a new chromosomal site by P-element-mediated transformation. Once that segment

Figure 3. *cis*-regulation of s15-1 gene expression. (*Top*) Broad analysis of the nucleotide sequence of the *D. melanogaster* s15-1 chorion gene and flanking regions. The distribution, in 100-bp intervals, of inverted repeats with 7/7 nucleotide matches is represented in the upper panel. The composition of the coding strand is represented in the lower panel (Wong et al. 1985). (*Bottom*) Schematic representation of constructs containing the hybrid s15-poly gene. (*Thin line*) *D. melanogaster* genomic sequences. The exons of the s15-1 gene are diagrammed by filled boxes, wide for coding sequences and narrow for untranslated regions; connecting lines represent the single intron of the gene. The positions of the putative TATA element of the promoter and of the conserved TCACGT hexanucleotide are indicated. Nucleotide positions are numbered with respect to the cap site, defined as +1. (*Open box*) Silk moth chorion gene coding sequences. The ends of the flanking s18-1 and s19-1 (left and right, respectively; see Fig. 2) are indicated. The table summarizes transcription of the indicated constructs when introduced into *Drosophila* through the Carnegie 20 vector or the amplifiable Ctc 201 vector. Quantitative expression is relative to the endogenous s15-1 gene. Checkmarks under temporal expression indicate correct developmental specificity of transcription.

is defined, in vitro mutagenesis can be used to define the critical regulatory elements at the nucleotide level. In turn, this knowledge should ultimately help in the detection and mechanistic analysis of *trans*-acting elements regulating chorion gene expression. Figure 3 shows schematically the DNA segments used in transformation experiments and summarizes the results obtained to date.

To permit discrimination between the transcripts produced by the endogenous and the introduced genes, the s15-1 gene used in our constructs was modified by the insertion, within the coding region, of a 288-bp segment of an unrelated silk moth chorion cDNA clone, derived from *A. polyphemus* (Jones et al. 1979). That DNA was inserted in frame, essentially resulting in augmentation of the *Drosophila* coding sequence by a moth-specific protein coding domain. This strategy was adopted to minimize potential problems with the marked gene such as mRNA instability or translational inefficiency, as well as to facilitate future analysis of the morphogenetic role of individual chorion proteins. The moth DNA insert does not cross-hybridize with *Drosophila* sequences, and thus can be used as a specific probe for the transduced gene and for its transcript (shown as s15-poly in Fig. 4 and 5, versus s15-E for the endogenous s15-1 transcript). The hybrid gene was incorporated into transformation vectors in constructs containing 320 bp of sequences downstream from the s15-1 termination codon and up to 815 bp of sequences upstream from the s15-1 putative cap site. These flanking sequences are bounded by convenient *Sal*I and *Bgl*II sites (see Fig. 3), and represent nearly a quarter of the s15-1 to s19-1 intergenic region and the entire s18-1 to s15-1 intergenic region.

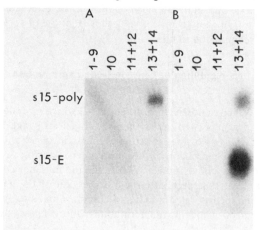

Figure 4. Developmental specificity of the s15-poly hybrid gene expression. RNA from follicles of the indicated developmental stages (1–14) of *Drosophila* oogenesis was isolated from transformed flies and analyzed as described in Methods. (*A*) RNA from transformants harboring the *Bgl*II–*Sal*I fragment within the Carnegie 20 vector was probed with an M13-derived single-stranded probe specific for the silk moth insert sequences. (*B*) RNA from transformants harboring the same fragment within the amplifiable Ctc 201 vector was probed with an M13-derived single-stranded probe specific for s15-1 sequences, which detects both the hybrid gene transcript (s15-poly) and the endogenous s15-1 transcript (s15-E).

Observed differences between the temporal specificities of individual chorion genes in normal flies (Griffin-Shea et al. 1982) had encouraged us to expect that the regulatory elements might be closely linked with each individual gene. However, to be able to detect elements possibly residing distally to the genes, each construct was tested in two parallel series of experiments. First, the construct was incorporated within the Carnegie 20 vector, in isolation from the remainder of the chorion gene cluster sequences. Second, it was placed "in context" within the gene cluster, by substituting the unmodified s15-1 gene present in the amplifiable transformation vector Ctc 201. The latter vector is a modified version of Carnegie 20, and includes a 10-kb segment of the chromosome III chorion locus (S1R3; see Fig. 2). The *Bgl*II–*Sal*I fragment carrying the unmodified s15-1 gene can be easily removed from this vector and replaced by modified s15-1 constructs. With the complete *Bgl*II–*Sal*I fragment, in either vector, transformant-specific RNA was detected (Fig. 4) at a level comparable on a per-gene copy basis with that attained by the endogenous gene within a factor of 2 to 4. The transcript is ovary specific, and is limited to follicles at stages 13 and 14 of oogenesis. Thus, the *cis*-regulatory elements controlling the developmental specificity and probably the quantitative levels of s15-1 expression are contained within the 1.6-kb *Bgl*II–*Sal*I fragment. Removal from the Carnegie 20 construct of the distal 429 bp of the same fragment, i.e., of sequences upstream of the *Xba*I site at −386 (with respect to the putative s15-1 cap site), reduces substantially the level of expression (Fig. 5), apparently without changing the developmental specificity (results not shown).

The formal argument that deletion of flanking sequences may increase the sensitivity of the transcriptional unit to position effects cannot be ruled out at this stage. Nevertheless, the simplest interpretation is that these experiments separated two functional domains of the s15-1 gene promoter: a distal one, upstream of −386, containing a quantitatively modulating element, and a proximal one containing elements responsible for both the developmental modulation as well as for the general transcriptional competency of the gene. Deletions, even closer to the cap site, at −305 and −280, do not abolish transcription; the developmental specificity has not been tested yet for these constructs.

cis- and *trans*-regulation of Moth Chorion Gene Expression

As stated earlier, moths and flies are quite distantly related, and the sequence and organization of their chorion genes reflect that evolutionary distance. However, in both groups the chorion genes share the same tissue and broad developmental specificity, being expressed in choriogenic follicle cells at the end of oogenesis. Since these regulatory properties seem rather fundamental and since, in general, development tends to be a conservative "tinkerer" during evolution, we won-

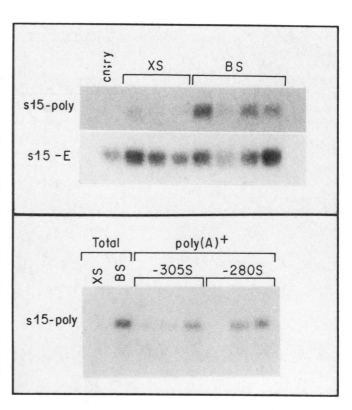

Figure 5. Effect of 5' deletions on s15-poly hybrid gene expression. (*Upper panel*) Each lane represents 20 μg ovarian RNA isolated from the indicated lines and analyzed as described in Methods. (*cn;ry*) Parental strain *Drosophila*; (XS and BS) lines transformed with Carnegie 20 vectors containing either the *Xba*I–*Sal*I fragment (XS) or the *Bgl*II–*Sal*I fragment (BS) of s15-poly. The filter was sequentially hybridized with an M13-derived single-stranded probe specific for the silk moth insert sequences and detecting the hybrid gene transcript (s15-poly), and a similar probe specific for s15-1 sequences and detecting the endogenous gene transcript (s15-E). (*Lower panel*) Total ovarian RNA (10 μg, XS and BS lanes) or ovarian poly(A)+ RNA (4 μg, −305S and −280S lanes) was analyzed. (−305S and −280S) Lines transformed with Carnegie 20 vectors containing BAL-31-generated 5' deletions of the s15-poly gene. −305 and −280 refer to the nucleotide positions of the 5' end points of the respective deletions in relation to the putative cap site of s15-1 (defined as +1). In these constructs, 3'-flanking sequences extended to the *Sal*I site, as usual (see Fig. 3). The filter was hybridized with the specific probe for the silk moth insert.

dered whether moth chorion genes could be expressed normally if transferred into flies. We found that indeed they can (Mitsialis and Kafatos 1985).

Two cloned genomic fragments of *B. mori* DNA (3.1 and 3.8 kb), each containing a single A/B gene pair, were inserted in Carnegie 20, in opposite orientations. The constructs were injected into *cn;ry Drosophila* embryos, and a total of 10 independent transformant lines were selected for both constructs. In every line, developmentally correct expression of the moth genes was detected. Typical results are shown in Figures 6 and 7. Transformant lines harboring similar constructs bearing a high-cysteine gene pair (Hc-A/Hc-B) from *B. mori* showed no significant accumulation of moth-specific transcripts, even when the Hc-A/Hc-B pair was introduced into *Drosophila* in amplifiable vectors. The reasons for this are most probably related to the fact that the Hc genes are recent evolutionary acquisitions of *B. mori*, and will be discussed in more detail elsewhere.

In Figure 6, it can be seen that silk moth chorion gene-specific transcripts, while absent from the parental fly line, are prominent in the transformed line and present only in females. Similar analyses proved that the transcripts are limited to the ovary (Mitsialis and Kafatos 1985). Thus, the genes are transcribed with correct sex and organ specificity. Transcripts are detected both with B-specific and with A-specific probes. They are polyadenylated, comparable in size to authentic moth chorion mRNAs, and surprisingly abundant. Position effects on expression are relatively minor (threefold variation in abundance for independent lines of the same construct). Unlike the endogenous fly chorion genes, the moth genes are not amplified in the fly ovaries, as they are not in the moth ovaries. Taking into account the differing gene dosage, the steady-state level of moth transcripts per gene copy in the most abundantly expressing line corresponds to double the level of the major endogenous chorion mRNA, s15-1.

To examine in more detail the developmental specificity of transcription, RNAs from staged follicles were analyzed. Typical results are presented in Figure 7. The moth gene transcripts are abundant mostly in late choriogenesis (stage 14), when each follicle consists only of follicular epithelial cells and the oocyte. Since the transcripts are absent from newly laid eggs, they are apparently produced in the follicular epithelial cells, which are shed at ovulation. Thus, the transcripts show marked temporal and tissue specificity, in addition to sex and organ restriction. The temporal specificity is identical for the paired A and B genes, in agreement with their coordinate expression in moths (results not shown).

An intriguing observation is that the moth genes show a bimodal profile of activation, mimicking the "late" *Drosophila* s15-1 gene (Fig. 7). Transcripts are present at a relatively low level in the early stages of oogenesis, then decline substantially in abundance and finally reappear at high levels toward the end of choriogenesis. The ratio of early-to-late transcripts is even higher for the moth genes than for s15-1, but the difference can be accounted for by the late amplification of s15-1 and the absence of moth gene amplification. In the case of s15-1, it has been shown (Thireos et al.

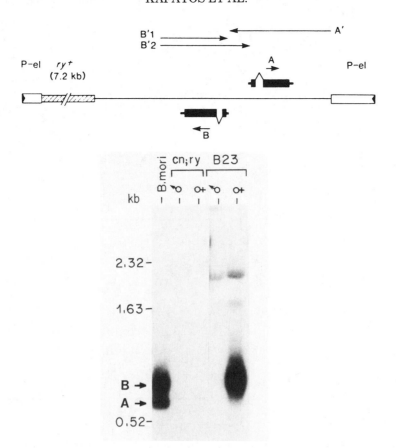

Figure 6. Sex specificity of silk moth chorion gene expression in transformed *Drosophila*. (*Top*) Structure of Carnegie 20-derived transformation vectors bearing *B. mori* chorion genes. The vector segments integrated into the *Drosophila* chromosomes is represented. (Open boxes) P-element sequences; (striped box) ry^+ sequences; (thin line) *B. mori* genomic sequences. The exons of the divergently transcribed chorion genes are diagrammed by filled boxes, wide for coding sequences and narrow for untranslated sequences; connecting lines represent the single intron of each gene. Arrows accompanying the A or B gene designation indicate the direction of transcription. Other arrows indicate the orientation and size of SP6-generated cRNA probes used for Northern analysis. (*Bottom*) Silk moth chorion gene-specific transcripts in transformants. RNA was isolated and analyzed with an A/B gene probe as described (Mitsialis and Kafatos 1985). (*B. mori*) 1 μg total silk moth ovarian RNA. Arrows indicate transcripts belonging to the A and B families of silk moth chorion genes; under the conditions of hybridization used here, transcripts of most members of the two families are detected. The remaining lanes contained 5 μg of poly (A)$^+$ RNA isolated from male (♂) or female (♀) *Drosophila* of the parental strain (*cn;ry*) or a transformant line harboring a *B. mori* chorion gene pair (B23). The migration and size in kilobases of single-stranded DNA markers is indicated at the left margin.

1980) that the early transcript is not translated, and it has been hypothesized that this premature and transient transcription might be related to and perhaps facilitates amplification. The similarity in the expression patterns suggest that moth A and B promoters are recognized by fly follicular cell factors as similar to those of s15-1, and are used accordingly.

We have searched for similarities between 5'-flanking sequences of moth and fly chorion genes, which might be expected by the similarities in expression patterns. All five *Drosophila* genes sequenced to date (Levine and Spradling 1985; Wong et al. 1985) share the hexanucleotide TCACGT at approximately the same position, 23-27 nucleotides 5' of the TATA element. In *B. mori*, the Hc-B genes (Iatrou and Tsitilou 1983), as well as the B genes (N. Spoerel et al., in prep.), both contain this sequence element at a similar position. In addition, patchy but recognizable similarities exist in the −20 to −110 region of the moth B gene and the fly s15-1 gene (Fig. 8). One patch encompasses the TATA element and its immediate vicinity (approximately −20 to −40). A second patch (approximately −57 to −93 in the B gene and −50 to −83 in s15-1) encompasses a series of short di- and trinucleotide matches, in addition to a heptanucleotide match; the latter includes the element TCACGT. A third patch of similarity encompasses nucleotides −94 to −110 in the B gene, and −93 to −112 in s15-1. It should be noted that short gaps must be introduced for alignment, within and between the similarity patches, but that these gaps cancel out, so that the patches and the entire region are similar in length between the B and s15-1 genes. Limited similarities are also seen further upstream, one centered at −195 and the other at −235 (B.12) or −275 (s15-1).

Our results indicate that some *trans*- and *cis*-acting regulatory elements responsible for the developmentally specific activation of chorion genes must have been conserved in *D. melanogaster* and *B. mori* since, for expression in the transformants, *trans* regulators of

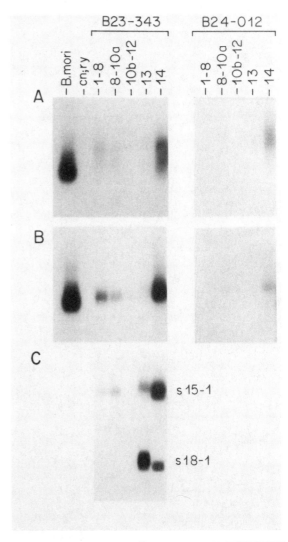

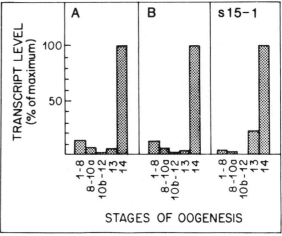

Figure 7. Temporal specificity of silk moth chorion gene expression in transformed *Drosophila*. (*Top*) RNA isolated from follicles representing the various stages (1–14) of *Drosophila* oogenesis was isolated from transformant lines and analyzed as described (Mitsialis and Kafatos 1985). (*B. mori*) 0.05 μg total ovarian silk moth RNA; (*cn;ry*) 15 μg total ovarian RNA from parental strain *Drosophila*; (B23-343 and B24-012) transformant lines harboring *B. mori* chorion genes. Lines of the B24 series contain a silk moth chorion gene pair that is very homologous but not identical to the one present in the B23 series; this pair has been inserted into the Carnegie 20 vector in the opposite orientation of that depicted in Fig. 6. Lanes 1-8 through 14 contained 15 μg of total RNA isolated from follicles at the corresponding developmental stages. Filters were hybridized with cRNA probes specific for the silk moth A gene (*A*), the silk moth B gene (*B*), or nick-translated probes specific for the *Drosophila* s15-1 or s18-1 genes (*C*). (*Bottom*) The relative abundance of silk moth chorion transcripts at the different developmental stages was quantitated and normalized as described in Methods. Values for the two silk moth genes (*A* and *B*) and the *Drosophila* s15-1 gene are expressed relative to respective transcript levels in stage-14 follicles. In lines of the B23 series, silk moth transcripts reach a steady-state level corresponding to 2.5% of the transcripts produced by the approximately 80-fold amplified s15-1 gene.

fly origin must have interacted with moth *cis* regulators residing within the transduced DNA fragments. It is attractive to speculate that this necessity of interaction between *cis* and *trans* elements imposes a constraint on their independent divergence and may thus account for their remarkable evolutionary conservation. Although some of the elements involved in the physiological heat shock response have been conserved between insects and mammals (Corces et al. 1981; Pelham 1982) and moderate tissue-specific restriction of transcription of the avian transferrin and skeletal actin genes in differentiated rodent cells has been reported

Figure 8. Sequence similarities in the 5'-flanking regions of chorion genes s15-1 of *D. melanogaster* and B.L12 of *B. mori*. The silk moth gene represents the particular member of the B gene family transduced into *Drosophila* in lines of the B23 series. Both sequences were searched for matches of at least 7/7 nucleotides and those occurring at similar positions are presented; sequences flanking the TATA and TCACGT elements in the region between −20 and −110 were aligned manually. Matching nucleotides are shown in capitals, mismatches in lower-case letters, and gaps in the alignment are shown as dots. Omitted sequences are represented by horizontal lines.

546

(McKnight et al. 1983; Nudel et al. 1985), the expression of silk moth chorion genes in *Drosophila* is, to our knowledge, the first example of such a prolonged evolutionary conservation for elements that confer clear-cut developmental regulation (sex, tissue, and temporal specificity of gene expression) in two species that have been separated by more than 200 million years.

ACKNOWLEDGMENTS

We wish to thank E. Fenerjian for dedicated secretarial assistance, B. Klumpar for excellent photography, and M. Yuk-See for artwork. This work was supported by grants from the National Institutes of Health, National Science Foundation, and American Cancer Society to F.C.K. B.M., S.A.M., and N.S. are recipients of postdoctoral fellowships from the Jane Coffin Child fund, the National Institutes of Health, and the Charles A. King Trust, respectively.

REFERENCES

CHURCH, G.M. and W. GILBERT. 1984. Genomic sequencing. *Proc. Natl. Acad. Sci.* **81:** 1991.

CORCES, V.A., A. PELLICER, R. AXEL, and M. MESELSON. 1981. Integration, transcription, and control of a *Drosophila* heat shock gene in mouse cells. *Proc. Natl. Acad. Sci.* **78:** 7038.

DE CICCO, D.V. and A.C. SPRADLING. 1984. Localization of a cis-acting element responsible for the developmentally regulated amplification of *Drosophila* chorion genes. *Cell* **38:** 45.

EICKBUSH, T.H. and F.C. KAFATOS. 1982. A walk in the chorion locus of *Bombyx mori*. *Cell* **29:** 633.

GOLDSMITH, M.R. and F.C. KAFATOS. 1984. Developmentally regulated genes in silk moths. *Annu. Rev. Genet.* **18:** 443.

GRIFFIN-SHEA, R., G.C. THIREOS, and F.C. KAFATOS. 1982. Organization of a cluster of four chorion genes in *Drosophila* and its relationship to developmental expression and amplification. *Dev. Biol.* **91:** 325.

GRIFFIN-SHEA, R., G. THIREOS, F.C. KAFATOS, W.H. PETRI, and L. VILLA-KOMAROFF. 1980. Chorion cDNA clones of *Drosophila melanogaster* and their use in studies of sequence homology and chromosomal location of chorion genes. *Cell* **19:** 915.

HONG, G.F. 1982. A systematic DNA sequencing strategy. *J. Mol. Biol.* **158:** 539.

IATROU, K. and S.G. TSITILOU. 1983. Coordinately expressed chorion genes of *Bomyx mori*: Is developmental specificity determined by secondary structure recognition? *EMBO J.* **2:** 1431.

JONES, C.W., N. ROSENTHAL, G.C. RODAKIS, and F.C. KAFATOS. 1979. Evolution of two major chorion multigene families as inferred from cloned cDNA and protein sequences. *Cell* **18:** 1317.

KAFATOS, F.C. 1983. Isolation of multigene families and determination of homologies by filter hybridization methods. In *Gene structure and regulation in development* (ed. S. Subtelny and F.C. Kafatos), p. 33. A.R. Liss, New York.

KAFATOS, F.C., C. DELIDAKIS, W. ORR, G. THIREOS, K. KOMITOPOULOU, and Y.-C. WONG. 1985. Studies on the developmentally regulated amplification and expression of *Drosophila* chorion genes. *Symp. Soc. Dev. Biol.* **43:** (in press).

KALFAYAN, L. and P.C. WENSINK. 1981. α-Tubulin genes of *Drosophila*. *Cell* **24:** 97.

KING, R.C. 1970. *Ovarian Development in* Drosophila melanogaster. Academic Press, New York.

KOMITOPOULOU, K., M. GANS, L.H. MARGARITIS, F.C. KAFATOS, and M. MASSON. 1983. Isolation and characterization of sex-linked female-sterile mutants in *Drosophila melanogaster* with special attention to eggshell mutants. *Genetics* **105:** 897.

LEVINE, J. and A.C. SPRADLING. 1985. DNA sequence of a 3.8 kb region controlling *Drosophila* chorion gene amplification. *Chromosoma* **92:** 136.

MCKNIGHT, G.A., R.E. HAMMER, E.A. KEUNZEL, and R.L. BRINSTER. 1983. Expression of the chicken transferrin gene in transgenic mice. *Cell* **34:** 335.

MCMASTER, G.K. and G.G. CARMICHAEL. 1977. Analysis of single and double-stranded nucleic acids on polyacrylamide and agarose gels by using glyoxal and acridine orange. *Proc. Natl. Acad. Sci.* **74:** 4835.

MELTON, D.A., P.A. KRIEG, M.R. REBAGLIATI, T. MANIATIS, K. ZINN, and M.R. GREEN. 1984. Efficient *in vitro* synthesis of biologically active RNA and RNA hybridization probes from plasmids containing a bacteriophage SP6 promoter. *Nucleic Acids Res.* **12:** 7035.

MITSIALIS, S.A. and F.C. KAFATOS. 1985. Regulatory elements controlling chorion gene expression are conserved between flies and moths. *Nature* **317:** 453.

NUDEL, U., D. GREENBERG, C.P. ORDAHL, O. SAXWL, S. NEUMAN, and D. YAFFE. 1985. Developmentally regulated expression of chicken muscle-specific gene in stably transfected rat myogenic cells. *Proc. Natl. Acad. Sci.* **82:** 3106.

ORR, W., K. KOMITOPOULOU, and F.C. KAFATOS. 1984. Mutants suppressing in *trans* chorion gene amplification in *Drosophila*. *Proc. Natl. Acad. Sci.* **81:** 3773.

OSHEIM, Y.N. and O.L. MILLER, JR. 1983. Novel amplification and transcriptional activity of chorion genes in *Drosophila melanogaster* follicle cells. *Cell* **33:** 543.

PELHAM, H.R.B. 1982. A regulatory upstream promoter element in the *Drosophila* hsp70 heat-shock gene. *Cell* **30:** 517.

PUSTELL, J. and F.C. KAFATOS. 1984. A convenient and adaptable package of computer programs for DNA and protein sequence management, analysis and homology determination. *Nucleic Acids Res.* **12:** 643.

———. 1986. A convenient and adaptable microcomputer environment for DNA and protein sequence manipulation and analysis. *Nucleic Acids Res.* (in press).

RUBIN, G.M. and A.C. SPRADLING. 1982. Genetic transformation of *Drosophila* with transposable element vectors. *Science* **218:** 348.

———. 1983. Vectors for P element-mediated gene transfer in *Drosophila*. *Nucleic Acids Res.* **11:** 6341.

SOUTHERN, E. 1975. Detection of specific sequences among DNA fragments separated by gel electrophoresis. *J. Mol. Biol.* **98:** 503.

SPRADLING, A.C. 1981. The organization and amplification of two chromosomal domains containing *Drosophila* chorion genes. *Cell* **27:** 193.

SPRADLING, A.C. and A.P. MAHOWALD. 1980. Amplification of genes for chorion proteins during oogenesis in *Drosophila melanogaster*. *Proc. Natl. Acad. Sci.* **77:** 1096.

SPRADLING, A.C. and G.M. RUBIN. 1982. Transposition of cloned P elements in *Drosophila* germ line chromosomes. *Science* **218:** 341.

SPRADLING, A.C., M.E. DIGAN, A.P. MAHOWALD, M. SCOTT, and E.A. CRAIG. 1980. Two clusters of genes for major chorion proteins of *Drosophila melanogaster*. *Cell* **19:** 905.

THIREOS, G., R. GRIFFIN-SHEA, and F.C. KAFATOS. 1980. Untranslated mRNA for a chorion protein of *Drosophila melanogaster* accumulates transiently at the onset of specific gene amplification. *Proc. Natl. Acad. Sci.* **77:** 5789.

WONG, Y.-C., J. PUSTELL, N. SPOEREL, and F.C. KAFATOS. 1985. Coding and potential regulatory sequences of a cluster of chorion genes in *Drosophila melanogaster*. *Chromosoma* **92:** 124.

The Molecular Basis of Differential Gene Expression of Two 5S RNA Genes

D.D. Brown and M.S. Schlissel
Department of Embryology, Carnegie Institution of Washington, Baltimore, Maryland 21210

We suggested that the differential expression of oocyte and somatic 5S RNA genes in *Xenopus* is caused by differences in affinity of a positive transcription factor TFIIIA for the internal control regions of the two kinds of 5S RNA genes and by the concentration of this factor in cells (Wormington et al. 1983; Brown 1984). The experiments summarized here (and detailed by Brown and Schlissel 1985) confirm the importance of these simple principles.

RESULTS

The Developmental Control That Needs to be Explained

The *Xenopus* genome contains a large oocyte 5S RNA multigene family (20,000 copies per haploid genome) and a smaller somatic 5S RNA multigene family (400 copies per haploid genome) (for review, see Brown 1982). In oocytes, both kinds of 5S RNA genes are active (Ford and Southern 1973). All RNA transcription ceases after meiosis and does not resume during the first 12 cleavage cycles of embryogenesis. RNA synthesis begins at midblastula (Brown and Littna 1964; Newport and Kirschner 1982a). At that time, there is synthesis of about equal amounts of oocyte and somatic 5S RNA (Wakefield and Gurdon 1983; Wormington and Brown 1983). Two cell divisions later, at gastrulation, only somatic 5S RNA genes are active; the oocyte 5S RNA genes are repressed. This remains the state of differential gene expression in somatic cells of *Xenopus*. We wish to know why the oocyte 5S RNA genes are expressed in oocytes, and why they are inactive in somatic cells. Furthermore, what is the molecular environment during early embryogenesis that gives rise to this differential gene expression?

TFIIIA and the Internal Control Region

TFIIIA interacts with an internal control region of about 50 nucleotides in the center of the 5S RNA gene (Bogenhagen et al. 1980; Engelke et al. 1980; Sakonju et al. 1980, 1981). This protein factor is specific for 5S RNA genes and is necessary but not sufficient for the accurate initiation of their transcription. At least two other factors (as yet poorly characterized), in addition to TFIIIA, are required (Segall et al. 1980). These factors form a stable transcription complex with the 5S RNA gene that in turn directs RNA polymerase III to initiate transcription faithfully (Bogenhagen et al. 1982).

Oocyte 5S RNA genes differ from somatic 5S RNA genes by six base changes within the gene; three of these are in the 5' part of the internal control region (Peterson et al. 1980). These three base changes weaken the binding of TFIIIA to the internal control region by about fourfold (Wormington et al. 1981). In experimental terms, it takes about four times more oocyte 5S RNA genes than somatic 5S RNA genes to compete for the binding of TFIIIA or to inhibit competitively the transcription of a second 5S RNA gene in a crude transcribing extract.

Abundance of TFIIIA Correlates with Differential Expression of 5S RNA Genes

In oocytes, TFIIIA is present in great excess and all 5S RNA genes are active. When TFIIIA is intermediate in concentration (at the midblastula transition [MBT]), most, but not all, of the oocyte 5S RNA genes are inactive. Complete inactivation of the oocyte 5S RNA genes occurs when TFIIIA concentration is low relative to gene number (Table 1).

The change in TFIIIA abundance with development and its correlation with differential expression is just that — a correlation. In this paper we summarize experiments that confirm the importance of TFIIIA concentration and its differential binding affinity to the two kinds of genes.

The Embryo Injection Assay

Previous studies on 5S RNA gene expression in *Xenopus* have used cell-free extracts (Birkenmeier et al. 1978) or injection into living oocytes (Brown and Gur-

Table 1. Factor Abundance and Transcription Discrimination

Developmental stage	Factor molecules per 5S RNA gene	Transcription efficiency (S/O)[a]
Oocyte	10^7	4
MBT	10	50
Gastrula	2	1000
Somatic cell	0.2	1000

Data from Pelham et al. (1981) and Wormington et al. (1983).
[a]Ratio of somatic 5S RNA to oocyte 5S RNA synthesized per gene. The gene ratio in the genome of *X. laevis* is 50:1 oocyte-to-somatic 5S RNA genes.

don 1977) as assay systems. In our experience, we have never seen preferential transcription of somatic over oocyte 5S RNA genes greater than 10-fold using either assay. Recently, Gargiulo et al. (1984) have enhanced somatic over oocyte 5S RNA gene transcription in *Xenopus* oocytes by injecting histones. The actual extent of preferential transcription was not stated.

We found that when a mixture of cloned somatic and oocyte 5S RNA genes was injected into *Xenopus* embryos, preferential expression of the cloned somatic 5S RNA genes was increased to as high as 200-fold over cloned oocyte 5S RNA genes (Brown and Schlissel 1985). Figure 1 is an example of such an experiment. A series of mutations were tested for their effect on this preferential transcription. Removal of DNA sequences flanking the two genes had no effect. Deletions up to the 5' or 3' borders of the internal control region resulted in the same relative synthesis of the altered gene compared with a control 5S RNA gene as is seen when the same gene mixture is assayed in vitro. However, deletions known to weaken the binding of TFIIIA to the internal control region are exaggerated by an order of magnitude in their effect when assayed by embryo injection.

There are important differences between the embryo injection assay and the other 5S RNA gene transcription assays. First, TFIIIA concentration is lower in embryos than it is in either living oocytes or in the oocyte nuclear extract used in the in vitro assay (Pelham et al. 1981). Second, genes injected into embryos (and oocytes) become assembled into a structure, presumably resembling chromatin (Wyllie et al. 1978). Increasing (or decreasing) TFIIIA concentration in an in vitro assay does not change the 4- to 10-fold somatic 5S RNA gene preference. Why is this small discrimination between oocyte and somatic 5S RNA genes in the in vitro assay so resistant to exaggeration? We believe the answer is related to the second point mentioned above, the assembly of injected genes into chromatin by embryos. The assembly of transcription complexes on naked DNA that occurs in vitro discriminates only slightly in favor of somatic 5S RNA genes (4- to 10-fold). However, the same genes injected into embryos are assembled into chromatin and in this configuration may bind factors with much different affinities than does naked DNA. The findings of Gargiulo et al. (1984) agree with this idea. Injection of cloned genes into oocyte nuclei resulted in their assembly into chromatin and an enhanced sensitivity to environmental changes that exaggerate somatic 5S RNA gene transcription. A difference between oocyte and embryo injection assays is the much higher concentration of TFIIIA in oocytes (Pelham et al. 1981). When the concentration of TFIIIA is increased in embryos, discrimination between somatic and oocyte 5S RNA genes is reduced to that seen in oocyte injection (Brown and Schlissel 1985).

TFIIIA Is Limiting for 5S RNA Gene Transcription in Embryos

The prediction that gradual inactivation of oocyte 5S RNA genes in embryos is due in part to a progressive decrease in the amount of TFIIIA per gene was tested by injecting TFIIIA into embryos just before or after MBT. This resulted in stimulation of endogenous 5S RNA transcription (Fig. 2). Analysis confirms that this enhanced 5S RNA synthesis is mainly oocyte-type 5S RNA. Embryos injected with aphidicolin to inhibit DNA replication still were able to respond to injected TFIIIA by increasing their transcription of oocyte 5S RNA genes.

DISCUSSION

Base changes in the 5S RNA gene internal control region that weaken TFIIIA binding can account for the preferential transcription of somatic over oocyte 5S RNA genes of greater than 100-fold when these genes are injected into embryos. Endogenous somatic 5S RNA genes are transcribed about 50-fold more efficiently than oocyte 5S RNA genes when RNA synthesis

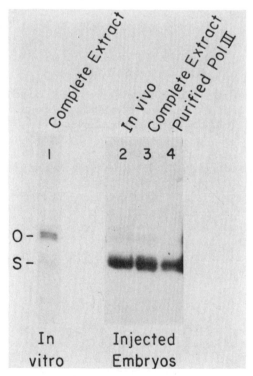

Figure 1. Differential expression of cloned oocyte and somatic 5S RNA genes assayed in vitro and by embryo injection. A 50:1 mixture of a cloned oocyte 5S RNA gene (pX1obs) and a cloned somatic 5S RNA gene (pX1s11) was either transcribed in vitro (lane *1*) or injected into embryos at first cleavage (lanes *2–4*). (Lane 2) The DNA was coinjected with [α-^{32}P]GTP and the RNA purified 3 hr later. Alternatively, the DNA was injected without the radioactive precursor and 3 hr later the embryos were homogenized and the DNA in the homogenate transcribed (lane *3*) in the in vitro extract or (lane *4*) with purified RNA polymerase III. The O indicates the position of oocyte 5S RNA, and S the position of somatic 5S RNA. (See Brown and Schlissel [1985] for experimental details.)

creases expression of the oocyte 5S RNA genes (Fig. 2). Thus, we conclude that two simple biophysical principles play key roles in oocyte-somatic 5S RNA gene discrimination—a difference in binding affinity of a positive transcription factor (TFIIIA) for two related DNA sequences (the internal control regions of the two kinds of 5S RNA genes) and the concentration of TFIIIA itself.

General Considerations of This Kind of Developmental Control

Genes or gene families that share the same *trans*-acting transcription factor(s) but bind to that factor with different affinity constants are candidates for differential control. When the factor concentration is high, all of the genes requiring it are active (Fig. 3). As the factor concentration drops, the gene that binds the factor most tightly (gene A in Fig. 3) is the last to be inactivated. When factor concentration rises with time, coordinated genes are not necessarily activated together. The gene that binds the factor most tightly is activated first and so on. The same principles hold for genes that share the same factor(s) when factor concentration forms a concentration gradient across an embryo. Marked differences in gene activity in neighboring cells could result from a gradient of a single *trans*-acting factor.

Rules for Turning Genes On

The repressed oocyte 5S RNA genes in somatic cells have been shown to lack transcription complexes. Instead, they are complexed with nucleosomes and histone H1 (Schlissel and Brown 1984). Removal of histone H1 enables factors to program the oocyte 5S RNA genes in vitro. Isolated somatic cell chromatin (including embryonic chromatin) supports transcription of the same ratio of somatic to oocyte 5S RNA genes that occurs in the living cells from which the chromatin was derived (Bogenhagen et al. 1982; Wormington and Brown 1983; Schlissel and Brown 1984). Yet, we have just shown that injection of TFIIIA into living embryos activates the oocyte 5S RNA genes (Fig. 2), suggesting that the inactive oocyte genes are not repressed stably in vivo. We presume that this apparent contradiction reflects the relative stability of repressed genes in vivo and in vitro; it may be explained by a dynamic exchange (Thomas and Rees 1983) of histone H1 binding to chromatin in living cells that enables a high level of TFIIIA to nucleate formation of a transcription complex. Indeed, activation of the endogenous oocyte 5S RNA genes requires a very high level of injected TFIIIA—50 ng or 500 times the endogenous level of TFIIIA, suggesting that the oocyte 5S RNA genes are repressed and not readily accessible to the factor. To activate such a gene, a high level of factor(s) is needed. Perhaps at the moment of DNA replication, when the nucleosome structure is perturbed momentarily, lower concentrations of factor will suffice. In

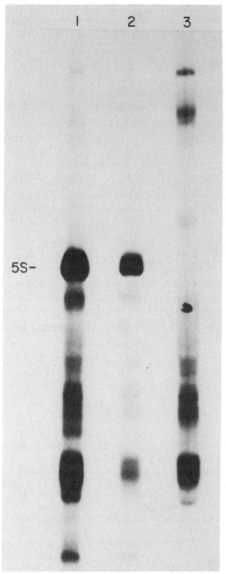

Figure 2. Stimulation of endogenous 5S RNA gene transcription in coenocytic (syncytical) embryos by injection of purified TFIIIA. Coenocytic embryos were prepared by centrifugation before first cleavage (Newport and Kirschner 1982b). Embryos received either one or two injections. The first injection was with 1 mg/ml of aphidicolin at MBT (stage 8) (lane 2). Then 1.5 hr later, embryos were injected either with a mixture of [α-^{32}P]GTP and 2.5 μg/μl of TFIIIA (lanes *1* and *2*) or [α-^{32}P]GTP alone (lane *3*). The position of 5S RNA is indicated.

begins at MBT (Table 1; Wormington and Brown 1983). Thus, injected genes mimic the differential expression of endogenous genes at first synthesis in embryogenesis. We have not been able to bring injected cloned genes under the exaggerated discrimination of greater than 1000-fold that characterizes gastrula-stage cells and adult somatic cells of *Xenopus*. The first discrimination that exists at MBT is clearly dependent upon the amount of TFIIIA, since injection of the factor in-

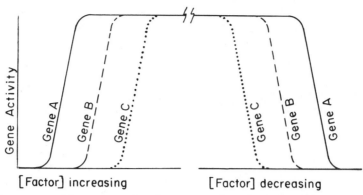

Figure 3. Diagram of the response of three genes to changing concentrations of a single positive *trans*-acting transcription factor required by all three for activity. The genes differ only in their affinity for the factor. The tightest binding is Gene A followed by Gene B; the weakest is Gene C.

the experiments reported here (Fig. 2), DNA replication was not required to activate the oocyte 5S RNA with TFIIIA. The prior injection of aphidicolin completely inhibits DNA replication but still permits activation of oocyte 5S RNA genes by TFIIIA injection (Fig. 2, lane 2). Maintenance of an active transcription complex requires a much lower concentration of factor(s) than formation of the complex de novo (Brown 1984).

Rules for Repressing a Gene

If stability of the 5S RNA gene transcription complex reflects characteristics of the activated state of other eukaryotic genes, then the activated (determined or committed) state is much more stable than the repressed state (Setzer and Brown 1985). The one biological event that is guaranteed to destabilize a transcription complex is replication. Perhaps one sister chromatid inherits the complex intact while the other is completely stripped of a complex. Alternatively, a partial complex might be partitioned to each sister chromatid. Such models for the epigenetic inheritance of a pattern of gene activity have been proposed (Brown 1984).

In the case of the gradual inactivation of oocyte 5S RNA genes in early embryos, we imagine that formation of a somatic 5S RNA gene transcription complex is favored by at least 100:1 over an oocyte 5S RNA gene transcription complex at MBT. This ratio is influenced by the amount of TFIIIA present at MBT. We propose that an additional feature of the system (not yet proven) is that a somatic 5S RNA gene transcription complex is stable to DNA replication whereas an oocyte 5S RNA gene complex is not. Thus each round of replication strips oocyte 5S RNA genes of their complexes requiring that they be reprogrammed again at the 50:1 ratio. The somatic 5S RNA genes in stable transcription complexes reform complexes on sister chromatids at each replication because of the strong cooperative interactions between factors.

ACKNOWLEDGMENTS

We thank E. Jordan for expert technical assistance. The work was supported in part by National Institutes of Health grant GM-22395. M. Schlissel is a trainee of the Medical Scientist Training Program, Johns Hopkins University School of Medicine.

REFERENCES

BIRKENMEIER, E.H., D.D. BROWN, and E. JORDAN. 1978. A nuclear extract of *Xenopus laevis* that accurately transcribes 5S RNA genes. *Cell* **15:** 1077.

BOGENHAGEN, D.F., S. SAKONJU, and D.D. BROWN. 1980. A control region in the center of the 5S RNA gene directs specific initiation of transcription. II. The 3′ border of the region. *Cell* **19:** 27.

BOGENHAGEN, D.F., W.M. WORMINGTON, and D.D. BROWN. 1982. Stable transcription complexes of *Xenopus* 5S RNA genes: A means to maintain the differentiated state. *Cell* **28:** 413.

BROWN, D.D. 1982. How a simple animal gene works. *Harvey Lect.* **76:** 27.

———. 1984. The role of stable complexes that repress and activate eucaryotic genes. *Cell* **37:** 359.

BROWN, D.D. and J.B. GURDON. 1977. High-fidelity transcription of 5S DNA injected into *Xenopus* oocytes. *Proc. Natl. Acad. Sci.* **74:** 2064.

BROWN, D.D. and E. LITTNA. 1964. RNA synthesis during the development of *Xenopus laevis*, the South African clawed toad. *J. Mol. Biol.* **20:** 95.

BROWN, D.D. and M.S. SCHLISSEL. 1985. A positive transcription factor controls the differential expression of two 5S RNA genes. *Cell* **19:** (in press).

ENGELKE, D.R., S.-Y. NG, B.S. SHASTRY, and R.G. ROEDER. 1980. Specific interaction of a purified transcription factor with an internal control region of 5S RNA genes. *Cell* **19:** 717.

FORD, P.J. and E.M. SOUTHERN. 1973. Different sequences for 5S RNA in kidney cells and ovaries of *Xenopus laevis*. *Nat. New Biol.* **241:** 7.

GARGIULO, G., R. RAZVI, and A. WORCEL. 1984. Assembly of transcriptionally active chromatin in *Xenopus* oocytes requires specific DNA binding factors. *Cell* **38:** 511.

NEWPORT, J. and M. KIRSCHNER. 1982a. A major developmental transition in early *Xenopus* embryos: I. Characterization and timing of cellular changes at the midblastula stage. *Cell* **30:** 675.

———. 1982b. A major developmental transition in early *Xenopus* embryos. II. Control of the onset of transcription. *Cell* **30:** 687.

PELHAM, H.R.B., W.M. WORMINGTON, and D.D. BROWN. 1981. Similar 5S RNA transcription factors in *Xenopus* oocytes and somatic cells. *Proc. Natl. Acad. Sci.* **78:** 1760.

PETERSON, R.C., J.L. DOERING, and D.D. BROWN. 1980. Characterization of two *Xenopus* somatic 5S DNAs and one minor oocyte-specific 5S DNA. *Cell* **20:** 131.

SAKONJU, S., D.F. BOGENHAGEN, and D.D. BROWN. 1980. A control region in the center of the 5S RNA gene directs specific initiation of transcription. I. The 5' border of the region. *Cell* **19:** 13.

SAKONJU, S., D.D. BROWN, D.R. ENGELKE, S.-Y. NG, B.S. SHASTRY, and R.G. ROEDER. 1981. The binding of a transcription factor to deletion mutants of a 5S ribosomal RNA gene. *Cell* **23:** 665.

SCHLISSEL, M.S. and D.D. BROWN. 1984. The transcriptional regulation of *Xenopus* 5S RNA genes in chromatin: The roles of active stable transcription complexes and histone H1. *Cell* **37:** 903.

SEGALL, J., T. MATSUI, and R.G. ROEDER. 1980. Multiple factors are required for the accurate transcription of purified genes by RNA polymerase III. *J. Biol. Chem.* **225:** 11986.

SETZER, D.R. and D.D. BROWN. 1985. Formation and stability of the 5S RNA transcription complex. *J. Biol. Chem.* **260:** 2483.

THOMAS, J.O. and C. REES. 1983. Exchange of histones H1 and H5 between chromatin fragments: A preference of H5 for higher-order structures. *Eur. J. Biochem.* **134:** 109.

WAKEFIELD, L. and J.B. GURDON. 1983. Cytoplasmic regulation of 5S RNA genes in nuclear-transplant embryos. *EMBO J.* **2:** 1613.

WORMINGTON, W.M. and D.D. BROWN. 1983. Onset of 5S RNA gene regulation during *Xenopus* embryogenesis. *Dev. Biol.* **99:** 248.

WORMINGTON, W.M., M. SCHLISSEL, and D.D. BROWN. 1983. Developmental regulation of *Xenopus* 5S RNA genes. *Cold Spring Harbor Symp. Quant. Biol.* **47:** 879.

WORMINGTON, W.M., D.F. BOGENHAGEN, E. JORDAN, and D.D. BROWN. 1981. A quantitative assay for *Xenopus* 5S RNA gene transcription in vitro. *Cell* **24:** 809.

WYLLIE, A.H., R.A. LASKEY, J. FINCH, and J.B. GURDON. 1978. Selective DNA conservation and chromatin assembly after injection of SV40 DNA into *Xenopus* oocytes. *Dev. Biol.* **64:** 178.

Symbiotic Nitrogen Fixation: Developmental Genetics of Nodule Formation

N. LANG-UNNASCH,* K. DUNN, AND F.M. AUSUBEL

Department of Genetics, Harvard Medical School and Department of Molecular Biology, Massachusetts General Hospital, Boston, Massachusetts 02114

Infection of alfalfa (*Medicago sativa*) with its bacterial symbiont *Rhizobium meliloti* results in the formation of root nodules, specialized organs within which the rhizobia fix nitrogen. The nodule tissues arise from a nodule meristem which forms when nonproliferating root cells are induced by the *Rhizobium* to dedifferentiate and begin dividing. The development of effective, nitrogen-fixing root nodules is a multistep process requiring genetic input from both the legume host and the bacterial symbiont. Both bacterial and plant mutants that block nodule formation at different morphological stages have been isolated (for reviews, see Vincent 1980; Verma and Long 1983). Plant and bacterial mutants also affect the biochemical development of the nodule as assayed by enzyme activities (Groat and Vance 1981), or by the accumulation of nodule-specific RNAs (Fuller et al. 1983) and polypeptides (Bisseling et al. 1983; Lang-Unnasch and Ausubel 1985). In this paper we describe the biochemical development of a class of ineffective nodules that do not contain bacteria. Our analysis suggests that the synthesis of nodule-specific proteins coded by the plant host involves at least two distinct mechanisms of regulation.

Nodule Development

Alfalfa root nodules are club shaped with a terminal meristem (embryonic growing region) and are similar in structure to nodules formed by many subtropical legumes (Dart 1977). Nodule formation begins with the attachment of *R. meliloti* to the root hairs; in response, the root hairs undergo a variety of characteristic deformations including branching and curling. Although many *Rhizobium* species will induce root hair deformations on a heterologous legume host, tight curls or "shepherd's crooks" are diagnostic of a compatible association. Root hair deformations are probably the result of changes in cell wall synthesis patterns elicited by the invading bacteria (Callaham and Torrey 1981).

Following attachment, the rhizobia become enclosed within a pocket in the root hair cell wall. An infection thread develops from this point, grows into the interior of the root, and ramifies. The infection thread is produced by the plant and consists primarily of cellulose. It serves as a conduit through which the rhizobia penetrate the interior of the root. As infection threads proliferate, root cortical cells dedifferentiate, divide, and form the nodule meristem. Within the nodule tissue resulting from outgrowth of this meristem, rhizobia are released from the infection threads into the host cell cytoplasm. The rhizobia, which are now termed bacteroids, are surrounded by a peribacteroid membrane which is host-derived (Brewin et al. 1985), although it may become modified with bacteroid proteins.

Differentiation of the bacteroids into the nitrogen-fixing state is characterized by the production of bacteroid-specific cytochromes and by the production of nitrogenase, the enzyme that catalyzes the reduction of free N_2 to ammonia. The plant also synthesizes nodule-specific proteins referred to as nodulins (Legocki and Verma 1980; for a review of the nomenclature, see van Kammen 1984). Leghemoglobin is a highly conserved nodulin, similar in structure to animal globins, which transports oxygen within the nodule and which controls the nodule oxygen tension (Wittenberg et al. 1974). Leghemoglobin is crucial to nodule function because nitrogenase is an oxygen-sensitive enzyme but, at the same time, bacteroids require a large and steady supply of oxygen for ATP production which is consumed rapidly by the energy-intensive nitrogen-fixing reaction.

Two nodulins involved in ammonia assimilation have been identified. Tropical legumes such as soybeans contain a nodule-specific uricase (Bergmann et al. 1983); both tropical and subtropical legumes contain a nodule-specific glutamine synthetase (Cullimore et al. 1983; K. Dunn and F.M. Ausubel, in prep.). Other nodulins, whose functions have not yet been determined, have been identified by cDNA cloning (Fuller et al. 1983; K. Dunn and F.M. Ausubel, in prep.) or immune assay (Legocki and Verma 1980; Bisseling et al. 1983; Lang-Unnasch and Ausubel 1985).

Mature alfalfa nodules contain a continuum of development. At the nodule tip, a zone of meristematic cells continues to divide throughout the life span of the nodule. Behind the meristem, a zone of cells is invaded by infection threads and newly released bacteroids. Still older cells form a late symbiotic zone containing differentiated bacteroids in the nitrogen-fixing state. Finally, the root proximal nodule cells form a senescent zone. As alfalfa nodules develop, these zones form sequentially and as they age, the proportion of cells in the senescent zone increases with respect to the other zones.

---

*Present address: Department of Biology, Case Western Reserve University, Cleveland, Ohio 44106.

Rhizobium Mutants That Alter Nodule Development

Many well-characterized symbiotically defective *R. meliloti* mutants have been isolated (for review, see Long 1984). These mutants fall into two major readily identifiable subgroups: those unable to induce visible nodule formation (Nod⁻) and those that induce ineffective, non-nitrogen-fixing nodules (Fix⁻). In *R. meliloti*, as in other so-called "fast-growing" *Rhizobium* species, symbiotic genes are located on large endogenous plasmids. *R. meliloti* contains two large plasmids, both greater than 500 kb, which carry symbiotic genes.

R. meliloti nod genes are located in two adjacent clusters approximately 15-20 kb from the *nif*HDK genes, which are *fix* genes coding for the subunits of nitrogenase (Long et al. 1982; Long 1984; Torok et al. 1984). *R. meliloti* strains mutated in the *nif* distal *nod* region do not induce root hair curling or the development of visible nodules. This region is referred to as the "common *nod*" region since genes from other *Rhizobium* species can complement these *R. meliloti nod* mutations (Banfalvi et al. 1981; Kondorosi et al. 1984; Marvel et al. 1985), and because these *nod* genes cross-hybridize with *nod* genes from other rhizobia (Schmidt et al. 1984). The *nif*-promixal *nod* region is believed to determine host specificity of nodulation (Torok et al. 1984); strains mutated in the host specificity region induce at least some root hair curling, but either induce no nodules or induce nodules 2-3 weeks later than wild-type strains.

Although the functions of the *nod* products have yet to be determined, it appears likely that only *nod* gene products are required to initiate nodule development. *Agrobacterium tumefaciens* strains carrying *R. meliloti* common and host-specific *nod* genes, but no other *Rhizobium* genes, induce Fix⁻ nodules on alfalfa (Wong et al. 1983; Truchet et al. 1984; Hirsch et al. 1984, 1985). The ultrastructural organization of these nodules is similar to wild-type nodules; however, these "pseudonodules" contain neither infection threads nor bacteroids. In addition to delineating the number of *Rhizobium* genes required to form a nodule, these experiments suggest that nodule morphogenesis can be separated from bacteroid formation.

R. meliloti mutants that elicit Fix⁻ nodules include those that lack or have a defective nitrogenase. *R. meliloti* strains have been isolated which contain Tn5 insertion mutations in each of the three structural genes of nitrogenase (*nif*H, *nif*D, and *nif*K; Ruvkun et al. 1982), in a regulatory gene required for transcriptional activation of the *nif*HDK operon (Szeto et al. 1984), and in a gene homologous to *Klebsiella nif*B (W. Buikema, pers. comm.). The ultrastructure of these Fix⁻ nodules is similar to wild-type except that they contain an enlarged zone of senescence (Hirsch et al. 1983). Using an immune assay to detect nodule-specific polypeptides (Lang-Unnasch and Ausubel 1985), we were unable to find any qualitative differences between these Fix⁻ nodules and wild-type nodules, but some specific polypeptides did differ quantitatively between these nodules. These differences indicated an accelerated aging process within the Fix⁻ nodules but suggested that the absence of nitrogenase was not itself a primary signal recognized by the plant.

In addition to *nif*⁻ mutants, *fix*⁻ *R. meliloti* mutants have been isolated that have no direct effect on nitrogenase synthesis. For example, Finan et al. (1985) recently described a class of *fix*⁻ *R. meliloti* mutants that initiate the formation of empty alfalfa nodules. These mutants are characterized by the lack of a particular acidic exopolysaccharide. Like the "pseudonodules" induced by the *A. tumefaciens* hybrids containing *R. meliloti nod* genes, the ultrastructures of these nodules are similar to wild-type except that these Fix⁻ nodules form no infection threads and contain no bacteria. The inducing *Rhizobia* are restricted to the intercellular spaces at the nodule epidermis. These exopolysaccharide-deficient mutants effectively uncouple the differentiation of the bacteria from the process of nodule development in the plant. However, unlike the nodules elicited by *A. tumefaciens-R. meliloti* hybrids, exopolysaccharide mutants elicit nodules in the same time frame and at the same frequency as wild-type *Rhizobium*. In this paper we have utilized *R. meliloti* exopolysaccharide mutants to study the regulation of nodulin biosynthesis.

EXPERIMENTAL PROCEDURES

Root nodules. Alfalfa (*M. sativa* cv. Iroquois) seeds were sterilized by soaking in concentrated sulfuric acid for 10 minutes. The seeds were washed thoroughly with sterile distilled water and sown on nitrogen-free agar slants (Meade et al. 1982). The seedlings were inoculated with a saturated culture of *R. meliloti* 3-4 days later. The exopolysaccharide-deficient *R. meliloti* mutants were isolated and characterized by Leigh et al. (1985) and were generously supplied by J. Leigh (Massachusetts Insitute of Technology) prior to publication. Seeds for alfalfa cultivars other than Iroquois were a gift from D.K. Barnes (University of Minnesota).

Nodules or roots were harvested by hand, immediately frozen in liquid nitrogen, and stored at $-80°C$ until needed.

Protein extraction and immune assay. Nodule and root extracts were prepared as previously described (Lang-Unnasch and Ausubel 1985), except that detergent was omitted from the homogenization buffer of extracts to be subjected to two-dimensional gel electrophoresis. Two-dimensional gel electrophoresis was performed according to O'Farrell (1974). SDS-polyacrylamide gel electrophoresis and the Western blotting procedures were described previously (Lang-Unnasch and Ausubel 1985).

Immune sera. Antiserum raised against plant-derived alfalfa nodule proteins was made "nodule-specific" by preadsorption with root proteins that were bound to CNBr-activated Sepharose (Lang-Unnasch and Ausubel 1985). Anti-glutamine synthetase immune serum was prepared against purified alfalfa glutamine synthetase and was a gift of E. Tischer and H. Goodman (Massachusetts General Hospital).

RNA isolation and hybridization. RNA was isolated from wild-type or mutant nodules by grinding frozen nodules to a fine powder in liquid nitrogen, adding two volumes of extraction buffer (0.2 M sodium acetate [pH 4.8], 1% SDS, 10 mM EDTA [pH 8.0]), and then extracting 3× with one volume of phenol/chloroform/isoamyl alcohol (25:24:1). RNA was precipitated from the aqueous phase by addition of one-third volume 8 M LiCl and incubation at 4°C for 16 hours followed by a second precipitation in ethanol at −20°C.

RNA dot blots were prepared from 4–6 µg of total RNA. The RNA was denatured in 6% formaldehyde, 1 M NaCl, 30 mM NaH_2PO_4 (pH 6.8), at 65°C for 15 minutes, then applied to nitrocellulose using a Schleicher and Schuell dot blotter. Hybridization was at 65°C in 50 mM Tris-HCl (pH 7.5), 0.1% sodium pyrophosphate, 0.1% SDS, 1 M NaCl, 1× Denhardt's solution, and 10% dextran sulfate.

RESULTS

Nodulins

Nodulins are nodule-specific proteins encoded by host plant genes. Although immunological and cDNA cloning techniques have been used to demonstrate the presence of approximately 15–30 nodulins in soybean (Legocki and Verma 1980), pea (Bisseling et al. 1983), and alfalfa (Lang-Unnasch and Ausubel 1985) nodules, the functions of only a few nodulins have been determined. Alfalfa nodulins of known function include leghemoglobin (Jing et al. 1982) and glutamine synthetase (K. Dunn and F.M. Ausubel, in prep.).

In soybeans, leghemoglobin is encoded by a small gene family (Sullivan et al. 1981; Hyldig-Nielsen et al. 1982) and in alfalfa there are probably five leghemoglobin genes whose protein products differ by only a few amino acids (Jing et al. 1982). Alfalfa leghemoglobin is readily visualized using the Western blotting technique in conjunction with a nodule-specific polyclonal immune serum (Lang-Unnasch and Ausubel 1985). We have also isolated a variety of alfalfa leghemoglobin cDNA clones based on their ability to cross-hybridize with previously cloned pea leghemoglobin genes (B. Burnett and F.M. Ausubel, pers. comm.). Glutamine synthetase is also encoded by small gene families in *Phaseolus* (J. Cullimore and B. Miflin, pers. comm.) and alfalfa (E. Tischer and H. Goodman, pers. comm.). We have isolated an alfalfa glutamine synthetase cDNA clone based on cross-hybridization with a previously isolated *Phaseolus* clone (Cullimore and Miflin 1983) and have shown that at least one glutamine synthetase gene is expressed exclusively in nodules (data not shown).

In addition to leghemoglobin, we have used the Western blotting technique to characterize 10 other potential alfalfa nodulins of unknown function (Lang-Unnasch and Ausubel 1985). Finally, in addition to leghemoglobin and glutamine synthetase, we have isolated three additional alfalfa nodulin cDNA clones (data not shown). One or more of the three unidentified nodulins may correspond to the nodule-specific polypeptides characterized by immune assay. Using these cDNAs as probes, we have shown that transcription of each of these nodulin genes is first detectable between 5 and 10 days after inoculation of alfalfa with *R. meliloti* (data not shown).

Nodule-specific Polypeptides in Effective Nodules

We have used the Western blotting technique to identify nodule-specific polypeptides in effective (nitrogen-fixing) nodules elicited by a wild-type strain of *R. meliloti*. Because there is a lot of genetic diversity between different alfalfa cultivars and because alfalfa seed tends to be genetically heterogeneous due to the fact that alfalfa is an outbreeding species, we compared the nodule-specific polypeptides from nine alfalfa cultivars. We were concerned that the genetic differences between the cultivars might be reflected in the expression of nodulins and that examination of a single cultivar might show a nonrepresentative pattern of nodulin expression. Figure 1 shows that nodules from all nine cultivars shared the same set of nodule-specific polypeptides that we described previously in nodules from the

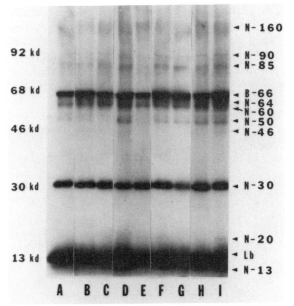

Figure 1. Nodule-specific polypeptides in alfalfa cultivars. A Western blot prepared from protein extracts of nodules was probed with "nodule-specific" immune serum. The nodules were harvested from the following alfalfa cultivars: (lane *A*) Iroquois, (*B*) Hairy Peruvian, (*C*) Indian, (*D*) African, (*E*) Chilean, (*F*) Ladak, (*G*) Turkestan, (*H*) Flemish, and (*I*) Grimm. The nodules of each cultivar shown were tested for acetylene reduction activity: A < B < C < D < E < F < G < H < I. The nodule-specific polypeptides are indicated by an N followed by their molecular weight in kilodaltons. Lb indicates leghemoglobin. A non-nodule-specific bacterial protein of 66 kD is labeled B-66.

alfalfa Iroquois cultivar (Lang-Unnasch and Ausubel 1985). On the other hand, the pattern of nodule-specific polypeptides did vary quantitatively between cultivars. For instance, nodules from alfalfa Grimm (Fig. 1, lane I) had more of two of the nodule-specific polypeptides, N-50 and N-20, than alfalfa Iroquois nodules in the same experiment (Fig. 1, lane A).

We also compared the nitrogen-fixing ability of the nine nodulated alfalfa cultivars by assaying their ability to reduce acetylene. All of the cultivars were inoculated with the same wild-type strain of R. meliloti, Rm1021, all were grown under the same conditions, and all were harvested at the same time, 21 days after inoculation. We found a range of nitrogen-fixing activities. Nodules from the cultivar Grimm were approximately five times more active than Iroquois nodules (data not shown).

The four nodule-specific polypeptides whose accumulation varied quantitatively in nodules from the different cultivars were N-50, N-46, N-20, and N-13 (Fig. 1). The concentration of these four polypeptides also varied in nodules grown under different conditions of heat and light (data not shown), in ineffective nodules produced by mutant alfalfa genotypes (data not shown), and in nodules induced by R. meliloti mutants lacking nitrogenase (Lang-Unnasch and Ausubel 1985). In alfalfa Iroquois nodules, three of these polypeptides (N-46, N-20, N-13) accumulated only during defined intervals in nodule development (Lang-Unnasch and Ausubel 1985). N-20 was only detectable for a short time after nodules first become visible (approximately 9 days postinoculation), N-13 was enriched in nodules during an interval 2-3 weeks after inoculation, and N-46 accumulated only in nodules 4-5 weeks after inoculation and then disappeared. The only other nodule-specific polypeptide whose concentration changed significantly within nodules between 11 and 42 days postinoculation was leghemoglobin (Lang-Unnasch and Ausubel 1985). The concentration of leghemoglobin in nodules from the nine alfalfa cultivars appeared equivalent (Fig. 1).

With the exception of leghemoglobin, the nodule-specific polypeptides in Figure 1 that did not differ quantitatively between cultivars (i.e., N-160, N-90, N-85, N-64, N-60, and N-30) also did not differ quantitatively within developing alfalfa Iroquois nodules or in ineffective nodules induced by a R. meliloti mutant lacking nitrogenase (Lang-Unnasch and Ausubel 1985).

The Induction of Empty Root Nodules

As described in the introductory section, under special conditions, alfalfa root nodules that do not contain infection threads or bacteria can be elicited. We made use of such "empty" nodules to determine whether the synthesis of certain alfalfa nodulins required penetration of R. meliloti into the nodule structure. In particular, we used exopolysaccharide-deficient mutants of R. meliloti which induce small, white, ineffective nodules (Finan et al. 1985; Leigh et al. 1985).

The exopolysaccharide mutants fall into six complementatation groups (A-F). We found that nodules could be seen on alfalfa roots 8-9 days after inoculation with complementation groups A, B, D, E, and F, the same time as nodules induced by the parental strain Rm1021. The appearance of root nodules induced by an exopolysaccharide-deficient mutant in group C, Rm7025, was delayed by at least a week.

We have used an immunological assay to examine nodulins present in empty nodules elicited by the exopolysaccharide mutants. The "nodule-specific" immune serum that we characterized previously binds to one R. meliloti protein, B-66, as well as to 11 polypeptides found only in the plant fraction of the nodule. B-66 is expressed at all stages of R. meliloti growth, including log- or stationary-phase cells of the free living bacterium and in bacteroids. Thus, the immunological detection of B-66 provides a marker for the presence of the rhizobia. As expected, the immune serum did not bind to the B-66 polypeptide in a Western blot of total nodule protein from root nodules induced by any of the exopolysaccharide-deficient R. meliloti strains that we tested (Figs. 2 and 3), although the B-66 polypeptide is very prominent in extracts of effective nodules induced by the wild-type R. meliloti. The free-living bacteria from all of these exopolysaccharide-deficient strains expressed the B-66 polypeptide (Fig. 4 and data not shown).

Expression of Nodule-specific Plant Polypeptides in Empty Nodules

Six of the 11 nodule-specific polypeptides that accumulate following inoculation with wild-type R. meliloti are also present in all of the ineffective nodules induced by the exopolysaccharide-deficient mutants (Figs. 2 and 3). These six nodule-specific polypeptides, N-160, N-90, N-85, N-64, N-60, and N-30, accumulate in virtually identical quantities in effective and all of the ineffective nodules. Each of these polypeptides was present in nodules harvested shortly after the nodules became visible and was essentially unchanged in relative concentration throughout the period of normal nodule development (Fig. 5).

In addition to the presence of these unidentified nodule-specific polypeptides, we have also found that the nodule forms of glutamine synthetase present in effective root nodules are also expressed in the empty nodules (Fig. 6). To distinguish the various forms of glutamine synthetase, Western blots were prepared following two-dimensional gel electrophoresis of the root and nodule extracts, and probed with antibodies against alfalfa glutamine synthetase. There are three major spots which have a molecular weight of approximately 41 kD and two spots of approximately 44 kD. We are not certain that all of these spots are actually glutamine synthetase, but there is precedent for multiple forms that vary in isoelectric point and to a lesser extent in molecular weight (Lara et al. 1985). The most electronegative, 41-kD spot is greatly enhanced in nod-

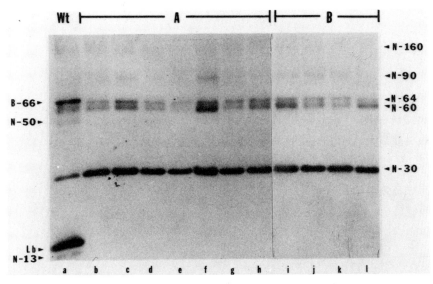

Figure 2. Nodule-specific polypeptides expressed in empty nodules. Nodules induced by a wild-type strain of *R. meliloti* (Wt) and by exopolysaccharide-deficient *R. meliloti* mutants in complementation group A (*A*) or complementation group B (*B*) were harvested 22 days after inoculation. A Western blot prepared from protein extracts of each of these nodules was probed with "nodule-specific" immune serum. The polypeptides are labeled as in Fig. 1. The *R. meliloti* strains used were: (lane *a*) Rm1021, (*b*) Rm7023, (*c*) Rm7034, (*d*) Rm7009, (*e*) Rm7011, (*f*) Rm7016, (*g*) Rm7031, (*h*) Rm7032, (*i*) Rm7002, (*j*) Rm7094, (*k*) Rm7013, and (*l*) Rm7014.

ule extracts and the 44-kD spots are present only in nodule extracts and not in root extracts. Presumably one of these three spots is the polypeptide encoded by the nodule-specific glutamine synthetase gene recently isolated in this lab (K. Dunn and F.M. Ausubel, in prep.). All five protein spots are present in empty nodules induced by the group-B mutant, Rm7094, though the 44-kD spots are less intense than those of the effective nodule control. Preliminary data suggest that RNA complementary to the 3' region of an alfalfa nodule-specific glutamine synthetase gene is also present in these exopolysaccharide-deficient nodules (data not shown).

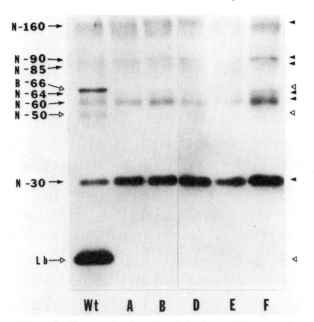

Figure 3. Characterization of nodules induced by exopolysaccharide-deficient *R. meliloti* mutants. Nodules induced by a wild-type strain of *R. meliloti* (Wt), and by representative exopolysaccharide-deficient *R. meliloti* mutants in five complementation groups (A, B, D, E, and F) were harvested 18 days after inoculation. A Western blot prepared from protein extracts of each of these nodules was probed with "nodule-specific" immune serum. The polypeptides are labeled as in Fig. 1. The *R. meliloti* strains used were: (lane *Wt*) Rm1021, (*A*) Rm7023, (*B*) Rm7094, (*D*) Rm7017, (*E*) Rm7022, and (*F*) Rm7055.

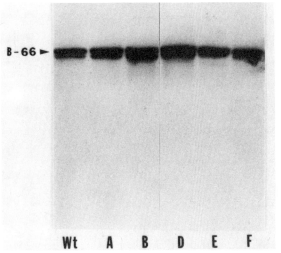

Figure 4. Exopolysaccharide-deficient *R. meliloti* mutants synthesize B-66. A Western blot was prepared from protein extracts of free-living *R. meliloti* wild-type (lane *Wt*) and exopolysaccharide-deficient mutants (lanes *A–F*). This blot was probed with total nodule immune serum. The position of the bacterial protein B-66 is indicated at left. The *R. meliloti* strains are: (lane *Wt*) Rm1021, (*A*) Rm7023, (*B*) Rm7094, (*C*) Rm7025, (*D*) Rm7017, (*E*) Rm7022, and (*F*) Rm7055.

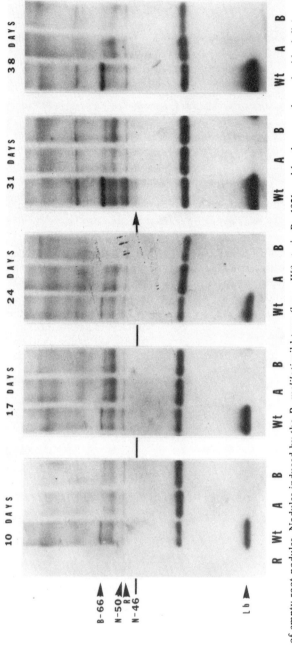

Figure 5. Development of empty root nodules. Nodules induced by the *R. meliloti* wild-type (lanes *Wt*) strain Rm1021 and by the exopolysaccharide-deficient strains Rm7023 (lanes *A*) and Rm7094 (lanes *B*) were harvested at weekly intervals beginning 10 days after inoculation. A Western blot prepared from protein extracts of these nodules and from uninfected roots (lane *R*) was probed with "nodule-specific" immune serum. The positions of those polypeptides present in the effective, wild-type-induced nodules, but not in the ineffective, mutant-induced nodules are indicated. Due to incomplete preadsorption of the immune serum with root proteins, a non-nodule-specific plant polypeptide of about 50 kD (R) is also visible.

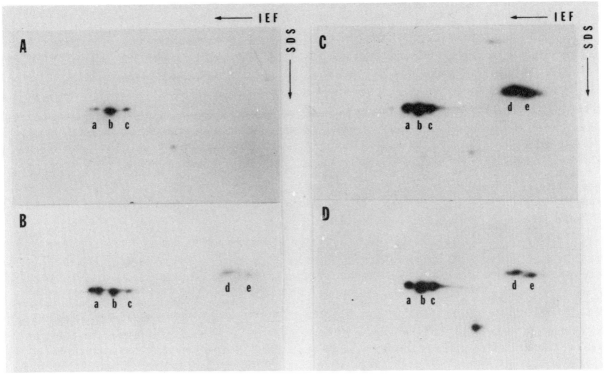

Figure 6. Heterogeneity of glutamine synthetase polypeptides from alfalfa roots and nodules. Protein extracts from uninfected roots (A) and from effective, Rm1021-induced nodules (B) were subjected to two-dimensional gel electrophoresis. A Western blot prepared from each gel was probed with an immune serum prepared against alfalfa glutamine synthetase. In a separate experiment, effective, Rm1021-induced nodules (C) and empty, ineffective Rm7094-induced nodules (D) were harvested 17 days after inoculation. Protein extracts from these nodules were also subjected to two-dimensional gel electrophoresis and a Western blot from each was probed with anti-glutamine synthetase immune serum.

Nodule-specific Polypeptides That Are Not Expressed in Empty Nodules

In effective root nodules, leghemoglobin is the major plant protein synthesized, comprising almost 20% of the polyadenylated mRNA (Auger et al. 1979). Leghemoglobin apoprotein is completely absent from the empty nodules induced by the exopolysaccharide-deficient mutant R. meliloti strains at all times throughout the nodule lifetime (Figs. 2, 3, and 5). Leghemoglobin mRNA was also not detected in total RNA prepared from empty nodules harvested either 17 or 24 days after inoculation (Fig. 7).

Three other nodule-specific polypeptides expressed in effective nodules appear to be missing from nodules induced by the exopolysaccharide-deficient R. meliloti. N-50 was clearly missing from the mutant-induced nodule extracts shown in Figures 2 and 3. This was much less clear in Figure 5 because an incomplete titration of the immune serum resulted in binding to a root protein that comigrates with N-50. The two other nodule-specific polypeptides missing from empty nodule extracts, N-46 and N-13, are normally expressed for only a short interval in effective nodules (Lang-Unnasch and Ausubel 1985). N-46 is normally expressed 4–5 weeks after inoculation with the wild-type R. meliloti (Fig. 5). N-13 is normally enriched in effective nodules 2–3 weeks after inoculation (Fig. 2). N-46 and N-13 did not accumulate in empty nodules at the same time as they accumulated in the effective nodules in these experiments.

DISCUSSION

Alfalfa seedlings inoculated with any of the exopolysaccharide-deficient R. meliloti mutants display the same set of responses. Ineffective nodules form which contain many of the plant proteins normally expressed in effective nodules, but lack immunologically detectable rhizobia. This consistent set of responses of the alfalfa to all of the R. meliloti mutants in the five complementation groups tested suggests that these mutants all lack the ability to signal the plant at the same step

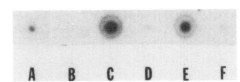

Figure 7. Leghemoglobin mRNA is not detected in empty nodules. RNA from uninfected root (A), effective nodules (C,E), empty, ineffective nodules (D,F), and yeast (B) was blotted on nitrocellulose. The blot was then probed with a nick-translated alfalfa leghemoglobin cDNA. Effective nodules were induced with R. meliloti 1021, and were harvested 17 (C) and 24 (E) days after inoculation. Ineffective nodules were induced with R. meliloti 7094, and were harvested 17 (D) and 24 (F) days after inoculation.

in the symbiotic process. The only known common feature of this collection of R. meliloti mutants is that they lack a particular acidic exopolysaccharide (Leigh et al. 1985). It is possible that this acidic exopolysaccharide is itself a signal or part of a signal recognized by the plant.

The lack of the exopolysaccharide signal only affects the synthesis of a subset of the normal complement of nodule-specific polypeptides. One group of nodule-specific polypeptides (e.g., N-30, glutamine synthetase) is expressed at about the same level in both ineffective nodules and effective nodules. However, a group of nodule-specific polypeptides (e.g., leghemoglobin) is apparently not synthesized at all in empty, ineffective nodules. To a lesser extent, synthesis of this second group of nodule-specific polypeptides also varies between alfalfa cultivars. It is aberrant in ineffective alfalfa nodules induced by R. meliloti mutants lacking nitrogenase (Lang-Unnasch and Ausubel 1985), in ineffective nodules produced by mutant alfalfa genotypes, and in nodules grown at reduced temperature and light (data not shown). In fact, leghemoglobin has been found to be diminished to some extent in all ineffective nodules studied to date (cf. Viands et al. 1979; Bisseling et al. 1983; Lara et al. 1983; Fuller and Verma 1984; Lang-Unnasch and Ausubel 1985).

In contrast to other ineffective nodules, empty nodules elicited by the exopolysaccharide-deficient R. meliloti mutants are unusual in that they appear not to synthesize detectable levels of leghemoglobin or other group two polypeptides. In the case of leghemoglobin, preliminary data suggest that transcription of the leghemoglobin genes is never activated in these nodules. In an effective symbiosis, leghemoglobin apoprotein is normally immunologically detectable 8–9 days after inoculation (Lang-Unnasch and Ausubel 1985). We have shown that leghemoglobin apoprotein is not present at any time from 10 through 38 days after inoculation with an exopolysaccharide-deficient R. meliloti mutant. Similarly, leghemoglobin RNA synthesis begins between 5 and 10 days after inoculation with a wild-type Rhizobium (data not shown) but was not detectable in the ineffective nodules at 17 or 24 days postinoculation. Although we cannot exclude the synthesis of unstable leghemoglobin mRNA that is not translated, it seems more likely that transcription of the leghemoglobin genes is never activated.

Since there are no obvious distinctions between the two groups of nodule-specific polypeptides that we have identified in this study, these groups may represent two distinct mechanisms of regulating a complex developmental system. It appears that expression of nodule-specific genes in the first group can be activated by the Rhizobium, even over a distance, in a single step. Further regulation of the products of these genes might be posttranslational, or it might be accomplished in response to specific negative signals from the bacterium or environment. In contrast, the second group of nodule-specific genes might require a close-range and/or continuous activation signal for their synthesis. The characterization of the plant's response to other well-characterized Rhizobium mutants should help to clarify these possibilities.

ACKNOWLEDGMENTS

We thank John Leigh for the exopolysaccharide-deficient R. meliloti mutants and Ed Tischer and Howard Goodman for the anti-glutamine synthetase immune serum, both provided prior to publication. This work was supported by a grant from Hoechst A.G.

REFERENCES

Auger, S., D. Baulcombe, and D.P.S. Verma. 1979. Sequence complexities of the poly(A)-containing mRNA in uninfected soybean root and the nodule tissue developed due to the infection by Rhizobium. Biochim. Biophys. Acta 563: 496.

Banfalvi, Z., V. Sakanyan, C. Koncz, A. Kiss, I. Dusha, and A. Kondorosi. 1981. Location of nodulation and nitrogen fixation genes on a high molecular weight plasmid of R. meliloti. Mol. Gen. Genet. 184: 318.

Bergmann, H., E. Preddie, and D.P.S. Verma. 1983. Nodulin-35: A subunit of specific uricase (uricase II) induced and localized in the uninfected cells of soybean nodules. EMBO J. 2: 2333.

Bisseling, T., C. Been, J. Klugkist, A. van Kammen, and K. Nadler. 1983. Nodule-specific host proteins in effective and ineffective root nodules of Pisum sativum. EMBO J. 2: 961.

Brewin, N.J., J.G. Robertson, E.A. Wood, B. Wells, A.P. Larkins, G. Galfre, and G.W. Butcher. 1985. Monoclonal antibodies to antigens in the peribacteroid membrane from Rhizobium-induced root nodules of pea cross-react with plasma membranes and Golgi bodies. EMBO J. 4: 605.

Callaham, D.A. and J.G. Torrey. 1981. The structural basis for infection of root hairs of Trifolium repens by Rhizobium. Can. J. Bot. 59: 1647.

Cullimore, J.V. and B.J. Miflin. 1983. Glutamine synthetase from the plant fraction of Phaseolus root nodules. Purification of the mRNA and in vitro synthesis of the enzyme. FEBS Lett. 158: 107.

Cullimore, J.V., M. Lara, P.J. Lea, and B.J. Miflin. 1983. Purification and properties of two forms of glutamine synthetase from the plant fraction of Phaseolus root nodules. Planta 157: 245.

Dart, P. 1977. Infection and development of leguminous nodules. In Treatise on dinitrogen fixation (ed. R.W.F. Hardy and W.S. Silver), p. 367. Wiley, New York.

Finan, T.M., A.M. Hirsch, J.A. Leigh, E. Johansen, G.A. Kuldau, S. Deegan, G.C. Walker, and E.R. Signer. 1985. Symbiotic mutants of Rhizobium meliloti that uncouple plant from bacterial differentiation. Cell 40: 869.

Fuller, F. and D.P.S. Verma. 1984. Appearance and accumulation of nodulin mRNAs and their relationship to the effectiveness of root nodules. Plant Mol. Biol. 3: 21.

Fuller, F., P.W. Kunstner, T. Nguyen, and D.P.S. Verma. 1983. Soybean nodulin genes: Analysis of cDNA clones reveals several major tissue-specific sequences in nitrogen-fixing root nodules. Proc. Natl. Acad. Sci. 80: 2594.

Groat, R.G. and C.P. Vance. 1981. Root nodule enzymes of ammonia assimilation in alfalfa (Medicago sativa L.). Plant Physiol. 67: 1198.

Hirsch, A.M., M. Bang, and F.M. Ausubel. 1983. Ultrastructural analysis of ineffective alfalfa nodules formed by nif::Tn5 mutants of Rhizobium meliloti. J. Bacteriol. 155: 367.

HIRSCH, A.M., D. DRAKE, T.W. JACOBS, and S. LONG. 1985. Nodules are induced on alfalfa roots by *Agrobacterium tumefaciens* and *Rhizobium trifolii* containing small segments of the *Rhizobium meliloti* nodulation region. *J. Bacteriol.* **161**: 223.

HIRSCH, A.M., K.J. WILSON, J.D.G. JONES, M. BANG, V.V. WALKER, and F.M. AUSUBEL. 1984. *Rhizobium meliloti* nodulation genes allow *Agrobacterium tumefaciens* and *Escherichia coli* to form pseudonodules on alfalfa. *J. Bacteriol.* **158**: 1133.

HYLDIG-NIELSEN, J.J., E.O. JENSEN, K. PALUDAN, O. WILBORG, R. GARRETT, Q.P. JORGENSEN, and K.A. MARCKER. 1982. The primary structure of two leghemoglobin genes from soybean. *Nucleic Acids Res.* **10**: 689.

KONDOROSI, E., Z. BANFALVI, and A. KONDOROSI. 1984. Physical and genetic map of a symbiotic region of *Rhizobium meliloti*: Identification of nodulation genes. *Mol. Gen. Genet.* **193**: 445.

JING, Y., A.S. PAAU, and W.J. BRILL. 1982. Leghemoglobins from alfalfa (*Medicago sativa* L. Vernal) root nodules. I. Purification and *in vitro* synthesis of five leghemoglobin components. *Plant Sci. Lett.* **25**: 119.

LANG-UNNASCH, N. and F.M. AUSUBEL. 1985. Nodule-specific polypeptides from effective alfalfa root nodules and from ineffective nodules lacking nitrogenase. *Plant Physiol.* **77**: 833.

LARA, M., H. PORTA, J. PADILLA, J. FOLCH, and F. SANCHEZ. 1984. Heterogeneity of glutamine synthetase polypeptides in *Phaseolus vulgaris* L. *Plant Physiol.* **76**: 1019.

LARA, M., J.V. CULLIMORE, P.J. LEA, B.J. MIFLIN, A.W.B. JOHNSTON, and J.W. LAMB. 1983. Appearance of a novel form of plant glutamine synthetase during nodule development in *Phaseolus vulgaris* L. *Planta* **157**: 254.

LEGOCKI, R.P. and D.P.S. VERMA. 1980. Identification of "nodule-specific" host proteins (nodulins) involved in the development of the *Rhizobium*-legume symbiosis. *Cell* **20**: 153.

LEIGH, J.A., E.R. SIGNER, and G.C. WALKER. 1985. Exopolysaccharide-deficient mutants of *Rhizobium meliloti* that form ineffective nodules. *Proc. Natl. Acad. Sci.* **82**: 6231.

LONG, S.R. 1984. Genetics of *Rhizobium* nodulation. In *Plant-microbe interactions* (ed. T. Kosuge and E.W. Nester), p. 256. Macmillan, New York.

LONG, S.R., W.J. BUIKEMA, and F.M. AUSUBEL. 1982. Cloning of *Rhizobium meliloti* nodulation genes by direct complementation of nod⁻ mutants. *Nature* **298**: 485.

MARVEL, D., G. KULDAU, A. HIRSCH, E. RICHARDS, J.G. TORREY, and F.M. AUSUBEL. 1985. Conservation of nodulation genes between *Rhizobium meliloti* and a slow growing *Rhizobium* strain that nodulates a non-legume host. *Proc. Natl. Acad. Sci.* **82**: 5841.

MEADE, H.M., S.R. LONG, G.B. RUVKUN, S.E. BROWN, and F.M. AUSUBEL. 1982. Physical and genetic characterization of symbiotic and auxotrophic mutants of *Rhizobium meliloti* induced by transposon Tn5 mutagenesis. *J. Bacteriol.* **149**: 114.

O'FARRELL, P.H. 1974. High resolution two-dimensional electrophoresis of proteins. *J. Biol. Chem.* **250**: 4007.

RUVKUN, G.B., V. SUNDARESAN, and F.M. AUSUBEL. 1982. Directed transposon Tn5 mutagenesis and complementation analysis of *R. meliloti* symbiotic nitrogen fixation genes. *Cell* **29**: 551.

SCHMIDT, J., M. JOHN, E. KONDOROSI, A. KONDOROSI, U. WIENEKE, G. SCHRODER, J. SCHRODER, and J. SCHELL. 1984. Mapping of the protein-coding regions of *Rhizobium meliloti* common nodulation genes. *EMBO J.* **3**: 1705.

SULLIVAN, D., N. BRISSON, B. GOODCHILD, and D.P.S. VERMA. 1981. Molecular cloning and organization of two leghemoglobin genomic sequences of soybean. *Nature* **289**: 516.

SZETO, W.W., J.L. ZIMMERMAN, V. SUNDARESAN, and F.M. AUSUBEL. 1984. A *Rhizobium meliloti* symbiotic regulatory gene. *Cell* **36**: 1035.

TOROK, I., E. KONDOROSI, T. STEPKOWSKI, J. POSFAI, and A. KONDOROSI. 1984. Nucleotide sequence of *Rhizobium meliloti* nodulation genes. *Nucleic Acids Res.* **12**: 9509.

TRUCHET, G., C. ROSENBERG, J. VASSE, J.-S. JULLIOT, S. CAMUT, and J. DENARIE. 1984. Transfer of *Rhizobium meliloti* pSym genes into *Agrobacterium tumefaciens*: Host-specific nodulation by atypical infection. *J. Bacteriol.* **157**: 134.

VAN KAMMEN, A. 1984. Plant genes involved in nodulation and symbiosis. In *Advances in nitrogen fixation research* (ed. C. Veeger and W.E. Newton), p. 587. Martinus Nijhoff/Dr. W. Junk, Boston.

VERMA, D.P.S. and S.R. LONG. 1983. Molecular biology of the *Rhizobium* legume symbiosis. In *Intracellular symbiosis* (ed. K. Jeon), p. 211. Academic Press, New York.

VIANDS, D.R., C.P. VANCE, G.H. HEICHEL, and D.K. BARNES. 1979. An ineffective nitrogen fixation trait in alfalfa. *Crop Sci.* **19**: 905.

VINCENT, J.M. 1980. Factors controlling the legume-*Rhizobium* symbiosis. In *Nitrogen fixation*, (ed. W.E. Newton and W.H. Orme-Johnson), vol. 2, p. 103. University Park Press, Baltimore.

WITTENBERG, J.B., F.J. BERGERSEN, C.A. APPLEBY, and G.L. TURNER. 1974. Facilitated oxygen diffusion, the role of leghemoglobin in nitrogen fixation by bacteroids isolated from soybean root nodules. *J. Biol. Chem.* **249**: 4057.

WONG, C.H., C.E. PANKHURST, A. KONDOROSI, and W.J. BROUGHTON. 1983. Morphology of root nodules and nodule-like structures formed by *Rhizobium* and *Agrobacterium* strains containing a *Rhizobium meliloti* megaplasmid. *J. Cell Biol.* **97**: 787.

Master Regulatory Loci in Yeast and Lambda

I. HERSKOWITZ

Department of Biochemistry and Biophysics, University of California, San Francisco, California 94143

There are two extreme starting points in studying the molecular basis of differentiation: One starting point is the specialized cell. In this approach, specializations are documented, and the goal is to work backward in time to determine how such differences arise. Another starting point is the fertilized egg. In this approach, the events occurring after fertilization (e.g., cleavage) are documented. The strategy is to work forward in time and determine how the early events ultimately lead to cellular specializations.

Regulatory proteins and genes responsible for cell specialization have been identified in yeast (Herskowitz 1983; Johnson and Herskowitz 1985). Their study provides an opportunity to look at the mechanisms responsible for maintenance of a stably differentiated eukaryotic cell. As discussed below, it is our premise that understanding the molecular basis of how cells *become* different requires understanding the ways in which regulatory proteins are themselves regulated. A further look at how regulatory proteins are regulated comes from studies of the growth decision of bacteriophage λ, which involves integration of multiple signals and takes us into the innermost recesses of cell physiology.

A Formal Description of Analysis of Cell-type Specialization

We first present a formal description of the "working backwards" approach, because it is a starting point for many molecular studies of differentiated cells and because below we describe yeast cell-type determination and the λ growth decision in this context. Specialized cell types, such as liver cells and brain cells or root cells and leaf cells, differ from each other in various observable ways, such as morphology and growth properties. They presumably differ from each other in their proteins. An initial goal is to determine why each type of cell exhibits its distinctive set of proteins. Do root cells contain precursors of the proteins unique to leaf cells? Do they contain an unusable form of the transcripts (unprocessed or unstable) or no transcripts at all for the proteins unique to leaf cells? Hybridization assays for stable RNA species show that in many cases the transcripts for the specialized proteins of one cell type are not produced in other cell types (see, e.g., Derman et al. 1981; Walker et al. 1983). The next step is to determine the mechanisms responsible for expression of a set of genes in one cell type and not in the other. Differences in RNA levels can in principle result from regulation of transcription itself or of transcript stability (perhaps governed by processing) (Darnell 1982). Whatever the molecular basis, the goal is to identify the proteins (or other agents) responsible for mediating cell-specific expression. As shown below, regulatory proteins responsible for cell-type-specific expression have been demonstrated in yeast.

The ultimate basis for differences between cells, we argue, results from the functioning of the regulatory agents. What then regulates activity of these agents? It is our premise that such key regulators exist and that modulation of their activity is responsible for cell diversity. One can imagine ways in which modulation might occur, for example, via regulating protein synthesis or stability, which might occur in response to extrinsic signals (such as hormones or cell contact) or intrinsic signals (such as a cell-division-counting mechanism). One of the main points of this paper is to give concrete examples of ways in which key regulators are regulated.

A Master Regulatory Locus Governing Cell Specialization in Yeast

Cell-specific differences and differentially regulated gene sets. In our studies of yeast, we have not made any assumptions about the molecular basis for cell specialization. Our starting point was simply that the three types of cells, **a**, α, and **a**/α, exhibit different properties (Roman et al. 1955). The fundamental properties are that **a** and α cells are specialized for mating, whereas **a**/α cells are specialized for meiosis and sporulation (Table 1). These macroscopic behaviors ultimately result from specialized proteins produced by the cell types. We have not identified such proteins by searching for proteins differences, for example, by two-dimensional gels or screening with monoclonal antibodies. Nor have we identified the genes coding for such proteins by differential hybridization methods (although such methods would have worked). Instead, we have studied mutants that affect the properties of

Table 1. Yeast Cell Types

Cell type	Mating	Pheromone production	Pheromone response	Meiosis and sporulation
a, **a**/**a**	with α	**a**-factor	α-factor	no
α, α/α	with **a**	α-factor	**a**-factor	no
a/α	no	no	no	yes

a/**a**, α/α, and **a**/α cells are diploids that differ only at the mating-type locus. (Their genotypes are *MAT***a**/*MAT***a**, *MAT*α/*MAT*α, and *MAT***a**/*MAT*α, respectively.)

the different yeast cell types and only later examined synthesis of the gene products. These mutants define approximately 20 genes termed *STE* genes (MacKay and Manney 1974; for review, see Klar et al. 1984), many (but not all) of which identify proteins that are found in one cell type but not in another. We can consider as examples the two genes *STE3* and *STE6*. *STE3* is necessary for mating by α cells but not for mating by **a** cells: **a** cells defective in the *STE3* gene mate normally (MacKay and Manney 1974). These observations require that *STE3* be expressed in α cells; it need not be expressed in **a** cells. Likewise, the *STE6* gene is required for mating only by **a** cells (J. Rine, cited in Wilson and Herskowitz 1984). Thus, **a** cells must express *STE6*; α cells need not express this gene. Because **a**/α cells do not mate, neither *STE3* nor *STE6* need be expressed in these cells. The mating properties of mutants defective in *STE3* or *STE6* are reflected in their likely roles. *STE3*, which is needed only by α cells, is likely to be a component of the receptor to **a**-factor, the mating factor produced by **a** cells (G.F. Sprague, Jr., pers. comm.). *STE6*, which is needed only by **a** cells, is necessary for synthesis of **a**-factor (Wilson and Herskowitz 1984).

We have cloned *STE3* and *STE6* in order to assay their transcripts in the different cell types. The results are simple: The *STE3* and *STE6* transcripts (assayed as stable RNA species by Northern hybridization) are present only in the cells in which the genes are needed (Table 2) (Sprague et al. 1983; Wilson and Herskowitz 1984). There are at least two genes in addition to *STE3* that have the α-specific expression pattern and thus are members of the α-specific gene set (Table 3). These are the duplicate genes, *MFα1* and *MFα2*, coding for the α mating factor (S. Fields, pers. comm.). There are four genes in addition to *STE6* whose transcripts are found only in **a** cells: *STE2*, *BAR1*, *MFA1*, and *MFA2* (see Johnson and Herskowitz 1985; V. MacKay; T. Manney; T. Brake; all pers. comm.). These genes constitute the **a**-specific gene set (Table 3). Our strategy of studying regulation of mutants with **a**- and α-specific defects thus identifies differentially regulated gene sets.

It should be noted that not all mutants that affect one cell type and not the other define differentially regulated genes: *STE13* is required for mating only by α cells but is expressed in all cell types (G.F. Sprague, Jr. and I. Herskowitz, unpubl.; see also Julius et al. 1983). Our strategy also has the limitation inherent in any mutant hunt: It fails to identify mutants in redundant genes. The two α-factor genes and the two **a**-factor genes were identified by a different route (Kurjan and Herskowitz 1982; Singh et al. 1983; Kurjan 1985; A. Brake et al., pers. comm.).

Expression of the *HO* gene differs from the **a**-specific and α-specific genes and defines a set of "haploid-specific" genes: Its transcript is present in both **a** and α cells but is absent in **a**/α cells (Table 3) (Jensen et al. 1983). A number of other genes (*STE5*, *RME1*, and *KAR1*) and the Ty1 element have a similar expression pattern (Table 3).

The mating-type locus. The mating-type locus governs both the gross phenotypes of yeast cells (mating and sporulation) and the molecular phenotypes, the cell-type-specific expression of genes just described. There are two alleles of *MAT*: *MATa* confers the **a** cell type and *MAT*α confers the α cell type. Cells with both alleles (*MATa/MAT*α), which result naturally by mating of the haploids, exhibit the **a**/α cell type (Roman et al. 1955). Given two facts—(1) that *MATa* and *MAT*α confer **a** and α cell types and (2) that there exist **a**-specific genes (such as *STE6*) and α-specific genes (such as *STE3*)—there are a number of specific molecular models that might be proposed to account for the way in which the mating-type locus controls expression of the specialized gene sets.

We first consider two models that are consistent with these observations (but which turn out not to be correct). In model I, each *MAT* allele codes for a positive regulator of expression. Thus, *MAT*α codes for an activator that turns on expression of α-specific genes in α cells. Similarly, *MATa* codes for a positive regulator that turns on expression of **a**-specific genes in **a** cells. Each type of cell thus expresses its distinctive set of genes because of the presence of a specific activator of that gene set. In model II, each *MAT* allele codes for a negative regulator of expression. *MAT*α codes for a negative regulator of expression of **a**-specific genes, and *MATa* codes for a negative regulator of α-specific genes. Although both of these models are consistent with the observations, the true situation is different.

*MAT*α codes for two regulatory activities (Fig. 1). One (α1, coded by *MAT*α1) is a positive regulator of expression of the α-specific genes. The second (α2, coded by *MAT*α2) is a negative regulator of expression of the **a**-specific genes. Thus, in α cells, the set of α-specific genes is turned on, and the set of **a**-specific genes is turned off. In **a** cells, the set of α-specific genes is silent because α1 is absent, and the set of **a**-specific genes is expressed because α2 is absent. (The role of the **a**1 product, coded by the *MATa* allele, is discussed below.) This hypothesis was derived from genetic and physiological studies (Strathern et al. 1981). Molecular analysis to determine the level at which the mating-type locus controls expression of the unlinked genes was carried out by Sprague et al. (1983) and by Wilson and Herskowitz (1984). These studies show that synthesis of the *STE3* transcript requires the α1 product and that synthesis of the *STE6* transcript is turned off by the α2 product.

Table 2. Differentially Expressed Sets of Genes

Cell type	α-specific genes	**a**-specific genes	Haploid-specific genes
α	+	−	+
a	−	+	+
a/α	−	−	−

+ or − indicates production of transcripts in α, **a**, or **a**/α cells. Members of the different gene sets are listed in Table 3. α/α diploids behave the same as α haploids; **a**/**a** diploids behave the same as **a** haploids.

Table 3. Members of the Cell-type-specific Gene Sets and Their Functions

α-Specific gene set (expressed only in α cells; activated by α1)
STE3	component of receptor to **a**-factor[a]
MFα1	structural gene for α-factor precursor
MFα2	structural gene for α-factor precursor

a-Specific gene set (expressed only in **a** cells; repressed by α2)
STE2	component of receptor to α-factor
STE6	synthesis of **a**-factor (processing enzyme?)
BAR1	degradation of α-factor (probably a protease)
MFA1	structural gene for **a**-factor precursor
MFA2	structural gene for **a**-factor precursor

Haploid-specific gene set (expressed in **a** and α cells but not in **a**/α cells; repressed by **a**1-α2)
STE5	necessary for mating by both **a** and α cells (function unknown)
HO	site-specific endonuclease that initiates mating-type switching
KAR1	component of the mitotic spindle, necessary for karyogamy
RME1	regulator of meiosis
Ty1	transposable element
MATα1[b]	structural gene for α1

[a]References are given in Johnson and Herskowitz (1985) and in the text.
[b]The MATα1 gene, although not expressed in **a** cells, can be considered to be a member of the haploid-specific gene set because it has the other defining characteristic of this set—its expression is repressed by **a**1-α2.

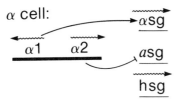

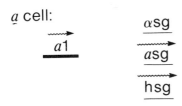

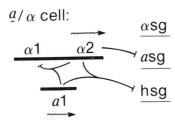

Figure 1. Control of expression of cell-type-specific gene sets by the mating-type locus. The mating-type locus is drawn at the left, and the cell-type-specific gene sets (αsg, α-specific genes; **a**sg, **a**-specific genes; hsg, haploid-specific genes) are drawn at the right. Wavy arrows indicate transcription. Pointed arrowheads indicate stimulation of gene expression; lines with blunt ends indicate inhibition of gene expression. (Adapted from Strathern et al. 1981.)

The studies above provide a formal description of α1 and α2, but they do not indicate the molecular mechanism by which α1 and α2 work or whether these products act directly on the regulated genes. The argument that α2 is a negative regulator of **a**-specific genes is deduced from the observations that cells with a functional MATα2 gene do not express **a**-specific genes, whereas cells lacking MATα2 function (due to a recessive mutation in the gene, or in a MA**T**a strain) do express **a**-specific genes. There are several molecular mechanisms by which a negative regulator might act. α2 might inhibit expression of the **a**-specific genes by inhibiting synthesis or activity of an activator protein that is necessary to turn on expression of the **a**-specific genes. There is a precedent for this type of negative regulation in yeast (Oshima 1982). Another mechanism is that the α2 protein might cause instability of **a**-specific gene transcripts in α cells. A final possibility and the one for which we have obtained strong evidence is that the α2 protein is a DNA-binding protein that acts directly on the members of the **a**-specific gene set to turn off their synthesis.

The α2 protein is a site-specific DNA-binding protein; it binds to a 32-bp site located 130 bp upstream of the transcript start point of the STE6 gene (Johnson and Herskowitz 1985; K.L. Wilson and I. Herskowitz, in prep.). This site is sufficient to place a gene not ordinarily under control by α2 under its control (Johnson and Herskowitz 1985). A nearly identical sequence (which we term the **a**-specific gene operator) is located upstream of all of the other members of the **a**-specific gene set (Johnson and Herskowitz 1985; Miller et al. 1985; K.L. Wilson and I. Herskowitz, in prep.). Our studies show that α2 is responsible for governing expression of an entire set of specialized genes by recognizing this sequence and binding to it to prevent expression.

We do not yet know whether the α1 protein binds to the DNA of genes whose expression it activates. Recent work indicates that α1 is not sufficient to activate expression of these genes: Transcription of the two α-factor genes also requires the *STE7*, *STE11*, and *STE12* genes (Fields and Herskowitz 1985); *STE7*, *STE11*, and *STE12* are also required for expression of the **a**-specific gene set. These proteins may define a requirement for multiple transcription factors.

A novel regulatory activity in a/α cells. We have seen that α cells have two regulatory activities coded by *MAT*α, α1 and α2. a/α cells exhibit a third regulatory activity, **a**1-α2. This activity, which requires both the **a**1 product of the **a** mating-type locus (Kassir and Simchen 1976) and the α2 product of *MAT*α, turns off expression of a wide variety of genes that are otherwise expressed in **a** and α cells. One such gene is *HO*, whose transcript is produced in both **a** and α cells but not in a/α cells (Jensen et al. 1983). Thus, the set of genes termed haploid-specific (Tables 2 and 3) is regulated by this activity. The molecular nature of the **a**1-α2 activity is not presently known. Given that α2 is a DNA-binding protein (Johnson and Herskowitz 1985), one can propose that **a**1-α2 is also a DNA-binding protein (Herskowitz 1982). **a**1 and α2 polypeptide chains might associate to form a species that is able to bind to a haploid-specific gene operator. Very different types of models can also be imagined. Because the *MAT***a**1 gene has two introns (Miller 1984), splicing of its RNA— perhaps influenced by α2— might generate different forms of the **a**1 product. Whatever the molecular explanation, it is clear that a novel regulatory activity results when the *MAT***a**1 gene is introduced into an α cell.

Although **a**1-α2 *inhibits* expression of a variety of genes, it *activates* meiosis and sporulation. Does this apparent dual action of **a**1-α2 reflect a single protein that can act as both a negative regulator and positive regulator, as is the case for the λcI protein (Ptashne et al. 1980)? Rine et al. (1981a) proposed that **a**1-α2 stimulates meiosis and sporulation by another negative action of **a**1-α2: It inhibits synthesis of an inhibitor of meiosis and sporulation encoded by the *RME1* gene. Recent studies show that the *RME1* gene is indeed negatively regulated by **a**1-α2 and that the presence of the *RME1* gene product inhibits meiosis and sporulation (A. Mitchell and I. Herskowitz, in prep.). Thus, **a**1-α2 stimulates meiosis and sporulation not by activating expression of the genes that are directly involved in these processes but rather by inhibiting synthesis of an inhibitor. The *RME1* product might carry out its inhibition by being a repressor, protease, or protein kinase that acts on a target gene or protein necessary for initiating meiosis.

Regulation of the Regulators Encoded by *MAT*

The above discussion documents that the mating-type locus is the central determinant governing cell specialization. As discussed above, regulation of such a master regulatory locus might itself play an important role in the course of development. For example, early events might influence the activity of the master regulatory locus so that cell specializations result. Having established that the cell-type-determining locus in yeast is the mating-type locus, we now discuss the ways in which it is regulated.

Turn off of **MAT**α1 *expression in a/α cells.* a/α cells do not mate and do not express either **a**-specific genes or α-specific genes. One of the mechanisms for turning off expression of these specialized gene sets in the a/α cell results from regulation of expression of the mating-type locus itself. In particular, the **a**1-α2 activity described above inhibits synthesis of the α1 product by turning off synthesis of the *MAT*α1 transcript (Klar et al. 1981; Nasmyth et al. 1981). Because α1 is absent, a/α cells do not express the α-specific gene set (see also Ammerer et al. 1985). A detailed analysis of the sites upstream of the *MAT*α1 gene required for this inhibition, presumably a recognition site for **a**1-α2, was described by Siliciano and Tatchell (1984). **a**1-α2 also inhibits expression of the *MAT*α2 gene approximately fivefold (Nasmyth et al. 1981; Hall et al. 1984). The functional significance of this inhibition is not known. These findings demonstrate that expression of the mating-type locus is regulated by products of the mating-type locus.

Changing the information at **MAT**. We have seen that the presence of the α allele at *MAT* causes a cell to exhibit the behaviors of an α cell; the presence of the **a** allele at *MAT* causes a cell to exhibit the behaviors of an **a** cell. Some yeast strains (those carrying the *HO* gene; Kostriken and Heffron 1984) have the remarkable ability to switch efficiently from **a** to α and from α to **a** (for review, see Herskowitz and Oshima 1981). They do this switching ("mating-type interconversion") by removing the allele at *MAT* and replacing it with the other allele, which is stored in the genome in a silent form. We have referred to movable segments of coding information as genetic cassettes (Hicks et al. 1977), because they are blocks of information expressed in one genomic position (in this case at *MAT*, analogous to the playback head of a tape recorder) and silent at another (in this case at two loci, one harboring a silent α cassette and another harboring a silent **a** cassette). Thus, yeast can switch from one specialized cell type to another by simply replacing the master regulatory locus.

Regulating the master regulators that govern cell specialization can generate a specific cell lineage. Studies of yeast cells capable of the high-frequency switching reveal a specific pattern of cell types in successive cell divisions (see Fig. 2) (Hicks and Herskowitz 1976; Strathern and Herskowitz 1979). We have established that cells in this population differ in their capacity to switch mating types: Mother cells are able to do so, whereas daughter cells cannot (Strathern and Herskowitz 1979; for an exhausting discussion, see Rine et

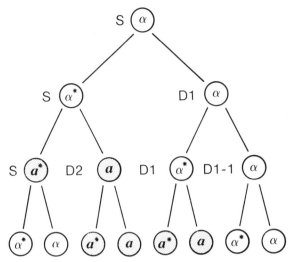

Figure 2. A mitotic cell lineage of yeast cells that are able to switch mating type. The initial cell, S, is a spore that is allowed to germinate and then undergo one cell-division cycle to yield two cells, S (the spore cell) and D1 (the spore's first daughter), both of which have the α mating type. In the next cell division, the S cell gives rise to two cells that have a mating type. This change in cell type occurs because the original mating-type locus (MATα in this case) has been removed and replaced with MATa, as a result of a genetic rearrangement. Asterisks indicate cells that are capable of switching mating types in their next cell divisions. As described in the text, only cells that have previously budded (cells that have been mothers) are competent. (Adapted from Strathern and Herskowitz 1979.)

progeny cells that are phenotypically **a** (Fig. 3). A cell lineage is thereby produced in which mother cells give rise to "specialized" **a** cells. In this lineage, one line of daughter cells maintains the α phenotype and thereby behaves as a stem cell line.

A Regulator in λ Responsible for the Decision between Two Different Modes of Growth

Studies of bacteriophage λ are instructive in thinking about the ways in which molecular decisions are made. As for cell-type determination in yeast, understanding the growth decision of bacteriophage λ ultimately requires understanding how a key regulatory protein is regulated. We argue that the phage cII protein acts as a molecular antenna that is used to assess the status of phage infection. We summarize some of the different types of inputs that it receives and how its outputs lead to a decisive commitment to one mode of growth or another (for review, see Herskowitz and Hagen 1980; Herskowitz and Banuett 1984). We introduce specific questions concerning bacteriophage λ by referring again to the general scheme of cell specialization presented above.

Outcome-specific differences. In the case of bacteriophage λ, we describe not specialized cell types but rather two different outcomes resulting from λ infection. Because λ is a temperate bacterial virus, infection does not always lead to cell death and production of phage (the lytic response). It may also lead to a re-

al. 1981b). Why mother cells but not daughter cells exhibit this potential is obviously intriguing. Studies by Nasmyth (1983) and Jensen and Herskowitz (1984) indicate that daughter cells fail to switch simply because they do not express the *HO* gene. Regulators of expression of the *HO* gene have been identified. There are five activator genes necessary for production of the *HO* transcript: *SWI1*, *SWI2*, *SWI3*, *SWI4*, and *SWI5* (Stern et al. 1984). Three genes (*SIN1*, *SIN2*, and *SIN3*) that may code for negative regulators of *HO* expression have been identified recently (P. Sternberg and M. Stern, pers. comm.). Distribution of these regulatory proteins between mother and daughter cells may hold the molecular answer to the difference between these cells in expressing *HO* and thus in switching cell types. We will return to this topic below.

Although the type of cell switching seen here for yeast involves a genetic rearrangement, the principle that regulation of regulators can generate specific cell lineages is a general one. To make this point explicit, we can produce a cell lineage in which the change in a cell's phenotype results not from gene rearrangement but simply from inactivating *MATα*. Inactivation of *MATα* causes the cell to be defective in both α1 and α2; such strains have the mating behavior of **a** cells. Let us imagine a gene that is expressed in precisely the same manner as *HO* and that inactivates *MATα*. Beginning with an α strain, inactivation of *MAT* in cell divisions of particular cells (mother cells) gives rise to

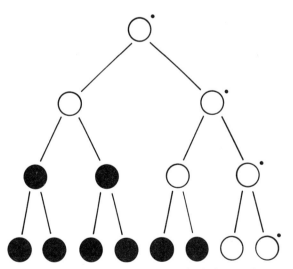

Figure 3. Mitotic cell lineage of a hypothetical yeast whose α mating-type locus is subject to inactivation. Cells derived from an initial spore cell are drawn as in Fig. 2 (mother cells to the left and daughter cells to the right). The initial cell has a functional mating-type locus (*MATα*) and exhibits the α cell type (○). In this hypothetical yeast, *MAT* is inactivated in cell divisions of mother cells, which yields cells that exhibit an **a** cell phenotype (●). (Although the molecular mechanism by which *MAT* is inactivated is not important here, yeast does have the ability to inactivate such blocks of genetic information; see Herskowitz and Oshima 1981; Klar et al. 1984.) The α cells marked with a dot comprise a stem cell lineage (see text).

sponse in which the infected cell survives (the lysogenic response). Survival does not occur because the phage takes pity on the host. Rather, the phage DNA becomes incorporated into the bacterial genome and is maintained in a silent (repressed) form. Thus, the phage replicates autonomously in the lytic pathway of viral development and as part of the bacterial genome in the lysogenic pathway.

These two outcomes reflect synthesis of two specialized sets of proteins after infection (Table 4). The lysogenic response requires production of the site-specific recombination enzyme (integrase) necessary for integrating the phage DNA into the host DNA and production of a repressor (the cI protein) that maintains the prophage in a repressed state. The lytic response requires synthesis of the proteins involved in phage morphogenesis (a group of 20 or so proteins for head and tail formation) as well as cell lysis (endolysin and others).

The regulatory genes responsible for synthesis of the specialized sets of proteins have been identified, and, as in the case of yeast, they act by regulating transcription of the two gene sets (Fig. 4). The phage Q gene product activates transcription of the large set of proteins essential for lytic development, all of which are read from the same promoter. Q is an antiterminator of transcription that allows transcription initiated upstream of the first gene to continue into the entire gene cluster (Forbes and Herskowitz 1982; Grayhack et al. 1985). The regulatory protein responsible for synthesis of the lysogenic proteins is the phage cII protein, which activates transcription of genes int and cI. cII stimulates transcription initiation by bacterial RNA polymerase at two promoters (pI and pRE) by binding to a site in the -35 region of these promoters (Ho et al. 1983; Wulff et al. 1984). cII also stimulates transcription from another promoter, termed paq, located within the phage Q gene (Hoopes and McClure 1985). Transcription from this promoter results in inhibition of lytic growth, probably by inhibiting synthesis of Q. cII thus stimulates lysogeny both by activating synthesis of the lysogenic set of proteins as well as by inhibiting lytic development.

cII as a decision-making protein. In the case of cell-type determination in yeast, the argument that the mating-type locus is the master regulatory locus for cell specialization came originally from the simple obser-

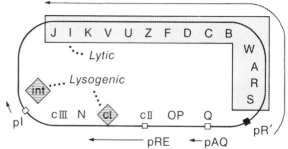

Figure 4. Gene sets of bacteriophage λ necessary for lytic and lysogenic pathways of growth. The DNA of bacteriophage λ is drawn after it has circularized. *int* and *cI* are genes necessary for the lysogenic response. Transcription initiated from promoters pI, pRE, and paq is stimulated by the phage cII protein, which binds to the sites indicated (□). Genes S through J are necessary for the lytic response. (S-R) Cell lysis; (A-F) head genes; (Z-J) tail genes. Transcription initiated from promoter pR' ordinarily terminates after 199 nucleotides (before it reaches the S gene). In the presence of the phage Q protein, which recognizes the site indicated (■), transcription continues into the entire cluster of genes S-J.

vation that there are two natural alleles at this locus, *MAT*a and *MAT*α. There are no analogous natural alleles involved in the decision of bacteriophage λ between lysis and lysogeny. Instead, analysis of a variety of mutants—both of the phage and of the host—has created a picture in which the cII protein is a key regulator that governs the decision between lysis and lysogeny: a high cII level leads to a lysogenic response, and a low cII level leads to a lytic response. Important observations that led to this view are as follows. First, there are two types of mutations of the *cII* gene: Null alleles result in a strong preference toward the lytic response (Kaiser 1957), and a dominant allele (*can1*) leads to a preference toward the lysogenic response (Jones and Herskowitz 1978). Thus, *cII* is a gene in which alleles with opposite effects on gene activity lead to opposite effects on a developmental choice. The use of such an argument for identifying decision-making genes has been developed by Horvitz and Hodgkin and colleagues in studies of nematode genes that govern cell lineages (Sternberg and Horvitz 1984) and sex determination (Hodgkin 1983). Second, a variety of mutations outside the *cII* gene—in both phage and host genes—that affect the efficiency of lysogeny all affect the level of cII. For example, mutants that exhibit a

Table 4. Outcome-specific Differences in Gene Expression in Cells Infected with Bacteriophage λ

	Cell lysis	Particle morphogenesis	Repression	Integration
Genes	S, R, Rz	A-F (head); Z-J (tail)	cI	int
Outcome				
lytic	+	+	−	−
lysogenic	−	−	+	+

+ or − indicates production of transcripts resulting from the lytic or lysogenic outcomes.

high frequency of lysogeny (*hfl* mutants) lead to increased levels of cII after phage infection (Epp 1978; Banuett et al. 1985).

We incorporate this role of cII into a view of infection by λ in which an explicitly programmed sequence of events leads to a metastable state that then becomes directed by cII toward lysis or lysogeny (Herskowitz and Hagen 1980). Immediately after infection, the first set of phage genes (including activator gene *N*) is transcribed. The N product next allows high-level expression of the second set of phage genes, which includes *c*II, *c*III, and genes for initiation of phage DNA replication. This is viewed as the decision-making period, in which the level of cII determines the outcome of infection. Execution of the decision then ensues: If cII levels are high, then cI and Int are synthesized, and lytic development is inhibited, both by cI and by cII-dependent inhibition of Q synthesis. If cII levels are not adequate, then Q synthesis ensues (to activate the morphogenetic and lysis proteins); synthesis of cI protein and Int does not occur.

Regulation of cII Activity by Phage and Host Factors

The factors that determine the level of cII activity during the uncommitted phase of growth and that lead to a commitment involve a variety of phage, host, and environmental factors (summarized in Fig. 5). The way in which the level of cII is affected introduces another molecular mechanism by which decision-making proteins are regulated—control of their stability.

Stability of cII. cII protein is unstable, having a half-life of approximately 1.5 minutes (Epp 1978; Hoyt et al. 1982). Mutations in the bacterial *hfl* genes, *hflA* (Belfort and Wulff 1971) and *hflB* (Banuett et al. 1985), result in enhanced lysogenization by the phage. They apparently do so by increasing the half-life of the cII protein to about twice its normal length and thereby increasing the level of cII. Thus, the wild-type *hfl* genes are responsible for specific degradation of the cII protein. We do not know at present whether they code for cII-specific proteases or whether their stimulation of cII degradation is indirect. Earlier, we discussed various possible mechanisms by which a negative regulator can act (as a repressor, a protease that inhibits an activator, etc.). We note that the Hfl proteins are negative regulators that are not repressors.

The phage *c*III gene is necessary for efficient lysogeny; it also appears to act by increasing stability of the cII protein (Epp 1978; Hoyt et al. 1982). cIII protein has been proposed (Gautsch and Wulff 1974) to be an antagonist of the Hfl proteins (because it is not needed in the absence of Hfl). Given that the Hfl proteins are involved in proteolysis of cII, then we can view cIII as a protease inhibitor. It is striking that cIII is quite small (54 amino acids; Knight and Echols 1983), which fits nicely with such a role.

Factors influencing the decision. In many cases, regulatory proteins are involved in responding to environmental signals. Well-known examples include monitoring nutrients in the growth medium and monitoring DNA damage (Walker 1984). The lysis-lysogeny decision is also influenced by external factors. In particular, the lysogenic response is favored over the lytic response under conditions of high multiplicity of infection and under conditions of nutritional starvation. This monitoring appears to occur via cII (summarized in Fig. 5). We next discuss how the multiplicity of infection and nutritional status of the infected cell are monitored and conclude with a discussion of ways in which commitment to lysis or lysogeny is made decisive.

The phage cIII protein appears to be responsible for assessing multiplicity of infection (Reichardt 1975). The level of cIII produced after infection is apparently limited by the number of copies of the *c*III gene. Thus, at high multiplicity of infection, a high level of cIII protein is produced, and Hfl is efficiently inhibited. Consequently, the level of cII is high, and lysogeny ensues. In contrast, cIII levels are low at low multiplicity of infection; hence, Hfl is relatively uninhibited and reduces cII levels. The manner in which the nutritional status of the infected cell is monitored is not fully understood but apparently involves cAMP and the *hfl* genes. In particular, the requirement of cAMP for efficient lysogenization is relieved in the absence of Hfl (Belfort and Wulff 1974). We have thus suggested that cAMP exerts a negative effect on Hfl in some manner (see Fig. 5). We know only that cAMP does not affect expression of the *hflA* gene (F. Banuett, pers. comm.); hence, cAMP might ultimately cause production of an inhibitor of Hfl *activity*, a bacterial analog of cIII.

Amplifying small differences to make a clean decision. One particularly intriguing aspect of the lysis-lysogeny decision is the question of how a "clean" decision can be made, so that an infected cell will yield either a successful lytic response or a successful lysogenic response. Failure to make a clean decision might

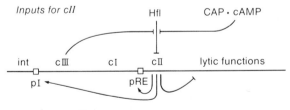

Figure 5. The central role of cII in the lysis-lysogeny decision. Lines with pointed arrowheads indicate stimulation of activity of targets (in this case, transcription initiation). Lines with blunt arrowheads indicate inhibitory interactions. Inhibition of lytic functions probably results both from an inhibition of DNA replication and from stimulation of transcription from p_{aq} (to inhibit synthesis of Q). (cAMP) Cyclic AMP; (CAP) catabolite gene activator protein. Additional details are given in the text. (Modified from Herskowitz and Hagen 1980.)

yield a mixed response in which, in the extreme, cells might establish repression during lytic development and thus result in a severely reduced yield of phage. The problem at a molecular level is one of how the various inputs that impinge on the cII protein are summed so that the cII activity is either above or below a threshold for activating the lysogenic response. In other words — How can infected cells that are nearly identical with respect to production of the cII polypeptide yield two different responses? There are a variety of molecular ways by which small initial differences in cII monomer levels can lead to much larger differences in cII activity. First, because cII is a tetramer (Ho and Rosenberg 1982), its activity is acutely dependent on the monomer concentration. Small initial differences in cII monomer levels can in principle be amplified into larger differences in cII activity if the tetramer is more resistant to proteolysis than the monomer or if the active form of cII can stimulate its own synthesis.

Another mechanism for ensuring commitment to one pathway or the other is to have components of the two pathways be mutually antagonistic (for discussion, see Monod and Jacob 1962; Johnson et al. 1981). An excellent example of such an antagonism is the well-characterized molecular switch seen in lysogens carrying a defective λ prophage (Eisen et al. 1970). These lysogens can produce either of two repressors, and these repressors inhibit each other's synthesis. Consequently, cells producing one repressor will remain producing that repressor. Does the lysis-lysogeny decision involve such mutual antagonisms? It would certainly make biological sense for cII and Q to be antagonists of each other. We have noted above the argument that cII inhibits synthesis of Q. cII also exerts an inhibition of DNA replication by an unknown mechanism (Smith and Levine 1964; see Schmeissner et al. 1980). To our knowledge, there is no evidence that a specific factor promoting lytic growth (such as Q) antagonizes cII. It would seem reasonable for DNA replication to influence the lysis-lysogeny decision. DNA replication per se does not necessarily reflect a commitment to the lytic response (inferred because the efficiency of lysogeny is enhanced by phage DNA replication). However, the *extent* of DNA replication might participate in setting the level of cII necessary for a lysogenic response. Thus, a high level of replication would necessitate a high cII level for lysogeny.

Regulation of Regulators during Development in Multicellular Eukaryotes

The most rudimentary cellular specialization that occurs during development results from cleavage of the fertilized egg to produce blastomeres of different constitutions. This results in identical nuclei residing in cytoplasms that have some type of difference. The difference between yeast mother and daughter cells with respect to their ability to switch mating types is an analogous situation (Strathern and Herskowitz 1979). Why do mother cells but not daughter cells express the gene necessary for switching (*HO*)? The answer may lie in the asymmetric distribution of proteins to the mother and daughter cells. As noted above, there are five activator genes necessary for production of the *HO* transcript. Perhaps these products are synthesized early in the cell cycle, before synthesis of the daughter bud has been initiated, and remain localized to a maternal structure. Returning to the fertilized egg, we can imagine that regulatory proteins such as the *SWI* products become asymmetrically distributed to sister blastomeres. They activate expression of genes analogous to *HO* that govern activity of a regulatory locus (analogous to *MAT*), which then ultimately triggers cell specialization. At any of these stages, a temporal or geographical signal might modulate activity of a key regulator in the same way as complex signals affect cII.

In this paper, we have described specific regulatory proteins that are responsible for cellular specialization in a eukaryotic cell and that are responsible for carrying out a complex integration of signals in a prokaryote. These proteins are themselves regulated by a diverse group of molecular mechanisms. On the basis of our experience in studying regulation of regulatory proteins, we anticipate that understanding the molecular basis for differentiation in multicellular eukaryotes will involve determinants of the early embryo that register their effects on regulatory proteins.

ACKNOWLEDGMENTS

I thank Alexander Johnson, Paul Sternberg, Flora Banuett, and Stan Fields for spirited discussion and invaluable comments during the preparation of this manuscript. The work from my laboratory described here was supported by research grants and predoctoral training grants from the National Institutes of Health. In addition, it is a great pleasure to acknowledge support to postdoctoral fellows from the Damon Runyon–Walter Winchell Cancer Fund, the American Cancer Society (National and California Divisions), the Jane Coffin Childs Fund, the U.S. Public Health Service, the Helen Hay Whitney Foundation, and the Weingart Foundation.

REFERENCES

Ammerer, G., G.F. Sprague, Jr., and A. Bender. 1985. Control of yeast α-specific genes: Evidence for two blocks to expression in *MATa/MATα* diploids. *Proc. Natl. Acad. Sci.* **82:** 5855.

Banuett, F., M.A. Hoyt, L. McFarlane, H. Echols, and I. Herskowitz. 1985. hflB, a new *E. coli* locus regulating lysogeny and the level of bacteriophage lambda cII protein. *J. Mol. Biol.* (in press).

Belfort, M. and D.L. Wulff. 1971. A mutant *Escherichia coli* that is lysogenized with high frequency. In *The bacteriophage lambda* (ed. A.D. Hershey), p. 739. Cold Spring Harbor Laboratory, Cold Spring Harbor, New York.

———. 1974. The roles of the lambda cIII gene and the *Escherichia coli* catabolite gene activation system in the establishment of lysogeny by bacteriophage lambda. *Proc. Natl. Acad. Sci.* **71:** 779.

DARNELL, J.E., JR. 1982. Variety in the level of gene control in eukaryotic cells. *Nature* **297:** 365.
DERMAN, E., K. KRAUTER, L. WALLING, C. WEINBERGER, M. RAY, and J.E. DARNELL, JR. 1981. Transcriptional control in the production of liver-specific mRNAs. *Cell* **23:** 731.
EISEN, H., P. BRACHET, L. PEREIRA DA SILVA, and F. JACOB. 1970. Regulation of repressor expression in λ. *Proc. Natl. Acad. Sci.* **66:** 855.
EPP, C. 1978. "Early protein synthesis and its control in bacteriophage lambda." Ph.D. thesis, University of Toronto, Canada.
FIELDS, S. and I. HERSKOWITZ. 1985. The yeast *STE12* product is required for expression of two sets of cell-type-specific genes. *Cell* (in press).
FORBES, D. and I. HERSKOWITZ. 1982. Polarity suppression by the Q gene product of bacteriophage λ. *J. Mol. Biol.* **160:** 549.
GAUTSCH, J.W. and D.L. WULFF. 1974. Fine structure mapping, complementation, and physiology of *Escherichia coli hfl* mutants. *Genetics* **77:** 435.
GRAYHACK, E.J., X. YANG, L.F. LAU, and J.W. ROBERTS. 1985. Phage lambda gene Q antiterminator recognizes RNA polymerase near the promoter and accelerates it through a pause site. *Cell* **42:** 259.
HALL, M.N., L. HEREFORD, and I. HERSKOWITZ. 1984. Targeting of *E. coli* β-galactosidase to the nucleus in yeast. *Cell* **36:** 1057.
HERSKOWITZ, I. 1982. The *MATα2* gene. *Rec. Adv. Yeast Mol. Biol.* **1:** 320.
———. 1983. Determination of yeast cell type. *Symp. Soc. Dev. Biol.* **41:** 65.
HERSKOWITZ, I. and F. BANUETT. 1984. Interaction of phage, host, and environmental factors in governing the λ lysis-lysogeny decision. *Proc. Int. Congr. Genet.* **15:** 59.
HERSKOWITZ, I. and D. HAGEN. 1980. The lysis-lysogeny decision of bacteriophage λ: Explicit programming and responsiveness. *Annu. Rev. Genet.* **14:** 399.
HERSKOWITZ, I. and Y. OSHIMA. 1981. Control of cell type in *Saccharomyces cerevisiae*: Mating type and mating type interconversion. In *The molecular biology of yeast* Saccharomyces: *Life cycle and inheritance* (ed. J.N. Strathern et al.), p. 181. Cold Spring Harbor Laboratory, Cold Spring Harbor, New York.
HICKS, J.B. and I. HERSKOWITZ. 1976. Interconversion of yeast mating types. I. Direct observations of the action of the homothallism (*HO*) gene. *Genetics* **83:** 245.
HICKS, J.B., J.N. STRATHERN, and I. HERSKOWITZ. 1977. The cassette model of mating type interconversion. In *DNA insertion elements, plasmids, and episomes* (ed. A. Bukhari et al.), p. 457. Cold Spring Harbor Laboratory, Cold Spring Harbor, New York.
HO, Y.M. and M. ROSENBERG. 1982. Characterization of the phage λ regulatory protein cII. *Ann. Microbiol.* **133A:** 215.
HO, Y., D.L. WULFF, and M. ROSENBERG. 1983. Bacteriophage λ protein cII binds promoters on the opposite face of the DNA helix from RNA polymerase. *Nature* **304:** 703.
HODGKIN, J. 1983. Two types of sex determination in a nematode. *Nature* **304:** 267.
HOOPES, B.C. and W.R. MCCLURE. 1985. A cII-dependent promoter is located within the Q gene of bacteriophage λ. *Proc. Natl. Acad. Sci.* **82:** 3134.
HOYT, M.A., D.M. KNIGHT, A. DAS, H.I. MILLER, and H. ECHOLS. 1982. Control of phage λ development by stability and synthesis of cII protein: Role of the viral cIII and host *hflA*, *himA*, and *himD* genes. *Cell* **31:** 565.
JENSEN, R.E. and I. HERSKOWITZ. 1984. Directionality and regulation of cassette substitution in yeast. *Cold Spring Harbor Symp. Quant. Biol.* **49:** 97.
JENSEN, R., G. SPRAGUE, JR., and I. HERSKOWITZ. 1983. Regulation of yeast mating-type interconversion: Feedback control of *HO* gene expression by the yeast mating-type locus. *Proc. Natl. Acad. Sci.* **80:** 3035.

JOHNSON, A.D. and I. HERSKOWITZ. 1985. A repressor (*MATα2* product) and its operator control expression of a set of cell-type specific genes in yeast. *Cell* **42:** 237.
JOHNSON, A.D., A.R. POTEETE, G. LAUER, R.T. SAUER, G.K. ACKERS, and M. PTASHNE. 1981. λ repressor and cro—Components of an efficient molecular switch. *Nature* **294:** 217.
JONES, M.O. and I. HERSKOWITZ. 1978. Mutants of bacteriophage λ which do not require the cIII gene for efficient lysogenization. *Virology* **88:** 199.
JULIUS, D., L. BLAIR, A. BRAKE, G. SPRAGUE, and J. THORNER. 1983. Yeast α factor is processed from a larger precursor polypeptide: The essential role of a membrane-bound dipeptidyl aminopeptidase. *Cell* **32:** 839.
KAISER, A.D. 1957. Mutations in a temperate bacteriophage affecting its ability to lysogenize *Escherichia coli*. *Virology* **3:** 42.
KASSIR, Y. and G. SIMCHEN. 1976. Regulation of mating and meiosis in yeast by the mating-type region. *Genetics* **82:** 187.
KLAR, A.J.S., J.N. STRATHERN, and J.B HICKS. 1984. Developmental pathways in yeast. In *Microbial development* (ed. R. Losick and L. Shapiro), p. 151. Cold Spring Harbor Laboratory, Cold Spring Harbor, New York.
KLAR, A.J.S., J.N. STRATHERN, J.R. BROACH, and J.C. HICKS. 1981. Regulation of transcription in expressed and unexpressed mating type cassettes of yeast. *Nature* **289:** 239.
KNIGHT, D.M. and H. ECHOLS. 1983. The cIII gene and protein of bacteriophage λ. *J. Mol. Biol.* **163:** 505.
KOSTRIKEN, R. and F. HEFFRON. 1984. The product of the *HO* gene is a nuclease: Purification and characterization of the enzyme. *Cold Spring Harbor Symp. Quant. Biol.* **49:** 89.
KURJAN, J. 1985. α-Factor structural gene mutations in *Saccharomyces cerevisiae*: Effects on α-factor production and mating. *Mol. Cell. Biol.* **5:** 787.
KURJAN, J. and I. HERSKOWITZ. 1982. Structure of a yeast pheromone gene (*MFα*): A putative α-factor precursor contains four tandem copies of α-factor. *Cell* **30:** 933.
MACKAY, V. and T.R. MANNEY. 1974. Mutations affecting sexual conjugation and related processes in *Saccharomyces cerevisiae*. I. Isolation and phenotypic characterization of nonmating mutants. *Genetics* **76:** 255.
MILLER, A.M. 1984. The yeast *MATa1* gene contains two introns. *EMBO J.* **3:** 1061.
MILLER, A.M., V.L. MACKAY, and K.A. NASMYTH. 1985. Identification and comparison of two sequence elements that confer cell-type specific transcription in yeast. *Nature* **314:** 598.
MONOD, J. and F. JACOB. 1962. Teleonomic mechanisms in cellular metabolism, growth, and differentiation. *Cold Spring Harbor Symp. Quant. Biol.* **26:** 389.
NASMYTH, K. 1983. Molecular analysis of a cell lineage. *Nature* **302:** 670.
NASMYTH, K.A., K. TATCHELL, B.D. HALL, C. ASTELL, and M. SMITH. 1981. A position effect in the control of transcription of yeast mating type loci. *Nature* **289:** 244.
OSHIMA, Y. 1982. Regulatory circuits of gene expression: The metabolism of galactose and phosphate. In *The molecular biology of the yeast* Saccharomyces: *Metabolism and gene expression* (ed. J.N. Strathern et al.), p. 159. Cold Spring Harbor Laboratory, Cold Spring Harbor, New York.
PTASHNE, M., A. JEFFREY, A.D. JOHNSON, R. MAURER, B.J. MEYER, C.O. PABO, T.M. ROBERTS, and R.T. SAUER. 1980. How the λ repressor and Cro work. *Cell* **19:** 1.
REICHARDT, L.F. 1975. Control of bacteriophage lambda repressor synthesis after phage infection: The role of the *N*, cII, cIII and cro products. *J. Mol. Biol.* **93:** 267.
RINE, J., G. SPRAGUE, JR., and I. HERSKOWITZ. 1981a. The *rme1* mutation of *Saccharomyces cerevisiae*: Map position and bypass of mating type locus control of sporulation. *Mol. Cell. Biol.* **1:** 958.
RINE, J., R. JENSEN, D. HAGEN, L. BLAIR, and I. HERSKO-

witz. 1981b. The pattern of switching and fate of the replaced cassette in yeast mating type interconversion. *Cold Spring Harbor Symp. Quant. Biol.* **45:** 951.

Roman, H., M.M. Phillips, and S.M. Sands. 1955. Studies of polyploid *Saccharomyces*. I. Tetraploid segregation. *Genetics* **40:** 546.

Schmeissner, U., D. Court, H. Shimatake, and M. Rosenberg. 1980. Promoter for the establishment of repressor synthesis in bacteriophage λ. *Proc. Natl. Acad. Sci.* **77:** 3191.

Siliciano, P.G. and K. Tatchell. 1984. Transcription and regulatory signals at the mating type locus in yeast. *Cell* **37:** 969.

Singh, A., E.Y. Chen, J.M. Lugovoy, C.N. Chang, R.A. Hitzman, and P.H. Seeburg. 1983. *Saccharomyces cerevisiae* contains two discrete genes coding for the α-factor pheromone. *Nucleic Acids Res.* **11:** 4049.

Smith, H.O. and M. Levine. 1964. Two sequential repressions of DNA synthesis in the establishment of lysogeny by phage P22 and its mutants. *Proc. Natl. Acad. Sci.* **52:** 356.

Sprague, G.F., Jr., R. Jensen, and I. Herskowitz. 1983. Control of yeast cell type by the mating type locus: Positive regulation of the α-specific *STE3* gene by the *MATα1* product. *Cell* **32:** 409.

Stern, M., R. Jensen, and I. Herskowitz. 1984. Five *SWI* genes are required for expression of the *HO* gene in yeast. *J. Mol. Biol.* **178:** 853.

Sternberg, P.W. and H.R. Horvitz. 1984. The genetic control of cell lineage during nematode development. *Annu. Rev. Genet.* **18:** 489.

Strathern, J.N. and I. Herskowitz. 1979. Asymmetry and directionality in production of new cell types during clonal growth: The switching pattern of homothallic yeast. *Cell* **17:** 371.

Strathern, J., J. Hicks, and I. Herskowitz. 1981. Control of cell type in yeast by the mating type locus: The α1-α2 hypothesis. *J. Mol. Biol.* **147:** 357.

Walker, G.C. 1984. Mutagenesis and inducible responses to deoxyribonucleic acid damage in *Escherichia coli*. *Microbiol. Rev.* **48:** 60.

Walker, M.D., T. Edlund, A.M. Boulet, and W.J. Rutter. 1983. Cell specific expression controlled by the 5′ flanking region of insulin and chymotrypsin genes. *Nature* **306:** 557.

Wilson, K.L. and I. Herskowitz. 1984. Negative regulation of *STE6* gene expression by the α2 product of yeast. *Mol. Cell. Biol.* **4:** 2420.

Wulff, D.L., M. Mahoney, A. Shatzman, and M. Rosenberg. 1984. Mutational analysis of a regulatory region in bacteriophage λ that has overlapping signals for the initiation of transcription and translation. *Proc. Natl. Acad. Sci.* **81:** 555.

Aspects of Dosage Compensation and Sex Determination in *Caenorhabditis elegans*

W.B. WOOD, P. MENEELY,* P. SCHEDIN, AND L. DONAHUE
Department of Molecular, Cellular, and Developmental Biology, University of Colorado, Boulder, Colorado 80309;
**Hutchinson Cancer Research Center, Seattle, Washington 98104*

The nematode *Caenorhabditis elegans* has two sexes, hermaphrodites and males. Hermaphrodites normally have two X chromosomes (XX or 2X) and males have only one (X0 or 1X). There is no Y chromosome, and sex is determined by the X/A ratio, i.e., the ratio of X chromosomes to sets of autosomes (Nigon 1949; Madl and Herman 1979). The X/A ratio acts to determine sex through a set of at least seven interacting autosomal genes, which have been defined, characterized, and shown to act as a regulatory pathway, primarily by Hodgkin and co-workers (Hodgkin and Brenner 1977; Hodgkin 1980; Doniach and Hodgkin 1984; Kimble et al. 1984; Hodgkin, this volume). The first gene in the pathway, *her-1*, acts through five intervening genes to regulate the major switch gene *tra-1*, whose activity determines somatic sexual development. At an X/A ratio of 1.0, *her-1* activity is low and *tra-1* activity is high, leading to hermaphrodite development. At an X/A ratio of 0.5, *her-1* activity is high and *tra-1* activity is low, leading to male development (see Hodgkin, this volume). Loss-of-function mutations in the *her-1* gene transform 1X animals into fertile hermaphrodites and have no effect on 2X hermaphrodites (Hodgkin 1980). Loss-of-function mutations in the *tra-1* gene transform 2X animals into fertile males and have no effect on 1X males (Hodgkin and Brenner 1977).

Like animals with XY systems of sex determination such as *Drosophila* and mammals, *C. elegans* compensates for the difference in X dose between 1X and 2X animals. We have shown this genetically with a technique first suggested by Muller (1950), using as rough measures of X-chromosome expression the severity of phenotypes that result from hypomorphic (partial loss-of-function) mutations on the X chromosome. Such phenotypes vary with the level of expression and therefore with the gene dosage of the mutant allele, in contrast to phenotypes resulting from null (amorphic, complete loss-of-function) alleles, which are customarily defined by their lack of allele dosage effects on phenotype. For X-linked hypomorphic alleles, the allele dosage differs by a factor of two between 1X and 2X animals. Therefore, if the resulting phenotypes are of the same severity in males and hermaphrodites, then there must be compensation for the difference in X dosage.

Table 1 (first line) shows the variations in phenotype that would be expected, assuming that dosage compensation is occurring, as the allele dosage and X-chromosome dosage are varied for an X-linked hypomorphic allele. The second line shows the expectations for a null allele of the same gene as a control. The phenotype resulting from the hypomorphic allele in the hemizygous male (with one copy) resembles that of a homozygous hermaphrodite (with two copies), rather than that of a heterozygous hermaphrodite carrying one copy of this allele and a deficiency of the locus.

Table 1. Gene-dosage Effects on Phenotypes Resulting from Null and Hypomorphic Alleles in *X*-linked Genes

Allele	*m/Df*[a]	*m/O* (males)	*m/m*	*m/m/m*
Hypomorph[b]	severely mutant	moderately mutant	moderately mutant	weakly mutant
Null[b]	severely mutant	severely mutant	severely mutant	severely mutant
let-2(mn114)[c]	embryonic lethal	adult	adult[d]	adult[e]
let-2(mn153)[f]	embryonic lethal	embryonic lethal	embryonic lethal	embryonic lethal

[a] *m* signifies the mutant allele in question; *m/Df* is a 2X hermaphrodite heterozygous for the mutant allele and a deficiency lacking the locus; *m/O* is a 1X male hemizygous for the mutant allele; *m/m* is a 2X hermaphrodite homozygous for the mutant allele; *m/m/m* is a 2A;3X animal carrying three copies of the mutant allele.
[b] See text.
[c] Hypomorphic allele.
[d] Sterile.
[e] Slightly fertile.
[f] Null allele.

We have demonstrated such behavior for hypomorphic and null alleles of five genes with quite different physiological functions in different regions of the X chromosome: *let-1*, *let-2*, *unc-3*, *lin-2*, and *lin-15*. These genes and their mutant phenotypes are: *let-1*, early embryonic lethality of unknown cause (Meneely and Herman 1981); *let-2*, later lethality or sterility (Meneely and Herman 1981) possibly resulting from basement membrane defects (J. Kramer, pers. comm.); *unc-3*, uncoordinated movement probably resulting from a nervous system defect (Herman 1984); *lin-2*, lack of a vulva resulting from a cell-lineage defect (Horvitz and Sulston 1980); and *lin-15*, production of multiple pseudovulvae resulting from a cell-lineage defect (Ferguson and Horvitz 1985). Results for the *let-2* alleles are shown in the lower half of Table 1. Null alleles such as *mn153* cause late embryonic lethality, regardless of dosage. In hermaphrodites, the hypomorphic allele *mn114* causes a similar phenotype when present in only one copy, but embryos with two copies develop into sterile adults, and those with three copies become adults with low fertility. However, 1X males with one copy develop to adulthood, like 2X hermaphrodites with two copies, indicating dosage compensation. Details of these experiments will be published elsewhere (P. Meneely et al., in prep.).

Further experiments of this kind using sexually transformed mutant animals suggest that dosage compensation occurs independently of sexual phenotype. Phenotypes resulting from X-linked hypomorphic mutations appear the same in sexually transformed *her-1* 1X hermaphrodites as in 2X hermaphrodites, indicating that compensation is occurring normally even when the 1X and 2X animals are both phenotypic hermaphrodites. Analogous observations have been made in comparisons of 1X males with *tra-1* 2X males (P. Meneely, unpubl.). Results of such an experiment with 1X and 2X hermaphrodites, using a hypomorphic allele of *lin-15* to compare levels of X expression, are shown in Table 2. The multiple pseudovulvae (Muv phenotype) characteristic of *lin-15* mutants can be seen as protrusions from the ventral hypodermis of the animal (Fig. 1, top panel). Hypomorphic alleles of this

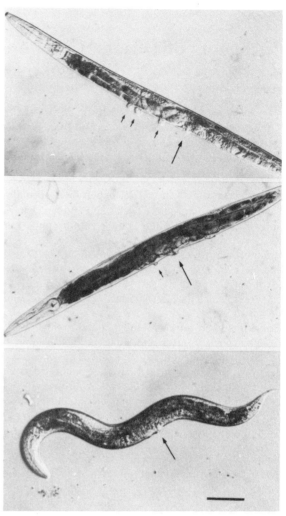

Figure 1. Bright-field photomicrographs of hermaphrodites homozygous for a hypomorphic allele of *lin-15* (see Fig. 2) that causes production of multiple pseudovulvae. (*Top*) 2X hermaphrodite; small arrows indicate the multiple pseudovulvae and a large arrow indicates the vulva. (*Middle*) 2A;3X hermaphrodite with a reduction in both number and size of pseudovulvae. (*Bottom*) 2X hermaphrodite that also carries the *dpy-21(e428)* allele and lacks pseudovulvae. Photographed with a Zeiss Universal Microscope. Bar, ~100 μm.

Table 2. Gene-dosage Effects on Penetrance of the Muv Phenotype for the X-linked Hypomorphic Allele *lin-15(n765)* in 2X and *her-1* 1X Hermaphrodites

Temp. (°C)	m/Df[a]	m/O	m/m	$m/m/m$
16	98 ± 2% (176/179)	14 ± 7% (15/108)	10 ± 3% (38/402)	1 ± 1% (1/137)
20	100% (83/83)	99 ± 2% (78/79)	100% (363/364)	65 ± 13% (32/49)
25	100% (90/90)	100% (130/130)	98 ± 6% (164/164)	100% (46/47)

Penetrance values are expressed as the percentage of animals with multiple vulvae (Muv phenotype). Errors are shown as 95% confidence limits. Below each value is shown in parentheses the number of Muv animals counted over the total number of animals scored.

[a] m signifies the hypomorphic allele *lin-15(n765)*; m/Df is heterozygous for this allele and a deficiency that removes the *lin-15* locus; m/O is a sexually transformed *her-1* 1X hermaphrodite hemizygous for the *lin-15* allele; m/m is homozygous for this allele; $m/m/m$ is a 2A;3X animal carrying three copies of the *lin-15* allele.

gene are particularly useful, because the percentage of animals exhibiting the Muv phenotype (penetrance) varies with temperature (Ferguson and Horvitz 1985), allowing quantitation of phenotypic expression over a wide range. As shown in Table 2, penetrance of the Muv phenotype increases with increasing temperature and decreases with increasing allele dosage. The penetrance in *her-1* hermaphrodites with one dose of the allele is similar to that in 2X hermaphrodites with two doses, indicating dosage compensation. In other experiments (not shown), we have demonstrated that the Muv phenotype resulting from hypomorphic *lin-15* alleles is the same in *her-1$^+$* 2X and *her-1* 2X hermaphrodites, indicating that the *her-1* mutation itself has no apparent effect on X-chromosome expression.

Preliminary evidence supports the view that dosage compensation occurs at the level of transcription of X-linked genes. We have made such measurements using gene-specific probes in a quantitative RNA dot-blot assay to determine transcript levels of three genes: the X-linked gene *act-4* (which probably codes for a cytoplasmic actin; M. Krause and D. Hirsh, pers. comm.) and the autosomal genes *act-1* (a muscle actin gene) and *unc-54* (a body-wall-muscle myosin gene). The gene-specific actin gene probes were kindly provided by M. Krause. The levels of *act-4* transcripts relative to the two autosomal transcripts were close to identical in adult hermaphrodites and males (hermaphrodite/male ratio 1.1 ± 0.03; weighted average of several experiments) and clearly not different by a factor of 2, as would be expected if there were no dosage compensation. Experiments are in progress to determine whether this compensation could involve gene-specific controls in addition to global X-chromosome effects. Further evidence for transcriptional compensation has been obtained in Northern blotting experiments with cloned probes for four other genes of unknown function in different regions of the X chromosome (B. Meyer, pers. comm.).

In summary, the X/A ratio appears to dictate two major decisions in embryonic development: determination of sexual phenotype and determination of the level of X-chromosome expression. Among the general questions raised by these findings are the following: (1) Through what mechanisms does the X/A ratio influence sex determination and level of X expression? (2) How are the decisions regarding sex determination and X expression related? (3) Are these decisions made once, in a manner that affects the entire embryo, or many times independently, by individual cells or groups of cells? (4) When in embryogenesis are these decisions made? Although none of these questions can yet be answered satisfactorily, some information pertinent to each of them is now available, as discussed in the following paragraphs.

Genes That Affect the Level of X Expression

The four genes *dpy-21* V, *dpy-22* X, *dpy-23* X, and *dpy-26* IV have been implicated in affecting X-chromosome expression. Mutations in these genes lead to the short phenotype known as Dumpy (Dpy) and other defects, which unlike the phenotypes of other Dpy mutants are dependent on the X/A ratio (Hodgkin 1983). The mutant phenotypes of these genes (see Table 3) have been postulated to result from inappropriate X expression because they resemble the phenotypes of X-chromosome aneuploids, i.e., animals with X/A ratios different from the normal values of 0.5 and 1.0 (Hodgkin 1983; Meneely and Wood 1984). Two additional genes with mutant phenotypes similar in many respects to those of *dpy-26* have been recently described as *dpy-27* III (J. Hodgkin, pers. comm.) and *dpy-28* III (J. Plenefisch and B. Meyer, pers. comm.). We refer to these six genes collectively as the X-dependent *dpy* genes.

Using phenotypes resulting from X-linked hypomorphic mutations to indicate levels of X expression as described above, we have obtained additional evidence that X-dependent *dpy* genes affect X-chromosome expression. Two probable loss-of-function alleles of *dpy-21* and a probable loss-of-function allele of *dpy-26* appear to cause *increases* in X expression. This conclusion is based on observations that they *suppress* the phenotypes resulting from hypomorphic alleles of several X-linked genes; i.e., they reduce the severity of the phenotypes in the same manner as does an increase in the dose of the hypomorphic allele. We have observed

Table 3. Influence of X/A Ratio on Phenotype for *dpy$^+$* Animals and Four X-dependent *dpy* Mutants

	Karyotype				
	3A;1X	2A;1X	2A;2X	2A;3X	2A;4X
X/A ratio	0.33	0.50	1.00	1.50	2.00
dpy$^+$	usually inviable	non-Dpy	non-Dpy	Dpy	inviable
dpy-21	—[a]	non-Dpy	Dpy	inviable	—
dpy-26	—	non-Dpy	Dpy[b]	inviable	—
dpy-22	—	usually inviable	Dpy, sickly	—	—
dpy-23	—	inviable	Dpy, sickly	—	—

[a] Not determined.
[b] Strong maternal effect on viability. Most progeny of homozygous hermaphrodites die as embryos, and homozygous stocks cannot be maintained.

this result for both *dpy-21* and *dpy-26* with three of the five X-linked genes listed in the preceding section. Similar results for *dpy-21* have been observed with a hypomorphic allele of *lin-14* X (L. DeLong and B. Meyer, pers. comm.), a gene that affects the timing of events in larval development (Ambros and Horvitz 1984). In control experiments, the *dpy-21* and *dpy-26* mutations did not suppress phenotypes resulting from null alleles of these X-linked genes. Moreover, the *dpy-21* mutations did not suppress phenotypes resulting from hypomorphic alleles in any of eight autosomal genes with mutant phenotypes similar to those of the X-linked genes. The *dpy-26* mutations also did not suppress one of these autosomal hypomorphs (the remainder were not readily testable due to the inviability of *dpy-26* hermaphrodites). As an additional control, mutations in several other *dpy* genes showed no effects on phenotypes resulting from either X-linked or autosomal hypomorphic mutations (P. Meneely et al., in prep.).

The results of such an experiment with a hypomorphic allele of *lin-15* X are illustrated in Figures 1 and 2. At 16°C, this allele causes about 30% of 2A;2X hermaphrodites to develop with a Muv phenotype; at 20°C, the penetrance is almost 100%. In 2A;3X hermaphrodites, the penetrance is significantly lower at both 16°C and 20°C. When 2A;2X hermaphrodites carry a mutation in either *dpy-21* or *dpy-26* in addition to the hypomorphic *lin-15* allele, they also show substantially decreased penetrance of the Muv phenotype at 16°C and 20°C. In fact, the presence of either the *dpy-21* or the *dpy-26* mutation suppresses the Muv phenotype somewhat more strongly than does the presence of an extra X chromosome (Fig. 2). The effects of the two mutations appeared to be additive in a *dpy-26;dpy-21* double mutant. In experiments similar to those shown in Figure 2, the penetrance of the Muv phenotype in *lin-15* animals carrying both these mutations was 0% and 45% at 20°C and 25°C, respectively, as compared with the values of about 20% and 95% at 20°C and 25°C, respectively, shown in Figure 2 for *lin-15* animals carrying either the *dpy-21* or the *dpy-26* mutation alone.

Consistent with this genetic evidence, gel-blot assays similar to those mentioned above have indicated increased levels of some X-linked transcripts in *dpy-21* mutants (B. Meyer, pers. comm.). In summary, *dpy-21* and *dpy-26* mutations appear to cause increased expression of several X-linked genes with quite different mutant phenotypes, suggesting a general effect on X-chromosome expression.

Similar genetic evidence suggests that a loss-of-function allele of *dpy-22* and a recessive allele of *dpy-23* both cause *decreases* in the level of X expression, based on the finding that they *enhance* the phenotypes resulting from hypomorphic alleles of X-linked genes. Data for a hypomorphic allele of *lin-15* X are presented in Figure 3. The penetrance of the Muv phenotype at 16°C, which in this experiment was about 37% in a 2A;2X homozygous animal carrying two copies of the *lin-15* allele, increases to 100% in a heterozygous animal carrying one copy of the *lin-15* allele and a deficiency (Df) that removes the *lin-15* locus. Penetrance in the homozygote at 16°C is increased to 67% by the *dpy-23* mutation and to 97% by the *dpy-22* mutation, which thus shows almost as great an effect as reducing the dosage of the *lin-15* allele from two copies to one. The *dpy-22* mutation also appeared to enhance the phenotypes resulting from hypomorphic alleles of three other X-linked genes, *let-1*, *lin-2*, and *unc-3*, although

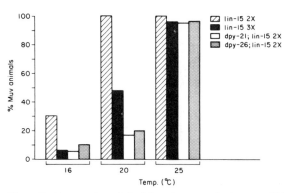

Figure 2. Suppression of the Muv phenotype by increased X dose and mutations in *dpy-21* or *dpy-26* in hermaphrodites homozygous for a synthetic hypomorphic allele of *lin-15* and a *him-5* mutation that causes increased frequency of X-chromosome nondisjunction at meiosis (*n833*; *him-5[e1490]*; *lin-15[n767]*; P. Meneely et al., in prep.). Penetrance of the mutant phenotype was compared at the indicated growth temperatures for animals of the following genotypes: 2X;3X (identified by their Dpy phenotype and small number of male progeny), 2X carrying the *dpy-21(e428)* mutation, and 2X carrying the *dpy-26(n199)* mutation.

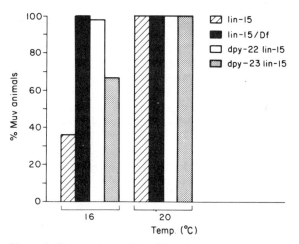

Figure 3. Enhancement of the Muv phenotype by decreased X dose and mutations in *dpy-22* or *dpy-23* in hermaphrodites homozygous for a synthetic hypomorphic allele of *lin-15* (*n833;lin-15[n767]*; see Fig. 2). Penetrance of the mutant phenotype was compared at the indicated growth temperatures for homozygous *lin-15* hermaphrodites, heterozygous hermaphrodites carrying one copy of the *lin-15* allele and a deficiency that removes the *lin-15* locus, and homozygous *lin-15* hermaphrodites carrying either the *dpy-22(e652)* or *dpy-23(e840)* mutations.

the results were not so clear-cut due to phenotypic effects of the *dpy-22* mutation itself. Unfortunately, control experiments are not feasible to determine whether the *dpy-22* mutation affects the phenotypes resulting from null alleles in these genes because these phenotypes are already fully penetrant or nearly so. No enhancement was observed with hypomorphic alleles in any of eight autosomal genes (P. Meneely et al., in prep.).

We have used quantitative RNA dot-blot assays (see above) to compare *dpy-22* and wild-type hermaphrodites for levels of transcripts from the X-linked *act-4* and autosomal *unc-54* genes. These experiments showed that the level of *act-4* RNA in the *dpy-22* mutant was slightly but significantly decreased to 0.85 ± 0.07 relative to a wild-type value of 1.0.

All the effects of mutations in X-dependent *dpy* genes described so far were observed in 2X animals. On the basis of less extensive evidence, 1X animals carrying these mutations also show apparent effects on X expression that are qualitatively similar to, although probably quantitatively different from, the effects seen in 2X animals. The *dpy-21* and *dpy-26* mutations suppress the Muv phenotype resulting from hypomorphic *lin-15* alleles in sexually transformed *her-1* 1X hermaphrodites, although not as strongly as in *her-1* 2X hermaphrodites (data not shown). The *dpy-22* mutation strongly enhances the Muv phenotype resulting from a *lin-15* hypomorphic allele in rare surviving *dpy-22 lin-15* 1X males, and the low viability of *dpy-22* and *dpy-23* 1X animals (Hodgkin 1983) suggests that these mutations affect 1X animals more strongly than 2X animals. Details of these experiments will be published elsewhere (P. Meneely et al., in prep.).

In summary, mutations in *dpy-21* appear to cause increased expression of X-linked genes. On the basis of somewhat less evidence, mutations in *dpy-26* appear to have similar effects, and recent experiments indicate that *dpy-27* and *dpy-28* mutations do as well (L. DeLong and B. Meyer, pers. comm.). These effects result from probable loss-of-function mutations, suggesting that these genes may normally act to decrease X-chromosome expression. The evidence is less complete for *dpy-22* and *dpy-23* effects; however, the results so far suggest that probable loss-of-function mutations in these genes cause decreased expression of at least some X-linked genes, with no apparent effects on the autosomal genes tested. Therefore, *dpy-22* and *dpy-23* may normally act to increase expression of X-linked genes. These observations are consistent with the earlier suggestion by Hodgkin (1983) that the X-dependent *dpy* genes could be involved in the process of dosage compensation. If so, the effects of both *dpy-21* and *dpy-22* mutations in both 1X and 2X animals further suggest that the compensation process in *C. elegans* could involve a combination of both positive and negative regulation of X-chromosome expression in both sexes. Such a mechanism would contrast with the situation in both *Drosophila*, which appear to compensate by hyperactivation of the single X chromosome in males (for review, see Baker and Belote 1983), and mammals, which compensate by inactivation of one of the two X chromosomes in females (for review, see Gartler and Riggs 1983).

In other experiments that may be relevant to the roles of the X-dependent *dpy* genes, we have introduced X duplications that do not include the *lin-15* locus into 2X animals carrying a hypomorphic allele of *lin-15*. One copy of an X duplication comprising about 25% of the X chromosome suppresses the Muv phenotype substantially, reducing the penetrance at 20°C from 100% without the duplication to 62% (283 animals examined) when it is present. Two copies of the same duplication suppress still more strongly, reducing the penetrance to 21% (101 animals examined). This effect, if also observed with other duplications and other X-linked hypomorphic alleles, would be consistent with interaction of X-chromosome material with a general negative regulator of X-chromosome expression present in limiting amount, such that its titration leads to increased X-chromosome expression. It would contrast with the results of similar experiments in *Drosophila*, where added X-chromosome material has been reported to cause apparent decreases in X-chromosome expression, consistent with limiting amounts of a positive regulator (Maroni and Lucchesi 1980; Williamson and Bentley 1983).

Relationship of Sex Determination and Dosage Compensation

There are three formal possibilities for the relationship between the decisions regarding sex determination and dosage compensation in response to the X/A ratio: (1) The X/A ratio could determine sexual phenotype, which in turn determines the level of X expression. (2) The X/A ratio could determine sexual phenotype and level of X expression independently. (3) The X/A ratio could determine the level of X expression, which in turn determines sexual phenotype. The first possibility seems unlikely on the basis of the finding that dosage compensation appears normal in sexually transformed mutants (Table 2). The second and third possibilities cannot be distinguished on the basis of existing information. However, it is of interest that a *dpy-21* mutation, which causes increased X-chromosome expression as discussed above, also causes hermaphroditization of 1X animals carrying X duplications (Meneely and Wood 1984). For duplications representing about 25% of the X chromosome, such animals are normally fertile males but show an intersexual phenotype if they carry a *dpy-21* mutation. A presumably similar phenomenon has been observed by J. Hodgkin (pers. comm.), who found that 3A;2X animals, normally males, are transformed to fertile hermaphrodites by a *dpy-27* mutation. These observations suggest that the *dpy-21* and *dpy-27* genes could normally play a role in sex determination, either directly or through effects on X-chromosome expression.

Is the Sex Determination Decision Made Once or Many Times?

Early experiments with *Drosophila* showed that 2X:3A aneuploids respond to the intermediate X/A ratio by developing as mosaic intersexes, with interspersed patches of male and female tissues (for review, see Baker and Belote 1983). These observations suggest that during embryogenesis, the decision to follow the male or female pathway of development can be made many times independently by individual cells or groups of cells, whose progeny may inherit their decision. In *Caenorhabditis*, 3A;2X animals are fertile males. However, Madl and Herman (1979) observed that 3A;2X animals that also carry a duplication of 25% of the X chromosome are intersexual, exhibiting a variety of phenotypes that range from superficially nearly normal male to superficially nearly normal hermaphrodite. These authors interpreted their observations to suggest that assessment of the X/A ratio could be made by more than one cell during development and that different cells could differ in their assessment.

We have examined such animals further for mosaicism at the organ and tissue levels. Some of these animals clearly show differences in the sex of the germ line and that of the somatic gonad based on morphology. We have observed gonads with male morphology but containing oocytes, as well as gonads with hermaphrodite morphology but containing only sperm. To test for mosaicism in somatic tissues, we have exploited the finding that vitellogenins are synthesized in the intestines of hermaphrodites, but not in males (Kimble and Sharrock 1983). Using as a probe a cloned fragment of DNA (kindly provided by T. Blumenthal) from one of the vitellogenin genes, *vit-5* (Blumenthal et al. 1984), we have assayed by in situ hybridization for the presence of vitellogenin mRNAs in dissected gonads and intestines of diploid, triploid, and aneuploid animals. A total of 100 2A;2X and 50 3A;3X hermaphrodites all showed strong hybridization of the labeled probe to all intestinal cells and none to the gonad, as illustrated in Figure 4a. In contrast, a total of 50 2A;1X males and 40 3A;2X males showed no signal over either tissue as illustrated in Figure 4b. These results confirm the specificity of synthesis observed for vitellogenin proteins by Kimble and Sharrock (1983) using different techniques. Therefore, we can interpret the presence of vitellogenin mRNAs as a characteristic of hermaphrodite intestinal cells, with the reservation that it need not necessarily imply an intrinsic decision with regard to sex determination in the intestinal cells themselves.

In 3A;2X animals that carry a partial X duplication and are therefore intersexual, we find clear examples of somatic mosaicism. Among 94 such intersexes that carry duplications representing about 25% of the X chromosome, 44 of the animals showed the hermaphrodite pattern of labeling over the entire intestine, and 33 showed the male pattern of no intestinal labeling. In some dissections, the morphology of the gonad could be scored as well. In five such dissections, the intestinal pattern did not correlate with the sex of the somatic gonad based on morphology. Among these cases, we observed three examples of animals with an apparently morphologically normal male gonad containing only sperm, and an intestine exhibiting the hermaphrodite pattern of vitellogenin expression in all cells (not shown). We also observed two examples of the converse: animals with an apparently normal hermaphrodite gonad containing abnormal-appearing oocytes and an intestine exhibiting the male pattern of no vitellogenin expression. These results argue against the suggestion of Kimble and Sharrock (1983) that the sexes of the gonad and intestine could normally be coordinately regulated. Instead, our observations indicate that for the somatic gonad, which derives from the MS founder-cell lineage, and the intestine, which derives from the E founder-cell lineage, the male/hermaphrodite decision can be made independently.

When Are Sex Determination Decisions Made?

The remaining 17 of the intersex animals described in the preceding section exhibited mosaicism within the intestine; i.e., some cells showed the hermaphrodite level of intense labeling and others showed the male level of no labeling. An example is shown in Figure 4 c and d. As described below, the patterns of vitellogenin expression observed in these mosaics are surprising in two respects. First, all the observed mosaics exhibited vitellogenin expression only in contiguous intestinal cells. On the basis of the known lineage of the intestinal cells (Sulston et al. 1983), this pattern is not consistent with the patterns expected if different male/hermaphrodite decisions were made independently by cells of the E lineage during intestinal development and then inherited by the progeny of these cells. Figure 5 (upper portion) shows the lineages of the intestinal cells on the left side of the animal, deriving from the cells Eal and Epl. The lineages on the right side, deriving from Ear and Epr, are identical and have been omitted for simplicity. The lower portion of the figure shows schematically the structure of the adult intestine, which consists of 20 cells arranged in 9 structural units: 1 quadruplet and 8 pairs of cells, 14 of which are binucleate. These units are designated int1 through int9, respectively. Cross-hatched cells are those deriving from the left side of the lineage. Because two cells on each side of the animal exchange places with their neighbors between the 16- and 20-E-cell stages of intestinal development, the descendants of a given intestinal precursor cell generally do not occupy adjacent positions in the adult intestine. For example, as can be seen from the lineage diagram, descendants of the Ea cell give rise to the units int1, int2, int3, and int5; the intervening unit, int4, is derived from descendants of Ep. Therefore, lineal inheritance of decisions regarding vitellogenin expression made during intestinal development would not give contiguous labeling patterns, but rather patches of labeled cells separated by unlabeled cells.

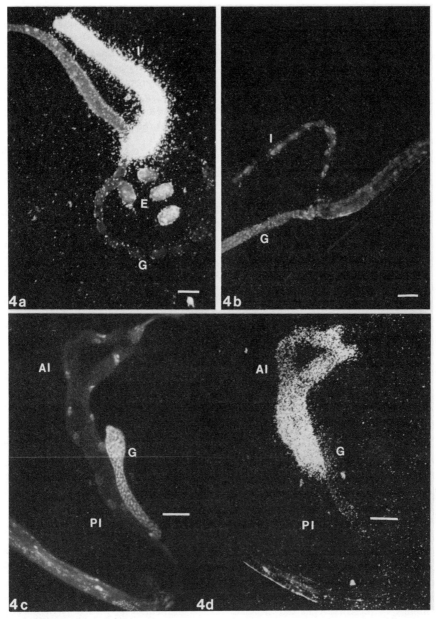

Figure 4. In situ hybridization of a vitellogenin gene probe to dissected intestines and gonads of hermaphrodite, male, and intersexual animals. Tissues dissected from adult animals according to the method of Kimble and Sharrock (1983) were mounted on glass slides, fixed, and hybridized to a nick-translated fragment of the *vit-5* gene cloned into the vector pUC8 (Blumenthal et al. 1984), using procedures described by Albertson (1984) with slight modifications, including use of ^{35}S-labeled probes (10^8 cpm/μg) rather than biotin-labeled probes. Slides were autoradiographed, stained with diamidinophenylindole hydrochloride (DAPI), and photographed as described by Edwards and Wood (1983), using a Zeiss Photomicroscope III equipped with dark-field and epifluorescence optics. (*a*) Dark-field/epifluorescence image of a dissected 2A;2X wild-type hermaphrodite, showing intense labeling of the intestine (I) and no labeling of the gonad (G). The preparation was photographed with visible dark-field and 365 nm epi-illumination in order to simultaneously visualize autoradiographic grains and DAPI-stained nuclei. (*b*) Dark-field/epifluorescence image of a dissected 2A;1X male, showing no labeling of either intestine (I) or gonad (G). Photographed as in *a*. (*c*) Epifluorescence image of a dissected 3A;2X intersex carrying a duplication of about 25% of the X chromosome, showing DAPI staining of intestinal (I) and gonadal (G) nuclei. (*d*) Dark-field image of the same preparation as in *c*, showing intense labeling of the anterior portion of the intestine (AI) and no labeling of the posterior intestine (PI) or gonad (G). Bar, ~50 μm.

The second surprising aspect is that in all intestinal mosaics, the cells expressing vitellogenins were anterior to the cells not expressing vitellogenins. The position of the junction between the labeled anterior and unlabeled posterior cells varied among different individuals but was never observed anterior to int3 or posterior to int6. A possibly related observation is that in some mosaics, the posterior portion of the intestine appeared thinner than the anterior portion, a difference that may be characteristic of males and hermaph-

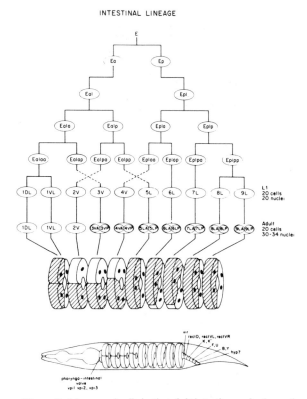

Figure 5. Lineage of cells in the adult intestine and schematic representation of the intestinal anatomy. Upper portion of the figure shows the lineage relationships and names of precursor cells on the left side of the animal that give rise to the intestinal cells (marked with cross-hatching on the diagram in the center of the figure). This diagram also shows schematically the arrangement of 20 cells into the 9 units that make up the intestine. Lower diagram shows these units in relation to associated structures in the adult animal (see text).

rodites. We often observed in dissections that normal male intestines appeared smaller in diameter than normal hermaphrodite intestines.

The observed mosaic patterns indicate first that the decision to express or not to express vitellogenin genes is not made randomly by intestinal cells but rather is positionally influenced. Second, the observation that labeled cells are always contiguous shows that this decision cannot be made irreversibly prior to the 20-E-cell stage, because an earlier decision would lead to vitellogenin expression in noncontiguous cells. The 20-E-cell stage is about midway through embryogenesis, after completion of the cell proliferation phase (Sulston et al. 1983). Synthesis of the vitellogenin proteins is not detected until much later, in the fourth stage of larval development when gonadogenesis is completed (Kimble and Sharrock 1983). Experiments are in progress to determine when the first vitellogenin transcripts appear.

The signal that turns on expression of the vitellogenin genes is not known. From our results, the decision to express or not to express these genes could be made at any time between mid-embryogenesis and the onset of this signal. Whether this decision reflects assessment of the X/A ratio in intestinal cells or in some other cells that generate the signal remains to be established. Nevertheless, the mosaics indicate that in animals with intermediate X/A ratios, there is some hermaphroditizing influence that can be stronger in the anterior than in the posterior of the animal. Experiments are in progress to investigate further the nature of this influence and its possible effects on other sexually dimorphic characteristics.

ACKNOWLEDGMENTS

This research was supported by a grant to W.B.W. from the National Institutes of Health (HD-11762) and by postdoctoral fellowships to P.M. from the American Cancer Sociaty (PF-1840) and the National Institutes of Health (GM-07775). Some of the strains used were provided by the *Caenorhabditis* Genetics Center, which is supported by contract NO1-AG-92113 between the National Institutes of Health and the Curators of the University of Missouri. We are grateful to J. Hodgkin, B. Meyer, and E. Ferguson for communication of unpublished results, to C. Trent for valuable discussion, and to P. Mains and C. Trent for comments on the manuscript.

REFERENCES

ALBERTSON, D. 1984. Localization of the ribosomal genes in *Caenorhabditis elegans* chromosomes by *in situ* hybridization using biotin-labeled probes. *EMBO J.* **3:** 1227.

AMBROS, V. and H.R. HORVITZ. 1984. Heterochronic mutants of the nematode *Caenorhabditis elegans*. *Science* **226:** 409.

BAKER, B.S. and J.M. BELOTE. 1983. Sex determination and dosage compensation in *Drosophila melanogaster*. *Annu. Rev. Genet.* **17:** 345.

BLUMENTHAL, T., M. SQUIRE, S. KIRTLAND, J. CANE, M. DONEGAN, J. SPIETH, and W. SHARROCK. 1984. Cloning of a yolk protein gene family from *Caenorhabditis elegans*. *J. Mol. Biol.* **174:** 1.

DONIACH, T. and J. HODGKIN. 1984. A sex-determining gene, *fem-1*, required for both male and hermaphrodite development in *Caenorhabditis elegans*. *Dev. Biol.* **106:** 223.

EDWARDS, M.K. and W.B. WOOD. 1983. Location of specific messenger RNAs in *Caenorhabditis elegans* by cytological hybridization. *Dev. Biol.* **97:** 375.

FERGUSON, E.L. and H.R. HORVITZ. 1985. Identification and characterization of 22 genes that affect the vulval cell lineages of the nematode *Caenorhabditis elegans*. *Genetics* **110:** 17.

GARTLER, S.M. and A.D. RIGGS. 1983. Mammalian X-chromosome inactivation. *Annu. Rev. Genet.* **17:** 155.

HERMAN, R.K. 1984. Analysis of genetic mosaics of the nematode *Caenorhabditis elegans*. *Genetics* **108:** 165.

HODGKIN, J. 1980. More sex-determination mutants of *Caenorhabditis elegans*. *Genetics* **96:** 649.

———. 1983. X chromosome dosage and gene expression in *C. elegans*: Two unusual dumpy genes. *Mol. Gen. Genet.* **192:** 452.

HODGKIN, J. and S. BRENNER. 1977. Mutations causing transformation of sexual phenotype in the nematode *Caenorhabditis elegans*. *Genetics* **86:** 275.

HORVITZ, H.R. and J.E. SULSTON. 1980. Isolation and genetic characterization of cell lineage mutants of the nematode *Caenorhabditis elegans*. *Genetics* **96:** 435.

KIMBLE. J. and W.J. SHARROCK. 1983. Tissue specific synthesis of yolk proteins in *Caenorhabditis elegans*. *Dev. Biol.* **96:** 189.

KIMBLE, J., L. EDGAR, and D. HIRSH. 1984. Specification of male development in *Caenorhabditis elegans*: The *fem* genes. *Dev. Biol.* **104:** 234.

MADL, J.E. and R.K. HERMAN. 1979. Polyploids and sex determination in *Caenorhabditis elegans*. *Genetics* **93:** 393.

MARONI, G. and J.C. LUCCHESI. 1980. X-chromosome transcription in *Drosophila*. *Chromosoma* **77:** 253.

MENEELY, P.M. and R.K. HERMAN. 1981. Suppression and function of X-linked lethal and sterile mutations in *Caenorhabditis elegans*. *Genetics* **97:** 65.

MENEELY, P.M. and W.B. WOOD. 1984. An autosomal gene that affects X-chromosome expression and sex determination in *Caenorhabditis elegans*. *Genetics* **106:** 29.

MULLER, H.J. 1950. Evidence of the precision of genetic adaptation. *Harvey Lect.* **43:** 165.

NIGON, V. 1949. Effets de la polyploidie chez un Nematode libre. *C.R. Acad. Sci.* **228:** 1161.

SULSTON, J.E., E. SCHIERENBERG, J.G. WHITE, and J.N. THOMSON. 1983. The embryonic cell lineage of the nematode *Caenorhabditis elegans*. *Dev. Biol.* **100:** 64.

WILLIAMSON, J.H. and M.M. BENTLEY. 1983. Dosage compensation in *Drosophila*: NADP-enzyme activities and cross-reacting material. *Genetics* **103:** 649.

The Sex Determination Pathway in the Nematode *Caenorhabditis elegans*: Variations on a Theme

J. HODGKIN, T. DONIACH, AND M. SHEN
MRC Laboratory of Molecular Biology, Hills Road, Cambridge, CB2 2QH, England

Sex determination in *Caenorhabditis elegans* is a problem that lends itself to investigation by a "top-down" approach. The primary sex-determining signal, the ratio of X chromosomes to autosomes, is known and can be subjected to experimental manipulation. It has also been possible to identify a set of regulatory genes that respond to this primary signal and act to direct development along one of two alternative developmental paths. Sexual dimorphism in this animal is extensive, involving sexually specialized differentiation in many different tissues, and thus a variety of events must be controlled correctly in order to achieve normal sexual differentiation. The same set of major regulatory genes appears to be involved in all of these events, but there are additional minor controls that seem to be necessary in some tissues but not in others. The interactions between these various genes have been extensively investigated, in the hope of understanding how the whole process of sexual differentiation is coordinated, and why it is organized in this way.

The two natural sexes of *C. elegans*, the self-fertilizing XX hermaphrodite and the XO male, exhibit many differences in development and final morphology (Fig. 1). For convenience, the main differences are summarized in Table 1 (there are additional differences that need not concern us here). The characteristics of a female are also summarized in Table 1: Females do not occur naturally in *C. elegans* but do occur in closely related species such as *Caenorhabditis remanei*; females can also be generated by a variety of mutations in *C. elegans*. It is likely that the only difference between a hermaphrodite and a true female in this species lies in the brief phase of spermatogenesis during larval growth of hermaphrodites. Most nematode species have male and female sexes, but no hermaphrodite sex, so we believe that hermaphroditism is a secondary specialization. If this is so, then the original nematode sex-determination mechanism must have been modified to permit spermatogenesis (a male function) to occur in the otherwise female body and gonad of the hermaphrodite. The genetic analysis of sex determination has provided some indication as to how this modification is achieved.

The powerful genetic methods that have been developed for investigating *C. elegans* have been used to analyze the nature of the primary sex-determining signal, which is the dosage of X chromosomes. Madl and Herman (1979) showed that the critical variable is the ratio of X chromosomes to autosomes, not the absolute number; thus, a 4A;2X tetraploid is male rather than hermaphrodite. They also showed that multiple X-chromosome sites contribute to this ratio, so there appears to be no single, major sex-determining gene on the X chromosome.

Several major autosomal sex-determining genes, however, have been identified (Hodgkin and Brenner 1977; Hodgkin 1980; Doniach and Hodgkin 1984; Kimble et al. 1984). These appear to respond to the X/A

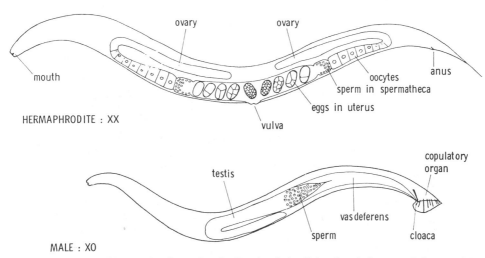

Figure 1. Schematic diagrams of hermaphrodite and male. For detailed cellular descriptions, see Sulston and Horvitz (1977), Kimble and Hirsh (1979), and Sulston et al. (1980, 1983). (Reprinted, with permission, from Hodgkin 1985b.)

Table 1. Sexual Dimorphism in *C. elegans*

	Male	Hermaphrodite	Female
Wild-type karyotype	XO	XX	—
Germ line	3000 sperm —	330 sperm 2000 oocytes	— 2000 oocytes
Somatic gonad	one arm, 56 nuclei	two arms, 143 nuclei	two arms
Vulva	absent	present	present
HSN cells	die	survive	survive
Tail anatomy	complex	simple	simple
Yolk protein synthesis in adult intestine	absent	present	present

ratio and to direct development along either the male or the hermaphrodite developmental pathway. Seven such genes have been identified so far, all of which affect sex determination in both germ line and soma, so that mutations in some of these genes can cause complete sexual transformation (e.g., Fig. 2). For each of these genes, at least four loss-of-function mutations have been isolated. Either amber alleles or genetic deficiencies, or both, have been obtained for six of the genes, and these have been used to infer the null phenotypes of these genes (see Table 2). Table 2 also lists the phenotypes of the rare, dominant mutations that have been obtained for several of the genes: In each case, the dominant mutations cause a sexual transformation opposite to that caused by the putative null alleles (Hodgkin 1983b; Trent et al. 1983). This is consistent with the dominant mutations causing some kind of constitutive or excess expression of an otherwise unaltered gene product.

Inspection of double-mutant phenotypes allows these seven genes to be placed in a hierarchy of epistasis (Hodgkin 1980, 1984; Doniach and Hodgkin 1984). We have interpreted these results in terms of a cascade of negative regulatory interactions, the essentials of which are shown in Figure 3. According to this basic model, only the first gene in the cascade, *her-1*, responds to the X/A ratio. When this ratio is high (as in an XX animal), the *her-1* gene (or its gene product) is repressed in some way, and thus the *her-1* gene is functionally "OFF." This leaves the *tra-2* and *tra-3* genes active so the *fem* gene activities are repressed, which in turn permits the *tra-1* gene to be active ("ON"). This gene activity alone can promote female somatic development. Alternatively, when the X/A ratio is low (as in an XO animal), an opposite set of activities is established. This means that the *fem* genes are ON and the *tra-1* gene is OFF, resulting in fertile male development.

This model explains most of the data on mutant phenotypes, both of single mutants and multiple mutants. It is also consistent with the time of action of some of the genes, as inferred from temperature-sensitive periods of temperature-sensitive alleles (for a summary, see Hodgkin 1984). However, it is only a formal model, which leaves many questions unanswered. These questions are of two kinds. First are the questions of molecular detail: What is the nature of the regulatory interactions between the different genes, or what precisely

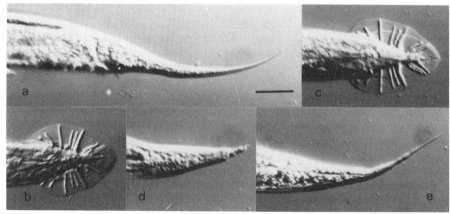

Figure 2. Complete sexual transformations effected by *tra-1* mutations, illustrated by tail phenotype. (*a*) Wild-type XX hermaphrodite; (*b*) wild-type XO male; (*c*) *tra-1*(−) XX male; (*d*) *tra-1(dom)*/+ XO female; (*e*) *tra-1(dom)* homozygous XO female. The *tra-1*(−) allele was *e1099*, and the *tra-1(dom)* allele was *e1575*. The animals in *d* and *e* were also homozygous for *dpy-26(n199)* in order to prove XO karyotype (Hodgkin 1983c). The animal in *d* has a truncated tail spike, characteristic of slightly incomplete female somatic development. Bar, 25 μm.

Table 2. Sex-determining Genes

Genotype	XX phenotype	XO phenotype
Wild type	hermaphrodite	male
(a) *Putative null phenotypes (recessive)*		
tra-1 III	male	male
tra-2 II	abnormal male	male
tra-3 IV	abnormal male	male
her-1 V	hermaphrodite	hermaphrodite
fem-1 IV	female	female
fem-2 III	female	female
fem-3 IV	female	female
(b) *Phenotype of dominant mutations*		
tra-1(dom)/+	female	female
tra-2(dom)/+	female	male
her-1(dom)	masculinized hermaphrodite	male

Table 3. Effect of *egl-41* and *fem-1* Genotype on HSN Cell Death in XX Animals

egl-41	fem-1(mat)	fem-1(zyg)	HSN phenotype
+	+	+	present (wild type)
−	+	+	absent (1/20)
−	+	−	absent (0/8)
−	−	+	present (9/14)
−	−	−	present (16/16)

of L1 larvae using Nomarski optics; the numbers record the fraction of animals with at least one surviving HSN cell.

is being measured in the assessment of the X/A ratio? Sure answers must await cloning and sequencing of these genes, which is now a feasible prospect thanks to advances in the molecular genetics of *C. elegans* (see, e.g., Eide and Anderson 1985).

The second kind are questions of overall organization and function: What is the purpose of this cascade of interactions? Does it serve a useful purpose, or is it merely a consequence of the way the system has evolved? A tentative answer can be provided, which is that the cascade permits more flexibility: It can be modified to function rather differently in different tissues. This is particularly necessary in the germ line of hermaphrodites, which must go through a phase of masculinization in order to permit spermatogenesis. As described in this paper, we have identified a modified control pathway that is necessary for spermatogenesis. We have also identified other minor controls that are required only for certain events during sexual development.

METHODS

The methods used were carried out essentially as described elsewhere (e.g., Doniach and Hodgkin 1984). For genetic nomenclature, see Horvitz et al. (1979). Photographs of live worms were taken using Nomarski optics. HSN cells were scored (Table 3) by inspection

RESULTS

Isolation of New Feminizing Mutations

Recessive mutations that transform XX animals (normally hermaphrodite) into functional females are readily obtained in *C. elegans*. Most of these, however, are mutations that affect spermatogenesis or sperm function (see, e.g., Argon and Ward 1980), and such mutations are unlikely to identify sex-determination genes. For this reason, feminizing mutations were sought using two new methods.

The first method was to screen for females in the F_1 progeny of mutagenized wild-type hermaphrodites, in the hope of identifying dominant feminizing mutations (T. Doniach, unpubl.). It was already known that at least one of the known sex-determination genes, *tra-1*, could yield dominant feminizing alleles (Hodgkin 1983b). Eleven mutations of this type were obtained, from approximately 75,000 F_1 progeny screened. One of these was a dominant mutation affecting spermatogenesis, resulting in arrest at the primary spermatocyte stage. Of the remainder, three were alleles of *tra-1* (resembling the dominant allele previously described); one was a semidominant allele of the gene *fem-3*, and six were dominant alleles of *tra-2* (see below).

The second method was to select for extragenic suppressors of a weak masculinizing mutation, *tra-3-(e1767)* (Hodgkin 1985a and unpubl.). A large number of suppressors were obtained and most of these have been characterized; 12 are dominant alleles of *tra-1*, 1 is a dominant allele of *tra-2*, 4 are recessive alleles of

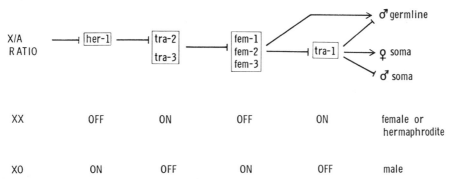

Figure 3. Basic model for interactions between sex-determining genes and proposed activity states for the various genes in the two sexes. The true activity states may really be "HIGH" and "LOW" rather than ON and OFF.

fem-1, 5 are recessive alleles of *fem-2*, and 16 are alleles of *fem-3*.

Germ-line-specific Modulation of *tra-2* Activity

Several dominant feminizing mutations of *tra-2* have been isolated as described above. The strongest of these mutations (e.g., *e2020* and *e2046*) cause complete feminization of XX animals in both heterozygotes and homozygotes, but they have little or no effect on XO animals, which are fertile males. Homozygous male/female strains of these mutations can therefore be propagated with no difficulty. They have been shown to be *tra-2* mutations by selecting for mutations conferring self-fertility on *e2020*/+ or *e2046*/+ XX animals. Intragenic revertants, e.g., *e2046e2113*, were obtained that have a recessive masculinizing phenotype. They resemble *tra-2(null)* mutations, such as the amber allele *e1425* (Hodgkin 1985a), and fail to complement *tra-2(null)* alleles, so they are assumed to be double mutants with a null mutation in *cis* to the dominant *e2020* or *e2046*. The *cis*-heterozygotes, for example, *e2046e2113*/+ +, have a different phenotype (hermaphrodite) from *trans*-heterozygotes such as *e2046*/*e1425* (female), demonstrating that *e2046* is a *tra-2* allele.

These dominant *tra-2* alleles have a major effect only on the XX germ line, and therefore differ from *tra-1(dom)* alleles, which cause complete feminization of germ line and soma in both XX and XO animals. The *tra-2* alleles appear to identify a germ-line-specific control required for spermatogenesis in the hermaphrodite. We therefore suggest that in the wild-type hermaphrodite, the activity of *tra-2* is initially repressed in the germ line, so that the first gametes to differentiate are sperm rather than oocytes. Subsequently, the repression of *tra-2* activity is alleviated, and the germ line switches to oogenesis. The net result is the production of both types of gametes in the XX hermaphrodite gonad: An initial (larval) phase of spermatogenesis is followed by sustained oogenesis in the adult. We interpret the *tra-2(dom)* alleles as being insensitive to the germ line repression, so the germ line in the XX animal makes only oocytes. However, these alleles are still responsive to the repression of *tra-2* activity by *her-1* activity in XO animals, and thus *tra-2(dom)* XO animals are fertile males.

At present, it is not clear how germ line repression is achieved (symbolized by ? in Fig. 6). If a particular gene activity is required for the transient repression, it should be possible to identify it by means of recessive mutations, which are predicted to feminize *tra-2*(+) XX animals but not *tra-2*(−) XX animals or *tra-2*(+) XO animals. Mutations in a gene *fog-2 V* seem to have such properties (T. Schedl and J. Kimble, pers. comm.). Note that this predicted gene activity need not be controlled by the X/A ratio (unlike other sex-determining genes), because in the XO animals, complete repression of *tra-2* activity is adequately achieved by the action of *her-1*. However, there may well be some form of temporal and spatial control of this activity, because it acts only early and only in (or on) the germ line.

The independent control of *tra-2* in germ line and soma raises the question of where these various control genes are acting. It is possible that the germ line phenotype is actually controlled by influences extrinsic to the germ line nuclei. It is already known that the mitosis/meiosis decision is controlled extrinsically by the distal tip cells of the somatic gonad (Kimble and White 1981). These cells might also influence the sperm/oocyte choice, in which case the necessary modulation of *tra* and *fem* activities could be taking place in the distal tip cells (or some other somatic cell), rather than in the germ line cells. Mosaic analysis, which is now feasible in *C. elegans* (see, e.g., Herman and Kari 1985), may provide answers to these questions.

The *fem* Genes

The *tra-3* reversion scheme described above yielded several isolates of each of the three *fem* genes, so it is possible that there are only three genes acting at this level in the control pathway. We have previously reported a detailed analysis of *fem-1* (Doniach and Hodgkin 1984), and Kimble et al. (1984) have described a temperature-sensitive allele of *fem-2*. The new alleles of *fem-2* probably include null alleles: They have stronger effects than those of the first allele *b245* but weaker than those of most *fem-1* alleles. The *fem-3* gene is a new discovery and is further described below.

Recessive mutations in the three *fem* genes have the following properties in common: They cause complete transformation of XX and XO animals into fertile females, are fully epistatic to masculinizing mutations of *tra-2* and *tra-3*, and are epistatic to masculinizing mutations of *tra-1* in the germ line but not in the soma. Consequently, the phenotype of *fem*(−);*tra-1*(−) is a male animal with oocytes formed inside a male gonad. In addition, all three *fem* genes show marked maternal effects. The common properties suggest that the three *fem* genes act together to repress *tra-1* activity and to promote spermatogenesis.

The mutations of *fem-3* range from weak alleles, which cause only a slight or temperature-sensitive feminization of XO animals and do not completely eliminate spermatogenesis from XX animals, to strong alleles (e.g., *e1996*), which cause complete feminization of XO and XX animals. These strong mutations behave similarly in complementation tests to two small LGIV deficiencies, *eDf18* and *eDf19*, which fail to complement *fem-3* and an adjoining gene, *daf-15* (Swanson et al. 1984). We therefore believe that the strong mutations are null alleles. It is important to establish this because the strong alleles show a slight dominant effect: *e1996*/+ XX animals are occasionally female rather than hermaphrodite and always show reduced self-progeny brood sizes (mean 252 for 35 animals as compared with 328 for 21 wild-type hermaphrodites), which probably reflects reduced spermatogenesis. The strong alleles are also weak dominant suppressors of *tra-3(e1767)*, which explains why *fem-3* mutations are found so frequently among *tra-3* suppressors. In contrast, null or putative null alleles of *fem-1* and *fem-2* are recessive. Probably, the partial dominance of *fem-3* alleles is a consequence of haplo-insufficiency for this locus. Another of the sex-determination genes, *tra-2*, also shows slight haplo-insufficiency (Trent et al. 1983).

A Paradoxical Maternal Effect

The *fem-3* gene shows a strong maternal rescue effect: *fem-3* XO progeny of *fem-3* XX mothers are female, but *fem-3* XO progeny of *fem-3/+* XX mothers are intersexual (Fig. 4a,b). A similar maternal effect has already been reported for *fem-1* and *fem-2* (Doniach and Hodgkin 1984; Kimble et al. 1984). Furthermore, the maternal contribution of the *fem-3(+)* gene product appears to be essential for normal male development. When *fem-3* XX females are crossed with wild-type males, many of the *fem-3/+* XO male progeny have abnormal gonads, and about 15% exhibit rudimentary vulval development (Fig. 4c). However, some *fem-3/+* XO males are normal (Fig. 4d), and those produced by *fem-3/+* or *+/+* mothers are always normal, indicating that the effect is not due to haplo-insufficiency.

These maternal effects are paradoxical because they indicate that in the wild-type hermaphrodite, the germ line is synthesizing substantial amounts of *fem* gene products and contributing these products to all oocytes. But the role of these gene products is to *masculinize* (by repressing *tra-1* activity and promoting spermatogenesis), and thus their presence in the adult hermaphrodite germ line (which is essentially female and oogenic) is inappropriate. It is possible that they are synthesized at a late stage or in an inactive form, and so they are unable to affect the type of gamete formed. The presence of *fem* gene products is also inappropriate in any oocyte that is fertilized to give an XX zygote, destined for female somatic development.

We rationalize the maternal effect by suggesting that it is important to keep the final sex-determination gene *tra-1* fully inactive during early embryogenesis, because it might otherwise be expressed at a sufficient level to cause irreparable feminization, eliminating spermatogenesis from the XX animal and causing intersexual development of the XO animal. The fact that there is an *essential* need for the maternally contributed *fem-3(+)* product is consistent with this explanation.

Posttranscriptional Control of *fem* Activity: Major and Minor Controls

We have argued previously (Doniach and Hodgkin 1984) that the maternal effect of *fem* genes means that

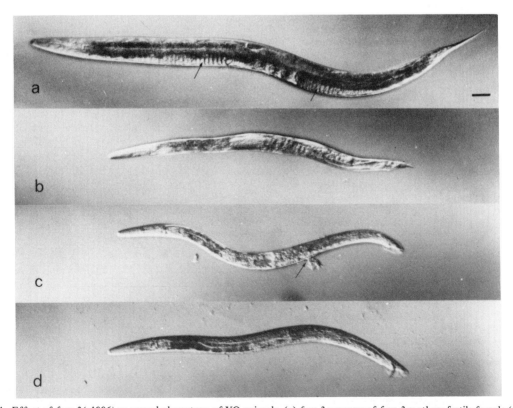

Figure 4. Effect of *fem-3(e1996)* on sexual phenotype of XO animals. (*a*) *fem-3* progeny of *fem-3* mother: fertile female (oocytes arrowed). (*b*) *fem-3* progeny of *fem-3/+* mother: intersex. (*c*) *fem-3/+* progeny of *fem-3* mother: sterile abnormal male (vulval structure arrowed). (*d*) *fem-3/+* progeny of *fem-3* mother: fertile male. Bar, 50 µm.

there must be a posttranscriptional mechanism for removing or inactivating *fem* gene products from the XX zygote, where they are inappropriate. The major control that achieves this is the activity of *tra-2* (see Fig. 1).

It is now clear that there is an additional minor control involving the gene *egl-41 V*. This activity seems to be required mainly for the very earliest event in hermaphrodite sex-specific development, which is the survival of the two HSN neurons. These two neurons undergo programmed cell death in male embryos (Sulston et al. 1983), but in hermaphrodites, they survive and differentiate, being necessary for normal egg laying. In *egl-41* hermaphrodites, the two neurons die as they do in males, and thus the adult *egl-41* hermaphrodite is egg-laying-deficient (hence the gene name). In contrast, the HSN deaths usually fail to occur in *egl-41* hermaphrodites if the maternally contributed *fem-1(+)* gene product is absent, and they never occur in the complete absence of the *fem-1(+)* gene product (Table 3).

Three mutations of *egl-41* were obtained by C. Desai and H.R. Horvitz (pers. comm.) on the basis of the egg-laying defect; one allele (*e2055*) was isolated independently by T. Doniach (unpubl.) as an extragenic suppressor of a weak, dominant feminizing mutation of *tra-2*. All four mutations have similar properties. It is not certain whether these mutations result in loss of function or gain of function in the *egl-41* gene, particularly since the mutations have some dominant effects, and therefore the present interpretation is tentative.

The existence of a minor feminizing pathway independent of *tra-2* was deduced previously (see Hodgkin 1980), mainly on the basis of the following argument. If the simple model (Fig. 1) were correct, then the phenotype of *tra-2* XX (and of *tra-2;tra-3* XX) should be identical to that of *tra-1* XX. This is not the case, because *tra-2* XX animals are incompletely male in tail anatomy, so there must be an additional feminizing influence acting upstream of *tra-1* and independent of *tra-2*. The *egl-41* activity appears to constitute such an activity, particularly since *tra-2;egl-41* XX animals are sometimes distinctly more masculinized than *tra-2* XX animals.

Note that the *egl-41* activity appears to be necessary only for one event in early hermaphrodite development. Consequently, *egl-41* XX animals are essentially hermaphrodite, apart from the lack of HSN cells, in contrast to *tra-2(-)* XX animals, which are completely male in germ line phenotype and almost completely male in the soma. The minor (*egl-41*) pathway can be regarded as a piece of "fine tuning" on the main mechanism.

Downstream Functions: Realizator Genes?

So far, we have discussed genes that regulate, directly or indirectly, the crucial control gene *tra-1*. The properties of *tra-1* mutations indicate that this activity alone, in the absence of any of the other regulatory activities and at any X/A ratio, can direct female development in both germ line and soma. Also, the absence of *tra-1* activity results in male somatic development. However, male germ line differentiation requires both the absence of *tra-1* activity and the presence of *fem* gene activity.

The model predicts the existence of particular somatic differentiation functions under the control of *tra-1*: In the presence of *tra-1* activity, the female functions are activated and the male functions are repressed. In the germ line, genes required for spermatogenesis are repressed by *tra-1* activity and activated by *fem* activity. Oogenesis appears to be a "default" program that is pursued if the *fem* genes are inactive or if *tra-1* is active.

One can draw a distinction between genes that are involved in choosing between two or more developmental possibilities and genes that are required to execute those possibilities. The distinction corresponds to that made by Garcia-Bellido (1975) between "selector" genes and "realizator" genes. Mutations in the two classes of genes are expected to have different consequences: Mutations in selector genes will result in developmental *transformations* (such as the sexual transformations described above), whereas mutations in realizator genes will result in defective development without transformation.

A number of sex-specific differentiation genes have been identified, which may be the targets for *tra-1* and *fem* gene control (i.e., realizator genes). Some of these are required for the final sexual differentiation of specific tissues such as gametes or intestinal cells. The major sperm-protein genes (Burke and Ward; 1983; Klass et al. 1984) and the yolk-protein genes (Blumenthal et al. 1984), which are expressed in the intestine (Kimble and Sharrock 1983), fall into this category. More interesting are genes that seem to be involved in earlier decisions. Mutants of this type have been sought by screening for male-specific abnormality (*mab* mutants). Of the ten *mab* genes so far described (Hodgkin 1983a), most appear to be required for normal development in both sexes, although they are more important for male development than for hermaphrodite development.

Two of the *mab* mutants have more striking phenotypes. One, *mab-9*, affects the early larval divisions of the male-specific blast cells B, U, and F (Sulston and Horvitz 1977; Sulston et al. 1983). These cells do not divide during wild-type hermaphrodite development, and *mab-9* mutant hermaphrodites appear to develop quite normally. In males, however, the B, U, and F cells contribute a total of 55 cells to the adult male tail, so the disruption of their division patterns leads to an extreme anatomical abnormality (Fig. 5a). Other events in male development, such as gonadal and germ line maturation, proceed normally, and therefore *mab-9* appears to be a good candidate for a realizator gene, executing one particular part of male development.

A second *mab* gene with an apparently male-specific phenotype is *mab-3*, which affects a number of male-

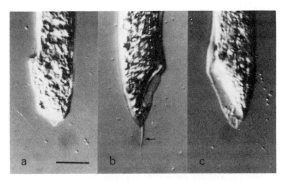

Figure 5. Tail phenotypes of male animals (cf. Fig. 2, b and c). (*a*) *mab-9(e1245)* XO; (*b*) *mab-3(e1240)* XO (tail spike arrowed); (*c*) *mab-3(e1240); tra-1(e1099)* XX. Bar, 25 μm.

specific developmental events. Mutant males usually fail to make the anterior six rays of the copulatory bursa (Fig. 5b) and also exhibit other abnormalities; mutant hermaphrodites are superficially wild type. A remarkable additional phenotype has been observed in three independently isolated *mab-3* alleles; adult *mab-3* XO males synthesize yolk proteins, which are observed to accumulate in the pseudocoelom of old adult males (M. Shen and C. Kenyon, unpubl.). Therefore, *mab-3* appears to have two rather different roles: (1) It is required for normal tail morphogenesis during larval male development and (2) it is required for the repression of yolk protein synthesis in the adult male intestine. If *mab-3* acts downstream from *tra-1*, then the *mab-3;tra-1* XX male is expected to have the same phenotype as the *mab-3* XO male. This is approximately what is observed, but the amount of yolk protein synthesized in the double mutant is much reduced. In addition, the hermaphroditic tail spike often seen in *mab-3* XO animals is almost never seen in the double mutant (Fig. 5c). These observations (and other unpublished results) indicate that the regulatory relationship between *tra-1* and *mab-3* is not simple and that there may be a feedback interaction between *mab-3* and the upstream *fem* genes, as indicated in Figure 6. Therefore, in the case of *mab-3*, the distinction between a selector gene and a realizator gene is blurred, because *mab-3* has characteristics of both.

A similar situation is observed with two genes that seem to have germ-line-specific functions, called *fog-1* I and *mog-3* II (*fog* stands for feminization of germ line and *mog* stands for masculinization of germ line, these being the respective mutant phenotypes). Mutations in these genes have been identified fortuitously during general screens for mutants with abnormal germ line differentiation (T. Doniach, unpubl.). The *fog-1* mutation eliminates spermatogenesis from XX animals and causes oogenesis in the testes of XO animals. The *mog-3* mutation eliminates oogenesis in XX animals, so that all gametes differentiate into sperm. Somatic tissues are not obviously affected by either mutation. The results of epistasis tests show that both of these genes appear to act downstream from the regulatory *fem* and *tra-1* genes. The double mutant *fog-1;mog-3* makes neither sperm nor oocytes; instead, abnormal gametes with an intermediate phenotype are formed. If these two genes had a purely regulatory role, one would expect that either sperm or oocytes or a mixture of both would be found in the double mutant, rather than the hybrid gametes observed. However, if the two genes had a purely differentiative role, then single mutants would be expected to make defective gametes of one class, rather than causing a transformation of one gamete class into the other. Thus, these genes, like *mab-3*, appear to have both a regulatory role and a differentiative role.

fog-1 and *mog-3* nevertheless conform to the generally hierarchical nature of sexual differentiation in *C. elegans* since they affect the germ line but not the soma. The *mab-3* and *mab-9* genes have a reciprocal role, affecting the soma but not the germ line. There are other

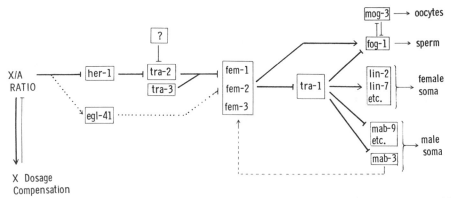

Figure 6. Summary of proposed gene interactions. (Heavy arrows) Major interactions; (pointed arrows) positive regulation; (barred arrows) negative regulation. Thin line from the question mark indicates an interaction that occurs only in the larval hermaphrodite germ line. Dotted line indicates an interaction (the minor pathway) that is required only in early hermaphrodite somatic development. Dashed line indicates an interaction required only in late male somatic development. Some of these minor regulative interactions are very speculative at present (for further explanations, see text). The interaction between X-chromosome dosage compensation and sex determination is not discussed in this paper, but is included in this figure for the sake of completeness.

candidates for genes in this class. For example, several of the *lin* genes (e.g., *lin-2* and *lin-7*) that affect vulva formation in the hermaphrodite (Ferguson and Horvitz 1985) may be required only for this sex-specific event and are not required for other parts of hermaphrodite development or for male development.

DISCUSSION

The main object of analyzing any developmental phenomenon in detail must be to relate it to other phenomena, in the hope of identifying general principles that apply to many different organisms. Every animal species seems to have quirks in some aspects of its development, peculiar processes that seem to be absent from other animals, even apparently close relatives. It is preferable to avoid the quirky, but lack of generality may only become apparent after a lot of analyses. Often, too, apparent similarities may turn out to be purely superficial. An example in the present context is the phenomenon of X-chromosome dosage compensation, which initially seemed to involve similar principles in *Drosophila* (Baker and Belote 1983; Cline 1984) and in *C. elegans* (Hodgkin 1983c), but now the mechanisms look increasingly different (Hodgkin 1985a; J. Hodgkin, unpubl.; B.J. Meyer, pers. comm.). Recent observations (J. Hodgkin, unpubl.) also suggest that the relationship between dosage compensation and sex determination is different in these two animals.

What generalities can be gleaned from the results described in this paper? First, and perhaps most important, is the fact that this kind of analysis is possible at all. We have identified key control genes, we have been able to mutate these genes so that the system becomes locked into one state or another, and we have been able to infer how the genes interact with each other. The sex-determination pathway involves a limited set of genes that can be considered and understood in isolation, and thus the system appears to be modular and comprehensible.

Second, the pathway is largely but not completely hierarchical: Genes that act at one level are mostly not affected by genes at a lower level, which act later temporally and/or in a more spatially restricted manner. An interesting exception to this is the gene *mab-3*, which may be involved in a feedback loop on the *fem* genes. This loop may be necessary to maintain the *tra-1* gene in a repressed state during later stages of development of XO animals. However, *mab-3* seems to be required only in a limited set of somatic tissues; other genes may play an analogous role in different tissues. Reinforcing feedback loops of this type have long been recognized as potentially useful mechanisms for maintaining a particular set of control gene activities (see, e.g., Kaufman 1973).

Third, maternal effects are important: The system cannot be understood solely on the basis of the gene activities in the zygote. It is clear that maternal gene products play vital roles in the early development of almost all animals. One may note also that maternal effects represent a special case of a more general principle: that the gene activities in any cell may be affected by antecedent gene activities in the mother of that cell. Posttranscriptional control mechanisms may be required to remove inherited products, as in the case of the *fem* gene products.

Fourth, regulatory interactions between different levels are mostly negative. This is surprising, because it seems to entail some rather cumbersome machinery to ensure that the key gene activity (*tra-1*) remains repressed and then to achieve derepression when required. Thus, the *fem* gene products (which we think are the proximate negative regulators of *tra-1*) are inherited maternally and then must be removed from XX zygotes, entailing the action of *egl-41*, a gene activity that otherwise seems dispensable. From a teleological point of view, we might argue that positive controls would be more efficient and economical.

Fifth, the same basic control pathway may be modified for different purposes in different tissues. The seven major genes affect sexual identity in all the sexually specialized parts of this animal, but there are additional minor genes that modify their action, so that the pathway has flexibility.

Finally, we have tried to understand the overall logic of the sex-determination system, treating it as if it had at least some elements of optimal design, but this may be a mistaken approach. Since animal development is the product of evolution, many of its features may be historical accidents. We have argued that some of the complexity of the gene interactions in *C. elegans* sex determination results from the need to modify a female sex into a hermaphrodite sex. Comparison of *C. elegans* with its close relative *C. remanei* (which has a male/female sex-determination system) should be possible once some of the sex-determining genes are cloned and may allow us to infer how much of this system is the result of a historical accident and how much represents an efficient solution to the general problems of sex determination.

ACKNOWLEDGMENTS

We are grateful to Judith Kimble, Tim Schedl, Chand Desai, Robert Horvitz, and Barbara Meyer for communicating unpublished results and to Cynthia Kenyon for discussion and for some of the observations on *mab-3*. T.D. is supported by a Thomas C. Usher Studentship. M.S. is supported by a National Science Foundation Graduate Fellowship.

REFERENCES

Argon, Y. and S. Ward. 1980. *Caenorhabditis elegans* fertilization-defective mutants with abnormal sperm. *Genetics* **96:** 413.

Baker, B.S. and J. Belote. 1983. Sex determination and dosage compensation in *Drosophila melanogaster*. *Annu. Rev. Genet.* **17:** 345.

Blumenthal, T., M. Squire, S. Kirtland, J. Cane, M. Donegan, J. Spieth, and W. Sharrock. 1984. Cloning of a yolk protein gene family from *C. elegans*. *J. Mol. Biol.* **174:** 1.

BURKE, D.J. and S. WARD. 1983. Identification of a large multigene family encoding the major sperm protein of *C. elegans*. *J. Mol. Biol.* **171**: 1.

CLINE, T. 1984. Autoregulatory functioning of a *Drosophila* gene product that establishes and maintains the sexually determined state. *Genetics* **107**: 231.

DONIACH, T. and J. HODGKIN. 1984. A sex determining gene, *fem-1*, required for both male and hermaphrodite development in *Caenorhabditis elegans*. *Dev. Biol.* **106**: 223.

EIDE, D. and P. ANDERSON. 1985. Transposition of Tc*1* in the nematode *Caenorhabditis elegans*. *Proc. Natl. Acad. Sci.* **82**: 1756.

FERGUSON, E.L. and H.R. HORVITZ. 1985. Identification and characterization of 22 genes that affect the vulval cell lineages of the nematode *Caenorhabditis elegans*. *Genetics* **110**: 17.

GARCIA-BELLIDO, A. 1975. Genetic control of wing disc development in *Drosophila*. *Ciba Found. Symp.* **29**: 161.

HERMAN, R.K. and C.K. KARI. 1985. Muscle-specific expression of a gene affecting acetylcholinesterase in the nematode *Caenorhabditis elegans*. *Cell* **40**: 509.

HODGKIN, J. 1980. More sex determination mutants of *Caenorhabditis elegans*. *Genetics* **96**: 649.

———. 1983a. Male phenotypes and mating efficiency in *Caenorhabditis elegans*. *Genetics* **103**: 43.

———. 1983b. Two types of sex determination in a nematode. *Nature* **304**: 267.

———. 1983c. X chromosome dosage and gene expression in *C. elegans*: Two unusual dumpy genes. *Mol. Gen. Genet.* **192**: 452.

———. 1984. Switch genes and sex determination in the nematode *C. elegans*. *J. Embryol. Exp. Morphol.* (suppl.) **83**: 103.

———. 1985a. Novel nematode amber suppressors. *Genetics* **111**: 287.

———. 1985b. Males, hermaphrodites, and females: Sex determination in *Caenorhabditis elegans*. *Trends Genet.* **1**: 85.

HODGKIN, J. and S. BRENNER. 1977. Mutations causing transformation of sexual phenotype in the nematode *Caenorhabditis elegans*. *Genetics* **86**: 275.

HORVITZ, H.R., S. BRENNER, J. HODGKIN, and R.K. HERMAN. 1979. A uniform genetic nomenclature for the nematode *Caenorhabditis elegans*. *Mol. Gen. Genet.* **175**: 129.

KAUFMAN, S.A. 1973. Control circuits for determination and transdetermination. *Science* **181**: 310.

KIMBLE, J. and D. HIRSH. 1979. Postembryonic cell lineages of the hermaphrodite and male gonads in *Caenorhabditis elegans*. *Dev. Biol.* **70**: 396.

KIMBLE, J.E. and W.J. SHARROCK. 1983. Tissue-specific synthesis of yolk proteins in *C. elegans*. *Dev. Biol.* **96**: 189.

KIMBLE, J.E. and J.G. WHITE. 1981. On the control of germ cell development in *Caenorhabditis elegans*. *Dev. Biol.* **81**: 208.

KIMBLE, J., L. EDGAR, and D. HIRSH. 1984. Specification of male development in *C. elegans*: The *fem* genes. *Dev. Biol.* **105**: 234.

KLASS, M.R., S. KINSLEY, and L.C. LOPEZ. 1984. Isolation and characterization of a sperm-specific gene family in the nematode *C. elegans*. *Mol. Cell. Biol.* **4**: 529.

MADL, J.E and R.K. HERMAN. 1979. Polyploids and sex determination in *Caenorhabditis elegans*. *Genetics* **93**: 393.

SULSTON, J.E. and H.R. HORVITZ. 1977. Postembryonic lineages of the nematode *Caenorhabditis elegans*. *Dev. Biol.* **56**: 110.

SULSTON, J.E., D.G. ALBERTSON, and J.N. THOMSON. 1980. The *Caenorhabditis elegans* male: Post-embryonic development of nongonadal structures. *Dev. Biol.* **78**: 542.

SULSTON, J.E., E. SCHIERENBERG, J.G. WHITE, and J.N. THOMSON. 1983. The embryonic cell lineage of the nematode *Caenorhabditis elegans*. *Dev. Biol.* **100**: 64.

SWANSON, M.M., M.L. EDGLEY, and D.L. RIDDLE. 1984. The nematode *Caenorhabditis elegans*. In *Genetic maps* (ed. S.J. O'Brien), vol. 3, p. 286. Cold Spring Harbor Laboratory, Cold Spring Harbor, New York.

TRENT, C., N. TSUNG, and H.R. HORVITZ. 1983. Egg-laying defective mutants of the nematode *Caenorhabditis elegans*. *Genetics* **104**: 619.

Sex-lethal, A Link between Sex Determination and Sexual Differentiation in *Drosophila melanogaster*

E.M. MAINE, H.K. SALZ, P. SCHEDL, AND T.W. CLINE
Biology Department, Princeton University, Princeton, New Jersey 08544

Early in this century it was discovered that a rather small change (50%) in the relative number of X chromosomes in *Drosophila melanogaster* is responsible for one of the more striking examples of differential gene expression observed in higher eukaryotes. This classical illustration of a fundamental puzzle in developmental biology involves fruit fly sexual differentiation. Diploid cells with two X chromosomes (X chromosomes:sets of autosomes = 1) differentiate as female, whereas diploid cells with only one X chromosome (X:A = 0.5) differentiate as male (Bridges 1925).

Since that time, considerable progress has been made toward understanding the nature of this quantitative developmental signal and the mechanism by which it is transduced into the remarkable qualitative and quantitative differences in development between the sexes (for review, see Baker and Belote 1983; Cline 1985). Recent experimental evidence suggests that although cells carry their sex-determination signal (the X:A balance) with them throughout development, they seem to read this signal only early in development, long before most overt aspects of sexual differentiation occur. They then initiate a sexual pathway commitment that seems to be maintained independent of the initiating signal (Cline 1985 and unpubl.). Thus, for *Drosophila*, the term sex determination seems to apply not only in its general sense, but also in the much more restricted sense of developmental pathway commitments maintained epigenetically by dividing cells. The classical paradigm for "determination" in this latter sense involves the process by which the precursor cells of the adult *Drosophila* integument acquire and maintain their segmental and subsegmental identities (Gehring 1972). Comparisons between the mechanisms of sex determination and segmental determination of *Drosophila* may lead to the discovery of common molecular mechanisms that govern developmental pathway commitments in metazoans.

Three broad aspects of *Drosophila* sexual development can be considered: initiation of the sexual pathway commitment, maintenance of the pathway commitment, and expression of the pathway commitment. We discuss here some of the ways that one can distinguish among genetic perturbations that disrupt different steps in this process, even when different perturbations involve the same gene. Awareness of these distinctions is likely to be important for understanding the way regulatory genes coordinate multicellular development.

The Regulatory Gene, Sex-lethal

Many regulatory genes are known to be involved in the programming of *Drosophila* sexual development; however, at this point only Sex-lethal (*Sxl*; 1-19.2) is known to be involved in all aspects of sexual dimorphism and in all three categories of sexual pathway steps listed above. Determination of the functional state of *Sxl* seems likely to be the most immediate response of cells to their perceived X-chromosome dose. Moreover, *Sxl* seems to be the most viable candidate for "carrier" of the sexually determined state: a gene (1) that maintains the developmental commitment in dividing cells independent of "upstream" genetic elements that initiate the pathway and (2) that directs "downstream" elements (including other regulatory genes) to express that commitment when sexual differentiation occurs. In light of this possibility, it is intriguing that the products of *Sxl* appear to be involved in the regulation of this gene, a rare example of *positive* autoregulation in higher eukaryotes (Cline 1984).

There is an important maternal component to *Drosophila* sex determination. Part of the biochemical machinery that the progeny need to read and respond properly to their X:A ratio is built into the egg during oogenesis under the direction of the maternal genome. The maternally synthesized product of a gene named daughterless (*da*; 2-41.5) is a positive regulator of *Sxl*. In the absence of this product, all progeny appear to develop as males, regardless of their chromosomal sex and regardless of their own genotype with respect to *da* (Cline 1983, 1984). Genes downstream from *Sxl* in the sense that their functioning depends on *Sxl* and that mutations in them are epistatic to mutations in *Sxl*, include regulatory elements such as *tra* and *dsx* that are involved solely in aspects of somatic sexual development and the autosomal male-specific lethal loci such as *mle* and *msl-1* that seem to be involved solely in dosage compensation (Baker and Belote 1983; Belote 1983). X-chromosome dosage compensation is a process necessitated by the nature of the sex-determination signal of *Drosophila* and the fact that approximately 20% of this fly's genes are X-linked. Males must compensate for the fact that they have half the dose of X-linked genes as females, yet have similar requirements for most of these genes' products. They hyperactivate their single dose of each compensated gene to match the total transcriptional output of the two gene doses in females (for review, see Stewart and Merriam 1980).

Dosage compensation is a cell-vital process; transcription of these genes at a level that is inappropriate for the X-chromosome dose will lead to a lethal imbalance of gene products. Analysis of *Sxl* benefits from, but is also complicated by, the fact that this gene is involved in both sex determination and dosage compensation. Special strategies must be used in order to avoid having perturbations in sexual development caused by interference with *Sxl* functioning masked by accompanying lethal perturbations in the vital process of dosage compensation.

Sxl is introduced here in the context of some of the ways that we believe can be used to distinguish the level at which various mutations in this master regulatory gene may act; i.e., whether they affect sex determination in the sense of sexual pathway choice and maintenance (the regulation of *Sxl*) or whether they affect sexual differentiation (the pathway expression functions of *Sxl* products). These examples also illustrate how one can study the effects on sexual development of perturbations affecting this female-cell-vital regulatory gene without those effects being masked by the death of the organism or the elimination of aberrant cells during its development. We then discuss the results of our recent molecular characterization of *Sxl* and its various mutant alleles (including mutants introduced in the illustrations). The complexity of the molecular results illustrates why it is so important to have specific assays for the different aspects of regulation and product functioning of *Sxl*.

Sauce for the Goose Is Poison for the Gander

Sxl products direct genetic females (XX AA) to differentiate as phenotypic females. In the absence of this gene activity, all individuals appear to develop as males, regardless of their X:A balance (Cline 1979a, 1983; Sanchez and Nöthiger 1982). Females homozygous for null alleles die as embryos, presumably from upsets in dosage compensation. In contrast, males who are hemizygous for loss-of-function alleles, or who are even deleted for the *Sxl* region, are viable and fertile (E. Maine et al., in prep.).

This recessive, female-specific (masculinizing) lethality of loss-of-function *Sxl* alleles contrasts with the dominant, male-specific (feminizing) lethality of gain-of-function alleles like Sxl^{M1}. The Sxl^{M1} lesion causes the gene to express its female-specific functions regardless of the normal sex-determination signals (Cline 1979b, 1983, 1984). This is clearly inappropriate for genetic males (they generally die as young larvae), but it has no adverse effect on genetic females. Indeed, such mutant lesions rescue females from the otherwise lethal effects of mutations in positive regulators of *Sxl*, such as *da*.

One of the strongest arguments that Sxl^{M1} kills males by imposing normal female-specific vital functions on them, i.e., functions that are inappropriate for haplo-X individuals, is based on the observation that intragenic suppressors of male lethality (Sxl^{fx}) of Sxl^{M1} are invariably associated with loss of female-specific vital functions in the doubly mutant allele, $Sxl^{M1,fx}$. The female-lethal phenotype of such male-viable Sxl^{M1} derivatives demonstrates the *cis*-dominant behavior of Sxl^{M1} (Cline 1981, 1984). The fact that the gain-of-function sex transformation is the reciprocal of the loss-of-function is especially significant in assigning a unique and central position for *Sxl* in the complex regulatory gene network that directs *Drosophila* sexual development.

Distinguishing Effects on Sexual Pathway Choice from Effects on Pathway Expression

Implications of diploid intersexes. With respect to effects on sexual development (ignoring viability considerations), elimination of any one of the three sexual pathway functions suggested for *Sxl* can be expected to generate the same basic phenotype: transformation of genetic females into phenotypic males. For that reason, situations that reduce, but do not eliminate, particular *Sxl* gene functions required for sexual development are especially useful in distinguishing the functioning of *Sxl* in sexual pathway initiation, maintenance, and expression.

There are two distinct classes of incomplete sexual transformation that can be distinguished easily in *Drosophila* because the sexual phenotype of individual cells in the adult integument can be assessed. The distinction between the two categories is most straightforward in the first tarsal segment of the foreleg, and for that reason, it is a favored region for the analysis of sexual perturbations. In the foreleg, the phenotype of the "true intersex" is intermediate between male and female, even at the level of individual cells. In contrast, the "mosaic intersex" phenotype is intermediate over all; but at the level of individual cells, the animal can be seen to be composed of a mixture of apparently normal male and normal female differentiated structures. The simplest situation that generates a true intersex phenotype is when all functions of the doublesex (*dsx*) gene are eliminated. In this case, genetic males and females appear to display nearly the same intersexual phenotype (see Baker and Ridge 1980). The most straightforward situation generating mosaic intersexes is when the primary sex-determination signal has a sexually ambiguous value of 0.67, as in the case for triploid fruit flies who have only two X chromosomes (see Cline 1983). In our discussion of intersexual tissue development in this paper, we are only considering animals in which all cells of the tissue in question are genetically identical. This contrasts with the situation of gynandromorphs, i.e., individuals composed of a mixture of XX and XO tissue generated by chromosome loss occurring during the first few nuclear divisions after fertilization. The phenotypic sexual mosaicism of gynandromorphs reflects their underlying genetic mosaicism. Gynandromorphs do serve to illustrate the point that sexual differentiation among the cells that form the adult integument is a cell autonomous process.

The class of intersex phenotype generated in a particular experimental situation can reflect the level at which the regulation of sexual development is perturbed, whether at the pathway-choice level (including initiation and maintenance aspects) for mosaic intersexes or at the pathway-expression level for true intersexes. As discussed below, different mutations in *Sxl* can generate both kinds of intersexuality among genetically identical cells.

Figure 1 illustrates the sexually dimorphic bristle pattern in the foreleg. Bristle shafts are the products of single polyploid cells and are generated when larvae metamorphose into adults (see Poodry 1980). Figure 1A shows almost normal female development, whereas Figure 1C shows normal male development. The most prominent difference between these patterns involves the sex comb, i.e., a row of distinctively thick, blunt, and dark male bristles that runs along the long axis of the leg at the distal end of the first tarsal segment. In the female pattern, this sex comb is replaced by additional transverse rows of long, thin, sharp, light bristles, like those in more proximal regions of the segment. The leg shown in Figure 1B is a sexual intermediate of the "true intersex" variety, although one in which cells display considerably more phenotypic variation than would be the case for a dsx^- leg. Note the morphologically intermediate sexual phenotype of the two bristles indicated in Figure 1B.

Figure 1 also illustrates the effects of one of the more recently generated genetic tools with which we are exploring the functioning of *Sxl*. All three sexual phenotypes illustrated (female, true intersex, and male) were generated in the *same* mutant genotype of a diploid male (XY AA) raised at different temperatures (29°C, 25°C, and 18°C, respectively). These genetic males carried a complex dominant mutant allele, $Sxl^{M1,fPa\text{-}ra}$, whose expression of female-specific *Sxl* functions depends on temperature, not on the X:A balance. This allele illustrates the effects of abnormally low but nonzero levels of the feminizing activities of *Sxl*. The molecular nature of two of the three lesions carried by this mutant allele is discussed below.

$Sxl^{M1,fPa\text{-}ra}$ was derived from the dominant, unconditionally male-lethal feminizing allele, Sxl^{M1}, in two rounds of PM hybrid dysgenesis (for review, see Engels 1983). In the first round, selection was for dysgenic derivatives of Sxl^{M1} that were viable and fertile in males. A male-viable derivative, $Sxl^{M1,fPa}$, was obtained in which a P-type mobile genetic element had inserted into Sxl^{M1} and thereby destroyed all female-specific functions of the gene. The null derivative had lost the ability of Sxl^{M1} to complement all loss-of-function alleles, and it could no longer support female differentiation in clones induced by somatic recombination. In the second round of dysgenesis, revertants were selected that had regained the ability to complement the female-specific *Sxl* allele, Sxl^{f9}, which we believe is defective only in a positive autoregulatory aspect of *Sxl* product function (see below). Although some revertants to full female function were obtained from $Sxl^{M1,fPa}$, several derivatives like $Sxl^{M1,fPa\text{-}ra}$ were recovered that have increased but sub-wild-type levels of female functions; moreover, in several of these cases, the level of the restored female functions depends on temperature, with higher expression at higher temperatures. Since the mutant alleles still carry the Sxl^{M1} lesion, they express female-specific functions in both ge-

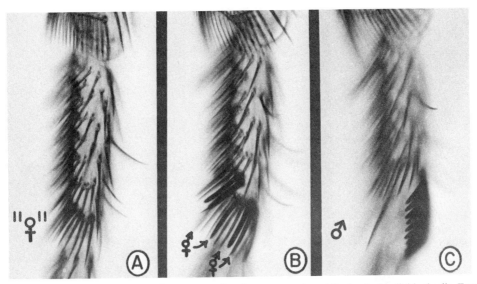

Figure 1. Effects of *Sxl* female-specific functions on sexual pathway expression at the level of individual cells. Female (*A*), true intersex (*B*), and male (*C*) foreleg adult sexual morphology are illustrated by three genetically identical males carrying the multiply mutant allele, $Sxl^{M1,fPa\text{-}ra}$, but grown at different temperatures (29°C, 25°C, and 18°C, respectively). This allele expresses the female-specific activities of *Sxl* as a function of temperature, rather than a function of the normal sex-determination signal, the X:A balance.

netic males and females. Thus, higher temperatures are nonpermissive for normal male development, even if the males also carry Sxl^+ (wild-type alleles actually exacerbate the situation, as expected from autoregulatory considerations).

The results with $Sxl^{M1,fPa-ra}$ show that low but nonzero levels of Sxl feminizing differentiation functions generate true intersex individuals, i.e., animals in which the differentiation of even individual cells is neither male nor female, but intermediate. The genotype illustrated in Figure 1 was the first demonstration of this for Sxl in males; true intersex development caused by low levels of Sxl expression had been shown previously in *females* carrying the temperature-conditional loss-of-function mutant allele, Sxl^{2593} (Cline 1984). Temperature-conditional and partial loss-of-function mutations allow one to manipulate the vital and nonvital functions of Sxl differentially and thereby help to avoid complications from viability effects. We interpret these intersexual phenotypes as reflecting aberrations in sexual pathway expression rather than pathway choice on a number of criteria, including both the appearance of the cells themselves and the nature of the situations in which the phenotype is known to arise.

Intersexes of the mosaic class—the phenotype we believe reflects effects on sexual pathway choice and/or maintenance, rather than pathway expression per se— can be generated by a different set of perturbations in Sxl and in genes that appear to be upstream of it in a regulatory sense. Figure 2A shows the mosaic intersex phenotype generated by the partial loss-of-function allele, Sxl^{fPb}, the allele used to isolate Sxl DNA (see below). Because females homozygous for this allele die before reaching the sexually dimorphic adult stage, we assayed the effect of this mutation on sexual development in clones produced by somatic recombination during the larval stages, a standard ploy to study mutations in vital genes. The mosaic intersexuality in this example is therefore displayed by a clone of homozygous mutant tissue in an otherwise nonmutant individual, rather than in the individual as a whole (for examples of the latter type, see Cline 1984). This mutant clone was generated by somatic recombination induced in a developing heterozygous female animal *after* the time we have previously shown that the sexual pathway commitment would have normally been made (see Cline 1985).

Figure 2, B and C, shows two other clones generated under identical conditions as those that produced the leg in Figure 2A. The clone in Figure 2B is entirely female, whereas that in Figure 2C is entirely male. Clearly, the proportions of phenotypically male and phenotypically female tissues varied from clone to clone, but cells with a sexually intermediate phenotype were never observed. We believe that this result reflects a defect in the sexual pathway maintenance aspect of Sxl functioning. Of 17 female-lethal Sxl alleles examined by such clonal analysis, only Sxl^{fPb} displayed a mosaic intersex phenotype. Homozygous diplo-X clones of null Sxl alleles induced under the same conditions are entirely male. Generating null mutant clones by removing Sxl^+ by somatic recombination just prior to the time when the cells must differentiate produces a true intersex phenotype (T. Cline, unpubl.); however, at no time in the development of a null allele heterozygote will somatic recombination generate a mosaic intersex clone.

The mosaic intersex phenotype can also be generated for diploid females when one reduces, but does not eliminate, the functioning of genes such as *da* known

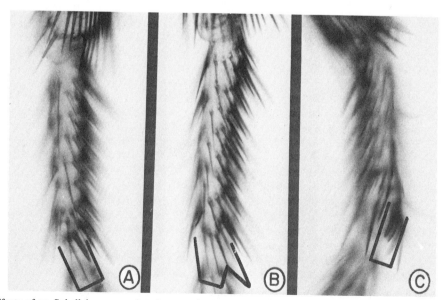

Figure 2. Effects of an Sxl allele on sexual pathway maintenance. Foreleg clones of tissue homozygous for the partial loss-of-function allele, Sxl^{fPb}, were founded by somatic recombination in an Sxl^{fPb}/Sxl^+ female between the late first and mid second larval instar stages. Three different sexual morphologies are exhibited by clones produced under identical conditions: (*A*) mosaic intersex; (*B*) entirely female; (*C*) entirely male.

to be required for the expression of *Sxl*. To observe this effect in such diploid situations, however, one must provide conditions that seem to suppress (at least in part) the upsets in dosage compensation that would normally accompany expression of *Sxl* that is inappropriate to the X:A balance (Cline 1984, 1985).

Dominance among daughterless daughters. In the previous section, assays for *Sxl* functioning were discussed that showed effects of mutant lesions on sexual pathway expression (Fig. 1) and pathway maintenance (Fig. 2). The experiment described in this section illustrates another situation that reduces, but does not eliminate, *Sxl* function, a situation that we believe can be exploited to assay the effects of lesions in *Sxl* on the first level of its functioning—on sexual pathway initiation—independent of effects on the subsequent two levels.

It was mentioned earlier that maternal *da*$^+$ gene activity is required for the subsequent expression of female-specific functions of *Sxl* in the progeny in response to their X:A balance. One *Sxl*$^+$ allele by itself is normally sufficient to direct all aspects of female development, *provided* mothers supply wild-type levels of *da*$^+$ activity. Thus, null *Sxl* alleles are *recessive* female-specific lethals. However, the situation is very different when maternal *da*$^+$ activity is reduced. Lowered levels that still suffice for the survival of all daughters with two doses of *Sxl*$^+$ can be lethal to daughters with only a single dose (Cline 1980). Thus, in this situation of reduced but nonzero *da*$^+$ activity, null *Sxl* alleles are dominant female-lethals.

Another situation in which fruit fly development is extremely sensitive to *Sxl*$^+$ dose is when the X:A balance is 0.67, as it is in triploid intersexes (Cline 1983). As described above, XX AAA individuals with two doses of *Sxl*$^+$ develop as mosaics of male and female cells; however, XX AAA flies with only a single functional copy of *Sxl* develop as nonmosaic males. As in the case of reduced maternal *da*$^+$ activity, loss-of-function *Sxl* alleles again are dominant.

These are two of the observations that led to the proposal that the effect of lowering maternal *da*$^+$ activity is equivalent to the effect of lowering the X:A balance and that both situations lower the probability of the cells' stably activating at least one *Sxl* allele at the point in development when sexual pathway initiation occurs (Cline 1984). The explanation for the extreme sensitivity of phenotype to *Sxl* dose in these two situations was twofold: a reduced number of targets for activation in situations where the probability of activation is low but nonzero, and a potentially reduced level of autoregulatory function of *Sxl* that may be important in stably activating the gene once it responds to the initial signal. The work that led to this proposal suggested that neither of these two factors necessarily involved the sexual pathway maintenance or expression functions of *Sxl*.

The partial loss-of-function alleles, *Sxl*^{f9} and *Sxl*^{fLS}, can provide a test for the hypothesis that these dominant effects of female-specific loss-of-function *Sxl* alleles involve aspects of *Sxl* functioning that are distinct from those involved in sexual pathway maintenance and/or expression. We propose that the functions assayed in these two *Sxl* dominance tests (a particularly easy assay in the case of the *da* effect) are functions involved in sexual pathway initiation, a necessary step in the generation of the female phenotype, but distinct from sexual differentiation per se.

Homozygous diplo-X clones of *Sxl*^{fLS} induced during early larval development grow poorly and are totally masculinized (Sanchez and Nöthiger 1982). In the same situation, homozygous *Sxl*^{f9} clones grow well and exhibit normal female differentiation (T. Cline, unpubl.). Thus, in this assay of female pathway maintenance and expression functions, *Sxl*^{fLS} looks much more defective than *Sxl*^{f9}. In contrast, with respect to their dominant-lethal effects on female viability under culture conditions that allow survival of more than 50% of the *Sxl*$^+$/*Sxl*$^+$ daughters of *da*/*da* mothers, *Sxl*^{fLS} is no more defective than *Sxl*$^+$, whereas *Sxl*^{f9} is nearly as defective as a null allele. By the hypothesis above, these results would indicate that *Sxl*^{f9} is considerably *more* defective than *Sxl*^{fLS} in sexual pathway initiation functions but is considerably *less* defective with respect to pathway maintenance and/or expression functions. The proposal that the mutations affect different functions of *Sxl* is supported by the fact that the two alleles fully complement each other. This proposal regarding *Sxl*^{f9}, *Sxl*^{fLS}, and the basis of *Sxl* dominance effects leads to what might seem like a rather unlikely prediction in the absence of the hypothesis: Even though the sexual phenotype of triploid intersexes is known to be exquisitely sensitive to genetic perturbations affecting *Sxl*, the strongly "masculinizing" allele, *Sxl*^{fLS}, should *not* masculinize triploid intersexes. Conversely, the allele, *Sxl*^{f9}, which has no effect on the sexual phenotype of clones in the adult integument induced during larval stages, *should* masculinize those same tissues in the triploid intersex situation.

The data in Table 1 show that this prediction is upheld. Intersexes with *Sxl*^{f9} are as masculine as *Sxl* null allele homozygotes. In contrast, the masculinizing allele, *Sxl*^{fLS}, has relatively little effect on intersex phenotype. The relative viability of the two genotypes of intersexes appears to be comparable (data not shown). These results support the statement that the various functions of *Sxl* required for female development can be affected differentially by mutations in ways that cannot be explained simply as effects on overall product levels. Moreover, they validate the proposal that clonal analysis and *da* dominance tests can be used to assay different aspects of *Sxl* functioning.

Molecular Analysis of *Sxl*

Analysis by the techniques of developmental genetics (see above) shows *Sxl* to be a functionally complex regulatory gene; however, the same analysis provides hope that its functional complexities can be understood, at least in part, through the use of a wide

Table 1. Effect of *Sxl* Genotype on the Sexual Phenotype of Triploid Intersexes (XXY AAA)

Sxl Genotype	No. animals scored	Phenotype of sexually dimorphic structures[a] (% of total cases with some [only] female tissue)						
		genitalia	analia	foreleg	hemitergite no. 7	hemisternite no. 6	hemisternite no. 7	overall sex index[b]
Masculinizing allele								
Sxl^{fLS}/Sxl^+	105[c]	90 [47]	91 [0]	— [0]	35 [9]	39 [20]	40 [17]	51 [14]
Nonmasculinizing allele								
Sxl^{f9}/Sxl^+	122[c]	2 [0]	6 [0]	— [0]	0 [0]	7 [0]	0 [0]	3 [0]
Homozygous null allele								
Sxl^{f1}/Sxl^{f1}	240[d]	0	0	— [0]	0	30 [0]	0	8 [0]
Wild-type control								
Sxl^+/Sxl^+	65[d]	89 [48]	89 [49]	— [0]	39 [17]	42 [17]	38 [21]	52 [21]

[a] Intersexes generated, raised, and scored exactly as described by Cline (1983).
[b] Calculated as the average for all structures scored.
[c] Nonbalancer live intersex adults from the following crosses: y^2; C(2L)RM,*dp*; C(2R)RM,*px*; C(3L)RM,h^2rs^2; C(3R), + ♂♂ × ♀♀ $Sxl^{f9}ct$ v/*Basc*,w^aB OR *y* Sxl^{fLS}/*Basc*,w^aB.
[d] Data from Cline (1983), crosses D and K.

variety of mutations affecting not only *Sxl* itself, but also the genes and gene products with which it interacts upstream and downstream in the regulatory network that directs sexual development. On the other hand, genetic analysis alone says very little about the specific molecular level at which gene interactions take place. For this, biochemical analysis is required.

Sxl region DNA was cloned by the P-element transposon-tagging method (Searles et al. 1982), starting with the P-element insertion mutant, Sxl^{fPb}. A molecular walk was then initiated in both directions from the site of this P-element insertion. Restriction fragments from this walk have been used as probes for the whole-genome Southern blot analysis of a wide variety of mutant *Sxl* alleles and their derivatives (for details of this work, see E. Maine et al., in prep.). Restriction fragments from this walk have been used for Northern blot analysis of poly(A)⁺ RNA as a function of developmental stage and genotype (E. Maine et al., unpubl.).

DNA sequence organization of wild-type and mutant Sxl alleles.

Figure 3 presents a restriction map of the region surrounding the site of the P-element insertion shown by reversion analysis to be responsible for the female-lethal phenotype of Sxl^{fPb}. A wide variety of mutant *Sxl* alleles are associated with gross changes in DNA within the cloned region. The centromere-distal breakpoint of $In(1)Sxl^{af}$ is shown in Figure 3; this male-viable inversion appears to eliminate all known *Sxl* functions. $Df(1)Sxl^{3G2}$ and $Df(1)Sxl^{ra}$ were derived from Sxl^{fPb} and Sxl^{M1}, respectively; they also eliminate all *Sxl* functions. $Df(1)Sxl^{3G2}$ is male viable and fertile by itself, whereas $Df(1)Sxl^{ra}$ is male viable and fertile, provided that the four other centromere-proximal vital genes that it deletes are covered.

DNA insertions have been identified in five loss-of-function (female-specific recessive-lethal) alleles, including Sxl^{fLS} (see above). It should be mentioned, however, that only the female-lethal phenotypes of Sxl^{fPa} and Sxl^{fPb} have been linked unequivocally with their DNA insertions by reversion analysis. This is not a trivial point, since two wild-type alleles, Canton S and Oregon R, differ with respect to the presence of a mobile element insertion (roo) in the second *Hin*dIII restriction fragment from the left end of the map.

The gain-of-function (male-specific dominant-lethal) allele, Sxl^{M1}, is associated with the insertion of the mobile DNA element, roo, at the location shown in Figure 3. At a very low frequency, Sxl^{M1} can revert to wild-type functioning; an incomplete excision of the roo element appears to suffice for the restoration of function. Four other dominant male-lethal alleles like Sxl^{M1} have been isolated. All are associated (functionally) with insertions very near Sxl^{M1}. One involves a roo element that can be distinguished from that in Sxl^{M1}. Two others have insertions of the mobile element, 297. The identity of the fifth insertion is not yet known. Sxl^{M1} and three other male-lethal insertions that have been mapped precisely fall within 1 kb of each other. The female-lethal inversion, the five female-lethal insertions, and the five male-lethal insertions all fall within the 11-kb region identified at the bottom of Figure 3. There appears to be no gross difference in DNA primary structure between the sexes over the region shown in Figure 3. The limits of the *Sxl* gene have not yet been determined; P-element-mediated transformation attempts are in progress to address this point.

Considerable excitement has been generated by the discovery of highly conserved DNA sequences (called homeo boxes) that appear to be characteristic of certain regulatory genes that appear to govern segmental pathway commitments in *Drosophila* (McGinnis et al. 1984; Poole et al. 1985). In view of this excitement, and in light of the many functional parallels between the genetic control of sex determination and segmentation, we could not resist the temptation to probe *Sxl* for homeo box sequence homologies. Two different

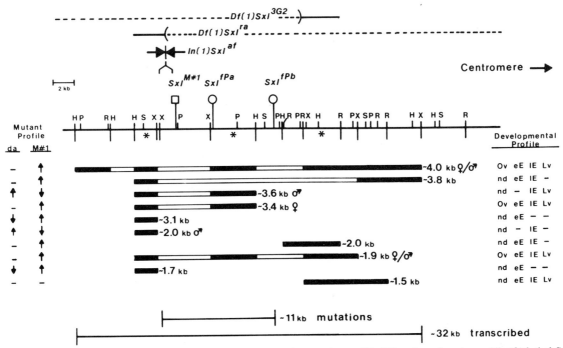

Figure 3. Molecular structure and functioning of *Sxl*. Locations of one male-specific (□) and two female-specific (○) lethal *Sxl* lesions and three chromosome rearrangements affecting *Sxl* function are given with respect to a restriction map of the cloned region. All ten pseudopoint lesions characterized so far fall within the 11-kb stretch indicated. Poly(A)+ transcripts that have homology with the cloned sequences are indicated beneath the restriction map in decreasing order of size. Restriction fragments used as probes in this Northern blot analysis are indicated by the vertical lines that extend below some sites on the map. Homologies between the various RNA species and the cloned DNA are indicated by the dark shading (exons); light stretches indicate a lack of homology (introns). The sex specificity of the transcripts is indicated in all cases where it is known. A crude developmental profile for the transcripts is shown at the right, with the following abbreviations indicating the presence of the particular RNA species from ovaries (Ov); early embryos, 0–5 hr after egg laying (eE); late embryos, 17–24 hr after egg laying (lE); and larvae of mixed stages (Lv). A dash signifies that the species was not present at substantial levels; nd indicates data not determined. Changes in the wild-type 0–12-hr embryo transcript pattern caused by the *da* maternal effect and *Sxl*^M1 are indicated at the left of the figure (increased, decreased, or unchanged for each transcript). Asterisks indicate restriction fragments that display weak homology with both of two rather divergent homeo box DNA sequences.

homeo-box-containing probes were used to screen for homology with cloned restriction fragments from the *Sxl* region. The restriction fragments used in this Southern blot analysis are indicated by the lines extending beneath the map in Figure 3. One probe was from the gene fushi tarazu and the other probe was from Antennapedia. Under the reduced stringency hybridization conditions described by McGinnis et al. (1984), both probes revealed homology with three nonadjacent restriction fragments from the *Sxl* region (Fig. 3, asterisks). Until these regions are sequenced, we cannot be certain that the homology is truly due to the presence of homeo boxes, although the fact that the three regions of homology correspond to what appear to be the major exons of this gene is certainly suggestive (see below).

Transcription from Sxl in wild-type males and females. Transcripts arrayed on Northern blots were probed for homology with the restriction fragments indicated in Figure 3 (vertical lines that extend beneath certain restriction sites on the map); horizontal bars under this map represent the collection of poly(A)+-containing transcripts that have been identified so far from this region, listed in order of decreasing size. The filled regions of the transcript bars signify homology between the transcript and the particular restriction fragment probe above it on the DNA map, while the empty regions signify that no hybridization was detected between the particular transcript and the corresponding DNA. Information on the developmental stages at which each transcript could be detected is indicated in the table at the right side of Figure 3 (data on pupal and adult stages are not yet available). The transcript pattern is clearly complex, both with respect to the structure and variety of species produced and with respect to their temporal pattern. There appear to be several introns in this gene, with a variety of splicing patterns being employed to generate the RNAs. The apparent structural complexity is certain to increase as the transcript sequence analysis proceeds; e.g., the 3.1-kb species obviously must have homology with regions besides the single 2-kb restriction fragment indicated. Preliminary analysis of cDNA clones (not shown) has already suggested the presence of additional introns and microexons.

Some transcripts such as the 4.0-, 3.4-, and 1.9-kb species appear to be present at all stages in development. At the other extreme are ephemeral embryonic transcripts such as the 3.1- and 1.7-kb species that are present by 2.5 hours after egg laying, increase several-fold between 2.5 and 5 hours after egg laying, but are almost gone between 5 and 8 hours, and are not detectable thereafter. The timing of these transcripts corresponds remarkably well to the only period in zygotic development when the temperature-conditional allele, Sxl^{2593}, fails to complement Sxl^{f9} at the nonpermissive temperature (T. Cline, unpubl.). As discussed above, Sxl^{f9} appeared to be defective specifically in its sexual pathway initiation function. Thus, these transcripts may be involved in Sxl regulation.

Most of the pseudopoint mutations mapped fall within the intervening sequences indicated on Figure 3, including all of the male-lethal regulatory variants. The fact that a functionally wild-type derivative of Sxl^{M1} still contains a fragment of the parental roo element insertion is consistent with the proposal that this region is not represented in mature mRNA. Of the five female-specific insertion alleles, only Sxl^{fPa} appears likely to be located in an exon. It is the only null allele in the group.

As might have been anticipated from the genetic analysis, there are female-specific transcripts (e.g., the 3.4-kb species). The 3.1- and 1.7-kb species whose sex specificity has not yet been determined are also candidates for female-specific transcripts on the basis of their behavior with respect to the masculinizing da maternal effect (see below). It is known that they are not male-specific. On the other hand, since males are viable and fertile without this region, the presence of some transcripts common to both sexes, and others even specific to males, came as somewhat of a surprise. One possibility is that the male-specific species provide a negative autoregulatory function. They appear to reach their highest levels late in the embryonic period, the point at which the sexual pathway choice may first become irreversible. The proposal of a vital negative function for the male-specific transcripts might seem at odds with the failure to recover loss-of-function Sxl alleles that are male-specific or non-sex-specific lethal (Nicklas and Cline 1983); however, the extensive overlap that may exist between some male-specific and female-specific transcripts (e.g., compare the 3.6- and 3.4-kb species) might make it difficult to destroy a negative autoregulatory Sxl function by mutation without simultaneously destroying the female-specific functions that are negatively regulated. The possibility that these male-specific transcripts serve nonessential functions, or even no function at all, must also be considered.

Changes in the Sxl transcript pattern associated with sex-specific lethals.

Sxl^{M1} and the da maternal effect have opposite effects on *Drosophila* development. In the absence of maternal da^+ activity, all progeny appear to develop as males as a consequence of a defect in the regulation of Sxl. This has lethal consequences on genetic females. Sxl^{M1} also appears to disrupt Sxl regulation, but in contrast to the da maternal effect, it causes individuals to develop as females, with lethal consequences for genetic males. The effects of these mutations on the Sxl transcript pattern is therefore of considerable interest. For the analysis of Sxl^{M1}, we wished to arrange the situation so that all sons (genetic males) would carry the Sxl^{M1} mutation. To accomplish this, we introduced the double-mutant allele, $Sxl^{M1,fm3}$, from the fathers, crossing them to attached-X mothers. Unlike Sxl^{M1} alone, this double-mutant allele is male-viable because it also contains a partial loss-of-function mutation (Cline 1984). Sxl^{fm3} maps centromere-proximal to Sxl^{M1} and does not appear to be associated with any gross change in the DNA.

The results of analysis of transcripts from 0–12-hour-old embryos of mixed genetic sex are summarized in Figure 3 (left side). In the absence of normal maternal da^+ activity, the level of the only two known male-specific transcripts (3.6 and 2.0 kb) is dramatically increased, whereas the level of two early embryo-specific transcripts (3.1 and 1.7 kb) that we know are normally present in females is reduced to undetectable levels. This is consistent in part both with masculinizing developmental consequences of this maternal effect and with the proposal that these early embryo transcripts are involved in the activation of Sxl (see above). On the other hand, the only species known to be female-specific, the 2.4-kb transcript, seems not to be affected during the early period examined. Moreover, the male-specific transcripts appear prematurely in this mutant situation. The da maternal effect does not simply impose a wild-type male pattern on genetic females, at least not at this very early period in development; the situation is a bit more complicated.

$Sxl^{M1,fm3}$ has a very different effect on the transcript pattern 0–12 hours after egg laying. As with da, its effects are consistent with expectations from genetic analysis but are not as simple as they might have been. This derivative of the feminizing allele, Sxl^{M1}, causes a 20-fold increase in the levels of seven of the ten transcripts shown. The three exceptions include the two transcripts known to be male-specific, the same two that are increased over this period in the absence of maternal da gene activity. They may be eliminated by this mutation; however, since in the wild-type situation in the best circumstances they are only just barely detectable by 12 hours after egg laying, this result may be consistent with the interpretation that they are not affected by $Sxl^{M1,fm3}$. Information on later time points is needed.

Recall that the regulatory abnormality of Sxl^{M1} is caused by a mobile DNA element insertion into what appears to be an intron of this gene. The fact that $Sxl^{M1,fm3}$ dramatically increases the level of transcripts that span this intron, as well as transcripts on either side of it which do not, suggests that Sxl^{M1} may be acting as an enhancer element (for review, see Khoury and Gruss 1983).

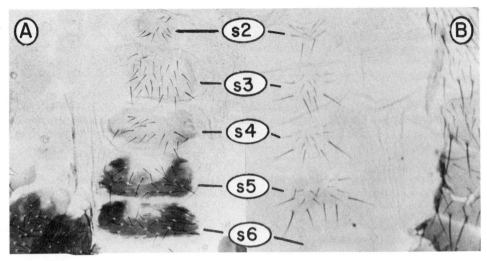

Figure 4. A possible ventral→dorsal transformation in adult abdominal development caused by the abnormal expression of a mutant *Sxl* allele. (*A*) "Tergite-like" sternites (S2-S6) of a $Sxl^{M1,fPa-ra}/Y;Dp(1;3)Sxl^+/+$ male raised at 22°C. (*B*) Normal male sternite development of a $Sxl^{M1,fPa-ra}/Y$ male raised at 18°C.

Sexual Relationships

In view of the parallels between the genetic strategy fruit flies appear to use to control sexual and segmental aspects of development, the similar kinds of complexities that exist in transcript patterns among major regulatory genes that govern these two processes and the possible presence of homeo boxes in sex-determining genes, one must wonder about the nature of the relationships that might exist between different regulatory networks in this organism. Do they use elements that are only evolutionarily related or might they use some elements in common? Could disruption of one system interfere with the operation of another? At this point, there is no hint from the literature of any direct overlap between these systems; however, in light of the cell lethality associated with disruption of *Sxl* functioning that can mask effects of sexual development, one must interpret the absence of such information with caution.

In this regard, the grossly abnormal phenotype shown in Figure 4 is intriguing. Figure 4B shows the normal development of the ventral male abdomen—the area of sternites (S2-S5 in males; females have an S6 and S7). In contrast, Figure 4A shows the same region exhibiting many of the developmental characteristics of the dorsal abdomen. Sternites appear to be developing as tergites. This apparent ventral-to-dorsal homeotic transformation is caused by $Sxl^{M1,fPa-ra}$ (see figure legend). The relationship of this abnormality to any aspect of sexual differentiation or dosage compensation is obviously unclear at this point. Much more analysis will be required to understand the phenotype more fully. Nevertheless, this observation does increase our conviction that unifying mechanistic principles governing development might be derived from comparisons among the different regulatory networks that govern what would seem to be fundamentally different aspects of *Drosophila* development.

ACKNOWLEDGMENTS

We thank N. Chapman for isolating $Sxl^{M1,fPa}$ and $Sxl^{M1,fPa-ra}$, J.M. Belote for drawing our attention to the stock that carried Sxl^{fPb}, L. Sanchez for providing us with Sxl^{fLS}, W.J. Gehring for providing the homeo box probes, and N. Lyons for technical assistance. This work was supported by grants from the National Institutes of Health, ACS, and March of Dimes.

REFERENCES

BAKER, B.S. and J.M. BELOTE. 1983. Sex determination and dosage compensation in *Drosophila melanogaster*. *Annu. Rev. Genet.* **17**: 345.

BAKER, B.S. and K.A. RIDGE. 1980. Sex and the single cell. I. On the action of major loci affecting sex determination in *Drosophila*. *Genetics* **94**: 383.

BELOTE, J.M. 1983. Male-specific lethal mutations of *Drosophila melanogaster*. II. Parameters of gene action during male development. *Genetics* **105**: 881.

BRIDGES, C.B. 1925. Sex in relation to chromosomes and genes. *Am. Nat.* **59**: 127.

CLINE, T.W. 1978. Two closely linked mutations in *Drosophila melanogaster* that are lethal to opposite sexes and interact with daughterless. *Genetics* **90**: 683.

―――. 1979a. A product of the maternally influenced Sex-lethal gene determines sex in *Drosophila melanogaster*. *Genetics* **84**: 723.

―――. 1979b. A male-specific lethal mutation in *Drosophila* that transforms sex. *Dev. Biol.* **72**: 266.

―――. 1980. Maternal and zygotic sex-specific gene interactions in *Drosophila melanogaster*. *Genetics* **96**: 903.

―――. 1981. Positive selection methods for the isolation and fine-structure mapping of *cis*-acting, homeotic mutations at the Sex-lethal locus in *D. melanogaster*. *Genetics* **97**: s23.

―――. 1983. The interaction between daughterless and Sex-lethal in triploids: A lethal sex-transforming maternal effect linking sex determination and dosage compensation in *Drosophila melanogaster*. *Dev. Biol.* **95**: 260.

―――. 1984. Autoregulatory functioning of a *Drosophila* gene product that establishes and maintains the sexually determined state. *Genetics* **107**: 231.

———. 1985. Primary events in the determination of sex in *Drosophila melanogaster*. In *Origin and evolution of sex* (ed. H. Halvorson and A. Munroy), p. 301. A.R. Liss, New York. (In press.)

ENGELS, W.R. 1983. The P family of transposable elements in *Drosophila*. *Annu. Rev. Genet.* **17:** 315.

GEHRING, W. 1972. The stability of the determined state in cultures of imaginal disks in *Drosphila*. In *The biology of imaginal disks* (ed. H. Ursprung and R. Nöthiger), p. 35. Springer-Verlag, New York.

KHOURY, G. and P. GRUSS. 1983. Enhancer elements. *Cell* **33:** 313.

MCGINNIS, W., M.S. LEVINE, E. HAFEN, A. KUROIWA, and W.J. GEHRING. 1984. A conserved DNA sequence in homoeotic genes of *Drosophila* Antennapedia and bithorax complexes. *Nature* **308:** 428.

NICKLAS, J.A. and T.W. CLINE. 1983. Vital genes that flank Sex-lethal, an X-linked sex-determining gene of *Drosophila melanogaster*. *Genetics* **103:** 617.

POODRY, C.A. 1980. Epidermis: Morphology and development. In *The genetics and biology of* Drosophila (ed. M. Ashburner and T.R.F. Wright), vol. 2d, p. 443. Academic Press, London.

POOLE, S.J., L.M. KAUVAR, B. DREES, and T. KORNBERG. 1985. The engrailed locus of *Drosophila*: Structural analysis of an embryonic transcript. *Cell* **40:** 37.

SANCHEZ, L. and R. NÖTHIGER. 1982. Clonal analysis of *Sex-lethal*, a gene needed for female sexual development in *Drosophila melanogaster*. *Wilhelm Roux's Arch. Dev. Biol.* **191:** 211.

SEARLES, L.L., R.S. JOKERST, P.M. BINGHAM, R.A. VOELKER, and A.L. GREENLEAF. 1982. Molecular cloning of sequences from a *Drosophila* RNA polymerase II locus by P-element transposon tagging. *Cell* **31:** 585.

STEWART, B. and J. MERRIAM. 1980. Dosage compensation. In *The genetics and biology of* Drosophila (ed. M. Ashburner and T.R.F. Wright), vol. 2d, p. 107. Academic Press, London.

Control of Sexual Differentiation in *Drosophila melanogaster*

J.M. Belote, M.B. McKeown, D.J. Andrew, T.N. Scott, M.F. Wolfner, and B.S. Baker
Department of Biology, University of California at San Diego, La Jolla, California 92093

Sex determination in *Drosophila melanogaster* provides an attractive system in which to study at both the genetic and molecular levels the hierarchies of genetic functions responsible for development in a higher eukaryote. The initial signal determining whether an individual will undergo female or male development is known. In addition, regulatory genes that control sexual differentiation in response to this signal have been identified by mutational analysis, and a model of how these genes interact with one another has been developed (Baker and Ridge 1980; Baker and Belote 1983; Cline 1983a, 1984). Finally, several of the terminal differentiation functions that are regulated in a sex-specific manner, and thus are the targets of these regulatory genes, are known, and these are also amenable to analysis at the genetic and/or molecular level (for review, see Baker and Belote 1983). Thus, the molecular cloning of both regulatory and structural genes now makes it possible to test particular aspects of the genetic model and to extend our understanding of how these genes, acting as parts of a regulatory circuit, initiate and maintain sexually dimorphic developmental events.

Sexual dimorphism in the adult fly is extensive. Sexual differences are manifest with respect not only to the morphology, bristle pattern, pigmentation, and segmentation of the external cuticle, but also to the internal reproductive systems, as well as several biochemical criteria. Innate behavioral differences between males and females that reflect a dimorphism in the structure and/or functioning of at least part of the fly's nervous system are also involved. Our focus here is on the regulatory genes that mediate the choice between the two disparate pathways of development responsible for this somatic sexual dimorphism.

The primary determinant of both somatic and germ line sex, as well as dosage compensation, is the relative numbers of X chromosomes and sets of autosomes, the X:A ratio (Bridges 1921; Maroni and Plaut 1973; van Deusen 1976; Schüpbach 1985). Although the Y chromosome is required in males for normal spermatogenesis, it plays no role in sex determination. Flies with a single X chromosome and a diploid complement of autosomes (i.e., an X:A ratio of 0.5) develop as males, and flies with equal numbers of X chromosomes and sets of autosomes (e.g., 2X:2A or 3X:3A, X:A = 1.0) develop as females. Significantly, individuals with an intermediate X:A ratio of .67 (2X:3A) develop as intersexual flies (referred to as triploid intersexes) that appear to be coarse-grained sexual mosaics composed of discrete patches of male and female tissue (Stern 1966). Distinct phenotypic boundaries between cells of opposite sexual genotype are possible in *Drosophila* because of the cell-autonomous nature of sexual differentiation (Morgan 1916; Cline 1979; Baker and Ridge 1980; Sanchez and Nöthiger 1982; Wieschaus and Nöthiger 1982); none of the identified steps in sex determination involve diffusible substances, such as sex hormones, that influence sexual development of the whole fly.

The mosaic sexual phenotype of triploid intersexes can be understood if at an early stage in development, a fly's cells monitor their X:A ratio and make an irreversible commitment to one sexual pathway or the other, and in 2X:3A individuals, the X:A ratio is sufficiently close to the border between the male- and female-determining signal that cells of the same genotype perceive this signal differently, and differentiate accordingly. This hypothesis of an early irreversible step in sex determination is supported by experiments in which X-chromosome loss was induced in XX:2A females at defined stages in development, and the resulting XO:2A clonal patches that encompassed sexually dimorphic regions of the cuticle scored for sexual phenotype in the adult (Baker and Belote 1983; Sanchez and Nöthiger 1983; B. Baker, in prep.). In these experiments, XO clones induced early in embryogenesis developed as male, whereas XO clones induced in the late larval or pupal stages differentiated as female, despite their male chromosome complement. This result is expected if the X:A ratio is monitored at a point in development between these two times and, as a result of that decision, the pathway of sexual differentiation is irreversibly determined. The further observation that the XO female clones (i.e., those XO clones produced late in development) grow poorly relative to XX cells is interpreted as meaning that the mechanism controlling the transcriptional activity of the X chromosome (i.e., dosage compensation) is irreversibly set concomitantly with the determination of sex. Thus, the reduced viability of XO female cells generated late in development is suggested to be a consequence of their single X chromosome being transcribed at the female rate; such cells would be synthesizing only half the normal amounts of X-encoded gene products. At both the genetic and molecular levels, the mechanism by which the X:A ratio is assessed is not known (for review, see Baker and Belote 1983), but it is likely that this irreversible step in the sex differentiation and dosage compensation pathways functions by setting the activities of a relatively small number of major regulatory loci.

The X:A ratio, acting in conjunction with the maternally contributed daughterless (da^+) gene product (Cline 1976, 1980), provides a signal to affect the expression of sets of regulatory loci that make up the second level in the sex-determination regulatory hierarchy. The Sex lethal (*Sxl*) locus is of particular importance in that it appears to control both somatic sexual differentiation and dosage compensation (Cline 1978, 1979, 1983a, 1984; Lucchesi and Skripsky 1981; Sanchez and Nöthiger 1982). The relationship of Sxl^+ to germ line sex determination is less clear, but it does appear to play some role in germ line differentiation (Cline 1983b; Schüpbach 1985). The effect of Sxl^+ on sex determination and dosage compensation is mediated through the functioning of sets of regulatory genes, including transformer (*tra*), transformer-2 (*tra-2*), intersex (*ix*), and doublesex (*dsx*), in the case of somatic sexual differentiation, and maleless (*mle*), male-specific lethal-1 (*msl-1*), male-specific lethal-2 (*msl-2*), and maleless-3 (*mle-3*), in the case of dosage compensation (Baker and Ridge 1980; Belote and Lucchesi 1980a,b; Baker and Belote 1983). Analogous regulatory genes governing germ line sex determination have not been identified.

The third level in the sex-determination regulatory hierarchy is composed of those genes whose products are directly responsible for the structure and function of sexually dimorphic tissues. These are structural genes whose activities are under the control of the sex-determination regulatory pathway. The yolk polypeptide genes (*Yp1*, *Yp2*, and *Yp3*), which are expressed in the adult fat body in a female-specific manner, are the best-characterized examples of sexually regulated terminal differentiation functions (Postlethwait et al. 1980; Bownes and Nöthiger 1981; Ota et al. 1981; Belote et al. 1985; Garabedian et al. 1985).

Regulatory Genes Controlling Somatic Sexual Differentiation

Several mutants, including *tra*, *ix*, *dsx*, and *tra-2*, have phenotypes which suggest that the wild-type functions of the loci they identify play important roles in somatic sexual differentiation. The observation that all aspects of somatic sex (i.e., morphological, biochemical, and behavioral) are determined by these genes suggests that they represent regulatory loci. These genes have important genetic properties in common with other well-studied developmentally important regulatory loci in *Drosophila*, e.g., the homeotic loci of the bithorax complex (Lewis 1978) and the Antennapedia complex (Kaufman et al. 1980), controlling body segment determination, and engrailed, controlling subsegment compartmentalization (Kornberg 1981). These genes act in a cell-autonomous manner, they are required at more than one stage, and perhaps continuously, in development, and they function as genetic switches mediating a choice between alternative paths of development. Thus, one may think of the sex-determination mutants as a special class of homeotic genes, with the pertinent developmental decision being between male and female differentiation.

The sex-determination mutants are autosomal and map to well-separated sites in the genome. Loss-of-function mutations at these loci are recessive, and none affect viability. The *tra*, *tra-2*, and *ix* mutations affect somatic sex determination only in females, with *tra* and *tra-2* transforming XX individuals into males and *ix* transforming XX individuals into intersexes. Null mutations at the *dsx* locus affect both XX and XY individuals, causing them to develop as phenotypically indistinguishable intersexes. In the intersexes characteristic of *dsx* and *ix* mutants, all sexually dimorphic cells exhibit intermediate sexual development, as if they simultaneously underwent both male-specific and female-specific differentiation. Thus, they are quite unlike 2X:3A triploid intersexes, which appear to be mosaics of male and female patches.

By defining the genetic properties of various alleles at four of the regulatory loci controlling somatic sexual differentiation, and characterizing the phenotypes of double-mutant combinations, Baker and Ridge (1980) developed a model for how these genes interact (Fig. 1). In this model, dsx^+ is a bifunctional locus that can be expressed actively in either of two ways: In males, dsx^+ functions to repress female-specific differentiation, and in females, dsx^+ acts in an opposite manner to repress male-specific differentiation. In chromosomally female individuals, the $tra-2^+$ and tra^+ genes are expressed in response to Sxl^+ function (see Cline 1983a, 1984), and their products act to maintain the dsx^+ gene in the female mode of expression. The product of the ix^+ gene is required in females in conjunction with that of the dsx^+ locus for the repression of male differentiation. In males, the $tra-2^+$ and tra^+ loci do not function (probably as a result of Sxl^+ inactivity) and, as a consequence, dsx^+ is expressed in the male mode.

Studies using temperature-sensitive *tra-2* alleles show that the functioning of this gene, and, by inference, the other genes in this regulatory pathway, is required at many different times during development, and perhaps continuously, for proper sexual differentiation. An important result from these studies was the observation that the sex-determination regulatory pathway acts at more than one time even within a given cell lineage to mediate different developmental decisions necessary for sexual development (Belote and Baker 1982; Belote et al. 1985).

One goal of our research is to isolate by molecular cloning the genetic components of this pathway, in order to test and extend this model, and to examine its functioning at the molecular level. As the first step toward this end, we have cloned the dsx^+ gene (B. Baker and M. Wolfner, in prep.). This gene is of particular interest in that it appears to play a central role in the regulatory pathway governing sexual differentiation and because it exhibits the unusual property of bifunc-

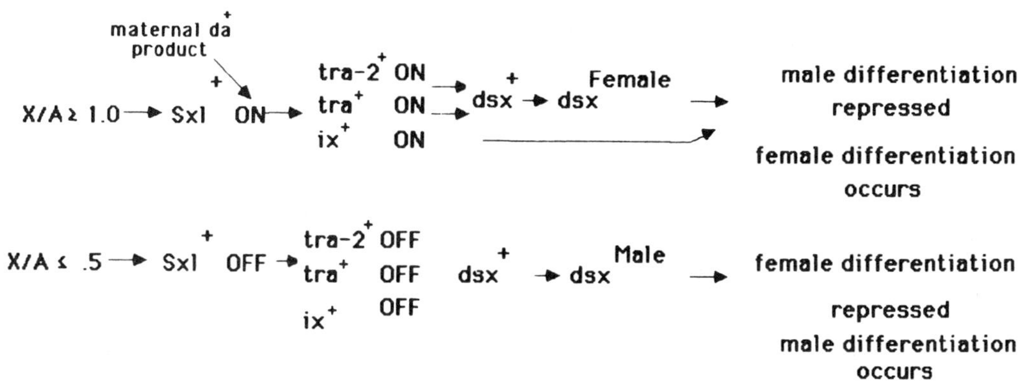

Figure 1. Model for the genetic control of somatic sexual differentiation. The bifunctional dsx^+ locus can express either of two functions: dsx^{Male} is expressed in males and represses female differentiation. The tra-2^+ and tra^+ loci function in response to Sxl^+ activity to allow the expression of dsx^{Female}. The ix^+ locus acts in conjunction with dsx^{Female} to repress male differentiation. Maternally supplied da^+ product is required for the activation of Sxl^+ in females.

tional active expression. More recently, we have cloned the tra^+ locus (J. Belote and M. McKeown, in prep.).

One major qualification of our genetic model is that it is, per force, limited to only those steps in the pathway that have been defined by the existence of mutants. Since most of the extant sex-determination mutants were discovered fortuitously, and saturation mutagenesis has not been attempted, it is reasonable to assume that there are some as yet unidentified genes that also play important roles in this process. Thus, another objective of our work is to fill in the gaps of our knowledge by identifying all of the genes involved in the regulatory pathway controlling sex. Both genetic and molecular approaches are being used to accomplish this.

RESULTS AND DISCUSSION

Molecular Analysis of Sex-determination Genes

The molecular cloning of both dsx^+ and tra^+ was achieved by following the same basic approach. First, X-ray mutagenesis and cytogenetic analyses were carried out (1) to determine as precisely as possible the cytological positions of these genes on the salivary gland chromosome map, (2) to define the number, distribution, and phenotypes of adjacent complementation groups, with particular interest in determining whether the regions contained other related regulatory genes, and (3) to obtain chromosome rearrangements (e.g., inversions, translocations, and deficiencies) that would provide molecularly detectable landmarks, allowing cloned DNA regions to be correlated with positions on the genetic and cytological maps. Some of the chromosome rearrangements recovered proved to be useful in allowing us to "jump" into the region of interest from previously cloned DNA regions (for a description of chromosome walking and jumping, see Bender et al. 1983a). Moreover, by determining the positions of selected breakpoints, we could delimit the maximum and minimum extents of the genes of interest.

The doublesex locus. The genetic studies of the dsx locus suggest that it is an unusual gene, capable of being actively expressed in either of two ways. It is proposed that dsx^+ acts as a negative regulator of sex-specific terminal differentiation functions, having opposite effects in the two sexes: In females, the dsx^+ locus acts to prevent the expression of male-specific terminal differentiation functions, whereas in males, dsx^+ acts to prevent the expression of female-specific terminal differentiation functions. The bifunctional nature of dsx^+ function is supported by several lines of evidence. That dsx^+ carries out active but opposite roles in the two sexes is suggested by the phenotypes of null mutants. These mutants convert both XX and XY individuals into morphologically identical intersexes whose phenotypes are interpreted as being the result of sexually dimorphic tissues trying to simultaneously undergo both male and female development (i.e., neither pathway being repressed). That dsx^+ functions differently in males and females is strongly supported by the existence of sex-specific dsx alleles. There are alleles of dsx that affect only XX flies (these include a class of dominant alleles) and there is one allele, dsx^{136}, that affects only XY individuals. It was to account for this genetic complexity that the bifunctional nature of dsx^+ was hypothesized. One major goal of the molecular studies of dsx is to determine the basis for this bifunctionality.

To ascertain whether dsx was part of a gene complex controlling sex determination and to collect the genetic material necessary to clone the dsx locus, we (B. Baker, T. Hazelrigg, T. Kaufman, and G. Hoff) carried out a saturation mutagenesis of the dsx region. The cytogenetic analysis of dsx was facilitated by the existence of dominant dsx alleles, dsx^D and dsx^{Mas}. X-ray mutagenesis of chromosomes carrying either of these alleles yields revertants of the dominant phenotype whenever the dsx^+ function associated with the dominant alleles (see below) is rendered nonfunctional (Denell and Jackson 1972; Duncan and Kaufman 1975). Such revertants can be scored in the F_1 generation following mutagenesis, allowing large numbers of irradiated chromosomes to be screened. Of 45,000 treated chromosomes examined, 32 revertants of dsx^{Mas} were recovered. Cytological examination of these revealed that 13 were inversions or translocations broken in salivary gland chromosome band 84E1·2. Chromosome deficiencies, obtained from this screen and from other sources, that delete this band are all dsx. These results unambiguously place the dsx locus at this position on the polytene chromosome map. Chemical mutagens were used to saturate this region, defining 23 complementation groups within a 26-band salivary gland chromosome region around dsx. Since none of the complementation groups, other than dsx itself, have alleles that exhibit a phenotype affecting sexual differentiation, we conclude that the dsx locus is not part of a gene cluster controlling sexual development. This is supported by molecular studies of the transcription units flanking dsx (see below), which show that they are expressed in both sexes and, in one case, correspond to an essential gene of unknown function. In the other case, transcripts are most abundant in adult females and embryos, suggesting that this locus may, in part, be active in maternal programming of the embryo.

One chromosome rearrangement (dsx^{Mas+R2}) recovered in the screen was a deficiency that deleted all the material between band 84C1·2 and 84E1·2, thus moving the dsx gene region near a previously cloned α-tubulin gene at 84B3-6 (Mischke and Pardue 1982). Starting from a clone of this region (provided by D. Mischke and M.L. Pardue, Massachusetts Institute of Technology), we (B. Baker and M. Wolfner) carried out a chromosome walk until we reached the dsx^{Mas+R2} breakpoint and then screened a library constructed from flies carrying this mutation to "jump" into the dsx region. DNA fragments from the dsx portion of the breakpoint-containing clone were then used to initiate a walk in a wild-type *Drosophila* library (Maniatis et al. 1978). This walk was continued for about 108 kb, extending beyond the breakpoints defining the maximal possible extent of the dsx^+ functional unit.

An analysis of the positions of chromosome breakpoints defining the dsx locus revealed that, like some of the other developmentally important regulatory genes that have been examined molecularly, e.g., Ultrabithorax (Bender et al. 1983b), Antennapedia (Garber et al. 1983; Scott et al. 1983), and Notch (Artavanis-Tsakonas et al. 1983; Kidd et al. 1983), the dsx^+ gene is rather large. An outside limit on the size of the dsx^+ locus is set by the positions of chromosome rearrangements on each side of 84E1·2. $Df(3R)Ns^{+R17}$ is a deficiency that deletes all known complementation groups immediately proximal to 84E1·2, but is dsx^+. The other relevant breakpoint is that of a translocation, $T(2;3)Es$, broken just distal to dsx^+ in a lethal complementation group one gene removed from dsx, and not affecting dsx^+ function. The distance between these breakpoints (~75 kb) provides a maximum estimate of the size of the dsx^+ genetic unit. The minimum estimate of dsx^+ is defined by the extent of chromosome breakpoints that inactivate dsx^+ function; the positions of these breakpoints span a 27-kb region. It should be pointed out that the available collections of such mutants were obtained as revertants of the dominant dsx alleles, dsx^D and dsx^{Mas}, which are thought to be irreversibly stuck in the male-specific mode of expression. Thus, these revertants identify only those sequences necessary for the male-specific function of dsx. Another estimate for the size of the dsx^+ gene is provided by an analysis of the transcripts from the region. These data (see below) are consistent with the size of dsx^+ being approximately 30 kb.

The sex-specific expression of dsx inferred from genetic studies could arise via regulation at the level of dsx^+ RNA (e.g., as a result of differential transcription initiation, transcription termination, and/or RNA processing), at the level of dsx^+ protein, or at the level of dsx^+ DNA (e.g., sex-specific rearrangements of the dsx^+ gene). This latter possibility would be formally analogous to the mechanism of mating-type control in *Saccharomyces cerevisiae* (Herskowitz and Oshima 1981).

Whole-genome Southern blot analysis of adult male and female DNAs failed to detect any differences in the gross DNA organization of the dsx^+ locus in the two sexes. In these experiments, 13 different restriction endonuclease digestion patterns were examined, using as probes DNA fragments that together encompass the entire dsx^+ region. Thus, the bifunctional nature of this locus is probably not due to sex-specific DNA rearrangements, unless such rearrangements at dsx^+ are restricted to a small number of cells or involve minor changes in the sequence at dsx^+.

Figure 2 summarizes the results of Northern blot analyses designed to look at the pattern of dsx^+ transcription during development. In this experiment, poly(A)$^+$ RNA was extracted from males or females of the indicated developmental stage and size-fractionated in a formaldehyde-agarose gel. After transferring the RNA to nitrocellulose filters, they were probed with radiolabeled DNA fragments from the dsx^+ region. Figure 2 (top) shows a depiction of the EcoRI fragments of the dsx^+ region that were used as probes; below are shown the fragments of the dsx^+ region that hybridize to various dsx^+ transcripts. No dsx^+ transcripts were detected in embryos. The levels of dsx^+ RNAs in all other stages were low, but detectable, by this technique.

One striking result of this analysis is that there are sex-specific dsx^+ transcripts. In pupae, and in adults, there are size classes of dsx^+ transcripts that are present in males but are not seen in females, and vice versa. In larvae, the major classes of dsx^+ transcripts are not detectably different in the two sexes. The exact molecular basis for these different dsx^+ transcripts is not known. This pattern could arise as the result of differential transcription termination and/or differential RNA splicing. Since the 5′ ends of the different dsx^+ transcripts have not been mapped precisely, it cannot be ruled out that different transcription start positions predispose different transcripts to undergo different termination or processing events.

Two features of these data are worthy of comment. First, the existence of sex-specific dsx^+ transcripts is

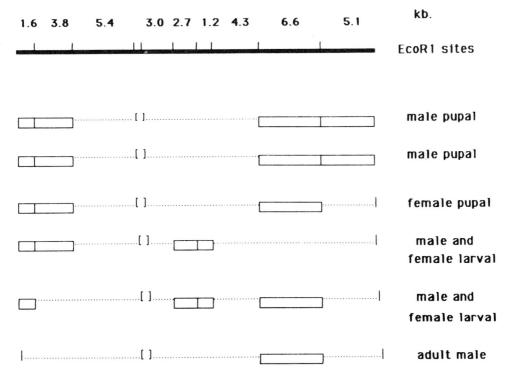

Figure 2. Transcription pattern of the dsx^+ gene. Bar at the top represents the EcoRI restriction fragments from the dsx^+ region. Each of these fragments (except the smallest fragment) was used to probe poly(A)$^+$ RNA isolated from males or females of the indicated stages. Open bars shown below correspond to the fragments that hybridize to dsx^+ transcripts of the sex and stage indicated.

satisfying in that it provides a potential basis for the bifunctional active expression of dsx^+, predicted from purely genetic considerations. Second, the stage-specific complexity of dsx^+ transcripts is surprising and not anticipated by the genetic studies. In larvae, there are two major transcripts that are not sex-specific; in pupae, new size classes of dsx^+ transcripts appear; and in adults, yet another pattern of dsx^+ transcripts is seen. Although the meaning of this developmental specificity of dsx^+ transcription pattern is not clear, it may be significant that the stages during development when there is little sexual dimorphism (i.e., the larval stages) are the stages when dsx^+ transcripts show little sex specificity.

The transformer locus. Null mutations at the *tra* locus, when homozygous, cause chromosomally female individuals (XX) to develop as phenotypic males (Sturtevant 1945). These diplo-X males have rudimentary gonads and are sterile, since the germ line cells of such flies are not transformed (Marsh and Wieschaus 1978). Nevertheless, as far as is known, all aspects of somatic sexual development resemble those of the male in transformed flies. There is no detectable effect of *tra* mutants in chromosomal males (XY). This sex specificity of the phenotype of *tra* mutants suggests that the *tra*$^+$ gene may be female-limited in its expression and is therefore of potential interest from the standpoint of its own regulation. One simple model of sex-determination regulatory gene interactions is that the product of the Sxl^+ gene functions in some manner in females to activate *tra*$^+$. A molecular understanding of these genes will provide an opportunity to approach these questions experimentally. A second facet of *tra*$^+$ gene function that is of interest derives from our model that the *tra*$^+$ and *tra-2*$^+$ loci act, either directly or indirectly, to specify that the dsx^+ gene be expressed in the female-specific mode of expression. As a first approach toward addressing these topics, we have cloned the *tra*$^+$ locus (M. McKeown and J. Belote, in prep.).

The initial step in the cloning of *tra*$^+$ was an X-ray mutagenesis screen and cytogenetic analysis of the chromosome region surrounding *tra*$^+$. Since previous work (Ashburner et al. 1981) placed the *tra* locus very near the scarlet (*st*) locus, our mutant hunt focused on this region of chromosome arm 3L (salivary gland chromosome region 73AB). Chromosome deficiencies (20), either induced in our screen or obtained from other sources, were tested for complementation with the *tra* mutation. The results place *tra* within the faintly banded region 73A5·9. Our collection of X-ray-induced mutants, together with a set of EMS-induced mutants obtained from M. Hoffmann (Harvard University), were used to define the number and distribution of complementation groups in the region flanking *tra* (73A2-73B1·2). This mutational analysis revealed 16 complementation groups within a 14-band polytene chromosome region. The good correlation between the number of bands and the number of complementation groups, together with the fact that many of the complementation groups were defined by more than one mutant allele, suggests that we have approached mutational saturation of the region. None of the mutants in the region, other than the *tra* alleles, appear to affect sexual differentiation. Thus, there does not appear to be a sex-determination gene cluster around *tra*. The complementation groups mapping nearest to *tra* are defined by lethal mutations with late larval or pupal lethal periods.

To clone the *tra*+ gene and the surrounding region, we carried out converging chromosome walks initiated on opposite sides of the 73A5·9 interval. One starting point was at the *Drosophila* cellular homolog of the v-*abl* oncogene (the Dash sequence; Hoffmann-Falk et al. 1983), at position 73B1·2. The starting clone containing this sequence was provided by M. Hoffmann (Harvard University). As the other starting point, we made use of a chromosome inversion ($In(3L)st^{a27}$) that we had induced with X-rays. This inversion has one breakpoint in the *st* locus at 73A3·4 and the other breakpoint in the previously cloned rosy (*ry*) gene region at 87D13·14 (Bender et al. 1983a). We constructed a genomic library of $In(3LR)st^{a27}$ fly DNA cloned into a bacteriophage λ vector and then used the *ry* region DNA clones (provided by W. Bender, Harvard University) to select recombinant phage that contained the *ry* region DNA fused, as a result of the inversion, to the *st* region DNA. Restriction fragments from the *st* portion of these clones were then used to initiate a chromosome walk in the 73A3·4 region. These two chromosome walks were continued until the entire region spanning the 73A3-73B1·2 interval was cloned (about 250 kb).

By using whole-genome Southern blot analysis to map the breakpoints of various deficiencies broken in the region, we were able to subdivide the molecular map into ten regions and to localize within these subdivisions the known complementation groups. In this way, the *tra* locus was mapped to within an approximately 40-kb region, approximately 100 kb proximal to the *st* locus (Fig. 3). $Df(3L)P,st^{B7}$ and $Df(3L)std11$ are both *tra*+, which means that everything required for *tra*+ function must lie between these deficiencies. In addition, $Df(3L)st^{g24}$ is *tra*, suggesting strongly that at least part of the DNA sequences necessary for *tra*+ activity lies to the left of this breakpoint.

As one approach toward identifying where the *tra* genetic unit lies, we analyzed, by whole-genome Southern blot analysis, the available *tra* mutants for molecular rearrangements within this region. Of four *tra* mutant alleles examined, one, the original *tra* mutation of Sturtevant (1945), showed a gross molecular rearrangement, a 1.1-kb deletion, in this region. Unfortunately, since the original *tra*+ chromosome from which the *tra* mutation was derived is not available for testing, it is impossible to know whether the *tra* mutation arose at the same time as, and thus presumably as a

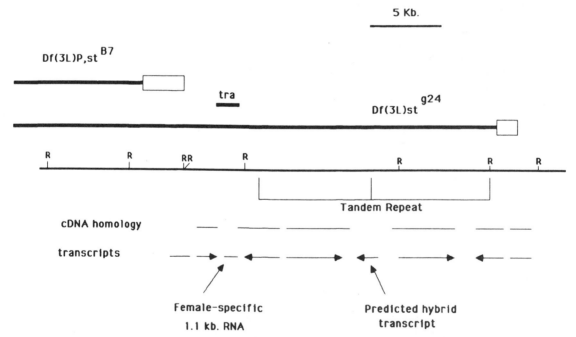

Figure 3. *Eco*RI map of the *tra* region. This region contains a 7.5-kb tandem duplication, with an *Eco*RI site difference between the two repeats. Heavy lines at the top represent regions deleted in the mutant chromosome indicated. Open boxes represent the limits of the deficiency breakpoint positions. Lines labeled cDNA homology represent the regions homologous to isolated cDNA clones; lines labeled transcripts show the map positions of poly(A)+ RNAs detected on Northern blots using either cDNA or genomic clone fragments as probes. Arrows indicate the direction of transcription. The 1.9-kb transcript that crosses the boundary between the tandem repeat is the predicted hybrid transcript formed by the fusion of the 5' end of one transcription unit with the 3' end of another transcription unit. The three leftmost transcripts are detected when Northern blots are probed with the 3.9-kb *Eco*RI genomic fragment. Of these, cDNA clones corresponding only to the center transcription unit have been isolated.

consequence of, this deletion or whether this deletion is actually unrelated to the *tra* mutant phenotype. Nevertheless, this deficiency provides one piece of circumstantial evidence suggesting where *tra* might lie within the cloned region. A completely independent line of evidence suggesting that this site is the *tra* locus comes from studies of the transcripts derived from this region.

For these studies, we made use of an excellent set of stage-specific cDNA libraries constructed by L. Kauvar, B. Drees, S. Poole, and T. Kornberg (University of California at San Francisco). These libraries were screened for clones that hybridize to DNA fragments that together span the entire region between the $Df(3L)P,st^{B7}$ and $Df(3L)st^{g24}$ right-hand breakpoints. The clones that were selected can be unambiguously classified into groups on the basis of where they map and whether or not they cross-hybridize (see Fig. 3). Northern blot analysis using cDNA clone fragments as probes showed that the sizes of the transcripts were approximately the same as the largest cDNA clone in each class.

From genetic studies of the sex-determination mutants, it is inferred that the tra^+ gene function is required in female larvae, pupae, and probably adults for normal sexual differentiation (Baker and Ridge 1980; Wieschaus and Nöthiger 1982; Baker and Belote 1983). In males, however, the tra^+ gene need not function at all for normal sexual development and fertility. The observation that there is a female-specific poly(A)$^+$ RNA (1.0 kb in size) found in larvae, pupae, and adults, taken together with the fact that this transcript derives from the position of the 1.1-kb deletion found in the *tra* mutant chromosome, provides strong evidence that this transcription unit corresponds to the tra^+ gene. None of the other transcripts that derive from this 20-kb region exhibit any sex specificity. Although conclusive proof of the position of the tra^+ locus must await the results of experiments in which DNA fragments from this region are transferred into mutant genomes via the P-mediated transformation technique (Rubin and Spradling 1982; Spradling and Rubin 1982), we believe it very likely that the 1.0-kb female-specific transcript corresponds to tra^+.

Not only is the putative tra^+ transcript relatively small, but the close proximity of the flanking transcription units implies that the tra^+ gene itself is small. Although the direction of transcription of tra^+ has not yet been determined, there is not much room on either side of the tra^+ transcribed region for an extensive stretch of nontranscribed regulatory sequences. The flanking transcription units (1.3 kb apart) are transcribed in both sexes, and the cDNA clones corresponding to these RNAs show no detectable cross-hybridization with the tra^+ transcript.

The tra^+ gene transcript not only is regulated in a sex-specific manner, but also exhibits stage-specific variation in abundance. Pupae and adults show the highest levels of tra^+ RNA, with the levels in larvae being lower. It is an interesting correlation that the stages during development when the tra^+ transcript is most abundant are also the stages when the dsx^+ gene exhibits its most pronounced sex-specific transcript patterns.

Searches for Additional Sex-determination Genes

Most of the extant sex-determination mutants were discovered by chance as spontaneous mutations and not as the result of systematic screening for sex-determination loci. The relatively low frequency with which such loci have been identified (about one new sex-determination locus every 10 years) suggests that the number of such genes is not high. Moreover, the mutants described to date may be a biased subset in that they either are completely viable (e.g., *dsx*, *tra*, *tra-2*, and *ix*) or exhibit a sex-specific lethal phenotype (e.g., Sxl^F, Sxl^M). Since a complete understanding of the sex-determination hierarchy requires that all components of the pathway be identified, we have undertaken two screens, one genetic and one molecular, to search for new sex-determination genes systematically. These approaches have the advantage that they circumvent potential problems associated with pleiotropic mutants for which it might otherwise be difficult to ascertain whether or not they are involved in sex determination.

A genetic screen for sex-determination mutants. During the course of studies designed to determine the epistatic relationships among sex-determination mutants (Baker and Ridge 1980), it was discovered that in XX flies simultaneously heterozygous for both *tra-2* and *tra*, the addition of a third sex-determination mutant in heterozygous condition frequently results in such flies developing as intersexes. The frequency of such intersexes varies between 10% and 100%, depending on the particular locus involved, and on other undefined genetic background effects. The degree of sexual transformation also varies, with the most extreme cases exhibiting a phenotype similar to that of *dsx/dsx* individuals. Here, we are not primarily concerned with the physiological basis for this interaction, but presumably, by reducing the levels of both the $tra-2^+$ and tra^+ gene products, the sex-determination regulatory pathway becomes sensitized to further perturbations such as those that might be caused by heterozygosity for a third gene in the pathway. Under such conditions, there is a frequent failure to repress fully the male-specific sexual differentiation functions, and intersexual development ensues.

Regardless of the basis for this interaction, the observation that every known sex-determination mutant, including *dsx*, *ix*, and Sxl^F, exhibits such a dominant phenotype when in the *tra-2/+*;*tra/+* genetic background suggests that such an interaction can be used to screen for recessive mutations in sex-determination genes without having to make them homozygous. One advantage of such a screen is that it can potentially allow the detection of pleiotropic mutations that affect not only sex determination, but also some vital process and thus are homozygous lethal. A second advantage is that one can initially screen a sizable fraction of the genome for sex-determination loci by introducing chromosome deficiencies in heterozygous condition into the $tra\text{-}2/+;tra/+$ background. If a deficiency removes an important sex-determination locus, then it might interact in this test to produce a high frequency of intersexes.

We (T.N. Scott and B. Baker) have carried out such a screen, testing about 70 deficiencies that together encompass just over 30% of the genome; 12 deficiencies tested positive in this assay, each producing $\geq 20\%$ intersexes among the $XX;tra\text{-}2/+;tra/+$ flies. Seven of these, however, were deficiencies that deleted a Minute locus. Minute loci are probably the structural genes for the ribosomal proteins (Huang and Baker 1976; Vaslet et al. 1980; Kongsuwan et al. 1985). Thus, the screen is probably not specific for sex-determination genes but may also detect mutations that disrupt certain general metabolic functions whose activities incidently modulate the expression of the sex-determination regulatory pathway.

The deficiency that gave the highest frequency of intersexes in the $tra\text{-}2/+;tra/+$ test was $Df(1)C246$, deleted for region 11D to 12A1·2 on the X chromosome. We have begun a genetic analysis of this region in an attempt to identify the gene responsible for this interaction. One approach is to saturate the region with mutations and to examine their phenotypes to determine whether they show defects in sexual differentiation. These mutant screens are currently in progress. Another way to study the role of this region in sex determination is to examine the interaction between $Df(1)C246$ and other genes involved in sex determination. Preliminary results of crosses involving $Df(1)C246$ and Sxl^F or da suggest that this region may contain a locus that functions in the early events controlling sex determination and dosage compensation. A striking sex-specific maternal-effect lethality is observed when $Df(1)C246/+$ females are crossed to Sxl^F/Y males. Such a cross yields very few daughters, and the female progeny that do survive often exhibit patches of male tissue. There is no such sex differential lethality among the progeny of $Df(1)C246/+$ females crossed to Sxl^+/Y males. These results suggest that $Df(1)C246/+$ females produce eggs that are deficient in some component necessary for female development and that this deficiency can be partially compensated for by Sxl^+ alleles in the zygote. This interaction is similar to one involving the Sxl^F allele and the second chromosome recessive maternal-effect mutant da. In that interaction, $da/+$ mothers mated with Sxl^F/Y fathers produce reduced numbers of female progeny (Cline 1980).

An interaction is also seen between $Df(1)C246$ and da. Females heterozygous for both $Df(1)C246$ and da when crossed to $da/+$ males yield many fewer da/da progeny than expected, suggesting that $Df(1)C246$ acts maternally to exacerbate the lethal effect of the da mutation (see Cline 1976).

Although these preliminary results demonstrating an interaction between $Df(1)C246$ and loci involved in both sex determination and dosage compensation are intriguing, a precise delimitation of the function of the sex-determination locus in this region of the X must await further data. Nevertheless, the discovery of interactions of $Df(1)C246$ with Sxl^F and da provides strong independent corroboration for the validity of the $tra\text{-}2/+;tra/+$ interaction screen.

A molecular screen for sex-determination loci. As another approach to the isolation of new sex-determination genes, we (D. Andrew and B. Baker) have searched for DNA sequences that share homology with the dsx^+ gene. The rationale for this approach is that genes with similar or parallel functions might share sequence homology either as a consequence of their evolutionary relationships and/or because similar sequences are important in their coordinate control or common function. A good candidate for this latter type of sequence is the recently identified homeo box region shared by several developmentally important regulatory genes (McGinnis et al. 1983, 1984; Laughon and Scott 1984; Levine et al. 1984; Scott and Weiner 1984; Fjose et al. 1985; Poole et al. 1985). This strategy has been used successfully to identify and isolate related genes of a variety of types, e.g., heat-shock protein genes (Craig et al. 1983), larval serum protein genes (Brock and Roberts 1981), chorion protein genes (see Kafatos 1983), and segmentation control genes (McGinnis et al. 1983). The fact that there is a simple genetic assay (i.e., the $tra\text{-}2/+;tra/+$ test) for the involvement of the cognates identified in this screen in the pathway controlling sex determination makes this approach feasible.

Low-stringency hybridizations ($5 \times$ SSPE, 29% formamide at 42°C) of dsx^+ region DNA probes to whole-genome Southern blots reveal 20–22 EcoRI restriction fragments in the genome (other than those at dsx) that hybridize the dsx^+ probes. Since several of these restriction fragments represent contiguous stretches of DNA, there appear to be a limited number of regions in the genome that share significant sequence homology with the dsx^+ region. Seven doublesex cognate sequences (dsc's) have been isolated as genomic clones from a standard *D. melanogaster* library (Maniatis et al. 1978) and mapped to salivary gland chromosome positions. Each dsc maps to a unique autosomal site. Although none of these exactly correspond to the position of a previously identified sex-determination locus, it is intriguing that two of them map very close to known sex-determination genes. The $dsc73$ cognate sequence is only 130 kb distal to the tra locus, and $dsc50$ maps just a few polytene chromosome bands (estimated to be ~150 kb) proximal to $tra\text{-}2$. The significance of the apparently nonrandom positions of these cognates is not known.

Several of the clones that contain the cognate sequences appear to have extensive homology with the dsx^+ region (i.e., the homology with dsx^+ is not limited to a single small region of the dsc clones). Their patterns of homology with dsx^+ DNA do not show the properties expected from simple gene duplication followed by deletions, insertions, and substitutions (i.e., homologous regions being colinear between the dsc's and dsx^+). Instead, the regions of homology appear to have undergone complex rearrangements involving large blocks of DNA.

All of the dsc clones hybridize to genomic DNA from *D. simulans, D. mauritiana, D. yakuba, D. pseudoobscura, D. erecta,* and *D. virilis* under high-stringency conditions (5 × SSPE, 50% formamide at 42°C). This indicates that some portions of the regions containing the dsc's are well conserved across the genus.

The clones containing the dsc sequences hybridize to poly(A)$^+$ RNA on Northern blots, and these RNAs are developmentally regulated. Several show pupal-specific expression, although none appear to be expressed in a sex-specific manner. It is not known at this time, however, whether the transcripts themselves contain the sequences that are homologous to dsx^+.

Of critical interest is the determination of whether the dsc's actually have a role in the control of sexual differentiation or whether they share homology with dsx^+ for other reasons. To examine this, chromosomal deficiencies that delete each of the dsc loci are being made simultaneously heterozygous with tra-2 and tra and the diplo-X flies examined for defects in sexual differentiation. One of the cognates, $dsc73$, has tested positive in this assay: A small deficiency that deletes $dsc73$, when heterozygous for tra-2 and tra, results in a high frequency (~30%) of diplo-X flies being intersexual. Moreover, a lethal mutant, *l(3L)133.37*, that maps very close to, if not at, the $dsc73$ sequence also interacts strongly with tra-2/+; tra/+ in females. Work is in progress to determine whether this mutant actually corresponds to the $dsc73$ locus, and, if so, what the function of this putative sex-determination gene might be.

CONCLUSION

The experiments described above are only the beginning steps in the genetic and molecular characterization of the regulatory hierarchy controlling somatic sexual differentiation. The molecular analyses of dsx^+ and tra^+ have just laid the groundwork for the interesting questions concerning their regulation, their interactions, and their (protein) products. The availability of dsx^+ and tra^+ recombinant DNA clones now makes it possible to examine their expression by RNA blot and in situ hybridization techniques. In this way, we have already established that the sex-specific regulation of both dsx^+ and tra^+ occurs at the RNA level (either transcription or processing). By examining the expression of these genes in flies carrying one or more of the sex-determination mutants, we can now test, and possibly extend, our genetic model of their interactions. The isolation by cloning of the other sex-determination loci (both known and previously unidentified) is also an important requirement for a genetic and molecular dissection of sex determination. It is our hope that by combining the techniques of genetics, cytology, and molecular biology, we can begin to understand the molecular mechanisms involved in controlling this developmental pathway.

ACKNOWLEDGMENTS

This work was supported by U.S. Public Health Service grants GM-23345, GM-07642, and GM-07199, National Science Foundation grant PCM-8202812, and a Helen Hay Whitney fellowship (to M.B.M.).

REFERENCES

ARTAVANIS-TSAKONAS, S., M.A.T. MUSKAVITCH, and B. YEDVOBNICK. 1983. Molecular cloning of *Notch*, a locus affecting neurogenesis in *Drosophila melanogaster*. *Proc. Natl. Acad. Sci.* **80:** 1977.

ASHBURNER, M., P. ANGEL, C. DETWILER, J. FAITHFULL, D. GUBB, G. HARRINGTON, T. LITTLEWOOD, S. TSUBOTA, V. VELISSARIOU, and V. WALKER. 1981. New mutants. *Drosophila Inform. Serv.* **56:** 186.

BAKER, B.S. and J.M. BELOTE. 1983. Sex determination and dosage compensation in *Drosophila melanogaster*. *Annu. Rev. Genet.* **17:** 345.

BAKER, B.S. and K. RIDGE. 1980. Sex and the single cell: On the action of major loci affecting sex determination in *Drosophila melanogaster*. *Genetics* **94:** 383.

BELOTE, J.M. and B.S. BAKER. 1982. Sex determination in *Drosophila melanogaster*: Analysis of *transformer-2*, a sex transforming locus. *Proc. Natl. Acad. Sci.* **79:** 1568.

BELOTE, J.M. and J.C. LUCCHESI. 1980a. Male-specific lethal mutations of *Drosophila melanogaster*. *Genetics* **96:** 165.

———. 1980b. Control of X chromosome transcription by the *maleless* gene in *Drosophila*. *Nature* **285:** 573.

BELOTE, J.M., A.M. HANDLER, M.F. WOLFNER, K.J. LIVAK, and B.S. BAKER. 1985. Regulation of yolk protein gene expression in *Drosophila melanogaster*. *Cell* **40:** 339.

BENDER, W., P. SPIERER, and D.S. HOGNESS. 1983a. Chromosome walking and jumping to isolate DNA from the *Ace* and *rosy* loci and the bithorax complex of *Drosophila melanogaster*. *J. Mol. Biol.* **168:** 17.

BENDER, W., M.A. AKAM, P.A. BEACHY, F. KARCH, M. PEIFER, E.B. LEWIS, and D.S. HOGNESS. 1983b. Molecular genetics of the bithorax complex in *Drosophila melanogaster*. *Science* **221:** 23.

BOWNES, M. and R. NÖTHIGER. 1981. Sex determining genes and vitellogenin synthesis in *Drosophila melanogaster*. *Mol. Gen. Genet.* **182:** 222.

BRIDGES, C.B. 1921. Triploid intersexes of *Drosophila melanogaster*. *Science* **54:** 252.

BROCK, H.W. and D.B. ROBERTS. 1981. Quantitative in situ hybridization reveals extent of sequence homology between related DNA sequences in *Drosophila melanogaster*. *Chromosoma* **83:** 159.

CLINE, T.W. 1976. A sex-specific, temperature-sensitive maternal effect of the daughterless mutation of *Drosophila melanogaster*. *Genetics* **84:** 723.

———. 1978. Two closely linked mutations in *Drosophila melanogaster* that are lethal to opposite sexes and interact with daughterless. *Genetics* **90:** 683.

———. 1979. A male-specific mutation in *Drosophila melanogaster* that transforms sex. *Dev. Biol.* **72:** 266.

———. 1980. Maternal and zygotic sex-specific gene interactions in *Drosophila melanogaster*. *Genetics* **96:** 903.

———. 1983a. The interaction between daughterless and sex-lethal in triploids: A novel sex-transforming maternal effect linking sex determination and dosage compensation in *Drosophila melanogaster*. *Dev. Biol.* **95:** 260.

———. 1983b. Functioning of the genes daughterless (*da*) and Sex-lethal (*Sxl*) in *Drosophila* germ lines. *Genetics* **104:** s16.

———. 1984. Autoregulatory functioning of a *Drosophila* gene product that establishes and maintains the sexually determined state. *Genetics* **107:** 231.

CRAIG, E., T.D. INGOLIA, and L.J. MANSEAU. 1983. Expression of *Drosophila* heat-shock cognate genes during heat shock and development. *Dev. Biol.* **99:** 418.

DENELL, R.E. and R. JACKSON. 1972. A genetic analysis of transformer-dominant. *Drosophila Inform. Serv.* **48:** 44.

DUNCAN, I.W. and T.C. KAUFMAN. 1975. Cytogenetic analysis of chromosome 3 in *Drosophila melanogaster*: Mapping of the proximal portions of the right arm. *Genetics* **80:** 733.

FJOSE, A., W.J. MCGINNIS, and W.J. GEHRING. 1985. Isolation of a homeo box-containing gene from the *engrailed* region of *Drosophila* and the spatial distribution of its transcripts. *Nature* **313:** 284.

GARABEDIAN, M.J., M.-C. HUNG, and P.C. WENSINK. 1985. Independent control elements that determine yolk protein gene expression in alternative *Drosophila* tissues. *Proc. Natl. Acad. Sci.* **82:** 1396.

GARBER, R.L., A. KUROIWA, and W.J. GEHRING. 1983. Genomic and cDNA clones of the homeotic locus *Antennapedia* in *Drosophila*. *EMBO J.* **2:** 2027.

HERSKOWITZ, I. and Y. OSHIMA. 1981. Control of cell type in *Saccharomyces cerevisiae*: Mating type and mating type interconversion. In *The molecular biology of the yeast* Saccharomyces: *Life cycle and inheritance* (ed. J.N. Strathern et al.), p. 181. Cold Spring Harbor Laboratory, Cold Spring Harbor, New York.

HOFFMANN-FALK, H., P. EINAT, B.-Z. SHILO, and F.M. HOFFMANN. 1983. *Drosophila melanogaster* DNA clones homologous to vertebrate oncogenes: Evidence for a common ancestor to the *src* and *abl* cellular genes. *Cell* **32:** 589.

HUANG, S.L. and B.S. BAKER. 1976. The mutability of the *Minute* loci in *Drosophila melanogaster* with ethyl methanesulfonate. *Mutat. Res.* **34:** 407.

KAFATOS, F.C. 1983. Structure, evolution, and developmental expression of the chorion multigene families in silkmoths and *Drosophila*. *Symp. Soc. Dev. Biol.* **41:** 33.

KAUFMAN, T.C., R.A. LEWIS, and B.T. WAKIMOTO. 1980. Cytogenetic analysis of chromosome 3 in *Drosophila melanogaster*: The homoeotic gene complex in polytene chromosome interval 84A-B. *Genetics* **94:** 115.

KIDD, S., T.J. LOCKETT, and M.W. YOUNG. 1983. The *Notch* locus in *Drosophila melanogaster*. *Cell* **34:** 421.

KONGSUWAN, K., Y. QUIANG, A. VINCENT, M.C. FRISARDI, M. ROSBASH, J.A. LENGYEL, and J.R. MERRIAM. 1985. A *Drosophila Minute* mutation encodes a ribosomal protein. *Nature* (in press).

KORNBERG, T. 1981. *engrailed*: A gene controlling compartment and segment formation in *Drosophila*. *Proc. Natl. Acad. Sci.* **78:** 1095.

LAUGHON, A. and M.P. SCOTT. 1984. Sequence of a *Drosophila* segmentation gene: Protein structure homology with DNA-binding proteins. *Nature* **310:** 25.

LEVINE, M., G. RUBIN, and R. TIJAN. 1984. Human DNA sequences homologous to a protein coding region conserved between homeotic genes of *Drosophila*. *Cell* **38:** 667.

LEWIS, E.B. 1978. A gene complex controlling segmentation in *Drosophila*. *Nature* **276:** 565.

LUCCHESI, J.C. and T. SKRIPSKY. 1981. The link between dosage compensation and sex determination in *Drosophila melanogaster*. *Chromosoma* **8:** 217.

MANIATIS, T., R.C. HARDISON, E. LACY, J. LAUER, C. O'CONNELL, D. QUON, G.K. SIM, and A. EFSTRATIADIS. 1978. The isolation of structural genes from libraries of eukaryotic DNA. *Cell* **15:** 687.

MARONI, G. and W. PLAUT. 1973. Dosage compensation in *Drosophila melanogaster* triploids: Autoradiographic study. *Chromosoma* **40:** 361.

MARSH, J.L. and E. WIESCHAUS. 1978. Is sex determination in germline and soma controlled by separate genetic mechanisms? *Nature* **272:** 249.

MCGINNIS, W., R.L. GARBER, J. WIRZ, A. KUROIWA, and W.J. GEHRING. 1984. A homologous protein-coding sequence in *Drosophila* homeotic genes and its conservation in other metazoans. *Cell* **37:** 403.

MCGINNIS, W., M.S. LEVINE, E. HAFEN, A. KUROIWA, and W.J. GEHRING. 1983. A conserved DNA sequence in homeotic genes of *Drosophila* Antennapedia and bithorax complexes. *Nature* **308:** 428.

MISCHKE, D. and M.L. PARDUE. 1982. Organization and expression of α-tubulin genes in *Drosophila melanogaster*. *J. Mol. Biol.* **156:** 449.

MORGAN, T.H. 1916. Mosaics and gynandromorphs in *Drosophila*. *Proc. Soc. Exp. Biol. Med.* **11:** 171.

OTA, T., A. FUKUNAGA, M. KAWABE, and K. OISHI. 1981. Interactions between sex transformation mutants of *Drosophila melanogaster*. I. Hemolymph vitellogenins and gonad morphology. *Genetics* **99:** 429.

POOLE, S.J., L.M. KAUVAR, B. DREES, and T. KORNBERG. 1985. The *engrailed* locus of *Drosophila*: Structural analysis of an embryonic transcript. *Cell* **40:** 37.

POSTLETHWAIT, J.H., M. BOWNES, and T. JOWETT. 1980. Sexual phenotype and vitellogenins in *Drosophila*. *Dev. Biol.* **79:** 379.

RUBIN, G. and A. SPRADLING. 1982. Genetic transformation of *Drosophila* with transposable element vectors. *Science* **218:** 348.

SANCHEZ, L. and R. NÖTHIGER. 1982. Clonal analysis of *Sex-lethal*, a gene needed for female sexual development in *Drosophila melanogaster*. *Wilhelm Roux's Arch. Dev. Biol.* **191:** 211.

———. 1983. Sex determination and dosage compensation in *Drosophila melanogaster*: Production of male clones in XX females. *EMBO J.* **2:** 485.

SCHÜPBACH, T. 1985. Normal female germ cell differentiation requires the female X chromosome to autosome ratio and expression of *Sex-lethal* in *Drosophila melanogaster*. *Genetics* **109:** 529.

SCOTT, M.P. and A.J. WEINER. 1984. Structural relationships among genes that control development: Sequence homology between Antennapedia, Ultrabithorax, and fushi tarazu loci of *Drosophila*. *Proc. Natl. Acad. Sci.* **81:** 4115.

SCOTT, M.P., A.J. WEINER, T.I. HAZELRIGG, B.A. POLISKY, V. PIRROTTA, F. SCALENGHE, and T.C. KAUFMAN. 1983. The molecular organization of the Antennapedia locus of *Drosophila*. *Cell* **35:** 763.

SPRADLING, A. and G. RUBIN. 1982. Transposition of cloned P elements into *Drosophila* germ line chromosomes. *Science* **218:** 341.

STERN, C. 1966. Pigmentation mosaicism in intersexes of *Drosophila*. *Rev. Suisse Zool.* **73:** 339.

STURTEVANT, A.H. 1945. A gene in *Drosophila melanogaster* that transforms females into males. *Genetics* **30:** 297.

VAN DEUSEN, E.B. 1976. Sex determination in germline chimeras of *Drosophila melanogaster*. *J. Embryol. Exp. Morphol.* **37:** 173.

VASLET, C.A., P. O'CONNEL, M. IZQUIERDO, and M. ROSBASH. 1980. Isolation and mapping of a cloned ribosomal protein gene of *Drosophila melanogaster*. *Nature* **285:** 674.

WIESCHAUS, E. and R. NÖTHIGER. 1982. The role of the transformer genes in the development of the genitalia and analia of *Drosophila melanogaster*. *Dev. Biol.* **90:** 320.

A Single Principle for Sex Determination in Insects

R. NÖTHIGER AND M. STEINMANN-ZWICKY
Zoological Institute, University of Zurich, CH 8057, Switzerland

On September 27, 1870, Gregor Mendel wrote to the famous botanist Nägeli in Zurich. In his letter, he suggested that sex determination might prove to be a phenomenon of heredity and segregation. But, as with his previous discovery of the principles of heredity, no attention was paid to his idea.

Today, nobody doubts that genes control the sexual pathway. This is clear for systems with *genetic* sex determination, where inherited genetic differences dictate maleness or femaleness; it is less obvious, at least at first glance, for systems with *environmental* sex determination, where there are no genetic differences between males and females. In the case of hermaphrodites, however, where the same genotype differentiates male and female organs within an individual, we realize that sex determination and differentiation are a matter of gene regulation; one group of cells can form a male organ, and another group can form a female organ. This is analogous to the situation in a fly, where one group of cells forms mesothoracic structures and another group forms metathoracic structures, depending on which developmental pathway is chosen. Such decisions are under the control of regulatory genes (Lewis 1978; Hodgkin 1984).

Textbooks and general reviews revel in emphasizing the bewildering variety of sex-determining mechanisms, with no attempt to look for an underlying principle. The phenomena are confusing even when we confine our observations to insects: To determine sex, *Drosophila* and *Sciara* use the ratio of X chromosomes to autosomes; in *Musca* or *Anopheles*, the Y chromosome can determine maleness; in some strains of *Musca* or *Chironomus*, dominant male determiners are found and in other strains, dominant female determiners are found; *Chrysomya* uses a maternal factor, the *Hymenoptera* a haplo-diplo mechanism; and in *Heteropeza*, the hemolymph of the mother determines the sex of the offspring.

We propose that this apparent multitude of mechanisms arises from minor variations of a common principle: a hierarchically built control system with a primary *signal*, present or absent, that is read by a *key gene* whose state of activity (OFF or ON) is used to control the sex differentiation genes through the action of a genetic *double switch* (Fig. 1). (The terms ON and OFF are used loosely throughout this paper to indicate that the functional product is required [ON] or not required [OFF]. We do not want to imply that the control must occur at the transcriptional level.)

This simple system forms the basis of sex determination in general. There are two pivotal factors: One is the *key gene* that must be ON to determine the female pathway or OFF to determine the male pathway; the other is the *double switch*, whose mutually exclusive functions, producing either M or F, result in expression of only the male or only the female differentiation genes. In systems with genetic sex determination, the *signal* is produced by the genes of the male embryo to repress the basically constitutive key gene. Mutations can occur in the signaling gene(s) and in the key gene, rendering the genes refractory to regulation, or the products ineffective or sensitive to environmental factors. These modifications create seemingly very different mechanisms of sex determination.

We believe, however, that all of these mechanisms serve one purpose: The key gene should be active in some animals, or cells, which then enter the female pathway, and inactive in other animals, or cells, which then enter the male pathway.

RESULTS AND ARGUMENTS

Our hypothesis is based on the analysis of examples found among insects, mainly of the order Diptera, whose sex-determination mechanisms have been studied in some detail. We show how these apparently very different mechanisms can be fitted into a single scheme and how the same principle is at work, even in very distant species.

The Paradigm of *Drosophila*

We begin with a brief description of *Drosophila melanogaster*, since the mechanism of sex determination is best analyzed in this species (Fig. 2). In *Drosophila*, the ratio of X chromosomes to sets of autosomes (X:A) acts as the discriminator between males and females. The key gene Sex-lethal (*Sxl*) is ON in animals with an X:A ratio of 1.0 and thus determines the female pathway; with an X:A ratio of 0.5, *Sxl* is OFF and thus determines the male pathway. A maternal product, da^+, is a prerequisite for *Sxl* to be active, but it has no discriminative effect in itself, since it is present in all eggs. The signaling genes, *R*, specifying maleness by repressing *Sxl* have not been identified, but they must be located on the autosomes, since a genotype with one X chromosome and one set of autosomes (X A) is female; but when another set of autosomes is added (X AA), male development results. For normal diploid males (X AA) and females (XX AA), a plausible hypothesis assumes that both sexes contain the same limited amount of product R but that two X chromosomes bind all re-

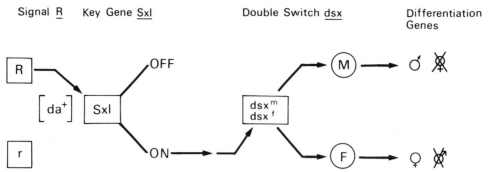

Figure 1. Basic scheme for the genetic control of sex determination. A signal gene, R, controls a key gene, Sxl, whose state of activity, ON or OFF, sets a genetic double switch, consisting of the two elements dsx^m and dsx^f, in such a way that it produces either M when dsx^m is ON or F when dsx^f is ON. The products, M or F, act to prevent either the female set or the male set of differentiation genes from being expressed. The allele R is only present in the male sex, where it prevents activity of Sxl. Females are homozygous r/r, which allows constitutive expression of Sxl. Activity of Sxl in the embryo appears to depend on the presence of the maternal gene product da^+.

pressor molecules (R) so that Sxl can be transcribed; when only one X chromosome is present, some R molecules are left to repress Sxl (Chandra 1985). The state of activity of Sxl, via a small number of regulatory genes ($tra-2$, tra, ix), is passed on to the genetic double switch of dsx. In genetic terms, dsx consists of two cistrons, dsx^m and dsx^f, only one of which functions in either sex. When both cistrons are deleted by mutation, the male set and the female set of differentiation genes appear to be active within a cell, producing an intersexual phenotype at the cellular level.

We conclude that the products of dsx^m (indicated by M in Fig. 2) are used to repress the female sex-differentiation genes, and the products of dsx^f (F) are used to repress the male-differentiation genes. It is important to realize that the simple binary code of the key gene, Sxl ON or OFF, cannot directly control the differentiation genes but must use a double switch as an intermediary. The two cistrons or functions of dsx are wired in such a way that when dsx^m is ON, dsx^f is OFF, and vice versa. Since the task to allow expression of only one of two sets of sex differentiation genes must be solved in every species, including those with hermaphroditism, we postulate that such a double switch, homologous to dsx of *Drosophila*, must be operative in all insects and probably far beyond this systematic group. For a review of sex determination in *Drosophila*, see Baker and Belote (1983) and Nöthiger and Steinmann-Zwicky (1985).

The Primitive State and Its Modifications

For future considerations, we restrict the discussion to the three elements R, Sxl, and da, which may be called the variables in the genetic system of sex determination (Fig. 3). We omit dsx because we know of no example where evolution played with the double switch. *Drosophila* is a highly evolved species of Diptera, with a specialized sex-determination mechanism that must have been acquired late in the phylogeny of this order. The primitive situation is probably represented by species where the male is heterozygous at a single sex-determining locus, usually symbolized by M/m = male, m/m = female (Lucchesi 1978). The simplest hypothesis is that M corresponds to our signaling gene(s) R in Figures 1 and 2 and specifies a repressor R for the key gene Sxl; the recessive m, or r in our terminology, does not code for a functional product, thus allowing Sxl to be active. Species with this mode of sex determination, e.g., *Culex*, may be characterized by the genotypic formula $R/r\ Sxl^+/Sxl^+$ for males and $r/r\ Sxl^+/Sxl^+$ for females. Sxl^+ is regulated by R versus r, i.e., by the presence or absence of product R. As we have seen in *Drosophila*, a maternal component produced by the

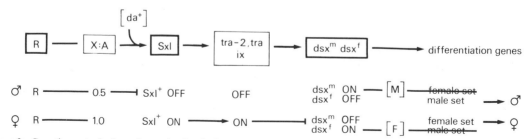

Figure 2. Genetic control of sex determination in *Drosophila* (see text: The Paradigm of *Drosophila*). The X:A ratio mediates between the (postulated) R genes and Sxl. The state of Sxl is communicated to the double switch $dsx^m\ dsx^f$ via the genes $tra-2$, tra, and perhaps ix as intermediaries. Although we do not yet understand why such intermediary genes are necessary, we expect them to exist in other insects, and they may also occur in other taxonomic groups (see e.g., *Caenorhabditis*; Hodgkin 1984).

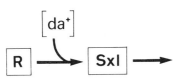

Figure 3. The three elements that form the variables for evolution of the genetic system controlling the sexual pathway: the signaling gene *R*, the key gene *Sxl*, and the maternally active gene *da*+. See Fig. 4 for variations.

gene *da*+ may be necessary for the key gene *Sxl*+ to be active.

Figure 4 illustrates how the primitive genetic system, represented by type 1, can be varied by mutations in its elements, *R*, *Sxl*, or *da* or by the evolution of heteromorphic chromosomes, so that sex-determining mechanisms result that appear very different from each other. Types 8 and 9 (Fig. 4) fall into the category of environmental sex determination and types 2 through 6 remain genetic; type 7 is difficult to classify because it is the genotype of the mother that determines the sex of the embryos. Of the systems with genetic sex determination, some have dominant male determiners with (type 2) or without (type 1) heteromorphic sex chromosomes, others have dominant female determiners that are epistatic (type 5) or hypostatic (type 6) to the dominant male determiners, and still others have a haplo-diplo (type 4) or a balance (type 3) mechanism. The common denominator of these various mechanisms is that they all achieve the same result, namely, an active product of *Sxl* in what is to be female and an inactive product in what is to be male. Table 1 gives concrete examples of sex determination in insects and an interpretation of how a particular mechanism could have arisen from a primitive type (compare corresponding numbers in Table 1 and Fig. 4). The evolution of heteromorphic sex chromosomes (types 2 and 3) can create a problem of gene dosage that may be solved by evolving mechanisms of dosage compensation (not discussed here; see Lucchesi 1978).

Some species, such as *Megaselia scalaris*, demonstrate that the dominant male determiner *R* can move around in the genome so that different chromosomes can acquire the status of sex chromosomes (Mainx 1964). *R* can also increase in number, as shown for some strains of *Musca domestica* with multiple *R* factors (Franco et al. 1982). This situation would lead to a dangerous surplus of males in a population unless evolution took countermeasures. When males from such a strain (type 5 in Fig. 4; male = $R/R\ Sxl^+/Sxl^+$) are crossed to standard females (type 2 in Fig. 4; female = $r/r\ Sxl^+/Sxl^+$), all offspring are male. In the strain with multiple *R* factors, however, the sex ratio is 1:1, due to a dominant female determiner that is epistatic, even over many *R* factors. This could be achieved by *Sxl* having mutated to a constitutive, nonrepressible state, Sxl^c, a type of mutation that is known in *Drosophila* (Cline 1978). Multiple *R* genes may also be present in *Drosophila*, where they appear to have spread on the autosomes. Parallel to this process, the X chromosomes have acquired binding sites that can capture and neutralize the *R* products, so that a delicate balance between genes producing repressors and DNA sequences (genes?) absorbing them is created.

We now briefly review some of the cases listed in

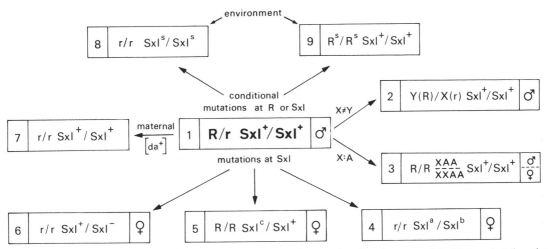

Figure 4. Variations in the mechanisms of sex determination. Examples are taken from insects, mainly Diptera, but the principle probably applies far beyond this systematic group. Variations for the primitive type (1, center), produced by mutations in *R*, *Sxl*, or *da* or by the evolution of heteromorphic sex chromosomes are shown (see text). *Mutations*: (Sxl^a, Sxl^b) alleles that yield nonfunctional, but complementing, products; (Sxl^c) allele that is not suppressible by *R*, dominant; (Sxl^-) loss of function, recessive; (Sxl^s, R^s) alleles whose products are sensitive to environmental influences. The nine boxes contain our interpretation of concrete examples listed in Table 1. Numbers *1* through *9* correspond to those in Table 1. Genotypes of the heterogametic sex are given. The opposite sex then is homozygous for the recessive allele: (*1*) female is $r/r\ Sxl^+/Sxl^+$; (*2*) female is $X(r)/X(r)\ Sxl^+/Sxl^+$; (*4*) male is $r\ Sxl^a$ or $r\ Sxl^b$ or $r/r\ Sxl^a/Sxl^a$; (*5*) male is $R/R\ Sxl^+/Sxl^+$; (*6*) male is $r/r\ Sxl^-/s\ r/r\ Sxl^-/s\ r/r\ Sxl^-/s\ r/r\ Sxl^-/Sxl^-$. For *7*, *8*, and *9*, the genotype of male and female is the same, sex being determined by a maternal factor (*7*) or environment (*8*,*9*).

Table 1. Sex-determining Mechanisms in Insects: Examples and Interpretation

Genotype, sex	Phenomenon	Interpretation R	Sxl allelic state	Sxl state of activity	Genus	References
(1) $M/m = ♂$	M, dominant male determiner	$M = R$			Aëdes	McClelland (1962)
$m/m = ♀$		$+/-$	$+/+$	inactive	Calliphora	Ribbert (1967)
	locus of M varies from strain to strain	$-/-$	$+/+$	active	Culex	Gilchrist and Haldane (1947)
		R on a transposable element?			Chironomus	Martin and Lee (1984); Hägele (1985)
					Megaselia	Mainx (1964)
					Musca	Franco et al. (1982)
(2) $Y/X = ♂$	Y, dominant male determiner	Y carrying R, heteromorphic	$+/-$	$+/+$	inactive	Anopheles	Baker and Sakai (1979)
$X/X = ♀$		$-/-$	$+/+$	active	Calliphora	Ullerich (1963)
					Musca	Franco et al. (1982)
(3) $X;AA = ♂$	X:A ratio	R is produced by autosomes, absorbed by X	$+/+$	$+/+$	inactive	Drosophila	Baker and Belote (1983)
$XX;AA = ♀$		$+/+$	$+/+$	active	Sciara	Nöthiger and Steinmann-Zwicky (1985); Metz (1938); Crouse (1960)
(4) haploid = ♂	multiple complementing alleles	mutations at Sxl	$-$	a or b	inactive product	Apis	Rothenbuhler (1957)
diploid = ♀		$-/-$	a/b	active product	Habrobracon	P.W. Whiting (1943a)
					Mormoniella	A.R. Whiting (1967)
(5) $MMM f/f = ♂$	several dominant male determiners M, one dominant female determiner F, F epistatic over M	$M = R$ on several chromosomes; $F = Sxl^c$, not repressible	$+/+$	$+/+$	inactive	Musca	Franco et al. (1982)
$MMM F/f = ♀$		$+/+$	c/+	active		
(6) $f/f = ♂$	F, dominant female determiner M of other strains epistatic over F	$f = Sxl^-$	$-/-$	$-/-$	inactive	Chironomus	Thompson and Bowen (1972)
$F/f = ♀$		$F = Sxl^+$	$-/-$	$+/-$	active		
(7) $f/f = ♀$ arrhenogenic	F', maternal factor; female determiner	$f = da^-$	$-/-$	$+/+$	inactive	Chrysomya	Ullerich (1973, 1984)
$F'/f = ♀$ thelygenic		$F' = da^+$	$-/-$	$+/+$	active		
(8,9) ♂ and ♀ have same genotype	environment determines sex —temperature —nutrition	R or Sxl conditional	s/s	s/s	depends on conditions	Aëdes	Horsfall and Anderson (1963)
						Heteropeza	Went and Camenzind (1980)

(R) Signal gene(s); (Sxl) key gene; (da) gene producing a maternal factor (see Fig. 1). Numbers 1 through 9 correspond to those in Fig. 4, where genetic symbols are explained. Cases and list of references represent a selected sample.

Table 1. For *Chironomus*, the normal situation is male heterogamety, corresponding to the primitive type 1 in Figure 4. The dominant male determiner M can occupy different chromosomal positions (Martin and Lee 1984). Of two geographically isolated races, A carries a dominant male determiner (M) on the left arm of chromosome 1 and B has a dominant female determiner (F) on the right arm of this chromosome. When A males were crossed to B females, M proved epistatic over F (Thompson and Bowen 1972). Assuming that M corresponds to R, and F to Sxl^+, the results suggest that B derived from the primitive type 1 by a null mutation in Sxl, accompanied by loss of R (see type 6 in Fig. 4), and thus females of B are heterozygous Sxl^+/Sxl^-, males are homozygous Sxl^-/Sxl^-, and both sexes are r/r.

The Hymenoptera (*Habrobracon*, *Apis*, and *Mormoniella*) are representatives of the classic haplo-diplo mechanism. P.W. Whiting (1943a), Rothenbuhler (1957), and A.R. Whiting (1967), however, describe diploid males obtained by inbreeding when egg and sperm carry the same allele of a sex-determining locus. At this locus, a series of several alleles exists. Female development ensues when animals are heterozygous for different alleles. We propose that this system resulted from mutations that produced a series of nonfunctional, but complementing, alleles at the key gene.

In *Chrysomya rufifacies*, sex is determined by a maternal effect (Ullerich 1973). Two kinds of females exist: (1) thelygenic females, which produce only daughters, and (2) arrhenogenic females, which produce only sons. Half of the daughters are thelygenic and half are arrhenogenic, indicating that a dominant female-determining factor, called F', segregates in thelygenic females. The case can be fitted into the standard scheme if we assume that all animals are homozygous r/r Sxl^+/Sxl^+, but Sxl^+ can only become active when the maternal product of F' is present in the egg (type 7 in Fig. 4; $da^+ = F'$). Females that are heterozygous F'/f are thelygenic, whereas females that are f/f are arrhenogenic. The genotype of the father with respect to F' or f is irrelevant, as shown by Ullerich (1984), who, by transplantation of pole cells, succeeded in fertilizing eggs of arrhenogenic females with sperm carrying the female-determining gene F'. The resulting zygotes with the genotype f/F' nevertheless developed as males, which shows that the paternal gene F' cannot impose female development on a zygote in which the product of F' was missing in the egg. Assuming that F' corresponds to da^+, the case is analogous to *Drosophila*. Cline (1978, 1983) showed that da has a maternal sex-determining effect, since its product is required for expression of the key gene Sxl in females. As in *Chrysomya*, the maternal effect of da/da mothers cannot be corrected by a da^+ sperm (Cline 1976). We conclude that the state of activity (ON or OFF) of Sxl is set early in development and is irreversible (Sánchez and Nöthiger 1983; Cline 1984).

Most mosquitoes, such as *Culex* or *Aëdes*, use the primitive mechanism with a dominant male-determining allele M, which corresponds to R in our terminology (Table 1 and type 1 in Fig. 4). For the subarctic species *Aëdes stimulans*, Horsfall and Anderson (1963) have shown that R/r (genetic males) develop as females at high temperatures and as males at low temperatures, whereas r/r remain females at all temperatures. The product of R is apparently thermolabile, so that it can repress Sxl^+ only at low temperature. A temperature-sensitive mutation at R (R^s) has partially transformed this system with genetic sex determination into a system with environmental sex determination. Complete environmental sex determination, however, would require homozygosity for R^s (type 9 in Fig. 4).

A system with environmental sex determination can also be obtained with mutations that render the product of Sxl thermosensitive (type 8 in Fig. 4). Conditional mutations at R or Sxl have different consequences: For R^s, the restrictive temperature is female-determining and for Sxl^s it is male-determining. This situation is not known among insects, but certain reptiles may fall into this category (Bull 1980).

A system with complete environmental sex determination is operative in the paedogenetic gall midge *Heteropeza* (Diptera). The nutritional status of the mother, i.e., her hemolymph, determines the sex of her offspring (Went and Camenzind 1980). This mechanism may correspond to either type 8 or type 9 in Figure 4, depending on whether the restrictive conditions are male-determining (type 8) or female-determining (type 9).

DISCUSSION

Transitions between Systems with Different Mechanisms of Sex Determination and Parallels to Homeotic Genes

Types 8 and 9 (Fig. 4) show that transitions between systems with genetic sex determination and systems with environmental sex determination are easily achieved by a single mutation in a control gene. As we pointed out earlier, such control genes must be present in all organisms whether they determine their sex genetically or environmentally. In the first case, the genes are regulated by an inherited allelic difference between the sexes and in the second case, they are regulated by environmental factors. We also define as "environmental" a difference in position within an individual, and thus include hermaphrodites in this class. Hermaphrodites produce male organs in one area of the body and female organs in another area of the body. This is formally analogous to the situation in a fly where a wing is produced in the second thoracic segment (T2) and a haltere is produced in the third thoracic segment (T3). The choice between wing or haltere is made by Ubx^+, which is present in every cell but is differentially regulated so that it is OFF in position T2 and ON in position T3. Mutations at Ubx exist that render the gene insensitive to positional information so that a wing is also made in T3 (Ubx^-) or a haltere in T2 (constitutive expression of *Ubx*⁺). A transition from positional (environmental) regulation to genotypic fixation of thoracic development has thus occurred (Lewis 1978).

Transitions from one mechanism of sex determination to another are not unique for insects—not even for animals. For the nematode *Caenorhabditis elegans*, Hodgkin (1983) has shown how this species that normally uses the X:A ratio to produce hermaphrodites and males can be transformed into one in which a single allelic difference at the control gene *tra-1* yields fertile males and females in equal numbers. Thus, new types of mechanisms for sex determination can be generated even by mutations of genes acting at a lower hierarchical level than the key gene.

Corn (*Zea mays*) is a classic hermaphrodite in which position determines sex: Male inflorescences are produced at the tip and female inflorescences are produced further down on the stem. As illustrated in Figure 5, two mutations are known that transform this environmental sex-determining mechanism into one that now operates genetically on the basis of a single allelic difference (Srb et al. 1965).

The Double Switch *dsx*

Good experimental evidence for the double switch exists only for *D. melanogaster*. As we reasoned before, however, such a device seems necessary whenever males and females express two nonoverlapping, different sets of differentiation genes. The function of the double switch then is to repress one of the two sets (Fig. 1). Mutations in the double switch characteristically have a sex-specific effect, turning either genetic males or genetic females into intersexes. The literature contains scarce evidence for the existence of such mutations. Newby (1942) described a dominant mutation causing intersexuality of XX flies in *Drosophila virilis*. This mutation has the genetic and developmental properties of *dsx*ᴰ of *D. melanogaster*. A few other cases of sex-specific intersexuality were reported by Laven (1955) for male *Culex*, by Whiting et al. (1934) and Whiting (1943b) for male and female *Habrobracon*, and by Milani (1967) for *Musca*.

The near lack of evidence, however, is no argument against the ubiquitous occurrence of a double switch but simply reflects the incompleteness or paucity of genetic studies.

CONCLUDING REMARKS

We summarize here the basic tenets of our proposition. To determine their sex, all insects use the same set of genetic elements that form a hierarchically built control system (Fig. 1). The different mechanisms that we observe at the phenomenological level (Table 1) all serve one purpose: A key gene, *Sxl*, makes an active product in what is to become a female and an inactive product or no product in what is to become a male. The differential activity of *Sxl* is achieved either by zygotic or maternal genes, *R* or *da*, by mutations of *Sxl* itself, or by environmental factors. The three elements, *R*, *da*, and *Sxl*, are the variables with which evolution can play. Single mutations and allelic differences in these elements can lead to apparently very different sex-determining mechanisms (Fig. 4). The state of activity of *Sxl* is used to control a double switch, *dsx*, consisting of two genetic elements. Their function is to repress either the male set or the female set of sex differentiation genes, thus achieving differentiation of one or the other sex (Fig. 1). We predict that a key gene and a double switch will be operative in all insects and probably in many other organisms.

As a generalization, we suggest that every individual has the complete genetic information to differentiate both sexes. A few control genes, differentially regulated in the two sexes, decide whether the male or the female set of differentiation genes is expressed, thus selecting either the male or female sexual pathway. In organisms with *environmental* sex determination, the state of activity of these control genes is regulated by external factors such as temperature and nutrition, whereas in organisms with *genetic* sex determination,

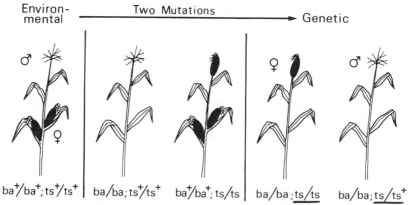

Figure 5. Transition from hermaphroditism with environmental sex determination to gonochorism with genetic sex determination, as exemplified by the transition from monoecious to dioecious corn (*Zea mays*). Two genes are involved, but in the end, sex is determined by a single allelic difference (underscored). (*ba*) Barren stalk; (*ts*) tassel seed.

an allelic difference at a control gene dictates the choice between the male or the female sexual pathway.

ACKNOWLEDGMENTS

We thank the Centro de Biologia Molecular in Madrid, in particular Drs. A. Garcia-Bellido, G. Morata, and P. Ripoll for their hospitality and interest. We are grateful to Dr. A Dübendorfer, Dr. D. Went, Ms. M. Eich, P. Gerschwiler, A. Kohl, and S. Schlegel for valuable help in the preparation of the manuscript. The work was supported by the Swiss National Science Foundation.

REFERENCES

BAKER, B.S. and J.M. BELOTE. 1983. Sex determination and dosage compensation in *Drosophila melanogaster*. *Annu. Rev. Genet.* **17**: 345.

BAKER, R.H. and R.K. SAKAI. 1979. Triploids and male determination in the mosquito, *Anopheles culicifacies*. *J. Hered.* **70**: 345.

BULL, J.J. 1980. Sex determination in reptiles. *Q. Rev. Biol.* **55**: 3.

CHANDRA, H.S. 1985. Sex determination: A hypothesis based on noncoding DNA. *Proc. Natl. Acad. Sci.* **82**: 1165.

CLINE, T.W. 1976. A sex-specific temperature-sensitive maternal effect of the daughterless mutation of *Drosophila melanogaster*. *Genetics* **84**: 723.

———. 1978. Two closely linked mutations in *Drosophila melanogaster* that are lethal to opposite sexes and interact with daughterless. *Genetics* **90**: 683.

———. 1983. The interaction between daughterless and Sex-lethal in triploids: A lethal sex-transforming maternal effect linking sex determination and dosage compensation in *Drosophila melanogaster*. *Dev. Biol.* **95**: 260.

———. 1984. Autoregulatory functioning of a *Drosophila* gene product that establishes and maintains the sexually determined state. *Genetics* **107**: 231.

CROUSE, H.V. 1960. The controlling element in sex chromosome behaviour in *Sciara*. *Genetics* **45**: 1429.

FRANCO, M.G., P.G. RUBINI, and M. VECCHI. 1982. Sex-determinants and their distribution in various populations of *Musca domestica* L. of Western Europe. *Genet. Res.* **40**: 279.

GILCHRIST, B.M. and J.B.S. HALDANE. 1947. Sex linkage and sex determination in a mosquito, *Culex molestus*. *Hereditas* **33**: 175.

HÄGELE, K. 1985. Identification of a polytene chromosome band containing a male sex determiner of *Chironomus thummi thummi*. *Chromosoma* **91**: 167.

HODGKIN, J. 1983. Two types of sex determination in a nematode. *Nature* **304**: 267.

———. 1984. Switch genes and sex determination in the nematode *C. elegans*. *J. Embryol. Exp. Morphol.* **83s**: 103.

HORSFALL, W.R. and J.F. ANDERSON. 1963. Thermally induced genital appendages on mosquitoes. *Science* **141**: 1183.

LAVEN, H. 1955. Intersexualität bei *Culex pipiens*. *Naturwissenschaften* **42**: 517.

LEWIS, E.B. 1978. A gene complex controlling segmentation in *Drosophila*. *Nature* **276**: 565.

LUCCHESI, J.C. 1978. Gene dosage compensation and the evolution of sex chromosomes. *Science* **202**: 711.

MAINX, F. 1964. The genetics of *Megaselia scalaris* Loew (Phoridae): A new type of sex determination in Diptera. *Am. Nat.* **98**: 415.

MARTIN, J. and B.T.O. LEE. 1984. A phylogenetic study of sex determiner locations in a group of australasian *Chironomus* species (Diptera, Chironomidae). *Chromosoma* **90**: 190.

MCCLELLAND, G.A.H. 1962. Sex-linkage in *Aëdes aegypti*. *Trans. R. Soc. Trop. Med. Hyg.* **56**: 4.

METZ, C.W. 1938. Chromosome behavior, inheritance and sex determination in *Sciara*. *Am. Nat.* **72**: 485.

MILANI, R. 1967. The genetics of *Musca domestica* and of other muscoid flies. In *Genetics of insect vectors of disease* (ed. J.W. Wright and R. Pal), p. 315. Elsevier/North-Holland, Amsterdam.

NEWBY, W.W. 1942. A study of intersexes produced by a dominant mutation in *Drosophila virilis*, Blanco stock. *Univ. Texas Publ.* **4228**: 113.

NÖTHIGER, R. and M. STEINMANN-ZWICKY. 1985. Sex determination in *Drosophila*. *Trends Genet.* **1**: 209.

RIBBERT, D. 1967. Die Polytänchromosomen der Borstenbildungszellen von *Calliphora erythrocephala*. *Chromosoma* **21**: 296.

ROTHENBUHLER, W.C. 1957. Diploid male tissue as new evidence on sex determination in honey bees *Apis mellifera* L. *J. Hered.* **48**: 160.

SÁNCHEZ, L. and R. NÖTHIGER. 1983. Sex determination and dosage compensation in *Drosophila melanogaster*: Production of male clones in XX females. *EMBO J.* **2**: 485.

SRB, A.M., R.D. OWEN, and R.S. EDGAR, eds. 1965. *General genetics*, 2nd edition, p. 380. W.H. Freeman, San Francisco.

THOMPSON, P.E. and J.S. BOWEN. 1972. Interactions of differentiated primary sex factors in *Chironomus tentans*. *Genetics* **70**: 491.

ULLERICH, F.H. 1963. Geschlechschromosomen und Geschlechtsbestimmung bei einigen Calliphorinen (Calliphoridae, Diptera). *Chromosoma* **14**: 45.

———. 1973. Die genetische Grundlage der Monogenie bei der Schmeissfliege *Chrysomya rufifacies* (Calliphoridae, Diptera). *Mol Gen. Genet.* **125**: 157

———. 1984. Analysis of sex determination in the monogenic blowfly *Chrysomya rufifacies* by pole cell transplantation. *Mol. Gen. Genet.* **193**: 479.

WENT, D.F. and R. CAMENZIND. 1980. Sex determination in the dipteran insect *Heteropeza pygmaea*. *Genetica* **52/53**: 373.

WHITING, A.R. 1967. The biology of the parasitic wasp *Mormoniella vitripennis* (Walker). *Q. Rev. Biol.* **42**: 333.

WHITING, P.W. 1943a. Multiple alleles in complementary sex determination of *Habrobracon*. *Genetics* **28**: 365.

———. 1943b. Intersexual females and intersexuality in *Habrobracon*. *Biol. Bull.* **85**: 238.

WHITING, P.W., R.J. GREB, and B.R. SPEICHER. 1934. A new type of sex-intergrade. *Biol. Bull.* **66**: 152.

Sex Determination in Mice

A. McLaren
MRC Mammalian Development Unit, Wolfson House, University College London, London NW1 2HE, England

Most of the sexual characteristics of mammals are controlled by sex hormones released from the gonads. The primary event of sex determination is thus the differentiation of the indifferent gonad into testis or ovary. An XY or XXY chromosome constitution is normally associated with the development of a testis, and an XO or XX constitution with an ovary. The simplest model of mammalian sex determination is to assume that the Y chromosome controls the production of a substance responsible for testis formation and that in the absence of this substance the indifferent gonad develops as an ovary. But what is this substance? Wachtel et al. (1975) proposed that the sex-determining substance was H-Y antigen, known to be controlled by the Y chromosome. This hypothesis is now cited as fact in some textbooks.

Sex Reversal

Thanks to the existence of an abnormal Y chromosome in the mouse, the location of the sex-determining region has been narrowed down to a relatively small portion of the Y. In this rearranged Y chromosome, a segment close to the centromere appears to have been duplicated and transferred to the distal end of the chromosome, beyond the X-Y pairing region (Singh and Jones 1982). As pointed out by Burgoyne (1982), any gene in this location has a 50% chance of being transferred to the X chromosome during the single obligatory X-Y crossover event postulated to occur during male meiosis. Of the progeny of male mice carrying the abnormal Y, 25% consist of XX individuals that develop as males instead of females. Evidently, these XX males have received the duplicated segment from their father, and since they develop as males, the testis-determining sequences must be included within this segment.

The occurrence of such XX male mice among the progeny of "carrier" males was first reported by Cattanach et al. (1971), and the mutation presumed to be responsible for it was termed Sex-reversal (*Sxr*). The "pseudoautosomal" pattern of inheritance is indistinguishable from true autosomal inheritance (Burgoyne 1982), and thus it was some years before the association of *Sxr* with the sex chromosomes was appreciated (Singh and Jones 1982; for review, see McLaren 1983a). The additional segment of chromosome has now been visualized cytologically, both on the X chromosome of X/X *Sxr* males and on the Y chromosome of X/Y *Sxr* carriers (Evans et al. 1982). Since it can be seen under the microscope, it must include a rather large stretch of DNA.

Sex Reversal Sex-reversed

Segments of autosome translocated to the X chromosome may become inactivated, although to a variable degree, when the X to which they are attached is inactivated (Cattanach 1974). If the segment of the Y chromosome carrying the testis-determining sequences were to become similarly inactivated when attached to an inactive X chromosome, no effect on sex ratio would be expected when inactivation is random, since we know from studies on XX/XY chimeras that considerably less than 50% of the cells in the gonad primordium need to be male-determining to ensure that the gonad develops as a testis (McLaren 1984). However, the situation is altered if the X chromosome carrying *Sxr* is partnered by an X-autosome translocation such as T(X;16)16H (hereafter termed T16H), which is expressed in all cells, with the X-chromosome partner silent. We crossed female mice heterozygous both for T16H and for an X-chromosome marker (*Pgk-1*) to distinguish the progeny carrying T16H, with males carrying *Sxr*. The T16H/*Sxr* progeny turned out to be of three types: male, female, and intersex (Cattanach et al. 1982; McLaren and Monk 1982).

Although it is at first sight surprising that mice of identical chromosomal constitutions should exhibit such very different phenotypes, the situation may be analogous to that presented by XX/XY chimeras, which also include some female and intersex individuals, as well as male individuals. We postulate that the X-chromosome inactivation event spreads to a variable degree, to include the attached testis-determining sequences in some cells but not all. The T16H/X *Sxr* mice would thus be mosaics, made up of a mixture of male- and female-determining cells, so that the sexual phenotype of any individual embryo would be determined by the relative proportion of the two cell populations that contributed to its gonadal primordium (McLaren 1983a). We have recently observed that the proportion of T16H/X *Sxr* mice developing as females rather than males depends on the origin (perhaps the *Xce* type) of the inactive X chromosome (A. McLaren, unpubl.).

The T16H/X *Sxr* females ("sex-reversal sex-reversed" mice) turn out to be fully fertile and to transmit *Sxr* to 50% of their progeny. By mating these females to *Sxr*-carrying males, we have obtained homozygous X *Sxr*/Y *Sxr* males (McLaren and Burgoyne 1983). When such males are crossed with normal

XX females, they sire only sons (apart from 1% of XO daughters, in which paternal nondisjunction has resulted in loss of the *Sxr*-carrying sex chromosome).

H-Y Histocompatibility Antigen

We have attempted to use the genetic system described above to shed some light on the identity of the mammalian testis-determining substance. H-Y antigen, central to the Wachtel et al. (1975) hypothesis, is a male-specific histocompatibility antigen, identified first by skin grafting (Eichwald and Silmser 1955). Skin-graft rejection is a T-lymphocyte-mediated response, which can now be tested in vitro, using either T-cell-mediated cytotoxicity tests or proliferative assays based on H-Y-specific T-cell clones (Simpson 1983). We have used skin grafting and both types of in vitro test to examine the H-Y status of X/X *Sxr* and T16H/X *Sxr* mice. X/X *Sxr* males proved without exception to react positively to H-Y antigen, provided that target and responder cells shared an H-2 haplotype, since the H-Y response is restricted by H-2 (Simpson et al. 1981, 1984). This result shows that the expression of H-Y histocompatibility antigen is controlled by DNA sequences contained within the *Sxr* fragment, rather than elsewhere on the Y chromosome. As expected, T16H/X *Sxr* males also proved to be H-Y-positive, and so too did T16H/X *Sxr* females, with one exception. If these individuals are indeed mosaics, containing a population of cells in which *Sxr* is not inactivated, it is perhaps not surprising that H-Y antigen should be present in females as well as in males. Such an observation does not in itself disprove the hypothesis of Wachtel et al. (1975): H-Y antigen might be a necessary but not a sufficient condition for male development, or the concentration of H-Y in the indifferent gonads of the T16H/X *Sxr* females might have been too low to induce testis formation.

H-Y-negative Males

The single exceptional T16H/X *Sxr* female was retested, using both cytotoxic and proliferative assays, and proved obdurately negative. Breeding revealed that she was not the extreme end of a mosaic spectrum, as we had initially suspected. Rather, she appeared to carry a variant *Sxr* fragment, since all her female T16H/X *Sxr* and male X/X *Sxr* or T16H/X *Sxr* descendants, irrespective of their H-2 haplotype, were similarly H-Y-negative (McLaren et al. 1984). Male descendants with a Y chromosome were, as expected, H-Y-positive. We term the variant fragment *Sxr'*; the pedigree has now been continued for eight generations.

We do not know whether the variant *Sxr'* fragment has lost, perhaps by unequal crossing over, the gene or genes controlling H-Y antigenicity while retaining those responsible for testis determination or whether the single locus assumed by Wachtel et al. (1975) to be responsible for both functions has suffered a mutational change such that the H-Y epitope is no longer expressed, but testis formation is still induced. We have recently shown (E. Simpson et al., in prep.) that H-Y negative X/X *Sxr'* males reject skin grafts from semi-syngeneic XY males as briskly as females do. This suggests that H-Y antigen expression is absent not only in adult T cells, but also in other adult tissues and probably also in the fetus, otherwise the X/X *Sxr'* males would develop immunological unresponsiveness to the antigen. Thus, any mutation that has occurred is probably a structural mutation, rather than a tissue-specific or temporal regulatory mutation.

The Wachtel/Ohno Hypothesis

The existence of males that do not express H-Y antigen argues strongly against the hypothesis of Wachtel et al. (1975). It is difficult to maintain that H-Y antigenicity is responsible for testis determination in the face of this evidence.

Female mice immunized with male tissue, however, mount not only a T-lymphocyte-mediated histocompatibility response, but also a B-lymphocyte-mediated antibody response. The anti-male antisera obtained from such females are low in titer and rather unstable, and serological assay systems used for their detection are notoriously temperamental and hard to repeat. Those who have studied the serological response (Goldberg et al. 1971; for review, see Wachtel 1983) have used the term "H-Y antigen" to denote the male-specific antigen that induces antibody formation in females. This begs the question as to whether the antigen(s) recognized by B lymphocytes is the same as or different from that recognized by T lymphocytes. Since there is no good evidence that the two antigens are the same, it has been suggested that the former should be termed serologically detected male (SDM) antigen (Silvers et al. 1982).

If our H-Y-negative X/X *Sxr'* mice turn out to be positive for SDM, it will show incontrovertibly that two different antigens are involved. SDM will then remain a strong candidate for the hypothetical testis-inducer postulated by Wachtel et al. (1975).

What Role for H-Y?

But if H-Y antigen is not determining maleness, what is it doing? A protein present in males but not in females could well be involved in male reproduction. H-Y-negative X/X *Sxr* males appear to copulate normally and show other indications of having normal testosterone levels. What about spermatogenesis? We cannot compare spermatogenesis in X/X *Sxr* and X/X *Sxr'* males, since in both, the germ cells are lost during the first few days of postnatal life. This is true of all situations in which germ cells with two X chromosomes develop as prospermatogonia within a testis (see McLaren 1983b). X *Sxr*/O males, on the other hand, do show spermatogenesis, although the spermatozoa that they produce are not normal (Cattanach et al. 1971;

Burgoyne 1984). We therefore compared the testes of X *Sxr*/O and X *Sxr'*/O males and observed a striking difference (P.S. Burgoyne et al., unpubl.). The X *Sxr*/O testes, as expected, contained spermatogonia, primary and secondary spermatocytes, spermatids, and spermatozoa. In contrast, the X *Sxr'*/O testes contained spermatogonia but only occasional patches of cells in meiotic prophase, and no spermatids or spermatozoa.

We therefore suspect that H-Y antigen may play a role in switching male germ cells from the spermatogonial stem cell lineage into the pathway of differentiation that normally leads through meiosis into spermiogenesis.

REFERENCES

BURGOYNE, P.S. 1982. Genetic homology and crossing-over in the X and Y of mammals. *Hum. Genet.* **61:** 85.

———. 1984. Meiotic pairing and gametogenic failure. *Symp. Soc. Exp. Biol.* **38:** 349.

CATTANACH, B.M. 1974. Position effect variegation in the mouse. *Genet. Res.* **23:** 291.

CATTANACH, B.M., C.E. POLLARD, and S.G. HAWKES. 1971. Sex-reversed mice: XX and XO males. *Cytogenetics* **10:** 318.

CATTANACH, B.M., E.P. EVANS, M. BURTENSHAW, and J. BARLOW. 1982. Male, female and intersex development in mice of identical chromosome constitution. *Nature* **300:** 445.

EICHWALD, E.J. and C.R. SILMSER. 1955. Untitled communication. *Transplant. Bull.* **2:** 148.

EVANS, E.P., M. BURTENSHAW, and B.M. CATTANACH. 1982. Cytological evidence for meiotic crossing-over between the X and Y chromosome of male mice carrying the sex reversing (*Sxr*) factor. *Nature* **300:** 443.

GOLDBERG, E.H., E.A. BOYSE, D. BENNETT, M. SCHEID, and E.A. CARSWELL. 1971. Serological demonstration of H-Y (male) antigen on mouse sperm. *Nature* **232:** 478.

MCLAREN, A. 1983a. Sex reversal in the mouse. *Differentiation* **23:** S93.

———. 1983b. Does the chromosomal sex of a mouse germ cell affect its development? *Symp. Br. Soc. Dev. Biol.* **7:** 225.

———. 1984. Chimeras and sexual differentiation. In *Chimeras in developmental biology* (eds. N. Le Douarin and A. McLaren), p. 381. Academic Press, London.

MCLAREN, A. and P.S. BURGOYNE. 1983. Daughterless X *Sxr*/Y *Sxr* mice. *Genet. Res.* **42:** 345.

MCLAREN, A. and M. MONK. 1982. Fertile females produced by inactivation of an X chromosome of *sex-reversed* mice. *Nature* **300:** 446.

MCLAREN, A., E. SIMPSON, K. TOMONARI, P. CHANDLER, and H. HOGG. 1984. Male sexual differentiation in mice lacking H-Y antigen. *Nature* **312:** 552.

SILVERS, W.K., D.L. GASSER, and E.M. EICHER. 1982. H-Y antigen, serologically detectable male antigen and sex determination. *Cell* **28:** 439.

SIMPSON, E. 1983. Immunology of H-Y antigen and its role in sex determination. *Proc. R. Soc. Lond. B. Biol. Sci.* **220:** 31.

SIMPSON, E., A. MCLAREN, P. CHANDLER, and K. TOMONARI. 1984. Expression of H-Y antigen by female mice carrying *Sxr. Transplantation* **37:** 17.

SIMPSON, E., P. EDWARDS, S. WACHTEL, A. MCLAREN, and P. CHANDLER. 1981. H-Y antigen in *Sxr* mice detected by H-2 restricted cytotoxic T cells. *Immunogenetics* **13:** 355.

SINGH, L. and K.W. JONES. 1982. Sex reversal in the mouse (*Mus musculus*) is caused by a recurrent nonreciprocal crossover involving the X and an aberrant Y chromosome. *Cell* **28:** 205.

WACHTEL, S.S., ed. 1983. *H-Y antigen and the biology of sex determination*. Grune and Stratton, New York.

WACHTEL, S.S., S. OHNO, G.C. KOO, and E.A. BOYSE. 1975. Possible role for H-Y antigen in the primary determination of sex. *Nature* **257:** 235.

Genetic and Molecular Analysis of Division Control in Yeast

S.I. REED, M.A. DE BARROS LOPES, J. FERGUSON, J.A. HADWIGER, J.-Y. HO, R. HORWITZ,
C.A. JONES, A.T. LÖRINCZ,* M.D. MENDENHALL, T.A. PETERSON,† S.L. RICHARDSON, AND
C. WITTENBERG

Biochemistry and Molecular Biology Section, Department of Biological Sciences, University of California, Santa Barbara, California 93106

In *Saccharomyces*, cell division is controlled both in response to nutrient limitation (Byers and Goetsch 1975; Johnston et al. 1977) and by mating pheromones in preparation for sexual conjugation (Bücking-Throm et al. 1973). In each case, control of division occurs late in the G_1 interval of the cell cycle. This has led to the proposal of an integrative gating event for cell division designated as "start" (Hartwell et al. 1974). It has been our goal, using genetic and molecular methods, to describe "start" in molecular terms, and in so doing, to define the biochemical mechanism of division control in a eukaryotic organism.

Mutations in four genes, *CDC28*, *CDC36*, *CDC37*, and *CDC39*, have been shown to confer conditional cell cycle arrest at "start" (Hartwell et al. 1974; Reed 1980). These genes have been isolated on recombinant plasmids (Reed et al. 1982; Breter et al. 1983), and the analysis of encoded products has been the primary focus of our investigations. Computer-based primary structure comparisons have detected homology between the predicted products of two of these genes and vertebrate oncogene products. The *CDC28* product shares homology with known protein kinases, including members of the *src* family of oncogenes (Lörincz and Reed 1984), whereas the *CDC36* product shares homology with a small portion of the polyprotein oncogene of avian erythroblastosis virus E26 (Peterson et al. 1984). Because the former homology relationship suggests a biochemical activity for the *CDC28* product often implicated in biological signalling, namely protein phosphorylation, we have attempted to gain entry into the biochemistry of division control by characterizing and then working outward from this point.

MATERIALS AND METHODS

Yeast strains, plasmids, and media. All yeast strains except those used in immunofluorescence experiments were congenic derivatives of *S. cerevisiae* strain BF264-15D (*MATα leu2 trp1 ade1 his3*) (Reed et al. 1985). Temperature-sensitive *cdc28* mutations were introduced by transplacement (Scherer and Davis 1979). *S. uvarum* used in immunofluorescent microscopy experiments was provided by J. Kilmartin and cultured as described (Kilmartin and Adams 1984). Unless stated otherwise, *S. cerevisiae* cells were cultured as described by Hartwell (1967).

YRp7[*CDC28.4*]HCN, a plasmid that maintains the *CDC28* gene at high intracellular copy in yeast has been described (Reed et al. 1982). Expression of *CDC28* at extremely high levels was achieved by cloning the gene into expression vector pMA91 (Mellor et al. 1983) to give pMA91[*CDC28.1*]. This construction places the *CDC28* coding region under control of the active promoter of the yeast phosphoglycerate kinase (PGK) gene. We have estimated, on the basis of immunoblotting assays, that overexpression is greater than 100-fold.

Immunoprecipitations and protein kinase assays. Immunoprecipitations, immune complex protein kinase assays, and phosphoamino acid analyses were performed as described by Reed et al. (1985). For cytoskeleton protein kinase assays, cytoskeletons were prepared from yeast spheroplasts essentially as described by Heuser and Kirschner (1980) for animal cells. For protein kinase assays, the cytoskeleton preparation was as described below for fluorescence microscopy. One milliliter (out of 10 ml) of Triton X-100 or potassium iodide-extracted cytoskeletons was transferred to an Eppendorf tube and washed by pelleting two times with kinase buffer (KB; 20 mM Tris-HCl [pH 7.2], 7.5 mM $MgCl_2$, 1 mM $ZnCl_2$, 150 mM NaCl). Cytoskeletons were finally resuspended in 15 μl of 1.33 × KB and 5 μl of [γ-^{32}P]ATP (ICN; >3000 Ci/mmole). After incubation at 23°C for 30 minutes, cytoskeletons were washed by addition of 300 μl of KB and collected by centrifugation. Analysis by SDS-polyacrylamide gel electrophoresis was as has been described (Reed et al. 1985).

Procedures for fluorescence microscopy of yeast cells. For immunofluorescent staining of *S. uvarum* cells, 10^9 log-phase cells were harvested by centrifugation, washed once in Sorenson's phosphate buffer (35 mM), and resuspended in 20 ml of the same buffer containing 3.7% formaldehyde. Fixation was for 1 hour at 23°C. Cells were washed two times again in Sorenson's

---

*Present address: Bethesda Research Laboratories, Gaithersburg, Maryland 20877.
†Present address: CSIRO-Division of Plant Industry, P.O. Box 1600, Canberra, A.C.T. 2601 Australia.

phosphate buffer and resuspended in 20 ml of the same buffer containing 0.5 mg zymolyase 60,000 (Kirin) and 25 µl of glusulase (Endo). Incubation was for approximately 30 minutes at 30°C. Spheroplasts were washed two times by centrifugation at low speed and resuspended in methanol at −20°C for 5 minutes. After centrifugation for 3 minutes at 3000 rpm in a Sorvall HB-4 rotor, cells were resuspended in acetone at −20°C followed by immediate centrifugation as above. Cells were resuspended in 4 ml of phosphate-buffered saline made 10 mg/ml bovine serum albumin (Sigma, Fraction V) (PBS-BSA); the suspension was divided into four Eppendorf tubes and washed two times by 1-second spins in a microfuge. Each tube then was used for a staining experiment. Pellets were resuspended in 25 µl of affinity-purified anti-*CDC28* IgG ($\sim$100 µg/ml) or preimmune IgG. Addition of 10 µl of goat IgG ($\sim$1 mg/ml) was found to reduce background. Incubation with mild agitation was for 1 hour at 37°C. Five 1-ml washes in PBS-BSA were performed by centrifugation for 1 second using a microfuge. Pellets were incubated with affinity-purified fluorescein-conjugated (FITC) goat anti-rabbit IgG (Miles) diluted appropriately and precentrifuged to remove aggregates. Then, 10–20 µl were added per pellet, again with normal goat IgG, and incubated as before, except that tubes were wrapped in aluminum foil to prevent photo-bleaching. Five washes were performed as above and pellets resuspended in 97% glycerol:3% PBS containing the antibleaching agent *p*-phenylenediamine as described by Adams and Pringle (1984). Samples were mounted and then observed and photographed using a Zeiss epifluorescence microscope fitted with a 100× objective.

For visualization of actin, procedures described by Adams and Pringle (1984) were followed. Rhodamine-phalloidin was obtained from Molecular Probes (Junction City, Oregon).

Cytoskeletons for immunofluorescent staining as well as for biochemical studies were prepared essentially as described by Heuser and Kirschner (1980). Cells (10^9) were washed once in stabilization buffer (SB; 0.15 M PIPES [pH 6.9], 0.5 mM $MgCl_2$, 0.1 mM EDTA) and then resuspended in 10 ml of SB and 1 M sorbitol. Spheroplasts were prepared by adding 25 µl of glusulase (Endo) and 200 µg zymolyase 60,000 (Kirin). The time course of spheroplast formation varies with the yeast strain used and must be monitored. The reaction is facilitated by addition of 2-mercaptoethanol to 0.5%. Spheroplasts were collected and washed one time in SB and 1 M sorbitol by centrifugation at low speed. Cytoskeletons were prepared by resuspending pellets in 10 ml SB, 0.5% Triton X-100. Incubation was at 23°C for 20 minutes. Cytoskeletons were collected by centrifugation at low speed. Further extraction was achieved in some cases by resuspending cytoskeletons in SB and 0.3 M potassium iodide and incubating for 3 hours at 4°C. For staining, cytoskeleton fractions were washed once with PBS and then stained with anti-*CDC28* IgG and FITC goat anti-rabbit IgG, as described for whole cells. Stained cytoskeletons were then fixed by resuspension in SB and 3.7% formaldehyde for 0.5 hour at 23°C. After one additional wash in SB, cytoskeletons were mounted and analyzed microscopically as has been described above.

Sequence analysis of yeast DNA. Sequence analysis of the start genes and flanking regions was performed using the dideoxy chain termination method (Sanger et al. 1977) essentially as has been described by Lorincz and Reed (1984). Isolation, subcloning, and identification of coding regions of start genes has been described (Reed et al. 1982; Breter et al. 1983). End points for sequencing were derived either by digesting double-stranded phage DNA with nuclease BAL-31 as we have described or by digestion with exonuclease III followed by S1 nuclease as has been described by Henikoff (1984).

RESULTS

The *CDC28* Protein

Lacking a specific functional assay for a protein, the development of immunological reagents is essential. We have raised antisera specific for the *CDC28* product by constructing an in-frame fusion between the *Escherichia coli lacZ* gene encoding β-galactosidase and the *CDC28* gene and using the resulting chimeric protein synthesized in *E. coli* as an immunogen (Reed 1982). The antisera immunoprecipitate a polypeptide of the expected molecular weight of the *CDC28* product on the basis of the DNA sequence of the gene (Fig. 1) (Reed et al. 1985). Immunoblotting using the antisera gives a similar result (J. Hadwiger et al., unpubl.). Furthermore, amplifying the *CDC28* gene by introducing it into a recipient strain on a plasmid maintained at high intracellular copy provides a significant increase in immunoprecipitable material (Fig. 1). We therefore conclude that, at least when reacted with yeast cell lysates, antisera prepared in the manner described immunoprecipitate predominantly the product of the *CDC28* gene.

Extracts prepared for immunoprecipitation experiments as shown in Figure 1 (lanes 1–4) were prelabeled using [^{35}S]methionine. Similar results were obtained prelabeling cells by growth in medium containing [^{32}P]orthophosphate (Fig. 1, lanes 5–8). The finding that the *CDC28* product is a phosphoprotein is consistent with the observation that many protein kinases are, themselves, phosphoproteins. The possible implications of this for regulation of protein kinase activity will be discussed.

The Associated Protein Kinase Activity

When immunoprecipitates were prepared in a manner similar to that used to isolate labeled *CDC28* product and incubated with [γ-^{32}P]ATP, a specific protein species of M_r 40,000 becomes labeled with phosphate (Fig. 2a) (Reed et al. 1985). We have demonstrated that

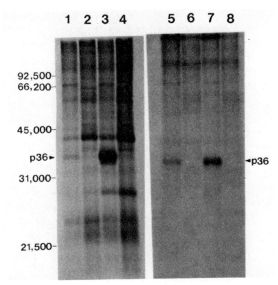

Figure 1. Analysis of in vivo-labeled *CDC28* product immune complexes by SDS-polyacrylamide gel electrophoresis. Lanes *1–4* correspond to ^{35}S-labeled samples whereas lanes *5–8* correspond to ^{32}P-labeled samples. (Lanes *1* and *5*) Wild-type cell lysate with anti-*CDC28* product immune serum; (lanes *2* and *6*) wild-type lysate with preimmune serum; (lanes *3* and *7*) high copy plasmid-containing cell lysate with immune serum; (lanes *4* and *8*) high copy plasmid-containing cell lysate with preimmune serum. Immunoprecipitated species (p36) is indicated. Molecular weight markers are muscle phosphorylase *b*, $M_r = 92,500$; serum albumin, $M_r = 66,200$; ovalbumin, $M_r = 45,000$; carbonic anhydrase, $M_r = 31,000$; soybean trypsin inhibitor, $M_r = 21,500$.

this in vitro reaction is absolutely labile for *CDC28* product prepared from one *cdc28*^{ts} mutant strain and thermolabile for *CDC28* product prepared from another *cdc28*^{ts} strain (Fig. 2b). Although the identity and function of p40, the phosphate recipient protein, are unknown, we feel, on the basis of these correlations in temperature-sensitive mutant strains, that it is likely that the *CDC28* product is the responsible protein kinase.

The question of the identity of p40 is obscured further by the results of a different in vitro assay. We have shown, as discussed below, that the *CDC28* product copurifies with yeast cytoskeletons. Protein kinase assays of these fractions show *CDC28*-specific phosphate incorporation into a protein of M_r 70,000 (Fig. 2c). This apparent discrepancy will be discussed below. In neither the immune complex assay nor the cytoskeleton assay does the *CDC28* protein kinase autophosphorylate.

Phosphoamino acid analysis of p40 labeled with phosphate in vitro indicates that the *CDC28* product possesses a protein kinase activity specific for threonine and serine residues (Fig. 2d). This result is consistent with an analysis of the evolutionary relationship of a number of protein kinases, including the *CDC28* product, based on primary structure. This study placed the *CDC28* kinase at a point in the proposed evolutionary hierarchy quite remote from emergence of tyrosine specificity (Feng et al. 1985).

The protein kinase activity observed in vitro required a divalent cation in addition to Mg^{++}. At 1 mM, Zn^{++} was found to be a more efficient stimulator than Ca^{++}, although an optimal activating concentration has been determined for neither cation (Reed et al. 1985). The effect of Zn^{++} is noteworthy since sites of

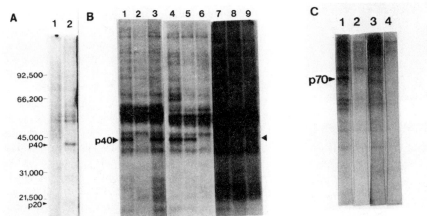

Figure 2. Protein kinase activity of mutant and wild-type *CDC28* product. (*A*) Wild-type immune complexes incubated with [γ-^{32}P]ATP. (Lane *1*) Preimmune serum; (lane *2*) anti-*CDC28* product serum. p40, the phosphate acceptor species, is indicated. (*B*) Comparison of wild-type and mutant immune complexes. (Lanes *1–3*) Wild-type, temperature-sensitive mutant *cdc28-4*, and mixture of the two, respectively, at permissive temperature (23°C). (Lanes *4–6*) Wild-type, temperature-sensitive mutant *cdc28-13*, temperature-sensitive mutant *cdc28-4*, respectively, at permissive temperature. (Lanes *7–9*) Same as lanes *4–6*, except that preincubation and reaction were performed at restrictive temperature (38°C). (*C*) Comparison of protein kinase activity in wild-type and mutant cytoskeletons. (Lanes *1* and *2*) Triton-extracted cytoskeletons from wild-type and temperature-sensitive mutant *CDC28-4* respectively, incubated with [γ-^{32}P]ATP at permissive temperature (23°C). (Lanes *3* and *4*) Potassium iodide-extracted cytoskeletons from wild-type and temperature-sensitive mutant *cdc28-4* incubated at permissive temperature. Phosphorylated species p70 is indicated. (*D*) Phosphoamino acid analysis of immune complex substrate p40 by two-dimensional thin-layer electrophoresis. Positions were determined using unlabeled standards: (S) phosphoserine; (T) phosphothreonine; (Y) phosphotyrosine.

possible zinc coordination were detected in the *CDC28* primary structure (Sulkowski 1985). However, the functional intracellular stimulator as well as its possible role in regulation of protein kinase activity remain to be established.

Intracellular Localization of the *CDC28* Product

Indirect immunofluorescence microscopy experiments have been performed utilizing anti-*CDC28* product antibody. The pattern observed is shown in Figure 3a. A granular, irregular cytoplasmic pattern is consistently obtained. A control experiment using preimmune serum at the same concentration gives no detectable staining (Fig. 3b). Gene amplification experiments utilizing the cloned *CDC28* gene on a plasmid (Reed et al. 1982) give the anticipated amplification of the immunofluorescence signal (S. Richardson and S. Reed, unpubl.). Consequently, we believe that the observed pattern represents the legitimate intracellular distribution of the *CDC28* product.

In the course of performing immunoprecipitation experiments, it was observed that the *CDC28* product was not freely extractable from cell lysates prepared using physiological concentrations of salt. A possible cytoskeletal association was therefore investigated. Detergent-extracted yeast cytoskeletons were found to retain the *CDC28* staining pattern observed using fixed cells (Fig. 3c). A rough quantitation of *CDC28* protein recovered in detergent-extracted cytoskeletons by immunoblotting suggests that it is completely accounted for in this fraction (S. Richardson et al., unpubl.). Further extraction by potassium iodide, a treatment found to be somewhat destructive to yeast cytoskeletons, still allows retention of a significant portion of *CDC28* protein in the rapidly sedimenting cytoskeletal fraction. As has been discussed above, a *CDC28*-specific protein kinase activity can be detected in both of these fractions. Under conditions where a major portion of the *CDC28* protein is not associated with the cytoskeletal fraction, notably where drastic overexpression has been engineered, the properties typical of a soluble protein have been observed in the course of a number of biochemical manipulations. We conclude then that the cytoskeletal association most likely reflects a true affinity and not a fortuitous surface aggregation.

Cytological Effects of *cdc28* Mutations

On the basis of its observed association with the yeast cytoskeleton, it was proposed that *CDC28* product may act by controlling an aspect of cytoskeletal structure or function. In yeast, it has been demonstrated that actin structures go through an elaborate series of organizational transitions during the course of the cell division cycle (Adams and Pringle 1984; Kilmartin and Adams 1984). Therefore, we sought to determine whether *CDC28* function played a role in actin organization by observing the effects of *cdc28* mutations. In the course of normal division in yeast, actin initially becomes concentrated at the site where a bud is to appear and then in the bud itself as it emerges (Fig. 4a) (Kilmartin and Adams 1984). In addition, a network of actin filaments is seen extending along the axis of the mother cell. In temperature-sensitive *cdc28* mutants, localization of actin occurs in buds normally at the permissive temperature (Fig. 4b). Following a shift to the restrictive temperature, delocalization of actin structures from buds to what appear to be random sites within the mother cell occurs. This spatial reorganization of nonfilamentous actin is accompanied by the disappearance of filamentous actin. Figure 4c shows mutant cells 1.5 hours after the shift to restrictive temperature. In this field, budded cells containing delocalized actin structures in the mother compartment are apparent. Other cells in the same field that have already become arrested in G_1 can be seen with similar actin structures. By 5 hours after the temperature shift (Fig. 4d), large cohesive actin structures have disappeared and residual actin is in the form of heterogeneously sized spots and masses. On the other hand, extreme overexpression of the *CDC28* product results in the aberrant actin structure typified in Figures 4, e and f. Under these circumstances, it appears that lateral aggregation of actin filaments has resulted in the appear-

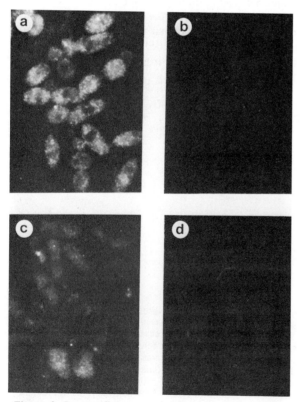

Figure 3. Immunofluorescent localization of the *CDC28* product in yeast cells and cytoskeletons. (*a*) Fixed cells, anti-*CDC28* product serum. (*b*) Fixed cells, preimmune serum. (*c*) Triton-extracted cytoskeletons, immune serum. (*d*) Cytoskeletons, preimmune serum.

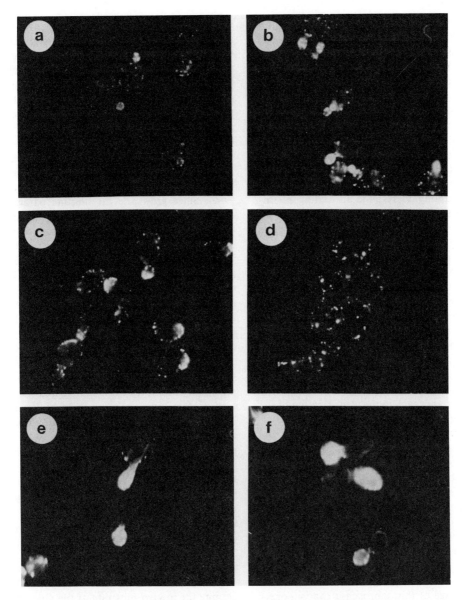

Figure 4. Effects of *cdc28* mutants on actin structure. (*a*) Wild-type cells stained with Rh-phalloidin. (*b*) Temperature-sensitive mutant *cdc28-4* at permissive temperature. (*c*) *cdc28-4*, 1.5 hr after shift to restrictive temperature (36°C). (*d*) *cdc28-4*, 5 hr after shift to restrictive temperature. (*e* and *f*) Wild-type cells containing plasmid (pMA91[*CDC28.1*]) with *CDC28* under control of phosphoglycerate kinase promoter.

ance of abnormally thick actin trunks extending into bud-like structures. In addition, the asymmetrical segregation of actin into buds is highly exaggerated. Although this condition is not lethal, growth is extremely poor. The micrographs shown in Figures 4, e and f, do not show the actin trunks because of the high-contrast film used. At the risk of overinterpreting these preliminary data, we suggest that underexpression of *CDC28* function results in disassembly of the actin cytoskeleton, as typified by functional mutants, whereas overexpression results in hyperassembly seen as thick trunks and extreme asymmetry of actin distribution. How these phenomena relate to protein kinase function remains to be determined.

Identifying Interacting Components

One of the most significant advantages exploitable as a result of the amenability of yeast to genetic analysis is the potential, having identified one genetic component of a system, to identify additional components that interact with the first. The strategy, initially described for the study of phage morphogenesis, is called pseudoreversion analysis (Jarvik and Botstein 1975). Starting with a missense mutation, second-site mutations are selected that can suppress the phenotype conferred by the first. The rationale is that second-site suppressors define genes and products that rescue on the basis of an interaction. Where allele-specific suppression can be demonstrated, direct interaction between gene products is suggested. Conveniently, second-site suppressors often confer phenotypes of their own, other than suppression, which are useful in their genetic analysis and which may provide clues to their function (Jarvik and Botstein 1975).

Preliminary pseudoreversion studies on *CDC28* have yielded second-site suppressors of temperature-sensitive *cdc28* mutations that also confer an inability to conjugate (M. Mendenhall et al., unpubl.). The simplest interpretation of these data is that the conjugation defect conferred by these second-site mutations is caused by an inability to respond to mating pheromone. Implicit with the relationship of *CDC28* to genes involved in pheromone-mediated cell cycle arrest is the suggestion that the *CDC28* product may be involved in this process as well. Consistent with this hypothesis is our isolation of a *cdc28* mutation as an extragenic suppressor of a *ste5* mutation. *STE5* is a gene shown to be required for the mating pheromone response in cells of both mating types, although the precise function of its product is not known (Hartwell 1980). Subsequently, we have determined that suppression of *ste5* mutations is a property of at least several *cdc28* alleles and that this suppression is mediated by reestablishment of the pheromone response.

Another similar approach to the identification of interacting components is suppression by gene dosage. The rationale here is that a labile protein may be stabilized, or an activity that it possesses amplified, by increasing the concentrations of interacting proteins. Operationally, dosage suppressors may be isolated in yeast by using libraries of yeast sequences maintained in plasmid vectors (Nasmyth and Reed 1980) which utilize a replicon derived from the episome 2μ circle (Broach et al. 1979). Derivatives of these vectors are maintained at relatively high intracellular copy. One then introduces the library into a mutant strain and selects sequences that can suppress the mutant phenotype. One advantage of this approach is that suppressor sequences are cloned simultaneously with their identification and therefore are available on plasmids for further characterization. Using this strategy on a temperature-sensitive *cdc28* mutant, we have obtained five different wild-type sequences that suppress when maintained on a high-copy plasmid. These are presently being analyzed.

DISCUSSION

The Physiological Role of the *CDC28* Protein Kinase

We have demonstrated that the yeast *S. cerevisiae* gene *CDC28* encodes a protein kinase as expected on the basis of DNA sequence data. The role of this protein kinase in the control of cell proliferation is not known. Our localization experiments suggest, however, that some aspect of cytoskeletal structure or function may be involved. Furthermore, pseudoreversion studies suggest that *CDC28* may play a role in the response to mating pheromone. There is reason to believe, however, that *CDC28* may be involved in other aspects of division control. A closely related gene, *CDC2*, has been isolated and characterized in the distantly related fission yeast *Schizosaccharomyces pombe* (Nurse et al. 1976). The homologous gene products share greater than 60% identity at the level of primary structure (P. Nurse, pers. comm.) and each can suppress corresponding mutations in the heterologous organism (Beach et al. 1982; D. Beach, pers. comm.). What is noteworthy about the *S. pombe* gene is that an allele, *wee2.1*, exists which causes loss of coordination between growth and division (Nurse and Thuriaux 1980). Since growth is the limiting parameter in cell doubling times, a method of delaying division in order to accommodate the necessary mass accumulation is essential. *wee2.1* mutants are incapable of this delay, with the result being a relaxation of the population to an abnormally small cell size. Due to the high degree of structural and functional homology between *S. cerevisiae CDC28* and *S. pombe CDC2*, it is conceivable that *CDC28* plays a similar role in division–growth coordination. We are seeking an allele of *CDC28* that confers a *wee2*-like phenotype.

Assuming that the *CDC28* product participates in some aspect of division control at the level of protein phosphorylation, then it is anticipated that the associated protein kinase activity must be itself regulated. The requirement for Zn^{++} or possibly Ca^{++} suggests that ion fluxes may serve as signals. Alternatively, or possibly in concert with ion signals, regulated phosphorylation of the *CDC28* product may play a role. Preliminary evidence suggests that the nonphosphorylated form of the protein is inactive in terms of protein kinase function. Furthermore, there is evidence for regulation of phosphorylation of the *CDC28* product (Reed 1985; J. Hadwiger et al., unpubl.). A detailed investigation of *CDC28* product phosphorylation is in progress.

At the molecular level, there is some ambiguity concerning the in vivo substrate of the *CDC28* protein kinase. Immune complex assays reveal a coprecipitated substrate protein (p40) of M_r 40,000. Assays of *CDC28* activity in purified cytoskeletons suggest a substrate of M_r 70,000. We feel that this discrepancy results from proteolysis during the immune complex assay. Immunoprecipitations for this assay must, of necessity, be performed for long periods of time using extremely concentrated cell lysates. Proteolytic artifacts under these conditions are a concern. The procedure employed for preparation of cytoskeletons results in rapid extraction of all soluble and membrane proteins which should include most proteases. It is therefore more likely that protease-sensitive molecules would remain intact under these conditions. Even so, the mobility of the phosphorylated species appears to increase slightly after extensive extraction with potassium iodide (Fig. 2c, lane 3), suggesting that this species has a high sensitivity to proteolysis even under these harsh conditions. This interpretation is consistent with an observation that has been difficult to reconcile with an in vivo substrate of M_r 40,000. Electrophoretic analysis of

CDC28 immunoprecipitates prepared from prelabeled yeast cell lysates has never revealed a coprecipitated species of M_r 40,000. On the other hand, a species of M_r 70,000 is observed using both ^{35}S- and ^{32}P-labeled lysates (Fig. 1), consistent with the results of cytoskeleton assays. The most likely explanation for the apparent discrepancy between the two types of immune complex assay is that immunoprecipitations from prelabeled cells are performed using considerably less concentrated (50-fold) lysates. Under these circumstances, proteolysis would be expected to be less severe. Finally, cytoskeletal preparations appear to be enriched for a M_r 70,000 species, although correspondence of this with the phosphorylated species has not yet been demonstrated (C. Jones and S. Reed, unpubl.). Should this be, in fact, the CDC28 substrate, additional fractionation of cytoskeletal preparations may allow its purification for further study.

Investigations of Other Start Genes

Similar investigations and approaches to those outlined above are in progress for three other yeast genes: CDC36, CDC37, and CDC39. Although CDC36 shows some homology at the level of product primary structure to a portion of the avian oncogene ets (Peterson et al. 1984), this relationship has been difficult to exploit, as the molecular activity of ets is not known. For CDC37 and CDC39, the predicted primary structure of products has not yielded a significant match with known proteins or gene products. The sequences are shown in Figure 5, a and b, respectively.

Immune sera prepared against the CDC36 product using an analogous strategy to that described for CDC28 immunoprecipitate a protein of M_r 21,000, as predicted from the DNA sequence of the gene (T. Peterson and S. Reed, unpubl.). The CDC36 product does not appear to be a phosphoprotein (T. Peterson and S. Reed, unpubl.). Immunofluorescent and biochemical localization experiments using these antisera have yet to be performed. Interestingly a deletion of the entire CDC36 coding region from the genome did not prove to be entirely lethal. This mutation confers, rather, extremely poor growth and a high degree of temperature sensitivity (T. Peterson and S. Reed, unpubl.). It has been possible to then isolate second-site suppressors of

(a)
```
               20            40            60            80           100
MYSHLNKRVDRILSNLPESSLTDLPAVTKFLNANFDKMEKSKGENVDPEIATYNEMVEDLFEQLAKDLDKEGKDSKSPSLIRDAILKHRAKIDSVTVEAK
              120           140           160           180           200
KKLDELYKEKNAHISSEDIHTGFDSSFMNKQKGGAKPLEATPSEALSSAAESNILNKLAKSSVPQTFIDFKDDPMKLAKETEEFGKISINEYSKSQKFLL
              220           240           260           280           300
EHLPIISEQQKDALMMKAFEYQLHGDDKMTLQVIHQSELMAYIKEIYDMKKIPYLNPMELSNVINMFFEKVIFNKDKPMGKESFLRSVQEKFLHIQKRSK
              320           340           360           380           400
ILQQEEMDESNAEGVETIQLKSLDDSTELEVNLPDFNSKDPEEMKKVKVFKTLIPEKMQEAIMTKNLDNINKVFEDIPIEEAEKLLEVFNDIDIIGIKAI
              420           440      449
LENEKDFQSLKDQYEQDHEDATMENLSLNDRDGGGDNHEEVKHTADTVD
```

(b)
```
               20            40            60            80           100
MAHRKLQQEVDRVFKKINEGLEIFNSYYERHESCTNNPSQKDKLESDLKREVKKLQRLREQIKSWQSSPDIKDKDSLLDYRRSVEIAMEKYKAVEKASKE
              120           140           160           180           200
KAYSNISLKKSETLDPQERERRDISGYLSQMIDELERQYDSLQVEIDKLLLLNKKKKTSSTTNDEKKEQYKRFQARYRWHQQQMELALRLLANEELDPQD
              220           240           260           280           300
VKNVQDDINYFVESNQDPDFVEDETIYDGLNLQSNEAIAHEVAQYFASQNAEDNNTSDANESLQDISKLSKKEQRKLEREAKKAAKLAAKNATGAAIPVA
              320           340           360           380           400
GPSSTPSPVIPVADASKETERSPSSSPIHNATKPEEAVKTSIKSPRSSADNLLPSLQKSPSSATPETPTNVHTHIHQTPNGITGATTLKPATLPAKPAGE
              420           440           460           480           500
LKWAVAASQAVEKDRKVTSASSTISNTSTKTPTTAAATTTSSNANSRIGSALNTPKLSTSSLSLQPDNTGASSSAATAAAVLAAGAAAVHQNNQAFYRNM
              520           540           560           580           600
SSSHHPLVSLATNPKSEHEVATTVNQNGPENTTKKVMEQKEEESPEERNKLQVPTFGVFDDDFESDRDSETEPEEEEQPSTPKYLSLEQREAKTNEIKKE
              620           640           660           680           700
FVSDFETLLLPSGVQEFIMSSELYNSQIESKITYKRSRDMCEISRLVEVPQGVNPPSPLDAFRSTQQWDVMRCSLRDIIIGSERLKEDSSSIYAKILENF
              720           740           760           780           800
RTLEMFSLFYNYYFAITPLEREIACKILNERDWKVSKDGTMWFLRQGEVKFFNEICEVGDYKIFKLDDWTVIDKINFRLDYSFLQPPVDTASEVRDVSVD
              820      834
NNNVNDQSNVTLEQQKQEISHGLQLLETIETGKN
```

Figure 5. Primary structure of CDC37 (a) and CDC39 (b) gene products predicted from DNA sequence.

this phenotype, completely bypassing the *CDC36* function (M. de Barros Lopes and S. Reed, unpubl.). These presumably define genes of related function, the analysis of which should prove useful in elucidating the role of *CDC36*.

Preparation of specific antisera for the *CDC37* and *CDC39* products is in progress. Although little is known concerning the genetic or biochemical attributes of the encoded products, there is evidence to suggest that the functions of *CDC37* and *CDC28* may be related. In analyzing the meiotic progeny of diploids heterozygous for temperature-sensitive mutations at both the *CDC28* and *CDC37* loci, we have found that the double mutants are barely viable and may be inviable for some alleles. Although the singly marked strains grow well at permissive temperature, the severely attenuated growth of double mutants suggests that the lesions behave in a cooperative manner. Further elucidation of the biochemical activities of the products of these genes should allow us to account for this phenomenon.

ACKNOWLEDGMENTS

This research was supported by U.S. Public Health Service Grant RO1 GM28005 and National Science Foundation Grant PCM84-02344 to S.I.R. S.I.R. acknowledges the support of American Cancer Society Faculty Research Award FRA-248. C.A.J. and M.D.M. acknowledge the support of EMBO and Jane Coffin Childs postdoctoral fellowships, respectively.

REFERENCES

Adams, A.E.M. and J.R. Pringle. 1984. Relationship of actin and tubulin distribution to bud growth in wild-type and morphogenetic-mutant *Saccharomyces cerevisiae*. *J. Cell Biol.* **98:** 934.

Beach, D., B. Durkacz, and P. Nurse. 1982. Functionally homologous cell cycle control genes in budding and fission yeast. *Nature* **300:** 706.

Breter, H.-J., J. Ferguson, T.A. Peterson, and S.I. Reed. 1983. The isolation and transcriptional characterization of three genes which function at start, the controlling event of the *S. cerevisiae* cell division cycle: *CDC36*, *CDC37* and *CDC39*. *Mol. Cell. Biol.* **3:** 881.

Broach, J.R., J.M. Strathern, and J.B. Hicks. 1979. Transformation in yeast: Development of a hybrid cloning vector and isolation of the *CAN1* gene. *Gene* **8:** 121.

Bücking-Throm, E., W. Duntze, L.H. Hartwell, and T.R. Manney. 1973. Reversible arrest of haploid yeast cells at the initiation of DNA synthesis by a diffusible sex factor. *Exp. Cell Res.* **76:** 99.

Byers, B. and L. Goetsch. 1975. Behavior of spindle plaques in the cell cycle and conjugation of *Saccharomyces cerevisiae*. *J. Bacteriol.* **124:** 511.

Feng, D.F., M.S. Johnson, and R.F. Doolittle. 1985. Aligning amino acid sequences: Comparison of commonly used methods. *J. Mol. Evol.* **21:** 112.

Hartwell, L.H. 1967. Macromolecular synthesis in temperature-sensitive mutants of yeast. *J. Bacteriol.* **93:** 1662.

———. 1980. Mutants of *Saccharomyces cerevisiae* unresponsive to cell division control by polypeptide mating hormone. *J. Cell Biol.* **85:** 811.

Hartwell, L.H., J. Culotti, J.R. Pringle, and B.J. Reid. 1974. Genetic control of the cell division cycle in yeast. *Science* **183:** 46.

Henikoff, S. 1984. Unidirectional digestion with exonuclease III creates targeted breakpoints for DNA sequencing. *Gene* **28:** 351.

Heuser, J.E. and M.W. Kirschner. 1980. Filament organization revealed in platinum replicas of freeze-dried cytoskeletons. *J. Cell Biol.* **86:** 212.

Jarvik, J. and D. Botstein. 1975. Conditional-lethal mutations that suppress genetic defects in morphogenesis by altering structural proteins. *Proc. Natl. Acad. Sci.* **72:** 2738.

Johnston, G.C., J.R. Pringle, and L.H. Hartwell. 1977. Coordination of growth with cell division in the yeast *Saccharomyces cerevisiae*. *Exp. Cell Res.* **105:** 79.

Kilmartin, J.V. and A.E.M. Adams. 1984. Structural rearrangements during the cell cycle of the yeast *Saccharomyces*. *J. Cell Biol.* **98:** 946.

Lörincz, A.T. and S.I. Reed. 1984. Primary structure homology between the product of yeast division control gene *CDC28* and vertebrate oncogenes. *Nature* **307:** 183.

Mellor, J., M.J. Dobson, N.A. Roberts, M.F. Tuite, J.S. Emtage, S. White, P.A. Lowe, A.J. Kingsman, and S.M. Kingsman. 1983. Efficient synthesis of enzymatically active calf chymosin in *Saccharomyces cerevisiae*. *Gene* **24:** 1.

Nasmyth, K.A. and S.I. Reed. 1980. Isolation of genes by complementation in yeast: Molecular cloning of a cell-cycle gene. *Proc. Natl. Acad. Sci.* **77:** 2119.

Nurse, P. and P. Thuriaux. 1980. Regulatory genes controlling mitosis in the fission yeast *Schizosaccharomyces pombe*. *Genetics* **96:** 627.

Nurse P., P. Thuriaux, and K. Nasmyth. 1976. Genetic control of the cell division cycle in the fission yeast *Schizosaccharomyces pombe*. *Mol. Gen. Genet.* **146:** 167.

Peterson, T.A., J. Yochem, B. Byers, M.F. Nunn, P.H. Duesberg, R.F. Doolittle, and S.I. Reed. 1984. A relationship between the yeast cell division cycle genes *CDC4* and *CDC36* and the *ets* sequence of oncogenic virus E26. *Nature* **309:** 556.

Reed, S.I. 1980. The selection of *S. cerevisiae* mutants defective in the start event of cell division. *Genetics* **95:** 561.

———. 1982. Preparation of product-specific antisera by gene fusion: Antibodies specific for the product of the yeast cell cycle gene *CDC28*. *Gene* **20:** 253.

———. 1985. Yeast cell cycle genes as proto-oncogenes? In *Viral and cellular oncogenes* (ed. G.M. Cooper). Martinus Nijhoff, Boston. (In press.)

Reed, S.I., J. Ferguson, and J.C. Groppe. 1982. Preliminary characterization of the transcriptional and translational products of the *Saccharomyces cerevisiae* cell division cycle gene *CDC28*. *Mol. Cell. Biol.* **2:** 412

Reed, S.I., J.A. Hadwiger, and A.T. Lörincz. 1985. Protein kinase activity associated with the product of the yeast cell division cycle gene *CDC28*. *Proc. Natl. Acad. Sci.* **82:** 4055.

Sanger, F., S. Nicklen, and A.R. Coulson. 1977. DNA sequencing with chain-terminating inhibitors. *Proc. Natl. Acad. Sci.* **74:** 5463.

Scherer, S. and R.W. Davis. 1979. Replacement of chromosome segments with altered sequences constructed *in vitro*. *Proc. Natl. Acad. Sci.* **76:** 4951.

Sulkowski, E. 1985. Purification of proteins by IMAC. *Trends Biotechnol.* **3:** 1.

Sexual Differentiation Is Controlled by a Protein Kinase Encoded by the *ran1+* Gene in Fission Yeast

D. BEACH

Cold Spring Harbor Laboratory, Cold Spring Harbor, New York 11724

Fission yeast is a haploid ascomycete that switches mating-type every few cell divisions during vegetative growth (Egel 1977; Beach 1983). Under conditions of nutrient deprivation, cells of opposite mating-type (h^+ or h^-), which have become arrested in the G_1 phase of the cell cycle, conjugate to form a diploid zygote (Egel 1971; Nurse and Bissett 1981). The zygote, now heterozygous at the mating-type locus (h^+/h^-), enters the premeiotic S phase and then undergoes two meiotic divisions that lead to the formation of a four-spored ascus.

Temperature-sensitive lethal mutations that allow haploid cells to sporulate have recently been described (Nurse 1985; Iino and Yamamoto 1985a,b). Loss of function of the *ran1+* gene bypasses two normally essential requirements for sporulation: (1) heterozygosity at the mating-type locus and (2) nutritional deprivation. In this paper we summarize experiments that characterize the transition from vegetative growth to meiosis in normal diploid strains and compare this process in haploids carrying a temperature-sensitive allele of the *ran1+* gene. In several key respects, entry into meiosis in *ran1.114* haploids precisely mimics the normal process of meiotic commitment. It is concluded that meiosis is initiated in h^+/h^- diploids by inhibition of the activity of the *ran1+* gene product through the combined activities of the *matP+* and *matM+* genes of the mating-type locus and a pathway that signals nutritional deprivation. The *ran1+* gene has been sequenced. Sequence homology between the predicted *ran1+* protein and known protein kinases suggest that the *ran1+* gene encodes a protein kinase.

Initiation of Premeiotic S Phase Marks Commitment to Meiosis

Meiosis is usually induced in cultures of fission yeast by nutritional shiftdown during log-phase growth, from complete medium to one lacking a nitrogen source (Egel 1973). After the nutritional shiftdown, high levels of sporulation are obtained but meiosis is preceded by at least two rounds of vegetative mitotic division. We had devised a protocol in which the mitotic-meiotic decision is taken in the virtual absence of mitotic cell division.

An $h^{-s}ade6.210/h^{+N}ade6.216$ diploid was grown to stationary phase in PM (Nurse 1975), a medium in which glucose rather than the nitrogen source (NH_4Cl) becomes limiting. At stationary phase the cells were transferred to PM lacking both carbon and nitrogen sources (glucose and NH_4Cl) for 15 hours. Glucose was then added to 0.2%. This led to a highly synchronous round of premeiotic DNA replication (Fig. 1) followed by meiotic division and sporulation in the virtual absence of mitotic cell division (a 20% increase in cell number followed addition of glucose).

Meiotic commitment was assayed by plating cells on PM medium supplemented with 10 μg/ml adenine. *ade6.210* and *ade6.216* haploid colonies arising after the completion of meiosis develop a pink color whereas *ade6.210/ade6.216* diploids remain white (Gutz et al. 1974). Comparison of the ratio of pink-to-white colonies after plating allowed assessment of the percentage of cells in the sporulating culture which had become irreversibly committed to meiotic haploidization. Prior to the addition of 0.2% glucose, the bulk of the culture was not committed to meiosis. Commitment to meiosis after addition of glucose very closely matched the time course of initiation of premeiotic S phase (Fig. 1).

The Cell-cycle "Start" Gene, *cdc2+* Is Not Essential for Premeiotic S Phase

Premeiotic DNA replication occurs under physiological conditions that do not support vegetative DNA replication in an $h^{-s}ade6.210/h^{-s}ade6.216$ strain (Fig. 2). There is considerable genetic evidence that the rate of progression through the mitotic cell cycle is controlled by the *cdc2+* gene. *cdc2+* is required in G_1 at "start" and also in G_2 prior to nuclear division (Nurse and Bissett 1981). *cdc2+* is homologous to the *cdc28+* cell-cycle "start" gene of *Saccharomyces cerevisiae* (Beach et al. 1982; Lorinz and Reed 1982; Hindley and Phear 1984) and is probably a protein kinase.

Since premeiotic DNA replication occurs under conditions that do not support vegetative DNA replication, it must be assumed that at least one key requirement for the mitotic cell cycle is bypassed during meiosis. We have found that a strain carrying the temperature-sensitive *cdc2.33* allele can undergo premeiotic S phase and sporulation at the restrictive temperature (Fig. 2). Asci containing two diploid spores rather than four haploid spores were formed. *cdc2+* is therefore not required either for premeiotic replication, for one of the two meiotic divisions, or for sporulation.

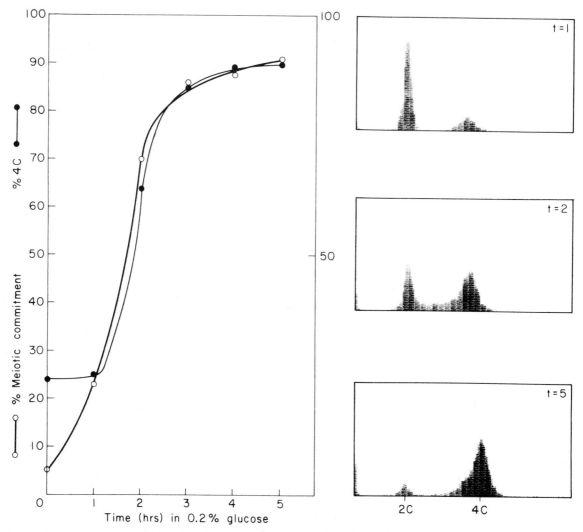

Figure 1. Time course of commitment to meiosis compared with the time of premeiotic DNA replication. The percentage of $h^{-s}ade6.210/h^{+N}ade6.216$ cells that had become committed to meiosis (○) is compared with the percentage of cells with a 4C DNA content (●). The culture was grown to stationary phase at 30°C in PM and transferred for 15 hr to a medium lacking both glucose and NH$_4$ Cl; the experiment was initiated by addition of glucose to 0.2%. The panels show the flow cytometric profiles at 1, 2, and 5 hr after glucose addition.

$ran1^+$ Activity Is Controlled by Mating-type and Nutritional Signals

At 34°C a haploid strain carrying the $ran1.114$ allele undergoes sporulation (Fig. 3A). The time course of sporulation was virtually independent of the initial cell density between 10^5–10^7 cells/ml. This shows that in the absence of $ran1^+$ sporulation occurs quite independently of normal mating-type and nutritional signals. It is proposed, therefore, that in h^+/h^- diploids these signals are responsible for inhibiting the activity of the $ran1^+$ gene, loss of which provokes meiosis and sporulation.

The observation that at 30°C the $ran1.114$ allele is partially functional has prompted the question of whether nutritional deprivation or heterozygosity at the mating-type locus can inhibit the residual $ran1^+$ activity and cause sporulation. An $h^{-s}ran1.114ade6.210$ strain does not sporulate rapidly at 30°C but only after stationary phase has been reached (Fig. 3B). It was possible to advance the onset of sporulation by culturing this strain in medium containing either one-tenth of the normal glucose concentration or one-twentieth of the normal nitrogen source (Fig. 3B).

At 30°C $ran1.114$ causes nutritional derepression of conjugation. A homothallic $h^{90}ran1.114ade6.210$ strain conjugated with high efficiency in complete minimal medium, whereas a wild-type h^{90} strain required nitrogen starvation (Fig. 3C). Comparison of the $h^{90}ran1.114ade6.210$ and $h^{-s}ran1.114ade6.210$ strains, cultured at 30°C, showed that sporulation occurred much earlier in diploid zygotes $(h^+ran1.114/h^-ran1.114)$ than in the nonmating heterothallic $h^{-s}ran1.114ade6.210$

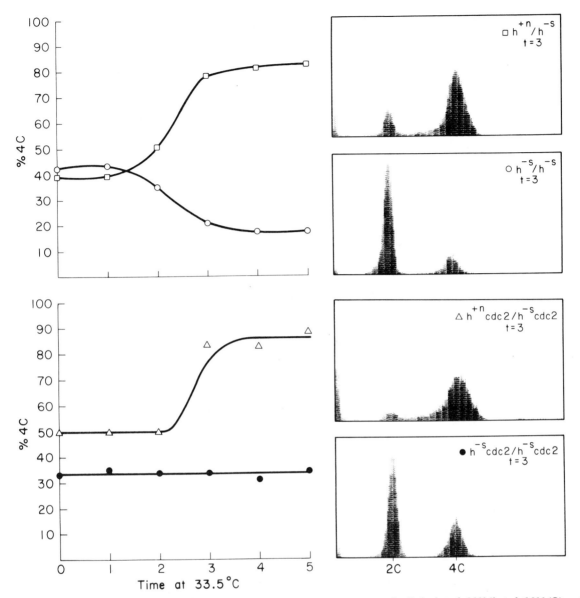

Figure 2. Premeiotic DNA replication in *cdc2.33* strains. (*Top panel*) Percentage of cells in $h^{-s}ade6.210/h^{-s}ade6.216$ (○) and $h^{-s}ade6.210/h^{+N}ade6.216$ (□) cultures with a 4C DNA content after addition of glucose to 0.2% at 33.5°C. The experimental protocol is similar to that in the previous experiment (Fig. 1). (*Bottom panel*) $h^{-s}ade6.210cdc2.33/h^{-s}ade6.216cdc2.33$ (●) and $h^{-s}ade6.210cdc2.33/h^{+N}ade6.216cdc2.33$ (△) cultures subjected to an identical experimental protocol. The inset panels show the flow cytometric profiles in each culture at 3 hr after addition of 0.2% glucose.

strain (Fig. 3C). These experiments suggest that both signals of nutrient depletion and heterozygosity at the mating-type locus can contribute toward inhibition of $ran1^+$ activity.

The observation that loss of $ran1^+$ leads to both conjugation (Fig. 3C) and sporulation (Fig. 3A) presents a paradox. Conjugation occurs from the G_1 phase of the cell cycle whereas sporulation is preceded by premeiotic DNA replication. Can inhibition of $ran1^+$ first cause G_1 arrest followed by release from this arrest as cells become committed to meiosis at the premeiotic S phase?

An $h^{-s}ran1.114ade6.210$ strain was cultured at 30°C and the percentage of cells with a G_2 DNA content was estimated as the cultures entered stationary phase (Fig. 4). As stationary phase was reached, the percentage of cells in the G_1 phase of the cell cycle was higher in the $ran1.114$ culture compared with a wild-type strain. However, at stationary phase a round of DNA replication took place without accompanying cell divison (Fig. 4). This DNA replication was followed by sporulation and has properties characteristic of premeiotic DNA replication (see below).

Vegetative DNA replication can be distinguished from premeiotic S phase by its dependence on $cdc2^+$ (Fig. 2). We have found that in a $ran1.114cdc2.33$ double-mutant, at the restrictive temperature, DNA replication immediately precedes sporulation (Fig. 5). This suggests that loss of *ran1*+ function does indeed provoke premeiotic DNA replication.

The following model for the control of the switch from vegetative cell division to meiosis is proposed. During growth of haploid cells in conditions of adequate nutrition the *ran1*+ gene product is expressed constitutively. Nitrogen starvation is thought to cause partial inhibition of *ran1*+ (Fig. 3B). This leads to accumulation of cells in the G_1 phase of the cell cycle in a state that encourages conjugation between cells of opposite mating-type (Fig. 3C). Conjugation necessarily leads to expression of both *matP*+ and *matM*+ mating-type information in the diploid zygote. In conjugation with continued nutritional deprivation, heterozygosity at the mating-type locus is presumed to lead to full inhibition of *ran1*+ activity (Fig. 3C). This causes the diploid zygote to be released from vegetative G_1 arrest and enter into premeiotic S phase (Fig. 4 and 5). Full loss of *ran1*+ marks the point of irreversible commitment to meiosis.

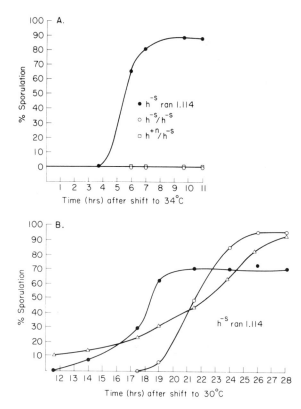

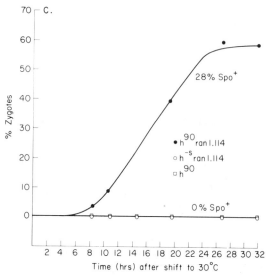

Figure 3. Expression of the *ran1.114* phenotype under different physiological conditions. (*A*) Cultures of $h^{-s}ade6.210/h^{-s}ade6.216$ (○), $h^{-s}ade6.210/h^{+N}ade6.216$ (□), and $h^{-s}ade6.210ran1.114$ (●) were incubated at 25°C to mid-log phase and then shifted to 34°C. The percentage of sporulated cells was counted at time intervals thereafter. (*B*) The $h^{-s}ade6.210ran1.114$ strain was cultured in PMA at 25°C until a density of 2×10^6 cells/ml was reached. The culture was split and incubated at 30°C, either in the same complete medium (○), one containing 10-fold less glucose (○), or one containing 20-fold less NH$_4$Cl (△). The percentage of sporulated cells was estimated at time intervals after the temperature and medium shift. (*C*) Cultures of $h^{90}ran1.114\ ade6.210$ (●), $h^{-s}ade6.210ran1.114$ (○), and $h^{90}ade6.216$ (□) were grown in PMA at 25°C to a density of 2×10^5 cells/ml, at which time the temperature was raised to 30°C. The percentage of zygotes was estimated in each culture at regular intervals thereafter. At the final time point, 28% of the $h^{90}ade6.210ran1.114$ culture had undergone zygotic sporulation. No spores were yet apparent in either of the other two cultures.

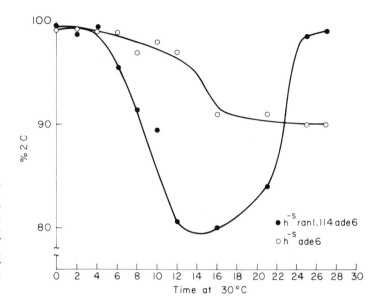

Figure 4. Cell-cycle analysis of wild-type and *ran1.114* strains entering stationary phase. $h^{-s}ade6.210$ (○) and $h^{-s}ade6.210ran1.114$ (●) strains were cultured in PMA at 25°C to a density of 2×10^6 cells/ml, after which the temperature was increased to 30°C. The percentage of cells in each culture with a 2C DNA content was determined by flow cytometry. Sporulation of the $h^{-s}ade6.210ran1.114$ culture began after 27 hr at 30°C.

3':5' cAMP Suppresses *ran1.114*

Sporulation in *ran1.114* strains is associated with total cessation of vegetative cell division. Addition of 3':5' cAMP to the culture medium at very high levels (50 mM) suppressed the *ran1.114* phenotype, allowing a culture to grow to saturation at 34°C without sporulation (Table 1A). This effect is apparently specific for 3':5' cAMP because it is abolished by preincubation with bovine brain phosphodiesterase and it is not mimicked by any probable impurity in a commercially available preparation of 3':5' cAMP. Furthermore, a similar effect was obtained with caffeine, a known phosphodiesterase inhibitor. The clearest evidence that that elevation of intracellular 3':5' cAMP suppresses the *ran1.114* phenotype was obtained by introduction of the S. cerevisiae gene encoding adenylate cyclase (*cyr1*) into fission yeast. This elevated the intracellular 3':5' cAMP level and also suppressed expression of the phenotype of *ran1.114* (Table 1). It is probable that the level of 3':5' cAMP required to suppress the *ran1.114* phenotype is above the usual physiological range.

Sequence Homology between *ran1*+, *cdc2*+, and the Family of Mammalian Protein Kinases

We have sequenced the *ran1*+ gene and undertaken a computer search for any proteins with homologies to the predicted *ran1*+ protein. To our surprise, we found that *ran1*+ has a region of approximately 300 amino acid residues that is 25% homologous to the product of the *cdc2*+ gene (Fig. 6). This is of particular interest because *cdc2*+ was previously shown to have homology to the *cdc28*+ gene of S. cerevisiae, to cAMP-dependent protein kinase, and to the family of mammalian protein kinase oncogenes (Hindley and Phear 1984).

ran1+ shares homology with cAMP-dependent protein kinase and the mammalian oncogenes, but the homology is greatest with *cdc2*+. This suggests that the product of *ran1*+ may itself be a protein kinase but is probably not a cAMP-dependent protein kinase. High levels of intracellular 3':5' cAMP may allow bypass of the requirement for the *ran1*+ gene product because 3':5' cAMP-dependent protein kinase and *ran1*+ protein kinase share certain key substrates for phosphorylation.

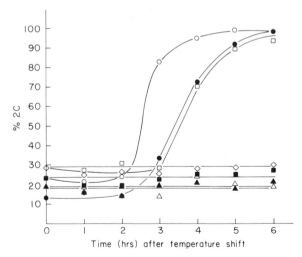

Figure 5. Assessment of DNA replication in *ran1cdc* double-mutants. Strains were cultured in PMA at 25°C to a density of 5×10^6 cells/ml. The cells were washed and resuspended in PM lacking NH$_4$Cl but supplemented with adenine (10 μg/ml) at 25°C. After 16 hr the cells were restored to complete medium (PMA) at 35.5°C. The percentage of cells with a 2C DNA content was determined at time intervals after the shift to 35.5°C. The strains used were: $h^{-s}ade6.216$ (●), $h^{-s}ade6.210ran1.114$ (○), $h^{-s}cdc2.33$ (■), $h^{+N}cdc10.129$ (△), $h^{+N}ade6.210$-*ran1.114cdc2.33* (□), $h^{+N}ade6.210ran1.114$-*cdc10.129* (▲), and $h^{-s}ade6.210ran1.114cdc2.33mei2.33$ (◇).

Table 1. Suppression of ran Mutants by 3′:5′cAMP

A.[a] Strain	Agent	Growth response	Final density (cells/ml)
(1) $h^{90}ran1.114ade6.216$	none	−	
	3′:5′ cAMP (50 mM)	+	6×10^7
	3′:5′- dibuturyl cAMP (50 mM)	+	4.3×10^7
	cAMP + phosphodiesterase (50 mM + 1 unit)	−	
	2′:3′ cAMP (50 mM)	−	
	5′ AMP (50 mM)	−	
	3′ AMP (50 mM)	−	
	adenine (25 mM)	−	
	adenosine (10 mM)	−	
	caffeine (5 mM)	+	1.1×10^7
	cAMP (2 mM)	−	
	caffeine (1 mM)	−	
	cAMP (2 mM), caffeine (1 mM)	+	3.4×10^7
(2) $ran1.\triangledown$	none	−	
	3′:5′ cAMP (50 mM)	+	5.5×10^7
	caffeine (5 mM)	+	9.8×10^6
(3) h^{+N} ran1.114 leu1.32 ade6	+ pDB248	−	
	+ pCYR1.2/pDB248	+	2.1×10^7
	+ pcAMPrK-YEP13	+	3×10^7

B.[b] Genotype	Temperature (°C)	cAMP (pmoles/mg protein)
$h^{+N}ran1.114$	25	5.9
$h^{+N}ran1.114$	34	6.8
$h^{+N}ran1.114mei2.24$	25	5.9
$h^{+N}ran1.114mei2.24$	34	7.4
$h^{90}ade6.216leu1.32$	25	10.8
$h^{90}ade6.216leu1.32$ + pCYR1.2/pDB248	25	51.5

[a](1) Spores of $h^{90}ran1.114$ ade6.216 were inoculated into PMA at 34°C at a density of 10^5/ml. After 40 hr the cell density was estimated. Chemical agents were added at the concentrations shown and the medium was readjusted to pH 5.6 if necessary. (2) Spores obtained from a diploid strain, $h^{-s}ade6.216leu1.32his3ran1.114/h^{+N}leu1.32ade6.210ran1.\triangledown$, were inoculated at 30°C into minimal medium, supplemented with adenine and histidine but not leucine. Only the 50% of spores that carry $ran1.\triangledown$ (Leu2+ insertion) have potential for growth under these conditions. Chemical agents were added at the concentrations given and the cell density was estimated after 40 hr of growth. (3) The $h^{+N}ran1.114leu1.32ade6.210$ strain was transformed to Leu+ with the yeast vector pDB248 (Beach and Nurse 1981) or cotransformed with this vector and pCYR1.2. pCYR1.2 carries the S. cerevisiae adenylate cyclase gene in the vector YEP50 (T. Kataoka, pers. comm.). Cotransformants were readily identified since pCYR1.2 suppressed the phenotype of ran1.114. pcAMPrK-YEP13 contains a gene, derived from S. cerevisiae, which shares 60% amino acid homology with mammalian cAMP-dependent protein kinase (T. Toda, pers. comm.). It is carried in the vector YEP13. Transformants were inoculated at 10^5 cells/ml into PMA at 34°C. The cell density was estimated after 40 hr.

[b]Assay of intracellular 3′:5′ cAMP. Cultures of the strains indicated were grown in minimal medium to a density of 2×10^7 cells/ml at 25°C. At this time the cultures were either harvested immediately or shifted to 34°C for 1 hr before harvesting.

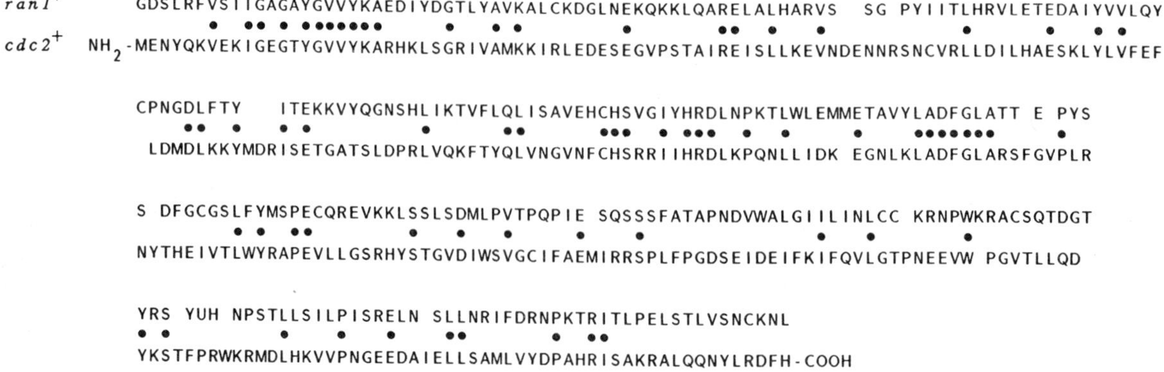

```
                                    ATP
ran1+      GDSLRFVSIIGAGAYGVVYKAEDIYDGTLYAVKALCKDGLNEKQKKLQARELALHARVS  SG PYIITLHRVLETEDAIYVVLQY
            • ••  •••••••••       •  ••                 •  ••            •          •    • ••
cdc2+  NH2-MENYQKVEKIGEGTYGVVYKARHKLSGRIVAMKKIRLEDESEGVPSTAIREISLLKEVNDENNRSNCVRLLDILHAESKLYLVFEF


           CPNGDLFTY   ITEKKVYQGNSHLIKTVFLQLISAVEHCHSVGIYHRDLNPKTLWLEMMETAVYLADFGLATT E PYS
            ••   ••      ••     • ••        •••  •••  •  •                  •••••••     •
           LDMDLKKYMDRISETGATSLDPRLVQKFTYQLVNGVNFCHSRRIIHRDLKPQNLLIDK EGNLKLADFGLARSFGVPLR


           S DFGCGSLFYMSPECQREVKKLSSLSDMLPVTPQPIE SQSSSFATAPNDVWALGIILINLCC KRNPWKRACSQTDGT
             • ••••   •                        •         •   •   •   •    •       •
           NYTHEIVTLWYRAPEVLLGSRHYSTGVDIWSVGCIFAEMIRRSPLFPGDSEIDEIFKIFQVLGTPNEEVW PGVTLLQD


           YRS YUH NPSTLLSILPISRELN SLLNRIFDRNPKTRITLPELSTLVSNCKNL
            ••  •••         •    •        ••   •   •  ••
           YKSTFPRWKRMDLHKVVPNGEEDAIELLSAMLVYDPAHRISAKRALQQNYLRDFH-COOH
```

Figure 6. Comparison of part of the predicted amino acid sequence of $ran1^+$ and the entire sequence of cdc^+. The lysine residue that is expected to bind ATP is indicated. The sequence of $cdc2^+$ is taken from Hindley and Phear (1984).

ACKNOWLEDGMENTS

I am grateful to Drs. P. Nurse and M. Yamamoto for providing yeast strains and to Dr. Louise Clarke for providing a DNA fragment containing the *ran1+* gene. Technical assistance of Linda Rodgers, Jane Gould, and Scott Silbiger is acknowledged. This work was supported by a grant from the National Institutes of Health.

REFERENCES

BEACH, D.H. 1983. Cell type switching by DNA transposition in fission yeast. *Nature* **305:** 682.

BEACH, D.H. and P.M. NURSE. 1981. High frequency transformation of the fission yeast, *S. pombe. Nature* **200:** 140.

BEACH, D.H., B. DURKACZ, and P.M. NURSE. 1982. Functionally homologous cell cycle control genes in budding and fission yeast. *Nature* **300:** 706.

EGEL, R. 1971. Physiological aspects of conjugation in fission yeast. *Planta* **98:** 89.

———. 1973. Commitment to meiosis in fission yeast. *Mol. Gen. Genet.* **121:** 277.

———. 1977. Frequency of mating-type switching in homothalic fission yeast. *Nature* **266:** 172.

GUTZ, H., H. HESLOT, U. LEUPOLD, and N. LOPRIENO. 1974. *Schizosaccharomyces pombe.* In *Handbook of genetics* (ed. R.C. King), vol. 1, p. 395. Plenum Press, New York.

HINDLEY, J. and G. PHEAR. 1984. Sequence of cell division gene, *cdc2,* from *Schizosaccharomyces pombe;* patterns of splicing and homology to protein kinases. *Gene* **31:** 129.

IINO, Y. and M. YAMAMOTO. 1985a. Mutants of *Schizosaccharomyces pombe* which sporulate in the haploid state. *Mol. Gen. Genet.* **198:** 416.

———. 1985b. Negative control for the initiation of meiosis in *Schizosaccharomyces pombe. Proc. Natl. Acad. Sci.* **82:** 2447.

LORINZ, A.T. and S.I. REED. 1982. Primary structure homology between the product of yeast cell division control gene *cdc28* and vertebrate oncogenes. *Nature* **307:** 183.

NURSE, P.M. 1975. Genetic control of cell size at division in yeast. *Nature* **256:** 547.

———. 1985. Mutants of the fission yeast *Schizosaccharomyces pombe* which alter the shift between cell proliferation and sporulation. *Mol. Gen. Genet.* **198:** 497.

NURSE, P.M. and Y. BISSETT. 1981. Gene required in G_1 for commitment to cell cycle and in G_2 for control of mitosis in fission yeast. *Nature* **292:** 558.

Regulation of the Yeast *HO* Gene

L. BREEDEN AND K. NASMYTH
MRC Laboratory of Molecular Biology, Cambridge, UK CB2 2QH, England

Saccharomyces cerevisiae has two haploid cell types (**a** and α) which can mate to form a third cell type: the **a**/α diploid. Cell type is primarily determined by the information encoded at the mating-type (*MAT*) locus. A haploid carries either the *MAT***a** or the *MAT*α allele, and the diploid carries both. Master regulator genes, encoded by these two *MAT* alleles, affect the expression of many genes and thereby cause cell type-specific differentiation. Two genes from *MAT*α (called α*1* and α*2*) and one gene from *MAT***a** (**a***1*) are involved in regulating cell-type-specific processes. The α*1* gene product induces transcription of α-specific genes, while α*2* turns off **a**-specific genes. In **a** cells, and indeed in cells lacking active *MAT* genes, **a**-specific genes are constitutively expressed. When the **a**/α diploid is formed, **a***1* and α*2* both act to turn off α*1* transcription, to repress haploid specific gene expression, and to induce diploid specific activities (Strathern et al. 1981; for review, see Sprague et al. 1983; Klar et al. 1984).

One such haploid specific process is the ability to switch mating type. At least three DNA repair genes, *RAD52* (Malone and Esposito 1980), *RAD51*, and *RAD54* (J. Game, pers. comm. in Game 1983) and a site-specific endonuclease coded by the *HO* gene (Kostriken and Heffron 1983), are involved in the mechanics of switching. The *HO* endonuclease initiates a switch by making a double-strand scission at a DNA sequence within the *MAT* locus (Strathern et al. 1982). This is followed by a gene conversion event (Klar and Strathern 1984) whereby silent **a** or α information coded by the *HML*α or *HMR***a** locus is copied into the *MAT* locus. Once moved into this location, the new mating-type information is expressed, and mating type is changed. Efforts to find other nonessential genes that are necessary for switching have led to the identification of five more genes, *SWI1,2,3,4*, and *5* (Haber and Garvik 1977; Stern et al. 1984). These genes promote switching indirectly, in that they are all required for the expression of the *HO* gene (Stern et al. 1984).

Switching is common in wild populations of yeast (Klar et al. 1984) and has probably evolved as a means of expediting diploidization. Diploids show higher DNA repair capabilities and are far less sensitive to mutagenesis (Zirkle and Tobias 1953; Mortimer 1958; Friis and Roman 1968). They can also sporulate, which allows them to survive long periods of starvation (Roman and Sands 1953). A clear disadvantage of mating-type switching is the potential destabilization of ploidy that would occur if diploids were able to switch, or if haploids in the act of mating were able to switch. In both of these cases the stable, nonmating diploid (**a**/α) would be converted to an **a**/**a** or α/α cell. These diploids, having only **a** or α mating-type information, would be able to mate and triploids would be produced. Given such consequences, it is not surprising that the ability to switch is restricted to haploids, and is very precisely controlled within the cell cycle of those haploids.

Studies of the switching potential of individual haploid cells by Hicks and Herskowitz (1976) and Strathern and Herskowitz (1979) have allowed these authors to formulate a set of switching rules. They found that haploid mother cells switch their mating type nearly every cell division. When a switch occurs, the changed mating type is inherited by both mother and daughter in the subsequent cell division. This suggests that the switch occurs before the completion of S phase. Newly divided daughter cells are incapable of switching. Only after these cells give rise to a daughter of their own do they gain the ability to switch. In summary, the capacity to switch is a haploid-specific trait. This capacity is distributed asymmetrically to mother cells at cell division, and is confined to a specific period within the cell cycle.

It has been known for some time that *HO* expression is required for switching to occur (Winge and Roberts 1949; Oshima and Takano 1971). More recently it has been suggested that *HO* expression triggers switching. *HO* is not transcribed in diploids (Jensen et al. 1983) or in haploid daughter cells (Nasmyth 1983). In addition, *HO* transcription and activity is limited to a short period late in the G_1 phase of the cell cycle (Nasmyth 1983). *HO* transcription occurs after the period when haploid cells are receptive to mating and before DNA synthesis, so that if a switch does occur, it would be inherited by both cells in the subsequent cell division.

These results suggest two possibilities. Either *HO* is one of several genes whose activity is regulated to control switching, or it is the only gene whose activity is controlled and as such it acts as a trigger for switching. Jensen and Herskowitz (1983) have addressed this question by constructing a *GAL10-HO* fusion in which the *HO* sequences upstream from −130 bp (from the mRNA start site) have been replaced by a 365-bp fragment from the *GAL10* regulatory region. When switching was analyzed in cells carrying this fusion, it appeared that two fundamental switching rules were violated: Both diploid and daughter yeast cells were able to switch when grown in galactose medium, where *GAL10* transcription is fully induced. These findings suggest that high levels of *HO* expression (probably 50- to 100-fold overproduction) can trigger mating type switching in two cell types that normally do not switch.

Whether *HO* expression is the trigger or simply a consequence of controlling mating type switching, it is clear that this gene is regulated in a cell type-specific and cell-cycle-regulated fashion. A molecular analysis of the control circuitry involved in its regulation will lead to important new insights into how the capacity to express a gene is asymmetrically distributed at cell division, and to how the expression of a gene can be limited to a specific phase in the cell cycle.

This paper is an account of the progress we have made toward determining which DNA sequences are involved in *HO* regulation. In particular we have identified two types of transcription activation elements within the 5′-flanking DNA of *HO*, one of which is cell cycle specific.

METHODS

The growth and genetic manipulations of yeast were done according to the methods of Mortimer and Hawthorne (1969). Yeast transformations were done as described by Beggs (1978), with modifications suggested by MacKay (1983). Oligonucleotide cloning and analysis of plasmids in bacteria were done as described by Nasmyth (1985b).

Two different β-galactosidase assays have been used in this work. A rapid screening method was devised to detect very low levels of activity. Colonies grown on selective or nonselective plates were replica-plated to nitrocellulose filters, then dipped in liquid nitrogen to permeabilize the cells (Casadaban et al. 1983). These filters were thawed and placed in a petri plate containing a disk of Whatmann 3MM paper saturated with 2 ml of Z buffer (Miller 1972) containing 0.6 mg of 5-bromo-4-chloro-3-indolyl-β-D-galactoside (X-Gal). This was covered and incubated at 30°C for 30 minutes to overnight. β-galactosidase activity is detected due to the blue color generated when the X-Gal substrate is cleaved.

When it was necessary to quantitate the amount of β-galactosidase activity present in yeast cells, colorimetric assays using O-nitrophenol-β-D-galactosides (ONPG) were performed as previously described. In this case yeast cells were permeabilized with 0.1% SDS, as suggested by Jensen (1983).

DNA SEQUENCES REQUIRED FOR *HO* EXPRESSION

Figure 1 depicts the 2 kb of DNA upstream from the *HO* structural gene. In order to ask whether or not this DNA was sufficient to confer all the regulatory properties known to be exerted on *HO*, the sequence extending from about -1915 bp down to the transcription start site was fused to the *MATα1* structural gene and transferred to a heterologous location (the *TRP1* locus on chromosome IV). Cells carrying this hybrid gene produce an α1 transcript at levels comparable to that of *HO* itself. This α1 transcription is still cell cycle regulated and is fully repressed in diploids (data not shown). This suggests that sequences between -1915 and $+1$ are sufficient for *HO* expression and for these two regulatory processes. Mother/daughter control was not observed with this construction. The reason for this loss of regulation is being investigated.

Within this 2-kb DNA sequence, two regions have been identified that are required for *HO* expression (Nasmyth 1985a). These regions are represented by striped boxes in Figure 1. Deletion of the DNA represented by these boxes virtually eliminates *HO* transcription. At -80 there is an AT-rich region that affects the efficiency and site of initiation of transcription. This has been called the TATA region, by analogy to other systems (Corden et al. 1980). More than 1 kb upstream from the TATA region, there is another region that is required for *HO* transcription. By analogy to other yeast genes, it is called an upstream activation site (UAS) (Guarente 1984). Such sequences have many but not all properties in common with the enhancer sequences found in higher eukaryotes (see below; for review, see Khoury and Gruss 1983; Guarente 1984; Guarente and Hoar 1984).

The region between the TATA and UAS elements can be eliminated without dramatically affecting the steady-state levels of *HO* mRNA (Nasmyth 1985a). For example, the three deletions -1274 to -788, -901 to -545, and -559 to -145 result in the successive removal of all sequences between the UAS and very near the TATA region. However, the level of *HO* transcription and its a/α and cell cycle regulation are not grossly perturbed by any of these deletions. Very large deletions (i.e., -1160 to -145) have little or no effect on the steady-state levels of *HO* mRNA attained, or the a/α repression of it, but such deletions do interfere with the cell cycle regulation of this transcript (see below).

This deletion analysis implies the following. First, the TATA and UAS elements are not only necessary but are sufficient for transcription of *HO*. No other sequence element between them is required. Second, the spacing between the -1200 UAS and the TATA element is not particularly important for their function. Though these two essential elements are more than 1 kb apart in the wild-type promoter, a deletion which brings them to within 200 bp of each other still produces apparently normal steady-state levels of *HO* mRNA. And third, if there are regulatory elements in the -1200 to -100 region that are involved in diploid repression and cell cycle regulation, then they must be redundant.

THE -1200 ELEMENT AT *HO* IS A UAS

The region 1200 bp upstream from the *HO* gene acts at a distance to promote *HO* transcription (Nasmyth 1985a). However, in order to show that it is an upstream activation sequence for the *HO* gene, it was necessary to show that it is required in *cis* to the *HO* gene, and that it was sufficient to provide UAS activity to a heterologous gene. The results described below show that the DNA sequence at −1200 from the *HO* gene fulfills these two requirements and is, therefore, a UAS.

To show that the −1200 DNA sequence was required in *cis* to the *HO* structural gene, a classic *cis-trans* test was done. *HO:lacZ* fusions were made in which the promoter and the first 38 amino acids of the *HO* gene were joined in frame to the *Escherichia coli* β-galactosidase gene (*lacZ*) from plasmid pMC1871 (Casadaban et al. 1983). This fusion produces an active and stable β-galactosidase protein in yeast cells (L. Breeden, unpubl.) that can be easily assayed. Three variants of this fusion were made. The first contained wild-type *HO* sequences fused to a defective *lacZ* gene (*HO+:lacZ−*). The second contained an *HO* promoter deletion lacking the DNA from −1362 to −995 (*hoΔUAS*) fused to an active *lacZ* gene (*hoΔUAS:lacZ+*), and the third contained the wild-type version of each (*HO+:lacZ+*). These fusions were inserted into the chromosomal *HO* locus of a *MATα* and a *mat* deletion strain (*matΔ*). These strains were mated to produce the *MATα/matΔ* pseudohaploids listed in Table 1, and assayed for β-galactosidase activity. (Note that this test is possible because *matΔ* strains mate as **a** cells, but do not produce the *a1* gene product which would, in concert with α2, act to repress transcription of *HO*). Table 1 shows the results of this *cis-trans* test. It is clear that the *HO+:lacZ+* fusion produces β-galactosidase in a pseudohaploid strain (*MATα/matΔ*). When the −1200 region of the *HO* promoter is deleted in *cis* and supplied in *trans* to the active *lacZ* gene, no β-galactosidase activity is observed. This result shows that the −1200 UAS is a *cis*-acting element that is required for *HO* gene expression.

To show that the −1200 sequence alone was sufficient to supply UAS activity to a heterologous gene lacking a UAS of its own, several promoter fusions were constructed. Both UAS and TATA elements required for expression of the yeast *CYC1* gene have been identified (Guarente and Mason 1983; Guarente et al. 1984). In addition, a set of *CYC1:lacZ* fusion plasmids have been made, which include varying amounts of the *CYC1* promoter region linked to the *E. coli lacZ* gene (Guarente and Mason 1983). When the whole *CYC1* promoter is linked to *lacZ*, yeast cells carrying this plasmid (pLGΔ-312) produce β-galactosidase, and this activity can be easily quantitated using colorimetric *lacZ* assays (see Methods). When the *CYC1* promoter is truncated so that it has lost its UAS sequences (plasmid pLGΔ-178), the fusion is no longer expressed. As a result, pLGΔ-178 is a convenient vector into which potential UAS sequences can be inserted and assayed for UAS activity. When the −1200 DNA from the *HO* promoter was inserted into this vector, and transformed into yeast cells, its ability to provide UAS activity to this heterologous gene was very apparent (see Table 2). Though not as efficient as the homologous *CYC1* UAS plasmid pLGΔ-312, which actually carries two UAS sequences (Guarente et al. 1984), the −1200 UAS promoted transcription nearly two orders of magnitude above background (pLGΔ-178).

Thus, by all criteria tested, the DNA sequence 1200 bp upstream from the *HO* mRNA start site is an essential, *cis*-acting sequence that is necessary for *HO* transcription and is sufficient to provide UAS activity to heterologous genes.

REDUNDANT REGULATORY ELEMENTS WITHIN THE *HO* PROMOTER

Although the −1200 UAS and the TATA sequences appeared sufficient for *HO* transcription, they were not sufficient for proper cell cycle regulation. However, only very large deletions resulted in the loss of cell cycle regulation, and none of the deletions resulted in loss of **a**/α repression. This suggested that these forms of control might act upon sequence elements that were repeated within the *HO* promoter. This possibility prompted a computer search for repetitive elements within the *HO* promoter. Two such elements were found.

a/α Repression

The first repeated sequence that was found was a palindromic sequence that was repeated 10 times near the *HO* gene. This sequence was also found at other diploid-repressed genes, and was shown to confer **a**/α repression on heterologous genes when placed between their UAS and TATA elements (Miller et al. 1985). The location of these sequence elements is depicted in Figure 1 by small boxes, and the derived consensus sequence is

```
T    A
  C TGTNN  NANNTACATCA.
C G      T
```

Table 1. *cis-trans* Test of −1200 UAS

Strain	β-Galactosidase activity
matΔ HO+:lacZ− / *MATα HO+:lacZ+*	+
matΔ HO+:lacZ− / *MATα hoΔUAS:lacZ+*	−

β-Galactosidase activity was assayed using the nitrocellulose filter assay (see Methods). In this experiment, + indicates that β-galactosidase activity was evident within 30 min and − denotes no activity after several hours at 30°C.

Table 2. The −1200 Sequence of *HO* Activates Transcription of a *CYC1:lacZ* Fusion

Plasmid	UAS inserted	β-Galactosidase activity
pLGΔ-178	none	0.48 ± 0.2
pKAN1200	−1200*HO*	30–60
pLGΔ-312	CYC 1	231 ± 79

Colorimetric assays using ONPG were performed as described (Miller 1972; see Methods). −1200 *HO* denotes insertion of *HO* DNA from −1516 to −1274 into the *Xho*I site of pLGΔ-178.

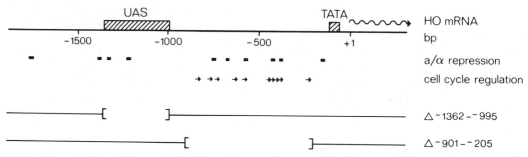

Figure 1. The *HO* promoter is depicted by a solid line. The approximate start site for transcription (+1) is denoted by a wavy line and arrow. This is 31 bp upstream from the translation start site. Sequence features denoted here are described in the text. Regions known to be required for *HO* expression are shown by striped boxes. Redundant sequences are shown below the line, either by solid boxes or arrows. Their proposed functions are described in the text. Two *HO* promoter deletions are also depicted: Δ −1362 to −995, and Δ −901 to −205. The DNA absent from these deletions is shown by bracketed gaps in the line.

Cell Cycle Control

The second repeated element found within the *HO* promoter was a short sequence (PuNNPy-CACGAAAA) that is repeated 10 times in a 600-bp region. The distribution and orientation of these sequences is shown with arrows in Figure 1. The location of $CACGA_4$ sequences coincides with a region that is missing from a set of three large deletions, all of which result in defects in the cell cycle regulation of *HO* (Nasmyth 1985b). The smallest of these deletions is depicted in Figure 1. Since all the $CACGA_4$ sequences fall within the region absent from this deletion, this repeated sequence was a good candidate for a cell cycle regulator.

To test this possibility, a chemically synthesized oligonucleotide bearing the $CACGA_4$ sequence was used to restore this sequence to the −901 to −205 deletion (see Fig. 1). The most striking effect of this sequence was its ability to activate transcription. The extent of activation increased with the numbers of $CACGA_4$ sequences included in the reconstruction, and this activation was only observed during late G_1 of the cell cycle. That is, synchronized cells showed a peak of transcription in late G_1. Cells arrested in early G_1 by α factor or a $cdc28^-$ block showed no activation (Nasmyth 1985b).

REPEATED $CACGA_4$ SEQUENCES HAVE UAS ACTIVITY

If the transcriptional activation apparent in the *HO* reconstruction described above was due to the $CACGA_4$ sequence itself, then these sequences alone should activate transcription from a heterologous yeast promoter. To test this prediction, we took advantage of the *CYC1-lacZ* vector pLGΔ-178, as described above. In this case, a synthetic oligonucleotide bearing the $CACGA_4$ sequence was cloned upstream from the *CYC1* TATA box in pLGΔ-178. Plasmids bearing one, two, or three $CACGA_4$ sequences were identified by sequence analysis and transformed into yeast cells, and these cells were assayed for β-galactosidase activity. The results of this experiment are summarized in Table 3. It is clear that the $CACGA_4$ sequence can activate transcription in the absence of any other *HO*-specific sequence.

It is also interesting to note that one $CACGA_4$ sequence leads to a low but discernible level of transcription. Two such sequences result in slightly higher β-galactosidase levels, and three elevate the β-galactosidase activity to a level comparable to that observed for the *HO* −1200 UAS cloned into the same site (compare Table 2, pKAN1200, to Table 3, entries pLB178-43 and pLB178-26). The significantly higher level of β-galactosidase activity promoted by three $CACGA_4$ sequences is apparent in both cases presented, despite the fact that the orientation of the sequences differ.

These observations were confirmed by looking directly at transcription from these plasmids. S1 nuclease protection experiments displayed in Figure 2 show that no transcript can be detected from the pLGΔ-178 parent plasmid (lane A). The addition of one box does not appear to increase transcription to a level detectable by this assay (lane B). Figure 2, lanes C–G, shows that transcription from plasmids carrying two $CACGA_4$ sequences can be detected and is quantitatively similar in these five examples, despite their different orienta-

Table 3. $CACGA_4$-activated Transcription of a *CYC1:lacZ* Fusion

Plasmid	Number and orientation of repeats	β-Galactosidase activity
pLGΔ-178 (A)	−	0.48 ± 0.2
pLB178-10 (B)	<	1.11 ± 0.5
pLB178-46 (C)	< >	2.35 ± 1
pLB178-27 (D)	< >	2.97 ± 1.2
pLB178-34 (E)	> <	2.91 ± 1.2
pLB178-12 (F)	> >	2.72 ± .52
pLB178-31 (G)	> >	1.87 ± 1.2
pLB178-26 (H)	< < <	20.80 ± 2.8
pLB178-43 (I)	< > >	17.45 ± 4

Results given are the average of at least five assays. − denotes no insert, > the insertion of an oligonucleotide into the *XhoI* site such that the orientation of the $CACGA_4$ sequence is the same as that at *HO*, and < a single insertion in the opposite orientation. The oligonucleotides used in this cloning had *XhoI* "sticky ends" with the sequence TCCACGAAAA in between. The capital letters following each entry correspond to the comparable lane in Fig. 2.

tions. With the plasmids containing three CACGA$_4$ repeats (lanes H and I), the level of transcription is about 10-fold higher and again appears to be independent of orientation.

It is clear that the CACGA$_4$ sequence can act as a UAS when placed upstream from the *CYC1* TATA box, and that this activation property is markedly enhanced by the presence of multiple copies of this sequence. This parallels the in vivo situation in that multiple copies of this sequence are also present within the *HO* promoter, and normal cell cycle regulation requires more than one copy (Nasmyth 1985b). However, it should be pointed out that CACGA$_4$ sequences are not tandemly repeated, but rather dispersed over a 600-bp region within the *HO* promoter. As such they are not sufficient for transcriptional activation. The -1362 to -995 deletion, which leaves all CACGA$_4$ sequences intact, abolishes *HO* transcription. So, it is formally possible that the transcriptional activation observed here is an artifact of the tandem duplication of CACGA$_4$ sequences, rather than a property intrinsic to that sequence. We believe that this is not the case for the following reasons. First, one CACGA$_4$ sequence is sufficient to cause a level of transcription which is detectably above background in both the colorimetric assays used in this work. Second, two CACGA$_4$ sequences cause more transcription and three cause even more, despite their different orientations in the plasmid. This means that the junction sequences differ in each of these constructions without affecting the level of transcription attained by them. And third, the CACGA$_4$-specific transcription that we observe from these plasmids is cell cycle regulated (see below).

CACGA$_4$-DEPENDENT ACTIVATION IS CELL CYCLE REGULATED

If the repeated CACGA$_4$ sequence is a cell cycle-specific activator of transcription, then the CACGA$_4$-driven expression of *CYC1:lacZ* on plasmids (e.g., pLB178-43) should be under cell cycle control. The experiment shown in Figure 3 was designed to see if that was true. In particular, we asked if the transcriptional activation by CACGA$_4$ sequences is, like that of *HO*, *CDC28* dependent. *CDC28* function is required for the cell to leave G$_1$ and undergo "start," which commits the cell to complete another mitotic cell cycle (Hartwell 1974). Temperature-sensitive mutations in *cdc28* cause cells (at high temperature) to arrest prior to "start." These cells are arrested in early G$_1$, as judged by the fact that they retain growth and mating capabilities that are normally associated with cells in this phase of the cell cycle (Reid and Hartwell 1977; Reed 1980). Cells arrested by *cdc28* mutations do not transcribe *HO* (Nasmyth 1983).

In order to ask whether or not CACGA$_4$-activated transcription would occur in *cdc28*-arrested cells, two of the CACGA$_4$-containing plasmids (pLB178-31 and pLB178-43) and the appropriate controls (pLGΔ-178 and pLGΔ-312) were transformed into a pair of strains that were *CDC28+* and *cdc28-6*. Cells carrying these plasmids were synchronized as unbudded G$_0$ cells by growth to saturation on uracil-deficient plates (to maintain selection for the plasmids) then inoculated into fresh medium and incubated at 37°C. Microscopic examination (data not shown) confirmed that the *cdc28-6* cells arrested with a typical *cdc28*-arrest phenotype as unbudded cells in early G$_1$ (Hartwell et al. 1973), and the *CDC28* cells resumed their cell cycle. Cells were harvested at 60, 90, and 120 minutes after inoculation, then transcription was analyzed using the S1 protection method (Berk and Sharp 1977, see Methods). For a control, *HO* transcription was monitored. As expected, the *CDC28* cells produced high levels of *HO* message, while the *cdc28-6* cells produced none (Nasmyth 1983). The *CYC1:lacZ* message was not made by either strain carrying the pLGΔ-178 plasmid, which lacks a UAS (Fig. 3a, e). When the normal *CYC1* UAS was present on the plasmid (pLGΔ-312; Fig. 3d, h), a high level of *CYC1:lacZ* transcript was made, and this transcription was *CDC28* independent. The central panels of this figure show transcription driven by two

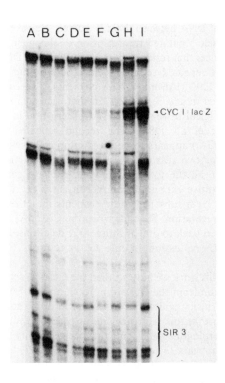

Figure 2. Transcriptional activation by the CACGA$_4$ sequence. The fragments protected by *CYC1:lacZ* mRNA are labeled, as are those protected by *SIR3* mRNA, which is used as an internal standard. The heavy bands flanking the *CYC1:lacZ*-specific band represent residual undigested probe. The *CYC1:lacZ* probe is a *Xho*I–*Hinc*II fragment of pLGΔ-178. This probe includes the transcription start site and can therefore distinguish faithful initiation from readthrough transcription. The *SIR3* probe is a fragment from the 3' end of the *SIR3* gene (Shore et al. 1984). (Lanes *A–I*) RNAs made from the yeast strain BY22 carrying the following plasmids: pLGΔ-178 (*A*), pLB178-10 (*B*), pLB178-46 (*C*), pLB178-27 (*D*), pLB178-34 (*E*), pLB178-12 (*F*), pLB178-31 (*G*), pLB178-26 (*H*), and pLB178-43 (*I*). BY22 is a *ura3–* strain, which allows these *URA3+* plasmids to be maintained by Ura+ selection.

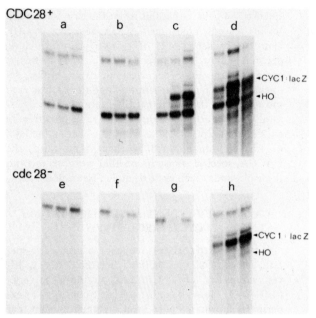

Figure 3. CACGA$_4$-mediated transcription is *CDC28* dependent. Transcription was analyzed from two strains, BY22 (*a–d*) and a *cdc28-6* derivative thereof (*e–h*) which carry one of four plasmids: pLGΔ-178 (*a* and *e*), pLB178-31 (*b* and *f*), pLB178-43 (*c* and *g*), or pLGΔ-312 (*d* and *h*). *HO* transcription was monitored using a probe of the *Bam*HI to *Hin*dIII region of the *HO* gene. The *CYC1:lacZ* probe is described in the legend to Fig. 2. The *SIR3* probe was also used in this experiment as an internal standard. Levels of *SIR3* message did not vary significantly in these RNA samples (data not shown). It may be noted that transcription of the *CYC1:lacZ* fusion begins at the same site in the CACGA$_4$-containing plasmids as it does in pLGΔ-312.

(Fig. 3b, f) or three (Fig. 3c, g) CACGA$_4$ sequences. It is clear that CACGA$_4$-activated transcription is absent in *cdc28*-6 arrested cells and present in *CDC28+* cells under the same conditions. One must therefore conclude that the UAS activity observed from repeated CACGA$_4$ sequences cannot occur when cells are arrested in early G$_1$ by the *cdc28*-6 mutation. By this criterion, the transcriptional activation exerted by this sequence is cell cycle regulated.

DISCUSSION

The *HO* gene is transcribed for only a brief period in the cell cycle of haploid mother cells. This period corresponds to late G$_1$, just after the *CDC28*-dependent "start" of the cell cycle (Nasmyth 1983). An extensive deletion analysis of the *HO* promoter sequence has been made that has allowed the identification of sequence elements that are essential for *HO* transcription, and others that are involved in the regulation of that transcription (Miller et al. 1985; Nasmyth 1985a,b). The aim of this work has been to characterize these important DNA sequences further with the hope of understanding the roles they play in the cell cycle-specific regulation of the *HO* gene.

The *HO* regulatory region is unusually long, by comparison to that of other characterized yeast genes (Guarente 1984). Sequences more than a kilobase apart are required for *HO* transcription. There is a TATA region 80 bp upstream from the mRNA start site that is required for transcription and that affects the site of initiation. At −1200 there is another essential sequence (Nasmyth 1985a). Here we have shown that the −1200 sequence is a transcription activator, in that it is a *cis*-acting element required for *HO* transcription. It can act at variable distances from the TATA box and can induce transcription of heterologous genes.

The −1200 UAS and −80 TATA box are sufficient to allow transcription of the *HO* gene when the kilobase of DNA between them is deleted. However, proper cell cycle-regulated transcription of this gene requires sequences that reside within that kilobase of DNA. Two sets of repeated sequences have been found between the −1200 UAS and TATA elements. The first set is involved in **a**/α repression, in that one such sequence is sufficient to repress *CYC1* transcription in diploids (Miller et al. 1985). The second repeated element (CACGA$_4$) appears to be involved in cell cycle regulation for two reasons. First, only large deletions that remove all the CACGA$_4$ sequences cause completely constitutive expression of *HO* in early G$_1$. Deletions that leave a subset of these elements intact retain cell cycle regulation. Second, when the CACGA$_4$ sequence is added back to one of those large deletions, the pulse of *HO* expression in late G$_1$ is restored (Nasmyth 1985b).

In this work we have studied the CACGA$_4$ element in isolation from the *HO* promoter in hopes of observing its intrinsic regulatory properties. In doing so we have shown that this sequence can provide UAS activity to a heterologous gene, and that its activating property only appears in late G$_1$, after the *CDC28*-dependent "start" of the cell cycle. This observation raises two important questions. First, what is the mechanism by which CACGA$_4$ mediates cell cycle-specific expression. Second, how is that mechanism of cell cycle specificity integrated into the wild-type *HO* promoter.

CACGA$_4$-mediated activation of transcription in late G$_1$ can be envisaged in two simple ways. The simplest explanation would be that there is an activating factor, required for CACGA$_4$-dependent transcription, that only functions during late G$_1$. The second possibility is that transcription activated from this sequence requires the transient removal of a repressor. Studies are currently underway to distinguish these two possibilities.

The second important question is how does this cell cycle-specific activator influence transcription from the intact *HO* promoter. CACGA$_4$ sequences alone, in conjunction with a TATA box, are sufficient to activate transcription, but within the context of the *HO* promoter they are not. Deletions that remove the −1200 UAS completely abolish *HO* transcription, so this sequence is clearly required, but deletions that remove all the CACGA$_4$ sequences do not abolish transcription. Thus, it would appear that CACGA$_4$ sequences are neither necessary nor sufficient for *HO* transcription. They are, however, necessary for proper cell cycle regulation of *HO* transcription and they appear to act as transcriptional activators in late G$_1$. This suggests that there are other elements involved in *HO* regulation, in particular some form of repression, that makes both the −1200 and CACGA$_4$ activation sequences necessary for transcription from the *HO* promoter. The target of repression must reside within the 600 bp that surround the CACGA$_4$ sequences because when that region is lost so is the need for the CACGA$_4$ sequences. Thus, the repression could be focused upon the CACGA$_4$ sequences themselves, or upon the sequences around them. Alternatively, the need for both the −1200 and CACGA$_4$ activators could simply be a result of the physical distance that separates the −1200 UAS from the *HO* TATA box.

The results of the experiments described here do not allow us to explain how *HO* expression is limited to a particular point in the cell cycle. However, they do provide compelling evidence that the CACGA$_4$ sequence plays an important role in this regulatory process. By studying the role of the CACGA$_4$ sequence in isolation from the *HO* promoter we have gained some understanding of how it affects transcription. This has allowed us to limit the number of possible mechanisms by which cell cycle-specific transcription could manifest itself, and has suggested the existence of some form of negative control that is also exerted upon the *HO* promoter.

CACGA$_4$ is a cell cycle-specific activator of transcription that is functionally separable from the *HO* gene. As such, it has an activity that can be assayed, and selected for or against. Clearly the most important consequences of this discovery is that we are now in a position to identify *trans*-acting genes that are required for this activity and to determine the pathway by which post-"start" gene expression is triggered.

ACKNOWLEDGMENTS

We would like to thank L. Guarente for providing plasmids, and S. Reed for sending the cdc28-6 strain. We are indebted to M. Squire for excellent technical assistance, and to A. Brand, A. Miller, and D. Shore for critically reading the manuscript. This work was supported by the Medical Research Council, U.K.

REFERENCES

BEGGS, J. 1978. Transformation of yeast by a replicating hybrid plasmid. *Nature* **275:** 104.

BERK, A.S. and P.A. SHARP. 1977. Sizing and mapping of early adenovirus mRNAs by gel electrophoresis of S1 endonuclease digested hybrids. *Cell* **2:** 721.

CASADABAN, M.J., A. MARTINEZ-ARIAS, S.K. SHAPIRA, and J. CHOU. 1983. β galactosidase gene fusions for analyzing gene expression in *E. coli* and yeast. *Methods Enzymol.* **100:** 293.

CORDEN, J., B. WASYLYK, A. BUSCHWALDER, P. CORSI, C. KEDINGER, and P. CHAMBON. 1980. Promoter sequences of eukaryote protein-coding genes. *Science* **209:** 1406.

FRIIS, J. and H. ROMAN. 1968. The effect of the mating type alleles on intragenic recombination. *Genetics* **59:** 33.

GAME, J.C. 1983. Radiation sensitive mutants and repair in yeast. In *Yeast genetics fundamental and applied aspects* (ed. J.F.T. Spencer et al.), p. 109. Springer Verlag, New York.

GUARENTE, L. 1984. Yeast promoters: Positive and negative elements. *Cell* **36:** 799.

GUARENTE, L. and E. HOAR. 1984. Upstream activation sites of the *CYC1* gene of *Saccharomyces cerevisiae* are active when inverted but not when placed downstream of the "TATA box." *Proc. Natl. Acad. Sci.* **81:** 7860.

GUARENTE, L. and T. MASON. 1983. Heme regulates transcription of the *CYC1* gene of *S. cerevisiae* via an upstream activation site. *Cell* **32:** 1279.

GUARENTE, L., B. LALONDE, P. GIFFORD, and E. ALANI. 1984. Distinctly regulated tandem upstream activation sites mediate catabolite repression of the *CYC1* gene of *S. cerevisiae*. *Cell* **36:** 503.

HABER, J.E. and B. GARVIK. 1977. A new gene affecting the efficiency of mating type interconversions in homothallic strains of *S. cerevisiae*. *Genetics* **87:** 33.

HARTWELL, L.H. 1974. *Saccharomyces cerevisiae* cell cycle. *Bacteriol. Rev.* **38:** 164.

HARTWELL, L.H., R.K. MORTIMER, J. CULOTTI, and M. CULOTTI. 1973. Genetic control of the cell division cycle in yeast. V. Genetic analysis of cdc mutants. *Genetics* **74:** 267.

HICKS, J.B. and I. HERSKOWITZ. 1976. Interconversion of yeast mating types. I. Direct observations of the action of the homothallism (HO) gene. *Genetics* **83:** 245.

JENSEN, R.E. 1983. "Control of mating type interconversion in S. cerevisiae." Ph.D. thesis, University of Oregon, Eugene.

JENSEN, R.E. and I. HERSKOWITZ. 1983. Directionality and regulation of cassette substitution in yeast. *Cold Spring Harbor Symp. Quant. Biol.* **48:** 97.

JENSEN, R., G.F. SPRAGUE, and I. HERSKOWITZ. 1983. Regulation of the yeast mating type interconversion: Feedback control of *HO* gene expression by the mating type locus. *Proc. Natl. Acad. Sci.* **80:** 3035.

KHOURY, G. and P. GRUSS. 1983. Enhancer elements. *Cell* **33:** 313.

KLAR, A.J.S. and J.N. STRATHERN. 1984. Resolution of recombination intermediates generated during yeast mating type switching. *Nature* **310:** 744.

KLAR, A.J.S., J.N. STRATHERN, and J.B. HICKS. 1984. Developmental pathways in yeast. In *Microbial development* (ed. R. Losick and L. Shapiro), p. 151. Cold Spring Harbor Laboratory, Cold Spring Harbor, New York.

KOSTRIKEN, R. and F. HEFFRON. 1983. The product of the *HO* gene is a nuclease: Purification and characterization of the enzyme. *Cold Spring Harbor Symp. Quant. Biol.* **48:** 89.

MALONE, R.E. and R.E. ESPOSITO. 1980. The RAD52 gene is required for homothallic interconversion of mating types and spontaneous mitotic recombination in yeast. *Proc. Natl. Acad. Sci.* **17:** 503.

MCKAY, V.L. 1983. Cloning of yeast STE genes in 2 μm vectors. *Methods Enzymol.* **101:** 325.

MILLER, A.M., V.L. MacKay, and K.A. Nasmyth. 1985. Identification and comparison of two sequence elements that confer cell-type specific transcription in yeast. *Nature* **314:** 598.

MILLER, J.H. 1972. *Experiments in molecular genetics*, p. 352. Cold Spring Harbor Laboratory, Cold Spring Harbor, New York.

MORTIMER, R.K. 1958. Radiobiological and genetic studies on a polyploid series (haploid to hexaploid) of *S. cerevisiae*. *Radiat. Res.* **9:** 312.

MORTIMER, R.K. and D.C. HAWTHORNE. 1969. Yeast genetics. In *The yeasts* (ed. A.H. Rose and J.S. Harrison), vol. 1, p. 385. Academic Press, New York.

NASMYTH, K. 1983. Molecular analysis of a cell lineage. *Nature* **302:** 670.

———. 1985a. At least 1400 bp of 5′ flanking DNA is required for correct expression of the *HO* gene in yeast. *Cell* **42:** 213.

———. 1985b. A repetitive DNA sequence which confers cell cycle START (CDC 28) dependent transcription on the *HO* gene in yeast. *Cell* **42:** 225.

OSHIMA, T. and I. TAKANO. 1971. Mating types in *Saccharomyces*: Their convertibility and homothallism. *Genetics* **67:** 327.

REED, S.I. 1980. The selection of *S. cerevisiae* mutants defective in the start event of cell division. *Genetics* **95:** 561.

REID, B. and L.H. HARTWELL. 1977. Regulation of mating in the cell cycle of *S. cerevisiae*. *J. Cell Biol.* **75:** 355.

ROMAN, H. and S.M. SANDS. 1953. Heterogeneity of clones of *Saccharomyces* derived from haploid ascospores. *Proc. Natl. Acad. Sci.* **39:** 171.

SHORE, D., M. SQUIRE, and K.A. NASMYTH. 1984. Characterization of two genes required for the position effect control of yeast mating type genes. *EMBO J.* **3:** 2817.

SPRAGUE, G.F., L.C. BLAIR, and J. THORNER. 1983. Cell interactions and regulation of cell type in the yeast *S. cerevisiae*. *Annu. Rev. Microbiol.* **37:** 623.

STERN, M., R. JENSEN, and I. HERSKOWITZ. 1984. Five SW1 genes are required for expression of the *HO* gene in yeast. *J. Mol. Biol.* **178:** 853.

STRATHERN, J.N. and I. HERSKOWITZ. 1979. Assymetry and directionality in production of new cell types during clonal growth: The switching pattern of homothallic yeast. *Cell* **17:** 371.

STRATHERN, J., J. HICKS, and I. HERSKOWITZ. 1981. Control of cell type in yeast by the mating type locus. The $\alpha 1$-$\alpha 2$ hypothesis. *J. Mol. Biol.* **147:** 357.

STRATHERN, J.N., A.J.S. KLAR, J.B. HICKS, J.A. ABRAHAM, J.M. IVY, K.A. NASMYTH, and C. MCGILL. 1982. Homothallic switching of yeast mating type cassettes is initiated by a double-stranded cut in the *MAT* locus. *Cell* **31:** 183.

WINGE, O. and C. ROBERTS. 1949. A gene for diploidization in yeast. *C. R. Trav. Lab. Carlsberg Ser. Physiol.* **24:** 341.

ZIRKLE, R.E. and C.A. TOBIAS. 1953. Effects of ploidy and linear energy transfer on radiobiological survival curves. *Arch. Biochem. Biophys.* **47:** 282.

The Role of Mitotic Factors in Regulating the Timing of the Midblastula Transition in *Xenopus*

J. NEWPORT, T. SPANN, J. KANKI, AND D. FORBES
Biology Department, University of California, San Diego, La Jolla, California 92093

In early *Xenopus* development, the cell cycle proceeds rapidly and synchronously until the twelfth cleavage (4000-cell stage) (Signoret and Lefresne 1971; Satoh 1977). At this time a transition occurs, called the midblastula transition (MBT), which involves the turn-on for the first time of detectable RNA transcription, the onset of motility, and an elongation of the cell cycle in all the cells of the embryo (Fig. 1) (Kobayakawa and Kubota 1981; Newport and Kirschner 1982a,b). The transition at the 4000-cell stage is induced because a particular nuclear/cytoplasmic ratio has been reached in the embryo (Fig. 1). We have proposed that this critical ratio causes the MBT through the titration of a cytoplasmic element by the newly made nuclei. In our model, at the MBT enough nuclei have been made to deplete the cytoplasm of this element and this in turn triggers the MBT. Since either new RNA transcription or cell motility can be blocked without preventing the MBT, these new cellular activities appear to be activated as a result of the MBT and do not regulate its occurrence. Therefore, we believe that the onset of the MBT is controlled by a fundamental change in the regulation of the cell cycle at this time in development. In this model, enzymatic activities which act to control the cell cycle are the components that are being titrated to achieve the particular nuclear/cytoplasmic ratio that induces the MBT. Experiments to test this prediction are described below.

The cell cycle prior to the MBT is biphasic, consisting of a 15-minute S phase followed by a 15-minute M phase. There are two periods during the early embryonic cell cycle that can be regulated such that they control the transition between the phases of the cell cycle. These switching periods regulate the S-to-M-phase transition and the M-to-S-phase transition. Any event that delays either of these two regulatory events would cause the overall cell cycle to elongate. After the MBT, examination of the embryonic cells clearly demonstrates that there is a decrease in the number of cells with mitotic figures relative to those with intact nuclei. Clearly an elongation of the cell cycle at the MBT has occurred, involving an expansion of the interphase periods of the cell cycle (G_1, S, and G_2). This strongly suggests that the elongation of the cell cycle which occurs at the MBT is the result of a delay in the molecular mechanisms that normally regulate the S-to-M-phase transition. Such a delay in effect creates a G_2 period within the cell cycle.

It has been demonstrated that a factor present in unfertilized eggs, called maturation promoting factor (MPF) (Wasserman and Smith 1978; Meyerhof and Masui 1979; Reynhout and Smith 1984), appears to play a pivotal role in inducing a cell to enter mitosis (Wu and Gerhart 1980; Miake-Lye et al. 1983; Newport and Kirschner 1984; Gerhart et al. 1984). This factor has been found to be a highly conserved factor in that mitotic cells from yeast or humans contain MPF activity which is active in inducing mitosis when injected

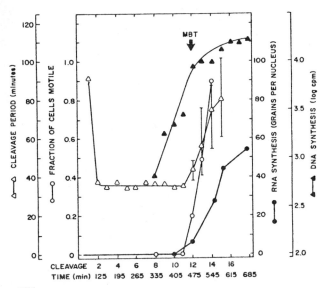

Figure 1. Various cellular parameters affecting the early cleavage cycle in *Xenopus*: the cleavage cycle (△), DNA replication cycle (▲), RNA synthesis per nucleus (●), and fraction of motile cells (○). The bars in the cleavage cycle after the twelfth cleavage indicate the heterogeneity in the cell cycle after the MBT.

into *Xenopus* eggs (Sunkara et al. 1979; Nelkin et al. 1980; Weintraub et al. 1982). We and others have presented evidence suggesting that the regulation of the early embryonic cell cycle in *Xenopus* is controlled by an oscillation of MPF activity (Gerhart et al. 1984; Newport and Kirschner 1984). MPF activity has been found to be present during mitosis and absent during S phase. It has been concluded that the cells in the embryo are in mitosis when MPF is present and in S phase when MPF is either destroyed or temporarily inactivated. A possible mechanism for the action of MPF in regulating the S-to-M-phase transition is one where MPF activates a number of enzymatic activities necessary for carrying out the initial events associated with mitosis (i.e., nuclear envelope breakdown and chromosome condensation). Such enzymatic activities, because of their close association with nuclei and as important regulators in initiating mitosis, would be prime candidates for titratable factors that regulate the onset of the MBT.

Because of the rapid nature of the cell cycle during early *Xenopus* cleavage, proteins and RNA molecules, such as the histones, nucleoplasmin, and the snRNAs, used at this time are produced and stored during oogenesis in amounts sufficient to supply 4000 or more cells (Laskey et al. 1977; Woodland and Adamson 1977; Forbes et al. 1983b). It would not be unexpected that the mitotic initiation activities described above would also be made and stored during oogenesis, such that the total amount of these activities would be the same in the unfertilized egg and the 4000-cell embryo. This last supposition is strongly supported by the following: We find that an egg arrested in mitosis, when injected with 1000–2000 rat thymus interphase nuclei, can induce the complete breakdown of the nuclei and condensation of their chromosomes. Such mitotic induction occurs even if protein synthesis is blocked (Newport and Kirschner 1984). Thus the egg does appear to contain stored mitotic initiation factors.

RESULTS

Nuclear Assembly In Vivo and In Vitro

In this report we describe experiments directed at establishing a molecular link between the regulation of the cell cycle and the nuclear/cytoplasmic ratio that triggers the MBT. It is hoped that this link will provide information about the molecular mechanism regulating the MBT. Previously we demonstrated that the MBT can be induced to occur prematurely when exogenous DNA was injected into a fertilized egg. The simplest model for initiation of the MBT would be a DNA-dependent titration of a cytoplasmic molecule. In subsequent experiments, it became apparent that a number of nuclear proteins might be the titrated component. This conclusion is based on our observation that when bacteriophage λ DNA is injected into *Xenopus* eggs, it undergoes a series of rearrangements, the end result of which is that the DNA is found encapsulated in structurally intact nuclei (Forbes et al. 1984a). These "syn-

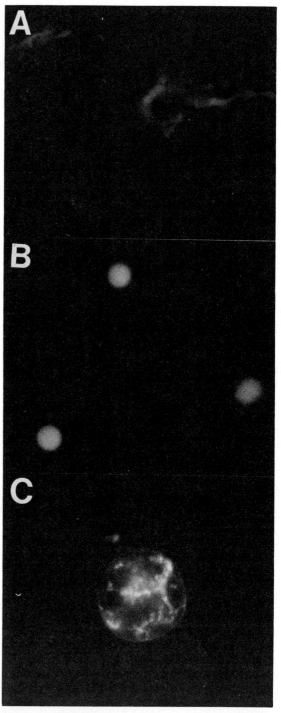

Figure 2. In vitro assembly of nuclei in extracts: 4 ng of λ DNA per egg equivalent of interphase egg extract was incubated for 30 (*A*), 60 (*B*), and 90 (*C*) min. The arrangement of the DNA was visualized by staining with the DNA-specific dye bisbenzimide. At 30 min the DNA is assembled into chromatin. By 60 min the DNA is highly condensed into spheres. At 90 min the DNA is encapsulated within a nuclear membrane and the DNA has decondensed.

thetic" nuclei are identical to "real" nuclei in that they are surrounded by a double bilayer membrane containing nuclear pores. The membranes are lined with a lamina composed of the major nuclear lamin proteins.

We further found that, when an egg containing such synthetic nuclei is arrested in mitosis, the nuclear membranes of the synthetic nuclei break down and the DNA becomes highly condensed, indicating that the nuclei can respond to cell cycle cues. The reconstituted nuclei are functional in that they carry out a limited amount of DNA synthesis (J. Newport, unpubl.), transcribe very efficiently, and transport nuclear-specific proteins (D. Forbes, unpubl.).

Extending the in vivo studies on nuclear assembly, we have made extracts from activated *Xenopus* eggs that are capable of reconstituting nuclei when naked DNA is added (Fig. 2). In such extracts the DNA is incorporated into nuclei through a series of discrete steps which include assembly into chromatin and condensation of the chromatin, followed by encapsulation by nuclear membranes and incorporation of nuclear pores. During this last step, the DNA decondenses and can be seen to be attached to the periphery of the nucleus (Fig. 2). The in vitro-reconstituted nuclei contain the double membrane system, nuclear pores, and nuclear lamin proteins found in normal nuclei. Using this system, we have been able to demonstrate that at least 1000 *Xenopus* nuclear equivalents of DNA per egg equivalent of extract can be assembled into nuclei. Thus these nuclear elements must be accumulated and stored within the egg during oogenesis. We are currently using this in vitro nuclear reconstitution system to examine how nuclei assemble and to identify nuclear components that could play a role in regulating the cell cycle.

Nuclear Disassembly in Mitotic Extracts

An unfertilized egg is stably arrested in second meiotic metaphase. We find that extracts made from these M-phase arrested eggs can induce exogenously added interphase nuclei to enter mitosis. Specifically, when isolated rat thymus nuclei are added to such M-phase or mitotic extracts, the nuclei undergo nuclear envelope breakdown and chromosome condensation (Fig. 3). The ability to make an M-phase extract from unfertilized eggs demonstrates that the factors necessary to initiate mitosis are present in excess in the cytoplasm of unfertilized eggs. These extracts are used to investigate the relationship between the initiation of mitosis and the timing of the MBT in the experiments described below.

To determine the capacity of mitotic extracts for converting interphase nuclei to mitotic nuclei, increasing amounts of rat liver nuclei were added to M-phase extracts. Aliquots of the extracts were removed at different times and analyzed for the extent of nuclear envelope breakdown and chromosome condensation. As shown in Figure 4, when the number of added nuclei was below 3000-6000 nuclei/egg-equivalent of M-phase extract, the time taken to bring about nuclear envelope breakdown and chromosome condensation was independent of the number of nuclei present. However, above 3000-6000 nuclei/egg-equivalent of extract the time required to bring about nuclear breakdown and chromosome condensation increased sharply.

This result indicates that biochemically the activities required to convert interphase nuclei to metaphase nuclei are in excess at a nuclear concentration below 3000-6000 nuclei/egg-equivalent of M-phase extract. Above this concentration of nuclei, the activities are rate-limiting, i.e., the rate of breakdown is linearly dependent on substrate concentration. Significantly the concentration of nuclei at which this break occurs is very close to the number of nuclei present within an embryo at the MBT (4000). This correlation strongly supports a model in which the slowing of the cell cycle at the MBT is due to the presence at this period in development of more nuclei than can be broken down in a short period of time by the limited number of mitotic factors present in the egg. As a result, there would be a progressively longer amount of time separating the completion of DNA synthesis and the initiation of mitosis each cell cycle past the MBT. Thus, a time gap between the end of S phase and mitosis (equal to a G_2 phase of the cell cycle) would become a permanent and increasingly important part of the cell cycle after the MBT.

To determine whether the mitotic activities which become rate-limiting for nuclear breakdown at high concentrations of nuclei are DNA-binding proteins, the following experiment was done: Nuclei were digested twice with DNAse I, followed by extraction with 2% Triton X-100. This treatment removed over 95% of the DNA and histone proteins from the nuclei (Dwyer and Blobel 1976). The major nuclear proteins left in these nuclear "ghosts" were the nuclear lamin proteins, A, B, and C, which represent approximately 50% of the remaining protein. When added to mitotic extracts, the DNA-depleted nuclei break down, demonstrating that they are able to respond to mitotic factors. Furthermore, when increasing concentrations of the "ghosts" are added to M-phase extracts containing 500 intact (DNA-containing) nuclei per egg equivalent, they inhibit nuclear breakdown of the intact nuclei at concentrations above 4000-6000 ghosts per egg equivalent (Fig. 4). Thus, the DNA-depleted nuclei appear to be as effective at inhibiting mitotic breakdown of nuclei as do high concentrations of DNA-containing nuclei. This strongly supports a model in which the factors that become rate-limiting for the initiation of mitosis regulate nuclear envelope breakdown rather than processes requiring binding to DNA.

The experiment described above suggests that the molecules contained within DNA-depleted nuclei (primarily nuclear lamins A, B, and C) are acting as a sink for the titration of important factors responsible for initiating mitosis within the nucleus. To test this hypothesis, M-phase extracts were first incubated with 15,000 DNA-depleted nuclei per egg-equivalent of extract for 2 hours. These ghosts, which did not break down during this time period, were then removed with a brief centrifugation. The extract was further incubated with 1000 intact nuclei per egg equivalent for 2 additional hours. In control extracts, which were treated identically but without the addition of ghosts, the added intact nuclei underwent nuclear envelope

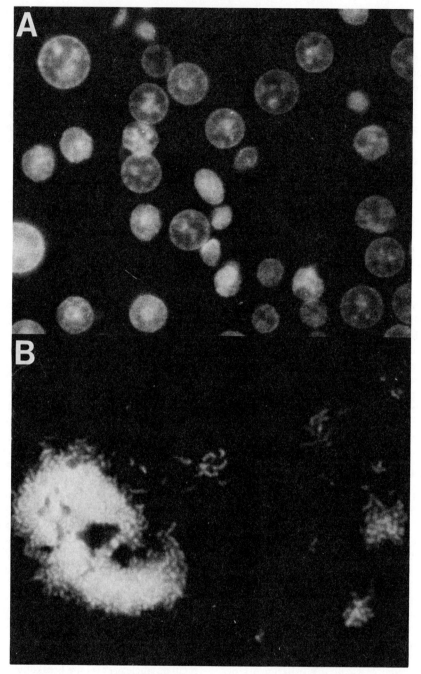

Figure 3. In vitro mitotic breakdown of nuclei: 2000 isolated rat thymus interphase nuclei per egg equivalent of mitotic extract were incubated for 0 (*A*) and 120 (*B*) min. By 120 min the nuclei have undergone nuclear envelope breakdown and chromosome condensation. The small fluorescent particles in *B* are individual condensed mitotic chromosomes.

breakdown and chromosome condensation during the second 2-hour period. However, the extracts preincubated with ghosts did not induce nuclear envelope breakdown and chromosome condensation of the intact nuclei during the second 2-hour period. This result demonstrates that factors responsible for the initiation of mitosis have an affinity for DNA-depleted nuclei and as such represent factors that can be titrated from the cytoplasm by the nucleus during the rapid cleavage period. It should be noted that such nuclear mitotic-inducing factors need not be and, probably are not, tightly bound to the nucleus. Rather, the roughly linear increase in time required for breakdown of nuclei when increasing levels of nuclei are added above 4000 nuclei per egg equivalent of extract (Fig. 4) suggests that these factors are enzymatic in function and bind weakly and transiently to nuclei while carrying out their function(s). From other experiments, it appears that when sufficient nuclear ghosts are present (15,000), the nuclear mitotic factors bind to the ghosts long enough to

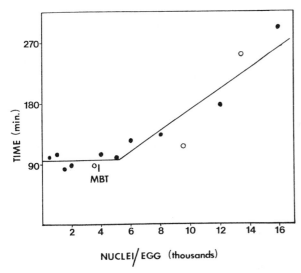

Figure 4. Nuclear breakdown in mitotic extracts: Different amounts of either intact nuclei (●) or DNA-depleted nuclei (○) were added to egg equivalents of mitotic extracts. Breakdown of nuclei was determined by visual examination at different times. Nuclei were considered to have undergone mitotic breakdown when the envelope was no longer present and the DNA was condensed.

be removed from an extract when the nuclear ghosts are pelleted.

DISCUSSION AND CONCLUSIONS

The rapid cleavage period following fertilization of *Xenopus* eggs is terminated after 12 cleavage cycles. At this time (MBT), new transcription and cell motility are activated. We propose that the activation of these processes is the result of a stabilization of the cell cycle resulting in an increasingly longer interphase period (Fig. 5). Specifically, we propose that this stable interphase period appears at the MBT because the proteins necessary to initiate mitosis become rate-limiting at this stage in development. We have demonstrated that mitotic extracts made from *Xenopus* eggs can break down up to 3000–6000 rat thymus nuclei per egg-equivalent of extract at a rate that is independent of the concentration of the nuclei. However, at concentrations of nuclei higher than 3000–6000, nuclear breakdown and chromosome condensation slows dramatically. Similar results are obtained when nuclei depleted of DNA are used. These results suggest that an unfertilized egg contains sufficient mitotic factors to initiate mitosis in nuclei until the MBT. Following this stage, a large enough number of nuclei have been produced such that the factors necessary for initiation of mitosis are rate-limiting and thus initiation of mitosis occurs at a slower rate. We predict that the new activities (RNA transcription and cell motility) present at the MBT result from the presence of a stable interphase period due to the slowdown in initiation of mitosis.

The finding that DNA-depleted nuclei are as effective as whole nuclei in titrating out the factor that converts nuclei into a mitotic conformation suggests that the mitotic initiation factors are not binding to chromatin. Rather they appear to be factors that could be regulating the nuclear lamina complex, since 50% of the protein in the DNA-depleted nuclei are the lamin proteins. Gerace and Blobel (1980) have demonstrated that the nuclear lamins are converted to a soluble form in CHO cells during mitosis and that this solubilization occurs coincident with a hyperphosphorylation of the lamins. Clearly the lamin kinase involved in this mitotic event would be a candidate for a nuclear mitotic initiation factor. Other investigators have shown that a topoisomerase II is a major protein of the nuclear lamina complex and propose that the DNA is attached to the nuclear membrane via this topoisomerase (Earnshaw et al. 1985; Earnshaw and Heck 1985). A mitotic regulator of the topoisomerase would be another candidate for a protein involved in initiation of mitosis in nuclei. We are currently attempting to purify the factor(s) responsible for the mitotic initiation event.

The experiments described above are consistent with the model that the timing of the MBT is dependent on a nuclear titration of factors that regulate the initiation of mitosis. The delay in the onset of mitosis and the increased duration of interphase would allow for the activation of a number of new cellular processes which are normally inactive during mitosis but become active for the first time in development at the MBT. The activation of RNA transcription at the MBT could occur simply as a result of the stabilization of the nucleus during this longer interphase. The establishment of a prolonged interphase at the MBT would also allow both microtubules and microfilaments to organize into the networks needed for cell motility and cell shape changes. During the rapid cleavage period before the MBT, the microtubules and microfilaments are used extensively for formation of the mitotic spindle and contractile ring. It has been reported that during mitosis in tissue culture cells, Golgi-mediated protein transport to the plasma membrane is blocked. We have preliminary evidence suggesting that the transport of plasma membrane proteins is blocked prior to the MBT. Thus, the establishment of a stable interphase through introduction of a rate-limiting step for the initiation of mitosis could result in the activation of a number of important cellular activities (Fig. 6). The activation of these cellular processes undoubtedly plays a major role

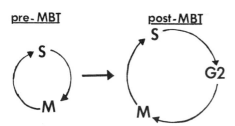

Figure 5. Diagram of the expansion of the cell cycle that occurs at the MBT due to a delay in the initiation of mitosis following completion of S phase.

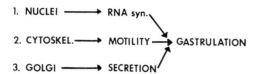

Figure 6. Predicted cellular processes activated at the MBT due to the establishment of a stable interphase period.

in controlling the next stage of development, gastrulation.

REFERENCES

DWYER, N. and G. BLOBEL. 1976. A modified procedure for the isolation of a pore complex lamina fraction from rat liver nuclei. *J. Cell Biol.* **70:** 581.

EARNSHAW, W. and M. HECK. 1985. Localization of topoisomerase II in mitotic chromosomes. *J. Cell Biol.* **100:** 1716.

EARNSHAW, W., B. HALLIGAN, C. COOKE, M. HECK, and L. LIU. 1985. Topoisomerase II is a structural component of mitotic chromosomes. *J. Cell Biol.* **100:** 1706.

FORBES, D.J., M.W. KIRSCHNER, and J.W. NEWPORT. 1983a. Spontaneous formation of nucleus like structures around bacteriophage DNA microinjected into *Xenopus* eggs. *Cell* **34:** 13.

FORBES, D.J., T.B. KORNBERG, and J.W. NEWPORT. 1983b. Small nuclear RNA transcription and assembly in early *Xenopus* development. *J. Cell Biol.* **97:** 62.

GERACE, L. and G. BLOBEL. 1980. The nuclear envelope lamina is reversibly depolymerized during mitosis. *Cell* **19:** 277.

GERHART, J.C., M. WU, and M. KIRSCHNER. 1984. Cell cycle dynamics of an M-phase specific cytoplasmic factor in *Xenopus laevis* oocytes and eggs. *J. Cell Biol.* **98:** 1247.

KOBAYAKAWA, Y. and H. KUBOTA. 1981. Temporal pattern of cleavages and the onset of gastrulation in amphibian embryos developed from eggs with reduced cytoplasm. *J. Embryol. Exp. Morphol.* **62:** 83.

LASKEY, R.A., A.D. MILLS, and N.R. MORRIS. 1977. Assembly of SV40 chromatin in a cell-free system from *Xenopus* eggs. *Cell* **10:** 237.

MEYERHOF, P.G. and Y. MASUI. 1979. Chromosome condensation activity in *Rana pipiens* eggs matured *in vivo* and in blastomeres arrested by cytostatic factor (CSF). *Exp. Cell Res.* **123:** 345.

MIAKE-LYE, R., J.W. NEWPORT, and M.W. KIRSCHNER. 1983. Maturation promoting factor induces nuclear envelope breakdown in cycloheximide-arrested embryos of *Xenopus laevis. J. Cell Biol.* **97:** 81.

NELKIN, B., C. NICHOLS, and B. VOGELSTEIN. 1980. Protein factor(s) from mitotic CHO cells induce meiotic maturation in *Xenopus laevis* oocytes. *FEBS Lett.* **109:** 233.

NEWPORT, J.W. and M.W. KIRSCHNER. 1982a. A major developmental transition in early *Xenopus* embryos. I. Characterization and timing of cellular changes at the midblastula stage. *Cell* **30:** 675.

———. 1982b. A major developmental transition in early *Xenopus* embryos. II. Control of the onset of transcription. *Cell* **30:** 687.

———. 1984. Regulation of the cell cycle during early *Xenopus* development. *Cell* **37:** 731.

REYNHOUT, J.K. and L.D. SMITH. 1984. Studies of the appearance and nature of a maturation-inducing factor in the cytoplasm of amphibian oocytes exposed to progesterone. *Dev. Biol.* **38:** 394.

SATOH, N. 1977. Metachromous cleavage and initiation of gastrulation in amphibian embryos. *Dev. Growth Differ.* **19:** 111.

SIGNORET, J. and J. LEFRESNE. 1971. Contribution a l'etude de la segmentation de l'oef d'axolotl. I. Definition de la transition blastulenne. *Ann. Embryol. Morphog.* **4:** 113.

SUNKARA, P.S., D.A. WRIGHT, and P.N. RAO. 1979. Mitotic factors from mammalian cells: A preliminary characterization. *J. Supramol. Struct.* **11:** 195.

WASSERMAN, W.J. and L.D. SMITH. 1978. The cyclic behavior of a cytoplasmic factor controlling nuclear membrane breakdown. *J. Cell Biol.* **78:** R15.

WEINTRAUB, H., M. BUSCAGLIA, S. FERREZ, A. WEILLER, F. BOULET, E. FABRE, and E.E. BAULIEU. 1982. Mise en evidence d'une activite "MPF" chez *Saccharomyces cerevisiae. C.R. Acad. Sci.* **295:** 787.

WOODLAND, H.R. and E.D. ADAMSON. 1977. The synthesis and storage of histones during the oogenesis of *Xenopus* oocytes. *Dev. Biol.* **57:** 118.

WU, M. and J.G. GERHART. 1980. Partial purification and characterization of maturation promoting factor from eggs of *Xenopus laevis. Dev. Biol.* **79:** 465.

Chromosome Replication in Early *Xenopus* Embryos

R.A. LASKEY, S.E. KEARSEY, M. MECHALI,* C. DINGWALL, A.D. MILLS, S.M. DILWORTH, AND J. KLEINSCHMIDT†

CRC Molecular Embryology Group, Department of Zoology, University of Cambridge, Cambridge CB2 3EJ, England

Free-living embryos are characterized by exceptional rates of chromosome replication. Table 1 illustrates the speed of cell division, and hence of chromosome replication of the most widely studied early embryos. By the time a mouse egg or a frog adult cell has divided once, the embryo of *Xenopus laevis* has already passed 20,000 cells. Each *Xenopus* chromosome is 40 times larger than the entire genome of *Escherichia coli*, and *Xenopus* replication forks move about 100 times more slowly than those of *E. coli*. Nevertheless, cleavage-stage embryos of *X. laevis* replicate their genomes faster than *E. coli* replicates its genome in log-phase growth. Much of the biochemistry of early development of free-living embryos is concerned with maintaining these exceptional rates of chromosome replication.

This paper considers two aspects of accelerated chromosome replication in embryos of *X. laevis*. First we consider the regulation of DNA replication. When purified DNA is microinjected into unfertilized eggs of *X. laevis*, it replicates. DNA equivalent to more than 250 diploid nuclei is replicated by each mononucleate egg under strict control of the egg's cell cycle. There have been conflicting reports of the DNA sequence requirements for replication in *Xenopus* eggs. This paper will attempt to resolve that confusion. The second aspect of accelerated chromosome replication to be considered in this paper is the assembly of chromatin from newly replicated DNA. Each cell cycle is completed in only 35 minutes. During this time the DNA is not only replicated completely, but also is assembled into chromatin, and condensed and divided at mitosis. The capacity for chromatin assembly is extraordinary. Each unfertilized egg can assemble DNA equivalent to more than 7000 diploid nuclei in only 1 hour in the absence of new protein synthesis. This paper will consider the process of chromatin assembly and the properties and functions of histone transer factors that mediate in the interaction of histones and DNA.

ACCELERATED DNA REPLICATION IN EARLY EMBRYOS OF *X. LAEVIS*

Adult cells of *X. laevis* have an S phase of the cell cycle of approximately 20 hours, yet the whole cell cycle of early embryos is only 35 minutes for cycle numbers 2–12 (Newport and Kirschner 1982). Part of the explanation of this acceleration lies in the spacing between consecutive initiations. Replication initiates at shorter intervals along the DNA in embryos than in adult cells (Callan 1972). This phenomenon has been documented most thoroughly in *Drosophila* embryos (Blumenthal et al. 1974), where the average distance between initiations in embryos is 7.9 kb compared with 40 kb in adult cells.

The observations that replication can initiate at closer intervals in embryos immediately raises questions about the DNA sequences involved in initiation at different developmental stages. Are there different sequence specificities, or is there selective masking of some potential initiation sites in adult cells so that only a subset is available for initiation? An approach to this question was provided by the observation of Gurdon et al. (1969) that injection of DNA into unfertilized eggs of *X. laevis* stimulates DNA synthesis, which was subsequently shown to be semiconservative replication (Laskey and Gurdon 1973; Ford and Woodland 1975).

It follows from these observations that microinjec-

---

*Present address: Institut Jacques Monod, Universite Paris 7, Tour 43, 2 Place Jussieu, 75251 Paris Cedex 05, France.
†Present address: Institut fur Zell und Tumorbiologie, Deutsches Krebsforschungszentrum, Im Neuenheimer Feld 280, 6900 Heidelberg, Federal Republic of Germany.

Table 1. Increase in Nuclear Numbers following Fertilization

Hours past fertilization	*Strongylocentrotus purpuratus*	*Xenopus laevis*	*Drosophila melanogaster* (nuclei)	*Mus musculus*
1	1	1	64	1
2	2	2	1500	
3	4	8	6000	
5	8	500		
10	100	20,000		
20	400	80,000		1

Data compiled from Nieuwkoop and Faber (1956), Hinegardner (1967), Zalokar and Erk (1976), Newport and Kirschner (1982).

tion of DNA templates into unfertilized eggs of *X. laevis* should provide an assay for DNA sequences that can serve as eukaryotic replication origins. Several laboratories tested this possibility by injecting various DNA templates into *Xenopus* eggs and assaying them for DNA replication. They reported conflicting results.

Sequence Specificity of DNA Replication in *Xenopus* Eggs

On the one hand Watanabe and Taylor (1980) and Chambers et al. (1982) reported that a 503-bp DNA insert from the *Xenopus* genome appears to act as a replication origin by increasing replication of pBR322 20-fold. In support of this, Hines and Benbow (1982) reported that an *Eco*RI fragment from the 26S gene region of the *Xenopus* rDNA repeat replicated more efficiently than either the other *Eco*RI fragment or the ColE1 cloning vector. They also reported preferential initiation in the same 26S coding region on the basis of electron microscopic data. In contrast, Harland and Laskey (1980) found that a range of prokaryotic templates including plasmids ColE1 and pBR322 or bacteriophages M13, φX174, and G4 replicated relatively efficiently after microinjection into *Xenopus* eggs. Furthermore, they found that the six *Hin*dIII fragments of SV40 DNA all replicated and that the fragment containing the viral replication origin did not replicate preferentially. There was a marked effect of template size, small templates supporting less incorporation than large. In the case of polyoma, the *Hin*dIII fragment containing the replication origin replicated two to three times more efficiently than the fragment that lacked the viral origin, but it was also 1.3 times larger and its preferential replication may have reflected the size difference. Similarly McTiernan and Stambrook (1980) concluded that replication of injected DNA does not require specific DNA sequences.

Although the aggregate of these reports is confusing, several consensus areas emerge. First, there appears general agreement that a specialized eukaryotic replication origin is not essential for replication in the egg. Second, none of the experiments described here excludes the possibility that initiation on the egg's own chromosomal DNA is subject to a sequence constraint that is not detected by microinjecting additional DNA templates.

Nevertheless, there is disagreement over the extent of replication of prokaryotic templates and the extent to which replication initiates at preferred sites on injected templates. Thus, McTiernan and Stambrook (1984) have mapped initiation sites on SV40 DNA after injection into eggs and concluded that initiations are scattered throughout the SV40 genome and not located at a small number of preferential sites. Mechali and Kearsey (1984) have attempted to resolve the confusion further by microinjecting the range of DNA templates examined in each of the studies referred to above using carefully matched, paired experiments. They confirmed the effect of size on efficiency of template replication and found that the amount of synthesis was proportional to the length of the template for all molecules tested between 4 and 11 kb. Above 11 kb the curve rises less steeply and below 4 kb the efficiency of replication decreases very sharply with decreasing template size. Mechali and Kearsey (1984) were unable to confirm the reports of preferential replication of cer-

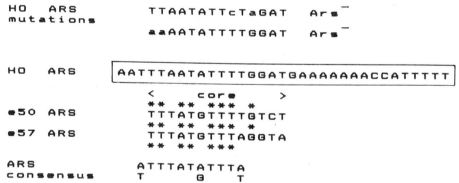

Figure 1. Evidence for high sequence specificity for autonomous DNA replication in yeast. Comparison of the *HO* ARS sequence from *S. cerevisiae* with two other ARS elements, one of which is selected from the genome of *X. laevis*, and with two mutant derivatives of the *HO* ARS which lack ARS function. The sequence of the *HO* ARS is shown boxed. It is important for autonomous replication as defined by the assay described in Kearsey (1984). The "core" region of this sequence is essential for ARS function, as shown by deletion mutations (see Kearsey 1984), and is largely composed of an AT-rich consensus sequence previously recognized as a conserved element in ARS elements (Broach et al. 1983). Also shown are the only regions of two ARS elements that show unambiguous homology to the ARS consensus. ARS activity has been located to a 49-bp region in the e50 ARS (Kearsey 1983) and to a 71-bp region in the e57 ARS. Some of the nucleotides conserved between the ARS elements in this alignment (indicated by *) have been mutated in the *HO* ARS: Two mutants are shown, each with double base substitutions, and in both cases autonomous replication is abolished (Kearsey 1984). Mutation of the conserved G (to A) alone reduces but does not abolish ARS activity (S.E. Kearsey, unpubl.) as measured by a quantitative sectoring assay similar to the one described by Hieter et al. (1985). The e50 ARS (from *S. cerevisiae* DNA) and the e57 ARS (from *X. laevis* DNA) do not appear to affect replication efficiency when injected into the *Xenopus* egg (Mechali and Kearsey 1984). This figure was compiled from data of Mechali and Kearsey (1984) and Kearsey (1984).

tain *Xenopus* DNA sequences even though they were using the same templates.

The ability of *Xenopus* eggs to initiate replication on such a wide range of templates contrasts sharply with the sequence requirements for autonomous replication of plasmids in yeast. Figure 1 summarizes results of mutagenesis studies on an autonomously replicating sequence (ARS element) which flanks the *HO* gene of *Saccharomyces cerevisiae* (Kearsey 1984). Deletion mutagenesis defines a core element that appears to be essential in *cis* for plasmid replication and is homologous to the ARS consensus of Broach et al. (1983). Furthermore, point mutants in either of the triple T clusters abolish replication (Kearsey 1984). When this assay was used to select sequences from the *Xenopus* genome that can replicate autonomously in yeast, a sequence homologous to the yeast consensus sequence was found in each of the isolates (Fig. 1 and Mechali and Kearsey 1984). This emphasizes the strict sequence constraint on replication of DNA molecules inserted into yeast and contrasts sharply with the results from *Xenopus* eggs. This contrast is further emphasized when the ARS elements isolated from the *Xenopus* genome are reintroduced into *Xenopus* eggs. The short ARS elements tested in this way replicated to the same extent as their M13 shuttle vector (Mechali and Kearsey 1984).

Experiments with injected DNA templates cannot prove that endogenous chromosomal DNA also initiates replication randomly with respect to DNA sequence. Indeed, in *Drosophila* embryos, which replicate even faster than *Xenopus* embryos, and which initiate replication at an average interval of only 7.9 kb (Blumenthal et al. 1974), random initiation of replication appears to be inadequate to ensure that replication is completed by the end of S phase. This follows from the observations that each S phase occupies about 4

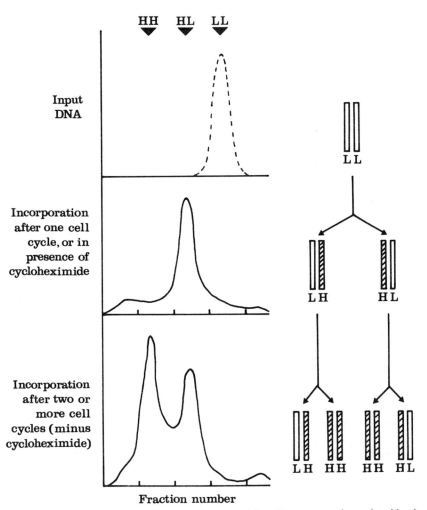

Figure 2. Semiconservative replication of bacteriophage λ DNA injected into *Xenopus* eggs is regulated by the mechanism that prevents reinitiation within a single cell cycle. The right-hand panel shows schematically the density of molecules synthesized semiconservatively from light DNA (LL) in the presence of the dense thymidine analog BrdUTP. The left-hand panel shows the density profiles observed when λ DNA is coinjected with BrdUTP and [^{32}P]dATP with or without cycloheximide, which arrests the egg's cell cycle clock. In the experiment depicted, the eggs were exposed to 50 ng/ml of the tumor promoter TPA, which increases replication efficiency about fivefold (Mechali et al. 1983). This figure was compiled from data from Mechali et al. (1983).

minutes, and replication forks proceed at 2.6 kb per minute, allowing 21 kb to be synthesized bidirectionally from each initiation in each S phase. This is sufficient for the average distance between initiations of 7.9 kb, provided that initiation occurs at a regular periodicity; but if initiation occurred at intervals that form a Poisson distribution around a mean of 7.9 kb, then 4% of the DNA would have failed to replicate by the time of mitosis. This would result in chromosome breakage and it argues that initiation is unlikely to occur at truly random intervals at least in the case of Drosophila embryos.

Although egg injection experiments do not allow the conclusion that the egg's chromosomes initiate at random DNA sequences, they force two important conclusions. First, they show that the enzymes of DNA replication in a Xenopus egg do not require specific DNA sequences to initiate replication. Second, they show that specialized eukaryotic replication origins are not required for the mechanism which ensures that DNA replicates only once in any cell cycle but does not reinitiate replication until the next cell cycle.

Coupling of DNA Replication to the Cell Cycle

The replication of injected DNA is highly regulated and tightly coupled to the egg's cell cycle (Harland and Laskey 1980). Thus, the capacity to replicate injected DNA can be induced by hormonal maturation of oocytes followed by activation by pricking. In addition, replication of injected DNA occurs only once within any cell cycle. This effect is seen most clearly when the autonomous cell cycle clock (Hara et al. 1980) is blocked by cycloheximide. Injected DNA molecules continue to initiate throughout a period equivalent to eight cell cycles, but reinitiation is completely and selectively blocked (Fig. 2). The mechanism that distinguishes replicated DNA from unreplicated DNA is not overwhelmed by injection of excess DNA and it continues to function when the overall efficiency of replication is increased by a factor of 5 by treating the egg with the tumor promoter TPA (Fig. 2 and Mechali et al. 1983). Note that the experiment illustrated in Figure 2 was performed with DNA from bacteriophage λ. Although replication in prokaryotes is not normally subject to the block that prevents reinitiation within a single cell cycle, prokaryotic DNA becomes regulated by this mechanism after it is injected into eggs. Therefore Xenopus egg injection experiments show that eukaryotic replication origins are not required either for the enzymes of replication to initiate or for the mechanism that regulates replication to only one round in any cell cycle.

There are several reasons to believe that specific DNA sequences are involved in replication initiation in other eukaryotic examples. The most convincing of these is yeast (see above). Therefore, it remains an interesting question to ask why replication is ever sequence-specific if it can occur so efficiently and under such rigorous control in a Xenopus egg without a de-

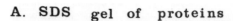
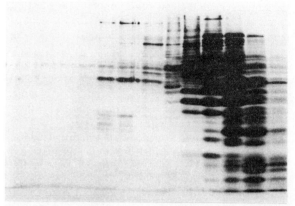
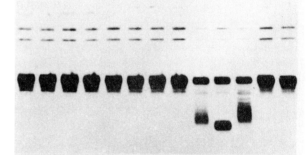
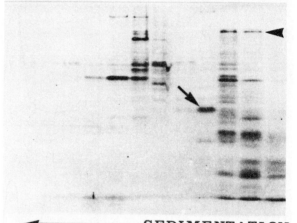

Figure 3. Cosedimentation of nucleosome assembly activity with nucleoplasmin (diagonal arrow in C). An egg supernatant was centrifuged through a 10–40% sucrose gradient and each fraction was analyzed in three ways (fractions nearest the top of the gradient are on the right in each panel): (A) Electrophoretic separation of the polypeptides in each gradient fraction; (B) assay of each fraction for insertion of superhelical turns into relaxed ^{3}H-labeled SV40 DNA; (C) as A, but each fraction was heated for 10 minutes at 80°C and centrifuged to remove precipitated proteins. One thermostable polypeptide (nucleoplasmin) copurifies with assembly activity (diagonal arrow). N1 and N2 are also heat-stable, but sediment more slowly than the main peak (horizontal arrowhead). (Reprinted, with permission, from Laskey et al. 1978a.)

tectable requirement for specific replication origins. The answer to this question is unknown, but it is possible to speculate that replication origins might have a role in building active chromatin, by defining the boundaries of active and inactive regions of chromatin. This alternative view of replication origins would argue that they are not required for DNA replication per se but for the regulation of gene expression. It would be consistent with the observation that early embryos of both *Xenopus* and *Drosophila* are transcriptionally quiescent at the time of rapid chromosome replication when replication initiates at close intervals on the DNA (McKnight and Miller 1976; Newport and Kirschner 1982). An interesting test of this alternative would be to assay for sequence specificity of replication after the midblastula transition when replication slows down and transcription initiates.

RAPID CHROMATIN ASSEMBLY IN *XENOPUS* EGGS AND EMBRYOS

To accelerate cell proliferation, it is not enough to replicate DNA rapidly, but it is also essential to assemble it into chromatin in time for mitosis. The *Xenopus* egg's capacity for chromatin assembly from purified DNA is even more remarkable than its capacity for DNA replication. The clarified homogenate from each unfertilized *Xenopus* egg can assemble nucleosomes from DNA equivalent to more than 7000 diploid nuclei in only 1 hour (Laskey et al. 1977; 1978a). This is achieved using a stored histone pool which is accumulated through oogenesis by uncoupled synthesis of histones in the absence of DNA synthesis (Adamson and Woodland 1974; Woodland and Adamson 1977). However, the conditions in which histones and DNA can self-assemble to form nucleosome cores are very limited (reviewed by Laskey and Earnshaw 1980). In particular, self-assembly is inhibited by excess histones and the *Xenopus* egg contains approximately a 7000-fold excess of histones over DNA. Fractionation of egg extracts revealed that the stored histones are bound to negatively charged factors to make negatively charged complexes. Furthermore, the complexes copurified with an assembly activity which facilitates the interaction of DNA with either the endogenous histones or with exogenous histones (Laskey et al. 1978b). Further fractionation led to the purification of nucleoplasmin (Fig. 3), an acidic, thermostable protein that facilitates nucleosome assembly in vitro by binding histones and transferring them to DNA (Laskey et al. 1978a; Earnshaw et al. 1980).

Until recently it has not been clear whether or not nucleoplasmin performs this role in vivo, as initial immunoprecipitation studies suggested that nucleoplasmin occurs only free in solution but not as a complex with histones (Krohne and Franke 1980a,b). In addition, clear evidence has emerged that part of the stored pool of histones H3 and H4 is complexed to two other acidic, nuclear polypeptides called N1 and N2 (Kleinschmidt and Franke 1982). This situation has been clarified by further work from the same laboratory (Kleinschmidt et al. 1985) that has demonstrated that two classes of acidic histone storage complex exist in *Xenopus* eggs. The first consists of histones H3 and H4 complexed to N1 and N2 as described by Kleinschmidt and Franke (1982). The second consists of nucleoplasmin bound to H2A, H2B, and two histones which coelectrophorese with H3 and H4, but have significantly different charges. The properties of nucleoplasmin and N1 and N2 are summarized in Table 2.

The reason why two classes of complex coexist remain to be elucidated. One possibility is illustrated in Figure 4. The observations that H2A and H2B are bound only to nucleoplasmin and not to N1 and N2, while H3 and H4 or proteins that electrophorese with H3 and H4, appear to be bound to either class of factor, suggests that these histone pairs might be transferred to DNA in separate histone transfer steps. This possibility is consistent with studies in somatic cells which suggest that histones H3 and H4 associate with nascent DNA before histones H2A and H2B (Worcel et al. 1978; Senshu et al. 1978). We stress, however, that Figure 4 is only a working hypothesis capable of integrating the existing data and it is not a proven pathway. Further work is required to determine the stoichiometry of the complexes and the sequence of their interactions. The possibility that histone H1 is also transferred to DNA via a similar complex also remains to be tested. Perhaps the most important question remaining is whether the histone transfer factors found

Table 2. Properties of N1, N2, and Nucleoplasmin from *Xenopus* Eggs

Properties	N1 and N2	Nucleoplasmin
Apparent molecular weight of polypeptide	110,000	30,000
Subunit number	~1-2	5
Isoelectric point	4.5-4.8	4.5-5.5
Heat stability	moderately stable	stable
Histone binding	binds H3 and H4 in vivo	binds core histones in vivo and in vitro
Concentration in *Xenopus* oocyte nucleus	4-6 mg/ml	5-8 mg/ml
Proportion of total protein in *Xenopus* oocyte nucleus	6-8%	8-10%

Compiled from data in Laskey et al. (1978a), Earnshaw et al. (1980), Krohne and Franke (1980a,b), Mills et al. (1980), Dabauvalle and Franke (1982), Kleinschmidt and Franke (1982), and Kleinschmidt et al. (1985).

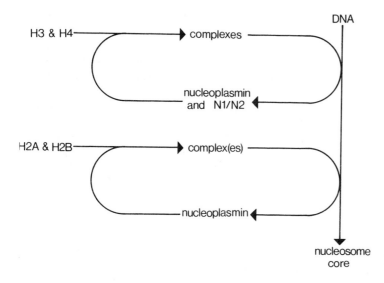

Figure 4. A hypothetical histone transfer pathway for nucleosome core assembly in *Xenopus* eggs.

in *Xenopus* eggs and oocytes are also involved in nucleosome assembly in somatic cells.

The extent to which the biochemistry of early *Xenopus* development is concerned with rapid chromosome replication can be seen from Figure 5, which shows the proteins of the *Xenopus* oocyte nucleus. N1, N2, and nucleoplasmin are three of the most abundant proteins contained in the oocyte nucleus. Nucleoplasmin is the most abundant protein, accounting for 8–10% of the total nuclear protein (Krohne and Franke 1980a,b; Mills et al. 1980), whereas N1 and N2 account for a further 6–8% (Dabauvalle and Franke 1982). Clearly, neither nucleoplasmin nor N1 and N2 are present at these high levels in somatic cells, and the extent of their presence or their involvement in nucleosome assembly in somatic cells remain to be determined.

SUMMARY

Early embryos of *X. laevis* achieve exceptional rates of DNA replication and chromatin assembly. These processes have been studied by microinjecting DNA templates into eggs or by incubating DNA in egg extracts. The egg is able to regulate replication of injected DNA without requiring specialized DNA sequences. Assembly of DNA into the nucleosome subunits of chromatin involves interaction of DNA with complexes containing two classes of acidic protein, namely nucleoplasmin and N1/N2.

ACKNOWLEDGMENTS

We are grateful to the Cancer Research Campaign for support and to Barbara Rodbard for help with the manuscript. M.M. and J.K. were supported by EMBO fellowships.

REFERENCES

Adamson, E.D. and H.R. Woodland. 1974. Histone synthesis in early amphibian development: Histone and DNA synthesis are not coordinated. *J. Mol. Biol.* **88:** 263.

Blumenthal, A.B., H.J. Kriegstein, and D.S. Hogness. 1974. The units of DNA replication in *Drosophila melanogaster* chromosomes. *Cold Spring Harbor Symp. Quant. Biol.* **38:** 205.

Broach, J.R., Y.-Y. Li, J. Feldman, M. Jayaram, J. Abraham, K.A. Nasmyth, and J.B. Hicks. 1983. Localization and sequence analysis of yeast origins of DNA replication. *Cold Spring Harbor Symp. Quant. Biol.* **47:** 1165.

Callan, H.G. 1972. Replication of DNA in the chromosomes of eukaryotes. *Philos. Trans. R. Soc. Lond. B.* **181:** 19.

Chambers, J.C., S. Watanabe, and J.H. Taylor. 1982. Dissection of a replication origin of *Xenopus* DNA. *Proc. Natl. Acad. Sci.* **79:** 5572.

Dabauvalle, M.-C. and W.W. Franke. 1982. Karyophilic proteins: Polypeptides synthesized *in vitro* accumulate in the nucleus on microinjection into the cytoplasm of amphibian oocytes. *Proc. Natl. Acad. Sci.* **79:** 5302.

Earnshaw, W.C., B.M. Honda, R.A. Laskey, and J.O. Thomas. 1980. Assembly of nucleosomes: The reaction involving *X. laevis* nucleoplasmin. *Cell* **21:** 373.

Ford, C.C. and H.R. Woodland. 1975. DNA synthesis in oocytes and eggs of *Xenopus laevis* injected with DNA. *Dev. Biol.* **43:** 189.

Gurdon, J.B., M.L. Birnstiel, and V.A. Speight. 1969. The replication of purified DNA introduced into living egg cytoplasm. *Biochim. Biophys. Acta* **174:** 317.

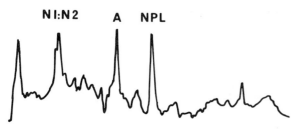

Figure 5. Three of the most abundant proteins in *Xenopus* oocyte nuclei are involved in histone binding and nucleosome assembly (namely nucleoplasmin [NPL], N1, and N2). Proteins from 10 nuclei of *Xenopus* oocytes were electrophoresed on a 15% SDS polyacrylamide gel. The peaks marked are N1 and N2, actin (A), and NPL.

HARA, K., P. TYDEMAN, and M. KIRSCHNER. 1980. A cytoplasmic clock with the same period as the division cycle in *Xenopus* eggs. *Proc. Natl. Acad. Sci.* **77:** 997.

HARLAND, R.M. and R.A. LASKEY. 1980. Regulated replication of DNA microinjected into eggs of *Xenopus laevis*. *Cell* **21:** 761.

HIETER, P., C. MANN, M. SNYDER, and R.W. DAVIS. 1985. Mitotic stability of yeast chromosomes: A colony colour assay that measures nondisjunction and chromosome loss. *Cell* **40:** 381.

HINEGARDNER, R.T. 1967. Echinoderms. In *Methods in developmental biology* (ed F.H. Wilt and N.K. Wessels), p. 139. Crowell, New York.

HINES, P.J. and R.M. BENBOW. 1982. Initiation of replication at specific origins in DNA molecules microinjected into unfertilized eggs of the frog *Xenopus laevis*. *Cell* **30:** 459.

KEARSEY, S.E. 1983. Analysis of sequences conferring autonomous replication in baker's yeast. *EMBO J.* **2:** 1571.

———. 1984. Structural requirements for the function of a yeast chromosomal replicator. *Cell* **37:** 299.

KLEINSCHMIDT, J.A. and W.W. FRANKE. 1982. Soluble acidic complexes containing histones H3 and H4 in nuclei of *Xenopus laevis* oocytes. *Cell* **29:** 799.

KLEINSCHMIDT, J.A., E. FORTKAMP, G. KROHNE, H. ZENTGRAF, and W.W. FRANKE. 1985. Co-existence of two different types of soluble histone complexes in nuclei of *Xenopus laevis* oocytes. *J. Biol. Chem.* **260:** 1166.

KROHNE, G. and W. FRANKE. 1980a. Immunological identification and localisation of the predominant nuclear protein of the amphibian oocyte nucleus. *Proc. Natl. Acad. Sci.* **77:** 1034.

———. 1980b. A major soluble acidic protein located in nuclei of diverse vertebrate species. *Exp. Cell Res.* **129:** 167.

LASKEY, R.A. and W.C. EARNSHAW. 1980. Nucleosome assembly. *Nature* **286:** 763.

LASKEY, R.A. and J.B. GURDON. 1973. Induction of polyoma DNA synthesis by injection into frog egg cytoplasm. *Eur. J. Biochem.* **37:** 467.

LASKEY, R.A., A.D. MILLS, and N.R. MORRIS. 1977. Assembly of SV40 chromatin in a cell-free system from *Xenopus* eggs. *Cell* **10:** 237.

LASKEY, R.A., B.M. HONDA, A.D. MILLS, and J.T. FINCH. 1978a. Nucleosomes are assembled by an acidic protein which binds histones and transfers them to DNA. *Nature* **275:** 416.

LASKEY, R.A., B.M. HONDA, A.D. MILLS, N.R. MORRIS, A.H. WYLLIE, J.E. MERTZ, E.M. DE ROBERTIS, and J.B. GURDON. 1978b. Chromatin assembly and transcription in eggs and oocytes of *Xenopus laevis*. *Cold Spring Harbor Symp. Quant. Biol.* **42:** 171.

MCKNIGHT, S.L. and O.L. MILLER. 1976. Ultrastructural patterns of RNA synthesis during early embryogenesis of *Drosophila melanogaster*. *Cell* **8:** 305.

MCTIERNAN, C.F. and P.J. STAMBROOK. 1980. Replication of DNA templates injected into frog eggs. *J. Cell Biol.* **87:** 45a.

———. 1984. Initiation of SV40 DNA replication after microinjection into *Xenopus* eggs. *Biochim. Biophys. Acta* **782:** 295.

MECHALI, M. and S.E. KEARSEY. 1984. Lack of specific sequence requirement for DNA replication in *Xenopus* eggs compared with high sequence specificity in yeast. *Cell* **38:** 55.

MECHALI, M., F. MECHALI, and R.A. LASKEY. 1983. Tumor promoter TPA increases initiation of replication on DNA injected into *Xenopus* eggs. *Cell* **35:** 63.

MILLS, A.D., R.A. LASKEY, P. BLACK, and E.M. DE ROBERTIS. 1980. An acidic protein which assembles nucleosomes *in vitro* is the most abundant protein in *Xenopus* oocyte nuclei. *J. Mol. Biol.* **139:** 561.

NEWPORT, J. and M. KIRSCHNER. 1982. A major developmental transition in early *Xenopus* embryos. I. Characterization and timing of cellular changes at the mid blastula stage. *Cell* **30:** 675.

NIEUWKOOP, P.D. and J. FABER. 1956. *Normal table of* Xenopus laevis *(Daudin)*. Elsevier/North-Holland, Amsterdam.

SENSHU, T., M. FUKADA, and M. OHASHI. 1978. Preferential association of newly synthesized H3 and H4 histones with newly synthesized replicated DNA. *J. Biochem.* **84:** 985.

WATANABE, S. and J.H. TAYLOR. 1980. Cloning an origin of DNA replication of *Xenopus laevis*. *Proc. Natl. Acad. Sci.* **77:** 5292.

WOODLAND, H.R. and E.D. ADAMSON. 1977. Synthesis and storage of histones during the oogenesis of *Xenopus laevis*. *Dev. Biol.* **57:** 118.

WORCEL, A., S. HAN, and M.L. WONG. 1978. Assembly of newly replicated chromatin. *Cell* **15:** 969.

ZALOKAR, M. and I. ERK. 1976. Division and migration of nuclei during early embryogenesis of *Drosophila melanogaster*. *J. Microsc. Biol. Cell.* **25:** 97.

Regulation of Histone Gene Expression

M. BUSSLINGER, D. SCHÜMPERLI, AND M.L. BIRNSTIEL
Institut für Molekularbiologie II der Universität Zürich Hönggerberg, CH-8093 Zürich, Switzerland

Regulation of Histone Gene Expression during Sea Urchin Development

The histone gene families, particularly those of the sea urchins, have proved ideal for the study of gene regulation during development. In these species, specific H1, H2A, and H2B variants of unique polypeptide sequence appear in a temporarily regulated way and are later diluted out in the chromatin, as new variant proteins emerge in rapid succession during ontogeny (Cohen et al. 1975; Newrock et al. 1978b; for review, see Maxson et al. 1983a).

Thus, egg and sperm cells have their own set of histone variants (Carroll and Ozaki 1979; for review, see von Holt et al. 1984). Upon entrance of the sperm into the egg, the sperm histones are rapidly replaced by the so-called cleavage-stage (CS) proteins (Newrock et al. 1978a), which are stored together with their mRNAs in the egg (Childs et al. 1979; Salik et al. 1981). The early histone proteins and their mRNAs predominate progressively from the eight-cell stage of the embryo up to the early blastula (Mauron et al. 1982; Maxson and Wilt 1982; Weinberg et al. 1983). Transcription of the early genes is terminated within one or two cell divisions at the hatching blastula stage (Childs et al. 1979). Concurrent with cessation of early histone gene transcription, there is a drastic restructuring of the chromatin of these genes (Bryan et al. 1983). The early mRNAs rapidly decay as a consequence of turnover (Maxson and Wilt 1982; Weinberg et al. 1983). Transcription of the late histone gene variants appears to initiate concurrently with that of the early genes, but late mRNA pools grow rapidly in size only in the mesenchyme blastula and at later developmental stages (Childs et al. 1979; Maxson et al. 1983b; Busslinger and Barberis 1985) as the number of nuclei and the stability of late mRNAs increase (Knowles and Childs 1984). The S1 mapping experiment shown in Figure 1 uses cloned cDNA and genomic histone gene variants of H2A and H2B subtypes and graphically displays the modulation of specific histone mRNA pools in the gonads and during the early sea urchin development.

Recent evidence suggests that late histone genes are a diverse family of genes that require further subdivision. Thus, some late H2A and H2B gene transcripts accumulate maximally during the gastrula, and others accumulate during the prism stage (Mohun et al. 1985; M. Busslinger, unpubl.). If one considers only the H2A and the H2B mRNA variants present in spermatocytes, eggs, and the different developmental stages up to the prism larvae, the developmental behavior of at least nine subtypes of genes (see Fig. 1) needs to be elucidated, and more diversity may as yet be discovered as later stages are investigated.

A question that remains to be resolved is to what extent, if at all, histone protein variants contribute to the establishment of cell lines and to the tissue-specific expression of other genes. The classical experiments of Grunstein and colleagues with yeast histones (Rykowski et al. 1981; Wallis et al. 1983) show that either of the two H2B variants alone guarantees all aspects of the life cycle so that, at least in this species, diversity of the H2B protein is not essential for survival in the laboratory.

Sea urchin H2A, H2B, and H1 proteins appearing during development show quite a varied and rich repertoire of sequences, and yet the primary structure of some variants is, curiously, conserved between species as if some cellular requirement demanded a specific sequence. The conserved aminoterminal pentapeptide repeats of the sperm H2B proteins are an example of this. von Holt has suggested that this basic repeated structure serves a function in condensation of the sea urchin sperm chromatin, which lacks the protamines found in vertebrates (von Holt et al. 1984). This concept is supported by the recent finding that the aminoterminal sequences of the sperm H2B proteins are rapidly phosphorylated when the sperm enters the egg, presumably facilitating chromatin decondensation and exchange of the sperm histone set with CS variants in the male pronucleus (Poccia et al. 1984; Green and Poccia 1985).

Developmental Studies Reveal Molecular Details of Posttranscriptional Regulation

The temporal pattern of synthesis of histone variants in the developing sea urchin can be traced, in the main, to changes in gene transcription (Knowles and Childs 1984) and in mRNA turnover (Maxson and Wilt 1982; Weinberg et al. 1983), but there is at least one case where nucleocytoplasmic transport of histone mRNAs (or the lack of it) is responsible for accurate timing of the synthesis of a class of histone variant proteins. The early histone variants are only synthesized during the early cleavage stages and in the blastula embryo, but the genes coding for these proteins are already activated, presumably transiently, in the sea urchin egg (Angerer et al. 1984). Relatively large pools, some 10^6 molecules (Mauron et al. 1982) of newly synthesized early histone mRNAs stay sequestered in the egg pronucleus, as detected by in situ hybridization (Venezky et al. 1981; Showman et al. 1982). As transcripts of the spacer sequences are not present, it is presumed

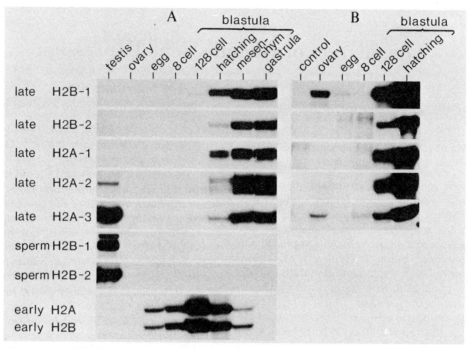

Figure 1. Developmental profile of histone mRNA accumulation during sea urchin embryogenesis (see Busslinger and Barberis 1985). S1 probes of different histone cDNA and genomic DNA clones were hybridized to 10 μg (A) or 40 μg (B) of total RNA isolated from testis, ovary, and developing embryos prior to S1 nuclease digestion and gel electrophoresis. Only the relevant part of the autoradiograph with the S1 signal of the correctly protected DNA sequence is shown for the different S1 mapping experiments. The autoradiographs in B were exposed for a longer time than those in A.

that mature histone mRNAs are stored in the nuclear compartment (DeLeon et al. 1983) which prevents their translation. During the period when early histone mRNAs are synthesized and stored in the nucleus, mRNAs for CS proteins are translated in the cytoplasm of the egg (Herlands et al. 1982). The blastula-type histone mRNAs are only released into the cytoplasm following breakdown of the nuclear membrane in the zygote (DeLeon et al. 1983) and are subsequently recruited onto the polysomes from the two-cell stage onward (Wells et al. 1981). It would be of great interest to delimit the sequences that allow such specific retention of the early histone mRNAs in the egg pronucleus.

Another interesting example documenting the diversity of molecular mechanisms that impinge on histone gene expression is the polycistronic transcription pattern of the histone gene quintet at the lampbrush chromosome stage of the amphibian oocyte. At this developmental stage, all five histone gene promoters appear to be involved in transcription initiation (Diaz and Gall 1985), but there is a pronounced lack of both transcription termination (Diaz et al. 1981) and of 3' processing distal to each gene (Diaz and Gall 1985), whereas these processes are clearly in evidence in other histone genes investigated. Instead, long transcripts are generated which, together with associated nuclear proteins, give the histone gene loci on the chromosome the typical lampbrush-loop structure. It remains to be established whether or not such polycistronic transcripts are processed to monocistronic histone mRNAs. These results are of general interest because they indicate that transcription termination and/or RNA 3' maturation are not constitutive, but are regulated processes.

Formally, these oocytes are in an extended prophase of the first meiotic division. In view of the large contribution by posttranscriptional mechanisms to regulated expression of histone genes, the question arises whether such an oocyte pattern of histone gene transcription could be found repeated at certain stages of other cell types.

Regulation of Replication Variants during the Cell Cycle of Mammalian Cells

The complexity of developmentally regulated histone genes is considerably smaller in species other than the sea urchin. In vertebrates, tissue-specific expression is restricted mainly to the family of H1 and H5 proteins, respectively, but there is also a diversity of core histone variants that can be classified into subtypes mainly according to their behavior relative to the cell cycle (for review, see Zweidler 1984).

Replacement variants of unique amino acid sequence are expressed constitutively throughout the cell cycle and in tissues that have ceased to divide. Their pattern of gene regulation is quite distinct from the classical replication variants, whose mRNA pools closely correlate with the S phase of the cell cycle. Histone gene variants with intermediate behavior have also been described (for review, see Zweidler 1984).

In tissue culture cells mRNA pools derived from replication-variant histone genes are down by a factor as large as 200 during G_1 and G_2 compared with their levels in S phase (DeLisle et al. 1983; Heintz et al. 1983; Lüscher et al. 1985). Nuclear run-on experiments suggest that the change in rate of gene transcription throughout the cell cycle makes a rather modest contribution to the cell cycle-dependent regulation of replication-type histone mRNAs, in that the number of nascent histone gene transcripts per nucleus varies by a factor of only 2-5 among G_1, S, and the G_2 phase (DeLisle et al. 1983; Sittman et al. 1983; Alterman et al. 1984; Lüscher et al. 1985). It must therefore be concluded that, as at some stages of development, posttranscriptional mechanisms make a major contribution to cell cycle regulation of histone gene expression.

To understand the molecular details of the regulated expression of histone genes during development and during the cell cycle, we must identify the relevant sequence elements and regulatory factors interacting with them.

Cell Cycle Regulation of H4 Histone mRNA in the Mouse Depends on Sequences at or near the 3' End of the Gene

Rapid progress in identifying the sequences imparting cell cycle regulation is being made by the use of cell transformation experiments. In a first such study by Lüscher et al. (1985), a temperature-sensitive cell cycle mutant of a mouse mastocytoma cell line (Zimmermann et al. 1981) was transformed with a cloned homologous H4 histone gene that was marked by insertion of a linker sequence within the coding body of the gene. The newly introduced histone gene becomes subject to much the same cell cycle regulation as the endogenous family of H4 histone genes. Cells arrested in G_1 phase at nonpermissive temperature contain less than 1/120 of the H4 mRNA of exponentially growing cells; this is true for both the endogenous and the transformed H4 gene. In experiments to delimit sequences essential for cell cycle regulation, it was found that a fusion gene consisting of the SV40 enhancer and promoter, the 5' end of the SV40 early gene, 233 bp of the 3' part of the H4 gene, and 230 bp of the adjacent spacer exhibited the same degree of cell cycle regulation as the endogenous H4 histone genes. It may therefore be concluded that sequences near or at the 3' end of the mouse H4 histone gene suffice for cell cycle regulation of histone mRNA levels in the absence of the H4 gene promoter. A detailed program of mutagenesis should reveal at what point(s) after transcription initiation cell cycle regulation acts.

A Novel Member of the U-snRNA Family Directs 3' Processing of Histone pre-mRNAs

It has been known for some time that the 3' terminus of most histone genes is flanked by two conserved sequences; i.e., by an inverted DNA repeat (which is nearly invariant in both size and sequence throughout evolution) and a GAAAGA core sequence at a short distance downstream of it (Busslinger et al. 1979; Birnstiel et al. 1985). Apart from the sequences immediately flanking the inverted DNA repeat, no other conserved sequences have been discovered so far at the 3' end. Whether any or all of these 3' conserved sequences are important for posttranscriptional regulation of histone genes is not known, but it is nevertheless interesting to note that all known cell cycle-dependent histone genes possess these terminal sequences, whereas cell cycle-independent histone genes do not (Brush et al. 1985).

The conserved DNA sequences, when transcribed into pre-mRNA, are known to serve as signals for the generation of the mature histone mRNA 3' ends by RNA processing. Mutations that destroy the capacity of the histone gene transcripts to form a terminal RNA palindrome are all processing deficient mutants (Birchmeier et al. 1984), as are partial, or total, deletion mutations of the downstream sea urchin CAAGAAAGA sequence (Georgiev and Birnstiel 1985). Insertion of short sequences between the palindrome and the downstream conserved RNA sequences are also null mutations (Georgiev and Birnstiel 1985). This suggests that the topology of the terminal histone pre-mRNA sequences is of major importance for 3' maturation (for review, see Birnstiel et al. 1985).

When injected into the mature frog oocyte, histone pre-mRNAs synthesized in vitro with an *Escherichia coli* (Birchmeier et al. 1984) or Sp6 transcription system (Georgiev et al. 1984; Krieg and Melton 1984) are readily processed; this allowed rapid screening of in vitro mutated genes and their transcripts. The sole exception so far is the H3 pre-mRNA of the sea urchin *Psammechinus miliaris* which is not processed by the frog oocyte system. However, the lesion in 3' maturation of the H3 pre-mRNA was found to be alleviated by injection of a 12S RNP "termination factor" (Stunnenberg and Birnstiel 1982). Later, the active principle was traced to a 60-nucleotide RNA (Galli et al. 1983), termed U7 RNA (Strub et al. 1984). Injection of this small RNA was found to complement the lack of 3' processing of the sea urchin H3 pre-mRNA in frog oocytes (Galli et al. 1983; Birchmeier et al. 1984; Georgiev and Birnstiel 1985).

The U7 RNA is about 58 nucleotides long and is present in the sea urchin egg at a molar concentration of only 1/30 to 1/50 (or less than 1% in mass) of that of the U1 RNA; hence it is a rare RNA species. Further characterization of the U7 RNA required its cloning as cDNA (Strub et al. 1984). With this analytical tool in hand, it has been established that the U7 RNA belongs to the Sm-type family of U-snRNPs, as U7 RNPs can be precipitated with polyclonal human and monoclonal murine antisera of the Sm serotype (gifts from E. De Robertis, University of Basel, and J. Steitz, Yale University). However, the efficiency of precipitation was less than that observed for the sea urchin U1-snRNP when monoclonal sera were used. The U7 RNA is quantitatively precipitable with antisera raised against the modified trimethyl-cap structure; hence the

U7 RNA is capped like other U-snRNPs (K. Strub and M.L. Birnstiel, unpubl.).

Characterization of the Chromosomal Genes Coding for the Sea Urchin U7 RNA

Sp6 RNA transcripts of the U7 cDNA clones were used to titrate the number of U7 RNA genes. This proved to be of the order of five genes per haploid genome. Five U7 RNA genes were found to be clustered in a 9.5-kb HindIII DNA restriction fragment that was cloned by standard procedures. As already suggested by the analysis of the cDNA clones (Strub et al. 1984), the U7 gene sequences were also found to be slightly heterogeneous in sequence, differing from one another by one or two nucleotides. Hence, the U7 RNA genes are a small, polymorphic gene family (M. De Lorenzi and M.L. Birnstiel, unpubl.). A representative U7 RNA sequence together with the 3'-flanking spacer (transcript) sequence is shown in the secondary structure model of Figure 2. The 3' two-thirds of the U7 RNA may fold into a stable stem-loop structure. The U7 RNA sequence is flanked on the 5' side by sequences that are conserved between the various copies of the U7 RNA genes and show weak homologies to the promoter of the sea urchin U1 RNA gene (Brown et al. 1985) near position −50. The 3' sequences contain, some 12 nucleotides downstream of the 3' terminus of the U-RNA (see Fig. 2), the conserved core sequence AAAGNNAGA which is highly characteristic of most, if not all, eukaryotic U-RNA genes (You et al. 1985; M. De Lorenzi and M.L. Birnstiel, unpubl.). This sequence has been shown to be necessary and sufficient for the generation of 3' ends of both human U1 and U2 RNA (Hernandez 1985; You et al. 1985).

The most interesting aspect of the U7 RNA sequence derived from an analysis of cDNA clones (Strub et al. 1984) and of the genomic DNA (M. De Lorenzi and M.L. Birnstiel, unpubl.) is the base complementarity within the 5' terminal 25 nucleotides to the conserved pre-mRNA sequences flanking the mature H3 mRNA terminus (Fig. 3). Thus, the core GAAAGA sequence of the conserved CAAGAAAGA motif and 13 out of 16 nucleotides of the conserved stem-loop structure of the histone pre-mRNA can be base-paired with the U7 RNA. In this hypothetical hybrid structure, nucleotides 7–12 of the U7 RNA and the less conserved spacer between the inverted repeat and the CAAGAAAGA sequence in the histone pre-mRNA are looped out, and it is here that the mature 3' end will ultimately be generated at a C or at an A residue, as indicated.

This model can be tested by preparing matched sets of base changes in both the pre-mRNA and the U7 RNA sequence to verify the essential contacts between them. As the U7 RNA gene can be expressed in the frog oocyte and is now available for in vitro mutagenesis experiments, we are following this line of attack with oocyte injection experiments. Another interesting question concerns the temporal pattern of expression

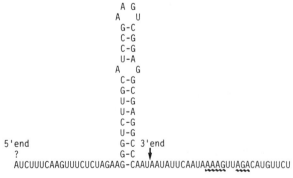

Figure 2. Sea urchin U7 RNA and 3'-flanking spacer sequences (M. de Lorenzi and M.L. Birnstiel, unpubl.). The U7 RNA and 3'-flanking sequences are derived from one of the five cloned U7 RNA genes. The 5' and 3' end of the U7 RNA as determined by analysis of the U7 cDNA clones (Strub et al. 1984) are indicated. The wavy line denotes the conserved downstream sequence motif present in most eukaryotic U-RNA genes (You et al. 1985).

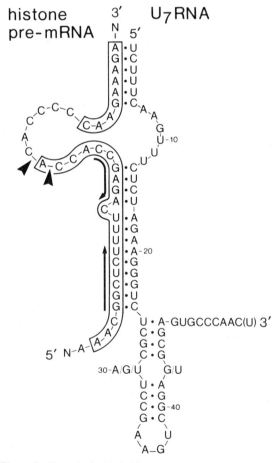

Figure 3. Hypothetical hybrid structure formed from sea urchin H3 pre-mRNA and U7 RNA during RNA 3' processing (see Strub et al. 1984). The RNA palindromic sequences are indicated by arrows. The conserved motifs of H3 pre-mRNA are boxed. (▶) Marks the 3' ends of mature sea urchin histone mRNAs.

of the U7 RNA variants in development and their role, if any, in cell cycle regulation of histone genes.

CONCLUSIONS

Sequence comparisons among a great many histone genes have identified several conserved structural features in both 5'- and 3'-flanking sequences unique to histone genes (Busslinger et al. 1979, 1980; for review, see Hentschel and Birnstiel 1981). The functional importance of many of these features was confirmed by oocyte injection and in vitro transcription experiments. In addition, roles for the bidirectional, far upstream histone modulator sequences for maximal transcription (Grosschedl and Birnstiel 1980a; Grosschedl et al. 1983), of the TATA box for selecting the cap site (Grosschedl and Birnstiel 1980b), and of the conserved 3' sequences for U7 RNA-dependent RNA maturation (for review, see Birnstiel et al. 1985) were established. The availability of tissue culture gene transfer experiments and the emergence of whole sea urchin transformation systems (Davidson et al. 1984; L. Vitelli, unpubl.) make it now possible to investigate the relevance of these sequences in directing the developmental and cell cycle regulation of histone gene expression.

ACKNOWLEDGMENTS

We would like to thank S. Oberholzer for operating the word processor and F. Ochsenbein for preparing the graphs. This work was supported by the State of Zürich and by the Swiss National Science Foundation, grants No. 3.484.83 and 3.542.83.

REFERENCES

ALTERMAN, R.M., S. GANGULY, D.H. SCHULZE, W.F. MARZLUFF, C.L. SCHILDKRAUT, and A.I. SKOULTCHI. 1984. Cell cycle regulation of mouse H3 histone mRNA metabolism. *Mol. Cell. Biol.* **4:** 123.

ANGERER, L.M., D.V. DELEON, R.C. ANGERER, R.M. SHOWMAN, D.E. WELLS, and R.A. RAFF. 1984. Delayed accumulation of maternal histone mRNA during sea urchin oogenesis. *Dev. Biol.* **101:** 477.

BIRCHMEIER, C., D. SCHÜMPERLI, G. SCONZO, and M.L. BIRNSTIEL. 1984. 3' Editing of mRNAs: Sequence requirements and involvement of a 60-nucleotide RNA in maturation of histone mRNA percursors. *Proc. Natl. Acad. Sci.* **81:** 1057.

BIRNSTIEL, M.L., M. BUSSLINGER, and K. STRUB. 1985. Transcription termination and 3' processing: The end is in site! *Cell* **41:** 349.

BROWN, D.T., G.F. MORRIS, N. CHODCHOY, C. SPRECHER, and W.F. MARZLUFF. 1985. Structure of the sea urchin U1 RNA repeat. *Nucleic Acids Res.* **13:** 537.

BRUSH, D., J.B. DODGSON, O.-R. CHOI, P.W. STEVENS, and J.D. ENGEL. 1985. Replacement variant histone genes contain intervening sequences. *Mol. Cell. Biol.* **5:** 1307.

BRYAN, P.N., J. OLAH, and M.L. BIRNSTIEL. 1983. Major changes in the 5' and 3' chromatin structure of sea urchin histone genes accompany their activation and inactivation in development. *Cell* **33:** 843.

BUSSLINGER, M. and A. BARBERIS. 1985. Synthesis of sperm and late histone cDNAs of the sea urchin with a primer complementary to the conserved 3' terminal palindrome: Evidence for tissue-specific and more general histone gene variants. *Proc. Natl. Acad. Sci.* **82:** 5676.

BUSSLINGER, M., R. PORTMANN, and M.L. BIRNSTIEL. 1979. A regulatory sequence near the 3' end of sea urchin histone genes. *Nucleic Acids Res.* **6:** 2997.

BUSSLINGER, M., R. PORTMANN, J.C. IRMINGER, and M.L. BIRNSTIEL. 1980. Ubiquitous and gene-specific regulatory 5' sequences in sea urchin histone DNA clone coding for histone protein variants. *Nucleic Acids Res.* **8:** 957.

CARROLL, A.G. and H. OZAKI. 1979. Changes in the histones of the sea urchin *Strongylocentrotus purpuratus* at fertilization. *Exp. Cell Res.* **119:** 307.

CHILDS, G., R. MAXSON, and L.H. KEDES. 1979. Histone gene expression during sea urchin embryogenesis: Isolation and characterization of early and late messenger RNAs of *Strongylocentrotus purpuratus* by gene-specific hybridization and template activity. *Dev. Biol.* **73:** 153.

COHEN, L.H., K.M. NEWROCK, and A. ZWEIDLER. 1975. Stage-specific switches in histone synthesis during embryogenesis of the sea urchin. *Science* **190:** 994.

DELEON, D.V., K.H. COX, L.M. ANGERER, and R.C. ANGERER. 1983. Most early-variant histone mRNA is contained in the pronucleus of sea urchin eggs. *Dev. Biol.* **100:** 197.

DELISLE, A.J., R.A. GRAVES, W.F. MARZLUFF, and L.F. JOHNSON. 1983. Regulation of histone mRNA production and stability in serum-stimulated mouse 3T6 fibroblasts. *Mol. Cell. Biol.* **3:** 1920.

DIAZ, M.O. and J.G. GALL. 1985. Giant readthrough transcription units at the histone loci on lampbrush chromosomes of the newt notophthalmus. *Chromosoma* **92:** 243.

DIAZ, M.O., G. BARSACCHI-PILONE, K.A. MAHON, and J.G. GALL. 1981. Transcripts from both strands of a satellite DNA occur on lampbrush chromosome loops of the newt notophthalmus. *Cell* **24:** 649.

GALLI, G., H. HOFSTETTER, H.G. STUNNENBERG, and M.L. BIRNSTIEL. 1983. Biochemical complementation with RNA in the *Xenopus* oocyte: A small RNA is required for the generation of 3' histone mRNA termini. *Cell* **34:** 823.

GEORGIEV, O. and M.L. BIRNSTIEL. 1985. The conserved CAAGAAAGA spacer sequence is an essential element for the formation of 3' termini of the sea urchin H3 histone mRNA by RNA processing. *EMBO J.* **4:** 481.

GEORGIEV, O., J. MOUS, and M.L. BIRNSTIEL. 1984. Processing and nucleo-cytoplasmic transport of histone gene transcripts. *Nucleic Acids Res.* **12:** 8539.

GREEN, G.R. and D.L. POCCIA. 1985. Phosphorylation of sea urchin sperm H1 and H2B histones precedes chromatin decondensation and H1 exchange during pronuclear formation. *Dev. Biol.* **108:** 235.

GROSSCHEDL, R. and M.L. BIRNSTIEL. 1980a. Spacer DNA sequences upstream of the TATAAATA sequence are essential for promotion of H2A histone gene transcription in vivo. *Proc. Natl. Acad. Sci.* **77:** 7102.

———. 1980b. Identification of regulatory sequences in the prelude sequences of an H2A histone gene by the study of specific deletion mutants in vivo. *Proc. Natl. Acad. Sci.* **77:** 1432.

GROSSCHEDL, R., M. MÄCHLER, U. ROHRER, and M.L. BIRNSTIEL. 1983. A functional component of the sea urchin H2A gene modulator contains an extended sequence homology to a viral enhancer. *Nucleic Acids Res.* **11:** 8123.

HEINTZ, N., H.L. SIVE, and R.G. ROEDER. 1983. Regulation of human histone gene expression: Kinetics of accumulation and changes in the rate of synthesis and in the half-lives of individual histone mRNAs during the HeLa cell cycle. *Mol. Cell. Biol.* **3:** 539.

HENTSCHEL, C.C. and M.L. BIRNSTIEL. 1981. The organization and expression of histone gene families. *Cell* **25:** 301.

HERLANDS, L., V.G. ALLFREY, and D. POCCIA. 1982. Translational regulation of histone synthesis in the sea urchin *Strongylocentrotus purpuratus*. *J. Cell Biol.* **94:** 219.

HERNANDEZ, N. 1985. Formation of the 3' end of U1 snRNA is directed by a conserved sequence located downstream of the coding region. *EMBO J.* **4:** 1827.

KRIEG, P.A. and D.A. MELTON. 1984. Formation of the 3' end of histone mRNA by post-transcriptional processing. *Nature* 308: 203.

KNOWLES, J.A. and G.J. CHILDS. 1984. Temporal expression of late histone messenger RNA in the sea urchin *Lytechinus pictus*. *Proc. Natl. Acad. Sci.* 81: 2411.

LÜSCHER, B., C. STAUBER, R. SCHINDLER, and D. SCHÜMPERLI. 1985. Faithful cell-cycle regulation of a recombinant mouse histone H4 gene is controlled by sequences in the 3' terminal part of the gene. *Proc. Natl. Acad. Sci.* 82: 4389.

MAURON, A., L. KEDES, B.R. HOUGH-EVANS, and E.H. DAVIDSON. 1982. Accumulation of individual histone mRNAs during embryogenesis of the sea urchin *Strongylocentrotus purpuratus*. *Dev. Biol.* 94: 425.

MAXSON, R.E. and F.H. WILT. 1982. Accumulation of the early histone messenger RNAs during the development of *Strongylocentrotus purpuratus*. *Dev. Biol.* 94: 435.

MAXSON, R., R. COHN, L. KEDES, and T. MOHUN. 1983a. Expression and organization of histone genes. *Annu. Rev. Genet.* 17: 239.

MAXSON, R., T. MOHUN, G. GORMEZANO, G. CHILDS, and L. KEDES. 1983b. Distinct organizations and patterns of expression of early and late histone gene sets in the sea urchin. *Nature* 301: 120.

MOHUN, T., R. MAXSON, G. GORMEZANO, and L. KEDES. 1985. Differential regulation of individual late histone genes during development of the sea urchin (*Strongylocentrotus purpuratus*). *Dev. Biol.* 108: 491.

NEWROCK, K.M., C.R. ALFAGEME, R.V. NARDI, and L.H. COHEN. 1978a. Histone changes during chromatin remodeling in embryogenesis. *Cold Spring Harbor Symp. Quant. Biol.* 42: 421.

NEWROCK, K.M., L.H. COHEN, M.B. HENDRICKS, R.J. DONNELLY, and E.S. WEINBERG. 1978b. Stage-specific mRNAs coding for subtypes of H2A and H2B histones in the sea urchin embryo. *Cell* 14: 327.

POCCIA, D., T. GREENOUGH, G.R. GREEN, E. NASH, J. ERICKSON, and M. GIBBS. 1984. Remodeling of sperm chromatin following fertilization: Nucleosome repeat length and histone variant transitions in the absence of DNA synthesis. *Dev. Biol.* 104: 274.

RYKOWSKI, M., J. WALLIS, J. CHOE, and M. GRUNSTEIN. 1981. Histone H2B subtypes are dispensable during the yeast life cycle. *Cell* 25: 477.

SALIK, J., L. HERLANDS, H.P. HOFFMANN, and D. POCCIA. 1981. Electrophoretic analysis of the stored histone pool in unfertilized sea urchin eggs: Quantification and identification by antibody binding. *J. Cell Biol.* 90: 385.

SHOWMAN, R.M., D.E. WELLS, D.A. ANSTROM, and R.A. RAFF. 1982. Message-specific sequestration of maternal histone mRNA in the sea urchin egg. *Proc. Natl. Acad. Sci.* 79: 5944.

SITTMAN, D.B., R.A. GRAVES, and W.F. MARZLUFF. 1983. Histone mRNA concentrations are regulated at the level of transcription and mRNA degradation. *Proc. Natl. Acad. Sci.* 80: 1849.

STRUB, K., G. GALLI, M. BUSSLINGER, and M.L. BIRNSTIEL. 1984. The cDNA sequences of the sea urchin U7 small nuclear RNA suggest specific contacts between histone mRNA precursor and U7 RNA during RNA processing. *EMBO J.* 3: 2801.

STUNNENBERG, H.G. and M.L. BIRNSTIEL. 1982. Bioassay for components regulating eukaryotic gene expression: A chromosomal factor involved in the generation of histone mRNA 3' termini. *Proc. Natl. Acad. Sci.* 79: 6201.

VENEZKY, D.L., L.M. ANGERER, and R.C. ANGERER. 1981. Accumulation of histone repeat transcripts in the sea urchin egg pronucleus. *Cell* 24: 385.

VON HOLT, C., P. DE GROOT, S. SCHWAGER, and W.F. BRANDT. 1984. The structure of sea urchin histones and considerations on their function. In *Histone genes and histone gene expression* (ed. G. Stein et al.), p. 65. Wiley, New York.

WALLIS, J.W., M. RYKOWSKI, and M. GRUNSTEIN. 1983. Yeast histone H2B containing large amino terminus deletions can function in vivo. *Cell* 35: 711.

WEINBERG, E.S., M.B. HENDRICKS, K. HEMMINKI, P.E. KUWABARA, and L.A. FARRELLY. 1983. Timing and rates of synthesis of early histone mRNA in the embryo of *Strongylocentrotus purpuratus*. *Dev. Biol.* 98: 117.

WELLS, D.E., R.M. SHOWMAN, W.H. KLEIN, and R.A. RAFF. 1981. Delayed recruitment of maternal histone H3 mRNA in sea urchin embryos. *Nature* 292: 477.

YOU, C.-J., M. ARES, JR., and A.M. WEINER. 1985. Sequences required for 3' end formation of human U2 small nuclear RNA. *Cell* 42: 193.

ZIMMERMANN, A., J.C. SCHAER, J. SCHNEIDER, P. MOLO, and R. SCHINDLER. 1981. Dominant versus recessive behavior of a cold- and a heat-sensitive mammalian cell cycle variant in heterokaryons. *Somat. Cell Genet.* 7: 591.

ZWEIDLER, A. 1984. Core histone variants of the mouse: Primary structure and differential expression. In *Histone genes and histone gene expression* (ed. G. Stein et al.), p. 339. Wiley, New York.

A Comparison of Several Lines of Transgenic Mice Containing the SV40 Early Genes

T. VAN DYKE, C. FINLAY, AND A.J. LEVINE
Department of Molecular Biology, Princeton University, Princeton, New Jersey 08544

Inoculation of simian virus 40 (SV40) into newborn hamsters can result in a broad spectrum of tumors derived from many tissue types (Tooze 1981). Subcutaneous injections of virus usually result in fibrosarcomas (Eddy et al. 1961); intracranial inoculations may produce gliomas, papillomas of the choroid plexus, and ependynomas (Gerber and Kirschstein 1962); and intravenous injections have given rise to leukemias, lymphomas, osteosarcomas, and reticulum cell sarcomas (Diamandopoulos 1972, 1978). In mice, attempts to induce tumors with SV40 have usually failed (see Tooze 1981). However, recent experiments suggest that small tumors may arise under special conditions of viral persistence and an altered host immune response (Ambramczuk et al. 1984). On the other hand, when the SV40 early-region genes were introduced into the germ line of mice, 65-90% of the transgenic mice developed tumors of a specific tissue type, the choroid plexus (Brinster et al. 1984 and pers. comm.). In these transgenic mice the expression of the SV40 large T antigen was generally limited to the choroid plexus tumors. Some mice expressed a low level of T antigen in the kidney and thymus, and cortical cysts of the kidney and thymic hyperplasia were observed occasionally (Brinster et al. 1984).

Transgenic mice carrying deletions in the SV40 early region have served to demonstrate that the large T antigen, but not the small t antigen, is required for the production of choroid plexus papillomas (Palmiter et al. 1985; Table 1). It also appears that the SV40 enhancer region, the 72-bp repeat (Gruss et al. 1981), plays a role in the tissue-specific induction of choroid plexus papillomas but it is not essential for tumorigenesis (Palmiter et al. 1985). Hybrid gene constructs with the T antigen gene have been used to demonstrate the general role of 5'-flanking sequences in controlling tissue specificity. Transgenic mice that carry the upstream sequences of the insulin gene (Hanahan 1985; Hanahan and Alpert, this volume) or the elastase I gene (Ornitz et al., this volume) linked to the SV40 T antigen coding sequence express the T antigen exclusively in the pancreas, and these mice develop pancreatic tumors.

The initial three experiments in which the SV40 early genes were introduced into embryos (Brinster et al. 1984) produced 25 mice that had detectable SV40 sequences present in tail DNA preparations. Sixteen of these mice died within a 3- to 6-month period of time. Several of them had confirmed tumors of the choroid plexus. However, nine of the 25 mice never developed tumors (Brinster et al. 1984). Representative lines of mice have been established from founder animals exhibiting both phenotypes (tumor positive or negative) and in each case the phenotype of the parent has been preserved in the offspring that carry SV40 DNA. Since the transgenes as well as the genetic background of the mice were similar or identical, the position of integration of the plasmid DNA could contribute to the difference in phenotypes. To study this phenomenon, the lines of transgenic mice exhibiting each phenotype have been compared for the timing of expression of the SV40 large T antigen. The brains of mice destined to develop tumors of the choroid plexus express low levels of SV40 tumor antigen at times prior to obvious tumor formation. In a mouse that will get a tumor and die at 15-16 weeks of life (SV11 mice), the SV40 T antigen is readily detectable by at least 8 weeks after birth. In contrast, during the same time interval, mice that carry SV40 DNA but do not have tumors (419 mice) do not express the SV40 T antigen in brain tissue. The differences in the expression of SV40 T antigen in SV11 mice (tumor positive) and 419 mice (tumor negative) could also be observed in cells derived from those mice and grown in culture, thus providing a model system for studying the basis of these differences. Finally, of 52 transgenic mice with SV40 DNA in the 419 line, a single mouse developed a T antigen-positive papilloma of the choroid plexus at 1.5 years of age. This animal represents a reversion of the tumor-minus phenotype. Southern blot analysis of the SV40 DNA from this tumor did not detect any rearrangements, translocations, or alterations at or near the integration site. This analysis did, however, detect differences in the degree of CpG methylation of cytosine residues in tumor DNA when compared with DNA from either tissues of a healthy 419 mouse or unaffected tissues of the same mouse.

Table 1. Tumor Incidence in Transgenic Mice Which Carry SV40 DNA

	Lines of mice			
	427[a]	SV11	419	SV8[b]
Number of generations	8	6	5	6
Total mice	76	80	101	117
Mice With SV40 DNA	38	35	52	59
Mice with tumors	33	26	1	0
Healthy mice sacrificed	5	9	9	8

[a]Experimental data reviewed from Brinster et al. (1984).
[b]Data combined for two separate lines, SV8-30 and SV8-33.

MATERIALS AND METHODS

Plasmid constructions and microinjection. Plasmids pSV-MK (Brinster et al. 1984), pSV-8, and pSV-11 (Colby and Shenk 1982) have been described. Microinjection of the linearized plasmids was as previously described (Brinster et al. 1984; Palmiter et al. 1985).

Development of mouse lines. The 427 founder transgenic mouse and the initial pedigree derived from this female have been described (Brinster et al. 1984). Two 419 male mice were obtained from R. Brinster and bred. In every case a 419 animal heterozygous for the SV40 DNA was crossed to a C57BL/6J inbred mouse (from Jackson Laboratory, Bar Harbor, Maine). Similar crosses have been carried out with mice containing the plasmid constructions pSV-11 (SV11 mice) and pSV-8 (SV8-30 and SV8-33 mice were derived from two independent founder animals that contained pSV-8 DNA at different integration sites). A summary of the number of generations and number of mice involved in these crosses is provided in Table 1.

Histological procedures. Histology was carried out by Dr. Douglas Miller in the pathology laboratory at the New Jersey–Rutgers School of Medicine. Tissues or tumors were fixed in 4% formaldehyde in 0.1% phosphate buffer (pH 7.2). Tissues were sectioned and stained by standard techniques.

Western blot analysis. Western blot analysis to detect the SV40 large T antigen was carried out as described by Brinster et al. (1984). T antigen present in whole tissue extracts was immunoprecipitated with Pab412, as described by Linzer et al. (1979). The resulting immunoprecipitates were then solubilized, and half of the sample was electrophoresed through a discontinuous 10% polyacrylamide gel. Proteins were electrophoresed from the gel onto nitrocellulose which was then incubated first with Pab412 and then with ^{125}I-labeled Protein A (New England Nuclear Corp.) to detect T antigen.

Isolation and analysis of DNA. Transgenic animals were detected by slot hybridization of tail nucleic acids to nick-translated SV40 viral or pSV-MK plasmid DNA. To isolate tail nucleic acid, 1/2-inch tail sections were incubated overnight at 55°C in 0.7 ml of 50 mM Tris-HCl (pH 8.0), 0.1 M EDTA, 0.1 M NaCl, 1% SDS, and 500 µg/ml Proteinase K. Samples were extracted once with phenol (equilibrated in 10 mM Tris-HCl [pH 8.0] and 1 mM EDTA [TE]), extracted once with a 1:1 mixture of phenol and chloroform/isoamyl alcohol (24:1) and once with chloroform/isoamyl alcohol (24:1). Nucleic acids were recovered by precipitation from 70% ethanol and were resuspended in TE. For slot blot analysis, 2 µg were added to 500 µl 2 M NaCl and 0.1 N NaOH. The sample was heated at 100°C for 3 minutes and filtered onto a nitrocellulose sheet.

Nucleic acid was isolated from other tissues as described by Brinster et al. (1984). Southern analysis was performed after electrophoresing 10 µg of digested DNA in a 0.8% agarose gel. Nick-translated probes ($>10^8$ cpm/µg) were as indicated in figure legends. Hybridizations of slot and Southern blots were carried out for 16–24 hours at 68°C in 5 × SSPE, 5 × Denhardt's reagent, 0.1% SDS, 100 µg/ml salmon sperm DNA, and 5×10^6 cpm/ml probe. Blots were washed twice in 2 × SSPE and 0.1% SDS and twice in 0.2 × SSPE and 0.1% SDS at 68°C.

Cell culture. Tissues were dissected from the mice and minced finely. The fragments were treated with trypsin at 37°C for 40 minutes, washed with Dulbecco's modified Eagles medium, and then plated into culture flasks in the same medium containing 15% fetal bovine serum and antibiotics. Following the first passage, cell cultures were routinely seeded at a standard density of 0.5×10^4 to 1.0×10^4 cells/cm² and passaged at confluency. Generation number (population doubling) was calculated as described by Cristofalo and Charpentier (1980).

Indirect immunofluorescence. Cells were plated on glass coverslips at a density of 1.0×10^4 cells/cm². Twenty-four hours after seeding, the cells were fixed in 1.5% formaldehyde in phosphate-buffered saline (PBS) for 20 minutes and solubilized in 0.1% Triton X-100 in PBS for 1 minute. The cells were incubated with Pab 412 for 30–60 minutes at 37°C, washed extensively in PBS, and then incubated with fluorescein isothiocyanate (FITC)-conjugated goat anti-mouse antiserum for 30 minutes at 37°C. After further washing in PBS, coverslips were mounted onto glass slides and examined in a fluorescence microscope.

RESULTS

Characteristics of the SV40-Transgenic Mouse Lines

Some transgenic mice with SV40 DNA in the germ line produce tumors of the choroid plexus while others have no apparent phenotype (Brinster et al. 1984). To investigate the nature of these differences, mice were obtained as representative examples of each of these two phenotypes. In each case a heterozygous (SV40 +/−) founder mouse was bred to a normal C57BL/6J mouse. The offspring were tested for the genetic transmission of SV40 DNA as described in Methods. Progeny that were SV40 DNA positive were then bred to normal C57BL/6J mice and the process was repeated. Four different lines of transgenic mice containing SV40 wild-type or mutant DNA in the germ line were established. Table 1 reviews the progress with each of these transgenic lines. Over several generations the 427 and SV11 mouse lines have transmitted the SV40 DNA in an expected Mendelian ratio (50%) and, without exception, there has been cosegregation of SV40 DNA and choroid plexus papillomas. Conversely, while the mouse lines 419 and SV8 transmitted SV40 DNA in an expected Mendelian fashion (50%), only one mouse in the 419 line developed a tumor (a T antigen-positive choroid plexus papilloma). The SV8 mice carry an

SV40 deletion mutation which encodes only the amino-terminal 311 amino acids of large T antigen, but leaves the small t sequence intact. Of nine independently derived SV8 transgenic mice, none have developed tumors (Palmiter et al. 1985) and none of their progeny have developed tumors (Table 1). Thus, it is likely the SV40 large T antigen gene product is required for the production of choroid plexus papillomas. The 419 mice apparently contain T antigen genes that are intact. An SV40 T antigen of the normal size (in SDS-polyacrylamide gels) and immunological properties is produced in the single 419 tumor (see Fig. 3 below) and in a cell line derived from a 419 mouse kidney (data not shown). Since a tumor can be produced and a transformed cell line can be obtained from 419 mice, this suggests that the T antigen gene is normal and that it fails to produce a tumor of the choroid plexus in most cases for reasons other than it carries a mutation.

Another interesting difference between these lines of mice is found when comparing two tumor-positive lines such as SV11 and 427. The time of death that results from a choroid plexus tumor in these lines is different (Table 2). Over six generations, SV11 mice have died with a mean time of 111 days, whereas over four generations, 427 mice have died at a mean time of 156 days. It is possible that these differences reflect the position of the SV40 genes in the mouse genome.

The Expression of SV40 T Antigen in SV11 and 419 Mice

The establishment of a colony of transgenic mice that contained SV40 DNA and developed tumors in a defined time interval (SV11 mice) permitted an analysis of the expression of the SV40 large T antigen as a function of time. When the tissues of tumor-bearing SV11 mice were examined for T antigen expression, T antigen levels were high in the tumors of all animals and lower amounts were detected in kidneys of some SV11 mice. T antigen was not detected in any other tissues (our unpublished results). To determine the timing of this tissue-specific expression, and to see if 419 mice expressed T antigen within a similar time frame, brains and kidneys of SV11 mice and 419 mice at a variety of ages were assayed for T antigen expression. Soluble protein extracts were prepared from these tissues and T antigen was immunoprecipitated and detected by Western blotting. The results of this experiment are presented in Figure 1. In the brains of SV11 mice, tumors (T, gross pathology) were detectable by 13.5 weeks and high levels of T antigen were present. Lower levels of SV40 T antigen were observed in brain tissue of SV11 mice as early as 8 weeks after birth when no (0, gross pathology) tumors were detectable by inspection of the tissue through a dissecting microscope. Kidneys of SV11 mice also expressed low levels of SV40 T antigen coincident with the onset of a tumor in the brain. Prior to this time, however, SV40 T antigen was not detected in kidneys by Western blotting procedures. The brain and kidney tissues of 419 mice (no tumors observed) were also negative for SV40 T antigen, as indicated in Figure 1. Longer exposures of this autoradiogram show very low levels of SV40 T antigen in brain tissue from older 419 mice (32 and 47 weeks of age) but not in kidney tissue at that time. Whether this is a general feature of the older 419 mice is presently under study.

Expression of the 419 and SV11 Phenotypes in Cell Culture

SV11 mice express SV40 T antigen in a tissue-specific fashion as early as 8 weeks after birth (Fig. 1), whereas 419 mice do not express those levels of SV40 T antigen at the same age. These phenotypic differences are reflected by similar observations with cells in culture that have been derived from SV11 and 419 mice. When tumor tissue from the choroid plexus of SV11 mice was placed in cell culture, the cells began growth quite rapidly and were T antigen-positive by indirect immunofluorescence following the first passage in culture (Fig. 2). Similarly, when brain tissue was obtained from SV11 mice prior to any detectable tumor, but at

Table 2. Time of Death of SV11 and 427 Mice

Mice	Generation	Mean age at death days	S.D.	Number of mice
SV11 Line	1	120		1
	2	109	30	2
	3	104	14	5
	4	114	6	6
	5	111	17	6
	6	104	12	6
	Average[b]	110	Total	26
427 Line[a]	1	155		1
	2	165		1
	3	153	11	5
	4	152	9	17
	Average[b]	156	Total	24

[a] Values from Brinster et al. (1984).
[b] Calculated as average of mean ages of death over all generations.

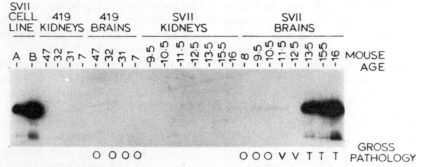

Figure 1. Detection of SV40 large T antigen in brains and kidneys of 419 and SV11 mice. SV40 large T antigen was isolated by immunoprecipitation from whole brains or kidneys of 419 or SV11 mice of different ages or from a cell line established from an SV11 tumor. The Western blotting procedure was used to detect T antigen. Lanes A and B contain T antigen from 1×10^6 and 2×10^6 cells, respectively, of an SV11 cell line in culture. The other lanes contain T antigen extracted from whole tissues. Kidneys and brains were dissected from the same mouse. Ages of the mice (in weeks) are indicated at the top of each lane and gross pathology is denoted at the bottom of brain lanes by 0 (no pathology observed), V (increased vascularization in one or more ventricles), or T (visible choroid plexus tumor). The X-ray film was exposed for 18 hr.

the time when T antigen was detected in the brain (8 weeks), the cells derived from this tissue were rapidly proliferating and were SV40 T antigen positive at the earliest time measured (Fig. 2). In contrast, brain cells from a 419 mouse or a control mouse with no SV40 DNA in the germ line (C57BL/6J) took much longer times in culture (80–100 days) before slowly proliferating fibroblast-like cells grew out. Whenever tested, in this experiment, 419-derived cells were SV40 T antigen negative (Fig. 2). Thus, the different phenotypes in vivo are reflected by the brain cells derived from these mice when grown in vitro.

A Papilloma of the Choroid Plexus in the 419-2 Mouse

At the age of 1.5 years, a single 419 mouse (419-2) developed a papilloma of the choroid plexus. Western blot analysis of T antigen levels in the tumor tissue, spleen, heart, kidney, liver, and brain tissue demonstrated T antigen in only the tumor tissue (Fig. 3). A related 419 mouse (419-22; 7 months of age) failed to express detectable levels of T antigen in similar tissues tested (Fig. 2). The brother of 419-2 (419-13), born in the same litter, is healthy at 1.7 years of age. This sug-

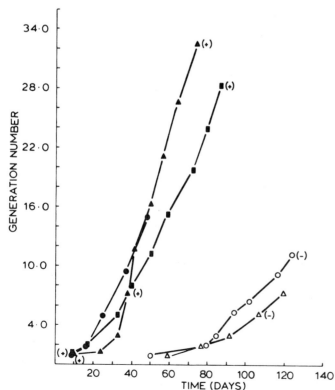

Figure 2. Analysis of 419 and SV11 brain cell growth in culture. Brain cultures from 419 (△), SV11 (■), and control (○) mice and tumors from SV11 (▲, ■) mice were initiated and passaged as described in Methods. T antigen expression was assayed by indirect immunofluorescence and the presence (+) or absence (−) of nuclear fluorescence in the culture is indicated. Generation number was calculated as described in Methods.

Figure 3. Western blot analysis of expression of SV40 large T antigen in the tumor of mouse 419-2. Tissues from two 419 mice (#2 and #22) were analyzed for the presence of T antigen as described in the legend to Fig. 1. 419-2 had a choroid plexus papilloma at 1.5 years of age and 419-22 was healthy at 7 months of age when sacrificed. Tissues are indicated as follows: (Br) brain; (Li) liver; (Ki) kidney; (He) heart; (Sp) spleen; (Tu) tumor; (Th) thymus. T antigen isolated from a 427 choroid plexus tumor was electrophoresed in the same gel as a reference. Markers are ^{14}C-labeled phosphorylase a and bovine serum albumin. The X-ray film was exposed for 14 hr.

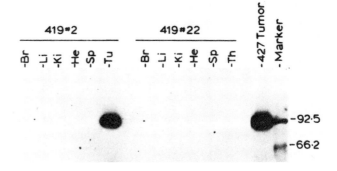

gests that the T antigen-positive papilloma in 419-2 was the result of a rare event (1 out of 52 mice to date) rather than another example of tumors arising at different times of age in a mouse family (Table 2). If the site of integration of SV40 DNA in 419 mice results in little or no expression of T antigen and, therefore, the absence of tumors, then a rare transposition or mutation event might give rise to T antigen expression and an SV40-induced tumor. To test this possibility, the DNA was extracted from the tumor tissue and kidney tissue of 419-2 and the kidney tissue of a related mouse 419-22 (negative for T antigen expression; Fig. 3). The DNA was digested with *Eco*RI, which cuts once, or *Bam*HI, which cuts twice in the pSV-MK sequences in 419 mice. The DNA was then analyzed by Southern blotting using pSV-MK[^{32}P]DNA as a probe. Figure 4 presents the autoradiogram of this Southern blot. The pSV-MK DNA in 419 animals appears to be located at a single integration site (only two flanking DNA fragments are observed) as a tandem repeated sequence of approximately 8–10 copies. Cutting the DNA with the enzymes *Bam*HI, *Eco*RI (Fig. 4), *Msp*I (Fig. 5), *Hin*fI, or *Sph*I (T. Van Dyke and A. J. Levine, unpubl.) failed to show any differences between DNA from tumor tissue and DNA from 419-2 or 419-22 kidney tissue. There is no evidence, therefore, for transposition, amplification, or mutations affecting pSV-MK and the surrounding sequences that are detectable by this analysis.

A difference between the DNA obtained from the 419-2 tumor and the DNAs from brain, liver, and kidney from 419-2 or 419-22 was observed, however, when the CpG cytosine methylation sites were examined using the *Hpa*II restriction enzyme. Comparison of the DNA from several tissues derived from 419 mice (419-2, 419-22) using the restriction enzymes *Msp*I (cuts at methylated CpG) and *Hpa*II (fails to cut at methylated CpG) indicates that the *Hpa*II sites near the SV40 DNA are extensively methylated in the 419 mouse family. The DNA extracted from the tumor tissue, however, is hypomethylated when compared with any other source of 419 DNA (Fig. 5).

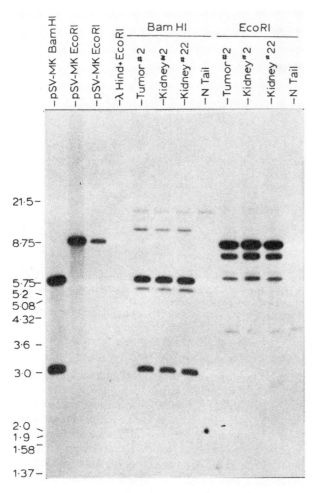

Figure 4. Analysis of pSV-MK DNA and the pattern of integration in 419-2 and 419-22 DNA. Southern blot analysis was performed on *Bam*HI- or *Eco*RI-digested 419-2 tumor and kidney and 419-22 kidney DNA, as described in Methods. The first four lanes from left to right contain markers as follows: eight copies of pSV-MK digested with *Bam*HI; seven copies of pSV-MK digested with *Eco*RI; one copy of pSV-MK digested with *Eco*RI; and λ DNA digested with *Hin*dIII and *Eco*RI. Tail DNA from nontransgenic mice digested with *Bam*HI or *Eco*RI was used as a negative control. The radioactive probe was nick-translated pSV-MK DNA. The X-ray film was exposed for 14 hr.

DISCUSSION

Transgenic mice have been produced that carry SV40 DNA in their germ line. Some of these mice (427, SV11) get tumors at characteristic times after birth and die within a remarkably reproducible time frame. Other families of transgenic mice (419), which contain apparently wild-type SV40 DNA, do not get tumors and express no detectable T antigen in the same time period. A low level of T antigen may be expressed in brains of

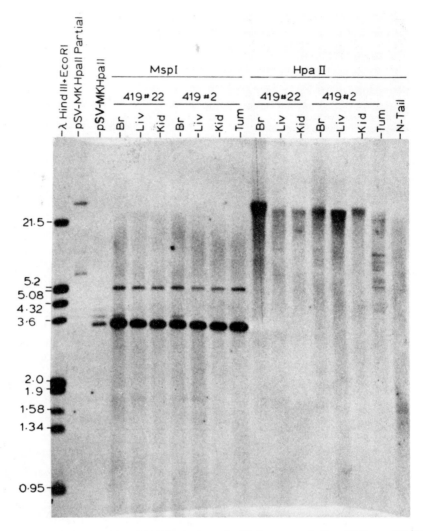

Figure 5. Comparison of methylation levels in pSV-MK DNA from the 419-2 tumor and other 419 tissues. Brain (Br), liver (Liv), and kidney (Kid) DNA from 419-2 and 419-22 and tumor (Tum) DNA from 419-2 were digested to completion with *Msp*I (methylation insensitive) or *Hpa*II (methylation sensitive) and then analyzed by Southern blotting. λ DNA digested with both *Hin*dIII and *Eco*RI and pSV-MK DNA digested partially or completely with *Hpa*II serve as markers. The radioactive probe was nick-translated SV40 DNA. The X-ray film was exposed for 5 days.

these mice and one older 419 mouse has developed a tumor of the choroid plexus. The pedigrees of different transgenic mice have been compared to determine why some mice get tumors while others do not. A comparison of the 427 and SV11 families of mice shows nonoverlapping times of death, with narrow standard deviations. Thus, the time of death is predictable and relatively constant through at least four to six generations. These data suggest that a timing mechanism may be operative in the development of these tumors.

Although the nature of this putative timing mechanism is presently unclear, these mice provide a model system for examining the time course of tumor development in vivo. The initial investigations into the time of T antigen expression in the brain tissue of SV11 mice revealed that these animals express low but detectable levels of T antigen in the youngest mice tested (8 weeks). These data demonstrate that T antigen is expressed prior to the formation of a visible tumor. This could be the result of a microscopic tumor, but if that is the case the kinetics of tumor growth and development would be quite slow because visible tumors do not appear for another 4-6 weeks in these mice. In addition, it should be pointed out that the studies described here do not localize the cell type expressing T antigen in the SV11 brain tissue, and so it remains a formal possibility that some as yet uncharacterized cell type produces T antigen in 8-week-old mice. The alternative to these hypotheses is that the cells of the choroid plexus produce T antigen prior to tumor production at 8 weeks of age. A second event, such as the induction of another gene(s) or increased levels of T antigen may then be required for the development of a choroid plexus papilloma. Stewart et al. (1984) have suggested that a second transforming event is necessary for the production of mammary adenocarcinomas in transgenic mice that express an MTV-*myc* fusion gene. This suggestion is based on the fact that the *myc* gene was expressed in other apparently normal tissues in these mice and that individual tumors were observed rather than a uniform neoplasia of entire mammary epithelium. Similarly, transgenic mice carrying the insulin-T antigen fusion gene expressed T antigen in all islets examined, but only a few tumor foci were detected within each pancreas (Hanahan 1985). Expression of T antigen in these mice was detectable in the islets of 16-day-old embryos (D. Hanahan and S. Alpert, pers. comm.), but tumors did not arise until mice were 10–20 weeks of age. This time lag between the onset of expression of T antigen and the time of tumor formation may be similar to that observed in the SV11 mice.

The 419 mice, with only one exception, failed to develop tumors of the choroid plexus or of any other tissue. There are a number of possible explanations for this phenotype. First, the lack of tumor formation could be the result of a mutant T antigen. Our initial investigations into this possibility, however, showed the molecular weight of the T antigen from the choroid plexus of the 419-2 mouse to be indistinguishable from that of a wild type. In addition, Southern analysis demonstrated that there were no rearrangements, insertions, or deletions in the 419 T antigen coding sequences. These analyses do not exclude the possibility that the T antigen carries a point mutation which affects its ability to cause a tumor. Transfection experiments designed to assess the transforming ability of the 419 T antigen gene are in progress. Second, 419 mice may simply not express sufficient levels of T antigen for tumorigenesis. The levels of T antigen visualized by Western analysis of a variety of 419 tissues were either very low or nondetectable. Fibroblast-like cells derived in culture from the tissues of a 14-week-old 419 brain and heart have also failed to express T antigen in vitro. In the initial passages of a 419 kidney culture, the cells grew slowly, had a nontransformed cell morphology, and were SV40 T antigen negative. In the eighth passage, however, a few SV40 T antigen-positive cells were detected by indirect immunofluorescence and foci quickly formed in the culture. The T antigen-positive cells rapidly overtook the cell culture and routinely achieved densities four- to fivefold higher than the initial culture. These results suggest that the 419 mice contain SV40 DNA that can produce a functional T antigen, but in spite of that, 419 mice fail to express T antigen and fail to produce tumors.

Elucidation of the mechanism(s) by which T antigen expression was activated in the tumor of 419-2 may lead to an understanding of how the T antigen genes have been rendered inactive in the 419 mouse line. Southern analysis of the tumor DNA showed that the formation of the 419-2 tumor was not associated with an amplification of SV40 DNA or a transposition, insertion, deletion, or rearrangement that would alter the pSV-MK or nearby host DNA. A transposition event involving the entire pSV-MK tandem array with the approximately 5 kb of 5'- and 3'-adjoining host DNA or mutations that do not produce major structural alterations in the DNA could not be ruled out by this analysis. The tumor-derived DNA, however, was found to be hypomethylated, when compared with other 419-derived DNAs. Alterations in gene methylation have been proposed as a possible mechanism involved in the regulation of gene transcription; CpG methylation patterns are known to be inherited and many examples of association between hypomethylation and increased gene expression have been demonstrated (for review, see Doerfler 1983). If 419 mice do not develop tumors as a result of a failure to express the required levels of T antigen, CpG methylation at or near the SV40 integration site could provide a reasonable explanation. Thus, a mistake in the inheritance pattern in CpG methylation, with increased age for example, could have resulted in hypomethylation of SV40 DNA, increased T antigen expression, and the development of the 419-2 tumor. Further experimentation is required to test this hypothesis.

ACKNOWLEDGMENTS

The authors gratefully acknowledge R. Brinster for providing the founder mice for these studies. C. Jackson and A. Teresky provided excellent technical assistance, L. Silver was instrumental in the organization of our animal colony records, and D. Miller and R. Trilstad provided excellent histological services. M. Clarke is gratefully acknowledged for her patient typing of this manuscript. This work was supported by a grant from the National Institute of Health, R01-CA38757.

REFERENCES

ABRAMCZUK, J.S., S. PAN, E. MAUL, and B.B. KNOWLES. 1984. Tumor induction by simian virus 40 in the mouse is controlled by long term persistence of the viral genome and the immune response of the host. *J. Virol.* **49:** 540.

BRINSTER, R.L., H.Y. CHEN, A. MESSING, T. VAN DYKE, A.J. LEVINE, and R.D. PALMITER. 1984. Transgenic mice harboring SV40 T-antigen genes develop characteristic brain tumors. *Cell* **37:** 367.

COLBY, W.W. and T. SHENK. 1982. Fragments of the simian virus 40 transforming gene facilitate transformation of rat embryo cells. *Proc. Natl. Acad. Sci.* **79:** 5189.

CRISTOFALO, V.J. and R. CHARPENTIER. 1980. A standard procedure for cultivating human diploid fibroblast-like cells to study cellular aging. *J. Tissue Cult. Methods* **6:** 117.

DIAMANDOPOULOS, G.T. 1972. Leukemia, lymphoma and osteosarcoma induced in the Syrian golden hamster by simian virus 40. *Science* **176:** 173.

———. 1978. Incidence, latency and morphological types of neoplasms induced by simian virus 40 inoculated intravenously into hamsters of three inbred strains and one outbred stock. *J. Natl. Cancer Inst.* **60:** 445.

DOERFLER. W. 1983. DNA methylation and gene activity. *Annu. Rev. Biochem.* **52:** 93.

EDDY, B.E., G.S. BERMAN, W.H. BERKELEY, and R.D. YOUNG. 1961. Tumors induced in hamsters by injection of rhesus monkey kidney cell extracts. *Proc. Soc. Exp. Biol. Med.* **107:** 191.

GERBER, P. and R.L. KIRSCHSTEIN. 1962. SV40-induced ependynomas in newborn hamsters. I. Virus-tumor relationships. *Virology* **18:** 582.

GRUSS, P., R. DHAR, and G. KHOURY. 1981. Simian virus 40 tandem repeated sequences as an element of the early promoter. *Proc. Natl. Acad. Sci.* **78:** 943.

HANAHAN, D. 1985. Heritable formation of pancreatic B-cell tumors in transgenic mice expressing recombinant insulin/simian virus 40 oncogenes. *Nature* **315:** 115.

LINZER, D.H., W. MALTZMAN, and A.J. LEVINE. 1979. The SV40 A-gene product is required for the production of a 54,000 MW cellular tumor antigen. *Virology* **90:** 308.

PALMITER, R.D., H.Y. CHEN, A. MESSING, and R.L. BRINSTER. 1985. The SV40 enhancer and large T-antigen are instrumental in development of choroid plexus tumors in transgenic mice. *Nature* **316:** 457.

STEWART, T.A., P.K. PATTENGAL, and P. LEDER. 1984. Spontaneous mammary adenocarcinomas in transgenic mice that carry and express MTV/myc fusion genes. *Cell* **38:** 627.

TOOZE, J., ed. 1981. *Molecular biology of tumor viruses*, 2nd edition, revised: *DNA tumor viruses*. Cold Spring Harbor Laboratory, Cold Spring Harbor, New York.

Persistence and Transmission in the Mouse of Autonomous Plasmids Derived from Polyoma Virus

M. RASSOULZADEGAN AND J. VAILLY

Laboratoire de Génétique Moléculaire des Papovavirus (INSERM U273), Centre de Biochimie, Université de Nice, 06034 Nice, France

By microinjecting DNA of the recombinant plasmid pPyLT1 into the male pronucleus of fertilized eggs (Gordon et al. 1980; Brinster et al. 1981; Costantini and Lacy 1981; Wagner et al. 1981), we created 13 transgenic strains of mice. The plasmid (Fig. 1) comprises bacterial vector sequences derived from pBR322, linked at their *Bam*HI site to a modified polyoma virus genome deleted of intron sequences (Treisman et al. 1981; Tyndall et al. 1981). Only one of the three viral T antigens, large T, is made, which, together with sequences at the origin of replication (also present in pPyLT1), represents all the virus-coded elements required for autonomous replication in mouse cells (for review, see Hand 1981). After transfection into permissive mouse 3T6 fibroblasts, pPyLT1 DNA was replicated, albeit with a limited efficiency as compared with similar constructs expressing the complete early region (Nilsson and Magnusson 1983; R. Kamen, pers. comm.). Two possible reasons for this relatively low efficiency are the limited rate of protein synthesis from the intronless large T-only gene and the absence of the small T protein, which exerts a still unexplained synergistic effect on autonomous replication (Nilsson and Magnusson 1983, 1984). Expression of pPyLT1 DNA was also shown to induce changes in growth control corresponding to an early stage of the transformation of rodent fibroblast cells in culture (immortalization, serum independence and reactivity to tumor promoters and to other oncogenes) (Rassoulzadegan et al. 1982, 1983; Land et al. 1983). These transformation-related functions and the replicative functions appear to be carried out by distinct domains of the multifunctional large T protein. A truncated form corresponding to its aminoterminal 40% was shown to be defective for initiation of DNA replication, but to confer immortality, and serum independence and to cooperate with other oncogenes (Rassoulzadegan et al. 1982, 1983; Land et al. 1983). Conversely, a large deletion in this aminoterminal region retained at least part of its activity in DNA replication (Nilsson and Magnusson 1984).

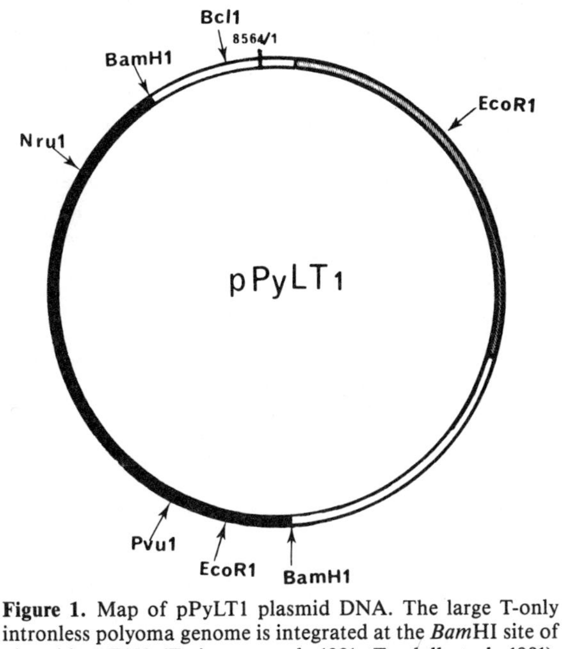

High Efficiency of Establishment of Transgenic Mice after Injection of pPyLT1 DNA

Circular DNA of plasmid pPyTL1 was injected into pronuclei of fertilized mouse eggs. Fifteen mice were born in three series of experiments (Table 1) and two of them died when they were 1 and 3 days old, respectively. Genomic DNA prepared at 4 weeks from the tail of the 13 living mice, or from the two dead mice, showed in all cases specific hybridization with nick-translated pPyLT1 DNA (Fig. 2), with polyoma DNA or with DNA of the bacterial vector alone (the latter control was necessary in view of the reported occurrence of polyoma virus contaminations in mouse colonies). Together with this unusually high frequency of establishment of transgenics, most mice maintained a large number of copies of the added genes (up to 100/cell), a situation also at variance with a number of previous observations on transgenic mice. All DNAs produced multiple bands by Southern blot hybridization with pPyLT1 probes after *Eco*RI cleavage. Still more unusual was the observation (not shown) that at least part of the DNA complementary to the probe was present as free form-I molecules, as judged from its apparent density in CsCl in the presence of ethidium bromide.

Both the establishment of the foreign genes with high

Figure 1. Map of pPyLT1 plasmid DNA. The large T-only intronless polyoma genome is integrated at the *Bam*HI site of plasmid pAT153 (Treisman et al. 1981; Tyndall et al. 1981). Nucleotide 1 of the recombinant DNA is that of the polyoma sequence (Soeda et al. 1980) and corresponds to the region of the origin of replication. (▨) Polyoma early; (☐) polyoma late; (■) pAT153.

680 RASSOULZADEGAN AND VAILLY

Table 1. Efficiency of Establishment of pPyLT1 DNA in Transgenic Mice

Experiment number	Plasmid DNA	Number of eggs transferred to foster mothers	Number of pups born	Number of transgenic pups
1	pPyLT1	10	1	1
2	pPyLT1	12	10	10
3	pPyLT1	8	4	4
4	pLT214	11	9	2

Form-I DNA of the indicated plasmid (6 ng/μl) was injected into fertilized eggs recovered from (C57BL/6 × DBA2)F_1 female mice that had been mated to (C57BL/6 × DBA2)F_1 males (Iffa-Credo, France). Transfer of the injected eggs into the oviduct of pseudopregnant females was carried out at the two-cell stage.

efficiency and their maintenance in high numbers of copies appeared to require the integrity of the large T coding region, or at least the presence of its carboxyterminal half. Microinjection of a deleted plasmid, pLT214 (Rassoulzadegan et al. 1982, 1983), which lacks all the coding region downstream of the viral *Eco*RI site (60% carboxyterminal part of the protein; see Fig. 1), produced only two transgenics out of nine mice born after reimplantation (Table 1), and both of them maintained a small number of integrated copies of the added genes (data not shown).

Hereditary Transmission and DNA Patterns in the Offspring

All the founder animals (designated F_0), when mated with the C57BL/6 parental strain, efficiently transmitted polyoma and pBR322 DNA sequences to their progeny (F_1). The same high efficiencies, often close to 100%, were subsequently observed in F_2 and F_3 backcross generations (Table 2). The number of copies in tail DNA of F_1 mice was in general lower than in the founders (only 1–10/cell genome), but animals with higher values (up to 100 copies/cell genome) were observed occasionally in various transgenic strains, with no obvious regularity.

Figure 3 shows results of Southern blot analysis performed comparatively on two representative founders and their progenies. In all 13 transgenic strains, only one type of DNA molecule was transmitted from F_0 to F_1. It was not one of the predominant forms in F_0 DNA, and the molecules maintained in the different strains exhibited different structures, none of them being identical with pPyLT1. Identical sizes and structures were observed, with variable copy numbers, in all the F_1 animals from a given founder (data not shown), and, in any given lineage, the simplified structure observed in F_1 animals was maintained without further changes in their F_2 and F_3 backcross progenies. In all the transgenic strains that we established with pPyLT1 DNA, the viral sequences had undergone rearrangements (Fig. 3), the bacterial vector sequences being usually conserved (see below).

The Transgenomic Recombinant Molecules Are Maintained Exclusively As Autonomous Plasmids in the Mouse

We observed (see above) that a fraction of the transgenomic sequences was present as free form-I DNA in F_0 mice. Further analysis was made difficult by the complexity of the profiles on Southern blots. As animals with low numbers of copies were generated in offspring, it became possible to investigate whether the recombinant DNA was present only in a nonintegrated form. One might alternatively consider that, as previously described for transformed fibroblast cells which contain free polyoma DNA (see Pellegrini et al. 1984 and references therein), these form-I molecules could originate by excision from an integrated head-to-tail

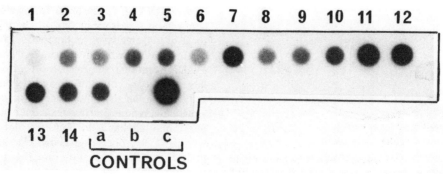

Figure 2. Dot blot hybridization of pPyLT1 DNA with tail DNA. (Dots *1–14*) 5 μg of DNA extracted from 2-cm sections of tails of mice born after injection of pPyLT1 DNA, using a protocol described by Davis et al. (1980), with the modification that one phenol extraction and one chloroform extraction were performed prior to ethanol precipitation. (Dots *a* and *c*) 5 and 50 μg, respectively, of DNA from the polyoma-transformed rat cell line FR3T3 MTT4 (Rassoulzadegan et al. 1982) which maintain about 10 integrated viral genomes (our unpublished results). (Dot *b*) Tail DNA from normal mouse (5 μg).

Table 2. Hereditary Transmission of Plasmid Sequences (F_1 to F_2)

Founder number	F_1 mouse number	Sex	Number of pups	Number of transgenic mice
3	15	male	3	2
3	15	male	5	5
3	15	male	6	6
12	51	male	12	10
6	26	female	12	5
8	36	female	12	11

F_1 mice derived from backcrosses of the indicated founder with C57BL/6 were again crossed with C57BL/6. Each line indicates the proportion of pups in each litter that were found positive with a pPyLT1 probe by dot blot and by Southern blot hybridization.

multimer, which might be, in fact, the only form transmitted hereditarily.

A series of independent experimental evidences excluded the second hypothesis, establishing that a small number of free molecules are stably maintained and regularly transmitted through mitosis and meiosis.

Molecules with the same size and structure were observed in total DNA and in selectively extracted low-molecular-weight DNA. Selective extraction was performed according to Hirt (1967) on primary cultures established from F_1 embryo fibroblasts. The same electrophoretic profiles were observed in these preparations, in total DNA from the same cultures, and in the DNA of adult F_1 mice from the same strain (data not shown).

The same bands were obtained on Southern blots after digestion with two enzymes which do not cleave pPyLT1 DNA, and only unit-length linear DNA was produced by cleavage with three "one-cut" restriction enzymes. Figure 4 shows blot hybridization profiles after cleavage of the DNA of one representative F_1 mouse with five restriction endonucleases. Two of them, *Bgl*II and *Hpa*I, which do not cleave pPyLT1 DNA, produced bands corresponding to the same electrophoretic mobility (Fig. 4D, E), a result that would be highly unlikely for DNA integrated in cellular sequences. Moreover, all the available enzymes that cleave pPyLT1 DNA only once (*Bcl*I, *Nru*I, and *Pvu*I; see Fig. 1) produced only one band, in all three cases at the same electrophoretic position. No additional fragment could be detected that might correspond to the insert-cellular DNA junctions of an integrated multimer, even after deliberate overexposure of the autoradiograms (Fig. 4, lanes A-C).

Polyoma-pBR322 recombinant molecules could be transferred efficiently from mouse DNA back into Escherichia coli. After transfection of total DNA prepared from various tissues of F_0, F_1, and F_2 animals, bacterial colonies resistant to ampicillin appeared with efficiencies in the range of 10-1000 resistant colonies per microgram of cellular DNA (a rather high value in view of the small fraction of plasmids in the total DNA). As shown in Figure 5, the structure of the DNA rescued in bacteria, and analyzed by restriction enzyme cleavage and gel electrophoresis, was in all cases

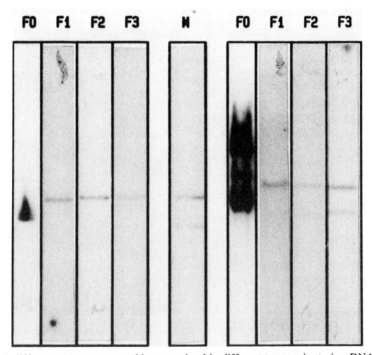

Figure 3. Plasmids with different structures are stably transmitted in different transgenic strains. DNA (20 µg) extracted from sections of the tails of two founders (F_0-8 and F_0-12), and of their respective F_1, F_2, and F_3 progenies was digested with *Bam*HI endonuclease, electrophoresed through 0.8% agarose gels, transferred to nitrocellulose, and hybridized with a ^{32}P-labeled pPyLT1 probe. Marker (M) is pPyLT1 DNA (10 pg) cleaved with *Bam*HI.

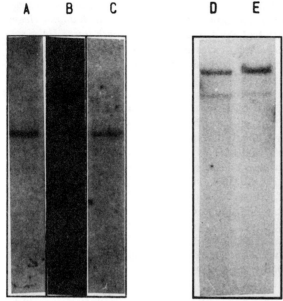

Figure 4. Absence of integrated pPyLT1 sequences. DNA was extracted from liver of a female F_1 mouse derived from founder number 8 (see Fig. 3). DNA (20 μg) was cleaved with the indicated endonuclease. One-cutters are *Bcl*I (lane *A*), *Pvu*I (lane *B*), and *Nru*I (lane *C*) (see Fig. 1). *Bgl*II (lane *D*) and *Hpa*I (lane *E*) have no cleavage sites on pPyLT1. Autoradiograms presented in *A* through *C* were voluntarily overexposed.

identical with that revealed in mouse DNA by blot hybridization.

Distribution in the Various Tissues

As exemplified in Figure 6, the plasmid DNA was found in all tissues analyzed so far (skin, muscle, brain, liver, kidney), usually in low copy numbers. Although higher numbers were observed in several instances (see brain DNA in Fig. 6), no single tissue consistently showed increased values in all the animals tested. A remarkable case is that of spermatozoons, where the same profile of fragments as in somatic cell DNA was observed, in all cases with a low number of copies (1–2/cell genome) (Fig. 7). Selective preparation of low-molecular-weight DNA confirmed the presence of free molecules in the germ line (data not shown). Since these males transmitted the plasmid DNA to their offspring with an efficiency close to 100% (see Table 2), the low average copy number in sperm is likely to reflect the presence of only one to two copies in every spermatozoon. This conclusion would in turn strongly suggest an active segregation mechanism operating during mitosis and meiosis.

The Phenotype of the Plasmid-bearing Transgenics

As already mentioned for two of the original F_0 mice, a number of baby mice born in F_1 to F_3 crosses died 1–3 days later. Detailed histological investigation is in progress, but so far, no clear cause of death has been found. No difference in either the state or the structure of polyoma and pBR322 sequences was observed between the dead and the surviving babies of the same

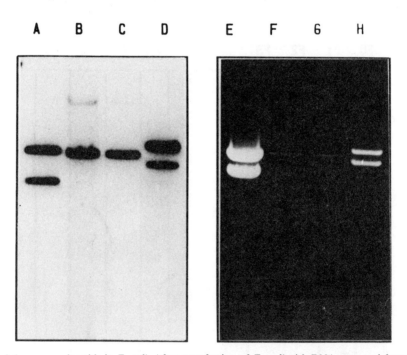

Figure 5. Rescue of the mouse plasmids in *E. coli*. After transfection of *E. coli* with DNA extracted from the tail of three F_2 mice from two transgenic strains, ampicillin-resistant colonies were selected. Plasmid DNA was prepared from bacteria using standard methods (Maniatis et al. 1982). All preparations were digested with *Bam*HI endonuclease. (Lanes *A–D*) Results of Southern blot analysis of pPyLT1 marker DNA (10 pg) (lane *A*), DNA of two mice of strain number 8 (lanes *B*, *C*), DNA of a mouse of strain number 12 (lane *D*). (Lanes *E–H*) Ethidium bromide-stained electrophoresis gels of DNA extracted from bacteria (0.3–1 μg/lane); (lane *E*) pPyLT1; (lanes *F–H*) plasmids rescued from mouse DNA shown in lanes *B–D*, respectively.

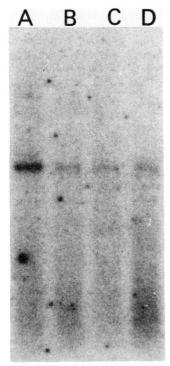

Figure 6. Plasmid DNA in various tissues. DNA was extracted as described in Fig. 2 legend from different tissues of an F_1 mouse (strain number 8) and further analyzed by Southern blot hybridization (BamHI cleavage, pPyLT1 probe). (Lane A) Brain; (lane B) liver; (lane C) kidney; (lane D) muscle.

or severely impaired pups.) Incidence of this early death syndrome appeared as characteristic of the progeny of one given founder: one of the F_0 males (F_0-2) did not produce any living progeny over a series of successive matings, five mice, including both males and females, produced repeatedly between 20 and 40% dead babies, and the remainder produced completely healthy offspring. Preliminary observations indicated that when the animals grew older, their progeny showed increased viability: for instance, viable progeny of male F_0-2 was obtained when he was 10 months old.

Animals who survived the first days after birth appeared perfectly healthy. Unlike what was observed for transgenics bearing either the SV40 large T (Brinster et al. 1984; Hanahan 1985) or rearranged *myc* oncogenes (Stewart et al. 1984), no incidence of spontaneous tumors was so far observed (the oldest living animal is 18 months old at the time of this report). This relative innocuousness of the polyoma sequences is in agreement with the absence in the original plasmid of coding capacity for the middle T viral protein, which is necessary for the terminal stages of oncogenic transformation (Treisman et al. 1981; Rassoulzadegan et al. 1983), but it may also be related to the deletions that occurred in the early viral sequences of the mouse plasmid, assuming that they would affect preferentially the aminoterminal part of the protein and its transforming functions.

ACKNOWLEDGMENTS

We thank F. Cuzin for useful discussions and for laboratory facilities and R.M. Catalioto for her partic-

litters. (These studies were made difficult by the cannibalizing habits of the mothers with respect to dead

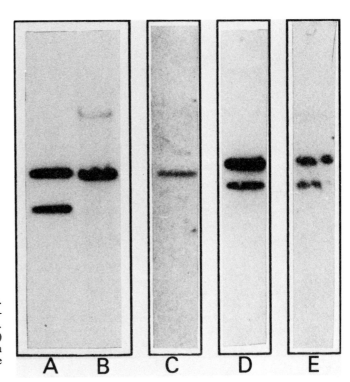

Figure 7. Autonomous DNA in haploid gametes. DNA of spermatozoa isolated from dissected epididymis was analyzed as described in the Fig. 7 legend. (Lane A) pPyLT1 marker (10 pg); tail DNA (lane B) and sperm DNA (lane C) from F_2 mouse of strain number 8; tail DNA (lane D) and sperm DNA (lane E) from F_2 mouse of strain number 12.

ipation in a series of experiments. We are indebted to L. Carbone and F. Tillier for skilled technical help. This work was made possible by grants from Centre National de la Recherche Scientifique, Association pour la Recherche sur le Cancer, and Foundation pour la Recherche Médicale, France.

REFERENCES

BRINSTER, R.L., H.Y. CHEN, A. MESSING, T. VAN DYKE, A.J. LEVINE, and R.J. PALMITER. 1984. Transgenic mice harboring SV40 T-antigen genes develop characteristic brain tumors. *Cell* **37:** 367.

BRINSTER, R.L., H.Y. CHEN, M. TRUMBAUER, A.W. SENEAR, R. WARREN, and R.A. PALMITER. 1981. Somatic expression of herpes thymidine kinase in mice following injection of a fusion gene into eggs. *Cell* **27:** 223.

COSTANTINI, F. and E. LACY. 1981. Introduction of a rabbit beta-globin gene into the mouse germ line. *Nature* **294:** 92.

DAVIS, R.W., M. THOMAS, J. CAMERON, T.P. STJOHN, S. SCHERER, and R.A. PADGETT. 1980. Rapid DNA isolation for enzymatic and hybridization analysis. *Methods Enzymol.* **65:** 404.

GORDON, J.W., G.A. SCANGOS, D.J. PLOTKIN, J.A. BARBOSA, and F.H. RUDDLE. 1980. Genetic transformation of mouse embryos by microinjection of purified DNA. *Proc. Natl. Acad. Sci.* **77:** 7380.

HANAHAN, D. 1985. Heritable formation of pancreatic B-cell tumors in transgenic mice expressing recombinant insulin-simian virus 40 oncogenes. *Nature* **315:** 115.

HAND, R. 1981. Functions of T antigens of SV40 and polyoma virus. *Biochim. Biophys. Acta* **651:** 1.

HIRT, B. 1967. Selective extraction of polyoma DNA from infected mouse cell cultures. *J. Mol. Biol.* **26:** 365.

LAND, H., L.F. PARADA, and R.A. WEINBERG. 1983. Tumorigenic conversion of primary embryo fibroblasts requires at least two cooperating oncogenes. *Nature* **304:** 596.

MANIATIS, T., E.F. FRITSCH, and J. SAMBROOK, eds. 1982. *Molecular cloning: A laboratory manual.* Cold Spring Harbor Laboratory, Cold Spring Harbor, New York.

NILSSON, S.V. and G. MAGNUSSON. 1983. T-antigen expression by polyoma mutants with modified RNA-splicing. *EMBO J.* **2:** 2095.

———. 1984. Activities of polyoma-virus large-T-antigen proteins expressed by mutant genes. *J. Virol.* **51:** 768.

PELLEGRINI, S., L. DAILEY, and C. BASILICO. 1984. Amplification and excision of integrated polyoma DNA sequences require a functional origin of replication. *Cell* **36:** 943.

RASSOULZADEGAN, M., A. COWIE, A. CARR, N. GLAICHENHAUS, R. KAMEN, and F. CUZIN. 1982. The roles of individual polyoma virus early proteins in oncogenic transformation. *Nature* **300:** 713.

RASSOULZADEGAN, M., Z. NAGHASHFAR, A. COWIE, A. CARR, M. GRISONI, R. KAMEN, and F. CUZIN. 1983. Expression of the large T protein of polyoma virus promotes the establishment in culture of "normal" rodent fibroblast cells. *Proc. Natl. Acad. Sci.* **80:** 4354.

SOEDA, E., J.R. ARRAND, N. SMOLAR, J.E. WALSH, and B.E. GRIFFIN. 1980. Coding potential and regulatory signals of the polyoma virus genome. *Nature* **283:** 445.

STEWART, T.A., P.K. PATTENGALE, and P. LEDER. 1984. Spontaneous mammary adenocarcinomas in transgenic mice that carry and express MTV/*myc* fusion genes. *Cell* **38:** 627.

TREISMAN, R., U. NOVAK, J. FAVALORO, and R. KAMEN. 1981. Transformation of rat cells by an altered polyoma virus genome expressing only the middle-T protein. *Nature* **292:** 595.

TYNDALL, C., G. LA MANTIA, C.M. THACKER, J. FAVALORO, and R. KAMEN. 1981. A region of the polyoma virus genome between the replication origin and late protein coding sequences is required in *cis* for both early gene expression and viral DNA replication. *Nucleic Acids Res.* **9:** 6231.

WAGNER, E.F., T.A. STEWART, and B. MINTZ. 1981. The human beta-globin gene and a functional viral thymidine kinase gene in developing mice. *Proc. Natl. Acad. Sci.* **78:** 5016.

The Ability of EK Cells to Form Chimeras after Selection of Clones in G418 and Some Observations on the Integration of Retroviral Vector Proviral DNA into EK Cells

M.J. Evans, A. Bradley, M.R. Kuehn, and E.J. Robertson
University of Cambridge, Department of Genetics, Cambridge CB2 3EH, England

The use of tumor-derived embryonal carcinoma (EC) cells and embryo-derived stem cells (EK cells) as models for studying aspects of early mouse embryogenesis is well established (Evans et al. 1983). The advent both of methods for establishing primary cultures of pluripotential cells directly from the embryo (Evans and Kaufman 1981; Martin 1981) and of the ability to maintain these cells in a fully totipotential and karyotypically unaltered form during in vitro culture (Robertson et al. 1983; Evans et al. 1985) reopens the issue of the practicality of attempting to assay in vivo, mutants produced in vitro.

We have shown that 15 EK cell lines of separate origins are able to contribute to chimeras and that the overall efficiency of chimera production is high (Table 1). Approximately two-thirds of injected blastocysts transferred to foster mothers in which pregnancy is established are recovered as live-born mice, and in recent experimental series over half of these are regularly chimeric. We have analyzed over 500 chimeric mice and have seen no evidence of abnormalities nor of teratocarcinoma-related neoplasms. The injected cells contribute to all tissues analyzed and most of the tissues are chimeric in any one mouse. The general distribution of both coat color contribution and contribution to the internal organs (as judged by glucose phosphate isomerase [GPI] analysis) is fine grained. These results and also the observation that contribution to the blood and to the germ cell line can be greatly increased by using host blastocysts that are genetically deficient in blood and germ line stem cells have been reported in greater detail elsewhere (Evans et al. 1985). Of particular importance is the observation that the injected tissue culture cells may contribute to functional sperm in the resultant chimeric mouse (Bradley et al. 1984).

One particularly attractive method for introducing novel genes and gene constructs is the use of retroviral vectors. Retroviruses have the additional feature of disrupting flanking sequences where they integrate and thereby mutagenizing chromosomal genes (Schnieke et al. 1983). The behavior of stem cells modified in culture may be studied during normal in vivo development in chimeric combination with a carrier embryo and via functional germ cell formation (Bradley et al. 1984) in subsequent transgenic generations.

We discuss here two aspects of studies designed to establish these techniques. First, we have assayed the developmental competence (by chimera formation) of EK cell clones that have been selected for the ability to grow in G418 after infection with the vector ZIPneoSVX (Cepko et al. 1984) and, second, we discuss some preliminary observations on the transformation of EK cells with retroviral vectors.

EXPERIMENTAL PROCEDURES

The EK cell line EK CC1.2 was maintained as previously described (Robertson et al. 1983). Psi-2 cell lines producing the ZIPneoSVX and the DO1 virus were provided by Dr. Richard Mulligan (MIT, Cambridge, Massachusetts). The DO1 transformant was derived following cocultivation of stem cells with inactivated

Table 1. Rates of Chimera Formation with Various EK Cell Lines

Description	Number of lines	Number of injected blastocysts	Live-born mice	Chimeras
Euploid XY lines derived from fertilized embryos	5 { series 1 }	656	454 (69%)	110 (24%)
	3 { series 2 }	650	427 (66%)	234[a] (55%)
Euploid XX lines derived from fertilized embryos	2	189	117 (62%)	41 (35%)
Diploid cell lines derived from parthenogenetically activated embryos	5	324	220 (68%)	40 (18%)

[a] From this series, of 47 phenotypically male animals that proved fertile, 15 were germ line chimeric.

(mitomycin C-treated) producer cells plated in the ratio of 1:2 with STO feeder cells. The virus was used as culture supernatant with the addition of 8 µg/ml Polybrene. Virus titers were measured as numbers of G418 (1 mg/ml)-resistant colonies formed after infection of an excess of 3T3 cells with serial dilutions of the viral suspension. G418 was obtained from Gibco Biocult as Genticin, and concentrations of G418 are expressed as milligrams dry weight of the product as supplied.

It was found that G418 at concentrations greater than 0.1 mg/ml completely abolished the progressive growth of clonal colonies of untransformed EK cell lines. Concentrations of either 0.2 or 0.3 mg/ml were used for selection in the transformation experiments. At these concentrations, it is possible to use normal STO cells as feeder cells but STO cells resistant to 1 mg/ml G418 were made by transformation with pSV2neo and these provide better feeder layers when G418 selection is used.

RESULTS

Following treatment of EK cells with ZIPneoSVX or DOI virus, both as cell suspensions and in monolayer, colonies able to grow in G418 were recovered. The effective viral titer of colony-forming units was approximately two orders of magnitude less on EK cells than on fibroblasts.

Single cell-derived colonies of G418-resistant EK cells were picked and cell lines established. These cells grew slowly in G418. Preliminary observations suggested that slow-growing cells colonize embryos less efficiently, therefore the cells were passaged without selection prior to being tested for their chimera-forming ability and the preparation of DNA for analysis.

Cells were injected into 3.5-day mouse blastocysts as previously described (Bradley et al. 1984). The results of these injections are presented in Table 2 and a typical chimeric animal pictured in Figure 1. The rate of chimera production was high (60%) and the resulting animals showed extensive coat color contribution from the injected cells. A total of 28 phenotypic male mice were test bred and of these 20 were fertile. One animal has been demonstrated to have a germ line contribution. These results are compared with those previously obtained for the parental line of untransformed EK.CC1.2 cells (Table 3).

The level of contribution of the EK cells to the various organs of the mice was determined by a semiquantitative analysis of the GPI isoenzyme analysis. The results of this analysis are presented for nine mice in Figure 2.

DNA from the cell lines and from various organs from chimeric mice was examined for the presence of the ZIPneoSVX genome. In the cell lines, the ZIPneoSVX genome was found at an average of less than a single copy per cellular genome. In DNA samples from the chimeric mice, practically no trace of the ZIPneoSVX genome was detected (C. Cepko, pers. comm.).

Using cocultivation of stem cells with cells producing the DO1 virus, a single EK-transformed colony was derived following G418 selection. This was maintained for eight passages in 0.2 mg/ml G418. At passage nine, subclones were derived that were maintained for three to four passages prior to analysis. Southern blotting performed on the parental line, DO11, demonstrated a pattern consistent with single-copy integration of proviral DNA. A single subclone maintained without selection for four passages showed an apparent reduction of the proviral genome to below an average of a single copy per cellular genome. Another subclone maintained in high levels of G418 (0.8 mg/ml) showed an increase above a single copy of proviral DNA per cellular genome.

DISCUSSION

Treatment of cell cultures with ZIPneoSVX suspensions allows clones of cells to be isolated that are able

Table 2. Chimera Formation by G418-resistant Clones of CC1.2 Following Infection with ZIPneoSVX

Clone	Blastocysts injected	Animals born	Chimeras	Female	Male
Series I[a]					
CC1.2.ZNB	16	10	7	2	5
CC1.2.ZNC	13	6	4	–	4
CC1.2.ZNG	15	12	10	5	5
Overall	44	28	21	7	14
Series II[b]					
CC1.2.G13	42	26	19	3	16
CC1.2.G14	30	22	17	3	14
CC1.2.G16	128	77	41	10	30
CC1.2.C12	93	51	31	13	18
CC1.2.H25	103	39	16	2	14
Overall	396	215	124	31	92
Totals	440	243	145(60%)	38(26%)	106(74%)

[a] Infected in suspension after trypsinization.
[b] Infected in monolayer culture.

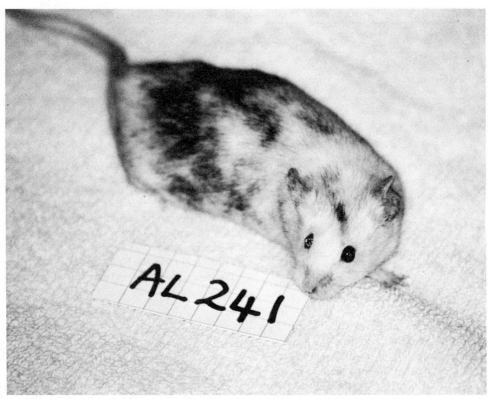

Figure 1. Chimeric phenotypic male mouse, G16.9. Upon autopsy this chimera proved to be an hermaphrodite. Clone G16 was isolated from the CC1.2 cell line following exposure to ZIPneoSVX virus. Pigmented cells in the coat and eyes are derived from the Black Agouti genotype of the CC1.2 cells following injection into an albino host blastocyst.

to grow in 0.2 mg/ml G418. G418-resistant clones can not be isolated from the untreated culture. This suggests that the viral vector is transforming the cells, but the level of G418 resistance conferred in EK cells by transformation with ZIPneoSVX is low, as is exhibited by the slow growth rate of these cells in G418. This probably reflects the low level of activity of transcription provided by the Moloney long terminal repeat (LTR) in cells of this phenotype (Linney et al. 1984). This has been clearly shown by studies with similar viruses by Rubenstein et al. (1984) where the level of the *neo* gene transcription is very greatly increased in EC cells by placing it under the control of an hybrid SV40/*tk* promoter. Wagner et al. (1985) have also shown good expression of the *neo* gene in EC cells using a retroviral vector with an internal *tk* promoter.

The *neo* gene in the vector DO1 is promoted by an internal SV40 promoter (R. Mulligan, pers. comm.) which might be expected to give a higher level of transcription and therefore better resistance and growth characteristics for the EK cells in selective media. Although we observe progressive growth in high levels of G418 in a subclone of the DO1 transformant, this appears to be accompanied by an amplification phenomena which might suggest that a single copy of the *neo* gene with the SV40 promoter is not sufficient to provide resistance at this level of selection. One interesting feature observed with both of these vectors is the instability of the introduced virus.

We do not have evidence of the state of integration in the other cell lines used but the observations suggest that one or other of two unusual situations may occur in EK cells. The viral genome is either maintained for long periods of cell growth as an episomal element, or if it integrates it is readily remobilized and lost from its integration site. In either case it is possible that the SV40 origin of replication present in the viral vector constructs is an instrumental factor. Willison et al. (1983) have shown that SV40 DNA injected into early normal embryos is maintained in an unintegrated form through early embryonic development, and Friedrich and Lehman (1981) showed that EC cells infected with SV40 still contain unintegrated full-length viral genome 10 days after infection.

Despite the instability of these proviral genomes it would appear that the transformation and selection protocol had not apparently affected the pluripotential phenotype of the cells, as they were able to form chimeras with the same efficiency and show the same sex distortion phenomena as the parental line. There was an extensive level of contribution by the introduced cells to all the somatic tissues of chimeric mice.

As a number of selected clonal lines were tested, limited numbers of male chimeras were obtained from each

Table 3. A Comparison of the Chimera-forming Ability of the Original Cell Line EK CC1.2 and of G418-selected Subclones

Cell line	Blastocysts injected	Animals born	Chimeras	Female	Male	Setup	Bred	Germ line
CC1.2	321	205	127	29	97	41	19	7
G418-selected subclones	440	243	145	38	106	28	20	1

clonal line. A single germ line chimera derived from the clone was observed, however, insufficient results are available to conclude that the other clonal lines are no longer able to form functional germ cells, particularly because some of the most heavily chimeric animals were not taken through to breeding.

As these results demonstrate that EK cells that have been subcloned and selected in the presence of 0.2 mg/ml G418 are still capable of undergoing normal embryonic development and contributing extensively to many tissues of an adult mouse, including the functional sperm, this selection protocol may be used to select genetically transformed cells in order that the activity of introduced genes may be followed in the developmental environment of a normal animal.

Clearly the possible influence of specific features in the construction of the retroviral vectors upon their efficacy for use with embryonic stem cells must be further investigated. We have preliminary evidence (T. Franz, unpubl.) for the stable integration of proviral DNA from a completely different retroviral vector (which does not contain the SV40 origin of replication) into the genome of EK cells and the successful incorporation of these transfected cells into a chimeric mouse.

ACKNOWLEDGMENTS

We would like to thank Pam Fletcher and Lesley Cooke for their excellent technical assistance and the Cancer Research Campaign for its generous and continuing support for this work. A.B. is a Beit Memorial Research Fellow. M.K. acknowledges a postdoctoral fellowship from the National Institute of General Medical Sciences, National Institutes of Health.

REFERENCES

Bradley, A., M. Evans, M.H. Kaufman, and E. Robertson. 1984. Formation of germ-line chimeras from embryo derived teratocarcinoma cell lines. *Nature* 309: 255.

Cepko, C.L., B.E. Roberts, and R.C. Mulligan. 1984. Construction and applications of a highly transmissible murine retrovirus shuttle vector. *Cell* 37: 1053.

Evans, M.J. and M.H. Kaufman. 1981. Establishment in culture of pluripotential cells from mouse embryos. *Nature* 292: 154.

Evans, M., A. Bradley, and E. Robertson. 1985. EK cell contribution to chimeric mice: From tissue culture to sperm. *Banbury Rep.* 20: 93.

Evans, M., E. Robertson, A. Bradley, and M.H. Kaufman. 1983. Relationship between embryonal carcinoma cells and embryos. Current problems in germ cell differentiation. *Symp. Br. Soc. Dev. Biol.* 7: 139.

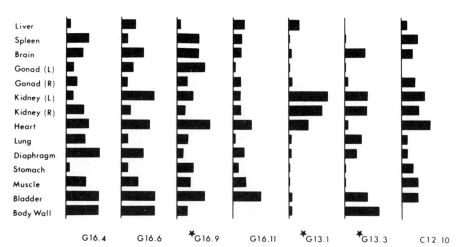

Figure 2. GPI analysis of nine chimeric phenotypic male mice constructed using clones isolated from CC1.2 cells following exposure to ZIPneoSVX virus. Host embryos carry the GPI isoenzyme type 1^b1^b, electrophoretically distinguishable from the transformed clones which carry the parental marker type 1^c1^c. Contributions by the injected cells ranged from 0% to 65%. G16.4, G16.6, G16.11, and C12.10 were normal breeding males. G16.9, G13.1, and G13.3 were nonbreeding phenotypic males. Upon autopsy all proved to be hermaphrodites.

FRIEDRICH, T.D. and J.M. LEHMAN. 1981. The state of simian virus 40 DNA in the embryonal carcinoma cell of the murine teratocarcinoma. *Virology* **110:** 159.

LINNEY, E., B. DAVIS, J. OVERHAUSER, E. CHAO, and H. FAN. 1984. Non-function of a Moloney murine leukemia virus regulatory sequence in F9 embryonal carcinoma cells. *Nature* **308:** 470.

MARTIN, G.R. 1981. Isolation of a pluripotent cell line from early mouse embryos cultured in medium conditioned with teratocarcinoma stem cells. *Proc. Natl. Acad. Sci.* **78:** 7634.

ROBERTSON, E.J., M.H. KAUFMAN, A. BRADLEY, and M.J. EVANS. 1983. Isolation, properties and karyotype analysis of pluripotential (EK) cell lines from normal and parthenogenetic embryos. *Cold Spring Harbor Conf. Cell Proliferation* **10:** 647.

RUBENSTEIN, J.L.R., J.-F. NICOLAS, and F. JACOB. 1984. Construction of a retrovirus capable of transducing and expressing genes in multipotential embryonic cells. *Proc. Natl. Acad. Sci.* **81:** 7137.

SCHNIEKE, A., K. HARBERS, and R. JAENISCH. 1983. Embryonic lethal mutation in mice induced by retrovirus insertion into the a1 (I) collagen gene. *Nature* **304:** 315.

WAGNER, E.F., M. VANEK, and B. VENNSTROM. 1985. Transfer of genes into embryonal carcinoma cells by retrovirus infection: Efficient expression from an internal promoter. *EMBO J.* **4:** 663.

WILLISON, K., C. BABINET, M. BOCCARA, and F. KELLY. 1983. Infection of preimplantation mouse embryos with simian virus 40. *Cold Spring Harbor Conf. Cell Proliferation* **10:** 307.

Gene Transfer into Murine Stem Cells and Mice Using Retroviral Vectors

E.F. WAGNER,* G. KELLER,† E. GILBOA,‡ U. RÜTHER,* AND C. STEWART*

*European Molecular Biology Laboratory, D-6900 Heidelberg, Federal Republic of Germany; †Basel Institute of Immunology, CH-4005 Basel, Switzerland; ‡Princeton University, Princeton, New Jersey 08544

Biological processes such as stem cell differentiation and embryonic development are believed to be controlled by differential gene expression. The control of gene expression during differentiation and development is poorly understood, thus the ultimate aim of developmental biology is to study the role of individual genes in these processes in vitro and in vivo.

We have decided to study gene regulation in the context of embryonal stem cell lines, hematopoietic precursor cells, and in early mouse embryos. All of these systems exhibit a distinct developmental program that can be followed in vitro and in vivo in the whole organism. Because of the lack of developmental mutants available to the mammalian embryologist, gene transfer is used to analyze gene regulation. The various ways of introducing genes into stem cells and mice are summarized in Figure 1A. Such approaches allow us to analyze elements controlling the correct cell-specific and temporal expression of genes as well as to investigate the consequences of gene expression on the developing organism.

However, an efficient delivery system is required to apply gene transfer techniques to the study of developmental questions. Retroviral vectors are the preferred method where the objective is to insert efficiently a specific, single gene into the chromosome of the target cell. The unique properties of the retrovirus-derived gene transfer system are based on the observation that retroviruses have evolved an efficient process for inserting their genes into the chromosome of the infected cells. In addition, it has been shown by Hwang and Gilboa (1984) that genes introduced into cells by retroviral infection are expressed 10- to 50-fold more efficiently as compared with genes introduced by other transfer methods, such as DNA-mediated transfection.

Several properties make the retroviral vectors particularly versatile: (1) A wide variety of cells can be infected using the same vectors; (2) the efficiency of gene transfer can reach a value of 100% and expression is very high; (3) genes up to 8 kb can be incorporated into the DNA vectors, which can be converted to viruses by transfection into an appropriate cell line (Mann et al. 1983); and (4) usually a single intact copy of the provirus which stably expresses the foreign gene(s) is integrated. Not only do retroviral vectors have advantages over other procedures, but for several systems this is the only method for successful gene transfer, e.g., for bone marrow cells (Joyner et al. 1983). However, there are also limitations, especially the size constraint (~8 kb), vector instability, and inefficient expression in early embryonic cells (Stewart et al. 1982).

However, we believe that the advantages of these vectors outweigh their disadvantages, and we will demonstrate the usefulness of these vectors in this paper. We set out to introduce by retrovirus infection foreign genes such as the neomycin (neo) resistance gene along with a transforming gene (v-myc) into established embryonal carcinoma (EC) cell lines and feeder-dependent stem cell lines (EK cells), which were subsequently used for chimera formation to study gene expression in vivo. Ways to overcome the inefficient and incorrect expression from viral regulatory elements in stem cells and early embryos will be discussed. In addition, the application of these vectors to genetically alter pluripotent precursor cells of the hematopoietic system and to study cell lineage relationships will also be reported.

METHODS

Cells and viruses. Cells were grown under conditions as described in Wagner et al. (1985a). The derivation of the EK cell lines was previously described by Evans and Kaufman (1981), except that primary mouse fibroblasts from 13-day-old embryos, instead of STO fibroblasts, were used as feeder cells. The construction and propagation of recombinant retroviruses was described previously (Wagner et al. 1985a).

Infection of cells with virus. EC cells were infected with various virus dilutions, as described by Wagner et al. (1985a), and EK cells were infected and selected in a similar way. Infection of bone marrow cells and reconstitution experiments were done according to G. Keller et al. (in prep.).

Infection of preimplantation embryos with virus. Four- to eight-cell-stage embryos were isolated from ICR females on the third day of pregnancy. The zona pellucida was removed using acidified Tyrode's solution (Stewart 1980) and the embryos were placed on confluent MMCV-neo (see Fig. 1B and Wagner et al. 1985a) virus-producing cells for 18 hours in a 1:1 mixture of DMEM with 10% fetal calf serum and Whitten's medium containing 2 μg/ml of Polybrene. Prior to transfer of the embryos to pseudopregnant recipients, the embryos were allowed to recover in Whitten's medium for 5-6 hours.

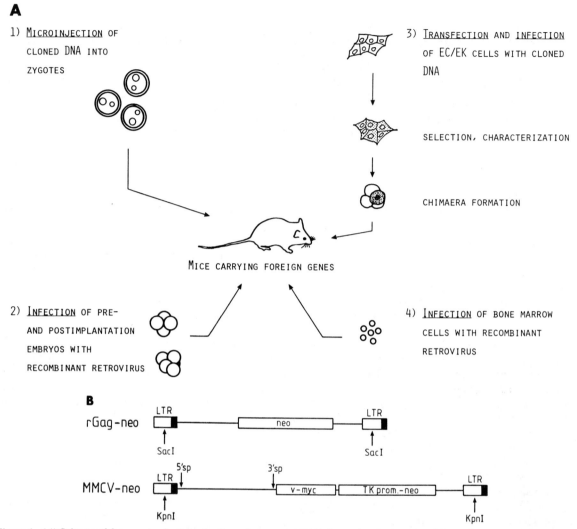

Figure 1. (*A*) Scheme of four routes for introduction of exogenous DNA into mice. Routes 1 and 2 are direct DNA injection or infection of early embryos with retroviruses; route 3 allows the characterization of cell clones in vitro before chimera formation, and in route 4 only specific cell lineages will receive the foreign gene. (*B*) Structure of retroviral vectors. The LTRs and adjacent viral sequences are derived from Mo-MLV. (5′ sp) 5′ splice donor site; (3′ sp) 3′ splice acceptor site.

Aggregation of embryos with EC and EK cells. The procedure to aggregate preimplantation embryos with EC and EK cells was previously described by Stewart (1982).

Mice. All mice were bred in our mouse colony at the European Molecular Biology Laboratory (EMBL).

Analysis of DNA and RNA. High-molecular-weight DNA was prepared, digested with restriction enzymes, and analyzed by Southern blotting with nick-translated probes, as described previously (Wagner et al. 1985a). Poly (A)⁺ RNA isolation and Northern and S1 analysis were also as described by Wagner et al. (1985a).

RESULTS AND DISCUSSION

Gene Transfer into EC Cells by Retroviral Infection:

Expression from the Viral Long Terminal Repeat

A difficulty of using retroviral vectors for infection of EC cells is that retroviral gene expression is inhibited in stem cells, even though infection results in the integration of proviral DNA into the host cell genome (Stewart et al. 1982; Gautsch and Wilson 1983; Niwa et al. 1983). The restriction of viral gene expression in EC cells may partly be due to the long terminal repeat (LTR) acting as an inefficient promoter (Linney et al. 1984), although additional blocks such as de novo methylation of proviral genomes (Stewart et al. 1982) may also exist. Despite these problems, viral vectors based on the genome of Moloney murine leukemia virus (Mo-MLV) were constructed, and consisted entirely of retroviral regulatory elements (promoters, splice sites, poly[A] addition signals) to express selectable (*neo* or *gpt*) and nonselectable (human β-interferon cDNA) genes following infection and selection of various EC cell lines (Fig. 1B). The purpose of these experiments was to test: (1) the frequency of obtaining stable cell clones following selection; (2) the efficiency of LTR transcription in EC cells compared with permissive 3T3 cells; and (3) the feasibility of expressing one or two genes from various positions in the retroviral genome.

The simplest of these vectors contained only the two viral LTRs and the selectable *neo* gene in the *gag* position of the proviral DNA (rGag-*neo*; Fig. 1B). Helper-competent viral stocks with high titers (10^6 neo-cfu/ml) were generated on NIH-3T3 cells and were used to infect EC cells such as F9, P19, and PC13. Overall, a frequency of 10^{-3} (0.1%) *neo*-resistant (neo^R) clones was obtained following G418 selection (Table 1). Similar results were obtained with an analogous Eco-*gpt* vector (Mann et al. 1983; C. Stewart et al., in prep.). The clones contained a single and intact proviral DNA copy of the recombinant genome and zero to five copies of helper virus DNA (Fig. 2A). The recombinant genome was usually found to be unmethylated when F9 cell clones were analyzed (Fig. 2A). Exceptions were found when P19 clones which expressed the *neo* or *gpt* gene, although being methylated in many sites of the genome, were tested (C. Stewart et al.; E. Wagner et al.; both in prep.). Expression of *neo*-specific RNA from the LTR was found to be largely reduced in EC cells when compared with the permissive NIH-3T3 cells (Fig. 2B). No apparent variation in the amount and size of RNA was seen when five individual clones were analyzed. These data show that LTRs can function in EC cells and they suggest that *cis*-acting elements at the site of integration, or mutations in the viral control elements, may be responsible for the expression in the EC cells. The same conclusions were reached by Sorge et al. (1984) and Taketo et al. (1985) using similar retroviral constructs.

The next series of vectors were constructed by inserting either the *neo* gene in the *env* position (rEnv-*neo*) or by engineering a double-expression vector containing the *neo* gene in the *gag* position and the human β-interferon cDNA in the *env* position (rDE-*neo-int*). Both types of vectors led to the derivation of stable EC cell clones with a frequency of 10^{-3}, as observed previously with rGag-*neo*. However, RNA analysis of selected cell clones showed that no correct genomic RNA

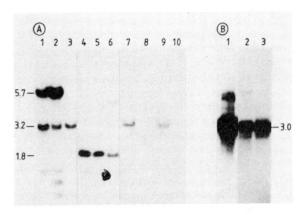

Figure 2. Analysis of DNA and RNA from EC and NIH-3T3 cells infected with rGag-*neo* and selected in G418-containing medium. (*A*) Southern analysis of genomic DNA from one individual NIH-3T3 clone and two F9 clones. Lanes *1-3* were digested with *Sac*I; lanes *4-6* with *Sma*I; lanes *7* and *9* with *Sac*I; and lanes *8* and *10* with *Sac*I/*Hpa*II. Lanes *1* and *4* are DNA from the NIH-3T3 clone; lanes *2, 5, 7,* and *8* are from one F9 clone; and lanes *3, 6, 9,* and *10* are from the second F9 clone. The filter was hybridized with a Mo-MLV-LTR specific probe (Müller and Müller 1984). (*B*) RNA analysis of virus-infected and -selected NIH-3T3 and F9 cell clones. Northern analysis of 3 μg of the poly(A)⁺ RNA and hybridization with a *neo*-specific probe. Lane *1* is again the NIH-3T3 clone and lanes *2* and *3* are two F9 clones.

was made with both vectors. In addition, the subgenomic, spliced RNA was heterogeneous in size on each EC cell clone, whereas the expected transcripts were observed when individual NIH-3T3 clones were examined (E. Wagner et al., in prep.). These results suggest that besides the inefficient transcription from the LTR in rGag-*neo* clones, additional problems with correct RNA processing or splicing are encountered in EC cells when genes are being expressed via the retroviral subgenomic RNA in rEnv-*neo* or in the double-expression vector system.

Gene Transfer into EC Cells by Retroviral Infection: Expression from Internal Promoters

One way to improve the frequency in obtaining selectable cell clones and the inefficient expression of foreign genes in EC cells following retroviral infection is the use of vectors that express the gene(s) of interest

Table 1. Relative Efficiencies of neo^R Colony Formation after Infection of EC and NIH-3T3 Cells with Retroviral Vectors

	Relative numbers of stable transformants			
Vector	NIH-3T3	F9	P19	PC13
rGag-*neo*[a]	1	~0.001	~0.001	~0.001
MMCV-*neo*	1	0.2-0.5	0.05-0.1	0.03-0.1
MMCV-*neo* (42)	1	0.2-0.5	—	—

[a]The same relative numbers of stable transformants were obtained with the rEnv-*neo* and rDE-*neo-int* vectors.

NIH-3T3 and EC cells were infected in parallel and subjected to selection for neo^R, as described (Wagner et al. 1985a). Virus titers were in the range of 1×10^6 *neo*-cfu/ml for rGag-*neo* and MMCV-*neo*, whereas the MMCV-*neo* (42) titer was 100-fold lower. The MMCV-*neo* constructs express the *neo* gene from the internal TK promoter.

from an internal promoter independent from the LTR. We have analyzed results obtained with two vectors that both express the *neo* gene from an internal thymidine kinase (TK) promoter (MV-4-*neo* and MMCV-*neo*; see Wagner et al. 1985a and Fig. 1B). One construct carries, in addition, a second gene, the v-*myc* oncogene, which has transforming properties in vitro (Vennström et al. 1984) and which can be expressed in NIH-3T3 cells from the viral LTR (Wagner et al. 1985a). Of the various EC cell lines infected with these internal promoter constructs, 10-50% of infected and selected cells gave rise to neo^R colonies when compared with NIH-3T3 cells (Table 1). The *neo* gene was expressed in both the nonpermissive EC and the permissive NIH-3T3 cells at similar levels, and transcription of *neo*-specific RNA was initiated at the proper site in the TK promoter (Fig. 3). No v-*myc* expression from the LTR was detectable in EC cells, even after induction of differentiation. In contrast, both transcription units are active in permissive NIH-3T3 cells and apparently escape the suppression of expression by an epigenetic mechanism, as was reported by Emerman and Temin (1984) for similar constructs. Thus, the use of internal promoters allows the isolation of many EC and EK cell clones carrying the foreign gene, which is efficiently expressed in stem cells in a possibly position-independent manner. A study by Rubenstein et al. (1984) showed that using a composite SV40-TK promoter in front of the *neo* gene also gives an increase in the frequency of obtaining neo^R colonies on EC cells such as PCC3 cells (up to 0.3%), when compared with NIH-3T3 cells. We are presently investigating the usefulness of various other internal promoters in retroviruses, especially the ones that can also be regulated in vitro and in vivo. It has already been shown that the SV40 promoter, the metallothionein promoter, and also the Mo-MLV promoter itself (E. Linney, pers. comm.) can function in this context.

The transfer of genes into various cells and animals, with the objective to study the regulation of those genes that are introduced by retroviral infection, is severely limited by the interactions between the LTR, which contains its own promoter and enhancer, and the introduced gene. We have developed several types of vectors that are derived from the genome of Mo-MLV and are designed to overcome most of these interferences. The main feature of these vectors, termed suicide retroviruses, is the use of a 3′ LTR fragment from which the enhancer and promoter sequences have been removed. Upon infection of target cells with the corresponding virus, the deletion is transferred to the 5′ LTR, resulting in a transcriptionally dead provirus. Thus virus transcription has "committed suicide," allowing uninhibited expression of the foreign gene(s) as well as minimizing the possibility of activating adjacent genes upon integration.

We have obtained results with several prototype vectors, which express the selectable *neo* gene from internal promoters such as TK, SV40, and mouse metallothionein; one construct contains, in addition, the c-*fos* gene under the control of the inducible human metallothionein promoter (MT-c-*fos*). Individual helper-free virus stocks were generated and were used to infect NIH-3T3 cells and various EC cells. DNA and RNA analysis proved that the 3′ deleted LTR was transferred to the 5′ LTR and that no LTR transcription was detectable (E. Gilboa et al., in prep.). The data on EC cells suggest that no enhancer function is required to express the *neo* gene. Moreover, the other vector carrying, in addition, a second transcription unit with MT-

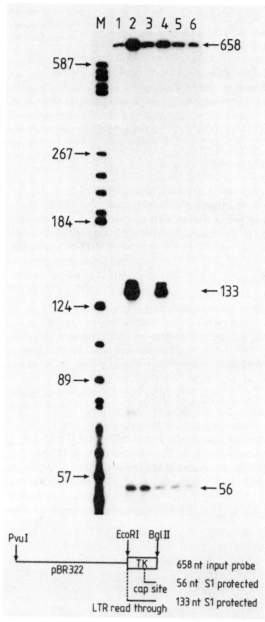

Figure 3. S1 nuclease analysis of RNA from rGag-*neo*-infected F9 cells (lane *1*), MV-4-*neo*-infected NIH-3T3 (lane *2*), and F9 cells (lane *3*); MMCV-*neo*-infected NIH-3T3 (lane *4*) and F9 cells before (lane *5*) and after (lane *6*) retinoic acid induction. Lane *M* contains ^{32}P-end-labeled *Hae*III fragments of pBR322. The TK-specific transcript is at position 56 and the LTR readthrough transcript at position 133. (Reprinted, with permission, from Wagner et al. 1985a.)

c-*fos*, which was inserted in the opposite orientation to the TK-*neo* fragment, still showed inducible c-*fos* expression (10- to 20-fold). Therefore, this approach of using suicide vectors will allow us to express genes from their own regulatory elements and from heterologous promoters independent from the retroviral LTRs and will be of great use for gene transfer studies in many systems, including application to somatic gene therapy in man.

Infection of EC Cells without Selection

For several purposes one would like to introduce a foreign gene in a given cell without selecting for a particular cell clone. Therefore, it is important to show that genes can be introduced into cells by infection with nonreplicating recombinant retroviruses and that they can be expressed without selection. Using EC cells we were able to show that almost every cell can be infected with a recombinant virus and that there is a high efficiency of proviral DNA integration even in the absence of selection (Wagner et al. 1985a). More importantly, some, but reduced, expression from the internal TK-promoter construct measured at the RNA and protein levels can be found after 3 weeks in infected but unselected cultures of EC cells, whereas no expression is detectable even 5 days after infection using a vector that uses the viral LTR to express the *neo* gene (Fig. 4). The reason why only a fraction of EC cells have the potential to express the *neo* gene in the absence of selection remains unknown. It appears that the block in expression is at the level of transcription, but additional translational effects cannot be ruled out. Specific methylation in the LTR seems to be an unlikely explanation for a mechanism, which is responsible for both the efficient expression of proviral genomes that were selected for expression and for the poor expression of proviruses obtained after infection without selection.

Infection of Feeder-dependent EC and EK Cells

Establishment of EK cells in vitro and formation of germ line chimeras. To use the EC/EK stem cell system for gene transfer experiments into the germ line of mice and for expression studies in chimeras, established cell lines are needed that yield high chimerism and germ line transmission after reintroduction of the cells into early embryos. Toward that aim, EK cells were established in vitro from 5-day-old blastocysts (day 1 = day of plug) according to the method of Evans and Kaufman (1981). The only modification used in the procedure was that primary mouse embryo fibroblasts, grown from 13-day-old mouse embryos, were used as a feeder cell layer rather than STO fibroblasts. We have succeeded so far in establishing 10 EK cell lines from blastocysts of the 129 (agouti-colored) inbred strain and have not yet obtained lines for C3H/J, C57B/6J, and F_1 hybrids between these strains.

One of these EK cell lines termed EKcs-1, derived from a 129 embryo, was karyotyped at the eighth passage in vitro as being male (X,Y), having the normal number of chromosomes in 76% of metaphases of the cultured cells; the remainder had lost one chromosome. To test the developmental totipotency of this cell line, cells from the eighth passage were aggregated with ICR (gpi^{bb}, Hbb^s) embryos to form chimeras as described by Stewart (1982). A total of 90 aggregates were transferred to pseudopregnant recipients and 33 mice were born; four males showed overt chimerism in the coat (see Fig. 5A). Three of the chimeras had a very extensive distribution of pigmented hair (more than 50%) and the fourth had only a small distribution of pigmented hair (on the head and some on the left rump) in its coat. All four males were mated with BALB/c females to test the germ line contribution of the cultured stem cells. All sired over 100 offspring; however, only the chimera with the weakest coat color distribution produced offspring that were all agouti (Fig. 5B). Such complete germ line transmission from the phenotypic male chimera is probably due to sexual mosaicism between the X/Y EK cells and X/X embryo cells, which are incapable of forming functional spermatozoa, as has been observed before (Bradley et al. 1984).

The lineage of the F_1 agouti offspring was confirmed by an independent marker at the hemoglobin *Hbb* locus. The chimeric father had only the *single* Hbb^s (ICR) variant in his blood lysate, whereas the mother had the *diffuse* Hbb^d (BALB/c) component (Fig. 6). All offspring had the typical Hbb^d pattern expected from crossing a BALB/c strain (Hbb^d) with a 129 strain (Hbb^d).

These results have confirmed those of Bradley et al. (1984) as to the feasibility of producing germ line chimeras from EK cells grown in vitro. Here, the germ line chimera was produced by the aggregation technique,

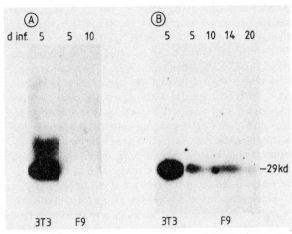

Figure 4. Neomycin-phosphotransferase (NPTII) activity in cell extracts from NIH-3T3 cells and F9 cells infected with rGag-*neo* (*A*) or MV-4-*neo* (*B*) without selection. The same amount of protein was loaded onto a nondenaturing acrylamide gel and the in situ assay was performed as described (Wagner et al. 1985b). (d. inf.) Days after infection without selection.

Figure 5. Chimeric mice and the germ line chimera derived from the karyotypically normal EK cell line, EKcs-1. (A) Two extensively chimeric males produced by aggregation of 2 ICR four- to eight-cell stage embryos with stem cells of the EK cell line. (B) Complete male germ line chimera produced as described above with minor chimerism in the coat (note the localization of pigment in the coat hair on the head region). All of its germ cells were of tumor-strain lineage, as seen in the all-agouti color of three F_1 offspring from a mating to a BALB/c mother (animal facing to the right).

which is technically simpler than the microinjection of blastocysts. We are at present isolating cell clones from the above-mentioned line as well as from other cell lines that carry a variety of different retroviral vectors. These clones will be karyotyped and then tested to see if they can also produce germ line chimeras.

Gene expression studies in chimeras. The first series of experiments using the EC/EK cell system in vitro in combination with gene expression studies in chimeras in vivo were undertaken to answer the following questions: (1) Can a gene that is expressed from an LTR and selected for activity in vitro at the stem cell level remain active following differentiation and development in vivo; and (2) are genes that are expressible from LTRs but silent in stem cells able to be reactivated in differentiated cells following chimera formation.

In one series of experiments, P10 EC cells were infected with the *neo*-containing vector rGag-*neo* (Wagner et al. 1985b). Clones were obtained at a frequency of 10^{-5} and one clone was isolated which carried a single intact proviral copy (C. Stewart et al., in prep.). The stability of expression of the *neo* gene from the LTR and the developmental capability of these cells was tested by reintroducing them into normal mouse embryos by the aggregation technique (Stewart 1982; Fuji and Martin 1983). Following manipulation, 34 mice were born; 11 of these showed overt chimerism in the coat. A wide range of chimerism was observed in various tissues assayed by the P10-specific allele glucose-phosphate-isomerase (*gpi-1b*). Expression of the LTR-driven *neo* gene at the RNA and protein levels measured by the neomycin-phosphotransferase (NPTII) assay (Reiss et al. 1984) was correlated with the presence of the P10-*neo*1 cells in various organs, thereby demonstrating that the gene, which was active at the stem cell level, maintained its activity during differentiation (C. Stewart et al., in prep.).

In a second series of experiments, Cp1 EK cells (obtained from Martin Evans, Cambridge University, England), which previously gave germ line chimeras (Bradley et al. 1984), were infected with helper-independent MMCV-*neo* virus (Wagner et al. 1985a). Several stem cell clones were isolated that expressed the selectable *neo* gene from the TK promoter and contained the silent v-*myc* gene, which can be expressed in fibroblasts from the LTR (Wagner et al. 1985a). Three characterized clones were used to form aggregation chimeras, and, currently, several mice are being analyzed for expression of the *neo* gene as well as for v-*myc* expression in various differentiated cells. It seems that the selectable gene remains active again throughout development and that the second gene, once being silent at the stem cell level, cannot be reactivated in various differentiated cell types (C. Stewart et al., in prep.).

These experiments demonstrate for the first time that it is possible to introduce genes by retroviral infection into pluripotent feeder-dependent stem cells, which, following characterization in vitro, are still able to undergo differentiation and development in the embryo and can be used for studying gene expression. It is still necessary, however, to demonstrate that the genetically altered stem cell clones can also produce germ line chimeras, which would then make this system an

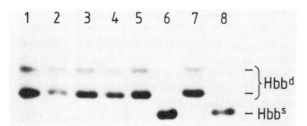

Figure 6. Electrophoretic separation of hemoglobin variants (*Hbb* locus) in cellulose acetate showing the characteristic pattern of the 129 strain *diffuse* type (Hbb^d) of four F_1 offspring (lanes *1–4*); the Hbb^d type of the BALB/c mother (lane *5*) and the chimeric father (lane *6*), which only showed the *single* type (Hbb^s) variant. Lane *7* is a Hbb^d 129 control and lane *8* shows the Hbb^s from an ICR mouse.

alternative to the routinely used microinjection of DNA into fertilized eggs to produce transgenic mice.

Retroviral Infection of Preimplantation Embryos

As an alternative to DNA injection into zygotes, genes can be introduced into mice by direct infection of preimplantation embryos with recombinant retroviruses (Fig. 1A). The feasibility of using infection of preimplantation embryos was demonstrated by experiments using wild-type Mo-MLV (Jaenisch et al. 1975), although retroviral gene expression was also shown to be repressed in these experiments (Jähner et al. 1982). We have set out to determine the efficiency of introducing foreign genes and the possibility of obtaining expression of the genes, when four- to eight-cell-stage embryos were infected.

In our initial experiments, the embryos were infected with the MMCV-*neo* virus (~5 × 10⁵ *neo* cfu/ml, Fig. 1B) carrying the v-*myc* gene and the *neo* gene, which is under the control of the internal TK promoter. Following infection of embryos by coculturing for 18 hours with virus-producing fibroblasts, the embryos were transferred to pseudopregnant foster mothers and allowed to develop to the 14th day of gestation. A total of 132 infected embryos were transferred. The 50 that developed normally were killed and analyzed for expression of the *neo* gene and for the presence of vector sequences. Expression of the *neo* gene was tested by explanting half of a fetal extract without yolk sac, placenta, and amnion into cultures in selective medium containing G418. Of these, 11 (22%) were found to have cells that proliferated under selective conditions, and, when RNA was prepared from these primary cultures, expression of the *neo*-specific RNA but not the v-*myc*-containing RNA originating from the LTR was found (Fig. 7B). The expression of the foreign gene was correlated with the presence of vector sequences, when DNA was analyzed either from the cell cultures or the corresponding second half of the extract including extraembryonic tissues (Fig. 7A). Thus, these results show that genes carried in retroviral vectors can readily be introduced into embryos by direct infection and that a gene from an internal promoter can be expressed in 14-day-old fetuses. Infected embryos have been allowed to develop to term and currently are being analyzed for the presence and expression of the foreign genes. DNA-positive animals are being mated to see if the newly introduced genes are transmitted to offspring and if expression can be found in subsequent generations.

Gene Transfer into Bone Marrow Cells by Retroviral Infection

The hematopoietic system is another well-characterized developmental cell system that recently has become accessible to gene transfer studies. Pluripotent precursors found in adult bone marrow can be manipulated in vitro and then transferred to hematopoietically deficient hosts, where they are able to generate

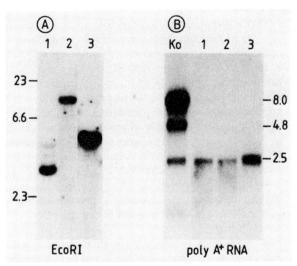

Figure 7. Analysis of DNA from fetal extracts (*A*) and RNA from primary cell cultures of fetuses (*B*) that had been infected with the MMCV-*neo* virus at the four- to eight-cell stage. (*A*) Southern analysis of genomic DNA from three fetuses. The DNA was digested with *Eco*RI, which cuts once in the provirus, and the filter was hybridized with a *neo*-specific probe. (*B*) RNA analysis of fetal extracts that had been grown in the presence of G418 in vitro. Poly(A)⁺ RNA (3 μg) was loaded in each lane and hybridization was performed using a ³²P-nick-translated *neo* probe. Lane *Ko* is a control RNA from MMCV-*neo*-infected and selected NIH-3T3 cells.

cells of the myeloid and lymphoid lineages (Abramson et al. 1977). These primitive pluripotent cells can be assayed only in long-term reconstitution experiments (Harrison 1980; Boggs et al. 1982), whereas their descendants, the more restricted multipotent and unipotent precursors, can be detected in vivo (Till and McCulloch 1961) and in vitro (for a review, see Burgess and Nicola 1983) by colony assays.

Initial in vitro experiments showed that precursors committed to neutrophil and macrophage differentiation can be infected with a retroviral vector carrying the bacterial *neo* resistance gene. Expression of the introduced gene was demonstrated by the growth of precursors in G418 and by the presence of *neo*-specific RNA in their progeny (Joyner et al. 1983). Recently, Williams et al. (1984) used a retrovirus to transfer the *neo* gene into spleen-colony forming cells (CFU-S) and Miller et al. (1984) obtained some evidence for human hypoxanthine phosphoribosyl transferase (*hprt*) expression in one mouse, reconstituted with bone marrow cells that had been infected with a replication-competent *hprt*-carrying retrovirus.

In our initial experiments using a *neo*-carrying retrovirus for bone marrow infections, we wanted to answer three basic questions: (1) What kind of precursor cells can be infected with the recombinant virus, and what efficiency of infection and expression can be obtained; (2) can the newly introduced gene be expressed in various hematopoietic lineages; and (3) can we study cell lineage relationships through the use of unique proviral integration sites.

The first experiments were performed to test the efficiency of infecting bone marrow cells with the *neo*-containing virus using the strategy outlined in Figure 8. In vitro colony assays showed that the majority of the precursors (50-95% of all colony-forming cells, CFC) survived the 24-hour coculture of fresh bone marrow cells with virus-producing cells and that 10-20% of these were able to grow and form colonies in the presence of G418 (Wagner et al. 1985b). Following a 48-hour selection step in liquid culture, 60-95% of the surviving CFC were G418 resistant. Short-term reconstitution experiments with virus-infected, nonselected cells for 12 days revealed the presence of intact vector DNA in a small fraction of the spleen cells of reconstituted mice. In contrast, the majority of spleen cells from mice reconstituted with preselected cells contained the recombinant virus and expressed the *neo* gene.

The next analysis was done on long-term reconstituted animals (6-17 weeks) to determine whether or not primitive precursors, capable of generating myeloid and lymphoid cells, have been infected with the recombinant viruses and to demonstrate expression of the introduced gene in the different lineages. Through analysis of unique viral integration sites of tissues and cell lines from several mice, we can show the infection of pluripotent precursor cells (G. Keller et al., in prep.). Expression of the newly introduced *neo* gene was found in all lineages tested. This allowed us to grow a variety of different cells (B- and T-cell hybridomas, mast cells, etc.) in the presence of G418 and thereby clearly demonstrate the presence of common lymphoid and myeloid integration sites.

The experiments described above are directed toward understanding early events involved in the commitment of pluripotent precursors to cells that display a more restricted developmental potential. One approach toward elucidating the mechanism involved at this stage of development in the hematopoietic system is to introduce genes that have the capacity to interfere with the normal pattern of differentiation of these primitive precursors. Experiments along these lines are now under way.

These experiments should also provide the basis for

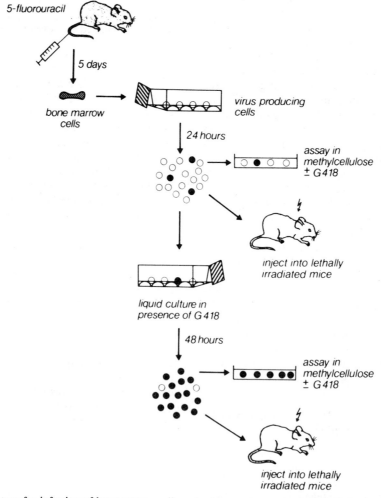

Figure 8. General strategy for infection of bone marrow cells and subsequent assay in methylcellulose or in reconstituted mice using either nonselected or G418-selected cells. The protocol for infection was identical to that described by G. Keller et al. (in prep.).

somatic gene therapy in man. However, additional information is needed from animal experiments concerning the efficiency of gene transfer, the possibility of regulating the gene of interest within a retrovirus, the consequences of proviral integration, and the stability and longevity of expression of the introduced gene.

CONCLUSIONS

The experiments discussed in this paper have shown that gene transfer by retroviral infection of embryonal stem cell lines, preimplantation embryos, and hematopoietic precursor cells offers a powerful tool for analyzing gene expression in vitro and in vivo. Retroviral vectors with internal promoters and suicide vectors with deleted LTRs that allow the efficient and regulatable expression of genes in stem cells independent from the viral LTR were described. With readily available EK cell lines that can form germ line chimeras, this stem cell system has been established to study gene expression and its consequences both in vitro and in vivo. It has yet to be proved that germ line chimeras can be obtained from selected cell clones and that this procedure could serve as an alternative to producing transgenic mice by DNA injection. In the hematopoietic system, we now have tools to alter various precursor cells genetically, to study cell lineage relationships, and to proceed to transfer foreign genes such as c-*onc* genes, whose functions in differentiation are still unknown. These approaches will provide us with novel information regarding the role of individual genes in normal development, in growth control, and in the process of neoplastic transformation.

ACKNOWLEDGMENTS

We thank Mirka Vanek and Christa Garber for technical assistance, Catherine Boulter for critically reading, and Ines Benner for typing the manuscript. U.R. is a recipient of an EMBO postdoctoral long-term fellowship and E.G. a recipient of a grant from ACS MV-116B.

REFERENCES

ABRAMSON, S., R.G. MILLER, and R.A. PHILLIPS. 1977. The identification in adult bone marrow of pluripotential and restricted stem cells of the myeloid and lymphoid systems. *J. Exp. Med.* **145:** 1567.

BOGGS, D.R., S.S. BOGGS, D.F. SAXE, L.A. GRESS, and D.R. CANFIELD. 1982. Hematopoietic stem cells with high proliferative potential. *J. Clin. Invest.* **70:** 242.

BRADLEY, A., M. EVANS, M.H. KAUFMAN, and E. ROBERTSON. 1984. Formation of germ-line chimaeras from embryo-derived teratocarcinoma cell lines. *Nature* **309:** 255.

BURGESS, A. and N. NICOLA. 1983. *Growth factors and stem cells.* Academic Press, New York.

EMERMAN, M. and H.M. TEMIN. 1984. Genes with promoters in retrovirus vectors can be independently suppressed by an epigenetic mechanism. *Cell* **39:** 459.

EVANS, M.J. and M.H. KAUFMAN. 1981. Establishment in culture of pluripotent cells from mouse embryos. *Nature* **292:** 154.

FUJI, J.T. and G.R. MARTIN. 1983. Developmental potential of teratocarcinoma stem cells in utero following aggregation with cleavage stage mouse embryos. *J. Embryol. Exp. Morphol.* **74:** 79.

GAUTSCH, J.W. and M.C. WILSON. 1983. Delayed de novo methylation in teratocarcinoma suggests additional tissue-specific mechanisms for controlling gene expression. *Nature* **301:** 32.

HARRISON, D.E. 1980. Competitive repopulation: A new assay of long-term stem cell functional capacity. *Blood* **55:** 77.

HWANG, L. and E. GILBOA. 1984. Expression of genes introduced into cells by retroviral infection is more efficient than that of genes introduced into cells by DNA transfection. *J. Virol.* **50:** 417.

JAENISCH, R., H. FAN, and B. CROKER. 1975. Infection of preimplantation mouse embryos and of newborn mice with leukemia virus: Tissue distribution of viral DNA and RNA and leukemogenesis in the adult animal. *Proc. Natl. Acad. Sci.* **72:** 4008.

JÄHNER, D., H. STUHLMANN, C.L. STEWART, K. HARBERS, J. LÖHLER, I. SIMON, and R. JAENISCH. 1982. De novo methylation and expression of retroviral genomes during mouse embryogenesis. *Nature* **298:** 623.

JOYNER, A., G. KELLER, R.A. PHILLIPS, and A. BERNSTEIN. 1983. Retrovirus transfer of a bacterial gene into mouse haematopoietic progenitor cells. *Nature* **305:** 556.

LINNEY, E., B. DAVID, J. OVERHAUSER, E. CHAO, and H. FAN. 1984. Non-function of a Moloney murine leukaemia virus regulatory sequence in F9 embryonal carcinoma cells. *Nature* **308:** 470.

MANN, R., R.C. MULLIGAN, and D. BALTIMORE. 1983. Construction of a retrovirus packaging mutant and its use to produce helper-free defective retrovirus. *Cell* **33:** 153.

MILLER, A.D., R.J. ECKNER, D.J. JOLLY, T. FRIEDMANN, and I.M. VERMA. 1984. Expression of a retrovirus encoding human HPRT in mice. *Science* **225:** 630.

MÜLLER, R. and D. MÜLLER. 1984. Co-transfection of normal NIH/3T3 DNA and retroviral LTR sequences: A novel strategy for the detection of potential c-*onc* genes. *EMBO J.* **3:** 1121.

NIWA, O., Y. YOKOTA, H. ISHIDA, and T. SUGAHARA. 1983. Independent mechanisms involved in suppression of the Moloney leukemia virus genome during differentiation of murine teratocarcinoma cells. *Cell* **32:** 1105.

REISS, B., R. SPRENGEL, H. WILL, and H. SCHALLER. 1984. A new sensitive method for qualitative and quantitative assay of neomycin phosphotransferase in crude cell extracts. *Gene* **30:** 211.

RUBENSTEIN, J.L.R., J.-F. NICOLAS, and F. JACOB. 1984. Construction of a retrovirus capable of transducing and expressing genes in multipotential embryonic cells. *Proc. Natl. Acad. Sci.* **81:** 7137.

SORGE, J., A.E. CUTTING, V.D. ERDMAN, and J.W. GAUTSCH. 1984. Integration-specific retrovirus expression in embryonal carcinoma cells. *Proc. Natl. Acad. Sci.* **81:** 6627.

STEWART, C.L. 1980. Aggregation between teratocarcinoma cells and preimplantation mouse embryos. *J. Embryol. Exp. Morphol.* **58:** 289.

———. 1982. Formation of viable chimaeras by aggregation between teratocarcinoma and preimplantation mouse embryos. *J. Embryol. Exp. Morphol.* **67:** 167.

STEWART, C.L., H. STUHLMANN, D. JÄHNER, and R. JAENISCH. 1982. De novo methylation, expression and infectivity of retroviral genomes introduced into embryonal carcinoma cells. *Proc. Natl. Acad. Sci.* **79:** 4098.

TAKETO, M., E. GILBOA, and M.J. SHERMAN. 1985. Isolation of embryonal carcinoma cell lines that express integrated recombinant genes flanked by the Moloney murine leukemia virus long terminal repeat. *Proc. Natl. Acad. Sci.* **82:** 2422.

TILL, J.E. and E.A. MCCULLOCH. 1961. A direct measurement of the radiation sensitivity of normal mouse bone marrow cells. *Radiat. Res.* **14:** 213.

VENNSTRÖM, B., P. KAHN, B. ADKINS, P. ENRIETTO, M.J. HAYMAN, T. GRAF, and P. LUCIW. 1984. Transformation of mammalian fibroblasts and macrophages in vitro by a murine retrovirus encoding an avian v-*myc* oncogene. *EMBO J.* **3**: 3223.

WAGNER, E.F., M. VANEK, and B. VENNSTRÖM. 1985a. Transfer of genes into embryonal carcinoma cells by retrovirus infection: Efficient expression from an internal promoter. *EMBO J.* **4**: 663.

WAGNER, E., U. RÜTHER, R. MÜLLER, C.L. STEWART, E. GILBOA, and G. KELLER. 1985b. Expressing foreign genes in stem cells and mice. *Banbury Rep.* **20**: (in press).

WILLIAMS, D.A., I.R. LEMISCHKA, D.G. NATHAN, and R.C. MULLIGAN. 1984. Introduction of new genetic material into pluripotent haematopoietic stem cells of the mouse. *Nature* **310**: 476.

The Regulation of Gene Expression in Murine Teratocarcinoma Cells

C.M. GORMAN,* D.P. LANE,† C.J. WATSON,‡ AND P.W.J. RIGBY§

Cancer Research Campaign, Eukaryotic Molecular Genetics Research Group, Department of Biochemistry, Imperial College of Science and Technology, London SW7 2AZ, England

The elucidation of the mechanisms that regulate gene expression during the early development of the mouse requires, at the very least, the availability of: (1) cloned genes that are appropriately regulated; (2) gene transfer systems in which the *cis*-acting DNA sequences that control expression and regulation can be analyzed, and (3) procedures for identifying the *trans*-acting factors, presumably proteins, that mediate the regulation.

There are a number of well-characterized genes that are regulated during early development, for example, those encoding α-fetoprotein (AFP) (Dziadek and Adamson 1978), the class-I major histocompatibility complex antigens (Rosenthal et al. 1984), and laminin (Cooper et al. 1983); in the case of AFP, considerable progress has been made in defining the *cis*-acting sequences required for proper regulation (Krumlauf et al., this volume). However, many of these genes are activated at relatively late times during development and are expressed in terminally differentiated cells. We wish to understand gene expression in stem cells, particularly those of the preimplantation embryo, and here we discuss experiments designed to identify and characterize genes that are expressed at this stage and repressed following differentiation. The isolation from early embryos of sufficient quantities of material remains a major problem and therefore we have begun by using an in vitro model system, namely embryonal carcinoma (EC) cells. These are the undifferentiated stem cells of teratocarcinomas and are thought to be analogous to the inner cell mass (ICM) cells of the blastocyst. EC cells can be obtained in the quantities required for various purposes including the preparation of in vitro transcription extracts and certain kinds of biochemical analyses. They also form the basis of the gene transfer system that we have employed (Gorman et al. 1984, 1985). To understand in detail the *cis*-acting sequences that regulate the expression of a particular gene, one must be able to create a large number of mutations in putative control regions and then assay the effects of such mutations when the genes are reintroduced into cells. The requisite number of mutations cannot be analyzed in transgenic mice and therefore it is necessary to develop an appropriate cell culture system. We have previously shown that it is possible to use calcium phosphate-mediated transfection to introduce exogenous genes into undifferentiated EC cells at high efficiency (Gorman et al. 1984). We have now used this system to analyze the regulation of viral control elements in these cells.

Several viruses, for example simian virus 40 (SV40), polyoma, and Moloney murine leukemia virus (Mo-MLV), cannot efficiently express their genomes in undifferentiated EC cells (Swartzendruber and Lehman 1975; Teich et al. 1977; Segal and Khoury 1979; Linnenbach et al. 1980; Stewart et al. 1982), whereas efficient expression occurs in differentiated cells. Differentiation is accompanied by changes in cellular gene expression and we assume that the viral control elements will enable us to probe the mechanisms that regulate cellular genes. We have used gene transfer to analyze papovavirus and retrovirus promoters in EC cells and have thus shown that undifferentiated cells contain a negative regulatory factor(s) that operates on enhancer sequences.

MATERIALS AND METHODS

Recombinant plasmids. pTSV3 contains the SV40 genome cloned as an *Eco*RI linear into the *Eco*RI site of pAT153 (Clayton et al. 1982). In pRSVcat, transcription of the chloramphenicol acetyl transferase (CAT) gene is driven by the long terminal repeat (LTR) of Rous sarcoma virus (RSV) (Gorman et al. 1982b); in pMSVcat the LTR is derived from Moloney murine sarcoma virus (Mo-MSV) (Laimins et al. 1984). pLTR10cat lacks the enhancer region whereas pLTR0cat lacks both the enhancer and elements of the promoter (Laimins et al. 1984). In pSV2cat transcription is driven by the early promoter of SV40; in pSV1cat the enhancer has been deleted (Gorman et al. 1982a,b). In pSrM2cat the SV40 enhancer has been replaced by that of Mo-MSV (Laimins et al. 1982). All recombinant plasmid DNAs were prepared according to Gorman et al. (1982a). DNAs to be used in transfection experiments were purified by two successive cesium chloride–ethidium bromide equilibrium density centrifugations because this improves the reproducibility of the experiments.

---

*Present address: Department of Molecular Biology, Genentech Inc., 460 Point San Bruno Boulevard, South San Francisco, California, 94080.

†Present address: Imperial Cancer Research Fund, Clare Hall Laboratories, South Mimms, Hertfordshire, England.

‡Present address: Chester Beatty Laboratories, Institute of Cancer Research, Fulham Road, London, SW3 6JB, England.

§Present address: Laboratory of Eukaryotic Molecular Genetics, National Institute for Medical Research, The Ridgeway, Mill Hill, London, NW7 1AA, England.

Cell culture and DNA-mediated gene transfer. Cells were grown in Dulbecco's modification of Eagle's medium supplemented with 10% (vol/vol) fetal calf serum and antibiotics. F9 cells were induced to differentiate by adding 10^{-6} M retinoic acid to the medium (Strickland and Mahdavi 1978). The cells were refed every 48 hours with fresh medium containing retinoic acid and differentiation was monitored by immunocytochemical staining (Gorman et al. 1985). PCC3 cells were induced to differentiate by treatment with 5×10^{-7} M retinoic acid for 10 days, the medium being changed every 2 days. Differentiation was again monitored by immunocytochemical staining.

DNA was introduced into cells by calcium phosphate transfection and expression was monitored by CAT assays or by immunocytochemical staining using the procedures described by Gorman et al. (1985).

Nucleic Acids. Polyadenylated cytoplasmic RNA was isolated from undifferentiated and differentiated PCC3 cells by the NP-40 lysis procedure (Favaloro et al. 1980). Cellular genomic DNA was isolated from PCC3 cells by the method given in Maniatis et al. (1982). cDNA inserts were isolated from λgt10 clones by digestion with *Eco*RI and preparative electrophoresis in 0.6% low-gelling-temperature agarose gels. They were labeled with ^{32}P using the procedure of Feinberg and Vogelstein (1984).

Transfer hybridization. Polyadenylated cytoplasmic RNA was fractionated by electrophoresis in 1.2% agarose gels containing formaldehyde and transferred to nitrocellulose filters (Thomas 1980). The filters were hybridized in $5 \times$ SSC, 0.01% (wt/vol) SDS, 20 mM NaH_2PO_4, $2 \times$ Denhardt's solution, 2 µg/ml sonicated salmon sperm DNA, and 50% (vol/vol) formamide at 42°C and washed in $0.2 \times$ SSC, 0.1% (wt/vol) SDS at 55°C. Genomic DNA was digested with restriction endonucleases, fractionated by electrophoresis in 0.6% agarose gels and, following depurination (Wahl et al. 1979), transferred to nitrocellulose filters (Southern 1975). These were hybridized in $5 \times$ SSC, 0.1% (wt/vol) SDS, $2 \times$ Denhardt's solution, and 2 µg/ml sonicated salmon sperm DNA at 65°C and washed in $0.1 \times$ SSC and 0.1% (wt/vol) SDS at 65°C.

Construction and analysis of cDNA libraries. Double-stranded cDNA was synthesized from polyadenylated cytoplasmic RNA from undifferentiated PCC3 cells by the procedures of Watson and Jackson (1985). This cDNA was subsequently cloned into λgt10 (Huynh et al. 1985) to generate a library comprising 6×10^5 independent events. The library was screened by plaque hybridization (Benton and Davis 1977) with ^{32}P-labeled cDNA synthesized from polyadenylated cytoplasmic RNA from either undifferentiated or differentiated PCC3 cells. Plaques that reproducibly hybridized more intensely to the undifferentiated probe were picked and purified, and the phage was grown in plate lysates, DNA was prepared, and the cDNA insert was isolated as described above.

RESULTS

Isolation and Characterization of cDNA Clones Corresponding to Genes Expressed in Undifferentiated EC Cells but Not in Differentiated Derivatives

We have constructed a λgt10-based cDNA library, containing 6×10^5 independent events, from the polyadenylated cytoplasmic RNA of undifferentiated PCC3 cells. This library was then differentially screened with ^{32}P-labeled cDNA probes prepared from undifferentiated and differentiated cell mRNA. Plaques (4×10^3) were screened and of these 15, which reproducibly showed more intense hybridization with the undifferentiated probe, were selected for further study.

Figure 1, A and B, shows a Northern blot analysis of mRNA from undifferentiated and differentiated PCC3 cells hybridized with one of the clones isolated as described above. The cDNA insert from λ22, which is 2.4 kb in length, detects a single RNA species of approximately 5.6 kb, which is present at much higher levels in undifferentiated cells than in their differentiated derivatives. We cannot tell whether the low level of expression seen in the differentiated cells results from unabated expression in a small percentage of cells that remain undifferentiated or from a low level of expression in all differentiated cells. When used to probe Southern blots of genomic DNA (Fig. 1C), the insert from λ22 hybridizes to a large number of fragments, suggesting either that it defines a multigene family or that it corresponds to a repetitive element.

The size of the mRNA detected by λ22 and the apparent moderate repetition of the corresponding genomic sequence suggest that λ22 might correspond to a transposable element or to an endogenous retrovirus. We know that this clone does not hybridize to an intracisternal A particle probe and we are presently characterizing it further.

We have isolated a number of other clones that have similar expression patterns and are currently analyzing them in detail.

Negative Regulation of Viral Enhancers in Undifferentiated EC Cells

If undifferentiated F9 cells are infected with SV40, only a small percentage of morphologically distinct, and presumably differentiated, cells synthesize the viral early gene product large T antigen, whereas if the cells are induced to differentiate immediately after infection then up to 50% of the nuclei stain 72 hours after infection (Fig. 2A,B). However, if undifferentiated cells are transfected with pTSV3, a plasmid that contains the entire SV40 genome, then 30-60% of the cells express both large T antigen and small t antigen, the two protein products of the viral early region (Fig. 2C-E). We have performed a large number of control experiments which show that the transfection procedure does not induce the differentiation of the EC cells (Gorman et al. 1985).

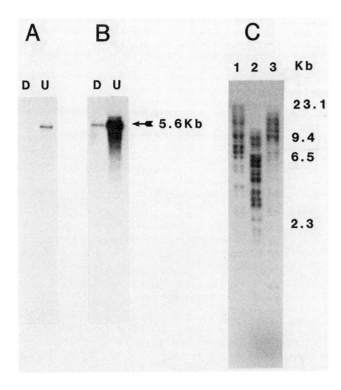

Figure 1. Transfer hybridization analysis of the cDNA clone λ22. (A) Polyadenylated cytoplasmic RNA from undifferentiated (U) and differentiated (D) PCC3 cells was fractionated by electrophoresis in a 1.2% agarose gel containing formaldehyde, transferred to a nitrocellulose filter, and hybridized with the ^{32}P-labeled cDNA insert from λ22. Autoradiography was for 2.5 hr. (B) The same filter as in A, exposed for 20 hr. (C) Genomic DNA from PCC3 cells was digested with the restriction endonucleases BamHI (lane 1), HindIII (lane 2), and EcoRI (lane 3), fractionated by electrophoresis in a 0.6% agarose gel, transferred to a nitrocellulose filter, and hybridized with the ^{32}P-labeled cDNA insert from λ22.

Why are the consequences of viral infection so different from those of the transfection of viral DNA? The latter procedure introduces into the cells many more copies of the viral genome than does the former, suggesting that the lack of expression following infection is due to a negative regulatory factor(s) which is titrated out by the large number of genomes introduced by transfection. In agreement with this idea, the number of undifferentiated cells expressing large T antigen increases as the amount of DNA transfected is increased (Gorman et al. 1985).

Table 1 gives the results of experiments in which the cat gene, which encodes the easily assayable bacterial enzyme CAT, driven by a variety of viral transcriptional control elements, has been introduced into undifferentiated F9 cells and expression has been monitored by assaying the amount of CAT enzyme synthesized in the transfected cells. The first striking observation is that pSV1cat, which lacks the viral enhancer, is expressed as efficiently as pSV2cat, which contains it, whereas in differentiated cells the enhancer-containing construct is much more efficiently expressed.

In agreement with previously published data, the LTR of MSV is inactive in undifferentiated cells but we were surprised to observe that pLTR10cat, in which the enhancer has been deleted, is expressed quite efficiently. This suggests that the MSV enhancer region is the target for a negative regulatory factor(s). To explore this idea further, we have used pSrM2cat, in which the SV40 enhancer has been replaced by that of MSV. We have shown that in these cells the SV40 early promoter functions without its homologous enhancer (pSV1cat) and thus if the MSV enhancer is the target for a negative regulatory factor, pSrM2cat should be nonfunctional. This expectation is fulfilled (Table 1).

These effects are specific to the undifferentiated state. The data of Table 1 show that in differentiated cells pSV2cat is expressed more efficiently than pSV1cat, i.e., the SV40 enhancer is functional, and that pMSVcat is more efficient than pLTR10cat, i.e., the activating function of the MSV enhancer is observed. Moreover, the dose-response effect seen when transfecting SV40 DNA into undifferentiated cells is not observed with differentiated cells (Gorman et al. 1985).

A final piece of evidence for a negative regulatory factor(s) comes from cotransfection experiments. If pMSVcat is cotransfected with pRSVneo, which contains the LTR of RSV driving the 3'-aminoglycoside phosphotransferase gene, then increasing the proportion of the latter DNA increases CAT activity (Table 1), indicating that the RSV LTR can compete for the negative regulatory factor(s). Similar competition effects are seen with pSV2cat but not with pSV1cat (Gorman et al. 1985).

DISCUSSION

The cDNA cloning experiments that we have described have led to the identification of a number of genes that are expressed at higher levels in undifferentiated embryonic stem cells than in their differentiated derivatives. We are presently isolating the corresponding genomic clones so that we can locate the sequences that specify this expression pattern by in vitro mutagenesis and DNA-mediated gene transfer. We also plan to attempt to use in vitro transcription systems derived

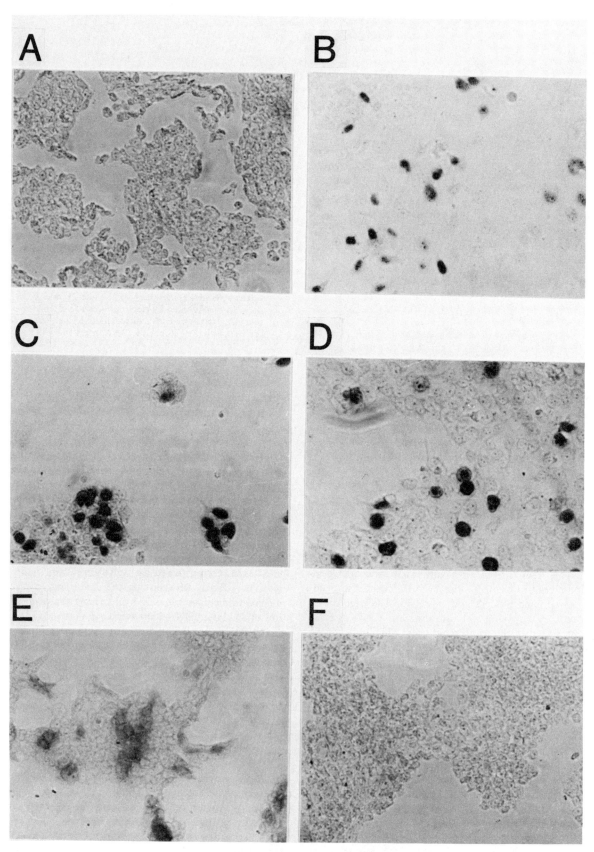

Figure 2. (*See facing page for legend.*)

Table 1. Activities of Viral Control Elements in Undifferentiated and Differentiated F9 Cells

Plasmid	Undifferentiated F9 cells	Differentiated F9 cells
pSV2cat	20	93
pSV1cat	19	3
pMSVcat	1.5	63
pLTR10cat	15	12
pLTR0cat	0	0
pSrM2cat	1.5	N.D.[a]
pMSVcat (4 µg) + pRSVneo (1 µg)	40	N.D.
pMSVcat (2.5 µg) + pRSVneo (2.5 µg)	75	N.D.

In all experiments, pRSVcat was used as a control and expression from this plasmid was taken as 100. CAT activities were determined by cutting out the spots from thin-layer chromatography (TLC) plates and counting them in a scintillation counter. The values given are the averages of at least three experiments.

Data taken from Gorman et al. (1985).

[a] N.D., Not done.

from EC cells (N.B. La Thangue and P.W.J. Rigby, unpubl.) to identify and then purify *trans*-acting factors involved in regulating the expression of these genes.

The use of well-characterized viral control elements to probe aspects of the regulation of cellular gene expression continues to be valuable. Our experiments with the SV40 early promoter/enhancer and with the LTR of MSV have demonstrated that undifferentiated EC cells contain a negative regulatory factor(s) that operates on sequences within the enhancer region of MSV. We think it likely that this molecule(s) is involved in maintaining the pattern of cellular gene expression characteristic of the undifferentiated state. We hope that by analyzing the activities of these viral sequences in the in vitro system we will be able to demonstrate the activity of this factor(s) and then purify and characterize it.

The E1a region of human adenoviruses encodes proteins that are required for the subsequent expression of all other viral transcription units (Berk et al. 1979; Jones and Shenk 1979) and can regulate the expression of cellular genes (Green et al. 1983; Kao and Nevins 1983; Treisman et al. 1983; Gaynor et al. 1984). Mutants defective in E1a function cannot grow in differentiated cells because the other viral genes are not expressed but they do grow on undifferentiated F9 cells (Imperiale et al. 1984), suggesting that these cells contain an activity that can substitute for E1a. It has recently been shown that, as well as acting as a *trans* activator of transcription, E1a products can act on enhancer regions to repress transcription (Borelli et al. 1984; Velcich and Ziff 1985). It is possible that the factor(s) we have detected corresponds to this E1a-like activity and experiments currently in progress should resolve this issue. It will also be of considerable interest to see if enhancers, which were originally recognized by their ability to increase the level of transcription in differentiated cells, normally perform the opposite function, or are nonfunctional in undifferentiated stem cells.

ACKNOWLEDGMENTS

C.M.G. was supported by a NATO Postdoctoral Fellowship and by a Long-term Fellowship from the European Molecular Biology Organisation. C.J.W. acknowledges a Research Studentship from the Science and Engineering Research Council. P.W.J.R. holds a Career Development Award from the Cancer Research Campaign, which also paid for this work.

REFERENCES

Benton, W.D. and R.W. Davis. 1977. Screening λgt recombinant clones by hybridization to single plaques in situ. *Science* **196:** 180.

Berk, A.J., F. Lee, T. Harrison, J. Williams, and P.A. Sharp. 1979. Pre-early adenovirus 5 gene product regulates synthesis of early viral messenger RNAs. *Cell* **17:** 935.

Borrelli, E., R. Hen, and P. Chambon. 1984. Adenovirus-2 E1a products repress enhancer stimulation of transcription. *Nature* **312:** 608.

Clayton, C.E., M. Lovett, and P.W.J. Rigby. 1982. Functional analysis of a simian virus 40 super T antigen *J. Virol.* **44:** 974.

Cooper, A.R., A. Taylor, and B.L.M. Hogan. 1983. Changes in the ratio of laminin and entactin synthesis in F9 embryonal carcinoma cells treated with retinoic acid and cyclic AMP. *Dev. Biol.* **99:** 510.

Dziadek, M. and E. Adamson. 1978. Localization and synthesis of alpha fetoprotein in post-implantation mouse embryos. *J. Embryol. Exp. Morphol.* **43:** 289.

Favaloro, J., R. Treisman, and R. Kamen. 1980. Transcription maps of polyoma virus-specific RNA: Analysis by two-dimensional nuclease S1 gel mapping. *Methods Enzymol.* **65:** 718.

Figure 2. Expression of SV40 T-antigen in infected and transfected EC cells. (*A*) Undifferentiated F9 cells were infected with SV40 virus at an m.o.i. of 20 pfu/cell. At 72 hr postinfection the cells were washed, fixed, and stained, using the immunoperoxidase technique, with 10 µg/ml PAb419, a monoclonal anti-T antibody that recognizes a determinant at or near the amino-termini of large T and small t antigens. (*B*) Undifferentiated F9 cells were infected with SV40 virus at an m.o.i. of 20 pfu/cell and then immediately induced to differentiate by the addition of retinoic acid to the medium. At 72 hr postinfection the cells were stained as described in *A*. (*C*) Undifferentiated F9 cells were transfected with 10 µg pTSV3 DNA; 40 hr later the cells were stained as described in *A*. (*D*) Undifferentiated F9 cells were transfected as in *C* and 40 hr later stained with PAb423, an anti-T monoclonal antibody that recognizes a determinant at or near the carboxyl terminus of large T antigen and does not react with small t antigen. (*E*) Undifferentiated F9 cells were transfected as in *C* and 40 hr later stained with PAb280, a monoclonal antibody that recognizes a determinant near the carboxyl terminus of small t antigen and does not react with large T antigen. (*F*) Undifferentiated F9 cells were transfected as in *C* and 40 hr later stained with TROMA-1, a monoclonal antibody that recognizes a determinant expressed on keratin in differentiated but not in undifferentiated cells. (Adapted from Gorman et al. 1985.)

FEINBERG, A.P. and B. VOGELSTEIN. 1984. A technique for radiolabelling DNA restriction endonuclease fragments to high specific activity. *Anal. Biochem.* **137:** 266.

GAYNOR, R., D. HILLMAN, and A. BERK. 1984. Adenovirus early region 1A protein activates transcription of a nonviral gene introduced into mammalian cells by infection or transfection. *Proc. Natl. Acad. Sci.* **81:** 1193.

GORMAN, C.M., D.P. LANE, and P.W.J. RIGBY. 1984. High efficiency gene transfer into mammalian cells. *Philos. Trans. R. Soc. Lond. B.* **307:** 343.

GORMAN, C.M., L.F. MOFFAT, and B.H. HOWARD. 1982a. Recombinant genomes which express chloramphenicol acetyl transferase in mammalian cells. *Mol. Cell. Biol.* **2:** 1044.

GORMAN, C.M., P.W.J. RIGBY and D.P. LANE. 1985. Negative regulation of viral enhancers in undifferentiated embryonic stem cells. *Cell* **42:** 519.

GORMAN, C.M., G. MERLINO, M. WILLINGHAM, I. PASTAN, and B.H. HOWARD. 1982b. Rous sarcoma virus long terminal repeat is a strong promoter when introduced into a variety of eukaryotic cells by DNA mediated transfection. *Proc. Natl. Acad. Sci.* **79:** 6777.

GREEN, M.R., R. TREISMAN, and T. MANIATIS. 1983. Transcriptional activation of cloned human β-globin genes by viral immediate-early gene products. *Cell* **35:** 137.

HUYNH, T.V., R.A. YOUNG, and R.W. DAVIS. 1985. Constructing and screening cDNA libraries in λgt10 and λgt11. In *DNA cloning: A practical approach* (ed. D.M. Glover), vol. 1, p. 49. I.R.L. Press, Oxford.

IMPERIALE, M.J., H.T. KAO, L. FELDMAN, J. NEVINS, and S. STRICKLAND. 1984. Common control of the heat shock gene and early adenovirus genes: Evidence for a cellular E1a-like activity. *Mol. Cell. Biol.* **4:** 867.

JONES, N. and T. SHENK. 1979. An adenovirus type 5 early gene function regulates expression of other early viral genes. *Proc. Natl. Acad. Sci.* **76:** 3665.

KAO, H.T. and J.R. NEVINS. 1983. Transcriptional activation and subsequent control of the human heat shock gene during adenovirus infection. *Mol. Cell. Biol.* **3:** 2058.

LAIMINS, L.A., P. GRUSS, R. POZZATTI, and G. KHOURY. 1984. Characterisation of enhancer elements in the long terminal repeat of Moloney murine sarcoma virus. *J. Virol.* **49:** 183.

LAIMINS, L.A., G. KHOURY, C. GORMAN, B. HOWARD, and P. GRUSS. 1982. Host-specific activation of transcription by tandem repeats from simian virus 40 and Moloney murine sarcoma virus. *Proc. Natl. Acad. Sci.* **79:** 6453.

LINNENBACH, A., K. HUEBNER, and C.M. CROCE. 1980. DNA transformed murine teratocarcinoma cells: Regulation of expression of simian virus 40 tumor antigens in stem versus differentiated cells. *Proc. Natl. Acad. Sci.* **77:** 4875.

MANIATIS, T., E.F. FRITSCH, and J. SAMBROOK. 1982. *Molecular cloning: A laboratory manual.* Cold Spring Harbor Laboratory, Cold Spring Harbor, New York.

ROSENTHAL, A., S. WRIGHT, H. CEDAR, R. FLAVELL, and F. GROSVELD. 1984. Regulated expression of an introduced MHC H-2K^{bm1} gene in murine embryonal carcinoma cells. *Nature* **310:** 415.

SEGAL, S. and G. KHOURY. 1979. Differentiation as a requirement for simian virus 40 gene expression in F9 embryonal carcinoma cells. *Proc. Natl. Acad. Sci.* **78:** 1100.

SOUTHERN, E.M. 1975. Detection of specific sequences among DNA fragments separated by gel electrophoresis. *J. Mol. Biol.* **98:** 503.

STEWART, C.L., H. STUHLMANN, D. JAHNER, and R. JAENISCH. 1982. De novo methylation, expression and infectivity of retroviral genomes introduced into embryonal carcinoma cells. *Proc. Natl. Acad. Sci.* **79:** 4098.

STRICKLAND, S. and V. MAHDAVI. 1978. The induction of differentiation in teratocarcinoma stem cells by retinoic acid. *Cell* **15:** 393.

SWARTZENDRUBER, D.E. and J.M. LEHMAN. 1975. Neoplastic differentiation: Interaction of simian virus 40 and polyoma virus with murine teratocarcinoma cells *in vitro*. *J. Cell. Physiol.* **85:** 179.

TEICH, N., R. WEISS, G. MARTIN, and D.R. LOWY. 1977. Virus infection of murine teratocarcinoma stem cell lines. *Cell* **12:** 973.

THOMAS, P.S. 1980. Hybridization of denatured RNA and small DNA fragments transferred to nitrocellulose. *Proc. Natl. Acad. Sci.* **77:** 5201.

TREISMAN, R., M. GREEN, and T. MANIATIS. 1983. Cis- and trans-activation of globin gene transcription in transient assays. *Proc. Natl. Acad. Sci.* **80:** 7428.

VELCICH, A. and E. ZIFF. 1985. Adenovirus E1a proteins repress transcription from the SV40 early promoter. *Cell* **40:** 705.

WAHL, G.M., M. STERN, and G.R. STARK. 1979. Efficient transfer of large DNA fragments from agarose gels to diazobenzyloxymethyl-paper and rapid hybridization by using dextran sulphate. *Proc. Natl. Acad. Sci.* **76:** 3683.

WATSON, C.J. and J.F. JACKSON. 1985. An alternative procedure for synthesising double stranded cDNA for cloning in phage and plasmid vectors. In *DNA cloning: A practical approach* (ed. D.M. Glover), vol. 1, p. 79. I.R.L. Press, Oxford.

Transformation of Embryonic Stem Cells with the Human Type-II Collagen Gene and Its Expression in Chimeric Mice

R.H. LOVELL-BADGE,* A.E. BYGRAVE,* A. BRADLEY,† E. ROBERTSON,† M.J. EVANS,†
AND K.S.E. CHEAH‡

*MRC Mammalian Development Unit, London NW1 2HE, England; †Genetics Department, University of Cambridge, Cambridge CB2 3EH, England; ‡Department of Biochemistry, Li Shu Fan Building, Hong Kong University, Hong Kong

Type-II collagen is the major extracellular component of developing cartilage, being synthesized in large amounts in the embryo. Its synthesis decreases as development progresses to very low levels in the adult. In the embryo, the correct regulation of type-II collagen is important not only for those structures that remain as cartilage, but also for the process of endochondral ossification whereby bone deposition occurs around a preformed cartilage model, and for the continuing growth of long bones. It is likely that primary defects in the type-II collagen gene will be found associated with some of the inherited growth disorders in man that affect cartilage or bone development (McKusick 1972; Cheah 1985; Stoker et al. 1985). Type-II collagen is made almost exclusively by chondrocytes which arise as condensations of mesenchymal cells in an inductive response with overlying ectoderm (Gumpel-Pinot 1981). These cells behave in a variety of ways throughout the body, and there is some evidence to suggest that the type-II collagen gene is differentially regulated in the different types of cartilage they produce (Kravis and Upholt 1985).

Suitable in vitro cell models for studying type-II collagen gene expression do not exist; indeed, isolated chondrocytes generally stop synthesizing the protein and become fibroblastic, making type I instead. Because of its fundamental role in the development of form and its interesting pattern and mode of regulation, we wished to establish a system in which we could begin to analyze the factors controlling the type-II collagen gene and the possible effects of its abnormal expression on morphogenesis.

The approach we have adopted is to introduce the entire human type-II collagen gene, contained within a cosmid clone CosHcol1 (Cheah et al. 1985) into embryonic stem cells (EK cells) (Evans and Kaufman 1981; Martin 1981; Robertson et al. 1983) by the calcium phosphate/DNA precipitate technique. Stably transformed sublines containing intact copies of the human gene were obtained and have been used to derive chimeric mice after injection into blastocysts. (Bradley et al. 1984; R.H. Lovell-Badge et al., in prep.). We describe here results obtained on the expression of the human collagen gene in chimeras derived with one of the transformed EK sublines, CP3C/N1.

MATERIALS AND METHODS

Cotransformation. The cosmid clone CosHcol1 contains the entire 30-kb human type-II collagen gene together with about 5 kb of 5'- and 2.2 kb of 3'-flanking sequence (see Fig. 1). CosHcol1 and pSV2neo were linearized with *Sal*I and *Eco*RI, respectively. CP3 EK cells were plated out onto 6-cm gelatinized dishes without feeders at 2.5×10^6 cells/dish in medium containing 15% fetal calf serum (FCS). The medium was changed after 24 hours to one containing only 10% serum and the calcium phosphate/DNA precipitate

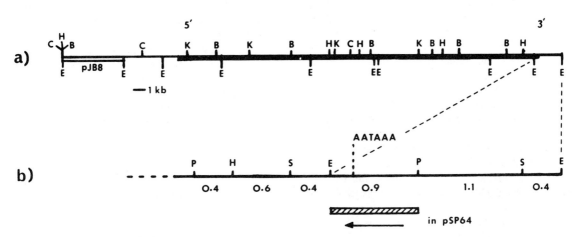

Figure 1. (*a*) Structure of the cosmid clone CosHcol1 containing the human type-II collagen gene. (*b*) The 3' end of CosHcol1 showing the fragment used to generate SP6 transcripts for RNase protection assays. (E) *Eco*RI; (P) *Pst*I; (H) *Hin*dIII; (B) *Bam*HI; (K) *Kpn*I; *Sac*I.

added (Wigler et al. 1977; Jami et al. 1983). Amounts of 5 μg of CosHcol1 and 0.5 μg of pSV2neo were used per dish. After 24 hours the medium was changed back to high serum. The following day the cells were passaged onto feeder layers in medium containing 0.3 mg/ml of G418 (Geneticin, Gibco Biocult). This concentration of G418 has been found to inhibit completely the growth of untransformed EK cells. pSV2neo-transformed STO fibroblasts were used as G418-resistant feeder cells. EK cell colonies growing in the selective medium were picked and expanded.

Preparation of total cellular RNA. Tissues were homogenized in 6 M urea and 3 M LiCl for 60 seconds with a Polytron. The RNA was left to precipitate at 4°C overnight, pelleted by centrifugation at 10,000 rpm for 20 minutes, and washed twice with urea/LiCl. The RNA pellet was taken up in 0.5% SDS, 10 mM Tris (pH 7.5), and 50 μg/ml of Proteinase K, and then incubated at 37°C for 1 hour, extracted with phenol/chloroform, and precipitated with ethanol.

RNase protection assays. A 900-bp *PstI/EcoRI* fragment at the 3' end of the collagen gene was subcloned into pSP64 (see Fig. 1) to give p64P2, which was used to generate [^{32}P]RNA probes for RNase protection assays. These were performed essentially as described by Zinn et al. (1983). Transcripts from pSV2neo were also detected by RNase protection. In this case [^{32}P]RNA probes were generated from the *Bam*HI–*Hin*dIII fragment of pSV2neo, containing the entire *neo* gene, in pSP65 (p65N).

RESULTS

CosHcol1 was introduced into CP3 EK cells by cotransformation with pSV2neo. The transformation frequency was about 10^{-6}, approximately 10-fold lower than that obtained with C1 1-D mouse fibroblasts cotransformed at the same time. Analysis of DNA in Southern blots demonstrated the presence of intact copies of cosHcol1 in about 70% of G418-resistant colonies. The transformation was shown to be completely stable, in terms of both *neo* expression and the presence of the human collagen gene when the cells were grown in the absence of selection for at least 10 passages. The developmental potential of a number of sublines has been examined in detail, but the results described here are for just one line, CP3C/N1, which contains about five copies of CosHcol1.

Chimeras were derived by injecting CP3C/N1 cells into 3.5-day blastocysts from MF1 albino mice. The frequency of chimera formation (47%) was similar to that described for untransformed EK cell lines (Evans et al., this volume) and they show often extensive contribution by the CP3C/N1 cells to all tissues looked at, by glucose phosphate isomerase (Gpi) isozyme analysis or by probing for the human collagen gene on Southern blots (R.H. Lovell-Badge et al., in prep.). The DNA analysis has shown no loss or rearrangement of CosHcol1 upon differentiation. The only abnormality noticed in these chimeras is a loss of *Agouti* gene function or action in hair follicle derivatives of the transformed EK cells. This peculiar effect is found in chimeras derived from all the other CosHcol1 transformants tested, and is presently under investigation.

Figure 2 shows an RNase protection assay to detect transcripts from the human type-II collagen gene. Figure 2a shows the specificity of the assay. The transcript from the human gene protects a 230-base fragment, which corresponds to the 3' end of the mRNA up to the poly(A) addition site. Smaller protected fragments, less than 80 bases, are found in tracks from samples containing RNA from mouse cartilage, and presumably demonstrate some sequence conservation between the mouse and human type-II collagen genes. RNA from the fibroblast line 1DC/N4, cotransformed with CosHcol1 and pSV2neo, protected a 230-base fragment, demonstrating that the cosmid contains a functional gene. Figure 2b shows a protection assay with RNA from CP3C/N1 cells in vitro, from tissues of an adult chimera derived from these cells, and from fetal chimeras dissected into two portions, soft tissues (viscera and brain) and the remaining carcass. There was a low level of transcription from CosHcol1 in the cells maintained in vitro. Apart from a very faint signal in sternum, we have been unable to detect transcripts from the human type-II collagen gene in any adult tissue shown here, and in a number of other analyses. The fetal chimeras show no expression of the human gene in the "soft tissues"; however, there were significant levels in the carcass samples. The presence of the smaller bands confirms that we expect endogenous type-II collagen gene expression only in the carcass sample. The variation in the level of the human transcript between the different embryo samples correlates very well with the extent of chimerism as judged by eye pigment and by Gpi isozyme analysis. Without knowing the proportion of chondrocytes derived from the transformed EK cells, it is difficult to estimate the precise level of transcription, but if one assumes that it is expressed tissue specifically, and that fetus 7 was 20% chimeric, it is possible to calculate a figure of about 200 copies per cell (from cutting out bands from the gel and determining cpm).

Figure 3 shows an RNase protection assay to detect transcripts from pSV2neo. The samples used were equivalent to those used in Figure 2b. There are protected fragments only in the tracks from CP3C/N1 cells maintained in vitro and in 1DC/N4. There is no clear expression of the *neo* gene in any of the chimera samples.

DISCUSSION

We have demonstrated that a cloned human type-II collagen gene can be stably introduced into mouse EK cells by cotransformation with pSV2neo and G418 selection. The selection procedure has not significantly affected the ability of the EK cells to contribute to chimeras, and the exogenous DNA can be detected in tis-

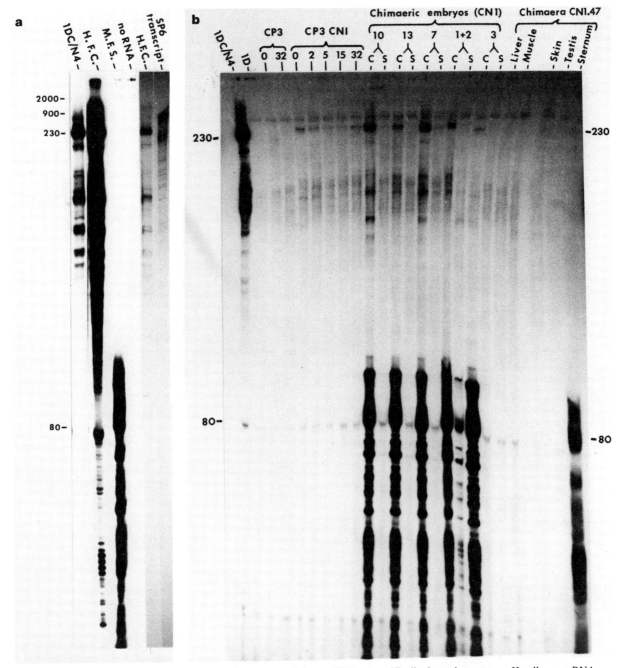

Figure 2. RNase protection assay using transcripts generated from p64P2 to specifically detect human type-II collagen mRNA. Total cellular RNA (20 μg) and 7 × 10⁵ cpm [³²P]RNA transcript were used for each track, except for samples from human fetal cartilage (5 μg) and CN1.47 sternum (14 μg). (*a, from left to right*) 1DC/N4, mouse fibroblast line cotransformed with CosHcol1 and pSV2neo; (H.F.C.) 14-week human fetal cartilage; (M.F.S.) 14.5-day mouse fetal sternum; (no RNA) control with no cellular RNA; (H.F.C.) brief exposure of human fetal cartilage track; (Sp6 transcript) from p64P2. (*b*) (CP3) Untransformed EK cells; (CP3/N1) CosHcol1/pSV2neo EK cotransformant; 0, 2, 5, 15, 32, days differentiation in vitro. The cells were maintained in the absence of selection. C and S refer to carcass and soft tissue portions from CP3C/N1 chimeric embryos. CN1.47 was an adult chimera showing 20–50% contribution from CP3/N1.

sues from these animals (R.H. Lovell-Badge et al., in prep.).

RNase protection assays demonstrate a degree of tissue specificity in the expression of the human collagen gene in chimeras derived from one transformed line, CP3C/N1. Transcripts are not found in a range of adult tissues, but they are found at significant levels in the appropriate "half" of 14.5-day fetal chimeras. The level found could be equivalent to about 200 copies/cell. This is likely to be less than the level of endogenous gene transcription. We cannot reliably estimate from the gels presented here, but chick limb chondrocytes are known to contain about 2000 copies of type-II collagen mRNA per cell (Kravis and Upholt 1985) and it would seem reasonable that mouse chondrocytes express at similar levels. This would give a figure of about 10% of endogenous expression for the introduced human gene. This low level could be attributed to the presence of vector sequences (Chada et al. 1985), to chromosomal position effects, or to species specificity. We are currently investigating the precise localization of transcripts from CosHcol, by in situ hybridization to tissue sections from chimeric embryos. This will also allow us to determine more accurately the proportion of cells in any tissue expressing the gene, and thus aid in the above calculations.

There is a very low level of transcription in the undifferentiated CP3C/N1 EK cells. Although representing less than one copy of RNA per cell, this expression is inappropriate. However, we need to verify that initiation of transcription is correct. There is some evidence for abnormal initiation of α-fetoprotein "minigenes" introduced into F9 embryonal carcinoma cells (Scott et al. 1984), and we could be seeing readthrough from adjacent pSV2neo sequences. pSV2neo is expressed in CP3C/N1 in vitro, in the absence of selection, both in undifferentiated cells and during differentiation in vitro. It is not, however, expressed in adult or fetal chimeras. This would suggest that gene regulation is under tighter control in vivo, or that the tissues produced in vitro are less differentiated than their counterparts in vivo.

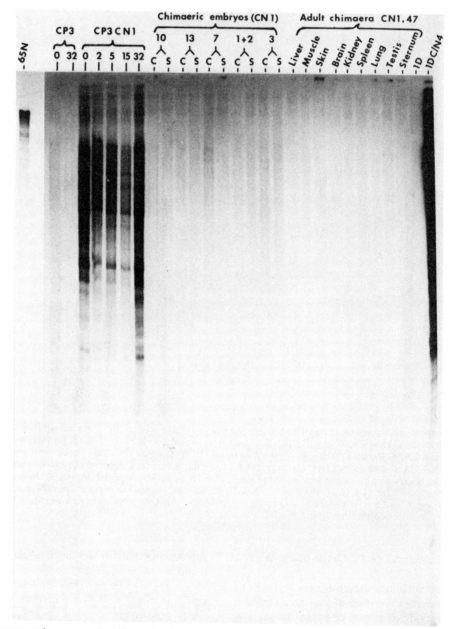

Figure 3. RNase protection assay using [^{32}P]RNA generated from p65N to detect specifically transcripts from pSV2neo. Total cellular RNA (20 μg) and 7×10^5 cpm [^{32}P]RNA were used for each track. The samples are equivalent to those in Fig. 2b.

The production of transgenic mice by direct injection into the fertilized egg is clearly a powerful technique for analyzing some of the factors regulating gene expression and for determining some functional aspects of particular gene products. However, we have shown here that genes may also be introduced into mice after transformation of EK cells in vitro, and that they may undergo appropriate regulation. There are a number of advantages to this approach, including that of being able to select particular characteristics in advance. It will be especially valuable where the introduced sequences are likely to have an effect on development, either by default or by design.

REFERENCES

BRADLEY, A., M. EVANS, M. KAUFMAN, and E. ROBERTSON. 1984. Formation of germ-line chimeras from embryo derived teratocarcinoma cell lines. *Nature* 309: 255.

CHADA, K., J. MAGRAM, K. RAPHAEL, G. RADICE, E. LACY, and F. COSTANTINI. 1985. Specific expression of a foreign β-globin gene in erythroid cells of transgenic mice. *Nature* 314: 377.

CHEAH, K.S.E. 1985. Collagen genes and inherited connective tissue disease. *Biochem. J.* 229: 287.

CHEAH, K.S.E., N.G. STOKER, J.R. GRIFFIN, F.G. GROSVELD, and E. SOLOMON. 1985. Identification and characterization of the human type II collagen gene (COL2A1) *Proc. Natl. Acad. Sci.* 82: 2555.

EVANS, M.J. and M.H. KAUFMAN. 1981. Establishment in culture of pluripotential cells from mouse embryos. *Nature* 292: 154.

GUMPEL-PINOT, M. 1981. Ectoderm-mesoderm interactions in relation to limb-bud chondrogenesis in the chick embryo: Transfilter cultures and ultrastructural studies. *J. Embryol. Exp. Morphol.* 65: 73.

JAMI, J., C. LASSERRE, D. BUCCHINI, R. LOVELL-BADGE, J. THILLET, M.-G. STINNAKRE, F. KUNST, and R. PICTET. 1983. Selection for transformed teratocarcinoma stem cells upon transfer of cloned genes. *Cold Spring Harbor Conf. Cell Proliferation* 10: 487.

KRAVIS, D. and W.B. UPHOLT. 1985. Quantitation of type II procollagen mRNA levels during chick limb cartilage differentiation. *Dev. Biol.* 108: 164.

MARTIN, G.R. 1981. Isolation of a pluripotent cell line from early mouse embryos cultured in medium conditioned with teratocarcinoma stem cells. *Proc. Natl. Acad. Sci.* 78: 7634.

McKUSICK, V.A. 1972. *Heritable disorders of connective tissue*, 4th edition. C.V. Mosby, Saint Louis, Missouri.

ROBERTSON, E.J., M.H. KAUFMAN, A. BRADLEY, and M.J. EVANS. 1983. Isolation, properties and karyotype analysis of pluripotential (EK) cell lines from normal and parthenogenetic embryos. *Cold Spring Harbor Conf. Cell Proliferation* 10: 647.

SCOTT, R.W., T.F. VOGT, M.E. CROKE, and S.M. TILGHMAN. 1984. Tissue-specific activation of a cloned α-fetoprotein gene during differentiation of a transfected embryonal carcinoma cell line. *Nature* 310: 562.

STOKER, N.G., K.S.E. CHEAH, J.R. GRIFFIN, F.M. POPE, and E. SOLOMON. 1985. A highly polymorphic region 3' to the human type II collagen gene. *Nucleic Acids Res.* 13: 4613.

WIGLER, M., S. SILVERSTEIN, L.S. LEE, A. PELLICER, Y.-C. CHENG, and R. AXEL. 1977. Transfer of purified herpes virus thymidine kinase gene to cultured mouse cells. *Cell* 11: 223.

ZINN, K., D. DiMAIO, and T. MANIATIS. 1983. Identification of two distinct regulatory regions adjacent to the human β-interferon gene. *Cell* 34: 865.

Introduction of Genes into Embryonal Carcinoma Cells and Preimplantation Embryos by Retroviral Vectors

J.F. NICOLAS, J.L.R. RUBENSTEIN, C. BONNEROT, AND F. JACOB
Unité de Génétique cellulaire du Collège de France et de l'Institut Pasteur, 75724 Paris Cedex 15, France

Recombinant retroviruses offer many advantages as vectors for introducing genes into eukaryotic cells. In particular, they permit the stable integration of new genes into the host chromosome. In addition, the genetic information required for production of a packageable genomic RNA (gRNA) is grouped at the 5' and 3' ends of the viral genome, thus enabling the *gag-pol-env* genes to be replaced with other DNA fragments up to 8 kb in length. To produce infectious virions carrying the recombinant gRNA, *trans*-complementing cell lines have been constructed (Mann et al. 1983; Cone and Mulligan 1984; Miller et al. 1985). These helper cell lines can produce viral stocks containing high titers of the recombinant virus, which are free of wild-type virions.

Our interest is to use these vectors to introduce and express genes in preimplantation mouse embryos. Although wild-type Moloney murine leukemia virus (Mo-MLV) can infect mouse embryos (Jaenisch et al. 1975), the proviral genes are not expressed in the embryonic cells (Pèriès et al. 1977; Teich et al. 1977). Furthermore, if the provirus integrates early in development, its gene expression often remains suppressed in the adult tissues (Jaenisch et al. 1981). This poses a serious problem regarding the utility of recombinant retroviruses for expressing genes during embryonic development. In an effort to overcome this block in gene expression, we have modified the retroviral structure. We found that inserting the composite SV40-thymidine kinase (SV*tk*) promoter in a Mo-MLV recombinant retrovirus greatly increased the efficiency of gene expression in embryonal carcinoma (EC) cells (Rubenstein et al. 1984a), a result that has been confirmed by Wagner et al. (1985).

In the first part of this paper we report that the internal SV*tk* promoter does not significantly alter the efficiency of producing a recombinant proviral structure. We show, however, that the site of proviral integration can have a significant effect on the level of expression, and that in EC cells rearrangements of the proviral structure occur in 10% of the tested clones. Finally, we demonstrate that recombinant retroviruses can be used to introduce genes into preimplantation mouse embryos.

In the second part of this paper, we show that genes which produce RNA complementary to a mRNA encoding an enzyme inhibit the production of that enzyme. We discuss the possible use of antisense RNA to study mammalian embryogenesis.

MATERIALS AND METHODS

Retroviral vectors and plasmids. Details on the structure of pMo-MLV-neo and pM-MLV-SVtk-neo can be found in Rubenstein et al. (1984a). The structure of pCH110 is described in Hall et al. (1983) and of pNSlacZ in Rubenstein et al. (1984b).

Blot hybridization. Restriction enzymes were obtained from Boehringer Mannheim Biochemicals and used under the conditions recommended by the suppliers. Cellular DNA purification and Southern blot hybridizations were performed as in Nicolas and Berg (1983). ^{32}P-Labeled nick-translated probes (specific activities of 2×10^8 cpm/µg) were made using materials from Amersham.

Cell culture, virus infection, and G418 selection. The properties of the EC cell lines are described in Nicolas et al. (1981). Viral production and infection are described in Rubenstein et al. (1984a). G418 selection of EC cells is described in Nicolas and Berg (1983). Cell fusions were performed using the methods of Pontecorvo (1975).

Embryos. Mice were obtained from the Pasteur Institute's animal facilities. Superovulation, harvesting of stage-2 and stage-4 embryos, and embryo reimplantation into pseudopregnant foster mothers are described in Wittingham (1975). Additional information will be described elsewhere (J.L.R. Rubenstein and J.F. Nicolas, in prep.).

RESULTS

Recombinant Retroviral Gene Expression in Embryonic Cells

Retroviruses are incapable of producing a productive infection in embryonic cells (Pèriès et al. 1977; Teich et al. 1977; Jaenisch et al. 1981). The block in their life cycle is due at least in part to inefficient expression of the promoter in the 5' long terminal repeat (LTR) (Linney et al. 1984). For instance, when a recombinant retrovirus is constructed in which the neomycin (*neo*) gene is placed under the transcriptional control of the 5' LTR (Mo-MLV-*neo*), expression of the *neo* gene in EC cells is extremely inefficient (Rubenstein et al. 1984a; Sorge et al. 1984).

We showed that a recombinant retrovirus modified by the insertion of the SV*tk* promoter between the 5' LTR and the *neo* gene—Mo-MLV-SV*tk*-*neo*—could efficiently express the *neo* gene in EC cells (Rubenstein et al. 1984a). Both the frequency of obtaining G418-resistant EC cells and the level of aminoglycoside 3′ phosphotransferase (APH[3′]II, the enzyme encoded by the *neo* gene) were shown to be increased. Figure 1 illustrates the difference in the level of APH(3′)II in G418-resistant PCC3 cells (a multipotential EC cell line; Nicolas et al. 1981) infected either with Mo-MLV-*neo* or with Mo-MLV-SV*tk*-*neo*.

A caveat in this system comes from the fact that retroviruses recombine at high frequency and that the insertion of internal promoters into retroviruses has sometimes led to structural rearrangements (Joyner and Bernstein 1983; Emerman and Temin 1984). Therefore, it was important to analyze whether the internal promoter affects the efficiency of viral production and whether the structure of the Mo-MLV-SV*tk*-*neo* provirus remains intact in both differentiated (NIH-3T3) and embryonic (LT-1, PCC3, PCC4) cells.

The SV*tk* Promoter Does Not Affect Viral Production in ψ2 Cells

DNA from either the pMO-MLV-neo or pMo-MLV-SVtk-neo plasmids was transfected into ψ2 cells by the calcium phosphate coprecipitation method. ψ2 is a *trans*-complementing cell line that makes the necessary gene products to produce infectious helper-free defective retroviruses (Mann et al. 1983). Forty-eight hours after transfection, the virus-containing supernatant was used to infect NIH-3T3 cells. These cells were grown in G418 to select for colonies of cells expressing the *neo* gene. The titer from transfections using pMo-MLV-neo (1.2×10^3/ml) was the same as from pMo-MLV-SVtk-neo (1.8×10^3/ml). Therefore, it is likely that the SV*tk* internal promoter does not significantly affect viral production. This is consistent with our earlier report that G418-resistant ψ2 clones transformed with pMo-MLV-neo or pMo-MLV-SVtk-neo produce essentially the same number of infectious virions (Rubenstein et al. 1984a).

Structure of Mo-MLV-SV*tk*-*neo* Provirus in EC Clones

To demonstrate that the virus produced by ψ2 was not rearranged, we infected EC cells with Mo-MLV-SV*tk*-*neo* and picked clones after G418 selection. Our restriction enzyme analysis on cellular DNA characterizes five parameters of the Mo-MLV-SV*tk*-*neo* provirus (Fig. 2). First, the *Xba*I digest tests the proviral length (4060 bp; the only *Xba*I sites are in the U_3 section of the two LTRs). Second, the *Xba*I-*Eco*RI double digest tests both the length between the 5′ LTR and the SV*tk* promoter (1179 bp, the *Eco*RI site is in the *tk* promoter) and the length between the *tk* promoter and the 3′ LTR (2881 bp). Finally, a *Sal*I-*Eco*RI digest tests both the size of the SV*tk* promoter (465 bp) and the number of integrated copies (the second *Eco*RI site is supplied by the genomic DNA). The probe was the SV*tk*-*neo* fragment from pMo-MLV-SVtk-neo.

In 18 out of 20 clones, the proviral structure was found to be unchanged. The two rearrangements (one PCC3 and one PCC4 clone) have alterations in the region 5′ to the SV*tk*-*neo* fragment. This was noted by the altered size of the *Xba*I-*Eco*RI fragment corresponding to this segment (Fig. 2, PCC3 ψA5). We do not know whether these alterations are due to recom-

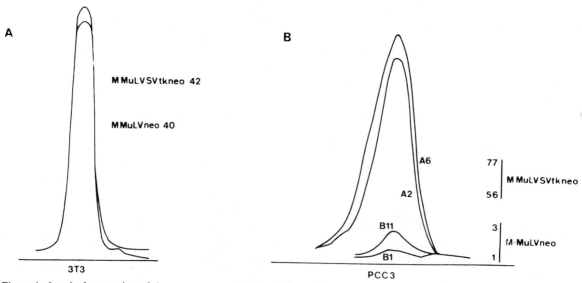

Figure 1. Level of expression of the *neo* gene-encoded phosphotransferase (APH[3′]II) in EC and in NIH-3T3 cells harboring one Mo-MLV-SV*tk*-*neo* or Mo-MLV-*neo* provirus. Protein extracts of cells grown in selective medium were electrophoresed and the in situ assay for APH(3′)II was performed as described by Reiss et al. (1984). The figure represents densitometer tracings of the band corresponding to APH(3′)II activity. The numbers correspond to arbitrary units proportional to the APH(3′)II activity. (*A*) NIH-3T3 clones, (*B*) PCC3 clones; B1 and B11 are two Mo-MLV-*neo*-infected clones; A2 and A6 are two Mo-MLV-SV*tk*-*neo*-infected clones. All four were selected in G418. (MMuLV) Mo-MLV.

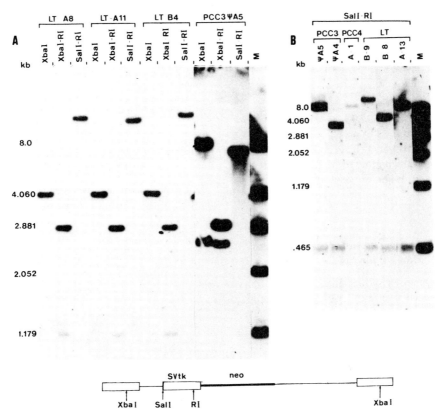

Figure 2. Southern gel analysis of total DNA extracted from EC clones infected with the Mo-MLV-SVtk-neo virus. DNA (10 μg) was digested with restriction enzymes for 5 hr, electrophoresed on an agarose gel, transferred to a nitrocellulose filter, and hybridized with a ^{32}P-labeled nick-translated SVtk-neo fragment. LT, PCC3, and PCC4 are three different EC cell lines (Nicolas et al. 1981). (A) Analysis of four EC clones using XbaI, XbaI-EcoRI, and SalI-EcoRI digestion. (B) Analysis of six EC clones using a SalI-EcoRI double digestion to identify the 465-bp SVtk promoter fragment. The size markers (M) are to the right of each figure, and their corresponding lengths in kilobases (kb) are to the left. Below is a schematic structure of the Mo-MLV-SVtk-neo provirus, which illustrates the positions of the two LTRs, the SVtk promoter, the neo gene, and the pertinent restriction enzyme sites.

bination between retroviruses or to a postintegration event.

Structure of Mo-MLV-SVtk-neo Provirus in Nonselected NIH-3T3 Clones

We also analyzed the structure of the Mo-MLV-SVtk-neo provirus in infected NIH-3T3 cells. These cells derived from clones that were picked at random prior to G418 selection. Using the same restriction enzyme analysis as for EC the clones, an unchanged structure was found in all cases (data not shown).

Rescue of the Mo-MLV-SVtk-neo Provirus from EC Cells

To determine if the viral elements of the Mo-MLV-SVtk-neo provirus were still functional, we showed that gRNA can be produced from these proviruses. This procedure was carried out by fusing G418-resistant PCC3 clones carrying a single provirus with mycophenolic acid-resistant ψ2 cells. The hybrid clones were then selected in medium containing G418, mycophenolic acid, and xanthine (Mulligan and Berg 1980). Three weeks later the supernatant was tested for the presence of neo-transducing virions by infecting NIH-3T3 cells. The hybrids of PCC3 × ψ2 cells produced 10^2–10^3 neo-transducing virions per milliliter. Restriction enzyme analysis revealed an unchanged structure of the provirus. Thus it is likely that no rearrangement of the LTRs had occurred during generation of virus or of infection of these EC cells.

In contrast, it was not possible to rescue any virus from PCC3 ψA5, one of the clones that has an alteration in the region 5′ to the SVtk-neo fragment.

The Level of neo Expression Varies in Different G418-resistant EC Clones

Cellular extracts from G418-resistant PCC3 and LT clones bearing a unique unrearranged provirus were assayed for APH(3′)II activity. Figure 3 illustrates variations in the level of expression between clones of both cell lines. A 10-fold variation was observed. Because the proviruses in these clones have the same structure, we suggest that these variations reflect interactions between some elements of the provirus and the state of the flanking DNA. This hypothesis may also account

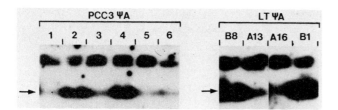

Figure 3. Autoradiograph of an in situ assay of aminoglycoside 3' phosphotransferase APH(3')II (Reiss et al. 1984). Extracts of different clones of LT or PCC3 infected with the Mo-MLV-SVtk-neo virus were assayed. The arrow marks the position of the APH(3')II activity.

for the fact that Mo-MLV-neo virus can express the neo gene in EC cells, although at a very reduced efficiency (see Rubenstein et al. 1984a; Sorge et al. 1984).

Recombinant Retroviruses to Introduce Genes into Mouse Embryos

Ten years ago, Jaenisch showed that wild-type Mo-MLV could infect early mouse embryos (Jaenisch et al. 1975) and thereby introduce a provirus into the mouse germ line. These results suggested that recombinant retroviral vectors could also be used to construct transgenic mice.

We have investigated this possibility by infecting preimplantation mouse embryos in tissue culture using a cell line producing a replication-defective recombinant retrovirus.

Stage-2 and stage-4 embryos from F_1(CBA/J × SJL/J) females mated with F_1(C57 B6 × DBA/2) males were obtained. The zona pellucida was removed using acid Tyrode's solution and the embryos were cocultivated in DMEM and 10% fetal calf serum with semiconfluent ψ2 fibroblasts producing the Mo-MLV-neo recombinant retrovirus. At the expanded blastocyst stage, the embryos were reimplanted into pseudopregnant foster mothers. Fourteen days after reimplantation, the fetuses were dissected and their DNA was purified. Southern analysis of the DNA revealed that two out of the 14 embryos tested contain the Mo-MLV-neo provirus (Fig. 4). When the DNA was digested with the restriction enzyme XbaI (which cuts once in each LTR), a band of the expected size (3450 bp) was obtained, suggesting that no major rearrangement in the virus had occurred. The DNA was also digested with the restriction enzymes BamHI and EcoRI to determine the number of proviral integration sites. BamHI cuts the provirus once just 3' to the neo gene, and EcoRI does not cut the provirus. Therefore, using a neo gene probe, each proviral integration site will produce only one fragment. Figure 4 demonstrates that both fetuses have only one provirus. These fragments have different sizes, demonstrating that the proviruses have integrated in different locations. Furthermore, the fact that there is only one provirus per fetus indicates that the infected cells have a clonal origin. The provirus was estimated to be present in 0.1–0.2 copies per genome equivalent. Therefore, it is likely that the provirus is integrated in a variety of different somatic tissues in the mouse.

Antisense RNA as a Tool Specifically to Inhibit Gene Expression

Our experiments involve cotransfection into cells of mixtures of two plasmids (Rubenstein et al. 1984b), one

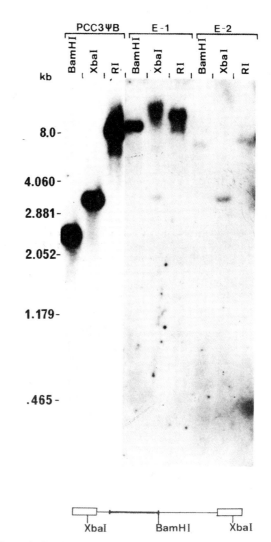

Figure 4. Southern gel analysis of total DNA extracted from 16-day-old fetuses. DNA (10 µg) was digested with either XbaI, BamHI, or EcoRI endonuclease. PCC3 ψB (an EC clone infected with Mo-MLV-neo) is used as control. Lanes E-1 and E-2 correspond to two different fetuses derived from embryos infected with Mo-MLV-neo at stages 2 and 4, respectively. Note that the XbaI digestion in E-1 is incomplete. The probe is a nick-translated SVtk-neo fragment. Below is a schematic of the Mo-MLV-neo provirus, showing the two LTRs (boxes), the neo gene (dark line), and the XbaI and BamHI sites.

that encodes an enzyme (either the bacterial β-galactosidase in pCH110 or APH[3']II phosphotransferase in pSVtk-neoβ) and one that expresses antisense RNA of either of these genes. In pCH110, the *E. coli lacZ* gene is under the transcriptional control of the SV40 early promoter (Hall et al. 1983). In the vector pNSlacZ, the *Hin*dIII–*Sac*I fragment of pCH110 has been inverted; hence it produces an RNA in which 2566 bp are complementary to the *lacZ* mRNA. This antisense RNA has, however, about 70 bp 5' and 1200 bp 3' to the inverted region that are the same as the *lacZ* mRNA.

To see whether the plasmid that expresses the antisense strand of *lacZ* affects β-galactosidase expression, the following experiment was done: pCH110 and pNSlacZ were mixed in varying proportions. The DNA (11 μg/5 × 10⁵ cells) was applied to mouse 3T6 cells, and β-galactosidase activity was measured after 48 hours. In the control experiment, pNSlacZ was replaced by pSVΔgpt (in this plasmid the *E. coli xgpt* gene is under the control of the SV40 promoter). This allows for a constant number of promoters in the two experiments. The percentage of inhibition was then plotted against the percent of pNSlacZ DNA in the transfection mixture (Fig. 5). The β-galactosidase activity diminished rapidly in presence of pNSlacZ: Even with a 1:1 mixture of the two plasmids a 75% inhibition is detected.

To determine whether the inhibition is specific for β-galactosidase, we included in each experiment 1 μg of the plasmid pSVneoβ. In this plasmid the *neo* gene is under the control of the SV40 early promoter. The level of the *neo* gene-encoded phosphotransferase was measured 48 hours after transfection (Fig. 5). Irrespective of the amount of pNSlacZ, no inhibition of the phosphotransferase was detected.

We presently are studying the level at which the inhibition by the antisense plasmid occurs. One possibility is that DNA recombination between the sense and antisense plasmids destroys the activity of the gene. Figure 6A shows that a recombination event would destroy the *lacZ* gene. To test whether recombination was actually taking place at a high enough frequency to account for the inhibition of the *lacZ* activity, we constructed the following plasmid. The molecule has no eukaryotic promoter, and the *lacZ* gene has a 4-bp deletion in its 3' domain at the *Sac*I site. Figure 6B shows that a recombinational event between this molecule and pCH110 would also inactivate the expression of the *lacZ* gene in eukaryotic cells. When the 3' deletion plasmid was transfected with pCH110, no inhibition of β-galactosidase activity was detected, thereby ruling out recombination as the mechanism of the inhibition.

We have also shown that expression from another gene can be inhibited by a plasmid that makes antisense RNA. In this case, we chose the *neo* gene. The antisense plasmid contained the entire *neo* gene in an inverted orientation. Using the same protocol as in the *lacZ* experiments, we observed a similar inhibition of the expression of the *neo* gene. Again, the inhibition was specific, as judged by the lack of inhibition of a cotransfected *lacZ* gene (J.L.R. Rubenstein, in prep.).

Therefore, our experiments show that a plasmid that produces an RNA antiparallel to an enzyme encoding mRNA inhibits the production of that enzyme in eukaryotic cells. The antisense plasmid does not inhibit a cotransfected plasmid which uses the same promoter

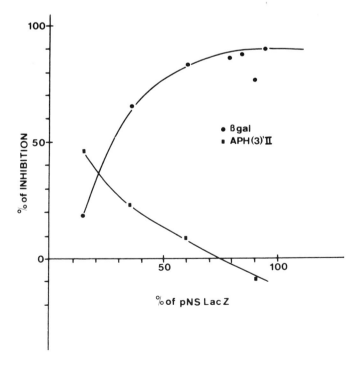

Figure 5. Inhibitory effect of pNS*lacZ* on β-galactosidase activity. Mixtures of increasing amounts of pNS*lacZ* relative to pCH110 (totaling 10 μg) were transfected into 3T6 mouse fibroblasts. pSVΔgpt replaced pNSlacZ in the control samples. All samples contained 1 μg of pSVneoβ. The β-galactosidase activity (or APH[3']II activity) was compared in samples containing pNSlacZ and in samples containing pSVΔgpt by calculating the percent inhibition. (●) Percent inhibition of β-galactosidase activity; (■) percent inhibition of APH(3')II phosphotransferase activity.

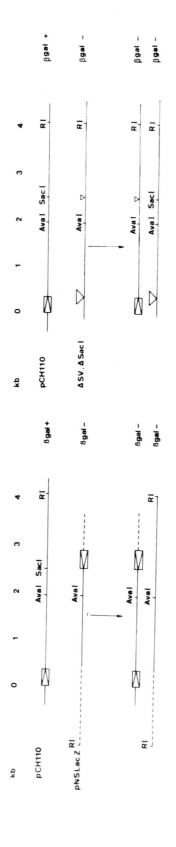

Figure 6. Homologous recombination between pCH110 and pNSlacZ vectors. (A) The regions of homology between the coding segment of pCH110 and the inverted region in pNSlacZ are aligned. The structure of the two products of a single crossover event within the homologous regions is represented. βgal+ corresponds to a structure expressing β-galactosidase and βgal− corresponds to a structure that does not express β-galactosidase. (λ) SV40 early promoter. The designations AvaI, SacI, and RI correspond to restriction enzyme sites. (B) The region of homology between the coding segment of pCH110 and the corresponding region of a promoterless deletion mutant of pCH110 (ΔSV, ΔSacI) are aligned. The structure of the two products of a single crossover event is also represented. (▽) A 4-bp deletion.

to express an unrelated gene. Furthermore, the inhibition is not caused by recombination. We suggest that the inhibition is due to an interaction between the mRNA and the antiparallel RNA.

CONCLUDING REMARKS

By several criteria, including structural analysis and biological assays, we have shown that recombinant retroviruses, with an additional internal promoter, can be propagated in producer cell lines and used to infect various cell types, including EC cells. The majority of infected NIH-3T3 and EC cells (LT, PCC3, PCC4) harbor only one recombinant provirus of the expected structure. In two out of 20 EC clones, however, rearrangements in the 5' portion of the provirus have occurred. Whereas one can rescue proviruses that have the normal structure, rescue of the rearranged proviruses has not been possible. Of note is the observation that the EC clones containing the rearranged provirus grow faster in selective medium than EC cells containing an unrearranged provirus. This fact, taken together with the observation that the site of integration influences the level of expression in EC cells suggests that the introduction of an internal promoter does not always overcome the difficulties of retroviral expression in EC cells. In addition, retroviral vectors in which the *neo* gene is under the control of the 5' LTR can express the *neo* gene in EC cells (Rubenstein et al. 1984a; Sorge et al. 1984), although the level of APH(3')II enzyme is very low (Fig. 1). Therefore, it is clear that the site of proviral integration can have a significant effect on the level and possibly the stability of gene expression.

We are now using these vectors to make transgenic mice. Recently, Stuhlmann et al. (1984) succeeded in introducing the *xgpt* gene into postimplantation mouse embryos by microinjecting cells producing a recombinant retrovirus (MSV-*gpt*) directly into the embryo. To spread the infection they coinjected fibroblasts producing wild-type Mo-MLV. This procedure, however, has a number of drawbacks, including the low percentage of embryonic cells infected, the presence of the leukemogenic wild-type virus, and the difficulty of infecting the germ line cells. As an alternative, we have cocultivated stage-2 and stage-4 mouse embryos with cells producing the Mo-MLV-*neo* virus. We found that about 20% of the fetuses harbor integrated copies of Mo-MLV-*neo* in 10–20% of their cells. It is interesting to note that the infected cells had a clonal origin. We do not know whether this is due to a heterogeneity in the susceptibility to infection of embryonic cells, or if the infection occurs at random in only a few cells of the embryos. In the latter case, retroviral vectors may become a powerful tool for cell lineage analysis. Other applications include insertional mutagenesis and tissue-specific gene expression by using appropriate promoters.

One of our goals is to use these vectors to analyze the early development of the mouse embryo. A potentially powerful technique would be to interfere specifically with mRNA metabolism of a given gene to produce phenotypic variants. We and others (Izant and Weintraub 1984; Rubenstein et al. 1984b) have shown that eukaryotic vectors producing antisense RNA of a given gene inhibit the expression of that gene. Recently, other laboratories have succeeded in inhibiting gene expression using antisense RNA. Melton has shown that injection of an antisense RNA of β-globin synthesized in vitro specifically inhibits β-globin synthesis in *Xenopus* oocytes (Melton 1985). Also phenocopies of the Krüppel developmental mutation have been produced after injection of antisense RNA into *Drosophila* embryos (Rosenberg et al. 1985). These results, combined with the fact that bacteria naturally utilize RNA complementary to mRNA to reduce gene expression of those mRNAs (Simons and Kleckner 1983; Mizuno et al. 1984), make it likely that the inhibition of the expression of endogenous eukaryotic genes will be practical when a means to amplify the production of antisense RNA is developed.

ACKNOWLEDGMENTS

We thank C. Babinet, H. Condamine, and J.L. Guénet for their help during our work with mouse embryos. We are indebted to Didier Rocancourt for excellent technical assistance and help in the preparation of the manuscript.

This work was supported by grants from the Centre National de la Recherche Scientifique (UA 269), the Foundation pour la Recherche Médicale, the Ligue Nationale Française contre le Cancer, the Ministère de l'Industrie et de la Recherche (82V1388), and the Fondation André Meyer. J.-F.N. is Maître de Recherche at Institut National de la Santé et de la Recherche Médicale. J.L.R.R. is a Fellow of the Association pour le Développement de l'Institut Pasteur.

REFERENCES

Cone, R.D. and R.C. Mulligan. 1984. High-efficiency gene transfer into mammalian cells: Generation of helper-free recombinant retrovirus with broad mammalian host range. *Proc. Natl. Acad. Sci.* **81:** 6349.

Emerman, M. and H.M. Temin. 1984. High-frequency deletion in recovered retrovirus vectors containing exogenous DNA with promoters. *J. Virol.* **50:** 42.

Hall, C.V., P.E. Jacob, G.M. Ringold, and F. Lee. 1983. Expression and regulation of *Escherichia coli lacZ* gene fusions in mammalian cells. *J. Mol. Appl. Genet.* **2:** 101.

Izant, J.G. and H. Weintraub. 1984. Inhibition of thymidine kinase gene expression by anti-sense RNA: A molecular approach to genetic analysis. *Cell* **36:** 1007.

Jaenisch, R., H. Fan, and B. Croker. 1975. Infection of preimplantation mouse embryos and of newborn mice with leukemia virus: Tissue distribution of viral DNA and RNA and leukemogenesis in the adult animal. *Proc. Natl. Acad. Sci.* **72:** 4008.

Jaenisch, R., D. Jähner, P. Nobis, I. Simon, J. Löhler, K. Harbers, and D. Grotkopp. 1981. Chromosomal position and activation of retroviral genomes inserted into the germ line of mice. *Cell* **24:** 519.

Joyner, A.L. and A. Bernstein. 1983. Retrovirus transduction: Generation of infectious retroviruses expressing dominant and selectable genes is associated with *in vivo* recombination and deletion events. *Mol. Cell. Biol.* **3:** 2180.

Linney, E., B. Davis, J. Overhauser, E. Chao, and H. Fan. 1984. Non-function of a Moloney murine leukaemia virus regulatory sequence in F9 embryonal carcinoma cells. *Nature* **308:** 470.

Mann, R., R.C. Mulligan, and D. Baltimore. 1983. Construction of a retrovirus packaging mutant and its use to produce helper-free defective retrovirus. *Cell* **33:** 153.

Melton, D. 1985. Injected anti-sense RNAs specifically block messenger RNA translation *in vivo. Proc. Natl. Acad. Sci.* **82:** 144.

Miller, A.D., M.F. Law, and I.M. Verma. 1985. Generation of helper-free amphotropic retroviruses that transduce a dominant-acting, methotrexate-resistant dihydrofolate reductase gene. *Mol. Cell. Biol.* **5:** 431.

Mizuno, T., M.Y. Chou, and M. Inouye. 1984. A unique mechanism regulating gene expression: Translational inhibition by a complementary RNA transcript (micRNA). *Proc. Natl. Acad. Sci.* **81:** 1966.

Mulligan, R.C. and P. Berg. 1980. Expression of a bacterial gene in mammalian cells. *Science* **209:** 1422.

Nicolas, J.F. and P. Berg. 1983. Regulation of expression of genes transduced into embryonal carcinoma cells. *Cold Spring Harbor Conf. Cell Proliferation* **10:** 469.

Nicolas, J.F., H. Jakob, and F. Jacob. 1981. Teratocarcinoma-derived cell lines and their use in the study of differentiation. In *Functionally differentiated cell lines* (ed. G. Sato), p. 185. A.R. Liss, New York.

Périès, J., E. Alves-Cardoso, M. Canivet, M.C. Debons-Guillemin, and J. Lasneret. 1977. Lack of multiplication of ecotropic murine C-type viruses in mouse teratocarcinoma primitive cells. *J. Natl. Cancer Inst.* **59:** 463.

Pontecorvo, G. 1975. Production of mammalian somatic cell hybrids by means of polyethylene glycol treatment. *Somatic Cell Genet.* **1:** 397.

Reiss, B., R. Sprengel, H. Will, and H. Schaller. 1984. A new sensitive method for qualitative and quantitative assay of neomycin phosphotransferase in crude cell extracts. *Gene* **30:** 217.

Rosenberg, U.B., A. Preiss, E. Seifert, H. Jäckle, and D.C. Knipple. 1985. Production of phenocopies by *Krüppel* antisense RNA injection into *Drosophila* embryos. *Nature* **313:** 703.

Rubenstein, J.L.R., J.F. Nicolas, and F. Jacob. 1984a. Construction of a retrovirus capable of transducing and expressing genes in multipotential embryonic cells. *Proc. Natl. Acad. Sci.* **81:** 7137.

———. 1984b. L'ARN non sens (nsARN): Un outil pour inactiver spécifiquement l'expression d'un gène donné *in vivo. C. R. Acad. Sci.* **299:** 271.

Simons, R.W. and N. Kleckner. 1983. Translational control of IS10 transposition. *Cell* **34:** 683.

Sorge, J., A.E. Cutting, V.D. Erdman, and J.W. Gautsch. 1984. Integration-specific retrovirus expression in embryonal carcinoma cells. *Proc. Natl. Acad. Sci.* **81:** 6627.

Stuhlmann, H., R. Cone, R.C. Mulligan, and R. Jaenisch. 1984. Introduction of a selectable gene into different animal tissue by a retrovirus recombinant vector. *Proc. Natl. Acad. Sci.* **81:** 7151.

Teich, N.M., R.A. Weiss, G.R. Martin, and D.R. Lowy. 1977. Virus infection of murine teratocarcinoma stem cell lines. *Cell* **12:** 973.

Wagner, E.F., M. Vanek, and B. Vennström. 1985. Transfer of genes into embryonal carcinoma cells by retrovirus infection: Efficient expression from an internal promoter. *EMBO J.* **4:** 663.

Whittingham, D.G. 1975. Fertilisation, early development and storage of mammalian ova. In *The early development of mammals* (ed. M. Balls and A.E. Wild), p. 1. Cambridge University Press, Cambridge, England.

Conservation and Divergence of *RAS* Protein Function during Evolution

C. BIRCHMEIER, D. BROEK, T. TODA, S. POWERS, T. KATAOKA, AND M. WIGLER
Mammalian Cell Genetics Section, Cold Spring Harbor Laboratory, P.O. Box 100, Cold Spring Harbor, New York 11724

The *ras* genes were first isolated as the transforming genes of Harvey and Kirsten sarcoma virus (Ellis et al. 1981). At least three different *ras* genes, Ha-*ras*, Ki-*ras*, and N-*ras*, exist in mammals and code for three very similar 21-kD proteins (Shimizu et al. 1983b). The *ras* proteins are localized in the plasma membrane (Willingham et al. 1980), bind guanine nucleotides (Shih et al. 1980, 1982), and have weak GTPase activity (Gibbs et al. 1984; McGrath et al. 1984; Sweet et al. 1984). A large number of tumor cells contain structurally mutated *ras* genes that are capable of tumorigenic transformation of NIH-3T3 cells upon DNA-mediated gene transfer (Reddy et al. 1982; Tabin et al. 1982; Taparowsky et al. 1982; Capon et al. 1983; Shimizu et al. 1983a; Yuasa et al. 1983). These oncogenic *ras* genes differ from their normal counterparts by single missense mutations which reduce the GTPase activity of the encoded proteins (Gibbs et al. 1984; McGrath et al. 1984; Sweet et al. 1984). Genes homologous to the mammalian *ras* have been identified in virtually every organism investigated, including yeast, slime molds, and fruit flies (Shilo and Weinberg 1981; DeFeo-Jones et al. 1983; Neuman-Silberberg et al. 1984; Powers et al. 1984; Reymond et al. 1984). In mice, the *ras* genes appear to be expressed in all cell types, and in all developmental stages (Mueller et al. 1982, 1983). The *ras* proteins are therefore presumed to be involved in a basic and ubiquitous system controlling cell proliferation.

In the yeast *Saccharomyces cerevisiae* there are two closely related but distinct genes, *RAS1* and *RAS2*, that encode proteins which are highly homologous to the mammalian *ras* proteins (DeFeo-Jones et al. 1983; Dhar et al. 1984; Powers et al. 1984). Although neither *RAS1* nor *RAS2* are by themselves essential genes, some *RAS* function is required for the continued growth and viability of haploid cells (Kataoka et al. 1984, 1985; Tatchell et al. 1984). This observation enabled us to demonstrate that yeast cells expressing the mammalian *ras* protein remain viable even when their endogenous *RAS* genes have been disrupted (Kataoka et al. 1985). Thus, it seems extremely likely that there has been considerable conservation of *ras* function throughout evolution.

RESULTS AND DISCUSSION

We have been examining the function of the yeast *RAS* genes. We have shown by combined biochemical and genetic experiments that activation of yeast adenylate cyclase is one of the essential functions of *RAS* (Toda et al. 1985; Broek et al. 1985). The dependence of yeast adenylate cyclase on *RAS* can also be demonstrated in vitro. Yeast *RAS2* protein, purified from an *Escherichia coli* expression system, will strongly activate the adenylate cyclase present in membranes prepared from yeast cells lacking endogenous *RAS* proteins (Fig. 1a). Human Ha-*ras* protein purified from an *E. coli* expression system will do the same (Fig. 1b), providing further evidence that there has been conservation of the biochemical function of *RAS* during evolution.

It was of considerable interest, therefore, for us to determine if vertebrate *ras* activates vertebrate adenylate cyclase. For this purpose we chose to examine the function of *ras* proteins in *Xenopus laevis* oocytes. Fully grown *Xenopus* oocytes are extemely large cells (diameter 1.1–1.4 mm). They can be easily injected and their size allows biochemical measurements on the injected cells. Moreover, oocytes contain an adenylate cyclase that, as in mammalian cells, is regulated by at least one GTP-binding protein, G_s, which can be ADP-ribosylated in response to exposure to cholera toxin (Olate et al. 1984). In addition, oocyte maturation, that is, the induction of meiosis, is modulated by cAMP. Oocytes, surgically removed from adult *Xenopus* ovaries, are arrested in the prophase of meiosis and can be triggered to undergo meiosis by treatment with progesterone, insulin, IGF-1, and a variety of other agents. Agents, such as cholera toxin or phosphodiesterase inhibitors, which increase cAMP levels, or microinjection of the catalytic subunit of the cAMP-dependent protein kinase (protein kinase A) inhibit maturation (O'Connor and Smith 1976; Maller and Krebs 1977; Schordaret-Slatkine et al. 1978). Injection of proteins that inhibit the activity of the protein kinase A, such as the heat-stabile kinase inhibitor or the regulatory subunit of protein kinase A, induce meiosis. Progesterone, the physiological inducer of maturation, lowers cAMP levels (Maller and Krebs 1977; Speaker and Butcher 1977; Mulner et al. 1979).

Mutant and wild-type human Ha-*ras* proteins were purified from *E. coli* carrying an expression plasmid (Gross et al. 1985). Upon injection of Ha-*ras* protein into *Xenopus* oocytes, maturation was induced, as judged by the appearance of a white spot in the pigmented half of the oocyte, by the breakdown of the germinal vesicle, and by the appearance of a meiotic

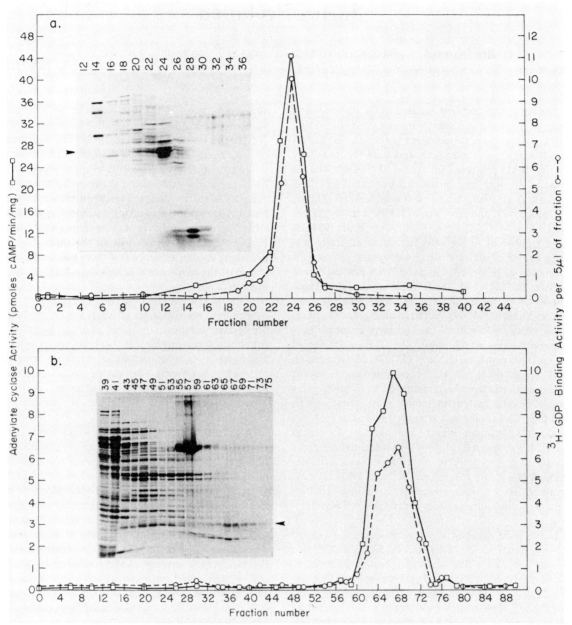

Figure 1. Copurification of *ras* proteins and adenylate cyclase-stimulating activity. (*a*) Partially purified fusion *RAS2* protein (f-*RAS2*) was applied to a Sephacryl S-300 column and 1-ml fractions were collected. [³H]GDP binding activity was determined with 5-μl aliquots of fractions. For reconstitution of adenylate cyclase activity, 5 μl of indicated fractions were incubated with 1 μl of 1 mM Gpp(NH)p for 30 min at 37°C. The samples were chilled and incubated with 30 μg of *bcy1 ras1⁻ ras2⁻* membrane proteins at 0°C for 30 min, followed by a 15-min incubation at 27°C in the presence of 0.5 mM ATP. The addition of a mixture containing [α-³²P]ATP started the 40-min reaction. Final concentrations of components in the 100-μl reaction mix were 20 units/ml creatine phosphokinase, 20 mM phosphocreatine, 20 mM MES (pH 6.2), 2.5 mM MgCl₂, 10 mM theophylline, 1 mM [³H]cAMP (20,000 cpm per reaction), 0.1 mg/ml bovine serum albumin, and 1 mM β-mercaptoethanol. SDS-polyacrylamide gel electrophoresis of Sephacryl S-300 column fractions (*inset*) shows a 42,000 molecular weight protein, identified as f-*RAS2* by immunoprecipitation, eluting at fractions containing maximal GDP binding activity and adenylate cyclase-stimulating activity. Determination of cAMP produced was carried out as previously described (Solomon et al. 1973). See Broek et al. (1985) for details. (*b*) Partially purified Ha-*ras*^Val12 protein was applied to a Sephadex G-75 gel filtration column (95 cm × 2.5 cm) and 4.5-ml fractions were collected. [³H]GDP binding activity and adenylate cyclase-stimulating activity were determined as described above. SDS-polyacrylamide gel electrophoresis of the Sephadex G-75 column fraction (*inset*) shows a 21,000 molecular weight protein, identified as Ha-*ras*^Val19 by immunoprecipitation, which elutes at fractions that contain GDP binding activity and a factor that stimulates adenylate cyclase.

spindle. When heat-denatured Ha-*ras* protein was injected, no maturation was observed (Fig. 2). Differences in the biological activity of the mutant and wild-type proteins were immediately apparent. Ha-*ras*^{Val12} protein, the product of the oncogenic gene with a missense mutation that specifies valine instead of glycine in position 12, induced maturation in 94% of the oocytes when 10 ng of the protein was injected (Fig. 2). The time course of induction of maturation for Ha-*ras*^{Val12}-injected cells was only slightly slower than for progesterone-treated cells (Fig. 2). The injected wild-type Ha-*ras* protein was much less efficient in the induction of maturation (Fig. 2). At the highest concentration tested, 400 ng of this protein induced maturation in only 55% of the injected oocytes. These experiments indicate that injected Ha-*ras* proteins induce meiosis in *Xenopus* oocytes, and that the *Xenopus* oocytes are much more sensitive to oncogenic Ha-*ras* protein than to wild-type protein.

A very early response of *Xenopus* oocytes to progesterone, the physiological inducer of meiosis, is a drop in cAMP levels, which is possibly the signal for induction of maturation (for review, see Maller 1983). We determined, therefore, whether we could observe a similar drop in *Xenopus* oocytes injected with Ha-*ras*^{Val12} protein. For this purpose, we injected [α-^{32}P]ATP into oocytes and then determined the amount of radioactivity converted into [^{32}P]cAMP after various treatments. In principle, measurements with this method reflect the equilibrium level of cAMP. Using this method, others have detected decreases in cAMP levels upon progesterone treatment, and increases upon cholera toxin treatment (Mulner et al. 1979). In our experiments, we detect a 70% drop in radioactive cAMP upon progesterone treatment and a sixfold increase upon cholera toxin treatment; however, we failed to detect any significant difference between oocytes injected with Ha-*ras*^{Val12} protein or injected with bovine serum albumin (Table 1). We conclude from this that the injected Ha-*ras*^{Val12} protein does not produce a sustained depression or elevation of adenylate cyclase activity in *Xenopus* oocytes. Changes of a transient nature would not be detected in our assay and can therefore not be completely excluded.

Cholera toxin, which raises cAMP production by activating adenylate cyclase, blocks oocyte maturation induced by progesterone (O'Connor and Smith 1976; Schorderet-Slatkine et al. 1978). Therefore, we sought to determine whether all the effects observed upon Ha-*ras*^{Val12} injections could be blocked by treatment of oocytes with cholera toxin. Cholera toxin inhibited the maturation induced by injected Ha-*ras*^{Val12} protein, as judged by the absence of germinal vesicle breakdown. However, the appearance of the white spot in the pigmented half of the oocyte could not be totally inhibited by this agent (data not shown). In many of the treated oocytes, we observed that the germinal vesicle had migrated to a position located directly under the white spot. The same concentration of cholera toxin inhibited completely germinal vesicle breakdown, migration, and the appearance of the white spot in progesterone-treated oocytes. These results also suggest that at least some of the effect of Ha-*ras* proteins on oocytes bypass the adenylate cyclase system, and they confirm the results described in the previous paragraph.

Our findings in *Xenopus* oocytes were unexpected in view of our results obtained in the yeast *S. cerevisiae*, where both yeast *RAS* and mammalian *ras* proteins strongly activate adenylate cyclase (Broek et al. 1985; Toda et al. 1985). We are therefore left with an apparent conflict, since yeast *RAS* and mammalian *ras* pro-

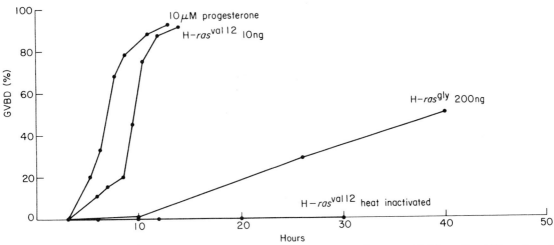

Figure 2. Time course of meiosis induced by injected Ha-*ras* protein. Ha-*ras* protein was purified from *E. coli* (Feramisco et al. 1984; Gross et al. 1985) and injected into the cytoplasm of *Xenopus laevis* oocytes (stage-VI oocytes, according to Dumont 1972). The time course of germinal vesicle breakdown (GVBD) is shown for oocytes injected with 10 ng of Ha-*ras*^{Val12}, 10 ng of heat-inactivated Ha-*ras*^{Val12}, 200 ng of Ha-*ras*^{Gly}, or for oocytes incubated in 10 μM progesterone, as indicated. Oocytes were fixed in 6% trichloroacetic acid after the indicated incubation times and examined for the presence of the germinal vesicle. For each experimental point a minimum of 20 oocytes were analyzed.

Table 1. cAMP Levels in Xenopus Oocytes after Various Treatments

Treatment	Time	[^{32}P]cAMP/total ^{32}P-labeled material ($\times 10^5$)		
		experiment 1	experiment 2	experiment 3
BSA injection	15 min	20	29	
	30 min		24	
	45 min		27	
	1 hr	25	26	18
	2 hr	33	26	
	4 hr			
BSA injection + progesterone treatment	1 hr	9	7	
Ha-ras^{Val12} protein injection	15 min	23	26	
	30 min		26	
	45 min		24	
	1 hr	28	21	
	2 hr	23	24	
	4 hr	31	24	
BSA injection + cholera toxin treatment	1 hr			109

Xenopus oocytes were injected with 10 ng of protein and 2 μCi [^{32}P]ATP each; after injection, the oocytes were transferred to medium containing 1 μM of the phosphodiesterase inhibitor, 3-isobutyl-1-methylxanthine; in addition, 10 μM progesterone or 50 ng/ml cholera toxin were added as indicated. After the indicated incubation times, the injected oocytes were transferred to 6% trichloroacetic acid, and unlabeled and [^{3}H]cAMP were added as marker and standard for recovery, respectively. The oocytes were homogenized, the insoluble material was removed, and the amount of soluble ^{32}P-labeled material was determined. The amount of [^{32}P]cAMP present in the sample was determined after the following purification steps: Dowex 40W-4X column (Solomon et al. 1973), two-dimensional thin-layer chromatography on polyethyleneimine cellulose plates; solvents for first and second dimension were 60 mM sodium acetate (pH 5.6) and 1.5 M (NH$_4$)$_2$SO$_4$, respectively (Boehme and Schultz 1976). After this step, half of the sample was treated with phosphodiesterase from bovine brain, and both aliquots were purified on neutral alumina columns. The amount of hydrolyzable [^{32}P]cAMP was calculated, and this value was adjusted for overall recovery of the [^{3}H]cAMP. The amount of [^{32}P]cAMP present in the total amount of soluble ^{32}P-labeled material was calculated. In experiment 3, oocytes were treated with collagenase to remove follicle cells.

teins appear to have a conserved biochemical function. Two major models can be considered that are consistent with our observations. First, the adenylate cyclase of *S. cerevisiae* may not be directly related to the adenylate cyclase of vertebrate cells. Rather, a domain of a protein which interacts with *RAS* proteins may have been conserved in evolution that has appeared as a regulatory domain on proteins with different function during the course of evolution. In the second model, *RAS* proteins interact with homologous effector proteins in both yeast and vertebrates, but the interactions of these proteins with subsequent cellular signaling systems have been mutated during the course of evolution. At the present time there is no compelling reason to prefer one model over the other. Clearly, we need to know whether the *RAS* proteins interact directly with the *S. cerevisiae* adenylate cyclase, or whether they act through other as yet unidentified proteins.

ACKNOWLEDGMENTS

This work was supported by grants from the National Institutes of Health and American Business for Cancer Research. C.B. is supported by the Swiss National Science Foundation, D.B. by the Damon Runyon-Walter Winchell Cancer Fund, and S.P. is a postdoctoral fellow of the Leukemia Society of America. We thank Patricia Bird for preparation of the manuscript.

REFERENCES

Boehme, E. and G. Schultz. 1976. Separation of cyclic nucleotides by thin layer chromatography on polyethyleneimine cellulose. *Methods Enzymol.* **38:** 27.

Broek, D., N. Samiy, O. Fasano, A. Fujiyama, F. Tamanoi, J. Northup, and M. Wigler. 1985. Differential activation of yeast adenylate cyclase by wild-type and mutant *RAS* proteins. *Cell* **41:** 763.

Capon, D., P. Seeburg, J. McGrath, J. Hayflick, U. Edman, A. Levinson, and D. Goeddel. 1983. Activation of Ki-*ras2* gene in human colon and lung carcinoma by different point mutations. *Nature* **304:** 507.

DeFeo-Jones, D., E. Scolnick, R. Koller, and R. Dhar. 1983. *ras*-related gene sequences identified and isolated from *Saccharomyces cerevisiae*. *Nature* **306:** 707.

Dhar, R., A. Niedo, R. Koller, D. DeFeo-Jones, and E. Scolnick. 1984. Nucleotide sequence of two *ras*^{H} related genes isolated from the yeast *Saccharomyces cerevisiae*. *Nucleic Acids Res.* **12:** 3611.

Dumont, J. 1972. Oogenesis in *Xenopus laevis*. Stages of oocyte development in laboratory maintained animals. *J. Morphol.* **136:** 153.

Ellis, R., D. DeFeo, T. Shih, M. Gonda, H. Young, N.

TSUCHIDA, D. LOWY, and E. SCOLNICK. 1981. The p21 *src* genes of Harvey and Kirsten sarcoma viruses originate from divergent members of a family of normal vertebrate genes. *Nature* 292: 506.

FERAMISCO, J., M. GROSS, T. KAMATA, M. ROSENBERG, and R. SWEET. 1984. Microinjection of the oncogenic form of the human H-*ras* (T24) protein results in rapid proliferation of quiescent cells. *Cell* 38: 109.

GIBBS, J., I. SIGAL, M. POE, and E. SCOLNICK. 1984. Intrinsic GTPase activity distinguishes normal and oncogenic *ras* p21 molecules. *Proc. Natl. Acad. Sci.* 81: 5704.

GROSS, M., R. SWEET, G. SATHE, S. YOKOYAMA, O. FASANO, M. GOLDFARB, M. WIGLER, and M. ROSENBERG. 1985. Purification and characterization of human H-*ras* proteins expressed in *E. coli*. *Mol. Cell. Biol.* 5: 1015.

KATAOKA, T., S. POWERS, S. CAMERON, O. FASANO, M. GOLDFARB, J. BROACH, and M. WIGLER. 1985. Functional homology of mammalian and yeast *RAS* genes. *Cell* 40: 19.

KATAOKA, T., S. POWERS, C. MCGILL, O. FASANO, J. STRATHERN, J. BROACH, and M. WIGLER. 1984. Genetic analysis of yeast *RAS1* and *RAS2* genes. *Cell* 37: 437.

MALLER, J. 1983. Interaction of steroids with cyclic nucleotide system in amphibian oocytes. *Adv. Cyclic Nucleotide Res.* 13: 295.

MALLER, J. and E. KREBS. 1977. Progesterone stimulated meiotic cell division in *Xenopus* oocytes. *J. Biol. Chem.* 252: 1712.

MCGRATH, J., D. CAPON, D. GOEDDEL, and A. LEVINSON. 1984. Comparative biochemical properties of normal and activated human *ras* p21 protein. *Nature* 310: 644.

MUELLER, R., D. SLAMON, J. TREMBLAY, M. CLINE, and I. VERMA. 1982. Differential expression of cellular oncogenes during pre- and postnatal development of the mouse. *Nature* 299: 640.

MUELLER, R., D. SLAMON, E. ADAMSON, J. TREMBLAY, D. MUELLER, M. CLINE, and I. VERMA. 1983. Transcription of c-onc genes c-*ras*^{ki} and c-*fms* during mouse development. *Mol. Cell. Biol.* 3: 1062.

MULNER, O., D. HUCHON, C. THIBIER, and R. OZON. 1979. cAMP synthesis in *Xenopus laevis* oocytes: Inhibition by progesterone. *Biochim. Biophys. Acta* 582: 179.

NEUMAN-SILBERBERG, F., E. SCHEJTER, F. HOFFMANN, and B.-Z. SHILO. 1984. The *Drosophila ras* oncogenes: Structure and nucleotide sequence. *Cell* 37: 1027.

O'CONNOR, C. and L. SMITH. 1976. Inhibition of oocyte maturation by theophylline: Possible mechanism of action. *Dev. Biol.* 52: 318.

OLATE, J., C. ALLENDE, J. ALLENDE, R. SEKURA, and L. BIRNBAUMER. 1984. Oocyte adenylate cyclase contains Ni, yet the guanine nucleotide dependent inhibition by progesterone is not sensitive to pertussis toxin. *FEBS Lett.* 175: 25.

POWERS, S., T. KATAOKA, O. FASANO, M. GOLDFARB, J. STRATHERN, J. BROACH, and M. WIGLER. 1984. Genes in *S. cerevisiae* encoding proteins with domains homologous to the mammalian *ras* proteins. *Cell* 36: 607.

REDDY, E., R. REYNOLDS, E. SANTOS, and M. BARBACID. 1982. A point mutation is responsible for the acquisition of transforming properties by the T24 human bladder carcinoma oncogene. *Nature* 300: 149.

REYMOND, G., R. GOMER, M. MEHDY, and R. FIRTEL. 1984. Developmental regulation of a *Dictyostelium* gene encoding a protein homologous to mammalian *ras* protein. *Cell* 39: 141.

SCHORDERET-SLATKINE, S., M. SCHORDERET, B. BOQUET, F. GODEAU, and E. BEAULIEU. 1978. Progesterone induced meiosis in *Xenopus laevis* oocytes: A role for cAMP at the maturation promoting level. *Cell* 15: 1269.

SHIH, T., A. PAPAGEORGE, P. STOKES, M. WEEKS, and E. SCOLNICK. 1980. Guanine nucleotide-binding and autophosphorylating activities associated with the p21^{src} protein of Harvey murine sarcoma virus. *Nature* 287: 686.

SHIH, T., P. STOKES, G. SMYTHES, R. DHAR, and S. OROSZLAN. 1982. Characterization of the phosphorylation sites and the surrounding amino acid sequences of the p21 transforming proteins coded for by the Harvey and Kirsten strains of murine sarcoma viruses. *J. Biol. Chem.* 257: 11767.

SHILO, B. and R. WEINBERG. 1981. DNA sequences homologous to vertebrate oncogenes are conserved in *Drosophila melanogaster*. *Proc. Natl. Acad. Sci.* 78: 6789.

SHIMIZU, K., D. BIRNBAUM, M. RULEY, O. FASANO, Y. SUARD, L. EDLUND, E. TAPAROWSKY, M. GOLDFARB, and M. WIGLER. 1983a. The structure of the K-*ras* gene of the human lung carcinoma cell line Calu-1. *Nature* 304: 497.

SHIMIZU, K., M. GOLDFARB, Y. SUARD, M. PERUCHO, Y. LI, T. KAMATA, J. FERAMISCO, E. STAVNESER, J. FOGH, and M. WIGLER. 1983b. Three human transforming genes are related to the viral *ras* oncogene. *Proc. Natl. Acad. Sci.* 80: 2112.

SOLOMON, Y., C. LANDOS, and M. RODBELL. 1973. A highly sensitive adenylate cyclase assay. *Anal. Biochem.* 58: 541.

SPEAKER, M. and F. BUTCHER. 1977. Cyclic nucleotide fluctuations during steroid induced meiotic maturation of frog oocytes. *Nature* 267: 848.

SWEET, R., S. YOKOYAMA, T. KAMATA, J. FERAMISCO, M. ROSENBERG, and M. GROSS. 1984. The product of *ras* is a GTPase and the T24 oncogenic mutant is deficient in this activity. *Nature* 311: 273.

TABIN, C., S. BRADLEY, C. BARGMANN, R. WEINBERG, A. PAPAGEORGE, E. SCOLNICK, R. DHAR, D. LOWY, and E. CHANG. 1982. Mechanism of activation of a human oncogene. *Nature* 300: 143.

TAPAROWSKY, E., Y. SUARD, O. FASANO, K. SHIMIZU, M. GOLDFARB, and M. WIGLER. 1982. Activation of the T24 bladder carcinoma transforming gene is linked to a single amino acid change. *Nature* 300: 762.

TATCHELL, K., D. CHALEFF, D. DEFEO-JONES, and E. SCOLNICK. 1984. Requirement of either of a pair of *ras*-related genes of *Saccharomyces cerevisiae* for spore viability. *Nature* 309: 523.

TODA, T., I. UNO, T. ISHIKAWA, S. POWERS, T. KATAOKA, D. BROEK, J. BROACH, K. MATSUMOTO, and M. WIGLER. 1985. In yeast, *RAS* proteins are controlling elements of the cyclic AMP pathway. *Cell* 40: 27.

WILLINGHAM, M., I. PASTAN, T. SHIH, and E. SCOLNICK. 1980. Localization of the *src* gene product of the Harvey strain of MSV to plasma membrane of transformed cells by electron microscopic immunocytochemistry. *Cell* 19: 1005.

YUASA, Y., S. SRIVASTAVA, C. DUNN, J. RHIM, E. REDDY, and S. AARONSON. 1983. Aquisition of transforming properties by alternative point mutations within c-*bas/has* human proto-oncogene. *Nature* 303: 775.

Proto-oncogenes of *Drosophila melanogaster*

J.M. BISHOP,*† B. DREES,‡ A.L. KATZEN,†‡ T.B. KORNBERG,‡ AND M.A. SIMON†‡

*Departments of ‡Biochemistry and Biophysics and *Microbiology and Immunology and †The Hooper Research Foundation, University of California Medical Center, San Francisco, California 94143*

The oncogenes of retroviruses arose by transduction of "proto-oncogenes" found in the genomes of vertebrates and perhaps of all metazoan organisms (Bishop 1983). It appears likely that proto-oncogenes represent a common keyboard on which carcinogens of many sorts can play. But what is the role of these genes in the daily affairs of normal cells and organisms? Why are they there? We do not know, but we suspect that they may figure in differentiation. This suspicion has at least three origins: the specificity with which retroviral oncogenes attack different tissues, as if the genes were designed to work only in certain types of cells; the fact that most retroviral oncogenes meddle with differentiation during the course of neoplastic transformation; and the emerging links between the expression of various proto-oncogenes and the course of differentiation in one or another developmental lineage (Bishop and Varmus 1982, 1985).

To explore the physiological function of proto-oncogenes, we have turned to *Drosophila melanogaster*. Shilo and Weinberg sounded the first alert that the fruit fly might possess at least some of the known proto-oncogenes (Shilo and Weinberg 1981), and as many as eight of these genes have now been isolated from *Drosophila* DNA. In this report, we summarize our efforts to explore the diversity and function of proto-oncogenes in *D. melanogaster*. Contrary to prior expectations, we find that at least one of these genes (c-*src*) is used principally in cells that are no longer dividing and thus appears not to be a mitotic signal.

EXPERIMENTAL PROCEDURES

Isolation of proto-oncogenes from the DNA of the Canton S strain of *D. melanogaster* has been described elsewhere (Simon et al. 1983; Katzen et al. 1985). The manipulation of molecular clones of DNA, the determination of nucleotide sequence, and the performance of hybridizations in situ with sections of tissue were all performed according to standard procedures (Simon et al. 1983; Katzen et al. 1985; Kornberg et al. 1985).

RESULTS AND DISCUSSION

Isolation of *Drosophila* Genes related to v-*src*

We first sought *Drosophila* genes from within the large family that encodes various forms of protein-tyrosine kinases (Hunter and Cooper 1985). The search began with genes related to v-*src*, the oncogene of Rous sarcoma virus. Three were found, located at positions 29A, 64B, and 73B on polytene chromosomes (Simon et al. 1983). The gene at 64B represents c-*src* as found earlier in avian and mammalian genomes (Stehelin et al. 1976; Spector et al. 1978; Parker et al. 1985); the gene at 73B is analogous to c-*abl* (Hoffman et al. 1983); and the gene at 29A is as yet unidentified. There is no doubt about the authenticity of these loci because each has been sequenced sufficiently to identify it as a member of the protein-tyrosine kinase family. We have since cast our net wider by isolating three additional genes that may also be members of the kinase family, but these are related more closely to c-*fps* than to c-*src* (A. Katzen, unpubl.).

It therefore appears that *D. melanogaster* resembles vertebrate organisms by possessing a substantial family of genes encoding protein-tyrosine kinases. The full size of this family and the physiological functions of its members are not yet apparent.

The Structure of the Protein Encoded by c-*src*(*Drosophila*)

A molecular clone of cDNA representing mRNA for c-*src*(*Drosophila*) was isolated and sequenced. The results revealed that the protein encoded by the *Drosophila* gene is remarkably similar to the product of avian and mammalian c-*src* (Fig. 1). The *Drosophila* and vertebrate proteins are virtually the same length; with the exception of a 10-kD domain at their aminoterminal extremes, they share ~50% of their amino acid sequences; they also share topographical landmarks defined previously on the vertebrate protein, including a tyrosine residue that is a substrate for phosphorylation, a lysine residue that marks the ATP binding site in the catalytic domain of the protein, and an aminoterminal glycine that is myristylated and apparently mediates attachment of the protein to membranes (see Hunter and Cooper 1985). Figure 2 illustrates the similarities between the *src* proteins of *Drosophila* and chickens in regions known to be archetypical of vertebrate c-*src*. It thus appears likely that the proteins encoded by *Drosophila* and vertebrate c-*src* have similar or identical enzymatic activities and biochemical functions.

Antibodies directed against the c-*src* protein of birds and mammals react specifically with a protein-tyrosine kinase found in tissues and cultured cells from *Drosophila* (Simon et al. 1983). We have not identified the responsible polypeptide, however, and further pursuit of the protein encoded by c-*src*(*Drosophila*) awaits the

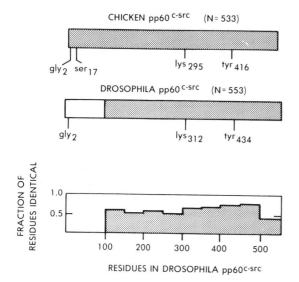

Figure 1. The proteins encoded by c-*src*(chicken) and c-*src*(*Drosophila*) are similar. The amino acid sequences of the proteins encoded by c-*src*(chicken) and c-*src*(*Drosophila*) were deduced from the nucleotide sequences of the genes (Takeya and Hanafusa 1982; Hoffman-Falk et al. 1983; Simon et al. 1983; M. Simon, unpubl.).

preparation of antibodies against peptides synthesized to represent portions of the *Drosophila* protein.

The Expression of c-*src*(*Drosophila*) Is Regulated during Development

How are the various protein-tyrosine kinase genes of *Drosophila* used? What is the purpose of the genetic diversification they represent? Might they have been fabricated for different developmental programs? As a first step toward answering these questions, we analyzed the occurrence of RNA transcribed from the kinase genes at positions 29A, 64B, and 73B during the course of *Drosophila* development. The three genes are expressed in distinctive patterns: 29A throughout development, but least prominently in larvae; 73B in abundance as maternal mRNA immediately after oviposition, and again during puparium; and 64B in a complex pattern of three mRNAs of different lengths (3.5, 5.0, and 5.5 kb), each abundant at distinctive periods of development (M. Simon, unpubl.). It is as if the different genes, and even different RNAs arising from a single gene, were used at different times and perhaps for different purposes during the course of development. To explore these thoughts further, we turned to hybridization in situ, using radioactive probes for the *src* gene at 64B and sections from embryos, larvae, and pupae. Table 1 summarizes the results.

At early times (0-2 hr following oviposition), the developing embryo is a syncytium containing synchronously dividing nuclei. Differentiation of form and function is not apparent, the nuclei remain totipotent, and transcription from the embryonic genome has yet to begin. RNA transcribed from c-*src* is abundant throughout the embryo and must have a maternal origin. Ten hours after the onset of embryogenesis, the expression of *src* has become specialized, being prominent only in cells that are the progenitors for the smooth muscle that will encase the primitive gut. By 16 hours after oviposition, organogenesis is well advanced, and the distribution of *src* RNA is again highly specialized—abundant only in the primitive brain and ventral nerve chord. Expression is no longer apparent in the gut. In late larvae and in pupae, expression of c-*src* is prominent in the imaginal disc that will give rise to the insect's head, and, within the disc, in two regions: the anlage for the individual facets of the com-

Conservation of Archetypical Amino Acid Sequences
in pp60^{c-src}

Dr: (308)-val-ala-<u>val</u>-lys-thr-leu- (315)
Ch: (291)-val-ala-<u>ile</u>-lys-thr-leu- (298)

Dr: (401)-his-arg-asp-leu- (406)
Ch: (383)-his-arg-asp-leu- (388)

Dr: (430)-asp-<u>asp</u>-glu-tyr-<u>cys-pro</u>- (437)
Ch: (412)-asp-<u>asn</u>-glu-tyr-<u>thr-ala</u>- (419)

Dr: (444)-lys-trp-thr-ala-pro-glu- (451)
Ch: (426)-lys-trp-thr-ala-pro-glu- (433)

Dr: (460)-ser-asp-val-trp-ser-<u>tyr</u>-gly-ile-leu-leu- (471)
Ch: (442)-ser-asp-val-trp-ser-<u>phe</u>-gly-ile-leu-leu- (453)

Figure 2. Archetypical amino acid sequences in the proteins encoded by c-*src*(chicken) and c-*src*(*Drosophila*). Conserved amino acid sequences have been identified that serve as signatures for the family of protein-tyrosine kinases (Ralston and Bishop 1984; Hunter and Cooper 1985). The figure aligns these sequences from the proteins encoded by c-*src*(chicken) and c-*src*(*Drosophila*).

Table 1. Expression of c-*src* during the Development of *D. melanogaster*

Stage of development	Areas of expression
Syncytial preblastoderm (0–2 hr after oviposition)	diffusely throughout preblastoderm (energids of preblastoderm nuclei)
Cellular blastoderm, including germ band extension (6 hr after oviposition)	all cells of blastoderm
Ten-hour embryo	anlage of smooth muscle cells in gut (visceral mesoderm)
Eighteen-hour embryo	brain and ventral nerve cord; *not* in gut
Third instar larvae (5.5 days)	brain, ventral ganglia, and eye-antennal imaginal disc
Early pupae (6–24 hr after puparium)	brain, ventral ganglia, eye-antennal imaginal disc, and gut (nature of cells uncertain)

pound eye and the brain. In addition, expression in the gut returns as of early puparium.

There is a surprise in these findings. With the exception of the early embryo, the regions where c-*src* is expressed in abundance represent mainly postmitotic tissues: cells are not dividing, but instead are committed to specific developmental lineages. The patterns of *src* expression in neural tissue of *Drosophila* are reminiscent of work with chickens, which has found expression of *src* in the brain and in terminally differentiated cells of the neuroretina (Cotton and Brugge 1983; Sorge et al. 1984; Fults et al. 1985). We look forward to a comparative study with other members of the kinase family. At stake is the hypothesis that the diversification of kinase genes is designed to provide different developmental pathways with individual access to the same biochemical function.

The *myb* Gene of *D. melanogaster*

Four major classes of proto-oncogenes have been perceived to date, according to the proteins they encode: growth factors that elicit tyrosine-phosphorylation; protein-tyrosine kinases; regulators of cAMP; and nuclear proteins whose functions are not known. The last of these have been particularly elusive in *Drosophila*. For example, we and others have failed repeatedly to identify an authentic representation of c-*myc* in the fruit fly. We therefore turned to another option.

Both v-*myb* and the cellular gene (c-*myb*) from which it arose encode proteins that are found in the nuclear matrix of the cell (Klempnauer et al. 1984; G. Ramsay and J.M. Bishop, unpubl.). We have identified and isolated a *Drosophila* gene that appears to be the homolog of vertebrate c-*myb*, residing at position 13E/F on the X chromosome. The conserved domain of c-*myb* comprises a minimum of 125 amino acids, 91 of which are identical in the chicken and *Drosophila myb* proteins, and 28 of which are conservative or neutral substitutions (see Fig. 3). The c-*myb* gene of *Drosophila* is expressed throughout the course of *Drosophila* development, but we have yet to determine the tissues within which the expression occurs.

Vertebrate c-*myb* is expressed principally in primitive hematopoietic cells (Gonda et al. 1982; Sheiness and Gardinier 1984), a fact that is thought to reflect some specialized function of the gene. It would be remarkable if c-*myb* should prove to have an analogous function in *Drosophila*, whose hematopoietic system is quite different from that of vertebrates. Our isolation of c-*myb* from *Drosophila* enlarges the variety of proto-oncogenes found in the fly, and it sets the stage for genetic analyses that may help to illuminate the principles that govern the genesis of developmental pathways during the course of evolution.

CONCLUSIONS

The use of *D. melanogaster* to study proto-oncogenes represents a compromise. The fruit fly is a holometabolous organism that undergoes several metamorphoses during its developmental cycle. Some adult tissues arise from cells that had been sequestered in a primitive state during previous stages of development. This is not the route by which *Homo sapiens* reaches its glory, so we must bear the compromise in mind. But the superficial distinctions between *Drosophila* and vertebrate organisms appear ever less threatening as the profusion of homeo boxes and other links between the humble and mighty of phylogeny continue to emerge (Gehring 1985).

The genetic map of proto-oncogenes in *Drosophila* has become quite busy (Fig. 4). There are three analogs of mammalian c-*ras*: c-*src* and several of its relatives; a gene analogous to c-*erb-B* and thus the receptor for epithelial growth factor (EGF); and c-*myb*. As a first approximation, we can say that virtually all of the major categories of proto-oncogenes are represented: protein kinases; regulators of cAMP; receptors for growth factors (and by inference, the growth factors themselves); and nuclear proteins that may bind to DNA, but whose function is presently unknown. Of these genes, only *ras* has been sighted in yeast. The prospects for genetic definitions of proto-oncogene function therefore still lie mainly with the fruit fly and we are therefore grateful for the cornucopia of proto-oncogenes the fly presents.

ACKNOWLEDGMENTS

Work summarized here was supported by the National Institutes of Health (grants CA12705, S07 RR05355, and GM 31286) and by funds from the G.W. Hooper Research Foundation. A. Katzen was supported by a National Science Foundation Graduate

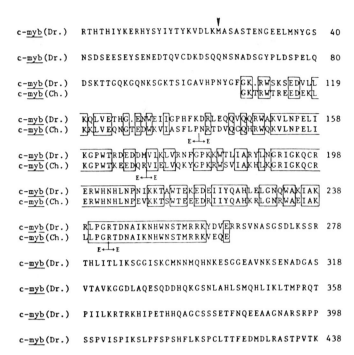

Figure 3. The protein encoded by c-*myb*(*Drosophila*). The alignment was compiled from published data (Klempnauer et al. 1982; Katzen et al. 1985) and from unpublished sequence provided by S. Gerondakis. (Arrowhead) Potential initiation codon; (E) locations of introns in the chicken gene.

Fellowship, and B. Drees and M. Simon by an Institutional Training Grant from the National Institutes of Health.

REFERENCES

BISHOP, J.M. 1983. Cellular oncogenes and retroviruses. *Annu. Rev. Biochem.* **52**: 301.

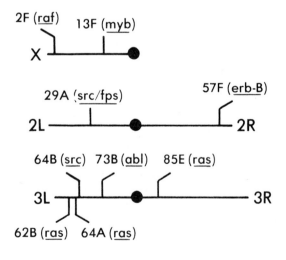

Figure 4. The proto-oncogenes of *D. melanogaster*. Proto-oncogenes isolated to date from *Drosophila* DNA are here assigned to their approximate positions on polytene chromosomes. (Details are from Hoffman-Falk et al. 1983; Simon et al. 1983; Neuman-Silberg et al. 1984; Katzen et al. 1985; Livneh et al. 1985; G. Mark, pers. comm.)

BISHOP, J.M. and H.E. VARMUS. 1982. Functions and origins of retroviral transforming genes. In *Molecular biology of tumor viruses, 2nd edition: RNA tumor viruses* (ed. R. Weiss et al.), p. 999. Cold Spring Harbor Laboratory, Cold Spring Harbor, New York.

———. 1985. Functions and origins of retroviral transforming genes. In *Molecular biology of tumor viruses*, 2nd edition, revised: *RNA tumor viruses* (ed. R. Weiss et al.), vol. 2, p. 249. Cold Spring Harbor Laboratory, Cold Spring Harbor, New York.

COTTON, P.C. and J.S. BRUGGE. 1983. Neural tissues express high levels of the cellular *src* gene product pp60[src]. *Mol. Cell. Biol.* **3**: 1157.

FULTS, D.W., A.C. TOWLE, J.M. LAUDER, and P.F. MANESS. 1985. pp60[c-src] in the developing cerebellum. *Mol. Cell. Biol.* **5**: 27.

GEHRING, W.J. 1985. The homeo box: A key to the understanding of development? *Cell* **40**: 3.

GONDA, T.J., D.K. SHEINESS, and J.M. BISHOP. 1982. Transcripts from the cellular homologs of retroviral oncogenes: Distribution among chicken tissues. *Mol. Cell. Biol.* **2**: 617.

HOFFMAN, F.M., L.D. FRESCO, H. HOFFMAN-FALK, and B.-Z. SHILO. 1983. Nucleotide sequence of the *Drosophila src* and *abl* homologs: Conservation and variability in the *src* family oncogenes. *Cell* **35**: 393.

HOFFMAN-FALK, H., P. EINAT, B.-Z. SHILO, and F.M. HOFFMAN. 1983. *Drosophila melanogaster* DNA clones homologous to vertebrate oncogenes: Evidence for a common ancestor to *src* and *abl* cellular genes. *Cell* **32**: 589.

HUNTER, T. and J.A. COOPER. 1985. Protein-tyrosine kinases. *Annu. Rev. Biochem.* **54**: 897.

KATZEN, A.L., T.B. KORNBERG, and J.M. BISHOP. 1985. Isolation of the proto-oncogene c-*myb* from *Drosophila melanogaster*. *Cell* **41**: 449.

KLEMPNAUER, K.-H., T.J. GONDA, and J.M. BISHOP. 1982. Nucleotide sequence of the retrovirus leukemia gene v-*myb* and its cellular progenitor c-*myb*: The architecture of a transduced oncogene. *Cell* **31**: 453.

KLEMPNAUER, K.-H., G. SYMONDS, G. EVAN, and J.M. BISHOP.

1984. Subcellular location of proteins encoded by viral and cellular *myb* genes. *Cell* **37**: 537.

KORNBERG, T., I. SIDEN, P. O'FARRELL, and M. SIMON. 1985. The engrailed locus of *Drosophila: In situ* localization of transcripts reveals compartment-specific expression. *Cell* **40**: 45.

LIVNEH, E., L. GLAZER, D. SEGAL, J. SCHLESSINGER, and B.-Z. SHILO. 1985. The *Drosophila* EGF receptor gene homolog: Conservation of both hormone binding and kinase domains. *Cell* **40**: 599.

NEUMAN-SILBERG, F.S., E. SCHEJTER, F.M. HOFFMANN, and B.-Z. SHILO. 1984. The *Drosophila ras* oncogenes: Structure and nucleotide sequence. *Cell* **37**: 1027.

PARKER, R.C., G. MARDON, R.V. LEBO, H.E. VARMUS, and J.M. BISHOP. 1985. Isolation of duplicated human c-*src* genes located on chromosomes 1 and 20. *Mol. Cell. Biol.* **5**: 831.

RALSTON, R. and J.M. BISHOP. 1984. Evolutionary relationships among oncogenes of DNA and RNA tumor viruses: *myc, myb*, and adenovirus E1a. *Cancer Cells* **2**: 165.

SHEINESS, D. and M. GARDINIER. 1984. Expression of a proto-oncogene (proto-*myb*) in hemopoietic tissues of mice. *Mol. Cell. Biol.* **4**: 1206.

SHILO, B.-Z. and R.A. WEINBERG. 1981. DNA sequences homologous to vertebrate oncogenes are conserved in *Drosophila melanogaster. Proc. Natl. Acad. Sci.* **78**: 6789.

SIMON, M.A., T.B. KORNBERG, and J.M. BISHOP. 1983. Three loci related to the *src* oncogene and tyrosine-specific protein kinase activity in *Drosophila. Nature* **302**: 837.

SORGE, L.K., B.T. LEVY, and P.F. MANESS. 1984. pp60$^{c\text{-}src}$ is developmentally regulated in the neural retina. *Cell* **36**: 249.

SPECTOR, D.H., H.E. VARMUS, and J.M. BISHOP. 1978. Nucleotide sequences related to the transforming gene of avian sarcoma virus are present in the DNA of uninfected vertebrates. *Proc. Natl. Acad. Sci.* **75**: 4102.

STEHELIN, D., H.E. VARMUS, J.M. BISHOP, and P.K. VOGT. 1976. DNA related to the transforming gene(s) of avian sarcoma viruses is present in normal avian DNA. *Nature* **260**: 170.

TAKEYA, T. and H. HANAFUSA. 1982. DNA sequence of the viral and cellular *src* gene of chickens. Comparison of the *src* genes of two strains of avian sarcoma virus and of the cellular homolog. *J. Virol.* **44**: 12.

Proto-Oncogene *fos* Is Expressed during Development, Differentiation, and Growth

J. Deschamps, R.L. Mitchell, F. Meijlink, W. Kruijer, D. Schubert, and I.M. Verma
The Salk Institute, P.O. Box 85800, San Diego, California 92138

Proto-oncogene *fos* is expressed during pre- and postnatal development of the mouse, cellular differentiation, and cell proliferation. Day-17 to -18 mouse amnion displays high levels of both *fos*-specific 2.2-kb transcripts as well as 55K *fos* protein. Bone marrow cells and macrophage cell lines express *fos* transcripts. Rapid induction of *fos* gene expression is witnessed when either a human promonocytic cell line, U937, or monomyelocytic cell line, HL-60, is treated with phorbol ester (TPA) to differentiate to mature macrophages. The highest levels of induction occur by 30 minutes but *fos* RNA is synthesized continuously, albeit three- to fourfold lower, for 96–108 hours. In contrast, *fos* protein can be detected for only 90–120 minutes postinduction. When PC12 cells differentiate to form neurites upon addition of nerve growth factor (NGF), the *fos* gene is rapidly but transiently induced, with maximal levels accumulating by 20–30 minutes. Unlike the monocytic cell differentiation, no *fos*-specific transcripts could be observed after 120–240 minutes of addition of NGF. Rapid and transient induction of the expression of the *fos* gene is also observed when quiescent mouse fibroblasts are stimulated to proliferate with mitogens. In an attempt to understand the regulation of transcription of the *fos* gene, a transcriptional enhancer element has been localized to reside between −60 to −400 nucleotides upstream of the 5′ cap nucleotide. We discuss the complex regulation of the *fos* gene in terms of its role in normal cells and its ability to induce cellular transformation in vitro.

INTRODUCTION

Delineation of the role of proto-oncogenes in the normal cell has become central to understanding their role in tumor induction (Bishop 1983; Hunter 1984). Structural similarities between the products of at least three proto-oncogenes and normal cellular proteins are well established. The *sis* oncogene encodes the β-chain of platelet-derived growth factor (PDGF) (Doolittle et al. 1983; Waterfield et al. 1983), the *erb-B* oncogene is the truncated form of epidermal growth factor (EGF) receptor gene (Downward et al. 1984; Ullrich et al. 1984), whereas the *fms* oncogene product is related if not identical to macrophage colony stimulating factor (M-CSF or CSF-1) receptor protein (Sherr et al. 1985). Extensive phylogenetic conservation and the tissue- and stage-specific expression of many proto-oncogenes lend support to the general belief that proto-oncogenes are critically involved in normal cellular processes (Shilo and Weinberg 1981; Müller et al. 1982). An extensive survey of the expression of proto-oncogenes has been compiled (Müller and Verma 1984). In this paper, we report the expression of proto-oncogene *fos* during mouse development and during differentiation and growth of a variety of cell types.

RESULTS

fos Gene Architecture

Elucidation of the complete molecular structure of the *fos* gene preceded knowledge of its expression in a variety of cell types. This was propitious because it is essential to know its architecture in order to grasp the subtle and complex regulation of its expression.

To date, two retroviruses containing the *fos* oncogene have been identified, namely Finkel-Biskis-Jinkins murine sarcoma virus (FBJ-MSV) and Finkel-Biskis-Reilly (FBR)-MSV (Finkel et al. 1966, 1973). The complete nucleotide sequences of their proviral DNA have been deduced (Van Beveren et al. 1983, 1984). Additionally, the nucleotide sequence of the cellular progenitor of the *fos* gene has also been determined (Van Beveren et al. 1983). Figure 1 is a diagram of the organization of viral and cellular *fos* genes and their deduced products. The salient features can be summarized as follows:

1. FBJ-MSV proviral DNA contains 4026 nucleotides, including two long terminal repeats (LTRs) of 617 nucleotides each, 1639 nucleotides of acquired cellular sequences (v-*fos*), and a portion of the envelope (*env*) gene.
2. Both the initiation and termination codons of the v-*fos* protein are within the acquired sequences that encode a protein of 381 amino acids, having a molecular weight of 49,601.
3. In cells transformed by FBJ-MSV, a phosphoprotein with an apparent M_r of 55,000 (p55) on SDS-polyacrylamide gel electrophoresis (SDS-PAGE) has been identified as the transforming protein (Curran et al. 1982). The discrepancy between the observed size and the size predicted by sequence analysis is likely due to the unusual amino acid composition of the *fos* protein (10% proline), since the v-*fos* protein expressed in bacteria has a similar relative mobility (MacConnell and Verma 1983).
4. The sequences in the c-*fos* gene that are homolo-

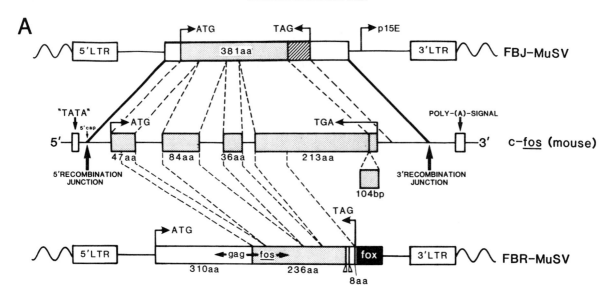

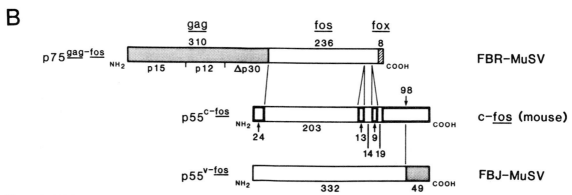

Figure 1. (A) Molecular architecture of FBJ-MSV (top) and FBR-MSV (bottom) proviral DNAs and the c-fos gene (middle). (Top) The stippled box indicates the acquired cellular sequences; arrows indicate the initiation and termination codons of v-fos proteins; the hatched region indicates the carboxyterminal 49 amino acids of the v-fos protein encoded in a different reading frame due to deletion of 104 bp of c-fos sequences. (Middle) The stippled boxes are the exons; the number of amino acids encoded by each exon is given. The 104-bp sequence that has been deleted in the v-fos sequence is indicated with a box below the line. Unlike the v-fos protein, the c-fos protein terminates at a TGA codon. (Bottom) Broken lines indicate the portions of the exons acquired from the c-fos gene; small, open triangles indicate deletion from FBR-MSV as compared with the c-fos gene. Details of the structure of FBR-MSV proviral DNA have previously been described (Van Beveren et al. 1984). (B) A schematic comparison of p75$^{gag\text{-}fos}$ (top), p55$^{c\text{-}cos}$ (middle), and p55$^{v\text{-}fos}$ (bottom) proteins. In p75 the gag-encoded portion is indicated with a stippled box, and that encoded by v-fos is shown by the hatched box. The region of p55$^{c\text{-}fos}$ indicated by thickened boxes and vertical arrows are those portions deleted in p75$^{gag\text{-}fos}$. The hatched region in p55$^{v\text{-}fos}$ is the carboxyterminal portion, which differs from that of p55$^{c\text{-}fos}$. The numbers refer to the number of amino acids encoded by each region.

gous to those in the v-fos gene are interrupted by four regions of nonhomology, three of which represent bona fide introns.

5. The 104-nucleotide-long fourth region, which is present in both mouse and human c-fos genes, represents sequences that have been deleted during the biogenesis of the v-fos gene. (The additional 104 nucleotides in the c-fos gene transcripts do not increase the predicted size of the c-fos proteins, because of a switch to a different reading frame.)

6. The c-fos protein has 380 amino acids, which is remarkably similar to the size of the v-fos protein (381 amino acids).

7. In the first 332 amino acids, the v-fos and mouse c-fos proteins differ at only five residues, whereas the remaining 48 amino acids of the c-fos protein are encoded in a different reading frame from that in the v-fos protein. Thus, the v-fos and c-fos proteins, though largely similar, have different carboxyl termini (Fig. 1B).

8. Despite their different carboxyl termini, both the v-fos and c-fos proteins are located in the nucleus. The c-fos protein undergoes more extensive modifications than the v-fos protein.

9. The mouse and human c-fos genes share greater than 90% sequence homology, differing in only 24

residues out of a total of 380 amino acids (van Straaten et al. 1983).
10. FBR-MSV proviral DNA contains 3791 nucleotides (specifying a genome of 3284 bases) and encodes a single *gag-fos* fusion product of 554 amino acids.
11. The *fos* portion of the gene lacks sequences that encode the first 24 and the last 98 amino acids of the 380-amino-acid mouse c-*fos* gene product (Fig. 1B). In addition, the coding region has sustained three small in-frame deletions, one in the $p30^{gag}$ portion and two in the *fos* region, as compared with sequences of AKR MLV and the c-*fos* gene, respectively (Van Beveren et al. 1984).
12. The gene product terminates in sequences termed *fox* (Fig. 1A), which are present in normal mouse DNA at loci unrelated to the c-*fos* gene. The c-*fox* gene(s) is expressed as an abundant class of polyadenylated RNA in mouse tissue.

Expression during Prenatal Development

We have previously shown that many proto-oncogenes are not only expressed in specific tissues during prenatal mouse development, but, moreover, their expression is temporally regulated as well (Müller et al. 1982; Müller and Verma 1984). Some oncogenes like *fos* and *fms* are expressed early during development, while others like *abl* are expressed at the highest levels during midgestation. Proto-oncogene Ha-*ras* is expressed at all stages of development. An extensive evaluation of the expression of *fos* and *fms* genes reveals that the highest levels of expression are confined to extraembryonal tissues (placenta, amnion, and yolk sac) (Müller and Verma 1984). The levels of *fms* expression are found to be low in day-10 to -12 extraembryonal membranes but increase approximately sixfold between days 12 and 18 to a level that is slightly higher than that observed in the placenta. A detailed analysis of the microsurgically isolated components of the late-gestation extraembryonal membranes showed that the levels of *fos* transcripts are higher in visceral yolk sac (endoderm and mesoderm), as compared with placenta. Furthermore, the highest levels of c-*fos* are observed in day-18 amnion. These levels are close to those of v-*fos* transcripts in FBJ-MSV-transformed cells. Figures 2, a and b, show the detection of *fos* expression in day-17 mouse amnion by in situ hybridization. It can be seen that *fos* transcripts are expressed in all cells of the amnion, which is a bilayered structure. The specificity of the hybridization was confirmed by the absence of a signal when only the vector (pBR322) DNA was used as a probe. The nuclear *fos* protein is detected in amnion cells, both by immunoprecipitation as well as by immunofluorescence with specific antisera (Curran et al. 1984).

Expression during Hematopoiesis

When expression of the *fos* gene was studied in postnatal tissues, we were surprised to find the highest expression in neonatal bone tissue (Müller and Verma 1984). Subsequent analysis, however, revealed that our neonatal bone preparations also contained some bone marrow tissue and that *fos* expression was confined to the latter. Initial experiments revealed c-*fos* transcripts in peritoneal exudate and macrophage cell lines. To determine the nature of the cell types expressing the c-*fos* gene, we analyzed the expression of c-*fos* during differentiation of hematopoietic cell lines.

We have previously shown that the expression of the *fos* gene is induced during TPA-induced differentiation of a promonocytic leukemia cell line, U937, or a monomyelocytic leukemia cell line, HL-60, to macrophages (Mitchell et al. 1985). No *fos* expression is detected when HL-60 cells are induced to differentiate into granulocytes (Mitchell et al. 1985). Figure 3A shows that *fos* mRNA transcripts are also expressed when a human erythroid cell line (HEL) or a chronic myelocytic leukemia cell line (K562) is treated with TPA, but that the response in a T-cell line, CCRF-CEM, is much less dramatic. In the case of U937 or HL-60, extensive kinetic analysis has been undertaken and the results can be summarized as follows: (1) Induction of *fos* expression is very rapid, occurring within 3 minutes of TPA addition; (2) maximal levels of induction are observed within 30 minutes and then the levels decline by four- to fivefold and remain essentially unchanged for at least 10 days, by which time over 99% of the viable cells in the culture are fully adherent macrophages; (3) another agent that induces the differentiation of U937 and HL-60 cells to macrophages, vitamin D (1,25-dihydroxy vitamin D_3), also induces the expression of *fos* (Fig. 3B). In the response of hematopoietic cells to TPA or vitamin D, the larger of the observed RNA transcripts is the unspliced *fos* mRNA, since it hybridizes specifically to a probe for intron I of the c-*fos* gene (Fig. 3C).

Although *fos* is expressed transiently in many cell types following treatment with TPA, the continuous expression of intermediate levels of *fos* mRNA in the population of differentiated adherent macrophages and the elevated levels of *fos* mRNA observed in peritoneal exudates enriched for macrophages may suggest that continuous expression of *fos* mRNA has a role during the full differentiation of macrophages. It remains unclear as to whether *fos* expression is required for early monocytic differentiation.

Expression during Neuronal Differentiation

To investigate further the role of the *fos* gene in differentiation, we have studied its induction by nerve growth factor (NGF) in the clonal rat pheochromocytoma cell line, PC12. In the presence of NGF, PC12 cells acquire properties of sympathetic neurons, including neurite outgrowth, increased electrical excitability, and changes in neurotransmitter synthesis (Green and Tischler 1976; Dichter et al. 1977). NGF may also act as a weak mitogen in these cells (Boonstra et al. 1983). Figure 4A shows that *fos* gene transcripts can be detected 5 minutes after addition of NGF, they are maximally abundant after 30 minutes, and their levels

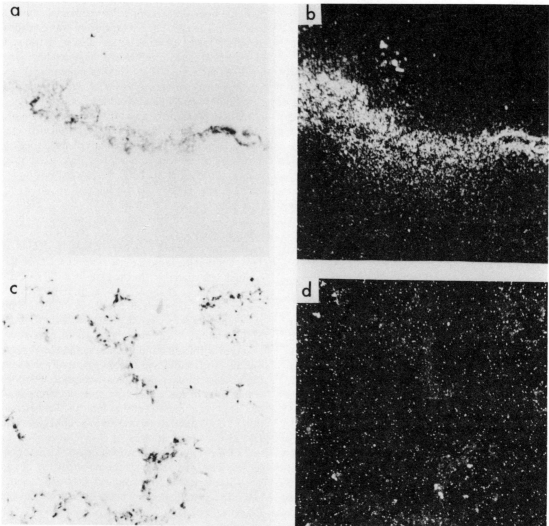

Figure 2. In situ hybridization of mouse amnion c-*fos* transcripts using a *fos*-specific probe. Day-17 mouse amnion was frozen in liquid nitrogen immediately after dissection, embedded in OCT, and sectioned using a cryomicrotome. Sections (0.7-μm) were collected on albumin-coated slides, fixed with ethanol (60%), acetic acid (10%), and chloroform (30%) at −20°C, and hybridized with ^{32}P-labeled nick-translated *fos*-specific probe (*a* and *b*), or with a ^{32}P-labeled nick-translated pBR322 fragment (*c* and *d*). Hybridization conditions were the following: slides were hybridized with 10 ng of ^{32}P-labeled probe per slide (specific activity 10^8 cpm/μg), the hybridization medium contained 50% deionized formamide, 10 mM Tris (pH 8), 1 mM EDTA, 300 mM NaCl, 500 μg/ml yeast carrier tRNA, 1× Denhardt. Hybridization solution (15 μl) was placed on the tissue section and covered with a 12-mm siliconized glass coverslip. The edges were sealed with rubber cement. After an overnight hybridization at 42°C, the slides were washed extensively at room temperature in 2× SSC, then at increasing temperature and decreasing salt concentration. The final washing was at 50°C in 0.1× SSC. Tissue was overlaid by a Kodak nuclear track photographic emulsion and exposed for 2 weeks at 4°C. Slides were then developed and stained with thionin before being mounted with permount.

decrease thereafter. The *fos* protein synthesis parallels the expression of *fos* mRNA (Fig. 4B), and the induced *fos* proteins are located in the nucleus (Fig. 4C).

Binding of NGF to its cell-surface receptor causes a rapid (within minutes) and transient increase in intracellular cAMP levels (Schubert et al. 1978; Traynor and Schubert 1984). Figure 5A shows that dibutyryl cAMP also induces *fos* gene transcription upon addition to PC12 cells, although with slightly slower kinetics than NGF. Exogenous K$^+$ induces neurite outgrowth without a detectable increase in the level of intracellular cAMP (Traynor and Schubert 1984). It causes an influx of Ca^{++} ions, which can directly stimulate neurite extension in PC12 cells. K$^+$ depolarization (50 mM) does induce *fos* expression although it does not lead to enhanced cAMP levels (Fig. 5B). In addition to NGF, cAMP and elevated K$^+$, TPA, and epidermal growth factor (EGF) also induce *fos* (Kruijer et al. 1985). All of these reagents also rapidly stimulate the phosphorylation of PC12 tyrosine hydroxylase at one or more of four distinct sites in the enzyme (McTigue et al. 1985). The phosphorylation of only one site is stimu-

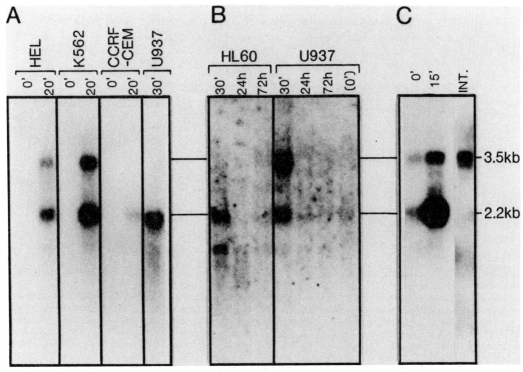

Figure 3. Expression of c-*fos* mRNA after treatment with TPA or vitamin D₃. (*A*) Cellular RNA was isolated from cells treated with TPA (20 ng/ml) for the times indicated in the figure. c-*fos*-specific transcripts were detected by hybridization to a ^{32}P-labeled human c-*fos* probe (*NcoI–XhoI* fragment; van Straaten et al. 1983). (*B*) c-*fos* transcripts detected in cellular RNA from HL-60 and U937 cells treated with vitamin D [1,25-(OH)₂; 10^{-7} M] for the times indicated in the figure. (*C*) Detection of c-*fos* transcripts in untreated U937 cells (0′) and in U937 cells treated with TPA (20 mg/ml) for 15 min using the *NcoI–XhoI* fragment as probe (15′) or using a probe specific for intron I of human c-*fos* (INT.). The unspliced primary transcript (3.5 kb) and the spliced mature *fos* mRNA (2.2 kb) are identified in the figure.

lated by TPA, but the TPA-stimulated site is also phosphorylated in response to cAMP, EGF, NGF, and K⁺ depolarization. Since TPA is a fairly specific activator of C-kinase, and since all conditions that cause the C-kinase-specific phosphorylation of tyrosine hydroxylase peptides also increase *fos* expression, it follows that C-kinase may be involved in *fos* induction.

PC12 cells have some degree of developmental plasticity in that they can be induced by NGF into cells with characteristics of sympathetic ganglion cells, and by corticosteroids into cells that resemble chromaffin cells (Schubert et al. 1980). Neither *fos* mRNA nor *fos* proteins were detected up to 5 days after the addition of dexamethasone, indicating that the expression of the *fos* gene is correlated with the pathway of differentiation induced by NGF but not corticosteroids (Fig. 5C).

Expression during Cell Growth

It has previously been shown that when quiescent mouse fibroblasts are treated with serum or growth factors like PDGF, EGF, or TPA, the proto-oncogene *fos* is rapidly induced (Cochran et al. 1984; Greenberg and Ziff 1984; Kruijer et al. 1984; Müller et al. 1984). The salient features of these observations are shown in a composite figure (Fig. 6) and can be summarized as follows: (1) Within 2–3 minutes of stimulation of growth, c-*fos* transcripts can be detected as measured by hybridization with ^{32}P-labeled cRNA (Kruijer et al. 1984). (2) Maximal levels of induction occur within 20 minutes (20-fold induction) of the exposure of cells to 0.83 nM purified platelet-derived growth factor (PDGF). The levels declined by 60 minutes and by 240 minutes little or no c-*fos* transcripts could be detected. (3) Addition of cycloheximide resulted in a 50-fold induction, suggesting stabilization of c-*fos* mRNA transcripts. (4) We estimate that after 20 minutes of exposure to PDGF, 0.0001% of NIH-3T3 cell RNA (0.0005% of mRNA) is c-*fos* mRNA. Assuming a cellular RNA content of 6 pg, this corresponds to about 5–10 copies of *fos* mRNA per cell. (5) Exposure to PDGF for as short as 30 minutes induces the synthesis of *fos* protein which can be detected by immunoprecipitation with *fos*-specific peptide antisera. (6) At least six to eight polypeptides are identified by immune precipitation, most of which represent modified forms of *fos* protein; however, some non-*fos* polypeptides are also precipitated. One possibility is that some of them may be related to *fos* and may react with peptide antisera (Cochran et al. 1984). (7) c-*fos* protein synthesis was maximal with PDGF concentrations that saturate PDGF binding sites at 37°C (1.0 nM) and half-maximal at 0.3–0.5 nM. (8) It appears that c-*fos* is transiently induced in response to a variety of mitogens in addition to inducers of differentiation.

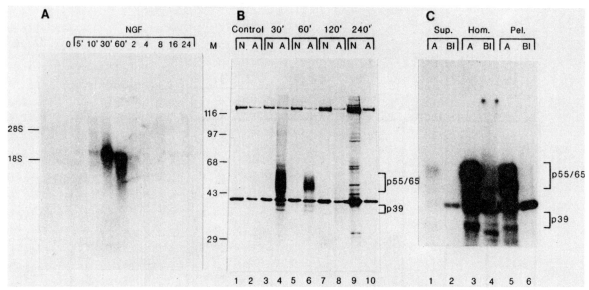

Figure 4. (A) fos expression in NGF-induced PC12 cell line. Northern blot analysis of total RNA (15 μg/lane) isolated at the indicated times after addition of 50 ng/ml of NGF. After electrophoresis and Northern transfer, fos-specific sequences were detected by hybridization to ^{32}P-labeled v-fos probe. (B) Characterization of fos proteins. Kinetics of fos synthesis. PC12 cells were induced for the indicated times followed by labeling with [^{35}S]methionine for 20 min. RIPA lysates were immunoprecipitated with normal rabbit serum (N, lanes 1, 3, 5, 7, and 9) or with affinity-purified M2 peptide antiserum (A, lanes 2, 4, 6, 8, and 10). The positions of fos and p39 proteins are indicated. (M) Molecular weight standards. (C) Subcellular localization of fos proteins. PC12 cells were induced with 50 ng/ml NGF for 30 min and labeled for 20 min in 5 ml of N2 (low methionine) medium with 0.1 mCi/ml of [^{35}S]methionine (New England Nuclear, 600-1000 Ci/mmole). Separations in a cytoplasmic and nuclear fraction were performed as described before (Kruijer et al. 1984). Briefly, after washing with Tris-saline the cells were scraped from the dish in 1.0 ml of 1 mM DTT, 10 mM HEPES (pH 7.4), and 10 units/ml Trasylol and homogenized in a Teflon/glass homogenizer. Half of the sample was removed and the remainder centrifuged at 600g for 5 min at 4°C. One volume of twice-concentrated RIPA buffer was added to each fraction. Half of the total homogenate (lanes 1 and 2), supernatant (lanes 3 and 4), and pellet (lanes 5 and 6) were immunoprecipitated with M2 peptide anti-serum (lanes 1, 3, and 5) or M2 peptide antiserum preincubated with excess M peptide (lanes 2, 4, and 6).

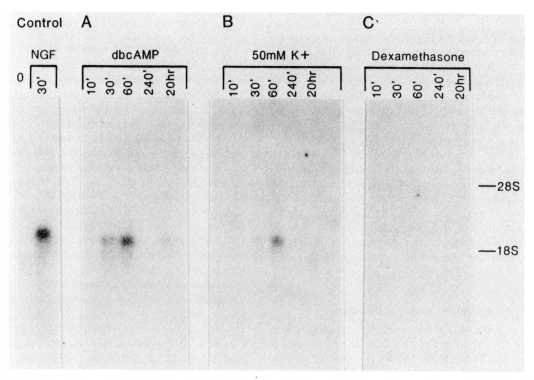

Figure 5. Induction of fos mRNA by dibutyryl cAMP (dbcAMP; A), 50 mM K$^+$ (B), and dexamethasone (C). Subconfluent PC12 cells were induced with NGF (50 ng/ml), dibutyryl cAMP (1 × 10^{-3} M), K$^+$ (50 mM), or dexamethasone (10^{-6} M) for the indicated times. Total RNA (15 μg/lane) was separated by electrophoresis and transferred to nitrocellulose. fos-specific sequences were detected by hybridization with a fos probe.

Transcriptional Regulation of the *fos* Gene

Because proto-oncogene *fos* is transiently induced in response to a variety of mitogens and differentiation inducers, we have begun characterization of its transcriptional control elements. A number of regulatory elements are required for efficient transcription of eukaryotic genes. These include, in addition to the proximal sequences like the TATA box and the CAT box, transcriptional enhancers that are generally located upstream from the cap site (Khoury and Gruss 1983). Enhancer sequences augment the level of transcription of a gene and in certain instances can act in a tissue-specific manner. We linked the 5′-flanking region of the *fos* gene including its promoter to the coding domain of the bacterial chloramphenicol acetyl transferase (CAT) gene (Gorman et al. 1982) and assayed for CAT enzymatic activity following transient expression in transfected cells. Figure 7 depicts the structure of the proto-oncogene *fos*, with its promoter and upstream sequences (van Straaten et al. 1983), the SV40-CAT transcription unit (pSV2CAT) (Gorman et al. 1982), and a variety of *fos*-CAT fusion constructs. Transient assays were performed in mouse NIH-3T3 and rat 208F fibroblasts as well as in the human amnion cell line AV-3 and in HeLa cells. The relative CAT activity observed for each construct in the different cell lines is tabulated in Figure 7.

Construct FC1 contains about 2250 nucleotides of upstream sequences linked to the CAT gene at the *Nae*I site (map position +41, the cap site being defined as +1; Fig. 7). The CAT activity assayed in extracts from NIH-3T3 cells transfected with FC1 DNA is about 60% of that measured after transfection with the control plasmid pSV2CAT in which the CAT gene is under the control of the SV40 early promoter and enhancer sequences. The enzymatic activity associated with all other *fos*-CAT hybrids is tabulated as a fraction of that exhibited by FC1. To identify sequences essential for transcription of the *fos* gene, we made progressive deletions in the upstream region. Constructs FC2, FC3, and FC4 in which the 3′ ends of the deletion map at positions −1450, −712, and −404 bp, respectively, were generated by utilizing unique restriction endonuclease sites. They display an activity similar to that observed with FC1. A series of plasmids containing more extensive deletions were obtained using BAL-31 exonuclease digestion from the *Sst*II site of FC4. The 3′ borders of all deletions were determined by restriction mapping, followed by direct nucleotide sequence analysis for FC5, FC9, FC10, and FC11. Constructs FC5, FC6, FC7, and FC8, in which the 3′ ends of the deletion map at positions −307 (FC5) to −220 (FC8) show about 50% of the CAT activity obtained with FC4. A further 2.5-fold drop in activity is observed with construct FC9 in which the 3′ end of the deletion maps at position −206. The CAT activity declines further in constructs FC10 and FC11 in which the deletion end points reach positions −124 and −64, respectively. As a negative control, construct FC20 was generated, in which the *Sst*II-*Nae*I fragment containing the promoter was reversed (Fig. 7). No CAT activity was observed, indicating that CAT gene transcription depends entirely upon the *fos* gene promoter. The results with the deletion constructs indicate that sequences between −64 and −404 are required for efficient utilization in the *fos* gene promoter.

To determine if the upstream sequences needed to augment the transcription of the *fos* gene are analogous to enhancers identified for other eukaryotic genes, we performed two types of experiments.

1. Since an enhancer element is functional in either orientation, we inserted the 690-bp *Bam*HI-*Nar*I fragment in both orientations in the *Sst*II site of construct FC10 generating constructs FC30 and FC40. As can be seen in Figure 7, both of these constructs give rise to much higher CAT activity than observed with FC10. CAT gene expression is, however, more efficiently restored if the *Bam*HI-*Nar*I fragment is inserted in the original orientation. Sequences upstream from the enhancer (map positions −2250 to −700) inserted in the same site, do not restore the CAT activity.

2. Removal of most of the SV40 enhancer sequences (*Acc*I-*Sph*I, Fig. 7) from pSV2CAT decreases the CAT synthesis 10-fold (Gorman et al. 1982). Insertion of *fos* enhancer sequences (*Nae*I-*Nar*I isolated from FC4) in place of the SV40 enhancer partially restores the CAT activity, whereas an internal fragment of the *fos* gene inserted at the same site does not. Both of these experiments indicate that the upstream sequences that we have identified have features of a prototypic transcriptional enhancer element.

As shown in Figure 6, a 20-fold induction of *fos* mRNA transcripts can be observed when growth factors are added to quiescent fibroblasts. To determine if we can localize the inducible sequences, we attempted two types of experiments: (1) Stable transfectants of *fos*-CAT constructs. The construct FC1 (Fig. 7) was transfected into rat 208F HPRT⁻ cells along with HPRT DNA with a ratio of 20:1. MAT-resistant colonies were selected and analyzed for the presence of integrated *fos*-CAT DNA. Several cell lines were generated which synthesized catalytically active CAT protein. R-FC1 cells from independent clones were grown until quiescent and then stimulated with serum-rich medium. Total cellular RNA was isolated at various time points after serum stimulation, and analyzed for expression of *fos* transcripts. Figure 8A shows the results obtained with one such clone, R-FC1-5. The endogenous *fos* mRNA transcripts show a 20-fold induction, but the *fos*-CAT mRNA is induced only 3.5-fold. The levels of endogenous c-*fos* transcripts drop by 60 minutes, whereas the *fos*-CAT mRNA does not. Figure 8B shows similar data obtained using another clone. Instead of being starved and serum stimulated, cells were treated with TPA for various periods of time. In

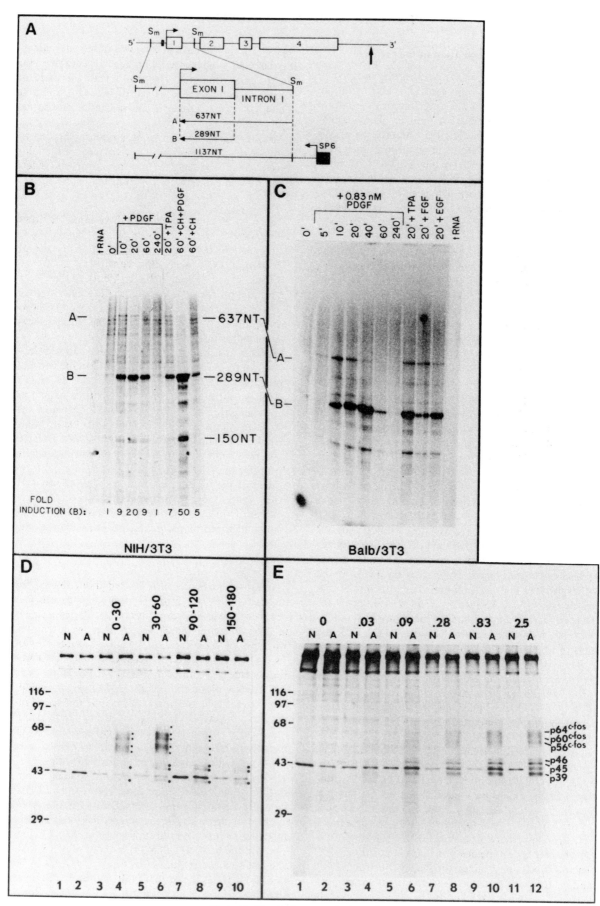

Figure 6. (*See facing page for legend.*)

control 208F cells the endogenous *fos* mRNA is induced 20-fold within 30 minutes of the addition of TPA. Only small amounts of *fos* transcripts are detected by 60 minutes. In R-FC1 cells the endogenous *fos* mRNA transcripts show 20-fold induction but the *fos*-CAT mRNA is induced only threefold. (2) Stable transfectants of c-*fos* gene constructs. We have previously described the construction of a recombinant construct MMV, which contains the c-*fos* promoter, 600 nucleotides of sequences upstream of the 5' cap site, the coding domain of mouse *fos* protein, and the poly(A) addition sequences and enhancer elements of FBJ-MSV (Miller et al. 1984). Cells stably transfected with this construct were phenotypically transformed and the c-*fos* protein could be detected. Such a cell line, R-MMV, was induced with TPA to determine if transcription from the exogenously added *fos* recombinant construct is inducible like that observed with the endogenous *fos* mRNA. Data in Figure 8C show that the exogenous *fos* gene is induced sixfold, compared with the 13-fold induction of the endogenous *fos* gene. At least two explanations can be readily advanced to account for the lower induction factor observed for exogenous *fos* sequences compared with the endogenous *fos* gene: (1) The exogenous *fos* DNA used in these studies might not carry all the sequences required for induction of transcription, and (2) sequences present in the c-*fos* coding domain (missing in FC1) or 3' from it (missing in MMV and FC1) might be crucial for the accumulation and/or stability of the induced *fos* mRNA. We are currently testing these hypotheses.

DISCUSSION

Regulation of *fos* Expression

fos is a multifaceted gene, the product of which may play a role during development, cellular differentiation, and cell growth. Since products of both the viral and cellular *fos* genes can induce the transformation of fibroblasts in vitro, it is puzzling that expression of the *fos* protein in vivo and the induced expression of *fos* in vitro do not result in transformation. It is possible that some cell types, such as peritoneal macrophages or macrophages in culture, are refractory to transformation by *fos*, even during sustained expression of *fos* mRNA. Perhaps fibroblasts and other cells that are normally susceptible to c-*fos*-induced transformation are not transformed because the expression of the *fos* protein is only transient.

The synthesis of the *fos* gene product displays an exquisite regulation. Our previous findings have led us to believe that *fos* protein synthesis may be regulated posttranscriptionally, or even more likely, at the translational level. Two sets of observations favor this notion. First, when the intact proto-oncogene c-*fos* is transcribed constitutively, it is unable to transform fibroblasts and only very low levels of the *fos* protein are detected (Miller et al. 1984). The *fos* gene becomes transforming when a stretch of only 67 bp located 527 bp downstream of the termination codon and 123 bp upstream of the poly(A) addition site is removed (Meijlink et al. 1985). We hypothesize that the *fos* protein may regulate its own synthesis, possibly by binding to the 67-bp region and altering the translational effi-

Figure 6. Analysis of PDGF-stimulated c-*fos* RNA and proteins. (*A*) Diagram of the molecular structure of the mouse c-*fos* gene, based on nucleotide sequence analysis. Expected sizes of the protected c-*fos* mRNA transcripts are indicated. Positions of putative 5' cap (rightward bent arrow), polyadenylated signal (↑), TATA box (■), exon (□), intron (———), SP6 phage promoter (filled box with leftward bent arrow), and vector (--------) are indicated. (*B* and *C*) Analysis of c-*fos* transcripts. Total RNA from PDGF-treated cells was used for RNA protection experiments. Times of induction (min) and types of treatment are indicated. As a control for self-hybridization of the probe, one hybridization contained 10 μg of tRNA in place of cellular RNA. The expected size of fragment A was 637 nucleotides and that of fragment B was 189 nucleotides. The size of the protected fragments was determined relative to denatured ^{32}P end-labeled pBR322 TaqI fragments as size markers. For experimental details, see Kruijer et al. (1984). (*D*) Time course of c-*fos* protein synthesis. BALB/c 3T3 cell cultures were treated with 0.67 nM pure PDGF for 0 min (lanes *3* and *4*), 30 min (lanes *5* and *6*), 90 min (lanes *7* and *8*), or 150 min (lanes *9* and *10*) before the addition of 100 μCi of [^{35}S]methionine. Another culture received an equivalent volume of bovine serum albumin (BSA) in 1 mM acetic acid for 30 min before labeling (lanes *1* and *2*). After a further 30-min incubation, cultures were washed, lysed, and one-third volumes were immunoprecipitated with 1 μg of IgG equivalent of nonimmune rabbit serum (N; lanes *1, 3, 5, 7,* and *9*) or 1 μg of affinity-purified IgG to M peptide (A; lanes *2, 4, 6, 8,* and *10*). Immunoprecipitates were analyzed by SDS-PAGE. (■) Proteins related to p55$^{c\text{-}fos}$; (○) unrelated proteins. A band of 43 kD, observed in all samples, is most likely actin. (*E*) Dose dependence of c-*fos* protein synthesis. NIH-3T3 cells were exposed to various concentrations of pure PDGF, diluted in 2 mM acetic acid containing 1 mg/ml BSA, for 30 min prior to addition of [^{35}S]methionine for 30 min. The final concentrations of PDGF were 0 nM (lanes *1* and *2*), 0.03 nM (lanes *3* and *4*), 0.09 nM (lanes *5* and *6*), 0.28 nM (lanes *7* and *8*), 0.83 nM (lanes *9* and *10*), or 2.5 nM (lanes *11* and *12*). Each lysate was immunoprecipitated with nonimmune rabbit serum (N; lanes *2, 4, 6, 7, 9,* and *11*) 1 μg of affinity-purified IgG to M peptide (A; lanes *2, 4, 6, 8, 10,* and *12*). Except where noted, confluent 35-mm dish cultures of 3T3 cells were incubated in 1 ml of Dulbecco's modified Eagle's (DME) medium containing 1% of the regular methionine concentration and 0.5% calf serum for 40–48 hr. Addition of PDGF (essentially homogeneous, purified from phenyl-Sepharose, dissolved in 1 mM acetic acid containing 1 mg/ml BSA) or an equal volume of 1 mg/ml BSA in 1 mM acetic acid, and [^{35}S]methionine (100 μCi of ~1000 Ci/mmole (Amersham/Searle) was made directly to the medium. Cultures were lysed after washing with cold Tris-buffered saline by adding 0.5 ml of RIPA buffer (0.15 M NaCl, 1% Nonidet PO-40, 5% sodium deoxycholate, 0.5% SDS, 2 nM EDTA, 100 units/ml Trasylol, 10 nM sodium phosphate, pH 7.0) and scraping. Lysates were clarified at 20,000*g* for 60 min at 4°C. IgG or antisera were added as indicated. After 1 hr at 0°C, 1 mg of Pansorbin (Calbiochem) was added for a further 1 hr. Immunoprecipitates were centrifuged through a solution of 10% sucrose in RIPA, then washed repeatedly by centrifugation in RIPA. Immunoprecipitations with rat antitumor serum or normal rat serum utilized goat antiserum to rat IgG added 30 min prior to Pansorbin. Immunoprecipitates were dissociated by incubation at 100°C for 2 min in 2% SDS, 20% s-mercaptoethanol, 10% glycerol, and 0.1 M Tris-HCl (pH 6.8), and one-half of each sample was analyzed on an SDS-polyacrylamide gel (12.5% acrylamide, 0.10% bis-acrylamide). Gels were stained to visualize the markers galactosidase, phosphorylase, BSA, ovalbumin, and carbonic anhydrase and impregnated with disphenyloxazole IPPO. Dried gels were exposed to presensitized film at −70°C. Exposure times: (*D*) 10 days; (*E*) 4 days.

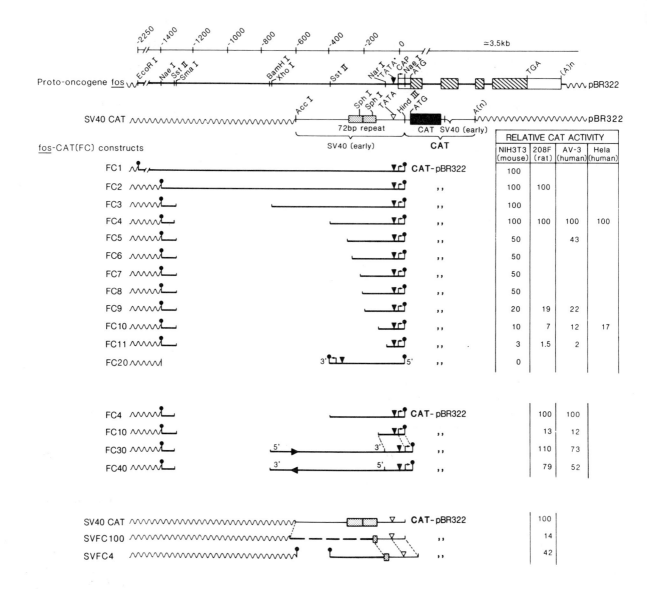

Figure 7. Schematic representation of the *fos*-CAT constructs. The structures of the upstream sequences of the human *fos* gene (van Straaten et al. 1983), SV40-CAT (Gorman et al. 1982), and various *fos*-CAT constructs are shown. The enzymatic activity measured in extracts of cells transfected with the *fos*-CAT (FC) and SV40-CAT constructs is tabulated. Standard recombinant DNA procedures were followed. To generate FC1, a 2.25-kb *Eco*RI-*Nae*I fragment containing upstream sequences and the promoter of the *fos* gene was inserted in place of the *Acc*I-*Hin*dIII fragment present in pSV2CAT using *Hin*dIII linkers (↑). FC2 was generated similarly after *Hin*dIII linker ligation to the isolated 4.2-kb *Nae*I fragment. *Sma*I-*Xho*I and *Sst*II-*Sst*II deletions in FC2 yielded constructs FC3 and FC4, respectively. After the deleted fragments corresponding to the residual 5′-flanking sequences and *fos* promoter were digested with *Hin*dIII and gel-purified, they were recloned in the *Sma*I-*Hin*dIII-digested FC2 DNA in place of the original 1.3-kb fragment. The extent of the deletions was determined by agarose gel electrophoresis followed in four cases by direct nucleotide sequence analysis. The 3′ borders of the deletions map at positions −307 (FC5), −206 (FC9), −124 (FC10), and −64 (FC11). FC20 was generated by inserting the 404-bp *Sst*II-*Nae*I fragment present in FC4, in reversed orientation into the *Acc*I-*Hin*dIII deleted pSV2CAT DNA. Constructs FC30 and FC40 were obtained by inserting the 671-bp *Bam*HI-*Nar*I fragment from FC2 in both orientations in the unique *Sst*II site of FC10 after blunt-end conversion. SVFC4 was constructed by replacing the *Acc*I-*Sph*I fragment from pSV2CAT with the *Hin*dIII-*Nar*I sequences isolated from FC4. SVFC100 was generated by inserting a 540-bp *Nae*I-*Sph*I fragment originating from the first intron of the *fos* gene, in the same *Acc*I-*Sph*I deleted vector as SVFC4. DNA transfection of NIH-3T3 and 208F cells was carried out by the calcium phosphate precipitation technique. DNA (5 µg) was used to transfect semiconfluent cells in a 10-cm petri dish. Duplicate plates were used for each DNA in every experiment, and two or three independent assays were performed for each construct. AV-3 and HeLa cells were transfected following the DEAE-dextran procedure, using 10 µg of DNA per 10-cm dish. Forty-eight hours after transfection, cell extracts were prepared and total protein concentration measured using the BioRad protein assay reagent. CAT assays were performed essentially as described by Gorman et al. (1982). The acetyl coenzyme A final concentration was 4 mM and 0.5 µCi of [^{14}C]chloramphenicol was used per assay. Under these conditions, all reactions were linear for at least 3 hr (not shown). As a positive control, pSV2CAT was included in every assay showing a variation of not more than 20% between experiments. CAT activity measured for constructs FC2 to FC40 is tabulated as a fraction of that exhibited by FC1. CAT activity observed using constructs SVFC100 and SVFC4 is expressed as a fraction of that corresponding to pSV2CAT. Although results from only one experiment are shown, an average of three assays were performed for each construct.

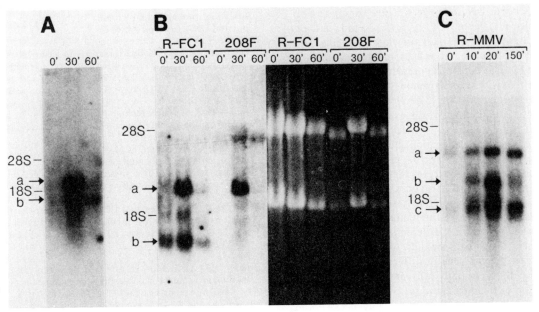

Figure 8. Mitogen and TPA stimulation of exogenous versus endogenous c-*fos* genes. (*A*) A stable rat fibroblast cell line carrying the FC1 *fos*-CAT constructs (see Fig. 7) was grown in Dulbecco and Vogt's modified Eagle's medium (DMEM) containing 10% fetal calf serum (FCS). When cells were confluent, the medium was replaced with DMEM plus 20% FCS and total RNA extracted at different time points after stimulation. Equal amounts of RNA (20 μg) were loaded on a formaldehyde-containing agarose gel and Northern analysis was performed, using equal amounts of a *fos* and a CAT-specific probe (independent CAT and *fos* hybridizations of a same blot had been previously performed). (a) Rat endogenous *fos* transcripts; (b) *fos*-CAT transcripts. (*B*) Another clone of R-FC1 cells was stimulated by TPA. Total cellular RNA was isolated after different periods of time and Northern analysis was performed, using *fos* and CAT-specific probes. (a) Endogenous *fos* transcripts; (b) *fos*-CAT hybrid transcripts. A picture of the ethidium bromide-stained blot is shown next to the autoradiogram, because not equal amounts of RNA were transferred in this case. (*C*) A stable transformed cell line (R-MMV) carrying the cellular mouse c-*fos* gene linked to a FBJ-MSV LTR (MMV, Miller et al. 1984) was stimulated by TPA. Total cellular RNA was isolated at various time points. Equal amounts (15 μg) of RNA were loaded on a formaldehyde gel, blotted onto nitrocellulose, and hybridized with a *fos*-specific probe. (a) MMV transcripts; (b) rat endogenous c-*fos* sequences; (c) transcripts from a presumably rearranged copy of the transfected MMV DNA. Induction factors were calculated by densitometric analysis of the hybridization signal obtained before, and 30 min (*A* and *B*) or 20 min (*C*) after stimulation.

ciency of the c-*fos* mRNA. Second, and perhaps related, is the observation that promonocytes induced to differentiate into macrophages continue to express *fos* mRNA for at least 10 days, even though the *fos* protein is only detected for 60–120 minutes following treatment with the inducer (Mitchell et al. 1985).

We have no firm grip on the mechanism of such a translational control. We have been unable to demonstrate any homologies or complementarity between the 67-bp stretch and the rest of the gene. In addition, there are at least two pitfalls in our hypothesis of a translational control mechanism. First, it is possible that the c-*fos* protein is made in cells carrying the nontransforming construct, but we are unable to detect the protein with our *fos* antisera because of extensive protein modifications. Second, the *fos* protein appears to be exclusively nuclear, at least by immunofluorescence, yet a direct translational control mechanism would likely require some cytoplasmic *fos* protein. Regardless of the molecular mechanism controlling *fos* at the transcriptional or posttranscriptional level, we believe that the natural expression of the c-*fos* protein does not transform cells because the expression of the protein is only transient. In contrast, v-*fos* escapes this regulation of *fos* protein synthesis due to its altered carboxyl terminus, and hence induces transformation.

Regulation of Transcription

Sequences required for efficient functioning of a gene are likely to be conserved during evolution. A transcriptional enhancer element of proto-oncogene *fos* has been identified in such a conserved region. It lies between −60 to −400 nucleotides upstream of the 5′ cap nucleotide. A dot matrix analysis of identities between the promoter and 5′-flanking regions of the human and mouse *fos* genes indicates two regions of striking similarity: one surrounding the cap site and "TATA" box and another stretching from −250 to −475, the region identified as the transcriptional enhancer. The same two regions also bear a DNase I hypersensitive site, another hallmark of a region involved in regulation of transcription. The precise nature of the essential sequences in the enhancer region and the molecular mechanism underlying their functioning remains to be established. The proto-oncogene *fos* could be a housekeeping gene, since it is transcribed at low levels in most if not all cells. One would thus expect the *fos* enhancer to be non-tissue specific, a notion borne out by our observations that the CAT assays, as well as the mapping of DNase I hypersensitive sites, give similar results in different cell types. At the moment, our investigations aim at explaining why we are unable to elicit a full inducible response to mitogens from exogenous c-*fos* gene constructs present in stable cell lines, as compared with the endogenous *fos* sequences.

What role does *fos* play in development, growth, and differentiation? Perhaps *fos* is expressed as a general anabolic response of cells to specific stimuli. If this is the case, such an explanation must account for the expression of *fos* in response to only a subset of differentiation inducers. Indeed, *fos* is expressed during differentiation of cells of the monocytic lineage to macrophages, but not granulocytes. Similarly, *fos* is induced when PC12 cells differentiate into neurites with NGF but not when they differentiate into chromaffin cells when treated with steroids. One way to establish the role of the *fos* gene in differentiation will be to introduce the gene into cells and determine if the expression of the introduced gene induces differentiation in the absence of other inducing agents. Another approach to studying the role of the *fos* gene will be to block *fos* protein synthesis in cells treated with differentiation inducers using eukaryotic vectors expressing anti-sense *fos* mRNA.

ACKNOWLEDGMENTS

We thank Liza Zokas for excellent technical assistance and Carolyn Goller for typing the manuscript.

This work was supported by grants from the National Institutes of Health, the American Cancer Society, and the Muscular Dystrophy Association of America. J.D. was supported by grants from the International Union Against Cancer (Geneva) and La Fondation Rose et Jean Hoguet (Brussels). R.L.M. was supported by a National Institutes of Health postdoctoral fellowship (08-F2GM 10161A). F.M. was the recipient of fellowships from the Netherlands Organization for the Advancement of Pure Research (Z.W.O) and the European Molecular Biology Organization (EMBO). W.K. was supported by a fellowship from the Dutch Queen Wilhemina Fund.

REFERENCES

Bishop, J.M. 1983. Cellular oncogenes and retroviruses. *Annu. Rev. Biochem.* **52**: 301.

Boonstra, J., W.H. Moolenaar, P.H. Harrison, P. Moed, P.T. van der Saag, and S.W. De Laat. 1983. Ionic responses and growth stimulation induced by nerve growth factor and epidermal growth factor in rat pheochromocytoma (PC12) cells. *J. Cell Biol.* **97**: 92.

Cochran, B.M., J. Zullo, I.M. Verma, and C.D. Stiles. 1984. Expression of c-*fos* oncogene and of a *fos*-related gene is stimulated by platelet-derived growth factor. *Science* **226**: 1080.

Curran, T., A.D. Miller, L.M. Zokas, and I.M Verma. 1984. Viral and cellular *fos* gene products are located in the nucleus. *Cell* **36**: 259.

Curran, T., G. Peters, C. Van Beveren, N.M. Teich, and I.M. Verma. 1982. FBJ murine osteosarcoma virus: Identification and molecular cloning of biologically active proviral DNA. *J. Virol.* **44**: 674.

Dichter, M.A., A.S. Tischler, and L.A. Greene. 1977. Nerve growth factor-induced increase in electrical excitability and acetylcholine sensitivity of a rat pheochromocytoma cell line. *Nature* **268**: 501.

Doolittle, R.F., M.W. Hunkapiller, L.E. Hood, S.G. Devare, E.C. Robbins, S.A. Aaronson, and H.N. Antoniades. 1983. Simian sarcoma virus *onc* gene, v-*sis*, is derived from the gene (or genes) encoding a platelet-derived growth factor. *Science* **221**: 275.

Downward, J., Y. Yarden, E. Mayes, G. Scrace, N. Totty, P. Stockwell, A. Ullrich, J. Schlessinger, and M.D. Waterfield. 1984. Close similarity of epidermal growth factor receptor and v-*erb*-B oncogene protein sequences. *Nature* **307**: 521.

Finkel, M.P., B.O. Biskis, and P.B. Jinkins. 1966. Virus induction of osteosarcomas in mice. *Science* **151**: 698.

Finkel, M.P., C.A. Reilly, Jr., B.O. Biskis, and I.L. Greco. 1973. Bone marrow viruses. *Colston Res. Soc. Proc. Symp.* **24**: 353.

Gorman, C.M., L.F. Moffat, and B.H. Howard. 1982. Recombinant genomes which express chloramphenicol acetyltransferase in mammalian cells. *Mol. Cell. Biol.* **2**: 1044.

Green, L.A. and A. Tischler. 1976. Establishment of a noradrenergic clonal line of rat adrenal pheochromocytoma cells which respond to nerve growth factor. *Proc. Natl. Acad. Sci.* **73**: 2424.

Greenberg, M.E. and E.B. Ziff. 1984. Stimulation of 3T3 cells induces transcription of c-*fos* proto-oncogene. *Nature* **311**: 433.

Hunter, T. 1984. The proteins of oncogenes. *Sci. Am.* **251**: 70.

Khoury, G. and P. Gruss. 1983. Enhancer elements. *Cell* **33**: 313.

Kruijer, W., D. Schubert, and I.M. Verma. 1985. Induction of the proto-oncogene *fos* by nerve growth factor. *Proc. Natl. Acad. Sci.* **82**: (in press).

Kruijer, W., J.A. Cooper, T. Hunter, and I.M. Verma. 1984. Platelet-derived growth factor induces rapid but transient expression of the c-*fos* gene and protein. *Nature* **312**: 711.

MacConnell, W.P. and I.M. Verma. 1983. Expression of FBJ-MSV oncogene (*fos*) product in bacteria. *Virology* **131**: 367.

McTigue, M., J. Cremins, and S. Halegoua. 1985. Nerve growth factor and other agents mediate phosphorylation and activation of tyrosine hydroxylase. *J. Biol. Chem.* **260**: 9047.

Meijlink, F., T. Curran, A.D. Miller, and I.M. Verma. 1985. Removal of a 67-base-pair sequence in the noncoding region of protooncogene *fos* converts it to a transforming gene. *Proc. Natl. Acad. Sci.* **82**: 4987.

Miller, A.D., T. Curran, and I.M. Verma. 1984. c-*fos* protein can induce cellular transformation: A novel mechanism of activation of a cellular oncogene. *Cell* **36**: 51.

Mitchell, R.L., L. Zokas, R.D. Schreiber, and I.M. Verma. 1985. Rapid induction of the expression of protooncogene *fos* during human monocytic differentiation. *Cell* **40**: 209.

Müller, R. and I.M. Verma. 1984. Expression of cellular oncogenes. *Curr. Top. Microbiol. Immunol.* **112**: 73.

Müller, R., R. Bravo, J. Buckhardt, and T. Curran. 1984. Induction of c-*fos* gene and protein by growth factors precedes activation of c-*myc*. *Nature* **312**: 716.

Müller, R., D.J. Slamon, J.M. Tremblay, M.J. Cline, and I.M. Verma. 1982. Differential expression of cellular oncogenes during pre- and postnatal development of the mouse. *Nature* **299**: 640.

Schubert, D., M. LaCorbiere, F.G. Klier, and J.H. Steinbach. 1980. The modulation of neurotransmitter synthesis by steroid hormones and insulin. *Brain Res.* **109**: 67.

Schubert, D., M. LaCorbiere, C. Whitlock, and N. Stallcup. 1978. Alterations in the surface properties of cells re-

sponsive to nerve growth factor. *Nature* **273:** 718.
SHERR, C.J., C.W. RETTENMIER, R. SACCA, M.F. ROUSSEL, A.J. LOUK, and E.R. STANLEY. 1985. The c-*fms* proto-oncogene product is related to the receptor for the mononuclear phagocyte growth factor, CSF-1. *Cell* **41:** 665.
SHILO, B.Z. and R.A. WEINBERG. 1981. DNA sequences homologous to vertebrate oncogenes are conserved in *Drosophila melanogaster*. *Proc. Natl. Acad. Sci.* **78:** 6789.
TRAYNOR, A. and D. SCHUBERT. 1984. Phospholipases elevate cyclic AMP and promote neurite extension. *Dev. Brain Res.* **14:** 197.
ULLRICH, A., L. COUSSENS, J.S. HAYFLICK, T.J. DULL, A. GRAY, A.W. TAM, J. LEE, Y. YARDEN, T.A. LIBERMANN, J. SCHLESSINGER, J. DOWNWARD, E.L.V. MAYES, N. WHITTLE, M.D. WATERFIELD, and P.H. SEEBURG. 1984. Human epidermal growth factor receptor cDNA sequence and aberrant expression of the amplified gene in A431 epidermoid carcinoma cells. *Nature* **309:** 418.
VAN BEVEREN, C., S. ENAMI, T. CURRAN, and I.M. VERMA. 1984. FBR murine osteosarcoma virus. II. Nucleotide sequence of the provirus reveals that the genome contains sequences acquired from two cellular genes. *Virology* **135:** 229.
VAN BEVEREN, C., F. VAN STRAATEN, T. CURRAN, R. MÜLLER, and I.M. VERMA. 1983. Analysis of FBJ-MuSV provirus and c-*fos* (mouse) gene reveals that viral and cellular *fos* gene products have different carboxy termini. *Cell* **32:** 1241.
VAN STRAATEN, F., R. MÜLLER, T. CURRAN, C. VAN BEVEREN, and I.M. VERMA. 1983. Complete nucleotide sequence of a human c-*onc* gene: Deduced amino acid sequence of the human c-*fos* protein. *Proc. Natl. Acad. Sci.* **80:** 3183.
WATERFIELD, M.D., G.T. SCRACE, N. WHITTLE, P. STROOBANT, A. JOHNSSON, A. WASTESON, B. WESTERMARK, C.-H. HELDIN, J.S. HUANG, and T.F. DEUEL. 1983. Platelet-derived growth factor is structurally related to the putative transforming protein p28^{sis} of simian sarcoma virus. *Nature* **304:** 35.

Viral Enhancer Activity in Teratocarcinoma Cells

P. SASSONE-CORSI, D. DUBOULE, AND P. CHAMBON

Laboratoire de Génétique Moléculaire des Eucaryotes du CNRS, U.184 de Biologie Moléculaire et de Génie Génétique de l'INSERM, Faculté de Médecine, 67085 Strasbourg Cédex, France

Enhancers were originally identified as long-range activators of transcription in eukaryotes. For instance, sequences upstream of the early promoter of the simian virus 40 (SV40) were characterized by their ability to stimulate transcription in *cis* from homologous, heterologous, and substitute promoter elements over considerable distances and in an orientation-independent manner (for references, see Banerji et al. 1981; Moreau et al. 1981; Hen et al. 1982; Wasylyk et al. 1983). Transcriptional enhancers were also the first DNA sequences found to confer tissue- or cell-specific gene expression, suggesting a positive interaction with specific regulatory molecules and a possible role in the control of gene expression during differentiation (Banerji et al. 1983; Gillies et al. 1983; Queen and Baltimore 1983). The mechanism by which enhancers stimulate initiation of transcription is not known (Chambon et al. 1984), but both in vivo and in vitro competition experiments suggest that *trans*-acting "positive" transcriptional factors interact with specific enhancer sequences (Scholer and Gruss 1984; Sassone-Corsi et al. 1984, 1985; Wildeman et al. 1984). In addition, it has been shown that the adenovirus E1A gene products can regulate negatively the transcriptional activity of several enhancers (Borrelli et al. 1984). Furthermore, it has been demonstrated by DNase I hypersensitivity assays and electron microscopy studies that the SV40 enhancer induces an alteration in chromatin structure over its own sequence (for references, see Jongstra et al. 1984). Thus, *trans*-acting "positive" and "negative" enhancer factors may play an important role in the combinatorial control of eukaryotic gene expression, particularly during the establishment of a given differentiated cell phenotype.

Here we will analyze the transcriptional activity of the SV40, polyoma virus, and murine sarcoma virus (MSV) enhancer elements in several undifferentiated and differentiated teratocarcinomas. Teratocarcinoma cell lines are useful because some of them correspond to specific tissues at specific stages of early mouse development. We show that, while the SV40 enhancer is active in all of the cell lines analyzed, the polyoma virus enhancer element exhibits "differentiation-specific" activity.

EXPERIMENTAL PROCEDURES

Tumors and cell lines. The various teratocarcinomas were transplanted by either subcutaneous (solid tumors, TDR 602, TDR 694, TDR 114, TDN 2283) or intraperitoneal (ascitic tumors, TDE 113, OTT 2158) injections into syngeneic mice from the strains 129/Sv-Sl^{J}C P and LT/Sv (hereafter 129 and LT, respectively) obtained from the Jackson Laboratory (Bar Harbor, Maine) and subsequently bred in our laboratory. In vitro cultures were established from the solid tumors by mild trypsinization of small pieces and attachment of the cells to the surface of tissue culture dishes in Dulbecco's modification of Eagle's medium supplemented with 10% fetal calf serum at 37°C in a humid 5% CO_2 atmosphere. The embryoid bodies contained in the ascitic tumors were plated under the same conditions, but without trypsinization; cells were allowed to attach and were grown overnight before removing the embryoid bodies by washing.

Cell transfection and RNA analysis by quantitative S1 nuclease mapping. All cell lines were transfected at 50-70% confluence by the calcium phosphate technique with 10 µg of recombinant plasmid per 10-cm Petri dish, as described by Banerji et al. (1981). Cells were washed after 24 hours, and 12 hours later cytoplasmic RNA was purified from cells lysed with 0.3% Nonidet-P40. Hybridizations were carried out with 30-50 µg of total cytoplasmic RNA dissolved in 20 µl of 10 mM PIPES (pH 6.5) and 0.4 M NaCl containing an excess of the appropriate single-stranded DNA probes (see figure legends) and hybridized at 68°C for 12 hours. The samples were then diluted into 200 µl of 30 mM NaOAc (pH 4.5), 3 mM $ZnCl_2$, and 400 mM NaCl, containing 40 units (BRL) of S1 nuclease, and then incubated for 2 hours at 25°C. Nuclease-resistant hybrids were analyzed on 8% acrylamide/8.3 M urea sequencing gels.

RESULTS

Teratocarcinoma Cells

Teratocarcinoma cells at various stages of differentiation were derived from seven teratocarcinomas propagated in vivo (Experimental Procedures): three teratocarcinoma-derived rhabdomyosarcomas (TDR), one teratocarcinoma-derived neuroblastoma (TDN), one teratocarcinoma-derived endodermal tumor (TDE), and two undifferentiated tumors. The solid tumor TDR 602 (Fig. 1a) led in vitro to the formation of only rounded or bipolar myoblasts (Fig. 1b) and was therefore considered as developmentally less advanced than tumor TDR 694 (Fig. 1c), which yielded elongated cells in vitro (not shown), and tumor TDR 114 (Fig. 1d), which led to multinucleated myotubes in vitro (Fig. 1e) and thus represents the most advanced muscle-like cell type among these groups of TDRs (TDR 602 and TDR 694 were derived from a spontaneous ovarian teratocarcinoma of LT origin, whereas TDR 114 was derived from a spontaneous testicular teratocarcinoma of strain 129 [Blüthmann et al. 1983]). TDN 2283, derived from the 129 tumor OTT 6050 B (Stevens 1970), was composed mainly of immature neural tissue (Fig. 1f). After a few

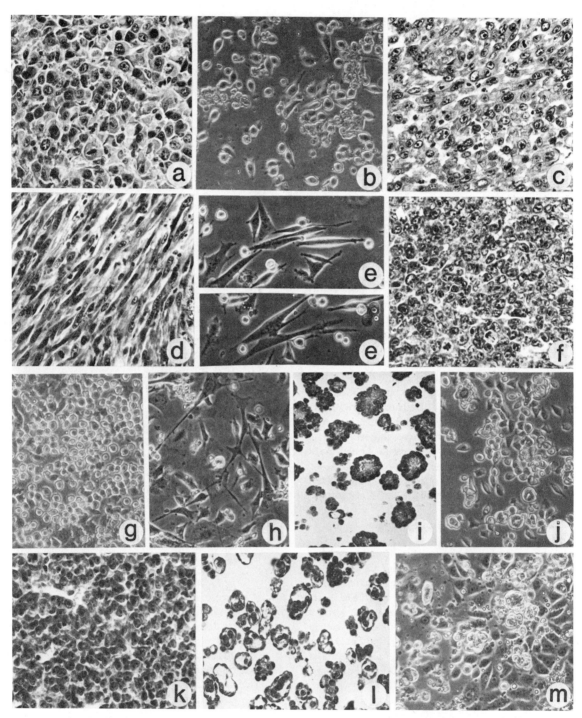

Figure 1. Histological sections and in vitro morphology of the different teratocarcinoma tumors. (*a, c, d*, and *k*) Histological sections of the solid teratocarcinoma tumors TDR 602, TDR 694, TDR 114, and OTT 2158, respectively; (*i* and *l*) histological sections of the embryoid bodies contained in the ascitic form of the teratocarcinoma tumors TDE 113 and OTT 2158, respectively. The morphology of the in vitro primary cultured cells derived from the teratocarcinoma tumors is shown in *b, e, g, h, j,* and *m*. TDR 602 after 1 week in culture (*b*), TDR 114 after 1 week in culture (*e*), TDN 2283 after 2 days (*g*) and 10 days (*k*) in culture, TDE 113 after 1 day in culture (*j*), and OTT 2158 after 3 days in culture (*m*).

days of in vitro culture, the TDN 2283 cells differentiated to form neural-like cells with long processes (compare Fig. 1g, h).

The ascitic form of endodermal tumor TDE 113 contains embryoid-like endodermal vesicles (Fig. 1i). In vitro, the tumor consists of epithelial-like cells (Fig. 1j) containing cytoplasmic granules. The solid form of this TDE 113 tumor from LT origin was characterized as parietal endoderm-like in view of the presence of hyaline material (Reichert's membrane) surrounding the cells (not shown). OTT 2158, derived from the 129 tumor OTT 6050 A (Stevens 1970), grows in vivo as a solid tumor consisting of undifferentiated embryonal carcinoma cells (Fig. 1k). The ascitic form of the tumor contains embryoid bodies composed of inner stem cells surrounded by a layer of primitive endodermal cells (Fig. 1l). After attachment in vitro, the overgrowth of endodermal-like cells was observed (Fig. 1m). The in vitro-established F9 undifferentiated embryonal carcinoma (EC) cell line (Strickland and Mahdavi 1980) was also used. A detailed description of most of these cell lines has been previously published by Blüthmann et al. (1983).

Activity of the SV40 Enhancer

The wild-type SV40 enhancer, which contains a 72-bp repeat, is located between SV40 coordinates 107 and 270 (Benoist and Chambon 1981) (coordinates following the BBB system; Tooze 1981). It has been previously shown that an enhancer fragment containing only one 72-bp sequence is also an efficient activator of transcription in vivo (Gruss et al. 1981; Moreau et al. 1981; M. Zenke et al., in prep.). To study the activity of the SV40 enhancer in teratocarcinoma cells, we have inserted the fragment containing one copy of the 72-bp sequence (between positions 179 and 270) upstream from position −109 of the rabbit β-globin gene (plasmid pG in Fig. 2A) (the β-globin promoter deleted to position −109 has been previously shown to be as efficient as the intact wild-type β-globin promoter [Dierks et al. 1983]). In HeLa or 3T6 cells, the presence of the SV40 enhancer in this pGB construction results in a 20- to 100-fold stimulation of transcription from the β-globin promoter (our unpublished observations). The stimulatory effect of the SV40 enhancer on β-globin transcription (compare pG to pGB in Fig. 2B) in the different teratocarcinoma cell lines (see Fig. 1) is always visible. In F9 undifferentiated EC cells, the stimulation is approximately 30-fold (Fig. 2B, compare lanes 1 and 2). A strong enhancer activity is also observed in the endodermal TDE 113 cells (lanes 9 and 10), and in the rhabdomyosarcomas TDR 602 (lanes 13 and 14) and TDR 114 (lanes 21 and 22) cells. In the other cell lines, OTT 2158 and TDR 694, the SV40 stimulatory effect is more moderate, but still very clear (Fig. 2B, lanes 5 and 6, and 17 and 18, respectively). We conclude that the SV40 enhancer is active in all the teratocarcinoma cell lines tested in this study, when linked to the heterologous β-globin promoter region.

Activity of the Polyoma Virus Enhancer

The polyoma virus enhancer, which does not contain a repeated sequence (de Villiers and Schaffner 1981), is located between the PvuII and BclI sites (coordinates 5262-5021 on the polyoma virus map; Tooze 1981). To study the polyoma virus enhancer activity in teratocarcinoma cells, we have used the recombinant pβ(244+)β constructed by de Villiers and Schaffner (1981), in which the PvuII-BclI fragment is inserted between two copies of a 4.5-kb chromosomal fragment containing the rabbit β-globin gene (see Fig. 2A). The presence of the polyoma virus enhancer in this recombinant results in a stimulation of transcription from the β-globin promoter of approximately 20- to 30-fold in mouse 3T6 fibroblasts (not shown). In contrast, no stimulation of transcription by the polyoma virus enhancer is visible in F9 undifferentiated EC cells (compare pβ2x and pβ[244+]β in Fig. 2B, lanes 3 and 4), nor in the other teratocarcinoma cell lines corresponding to early embryonic or extraembryonic tissues, such as the endodermal OTT 2158 (lanes 7 and 8) or the mesodermal TDR 602 (lanes 15 and 16). However, when phenotypically more differentiated cells, like the endodermal TDE 113 (lanes 11 and 12), were transfected, the polyoma virus enhancer activity became visible, being in some cases (for example in the mesodermal TDR 694 and TDR 114) similar to that of the SV40 enhancer (Fig. 2B, lanes 19 and 20, 23 and 24, respectively). We conclude from these results that, in contrast to the SV40 enhancer, the polyoma virus enhancer is active only in the more developmentally advanced cells.

Activity of the SV40 21-bp Repeat Upstream Promoter Element and of the MSV Enhancer in F9 Undifferentiated EC Cells

The SV40 enhancer stimulates transcription in mouse 3T3 fibroblasts at approximately the same extent, whether it is located in its homologous SV40 early promoter environment or linked to a heterologous promoter element (data not shown). To test the generality of this observation, we transfected recombinants pA0 and pA56 (see Fig. 2A) in F9 EC cells. Surprisingly, the transcription of pA56 (in which the enhancer is deleted from the SV40 early promoter) is markedly stronger in F9 EC cells (Fig. 2B, lanes 25 and 26) than in mouse 3T3 fibroblasts (Augereau and Wasylyk 1984; and our unpublished observations). In addition, although the transcriptional levels of the enhancerless recombinants pG and pA56 are the same in mouse 3T3 cells (data not shown), pA56 is transcribed more efficiently than pG in F9 EC cells (Fig. 2B, lanes 2 and 26). These results strongly suggest that the upstream SV40 early promoter element, the 21-bp repeat region, is more efficient at stimulating transcription in F9 undifferentiated EC cells than the β-globin upstream promoter region.

To investigate whether the SV40 21-bp repeat region is also activated in F9 undifferentiated cells, when

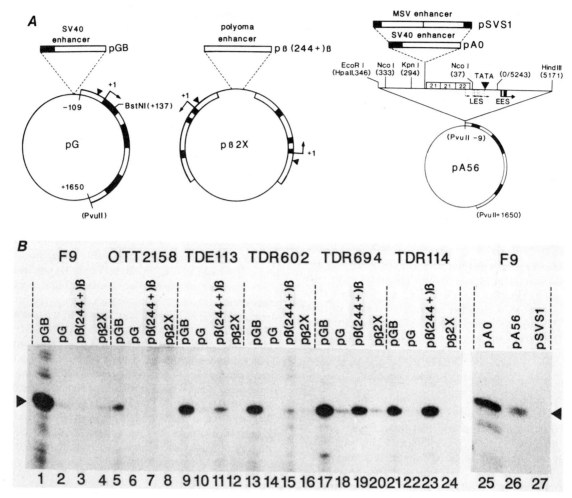

Figure 2. (A) Structure of the recombinant plasmids. pG contains the rabbit β-globin genomic sequence between positions −109 to +1650 inserted in pBR322 (Dierks et al. 1983). pGB was constructed by the insertion of the SV40 enhancer containing a single 72-bp sequence (positions 179–270) into a polylinker previously added upstream from position −109 of pG. pβ2x contains two copies of the rabbit 4.5-kb chromosomal fragment containing the rabbit β-globin gene; pβ(244+)β contains the polyoma virus enhancer (fragment PvuII [5262]–BclI [5021]) inserted between the two β-globin sequences of pβ2x (de Villiers and Schaffner 1981). pA56 contains the SV40 early promoter region (346–5171), deleted of the enhancer element (between 270 and 101) linked to promoterless rabbit β-globin sequences from −9 to +1650 (Van Ooyen et al. 1979). pA0 is derived from pA56 by inserting the SV40 enhancer containing a single 72-bp sequence. pSVS1 contains the Mo-MSV enhancer Sau3A–XbaI fragment (positions 327–529; Levinson et al. 1982) inserted in pA56 at the same position as the SV40 enhancer in pA0. The BstNI site at +137 of the β-globin sequence has been used to prepare a single-stranded 5′-labeled probe for S1 nuclease mapping experiments (Borrelli et al. 1984). Solid triangles represent the TATA box elements; solid boxes in the β-globin sequence indicate the exons. None of the recombinants is drawn to scale. (B) Quantitative S1 nuclease mapping analysis of the RNA from transfected teratocarcinoma cells. The recombinants were transfected into the different teratocarcinoma cells by the calcium phosphate technique. Cytoplasmic RNA was analyzed as described in the Experimental Procedures. The autoradiographic bands (arrowhead) correspond to S1 nuclease-resistant hybrids of RNA initiated at the β-globin gene cap site (lanes 1–24) and the rabbit β-globin probe 5′-labeled at the BstNI site (+137), as described (Borrelli et al. 1984). The bands in lanes 25–27 correspond to the end point of homology (position −9 of the β-globin sequence, see A) between the transcripts initiated from the SV40 early start sites and the β-globin probe (similar results were obtained when a probe fully homologous to pA0 was used). The results shown in the figure are representative of a series of transfection experiments.

linked to another enhancer element, we constructed the plasmid pSVS1 described in Figure 2A. This hybrid promoter contains the Moloney murine sarcoma virus (Mo-MSV) enhancer (Levinson et al. 1982) inserted in the SV40 early promoter in place of the 72-bp repeat region. This Mo-MSV enhancer has been described to be efficient at stimulating heterologous promoter elements in mouse LMTK⁻ and 3T3 cells (Levinson et al. 1982; Augereau and Wasylyk 1984). Unexpectedly, its presence completely abolishes transcription from the SV40 early promoter in F9 EC cells (Fig. 2B, lanes 25–27). This striking result has also been observed by other investigators (P. Rigby, pers. comm.).

DISCUSSION

The study of promoter activities in teratocarcinoma cells is interesting since it could give some indications regarding the presence or the absence of *trans*-acting transcriptional regulatory proteins during differentiation. Because of the transcriptional characteristics of enhancer elements, it has been proposed that they can be targets for cellular regulatory molecules at specific stages of development and cellular differentiation. We have reported here results concerning the SV40 and polyoma virus enhancer activity in a series of teratocarcinoma cell lines that represent specific stages of differentiation of mouse embryonic or extraembryonic tissues. Whereas the SV40 enhancer element is active in all the cell lines tested in this study, the polyoma virus enhancer exhibits a differentiation-specific activity. In particular, there is no stimulation of β-globin gene expression by polyoma virus enhancer in F9 undifferentiated EC cells and in "early" endodermal (OTT 2158) or mesodermal (TDR 602) cells. Similar results have been obtained by others using undifferentiated PCC3 EC cells and an enzymatic chloramphenicol acetyl transferase (CAT) assay: following transfection of hybrid recombinants containing the polyoma virus enhancer linked to the α2-collagen promoter, the stimulation of the transcription was only threefold, whereas it was 12-fold in 3T6 fibroblasts (Herbomel et al. 1984). In contrast, the polyoma virus enhancer is active in the parietal endoderm-like TDE 113 teratocarcinoma and in the more developmentally advanced TDR 694- and TDR 114-derived rhabdomyosarcomas (see Fig. 2B). The activity of both polyoma virus and SV40 enhancers is also high in the neural-like cells derived from TDN 2283 (Fig. 1) (data not shown).

It is interesting to note that an enhancerless SV40 early promoter is more active in undifferentiated F9 EC cells than in mouse fibroblasts (pA56 in Fig. 2). Since the enhancerless β-globin promoter is not active in F9 EC cells (pG in Fig. 2A), it appears that the SV40 21-bp repeat upstream promoter element may be specifically activated in these undifferentiated cells. It is very striking that this activation is completely abolished when the MSV enhancer is inserted in the place of the SV40 72-bp repeat enhancer (pSVS1 in Fig. 2, lanes 26 and 27). This result is in contrast to what has been observed in mouse fibroblasts, in which the same hybrid promoter is active (Levinson et al. 1982; Augereau and Wasylyk 1984).

That both polyoma virus and MSV enhancer elements are nonfunctional in undifferentiated EC cells may be related to the observation of Imperiale et al. (1984), who have reported the presence in F9 EC cells of an adenovirus E1A-like protein that allows the growth of an adenovirus 2 mutant deficient in E1A (*dl*312). In fact, it has been shown that the E1A gene products repress the stimulatory activity of several enhancer elements (Borrelli et al. 1984). Thus the nonfunction of the polyoma virus enhancer in F9 EC cells could be related to the presence of an E1A-like protein rather than to the lack of a positive *trans*-acting stimulatory factor. This putative E1A-like protein with enhancer repressor activity may have different affinities for different enhancer elements; that is, lower for the SV40 enhancer and higher for the MSV enhancer, in which case it completely abolishes transcription. In this respect, it is interesting to note that the Moloney murine leukemia virus (Mo-MLV) enhancer, closely related to MSV, is also nonfunctional in undifferentiated F9 EC cells (Linney et al. 1984).

In conclusion, using several teratocarcinoma cell lines, we have demonstrated that the activity of enhancer elements is subjected to striking regulations during cellular differentiation. Such regulation may be caused by the selective presence of *trans*-acting stimulatory enhancer factors appearing during differentiation or by the disappearance of enhancer-specific repressor proteins, or by a combination of the two processes. In vivo competition experiments are in progress to distinguish between these two possibilities of control of enhancer activity in teratocarcinoma cells.

ACKNOWLEDGMENTS

We thank H. Blüthmann and L.C. Stevens for gift of the teratocarcinoma cell lines; W. Schaffner, C. Weissman, P. Augereau, and P. Jalinot for recombinants; and C. Aron, C. Werlé, and B. Boulay for illustrations and preparation of the manuscript. P. Sassone-Corsi and D. Duboule are recipients of a fellowship from the Fondation pour la Recherche Médicale Française and a long-term fellowship of the European Molecular Biology Organization, respectively. This work was supported by CNRS (ATP 3582), INSERM (PRC 124026), the Fondation pour la Recherche Médicale, the Ministère de l'Industrie et de la Recherche (82V1283), and the Association pour le Développement de la Recherche sur le Cancer.

REFERENCES

AUGEREAU, P. and B. WASYLYK. 1984. The MLV and SV40 enhancers have a similar pattern of transcriptional activation. *Nucleic Acids Res.* **12**: 8801.

BANERJI, J., L. OLSON, and W. SCHAFFNER. 1983. A lymphocyte specific cellular enhancer is located downstream of the joining region in immunoglobulin heavy chain genes. *Cell* **33**: 729.

BANERJI, J., S. RUSCONI, and W. SCHAFFNER. 1981. Expression of a β-globin gene is enhanced by remote SV40 DNA sequences. *Cell* **27**: 299.

BENOIST, C. and P. CHAMBON. 1981. *In vivo* sequence requirements of the SV40 early promoter region. *Nature* **290**: 304.

BLÜTHMANN, H., E. VOGT, P. HÖSLI, L.C. STEVENS, and K. ILLMENSEE. 1983. Enzyme activity profiles in mouse teratocarcinoma. A quantitative ultra microscale analysis. *Differentiation* **24**: 65.

BORRELLI, E., R. HEN, and P. CHAMBON. 1984. The adenovirus-2 E1A products repress stimulation of transcription by enhancers. *Nature* **312**: 608.

CHAMBON, P., A. DIERICH, M.P. GAUB, S. JAKOWLEV, J. JONGSTRA, A. KRUST, J.P. LEPENNEC, P. OUDET, and T. REUDELHUBER. 1984. Promoter elements of genes coding for proteins and modulation of transcription by oestrogens and progesterone. *Recent Prog. Horm. Res.* **40**: 1.

DE VILLIERS, J. and W. SCHAFFNER. 1981. A small segment of polyoma virus DNA enhances the expression of a cloned β-globin gene over a distance of 1400 base pairs. *Nucleic Acids Res.* **9**: 6521.

DIERKS, P., A. VAN OOYEN, M. COCHRAN, C. DOBKIN, J. REISER, and C. WEISSMAN. 1983. Three regions upstream from the cap site are required for efficient and accurate transcription of the rabbit β-globin gene in mouse 3T6 cells. *Cell* **32:** 695.

GILLIES, S.D., S.L. MORRISON, V.T. OI, and S. TONEGAWA. 1983. A tissue-specific transcription enhancer element is located in the major intron of a rearranged immunoglobulin heavy chain gene. *Cell* **33:** 717.

GRUSS, P., R. DHAR, and G. KHOURY. 1981. Simian virus 40 tandem repeated sequences as an element of the early promoter. *Proc. Natl. Acad. Sci.* **78:** 943.

HEN, R., P. SASSONE-CORSI, J. CORDEN, M.P. GAUB, and P. CHAMBON. 1982. Sequences upstream from the TATA box are required *in vivo* and *in vitro* for efficient transcription from the adenovirus serotype 2 major late promoter. *Proc. Natl. Acad. Sci.* **79:** 7132.

HERBOMEL, P., B. BOURACHOT, and M. YANIV. 1984. Two distinct enhancers with different cell specificities coexist in the regulatory region of polyoma. *Cell* **32:** 319.

IMPERIALE, M.J., H.T. KAO, L.T. FELDMAN, J.R. NEVINS, and S. STRICKLAND. 1984. Common control of the heat shock gene and early adenovirus genes: Evidence for a cellular E1A-like activity. *Mol. Cell. Biol.* **4:** 867.

JONGSTRA, J., T. REUDELHUBER, P. OUDET, C. BENOIST, C.B. CHAE, J.M. JELTSCH, D.J. MATHIS, and P. CHAMBON. 1984. Induction of altered chromatin structures by the SV40 enhancer and promoter elements. *Nature* **307:** 708.

LEVINSON, B., G. KHOURY, G. VANDE WOUDE, and P. GRUSS. 1982. Activation of SV40 genome by 72-base pair tandem repeats of Moloney sarcoma virus. *Nature* **295:** 568.

LINNEY, E., B. DAVIES, J. OVERHAUSER, E. CHAO, and H. FAN. 1984. Non-function of a Moloney murine leukaemia virus regulatory sequence in F9 embryonal carcinoma cells. *Nature* **308:** 470.

MOREAU, P., R. HEN, B. WASYLYK, R. EVERETT, M.P. GAUB, and P. CHAMBON. 1981. The 72 bp repeat has a striking effect on gene expression both in SV40 and other chimeric recombinants. *Nucleic Acids Res.* **9:** 6047.

QUEEN, C. and D. BALTIMORE. 1983. Immunoglobulin gene transcription is activated by downstream sequence elements. *Cell* **33:** 741.

SASSONE-CORSI, P., A. WILDEMAN, and P. CHAMBON. 1985. A *trans*-acting factor is responsible for the simian virus 40 enhancer activity *in vitro*. *Nature* **313:** 458.

SASSONE-CORSI, P., J.P. DOUGHERTY, B. WASYLYK, and P. CHAMBON. 1984. Stimulation of *in vitro* transcription from heterologous promoters by the SV40 enhancer. *Proc. Natl. Acad. Sci.* **81:** 308.

SCHÖLER, H.R. and P. GRUSS. 1984. Specific interaction between enhancers and cellular components. *Cell* **36:** 403.

STEVENS, L.C. 1970. The development of transplantable teratocarcinomas from intratesticular grafts of pre- and postimplantation mouse embryos. *Dev. Biol.* **21:** 364.

STRICKLAND, S. and V. MAHDAVI. 1980. The induction of differentiation in teratocarcinoma stem cells by retinoic acid. *Cell* **15:** 394.

TOOZE, J. 1981. *Molecular biology of tumor viruses*, 2nd edition, revised: *DNA tumor viruses*. Cold Spring Harbor Laboratory, Cold Spring Harbor, New York.

VAN OOYEN, A., J. VAN DEN BERG, N. MANTEI, and C. WEISSMAN. 1979. Comparison of total sequence of a cloned rabbit β-globin gene and its flanking regions with a homologous mouse sequence. *Science* **206:** 337.

WASYLYK, B., C. WASYLYK, P. AUGEREAU, and P. CHAMBON. 1983. The SV40 72 bp repeat preferentially potentiates transcription starting from proximal natural or substitute promoter elements. *Cell* **32:** 503.

WILDEMAN, A., P. SASSONE-CORSI, T. GRUNDSTROM, M. ZENKE, and P. CHAMBON. 1984. Stimulation of *in vitro* transcription from the SV40 early promoter by the enhancer involves a specific trans-acting factor. *EMBO J.* **3:** 3129.

Coordinate Expression of Myogenic Functions and Polyoma Virus Replication

A. Felsani, R. Maione, L. Ricci, and P. Amati
Dipartimento di Biopatologia Umana, Sezione di Biologia Cellulare, Università di Roma La Sapienza, Policlinico Umberto I, 00161 Rome, Italy

Polyoma virus has been extensively used to study degrees of cell differentiation (Georges et al. 1982; Tanaka et al. 1982) and tissue specificity (de Villiers et al. 1984; De Simone et al. 1985; Maione et al. 1985) by means of analyses of cell permissivity to viral growth. The viral nontranscribed regulatory region that directs early and late mRNA transcription, as well as DNA replication (Tyndall et al. 1981), has also been used as an enhancer for heterologous genes in chimeric plasmids (de Villiers and Shaffner 1981) to analyze the regulatory interactions of putative cell functions in different tissue cell lines (Khoury and Gruss 1983; Herbomel et al. 1984).

The overall picture of the results of these studies clearly lends itself to the notion that the polyoma regulatory region interacts in a very composite fashion with cell factor(s). These cell factors may play a positive or negative role on viral gene expression (Borrelli et al. 1984) and replication (Maione et al. 1985).

With the aim of correlating the activity of these putative cell factors with their role in regulating endogenous gene expression, we have analyzed the ability of polyoma to replicate in established differentiated cell lines that are able to undergo further differentiation in vitro.

In the present work, we describe the activation of polyoma gene expression and replication in the mouse myoblasts C2 cell line (Yaffe and Saxel 1977). These cells, under determined growth conditions, undergo the myogenic differentiative step from myoblast (Mb) to myotubes (Mt). Several cell-specific functions that are induced during the myogenic process have been identified (for a review, see Pearson 1980) and special attention has been paid to the regulation of contractile protein synthesis (for a review, see Buckingham 1985).

The results that we obtained show that polyoma virus is unable to replicate in C2 cells in the Mb condition and that its genome appears diluted in progeny cells. However, if myogenic differentiation is induced, viral functions are expressed and replication, as well as maturation, takes place. This property of infected C2 cells is maintained during active growth for a length of time that is directly proportional to multiplicity of infection (moi). The appearance of α-actin mRNA (a characteristic Mb differentiative function (Caravatti et al. 1982; Minty et al. 1982) precedes polyoma early mRNA synthesis, suggesting a coordinate control of expression.

A comparison of the nucleotide sequences of the regulatory regions α-actin and the myosin light chain (MLC1) with that of the polyoma enhancer has been reported recently (Daubas et al. 1985) and will be discussed in the light of our results.

METHODS

Cell Culture Conditions and Viral Infection

The mouse C2 myoblast line (Yaffe and Saxel 1977) obtained from M. Buckingham, was grown in Dulbecco's modified Eagle's medium supplemented with 10% fetal calf serum. To maintain their active growth, the cells were diluted 1:10 when they reached a concentration of 1×10^6 cells/10-cm-diameter plate. The generation time in these conditions is approximately 18 hours. To induce differentiation, cells were allowed to grow to confluence in plates whose bottoms had been coated with collagen (Vitrogen 100, Flow Lab) and the medium was changed every day. Under these conditions, cells reach confluence after 2–3 days, myotubes appear on day 5, and complete differentiation occurs on days 9 or 10.

The polyoma A2 wild-type (wt) strain (Ruley and Fried 1983), propagated on 3T6 cells, was used for standard myoblast infection at the reported multiplicities. Virus production was titrated, after cell destruction by freezing and thawing, using the sheep red blood cells agglutination test (Tooze 1981).

DNA and RNA Extraction, Restriction Endonuclease Digestion, and Blot Hybridization

Viral DNAs were extracted from cell cultures using the Hirt selective procedure (Hirt 1967). Digestion with restriction endonucleases was carried out under the conditions specified by the manufacturer (New England Biolabs, Beverly, Mass.). To detect the amount and constitution of polyoma virus genomes in the Hirt extracts, Southern blots (Southern 1975) and dot blots were performed essentially according to Maniatis et al. (1982). The probes used to detect viral DNA were the pAT153 plasmid carrying whole polyoma genome inserted in the *Bam*HI restriction site (pPy53A6.6; Treisman et al. 1981) or the uncloned polyoma genome.

Messenger and ribosomal RNA determination was performed by dot hybridization analysis of cellular cytoplasmic preparation as described by White and Bancroft (1982). The probes used in the hybridization were: for polyoma early mRNA, an M13mp8 vector carrying the polyoma fragment PvuII-HindIII (nt 1144-1659); for cell rRNA, the pS10R plasmid (La Volpe et al. 1985); and, for α-actin-specific mRNA, the pAM91-200 plasmid (Minty et al. 1981).

Cell Transfection

Polyoma DNA from pPy53A6.6 plasmid excised by BamHI endonuclease was transfected into C2 cells at the Mb stage using the calcium phosphate standard method (Wigler et al. 1977) at 20 µg/ml per plate.

RESULTS

Viral Replication

The C2 cells were infected at different moi ranging from 20 to 200 and were kept in active growth by diluting them 1:5 every 2 days. At each passage a fraction of the infected cells was subjected to Hirt extraction (Hirt 1967) to determine the level of polyoma genome presence; another fraction was kept undiluted, to induce cell differentiation. The undiluted samples were also Hirt-extracted or lysed for mature virus titration after 9 and 11 days of further incubation.

Results obtained in several independent experiments at different moi, using Southern blots of Hirt extracts, show that the viral genomes in actively growing cells (diluted samples) had been segregated without replication in progeny cells. Moreover, mature viral particles were never recovered. In the fractions of cells that had been kept undiluted at different passages, viral replication was observed until about 50-80 polyoma genomes per plate were present. Instead, mature viral particles were observed only at stages when at least 10^3 genomes per plate were calculated to be present. This rough calculation was based on the original virus genomes' number per plate (dependent on moi and cell number) and the dilution factor of the serial passages. Results of an experiment performed in parallel at moi 20 and 200 are shown in Figure 1, where it can be seen that the input virus is detectable over time in ways proportional to the moi. Viral replication takes place up to passage 6, but with evident differences between the two moi. At this passage the number of polyoma genomes present per plate was calculated to be 60 for moi 20 and 600 for moi 200. At passage 8, the genomes per plate were calculated to be 2 and 20, respectively, both under the productive threshold observed in experiments at moi 50 and 100 (data not shown).

Mature viral genomes detectable by hemoagglutination (at least 10^6 particles/ml) were observed in cells from passage 3 at moi 20 and passage 5 at moi 200.

Timing of Polyoma Induction during Cell Differentiation

To analyze the sequence of events that are correlated to polyoma replication during myogenic differentiation, we followed the time course of the appearance of α-actin mRNA, polyoma early mRNA, and polyoma DNA in C2 myoblast cells infected at moi 50 and allowed growth without dilution.

Results reported in Figure 2 show that α-actin mRNA precedes the synthesis of polyoma early mRNA by approximately 1 day and that polyoma DNA synthesis follows the early mRNA synthesis. These results indicate that the induction of cell differentiative functions is coordinated with the expression of polyoma genomes.

Molecular Fate of Input Viral Genome

To analyze the fate of the input viral genome in relation to the lack of replication observed in the experiments described above, we transfected actively growing C2 cells with BamHI-linearized polyoma genome excised from the pPy53 plasmid. To distinguish replicated viral DNA, MboI restriction endonuclease sensitivity was utilized, since this enzyme can discriminate unreplicated prokaryotic methylated DNA from the replicated eukaryotic unmethylated kind. The amount of polyoma genomes in a constant number of originally infected cells and their sensitivity to digestion by the MboI restriction endonuclease were measured both during myogenic differentiation and in the myoblast condition. To obtain homogeneity, cells from several transfected plates were pooled together 12 hours after transfection and replated (10^5 cells/plate). The cells in which differentiation was to be induced were kept undiluted, and those used for the analysis of the viral genome in the Mb stage were passed as described in Methods. Hirt extracts were performed from all the progeny cells deriving from each original plate.

Results reported in Figure 3 show that all polyoma molecules become sensitive to MboI digestion 1 day after transfection. The total amount of polyoma genomes remained unaltered during myogenesis until day 5 and afterwards increased drastically. Conversely, during growth in the Mb conditions, the amount of polyoma DNA remained constant. Since conversion to MboI susceptibility requires viral DNA replication, we must assume that viral DNA in Mb cells performed one but not more than a few rounds of replication before early functions could be expressed.

DISCUSSION

The results of this study demonstrate that actively growing myoblast C2 cells are not permissive to polyoma growth and that the viral genome is not integrated because it is diluted in progeny cells. The virus seems to replicate when cells reach confluence and before extensive myotube formation. The expression of early viral mRNA is preceded by α-actin mRNA induction. This coordinate expression of viral functions necessary for viral replication, and of cellular functions characterizing myotube differentiation, strongly suggests that

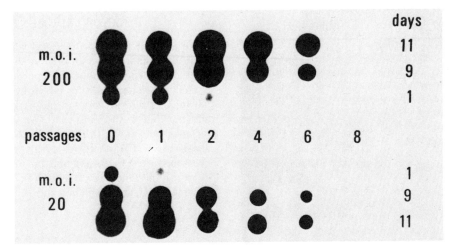

Figure 1. Dot blot hybridization with ^{32}P-labeled pPy53 plasmid of Hirt extracts from polyoma-infected C2 myoblast cells at moi 20 and 200. Plates of C2 myoblasts were Hirt-extracted at each passage, before and after myogenic differentiation. Days denote further incubation of undiluted cell samples.

a common factor(s) must be involved. This fact has induced Daubas et al. (1985) to search for sequence homologies in the 5′ regulatory region of cardiac myosin L C genes and in the polyoma regulatory region. They found a similarity between the consensus sequence of the A domain of polyoma and sequences upstream of the MLC1 promoter.

The analysis of sequence homologies in the core of DNA domains that have been shown to be involved in the regulation of gene expression has led Hearing and Shenk (1983) to propose a consensus sequence common to a variety of enhancers. In Table 1, we compare this proposed consensus sequence with the core sequence of the polyoma A domain, and with the se-

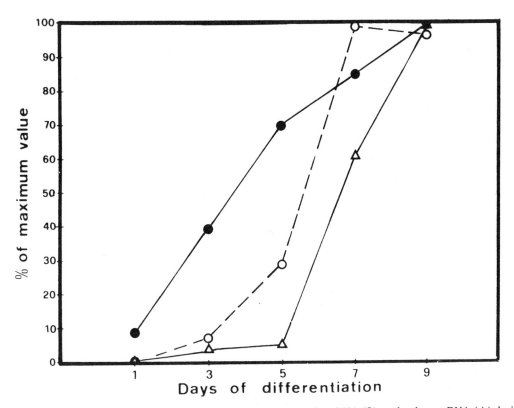

Figure 2. Time course of the synthesis of α-actin mRNA (●), polyoma early mRNA (○), and polyoma DNA (△) during myogenic differentiation of polyoma-infected C2 cells. Hybridization with ^{32}P-labeled specific probes was performed with cytoplasmic extracts for RNAs and with nuclear extracts for DNA determination. The values were calculated (as percent of maximum expression) from the rectilinear part of the curve given by six serial dilutions for each sample immobilized by dot blot on nitrocellulose filters. The mRNA values were normalized to rRNA present in each sample.

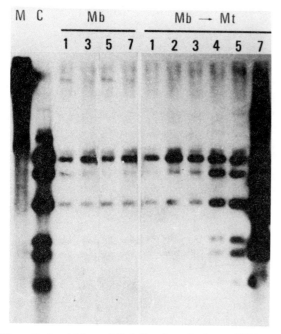

Figure 3. Sensitivity to *Mbo*I digestion of polyoma DNA extracted from transfected C2 cells during myogenic differentiation and Mb growth. Linearized polyoma genomes from pPy53A6.6 plasmid with *Bam*HI digestion were transfected into C2 myoblast cells. At different days of sparse or confluent growth, Hirt extracts were digested by *Mbo*I endonuclease, run on a 1.4% agarose gel, and blot-hybridized with ^{32}P-labeled viral genome. Numbers denote days from transfection. (M) Methylated polyoma DNA from plasmid; (C) unmethylated polyoma DNA from 3T6.

quences in MLC1 and the α-actin regulatory region examined by Daubas et al. (1985). The proposed consensus core sequences of these last two genes are included in stretches of 29 bp which are 76% homologous. This fact lends further support to our hypothesis of a common cell factor that regulates the expression of polyoma early mRNA as well as the differentiation-specific gene functions.

We are inclined to believe that the sensitivity to *Mbo*I digestion of polyoma genomes in cells in the Mb state may be explained in one of two ways: as a demethylation that is not related to replication or, alternatively, as an extremely limited replication that is independent of the expression of early viral functions (as in the case of polyoma wild-type infecting embryonal carcinoma F9 cells observed by Dandolo et al. [1984]).

ACKNOWLEDGMENTS

We are grateful to Laura Pozzi who has done the basic preliminary observations on the differentiative stage dependence of polyoma replication. We are indebted to Arlette Cohen for all her helpful advice and to Margaret Buckingham for the gift of the C2 cell line and the α-actin-specific probe. We acknowledge the technical help of Mr. L. De Angelis. A.F. is on leave of absence from I.E.M., C.N.R., Naples. This work has been supported by Programmi Finalizzati Ingegneria Genetica and Oncologia, CNR; Associazione Italiana Ricerca sul Cancro, Milano; and Ministero Pubblica Istruzione, Italy.

REFERENCES

BORRELLI, E., R. HEN, and P. CHAMBON. 1984. Adenovirus-2 E1A products repress enhancer-induced stimulation of transcription. *Nature* **312**: 608.

BUCKINGHAM, M. 1985. Actin and myosin multigene families: Their expression during the formation of skeletal muscle. *Essays Biochem.* **20**: 77.

CARAVATTI, M., A. MINTY, B. ROBERT, D. MONTARRAS, A. WEYDERT, A. COHEN, P. DAUBAS, and M. BUCKINGHAM. 1982. Regulation of muscle gene expression: The accumulation of messenger RNAs coding for muscle specific protein during myogenesis in a mouse cell line. *J. Mol. Biol.* **160**: 59.

DANDOLO, L., J. AGHION, and D. BLANGY. 1984. T-antigen-independent replication of polyomavirus DNA in murine embryonal carcinoma cells. *Mol. Cell. Biol.* **4**: 317.

DAUBAS, P., B. ROBERT, I. GARNER, and M. BUCKINGHAM. 1985. A comparison between mammalian and avian fast skeletal muscle alkali myosin light chain genes: Regulatory implications. *Nucleic Acids Res.* **13**: 4623.

DE SIMONE, V., L. LANIA, G. LA MANTIA, and P. AMATI. 1985. Polyomavirus mutation that confers a cell-specific *cis* advantage for viral DNA replication. *Mol. Cell. Biol.* **5**: 2142.

DE VILLIERS, J. and W. SCHAFFNER. 1981. A small segment of polyoma DNA enhances the expression of a cloned β-globin gene over a distance of 1,400 base pairs. *Nucleic Acids Res.* **9**: 6251.

DE VILLIERS, J., W. SCHAFFNER, C. TYNDALL, S. LUPTON, and R. KAMEN. 1984. Polyoma virus DNA replication requires an enhancer. *Nature* **312**: 242.

GEORGES, E., M. VASSEUR, and D. BLANGY. 1982. Polyoma virus mutants as probes of variety among mouse embryonal carcinoma cell lines. *Differentiation* **22**: 62.

HEARING, P. and T. SHENK. 1983. The adenovirus type 5 E1A transcriptional control region contains a duplicated enhancer element. *Cell* **33**: 695.

HERBOMEL, P., B. BOURACHOT, and M. YANIV. 1984. Two distinct enhancers with different cell-specificities coexist in the regulatory region of polyoma. *Cell* **39**: 653.

Table 1. Sequence Homology between Consensus Sequences

	A A - G G A A G T G A - C C		Consensus core sequence	
5103	...T A A G C A G G A A G T G A C...	5117	Py	
	...T g A G C c G G A A - T G c C...	−24	MLC1	11/15
	....C A a G A A - T G g C...	−31	α-actin	8/11

The consensus core sequence proposed by Hearing and Shenk (1983) is compared with the polyoma core sequence of the A domain (Herbomel et al. 1984), with the proposed one for MLC1 (Daubas et al. 1985) and with that of α-actin (I. Garner, unpublished results reported in Daubas et al. 1985). The numbers refer, for polyoma, to the nucleotide numbering of Tyndall et al. (1981), and, for MLC1 and α-actin, to the distance to CAT box.

Hirt, B. 1967. Selective extraction of polyoma DNA from infected mouse cell cultures. *J. Mol. Biol.* **26**: 365.

Khoury, G. and P. Gruss. 1983. Enhancer elements. *Cell* **33**: 313.

La Volpe, A., A. Simeone, M. D'Esposito, L. Scotto, V. Fidanza, A. De Falco, and E. Boncinelli. 1985. Analysis of the heterogeneity region of the human ribosomal spacers. *J. Mol. Biol.* **183**: 223.

Maione, R., C. Passananti, V. De Simone, P. Delli Bovi, G. Augusti-Tocco, and P. Amati. 1985. Selection of mouse neuroblastoma cell specific polyoma virus mutants with stage differentiative advantages of replication. *EMBO J.* (in press).

Maniatis, T., E.F. Fritsch, and J. Sambrook. 1982. *Molecular cloning: A laboratory manual.* Cold Spring Harbor Laboratory, Cold Spring Harbor, New York.

Minty, A., S. Alonso, M. Caravatti, and M. Buckingham. 1982. A fetal skeletal muscle actin mRNA in the mouse and its identity with cardiac actin mRNA. *Cell* **30**: 185.

Minty, A., M. Caravatti, B. Robert, A. Cohen, P. Daubas, A. Weydert, F. Gros, and M. Buckingham. 1981. Mouse actin messenger RNAs–Construction and characterization of a recombinant plasmid molecule containing a complementary DNA transcript of mouse actin mRNA. *J. Biol. Chem.* **256**: 1008.

Pearson, M.L. 1980. Muscle differentiation in cell culture: A problem in somatic cell and molecular genetics. In *The molecular genetics of development* (ed. T. Leighton and W.F. Loomis), p. 361. Academic Press, New York.

Ruley, E.H. and M. Fried. 1983. Sequence repeats in a polyoma virus DNA region important for gene expression. *J. Virol.* **47**: 233.

Southern, E. 1975. Detection of specific sequences among DNA fragments separated by gel electrophoresis. *J. Mol. Biol.* **98**: 503.

Tanaka, K., K. Chowdhury, K.S.S. Chang, M. Israel, and Y. Ito. 1982. Isolation and characterization of polyoma virus mutants which grow in murine embryonal carcinoma and trophoblast cells. *EMBO J.* **1**: 1521.

Tooze, J. 1981. *Molecular biology of tumor viruses*, 2nd edition, revised: *DNA tumor viruses.* Cold Spring Harbor Laboratory, Cold Spring Harbor, New York.

Treisman, R., U. Novak, J. Favaloro, and R. Kamen. 1981. Transformation of rat cells by an altered polyoma virus genome expressing only the middle-T protein. *Nature* **292**: 595.

Tyndall, C., G. La Mantia, C.M. Thacker, J. Favoloro, and R. Kamen. 1981. A region of the polyoma virus genome between the replication origin and late protein coding sequences is required in *cis* for both early gene expression and viral DNA replication. *Nucleic Acids Res.* **9**: 6231.

White, B.A. and F.C. Bancroft. 1982. Cytoplasmic dot hybridization. Simple analysis of relative mRNA levels in multiple small cell or tissue samples. *J. Biol. Chem.* **257**: 8569.

Wigler, M., S. Silverstein, L.S. Lee, A. Pellicer, Y.C. Cheng, and R. Axel. 1977. Transfer of purified herpes virus thymidine kinase gene to cultured mouse cells. *Cell* **11**: 223.

Yaffe, D. and O. Saxel. 1977. Serial passaging and differentiation of myogenic cells isolated from dystrophic mouse muscle. *Nature* **270**: 725.

Structure and Regulated Transcription of DIRS-1: An Apparent Retrotransposon of *Dictyostelium discoideum*

J. CAPPELLO,* K. HANDELSMAN,* S.M. COHEN,* AND H.F. LODISH*†
*Whitehead Institute for Biomedical Research, Cambridge, Massachusetts 02142;
†Department of Biology, Massachusetts Institute of Technology, Cambridge, Massachusetts 02139

Transposon- and retrovirus-like DNA segments form a major part of the repetitive DNA in many eukaryotes. They are a cause of many "spontaneous" mutations in *Drosophila* and other organisms, and many catalyze the rearrangements of large segments of DNA that have occurred during evolution (Flavell et al. 1981; Kleckner 1981; Spradling and Rubin 1981; Fedoroff 1983; Roeder and Fink 1983). In the course of cloning segments of *Dictyostelium discoideum* genomic DNA that are expressed preferentially during early differentiation, we isolated the DIRS-1 transposon (Zuker and Lodish 1981; Chung et al. 1983). DIRS-1 is unusual in that its transcription is induced either by heat shock or by the high cell density obtained during the early stages of differentiation. It is also novel in that its structure is unlike the structure of any transposon in prokaryotes or eukaryotes identified to date. DIRS-1 has been independently isolated by Rosen et al. (1983) who call it Tdd-1.

Structure of DIRS-1

The DIRS-1 genomic family has several structural features in common with well-characterized prokaryotic and eukaryotic transposable elements. There are multiple copies of DIRS-1-related sequences in the genome. The major component of these (about 40 copies) is a homologous 4.7-kb element that consists of 4.4-kb of internal unique sequence flanked by 330-bp inverted terminal repeats (ITRs). The elements are inserted at dispersed genomic locations, and the genomic locations of the elements vary in different *Dictyostelium* genetic stocks. Elements that are inserted at different chromosomal locations have similar terminal repeats and internal 4.1-kb segments (Chung et al. 1983).

DIRS-1 is an unusual eukaryotic transposon because it has long terminal repeats (330 nucleotides) that are inverted (Fig. 1) (Zuker et al. 1984). A great majority of eukaryotic transposons and transposable-like elements (e.g., retroviruses) have direct terminal repeats (Majors et al. 1981; Spradling and Rubin 1981). Only four eukaryotic transposable elements are known to bear ITRs: the P and FB elements of *Drosophila* (Spradling and Rubin 1981; O'Hare and Rubin 1983), the σ element of yeast (Del Rey et al. 1982), and the Tc1 element of *Caenorhabditis elegans* (Rosenzweig and Hirsh 1983). The ITRs flanking the P and σ elements are very short (31 or 15 bp and 8 bp, respectively), those at the ends of the FB element contain periodic tandem repeats of a 10-, 20-, and 31-bp sequence (Potter 1982), and those of the Tc1 element are 54 bp in length (Rosenzweig et al. 1983). DIRS-1 is also unusual in that the lengths of the left and right ITRs differ by 27 bp; to our knowledge, no other transposon has left and right ITRs of different lengths.

Both the left and right ITRs flanking a single DIRS-1 element frequently contain nucleotide sequence alterations that are not found in other cloned copies of ITRs that we have sequenced; this suggests that the left and right ITRs of a DIRS-1 element may be copied or corrected from a single DNA sequence during transposition (Zuker et al. 1984). The direct repeats of Ty1 and copia elements may be copied or corrected by an analogous process (Eibel et al. 1981).

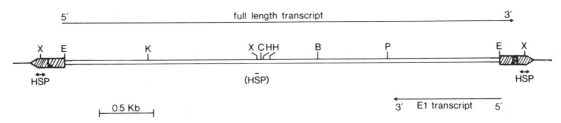

Figure 1. Structure of DIRS-1 and its transcripts. This restriction map is a consensus of six cloned DIRS-1 elements. DIRS-1 is 4.7 kb in length and contains 330-bp ITRs (hatched boxes). The unique internal 4.1-kb sequence is indicated by the open box. The ITRs contain heat-shock promoters (HSP): The heat-shock promoter in the left ITR is responsible for directing rightward transcription of the full-length 4.5-kb RNA transcript, and the heat-shock promoter in the right ITR directs leftward transcription of the E1 RNA. Arrows above and below the element indicate the extent and direction of these transcripts. An internal heat-shock promoter has been shown to be functional in yeast (J. Cappello, unpubl.). Restriction enzymes: (B) (*Bgl*II); (C) *Cla*I; (E) *Eco*RI; (H) *Hin*dIII; (K) *Kpn*I; (P) *Pvu*II; (X) *Xba*I.

Insertion of DIRS-1

The majority of DIRS-1-related sequences in the genome are not part of an intact 4.7-kbp DIRS-1 element (Chung et al. 1983). Most of these vary in size and pattern of restriction enzyme digestion from the canonical DIRS-1. It is possible that some of these fragments were derived by deletions within DIRS-1. However, recent data suggest that many of these partial DIRS-1 sequences are generated by a novel preference for insertion sites exhibited by the transposition of DIRS-1: DIRS-1 inserts preferentially into itself, but the target site within DIRS-1 is more or less random (Cappello et al. 1984a). To establish this, the structures of several genomic clones containing intact 4.7-kb DIRS-1 elements were examined. Except for Cpl9-5 (not shown in Fig. 2), all five cloned genomic DIRS-1 elements are flanked by DIRS-1 internal sequences (Fig. 2). Cloned cDNA pCCA5 is a partial transcript of an apparent DIRS-1-into-DIRS-1 insertion (Cappello et al. 1984a). The 50–100 nucleotides immediately flanking the terminal repeats of each cloned DIRS-1 element (except one) are 80–100% homologous to internal sequences of DIRS-1. By using blot hybridization, Zuker (1983) showed that, in several cases, the homology of the flanking regions of intact DIRS-1 elements was longer than these immediate flanking nucleotides (Zuker 1983).

In one clone, SB41, the DIRS-1-related flanking sequences are those that would result from a precise DIRS-1-into-DIRS-1 insertion: The EcoRI fragments flanking the intact DIRS-1 element in SB41 equal the total 4.1-kb internal EcoRI fragment of DIRS-1 (Cappello et al. 1984a).

At least for the DIRS-1-into-DIRS-1 insertion in SB41, there is no duplication of the target nucleotide sequence at either side of the inserted DIRS-1 element. Flanking sequences on both sides of the DIRS-1 element can be rejoined and aligned perfectly with the internal sequence of DIRS-1. The alignment accounts for every nucleotide present in the DIRS-1 sequence before insertion except one. This possible deletion of a T residue could be explained merely by an ambiguity in the definition of the final residue of the terminal repeat sequences, which are also T residues. The transposable elements of Tc1 from *C. elegans* and Tn554 from *Staphylococcus aureus* also transpose without an apparent duplication of host sequences at the insertion site (Rosenzweig et al. 1983; Murphy and Lofdahl 1984).

The locations of the flanking-region homologies, which define the proposed sites of insertion in each of our clones, are distributed throughout DIRS-1 (Fig. 3). This result indicates that within the DIRS-1 sequence, there is no absolute preference for insertion sites. In addition, because the extent of the DIRS-1 sequence in each of the flanking regions is variable, ranging from a full uninterrupted complement of DIRS-1 sequences in SB41 to probably no more than a few hundred bases on either side of the intact DIRS-1 element in another genomic clone, CP19-1, it is possible that a preexisting DIRS-1-related sequence need not be a complete DIRS-1 in order to function as a target for insertion by a second DIRS-1.

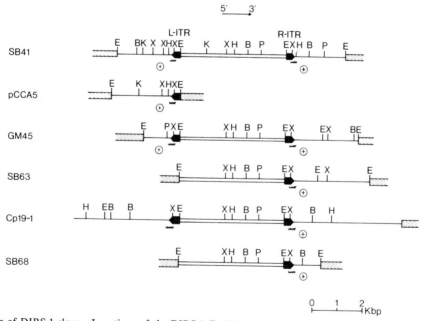

Figure 2. Structure of DIRS-1 clones. Locations of the DIRS-1 flanking sequences homologous to DIRS-1 are shown. Boxed regions of each clone are the intact DIRS-1 elements. L-ITR and R-ITR refer to the positions of the ITRs at both ends of DIRS-1. Wavy lines below the restriction maps denote the positions of the flanking nucleotides we have sequenced that are homologous to internal DIRS-1 sequences. Additional segments of flanking regions that have been shown to be homologous to DIRS-1 by hybridization are indicated by a plus sign below the maps.

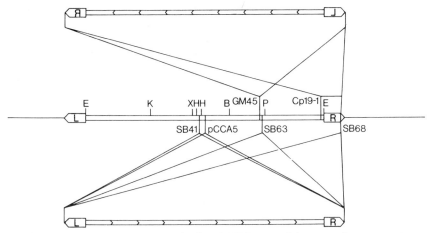

Figure 3. Model for the generation of sequences present in cloned DIRS-1 containing genomic segments. Intact DIRS-1 elements are flanked by internal DIRS-1 sequences. We interpret this to be the result of an insertion of an intact DIRS-1 element into a second DIRS-1 or DIRS-1-related sequence. This is illustrated for five cloned genomic DIRS-1 sequences and for the cDNA clone pCCA5. The position along the DIRS-1 element denoted by the homology present in the flanking sequences is the putative point of insertion of the incoming element. In this model, the preexisting target element (the center element of the diagram) generates the DIRS-1-related flanking sequences on both sides of the incoming element (those elements depicted above and below). As a generalization, the preexisting target sequence is depicted as an intact element. However, in at least two cases (GM45 and Cp19-1), the DIRS-1-related sequences flanking the intact element are not those expected from insertion at a single point in an intact DIRS-1 element. Only cloned SB41 is known to contain both parts of the intact 4.1-kb EcoRI fragment of DIRS-1 flanking the central element. The insertion events associated with clones GM45 and Cp19-1 are depicted as inversions; i.e., the incoming element is inverted with respect to the DIRS-1-related sequence at the target site. Those events depicted for SB41, pCCA5, SB63, and SB68 are depicted as colinear insertions.

The orientation of the incoming element can be either colinear or inverted with respect to the preexisting DIRS-1-related sequence. The DIRS-1-related flanking sequences in four cases are in the same orientation as the central complete DIRS-1 element. In the remaining two cases, they are inverted with respect to the central complete element.

Several hypotheses can explain the preference of DIRS-1 for self-insertion. The simplest involves the recognition of internal DIRS-1 sequences as insertional "hot spots." Since the exact location of insertion sites within a DIRS-1 sequence can vary, a level of recognition more complex than the actual nucleotide sequence must be involved. The secondary structure of an integrated element, possibly in conjunction with specifically bound proteins, might serve as a recognition signal. Alternatively, it is possible that a specific nucleotide sequence in DIRS-1 is initially recognized by the incoming element; this could define a region into which the incoming element might insert that could span several kilobases of target DIRS-1 sequence. A possible analogy is provided by the KB restriction system of *Escherichia coli*. The K and B restriction nucleases recognize a specific target sequence within the DNA but, after binding to the DNA, move a random distance along it before cleaving (Yvan 1981).

DIRS-1 Transcription

A rather heterogeneous set of polyadenylated transcripts of DIRS-1-related sequences accumulate in response to a variety of cellular stresses (Fig. 4) (Zuker et al. 1983). DIRS-1 transcripts are almost undetectable in low-density exponentially growing cells. Their abundance increases very rapidly after the cells are plated for development. The trigger for this is the high cell density required for differentiation (Zuker et al.

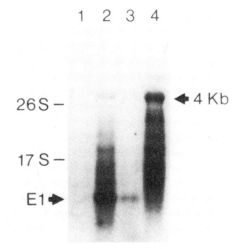

Figure 4. Developmental and heat-shock regulation of DIRS-1 RNAs. RNA filter hybridization of *Dictyostelium discoideum* RNAs with labeled DIRS-1 DNA. Poly(A)⁺ RNA (4 μg) isolated from vegetative cells (lane *1*), heat-shock-treated vegetative cells (lane *2*), and 1-hr (lane *3*) and 15-hr (lane *4*) filter-developed cells were size-fractionated on a 1% agarose-formaldehyde gel and transferred to Gene Screen. The filter was hybridized for 48 hr at 65°C with 10⁶ cpm/ml of nick-translated DNA from clone pB41.6. Positions of the ribosomal RNAs are indicated as 26S and 17S. Positions of the ~4.5-kb and E1 RNAs are indicated.

1983). Growing cells express DIRS-1 RNAs when subjected to heat shock or when grown to very high densities. Individual species of DIRS-1 RNAs that are induced very early in development (0.5–2.5 hr after plating) are also induced by heat shock (Zuker et al. 1983). During both differentiation and heat shock, the increase in the level of DIRS-1 RNAs is due, in part, to a four- and fivefold increase in the rate of RNA transcription (Zuker 1983). The maximum accumulation of DIRS-1 RNA can be achieved by any single stress; it is not enhanced by the application of a second stress condition (i.e., development and heat shock together).

Transcription of DIRS-1 and related sequences generates two predominant discrete polyadenylated RNA species: a "full-length" 4.5-kb transcript and E1 (Zuker et al. 1983). As depicted in Figure 1, E1 is transcribed from right to left on the "standard" DIRS-1 (Cohen et al. 1984; Zuker et al. 1984). In contrast, the majority of DIRS-1 RNAs, including the 4.5-kb transcript, are transcribed from the opposite strand, from left to right (Fig. 1).

E1 is unusual in several respects. It is transcribed exclusively during early development and, unlike the majority of DIRS-1 RNAs, reaches its maximal accumulation by 1 hour of development and is almost undetectable by 2 hours of development (Zuker et al. 1983). In contrast to other DIRS-1 RNAs, E1 is transcribed from the opposite strand, initiating near the right terminal repeat and extending leftward into DIRS-1 (Cohen et al. 1984), and transcription of E1 is induced to a much greater extent by heat shock (Fig. 4). The reasons for the differences in the regulation of E1 and the other DIRS-1 RNAs are not known. In particular, we do not know whether E1 is transcribed from complete copies of DIRS-1 or from some of the large number of partial copies present in the genome. As noted above, other heterogeneously sized DIRS-1-related RNAs may also be transcripts of partial copies of DIRS-1. However, we cannot exclude the possibility that some of these may be processed from the full-length 4.5-kb transcript. Most cytoplasmic DIRS-1 RNAs are polyadenylated and are associated with polysomes, suggesting that they may encode protein(s) (Zuker 1983). However, the longest reading frame in E1 is only 294 bases, suggesting that E1 is not translated (see below).

As is the case for many developmentally induced Dictyostelium mRNAs (Mangiarotti et al. 1982), the expression of DIRS-1 RNAs is regulated at the levels of RNA transcription and also at the stability of the transcripts. DIRS-1 transcripts are in very low abundance in vegetative cells grown at low density, but they appear immediately upon plating at high density for development and continue to accumulate throughout differentiation. DIRS-1 transcription rates, determined by in vitro runoff transcription in isolated nuclei, indicate that DIRS-1 transcription increases four- and fivefold both in early developing cells and in heat-shocked cells (Zuker 1983). Some DIRS-1 transcripts are synthesized in vitro by nuclei isolated from vegetative cells; however, proportionally much less DIRS-1 RNA is found in the polyadenylated nuclear or cytoplasmic RNA isolated from growing cells. Thus, DIRS-1 RNAs transcribed in growing cells may be rapidly degraded. Alternatively, the apparent transcription of DIRS-1 RNA in nuclei from growing cells may reflect an artifactual induction of DIRS-1 transcription caused by stress to the cells during the preparation of nuclei.

The initial burst of DIRS-1 transcription at about 3 hours of development is followed by a drop to the approximate rate observed in vegetative cells; however, cytoplasmic DIRS-1 RNAs continue to accumulate to increasing levels until about 15 hours of development. This accumulation is due, in large part, to an increase in the processing or stability of DIRS-1 transcripts.

Processing and nuclear transit times of DIRS-1 transcripts in developing cells was assayed by continuous labeling of cells with [^{32}P]PO$_4$, followed by measuring the accumulation of DIRS-1-related poly(A)$^+$ RNA and poly(A)$^-$ RNA in the nucleus and the cytoplasm. Results indicate that polyadenylation of DIRS-1 RNA occurs within 45 minutes after synthesis, as is the case with other Dictyostelium RNAs (Mangiarotti et al. 1983); however, DIRS-1 RNAs begin to accumulate in the cytoplasm only after a lag of 135 minutes. This nuclear transit time is much longer than that of constitutively expressed mRNAs or most other developmentally induced Dictyostelium RNAs, which range from 25 to 60 minutes and from 50 to 100 minutes, respectively. DIRS-1 RNAs are also unusual in that only 25% of the nuclear transcripts appear in the cytoplasm. The remaining 75% are apparently degraded within the nucleus or immediately after export to the cytoplasm (Mangiarotti et al. 1983). The processing efficiency of most other Dictyostelium mRNAs ranges from 60% to 70%, although some RNAs unrelated to DIRS-1 are processed at about 20% efficiency.

Transcription and processing of DIRS-1 RNAs has several features in common with those of the Drosophila copia-like transposable elements (Stanfield and Lengyel 1980; Falkenthal et al. 1981). Of copia transcripts, 90% are also rapidly degraded in the nucleus. The 10% of the poly(A)$^+$ copia RNA that is exported to the cytoplasm is stable. In both cases, the reasons for the low efficiency of processing and transport of nuclear RNAs are not known.

Sites of Transcription Initiation

Using S1-nuclease mapping and primer extension, we have mapped several sites within DIRS-1 where transcription initiation occurs. These are located at or near the junctions of both ITRs with the internal sequence (Cohen et al. 1984).

Transcription of the majority of the DIRS-1-related RNAs is directed inward (rightward) from the left repeat. Transcription initiates predominantly at a site 9 bases upstream of the center of the EcoRI site that defines the end of the terminal repeat (Cohen et al. 1984). This initiation site is located within a short oligo(dT) sequence. An oligo(dT) sequence preceding transcriptional initiation sites is a characteristic feature of *Dictyostelium* protein-coding genes and may be required for precise initiation (Kimmel and Firtel 1983). A sequence identical to the *Dictyostelium* consensus TATA box sequence begins at position −27 (Fig. 5).

Initiation of transcription of E1 RNA is more heterogeneous than that of the 4.5-kb RNA: E1 transcripts initiate over a range of 300 bases (Cohen et al. 1984). The longest E1 RNAs initiate near the boundary of the right ITR and the internal DIRS-1 sequence. This position in the right ITR is analogous to that of the initiation of the 4.5-kb transcript in the left ITR and may be directed by a TATA box located at position −42 within the right terminal repeat. The sequence of the first 120 nucleotides downstream (leftward) from the *Eco*RI restriction site defining the right ITR is unusual in that it contains a 100-base stretch consisting almost entirely of adenine and thymine residues. Consequently, other perfect matches for the consensus TATA box are located within this sequence.

The locations of the sites for initiation of transcription thus support the idea that DIRS-1 contains within its terminal repeats promoters or enhancers for RNA synthesis. All ITRs contain a 25-bp sequence that possesses two tandem and near-perfect matches with the consensus *Drosophila* heat-shock promoter sequence (Cohen et al. 1984; Zuker et al. 1984) (see below). We presume that this putative heat-shock promoter sequence in DIRS-1 is responsible for regulating DIRS-1 transcription during both heat shock and development in *Dictyostelium*. This suggestion is supported by the finding that the same transcription initiation sites are used during both heat shock and developmental induction of DIRS-1 transcription (Cohen et al. 1984).

DIRS-1 Transcription in Yeast

Although the major initiation sites of DIRS-1 transcription are located an appropriate distance downstream from the putative promoters contained in the terminal repeats, the actual role of these sequences in the transcription of DIRS-1 is difficult to assess in *Dictyostelium*. For this reason, we transformed *Saccharomyces cerevisiae* (yeast) cells with a plasmid containing an isolated terminal repeat (Cappello et al. 1984b). This

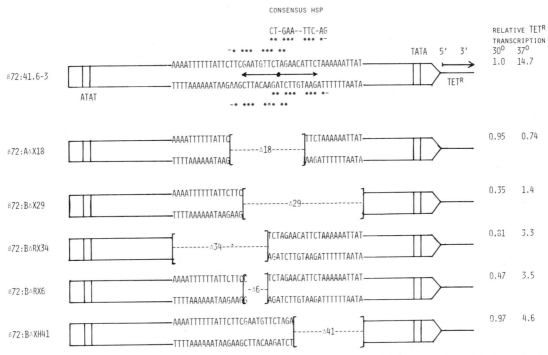

Figure 5. Organization of the DIRS-1 terminal repeat. A schematic representation of the organization of the heat-shock promoter sequences and TATA box sequences contained in the terminal repeats is presented. The consensus sequence of the region of the terminal repeat containing the *Dictyostelium* heat-shock promoters is shown. Asterisks indicate homology with the *Drosophila* consensus heat-shock promoter (HSP) "CT-GAA--TTC-AG." Dash lines indicate mismatched nucleotides. Two tandem overlapping copies of the heat-shock promoter spanning the 18-bp palindrome (indicated by divergent arrows) are present on each strand of the terminal repeat. The positions of TATA boxes at each end of the repeat are indicated. Previously, we showed that in yeast, the right ITR of clone SB41 was both capable of and sufficient for directing the heat-shock inducible transcription of plasmid flanking sequences, in this case, the tetracycline resistance gene of pBR322. To identify the sequences responsible for this regulation, we generated a series of deletions of the ITR. By S1-nuclease protection and Northern blot analysis, we assayed their ability to regulate transcription of the Tet^R gene in yeast (J. Cappello, unpubl.). The wild-type DIRS-1 ITR (β72:41.6-3) and five of the deletion constructs are shown. The relative accumulation of transcripts with respect to the wild-type ITR 30°C level was measured both at 30°C and after a 1-hr heat shock at 37°C. HIS3 RNA, which remains constant throughout the 1-hr heat shock (J. Cappello, unpubl.), was used as an internal control for the amount of total RNA per sample.

ITR induced transcription of flanking DNA sequences; as in *Dictyostelium,* transcription increased severalfold after heat shock. An essential component of the DIRS-1 heat-shock promoter, identified by deletion mapping, is the 18-bp palindromic sequence (Fig. 5) located 120 bp upstream of the initiation sites of transcription. Each half of this palindrome contains an apparent functional heat-shock promoter, as judged from the similarity of 9 of 10 and 10 of 10 bases with the consensus *Drosophila* heat-shock promoter (Fig. 5) (Pelham 1982; Pelham and Bienz 1982; Zuker et al. 1983; Cohen et al. 1984). Deletion of the entire palindrome (mutant AΔX18 and BΔX29) abolished heat-shock-inducible promoter activity. Deletions that remove either half of the palindrome, leaving intact only a single copy of the consensus heat-shock promoter, achieved about one third of the normal heat-shock induction (Fig. 5; mutants BΔRX34 and BΔXH41). These results establish that the 18-base palindrome containing two near-perfect matches with the *Drosophila* consensus heat-shock promoter is an upstream activator transcription that responds to heat shock. These data suggest that both halves of the palindrome function additively to produce full heat-shock induction. Whether this cooperativity is due to the structural symmetry of the palindrome or to the divalence of the promoter sequence is not clear.

Although both terminal repeats contain functional heat-shock promoters, transcription of the intact DIRS-1 element in yeast occurs almost exclusively from the left promoter (J. Cappello, unpubl.). As in *Dictyostelium*, a transcript of 4.5 kb is produced that spans most of the element. In contrast to *Dictyostelium*, very little E1-like RNA (transcription inward from the right terminal repeat promoter) is evident. Clearly, the transcripton of DIRS-1 is complex and involves more than merely the function of the heat-shock promoters contained in both left and right ITRs. Although a full understanding of DIRS-1 transcription will eventually arise only from its study in *Dictyostelium*, the functional data acquired from the mapping of these promoters in yeast cells have provided the first definition of a *Dictyostelium* promoter sequence.

Nucleotide Sequence of DIRS-1

As mentioned above, DIRS-1 RNAs are polyadenylated and associated with polysomes (Zuker 1983). These observations are consistent with the notion that at least some DIRS-1 transcripts are mRNAs that encode polypeptides. These protein products might function in the transposition of DIRS-1. To characterize these products, we have determined the complete nucleotide sequence of DIRS-1.

The unique internal sequence of DIRS-1 (between the *Eco*RI sites) is 4158 bp long and contains three long open reading frames (Fig. 6, ORF1, 2, and 3) (Cappello et al. 1985). ORF1 extends through the first 1150 bases of DIRS-1 and could encode a polypeptide of 36,000 daltons. ORF2 overlaps the final 170 nucleotides of ORF1 and extends for 2800 bases. Using the first ATG codon of ORF2 for initiation, a translation product of over 100,000 daltons is predicted.

The most surprising and remarkable feature of the sequence of DIRS-1 is ORF3. This reading frame also overlaps the final 55 nucleotides of ORF1, extends for 2000 bases, and could encode a polypeptide of about 69,000 daltons. Consequently, ORF3 completely and co-directionally overlaps ORF2. For more than 2000 bases, DIRS-1 potentially encodes two different polypeptides simultaneously using one nucleotide sequence.

The polarity of all three open reading frames of DIRS-1 is such that they could be encoded by the 4.5-kb RNA. No open reading frames longer than 100 amino acids are encoded on the opposite strand; thus, it is doubtful that the DIRS-1 transcript E1 encodes a protein. Because ORF1 initiates at the first AUG codon from the 5' end of the 4.5-kb transcript, initiation of its translation is straightforward. However, the utilization of ORF2 and ORF3 would require either the use of an internal AUG codon or the elimination of the 5' region of the 4.5-kb RNA, including most of ORF1. This could be accomplished by RNA splicing or by transcriptional initiation within DIRS-1. Evidence for or against both of these possibilities is under continued investigation.

The occurrence of the two long overlapping reading frames is totally unexpected. All three long overlapping open reading frames are found in a second genomic copy of DIRS-1 that we have sequenced (Cappello et al. 1985). The reason for the conservation of the coding capabilities for two very different polypeptides in such a small space is of great interest. Conservation of these unusual DIRS-1 open reading frame sequences in *Dictyostelium* might indicate that DIRS-1 proteins are important for the transposition and/or maintenance of DIRS-1 sequences in the *Dictyostelium* genome.

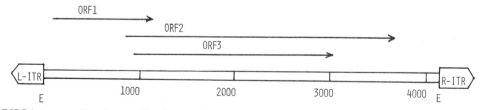

Figure 6. DIRS-1 open reading frames. The diagram depicts the nucleotide sequence of DIRS-1 and the open reading frames longer than 100 amino acids encoded. The large boxes at each end denote the positions of the left and right inverted terminal repeats (L-ITR and R-ITR). The numbering begins at the *Eco*RI site (E), which marks the boundary of the L-ITR and the unique internal region. ORF1 begins with an ATG, and ORFs 2 and 3 do not.

ORF3 Contains Homology with Reverse Transcriptase

The accumulation of nucleotide sequences from the vertebrate retroviruses, animal DNA viruses, and some plant RNA and DNA viruses has revealed that many of these encode an RNA-dependent polymerase involved in their replication (Kamer and Argos 1984). A comparison of these "*pol*" gene products has shown that a region of about 200 amino acids in these polypeptides can be aligned to display a significant homology (Toh et al. 1983). At the core of this homology region are ten amino acid positions invariant in all known reverse transcriptase genes. Recently, a reverse transcriptase activity has been implicated in the transposition of the yeast transposon, Ty*1* (Boeke et al. 1985). The nucleotide sequence of Ty*1*, as well as the sequences of two *Drosophila* transposons, *copia* and the *copia*-like element, 17.6, has a similar homology with reverse transcriptase (Saigo et al. 1984; Clare and Farabaugh 1985; Mount and Rubin 1985).

A comparison of the DIRS-1-encoded polypeptides with the amino acid sequence deduced for the *pol* genes of Rous sarcoma virus (RSV) (Schwartz et al. 1983) and Moloney murine leukemia virus (Mo-MLV) (Shinnick et al. 1981) revealed that ORF3 is homologous to this conserved reverse transcriptase region. A dot matrix analysis of the amino acid sequence of ORF3 compared with that of Mo-MLV *pol* gene (data not shown) shows that a region of 200 amino acids of ORF3 is nearly as homologous to the retroviral reverse transcriptase genes as the two retroviral *pol* genes are to themselves. The location of the reverse transcriptase homology region in ORF3 is similar to that in the retroviral *pol* genes; it begins 103 amino acids from the 5' end of ORF3 as compared with 86 and 222 amino acids from the 5' ends of the *pol* genes of RSV and Mo-MLV, respectively.

A region of ORF3 is aligned in Figure 7 with the amino acid sequence of the Mo-MLV reverse transcriptase. Also shown is the sequence of the RNA-depen-

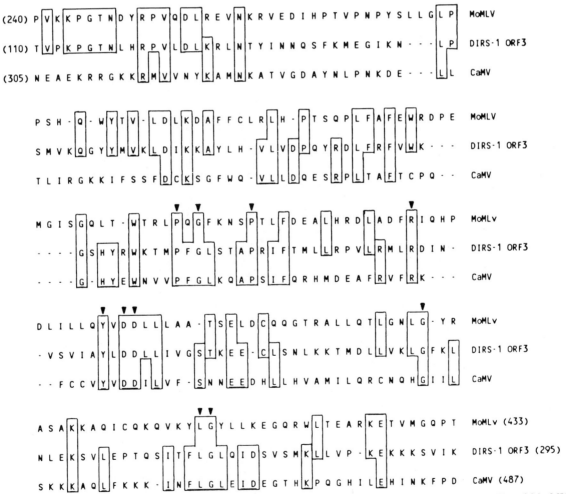

Figure 7. ORF3 homology with reverse transcriptase. The 200 amino acid regions of homology between ORF3 and Mo-MLV are aligned with the sequence of the RNA-dependent polymerase of CaMV. Amino acid positions for the first and last residues displayed are given for each sequence. Amino acid positions identical with the DIRS-1 ORF3 sequence are boxed. Toh et al. (1983) have shown that all known reverse transcriptase sequences contain ten invariant amino acid positions. These positions are denoted by arrowheads above the sequence.

dent polymerase encoded by cauliflower mosaic virus (CaMV) (Gardner et al. 1981). The DIRS-1 ORF3 product is identical at 56 out of 240 amino acid positions (25%) with both Mo-MLV and CaMV *pol* proteins. The ORF3 product contains all ten of the invariant reverse transcriptase residues. Many of the nonidentical positions have conservative amino acid substitutes. On the basis of this homology, we classify DIRS-1 as a possible "retrotransposon."

If the ORF3-encoded polypeptide is a reverse-transcriptase-like enzyme involved in the replication of DIRS-1, an interesting question arises concerning the mechanism by which a transposon with nonidentical ITRs can regenerate these sequences using the information encoded in the nearly genomic-size 4.5-kb DIRS-1 RNA. We hope to be able to answer this question by determining the functions of the DIRS-1-encoded polypeptides and by assaying the transposition of DIRS-1 by reintroducing modified DIRS-1 elements into yeast and *Dictyostelium* by transformation.

By using antibodies prepared against the hypothetical polypeptides encoded by ORF1, ORF2, and ORF3, we are trying to detect these proteins in *Dictyostelium*, as well as in DIRS-1-transfected yeast. A functional assay for DNA transposition developed in yeast containing DIRS-1 sequences will facilitate the study of the DIRS-1 translation products and their functions.

REFERENCES

BOEKE, J.D., D.J. GARFINKLE, C.A. STYLES, and G.R. FINK. 1985. Ty elements transpose through an RNA intermediate. *Cell* **40:** 491.

CAPPELLO, J., S.M. COHEN, and H.F. LODISH. 1984a. *Dictyostelium* transposable element DIRS-1 preferentially inserts into DIRS-1 sequences. *Mol. Cell. Biol.* **4:** 2207.

CAPELLO, J., K. HANDELSMAN, and H.F. LODISH. 1985. Sequence of *Dictyostelium* DIRS-1: An apparent retrotransposon with inverted terminal repeats and an internal circle junction sequence. *Cell* (in press).

CAPPELLO, J., C. ZUKER, and H.F. LODISH. 1984b. Repetitive *Dictyostelium* heat-shock promoter functions in *Saccharomyces cerevisiae*. *Mol. Cell. Biol.* **4:** 591.

CHUNG, S., C. ZUKER, and H.F. LODISH. 1983. A repetitive and apparently transposable DNA sequence in *Dictyostelium discoideum* associated with developmentally regulated RNAs. *Nucleic Acids Res.* **11:** 4835.

CLARE, J. and P.J. FARABAUGH. 1985. Nucleotide sequence of a Ty*1* element: Evidence for a novel mechanism of gene expression. *Proc. Natl. Acad. Sci.* **82:** 2829.

COHEN, S.M., J. CAPPELLO, and H.F. LODISH. 1984. Transcription of *Dictyostelium discoideum* transposable element DIRS-1. *Mol. Cell. Biol.* **4:** 2332.

DELREY, F.J., T.F. DONAHUE, and G.R. FINK. 1982. Sigma, a repetitive element found adjacent to tRNA genes of yeast. *Proc. Natl. Acad. Sci.* **79:** 4138.

EIBEL, H., J. GAFNER, A. STOTZ, and P. PHILIPPSEN. 1981. Characterization of the yeast mobile element Ty*1*. *Cold Spring Harbor Symp. Quant. Biol.* **45:** 609.

FALKENTHAL, S., M.L. GRAHAM, E.L. KORN, and J.A. LENGYEL. 1981. Transcription, processing and turnover from the *Drosophila* mobile genetic element copia. *Dev. Biol.* **92:** 294.

FEDOROFF, N.V. 1983. Controlling elements in cell lineages. In *Mobile genetic elements* (ed. J.A. Shapiro), p. 1. Academic Press, New York.

FLAVELL, R.B., M. O'DELL, and J. HUTCHINSON. 1981. Nucleotide sequence organization in plant chromosomes and evidence for sequence translocation during evolution. *Cold Spring Harbor Symp. Quant. Biol.* **45:** 501.

GARDNER, R.C., A.J. HOWARTH, P. HAHN, M. BROWN-LEUDI, R.J. SHEPHERD, and J. MESSING. 1981. The complete nucleotide sequence of an infectious clone of cauliflower mosaic virus by M13 mp7 shotgun sequencing. *Nucleic Acids Res.* **9:** 2871.

KAMER, G. and P. ARGOS. 1984. Primary structural comparison of RNA-dependent polymerases from plant, animal and bacterial viruses. *Nucleic Acids Res.* **12:** 7269.

KIMMEL, A.R. and R.A. FIRTEL. 1983. Sequence organization in *Dictyostelium*: Unique structure at the 5' ends of protein coding genes. *Nucleic Acids Res.* **11:** 541.

KLECKNER, N. 1981. Transposable elements in prokaryotes. *Annu. Rev. Genet.* **15:** 341.

MAJORS, J.E., R. SWANSTROM, W.J. DELORRE, G.S. PAYNE, S.H. HUGHES, S. ORTIZ, N. QUINTRELL, J.M. BISHOP, and H.E. VARMUS. 1981. DNA intermediates in the replication of retroviruses are structurally (and perhaps functionally) related to transposable elements. *Cold Spring Harbor Symp. Quant. Biol.* **45:** 731.

MANGIAROTTI, G., P. LEFEBVRE, and H.F. LODISH. 1982. Differences in the stability of developmentally regulated mRNAs in aggregated and disaggregated *Dictyostelium discoideum* cells. *Dev. Biol.* **89:** 82.

MANGIAROTTI, G., C. ZUKER, R. CHISHOLM, and H.F. LODISH. 1983. Different mRNAs have different nuclear transit times in *Dictyostelium discoideum* aggregates. *Mol. Cell. Biol.* **3:** 1151.

MOUNT, S. and G.M. RUBIN. 1985. Complete nucleotide sequence of the *Drosophila* transposable element copia; homology between copia and retroviral proteins. *Mol. Cell. Biol.* **5:** 1330.

MURPHY, E. and S. LOFDAHL. 1984. Transposition of Tn554 does not generate a target duplication. *Nature* **307:** 292.

O'HARE, K. and G.M. RUBIN. 1983. Structures of P transposable elements and their sites of insertion and excision in the *Drosophila melanogaster* genome. *Cell* **34:** 25.

PELHAM, H.R.B. 1982. A regulatory upstream promoter element in the *Drosophila* HSP 70 heat shock gene. *Cell* **30:** 517.

PELHAM, H.R.B. and M. BIENZ. 1982. A synthetic heat-shock promoter confers heat inducibility on the herpes simplex thymidine kinase gene. *EMBO J.* **1:** 1473.

POTTER, S.S. 1982. DNA sequence of a foldback transposable element in *Drosophila*. *Nature* **297:** 201.

ROEDER, G.S. and G.R. FINK. 1983. Transposable elements in yeast. In *Mobile genetic elements* (ed. J.A. Shapiro), p. 250. Academic Press, New York.

ROSEN, E., A. SIVERTSEN, and R.A. FIRTEL. 1983. An unusual transposon encoding heat shock inducible and developmentally regulated transcripts in *Dictyostelium*. *Cell* **35:** 243.

ROSENZWEIG, B., L.W. LIAO, and D. HIRSH. 1983. Target sequences for the *C. elegans* transposable element Tc*1*. *Nucleic Acids Res.* **11:** 7137.

SAIGO, K., W. KUGIMIYA, Y. MATSUO, S. INOYE, K. YOSHIOKA, and S. YUKI. 1984. Identification of the coding sequence for a reverse transcriptase-like enzyme in a transposable genetic element in *Drosophila melanogaster*. *Nature* **312:** 659.

SCHWARTZ, D., R. TIZARD, and W. GILBERT. 1983. Nucleotide sequence of Rous sarcoma virus. *Cell* **32:** 853.

SHINNICK, T.M., R.A. LERNER, and J.G. SUTCLIFFE. 1981. Nucleotide sequence of Moloney murine leukemic virus. *Nature* **293:** 543.

SPRADLING, A.C. and G.M. RUBIN. 1981. *Drosophila* genome organization: Conserved and dynamic aspects. *Annu. Rev. Genet.* **15:** 219.

STANFIELD, S.W. and J.A. LENGYEL. 1980. Small circular deoxyribonucleic acid of *Drosophila melanogaster*: Homologous transcripts in the nucleus and cytoplasm. *Biochemistry* **19:** 3873.

TOH, H., H. HAYASHIDA, and T. MIYATA. 1983. Sequence homology between retroviral reverse transcription and putative polymerases of hepatitis B virus and cauliflower mosaic virus. *Nature* **305:** 827.

YVAN, R. 1981. Structure and mechanism of multifunctional restriction endonucleases. *Annu. Rev. Biochem.* **50:** 285.

ZUKER, C. 1983. "Repetitive developmentally regulated genes in *Dictyostelium*." Ph.D. thesis, Massachusetts Institute of Technology, Cambridge.

ZUKER, C. and H.F. LODISH. 1981. Repetitive sequences cotranscribed with developmentally regulated *Dictyostelium discoideum* mRNAs. *Proc. Natl. Acad. Sci.* **78:** 5386.

ZUKER, C., J. CAPPELLO, R.L. CHISOLM, and H.F. LODISH. 1983. A repetitive *Dictyostelium* gene family that is induced during differentiation and by heat shock. *Cell* **34:** 997.

ZUKER, C., J. CAPPELLO, H.F. LODISH, P. GEORGE, and S. CHUNG. 1984. *Dictyostelium* transposable element DIRS-1 has 350-base-pair inverted terminal repeats that contain a heat shock promoter. *Proc. Natl. Acad. Sci.* **81:** 2660.

Regulation of Cell-type-specific Differentiation in *Dictyostelium*

W.F. LOOMIS
Department of Biology, University of California, San Diego, La Jolla, California 92093

Development of *Dictyostelium* has attracted attention for many years because large numbers of cells synchronously proceed through a series of morphological stages to produce fruiting bodies consisting of two quite different cell types: spores and stalk cells (Loomis 1975). Exponentially growing amebae are induced to develop by removal of exogenous nutrients. This triggers the expression of several dozen genes that were not active during growth. Several of these genes are responsible for the components of a chemotactic signaling system that is used several hours later for aggregation of cells into mounds containing up to 10^5 cells. An extracellular surface sheath is produced that integrates the cells within each aggregate such that subsequent differentiation is synchronized and internally regulated. Aggregates take on elongated slug shapes and proceed to develop as independent organisms. Under standardized laboratory conditions, slugs will proceed through subsequent stages in lock-step, allowing molecular analyses of large populations to reflect the morphological steps. The first signs of divergence of cell types occur just as aggregation is completed. At this time, a new set of genes is activated in some, but not all, cells. These cells will differentiate into spores several hours later and so are referred to as prespore cells. In migrating slugs, they are found in the posterior portions but not in the anterior quarter. Thus, expression of developmental genes is subject to both temporal and spatial regulation in *Dictyostelium* (Loomis 1982).

Varying the conditions under which cells are allowed to develop shows that progression through the temporal stages is dependent on both internal and external signals. These characteristics are reminiscent of processes that occur during embryogenesis of metazoan organisms. Resolution of the mechanism of cell-type regulation in *Dictyostelium* will have bearing on similar processes in embryos.

RESULTS

Temporal Control

The pattern of synthesis of several hundred individual proteins has been analyzed by pulse labeling with [^{35}S]methionine at hourly intervals throughout development of *Dictyostelium discoideum*. The proteins were distinguished by two-dimensional gel electrophoresis. These studies have been carried out in several independent laboratories over the last few years (Alton and Lodish 1977; Alton and Brenner 1979; Coloma and Lodish 1981; Morrissey et al. 1981, 1984; Ratner and Borth 1983; Cardelli et al. 1985; Finney et al. 1985). When the results of these studies are analyzed together, it can be seen that 150–200 new proteins are rapidly synthesized throughout development (Fig. 1). There are bursts of activity (1) immediately following the induction of development ($T = 0$ hr), (2) when chemotactic signaling is maximal ($T = 8$ hr), (3) as the slug is integrated ($T = 12$ hr), and (4) at culmination ($T = 22$ hr). However, proteins begin to be synthesized at other times, which suggests that there are at least a dozen stages of biochemical differentiation during development of *Dictyostelium*. Detailed studies on specific enzymes, lectins, and membrane proteins have shown that the cells are continuously modifying their biochemical composition and their physiological functions (Loomis 1975; Siu et al. 1977; West and MacMahon 1979; Barondes et al. 1982). A similar view of differentiation in *D. discoideum* comes from analysis of newly transcribed sequences either by global hybridization techniques or by detailed studies on cloned cDNAs (Firtel 1972; Kimmel and Firtel 1982; Lodish et al. 1982; Barklis and Lodish 1983).

The external signal that triggers expression of the early developmental genes is a combination of nutrient exhaustion and high cell density (Sussman and Sussman 1969; Marin 1976; Chisholm et al. 1984). Cells respond to the absence of exogenous amino acids by altering their pattern of macromolecular biosynthesis. Many of the abundant proteins synthesized during growth are no longer synthesized as a consequence of competition for ribosomes and turnover of their mRNA (Cardelli et al. 1985). Moreover, they start to synthesize about two dozen major new proteins. Synthesis of at least some of these proteins is dependent on a critical cell density, which is usually achieved when 3×10^7 to 10^8 cells are deposited on a filter support saturated with a few milliliters of buffer. However, if the cells are suspended at low cell density (5×10^6/ml), specific early proteins are not made (Grabel and Loomis 1978). The requirement for a threshold of cell density can be overcome for an early enzyme (*N*-acetyl glucosaminidase) and a set of cloned sequences expressed at 6 hours of development by incubating the cells in conditioned medium (Grabel and Loomis 1978; Mehdy and Firtel 1985). These results indicate that the cells monitor the environment to ensure that there is a sufficient number of cells for multicellular development before embarking on the program of differentiation.

Starting 2 hours after the initiation of development, the components for chemotaxis accumulate for several hours (Loomis 1979; Devreotes 1982). In *D. discoideum*, cAMP is used as the chemoattractant; adenyl cyclase is initially very low but increases to a peak of activity at 6 hours. During the same period, cAMP phosphodiesterase is synthesized and secreted to keep the cAMP levels in bounds. The phosphodiesterase is regulated, in turn, by a protein inhibitor that is synthesized when exogenous cAMP is low but not when cAMP is added to the environment (Yeh et al. 1978). Thus, cells in dense populations where the cAMP concentration is high have full phosphodiesterase activity so as to lower the background noise. A cAMP-binding protein of 40 kD is also synthesized during this period and inserted into the membrane. Only after these components are in place do the cells become responsive to cAMP (Devreotes 1982).

Eight hours after initiation of development, a fairly large number of new proteins appear (Fig. 1). One of these is a surface glycoprotein of 80,000 daltons, gp80, that appears to play a role in cell-cell adhesion (Muller and Gerisch 1978; Murray et al. 1981; Loomis et al. 1985). It is synthesized in suspensions of axenically grown cells only if pulses of 10^{-9} M cAMP are given every 5 minutes. Due to the relay system of *D. discoideum* in which a pulse of cAMP stimulates a transient increase in adenyl cyclase activity and resultant cAMP secretion, the cells are normally exposed to cAMP in pulses about every 5 minutes. Cells respond to such pulses not only by chemotactic migration, but also by activating a specific set of genes. The gene for gp80 is transcribed into a 1.8-kb mRNA when cells developed for 8 hours are exposed to cAMP pulses (C. Hong and W.F. Loomis, in prep.). The cells then acquire an EDTA-resistant adhesiveness and form tight aggregates.

A few hours later, another group of genes is activated, some of which are expressed only in prespore cells (Alton and Brenner 1979; Morrissey et al. 1981, 1984; Ratner and Borth 1983). About a dozen cloned cDNAs have been studied that recognize mRNAs that first appear at 12 hours of development (Barklis and Lodish 1983; Mehdy et al. 1983). If cells are dissociated from aggregates shortly thereafter, these mRNAs disappear rapidly but reappear in disaggregated cells if cAMP is added to 10^{-4} M. Thus, cAMP appears to play a role in regulating those genes normally expressed in aggregates. However, it is important to be aware that the inducing concentration of cAMP is about 100 times higher than the concentration of cAMP that cells are normally exposed to under developmental conditions (Brenner 1978; Merkle et al. 1984). The measured concentration of cAMP in developed cells of *Dictyostelium* is only 1–5 μM, and the cAMP-binding protein saturates at 1 μM (Devreotes 1982); however, maximal induction of gene expression requires at least 100 μM cAMP (Mehdy and Firtel 1985). It appears that high levels of exogenous cAMP bypass a normally required signal that is provided by multicellularity within an aggregate. It is possible that the high levels of exogenous cAMP are required in order to raise the internal cAMP level to an inducing threshold. During normal development, multicellular conditions would trigger a rise in internal cAMP.

The final burst of new gene expression occurs as cytodifferentiation of spores and stalk cells begins at 22 hours of development (Fig. 1). The decision to undergo terminal differentiation is regulated by the pNH_3 level (Sussman 1982). Throughout development, protein turnover provides much of the energy source and generates ammonia. As the reserves dwindle, the rate of NH_3 release decreases until a threshold is crossed and the final set of genes is activated. Many of these genes are responsible for proteins synthesized exclusively in stalk cells (Coloma and Lodish 1981; Morrissey et al. 1984).

The temporal sequence is accelerated in strains carrying *rde* mutations but cannot be speeded up in wild-type strains (Loomis 1975). Thus, time elapsed since the initiation of development is an essential component of the sequence. There is considerable genetic evidence for a central dependent sequence occurring throughout the course of development of *D. discoideum* in which early stages are essential prerequisites for later stages (Loomis et al. 1976; Blumberg et al. 1982). The temporal sequence is likely to depend on such internal events as the synthesis, accumulation, and membrane insertion of cAMP-binding sites, as well as the external signals of induction, multicellularity, and reduced pNH_3. The cascade of internal and external signals that may regulate the temporal progression are shown schematically in Figure 2.

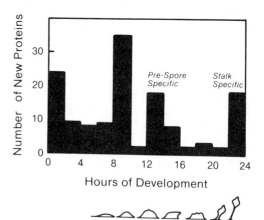

Figure 1. Stage-specific proteins. Individual proteins were recognized on two-dimensional electropheretograms after a 2-hr in vivo pulse-labeling of cells with [^{35}S]methionine at intervals during development. Cell-type-specific proteins were analyzed in cells manually dissected from migrating slugs or separated on Percoll gradients. Counts of new proteins were made from over a dozen determinations at each stage in our laboratory and compared with those found at the same stage in several other laboratories (see text). The stages of morphogenesis are indicated below the time axis.

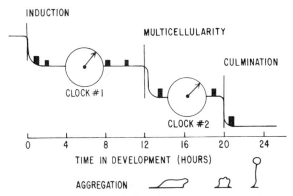

Figure 2. Conceptual model for temporal regulation. The external conditions of induction, multicellularity, and culmination are necessary for continuation beyond the time indicated. Blocks of gene expression (■) are separated by invariant temporal periods, indicating internal timing mechanisms before and after the external multicellular signal.

Cell-type Specificity

Stalk cells are derived from cells in the anterior of slugs, whereas spores are derived from posterior cells (Raper 1940). Prespore and prestalk cells can be separated manually either by microdissection of slugs or by density separation on Percoll gradients (Tsang and Bradbury 1981; Borth and Ratner 1983; Devine and Loomis 1985), and the biochemical composition of the two cell types can then be analyzed. The first difference found was in the activity of UDP galactose polysaccharide transferase, an enzyme involved in the biosynthesis of a spore-specific polysaccharide (Newell et al. 1969). It is localized exclusively in prespore cells in the posterior of slugs. Two-dimensional gel electrophoresis of [^{35}S]methionine-labeled proteins in prespore and prestalk cells uncovered another score of prespore-specific protein (Alton and Brenner 1979; Morrissey et al. 1981, 1984; Ratner and Borth 1983). The accumulation of these proteins was also investigated in the separated cell types by silver staining of two-dimensional gels (Morrissey et al. 1984). Most of the cell-type-specific proteins accumulated as expected from their synthetic periods; however, we found that one protein (PSP 59) was synthesized and accumulated from 8 to 12 hours in all cells and then preferentially repressed in prestalk cells. The other prespore proteins were synthesized and accumulated only in prespore cells starting at 12 hours of development (Fig. 3); synthesis of some ceased at 18–22 hours, whereas synthesis of others continued into spores.

The only abundant prestalk protein (PST 71) was also a case of differential repression, since it was synthesized and accumulated in all cells from 8 to 12 hours and then preferentially repressed in prespore cells. During culmination, 12 stalk-specific proteins were observed (Fig. 3).

cDNA sequences prepared from mRNA from 15- and 22-hour cells have yielded several prestalk-specific clones as well as a dozen prespore-specific clones. Accumulation of these mRNAs occurs predominantly after 12 hours of development (Barklis and Lodish 1983; Mehdy et al. 1983). One of the prestalk-specific clones codes for *Dictyostelium ras* protein (Reymond et al. 1984). Although the *ras* mRNA is present almost exclusively in prestalk cells, the *ras* protein is present in all cells due to expression of the gene during growth (Reymond et al. 1984). This indicates that cell-type-specific localization of a specific mRNA need not imply that the protein product is localized exclusively to that cell type. The proteins coded by the other cloned sequences are not yet known.

Stalky Mutants

Several years ago, we isolated a set of mutations that were located in a single locus, *stkA*, and resulted in differentiation of all cells into stalk cells (Morrissey and Loomis 1981). The mutations appear to affect a fairly late step during culmination, since the pattern of cell-type-specific protein synthesis was not affected at the slug stage; i.e., posterior cells still synthesized prespore-specific proteins. However, at culmination, these cells as well as the anterior cells shifted over to synthesis of stalk-specific proteins and differentiated uniformly into stalk cells (Fig. 3). Morever, C. West (in prep.) recently analyzed a temperature-sensitive mutation in the *stkA* locus and showed that the critical period extends to 18 hours of development, just as culmination starts. These mutants clearly show that accumulation of prespore proteins does not restrict cells to spore differentiation and indicate that a conversion of prespore cells to stalk cells can occur rapidly late in development unless the *stkA* gene is active. The converse, differentiation of prestalk cells into prespore cells, takes a period of several hours for regulation (Raper 1940; Sakai 1973). These results indicate that prespore cells are totipotent, whereas prestalk cells have not undergone the molecular processes necessary to form spores.

Spore Coat Proteins

As the posterior cells move up the elongating stalk during culmination, they release the components of spore coats from prespore vesicles and encapsulate (Fig. 4). When spores germinate, the spore coats are left behind as empty shells that can be easily isolated (Orlowski and Loomis 1979). When reduced and denatured, the proteins of the spore coats can be separated on two-dimensional gels (Fig. 5). Only three major proteins, SP96, SP70, and SP60, are extracted from purified spore coats. They form a disulfide cross-linked complex in the coat itself (Devine et al. 1983). The spore coat proteins migrate as diffuse spots on two-dimensional gels; however, analysis of proteolytic cleavage fragments taken from various positions within the spots indicates no heterogeneity. Moreover, amino acid analyses of gel-separated SP70 and SP60 give a unique sequence at the amino terminus for each spore coat pro-

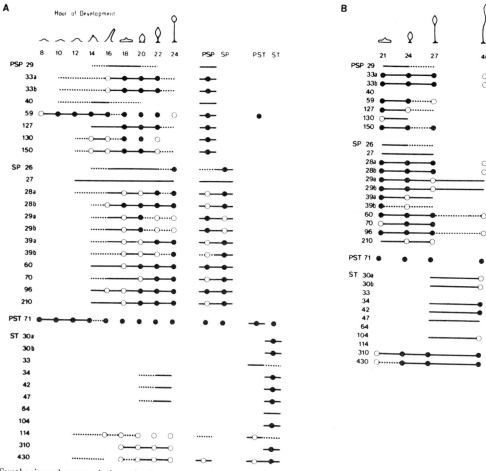

Figure 3. Synthesis and accumulation of cell-type-specific proteins. (*A*) Synthesis of specific proteins recognized on two-dimensional gels are represented by a line during the synthetic period. A dashed line indicates low levels of incorporation of [^{35}S]methionine into that protein. Accumulation was measured by silver staining the gels. (●) Strong staining of the specific protein; (○) weak staining. Samples were taken during synchronous development on filters. Isolated cell types were prepared from migrating slugs (PSP, prespore cells; PST, prestalk cells) or fruiting bodies (SP, spores; ST, stalks). (*B*) An identical analysis was carried out on a strain carrying the *stkA508* mutation in which the proportion of spores to stalks is altered such that only stalk cells are formed (Morrissey et al. 1984).

tein (Table 1) (B. Dowds and W.F. Loomis, in prep.). No sequence of SP96 could be determined, since the amino terminus appears to be blocked. The aminoterminal sequence of SP60 is of particular interest in that the six-amino-acid sequence, Gly-Asp-Trp-Asn-Asn-Asx, occurs at least three times. Such a repeating structure may facilitate self-assembly of the protein. These sequences will be useful in isolating and confirming the identity of cloned DNA fragments coding for SP70 and SP60.

A cloned cDNA derived from SP96 mRNA has been recovered by screening a pBR322 cDNA bank made from mRNA isolated from cells that had developed for 15 hours (Dowds and Loomis 1984). The plasmids were screened by in vitro translation of hybrid-selected mRNA followed by gel electrophoresis of proteins immunoprecipitated by antiserum specific to SP96. The cDNA clone, pSP96, hybridizes to a 2.4-kb mRNA that first appears at 12 hours of development and is specific to prespore cells (Fig. 7). It is not found in prestalk

Table 1. Aminoterminal Amino Acid Sequence of Spore Coat Proteins

SP70: (1) Ile Asn X Asp Gly Leu Ser Lys Asp Gln (10) X Glu Gln Asn Phe Pro Gln Ala Gln Ile Leu (20)

SP60: (1) Gly Asp Trp Asn Asn Asn Gly Asp Trp Asn (10) Asn Gly Asp Trp Asn Asn Asp Gly Asp Trp Asn (20)

X indicates modified amino acid.

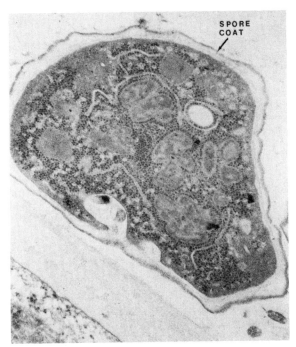

Figure 4. Encapsulating spore. A culminating fruiting body was fixed, sectioned, and observed in an electron microscope. The spore coat is clearly visible on the surface and is still discontinuous at this early stage in encapsulation. The stalk tube is visible at the lower left.

cells and disappears following encapsulation of spores. We are now using this cDNA to screen genomic banks for the complete gene sequence.

Signals Regulating Spore Coat Genes

The genes coding for spore coat proteins SP96, SP70, and SP60 are not expressed until after 12 hours of development, when the cells have aggregated into mounds (Figs. 4 and 6). One possible signal considered as a potential requirement for expression of genes at this stage is the presence of contact sites A (csA) (Chisholm et al. 1984), which is defined as EDTA-resistant cell-cell adhesion (Beug et al. 1970). A glycoprotein of 80,000 daltons, gp80, has been implicated in this adhesion mechanism by immunological techniques (Muller and Gerisch 1978). Monoclonal antibodies reactive with the carbohydrate side groups found on gp80 block adhesion (Murray et al. 1983; Loomis et al. 1985). gp80 accumulates between 4 and 12 hours of development concomitantly with the appearance of csA. We have isolated mutations in the *modB* locus (which maps to linkage group VI) that result in a block to the posttranslational modification of gp80 (Murray et al. 1984; Loomis et al. 1985). These mutations result in cells that fail to develop csA and form fewer spores. The fruiting bodies of *modB* strains are less than half the size of those of wild-type (*modB*⁺) strains. However, the timing of morphogenetic stages is unaffected in *modB* strains. Likewise, accumulation of SP60, SP70, and SP96 as judged from silver-stained material on two-dimensional gels occurred normally in cells lacking csA (W.F. Loomis, unpubl.). These results appear to rule out a requirement for csA in the expression of the spore coat protein genes.

Although *modB* mutant cells lack EDTA-resistant adhesion, they nevertheless aggregate and form mounds while adhering with the weaker EDTA-sensitive adhesion that appears during early development. To determine whether this multicellular condition was necessary for expression of the spore coat genes, cells were kept from aggregating by inducing development under sparse conditions where the cells were separated by distances too great for chemotactic signaling. Cells submerged in phosphate buffer will attach to the bottom of tissue-culture flasks. At a density of 4×10^3 cells/cm² or less, the cells fail to aggregate; at higher densities, they aggregate in a manner similar to that

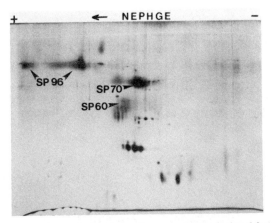

Figure 5. Spore coat proteins. Spore coats were isolated from germinated spores and analyzed by two-dimensional gel electrophoresis. Silver staining shows three major proteins: SP96 (96,000 daltons), SP70 (70,000 daltons), and SP60 (60,000 daltons). The other spots below SP60 are RNase and DNase added to the sample.

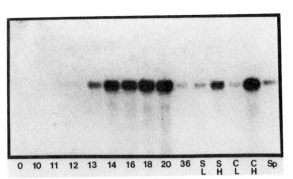

Figure 6. Developmental Northern blots of pSP96. Constant amounts of RNA isolated at various times in development (indicated in hours) were separated on a gel, transferred to nitrocellulose, and hybridized to ³²P-labeled pSP96. Prestalk cells (L) and prespore cells (H) were isolated from slugs (S) and culminants (C). RNAs from these cells as well as from spores were also analyzed by hybridization to pSP96 (Dowds and Loomis 1985).

Table 2. Expression of Spore Coat Proteins in Unaggregated Cells

Condition	Density (cells/cm²)	SP96/SP70
Submerged	4×10^5	+ + +
Submerged	4×10^3	−
Submerged + cAMP	4×10^5	+ + +
Submerged + cAMP	4×10^3	+ +

cAMP (1 mM) was added at 6 hr of development. Cells were collected at 24 hr, and extracts were separated electrophoretically on SDS-acrylamide gels. The separated proteins were immunologically recognized by the Western procedure.

seen on an air-water interface. Samples were taken at various times up to 2 days following incubation at 4×10^3 cells/cm². Accumulation of SP96 and SP70 was never observed, although accumulation of earlier gene products occurred normally (Table 2). Thus, these sparse conditions are not permissive for triggering these late genes. However, if 1 mM cAMP was added to the buffer, the cultures at 4×10^3 cells/cm² accumulated almost as much SP96 and SP70 as did cells under normal developmental conditions (Table 2). The addition of cAMP did not affect the behavior of the cells that remained as isolated single cells attached to the plastic surface of the flask. The stimulation of late gene expression by cAMP has been previously observed using cloned cDNA probes for unassigned developmentally regulated mRNAs (Barklis and Lodish 1983; Mehdy et al. 1983; Mehdy and Firtel 1985).

Anterior-like Cells and Regulation

The presence of SP96 and SP70 can be recognized in individual cells by fluorescently labeled antibodies (Fig. 7). The anterior portions of slugs (prestalk cells) are devoid of these spore coat proteins, whereas most cells in the posterior (prespore cells) have accumulated them. However, there are an appreciable number of cells in the posterior region that lack these spore coat proteins (Fig. 7). Sternfeld and David (1982) first brought attention to "anterior-like" cells in the posterior that they recognized by neutral red staining properties. We found that anterior-like cells could be separated from posterior fragments by density gradient centrifugation (Devine and Loomis 1985). Two-dimensional gel electrophoretic analyses of newly synthesized proteins confirmed that, unlike other cells in the posterior, anterior-like cells do not synthesize any of the spore coat proteins (Fig. 8). They do synthesize a prestalk-specific protein (ST430), although at a lower relative rate than prestalk cells isolated from the anterior of slugs. Anterior-like cells also contain prestalk-specific transcripts, C1 and PL1, but at lower levels than are found in true anterior cells (Devine and Loomis 1985). Unlike prestalk cells, anterior-like cells synthesize a protein (PSP 59) that is synthesized at a high rate in prespore cells (Fig. 8). On this basis, anterior-like cells can be distinguished from anterior cells. Sternfeld and David (1982) have shown that when the anterior cells of a slug are removed, the anterior-like cells migrate to the front of the decapitated slug and ultimately differentiate as stalk cells.

By injecting small numbers of anterior cells labeled with the fluorescent dye XR1TC into the anterior of unlabeled slugs, we have been able to follow them during subsequent migration. Within a few hours, some of the labeled cells were found to have moved into the posterior half; when these slugs formed fruiting bodies, a few of the spores were found to be fluorescent (W.F. Loomis, unpubl.). Thus, there is redistribution and respecification of some of the anterior cells. The anterior-like cells appear to be those undergoing respecification.

Cells in an isolated fragment taken from the anterior of a slug differentiate only into stalk cells unless they

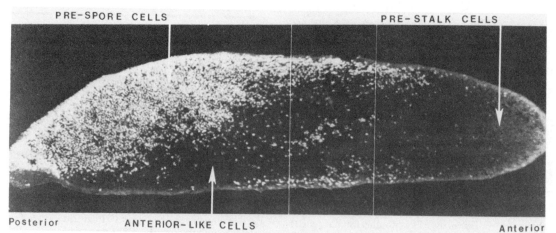

Figure 7. Localization of SP96. Sections (10 μm) of a migrating slug were stained with antiserum specific to SP96. Bound antibody was recognized with fluorescently labeled goat anti-rabbit IgG. Prestalk cells (unlabeled) were isolated by density of Percoll gradients from the anterior quarter; prespore cells (labeled) and anterior-like cells (unlabeled) were isolated from the posterior half by density on Percoll gradients (Devine et al. 1983).

are allowed 1 day to regulate before undergoing terminal differentiation (Raper 1940; Sakai 1973). During the period of regulation, the isolated anterior fragments migrate around as small slugs. There is a lag of several hours before prespore vesicles and spore coat proteins are made. During the next 8 hours, they accumulate in the posterior of regulating slugs. The presence of prespore cells then inhibits cells at the anterior from expressing the genes for spore coat proteins.

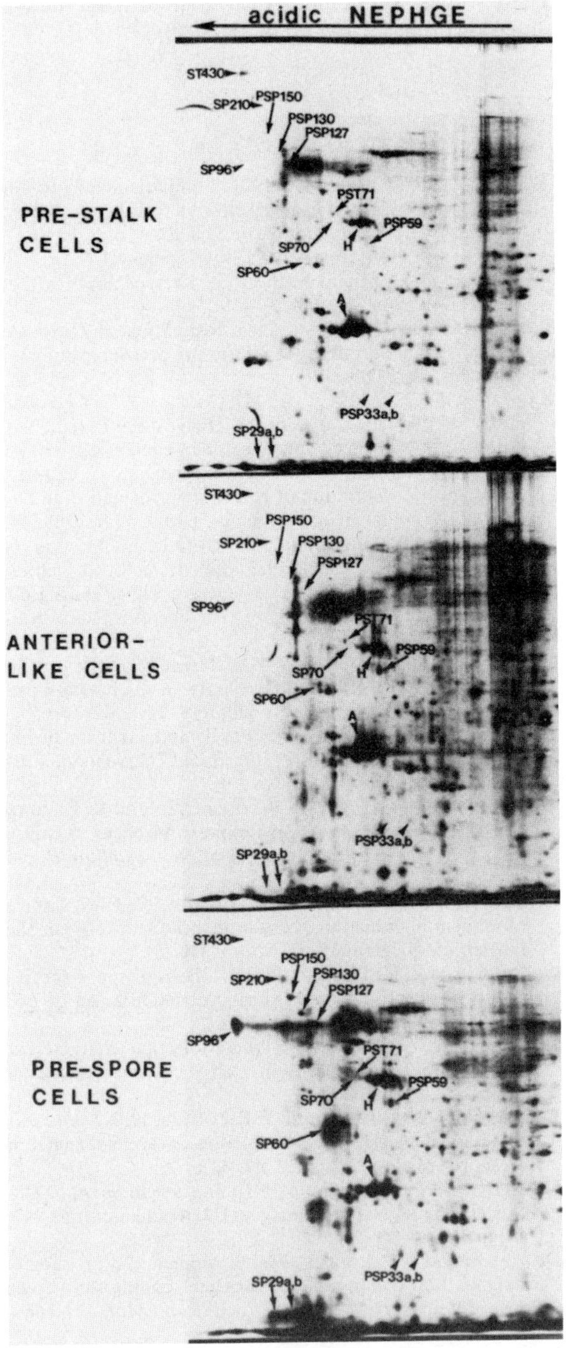

DISCUSSION

The expression of most developmental genes in *D. discoideum* occurs in invariant order at discrete stages spread throughout the 24-hour cycle. Conditions that are permissive for biochemical differentiation but preclude morphogenesis (such as sparse cultures) still result in an ordered progression: Early genes are expressed before late genes. These results indicate a causal sequence that regulates the temporal aspects of development.

Cells appear to signal population density by secreting, shortly after the initiation of development, factors that must increase above threshold levels before later genes are expressed. Several years ago, we showed that accumulation of an early developmental gene product, N-acetylglucosaminidase, responds to a heat-stable, low-molecular-weight effector secreted by developing cells (Grabel and Loomis 1978). Although addition of the effector was sufficient for expression of this early gene, it was not sufficient for expression of later genes; they would only be expressed when the cells were allowed to form multicellular structures (Grabel and Loomis 1978). One of the possible signals found in multicellular aggregates is tight cell association mediated by csA. We examined this question by inactivating csA with mutations in the *modB* locus. Although mutant cells lack EDTA-resistant cell-cell adhesion, they do proceed through the temporal stages of development and express late developmental genes (Loomis et al. 1985). Thus, the tight cell contacts provided by csA are not essential genetic regulatory signals. This conclusion is further supported by the expression of spore coat genes in sparse cultures that preclude aggregation (Table 2). However, exogenous cAMP is required for late gene expression in sparse cultures. Since the level of cAMP that is effective is at least an order of magnitude greater than the levels found in aggregates, it appears that we are bypassing a normally required signal, rather than mimicking the endogenous conditions. The multicellular state may be recognized by mechanisms that raise the internal cAMP levels such that cells proceed through later steps in development. In the sparse cultures, high exogenous cAMP may bypass the requirement for multicellularity by directly raising the internal cAMP levels. These results show that temporal control is mediated, in part, by a diffusible mass effector that is essential for early gene expression and a subsequent requirement that can be filled by addition of high concentrations of cAMP.

Two cell types, prespore and prestalk cells, can be seen to diverge as aggregates are formed 12 hours after the initiation of development (Fig. 4). In migrating slugs, these cells are found in the posterior and anterior, respectively. It is an open question whether the

Figure 8. Protein synthesis in isolated cell types. Migrating slugs were labeled with [^{35}S]methionine for 1 hr before being manually dissected into anterior quarters and posterior halves. Light anterior cells (prestalk), light posterior cells (anterior-like), and heavy posterior cells (prespore) were separated on preformed Percoll gradients, and their newly synthesized proteins were analyzed by NEPHGE/SDS two-dimensional electrophoresis. Cell-type-specific proteins are indicated. Note that anterior-like cells synthesize PSP59, whereas prestalk cells do not (Devine and Loomis 1985).

cell types diverged in randomly positioned cells that subsequently sorted out to the anterior and posterior or whether the cells responded to position-dependent signals and differentiated in place (MacWilliams and David 1984). The timing of appearance of cell-type-specific markers is consistent with either mechanism, since it starts in aggregates and is not complete until the spatial organization is established. A few gene products become cell-type-specific by preferential removal from one or the other of the cell types (Morrissey et al. 1984). These markers accumulate earlier in all cells but only become localized to a specific cell type following aggregation. It may be necessary to define the informational signaling system fully before these alternatives can be distinguished.

The phenotype of strains carrying mutations in the stkA locus clearly shows that cells that have undergone prespore-specific differentiations can nevertheless rapidly differentiate into stalk cells (Fig. 5). Thus, prespore cells are competent to form either spores or stalk cells. On the other hand, prestalk cells that have not undergone prespore-specific differentiation form stalk cells but not spores. They appear to have been restricted to prestalk differentiation by signals emanating from the posterior compartment, since they do regulate when the posterior is removed. This suggests that a long-range inhibitor is functioning in a fieldwide manner encompassing the whole slug.

SUMMARY

During development of Dictyostelium, there are at least a dozen discrete stages of differentiation that can be distinguished by the expression of specific genes. The early stages are triggered by amino acid starvation and are dependent on a small heat-stable effector secreted by the cells to indicate a critical cell density. After development has proceeded for 12 hours, late genes are expressed that are dependent on the conditions found in multicellular aggregates. Cells monitor these conditions and appear to respond by raising their internal cAMP levels to act as a second messenger. Multicellularity can be bypassed as an essential condition if high levels of cAMP are added to the environment. The EDTA-resistant cell-adhesion mechanism that develops by 12 hours is not a required aspect of multicellularity. A final set of genes can be induced in 18-hour-developed cells by lowering the pNH$_3$.

The spore coat proteins are well-characterized markers for prespore differentiation; their genes are first expressed at 12 hours. Prestalk cells do not express these genes. A small number of prestalk cells become redistributed in the posterior during slug migration and appear to undergo respecification when their position is changed. Prestalk genes become repressed in these "anterior-like" cells and prespore genes are activated. These results clearly indicate that a fieldwide system of positional information regulates cell-type differentiation in Dictyostelium.

ACKNOWLEDGMENTS

Studies in my laboratory were carried out in collaboration with John Bergmann, Barbara Dowds, Kevin Devine, Paul Farnsworth, Danny Fuller, Choo Hong, David Knecht, James Morrissey, and Steve Wheeler. These studies were supported by a grant from the National Institutes of Health (GM-23822).

REFERENCES

ALTON, T. and M. BRENNER. 1979. Comparison of proteins synthesized by anterior and posterior regions of Dictyostelium discoideum pseudoplasmodia. Dev. Biol. 71: 1.

ALTON, T. and H. LODISH. 1977. Developmental changes in mRNAs and protein synthesis in Dictyostelium discoideum. Dev. Biol. 60: 180.

BARKLIS, E. and H. LODISH. 1983. Regulation of Dictyostelium discoideum mRNAs specific for prespore and prestalk cells. Cell 32: 1139.

BARONDES, S., W. SPRINGER, and D. COOPER. 1982. Cell adhesion. In The development of Dictyostelium (ed. W.F. Loomis), p. 195. Academic Press, San Diego.

BEUG, H., G. GERISCH, S. KEMPF, V. RIEDEL, and G. CREMER. 1970. Specific inhibition of cell contact formation in Dictyostelium by univalent antibodies. Exp. Cell Res. 63: 147.

BLUMBERG, D., J. MARGOLSKEE, S. CHUNG, E. BARKLIS, N. COHEN, and H. LODISH. 1982. Specific cell-cell contacts are essential for induction of gene expression during differentiation of Dictyostelium discoideum. Proc. Natl. Acad. Sci. 79: 127.

BORTH, W. and D. RATNER. 1983. Different synthetic profiles and developmental fates of prespore versus prestalk proteins of Dictyostelium. Differentiation 24: 213.

BRENNER, M. 1978. Cyclic AMP levels and turnover during development of the cellular slime mold Dictyostelium discoideum. Dev. Biol. 64: 210.

CARDELLI, J., D. KNECHT, M. WUNDERLICH, and R. DIMOND. 1985. Major changes in gene expression occur during at least four stages of development of Dictyostelium discoideum. Dev. Biol. 110: 147.

CHISHOLM, R., E. BARKLIS, and H. LODISH. 1984. Mechanism of sequential induction of cell-type specific mRNAs in Dictyostelium differentiation. Nature 310: 67.

COLOMA, A. and H. LODISH. 1981. Synthesis of spore-specific and stalk-specific proteins during differentiation of Dictyostelium discoideum. Dev. Biol. 81: 238.

DEVINE, K. and W.F. LOOMIS. 1985. Molecular characterization of anterior-like cells in Dictyostelium discoideum. Dev. Biol. 107: 364.

DEVINE, K., J. BERGMANN, and W.F. LOOMIS. 1983. Spore coat proteins of Dictyostelium discoideum are packaged in prespore vesicles. Dev. Biol. 99: 437.

DEVREOTES, P. 1982. Chemotaxis. In The development of Dictyostelium (ed. W.F. Loomis), p. 117. Academic Press, San Diego.

DOWDS, B. and W.F. LOOMIS. 1984. Cloning and expression of a cDNA that comprises part of a gene coding for a spore coat protein of Dictyostelium discoideum. Mol. Cell. Biol. 4: 2273.

FINNEY, R., C. LANGTIMM, and D. SOLL. 1985. Regulation of protein synthesis during the preaggregative period of Dictyostelium discoideum development: Involvement of close cell association and cAMP. Dev. Biol. 110: 171.

FIRTEL, R. 1972. Changes in expression of single-copy DNA during development of the cellular slime mold Dictyostelium discoideum. Dev. Biol. 66: 363.

GRABEL, L. and W.F. LOOMIS. 1978. Effector controlling accumulation of N-acetylglucosaminidase during development of Dictyostelium discoideum. Dev. Biol. 64: 203.

KIMMEL, A. and R. FIRTEL. 1982. The organization and expression of the *Dictyostelium* genome. In *The development of* Dictyostelium (ed. W.F. Loomis), p. 234. Academic Press, San Diego.

LODISH, H., D. BLUMBERG, R. CHISHOLM, S. CHUNG, A. COLOMA, S. LANDFEAR, E. BARKLIS, P. LEFEBVRE, C. ZUKER, and G. MANGIAROTTI. 1982. Control of gene expression. In *The development of* Dictyostelium (ed. W.F. Loomis), p. 325. Academic Press, San Diego.

LOOMIS, W.F. 1975. *Dictyostelium discoideum*, a developmental system. Academic Press, San Diego.

———. 1979. Biochemistry of aggregation in *Dictyostelium*. *Dev. Biol.* **70:** 1.

———. 1982. *The development of* Dictyostelium. Academic Press, San Diego.

LOOMIS, W.F., S. WHITE, and R. DIMOND. 1976. A sequence of dependent stages in development of *Dictyostelium discoideum*. *Dev. Biol.* **53:** 171.

LOOMIS, W.F., S. WHEELER, W. SPRINGER, and S. BARONDES. 1985. Adhesion mutants of *Dictyostelium discoideum* lacking the determinant recognized by two monoclonal adhesion blocking antibodies. *Dev. Biol.* **109:** 111.

MARIN, F. 1976. Regulation of development in *Dictyostelium discoideum*. I. Initiation of the growth to development transition by amino-acid starvation. *Dev. Biol.* **48:** 110.

MACWILLIAMS, H. and C. DAVID. 1984. Pattern formation in *Dictyostelium*. In *Microbial development* (ed. R. Losick and L. Shapiro), p. 255. Cold Spring Harbor Laboratory, Cold Spring Harbor, New York.

MEHDY, M. and R. FIRTEL. 1985. A secreted factor and cAMP jointly regulate cell-type specific gene expression in *Dictyostelium discoideum*. *Mol. Cell. Biol.* **5:** 705.

MEHDY, M., D. RATNER, and R. FIRTEL. 1983. Induction and modulation of cell-type specific gene expression in *Dictyostelium*. *Cell* **32:** 763.

MERKLE, R., K. COOPER, and C. RUTHERFORD. 1984. Localization and levels of cyclic AMP during development of *Dictyostelium discoideum*. *Cell Differ.* **14:** 257.

MORRISSEY, J. and W.F. LOOMIS. 1981. Parasexual genetic analysis of cell proportioning mutants of *Dictyostelium discoideum*. *Genetics* **99:** 183.

MORRISSEY, J., K. DEVINE, and W.F. LOOMIS. 1984. The timing of cell-type specific differentiation in *Dictyostelium discoideum*. *Dev. Biol.* **103:** 412.

MORRISSEY, J., P. FARNSWORTH, and W.F. LOOMIS. 1981. Pattern formation in *Dictyostelium discoideum*: An analysis of mutants altered in cell proportioning. *Dev. Biol.* **83:** 1.

MULLER, K. and G. GERISCH. 1978. A specific glycoprotein as the target site of adhesion blocking Fab in aggregating *Dictyostelium* cells. *Nature* **274:** 445.

MURRAY, B., H. NIMAN, and W.F. LOOMIS. 1983. Monoclonal antibody recognizing gp80, a membrane glycoprotein implicated in intercellular adhesion of *Dictyostelium discoideum*. *Mol. Cell. Biol.* **4:** 863.

MURRAY, B., L. YEE, and W.F. LOOMIS. 1981. Immunological analysis of a glycoprotein (contact sites A) involved in intercellular adhesion of *Dictyostelium discoideum*. *J. Supramol. Struct. Cell. Biochem.* **176:** 387.

MURRAY, B., S. WHEELER, T. JONGENS, and W.F. LOOMIS. 1984. Mutations affecting a surface glycoprotein, gp80, of *Dictyostelium discoideum*. *Mol. Cell. Biol.* **4:** 514.

NEWELL, P., J. ELLINGSON, and M. SUSSMAN. 1969. Synchrony of enzyme accumulation in a population of differentiating slime mold cells. *Biochim. Biophys. Acta* **177:** 610.

ORLOWSKI, M. and W.F. LOOMIS. 1979. Plasma membrane proteins of *Dictyostelium*: The spore coat proteins. *Dev. Biol.* **71:** 297.

RAPER, K. 1940. Pseudoplasmodium formation and organization in *Dictyostelium discoideum*. *J. Elisha Mitchell Sci. Soc.* **56:** 241.

RATNER, D. and W. BORTH. 1983. Comparison of differentiating *Dictyostelium discoideum* cell types separated by an improved method of density gradient centrifugation. *Exp. Cell Res.* **143:** 1.

REYMOND, C., R. GOMER, M. MEHDY, and R. FIRTEL. 1984. Developmental regulation of a *Dictyostelium* gene encoding a protein homologous to mammalian *ras* protein. *Cell* **39:** 141.

SAKAI, Y. 1973. Cell type conversion in isolated prestalk and prespore fragments of the cellular slime mold *Dictyostelium discoideum*. *Dev. Growth Differ.* **15:** 11.

SIU, C.-H., R.A. LERNER, and W.F. LOOMIS. 1977. Rapid accumulation and disappearance of plasma membrane proteins during development of wild-type and mutant strains of *Dictyostelium discoideum*. *J. Mol. Biol.* **116:** 469.

STERNFELD, J. and C. DAVID. 1982. Fate and regulation of anterior-like cells in *Dictyostelium* slugs. *Dev. Biol.* **93:** 111.

SUSSMAN, M. 1982. Morphogenetic signaling, cytodifferentiation, and gene expression. In *The development of* Dictyostelium (ed. W.F. Loomis), p. 353. Academic Press, San Diego.

SUSSMAN, M. and R. SUSSMAN. 1969. Patterns of RNA synthesis and of enzyme accumulation and disappearance during cellular slime mold cytodifferentiation. *Symp. Soc. Gen. Microbiol.* **19:** 403.

TSANG, A. and J. BRADBURY. 1981. Separation and properties of prestalk and prespore cells of *Dictyostelium discoideum*. *Exp. Cell Res.* **132:** 433.

WEST, C. and D. MACMAHON. 1979. The axial distribution of plasma membrane molecules in pseudoplasmodia of the cellular slime mold *Dictyostelium discoideum*. *Exp. Cell Res.* **124:** 393.

YEH, R., F. CHAN, and M. COUKELL. 1978. Independent regulation of the extracellular cyclic AMP phosphodiesterase inhibition system and membrane differentiation by exogenous cyclic AMP in *Dictyostelium discoideum*. *Dev. Biol.* **66:** 361.

Two Feedback Loops May Regulate Cell-type Proportions in *Dictyostelium*

H. MacWilliams, A. Blaschke, and I. Prause
Zoolgisches Institut der Ludwig-Maximilians-Universität, 8000 München 2, Federal Republic of Germany

The Dictyosteliaceae are "social amebae" whose life cycle includes both unicellular and multicellular phases. Vegetative growth is strictly unicellular. Upon exhaustion of the food supply, the cells aggregate, forming at first a loose mound, which elongates, falls over, and migrates for a variable period as a "slug" before transforming into a fruiting body in which each ameba becomes either a stalk cell or a spore. The question of how individual amebae decide for the stalk or spore pathway has received increasing attention in recent years (Morrissey 1982).

Two kinds of amebae can be distinguished in the slug stage; in normal development, these "prestalk" and "prespore" cells form the stalk cells and the spores, respectively. The precursor cell types are partially differentiated: Prestalk cells contain autophagic vacuoles (Yamamoto and Takeuchi 1983) that hypertrophy and fuse to form the central vacuole of stalk cells (Maeda and Takeuchi 1969; Tasaka and Maeda 1983); prespore cells are dehydrated (Schaap 1983) and contain prespore vesicles filled with spore coat proteins (Devine et al. 1982, 1983).

The precursor cells can switch differentiation pathways. Switching can be observed in isolated prestalk or prespore cells: Cells of the missing type appear within the initially pure population, and in approximately 6 hours, the original ratio of the two cell types is restored (Weijer and Durston 1985). This process is often called "proportion regulation."

The prestalk and prespore cells are normally found in separate prestalk and prespore zones. It is not absolutely clear whether spatial patterning causally precedes or follows differentiation into prestalk and prespore cells. The hypothesis of "position-directed cell differentiation" states that a spatial pattern of differentiation-directing substances is created within the slug and that amebae that happen to lie in prestalk and prespore zones receive instructions to become prestalk and prespore cells, respectively (for review, see MacWilliams and Bonner 1979). The hypothesis of "sorting out" states that amebae first decide to become prestalk or prespore cells and then establish the prestalk and prespore zones.

Sorting is a sufficient explanation for the formation of spatial pattern. Preaggregation amebae are heterogeneous, with certain subgroups preferentially forming prestalk and prespore cells (Weijer et al. 1984). Once formed, prestalk and prespore cells can sort out from one another to establish the prestalk and prespore zones (Sternfeld and David 1981). There is no unequivocal evidence for position-directed differentiation (MacWilliams and David 1984). For the sake of simplicity and clarity, we take a "pure sorting" point of view in this paper.

CELL-TYPE PROPORTIONS ARE CONTROLLED BY NEGATIVE FEEDBACK

The existence of proportion regulation has long been taken to indicate that the choice of differentiation pathway is controlled at the supracellular level. This is not a strong inference. Suppose that prestalk cells spontaneously switch to prespore with a probability of p (per cell per unit time) and that prespore cells switch to prestalk with a probability of q. A prestalk/prespore ratio of q/p will be established by mass action without any sort of cell communication. If this ratio is disturbed, it will be automatically reestablished. Recent results, however, allow one to make a strong argument for supracellular control of cell-type proportions. This argument is constructed below.

Shifts in Proportioning

Different kinds of amebae form slugs with different proportions of prestalk and prespore cells. Proportions are influenced by the growth medium; amebae raised without glucose (G⁻ cells) form more prestalk cells than G⁺ amebae raised normally (Leach et al. 1973; Forman and Garrod 1977). In addition, mutants have been described in which the cell-type proportions are shifted (MacWilliams 1982) (see also Table 1). Proportions are affected by the cell-cycle phase at which development starts; immediately after the mitotic phase, amebae form more prestalk cells than amebae of other phases (Weijer et al. 1984; Durston et al. 1985).

Proportioning Is Correlated with "Differentiation Preference"

When amebae of the types discussed above are mixed with "normal" cells (G⁻ with G⁺, mutants with wild-type, and postmitotic with premitotic), then, without exception, the ameba that forms more prestalk cells when differentiating alone is enriched in the prestalk zone of the chimera, whereas the ameba that forms more prespore cells is enriched in the prespore zone.

Feedback Model

The above correlation is expected if proportioning is regulated by negative feedback and if glucose, cell-cycle-phase, and the mutants so far described affect sensitivity to the feedback signal (MacWilliams 1982). Imagine, for example, a feedback system in which the prestalk cells produce a "PSP→PST inhibitor," a substance that tends to hold cells in the prespore state. In a mutant whose *sensitivity* is increased, less inhibitor is required; the equilibrium proportion of (inhibitor-producing) prestalk cells decreases. Moreover, in a mixture of mutant and wild-type cells, more of the (hypersensitive) mutant cells than wild-type cells will form prespore, and the mutant will be enriched in the prespore zone. Thus, sensitivity shifts produce correlated shifts in proportioning and differentiation preference.

Alternative Stochastic Model

A correlation between shifts in proportioning and differentiation preference is also expected if the ratio of prestalk to prespore cells is determined by cell autonomous switching. Consider, for instance, two strains: A normal one that makes 20% prestalk and a mutant with changed q/p that makes 5% prestalk when differentiating alone. Without supracellular control, each partner strain will form the same proportions of prestalk and prespore cells in a chimera as it forms alone. In a 1:1 mixture of the two strains, the ratio of mutant to wild-type cells in the prespore zone will be 95:80. As in the feedback model, the reduced-prestalk mutant shows prespore differentiation preference.

Cell-fate Calculations Demonstrate Supracellular Control

In the stochastic model, each strain behaves the same in the mixture as it behaves alone. In contrast, in the negative-feedback model, admixture of inhibitor-hypersensitive mutant cells will affect the inhibitor level and influence the differentiation of wild-type cells. Pure mutant slugs have a decreased level of PSP→PST inhibitor, and in chimeras, the equilibrium inhibitor level decreases with an increasing fraction of mutant cells (see A. Blaschke et al., in prep.). Thus, the fraction of cells forming prestalk will increase with an increasing fraction of mutant cells.

In mixtures of Hs2 (a strain with reduced prestalk zone apparently due to hypersensitivity) and its parent, we have shown that cell fate does depend on mixing proportions; as expected, the fraction forming prestalk increases with increasing fraction Hs2 (Fig. 1) (A. Blaschke et al., in prep.). We have also found such shifts in G$^-$/G$^+$ mixtures (A. Blaschke et al., in prep.) (data not shown).

THE BEHAVIOR OF ANTERIOR-LIKE CELLS SUGGESTS THAT THERE ARE TWO FEEDBACK SYSTEMS

In addition to prestalk and prespore cells, *Dictyostelium* slugs contain a small fraction of "anterior-like" cells; these have many characteristics of prestalk cells but are found (scattered) in the prespore zone (Sternfeld and David 1982; Ratner and Borth 1983; Devine and Loomis 1985). In Hs1/Hs2 chimeras, the anterior-like cells show a ratio of Hs1 to Hs2 which is the same as that among prespores and different from that among prestalk cells. In G$^-$/G$^+$ chimeras, in contrast, the G$^-$/G$^+$ ratio among anterior-like cells resembles that among prestalk cells rather than prespores (Table 1). The significance of this is more apparent if the result is stated in a different way: In Hs1/Hs2 chimeras, the ratio of anterior-like cells to prespore cells is the same as that in Hs1 and Hs2, and in G$^-$/G$^+$ chimeras, the ratio of anterior-like cells to prestalk cells is the same as that in G$^+$ and G$^-$. This suggests that a specific

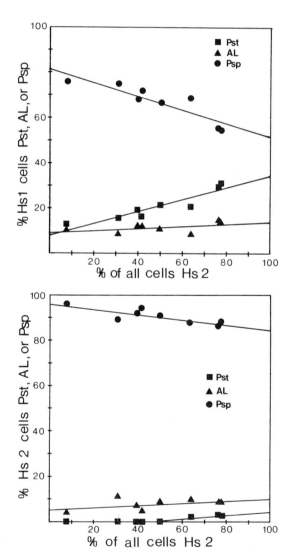

Figure 1. Fate of Hs1 (*top*) and Hs2 (*bottom*) in chimeric slugs formed by various proportions of the two strains. With an increasing fraction of Hs2 in the chimera as a whole, the fraction of both Hs1 and Hs2 cells that form prespore (Psp) decreases and the fraction forming prestalk (Pst) increases. The ordinate values were calculated from measurements of the fraction of Hs2 cells in each cell type and measurements of the overall cell-type proportions of the chimeras. For further discussion, see A. Blaschke et al. (in prep.).

Table 1. Cell-type Proportions and Differentiation Preference in Some Strains of D. discoideum

Strain	Parent	Cell-type proportions			% Mutant cells in various types in a 50:50 mixture with parent			
		% PST	% AL	% PSP	PST	AL	PST+AL[a]	PSP
Hs1[b]	—	11	9	80				
Hs2[b]	Hs1	6	10	84	3	55	—	65
G−[b]	—	25	3	72				
G+[b]	G−[c]	12	6	82	28	24	—	58
Hs1[d]	—	7	9	84				
Hs3[d]	Hs1	18	23	59	—	—	83	40
NP187[d]	—	7	8	85				
NP429[d]	NP187	2	9	89	—	—	21	57

[a] Anterior-like and prestalk cells were not determined separately in these experiments.
[b,d] Slugs prepared from cells suspended in 17 mM potassium phosphate buffer[b] or water[d] (and migrating on 1% agar) were dissected into prestalk and prespore zones and disaggregated with cellulase (Weijer et al. 1984). The proportions of prespore and non-prespore cells in each zone were determined with the aid of the antiprespore monoclonal antibodies MUD-1 and MUD-50 (Krefft et al. 1983). In mixing experiments, one cell type was stained using XRITC (Weijer et al. 1984) before mixing. For a detailed description of the methods, see A. Blaschke et al. (in prep.). The figures represent the average of at least three independent determinations, in each of which several thousand cells were counted; the individual determinations varied by at most 2-3%. The cell type proportions of Hs1 differ in slugs prepared from water- and buffer-suspended cells. The different preparation methods are required to obtain usable slugs from the different mutants.
[c] G+ and G− cells are the G and NS cells discussed by Leach et al. (1973) and are both prepared from the strain Ax2; G− is formally considered the parent here to facilitate comparison with the other data.

mechanism regulates the ratio of anterior-like cells to prespore cells and that a second, independent mechanism regulates the ratio of anterior-like cells to prestalk cells. These two mechanisms would suffice to regulate the mutual proportions of all three types of cells in *Dictyostelium* slugs. Thus, there may not be direct interaction between prestalk and prespore cells.

Alteration of the Prestalk/Anterior-like System in Hs2

In Hs1/Hs2 chimeras, the ratio of prestalk cells to anterior-like cells is higher among Hs1 cells than among Hs2 cells in all mixing proportions, suggesting that Hs2 cells are more sensitive to an AL→PST feedback inhibitor. In fact, calculations show that this idea can account, quantitatively, for all of the results of Hs1/Hs2 mixing experiments (A. Blaschke et al., in prep.).

Alteration of the Prespore/Anterior-like System in G−

In G+/G− mixtures, the ratio of anterior-like cells to prespore cells is higher among the G− amebae than in the G+ cells at all mixing ratios, suggesting that G− is more sensitive than G+ to a AL→PSP feedback inhibitor. This idea can account for many but not all of the experimental observations. Although the ratio of prestalk cells to anterior-like cells is the same among the G+ and G− subpopulations in any given mixture, suggesting that G+ and G− cells have the same sensitivity to the AL→PST inhibitor, the ratio shifts with the fraction of G− cells in the mixture from about 2 (in pure G+) to more than 5 (in pure G−). This behavior can be interpreted by assuming that G− cells have a second alteration: The rate of production (or breakdown) of the AL→PST inhibitor is changed (A. Blaschke et al., in prep.).

CELL-TYPE PROPORTIONS IN OTHER MUTANTS SUPPORT THE TWO-FEEDBACK-SYSTEM CONCEPT

The short-prestalk mutant NP429 (Newell and Ross 1982a) shows a dramatic reduction of the prestalk cells (from 12% in the parent strain NP187 to less than 3% in the mutant), but the ratio of anterior-like cells to prespore cells remains 0.1 (I. Prause and H. MacWilliams, in prep.). This mutant may, like Hs2, be specifically affected in sensitivity to the AL→PST inhibitor.

The long-prestalk mutant Hs3 (MacWilliams 1982) shows an increased fraction prestalk (from 7 to 18, a factor of 2.6) and an increased fraction anterior-like (from 9 to 22, a factor of 2.4). The ratio of prestalk cells to anterior-like cells remains close to constant. This mutant may specifically affect the prespore/anterior-like system.

The concept that cell-type proportions in *Dictyostelium* slugs is controlled by two independent feedback systems allows explanation of the puzzling results in two related fields of investigation: factors influencing *Dictyostelium* differentiation and transplantation properties of *Dictyostelium* slug tissue.

THE BIOLOGY OF SLIME MOLD "MORPHOGENS" SUPPORTS THE TWO-FEEDBACK-SYSTEM CONCEPT

Two substances can be isolated from *Dictyostelium* that affect *Dictyostelium* cell-type proportions at physiological concentrations: differentiation-inducing factor ("DIF," whose structure is not yet known) and 3′,5′-cAMP.

DIF

DIF activity can be demonstrated in an in vitro culture system in which prestalk cells (of appropriate strains) form stalk and prespore cells form spores (Tsang and Bradbury 1981). Vegetative amebae of these strains placed into the in vitro system without DIF form prespores and spores; in the presence of DIF, however, they differentiate as stalk cells (Town and Stanford 1979).

It has been suggested that amebae in the presence of DIF become prestalk cells before differentiating as stalk (Kay and Jermyn 1983; Kopachik et al. 1983), but the markers cited to support this claim ("prestalk-specific" acid phosphatase isozyme, lack of reaction with antiprespore serum) do not distinguish between prestalk and anterior-like cells (see Devine and Loomis 1985) and could as well indicate that the cells become anterior-like cells. Available evidence suggests that anterior-like cells as well as prestalk cells form stalk in the in vitro system: "Light cells" isolated from disaggregated slugs on Percoll gradients form stalk as efficiently as cells from the prestalk zones (Tsang and Bradbury 1981). Light cells consist of about half prestalk cells and half anterior-like cells (Ratner and Borth 1983).

cAMP

Weijer system. Two laboratories have shown that cAMP can dictate the cell-differentiation pathway. Weijer and Durston (1985) isolated prespore cells from disaggregated slugs and placed them in roller tubes with and without cAMP. In controls, about 40% of the prespore cells convert to prestalk cells or anterior-like cells (it is not known which) within 6 hours. cAMP blocks this conversion. Weijer further found that phosphodiesterase accelerates the conversion and reduces the equilibrium proportion of prespore cells. Adenosine, which interferes with cAMP binding in *Dictyostelium* (Newell and Ross 1982b; Van Haastert 1983), had an effect similar to that of phosphodiesterase.

Ishida system. In the mutant dev-1510 (Ishida 1980a), slugs spontaneously transform into heaps of terminally differentiated cells, the prestalk zone giving rise to stalk cells while the prespore zone yields spores (Ishida 1980b). When aggregation-competent amebae of dev-1510 are embedded in agar without cAMP, they differentiate predominantly into stalk, but in the presence of cAMP (10^{-5} M or higher), stalk formation is suppressed and spore formation occurs.

DIF, cAMP, and the Two-Feedback-System Model

For a variety of reasons, it is attractive to assume that cAMP plays the role of the AL→PST inhibitor discussed above; DIF could play the role of the AL→PSP inhibitor.

cAMP and the Prestalk/Anterior-like Equilibrium

Caffeine effect. In *Dictyostelium*, caffeine blocks the cAMP-stimulated activation of adenylate cyclase (Brenner and Thomas 1984; Theibert and Devreotes 1984) and might thereby be expected to reduce extracellular levels of cAMP. In slugs placed on caffeine, the anterior-like cells suddenly begin to attract each other, forming a series of aggregation foci within the prespore zone (C.J. Weijer, in prep.). This suggests that the anterior-like cells have converted to prestalk cells (which attract each other chemotactically; see below). The time required for aggregate formation (1-2 hr) is the same as the time required for anterior-like/prestalk conversion (Sternfeld and David 1982). When one removes the slugs from caffeine, each individual aggregation focus becomes a prestalk zone, and the slug splits up into as many small slugs.

5'-nucleotidase as a homeostatic regulator. Histochemical staining for 5'-nucleotidase clearly shows that this activity is restricted to anterior-like cells (C.J. Weijer, unpubl.). Armant and Rutherford (1979) have described a peak of activity at the prestalk-prespore boundary, which is apparently due to the presence (in this position) of prestalk cells that recently were anterior-like cells (MacWilliams 1984). Since 5'-nucleotidase produces adenosine from 5'-cAMP, the anterior-like cells could be the major source of adenosine in slugs. If cAMP converts prestalk cells to anterior-like cells, production of adenosine by anterior-like cells will be homeostatic; when many anterior-like cells are present, the resulting adenosine interferes with cAMP binding and opposes the production of further anterior-like cells.

The cell-type distribution of phosphodiesterase. Schaap and Spek (1984) have found that phosphodiesterase is highly enriched in the light cells isolated on density gradients. Brown and Rutherford (1980) have reported that phosphodiesterase is fairly uniformly distributed in the *Dictyostelium* slug; i.e., it is only moderately enriched in the prestalk zone. These results do not suggest that phosphodiesterase is prestalk-specific but are consistent with the idea that it is anterior-like-specific, if one supposes that some anterior-like cells are present in the prestalk zone. This idea is supported by Rutherford's finding of very high phosphodiesterase activity in a tiny zone at the rear of the slug; this could represent the "rearguard zone," an accumulation of anterior-like cells (Sternfeld and David 1982). If cAMP is the AL→PST inhibitor, the association of phosphodiesterase with anterior-like cells will (as 5'-nucleotidase) have a homeostatic effect.

cAMP and prespore/spore formation. The results with the Ishida and Weijer systems, which have been interpreted as suggesting that cAMP stimulates the formation of prespore cells, are consistent with the idea that cAMP primarily converts prestalk cells to anterior-like cells. The prespore/anterior-like proportioning system would then convert the "excess" anterior-like cells to prespores. Both the Weijer and Ishida effects are observed at high cell densities, at which the prespore/anterior-like system could be intact. Neither the Weijer system nor the Ishida system achieves 100% prespores/spores; a fraction of "other cells," which may be anterior-like cells, remain even at high cAMP concentrations.

DIF and the Anterior-like/Prespore Equilibrium

Effect on stalk formation. cAMP is present at millimolar concentrations in the in vitro system used to test for DIF activity. If cAMP converts prestalk cells to anterior-like cells, it is likely that only anterior-like cells and prespore cells exist under these conditions. If DIF is the AL→PSP inhibitor, it should stabilize the anterior-like state. If, as suggested above, the anterior-like cells differentiate as stalk in the in vitro system, DIF will stimulate the formation of stalk cells in this system.

Differentiation preference of a DIF-sensitivity mutant. Our preliminary evidence supports the idea that DIF acts on the anterior-like/prespore equilibrium. Strain HM2 shows threefold higher sensitivity to DIF than its parent V12M2 (R. Kay, pers. comm.) and an increase in the fraction of non-prespore cells (H. MacWilliams et al., in prep.). Our group has found that in HM2/V12M2 mixtures, HM2 preferentially forms prestalk cells and V12/M2 forms prespore cells, as expected if DIF sensitivity is relevant to cell differentiation in vivo. In the mixtures so far examined, the ratio of HM2 to V12M2 is the same in anterior-like and prestalk cells, suggesting that there is no change in sensitivity to the AL→PST inhibitor.

A PARADOXICAL TRANSPLANTATION PHENOMENON IS EXPLAINED BY THE TWO-FEEDBACK-SYSTEM CONCEPT

Slug Morphogenesis and the "Tip"

Morphogenesis in *Dictyostelium* slugs is controlled by a small region of cells at the anterior extreme called the "tip." Tip formation immediately precedes the elongation of the aggregate into a slug. A tip transplanted into the side of a slug causes the tissue to elongate in its direction, ultimately "organizing" a second slug. The tip also influences slug shape (MacWilliams 1984). Like "dominant regions" in other organisms (see MacWilliams 1983a,b), tips inhibit tip formation, and transplantation experiments demonstrate gradients of tip inhibition and of resistance to it along the slug axis (Durston 1976; Durston and Vork 1977, 1979; MacWilliams 1982).

Process of Tip Formation

The tip itself is inevitably composed of prestalk cells; strains with very small prestalk zones (such as HS2 and NP429) have very small tips. A local assembly of prestalk cells appears necessary and (at least sometimes) sufficient for tip formation. Thus, an injected mass of prestalk cells often becomes a tip. In stirred slugs, in which all cell types are randomly mixed, the prestalk cells can be induced to coalesce about the tip of an electrode which iontophoretically releases cAMP. If the electrode is then withdrawn, the aggregated prestalk cells become a tip (Matsukuma and Durston 1979). Since prestalk cells themselves appear to produce cAMP (Pan et al. 1974; Brenner 1977; Town and Stanford 1977), tip formation can be plausibly described as an autocatalytic process (see MacWilliams and David 1984), in which an aggregate of prestalk cells produces a stimulus that brings about further aggregation. If, in a mass with no tip, random factors or weak environmental gradients lead to a local increase in the concentration of prestalk cells, a local maximum in the cAMP concentration will result. This will attract further prestalk cells, which in turn will further strengthen the cAMP peak. The process can continue until all prestalk cells have aggregated in one place.

In transplantation experiments, one can show that the tip inhibits tip formation; this inhibition is altered in some proportioning mutants. Here, we suggest that tip inhibition is identical to AL→PST inhibition.

Tip Inhibition and Resistance to Inhibition in Short-prestalk Mutants

Properties of Hs2. This mutant shows both a drastically reduced ability to form tips (transplanted Hs2 prestalk cells fail to form tips under most circumstances) and a drastically reduced ability to inhibit tip formation (transplants too small to form tips in Hs1 hosts do so readily in Hs2 hosts) (MacWilliams 1982). Both properties are expected if Hs2 has increased sensitivity to the AL→PST inhibitor as suggested above. Hs2 prestalk cells should be readily converted to anterior-like cells when transplanted into a wild-type slug; the anterior-like cells will fail to attract each other. Since the AL→PST inhibition level in Hs2 slugs is lower than that in Hs1 slugs, transplants to Hs2 should be less effectively converted to anterior-like cells than transplants to Hs1. Assuming that the AL→PST inhibitor is cAMP, tip formation in Hs2 hosts will also be enhanced by low background cAMP levels; transplants to Hs2 can more easily produce a local cAMP "peak."

Other short-prestalk mutants. The transplantation properties of Hs2 are examples of a general phenomenon: In two other mutants with reduced prestalk zones, NP429 and KY3 (Yanigasawa et al. 1967), the same combination of transplantation properties is observed (H. MacWilliams, in prep.). NP429 seems likely to be affected, like Hs2, in the sensitivity to AL→PST inhibitor (the cell-differentiation preferences of KY3 have not been investigated).

Paradoxical properties of a long-prestalk mutant. In the mutant Hs3, which has an enlarged prestalk zone, the resistance to tip inhibition is unaltered, and the tip-inhibition level is changed only in that the slope of its gradient is increased (MacWilliams 1982). This suggests that Hs3 is not an opposite of Hs2 but is a mutant of a different kind.

The idea that two independent mechanisms regulate proportioning allows one to understand how this could be. If, as suggested above, the proportioning shift in Hs3 is due to an increased sensitivity to the AL→PSP inhibitor, rather than to a decreased sensitivity to the AL→PST inhibitor, Hs3 prestalk cells will not resist conversion to anterior-like cells more than Hs1, and Hs3 hosts will not suppress tip formation via increased background levels of cAMP. One can even explain the observed steepening of the inhibition gradient. The increase in sensitivity to AL→PSP inhibitor leads to an increase in anterior-like cells (as observed; see above); the resulting increased phosphodiesterase activity steepens the cAMP (tip inhibition) gradient.

REFERENCES

ARMANT, D.R. and C.L. RUTHERFORD. 1979. 5-prime AMP nucleotidase is located in the area of cell-cell contact of prespore and prestalk regions during culmination of *Dictyostelium discoideum*. *Mech. Ageing Dev.* **10:** 199.

BRENNER, M. 1977. The cyclic AMP gradient in migrating pseudoplasmodia of the cellular slime mold *Dictyostelium discoideum*. *J. Biol. Chem.* **252:** 4073.

BRENNER, M. and S.D. THOMAS. 1984. Caffeine blocks activation of cAMP synthesis in *Dictyostelium discoideum*. *Dev. Biol.* **101:** 136.

BROWN, S.S. and C.L. RUTHERFORD. 1980. Localization of cyclic nucleotide phosphodiesterase in the multicellular stages in *Dictyostelium discoideum*. *Differentiation* **16:** 173.

DEVINE, K.M. and W.F. LOOMIS. 1985. Molecular characterization of anterior-like cells in *Dictyostelium discoideum*. *Dev. Biol.* **107:** 364.

DEVINE, K.M., J.E. BERGMANN, and W.F. LOOMIS. 1983. Spore coat proteins of *Dictyostelium discoideum* are packaged in prespore vesicles. *Dev. Biol.* **99:** 437.

DEVINE, K., J. MORRISSEY, and W. LOOMIS. 1982. Differential synthesis of spore coat proteins in prespore and prestalk cells of *Dictyostelium*. *Proc. Natl. Acad. Sci.* **79:** 7361.

DURSTON, A. 1976. Tip formation is regulated by an inhibitory gradient in the *Dictyostelium discoideum* slug. *Nature* **263:** 126.

DURSTON, A. and F. VORK. 1977. The control of morphogenesis and pattern in the *Dictyostelium discoideum* slug. In *Development and differentiation in the cellular slime moulds* (ed. P. Cappuccinelli and J.M. Ashworth), p. 17. Elsevier/North-Holland, Amsterdam.

———. 1979. A cinematographical study of the development of vitally stained *Dictyostelium discoideum*. *J. Cell Sci.* **36:** 261.

DURSTON, A., C. WEIJER, H.F. JONGKIND, T. VERKERK, A. TIMMERMANS, and W. TE KULVE. 1985. A flow fluorimetric analysis of the cell cycle during growth and differentiation in *Dictyostelium discoideum*. *Wilhelm Roux's Arch. Dev. Biol.* **194:** 18.

FORMAN, D. and D.R. GARROD. 1977. Pattern formation in *Dictyostelium discoideum*. I. Development of prespore cells and its relationship to the pattern of the fruiting body. *J. Embryol. Exp. Morphol.* **40:** 215.

ISHIDA, S. 1980a. A mutant of *Dictyostelium discoideum* capable of differentiating without morphogeneis. *Dev. Growth Differ.* **22:** 143.

———. 1980b. The effects of cyclic AMP on differentiation of a mutant *Dictyostelium discoideum* capable of developing without morphogenesis. *Dev. Growth Differ.* **22:** 781.

KAY, R. and K. JERMYN. 1983. A possible morphogen controlling differentiation in *Dictyostelium*. *Nature* **303:** 242.

KOPACHIK, W., A. OOHATA, J. DHOKIA, J. BROOKMAN, and R. KAY. 1983. *Dictyostelium* mutants lacking DIF, a putative morphogen. *Cell* **33:** 397.

KREFFT, M., L. VOET, H. MAIRHOFER, and K. WILLIAMS. 1983. Analysis of proportion regulation in slugs of *Dictyostelium discoideum* using a monoclonal antibody and a FACS-IV. *Exp. Cell Res.* **147:** 235.

LEACH, C.K., J.M. ASHWORTH, and D.R. GARROD. 1973. Cell sorting out during differentiation of mixtures of metabolically distinct populations of *Dictyostelium discoideum*. *J. Embryol. Exp. Morphol.* **29:** 647.

MACWILLIAMS, H.K. 1982. Transplantation experiments and pattern mutants in cellular slime mold slugs. In *Developmental order: Its origin and regulation* (ed. S. Subtelny and P. Green), p. 463. A.R. Liss, New York.

———. 1983a. *Hydra* transplantation properties and the mechanism of *Hydra* head regeneration. I. Properties of the head inhibition. *Dev. Biol.* **96:** 217.

———. 1983b. *Hydra* transplantation properties and the mechanism of *Hydra* head regeneration. II. Properties of the head activation. *Dev. Biol.* **96:** 239.

———. 1984. Cell-type ratio and shape in slugs of the cellular slime molds. In *Pattern formation* (ed. G. Malacinski), p. 132. MacMillan, New York.

MACWILLIAMS, H.K. and J.J. BONNER. 1979. The prestalk-prespore pattern in cellular slime molds. *Differentiation* **14:** 1.

MACWILLIAMS, H.K. and C.N. DAVID. 1984. Pattern formation in *Dictyostelium*. In *Microbial development* (ed. L. Shapiro and R. Losick), p. 255. Cold Spring Harbor Laboratory, Cold Spring Harbor, New York.

MAEDA, Y. and I. TAKEUCHI. 1969. Cell differentiation and fine structures in the development of the cellular slime molds. *Dev. Growth Differ.* **11:** 232.

MATSUKUMA, S. and A. DURSTON. 1979. Chemotactic cell sorting in *Dictyostelium discoideum*. *J. Embryol. Exp. Morphol.* **50:** 243.

MORRISSEY, J. 1982. Cell proportioning and pattern formation. In *Development of* Dictyostelium discoideum (ed. W.F. Loomis). Academic Press, New York.

NEWELL, P.C. and F.M. ROSS. 1982a. Genetic analysis of the slug stage of *Dictyostelium discoideum*. *J. Gen. Microbiol.* **128:** 1639.

———. 1982b. Inhibition by adenosine of aggregation centre initiation and cyclic AMP binding in *Dictyostelium*. *J. Gen. Microbiol.* **128:** 2715.

PAN, P., J.T. BONNER, H. WEDNER, and C. PARKER. 1974. Immunofluorescence evidence for the distribution of cyclic AMP in cells and cell masses of the cellular slime molds. *Proc. Natl. Acad. Sci.* **71:** 1623.

RATNER, D. and W. BORTH. 1983. Comparison of differentiating *Dictyostelium discoideum* cell types separated by an improved method of density gradient centrifugation. *Exp. Cell Res.* **143:** 1.

SCHAAP, P. 1983. Quantitative analysis of the spatial distribution of ultrastructural differentiation markers during development of *Dictyostelium discoideum*. *Wilhelm Roux's Arch. Dev. Biol.* **192:** 86.

SCHAAP, P. and W. SPEK. 1984. Cyclic AMP binding to the cell surface during development of *Dictyostelium discoideum*. *Differentiation* **27:** 83.

STERNFELD, J. and C. DAVID. 1981. Cell sorting during pattern formation in *Dictyostelium*. *Differentiation* **20:** 10.

———. 1982. Fate and regulation of anterior-like cells in *Dictyostelium* slugs. *Dev. Biol.* **93:** 111.

TASAKA, M. and Y. MAEDA. 1983. Ultrastructural changes of the two types of differentiated cells during the migration and early culmination stages of *Dictyostelium discoideum*. *Dev. Growth Differ.* **25:** 353.

THEIBERT, A. and P.N. DEVREOTES. 1984. Cyclic 3′,5′ AMP relay in *Dictyostelium discoideum*: Adaptation is independent of activation of adenylate cyclase. *J. Cell Biol.* **106:** 166.

TOWN, C. and E. STANFORD. 1977. Stalk cell differentiation by cells from migrating slugs of *Dictyostelium discoideum*: Special properties of the tip cells. *J. Embryol. Exp. Morphol.* **42:** 105.

———. 1979. An oligosaccharide-containing factor that induces cell differentiation in *Dictyostelium discoideum*. *Proc. Natl. Acad. Sci.* **76:** 308.

TSANG, A. and J. BRADBURY. 1981. Separation and properties of prestalk and prespore cells of *Dictyostelium discoideum*. *Exp. Cell Res.* **132:** 433.

VAN HAASTERT., P.J.M. 1983. Binding of cAMP and adenosine derivatives to *Dictyostelium discoideum* cells. *J. Biol. Chem.* **258:** 9643.

WEIJER, C.J. and A.Y. DURSTON. 1985. Influence of cyclic AMP and hydrolysis products on cell type regulation in *Dictyostelium discoideum*. *J. Embryol. Exp. Morphol.* **86:** 19.

WEIJER, C.J., G. DUSCHL, and C.N. DAVID. 1984. Dependence of cell-type proportioning and sorting on cell cycle phase in *Dictyostelium discoideum*. *J. Cell Sci.* **70:** 133.

YAMAMOTO, A. and I. TAKEUCHI. 1983. Vital staining of autophagic vacuoles in differentiating cells of *Dictyostelium discoideum*. *Differentiation* **24:** 83.

YANIGASAWA, K., W. LOOMIS, and M. SUSSMAN. 1967. Developmental regulation of the enzyme UDP galactose polysaccharide transferase. *Exp. Cell Res.* **46:** 328.

cAMP Receptors Controlling Cell-Cell Interactions in the Development of *Dictyostelium*

P. KLEIN, D. FONTANA, B. KNOX, A. THEIBERT, AND P. DEVREOTES
Department of Biological Chemistry, Johns Hopkins University, School of Medicine, Baltimore, Maryland 21205

Cell-Cell Communication in Development

Dictyostelium is becoming increasingly recognized as a genetically and biochemically accessible system for studies of transmembrane signaling and cell-cell communication in development. The life cycle is characterized by a striking transition whereby single cells spontaneously form a multicellular organism that undergoes morphogenesis and differentiates into two cell types (Bonner 1982). Early in development, an intercellular communication system appears that coordinates the highly organized aggregation of several million cells. Spiral and concentric waves, triggered at 6-minute intervals, sweep through the cell monolayer, reaching the edge of the 1-2-cm aggregation territories in about 1 hour (Alcantara and Monk 1974). The elegant wave patterns, which resemble those seen in the Zabotinsky-Zaikin chemical reactions (Fig. 1A), are organized by cAMP. In this system, in addition to its role as an intracellular second messenger, cAMP acts extracellularly as a cell-cell-signaling molecule and chemoattractant. The distribution of extracellular cAMP was revealed to form spiral and concentric wave patterns identical to those formed by the cells (Fig. 1B) (Tomchik and Devreotes 1981). Aggregation is brought about by a precisely timed sequence of events (Fig. 1C): At aggregation centers, cAMP levels spontaneously oscillate and each peak initiates one of the propagated cAMP waves. The leading edge of each passing cAMP wave provides a gradient that orients the chemotactically sensitive cells toward the aggregation center. Cells move up the gradient for several minutes until the peak of the wave reaches the position of the cell. The cells then move randomly until the next wave elicits another coordinated movement step. About 30 waves are required to attract the cells into the multicellular structure (Devreotes 1982).

The two responses essential to this coordinated aggregation are chemotaxis (Konijn 1970) and cAMP signaling (Roos et al. 1975; Shaffer 1975). Both are mediated by cell-surface receptors and provide excellent models for transmembrane signaling in eukaryotic cells. The chemotactic response appears to differ from that in leukocytes and macrophages only in minor details. Recent evidence suggests that the cAMP-signaling response is a receptor/G-protein-mediated activation of adenylate cyclase akin to those mediated by hormones and neurotransmitters. The adaptation or "desensitization" properties of these responses have been investigated extensively in *Dictyostelium* (Devreotes and Steck 1979; Dinauer et al. 1980b,c; Fontana et al. 1985).

In *Dictyostelium*, most of the known components of these response systems are under tight developmental regulation. The periodic cAMP signaling, in turn, regulates the expression of these components, as well as a host of other developmentally controlled gene products (Chisholm et al. 1984). Evidence for this is the observation that low, constant concentrations of exogenous cAMP profoundly inhibit the early developmental program, whereas repeated applications of cAMP at 6-minute intervals accelerate the process. In this paper we report our recent observations on chemotaxis and cAMP signaling, identification of the receptors mediating these responses, and the discovery of cAMP-induced modification of the receptors that may be the biochemical mechanism of sensory adaptation.

Chemotaxis and cAMP Signaling

The chemotactic response of *Dictyostelium discoideum* has been studied previously with the small population assay (Konijn 1970). With this assay, Pan et al. (1972) showed that growing amebae are insensitive to cAMP but are attracted to folic acid. During development, amebae lose sensitivity to folic acid as they become chemotactically responsive to cAMP. To observe more rapid chemotactic responses, we have recently developed a perfusion assay whereby responses are carried out on a time scale of a few seconds. The perfusion chamber is a stainless steel block the size of a microscope slide and contains a diamond shape hole. Glass coverslips form the top and floor of the chamber, and amebae are attached to the top coverslip. A pump draws fluid through inlet and outlet needles, exchanging the solution surrounding the amebae every 1-2 seconds. The behavior of the amebae is time-lapse-recorded.

When amebae are perfused with buffer, they move about randomly with a speed of about 7-10 μm/min; however, when an appropriate stimulus is introduced, the amebae retract their pseudopodia. After 15-30 seconds, the pseudopodia are reextended in all directions; the amebae become immobilized and appear flattened. About 2-3 minutes later, the cells resume their random motion even in the continued presence of stimulus (Fig. 2). We term this behavior the shape-change response. The shape-change response appears to be an expression of the chemotactic response in general. Growing amebae, which are chemotactic to folic acid but not to

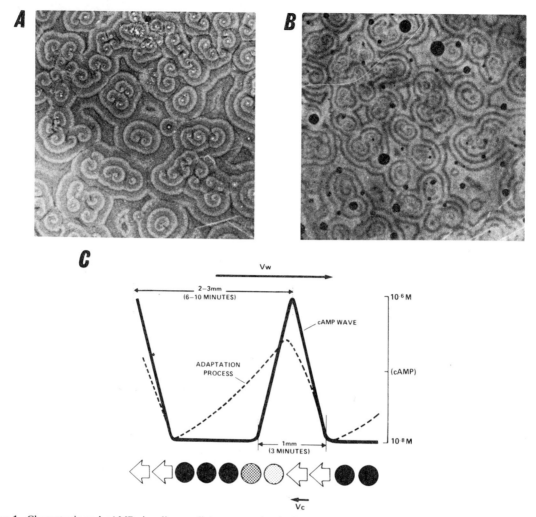

Figure 1. Chemotaxis and cAMP signaling mediate aggregation in *Dictyostelium*. (*A*) Dark-field photograph of aggregating cells at 5 hr in the developmental program. Territories containing about one million cells are about 1–2 cm in diameter. (*B*) Fluorographic image of cAMP waves within monolayer of aggregating cells. Waves were detected by the solid-phase isotope dilution technique referred to in the text. (*C*) Dynamics of signal relay and chemotaxis. The heavy line representing the cAMP concentration is drawn from analyses of the scans of the optical density of fluorographic images of cAMP waves. Symbols at the lower part of the diagram represent a single radial line of cells: (open arrows) cells moving toward center; (shaded circles) randomly oriented cells; (arrow vectors) speed and direction of motion of the cAMP wave ($V_w = 300$ μm/min) and the moving cells ($V_c = 20$ μm/min).

cAMP, show the same shape-change response to folic acid but not to cAMP. Developed amebae show the shape-change response to cAMP but not to folic acid. This is consistent with the previously observed developmental relationship between these two chemotactic responses.

The resumption of motion in the presence of a continuous level of stimulus suggests that the chemotactic response adapts. An adaptation process for chemotaxis was also detected in the small-population assay. Cells will orient in a gradient formed by a drop of 10^{-6} M cAMP on a background of 10^{-5} M cAMP (Van Haastert 1983a). The observation that amebae that have adapted to a low cAMP concentration still respond to higher concentrations (seen in both assays) suggests that adaptation is not an all-or-none process but that the degree of adaptation is dependent on the final concentration of cAMP to which the amebae are exposed.

After removal of the stimulus for 10–15 minutes, amebae regain responsiveness.

Very early in development, amebae display the shape-change response to both folic acid and cAMP. With these amebae, it can be demonstrated that adaptation to folic acid is independent of adaptation to cAMP. Furthermore, deadaptation to one of the stimuli can occur in the presence of the other. This result was suggested, but could not be directly demonstrated, with the small-population assay. With the perfusion assay, we hope to establish the kinetics of chemotactic excitation, adaptation, and deadaptation in *Dictyostelium* that will enable us to evaluate biochemical changes involved in the chemotactic response.

In another type of perfusion assay designed to study the cAMP-signaling response, amebae are labeled with [³H]adenosine so that [³H]cAMP will be secreted. It has been shown previously that the rate of [³H]cAMP

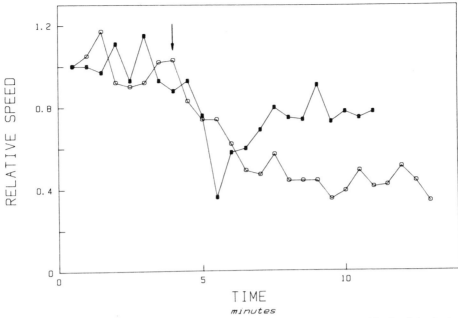

Figure 2. Rapid cell-shape changes in response to cAMP. Shown is the relative speed (averaged for 7 cells) prior to and following addition of cAMP stimuli. (●) Response to a stimulus jump from 0 μM to 1 μM cAMP. Stimulus was added at arrow and held constant. Cells stop momentarily and then resume random motion with the speed nearly returning to the prestimulus levels. (○) Response to the four successive increments in the stimulus concentration. Stimulus was 0→1 nM→10 nM→100 nM→1 μM. Each increment was applied for 3 min. Cells gradually slow down and speed remains at about 40% of the prestimulus level.

secretion is proportional to the intracellular concentration of cAMP and reflects an increase in synthesis, not release of cAMP from intracellular stores (Dinauer et al. 1980a). When the amebae are perfused with buffer, no cAMP is secreted, whereas when amebae are perfused with cAMP, they respond by synthesizing and secreting [³H]cAMP (Fig. 3, left) for a few minutes. A further increase in the occupancy of the cAMP surface receptors elicits another transient secretion response. The magnitude of the response is proportional to the change in the receptor occupancy. Therefore, like the adaptation of chemotaxis, the adaptation of the signaling response is not an all-or-none process but depends on the stimulus level. With the use of the inhibitor caffeine, which blocks activation of the adenylate cyclase (Brenner and Thoms 1984), but not adaptation (Theibert and Devreotes 1983), it has been shown that adaptation is independent of the rise in intracellular cAMP. Upon removal of the cAMP stimulus, the amebae regain their responsiveness to cAMP (Fig. 3, right). The kinetics of deadaptation of the cAMP-signaling response is similar to that observed for the cAMP chemotactic response.

The transient nature of the signaling and chemotactic responses is necessary if they are to mediate the aggregation of a million individual amebae. The adaptation component of the signaling response results in the orderly movement of the cAMP wave from the aggregation center to the edge of the aggregation territory, and deadaptation is necessary if another wave is to be propagated. The adaptation to the chemotactic response keeps the amebae moving toward the aggregation center and prevents them from turning and following the wave as the peak cAMP concentration passes them. The deadaptation of the chemotactic response allows them to move toward the center in response to every wave of cAMP. Evidence presented below suggests that the response and the adaptation of both the signaling and chemotactic responses are mediated by receptors of the same affinity and specificity for cAMP.

Photoaffinity Labeling of the Surface cAMP Receptor

The cAMP receptor on the surface of *D. discoideum* has been extensively characterized by binding studies with radioactive cAMP and cAMP analogs. The receptor shows saturable binding (K_D = 30–50 nM) with approximately 100,000 sites per cell (Green and Newell 1975; Henderson 1975; Van Haastert 1983b). For a variety of cAMP analogs, an order of potency has been established for stimulation of chemotaxis, cAMP signaling, and cGMP accumulation and for inhibition of [³H]cAMP binding (Van Haastert and Kien 1983; A. Theibert et al., in prep.). The close correlation of analog binding with stimulation of these physiological responses suggests that the various responses are mediated by the same receptor. Furthermore, the order of potency distinguishes this receptor from other cAMP-binding proteins such as phosphodiesterase and the regulatory subunit of cAMP-dependent protein kinase (Van Haastert and Kien 1983). Developmental regulation of [³H]cAMP-binding parallels the regulation of cAMP-mediated chemotaxis and signaling with a maximum level as cells are entering the aggregation phase

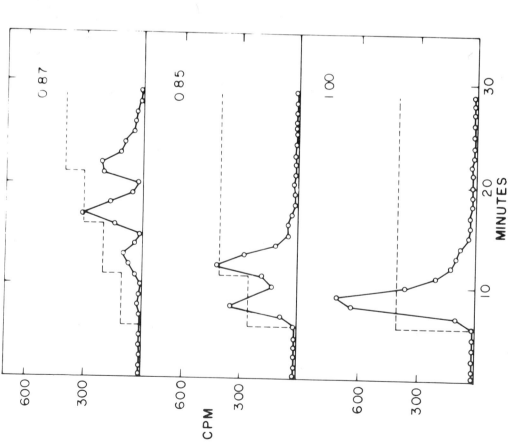
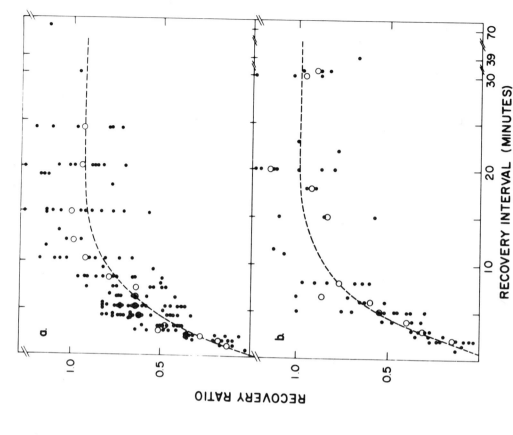

Figure 3. Adaptation and deadaptation properties of cAMP-mediated cAMP secretion. (*Left*) Response to sequential increments in cAMP stimulus concentration. Stimuli: (*Bottom*) 0→1 μM; (*middle*) 0→50 nM→1 μM; (*top*) 0→1 nM→10 nM→100 nM→1 μM. Dashed lines indicate the approximate changes in the receptor occupancy, and solid lines indicate rate of [³H]cAMP secretion. Number in upper right of each panel indicates the total amount of [³H]cAMP secretion in each case, normalized to that secreted in response to the 0→1 μM stimulus. (*Right*) Recovery of the cAMP-signaling response after adaptation to cAMP. Two identical stimuli were applied separated by the indicated recovery interval. The magnitude of the second response normalized to that of the first is plotted. (*a*) 10 nM cAMP; (*b*) 10 μM cAMP.

of development (Green and Newell 1975; Henderson 1975).

Although the cAMP receptor has a rapid dissociation rate ($t_{1/2} < 2$ sec) in most assays (Mullens and Newell 1978), binding can be greatly stabilized by ammonium sulfate (see Van Haastert and Kien 1983). This stabilization has made it possible to photoaffinity-label the receptor with 8-N_3-[^{32}P]cAMP with high specificity (Theibert et al. 1984). On SDS-PAGE, the predominantly labeled protein appeared as a closely spaced doublet with a molecular weight of 40,000–43,000 (Fig. 4); labeling was completely inhibited with 1 μM cAMP.

The photolabeled doublet has the characteristics of the cAMP receptor with respect to saturable binding, affinity for cAMP analogs, and developmental regulation. 8-N_3-[^{32}P]cAMP binding and photoaffinity labeling are saturable ($K_D = 0.3$ μM). As shown in Figure 4B, the intensity of the photoaffinity-labeled band increased with increasing concentrations of ligand, reaching saturation at about 1 μM. Lanes 1, 4, 7, 10, and 13 (Fig. 4B) show nonspecific labeling in parallel sam-

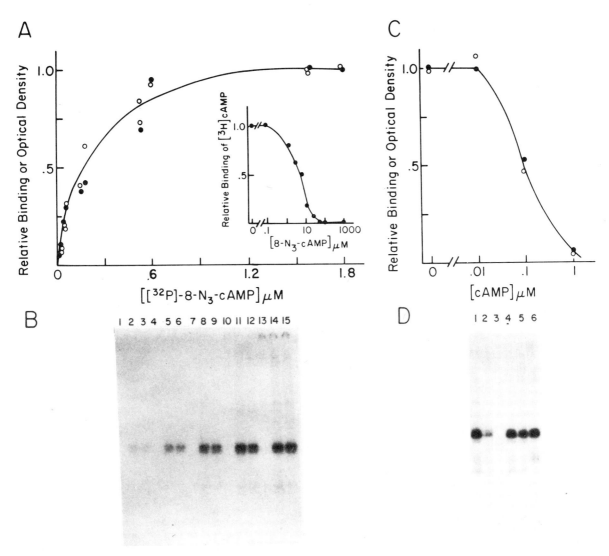

Figure 4. Saturation curve: (*A*) Cells were incubated with 8-N_3-[^{32}P]cAMP at 18.5 nM to 1.8 μM. Binding was normalized to maximal binding (●). Maximal binding was typically about 10^5 sites/cell (80,000 cpm/10^7 cells). (*B*) Cells were then irradiated, and membranes were isolated and run on 10% SDS-PAGE. A representative autoradiogram is shown with ligand present at 20 nM (lanes *1–3*), 55.6 nM (lanes *4–6*), 0.167 μM (lanes *7–9*), 0.5 μM (lanes *10–12*), and 1.5 μM (lanes *13–15*). The major band migrates as a doublet with an apparent molecular weight of 40,000–43,000. Lanes *1, 4, 7, 10,* and *13* show photolabeling when 100 μM cAMP was included in the binding assay. Autoradiograms were scanned, and the optical density in the doublet (m.w. = 40,000–43,000) was normalized as a fraction of maximal optical density (open circles in *A*). (*C*) Cells were incubated with 8-N_3-[^{32}P]cAMP at 0.2–0.3 μM final concentration with nonradioactive cAMP present at 10 nM, 0.1 μM, and 1 μM, and binding was determined and plotted as described above (●). (*D*) Cells were irradiated, and membranes were analyzed on SDS-PAGE as above. The autoradiogram shows photolabeling with cAMP present at 10 nM (lane *1*), 0.1 μM (lane *2*), and 1 μM (lane *3*). Lanes *4–6* are controls in the absence of cAMP. The autoradiograms were scanned and plotted (○) as the fraction of binding in the absence of cAMP. (*Inset*) The competition of [^{3}H]cAMP binding by 8-N_3-cAMP.

ples when excess cAMP is included. As shown in Figure 4A, photoaffinity labeling correlated well with noncovalent binding of 8-N_3-[^{32}P]cAMP. The specificity for cAMP is demonstrated in Figure 4, C and D, in which increasing concentrations of unlabeled cAMP competed 8-N_3-[^{32}P]cAMP binding and photoaffinity labeling of the doublet band with a K_I (70 nM) near the K_D for cAMP.

Inhibition of 8-N_3-[^{32}P]cAMP binding and photoaffinity labeling by several cAMP analogs showed the following order of potency: 2'-deoxy-cAMP > 6-chloro-cAMP > 8-bromo-cAMP > N^6-monobutyryl-cAMP. This parallels the order established for inhibition of [^{3}H]cAMP binding and for stimulation of chemotaxis and cAMP signaling (Van Haastert and Kien 1983; A. Theibert et al., in prep.). Furthermore, this order of potency is clearly distinct from that found for other known cAMP-binding proteins in *Dictyostelium*.

Binding and photoaffinity labeling by 8-N_3-[^{32}P]cAMP closely follows the pattern of cAMP binding during early development. AX-3 cells were starved for various times between 0 and 10 hours and then photoaffinity-labeled. Figure 5 shows that both binding and photoaffinity labeling reached a maximum at 5-6 hours of development, as cells were beginning to form tight aggregates, and declined to less than 10% of maximum by 9 hours of development.

Several investigators have reported photoaffinity labeling of intact *Dictyostelium* with 8-N_3-[^{32}P]cAMP (Hahn et al. 1977; Wallace and Frazier 1979; Juliani and Klein 1981). In most of these reports, radioactivity was incorporated into many bands on SDS-PAGE. cAMP competition of specific bands was reported by Wallace and Frazier (1979; m.w. = 40,000), Juliani and Klein (1981; m.w. = 45,000), and Hahn et al. (1977; m.w. = 36,000 and 33,000). Each of these bands may be related to the doublet we have identified. However, in addition to the high degree of nonspecific labeling in these studies, the bands were not distinguished from the regulatory subunit of cAMP-dependent protein kinase, which is also labeled by 8-N_3-[^{32}P]cAMP and has a similar molecular weight (m.w. = 41,000; Leichtling et al. 1984). Our results differ in that the doublet was specifically labeled with high efficiency and was distinguished from the regulatory subunit of protein kinase by multiple criteria: Photoaffinity labeling of the regulatory subunit was markedly inhibited by N^6-monobutyryl-cAMP at 1 µM but not by 2'-deoxy-cAMP even at 100 µM. This is opposite the potency of these analogs for inhibition of binding of both [^{3}H]cAMP and 8-N_3-[^{32}P]cAMP to the surface cAMP receptor. The developmental regulation of the doublet was also distinct from the regulatory subunit. When the cytosols from the cells used in Figure 5 were photoaffinity-labeled using the protocol of Leichtling et al. (1984), the intensity of the photoaffinity-labeled regulatory subunit rose to a plateau at 9 hours, as previously reported by these authors. Furthermore, the two proteins had different subcellular distributions. Using antiserum against the regulatory subunit (a gift from B. Leichtling) for Western blot analysis, we found that this protein was localized to the cytosolic fraction and was absent from membranes, whereas 80% of the surface cAMP recep-

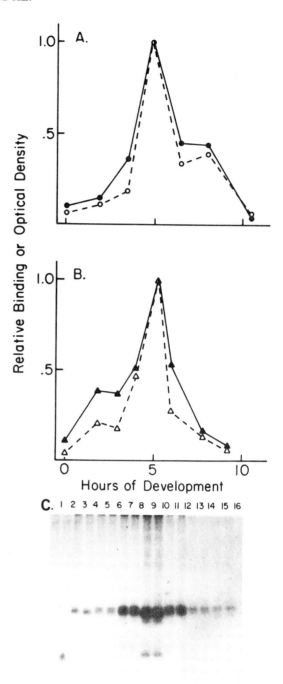

Figure 5. Developmental regulation: Development was initiated by transferring cells to development buffer (5 mM Na_2HPO_4, 5 mM KH_2PO_4, 200 µM $CaCl_2$, 2 mM $MgCl_2$) at 1-1.5-hr intervals. After 9-11 hr, all cells were harvested and, binding and photolabeling were carried out. (*A, B*) Comparison of noncovalent binding with photolabeling (as determined by optical scanning of autoradiograms) as a function of developmental age in two different experiments. (*C*) Representative autoradiogram (corresponding to the data shown in *B*) with each time point run in duplicate, for cells at 0-9.5 hr of development. The origin of the variation in electrophoretic mobility is discussed in the text.

tor was recovered in membranes and none were present in the cytosolic fraction.

Meyers-Hutchins and Frazier (1984) have purified a cAMP-binding protein (m.w. = 70,000) that can be labeled with 8-N_3-[^{32}P]cAMP, but they were not able to label it on intact cells. We have observed photoaffinity-labeled doublets with molecular weights of 55,000 and 70,000 that are present at less than 3% of the labeled protein.

Despite the differences in specificity, developmental regulation, and subcellular location, the cAMP receptor does share some characteristics with the regulatory subunit of protein kinase. Both are cAMP-binding proteins that either directly or indirectly regulate kinase activities. In both cases, cAMP binding can be stabilized by ammonium sulfate (as in the CAP protein of *Escherichia coli*). In general, regulatory subunits of protein kinase can be phosphorylated (although this has not been reported for the regulatory subunit in *Dictyostelium*), and in some cases, the phosphorylation is associated with a shift in electrophoretic mobility (Robinson-Steiner et al. 1984), as is also seen with the cAMP receptor (see below). The ability of cAMP to induce certain genes in *D. discoideum* is reminiscent of the CAP protein of *E. coli*, which induces several genes when cAMP levels rise. It is possible that *Dictyostelium* has a class of cAMP-dependent regulatory proteins with different functions and subcellular localizations.

Receptor Modification and Sensory Adaptation

The relative intensities of the bands composing the receptor doublet varied in independent cultures of cells (see Figs. 4 and 5). It is well known that *Dictyostelium* undergoes spontaneous oscillations in cAMP synthesis (Gerisch and Wick 1975). We hypothesized that there are two forms of the receptor and that the physiological state of the cells at the moment of sampling determines which form is predominant. To test our hypothesis, spontaneous oscillations (Fig. 6A) were monitored by continuous measurement of light scattering, and samples were taken at 1-minute intervals. The cells were washed at 0°C, and the surface receptors were photoaffinity-labeled. As shown in Figure 6, B through D, receptors were found in either a high (m.w. = 40,000) or a low (m.w. = 43,000) mobility form, and the relative distribution between the two forms varied at the same frequency as the cAMP oscillations (the higher mobility form was designated R and the lower mobility form was designated D). Just before the phase of active cAMP synthesis, the R form was predominant, whereas during the phase of cAMP synthesis, the D form increased in relative intensity. The fraction of intensity in the D form reached a peak of about 65% and then gradually returned to a value of about 30% just before the cycle repeated (P. Klein et al. 1985).

To examine whether the shifts between the R and D forms of the receptor depended on exogenous cAMP, cells were pretreated with caffeine, which blocks the spontaneous cAMP oscillations (Brenner and Thoms 1984). The kinetics of the cAMP-induced shift from the R form to the D form of the receptor was examined after addition of exogenous cAMP. As shown in Figure 7A, prior to addition of cAMP, the fraction of receptors in the D form was about 0.1. The addition of 1 μM cAMP triggered a time-dependent redistribution in the fraction of receptors in each of the receptor forms. By 15 minutes, the fraction of receptors in the D form had increased to a plateau value of 0.80. This ratio remained constant after 26 minutes of continuous stimulation. A detectable shift was observed within 15 seconds of addition of cAMP. The transition occurred with a half-time of about 2.5 minutes (Devreotes and Sherring 1985).

After 10 minutes of stimulation with 1 μM cAMP, a portion of the cells was removed, washed free of cAMP at 0°C, and incubated at 0°C or 22°C. As shown in Figure 7A, at 22°C, there was a time-dependent return toward the basal state until the fraction of receptors in the D form was about 0.12 after 32 minutes. A detectable decrease in the fraction of receptors in the D form was observed within 30 seconds of warming to 22°C. The complete transition from the D form to the R form occurred with a half-time of 5–6 minutes. At 0°C, no redistribution occurred, and the fraction of receptors in the D form remained at 0.8 for as long as 32 minutes (Devreotes and Sherring 1985).

The kinetics of cAMP-induced shift from the R form to the D form of the receptor was also examined at 50 nM cAMP, a concentration that occupies about 50% of the surface receptors. The lower concentration of cAMP induced a rapid increase in the fraction of receptors in the D form which reached a plateau value of about 0.4 after about 10 minutes. No further increase occurred between 10 and 26 minutes. A detectable shift was observed within 15 seconds of addition of the stimulus. The transition occurred with a half-time of about 2 minutes. Again, when cells that had been pretreated with cAMP for 10 minutes were washed free of cAMP, there was a gradual return to the basal state in which the fraction of receptors in the D form was again about 0.10. A detectable decrease in the fraction of receptors in the D form was observed within 30 seconds of warming to 22°C, and the half-time of the decline was about 5 minutes, similar to that which occurred upon removal of 1 μM cAMP (Devreotes and Sherring 1985).

These data indicate that the steady-state distribution between R and D forms of the receptor is dose-dependent. The steady-state distribution of R and D forms was examined as a function of the cAMP stimulus concentration. As shown in Figure 7B, a detectable increase in the fraction of receptors in the D form was observed with 0.5 nM cAMP, and the maximal fraction in the D form, about 0.8, occurred with 1 μM cAMP. About 50% of the maximal shift occurred with about 30 nM cAMP. After 15 minutes of persistent stimulation with these concentrations of cAMP, each set of cells received an increment in the cAMP stimulus concentration from the indicated concentration to about 10 μM cAMP, and incubation was continued for an additional 15 minutes. As shown in Figure 7B, the frac-

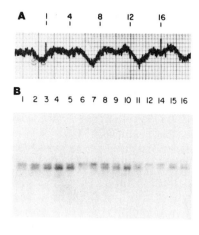
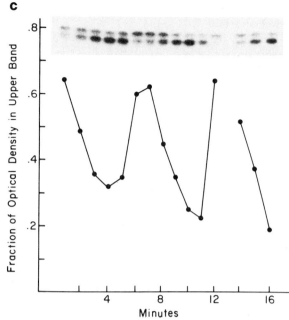

Figure 6. Photolabeling during spontaneous oscillations in cAMP synthesis: Cells were developed for 3.5 hr. (*A*) Oscillations in cAMP synthesis were monitored by measuring light scattering using a modification of the procedure described by Gerisch and Hess (1974). (*B*) Samples were taken at 1-min intervals (sample 13 was lost) during the oscillations and photolabeled. Isolated membranes were run on 10% SDS-PAGE (10% acrylamide, 0.8% *bis*-acrylamide), and autoradiograms were exposed without intensifying screens. (*C*) The same samples were run on SDS-PAGE with 10% acrylamide and 0.05% *bis*-acrylamide, which enhanced the separation of the two bands of the doublet. (*D*) The autoradiogram in C was scanned and the optical density of the upper band was plotted as a fraction of total optical density in the doublet.

tion of receptors in the D form rose to about 0.8 in all cases (Devreotes and Sherring 1985).

These observations suggest that cAMP induces a reversible modification of surface cAMP receptors that alters electrophoretic mobility in SDS-PAGE. The kinetics and cAMP concentration dependence of this modification correlate closely with those of the adaptation process that gradually extinguishes the cAMP-induced activation of guanylate and adenylate cyclases (Van Haastert and Van der Heijden 1983; Devreotes 1982), cell-shape changes (Fontana et al. 1985), and myosin heavy- and light-chain kinase activation (Berlot et al. 1985). As described above, cells only respond to increases in the fractional occupancy of surface receptors. When occupancy is held constant, responses subside within 10-20 minutes. The magnitude of the elicited response is proportional to the fractional increase in receptor occupancy. This holds for the initial challenge with cAMP or for any subsequent increment in the stimulus level. Several theoretical treatments have shown that receptor modification could account for these features of adaptation (Goldbeter and Koshland 1982; Block et al. 1983; Martiel and Goldbeter 1984; B. Knox et al., in prep.). The pattern of modulation of receptor modification theoretically required to bring about adaptation is consistent with the data presented in Figure 7 A and B.

The rates of adaptation and deadaptation of the adenylate cyclase have been measured. Detectable adaptation occurs within 20 seconds of stimulation with 1 μM cAMP, and adaptation is nearly complete after 12 minutes. Complete adaptation occurs slightly sooner with lower concentrations of cAMP (Dinauer et al. 1980c). This agrees closely with the kinetics of the cAMP-induced shifts in the distribution of the R and D forms of the receptor. Deadaptation is a first-order process, since it occurs with the same half-time (about 4 min) following removal of either a 10 nM or 10 μM cAMP stimulus (Dinauer et al. 1980b). Consistent with this, the reversals in the distribution of receptors following removal of 1 μM or 50 nM cAMP occurred with similar half-times (6 and 5 min, respectively). The cAMP-induced shifts in receptor distribution can been understood in terms of the diagram shown below.

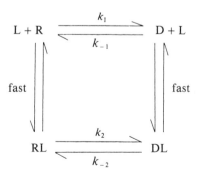

Here, R and D represent free states of the receptor that are unmodified or modified, respectively. The forma-

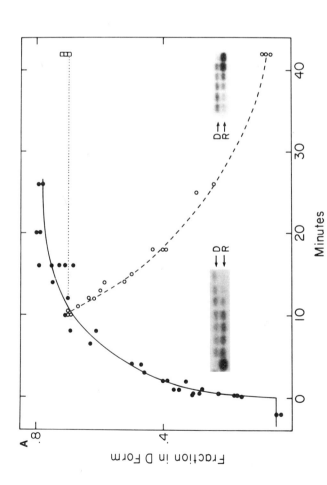
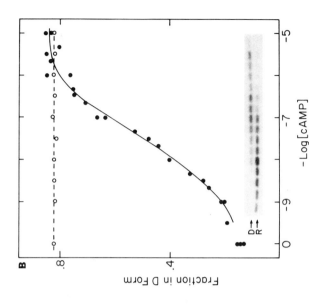

Figure 7. (A) Kinetics of reversible modification of the receptor induced by a cAMP stimulus concentration that saturates binding. Cells were stimulated with 1 μM cAMP, and samples were taken at the indicated intervals, photoaffinity-labeled, and analyzed by SDS-PAGE. The results of four independent experiments are combined for kinetics of modification following addition of the stimulus (●). (*Left inset*) Portion of a typical gel. It has been previously demonstrated that 90% of the radioactivity in the gel is in the doublet bands shown here. Sequential lanes show samples taken at 0, 0.25, 0.5, 1, 2, 4, 8, and 16 min after stimulus addition. The results of two independent experiments are combined for the kinetics of reversal following removal of the stimulus (○). (*Right inset*) Portion of a typical gel. Sequential lanes are samples taken at 0.5, 2, 4, 8, 16, and 32 min after stimulus removal. In two experiments, cells were held at 0°C for 32 min following the removal of cAMP. After 15 min of persistent stimulation, samples were taken, photoaffinity labeled, and analyzed by SDS-PAGE. (B) Concentration dependence of cAMP induced receptor modification at steady state. Cells were stimulated with the indicated concentrations of cAMP. After 15 min of persistent stimulation, samples were taken, photoaffinity labeled, and analyzed by SDS-PAGE. Two independent experiments were combined for the initial challenge experiment (●). (*Inset*) Portion of a typical gel. The first two lanes are controls not stimulated with cAMP. The remaining 14 lanes (sequentially from right to left) are for stimulation with a twofold dilution series starting at 10 μM cAMP. In one experiment, after aliquots were taken at 15 min, 10 μM was added to each set of cells. After an additional 15 min, aliquots were taken for analysis (○).

tion of complexes with cAMP gives two more states designated RL and DL, where L represents cAMP. In the absence of cAMP, only the free states of the receptor, R and D, exist, and the equilibrium between these two states heavily favors the R state. When a concentration of cAMP that saturates the binding sites is added, only liganded states of the receptor, RL and DL, exist, and the equilibrium between these states heavily favors the DL state. The experimentally observed shift in receptor forms (Fig. 7A) represents the time-dependent reequilibration between the RL and DL states. When the stimulus is removed, there are again only free states of the receptor, and the experimentally observed reversal in the distribution of receptor forms (Fig. 7A) represents the reequilibration between the D and R states. With cAMP concentrations that partially saturate binding sites, all four receptors states are present, and the relative amount of unmodified and modified forms depends on the cAMP concentration (Fig. 7B).

As illustrated in Figure 7, A and B, sets of cells could be prepared in which receptors were predominantly in the R or D form. The affinities of receptors in the two forms were measured by binding of [^{3}H]cAMP. The two forms of the receptor were found to have nearly identical affinities. The K_D of the preparation predominantly in the R form was about 15 nM and that of the preparation predominantly in the D form was about 30 nM (Devreotes and Sherring 1985). This observation, taken together with the kinetic data for receptor modification, suggests that the four receptor states are not in thermodynamic equilibrium. Rather, steady-state distribution is attained. Quantitative analysis of these data indicates that, starting with state R, the product of equilibrium constants for the clockwise direction arriving at state DL is about 70-fold less than that of the product in the counterclockwise direction. This suggests that an energy input of several kilocalories is required to maintain the steady-state distribution.

In Vivo and In Vitro Phosphorylation of the Surface cAMP Receptor

As discussed above, the cAMP-induced modification of the cAMP receptor was shown to correlate closely with adaptation of the cAMP-signaling response, suggesting that a reversible covalent modification of the receptor may be the molecular basis of adaptation. Covalent modification of chemotactic and signal-transducing receptors has been widely documented. Methylation of the MCPs in chemotactic bacteria, phosphorylation of the β-adrenergic receptor in turkey erythrocytes, and phosphorylation of rhodopsin in the rod outer segment all correlate with the adaptation (desensitization) of the physiological responses coupled to these receptors. In *Dictyostelium*, phosphorylation of a band that comigrates with photoaffinity-labeled cAMP receptors has been reported (Lubs-Harkness and Klein 1982; C. Klein et al. 1985). Here, we describe our preliminary studies on the phosphorylation of the cAMP surface receptor that we have previously identified and characterized by cAMP photoaffinity-labeling studies.

Cells were developed for 4-5 hours (the peak in receptor binding activity), labeled with [^{32}P]orthophosphate for 30 minutes, and then incubated with or without cAMP for an additional 15 minutes. Membranes were prepared from these cells exactly as in the photolabeling studies (by osmotic shock) and separated by SDS-PAGE. The autoradiograph and corresponding scan are shown in Figure 8A. Under these labeling conditions, the major phosphorylated proteins appear in the 40,000-43,000-molecular-weight region (bracket), although many minor phosphorylated bands are observed. Before addition of cAMP to the cells, there is a major phosphorylated band (m.w. = 40,000) that comigrates with the photoaffinity-labeled receptor (not shown). After stimulation with cAMP for 15 minutes, this band disappears and a new band (m.w. = 43,000) appears that is more heavily phosphorylated. This band comigrates with the 8-N_3-[^{32}P]-cAMP-labeled receptor from cAMP-treated cells. Both the 40,000- and 43,000-molecular-weight phosphorylated proteins are equivalent to the respective photolabeled proteins, since they copurify to homogeneity. Furthermore, these appear to be the only phosphorylated proteins whose mobility and intensity change upon addition of cAMP.

We have not yet demonstrated directly that the phosphorylation of the receptor is the basis for its change in electrophoretic mobility, but several lines of evidence strongly support this mechanism. The kinetics of in vivo phosphorylation upon stimulation with cAMP closely follows the kinetics of the shift to the D form with half-maximal phosphorylation occurring after approximately 2 minutes of stimulation. Furthermore, both the phosphorylation and the shift in mobility are reversible. The cAMP concentration dependence is also very similar for phosphorylation and shift to the D form, with half-maximal phosphorylation occurring at approximately 50 nM cAMP. It is conceivable that the shift in mobility is due to another modification or conformational change in the receptor that then allows phosphorylation to occur. To determine whether phosphorylation causes the shift, treatment of the phosphorylated receptor with purified phosphatase and kinase to produce a shift in electrophoretic mobility must be performed.

The increase in phosphorylation of the cAMP receptor after cAMP stimulation of intact cells is clearly independent of intracellular cAMP, which was demonstrated by inhibiting the rise in intracellular cAMP with caffeine. Phosphorylation of the receptor was unaffected. Adaptation of the signaling response and the shift in mobility of the surface receptor are also independent of cAMP. Two different molecular mechanisms that could lead to the increase in phosphorylation of the receptor upon binding cAMP are (1) an activation of a receptor-specific kinase or (2) activation of the receptor (e.g., through a conformational change after binding cAMP) that would allow it to be phosphorylated. The latter appears to be the case in

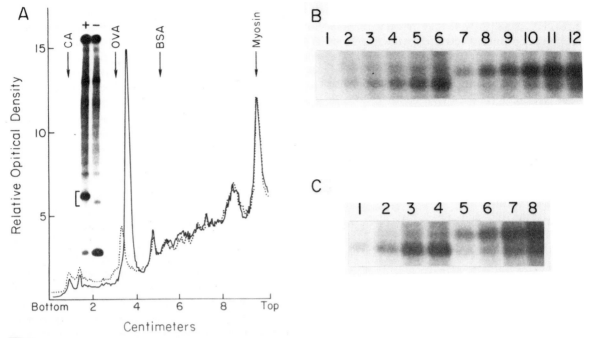

Figure 8. In vivo and in vitro receptor phosphorylation. (*A*) Developed, intact cells were labeled with [^{32}P]orthophosphate at 10^8 per milliliter and 1 mCi ^{32}P per milliliter. After 30 min, cells were split and incubated either in the absence (−) or presence (+) of 10^{-5} M cAMP and 10 mM DTT for an additional 15 min. Membranes were prepared and separated by SDS-PAGE as described by Klein et al. (1985). Each lane represents membranes from 2×10^6 cells. The two lanes were scanned with a densitometer, and the scan is shown with the corresponding molecular-weight markers (− is the dashed line and + is the solid line). Bands at the dye front, which represent the ^{32}P-labeled phospholipids, were omitted from the densitometric scan. (*B*) Cells were lysed with 10 mM CHAPS at a density of 7×10^6 per milliliter, and the lysate was incubated with reaction mix (2 mM MgSO$_4$, 7.5 mM Tris [pH 7.5], 10 μM ATP, and [^{32}P]ATP; 2×10^7cpm/300 μl assay). (Lanes *1–6*) Cells were removed prior to stimulation, lysed, and incubated for 10, 20, and 30 sec, and 1, 2, and 5 min, respectively. (Lanes *7–12*) Intact cells were stimulated with 10^{-6} M cAMP and 10 mM DTT for 1 min prior to lysis and incubated with reaction mix for 10, 20, and 30 sec and 1, 2, and 5 min, respectively. The reaction was stopped by addition of 0.5 M ATP. The receptor, which is localized to the detergent-resistant membranes, was retained in the pellet after centrifugation at 12,000g for 10 min. Each lane represents the pellet from 3×10^6 cells. The 38-45-kD region of the gel is shown. (*C*) Cells were lysed with 10 mM CHAPS in the presence of protease inhibitors. The lysate was centrifuged at 12,000g for 10 min, and the supernatant was removed. The pellet was reextracted with CHAPS and centrifuged at 12,000g for 10 min; it was then resuspended in CP buffer (10% glycerol, 10 mM HEPES [pH 7.5], and 1 mM EDTA) and sonicated. The detergent-resistant membrane fraction was then incubated with reaction mix (as in *B*, except with 5 mM Mg). (Lanes *1–4*) Membranes were prepared from cells removed prior to cAMP addition and incubated for 1, 2, 5, and 10 min, respectively. (Lanes *5–8*) Membranes were prepared from cells that had been stimulated with 10^{-6} M cAMP and 10 mM DTT for 15 min prior to lysis and incubated for 1, 2, 5, and 10 min, respectively. Each lane represents membranes from 2×10^6 cells. After incubation, the samples were prepared and electrophoresed as described in *B*. Again, the 38-45-kD region of the gel is shown.

the rod outer segment, in which there is strong evidence that after light is absorbed, a conformational change in the rhodopsin molecule makes it available as a substrate for rhodopsin kinase which is independent of cyclic nucleotides. To elucidate the molecular mechanism of both the basal and the cAMP-induced increase in phosphorylation of the cAMP receptor, a cell-free system is required. We have therefore devised an in vitro assay to monitor receptor phosphorylation.

Intact cells, with or without a cAMP stimulus, were lysed using the zwitterionic detergent CHAPS and incubated with MgATP and [γ-^{32}P]ATP. Membranes were prepared and separated by SDS-PAGE, and the gel was autoradiographed. Figure 8B (lanes 1–6; increasing time of reaction) shows phosphorylated membrane proteins in the 40,000-molecular-weight region from unstimulated cells; lanes 7–12 (increasing time of in vitro reaction) were taken from stimulated cells. As observed in the in vivo studies, the phosphorylated band (m.w. = 40,000) is decreased after stimulation of cells with cAMP, whereupon a new band with an apparent molecular weight of 43,000 is observed. This doublet band is again the only protein that changes upon cAMP stimulation. In this experiment, cells were stimulated with cAMP prior to lysis. Addition of cAMP to broken cells does not result in the shift or increased phosphorylation of any bands.

The receptor is apparently localized to a detergent-insoluble membrane fraction. Since dilution of the extract did not diminish the receptor phosphorylation, we predicted that the receptor kinase may be localized in this fraction as well. To test this, CHAPS-insoluble membranes were incubated with the [γ-^{32}P]ATP reaction mix, and reaction time courses are shown in Figure 8C (lanes 1–4 and 5–8). Nearly all the receptor kinase activity is recovered in this fraction. In vitro phosphorylation can also be observed in lysates of cells prepared by other techniques (e.g., lysed through membrane filters or in the detergent Triton X-100). It should be possible, using exogenous phosphate acceptors such as kemptide or histones and through the reconstitution of activated and unactivated cells, to determine whether the mechanism of activation is through activation of the receptor or its kinase.

The increase in phosphorylation after cAMP stimulation observed both in vivo and in vitro can be as high as tenfold over unstimulated cells (Fig. 8A scan). One-dimensional peptide maps demonstrate that the basal and stimulated phosphorylations occur on different fragments; one distinct basal peptide and two cAMP-stimulated peptides were observed. The same distinct peptides are observed both in vivo and in vitro, indicating that the physiologically relevant sites are accessible in vitro. In vivo, the basal and cAMP-stimulated phosphorylation sites can also be distinguished by their relative rates of phosphorylation. The basal phosphorylation (of the R form) increases slowly over a period of several hours during incubation of the cells with [^{32}P]phosphate, whereas the cAMP-stimulated phosphorylation (appearing as the D form) is nearly maximal within 15 minutes of [^{32}P]phosphate and cAMP addition. Phosphoamino acid analysis shows that the amino acids modified in vivo are predominantly serine and threonine. Phosphotyrosine was not detected.

The in vivo phosphorylation kinetics closely resembles the observed kinetics of the shift in electrophoretic mobility detected by photoaffinity labeling. In contrast, when assayed in vitro, the kinetics of phosphorylation appear to be slightly faster. Within 30 seconds of addition of the cAMP stimulus, less than half the maximal mobility shift or phosphorylation observed in vivo has occurred. However, there is no difference in the pattern of in vitro phosphorylation of lysates or membranes taken between 30 seconds and 16 minutes after addition of the stimulus. One interpretation of these observations is that the substrate that is phosphorylated in the in vitro assays is an activated form of the receptor, which, at any time, represents a small fraction of total receptors. Future studies will be focused on the mechanism and role of the stimulated phosphorylation, with the ultimate goal of understanding adaptation in *Dictyostelium*.

SUMMARY AND PERSPECTIVES

D. discoideum offers an experimentally accessible system for the study of cell-cell communication, signal transduction, chemotaxis, and the control of gene expression by intercellular signals. Several generalizations become evident from these studies: (1) Multiple responses apparently can be mediated by one surface receptor. (2) The responses to a chemoattractant are transient in the presence of a persistent stimulus and therefore must have an adaptation mechanism. (3) Covalent modification of receptors may suggest a mechanism for adaptation with implications for adapting systems in general.

Identification of the surface cAMP receptor has revealed that the receptor is modified in a manner that correlates closely with the kinetics and dose-dependence of the transient responses to cAMP. This modification, reflected in a change in electrophoretic mobility, is associated with a change in the level of phosphorylation of the receptor in vivo and in vitro. Stimulus-induced phosphorylation of receptors has been observed in a wide variety of systems. Phosphorylation is correlated with adaptation for the β-adrenergic receptor and rhodopsin, as shown here for the cAMP receptor. Definitive proof that phosphorylation is responsible for adaptation is still lacking. We also have not definitively shown that phosphorylation of the cAMP receptor is the modification underlying the shift in electrophoretic mobility. It should now be possible to purify the cAMP receptor and determine the structural differences causing the shift in mobility. Reconstitution experiments will then allow for determination of the role of each form of the receptor in stimulation of adenylate cyclase and other cAMP-mediated responses.

As stated at the outset, the cAMP receptor influences the developmental regulation of several gene products. For example, the induction of the cAMP receptor itself, adenylate cyclase, and the membrane phosphodiesterase (for review, see Devreotes 1982) requires pulses of cAMP at 6-minute intervals during early development. This regulation is adaptive, since cAMP supplied at a constant level, which maintains the receptor in the D form, suppresses the expression of each of these proteins. Perhaps the R form is responsible for the induction of these early genes and the D form is responsible for their suppression. Further knowledge of the receptor and its regulation by covalent modification may give insights to these questions.

REFERENCES

ALCANTARA, F. and M. MONK. 1974. Signal propagation in the cellular slime mold *Dictyostelium discoideum*. *J. Gen. Microbiol.* **85:** 321.

BERLOT, C., J. SPUDICH, and P. DEVREOTES. 1985. Chemoattractant-elicited increases in myosin phosphorylation in *Dictyostelium*. *Cell* (in press).

BLOCK, S., J. SEGALL, and H. BEUG. 1983. Adaptation kinetics in bacterial chemotaxis. *J. Bacteriol.* **154:** 312.

BONNER, J. 1982. Comparative biology of cellular slime molds. In *The development of* Dictyostelium discoideum (ed. W.F. Loomis), p. 1. Academic Press, New York.

BRENNER, M. and S. THOMS. 1984. Caffeine blocks activation of cAMP synthesis in *Dictyostelium discoideum*. *Dev. Biol.* **101:** 136.

CHISHOLM, R., D. FONTANA, A. THEIBERT, H.F. LODISH, and P.N. DEVREOTES. 1984. Development of *Dictyostelium discoideum*: Chemotaxis, cell-cell adhesion, and gene expression. In *Microbial development* (ed. R. Losick and L. Shapiro), p. 219. Cold Spring Harbor Laboratory, Cold Spring Harbor, New York.

DEVREOTES, P.N. 1982. Chemotaxis. In *The development of* Dictyostelium discoideum (ed. W.F. Loomis), p. 117. Academic Press, New York.

DEVREOTES, P. and J. SHERRING. 1985. Kinetics and concentration dependence of reversible cAMP-induced modification of the surface cAMP receptor in *Dictyostelium*. *J. Biol. Chem.* **260:** 6378.

DEVREOTES, P.N. and T.L. STECK. 1979. Cyclic 3′,5′-AMP relay in *Dictyostelium discoideum*. II. Requirements for initiation and termination of the response. *J. Cell Biol.* **80:** 300.

DINAUER, M., S.A. MCKAY, and P.N. DEVREOTES. 1980a. Cyclic 3′,5′-AMP relay in *Dictyostelium discoideum*. III. The relationship of cAMP synthesis and secretion during the signaling response. *J. Cell Biol.* **86:** 537.

DINAUER, M., T. STECK, and P. DEVREOTES. 1980b. Cyclic 3′,5′-AMP relay in *Dictyostelium discoideum*. IV. Recovery of the cAMP signaling response after adaptation to cAMP. *J. Cell Biol.* **86:** 545.

———. 1980c. Cyclic 3′,5′-AMP relay in *Dictyostelium discoideum*. V. Adaptation of the cAMP signaling response during cAMP stimulation. *J. Cell Biol.* **86:** 554.

FONTANA, D., A. THEIBERT, T.-Y. WONG, and P. DEVREOTES. 1985. Cell-cell interactions in the development of *Dictyostelium*. In *The cell surface in cancer and development* (ed. M. Steinberg). Plenum Press, New York. (In press.)

GERISCH, G. and B. HESS. 1974. Cyclic AMP controlled oscillations in suspended *Dictyostelium* cells: Their relation to morphogenetic cell interactions. *Proc. Natl. Acad. Sci.* **71:** 2118.

GERISCH, G. and U. WICK. 1975. Cyclic AMP oscillations in cultures of *Dictyostelium discoideum*. *Biochem. Biophys. Res. Commun.* **65:** 364.

GOLDBETER, A. and D. KOSHLAND. 1982. Simple molecular model for sensing and adaptation based on receptor modification with application to bacterial chemotaxis. *J. Mol. Biol.* **161:** 395.

GREEN, A. and P. NEWELL. 1975. Evidence for the existence of two types of cAMP binding sites in aggregating cells of *Dictyostelium discoideum*. *Cell* **6:** 129.

HAHN, G., K. METZ, R. GEORGE, and B. HALEY. 1977. Identification of cAMP receptors in the cellular slime mold *Dictyostelium discoideum* using a photo affinity analog. *J. Cell Biol.* **75:** 41A.

HENDERSON, E. 1975. The cAMP receptor of *Dictyostelium discoideum*. Binding characteristics of aggregation competent cells and variation of binding levels during the life cycle. *J. Biol. Chem.* **250:** 4730.

JULIANI, M. and C. KLEIN. 1981. Photoaffinity labeling of the cell surface adenosine 3′,5′-monophosphate receptor of *Dictyostelium discoideum* and its modifaction in down-regulated cells. *J. Biol. Chem.* **256:** 613.

KLEIN, C., J. LUBS-HARKENESS, and S. SIMONS. 1985. cAMP induces a rapid and reversible modification of the chemotactic receptor in *Dictyostelium discoideum*. *J. Cell Biol.* **100:** 715.

KLEIN, P., A. THEIBERT, D. FONTANA, and P. DEVREOTES. 1985. Identification and cyclic AMP-induced modification of the cyclic AMP receptor in *Dictyostelium discoideum*. *J. Biol. Chem.* **260:** 1757.

KONIJN, T. 1970. Microbiological assay of cyclic 3′,5′-AMP. *Experientia* **26:** 367.

LEICHTLING, B., I. MAJERFELD, E. SPITZ, K. SCHALLER, C. WOFFENDIN, S. KAKINUMA, and H. RICKENBURG. 1984. A cytosolic cyclic AMP-dependent protein kinase in *Dictyostelium discoideum*. II. Developmental regulation. *J. Biol. Chem.* **259:** 662.

LUBS-HARKNESS, J. and C. KLEIN. 1982. Cyclic nucleotide-dependent phosphorylation in *Dictyostelium discoideum* amoebae. *J. Biol. Chem.* **257:** 12204.

MARTIEL, J. and A. GOLDBETER. 1984. Oscillations et relais des signaux d'AMP cyclique chez *Dictyostelium discoideum*: Analyse d'un modele fonde sur la modification du recepteur 1-AMP cyclique. *C.R. Acad. Sci.* **298:** 549.

MEYERS-HUTCHINS, B. and W. FRAZIER. 1984. Purification and characterization of a membrane-associated cAMP-binding protein from developing *Dictyostelium discoideum*. *J. Biol. Chem.* **259:** 4379.

MULLENS, I and P. NEWELL. 1978. cAMP binding to cell surface receptors of *Dictyostelium*. *Differentiation* **10:** 171.

PAN, P., E.M. HASS, and J.T. BONNER. 1972. Folic acid as second chemotactic substance in the cellular slime molds. *Nat. New Biol.* **237:** 181.

ROBINSON-STEINER, A.M., S.J. BEEBE, S.R. RANNELS, and J.D. CORBIN. 1984. Microheterogeneity of type II cAMP-dependent protein kinase in various mammalian species and tissues. *J. Biol. Chem.* **259:** 10596.

ROOS, W., V.D. NANJUNDIAH, and G. GERISCH. 1975. Amplification of cAMP signals in aggregating cells of *Dictyostelium discoideum*. *FEBS Lett.* **53:** 139.

SHAFFER, B.M. 1975. Secretion of cAMP induced by cAMP in the cellular slime mold *Dictyostelium discoideum*. *Nature* **255:** 549.

THEIBERT, A. and P.N. DEVREOTES. 1983. Cyclic 3′,5′-AMP relay in *Dictyostelium discoideum*: Adaptation is independent of activation of adenylate cyclase. *J. Cell Biol.* **97:** 173.

THEIBERT, A., P. KLEIN, and P.N. DEVREOTES. 1984. Specific photoaffinity labeling of the cAMP surface receptro in *Dictyostelium discoideum*. *J. Biol. Chem.* **259:** 12318.

TOMCHIK, K.J. and P.N. DEVREOTES. 1981. cAMP waves in *Dictyostelium discoideum*: Demonstration by a novel isotope dilution fluorography technique. *Science* **212:** 443.

VAN HAASTERT, P. 1983a. Sensory adaptation of *Dictyostelium discoideum* cells to chemotactic signals. *J. Cell Biol.* **96:** 1559.

———. 1983b. Binding of cAMP and adenosine derivatives ot *Dictyostelium discoideum* cells. *J. Biol. Chem.* **258:** 9643.

VAN HAASTERT, P. and E. KIEN. 1983. Binding of cAMP derivatives to *Dictyostelium discoideum* cells. *J. Biol. Chem.* **258:** 9636.

VAN HAASTERT, P. and P. VAN DER HEIJDEN. 1983. Excitation, adaptation, and deadaptation of the cAMP-mediated cGMP response in *Dictyostelium discoideum*. *J. Cell Biol.* **96:** 347.

WALLACE, L. and W. FRAZIER. 1979. Photoaffinity labeling of cyclic-AMP and AMP-binding proteins of differentiating *Dictyostelium discoideum*. *Proc. Natl. Acad. Sci.* **76:** 4250.

Regulation of Cell-type-specific Gene Expression in *Dictyostelium*

R.H. GOMER, S. DATTA, M. MEHDY,* T. CROWLEY, A. SIVERTSEN, W. NELLEN,
C. REYMOND, S. MANN, AND R.A. FIRTEL
Department of Biology, University of California, San Diego, La Jolla, California 92093

We are interested in the problems of how gene expression is regulated during development and how an initially homogeneous population of cells is partitioned into two distinct cell types. To approach these questions, we are studying a simple eukaryote, *Dictyostelium discoideum*. Approximately 10 hours after exhaustion of a food source, *Dictyostelium* amebae form a multicellular aggregate. Decisions are made at this time (or earlier according to some models [MacWilliams and Bonner 1979; Tasaka and Takeuchi 1981, McDonald and Durston 1984]) as to whether individual cells will differentiate into prespore or prestalk cells, the precursors to the terminally differentiated spore and stalk cells. As development proceeds, these cell types become located in distinct regions of the aggregate. This is most apparent in the migrating slug or pseudoplasmodium, in which the anterior 15–20% of the slug consists of prestalk cells and the remainder consists predominantly of prespore cells. In contrast to development in many eukaryote systems, the differentiation of prestalk and prespore cells is plastic until very late in development, and the two cell types can dedifferentiate and then redifferentiate into the other cell type. This presumably depends on the position of the cell within the multicellular aggregate and on physiological factors involved in this determination process (see Loomis 1982; Mehdy and Firtel 1985).

We are trying to identify the mechanisms underlying the differentiation of prestalk and prespore cells. This involves identifying first the biochemical signals and then the receptors and defining when they are present in development. Related to these questions are (1) how spatial differentiation is established within the multicellular aggregate and (2) the nature of the intracellular linkage between extracellular physiological signals and expression of cell-type-specific genes.

To approach these goals, we have identified genes that are preferentially expressed in either prespore or prestalk cells and a cell-type-nonspecific gene that is developmentally regulated and expressed equally in both cell types. The expression of cell-type-specific genes serves as a molecular marker for the differentiation of the individual cell types (Mehdy et al. 1983). This method has been used to identify two physiological factors, cAMP and a low-molecular-weight diffusible factor called conditioned medium factor (CMF), that regulate cell-type-specific gene expression. Genomic clones for prespore and prestalk genes, including the *Dictyostelium ras* gene, have been isolated, and DNA-mediated transformation has been used to examine their regulation in transfected *Dictyostelium* cells and to identify a region containing the *cis*-acting DNA regulatory sequences (Mehdy et al. 1983; Nellen et al. 1984a,b; Reymond et al. 1984; Mehdy and Firtel 1985). To approach the question of spatial differentiation, antibodies against proteins encoded by the cell-type-specific mRNAs have been made and used to examine the localization of the individual cell types during morphological differentiation. In this paper we review and describe experiments aimed at understanding the process by which cells choose between two alternate developmental pathways.

RESULTS

Identification of Physiological Factors Controlling Cell-type-specific Gene Expression

We have isolated cDNA clones complementary to mRNAs that are preferentially expressed in either prestalk or prespore cells. In addition, a cell-type-nonspecific gene that is developmentally regulated and equally expressed in both cell types has been identified (Mehdy et al. 1983, 1984; Reymond et al. 1984). Most prestalk mRNAs can first be detected late in aggregation as the multicellular aggregate forms. The level of expression of these genes peaks at the tight aggregate stage and then falls off during the next few hours of development. Most prespore genes are not expressed until the tight aggregation stage, when the level of prestalk RNAs begins to decline. The level of prespore RNAs peaks at the beginning of culmination and then decreases dramatically during the next few hours. Comparison of the expression of these genes with morphological differentiation indicates that cell-type-specific gene expression is tightly coupled to cellular differentiation. For example, in mutant FR17 (a temporally deranged mutant that forms mature stalk cells and spores in 16 hours instead of 25 hours), the timing of prestalk and prespore gene expression is accelerated and follows the timing of cell differentiation (C.L. Saxe and R.A. Firtel, in prep.). A developmental time course of representative mRNAs is shown in Figure 1. The ma-

---
*Present address: Plant Biology Laboratory, The Salk Institute, San Diego, California 92138.

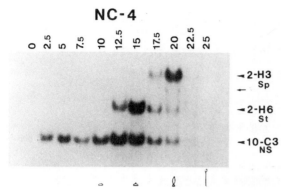

Figure 1. Expression of prespore, prestalk, and cell-type-nonspecific genes during the development of wild-type strain NC-4. Cells were developed on filter pads, and the resulting total cellular RNAs were analyzed by blot hybridization as described previously (Mehdy et al. 1983). Hours in development and approximate developmental morphologies are indicated. (2-H3) Prespore mRNA; (2-H6) prestalk mRNA; (10-C3) cell-type-nonspecific mRNA (for details, see Mehdy et al. 1984).

jority of the prestalk and prespore genes are coordinately regulated, although several cell-type-specific genes have been identified that show individual patterns of expression. Work similar to that described in this paper has been carried out in other laboratories (Barklis and Lodish 1983; Chisholm et al. 1984). The cell-type-nonspecific gene, 10-C3, is not expressed in vegetative cells. The RNA can be detected at low levels early during development and then parallels the kinetics of prestalk gene expression in wild-type strain NC-4 (see Fig. 1).

To identify the physiological and biochemical factors affecting cell-type-specific gene regulation, it was necessary to establish in vitro culture systems in which the regulation of the prestalk- and prespore-specific genes can be examined in the absence of multicellular differentiation. This allows us to control differentiation by the addition of exogenous factors. Two in vitro culture systems were established (Mehdy et al. 1983; Mehdy and Firtel 1985). In the first, starved, washed cells were suspended at a high density in buffered salts and shaken either at 70 rpm (which allows formation of multicellular agglomerates similar to those seen in normal development) or at 230 rpm, in which the cells remained isolated with an absence of sustained cell-cell contacts (Mehdy et al. 1983; Reymond et al. 1984). These conditions allowed us to determine whether cAMP and cell-cell or cell-surface interactions are necessary for cell-type-specific gene expression. The cell-type-nonspecific gene, 10-C3, is induced in both slow- and fast-shaking cultures independently of exogenous cAMP. Addition of cAMP to a fast-shaking culture after 6 hours is sufficient to induce prestalk (but not prespore) gene expression. In slow-shaking cultures, prestalk gene expression occurs independently of exogenous cAMP, as would be expected for multicellular aggregates, whereas prespore gene expression is enhanced by cAMP. Thus, prestalk gene expression requires *either* cAMP or cell aggregation, and prespore gene expression requires *both* cAMP and cell aggregation. Cell-type-nonspecific gene expression does not require either condition.

Earlier reports showed that single wild-type cells could differentiate into stalk cells under specific conditions (Bonner 1970; Town et al. 1976). Using a mutant strain, Kay (1982) was able to show that single cells could differentiate into mature spores. We were interested in examining the requirements for prestalk and prespore gene expression in single-cell monolayers plated in submerged culture on petri dishes (Mehdy et al. 1984; Mehdy and Firtel 1985). Cells were plated from 900 cells/cm² ($1\times$ cell density) to 360,000 cells/cm² ($400\times$ cell density). At $1\times$, $2\times$, and $5\times$ cell densities, all cells are single, with distances of many cell diameters between the individual cells. At a $20\times$ density, approximately 90% of the cells are single cells and the remainder are present in small aggregates of two to five cells. At higher cell densities, the cells are predominantly in aggregates. In the absence of cAMP, cells at $1-20\times$ cell densities did not express prestalk- or prespore-specific genes. At higher cell densities (e.g., $400\times$), both classes of cell-type-specific genes were expressed at very low levels, probably in response to endogenous cAMP produced in the aggregates. When cAMP was added to identical cultures, neither prestalk nor prespore genes were induced in $1\times$ and $5\times$ cultures, but they were induced in $20\times$ and $400\times$ cultures at levels comparable to those in normal multicellular aggregates.

These results suggested that a density-dependent secreted factor was necessary for prestalk and prespore gene expression; $20\times$ cultures contain predominantly (>90%) single cells, yet the level of prespore and prestalk gene expression is similar to levels found in $400\times$ cultures with cAMP or normally developing aggregates. To determine whether cells are secreting a soluble factor, culture medium was isolated from high-density $100\times$ cultures after 20 hours, filtered to remove any possible cell debris, and then assayed on cells at a $2\times$ cell density for the ability to promote prestalk and prespore gene expression. Cultures containing fresh medium (buffered salts; FM) showed no prestalk or prespore gene expression in the absence of cAMP (data not shown) (Mehdy and Firtel 1985). In the presence of cAMP, cell-type-specific genes are expressed in cultures containing FM (Fig. 2, lanes 0–20 FM) but at a highly reduced level relative to cells containing conditioned medium (Mehdy and Firtel 1985). Low-density cells in the presence of CMF, however, expressed prestalk and prespore genes at a high level in the presence (Fig. 2, lanes 0–20 CM), but not the absence of, exogenous cAMP (data not shown). These results indicate that cAMP is necessary for prespore and prestalk gene expression and that prespore gene expression can be induced in the absence of multicellular aggregate formation. The results do, however, suggest that a cell-surface interaction is necessary for the expression of prespore genes. Thus, a soluble factor produced by cells at a high density is necessary for prestalk and prespore

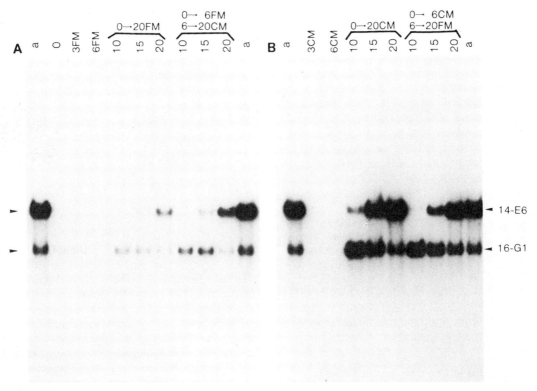

Figure 2. The period of CMF action on cell-type-specific gene expression. Vegetative Ax-3 cells were resuspended in conditioned or fresh medium. The suspensions were plated on petri dishes at a density of 1800 cell/cm². Half of the plates in each culture were treated with 300 μM cAMP after 6 hr. At this time, the medium was decanted from the remaining plates. Those cells previously incubated in fresh medium were now incubated in conditioned medium containing 300 μM cAMP. Those cells previously incubated in conditioned medium were now incubated in fresh medium containing 300 μM cAMP. A fraction of the plates from each culture was harvested at the times indicated. Total RNA was purified, size-separated by electrophoresis on formaldehyde gels, and blotted onto Gene Screen. Lane a is RNA from culminating aggregates as a reference. The filter was hybridized with a labeled prespore-specific cDNA clone, 16-G1, and 14-E6.

gene expression. Preliminary biochemical characterizations indicate that the factor has a low molecular weight and is hydrophobic: Approximately half of the biologically active material binds to a C12 hydrophobic column (M.C. Mehdy and R.A. Firtel, unpubl.).

We have identified the developmental period of CMF production by cells (Mehdy and Firtel 1985). CMF production is maximal during aggregation and then continues at a lower rate after the formation of multicellular aggregates. These kinetics suggested that CMF is important for a step in cell differentiation prior to the induction of prespore- or prestalk-specific genes. To examine this more directly, we determined the effect of CMF on early gene expression. These early genes are not expressed in vegetative cells and are induced during early development (Rowekamp and Firtel 1980; Poole et al. 1981; S. Mann et al., in prep.). Expression of these genes can be induced in vegetative cells by pulses of cAMP, which mimics the cAMP pulsing normally occurring during aggregation. The results of these experiments show that CMF stimulates the expression of the preaggregation genes but has no detectable effect on the expression of a set of genes, including actin, that are expressed in vegetative growth and during early development.

Since CMF is produced maximally during aggregation and stimulates the expression of preaggregation genes, it appeared that CMF could be regulating events during early development and that these events, in turn, affect prespore and prestalk gene expression. We therefore examined the requirement of CMF for prestalk and prespore gene expression at various stages of *Dictyostelium* development. In the experiment shown in Figure 2, cells were plated in conditioned medium for 6 hours. The medium was then removed and replaced with fresh medium containing cAMP. After incubation for an additional 14 hours, prestalk and prespore gene expression was assayed. The results showed that CMF need not be present during the actual induction of prespore or prestalk gene expression as long as it is present during the first 6 hours after starvation, the time at which preaggregation genes are expressed (Fig. 2, lanes 0–6 CM and 6–20 FM). In the same experiment, CMF was added at 6 hours, along with the cAMP. Under these conditions, prestalk and prespore genes are expressed at substantially lower levels (Fig. 2, lanes 0–6 FM and 6–20 CM). These experiments suggest that CMF is required for a step in early differentiation necessary for the induction of the cell-type-specific genes by cAMP. We have defined the developmental state at which cells are capable of being induced by cAMP as "competence" (see Discussion).

We have examined the inducibility of prestalk and prespore gene expression by cAMP added at various times in low-density cultures to cells incubated in conditioned medium (Mehdy and Firtel 1985). When cAMP is added shortly after plating the cells (<2 hr), there is no induction of late gene expression. As cAMP is added at successively later times, the level to which prestalk and prespore mRNAs accumulate rises and reaches a maximum when cAMP is added 6-8 hours after starvation; when cAMP is added more than 12 hours after starvation, there is negligible prespore gene expression, and prestalk gene expression is reduced. In all cases, prespore gene expression occurs between 4 and 12 hours after addition of cAMP, whereas prestalk gene expression occurs within 1 hour. The results indicate that during the first 6 hours after plating in conditioned medium, the level of prestalk and prespore gene expression is proportional to the level of preaggregation gene expression. There is thus a relatively narrow window for the induction of prespore gene expression under these conditions.

Regulation of Cell-type-specific Genes in Transfected Cells

We have used a DNA-mediated transformation system established in our laboratory to reintroduce gene constructs containing prestalk-specific promoters back into *Dictyostelium* cells. The transformation system is based on a gene fusion (denoted B10S) containing a *Dictyostelium* Actin 6 promoter fused to the neomycin resistance gene from Tn5 (Nellen et al. 1984a,b; Nellen and Firtel 1985). This gene fusion encodes a functional kanamycin phosphotransferase that confers resistance to the drug G418. The fusion carries approximately 600 nucleotides of the 5'-flanking sequences, the 5'-untranslated region, and eight codons of the coding region of Actin 6. This is fused in-frame to the neomycin-resistant gene-coding sequences. Vectors also carry a 3' polyadenylation and/or termination signal from the *Dictyostelium* Actin 8 gene (Nellen and Firtel 1985). When the gene fusion, cloned into pBR322, is transfected into *Dictyostelium* cells, the DNA integrates as a tandem array. In a transformed population, it is present at an average of approximately 5 copies per cell, but cells that have copy numbers ranging from 50 to 150 copies can be selected by growth on G418. We have also recently established extrachromosomal vectors on the basis of a cloned endogenous plasmid present in several *Dictyostelium* strains (Firtel et al. 1985). The Actin 6–neomycin-resistance gene (Actin 6–*neo*^{R}) fusion ligated into the cloned *Dictyostelium* plasmid can be used to directly transform *Dictyostelium* cells. This vector replicates extrachromosomally in *Dictyostelium* and is present at approximately 50 copies per cell. Transformation frequencies are approximately 10^3 per 10^7 cells for the integrating plasmids and approximately tenfold higher with the extrachromosomal vectors. We have used this DNA-mediated transformation to examine the regulation of two prestalk genes, the *Dictyostelium* proto-oncogene homolog of mammalian *ras* (*Dd-ras*) and a gene encoding a cysteine protease related to cathepsin B, denoted *pst-cath*.

The *ras* gene is single copy and encodes a low-abundance 1.2-kb mRNA in vegetative cells, which, upon the onset of development, rapidly disappears (Reymond et al. 1984). *ras* gene expression is then reinduced in concert with other prestalk genes late in aggregation. Two mRNAs (0.9 kb and 1.2 kb) are specifically expressed in prestalk cells but not in prespore cells. *ras* protein levels remain constant or decrease slightly from vegetative cells through early development and then decrease rapidly at about the time of prespore gene expression. *ras* protein is found in prestalk cells but not in prespore cells in pseudoplasmodia (R.H. Gomer et al., in prep.). *ras* gene expression is therefore unusual for a prestalk gene since there is expression in vegetative cells. The analysis of DNA sequence data in the 5'-flanking region of the *ras* gene and *pst-cath* shows a GC-rich region lying approximately 200 nucleotides upstream of each cap site. Within this is a core sequence that is homologous between the two prestalk-specific genes. We feel that this is the regulatory region associated with prestalk-specific regulation (S. Datta et al., unpubl.).

As described above, cAMP can induce prestalk gene expression in high-density, fast-shaking cultures (no sustained cell contact) or in low-density, single-cell culture containing CMF. When normal developing aggregates are dissociated, the level of prestalk and prespore mRNAs rapidly decreases to levels less than 10% of the original level in the aggregate by 3–4 hours after dissociation (Barklis and Lodish 1983; Mehdy et al. 1983; Reymond et al. 1984; Mehdy and Firtel 1985). In vitro nuclear run-on experiments and pulse-labeling experiments indicate that these genes are not transcribed under these conditions (Landfear et al. 1982; A. Sivertsen and R.A. Firtel, unpubl.). When cAMP is added to disaggregated cells in fast-shaking culture, the levels of both prestalk and prespore mRNAs rapidly rise to between 50% and 100% of the initial level in the multicellular aggregate within 1 hour. In the case of the *Dictyostelium ras* gene, the 0.9-kb mRNA rises to similar levels and the 1.2-kb mRNA increases approximately five- to tenfold higher than that in multicellular aggregates. A similar "superinduction" of the 1.2-kb mRNA is seen in fast-shaking cultures in the presence of cAMP.

We have utilized DNA-mediated transformation to examine the regulation of the *Dictyostelium ras* gene and *pst-cath* under a number of developmental conditions. A series of constructions were made that contain various regions of the 5' portion of the *Dd-ras* gene fused to a *Dd-ras* cDNA clone (see Fig. 3A). All constructs have been transformed into *Dictyostelium* cells and are expressed during vegetative growth (Reymond et al. 1985). The *ras* gene constructs lack proper transcription termination and/or polyadenylation signals. 3'-end processing of the *ras* gene transcripts occurs

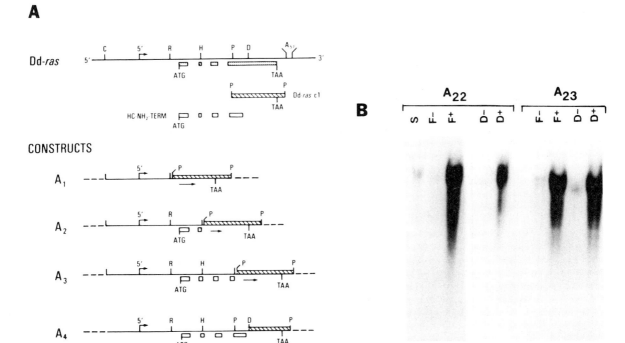

Figure 3. Expression of *ras* gene constructs in transformants. (*A*) A description of the construction of the A series vectors was published recently (Reymond et al. 1985). A map of the cloned endogenous *ras* gene (Reymond et al. 1984) is shown with the regions coding for the *ras* protein (stippled area). Open boxes indicate the regions coding for the highly conserved amino terminus of the protein. The map of the *Dd-ras* cI cDNA is shown in relationship to the genomic sequence (hatched area). The putative 5′ start of the mRNA (→) and the ATG (initiation) and TAA (termination) codons of the proteins are indicated. (*B*) Regulation of expression of *ras* gene constructs. Cells from single colonies transformed with construct A_2 were placed in fast-shaking cultures under starvation (Mehdy et al. 1983). After 6 hr, EDTA was added to cultures to a final concentraton of 1 mM and cAMP to 300 μM was added to half the cultures. After an additional 13 hr, RNA was extracted and analyzed (Mehdy et al. 1983). A large increase in the RNA level can be seen in samples from the plus cAMP cultures (F^+ lanes) as compared with levels found in cultures without exogenously added cAMP (F^- lanes) or in slowly shaken cultures (lanes S). Vegetative cells from the two single-colony isolates were plated for development. At 18 hr (between the tight aggregate and early cumulation stage), the cells were disaggregated into single cells and placed in fast-shaking culture for 13 hr. cAMP was given to half the culture, and RNA was isolated from cells 2 hr later (see Mehdy et al. 1983). (S) Slow-shaking culture; (F^+ and F^-) fast-shaking culture with and without added cAMP, respectively; (D^+ and D^-) disaggregated cells with and without cAMP, respectively. Total RNA (7.5 μg) was loaded per lane.

within the 5′-flanking region of the Actin 6-*neo*[R] gene fusion immediately downstream from the *ras* gene fusion. These fusions therefore express RNAs that are larger than the 1.2-kb endogenous *Dd-ras* mRNA and thus can be distinguished from the endogenous *ras* gene transcript.

To determine whether the *ras* constructs are regulated by cAMP, starved vegetative cells were placed in fast-shaking culture for 5 hours and then cAMP was added to half the culture. The cultures were then shaken for an additional 15 hours. Figure 3B shows the hybridization of a *Dd-ras* probe to RNA isolated from these cultures. A significant amount of hybridization is observed to RNA isolated from cultures containing exogenous cAMP but not in cultures lacking exogenous cAMP. We also examined the regulation of *ras* gene expression in disaggregated cells with and without added cAMP. The level of *ras* mRNA is very low in disaggregated cells lacking cAMP but is very high in cultures containing cAMP. These results are consistent with those obtained with the wild-type gene, indicating that these constructs are properly regulated by cAMP after transfection into *Dictyostelium* cells and that the *cis*-acting regulatory sequences are localized within approximately 500 nucleotides upstream of the cap site or within a portion of the 5′-untranslated region or protein-coding region contained within the cDNA clone (Reymond et al. 1985).

Spatial and Temporal Regulation of Prestalk and Prespore Genes

We have shown that cAMP is sufficient to induce prestalk gene expression in competent cells (i.e., fast-shaking cells at high density or low-density cells in cultures containing CMF). However, to understand the mechanisms by which patterning is established in a multicellular organism, one must also understand how spatial differentiation is established between two separate cell types. Eventually, we would like to know how physiological factors such as cAMP help to determine this patterning and the developmental decision of cells to differentiate into either prespore or prestalk cells.

To examine the spatial pattern of expression of individual genes within the developing multicellular aggregates, we have made antibodies against *Dictyostelium* proteins encoded by prespore and prestalk genes. DNA fragments carrying open reading frames from these genes were cloned into λgt11 with EcoRI linkers in-frame within the carboxyterminal region of the β-galactosidase gene to encode a fusion protein (Young and Davis 1983). Fusion proteins were purified from *Escherichia coli* infected with the recombinant λgt11 phage and used as antigens to immunize rabbits (Reymond et al. 1984; R.H. Gomer et al., in prep.).

We have used these antibodies produced against prespore or prestalk proteins in immunofluorescence studies to examine the distribution of cells within the multicellular aggregate expressing a specific gene (Gomer and Lazarides 1981, 1983). In addition, they have been used to assay cells in the in vitro culture systems to determine whether specific genes are being expressed. Figure 4, a and b, shows immunofluorescence staining of migrating *Dictyostelium* pseudoplasmodia using antibodies against the protein encoded by *Dd-pst-cath* (see Fig. 4a), a prestalk gene, and against *Dd-beejin* (Fig. 4b), a prespore gene (S. Datta et al., in prep.). As is expected from the known spatial distribution of prespore and prestalk cells within the slug, the anti-*pst-cath* antibody preferentially stains the anterior portion of the slug. In addition, there is some staining along the edges of the slug, suggesting that cells expressing this gene may represent an exterior layer around a majority of the slug. It should also be noted that the staining is more intense in the tip region and on the outer edges than in the central portion of the anterior of the slug. In contrast, the anti-*beejin* antibody preferentially stains the posterior region. As has been observed using vital dyes as stains and cell-type-specific monoclonal antibodies, there is a fairly discrete partitioning between the prespore and the prestalk cells in slugs (Bonner 1952; Tasaka et al. 1983; Krefft et al. 1984).

Spatial and Temporal Regulation of *pst-cath* Gene Fusions

We have examined the regulation of a *pst-cath* gene in transformants to determine whether it is properly regulated in both a temporal and spatial manner and whether the expression was regulated in in vitro cultures by cAMP (S. Datta et al., in prep.). To tag the *pst-cath* gene in transformants and to enable us to distinguish the expression of the transfected gene from that of the endogenous gene, we made a fusion in which the first approximately 300 nucleotides of the *pst-cath*-coding sequence, the 5′-untranslated region, and approximately 1 kb of the 5′-flanking region was fused in-frame to the *E. coli* β-glucuronidase-coding region. The *pst-cath*/β-gluc fusion was inserted into the B10S transformation vector and transfected into *Dictyostelium* cells. A cell population was then selected that had a high average copy number of vector DNA (∼150 copies/cell) and single clones were isolated (Nellen and Firtel 1985; S. Datta et al., in prep.). Figure 5 (top) shows a map of the constructs and the expression of the *pst-cath*/β-gluc gene in transformants during development. RNA isolated from transformants at various developmental stages was also assayed for expression of the endogenous *pst-cath* gene, using a probe

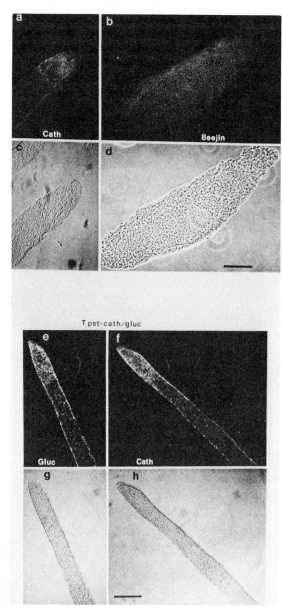

Figure 4. Cryosections of fixed *Dictyostelium* slugs stained by indirect immunofluorescence. (*a–d*) Kessin Ax-3 slugs; (*e–h*) slugs from the transformant *pst-cath*/β-gluc. Phase images of the sections shown in *a*, *b*, *e*, and *f* are shown in *c*, *d*, *g*, and *h*, respectively. Primary antibodies for the indirect immunofluorescence were rabbit anti-*Dd-cath*/β-galactosidase fusion protein (*a*, *f*), rabbit anti-*beejin*/β-galactosidase fusion protein (*b*), and rabbit anti-β-glucuronidase (*e*). The latter was a gift from R. Jefferson and D. Hirsch (University of Colorado, Boulder). Bars, 100 μm.

that recognizes the 3' end of the endogenous *pst-cath* gene and does not cross-hybridize to the gene fusion (Fig. 5, bottom, part A and B). The *pst-cath*/β-gluc gene does not have an appropriate *Dictyostelium* 3'-termination signal, so that the resulting RNA is transcribed from the *pst-cath* cap site through a majority of the clone and terminates in a region in the 5'-flanking sequences of the Actin 6 promoter. Termination of transcription within this region has been noted previously (see above; Nellen and Firtel 1985). In addition, a smear of lower-molecular-weight RNA is observed that results from either an instability of this transcript or random termination within the plasmid sequences 3' to the β-glucuronidase gene. As shown in Figure 5, the

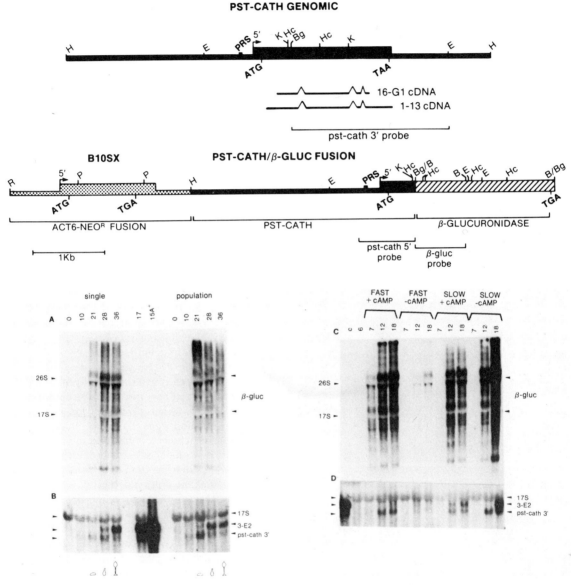

Figure 5. Regulation of *pst-cath* in transformants. (*Top*) Restriction map of a genomic *Hin*dIII fragment containing the *pst-cath* gene and an in-frame fusion of the *pst-cath* gene to the coding region of *E. coli* β-glucuronidase gene (S. Datta et al., in prep.). This plasmid, pSD1, carrying the *pst-cath*/β-gluc gene also carries the Actin 6–*neo*R fusion gene from B10SX (Nellen and Firtel 1985) as a selectable marker for selection of G418-resistant transformed cells. PRS is the putative regulatory sequence. (*Bottom*) *Dictyostelium* cells were transformed with pSD1. Populations capable of growth in G418 were selected and single colonies were isolated. Cells were shown to have about 100 copies of the vector per cell. (*A*) Cells from a single colony (single) or a population (population) were plated for development and harvested at the appropriate morphological stages. RNA was extracted and then sized on denaturing gels and blotted to Gene Screen (see Mehdy et al. 1983). Duplicate blots were hybridized to either a β-gluc-specific probe (*A*) or a mixed probe complementary to the 3' part of *pst-cath* and a prespore-specific gene probe, 3E-2 (*B*). See upper panel for location of specific probes. Lanes show the times at which cells were harvested during development. Note that cells transformed with *pst-cath* complete development in approximately 35 hr instead of 25 hr. Prestalk and prespore genes, however, are expressed at the proper morphological stage. To observe cAMP regulation, transformed cells were fast shaken for 6 hr and then separated into either fast- or slow-shaking cultures with or without added cAMP (Mehdy et al. 1983). Cells were harvested at various times and RNA was extracted. A Northern blot of the RNA was hybridized with the same probes as in *A* and *B*. (*C*) β-gluc probe; (*D*) 3'-*pst-cath* and 3E-2 probes.

developmental time course of the endogenous gene (part B) and the transformed *pst-cath*/β-gluc gene fusion (part A) are similar. Also shown is the hybridization of a prespore-specific gene that is unchanged relative to its pattern in untransformed normal developing cells (see Fig. 5, part B). It can be seen that the *pst-cath* endogenous mRNA and the prespore mRNA are not degraded.

To determine whether the gene is also regulated by cAMP, *pst-cath*/β-gluc transformants were assayed in fast- and slow-shaking cultures with and without cAMP. Figure 5 (part C) shows that the *pst-cath*/β-gluc gene fusion is induced in fast-shaking culture with cAMP and in slow-shaking culture in a manner identical with that of the endogenous *pst-cath* gene (Fig. 5, part D). Expression of the fusion gene was also assayed in lower-density cell cultures containing CMF with or without exogenous cAMP (Mehdy and Firtel 1985; see Fig. 2). Expression of the fusion gene was assayed by immunofluorescence using the anti-β-glucuronidase antibody. As shown in Figure 6, cultures containing exogenous cAMP show a high level of expression of the gene (50% of cells), whereas those lacking cAMP do not. These results suggest that the *pst-cath*/β-gluc gene fusion is being properly regulated by the same signals as the endogenous gene.

We have also examined the spatial expression of the gene fusion within developing *Dictyostelium* cells. The *pst-cath*/β-gluc gene encodes a fusion protein that can be recognized by anti-β-glucuronidase antibodies or antibodies against that portion of *pst-cath* present in the gene fusion. We have used these antibodies to examine the expression of the gene fusion in migrating slugs. Figure 4, a, e, and f, shows immunofluorescent staining of untransformed slugs and slugs carrying the gene fusion stained with either an anti-*pst-cath* antibody or an anti-β-glucuronidase antibody. As can be seen, both antibodies stained cells within the anterior region of the slug, in agreement with the prestalk-specific expression of the RNA encoding the antigen. Moreover, the pattern of expression of the endogenous gene in wild-type untransformed cells (Fig. 4a) and the gene fusion in transformants is similar (Fig. 4e,f), again suggesting a spatial as well as temporal regulation of the gene.

Use of Anti-sense Constructs to Block Gene Expression in *Dictyostelium*

To examine the function of individual developmentally regulated genes in *Dictyostelium*, it would be useful to specifically block expression of these genes. One approach to achieve this in the absence of specific mutations is to express an anti-sense RNA complementary to the mRNA encoded by the endogenous gene (Coleman et al. 1984; Izant and Weintraub 1984; Mizuno et al. 1984). This approach has the potential of specifically blocking the expression of a single-copy gene or members of a multigene family. We have tested this approach in *Dictyostelium* using the developmentally regulated discoidin I multigene family. Discoidin I is encoded by a three-member multigene family. The protein, which is maximally expressed during the preaggregation stage, has been implicated in cell-substratum attachment, is involved in ordered cell migration, and is analogous to fibronectin (Springer et al. 1984). Lack of expression of discoidin I is not lethal for *Dictyostelium* (Alexander et al. 1983). We have taken advantage of the fact that the DNA sequences of the protein-coding region of all three members of the discoidin I multigene family are very homologous (Rowekamp et al. 1980; Poole et al. 1981). We constructed a vector carrying a piece of the coding sequence in an anti-sense orientation downstream from the 5′ end of the Disc I-α gene (see Fig. 7A) (Poole and Firtel 1984). This vector, Disc I-α Rev, was cotransformed into *Dictyostelium* cells with vector B10SX, a derivative of B10S, selecting for G418 resistance (Crowley et al. 1985). Cells were then selected for growth in G418. Analysis of vector DNA within these cells indicated that the Disc I-α Rev vector and B10SX were both present at about 100 copies per cell.

We have investigated the RNA expression within a series of single-colony isolates derived from these transformants. Figure 7B shows that the level of discoidin I RNA is reduced at least 10–20-fold in comparison with the level in B10SX transformants not carry-

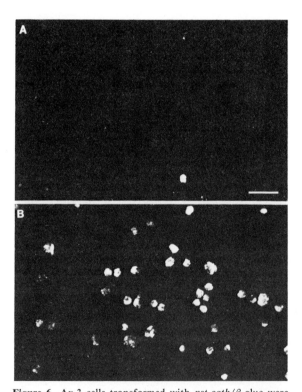

Figure 6. Ax-3 cells transformed with *pst-cath*/β-gluc were grown in fast-shaking culture to inhibit formation of cell-cell contacts. After 6 hr, the culture was divided in half; cAMP was then added to half of the culture every 2 hr. Samples of the cells were removed, allowed to adhere to coverslips, and fixed and stained for β-glucuronidase by indirect immunofluorescence. (*A*) Cells from cultures lacking cAMP; (*B*) cells from cultures with added cAMP. Approximately 70 cells were in each field, determined by counting cells in a phase image of the field (not shown). Bar, 30 μm.

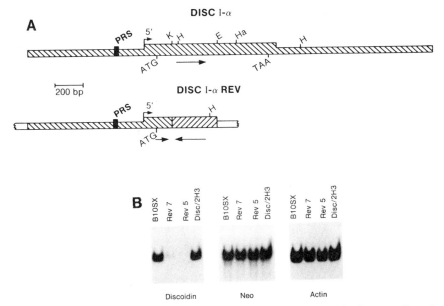

Figure 7. Analysis of discoidin I expression in transformants. (*A*) Restriction map of the Disc I-α gene. To make the Disc I-α Rev construction, the *Kpn*I/*Bgl*II fragment containing the 5' end of the gene was inserted into pUC18cm that had been linearized with *Kpn*I and *Bam*HI. The resulting plasmid, pDdDisc 5', was linearized with *Kpn*I and digested with S1 nuclease to give blunt ends. The 310-bp *Kpn*I/*Eco*RI fragment from within the gene was purified, treated with S1 nuclease, and then ligated to the linear pDdDisc 5'. The DNA was used to transform *E. coli*. Plasmid DNA made from single colonies was then screened by restriction enzyme digests to select the appropriate clone. In Disc I-α Rev, the *Kpn*I/*Eco*RI portion of the coding region is inverted with respect to the promoter. PRS is putative regulatory sequence (Poole and Firtel 1984). (*B*) Assay of discoidin RNA expression in Disc I-α Rev *Dictyostelium* transformants. RNA was extracted from vegetatively growing cells, separated on an agarose formaldehyde gel, and then transferred to Gene Screen. The filter was then hybridized with a nick-translated discoidin-I-specific probe. After autoradiography, the probe was removed by sequential washes of 5 mm EDTA (pH 7.2) at 95°C. The filter was then probed with a Tn5 *neo*R gene-specific fragment to assay for the level of Actin 6–*neo*R fusion gene mRNA. The filter was then stripped and probed with a *Dictyostelium* actin-specific probe. Control lanes: (B10SX) RNA from cells transformed with the B10SX transformation vector; RNA from cells transformed with a B10 construction containing the Disc I-α promoter linked to the coding region of the prespore-specific gene, 2H3 (Disc/2H3). (Rev5 and Rev7) RNA from two single-colony isolates carrying approximately 100 copies of the Disc I-α Rev construct (Crowley et al. 1985).

ing Disc I-α Rev DNA. We have also examined the level of discoidin I mRNA in transformants carrying a high copy number of the Disc I-α promoter fused out of reading frame to the 3' end of a prespore gene (Disc/2H3) (Crowley et al. 1985). The level of discoidin I mRNA in these cells is the same as that in control cells, suggesting that the decreased level of the discoidin I mRNA in Disc I-α Rev transformants is not due to promoter competition by the high copy number of the Disc I-α promoter.

Discoidin I has been proposed to be analogous in function to fibronectin in metazoan cells. Both proteins have the sequence Gly-Arg-Gly-Asp, which has been implicated as important for cell attachment by fibronectin (see Springer et al. 1984). When control *Dictyostelium* cells are plated at a high density in submerged culture under starvation conditions, the cells form streams on the surface of the petri dish. When synthetic peptides containing the four-amino-acid sequence Gly-Arg-Gly-Asp are added to these cultures, the cells do not stream, presumably because the peptide competes with discoidin I for receptor sites. We have examined streaming in wild-type axenic cultures, in cultures carrying the Disc/2H3 gene fusion and the Disc Rev gene. Ax-3 cells and cells transformed with Disc/2H3 stream normally, whereas the cells carrying a high copy number of Disc I-α Rev do not; immunofluorescent staining using anti-discoidin I antibodies shows that more than 95% of the cells are not expressing discoidin I (data not shown). Presumably, this is a phenocopy of a discoidin I⁻ phenotype (Springer et al. 1984) and results from a blockage of discoidin I expression (Crowley et al. 1985).

DISCUSSION

We have established in vitro culture systems that enable us to examine the requirements for prestalk and prespore gene expression. Experiments using cultures in which the cells are either in suspension or plated on a plastic surface suggest that a cell-surface interaction is necessary for prespore gene expression, whereas prestalk gene expression does not require such an interaction. Addition of cAMP can induce prestalk gene expression in cells made competent by CMF (see below). Prespore gene expression is similarly induced by cAMP, but the expression is temporally displaced within 4–12 hours from the time of induction of prestalk genes. These results suggest that cAMP induces other cellular events that then result in the induction of prespore gene expression. Prespore gene expression appears to require other factors, including unstable protein molecules, as evidenced by the sensitivity of prespore gene expression to cycloheximide.

A second factor, CMF, must be present during the first 6 hours after starvation to allow the cells to reach a state of competence for prespore and prestalk gene expression by cAMP. CMF is secreted rapidly upon the onset of starvation and continues to be secreted, although at a lower level, throughout later development. Prestalk genes are induced rapidly in response to permissive CMF and cAMP conditions, whereas prespore genes are delayed in their induction until about 18 hours after starvation. It seems probable that additional developmental parameters besides cAMP and CMF affect prespore gene induction and that these somehow result in a fixed timing of prespore gene induction relative to prestalk gene expression.

CMF also affects the expression of preaggregation genes but does not have a detectable effect on actin genes or a gene maximally expressed in vegetative cells. We do not know whether the same or different components in the conditioned medium influence early and late gene expression. A dependence of late gene expression on early gene expression is suggested in our experiments by varying the time of cAMP treatment of starving cells. The expression of three early genes is reduced by micromolar concentrations of cAMP (Williams et al. 1980; S. Mann et al., in prep.). Immediate application of cAMP blocks nearly all expression of these genes, whereas progressively later cAMP applications allow greater accumulation of these mRNAs. Prespore and prestalk genes are also not induced when micromolar levels of cAMP are present starting from the initiation of development. As with the early genes, cells starved 2-6 hours prior to cAMP exposure show increased induction of prespore and prestalk genes. Since the level of late gene expression is roughly proportional to the level of early gene expression prior to cAMP addition, it is possible that the expression of certain early genes is necessary for late gene expression.

Figure 8 shows a model of the interrelationship between CMF, cAMP, and preaggregation gene expression to prespore and prestalk gene expression. In our model, CMF is required for preaggregation gene expression. Expression of these genes is stimulated by nanomolar pulses of cAMP that mimic those which occur in vivo during aggregation (S. Mann et al., in prep.). This can also result in the induction of prestalk and prespore genes (Chisholm et al. 1984). Preaggregation gene expression is necessary for the cells to become competent for induction of prestalk and prespore gene expression. cAMP added in micromolar levels induces prestalk and prespore gene expression in competent cells. When micromolar levels of cAMP are added during the first 6 hours, preaggregation gene expression is turned off, and the level of prestalk and prespore gene expression is proportional to the level to which the preaggregation gene was induced. Prespore gene expression is always delayed relative to prestalk gene expression, possibly because other cellular processes must be expressed prior to prespore gene induction. Our model would suggest that during normal development, preaggregation genes are induced in starved cells that are exposed to a sufficiently high concentration of CMF, i.e., under conditions of high cell density. Some genes expressed during this period are necessary for subsequent prestalk and prespore gene induction by cAMP. Proteins expressed during this period that are required for late gene expression may include the cAMP-dependent protein kinase system (Leichtling et al. 1984) and/or other cAMP-binding proteins that may interact directly with DNA regulatory sequences. We do not feel that the cell-surface cAMP receptor plays a direct role in mediating induction of prestalk or prespore gene expression, since the level of receptor has decreased at the time late expression is induced (Devreotes 1982). As the cells form multicellular aggregates, the level of cAMP may rise, resulting in an inhibition of preaggregation gene transcription and induction of prespore and prestalk gene expression.

The molecular basis of the linkage between physiological factors and cell-type-specific gene expression can be examined using the *Dictyostelium* DNA-mediated transformation system. Gene fusions made with either of two different prestalk promoters, *Dd-ras* or *pst-cath*, are properly regulated by cAMP in fast- and slow-shaking culture and in disaggregated cells. In addition, the *pst-cath/β*-gluc gene fusion shows proper spatial regulation. Our laboratory is in the process of identifying the *cis*-acting DNA sequence necessary for the induction of the *Dd-ras* gene with cAMP. Preliminary results suggest that the GC-rich sequence lying approximately 200 nucleotides upstream of the cap site is necessary. In similar experiments, we have used the frequency of transformation as an assay for *cis*-acting sequences necessary for transcription from the Actin 6 promoter that controls expression of the neomycin resistance gene in the Actin 6–neo[R] gene fusion. Deletion analysis indicates that a region approximately 400 nucleotides 5' to the Actin 6 cap site is necessary for the expression of the Actin 6–neo[R] gene fusion. Eventually, we hope to identify the *cis*-acting sequences controlling spatial and temporal regulation of gene expression. We also hope to identify the *trans*-acting factors and reg-

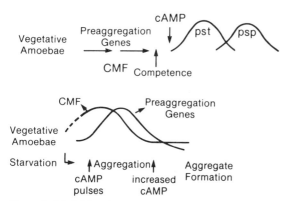

Figure 8. Model for the control of prestalk and prespore cell differentiation (see text for details).

ulatory pathways associated with regulation by cAMP, CMF, and other factors.

The regulation of prespore and prestalk gene expression can also be studied by examining the localization of the gene products within developing aggregates. The use of antibodies directed against fusion proteins produced in bacteria has several advantages. First, the antibodies are polyclonal against a multiple number of epitopes, an advantage over monoclonal antisera which can give incorrect results due to modification or inaccessibility of the binding sites (Blose et al. 1982; Franke et al. 1983; Danto and Fischman 1984). Second, the antibodies are directed against the polypeptide backbone, eliminating the possibility that the antibody is binding to a posttranslational modification (e.g., glycosylation) that might be found on products of more than one gene or that might be regulated in a manner other than that of the gene transcription (and thus hopefully translation) itself (Grant and Williams 1983; Loomis et al. 1983; Knecht et al. 1984).

These antibodies will allow us to examine where prespore and prestalk genes are first expressed in developing aggregates and the spatial distribution of the differentiated cells at various developmental stages in addition to the slug stage (Fig. 4). Immunofluorescence using these antibodies is also being used to assay for the effect of CMF on a relatively small number of cells growing at low cell density. Since this assay requires literally only a few drops of medium containing CMF and takes about 5 hours, it is being used in an attempt to purify and identify CMF.

Finally, we are using the *Dictyostelium* transformation system to examine the function of developmentally regulated genes such as discoidin. In the experiments shown in Figure 6, the inhibition of discoidin I expression is presumably due to formation of a RNA-RNA hybrid between the anti-sense and sense strands. This suggests that a portion of a common coding sequence within a multigene family can be used to block the expression of all gene members. Thus, this approach represents a potential method of blocking the expression of an entire gene family that would not be possible using classic genetic approaches. Using the molecular, cell biological, and physiological methods described in this paper, we now are in a strong position to examine further the molecular basis of cell-type differentiation in *Dictyostelium*.

ACKNOWLEDGMENTS

We thank Richard Jefferson and David Hirsh for recommending the *E. coli* β-glucuronidase gene and for generously giving us affinity-purified antibodies against the β-glucuronidase gene. M.C.M. and S.M. were recipients of National Science Foundation predoctoral fellowships and have been supported by U.S. Public Health Service fellowships. R.H.G. and T.C. were supported by National Institutes of Health postdoctoral fellowships. C.R. and W.N. were supported by postdoctoral fellowships, respectively, from the Swiss National Science Foundation (No. 83.123.0.83) and from the Deutsche Forschungsgemein-schaft (Ne 285-1). C.R. is a Fogharty Fellow International (1-FO5TW03433). A.S. was supported by a fellowship from the Danish Natural Science Research Council. This work was supported by U.S. Public Health Service grants GM-24279 and GM-30693 from the National Institutes of Health to R.A.F.

REFERENCES

ALEXANDER, S., T.M. SHINNICK, and R.A. LERNER. 1983. Mutants of *Dictyostelium discoideum* blocked in expression of all members of the developmentally regulated discoidin multigene family. *Cell* **34:** 467.

BARKLIS, E. and H.F. LODISH. 1983. Regulation of *Dictyostelium discoideum* mRNAs specific for prespore and prestalk cells. *Cell* **32:** 1139.

BLOSE, S.H., F. MATSUMARA, and J.C. LIN. 1982. Structure of vimentin ten nanometer filaments probed with a monoclonal antibody that recognizes a common antigenic determinant on vimentin and tropomyosin. *Cold Spring Harbor Symp. Quant. Biol.* **46:** 455.

BONNER, J.T. 1952. The pattern of differentiation in amoeboid slime molds. *Am. Nat.* **86:** 79.

———. 1970. Induction of stalk cell differentiation by cyclic AMP in the cellular slime mold *Dictyostelium discoideum*. *Proc. Natl. Acad. Sci.* **65:** 110.

CHISHOLM, R.L., E. BARKLIS, and H.F. LODISH. 1984. Mechanism of sequential induction of cell-type specific mRNAs in *Dictyostelium* differentiation. *Nature* **310:** 67.

COLEMAN, T., P.J. GREEN, and M. INOUYE. 1984. The use of RNAs complementary to specific mRNAs to regulate the expression of individual bacterial genes. *Cell* **37:** 429.

CROWLEY, T., W. NELLEN, R.H. GOMER, and R.A. FIRTEL. 1985. Phenocopy of discoidin I-minus mutants by antisense transformation in *Dictyostelium*. *Cell* (in press).

DANTO, S.I. and D.A. FISCHMAN. 1984. Immunocytochemical analysis of intermediate filaments in embryonic heart cells with monoclonal antibodies to desmin. *J. Cell Biol.* **98:** 2179.

DEVREOTES, P.N. 1982. Chemotaxis. In *The development of Dictyostelium discoideum* (ed. W.F. Loomis), p. 117. Academic Press, New York.

FIRTEL, R.A., C. SILAN, T.E. WARD, P. HOWARD, B.A. METZ, W. NELLEN, and A. JACOBSON. 1985. Extra chromosomal replication of shuttle vectors in *Dictyostelium discoideum*. *Mol. Cell. Biol.* (in press).

FRANKE, W.W., E. SCHMID, J. WELLSTEED, C. GRUND, O. GIGI, and B. GEIGER. 1983. Change of cytokeratin filament organization during the cell cycle: Selective masking of an immunological determinant in interphase Pt K2 cells. *J. Cell Biol.* **97:** 1255.

GOMER, R.H. and E. LAZARIDES. 1981. The synthesis and deployment of filamin in chicken skeletal muscle. *Cell* **23:** 524.

———. 1983. Highly homologous filamin polypeptides have different distributions in slow and fast muscle. *J. Cell Biol.* **97:** 818.

GRANT, W.N. and K.L. WILLIAMS. 1983. Monoclonal antibody characterization of slime sheath: The extra-cellular matrix of *Dictyostelium discoideum*. *EMBO J.* **2:** 935.

IZANT, J.G. and H. WEINTRAUB. 1984. Inhibition of thymidine kinase gene expression by anti-sense RNA: A molecular approach to genetic analysis. *Cell* **36:** 1007.

KAY, R.R. 1982. cAMP and spore differentiation in *Dictyostelium discoideum*. *Proc. Natl. Acad. Sci.* **79:** 3228.

KNECHT, D.A., R.L. DIMOND, S. WHEELER, and W.F. LOOMIS. 1984. Antigenic determinants shared by lysosomal proteins in *Dictyostelium discoideum*: Characterization using monoclonal antibodies and isolation of mutations affecting the determinant. *J. Biol. Chem.* **259:** 10633.

KREFFT, M., L. VOET, J.H. GREGG, H. MAIRHOFER, and K.L.

WILLIAMS. 1984. Evidence that positional information is used to establish the prestalk-prespore pattern in *Dictyostelium discoideum* aggregates. *EMBO J.* **3:** 201.

LANDFEAR, S.M., P. LEFEBVRE, S. CHUNG, and H.F. LODISH. 1982. Transcriptional control of gene expression during development of *Dictyostelium discoideum*. *Mol. Cell. Biol.* **2:** 1417.

LEICHTLING, B.H., I.H. MAJERFELD, K.L. SCHALLER, C. WOFFENDIN, S. KAKINUMA, and H.V. RICKENBERG. 1984. A cytosolic cyclic AMP-dependent protein kinase in *Dictyostelium discoideum*. II. Developmental regulation. *J. Biol. Chem.* **259:** 662.

LOOMIS, W.F., ed. 1982. *The development of* Dictyostelium discoideum. Academic Press, New York.

LOOMIS, W.F., B.A. MURRAY, L. YEE, and T. JONGENS. 1983. Adhesion-blocking antibodies prepared against gp 150 react with gp 80 of *Dictyostelium*. *Exp. Cell Res.* **147:** 231.

MCDONALD, S.A. and A.J. DURSTON. 1984. The cell cycle and sorting behavior in *Dictyostelium discoideum*. *J. Cell Sci.* **66:** 195.

MACWILLIAMS, H.K. and J.T. BONNER. 1979. The prestalk-prespore pattern in cellular slime molds. *Differentiation* **14:** 1.

MEHDY, M.C. and R.A. FIRTEL. 1985. A secreted factor and cyclic AMP jointly regulate cell-type specific gene expression in *Dictyostelium discoideum*. *Mol. Cell. Biol.* **5:** 705.

MEHDY, M.C., D. RATNER, and R.A. FIRTEL. 1983. Induction and modulation of cell-type specific gene expression in *Dictyostelium*. *Cell* **32:** 761.

MEHDY, M.C., C.L. SAXE III, and R.A. FIRTEL. 1984. The regulation of cell-type specific genes in *Dictyostelium*. *UCLA Symp. Mol. Cell. Biol. New Ser.* **19:** 293.

MIZUNO, T., M.-Y. CHOU, and M. INOUYE. 1984. A unique mechanism regulating gene expression: Translational inhibition by a complementary RNA transcript (micRNA). *Proc. Natl. Acad. Sci.* **81:** 1966.

NELLEN, W. and R.A. FIRTEL. 1985. High copy number transformants and cotransformants in *Dictyostelium*. *Gene* (in press).

NELLEN, W., C. SILAN, and R.A. FIRTEL. 1984a. DNA-mediated transformation in *Dictyostelium*. *UCLA Symp. Mol. Cell. Biol. New Ser.* **19:** 663.

―――. 1984b. DNA-mediated transformation in *Dictyostelium discoideum*: Regulated expression of an actin gene fusion. *Mol. Cell. Biol.* **4:** 2890.

POOLE, S.J. and R.A.FIRTEL. 1984. Conserved structural features are found upstream from the three co-ordinately regulated discoidin I genes of *Dictyostelium discoideum*. *J. Mol. Biol.* **172:** 203.

POOLE, S., R.A. FIRTEL, E. LAMAR, and W. ROWEKAMP. 1981. Sequence and expression of the discoidin I gene family in *Dictyostelium discoideum*. *J. Mol. Biol.* **153:** 273.

REYMOND, C.D., W. NELLEN, and R.A. FIRTEL. 1985. Regulated expression of *ras* gene constructs in *Dictyostelium* transformants. *Proc. Natl. Acad. Sci.* (in press).

REYMOND, C.D., R.H.GOMER, M.C. MEHDY, and R.A. FIRTEL. 1984. Developmental regulation of a *Dictyostelium* gene encoding a protein homologous to mammalian *ras* protein. *Cell* **39:** 141.

ROWEKAMP, W. and R.A. FIRTEL. 1980. Isolation of developmentally regulated genes from *Dictyostelium*. *Dev. Biol.* **79:** 409.

ROWEKAMP, W., S. POOLE, and R.A. FIRTEL. 1980. Analysis of multigene family coding the developmentally regulated carbohydrate-binding protein discoidin-I in *D. discoideum*. *Cell* **20:** 495.

SPRINGER, W.R., D.N.W. COOPER, and S.H. BARONDES. 1984. Discoidin I is implicated in cell-substratum attachment and ordered cell migration of *Dictyostelium discoideum* and resembles fibronectin. *Cell* **39:** 557.

TASAKA, M. and I. TAKEUCHI. 1981. Role of cell sorting in pattern formation. *Differentiation* **18:** 191.

TASAKA, M., T. NOCE, and I. TAKEUCHI. 1983. Prestalk and prespore differentiation in *Dictyostelium* as defined by cell-type specific monoclonal antibodies. *Proc. Natl. Acad. Sci.* **80:** 5340.

TOWN, C.D., J.D. GROSS, and R.R. KAY. 1976. Cell differentiation without morphogenesis in *Dictyostelium discoideum*. *Nature* **262:** 717.

WILLIAMS, J.G., A.S. TSANG, and H. MAHBUBANI. 1980. A change in the rate of transcription of a eucaryotic gene in response to cyclic AMP. *Proc. Natl. Acad. Sci.* **77:** 7171.

YOUNG, R.A. and R.W. DAVIS. 1983. Efficient isolation of genes by using antibody probes. *Proc. Natl. Acad. Sci.* **80:** 1194.

Early *Dictyostelium* Development: Control Mechanisms Bypassed by Sequential Mutagenesis

G. GERISCH, J. HAGMANN, P. HIRTH, C. ROSSIER,* U. WEINHART, AND M. WESTPHAL
Max-Planck-Institut für Biochemie, D-8033 Martinsried bei München, Federal Republic of Germany

Dictyostelium discoideum is a microorganism that has chosen an unusual way to reach multicellularity: Its ameboid cells stay single during growth and, after nutrient exhaustion, aggregate into a multicellular body, the slug. Eventually, the cells differentiate into spores and stalk cells, the two major constituents of the fruiting body formed at the end of development. Although *Dictyostelium* represents a side branch of evolution that has not given rise to higher multicellular organisms, the mechanisms of its development may help to define general principles by which the interaction of cells is controlled and by which their differentiation is programmed in time and ordered in space. In this paper we concentrate on the early stages of development.

The best-studied factor that serves as a signal during the development of *D. discoideum* is cAMP. Originally identified as the chemoattractant that helps aggregating cells to accumulate in aggregation centers (Konijn et al. 1967), cAMP turned out to be a multipurpose signal. During early development, it is produced rhythmically by the periodic activation of adenylate cyclase at intervals of 5–9 minutes and is released from the cells in the form of pulses (Gerisch and Wick 1975). cAMP begins to act during the interphase between the growth phase and the aggregation-competent stage by stimulating cellular development, as shown by the enhanced formation of EDTA-stable cell contacts, a characteristic of aggregation-competent cells (Darmon et al. 1975; Gerisch et al. 1975; Chisholm et al. 1984). The following are the main features of cAMP action during the early phase of development:

1. Pulses of extracellular cAMP activate cell-surface receptors and elicit at least three fast responses, i.e., responses that occur within intervals of a few seconds to 1 minute after stimulation—the activation of guanylate cyclase, a net Ca^{++} influx into the cells, and the activation of adenylate cyclase (for review, see Gerisch 1982).
2. Two of the fast responses, activation of guanylate and adenylate cyclases, but not the Ca^{++} influx (Bumann et al. 1984), adapt to constant cAMP levels. Adaptation in the adenylate-cyclase-activating pathway is paralleled by a modification of the cAMP receptors (Juliani and Klein 1981; P. Klein et al. 1985), probably a phosphorylation (C. Klein et al. 1985). The half-time of receptor modification is 2.5 minutes and that of its reversal is 6 minutes (Devreotes and Sherring 1985). Adaptation has two consequences. First, periodic pulses of cAMP are the most efficient stimuli for triggering intracellular responses mediated by cGMP, cAMP, or any other factor coupled to the adapting system. Second, hydrolysis of cAMP by cell-surface and extracellular phosphodiesterases is important for repetitive stimulation (Brachet et al. 1979).
3. Slow responses, probably mediated through the activation or deactivation of specific genes, include the regulation of components of the cAMP signal system, such as the cell-surface receptors, adenylate cyclase, plasma-membrane-bound and extracellular phosphodiesterases, and a glycoprotein released into the medium that acts as an inhibitor of the latter enzyme (Gerisch 1979). In addition, the expression of a cell-surface glycoprotein called contact site A (csA) is regulated by cAMP. This 80-kD glycoprotein was originally purified as the target antigen of polyclonal Fab that blocks the EDTA-stable type of intercellular adhesion (Müller and Gerisch 1978). csA expression is stimulated by pulses and is suppressed by low steady-state concentrations of cAMP (Gerisch et al. 1984), indicating a link to the adapting pathway. In this present study, the 80-kD glycoprotein serves as a cell differentiation marker that is easily recognizable on the surface of living cells and allows the selection of mutants by monoclonal antibody labeling and fluorescence-activated cell sorting.

cAMP receptors, membrane-bound phosphodiesterase, and adenylate cyclase are constituents of a system that controls development and are also subject to control by that system. Expression of these membrane proteins is enhanced through the adapting pathway (Roos et al. 1977). The control network is complicated by the fact that it contains not only these positive interactions that drive cell differentiation toward the aggregation-competent stage, but also negative feedback control of extracellular cAMP (Fig. 1). Extracellular phosphodiesterase and its inhibitor are involved in this negative control, since the phosphodiesterase is induced and the inhibitor is suppressed by extracellular cAMP. These two responses do not depend on pulsatile signals (see Yeh et al. 1978), suggesting that not all responses of the slow type are subject to adaptation.

---

*Present address: Laboratoire de Microbiologie générale, Université de Genève 1211 Genève 4, Switzerland.

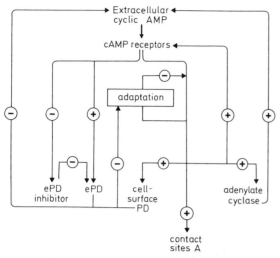

Figure 1. Network showing positive (right loops) and negative (left loops) interactions in the cAMP signal system controlling early development. Evidence for the positive interactions is provided by the effects of pulses of 20 nM cAMP applied every 6 min. These pulses enhance the expression of cAMP receptors as determined by cAMP binding to the cell surface, of adenylate cyclase determined as the basal activity of the enzyme in cell homogenates (the basal activity is the activity measured between pulses that activate the cyclase), and of the membrane-bound phosphodiesterase that represents primarily the cell-surface enzyme (Roos et al. 1977). A continuous flux of the same average amount of cAMP applied per unit time as applied in the form of pulses has no effect, or a very weak stimulating effect, on the expression of these components, indicating that a process responsible for adaptation is implemented in the regulatory pathways of these components. The extracellular (ePD) and cell-surface phosphodiesterases have a dual function in the network. They are essential for the pulse shape of the signals and therefore allow the cells to de-adapt after stimulation. This effect is indicated by an arrow from phosphodiesterase to the boxed adaptation system. On the other hand, reduction of the half-life of extracellular cAMP by the phosphodiesterase also decreases the amplitudes of the pulses, which is indicated by the negative feedback loop.

We have undertaken an analysis of the network controlling the early development of *D. discoideum* (1) to approach an understanding of the control system by combining biochemical and genetic methods with the use of cDNA probes for the expression of specific genes and (2) to uncouple by mutation the production of proteins directly involved in the aggregation process from the control system. Uncoupling is important for the selection of mutants defective in structural genes for proteins that are required in the chemotactic response or in the intercellular adhesion of aggregating cells. These genes are often developmentally regulated. If the wild type is mutagenized, most of the aggregation-deficient mutants obtained are defective in regulatory genes and are consequently blocked before the cells reach the aggregation stage and express the structural genes in question. Mutants defective in the structural genes will occur more often among the aggregation-deficient mutants if expression of these genes is uncoupled, in a first step of mutagenesis, from the control system that governs wild-type development.

METHODS

Cell culture and mutagenesis. Cells of AX2 clone 214 and of mutants derived from it were cultivated at 23°C in nutrient medium containing yeast extract, peptone, and maltose as described previously (Malchow et al. 1972). For the assay of csA expression by immunoblotting, AX2 and HG302 cells were grown up to a density of not more than 5×10^6/ml. To initiate development, these cells were washed and shaken at 150 rpm in 17 mM Soerensen phosphate buffer (pH 6.0; "nonnutrient buffer") at a cell density of 1×10^7/ml. HG518 cells, which do not grow in suspension, were cultivated in polystyrene petri dishes; 15 ml of nutrient medium per dish (85 mm dia) was inoculated with 7.5×10^5 cells of this mutant or, in parallel, with cells of AX2 and HG302. At various times, cells attached to the bottom of the dish were photographed to document aggregation and then gently detached from the plastic surface for counting. For the assay of adenylate cyclase, of EDTA-stable cell adhesion, and of the expression of the csA glycoprotein, the suspended cells were washed in nonnutrient buffer.

For colony blotting, mutants were cultivated with *Escherichia coli* B/2 on nutrient agar containing 0.1% Bacto-peptone (Difco), 0.1% glucose, and 2% Bacto-agar (Difco) in 17 mM Soerensen phosphate buffer (pH 6.0). For mutagenesis, spores were UV-irradiated or cells were treated with *N*-methyl-*N'*-nitro-*N*-nitrosoguanidine (MNNG). The mutagenized spores were plated onto nutrient agar to screen for morphogenetic mutants. The MNNG-treated cells were cultivated for several generations in nutrient medium before they were starved in nonnutrient buffer for the selection of csA-negative mutants by fluorescent antibody labeling and cell sorting.

Immunological procedures. The csA-specific monoclonal antibodies used, mAb 33-294-17 and mAb 41-71-21, were prepared and labeled with ^{125}I as described by Bertholdt et al. (1985). These antibodies are hereafter referred to as mAb 294 and mAb 71, respectively. For immunoblotting, total homogenates of 1×10^6 cells per lane were subjected to SDS-PAGE in 10% gels (Laemmli 1970), and the proteins were transferred to BA 85 nitrocellulose (Schleicher and Schüll, Dassel) according to the method of Towbin et al. (1979) and labeled with ^{125}I-mAb 294 as described recently (Bertholdt et al. 1985). For cell sorting, mutagenized cells were incubated with mAb 71 and labeled with FITC-conjugated sheep anti-mouse IgG (Institute Pasteur Production). Unlabeled cells were selected and cloned onto nutrient agar plates using a FACS IV cell sorter equipped with a single-cell deposition system as described by Francis et al. (1985). For colony blotting, a nitrocellulose filter was placed on the surface of an agar plate culture, removed, and immediately put on a metal plate cooled by dry ice until the cells were frozen. When thawed, the cells became homogenized and their proteins bound to nitrocellulose. The filter was incubated with ^{125}I-labeled mAb 71 and autoradiographed. Subsequently, the proteins of the blotted colonies were stained with a solution of 0.2% Ponceau S (Serva) in 3% TCA and washed in 3% TCA.

Assays of adenylate cyclase and EDTA-stable cell adhesion. Adenylate cyclase activity was determined in cell homogenates obtained by freezing and thawing 1.5×10^7 cells in 250 µl of buffer containing 10 mM Tris-HCl (pH 8.0), 1 mM EDTA, 1 mM DTT, 5 mM benzamidine, and 0.2 M sucrose. Homogenates (30 µl) were immediately mixed with 70 µl of an assay mixture, giving final concentrations of 50 mM Tris-HCl (pH 8.0), 5 mM cAMP, 20 mM phosphocreatine, 20 µg per 100 µl of creatine kinase, 10 mM DTT, 5 mM $MgCl_2$, 2 mM $MnCl_2$, and 0.2 mM [α-^{32}P]ATP (50 mCi/mmole). After 10 minutes at 24°C, the reactions were terminated by SDS, and [^{32}P]cAMP was isolated as described by Salomon et al. (1974).

Cell adhesion in the presence of 10 mM EDTA was determined by subjecting suspensions of 1×10^7 washed cells per milliliter of nonnutrient buffer to a microprocessorized version of the agglutinometer described by Beug and Gerisch (1972). Decreased light scattering as a measure of cell agglutination was recorded automatically under conditions of constant shear.

Preparation of poly(A)+ RNA from membrane-bound polysomes and cDNA synthesis. AX2 cells were starved for 6 hours in nonnutrient buffer. Rough endoplasmic reticulum was prepared according to the method of Cardelli et al. (1981), and the RNA was extracted with phenol-chloroform and chromatographed on oligo(dT)-cellulose. First-strand synthesis of cDNA was primed with oligo(dT)$_{12-18}$. The second strand was synthesized according to the method of Gubler and Hoffman (1983) using RNase H, *E. coli* DNA polymerase, and *E. coli* DNA ligase. The 3' ends of the cDNA were tailed with an average of 12 dGs by terminal transferase.

cDNA cloning and selection of probes for developmentally regulated messages. For cDNA cloning, plasmid p2732B constructed by J.D. Monahan (Roche Institute) was used. To prevent religation, the vector DNA was cut with *Bgl*II and *Cla*I. Subsequently, the 3' ends were tailed with an average of 12 dCs by terminal transferase. Vector and cDNA were annealed according to the method of Roewekamp and Firtel (1980). A RecBC$^-$ strain, *E. coli* BJ5183, was transformed with the products of annealing, and colonies of transformants were selected with ampicillin and transferred to nitrocellulose filters in duplicate as described by Grunstein and Hogness (1975). One set of filters was hybridized with a first-strand [^{32}P]cDNA probe from AX2 cells starved for 6 hours. The cDNA probe was synthesized using membrane-bound polysomal poly(A)+ RNA and an oligo(dT) primer. The second set of nitrocellulose filters was probed with a ^{32}P-labeled first strand of cDNA from total cytoplasmic poly(A)+ RNA of growth-phase cells. Nick-translated DNA was prepared from clones that hybridized preferentially with the 6-hour probe and then used for Northern blots with total cytoplasmic RNA of growth-phase cells and of cells starved for 5 hours. Clones that gave a clearly stronger signal with the latter RNA than with the RNA from growth-phase cells were used for the analysis of gene expression during early development of wild-type and mutant cells.

RESULTS

Probes for Gene Expression in Early Development

Membrane proteins play a major role in the interaction of aggregating cells, in cAMP-induced responses, as well as in cell adhesion. Two types of probes were applied for investigating the expression of genes coding for membrane proteins: cloned cDNA and monoclonal antibodies. A cDNA library was obtained from poly(A)+ RNA isolated from membrane-bound polysomes. The library was differentially screened with RNA from growth-phase cells and from cells at the 6-hour stage of development. Sixteen clones that in Northern blots reacted more strongly with the RNA from the 6-hour stage were selected. These clones were distinguishable from each other by their labeling patterns in genomic Southern blots. Two classes of genes that differed in the time course of their expression were represented by these clones (Fig. 2). The levels of transcripts of class-A genes increased steadily over the first 4-6 hours of development. The transcripts of class-B genes were very weakly and often not detectably expressed during the first 2 hours of development, and the levels of their transcripts increased sharply thereafter. The latter class of genes, represented by five of our clones, was strongly induced by pulses of cAMP and therefore seems to be coupled to the positive-feedback branch of the network shown in Figure 1.

Monoclonal antibodies have been raised as probes for expression of the csA glycoprotein. This glycoprotein is an integral membrane protein that carries several modifications (Fig. 3). Because of the usual cross-reactivity of anticarbohydrate antibodies with other glycoproteins, only antiprotein antibodies are reliable probes for the expression of the csA glycoprotein. We have chosen mAb 294, which efficiently labels the SDS-treated protein, for identification of the glycoprotein by immunoblotting after SDS-PAGE and mAb 71, which recognizes the external part of the protein, for labeling the surfaces of living cells.

Mutants in which the Requirement for cAMP Pulses Is Bypassed

When cAMP is applied in the form of periodic pulses of 20 nM amplitude to a suspension culture of wild-type AX2 cells, csA expression is stimulated. When the same average amount of cAMP per unit time is applied as a continuous flux, a quasi-steady-state concentration of cAMP is maintained that is shifted only slowly with changes in the extracellular and cell-surface phos-

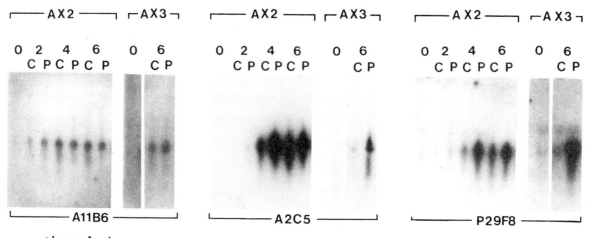

Figure 2. Regulation of three RNA species in control cells (C) and in cells stimulated by cAMP pulses (P) of two axenic strains of *D. discoideum*, AX2 and AX3. The designation of the cDNA probes used for labeling of the Northern blots is given at the bottom. (0-6) Hours of starvation of the cells in nonnutrient buffer. The class-A cDNA clone A11B6 hybridized with an RNA species that was weakly expressed at the 0-hr stage, was already significantly expressed after 2 hr of starvation in AX2, and was not strongly induced in AX2 or AX3 by cAMP pulses. The two class-B cDNA clones, A2C5 and P29F8, hybridized with RNA species that were barely expressed at 2 hr of starvation in the AX2 strain and were strongly induced by cAMP pulses both in AX2 and AX3. The effect of cAMP pulses is most clearly seen in AX3, and at the 4-hr stage in AX2.

phodiesterase activities. Under these conditions, csA expression is inhibited (Gerisch et al. 1984). The inhibition of development, as monitored by csA expression, by a continuous flux of cAMP is apparently due to adaptation of the cells, i.e., to interference with the self-stimulation of cells by their own periodic production of cAMP pulses. In principle, this inhibition could be used for the selection of mutants that do not require cAMP pulses for their development, and two types of mutants might be expected: mutants that do not need cAMP at all and mutants that do not need pulsatile signals. Mutants of the first type would be called bypass mutants because they proceed with development without requiring cAMP signals, and mutants of the second type would include those that do not adapt to cAMP (Fig. 4).

Since cAMP induces extracellular phosphodiesterase, it is rapidly hydrolyzed when added to agar plate cultures. Mutant selection is only possible with a phosphodiesterase-resistant cAMP analog that acts as an agonist on the receptors. A thioanalog, 3',5'-cyclic adenosine phosphorothioate (cAMPS), meets these requirements. Wild-type AX2 cells grow normally but do not aggregate in the presence of 5×10^{-7} M cAMPS (Rossier et al. 1978). Thirty-five mutants have been isolated that aggregate and form fruiting bodies under these conditions. The aggregates formed in the presence of cAMPS are numerous but small. Although chemotactic orientation along cAMP gradients is suppressed by high uniform concentrations of cAMPS, the cells can aggregate by random movement and adhesion of colliding cells (E. Wallraff, pers. comm.). This means that only the adhesion system needs to be established in the presence of cAMPS for aggregation to occur.

Most of the cAMPS-resistant mutants form normal aggregates in the absence of cAMPS, suggesting that their cAMP signal system develops normally. One mutant, HG302, forms small aggregates both in the absence and in the presence of cAMPS, indicating that its cAMP signal system does not acquire its normal function. This mutant is apparently defective in a control element that plays a central role in the establishment of the signal system during development. HG302 produced almost no extracellular phosphodiesterase and phosphodiesterase inhibitor, and it also had low membrane-bound phosphodiesterase activity and a small

CONTACT SITES A: 80 KD GLYCOPHOSPHOLIPOPROTEIN

Developmental regulation	Absent from growth phase cells - maximally expressed at the aggregation stage
Modifications	Acylated with palmitic acid
	Phosphorylated at serine residues
Glycosylation	Type I carbohydrate:
	N-glycosidically linked, sulfated
	Type II carbohydrate:
	recognized by wheat germ agglutinin
	and many monoclonal antibodies
Partially glycosylated or unglycosylated forms	68 kd lacks type II carbohydrate
	66 kd lacks type I carbohydrate
	53 kd lacks both type I and II carbohydrate

Figure 3. Modifications of the csA glycoprotein. Evidence for phosphorylation was provided by Coffman et al. (1981) and Schmidt and Loomis (1982); sulfation was demonstrated by Stadler et al. (1983) and acylation by Stadler et al. (1984). Data on glycosylation are based on results of Ochiai et al. (1982) and Bertholdt et al. (1985).

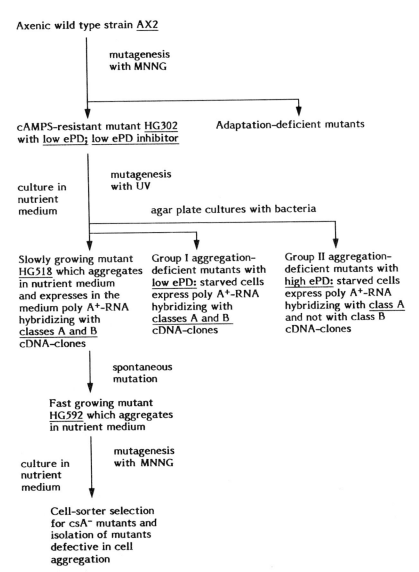

Figure 4. Sequential mutagenesis as used for the selection of bypass mutants or other mutants in the system controlling early development and for the isolation of mutants defective in proteins required for aggregation.

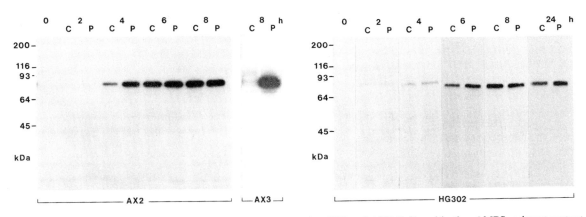

Figure 5. Expression of the csA glycoprotein in the wild-type strains AX2 and AX3 (*left*) and in the cAMPS-resistant mutant HG302 (*right*). Hours of starvation in nonnutrient buffer are indicated at top of the panels. Control cells (C) and cells stimulated every 6 min with pulses of 20 nM cAMP (P) were run in parallel.

number of cAMP receptors. Its adenylate cyclase activity was about half that of the wild type. Despite these defects in the cAMP signal system, the csA glycoprotein was under developmental control; it was not expressed during growth and appeared after several hours of starvation (Fig. 5). Expression of the csA glycoprotein was not reduced by cAMPS in HG302, in contrast to wild type, where it was inhibited. The mutant was analyzed by parasexual genetics and proved to be defective in a single gene, or in a group of closely linked genes, that mapped on linkage group II (Wallraff et al. 1984).

Second Mutagenesis: Screening for Aggregation-deficient Mutants

Developmental regulation of the csA glycoprotein in HG302 indicates that factors other than cAMP pulses maintain developmental control when the cAMP signal system is bypassed by mutation. To cause defects in the other control mechanisms, HG302 was mutagenized and aggregation-deficient mutants were isolated. These mutants fell into two groups: Group I mutants had retained the low extracellular phosphodiesterase production of the parent HG302 mutant; in group II mutants,

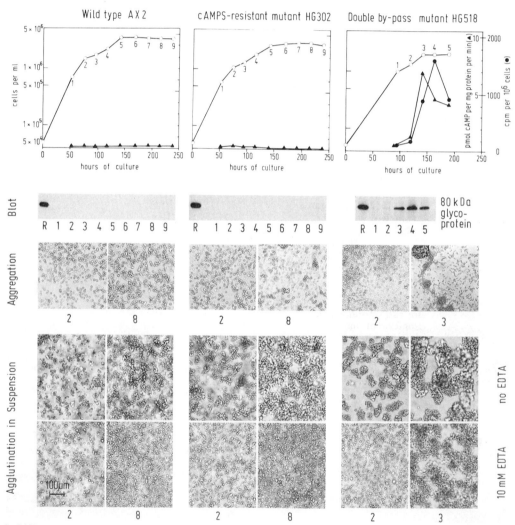

Figure 6. Differentiation of a double bypass mutant, HG518, in nutrient medium. Expression of two membrane proteins of developing cells, adenylate cyclase and the csA glycoprotein, and formation of EDTA-stable contacts were assayed in the parent strain AX2 (*left*), in mutant HG302 (*center*), and in HG518 derived from it (*right*). Graphs on top show growth curves in a semilogarithmic plot (○), the activities of adenylate cyclase (▲), and, for HG518, quantitation of the csA glycoprotein by immunoblotting with ^{125}I-labeled mAb 294 (●). From the immunoblot shown below the graph, the 80-kD band was cut out and counted in a γ-counter, and background was subtracted. In the blots of the AX2 and HG302 strains, no label was seen. A reference (R) of AX2 cells starved for 8 hr in nonnutrient buffer was run in each gel. (*1–9*) Lane numbers corresponding to the respective points on the growth curves. Cell aggregation in nutrient medium was followed in petri dishes with the cells adhering to the plastic surface. Numbers below the photographs refer to the points on the growth curves. Only the stationary phase of HG518 showed stream formation, cell elongation, and chemotactic orientation. At the same time points as for aggregation, cell adhesion was assayed in an agglutinometer in the absence and presence of EDTA. In the absence of EDTA, all cells agglutinated. Only stationary phase HG518 cells formed EDTA-stable cell contacts.

the extracellular phosphodiesterase activities were higher and resembled wild-type levels.

Group I and II mutants are distinguishable by the developmentally regulated genes they express. Group I mutants expressed both class-A and class-B genes, whereas group II mutants expressed only class-A genes. These results are consistent with the hypothesis that the group I mutants are deficient in the aggregation process, rather than in development to the aggregation-competent stage. The group II mutants are clearly regulatory mutants. They indicate that class-A gene expression is independent of class-B gene expression and that class-B genes have a control mechanism in common.

Selection of a Double Bypass Mutant

In addition to the selection of aggregation-deficient mutants, progeny of mutagenized HG302 were cultivated in nutrient medium, and aggregation of cells within the medium was scored to detect mutants in which, in addition to the cAMP signal system, other control mechanisms were bypassed. Neither wild-type AX2 nor HG302 cells aggregated in the nutrient medium, even when they remained in the stationary phase for a long time, and they did not form the EDTA-stable contacts typical of aggregation-competent cells (see Fig. 6).

Screening for double bypass mutants has yielded HG518, a mutant that aggregates in nutrient medium. The capacity of aggregating is reflected in the formation of EDTA-stable cell contacts that are detectable in the nutrient medium after HG518 has finished growth. The development of HG518 under these conditions was confirmed by assaying two developmentally regulated membrane constituents, adenylate cyclase and csA. Both of these constituents were not expressed in AX2 and HG302 as long as the cells remained in the nutrient medium (Fig. 6). In HG518, the adenylate cyclase activity sharply increased when the cells entered the stationary phase, and csA glycoprotein appeared somewhat later. This order of increase is similar to what is observed during development of wild-type cells in nonnutrient buffer, where the adenylate cyclase sharply increases after 2 hours of starvation (Klein 1976) and the csA glycoprotein increases after 4 hours of starvation (Fig. 5). The conclusion from these results is that in HG518, two developmental controls have been bypassed, that of cAMP pulses and a control that suppresses development in the presence of nutrient medium. A third control is still functioning in HG518; this control is responsible for suppression of developmentally regulated proteins during growth and is switched off when the cells reach the stationary phase.

Expression of developmentally regulated genes in HG518 has also been investigated with cDNA probes. Figure 7 shows that in the single bypass mutant HG302, three RNA species tested were more or less strictly suppressed during growth and were expressed after star-

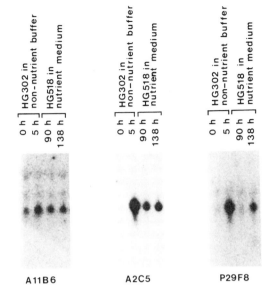

Figure 7. Northern blots showing developmental regulation of three RNA species in HG302 and their expression in HG518 at two different stages in nutrient medium. The same cDNA clones as in Fig. 2 were applied as probes.

ing the cells in nonnutrient buffer. These results are similar to those obtained with AX2 wild-type cells (Fig. 2). In addition, Figure 7 shows that although all three RNA species were expressed in HG518 after 138 hours of culture, only two of them were expressed after 90 hours. (Since an independent culture has been used for this experiment, the stages of development at a given time were not precisely the same as in Fig. 6.) It remains to be examined whether RNA species hybridizing with clones A11B6 or A2C5 are constitutively expressed in HG518 during growth or whether they are expressed early during development (see adenylate cyclase in Fig. 6). The RNA hybridizing with P29F8 might be regulated with a similar time course as the csA glycoprotein.

Third Mutagenesis: Screening for Mutants Defective in csA and in Other Proteins Required for Development

For unknown reasons, the growth of HG518 proved to be anchorage-dependent, i.e., the mutant grew axenically on plastic surfaces but not in suspension. The anchorage dependence was associated with slow growth on agar plate cultures with bacteria. Consequently, fast-growing spontaneous mutants were selected that otherwise resembled HG518. These mutants also grew in suspension. One of them, HG592, served as starting material for selecting mutants defective in the csA glycoprotein and for screening for mutants with various morphogenetic defects.

The colony immunoblotting technique (Francis et al. 1985) proved to be of great help in the identification and characterization of mutants. Colonies were directly blotted from agar surfaces onto nitrocellulose filters and were labeled with a csA-specific antibody. Figure 8 illustrates blotted colonies of HG592 and of two mutants derived from it. After antibody labeling and autoradiography, the blotted proteins were stained to reveal the size and details of the colonies, e.g., shape and positions of aggregates. Colonies of HG592 show, like wild-type colonies, a ring-shaped csA label coinciding with the zone of aggregation (Fig. 8, left panel). The colonies extend beyond the circumference of the csA ring, reflecting the developmental regulation of csA. The weak csA label in the inner circle of the colonies is due in part to down-regulation of csA at the multicellular stage and in part to the fact that after aggregation, cells are concentrated in small areas. Therefore, ring-shaped csA labeling indicates development proceeding to the multicellular stage.

Mutants blocked at the aggregation stage as well as mutants that express csA, but, for some reason, are incapable of aggregating, retain the csA label up to the center of the colonies (Fig. 8, right panel). csA-negative mutants are also easily detected by colony blotting, as shown for a nonaggregating mutant (Fig. 8, center panel). Current work is concentrating on the identification of the regulatory or structural gene defects in these types of mutants.

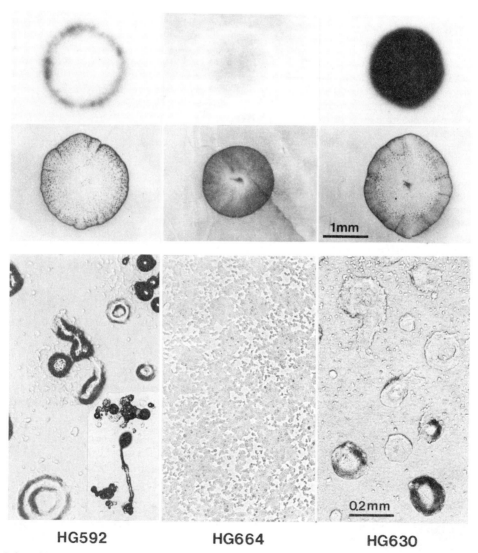

Figure 8. Colony blots and aggregation behavior of the double bypass mutant HG592 and of two morphogenetic mutants derived from it. Colonies were either blotted onto nitrocellulose (*top* and *center*) or photographed on the agar surface to show details of development (*bottom*). The blots were labeled with csA-specific mAb 71 (*top*) and subsequently stained with Ponceau S (*center*). HG592 shows a ring of label where the cells aggregated, HG664 did not react at all with the antibody, and HG630 retained the label over the whole colony. HG592 formed small aggregates that developed into minute, but almost normally organized, fruiting bodies. HG664 did not aggregate, and HG630 was blocked after aggregation and before tip formation. Tips would indicate the beginning of slug formation.

DISCUSSION

Development in higher organisms is probably controlled by complicated interactions of regulatory elements, rather than by factors that act in a single linear sequence such that the gene-A product induces gene B, the product of that gene induces gene C, and so forth. Sequential gene activation has been suggested to govern the developmental program in *D. discoideum* (Loomis et al. 1976; Chisholm et al. 1984). However, the control of development by cAMP provides, in our opinion, evidence for the organization of regulatory elements into a network.

The relief of gene expression from feedback control by the cAMP signal system, as in the cAMPS-resistant mutants, and the sequential elimination of other regulatory mechanisms through bypassing facilitate investigation of the following topics: (1) Because two control mechanisms implemented in the early development of wild-type AX2 are not required in HG592, mutations in genes specifically required for these controls will not result in the inability of HG592-derived mutants to aggregate. In consequence, among aggregation-deficient mutants isolated from HG592, a high percentage should be defective in proteins immediately involved in aggregation, whereas aggregation-deficient mutants isolated from the wild type are mainly regulatory ones. (2) By the same argument, it is feasible to select from HG592 mutants defective in the structural gene for the csA glycoprotein, using protein-specific monoclonal antibodies and a fluorescence-activated cell sorter. (3) Assuming the cAMP signal system is involved in both early and postaggregative development, the consequence would be that mutants defective in that system are blocked at an early stage. If obtained in the genetic background of a bypass mutant that does not require the signal system for early development, the same mutation might cause a block at the postaggregative stage. For example, a mutant defective in the production or recognition of cAMP would be a nonaggregating one if obtained from wild type; if obtained from a bypass mutant, it would aggregate and would thus be amenable to the investigation of cAMP functions in later stages of development. (4) Mutants that express the csA glycoprotein already during exponential growth may be selected from HG592 using fluorescent antibody labeling and a cell sorter. In these mutants constitutive for csA, the last control mechanism still functional in HG592 should be bypassed. These examples illustrate the potential inherent in the combined analysis of regulatory networks by sequential mutagenesis and by a biochemical approach toward the identification of the components involved.

ACKNOWLEDGMENTS

We thank Dipl.-Ing. Rainer Merkl for cooperation in mutant selection with the cell sorter, Barbara Fichtner for providing us with the antibodies, and Dr. Jeffrey Segall for his comments on the manuscript.

REFERENCES

BERTHOLDT, G., J. STADLER, S. BOZZARO, B. FICHTNER, and G. GERISCH. 1985. Carbohydrate and other epitopes of the contact site A glycoprotein of *Dictyostelium discoideum* as characterized by monoclonal antibodies. *Cell Diff.* **16:** 187.

BEUG, H. and G. GERISCH. 1972. A micromethod for routine measurement of cell agglutination and dissociation. *J. Immunol. Methods* **2:** 49.

BRACHET, P., E.L. DICOU, and C. KLEIN. 1979. Inhibition of cell differentiation in a phosphodiesterase defective mutant of *Dictyostelium discoideum*. *Cell Differ.* **8:** 255.

BUMANN, J., B. WURSTER, and D. MALCHOW. 1984. Attractant-induced changes and oscillations of the extracellular Ca^{++} concentration in suspensions of differentiating *Dictyostelium* cells. *J. Cell Biol.* **98:** 173.

CARDELLI, J.A., D.A. KNECHT, and R.L. DIMOND. 1981. Membrane-bound and free polysomes in *Dictyostelium discoideum*. *Dev. Biol.* **82:** 180.

CHISHOLM, R.L., E. BARKLIS, and H.F. LODISH. 1984. Mechanism of sequential induction of cell-type specific mRNAs in *Dictyostelium* differentiation. *Nature* **310:** 67.

COFFMAN, D.S., B.H. LEICHTLING, and H.V. RICKENBERG. 1981. Phosphoproteins in *Dictyostelium discoideum*. *J. Supramol. Struct. Cell. Biochem.* **15:** 369.

DARMON, M., P. BRACHET, and L. PEREIRA DA SILVA. 1975. Chemotactic signals induce cell differentiation in *Dictyostelium discoideum*. *Proc. Natl. Acad. Sci.* **72:** 3163.

DEVREOTES, P.N. and J.A. SHERRING. 1985. Kinetics and concentration dependence of reversible cAMP-induced modification of the surface cAMP receptor in *Dictyostelium*. *J. Biol. Chem.* **260:** 6378.

FRANCIS, D., K. TODA, R. MERKL, T. HATFIELD, and G. GERISCH. 1985. Mutants of *Polysphondylium pallidum* altered in cell aggregation and in the expression of a carbohydrate epitope on cell surface glycoproteins. (in press).

GERISCH, G. 1979. Control circuits in cell aggregation and differentiation of *Dictyostelium discoideum*. In *VIIIth Congress of the International Society for Developmental Biology*, Tokyo (ed. J.D. Ebert and T. Okada), p. 225. Wiley, New York.

———. 1982. Chemotaxis in *Dictyostelium*. *Annu. Rev. Physiol.* **44:** 535.

GERISCH, G. and U. WICK. 1975. Intracellular oscillations and release of cyclic AMP from *Dictyostelium* cells. *Biochem. Biophys. Res. Commun.* **65:** 364.

GERISCH, G., H. FROMM, A. HUESGEN, and U. WICK. 1975. Control of cell-contact sites by cyclic AMP pulses in differentiating *Dictyostelium* cells. *Nature* **255:** 547.

GERISCH, G., A. TSIOMENKO, J. STADLER, M. CLAVIEZ, D. HÜLSER, and C. ROSSIER. 1984. Transduction of chemical signals in *Dictyostelium* cells. In *Information and energy transduction in biological membranes* (ed. E. Helmreich), p. 237. A.R. Liss, New York.

GRUNSTEIN, M. and D.S. HOGNESS. 1975. Colony hybridization: A method for the isolation of cloned DNAs that contain specific genes. *Proc. Natl. Acad. Sci.* **72:** 3961.

GUBLER, U. and B.J. HOFFMAN. 1983. A simple and very efficient method for generating cDNA libraries. *Gene* **25:** 263.

JULIANI, M.H. and C. KLEIN. 1981. Photoaffinity labeling of the cell surface adenosine 3′:5′-monophosphate receptor of *Dictyostelium discoideum* and its modification in down-regulated cells. *J. Biol. Chem.* **256:** 613.

KLEIN, C. 1976. Adenylate cyclase activity in *Dictyostelium discoideum* amoebae and its changes during differentiation. *FEBS Lett.* **68:** 125.

KLEIN, C., J. LUBS-HAUKENESS, and S. SIMONS. 1985. cAMP induces a rapid and reversible modification of the chemotactic receptor in *Dictyostelium discoideum*. *J. Cell Biol.* **100:** 715.

KLEIN, P., A. THEIBERT, D. FONTANA, and P.N. DEVREOTES.

1985. Identification and cyclic AMP-induced modification of the cyclic AMP receptor in *Dictyostelium discoideum*. *J. Biol. Chem.* **260:** 1757.

Konijn, T.M., J.G.C. van de Meene, J.T. Bonner, and D.S. Barkley. 1967. The acrasin activity of adenosine-3',5'-cyclic phosphate. *Proc. Natl. Acad. Sci.* **58:** 1152.

Laemmli, U.K. 1970. Cleavage of structural proteins during the assembly of the head of bacteriophage T4. *Nature* **227:** 680.

Loomis, W.F., S. White, and R.L. Dimond. 1976. A sequence of dependent stages in the development of *Dictyostelium discoideum*. *Dev. Biol.* **53:** 171.

Malchow, D., B. Nägele, H. Schwarz, and G. Gerisch. 1972. Membrane-bound cyclic AMP-phosphodiesterase in chemotactically responding cells of *Dictyostelium discoideum*. *Eur. J. Biochem.* **28:** 136.

Müller, K. and G. Gerisch. 1978. A specific glycoprotein as the target site of adhesion blocking Fab in aggregating *Dictyostelium* cells. *Nature* **274:** 445.

Ochiai, H., J. Stadler, M. Westphal, G. Wagle, R. Merkl, and G. Gerisch. 1982. Monoclonal antibodies against contact sites A of *Dictyostelium discoideum*: Detection of modifications of the glycoprotein in tunicamycin-treated cells. *EMBO J.* **1:** 1011.

Roewekamp, W. and R.A. Firtel. 1980. Isolation of developmentally regulated genes from *Dictyostelium discoideum*. *Dev. Biol.* **79:** 409.

Roos, W., D. Malchow, and G. Gerisch. 1977. Adenylyl cyclase and the control of cell differentiation in *Dictyostelium discoideum*. *Cell Differ.* **6:** 229.

Rossier, C., G. Gerisch, and D. Malchow. 1978. Action of a slowly hydrolysable cyclic AMP analogue on developing cells of *Dictyostelium discoideum*. *J. Cell Sci.* **35:** 321.

Salomon, Y., C. Londos, and M. Rodbell. 1974. A highly sensitive adenylate cyclase assay. *Anal. Biochem.* **58:** 541.

Schmidt, J.A. and W.F. Loomis. 1982. Phosphorylation of the contact site A glycoprotein (gp80) of *Dictyostelium discoideum*. *Dev. Biol.* **91:** 296.

Stadler, J., G. Bauer, and G. Gerisch. 1984. Acylation in vivo of the contact site A glycoprotein and of other membrane proteins in *Dictyostelium discoideum*. *FEBS Lett.* **172:** 326.

Stadler, J., G. Gerisch, G. Bauer, C. Suchanek, and W.B. Huttner. 1983. In vivo sulfation of the contact site A glycoprotein of *Dictyostelium discoideum*. *EMBO J.* **2:** 1137.

Towbin, H., T. Staehelin, and J. Gordon. 1979. Electrophoretic transfer of proteins from polyacrylamide gels to nitrocellulose sheets: Procedure and some applications. *Proc. Natl. Acad. Sci.* **76:** 4350.

Wallraff, E., D.L. Welker, K.L. Williams, and G. Gerisch. 1984. Genetic analysis of a *Dictyostelium discoideum* mutant resistant to adenosine 3':5'-cyclic phosphorothioate, an inhibitor of wild-type development. *J. Gen. Microbiol.* **130:** 2103.

Yeh, R.P., F.K. Chan, and M.B. Coukell. 1978. Independent regulation of the extracellular cyclic AMP phosphodiesterase-inhibitor system and membrane differentiation by exogenous cyclic AMP in *Dictyostelium discoideum*. *Dev. Biol.* **66:** 361.

Cell Interactions Govern the Temporal Pattern of *Myxococcus* Development

D. KAISER, L. KROOS, AND A. KUSPA

Department of Biochemistry, Stanford University School of Medicine, Stanford, California 94305

Myxobacteria are simple cells that exhibit multicellular development (see Fig. 1) (Reichenbach and Dworkin 1981; Rosenberg 1984). The growing cells (0 hr) of *Myxococcus xanthus* are long thin rods, 0.5 μm by 5 μm. When a culture is starved, growth stops, but development proceeds if the cells are at high density and are located on a solid surface (Shimkets 1984). As shown in Figure 1, by 4 hours, centers have been cho-

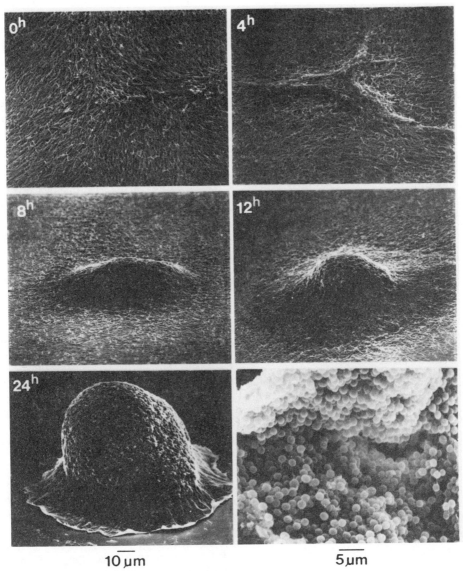

Figure 1. Fruiting body development in *M. xanthus*. Scanning electron micrographs were made in submerged culture by Kuner and Kaiser (1982). The time marked at the upper left corner of each frame shows the interval measured from the beginning of starvation at which the equivalent stage is reached in cultures developing on TPM plates as described in Methods. The 10-μm scale applies to these micrographs. The lower right frame shows a mature fruiting body that has cracked open, revealing its spores (5-μm scale).

sen into which cells begin to migrate. The centers are asymmetric at first, but as more cells accumulate there, they become circular mounds. The mounds enlarge as more cells enter them, until each contains about 100,000 cells and the surrounding region is depleted of cells. Sporulation starts within the mounds at around 20 hours. Eventually, all cells either lyse or become exospores. More than half the cells in a mound lyse (the number depends on the nutritional conditions), perhaps providing signals, nutrients, or structural components for the developing exospores (Wireman and Dworkin 1977; Teintze et al. 1985). The result—a mature fruiting body—is a dense mound of spores (Fig. 1, lower right frame), 0.1 mm high. The fruiting body is covered and protected by an acellular skin of hardened slime, evident in the lower left frame of the figure.

Myxobacteria, which represent one of the 15 or more independent occasions that multicellularity has evolved (Valentine 1978), live by secreting hydrolytic enzymes with which they degrade particulate organic matter in soil (Rosenberg 1984). The feeding process has evolved to become cooperative through the close association of many cells with each other, allowing them to share extracellular enzymes (Rosenberg et al. 1977). Myxobacteria move by gliding on surfaces, which facilitates their feeding on particles and permits them to climb upon each other when they construct a fruiting body. As an experimental system, myxobacteria are amenable to genetic transduction, plasmid transfer from *Escherichia coli*, transposon mutagenesis, and other genetic techniques applicable to gram-negative bacteria (for summary, see Kaiser 1984). They have a genome consisting of 5700 kb of unique DNA, only one-fourth larger than *E. coli* (Yee and Inouye 1982), and can be grown to high density in the laboratory on a tryptic digest of casein with a generation time of 4 hours.

Myxobacteria present, in simple form, a general development problem: What coordinates the behavior of cells during multicellular development? Fruiting body development occurs rather synchronously; e.g., at 4 hours after starvation, small irregular aggregates appear all over a culture plate. New proteins are synthesized according to a reproducible time schedule (Inouye et al. 1979a), and sporulation occurs only after a full quota of cells have accumulated in a nascent fruiting body. We are using the experimental simplicity of *Myxococcus* to test the hypothesis that coordination arises from the passage of regulatory signals between cells.

METHODS

Selection of extracellularly complementable mutants. Fruiting competent bacteria, after exposure to a mutagen, were plated on starvation medium at high cell density (Hagen et al. 1978). Mutant cells will thus be surrounded by normal cells and, after 24 hours, fruiting bodies form on the culture plate. After fruiting, the entire plate is heated to 50°C for 2 hours. Spores survive the heating, but rod-shaped cells are killed. The fruiting bodies are scraped from the plates; the spores are dispersed and then plated at very low density on CF medium (Hagen et al. 1978). At low density, each spore germinates and then grows to form a separate colony. Because CF medium contains a low nutrient level (0.015% casein hydrolysate), the nutrient is soon exhausted by cell growth, and the colonies are thin. This starvation induces fruiting, and fruiting bodies develop on the surface of the thin colonies. However, colonies formed by the desired mutants lack normal fruiting bodies and are easily recognized. An important feature of development in *Myxococcus* is that mutants unable to develop can nevertheless be propagated by vegetative growth.

Transposition of Tn5 lac. P1::Tn5 *lac* was used to obtain transpositions of Tn5 *lac* in *M. xanthus* strains DK101 (Hodgkin and Kaiser 1977) and DK1622 (Kaiser 1979) as described previously (Kroos and Kaiser 1984). Since bacteriophage P1 is not maintained in *Myxococcus* (Kaiser and Dworkin 1975), P1::Tn5 *lac* acts as a suicide vector for introduction of Tn5 *lac*. Selection for kanamycin resistance permits isolation of Tn5 *lac* insertions in the *Myxococcus* genome. Insertions of Tn5 *lac* in the protein S genes were obtained by infecting *E. coli* strain HB101 (Boyer and Roulland-Dussoix 1969) containing pSI003 (pBR322 with the protein S gene, *tps*, cloned in the *Bam*HI site [kindly provided by Inouye et al. 1983]) with P1::Tn5 *lac* (details of the isolation procedure will be published elsewhere). Plasmids with Tn5 *lac* inserted in either orientation in the protein S gene were identified by restriction mapping.

***Expression of β-galactosidase in* Myxococcus.** For rapid screening of β-galactosidase (β-Gal) activity during growth and development, cells were transferred to CTT nutrient agar (Hodgkin and Kaiser 1977) containing 40 μg/ml of 5-bromo-4-chloro-3-indolyl-β-D-galactoside (X-Gal) and to TPM starvation agar (Bretscher and Kaiser 1978) containing 20 μg/ml of X-Gal and were examined after 3 days of incubation at 32°C. CTT contains 1% hydrolyzed casein, and TPM contains only buffer and salts. More X-Gal was used in CTT agar because the blue color of hydrolyzed X-Gal was less easily seen against the yellow background of nutrient agar than against the white background of starvation agar.

To quantitate β-Gal, cells growing exponentially in CTT liquid were sedimented and resuspended in TPM liquid at 5×10^9 cells/ml. An aliquot was stored at $-20°C$ for later determination of β-Gal activity in growing cells ($t = 0$ sample). For development, aliquots were spotted on TPM agar, incubated at 32°C, scraped from the agar at various times into TPM liquid, and stored at $-20°C$ until all samples were collected. β-Gal specific activity was determined by measuring *o*-nitrophenyl-β-D-galactoside hydrolysis (Miller 1972) and protein content (Bradford 1976) of supernatants after rod-shaped cells were disrupted by sonication and debris was removed by centrifugation. Spores were disrupted by sonication with glass beads (Teintze et al. 1985).

Restriction maps of **Myxococcus** *DNA adjacent to* Tn*5* lac *insertions.* Myxococcus DNA was prepared as described previously (Avery and Kaiser 1983). For each strain, aliquots of DNA were digested with each of six restriction enzymes, SmaI, XhoI, EcoRI, SalI, PstI, and HindIII, that cut Tn5 lac at or to the right of the single EcoRI site in Tn5 lac shown in Figure 3. The digests were electrophoresed on 0.4% agarose gels and analyzed by Southern blot filter hybridization (Southern 1975), using pLRK31 as a nick-translated probe. Since pLRK31 contains only sequences to the left of the EcoRI site in Tn5 lac, subcloned into pBR322 (Bolivar et al. 1977), it detects a single fragment in each of the Myxococcus chromosomal digests. The size of the fragment gives the distance from the known restriction site in Tn5 lac to the first chromosomal restriction site to the left of Tn5 lac (oriented as drawn in Fig. 3).

Construction and extracellular complementation of spoA⁻ Tn*5* lac *mutants.* Tn5 lac insertions were introduced into the spoA mutants DK476 and DK480 (Hagen et al. 1978) and their wild-type parent DK101 by generalized transduction with myxophage Mx8 (Martin et al. 1978) according to the method of Hodgkin and Kaiser (1977). Kanamycin-resistant colonies were selected, and their Spo phenotype was confirmed. For extracellular complementation, spoA⁻ Tn5 lac strains were allowed to develop on TPM agar as described above, but at $t = 0$, they were overlaid with an equal number of spoA or wild-type cells (which do not harbor a Tn5 lac insertion). Samples were collected and assayed as described above, but the measured specific activity of β-Gal was corrected, taking into account that only half of the cells in these mixtures could produce β-Gal.

RESULTS

Extracellular Complementation

To test whether signals must be passed between Myxococcus cells during fruiting body development, we searched for mutants defective in developmental signal production. Such mutants are expected to grow normally and be conditionally defective in development. Alone, these mutants are not able to develop, but when they are mixed with wild-type cells, development occurs. The idea is that developing wild-type cells would provide the missing signal and allow the mutants to develop.

A selection procedure for extracellularly complementable mutants has been worked out (see Methods). Even without selection, about half the mutants that fail to develop spores in fruiting bodies are rescued by the addition of wild-type cells (Hagen et al. 1978). A total of 51 extracellularly complementable mutants, some isolated by the selection procedure described above and some without selection, have been studied. When pairs of mutants are mixed with each other, four (extracellular) complementation classes, spoA, spoB, spoC, and spoD, can be defined (Fig. 2). Spores isolated from

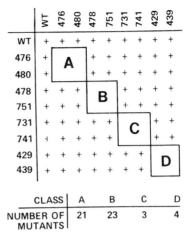

Figure 2. Strains were mixed and allowed to develop as described by Hagen et al. (1978). WT is wild-type strain DK101 and the three digit numbers along the two axes are strain numbers (DK) of spo⁻ mutants derived from DK101. + indicates that sporulation occurred in the mixture; boxed regions indicate no sporulation. Boxes labeled A through D define spoA, spoB, spoC, and spoD mutants, respectively.

fruiting bodies formed in these mixtures of mutants are of both input types (Hagen et al. 1978; K. Mayo, unpubl.). Fruiting is believed to be an asexual process, and it has been found experimentally that recombinants are not present in the fruiting bodies formed by complementation (K. Mayo, unpubl.). The extracellularly complementable mutants are not simply auxotrophs cross-fed in mixtures with wild type or other mutants, because the mutants grow on minimal medium like the wild type (LaRossa et al. 1983).

The four classes are unequally represented in the sample of 51 mutants shown in Figure 2. Class-C mutants belong to a single genetic complementation group (Shimkets et al. 1983), and transposon mutagenesis of a 2-kb cloned DNA fragment containing the spoC locus has produced more C mutants (L. Shimkets, unpubl.). Finding a larger number of spoA and spoB mutants than spoC mutants may reflect a multiplicity of loci responsible for each of those functions, and preliminary linkage mapping supports this view.

Although all of these mutants were selected for their failure to sporulate, these mutants are apparently not defective in sporulation per se because they can be complemented by wild-type cells. The classes arrest development earlier than the time of sporulation in wild type and at different times as illustrated in Table 1. Developmentally induced protein S is normally formed at 6 hours (Inouye et al. 1979a), and protein H, a lectin, is formed at about 12 hours (Cumsky and Zusman 1979). The failure of mutants to exhibit early developmental markers suggests that sporulation is at the end of a dependent developmental pathway. Since the behavior of the mutants is consistent with their being blocked at different points on a pathway, it is possible that the putative signals missing in these mutants help to establish the time sequence.

Table 1. The Four Mutant Classes Arrest Development at Different Times

Starvation high cell density ======> solid surface	prot S (6 hr)	prot H (12 hr)	lysis	spores (24 hr)
Wild type	+	+	+	+
spoA	delayed	−	−	−
spoD	+	−	−	−
spoB	+	+	−	−
spoC	+	+	−	−

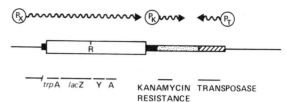

Figure 3. Structure of Tn5 lac. Transcripts (wavy line) are shown above the DNA, originating from P_K, the promoter of the kanamycin resistance gene; P_T, the promoter of the transposase gene; and P_X, the promoter of a transcript into which coding sequence Tn5 lac has inserted in the correct orientation to make a fusion. Tn5 lac consists of sequences from IS50L (■), the trp-lac fusion segment (□), the central region of Tn5 (stippled area), and IS50R (hatched area). The single EcoRI site (R) in Tn5 lac is shown. Proteins (solid line), shown below the DNA, include a truncated polypeptide (broken line) encoded by the gene into which Tn5 lac has inserted.

Tn5 lac

Because the time sequence implied by Table 1 is based on four markers, it is known only at low resolution. To increase the resolving and discriminating power of this experiment, more markers and single-gene markers (as opposed to process markers such as lysis and sporulation) are needed. To identify single gene markers, one of us (L.K.) has constructed a promoter probe, called Tn5 lac (Kroos and Kaiser 1984).

As illustrated in Figure 3, Tn5 lac retains the ends of Tn5 and the transposase encoded in the right inverted repeat of Tn5, which enable it to transpose (Berg and Berg 1983). The neomycin phosphotransferase gene of Tn5, which confers resistance to neomycin and kanamycin, is present in Tn5 lac, allowing selection for drug resistance. A promoter-less trp-lac segment (Casadaban et al. 1980) near the left end of Tn5 lac fuses the transcription of lacZ, which encodes β-Gal, to the promoter of any transcription unit into which Tn5 lac inserts in the correct orientation. Tn5 lac can make transcriptional, but not translational, fusions because translation stop codons in all three reading frames in the first 80 bp of the trp-lac segment (Yanofsky et al. 1981) prevent the formation of fusion proteins.

Tn5 lac Detects Known Developmental Regulation in Myxococcus

Protein S forms one of the outer coats of Myxococcus spores (Inouye et al. 1979b), and together with the mRNA that encodes it, it is first detected at about 6 hours after starvation (Inouye et al. 1979a; Downard et al. 1984). Plasmids containing a Tn5 lac insertion in the protein S gene (see Methods) were introduced into M. xanthus strain DK1622 from E. coli using generalized P1 transduction (O'Connor and Zusman 1983) and selecting for kanamycin resistance. Strains were identified in which the normal protein S gene was replaced by the Tn5 lac-containing gene, and their structures were verified by Southern blot hybridization (Southern 1975) with Tn5 lac as probe (data not shown). A strain that contains the protein S gene with Tn5 lac in the correct orientation to fuse lacZ expression to the protein S promoter increases its specific activity of β-Gal beginning at about 6 hours after starvation. The same result was observed with a protein fusion of protein S to β-Gal (generated in vitro by Downard and Zusman [1985]). The increase in β-Gal expression observed with the Tn5 lac fusion during development was evident upon comparing the color of colonies on nutrient and starvation agar plates containing X-Gal (which turns blue when hydrolyzed by β-Gal). The strain gives yellow colonies on the nutrient plate, due to endogenous Myxococcus pigments (Reichenbach and Kleinig 1984), and green colonies on the starvation plate, due to the mixture of yellow and blue pigments. A strain with Tn5 lac in the opposite orientation in the protein S gene remains yellow on the starvation plate, indicating that β-Gal is not expressed during development when Tn5 lac is in the wrong orientation to make a fusion, and this was confirmed by specific activity measurements. This reconstruction experiment with a gene of known developmental regulation showed that Tn5 lac could be used to identify developmentally regulated genes in Myxococcus.

Transposition of Tn5 lac into Myxococcus

We have generated 2374 Tn5 lac insertion-containing strains and screened them for expression of β-Gal on nutrient and starvation agar plates containing X-Gal. Table 2 shows that nearly one fourth of these strains made β-Gal on nutrient agar plates, indicating that Tn5 lac inserted frequently in transcription units in the correct orientation to fuse lacZ expression to promoters active during growth. Most of the strains appeared to express the same level of β-Gal on starvation agar as on nutrient agar. No strains appeared to decrease their synthesis of β-Gal during development, although a reduction in enzyme activity may have been difficult to detect by colony color due to the stability of β-Gal (Lin and Zabin 1972). Of the 2374 strains, 94 appeared to

Table 2. Transposition of Tn5 lac into Myxococcus

Characteristic	No. of strains (%)
Kanamycin resistant	2374 (100)
Made β-Gal during growth	548 (23)
Increase β-Gal during development as judged from colony color	94 (4)
Increase β-Gal specific activity more than threefold during development	36 (1.5)
Developmentally defective insertion mutants	8 (0.3)

produce more β-Gal during development than during growth, because the colonies were more green on starvation agar than on nutrient agar.

The specific activity of β-Gal was measured in sonic extracts of cells harvested from starvation agar at 1, 2, and 3 days after plating. Thirty-six strains were found to increase β-Gal specific activity more than threefold during development. As a control, ten strains that did not appear to increase β-Gal expression during development in the plate screens were also tested for β-Gal in sonic extracts; these strains showed the same or less-specific activity during development than during growth. Thus, the 36 strains that increase β-Gal specific activity more than threefold during development contain Tn5 lac in regions of the Myxococcus chromosome that are selectively expressed during development.

If Tn5 lac inserts in an operon essential for development, it may produce a developmental mutant. All 2374 Tn5 lac insertion-containing strains were examined for the formation of darkened, spore-filled fruiting bodies after 3 days of development. At this level of analysis, only eight strains had a defect that reduced or abolished aggregation and/or sporulation. Of these eight mutant strains, three are among the 36 strains that increase β-Gal more than threefold during development (strains with insertions Ω4408, Ω4491, and Ω4414). The other five strains did not appear to change their expression during development in the plate screens, and three of the five made β-Gal during growth. When the 36 strains that increase β-Gal more than threefold during development were examined at 4-6-hour intervals throughout development, four additional strains appeared to be slightly delayed for aggregation (Ω4425, Ω4427, and Ω4442) or reduced for sporulation (Ω4473). Thus, 7 of the 36 strains that increase β-Gal strongly during development appear to have a defect in aggregation or sporulation.

Characteristics of the Strongly Expressing Strains

The specific activity of β-Gal was examined at 4-6-hour intervals during development for the 36 strains that increase β-Gal more than threefold overall. Figure 4 shows the timing of enzyme synthesis for three of these strains. If the linear portion of each curve is extrapolated down to the level of β-Gal synthesized during growth, the time at which β-Gal expression begins to increase during development can be estimated (which we will call the "expression time"). Table 3 shows the specific activity of β-Gal observed during exponential growth, the maximum specific activity observed during development, and the estimated expression time for the 36 strains. For many strains with late expression times, the maximum β-Gal-specific activity was detected only when the spores were disrupted. The strains can be separated into three groups on the basis of their expression times: (1) 0-5 hours, during which starvation initiates development and aggregation begins; (2) 9-15 hours, when much of aggregation has been completed; and (3) 18-24 hours, when sporulation begins. The specific activity profiles of the three strains shown in Figure 4 are examples of the three groups, called initiation/aggregation, postaggregation, and sporulation. The number of strains in each group is indicated in Figure 4 and in Table 3.

It is obvious from Table 3 that not all strains in a given group have the same specific activity profile. For example, different strains express different levels of β-Gal during growth, reach different maxima at different times, and have different estimated expression times. Yet, some of the profiles for different strains are similar. To determine whether the 36 strains have Tn5 lac inserted at different places in the Myxococcus chromosome, restriction maps of the DNA adjacent to each Tn5 lac insertion (in the direction from which transcription must originate to express lacZ) were determined by Southern blot filter hybridization experiments (see Methods).

Most of the insertions are in different regions of the Myxococcus genome, because the restriction maps show no resemblance (data not shown). However, in the five cases indicated by brackets in column 1 of Table 3, two or more insertions have related restriction maps and thus may be in the same region. In all five cases, strains with related maps express β-Gal at about

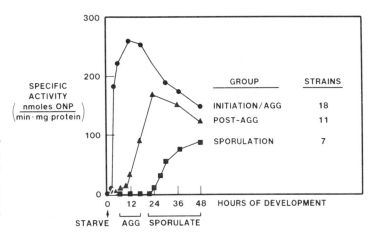

Figure 4. β-Gal synthesis in developing Tn5 lac strains. The specific activity of β-Gal in sonic extracts of cells harvested at different times during development is shown for strains representing the initiation/aggregation group (●; Ω4521), the postaggregation group (▲; Ω4514), and the sporulation group (■; Ω4480). The number of strains in each group is shown.

Table 3. Developmental β-Gal Expression from Tn5 lac Insertions

Ω	Sp. act.[a] at $t = 0$[b]	Maximum sp. act.[a]	Expression time[c]
4523	2	15	5
4454[d]	5	19	0
4491	6	21	0
4427	6	43	2
4408[d]	7	45	1
4530	7	46	4
4531	5	67	4
4521	7	110	2
4499	7	130	4
4400[d]	3	250	5
4494	19	100	0
4457[d]	41	150	1
4455[d]	25	240	3
4411	27	230	5
4442[d]	40	220	0
4425	75	320	2
4469	55	330	4
4445[d]	22	490	3
4474	2	140	13
4514	6	170	9
⌈ 4414	14	2200	13
⌊ 4473	14	1500	12
⌈ 4500	3	200	13
⌊ 4403[d]	3	210[e]	11
4492	17	200[e]	10
4506	7	230[e]	11
4406[d]	3	540[e]	13
⌈ 4451[d]	9	1100[e]	12
⌊ 4459[d]	11	1100[e]	15
⌈ 4435	3	2000[e]	22
4480	3	620[e]	22
4495	7	1000[e]	20
⌊ 4497	5	630[e]	20
⌈ 4511	3	270[e]	18
⌊ 4529	4	190[e]	21
4401[d]	5	590[e]	24

Site of Tn5 lac insertion is designated by Ω followed by a number. Brackets at the left indicate that the insertions are within 1 kb of each other, oriented to detect transcription from the same direction. Only Ω4480 and Ω4495 could possibly be siblings.

[a] Units are nmoles o-nitrophenol produced/min/mg of protein in sonic extracts (see Methods).

[b] $t = 0$ means that growing cells were sedimented, resuspended, and frozen immediately (see Methods).

[c] Expression times, estimated as described in the text, are plus or minus 3 hr.

[d] These insertions were transduced into DK1622 from DK101 using Mx8 (Martin et al. 1978) for comparison of their expression times with the insertions isolated in DK1622.

[e] Maximum specific activity was observed when spores were disrupted.

the same time and level during development, suggesting that some pairs of insertions are in the same transcript.

None of the 18 insertions in the initiation/aggregation group are in related genomic regions. Taking restriction map similarities into account, the postaggregation and sporulation groups reveal 11 different transcripts. Thus, Tn5 lac transposition into *Myxococcus* identified 29 different developmentally regulated genes. These insertions can be used as markers to test the hypothesis that extracellular signals coordinate gene expression during development.

β-Gal Expression in *spoA⁻* Tn5 lac Strains

Because the synthesis of protein S is delayed and later developmental markers are not expressed in *spoA* mutants, they appear to be blocked early in development (Table 1). Using the set of developmentally regulated Tn5 lac insertions just described, we could investigate more precisely the role of *spoA* in development. Twenty of the insertions in Table 3 and an insertion in the S gene were introduced into two *spoA* mutants and their parent (which develops normally). β-Gal expression during development was determined for each strain. A typical result is illustrated in Figure 5. The normal developmental increase in β-Gal activity seen in the wild-type background for this insertion (Tn5 lac-4514) is blocked in the *spoA* mutant. Thus, the β-Gal expression from Tn5-lac-4514 is A-dependent. A similar result was obtained for 18 of the 21 insertions tested. All 15 Tn5 lac insertions that have expression times after 5 hours of development were A-dependent. Three of six insertions that express β-Gal earlier than 5 hours were also A-dependent. Of these, the earliest effect was seen for two insertions with expression times of 1.5 hours in wild type. The three A-independent insertions begin to express β-Gal at 0, 3, and 5 hours of development. For each A-dependent and A-independent insertion tested, similar β-Gal expression was observed in both *spoA* mutants. Thus, *spoA* genes may be required as early as 1.5 hours for the proper expression of developmentally regulated genes.

Rescue of A-dependent β-Gal Expression

Wild-type cells rescue the sporulation defect of *spoA* mutants (Fig. 2). To test whether the expression of A-

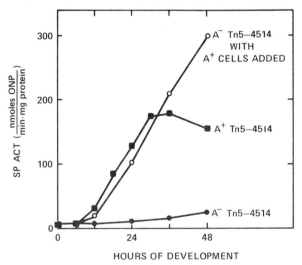

Figure 5. Developmental expression of β-Gal in a *spoA⁻* Tn5 lac strain. Tn5 lac-4514 was transduced into *spoA* mutant and wild-type strains (see Methods). The developmental expression of β-Gal for the insertion in the wild-type (■) and *spoA* mutant (●) strains is shown. The abolished β-Gal expression in the *spoA⁻* derivative can be rescued by the addition of wild-type cells (○). When A⁺ cells were added to A⁺ Tn5 lac-4514, β-Gal expression was similar to that seen with the A⁺ Tn5 lac-4514 strain alone.

dependent Tn5 lac insertions could also be rescued, wild-type cells were added to spoA⁻ Tn5 lac strains. The cell mixtures were allowed to develop, and β-Gal activity was determined at various times. Indeed, developmental expression of β-Gal was restored in all eight A-dependent Tn5 lac insertions tested (e.g., Fig. 5). The expression times of the eight insertions ranged from 1.5 to 22 hours in wild type. The timing of the rescued expression was the same as that seen in wild type for each insertion tested. This is illustrated by Tn5 lac-4514 in Figure 5. These observations indicate that the gene expression blocked in the spoA mutants can be rescued by an extracellular factor supplied by normal cells.

DISCUSSION AND SUMMARY

Myxococcus development depends on several cell-to-cell interactions, possibly four. Differences between the interactions are implied by the existence of four extracellular complementation groups. However, two complementation groups do not necessarily imply distinct signaling systems because, for example, an extracellular signal might be processed by extracellular enzymes.

The cell-to-cell interactions defective in *spoA*, *spoB*, *spoC*, and *spoD* mutants are required at different times in development because the mutant classes arrest at different stages. In particular, the cell interaction defective in the *spoA* mutants occurs early. Three types of experiments support the idea that the *spoA* interaction is early. First, when wild-type cells are starved, they arrest growth and initiate development. *spoB*, *spoC*, and *spoD* mutants arrest growth but are unable to develop; *spoA* mutants fail to arrest growth when subjected to a mild starvation that is sufficient to initiate development of wild-type strains, suggesting that their defect lies near the choice between growth and development (LaRossa et al. 1983). Second, of the four developmental markers in Table 1, only the first is expressed in *spoA* mutants and that expression is delayed. Finally, 18 of 21 developmentally regulated Tn5 lac insertions depend on *spoA* for expression of β-Gal and the three that do not are expressed early.

Tn5 lac insertions provide a new set of developmental markers in which gene expression is indicated by the synthesis of β-Gal. The set of strongly developmentally regulated Tn5 lac insertions described here represents 29 different transcription units, with a range of expression times from minutes after being starved to the time of sporulation. Although this set is not expected to include insertions in all developmentally regulated genes, it does offer a fair sample of such genes. The pattern of Tn5 lac transpositions to bacteriophage λ (Kroos and Kaiser 1984) and into the genome of *Myxococcus* is consistent with the low target-sequence specificity observed for Tn5 (Berg and Berg 1983). Among the 36 strongly developmentally regulated Tn5 lac insertions, five cases of two or more insertions located within 1 kb of each other were observed. Keeping in mind that these 36 Tn5 lac insertions had been selected for strongly developmentally regulated expression, at least five such cases would have been expected by chance alone in one out of four experiments, assuming random transposition of Tn5 lac.

In addition to the 36 strongly developmentally regulated Tn5 lac insertions, 58 appeared to increase β-Gal expression weakly during development, as judged from colony color (Table 2). If Tn5 lac insertion is random and since Tn5 lac can insert in two orientations in a transcription unit (only one of which is expected to allow transcription of the lacZ gene), it can be estimated that 8% of the *Myxococcus* genome augments its transcription during development. Since 23% of the Tn5 lac insertion strains made β-Gal during growth, at least 46% of the genome is transcribed during growth. Insertions in genes essential for growth would not have been recovered.

Most Tn5 lac insertions into a coding region would be expected to inactivate the target gene, yet few developmentally regulated Tn5 lac insertions in *Myxococcus* were found to have a developmental defect. This suggests that many of the products of developmental genes are not essential for the completion of development and do not control development. This is apparently also the case in *Dictyostelium* (Loomis 1975). A well-studied example of *Myxococcus* is protein S, of which a million molecules are synthesized per cell, yet its absence has no effect on aggregation or the production of heat-resistant spores (Komano et al. 1984; Zusman 1984).

In contrast, the *spoA*, *spoB*, *spoC*, and *spoD* gene products do appear to regulate the main developmental pathway in *Myxococcus*, since mutations in these genes reduce or abolish aggregation and/or sporulation. When combined with a *spoA* mutation, 18 of 21 developmentally regulated Tn5 lac insertions are not expressed and thus are A-dependent. β-Gal expression can be rescued in A-dependent Tn5 lac cells if they are mixed with wild-type cells that supply the missing interactions. Since β-Gal expression from the earliest A-dependent insertion can be rescued by addition of wild-type cells, and since the timing of the rescued synthesis is the same as that seen in normal development, it is likely that rescue reflects a normal developmental process. From this point of view, wild-type cells produce a signal (A factor) that is required for normal development. The rescue of β-Gal expression in a *spoA*⁻ mutant containing an A-dependent Tn5 lac should serve as an assay for the purification of A factor from wild-type cells. Introduction of developmentally regulated Tn5 lac insertions into extracellularly complementable mutants of the other classes may provide simple assays for the other developmental signals and will further test the hypothesis that extracellular signals coordinate gene expression during *Myxococcus* development.

ACKNOWLEDGMENTS

Unpublished work from the authors' laboratory was supported by grants from the National Institutes of Health (GM-23441 and AG-02908). Stipend support for

A.K. and L.K. was provided by a National Institutes of Health grant (GM-07599).

REFERENCES

AVERY, L. and D. KAISER. 1983. In situ transposon replacement and isolation of a spontaneous tandem genetic duplication. *Mol. Gen. Genet.* **191**: 99.

BERG, D.E. and C.M. BERG. 1983. The prokaryotic transposable element Tn5. *Biotechnology* **1**: 417.

BOLIVAR, F., R.L. RODRIGUEZ, P.J. GREENE, M.C. BETLACH, H.L. HEYNEKER, H.W. BOYER, J.H. BROSA, and S. FALKOW. 1977. Construction and characterization of new cloning vehicles. II. A multipurpose cloning system. *Gene* **2**: 95.

BOYER, H.W. and D. ROULLAND-DUSSOIX. 1969. A complementation analysis of the restriction and modification of DNA in *Escherichia coli*. *J. Mol. Biol.* **41**: 459.

BRADFORD, M. 1976. A rapid and sensitive method for the quantization of microgram quantities of protein utilizing the principle of protein-dye binding. *Anal. Biochem.* **72**: 248.

BRETSCHER, A.P. and D. KAISER. 1978. Nutrition of *Myxococcus xanthus*, a fruiting myxobacterium. *J. Bacteriol.* **133**: 763.

CASADABAN, M.J., J. CHOW, and S.N. COHEN. 1980. In vitro gene fusions that join an enzymatically active β-galactosidase segment to amino-terminal fragments of exogenous proteins: *Escherichia coli* plasmid vectors for the detection and cloning of translational initiation signals. *J. Bacteriol.* **143**: 971.

CUMSKY, M. and D.R. ZUSMAN. 1979. Myxobacterial hemagglutinin: A development-specific lectin of *Myxococcus xanthus*. *Proc. Natl. Acad. Sci.* **76**: 5505.

DOWNARD, J.S. and D.R. ZUSMAN. 1985. Differential expression of protein S genes during *Myxococcus xanthus* development. *J. Bacteriol.* **161**: 1146.

DOWNARD, J.S., D. KUPFER, and D.R. ZUSMAN. 1984. Gene expression during development of *Myxococcus xanthus*: Analysis of the genes for protein S. *J. Mol. Biol.* **175**: 469.

HAGEN, D.C., A.P. BRETSCHER, and D. KAISER. 1978. Synergism between morphogenetic mutants of *Myxococcus xanthus*. *Dev. Biol.* **64**: 284.

HODGKIN, J. and D. KAISER. 1977. Cell-to-cell stimulation of movement in nonmotile mutants in *Myxococcus*. *Proc. Natl. Acad. Sci.* **74**: 2938.

INOUYE, S., Y. IKE, and M. INOUYE. 1983. Tandem repeat of the genes for protein S, a development specific protein of *Myxococcus xanthus*. *J. Biol. Chem.* **258**: 38.

INOUYE, M., S. INOUYE, and D.R. ZUSMAN. 1979a. Gene expression during development of *Myxococcus xanthus*: Pattern of protein synthesis. *Dev. Biol.* **68**: 579.

———. 1979b. Biosynthesis and self-assembly of protein S, a development-specific protein of *Myxococcus xanthus*. *Proc. Natl. Acad. Sci.* **76**: 209.

KAISER, D. 1979. Social gliding is correlated with the presence of pili in *Myxococcus xanthus*. *Proc. Natl. Acad. Sci.* **76**: 5952.

———. 1984. Genetics of myxobacteria. In *Myxobacteria development and cell interactions* (ed. E. Rosenberg), p. 166. Springer-Verlag, New York.

KAISER, D. and M. DWORKIN. 1975. Gene transfer to a myxobacterium by *Escherichia coli* phage P1. *Science* **187**: 653.

KOMANO, T., T. FURUICHI, M. TEINTZE, M. INOUYE, and S. INOUYE. 1984. Effects of deletion of the gene for the development-specific protein S on differentiation in *Myxococcus xanthus*. *J. Bacteriol.* **158**: 1195.

KROOS, L. and D. KAISER. 1984. Construction of Tn5 *lac*, a transposon that fuses *lacZ* expression to exogenous promoters, and its introduction into *Myxococcus xanthus*. *Proc. Natl. Acad. Sci.* **81**: 5816.

KUNER, J. and D. KAISER. 1982. Fruiting body morphogenesis in submerged cultures of *Myxococcus xanthus*. *J. Bacteriol.* **151**: 458.

LAROSSA, R., J. KUNER, D. HAGEN, C. MANOIL, and D. KAISER. 1983. Developmental cell interactions of *Myxococcus xanthus*: Analysis of mutants. *J. Bacteriol.* **153**: 1394.

LIN, S. and I. ZABIN. 1972. β-galactosidase: Rates of synthesis and degradation of incomplete chains. *J. Biol. Chem.* **247**: 2205.

LOOMIS, W.L. 1975. *Dictyostelium discoideum: A developmental system*. Academic Press, New York.

MARTIN, S., E. SODERGREN, T. MASUDA, and D. KAISER. 1978. Systematic isolation of transducing phages for *Myxococcus xanthus*. *Virology* **88**: 44.

MILLER, J., ed. 1972. *Experiments in molecular genetics*. Cold Spring Harbor Laboratory, Cold Spring Harbor, New York.

O'CONNOR, K.A. and D.R. ZUSMAN. 1983. Coliphage P1-mediated transduction of cloned DNA from *Escherichia coli* to *Myxococcus xanthus*: Use for complementation and recombinational analyses. *J. Bacteriol.* **155**: 317.

REICHENBACH, H. and M. DWORKIN. 1981. The order Myxobacterales. In *The prokaryotes: A handbook on habitats, isolation, and identification of bacteria* (ed. M.P. Starr et al.), p. 328. Springer-Verlag, Berlin.

REICHENBACH, H. and H. KLEINIG. 1984. Pigments of myxobacteria. In *Myxobacteria development and cell interactions* (ed. E. Rosenberg), p. 128. Springer-Verlag, New York.

ROSENBERG, E. 1984. *Myxobacteria development and cell interactions*. Springer-Verlag, New York.

ROSENBERG, E., K.H. KELLER, and M. DWORKIN. 1977. Cell density-dependent growth of *Myxococcus xanthus* on casein. *J. Bacteriol.* **129**: 770.

SHIMKETS, L. 1984. Nutrition, metabolism, and the initiation of development. In *Myxobacteria development and cell interactions* (ed. E. Rosenberg), p. 92. Springer-Verlag, New York.

SHIMKETS, L.J., R.E. GILL, and D. KAISER. 1983. Developmental cell interactions in *Myxococcus xanthus* and the *spoC* locus. *Proc. Natl. Acad. Sci.* **80**: 1406.

SOUTHERN, E.M. 1975. Detection of specific sequences among DNA fragments separated by gel electrophoresis. *J. Mol. Biol.* **98**: 503.

TEINTZE, M., T. FURUICHI, R. THOMAS, M. INOUYE, and S. INOUYE. 1985. Differential expression of two homologous genes coding for spore-specific proteins in *Myxococcus xanthus*: In *Spores IX: The molecular biology of microbial differentiation* (ed. J. Hoch and P. Setlow), p. 253. American Society of Microbiology, Washington, D.C.

VALENTINE, J.W. 1978. The evolution of multicellular plants and animals. *Sci. Am.* **239**: 140.

WIREMAN, J.W. and M. DWORKIN. 1977. Developmentally induced autolysis during fruiting body formation by *Myxococcus xanthus*. *J. Bacteriol.* **129**: 796.

YANOFSKY, C., T. PLATT, I.P. CRAWFORD, B.P. NICHOLS, G.E. CHRISTIE, H. HOROWITZ, and M. VANCLEEMPUT. 1981. The complete nucleotide sequence of the tryptophan operon of *Escherichia coli*. *Nucleic Acids Res.* **9**: 6647.

YEE, T. and M. INOUYE. 1982. Two-dimensional DNA electrophoresis applied to the study of DNA methylation and the analysis of genome size in *Myxococcus xanthus*. *J. Mol. Biol.* **154**: 181.

ZUSMAN, D.R. 1984. Developmental program of *Myxococcus xanthus*. In *Myxobacteria development and cell interactions* (ed. E. Rosenberg), p. 185. Springer-Verlag, New York.

Temporal and Spatial Control of Flagellar and Chemotaxis Gene Expression during *Caulobacter* Cell Differentiation

R. CHAMPER, R. BRYAN, S.L. GOMES,* M. PURUCKER, AND L. SHAPIRO

Department of Molecular Biology, Division of Biological Sciences, Albert Einstein College of Medicine, Bronx, New York 10461

The appearance of asymmetry in a single cell is among the earliest events in a developmental program. Early events include the establishment of cellular polarity by the programmed spatial distribution of specific gene products. These products can result from the temporally regulated transcription and translation of specific sets of genes. We demonstrate here that such events occur during the *Caulobacter crescentus* cell cycle. Thus, genetic mechanisms that can measure time and can arrange cell constituents in three-dimensional space mediate the establishment of asymmetry in *C. crescentus*. To learn more about these mechanisms, it is essential to study the organization and function of the relevant genetic sequences in the context of the physiology and architecture of the entire cell. *C. crescentus* carries out a defined and relatively simple set of subcellular morphological changes prior to cell division. The predivisional cell assembles a flagellum, pili, and the components of a chemosensory system at one cell pole. The asymmetry thus established in the predivisional cell results in the production of two daughter cells that differ both structurally and functionally. Because this organism is amenable to biochemical and genetic analysis, it is feasible to define completely the molecular basis of its cellular differentiation.

The cellular changes that occur at defined times in the *C. crescentus* cell-division pathway are shown schematically in Figure 1. For a portion of the cell cycle, this bacterium bears a polar flagellum and several pili and is motile. The motile swarmer cell sheds its flagellum (Poindexter et al. 1967; Shapiro and Maizel 1973) and loses its pili (Shapiro and Agabian-Keshishian 1970; Smit and Agabian 1982a) after a period of time equivalent to one third of the cell cycle. A slender tube-like stalk is then formed by the localized synthesis of cell envelope at the site previously occupied by the flagellum (Schmidt and Stanier 1966; Smit and Agabian 1982b). The swarmer cell is unable to replicate its chromosome, but once stalk formation begins, DNA replication is initiated (Degnan and Newton 1972). Upon completion of DNA replication, a new flagellum and several pili are assembled at the cell pole opposite the stalk in the predivisional cell. These events each initiate at a defined temporal fraction of the cell cycle and are independent of environmental fluctuations and cell contact (Poindexter 1964). The replication of the chromosome and the distribution of the duplicated chromosomes to the two cell poles are required for subsequent cell division (Osley and Newton 1980). Binary fission of the predivisional cell, with a flagellum at one pole and a stalk at the other, yields a swarmer cell and a stalked cell. Chromosome replication is initiated in the daughter stalked cell, but the daughter swarmer cell cannot initiate replication until it has lost its flagellum and has become a stalked cell later in the cell cycle. Thus, the asymmetry apparent in the predivisional cell yields progeny cells that differ structurally and have different programs of temporally regulated gene expression and DNA replication.

The experiments described in this paper are aimed at defining the mechanisms that control the timed expression and spatial distribution of the *C. crescentus* flagellar (*fla*) and chemosensory (*che*) proteins. We present evidence here that the *che* gene products are sequestered to one portion of the predivisional cell. The temporal order of *fla* and *che* gene expression appears to be controlled by a cascade of *trans*-acting factors that function at the level of transcription.

METHODS

Generation of the Tn5 β-galactosidase promoter probe. Construction of the transposon of Tn5-RC96 (Fig. 2, panel I) utilized a strategy first employed in the construction of the neomycin phosphotransferase II (NPT-II) promoter probe Tn5-VB32 (Bellofatto et al. 1984). pRC96 was generated by inserting a promoterless β-galactosidase gene into the *Escherichia coli* plasmid pRZ341 (*Col*E1::Tn5-341) (Johnson and Reznikoff 1983). The gene encoding NPT-II and most of IS50L is deleted from the Tn5 insert in pRZ341 and carries a restriction fragment from Tn*10* encoding tetracycline resistance (*Tet*[R]) (Fig. 2, panel I). A 7.1-kb restriction fragment of pMC871 (Casadaban et al. 1980) containing internal sequences of the *trp* operon (with translational stops in all three reading frames to prevent the formation of fusion proteins), followed by a promoterless *lacZ* gene and a *lacY* gene, was inserted adjacent to the remaining portion of IS50L in pRZ341. To facilitate transfer of Tn5-RC96 into *C. crescentus*, it was inserted into pJB9JI, a self-transmissible P-type R plasmid encoding gentamycin resistance (Gm[R]) and spectinomycin resistance (Sp[R]) and containing bacteriophage Mu (Beringer et al. 1978). The plasmid pRC96

---

*Present address: Instituto de Quimica, Universidade de Sao Paulo, Sao Paulo, Brasil.

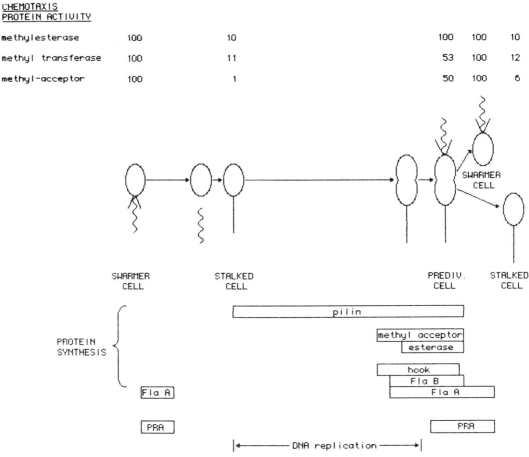

Figure 1. Schematic diagram of the *C. crescentus* cell division cycle showing the time of synthesis (□) of pilin (Smit and Agabian 1982a), flagellar proteins (Osley et al. 1977; Agabian et al. 1979), and chemotaxis proteins (Gomes and Shapiro 1984). The time of appearance of DNA phage receptor activity PRA is also indicated (Agabian-Keshishian and Shapiro 1971; Lagenaur et al. 1974; Huguenel and Newton 1982; E. O'Neill and R. Bender, unpubl.). The activity of several proteins involved in chemotaxis methylation function, as a percentage of that observed in the swarmer cell, is shown above the appropriate cell stage in the cell-cycle diagram. (Reprinted, with permission, from Shapiro 1985.)

(pJB9JI::Tn5-RC96) was transferred from *E. coli* to *C. crescentus* SC1140 (*lacA*) by conjugation. The plasmid vector pJB9JI containing Tn5-RC96 was not stable in *C. crescentus* because of its Mu sequences (Ely and Croft 1982). Therefore, any Tet^R, Gm^S colonies were the result of transposition of Tn5-RC96 into the chromosome. SC1140 Tet^R exconjugants were tested for their level of β-galactosidase production on X-Gal (5-bromo-4-chloro-3-indolyl-β-D-galactoside) plates.

Generation and mapping of **flaY**::*Tn5-VB32.* To generate *fla*::Tn5-VB32 flagellar mutants by transposon insertional inactivation, Tn5-VB32 (see Fig. 2, panel II) was transferred from *E. coli* AEE431 to wild-type *C. crescentus* AE5000 by conjugation, as described previously (see Bellofatto et al. 1984). Nonmotile, Tet^R conjugants were selected and then screened for kanamycin resistance (Kn^R). Transfer was confirmed by transduction and Southern blot analysis. Motility was assayed as the ability to form swarms on semisolid PYE (Poindexter 1964) plates (0.3% agar). The map location of the *flaY*::Tn5-VB32 was determined by conjugation and generalized transduction us-

ing the *C. crescentus* phage φCr30, as described previously (Hodgson et al. 1984). One of the insertions (AE8001) mapped to the *flaYEF* region of the chromosome. Southern blot analysis localized the insertion to a restriction fragment containing *flaY*. The Tn5-VB32 insert carries a promoter-less NPT-II gene. The fact that *flaY*::Tn5-VB32 is resistant to kanamycin argues that the *flaY C. crescentus* promoter has been accessed.

Preparation of NPT-II antibody and immunoprecipitation of cellular proteins. Tn5 encodes the periplasmic enzyme NPT-II that confers kanamycin resistance to its host cell. The enzyme was purified to homogeneity from *E. coli* HB101 cells containing a high copy number plasmid carrying a wild-type Tn5. Cultures were osmotically shocked according to the method of Davies (Williams and Northrop 1976) to release periplasmic enzymes. Nucleic acids were precipitated by bringing the supernatant to 1.6% (w/w) streptomycin sulfate and centrifuging at 15,000g for 15 minutes. The streptomycin sulfate supernatant was brought to 65% saturation with ammonium sulfate,

I. β-galactosidase promoter probe Tn5-RC96

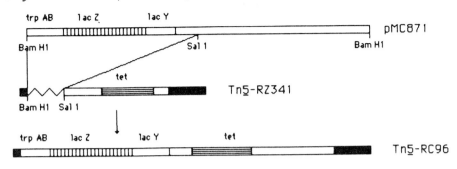

II. Neomycin phosphotransferase promoter probe Tn5-VB32

Figure 2. (Panel I) Diagram of the construction of a transcription fusion probe, Tn5-RC96, in which a promoter-less β-galactosidase gene is inserted into an altered Tn5 transposon. The Tn5-RC96 insert contains the 53 bases of the left end of IS50L joined to the 275 bp of pBR322, an internal segment of the *trp* operon carrying translation stop signals, the *lacZ* gene (lacking its promoter but beginning with its translation initiation sequence), and *lacY*. For *lacZ* to express β-galactosidase, it must insert 3' to a chromosomal promoter. (Panel II) Diagram of the transcription fusion probe Tn5-VB32 that contains the NPT-II gene (*neo*) from Tn5 lacking its promoter sequence (Bellofatto et al. 1984). Translational stops in all three reading frames prevent a protein fusion with the *neo* gene product. Both of these transcription fusion probes contain intact tetracycline resistance (*tetA*) and tetracycline repressor (*tet*R) genes from Tn10. Black boxes are IS50 sequences; horizontal lines delineate the *tet* genes; vertical lines delineate the β-galactosidase structural gene (*lacZ*); the stippled area delineates the gene encoding NPT-II (*neo*); and the zigzag line is a pBR322 fragment. Transcription of *lacZ* and the *neo* gene is left to right (5'–3').

and after centrifugation, the pellet was resuspended in 0.2 mM KPO₄, 10 mM MgCl₂, 0.1 mM EDTA, 1 mM DTT, and 1 M KCl. An affinity resin was prepared by coupling a twofold molar excess of neomycin to activated agarose (Affi-Gel 10), according to the method described by Bio-Rad Laboratories. The procedure followed was a modification of the method designed by D. Gelfund (pers. comm.). The resuspended ammonium sulfate precipitate was dialyzed and applied to the affinity column (equilibrated with the same buffer) and washed serially with 200-ml volumes of 0.1 M KCl, 0.5 M KCl, and 1.0 M KCl. The NPT-II enzyme was eluted from the column with 25 ml of 1.5 M KCl and 0.3 mM neomycin and analyzed for purity by electrophoresis through a 10% SDS-polyacrylamide gel. Antibody was raised by inoculating a rabbit with 500 μg of NPT-II and boosting at 4 and 6 weeks. Cellular proteins were immunoprecipitated with anti-NPT-II antibody and visualized by autoradiography following electrophoresis through 10% SDS-polyacrylamide gels, as described previously (Bellofatto et al. 1984).

RESULTS AND DISCUSSION

Differential Protein Synthesis and Placement as a Function of the Cell Cycle

Structure of the polar flagellum and time of synthesis of its protein components. The definition of the molecular events that lead to the localized biogenesis of the flagellum in *C. crescentus* requires an understanding of its three-dimensional structure, the manner in which it is anchored to the cell, and its protein composition. Analysis of the flagellum is facilitated by the fact that it is released into the medium during each cell cycle and is thus easily isolated. A diagram of the *C. crescentus* flagellum is shown in Figure 3.

The flagellum can be viewed as being composed of three subassemblies: the basal body complex, the hook, and the filament (Shapiro and Maizel 1973; Johnson et al. 1979; Wagenknecht et al. 1981). The basal body serves to attach the flagellum to the cell envelope and to drive flagellar rotation. The *C. crescentus* basal body has five rings threaded on a rod that spans the layers of the outer membrane, the peptidoglycan, and the inner membrane (Johnson et al. 1979; Stallmeyer et al. 1985). Purified basal bodies have been shown to be composed of five major proteins, in addition to the 32-kD rod protein (K. Hahnenberger and L. Shapiro, unpubl.).

The hook subassembly is composed of a single protein of 70 kD (Lagenaur et al. 1978; Johnson et al. 1979; Sheffery and Newton 1979) arranged in a right-handed helix with intersecting families of 5-, 6-, and 11-start helices (Wagenknecht et al. 1981). Based on the presence of 16 6-start grooves, each hook contains 295 ± 13 elongated 70-kD monomers.

The filament, like the hook, has its component protein monomers arranged in a right-handed helix (Koyasu and Shirakihara 1984). Each flagellar filament is composed of two major flagellin monomers, a 27.5-kD flagellin-B monomer that forms a relatively short por-

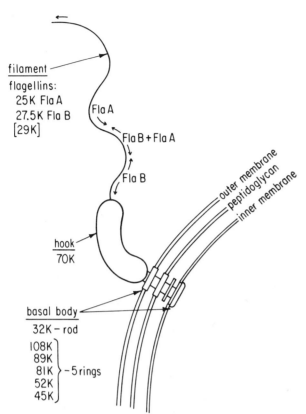

Figure 3. Diagram of the *C. crescentus* flagellum and the placement of its basal body in relation to the cell envelope. The flagellar protein components are indicated and their molecular masses are given in kilodaltons (K). (Reprinted, with permission, from Ely and Shapiro 1984.)

tion of the hook-proximal filament and a 25-kD flagellin-A monomer that forms the rest of the filament (Lagenaur and Agabian 1976, 1977; Marino et al. 1976; Sheffery and Newton 1977; Fukuda et al. 1978; Koyasu et al. 1981; Weissborn et al. 1982). These two flagellins, in addition to a third 29-kD flagellin monomer, form a gene family that appears to have evolved from a common ancestral gene (Gil and Agabian 1982, 1983; Weissborn et al. 1982). Mutant analysis has shown that both flagellin A and flagellin B are required to assemble a functional filament (Fukuda et al. 1981; Koyasu et al. 1981; Weissborn et al. 1982; Johnson et al. 1983). The 29-kD flagellin monomer, which is detected intracellularly in wild-type cells and assembles into stub-like filaments in some nonmotile mutants, appears to be required for the normal filament assembly process (Weissborn et al. 1982; Gil and Agabian 1983).

Several protein components of the flagellum have been shown to be synthesized during a specific time period in the cell cycle (Fig. 1). The synthesis of flagellin A, flagellin B, and hook proteins is initiated just prior to their assembly into a flagellum in the predivisional cell (Shapiro and Maizel 1973; Lagenaur and Agabian 1978; Agabian et al. 1979; Osley and Newton 1980; Sheffery and Newton 1981). The synthesis of the major component of the filament, flagellin A, continues in the swarmer cell following cell division, whereas the synthesis of both the hook protein and flagellin B is completed in the predivisional cell (Osley et al. 1977; Lagenaur and Agabian 1978). It may be that the organization of the flagellin monomers in the filament, with flagellin B comprising a short, hook-proximal region, reflects the fact that flagellin-A synthesis continues long after flagellin-B synthesis has terminated.

Time of synthesis and placement of proteins involved in chemosensory transduction. The direction of flagellar rotation in a chemical gradient is mediated in part by the methylation state of chemotaxis membrane proteins. Those proteins involved in the chemotaxis methylation machinery in *C. crescentus* (the methyl-accepting chemotaxis proteins [MCPs], the carboxylmethyltransferase, and the carboxylmethylesterase) are under cell-cycle control (Shaw et al. 1983; Gomes and Shapiro 1984). The appearance of all three activities coincides with the presence of the polar flagellum (Fig. 1). Upon the release of the flagellum during the transition from swarmer cell to stalked cell, all three chemotaxis methylation activities are lost. To determine whether these cell-cycle-specific activities are due to the differential synthesis of the proteins, antibody to an MCP protein from *Salmonella*, the TAR protein (A. Rousso and D. Koshland, unpubl.), and antibody to *Salmonella* carboxylmethylesterase (Stock and Koshland 1978) were shown to cross-react with the comparable proteins from *C. crescentus* and were then used to immunoprecipitate pulse-labeled proteins from synchronized cell populations (Gomes and Shapiro 1984). As shown schematically in Figure 1, both the MCP and the carboxylmethylesterase are synthesized in the predivisional cell coincident with the time of synthesis of the flagellar proteins.

The flagellar proteins are localized to the incipient swarmer cell pole of the predivisional cell upon their assembly into a flagellum. The chemotaxis methylation proteins are not part of the flagellum (K. Hahnenberger, unpubl.), yet upon cell division, the MCPs and the soluble methyltransferase and methylesterase are present only in the daughter swarmer cell. We investigated the manner in which this occurs in the case of the integral membrane MCP protein. Synchronized cultures were pulse-labeled with ^{14}C-labeled amino acids at the stages of the cell cycle indicated in Figure 4 (panel I). Labeled MCP was detected predominantly in the predivisional cell, whereas upon cell division, neither the daughter swarmer cell nor the daughter stalked cell was able to synthesize MCP. However, in vitro assays of MCP methylation, using membrane substrate from the progeny swarmer and stalked cells, as well as methyltransferase from a cytoplasmic extract of a mixed-cell population, showed that only the daughter swarmer cell contained MCPs (Fig. 4, panel II). Furthermore, in vivo methylation in the presence of [^{3}H]methionine showed that only the progeny swarmer cell methylated its MCPs (Fig. 4, panel III). These results suggest that the MCPs synthesized in the predivisional cell are localized in the incipient swarmer cell portion of the predivisional cell, and upon equitorial cell division, they are sequestered in the swarmer cell. Consequently, the daughter stalked cell is devoid of MCPs. This conclusion is supported by the observation that ^{14}C-labeled amino acids incorporated into MCP in the predivisional cell are specifically chased into the daughter swarmer cell (Fig. 4, upper portion of panel I). Therefore, one manifestation of the generation of polarity in the predivisional cell is the localized placement of newly synthesized membrane proteins.

To determine the molecular signals that dictate the timed "turn-on" of specific protein synthesis and protein localization, the genes encoding several of the flagellar and chemotaxis proteins have been identified and isolated. The biogenesis, rotation, and chemosensory response of the C. crescentus flagellum requires the expression of at least 27 flagellar (fla) genes, three motility (mot) genes, and at least eight chemotaxis (che) genes (Johnson and Ely 1979; Johnson et al. 1983; Ely and Shapiro 1984; Ely et al. 1984; Shapiro et al. 1985). These genes must include those that encode a minimum of 12 flagellar structural proteins and the regulatory sequences and proteins required for flagellar assembly, localization, and rotor function. The structural genes for the hook protein (flaK) (Ohta et al. 1982, 1984), the 29-kD flagellin (Milhausen et al. 1982; Purucker et al. 1982), the motC protein (K. Hahnenberger and L. Shapiro, unpubl.), and the chemotaxis methylation proteins, including the MCP, methyltransferase (cheR), and methylesterase (cheB) (W. Alexander and L. Shapiro, unpubl.), have been identified and cloned (Fig. 5). Tn5 insertion mutants in the MCP, cheR, and cheB are polar, suggesting that this gene cluster forms an operon (W. Alexander and L. Shapiro, unpubl.). N. Agabian and co-workers (unpubl.) have found that the sequences encoding the 25-kD flagellin A and the 27.5-kD flagellin B are grouped with the 29-kD flagellin adjacent to the flaYE gene cluster, but the organization of their transcription units has not yet been reported.

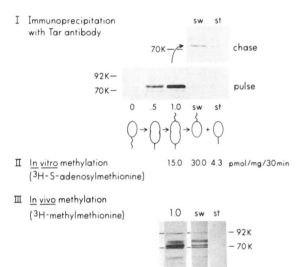

Figure 4. Differential activity and synthesis of MCPs during the C. crescentus cell cycle. C. crescentus CB13 swarmer cells were isolated and allowed to proceed synchronously through the cell cycle. (Panel I) Cells were pulse-labeled with ^{14}C-labeled reconstituted protein hydrolysate at the indicated stages in the cell cycle to measure the synthesis of the MCPs. The ^{14}C-labeled cells were immunoprecipitated with anti-MCP antibody as described in Methods. Autoradiographs of SDS–10% polyacrylamide gel electrophoretograms are shown for the indicated points in the cell cycle. The arrow indicates an autoradiograph of a pulse-chase experiment in which counts incorporated into the MCPs in predivisional cells were chased specifically into the daughter swarmer cell. Counts chased into the daughter cells were immunoprecipitated with anti-MCP antibody. Numbers below the autoradiograms represent fractions of the cell division time. (Panel II) In vitro methylation of MCPs was measured in membrane fractions isolated from the indicated cell types. Membrane fractions were incubated with C. crescentus carboxylmethyltransferase and S-adenosyl-L-[methyl-^{3}H]methionine (4 × 10^{3} cpm/pmole) as the methyl donor. The incorporation of 1 pmole of methyl-^{3}H from S-adenosylmethionine per milligram of methyltransferase is shown per milligram of membrane protein. (Panel III) Cell-cycle samples were pulse-labeled with [methyl-^{3}H]methionine to measure in vivo methylation of the MCPs. Protein molecular weights (in kD) are indicated. (Adapted from Gomes and Shapiro 1984.)

Temporal Regulation of Flagellar Gene Expression

We have studied the transcriptional regulation of flagellar protein synthesis using a cloned flagellin gene to detect differential mRNA synthesis as well as a transposon insertion in which the promoter region of a flagellar gene drives the expression of NPT-II. The gene encoding the 29-kD flagellin (Fig. 5) (Purucker 1985) was nick-translated and used as a probe with Northern blots of mRNA isolated from various stages of the cell cycle (Fig. 6). Because the nucleotide se-

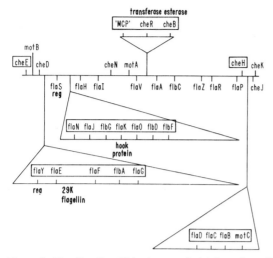

Figure 5. The flagellar (*fla*), chemotaxis (*che*), and motility (*mot*) genes of *Caulobacter crescentus* CB15. The map positions of these genes were determined by Ely et al. (1984). Cloned genes are enclosed in boxes. *cheE* and *cheH* were isolated by S.L. Gomes (unpubl.), *flaYE* by Purucker et al. (1982), the 29 kD flagellin gene by Purucker et al. (1982) and Milhausen et al. (1982), *flaF-flaG* by P. Schoenlein and B. Ely (unpubl.), the *flaN-flbF* gene cluster by Ohta et al. (1982, 1984), and the *flaD-motC* gene cluster by K. Hahnenberger and L. Shapiro (unpubl.).

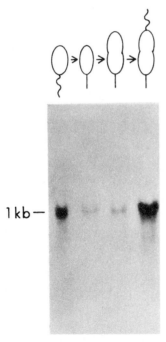

Figure 6. Northern blot hybridization of nick-translated pFB1 DNA (Purucker 1985), which contains the 29-kD flagellin gene, to *C. crescentus* RNA isolated from synchronized cell populations at the indicated stages of the cell cycle. Swarmer cells were isolated in Ludox gradients, as described by Evinger and Agabian (1977), and allowed to continue synchronously through the cell cycle. Aliquots of cell culture were taken at the indicated stages, and RNA was isolated as described by Purucker et al. (1982). RNA samples were electrophoresed through 1.5% agarose and 2.2 M formaldehyde, transferred to nitrocellulose, and treated as described by Derman et al. (1981).

quences of the three flagellin genes share a great deal of homology (Gil and Agabian 1982; Weissborn et al. 1982), the 29-kD flagellin probe is expected to hybridize to mRNA encoding all three flagellin proteins. A single band of approximately 1 kb was detected primarily in the RNA from flagellum-bearing swarmer cells and predivisional cells. The presence of flagellin mRNA is sharply decreased during the transition from swarmer cell to stalked cell. This mRNA is again detected in the predivisional cell coincident with the time of flagellin protein synthesis. As a control, the Northern blot that had been probed with the 29-kD flagellin gene was washed and rehybridized to nick-translated ribosomal DNA from *C. crescentus*. Ribosomal RNA, which is present throughout the cell cycle, was detected in equal intensity in all the stages of the cell cycle (data not shown). These results show that the level of flagellin mRNA is regulated as a function of the cell cycle.

Additional evidence that the turn-on of flagellar gene expression occurs at the transcriptional level was obtained using *fla* insertion mutants generated with the transcription fusion vector Tn5-VB32 (Fig. 2) (Bellofatto et al. 1984). As described in Methods, the gene that encodes NPT-II, and thus confers resistance to kanamycin, is carried on Tn5-VB32 but lacks its own promoter (Fig. 2, panel II). In nonmotile insertion mutants, the expression of NPT-II is under the control of a given *fla* gene promoter. Fusion proteins cannot be formed because of translational stop codons in all three reading frames 5' to the start of the *neo* gene. We have determined that when the synthesis of NPT-II mRNA is driven by a *fla* promoter, NPT-II protein synthesis is temporally regulated in a manner analogous to the synthesis of flagellar proteins (R. Champer and L. Shapiro, unpubl.). In one case, NPT-II synthesis occurred predominantly in the predivisional cell at a time in the cell cycle coincident with the synthesis of flagellar and chemotaxis proteins (see Fig. 1). This observation suggests that the information signaling the time of gene turn-on is regulated at the level of transcription. It remains to be determined whether the temporal control of transcription occurs by the modulation of mRNA initiation, by a mechanism analogous to attenuation, or by changes in mRNA stability.

Trans-acting Control of *fla* and *che* Gene Expression

There are at least three sets of regulatory signals (not necessarily mutually exclusive) that must act upon the *C. crescentus fla* and *che* genes: those that control the temporal turn-on of transcription, those that control the spatial distribution of the transcription and/or translation products, and those that order the flagellar assembly process. The coordinate expression of a large number of *fla* and *che* genes could be controlled by a common activation factor that works simultaneously, but separately, on those genes that are widely scattered on the chromosome. It is equally feasible that these genes are controlled by a cascade mechanism in which the product of a gene at the top of a hierarchy func-

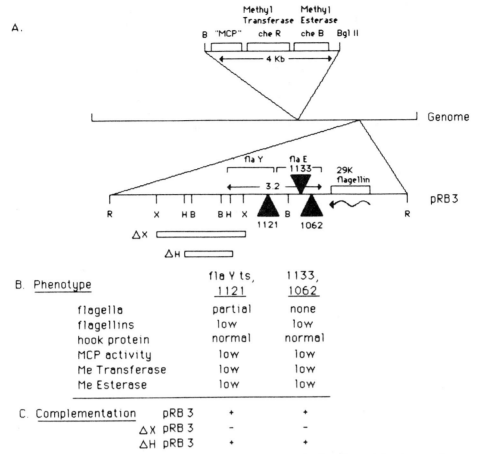

Figure 7. (A) Portion of the C. crescentus CB15 genome showing the relative locations of the chemotaxis genes MCP, cheR, and cheB, the flagellar genes flaY, flaE, and the 29-kD flagellin gene. A restriction map of the flaYE region of CB15 and the location of Tn5 insertions (▲) is also shown. Wavy arrow beneath the 29-kD flagellin gene shows the direction of transcription (Gil and Agabian 1983; Purucker 1985). The cloned EcoRI fragment carrying flaY, flaE, and the 29-kD flagellin is contained on a pRK291 derivative, pRB3. The open bars beneath the restriction map indicate the extent of two deletions of pRB3 constructed in vitro. ΔX indicates that the DNA between the two Xho (X) sites is deleted and ΔH indicates that the DNA between the two HindIII (H) sites is deleted. (B) Phenotypes of two mutations in flaY (a flaY temperature-sensitive [SC274] and a flaY Tn5 insertion [SC1121]) and two mutations in flaE (flaE Tn5 insertion mutants [SC1133 and SC1062]). Arrowheads indicate the approximate boundaries of flaY and flaE, which are separated by 3.2 kb of DNA. The flagellar structure was analyzed by electron microscopy, the levels of the flagellins and the hook protein were analyzed by immunoprecipitation with the relevant antibody, as described previously (Gomes and Shapiro 1984), and the activities of the methyl-accepting protein (MCP), the methyltransferase, and the methylesterase were determined as described previously (Gomes and Shapiro 1984). (C) Ability of pRB3 and its derivatives, ΔXpRB3 and ΔHpRB3, to complement SC274, SC1121, SC1133, and SC1062, as described in text.

tions in *trans* to allow the ordered expression of other *fla* and *che* genes. This type of cascade relationship has been shown to order the expression of flagellar and chemotaxis genes in *E. coli* (Komeda 1982).

We have recently found that in *C. crescentus*, the products of one group of *fla* genes are required for the expression of other *fla* and *che* genes that map elsewhere on the chromosome (Bryan et al. 1984). Initially, we observed that Tn5 insertions in *flaY* and *flaE* genes that map 3′ to the 29-kD flagellin gene result in the loss of motility (Fig. 7) (Bryan et al. 1984). Analysis of these Tn5 insertion mutants and a temperature-sensitive mutant in *flaY* showed that the level of synthesis of all three flagellins and the level of MCP, methyltransferase, and methylesterase activity were all low. The only known phenotypic difference between *flaY* and *flaE* is that *flaY* mutants are able to assemble a partial, stub-like filament, whereas *flaE* mutants are unable to assemble a flagellum. Complementation by an intact *flaY*, *flaE* nucleotide sequence carried on plasmid pRB3 restored normal levels of flagellin synthesis and normal activities of the chemotaxis methylation proteins in all of the mutants tested (Fig. 7). These results suggest that the products of the *flaY* and *flaE* genes function in *trans* to regulate the level of the 3′-proximal flagellin genes and the distal chemotaxis genes. Another component of the flagellum, the hook protein, was found not to be under *flaYE* control.

To determine whether the *trans*-acting *flaYE* products are functioning in a regulatory cascade, and whether this regulation occurs at the level of transcription, double mutants carrying Tn5 flagellar gene fusions were analyzed. For example, a *flaY* mutant was generated by insertional inactivation with a Tn5-VB32 transposon (as described in Methods) and found to be resistant to kanamycin at concentrations of 200 μg/ml. The kanamycin resistance of this nonmotile insertion mutant was shown to be due to the fact that the expression of the promoter-less NPT-II gene in the insert was driven by the upstream *C. crescentus* flaY promoter (Fig. 8, panel II). To determine whether the products of fla genes that map elsewhere on the chromosome are required for the expression of normal levels of flaY, flaY::Tn5-VB32 was transduced into several different fla⁻ strains and assayed for NPT-II synthesis and kanamycin resistance. The rationale of these experiments is that a *trans*-acting function for a given fla gene product would be detected if a mutation in that fla gene failed to allow the normal level of expression of NPT-II driven by the flaY promoter. Because a protein fusion between the product of the gene encoding NPT-II and the interrupted flaY gene cannot occur, any *trans*-

I.

	Double fla mutants		Promoter Activity	
	host strain mutant	fla::Tn5-VB32	NPT II synthesis	drug resistance
	CB 15 (wild type)	-	none	Kn^s
	CB 15 (wild type)	flaY (AE8001)	high	Kn^r
	flaYE (SC520)	flaY (AE8001)	low	Kn^s
	flaS (SC508)	flaY (AE8001)	none	Kn^s
	flaD (SC252)	flaY (AE8001)	high	Kn^r

II. Trans-acting regulation

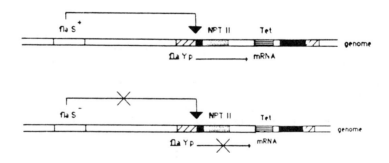

III. Hierarchy of Control

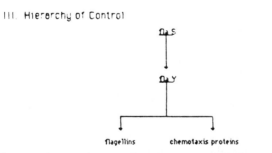

Figure 8. Effect of fla⁻ mutations on the expression of kanamycin resistance by flaY::Tn5-VB32 fusions. (Panel I) Expression of a promoter-less NPT-II gene carried by Tn5-VB32 when it is inserted downstream from the flagellar gene flaY. The flaY gene, interrupted by Tn5-VB32, was transduced into wild-type cells (CB15) and into several different fla⁻ strains, as indicated. The double mutants were each tested for their ability to synthesize NPT-II by pulse-labeling the cultures with ¹⁴C-labeled reconstituted protein hydrolysate and immunoprecipitating the labeled proteins with antibody to NPT-II (see Methods). The cultures were also tested for their ability to grow in the presence of kanamycin. (Panel II) Schematic representation of the flaS gene product functioning in *trans* to control the level of the flaY-NPT-II transcript. When the flaS gene product is lacking, the transcript encoding NPT-II is not available. (Panel III) Hierarchy of *trans*-acting control based on complementation with cloned genes (Fig. 5) and the kanamycin resistance phenotypes observed in double mutants carrying Tn5-VB32 insertionally inactivated fla genes.

acting effect must occur at the level of the transcript. The phenotypes of several double mutants carrying the *flaY*::Tn5-VB32 mutation are shown in Figure 8 (panel I). Two *fla*⁻ mutants, *flaS* (SC508) and *flaYE* (SC520), reduced or abolished the synthesis of NPT-II. The general map position of the Tn5-VB32 insertion in AE8001 was determined by conjugation and shown to lie within the *flaYEF* region (Fig. 5). Southern blot analysis has shown that its position is within the *flaY* gene.

Several lines of evidence argue for the presence of an hierarchial system of control (Fig. 8, panel III) that operates at the level of transcription in *C. crescentus* flagellar biogenesis. Mutations in the *flaYE* genes were shown to result in the pleiotropic down-regulation of distal flagellin and chemotaxis genes (Fig. 7) (Bryan et al. 1984). Correction of this phenotype by plasmid complementation with an intact *flaYE* region (and the failure of *flaY* deletions to complement) showed that the product of the *flaY* gene functions in *trans* to allow the normal level of expression of several *fla* and *che* genes. In addition, expression of *flaS* is required for normal *flaY* function. These results have allowed the construction of a segment of the hierarchy shown in Figure 8 (panel III). It remains to be determined in what manner the turn-on and turn-off of flagellar synthesis is coupled to the periodicity of the *C. crescentus* cell cycle.

The β-galactosidase transcription fusion vector pRC96 allows us to ask the same kinds of questions as were asked for the Tn5-VB32 NPT-II promoter probe, with several distinct advantages. Simple biochemical assays of β-galactosidase production may be performed quickly and quantitatively. Should immunoprecipitations be required, antibody is commercially available. In *C. crescentus*, many of the existing *fla* mutants are the result of wild-type Tn5 insertion (NPT-II promoter intact) and cannot be used with Tn5-VB32. With the β-galactosidase vector, we will be able to transduce *fla*::Tn5-RC96 mutants into *fla*::Tn5 mutants and assay their impact on β-galactosidase production. Finally, the presence of an intact *Bam* site near the 5' end of the Tn5-RC96 promoter probe will provide easy access to cloning, sequencing, and performing S1-nuclease studies on putative *C. crescentus* flagellar promoter regions.

SUMMARY

Each *Caulobacter* cell division yields daughter cells that differ from one another both structurally and functionally. By focusing on the biogenesis of the polar flagellum and the proteins of the chemosensory system, several laboratories have now defined an extensive network of genes whose temporal expression is controlled in the predivisional cell. The differential turn-on of these genes contributes to the generation of asymmetry in the predivisional cell in that the products of these genes are targeted to specific cellular locations. To define the mechanisms that mediate this temporal and spatial control, *fla* genes whose products are not known were accessed by the insertion of transposon-carried drug resistance markers. The transposons were altered so that upon insertion into the chromosome, transcription fusions are formed in which the promoter regions of *fla* genes drive the expression of the downstream promoter-less drug resistance genes. Assays of the differential placement of the promoter-less drug resistance proteins (encoded within the interrupted *fla* genes) allow us to determine whether the positioning of the *fla* gene products is controlled by signal sequences in their proteins, by specific mRNA-targeting sequences in the 5'-regulatory regions of these genes, or by specific transcription from only one of the two newly replicated chromosomes in the predivisional cell.

ACKNOWLEDGMENTS

This investigation was supported by U.S. Public Health Service grants GM-1130 and GM-32506-02 from the National Institutes of Health and Core Cancer Center grant NIH/NCI P30-CA-13330. M.P. and R.C. are Medical Science Training Program Fellows (grant 5T5-GM1674). S.L.G. was an International Research Fellow of the Fogarty International Center (NIH) and was partially supported by the Consellio Nacional de Desenvolvimento Cientifico e Technologico (CNPq), Brasil. The authors are grateful to P. Schoenlein for her assistance in mapping *flaY*::Tn5-UB32.

REFERENCES

AGABIAN, N., M. EVINGER, and E. PARKER. 1979. Generation of asymmetry during development. *J. Cell Biol.* **81:** 123.

AGABIAN-KESHISHIAN, N. and L. SHAPIRO. 1971. Bacterial differentiation and phage infection. *Virology* **44:** 46.

BELLOFATTO, V., L. SHAPIRO, and D. HODGSON. 1984. Generation of a Tn5 promoter probe and its use in the study of gene expression in *Caulobacter crescentus*. *Proc. Natl. Acad. Sci.* **81:** 1035.

BERINGER, J., J. BEYNON, A.V. BUCHANAN-WOLLASTAN, and A. JOHNSTON. 1978. Transfer of the drug resistance transposon Tn5 to *Rhizobium*. *Nature* **276:** 633.

BRYAN, R., M. PURUCKER, S.L. GOMES, W. ALEXANDER, and L. SHAPIRO. 1984. Analysis of the pleiotropic regulation of flagellar and chemotaxis gene expression in *Caulobacter crescentus* using plasmid complementation. *Proc. Natl. Acad. Sci.* **81:** 1341.

CASADABAN, M., J. CHOU, and S. COHEN. 1980. *In vitro* gene fusions that join an enzymatically-active β-galactosidase segment to amino-terminal fragments of exogenous proteins. *J. Bacteriol.* **143:** 971.

DEGNAN, S.T. and A. NEWTON. 1972. Chromosome replication during development in *Caulobacter crescentus*. *J. Mol. Biol.* **64:** 671.

DERMAN, E., K. KRAUTER, L. WALLING, C. WEINBERGER, M. ROY, and J.E. DARNELL. 1981. Transcriptional control in the production of liver-specific mRNAs. *Cell* **32:** 731.

ELY, B. and R.H. CROFT. 1982. Transposition mutagenesis in *Caulobacter crescentus*. *J. Bacteriol.* **149:** 620.

ELY, B. and L. SHAPIRO. 1984. Regulation of cell differentiation in *Caulobacter crescentus*. In *Microbial development* (ed. R. Losick and L. Shapiro), p. 1. Cold Spring Harbor Laboratory, Cold Spring Harbor, New York.

ELY, B., R.H. CROFT, and C.J. GERARDOT. 1984. Genetic mapping of genes required for motility in *Caulobacter crescentus. Genetics* **108:** 523.

Evinger, M. and N. Agabian. 1977. Envelope-associated nucleoid from *Caulobacter crescentus* stalked and swarmer cells. *J. Bacteriol.* **132:** 294.

Fukuda, A., S. Koyasu, and Y. Okada. 1978. Characterization of two flagella-related proteins from *Caulobacter crescentus. FEBS Lett.* **95:** 70.

Fudaka, A., M. Asada, S. Koyasu, H. Yoshida, K. Yginuma, and U. Okada. 1981. Regulation of polar morphogenesis in *Caulobacter crescentus. J. Bacteriol.* **145:** 559.

Gil, P.R. and N. Agabian. 1982. A comparative structural analysis of the flagellin monomers of *Caulobacter crescentus* indicates that these proteins are encoded by two genes. *J. Bacteriol.* **150:** 925.

———. 1983. The nucleotide sequence of the M_r = 28,500 flagellin gene of *Caulobacter crescentus. J. Biol. Chem.* **258:** 7395.

Gomes, S.L. and L. Shapiro. 1984. Differential expression and positioning of chemotaxis methylation proteins in *Caulobacter. J. Mol. Biol.* **178:** 551.

Hodgson, D.A., P. Shaw, and L. Shapiro. 1984. Isolation and genetic analysis of *Caulobacter* mutants defective in cell shape and membrane lipid synthesis. *Genetics* **108:** 809.

Huguenel, E.D. and A. Newton. 1982. Localization of surface structures during prokaryotic differentiation: Role of cell division in *Caulobacter crescentus* differentiation. *Differentiation* **21:** 71.

Johnson, R.C. and B. Ely. 1979. Analysis of non-motile mutants of the dimorphic bacterium *Caulobacter crescentus. J. Bacteriol.* **137:** 627.

Johnson, R.C. and W. Reznikoff. 1983. DNA sequences at the ends of transposon Tn5 required for transposition. *Nature* **304:** 280.

Johnson, R.C., D.M. Ferber, and B. Ely. 1983. Synthesis and assembly of flagellar components by *Caulobacter crescentus* motility mutants. *J. Bacteriol.* **154:** 1137.

Johnson, R.C., J.P. Walsh, B. Ely, and L. Shapiro. 1979. Flagellar hook and basal complex of *Caulobacter crescentus. J. Bacteriol.* **138:** 984.

Komeda, Y. 1982. Fusions of flagellar operons to lactose genes on a Mu *lac* bacteriophage. *J. Bacteriol.* **150:** 16.

Koyasu, S. and Y. Shirakihara. 1984. *Caulobacter crescentus* flagellar filament has a right-handed helical form. *J. Mol. Biol.* **173:** 125.

Koyasu, S., M. Asada, A. Fukuda, and Y. Okada. 1981. Sequential polymerization of flagellin A and flagellin B into *Caulobacter* flagella. *J. Mol. Biol.* **153:** 471.

Lagenaur, C. and N. Agabian. 1976. Physical characterization of *Caulobacter crescentus* flagella. *J. Bacteriol.* **128:** 435.

———. 1977. *Caulobacter* flagellins. *J. Bacteriol.* **132:** 731.

———. 1978. *Caulobacter* flagellar organelle: Synthesis, compartmentation and assembly. *J. Bacteriol.* **135:** 1062.

Lagenaur, C., M. De Martini, and N. Agabian. 1978. Isolation and characterization of *Caulobacter crescentus* flagellar hooks. *J. Bacteriol.* **136:** 795.

Lagenaur, C., S. Farmer, and N. Agabian. 1974. Absorption properties of stage-specific *Caulobacter* ϕCbK. *Virology* **77:** 401.

Marino, W., S. Ammer, and L. Shapiro. 1976. Conditional surface structure mutants of *Caulobacter crescentus* temperative-sensitive flagella. *J. Mol. Biol.* **107:** 115.

Milhausen, H., P.R. Gil, G. Parker, and N. Agabian. 1982. Cloning of developmentally-regulated flagellin genes from *Caulobacter crescentus* via immunoprecipitation of polyribosomes. *Proc. Natl. Acad. Sci.* **79:** 6847.

Ohta, N., L.-S. Chen, and A. Newton. 1982. Isolation and expression of cloned hook protein gene from *Caulobacter crescentus. Proc. Natl. Acad. Sci.* **79:** 4863.

Ohta, N., E. Swanson, B. Ely, and A. Newton. 1984. Physical mapping and complementation analysis of Tn5 mutations in *Caulobacter crescentus*: Organization of transcriptional units in the hook cluster. *J. Bacteriol.* **158:** 897.

Osley, M.A. and A. Newton. 1980. Temporal control and the cell cycle in *Caulobacter crescentus*: Roles of DNA chain elongation and completion. *J. Mol. Biol.* **138:** 109.

Osley, M.A., M. Sheffery, and A. Newton. 1977. Regulation of flagellin synthesis in the cell cycle of *Caulobacter*: Dependence on DNA replication. *Cell* **12:** 393.

Poindexter, J.S. 1964. Biological properties and classification of the *Caulobacter* group. *Bacteriol. Rev.* **28:** 231.

Poindexter, J.S., P.R. Hornack, and P.A. Armstrong. 1967. Intracellular development of a large DNA bacteriophage lytic for *Caulobacter crescentus. Arch. Microbiol.* **59:** 237.

Purucker, M. 1985. "Structure and function of a flagella gene cluster." Ph.D. thesis, Albert Einstein College of Medicine, New York.

Purucker, M., R. Bryan, K. Amemiya, B. Ely, and L. Shapiro. 1982. Isolation of a *Caulobacter* gene cluster specifying flagellum production by using non-motile Tn5 insertion mutants. *Proc. Natl. Acad. Sci.* **79:** 6797.

Schmidt, J.M. and R.Y. Stanier. 1966. The development of cellular stalks in bacteria. *J. Cell Biol.* **28:** 423.

Shapiro, L. 1985. Generation of polarity during *Caulobacter* cell differentiation. *Annu. Rev. Cell Biol.* **1:** (in press).

Shapiro, L. and N. Agabian-Keshishian. 1970. Specific assay for differentiation in the stalked bacterium *Caulobacter crescentus. Proc. Natl. Acad. Sci.* **69:** 200.

Shapiro, L. and J.V. Maizel, Jr. 1973. Synthesis and structure of *Caulobacter crescentus* flagella. *J. Bacteriol.* **113:** 478.

Shapiro, L., W. Alexander, R. Bryan, R. Champer, P. Frederikse, S.L. Gomes, K. Hahnenberger, and B. Ely. 1985. Biogenesis of a polar flagellum and a chemosensory system during *Caulobacter* cell differentiation. In *Motility and chemosensory transduction in microorganisms*, Katzir-Katchalsky Conference (ed. M. Eisenback). Weizmann Institute, Rehovat, Israel. (In press.)

Shaw, P., S.L. Gomes, K. Sweeney, B. Ely, and L. Shapiro. 1983. Methylation involved in chemotaxis is regulated during *Caulobacter* differentiation. *Proc. Natl. Acad. Sci.* **80:** 5261.

Sheffrey, M. and A. Newton. 1977. Reconstitution and purification of flagellar filaments from *Caulobacter crescentus. J. Bacteriol.* **132:** 1027.

———. 1979. Purification and characterization of a polyhook protein from *Caulobacter crescentus. J. Bacteriol.* **138:** 575.

———. 1981. Regulation of periodic protein synthesis in the cell cycle: Control of initiation and termination of flagellar gene expression. *Cell* **24:** 49.

Smit, J. and N. Agabian. 1982a. *Caulobacter crescentus* pili: Analysis of production during development. *Dev. Biol.* **89:** 237.

———. 1982b. Cell surface patterning and morphogenesis: Biogenesis of a periodic surface array during *Caulobacter* development. *J. Cell Biol.* **95:** 41.

Stallmeyer, M.J.B., D.J. De Rosier, S.-I. Aizawa, R.M. Macnab, K. Hahnenberger, and L. Shapiro. 1985. Structural studies of the basal body of bacterial flagella. *Biophys. J.* **47:** 48a (Abstr.).

Stock, J.B. and D.E. Koshland. 1978. A protein methylesterase involved in bacterial sensing. *Proc. Natl. Acad. Sci.* **75:** 3659.

Wagenknecht, T., D. DeRosier, L. Shapiro, and A. Weissborn. 1981. Three-dimensional reconstruction of the flagellar hook from *Caulobacter crescentus. J. Mol. Biol.* **151:** 439.

Weissborn, A., H.M. Steinman, and L. Shapiro. 1982. Characterization of the proteins of the *Caulobacter crescentus* flagellar filament: Peptide analysis and filament organization. *J. Biol. Chem.* **257:** 2066.

Williams, J. and D. Northrop. 1976. Purification and properties of gentamicin acetyltransferase I. *Biochemistry* **15:** 125.

Molecular Genetics of *Drosophila* Neurogenesis

B. YEDVOBNICK,*‡ M.A.T. MUSKAVITCH,† K.A. WHARTON,* M.E. HALPERN,* E. PAUL,*
B.G. GRIMWADE,* AND S. ARTAVANIS-TSAKONAS*

Department of Biology, Yale University, New Haven, Connecticut 06520; †Department of Biology, University of Indiana, Bloomington, Indiana 47405

The central nervous system in *Drosophila* is derived from a set of embryonic precursor cells called neuroblasts. Morphogenetic movements are initiated immediately after blastoderm formation, resulting in the establishment of the ectodermal and endodermal germ layers. Subsequently, the morphologically homogeneous ectodermal layer exhibits the first visible signs of differentiation (Poulson 1950; Hartenstein and Campos-Ortega 1984). The region of the ectoderm below the mesoderm and lying symmetrically on both sides of the ventral midline consists of approximately 1600 cells and is designated the neurogenic region. As development proceeds, cells within this medial ectodermal region uniformly increase in size, rendering them morphologically distinct from the remainder of the ectodermal cells (Hartenstein and Campos-Ortega 1984). About one-fourth of the cells in the neurogenic region then segregate toward the interior of the embryo. These cells are the neuroblasts, and will eventually give rise to the central nervous system. The rest of the cells, designated dermoblasts, will give rise to the ventral epidermis. Embryological analyses indicate that cells in the neurogenic region are actually a mixture of precursors that will give rise to epidermal and neural structures (Hartenstein and Campos-Ortega 1984). The procephalic ectoderm also appears to be a mixture of epidermal and neural precursor cells. Although less rigorously characterized, neuroblast segregation in the procephalic lobe appears to be analogous, though not identical, to that in the germ band.

We are interested in gaining insight into the molecular mechanisms underlying the initial events in the differentiation of the central nervous system. How is the neurogenic region defined? What are the factors responsible for the differentiation of the neuroblasts in the neurogenic region? What molecular differences distinguish neuroblasts from their neighboring dermoblasts? Virtually nothing is known about mechanisms that govern these early differentiation events in the central nervous system. However, it was recognized very early that these events are under genetic control. Poulson was the first to show that lesions at the Notch locus resulted in a hypertrophied nervous system at the expense of hypodermal structures, suggesting that wild-type Notch activity is necessary for the correct differentiation of neuroblasts and dermoblasts within the ectoderm (Poulson 1937; Wright 1970). The phenotypic description of Notch mutants by Poulson led Wright to propose that the developmental role of the Notch gene product is to suppress the potential of a subset of ectodermal cells to become neuroblasts (Wright 1970). In the absence of such a function, cells in the neurogenic region would be channeled into a neural rather than an epidermal pathway. The initial observations of Poulson and the interpretation proposed by Wright have been confirmed and extended by Campos-Ortega and his colleagues. They have studied the phenotype resulting from Notch mutations, as well as the phenotype resulting from other mutations that apparently affect these developmental events (Lehman et al. 1983).

Exhaustive mutagenic screens for zygotic mutations have revealed the existence of six genes in the *Drosophila* genome that when mutated may result in embryonic phenotypes similar to Notch mutations (Lehman et al. 1983; Jurgens et al. 1984; Nusslein-Volhard et al. 1984; Wieschaus et al. 1984). These genetic loci, Notch (*N*), Delta (*Dl*), Enhancer of Split (*E[spl]*), mastermind (*mam*), big brain (*bib*), and neuralized (*neu*), are collectively known as the zygotic neurogenic loci. In addition, a few more maternally acting neurogenic genes have been identified: almondex (Shannon 1973; Lehman et al. 1983), pecanex (Perrimon et al. 1984), and others (T. Schuepbach; A. Mahowald; both pers. comm.). Females homozygous for such mutations yield progeny that exhibit hypertrophy of the nervous system, as has been observed in embryos carrying mutations in the zygotic neurogenic loci. The genetic analysis of neurogenic gene function to date suggests that it is the null state, within the context of either zygotic or maternal gene expression, that results in the hyperplasy (Lehman et al. 1983). Finally, a third group of genes affecting neurogenesis has been identified (N. Perriman, K. Konrad, L. Engstrom, and A. Mahowald, pers. comm.). Germ line mosaics of zygotic, postembryonic lethal mutations mapping within a given X chromosomal interval, revealed the existence of a maternal component apparently essential for neurogenesis in several of these mutations. Elimination of this component during oogenesis via germ line mosaics leads to a neurogenic phenotype. If one assumes that such genes are randomly distributed throughout the genome, the analysis of Mahowald and his colleagues raises the possibility that more than 50 loci having a maternal neurogenic component may exist in the genome.

‡Present address: Department of Biology, Emory University, Atlanta, Georgia 30322.

The existence of a group of genes, which when mutated give rise to a similar phenotype, raises the possibility that they define a developmental pathway. The concept of a developmental pathway implies some hierarchical order of the various genetic units involved in it as well as functional interactions between them. Notwithstanding the fact that the phenotype of most neurogenic mutants have not been studied in detail, the embryological analysis indicates that they all affect the developmental events that lead to the differentiation of the ectoderm into epidermal and neural precursors. One possibility, therefore, is that the activity of all neurogenic loci in concert defines a developmental pathway. According to that hypothesis, one or more neurogenic loci would control each step in the pathway. Alternatively, the differentiation of the neurogenic region may be dependent on the correct expression of a gene ensemble which is functionally unrelated. For instance, it is conceivable that for an ectodermal cell to become a neuroblast, the level of specific molecules in the cell, which may be controlled either by genetic units or epigenetic mechanisms, must reach certain values. Changes of those values may result in channeling the ectodermal cell into a neural rather than an epidermal pathway. Another way of describing such an event would be to say that the state of the neuroblast in the neurogenic region reflects a developmental "depression" or "sink" which can be reached in many developmentally unrelated ways, rather than a state reached by distinct determinative steps, each of which is controlled by one or more neurogenic genes.

In our attempts to explain the existence of multiple neurogenic loci, a third hypothesis can be formulated which is conceptually a composite of the two previous ones. Namely, that the phenotypic manifestation of one neurogenic locus, locus A, does not reflect the direct participation of that locus in the developmental pathway, but is rather the indirect result of the interaction of locus A with another neurogenic locus, say locus B, which itself is directly involved in a determination of the neurogenic pathway. To gain insight into the mechanisms underlying neurogenesis, we have initiated a molecular study of neurogenic loci. Our attention is drawn specifically to the zygotic neurogenic loci for two reasons. As it has been argued before (e.g., Garcia-Bellido 1975), genes with functions involved in binary developmental decisions in the embryo are likely to be expressed during zygotic development. The zygotic neurogenic loci are therefore the best candidates for genes that encode functions affecting the commitment of ectodermal cells into a neural or epidermal fate. Moreover, it is the zygotic loci that have been best characterized in developmental and genetic terms (Lehman et al. 1983). We are seeking to determine the biochemical identity of their products, the regulation of their activity during development, and the nature of the interactions between these unlinked loci. We have initiated our study with the analysis of the Notch locus, which, genetically and phenotypically, is the best understood neurogenic locus.

EXPERIMENTAL PROCEDURES

Nucleic acids. Preparation and analysis of nucleic acids was performed as described previously (Grimwade et al. 1985).

Drosophila *strains*. The fly stocks used throughout this work were maintained as described previously (Grimwade et al. 1985). Except when otherwise mentioned, the descriptions of the mutants can be found in Grimwade et al. (1985).

Construction of cDNA clones. Cloning of cDNAs was performed according to the protocol of Huynh et al. (1985), kindly provided to us prior to publication, with minor modifications. First-strand synthesis was completed using Sephacryl S-200 repurified reverse transcriptase (Life Sciences) at 42°C for 90 minutes in 50 mM HEPES (pH 8.5), 40 mM KCl, 8mM $MgCl_2$, 4 mM dithiothreitol (DTT), 1 mM d(ATGC)TP, 12 μg/ml of oligo(dT)$_{12-18}$, and 2 μg poly(A)$^+$ RNA. The RNA:DNA duplexes were denatured by boiling for 90 seconds, cooled, and second-strand synthesis was initiated through addition of: 50 μl of 2× buffer (100 mM HEPES [pH 6.9], 100 mM KCl, 20 mM $MgCl_2$, 10 mM DTT), 1 μl of 10 mM d(ATGC)TP, and 60 units of DNA polymerase I (New England Biolabs). Second-strand synthesis occurred at 15°C for 5 hours. After second-strand synthesis, 400 μl of S1 nuclease digestion buffer (30 mM NaAc [pH 4.4], 250 mM NaCl, 1 mM $ZnCl_2$) and 2500 units of S1 nuclease (Boehringer Mannheim) were added and the mixture was incubated at 37°C for 30 minutes. The reaction was terminated through addition of 5 μl of 1 mM EDTA and phenol-chloroform extraction. The aqueous phase was ethanol-precipitated and pelleted at 4°C in a microfuge. The pellet was resuspended in 20 μl of EcoRI methylase buffer (50 mM Tris-HCl [pH 7.5], 1 mM EDTA, 5 mM DTT), to which 2 μl of 100 mM S-adenosyl-L-methionine and 30 units of EcoRI methylase (New England Biolabs) were added. The methylase reaction was run at 37°C for 30 minutes, and heat-inactivated at 70°C for 10 minutes. After cooling, the cDNA ends were filled in through addition of 2.5 μl of 100 mM $MgCl_2$, 2.5 μl of 0.2 mM d(ATGC)TP, 10 units of DNA polymerase I, and incubation at room temperature for 30 minutes. At that time, 10 μl of 50 mM EDTA and 1 μg of λgt10 DNA were added and the mixture was phenol-chloroform-extracted. The organic phase was extracted twice more, and the aqueous phases were then pooled and ethanol-precipitated. The cDNA pellet was resuspended in 8 μl of kinased EcoRI linkers (100 μg/ml), and 400 units of DNA ligase (New England Biolabs) were added; the ligation was run at 13°C for 20 hours. An equal volume of 2× digestion buffer (50 mM Tris-HCl [pH 7.5], 10 mM $MgCl_2$, 200 mM NaCl, 4 mM DTT) was added, and the cDNAs were digested with 40 units of EcoRI (New England Biolabs) at 37°C for 60 minutes. An additional 20 units of enzyme was then added and the digestion continued for 60 minutes. The enzyme was heat-inactivated at 70°C for 10 minutes, and the digestion products were separated on a 1% agarose gel. The molecular weight range of 500–7000 bp was excised from the gel and electroeluted into dialysis tubing. The eluate was purified by passage through an Elu-Tip column (Schleicher & Schuell) and ethanol-precipitated with 1 μg of EcoRI-cut λgt10. The pellet was resuspended in 4 μl of 10 mM Tris-HCl (pH 7.5) and 10 mM MgCl$_2$, and heated at 42°C for 15 minutes. After cooling, 0.5 μl of 10 mM ATP, 0.5 μl of 100 mM DTT, and 0.5 μl of (200 units) ligase were added and the mixture incubated at 13°C for 20 hours. Aliquots of the ligation mixture were in vitro-packaged and plated on c600 Hf1.

The production of specifically primed Notch cDNAs was performed as follows. A Pst–BglII restriction fragment derived from the 5' region of the 7-kb exon (exon E) was subcloned into the M13 vectors mp8 and mp9. Single-stranded mp9 subclones were primed and the insert copied into double strands using the Klenow fragment. Because the copied insert is susceptible to restriction digestion, the products of the Klenow reaction were digested, denatured, and a single-stranded DNA purified according to the protocol of Burke (outlined in Akam 1983). The single-stranded primer was hybridized for 60 minutes with 5 μg of poly(A)$^+$ at 40°C in 80% formamide, 40 mM PIPES (pH 6.4), 400 mM NaCl, and 1 mM EDTA. The primed RNA was precipitated and then dissolved in first-strand synthesis buffer (see above).

Hybridization to embryo tissue sections. In situ hybridization to tissue sections was performed using the method of Hafen et al. (1983) with the exception that aged Oregon-R embryos were collected, dechorionated, and placed directly in embedding compound for sectioning. The preparation of tritiated nick-translated DNA fragments and the conditions for hybridization and autoradiography were as described in Hafen et al. (1983). The fushi tarazu plasmid p523B, containing a 3- to 4-kb genomic EcoRI fragment, was supplied by W. McGinnis (Yale University).

RESULTS AND DISCUSSION

Genetics of the Notch Locus and Its Interactions

The Notch locus is located at band 3C7 of the salivary gland chromosomes and is genetically defined by an array of noncomplementing amorphic embryonic lethal Notch (N) alleles. Hemizygous or homozygous Notch embryos show the neurogenic phenotype, namely hypertrophy of the nervous system at the expense of hypodermal structures. Heterozygous Notch females of the type N/+ display a dominant phenotype consisting of variably notched wings, thickened wing veins, and minor bristle abnormalities (Welshons 1965, 1974; Welshons and Keppy 1981). Interspersed within the array of dominant Notch alleles is a second class of mutations that behave as recessive visibles and can affect either eye (e.g., *fa, spl, fa^{swb}*) or wing (e.g., *nd, nd^2*) morphology. In general, eye mutations complement wing mutations whereas heterozygous combinations between eye mutants and between wing mutants are usually, although not always, noncomplementing. When dominant Notch mutations are combined with recessive visibles, a pseudodominant behavior of the recessive is observed (Welshons 1965). Finally, a third class, the Abruptex (*Ax*) mutations, maps within the limits of the Notch locus. These are dominant alleles affecting wing venation as well as bristle and hair formation, and in the heterozygous form they show complex interactions with Notch alleles. In spite of the various mutant classes and the complex mutant interactions, the genetic analysis indicates that all mutations affect a single functional unit (Welshons 1971; Foster 1975; Portin 1975).

The extensive genetic analysis of the Notch locus has revealed a number of phenotypic interactions between Notch and other genetic loci. Welshons and his colleagues have described experiments that reveal the *cis* interaction of Notch with sequences mapping a few salivary chromosome bands distal to 3C7 (Keppy and Welshons 1977), and have identified at least one other locus that seems to affect in *trans* the activity of Notch (W. Welshons, pers. comm.). Interaction between cubitus interruptus and Notch has been documented (House and Lutes 1975), and Hairless appears also to influence Notch functions (Lindsley and Grell 1968). The nature or the developmental significance of these interactions remains obscure. The phenotypic interactions between genes must be interpreted with caution, since they may be the result of secondary effects rather than the manifestation of a direct developmental or molecular relationship between the two loci.

Three of the zygotically acting neurogenic loci, Notch, Delta, and Enhancer of split, exhibit a particularly interesting set of interactions (Lindsley and Grell 1968; Dietrich and Campos-Ortega 1984; Harris 1985). In contrast to the other neurogenic loci, hemizygosity for any one of these three loci yields a haplo-insufficient wing phenotype that involves nicking of the wing and/or disruption of wing venation. This wing venation phenotype is suppressible by the mutation Hairless (*H*, 3−69.5, 92F-93 AB) for all of these loci.

Enhancer of split (*E[spl]*, 3-89.1 96F5-7) is a dominant mutation that leads to enhancement of the eye phenotype associated with split, a recessive visible mutation within the Notch locus (Welshons 1956). The strength of the effect of *E(spl)* is evidenced by the observation that "roughening" of the eye occurs when *E(spl)* is present in the third chromosome even if split is heterozygous with a wild-type Notch allele. However, the *E(spl)* mutation yields no visible phenotype in the absence of the split mutation. Thus, the *E(spl)* locus was first identified on the basis of a dominant, presumably neomorphic allele that acts in *trans* to influence the phenotype of a mutation within Notch. Reversion of the dominant *E(spl)* phenotype generally leads to a nonfunctional copy of the locus, thus eliminating the neomorphic as well as the normal function. Revertants of Enhancer of split (*E[spl]^R*) have therefore lost their ability to enhance the split phenotype. In the homozygous condition, E(spl)^R leads to embryonic lethality, and the resulting embryos exhibit a neurogenic phenotype (Lehman et al. 1983). Therefore E(spl) initially identified on the basis of its interaction with Notch, is itself required for the proper completion of neurogenesis.

Revertants of E(spl), in turn, exhibit a curious interaction with alleles of Delta (Dl, 3-66.2, 92A2). Although Dl is far separated from E(spl) on the meiotic recombination map, an animal heterozygous for null mutations at both Dl and E(spl) dies before the first larval instar (Lehman et al. 1983). A number of Dl alleles, including a deficiency (Df[3R] Cha^{M8}, 91C7-92A8; L. Myers and W. Gelbart, unpubl.) which eliminates Delta, demonstrates this interaction in combination with E(spl)^{RS1}, a spontaneous revertant of E(spl). Numerous E(spl)^R alleles tested in combination with Df (3R) Cha^{M8} and Dl^3, an apparent point mutation, also exhibit early lethality. At this stage in the analysis, there appears to be no allele specificity for this interaction. However, one would expect that if these interactions are dosage-dependent then certain hypomorphic, as opposed to amorphic, alleles might be viable in the heterozygous combination. What is clear is that embryogenesis is extremely sensitive to the combined dosage of the products from these two loci.

This brief account of the phenotypic interactions of E(spl)^R, Dl, and Notch, which have been reported in the literature, raises the possibility that the gene products may be interacting at the molecular level. The mechanisms underlying these interactions, however, must await the characterization of the regulation and products of these genes. Trans interactions may provide the key in understanding the broad developmental questions we pose, but as already mentioned, caution must be exercised in attempting to use phenotypic interactions as a tool in identifying pathways that are part of the developmental logic of morphogenesis.

In spite of the extensive genetic and embryological characterization of the Notch locus, as well as efforts to unravel its molecular nature (Thorig et al. 1981), the biochemical characteristics and the mode of action of its products during development remain unknown. We approached the molecular analysis of the Notch locus by cloning the entire chromosomal region in which Notch was known to reside (Artavanis-Tsakonas et al. 1983). Using N^{76B8}, a chromosomal inversion between 3C7 (Notch) and 3C10-11, and cloned sequences spanning the 3C10-11 breakpoint, we were able to "jump" into the Notch locus. Once sequences in the 3C7 region were acquired, a chromosomal walk was initiated and 150 kb of contiguous DNA was isolated. We sought to localize the Notch locus within this chromosomal region by whole genome Southern analysis comparing the Notch mutant and wild-type chromosomes. Molecular lesions corresponding to specific mutations that have been localized by recombination on the Notch genetic map should fall in an array along the physical map as predicted by the respective genetic map positions (Artavanis-Tsakonas et al. 1983, 1984; Grimwade et al. 1985). We were thus able to correlate the physical map with the genetic map as shown in Figure 1, which summarizes our current knowledge regarding the molecular and genetic positions of all known point mutations in the Notch locus. Numerous chromosomal rearrangements with breakpoints along the Notch map have also been identified. This analysis revealed that all identified mutations in Notch map within an approximately 40-kb-long interval, indicating that the physical length of the locus corresponds to that length. Moreover, a correlation between physical and genetic map distances could be established. In the Notch locus, 0.01 cM corresponds to ~2.8 kb of DNA.

Although the molecular analysis of mutations in the Notch locus argues that it is physically limited within 40 kb of DNA, the experimental arguments supporting this contention are indirect (Grimwade et al. 1985). Direct proof regarding the distal limit of the gene was obtained from a deletion analysis, since the relationship between the lack of specific DNA sequences to mutant or wild-type phenotype can be unambiguously established. Df(1)w^{67k30}, deficient for salivary bands 3C2-3C6, is a recessive lethal and lacks the normal functions of white, roughest, and verticals, but does not affect Notch. Comparative Southern blot analysis of Df(1)w^{67k30} and wild-type chromosomes shows that the deficiency breakpoint maps within the BglII-HindIII fragment between coordinates -28.5 and -27.3. Since animals carrying the deficiency are phenotypically wild type for Notch, we conclude that the sequences mapping distal to -28.5 are not necessary for normal Notch function. There are no analogous deletions at the proximal end of the locus, and hence we cannot use the same type of rigorous argument to define the proximal limits. Nevertheless, the transcriptional analysis of the locus strongly suggests that sequences lying proximal to the EcoRI-BglII fragment between coordinates $+10.3$ and $+11.3$ code for transcripts that show developmental regulation very different from that expected from a Notch product. It should be noted, however, that such arguments only indirectly define the proximal limits (Grimwade et al. 1985). We are currently seeking direct proof of the physical limits of the Notch locus by attempting to rescue the phenotype of mutant animals via DNA-mediated transformation using the DNA sequences mapping between coordinates -28.5 and $+11.3$. If such experiments succeed, they will provide direct proof of the limits of the locus.

The Transcriptional Activity of Notch

We first examined the transcriptional activity of the Notch locus by Northern analysis. Radioactively labeled contiguous fragments deriving from within as well as from outside the 40-kb region were used as probes to blots of agarose gels containing electrophoretically fractionated poly(A)$^+$ RNA from various developmental stages. Our results can be summarized as follows: fragments derived from within the 40-kb region are complementary either to a predominant 10.5-kb RNA size class, or alternatively, appear not to be transcribed. The only exception to this general hybridization profile is provided by fragments containing sequences between coordinates +8.5 and +9.0. Such probes detect a heterogeneous group of RNA species ranging in size from approximately 0.7 kb to 11.0 kb due to the presence of a small repetitive sequence discussed in more detail below. The 40-kb region is bordered distally by fragments complementary to a 7-kb transcript and proximally to 0.9-kb and 0.7-kb transcripts (Grimwade et al. 1985). The deletion analysis mentioned above excludes the possibility that the 7-kb transcript is part of the Notch locus, while it leaves open the question of whether or not the 0.7-kb and 0.9-kb transcripts are relevant for wild-type Notch function.

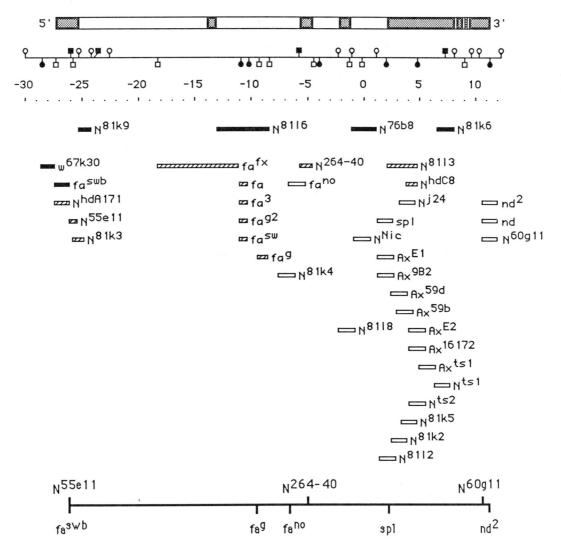

Figure 1. Correlation between physical, transcriptional, and genetic maps of the Notch locus: A simplified restriction map is shown above the scale (in kilobase pairs), which represents the system of coordinates used in the text, and has been discussed in Grimwade et al. (1985). The centromere maps to the right of the map. (♀) EcoRI; (♦) BglII; (♀) HindIII; (■) XhoI. Above the restriction map is a schematic representation of the Notch transcription unit (bar) which spans the 40-kb interval within which the Notch locus resides. The direction of transcription is from left to right. Restriction fragments homologous to the mature 10.5-kb poly(A)+ RNA are depicted as shaded areas. Below the scale, the various bars indicate the area within which particular mutations map. Given that some mutations have been mapped both genetically as well as molecularly, the physical and genetic maps (a rudimentary genetic map is shown at the bottom of the figure) could be aligned. Most known alleles with normal cytology are listed in the figure. The positions of alleles with known molecular lesions are shown by hatched bars while those with no known restriction pattern abnormalities but known genetic positions are shown by open bars. Such mutants have been placed in the map using their recombination frequencies from known markers within the locus, assuming that 0.01 cM equals 2.8 kb of DNA. The breakpoints of many chromosomal rearrangements have also been molecularly mapped (Grimwade et al. 1985) but only a few are shown here (solid bars) to illustrate that chromosomal breaks mapped throughout the locus confer a null phenotype. The various alleles have been described previously (Portin 1975; Welshons and Keppy 1981; Grimwade et al. 1985).

The direction of transcription of the 10.5-kb RNA was determined (Artavanis-Tsakonas et al. 1984), and the mapping of the transcribed sequences suggested that the entire 40-kb region, defined by the molecular genetics as being the Notch locus, is transcribed as a single transcription unit and spliced into a 10.5-kb mature poly(A)+ RNA. A combination of Northern analysis, S1 nuclease mapping, and cDNA and genomic sequence analysis allowed the mapping of the exonic regions within the 40-kb region. Our current knowledge is summarized in Figure 1. The fact that the Notch locus seems to give rise to a single large transcription unit provides a plausible explanation to the genetic data which suggest that, in spite of the many different mutant classes and its size, Notch behaves as a single genetic unit. Given such an organization of the Notch sequences, it is expected that, in general, rearrangements breaking anywhere within the 40-kb region will result in an aberrant and presumably nonfunctional product. Indeed, as examination of the phenotypes reveals, rearrangement invariably gives Notch phenotypes that are indistinguishable from that of a mutant lacking the entire locus. Finally, the distribution of exonic sequences is clearly reflected by the distribution of mutations in the locus. Most mutants appear to map toward the proximal end, and the RNA mapping shows that most of the exonic regions map to the same location. This can be seen diagramatically in Figure 1.

The existence of mutant classes in the Notch locus that affect various structures in the animal suggests that Notch activity is needed at more than just one developmental stage. In agreement with this contention, the analysis of temperature-sensitive alleles of Notch has shown several temperature-sensitive periods (Shellenbarger and Mohler 1975, 1978). Moreover, the existence of a maternal component of Notch expression has been revealed by the construction of germ line mosaics homozygous for a Notch mutation (Jimenez and Campos-Ortega 1982). We sought to correlate these observations with the developmental regulation of Notch transcription. The profile of the transcriptional activity of Notch during development was determined by preparing RNA from different developmental stages and examining it by Northern analysis for the presence of Notch-specific transcripts. Figure 2 shows a summary of these results. The timing of the transcriptional activity of the Notch locus coincides rather closely with the temperature-sensitive periods, as determined by the genetic analysis (Artavanis-Tsakonas et al. 1983). Transcripts are also found in unfertilized eggs, as predicted by the germ line mosaic analysis. The presence of transcripts in adult females shown in Figure 2 is presumably due to expression of Notch in their ovaries. It seems therefore that the expression of Notch as revealed by the genetic analysis coincides with periods of transcript accumulation.

Superficially, Northern analysis indicates that throughout development the size of the Notch transcripts remains constant, suggesting that the locus produces only one transcriptional product. There are sev-

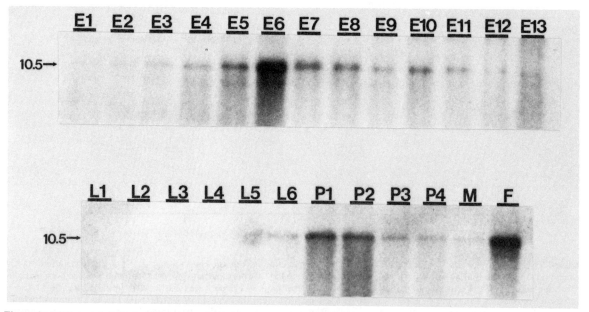

Figure 2. RNA isolated from different developmental stages was electrophoretically separated on formaldehyde-agarose gels and transferred onto nitrocellulose filters as described previously (Artavanis-Tsakonas et al. 1984). The blots were hybridized with a ^{32}P-labeled probe prepared by nick-translation of a genomic fragment mapping between +1.1 and +8.1 (exonic region b in Fig. 3). The autoradiograph shows only the region containing the 10.5-kb Notch RNA. Each lane in the lower panel contains 12 µg of poly(A)+ RNA whereas each lane in the gel shown above contains 30 µg of total RNA. (E) Embryo; (L) larva; (P) pupa; (M) adult male; (F) adult female. Ages of embryos in hours after oviposition at 25°C: (E1) 1–2; (E2) 2–3; (E3) 3–4; (E4) 4–5; (E5) 5–6; (E6) 6–7; (E7) 7–8; (E8) 8–9; (E9) 9–10; (E10) 10–11; (E11) 11–12; (E12) 13–16; (E13) 16–19. Ages of larvae in hours after hatching at 25°C: (L1) 27–30 (first instar); (L2) 48–51 (second instar); (L3) 71–73 (early third instar); (L4) 99–101 (mid-third instar); (L5) 120–125 (late third instar); (L6) climbing third instar. Pupal ages in hours after puparium formation: (P1) 0–24; (P2) 24–48; (P3) 48–72; (P4) 72–96.

eral reasons preventing us from reaching such a conclusion with confidence. Small size differences (<200 bp) between transcripts would escape detection by Northern blot analysis and yet we know from our mapping data, as well as that of other workers (Kidd et al. 1983), that small exons exist. In addition, apparently identical-size classes of Notch transcripts do not necessarily imply identical RNA species. Finally, and perhaps more importantly, long exposures of autoradiograms reproducibly reveal the existence of minor yet distinct species of transcripts with sizes different from that of the main 10.5-kb size class. At least one more size class, approximately 9-kb in length, is also weakly detectable in the early embryonic stages with the probe used in the experiments depicted in Figure 2. Kidd et al. (1983) have also detected a 9-kb size class in embryonic RNA. These considerations raise the possibility that Notch may produce multiple transcription products derived from a single transcription unit. Such multiplicity of products may explain the existence of multiple mutant classes as well as the complex complementation patterns observed between these various classes.

The Analysis of Embryonic Notch cDNA Clones

In order to unambiguously identify different Notch transcripts, to determine the exact number and sizes of exons, and to assess the importance of the minor RNA species, we undertook the construction and analysis of cDNA clones from peak stages of Notch expression. Given the length of the mature Notch transcript(s), rather than attempt to generate full-length products, the cDNA analysis was divided into two parts. The exons in the 3' region (between coordinates +10 and +12) were studied using cDNAs generated by standard oligo(dT) priming of total poly(A)+ RNA. The 5' exonic regions (between coordinates +2.5 and −28.5) were studied using cDNAs produced by specific priming of total poly(A)+ RNA using a single-strand primer homologous to sequences approximately 8 kb from the 3' end of the Notch RNA.

Poly(A)+ RNA was isolated from the peak stages of Notch RNA accumulation as indicated by Northern blot analysis: unfertilized eggs, 4- to 7-hour embryos, climbing third instar larvae, 24- to 48-hour pupae, and adult females. These RNA samples were used as templates for the construction of oligo(dT)-primed cDNA clones, as described in the Experimental Procedures. Approximately 500,000 cDNA clones in λgt10 phage were obtained from each stage. After amplification, each staged library was screened with a probe homologous to the 3' end (mapping between coordinates +10 and +12) of the Notch transcription unit. Approximately 500,000 cDNAs from each stage were screened and multiple positive phage were isolated from every stage. The number of positives from each stage reflected the relative abundance of Notch RNA as determined by Northern analysis. A number of other cDNA clones were also selected from two embryonic cDNA libraries constructed and kindly provided to us by M. Goldschmidt-Clermont and D.S. Hogness (Stanford University) and by L. Kauvar and T. Kornberg (University of California, San Francisco).

Specifically primed cDNA clones were constructed as described in the Experimental Procedures using as a primer a PstI-BglII fragment mapping between +2.5 and +5.0. Approximately 50 Notch cDNAs were isolated from the embryonic library using as a probe a mixture of fragments that are complementary to exonic regions a, b, c, and d (see Fig. 3). Apart from these clones, two other populations of phage were identified in this library. First, phage consisting only of λgt10 vector sequences and, second, recombinant phage containing inserts complementary exclusively to the primer. Recombinant phage containing only primer sequences were not expected, and were presumably the result of self-priming caused by the secondary structure of the 2.5-kb-long single-strand primer.

We started the cDNA analysis with the systematic examination of the embryonic stage because this stage is most relevant to neurogenesis. DNA was prepared from several clones, radioactively labeled, and hybridized to blots containing a series of electrophoretically separated genomic restriction fragments that span the entire locus. The cross-hybridization results for 33 cDNA clones are schematically summarized in Figure 3. They show that cDNA clones spanning the entire length of the 10.5-kb RNA have been isolated. The six cDNAs listed at the bottom of the figure, cDNAs 2-5, 1-7, B, F, E2, and E6, were constructed by oligo(dT) priming, whereas the rest are the result of specific priming.

At the level of resolution offered by the cross-hybridization analysis, it appears that there is only one major embryonic Notch transcript, in agreement with the Northern analysis. With two exceptions, discussed below, the exonic sequence arrangement of all cDNA clones reflects the order depicted in Figure 2.

The structure of the Notch embryonic transcript was further characterized by the determination of the entire nucleotide sequence of cDNA 0 which is complementary to all the exonic regions mapping 5' to coordinate +3, as well as the determination of sequences from selected oligo(dT)-primed cDNA clones to determine the structure of 3' exonic regions. We were thus able to define accurately all the intron–exon junctions. The results of the sequence analysis are beyond the scope of this paper and will be reported elsewhere. However, certain salient features can be summarized as follows.

The genomic regions a, e, f, g, and h indicated in Figure 3 contain one exonic region each whereas c contains two. The smallest exonic region of 90 bp in length maps within region b, and the largest, approximately 6.5 kb in length, maps within e. Exons e, f, g, and h are separated in the genome by three small intronic sequences, 88 bp, 77 bp, and 75 bp in length, respectively. Although our mapping data are in general agreement with the S1 nuclease analysis utilizing pupal RNA, re-

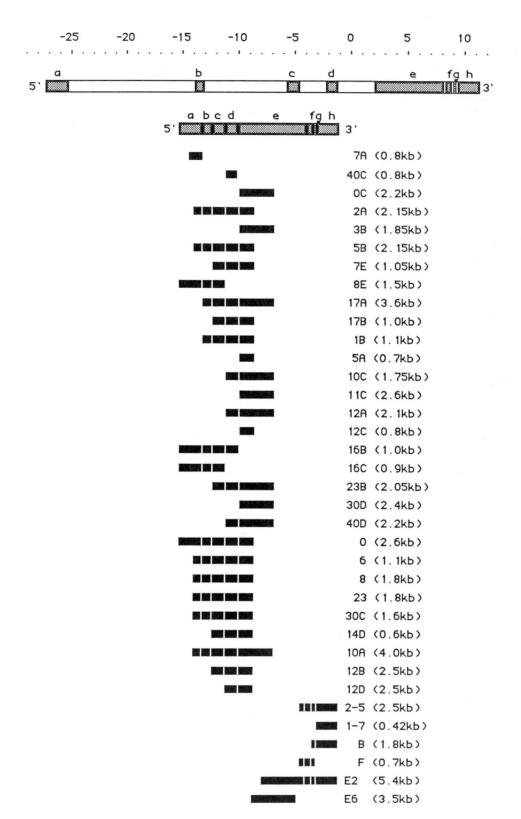

Figure 3. Cross-hybridization summary of Notch cDNA and the genomic region defining the Notch locus: The scale is the same one used in Fig. 1. Underneath the block diagram is the transcriptional map of Notch. Stippled blocks indicate genomic areas within which exonic regions can be found and open blocks indicate intronic regions. Each exonic region is designated a, b, c, d, e, f, g, and h (also see text). Below this diagram a schematic representation of the mature 10.5-kb transcription unit is shown. The name and size (in parentheses) of each cDNA is given adjacent to blocks that illustrate the approximate extent of cross-hybridization of a given cDNA to various exonic regions.

ported by Kidd et al. (1983), we differ significantly in three points. First, their analysis suggests that the region mapping approximately between coordinates −25 and −6 is entirely intronic. According to our data, the region between −25 and −6 has at least one exon (exon b). Second, their data indicate the existence of multiple termination sites at the 3′ end of the RNA. Our cDNA sequence analysis based on several embryonic cDNA clones has so far not identified multiple terminations at this stage. The indication of multiple termination sites may reflect the fact that exonic region h, which defines the 3′ end of the 10.5-kb RNA, has, according to our sequence analysis, several AT-rich stretches that could be misidentified as termination sites by the S1 nuclease analysis. Finally, Kidd et al. (1983) suggest that the exon in region e extends the entire 7-kb EcoRI fragment between coordinates +1.1 and +8.1. Our data fail to detect any exonic sequences in the approximately 1.2-kb region between coordinates +1.1 and +2.3. We have not yet analyzed pupal cDNAs in this area, and since their S1 nuclease analysis was carried out with pupal RNA it is conceivable that the difference between these two sets of data may reflect stage-specific differences.

Comparison of genomic and cDNA sequences at the 5′ end of the transcription unit reveals that cDNA 0 extends to within 26 bp of what we believe is the relevant TATAA sequence, which is 300 nucleotides 5′ of the XhoI site at coordinate −26.0. Preliminary data regarding open reading frames in the cDNA 0 sequence suggest that the 10.5-kb Notch transcript must have an untranslated leader sequence in the order of 1 kb in length. As reported previously (Wharton et al. 1985), the sequence analysis also revealed the existence of a 93-bp-long repetitive sequence within exon g. It was shown that this repetitive sequence element is composed of 30 CAX triplets with X being either A or G. This sequence is a member of a large family of dispersed cross-hybridizing elements of similar length and sequence. We have termed those elements the *opa* repetitive family and have shown that *opa* sequences can be found in a large number of developmentally regulated transcripts.

Inspection of *opa* and sequences adjacent to it in the Notch locus raises the possibility that this repetitive element is not only transcribed, but may also be translated. Within *opa*, with the exception of one CAT triplet, CAA and CAG both code for glutamine. The sequence information gathered to date does not allow us to determine with certainty if the *opa* sequence is translated at Notch. If it were, it would give rise to a 30-amino-acid-long stretch of almost pure glutamines. Sequence analysis of four other transcribed *opa* elements revealed the existence of stop codons in two *opa* sequences, indicating that translation cannot be a general property of *opa*. Interestingly, *opa* sequences are found in several transcripts of homeotic loci. These repetitive sequences were first observed in the Antennapedia region, and were termed the M repeat. Direct comparisons of the M repeat with *opa* suggest that they are equivalent (McGinnis et al. 1985). Sequence analysis from several homeotic loci indicates that these repetitive elements may indeed be translated into polyglutamine-rich regions (see, for example, M. Scott et al., this volume). The functional role of such sequences is unknown, and thus several possibilities exist. It is conceivable, for instance, that they confer to the protein stability properties that are critical to the control of protein expression or that they serve to link between functional domains of a given protein.

It has already been mentioned that using cross-hybridization of cDNA to genomic regions as the tool for detection of heterogeneous Notch RNA species in the embryo, two out of the 33 cDNA clones examined revealed differences. The first, cDNA 7A, is a 0.8-kb cDNA which cross-hybridizes with the exon residing in area a (Fig. 3). The homology of cDNA 7A, however, although not reaching the exonic sequences of area b, does extend beyond the exon a. Sequence analysis of cDNA 7A and the relevant genomic region shows that cDNA 7A sequences extend beyond the splicing junction into the intron between a and b. We cannot distinguish from these data if cDNA 7A reflects a different embryonic RNA species or an unprocessed Notch RNA template.

The sequence arrangement of 14D, however, raises the possibility that more than one kind of Notch RNA may be present in the embryo. Comparison of cDNA 14D sequences with cDNA 0 shows some intriguing differences. cDNA 14D shows homology to the genomic region containing exons c, d, and e, as does cDNA 0. Yet it appears that although cDNA 14D and cDNA 0 share sequences of exon e, the splicing points into this exon differ between these two clones. Moreover, while both cDNA 14D and cDNA 0 show homology to the genomic areas c and d, the two cDNAs do not share sequences except those defined by the exonic region e. Given that only one clone with the sequence arrangement of cDNA 14D was isolated, and that the sequence analysis of the Notch transcription unit is not yet complete, we cannot exclude the possibility of a cloning artifact. The distribution of cDNA 14D homologies to three noncontiguous genomic regions renders such a possibility unlikely. The relative abundance of the minor RNA species to the 10.5-kb transcript is one to two orders of magnitude smaller. Therefore, if cDNA 14D represents the arrangement of RNA sequences in one of the minor embryonic RNA species, it is not surprising that only one out of 33 cDNAs shows that sequence arrangement. Moreover, given that the criteria in classifying different cDNA classes are based so far on low-resolution cross-hybridization data, cDNA clones with sequence arrangements similar to cDNA 14D may have escaped detection.

Tissue Distribution of Notch Transcripts in the Embryo

Poulson's phenotypic description of Notch embryos shows that the absence of the Notch product(s) results in the neurogenic phenotype. Based on Poulson's observations (1937), Wright proposed that the function of the Notch product(s) is to repress the developmental potential of a cell to become a neuroblast (Wright 1970). In an extension of Wright's hypothesis, similar roles have been suggested for the other zygotic neurogenic loci by Campos-Ortega and his colleagues (Campos-Ortega et al. 1984). The simplest way repressor action could be realized is to have the Notch locus expressed selectively in the ectoderm or more specifically in the dermoblasts within the neurogenic region. The availability of Notch-specific probes allows us to test directly such hypotheses.

The 7-kb EcoRI fragment mapping between coordinates +1 and +8 is, with the exception of the 1.1-kb-long sequence mapping between the EcoRI site at +1 and the BglII site at +2.1, entirely complementary to exon e. This fragment was nick-translated with [³H]thymidine, and hybridized to tissue sections from wild-type embryos, as described in the Experimental Procedures. Figure 4 shows the results of such an experiment. The distribution of the autoradiographic grains appears to be uniform over a gastrulating embryo. Not only is it uniform over the entire ectodermal region, but Notch-specific transcripts are also detected in the mesoderm.

Two kinds of controls suggest that the uniform distribution of grains reflects the presence of Notch transcripts rather than nonspecific hybridization. The sections above and below the one shown in Figure 4 were hybridized, respectively, to nick-translated, double-stranded probes of the 3-kb-long bacterial plasmid pAT153 (a derivative of pBR322) and a 3.4-kb fragment containing sequences complementary to fushi tarazu (ftz), a gene expressed in a characteristically "striped" pattern during early embryogenesis (Hafen et al. 1984). In the first case no significant hybridization could be detected while in the second the typical hybridization pattern of ftz could clearly be seen (data not shown).

Figure 2 shows the Northern analysis using the same 7-kb-long Notch-specific probe used for the in situ hybridization. It is apparent that the transcription of the locus does not remain constant during embryogenesis. Preliminary in situ hybridization results to embryos of different stages suggest that the grain density parallels the pattern of Notch RNA accumulation observed by Northern blots. These observations also indicate that the signal we observe in the in situ hybridization is due to specific hybridization of the probe to Notch transcripts.

In spite of the preliminary nature of our in situ hybridization results, it seems that Notch transcripts are not obviously localized in a tissue-specific manner, and thus the hypothesis that Notch is expressed exclusively in the ectoderm or only in the neurogenic region is not supported. Such conclusions must be qualified by several considerations: our data regarding Notch transcription raise the possibility of more than one embryonic transcript. The in situ hybridization experiments were performed using a probe complementary only to exon e. Probes containing other exonic regions may well reveal a differential distribution of certain Notch transcripts. In addition, the resolution of our histological preparations is not sufficient for us to state with confidence that the detected transcripts are present in all cells. It is therefore possible that the cells in the neurogenic region that are to become neuroblasts do not express the Notch locus. Since in the early embryo the neuroblasts are mixed with cells destined to give rise to epidermal structures, the signal in the neurogenic re-

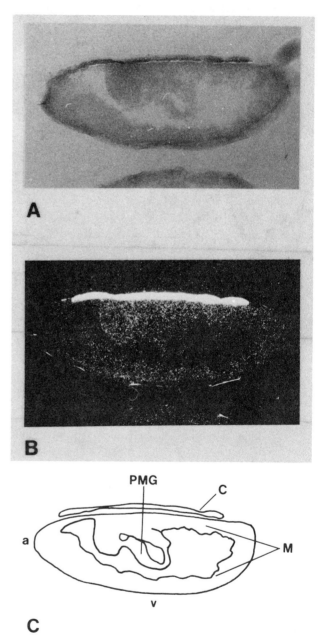

Figure 4. Detection of Notch transcripts in situ: (A) A sagittal section of a gastrulating embryo: (B) The corresponding dark-field photomicrograph. The autoradiograph exposure was for 3.5 weeks: (C) A diagramatic representation of the same embryo. (a) anterior; (v) ventral; (c) chorion remnants; (m) mesoderm; (PMG) posterior midgut invagination.

gion may arise from only the dermoblasts. It should also be remembered that unfertilized eggs contain Notch transcripts (Artavanis-Tsakonas et al. 1983). This maternally derived RNA can contribute to the in situ hybridization signal we observe in the zygote. It is, however, unlikely that our conclusions regarding the non-tissue-specific expression of Notch using the exonic region e as a probe suffers from that complication. Northern analysis shows that the zygotic expression of Notch results in approximately two orders of magnitude higher transcript levels than those found in unfertilized eggs. Finally, although the Notch transcripts are uniformly distributed throughout the embryo, it does not necessarily follow that the Notch protein products, assuming such products exist, will show identical distributions.

Generation of Isogenic Alleles of *Dl* and *E(spl)*

The combination of genetic, embryological, and molecular analysis provides a powerful tool in elucidating the role the neurogenic genes play in neurogenesis. Although many questions regarding the activity of the Notch locus and its relation to neurogenesis can be addressed with the available probes, it is necessary to obtain more information regarding the molecular characteristics of the other neurogenic loci. For this reason, we are characterizing in more detail four other zygotic neurogenic genes: Delta (*Dl*), Enhancer of Split (*E(spl)*), mastermind (*mam*), and big brain (*bib*). Initially we have two objectives in mind: to isolate more alleles of these genes, and to isolate the nucleic acid sequences defining them.

The molecular analysis of each neurogenic gene is predicated on the definition of the chromosomal DNA interval required to provide the regulated function of each locus. The first stage of such analysis is a comparative study of the structures of chromosomes bearing mutations that affect the locus of interest and wild type or parental chromosomes. The analysis of the *Dl* and *E(spl)* loci has been initiated by the generation of a set of isogenically derived alleles for each locus.

A set of *Dl* alleles has been isolated on the basis of a first-generation screen for induced mutations which yield the dominant wing venation phenotype associated with *Dl* haplo-insufficiency. Mutagenized males were isogenic for a third chromosome bearing the markers spineless, ebony, and rough. A screen of 73,600 ethylmethane sulfonate (EMS)-mutagenized chromosomes yielded 16 mutations and analysis of 35,900 X-ray-mutagenized chromosomes resulted in the isolation of 15 additional mutations. The frequency of occurrence of X-ray-induced lesions is comparable to that observed for Notch (Grimwade et al. 1985), suggesting that the target lengths may be comparable. Individual mutations vary with respect to the severity of the associated venation phenotypes, presumably reflecting variable reduction in the expression of Delta for different alleles.

New alleles of the *E(spl)* locus have been selected in a first-generation screen for apparent reversion of the enhancing effect of the *E(spl)* allele on split. These mutations were induced in males isogenic for a third chromosome bearing *E(spl)* and taxi, a wing mutation closely linked to *E(spl)* (Lindsley and Grell 1968). Analysis of 46,100 EMS-mutagenized chromosomes has yielded a minimum of 18 mutations and screens of 24,900 chromosomes following X-ray mutagenesis led to the isolation of at least 26 mutations. The relatively high frequency of isolation of X-ray-induced revertants could reflect a large target length, greater than twice that of Notch, or may indicate the occurrence of mutations at loci other than *E(spl)* which can dominantly suppress the enhancing effect of the *E(spl)* allele. A range of wing phenotypes is observed among these mutations, as for the set of *Dl* alleles described above, suggesting a similar range of effects on the function of the locus for different alleles. Indeed, some revertants without wing effects could carry lesions in *trans*-acting (second site) suppressors of *E(spl)* which do not yield haplo-insufficient wing phenotypes.

Analysis of the segregation pattern of these apparent revertants has revealed that all except eight map to the third chromosome, consistent with linkage to *E(spl)*. These data also suggest that the mutations responsible for the remaining reversions are linked to the second or fourth chromosomes. Two of these mutations have been shown to segregate with the second chromosome, demonstrating that this screen leads to identification of loci that can be mutated to yield *trans*-acting suppressors of the *E(spl)* mutation. Given the synthetic lethality of transheterozygotes for *Dl* and *E(spl)* null mutations and its implication for the combinatorial action of different neurogenic genes, we will test any *trans*-acting suppressors among the mutations thus far isolated to determine whether they are allelic to previously defined neurogenic loci or constitute additional genes required for the completion of embryonic neurogenesis.

Hybrid Dysgenesis and Molecular Cloning of Neurogenic Genes

Molecular analysis of the functions of neurogenic loci depends on the isolation of the chromosomal sequences that correspond to these genes. As was the case with the Notch locus before cloning it, the transcriptional or translational products of the other neurogenic loci are unknown and therefore it is not possible to use conventional hybridization procedures to obtain the genes. Alternative gene isolation methods that overcome this restriction have to be sought. The approach we decided to take initially was to attempt to "tag" the locus of interest with a transposon. Generation of mutation by hybrid dysgenesis often, although not always, leads to the insertion of the P-transposable elements into the gene of interest (Kidwell et al. 1977; Simmons and Lim 1980). Since the P-element sequences have been isolated, such hybrid dysgenesis-induced insertion mutants provide a tool for cloning other neurogenic genes (O'Hare and Rubin 1983).

There are two limitations to this procedure, however. First, not every gene is capable of mutating during hybrid dysgenesis, and second, not all mutations generated are caused by P-element insertions. Some may be caused by chromosomal rearrangements, or by the insertion of other transposable elements. Here we will describe the results of two hybrid dysgenesis experiments designed to isolate mutations in $E(spl)$, Dl, bib, and mam.

The phenotype associated with heterozygosity for Dl or $E(spl)$ constitutes the basis for a first-generation screen for mutations affecting these loci (Fig. 5). A screen of 70,000 chromosomes exposed to dysgenic conditions has yielded a set of six independent third-chromosome mutations (HD9, HD28, HD30, HD40, HD62, HD82) that yield a dominant wing venation phenotype in heterozygous condition and result in embryonic lethality when homozygous. Assignment of these mutations to either Dl or $E(spl)$ is being attempted on the basis of complementation tests which should delineate alleles at these two loci.

The scheme for the generation and screening of hybrid dysgenic mutations in bib and mam mapping, respectively, in positions 34.7 and 70.3 on the second chromosome is outlined in Figure 5. Unlike mutations in $E(spl)$ and Dl, which display a dominant phenotype and can therefore be scored in a first-generation screen, a second-generation screen based on the homozygous lethality of bib and mam had to be devised to score mutations in these loci. Approximately 6000 chromosomes were screened and scored over a double-mutant bib mam chromosome. We thus isolated six mutations that were lethal over the double-mutant chromosome. Further complementation tests showed that all six mutations did complement bib mutation, but failed to complement mam mutants. It seems likely, therefore, that HD 2/3, HD 3/1, HD 13/6, HD 11/2, HD 10/1, and HD 6/4 are alleles of the mam locus. Given the high number of mutations recovered in mam it appears that this locus may be a "hot spot" for hybrid dysgenesis-induced mutations.

We are currently testing for the insertion of P-element sequences at the appropriate cytological locations in chromosomes bearing the hybrid dysgenesis-induced mutations. Identification of P-element sequences within each locus will provide a tool for the physical isolation of the chromosomal segments corresponding to the mutated gene.

CONCLUDING REMARKS

Until the development of recombinant DNA technology and DNA-mediated transformation of whole animals, the best and usually only functional assay of a gene was its mutational analysis. In most cases, especially for genes involved in complex developmental events, this is still true. The assumption, or the hope, of such analyses is that the phenotypic description of a particular mutant provides clues as to the wild-type function of the gene in question, and hence the mode of action and topological localization of gene activity during development. In the cases where such conten-

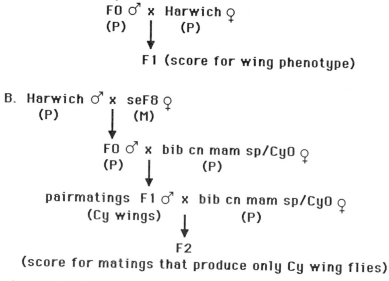

Figure 5. Scheme for the generation of mutants by hybrid dysgenesis in Delta (Dl) and Enhancer of split ($E[spl1]$) (A) or in big brain (bib) or mastermind (mam) (B). The strains Harwich (P cytotype) and seF8 (M cytotype) were obtained from M. Kidwell (Brown University).

tions could be tested, the most notable example being the bithorax locus, they were shown to be largely justified (Beachy et al. 1985). Yet the validity of extrapolating from mutant phenotypes to wild-type functions is dependent on the developmental event, the accuracy of the phenotypic description, and, of course, the extent of the genetic analysis. In spite of the substantial advances made in describing neurogenesis in *Drosophila* (Hartenstein and Campos-Ortega 1984), we believe that the absence of clear-cut landmarks or neuroblast-specific molecular markers in the early embryos still poses an inherent difficulty in accurately describing the neurogenic phenotypes. Moreover, mutations in neurogenic loci cause pleiotropic effects (Cross and Sang 1978). These facts must be kept in mind while attempting to use the genetic and molecular analysis of neurogenic loci to unravel the developmental mechanisms underlying neurogenesis. We have argued in the beginning of this paper that we cannot be certain how many of the neurogenic loci are part of the primary developmental logic of neurogenesis, versus how many affect the process only indirectly. We have described experiments designed to characterize the molecular properties of some of the zygotically acting neurogenic genes. Notwithstanding the arguments put forward above, only rigorous molecular characterizations, similar to that being carried out for the Notch locus, will provide the necessary tools to address and resolve the issues raised by the genetic and embryological analyses. Such an approach, although laborious and not without pitfalls, is perhaps the only way to determine the function of the neurogenic gene products and their relationship to neurogenesis.

Regarding the possible role the neurogenic loci play in development, the work of Doe and Goodman (1985a,b) on neuroblast differentiation in the grasshopper provides some very interesting clues. These workers have shown that the region of the embryonic ectoderm that becomes neuroepithelium is composed of a mixture of cells, some of which will become neuroblasts and the rest of which will either die or give rise to nonneuronal structures. The neuroblasts in the neuroepithelium are surrounded by cells that will acquire nonneuronal phenotypes. Laser ablation experiments provide convincing evidence that as a given ectodermal cell takes the neural pathway it inhibits its neighbors from becoming neuroblasts. There are many similarities between *Drosophila* and grasshopper both in the pattern of identified neurons as well as that of neuronal precursor cells (Hartenstein and Campos-Ortega 1984; Thomas et al. 1984). It is therefore probable that analogous cellular mechanisms drive the differentiation of the *Drosophila* neuroblasts. If this is the case it is, as they argue, conceivable that the neurogenic loci are involved in the control of the interaction between neuroblasts and dermoblasts (Doe and Goodman 1985b).

Utilizing the sophisticated genetic and molecular analyses that *Drosophila* allows as an experimental system, we hope that we will be able to elucidate at the molecular level the cellular mechanisms involved in neurogenesis. Such information may not only elucidate the embryonic origins of the neurons, but may also provide the basis for addressing questions involving higher-order structures of the nervous system.

ACKNOWLEDGMENTS

This work was supported by grants GM29093 and GM 33291 from the National Institutes of Health. B.Y. is a National Institutes of Health postdoctoral fellow, and M.E.H. is a recipient of a Canadian NSERC fellowship. We thank our colleagues, K. Johansen, K. Markopoulou, D. Hartley, and R.G.P.-Ramos for helpful discussions, M. Levine and M. Akam for advice on in situ hybridization, and T. Huynh for providing the cDNA construction protocol prior to publication. We also wish to acknowledge the expert technical assistance of R. Schlesinger-Bryant and A. Terry. Finally, we would like to thank Dr. W. Welshons for his invaluable help and advice in regard to many aspects of this work.

REFERENCES

AKAM, M.E. 1983. The location of ultrabithorax transcripts in *Drosophila* tissue sections. *EMBO J.* **2**: 2075.

ARTAVANIS-TSAKONAS, S., B.G. GRIMWADE, R.G. HARRISON, K. MARKOPOULOU, M.A.T. MUSKAVITCH, R. SCHLESINGER-BRYANT, K. WHARTON, and B. YEDVOBNICK. 1984. The Notch locus of *Drosophila melanogaster*: A molecular analysis. *Dev. Genet.* **4**: 233.

ARTAVANIS-TSAKONAS, S., M.A.T. MUSKAVITCH, and B. YEDVOBNICK. 1983. Molecular cloning of Notch, a locus affecting neurogenesis in *Drosophila melanogaster*. *Proc. Natl. Acad. Sci.* **80**: 1977.

BEACHY, P.A., S.L. HELFAND, and D.S. HOGNESS. 1985. Segmental distribution of bithorax complex proteins during *Drosophila* development. *Nature* **313**: 545.

CAMPOS-ORTEGA, J.A., R. LEHMAN, F. JIMENEZ, and U. DIETRICH. 1984. A genetic analysis of early neurogenesis in *Drosophila*. In *Organizing principles of neural development* (ed. S.C. Sharma), p. 129. Plenum Press, New York.

CROSS, D. and J. SANG. 1978. Cell culture of individual *Drosophila* embryos. II. Culture of X-linked embryonic lethals. *J. Embryol. Exp. Morphol.* **45**: 173.

DIETRICH, U. and J.A. CAMPOS-ORTEGA. 1984. The expression of neurogenic loci in imaginal epidermal cells of *Drosophila melanogaster*. *J. Neurogenet.* **1**: 315.

DOE, C.Q. and C.S. GOODMAN. 1985a. Early events in insect neurogenesis. I. Development and segmental differences in the pattern of neuronal precursor cells. *Dev. Biol.* **111**: 193.

―――. 1985b. Early events in insect neurogenesis. II. The role of cell interactions and cell lineage in the determination of neuronal precursor cells. *Dev. Biol.* **111**: 206. (in press).

FOSTER, G.G. 1975. Negative complementation at the Notch locus of *Drosophila melanogaster*. *Genetics* **81**: 99.

GARCIA-BELLIDO, A. 1975. Genetic control of wing disc development in *Drosophila*. *Ciba Found. Symp.* **29**: 161.

GRIMWADE, B.G., M.A.T. MUSKAVITCH, W.J. WELSHONS, B. YEDVOBNICK, and S. ARTAVANIS-TSAKONAS. 1985. The molecular genetics of the Notch locus in *Drosophila melanogaster*. *Dev. Biol.* **107**: 503.

HAFEN, E., A. KUROIWA, and W.J. GEHRING. 1984. Spatial distributions of transcripts from the segmentation gene fushi tarazu during *Drosophila* embryonic development. *Cell* **37**: 833.

HAFEN, E., M. LEVINE, R.L. GARBER, and W.J. GEHRING. 1983. An improved *in situ* hybridization method for the detection of cellular RNAs in *Drosophila* tissue sections and its application for localizing transcripts of the homeotic Antennapedia gene complex. *Embo J.* **2**: 617.

HARRIS, W.A. 1985. *J. Neurogenet.* **2**: 179.

HARTENSTEIN, V. and J.A. CAMPOS-ORTEGA. 1984. Early neurogenesis in wild-type *Drosophila melanogaster*. *Wilhelm Roux's Arch. Dev. Biol.* **193**: 308.

HOUSE, V.L. and C.M. LUTES. 1975. Interactions of the Abruptex (Ax) and cubitus-interruptus-recessive (ci^D) mutants in *Drosophila melanogaster*. *Genetics* **80**: s42.

HUYNH, T.V., R. YOUNG, and R. DAVIS. 1985. Constructing and screening cDNA libraries in λgt10 and λgt11. In: *DNA cloning techniques: A practical approach* (ed. D. Glover). IRL Press, Oxford. (In press.)

JIMENEZ, F. and J.A. CAMPOS-ORTEGA. 1982. Maternal effects of zygotic mutants affecting early neurogenesis in *Drosophila*. *Wilhelm Roux's Arch. Dev. Biol.* **191**: 191.

JURGENS, G., E. WIESCHAUS, C. NUSSLEIN-VOLHARD, and H. KLUDING. 1984. Mutations affecting the pattern of the larval cuticle in *Drosophila melanogaster*. II. Zygotic loci on the third chromosome. *Wilhelm Roux's Arch. Dev. Biol.* **193**: 283.

KEPPY, D.O. and W.J. WELSHONS. 1977. The cytogenetics of a recessive visible mutant associated with a deficiency adjacent to the *Notch* locus in *Drosophila melanogaster*. *Genetics* **85**: 497.

KIDD, S., T.J. LOCKETT, and M.W. YOUNG. 1983. The *Notch* locus of *Drosophila melanogaster*. *Cell* **34**: 421.

KIDWELL, M.G., J.F. KIDWELL, and J.A. SVED. 1977. Hybrid dysgenesis in *D. melanogaster*: a syndrome of aberrant traits including mutation, sterility and male recombination. *Genetics* **86**: 813.

LEHMAN, R., F. JIMENEZ, W. DIETRICH, and J.A. CAMPOS-ORTEGA. 1983. On the phenotype and development of mutants of early neurogenesis in *D. melanogaster*. *Wilhelm Roux's Arch. Dev. Biol.* **192**: 62.

LINDSLEY, D.L. and E.H. GRELL. 1968. Genetic variations of *Drosophila melanogaster*. Publ. Carnegie Inst. Wash. Publ. no. 627.

MCGINNIS, W., K. HARDING, F. KARCH, and M. LEVINE. 1985. Homeobox genes of the Antennapedia and Bithorax complexes of *Drosophila*. *Cell* (in press).

NUSSLEIN-VOLHARD, C., E. WIESCHAUS, and H. KLUDING. 1984. Mutations affecting the pattern of the larval cuticle in *Drosophila melanogaster*. I. Zygotic loci on the second chromosome. *Wilhelm Roux's Arch. Dev. Biol.* **193**: 267.

O'HARE, K. and G.M. RUBIN. 1983. Structures of P transposable elements and their sites of insertion and excision in the *Drosophila melanogaster* genome. *Cell* **34**: 25.

PERRIMON, N., L. ENGSTROM, and A.P. MAHOWALD. 1984. Developmental genetics of the 2E-F region of the Drosophila X chromosome: A region rich in "developmentally important" genes. *Genetics* **108**: 559.

PORTIN, P. 1975. Allelic negative complementation at the *Abruptex* locus of *Drosophila melanogaster*. *Genetics* **81**: 121.

POULSON, D.F. 1937. Chromosomal deficiencies and the embryonic development of *Drosophila melanogaster*. *Proc. Natl. Acad. Sci.* **23**: 133.

———. 1950. Histogenesis, organogenesis and differentiation in the embryo of *Drosophila melanogaster* Meigen. In *Biology of* Drosophila (ed. M. Demerec), p. 168. Wiley, New York.

SHANNON, M.P. 1973. The development of eggs produced by the female sterile mutant *almondex* of *Drosophila melanogaster*. *J. Exp. Zool.* **183**: 383.

SHELLENBARGER, D.L. and J.D. MOHLER. 1975. Temperature-sensitive mutations of the *Notch* locus of *Drosophila melanogaster*. *Genetics* **81**: 143.

———. 1978. Temperature-sensitive periods and autonomy of pleiotropic effects of $1(1)N^{ts1}$, a conditional *Notch* lethal in *Drosophila*. *Dev. Biol.* **62**: 432.

SIMMONS, M.J. and J.K. LIM. 1980. Site specificity of mutations arising in dysgenic hybrids of *Drosophila melanogaster*. *Proc. Natl. Acad. Sci.* **77**: 6042.

THOMAS, J.B., M.J. BASTIANI, M. BATE, and C.S. GOODMAN. 1984. From grasshopper to *Drosophila*: A common plan for neuronal development. *Nature* **310**: 203.

THORIG, G.E.W., P.W.H. HEINSTRA, and W. SCHARLOO. 1981. The action of the *Notch* locus in *D. melanogaster*. II. Biochemical effects of recessive lethals on mitochondrial enzymes. *Genetics* **99**: 65.

WELSHONS, W.J. 1956. Dosage experiments with *split* mutants in the presence of an enhancer of *split*. *Drosophila Inform. Serv.* **30**: 157.

———. 1965. Analysis of a gene in *Drosophila*. *Science* **150**: 1122.

———. 1971. Genetic basis for two types of recessive lethality at the *Notch* locus of *Drosophila*. *Genetics* **68**: 259.

———. 1974. The cytogenetic analysis of a fractured gene in *Drosophila*. *Genetics* **76**: 775.

WELSHONS, W.J. and D.O. KEPPY. 1981. The recombination of analysis of aberrations and the position of the *Notch* locus on the polytene chromosome of *Drosophila*. *Mol. Gen. Genet.* **181**: 319.

WHARTON, K.A., B. YEDVOBNICK, V.G. FINNERTY, and S. ARTAVANIS-TSAKONAS. 1985. opa: A novel family of transcribed repeats shared by the *Notch* locus and other developmentally regulated loci in *D. melanogaster*. *Cell* **40**: 55.

WIESCHAUS, E., C. NUSSLEIN-VOLHARD, and G. JURGENS. 1984. Mutations affecting the pattern of the larval cuticle in *Drosophila melanogaster*. III. Zygotic loci on the X-chromosome and fourth chromosome. *Wilhelm Roux's Arch. Dev. Biol.* **193**: 296.

WRIGHT, T.R.F. 1970. The genetics of embryogenesis in *Drosophila*. *Adv. Genet.* **15**: 261.

Gene Expression in Differentiating and Transdifferentiating Neural Crest Cells

D.J. ANDERSON,* R. STEIN, AND R. AXEL

Howard Hughes Medical Institute, Columbia University College of Physicians and Surgeons, New York, New York 10032

The neural crest is a transient population of cells that detach from the top of the neural tube and then disperse throughout the embryo giving rise to the peripheral nervous system, melanocytes, and the adrenal medulla, as well as a diverse array of other cell types. Transplantation experiments suggest that most premigratory cells throughout the neural crest have the same developmental potential, regardless of their actual fate (LeDouarin 1982; Weston 1982.) Thus, the environment encountered by crest cells during or after migration appears to be critical in determining their final phenotype.

One neural crest developmental decision that shows considerable environmental influence is the choice between an endocrine and a neuronal phenotype. Both adrenergic endocrine cells (chromaffin cells) of the adrenal medulla and noradrenergic neurons of the paravertebral sympathetic ganglia derive from the so-called sympathoadrenal (Landis and Patterson 1981) region of the crest, which originates in the caudal thoracic region (LeDouarin 1982). Transplantation experiments have shown that when this region is removed and replaced by a graft of more rostrally originating crest, histologically distinguishable donor-derived chromaffin cells develop in the adrenal medulla, despite the fact that the donor crest normally never gives rise to these cells but only to neurons (LeDouarin 1980). Thus, the embryonic environment ventral to the adrenomedullary crest appears to elicit the chromaffin phenotype from cells that migrate into it.

The effect of environment on the sympathoadrenal decision is apparent not only during development, but also in the "plasticity" of the fully differentiated derivatives. Thus, chromaffin cells, although expressing a terminally differentiated phenotype, retain the capacity to transdifferentiate into neurons both in vitro and in vivo, when exposed to nerve growth factor (NGF) (Unsicker et al. 1978; Aloe and Levi-Montalcini 1979; Doupe et al. 1985a,b). The in vitro conversion to a neural phenotype is blocked by glucocorticoids, suggesting that maintenance of the chromaffin phenotype may be a consequence of the ultimate location of the adrenal medullary cells, which are surrounded by adrenal cortical cells expressing large quantities of glucocorticoids (Unsicker et al. 1978). Taken together, these and other observations have suggested that NGF and glucocorticoids may play opposing roles in promoting the neural and chromaffin phenotypes, respectively, from a common precursor cell (Landis and Patterson 1981; Doupe et al. 1985b). This developmental decision therefore provides an opportunity to understand the molecular mechanisms that underlie the expression of alternative phenotypes by a multipotential precursor, and the subsequent plasticity of these phenotypes, in response to defined environmental signals.

We have isolated cDNA clones for several mRNAs that are abundant in adult sympathetic neurons but not in adrenal chromaffin cells. Using RNA blot and in situ hybridization, we have analyzed the pattern of emergence of these messengers in embryonic development and their plasticity in cell culture. We find that these neural-specific genes are not synchronously induced but rather appear at recognizable stages in neurogenesis. Moreover, one of the genes is induced by NGF in PC12 cells and in adrenal medullary primary cultures, suggesting that the hormone may control the transcriptional activation of this gene early in development. The results suggest a mechanism in which the developmental history of some of these neural-specific genes could account for their plasticity in mature chromaffin cells.

RESULTS

The Appearance of Neural-specific mRNAs during Development

We have isolated several cDNA clones encoding mRNA transcripts expressed in ganglionic neurons but not in adrenal chromaffin cells by performing differential hybridization to a rat sympathetic ganglion cDNA library. The tissue-specificity of these clones was established by both Northern blot analysis and in situ hybridization. Clones SCG4, SCG5, and SCG10 detect abundant RNA species in both sympathetic ganglia and brain, which are either absent or detectable at much lower levels in the adrenal medulla and in nonneural tissues (Fig. 1). In situ hybridization to sections and cultures of superior cervical ganglion, using single-stranded RNA probes (Cox et al. 1984), revealed grain accumulations over the cell bodies of principal neurons, but not over the nonneural cells within the ganglion (data not shown). Consistent with the Northern analysis, no hybridization was observed to sections of adrenal chromaffin cells (Anderson and Axel 1985).

---

*Present address: Division of Biology, California Institute of Technology, Pasadena, California 91125.

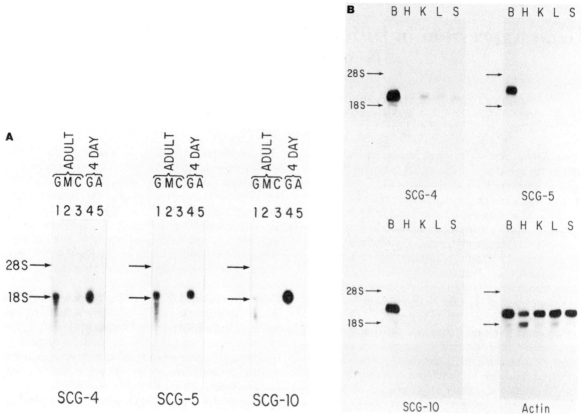

Figure 1. Tissue-specificity of clones SCG4, SCG5, and SCG10 expression. (*A*) Northern blots containing 10 μg/lane of total RNA from adult SCG (G, lanes *1*), adrenal medulla (M, lanes *2*), adrenal cortex (C, lanes *3*) or 4-day postnatal SCG (G, lanes *4*), and whole adrenal (A, lanes *5*) were hybridized with nick-translated probes as indicated below the panels. (*B*) As in *A*, except the lanes contained RNA from 6-day postnatal tissues: (B) brain; (H) heart; (K) kidney; (L) liver; (S) spleen. The probes used are indicated below the panels; "Actin" was a human β-actin probe that detects a smaller cardiac actin transcription in the heart RNA. All autoradiograms represent overnight exposures; we estimate the relative abundance of SCG4 and SCG10 in 4-day ganglia as 0.25 and 0.6%, respectively.

The availability of neuron-specific clones that detect abundant mRNAs now permitted us to relate the appearance of neural-specific mRNAs to the migratory events in neural crest differentiation. In the rat, neural crest cell migration begins around the ninth day of gestation (E9), proceeds along the embryo in a rostro-caudal gradient, and ends about E11.5–E12. At this stage, the sympathetic ganglia primordia have largely formed and consist of collections of primitive neuroblasts that continue to divide throughout gestation, enlarging the ganglia (Hendry 1977; Rothman et al. 1978). The formation of the adrenal medulla occurs somewhat later via a secondary ventral migration of crest cells from the caudal thoracic region to a position above the kidney tubules where the cells coalesce with the mesodermally derived cortical cells. This secondary migration occurs beginning around E13 and continues until birth. A discrete adrenal medulla is visible by E17.5 (Bohn et al. 1981; Teitelman et al. 1982).

In initial experiments we examined the developmental appearance of the neural-specific RNAs by performing Northern blot analyses on total embryo RNA. SCG5 mRNA is barely detectable between E10.5 and E13.5, but undergoes at least a 20-fold induction at birth. Thus, significant levels of SCG5 mRNA do not appear to accumulate in sympathetic neurons or the CNS until long after the cessation of neural crest migration (Anderson and Axel 1985). In contrast to SCG5 mRNA, SCG4 and SCG10 mRNAs were both detectable at significant levels in midgestational embryos, although their patterns of expression during development are distinct (Fig. 2). SCG10 mRNA is barely detectable at E10.5, but undergoes a 30-fold induction over the next 48 hours (Fig. 2A). During this period, SCG4 is expressed at fairly constant levels but then declines significantly beginning on about E13.5 (Fig. 2B).

Regulated Expression of SCG10

The temporal appearance of SCG10 mRNA suggests that this gene is not expressed at high levels in the neural crest, but rather is induced in crest cells following migration and ganglion formation. To confirm this interpretation, we performed in situ hybridizations with an SCG10 probe to embryo sections at various stages of development. In very caudal sections of E11.5 em-

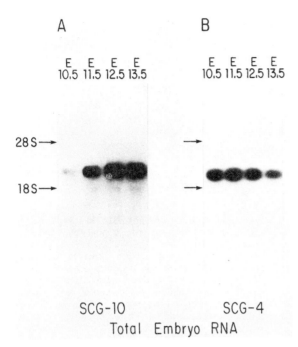

Figure 2. Early expression of SCG10 and SCG4 mRNAs. Total cellular RNA (10 μg) isolated from embryos of the stages indicated was subjected to Northern analysis as described in Methods. Blots were hybridized with ^{32}P-labeled SP6 RNA probes labeled to a specific activity of 4×10^9 dpm/μg. Blots A and B were exposed for different times and therefore the signal intensities are not directly comparable. (A) SCG10 probe; (B) SCG4 probe.

bryos, significant hybridization is observed to the neurons of the ventrolateral neural tube, but not to cells in the migrating neural crest (Fig. 3D). In more rostral sections of the same embryo (Fig. 3B,C), much higher grain densities were apparent in the ventral and lateral margins of the neural tube, suggesting that expression of SCG10 mRNA occurs in the neural tube in a rostro-caudal gradient. Later, at E13.5, SCG10 transcripts were detectable at high levels in the sympathetic and sensory ganglia (Fig. 4A) and in parasympathetic primordia in the arterial walls (Fig. 4B). However, no hybridization is observed with the ependymal zone of the neural tube which contains mitotic undifferentiated neuroblasts (Fig. 4A). Thus, in both the central and peripheral nervous system, expression of SCG10 mRNA correlates with the earliest overt signs of commitment to a neuronal phenotype, occurring first in the ventrolateral neural tube and later in the peripheral ganglion primordia.

Comparison of the intensity of grains over developing ganglia suggest that a large component of the induction seen on Northern blots is likely to be due to an increase in the proportion of embryonal cells which are neurons as the consequence of either mitotic expansion (Rothman et al. 1978) or conversion of cells. Quantitatively, however, a 30-fold relative increase in the neural population (needed to account for the magnitude of SCG10 mRNA induction) would exceed that actually generated in this 48-hour interval, so that induction is probably occurring on a per cell basis as well. In support of this, we observe an increase in grain density between E11.5 and E13.5 in the ventrolateral neural tube.

The observation that SCG10 mRNA appears in sympathetic neurons subsequent to neural crest migration next led us to ask whether any known environmental factors could influence the expression of this gene. Previous studies have suggested that the choice of a neuronal phenotype by sympathoadrenal precursors may be influenced by NGF in the cell's local environment (Unsicker et al. 1978; Aloe and Levi-Montalcini 1979; Landis and Patterson 1981; Doupe et al. 1985b). We therefore examined the ability of this polypeptide hormone to influence levels of SCG10 mRNA in vitro. We observed that SCG10 mRNA was induced almost 10-fold by NGF in PC12 cells (Fig. 5A), an adrenal medullary pheochromocytoma tumor line (Greene and Tischler 1976) that has many properties of a committed adrenergic precursor and which responds to NGF with neurite outgrowth. This induction is also observed in primary cultures of dissociated neonatal adrenal chromaffin cells exposed to NGF. In situ hybridization to adrenal chromaffin cell cultures in the presence of NGF revealed that those cells that converted to a neuronal morphology (Unsicker et al. 1978; Naujoks et al. 1982; Ogawa et al. 1984; Doupe et al. 1985a) also expressed SCG10 mRNA at high levels (Fig. 6A,B).

Thus, SCG10 mRNA is induced by NGF exposure in both primary cultures of chromaffin cells and in an established pheochromocytoma cell line. These results suggest that the dramatic appearance of this RNA early in the development of sensory and sympathetic neurons may result from the migration of these cells to an environment containing high local concentrations of NGF, or from the synchronous appearance of NGF in the embryo.

The Regulated Expression of SCG4

The pattern of expression of SCG4 mRNA during development is distinct from that of SCG10. SCG4 mRNA is highly enriched in adult neurons. Early in development, however, this RNA is abundant in virtually all tissues of the embryo. Expression of SCG4 mRNA declines precipitously in all tissues during late gestation but then returns to high levels after birth, specifically in the nervous system. Northern blot analysis of total embryo RNA reveals that SCG4 mRNA is expressed as an abundant species in midgestational embryos and is already quite abundant at E10.5, a time when SCG10 mRNA is barely detectable (Fig. 2B). This RNA is maintained at high levels (~10 times the abundance of SCG10) until E14 and then declines gradually to low levels at birth. In situ hybridizations yielded an unexpected result: although the SCG4 mRNA was clearly abundant in sympathetic ganglia, dorsal root ganglia, and the neural tube, hybridization was also detected in several other nonneural tissues, particularly

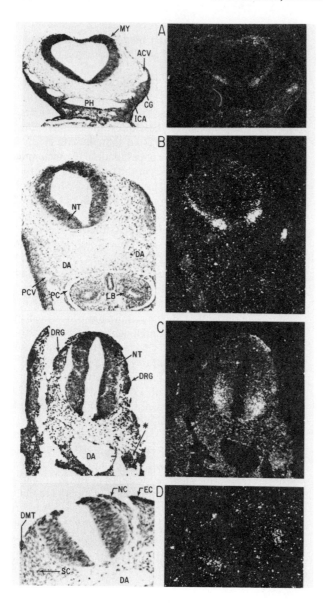

Figure 3. In situ distribution of SCG10 mRNA at stage E11.5. Sections obtained from a single embryo series were hybridized in situ with ^{35}S-labeled SP6 RNA transcripts from clone 10 and are displayed as brightfield–dark-field pairs. The sections proceed from A to D in a rostral to caudal direction. Sections are transverse but slightly oblique, thus structures ventral to the neural tube in a given section appear slightly more caudal in characteristics. (A) Region of the myelencephalon (MY). A cranial ganglion (CG) is visible ventral to the anterior cardinal vein (ACV) and dorsolateral to the internal carotid artery (ICA), which shows strong hybridization signal in dark field. Note also the high grain density in the ventrolateral neural tube. (PH) Pharynx. (B) Region of the posterior myelencephalon. (DA) Dorsal aorta; (NT) neural tube; (PCV) posterior cardinal vein; (PC) pleural cavity; (LB) lung bud. (C) Trunk region. Due to the curvature of the embryo, this section shows a more tangential appearance. (NT) Neural tube; (DRG) dorsal root ganglia; (DA) dorsal aorta; (*) putative sympathetic ganglion primordium. Note the relative absence of grains over this structure relative to the ventrolateral neural tube and the lateral aspect of the DRG. (D) More caudal region. Migrating neural crest cells (NC) are visible between the neural tube and a somite, which is divided into dermomyotome (DMT) and scelerotome (SC) regions. (DA) Dorsal aorta; (EC) ectoderm. B and D were hybridized in parallel and thus the grain densities are directly comparable. Exposure times were 3–4 days. Magnifications: A, 36×; B and C, 45×; D, 93×.

the lung (not shown). Northern blot analysis however confirmed this observation: SCG4 mRNA is abundant in heart, kidney, liver, lung, adrenal, and brain from E16 embryos (Fig. 7), declines neonatally, and reappears postnatally in neurons.

If, as we have suggested previously, the induction of clone 10 is in part mediated by the appearance of NGF, we would not predict that the expression of clone 4 is also mediated by NGF, as it is not coordinately expressed in time with clone 10 and, more importantly, it is expressed at high levels in tissues such as lung and heart which are not thought to contain NGF receptors. Indeed, we observed that SCG4 mRNA is expressed at high levels in naive P12 cells and is not affected by NGF treatment (Fig. 5B). This finding was therefore consistent with the presumed embryonic origin of this cell line. A different result is obtained when primary adrenal chromaffin cells are exposed to NGF. Chromaffin cells cultured from 7-day postnatal animals exhibited little or no in situ hybridization signals over background with an SCG4 probe (Fig. 6A, bottom), a finding consistent with previous data from Northern blot analyses (cf. Fig. 1). When similar cultures were then exposed to nerve growth factor for 1 month, those cells that morphologically converted to a neural phenotype showed strong hybridization signals (Fig. 6B, bottom) similar in intensity to that observed for bona fide sympathetic neurons treated in parallel (not shown).

These results suggest that the expression of clone 4 mRNA may be regulated by distinct mechanisms at different times in development. Early in embryogenesis, a high level of constitutive synthesis of clone 4 RNA is observed in a host of neural and nonneural tissues and this synthesis is unlikely to be regulated by NGF. Postnatally, the expression of clone 4 declines in all tissues and increases solely in tissues of neural origin where its expression may be regulated by exposure to NGF.

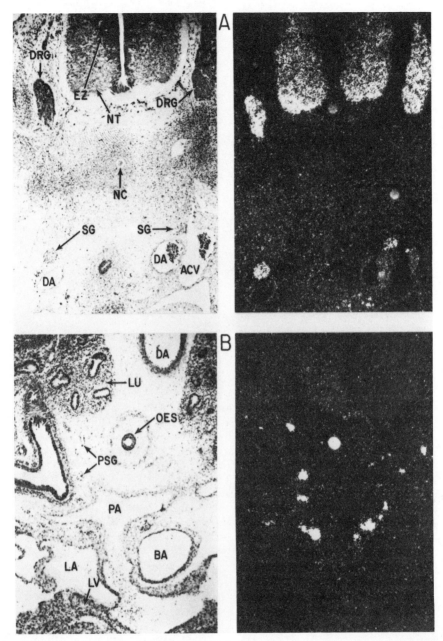

Figure 4. In situ distribution of SCG10 at later stages of development. (*A*) Section through trunk region E13.5. Note intense hybridization to neural tube (NT), dorsal root ganglia (DRG), and sympathetic ganglia (SG), while the ependymal zone (EZ) containing undifferentiated neuroblasts is negative. (NC) Notochord; (DA) dorsal aorta; (ACV) anterior cardinal vein. Magnification, 40×. (*B*) Section through ventral trunk E15. (DA) Dorsal aorta; (LU) lung; (BA) bulbus arteriosus; (LA) left atrium; (LV) left ventricle. Note strong signal to putative parasympathetic ganglia (PSG) surrounding the pulmonary artery (PA). The circular white spot in dark field (*right panel*) is an artifact.

DISCUSSION

Regulated Expression of Specific RNAs during Neural Development

We have followed the emergence of neurons during developmental time by performing in situ and blot hybridization with three neural-specific cDNA probes. Our data suggest that the induction of neural-specific genes does not occur synchronously but rather occurs in at least two major stages in vertebrate neurogenesis, one coinciding with the earliest overt signs of commitment to a neural phenotype (SCG10) and the other corresponding to maturation events that occur early in postnatal life (SCG4 and SCG5).

The observation that each clone appears at different times in development raises the question of what factors control these patterns of expression. NGF has previously been implicated in the chromaffin-neuron decision that occurs early in development (Aloe and Levi-Montalcini 1979; Landis and Patterson 1981). Clone SCG10 was inducible by NGF both in primary cultures of neonatal chromaffin cells and in the NGF-responsive pheochromocytoma tumor cell line PC12. It is therefore possible that NGF may be responsible for

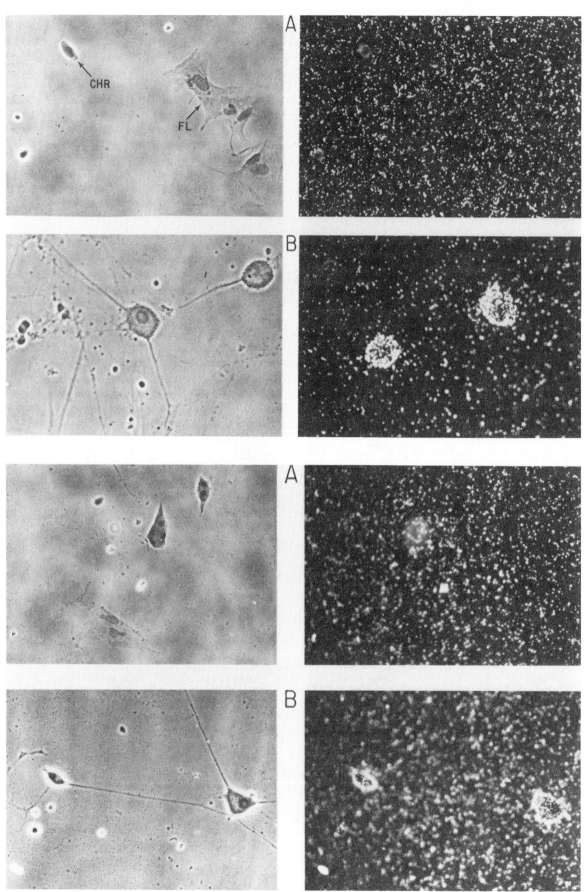

Figure 6. (*See facing page for legend.*)

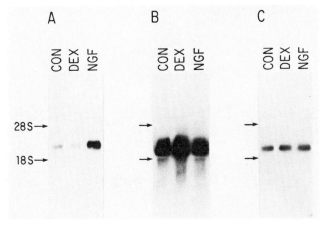

Figure 5. SCG10 mRNA is induced by NGF in PC12 cells. PC12 cells, at ~0.5–1.0 × 10⁶ cells/10-cm dish, were treated for 4 days with medium containing no additives (lanes CON), 5 μM dexamethasone (lanes DEX), or 50 μg/ml β-NGF (lanes NGF). Shown are three identical Northern blots containing 10 μg/lane of total RNA prepared from the indicated cultures. Probes used were SCG10 (*A*), SCG4 (*B*), or human γ-actin (*C*). *A* and *B* were hybridized with ³²P-labeled SP6 probes of the same specific activity, and the blot was exposed for the same length of time to show the relative abundance of the two mRNAs.

the transcriptional activation of SCG10 mRNA in sensory and sympathetic neurons during early development. This induction may be a consequence of the appearance of NGF, NGF receptors, or alternatively, the migration of crest cells to an environment enriched in NGF. However, we also observe ubiquitous expression of SCG10 mRNA in parasympathetic and central neurons, most of which are not responsive to NGF by classical criteria. We therefore think it unlikely that NGF is the sole mediator of SCG10 mRNA induction in the central and peripheral nervous system. Rather, expression of SCG10 mRNA may be controlled by different factors in different types of neurons.

SCG4 mRNA was also induced by NGF in postnatal chromaffin cells, but was constitutively expressed by PC12 cells. Our in situ hybridization data for SCG4 indicated that in the embryo it is abundant in both neural and nonneural cells, but is then extinguished and reexpressed selectively in neurons after birth. The NGF induction of SCG4 in chromaffin cells may thus reflect the reexpression of the gene that normally occurs in neurons of the same age. Likewise, the high level of expression of SCG4 mRNA in PC12 cells is consistent with the presumed embryonic nature of this tumor cell line (Greene and Tischler 1976).

The observation that glucocorticoids inhibit the NGF-induced conversion of chromaffin cells to neurons has led to the suggestion that these hormones may promote endocrine differentiation by blocking the NGF-induced expression of the neuronal phenotype by embryonic adrenal medullary pheochromoblasts. Preliminary in situ hybridization data suggest that, at E14, SCG10 mRNA is not expressed by, or is present at low levels in, immature chromaffin cells compared to neighboring sympathetic neurons (not shown.) Yet, the adrenal cortex does not secrete glucocorticoids until E17 (Bohn et al. 1981). The implication is that adrenal environmental factors other than glucocorticoids may suppress (or permit only transient, low-level) expression of SCG10 mRNA by E14 pheochromoblasts. Consistent with this, dexamethasone only weakly (~twofold) inhibited NGF induction of SCG10 mRNA in PC12 cells (not shown), and had no apparent effect on neurite outgrowth from organ-cultured E17 adrenal glands (Unsicker et al. 1985). In vivo, then, glucocorticoids may serve to maintain the differentiated chromaffin phenotype rather than to prevent initial differentiation of pheochromoblasts along the neuronal pathway.

Molecular Mechanism of Chromaffin Cell Plasticity

Chromaffin cells, expressing the phenotype of a mature endocrine cell, convert at high frequency to a neuronal phenotype both in vitro and in vivo in response to NGF (Unsicker et al. 1978; Aloe and Levi-Montalcini 1979; Doupe et al. 1985a). Our data illustrate this phenotypic plasticity at the level of RNA: mRNAs whose expression is normally restricted to neurons are induced to high levels in chromaffin cells exposed to NGF. Although the induction of neural-specific mRNAs by NGF in primary chromaffin cells may differ mechanistically from their initial induction in vivo, we believe that the plasticity of these genes in response to NGF nevertheless is a consequence of developmental changes in their chromatin structure in the sympathoadrenal precursor.

Specifically, the common precursor may maintain a battery of genes, specific to either sympathetic neurons or chromaffin cells, in an "open" chromatin configuration that is poised for transcription (Groudine and Weintraub 1982). The selection of those genes to be expressed would depend upon environmental signals, such as NGF or glucocorticoids, encountered by the

---

Figure 6. SCG10 and SCG4 mRNAs are induced by NGF in primary chromaffin cells. Shown are phase/dark-field pairs of in situ hybridizations to cultured cells. (*Top*) SCG10 probe; (*bottom*) SCG4 probe. (*A*) Chromaffin cells (CHR) grown in 5 μM dexamethasone. Note fibroblastic flat cells (FL) that are also negative with both probes. (*B*) Chromaffin cells switched after 1 week in dexamethasone to medium containing 50 ng/ml NGF and 10⁻⁵ M each of cytosine arabinoside and fluorodeoxyuridine. Cultures were maintained in this medium for 3 weeks prior to fixation. The photomicrographs are focused on the upper surface of the cells and thus the background is out of focus. Magnification, 215×.

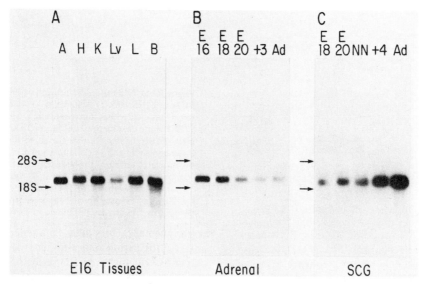

Figure 7. Developmental regulation of SCG4 mRNA. (*A*) Expression of SCG4 mRNA in nonneural tissues from E16 embryos. Each lane contains 10 μg of total RNA from adrenal gland (lane *A*), heart (lane *H*), kidney (lane *K*), liver (lane *Lv*), lung (lane *L*), and brain (lane *B*). Compare with Fig. 1B. (*B* and *C*) SCG4 mRNA declines in late gestation and is reexpressed selectively in nervous tissue. (*B*) Each lane contains 10 μg of total RNA from adrenal glands of the ages indicated. The amount detectable in adult (Ad) tissue corresponds to ~30–40 copies of mRNA/cell, assuming 2×10^5 mRNA molecules/cell. (*C*) RNA from developing superior cervical ganglion (SCG). Adult levels correspond to about 1400 copies/cell.

cell during or after migration. The inducibility of neural-specific genes in chromaffin cells may therefore reflect the persistence of these genes in such a transcriptionally competent configuration, despite the fact that they are not expressed as part of the mature chromaffin phenotype. In support of this model, studies of the structure of several tissue-specific genes have demonstrated that a nuclease-sensitive, "open" chromatin configuration developmentally precedes active transcription (Hofer et al. 1982; Sheffery et al. 1982; Burch and Weintraub 1983). In this way, the precursor may actively differentiate along one pathway while retaining the potential to switch to another pathway in response to a changed environment.

Why Have Plasticity?

The cells of the neural crest undergo extensive migrations before terminally differentiating into derivatives of widely varying phenotypes. The migratory pathways followed by neural crest cells are complex and may depend largely upon the cell's starting position (LeDouarin 1982; Thiery et al. 1982). While "homing" mechanisms may play a role in specifying a crest cell's final destination in advance of migration, the available data nevertheless suggest the process involves a fair amount of chance and randomness. If the ultimate environment of a crest cell cannot be completely determined in advance of migration, and this environment determines the cell's fate, then premigratory neural crest cells must face considerable developmental uncertainty. The multipotentiality apparent within the sympathoadrenal lineage, at least, constitutes one evolutionary strategy for accommodating this uncertainty.

This multipotentiality may be exploited to generate the phenotypic diversity displayed by the nervous system, and its persistence may permit changes in neuronal phenotypes in the mature organism (Furshpan et al. 1976; Patterson 1978; Black 1982; Landis and Keefe 1983; Black et al. 1984; Doupe et al. 1985a).

ACKNOWLEDGMENTS

We are grateful to Drs. Allison Doupe, Paul Patterson, Tom Jessell, and Story Landis for their helpful suggestions. We also thank Drs. Robert Angerer, Jim Roberts, and Eva Dworkin for providing detailed procedures for in situ hybridization, and Dr. John Pintar for providing some of the embryo sections used in this study and helping us to interpret the embryonic anatomy. This work was supported by the Howard Hughes Medical Institute, and a fellowship to D.J.A. from the Helen Hay Whitney Foundation.

REFERENCES

Aloe, L. and R. Levi-Montalcini. 1979. Nerve growth factor-induced transformation of immature chromaffin cells *in vivo* into sympathetic neurons: Effect of antiserum to nerve growth factor. *Proc. Natl. Acad. Sci.* **76:** 1246.

Anderson, D.J. and R. Axel. 1985. Molecular probes for the development and plasticity of neural crest derivatives. *Cell* **42:** 649.

Black, I.B. 1982. Stages in neurotransmitter development in autonomic neurons. *Science* **215:** 1198.

Black, I.B., J.E. Adler, C.F. Dreyfus, G.H. Jonakait, D.H. Katz, E.F. LaGamma, and K.H. Markey. 1984. Neurotransmitter plasticity at the molecular level. *Science* **225:** 1266.

Bohn, M.C., M. Goldstein, and I. Black. 1981. Role of glucocorticoids in expression of the adrenergic phenotype in rat embryonic adrenal gland. *Dev. Biol.* **82:** 1.

BURCH, J.B.E. and H. WEINTRAUB. 1983. Temporal order of chromatin structural changes associated with activation of the major vitellogenin gene. *Cell* **33:** 65.

COX, K.M., D.V. DELEON, L.M. ANGERER, and R.C. ANGERER. 1984. Detection of mRNAs in sea urchin embryos by *in situ* hybridization using asymmetric RNA probes. *Dev. Biol.* **101:** 485.

DOUPE, A.J., S.C. LANDIS, and P.H. PATTERSON. 1985a. Environmental influences in the development of neural crest derivatives: Glucocorticoids, growth factors and chromaffin cell plasticity. *J. Neurosci.* **5:** 2119.

DOUPE, A.J., P.H. PATTERSON, and S.C. LANDIS. 1985b. Small intensely fluorescent (SIF) cells in culture: Role of glucocorticoids and growth factors in their development and phenotypic interconversions with other neural crest derivatives. *J. Neurosci.* **5:** 2143.

FURSHPAN, E., P. MACLEISH, P. O'LAGUE, and D. POTTER. 1976. Chemical transmission between rat sympathetic neurons and cardiac myocytes developing in microcultures: Evidence for cholinergic, adrenergic and dual-function neurons. *Proc. Natl. Acad. Sci.* **73:** 4220.

GREENE, L.A. and A.S. TISCHLER. 1976. Establishment of a noradrenergic clonal line of rat adrenal phaechomocytoma cells which respond to nerve growth factor. *Proc. Natl. Acad. Sci.* **73:** 2424.

GROUDINE, M. and H. WEINTRAUB. 1982. Propagation of globin DNAase I-hypersensitive sites in absence of factors required for induction: A possible mechanism for determination. *Cell* **30:** 131.

HENDRY, I.A. 1977. Cell division in the developing sympathetic system. *J. Neurocytol.* **6:** 299.

HOFER, E., R. HOFER-WARBINEK, and J.E. DARNELL, JR. 1982. Globin RNA transcription: A possible termination site and demonstration of transcriptional control correlated with altered chromatin structure. *Cell* **29:** 887.

LANDIS, S.C. and D. KEEFE. 1983. Evidence for neurotransmitter plasticity *in vivo*: Developmental changes in properties of cholinergic sympathetic neurons. *Dev. Biol.* **98:** 349.

LANDIS, S.C. and P.H. PATTERSON. 1981. Neural crest cell lineages. *Trends Neurosci.* **4:** 172.

LEDOUARIN, N.M. 1980. The ontogeny of the neural crest in avian embryo chimeras. *Nature* **286:** 663.

———. 1982. *The neural crest*. Cambridge University Press, Cambridge, England.

NAUJOKS, K.W., S. KORSCHING, H. ROHRER, and H. THOENEN. 1982. Nerve growth factor-mediated induction of tyrosine hydroxylase and of neurite outgrowth in cultures of bovine adrenal chromaffin cells: Dependence on developmental stage. *Dev. Biol.* **92:** 365.

OGAWA, M., T. ISHIKAWA, and A. IRIMAJIRI. 1984. Adrenal chromaffin cells form functional cholinergic synapses in culture. *Nature* **307:** 66.

PATTERSON, P.H. 1978. Environmental determination of autonomic neurotransmitter functions. *Annu. Rev. Neurosci.* **1:** 1.

ROTHMAN, T.P., M.D. GERSHON, and H. HOLTZER. 1978. The relationship of cell division to the acquisition of adrenergic characteristics by developing sympathetic ganglion cell precursors. *Dev. Biol.* **65:** 322.

SHEFFERY, M., R.A. RIFKIND, and P.A. MARKS. 1982. Murine erythroleukemic cell differentiations: DNAase I hypersensitivity and DNA methylation near the globin genes. *Proc. Natl. Acad. Sci.* **79:** 1180.

TEITELMAN, G., T.H. JOH, D. PARK, M. BRODSKY, M. NEW, and D.J. REIS. 1982. Expression of the adrenergic phenotype in cultured fetal adrenal medullary cells: Role of intrinsic and extrinsic factors. *Dev. Biol.* **80:** 450.

THIERY, J.P., J.L. DUBAND, and A. DELOUVEE. 1982. Pathways and mechanisms of avian trunk neural crest cell migration and localization. *Dev. Biol.* **93:** 324.

UNSICKER, K., B. KRISCH, J. OTTEN, and H. THOENEN. 1978. Nerve growth factor-induced fiber outgrowth from isolated rat adrenal chromaffin cells: Impairment by glucocorticoids. *Proc. Natl. Acad. Sci.* **75:** 3498.

UNSICKER, K., T.J. MILLAR, T.H. MÜLLER, and H.-D. HOFMANN. 1985. Embryonic rat adrenal glands in organ culture: Effects of dexamethasone, nerve growth factor and its antibodies on pheochromoblast differentiation. *Cell Tissue Res.* **241:** 207.

WESTON, J.A. 1982. Motile and social behavior of neural crest cells. In *Cell behavior* (ed. R. Bellairs et al.), p. 429. Cambridge University Press, Cambridge, England.

A Biological Clock in *Drosophila*

M.W. Young,* F.R. Jackson,*‡ H.-S. Shin,†§ and T.A. Bargiello*
The Rockefeller University, New York, New York 10021; †Memorial Sloan Kettering Cancer Center, New York, New York 10021

A number of developmental and behavioral processes in multicellular animals are timed by chemical and/or electrical signals that originate in discrete regions of the brain. These processes range from developmental cycles, such as cycles of ovulation in mammals, to behavioral cycles, such as our own sleep/wake rhythms. Elegant physiological studies have shown that pacemakers can be composed of less than a few thousand cells (for review, see Takahashi and Zatz 1982; J.S. Takahashi; G.D. Block; both pers. comm.), but little is known about the molecular workings of any of these biological clocks.

An understanding of clock biochemistry can be approached in *Drosophila* by taking advantage of mutations affecting biological rhythms. Several mutations affecting the *Drosophila* clock have been recovered and characterized in considerable detail (Konopka and Benzer 1971; Jackson 1983). Five clock genes have now been defined in *D. melanogaster*. Mutations in some of these genes seem to affect the pace of a central neural clock, while others appear to affect the coupling of timed behavioral and developmental activities to the clock. Screens for clock mutations in *Drosophila* usually involve two kinds of assays. Wild-type flies have locomotor activity rhythms (analogous to sleep/wake rhythms) with a 24-hour period, and eclosion (hatching of adults from pupal cases) occurs with a circadian rhythm in populations of *Drosophila*. Flies that show heritable variations in the timing of these activities should carry mutations in genes required for the maintenance or function of a biological clock.

Period (*per*) was the first clock gene to be recognized in *D. melanogaster* (Konopka and Benzer 1971). Several mutant alleles of this X-linked gene have been recovered and these fall into three phenotypic classes. per^0 alleles cause arrhythmia with respect to both locomotor activity and eclosion. per^l mutants are rhythmic, but the period of locomotor activity rhythms and eclosion rhythms has been lengthened to 29 hours. per^s mutants have short period rhythms of 19 hours.

The tissue focus of *per*'s control over circadian rhythms has been mapped to the brain. The brain of a per^s fly can be transferred to the abdomen of a per^0 host with the result that a short-period locomotor activity rhythm is restored in the recipient (Handler and Konopka 1979). Analyses of flies that are genetically mosaic for *per* also prove that the head is the source of *per*'s control over these rhythms (Konopka et al. 1983).

Mutations at the *per* locus also affect noncircadian rhythms (Kyriacou and Hall 1980). Certain components of the courtship song of *D. melanogaster* are rhythmic with a periodicity of about 55 seconds in wild-type flies. Flies that carry the per^l mutation sing a song with an 80-second period, the per^s song rhythm has a 40-second period, and per^0 flies sing an arrhythmic courtship song. Interestingly, the tissue focus for *per*'s control over these rhythms may be nervous tissue in the thorax rather than the brain (for review, see Hall 1984). Evidently *Drosophila*'s ability to time can be affected in a very fundamental way by this set of mutations.

Isolation of a Clock Gene

Genomic DNA containing the *per* locus has been cloned by two laboratories (Bargiello and Young 1984; Reddy et al. 1984). A restriction map of part of a chromosomal walk through the *per* region is shown at the bottom of Figure 1. The structures of three chromosomal rearrangements are also shown (Fig. 1, top). For each rearrangement, chromosomal sequences that are present and in the wild-type configuration are stippled. The location of a breakpoint in each rearrangement is indicated by a vertical jagged line. Df(1)TEM202 (TEM) and Df(1)64j4 (j4) are deficient for a common DNA region covering the central two-thirds of the restriction map. Both of these deletions fail to complement mutations of the *per* locus. Moreover, Df(1)64j4/ Df(1)TEM202 heterozygous females, which are homozygously deficient for the central region of the cloned DNA, are arrhythmic (Young and Judd 1978; Smith and Konopka 1981). From all of this it can be concluded that loss of *per* locus function must be associated with arrhythmicity, whereas long- and short-period phenotypes should be linked to altered expression of the gene or to changes in the structure of the gene product.

The distal element of T(1;4)JC43 (Fig. 1, top) has been used as a *per* locus duplication. This duplication partially complements the arrhythmic phenotype of per^0 mutants; long-period (35-hr) rhythms can be produced by these JC43/per^0 flies (Smith and Konopka 1981). Apparently the JC43 break has damaged but not deleted *per*, and the chromosomal interval lying between the breakpoints of TEM and JC43 includes *per* locus DNA.

Figure 1 shows that four poly(A)⁺ transcripts are homologous to the cloned DNA. Slightly different lengths

---

‡Present address: Worcester Foundation for Experimental Biology, Shrewsbury, Massachusetts 01545.
§Whitehead Institute for Biomedical Research, Cambridge, Massachusetts 02142.

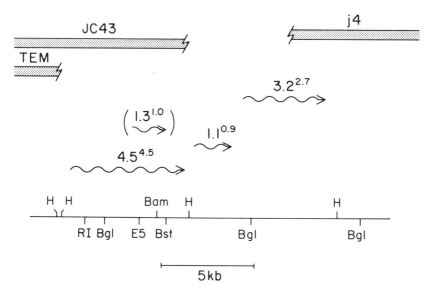

Figure 1. Physical map of *per* region DNA. The positions of three chromosomal rearrangements, T(1;4)JC43, Df(1)64j4, and Df(1)TEM202, are shown relative to the cloned DNA. Stippled bars indicate DNA still present and in the wild-type configuration. A vertical jagged line represents the location of a breakpoint. Only the distal element of T(1;4)JC43 is shown. The sizes (see also text), locations, and orientations of several transcripts are indicated. Symbols for restriction endonuclease cleavage sites are: (H) *Hin*dIII; (RI) *Eco*RI; (Bgl) *Bgl*II; (E5) *Eco*RV; (Bam) *Bam*HI; (Bst) *Bst*EII. All cleavage sites for these enzymes are not shown. For a more complete map see Bargiello and Young (1984).

have been reported for these transcripts, so to avoid confusion sizes given in Figure 1 (from Bargiello and Young 1984) are shown with superscripts indicating sizes measured by Reddy et al. (1984). Two of these transcripts, 1.1 and 3.2 kb, are poor candidates for *per* locus products because they are transcribed from DNA outside the interval covered by JC43 (Bargiello and Young 1984). Nevertheless, the abundance of one of these transcripts, the 1.1-kb RNA, has been found to oscillate with a circadian (24-hr) period (Reddy et al. 1984). Experiments described in a later section of this paper have ruled out a role for this oscillating transcript in *per*'s control of biological rhythms. In this regard it should be mentioned that over the years a great number of macromolecules have been shown to oscillate in this way in a variety of organisms (cf. Aschoff 1981), and it seems likely that the syntheses of most of these are dependent on, rather than part of, biological clocks.

To the left of the region coding for the 1.1- and 3.2-kb RNAs, two transcripts homologous to the DNA contained within the JC43 duplication are indicated. One of these RNAs, a 1.3-kb RNA in Figure 1, is shown in parentheses because, although it is homologous to the cloned DNA region, it is transcribed from another chromosomal location. Homology of this RNA to the *per* region appears to be restricted to less than 0.5 kb, and this transcript is detected with *per* locus probes in poly(A)+ RNA derived from Df(1)TEM202/Df(1)64j4 heterozygous females, which are homozygously deficient for the *per* region (see also Fig. 2b and related text). These results make it clear that the RNA cannot be transcribed from this chromosomal interval (Bargiello and Young 1984). The remaining RNA in the JC43 region, 4.5 kb (Fig. 1), is transcribed from an approximately 7-kb segment of the DNA. This RNA appears to be a good candidate for a *per* locus transcript.

The transcripts formed by flies carrying the chromosomal rearrangements shown in Figure 1 have been compared with those produced by wild-type flies (Bargiello and Young 1984; Reddy et al. 1984). As expected, the 4.5-kb RNA is not synthesized in Df(1)TEM202/Df(1)64j4 heterozygotes (Fig. 2a). The 4.5-kb transcript is also missing in T(1;4)JC43 flies, but a new, 11.5-kb transcript is formed (Fig. 2a). DNA containing the T(1;4)JC43 rearrangement breakpoint has been cloned (Bargiello and Young 1984), and the break maps to the 3′ nontranslated region of the transcription unit, giving rise to the 4.5-kb RNA in *per*+ flies (F.R. Jackson et al., in prep.). Apparently, the transcription unit coding for the 4.5-kb RNA is still functional in T(1;4)JC43, but a larger fusion transcript is now formed. This includes all of the protein coding sequence carried by the 4.5-kb RNA in wild-type flies, and 7 kb of RNA derived from the fourth chromosome. The residual *per* activity in JC43 flies is probably due to inefficient expression or utilization of the fusion transcript.

DNA from *per* Is Homologous to Other *Drosophila* Genes

Figure 2, b and c, shows that the DNA region transcribed to form the 4.5-kb RNA in wild-type flies is homologous to a number of transcripts synthesized by other chromosomal regions in *Drosophila*. Poly(A)+ RNA from wild-type flies (Fig. 2b, left), and RNA from Df(1)TEM202/Df(1)64j4 flies (Fig. 2b, right), has been hybridized with a portion of the 4.5-kb RNA coding

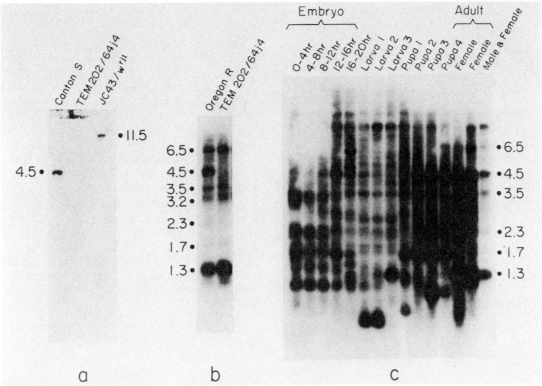

Figure 2. Characterization of *per* locus transcription in wild-type and mutant *Drosophila*. (*a*) Female, adult, poly(A)+ RNA (20 µg) was separated on a 1.0% agarose/2.2 M formaldehyde gel, transferred to Gene-Screen Plus (New England Nuclear), and subsequently hybridized with a ^{32}P-labeled, 2.7-kb *Eco*RI fragment (includes left end of Fig. 1 map). Df(1)w^{rj1} (w^{rj1}) removes all DNA in the *per* region (Bargiello and Young 1984). (*b*) Female, adult, poly(A)+ RNA (5 µg) was electrophoresed as above, transferred to Biodyne A membrane (Pall Ultrafine Filtration), and hybridized with a ^{32}P-labeled Riboprobe transcribed from the 1.5-kb *Bam*HI-*Hin*dIII fragment of Fig. 1 (see Bargiello et al. [1984], for construction of vectors for Riboprobe synthesis and related methods). (*c*) Total poly(A)+ RNA (5 µg) from the indicated stages was electrophoresed through a 0.8% agarose/2.2 M formaldehyde gel, transferred to Biodyne A membrane, and hybridized with Riboprobe transcribed from the 2.5-kb *Eco*RV-*Hin*dIII fragment (Fig. 1). For *a–c*, following hybridization, washes were preformed at 65°C in 5 mM NaCl, 2 mM NaPO$_4$ (pH 7), 0.1 mM EDTA, and 0.1% SDS. Sizes (in kb) indicated were measured from single-stranded RNA standards transcribed from DNA segments of known lengths cloned into the Riboprobe vector pSP64 (Promega Biotec).

region (see Fig. 2 legend for a description of the probe). The 4.5-kb transcript is detected only in the preparation of wild-type RNA. In contrast, several transcripts are detected in both RNA preparations, the predominant RNA being 1.3 kb, as discussed earlier. Because Df(1)TEM202/Df(1)64j4 flies are homozygously deficient for all of the DNA encoding the 4.5-kb RNA, it can be concluded that a number of actively transcribed genes are homologous to *per* locus DNA.

In Figure 2c a much larger spectrum of RNAs is detected with a slightly longer DNA probe. Homology to the new transcripts is limited to a 1-kb region of the *per* locus (Fig. 2, legend), so most of these transcripts must also come from other chromosomal intervals. Some of these transcripts appear to be homologous to portions of the *per* locus that are conserved in a wide variety of vertebrate and invertebrate species, and the nature of these homologies will be covered in a later section of this paper.

It will be of considerable interest to determine whether any of the *per*-homologous transcripts detected in Figure 2, b and c, are, like *per*, centrally involved in functions of the *Drosophila* clock. One DNA homology, probably that coding for the 1.3-kb transcript of Figure 2b, has been located by in situ hybridization of *per* locus DNA to polytene chromosomes. This homology maps to X chromosome region 1B (Bargiello and Young 1984; T.A. Bargiello and M.W. Young, unpubl.). Clock mutants have not been recovered in this chromosomal interval.

Restoration of Biological Rhythms in Transgenic *Drosophila*

The studies reported so far are consistent with the conclusion that the *per* locus product is the 4.5-kb RNA. To test further the ability of this transcription unit to supply *per* locus function, this DNA has been introduced into the germ line of *per^0* flies by P-element-mediated transformation (Bargiello et al. 1984). A 7.1-kb *Hin*dIII fragment containing only the 4.5-kb RNA coding region (Fig. 1) restores circadian rhythms in arrhythmic flies, with high penetrance of the rhythmic phenotype, as judged by analyses of two normally periodic phenomena, eclosion and locomotor activity. A related DNA segment containing most but not all of this transcription unit also transforms (Zehring et al. 1984). Transformants carrying the latter DNA (e.g., segment 14.6 from Zehring et al. 1984) produce a new 4.7-kb RNA in place of the usual 4.5-kb *per* transcript (J. Hall and M. Rosbash, pers. comm.).

In Figure 3c the temporal profile of eclosion (emergence of adults from pupal cases) is followed in wild-type flies. Eclosion is restricted to a few hours each morning in wild-type populations of *D. melanogaster*. Eclosion of a large population of flies can be synchronized in the laboratory by supplying a cycle of light and dark to developing larvae or pupae. An example of this is provided in Figure 3c. Eclosion was followed for a 9-day period. For the first 2 days of the experiment, a cycle of 12 hours light followed by 12 hours dark was provided. Flies entrain to this cycle, showing rhythmic hatching with a 24-hour periodicity. At the end of the second day the light/dark cycle was discontinued, and larvae and pupae were maintained for another 7 days in constant darkness. Flies continued to eclose rhythmically indicating that the rhythm can be sustained in the absence of further environmental signals.

Figure 3b shows the eclosion pattern for a population of *per^0* flies. Even in the presence of a light/dark cycle no rhythmicity is detected. Flies tested in Figure 3a are also *per^0*, but differ from those examined in Figure 3b in that they carry a single, autosome-linked copy of a *per$^+$* transposon. The *per$^+$* DNA carried by this transposon insertion (designated P1.48C) is sufficient to restore rhythmicity during the light/dark cycle and subsequently during constant darkness. The period of eclosion rhythms during entrainment is 24 hours, whereas the period of free-running rhythms (during constant darkness) is 27 hours.

Locomotor activity recordings for individual wild-type flies and for *per^0* flies carrying the P1.48C transposon are presented in Figure 4. Bouts of activity are registered as vertical deflections from an otherwise uninterrupted horizontal line. Each horizontal line represents a 48-hour trace of the fly's behavior and consecutive 48-hour intervals are presented with the first

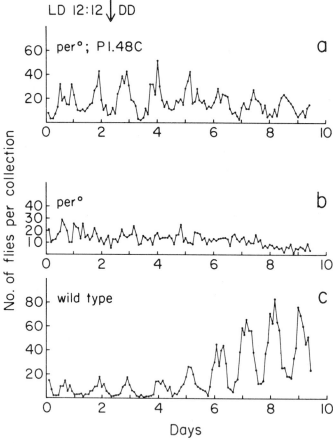

Figure 3. Eclosion patterns for wild-type, mutant, and transformed *Drosophila*. Eclosion (emergence from pupal cases) was followed simultaneously in three *Drosophila* populations. Flies were collected manually every 2 hr from culture bottles at 25°C. Cultures were entrained to a cycle of 12 hr light/12 hr dark for 3 days prior to the first collection. The light/dark cycle was maintained for the first 2 days of the experiment (LD 12:12). Cultures then were kept in constant darkness (DD) for the remaining 7 days of the experiment. The period of each rhythm was measured during DD by determining the time elapsed between medians of successive peaks of eclosion. (*a*) *per^0*; P1.48C, flies have rhythms with a 27.2-hr period. (*b*) *per^0* flies are arrhythmic. (*c*) Flies designated "wild-type" are heterozygous females of the genotype *per^0/+*. These have a period of 25.0 hr. For details of construction of strain *per^0*; P1.48C by P-element-mediated transformation see Bargiello et al. (1984).

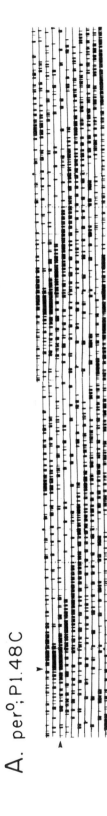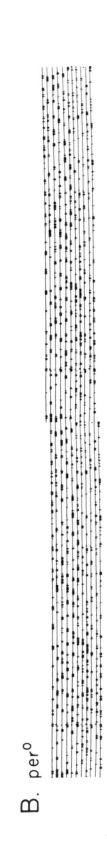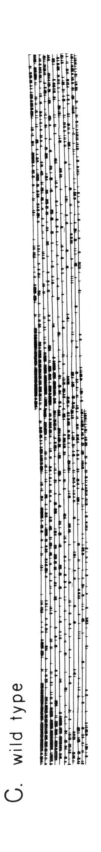

Figure 4. *(See next page for legend.)*

day at the top and last day at the bottom of each record. Figure 4b shows the activity record of a per^0 fly. Aperiodic activity is evident for the duration of the experiment. Figure 4c shows the rhythmic behavior of a wild-type (Oregon-R) fly. In contrast to the behavior of the per^0 fly, bouts of activity are seen in the per^+ fly every 24 hours. Figure 4, a and d, shows behavior traces for transformed flies. In both cases the same transposon insertion (P1.48C) complements a *per* locus mutation. In Figure 4a the arrhythmic mutation being complemented is the ethylmethanesulfonate (EMS)-induced allele per^{01} (Konopka and Benzer 1971). This allele is not associated with a detectable deletion of DNA in the *per* region (Bargiello and Young 1984). Figure 4d demonstrates that the transposon complements the 10-kb homozygous deficiency in a Df(1)TEM202/Df(1)64j4 fly. This deletion removes DNA coding for the 4.5-kb transcript, the 1.1-kb oscillating RNA, and the 3.2-kb RNA mapped in Figure 1. As the P1.48C transposon only replaces DNA encoding the 4.5-kb RNA, it can be concluded that, in this chromosomal region, only a single gene is required for the production of these biological rhythms (Bargiello et al. 1984). Consistent with this conclusion, Zehring et al. (1984) have found that DNA neighboring that coding for the 4.5-kb transcript does not restore rhythmicity to per^0 flies.

The period of the locomotor activity rhythms of the transformed flies in Figure 4 is about 27 hours. Similarly, the eclosion rhythm of P1.48C-transformed flies has a period of 27 hours (Fig. 3a). Thus, the transformed flies, after receiving wild-type transforming DNA, have longer than wild-type periods. The finding that this strain of transformed flies has long-period rhythms is not particularly surprising, because earlier genetic analyses of *per* have shown that period length is sensitive to *per* locus dosage. Females carrying only one dose of this X-linked gene have circadian periodicities of locomotor activity and eclosion rhythms that are about 1 hour longer than females carrying two doses of the gene (Smith and Konopka 1982). If a *per* product were underproduced by the transforming DNA, long-period rhythms would be expected. Consistent with this expectation, the 4.5-kb transcript is significantly underproduced in flies carrying the P1.48C transposon. It has been estimated that each copy of *per* in the Oregon-R (wild type) strain produces approximately five times as much 4.5-kb RNA as that generated by the transforming DNA in Df(1)TEM202/Df(1)64j4;P1.48C flies (Bargiello et al. 1984).

Importantly, not all flies transformed with wild-type *per* locus DNA have 27-hour rhythms. Certain strains of *Drosophila* independently transformed with the 7.1-kb HindIII fragment have rhythms with periodicities in the wild-type range (e.g., see Fig. 5), while other strains have rhythms with periodicities even longer than those seen in the P1.48C-transformed flies discussed above. Variations in period length may be governed by the specific chromosomal location of this transforming DNA. The results of a study of chromosomal position, period length, and RNA abundance in several independently transformed strains will be reported elsewhere (F.R. Jackson et al., unpubl.).

Temporal Control of *per* Transcription

Since we have suggested that the titer of the 4.5-kb RNA determines the period length for *Drosophila* circadian rhythms, the synthesis of this RNA has been followed in more detail under a variety of environmental conditions and over the course of a daily cycle of light and dark. Figure 6 shows that the 4.5-kb RNA is produced at a constant level around the clock. For RNA preparations labeled Oregon-R and per^0, adults were collected and RNA extracted at 4-hour intervals. Equal amounts of total poly(A)$^+$ RNA were slot-blotted to nitrocellulose and subsequently hybridized with a *per* locus probe (see Fig. 6, legend). Flies had been exposed to a cycle of 12 hours light/12 hours dark for 3 days prior to these collections and the first and fourth collections were made at lights on and lights off, respectively. No change in the production of the 4.5-kb transcript can be detected throughout a daily cycle in either Oregon-R (per^+) or per^0 flies. per^0 flies apparently produce wild-type levels of a nonfunctioning 4.5-kb RNA (see also Bargiello and Young 1984).

Figure 6 (right) shows that the level of 4.5-kb RNA detected in Oregon-R flies is not affected by the absence of a light/dark cycle. Poly(A)$^+$ RNA from a population of flies grown for three generations in constant light, and poly(A)$^+$ RNA extracted from flies grown for three generations in continuous darkness were blotted to nitrocellulose and probed as before. Levels of 4.5-kb RNA were, in both cases, indistinguishable from preparations made during the light/dark cycle experiments.

Developmental Control of *per* Transcription

Circadian rhythms of locomotor activity and the ultradian rhythms of courtship song are measured in

---

Figure 4. Locomotor activity records. Individual male (*A–C*) or female (*D*) flies of the indicated genotypes (wild type = Oregon-R) were maintained in clear plastic tubes positioned between a light-emitting diode (900-nm peak emission) and a phototransistor. In this arrangement movement of the fly was registered as an all-or-none signal causing a pen to deflect on an event recorder. Records were formed by plotting successive days (24 hr) of activity beneath each other, then plotting twice the resulting chart of activity. Thus, each horizontal line shows a continuous 48-hr interval. For *A*, 3 days of the record show entrainment to a light/dark cycle (LD 12:12), with lights on at 0 and off at 12. Lights off at the end of the LD cycle is indicated by the two arrowheads (12:00 on day 3). After this the fly remained in constant darkness. For *B–D*, records include only activity measured in constant darkness.

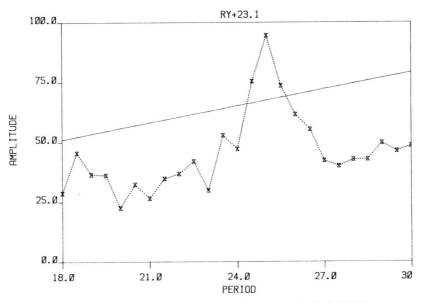

Figure 5. Period spectrum analysis of locomotor activity from a single P1/23 male. Strain P1/23 was generated by P-element-mediated transformation using pCP1 (also used to form P1.48C) as described by Bargiello et al. (1984). Locomotor activity was assayed as in Fig. 4 except that data were automatically collected and analyzed by an Apple IIe computer. The program used for spectral analysis of the data was provided by Sulzman (Sulzman 1982). This program evaluates the significance of spectral components using a t test, and the $p < 0.05$ level is indicated in the figure as a diagonal line. Four individuals have been tested from this strain with an average periodicity of 24.6 hr. The individual record shown indicates rhythmic locomotor activity with a period of 25.0 hr.

Drosophila adults. Eclosion rhythms are also measured at the earliest stages of the adult's life. However, it is clear that the time of emergence of the adult from the pupal case can be dictated by changes in the light/dark cycle administered early in pupation or even during the larval instars. For example, Figure 3c shows that rhythmic, synchronous eclosion continues to occur within a wild-type *Drosophila* population for at least 7 days following cessation of a light/dark cycle. Under the conditions used to culture *Drosophila* in this eclosion experiment (see Fig. 3, legend), flies emerging on the last day of the experiment (day 9 in Fig. 3c) would have been entrained to a light/dark cycle last seen when they were second or third instar larvae (day 2 in Fig. 3c). If the time of eclosion can be set during the larval and pupal stages of development, a clock under the influence of the *per* locus might be running long before adulthood is reached.

Figure 7 shows that *per* is transcribed from the middle of embryogenesis through adulthood. The 4.5-kb transcript is most abundant in adults and during pupation. In fact, this RNA was originally detected only during these two late stages of development (Bargiello and Young 1984; Reddy et al. 1984). Detection of the less abundant 4.5-kb transcripts in earlier developmental stages is probably only afforded by the use of more sensitive RNA probes (Fig. 7, legend). The finding that *per* is expressed very early in *Drosophila* development raises the possibility that a clock may be formed and running within hours of fertilization.

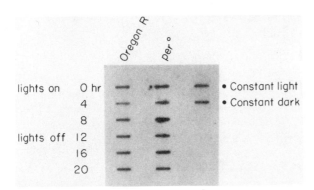

Figure 6. Temporal control of *per* locus transcription. Poly(A)$^+$ RNA was extracted from adult Oregon-R or per^0 flies at the indicated phases of a light/dark cycle (LD 12:12). RNA was also obtained from adult, Oregon-R flies after three generations of culture in constant darkness or constant light. RNA (5 μg) from each sample was slot-blotted to nitrocellulose and hybridized to a ^{32}P-labeled, 2.7-kb *Eco*RI fragment (covers left end of Fig. 1 map). This probe hybridizes only with the 4.5-kb *per* transcript in adult RNA preparations (Bargiello and Young 1984; see also Fig. 7).

DNA Homologies in Other Species

DNA homologous to the *per* locus has been found in a number of vertebrates (Shin et al. 1985). In Figure 8, a *per* locus probe is hybridized to total genomic DNA from yeast, *Drosophila*, chickens, cat, mouse, and humans. At least two homologies are detected in *Drosophila*, indicating, as before, that *per* is related to other *Drosophila* genes. Multiple homologies are detected in chickens and mice. A single, weaker homology is observed in human DNA. Several cloned DNA segments having homology to *per* were isolated from a mouse cosmid library, and each contains a discrete region of DNA homology to the *per* locus. Such mouse clones also hybridize with each other by virtue of those DNA regions homologous to *per* (Shin et al. 1985).

Figure 9 shows that a restriction fragment carrying the *per*-related DNA from one of the mouse cosmids is homologous to poly(A)$^+$ RNA synthesized in a number of mouse tissues. The mouse DNA is also homologous to poly(A)$^+$ RNA from *Drosophila* (Fig. 9, left). One of the homologous *Drosophila* transcripts is the 4.5-kb *per* RNA; two additional transcripts apparently come from other chromosomal regions bearing *per* locus homologies (Shin et al. 1985).

The DNA sequence of that portion of *per* bearing homology to the cloned mouse DNA has been determined and appears in Figure 10. Also shown in that figure is a portion of the mouse DNA sequence (from clone cp2.2) which has been aligned with respect to the fly sequence according to DNA and predicted protein homology. The mouse homologous sequence at the *per* locus covers about 0.3 kb of a 0.7-kb RNA coding region. This coding region contains a single open reading frame and a conceptual translation of this sequence is designated D protein in Figure 10. The *Drosophila* DNA sequence is designated D DNA, whereas the DNA sequence of the homologous mouse DNA is designated M DNA in the figure. Both the mouse and fly DNA segments appear to encode long stretches of a very simple amino acid sequence. The encoded protein sequence is composed of only two amino acids, threonine and glycine, and codons for the two amino acids alternate with each other. Up to 48 alternating threonine and glycine codons can occur in a stretch in *Drosophila*, while stretches up to 40 codons long occur in the mouse. Shorter stretches of these alternating codons occur in both the fly and mouse DNA segments, but these are separated from each other by codons for unrelated amino acids.

The sequence homology detected in the mouse is part of a single 2.1-kb open reading frame. Codons for alternating threonine and glycine are not found outside of this open reading frame. The alternating codons for threonine (ACN) and for glycine (GGN) are never in-

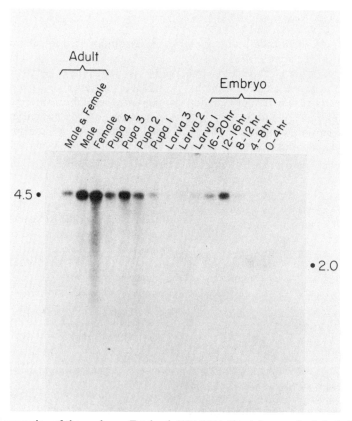

Figure 7. Developmental expression of the *per* locus. Total poly(A)$^+$ RNA (5 µg) from each of the indicated stages was electrophoresed through a 0.8% agarose/2.2 M formaldehyde gel, transferred to Biodyne A membrane, and hybridized with a ^{32}P-labeled Riboprobe transcribed from the 1.5-kb *Hin*dIII–*Eco*RI fragment of Fig. 1. Weak hybridization to a 2.0-kb transcript has been observed in some preparations of 0- to 8-hr embryos. This may reflect hybridization to *per*-homologous RNA transcribed from another chromosomal region as described previously (see Fig. 2 and related text). For additional procedural details see Fig. 2c legend.

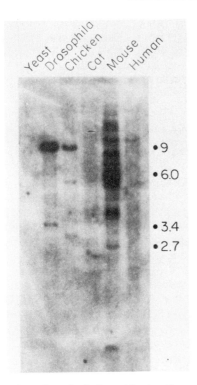

Figure 8. *per* locus homologies in vertebrates. Genomic DNA from yeast, *Drosophila*, chicken, cat, mouse, and human was digested with BamHI, separated on a 0.75% agarose gel, and transferred to nitrocellulose. A ^{32}P-labeled, 1.3-kb PstI-BamHI fragment (contains the 1-kb EcoRV-BamHI fragment of Fig. 1), was hybridized at 40°C in 5× SSC, 50% formamide, 7 mM Tris (pH 7.5), 1× Denhardt's solution, 25 µg/ml salmon sperm DNA, and 10% dextran sulfate. Washes were in 0.2× SSC, 0.1% SDS at 45°C. Numbers show sizes (in kb) of some hybridizing DNA segments as measured against restriction fragments of λ cI857.

of the Dayhoff and GenBank sequence libraries, poly serine-glycine protein segments have been found in some proteoglycans. In these molecules the repeated protein segment acts as a site for glycosylation (Bourdon et al. 1985). Thus, serine-glycine and threonine-glycine repeats may be substrates for posttranslational modification of the *per* protein.

In summary, transcripts from mouse and *Drosophila* genes appear to code for proteins with homologous parts. We know that one of these proteins, the *per* product, is essential for the construction or maintenance of a biological clock in *Drosophila*. Finding an unusual DNA sequence coding for alternating amino acids at the *per* locus that is also associated with genes in the mouse raises the attractive possibility that homologous proteins play a role in the generation of biological rhythms in many species.

terrupted by inversions or DNA insertions that might shift the reading frame, although clusters of these codons are often separated from each other by insertions of 3 bp or multiples of 3 bp. These constraints on the mouse DNA sequence make it probable that the cloned DNA encodes a protein segment containing long stretches of alternating threonine and glycine. The polarity of transcripts homologous to this mouse DNA is also consistent with this conclusion (see Fig. 9, legend).

The mouse and fly DNA segments are also related by a second homology. Stretches of alternating serine and glycine codons are found in both cloned DNAs. The longest of these stretches occurs in the mouse sequence, 22 alternating serine-glycine codons (not shown, but see Shin et al. 1985). Serine-glycine codons, as they occur in the fly DNA sequence, are underlined in Figure 10. The significance of this second homology may lie in the close chemical relationship of serine and threonine. Since both are uncharged, hydroxylated, polar amino acids that can be phosphorylated or glycosylated, serine-glycine and threonine-glycine protein segments could serve related functions.

While we have not detected DNA or protein homologies to threonine-glycine repeats in a computer search

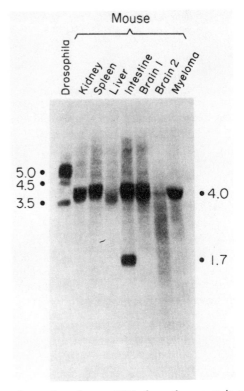

Figure 9. *per* homologous DNA from the mouse is transcribed. Poly(A)$^+$ RNA (5 µg) from Oregon-R, adult *Drosophila* (*left*) or from the indicated mouse tissues, was separated on a 1% agarose/2.2 M formaldehyde gel and transferred to Biodyne A membrane. ^{32}P-labeled Riboprobe was transcribed from a 2.5-kb EcoRI fragment derived from mouse clone cp2.2. This restriction fragment is extensively homologous to the *per* locus and its complete DNA sequence has been determined (Shin et al. 1985). Hybridizations using this single-stranded RNA probe were at 65°C in 50% formamide, 50 mM NaPO$_4$ (pH 7), 0.8 M NaCl, 1 mM EDTA, and 0.1% SDS. Washes were at 65°C in 5 mM NaCl, 2 mM NaPO$_4$ (pH 7), 0.01 mM EDTA, and 0.01% SDS. Brain 1 is from a fetal mouse (16 day), and Brain 2 is from a 1-day-old mouse.

```
D protein      GLY THR CYS VAL SER GLY ALA SER GLY PRO MET SER PRO VAL HIS
D DNA          GGC ACG TGT GTC AGT GGC GCC AGT GGT CCG ATG AGT CCC GTC CAC
                           --- ---         --- ---


D protein      GLU GLY SER GLY GLY SER GLY SER SER GLY ASN PHE THR THR ALA
D DNA          GAG GGC AGC GGG GGC AGT GGC TCC TCG GGC AAC TTC ACC ACC GCC
                           --- ---         --- ---         --- ---


D protein      SER ASN ILE HIS MET SER SER VAL THR ASN THR SER ILE ALA GLY
D DNA          AGT AAC ATA CAC ATG AGC AGT GTG ACA AAT ACG AGC ATT GCC GGC
M DNA                                  GTC ACA ... ... ... ... ... ... GGC


D protein      THR GLY     GLY THR                 GLY THR GLY THR GLY THR
D DNA          ACT GGT ... GGC ACG ... ... ... ... GGC ACT GGT ACA GGT ACA
M DNA          ACA GGG ACA GGC ACA GCC AAA GTC ACA GGC ACA GGG ACA GGC ACA


D protein      GLY THR GLY THR GLY THR GLY THR GLY THR GLY THR         GLY
D DNA          GGT ACT GGA ACT GGA ACT GGA ACC GGG ACA GGA ACT ... ... GGA
M DNA          GGC ACA GGC ACA GGC ACA GGC ACA GGC ACA GGC ACA GAC ACA GGC


D protein      THR GLY THR                 GLY THR GLY THR GLY THR GLY THR
D DNA          ACC GGG ACA ... ... ... ... GGA ACT GGA ACC GGG ACA GGA ACT
M DNA          ACA GGC ACA GCC AAA GTC ACA GGC ACA GGC ACA GGC ACA GGT ACA


D protein      GLY THR GLY THR GLY THR GLY THR GLY THR GLY THR GLY THR GLY
D DNA          GGA ACG GGA ACA GGT ACA GGC ACA GGC ACA GGC ACT GGA ACA GGC
M DNA          GGC ACA GGC ACA GGC ACA GGT ACA GGC ACA GGC ACA GGC ACA GGC


D protein      ASN GLY THR ASN SER                 GLY THR GLY THR
D DNA          AAT GGA ACA AAT TCC ... ... ... ... GGC ACC GGA ACC
M DNA          ... ... ACA ... ... GCC AAA GTC ACA GGC ACA GGC ACA
```

Figure 10. Sequence comparison of a portion of the *per* locus and homologous mouse DNA. *per* locus DNA having homology to the mouse is shown as D DNA. D protein is the predicted translation product of the *per* DNA. M DNA designates homologous DNA from mouse clone cp2.2. Gaps in the alignment of the mouse and fly sequences allow maximum DNA homology to be presented. Gaps always involve multiples of three nucleotides such that a shift in the reading frame for protein does not occur.

Summary

The *per* locus plays a central role in the organization and function of the *Drosophila* biological clock. The gene has been mapped to a 7-kb DNA segment by physically locating the breakpoints of several chromosomal rearrangements that disrupt *per* function. This DNA contains a single transcription unit which produces a 4.5-kb poly(A)$^+$ RNA. No oscillation in the synthesis of this transcript is detected when *per* expression is followed over a 24-hour cycle of light and dark. When wild-type DNA containing only this transcription unit is transferred to the genome of a *per*^{0} (arrhythmic) fly by P-element-mediated transformation, the 4.5-kb RNA is produced and rhythmic behavior is restored with high penetrance of the rhythmic phenotype. This transforming DNA will also complement chromosomal deletions that include the *per* locus and adjoining transcription units, indicating that only one gene in this chromosomal interval plays a measureable role in the production of circadian rhythms.

Transformed flies having rhythms with longer than wild-type periodicity underproduce the 4.5-kb RNA. We suggest that the periodicity of circadian rhythms in *Drosophila* is determined by the level of expression of a *per* locus protein encoded by the 4.5-kb transcript.

We have found DNA homologous to the *per* locus in several species of vertebrates. DNA exhibiting very high homology to a portion of the *per* locus has been cloned from the mouse and the DNA sequence of the conserved segment has been determined. Mouse and fly DNAs both appear to code for long protein segments composed exclusively of alternating threonine and glycine or serine and glycine residues.

ACKNOWLEDGMENTS

We thank Suk-Hyeon Yun and Julie Pan for technical assistance. This work was supported by grants from the Andre and Bella Meyer Foundation and the National Institutes of Health to M.W.Y.

REFERENCES

ASCHOFF, J., ed. 1981. *Handbook of behavioral neurobiology*, vol. 4. Plenum Press, New York.

BARGIELLO, T.A. and M.W. YOUNG. 1984. Molecular genetics

of a biological clock in *Drosophila. Proc. Natl. Acad. Sci.* **81:** 2142.

BARGIELLO, T.A., F.R. JACKSON, and M.W. YOUNG. 1984. Restoration of circadian behavioural rhythms by gene transfer in *Drosophila. Nature* **312:** 752.

BOURDON, M.A., A. OLDBERG, M. PIERSCHBACHER, and E. RUOSLAHTI. 1985. Molecular cloning and sequence analysis of a chondroitin sulfate proteoglycan. *Proc. Natl. Acad. Sci.* **82:** 1321.

HALL, J.C. 1984. Complex brain and behavioral functions disrupted by mutations in *Drosophila. Dev. Genet.* **4:** 355.

HANDLER, A.M. and R.J. KONOPKA. 1979. Transplantation of a circadian pacemaker in *Drosophila. Nature* **279:** 236.

JACKSON, F.R. 1983. The isolation of biological rhythm mutations in the autosomes of *Drosophila melanogaster. J. Neurogenet.* **1:** 3.

KONOPKA, R.J. and S. BENZER. 1971. Clock mutants of *Drosophila melanogaster. Proc. Natl. Acad. Sci.* **68:** 2112.

KONOPKA, R.J., S. WELLS, and T. LEE. 1983. Mosaic analysis of a *Drosophila* clock mutant. *Mol. Gen. Genet.* **190:** 284.

KYRIACOU, C.P. and J.C. HALL. 1980. Circadian rhythm mutations in *Drosophila melanogaster* affect short-term fluctuations in the male's courtship song. *Proc. Natl. Acad. Sci.* **77:** 6929.

REDDY, P., W.A. ZEHRING, D.A. WHEELER, V. PIRROTTA, C. HADFIELD, J.C. HALL, and M. ROSBASH. 1984. Molecular analysis of the *period* locus in *Drosophila melanogaster* and identification of a transcript involved in biological rhythms. *Cell* **38:** 701.

SHIN, H.S., T.A. BARGIELLO, B.T. CLARK, F.R. JACKSON, and M.W. YOUNG. 1985. An unusual coding sequence from a *Drosophila* clock gene is conserved in vertebrates. *Nature* **317:** 445.

SMITH, R.F. and R.J. KONOPKA. 1981. Circadian clock phenotypes of chromosome abberations with a breakpoint at the *per* locus. *Mol. Gen. Genet.* **183:** 243.

———. 1982. Effects of dosage alterations at the *per* locus on the period of the circadian clock of *Drosophila. Mol. Gen. Genet.* **185:** 30.

SULZMAN, F.M. 1982. Microcomputer monitoring of circadian rhythms. *Comput. Biol. Med.* **12:** 253.

TAKAHASHI, J.S. and M. ZATZ. 1982. Regulation of circadian rhythmicity. *Science* **217:** 1104.

YOUNG, M.W. and B.H. JUDD. 1978. Nonessential sequences, genes, and the polytene chromosome bands of *Drosophila melanogaster. Genetics* **88:** 723.

ZEHRING, W.A., D.A. WHEELER, P. REDDY, R.J. KONOPKA, C.P. KYRIACOU, M. ROSBASH, and J.C. HALL. 1984. P-element transformation with *period* locus DNA restores rhythmicity to mutant, arrhythmic *Drosophila melanogaster. Cell* **39:** 369.

Cell Adhesion Molecule Expression and the Regulation of Morphogenesis

G.M. EDELMAN

The Rockefeller University, New York, New York 10021

A number of empirical observations suggest that the problem of morphogenesis must be viewed in terms of developmental genetic and mechanochemical factors acting jointly to yield a common phenotype within a taxon. The requisite combination of molecular genetics and macroscopic mechanism implies that regulation of pattern formation must range across a wide variety of levels of organization from the gene and functional gene products, to cells, to tissues and organs, and back to the gene. This multilevel control is exercised by regulatory primary processes of development (cell adhesion, differentiation) interacting with a number of other primary processes (cell division, cell motion, cell death) that act as driving forces for the emergence of form.

In regulative development, cells of different history are brought together by morphogenetic movements to result in embryonic induction, or milieu-dependent differentiation. Embryonic induction can be reciprocal or asymmetric, and can occur in a series of stages that are either stimulatory or inhibitory, depending on the prior paths and local contacts of the induced and inducing cell populations (Jacobson 1966). There is an apparent paradox here: The genome itself cannot contain specific information about the exact position of the participating cells in time and space and yet morphogenesis is under genetic control.

These observations allow us to cast the basic morphogenetic problem in the form of two linked questions: (1) How does a one-dimensional genetic code specify a three-dimensional animal? (2) How is the answer to this question consistent with the possibility of relatively rapid morphological change in relatively short periods of evolutionary time?

This latter problem of morphologic evolution involves convergent paths, and no single ontogenetic mechanism is likely to provide a solution for all taxa. Nonetheless, a consideration of the genetic and mechanochemical constraints in regulative development suggests that an understanding of cell adhesion at the molecular level might be particularly revealing. Work that led to the discovery and chemical analysis of cell adhesion molecules (CAMs) has been reviewed in detail elsewhere (Edelman 1983, 1984a, 1985); these reviews contain bibliographic references to work in our own and other laboratories on CAMs in various species. Here, I will focus on evidence obtained in our laboratory indicating: (1) that CAMs have definite sequences of expression that correlate strongly with major morphogenetic events (Thiery et al. 1982, 1984; Edelman et al. 1983), (2) that one of their main functions is to regulate morphogenetic movements either directly or indirectly (Edelman 1984b; Grumet et al. 1984), and (3) that their expression can be cyclic in time and periodic in space (Chuong and Edelman 1985a,b; Crossin et al. 1985).

Types and Properties of CAMs

So far, three CAMs of different specificity and structure have been isolated and characterized (Fig. 1). The first two, liver cell adhesion molecule (L-CAM) (Cunningham et al. 1984) and neural cell adhesion molecule (N-CAM) (Edelman 1983), are called primary CAMs and appear early in embryogenesis upon derivatives of multiple germ layers. The third, neuron-glia CAM (Ng-CAM), is a secondary CAM that is not seen in early embryogenesis and that appears only on neuroectodermal derivatives, specifically on postmitotic neurons (Grumet and Edelman 1984; Grumet et al. 1984).

All of the CAMs are glycoproteins synthesized by the cells on which they function. N-CAM and L-CAM are intrinsic membrane proteins; this appears also to be true of Ng-CAM, but has not been as firmly established. N-CAM and probably L-CAM bind by homophilic mechanisms, that is, CAM on one cell to another identical CAM on an apposing cell. While N-CAM binding is calcium independent, L-CAM depends on this ion both for its integrity and its binding; these two primary CAMs show no binding cross-specificity for each other. In contrast, the secondary Ng-CAM on neurons appears to bind by a heterophilic mechanism, that is, Ng-CAM on a neuron to another CAM or a chemically different receptor on glia (Grumet and Edelman 1984).

It has been suggested (Edelman 1984b) that CAMs act to regulate binding via a series of cell-surface modulation mechanisms including changes in their prevalence at the cell surface, in their position or polarity, and in their chemistry of binding. All of these mechanisms have been shown to occur for one CAM or another at different developmental times (Edelman 1984a). The known case of chemical modulation is related to the presence (Hoffman et al. 1982) on N-CAM of α-2-8-linked polysialic acid (Finne et al. 1983) at three sites (Crossin et al. 1984) present in the middle domain of the molecule. In the microheterogeneous embryonic (E) form of N-CAM, there are 30 g/100 g of polypeptide, and in the discrete adult (A) forms this is reduced to 10 g/100 g polypeptide. Recent in vitro

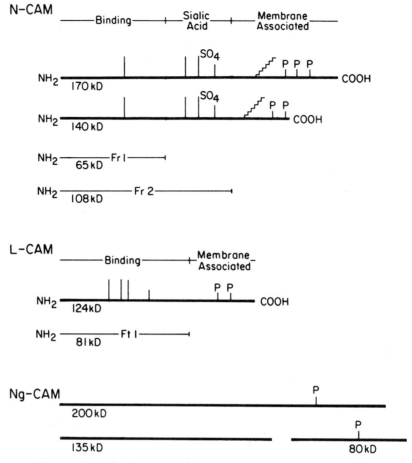

Figure 1. Diagrams of the linear chain structure of two primary CAMs (N-CAM and L-CAM) and of the secondary Ng-CAM. N-CAM is seen as two closely related polypeptide chains, one shorter than the other. Below them are the fragments Fr1 and Fr2 derived by limited proteolysis. As indicated by vertical lines, most of the carbohydrate is covalently attached in the middle domain at three sites (Crossin et al. 1984) and it is sulfated, although the exact sulfation site is unknown. Attached to these carbohydrates is polysialic acid. There are phosphorylation sites as well in the COOH terminal domains (Sorkin et al. 1984). The diagonal stepped bars refer to covalent attachment of palmitate. L-CAM yields one major proteolytic fragment (Ft1) and has four attachment sites for carbohydrate (vertical lines) but lacks polysialic acid (Cunningham et al. 1984). It is also phosphorylated in the COOH terminal region. Ng-CAM is shown without polarity but it is likely that the NH_2 terminus is to the left. There are two components (135 kD and 80 kD) that are probably derived from a posttranslationally cleaved precursor. Each is related to the major 200-kD chain (which may be this precursor) and the smaller is arranged as shown on the basis of a known phosphorylation site.

studies (Friedlander et al. 1985) suggest that N-CAM turns over at the cell surface and that the E form is replaced by newly synthesized A forms. Although the carbohydrate is not directly involved in binding, kinetic studies (Hoffman and Edelman 1983) of CAM vesicle binding suggest that E-to-A conversion can result in a fourfold increase in binding rates. It seems likely that the charged polysialic acid either modulates the conformation of the neighboring CAM binding region or directly competes with homophilic binding from cell to cell by charge repulsion (Edelman 1983). Even more striking than the effects of E-to-A conversion is the dependency of homophilic binding on changes in CAM prevalence or surface density: a twofold increase in E forms leads to greater than 30-fold increases in binding rates (Hoffman and Edelman 1983). This is likely to result from increases in valency by *cis* association at the cell surface of two or more CAM polypeptides.

Similar rate studies across a variety of vertebrate species (Hoffman et al. 1984) suggest that the N-CAM binding mechanism is conserved during evolution. The existence of such nonlinear binding effects accompanying surface modulation is compatible with the idea that CAMs act as sensitive regulators of cell aggregation and cell motion.

The CAM Expression Sequence

Correlation of the time and place of expression of each CAM with sites of embryonic induction and with key events of histogenesis has been observed using immunocytochemistry. I first describe the major sequence for the whole embryo (Fig. 2) and then consider some more detailed examples of histogenetic sequences as seen in the brain and in the feather.

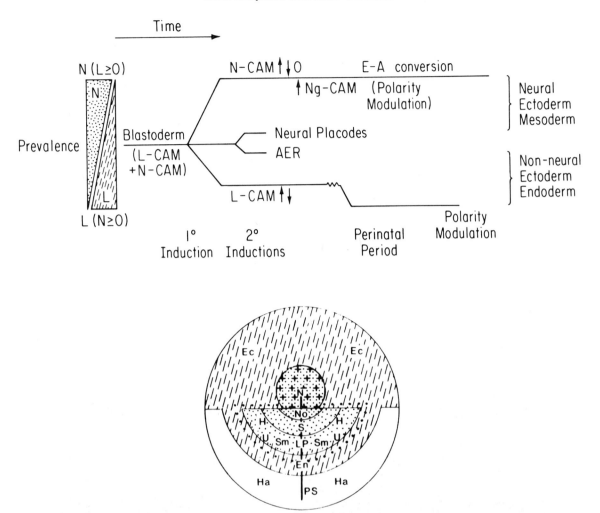

Figure 2. Major CAM expression sequence and composite CAM fate map in the chick. (*Top*) Schematic diagram showing the temporal sequence of expression of CAMs during embryogenesis. Vertical wedges at the left refer to relative amounts of each CAM in the different parts of the embryo; i.e., the line referring to blastoderm has relatively large amounts of each CAM whereas that for neural ectoderm has major amounts of N-CAM but little or no L-CAM. After an initial differentiation event, N-CAM and L-CAM diverge in cellular distribution and are then modulated in prevalence (↑↓) within various regions of inductions or actually decrease greatly (0) when mesenchyme appears or cell migration occurs. Note that placodes which have both CAMs echo the events seen for neural induction. Just before appearance of glia, a secondary CAM (Ng-CAM) emerges; unlike the other two CAMs, this CAM would not be found in the fate map (*bottom*) before 3.5 days. In the perinatal period, a series of epigenetic modulations occurs: E-to-A conversion for N-CAM and polar redistribution for L-CAM. (*Bottom*) Composite CAM fate map in the chick. The distribution of N-CAM (stippled), L-CAM (slashed), and Ng-CAM (crossed) on tissues of 5- to 14-day embryos is mapped back onto the tissue precursor cells in the blastoderm. Additional regions of N-CAM staining in the early embryo (5 days) are shown by larger dots. In the early embryo, the borders of CAM expression overlap the borders of the germ layers, i.e., derivatives of all three germ layers express both CAMs. At later times overlap is more restricted: N-CAM disappears from somatic ectoderm and from endoderm, except for a population of cells in the lung. L-CAM is expressed on all ectodermal and endodermal epithelia but remains restricted in the mesoderm to epithelial derivatives of the urogenital system. The vertical bar represents the primitive streak (PS); (Ec) intraembryonic and extraembryonic ectoderm; (En) endoderm; (N) nervous system; (No) precordal and chordamesoderm; (S) somite; (Sm) smooth muscle; (Ha) hemangioblastic area.

The two characterized primary CAMs, N-CAM and L-CAM, are both present at low levels on the chick blastoderm before gastrulation. In this species, CAM appearances have not been documented at very early times but in the mouse, L-CAM appears at the two-cell stage (Shirayoshi et al. 1983) and N-CAM appears at later times preceding neural induction. As gastrulation occurs in the chick and cells ingress through the primitive streak, the amount of detectable CAMs decreases (Thiery et al. 1982, 1984, 1985; Edelman et al. 1983; Edelman 1984b; Crossin et al. 1985), presumably reflecting the fact that they have been downregulated or masked. This phenomenon is seen particularly in mesoblast cells that will ultimately give rise to the mesoderm.

Following gastrulation and coincident with neural induction, there is a marked change in the distribution of the two primary CAMs. An increase in immunofluorescent N-CAM staining appears in the region of the neural plate and groove as L-CAM staining disappears.

In conjugate fashion, L-CAM staining is enhanced in the surrounding somatic ectoderm as N-CAM staining slowly diminishes (Fig. 3). The placodes that will give rise to neural structures first express both CAMs but eventually lose L-CAM. The apical ectodermal ridge of the limb bud expresses both CAMs as the limb is induced. Somewhat later, after neurulation, all sites of secondary induction show changes in cell-surface prevalence of N-CAM, L-CAM, or both (Fig. 2). Neural crest cell migration is accompanied by downregulation or masking of cell surface N-CAM (Thiery et al. 1982).

At about E 3 to E 4, the secondary Ng-CAM appears for the first time in the central nervous system (CNS) and peripheral nervous system (PNS) (Grumet and Edelman 1984; Grumet et al. 1984; Thiery et al. 1985) and can be present on the same neuronal surfaces as N-CAM (Grumet et al. 1984). Following the formation of tracts, and particularly in the perinatal period, N-CAM undergoes E-to-A conversion (Chuong and Edelman 1984), Ng-CAM diminishes greatly in amount in all myelinating areas of the CNS (Daniloff et al. 1985), and L-CAM shows polarity modulation in epithelia such as the exocrine pancreas (Gallin et al. 1983).

As already mentioned, when neural crest cells migrate as an ectomesenchyme, they lack detectable surface N-CAM. The N-CAM reappears or is unmasked at sites where these cells form ganglia (Thiery et al. 1982). Similarly, the moving secondary mesenchymal cells destined to induce feather placodes lose N-CAM but regain it in the vicinity of the L-CAM-positive ectoderm as dermal condensations are formed (Chuong and Edelman 1985a,b).

As seen for the primary CAMs in blastoderm and for Ng-CAM and N-CAM on neurons, at certain times, one cell can express two CAMs. In feather formation, and in the induction of pharyngeal and gut appendages, cells from ectoderm or endoderm can also express both L-CAM and N-CAM simultaneously; this is also true of kidney mesoderm (Chuong and Edelman 1985a,b; Crossin et al. 1985). In general, the expressions are dynamic and change so that one or the other CAM disappears during maturation to the adult state. Epithelia and mesenchyme show two different modes (Crossin et al. 1985) of modulation of CAM expression (Table 1). Thus, at early times a mapping of CAM distribution together with a classical map of tissues fates onto the

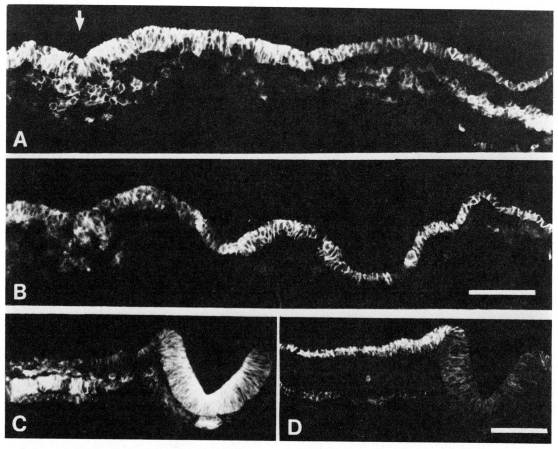

Figure 3. Immunocytochemical staining of the early ectodermal epithelial structures showing transition from distribution of N and L together on the blastoderm to their segregation upon neural induction. (*A,B*) The head fold-stage embryo (stage 6, Hamburger and Hamilton 1951) was sectioned transversely through the primitive streak (arrow) and was stained with affinity-purified anti-N-CAM (*A*) or anti-L-CAM (*B*). (*C,D*) Transverse sections through a five-somite embryo posterior to the last formed somite were stained with affinity-purified anti-N-CAM (*C*) or anti-L-CAM (*D*). Note loss of L-CAM in the neural plate.

Table 1. Modulation Modes of CAM Expression during Chicken Embryogenesis

Mode I: Mesenchyme[a]		Mode II: Epithelia[b]
Ectodermal		**Ectodermal**
$\underline{N \longrightarrow O \longrightarrow N}$		NL→N
neural crest	peripheral nerve ganglia	neural tube placode-derived ganglia
		NL→L
		somatic ectoderm stratum germinativum apical epidermal ridge branchial ectoderm
Mesodermal		NL→N→*
$\underline{N \longrightarrow O \longrightarrow N}$		lens marginal and axial plate of feather
somite	skeletal muscle (end plate only) dermal papilla (feather)	
		NL→L→*
nephrotome	germinal epithelium of gonad gonadal stroma	stratum corneum feather barbule, rachis
splanchnopleure	spleen stroma lamina propria of gut some mesenteries	**Mesodermal**
		N→NL→L
		Wolffian duct mesonephric tubules Mullerian duct
$\underline{N \longrightarrow O \longrightarrow N \longrightarrow *}$		**Endodermal**
somite	chondrocytes	
		NL→L
lateral plate	smooth muscle	trachea gastrointestinal epithelium hepatic duct gall bladder thyroid pharyngeal derivatives
		$\underline{NL}$ parabronchi (lung epithelia)

Mode I shows cyclic changes in N-CAM or disappearance. These transitions occur with movement (see Crossin et al. 1985). Mode II shows replacement of one CAM by another or disappearance.

[a] O indicates low levels of CAM. The original and terminal tissues containing high states of N-CAM are listed in the table; in some cases, the terminal state can be replaced by a differentiation product.

[b] * indicates differentiation products (e.g., keratin, crystallin) with disappearance of CAM.

blastoderm will show a less detailed but congruent map for N-CAM and L-CAM with overlapping CAM distributions at the tissue fate borders. The overlap disappears (Crossin et al. 1985) as the adult state is achieved (Fig. 2).

Neural Histogenesis

Within the overall expression sequence described here, a set of microsequences can be discerned as histogenetic events characterized by cellular differentiation occur. Following neural induction and neurulation, N-CAM is found distributed throughout the nervous system. At E 3.5 in the chick, the secondary Ng-CAM appears on neurons already displaying N-CAM (Daniloff et al. 1985; Thiery et al. 1985). In the CNS, it is seen mainly on extending neurites and is seen only faintly or not at all on cell bodies. This polarity modulation is not seen on cells known to be undergoing migration along guide glia, for example, in the cerebellum or spinal cord (Fig. 4). In regions containing such cells, Ng-CAM is found both on leading processes of the cell bodies and on neurites. A contrasting picture is seen in the PNS: after its appearance, Ng-CAM is found at all times on neurites and cell bodies (Daniloff et al. 1985; Thiery et al. 1985).

Ng-CAM thus shows a detailed microsequence of appearance that follows known sequences of neurite extension and cell migration (Daniloff et al. 1985; Thiery et al. 1985). The order of appearance shown in Table 2 is reproducible from animal to animal. This suggests that local signals related to cellular maturation and possibly to growth factors produced by glial precursors are responsible for both the appearance and the remarkable prevalence modulation of Ng-CAM at the cell surface.

Over the considerable period of time that these Ng-CAM appearances occur, alterations of a lesser degree and with longer time courses can be seen in the prevalence of N-CAM on different cell surfaces. At a time in the macrosequence (Fig. 2) when many neural tracts have been established and myelination is to begin, Ng-

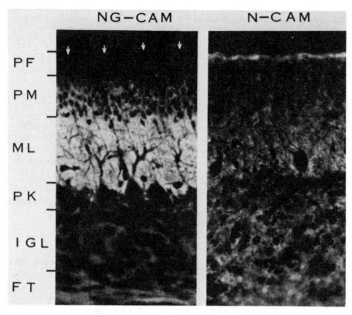

Figure 4. Localization of Ng-CAM and N-CAM on 17-day embryonic chicken cerebellum. Frozen sections were allowed to react sequentially with rabbit anti-Ng-CAM or anti-N-CAM IgG and fluorescein-conjugated goat anti-rabbit IgG. Fluorescence micrographs of comparable fields are shown; treatment with preimmune sera gave no staining. Note that staining for N-CAM but not for Ng-CAM (arrows indicate the cerebellar surface) was visualized in the proliferative zone (PF) of the external granular layer. In contrast, in the molecular layer (ML), staining for Ng-CAM is more intense than for N-CAM. (PM) Premigratory zone; (PK) Purkinje cell; (IGL) internal granular layer; (FT) fiber tract. Bar, 10 μm.

CAM as detected by immunohistochemical methods is downregulated at the cell surface in all CNS tracts that are to become white matter (Table 2). No such downregulation is seen in the PNS. At roughly similar times, N-CAM undergoes E-to-A conversion (Chuong and Edelman 1984; Daniloff et al. 1985) which, as discussed above, leads to increases in binding rates in vitro.

The net result of the various types of cell-surface modulation occurring developmentally for the two neuronal CAMs is a striking change in their relative distributions in most areas of the CNS (Fig. 4; Table 2). But in areas of the CNS that remain capable of forming new connections into adult life as well as in the PNS, the relative distribution of the two CAMs does not change. A set of maps has been constructed (Daniloff et al. 1985) that display the changes and constancies for many brain areas in the chick. At this point, the data strongly suggest that surface modulation events and CAM expressions can occur in relatively small cell populations in a defined order. This conclusion is reinforced by results obtained on microsequences of CAM expression in feather histogenesis which also suggest that the successive signals for CAM expression are sharply localized in time and space.

Periodic Expression Sequences and CAM Cycles in Feather Histogenesis

At all sites of secondary embryonic induction so far examined (Crossin et al. 1985), collections of cells linked by L-CAM (or cells expressing both L-CAM and N-CAM) are found adjacent to collectives of cells linked by N-CAM (Table 1). The potential significance of these primary "CAM couples" in morphogenesis is not yet fully understood, but examination of periodic and hierarchically arranged structures such as the feather shows unequivocally that they are correlated with other cytodifferentiation events.

The induction of feathers occurs through periodic accumulations in the skin of dermal condensations of mesodermal mesenchyme (Sengel 1976). Each of these condensations acts upon ectodermal cells to induce placodes. Feather induction proceeds in major feather fields in rows from the medial to the lateral aspects of the chick skin beginning just lateral to the midline. Periodicity is also seen within each feather site: after placode formation, a dermal papilla is formed involving another couple containing mesodermal (N-CAM) and ectodermal (L-CAM) elements. Afterwards, L-CAM-positive papillar ectoderm produces collar cells that express both CAMs (Chuong and Edelman 1985a, b). These cells will provide the basis for formation of barb ridges and barbule plates through alternating CAM couples, ultimately yielding three hierarchical levels of branching: rachis, ramus, and barbules (Fig. 5).

There is an extraordinary sequence (Chuong and Edelman 1985a,b) of CAM couple expression linked first to cell movement and then to cell division in the formation (Fig. 5) of this hierarchy: (1) Initially, L-linked ectodermal cells are approached by CAM-negative mesenchyme cells moving into the vicinity. Just beneath the ectoderm, the mesenchyme cells become N-

CAM positive and accumulate in lens-shaped aggregates that induce placode formation in the L-CAM linked ectodermal cells. (2) Subsequently, invagination occurs and another CAM couple of mesodermal (N) and ectodermal (L) collectives is seen. (3) A laminin- and fibronectin-positive basement membrane forms between the pulp and the ectoderm in the potential feather filament. This excludes mesodermal mesenchyme from further participation in the processes of barb ridge formation. Papillar ectoderm cells express L-CAM, providing progenitors for collar cells that express both N-CAM and L-CAM. (4) Derivatives of these cells lose N-CAM while retaining L-CAM and form barb ridges by division. In the valleys between the ridges, single or small numbers of basilar cells then express N-CAM while losing L-CAM. This process extends cell by cell up each ridge resulting in the formation of the N-CAM-positive marginal plate. The net result is alternating barb ridges (L-CAM linked) and marginal plates (N-CAM linked). (5) As ridge cells organize into barbule plates linked by L-CAM, a similar process recurs—N-CAM is expressed in cells lying be-

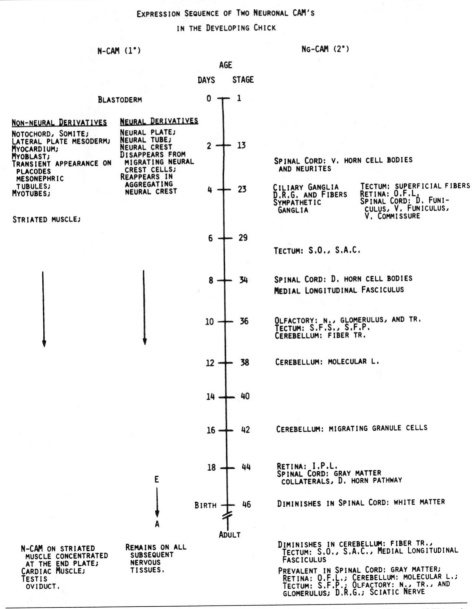

Table 2. Expression Sequence of Two Neuronal CAMs in the Developing Chick Nervous System

(D) Dorsal; (E to A) embryonic to adult conversion of N-CAM; (IPL) inner plexiform layer; (L) layer; (OFL) optic fiber layer; (SAC) stratum album centrale; (SFP) stratum fibrosum superficiale; (SO) stratum opticum; (TR) tract; (V) ventral.
(Reprinted, with permission, from Daniloff et al. 1985.)

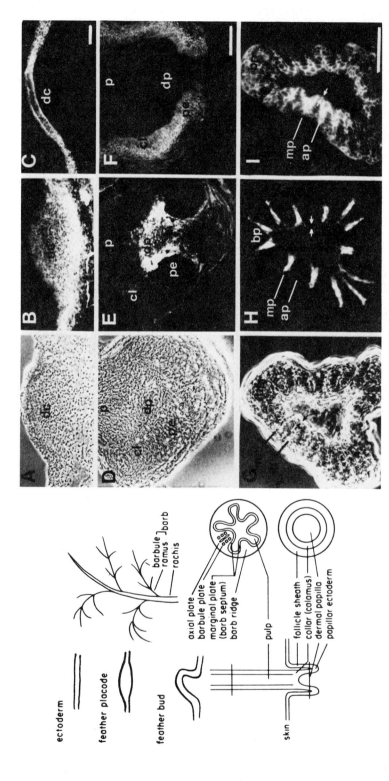

Figure 5. CAM collectives in feather development. Diagram at left shows successive stages and nomenclature in feather development. At the right are comparisons of anti-N-CAM and anti-L-CAM fluorescence at some of these stages. Feather placode from back skin of a stage-33 embryo photograph: (*A*) phase photograph, (*B*) anti-N-CAM fluorescence, and (*C*) anti-L-CAM fluorescence. (DC) Dermal condensation. Feather follicles (*D*) from the wing skin of a newly hatched chicken show intense N-CAM staining in the dermal papilla (*E*) and L-CAM staining in the papillar ectoderm (*F*). The collar epithelium stains for both N-CAM and L-CAM. Feather filaments from the skin of the back of a stage-44 embryo (*G*) show staining for both N-CAM (*H*) and L-CAM (*I*). N-CAM staining occurs in marginal and axial plates and strong L-CAM staining occurs in the barb ridge epithelium; there is thus a periodic appearance of the two CAMs. Arrows point to the basilar layer. (ap) Axial plate; (bp) barbule plate; (mp) marginal plate.

tween each of the future barbules resulting in yet another level of periodically expressed CAM couples. The net result is a series of cellular patterns in which cell collectives expressing L-CAM alternate with those expressing N-CAM at both the secondary barb level and the tertiary barbule level. (6) Finally, after further growth of these structures and extension of the barb ridges into rami, the L-CAM-positive cells keratinize and the N-CAM-positive cells die without keratinization. This leaves extracellular N-CAM, which appears to be resorbed or dispersed, leaving alternate spaces between rami and between barbules yielding the characteristic feather morphology.

A key feature of this histogenetic CAM expression sequence is periodic CAM modulation in a cycle on particular cells, such as those of the inducing mesenchyme. Another important feature is the periodic and successive formation of adjacent N-linked and L-linked cell collectives (CAM couples, Table 1). Finally, there is a striking association of particular cytodifferentiation events with one or another member of a couple as shown, for example, by keratin expression in the L-linked collectives (Table 1; Fig. 5). This association clearly illustrates some modes by which the primary processes of development constituting the driving forces for morphogenesis are linked to the regulatory processes of adhesion. Morphogenetic movement is coupled to expression of a CAM cycle in mesenchymal cells in the original induction. Cell division is associated with the formation of papillar ectoderm and barb ridge formation. Cell death is linked to the existence of N-linked collectives in barb and barbule formation and it comprises the terminal stages of hierarchical pattern formation in the feather; the conjugate process, cell differentiation as marked by keratin expression, is linked to prior morphoregulatory differentiation in L-linked cells.

Causal Significance and CAM Perturbation

While expression sequences reveal correlations between CAM expression, surface modulation, and key morphogenetic events, they do not provide any direct indication of the causal role of the CAMs. Considered a priori, CAM expression could be a cause or an effect; indeed, if CAM expression occurs in cycles and in parallel with other processes it could be both cause and effect. By means of perturbation experiments and by identifying and characterizing CAM genes, one may provide a basis of a search for the appropriate initiating signals and analysis of the causal sequences of cellular controls. In the remaining parts of this paper, I will briefly summarize some observations on CAM perturbation, provide evidence on the nature of CAM genes, and suggest a theoretical model of CAM regulation that can be tested in systems of embryonic induction.

Experiments on perturbation of CAM function or alterations of CAM expression have been carried out in a number of systems (Edelman 1983, 1984a, 1985). Addition of anti-L-CAM antibodies leads to failure of histotypic aggregate formation by liver cells (Bertolotti et al. 1980). Anti-N-CAM antibodies disrupt neural fasciculation in vitro and greatly disrupt layer formation in the chick retina during organ culture (Buskirk et al. 1980). Implanted anti-N-CAM antibodies disrupt retinotectal map formation in the frog (Fraser et al. 1984). When neural cells were transformed by a temperature-sensitive mutant of Rous sarcoma virus (RSV) (Brackenbury et al. 1984; Greenberg et al. 1984), they retained normal morphology, adult N-CAM levels, and normal aggregation behavior at the nonpermissive temperature. At the permissive temperature, the cells transformed and within hours, they downregulated their surface N-CAM and became more mobile.

Perturbation of normal cell–cell interactions in vivo also leads to alteration of CAM expression and distribution. N-CAM is present at the end plate of striated muscles (Rieger et al. 1985) but is absent from the rest of the surface of the myofibril. After cutting the sciatic nerve, the N-CAM disappears from the end plate, anti-N-CAM staining is increased in the cytoplasm and the molecule appears diffusely at the cell surface (Fig. 6). Thus, perturbation of morphology can be accompanied by altered CAM modulation. Alteration of CAM modulations have also been seen in genetic defects. In the mouse mutant *staggerer*, which shows connectional defects in the cerebellum between parallel fibers and Purkinje cells accompanied by extensive granule cell death, E-to-A conversion of N-CAM is greatly delayed in the cerebellum (Edelman and Chuong 1982).

All of these examples indicate that perturbations in CAM binding can lead to altered morphogenesis and that altered morphogenesis can lead to changes in CAM modulation patterns.

CAM Genes

To delineate the causal role of CAMs, it is particularly important to analyze CAM gene expression in relation to that of other products of differentiation. So far, cDNA clones for the two primary CAMs have been obtained (Murray et al. 1984; Gallin et al. 1985) and are being employed for these analyses.

Two independent cDNA clones were derived from enriched mRNA coding for chick N-CAM that had been prepared by immunoprecipitation of polysomes with antibodies to N-CAM (Murray et al. 1984). The plasmids hybridized to two discrete 6- to 7-kb-long RNA species in poly(A)$^+$ mRNA from embryonic chick brain but not to comparably prepared RNA from liver. The two components detected by the cDNA probes for N-CAM appear to correspond separately to the two known polypeptide chains existing for N-CAM; recently, a probe has been found (B.A. Murray et al., unpubl.) that detects only one of the two mRNA species. A cDNA clone for L-CAM has been obtained from poly(A)$^+$ RNA of embryonic liver using the λgt11 expression vector (Gallin et al. 1983). The clone was

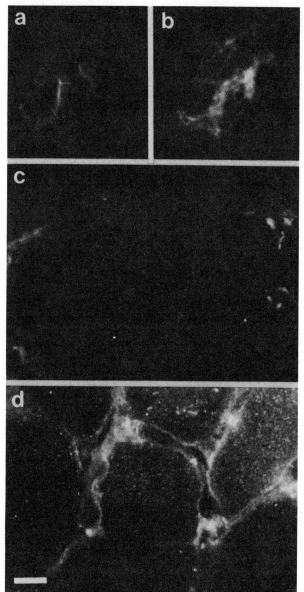

Figure 6. N-CAM at the motor end plate and changes in N-CAM prevalent in muscle after denervation (Rieger et al. 1985). Staining of motor end plate with fluoresceinated α-bungarotoxin (*a*) and rhodamine-labeled anti-N-CAM (*b*). Two weeks after section of the sciatic nerve in one thigh of the mouse, the gastrocnemius muscles from the unperturbed side (*c*) and the denervated side (*d*) were dissected, frozen, and cut in cross section. The sections were stained with fluorescein-labeled anti-N-CAM IgG. Bar, 10 μm. Magnification, 800×.

complementary to a single 4-kb mRNA which was found in tissues from all organs expressing L-CAM but not from those that lacked this CAM.

All of these cDNA probes have been used to explore the multiplicity of genes coding for primary CAMs (Fig. 7). In neither case of the two primary CAMs was there evidence of a large family of genes. Southern blotting analysis with one of the probes for N-CAM detected only one fragment in chicken genomic DNA digested with several restriction enzymes (Murray et al. 1984), suggesting that sequences corresponding to those of the probe are present at most a few times and possibly only once in the chicken genome. Southern blotting analysis of the cDNA probe for L-CAM (Gallin et al. 1985) showed components consistent with the presence of from one to three L-CAM genes.

The results support the notion that the complexity of CAM expression during development is not a reflection of structural gene complexity. Moreover, there does not appear to be a complex processing pathway for message during synthesis of the CAMs. Instead, it appears that initial control of CAM expression is seated in early steps of control of expression of the respective structural genes and possibly of control of the rate of mRNA turnover. The system of perturbation using temperature-sensitive RSV mutants to transform rat cerebellar cell lines (Brackenbury et al. 1984; Greenberg et al. 1984) has been used to analyze the level of the control of expression of N-CAM. The data (R. Brackenbury and G.M. Edelman, unpubl.) are consistent with the notion that most of the control occurs at the transcriptional level.

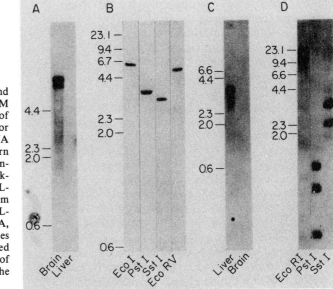

Figure 7. Hybridization analyses of mRNA species and genomic sequences for N-CAM (*A,B*) and L-CAM (*C,D*). (*A*) Hybridization analysis (Northern blot) of poly(A)+ RNA from 9-day embryonic chicken brain or 14-day chicken liver probed with the N-CAM cDNA clone pEC001. (*B*) Hybridization analysis (Southern blot) of adult chicken liver DNA digested with the indicated restriction endonucleases and probed with nick-translated pEC001. (*C*) Hybridization analysis of L-CAM mRNA. Poly(A)+ RNAs of liver and brain from 11-day embryos were hybridized with the pEC301 L-CAM cDNA clone. (*D*) Embryonic chicken liver DNA, digested with the indicated restriction endonucleases and probed with pEC301. Hybridized ^{32}P-labeled probes were detected by autoradiography. Sizes of marker fragments (in kilobase pairs) are shown to the left of each group of lanes.

The Regulator Hypothesis

The evidence on the structure, function, and sequences of CAM expression and that from perturbation experiments suggests that these molecules play a major regulatory role in morphogenesis. This conclusion provides the basis for the regulator hypothesis (Edelman 1984b) which states that, by means of cell-surface modulation, CAMs are key regulators of morphogenetic motion, epithelial integrity, and mesenchymal condensation. The evidence suggests that the genes affecting CAM expression (morphoregulatory genes) act independently of and prior to those controlling tissue-specific differentiation (historegulatory genes) inasmuch as CAM expression in most induced areas does initially precede the expression of cytodifferentiation products. This is consistent with the observation that the expression of CAM types in tissues overlaps different tissue types as seen in a classical fate map (Fig. 2).

At the level of a given kind of cell and its descendants, CAM expression may be viewed as occurring in a cycle (Fig. 8). Traversals of the outer loop of this cycle lead either to the switching on of one or another of the CAM genes or to their switching off. Switching on and off of the same genes is suggested in the case of mesenchymal cells contributing to dermal condensations as well as in the case of neural crest cells. Switching to a different CAM gene is suggested in epithelia (Table 1). The subsequent action of historegulatory genes (inner loop, Fig. 8) is pictured to be the result of signals arising in the new milieux that occur through CAM-dependent cell aggregation, motion, and tissue folding. If the expression of historegulatory genes led to altered cell motion or shape or altered posttransational events, this would alter the effects of subsequent traversals of the outer loop on morphogenesis. One key example directly affecting cells containing N-CAM concerns the genes specifying the enzyme responsible for E-to-A conversion.

Combination of the outer and inner loop of the cycle and the linkage of two such cycles by the formation of CAM couples (see Table 1) could lead to a rich set of effects altering the path of morphogenesis. In considering how two such cycles (each related to a different CAM) might interact, we are brought back to the nature of signals during induction that activate morpho-

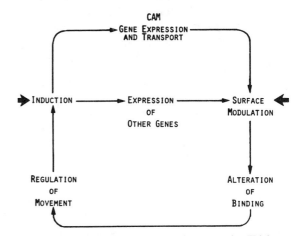

Figure 8. Hypothetical CAM regulatory cycle (Edelman 1984b). Early induction signals (heavy arrow at left) lead to CAM gene expression. Surface modulation (by prevalence changes, polar redistribution on the cell, or chemical changes such as E-to-A conversion) alters the binding rates of cells. This regulates morphogenetic movements which in turn affect embryonic induction or milieu-dependent differentiation. The inductive changes can again affect CAM gene expression as well as the expression of other genes for specific tissues. The heavy arrows at left and right refer to candidate signals for initiation of induction which are still unknown. These signals could result from global surface modulation as a result of CAM binding (*right*) or from release of morphogens affecting induction (*left*).

regulatory and histororegulatory genes. It remains unknown whether these signals are morphogens released by cells linked by a particular CAM (Fig. 8, left large arrow) or whether they are derived from mechanical alterations of the cell surface or cytoskeleton through global cell-surface modulation (Fig. 8, right large arrow).

On the assumption that both types of signals are involved, a concrete testable model for initial feather induction that leads to periodicity can be formulated. The assumption of this model is that CAM morphoregulatory genes act prior to histororegulatory genes and that, at least early in feather formation, the two kinds of genes are independent. Signals to express N-CAM in mesodermal mesenchyme cells are assumed to be triggered by the release of signals from L-CAM-linked cell collectives in the ectoderm (Fig. 5C). Reciprocal signals arising from the resultant N-CAM-linked mesodermal collectives (Fig. 5C) locally turn off the signals from L-linked cells while inducing placode formation in these cells. These reciprocal signals would reach threshold levels only when the N-linked mesodermal collective attained a sufficient size. This model would lead to the emergence of periodic feather induction sites and it prompts the prediction that the blockage of cell linkage by L-CAM in the ectoderm and of linkage by N-CAM in the dermal condensation would block periodic placode formation.

The regulator hypothesis assumes that the cell is the unit of control, that the cell surface is the nexus of control events, that cell adhesion and differentiation order the driving forces of the processes of cell movement and cell division, and that adhesion acts by generating local signals affecting collectives in CAM couples during epigenesis. These signals are assumed to consist of both morphogens and mechanochemical factors such as CAM cross-linkage affecting global cell surface modulation. While the cell is considered to be the unit of control, the unit of induction is considered to be a cell collective of sufficient size linked by a particular primary CAM or combination of CAMs. CAMs are thus hypothesized to be molecules that provide the linkage between the genes and the mechanochemical requirements of epigenetic sequences. Linked CAM cycles occurring in various contexts provide a potential solution to the problem of mechanochemical control of pattern ranging over the various levels from gene through organ back to gene during regulative development. Experimental test of this proposal is now feasible using perturbation of in vitro induction systems by antibodies to CAMs and analysis of gene expression with cDNA probes.

ACKNOWLEDGMENTS

This work was supported by U.S. Public Health Service grants HD-09635, HD-16550, and AM-04256. I wish to thank Dr. Robert Brackenbury for valuable criticism and advice.

REFERENCES

BERTOLOTTI, R., U. RUTISHAUSER, and G.M. EDELMAN. 1980. A cell surface molecule involved in aggregation of embryonic liver cells. *Proc. Natl. Acad. Sci.* **77:** 4831.

BRACKENBURY, R., M.E. GREENBERG, and G.M. EDELMAN. 1984. Phenotypic changes and loss of N-CAM mediated adhesion in transformed embryonic chicken retinal cells. *J. Cell Biol.* **99:** 1944.

BUSKIRK, D.R., J.-P. THIERY, U. RUTISHAUSER, and G.M. EDELMAN. 1980. Antibodies to a neural cell adhesion molecule disrupt histogenesis in cultured chick retinae. *Nature* **285:** 488.

CHUONG, C.-M. and G.M. EDELMAN. 1984. Alterations in neural cell adhesion molecules during development of different regions of the nervous system. *J. Neurosci.* **4:** 2354.

———. 1985a. Expression of cell adhesion molecules in embryonic induction. I. Morphogenesis of nestling feathers. *J. Cell Biol.* **101:** 1009.

———. 1985b. Expression of cell adhesion molecules in embryonic induction. II. Morphogenesis of adult feather. *J. Cell Biol.* **101:** 1027.

CROSSIN, K.L., C.-M. CHUONG, and G.M. EDELMAN. 1985. Expression sequences of cell adhesion molecules. *Proc. Natl. Acad. Sci.* (in press).

CROSSIN, K.L., G.M. EDELMAN, and B.A. CUNNINGHAM. 1984. Mapping of three carbohydrate attachment sites in embryonic and adult forms of the neural cell adhesion molecule (N-CAM). *J. Cell Biol.* **99:** 1848.

CUNNINGHAM, B.A., Y. LEUTZINGER, W.J. GALLIN, B.C. SORKIN, and G.M. EDELMAN. 1984. Linear organization of the liver cell adhesion molecule L-CAM. *Proc. Natl. Acad. Sci.* **81:** 5787.

DANILOFF, J.K., C.-M. CHUONG, G. LEVI, and G.M. EDELMAN. 1985. Differential distribution of cell adhesion molecules during histogenesis of the chick nervous system. *J. Neurosci.* (in press).

EDELMAN, G.M. 1983. Cell adhesion molecules. *Science* **219:** 450.

———. 1984a. Modulation of cell adhesion during induction, histogenesis, and perinatal development of the nervous system. *Annu. Rev. Neurosci.* **7:** 339.

———. 1984b. Cell adhesion and morphogenesis: The regulator hypothesis. *Proc. Natl. Acad. Sci.* **81:** 1460.

———. 1985. Cell adhesion and the molecular processes of morphogenesis. *Annu. Rev. Biochem.* **54:** 135.

EDELMAN, G.M. and C.-M. CHUONG. 1982. Embryonic to adult conversion of neural cell adhesion molecules in normal and staggerer mice. *Proc. Natl. Acad. Sci.* **79:** 7036.

EDELMAN, G.M., W.J. GALLIN, A. DELOUVÉE, B.A. CUNNINGHAM, and J.-P. THIERY. 1983. Early epocal maps of two different cell adhesion molecules. *Proc. Natl. Acad. Sci.* **80:** 4384.

FINNE, J., U. FINNE, H. DEAGOSTINIBAZIN, and C. GORIDIS. 1983. Occurrence of $\alpha 2$-8 linked polysialosyl units in a neural cell adhesion molecule. *Biochem. Biophys. Res. Commun.* **112:** 482.

FRASER, S.E., B.A. MURRAY, C.-M. CHUONG, and G.M. EDELMAN. 1984. Alteration of the retinotectal map in *Xenopus* by antibodies to neural cell adhesion molecules. *Proc. Natl. Acad. Sci.* **81:** 4222.

FRIEDLANDER, D., R. BRACKENBURY, and G.M. EDELMAN. 1985. Conversion of embryonic form to adult forms of N-CAM *in vitro* results from *de novo* synthesis of adult forms. *J. Cell Biol.* **101:** 412.

GALLIN, W.J., G.M. EDELMAN, and B.A. CUNNINGHAM. 1983. Characterization of L-CAM, a major cell adhesion molecule from embryonic liver cells. *Proc. Natl. Acad. Sci.* **80:** 1038.

GALLIN, W.J., E.A. PREDIGER, G.M. EDELMAN, and B.A. CUNNINGHAM. 1985. Isolation of a cDNA clone for the liver cell adhesion molecule (L-CAM). *Proc. Natl. Acad. Sci.* **82:** 2809.

GREENBERG, M.E., R. BRACKENBURY, and G.M. EDELMAN. 1984. Alteration of neural cell adhesion molecule (N-CAM) expression after neuronal cell transformation by Rous sarcoma virus. *Proc. Natl. Acad. Sci.* **81:** 969.

GRUMET, M. and G.M. EDELMAN. 1984. Heterotypic binding between neuronal membrane vesicles and glial cells is mediated by a specific neuron-glial adhesion molecule. *J. Cell Biol.* **98:** 1746.

GRUMET, M., S. HOFFMAN, C.-M. CHUONG, and G.M. EDELMAN. 1984. Polypeptide components and binding functions of neuron-glia adhesion molecules. *Proc. Natl. Acad. Sci.* **81:** 7989.

HAMBURGER, V. and H.L. HAMILTON. 1951. Series of normal stages in the development of the chick embryo. *J. Morphol.* **88:** 49.

HOFFMAN, S. and G.M. EDELMAN. 1983. Kinetics of homophilic binding by E and A forms of the neural cell adhesion molecule. *Proc. Natl. Acad. Sci.* **80:** 5762.

HOFFMAN, S., C.-M. CHUONG, and G.M. EDELMAN. 1984. Evolutionary conservation of key structures and binding functions of neural cell adhesion molecules. *Proc. Natl. Acad. Sci.* **81:** 6881.

HOFFMAN, S., B.C. SORKIN, P.C. WHITE, R. BRACKENBURY, R. MAILHAMMER, U. RUTISHAUSER, B.A. CUNNINGHAM, and G.M. EDELMAN. 1982. Chemical characterization of a neural cell adhesion molecule purified from embryonic brain membrane. *J. Biol. Chem.* **257:** 7720.

JACOBSON, A. 1966. Inductive processes in embryonic development. *Science* **152:** 25.

MURRAY, B.A., J.J. HEMPERLY, W.J. GALLIN, J.S. MACGREGOR, G.M. EDELMAN, and B.A. CUNNINGHAM. 1984. Isolation of cDNA clones for the chicken neural cell adhesion molecules (N-CAM). *Proc. Natl. Acad. Sci.* **81:** 5584.

RIEGER, F., M. GRUMET, and G.M. EDELMAN. 1985. N-CAM at the vertebrate neuromuscular junction. *J. Cell Biol.* **101:** 285.

SENGEL, P. 1976. *Morphogenesis of skin.* Cambridge University Press, New York.

SHIRAYOSHI, Y., T.S. OKADA, and M. TAKEICHI. 1983. The calcium-dependent cell-cell adhesion system regulates inner cell mass formation and cell surface polarization in early mouse development. *Cell* **35:** 631.

SORKIN, B.C., S. HOFFMAN, G.M. EDELMAN, and B.A. CUNNINGHAM. 1984. Sulfation and phosphorylation of the neural cell adhesion molecule N-CAM. *Science* **225:** 1476.

THIERY, J.-P., A. DELOUVÉE, M. GRUMET, and G.M. EDELMAN. 1985. Initial appearance and regional distribution of the neuron-glia cell adhesion molecule (Ng-CAM) in the chick embryo. *J. Cell Biol.* **100:** 442.

THIERY, J.-P., J.-L. DUBAND, U. RUTISHAUSER, and G.M. EDELMAN. 1982. Cell adhesion molecules in early chick embryogenesis. *Proc. Natl. Acad. Sci.* **79:** 6737.

THIERY, J.-P., A. DELOUVÉE, W.J. GALLIN, B.A. CUNNINGHAM, and G.M. EDELMAN. 1984. Ontogenetic expression of cell adhesion molecules: L-CAM is found in epithelia derived from the three primary germ layers. *Dev. Biol.* **102:** 61.

Neurogenesis in Grasshopper and fushi tarazu *Drosophila* Embryos

C.Q. DOE AND C.S. GOODMAN
Department of Biological Sciences, Stanford University, Stanford, California 94305

Neurogenesis in insect embryos begins shortly after gastrulation as an undifferentiated two-dimensional epithelial sheet is transformed into a highly differentiated three-dimensional central nervous system (CNS). Here we discuss our studies aimed at understanding the mechanisms underlying this early phase of neuronal development during which time hundreds of neurons are born in each segmental neuromere and acquire their individual determination. During the next phase of development, each of these neurons begins to express its unique specificity as it extends a process, called a growth cone, which finds and recognizes its appropriate targets (e.g., Goodman et al. 1984; Bastiani et al. 1985).

The generation of the stereotyped pattern of identified neurons from the neurogenic region in each segment involves two major stages. In the first stage, a morphologically uniform two-dimensional epithelial sheet develops into a stereotyped array of neuroblasts (NBs; Wheeler 1893; Bate 1976) and nonneuronal cells. In the second stage, each NB within this array divides asymmetrically to generate a characteristic chain of progeny, called ganglion mother cells (GMCs), each of which divide once more symmetrically to generate pairs of sibling cells (which differentiate into neurons). Thus, each NB contributes an essentially one-dimensional string of paired progeny which transforms the two-dimensional plate of NBs into a three-dimensional CNS (Doe and Goodman 1985a).

In this paper we describe the results of experiments on the grasshopper and *Drosophila* embryos aimed at understanding the mechanisms underlying these two stages of neurogenesis. We begin by discussing the results of laser ablations of individual identified cells in the grasshopper embryo. These results give rise to a model for insect neurogenesis. After briefly describing neurogenesis in *Drosophila*—quite similar to that of the grasshopper—we discuss a genetic manipulation, using the mutant fushi tarazu, aimed at testing one of the predictions of this model.

Neurogenesis in the Grasshopper Embryo

The grasshopper CNS is composed of a brain and a chain of segmental ganglia. Each hemiganglion contains about 1000 neurons, most of which can be individually identified by their unique morphology and synaptic connectivity. Development of these neurons is initiated shortly after gastrulation as the midventral strip of ectoderm becomes a neurogenic region. Little is known about when or how the ventral neurogenic ectoderm becomes separated from the lateral and dorsal ectoderm (which gives rise to epidermis and the sensory neurons of the peripheral nervous system [PNS]); however, several studies in grasshopper and *Drosophila* suggest that by the time neurogenesis begins, a boundary does exist between these regions of ectoderm (e.g., Campos-Ortega 1985; Doe and Goodman 1985b).

In each hemisegment, ~150 neurogenic ectodermal cells (nECs) give rise to a stereotyped pattern of 30 identified NBs (Fig. 1A); the nECs that do not become NBs instead differentiate into nonneuronal cells or die. Each NB generates a stereotyped chain of from 3 to 50 GMCs and then dies; each GMC divides once to generate a pair of sibling identified neurons. In this way, NBs generate characteristic families of from 6 to 100 neurons (e.g., Fig. 1B). A second class of neuronal precursors, called midline precursors (MPs; Bate and Grunewald 1981), arise along the dorsal midline; each of the seven MPs per segment divides symmetrically only once to produce a pair of sibling neurons, much like GMCs.

Each of the NBs and MPs can be individually identified according to its time of formation and position within the array, and according to the identified neurons that it generates (e.g., Goodman and Spitzer 1979; Raper et al. 1983; Taghert and Goodman 1984). Many of our experiments described here focus on a specific lineage: NB 1-1 generates GMC-1 which divides into a pair of sibling cells which differentiate into the identified aCC and pCC neurons (Fig. 1B) (Goodman et al. 1982, 1984). These two neurons have many cell-specific characteristics including their distinctive cell body location and axonal morphology. NB 1-1, GMC-1, and the aCC and pCC neurons are all highly accessible and easily identified, thus providing excellent assays for many of the cell ablation experiments described below (Doe and Goodman 1985b; Kuwada and Goodman 1985).

In summary, the hemisegmental pattern unfolds as ~150 epithelial cells generate 30 NBs; these NBs then produce ~500 GMCs, which divide into ~1000 neurons. What is most remarkable is that, although the epithelial sheet appears morphologically uniform, the NBs, GMCs, and neurons are all uniquely identified cells. Thus, the problem of how neuronal diversity and specificity is generated can be reduced to two more accessible questions, as discussed below.

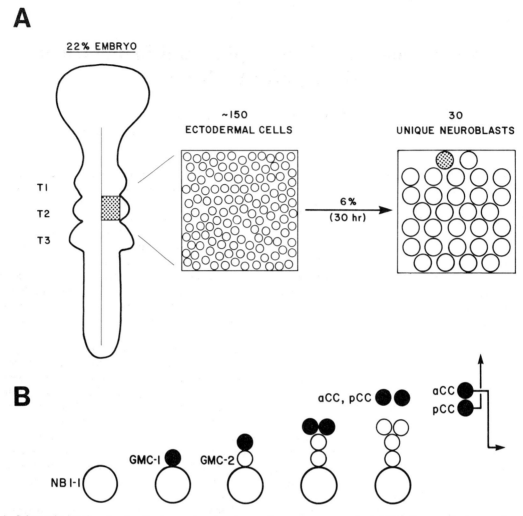

Figure 1. Schematic drawing showing the two major steps in grasshopper neurogenesis. (*A*) The first step is the transformation of a sheet of ectodermal cells into a stereotyped pattern of unique neuronal precursor cells, NBs. (*B*) The second step of neurogenesis is the production of a characteristic family of neurons from each NB. Each NB divides asymmetrically to generate a chain of ganglion mother cells (GMCs), which divide once symmetrically to produce a pair of postmitotic neurons. The lineage of NB 1-1 (shaded in *A*) is illustrated; the identified aCC and pCC neurons are always derived from the first GMC (GMC-1) of NB 1-1, and are used here as indicators of NB 1-1 development (see Goodman et al. 1984).

Alternative Hypotheses for Insect Neurogenesis

How does a seemingly uniform epithelial sheet generate a stereotyped pattern of unique NBs? Several alternative hypotheses could be proposed to answer this question (Fig. 2). (1) The process could be governed by a highly invariant cell lineage (Fig. 2A), such as in the nematode (e.g., Sulston et al. 1983). This alternative would imply that from early in development, individual cells within the epithelial sheet are committed to become specific NBs (Fig. 2A). (2) The epithelium could be subdivided into equivalence groups (e.g., Sulston and White 1980; Kimble 1981). Each equivalence group could thus be commited to generate a specific NB (the dominant fate) and its surrounding nonneuronal support cells, with the fate of each cell within the group being specified by cell interactions (Fig. 2B). (3) The process could be governed entirely by cell interactions within a sheet of equivalent ectodermal cells (Fig. 2C), as proposed for the formation of chaetae in insects (e.g. Wigglesworth 1940; Richelle and Ghysen 1979). All three alternatives can be directly tested by single cell ablation experiments.

How does each NB generate its characteristic family of neurons? The same three alternatives equally apply to this question. Each neuron could be uniquely determined entirely by its lineage from a specific GMC and NB. Alternatively, each neuron could be determined by both its lineage and interactions within an equivalence group. Finally, each neuron could be determined entirely by positional interactions. Here too, all three alternatives can be tested directly.

Differentiation of Neuroblasts

Before describing the experimental tests of the above

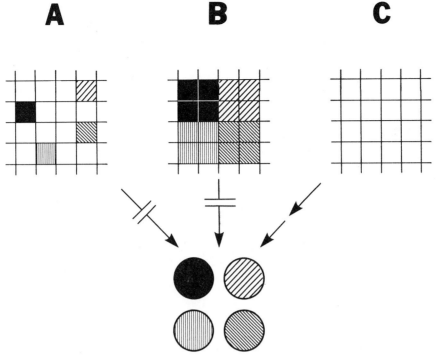

Figure 2. Three possible mechanisms of NB determination. nECs are represented as small squares; NBs derived from these nECs are represented below as circles. Shading of nECs indicates a restriction in fate to the similarly shaded NB. (*A*) Early determination: Individual nECs are determined early in development to produce specific NBs. (*B*) Equivalence groups: Groups of nECs are restricted to form a specific NB and its support cells; interactions amongst the cells within a group allow only one cell to enlarge into the NB, the rest forming nonneuronal support cells around the NB (not shown). (*C*) Cell interactions: Initially all nECs are equivalent; a positional mechanism triggers local interactions among nECs (first arrow), and may provide the information for NB fate specification. Local cell interactions result in only one cell forming a NB (second arrow).

hypotheses, it is important to describe in detail the development of the NBs and their cellular environment. Neurogenesis begins with a sheet of undifferentiated neurogenic ectodermal cells (nECs). The nECs differentiate into a precise number and spatial arrangement of NBs per hemisegment (e.g., 30 in the mesothoracic) as well as a variable number of nonneuronal cells surrounding each NB; in addition, some nECs die. There are two types of nonneuronal cells around each NB: cap cells (Kawamura and Carlson 1962), which tightly cap the ventral surface of each NB, and sheath cells, which surround each NB and extend end feet to both the dorsal and ventral surfaces of the neuroepithelium (Fig. 3). Each NB, its cap cell, and the adjacent sheath cells differentiate concurrently from a group of nECs.

The first sign of NB differentiation from a nEC is the dorsal migration of the nucleus and cytoplasm (Fig. 3B). Often more than one cell in a group begins to shift dorsally, but ultimately only one differentiates into a NB. As one of the nECs becomes a NB, further changes occur. The cell delaminates dorsally, the nucleus assumes a bilobed shape, and the cell enlarges from its initial diameter of 5–10 μm to a diameter of 20–30 μm (Fig. 3C). Unlike NBs in *Drosophila* (Poulson 1950; Hartenstein and Campos-Ortega 1984), NBs in the grasshopper maintain their large size until they degenerate. Approximately 1% (~5 hr) after a NB has en-

larged, it begins its stereotyped series of asymmetric divisions producing a chain of smaller GMCs extending dorsally (Fig. 3D).

It is important to note that all NBs do not develop simultaneously (Fig. 4). While some groups of nECs are differentiating into early arising NBs and their adjacent sheath and cap cells, other nECs continue to divide, ultimately differentiating into the later arising NBs and their nonneuronal neighbors. About 6% of embryonic development (30 hr in the grasshopper) after neurogenesis begins, the neurogenic region has been wholly transformed from a sheet of nECs to a stereotyped pattern of NBs surrounded by sheath and cap cells. Although each NB forms at a particular time, for simplicity we have shown the NBs as early-, middle-, and late-arising groups (Fig. 4). Clearly, the NB pattern does not form in either a simple anterior–posterior or medial–lateral gradient. The important temporal information for most of the experiments described below is that NB 1-1 appears about 15 hours earlier than NB 1-2 (Doe and Goodman 1985a). Thus, for many hours after NB 1-1 forms, a cluster of nECs lies in the adjacent 1-2 location. In addition, we will discuss experiments involving NB 7-3, the final NB to appear, which arises from the last cluster of nECs in the hemisegment, surrounded on all sides by differentiated cells (NBs, sheath, and cap cells).

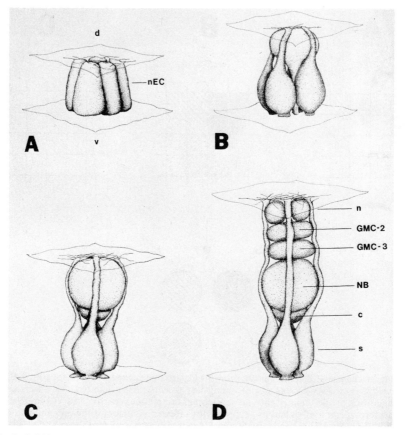

Figure 3. Morphological differentiation of neuroblasts and support cells in the neurogenic region. Drawings based on embryos viewed by transmission and scanning electron microscopy and Lucifer Yellow dye injections of single cells. Dorsal (d) and ventral (v) surfaces of the neuroectoderm are indicated. (*A*) A group of neurogenic ectodermal cells (nECs). (*B*) One nEC has begun differentiating into a NB; the cytoplasm and nucleus have shifted dorsally and delamination of the cell from the ventral surface of the embryo has begun. (*C*) The new NB is fully delaminated from the ventral surface. The surrounding cells, formerly nECs, have differentiated into a cap cell (c; drawn smaller than scale for clarity); and sheath cells (s) extending dorsally around the NB. (*D*) The NB has divided three times, generating three ganglion mother cells (GMCs). The first of the three GMCs (GMC-1) has divided into two postmitotic neurons (n). The second and third GMCs (GMC-2 and GMC-3) have yet to divide.

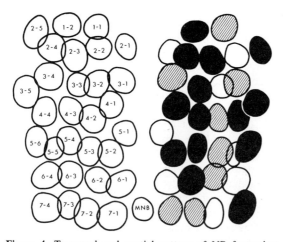

Figure 4. Temporal and spatial pattern of NB formation. Camera lucida drawing of a mature NB pattern; NB identities are indicated in the left hemisegment, time of formation of each NB is indicated by shading in the right hemisegment. Although each NB enlarges at a specific time, for simplicity only three groups are depicted: black NBs form first, cross-hatched NBs form second, and clear NBs form last.

Neuroblast Pattern Is Generated by Cell Interactions

How is the sheet of nECs transformed into a stereotyped pattern of NBs? One hypothesis is that individual nECs are determined to become specific NBs (Fig. 2A). To test this possibility, we ablated one or more nECs in the group at the 7-3 position both in vitro and in ovo at a time before NB 7-3 had appeared (Doe and Goodman 1985b). When only the enlarging nEC is killed (i.e., the cell that is differentiating into NB 7-3), a new NB 7-3 always forms. When all the nECs at the 7-3 position are ablated, NB 7-3 never develops. Three conclusions can be drawn from this and similar experiments (Doe and Goodman 1985b). (1) More than one nEC can become a particular NB. In fact, all nECs in the NB 7-3 group, and probably all nECs, have the potential to form a NB. (2) The enlarging NB 7-3 inhibits the adjacent nECs from assuming a NB fate; when the enlarging NB is ablated, the nECs are released from inhibition and a new NB will form. Lateral inhibition of NB fate around an enlarging NB ensures that only one NB develops per group of nECs, and may play a role in the concomitant differentiation of the surrounding pattern of nonneuronal cells. (3) The fact that NB 7-3 is never made when all nECs are ablated, despite the proximity of many other differentiated cell types, indicates that only nECs can form NBs. This conclusion is supported by many additional experiments, each indicating that only undifferentiated nECs have the potential to form a NB; other differentiated cell types within the neurogenic region (such as NB progeny, cap cells, and sheath cells), as well as lateral ectoderm and mesoderm, all lack the potential to form NBs (Doe and Goodman 1985b).

Neuroblasts Are Determined by Their Position

We know that nECs can replace an ablated neighboring NB, as well as the enlarging NB of their own position (Doe and Goodman 1985b). When nECs from one NB position replace an adjacent (different) NB, what fate does the new NB have? If it is the same as the NB normally made by the nECs (donor fate), then the nECs are probably determined to produce a specific NB (Fig. 2B). If, on the other hand, the new NB has the fate of the ablated NB (host fate), then the regulating nEC was not determined and has assumed the fate of the position in which it enlarged into a NB (Fig. 2C). To distinguish between these possibilities, we performed in ovo laser ablation experiments of NB 1-1. We chose this NB because we could unambiguously assay its first two progeny, the aCC and pCC neurons (Goodman et al. 1982, 1984). In addition, NB 1-1 forms at least 15 hours ahead of NB 1-2, and thus while NB 1-1 produces its first three progeny there exists a group of undifferentiated nECs at the adjacent 1-2 position.

When NB 1-1 is ablated, one of the adjacent nECs from the 1-2 position enlarges to form a NB in the 1-1 position. Thus, two NBs are produced from a single cluster of nECs: one in the initial position (1-2) and one in the new position (1-1). If nECs are restricted to one NB fate (Fig. 2B), we would expect loss of the progeny derived from NB 1-1; if position of enlargement determines NB fate, we would expect the regulated NB to have the 1-1 identity and produce the aCC and pCC neurons. The results of such experiments confirm the latter prediction: the regulated NB produces the aCC and pCC neurons (Figs. 5 and 6C; Doe and Goodman 1985b). The only difference from the control side is that the aCC and pCC neurons from the regulated NB are delayed about 10 hours, approximately the time it takes for NB regulation to occur. These results show that groups of nECs are not determined to form specific NBs (Fig. 2B); rather the fate of the regulated NB is determined shortly after its enlargement, probably due to its position in the neuroectoderm.

Ganglion Mother Cells Are Determined by Their Lineage

Given that NBs are positionally determined within the two-dimensional epithelial sheet, we wondered whether the NB progeny in the third dimension, the GMCs, are determined by position or lineage. We already had a good indication that the NB of origin was critical to neuronal determination; when NB 7-3 was permanently ablated (i.e., later in development when no nECs remained to replace it), two of its characteristic progeny, the identified neurons S1 and S2, were not generated by any other NB (Taghert and Goodman 1984; Taghert et al. 1984). Thus, the two sets of experiments described below were designed to test whether GMCs are determined not only by their NB of origin, but more specifically by their particular cell division of origin.

In the first set of experiments, we used a microelectrode to ablate GMC-1 from NB 1-1 in vitro immediately after its birth, but before GMC-2 had been born (Doe and Goodman 1985b). Cell assays revealed that GMC-2 (born about 5 hr after GMC-1) did not regulate to replace GMC-1; the aCC and pCC neurons, normally derived from GMC-1, were absent (Fig. 6E). As described above, a regulated NB 1-1 will produce a normal GMC-1, and consequently the aCC and pCC neurons, up to 15 hours later than normal. The fact that GMC-2 is born less than 5 hours after ablation of GMC-1 and yet fails to make the aCC and pCC neurons suggests the control of GMC-1 fate by lineage rather than temporally regulated interactions.

In the second set of experiments, we wanted to know whether a regulated NB would begin its lineage anew, even if the ablated NB had already divided, and despite the presence of progeny from the ablated NB. To answer this question, we ablated NB 1-1 after it had produced GMC-1 and GMC-2 (Figs. 5B and 6D). We assayed for duplicated aCC and pCC neurons. When regulation of NB 1-1 occurred, we observed duplication of the aCC and pCC neurons (Fig. 5B). The duplicated neurons were about 20 hours delayed relative to their control contralateral and ipsilateral homologs. In all experiments, the duplicated pCC extended its growth cone anteriorly as normal along the MP1/dMP2 fascicle, whereas the duplicated aCC extended its growth cone anteriorly and laterally along the U fascicle from the next anterior segment (instead of out the U fascicle from its own segment as normal). This departure from the aCC's normal pathway can be attributed to its 20-hour delay in initiating axonogenesis, an interesting experimental result outside the scope of this paper (see C.Q. Doe et al., in prep.). Taken together, these two sets of experiments indicate that GMCs are determined by their cell division of origin from a particular NB.

Neurons Are Determined by Interactions and Lineage

Having shown that GMC-1 from NB 1-1 is determined by its lineage, we wondered what causes its two progeny, the aCC and pCC neurons, to differ from one another (Kuwada and Goodman 1985). When either one of these two GMC-1 progeny is ablated within 5 hours after their birth, the remaining cell differentiates into the pCC (Fig. 6F). However, when ablations are made between 5 and 10 hours after their birth (yet still before their axonogenesis), the remaining cell becomes either the aCC or pCC with equal probability. These results suggest that the sibling progeny from GMC-1: (1) are initially equivalent, (2) become uniquely determined by early interactions, and (3) exhibit a hierarchy of fates whereby the pCC is dominant. The same results were obtained for a pair of sibling progeny produced by another neuronal precursor, MP3 (Kuwada and Goodman 1985). Thus identified neurons are determined by their lineage from a specific NB and GMC, and by their interactions with their sibling from the same GMC; the pairs of GMC progeny form equivalence groups within which cell interactions determine their ultimate fate according to a hierarchy.

Conclusions from Cell Ablation Studies

These cell ablation studies of neurogenesis in the grasshopper embryo lead to the following conclusions:

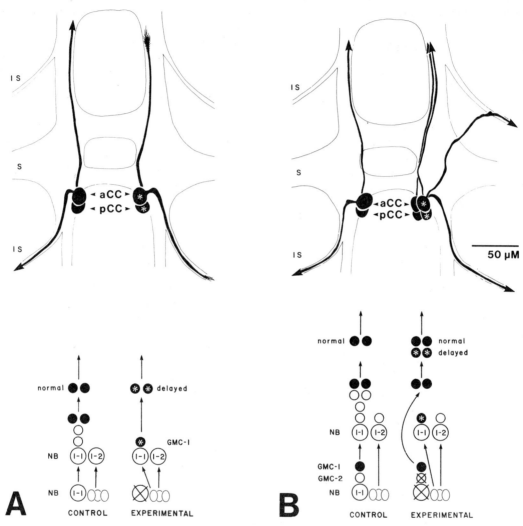

Figure 5. Regulation of neuroblasts and duplication of identified neurons. (*A*) The identity of a developing NB is apparently determined by its position of enlargement in the neuroectoderm. When NB 1-1 is ablated before it begins to divide, one nEC from the adjacent 1-2 position moves over and enlarges into a NB (shown schematically at the bottom of the figure). This new NB is shown to have the determination of NB 1-1 (the host position), rather than NB 1-2 (the donor position) by the development of aCC and pCC neurons (blackened with asterisk on the experimental side), which are derived from GMC-1 of NB 1-1. As expected, the aCC and pCC neurons on the experimental side are about 10 hr delayed relative to the contralateral control aCC and pCC neurons, due to the time it takes for regulation of the ablated NB 1-1. (*B*) Duplicated aCC and pCC neurons produced by a regulated NB 1-1. Schematic at the bottom of the figure illustrates the ablation protocol: on the experimental side NB 1-1 was ablated (as was GMC-2) but GMC-1 (blackened) was left untouched. GMC-1 ultimately develops into normal aCC and pCC neurons (blackened), as does the GMC-1 in the control hemisegment. Regulation of the killed NB 1-1 occurs as a nEC from the adjacent 1-2 position enlarges into a NB. This new NB begins the NB 1-1 lineage anew, despite the presence of GMC-1 progeny from the ablated NB 1-1, as seen by the existence of a duplicate pair of aCC and pCC neurons (blackened with asterisk). The camera lucida (top of figure) shows only the proximal part of each neuron; their axons extend beyond the figure borders. Both pCCs extend anteriorly as normal in the MP1/dMP2 fascicle; the normal aCC follows its usual path out the intersegmental nerve (IS) of its own segment, whereas the duplicate aCC grows out the IS nerve of the adjacent anterior segment. (S) segmental nerve. Thin line represents outline of neuropil region. Scale, 50 μm.

(1) The neuroblast pattern is generated by two types of cell interactions, one positional giving rise to the enlargement of NBs in a specific pattern, and the other local giving rise to the inhibition of neighboring nECs; (2) NBs are determined by their position; (3) GMCs are determined by their lineage; (4) neurons are determined by their lineage and interactions with their sibling. Each of these conclusions is discussed below.

Within the sheet of undifferentiated nECs, any cell can become a NB. Cell interactions between the ~150 equivalent nECs allow 30 cells to enlarge into NBs. Each NB enlarges at a precise location at a specific

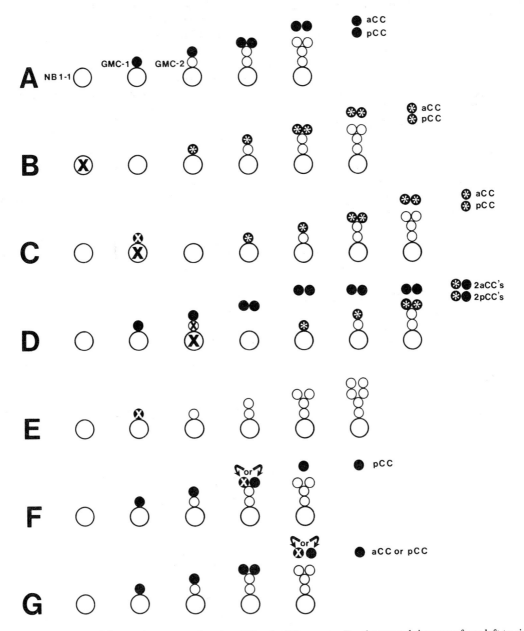

Figure 6. Cell lineage and cell interactions generating the aCC and pCC neurons. Developmental time goes from left to right in approximately 5-hr intervals between drawings. The aCC and pCC, as well as their precursor GMC-1, are blackened; when derived from a regulated NB 1-1 these cells are marked with an asterisk (*). (A) Normal lineage of aCC and pCC neurons. (B) Ablation of NB 1-1. An adjacent nEC enlarges to replace the NB; it produces normal aCC and pCC neurons, despite being delayed up to 10 hr. (C) Ablation of both NB 1-1 and GMC-1 results in regulation and the production of normal aCC and pCC neurons delayed 5 hr more than those in B. (D) Duplication of aCC and pCC neurons following ablation of NB 1-1 and GMC-2, but not GMC-1. The original GMC-1 produces normal aCC and pCC neurons, whereas the regulated NB 1-1 begins the lineage anew and produces a second duplicate set of aCC and pCC neurons delayed 15-20 hr relative to the normal aCC and pCC neurons. (E) Ablation of GMC-1 immediately after its birth; GMC-2 does not make aCC and pCC neurons, despite being born only 5 hr later than GMC-1. (F) Ablation of either of the progeny of GMC-1 within 5 hr of their birth. The remaining cell differentiates into the pCC neuron. (G) Ablation of either of the progeny of GMC-1 5-10 hr after their birth; the remaining cell shows an equal probability of forming an aCC or pCC neuron. See text for discussion of these results.

time. Cell ablation experiments suggest that the NB fate is dominant in a fate hierarchy. Evidently, once a cell enlarges to become a NB, it inhibits the cells around it from becoming NBs (Fig. 7A,B); they then differentiate into various nonneuronal cells. Released from inhibition by NB ablation, one of the nearby nECs will differentiate as a NB (Fig. 7C,D); in contrast, the neighboring differentiated nonneuronal cells cannot become NBs. Although our results suggest that many if not all nECs are equivalent, we cannot rule out the possibility that boundaries exist within the ~150 nECs restricting their fate, for example, either at segment borders or at other borders within a segment. We have observed regulation of six NBs from five groups of neighboring nECs, and it is only in these regions that we can be sure of nEC equivalence: from 1-2 to 1-1; from 3-4 to 3-5; from 3-3 to 4-4; from 7-2 to 7-1; from 7-3 to 7-2; and from 7-3 to 7-4.

Similar short-range inhibition of cell differentiation has been proposed in both plants (Schoute 1913; Wilcox et al. 1973; Mitchison 1977) and animals (Wigglesworth 1940; Moscoso del Prado and Garcia-Bellido 1984). It is interesting that mechanisms proposed for the development of insect macrochaetae mother cells (Richelle and Ghysen 1979), which generate peripheral sensory neurons, can also be used to explain the development of insect neuroblasts, which generate central neurons, as discussed below. (1) Positional interactions within an epithelial sheet give groups of cells in specific positions a high probability of forming a NB or chaeta mother cell. (2) Cell interactions amongst one such group of cells results in one cell differentiating into a NB or chaeta. (3) The differentiated cell inhibits its neighbors from undertaking an identical developmental pathway; these cells assume nonneuronal fates. (4) Ultimately, a highly stereotyped pattern of NBs or chaetae is formed. An apparent difference is that, in contrast to the pattern of macrochaetae, the NB pattern is more densely packed and the observed inhibition confined to nearest neighbors. Interestingly, when the scute gene is "derepressed" (Moscoso del Prado and Garcia-Bellido 1984), a greater number of macrochaetae are produced. Clonal analysis suggests that the density of macrochaeta mother cells at their time of determination is one per one to two cell diameters (Moscoso del Prado [1982], cited in Moscoso del Prado and Garcia-Bellido 1984), quite similar to the density of NBs.

It appears that each NB is assigned its unique identity according to its position of enlargement within the neurogenic epithelium. A nEC can be "transplanted" to an adjacent NB position by in ovo ablation of that NB. The nEC then enlarges in place of the ablated NB and differentiates into the positionally correct NB, producing neurons characteristic of the host rather than donor position. That each NB is uniquely determined to generate a specific family of neurons is clear. How such a pattern of 30 unique NBs is specified is currently unknown.

Determined by its position, each NB goes on to generate its characteristic chain of GMCs by an invariant cell lineage. For example, our results indicate that GMC-1 from NB 1-1 is born intrinsically determined, i.e., as a consequence of its lineage, rather than as a

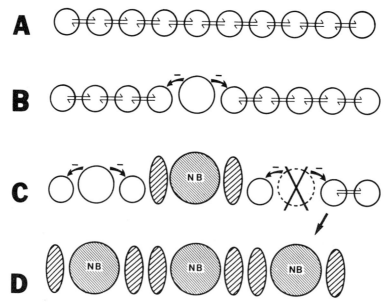

Figure 7. Schematic drawing illustrating the local inhibition of adjacent nECs by an enlarging NB. (*A*) All nECs are equivalent and interact with each other (half-arrows). (*B*) One cell begins to enlarge into a NB and inhibits the adjacent cells from enlarging (arrows). (*C*) The middle cells have differentiated (NB, narrow cross-hatching; support cells, wide cross-hatching); the lateral NBs have begun enlarging and inhibit the adjacent nECs (arrows). If an enlarging NB is ablated (dashed circle on the right), the adjacent nECs are released from inhibition and one will enlarge to replace it. (*D*) Ultimately all cells differentiate into either NBs or nonneuronal support cells.

consequence of its spatiotemporal environment (Doe and Goodman 1985b). One of the best confirmations of this conclusion is the observation of duplicated neurons (Fig. 5). Even though the ablated NB 1-1 has already generated GMC-1, the regulated NB begins its lineage anew. Although delayed some 20 hours relative to normal, the regulated NB 1-1 nevertheless generates a new GMC-1, thus giving rise to two pairs of aCC and pCC neurons, both of which survive and differentiate.

Finally, each GMC generates a pair of equivalent progeny, the fate of each individual neuron being determined by both its GMC of origin and interactions with its sibling (Kuwada and Goodman 1985). The pairs of GMC progeny act as equivalence groups in which each of the two neurons becomes uniquely determined according to a fate hierarchy. For example, if GMC-1 from NB 1-1 is ablated, the aCC and pCC neurons never form, indicating that only the progeny from GMC-1 can become these two neurons. However, shortly after their birth, if either one of the two GMC-1 progeny is ablated, the remaining cell becomes the pCC neuron, indicating that pCC is the dominant fate. Thus, a complex pattern of both cell lineage and cell interactions is involved in determining each individual identified neuron.

Neurogenesis in Wild-type *Drosophila* Embryos

Whereas the grasshopper embryo is ideal for the cellular studies described above, the *Drosophila* embryo has obvious attributes for a genetic approach. Fortunately, there exist many similarities between *Drosophila* and grasshopper embryos, including many of the same identified neurons (Goodman et al. 1984; Thomas et al. 1984; Bastiani et al. 1985).

Given the remarkable homologies in the patterns of identified neurons and axon fasciculation in grasshopper and *Drosophila*, it is not surprising that there is also considerable similarity in the earlier events of neurogenesis giving rise to these neurons. The *Drosophila* embryo has a neurogenic region which initially contains 60 cells per hemisegment (Campos-Ortega 1985), but several rounds of mitosis during the formation of the NB pattern increase this number of cells at least twofold (Hartenstein and Campos-Ortega 1985). The stereotyped pattern of NBs per hemisegment (13–16 NBs, Hartenstein and Campos-Ortega 1984; ∼20, C.Q. Doe and C.S. Goodman, in prep.) develops from these neurogenic ectodermal cells (nECs). Many of the remaining cells of the neurogenic region become sheath cells which later contribute to the ventral epidermis (Technau and Campos-Ortega 1985).

As in the grasshopper, the basic neuromere in the *Drosophila* embryo is segmental (not parasegmental) as identified by the stereotyped pattern of NBs that produces it. The posteriormost NB in the pattern (the median NB or MNB) lies directly between the tracheal pits when they first become visible (C.Q. Doe and C.S. Goodman, unpubl.); the tracheal pits apparently demarcate the future segment border (Martinez-Arias and Lawrence 1985).

In summary, neurogenesis in the *Drosophila* mesothoracic hemisegment unfolds as initially ∼60 nECs generate ∼20 NBs; these NBs then produce ∼125 GMCs, which divide into ∼250 neurons.

Although there are many similarities between grasshopper and *Drosophila* neurogenesis, there are also a few small differences. For example, there are fewer cells in the neurogenic region in *Drosophila*. Furthermore, the NBs appear to shrink in size as development proceeds (Poulson 1950), rather than maintain their original size as in grasshopper (Doe and Goodman 1985a). However, because of the striking similarities in neurogenesis between the two insects, we have begun to use *Drosophila* genetics to help understand the mechanisms uncovered by our cellular analysis of the grasshopper embryo.

Neurogenesis in fushi tarazu *Drosophila* Embryos

The cellular analysis of insect neurogenesis described above suggests that the precise pattern of NBs in both grasshopper and *Drosophila* is controlled by two kinds of cell interactions among the nECs in each hemisegment: the first positional to determine the spatial and temporal pattern of NB enlargement, and the second local to inhibit cells immediately adjacent to each NB from also becoming NBs.

One way to test this model is to manipulate the size of the segmental primordium, and examine the consequences on pattern formation in the nervous system as compared with the epidermis. The model makes two predictions about how the pattern of neurogenesis will respond to altered segmental primordia. First, the model predicts that the epidermis, CNS, and PNS all use the same positional cues for their pattern formation, and thus the nervous system and epidermis will show parallel shifts in their patterns. Second, the model predicts that as the segmental field increases, the local inhibition around each NB will no longer include all cells in the neurogenic region and thus extra (duplicate) NBs will be generated.

The first prediction, for example, can be tested by simply examining different sized segments in the grasshopper embryo. The NB pattern can compensate for these slight size alterations, as the same pattern and number of NBs is produced in both the relatively smaller (∼20%) A8 segment as compared with the T2 segment (Doe and Goodman 1985a).

This first prediction can be better tested, however, by examining *Drosophila* mutants that more dramatically alter the size of the segmental primordium. For example, if a segment is twice its normal size, there are three possible consequences for the pattern of neurogenesis. First, the model predicts that the pattern will be relatively normal but simply stretched out over twice the distance. Alternatively, the pattern in the double-sized segment could contain two segments worth of precursors in tandem array. Third, the pattern could be totally disrupted.

To test these alternatives, we used a null allele of the mutant fushi tarazu (ftz^{w20}; Wakimoto et al. 1984) to create double-sized embryonic segments. If the pattern of neurogenesis is stretched out over twice the distance as predicted, then the model's second prediction can also be tested by looking for duplications interspersed in the normal pattern.

The embryonic lethal phenotype of ftz^{w20}/ftz^{w20} homozygotes (*ftz* embryos) is a loss of alternate segments, with a concomitant approximately twofold enlargement of the remaining segments (Wakimoto et al. 1984). These expanded segments exhibit the phenotype of a specific single segment, not the composite features of two segments (Wakimoto et al. 1984). Interestingly, cuticular analysis shows one enlarged ventral denticle belt and one pair of tracheal pits per double-sized segment (Wakimoto et al. 1984), suggesting that both wild-type and double-sized segments are using the same positional cues to control cuticular patterns.

We examined young *ftz* embryos (0–8 hr) with the DNA stain Hoechst 33258 and observed no signs of cell death throughout the formation of the NB pattern. The first obvious sign of segmentation (ectodermal grooves) reveals double-sized segments with twice the number of cells; for example, at 7 hours, the length of the neurogenic region in a wild-type segment is ~12 cells whereas the length in a *ftz* segment is 22–24 cells. At no point do we see further subdivision of the double-sized segment in *ftz* embryos.

We have examined neurogenesis in the CNS and PNS of both wild-type and *ftz* embryos using Nomarski interference contrast optics and the monoclonal antibody (MAb) SOX2, which stains a subset of identified neurons in the CNS and all neurons in the PNS (Goodman et al. 1984) (Fig. 8).

In the CNS, cell bodies of identified neurons are located in the same relative positions within both wild-type (i.e., normal-sized) and double-sized *ftz* segments (Fig. 8). The segmental pattern of NBs also appears to be stretched out to cover twice the distance in *ftz* embryos. For example, the single median NB (MNB) can be observed at the posterior end of each neuromere, between the pair of tracheal pits. However, the pattern is not totally normal. We often observe extra NBs interposed in the pattern; instead of the normal number of six NBs in the medial column, *ftz* segments often have seven to eight NBs in this column (Fig. 8). Furthermore, we have occasionally observed duplication of certain neurons in the CNS. Thus, consistent with the predictions of the model, the most common phenotype in a *ftz* neuromere is a stretched out segmental pattern of NBs and neurons, with occasional duplications.

The same phenotype is observed in the PNS. The same relative positions, and, in most cases, the same number of sensory neurons can be observed in both wild-type and double-sized *ftz* segments (Fig. 8). Because the segment is almost twice as long as wild-type, the groups of sensory neurons in *ftz* segments are spaced twice as far apart. However, within specific groups of sensory neurons, the individual neurons are typically clustered with their normal number and density. For example, the lateral group of five chordotonal neurons (lch 5; A. Ghysen et al., in prep.) occur as a group in their same relative segmental position in *ftz* embryos. Even in *ftz* segments, this lch 5 group always contains the normal number of five neurons spaced at the wild-type density (rather than stretched out), suggesting that these sensory neurons may be initially determined as a unit rather than as individual neurons. A different group of dorsal sensory neurons, the des cluster, occasionally contains extra neurons. Finally, there is variability in the phenotype of sensory neurons; some *ftz* embryos show less spacing (along the anterior-posterior axis) between sensory cells.

Thus in both the CNS and PNS of *ftz* segments, we find a single segmental pattern of neurons stretched out over a double-sized segment, similar to the stretched out cuticular pattern (Wakimoto et al. 1984).

Although neurons in the CNS and PNS of *ftz* embryos appear to be born in their appropriate segmental positions, their axons often extend along highly abnormal pathways. This aspect of the *ftz* phenotype may be due to the greater spacing between developing neurons in both the CNS and PNS, resulting in abnormal cellular interactions and consequently abnormal axonal pathways.

It appears that the positional cues used by wild-type segments are also used by the double-sized *ftz* segments. These positional cues are used by neuronal precursor cells in both the CNS and PNS, as well as the epidermal cells from which the cuticle derives. The fact that the CNS and PNS patterns develop fairly accurately in double-sized segments indicates that either the precursors are responding to an expanded set of positional cues, or that the cell interactions leading to the ultimate precursor cell pattern can occur relatively normally despite a twofold increase in the field of cells.

Genetic Predictions

The model for insect neurogenesis described in this paper, based on cellular and genetic analysis of grasshopper and *Drosophila* embryos, leads to the prediction of four kinds of genes involved in neurogenesis:

(1) *Positional information:* It is likely that a common genetic mechanism is used to generate positional determination throughout the ectoderm, be it by some segmental pattern of positional information or alternatively by cell interactions leading to positional determination (e.g., Garcia-Bellido 1981). In either case, the nervous system appears to use the same positional cues as does the epidermis. These different tissues respond to this positional determination in tissue and cell-specific ways.

(2) *Local interactions:* The model would lead to the prediction of genes involved in the inhibition of neigh-

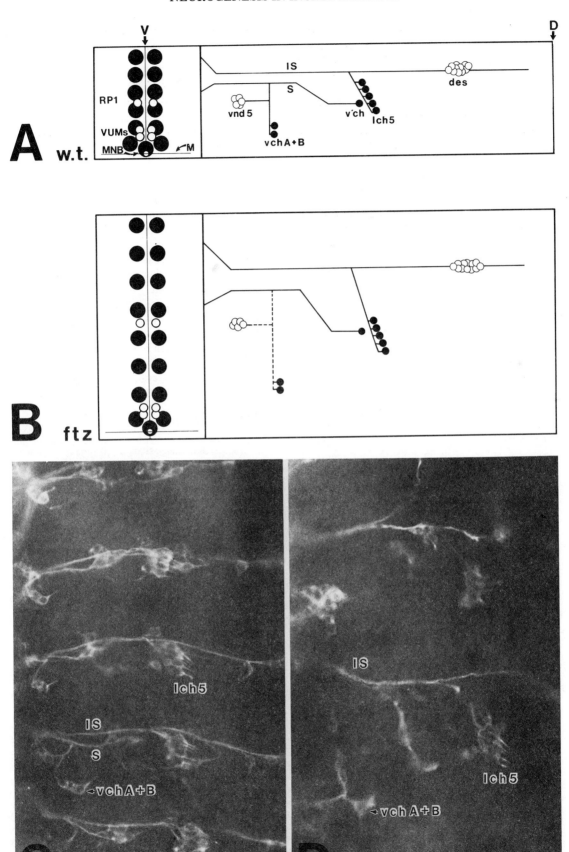

Figure 8. (See next page for legend.)

boring nECs by NBs leading to their subsequent differentiation into nonneuronal fates. The neurogenic genes in *Drosophila* (e.g. Lehmann et al. 1983; Campos-Ortega 1985) are likely candidates for these genes; mutation in any of these genes results in an increase in NBs at the expense of nonneuronal cells. A similar inhibitory interaction has been observed in the development of the precursor cells for the sensory neurons in the PNS (e.g., Richelle and Ghysen 1979). Interestingly, we observe hypertrophy of neurons in both the PNS and CNS in a number of alleles of one of these neurogenic genes, Delta (C.Q. Doe et al., in prep).

(3) NB and GMC determination: For some of the most important events of insect neurogenesis, the determination of the ~20 unique NBs in each hemisegment and the lineal determination of their chain of GMC progeny, there have been no candidate genes. One hypothesis would predict that, just as the bithorax and Antennapedia gene complexes control the determination of 14 segments (e.g., Lewis 1978; Bender et al. 1983; Lawrence and Morata 1983), so an analogous set of gene complexes might control the determination of the ~20 NBs in each hemisegment, and the subsequent determination of their ~125 GMC progeny. It will be interesting to see if this prediction holds true.

(4) Segmental specializations: Although little has been said here about the development of segmental specializations in the insect CNS, there are striking differences between segments in both the number and phenotypes of the differentiated neurons. The differences in the neurons in each segmental neuromere are controlled by differences in the number of neuronal precursor cells (Doe and Goodman 1985a), their number of divisions, the pattern of cell death of their progeny (Goodman and Bate 1981), and the differentiation of their progeny (e.g., Bastiani et al. 1984). It has been known for many years that the homeotic genes of the bithorax and Antennapedia complexes control segmental determination in the epidermis (e.g., Lewis 1978; Kaufman et al. 1980). Analysis using in situ hybridization (Akam 1983; Levine et al. 1983) and more recently using antibodies against fusion proteins (White and Wilcox 1984; Beachy et al. 1985) has revealed substantial homeotic gene expression in a segment-specific fashion in the CNS. It is quite likely that the homeotic genes are directly responsible for segmental specializations in the development of the CNS.

ACKNOWLEDGMENTS

We thank Dr. Thomas Kaufman for the mutant ftz^{w20}. This work was supported by NICHHD predoctoral traineeship to C.Q.D. and National Institutes of Health grant to C.S.G.

REFERENCES

Akam, M. 1983. The location of *Ultrabithorax* transcripts in *Drosophila* tissue sections. *EMBO J.* **2**: 2075.

Bastiani, M.J., K.G. Pearson, and C.S. Goodman. 1984. From embryonic fascicles to adult tracts: Organization of neuropile from a developmental perspective. *J. Exp. Biol.* **112**: 45.

Bastiani, M.J., C.Q. Doe, S.L. Helfand, and C.S. Goodman. 1985. Neuronal specificity and growth cone guidance in grasshopper and *Drosophila* embryos. *Trends Neurosci.* **8**: 257.

Bate, C.M. 1976. Embryogenesis of an insect nervous system. I. A map of the thoracic and abdominal neuroblasts in *Locusta migratoria*. *J. Embryol. Exp. Morphol.* **35**: 107.

Bate, C.M. and E.B. Grunewald. 1981. Embryogenesis of an insect nervous system. II. A second class of neuron precursor cells and the origin of the intersegmental connectives. *J. Embryol. Exp. Morphol.* **61**: 317.

Beachy, P.A., S.L. Helfand, and D.S. Hogness. 1985. Segmental distribution of bithorax complex proteins during *Drosophila* development. *Nature* **313**: 545.

Bender, W., M. Akam, F. Karch, P.A. Beachy, M. Peifer, P. Spierer, E.B. Lewis, and D.S. Hogness. 1983. Molecular genetics of the bithorax complex in *Drosophila melanogaster*. *Science* **221**: 23.

Campos-Ortega, J.A. 1985. Genetics of early neurogenesis in *Drosophila melanogaster*. *Trends Neurosci.* **8**: 245.

Doe, C.Q. and C.S. Goodman. 1985a. Early events in insect neurogenesis. I. Development and segmental differences in the pattern of neuronal precursor cells. *Dev. Biol.* **111**: 193.

———. 1985b. Early events in insect neurogenesis. II. The role of cell interactions and cell lineage in the determination of neuronal precursor cells. *Dev. Biol.* **111**: 206.

Garcia-Bellido, A. 1981. From the gene to the pattern: Chaeta differentiation. In *Cellular controls in differentiation* (ed. C.W. Lloyd and D.A. Rees), p. 281. Academic Press, New York.

Goodman, C.S. and M. Bate. 1981. Neuronal development in the grasshopper. *Trends Neurosci.* **4**: 163.

Goodman, C.S. and N.C. Spitzer. 1979. Embryonic devel-

Figure 8. The pattern of neuroblasts (NBs) and neurons in the CNS and PNS of wild-type and fushi tarazu (*ftz*) *Drosophila* embryos. (*A,B*) Schematic diagrams of the CNS and PNS in a wild-type segment (*A*) and double-sized *ftz* segment (*B*). (*A*) To the left are both hemisegments of the ventral neurogenic region (V marks ventral midline; anterior at top) which shows the medial column of NBs (blackened), the median NB (MNB), several identified neurons (RP1, VUMs), and the two muscle pioneers (M). To the right of the neurogenic region are shown some of the ventral (v), lateral (l), and dorsal (d) sensory neurons (D marks dorsal midline) (nomenclature for sensory neurons from A. Ghysen et al., in prep.). (*B*) The NBs and neurons in the double-sized *ftz* segment appear to be stretched out to cover twice the distance in a relatively normal pattern. However, we often observe extra NBs interposed in the pattern, and occasionally observe duplicate neurons in both the CNS and PNS. The dotted lines refer to typically abnormal axon pathways in the *ftz* embryos. (*C,D*) Photographs of sensory neurons in whole-mount preparations of a wild-type (*C*) and *ftz* (*D*) embryo stained with the SOX2 MAb and rhodamine-conjugated second antibody. The cluster of five lateral chordotonal neurons (lch 5) and the two ventral chordotonal neurons (vch A and B) are located in relatively normal positions in the stretched out *ftz* segments. For further discussion, see text. (IS) Intersegmental nerve; (S) segmental nerve. Scale bar, 20 μm.

opment of identified neurones: Differentiation from neuroblast to neurone. *Nature* **280:** 208.
GOODMAN, C.S., J.A. RAPER, R.K. HO, and S. CHANG. 1982. Pathfinding of neuronal growth cones in grasshopper embryos. In *Developmental order: Its origin and regulation* (ed. S. Subtelny and P.B. Green), p. 275. A.R. Liss, New York.
GOODMAN, C.S., M.J. BASTIANI, C.Q. DOE, S. DU LAC, S.L. HELFAND, J.Y. KUWADA, and J.B. THOMAS. 1984. Cell recognition during neuronal development. *Science* **225:** 1271.
HARTENSTEIN, V. and J.A. CAMPOS-ORTEGA. 1984. Early neurogenesis in wild-type *Drosophila melanogaster*. *Wilhelm Roux's Arch. Dev. Biol.* **193:** 308.
———. 1985. Fate-mapping in wild-type *Drosophila melanogaster*. I. The spatio-temporal pattern of embryonic cell divisions. *Wilhelm Roux's Arch Dev. Biol.* **194:** 181.
KAUFMAN, T.C., R. LEWIS, and B. WAKIMOTO. 1980. Cytogenetic analysis of chromosome 3 in *Drosophila melanogaster*: The homoeotic gene complex in polytene interval 84A-B. *Genetics* **94:** 115.
KAWAMURA, K. and J.G. CARLSON. 1962. Studies on cytokinesis in neuroblasts of the grasshopper, *Chortophaga viridifasciata* (de geer). III. Factors determining the location of the cleavage furrow. *Exp. Cell Res.* **26:** 411.
KIMBLE, J. 1981. Alterations in cell lineage following laser ablation of cells in the somatic gonad of *Caenorhabditis elegans*. *Dev. Biol.* **87:** 286.
KUWADA, J.Y. and C.S. GOODMAN. 1985. Neuronal determination during embryonic development of the grasshopper nervous system. *Dev. Biol.* **110:** 114.
LAWRENCE, P.A. and G. MORATA. 1983. The elements of the bithorax complex. *Cell* **35:** 595.
LEHMANN, R., F. JIMENEZ, U. DIETRICH, and J.A. CAMPOS-ORTEGA. 1983. On the phenotype and development of mutants of early neurogenesis in *Drosophila melanogaster*. *Wilhelm Roux's Arch. Dev. Biol.* **192:** 62.
LEVINE, M., E. HAFEN, R.L. GARBER, and W.J. GEHRING. 1983. Spatial distribution of Antennapedia transcripts during *Drosophila* development. *EMBO J.* **2:** 2037.
LEWIS, E.B. 1978. A gene complex controlling segmentation in *Drosophila*. *Nature* **276:** 565.
MARTINEZ-ARIAS, A. and P.A. LAWRENCE. 1985. Parasegments and compartments in the *Drosophila* embryo. *Nature* **313:** 639.
MITCHISON, G.J. 1977. Phyllotaxis and the Fibonacci series. *Science* **196:** 270.
MOSCOSO DEL PRADO, J. and A. GARCIA-BELLIDO. 1984. Cell interactions in the generation of chaetae pattern in *Drosophila*. *Wilhelm Roux's Arch. Dev. Biol.* **193:** 246.

POULSON, D.F. 1950. Histogenesis, organogenesis and differentiation in the embryo of *Drosophila melanogaster*. In *Biology of* Drosophila (ed. M. Demerec), p. 168. Wiley, New York.
RAPER, J.A., M. BASTIANI, and C.S. GOODMAN. 1983. Pathfinding by neuronal growth cones in grasshopper embryos. I. Divergent choices made by the growth cones of sibling neurons. *J. Neurosci.* **3:** 20.
RICHELLE, J. and A. GHYSEN. 1979. Determination of sensory bristles and pattern formation in *Drosophila*. I. A model. *Dev. Biol.* **70:** 418.
SCHOUTE, J.C. 1913. Beitrage zur Blattstellungslehre. *Recl. Trav. Bot. Neerl.* **10:** 153.
SULSTON, J.E. and J.G. WHITE. 1980. Regulation and cell autonomy during postembryonic development of *Caenorhabditis elegans*. *Dev. Biol.* **78:** 577.
SULSTON, J.E., E. SCHIERENBERG, J.G. WHITE, and J.N. THOMSON. 1983. The embryonic cell lineage of the nematode *Caenorhabditis elegans*. *Dev. Biol.* **100:** 64.
TAGHERT, P.H. and C.S. GOODMAN. 1984. Cell determination and differentiation of identified serotonin-immunoreactive neurons in the grasshopper embryo. *J. Neurosci.* **4:** 989.
TAGHERT, P.H., C.Q. DOE, and C.S. GOODMAN. 1984. Cell determination and regulation during development of neuroblasts and neurones in grasshopper embryos. *Nature* **307:** 163.
TECHNAU, G.M. and J.A. CAMPOS-ORTEGA. 1985. Fate-mapping in wildtype *Drosophila melanogaster*. II. Injections of horseradish peroxidase in cells of the early gastrula stage. *Wilhelm Roux's Arch. Dev. Biol.* **194:** 196.
THOMAS, J.B., M.J. BASTIANI, M. BATE, and C.S. GOODMAN. 1984. From grasshopper to *Drosophila*: A common plan for neuronal development. *Nature* **310:** 203.
WAKIMOTO, B., F.R. TURNER, and T.C. KAUFMAN. 1984. Defects in embryogenesis in mutants associated with the Antennapedia gene complex of *Drosophila melanogaster*. *Dev. Biol.* **102:** 147.
WHEELER, W.M. 1893. A contribution to insect embryology. *J. Morphol.* **8:** 1.
WHITE, R.A.H. and M. WILCOX. 1984. Protein products of the bithorax complex in *Drosophila*. *Cell* **39:** 163.
WIGGLESWORTH, V.B. 1940. Local and general factors in the development of pattern in *Rhodnius prolixus*. *J. Exp. Biol.* **17:** 180.
WILCOX, M., G.J. MITCHISON, and R.J. SMITH. 1973. Pattern formation in the blue-green alga, *Anabaena*. I. Basic mechanisms. *J. Cell. Sci.* **12:** 707.

Summary

G.M. RUBIN

Department of Biochemistry, University of California, Berkeley, California 94720

After hearing the presentations at this Symposium, I believe we can be optimistic about the possibility of understanding the basic features of development at the molecular level, certainly before the 100th Symposium, and perhaps by the 60th. Such optimism is not new. I would like to quote from the Foreward to the 1956 Symposium volume written by Mislav Demerec, then director of Cold Spring Harbor Laboratory:

> There is very little doubt that genetic mechanisms play a cardinal role in development in higher organisms—development from a single, undifferentiated cell into the aggregate of billions of highly diversified cells that make up an individual. But we are still far from understanding the processes through which this differentiation is accomplished. Our knowledge about the structure of genetic components is considerable, and is being rapidly advanced by the utilization of new research methods. We also know a good deal about the primary function of genes, and have reason to believe that each individual gene controls a specific chemical process in a living cell. Furthermore, there is reliable evidence to indicate a complex interaction between these processes, and to suggest an intricate system of interrelationships among the units that constitute the genome of a cell. But our information is still very meager regarding the function of that system which extends to a group of cells and is responsible for their differentiation.
> Recent developments, however, indicate that we may be on the verge of bridging the gap that now separates genetics and experimental embryology.

A major conclusion to be drawn from this year's Symposium is that we have now bridged this gap. But why has it taken nearly 30 years to do so? Most of us would agree that the experimental techniques available in 1956 were not up to the task and that the many advances in methodology of the past decades, such as the development of recombinant DNA methods, have built that bridge for us.

A major lesson of this Symposium must be the recognition of the essential contributions genetic analysis has made to many of the exciting advances described in this volume. An appreciation of the power of genetic analysis for the study of development is also not new; as T.H. Morgan wrote in 1934, "The embryologist is coming to realize his dependence on the evidence from genetics." This dependence was emphasized by J. Monod and F. Jacob in their concluding remarks to the 1961 Symposium volume. In discussing the many technical difficulties then existing for the study of differentiation in mammalian systems, they wrote: "The greatest obstacle is the impossibility of performing genetic analysis, without which there is no hope of ever dissecting out the mechanisms of differentiation." Increasingly in the years since that was written, the tools of genetic analysis have been focused on a small number of model organisms in which these difficulties did not exist or could be overcome. I do not wish to imply that organisms in which genetic analysis is not now possible have not, and will not, continue to make major contributions to our understanding of development. Often, a particular organism will be best suited for a specific question or experiment independent of its genetic facility, and indeed several elegant examples are presented in this volume. Moreover, new methods of nonclassical genetic manipulation, such as gene transfer, in vitro mutagenesis, and antisense RNA production, are beginning to fill some of the void left by the inability to carry out genetic analysis. Such alternatives were also apparent as early as 1961: "But it should be noted that actual genetic mapping may not necessarily be required. Adequate techniques of nuclear transfer, combined with systematic studies of possible inducing or repressing agents, and with the isolation of regulatory mutants, may conceivably open the way to the experimental analysis of differentiation at the genetic-biochemical level" (Monod and Jacob 1961). We have certainly seen the beginning of the successful applications of such approaches at this Symposium. The extent to which the continued development and improvement of methods for nonclassical genetic manipulation are successful will determine in large part the degree of success of efforts to understand development in organisms in which conventional genetic analysis is not possible. I think we can be optimistic on this point, but it is important to realize that such techniques have limitations. They do not offer a substitute for the ability to screen or select among large numbers of organisms for mutations affecting a particular developmental process. It is through such screens that most of the mutations in yeast, flies, and worms described in this volume were isolated.

I also do not wish to imply that genetics alone will provide the answers to development. Rather, I wish to emphasize that at this Symposium, we once again witnessed the extraordinary power of the combination of genetic analysis and biochemical techniques—the same combination that elucidated the molecular basis of gene regulation in bacteria. I disagree with those who argue

---

All references cited without a date refer to papers in this volume.

that modern molecular biology and biochemistry could now decipher bacterial regulation without the help of genetics; the *lac* operon, maybe, but not the life cycle of bacteriophage λ or the mechanism of DNA replication. The more complex the system, the greater the number of interacting parts, the greater the requirement for genetic dissection. A different perspective, with more emphasis on nongenetic approaches to problems of early development, can be found in J. Gurdon's Introduction to this volume. With my bias clearly stated, I would like to review some of the themes, highlights, and different approaches presented in this volume.

A result with widespread implications was the observation that functional reversible changes occur in the genome during gametogenesis in the mouse and that these changes result in the male and female haploid genomes being functionally distinct (Solter et al.). Zygotes containing either two female or two male pronuclei cannot develop normally. This observation is reminiscent of a phenomenon in maize called "presetting" (McClintock 1964).

A number of approaches to the problem of how positional information is established in the embryo during oogenesis are presented (Weeks et al.; Jamrich et al.; Wylie et al.). A direct biochemical approach was used to search for maternal mRNA localized in either the animal or vegetal pole of the *Xenopus* egg (Weeks et al.). Four sequences meeting this criterion were isolated, leading to the estimate that only 1 mRNA in 10,000 is specifically localized. For one of these RNA species, it was demonstrated that injection of an antisense RNA leads to defective gastrulation.

The genetic approach to this problem involves the isolation of mutations in maternal-effect genes, genes whose expression by the mother during oogenesis are essential for subsequent normal embryonic development. This approach has been applied in the nematode (Schierenberg et al.; Hirsch et al.) and to a greater extent in *Drosophila*, where mutations affecting either anterior-posterior polarity, such as bicaudal (Mohler and Wieschaus), or dorsal-ventral polarity, such as dorsal (Steward et al.), have been isolated and studied. The dorsal gene was shown to be transcribed during oogenesis. dorsal transcripts are present in the early embryo but apparently are degraded prior to blastoderm formation.

Once axes of polarity are established in the *Drosophila* embryo, other genes somehow read this positional information and divide the embryo into a series of segments. The mechanism by which these segmentation genes carry out their function is under intense investigation. Most, if not all, of the genes involved in this process have been identified, primarily through exhaustive genetic screens (Nüsslein-Volhard and Wieschaus 1980; Nüsslein-Volhard et al.) One class of mutations, called gap mutants, are required for proper development of one region of the embryo but are not required in other parts of the embryo (Jäckle et al.). The mutant phenotype for one of these genes, Krüppel, could be partially rescued by injection of the cloned gene into the embryo and the Krüppel phenotype could be mimicked by injection of antisense RNA (Jäckle et al.). Other genes, so-called pair-rule genes, appear to be required to set up the segmental boundaries and subboundaries. Three of these genes, hairy, fushi tarazu, and engrailed, are discussed in a number of papers in this volume (Ish-Horowicz et al.; Ali et al., O'Farrell et al.; Laughon et al.; Gehring). These genes begin to be expressed quite early, prior to the cellular blastoderm stage. Initially, expression is uniform over all or a portion of the embryo, and then the pattern of expression begins to sharpen up, leading to "zebra-like" arrays. The stripes of expression of these genes partially overlap.

The mechanism by which the expression of such genes is progressively localized remains unknown. Correct localization depends on the prior correct action of maternal-effect genes. For example, we saw that fushi tarazu gene expression was altered in a predictable way in bicaudal embryos (Mohler and Wieschaus). The patterns of expression of these genes and the results of extensive genetic analysis are consistent with the pattern being established by interaction between the various pair-rule genes and their products to translate the initial continuous asymmetries of the egg into a series of discrete metameric units.

The next step in *Drosophila* development involves the specifying of segment identity. This function is performed by the products of the homeotic genes. Mutations in these genes lead to the apparent transformation of one segment, or part of a segment, into another. Lewis, over the past 30 years, has identified a number of loci that are all required for the formation of diverse body segments in the posterior half of the fly (Lewis). These closely linked genes are known as the bithorax complex. Other homeotic genes, those in the Antennapedia complex, affect differentiation in the anterior half of the fly (Kaufman et al. 1980).

On the basis of embryonic lethality, the bithorax complex can be divided into three functional domains (Sánchez-Herrero et al.), and these domains themselves can be further divided on the basis of other phenotypes (Lewis). This view of three functional domains is supported by detailed molecular analysis of chromosome rearrangements that disrupt the complex (Bender et al.). For the most part, chromosome breakpoints within a domain appear to inactivate the entire domain. It is worth noting that there are three homeo boxes in the bithorax complex and they each appear to be associated with one of these domains. Each domain produces a *trans*-acting product that appears to instruct a cell to follow a particular developmental pathway. Contained within each domain are *cis*-acting genes that help limit the expression of *trans*-acting products to their proper spatial domains. The Ultrabithorax (*Ubx*) domain has been analyzed in the most detail. The *Ubx* transcription unit itself encodes several overlapping RNA species, and a *Ubx* protein product that contains a homeo box and is found in nuclei in embryonic and later stages

(Hogness et al.). Surprisingly, the other transcription unit detected within the *Ubx* domain, bithoraxoid, does not appear to have an open reading frame for translation, although bithoraxoid mutations were shown to affect the distribution of *Ubx* proteins in the embryo (Hogness et al.). The description of the patterns of accumulation of *Ubx* transcripts illustrates the dynamic nature of *Ubx* transcription during embryogenesis (Akam et al.). As in the case of the fushi tarazu, hairy, and engrailed transcripts, the patterns of transcript accumulation change with time. Interaction between the products of the various homeotic genes appears to be important in setting up these patterns (Ingham; Levine et al.). This view of dynamic interaction contrasts with the apparent lineage specificity of expression of genes reported for the sea urchin embryo (Davidson et al.).

The patterns of expression of the segmentation and homeotic genes described by molecular techniques is remarkably consistent with the predictions made solely on the basis of the analysis of the phenotypes of mutants lacking the products of the corresponding wild-type genes. This congruence is a tribute both to the usefulness of genetic analysis and to the skill of several *Drosophila* geneticists. This success gives us confidence that the function of a wild-type gene product generally can be inferred from the phenotype of a mutant lacking this product, even in complex organisms.

Genetic analysis of nematode development has led to the discovery of genes affecting cell lineage, cell death, induction, and differentiation (Fixen et al.; Hirsh et al.). A wealth of information has already been gleaned from the genetic analysis of these mutations. Until recently, however, cloning of specific genes from nematode genomes has been very difficult. Due to strategies for cloning genes based on the use of Tc*1* elements (Emmons et al.) as genetically mappable and biochemically identifiable landmarks and other emerging methods, we can look forward to a wealth of molecular information concerning these genes and their products in the near future.

Several examples of mutations produced by DNA insertion in transgenic mice are described (Covarrubia et al.; Jaenisch et al.; Mark et al.). Given the large number of transgenic mice being produced, this approach should provide a useful source of new mutations which have the advantage that the affected gene can easily be isolated by recombinant DNA methods. The major limitation of this approach is that only a small minority of embryonic lethal mutations will be in genes with functions of developmental interest. As Mary Lyon put it, "A lot of things can cause a mouse embryo to go wrong early." Extrapolating from *Drosophila*, one might expect 1 in 30 embryonic lethal mutations to be in genes with specific and controlling functions in embryogenesis. However, one case that might fall into this class is the isolation of a gene allelic to the previously isolated mutation, limb deformity (P. Leder et al., pers. comm.).

Much progress has been made in the analysis of sex determination in yeast, worms, flies, and mice. The details of sex determination in yeast have been worked out in exquisite detail (Herskowitz et al.; Breeden and Nasmyth; Beach). The genes have been identified and cloned, and their expression and mechanism of control have been analyzed. The biochemical functions of many of the products have been defined.

In *Drosophila* and nematodes (Hodgkin et al.; Wood et al.), the genetic circuits controlling sex determination have been worked out by elegant genetic analysis. In *Drosophila*, two of the most important genes in the circuit, Sex-lethal (Maine et al.) and doublesex (Belote et al.), have been cloned and their molecular analysis is proceeding rapidly. Each gene is complex, having multiple overlapping transcripts, of which some are male-specific, some are female-specific, and some are present in both sexes. The biochemical function is not known for the products of any of the *Drosophila* or nematode genes.

In mice, the long-standing hypothesis of the role of the H-Y antigen in sex determination has been disproved, and the testes-determining factor has been mapped to a small region of the Y chromosome (McLaren).

The surprising finding of evolutionary conservation of a number of DNA sequences known to encode important functions has provided a way to relate observations made in diverse systems. The use of sequence cross-homology as a tool to utilize the advantages or advances of one experimental system in another was a recurring theme in this year's Symposium. In the case of oncogenes, genes first identified in mammalian systems are being studied in *Drosophila* and yeast to take advantage of genetic methods (Bishop et al.). Conservation of function was also observed in some cases—the mammalian *ras* gene functioned in yeast (Wigler et al.); *cis* and *trans* signals regulating chorion gene expression were conserved over the 200 million years separating moths and *Drosophila* (Kafatos et al.).

By far the most widely noted cross-homology, however, was due to the homeo box. Homeo box homology has been detected in a large number of phyla, including annelids, arthropods, echinoderms, and chordates (Gehring; McGinnis). Homeo box homologous sequences from frogs (De Robertis et al.), mice (Colberg-Poley et al.; Joyner et al.; Ruddle et al.), and man (Joyner et al.; Ruddle et al.; Boncinelli et al.) have been isolated and characterized. The chromosomal locations of several mouse and human homeo boxes have been determined. Clusters of closely linked homeo boxes have been observed, as is the case in *Drosophila*. Although the conservation of sequences outside the homeo box among vertebrates is sufficient to identify homologous genes between species, it is not yet possible to assign unambiguously a vertebrate gene to a particular *Drosophila* gene. Transcription of many of these homeo box sequences has been detected in *Xenopus* embryos, in mouse embryos, and in certain cell lines in culture. The striking conservation of the homeo box strongly suggests that some aspects of its function are conserved. However, there is at present no evidence that any of the homeo-box-containing genes in vertebrates have a controlling function in embryonic development. Given the intense effort being put into the study of these genes, a clue to their function in vertebrates should be forthcoming soon.

The analysis of mechanisms used to control the expression of various genes in a wide variety of organisms indicates that many basic features of gene regulation are conserved from yeast to man. For example, upstream activating sequences with tissue specificity and properties similar to viral enhancer sequences are reported in several papers (Struhl et al.; Shepherd et al.; Posakony et al.; Bourouis and Richards). Examples of control at the level of RNA processing and translation are also presented (Rubin et al.; Busslinger et al.). A complex pattern of transcription was observed for many *Drosophila* genes, especially the homeotic genes and genes with controlling functions in sex determination, suggesting that control at the level of differential RNA processing may be more common than expected.

Much progress has been made in the development and improvement of gene-transfer techniques in a number of organisms, although none can match the perfection of the yeast methodology. In *Drosophila*, single-copy transformants are readily obtained, and in the vast majority of cases, the transferred gene is properly regulated both qualitatively and quantitatively (see Shepherd et al.; Kalfayan et al.; Meyerowitz et al.). In the mouse, it is now possible in a large number of cases to achieve correct tissue and temporal expression, but levels of expression are highly variable, especially if normalized to gene copy number (Hammer et al.; Ornitz et al.; Westphal et al.; Krumlauf et al.; Lovell-Badge et al.). The basic feasibility of gene transfer in plants (Horsch et al.; Schell et al.), nematodes (Hirsch), sea urchins (Davidson et al.), and *Dictyostelium* (Gomer et al.) has also been demonstrated. Continued improvements in methodology can be expected in all these systems, notably in the case of viral vectors in mice (Nicolas et al.; Evans et al.; Rassoulzadegan and Vailly; Felsani et al.; Wagner et al.; Jaenish et al.). The observation of presetting (see above) emphasizes the importance of germ line transformation methods, since only these methods allow the transferred gene to pass through the entire developmental history of the organism.

Considerable progress has also been made in the development of other surrogate genetic techniques. Antisense RNA has been used to inhibit gene function (Weeks et al.), and promoter fusions have been used to create dominant gain-of-function mutations (D. Hanahan, pers. comm.). Such methods provide valuable tools in the study of the function of genes that have been cloned but cannot be studied by conventional genetic analysis. Their major limitation is that they require the prior identification and isolation of the gene to be studied.

Where can we expect rapid progress over the next few years? Classical embryology has defined the problem of development in general descriptive terms. "It is known that the protoplasm of different parts of the egg is somewhat different, and that the differences become more conspicuous as the cleavage proceeds, owing to the movements of materials that then take place. From the protoplasm are derived the materials for the growth of the chromatin and for the substances manufactured by the genes. The initial differences in the protoplasmic regions may be supposed to affect the activity of the genes. The genes will then in turn affect the protoplasm, which will start a new series of reciprocal reactions. In this way we can picture to ourselves the gradual elaboration and differentiation of the various regions of the embryo" (Morgan 1934). But how can we reach an understanding of development in molecular detail? I believe genetics can both define the problem in specific terms and help dissect it into manageable bits. Genetic analysis has and will continue to be the primary, and perhaps the only, way in many complex systems to identify the important genes and gene products. Recombinant DNA methods will make it possible to obtain sufficient quantities of those genes and their products for biochemical analysis. Gene-transfer methods should allow us to manipulate their structure and time and place of expression. Biochemistry, cell biology, and sophisticated genetic analyses will be required to understand how these gene products execute the processes of determination, differentiation, and morphogenesis. The required experiments will not be as straightforward as the identification and isolation of the genes. In this volume, we see just the very beginnings of this type of detailed biochemical (Edelman et al; Brown and Schlissel; Wassarman et al.; Klein et al.) and genetic (D. Botstein, pers. comm.) analyses. It is clear, however, that it is the gene products rather than the genes themselves that drive development, and increasingly we will have to focus on identifying the biochemical activities of these proteins. Likewise, the interactions that occur between cells to execute morphogenesis await our attention. I see the major conceptual breakthroughs in developmental biology over the next few years coming from such molecular and biochemical analyses of the *products* of genes that have already been identified by genetic analysis in flies and worms.

REFERENCES

KAUFMAN, T.C., R. LEWIS, and B. WAKIMOTO. 1980. Cytogenetic analysis of chromosome 3 in *Drosophila melanogaster*. The homeotic gene complex in polytene chromosome interval 84A-B. *Genetics* **94**: 115.

MCCLINTOCK, B. 1964. Aspects of gene regulation in maize. *Carnegie Inst. Wash. Year Book* **63**: 592.

MORGAN, T.H. 1934. *Embryology and genetics*. Columbia University Press, New York.

NÜSSLEIN-VOLHARD, C. and E. WIESCHAUS. 1980. Mutations affecting segment number and polarity in *Drosophila*. *Nature* **287**: 795.

Author Index

A

Acampora, D., 301
Akam, M.E., 195
Alberts, B.M., 79
Ali, Z., 229
Amati, P., 753
Ambrosio, L., 223
Anderson, D.J., 855
Andrew, D.J., 605
Aronson, J., 45
Artavanis-Tsakonas, S., 841
Ashburner, M., 505
Ausubel, F.M., 555
Awgulewitsch, A., 277
Axel, R., 855

B

Babinet, C., 51
Baker, B., 421
Baker, B.S., 605
Baltimore, D., 417
Banks, J., 307
Bargiello, T.A., 865
Baumgartner, S., 127
Beach, D., 635
Beachy, P.A., 181
Belote, J.M., 605
Bender, W., 173
Bernstein, K.E., 411
Bingham, P.M., 337
Birchmeier, C., 721
Birnstiel, M.L., 665
Bishop, J.M., 727
Blackman, R.K., 119
Blanchet, P., 51
Blaschke, A., 779
Bleil, J.D., 11
Boncinelli, E., 301
Bonnerot, C., 713
Bopp, D., 127
Bourouis, M., 355
Bradley, A., 685, 707
Breeden, L., 643
Breindl, M., 439
Brinster, R., 371, 379, 399
Broek, D., 721
Brown, D.D., 549
Brûlet, P., 51
Bryan, R., 831
Burri, M., 127
Busslinger, M., 665
Bygrave, A.E., 707

C

Cappello, J., 759
Carroll, S.B., 253
Casanova, J., 165
Chada, K., 361
Chambon, P., 747
Champer, R., 831
Chapman, C.H., 337
Chapman, V.M., 371

Cheah, K.S.E., 707
Chen, W., 489
Chepelinsky, A.B., 411
Cline, T.W., 595
Cohen, S.M., 759
Colberg-Poley, A.M., 285
Coleman, K.G., 229
Condamine, H., 51
Costantini, F., 361, 417
Covarrubias, L., 447
Crosby, M.A., 347
Crowley, T., 801

D

Dalton, D., 277
Datta, S., 801
Davidson, E.H., 321
Dawid, I.B., 31
de Barros Lopes, M.A., 627
de Cicco, D., 527
Decker, G.L., 91
de Crombrugghe, B., 411
Delidakis, C., 537
De Robertis, E.M., 271
Deschamps, J., 733
Desplan, C., 235
Devreotes, P., 787
Dilworth, S.M., 657
DiNardo, S., 235
Dingwall, C., 657
Doe, C.Q., 891
Donahue, L., 575
Doniach, T., 585
Doyle, H., 209
Drees, B., 229, 727
Duboule, D., 747
Dunn, K., 555
Duprey, P., 51

E

Eckes, P., 421
Edelman, G.M., 877
Edström, J.-E., 127
Ellis, H., 99
Emmons, S.W., 313
Evans, M.J., 685, 707
Evans, R.M., 389

F

Faiella, A., 301
Fainsod, A., 277
Farach, H.A., Jr., 91
Farach, M.C., 91
Fedoroff, N., 307, 421
Felsani, A., 753
Ferguson, J., 627
Ferguson-Smith, A., 277
Fidanza, V., 301
Fienberg, A., 277
Finlay, C., 671

Firtel, R.A., 801
Fischer, J.A., 515
Fixsen, W., 99
Florman, H.M., 11
Flytzanis, C.N., 321
Fontana, D., 787
Forbes, D., 651
Fraley, R.T., 433
Frei, E., 127
Fritz, A., 271

G

Garabedian, M.J., 521
Garfinkel, M.D., 347
Gavis, E.R., 181
Gehring, W.J., 243
Gelbart, W.M., 119
Gerisch, G., 813
Gilbert, S.F., 45
Gilboa, E., 691
Goetz, J., 271
Goldschmidt-Clermont, M., 181
Gomer, R.H., 801
Gomes, S.L., 831
Goodman, C.S., 891
Gorman, C.M., 701
Grant, S.R., 91
Greve, J.M., 11
Grimaila, R., 119
Grimwade, B.G., 841
Grosschedl, R., 417
Gruss, P., 285
Gurdon, J.B., 1
Gustavson, E., 229

H

Hadwiger, J.A., 627
Hagmann, J., 813
Halpern, M.E., 841
Hammer, R.E., 371, 379, 399
Handelsman, K., 759
Harbers, K., 439
Handelsman, K., 759
Harbers, K., 439
Harding, K., 209
Hart, C.P., 277
Harte, P.J., 181
Harvey, R.P., 21
Hauser, C., 291
Hazelrigg, T., 329
Heasman, J., 37
Helfand, S.L., 181
Herrera-Estrella, L., 421
Herskowitz, I., 565
Hiatt, A.C., 475
Hill, D.E., 489
Hirsh, D., 69
Hirth, P., 813
Ho, J.-Y., 627
Hodgkin, J., 585

Hoey, T., 209
Hoffmann, F.M., 119
Hogness, D.S., 181
Hollenberg, S., 389
Holwill, S., 37
Hope, I.A., 489
Horsch, R.B., 433
Horvitz, R., 99
Horwitz, R., 627
Howard, K.R., 135
Hung, M.-C., 521

I

Imanishi-Kari, T., 417
Ingham, P.W., 135, 201
Irish, V.F., 119
Ish-Horowicz, D., 135

J

Jäckle, H., 465
Jackson, F.R., 865
Jacob, F., 51, 713
Jaenisch, R., 439
Jähner, D., 439
Jamrich, M., 31
Jefferson, R., 69
Jones, C.A., 627
Joyner, A., 291
Jürgens, G., 145

K

Kafatos, F.C., 537
Kaghad, M., 51
Kaiser, D., 823
Kalfayan, L., 527
Kanki, J., 651
Karch, F., 173
Karess, R.E., 329
Karr, T.L., 79, 229
Kassis, J.A., 235
Kataoka, T., 721
Katzen, A.L., 727
Kaulen, H., 421
Kauvar, L.M., 229
Kearsey, S.E., 657
Keller, G., 691
Kellogg, D.R., 79
Kemphues, K.J., 69
Kerk, N., 277
Kerridge, S., 165
Khillan, J.S., 411
Kienlin, A., 465
Kingsbury, J., 307
Klein, P., 787
Kleinschmidt, J., 657
Kluding, H., 145
Knipple, D.C., 465
Knox, B., 787
Kornberg, T., 229, 291, 727
Kreuzaler, F., 421

AUTHOR INDEX

Kroos, L., 823
Kruijer, W., 733
Krumlauf, R., 371
Kuehn, M.R., 685
Kuner, J.M., 235
Kuspa, A., 823

L

Lacy, E., 453
Lane, D.P., 701
Lang-Unnasch, N., 555
Laskey, R.A., 657
Laski, F.A., 329
Laughon, A., 253
Laverty, T., 329
La Volpe, A., 301
Lawrence, P.A., 105
Lee, J.J., 321
Lehmann, R., 465
Lennarz, W.J., 91
Levine, A.J., 671
Levine, J., 527
Levine, M., 209
Levis, R., 329
Levitt, A., 313
Lewis, E.B., 155
Lingappa, J.R., 537
Lipshitz, H.D., 181
Lodish, H.F., 759
Löhler, J., 439
Loomis, W.F., 769
Lörincz, A.T., 627
Losick, R., 483
Lovell-Badge, R.H., 707
Lowe, B.A., 475

M

MacWilliams, H., 779
Magram, J., 361
Mahon, K.A., 411
Maine, E.M., 595
Maione, R., 753
Malmberg, R.L., 475
Maniatis, T., 515
Mann, S., 801
Mariani, B., 537
Mark, W.H., 453
Martin, C.H., 347
Martin, G., 271
Martin, G., 291
Martinez-Arias, A., 195
Mathers, P.H., 347
Mattaj, I.W., 271
McGinnis, W., 263, 277
McGrath, J., 45
McIndoo, J., 475
McKeown, M.B., 605
McLaren, A., 623
Mechali, M., 657
Mehdy, M., 801
Meijlink, F., 733
Melton, D.A., 21
Mendenhall, M.D., 627
Meneely, P., 575
Messing, A., 399
Meyerowitz, E.M., 347
Miller, K.G., 79

Mills, A.D., 657
Mintz, B., 447
Mitchell, R.L., 733
Mitsialis, S.A., 537
Mohler, J., 105
Mohr, I.J., 79
Morata, G., 165
Morello, D., 51
Muskavitch, M.A.T., 841

N

Nasmyth, K., 643
Nellen, W., 801
Nelson, C., 389
Nelson, O., 307
Newport, J., 651
Nicolas, J.F., 713
Nishida, Y., 447
Noll, M., 127
Nöthiger, R., 615
Nüsslein-Volhard, C., 145

O

O'Driscoll, M., 37
Oettinger, M.A., 489
O'Farrell, P.H., 235
Ornitz, D.M., 399
Orr-Weaver, T., 527
Overbeek, P.A., 411

P

Palmiter, R.D., 379, 399
Papayannopoulou, T., 361
Parks, S., 527
Paul, E., 841
Peattie, D.A., 181
Peifer, M., 173
Peterson, T.A., 627
Piatigorsky, J., 411
Pinchin, S.M., 135
Pinkert, C.A., 399
Poole, S.J., 229
Posakony, J.W., 515
Posakony, L.M., 119
Powers, S., 721
Prause, I., 779
Preiss, A., 465
Purucker, M., 831

R

Rabin, M., 277
Raboy, V., 307
Radice, G., 361
Radomska, H., 209
Rassoulzadegan, M., 679
Rebagliati, M.R., 21
Reed, S.I., 627
Reymond, C., 801
Ricci, L., 753
Richards, G., 355
Richardson, S.L., 627
Rigby, P.W.J., 701
Riley, P.D., 253

Rio, D.C., 329
Robertson, E.J., 685, 707
Robinson, J.J., 321
Rogers, S.G., 433
Roller, R.J., 11
Rosahl, S., 421
Rose, S.J., III, 321
Rosenberg, U.B., 465
Rosenfeld, M.G., 389
Rossier, C., 813
Ruan, K.S., 313
Rubenstein, J.L.R., 713
Rubin, G.M., 329, 905
Ruddle, F.H., 277
Rüther, U., 691

S

Saint, R.B., 181
Salo, E., 271
Salz, H.K., 595
Salzmann, G.S., 11
Samuels, F.G., 11
Sánchez-Herrero, E., 165
Sargent, T.D., 31
Sassone-Corsi, P., 747
Savakis, C., 505
Schedin, P., 575
Schedl, P., 223, 595
Schell, J., 421
Schiefelbein, J.W., 307
Schierenberg, E., 59
Schlissel, M.S., 549
Schmidt, A., 411
Schubert, D., 733
Schümperli, D., 665
Scott, M.P., 253
Scott, T.N., 605
Scotto, L., 301
Segal, D., 119
Seifert, E., 465
Shapiro, L., 831
Shen, M., 585
Shepherd, B., 521
Sher, E., 235
Shin, H.-S., 865
Signorelli, K., 453
Simeone, A., 301
Simon, M.A., 727
Sivertsen, A., 801
Smith, G.D., 271
Snape, A., 37
Soeller, W., 229
Solter, D., 45
Spann, T., 651
Spena, A., 421
Spencer, F.A., 329
Spoerel, N., 537
Spradling, A., 527
Stamatoyannopoulos, G., 361
Stein, R., 855
Steinmann-Zwicky, M., 615
Sternberg, P., 99
Steward, R., 223
Stewart, C., 691
Stinchcomb, D.T., 69
St. Johnston, R.D., 119
Storfer, F.A., 253
Struhl, K., 489

Sucov, H.M., 321
Swanson, L., 389

T

Theibert, A., 787
Theis, J., 235
Tilghman, S.M., 371
Tjian, R., 291
Toda, T., 721

U

Utset, M., 277

V

Vailly, J., 679
Van Dyke, T., 671
Vasseur, M., 51
Verma, I.M., 733
VijayRaghavan, K., 347
Voss, S.D., 285

W

Wagner, E.F., 691
Wakimoto, B., 527
Walter, M., 79
Wassarman, P.M., 11
Watson, C.J., 701
Weaver, D., 417
Wedeen, C., 209
Weeks, D.L., 21
Weiffenbach, B., 173
Weinberger, C., 389
Weinhart, U., 813
Weinzierl, R., 195
Weir, M.P., 229
Wensink, P.C., 521
Westphal, H., 411
Westphal, M., 813
Wharton, K.A., 841
Wieschaus, E.F., 105
Wigler, M., 721
Wilde, C.D., 195
Willmitzer, L., 421
Wittenberg, C., 627
Wolfner, M.F., 605
Wood, W.B., 575
Woodward, H.D., 91
Wright, C., 271
Wright, D., 235
Wylie, C.C., 37

Y

Yedvobnick, B., 841
Yesner, L., 313
Young, M.W., 865

Z

Zachar, Z., 337
Zeller, R., 271
Zuber, P., 483
Zuker, C.S., 329

Subject Index

A

Abelson murine leukemia virus (Ab-MLV), 418
Ac controlling elements, maize, 307–311
Acrosome reaction, 13–14
Actin
 -binding proteins in the embryo, 83–85
  monoclonal antibodies to, 86–87
 CyIII genes, 321–327
 gene family in sea urchin, 322
 lineage-specific expression, 321
 and microtubules in developing embryos, 81–82
Actinomycin D
 blocks spicule formation in sea urchin embryos, 91
Activator-Dissociator (*Ac-Ds*)
 elements in maize, 307–311
 -inhibitor model, gradient of morphogen, 110
 transposition in heterologous plants, 429–430
Adenovirus
 E1a function in undifferentiated F9 cells, 705
Adenylate cyclase, 813
 and cAMP signaling in *Dictyostelium*, 789
 is regulated by G protein in oocytes, 721
Agrobacterium tumefaciens
 Ti vectors, 421, 433–436
Albumin gene expression in fetal liver, 371
Alcohol dehydrogenase (ADH)
 fusion genes (*Sgs3*), 356, 358
 gene transcription in *Drosophila*, 505–513, 515–519
Alfalfa (*Medicage sativa*), 555–562
Amino acid biosynthesis, control of in yeast, 501–502
Amorphic alleles, 575
Amplification, gene,
 of *Drosophila* chorion, 527–534
 of insect chorion, 537–547
Androgenones, diploid parental, 46
Anopheles, sex determination in, 615
Antennapedia complex (ANT-C), 113, 209–221. See also *Drosophila*
 Antp gene, 201, 243
  GC-rich repeats in cDNA for, 255
  conservation of homeo domain structure of, 263–269
  homeo boxes of, 291
  homeotic mutants (*Antp*), 243, 253–260
  molecular organization, 254
  proteins, 253–260
Aphidicolin, 550

B

Bacillus subtilis, promoter utilization in, 483–488
Bacteriophage λ regulatory loci, 565–572
Bicaudal mutations, 223. See also *Drosophila*
Biological clock, *Drosophila,* 865–875
Bithorax complex (BX-C), 113. See also *Drosophila*
 anatomy and function, 165–171
 bithoraxoid (*bxd*), 173
 complementation and the zeste locus, 342
 domains of *cis*-interaction, 173–179
 gene regulation, 155–163, 244
 homeo box of, 291
 Ubx domain, 181–194, 201
B lymphoid cells, differentiation of, 417
Blastocysts, transferred to foster mothers, 685, 708

C

c-abl, 727
Caenorhabditis elegans (*C. elegans*)
 cell determination during embryogenesis, 59–67
 cell fates in, 99–104
 cell lineage of the intestinal anatomy, 582
 Dumpy (*Dpy*) mutations, 577–579
 feminization of germ line (*fog*), 591
 genes affecting early development, 69–77
 her-1 gene, 575, 587
 hermaphrodite sex, 575–592, 620
 masculinization of germ line (*mog*), 591
 nematode gene functions, 75
 regulation of Tc1 transposable elements in, 313–319
 sex determination in, 575–582, 585–592, 620
 sexually transformed mutant animals, 576, 585–586
 tra-1 gene, 575, 588, 590–592, 620
cAMP
 -dependent protein kinase regulatory subunit in *Dictyostelium,* 792
 in *Dictyostelium,* 770, 782, 787–798
 as a chemoattractant, 787
 induces prestalk gene expression in, 802–811
 phosphorylation of the surface receptor for, 796
 receptors, 789, 813
 modification and sensory adaptation, 793
 -resistant mutants, 816
 S. cerevisiae (*cyr1*) gene, 639
 signaling in *Dictyostelium,* 813–821
CAP protein of *E. coli,* 793
Cartilage, 707
Caulobacter, cell differentiation in, 831–839
cDNA, maternal mRNA, 21–24
Cell adhesion molecules (CAMs)
 expression of, 877–888
 genes, 885
Cell cycle, 627–670
 coupling of DNA replication to, 660
 cytoplasmic control of, 60–63, 65
 expression of *HO* gene, 643
 gene regulation in *Caulobacter,* 831–839
 regulation of cell type in *Dictyostelium,* 779–784, 801–811
 regulation of histone mRNA, 667
 -specific activator (of *HO* transcription), 648
 in *Saccharomyces,* 627–634, 635–640, 648
 in *Xenopus,* 651–656
Cell death, programmed, 103
Cell division
 asymmetric, 2–4
 control of in yeast, 627–634
 and genes that affect cell fates, 100–101
Cell fates
 alteration of blastoderm, 105–111
 specification during development, 99
Cell fusion, 62
Cell interactions, and embryonic induction, 4–6
Cell lineage(s). See also *Caenorhabditis elegans;* Embryonic lineages
 neural lineage mutant *unc-86*, 101
 -specific gene expression, 321–327
 sublineages, 100
Cell-type specialization in yeast, 565
Central nervous system (CNS) development
 and *bxd*, 187
 in *Drosophila,* 841–853
 and expression of engrailed (*en*), 237
 and *Ubx* proteins, 193, 211

911

SUBJECT INDEX

c-*fps*, related oncogenes in *Drosophila*, 727–730
Chalcone synthase, 421
Chemosensory transduction, in *Caulobacter*, 834–835
Chemotactic orientation, in *Dictyostelium*, 818
Chemotaxis
 and cAMP signaling in *Dictyostelium*, 787–798
 and gene expression in *Caulobacter*, 831–840
Chimeric genes. *See also* Chloramphenicol acetyl transferase; Fusion genes; Transgenic mice
 regulation in plants, 421–430
 in transgenic mice, 411–415
Chimeric mice, 685–688. *See also* Transgenic mice
 expression of human type-II collagen in, 707–710
 gene expression studies in, 696–699
Chironomus, dominant male determiners in, 615, 619
Chloramphenicol acetyl transferase (CAT) chimeric genes, 411–415
Chondrocytes, 707–710
Chorion gene amplification
 in *Drosophila*, 527–534
 in insects, 537–547
Choroid plexus papillomas, and T-antigen expression, 671–677
Chromaffin cells, 855
 plasticity, 862
Chromatin
 assembly of genes into, 550
 assembly in early *Xenopus* embryos, 657
 conformation of the $\alpha 1(I)$ collagen gene, 440–441
 structure in TATA regions, 498–500
Chromosome
 aberrations
  and hybrid dysgenesis, 329
  of pair-rule genes (*Drosophila*), 147
  within bithorax, 156
 breakage, at *Ac* and *Ds*, 307–311
 dissection of polytene, 465
 localization of homeo box loci in mouse, 278
 polytene, 347
 rearrangements and *Ubx* mutations, 196
 replication in *Xenopus* embryos, walking
  and bithorax (BX-C), 244
  in *Drosophila*, 127
  using a *Krüppel* cDNA, 465
Circadian rhythms in *Drosophila*, 867
cis-acting regulatory elements, 906
 bxd on *Ubx*, 187, 192–194
 cis-inactivation (CIN) rule, 160, 174–179
 cis-overexpression (COE) rule, 160, 163
 and cisvection, 155, 157–159
 in *Dictyostelium*, 801, 810
 domains in bithorax, 173–179
 in *Drosophila* and moths, 538, 541–547
 of ADH gene, 505–513, 515–519
 and 68C glue puff regulation, 350
 regulate gene expression, 8–9
 and tissue-specific activation of AFP genes, 371
 and wing posture development, 125
 and yolk protein 1 gene in *Drosophila*, 521–526
Cisvection, 155, 157–159
Cleavage
 asymmetric, 65, 69
 cytoplasmic control of early, 63–67
 positioning of furrows, 69
$\alpha 2(I)$ collagen gene fusions, promoter activity, 411–415
Collagen
 type I
  and mechanical stability of the circulatory system, 443
  role in embryonic development, 439–444
  provirus insertion into, 439
 type II, transformation of EK cells with human, 707–710
Compartmentation, and the selector-gene hypothesis, 113–116
Contrabithorax (*Cbx*), 156. *See also* *Drosophila*
Controlling elements, *Ac-Ds*, *Spm* in maize, 307–311
αA crystallin chimeric genes, 411–415
c-src, 727
Cytokeratin EndoA
 gene regulation, 51–56
 of the trophectoderm, 51
Cytoplasm, -nuclear interactions in early development, 11–98
Cytoplasmic determinants
 and embryogenesis, 271
 partitioning of, 2–4
Cytoplasmic localization, and asymmetric cell division, 2–4
Cytoplasmic organization, *Drosophila* embryos, 78–89
Cytoplasts, 73
 transfer from cleavage-stage embryos, 47
Cytoskeleton
 and *cdc*28 mutants, 630
 filament systems in embryo(s), 79, 89
 and pattern formation in *Drosophila*, 83–85
 protein kinase assays, 627

cDNA, 254, 267
 protein sequence, 255
 transcripts in early gastrula, 248, 254
Determinants, cytoplasmic, 3
Determination, 1
 of cell fate, 65–67
Determinative switching and homeodomain-containing proteins, 257–260
Development. *See also* Cell lineage; *cis*-acting regulatory elements; Embryos; Gene regulation; Homeo box; Pattern; Segmentation; Tissue-specific elements; *names of specific organisms*
 and cAMP receptors in *Dictyostelium*, 787–799
 and cell interactions in *Myxococcus*, 823–830
 cytoplasm-nuclear interactions in early, 11–98
 and early mutations due to DNA rearrangements, 447–452
 genetic control of, 243–251
 and genetics of nodule formation, 555–563
 and hemoglobin switching, 361
 and histone gene expression, 665–669
 induced defects in, 439–482
 in vitro of preimplantation embryos, 460–462
 mechanisms of, cooperativity, 6–9
 nervous system, 841–908
 network regulation in *Dictyostelium*, 814
 proto-oncogene *fos* expression during, 733–744
 regulation of *c-src* in *Drosophila*, 728
Developmental
 arrest, insertional mutation causes, 453
 control of *Drosophila* yolk protein genes, 521–526
 expression of localized RNAs, 27
 regulation
  of the AFP minigene in liver, 374
  in *B. subtilis*, 483–487
  of globin genes in transgenic mice, 361–369
  of the growth hormone gene, 391–398
  of neuroendocrine genes, 389–396
 switching in tobacco, 475–480
Dictyostelium
 aggregation-deficient mutants, 818–819
 cAMP-resistant mutants, 816
 cell-type-specific
  differentiation in, 769–776
  expression, 779–784, 801–811
 discoidin I multigene family, 808
 early development, 813–821
 EDTA-stable contacts, 818–819
 feedback loops regulate cell-type proportions, 779–784
 regulation of *pst-cath* fusion gene, 806
 retrotransposon DIRS-1 in, 759–766
 slug morphogenesis, 783

D

Decapentaplegic (DPP-C) gene complex, 119–125. *See also* *Drosophila*
 mutations in, 120, 122–125
 organization of, 121
 synapsis-dependent complementation of, 342
Deformed (*Dfd*) allele

SUBJECT INDEX

Differentiation, 1, 759-840. *See
 also* Tissue-specific elements;
 names of specific organisms
 cAMP effects in *Dictyostelium*, 782
 cell-type-specific, in *Dictyostelium*, 769-776
 -inducing factor (DIF), 781
 endodermal, role of homeo box sequences in, 285-290
 first in mouse, 51-57
 gene expression during in *Caulobacter*, 831-840
 immune in transgenic mice, 417-420
 neural crest cells, 855-863
 "proportion regulation," 779-784
 proto-oncogene *fos* expression during, 733-744
 sexual, *Drosophila*, 595-604, 605-614
Diploid parental androgenones, 46
DIRS-1 retrotransposon
 of *Dictyostelium discoideum*, 759-766
 transcription in yeast, 763-764
DNA
 -binding protein(s)
  engrailed (*en*), 235-241
  and homeo box function, 274
  homeo box homology with *MAT*, 249
  and transcription of yeast genes, 489-502
 cis-acting elements, 521-526
 integration patterns in transgenic mice, 448-449
 rearrangements cause early developmental mutations in transgenic mice, 447-452
 "recognition helix" of homeo domains, 257
 repair, and integration of foreign DNA, 450
 replication in *Xenopus* embryos, 657-662
 strain polymorphisms, Tc-1 induced, 313
 transformation of *C. elegans*, 74-75
 and genetic complementation, 76
DNase I hypersensitive sites, 365-367, 401, 404-405
 in α1(I) collagen genes, 440
 in *gal* regulatory region, 492
Dolichol synthesis, in sea urchin embryos, 92-93
Dorsal (*dl*) gene expression, 223-228
Dosage compensation
 by ADH gene, 510
 in *C. elegans*, 575-582
 X chromosome in *Drosophila*, 595, 605, 615. *See also* Sex determination
Drosophila arizonensis, 510
Drosophila mauritania, 505
Drosophila melanogaster. See also specific locus names
 68C glue puff, 347-352
 alcohol dehydrogenase (ADH) gene transcription, 505-513, 515-519

Antennapedia (ANT-C) complex, 113, 201, 243, 253-260
bicaudal mutations in, 105-111
big brain (*bib*), 841, 851
biological clocks in, 865-874
bithorax complex (BX-C), 113, 155-163, 165-171, 173-179, 181-194, 243-244
contrabithorax (*cbx*), 156
cytoplasmic organization of early embryo, 79-90
daughterless (*da*) regulates sex-lethal (*Sxl*), 595, 599, 606, 615
Decapentaplegic (DPP-C), 119-125
Deformed (*Dfd*) loci, 209
Delta (*Dl*) mutations, 841, 851-852
dorsal (*dl*), 223-228
double-sex (*dsx*), 606-609, 616, 620
embryos, cytoplasmic organization of, 79
engrailed (*en*) gene locus, 113, 229-232, 235-241, 253
 en-related, invected (*inv*), 232
Enhancer of split (*E*[*spl*]), 841, 851-852
even-skipped (*eve*), 145-153, 238
extra sex combs (*esc*) gene, 127-133
fushi tarazu (*ftz*), 108, 135-136, 141, 151-153, 210, 238, 244, 254
 embryos, 906-907
 neurogenesis in, 891, 899-902
hairy (*h*) locus, 135-143, 151
homeo box gene families, 209-221, 243-250
hsp70 promoter, 427
invected (*inv*) gene locus, 232
Krüppel (*Kr*) segmentation genes, 465-472
maleless (*mle*), 606
male-specific lethal (*msl*), 606
Malpighian tubules, 506, 508, 516
master-mind (*mam*), 841, 852
mosaic intersexes, 580
mosaic at white locus, 332
neuralized locus (*neu*), 841
neurogenesis in, 841-853
Notch (*N*) mutations, 841, 843-853
odd-skipped (*odd*), 145-153
oogenesis in, 527-534
opa repetitive elements in, 263
paired (*prd*), 145-153, 238
"pair-rule" genes, 136, 145-153, 238
pattern formation in muscle, 113-116
P-element transposition, 329-335, 867
per locus, 865-874
Polycomb⁻ embryos (Pc⁻), 213
polytene chromosomes, 347
proto-oncogenes of, 727-730
runt mutant embryos, 238
salivary gland glue, 347-352, 355-359
segmentation loci, 136, 141, 145-153, 155-171, 209, 243. *See also* Bithorax
selector gene hypothesis, 201-208
sex combs reduced (*Scr*), 209

sex determination in, 595-603, 605-613, 615-620
Sex-lethal (*Sxl*) gene, 595-596, 606, 615
Sgs genes, 348, 352, 355
Sgs-4 locus, 342
super sex combs (*sxc*) loci, 201-208
transformer (*tra*) genes, 606, 609-611
transvection effects at the white locus, 337
trithorax (*trx*) mutations, 201-208
Ultrabithorax (*Ubx*)
 function and expression, 195-200, 209-221
 mutations, 136, 167-171, 201. *See also* Ultrabithorax
white gene, 329-335, 337-345
xanthine dehydrogenase rosy gene, 506, 511
X hyperactivation in, 579
yolk protein (*yp1*) gene, 521-526
zeste locus, 339
Drosophila mohavensis, 510
Drosophila mulleri, 510, 517-519
Drosophila orena, 505, 512-513
Drosophila simulans, 505

E

Ecdysteroid hormone, and ADH levels, 506
Egg. *See also* Oocyte; Oogenesis
 -laying-deficient (*egl*) mutants in *C. elegans*, 590-592
 localized maternal mRNA in (*Xenopus*), 21-30
 localization in amphibians, 2
 receptor for sperm (mouse), 11-19
EK cells, chimera formation of, 685-689
Elastase I
 gene regulation in transgenic mice, 399-408
 -LGH pedigrees, 402
Embryo
 ADH transcription in *Drosophila*, 509, 519
 in *C. elegans*, 59-67, 69-77
 cell-surface (glyco)protein expression in sea urchin, 91-98
 chromosome replication in *Xenopus*, 657-662
 cleavage-blocked, 60
 collagen, type-I, in early, 442
 cytoplasmic organization of *Drosophila*, 79-89
 developmental arrest is caused by insertional mutation (transgenic mouse), 453-467
 early developmental mutations in due to DNA rearrangments (transgenic mice), 447-452
 E-to-A conversion of CAM glycoproteins, 878
 first differentiations in (mouse), 51-57

Embryo *(continued)*
  gene activation in mammalian, 45–49
  genes affecting segmental subdivision of (*Drosophila*), 145–154
  germ line integrations in preimplantation, 439
  injection assay, 549
  lineage-specific gene expression in (sea urchin), 321–328
  neurogenesis in (grasshopper and *Drosophila*), 891–903
  parasegments, 113. *See also* Segmentation
  parthenogenic, 45
  postimplantation developmental arrest, 448
  preimplantation introduction of genes into with retroviral vectors, 713–720
  preimplantation stage, 456–462
  spatial organization and homeo boxes, 243–250
  -specific RNAs, accumulation, 32
  UV irradiation of, 31
Embryonal carcinoma (EC) cells,
  activity of SV40 21-bp repeat in, 749–450
  F9 cells, 285
  feeder-dependent, 695
  introduction of genes into with retroviral vectors, 713–720
  lines, 685, 691, 701, 713
  role of homeo box sequences in differentiation of, 285–290
Embryonic
  genes, onset of activation of, 48
  induction
    and cell interactions, 4–6
    permissive, 5
  lethality, 906
  and dominance of RSV-CAT transgenic strains, 415
  lineages
    in *C. elegans*, 70
    and transformed DNA, 75
Endocrine system, developmental gene expression, 389–396
Engrailed (*en*) locus
  and development, 229–232, 235–241
  en-related, invected (*inv*), 232
  and gastrulation, 253
  homeo box, 291–292, 294–299
    amino acid substitution, 245
  late blastoderm expression of, 248
  protein, DNA-binding, 235–241, 253
Enhancer activity, viral, in teratocarcinoma cells, 747–752
Enhancer sequences, 355, 358, 908
  in *Drosophila*, *cis*-acting DNA sequences, 526, 534
  function and transvection, 337–345
  -like, in *gal* regulatory regions, 492
  negative regulation in undifferentiated EC cells, 702–705
  pituitary-specific, 392
  polyoma, 749
  and *Sgs3* transcription, 358
  SV40, 356, 747, 749
    and tumorigenesis, 671
  in teratocarcinoma cells, 747–756
  and tissue-specific gene expression, 389
  and transvection effects, 342
    hypothesis, 343–345
  and upstream activation sites (UAS) in yeast, 644
Epidermal keratin genes, 31–34
Epitope selection, 390
Equivalence groups and equipotential cells in nematodes, 99
Escherichia coli
  gene fusions, 75. *See also* Fusion genes
  hygromycin phosphotransferase fusion proteins, 77. *See also* Fusion proteins
Evolution
  and conservation
    of *cis*- and *trans*-regulatory elements, 545
    of RAS protein function, 721–724
    of globin genes, 368
    of homeo box homology, 249–250, 268
  morphologic, 877
  and RAS protein function, 721–725
Exogastrulation, 31–34
Expression, gene. *See* Gene expression
Extracellular matrix, and type I collagen, 442
Extra sex combs (*esc*) gene, 127–133

F

Fate mapping, and neurogenesis, 881
Feather histogenesis and CAM cycles, 822
Feedback loops, and regulation of cell-type proportion in *Dictyostelium*, 779–785
Fertilization
  receptor for sperm, 11–18
  and regulation by ZP3, 17
α-Fetoprotein (AFP)
  gene expression in transgenic mice, 371–377
Fibronectin, and discoidin I gene expression in *Dictyostelium*, 808
Flagellar (*fla*) gene expression in *Caulobacter*, 835–839
Flanking DNA sequences of transgenes, 449
Flower development, abnormal, 479
fos, gene expression in mice, 733–744
Founder-cell lineage, 580
fushi tarazu (ftz)
  embryos, neurogenesis in, 891–903
  pattern elements, 135–136, 141, 151–153, 210, 238, 254
  transcripts, 108, 244, 247
Fusion genes
  β-globin-*Sgs3*, 356
  chloramphenicol acetyl transferase (CAT), 411–415
  elastase I-growth hormone, 399
  MTrGH (growth hormone), 379
  Sgs3-alcohol dehydrogenase (glADH), 356
Fusion proteins
  β-Gal-*esc*, 131
  β-Gal-*ftz*, 248, 255
  β-Gal-*Ubx*, 183
  CyIIId-CAT (actin-chloramphenicol acetyl transferase), 326
  engrailed (*en*), 238–239
  globin-, 361–369
  spoVG-lacZ, 484–487

G

Gain-of-function alleles, 596. *See also* Sxl alleles in *Drosophila*
gal1–gal10, control of transcription in yeast, 489–502
Gamete interactions, 11–12
Gastrulation
  of bicaudal embryos, 106–108
  and expression of cell surface glycoproteins, 91–98
Gene(s) (ϵ). *See also* specific gene names
  ADH transcription in *Drosophila*, 505–514
  α-fetoprotein in transgenic mice, 371–378
  affecting cell fates in *C. elegans* development, 99–104
  affecting early development in *C. elegans*, 69–78
  affect segmental subdivision *Drosophila*, 155–164
  chimeric in plants, 421–431
  chorion regulation in *Drosophila*, 527–535
  Dorsal in *Drosophila*, 223–228
  extra sex combs in *Drosophila*, 127–134
  glue gene regulation in *Drosophila*, 255–260
  HO gene in yeast, 643–650
  human globin in transgenic mice, 361–370
  human growth hormone in transgenic mice, 399–409
  hybrid in *Drosophila*, 355–360
  Krüppel in *Drosophila*, 465–473
  neuroendocrine, 389–397
  ran1$^+$ in fission yeast, 635–641
  regulation of transcription of in yeast, 483–488
  selector, *Drosophila*, 201–208
  spo0H in *B. subtilis*, 483–488
  T antigen in transgenic mice, 399–409
  yolk protein I, developmental control of in *Drosophila*, 521–526
Gene amplification
  cis-regulation of, 541
  of *Drosophila* chorion genes, 527–534
  of insect chorion genes, 537–547
  trans-regulation of, 539

SUBJECT INDEX

Gene-controlling molecules, nuclear, 8–9
Gene expression
and activation of embryonic genome(s), 45–49
ADH in *Drosophila*, 515–520
altered morphogenesis and, 31–34
cell adhesion molecule, 877–889
cell-type-specific
in *Dictyostelium*, 801–812
markers of, 7
of chimera in plants, 421–430
cisvection, 155, 157–159
constitutive expression and poly-(dT:dT) sequences, 493
control of, 483–563
α-fetoprotein in transgenic mice, 371–378
flagellar during *Caulobacter* cell differentiation, 831–840
hemizygous, 47
histone, 655–670
historegulatory genes, 887. *See also* Tissue-specific elements
homeo box sequences during murine endodermal differentiation, 285–290
lineage-specific in sea urchin, 321–328
mammalian homeo-box-containing genes, 291–300
model for *cis*-regulation, 161–163. See also *Cis*-acting regulatory elements
morphoregulatory genes (CAM), 887
and mother-daughter decisions, 101
neural crest cells, 855–863
in plants, 421–437
promoter-directed in transgenic mice, 399–409, 411–416
regulation of histone, 665–669
regulation of in murine teratocarcinoma cells, 701–706
5S RNA genes, 549–553
and sister-sister decisions, 100
spatial of the *ftz* gene, 108
temporal and spatial control
of CAT fusions, 413
of *Krüppel*, 468–469
tissue-specific patterns of in transgenic mice, 399–409, 411–416
in transgenic mice, 361–420
Gene families. *See also* Multigene families
histone, 665–669
homeo box (*Drosophila*), 209–222
Gene fusions between nematode and bacterial genes, 75. *See also* Fusion genes
Gene rearrangements, immunoglobulin, 420
Gene transfer, 908
by *Agrobacterium tumefaciens*, 421
to increase animal growth, 379–386. *See also* Growth hormone
into mature plants, 429
in mice. *See* Transgenic mice
microinjection, sea urchin, 325
and P elements, 355
in plants, 421, 433–436

transgenic rabbits, 383–384
using retroviral vectors, 691–699
Germ cell
determination in *X. laevis*, 37–43
formation of functional, 685, 695
Germination (*ger*) genes, in *B. subtilis*, 483–487
Germ line
inversion of polarity in, 59
transformation
by *ftz*, 247–248
using P elements, 355
transposition, 313, 330
Germ plasm
of amphibia, 2
and formation of primordial germ cells, 37–43
Globin gene(s)
developmental regulation in transgenic mice, 361–369
embryonic expression of fetal in transgenic mice, 367–369
fusion with *Sgs3*, 356
human in transgenic mice, 361, 454
Glucocorticoids
and metallothionein gene expression, 382
receptor, cDNA cloning of, 389–391
β-Glucuronidase (*uidA*) gene fusions in *E. coli*, 75–76
Glucose phosphate isomerase (GPI-1) isozymes, 46
Glycoproteins
cell adhesion molecules (CAMs), 877–888
contact site A (csA) in *Dictyostelium*, 813, 816
mutants, 819–821
and gastrulation and spicula formation in sea urchin, 91–98
mRNAs, 93
sperm receptor ZP3, 11–18
ZP1-3 of zona pellucida, 12
Gradient of morphogen, concentration of, 110
Grasshopper, neurogenesis in, 891–903
Growth, *fos* proto-oncogene expression during, 733–745
Growth hormone
-elastase I fusion genes, 399–408
expression in transgenic mice, 379–384
and prenatal lethalities, 447
releasing factor (GRF), 379
Gynandromorphs, 115
Gynogenones, diploid biparental, 46

H

Haplo-insufficient
dominant phenotype, 146
region, 119
Haplo-lethal (*Hin-d*) alleles, 120
Heat shock
gene promoter, 427
induces transcription of DIRS-1, 759, 761–764
Hematopoietic precursor cells, gene

regulation in, 691, 697–699, 708
Heterochromatin, 332
Heteropezea, sex determination in, 615, 619
his3
gene transcription, 496–498
upstream regulatory elements, 494
Histone
developmental gene expression, 665–669
H1 binding to chromatin, 551
H3 and H4 are coupled to nuclear polypeptides N1 and N2, 661
pre-mRNA, 668
transfer pathway for nucleosome core assembly, 662
HL-60 monomyelocytic cell line, 733, 735
(HMG)CoA reductase, β-hydroxymethyl glutaryl, 92
HO endonuclease,
gene regulation of, 643–649
negative regulators of, 569. *See* Mating-type allele
Homeo box(es), 243–306
and adult lethality, 157, 161
cDNA clones containing, 301–305
conservation of sequence of, 303–305
as controlling circuits, 250
cross-hybridizing loci, 215–223
cytogenetic locations of, 215
domain structure, 254
conservation of, 263–269
Drosophila and *Xenopus* homologies, 273
open reading frame, 245
in 3'-exon regions of *Ubx*, 184, 209, 291–299
expression of genes containing, 271, 291–299
expression in transformed fibroblasts, 305
genes, combinatorial regulation by, 257
and genetic control of development, 243–250
homology in animal genomes, 209–221, 240, 245, 265, 908
human loci, Hu1, Hu2, 293–294, 301
linkage relationships, 278
in mammalian genes, 277–283, 285–289, 291–299, 301–305
sequences conserved in higher genomes, 263–269
sequences in mouse, 278–283, 285–289, 291–299, 301
in sex-lethal, 601
specificity of induced transcripts, 287
in *Ubx*, 174, 209
in *Xenopus laevis*, 271–274
Homeotic gene complexes
characteristics of, 253
and control of development, 243–250
mutations
of bithorax, 173, 181, 201, 209. *See also* Bithorax; Ultrabithorax
nuclear localization of proteins encoded by, 255–260

Homeotic gene complexes *(continued)*
  and segmentation in *Drosophila*, 209, 243–250
  and sex determination, 619
Homeotic mutants, 155–242
Homeotic transformation, caused by sex-lethal (*Sxl*), 603, 619
Hox (homeo box) loci in mouse, 278–283, 285–289, 291–299
  Mu1, 293
Human growth hormone gene, and prenatal lethalities in transgenic mice, 447–451
H-Y histocompatibility antigen, 623, 907
Hyaline gene, 324
Hybrid dysgenesis, 329
  and molecular cloning of neurogenic genes, 851–853. *See also* P elements
Hydroxy-proline-containing proteins and spiculogenesis, 91
Hymenoptera, sex determination in, 615, 619
Hypomorphic alleles, 575
Hypothalamus expression of GH fusion genes in, 394

I

Imaginal disk regeneration, 119
  disk-specific *dpp* alleles, 120
Immunodifferentiation, in transgenic mice, 417–420
Immunoglobulin heavy-chain gene, expression in transgenic mice, 417–420
Imprecise excision and transposition, 316–317
Induction
  embryonic, 4–6
  instructive, 5–6
  permissive, 5
Inner cell mass (ICM) cells, 47
Insect. *See also Drosophila*; Grasshopper
  expression and amplification of chorion genes in, 537–547
  sex determination in, 605–614
In situ hybridization
  of *Dfd+* and *Antp* in early gastrula, 249, 254
  of homeo box *en* probe, 248
  of *Krüppel* gene, 469
  of mouse homeo box (*Hox*) loci, 280
  of vitellogenin mRNA, 580
  of *Ubx* transcripts, 187, 201, 209
Insertional mutagens
  in transgenic mice, 453–462
  retroviruses as, 439–444
Intermediate filament patterns in *Drosophila* embryos, 82

K

Karyoplasts, introduced into enucleated zygotes, 47

Krüppel (*Kr*) alleles
  central domain, 469
  segmentation genes of *Drosophila*, 465–472

L

λ bacteriophage, 569–572
  cII as a developmental regulator, 570
Lineage markers, 39
Lineage-specific cloned genes, 323
Liver cell adhesion molecules (L-CAM), 877

M

Maize, transposable elements in, 307–311
Malpighian tubules, 465, 506, 508, 516
Mammalian homeo box genes, 277–283, 285–289, 291–299, 301. *See also* Homeo box(es)
Maternal
  control of early development, 60
  dosage effects of *Kr* (*Krüppel*), 465
Maternal effect(s)
  daughterless (*da*) regulates sex-lethal (*Sxl*), 595
  of *esc* gene, 127–133
  of *fem* genes in *C. elegans*, 589
  genes, 60, 906
  and embryonic polarity, 223
  mutants in *C. elegans*, 70–74
  of segmentation genes (*Drosophila*), 147–148
  and sex determination, 615
  in *Chrysomya*, 615, 619
  and spatial orientation of embryo, 243
Maternal factors, localized mRNAs in *Xenopus* eggs, 21–29
Maternal genomes, and nonequivalence of paternal, 45
Mating-type allele(s) (*MAT*), 566–572
  a/α repression, 645, 568
  and *Drosophila* and *Xenopus* homeo boxes, 273
  genes, 245
  and HO expression, 643–649
  model for gene expression, 257–260
Maturation promoting factor (MPF), 63
  in *Xenopus* embryos, 651–656
Medicago sativa, nodule formation in, 555–562
Megaselia scalaris, sex determination in, 617
Methotrexate resistance in transformed plant cells, 436
Methylation
  of α1(I) collagen in MOV13, 440
  de novo of flanking host sequences, 441
  of MCPs in chemotactic bacteria, 796

Meiosis
  in fission yeast, 635–640
  Ha-*ras* proteins induce, 723
Meiotic spindles in *zyg-9* mutants, 73
Mesoderm
  inhibition of formation of, 34
  and muscle formation, 3–4
  and the selector gene hypothesis, 114
Metameric phase(s)
  and cellularization of blastoderm, 141, 152
  metasegmental patterns, 152, 187
  pattern elongation, 269
Microinjection
  of cloned globin genes, 362
  into mouse germ line, 411
  of retroviruses into mouse zygotes, 439–444
Microtubule-associated protein (MAP 2), in developing neurons, 83
Midblastula transition
  and activation of zygotic genome of *Xenopus*, 31
  and elongation of cell cycle, 651–656
Midgestation embryos, permissive for retrovirus expression, 441
Mitochondrial cloud of *Xenopus* oocytes, 37
Mitosis, induced by maturation promoting factor (MPF), 651–656
Mitotic recombination and BX-C, 166
Mobile genetic elements. *See also* Transposons
  roo in *Drosophila*, 600–602
Moloney murine leukemia virus (Mo-MLV), 713
  MOV substrains, 439–444
Morphogenesis, and embryonic induction, 877–888
Morphogenetic
  determinants in *Drosophila* embryos, 110
  patterns during embryogenesis, 223
Morphogens and mechanochemical factors (CAMs), 888
Mouse. *See also* Chimeric mice; *Musca domestica*; Transgenic mice
  autonomous plasmids derived from polyoma virus, 679–684
  chimera formation in EK cells of, 685–689
  expression of human type II collagen gene in, 707–711
  α-fetoprotein gene expression in, 371–378
  first differentiations in embryo of, 51–57
  gene transfer in, 691–700
  homeo box loci (*Hox-1-4*), 278–283, 285–289, 291–299
  human globin genes in, 316–370
  nuclear transfer in embryo, 45–50
  regulating gene expression in teratocarcinoma cells of, 701–706
  sex determination in, 623–625
  sperm receptor in egg of, 11–19
MOV substrains, carry Mo-MLV pro-

SUBJECT INDEX

viruses, 439–444
MOV-13, 439
M-strain females, 329
MTV-*myc* fusion gene in transgenic mice, 677
Musca domestica. See also Transgenic mice
  multiple R factors in, 617
  sex determination in, 615, 620
Multigene families
  actin CyIII genes, 321–327
  of chorion genes, 527–534, 537–547
  discoidin I in *Dictyostelium*, 808
  globin genes, 361–369
  of organ-specific cDNAs, 425–426
Muscle
  developing striated, 41
  myosin heavy-chain gene, 324
  selector genes and development of, 114–116
  vegetal induction of, 6
Mutagenesis
  insertional
    by *Ac-Ds*, 307–311, 429–430
    by retroviruses, 439–444
  sequential and *Dictyostelium* development, 813–822
Mutants, homeotic, 155–242
Mutations
  affecting cell fates, 100
  bicaudal (*bic, BicC, BicD*) loci, 105
  in bithorax, 113
    Mcp, 115
    Ubx domains, 181–194
  bypass in *Dictyostelium*, 815, 819
  cell cycle (*cdc*) in yeast, 630–634
  in decapentaplegic (DPP-C), 119–125
  and DNA rearrangements in transgenic mice, 447–451
  dominant gain-of-function, 908
  embryonic-lethal (*EL*) alleles (in DPP-C), 120
  engrailed (*en*), 229–232
  extracellular complementation in myxobacteria, 825
  feminizing (*fem*) in *C. elegans*, 587–592
  fushi tarazu (*ftz*), 108, 135–136, 141, 151–153, 210, 238, 244, 254
  homeotic, 155–242, 173, 209
  maternal-effect, 7
  in polyamine synthesis in plants, 477–480
  and reiterative expression of specific cell types (*lin-4, lin-14*), 101
  retrovirus-induced, 439–444
  screening for sex determination, 611–613, 618
  segmentation, 135, 145–153, 209, 243. See also Bithorax
  shv alleles (in DPP-C), 120
  transcription "initiator," 492
  in transgenic mice, 907
  ts maternal gene, 60
  Ultrabithorax (*Ubx*), 136, 167–171, 195–201
  in zygotic genes in *Drosophila*,
136, 145. See also *Drosophila*
  zygote-defective in *C. elegans*, 70–74
myb gene of *Drosophila*, 729
Myoblast C2 cell line, 753–756
Myxococcus, development in, 823–829

N

Nematodes, development, 907. See *Caenorhabditis elegans*
Neomycin phototransferase assays, 427–428
Nerve growth factor (NGF)
  induces neural-specific genes in PC12 cells, 855
  induces neurite formation, 733, 735–737
Nervous system, development in, 841–851
Neural
  cell adhesion molecules (N-CAM), 877
  crest cells, expression in differentiation, 855–862
  histogenesis and cell migration, 881
  induction and CAM distribution, 879
  -specific expression and the metallothionein promoter (MT-I), 396
  -specific genes, 855–862
Neuroblasts, differentiation of, 892–895
Neurogenesis
  in fushi tarazu embryos, 891, 899–902
  in grasshopper embryos, 891–902
  molecular genetics of in *Drosophila*, 841–854
  neuron-glia adhesion molecules (Ng-CAM), 877
Nicotiana tabacum, genetics of polyamine synthesis, 475–480
Nitrogen fixation, 555–562
N-linked glycosylation
  function in gastrulation, 94
  tunicamycin inhibits spiculogenesis, 91
  of zona pellucida glycoproteins, 14. See also Glycoproteins
Nodulins, 557–562
Nuclear
  extrusion, 62
  polypeptides N1 and N2, 661
  transfer
    and activation of embryonic genomes, 45–49
    in amphibians, 2
    interspecies, 47
    transplantation of mutants in bithorax, *Mcp*, 115
Nucleoplasmin, 652
  and histone(s) H2A and H2B, 661–662
Null alleles of X-chromosome genes, 575–576

O

Oligosaccharide assembly, 93
  N-linked, 14. See also *N*-linked glycosylation
  O-linked, 16
  role in sperm receptor activity, 14–18
Oncogenes, 727–730. See also Proto-oncogenes
  homology with CDC products, 627, 633, 639
  RAS, 721–725
Oocyte
  mitochondria in previtellogenic, 38
Oogenesis
  BicD acts during, 105–106
  and dorsal (*dl*) gene expression, 223–228
  in *Drosophila*, 527–534
  embryonic polarity in, 223
opa elements in engrailed (*en*), 263
Organogenesis, and type I collagen, 442
Outgrowth (*out*) genes in *Bacillus subtilis*, 483–487

P

Pair-rule genes of *Drosophila*, 136, 906
  even-skipped (*eve*), 145–153, 238
  odd-skipped (*odd*), 145–153
  paired (*prd*), 145–153, 238
Pancreatic
  acinar cells express elastase I gene, 399
  carcinomas, induced by elastase-SV40 T antigen genes, 404
Parietal endoderm, and differentiation of F9 EC cells, 285
Parthenogenesis, mammalian, 45
Patatin-encoding cDNA clones, 426
Paternal genomes, 45
Pattern
  formation and pigmentation, 335
  genes controlling, 267, 291. See also *ftz; Dfd; Scr; Antp; Ubx*
  role of *BicD*, 109
PC12 cells
  differentiation, 733, 735–737
  neural-specific genes in, 855, 861
P element
  in 68C glue puff genes, 350
  -mediated gene transfer, 129, 313
  transformation at *per* locus, 867
  transposition, germ-line specificity of, 329–335
  vectors, 355
P granule(s), 71
  segregation, 79
Pheochromocytoma tumor cell line PC12, 855, 861
Phosphodiesterase, inhibition by contact site A (csA)/cAMP, 823
Phytochrome, photoreceptors, 421
Photomorphogenesis, 421
Plants. See also *Dictyostelium discoideum*; Nitrogen fixation;

SUBJECT INDEX

Plants *(continued)*
    Saccharomyces; Tobacco; Yeast
    chimeric genes in, 421–431
    controlling elements in, 307–311
    developmental switches in, 475–482
    master regulatory loci in yeast, 565–574
    transgenic, 433–437
    yeast gene transcription regulation, 489–503
Plasmids, autonomous, transgenic, recombinant, 680
Pluripotential cells
    early transcription of E.Tn. in, 54
    primary cultures of, 685, 691, 701
Polar lobe formation, 2
Polyamine synthesis
    in tobacco, 475–480
    variants, 480
Polyoma virus
    enhancer, 749, 753
    replication, 753–756
    in transgenic mice, 679–684
Positional information. *See also* Determinants
    in *Drosophila* embryos, 79–89
    and gene expression, 332
    of transformed genes, 534
Preimplantation embryos, introduction of genes into by retrovirus vectors, 713–719
Prespore cells, differentiation in *Dictyostelium,* 779–784, 801–811
Prestalk cells, differentiation in *Dictyostelium,* 769, 779, 801–811
Primordial germ cells (PGCs)
    formation and migration of, 37–43
    infected with retroviruses, 439
Promoter sequences
    and CAT expression in transgenic mice, 411–416
    of αA crystallin gene, 411–415
    of $\alpha 2(I)$ collagen gene, 411–415
    of avian sarcoma virus genes, 411–415
    elastase I and gene expression in transgenic mice, 399–409
    exclusion, 513
    and gene transcription in yeast, 489–503
    papovavirus and retrovirus in EC cells, 688, 701
    regulatory elements within *HO,* 645
    and regulatory gene for utilization of in *B. subtilis,* 483–488
    SV40 early, 747
Protein kinases
    control of sexual differentiation in fission yeast, 635–640
    homology with CDC28, 627, 632–633, 639
Proto-oncogenes
    of *Drosophila,* 727–730
    fos expression in mouse embryos, 733–744
    ras gene in *Dictyostelium,* 801–811
Puffing
    and 68C glue puff, 347–352
    and high-level transcription, 352

R

ras
    gene in *Dictyostelium,* 801–811
    proteins in *Dictyostelium,* 771
    protein function, 721–724
Recessive lethal mutations
    linked to T-DNA, 435
    in MOV-13 mice, 439
Regulation, genetic. *See also* Homeotic mutants
    ADH expression in *Drosophila,* 515–520
    CAM and morphogenesis, 877–889
    chorion gene amplification and expression, 527–535
    developmental gene in *B. subtilis,* 483–488
    gene expression in murine teratocarcinoma cells, 701–706
    gene transcription in yeast, 489–503
    glue gene in *Drosophila,* 355–360
    histone gene expression in sea urchin, 665–670
    HO gene in yeast, 643–650
    loci in yeast and phage λ, 565–574
    sites in yeast, 489–503
    of transcription, 337–346
    of transposable elements in *C. elegans,* 313–320
Repeated $CACGA_4$ sequences, and cell cycle control in yeast, 646–649
Replication
    destabilized transcription complexes, 552
    mutants are amplification-defective, 541
    and polyoma large T antigen, 679
    polyoma virus, 753–757
    sequence specificity for autonomous, 658
    in *Xenopus* embryos, 657–662
Rescue, of *Kr* embryos, 467–472
Retrotransposon, in *D. discoideum,* 759–767
Retroviruses, 418
    and insertional mutagenesis, 439–444
    tRNA binding site, 54
    vectors, 685–688, 691–699, 713–719
Reverse transcriptase, homology with DIRS-1 transposon, 765
Rhizobium, mutants lack exopolysaccharide, 556
RNA
    antisense, 8, 28, 162, 717, 801
    Kr, 468
    differential
        processing, 908
        transcript mapping, 128, 132
    localized maternal mRNA, 21–29
    maternal dorsal (*dl*), 226
    polymerase II transcription of yeast mRNA, 489
    pre-mRNA 3'-processing of histone transcripts, 665–669
    -RNA hybrid (sense-antisense), 811
    U7-snRNA, 665–669
        genes encoding, 668

Root nodule formation, 555–562
Rous sarcoma virus (RSV)
    promoter sequences, 411–415
    RSV-CAT fusions, 414

S

Saccharomyces. See also Yeast
    cell cycle controls, 627–634, 635–640
    mating-type allele (*MAT*) and *HO,* 643–649. *See also* Mating-type allele
Salivary gland intermolt puffs, 347
Sciara, sex determination in, 615
Sea urchin embryos
    and developmental expression of glycoproteins, 91–98
    histone gene families, 665–669
    lineage-specific gene expression in, 321–327
    Strongylocentrotus purpuratus, 321
Segmentation, 906. *See also Drosophila*
    and compartment formation, 248
    in *Drosophila,* 135–143, 145–153, 165–171, 173, 209
    effect of *sxc* mutations, 202–203
    effect of *trx,* 205–208
    and engrailed (*en*), 229–232, 235–241
    genes and spatial organization, 243
    and homeo boxes in vertebrates, 274
    and localization of *ftz* and *en* transcripts, 248, 254
    nuclear localization of proteins encoded by, 255–260
    parasegments, 152, 182, 195
    pattern of *Krüppel* (*Kr*) genes, 465–472
Selector genes, 113
    spatial pattern of expression, 201–208
Serine protease genes, 399
Serologically detected male (SDM) antigen, 624
Sex combs reduced (*Scr*) gene, cDNA, 254, 267
Sex chromosomes, and the dominant male determiner, *R,* 617
Sex determination, 565–625, 907
    and dosage compensation in *C. elegans,* 579–582, 585–592
    in *Drosophila,* 595–603, 605–613, 615–620
    in insects, 615–620
    mechanisms, 618
    in mice, 623–625. *See also Musca*
Sex-lethal (*Sxl*)
    and daughterless (*da*), 595, 599, 606
    in *Drosophila,* 595–602, 605–613, 615
    R signaling genes repress, 615
Sex-reversal (*Sxr*), sex-reversed, 623
Sexual differentiation, in fission yeast, 635–640
Signal transducing receptors
    β-adrenergic receptor, 796
    rhodopsin, 796

SUBJECT INDEX

Silk moth follicular epithelial cells, 537
Sleep/wake rhythms in *Drosophila*, 865
Slime mold. *See Dictyostelium discoideum*
Solanum tuberosum, developmentally controlled genes in, 422–430
Somatic cell lineages, mapping in *C. elegans*, 69
Somatic excision, transposition of Tc1, 314
Somatic sexual development
  in *C. elegans*, 575, 591
  in *Drosophila*, 606. *See also* Sex-lethal
Somatic synapse and the enhancer hypothesis, 344
Sperm
  receptor in mouse, 11–18
  species-specific receptors, 12
Spermatogenesis in *Sxr* males, 624
Spicula matrix protein gene, 324
Spiculogenesis
  antibodies block, 95
  is inhibited by tunicamycin, 91
  in sea urchin embryos, 91–98
  in vitro, 94
Spm controlling elements, maize, 307–311
Spore coat proteins of *Dictyostelium*, 771–776
Sporulation
  genes (*spo*) in *B. subtilis*, 483–487
    spo0 genes, 484
    spoVG genes, 483
  in myxobacteria, 823–829
Stalk cell
  differentiation in *Dictyostelium*, 769
  mutants, 771
  embryo-derived, 685, 691, 701
  transformation with type II collagen, 707–710
Stem cells
  gene transfer into, using retroviral vectors, 691–700
  transformation of with type II collagen gene, 707–711
Steroid hormone
  ecdysterone, 347, 349
  -receptor interactions, 389
Super sex combs (*sxc*) loci, 201–208
Suppressor-mutator (*Spm*) elements in maize, 307–311
SV40
  early genes expressed in transgenic mice, 671–677
  T antigen
    -elastase fusion genes, 399–408
    in EC cells, 702–705
    induces pancreatic carcinomas, 404
Synapsis "transvection effects," 337

T

T antigen. *See also* SV40
  in choroid plexus tumors, 671
  of polyoma virus in transgenic mice, 679–684
TATA elements, 498–500, 532–534. *See also* Transcription
Tc1 transposons and polymorphisms, 74, 313–319
Tenebrio, 114
Teratocarcinoma cells
  gene expression in, 701–705
  viral enhancer activity in, 747–756
Teratoma callus, 422
Terminal differentiation processes, 417
Testes-inducer, and SDM antigen, 624
Ti plasmid
  of *Agrobacterium tumefaciens*, 421, 433–436
  -derived plant gene vectors, 427
Tissue-specific elements. *See also* Gene expression; Transgenic mice
  control yolk protein synthesis in *Drosophila*, 521–526
  expression of hybrid SV40-T-antigen chimera, 671
  gene expression in transgenic mice, 399–409, 411–416
  gene transcription of ADH gene in *Drosophila*, 508–509, 516
  of transgenes, 418
Tobacco, genetics of polyamine synthesis in, 475–482
tRNA, gene expression in DPP-C mutants, 125
trans-acting regulatory elements, 8–9, 906
  and *Drosophila* chorion genes, 539–541
  interact
    with *cis*-acting sequences, 351
    with enhancers, 747
  and regulation by cAMP, 810
  Spm suppressor element, 307
  transcription factors, 551
  white-spotted alleles, 339
trans-regulation by bithorax (BX-C), 163, 193
trans-regulatory products (TRPs), 161
Transcription
  accurate initiation, 491–492
  activation
    by CACGA$_4$ sequences, 646
    by enhancers, 747
  activator sequences, 359
  of alcohol dehydrogenase gene in *Drosophila*, 505–513, 515–519
  autoregulation of ADH gene, 510
  cell cycle control, 646–649
  of chorion gene clusters
    in *Drosophila*, 527–534
    in insects, 537–547
  and chromatin structure, 361
  complexes on naked DNA, 550
  constitutive and coordinately regulated of yeast genes, 489–502
  early embryonic of E.Tn., 54–56
  factor TFIIIA, 549–552
  heat-inducible expression from hsp70 promoter, 427
  of heterologous fusions of *Sgs3*, 357
  of histone gene(s) in the cell cycle, 667
  hormonal regulation of 68C gene, 347–352
  initiation in yeast, 500–501
  light-inducible in plants, 421–424
  and methylation, 677
  nuclear assays, 400
  organ-specific control of, 425
  posttranscriptional mechanisms, 425
  pre-mRNA 3' processing, 667–669
  positive/negative regulatory elements, 491
  postnatal repression of AFP, 374
  promoter elements in yeast, 489–502
  regulation of DIRS-1 in *Dictyostelium*, 759–766
  regulation of *fos* gene, 739–744
  repression, 552
  of retrotransposon DIRS-1 in *Dictyostelium*, 759–767
  RNA polymerase sigma factors, 484
  sequence-specific *en* binding, 238–241
  species differences in ADH gene, 517–519
  spo0H-dependent promoter utilization, 484
  tissue-specific, 371
    of β-globin genes, 363–365
  trans-acting factors, 551
  transvection effects, 337–345
  of X-linked genes in *C. elegans*, 577
Transduced genes, 332
Transformation
  of ADH gene in *Drosophila*, 506, 519
  of *C. elegans*, 74–75
  of *Drosophila*, 9
  P-element mediated, 129
  of chorion genes, 528, 541
  of *yp* genes, 522
  plant. *See also* Ti plasmid
    leaf disc system, 434–436
    split end vector system, 433–436
Transgenic mice, 361–420
  AFP minigenes in, 371–377
  developmental mutations in, 447–451, 453–462
  DNA rearrangements in, 447–451
  elastase I promoter in, 399–408
  express human globin genes, 361–369
  express mGH fusion genes, 392–396
  express a rearranged immunoglobulin heavy-chain gene, 418–420
  harboring inducible growth hormone fusion genes, 379–386
  immunodifferentiation studies in, 417–420
  mammary carcinomas in, 677
  methylation in, 442
  pancreatic carcinomas in, 404–408
  polyoma virus genomes in, 679–684
  prenatal lethalities, 447
  recessive prenatal lethal mutations in, 455
  and retroviral vectors, 685, 713, 716. *See also* Retroviral vectors
  SV40 early genes, 671–677
  tissue-specific CAT expression, 411–415
  tissue-specific gene expression in,

SUBJECT INDEX

Transgenic mice *(continued)*
399–409
Transgenic plants, 433–436
Transgenic rabbit(s), and growth hormone, 383
Transplantation
by nuclear extrusion, 62
pole cells, 145
single-cell in *Xenopus*, 39
Transposable element(s)
maize, 307–311
Tc1 in *C. elegans*, 69, 313–319
Transposition
of *Ac* in heterologous plants, 429–430
DIRS-1 of *D. discoideum*, 759–766
early, E.Tn., 51, 54–56
and extrachromosomal Tc1 DNA, 316
in maize, 307–311
P-element mediated, 329–335
tissue-specific regulation, 314
Tn5
in *Caulobacter*, 831–839
in myobacteria, 824, 826
Transvection, 163
synapsis-dependent complementation, 155, 159–160
and transcription at the white locus, 337–345
in vertebrates, 344
Trithorax (*trx*) loci, 201–208
Trophoectoderm
intermediate filaments of, 51
positional gene activation, 56
Tubulin, in *Drosophila* embryos, 82
Tumorigenesis, and the SV40 enhancer, 671
Tunicamycin, prevents gastrulation, 91

U

U937 promonocytic cell line, 733, 735
Ultrabithorax (*Ubx*), 906. *See also* Bithorax; Segmentation
anterobithorax (*abx*), 182
PPX function, 183
antibodies, 202
bithorax (*bx*) gene, 182
bithoraxoid (*bxd*) gene, 182
BXD-like functions, 183
transcripts, 188–194
function and expression, 195–200, 209–221
homeo box homology with *en*, 263
mutations, 136, 167–171, 201
abd-A, *Abd-B* genes, 168–171, 181, 209
postbithorax (*pbx*) gene, 182
postprothorax (*ppx*) alleles, 195–196
proteins, 185–188, 253
regulation and products of, 181–194
RNAs, 183, 209
sequence complementarity with *bxd* transcripts, 190–191, 203–205

V

Vectors, retroviral
and gene transfer into murine stem cells, 691–700
in mouse EK cells, 685–689
use of to introduce genes into embryonal carcinoma cells, embryos, 713–720
Vegetal pole blastomeres, donors in transplantation, 40
Virus. *See* Enhancer activity; Polyoma; Retrovirus; SV40; Vectors
Visceral endoderm, AFP in, 371
Vitellogenesis, 38, 580–582
v-*src*, -related oncogenes in *Drosophila*, 727

W

White locus, transvection effects at, 337–345

X

X/A ratio determines sex
in *C. elegans*, 575
in *Drosophila*, 595, 606, 615
Xanthine dehydrogenase, rosy gene in *Drosophila*, 506, 511, 516
X chromosome
aneuploids, 577
and *dpy* gene expression, 577
expression in *C. elegans*, 575, 620
-linked AFP genes, 372, 374–377
inactivation in mammals, 579
Xenopus
activation of zygotic genome, 31
chromosome replication in embryos of, 657–662
germ cell determination in, 37–43
Ha-*ras* induces meiosis in, 723
homeo boxes in, 271–275
maternal mRNAs in, 21–29
5S RNA genes, 549–552

Y

Y chromosome, and sex determination in mammals, 623
Yeast. *See also* Saccharomyces; Mating type (*MAT*)
cell cycle control in, 627–634, 635–640
developmental regulatory proteins, 565–572
fluorescence microscopy of, 627
HO gene regulation in, 643–650
mating-type switching, 568
RAS protein function in, 721–725
regulation of gene transcription in, 489–503
sequence specificity for autonomous DNA replication, 658
upstream activation sites (UAS), 644
Yolk protein I gene, developmental control of in *Drosophila*, 521–526

Z

Zea mays, 307–311. *See also* Maize
hermaphrodites in, 620
Zona pellucida glycoproteins, 11–18
Zyg-9 (zygote-defective)
alleles, 73
mutants, 70–74
Zygotes, enucleated, 47
Zygotic genes, and polarity, 223